Asdell's Patterns of Mammalian Reproduction

A Compendium of Species-Specific Data

VIRGINIA HAYSSEN
SMITH COLLEGE
ARI VAN TIENHOVEN
CORNELL UNIVERSITY
ANS VAN TIENHOVEN
ITHACA, NEW NORK

COMSTOCK PUBLISHING ASSOCIATES
A DIVISION OF
CORNELL UNIVERSITY PRESS
ITHACA AND LONDON

This book was printed from camera-ready pages provided by the authors.

First and second editions of *Patterns of Mammalian Reproduction* published 1946, 1964. *Asdell's Patterns of Mammalian Reproduction: A Compendium of Species-Specific Data* published 1993 by Cornell University Press.

Library of Congress Cataloging-in-Publication Data

Hayssen, Virginia Douglass.
Asdell's patterns of mammalian reproduction : a compendium of species-specific data / Virginia Hayssen, Ari van Tienhoven, Ans van Tienhoven.
p. cm.
Rev. ed. of: Patterns of mammalian reproduction / S.A. Asdell. 2nd ed. 1964.
Includes bibliographical references and index.
ISBN 0-8014-1753-8 (alk. paper)
1. Reproduction. 2. Mammals—Physiology. I. Van Tienhoven, Ari. II. Van Tienhoven, Ans. III. Asdell, S. A. (Sydney Arthur), 1897–1987. Patterns of mammalian reproduction. IV. Title. V Title: Patterns of mammalian reproduction.
QP251.H33 1993
599′ .016—dc20 92-56775

Printed in the United States of America

♾ The paper in this book meets the minimum requirements of the American National Standard for Information Sciences — Permanence of Paper for Printed Library Materials, ANSI Z39.48-1984.

We dedicate this book to Kenneth E. Wing, without whose administrative support it would not have survived beyond an early embryonic stage. We also dedicate this book to our teachers, who represent the past; our colleagues, who represent the present; and our students, who represent the future. Their inspiration, knowledge, and labor provided the foundation on which this book is based.

VH adds a personal dedication to the memory of her brother, Sandy.

Contents

Preface

More than fifty years ago Sydney A. Asdell, one of the great minds of the early years of reproductive physiology, undertook the difficult task of compiling the basic facts of reproductive anatomy and physiology for all mammals. The results appeared in his much praised and frequently consulted *Patterns of Mammalian Reproduction*, published in 1946. The discerning intellect of this man is apparent in the introductory section to the book, a section he titled "The Scaffolding".

For the second edition, published in 1964, Dr. Asdell changed the book from an introduction to mammalian reproductive physiology to an encyclopedia of reproductive data. The introductory scaffolding was razed to clear the way for a more extensive compilation of species-specific data on mammals. Interpretive accounts detailing physiological mechanisms were still included for a diverse array of mammals, but the overall goal was to provide the basic information available on the reproduction of individual species.

In 1976 Dr. Asdell asked Ari van Tienhoven to prepare a revision of the book. Dr. van Tienhoven initially declined, intending to find a replacement, then accepted in 1984 when 1) a sabbatical leave provided the opportunity to initiate the collection of data, 2) his efforts to find someone to revise the book failed, and 3) his two coauthors — Virginia Hayssen and Ans van Tienhoven — as well as Kenneth E. Wing, Associate Dean of the College of Agriculture and Life Sciences at Cornell University, assured him that the task of collecting the data was not insurmountable. When it proved to be more formidable than originally imagined, Dr. van Tienhoven had due cause to doubt the veracity of his coauthors. But with the book's final phases under way eight years later, their encouragement seems justified.

As the scientific literature has expanded, the scope of each edition has contracted. Empirical analyses have replaced physiological interpretations, and the emphasis is on a neutral presentation of the published facts of mammalian reproduction. The scope of this revision is similar in two main ways. First, we do not provide extensive interpretations of physiological mechanisms for each taxon as Dr. Asdell did. Second, we have not included data on domestic and laboratory species, as a voluminous literature is readily available. For instance, J. Vaissaire's review of the reproduction of laboratory and domestic animals summarizes some 6000 references (Sexualité et reproduction des mammifères domestiques et de laboratoire [Paris: Maloine, 1987]).

To compile this revision we examined over 150 core journals (see Appendix 2), gleaning data as well as referrals to other sources. We chose articles whose titles and/or abstracts suggested the presence of data on reproductive parameters. In all, over 20,000 sources were reviewed. We used primary sources except when they were not accessible or when we could not translate the language in which they were published. For Russian, Finnish, Chinese, and Japanese publications, we read English abstracts, summaries, or translated secondary sources. Even though some review articles (e.g., field guides, regional accounts) do not cite sources, we included data from such articles if these data were the sole information we could find or if the information differed from that in the primary sources we cite.

Data on a selected set of physical, ecological, and reproductive parameters were culled from each article. These data were coded for computer entry. After extensive checking for errors, we sorted the files and reformatted the appropriate data into tables or summarized them for the ordinal and familial accounts. Bibliographic information for each reference was also computer coded and entered into bibliographic data files. Although we proofread the manuscript at every stage of its compilation and endeavored to be thorough in our survey of the field, errors and omissions certainly exist. We would appreciate notification of any errors you find, so that they may be corrected in future printings.

This book has been a labor of love not only for us but also for the many who assisted us. We thank Kenneth E. Wing, Associate Dean of the College of Agriculture and Life Sciences of Cornell University[1], for his important support in the initial stages of our undertaking and for his sustained interest.

[1] As of 1 December, 1992, President of the State University of New York at Cobleskill, NY.

The collection of the data would have been less thorough if it were not for the crucial support we received from a multitude of librarians. The reference librarians at Mann Library, Cornell University, and the staff librarians at the Smith College Science Library were always willing to go the extra mile to find articles for which we had incomplete citations or to find information on obscure species. The collegiality of these librarians has been an example of professionalism and a source of friendship. The librarians handling interlibrary loans at both Cornell University and Smith College were equally helpful and patient with us, as were the desk librarians, who were always ready to search for publications which were not in their appointed places. J. Craig of the Morrill Science Library, University of Massachusetts at Amherst, spent many, many hours of his time finding and copying references. His generosity and dedication were boundless and are admired as well as appreciated. We are also grateful to J.P. Hearn and the librarians of the Zoological Society of London for making the van Tienhovens' sabbatical stay in the society's library a pleasant and profitable one.

We thank the librarian of the Library of the Royal Society of London for his cordial welcome and the free use of the collection. Use of the research library at the Brookfield Zoo, Brookfield, Ill., as well as much appreciated desk space, was made possible by G. Rabb, M. Rabb, P. Parker, and R. Lacy. A. Petric of the Brookfield Zoo supplied gestation lengths for some hoofed stock from the zoo's records. M. Jones and K. Benirschke provided access to data from the San Diego Zoological Society's records as well as helpful advice. L. Gordon smoothed our use of the mammal collection at the National Museum of Natural History. We are indebted to C. and E. James, Adelaide, Australia, J. Wright, Carlisle, Penn., and the late W. Wimsatt, Ithaca, N.Y., for access to their personal libraries.

We are grateful to the following colleagues who graciously sent us compilations of their reprints for review or provided data from obscure sources: K. Armitage, R.T.F. Bernard, K. Bhatnagar, H.E. Broadbooks, G.N. Cameron, L.N. Carraway, J.A. Chapman, D. Dewsbury, J.E.C. Flux, M. Gosling, P.D. Heideman, N.R. Holler, D.D. Hopkins, B.E. Horner, H. Ikeda, J.U.M. Jarvis, U. Judes, J. Kenagy, M.L. Kennedy, D. Klingener, J.A. Lackey, J. Murie, D.K. Odell, R.H. Pine, A.I. Roest, L. Selwood, M. Tuttle, T.A. Uchida, T. Vaughan, B.J. Verts, and D.W. Walton.

Our thanks to P.J. Carden, Cornell University, for translating Russian figure legends and table headings and to V.W. Hayssen, Hartland, Wisc., and G. Erf, Smith College, for help with French and German idioms.

The scientific names of mammals change over time. We are in debt to M. Carleton, National Museum of Natural History, Smithsonian Institution, K.A.R. Kennedy, Section of Ecology and Systematics, Cornell University, K. Koopman, American Museum of Natural History, and D.K. McClearn, Section of Ecology and Systematics, Cornell University, for their help in resolving many of our most intransigent taxonomic synonymies. Their professional courtesy and gracious willingness to decipher the systematic lexicon are a credit to their field.

The patience and help of T. Ferrara, Cornell University, and S. Beaumier, Smith College, in word processing and data entry are gratefully acknowledged. The many work-study students who helped code and enter bibliographic and scientific data also deserve our sincere thank-you.

The financial support of Cornell University, Smith College, and the Friends of the Smith College Library for clerical assistance, work-study students, library fees, and supplies are gratefully acknowledged.

VH would like to acknowledge the personal support and encouragement of her brother Sandy who had as many views of the bright side of life as there are numbers in this book.

Finally, we owe a debt to Cornell University Press and our editor, Robb Reavill, for the years of commitment to this book.

VIRGINIA HAYSSEN
ARI VAN TIENHOVEN
ANS VAN TIENHOVEN

Northampton, Massachusetts
Ithaca, New York

Organization of the Book

In the almost thirty years since the appearance of the second edition of *Patterns of Mammalian Reproduction*, the field of reproductive physiology has expanded dramatically. Because we could not cover in detail all the developments and still thoroughly review the basics of reproduction across mammals, we have chosen the latter course. To compensate we have provided lists of citations arranged by topic to the literature on the endocrinology, reproductive anatomy, and reproductive physiology of each family. In addition, citations pertaining to domestic and common laboratory species have been curtailed.

As in previous editions, the data are organized taxonomically. The pattern of reproduction is summarized for each mammalian order and family. The familial accounts also include data on neonatal development and number of mammae, summarized by genus. Following each familial account is a tabular listing of species-specific data for neonatal mass and size, weaning mass and size, litter size, age at sexual maturity, estrous cycle length, gestation length, lactation length, number of litters per year and/or interbirth (interlitter) interval, and seasonality of reproduction. Each reproductive variable is listed separately. Genera, and species within each genus, are arranged alphabetically. Where appropriate, data are ordered by sequence of the reproductive cycle; for example, the listings for sexual maturity provide estimates of the age at first conception before those for the age at first birth. Data on length, size, and mass are ordered numerically from smallest to largest. Literature sources for each entry are provided. In sum, the published data for a given species are listed by reproductive variable for each family. This organization makes finding information for any one species more difficult but allows easier comparisons among species as well as easier extrapolations to species for which no data exist.

What taxonomy was used?

Whenever possible, our taxonomy follows *Mammal Species of the World* by J.H. Honacki, K.E. Kinman, and J.W. Koeppl. M. Carleton (National Museum of Natural History, Smithsonian Institution), K.A.R. Kennedy (Section of Ecology and Systematics, Cornell University), K. Koopman (American Museum of Natural History), and D.K. McClearn (Section of Ecology and Systematics, Cornell University) provided expert assistance in resolving taxonomic synonymies as did colleagues at the 1991 meeting of the American Society of Mammalogists. Notwithstanding these efforts, nomenclatural errors may still exist, especially for species described in older references. Questionable taxonomy is indicated by a question mark after the species name.

Medical journals often do not provide the scientific name of the study subjects. When such publications contained useful data, we made the following nomenclatural assignments: baboon (*Papio hamadryas*), fox (*Vulpes vulpes*), ferret (*Mustela putorious*), mouse (*Mus musculus*), rat (*Rattus norvegicus*), and rabbit (*Oryctolagus cuniculus*).

To include as much information on little-known species as possible, we generally do not list data for domesticated species — dog (*Canis familiaris*), cat (*Felis catus*), horse (*Equus caballus*), pig (*Sus scrofa*), cow (*Bos taurus*), sheep (*Ovis aries*), goat (*Capra hircus*) — and common laboratory animals — rhesus macaque (*Macaca mulatta*), lab mouse (*Mus domesticus*), white rat (*Rattus norvegicus*), Mongolian gerbil (*Meriones unguiculatus*), golden hamster (*Mesocricetus auratus*), Siberian hamster (*Phodopus sungorus*), guinea pig (*Cavia porcellus*), and white rabbit (*Oryctolagus cuniculus*). Data from their wild counterparts are included.

What if you do not know the family of the animal?

An index of mammalian species provides the page number for the appropriate family.

What if you do not know the scientific name of the animal?

An index of the common names of over 300 well-known mammals provides the page number for the appropriate family.

What data are in the familial summaries?
Each familial account includes the following information when available:

- Common name of family
- Alphabetic listing of all the genera in the family with an estimate of their adult mass or head-body length if mass data were not available. For a few genera, neither estimate of adult size was found.
- General habitat, diet, and geographical distribution of family
- Sexual dimorphism of mass (by genus if appropriate and available)
- Presence of specialized reproductive features
    - anatomical or functional asymmetry
    - induced ovulation
    - copulatory plug, copulatory tie
- Placental morphology (layers and shape)
- Implantation (location and depth)
- Uterine shape
- Presence of placentophagia or postpartum estrus
- Developmental state of neonate (by genus if appropriate)
    - furred or naked at birth
    - blind or time to eye opening
- Number of mammae (by genus if appropriate)
- Male reproductive anatomy
- Citations, by topic, of publications which detail reproductive structure and/or function

What do the citations grouped at the end of each familial account cover?
In reviewing the literature we read many papers which detailed the structure or function of reproduction in various species, but which did not have data appropriate for inclusion in the tables. We grouped citations to such papers by topic and listed them at the ends of the familial accounts. Thus these bibliographical listings are not exhaustive for a given topic, but instead cite the literature we encountered in our search for reproductive data. In general, we categorized papers into the following topics: **female anatomy** (histology, fine structure, or development of ovary, ovarian follicles, corpora lutea, uterus, ducts, reproductive glands including mammary glands); **male anatomy** (histology, fine structure, or development of testes, accessory glands, ducts, or secondary sex characteristics); **endocrinology** (reproductive hormone levels, pharmacological experiments, gonadectomy experiments, hypophysectomy experiments, binding proteins, uterine reactions, corpora lutea function); **spermatogenesis** (spermiogenesis, testicular blood flow, sperm morphology, seasonal changes in testicular morphology, physiology, or function); **copulation** (courtship, sexual behavior, effects of hormones on sexual behavior); **milk composition; reproductive energetics** (energetic cost of lactation or gestation, metabolic rates during reproduction); and **light** (effects of photoperiod, melatonin levels, and pinealectomy on reproductive processes). For some families other topics (or subtopics) are included.

What are the sources of the unannotated information?
Unannotated information came from the following sources or from reviews cited in them — common names: Corbet and Hill 1980, Nowak and Paradiso 1983, Nowak 1991; character synopses: Anderson and Jones 1984; natural history: *Grzimek's Encyclopedia of Mammals* 1990, Nowak and Paradiso 1983, Nowak 1991, accounts in Mammalian Species and regional compilations; sex chromosomes: Fredga 1970. The American Society of Mammalogists has synthesized the literature on over 400 mammalian species in separate short pamphlets in their series Mammalian Species. These are listed by taxon in Appendix 1.

What species-specific data tables are provided for each family?
Tables of neonatal mass, neonatal length, weaning mass, weaning length, age at sexual maturity, estrous cycle length, gestation length, lactation length, number of litters per year and/or interbirth/interlitter interval, and seasonality of reproduction are provided for each family if appropriate data were available.

How are the data organized within each table?

For each reproductive category, the data are organized alphabetically by genus, and alphabetically by species within each genus. For each species, independent estimates are listed separately and the citations given. All data are numerically ordered for each species from smallest to largest, but some are grouped into subcategories, as follows:

NEONATAL MASS, WEANING MASS

- Neonate mass refers to mass at birth
- Weaning-mass data are accompanied by an age when reported
- Sex is noted when appropriate

NEONATAL LENGTH, WEANING LENGTH

- Neonate length refers to length at birth
- Weaning-length data are accompanied by an age when reported
- Sex is noted when appropriate
- Crown-rump (CR) or head-body (HB) lengths are indicated if known (unfortunately, many published data do not indicate what was measured: CR, HB, or total length)

LITTER SIZE

- Units (e.g., embryos, corpora lutea, or young) provided when available

AGE AT SEXUAL MATURITY

- Female (f) data first, then male (m) data, then unknown or both (f/m)
- For each sex data are grouped by sequence of occurrence
  - for females:
    - 1st estrus
    - 1st mating
    - 1st conception
    - 1st pregnant
    - 1st birth
  - for males:
    - large testes
    - 1st spermatogenesis
    - 1st sperm
    - 1st mating
    - 1st sire offspring

ESTROUS CYCLE LENGTH

- Portions of the cycle are listed first (i.e. follicular phase, luteal phase, menses, proestrus, estrus, metestrus, diestrus, anestrus) followed by length of entire cycle

GESTATION LENGTH

- Time of implantation first, followed by full gestation length with gestational delays noted

LACTATION LENGTH

- The following measures of lactation length are reported in an approximate sequence of occurrence (unfortunately, many sources did not state the measure used):
  - teat attachment
  - den emergence
  - 1st solid food
  - pouch exit
  - duration of nursing, suckling, lactation
  - weaning

NUMBER OF LITTERS PER YEAR or INTERBIRTH/INTERLITTER INTERVAL

- Interbirth/interlitter intervals with and without lactation are indicated if appropriate

SEASONALITY OF REPRODUCTION

- The months or time of year when specific reproductive stages occur are presented along with where the data were collected
- The following measures of seasonality are used:
  - testes large
  - mating season
  - pregnant females present
  - birth season
  - lactating females present
  - young found
  - reproductive period

- Ambiguous citations to "breeding seasons" were assumed to mean "reproduction occurs" and are listed as "repro" in the tables
- For each species, the above measures are listed by sequence of occurrence
- Within each measure, data are listed chronologically by month or time of year

How is each line of data organized?

For each line of data the mean (numerical average) is presented first, followed by the standard deviation (sd), the standard error (se), the median or mode, the range (r), and the sample size (n). Unspecified estimates of variability are indicated by ±. Confidence intervals are indicated by ci and reported verbatim from the references, i.e. if only one number rather than an interval was provided, only one number is reported here.

English measures were converted to metric and usually rounded in such a way as to retain the accuracy of the original data.

Data from the same reference may be given for more than one entry if they represent different measures, treatments, populations, or subspecies, and if the samples were large for all groups or the data appeared to be, or were, statistically different.

Identical data from different references may be merged to one entry if the data were from the same author, the data were from secondary sources, or the data were from different authors who used each other's work. Similar data from different references may be merged in one entry if they were from the same author and not significantly different. The largest sample size is reported if similar data were merged.

Some data were not used. Data published before World War I which conflicted with those published after World War II were deleted. Data from review articles were not used if they duplicated data from the primary literature. If many more accurate data were available, the following may have been deleted: anecdotal data, data with small sample size (n1-2), data given without methods, data given without estimate of variability, data given without sample size.

Some words of caution

The data are only as good as their sources, and the sources are quite variable. We accurately recorded what has been published, but what has been published may not be accurate. We spent eight years collecting and formatting the data. Resolving all the controversy within them could take another eight years, at least. If a particular value is critically important, please establish its veracity by direct observation.

The adult mass data may be for only the best known species or for only a few specimens and should be considered approximate. Most adult-mass data are from the publications we reviewed, augmented by general sources, such as *Walker's Mammals of the World*, *Grzimek's Encyclopedia of Mammals*, or regional accounts.

Finally, we would like to make a plea to all those readers who will write, edit, or review manuscripts which have data suitable for a compendium such as this. Please make sure that units are included with each measure and that an indication is clearly made as to which estimate of variation was used.

Abbreviations

ca	corpus albicans
cl	corpus luteum
cm	centimeter
CR	crown-rump length
d	day
ea	each
emb	embryo or fetus
est	estimate
f	female
FA	forearm length (Chiroptera)
g	gram
gest	gestating, gestation
h	hour
HB	head-body length
ibi	interbirth or interlitter interval
kg	kilogram
lact	lactating, lactation
lt	litter
m	male or meter
mm	millimeter
mo	month
ppe	postpartum estrus
preg	pregnant
ps	placental scars
repro	reproduction
rpy	removal of pouch young
TL	total length
vag perf	vaginal perforation
wk	week
yg	young
yr	year

<u>statistics</u>

~	approximately
<	less than
≤	equal to or less than
>	greater than
≥	equal to or greater than
±	unknown variance estimate
ci	confidence interval
est	estimate
min	minimum
max	maximum
n	sample size
r	range
sd	standard deviation
se	standard error

<u>seasonality</u>

fall	fall, autumn
repro	reproduction occurs
spr	spring
sum	summer
win	winter
yr rd	year round

<u>months</u>

Jan	January
Feb	February
Mar	March
Apr	April
May	May
June	June
July	July
Aug	August
Sept	September
Oct	October
Nov	November
Dec	December

<u>location</u>

Co	county
E, e	East (eastern)
I	island
N, n	North (northern)
R	river
S, s	South (southern)
W, w	West (western)

<u>Canadian Provinces/Territories</u>

ALB	Alberta
BC	British Columbia
MAN	Manitoba
NB	New Brunswick
NEW	Newfoundland
NS	Nova Scotia
NWT	Northwest Territories
ONT	Ontario
PEI	Prince Edward Island
QUE	Quebec
SAS	Saskatchewan
YUK	Yukon

Australian States/Territories

ACT	Australian Capital Territory
NSW	New South Wales
NT	Northern Territory
QLD	Queensland
SA	South Australia
TAS	Tasmania
VIC	Victoria
WA	Western Australia

Countries

UK	United Kingdom
US	United States of America
USSR	Union of Socialist Soviet Republics (used when we were uncertain as to which new republic was appropriate)

US States

AK	Alaska
AL	Alabama
AR	Arkansas
AZ	Arizona
CA	California
CO	Colorado
CT	Connecticut
DC	District of Columbia
DE	Delaware
FL	Florida
GA	Georgia
HI	Hawaii
IA	Iowa
ID	Idaho
IL	Illinois
IN	Indiana
KS	Kansas
KY	Kentucky
LA	Louisiana
MA	Massachusetts
MD	Maryland
ME	Maine
MI	Michigan
MN	Minnesota
MO	Missouri
MS	Mississippi
MT	Montana
NC	North Carolina
ND	North Dakota
NE	Nebraska
NH	New Hampshire
NJ	New Jersey
NM	New Mexico
NV	Nevada
NY	New York
OH	Ohio
OK	Oklahoma
OR	Oregon
PA	Pennsylvania
RI	Rhode Island
SC	South Carolina
SD	South Dakota
TN	Tennessee
TX	Texas
UT	Utah
VA	Virginia
VT	Vermont
WA	Washington
WI	Wisconsin
WV	West Virginia
WY	Wyoming

Class Mammalia

Lactation is the quintessence of mammals and the production of milk from mammary glands distinguishes mammals from all other animals. The three major taxa of mammals, Monotremata, Eutheria, and Marsupialia, are also characterized by their reproductive biology. The monotremes (Prototheria or Monotremata) are the only mammals which lay and incubate eggs, while both the Eutheria and the Marsupialia are viviparous. After very short gestation lengths, marsupials give birth to tiny offspring which depend on milk for most of their growth and development. Consequently, lactation in marsupials is unusually complex and its physiological control is highly evolved. In contrast, eutherians use gestation to provide a large portion of their maternal investment and the placental structures of eutherians are highly derived.

The length of lactation varies almost three orders of magnitude from 4 d (*Cystophora*, hooded seal) to over 900 d (Pongidae). In utero development (gestation or pregnancy) is less variable and ranges from 10-14 d in several marsupials to over 650 d in elephants. Most mammals give birth to between 1 and 15 young at intervals ranging from 3-4 wk for many rodents to 3-4+ years for sirenians, elephants, and rhinos among others. The highly specialized monotremes lay 1-3 eggs every 1-2 years.

Mammalian blastocysts are bilaminar with a trophoblast. Embryo formation is either from the outer layers of the blastocyst (monotremes and marsupials) or from the inner cell mass (eutherians). Embryonic Wolffian ducts differentiate into the male reproductive system, while Müllerian ducts form the female reproductive tract and are medial to the ureters in eutherians. Both sexes have paired gonads which secrete steroid hormones and produce gametes. The uterine lining (endometrium) and the placenta have endocrine as well as nutritive functions. Reproductive products may exit the body directly or in combination with renal metabolites via a urogenital sinus or with both renal and digestive material via a cloaca.

Females have 1 (probably 2, *Distoechurus*) to 29 (*Tenrec*) mammary glands as well as a pair of oviducts and uteri. Uteri may be fused along some (bipartite) or all (simplex) of their length. The utero-tubal junction separates the oviducts from the uteri, while the cervix separates the uterus from the terminal portion of the female tract (the vagina). Female monotremes and eutherians have a single vagina, whereas the structure is paired in the Marsupialia.

Male epididymides and deferent ducts are paired, but fuse before they pass through the penis which may be adorned with diverse protruberances, split along some or most of its length, or supported with a bone (os penis or baculum). The number and type of accessory glands associated with the male reproductive tract vary, although most males have at least a pair of bulbourethral glands and a single prostatic gland (only additional glands are noted in the familial accounts). Males are heterogametic structure. Sex chromosomes and sex differentiation vary (3492, 4013), although the female sex chromosome is highly conserved (10261a, 11474a, 11474b, 11474c). Male monotremes, eutherians, and didelphid marsupials have mammary glands (9045a, 9190a).

Several recent synthetic reviews of mammalian reproductive biology exist (464, 1317, 5891, 6165, 11190a). Comparative reviews exist for a few topics: implantation (505, 3010, 9461, 11803), sperm (710, 2169), environmental control (1316), fetal growth rates (3490), gestation (5726a), corpus luteum regulation (9312), birth (7775, 10804), litter-size variability (7698, 12062), origin of lactation (971), lactation (4523a, 5291, volumes 41 and 51 Symp Zool Soc London), and milk composition (5294, 5295).

Order Monotremata

Oviparity (egg-laying) is the distinctive reproductive feature of the highly specialized monotremes. The yolk-filled ova, already large at fertilization (4 mm in diameter), acquire a permeable shell membrane and shell through which absorption of nutritive, uterine fluids occurs (5078). Cleavage is meroblastic initially (1550). The 15x17 mm, ovoid eggs are laid when the embryos have 19-20 somites (3358, 3360, 3361, 4178, 4738, 4768, 5079, 11761, 11901). Embryogenesis occurs mainly during incubation. Although gestation and lactation lengths appear to be longer in the larger echidna than in the platypus, incubation lengths are similar, as is the time to sexual maturity.

Hatchlings are tiny, naked, and blind with a specialized, deciduous, egg-tooth bone and well-developed olfaction (1055, 4069). They suckle from mammary areolae which are similar to those of humans (4064). Echidna females possess a pouch, while platypus females do not. Lactation is long and milk composition is variable, but usually high in total solids, lipids, iron, and copper (4065, 4067, 4069, 5288, 7247a, 8245a, 10724).

The name of the order refers to the presence of a cloaca which communicates via a single vagina with a pair of uteri in females. Males have abdominal testes and epididymides; two types of accessory glands; a bifid or lobate penis within the cloaca; a subdivided urethra with multiple exits; and filiform spermatozoa (709, 1621, 2641, 4064). Urine does not pass through the penis (5431a). A spur is often present on the hind ankle of adult males and may be present on juveniles of both sexes (4064, 10474, 11152). The presence of a Y-chromosome is controversial (4013). Monotremes do not have antral follicles, bacula, scrota, or seminal vesicles.

Reviews of the general biology of monotremes (449a, 449b, 4064, 4064a) and captive breeding (1987, 3721) are available.

Order Monotremata, Family Tachyglossidae

Echidnas (*Tachyglossus* 2.5-6 kg; *Zaglossus* 5-10 kg) are specialists on ants and termites or earthworms in Australia and New Guinea. Both uteri and ovaries are functional (3358, 3360, 4768, 5078, 8164). Penial spines are not present (1031; cf 1621). Females have one more chromosome than males (4013). Details of female anatomy (4065, 4069, 8162, 8164), male anatomy (709, 2638, 2639, 2640, 4945, 5431a, 11101), endocrinology (9789a), sperm maturation (2641, 5431a), mating (4512b), lactation (4028, 4063, 4068), and milk (5445a, 5505, 6865, 7243a, 10725) are available.

HATCHLING MASS[1]		
Tachyglossus aculeatus	378 mg	4069
	380 mg	321
HATCHLING SIZE		
Tachyglossus aculeatus	12 mm	10698
	12.5 mm	6664
	14.7 mm	4069
Zaglossus bruijni	17 mm	11151
WEANING MASS		
Tachyglossus aculeatus	400 g; n1; pouch exit	6063
Zaglossus bruijni	~400 g; pouch exit	11151
WEANING SIZE		
Zaglossus bruijni	305 mm; pouch exit	11151
LITTER SIZE		
Tachyglossus aculeatus	1 egg	11901
	1 egg	9763
	1.03; r1-2; n140 lt	3358, 3360
	2 eggs; n1 f	1338
	2 eggs; r1-3; n2 f	8164
Zaglossus bruijni	mode 1 egg; r1-3	11151

[1] The left column lists, in CAPITAL letters, the reproductive variables and, in *italics*, the scientific names of the species. The middle column provides the data, and the right column gives the citations to the publications in which the data were found.

SEXUAL MATURITY		
Tachyglossus aculeatus	~1 yr	8798
	est 2 yr	777
CYCLE LENGTH		
Tachyglossus aculeatus	est 9-27 d	10698
	>16 d	4069
	est 18-28 d; n1	1338
	1 mo	8798
GESTATION		
Tachyglossus aculeatus	min 17 d; removal of male to birth	4064, 4064a
	est 18-27 d	9142a
	27 d	1338
INCUBATION		
Tachyglossus aculeatus	10 d	9142a
	10-10.5 d	4069
Zaglossus bruijni	~10 d	11151
LACTATION		
Tachyglossus aculeatus	pouch exit: ~55 d	4064
	pouch exit: 2 mo	10698
	pouch exit: 63 d	4027
	pouch exit: 3 mo	10474
	wean: 3+ mo after pouch exit	4064
	independent: 7-9 mo	10698
	wean: 195-200 d; n1	17
	wean: ~200 d	4068
Zaglossus bruijni	pouch exit: 8 wk	11151
	maternal care: 3 mo	11151
	wean: 5 mo	10809
INTERLITTER INTERVAL		
Tachyglossus aculeatus	est 2 yr	777
SEASONALITY		
Tachyglossus aculeatus	estrus: end July; Australia	9763
	mate: June-early Sept (last egg); TAS, Australia	3360
	mate: July-Aug; Australia	10474
	mate: July-Aug; Snowy Mt, NSW, Australia	680a
	mate: est early Sept; n1; Australia	1338
	mate: mid Sept; SA, Australia	11901
	mate: start rainy season; n Australia	2214
	eggs in utero: end July; se Australia	1550
	eggs in utero: Sept-Oct; QLD, Australia	777
	eggs laid: early Oct; n1; Australia	1338
	pouch with egg: end Aug; Australia	9763

Order Monotremata, Family Ornithorhynchidae, Genus *Ornithorhynchus*

The platypus (*Ornithorhynchus anatinus* 0.5-2 kg) is a highly specialized, aquatic, nocturnal predator of invertebrates in streams of temperate Australia. Males may be larger than females (8798). Penial spines are present (1031). The left ovary and uterus are prominent and preferentially used (775, 3360, 3607, 4768, 5078, 8157, 8158, 8159, 9763). Eyes open at 11+ wk (3302). Details of anatomy (3615, 3627, 4767, 4778, 4944, 8157, 11101), endocrinology (1618, 1619, 1620, 1621, 1767, 4315a, 7049, 7059a), courtship (10474a), milk composition (222, 3764, 7247), and reproductive energetics (4028a) are available.

HATCHLING SIZE		
Ornithorhynchus anatinus	25 mm	10698
	~25 mm	8798
LITTER SIZE		
Ornithorhynchus anatinus	1.83 eggs in utero; r1-2; n6 f	4737
	2 oocytes	4768
	2 eggs	11901
	2 eggs; r1-3	8798
	mode 2; r1-4	775
	2 eggs in utero; n1 f	1550

	2 eggs in utero; n1 f	4772
	2 eggs in nest?	1467
	2 yg ea; n2 nest	776
	2.0; n2 lt	11761
	2 cl ea; n2 f	8158
	2.5; r2-3; n2	775
	3 cl; n1 ovary	8159
	3 emb; n1 lt	8159
SEXUAL MATURITY		
Ornithorhynchus anatinus	f: smallest lact f: 670-750 g; n4 areas	4010
	f: min 2nd season	4009a
	f/m: 1st mate: 1-2 yr	4010
	f/m: 1 yr	8798
	f/m: 2 yr	9763
GESTATION		
Ornithorhynchus anatinus	>10 d	5078
	12-14 d	10698
	est 14 d	3302, 3303
	1 mo	8798
INCUBATION		
Ornithorhynchus anatinus	incubate: est 7-10 d	3302
	incubate: 1-2 wk	10474
	incubate: 12 d	3303
LACTATION		
Ornithorhynchus anatinus	wean: 3-4 mo; field data	4009, 4009a, 4010
	den emergence: 4 mo	8798
	nurse: 4-5 mo	10474
INTERLITTER INTERVAL		
Ornithorhynchus anatinus	1 yr	4010
SEASONALITY		
Ornithorhynchus anatinus	quiescent: May-July; NSW, Australia	4010
	estrus: mid Aug; Australia	9763
	mate: Aug; QLD, Australia	10474
	mate: Aug-Sept; Australia	1467
	mate: Aug-Sept; NSW, Australia	3607
	mate: Sept; NSW, VIC, Australia	10474
	mate: Oct; TAS, Australia	10474
	mate: Oct; 1 pr; captive	3303
	f with ova: Oct; n3; Australia	8158
	uterine eggs: end Aug; Australia	9763
	uterine eggs: Aug-Sept; n6; Australia	4737
	uterine eggs: mid Aug; n1; NSW, Australia	1550
	uterine eggs: Oct; NSW, Australia	775
	uterine eggs: Oct; n1; Australia	4772
	lay eggs: mid Aug-Oct; Australia	1467
	lay eggs: end July-early Sept; TAS, e Australia	8798
	hatch: Oct; VIC, Australia	5701
	lact f: Oct-Mar, peak Dec; NSW, Australia	4010
	lact f: Dec; Australia	8162
	young found: Oct-Nov; n2 nests; Australia	776

Order Marsupialia

The taxonomically and ecologically diverse marsupials give birth to extremely altricial neonates with perinatal specializations of the pectoral girdle and digits (5863). Lactation is long and biphasic while gestation is short. Neontates are an exceptionally small proportion of maternal mass and embryogenesis is completed during the first phase of lactation, when neonates are continuously attached to a teat. Marsupial development at teat detachment (first pouch exit for those marsupials with pouches) is often comparable to that of eutherian neonates at birth. A pouch encloses the mammary teats in many, but not all, families. Pouches are not an ancestral trait for marsupials and probably have evolved independently several times. Females have a duplex uterus with two cervices and a median birth canal between the lateral vaginae. The birth canal in some families is reformed with each pregnancy. Only peramelids use chorioallantoic placentation, all other marsupials use choriovitelline placentation only. Testicular vasculature includes a rete mirabile (4136). The prostate is disseminate. The penis is often bifid except in Notoryctidae and Tarsipidae and lacks a baculum. When present, the scrotum is prepenial. The spermatazoan acrosome is exceptionally stable (6941).

Unlike eutherians, X-chromosome inactivation in marsupials is incomplete and non-random (the paternal X is inactivated; 4013). Scrotal development occurs when a single X-chromosome is present (even in the absence of a Y-chromosome), while pouch development proceeds with a pair of X-chromosomes (even in the presence of a Y-chromosome; 4013).

Reviews exist on the following topics: reproductive physiology (9852, 11008, 10998a), captive breeding (1987, 3721, 9205), anatomy (8292, 11098, 11100), sperm (2169), sperm pairing (934), sex differentiation (9041a), placentation (5075, 9038), parental investment (9378, 9379), and milk composition (4026). More general reviews exist for several taxonomic or ecological groups: Australian mammals (10474); carnivorous marsupials (356); possums and opossums (357); bandicoots and bilbies (9719); possums and gliders (10088); kangaroos, wallabies, and rat-kangaroos (4071).

Order Marsupialia, Family Didelphidae

New World opossums (*Caluromys* 130-500 g; *Caluromysiops* 250-300 mm; *Chironectes* 600-800 g; *Didelphis* 1-5.5 kg; *Glironia* 160-205 mm; *Lestodelphys* 132-144 mm; *Lutreolina* 200-540 g; *Marmosa* 10-150 g; *Metachirus* 400-800 g; *Monodelphis* 20-155 g; *Philander* 222-1500 g) are nocturnal, terrestrial to semiarboreal (*Chironectes* is semiaquatic) omnivores of forested areas primarily in Central and South America (*Didelphis* is found in North America). Data on *Caluromysiops*, *Glironia*, and *Lestodelphys* are limited.

Didelphid males are often larger than females (439, 3024, 4976, 8555, 11007; cf 5210). Copulation can last over 2 h (*Marmosa* 604, 606, 2956). As large numbers of eggs may be ovulated (3123, 4443, 8850), litter size is limited by the number of mammae: 4 (*Glironia* 6859), 4-5 (*Chironectes* 6860), 4-13 (*Marmosa* 2956, 3024, 3840, 6022), 5-9 (*Philander* 7983), 5-11 (*Metachirus* 3024, 6021), 6-13 (*Monodelphis* 8552, 8555), 7-17 (*Didelphis* 630, 3024, 4284, 4437, 4438, 5146, 6022, 7097, 7497, 9074, 9745), and 9 mammae (*Lutreolina* 6857).

Neonates open their eyes at 2-5 wk (*Monodelphis* 6009, 6744), 35-40 d (*Marmosa* 608, 2937, 2956), or 50-72 d (*Didelphis* 4444, 9074). *Caluromys*, *Caluromysiops*, *Didelphis*, and *Philander* females have pouches, while *Glironia*, *Lutreolina*, *Marmosa*, *Metachirus*, and *Monodelphis* do not. Both male and female *Chironectes* have a posteriorly opening pouch (3024, 9292). Most didelphids are polyestrous, although *Didelphis albiventris* may be monoestrous (10494) and *Monodelphis dimidiata* may be semelparous (8555). *Marmosa* (3840) and *Monodelphis* (3139) have bimodally distributed estrous cycle lengths. The shorter cycles (about equal to the gestation length) are sterile. At 2 wk the length of gestation is about half the usual estrous cycle length, both of which are relatively constant across the family. Lactation is much longer (1.5-3 mo) and more variable. Earliest sexual maturity is at about twice the lactation length (3-6 mo). Polyovular follicles are common for *Didelphis* (4443).

Didelphid sperm become paired in the epididymides (934a). The bifid penis is posterior to the scrota in *Philander* and *Caluromys* (1423, 9088a).

Details of placentation (6009b, 6010, 6011), shell (6009a), anatomy (484, 490, 604, 3315, 3352, 3845, 4439, 4766, 5201, 6154, 6916, 7830, 7948, 7948a, 8484, 9204, 9746, 11012), endocrinology (2032, 3108a, 3138, 3722, 4363, 4364, 4365, 6155, 6916, 8083a, 8380, 11309, 12093), sex differentiation (1463a, 1463b, 1463c, 1463e, 3141, 4436, 7491c, 7491d, 7507b, 9045a, 9355a, 9355b), ovulation/conception (606, 710, 3136, 3137, 3139, 3840, 4439, 6916, 9206, 9207), spermatogenesis (710, 934, 6694, 8129), and milk composition (807a, 2143) are available.

NEONATAL MASS		
Caluromys philander	200 mg	438, 439
Didelphis marsupialis	~60 mg	630
	150 mg	4452
	160 mg	10994
	200 mg	439
	~390 mg	5146
Didelphis virginiana	130 mg	4444
	150 mg; r140-190; n4	7492
	160 mg	4284
Marmosa murina	90 mg	2956
Marmosa robinsoni	~100 mg	608
Monodelphis dimidiata	70-80 mg	2222
	r80-110 mg; n3	8555
Monodelphis domestica	100 mg	3142, 11215
Philander opossum	200 mg	439
NEONATAL SIZE		
Caluromys philander	10 mm	438, 439
Didelphis marsupialis	10 mm	439
	~10 mm	5146
Didelphis virginiana	HB: 13.8 mm; r13.0-15.4; n28 yg, 9 lt, 1 d	9074
	CR: 14 mm	4284
	14 mm	7492
Marmosa cinerea	10 mm	439
Marmosa murina	10 mm	439
Marmosa robinsoni	6-14 mm	451
	8-12 mm	1987
Monodelphis dimidiata	CR: ~10 mm	8555
Monodelphis domestica	10 mm	11215
Philander opossum	10 mm	439
WEANING MASS		
Caluromys philander	11 g; 75-80 d pouch exit	438
Didelphis spp	75-105 g	11007
Didelphis marsupialis	25 g; controls; 39 g; food supplemented	461
Didelphis virginiana	120-150 g	1464
	150 g	5473
Marmosa robinsoni	10-12 g	608
Monodelphis domestica	3-5 g; 3 wk	6744
	r16.53-24.05 g; n4 groups, n35 lt, 56 d	2063
Philander opossum	100-120 g	1987
WEANING SIZE		
Caluromys philander	75 mm; 75-80 d pouch exit	438
Didelphis virginiana	HB: 195.2 mm; r171.6-217.7; n41 yg, 10 lt, 95 d	9074
LITTER SIZE		
Caluromys spp	mode 3-4; r1-6	3027
Caluromys derbianus	2 emb; n1 f	931
	2.75 yg; r2-3; n4 f	931
	3 yg; n1 lt	4228
	3.3; r2-4; n13 lt	8482
	4 yg; n1 f	5423c
Caluromys philander	4 yg; n1 lt	4766
	4.1 yg; r2-7; n42 lt	439
	4.2 yg; r2-7; n28	438
	6 yg ea; n3 lt	7994
Caluromysiops irrupta	2 yg; n1 f; via photo	5201

Chironectes minimus	2-3	3027
	4.67; r4-5; n3 lt	9292
	max 5	7465
Didelphis albiventris	4.2; sd1.40; mode 4-5; r2-7; n10 lt	11007
	6.2; sd2.2; r3-9; n12 lt	10494
	6.5 emb; ±2.4; r3-10; n35 f	1675
Didelphis aurita	19-31 oocytes	4756
Didelphis marsupialis	4-6 yg	5111
	4.49; sd1.36; mode 5; r1-7; n41 lt	11007
	4.5; r2-7; n42 lt	439
	4.67 yg; r2-7; n3 f	3022, 3024
	5-10 yg	5146
	5 yg; n1 f	11751
	5.9; r2-10; n15	7994, 7998
	6.0 yg; r2-9; n29	3320
	6.54; sd2.19; mode 7; r1-11; n37 lt	11007
	7 yg; n1 lt	9029
	7.5; r3-10	7994, 7995, 7998
	7.5; r6-9; n2 lt	12128
	7.52; se0.36; n18 f	461
	7.61; se0.34; n20 f	461
	8 yg; n1 lt, <7 wk	2064
	mode 8-9; r7-10	6021
	9.78 yg; r5-17; n9 f	7624
	10 yg; n1 f	6084a
	11.7; r11-12; n3	7576
Didelphis virginiana	6 yg; n1 f	5423c
	6.26; mode 7; r3-11; n50 lt	1464, 1465
	6.7; r1-14	7492
	6.8 yg; n65 lt	6295
	6.8 yg; r1-10; n68 lt	2899
	7 yg ea; n2 f	1290
	7.1 yg; se0.18; r1-11; n143 f	7077
	7.2 yg; mode 7; r4-11; n44 lt	9074
	7.33 yg; r6-9; n3 f	10471
	7.4 yg; mode 7; r1-12; n28 lt	3286
	7.74 yg; n57 lt; 8.5 yg ≤40 mm; 6.6 yg >40 mm	6514
	7.9 yg; n85 lt	4930
	7.91; min 1; n58 lt	9465
	8 yg; n1 f	2191
	8.0 yg; r6-9; n4 lt	3212
	8.2 yg; r5-11; n5 lt; not at birth	4976
	8.6; n23 lt	9074
	8.65; ±1.83; mode 8-10; r6-12; n31 lt	2190
	8.66; mode 9; r2-15; n346 lt	4284
	8.8 yg; r3-13; n10 lt	6326
	8.9 yg; mode 8; r5-13; n42 lt	9073
	9 yg; r6-12; n7 lt	11834
	9.67 yg; r7-13; n3 f	8907
	mode 12-16 emb; r7-27	9745
	13 yg; n1 f	970
	14 yg; n1 f	1290
	19.5 yg; r18-21; n2 f; max 13-14 attach	4438, 4446
	22 ova; n1	10283
	22 ova; max 43	4438
	23 emb; n1 f	4446
	max 25 emb	4284
	29.6 oocytes; r19-40	9206
	r30-56 ova, blastocysts; n11 f	4446
	max 50 ovulations	1464
	max 70 ovulations	4444
Lutreolina crassicaudata	7 yg; n1 lt	6372
	mode 8-9; r7-10	6021

Marmosa spp	2; n1 lt	3021
	6 yg; n1 f	5063
	9; n1 lt	8692
	9.0; r8-10; n2 lt	2937
Marmosa canescens	8 yg ea; n2 f	5410
	13 emb; n1 f	5410
Marmosa cinerea	2 yg; n1	10683
	5 yg; n1 f	676
	6.2; max 11; n4 lt	439
Marmosa constantiae	8 yg; r7-9; n2	10683
Marmosa elegans	17 emb; n1 lt	6022
Marmosa fuscata	6	7995, 7998
Marmosa lepida	5 yg; n1 f	10683
Marmosa marica	6 yg	10683
Marmosa mexicana	10 yg; n1 f	1475
	11 yg; n1 lt	3750
	13 yg; n1 f	3212
Marmosa murina	5.5 yg; r4-7; n2 f	5111
	5-6 yg	5111
	6 yg	5111
	8.4; max 11; n33 lt	439
	12; r11-13; n3 lt	2956
Marmosa parvidens	6.5 emb; r6-7; n2 f	8553
Marmosa robinsoni	4.8-7.07 yg; n3 samples	604
	6.0; n70 lt	606
	6.91 yg; r1-12; n11 lt	10813
	7.8; sd3.1; n18 lt parental generation; 8.2; sd3.6; n15 lt F_1; 9.1; sd2.7; n16 lt F_2; r1-13	3840
	9.0; r8-10; n2 f	2937
	10.0 yg; r6-13; n16	3320
	12 yg; n1 lt	605
	12.25 yg; r3-16; n4 f; not birth	451
	14.0; r13-15; n13	7995, 7998
	19.8 cl; ±0.9; mode 21; n23	3840
Marmosa tyleriana	3 enlarged teats; n1	10683
Metachirus nudicaudatus	2 yg; r1-3; n2 lt	3024
	max 9	1987
Monodelphis brevicaudata	6 yg	5111
	7.5; r7-8; n5 lt	7995, 7998
Monodelphis dimidiata	12; r8-16; n2 lt	2222, 8555
Monodelphis domestica	1-11	10493
	3.79-5.55 wean	2063
	5-13; 13 rare	11215
	6 ovulations/ovary	490
	6.00-7.45 yg	2063
	7 yg; r3-14	3142
	8	10797
	8.4; sd2.6; r6-11	10494
	8.6 yg; sd2.6; r2-13	3136, 6009
Philander spp	1	9746
Philander opossum	3.3; n11, 3 lt	3024
	3.43 yg; r1-7; n7 f	5111
	3.67 yg; r2-5; n3 lt	3024
	4.2; r2-7; n179 lt	439
	4.6 yg; r2-7; n34	3320
	4.75 yg; r4-5; n4 lt	3212
	5.0 yg; r4-6; n2 lt	4228
	5.75 yg; r4-7; n4 f	7624
	6 yg; n1 f	5423c
	6.0 yg; r5-7; n2 f	2341
	6.05; mode 7; r3-7; n21 lt	8483
	7 yg; n1 lt	4766
	8 emb; n1 lt	3016

SEXUAL MATURITY

Caluromys derbianus	f/m: 7-9 mo	1987
Caluromys philander	f: 1st lact: 10 mo	439
Chironectes minimus	f: 1st estrus: ~10 mo	8093
Didelphis spp	f/m: 7 mo	11007
Didelphis albiventris	f/m: ~1 yr	10494
Didelphis marsupialis	f: 1st mate: 6 mo	5146
	f: 1st mate: 800 g	4452
	f: 1st lact: 6 mo	439
	m: 1st mate: 8 mo	5146
	f/m: 6 mo; n26	7998
Didelphis virginiana	f: 1st birth: 186 d	9074
	f: 6 mo	9074
	m: testes descend: 77 d	3256
	m: glands mature: 100 d	3256
	m: 8 mo	9074
Marmosa fuscata	f/m: 6 mo; n10	7998
Marmosa robinsoni	f: 1st estrus: 40 wk; 40 g; n200	608
	f: 1st mate: 278 d	604
	m: 1st mate: 1 yr	608
	f/m: 6 mo; n13	7998
Monodelphis brevicaudata	f/m: 6 mo; n5	7998
Monodelphis domestica	f/m: 3-5 mo	10921
	f/m: 3.5 mo	6744
	f/m: 4-5 mo	11215
	f/m: 5-7 mo	10494
Philander opossum	f: 1st estrus: 15 mo	1987
	f: 1st lact: 8 mo	439

CYCLE LENGTH

Caluromys derbianus	mode 27-29 d; r16-39; n1 f	1422
Caluromys lanatus	mode 27-29 d; r20-31; n3 f	1422
Caluromys philander	38.6 d; se1.4; r28-45; n14 cycles, 4 f	8380, 8380a
Chironectes minimus	estrus: max 24 h	8093
Didelphis marsupialis	7 d	6915
Didelphis virginiana	estrus: 3-5 h	9745
	estrus: 36 h	9074
	23-28 d	3174
	25.5 d; r17-38	5473
	28 d; r22-45	4439
	4-6 wk	9745
	29.2 d; se1.0; r22-42; n24 cycles, 18 f	3315
	29.5 d; r22-38; n8-10 f; increases throughout season	9074
	~1 mo	4438
Marmosa robinsoni	estrus: ~3 d	3840
	15.6 d; se0.7; mode 15; r13-19; n10 cycles, 36 f	3840
	23 d; r18-31	604
	25.5 d; se0.5; mode 24; r23-30; n23 cycles, 21 f	3840
	27 d	605
	mode 28 d; n500 observations	608
Monodelphis domestica	estrus: 5.9 d; sd2.11; r3-12	3139
	estrus: r4-8 d; n6 f; 12 d; n1 f	3136
	estrus occurred: 8.5 d after m introduced; 95%ci 7.56-9.2; r8.5-28; n80 f;	490
	14.4; sd2.8; r11-17; n5	3139
	28 d	11215
	~1 mo	10921
	~30 d	3140
	32.3; sd3.4; r28-39; n10	3139

GESTATION		
Caluromys philander	24 d; se0.9; n13	8380a
Didelphis spp	13 d	2937
Didelphis albiventris	12 d; n1	11007
Didelphis marsupialis	12-13 d	5146
	13 d	6916
	2 wk	6021
Didelphis virginiana	11 d	4438
	12.6 d; n1	4439
	12 d, 18+ h	1465
	12.8 d	9745
	12-13 d	7492
	12-13 d	4439, 4444
	13 d	4284
	13 d	4452
	13 d; n1	3594
	13 d; r12.85-13.19; n18 f	9074
Lutreolina crassicaudata	2 wk	6021
Marmosa spp	14 d	2937
Marmosa murina	≤13 d; last paired to birth	2956
Marmosa robinsoni	min 13 d; pairing to birth	10813
	min 14 d; pairing to birth	605
	13.5-14.5 d	3840
	14 d	2937
	14 d	608
	~15 d	604
Monodelphis domestica	14 d	10797
	14-15 d	3136, 3140, 3142, 11215
	15 d	10921
LACTATION		
Caluromys philander	pouch life: 75-80 d	438, 439
	wean: ~100 d (4-6 wk after teat detachment)	438
Chironectes minimus	teat attachment: 48 d	9292
Didelphis spp	wean: ~100 d	11007
Didelphis marsupialis	pouch life: 80 d	439
	wean: ~60 d	461
	wean: 90-100 d	5146
Didelphis virginiana	teat attachment: 50-65 d	7097
	solid food: 75 d	11834
	solid food: 85 d	2190
	solid food: 92 d	9074
	pouch exit: 80-87 d	9074
	pouch exit: 90-100 d	5473
	wean: 80-90 d	4444
	wean: 96-104 d	9074
	wean: 4 mo	1464
	forage: 90 d	11834
Marmosa spp	teat attachment: 20 d	2937
Marmosa murina	wean: 62 d	2956
Marmosa robinsoni	teat attachment: 20 d	2937
	teat attachment: 21 d	608
	solid food: 50-58 d	1987
	wean: 60-70 d	604, 608
	wean: 65-75 d	10814
	wean: 70 d	605
Monodelphis dimidiata	teat attachment: <2 wk	8555
Monodelphis domestica	teat attachment: <14 d	4522
	teat attachment: 14-17 d	6009
	teat attachment: 3 wk	6744
	den emergence: 4 wk	6744
	solid food: 4-5 wk	3142
	nurse: 6-8 wk	10494
	wean: 49-56 d	6009
	wean: 50 d	11215

	independent: 7 wk	3142
	wean: 56 d	10797
Philander opossum	pouch life: 80 d	439
INTERLITTER INTERVAL		
Didelphis virginiana	4 mo	9074
Monodelphis domestica	min 7-8 wk	10494
LITTERS/YEAR		
Caluromys philander	2	439
Didelphis albiventris	1	10493, 10494
Didelphis marsupialis	1.4	7998
	2	5146
	2 4	39
	est 2	3320
	est 2+	630
Didelphis virginiana	1	9745
	1-2	4930
	1-2	9074
	1-3	7492
	2	6295
	2	4284
	2	3286
	2	5473
	2; r1-3?; no good evidence for 3	4439, 4444
Marmosa fuscata	1.6	7998
Marmosa robinsoni	3.6	7998
Monodelphis brevicaudata	1.5	7998
Monodelphis dimidiata	1	8555
Monodelphis domestica	max 4	11215
	max est 5-6	10494
Philander opossum	3	439
SEASONALITY		
Caluromys derbianus	repro: est yr rd; Nicaragua	931, 8482
Caluromys philander	preg f: Apr-June, Oct-Dec; French Guyana	438
	lact f: low July-Sept; French Guyana	8380
Chironectes minimus	birth: Dec-Jan; Brazil	1519
Didelphis albiventris	m fertile: yr rd; ne Brazil	1675
	mate: Mar-Sept; Colombia	11007
	birth: Nov-Dec; ne Brazil	10494
	birth: peak Nov-Dec; ne Brazil	6808
Didelphis aurita	preg/lact f: June-July, Oct; Brazil	4756
Didelphis marsupialis	sperm: yr rd; Panama	3320
	sperm: yr rd; Nicaragua	930
	anestrus: Oct-Dec; Panama	3320
	mate: Jan-Aug; e Colombia	11007
	mate: Apr-Sept; not sampled later; w Colombia	11007
	mate: Jan-Sept; Panama	3320
	mate: Oct-Nov, again when yg independent; Argentina	6021
	mate: peak end Dec; Marago I, Brazil	2064
	conceive: peaks Jan-early June, Aug; Nicaragua	930
	lact f: Mar, May, Aug; n3; Barro Colorado I, Panama	3024
	lact f: Mar-Oct, n Venezuela	7994
Didelphis virginiana	sperm: yr rd; TX	4444
	anestrus: Oct-Dec, lab, PA	3174
	uteri large, cl: Feb-July; DC	7508
	estrus: mid Jan-mid Mar; FL	1465
	estrus: end Feb-mid Apr; captive	9745
	mate: Jan-?; lab, PA	3174
	mate: peak mid Jan-Apr, none Nov-Dec; TX	4439, 4444
	mate: peak Feb-Mar; NE	5405b$_1$
	ovulate: Dec-July; captive	5473
	conceive: mid Jan-June; central CA	9074
	conceive: Oct-July, peaks Jan-Feb, mid Apr-early May, Oct; OR	4976

	birth: Jan; FL, LA, AK	7492
	birth: Feb; MI	7492
	birth: end Feb-mid June; IL	7492
	birth: Jan-July; captive	5473
	birth: end Jan-Mar, peak Feb 31/33; FL	1464
	birth: end Jan-mid Mar, peak end Jan-mid Feb; n47 lt; FL	1465
	birth: Feb-June; MO	9073
	birth: end Feb-early July; MD	6514
	birth: Mar-July; KS	3286
	lact f: Jan-Sept, none Oct-Dec; TX	4438, 4444
	lact f: end Jan-mid Oct; central CA	9074
	lact f: Feb-?; IA	11834
	lact f: Feb-Mar, Aug; n3; NC	1290
	lact f: mid Feb-July; GA, FL	7077
	lact f: Mar-June; El Salvador	3212
	lact f: Aug; n1; Cozumel I, Qunitana Roo, Mexico	5423c
Lutreolina crassicaudata	mate: Oct-Nov, again when yg independent; Argentina	6021
Marmosa alstoni	preg/lact f: Aug, Oct; n3; Mexico	10683
Marmosa cinerea	preg/lact f: Jan-Mar, Sept-Nov; n10; n S America	10683
Marmosa constantiae	preg/lact f: Jan; n2; Bolivia	10683
Marmosa elegans	preg/lact f: Feb, Sept-Oct; n3; Chile, Bolivia	10683
Marmosa fuscata	lact f: May-Nov; few data; n Venezuela	7994
Marmosa germana	preg/lact f: Jan, Mar-Apr, Sept-Oct; n9; Ecuador, Peru	10683
Marmosa grisea	preg/lact f: Mar; n1; Bolivia	10683
Marmosa juninensis	preg/lact f: Nov; n1; Peru	10683
Marmosa lepida	preg/lact f: Jan, Mar; n2; Ecuador, Bolivia	10683
Marmosa marica	preg/lact f: Feb, July; n2; Venezuela	10683
Marmosa mexicana	lact f: June; El Salvador	3212
Marmosa murina	preg/lact f: Feb-June, Aug, Oct; n10; n S America	10683
Marmosa noctivaga	preg/lact f: Jan, Apr, Oct; n4; nw S America	10683
Marmosa parvidens	preg f: June-July; n2; Brazil	8553
Marmosa phaea	preg/lact f: May; n1; Colombia, Ecuador	10683
Marmosa quichua	preg/lact f: July, Oct; n3; Peru	10683
Marmosa rapposa	preg/lact f: Nov-Dec; n3; Peru	10683
Marmosa robinsoni	sperm: yr rd; Panama	3320
	anestrus: Oct-Feb; Panama	3320
	mate: Mar-Sept; Panama	3320
	preg f: none June, Aug, Nov-Apr; n40 f; Barro Colorado I, Panama	3024
	preg/lact f: Apr-June; Venezuela	451
	preg/lact f: May-Sept, Dec; n28; s Central America, n S America	10683
	lact f: May-Nov; few data; n Venezuela	7994
Marmosa rubra	preg/lact f: Mar-Apr, Dec; n4; Ecuador	10683
Marmosa tyleriana	preg/lact f: Feb; n1; s Venezuela	10683
Monodelphis brevicaudata	lact f: May-Aug, Nov; few data; n Venezuela	7994
Monodelphis dimidiata	birth: Dec-Jan; Argentina	8555
Monodelphis domestica	mate: yr rd; lab	10921
	mate: yr rd; lab	6744
	birth: yr rd; lab	11215
	lact f: min 8 mo of yr; ne Brazil	10493, 10494
	repro: aseasonal; ne Brazil	6808
Philander opossum	sperm: est yr rd; Panama	3320
	anestrus: Oct-Dec; Panama	3320
	mate: Jan-Sept; Panama	3320
	lact f: Feb-Oct; Nicaragua	930
	lact f: Mar; n1; Campeche, Mexico	5423c
	lact f: Mar-July; Nicaragua	8483
	lact f: Mar-Aug; El Salvador	3212
	lact f: Apr-July; Barro Colorado I, Panama	3024
	lact f: June, Sept-Oct; Venezuela	1987
	repro: est not seasonal; Veracruz, Mexico	4228

Order Marsupialia, Family Microbiotheriidae, Genus *Dromiciops*

The 20-40 g, Chilian microbiothere, *Dromiciops australis*, is a nocturnal, scansorial inhabitant of cool, moist forests where it may live in pairs (6773, 6777, 8556). Arthropods predominate in its diet. Males have scrotal testes and females have 4 teats in a small pouch (6777). Both scrotum and pouch are covered with rufous hair. Few accounts of its biology exist (3265, 6858).

LITTER SIZE		
Dromiciops australis	1 yg; n1 f	8305a
	2 yg; n1	6777
	2 yg; n1	4054
	3 yg; n1	8138
	4 yg; n1	6022
	5 yg; n1 nest	8478
SEXUAL MATURITY		
Dromiciops australis	f/m: 2nd yr	6777
SEASONALITY		
Dromiciops australis	lact f: Nov; n1; Chile	8138
	lact f: Dec; n1; Patagonia, Argentina	8305a
	lact f: Dec; n1; Malleco, Chile	4054
	repro: spring; Chile	6858

Order Marsupialia, Family Caenolestidae

The 15-45 g caenolestids (*Caenolestes, Lestoros, Rhyncholestes*) are terrestrial, nocturnal insectivores of South American cold and wet forests (5821, 6861). Females have 4 (*Caenolestes, Lestoros*) to 7 (*Rhyncholestes*) mammae, but no pouch (5821, 8257). *Caenolestes* males have a prostate gland 6 times the volume of their testes, 3 pairs of bulbourethral glands, a deeply forked glans penis (females have a deeply bifid clitoris; 8136, 9204). Distinctive sperm become paired in the epididymides (934a). Details of caenolestid anatomy are available (672, 9204). The reported reproductive biology of caenolestids is extremely limited.

LITTER SIZE		
Caenolestes spp	3 emb; n1 f	8136
Caenolestes caniventer	2 emb; n1 f	609
Caenolestes obscurus	3-4 enlarged teats; n4 f	5821
SEASONALITY		
Caenolestes obscurus	lact f: Aug; s Colombia	5821
Rhyncholestes raphanurus	repro f: Dec; n1; Chile	7242

Order Marsupialia, Family Dasyuridae

Dasyurids (*Antechinus* 10-120 g; *Dasycercus* 60-175 g; *Dasyuroides* 70-170 g; *Dasyurus* 0.4-4 kg; *Murexia* 105-285 mm; *Myoictis* 170-250 mm; *Neophascogale* 170-230 mm; *Ningaui* 8-13 g; *Phascogale* 45-200 g; *Phascolosorex* 117-226 mm; *Planigale* 5-15 g; *Sarcophilus* 4-18 kg; *Sminthopsis* 10-50 g) are terrestrial, nocturnal carnivores and insectivores from Australia and New Guinea. The New Guinean genera *Murexia, Myoictis, Neophascogale*, and *Phascolosorex* are poorly known.

Dasyurid males are often larger than females (423, 732, 1540, 2543, 2837, 3304, 3307, 4989, 5077, 6344, 7552, 7750, 9595, 10699, 11646, 11914; cf 1540, 4989). Copulation may be lengthy, e.g., 5-12 h in *Antechinus* (1194, 3131, 3163, 3305, 3439, 3601, 6344, 6825, 7280, 10642, 11646, 11913, 11915, 11922a; cf 423).

Sperm storage (710, 9757, 9761) and delayed development (9755) occur in *Antechinus*. Estrus has not been recorded post partum (425, 10126). In some species (e.g., *Antechinus, Dasyurus, Phascogale*) males die immediately after the mating season (589b, 609a, 731, 1190, 1191, 1192a, 1193, 1194, 1195, 1209, 1768a, 1936, 2174, 2189, 2543, 2553a, 4029a, 6307a, 7046a, 7047, 7455, 9699a, 10698, 11392, 11748, 11886, 11910b, 11913, 11917b, 11922a).

Pouch young open their eyes at 34-43 d (*Planigale* 424, 11646), 48-50 d (*Ningaui* 3163; *Sminthopsis* 3131, 3439, 3843; cf 10642), 58-70 d (*Antechinus* 6825, 9804, 11721, 11746, 11912), 60-74 d (*Dasycercus* 3305,

7280), 67-78 d (*Dasyuroides* 423, 425, 3664, 7181), 72-91 d (*Dasyurus* 2028, 3297, 7221, 9775; cf 3301, 9803), 87-105 d (*Sarcophilus* 3298, 3304, 4132).

Females have 4 (*Sarcophilus* 3304, 4040, 4132, 5076, 5077; *Phascolosorex* 7983; *Murexia* 7983), 4-12 (*Antechinus* 610, 731, 1540, 1935, 1936, 2554, 2838, 2841, 4043, 9757, 10108, 10703, 11392, 11395, 11396, 11425, 11721, 11746, 11748, 11886), 5-7 (*Dasyuroides* 423, 425, 11914), 5-15 (*Planigale* 95, 7983, 8992, 10699, 11646), 6 (*Myoictis*, 7983), 6-7 (*Ningaui* 354, 3163), 6-8 (*Dasycercus* 3305, 10698; *Dasyurus* 732, 3297, 4040, 11511, 11512), 6-11 (*Sminthopsis* 3838, 3843, 5836, 6436, 7552, 11923, 11927), or 8-13 mammae (*Phascogale* 2189).

Dasyurid marsupia vary from relatively undeveloped, flat, or non-existent (*Antechinus, Dasycercus, Dasyuroides, Dasyurus, Myoictis, Phascogale*) to extensive, deep pouches (*Antechinomys, Planigale, Sarcophilus, Sminthopsis*; 2189, 3163, 3304, 3439, 4989, 8617, 9852, 11916). An accessory erectile structure ventral to the urethral penis is present in some species (353, 11917a).

Details of anatomy (3351, 3352, 4754, 5077, 8292, 8297, 9761a, 11098, 11100, 11903, 11917), endocrinology (1388, 3326, 3327, 4803, 4808, 9699a, 9790, 11747, 11748, 11921, 11922a), sex differentiation (11029), ovulation/conception (1274, 4755, 4773, 5076, 7999, 9470, 9756, 9758, 9759, 9760, 10129), spermatozoa (10619), placentation (4754, 5075, 5076a), birth (5116), milk composition (4026, 4029, 7246), and light (2553b, 3839, 7006, 10126) are available.

NEONATAL MASS		
Antechinus macdonnellensis	12.3 mg; n1	11922
Antechinus stuartii	15 mg	11912
	16.4 mg	6825
Dasyuroides byrnei	26.8 mg; sd4.3; n62	3326
Dasyurus geoffroii	11 mg; r9-13; n2 in alcohol	9775
	14.5 mg; n1	376
Dasyurus viverrinus	20 mg	7218
Ningaui spp	5 mg	3163
Planigale gilesi	9.7 mg; n1	11646
Sarcophilus harrisii	18-29 mg	4132
Sminthopsis macroura	10 mg; n1	3838
NEONATAL SIZE		
Antechinus bilarni	CR: 7.2 mm; sd1.3	731
Antechinus macdonnellensis	CR: 3.8 mm; n1	11922
Antechinus minimus	3-5 mm	11746
Antechinus rosamondae	CR: 3.0-3.5 mm	11922a
Antechinus stuartii	CR: 4.9 mm	6825
Antechinus swainsonii	4.6 mm; r4.5-4.7	11721
Dasyuroides byrnei	4 mm	423
	4 mm	6697
	5.82 mm; sd0.37; n62	3326
Dasyurus geoffroii	4.4 mm; n2	9775
Dasyurus viverrinus	4.7-4.9 mm	1387
	7.6 mm	4754
Ningaui spp	3.5-4.5 mm	3163
Planigale gilesi	2.0 mm	11646
Planigale maculata	f: 5.7 mm	10699
Sarcophilus harrisii	12.5 mm; n4	3304
Sminthopsis crassicaudata	CR: 3.5-4 mm; from graph	7552
Sminthopsis longicaudata	CR: 3.0 mm	11927
Sminthopsis macroura	f: CR: 4.1 mm; n1	3838
Sminthopsis murina	5.4 mm; 3 d	3439
WEANING MASS		
Antechinus rosamondae	8.2 g; n1. 90 d	11922a
Dasycercus cristicauda	~22 g; 100-110 d	7280
Dasyuroides byrnei	~45 g	423
Dasyurus geoffroii	f: 35.2 g; sd2.0; n8, 72 d	9775
	m: 37.0 g; sd2.0; n5, 72 d	9775
Dasyurus viverrinus	f: 379.2 g; n19, 135 d	7221
	400 g	7218
	m: 465.8 g; n26, 135 d	7221

	f: 472.1 g; n17, 155 d	7221
	m: 652.9 g; n19, 155 d	7221
	f: 713.8 g; n10, 205 d	7221
	m: 1131.4 g; n11, 205 d	7221
Ningaui spp	2.5 g	3163
Planigale gilesi	f: 5.81 g; n5; independence	6519
	m: 6.03 g; n3; independence	6519
Planigale maculata	4.6-5.1 g	424
Planigale tenuirostris	f: 2.67 g; n4; independence	6519
	m: 2.93 g; n6; independence	6519
Sarcophilus harrisii	~200 g	4132
Sminthopsis crassicaudata	5-8 g	3843
Sminthopsis leucopus	f: 9 g	11923
	m: 11 g	11923
Sminthopsis macroura	7 g; r5.3-8.2	3838
Sminthopsis virginiae	18-20 g	10642
WEANING SIZE		
Ningaui spp	21.5 mm	3163
LITTER SIZE		
Antechinus apicalis	7; n1 lt	10917
	8; n1 lt	11915
Antechinus bellus	est 12+ ova	10703
	14 emb; r11-16	1540
	15 cl; r14-17	1540
Antechinus bilarni	2-6	10698
	r4.4-4.6 yg; n3 yr data; at pouch exit	734
	4.7; sd1.2; r3-6; n23 lt	731
	5.2; sd0.9; r3-6; n15 lt	731
	7 cl	1540
	7.5 emb	1540
Antechinus flavipes	4 yg	4989
	6.2 born; r3-12; n5 f	11912
	8.7 yg; sd1.44; r4-10; n13 lt	10108
	9.0 yg; sd1.56; r4-10; n20 lt	10108
	max 10	2841
Antechinus godmani	1-6 yg; n3 lt	11161
Antechinus leo	7 yg; n1 lt	11160
Antechinus macdonnellensis	mode 5	6344
	5.88 yg; r5-6; n8	11922
	max 6	11915a
	20.5 cl; r12-30; n10	11922
Antechinus melanurus	3.5 yg; r3-4; n4	2838
Antechinus minimus	6 yg; n1 lt	11425
	6 yg; n1 lt	4043
	6-8 yg	11746, 11748
	7.4; n8 f	11392
	8 yg ea; n12 lt	1935
Antechinus naso	3 yg, 4 swollen teats; n2 f	2838
Antechinus rosamondae	mode 6-8	6344
	6.3 yg; mode 8; r1-8; n10 f	11922a
Antechinus stuartii	3-5	11913
	5.025; r2-8; n40 lt	9804
	5.8 cl/side, 3.5 emb/side; n67 f; captive	9757
	6.1 cl/side; sd1.4; n100 sides, 53 f	9758
	6.4 cl/side, 4.8 emb/side; n22 f; wild mated	9757
	6.8; r3-8; n4 lt	6825
	6-8 yg	2554
	7.16; r0-10; n12 lt	11912
	7.17; r2-10; n107 lt	1935
	7.5	11886
	7.7 yg; se0.18; r7-8; n7 lt	610
	8 yg ea; n8 f	9757
	9-17 cl	9761

Antechinus swainsonii	7.5; mode 8; r3-8; n90 lt	1935
	8 yg; n1 lt	11395
	8.5; r8-9; n2 lt	11721
	9.33; r8, 8 yg, 12 emb; n3 lt	4043
Dasycercus cristicauda	5.2; r5-6; n3 lt	7280
	6-8	3305
	7-8	11903
Dasyuroides byrnei	3.4 wean; n10 lt	3601, 7181
	4.3 yg; r2-6; n10 lt	3601, 7181
	4.8; mode 6; r1-7; n79 lt	425
	mode 5	6697
	5.3 yg; n15 lt	423
	5.5 yg; sd1.1; n35 lt	3326
	6 yg; n1 lt	5116
	max 17 born	11914
Dasyurus geoffroii	mode 6; r3-6; n9 lt, 6 f	9775
Dasyurus hallucatus	r4.4-5.6; n3 yr data; at pouch exit	734
	6.1 yg; sd2.2; n9 f	9618
	6.3 yg; sd2.2; n4 f	9618
	6.4 yg; sd1.1; n21 lt	732
	7.8 yg; sd0.4; n22 f	9618
	8 yg ea; n2 lt	3306
	17.0 blastocysts; sd5.4; n4 f	9618
Dasyurus maculatus	r1-3 yg; n6 lt	2028
	3.5 yg; r2-5; n2 lt	9802a, 9803
	4-6	10917
Dasyurus viverrinus	1-6	7221
	4.3 yg; r3-5; n3 lt	11511
	4.7 yg; mode 6; r1-6; n7 lt	3325
	5.8 yg; mode 6; r5-6; n12 lt	4040
	6	4754
	6 yg	1387
	6 yg; n1 f	3297
	20-35 oocytes	4755
Ningaui spp	5 yg; n1 f	3313
	6.2; mode 7; r2-7; n6 lt	3163
	7 yg; n1 lt	5348
Ningaui yvonneae	7 yg ea; n2 lt	2082a
Phascogale calura	mode 6-8	6344
Phascogale tapoatafa	6.4 yg; r3-8; n13 lt	2189
Planigale gilesi	4.5; r3-7; n4 lt	11646
	6.1; r3-10; n13 lt	8992
Planigale ingrami	3.0 wean; n3 lt	7072a
	4-6 yg	355
	4-12 yg	4587
	7 yg; n1 lt	3307
	10; r8-12; n2 lt	2309
	15 yg; n1 nest	3308
Planigale maculata	5 yg; n1 f	6826
	8.0 yg; r4-12; n21 lt	10699
	8.1 yg; r5-11; n9 lt	424
Planigale tenuirostris	6 yg; n1 f	11498
	6.0; r4-9; n10 lt	8992
Sarcophilus harrisii	2 yg; n1 lt	4462
	2 yg; n1 f	3751
	2; r1-3; n2 lt	4040
	2.88 yg; mode 4; r1-4; n53 lt	4132
	4 yg; n1 lt	3298, 3304
	mode 4 71%; r1-4; n21 lt	5077
	21 oocytes; n1 f	3353
	max 23 ovulations	5076
	39 ovulations; r11-56; n6 f	5077
Sminthopsis crassicaudata	r1-9 born; n9 lt	11928
	5.4; se0.3; r3-8; n24 lt, 50-70 d	7552

	5.47 yg; n105 lt	3843
	6-8; n8 lt	6886
	6.09; r3-10; mode 5-6; n11 lt	3131
	7.0 wean; sd5.7; n120 lt	778
	7.3 yg or swollen teats; mode 8; r3-10; n23 museum f	3843
	7.5; se0.5; r5-10; n14 lt, 0-9 d	7552
	8.1; r7-9; n10 lt, <10 d	7552
	8.5 emb; r7-10; n2 f	7552
Sminthopsis laniger	4.5 yg; r3-6; n2 f	6436
	6 yg ea; n3 lt	11918
	7; n1 lt	10917
Sminthopsis leucopus	8 yg; n1	11923
	13 cl; r11-14; n3	11923
Sminthopsis longicaudata	3.0; r1-5; n2	11927
Sminthopsis macroura	1-8	3838
	6.0 yg; r1-8; n89 lt	11920
	14.7 emb; 8.4; sd2.8; n10 right horn, 6.3; sd4.5; n10 left horn	9761a
	20.7 cl	11920
Sminthopsis murina	6.75 yg; r1-10; n8 lt	3439
	10 yg; n1 lt	2841
Sminthopsis ooldea	7; r5-8; n7 lt	426
SEXUAL MATURITY		
Antechinus apicalis	f/m: 10-11 mo	11915
Antechinus bilarni	f/m: 1st mate: 11 mo	731
Antechinus flavipes	f/m: 1st mate: 11 mo	11913
Antechinus macdonnellensis	f/m: ~11 mo	11915a
	f/m: 1 yr	10698
Antechinus minimus	f/m: 1st mate: 11 mo	11392
Antechinus ningbing	f/m: 10-11 mo	11919
Antechinus rosamondae	f/m: ~10 mo	11922a
Antechinus stuartii	m: 9-10 mo	11912
	f/m: 1st mate: 11 mo	11913
	f/m: 320 d	6825
Antechinus swainsonii	f/m: 1st mate: 10-11 mo	11721
Dasycercus cristicauda	f/m: 1 yr	10698
Dasyuroides byrnei	f: 1st mate: 238 d	423, 425
	m: 1st mate: 216 d	425
	f/m: 8-11 mo	10698
	f/m: 10-11 mo	7181
Dasyurus geoffroii	f: 12 mo	376
Dasyurus maculatus	f/m: 1st mate: 11 mo	6346
	f/m: 1 yr	2028
Dasyurus viverrinus	f/m: 1st mate: 11 mo	6346
Phascogale tapoatafa	f/m: 1 yr	10698
Planigale maculata	f: 7.0 g	10699
	m: 8.5 g	10699
	f/m: 1st mate: 290 d	424
Sarcophilus harrisii	f: ~4-4.9 kg; n19	5077
	f/m: 1st mate: 2 yr	4134
Sminthopsis crassicaudata	f: 1st estrus: 100-120 d	1247
	f: 1st estrus: 115 d	3131
	f: 1st mate: 6 mo	778
	f: 1st birth: 18.5 wk; n1	3131
	f: 1st birth: 158.5 d; r152-165; n2	3843
	m: 1st mate: 159 d	3131
	m: 160 d	1247
Sminthopsis laniger	f: 1st estrus: 11.5 mo	11918
Sminthopsis leucopus	f/m: 11 mo; n7	11923
Sminthopsis macroura	f: 1st estrus: 119 d	3838
	f: 1st mate: bimodal 86-159 d, 12.5-16.0 g; or 185-262 d, r18.0-22.5 g	11920
	m: 1st sperm: 141-350 d; n60	11921
	f/m: 1st mate: 135 d	3838
Sminthopsis murina	m: 90-100 d	3439

Sminthopsis ooldea	f/m: 10 mo	426
Sminthopsis virginiae	f: 1st estrus: 230 d	10642
CYCLE LENGTH		
Antechinus apicalis	estrus: 4 d	11915
Antechinus macdonnellensis	estrus: 18.7 d; r10-27; n15	11922
Antechinus stuartii	estrus behavioral: 6.2 d; max 13	11912
	estrus histological: 19.33 d; sd4.42; n41 f	9757
Dasyuroides byrnei	60.3 d; sd7.1; n15	3326
Dasyurus maculatus	21 d	9803
Dasyurus viverrinus	37.4 d; se0.65; r28-43; n27	4803
	37.5 d; se0.7; r32-43; n20	4803
Planigale gilesi	estrus: 3 d	7983
	mode 21 d; r15-22; n7 cycles, 3 f	8992
Planigale tenuirostris	estrus: 1 d	7983
	mode 33 d; r31-34; n4 cycles, 4 f	8992
Sminthopsis crassicaudata	estrus: 1-3 d	1247
	estrus: 1-3 d	3131
	~28 d	1247
	mode ~30 d; r29-45	3131
	31.1 d; ±0.7; r25-37; n25 cycles; 17 f	3843
Sminthopsis laniger	30-51 d	10698
Sminthopsis leucopus	43 d; r40-46; n2 cycles, 2 f	11925
Sminthopsis longicaudata	34.4 d; r29-45; n5	11927
Sminthopsis macroura	23.25 d; r19-30; n25	11920
	26.2 d; ±0.5; r21-38; n32 cycles, 18 f	3838
Sminthopsis murina	24 d; r20-30; n5 f	3439
Sminthopsis virginiae	r29-40 d; n3 f	10642
GESTATION		
Antechinus apicalis	44-53 d	11915
Antechinus flavipes	23-27 d	11912, 11913
	~1 mo	2841
Antechinus macdonnellensis	> 38 d	11915a
	40 d; r36-44; n3	11922
	51 d; r45-55; n3; with sperm storage?	11922
Antechinus minimus	27.8 d; sd2.2; r26-30; n7; field capture to birth	11746
	30.6 d; sd1.5; r29-32; n3; lab	11746
Antechinus ningbing	est 45-52 d	11919
Antechinus rosamondae	40-50 d	10698
	49.7 d; r38-62; n10	11922a
Antechinus stuartii	diapause: 1-5 d	9755
	26-35 d; last mating to birth	11912, 11913
	27.2 d; se0.55; sd1.83; mode 26-28; r25-31; n26 f	9754, 9755, 9757
	31 d	2553
	32-34 d	11396
Antechinus swainsonii	28 d	2553
	31.5 d; r28-35; n2	11721
Dasycercus cristicauda	~30 d	7280
	~30 d; n1	3305
	35-44 d; n6	11914
Dasyuroides byrnei	r30-35 d; n32	11914
	r30-36 d; n10	423, 425
	32 d	6697
	32.5 d; sd1.8; r28-36; n35	3326
	33 d 18 h 58 min; n1	5116
	median ~35 d; r34-36; n10	7181
Dasyurus geoffroii	16-23 d	376
Dasyurus maculatus	21 d	9803
	~3 wk	3301
Dasyurus viverrinus	20 d; ±2	1387
	20.3 d; se0.63; r19-22; n4	4803
	20.5 d; se0.8; n6	4803
	21.02; sd2; r20-24; n4	3325
	est 3 wk; n3 lt	11511
Ningaui spp	14.7 d; r13.5-21.25; n5	3163

Phascogale tapoatafa	27.5 d; r26-29; n2; capture to birth	2189
	30 d	10698
Planigale gilesi	15.5 d; r14.5-16.5; n3	11646
Planigale maculata	19-20 d; n2	424
	20.5 d; r20-21; n2	11158
Planigale tenuirostris	mode 19 d; r18.5-19.5; n5 f	8992
Sarcophilus harrisii	~3 wk	5077
	31 d	4134
Sminthopsis crassicaudata	13 d	7552
	13.9 d; r13-16; n10	3843
	~16 d	3131
	16-18 d	1247
Sminthopsis laniger	11-12 d	10698
	est ≤12 d; weight changes	11918
Sminthopsis leucopus	est 20 d; no epithelial cells in urine to milk secretion; 3 wk	11923
Sminthopsis longicaudata	est <15 d; young present 17, 19 d after mating	11927
Sminthopsis macroura	10.7 d; sd0.67; most 10.5-11.0 d; r9.5-12.0	9761a
	r11-20 d; n74 lt	11920
	12-13 d; n8	3838
Sminthopsis murina	12.5 d; se0.28; r11-13.5	3439
Sminthopsis ooldea	≤ 21 d; pairing to birth	426
Sminthopsis virginiae	13-16, 16-19, 17-20 d; n3	10642
LACTATION		
Antechinus apicalis	wean: 4 mo	11915
Antechinus bellus	teat attachment: 4-5 wk	3520
Antechinus bilarni	wean: min 4 mo	731
Antechinus flavipes	teat attachment: 31 d	11912
	pouch exit: 30-36 d	10108
	nurse: 3-4 mo	10108
	wean: 4 mo	11912
Antechinus macdonnellensis	wean: 3-4 mo	11915a
	nurse: 14 wk	11922
Antechinus minimus	teat attachment: 2 mo	11746, 11748
	solid food: 8 wk	11425
	nurse: 12 wk	11425
	nurse: 80-90 d	11746
Antechinus ningbing	wean: ~16 wk	11919
Antechinus rosamondae	solid food: 70 d	11922a
	wean: 90-120 d	11922a
	wean: 4 mo	10698
Antechinus stuartii	teat attachment: 30-40 d	11912
	teat attachment: ~5 wk	11396
	teat attachment: 40 d	6825
	nurse: 90 d	6825
	nurse: min 3 mo	11396
	wean: 3 mo	11886
	wean: 3-4 mo	11912
Antechinus swainsonii	teat attachment: 8 wk	11721
	solid food: 12 wk	11721
	nurse: 14 wk	11721
Dasycercus cristicauda	teat attachment: 55-60 d	7280
	teat attachment: ~58 d; n1	3305
	teat attachment: ~1 mo	10917
	solid food: 108 d	7280
	wean: 121 d	7280
	independent: 3.5 mo	3305
Dasyuroides byrnei	teat attachment: 56 d	423, 425
	teat attachment: 56 d; r50-59	7181
	solid food: 95-96 d	7181
	wean: 100-105 d	423, 425
	nurse: 127 d; r115-144	7181
Dasyurus geoffroii	pouch exit: 62-72 d	9775
	pouch exit: 70-90 d	9378
	wean: 120+ d	9378

Dasyurus hallucatus	teat attachment: 8 wk	3306
	nurse: 6-7 mo	732
Dasyurus maculatus	teat attachment: 47-50 d	9803
	solid food: 17 wk	2028
	wean: 120-150 d; n2 lt	9803
	independent: 4.5 mo	3301
Dasyurus viverrinus	teat attachment: 9 wk; n1	3297
	teat attachment: 60-65 d	1387
	solid food: 107-112 d	7221
	wean: 102-142 d singletons; -200 d larger lt	7221
	nurse: 4 mo	9470
	independent: 4.5 mo	3297
	wean: 5.5-6 mo	1387
Ningaui spp	teat attachment: 42-44 d	3163
	solid food: 70 d	3163
	wean: 76-81 d	3163
Phascogale tapoatafa	teat attachment: 7-8 wk	2189
	wean: ~16 wk	2189
Planigale gilesi	teat attachment: 37 d	11646
	solid food: 65 d	11646
	solid food: 66 d; n1 lt	6519
Planigale maculata	teat attachment: 26 d; r24-28; n2	11158
	teat attachment: 28 d	424
	solid food: 55 d	424
	solid food: 57 d	11158
	independent: 70 d	424
	independent: 72 d	11158
Planigale tenuirostris	teat attachment: 36 d	8992
	nurse: 95 d	8992
Sarcophilus harrisii	teat attachment: 15 wk	3304
	pouch exit: 105 d	4132, 4134
	solid food: 5 mo	3304
	wean: ~8 mo	4132, 4134
Sminthopsis crassicaudata	teat attachment: 40 d	3843
	teat attachment: 40-43 d; n4 lt	3131
	solid food: 60 d	3843
	wean: 65-69 d; n3 lt	3131
	wean: ~70 d	1247
	wean: ~70 d	3843
Sminthopsis laniger	wean: 3 mo	11918
Sminthopsis leucopus	wean: 2.5 mo	11923
Sminthopsis macroura	teat attachment: 40 d	3838
	solid food: 60 d	3838
	wean: 70 d	3838
	wean: 70 d	11920
Sminthopsis murina	teat attachment: 34 d; n1 lt	3439
	solid food: 63 d; n1 lt	3439
Sminthopsis ooldea	wean: ~70 d	426
Sminthopsis virginiae	teat attachment: 55-56 d	10642
	solid food: 74-78 d	10642
	wean: 80-90 d	10642

INTERLITTER INTERVAL

Dasyuroides byrnei	161-173 d	425
Planigale maculata	40 d; n1 with loss of lt	424
Planigale tenuirostris	min 91 d	8992
Sminthopsis crassicaudata	mode 85 d; r82-210	778
	~12 wk	3131
	min 12 wk	3843
	~97 d; min 12 wk	3131
	4 mo	6886

LITTERS/YEAR		
Antechinus apicalis	1	11915
Antechinus bilarni	1	731
Antechinus flavipes	1	10108
Antechinus minimus	1	11748
Antechinus stuartii	1	11912
	1	11886
Dasycercus cristicauda	1/season	10698
Dasyuroides byrnei	max 2 lt/season	425
Dasyurus hallucatus	1	732
Ningaui spp	1-2	3163
Phascogale tapoatafa	1	2189
Planigale maculata	several	424
Sarcophilus harrisii	1	5077
Sminthopsis crassicaudata	1-3	10129
	2	7551
	est 2	7552
Sminthopsis laniger	>1/season	11918
Sminthopsis murina	max 2	3439
Sminthopsis macroura	2	11920
SEASONALITY		
Antechinus apicalis	sperm in urine: Jan-mid Apr; captive	11915
	mate: Mar-Apr; captive	11915
	birth: Apr; sw Australia	11915
Antechinus bellus	fertile m: mid June-Aug; NT, Australia	1540
	fertile m: mid June-mid Aug; NT, Australia	10703
	mate: mid-end Aug; NT, Australia	3520
	birth: end Sept-early Oct; NT, Australia	1540
	birth: mid-end Sept; NT, Australia	3520
Antechinus bilarni	pouch/scrotal development: June-Aug; NT, Australia	731
	mate: May-mid July; NT, Australia	1540
	mate: June-July; NT, Australia	731
	birth: Aug-Sept; NT, Australia	731
	lact f: Aug-Dec; NT, Australia	731
Antechinus flavipes	m death: Sept; QLD, Australia	10108
	m death: end Sept; QLD, Australia	2841
	mate: June-July; lab	11913
	mate: end July; ACT; mid Aug; NWS, Australia	2552a
	mate: end Aug-early Sept; QLD, Australia	10108
	mate: end Sept; QLD, Australia	2841
	birth: July-Aug; lab	11913
	birth: end Sept-early Oct; QLD, Australia	10108
	birth: end Oct; QLD, Australia	2841
	lact f: Oct-Jan, QLD, Australia	10108
Antechinus godmani	birth: July-Aug; ne QLD, Australia	11161
Antechinus leo	birth: early Nov; QLD, Australia	11160
Antechinus macdonnellensis	birth: late July-early Sept; central Australia	11922
	birth: Aug-Sept; Australia	11915a
Antechinus melanurus	mate: yr rd; few data; Mt Erimbari, New Guinea	2838
Antechinus minimus	prostate, cowpers: peak June-July; VIC, Australia	11747
	spermatogenesis: Feb-May; VIC, Australia	11747
	testes: peak May; VIC, Australia	11747
	m death: end July-early Aug; VIC, Australia	11747
	mate: May; VIC, Australia	11392
	preg f: Dec; n1; TAS, Australia	4043
	birth: July; VIC, Australia	11392
	birth: July-Aug; VIC, Australia	11746, 11748
	lact f: Aug; n1; captive	11425
Antechinus naso	mate: yr rd; few data; Mt Erimbari, New Guinea	2838
Antechinus ningbing	mate: June; WA, Australia	11919
Antechinus rosamondae	mate: Sept; nw WA, Australia	11922a
	birth: Oct-Nov; nw Australia	6346
	birth: Nov; nw WA, Australia	11922a

Antechinus stuartii	spermatogenesis: fails May; VIC, Australia	5694
	testes: large June-Aug; ACT, Australia	11913
	mate: July-Aug or Aug-Sept; 2 stocks, lab	9757
	mate: July-Sept; lab; ACT, Australia	11912, 11913
	mate: mid July-Sept; ACT, Australia	2553
	mate: early Aug; NSW, Australia	6825
	mate: early-mid Aug; NSW, Australia	2552a
	mate: mid-end Sept; QLD, Australia	11886
	birth: Aug-Sept; NSW, Australia	4989
	birth: Aug-Oct; ACT, Australia	11913
	birth: end Aug-early Sept or end Sept-early Oct; 2 stocks, lab	9757
	birth: end Sept; se Australia	6384
	birth: 3rd wk Sept-early Oct; e Australia	1194
	wean: Feb; QLD, Australia	11886
Antechinus swainsonii	mate: May-Sept; mainland, Australia	11721
	mate: May-Sept; ACT, Australia	2553
	mate: mid July; VIC, Australia	7047
	mate: Sept-Oct; TAS, Australia	11721
	birth: mid Aug; VIC, Australia	11395
	birth: end Aug; se Australia	6384
	lact f: ?-Feb; w TAS, Australia	4043
	wean: Dec; se Australia	6384
Dasycercus cristicauda	mate: end June-July; captive	3305
	birth: June-July; central Australia	6346
	repro: June-Sept; central Australia	11903
Dasyuroides byrnei	mate: Apr-Dec, 75% May-July; lab	423, 425
	mate: Dec-Apr; captive	7181
	birth: May-Oct; captive	6697
	birth: June; e central Australia	11495
Dasyurus geoffroii	mate: Apr; captive	353
	birth: May-Sept, peak June-July; n58; WA, Australia	10196
Dasyurus hallucatus	lact f: July-Sept; NT, Australia	732
	lact f: July-Jan; WA, Australia	9618
Dasyurus maculatus	mate: June-July; VIC, Australia	3301
	birth: ~May; Australia	10917
	birth: July-Aug; VIC, Australia	3301
Dasyurus viverrinus	testes: large Mar-June, small Nov-Feb; TAS, Australia	1388
	f cycles: May-Aug; se Australia	4803
	mate: peak end May-early June; TAS, Australia	3325
	mate: May-July; Australia	9470
	mate: May-June; TAS, Australia	3846
	lact f: July 5-26; ne TAS, Australia	4040
Ningaui spp	birth: end Oct-early Dec; n5; captive	3163
Ningaui ridei	repro: Sept-Feb; NT, Australia	7983
Ningaui timealeyi	repro: Sept-Mar, peak Nov-Jan; Australia	7983
Phascogale calura	mate: last 3 wk July; WA, Australia	1190
Phascogale tapoatafa	birth: end July-early Aug; VIC, Australia	2189
	lact: Aug-Nov; VIC, Australia	2189
Planigale gilesi	mate: end Nov-Dec; captive	11646
	birth: end Aug-mid Jan; NSW, Australia	8992
Planigale ingrami	birth: Dec-Mar; NT, QLD, Australia	4587
	lact f: Nov-Dec, Feb-Mar; n4; QLD, Australia	3307
	repro: Feb-Apr; NT, Australia	355
	repro: summer; WA, Australia	11916
Planigale maculata	mate: est yr rd; NT, Australia	424
	birth: est yr rd; NSW, QLD, Australia	11158
	pouch young: yr rd; NT, Australia	10699
Planigale tenuirostris	birth: Sept-Jan, NSW, Australia	8992
	lact f: Oct; n1; SA; Australia	11498
Sarcophilus harrisii	spermatogenesis: Apr-Aug; TAS, Australia	5077
	mate: Mar; TAS, Australia	4132
	mate: May-June; TAS, Australia	3304
	preg f: Mar-May; TAS, Australia	5076
	birth: Mar-May; TAS, Australia	5076

	birth: Mar-Apr; TAS, Australia	5077
	birth: Apr-Sept; TAS, Australia	6346
	birth: end May-early June; TAS, Australia	3304
	lact f: July; n2; ne TAS, Australia	4040
Sminthopsis crassicaudata	mate: June-July; SA, Australia	11901
	birth: May-Jan; captive	3131
	birth: est June-Dec; arid, se Australia	7552
	birth: July-Aug; VIC, Australia	3296
	birth: end July-Feb; mesic, se Australia	7552
	birth: yr rd, peak July-Jan; captive	3843, 10129
	birth: est yr rd; captive	6886
Sminthopsis laniger	birth: Aug-Dec; sw QLD, Australia	11918
Sminthopsis leucopus	sperm in urine: June-Oct; VIC, Australia	11923
	mate: Aug; VIC, Australia	11923
	birth: Aug-Sept; VIC Australia	11923
	wean: end Nov; VIC, Australia	11923
	lact f: end Aug-Oct; se Australia	8994
Sminthopsis macroura	birth: June-Feb, peak Aug-Dec; captive	11920
	birth: July-Feb; captive	3838
Sminthopsis murina	birth: Sept-Jan; NSW, Australia	3439
Sminthopsis ooldea	birth: Sept-Jan; SA, Australia	426
Sminthopsis virginiae	lact f: July-Nov; NT, Australia	7553

Order Marsupialia, Family Myrmecobiidae, Genus *Myrmecobius*

The 300-700 g, diurnal, solitary, and endangered numbat, *Myrmecobius fasciatus*, feeds on ants and termites in Western Australia (10809). Females have 4 teats and lack a well-developed pouch (1539, 3521, 9852, 10698). Details of milk composition (4066) are available.

LITTER SIZE

Myrmecobius fasciatus	3.4 yg; mode 4; r2-4; n8 lt	1538, 1539
	4 yg; n1 f	1827
	4 yg; n1 f	3521

LACTATION

Myrmecobius fasciatus	independent foraging: 6 mo	1538, 1539
	wean: 7 mo	10698
	wean: 9 mo	4066

SEASONALITY

Myrmecobius fasciatus	birth: Jan-May; sw WA, Australia	1538, 1539
	birth: end Jan; sw WA, Australia	4066
	lact f: Mar-Oct; few data; WA, Australia	3521
	lact f: May; n1; WA, Australia	1827
	repro: June-July; WA, Australia	11901

Order Marsupialia, Family Thylacinidae, Genus *Thylacinus*

The Tasmanian wolf, *Thylacinus cynocephalus*, is either extinct or restricted to a small part of Tasmania, Australia (10809). This nocturnal, possibly solitary, carnivore ranges from 15 to 25 kg. Females have 4 teats (8617). A pouch is present in both sexes (8617). Anatomical details are available (8298).

LITTER SIZE

Thylacinus cynocephalus	mode 2-3; max 4	10698
	3-4	4130

SEASONALITY

Thylacinus cynocephalus	small juveniles: yr rd, peak May-Aug; TAS, Australia	4130

Order Marsupialia, Family Notoryctidae, Genus *Notoryctes*

The solitary and fossorial *Notoryctes typhlops* is a 40-70 g, Australian insectivore. The pouch, present in both sexes, opens posteriorly (10423). Females have 2 mammae. Nothing has been recorded about its reproduction, but its anatomy has been described (10424, 10425, 10601).

There are no tabular data for *Notoryctes*.

Order Marsupialia, Family Peramelidae

Bandicoots (*Chaeropus, Echymipera, Isoodon, Microperoryctes, Perameles, Peroryctes, Rhynchomeles*) are 200-2300 g, terrestrial, nocturnal omnivores from Australia, Indonesia, and New Guinea. *Chaeropus* may be extinct. The New Guinea and Indonesian genera *Echymipera, Microperoryctes, Peroryctes,* and *Rhynchomeles* are little known. Males are generally larger than females (286, 6670). In addition to the usual choriovitelline placenta, bandicoots have a chorioallantoic placenta (3354, 3356, 4750, 4751, 4753, 4763, 5075, 8195, 8294, 10710, 10711).

The pouch opens posteriorly (6663, 10917). Neonates have deciduous claws and open their eyes at 44-49 d (3685, 4584, 6663, 6665, 6666, 6699, 6825). Females have 6-10 teats (4233, 8798, 10698, 10917).

Details of anatomy (4752, 4753, 4923, 5073, 6671, 8000, 8291, 10858), endocrinology (1618, 1619, 2214, 3669, 3670, 3672, 3676, 3679, 3684, 3687, 3688, 3689, 3690, 5073, 5351), sex-chromosome mosaicism (1906a, 1908, 3492, 4519, 4520, 11440), primordial germ cells (11028), spermatogenesis (6931, 9793a), ovulation or conception (6673), and milk composition (7220) are available.

NEONATAL MASS		
Isoodon macrourus	188.3 mg; r148-220; n6	6699
Isoodon obesulus	350 mg	4584
Perameles gunnii	250 mg	4584
Perameles nasuta	237 mg; r209-254; n15	6663, 6665
NEONATAL SIZE		
Isoodon macrourus	CR: 13.83 mm; r13.5-14.5; n3	6699
Isoodon obesulus	11.0 mm	4584
Perameles nasuta	CR: 12.8 mm; sd0.75; n13	6663, 6665
WEANING MASS		
Isoodon macrourus	138-212.5 g	6699
LITTER SIZE		
Chaeropus ecaudatus	1 yg; n1	6662
	2 yg	10698
Isoodon macrourus	2.16 wean; ±0.5; n87 lt	3680
	2.2 yg; r1-3; n9 lt	2841
	2.5 yg; sd1.03; n16 lt, 12 f	5646
	2.6; se0.2; r1-5; n33 1st lt	3674
	2.9 yg; r1-4; n12 lt	10443
	2.9; r1-5; n44 lt	4233
	3.0 yg; ±0.15; mode 3; r1-6; n136 lt, 0-10 d	3680
	3.03 yg; 2.31 wean; n32 lt	3691
	3.07 yg; se0.11; r1-6; n112	6670, 6671
	3.07 yg; se0.12; n222 lt, 0-10 d	3681
	3.19 yg; 2.56 wean; r1-5; n27 lt	3686
	3.38 yg; r1-7; n21 f	3946
	3.4 yg; se0.2; n36, 1-5 d	3671
	3.6 yg; se0.3; n14, 1-10 d	3671
	4.0; r1-7; n8 lt	6699
	4.0 yg; se0.4; n14 births	3671
	5.05 cl; se0.23; r3-7; 4.05 emb; se0.29; r1-6; n20	6671
	5.13 cl; r3-7; n39	6671
Isoodon obesulus	1.7 yg; r1-3; n6 f	11492
	2.3 yg; r1-5; n9 lt	6663
	2.8 yg; mode 4; r1-4; n15 lt	4584
	3.07 yg; mode 3-4; r1-6; n56 lt	10448

Perameles bougainville	mode 1-2; r1-3	10809
	2 yg; n1	9194
	2-4	10917
Perameles eremiana	mode 2	10917
	2	11901
Perameles gunnii	2.1 yg; r1-5; n25 lt, 9 f	9717
	2.23 yg; r1-4; n22 lt	6663
	2.33 yg; mode 2; r1-4; n54 lt, 1-20 d	4584
Perameles nasuta	1.83 yg; n6 lt	6665
	2-3 yg	6675
	2.44 yg; r1-5; n52 lt	6663
	2.6; r1-5	6670
	2.65 yg; se0.07; r1-5; n196	6671
	3.31 cl; r2-6; n45	6671
	3.33 cl; se0.22; r2-6; 2.63 emb; se0.26; r1-5; n24	6671
SEXUAL MATURITY		
Isoodon macrourus	m: high testosterone: 179.8 d; se6.8; r127-243; n17	3677
	m: 1st sperm: 199.5 d; se8.2; r201-224; n10	3677
	m: 1st sire: 349.7 d; se14.0; r293-383; n7	3677
	f: 1st birth: 249.6 d; se11.4; r161-399; n33	3674
	f: 1st repro: ~4 mo	4584
Perameles gunnii	f: 1st mate: ~2.5-3.5 mo	4584
	m: epididymal sperm: 4-6 mo	4584
Perameles nasuta	f: 1st birth: ~4 mo	6663
	m: epididymal sperm: 150 d	6663
CYCLE LENGTH		
Isoodon macrourus	rpy to ovulation: 6 d	3678
	rpy to estrus: ~20 d	3670
	20 d; mode 18; r9-34; n39 cycles, 15 f	6668
	22.1 d; se1.6; r14-30; n10 f	3678
Perameles nasuta	rpy to estrus: 17-26 d	1907
	~20 d; n3 cycles	1907
	21 d; r10-34; n9 cycles, 8 f	6668
	26 d; r17-34; n8 cycles	5073
GESTATION		
Isoodon macrourus	12 d, 8-11 h; n1	6666
Perameles nasuta	11-13 d; n1	5073
	12.5 d; n3	10443
LACTATION		
Isoodon macrourus	teat attachment: 42.4 d; se0.5; n10 lt	3685
	pouch exit: 48.7 d; se0.7; n8 lt	3685
	pouch exit: 50 d; ±7; n6 lt	3946
	pouch exit: 8 wk	6699
	wean: 59.8 d; se0.9; n20 lt	3685
Isoodon obesulus	wean: 58 d	10448
Perameles gunnii	wean: 59-61 d	4584
Perameles nasuta	pouch exit: 48-63 d	6663
	pouch exit: 50-54 d	10443
	wean: ~67 d	5073
	wean: ~75 d	6665
INTERLITTER INTERVAL		
Isoodon macrourus	56 d; n6 lt, 1 f	3946
	61 d	3685
Isoodon obesulus	89.5 d; r51-108; n3 f	5646
Perameles gunnii	~65 d	9717
LITTERS/YEAR		
Isoodon obesulus	1	11492
	2.61/season	10448
	max 3	5646
Perameles gunnii	3.8; r3-4	4584

SEASONALITY		
Chaeropus ecaudatus	mate: May-June; Australia	10917
Isoodon macrourus	prostate: large July-Sept, small Mar-May; captive	10858
	birth: end July-early Sept; n6; captive	3676
	birth: Sept-Apr; WA, Australia	5646
	birth: yr rd, low Apr-June; QLD, Australia	3671
	birth: yr rd; QLD, Australia	4233
	birth: yr rd; captive	6699
	birth: yr rd, peak Aug-Oct, low Mar-June; captive	3681
	lact f: Sept-Feb; QLD, Australia	2841
	lact f: peak end Sept-Nov; TAS, VIC, NSW, QLD, Australia	597
	lact f: yr rd; QLD, Australia	3946
Isoodon obesulus	mate: July-Feb; TAS, Australia	4584
	birth: Sept-Oct; SA, Australia	11492
	lact f: Aug-Jan; VIC, Australia	10448
Perameles bougainville	mate: May-Aug; Bernier I, w SA, Australia	10809
Perameles gunnii	epididymal sperm: yr rd; TAS, Australia	4584
	mate: end May-Dec; TAS, Australia	4584
	birth: peak July-Nov; VIC, Australia	9717
Perameles nasuta	mate: peak spring-summer; captive	6668, 6675
	mate: yr rd; captive	1907
	lact f: May; n1; QLD, Australia	2841
	lact f: yr rd; n~50; NSW, Australia	6663

Order Marsupialia, Family Thylacomyidae, Genus *Macrotis*

The nocturnal bilbies, *Macrotis lagotis* (800-2500 g) and *M. leucura* (310-435 g), from arid Australian grasslands are fossorial when inactive (5337). These colonial omnivores feed preferentially on ants and termites in parts of their range (10174, 11491). Males are larger than females (5337) and have an extra chromosome (4013). Eyes open 43-64 d (5088). Females have 6-8 teats (10474, 10917). *Macrotis leucura* may be extinct. Details of ovarian anatomy (7037) are available.

LITTER SIZE		
Macrotis lagotis	1 yg; n1 lt	6662
	1.6 yg; r1-2; n5 lt	5088, 5089
	2 yg; n1 lt	5337
CYCLE LENGTH		
Macrotis lagotis	20.6 d; se7.3; r12-37; n14 cycles, 6 f	7037
GESTATION		
Macrotis lagotis	14.0 d; se1.4; r13-16; n4; cornified cells to birth	7037
	21 d; n1	5337
LACTATION		
Macrotis lagotis	pouch exit: 67-68 d	5088
	pouch exit: 80 d; se2; n6	7037
	nurse: r21-88 d after pouch exit; n4	7037
SEASONALITY		
Macrotis lagotis	birth: Mar-May; Australia	11901

Order Marsupialia, Family Phalangeridae

Phalangers (*Phalanger, Trichosurus, Wyulda*) are 1-5 kg, arboreal herbivores in forests of Australia and New Guinea. Males are occasionally reported larger than females (3785, 5097, 8746, 10081) and may have an extra chromosome (7659). Females have larger adrenal glands than males with a distinctive zone (1553, 5696a, 9141, 11573). Estrus does not occur after birth (3784, 5610, 5613, 10127, 10130). The eyes of *Trichosurus* neonates open at 100-120 d (5014, 6676, 10998). Female *Trichosurus* have 2 mammae within a pouch (4973, 10917, 10998, 11902).

Details of anatomy (378, 3787, 5613, 8292, 10735, 11098, 11100), endocrinology (195, 1060a, 1065, 1556, 1618, 1619, 1766, 2030, 2174, 2179, 2180, 2182, 3675, 3688, 4807, 5097, 5405, 6824a, 9790, 9791, 9793, 9854, 9929, 9930, 10395a, 10397, 10398, 10805, 10990, 10992, 11308, 11571, 11572), intersexes (1059, 1060, 1061, 1062, 1063, 1064, 1067, 9863), spermatogenesis (2167, 4369a, 9208, 9746, 9793a, 10735a), prostate function (2030), semen sugars (9209), superfetation (5610), egg membranes/placentation (5082, 9853), implantation (5075), mammary glands (9854), milk composition (1058, 2141, 4072, 4100, 4101), and day length (3682) are available.

NEONATAL MASS		
Trichosurus vulpecula	180 mg; r170-190; n2	5612
	205 mg; r200-210; n2	6674
	est 230 mg from growth regression	5612
	500 mg; 2 d	8746
NEONATAL SIZE		
Trichosurus vulpecula	CR: 13 mm; n1	6674
WEANING MASS		
Trichosurus arnhemensis	470 g	10081
Trichosurus caninus	m: 666.67 g; se92.80; n3	5014
	f: 881.00 g; se42.56; n5	5014
Trichosurus vulpecula	321 g; r229-439; 170 d	750
	~600 g; 170 d	6676
	f: 1300 g	10081
LITTER SIZE		
Phalanger carmelitae	1 yg	7210
Phalanger gymnotis	1 yg	7210
Phalanger maculatus	1 yg; n1	7207
	2-4	10917
Trichosurus arnhemensis	1 10081	
Trichosurus caninus	1.08; r1-2; n13 lt	10130
Trichosurus vulpecula	1 ea; n1067 lt	750
	1.03; r1-2; n36 f	10985
	1.09 cl; r1-2; n99 lt	10127
	1.13; mode 1; r1-3; n60 lt	8798, 12128
Wyulda squamicaudata	1 ea; n4 lt	5097
SEXUAL MATURITY		
Phalanger maculatus	m: testes descend: est 8 mo; n1	7207
Trichosurus caninus	f: 1st ovulate: end 2nd yr	10130
	f: 1st birth: end 3rd yr	10130
Trichosurus vulpecula	f: large follicles: 8-10 mo	8746
	f: 1st estrus: 9 mo	8545
	f: 1st estrus: 12 mo	5613
	f: 1st birth: mode 2 yr	750
	f: 1st birth: 2 yr	2120
	f: 1st wean: 2.6 yr	5610
	f: end 1st yr	10127
	f: early 2nd yr; a few end 1st yr	3784
	m: testes large: 1925 g; n1	2810
	m: sperm: 2268 g	10985
	m: 24 mo	10998
	m: ~6 mo later than f	3784
	f/m: 15 mo	4973
	f/m: 1st mate: yr following birth	11900
Wyulda squamicaudata	f: ~2 yr; ~1.1 kg	5097
CYCLE LENGTH		
Phalanger maculatus	est 28 d	7210
Trichosurus caninus	rpy to estrus: 10-11 d	10130
	26.4 d; ±1.0; mode 23-24 47%; r23-38; n17 cycles	10130
Trichosurus vulpecula	rpy to estrus: 8.02 d; ±0.18	8545
	luteal phase: 8 d	9930
	22-23 d	5610
	mode 22-32 d; r22-58; n25 f	8545
	24.5 d; r21-30; n13 f	6674
	25.69 d; se0.31	9859

	26 d	10805
	26 d; r24-29; n4 cycles, 4 f	1858
	26.3 d; se1.46; abstract; 27.2 d; se0.84; text; n39 cycles	2181
	26.59 d; ±0.31; n49 cycles, 23 f	8545
GESTATION		
Trichosurus caninus	16.2 d; ±0.2; mode 16 60%; r15-17; n10 lt	10130
Trichosurus vulpecula	16-18 d; n10 f	8545
	17.1; se0.14; n21 lt	2181
	17.5 d; r17-18; n2	5610
	~17.5 d; r17-21; n6	6674
	≤21 d	6676
LACTATION		
Phalanger maculatus	pouch exit: est 6-7 mo	7207
Trichosurus arnhemensis	wean: 165-195 d	10998
	wean: 6 mo	10081
Trichosurus caninus	teat attachment: 81-112 d	5014
	pouch exit: most 175-200 d	5014
	wean: 240-270 d	5014
Trichosurus vulpecula	teat attachment: 94 d	2809
	pouch exit: 112-144 d	6676
	pouch exit: 121 d	2809
	pouch exit: 140 d	10127
	pouch exit: 170 d	750
	pouch exit: ~5 mo	2810
	wean: 5-6 mo	4973
	wean: 175 d	5016
	wean: 220-290 d	10127
	nurse: 6-7 mo	2810
Wyulda squamicaudata	wean: >8 mo	5097
INTERLITTER INTERVAL		
Trichosurus vulpecula	9.0 mo; se0.46; abstract; 9.4 mo; se0.36; text; n41	2181
LITTERS/YEAR		
Trichosurus arnhemensis	1.8	10081
Trichosurus caninus	1	5014
Trichosurus vulpecula	1	750
	1-2	6674
	1-2	10985
	1-2	5610, 5611
	2	2810
Wyulda squamicaudata	1	5097
SEASONALITY		
Phalanger carmelitae	lact f: Feb-Apr, Aug-Sept; New Guinea	7210
Phalanger gymnotis	repro: yr rd; New Guinea	7210
Phalanger orientalis	repro: June-Oct; New Guinea	7210
Trichosurus arnhemensis	mate: yr rd; ne Australia	5690
Trichosurus caninus	birth: peak Mar-May; NSW, Australia	5014
Trichosurus vulpecula	sperm: few summer, some aspermy each mo; New Zealand	5610
	sperm: yr rd; NSW, Australia	10127
	testes large, sperm: yr rd; New Zealand	3784, 3787
	estrus: Apr-May; New Zealand	5613
	mate: Mar; SA, Australia	6674
	mate: June; NSW, Australia	6674
	mate: June; n1; SA, Australia	11900
	mate: peak Sept-Oct, minor peak Apr; New Zealand	3784
	mate: yr rd; n Australia	2214
	mate/birth: May; New Zealand	1298
	preg f: Mar-Apr, Aug-Sept; NSW, Australia	1058
	birth: Mar-Apr, Sept; ACT, Australia	2809
	birth: peak Mar-Apr, none Jan-Feb; SA, Australia	9852
	birth: Mar-May, Aug-Oct; VIC, Australia	8746
	birth: Mar-June, peak Apr; NSW, Australia	10127
	birth: end Mar-mid Apr; n4; captive	3675
	birth: peak Apr-May, some mid Aug-Nov; ACT, Australia	2810
	birth: Apr-June 99%; New Zealand	750

	birth: Apr-June, peak July-Sept; New Zealand	2120
	birth: yr rd, 86% Mar-May; New Zealand	5612
	birth: yr rd, peak Mar-May; zoo	4973
	birth: yr rd, peak Nov-Mar; SA, Australia	9469
	birth: yr rd; New Zealand	10985
	birth: yr rd; n60; zoo	12128
	pouch exit: ~end Sept; SA, Australia	11901
Wyulda squamicaudata	birth: mid Feb-mid Aug; n9; WA, Australia	5097

Order Marsupialia, Family Burramyidae

Pygmy possums (*Acrobates*, *Burramys*, *Cercartetus*, *Distoechurus*) are 6-100 g, nocturnal, arboreal omnivores from heath, forests, and shrublands of New Guinea and Australia. Sexual dimorphism is variable (3314, 10081). Embryonic diapause and postpartum estrus occur (1164, 1856, 10483, 11453, 11455, 11456). Neonates open their eyes at 37-48 d (*Burramys* 2582, 5691, 10998), 41 d (*Cercartetus* 11453), or 60-76 d (*Acrobates* 3314, 10998, 11454). Pygmy possums usually have 4-6 teats in a forward opening pouch (2582, 2837, 3314, 10123, 10698, 10917, 10998, 11393, 11394, 11901), although *Distoechurus* has been reported to have 1 or 2 (10684, 11455). Details of anatomy (1856, 11455, 11456, 11457) are available.

NEONATAL MASS		
Acrobates pygmaeus	17.8 mg	11454
Cercartetus concinnus	12 mg	10998
Cercartetus nanus	19.8 mg	11453
NEONATAL SIZE		
Acrobates pygmaeus	CR: 5.5 mm	11454
Burramys parvus	CR: 7.2-7.5 mm	6785
Cercartetus nanus	CR: 5-7 mm	11453
WEANING MASS		
Acrobates pygmaeus	1.7 g; pouch exit, ~65 d	11454
	7 g	3314
	8-8.5 g; 100 d	11454
Burramys parvus	~22 g; 7-8 wk	5691
Cercartetus caudatus	11-12 g	435
	11-13 g	2838
Cercartetus nanus	2.2-2.8 g; pouch exit, 33-37 d	11453
	8-12 g	10474
	12-13 g; 60-65 d	11453
LITTER SIZE		
Acrobates pygmaeus	2.52 wean; mode 4; n33	11454
	2.7 yg; r1-4; n32	3314
	2.81 yg; mode 2-3; r1-4; n59	11454
	3.0 enlarged teats; r2-4; n28	3314
	3.27 enlarged teats; mode 4; r1-4; n51	11454
	3.5 emb; r2-5; n15 f	11455, 11456
Burramys parvus	2.25; r1-4; n4 lt	5691
	3.6 yg; sd0.9; n441	6785
	4 yg ea; n2 lt	2582
Cercartetus caudatus	2.0 yg; mode 2; r1-3; n8 lt	2838
	2.7 yg; r1-4; n7	435
Cercartetus concinnus	3.33 wean; r1-6; n7 nests	11455
	4 yg; r3-6; n4 f	1856
	4.25 emb; r4-5; n4 f	1856
	4.62 yg; mode 5; r3-6; n21 lt; lt of 10 excluded	11455
	5 yg; r1-6	10917
	5.29 enlarged teats; mode 6; r4-6; n7 lt	11455
	6 yg; n1	1164
	6 yg; n1 f	1650
	8 emb; r6-12; n7 f	11455
Cercartetus lepidus	3.25 yg; r2-4; n4 lt	11455, 11455a
	3.33 yg; r2-4; n3 lt	4043a
	4 yg; n1 f	2082a
Cercartetus nanus	3.75 yg; mode 4; r2-5; n53 lt	11453

	3.90 wean; mode 4; r3-4; n10 nests	11453
	3.92 enlarged teats; mode 4; r3-6; n55 lt	11453
	5 yg; n1	11394
Distoechurus pennatus	1 yg	10684
	1 yg; n1	11924
	1 yg; n1 f	11455
	2 emb; n1 f	11455
SEXUAL MATURITY		
Acrobates pygmaeus	f: 1st mate: 8 mo	3314
	m: testes large: 7-12 mo	3314
	m: est 9-18 mo; n2	11457
Burramys parvus	f: pouch development: 2nd yr	5691
	f: 1st mate: ~10-11 mo	6785
	m: large scrota: 8 mo	5691
	m: 1st mate: ~10-11 mo	6785
Cercartetus caudatus	f: 1st birth: 460-470 d; n2	435
Cercartetus nanus	f: 1st preg: 4.5-9 mo	11453
	m: large scrota: 4.5-9 mo	11453
CYCLE LENGTH		
Burramys parvus	20.3 d; r17-26.5	6785
GESTATION		
Burramys parvus	14-16 d; n1	5691
Cercartetus concinnus	min 51 d	1164
LACTATION		
Acrobates pygmaeus	pouch exit: 50-60 d	3314
	pouch exit: ~65 d	11454
	wean: 95-100 d	3314
	wean: 100 d	11454
Burramys parvus	pouch exit: 3-4 wk	5691
	pouch exit: min 24 d	2582
	pouch life: est ~35 d	6785
	wean: 55-60 d (table), 8-9 wk (abstract)	5691
	1st trapped: ~71 d	6785
Cercartetus caudatus	1st pouch exit: ~34 d; ±5	435
	independent: 92 d; ±10	435
Cercartetus concinnus	wean: ~50 d	1164
Cercartetus lepidus	independent: 3 mo	10474
Cercartetus nanus	pouch exit: 33-37 d	11453
	wean: 60-65 d	11453
INTERLITTER INTERVAL		
Acrobates pygmaeus	r68-161 d; n11	11454
Cercartetus nanus	r56-135 d; n20	11453
LITTERS/YEAR		
Acrobates pygmaeus	2	3314
	2	11456
Burramys parvus	1	2582
	1	6785
Cercartetus caudatus	est 2	435
Cercartetus concinnus	>1	1856
SEASONALITY		
Acrobates pygmaeus	birth: July-Jan, peak Aug, Nov; VIC, Australia	3314
	birth: mid July-Feb; VIC, Australia	11454, 11456
Burramys parvus	birth: Jan, Apr, Sept, Nov; n4; captive	5691
	birth: end Oct-mid Nov; VIC, Australia	6785
	birth: Nov-Dec; n4; NSW, Australia	2582
Cercartetus caudatus	lact f: May-Dec, perhaps yr rd, low mid May-July; New Guinea	2837
Cercartetus concinnus	preg/lact f: est yr rd; WA, Australia	1856
	lact f: Jan, July, Sept; SA, Australia	10917, 11901
	lact f: May-Feb; SA, VIC, Australia	11455
	lact f: July-May; WA, Australia	11455
Cercartetus lepidus	birth: Sept-Jan; TAS, Australia	10474
	birth: yr rd; VIC, Australia	11455a
Cercartetus nanus	birth: Apr-Aug; Australia	10081
	birth: yr rd, peak Oct/Nov-Mar/Apr; VIC, Australia	11453

Order Marsupialia, Family Petauridae

Petaurids (*Dactylopsila* 250-400 g; *Gymnobelideus* 100-165 g; *Petaurus* 100-710 g; *Pseudocheirus* 440-2000 g; *Schoinobates* 900-1800 g) are nocturnal, arboreal, gliding omnivores from Australian and New Guinean forests. *Gymnobelideus* is endangered (10080). Males may be larger than females in *Petaurus* and *Pseudocheirus*, but not in *Gymnobelideus* (10080, 10081). Males have a large prostate, 2 pairs of bulbourethrals, and 3 pairs of paracloacal glands (10124, 10735). A copulatory plug is present after mating in *Petaurus* (10120). Females have 2-4 teats (3353, 4192, 10120, 10698, 10917, 10998). *Schoinobates* females retain the entire Wolffian duct (10141). Neonates open their eyes 74-82 d (*Petaurus* 10120), 91-106 d (*Pseudocheirus* 4192, 10802), or 121 d (*Schoinobates* 10998). Details of anatomy (363, 5083, 8293, 10124), endocrinology (1618, 1619, 9790), spermatogenesis (534), placentation (549, 9853), and milk composition (7880) are available.

NEONATAL MASS		
Petaurus breviceps	194 mg; n2	1856, 10120
Pseudocheirus herbertensis	180 mg; n1	4192
Pseudocheirus peregrinus	300 mg; n1	10802
Schoinobates volans	273 mg; n1 preserved	549
NEONATAL SIZE		
Pseudocheirus peregrinus	TL: ~10 mm; n3	9169a
	CR: 15 mm; n1	10802
Schoinobates volans	CR: 14.6 mm; n1 preserved	549
WEANING MASS		
Gymnobelideus leadbeateri	30-38 g; 120 d	10080
Petaurus australis	285 g	10081
Petaurus breviceps	4.5-5 g; 40 d	10120
	45 g; 4 mo	10081
Pseudocheirus herbertensis	115-120 g; pouch exit	4192
	200 g	10081
Pseudocheirus peregrinus	~100 g; 1st solid food	8205
	f: ~310 g; 6 mo	8204
	m: ~350 g; 6 mo	8204
	400 g	10081
Schoinobates volans	24.2 g; teat detachment	10141
	150 g; pouch exit	10141
	500-600 g; independent	10141
LITTER SIZE		
Dactylopsila trivirgata	1-2 yg; n3 lt	11159
	1.5 yg	10081
	2 yg; n1 lt	11159
Gymnobelideus leadbeateri	1.5 yg; n24 lt	10080
Petaurus australis	1 yg ea; n2 lt	4636
	1 yg ea; n4 lt	2103
	1 yg; n10 lt	3862
	mode 1; r1-2	2104
Petaurus breviceps	1.5; r1-2; n22 lt	12128
	1.57; r1-2; n14 lt	10120
	1.8 yg; 81% twins; r1-2; n116 lt	10533
Petaurus norfolcensis	1.5; r1-2; n15 lt	12128
Pseudocheirus herbertensis	1.8; r1-3; n12 lt	4192
Pseudocheirus peregrinus	1-3 ova	5083
	1 yg; n1 lt	2841
	1.7 yg; r1-3; n22 lt, 6 f	9169a
	1.94; mode 2; r1-3; n84 lt	5017
	2-3 emb	9853
	mode 2; r1-4	8205
	2 enlarged teats; r1-3; n58 f	5083, 10802
	4 emb; n1 lt	3353
Schoinobates volans	1 emb ea; n17 f	549

SEXUAL MATURITY

Gymnobelideus leadbeateri	f: 1st repro: 1.3 yr	10080
	m: change scrotal color: 378-550 d	10080
Petaurus breviceps	f: 1st mate: 236.5 d; r227-246; n2	9669
	f: 1st birth: 15 mo; n1	10120
Pseudocheirus herbertensis	f: 1st birth: 16 mo; n2	4192
Pseudocheirus peregrinus	f: 1st conceive: 11 mo	9169a
	f: 12.5 mo; n50	8205
	m: testes descend: ~10-12 mo	8205
	m: 1st sire: 11 mo	9169a
	f/m: 12 mo	5017
Schoinobates volans	f/m: 1.5-2 yr	10141

CYCLE LENGTH

Petaurus breviceps	rpy to estrus: 11-13 d; n3	10120
	29 d; n1	10120
Pseudocheirus peregrinus	28 d	5083
Schoinobates volans	rpy to estrus: 0 d	10998

GESTATION

Gymnobelideus leadbeateri	<20 d	10080
Petaurus breviceps	16 d; r15-17; n3	10120
	16 d; n11 lt	9669

LACTATION

Gymnobelideus leadbeateri	pouch exit: 87.4 d; r80-93; n5	10080
	independent: est 120 d	10080
Petaurus australis	pouch life: 100 d	4636
	wean: 5 mo	9383
Petaurus breviceps	teat attachment: 40 d	10120
	pouch exit: 74 d	9669
	wean: 4 mo	10120
Petaurus norfolcensis	wean: 4.0 mo	10081
Pseudocheirus herbertensis	pouch exit: 115-120 d	4192
	wean: est 150-160 d	4192
Pseudocheirus peregrinus	teat attachment: 42 d	5017
	pouch exit: 100-120 d	9169a
	pouch exit: ~120 d	10802
	pouch exit: 125-130 d	5017
	solid food: 4-5 mo	8205
	wean: 145-220 d	5017
	wean: 5-7 mo	8204
	nurse: 5-8 mo	8205
	wean: 160 d, (4-6 wk after pouch exit at 110-120 d)	9169
	wean: 6-7 mo	10802
Schoinobates volans	teat attachment: 6 wk	1987
	teat attachment: 2-3 mo; field data	10141
	teat attachment: 93 d	10998
	pouch exit: 170 d	10998
	pouch exit: 6 mo	4635
	wean: 7.5 mo	10081
	wean: 230 d	10998

INTERLITTER INTERVAL

Petaurus australis	2 yr	2103
Pseudocheirus peregrinus	3-10.25 mo; n16 intervals, 6 f	9169a

LITTERS/YEAR

Gymnobelideus leadbeateri	1-3	10080
Petaurus australis	1	4636
Petaurus breviceps	1-2	10120
Pseudocheirus herbertensis	mode 1; r1-2	4192
Pseudocheirus peregrinus	mode 1; r1-2	5083
	1-2	5017
	1-2	8205
Schoinobates volans	1	11010

SEASONALITY

Dactylopsila trivirgata	mate: Aug, perhaps Mar; few data; QLD, Australia	11159
	birth: Feb-Aug; QLD, Australia	10081
Gymnobelideus leadbeateri	birth: Mar-Dec, peaks Apr-June, Oct-Dec; VIC, Australia	10080
Petaurus australis	birth: Apr, June, Aug-Dec; few data; VIC, Australia	4993
	birth: Apr, Nov-Dec; n3; NSW, Australia	4993
	birth: June-Dec, peak July-Sept; n10; NSW, Australia	3862
	birth: yr rd, except Nov; QLD, Australia	4993
	birth: Aug-Oct; n4; VIC, Australia	2103
	birth: yr rd; QLD, Australia	9383
Petaurus breviceps	mate: June-Nov; n14; captive	10120
	mate: peak Sept-Dec; VIC, Australia	10533
	birth: Mar-Sept, Dec, peaks Apr, June; n22; zoo	12128
	birth: July; VIC, Australia	10917
	birth: Aug-Dec; Australia	10081
	birth: Sept-Nov; VIC, Australia	10533
Petaurus norfolcensis	birth: Jan-Apr, June-Oct; n15; zoo	12128
	birth: June-Jan; Australia	10081
Pseudocheirus dahli	mate: yr rd; Arnhem land, Australia	2214
Pseudocheirus herbertensis	birth: Apr-Dec, peak May-July 74%; lab	4192
	birth: June-July, Dec; Australia	10081
Pseudocheirus peregrinus	mate: May-Aug; TAS, Australia	3353
	birth: Apr-Aug, Oct-Feb; n22; captive	9169a
	birth: Apr-Dec, peak end May-early June, mid Oct; VIC, Australia	5017
	birth: Apr-Dec, peak May-June; VIC, Australia	8205
	birth: peaks May-Aug, Nov-Dec; VIC, Australia	4822
	lact f: May-Dec; VIC, Australia	5083, 10802
Schoinobates volans	spermatogenesis: Dec-July, peak Mar; NSW, Australia	534
	testes: large Mar, small Aug-Sept; NSW, Australia	534
	birth: Apr-May; NSW, Australia	10141
	birth: July-Aug; VIC, Australia	10917

Order Marsupialia, Family Macropodidae

Kangaroos and wallabies (*Aepyprymnus* 1-4 kg; *Bettongia* 1-2 kg; *Caloprymnus* .65-1.5 kg; *Dendrolagus* 6-10 kg; *Dorcopsis* 2-7 kg; *Hypsiprymnodon* 400-700 g; *Lagorchestes* 2-5 kg; *Lagostrophus* 1-2.5 kg; *Macropus* 3-90 kg; *Onychogalea* 5 kg; *Peradorcas* 1-1.5 kg; *Petrogale* 1-9 kg; *Potorous* 0.7-2 kg; *Setonix* 2-5 kg; *Thylogale* 3.5-6.5 kg; *Wallabia* 11-17 kg) are terrestrial herbivores in Australia and New Guinea. *Caloprymnus* is presumed extinct.

Female *Lagorchestes conspicillatus* have one more chromosome than males, while male *Potorous tridactylus* and *Wallabia bicolor* have one more than females (4013). *Macropus* males are often larger than females (799, 7646, 9861). A postpartum estrus (prepartum in *Wallabia*) and embryonic diapause are usual (590, 791, 1390, 1410, 1861, 2857, 3328, 3329, 3333, 3873, 3947, 4366, 4537, 4538, 4539, 5159, 5359, 5360, 5362, 5814, 6449, 6995, 7217, 7219, 7834, 8544, 8672, 9045, 9047, 9275, 9277, 9280, 9377, 9419, 9850, 9851, 9855, 9857, 9858, 9859, 9860, 9861, 9870, 10125, 10442, 10987, 10991, 11006, 11009, 11030, 11412, 11417, 11982; cf 6997, 6998, 7504, 9859).

Neonates open their eyes at 78-92 d (*Aepyprymnus* 5359; *Bettongia* 10991), 90-105 d (*Potorous* 3683, 4129, 4585), 107-120 d (*Petrogale* 5360), 115-150 d (*Macropus* 2859, 5814, 6994, 7216, 7646, 9861, 11426; *Thylogale* 9281). Females have 2-4 mammae (591, 1842, 3311, 3353, 5363, 8089, 8295, 10991).

Macropus eugenii is the "laboratory rat" of macropod reproductive physiology as its extensive literature attests (see 11008 for pre-1987 review). Recent descriptions of endocrinology (1665, 3330, 4804, 4812, 5431, 9867, 10395a, 11005), sex differentiation (7990, 9863, 11004), sperm and testicular maturation (5117, 5428, 5431, 6941, 9793a, 11732), milk and lactation (2049, 2705, 7245, 7248, 7879) are available.

For other macropods, details of anatomy (3352, 3359, 4411a, 6968, 7504, 8291, 8293, 8296, 9208, 9851, 9856, 10989, 11413), endocrinology (1200, 1537, 1618, 1619, 1857, 2184, 3169, 3170, 6454, 9279, 9790, 9860, 11001a, 11412), sperm (5074, 7871, 9209, 9210, 9746, 9793a), gestation (3357, 9861), placentation (1200a, 1201, 1549, 9853), birth (718, 1588, 3299, 3355, 4437, 9858, 10986), marsupium (1745, 1945, 6080, 11983), lactation (2184, 6449, 9857), milk composition (497, 1066, 1945, 1985, 2160, 4026, 4070, 5445, 5499, 5500, 6374, 6375, 7215, 7247b, 7249, 8674, 10162, 11982a), weaning (9276), and seasonal quiescence (2176, 2177, 2183, 6590, 6592, 6593) are available.

NEONATAL MASS		
Bettongia gaimardi	307 mg; sd15; n5	9278
Bettongia lesueuri	317 mg; r267-344; n9	10991
Macropus agilis	634 mg; r580-660; n5	7216
Macropus eugenii	413.8 mg; r363-492; n4	11009
	444.2 mg; sd61; n25	11451
Macropus fuliginosus	934.33 mg; r828-1113; n6	8666
Macropus giganteus	898.67 mg; r740-1000; n3	8666
Macropus parma	506.7 mg; se10; r436-569; n17	6994, 6999
Macropus rufogriseus	537 mg; r453-690; n4	7219
	558 mg; r493-600; n6	11411
	700 mg; n6	6597
Macropus rufus	817 mg; r745-900; n5	9861, 9862
Petrogale xanthopus	508 mg; r460-572; n3	8672
Potorous tridactylus	333 mg; n1	5072
Setonix brachyurus	310-350 mg	11981
	382.5 mg; se43.6; r309.4-477.4; n22	9910
	500 mg	591
Thylogale billardieri	420 mg; n3 yg	9281
NEONATAL SIZE		
Bettongia lesueuri	CR: 13.75 mm; r13.00-14.30; n9	10991
Macropus fuliginosus	CR: 22.1 mm; n1; near term	8671
Macropus parma	CR: 18.33 mm; se0.21; r16.5-19.7; n17	6994, 6999
Macropus rufogriseus	CR: 16.6 mm; r15.8-17.5; n6	11411
Potorous tridactylus	CR: 14.7-16.1 mm	4129
	CR: 18 mm; r18-19; n4	5072
WEANING MASS		
Bettongia gaimardi	354.6 g; sd50; r291-460; n17; pouch exit	9277
	1046-1400 g; 23-24 wk	9278
Bettongia penicillata	min 145 g; pouch exit	10125
Macropus agilis	2286.7 g; sd256.8; n12; pouch exit	7216
Macropus parma	754 g; se25; r621-868; n10; 210 d, pouch exit	6994, 6999
Macropus rufus	~5 kg; pouch exit	9862
Peradorcas concinna	300 g	7834
Petrogale penicillata	735 g; sd26; r705-750; n3; pouch exit	1909
Potorous tridactylus	90-145 g; 1st pouch exit	3683
Setonix brachyurus	mode ~500 g; pouch exit	9910
Thylogale billardieri	1029 g; sd112; r857-1170; n6; pouch exit	9281
WEANING SIZE		
Macropus parma	276.9 mm; se2.9; r264-291; n10; 210 d, pouch exit	6994, 6999
LITTER SIZE		
Aepyprymnus rufescens	1 yg ea; n27 lt	5359
	1.07; r1-2; n15 lt	12128
Bettongia gaimardi	1 emb	3357
	1 yg ea; n23 lt	9278
	1 emb ea; n70 f	3359
	1.04; r1-2; n26 lt	12128
Bettongia lesueuri	1 ea; n8 lt	10991
	1 yg; n1	9194
Bettongia penicillata	1 yg; n~10 lt	11310
Caloprymnus campestris	1 yg; n7 f	3257
Dendrolagus matschiei	1 yg	8089
	1 yg; n6 lt	5111a
	1 ea; n16 lt	2152
Dendrolagus ursinus	1; n1 lt	12128
Dorcopsis vanheurni	1 yg	7983
Hypsiprymnodon moschatus	2; n1 lt	8901
	2 yg	5363
Lagorchestes conspicillatus	1 yg	5362

Lagostrophus fasciatus	1 yg ea; n9 f	10989
	mode 1 cl; r1-2	10989
Macropus agilis	1 yg ea; n15 lt	7763
	1 yg ea; n64 lt	7216
	1.01 yg; r1-2; n134 lt	5814
Macropus dorsalis	1 ea; n3 lt	12128
Macropus eugenii	1.01 yg; r1-2; n262 births	5154
	1.07 yg; r1-2; n14 lt	12128
Macropus fuliginosus	1.00; r1-2; n280 lt	7955
	1.01 yg; r1-2; n~170 lt	8666
Macropus giganteus	1.0 yg ea; n~118 lt	8666
	1.00 yg; r1-2; n201 lt	8286
	1.02 yg; r1-2; n41 lt	12128
Macropus irma	1 yg	9120
Macropus parma	1 cl; 1 yg	6997
Macropus parryi	1 yg ea; n16 lt	6996
Macropus robustus	1 yg; 0.2% twins	2859
Macropus rufogriseus	1 yg ea; n9 lt	9426
	1.01 yg; r1-2; n103 lt	3311
	1.16 yg; r1-3; n51 lt	12128
Macropus rufus	1 yg ea; n44 lt	12128
	1 yg; 1/1000 twins	9859
Onychogalea fraenata	1 yg; n1 lt	12128
Petrogale penicillata	1 ea; n7 lt	12128
	1 yg ea; n~12 lt	5360
Petrogale xanthopus	1.06 yg; r1-2; n36 lt	12128
Potorous longipes	1 yg; n1	9720
Potorous tridactylus	1 yg ea; n5 f	4585
	1 yg ea; n30+ lt	5072
	1 yg ea; n38 lt, 13 f	648
	1 yg ea; n42 lt, 21 f	4129
Setonix brachyurus	1 yg; n1	3299
	1 yg ea; n23 lt	9911
Thylogale billardieri	1 yg	10474
Wallabia bicolor	1.1 yg; r1-2; n7 lt	12128
SEXUAL MATURITY		
Aepyprymnus rufescens	f: 1st mate: 313 d; n1	5359
	m: 1st mate: 379 d; n1	5359
Bettongia gaimardi	f: 1st birth: 272 d; ±53	9277
	m: testosterone surges: ~30 wk; 1.5 kg	9279
	m: 1st mate or sperm in urine: 272 d; ±53	9277
Bettongia lesueuri	f: 1st mate: 217 d; r207-227; n2	10991
	f: 1st birth: 200 d	10991
Bettongia penicillata	f/m: end 1st yr	10809
Hypsiprymnodon moschatus	f/m: 1+ yr	5363
Lagorchestes conspicillatus	f/m: ~1 yr	5362
Lagostrophus fasciatus	f/m: est 1 yr	10989
Peradorcas concinna	f: 1st birth: 470 d; n1	7834
Macropus agilis	f: 1st estrus: 330-480 d	5814
	m: epididymal sperm: 385-401 d	5814
Macropus eugenii	f: 1st mate: < 1 yr	301, 590
	m: rapid testes growth: 18-20 mo	5155
	m: 2 yr	590
Macropus fuliginosus	f: teat eversion: 657.4 d; r426-949; n9	8671
	f: 1st mate: 647.1 d; r493-791; n9	8671
	f: 1st birth: 749.1 d; r524-1096; n11	8671
	f: 1st birth: 20 mo	8667
	m: tubular sperm: 20 mo	8667
	m: active sperm: 31 mo	8671
Macropus giganteus	f: teat eversion: 636 d; r543-751; n8	8671
	f: 1st mate: 639.3 d; r579-707; n3	8671
	f: 1st mate: mode 21 mo; r17-28	5810
	f: 1st birth: 812.6 d; r616-1446; n9	8671
	f: 1st birth: 20-36 mo	8669

	m: epididymal sperm: 20-30 mo	5810
	m: active sperm: 48 mo	8671
	m: 1st sperm: 20-36 mo	8669
	f/m: 1st repro: 16 mo	8286
Macropus parma	f: 1st mate: 357-491 d	6997, 6999
	f: 1st mate: est 19 mo	7001
	m: 1st sperm: 19-20 mo	6997
	m: 1st sire: 24-25 mo	6997
Macropus parryi	f: 1st mate: 24-25 mo	5581a
Macropus robustus	f: 1st mate: 2 yr	2859
Macropus rufogriseus	f: 1st mate: 13 mo	7054
	f: 1st lact: 14-18 mo	1665
	m: testes size: 18-22 mo	1665
Macropus rufus	f: 1st mate: min 14-20 mo	3525, 9858, 9861
	m: 1st sperm: 2-3 yr	3525, 9858, 9861
Petrogale penicillata	f: 1st mate: 540 d	5360
	m: 1st mate: 590 d	5360
Petrogale xanthopus	f: 1st conceive: ~16 mo	4993
	f: 1st birth: 541.3 d; sd112.6; r321-675; n10	8672
Potorous tridactylus	f: 1st birth: 11.5-17 mo	9718
	m: 600 g	5074
	f: 1st mate: ~1 yr	4128
Setonix brachyurus	f: 1st mate: 16-23 mo; ~2.5 kg	2811
	f: 1st birth: 8-10 mo; n5	9910
	m: 1st sire: 389-455 d; n7 d	9910
Thylogale billardieri	f: 1st birth: 14-15 mo	9280
	m: 14 mo; 3.9-4.75 kg	9280
Wallabia bicolor	f/m: 15 mo	5813
CYCLE LENGTH		
Aepyprymnus rufescens	rpy to estrus: 8.5 d; sd4.2; r4-14; n4	7504
	r21.0-25.0 d; n5 cycles, 4 f	5359
	33.8 d; sd2.1; mode 33-36; r28-36; n11 cycles, 4 f	7504
Bettongia gaimardi	rpy to estrus: 18.9 d; sd1.3; n16	9277
	22.6 d; sd3.5; r17-37; n46 cycles, 10 f	9277
Bettongia lesueuri	rpy to estrus: 21.50 d; mode 22; r16-28; n14	10991
	22.17 d; mode 23; r11-35; n18 cycles	10991
Macropus agilis	r29-34 d; n8 cycles, 4 f	5814
	32.4 d; sd2.2; r28-41; n23 cycles, 6 f	7216
Macropus eugenii	estrus: 12 h; n7 cycles, 7 f	6373
	rpy to estrus: 26.4 d; sd0.57; r25.5-28.5; n10 cycles with gestation, 10 f	7217
	rpy to estrus: 30.4 d; sd0.99; r28.5-31.5; n10 cycles, 10 f	7217
	28.4 d; se0.1	799
	29.4 d; sd1.26; r28-30; n10 cycles with gestation, 10 f	7217
	29.6 d; se0.45; r27-32; n12 cycles with gestation, 6 f	8674a
	30.6 d; sd1.17; r29-33; n10 cycles, 10 f	7217
	32.1 d; se0.58; r30-37; n16 cycles, 6 f	8674a
Macropus fuliginosus	rpy to estrus: 6.3 d; sd0.1	8667
	34.6 d; sd3.2; r25-48; n165 cycles, 8 f	8667
	34.85 d; sd4.42; r25-55; n292 cycles, 31 f	8671
Macropus giganteus	r41-43 d; n4 f	5810
	45.58 d; sd9.82; r26-81; n105 cycles, 19 f	8671
Macropus parma	rpy to estrus: 6-15 d	6997
	40.34 d; ±0.47; r36-48; n51 cycles	6995, 6998
	41.78 d; se0.72; r36-59; n58 cycles, 16 f	6997
Macropus parryi	rpy to estrus: 6 d; n1	5581a
	42.2 d; r41-44; n5	5581a
Macropus rufogriseus	rpy to estrus: 26 d	11459
	32.9 d; sd2.3; r28-40; n38 cycles, 8 f	7219
	33.4 d; sd2.3; r30-38; n14 cycles, 4 f	7219
Macropus rufus	rpy to estrus: 34.0 d; se0.89; n8	9855
	rpy to estrus: 34.46 d; sd1.92; n13 cycles, 10 f	9857
	34.30 d; se0.50; n23 cycles, 9 f	9858
	34.64 d; sd2.22; n42 cycles, 13 f	9857

	34.81 d; se0.63; n16 nonpreg	9855
	35.53 d; se0.52; n17 preg	9855
Peradorcas concinna	33.73 d; sd1.65; mode 35; r31-36; n52 cycles	7834
Petrogale penicillata	rpy to estrus: 31 d; n2	5360
	31.06 d; r30.2-32.0; n5 cycles, 5 f	5360
Petrogale xanthopus	rpy to estrus: 34.375 d; n1	8672
	r32-37 d; n3	8672
Potorous tridactylus	41.5 d; r39-44; n5 cycles	5072
Setonix brachyurus	luteal phase: 20 d	9851
	rpy to estrus: 26 d	9859
	28 d	9851, 9859
Thylogale billardieri	rpy to estrus: 29.2 d; r27-31; n6	9280
	remove cl to estrus: 11 d; n2	9280
	30.3 d; r30-32; n12 cycles	9280
Wallabia bicolor	rpy to estrus: 26 d; r23-28; n3	9859
	32 d; r27-44; n5	9859
GESTATION		
Aepyprymnus rufescens	rpy to birth: 18.7 d; r18-20.8	5359
	r22.0-23.6 d; n5	5359
	25.5 d; sd9.8; r21-30; n2	7504
Bettongia gaimardi	rpy to birth: 17.6 d; sd0.9; r16-19	9277
	19.5 d	9746
	21.3 d; sd0.8; 20-22; n15	9275, 9277
Bettongia lesueuri	rpy to implantation: 21-22 d	11459
	rpy to birth: 20.00 d; mode 20; r17-24; n14	10991
	21.33 d; mode 21; r22-23; n6	10991
	21-21.5 d; n1	10442
Bettongia penicillata	rpy to birth: 17.75 d; sd1.12; n20	10125
Dendrolagus matschiei	~32 d	8089
Lagorchestes conspicillatus	rpy to birth: 28-30 d	5362
	29-31 d; no lactation	5362
	5 mo with diapause	5362
Macropus agilis	rpy to birth: 25.85 d; r24.2-27.7; n4	5814
	29.4 d; sd0.9; r28-31; n16 lt, 8 f	7216
	30.52 d; r29.0-33.1; n4 lt, 4 f	5814
Macropus eugenii	rpy to birth: 26.2 d; se0.67; r23.5-26.5; n10	7217
	rpy to birth: 26.9 d; sd0.56; n10	12041
	rpy to birth: 27.30 d; sd0.95; n30	11451
	rpy to birth: 27.6 d; sd1.2; n16	11009
	28.4 d; se0.2; n12	791, 799
	12 mo with diapause	791, 799
Macropus fuliginosus	30.56 d; sd2.55; mode 30; r27-46; n115 lt, 30 f	8666
	30.8 d; sd1.8; r28-35; n25 lt, 6 f	8667
Macropus giganteus	36.44 d; sd1.63; mode 36; r33-45; n53 lt, 23 f	8666
	36.7 d; r35-38; n14	5810
Macropus parma	rpy to birth: 31.3 d; r30.5-32; n6	6998
	34.54 d; se0.13; mode 34-35; r33-36; n28 lt, 16 f	6995, 6997
Macropus parryi	36.3 d; r34-38; n3	5581a
Macropus robustus	rpy to birth: 33-34 d	5812
	35 d; mating to birth; 8-9 mo with diapause	5812
Macropus rufogriseus	rpy to birth: 22.0 d; sd4.5	7219
	rpy to birth: 26.8 d; sd0.6; n6	11412
	rpy to birth: 27 d; se0.2; n5	2183
	rpy to birth: 27.3 d; r26-28; n3	3311
	mating to ppe: 30 d; sd1.4; r28.5-34; n12 lt, 5 f	7219
	mating to ppe: 30.5 d; sd1.7; r29-41; n19 lt, 8 f	7219
	max 12 mo with diapause	3311
Macropus rufus	rpy to birth: 31.34 d; se0.35; n8	9855
	rpy to birth: 31.64 d; sd0.65; n7 f	9857
	33.00 d; sd0.32; n20 lt, 14 f	9857
	33.17 d; se0.16; n14	9855
Peradorcas concinna	30 d; n1	7834
Petrogale penicillata	rpy to birth: 29.1 d; r27.6-30; n6	5360
	30.8 d; r30-32; n5	5360

	199.5 d; r193-206; n4 with diapause	5360
Petrogale xanthopus	rpy to birth: 31.5 d; n1	8672
	31.8 d; sd0.76; 31-33 d; n5 birth, 3 f	8672
Potorous tridactylus	rpy to birth: 29 d; n4	9870
	30, 32-43, 34, 37.5 d; n4	5072
Setonix brachyurus	rpy to birth: mode 25 d; r24-27; n26	9910
	rpy to birth: 25.8 d; r25-26; n17	9911
	25-26 d; n1	9910
	27 d	9851, 9853, 9859
Thylogale billardieri	rpy to birth: 28.7 d; r27-31; n6	9280
	30.2 d; mode 30; r30-31; n5	9280
Wallabia bicolor	35 d; r33-38; n4 lt, 2 f	9859
	36.8 d; r35-38	7983
LACTATION		
Aepyprymnus rufescens	teat attachment: 87.3 d; r84-92	5359
	pouch exit: 114 d; r105-119; n5	5359
	nurse: ~164 d	5359
Bettongia gaimardi	teat attachment: ~5 wk	9278
	pouch exit: 106.1 d; sd2.5; r101-111	9275, 9277, 9278
	wean: 40-60 d after pouch exit	10474
	wean: 8-9 wk after pouch exit	9278
Bettongia lesueuri	pouch exit: 116.25 d; mode 115; r113-122; n4	10991
	nurse: 23-74 d after pouch exit	10991
Bettongia penicillata	pouch exit: 97.9 d; sd3.28; n25	10125
Dendrolagus matschiei	solid food: 266 d	8089
	pouch exit: 305 d	8089
Hypsiprymnodon moschatus	pouch exit: 21 wk	5363
Lagorchestes conspicillatus	pouch exit: 5 mo	5362
Macropus agilis	pouch exit: 208.5 d; r200-220; n9	5814
	pouch exit: 219.3 d; sd9.7; r207-235; n15	7216
	nurse: 321.5 d; r288-340; n6	7216
Macropus eugenii	teat attachment: 100-110 d	7646
	solid food: 190-220 d	7646
	pouch exit: ~6 mo	590
	pouch exit: 241 d; se8; r222-259; n4	4811
	pouch exit: 35-40 wk	11699
	pouch exit: 250 d	799
	wean: 269 d; se4; r258-282; n6	4811
	wean: 45 wk	4029b
Macropus fuliginosus	1st pouch exit: 298.40 d; sd34.29; r236-359; n25	8666
	pouch exit: 300 d	8667
	permanent pouch exit; 323.12 d; sd22.92; r280-359; n24	8666
Macropus giganteus	pouch exit: 297 d; r277-308; n12	5810
	1st pouch exit: 283.86 d; sd24.70; r247-329; n21	8666
	permanent pouch exit: 319.23 d; sd18.36; r297-359; n22	8666
	nurse: 18 mo	5810
Macropus parma	teat attachment: ~110 d	6994
	1st pouch exit: 160-175 d	6997
	permanent pouch exit: 211.9 d; se1.0; r207-218; n10	6997
	wean: 2.5-3.5 mo after pouch exit	6997
Macropus parryi	pouch exit: 9 mo	5581a
	nurse: 10-15 mo	5581a
Macropus robustus	teat attachment: 120 d	2859
	pouch exit: 7-8 mo	2857, 2859
	wean: 13 mo	2859
Macropus rufogriseus	pouch exit: 77-204 d	3311
	pouch exit: 280 d	7219
	pouch exit: 9-10 mo	5335
	wean: 4-5 mo after pouch exit	5335
Macropus rufus	pouch exit: 236.50 d; se2.66; n10	9855
	wean: 360 d	9379
Peradorcas concinna	pouch exit: 160 d	7834
	independent: 175 d	7834, 9485

Petrogale penicillata	pouch exit: 189-227 d	5360
	pouch exit: 194 d; sd8; r188-205; n4	1909
	independent: 59-86 d after pouch exit	1909
Petrogale xanthopus	pouch exit: 194.9 d; sd3.9; 190-201; n7	8672
Potorous tridactylus	teat attachment: 64 d; n1	5072
	1st pouch exit: 103.2 d; se2.0; r100-106; n6	3683
	pouch exit: 17 wk	4585
	permanent pouch exit: 119.7 d; se1.6; r115-124; n6	3683
	pouch exit: 140 d	4133
	wean: 21-23 wk	4585
Setonix brachyurus	pouch exit: 185-195 d; r177-196; n19	9910
Thylogale billardieri	pouch exit: 202 d; sd7; r196-212; n6	9281
	nurse: 3 mo after pouch exit	9281
Wallabia bicolor	pouch exit: 256 d	9859
INTERLITTER INTERVAL		
Lagorchestes conspicillatus	5 mo	5362
Macropus agilis	potentially 7-8 mo	5814
Macropus eugenii	1 yr with diapause	791
Macropus fuliginosus	372.02 d; sd61.39; r299-559; n45 intervals, 24 f	8666
	398.2 d; sd82.9; r278-559	8667
Macropus giganteus	306 d; r299-312; n4 with diapause	5810
	356 d; r338-368; n5	5810
	362.59 d; sd56.90; r283-523; n27 intervals, 15 f	8666
Macropus robustus	8-9 mo	5812
Macropus rufogriseus	9 mo	7054
Petrogale penicillata	~200 d	5360
Potorous tridactylus	122.8 d; se2.2; r118-129; n5	3683
	min 5-6 mo; yg lives	4585
Wallabia bicolor	~8 mo	7983
SEASONALITY		
Aepyprymnus rufescens	mate: est yr rd; QLD, Australia	7504
	birth: yr rd; few data; captive	5359
	birth: yr rd; n15; zoo	12128
Bettongia gaimardi	estrus: fall-spring; captive	9746
	mate: winter; TAS, Australia	3353
	birth: none Jan, Sept; n26; zoo	12128
	birth: yr rd; captive	9275
Bettongia lesueuri	mate: yr rd, peaks Apr-May, Nov-Dec; captive	10991
Bettongia penicillata	birth: yr rd; captive	10125
Caloprymnus campestris	pouch yg: June, Aug, Dec; s central Australia	3257
Hypsiprymnodon moschatus	mate: Feb-May, rainy season; few data; NSW, Australia	8901
	conceive: Feb-June; QLD, Australia	5363
Lagorchestes conspicillatus	birth: yr rd; QLD, Australia	5362
	birth: yr rd, peaks Mar, Sept; Barrow I, WA, Austraila	10474
Lagostrophus fasciatus	birth: peak est before July; WA, Australia	5071, 10989
Macropus agilis	birth: yr rd; n135; QLD, Australia	5814
	birth: yr rd; captive	7216
Macropus eugenii	spermatogenesis: yr rd; captive	4539
	prostate: peak Jan; captive	5155
	mate: Jan-Feb; captive	1664
	mate: end Jan; SA, Australia	3333
	mate: Feb-May; captive	9047
	ovulate: Dec-Jan; SA, Australia	3333
	obligate diapause: June-Dec; captive	11003
	implant: end Dec; SA, Australia	3333
	birth: Jan-Aug, most Jan-Feb; SA, Australia	301
	birth: end Jan-early May; n10; captive	4811
	birth: Apr-Oct; n14; zoo	12128
Macropus fuliginosus	birth: yr rd; captive	8671
	birth: yr rd, peak summer; SA, Australia	8667
	birth: yr rd, peak Nov; VIC, Australia	7956
Macropus giganteus	epididymal sperm: yr rd; s QLD, Australia	5810
	mate: yr rd, peak Oct-Dec; s QLD, Australia	5810
	birth: yr rd; captive	8666, 8671

	birth: yr rd, 70% Dec-Feb; TAS, Australia	8286
	birth: yr rd, low winter; e Australia	8669
Macropus parma	mate: yr rd, low July-Dec; captive	6998
	birth: peak Feb-June; NSW, Australia	7000
	birth: yr rd; n1 yr; Mar-July; n1 yr; New Zealand	7001
Macropus parryi	birth: yr rd, peak Mar-July; NSW, Australia	5581a
Macropus robustus	mate: yr yr, peak Sept-Nov; WA, Australia	2857
	birth: yr rd; WA, Australia	9421
Macropus rufogriseus	prostate: max Feb-Mar, low Aug-Oct; New Zealand	1665
	mate: peak end summer; TAS, Australia	4128
	birth: Jan-July; TAS, Australia	7219
	birth: Jan-Aug; se Australia	7219
	birth: Feb-Mar; New Zealand	1665
	birth: July-Dec, peak Aug; captive	3311
	birth: yr rd, peak Jan; QLD, Australia	7054
Macropus rufus	estrus: yr rd; captive	9858
	birth: yr rd; NSW Australia	3525
	birth: yr rd; WA, Australia	9421
	birth: yr rd; n44; zoo	12128
Peradorcas concinna	small pouch young: Aug; NT, Australia	9485
Petrogale penicillata	birth: yr rd; captive	5360
Petrogale xanthopus	mate: yr rd; captive	8672
	birth: yr rd; n36; zoo	12128
Potorous longipes	birth: est yr rd; se Australia	10474
Potorous tridactylus	mate: winter; TAS, Australia	3353
	mate: yr rd, peak Dec-Feb; TAS, Australia	6733
	mate: yr rd, peaks end Aug-early Oct, Jan; TAS, Australia	4127
	mate: est yr rd; TAS, Australia	4128
	birth: not May-June; TAS, Australia	5072
	birth: yr rd; TAS, Australia	4585
	birth: yr rd, peak Sept-Nov; captive	9718
	birth: yr rd; captive	4133
	birth: yr rd; captive	648
Setonix brachyurus	anestrus: Aug-Jan; WA, Australia	9849
	mate: yr rd; WA, Australia	9912
	birth: Jan-June; Rottnest I, Australia	9912
	birth: yr rd; mainland, Australia	9909, 9912
	birth: peak Feb-Mar; WA, Australia	9849
	birth: peak Mar; WA, Australia	9910
	birth: yr rd, peak Feb-May; WA, Australia	2811
Thylogale billardieri	mate: yr rd; TAS, Australia	9280
	birth: peak Apr-June; TAS, Australia	9280
Wallabia bicolor	birth: yr rd; n18; captive	12128
	birth: yr rd; QLD, Australia	5813

Order Marsupialia, Family Phascolarctidae, Genus *Phascolarctos*

Koalas (*Phascolarctos cinereus* 4-14 kg) are solitary, arboreal folivores of Australian forests. Males are larger than females (7107, 10081). Their Sertoli cells have crystalloid inclusions (4370a, 5696) and their sperm are structurally heterogeneous (11681). A copulatory plug is produced (10122). Females have 2 teats and a posteriorly opening pouch (10698, 11901). Young are coprophagous (7358). Two symposium volumes review koala biology (807, 6345). Details of semen (11681) and placentation (1549, 8294) are available.

NEONATAL MASS		
Phascolarctos cinereus	360 mg	10917
WEANING MASS		
Phascolarctos cinereus	0.82-1.49 kg; n14; 12 mo	10122
LITTER SIZE		
Phascolarctos cinereus	1 yg; n4 lt	3300
	1; 2 occasional	10698
	1 yg ea; n31 lt, 25 f	10122

SEXUAL MATURITY		
Phascolarctos cinereus	f: 1st conceive: 2 yr; n3	10122
	f: 12-14 mo	8706
	f: end 2 yr	2884
	f: end 2 yr	6894
	m: 1st sire: 3 yr	10122
CYCLE LENGTH		
Phascolarctos cinereus	27-30 d	2884, 10122
GESTATION		
Phascolarctos cinereus	25-30 d	8700, 8706
	34.6 d; r34-36; n7 lt	10122
LACTATION		
Phascolarctos cinereus	teat attach: ~5.5 mo	8700
	pouch exit: 5 mo; n1	3300
	pouch exit: ~5-6 mo	7107
	pouch exit: 6-7 mo	2884
	pouch exit: 209-224 d	10122
	pouch exit: 240-270 d	9379
	wean: ~10 mo	8700, 8704, 8706
	wean: 11 mo	6898
	wean: ≤12 mo	2884
	wean: 12 mo	10122
	independent: 1 yr	7107
INTERLITTER INTERVAL		
Phascolarctos cinereus	9-12 mo	2884
	1 yr	7107
	every other yr	10917
SEASONALITY		
Phascolarctos cinereus	mate: ~Oct; se Australia	1550
	birth: Jan-Mar; VIC, Australia	6898
	birth: Feb-Sept; VIC, Australia	7107
	birth: Oct-Jan 80%; QLD, Australia	10122
	birth: Nov-Mar; QLD, Australia	6899

Order Marsupialia, Family Vombatidae

Wombats (*Lasiorhinus krefftii*, *L. latifrons*, *Vombatus ursinus*) are 20-40 kg of muscle inside a tough integument. These crepuscular and nocturnal, solitary herbivores burrow extensively in the semi-arid regions of Australia. Females are larger than males (5336). A copulatory plug occurs (2158). Details of anatomy (1329, 8000), endocrinology (8400, 9790), and placentation (8294) are available.

NEONATAL MASS		
Lasiorhinus latifrons	500 mg; n3	3632
NEONATAL SIZE		
Lasiorhinus latifrons	2 cm; n3	3632
WEANING MASS		
Lasiorhinus latifrons	1.4-3.4 kg; 8 mo	3632
	2 kg; solid food	11587
	5.5-7.5 kg; 1 yr	3632
	13.55 kg; ±0.79; n10; 1st capture	11587
Vombatus ursinus	5.7 kg; n2	8747
LITTER SIZE		
Lasiorhinus latifrons	1 yg; n4 lt	2158
	1 yg; n8 f	11587
Vombatus ursinus	1 yg; n1 lt	2021
	1; twins recorded	10698
SEXUAL MATURITY		
Lasiorhinus latifrons	f: 18 mo	3632
	f: lact: 18.75 kg	11587
	m: spermatogenesis: 18 mo	3632
Vombatus ursinus	f/m: 2 yr	10698

CYCLE LENGTH		
Vombatus ursinus	33 d; r32-34; n8 cycles, 4 f	8400
GESTATION		
Lasiorhinus latifrons	20-21, 21-22 d; n2	2158
LACTATION		
Lasiorhinus latifrons	solid food: 265 d; n1	2158
	pouch exit: 6 mo	11587
	pouch exit: 9 mo	3632
	wean: 8-9 mo	11587
	wean: 12 mo	3632
Vombatus ursinus	pouch exit: 9.5 mo	8747
	independent: 12-15 mo	10698
SEASONALITY		
Lasiorhinus latifrons	mate: summer; SA, Australia	9852
	birth: Oct-Dec; SA, Australia	11587
Vombatus ursinus	mate: yr rd, peak summer-fall; Australia	10698
	birth: June; s VIC, Australia	8747

Order Marsupialia, Family Tarsipedidae, Genus *Tarsipes*

Honey possums (*Tarsipes spenserae*) are small (10-20 g), gregarious, arboreal inhabitants of southwest Australian heath where they subsist on pollen and nectar. Females are larger than males (2168, 9046, 10081, 11911) and have 4 teats (2373, 9046). A postpartum estrus and embryonic diapause are present (9037). The penis is not bifid (11926). Neonates open their eyes at 56 d (9046). Details of anatomy (2168, 2373, 9305) and sperm structure (4370, 4371, 10735) are available.

NEONATAL MASS		
Tarsipes spenserae	4.3 mg; r3-6; n3	9046
WEANING MASS		
Tarsipes spenserae	2 g; pouch exit	11911
	2.5 g; n2; pouch exit	2804
	4-5 g; wean	2804, 11911
LITTER SIZE		
Tarsipes spenserae	2.47 yg; mode 2 55%; r1-4; n154 lt	2804, 11911
	2.67 yg; n15 f	9037
	2.8 yg; r1-4; n20 lt	9526
	5 emb	9039
SEXUAL MATURITY		
Tarsipes spenserae	f: 1st lact: 8-10 g; n46	2804
	f: 5-6 mo, 8-10 g	11911
	m: 5-6 mo, 7-9 g	11911
GESTATION		
Tarsipes spenserae	21-28 d	7983
	~3 mo	11911
LACTATION		
Tarsipes spenserae	pouch exit: 63-69 d	9380
	wean: 90 d; n1 lt	9380
LITTERS/YEAR		
Tarsipes spenserae	est 3	9046
SEASONALITY		
Tarsipes spenserae	mate: yr rd; s WA, Australia	11911
	birth: yr rd; WA, Australia	9526
	birth: yr rd, peak spring, low Dec-Jan; WA, Australia	9046
	lact f: peaks Feb-Mar, June-July, Sept-Oct, low Dec; s WA, Australia	11911

Order Edentata=Xenarthra

Anteaters, sloths, and armadillos are the highly specialized remnants of a diverse heritage. Males have abdominal testes. Females are monotocous and have a simplex uterus. Precocial offspring are usual except for armadillos. Edentate biology has been reviewed (7482). Details of anatomy (5574, 11840), physiology (9350), and placentation (2396, 10958) are available.

Order Edentata, Family Myrmecophagidae

Central and South American anteaters (*Cyclopes, Myrmecophaga, Tamandua*) inhabit neotropical forests and savannas. The solitary, giant anteater (*Myrmecophaga tridactyla* 20-40 kg) is diurnal and terrestrial, while the silky anteater (*Cylcopes didactylus* 175-350 g) is nocturnal and arboreal. Tamanduans (*T. mexicana, T. tetradactyla* 2-7 kg) are active on the ground or in trees, during the day or at night. Females have a terminally, bipartite vagina and a urogenital sinus (3394). The deciduate, hemochorial, discoidal, or cotyledonary placenta implants in the fundus of a simplex uterus (231, 695, 3394, 4953, 7563, 10310, 10958, 11840, 11841) and is occasionally eaten (628, 4953, 7922). A postpartum estrus occurs at 5-8 d (*Myrmecophaga* 4953, 5965). After birth, neonates suckle from 2 pectoral mammae and can climb without assistance onto their mother's back (3024, 8798).

NEONATAL MASS		
Myrmecophaga tridactyla	790 g, stillborn	5965
	f: 1480 g; n1	4953
	m: 1720 g; n1	628
LITTER SIZE		
Cyclopes didactylus	1	3024
	1 yg	5111
Myrmecophaga tridactyla	1 emb; n1	7324
	1 ea; n2 lt	4953
	1 ea; n2 lt	628
Tamandua tetradactyla	1 emb ea; n10 f	695
	1 yg; r1-2	7983
CYCLE LENGTH		
Myrmecophaga tridactyla	estrus: 2-3 d	5965
GESTATION		
Myrmecophaga tridactyla	~6 mo	4953
	181 d; r178-184; n2	5965
	183 d; r168-197	628
	~190 d; n8 lt, 1 f	7922
Tamandua tetradactyla	130-190 d	7983
LACTATION		
Myrmecophaga tridactyla	solid food: 12 wk	628
	independent: 5-6 mo	8798
SEASONALITY		
Myrmecophaga tridactyla	mate: Apr; zoo	5965
	birth: Mar, Dec; n3; captive	628
	birth: Nov-Dec; n2; zoo	4953
	birth: not seasonal; n8; zoo	7922

Order Edentata, Family Bradypodidae, Genus *Bradypus*

Three-toed sloths (*Bradypus torquatus, B. tridactylus, B. variegatus*) are 3-6 kg, solitary, arboreal folivores in tropical forests of Central and South America. Males are smaller than females (717). Mating may extend over 2 d (3024). The 60 g, deciduate, labyrinthine placenta is partially endotheliolchorial and partially hemochorial, and is variously described as diffuse, discoidal, bidiscoidal, or cotyledonary (231, 694, 770, 1022, 2395, 2396, 2397, 4906, 5384, 5736, 7563, 10310, 10957, 11836, 11838, 11839, 11840). Implantation into the simplex uterus is superficial, central and/or antimesometrial (231, 770, 7563, 11838, 11840). Placentophagia occurs (4657). The furred and open-eyed neonates suckle from 2 mammae (717, 4657, 11840). The Sertoli cells of males have crystalloid inclusions (10883). Details of endocrinology (3786) and placental fine structure (5736) are available.

NEONATAL MASS		
Bradypus tridactylus	m: 185 g; n1, term fetus	770
Bradypus variegatus	m: 340 g; n1	4657
NEONATAL SIZE		
Bradypus variegatus	m: 19.0 cm; n1	4657
	TL: 31 cm	11840
LITTER SIZE		
Bradypus tridactylus	1 yg	5111
	1	717
Bradypus variegatus	1	3024
	1	7483
	1 emb ea; n31 f	11840
GESTATION		
Bradypus tridactylus	106 d	4910
Bradypus variegatus	4-6 mo	11838
	5-6 mo	7483
LACTATION		
Bradypus tridactylus	solid food: 5 wk	717
Bradypus variegatus	solid food: 4 d	4657
	solid food: 2 wk	7749
	nurse: 3-4 wk	7483
	nurse: 2 mo	7749
INTERLITTER INTERVAL		
Bradypus variegatus	1 yr	7483
SEASONALITY		
Bradypus tridactylus	courtship: Mar-Apr; British Guyana	717
	birth: July-Sept, dry season; British Guyana	717
Bradypus variegatus	preg f: Apr-July; Panama	11838
	preg/lact f: Jan; Panama	3024
	birth: Nov, early dry season; n1; zoo	4657

Order Edentata, Family Choloepidae, Genus *Choloepus*

Two-toed sloths (*Choloepus didactylus, C. hoffmanni*) are 3-8 kg, long-lived, arboreal folivores from the Neotropics. They have 100 g, deciduate, labyrinthine, discoidal or lobular placentae that are transitional between syndesmochorial and endotheliochorial (770, 2396, 2397, 9400, 10957, 10963, 11839) and eaten after birth (11155). Estrus occurs 1 d after removal of young (2940). Neonates are furred and mobile with open eyes (7230, 9067, 11282). Females have an additional chromosome (3492) and menstruate (770). Males have seminal vesicles, but no expressable penis (7230, 11840). Details of anatomy (7358a, 9400) and birth (7038) are available.

NEONATAL MASS		
Choloepus didactylus	300 g; 3 d after death	9067
	364.3 g; r309-408; n3	11282
Choloepus hoffmanni	r350-454 g; n7	7230
	f: 375 g; n1	8479
NEONATAL SIZE		
Choloepus didactylus	16-18 cm	9067
LITTER SIZE		
Choloepus didactylus	1 emb; n1 f	9400
	1 yg; n1	7038
	1 yg; n1	10452
	1 yg; n1 lt	9067
	1 ea; n20 lt	2940
Choloepus hoffmanni	1 yg ea; n2 f	7232
SEXUAL MATURITY		
Choloepus didactylus	f: >3 yr	2940
	m: >4.5 yr	2940
Choloepus hoffmanni	f: 1st preg: 2 yr; n1	7230, 7232
CYCLE LENGTH		
Choloepus hoffmanni	menses: 1-3 d	7230
	5-7 d	7230

GESTATION		
Choloepus didactylus	5 mo 20 d; n1	11282
	>263 d; n1; death of m to birth	10452
	309-325 d; n1; pairing to birth; >7 mo; n1	2940
Choloepus hoffmanni	>225 d	7232
	≥8-9 mo; delay?	11155
LACTATION		
Choloepus didactylus	solid food: 1st wk	2940
	wean: 5 mo	11282
Choloepus hoffmanni	solid food: 15-27 d; n2	7230
	maternal care: 6 mo	7232
INTERLITTER INTERVAL		
Choloepus didactylus	423 d; r361-622; n14; yg dies	2940
	800 d; r472-1093; n6; yg lives	2940
SEASONALITY		
Choloepus didactylus	birth: Jan-Mar, May-Aug, Oct, Dec; n21; captive	7038

Order Edentata, Family Dasypodidae

Armadillos (*Cabassous* 2-6 kg; *Chaetophractus* 800-4600 g; *Chlamyphorus* 85 g; *Dasypus* 1-10 kg; *Euphractus* 3-6 kg; *Priodontes* 29-60 kg; *Tolypeutes* 1.25-1.5 kg; *Zaedyus* 1-2 kg) are solitary, terrestrial, predominantly nocturnal predators of ants and termites in the New World. Male *Chaetophractus* are reported larger than females (4055, 11610). *Cabassous* sperm form rouleaux in the epididymis (4553). The penis can be 30-40% of the head-body length (2944, 7224a, 11480).

The hemochorial, villous, discoid or zonary (ring shaped) placenta has a superficial, antimesometrial implantation (231, 273, 1717, 2396, 3007, 3008, 3011, 3232, 6174, 7563, 7868, 9707, 10310, 10478, 11345, 11968) into a simplex uterus (762, 3014, 3231, 3582, 3586, 6174, 8260, 10640; cf 762, 3231, 3582). Armadillos are monoestrous and monoovular, although polyembryony and the subsequent production of genetically identical siblings are common (3009, 3231, 3582, 7867, 7868, 8259, 8260). Implantation is often delayed (1415, 3012, 3015, 4288, 10641). Neonates are blind (eyes open: 16-30 d *Chaetophractus* 6085, 7983; 22-25 d *Euphractus* 4121; 3-4 wk *Tolypeutes* 7222, 7224a) and generally naked, although *Chaetophractus* has some ventral hair at birth (4121, 7222, 7223, 7229, 9169). Females have 2-4 mammae (2944, 2952, 6174).

Details of anatomy (210, 3014, 7358a, 10640, 11503), endocrinology (1292, 1293, 2205, 3013, 4294, 6143, 7756, 7757, 8364, 8365, 8366), male function (708, 1593, 2200, 3586, 5574, 7747, 7748, 7949, 8367, 9649, 10874, 11100, 11480, 11503, 11840), follicular development (7866, 8362), spermatozoan fine structure (7748), and enlarged fetal adrenals (762, 7558) are available.

NEONATAL MASS		
Cabassous spp	f: 113 g; n1	7229
	m: 100 g; n1	7229
Chaetophractus villosus	f: 102 g; n1	700
	m: 113 g; n1	700
	142 g	6085
	~155 g	7983
Dasypus novemcinctus	r28.6-111.4 g; n65; r48.25-111.4; without 28.6	10466
	133 g; n1	502
Euphractus sexcinctus	95-115 g	4121
Priodontes maximus	max 113 g	7983
Tolypeutes matacus	113 g; 3 d	7222
Zaedyus pichiy	95-115 g	7983
NEONATAL SIZE		
Dasypus novemcinctus	25 cm	6174
LITTER SIZE		
Cabassous spp	1 ea; n2 lt	7229
Cabassous centralis	1 yg, emb; n8 f	3582
Chaetophractus villosus	1-2	12128
	1.62; r1-2; n8 lt	700
	1.62; r1-2; n8 lt	3712
	2 yg; n1 lt	9169
Dasypus hybridus	4; n1 lt	12128

	r7-12 emb; n60 f	3231
	8-12	762
	8 emb ea; n2 f	11345
	8.4; mode 9; r7-9; n9 lt	3231
Dasypus kappleri	8-12 yg	5111
	2 ea; n4 lt	3582
Dasypus novemcinctus	3.25 emb; r1-4; n4 f	3232
	mode 4	3012
	mode 4; r2-5	3582
	mode 4; r2-6	762
	4 yg; 1 cl	8260
	4 emb; 1 ovum	8259
	4 emb; 1 ovulation	8362
	4; n1 lt	10478
	4 emb; n1 f	7624
	4 emb; n1 f	9029
	4 yg; n1 lt	5410
	4 ea; n2 lt	4228
	4 emb ea; n2 f	6174
	4 emb ea; n3 f	502
	4.0; mode 4 95%; r3-6; n60 f	1411
	4-5 emb; 1 chorion	8259
	6 emb; 7th resorbing; n1 f	1411
Dasypus sabanicola	4	8183
	4; mode 4	3582
Dasypus septemcinctus	4-8; seldom 12	8798
Euphractus sexcinctus	1-3	4121
	1; n1 lt	1717
	1.5; r1-2; n4 lt	12128
	3 emb; n1 f	9456a
Priodontes maximus	1-2 yg	7983
Tolypeutes matacus	1	7223
	1 ea; n50 lt	3582
	1 yg ea; n18 lt	7224a
	mode 1	2944
Tolypeutes tricinctus	1 emb ea; n2 f	9456a
Zaedyus pichiy	2; r1-3	7223
SEXUAL MATURITY		
Chaetophractus villosus	f/m: 9 mo; n2	3712
Dasypus novemcinctus	f: ~1 yr	10641
	f: 1st mate: 1.5 yr	8260
	f: progesterone high: 17-20 mo	8363
Euphractus sexcinctus	f/m: 9 mo	8999
Priodontes maximus	f/m: 9-12 mo	7983
Zaedyus pichiy	f/m: 9-12 mo	7983
GESTATION		
Chaetophractus villosus	60-75 d	9169
Dasypus hybridus	~4 mo	4292
Dasypus novemcinctus	3-5 mo delay + 4-4.5 mo development	1415, 3012, 6143, 6144, 10641
	3.5 mo delay + 4.5 mo gestation	8361, 8364
	134 d	5721
	delay 3 wk; ~140 d mating to birth season	8260
	~4 mo delay + ~4-4.5 mo development	4288
	12.6-24.3 mo; n21; capture to birth	10464a, 10465
Dasypus sabanicola	2-4 mo	8183
Dasypus septemcinctus	120 d	8798
Euphractus sexcinctus	60-65 d	4121
	74 d	9169
Priodontes maximus	4 mo	7983
Tolypeutes matacus	120 d; pairing to birth	7224a
	5-6 mo; field data	2944
Zaedyus pichiy	60 d	7983

LACTATION

Chaetophractus villosus	solid food: 55 d; ~13 cm	9169
Dasypus novemcinctus	wean: 4-5 mo	762
Euphractus sexcinctus	solid food: 1 mo	4121
	wean: 4 wk	8798
Priodontes maximus	wean: 4-6 wk	7983
Tolypeutes matacus	solid food: 59 d	7222
	wean: est 5-6 wk	2952
	wean: 72 d	7222
Zaedyus pichiy	wean: 6 wk	7983

SEASONALITY

Chaetophractus villosus	conceive: Sept; Santa Fe, Argentina	6023
	birth: Mar-Sept; n9; zoo	12128
	birth: Nov; Santa Fe, Argentina	6023
Dasypus hybridus	implant: ~early June; Argentina	4292
	preg f: June-?; Buenos Aires, Argentina	3231
	birth: Oct; Argentina	4292
Dasypus novemcinctus	prostate: large Dec-Jan, small June-Aug; se Brazil	1593
	sperm: yr rd; TX	7045
	sperm: yr rd; TX, FL	3582
	testes: large Apr-Sept, small Jan-Mar; TX	7045
	testes: large Nov, Jan; Bela Horizonte, Brazil	10874
	testosterone: high Mar-Aug, low Oct-Dec; captive	8367
	mate: Oct-Nov; TX	8260
	estrus: June-July; TX	3012
	ovulate: June-Aug; TX	4288
	ovulate: July; TX	4294
	ovulate: July-Aug, Oct-Dec; lab	8362
	ovulate: July-Sept; TX	10641
	implant: Oct-June; TX	4288
	implant: Nov-Dec; lab	6144
	implant: end Nov-early Dec; TX	3012
	birth: Feb-Apr; Mexico	11303
	birth: Mar-Apr; TX, FL	3582
	birth: Mar-Apr; TX	4288
	birth: Mar-Apr; TX	8260
	birth: mid Mar-mid Apr; lab	8366
	birth: end Mar-early Apr; TX	3012
Dasypus sabanicola	conceive: Apr-May; Apure, Venezuela	8183
	birth: Aug-Sept; Apure, Venezuela	8183
Euphractus sexcinctus	spermatogenic arrest: May; Brazil, Chile, lab?	7949
	preg f: Jan; few data; Uruguay	8999
	preg f: July; Matto Grasso, Brazil	9456a
	preg f: Sept-Oct; few data; central Brazil	8999
Tolypeutes matacus	mate: May-June; Bolivian Chaco	2944, 2952
	birth: Nov; Bolivian Chaco	2944, 2952
	birth: Nov-Jan; Asuncion, Paraguay	7223
Tolypeutes tricinctus	preg f: July-Aug; n2; Matto Grasso, Brazil	9456a
Zaedyus pichiy	birth: Jan-Feb; Mendoza, San Juan, Argentina	7223

Order Insectivora

The six families of the order Insectivora are a taxonomic hodge-podge with little close phylogenetic relationship. A 19th century monograph on the order is available (2662) as is a review of reproductive physiology (9350).

Order Insectivora, Family Solenodontidae, Genus *Solenodon*

The 1 kg, nocturnal solenodonts (*Solenodon cubanus, S. paradoxus*) from Cuba, Santa Domingo and Haiti are insectivores with antimesometrial implantation of a hemochorial, discoidal, labyrinthine placenta into a uterus with a large body and two blunt horns (176, 11853). Naked and blind young (176, 1285, 11269) suckle from two teats near the base of the tail (176, 310). The teats are also used to transport young (2930). Females have a urogenital sinus (176, 11853). Testes are abdominal and the penis retracts into the abdominal wall (176, 2662, 8967). A German monograph on *S. paradoxus* is available (7426, 7427, 7428, 7429; cf 2933).

NEONATAL MASS		
Solenodon paradoxus	est 100 g	2930
LITTER SIZE		
Solenodon cubanus	mode 1-2	10809
Solenodon paradoxus	1 emb; n1 f	11853
	1 emb; n1 f	176
	1 yg; n1 lt	1285
	mode 2; r1-3	7428
	3; n1 lt	11269
CYCLE LENGTH		
Solenodon paradoxus	9-13 d	2930
GESTATION		
Solenodon paradoxus	est >50 d	2930
LACTATION		
Solenodon paradoxus	wean: est 75 d	2930

Order Insectivora, Family Tenrecidae

The solitary, nocturnal tenrecs and their relatives (*Dasogale, Echinops* 110-250 g; *Geogale* 71 mm; *Hemicentetes* 70-170 g; *Leptogale, Limnogale* 50-90 g; *Microgale* 34-85 g; *Micropotamogale* 135 g; *Oryzorictes* 42-48 g; *Potamogale* 1 kg; *Setifer* 125-365 g; *Tenrec* 830-2400 g) are terrestrial, semiaquatic, or semifossorial predators on invertebrates in Madagascar and Africa. The reproductive biology of *Dasogale, Geogale, Leptogale, Limnogale*, and *Oryzorictes* is not known.

Male *Echinops* have an additional chromosome (3492). *Potamogale* has a simplex uterus with two nonfunctional cornu (4762), while other genera have bicornuate uteri (3218, 3855, 9236, 10484). Females have well-developed prostate and bulbourethral glands (6364, 8436). *Hemicentetes* (3218, 3857) and *Setifer* (8620) are reportedly polyestrous, although tenrecs may be induced ovulators (7984). Ovarian follicles lack antra, and fertilization occurs within the ovary (3218, 3219, 6169, 7892, 10484, 10485). Polyovular follicles are common (*Tenrec* 7890). *Hemicentetes* has no morula (3855), but *Tenrec* does (7892). The labyrinthine, hemochorial, discoidal or diffuse placenta has a circumplacental chorion, a hemophagous organ, but no consistent implantation site (1025, 3218, 3854, 3855, 3856, 4104, 4762, 4814, 7563, 7891, 10310, 10486). Placentophagia occurs (3975, 4949, 8620).

The naked neonates open their eyes at 7-9 d (*Echinops* 2942, 3844, 3975; cf 9-15 d 4949, 9138), 7-10 d (*Hemicentetes* 2934, 2942, 3454, 3975), 9-14 d (*Setifer* 2934, 2942, 3975, 8620, 8798, 9138), 9-15 d (*Tenrec* 2934, 2942; cf 4-6 d 6604), or 18 d (*Microgale* 2939).

Females have 6 (*Limnogale* 2934), 10 (*Setifer* 2934, 8798), or 21-29 mammae (*Tenrec* 6604, 7890, 9236). Males have abdominal testes (2662, 2934, 7890a, 8436) and the penis retracts into the cloaca (176). Details of endocrinology (7891) and birth (8622) are available.

NEONATAL MASS		
Echinops telfairi	3.8-9.6 g; smallest die	3975
	7.7 g; n8	3844
	8.6 g; n9	3844
Hemicentetes semispinosus	11.5 g; r9.9-12.9; n4 yg, 1 lt	3975
Microgale talazaci	3.6 g; n1, 1 d	2939
Setifer setosus	24.7 g	3975
Tenrec ecaudatus	10-18 g	6604
	r22.8-27.4 g; n4	2934
	est 24 g	7891
NEONATAL SIZE		
Hemicentetes semispinosus	25 mm	3454
	60-66 mm	2934
Tenrec ecaudatus	r84-92 mm; n4	2934
WEANING MASS		
Tenrec ecaudatus	150-300 g; 5 wk	6604
LITTER SIZE		
Echinops telfairi	r1-10 yg; n3 lt	2942
	1.7; r1-3; n3 lt, 1 f	8622
	3.4 yg; r2-5; n9 lt	9138
	3.5; r2-12; n12 lt	9523
	mode 5 yg; r1-9; n25 lt	3844
	5.4; r3-8; n5 lt	4949
	5.85; r2-10; n13 lt, 12 f	3975
Hemicentetes semispinosus	2; n1 lt	3454
	3.6 yg; r1-11; n~36 lt	2934, 2942, 3975
	max 10 emb	8436
Microgale dobsoni	2.67 yg; r2-3; n3 lt	2939
Microgale talazaci	r1-3; n4 lt	2934, 2939
Micropotamogale ruwenzorii	2 emb; n1 f	8852
Potamogale velox	1-2 yg	4354a
	1.4 emb, yg; r1-2; n5 lt	2753
Setifer setosus	1-4 emb	2942
	1-6	8620
	2.5 yg; r1-4; n2 lt	2942
	2.6; r2-3; n5 lt	5573
	2.6; r1-5; n8 lt	9523
	2.7 yg; r1-5; n13 lt	9138
	max 6 emb	10484
	max 10 emb	8436
Tenrec ecaudatus	r1-4 yg; n3 lt	2942
	10-32 emb	2942
	10.5 cl; ±0.5; r3-18	7892
	12 emb; n1 f; 15 ovulations	9236
	12.3 emb; r10-15; n3 f	5573
	19.75 emb; r18-25; n4 f	8906
	21 yg; n1 lt	3854
	25 yg; r19-31; n2 lt	6604
	24 emb; r16-32; n2 f	3854
	max 42 ovulations; max 32 emb	1026
SEXUAL MATURITY		
Echinops telfairi	f: 1st conceive: 6 mo; n18	2942
	f/m: 1 yr	9523
	f/m: 14 mo	4949
Hemicentetes semispinosus	f: 1st conceive: 30-35 d; n15	2934, 2942
	f: 8 wk	3975
Microgale dobsoni	f/m: min 22 mo	2939
Microgale talazaci	f/m: ≥ 21 mo	2939
Setifer setosus	f: 1st conceive: 2 mo	9138
	f: 1st conceive: 6 mo; n2	2942
	f: 1st birth: 14 mo; n1	8798
	m: 8.5 mo; n1	8798
Tenrec ecaudatus	f: 1st conceive: 6 mo; n1	2942

CYCLE LENGTH		
Echinops telfairi	6-7 d; r4-13; n7 cycles, 2 f	3844
GESTATION		
Echinops telfairi	49 d; n1	3844
	est 7 wk	9523
	60 d	10797
	62-65 d; n5	2942
	66 d	2934
Hemicentetes semispinosus	~4 wk; field data	3218
	~45-55; n2	2942
	>50 d; capture to birth	3454
	55-58 d	2934
Microgale dobsoni	58-64 d	2930, 2939
Microgale talazaci	58-64 d	2930, 2939
Setifer setosus	~5 wk; field data	3218
	~7 wk; field data; not many	5573
	51-56 d	9138
	51-61 d; pairing to birth	9523
	54-57 d; ibi or pairing to birth	8620
	63 d	2934
	65-69 d; n1	2942
Tenrec ecaudatus	implant: 9.7 d; ±0.5	7892
	53 d	6604
	r58-64 d; n3	2934, 2942
	est 60 d	7891
	~9 wk; field data, not many	5573
LACTATION		
Echinops telfairi	solid food: 12-14 d	9138
	solid food: 13 d	3844
	solid food: 16 d	4949
	wean: 30 d	9138
	independent: 30-35 d	3844
	wean: r18-22 d; n18 lt	2942
	wean: 23 d	10797
	wean: 32 d	3975
	wean: 33 d	4949
	wean: ~5 wk; field data	3218
Hemicentetes semispinosus	den emergence: 9 d	2934
	solid food: 13 d	2934
	wean: r18-25 d; n15 lt	2942, 3975
Microgale dobsoni	wean: 28-30 d	2939
Microgale talazaci	wean: 28-30 d	2939
Setifer setosus	solid food: 12-14 d	9138
	wean: 15-20 d; n2 lt	2942
	wean: 30 d	9138
Tenrec ecaudatus	nest emergence: est 15-20 d	7891
	solid food: 1 wk	6604
	wean: 25-30 d; n1 lt	2942
LITTERS/YEAR		
Potamogale velox	2	2753
Setifer setosus	2	5573
Tenrec ecaudatus	1; 2 rare	2934
SEASONALITY		
Echinops telfairi	mate: est Nov-Dec; Madagascar	3975
	mate: all yr; captive	4949
	birth: Jan-Mar; Madagascar	9523
	birth: July-Aug; captive	9523
	birth: June-Sept, peak July; n15; zoo	9138
Hemicentetes semispinosus	mate: Sept-Dec; Madagascar	2934
	mate: Nov-May; Madagascar	3975
	birth: Nov-Jan; Madagascar	2934
	young found: yr rd; anecdotal reports; Madagascar	8436
Microgale dobsoni	mate: Dec-Aug; captive	2934
	birth: Feb-May; captive	2934

Potamogale velox	preg f: Apr-May; few data; Cameroun	9468
	preg f: June; n2; s Cameroun, Benito, equatorial Guinea	647
	birth: early rainy season; Zaire	11253
Setifer setosus	mate: Sept, after Nov; ne Madagascar	5573
	mate: Sept-mid Oct; Madagascar	2934
	ovulate: Sept-Oct, no data other mo; Madagascar	10484
	preg f: Nov-Dec; n5; Madagascar	5575
	preg f/nearly full grown yg: Nov-Dec; ne Madagascar	5573
	birth: Jan-Feb, Apr-June, Aug-Nov, none Mar, July, Dec; zoo	9138
	birth: July, Dec, none Mar; few data?; captive	8620
	young found: yr rd; according to natives; Madagascar	8436
Tenrec ecaudatus	mate: Aug-Oct; Seychelles	7892
	mate: Oct-Nov; Madagascar	2934
	mate: annually; Seychelles	7891
	preg f: Dec-Jan; n3; Madagascar	5573
	preg f: Oct-Nov; n4; Madagascar	8906
	birth: Dec-Feb; Madagascar	8436
	birth: Jan-Feb; Madagascar	2934
	lact f: Feb; n1; Madagascar	5573

Order Insectivora, Family Chrysochloridae

Golden moles (*Amblysomus* 40-100 g, *Calcochloris* 20-30 g, *Chlorotalpa* 50-75 g, *Chrysochloris* 40-90 g, *Chrysospalax* 350-550 g, *Cryptochloris* est 20 g, *Eremitalpa* 15-20 g) are blind, subterranean carnivores from equatorial and southern Africa. Reproductive data for *Calcochloris, Chlorotalpa*, and *Cryptochloris* are lacking. Male *Amblysomus* are slightly larger than females (6136). The endothelial, discoidal, labyrinthine placenta has a superficial, mesometrial implantation into a bicornuate uterus (2394, 2662, 7563, 10310, 11108). Testes are abdominal and the penis is retractile (6136). A urogenital sinus is present (6136). Neonates are naked (6135, 6136, 10158) and suckle from 4 teats (6135, 10158).

NEONATAL MASS		
Amblysomus hottentotus	~4.5 g	6136
NEONATAL SIZE		
Amblysomus hottentotus	HB: 47 mm	6135
Chrysochloris stuhlmanni	~47 mm	5759
LITTER SIZE		
Amblysomus hottentotus	2 emb; n1 f	10158
Chrysospalax villosus	2 emb; n1 f	10158
Eremitalpa granti	1 emb ea; n2 f	4927
LACTATION		
Chrysochloris stuhlmanni	nurse: 2-3 mo	5759
SEASONALITY		
Amblysomus hottentotus	repro: yr rd; S Africa	6136
	birth: perhaps summer; S Africa	10158
Chrysochloris asiatica	birth: winter; S Africa	10158
Eremitalpa granti	mate: Oct-Nov; S Africa	11108

Order Insectivora, Family Erinaceidae

Gymnures and hedgehogs (*Echinosorex* 0.5-2 kg; *Erinaceus* 140-1400 g; *Hemiechinus* 220-400 g; *Hylomys* 15-80 g; *Neohylomys* 50-59 g; *Neotetracus* 101-111 mm; *Paraechinus* 150-450 g; *Podogymnura* 50-70 mm) are solitary, nocturnal, terrestrial insectivores in Eurasia and Africa. Male *Erinaceus* may be larger than females (2045; cf 4688), while the reverse is reported for *Echinosorex* (2313). Coitus may induce ovulation in *Erinaceus* (10416; cf 3795) and a vaginal plug is present after mating (*Erinaceus* 1570, 5759, 10416). The hemoendothelial, labyrinthine, discoid placenta has a decidica basalis and an eccentric, antimesometrial, interstitial implantation into a bicornuate or bipartite uterus with a massive cervix (2662, 6727, 7178, 7563, 8219, 9678, 9839, 10310). A postpartum estrus is not usual (2045, 2417, 6439, 6440, 7527, 7529, 7530, 12094).

Neonates are blind, but generally have quills (2951, 4158, 4690, 5224, 7226, 8049, 8718, 9175, 11438). Eyes open at 8-22 d in *Erinaceus* (69, 1301, 2045, 2384, 3571, 4687, 5759, 6440, 7529, 7530, 10153) and at 23-29 d in *Paraechinus* (2951). Females have 4-10 mammae (179, 2662, 2951, 6369, 7621, 10158). Testes are abdominal (2662, 6727). Males have seminal vesicles and females have homologous glands and a large clitoris (2662, 3984, 7454). A urogenital sinus is present (3985). Details of ovarian function (2417, 3436, 3795, 8219, 9839, 9842) and testicular function (165, 912, 1570, 1839, 2075, 2823, 2957, 3436, 3794, 3984, 3985, 9410, 9411, 9412, 9413, 9414, 10747) are available.

NEONATAL MASS		
Echinosorex gymnurus	14.5 g; r14-15; n2	7161
Erinaceus albiventris	6-6.5 g	6440
	7-9 g	1301
	8-11 g	7226
Erinaceus concolor	22 g	9632
Erinaceus europaeus	10.7 g	69
	15 g; r8-22	7529, 7530
	m: 16.1 g; r12.8-20.12; n7	4686, 4687, 4688
	f: 17.17 g; r11.9-25.14; n5	4686, 4687, 4688
Erinaceus frontalis	9.8-11.3 g	10154
	m: 10.67 g; r9.5-12.0; n3	5224
	f: 10.8 g; r10.5-11.5; n3	5224
Hemiechinus auritus	8 g; n1	9632
	8.325 g	8718
	11 g; n1	11438
	13.15 g; n1	4690
Paraechinus aethiopicus	8-9 g; n3, 1 d	2951
NEONATAL SIZE		
Echinosorex gymnurus	80.5 mm; r80-81; n2	7161
Erinaceus europaeus	25 mm	10338
	56.4 mm	69
	m: 68 mm	4686
	f: 91.5 mm; r87-96; n2	4686
Erinaceus frontalis	f: HB: 52.3 mm; r51-53; n3	5224
	m: HB: 52.5 mm; r52.5-53; n3	5224
Hemiechinus auritus	47 mm	9175
	50 mm; n1	4690
	HB: 61 mm; n1	11438
WEANING MASS		
Erinaceus spp	306.8 g; ±57.4; n6	6042
Erinaceus europaeus	125-345 g; 40 d	7529
WEANING SIZE		
Erinaceus europaeus	160 mm	2045
LITTER SIZE		
Echinosorex gymnurus	1.9; r1-2; n7 lt	7161
	2 emb; n1 f	7178
	2 emb; n1 f	557
	2 emb; n1 f	2313
Erinaceus albiventris	2-4 yg	4354a
	r2-3; n6 lt	7226
	3	4693
	3.0; r1-6	1301
	4 emb; n1 f	4924
	9; n1 lt	6440
Erinaceus concolor	3 yg; n1 nest	9632
Erinaceus europaeus	2 yg; n1 f	2109
	2-3 cl	3795
	mode 3-8; r2-10	3571
	3.0; n113 lt	7529, 7530
	3.86 yg; r3-5, n7 lt	8922
	4-5; max 7; n67 lt	6439
	4-7 emb	2417
	4-8	69
	4.25 yg; r2-6; n76 nests	7538

	4.4 healthy emb, 4.6 implanted emb; n53 f	7527
	4.5; r4-5; n2 f	11255
	4.7; r3-7; n3 lt	4686, 4687, 4688
	4.75; mode 4-6; r2-9	11399
	5 emb; n1 f	1590
	5.2 wean; sd2.0; se0.2; n85	6035
	6.0 yg; r5-8; n3 f	8049
	6.8 emb, yg; r5-8; n4 lt	10338
Erinaceus frontalis	3-6	8965
	4; n1 lt	4180
	~4; r1-9	10158
	5; r2-10	2384
	9; n1 lt	5224
Hemiechinus auritus	1-6	9175
	2; n1 lt	11438
	2.33; r1-5; n12 lt	6727
	2.4 yg; r2-4; n5 lt	8718
	2.5 yg; r2-3; n2 nests	9632
	3 emb; n1 f	7716
	4; n1 lt	4690
	4 yg; n1 f	8049
	4-7	4645
	5; r4-6; n3 lt	8049
	5.8; r2-8	910
	6 yg; 1 lt	4158
Hylomys suillus	2 emb; n1 f	7161
	2-3 yg	6369
Neotetracus sinensis	4.3 emb; r4-5; n3 f	179
Paraechinus aethiopicus	2 emb; n1 f	2951
	4 yg; n1 lt	2951
Paraechinus hypomelas	3-4	9175
	4 yg; n1 f	8049
Paraechinus micropus	1.5 yg; r1-2; n2 f	8718
	4; n1 lt	11438
	5 yg; 1 lt	4158
SEXUAL MATURITY		
Erinaceus albiventris	f/m: 1st mate: 4 mo	6440
	f/m: 61-68 d	1301
	f/m: yr after birth	5759
Erinaceus europaeus	f: 1st birth: 11 mo	7529, 7530
	f: 1st birth: 1 yr	4688
	f/m: 5 mo; captive; 9 mo; wild	6439
	f/m: 10-12 mo	3571
CYCLE LENGTH		
Hemiechinus auritus	7.9 d; n16 cycles, 25 f	9839
	proestrus: 1.90 d; r1-3; metestrus: 1.8 d; r1-2.5	9839
	estrus: 1.2 d; r0.5-2.2; diestrus: 2.5 d; 1.5-4.5	9839
GESTATION		
Echinosorex gymnurus	35-40 d	6369
Erinaceus albiventris	28-40 d; pairing to birth	7226
	5-6 wk	4693
	mode 36 d; r34-44; pairing to birth	1301
	37 d	6440
	r37-38 d; n3	4694
Erinaceus europaeus	~1 mo	2662
	34-35 d; r31-39; n13 lt	7527, 7529, 7530
	35-40 d	8922
	5-6 wk; r34-49 d	4687, 4688
	37 d; min ibi	6439
Erinaceus frontalis	30-40 d	2384
	~40 d	10153
Hemiechinus auritus	30-32 d	12094
	35-42 d	9175
	36-37 d; n1	4690

	~7 wk	4645
Hylomys suillus	30-35 d	6369
Paraechinus micropus	est 35-39 d	11438
LACTATION		
Erinaceus spp	solid food: 21-22 d	6042
	wean: 42.5 d; r42-43; n2	6042
Erinaceus albiventris	solid food: ~1 mo	5759
	wean: ~40 d	5759
	solid food: 44 d	8798
	wean: 2 mo	8798
Erinaceus europaeus	den emergence: 3-4 wk	7527, 7529, 7538
	den emergence: 3-6 wk	6035
	solid food: 18-20 d	3571
	solid food: 23-24 d	4687
	solid food: 25-26 d; n2 lt	7527, 7529
	wean: 38-44 d	7527, 7529, 7538
	wean: 38-45 d; n5 lt	8922
	wean: 40+ d	6439
	wean: 45 d	4687
Erinaceus frontalis	forage: 6 wk	10158
Hemiechinus auritus	solid food: ~1 mo	8049
	den emergence: ~3 wk	8049
Paraechinus aethiopicus	solid food: 40 d	2951
Paraechinus micropus	wean: est ~5 wk	11438
LITTERS/YEAR		
Echinosorex gymnurus	2	6369
Erinaceus europaeus	1	11399
	1-2	2417
	1-2; 2 may not be exceptional	10338
	1.5; r1-2; n2	4687
	2; n1 f	7527
Hemiechinus auritus	1	9175
	1	4645
Paraechinus hypomelas	1	9175
SEASONALITY		
Echinosorex gymnurus	preg f: May-June, Sept, Nov; Madagascar	7161
	preg f: mid July-mid Aug; n1/5; N Borneo, Indonesia	2313
	repro: est yr rd; Thailand	6369
Erinaceus albiventris	estrus: yr rd; captive	4693
	mate: Apr; n1; captive	6440
	preg f: May; n1; Somalia, Ethiopia, Sudan	4924
	birth: May; n1; captive	6440
	birth: June, Sept, Oct; n3; captive	4694
Erinaceus europaeus	spermatogenesis: Jan-May; Europe	2075
	spermatogenesis: Apr-Aug, peak mid Apr-mid June; captive	7527
	spermatogenesis: May-Aug; central France	3794
	testes: large Mar-Apr, small Nov-Dec; France	9413
	testes: large Apr, small July-Jan; Europe	2075
	testes: large Apr-May; British Isles	2662
	testes: large Apr-Aug; Wales, England, UK	165
	testes: large May, small Sept, increase Jan; captive	10747
	testes: active Dec-Sept; France	9413
	cl present: Apr-Sept; Wales, England, UK	2417
	estrus: Apr-Sept; Germany	9678
	estrus: Oct, Mar; captive	4693
	mate: Mar; captive	7530
	mate: Mar-July; Switzerland	850
	mate: Mar-Aug; central Europe	3571
	mate: Apr-Sept; France	3795
	mate: end Apr-Aug/early Sept; captive	7527
	mate: end July-Sept; Denmark	11399
	mate: spring-summer; captive	4687
	preg f: May-June, Oct; British Isles	2662
	preg f: May-Sept; Great Britain	7529

	birth: May-June, Sept-Oct; captive	9410
	birth: May/June-Sept; central Europe	3571
	birth: end May; captive	7530
	birth: end May-early June; 1 lt 23 June; USSR	69
	birth: June; n1; lab	4686
	birth: end July; captive	4687
	birth: peak Aug; Denmark	11399
	birth: Sept; n1; zoo	2109
	birth: mid Sept; captive	7527
	birth: end Sept; France	3795
	birth: spring; Britain, Ireland	2045
	birth: summer; central Europe	4688
	birth: winter; captive, Switzerland	6439
	repro: Apr-Aug; e Europe	8798
Erinaceus frontalis	mate: Nov; S Africa	5224
	birth: Oct-Mar; Botswana	10154
	birth: Oct-Mar/Apr; S Africa	10158
	birth: Nov; Zimbabwe	10153
Hemiechinus auritus	sperm: Mar-Aug; Rajasthan, India	3985
	testes: large Jan-Aug, small Sept-Dec; Jaipur, India	402
	testes: large summer; Agra, India	6727
	estrus: mid Feb-May; captive	4693
	estrus: Apr-Aug; Rajasthan, India	9839
	mate: after hibernation; Pakistan	9175
	birth: July-Sept; Rajasthan, India	8720
	birth: July; Stavropol, Russia	910
	birth: peak Aug-Sept; Pakistan	9175
	birth: early summer; Europe, w USSR	4645
	lact f: June; n1; e Europe, n Asia	8049
Hylomys suillus	preg f: Mar; n1; Malaysia	7161
	repro: est yr rd; Thailand	6369
Paraechinus aethiopicus	lact f: June; n1; Egypt	11468
Paraechinus hypomelas	birth: end spring-early summer; Pakistan	9175
	lact f: May; n1; s Turkmenia	8049
Paraechinus micropus	repro: seasonal; India	912

Order Insectivora, Family Soricidae

Shrews (*Anourosorex* 14-35 g; *Blarina* 7-30 g; *Blarinella, Chimarrogale* 28-32 g; *Crocidura* 2-66 g; *Cryptotis* 3-9 g; *Diplomesodon* 7-13 g; *Feroculus* 35 g; *Megasorex* 9.5-11.7 g; *Myosorex* 7-28 g; *Nectogale, Neomys* 6-23 g; *Notiosorex* 3-6 g; *Paracrocidura* 6-16 g; *Scutisorex* 70-113 g; *Solisorex* 59-66 mm; *Sorex* 2-25 g; *Soriculus* 4-14 g; *Suncus* 1-110 g; *Sylvisorex* 3-92 g) are terrestrial to semiaquatic, solitary insectivores from Eurasia, Africa, North and Central America, and northern South America. Males have been reported to be larger than females (*Blarina* 8299, *Crocidura* 4601, *Myosorex* 9329; cf 673, *Sorex* 716, 5789, 9159; cf 1912, 3423, 7893, and *Suncus* 572, 1331, 8501, 8903, 11999a).

Ovulation may be induced by coitus (1210, 1217, 2001, 2737, 6728, 7410, 7420, 7524, 8299, 8757, 10677). A copulatory tie is reported in *Blarina* (8299), *Cryptotis* (7409, 7410), and perhaps *Neomys* (7273), but is not present in *Sorex* (2155) or *Suncus* (3374). A vaginal plug is reported for *Neomys*, as is intrauterine insemination (3630, 8757). Conception occurs within the follicle and corpora lutea development begins before ovulation (2737, 8299, 10482).

The endothelio-endothelio, endotheliochorial, or hemochorial, labyrinthine, villous, discoid placenta has an antimesometrial, superficial implantation into a bipartite or bicornuate uterus (231, 373, 1210, 1221, 2737, 4480, 5734, 5823, 7410, 7419, 7563, 7573, 8171, 8299, 9481, 9483, 9840, 9841, 10019, 10310, 11814) and may be eaten after birth (335, 5118). A postpartum estrus occurs (1210, 1217, 2004, 2045, 2155, 2156, 2737, 3373, 4600, 4601, 4604, 4894, 5244, 5373, 7273, 7409, 7410, 8757, 9175, 9916, 10017, 10019, 10482, 11313, 11315, 11316; cf 10339) and may be accompanied by delayed implantation (1210, 5039, 8058a).

Neonates are blind and naked (187, 1013, 1474, 2347, 2437, 3375, 3455, 3571, 4266, 4279, 4901, 5146, 7161, 7739, 8049, 9175; cf 335, 2589, 2736, 9914, 11314). Their eyes open at 5 d (*Blarina* 7333), 9-13 d (*Crocidura* 335, 2589, 3458, 3571, 4602, 4603, 10158, 11313, 11314, 11391; cf 11473), 1-2 wk (*Cryptotis* 2004,

7410; *Suncus* 2736, 2744, 3373, 9175, 9914, 11315), 15-18 d (*Myosorex* 673), 15-21 d (*Sorex* 2045, 2156, 3423, 3571, 5118, 5373, 9175, 11314, 11390; cf 1474), 20-24 d (*Neomys* 2156, 7272, 7275, 7276).

Females have 4-6 mammae (*Sylvisorex* 7260), 6 mammae (*Blarina* 1474, 4266; *Crocidura* 331, 2575, 3571, 4602, 4924, 5405a, 7167a, 7621, 8049, 9945; cf 8798; *Cryptotis* 4494; *Sorex* 502, 1474, 2548, 7621, 9175), or 10-14 mammae (*Neomys* 7274, 8757). The vagina may be sigmoid during estrus (*Blarina* 3725).

Males have a sigmoid penis which may have grooves or spicules, abdominal testes which may migrate seasonally (*Blarina* is ascrotal), and seminal vesicles which may not be present in all genera (373, 1211, 2742, 2845, 3984, 4277, 6888, 6945, 7438, 7441, 8299, 8774). *Sorex araneus* males have an additional chromosome (3492).

Details of anatomy (1210, 4827, 4828, 6019, 8757, 10558, 10676, 11677, 11678), intersex (1216), endocrinology (2743, 3565, 4605, 11807; *Suncus* 494, 2738, 2739, 3161, 4476, 4480, 4481, 6728, 7170, 9146, 9147, 9149), spermatogenesis (2742, 3603, 3604, 9995), sexual behavior (1900, 9143, 9144, 9145, 9146, 9148, 9153), milk composition (2740, 7581), photoperiod (9151, 11499), and reproductive energetics (3698, 7581, 7582) are available.

NEONATAL MASS

Blarina brevicauda	1.34 g; n1, 36-48 h	4266
Crocidura bicolor	0.25 g ea; n2	11473
	1.4 g; n1, 1 d	2589
Crocidura flavescens	2.26 g; sd0.20; n2	3698
Crocidura hirta	~1 g	7167a
Crocidura leucodon	0.8 g	3455
	1.0 g; n2	3458
	1 g	3571
Crocidura russula	0.78 g; r0.5-1.1; n39	4602, 4603
	1.11 g; sd0.23; n40	3698
Crocidura suaveolens	0.53 g; r0.42-0.67	11313, 11314
Crocidura viaria	1.39 g; sd0.23; n13	3698
Cryptotis parva	0.3-0.4 g	7409
	0.32 g	4279
Myosorex varius	~1 g	673
Neomys anomalus	0.52-0.61 g	7272
Neomys fodiens	0.53-0.89 g; n72 yg, 9 lt	7275
	0.62 g; n34 yg, 3 lt before nursing	7275
	1.0 g	2156
Sorex araneus	0.391 g; r0.345-0.431; n16	11314
	0.4-0.5 g	3571
	0.42 g	1838
	0.5 g	2156
Sorex cinereus	~0.1 g	5146
	0.28 g; se0.02; r0.23-0.36; n15	3423
Sorex coronatus	0.57 g; sd0.10; n12	3698
Sorex minutus	0.22 g; n1	3698
	0.25 g	2156
Sorex sinuosus	0.5 g	7869
Sorex vagrans	~0.5 g	5373
Suncus etruscus	0.18-0.27	3373
	0.19 g; r0.17-0.21; n2	11315
	0.20 g	3374, 3375
Suncus murinus	f: 2.0 g; r1.4-3.4; n32 some with 0.2 g milk	2736, 2744
	2.1 g; se0.1; r1-4; n211 lt	4477
	f: 2.2 g; sd0.4; n7	9914
	m: 2.3 g; sd0.3; n6	9914
	m: 2.3 g; r1.6-3.1; n28 some with 0.2 g milk	2736, 2744
	2.7 g; n40 lt, 82 yg	2741
	2.8 g; se0.2; r1-6; n51 lt	4477
	3.1 g; n1 lt, 4 yg	7161
	m: 3.6 g; sd0.6; n4	9914
	f: 3.7 g; sd0.4; n2	9914

NEONATAL SIZE		
Blarina brevicauda	TL: 31 mm; 2 d	3725
Crocidura bicolor	HB: 18.7 mm ea; n2	11473
	HB: 20 mm; 1 d	2589
	HB: ~20 mm; n3	335
Crocidura hirta	HB: ~30 mm	7167a
Crocidura russula	HB: 29 mm; n311	4603
Myosorex varius	HB: 30 mm	673
Neomys fodiens	20.5 mm; r17.0-22.5; n8	7275
Sorex araneus	18 mm	1838
Sorex cinereus	CR: 12-14 mm	3423
Sorex minutus	16 mm	5118
Suncus etruscus	HB: 14.2-15.4 mm	3373, 3375
WEANING MASS		
Crocidura bicolor	3.0 or 3.4 g; 23 d	2589
Crocidura flavescens	20.7 g; sd1.4; n3	3698
Crocidura hirta	~10 g; 15-20 d, from graph	7167a
Crocidura russula	f: 6.8 g; r5.01-9.0; n49	4603
	m: 7.3 g; r4-10; n52	4603
	9.0 g; sd0.7; n19	3698
Crocidura suaveolens	≤4 g	9253
Crocidura viaria	13.7 g; sd1.9; n13	3698
Cryptotis parva	at adult size	7409
Neomys fodiens	10 g	2156
Sorex araneus	~7 g; 18 d	2045
	10 g	2156
Sorex cinereus	3.5 g; ±0.02; r3.5-3.6; n6	3423
Sorex coronatus	8.6 g; sd0.9; n9	3698
Sorex fumeus	4 g; ~1 mo	4277
Sorex minutus	3.2 g; n1	3698
	7 g	2156
Sorex vagrans	5-6 g; ~1 mo	5373
Suncus etruscus	~2 g	11315
	2.05-2.21 g; 19-21 d	3373
Suncus murinus	f: 13.2 g; n19, 20 d	2736
	m: 15.0 g; n19, 20 d	2736
	f: 28.2 g; sd2.3; n9	9914
	m: 34.2 g; sd5.5; n11	9914
	f: 46.6 g; sd8.2; n14	9914
	m: 62.9 g; sd15.7; n11	9914
WEANING SIZE		
Crocidura russula	66.1 mm	4603
LITTER SIZE		
Anourosorex squamipes	3.2 emb; r2-4; n4 f	148
Blarina brevicauda	4 emb; n1 f	2547
	4 emb; n1 f	5798
	4.5 emb; r4-5; n2 f	4280
	4.7 born; se0.2; r1-8; 3.41 weaned; n81 lt	1028
	4.75 emb; r4-6 n2 f	5405b,
	5 emb; n1 f	495
	5; n1 lt	7333
	5 yg; n1	6882
	5 emb; n1 f	7235
	5.3 emb; r2-8; n32 f	8299
	6-7 emb, yg; r5-10; n29 f	4266
	mode 6; r4-8	9164
	6.1 cl; n23 f	8299
	6.3 emb; r5-8; n8 f	1426
	6.6 ps, emb; r5-8	10839
	6.8; r5-8; n6 lt	11956
Blarina carolinensis	3 emb; n1 f	6298
	3.7 emb; r3-5; n3 f	1290
	4 emb; n1 f	7498
Crocidura allex	4 emb; n1 f	4924

Crocidura attenuata	1 emb; n1 f	179
Crocidura bicolor	2; n1 lt	11473
	2.0; r1-3; n2 lt	4924
	3; n1 lt	335
	3 yg; n1 lt	2589
	4 emb; n1 f	7167a
	4 emb; n1 f	10154
Crocidura bottegi	3 emb; n1 f	4354a
Crocidura buttikoferi	3 emb; n1 f	2955
Crocidura crossei	2-4 yg	4354a
Crocidura cyanea	2 emb; n1 f	10154
	3.6 emb; r2-6; n5 f	10158
	4.7 emb; r4-6; n3 f	7167a
Crocidura dracula	2 emb; n3 f	179
Crocidura dsinezumi	4.3 emb; r3-5; n3 f	5405a
Crocidura eisentrauti	2 emb; n1 f	2955
Crocidura flavescens	2 emb; n1 f	1948
	2 yg; n1 f	6828a
	2.28 emb; mode 2; r1-4; n29 f	2575
	3.0 emb; mode 3; r2-4; n6 f	2444
	3.8 yg; sd1.5; n5 lt	3698
	mode 4 yg	4354a
	4 emb; n1 f	4924
	4 emb; n1 f	6828a
	4 emb; n1 f	9945
	4.2; r3-5; n12 lt	9893
Crocidura foxi	2 emb; n1 f	4354a
Crocidura fuliginosa	r1-2; n3 lt	7161
Crocidura fumosa	3.75 emb; r1-6; n4 f	4924
Crocidura gracilipes	2 emb; n1 f	2575
	3 emb; n1 f	2443
	3 emb; n1 f	2444
	3.25 emb; r3-4; n4 f	4924
Crocidura hirta	2; r1-2; n3 lt	187
	2.5 yg; r1-4; n4 lt	7167a
	3.4; r1-9; n9	10158
	3.6; r3-4; n5	10158
	3.8; r2-5; n7 lt	8965
	mode 4; r1-5	2449
	4 emb; n1 f	6272
	4.0 emb; r3-5; n2 f	9945
	4.2 emb; r4-5; n5 f	7167a
	4.25 emb; r2-6; n4 f	331
	4.3; r2-6; n13 lt	9892, 9893
Crocidura horsfieldi	3 emb, yg ea; n3 lt	8501
	3.3 emb; r3-4; n3 f	7382
Crocidura jacksoni	3.5 emb; r3-4; n2 f	4924
Crocidura kivuana	1.67 emb; r1-2; n3 f	2575
Crocidura lanosa	3.3 emb; r2-6; n3 f	2575
Crocidura leucodon	2; n1 lt	3458
	3; n1 lt	3455
	mode 3-6	3571
	3.37 yg; n19 lt	4602
	5-10	8798
Crocidura littoralis	3.0 emb; r2-4; n2 f	2575
Crocidura luna	2 emb; n1 f	331
Crocidura mariquensis	3.3 emb; r2-5; n22 f	10154
	3.7 emb; r2-5; n14 f	9892, 9893
	4.2 emb; r2-7	8965
	4.5 emb; r4-5; n2 f	7167a
Crocidura nigricans	4 emb; n1 f	9893
Crocidura niobe	3 emb; n1 f	2575
Crocidura planiceps	3 yg; n1 nest	4354a

Crocidura poensis	2 emb; n1 f	1948
	3 emb; n1 f	2955
Crocidura russula	2-6	5485
	3-10	3571
	3.0 yg; mode 4; r1-7; n937 lt	4600, 4601, 4602
	3.5; r1-7; n86 lt	11316
	3.65 emb	6574
	4.1; r3-6; n9 lt	955
	4.4 yg; sd1.3; n49 lt	3698
	4.5; r4-5; n2 lt	7898
	5-10	8049
Crocidura silacea?	4 emb; n1 f	334
Crocidura suaveolens	2 emb; n1 f	5096
	2.97 emb; r1-5; n31 f	9253
	3.93; r2-5	11313
	5-10	8049
Crocidura turba	2; n1 lt	2372
Crocidura usambarae	2 emb	5032
Crocidura viaria	3.5 yg; sd1.0; n24 lt	3698
Crocidura zaodon	2.5 emb; r2-3; n2 f	4924
Crocidura zarudnyi	6 emb; n1 f	2404a
Cryptotis magna	3 emb ea; n2 f	9177
Cryptotis mexicana	3 emb; n1 f	4228
Cryptotis parva	2.67 emb; r2-3; n3 f	6303
	3 yg; n1 nest	1924a
	3.33 emb, yg; r3-4; n3 f	2347
	4 emb; r3-5; n3 f	1290
	4.3 yg; r1-9	7409
	4.56	7410
	4.9 emb; r2-7	11626
	5-10	8798
	5 emb; n1 f	1812a
	5.0 yg; r4-6; n5 f	2004
	6 yg; n1 nest	11609
	8 emb; n1	6298
Diplomesodon pulchellum	5 yg	4645
Myosorex babaulti	3 or 4 emb; n1 f	2575
Myosorex cafer	3.0 emb; r2-4; n3 f	10158
	3 emb; n1 f	2096
Myosorex geata	4 yg	5759
Myosorex norae	1; r1-2	5759
	2 emb; n1 f	4924
Myosorex polulus	1-2	5759
	1.5 emb; r1-2; n2 f	4924
Myosorex varius	2.8 emb; r1-4; n19 f	9329
	2.9; mode 3; n10 lt	673
	3 yg; n1 f	3977
Neomys anomalus	mode 5-6; r3-11	3571
	4.0; r2-5; n4 lt	7272
	13 emb; n1 f	5998
Neomys fodiens	3; n1 lt	7898
	3-8	1123
	3.5; r2-5; n2 lt	3630
	6.1 yg; max 10	11316
	6.28 yg; sd2.78; mode 6; r3-15; n18 lt	7273
	6.38 yg; n29 lt	7275
	6.7 yg; r3-15	7276
	6.8 emb; r3-8	8757
	8.0 emb; r5-11; n2 f	1590
	11 emb; n1 f	11508
	11 emb; n1 f	6438
Notiosorex crawfordi	3 emb; n1 f	8754
	3 emb; n1 f	2617
	3 yg; n1 nest	8754

	3.5 emb, yg; r3-5; n4 f	4901
	5 emb; n1 f	529
	5 emb; n1 f	1871b
	5 emb; n1 f	5410
Scutisorex somereni	1.86 emb; mode 2; r1-3; n14 f	2575
Sorex alpinus	6	3571
Sorex araneus	1-11 emb	10677
	2-8	1123
	3-8 emb	9771
	4; r1-6	9916
	4.33 yg; r2-7; n6 lt	3190
	4.5 emb; r2-8; n4 f	3190
	4.8-5.4	9712
	mode 5-7; r2-10	3571
	5-10 emb	10482
	5 emb; n1	1590
	5.14; n29 lt	9712
	5.25 cl; r3-8; n4 f	3190
	5.3 emb; r4-7; n3 f	7901
	5.40; r4-11; n30 young f	10346
	5.46 emb; r1-9	1210
	5.8; r4-10; n8 lt	11314
	5.9; r2-8; n12 lt	11316
	5.92 yg; r2-11; n72 lt	10677
	6-7 cl	10482
	6; n1 lt	11390
	6.0 emb; r4-8; n6 yg f	8773
	6.45 emb; r1-9; n51 f	1210
	6.91; r2-8; n82 adult f	10346
	7.1 emb; r4-11; n66 f	4577
	7.3 emb; r4-9; n9 f	1838
	7.3; r5-9; n9 lt	10339
	7.35 cl; mode 7; r4-12; n97 f	1210
	7.5 emb; n15 f	8372
	7.5-8.3; mode 7-9; r4-11; across populations	10017
	7.7; mode 7; r1-10	2156
	7.75 emb; mode 8; n4 f	5039
	7.8; r6-10	9159
	7.9 emb; r6-10; n8 f	10344
Sorex arcticus	6.0 emb; r5-7; n2 f	10144
	6.0 emb; r4-8; n5 f	1912
	6.0 emb; r5-8; n12 f	1426
	6.25 emb; r3-9; n4 f	506
	7.25 emb; r6-9; n4 f	5210
	7.3 cl; se0.36; r5-9; n15 f	506
	7.5 emb; r6-9; n2 f	495
	7.7 emb; se0.42; r5-9; n10 f	506
	10 emb; n1 f	716
Sorex caecutiens	4.0 cl; r3-5; n2 f	8058a
	5.9-8.9	8058a
	7.0 emb; r5-9; n3 f	8058a
	7.1 emb; n8 f	8058a
Sorex cinereus	2-10	1474
	3 emb; n1 f	574
	4-8 yg	502
	4 emb; n1 f	6166
	4.6 yg; r3-12	1871a
	6.0 emb; r5-7; n2 f	495
	6.5 emb; r6-7; n2 f	12049
	6.5 emb; r6-7; n2 f	10839
	6.7; mode 7; r5-9	3423
	6.8 emb; r5-7; n23 f	1426
	7 emb; n1 f	8808
	7 yg; n1 nest	8766

	7.91 emb; se0.4; r3-12; n32 f	5147a
	8 emb; n1 f	2545
	8 emb; n1 f	11956
	9.0 emb; r7-12; n3 f	6553
Sorex coronatus	3.692 emb; ±1.109; r2-6; n13	6573
	5.1 yg; sd1.5; n19 lt	3698
Sorex cylindricauda	5 emb; n1 f	7382
Sorex daphaenodon	6 emb; n1 f	12105
Sorex dispar	2 emb; n1 f	5799
	2 emb; n1 f	10683a
	5 emb ea; n2 f	9111a
Sorex fontinalis	5; r4-6	9164
Sorex fumeus	5.1 emb	8155
	5.5 emb; r2-8; n42 f	4277
	6.6; r6-8	11956
	9 emb; n1 f	10144
Sorex gracillimus	3.5 cl; r3-4; n2 f	8058a
	6.0 live emb ea; n2 f (1 f with an additional resorbing emb)	8058a
Sorex haydeni	4-10 yg	1871a
Sorex hoyi	5.3 emb; r3-8; n3 f	6554
	6 emb; n1 f	4525
	7 emb; n1 f	9700
Sorex isodon	7.1 emb; se0.22; mode 7; r1-10; n61 f	10019
Sorex longirostris	3.9 emb; sd1.36; r1-6; n15 f	3512
	4 emb; n1 f	2821b
	mode 4	1474
	6-10 yg	3511
Sorex merriami	6.0 emb, swellings; r5-7; n3 f	5352
Sorex minutus	2 emb; n1 f	9771
	mode 5-6; r2-9	1123
	5; n1 lt	7901
	5.57 cl; mode 6; max 7; n21 f	4001
	6 emb; n1 f	7382
	6.0; r4-9; n9 f	8773
	6.0 emb; r2-10; n3 f	9159
	6.23 emb; r2-8; n31 f	1217
	6.5 emb; r6-7; n2 f	10339
	6.77 cl; r5-8; n39 f	1217
	9 yg; n1 lt	3698
Sorex monticolus	4-7	1474
	7 emb; r6-8; n2 f	1871a
Sorex nanus	6-7 yg	1871a
	6.5 emb; r6-8; n4	4894
Sorex ornatus	4-6 emb	8156
Sorex pacificus	4.2 emb; se0.2; r3-5; n20 f	1613, 1614
	4.8 ps; se0.5; r3-6; n6 f	1614
Sorex palustris	3 emb	2001
	4.7; r3-6; n3 f	7809
	5 emb; n1 f	6560
	5.43 emb, ps; r1-8; n14 f	2001
	mode 6 emb; r3-10	760
	6 emb; n1 f	1119
	8 emb; n1 f	6553
Sorex raddei	2.7 emb; n3 f	10274
Sorex trowbridgii	3.77 ps; mode 4; r3-4; n13 f	3623
	3.89 emb; mode 4; r3-5; n28 f	3623
	3.94 cl; mode 4; n34 f	3623
	5; r3-6; n8 f	5244
Sorex tundrensis	12 emb; n1 f	8808
Sorex unguiculatus	5.0 cl; sd1.5; r2-8; n12 non-preg f	8058a
	5.5 emb >8 mm; sd0.5; r5-6; n11 f	8058a
	5.6 emb <8 mm; sd1.0; r4-8; n20 f	8058a
Sorex vagrans	5 emb; n1 f	6488
	5.16 emb, yg; r2-9; n39 f; 5.55 emb; 4.74 yg	5373

	5.6 emb; se0.43; r4-7; n7 f	11230
	6.4 emb; r2-9; n32 f	1910
	6.5 emb; r5-8; n2 f	6553
	9 emb; n1 f	4896
Soriculus caudatus	4.75 emb; r3-6; n8 f	7382
Soriculus nigrescens	4.89 emb; r3-7; n9 f	7382
Suncus etruscus	3.76 yg; mode 4; r2-5; n45 lt, 14 f	3373, 3374
	4.0; r2-5; n5 lt	11315
	4-6	9175
	5 emb; n1 f	7923
Suncus lixus	3 emb; n1 f	328a, 331
Suncus murinus	1-4 cl	2737
	1.84; r1-3; n45 lt	2736, 2740, 2741, 2744
	2 emb ea; n3 f	8501
	2.7 emb; se0.14	572
	2.7; mode 3; r1-5; n71 lt	4406
	2.7; sd1.3; mode 2-3; r1-5; n30	9914
	2.99 emb; sd1.2; r1-7	1331
	3.0; sd1.2; mode 2-3; r1-5; n23	9914
	3.21 yg; mode 3; n24 lt	7524
	3.5 emb; n2 f	9265
	3.5; ±0.16	11999a
	3.67 emb; mode 3; n67 f	7524
	3.8 emb; sd0.286; est n44 f	6589
	3.85 emb; r1-8; n7 f	7382
	4-5 yg	8410a
	4.71 emb; r3-6; n21 f	8903
Suncus varilla	3 emb; n1 f	8965
	3.3 yg; ±1.2; r2-7; n16 lt	6658
	3.7 emb; ±1; r2-6; n30 f	6658
	4 emb; n1 f	6272
Sylvisorex granti	1.6 emb; r1-2; n5 f	2575
	2; n1 lt	187
Sylvisorex lunaris	2.7 emb; r2-4; n3 f	2575
Sylvisorex megalura	1.8 emb; mode 2; r1-2; n5 f	2575
	2 emb; n1 f	330, 331
Sylvisorex morio	2 emb; n3 f	2955

SEXUAL MATURITY

Blarina brevicauda	f: receptive: 47 d; n1	8299
	f: 1st conceive: 1-2 mo	2261
	f: 1st birth: 64 d; n1	1028
	m: 1st sperm 1-2 mo	2261
	m: 1st mate: 85 d; n1	8299
	f/m: fall born: 5 mo; spring born: <5 mo	8299
	f/m: mate: 10 wk	7333
Crocidura leucodon	f/m: 3 mo	8798
	f/m: 4 mo	3571
Crocidura russula	f: 1st mate: 58-416 d; n13; 9/13 <94 d	11316
	f: 1st mate: 7-15 wk	4600, 4601, 4602, 4604
	f: 1st birth: 13.6 wk; r11-19; n15	4601
	m: 1st sperm: 30-32 d	4600, 4601
	m: 1st mate: 71-642 d; n9; 4/9 <114 d	11316
	f/m: 1-3 mo	5276a
Crocidura suaveolens	f: 1st mate: 2 mo; n1 f	9253
	f/m: 1st mate: 1 yr	11313
Cryptotis parva	f: 1st mate: 31 d	7410
	f: 31 d	7409
	m: 1st sire: 36 d	7409, 7410
Neomys anomalus	f/m: 1st mate: 3-4 mo	3571
Neomys fodiens	f/m: 1st mate: rarely yr of birth; most 2nd season	8757
	f/m: 1st mate: 3-4 mo	3571
Sorex araneus	f: 3-5 mo; n "a few"	3571
	f: 4.6% 4-6 mo; mode 10-11 mo	4577
	f: ~9 mo-1 yr	10339
	m: testes large: est yg of yr; n1; 6 g	5982

	m: testes large: 10-11 mo	4577
	m: testicular sperm: ~1 yr	1211
	m: ~11 mo-1 yr	10339
	f/m: 1st mate: 9-12 mo	3571
	f/m: 10-11 mo old	9916
	f/m: ~1 yr; occasionally yr of birth	1123
Sorex cinereus	f: 4-5 mo	1474
Sorex fumeus	f/m: 1st mate: 8-12 mo	4277
Sorex isodon	m: testes large: 9 mo-1 yr	10019
Sorex minutus	f: 1st mate: 9-12 mo; a few f already 3-5 mo	3571
	f: 1st birth: 9-10 mo	9175
	m: ~4 g; min 3.4 g	4001
	m: > 4 g	1217
	f/m: not 1st season	1217
Sorex mirabilis	f/m: 11 mo	8079
Sorex palustris	f: 3+ mo	1474
	f/m: 1st mate: occasionally 4 mo	2001
Sorex vagrans	m: 1st mate: 8.4 g	5373
	f: 1st mate: 7.7g	5373
	f: 1st preg: yr of birth	5373
	f: 1 yr	11230
Suncus etruscus	f/m: prob 1 yr	3373
Suncus murinus	f: large follicles: 3 wk	5506
	f: 1st cl: 20 g; 50% preg or lactating 32 g	1331
	f: 1st mate: 30 d	2737
	f: 1st mate: 5 mo	7524
	f: 1st mate: before fully adult	8501
	f: 1st preg: 27 g	4406
	f: 1st repro: 60 d	2737
	m: spermatogenesis: 5 wk	5506
	m: testes anatomy: 44.5 g	1331
	m: testicular sperm: 27 d	4476
	m: testicular sperm: 31 d	2737
	m: 1st sire: 51 d	2737
	m: 50% fertile 31.5 g, 95 % fertile 46.1 g	4406
Suncus varilla	m: epididymal sperm: >9 mo	6658
CYCLE LENGTH		
Blarina brevicauda	estrus: 2-4 d	8299
Sorex araneus	estrus: < 1 d	2045
GESTATION		
Blarina brevicauda	15 d	7333
	18-20 d	8299
	≥21 d; n1	4266
Crocidura flavescens	r15-35; n12	2955
Crocidura hirta	18 d	2449, 5759, 10158
	~18 d	7167a
Crocidura leucodon	est 20-35 d	11391
	31 d	3455
	31-33 d	3571, 8798
Crocidura russula	28-31 d	3571
	28.5 d; mode 29; r24-32	4600, 4601, 4604
	30 d	11316
Crocidura suaveolens	26.8 d; mode 26-27; r26-29 d; n13	11313
	28 d	9253
Cryptotis parva	12-16 d	8798
	r21-23 d; n5	2004, 7409, 7410
Neomys anomalus	26 d	7272
Neomys fodiens	20.5 d; r19-22; n8 lt, 5 f; pairing to birth	7273, 7276
	20 d; r20-22	11316
	~24 d; non-lactating f	8757
	est 24 d	3630

Sorex araneus	13-19 d	1210
	20 d	2437
	21 d; n1	2156
	23-28 d	11316
Sorex coronatus	24.3 d; r24-25; n3	3841
Sorex haydeni	19-22 d	1871a
Sorex fumeus	~3 wk	4277
Sorex minutus	~3 wk	3571
	~22 d	9175
	~4 wk	8049
Sorex vagrans	20 d	5373
Suncus etruscus	est 27-28 d; pairing to birth, min ibi	3373
	27.5; 4-5 d longer with lactation	11315
Suncus murinus	implant: 7 d	2737
	implant: 7-8 d	4480
	placentation: 12 d	4480
	28-32 d	9914
	29.6 d; se0.1; n53	4477
	30-31 d	7524
	30.2 d; mode 30-31; r28.5-31.5; n20	2737
	30.3 d; se0.1; n156	4477

LACTATION

Blarina brevicauda	leave nest: 12 d	7333
	solid food: 21 d	1028
	wean: mode 25 d; r22-30	1028
	wean: 22 d	6882
Crocidura bicolor	solid food: min 11 d; ~15 d	335
	solid food: 19 d	2589
	nurse: 15 d	335
Crocidura hirta	wean: 15-20 d	7167a
	wean: 18 d	5759
Crocidura leucodon	solid food: 16 d	11391
	wean: 18-19 d	11391
	wean: 26 d	8798
	wean: 16-26 d	3571
Crocidura russula	solid food: 2 wk	4602, 4603
	wean: 22-24 d	4602, 4603
	independent: 6 wk	8049
	wean: ~16 d	3571
Crocidura suaveolens	independent: 6 wk	8049
Cryptotis parva	wean: 18-21 d	2004, 7409, 7410
Myosorex varius	wean: 20-22 d	673
Neomys anomalus	wean: 29-30 d	7272
Neomys fodiens	solid food; 27.8 d; sd1.0; r27-30	7273, 7275, 7276, 7276a
	nurse: 38 d; r34-41; n7 lt	7273, 7275, 7276, 7276a
	wean: est 21 d	3630
	wean: 4 wk	2156
	independent: 6 wk	8049
	wean: ~37 d	8757
	wean: 6 wk; r5-6 wk	3571
Notiosorex crawfordi	independent: 40 d	4901
Sorex araneus	den emergence: 17 d	2437
	den emergence: 18-19 d	12110
	den emergence: 3 wk	3571
	feeding by themselves: 22 d	2437
	wean: 21 d	11314
	wean: 4 wk	2156
Sorex cinereus	wean: 20-27 d	3423
Sorex minutus	wean: 22 d	9175
	wean: 23 d	2156
Sorex monticolus	wean: <3 wk	1871a
Sorex vagrans	wean: ~16-25 d	5373
Suncus etruscus	wean: 19-21 d	3373
	wean: 20 d	11315

Suncus murinus	solid food: 20 d	2736
	wean: 2+ wk	7524
	wean: ~20 d	2736
	wean: 20 d	9914
INTERLITTER INTERVAL		
Crocidura hirta	59 d; n1	7167a
Crocidura russula	28.5 d; mode 29; r24-32	4600, 4601, 4602, 4604
Neomys fodiens	20 d-4 mo	11316
Sorex araneus	28-43 d	11316
	43-44 d	9712
Sorex nanus	1-2 mo?	4894
Suncus murinus	56.2 d; r30-101; n12; yg survive	2737
SEASONALITY		
Anourosorex squamipes	testes: large June-Aug; small Nov-Dec; Taiwan	148
Blarina brevicauda	sperm: none Nov-early Jan; e US	8299
	testes: large end Apr-early Sept; QUE, Canada	11956
	mate: est by 10 Feb; PA	1831
	mate: est Mar-Aug; se MAN, Canada	1426
	preg f: Apr-May; n4; NE	5405b$_1$
	preg f: May; n1; MN	495
	birth: by end Feb; PA	1831
	birth: Mar-May, Aug-Sept; US	1474
	birth: Apr-Sept; e US	8299
	birth: Apr-Sept; PA	9164
	birth: peak May-June; NY	2261
	birth: spring, late summer; NY	4266
	lact f: May, July-Aug; NE	5405b$_1$
	lact f: Oct; FL	7498a
	repro: peak Mar-July, 2nd peak Sept; IL	3743
	repro: Apr, May (and Nov?); NY	7235
Blarina carolinensis	preg f: Mar-Apr, July; n3; NC	1290
Crocidura spp	preg f: yr rd except May, July, Dec; Zambia	9893
Crocidura allex	preg f: Sept; n1; Aberdare Mt, Kenya	4924
Crocidura bicolor	preg f: Nov-Dec; s Africa	7167a
	birth: Oct; few data; n Zimbabwe	335
	lact f: Apr; n1; n Zimbabwe	331
Crocidura bottegi	preg f: Mar; n1; Nigeria	4354a
Crocidura crossei	preg f: Apr-May, Oct-Nov; Nigeria	4354a
	preg f: May, July-Aug, Oct; Nigeria	5190a
Crocidura cyanea	birth: est warm, wet summer; S Africa	10158
	yg found: Nov-Jan; s Africa	7167a
Crocidura flavescens	preg f: Jan; n1; Kenya, Uganda	4924
	preg f: Jan; n1; Namibia	9945
	preg f: Jan; n2; captive	6828a
	preg f: Oct-Nov; Nigeria	5190a
	preg f: yr rd; Uganda, Rwanda, Burundi	2575
	young found: most mo; e Africa	5759
Crocidura foxi	preg f: July; n1; Nigeria	4354a
Crocidura fuliginosa	preg/lact f: Mar-Apr, Sept-Nov; n4; Malaysia	7161
	juveniles present: Aug; Malaysia	7161
Crocidura gracilipes	sexually active f: Mar, June, Nov; Uganda, Rwanda, Burundi	2575
	lact f: Aug; n1; e Africa	5759
Crocidura grandiceps?	preg f: Mar, May-June, Sept; Nigeria	5190a
Crocidura hirta	preg f: Nov; n1; Namibia	9945
	preg f: Nov-May; Transvaal, S Africa	8965
	preg f: Dec-Feb; n5; s Africa	7167a
	birth: Jan-Apr; n4; captive	7167a
	birth: Sept-May; S Africa	10158
	lact f: Mar-May; s Tanzania	187
	lact f: Apr; n1; s Africa	7167a
Crocidura horsfieldi	mate: est July-Aug; Sri Lanka	8501
	preg f: Apr; n3; Nepal	7382
Crocidura jacksoni	preg f: Nov, June; n2; Kenya	4924

Crocidura leucodon	birth: spring-fall; Germany	3571
	repro: not seasonal; Germany	8798
Crocidura luna	preg f: Mar-Apr; Zambia	7167a
	preg f: Apr; n1; s Africa	7167a
	lact f: Apr; n1; s Africa	7167a
Crocidura mariquensis	mate: end dry season-end rains; Zambia	9893
	preg f: Aug-Apr; Botswana	10154
	preg f: Aug-Dec, Feb-Apr; S Africa	10158
	preg/lact f: Jan-Feb, Oct; Transvaal, S Africa	8965
Crocidura russula	mate: Feb-Sept, but low Aug; Spain	6571
	mate: end Feb-Oct; Channel I, UK	955
	mate: yr rd; Germany	3571
	mate: yr rd, peak Mar-May; captive	4600
	preg f: Feb-Sept; Spain	6574
	preg f: Mar-Sept; Italy	5485
	preg f: peak Apr, low Dec; lab	4601
	birth: yr rd; lab	11316
Crocidura suaveolens	birth: Apr-Sept; lab	11313
	birth: almost yr rd; e Europe, n Asia	8049
Crocidura turba	preg f: July; n1; ne Zaire	2372
Crocidura zarudnyi	preg f: May; n1; Iran	2404a
Cryptotis magna	preg f: May, Oct; Oaxaca, Mexico	9177
Cryptotis mexicana	preg f: Dec; n1; Veracruz, Mexico	4228
	lact f: Oct; Veracruz, Mexico	4228
Cryptotis parva	mate: Mar-Nov; north; yr rd; south; US	4279
	preg f: Feb, May, Sept; n3; NC	1290
	preg f: June, Aug, Oct; n3; FL	6303
	preg f: Oct; n1; KS	1812a
	preg/lact f: Apr-Oct; FL	6303
Diplomesodon pulchellum	birth: Apr-Aug; USSR	4645
Megasorex gigas	lact f: June; n1; Guerrero, Mexico	368
Myosorex cafer	preg f: Oct-Nov; n3; Zimbabwe	10158
Myosorex norae	preg f: Oct; n1; Kenya	5759
	preg f: Oct; n1; Kenya	4924
Myosorex polulus	preg f: Oct; n2; Kenya	4924
Myosorex varius	testes: small May-June; S Africa	9329
	mate: Sept-Mar; S Africa	9329
	lact f: Sept-Oct, Feb, Apr; n9; Transvaal, S Africa	8965
Neomys anomalus	mate: Mar-Apr?; Germany	3571
	birth: June-Aug; n4; captive	7272
Neomys fodiens	testes: large Apr-May; lab	8757
	mate: Apr-May; Baltic area, Europe	8049
	mate: Apr-Sept; central Europe	3571
	mate: mid Apr-Sept, peak May-June; lab	8757
	birth: May-June; Europe, w USSR	4645
	birth: May-Oct; central Europe	3571
	repro: low Aug; lab	11316
	repro active f: end Mar-Oct; Germany	3630
	repro: Mar-?; Germany	10347
	sexually active: Apr-Sept; Poland	1123
Notiosorex crawfordi	preg f: Apr; n1; CA	369
	preg f: Nov; n2; OK	529
	birth: summer; AZ, CA	4901
Scutisorex somereni	sexually active f: Jan, Mar-Apr, June-Aug, Dec; central Africa	2575
	lact f: May; n3; Uganda, Zaire	5759
Sorex alpinus	birth: May, Aug; Germany	3571
Sorex araneus	sperm: mid Mar-early Nov; Wales, England, UK	1211
	spermatogenesis: end Feb-early Mar; s Britain	2045
	testes: large Feb-July; UK	2156
	testes: large end Feb-Apr; n Finland	4577
	testes: large Apr-Sept; England, UK	1838
	uteri: large Apr-July; England, UK	1838
	uteri: large May-July; UK	2156
	estrus: mid-end Apr, mid-end June; Germany	9159

	mate: Mar-?; England, UK	1427
	mate: Mar-Apr; Europe, USSR	4645
	mate: est Apr/May-?; Finland	10017
	preg f: Apr-Aug; n Finland	4577
	preg f: Apr-early Sept; e Germany	10339
	preg f: Apr-Sept; Poland	2436
	preg f: peak May; few data; Britain	8372
	preg f: May-June; Germany	10344
	preg f: May-Oct; Great Britain	1210
	preg f: May-Oct, none Nov; Poland	10677
	preg f: May-Oct; Poland	2436
	preg f: May-Oct; Wales, Scotland, UK	1210
	birth: (Apr) May-Oct; Germany	3571
	birth: end Apr-Aug; England, UK	9916
	birth: end May-mid Sept; n9; lab	11314
	birth: May-Oct; Wales, Scotland, UK	1210
	birth: mid May-Oct; Poland	10677
	birth: spring-summer; s Britain	2045
	sexually active: Apr-Oct; Poland	1123
Sorex arcticus	active testes: Feb-June; WI	1912
	mate: est Mar-Aug; se MAN, Canada	1426
	preg f: May, July; n2; NS, Canada	10144
	preg f: Apr, June; n2; MN	495
	preg f: Sept; n1; none end July-early Sept; AK	716
	preg/lact f: Apr-Sept; MN	506
	birth: early-mid May; BC, Canada	2088
Sorex cinereus	testes: large spring; few data; s QUE, Canada	11956
	estrus: spring; few data; s QUE, Canada	11956
	mate: est May-Aug; se MAN, Canada	1426
	preg f: Jan, Apr-May, Sept; N America	1474
	preg f: Apr, June; n2; MN	495
	preg f: mid June-mid July; n12; SD, WY	10954
	preg f: July; n1; se MT	6166
	preg f: July-Aug; n2; YUK, Canada	12049
	preg f: none mid July-early Sept; n39; AK	716
	lact f: July; n1; ND	3703
	lact f: July; n1; NE	5405b,
Sorex coronatus	preg f: Feb-Sept; n Iberian Peninsula	6573
	preg f: May-July; ne Iberian Peninsula	6572
Sorex cylindricauda	preg f: Sept; n1; Nepal	7382
Sorex dispar	testes: large Apr-Aug; VT	5799
	preg f: May; n2; PA	9111a
	preg f: Aug; n1; NY	10683a
	preg/lact f: May-Aug; VT	5799
	birth: May; ne US	1474
Sorex fontinalis	f fertile: end Feb-?; PA	9164
	preg f: end Mar-Sept; PA	9164
Sorex fumeus	testes: large Apr-Aug; NY	4277
	mate: mid Apr-Aug; QUE, Canada	11956
	preg f: Mar-Apr; TN, NC	8155
	preg f: May; n1; NS, Canada	10144
	birth: Apr-Oct; e US, Canada	1474
Sorex gaspensis	testes: large: end Apr-Aug; NB, Canada	5799
Sorex hoyi	preg f: July; n1; IA	6554
	preg f: July; n1; IA	9700
	lact f: July-Aug; n2; WI	6554
	lact f: Aug; n1; CO	6554
	lact f: Aug; n1; AK	6554
Sorex isodon	preg f: mid May-early Oct; e Finland	10019
Sorex longirostris	preg f: Apr-Sept, peak Apr; se US	3512
	birth: Apr; e US	1474
Sorex merriami	preg f: Apr, May, July; WA	5352
	large mammary glands: Mar, Oct; WA	5352

Sorex minutus	spermatogenesis: Mar-Aug; Wales	1217
	mate: Apr-Aug; mid Europe	3571
	mate: Apr-Oct; Poland	1123
	mate: May?; USSR	8049
	preg f: mid Apr-early Oct, peak June; Wales	1217
	preg f: end Apr-early May, early July; Germany	9159
	preg f: May; n1; Nepal	7382
	preg f: May-June; few data; Britain	8372
	preg, lact f: Apr-Oct; Poland	2436
	birth: ?-Aug; Pakistan	9175
	birth: May-Sept; mid Europe	3571
	birth: June; USSR	8049
	birth: June-Oct; Ireland	4001
Sorex monticolus	preg f: June, Aug; n2; WY	1871a
	birth: July; few data; nw N Am	1474
Sorex nanus	birth: July-Sept; WY	1871a
	birth: end July-early Sept; alpine US	4894
Sorex ornatus	repro: end Feb-Oct; CA	8156
Sorex pacificus	preg f: Mar, May-Sept; CA	1614
	preg f: mid Apr-Aug, Nov; OR	1614
Sorex palustris	sperm present; Jan-July; MT	2001
	preg f: Mar; WY	1871a
	birth: end Feb-June; n US	1474
Sorex raddei	mate: Mar; Turkey	10274
	birth: mid-end May; Turkey	10274
Sorex trowbridgii	mate: Feb-early June; CA	5244
	mate: Feb-Oct; OR	3623
	birth: Mar-May, occasionally July; w US	1474
	birth: end Apr/early May-?; BC, Canada	2088
Sorex vagrans	preg f: Apr-early Aug; MT	1910
	preg f: mid June-mid Aug; few data; CO	11230
	preg f: Sept, mid Feb-early June; CA	5373
	birth: Feb-May; Oct-Nov; nw US	1474
	birth: peak end Feb-early June, sm peak fall; CA	5373
	birth: Mar-July; WY	1871a
	birth: Mar-summer; s BC, Canada	2088
	birth: spring; w US	5146
Soriculus caudatus	preg f: May, Aug; n8; Nepal	7382
Soriculus fumidus	testes: large Mar-Apr; Taiwan	148
Soriculus nigrescens	preg f: May, July-Aug, Oct; n9; Nepal	7382
Suncus etruscus	preg/lact f: May-Sept; France	3373, 3374
	birth: yr rd, peak Aug-Oct; Pakistan	9175
	repro: Apr-May; lab	11315
Suncus lixus	preg f: Jan; n1; Zambia	328a
Suncus murinus	epididymal sperm: yr-rd; w Rajasthan, India	8903
	sperm: peak Jan-June; Malaysia	4406
	sperm: yr rd; Pakistan	494
	mate: yr rd, peak June-Sept; Japan	7524
	preg f: peak Jan-Mar, Apr-June; Burma	1331
	preg f: Feb-Aug; Pakistan	6728
	preg f: Mar-Sept, peak June-July; w Rajasthan, India	8903
	preg f: Apr, June-July, Sept; n7; Nepal	7382
	preg f: June, July, Dec; n3; Sri Lanka	8501
	preg f: yr rd, peak Mar-Apr, July-Aug; Zhejiang, China	11999a
	preg f: yr rd, peak Apr-June, Oct-Dec; Guam	572
	preg f: yr rd, peak 42% Oct-Dec; Malaysia	4406
	preg f: yr rd; w Bengal, India	6589
	birth: spring & end of monsoon (end Sept); Pakistan	9175
	young found: yr rd; Guam	8410a
Suncus varilla	testes: peak Aug-Feb; S Africa	6658
	mate: Aug-Mar; S Africa	6658
	preg f: peak Aug-Oct; S Africa	6658

Order Insectivora, Family Talpidae

Moles (*Condylura* 35-85 g; *Desmana* 100-220 g; *Galemys* 50-80 g; *Mogera* 87-220 mm; *Neurotrichus* 8-11 g; *Parascalops* 40-85 g; *Parascaptor* 110-115 mm; *Scalopus* 40-140 g; *Scapanulus* 35-38 mm; *Scapanus* 50-170 g; *Scaptochirus* ~140 mm; *Scaptonyx* 65-90 mm; *Talpa* 60-150 g; *Uropsilus* 12-20 g; *Urotrichus* 15 g) are subterranean or semiaquatic insectivores from Eurasia and North America. Males may be larger than females (*Talpa* 3842, 7186, 10341; *Parascalops* 2844).

Copulation may trigger ovulation (*Talpa* 2045, 2423; *Scalopus* 2005). A copulatory plug is present in *Condylura* and *Parascalops* (2847, 2848). The villous, labyrinthine, diffuse, discoid placenta invades the uterine wall (231, 7526, 7563, 7564, 7567, 8736, 10310), has an antimesometrial, superficial implantation into the bipartite or bicornuate uterus (2662, 7526, 7563, 8460, 8736, 8789, 10341), and is absorbed after birth (2423, 6955; cf 8462). Delayed implantation may occur (*Talpa* 486) as well as a postpartum estrus (*Galemys* 8209).

Neonates are blind and naked (2231, 2844, 4268, 4455, 5209, 6088, 11897). Their eyes open at 20-22 d (*Talpa* 2045, 3571, 3842, 8049). Females have 6-8 mammae (63, 1474, 2662, 7621, 8049), several accessory reproductive glands, and a long, penis-like clitoris (3833, 7186, 8049, 8459, 8460, 10046, 11897). A vaginal opening exists only during the breeding season (*Talpa* 63, 6955, 11897; *Scalopus* 2005; *Parascalops* 2844).

Males have abdominal testes (2045, 2846, 2848, 3833). *Neurotrichus* males have ampullary glands (2849), while *Condylura*, *Neurotrichus*, and *Parascalops* males have inguinal and perineal glands (2846, 2848, 2849). The penis is usually decorated (2848, 2849, 8789), although not in *Parascalops*. Both *Parascalops* and *Talpa* have bacula (2846, 3833). The male reproductive system can reach 14% of body mass in *Condylura* (2848, 4268).

Details of anatomy (211, 2005, 2423, 2775, 4105, 4530, 6334, 8678, 8789, 10566, 10567), intersex (1360), and endocrinology (3831, 8049, 8832, 9007, 10626, 10668) are available.

NEONATAL MASS		
Neurotrichus gibbsii	0.67 g; n1	2231
Scalopus aquaticus	5.355 g; r5.35-5.36; n2	4455
Scapanus townsendii	est ~5 g	6088
Talpa europaea	3 g	8274
	3.5 g	63
NEONATAL SIZE		
Galemys pyrenaicus	30-32 mm; large emb	8209
Neurotrichus gibbsii	HB: 21 mm; n1	2231
Scapanus townsendii	50 mm	6088
Talpa europaea	30 mm	8274
	33-34 mm	3834
	35 mm	7186
	35-40 mm	3842
	44 mm	63
WEANING MASS		
Scapanus townsendii	60-80 g; 30 d	6088
Talpa europaea	44.2 g; r33-67; n5, 33 d	3837
	~60 g; 3 wk	2045
WEANING SIZE		
Scapanus townsendii	≥115 mm	6088
LITTER SIZE		
Condylura cristata	2 yg; n1 nest	11667
	5 emb; n1 f	11956
	5 ps; n1 f	10839
	5 yg; r3-7; n9 lt	2851
	5.4 ps; n9 f	2851
	5.44 emb; r2-7; n25 f	2851
	5.6 yg; r5-7; n5 nests	2322
	6.3 emb; r6-7; n7 f	4268

Desmana moschata	3-5	4645
	3-5	8049
Galemys pyrenaicus	3.6 emb; r1-5; n53 f	8209
	mode 4 yg; r1-5	8461
Neurotrichus gibbsii	2.8 emb, yg; r1-4; n5 lt	2231
	4 emb; n1 f	8825a
Parascalops breweri	4	2844
	4-5 emb; max 8	4247
Scalopus aquaticus	1 emb; n1 f	359a
	2-4 emb; n5 f	1290
	2 emb; n1	8026
	3 emb; n1 f	5405b,
	3 emb; n1 f	5803
	3 emb; n1 f	12006a
	3.9; mode 4	8736
	3.91 emb; mode 4; r3-5; n33 f	2005
	mode 4; r2-5	5209
	4-5 emb	359a
	4 ps; n1 f	9164
	4.33 cl; r3-7; n15 f	2005
Scapanus latimanus	2-5	1474
Scapanus orarius	2-4 emb	3816
	2-5	2088
Scapanus townsendii	2.5 emb; mode 3; r1-3; n18 f	6088
	2.87 yg; mode 3; r1-4; n94 nests	6088
	3.5 emb, yg; mode 3; r2-6; n8 f	7490
Talpa altaica	est 4.6; r3-6	5463
Talpa europaea	2-6	3834
	2-6 emb	3833
	2.99 emb text, 3.0 emb table; r2-4; n10 f	11255
	3-4	6955
	3-4	2423
	mode 3-7; r3-9	3571
	3 emb; 1 f	1590
	3.12 emb; r2-4; n17 f	6573a
	3.5; se0.17; mode 4; n48 lt	3842
	3.63; r1-7; n60 lt	63
	3.65 emb; r1-6; n26 f	10344
	3.80; se0.11; mode 4; n45 lt	6216
	3.82 emb; ±0.13; r2-6	3836
	4 emb; n1 f	7902
	4.54 emb; r3-7; n46 f	10341
	4.8; mode 4 90%; r2-7; n68 lt	545
	5.08; se0.39; mode 5; n13 lt	635
	5.73; se0.06; mode 6; n11 lt	7357
Urotrichus talpoides	3-4	11051
	3.5 emb; mode 4; r2-5; n19 f	12023
	3.6; n8 lt	5168
	3.75 emb; se0.44	5137
	3.9 emb; mode 3; r3-5; n7 f	11050

SEXUAL MATURITY

Condylura cristata	f/m: ~10 mo	1474
Parascalops breweri	f/m: 10 mo	2844
Scalopus aquaticus	f/m: 1st mate: 10 mo	1871a
	f/m: 1st mate: 1 yr	1474
	f/m: not yr of birth	2005
Scapanus orarius	f/m: 9-10 mo	4456

Talpa europaea	f: 1st mate: 1 yr	10341
	f: 1st conceive: 8 mo	3834
	f: 1st birth: spring of yr after birth	2045
	f: 1st birth: 12 mo	3571
	f: 1 yr	6573a
	m: 1st mate: 1 yr	7186
	m: 1 yr	3842
	m: 1 yr	6573a
Urotrichus talpoides	f/m: est 1 yr	5137
CYCLE LENGTH		
Talpa europaea	estrus: est 20-30 h	2045
	<1 mo	6955
GESTATION		
Desmana moschata	45-50 d	4645
Galemys pyrenaicus	~30 d	8209
Neurotrichus gibbsii	>15 d; capture to birth	2231
Parascalops breweri	est 4-6 wk	2844
Scalopus aquaticus	est ~4 wk	2005
	40-45 d	1871a
	~6 wk	359a
	~6 wk	1474
	~45 d	5210
Talpa altaica	9-10 mo	1122
Talpa europaea	~3 wk; field data	10341
	30-40 d	4645
	~4 wk	6955
	est 4+ wk	63
	1 mo	2423
	30 d	3833, 3834
	40 ? d; r3-4 ? wk	3571
LACTATION		
Condylura cristata	independent: 3 wk	1474
Desmana moschata	independent: 40 d	4645
Parascalops breweri	leave nest: ~1 mo	1474
Scalopus aquaticus	independent: ~1 mo	1474
	independent: 5-6 wk	1871a
Scapanus townsendii	leave nest: 30 d	6088
Talpa europaea	wean: 1 mo	8274
	wean: 4-5 wk	2423
	wean: 33 d	3837
	independent: 4-6 wk	3571
	independent: 5-6 wk	4645
	independent: 5-6 wk	8049
	wean: 3 wk	6955
Urotrichus talpoides	independent: 4 wk	5168
LITTERS/YEAR		
Condylura cristata	1	4268
	min 2	7235
Desmana moschata	est 2	8049
Neurotrichus gibbsii	>1	1474
Parascalops breweri	1	2844
	1; possibly 2	1474
Scalopus aquaticus	1	2005
	est 1	5209
Scapanus latimanus	1	1474
Scapanus orarius	1	1474
Scapanus townsendii	1	1474, 2088, 9575
Talpa europaea	1	6955
	1	63
	1	10341
	1	2423
	1+; mode 1; 3% 2; r1-2	3842
	mode 1; 3% 2; r1-2	7186
Urotrichus talpoides	1-2	5137

SEASONALITY

Condylura cristata	mate: end Mar-Apr; NY	2847, 2848
	birth: end Mar-early Aug; NY	2851
	birth: Apr-June; NC, ne US, e Canada	1474
	birth: Apr-June, peak early May; NY	4268
	preg f: end Mar-early June; NY	4268
Desmana moschata	birth: yr rd, peak spring; USSR	4645
Galemys pyrenaicus	m fertile: Nov-May; France	8461
	preg f: Feb-June; France	8459, 8461, 8462
Neurotrichus gibbsii	m fertile: Feb; n1; WA	2231
	preg f: Feb; n1; WA	2231
	preg f: Apr; n1; BC, Canada	8825a
	birth: Mar; BC, Canada	2088
	lact f: Sept; n2; WA	2231
	lact f: Sept; BC, Canada	2088
Parascalops breweri	mate: end Mar-early Apr; NH	2844
	birth: early May; ne US, ne Canada	1474
Scalopus aquaticus	mate: Mar-Apr; s WI	8736
	mate: end Mar-early Apr; WI	2005
	mate: Feb-early Mar; MO	2005
	preg f: Feb; n1; TX	12006a
	preg f: Mar-May; n3; NC	1290
	preg f: May; n1; NE	5405b$_1$
	birth: Mar; south; May; north; central, e US	1474
	birth: Mar-Apr, n US; earlier s US	5209
Scapanus latimanus	birth: Mar-Apr; US	1474
Scapanus orarius	mate: Jan-early Mar; BC, Canada	3816
	birth: Mar-Apr; BC, Canada	3816
Scapanus townsendii	m fertile: Feb; US	1474
	mate: Feb; BC, Canada	2088
	birth: Mar-Apr; US	1474
	birth: mid-end Mar; WA	9575
Talpa altaica	implant: spring; Siberia	1122
	birth: spring; Siberia	1122
Talpa europaea	spermatogenesis: Jan-Mar; Austria	10668
	spermatogenesis: Jan-mid Apr; Europe	2075
	spermatogenesis: peak Feb-Mar, low May; Britain	2045
	spermatogenesis: Feb-May; England, UK	6543
	spermatogenesis: peak Mar, none July-Jan; England, UK	10567
	spermatogenesis: Dec; location not given	9007
	epididymal sperm: Mar-Apr, none May; Austria	10668
	testes: large Jan-Feb, small May-June; France	3832
	testes: enlarged Feb; Germany	10341
	testes: large Feb-Mar, small rest of yr; England, UK	3842
	testes: large Feb-Apr; Britain, Netherlands	7186
	testes: large Feb-Apr; England, UK	63
	testes: large Feb-Apr; England, UK	6543
	testes: large end Feb-early Apr, small Aug; Austria	10668
	testes: large Mar, small Nov; Lancashire, England, UK	8832
	testes: large early Apr; central, s Ural, USSR	8274
	testes/acc gland: large mid Mar-end Apr; England, UK	2423
	vaginal clefts open: mid Apr; central, s Ural, USSR	8274
	ovarian medullary cords: max Mar-Apr; France	3832
	ovary: quiescent June-Feb; France	3831
	mate: mid Feb-mid Mar; France	3831, 3833, 3834
	preg f: Jan-Apr; Spain	6573a
	preg f: Feb-June; England, UK	3836
	preg f: mid Feb-?; Spain	11255
	preg f: Mar-June; Germany	10341, 10344
	preg f: Aug; n1; odd, usually none after July; Germany	7902
	preg/lact: Feb-June; France	3831
	birth: end Mar-Apr; Austria	10668
	birth: mid-end Apr; Britain, Netherlands	7186
	birth: mid Apr-May; central European Russia, w Siberia	8049

	birth: mid Apr-mid May; sw European USSR	4645
	birth: mid Apr-June, peak May; Ukraine w of Dnieper	7357
	birth: mid Apr-June; England, UK	63
	birth: end Apr-early May; England, UK	6955
	birth: May-June; Britain	2045
	birth: end May-early June, 25% preg summer; central, s Ural, USSR	8274
	lact f: Feb-June; Spain	6573a
Talpa micrura	lact f: Mar; n1; Malaysia, Cameron Highlands	7161
Urotrichus talpoides	active cauda epididymis: Feb-May; Chiba, Japan	5168
	testes: large Jan-May; Fukucka, Japan	12023
	testes: large Mar, small Oct; Japan	11051
	mate: Feb-May, small peak Aug-Sept; Niigata, Honshu, Japan	5137
	preg f: mid Jan-?; Fukucka, Japan	12023
	preg f: Mar-Apr; Chiba, Japan	5168
	preg f: Mar-June; Japan	11051
	lact f: mid Feb-?; Fukucka, Japan	12023
	lact f: end Apr-mid June; Japan	11051

Order Scandentia, Family Tupaiidae

Tree shrews (*Anathana* 175-200 mm; *Dendrogale* 35-55 g; *Ptilocercus* 23-60 g; *Tupaia* 30-320 g; *Urogale* 350 g) are arboreal omnivores from tropical rainforests of southeast Asia. *Tupaia* males may be larger than females (4626). Ovulation may be induced (*Tupaia* 2011, 4474, 6890, 9917). The labyrinthine, bidiscoidal placenta has been described variously as hemochorial, endotheliochorial, or epitheliochorial, depending upon fetal age, and it has a superficial, bilateral, antimesometrial implantation at areas of the bicornuate uterus that are free of endometrial glands (231, 1491, 4764, 6087, 6622, 6623, 6890, 7179, 7180, 9684, 9685, 10310, 11111, 11256). Placentophagia occurs (*Tupaia* 10281) as does a postpartum, occasionally prepartum, estrus (*Tupaia* 3709, 4474, 4626, 6087, 6890, 7986, 9682, 9685, 9917, 10224, 10281, 10482, 11164; *Urogale* 2109). Delayed implantation may occur (6087, 6622, 6890; cf 9684). *Tupaia* young are suckled once every 48 hours (6889, 6892, 7986, 9682).

Tupaia neonates are naked and open their eyes at 13-25 d (398, 3706, 6889, 7161, 7535, 7986, 9217, 9917, 10223, 10280, 10281). Females menstruate (2011; cf 11164) and have 2-6 mammae (1990, 2313, 6678, 7161, 10279, 10281, 11899). Males have prepenial testes, seminal vesicles, and an ampullary gland (1491, 1990, 2310, 6890, 11256, 11899). *Anathana* males have abdominal testes (11256). Details of anatomy (136, 2777, 6311, 10482, 11256, 11899), endocrinology (1989, 7179, 9685), spermatogenesis (1591), and milk composition (2209, 6890, 9682, 11716b) are available.

NEONATAL MASS		
Tupaia glis	r10-12 g; n5	8798
	11.0-19.0 g	10223
	11.2 g; n8	10281
	12.6 g; suckled once	10280
	m: 12.9 g; r9.1-19.3	9682
	m: 14.5 g; n10	9917
	15 g; with ~6 g milk; r~5-20; n8	6889, 6890
Tupaia montana	11.0 g; r8.0-15.7; n3	10224
Urogale everetti	19.15 g; n1, 2 d	2109
	20.8 g; n1	2109
NEONATAL SIZE		
Tupaia glis	79 mm; n8	10281
	100-140 mm	10223
Urogale everetti	130 mm; 2 d	2109
WEANING MASS		
Tupaia spp	80 g; r~50-110; n8	6889
Tupaia glis	~100 g	9682
LITTER SIZE		
Anathana wroughtoni	5 emb; n1 f	11256
Tupaia glis	1.83 emb; mode 2; r1-2; n6 f	4626
	1.9; r1-2; n10 lt	5581
	mode 2; r1-4	11344
	2 emb	6201
	2 emb; n1 f	6678
	2 emb ea; n2 f	2313
	2 yg ea; n4 lt, 1 f	398
	2 yg ea; n9 lt	4474
	2.0; n5 lt	4406
	2.0 yg; r1-3; n57 lt	7535
	2.09 live yg; mode 2; r1-3; n21 lt	6890
	2.23; r1-4; mode 2; n56 lt	9682
	2.25 emb; r1-3; n4 f	6678
	2.37; r2-3; n8 lt	3706
	2.38; r1-3; n16 lt	10281
	2.5 yg; n54 lt	9683
	2.7; r2-3; n3 lt	7986
	2.83; r2-3; n6 lt	3706
	3 emb; n1 f	10824
	3-5 yg	4813a
Tupaia javanica	max 2	4764
	4-5 cl; 2 emb	10482

Tupaia minor	1-3	2209
	1.9 yg; r1-2; n8 lt	7161
	2 emb ea; n4 f	2313
	2; n48 lt	4406
Tupaia montana	r1-3 yg; n19	10224
Tupaia nicobarica	1 emb; n1 f	6678
Tupaia tana	1-3	2209
	2 emb ea; n3 f	2313
	2.5 yg; r2-3; n2 lt	9217
	4 emb; n1 f	6678
Urogale everetti	1.67; r1-2; n3 lt	2109
	1.86 yg; r1-2; n7 lt, 3 f	11612
SEXUAL MATURITY		
Tupaia glis	f: 1st birth: 150-300 d	9683
	f: 1st birth: 6 mo	10281
	f: 1st preg: 130 g	4406
	m: fertile: 50% 114.5 g, 95% 190.6 g	4406
	m: sperm: 90 d	1989
	m: 5 mo	398
	m: 6 mo	10281
	f/m: 90 d	5591
	f/m: 90-100 d	9917
	f/m: est 3-4 mo	9913
	f/m: 5-6 mo	7535
Tupaia minor	f: 1st preg: 46 g	4406
	m: fertile: 50% 16.3 g, 95% 449.7 g	4406
CYCLE LENGTH		
Tupaia glis	estrus: 2-4 h	2011
	10-12 d or 18-22 d	2011, 10223
Tupaia montana	r9-12 d; n16; intercopulation interval	10224
GESTATION		
Tupaia glis	implant: 7 d	9684
	43-45 d	9917
	43-46 d	10281
	43-46 d; n7 lt	4474
	43.2 d	9682
	44 d; n1	398
	44 d; r40-47	6087
	45 d, r40-52; n3	3706, 3709
	~45 d	7535
	est 46-50 d	10824
	46-50 d	4626
	~47 d; mode 49; r41-50; n39	2011, 10223
Tupaia montana	45-53 d; n20 lt	10224
Urogale everetti	~56 d	11612
LACTATION		
Tupaia glis	leave nest: 23 d	3706
	leave nestbox: 32 d	781a
	den emergence: 33 d	6890, 6892
	wean: few d after den emergence	6892
	solid food: 35 d	398
	wean: ~30 d	9682
	wean: ~30 d	11344
	wean: 6 wk	7535
Tupaia tana	solid food: 4 wk	9217
	solid food: 31 d	10223
	attempt nurse: 53 d	10223
Urogale everetti	den emergence: ~1 mo	2109
INTERLITTER INTERVAL		
Tupaia glis	44 d; mode 43-45; r41-120	9682, 9683, 9685
	45 d	11344
	min 45 d	6890, 6892
	45.3 d; r43-47	7986
	mode 49 d; r41-50	2011

SEASONALITY

Tupaia glis	testes: large Feb, small Sept; lab	1990
	mate: yr rd; lab	2011
	mate: Jan-May; captive	4626
	preg f: end Apr-mid Aug; n2/10; N Borneo, Indonesia	2313
	preg f: Aug; n1; Phu-Guy, Vietnam	10824
	birth: yr rd; Malaysia	7161
	lact/preg f: Jan-June, none July-Aug; Malaysia	6200
	repro: est yr rd; captive	4474
	repro: Feb-June; Penany I, Malaysia	6201
Tupaia minor	birth: yr rd, peak May-July; Malaysia	7161
Tupaia montana	birth: yr rd; captive	10224
Tupaia tana	preg f: July-Aug, Nov; no Feb-June data; Java, Banka, Indonesia	12120
	preg f: mid-end July; n3/15; N Borneo, Indonesia	2313
Urogale everetti	birth: Aug, Oct, Dec; n1 pr; zoo	2109

Order Dermoptera, Family Cynocephalidae, Genus *Cynocephalus*

The 1-2 kg, nocturnal flying lemurs (*Cynocephalus variegatus*, *C. volans*) glide through the tropical rainforest of southeast Asia seeking fruit, young shoots, and leaves. The sexes are often dimorphic with respect to color. The hemochorial, labyrinthine or discoidal (7563, 10310) placenta has a superficial, antimesometrial, eccentric (7563) implantation into a duplex uterus (11164). Females have 2 pair of auxillary teats (7161, 11611). Copulation can occur during lactation (11833a) and lactation can occur concurrently with gestation (7161).

NEONATAL MASS		
Cynocephalus volans	35.8 g; near term	11833a
NEONATAL SIZE		
Cynocephalus volans	max 254 mm	281a
LITTER SIZE		
Cynocephalus variegatus	1	7161
	1 yg; n1 f	557
	1 yg; n1 f	2313
Cynocephalus volans	1	11611
	1	11833a
	1 yg	988a
GESTATION		
Cynocephalus variegatus	~8 wk	6369
Cynocephalus volans	est 150 d	11833a
LACTATION		
Cynocephalus volans	dependent: ~200 d	11833a

Order Chiroptera

With at least 17 families, 175 genera, and over 900 species, bats are second only to rodents in phylogenetic diversity. This taxonomic diversity is paralleled by an equally varied reproductive biology. Bats are extraordinarily flexible with respect to the timing of the major events of the reproductive cycle. Sperm storage in both males and females dissociates the timing of copulation from that of conception. Similarly, embryonic diapause and delayed implantation release the timing of birth from dependence on the timing of conception.

In general, females produce one offspring each breeding season. Testes are often abdominal during non-reproductive seasons.

Ordinal comparisons of anatomy (419, 2071, 6720, 6966, 9178), reproductive asymmetry (11804), ovum development (11095), physiology (9350), sperm storage (4449, 8830, 8834, 11018, 11792), implantation (8947), gestational delays (1460, 8827), birth (11801), placentation (1228, 3926, 11799), endocrinology (6139), male reproductive cycles (4160, 6065), reproductive energetics/ecology (6110, 8176, 8835), and methodology (6110a) are available. One journal, Myotis, focuses on bats.

Order Chiroptera, Family Pteropodidae

Flying foxes and fruit bats (*Acerodon* 450-565 g; *Aethalops* 19 g; *Alionycteris* 64 mm; *Aproteles, Balionycteris* 9-15 g; *Boneia* 150-194 g; *Casinycteris* 40 g; *Chironax* 12-22 g; *Cynopterus* 26-100 g; *Dobsonia* 164-827 g; *Dyacopterus* 70-85 g; *Eidolon* 215-390 g; *Eonycteris* 35-80 g; *Epomophorus* 40-310 g; *Epomops* 60-220 g; *Haplonycteris* 16-20 g; *Harpyionycteris* 83-142 g; *Hypsignathus* 218-450 g; *Latidens* FA 67.5 mm, n1; *Macroglossus* 10-23 g; *Megaerops* 20-38 g; *Megaloglossus* 8-20 g; *Melonycteris* 77-106 mm; *Micropteropus* 10-35 g; *Myonycteris* 23-61 g; *Nanomycteris* 16-35 g; *Neopteryx* 190-250 g; *Notopteris* 50-60 g; *Nyctimene* 25-50 g; *Otopteropus* 17-21 g; *Paranyctimene* 22 g; *Penthetor* 35-65 g; *Plerotes* 87 mm; *Ptenochirus* 86-105 mm; *Pteralopex* 255-280 mm; *Pteropus* 45-1600 g; *Rousettus* 55-910 g; *Scotonycteris* 20-64 g; *Sphaerias* 26-30 g; *Styloctenium* 175-220 g; *Syconycteris* 11-25 g; *Thoopterus* 67-99 g) are primarily frugivorous or nectivorous inhabitants of the old-world tropics and subtropics. Males are often larger than females (191, 696, 1185, 3215, 5226, 6845, 7717, 8494, 8817, 9175, 10780, 11150, 11270; cf 3903, 5759, 8494, 9297). *Hypsignathus* mates in leks (1185).

The hemochorial or endotheliochorial, labyrinthine, discoid or diffuse, deciduate placenta has a mesometrial or circumpherential implantation of intermediate depth into a duplex or bicornuate uterus (181, 2198, 2948, 3914, 3917, 3924, 3926, 4573, 5527, 5539, 5617, 5924, 6627, 6846, 6847, 7416, 7417, 7563, 7677, 7679, 9178, 9746, 10310, 11799, 11898; cf 6627) and is eaten after birth (6093, 6094, 9365). A postpartum estrus occurs (252, 567, 696, 3914, 3916, 3921, 4572, 6031, 8081, 8082, 8871, 8874, 8875, 10780) as do gestational delays (*Eidolon* 7676, 7677; *Haplonycteris* 4571, 4572, 4573).

Dobsonia may be born with open eyes (2835), but most pteropids open their eyes 1-10 d after birth (334, 898, 3903, 5759, 6093, 6094, 7679, 7859, 9175, 9297, 10158; cf 9365). Neonates usually have some fur at birth (331, 334, 2835, 2948, 5759, 6093, 7679, 9297, 10158; cf 898, 1342, 9175). Females have 2 mammae (331, 6720, 9175, 9297, 11898) and an anatomical bridge linking the ovary and the uterus (3937, 3940, 6847). Males have seminal vesicles, ampullary glands, urethral glands, and a baculum (6065, 7441; *Cynopterus* 7439, 7440, 11071; *Harpyia* 9178; *Pteropus* 4097, 7665, 8863; *Rousettus* 3938, 6721).

Details of female anatomy (419, 3937, 5617, 6720, 6846, 6847, 9178, 9490, 10878, 11898), male anatomy (568, 5599), spermatogenesis (7064), and endocrinology (252, 567, 5128, 5526, 7065, 10878) are available.

NEONATAL MASS

Cynopterus sphinx	10.1 g; n1 stillborn	8873
	11.0 g; r10.8-11.2; n2	8873
	11 g	9471
	12-14 g	3915
Dobsonia moluccensis	50-60 g	2835
Dobsonia peroni	48 g; n1 emb	3903
Eidolon helvum	min 39 g	3566
	40-50 g	5759
	40-53 g; term emb	7677
	45-55 g; term emb	3184
	m: ~50 g; n1	2948
	f: 58.5 g	3566

	m: 58.7 g	3566
Eonycteris spelaea	9.0 g; r8.2-9.8; n2 smallest yg	898
	14.6 g; term emb	898
Epomophorus labiatus	11 g	8082
Epomops buettikoferi	24 g; n1 term emb	11273
Epomops franqueti	~19 g	8081
	20 g; near term	5127
Hypsignathus monstrosus	~40 g	1185
Myonycteris torquata	11 g; n1 term emb	11273
Pteropus spp	45 g; n1	9365
Pteropus poliocephalus	~70 g; r46-92	7833a
	90 g	10879
Rousettus aegyptiacus	17.9 g; n1	567
	18.7 g	7679
	22.5 g; n1; 2 d	6093, 6094
	22.7 g; se2.5; n11	7951
	23-24 g	5226
Rousettus leschenaulti	12 g	3915, 3921
NEONATAL SIZE		
Dobsonia beauforti	FA: 42.7 mm; n1	807b
Epomophorus crypturus or *Epomophorus gambianus*	HB: ~70 mm; FA: 32.5 mm	334
Myonycteris torquata	greatest length: 24 mm; n1 largest emb	807c
	greatest length: ~34 mm; n1 smallest yg	807c
Pteropus spp	130 mm; n1	9365
Pteropus giganteus	FA: 72 mm; n1	7013b
Pteropus poliocephalus	FA: 60 mm; n1	7833a
Rousettus aegyptiacus	56 mm; n1	6093
WEANING MASS		
Cynopterus sphinx	21.2 g; r19.8-22.3; n3 juveniles	10286
	~35 g	9471
Eidolon helvum	120-150 g	3566
Rousettus aegyptiacus	63.1 g; n1; 3 mo	567
	84.5 g; r67-98; 3 mo	5226
LITTER SIZE		
Acerodon celebensis	1 emb; n1 f	809
Acerodon mackloti	1 emb ea	3903
Boneia bidens	1 emb ea; n2 f	809
Casinycteris argynnis	1 emb	252
Chironax melanocephalus	1 emb ea; n2 f	809
Cynopterus brachyotis	1	7161
	1	8494
	1 yg; n1 lt	6394
	1 emb ea; n5 f	809
Cynopterus sphinx	1	9471
	1 cl	8874
	1 emb	5617
	1 emb	10286
	1 yg	3915
	1; n1 lt	12128
	1 emb; n1 f	3903
	1 emb ea; n165 f	3927
	1-2	1340
	2 cl; n1 f	3903
	2 yg; n1 lt	3903
Dobsonia beauforti	1 emb ea; n6 f	807b
Dobsonia inermis	1 emb; n1 f	7073
Dobsonia minor	1 emb ea; n3 f	7073
Dobsonia moluccensis	1 emb; n1 f	7073
	1 yg ea; n9 f	2835
Dobsonia peroni	1 emb; n1 f	3903
Eidolon helvum	1	7677
	1	3566
	1 yg	972

	1; 2 rare	252
	1.2 emb; r1-2; n5 f	191
	1; 2 rare, 1 pr twins in > 100 pregancies	5759
Eonycteris spelaea	1 mode; 2 rare	696
	1.13 emb; mode 1; r1-2; n23 f	898
Epomophorus angolensis	1	9945
Epomophorus crypturus	1	252
	1; twins reported	10158
	1 emb; n1 f	331, 337
Epomophorus gambianus	1 emb ea; n2 f	71
	1 emb ea; n5 f	8611
Epomophorus labiatus	1	252
	1 emb; n1 f	191
	1 emb; n1 f	185
	1 emb ea; n2 f	11272
Epomophorus wahlbergi	1	10158
	1 emb; n1 f	331, 337
	mode 1; r1-2	252
Epomops buettikoferi	1 emb; n1 f	11273
	1 emb ea; n8 f	1948
Epomops franqueti	1	252
	1	1346
	1	191
	1 yg	972
	1 ea; n2 lt	71
Haplonycteris fischeri	1	4572, 4573
Harpionycteris celebensis	1 emb ea; n2 f	809
Hypsignathus monstrosus	1	252
	1	191
	1; twins recorded	5759
	2 emb; n1 f	1347
Macroglossus minimus	1	10698
	1; n1 lt	7075
	1 emb; n1 f	6447
	1 emb ea; n3 f	809
	1 emb ea; n12 f	7073
Megaloglossus woermanni	1 emb ea; n2 f	8084
	1 emb ea; n3 f	5387
	2 emb, different sizes; n1 f	2198
Melonycteris woodfordi	1 emb ea; n2 f	7073
Micropteropus pusillus	1 emb; n1 f	4747
Myonycteris torquata	1	1346
	1 emb ea; n2 f; case of twins	1948
	1 emb ea; n2 f	11881
	1 emb ea; n3 f	5387
	2 emb, different sizes; n1 f	11273
Nanonycteris veldkampi	1 emb ea; n9 f	1948
Neopteryx frosti	1 emb; n1 f	809
Notopteris macdonaldi	1 yg	9457
	1 yg ea; n4 f	7835
	1 emb ea; n11 f	7835
Nyctimene aello	1 emb ea; n4 f	7073
Nyctimene albiventer	1 emb ea; n6 f	7073
Nyctimene cephalotes	1 emb ea; n1 f	809
Nyctimene robinsoni	1 emb ea; n2 f	7075, 7076
Paranyctimene raptor	1 emb ea; n7 f	7073
Penthetor lucasi	1 emb; n1 f	10314
Pteropus alecto	1; n many	7075
	2 yg; n1 lactating f	2214
Pteropus anetianus	1 emb ea; n8 f	517
Pteropus conspicillatus	1; n many	7075
Pteropus giganteus	1	3915, 3927
	1	1340
	1	8494, 8501

	1	6845
	1; n1 lt	12128
	1 emb ea; n3 f	7013b
	1 ea; n3 lt	3370
	1 emb ea; n70 f	3927
	1 emb ea; n~100	7416
Pteropus howensis	1 emb ea; n2 f	9457
Pteropus ornatus	1 emb ea; n14 f	9457
Pteropus poliocephalus	1	10878, 10879
	1	7075
	1; n1 lt	12128
Pteropus rayneri	1 emb; n1 f	9457
Pteropus rodricensis	1	1625
	1 yg ea; n3 f	8652
Pteropus rufus	2 emb ea; n3 f	8906
Pteropus scapulatus	1	7075
	1 yg; n2 lt	7833a
Pteropus tokudae	1 yg; n1 f	8371
Pteropus tonganus	1 yg	517
	1 yg	9457
Pteropus vampyrus	1	7161
	1 yg	6369
	1 emb; n1 f	2313
Rousettus spp	1; n1 lt	6093
Rousettus aegyptiacus	1	5226
	1 emb	7679
	1 emb; n1 f	11272
	1 emb; n1 f	11273
	1 emb; n6 f	7717
	1 emb ea; n11 f	1948
	1 emb; n12 f	331
	1+; r1-2; occasionally 2	4698
	1; r1-2	6094
	1 yg; 1.3% twins	8084
	1.25; n73 lt	12128
Rousettus angolensis	1	252
	1 emb ea; n2 f	11273
	1 emb ea; n2 f	11881
Rousettus amplexicaudatus	1 emb ea; n2 f	809
	1 emb ea; n2 f	7073
Rousettus celebensis	1 emb or yg ea; n5 f	809
Rousettus leschenaulti	1	8494
	1	9175
	1 yg	1340, 1342
	1 emb; n1 f	10872
	1 emb ea; n636 f	3914, 3915, 3916, 3921 3922, 3927, 3931
	1-2; twins rare	5537
Scotonycteris ophiodon	1 emb	252
	1 emb; n1 f	1948
	1 emb; n1 f	2955
Scotonycteris zenkeri	1	252
	1 emb; n1 f	1948
Syconycteris australis	1 yg; n3 lt	7833
	1 emb ea; n12 f	7073
Thoopterus nigrescens	1 emb ea; n2 f	809
SEXUAL MATURITY		
Cynopterus sphinx	f: ~5 mo; ~50 g	9471
	m: 15-20 mo	9471
Eidolon helvum	f: 200 g	2430
	m: testes large: 200-249 g	7677
Eonycteris spelaea	f: 1 yr	696
	m: 1 yr	696

Epomops buettikoferi	f: 1st mate: 6 mo	10780
	m: 11 mo	10780
Haplonycteris fischeri	f: 1st conceive: 3-5 mo	4571, 4573
Hypsignathus monstrosus	f: 1st preg: 6-7 mo	1185
	m: epididymal sperm: 12-18 mo	1185
Micropteropus pusillus	f: 1st mate: 6 mo	10780
	m: 7 mo	10780
Pteropus spp	f: 1st birth: 26 mo; n1	9365
Pteropus giganteus	f/m: 1 yr	8798
	f/m: 2 yr	1346
Pteropus poliocephalus	m/f: 18 mo	7833a
	m: spermatogenesis: 16 mo; 500-600 g	10879
Pteropus scapulatus	m/f: 18 mo	7833a
Pteropus tonganus	m: testes large: 600 g	517
	f: 1st preg: 450 g	517
Rousettus aegyptiacus	f/m: ~1 yr; ~90 g	7679
Rousettus leschenaulti	f: 7-8 mo; 55 g	3921
	m: 73 g	3921
GESTATION		
Cynopterus sphinx	3 mo	10286
	115-125 d	7417
	~120 d	9471
	~5 mo	8871
	150 d 1st pregnancy; 120 d 2nd pregnancy	6031
Eidolon helvum	implant: 3 mo	7676, 7677
	implant: 4-6 mo; field data	3566
	postimplantation growth: ~4 mo	7676, 7677
	~4 mo	3184
	~4 mo; field data	252, 254
	7 mo	7676, 7677
	9-12 mo with delay; field data	3566
Eonycteris spelaea	> 6 mo; poss 200 d; n35 recaptured f	696
Epomophorus crypturus	est ~3 mo; field data	252
Epomophorus gambianus	3? mo; field data	252
	5-6 mo	10780
Epomops buettikoferi	5-6 mo	10780
Epomops franqueti	5-6 mo	8081
Haplonycteris fischeri	delay: 2-6 mo primiparous f; max 8 mo	4573
	delay: 3-8 mo	4572
	slow growth: 3-9 mo; rapid growth: 2.5-3 mo	4571
	11.5 mo	4573
	1 yr	4571
Micropteropus pusillus	5-6 mo	10780
Pteropus spp	6.5 mo	9365
Pteropus giganteus	~5 mo; field data	7416
	140-150 d	1340, 1342, 3915
	6 mo; field data	6845
	199 d; r198-200; n3	3370
Pteropus poliocephalus	6 mo	10878, 10879
Pteropus tonganus	5-6 mo	517
Rousettus aegyptiacus	4 mo; field data	7679
	est min 4 mo	6093
	105-107 d	5759
Rousettus angolensis	4 mo; field data	255
Rousettus leschenaulti	15 wk	8494
	3.5 mo	1640
	4 mo; field data	3916
	125 d	3915, 3921

LACTATION		
Cynopterus brachyotis	1st flight: 3 mo; n1	6394
Cynopterus sphinx	wean: 40-45 d	9471
Dobsonia moluccensis	independent: 6 mo	10698
Eidolon helvum	wean: ~1 mo	7677
	wean: 2 mo	254
	wean: 3-4 mo	3566
Eonycteris spelaea	teat attachment: 4-6 wk	696
	nurse: est 3 mo	696
Epomophorus crypturus	wean: est ~2 mo	252
Epomops buettikoferi	wean: 7-8 wk	10780
Haplonycteris fischeri	wean: 2.5 mo	4572
Micropteropus pusillus	wean: 7-8 wk	10780
Pteropus giganteus	solid food: ~4 mo	8798
	wean: 3 mo	7859
	wean: 5 mo	1346
Pteropus rodricensis	solid food: 11 wk	8652
Rousettus aegyptiacus	den emergence: 70 d	6093
	nurse: 6 wk	7679
	independent: 3 mo	5226
	independent: by 4 mo	567
	wean: min 4 mo	6093
LITTERS/YEAR		
Cynopterus sphinx	2	10286
	2	8871
	2	3915
Eidolon helvum	1	7677
	1	3566
Eonycteris spelaea	2	696
Epomophorus crypturus	est 2	252
Epomophorus gambianus	1	252
	2	10780
Epomops buettikoferi	2	10780
Micropteropus pusillus	2	10780
Pteropus giganteus	1	1340, 1342
Rousettus aegyptiacus	1	331
	1	9175
	2	567
	2	7679
Rousettus leschenaulti	1	9175
	2	3914, 3915, 3916, 3921, 3922
SEASONALITY		
Acerodon celebensis	preg f: Feb?; n1; Sulawesi, Indonesia	809
Acerodon mackloti	preg f: Mar, May; Timor, Indonesia	3903
Aethalops alecto	preg f: Feb-Mar; Malaysia	7161
	preg f: Feb, Apr, June; n7; Malaysia	6446
Balionycteris maculata	preg f: Jan-Mar; Malaysia	7161
	preg f: Feb, Apr; n2; Malaysia	6446
	preg f: Aug; Malaysia	7161
Boneia bidens	preg f: June-July; n2; n Sulawesi, Indonesia	809
Casinycteris argynnis	birth: May; central Africa	252
	lact f: May-July; central Africa	252
Chironax melanocephalus	preg f: Jan-Mar; Malaysia	7161
	preg f: Feb, Apr; n2; Malaysia	6446
	preg f: Mar-Apr; n2; Sulawesi, Indonesia	809
Cynopterus brachyotis	preg f: Jan-Feb, Apr-May, Sept; n117; Malaysia	6445a, 6446
	preg f: Jan, Mar; n5; Sulawesi, Indonesia	809
	preg f: May, Aug; Sri Lanka	8494
	preg f: yr rd; Malaysia	7161
	preg/lact f: peak Feb; Philippines	135
	preg/lact f: May, Aug, Oct; Philippines	7586
	lact f: June; n1; Sarawak, Malaysia	6447
Cynopterus horsfieldi	preg f: yr rd; Selangor, Malaysia	7161

Cynopterus sphinx	breeding condition: Mar-May; no data May-Jan; Timor, Indonesia	3903
	reproductive tract: enlarged Dec; India	11071
	sperm: yr rd; India	10286
	anestrus July-Oct; India	9471
	mate: Mar; India	8874
	mate: early Oct; India	3937
	mate: yr rd; Sri Lanka	8501
	conceive: Mar; India	8874
	preg f: Feb-July; India	10286
	preg f: Oct-July; India	3915
	birth: Feb-Apr, June-July; India	9471
	birth: Mar-Aug; India	1340
Dobsonia inermis	preg f: Sept; n1; Bougainville, Solomon I	7073
Dobsonia minor	preg f: Jan, May-June; n3; Papua, New Guinea	7073
Dobsonia moluccensis	preg f: Mar; n1; Papua, New Guinea	7073
	preg/lact f: June-Aug; New Guinea	10114
	birth: peak mid Aug-mid Nov; New Guinea	2835
	birth: end winter-spring; Australia	10698
Dobsonia peroni	testes: large Mar-Apr; no data June-Jan; Timor, Indonesia	3903
Eidolon helvum	spermatogenesis: Apr-June; e Africa	5759
	testes: large Apr-June, regress July-Mar; Uganda	7676, 7677
	testes: large May-June; W Africa	5388
	mate: Apr-June; Uganda	7676, 7677
	mate: May-July; Nigeria	3566
	mate: June-July, Sept-Oct; central Africa	254
	implant: Oct; e Africa	5759
	implant: Oct-Nov; Uganda	7676, 7677
	implant: Nov-Dec; Nigeria	3566
	preg f: Feb; n1; Liberia	11273
	preg f: Oct-Feb; Uganda	7677, 7678, 7680
	preg f: Nov; n5/12; Zaire	191
	preg/lact f: Jan-Apr; n4; Cameroun	2948
	birth: Feb-Mar; Uganda	7676, 7677, 7678
	birth: mid Mar-May, peak Apr-May; Nigeria	3566
	birth: Apr; e Africa	7684
	birth: Oct-Dec, Feb-Mar; central Africa	254
	birth: Dec-Feb; e Africa	5759
Eonycteris spelaea	preg f: Feb, Apr; n5; Malaysia	6446
	preg f: yr rd, peaks May-June, Oct-Nov; Malaysia	696
	preg f: yr rd; Selangor caves, Malaysia	7161
Epomophorus crypturus	preg f: end July-Mar; S Africa	10158
	birth: peak Sept; S Africa	10158
	lact f: Mar; n1; n Zimbabwe	331
Epomophorus gambianus	mate: Sept-Oct; few data; central Africa	252
	preg f: Feb-Mar, May; Niger	8611
	preg f: Mar, May, Sept; Burkina Faso	1139
	birth: Jan-Feb; few data; central Africa	252
	birth: Apr, Oct; Sudan	10780
	birth: Dec-Jan; few data; W Africa	5903
Epomophorus labiatus	testes: large Sept-Oct,; Uganda	8082
	mate: end Mar-early Apr?, end Sept-early Oct?; Uganda	8082
	mate: peak Oct-June; central Africa	252
	preg f: Feb; n1; Kenya	185
	preg f: May, Dec; n2; Zaire	11272
	birth: Feb-Mar; central Africa	11270
	birth: Feb-Apr; Sept-Oct; Uganda	8082
	birth: Mar; Kenya, Tanzania	252
	birth: Oct; Rwanda, Tanzania	252
	birth: Nov; Zambia	252
	young found: Nov-Jan; Zaire	191
Epomophorus wahlbergi	preg f: Feb; n1; Zambia	331
	birth: Mar; Kenya	252
	birth: Mar, Nov; Zambia	252

	birth: Oct; Tanzania	252
Epomops buettikoferi	preg f: Dec; n6; Mt Nimba, Liberia	11273
	birth: peaks Mar-Apr, Sept-Oct; W Africa	10780
	birth: peaks Aug, Dec-Jan; Mt Nimba, Liberia	1948, 11881
Epomops franqueti	spermatogenesis: Jan-Apr; June-Sept; Gabon	252
	mate: yr rd; Uganda; e Africa	5759
	preg f: Feb-Mar; n9; Cameroun	2955
	preg f: Mar-July, Sept-Jan; Kisangani, Zaire	5127
	preg f: peak Apr-Sept, Sept-Feb; Uganda	8081
	preg/lact f: yr rd; central Africa	252
	embryo weight: peaks Jan, July; Kisangani, Zaire	5127
	birth: early rainy season; Uganda	8081
	lact f: Mar-Apr; Rio Muni, W Africa	5387, 5388
	young found: yr rd; Zaire	191, 1346
Haplonycteris fischeri	testes: large May-July; Philippines	4573
	mate: June; multiparous; Sept-Dec; nulliparous; Philippines	4571, 4573
	mate: yr rd; Philippines	4572
	implant: Sept-Dec; Philippines	4571
	preg/lact f: May, Aug, Oct-Nov; Philippines	7586
	birth: May-June; Philippines	4571, 4572
	wean: end Sept-Oct; Philippines	4572
Harpionycteris celebensis	preg f: Jan, Sept; n2; Sulawesi, Indonesia	809
	lact f: Jan; n1; Sulawesi, Indonesia	809
Hypsignathus monstrosus	m calling: mid Dec-early Feb, June-Aug; Gabon	1185
	preg f: May, Dec; Zaire	191
	preg f: July, Oct-Sept; n4; Mt Nimba, Liberia	11881
	preg f: Nov; n1; Cameroun	2948
	birth: Jan-Mar; June-July; Cameroun?, central Africa	252
	birth: est peaks Feb, July; e Africa	5759
	birth: Dec-Mar; Gabon	1346
Macroglossus minimus	preg f: Jan-Mar, July-Sept; n12; Papau, Newguinea, Bougainville, Solomon I	7073
	preg f: Feb, Mar, May; n3; Indonesia	809
	preg f: Feb, Apr, June; n4; Malaysia	6446, 6447
	preg f: Aug; n1; Australia	10698
	preg f: yr rd; New Guinea, Solomon I	7073
	preg/lact f: peak Jan-Aug; Philippines	135
	preg/lact f: June-Aug; New Guinea	10114
	preg/lact f: Aug, Oct; Philippines	7586
	lact f: Dec-Jan, Mar; n6; Solomon I	8481
	repro: yr rd; Bismarck Archipelago	10114
Magaerops ecaudatus	preg f: Feb-Mar, June; Malaysia	7161
	preg f: Feb, Apr, June; n10; Malaysia	6446
Megaloglossus woermanni	preg f: Mar; Zaire	808
	preg f: Aug; n2; Gabon	5387
	preg f: Aug; n4; Mt Nimba, Liberia	11881
	preg f: Sept; n1; Rio Muni, W Africa	5387
	preg/lact f: July-Aug; Liberia	11881
	lact f: Jan, July; Gabon	252
	lact f: Jan, Mar; Cameroun	252
	lact f: Feb, May; Zaire	808
	lact f: Apr; n1; Rio Muni, W Africa	5387
	lact f: July-Aug; n2; Mt Nimba, Liberia	11881
	lact f: Oct; Fernando Po, central Africa	252
Melonycteris aurantius	lact f: Oct; Solomon I	8481
Melonycteris woodfordi	preg f: July, Sept; n2; Bougainville, Solomon I	7073
Micropteropus pusillus	conceive: Oct-early Nov; Zaire	808
	preg f: May, Nov; n2; W Africa	5387
	preg f: Aug; n1; Ethiopia	4747
	birth: Mar-Apr, Sept-Oct; Sudan	6844, 10780
	lact f: Jan; n2; s Burkina Faso	5903
	lact f: May; n1; W Africa	5387
	young found: Mar-Apr; W Africa	5387, 5388
	repro: Dec-Jan; tropical Africa	1346

Myonycteris torquata	preg f: Jan-Mar; n21; Mt Nimba, Liberia	11273
	preg f: Oct; n3; W Africa	5387
	birth: Aug-Sept, Feb-Mar; Ivory Coast	10778
	birth: peak Sept; Mt Nimba, Liberia	11881
	birth: Dec-Feb; n16; Gabon	1346
Nanonycteris veldkampi	preg f: mid Nov-mid Dec; n12; Mt Nimba, Liberia	11881
	preg f: Dec-Mar; n7; Mt Nimba, Liberia	11273
	birth: May-June; Mole Nat Park, w Africa	6844
	lact f: Jan; n1; Mt Nimba, Liberia	11273
Neopteryx frosti	preg f: Mar; n1; Sulawesi, Indonesia	809
Notopteris macdonaldi	preg/lact f: Dec-Jan; New Caledonia	7835
	preg f: July; New Caledonia	9457
	lact f: Aug; New Caledonia	9457
Nyctimene aello	preg f: Jan-Feb; n4; Papua, New Guinea	7073
Nyctimene albiventer	preg f: Jan, July-Aug; n6; Papua, New Guinea	7073
	lact f: Feb, May, Aug; n8; Papua, New Guinea	7073
	lact f: July; n1; Bougainville, Solomon I	7073
Nyctimene cephalotes	preg f: Jan, Oct; n2; Sulawesi, Indonesia	809
Nyctimene robinsoni	preg f: Aug; n2; QLD, Australia	7076
Otopteropus cartilagonodus	preg f: Apr; n1; Luzon, Philippines	7586
Paranyctimene raptor	preg f: Jan-Feb, May; n7; New Guinea	7073
Penthetor lucasi	preg f: Jan; n1; Sarawak, Malaysia	10314
	preg f: Feb, Apr, June; n4; Malaysia	6446
	lact f: May; n1; Sarawak, Malaysia	10314
	preg f: June, Sept; none Jan-Mar, July; Malaysia	7161
Ptenochirus jagori	preg/lact f: peak Jan-Mar; Philippines	135
	preg/lact f: May, Aug, Oct; Luzon, Philippines	7586
Pteralopex anceps	lact f: July; n1; Solomon I	8481
Pteropus alecto	mate: Mar-Apr; n Australia	2214
	preg f: Jan; n1; Sulawesi, Indonesia	809
Pteropus anetianus	testes large: Oct-Jan; New Hebrides	517
	preg f: May; n2; New Hebrides	9457
Pteropus giganteus	mate: July-Aug; Madras, India	7859
	mate: early Sept; India	7416
	mate: early Oct; India	3937
	mate: mid Oct; zoo	3370
	mate: Dec-Jan; India, Burma, Sri Lanka	6845, 6847
	preg f: Nov-Apr; India	3915
	birth: Jan-Mar; w India	1340, 1342, 1346
	birth: May-June; Sri Lanka	1340, 1342, 1346
	birth: end Jan-early Feb; India	7416
	birth: Mar; Madras, India	7859
	birth: end Apr-early May; zoo	3370
	birth: May-June, Dec-Mar; zoo	517
	birth: end May (perhaps early June); Sri Lanka	6845
	birth: end May-early June; Sri Lanka	8494, 8501
	wean: May; Madras, India	7859
	young found: yr rd; Africa	8798
Pteropus howensis	preg f: Aug; Solomon I	8481
	preg f: Aug; n2; Solomon I	9457
Pteropus neohibernicus	preg/lact f: June-Aug; New Guinea	10114
Pteropus ornatus	preg f: Apr, June-Oct; n14; New Caledonia	9457
Pteropus poliocephalus	testes large: Feb-Apr; QLD, Australia	7065
	mate: peak Mar-Apr; QLD, Australia	10878, 10879
	birth: Sept-Oct; e Australia	7833a
Pteropus rayneri	preg f: July; n1; Solomon I	9457
	preg f: July or Nov; Solomon I	8481
Pteropus rodricensis	mate: June, Nov-Dec; Rodriques I, UK	1625
	birth: Nov-Dec, Mar; Rodriques I, UK	1625
Pteropus rufus	preg f: Oct; n3; Madagascar	8906
Pteropus scapulatus	mate: Oct; n1; QLD, Australia	8954a
	preg f: Feb; NSW, Australia	8954a
	birth: Apr-May; n21; e Australia	7833a
	birth: May; Australia	623

Pteropus temmincki	f with yg: June, July, or Aug; n1; New Guinea	10114
Pteropus tonganus	testes: large Nov-Feb; New Hebrides	517
	Leydig cells: large Oct-Feb; New Hebrides	4097
	mate: Feb; New Hebrides	4097
	preg f: Feb-Oct; New Hebrides	517
	preg f: May; New Hebrides	9457
	preg f: June; New Caledonia	9457
	birth: peak end Aug-early Sept; New Hebrides	517
Pteropus vampyrus	preg f: Apr; n1; Philippines	7586
	preg f: Nov-Jan; few data; Selangor, Malaysia	7161
	birth: end Mar-early Apr; Thailand	6369
Rousettus aegyptiacus	sperm: yr rd, peak Feb-Apr; Egypt	6721
	testes: peak Apr, Aug, min Nov, Feb, June-Aug; Uganda	7679
	mate: end Apr; not seasonal; Egypt	6093
	mate: June-mid Sept; Cape Province, S Africa	6549
	mate: est Nov-Jan; Egypt	8817
	mate: yr rd; Egypt	6721
	preg f: Mar; n1; Mt Nimba, Liberia	11273
	preg f: Aug; n1; Mt Nimba, Liberia	11881
	preg f: Oct 90%; Transvaal, S Africa	5226
	preg f: end Oct-early Dec; n11; Mt Nimba, Liberia	1948
	preg f: Dec-Mar, July-Sept; Uganda	7678
	birth: Mar, Sept, before peak rain; Uganda	7678, 7679
	birth: peak Mar, Sept; e Africa	8084
	birth: Oct-June; S Africa	4698
	birth: Nov-Dec; Transvaal, S Africa	5226
	birth: yr rd; se Zaire, Rwanda	255
	birth: yr rd; n73; zoo	12128
	birth: yr rd; zoo	517
	lact f: Feb; W Africa	5387
	lact f: Mar-May; Egypt	8817
	lact f: Aug; n1; Mt Nimba, Liberia	
Rousettus amplexicaudatus	mate: May-July; Java	5924
	preg f: Feb, Apr, June, Dec; n10; Malaysia	6446
	preg f: Mar; n2; Sulawesi, Indonesia	809
	preg f: July, Sept; n2; Bougainville, Solomon I	7073
	preg/lact f: peak Feb-Apr; Philippines	135
	preg/lact f: June-Aug; New Guinea	10114
	birth: Mar-May; no data June-Jan; Timor, Indonesia	3903
	lact f: Jan; n1; New Guinea	7073
	lact f: Dec; n1; Sulawesi, Indonesia	809
	lact f: Dec-Jan; n11; Solomon I	8481
Rousettus angolensis	preg f: Feb, Dec; n2; Mt Nimba, Liberia	11273
	preg f: July-Sept; Mt Nimba, Liberia	11881
	birth: June, Nov-Feb; se Zaire, Rwanda	255
	birth: Dec-Jan; Cameroun	2948
Rousettus celebensis	preg f: Mar; n1; Sulawesi, Indonesia	809
	birth: Mar; n1; Sulawesi, Indonesia	809
	lact f: Apr; n3; Sulawesi, Indonesia	809
Rousettus lanosus	mate: yr rd; e Africa	5759
Rousettus leschenaulti	anestrus: July-Nov; India	3916
	seminal vesicles enlarged: Oct-Apr; India	3938
	spermatogenesis: Nov-Mar; India	1342
	testes: abdominal May-Oct; India	3938
	mate: Nov-early Dec; Pakistan	9175
	conceive: mid Nov-mid Mar; India	3916
	preg f: Nov-July; India	3915, 3921
	birth: mid Feb-early Mar; Pakistan	9175
	birth: Mar, July; India	3916
	birth: Mar, July-Aug; India	1340, 1342
Scotonycteris ophiodon	preg f: end Aug-early Sept, Nov; n3; Mt Nimba, Liberia	11881
	birth: end Dec-early Jan; central Africa	252
	lact: Feb-Apr; central Africa	252

Scotonycteris zenkeri	mate: Aug-Sept; central Africa	252
	preg f: Aug; n1; Mt Nimba, Liberia	11881
	preg f: Nov; n1; Mt Nimba, Liberia	1948
	birth: Dec-Jan; central Africa	252
	lact: Dec-Mar/Apr; central Africa	252
	lact f: Dec; n1; Mt Nimba, Liberia	11273
Syconycteris australis	preg f: Jan, Mar-May, July, Sept; n12; Papua, New Guinea	7073
	preg f: Feb; n4; Beerwah, QLD, Australia	2841
	lact f: Jan; n2; Papua, New Guinea	7073
	lact f: Apr; n1; New Guinea	7073
Thoopterus nigrescens	preg f: Jan; n2; Sulawesi, Indonesia	809

Order Chiroptera, Family Rhinopomatidae, Genus *Rhinopoma*

The insectivorous, mouse-tailed bats (*Rhinopoma* 10-45 g) range from western Africa to southeast Asia. The zygote enters the uterus in the morula stage (5536). The labyrinthine, endotheliochorial, discoidal placenta has a superficial, mesometrial, circumferential, circumpuntial implantation into a symmetrical bicornuate uterus (3913, 3917, 5532, 5534, 5535, 5536, 10288, 11799) and is not normally eaten (248). Birth lasts 20-25 min and the naked, blind neonate is born head first (248, 5534, 9175). Females have 2 pectoral mammary glands, 2 pubic false teats (248, 5534, 5759, 6720), and are monoestrous (248, 5534). Males have ampullary glands, which store sperm, abdominal testes (248, 5533, 6065, 9999), urethral glands (7440), and seminal vesicles (9178). Details of female anatomy (553, 3937, 5534, 6161, 10288) and endocrinology (553) are available.

NEONATAL MASS		
Rhinopoma hardwickei	2-3 g; n6	5534
WEANING MASS		
Rhinopoma hardwickei	7-8 g; 20 d	5534
	12-13 g; 2 mo	5534
LITTER SIZE		
Rhinopoma hardwickei	1 emb	553, 5534
Rhinopoma microphyllum	1	248
	1 yg	8720
	1 emb; r1-2; 1 set twins; n237 (table), n273 (text) f	6161
SEXUAL MATURITY		
Rhinopoma hardwickei	f: 8.5-9 mo	5534
	f: 9-9.5 mo	553
	m: 16-17 mo	5534
Rhinopoma microphyllum	f/m: 2nd yr	248
GESTATION		
Rhinopoma hardwickei	95-100 d	553, 5534
Rhinopoma microphyllum	70-80 d	1340
	4 mo	6161
	123 d	3915
	min 123 d	248
LACTATION		
Rhinopoma hardwickei	teat attachment: 20 d	5534
	wean: 2 mo	553
Rhinopoma microphyllum	wean: 4-6 wk	248
INTERLITTER INTERVAL		
Rhinopoma hardwickei	1 yr	553, 5534
Rhinopoma microphyllum	1 yr	248
SEASONALITY		
Rhinopoma hardwickei	testes: regress Mar; India	5533
	spermatogenesis: Nov-Dec; India	5533
	mate: mid Feb-early Mar; India	553
	mate: mid Feb-mid Apr; India	5533
	mate: end Feb-mid Apr; India	5534
	ovulate: mid Mar; India	553
	ovulate: mid Mar-Apr; India	5534
	conceive: Apr; India	5533
	preg f: Apr; Egypt	8818

	birth: June; India	1340
	birth: June; Pakistan	9175
	birth: mid June-early July; Agra, India	553
	birth: mid June-July; India	5534
	birth: est July; Sudan	5759
	lact f: July; n8; India	9062
	half grown juveniles: mid July; n4; W Africa	5903
Rhinopoma microphyllum	torpor: end Oct-Feb/early Mar; India	248
	testes size: increase Oct, large end Jan, small end Apr; India	248
	mate: Feb-Mar; India	6161
	mate: Mar; India	248
	mate: Mar; India	3937
	mate: Apr-early July; Bengal, India	10288
	birth: June; India	1340
	birth: end June; India	6161
	birth: end July-early Aug; India	248
	half grown yg: Aug; W Africa	5903

Order Chiroptera, Family Emballonuridae

Sheath-tailed, sac-winged, and ghost bats (*Balantiopteryx* 5-9 g; *Centronycteris* 5 g; *Coleura* 9-11 g; *Cormura* 10 g; *Cyttarops* 50-55 mm; *Diclidurus* 19 g; *Emballonura* 4-40 g; *Peropteryx* 6-19 g; *Rhynchonycteris* 2-4 g; *Saccolaimus* 55-105 g; *Saccopteryx* 3-11 g; *Taphozous* 10-30 g) are insectivores from the tropics and subtropics. Females may be larger than males (1186, 1187, 6568, 7122, 8494, 10114; cf 5759). *Taphozous* exhibits reproductive asymmetry as the right ovary and uterus are used preferentially (4399, 5261, 5262, 5715, 5716, 5829, 9491; cf 3911, 3912, 6030, 9490). *Balantiopteryx* and *Peropteryx* use both ovaries (6568, 11804). The labyrinthine, endotheliochorial, discoidal placenta has a placental hematoma and a superficial, circumferential or lateral implantation into a bipartite or bicornuate uterus (3912, 3913, 3917, 5715, 5716, 6568, 6627, 6966, 8948, 9178, 9491, 11799, 11808). Development may be delayed (*Taphozous* 6029). Placentophagia is reported (9175), as is estrus postpartum (1340, 1346, 4399, 6029, 6030, 8948, 9491; cf 5831), but monoestry also occurs (*Balantiopteryx* 383).

Neonates are 25-30% of adult mass and most are born blind and naked (5759, 7161, 8501, 9175, 9491, 10669; cf 5903); however, *Taphozous* is born furred with open eyes (5380). *Taphozous* females have 2 mammae (6720). Males have penial spines and ampullary glands or seminal vesicles (5829, 6065, 6966, 7440, 7441, 7662, 9178). Details of anatomy (419, 6721, 9490), endocrinology (5383), spermatogenesis (5382, 7439), and social organization (1186, 1187, 1188, 1189) are available.

NEONATAL MASS		
Balantiopteryx plicata	~2 g	2341
	2 g	6568
Saccopteryx bilineata	3.5 g; r3.2-4.0	10669
Taphozous georgianus	7.8 g; r7.3-8.1; n4	5380
Taphozous melanopogon	7.5 g; r7-8; n2 neonates; 7-8 g full term	9491
NEONATAL SIZE		
Taphozous georgianus	HB: 50.2 mm; r46-56; n4; FA: 35.5 mm; r34-36; n4	5380
Taphozous nudiventris	47-51 mm	111
WEANING MASS		
Taphozous georgianus	13-16 g; 18-33 d	5380
Taphozous melanopogon	~20 g	9491
Saccopteryx bilineata	m: 6.6 g; se0.41; n13	10669
	f: 7.0 g; se0.31; n8	10669
WEANING SIZE		
Taphozous georgianus	HB: 62-68 mm; FA: 54-61 mm; 18-33 d	5380
LITTER SIZE		
Balantiopteryx io	1 yg	384
Balantiopteryx plicata	1 yg	2341
	1 emb; n1 f	2350, 2351
	1 emb ea; n3 f	3204
	1 emb ea; n10 f	1937
	1 emb ea; n21 f	5409

Centronycteris maximiliani	1 emb; n1 f	6260
Coleura afra	1	10158
	1	5759
	1	252
Diclidurus albus	1 yg	1671
Emballonura nigrescens	1 emb ea; n9 f	7073
Emballonura monticola	1	7161
Peropteryx macrotis	1 emb; n1 f	9113
	1 emb ea; n3 f	10974
	1 emb, yg ea; n14 f	7624
	1 emb ea; n20 f	3999a
Rhynchonycteris naso	1 emb; n1	11526
	1 yg; n1	452
	1 emb ea; n2 f	10974
	1 emb ea; n3 f	1475
	1 emb ea; n3 f	5405e
	1 emb ea; n4 f	1635
	1 emb ea; n6 f	2230, 4228
	1 yg ea; n9 lt	3999a
	1 emb ea; n15 f	7624
	1 emb, yg ea; n17 f	7624
Saccolaimus peli	1	191
Saccolaimus saccolaimus	1	8494, 8501
	1 yg; n1 f	7161
Saccopteryx bilineata	1 emb; n1 f	1162
	1 yg; n1 f	3024
	1 emb ea; n2 f	4228
	1 emb ea; n2 f	11471a
	1 emb ea; n4 f	10974
	1 emb, yg ea; n4 f	9113
	1; n5	452
	1 emb ea; n7 f	7624
	1 emb ea; n38 f	3999a
	1 ea; n41 f	1186
Saccopteryx canescens	1 ea; n2 lt	452
Saccopteryx leptura	1 emb; n1 f	3999a
	1 emb; n1 f	5425
	1 ea; n2 lt	452
	1 emb ea; n2 f	5405c
	1 emb ea; n3 f	2345
Taphozous georgianus	1 ea; n3 lt	7075
	1 yg; n282	5380
Taphozous longimanus	1	1340, 1346
	1	3931
	1	6030
	1 yg	3912, 3927
	1 emb; n1 f	8501
	1 emb ea; n129 f	3927
Taphozous mauritianus	1 yg	5759
	1	252
	1	10158
	1	1346
	1	191
Taphozous melanopogon	1	3927
	1	1340, 1342, 1346
	1	7161
	1 cl	5261, 5262
	1 emb ea; n57 f	3927
	1 emb ea; n89 f	9491
	1 emb, yg ea; n146 lt	5714, 5716
Taphozous nudicluniatus?	1; n1 lt	7075

Taphozous nudiventris	est 1	10006
	1	1340
	1	9175
	1	111
Taphozous perforatus	1	1340
	1 emb ea, n2 f	8611
	1 emb ea; n2 f	10158
SEXUAL MATURITY		
Balantiopteryx plicata	f: 1st birth: 1 yr	1187
Coleura afra	f: 1st birth: 1 yr	7122
Rhynchonycteris naso	f: 1st birth: max 18 mo	1187
Saccopteryx bilineata	f: 1st birth: 1 yr	1187
Saccopteryx leptura	f: 1st birth: max 18 mo	1187
Taphozous georgianus	f: 1st mate: 9 mo	5380
	m: 1st sperm: 8-9 mo	5380
	m: 1st mate: 21 mo	5380
Taphozous melanopogon	f: 1st conceive: 8 mo	5716
	f/m: not w/in 1st yr	9491
GESTATION		
Balantiopteryx plicata	~4.5 mo; r4.5-8 mo	6568
Peropteryx kappleri	~5 mo; field data	8948
Peropteryx macrotis	4-4.5 mo	6568
Saccopteryx leptura	4-4.5 mo	6568
Taphozous georgianus	~3 mo	5379, 5380
	est 4 mo	5829
Taphozous longimanus	105 d; 1st preg; 86 d; 2nd preg; indirect	6029, 6030
Taphozous mauritianus	~6 wk; field data; not many	5573
Taphozous melanopogon	120-125 d	9491
	~4 mo	5714
Taphozous perforatus	3 mo	255
LACTATION		
Balantiopteryx plicata	1st fly: 2 wk	383
	wean: 9 wk	383
	wean: 3 mo	1189
Rhynchonycteris naso	wean: 3 mo	1189
Saccopteryx bilineata	solid food: 2 wk	1186
	solid food: 18-30 d	10669
	wean: 8-10 wk	1186
	wean: ~2 mo	10669
Saccopteryx leptura	wean: 2.5 mo	1186
Taphozous longimanus	wean: 2-3 wk	6030
Taphozous georgianus	independent: 18-33 d	5380
Taphozous melanopogon	wean: 5-6 wk	9491
	nurse: 2 mo	1342
	wean: 2 mo	7161
LITTERS/YEAR		
Balantiopteryx plicata	1	1187
	1	6568
Coleura afra	1	5759
	2; r1-2	7122
Peropteryx kappleri	3	10785
Rhynchonycteris naso	max 2	1187
	>1	2230
Saccopteryx bilineata	1	1187
	1	10669
Saccopteryx leptura	max 2	1187
Taphozous longimanus	1	1340
	2	6029
	2	3912, 3937
Taphozous melanopogon	1	5714, 5716
	1	1342
Taphozous nudiventris	1	9175
	min 2	10006
Taphozous perforatus	2	9297

SEASONALITY

Balantiopteryx io	preg f: Apr; n2; Veracruz, Mexico	2230a, 4228
	lact f: July; n3; Oaxaca, Mexico	523
Balantiopteryx plicata	sperm: yr rd, low May, Aug; Central America	6568
	mate: end Jan-mid Feb; Central America	6568
	ovulate: Mar; Central America	6568
	preg f: May; n3; El Salvador	3204
	preg f: May-July; n10; Mexico	1937
	preg f: June; n21; Sinaloa, Mexico	5409
	preg f: June-July; n14; Sinaloa, Mexico	5413a
	birth: June-mid July; Central America	6568
	birth: synchronous end June; Costa Rica	1187
	birth: end June-early July; Mexico	2341
	lact f: July; Oaxaca, Mexico	523
	lact f: Sept; n1; Jalisco, Mexico	11471a
Centronycteris maximiliani	preg f: May; n1; Nicaragua	4048,
	preg f: May; n2; Costa Rica	6260
Coleura afra	mate: end June, end Dec; Kenya	7122
	ovulate: July, Dec; Kenya	7122
	preg f: end Dec; Tanzania	5759
	birth: end Jan; central, e Africa	252
	birth: Mar-June, Oct-Dec; Kenya	7122
	birth: Oct, end of rains; Sudan	5759
Coleura seychellensis	birth: Nov-Dec; Seychelles	7893
Cormura brevirostris	preg f: Aug-Sept; French Guiana	1350a
Diclidurus albus	preg f: Jan-June; Jalisco, Mexico	4671
	preg f: Mar, May; n2; Michoacan, Mexico	4671
Emballonura monticola	preg f: Feb-Mar, Oct-Nov; Selangor, Malaysia	7161
	preg f: Oct, Dec; n2; Malaysia	6446
	birth: Mar; Kuala Lumpur, Malaysia	3974
Emballonura nigrescens	preg f: Feb, May-July; n9; Papua, New Guinea	7073
Peropteryx kappleri	conceive: May-June, Dec-Jan; Colombia	8948
	preg f: Apr; Costa Rica	6260
	birth: peak Mar-May, Oct-Dec; Colombia	8948
	birth: mid May-early June; Costa Rica	1187
Peropteryx macrotis	preg f: Feb-Mar; n3; Guatemala	5405e
	preg f: Apr; n16; Yucatan, Mexico	5424f
	preg f: Apr, Aug-Sept; n20; Peru	3999a, 10974
	preg f: Oct; French Guiana	1350a
	lact f: Aug; n1; Quintana Roo, Mexico	5424f
Rhynchonycteris naso	preg f: Jan-Feb, Apr; Veracruz, Mexico	2230, 4228
	preg f: Feb; n3; Campeche, Mexico	5424f
	preg f: Mar; n2; Trinidad	3900a
	preg f: Mar; n3; Guatemala	5405e
	preg f: May; n3; El Salvador	1475
	preg f: May, July-Sept; n9; Peru	3999a, 10974
	preg f: mid July-mid Aug; n4; Trinidad	1635
	preg f: Aug; n1; Guatemala	2552
	preg f: Oct-Dec; French Guiana	1350a
	birth: Apr; none Nov-Mar; Costa Rica	1187
	lact f: July; Chiapas, Mexico	1641
	lact f: Oct; Veracruz, Mexico	4228
Saccolaimus peli	epididymal sperm: Aug; Gabon	252
	preg f: June, Dec; n3; ne Zaire	191, 252, 1207
	birth: Jan; Gabon	252
	birth: Jan, July; ne Zaire	252
Saccolaimus saccolaimus	preg f: Sept; n1; Malacca, se Asia	7161
	preg f: Sept-Nov; Sri Lanka	8501
	lact f: Sept-Nov; Sri Lanka	8493, 8494
Saccopteryx bilineata	mate: Dec; Panama	10669
	preg f: Feb; n2; Veracruz, Mexico	4228
	preg f: Feb, Apr; n2; Guatemala	5405e
	preg f: Mar; n2; Jalisco, Mexico	11471a
	preg f: Mar; n1; El Salvador	3204

	preg f: Mar-May, July; Trinidad	3900a
	preg f: Apr; n1; Panama	4228a
	preg f: June; n4; Veracruz, Mexico	6148a
	preg f: June-Aug, Oct; n38; Peru	3999a, 10974
	birth: May-June; Panama, Costa Rica	3322
	birth: synchronous early May; Panama	10669
	birth: end May-mid June; Trinidad	1186
	lact f: June-July; Trinidad	3900a
	lact f: July; n1; Guatemala	2552
Saccopteryx leptura	preg f: Feb; n2; Nicaragua	5405c
	preg f: Feb; n1; Panama	2345
	preg f; Mar; n1; Nicaragua	5425
	preg f: Apr-May; n2; Nicaragua	2345
	preg f: May; Trinidad	3900a
	birth: May; Costa Rica	1187
	birth: May, Nov; Costa Rica	1187
	birth: June-July; Trinidad	1186
Taphozous georgianus	accessory glands: enlarge spring; central QLD, Australia	5381
	epididymal sperm: yr rd; central QLD, Australia	5381
	sperm: yr rd, low mid fall-mid winter; WA, Australia	5829, 5831
	scrotal testes: summer; WA, Australia	5829
	scrotal testes: Nov-Feb; QLD, Australia	5380, 5382
	birth: Oct-Feb; WA, Australia	5829, 5831
	birth: end Nov-early Dec; QLD, Australia	5380
Taphozous longimanus	quiescent: Sept-Dec; India	6030
	mate: mid Jan-early Feb, May-early June; Varanasi, India	6029
	mate: mid Jan, end Apr-mid May; India	6030
	mate: yr rd; India	3912
	preg f: Sept; n1; Sri Lanka	8501
	preg/lact f: yr rd; India	3911
	birth: end Apr-early May; Varanasi, India	6029
	birth: yr rd; India	1340, 1345, 3912
Taphozous mauritianus	testes: large Dec; ne Zaire, Tanzania	252
	mate: July; ne Zaire, Angola, Tanzania, s Sudan	252
	mate: Oct-Nov; Katanga, Zambia	252
	mate: Dec-Jan; ne Zaire; Tanzania	252
	birth: Feb; Katanga, Zambia	252
	birth: est Apr-May; ne Zaire., Tanzania	252
	birth: Oct-Nov; ne Zaire, Angola, Tanzania, s Sudan	252
	young found: yr rd; Zaire	191
	repro: yr rd; equatorial Africa	253
Taphozous melanopogon	anestrus: Dec; India	5262
	regression genital system: Oct-Nov; India	5262
	preg f: Jan-May; India	9491
	preg f: Jan-June, peak Feb-Apr; n522; India	5716
	preg f: Jan-Mar, Oct-Nov; no data other mo; Malaysia	7161
	preg f: only Feb-May; India	5715
	preg f: May-June; India	5262
	preg/lact f: Sept; Sri Lanka	8501
	birth: mid Apr-mid May; India	1340, 1342
	birth: end May-early June; India	9491
	birth: end July-early Aug; India	5262
	lact f: Sept; India	5262
	wean: June-mid July; India	9491
	young of all ages found: Sept; Sri Lanka	8494, 8501
Taphozous nudiventris	hibernate: Dec?-Mar; Iraq	111
	epididymal sperm: Jan-Feb, Oct; Cairo, Egypt	6721
	spermatogenesis: Feb-May; Cairo, Egypt	6721
	mate: Sept; Pakistan	9175
	mate: fall; Iraq	110, 111
	ovulate: Mar; Iraq	111
	preg/lact f: July; few data; Burkina Faso, w Africa	5903
	birth: mid Apr; Pakistan	9175
	birth: end May; Iraq	111

	birth: mid June-mid July; India	1340
Taphozous perforatus	mate: Jan-Mar; Africa	255
	preg f: Feb-Mar; Niger	8611
	birth: Apr-May; India	1340
	birth: Apr-May; Kenya	252, 255
	birth: May-June; Nigeria	252, 255
	birth: end May-early June; few data; Nigeria	4399
	wean: mid June; Africa	252
	nearly full grown: mid Aug; Burkina Faso, w Africa	5903
	repro: est not seasonal; equatorial Africa	253

Order Chiroptera, Family Craseonycteridae, Genus *Craseonycteris*

Few reproductive data have been reported for *Craseonycteris thonglongyai*, a 1.7-2 g bat endemic to Thailand. Females have 2 pectoral mammae and 2 inguinal false teats (4746). Details of its natural history are available (3576, 4748, 7109).

LITTER SIZE		
Craseonycteris thonglongyai	1 yg	2747
INTERLITTER INTERVAL		
Craseonycteris thonglongyai	1 yr	2747
SEASONALITY		
Craseonycteris thonglongyai	repro: end Apr-?; Thailand	2747

Order Chiroptera, Family Nycteridae, Genus *Nycteris*

Nycterids (*Nycteris*) are 5-43 g insectivores primarily from Africa, but also from Malaysia and Indonesia. Implantation may preferentially occur in the right horn of the bicornuate or duplex uterus (824, 6966, 10154). A postpartum estrus is present in tropical species (255, 6962, 6966) after the birth of a furred neonate (11270). Temperate forms are monoestrous, but have two periods of follicular development (824). Females have 2-4 mammae (331, 502, 4079, 6720, 7557). Males have ampullary glands and a baculum (6065, 6966).

NEONATAL MASS		
Nycteris arge or *N major*	4 g; n1 emb	11273
LITTER SIZE		
Nycteris arge or *N major*	1 emb; n1 f	11273
Nycteris grandis	1 emb; n1 f	11273
	1.5 emb; r1-2; n4 f	1347
Nycteris hispida	1	11270
	1 emb; n1 f	191
Nycteris javanica	1 emb, yg; n2 lt	7161
Nycteris macrotis	1 emb; n1 f	252
	1 emb; n1 f	8611
	1 yg; n1 f	8611
	1 emb ea; n2 f	4924
Nycteris nana	1	11270
Nycteris thebaica	1	10158
	1	8965
	1	10154
	1 emb; n1 f	331
	1 emb; n1 f	9945
	1 emb ea; n3 f	7719
	1.5 emb; r1-2; n2 f	185, 186
GESTATION		
Nycteris hispida	~3 mo	11270
Nycteris macrotis	2.5 mo	255
Nycteris nana	~3.5 mo; field data	11270
Nycteris thebaica	2.5 mo	255
	5 mo	824

LACTATION

Nycteris arge	nurse: ~2 mo	5759
Nycteris macrotis	wean: 2 mo	255
Nycteris nana	wean: 45-60 d	252
	wean: 45-60 d	11270
Nycteris thebaica	wean: 2 mo	824
	wean: 2.5 mo	255

LITTERS/YEAR

Nycteris hispida	1 or 2	11270
Nycteris macrotis	2-3	252, 255
Nycteris nana	1	11270
Nycteris thebaica	1-2?	255

SEASONALITY

Nycteris arge	mate: Jan-Feb; Zaire	252
	mate: Oct; Zaire	252
	birth: Jan; Zaire	252
	birth: Apr-May; Zaire	252
	preg f: Mar, Dec; n3; Albert Park, Zaire	11272
	lact f: Apr; n1; Albert Park, Zaire	11272
	wean: Mar-June/July; Garamba Nat Park, Zaire	252
Nycteris grandis	birth: Jan; Cameroun	252
	birth: mid Apr, early Aug, mid Dec; Gabon	252
	birth: May; Garamba Nat Park, Zaire	252
	lact f: July-Aug; n2; Mt Nimba, Liberia	11811
Nycteris hispida	mate: early Jan; Zaire	11270
	mate: Jan; Tanzania, Zaire	252
	mate: June-July; Gabon	252
	mate: Dec; Cameroun	252
	preg f: Aug; n1; Rio Muni, W Arfica	5387
	preg f: Aug; n1; Zaire	191
	birth: Jan, June; Zaire	11270
	birth: Feb-Mar; Cameroun	252
	birth: Mar-Apr; Tanzania, Zaire	252
	birth: end Mar-early Apr; Zaire	11270
	birth: Sept-Oct; Gabon	252
	birth: Dec; n3; w Great Rift Valley, e Africa	6966
	lact f: Oct, Dec; Rio Muni, W Arfica	5387
	wean: May; Cameroun	252
	wean: June; Garamba Nat Park, Zaire	11270
	wean: June-July; Tanzania, Zaire	252
	wean: Nov; Gabon	252
	wean: Dec; Rio Muni, Guinea	252
Nycteris javanica	preg f: Mar; n1; Perlis, Malaysia	7161
	birth: Apr; n1; Kuala Lumpur, Malaysia	3974
	lact f: Nov; n1; Selangor, Malaysia	7161
Nycteris macrotis	mate: Aug, Oct; central Africa	252, 255
	mate: Dec; Garamba Nat Park, Zaire	252
	preg f: May; n1; Niger	8611
	preg f: end Dec; n1; Tanzania	5759
	birth: Jan; Kenya	252
	birth: Jan-Mar; Rwanda	252, 255
	birth: Mar; Garamba Nat Park, Zaire	252
	birth: Oct-Nov, early Jan; Rwanda	252
	birth: Oct-Dec; Liberia	255
	lact f: Dec; few data; Tanzania	186
Nycteris nana	mate: early Dec; Zaire	11270
	mate: end Dec-early Jan; central Africa	252
	birth: simultaneously mid Mar; central Africa	252
	birth: mid Mar; Zaire	11270
	wean: May-mid June; Zaire	11270
	wean: mid May; central Africa	252
Nycteris thebaica	estrus: July; Natal, S Africa	824
	mate: Apr; Rwanda	252
	mate: June; Natal, S Africa	824

	mate: July; se Zaire, Tanzania	252
	mate: Aug; Zambia	252
	mate: Oct-Nov; Kenya, Tanzania	252
	conceive: early Nov, early June; Natal, S Africa	824
	preg f: end Jan-early Feb; n2; e Africa	185
	preg f: May; n3; Saudi Arabia	7719
	preg f: Aug; n1; Zambia	331
	preg f: Aug-Oct only; Zimbabwe	10158
	preg f: Oct; n1; Namibia	9945
	preg f: Dec; n1; Tanzania	186
	birth: Jan-Feb; Kenya, Tanzania	252
	birth: July; Rwanda	252
	birth: ~Sept-Feb; few data; Botswana	10154
	birth: mid Oct; Katanga, Zaire	252
	birth: early Nov; Zambia	252
	birth: early Nov; Natal, S Africa	824
	birth: early summer; Transvaal, S Africa	8965
	birth: yr rd; se Zaire, Rwanda	255
	wean: Jan-Feb; Zambia	252
	wean: Apr; Kenya, Tanzania	252
	wean: end Sept; Rwanda	252
	wean: end Dec; se Zaire, Tanzania	252

Order Chiroptera, Family Megadermatidae

False vampire and yellow-winged bats are 20-170 g predators on insects and small vertebrates in equatorial Africa (*Cardioderma*, *Lavia*), tropical Australia (*Macroderma*), India, and southern and southeastern Asia (*Megaderma*). Females may be larger than (*Cardioderma* 7121; *Lavia* 4924) or equal to (*Megaderma* 8494) males in size. Ovulation in *Megaderma* is predominantly from the left ovary (8872, 8876, 8890), while *Cardioderma* uses both ovaries (6966). Multiovular follicles are common (3937). Connective tissue links the ovaries with the uterus (3937). Embryos enter the bipartite or bicornuate uterus in the morula stage where they develop endotheliochorial, discoid placentae with superficial, circumferential, implantation that can be either mesometrial, antimesometrial, or lateral (3910, 3917, 3928, 3929, 6966, 11799). The left uterine horn is used preferentially (3928, 8872, 8876, 8887, 9490; cf 3910).

Placentophagia occurs occasionally (3930, 11270) after the birth of either naked and blind (*Cardioderma* 4924, 5759) or furred (*Lavia* 11270) and open-eyed (*Megaderma* 3930) neonates. Females have 2 functional, pectoral teats and both sexes have 2 nonfunctional pubic teats (2698, 4669, 6966, 7161, 9297). Males have urethral glands, ampullary glands or seminal vesicles, and a baculum (5704, 6065, 6966, 7663, 9178). Details of birth (3930) are available.

NEONATAL MASS		
Megaderma lyra	7-8 g	3915
	10.25 g; r8.6-11.38; n11	3930
LITTER SIZE		
Lavia frons	est 1	11663
	1	9297
	1	11270
	1 yg	11233, 11234
	1 emb; n1 f	191
Macroderma gigas	1 yg	7075, 7076
Megaderma lyra	1	8890
	1	8876
	1 yg	1340, 1343
	mode 1 yg; stillborn twins; n1 lt	3930, 3931
	1 emb; 0.3% twins; r1-2; n330 f	3927
	2; n1 lt	8887
Megaderma spasma	1	8494, 8501
	1 yg; 2 rare	1340, 1343, 1346
	mode 1; r1-2	7161

SEXUAL MATURITY		
Megaderma lyra	f: 1st mate: 1.5 yr	8876
	m: 15 mo	8890
	f: 18 mo	8890
GESTATION		
Cardioderma cor	3 mo	252
	~3 mo	5759
Lavia frons	~3 mo	5759
	~3.5 mo; field data	11270
Megaderma lyra	5 mo; field data	3915, 3928
	~150 d: field data	8876
	150-160 d	7161
	150-160 d	8890
Megaderma spasma	70-90 d	1340
	~5 mo	7161
LACTATION		
Lavia frons	wean: >50 d	11234
	wean: 2.5 mo	252
Megaderma spasma	wean: 4-5 wk; few data	8876
Megaderma lyra	wean: 2 mo	8890
LITTERS/YEAR		
Cardioderma cor	min 2	252
Lavia frons	1	11270
SEASONALITY		
Cardioderma cor	mate: Apr-May; Kenya	252
	mate: Sept-Oct; Tanzania	252
	preg f: mid-May-mid-Aug, Nov-Jan; Kenya	252
	preg f: mid Oct-mid Jan; Tanzania	252
	preg/lact f: yr rd; few data; Kenya	7121
	birth: early Jan; Tanzania	252
	birth: Jan, Aug; Kenya	252
	birth: Aug+; Eritrea to n Zambia	5759
	lact f: Aug; n1; e Africa	4924
	wean: mid Apr, end Oct; Kenya	252
	wean: end Mar; Tanzania	252
Lavia frons	mate: Jan, n1; Seronera, Tanzania	11663
	mate: end May; se Zaire	252
	mate: end Dec; Rwanda	252
	mate: end Dec; Zaire	11270
	preg f: Mar; n1; Zaire	191
	birth: end Mar; Rwanda	252
	birth: end Aug; se Zaire	252
	birth: early Apr; Zaire	11270
	birth: early Apr; n2; Kenya	11233
	birth: mid-Apr, mid-Aug, mid-Nov, mid-Jan; Tanzania	252
	repro: yr rd; equatorial Africa	253
Macroderma gigas	mate: end Oct-early Nov; n Australia	10809
	preg f: June; n1; nw Australia	2697a, 2698
	preg f: Aug; n1; QLD, Australia	7076
	birth: Oct-Nov; Pilbara, WA, Australia	2698
	birth: Sept; n1; nw Australia	2698
	young seen: Sept-Nov; Australia	10698
Megaderma lyra	spermatogenesis: July-Dec; Maharashtra, India	8890
	sperm: July-Jan; India, Sri Lanka	8876
	testes: large Oct, small Dec; India, Sri Lanka	8876
	mate: early Oct; India	3937
	mate: early-mid Nov; India	3928
	mate: end Nov; India, Sri Lanka	8876
	preg f: Feb-Apr; central India	5705
	preg f: end Nov-mid Apr; Maharashtra, India	8890
	preg f: end Nov-Apr; India, Sri Lanka	8876
	preg f: Nov-May; India	3915
	birth: Mar; n1; Kuala Lumpur, Malaysia	3974
	birth: end Mar; India	1340

	birth: Apr; n1; India	8887
	birth: mid Apr; India	3928
	birth: end Apr; India, Sri Lanka	8876
	lact f: May-June; Maharashtra, India	8890
Megaderma spasma	mate: est Nov; few data; India, Sri Lanka	8876
	mate: Dec-Jan; India	1343
	preg f: Feb-Apr, Aug; no data other mo; Selangor, Malaysia	7161
	birth: Feb; n16/21; Kuala Lumpur, Malaysia	3974
	birth: Mar-May; India	1340, 1343
	birth: est end Apr; few data; India, Sri Lanka	8876
	birth: May; Sri Lanka	8493
	lact f: May; few data; Sri Lanka	8501

Order Chiroptera, Family Rhinolophidae

Horseshoe, leaf-nosed, and trident bats (*Anthops* ~50 mm; *Asellia* 6-10 g; *Aselliscus* 3-8 g; *Cloeotis* 3-8 g; *Coelops* 6-9 g; *Hipposideros* 5-180 g; *Paracoelops* 7 g; *Rhinolophus* 2-32 g; *Rhinonycteris* 8-10 g; *Triaenops* 8-25 g) are old-world insectivores. Females are often larger than males (4968, 8494, 9297, 9945, 11270; cf 191).

Ovulation may be induced (1640). Asymmetry of female reproductive function is common and either side may predominate (202, 826, 828, 832, 906, 908, 3573, 3918, 3919, 3925, 3934, 3936, 3941, 3942, 5527, 5528, 5529, 5538, 6627, 6718, 6956, 7161, 7208, 7681, 8057, 8875, 8882, 9241, 9489, 10068; cf 6966). Vaginal plugs (3573, 6956, 8055, 8830, 8918, 9238, 9239, 9241, 9437, 11318), sperm storage (826, 828, 830, 831, 3573, 3574, 6956, 7208, 7519, 8918, 9437, 11018, 11019), delayed development (832, 3942, 8882, 9489), delayed implantation (3942, 7208, 8882), monoestry (832, 1348, 8875), and polyovular follicles (11318) are reported.

The endotheliochorial or vasochorial, labyrinthine, bidiscoidal placenta has a mesometrial or circumferential, superficial implantation into a bicornuate or bipartite uterus (202, 828, 832, 905, 906, 907, 908, 1228, 1524, 3574, 3913, 3917, 3918, 3919, 3925, 3934, 3936, 3942, 5527, 5528, 5529, 5538, 6718, 6966, 7563, 7681, 9178, 10068, 11318, 11799) and is often eaten after birth (8873, 8877, 11270). A bicornuate vagina is reported (5529, 5538).

Neonates, up to 25% of adult mass (7681), are blind and either naked or covered with downy hair (897, 1343, 1349, 3571, 6718, 7681, 8798, 8873, 8875, 8877, 9175, 10285, 11270, 11623, 12021). Eyes open at 3-15 d (*Rhinolophus* 3571, 12021). Most females have 2 non-functional pubic teats as well as 2 pectoral mammary glands (331, 3571, 4669, 6720, 6956, 7161, 9297, 10158). Seminal vesicles, ampullary or urethral glands and a baculum are present in males (828, 3574, 5704, 6065, 6957, 6966, 7440, 9178).

Details of female anatomy (832, 3937, 6718, 6956, 8878, 9178, 11093, 11318), male anatomy (831a, 2830, 6957, 8054, 11293), endocrinology (830, 3936, 4661), spermatogenesis (8056), spermiophagy/abnormal sperm (827, 829, 11019), and conception (8057) are available.

NEONATAL MASS		
Hipposideros ater	1-1.5 g	3915, 3934
Hipposideros bicolor	1.2 g	3915
Hipposideros caffer	2.8 g; ±0.9; n8 term emb	7681
Hipposideros commersoni	28 g; n1	1349
Hipposideros fulvus	2.2 g	3915, 6718
Hipposideros lankadiva	8 g; n1; smallest neonate	9489
	8-11 g; term emb	897
Hipposideros larvatus	r3.5-4.0 g; n3	10872
Hipposideros ruber	3.0 g; term emb	11273
Hipposideros speoris	2.1 g; n1	8873
Rhinolophus cornutus	2.4 g	6115
	2.55 g; r2.4-2.7; n2	12021
Rhinolophus euryale	3.9 g; r3-4.5; n11	2402
Rhinolophus ferrumequinum	5.8 g	6115
Rhinolophus rouxi	1.8-2.1 g	8877
	4.2 g; r3.7-4.7; n4	10285
NEONATAL SIZE		
Asellia tridens	27.7-29.8 mm	110
Hipposideros caffer	FA: 26 mm; ±5.0; n17	7681

WEANING MASS		
Rhinolophus cornutus	7.1 g; r6.5-7.9; 55 d	12021
Rhinolophus rouxi	11.0 g; 3 wk	10285
LITTER SIZE		
Aselliscus tricuspidatus	1 yg	10114
	1 emb ea; n12 f	7073
Cloeotis percivali	1 emb	10158
Coelops frithi	1 emb ea; n2 f	7983
Hipposideros abae	1 or 2; large number of young suggests 2	252
	1-2	11270
Hipposideros ater	1 yg	3915, 3927, 3931, 3934
	1 emb or yg; n2 lt	5340
	1 emb ea; n208 f	3927
Hipposideros bicolor	1 yg; n many	1340, 1343
	1 emb; n1 f	10872
	1 emb ea; n24 f	3915, 3936
Hipposideros caffer	1	11270
	1 emb	10158
	1 yg	1348, 1351
	1 emb ea; n3 f	8611
	1 emb ea; n>60 f	7681
Hipposideros cineraceus	1 emb; n1 f	10872
	1 emb; n1 f	5705
Hipposideros commersoni	1 emb; n1 f	331
	1 yg	1351
Hipposideros cyclops	1	11270
Hipposideros diadema	1 yg	7161
	1 yg ea; n6 lt	7075
Hipposideros fulvus	1 emb ea; n3 f	8494, 8501
	1 emb ea; n12 f	5527
	1 emb ea; n123 f	3927
	1.03 emb; r1-2; n64 f	3915, 3927, 3931, 5528, 6718
	1.25 emb; r1-2; n4 f	10872
Hipposideros galeritus	1 yg	8494, 8501
	1 emb ea; n3 f	7073
	1 emb ea; n4 f	6447
	1 yg ea; n5 lt	7075
Hipposideros lankadiva	1	3927
	1 emb ea; n66 f	3927
	1 emb ea; n169 f	9489
Hipposideros larvatus	1 emb ea; n3 f	10872
Hipposideros ridleyi	1 emb; n1 f	3974
Hipposideros ruber	1 emb, yg; n4 lt	11273
Hipposideros speoris	1	3918, 3927, 3931
	1; n "several"	8494, 8501
	1 yg ea; n200 lt	1340, 1343
	1 emb ea; n280 f	3927
Rhinolophus affinis	1 emb	7161
Rhinolophus alcyone	1 emb; n1 f	11273
Rhinolophus capensis	1	10158
Rhinolophus clivosus	1 emb, yg	10158
	1 yg	9945
	1 emb	8965
	1; seldom 2	252
Rhinolophus cornutus	1	6115
Rhinolophus darlingi	1	8965
	2 emb; n1 f	10158
Rhinolophus euryale	1	436
Rhinolophus ferrumequinum	1	6115
	1	11318
	1; seldom 2	3571
	1 ovum	11019
	1 emb	8049

Rhinolophus fumigatus	1	11270
	1 emb	10158
	est 1 emb ea; n2 f	8611
Rhinolophus hildebrandti	mode 1; 2 emb; n1 f	252
Rhinolophus hipposideros	mode 1; r1-2; 35% twins based on population dynamics	10068
	1-3; 3 eggs exceptional	436
	1; r1-2	3571
	1 emb ea; n17 f	1524
	1.02 emb; r1-2; n46 f	3573, 3574
Rhinolophus landeri	1	252, 255
	1 emb; n1 f	10158
Rhinolophus lepidus	1	1340
	mode 1 emb	7161
	2 emb; n1 f	5705
Rhinolophus luctus	1	1340
	1 emb; n1 f	8494, 8501
Rhinolophus macrotis	mode 1 emb	7161
Rhinolophus maclaudi	1 emb; n1 f	73
Rhinolophus megaphyllus	1 yg; n many	7074, 7075
Rhinolophus rouxi	1	8882
	1 yg	8494, 8501
	1 yg; n150 lt	1340, 1343
	1 emb; 0.4 % twins; r1-2; n285 f	3927
	1 yg; 2 rare	3927, 3941, 3942
Rhinolophus simulator	1 emb; n1 f	331
Rhinolophus stheno	mode 1 emb	7161
Rhinolophus swinnyi	1 emb	252
SEXUAL MATURITY		
Hipposideros bicolor	f/m: 18-20 mo	1341
Hipposideros caffer	f/m: 2nd yr	5759
	f/m: 2 yr	1349
Hipposideros commersoni	f/m: 2 yr	1349
	f/m: ~2 yr	5759
Hipposideros lankadiva	f/m: 16-17 mo min	9489
Hipposideros speoris	f: 7.5-8 mo	3918
	f/m: 18-20 mo	1341
Rhinolophus capensis	f/m: > 1 yr	828
Rhinolophus clivosus	f: 1st preg: 2 yr 20%; 3 yr 67%	11818
	f: 1st estrus: 6 mo	826
	m: 1st mate: 2nd yr	826
Rhinolophus euryale	f: 1st birth: 2-3 yr	2586
	m: spermatogenesis: 2-3 yr	2586
Rhinolophus ferrumequinum	f: ovary development: 18 mo	9239
	f: 1st mate: 15 mo	6956
	f: 1st mate: 2 yr	9241
	f: 1st birth: 3-4 yr	2586
	f: 1st birth: 5 yr; r3-7	8835
	m: testes development: 18 mo	9239
	m: spermatogenesis: 3-4 yr	2586
	m: urethral gland size: 3 yr	9241
Rhinolophus hipposideros	f: 1st estrus: some 4 mo; most end 2nd yr	3573
	f: large follicle, sperm in vagina: 1.5 yr	10068
	f: 1st birth: 1-2 yr	2586
	f: 1st birth: 15 mo	6956
	f: 10-15 mo	3574
	f: 2 yr	9241
	m: sperm: 12-14 mo	3574, 3580
	m: spermatogenesis: 1-2 yr	2586
	f/m: 1st mate: 1-1.5 yr	3571
	f/m: 1st mate: 2nd yr	1524
Rhinolophus rouxi	f: min 19 mo	3942
	m: min 16 mo	3942
	f/m: 1 yr	1343

GESTATION

Asellia tridens	9-10 wk	110
Cloeotis percivali	3 mo	255
Hipposideros abae	3 mo	11270
Hipposideros ater	155-160 d	3915
	190-200 d	3934
Hipposideros bicolor	40-45 d	3915
Hipposideros caffer	preimplant: ~21 d; r14-28	832
	postimplant: 199 d	832
	3 mo; field data	7681
	80-100 d; field data	1349
	220 d	832
Hipposideros commersoni	5 mo with 2 mo delay	252, 254, 255
	5 mo; field data	1349
Hipposideros cyclops	3.5 mo	11270
Hipposideros fulvus	150-160 d	3915, 6718
	150-160 d	5527
Hipposideros lankadiva	~260 d with delay	9489
Hipposideros ruber	3 mo	252, 255
Hipposideros speoris	~90 d	1340
	135-140 d	3918
Rhinolophus blasii	3 mo	255
Rhinolophus capensis	3-4 mo; field data	828
Rhinolophus clivosus	3 mo	255
	3.5-4 mo	826
Rhinolophus euryale	est 3 mo; field data	2402
Rhinolophus ferrumequinum	10-11 wk	9238
	est ~3 mo	9437
	3-4 mo; est ovulation to birth	8918
Rhinolophus fumigatus	3 mo	255
	3 mo; field data	11270
Rhinolophus hildebrandti	~3 mo	252, 255
Rhinolophus hipposideros	~7 wk	9175
	~75 d	3571
	~2.5 mo	3573, 3574
Rhinolophus landeri	delay: 2 mo	252, 255
	postdelay gestation: 3 mo	252, 255
	5 mo	252, 255
Rhinolophus rouxi	implant: 40-45 d	8882
	3 mo; visible emb to birth	10285
	150 d; ±8; with delay; field data	3942, 8875, 8882
Rhinolophus simulator	3 mo	255
Rhinolophus swinnyi	~3 mo	252, 255

LACTATION

Hipposideros ater	wean: 2.5-3 mo	3934
Hipposideros caffer	nurse: 2+ mo	832
	wean: 3+ mo	1349
Hipposideros commersoni	wean: 5+ mo	1349
Hipposideros cyclops	solid food: ~2 mo	252, 11270
Hipposideros fulvus	wean: 22 d	5527
	wean: ~2 mo	6718
Hipposideros lankadiva	wean: ~3 mo	9489
Hipposideros speoris	wean: 50-55 d	3918
Rhinolophus capensis	wean: 2 mo	828
Rhinolophus clivosus	nurse: 2 mo	826
Rhinolophus cornutus	independent: ~55 d	12021
Rhinolophus euryale	wean: est 3.5 mo	9241
Rhinolophus ferrumequinum	wean: est ~2 mo	9437
Rhinolophus fumigatus	wean: 2 mo	11270
Rhinolophus hipposideros	wean: 4-6 wk	3573, 3574
Rhinolophus landeri	wean: 2 mo	255
Rhinolophus megaphyllus	wean: 2 mo	2830
Rhinolophus rouxi	nurse: 1-3 mo	1640
	nurse: ~2 mo; field data	3942

LITTERS/YEAR

Hipposideros abae	1	11270
Hipposideros caffer	1	1348, 1351
	1	11270
Hipposideros commersoni	1	1351
Hipposideros cyclops	1	11270
Hipposideros galeritus	1	8494
Hipposideros speoris	1	1343
Rhinolophus euryale	1	9241
Rhinolophus ferrumequinum	1	3571
	1	9241
Rhinolophus fumigatus	1	11270
Rhinolophus hildebrandti	1	252
Rhinolophus hipposideros	1	9241
	1	3571
Rhinolophus rouxi	1	1343

SEASONALITY

Asellia tridens	mate: fall; Iraq	110
	birth: May-June; few data; Burkina Faso, w Africa	5903
	birth: June-July; Iraq	110
	young found: July, Nov; Burkina Faso, w Africa	5903
Aselliscus tricuspidatus	preg f: July; n12; New Guinea	7073
	preg f: Oct; n1; New Hebrides	518
	f with yg; June, July, or Aug; New Guinea	10114
Cloeotis percivali	birth: Oct; se Zaire, Zambia	252, 255
	birth: est Oct-Dec; S Africa	10158
Coelops frithi	preg f: Jan; n2; Java, Indonesia	7983
Hipposideros abae	mate: Dec; Cameroun, Zaire, Uganda	252, 11270
	birth: Mar; Cameroun, Zaire, Uganda	252, 11270
Hipposideros armiger	preg f: Feb-May, Sept-Oct; Malaysia	7161
	preg f: Feb, June; n2; Malaysia	6446
	birth: Mar; few data; Kuala Lumpur, Malaysia	3974
Hipposideros ater	preg f: mid Nov-June; India	3915, 3934
	lact f: Jan; QLD, Australia	7076
Hipposideros beatus	birth: mid Mar; boreal; Nov; austral, Gabon	252
	birth: mid Mar; central Africa	11270
	wean: end May; central Africa	11270
Hipposideros bicolor	mate: Mar; India	3937
	mate: Oct; Selangor, Malaysia	7161
	preg f: Mar-May; India	3915
	birth: Apr; few data; Kuala Lumpur, Malaysia	3974
	birth: Apr; Selangor, Malaysia	7161
	birth: end Apr-early May; India	1340, 1343
Hipposideros caffer	sperm: only Nov; Nigeria	7208
	testes: peak Nov; Nigeria	7208
	mate: end Apr-early May; Natal, S Africa	832
	mate: Oct-Jan; central Africa	11270
	mate: Nov; e Africa	5759
	mate: Nov-Dec; boreal, central Africa	252
	mate: July-Aug; austral, central Africa	252
	preg f: Feb, May; n3; Niger	8611
	preg f: Mar; n15; Nigeria	7208
	preg f: Mar-Apr, none Nov; Zaire	11272
	preg f: Aug; n1; Rio Muni, W Africa	5387
	preg f: Aug; Zimbabwe	331
	preg f: Dec-Mar; Uganda	7678, 7681
	birth: Feb-Mar; Uganda	5759
	birth: Mar; Uganda	7681
	birth: Mar, Oct; ne Gabon	1348, 1349, 1351
	birth: Mar-Apr; Nigeria	255
	birth: Apr; Uganda	7678
	birth: Oct; Gabon	255
	birth: Oct/Nov-early Dec; Natal, S Africa	832

	birth: Oct-Dec, Feb-Apr; central Africa	11270
	birth: Oct-Dec; S Africa	10158
	lact f: Nov; Rio Muni, W Africa	5387
Hipposideros camerunensis	preg f: Oct; n12; Cameroun	2955
	lact f: Oct; n2; Cameroun	2955
Hipposideros cineraceus	mate: Oct; Malaysia	7161
	birth: end Apr; Malaysia	7161
	preg f: Mar; n1; central India	5705
Hipposideros commersoni	mate: May; Gabon	254
	mate: May; ne Gabon	1349
	emb development: June-?; central Africa	252
	birth: Oct; ne Gabon	1349, 1351
	birth: Oct; Gabon	254
	birth: Oct; se Zaire, Rwanda	255
Hipposideros coxi	lact f: Jan; Sarawak, Malaysia	10314
Hipposideros cyclops	mate: early Dec; central Africa	11270
	preg f: Feb-Mar; n3; Zaire	11272
	birth: mid Mar; central Africa	11270
	wean: mid May; central Africa	11270
Hipposideros diadema	mate: Oct; Kuala Lumpur, Malaysia	3974
	preg f: Apr, Oct; n3; Malaysia	6446
	preg f: May; n1; Philippines	7586
	birth: Mar; Kuala Lumpur, Malaysia	3974
	lact f: Mar-Apr; Selangor, Malaysia	7161
Hipposideros fuliginosus	birth: Mar; central Africa	252
	young found: Apr; few data; Cameroun	71
Hipposideros fulvus	mate: early Nov; India	6718
	mate: mid Nov; Maharashtra, India	5527
	ovulate: mid Nov; Maharashtra, India	5527
	conceive: mid Nov; India	5528
	preg f: Mar; n3; Sri Lanka	8501
	preg f: Mar-Apr; n7; India	5705
	preg f: mid Nov-Apr; India	6718
	preg f: Nov-May; India	3915
	birth: end Apr-early May; India	5528
	birth: May; n1; central India	5705
Hipposideros galeritus	preg f: Apr; n1; Sri Lanka	8494, 8501
	preg f: June; n4; Malaysia	6447
	preg f: July, Sept; n3; Papua, New Guinea	7073
	birth: Feb; few data; Kuala Lumpur, Malaysia	3974
Hipposideros lankadiva	mate: Aug; Chandrapur, Maharashtra, India	908
	conceive: Aug; Chandrapur, Maharashtra, India	908
	preg f: Mar-May, Feb-Apr; 2 areas; Chandrapur, Maharashtra, India	897
	implant: end Sept; Chandrapur, Maharashtra, India	908
	birth: mid-end May; Chandrapur, India	9489
	birth: early June; Chandrapur, Maharashtra, India	908
Hipposideros ridleyi	birth: end Apr; Kuala Lumpur, Malaysia	3974
Hipposideros ruber	mate: Dec; Uganda; early July; central Africa	252
	birth: Mar; Liberia	11881
	birth: Mar; Zaire	252
	birth: Oct; se Zaire	252, 255
Hipposideros speoris	conceive: end Dec-mid Mar; Chandrapur, Maharashtra, India	8878
	preg f: Jan-July; Chandrapur, Maharashtra, India	3918
	birth: Jan; n1; s India	11507b
	birth: May; India	1340, 1343
	birth: early Aug; Sri Lanka	8494, 8501
Rhinolophus affinis	preg f: Feb-Mar; Pahang, Malaysia	7161
	preg f: peak Apr-May, Oct; Selangor, Malaysia	7161
Rhinolophus alcyone	mate: Aug; central Africa	252
	preg f: June; n3; w Uganda	5759
	birth: Sept; n1; Uganda	5759
	birth: end Nov-early Dec; central Africa	252
	lact f: June; n1; w Uganda	5759

Rhinolophus blasii	birth: June; n Africa	72
	birth: Dec; se Zaire	255
	lact f: May; n1; Iran	2404a
Rhinolophus capensis	spermatogenesis: Oct-May; Cape Province, S Africa	828
	spermatogenesis: Jan-Apr; inactive: June-Sept; Cape Province, S Africa	830
	epididymal sperm: Apr-May; Cape Province, S Africa	828
	mate: Aug-Sept; Cape Province, S Africa	828, 830
	ovulate: Aug-Sept; Cape Province, S Africa	828
	birth: Nov-Dec; Cape Province, S Africa	828
	birth: Dec; S Africa	831
	birth: mid Dec-?; Cape Province, S Africa	10158
	lact f: Dec-Jan; Cape Province, S Africa	828
Rhinolophus clivosus	mate: Apr-May, Sept-Oct; Natal, S Africa	826
	mate: July-Aug; se Zaire, Rwanda	255
	preg f: Nov; Transvaal, S Africa	8965
	birth: Oct-Nov; se Zaire, Rwanda	252, 255
	birth: end Nov-early Dec; Natal, S Africa	11818
	birth: Dec-Feb; Natal, S Africa	826, 831
Rhinolophus cornutus	birth: early July; Japan	12021
Rhinolophus darlingi	preg f: Oct; n1; Transvaal, S Africa	8965
	birth: est early summer; S Africa	10158
Rhinolophus euryale	mate: Oct?; France	9241
	birth: end June-July; France	9241
	wean: mid Sept; France	9241
Rhinolophus ferrumequinum	sperm in uterus: Nov-Mar; Gloustershire, England	6956
	epididymal sperm: yr rd; France	9437
	spermatogenesis: May-Oct; France	9437
	spermatogenesis: May-Oct; Bologna, Italy	11293
	m: sex organs: developed Sept-May, regressed May-Aug; France	9241
	estrus: ?-Oct; Gloustershire, England	6956
	mate: Aug, early winter; Gloustershire, England	8918
	mate: mid-end Oct; Japan	7519, 8057
	mate: Oct; France	9241
	mate: Oct-Nov; France	9437
	copulatory plug: ?-Mar; France	9437
	ovulate: early Apr; Japan	7519, 8057
	ovulate: early-mid Apr; Gloustershire, England	6956
	ovulate: Apr; France	9437
	ovulate: ~Apr; Gloustershire, England	8918
	implant: before Mar 15; Sardinia, Italy	3517
	preg f: Mar-Apr; Sardinia, Italy	3517
	birth: May-July; Germany	3571
	birth: June; Algeria	5988
	birth: June; France	9238, 9241
	birth: June-July; Gloustershire, England	6956
	birth: end June-early July; France	9437
	birth: July-Aug; Gloustershire, England	8918, 8919
Rhinolophus fumigatus	mate: Dec: Zaire, Ethiopia	255
	preg f: Feb; n2; Niger	8611
	birth: Mar-Apr; Ethiopia, Zaire	252, 255
	birth: Oct-Nov; Kenya	252, 255
	birth: est Oct-Dec; S Africa	10158
Rhinolophus hildebrandti	mate: Aug-Sept; central Africa	252
	preg f: est Oct-early Mar; Uganda	7680
	birth: Dec-early Jan; central Africa	252
	birth: Dec-Jan; se Zaire, Rwanda	255
	birth: est summer Oct-Dec; S Africa	10158
Rhinolophus hipposideros	sperm in uterus: Nov-Mar; Gloustershire, England	6956
	spermatogenesis: May-Sept; Czechoslovakia	3574
	spermatogenesis: July-Oct; Europe	2075
	testes: developed Sept; France	9241
	estrus: Oct; Gloustershire, England	6956

	mate: Jan; Netherlands	10068
	mate: Oct-Apr; Moravia, Czechoslovakia	3573, 3580
	mate: Oct; France	9241
	ovulate: Apr; Moravia, Czechoslovakia	3573, 3574
	ovulate: Apr; Gloustershire, England	6956
	birth: (May) June-July; Europe	3571
	birth: June-July; Gloustershire, England	6956
	birth: June-July; France	9241
	birth: June-Aug; Netherlands	10068
	birth: July; Czechoslovakia	3574
Rhinolophus landeri	sperm: July-Dec; Nigeria	7208
	mate: June; central Africa	252
	mate: Nov; Nigeria	255
	preg f: Mar; Nigeria	7208
	preg f: Oct; n1; Zimbabwe	10158
	unimplanted trophoblasts: Nov; Nigeria	7208
	implant: Feb; Nigeria	255
	implant: June; Tanzania	255
	birth: end Apr; Nigeria	255
	birth: Sept-Oct; Tanzania	252, 255
	birth: Oct-Jan; Cameroun	252
	lact f: May; Nigeria	7208
Rhinolophus lepidus	preg f: Feb-Apr; Pahang, Malaysia	7161
	birth: mid Apr-mid May; India	1340
Rhinolophus luctus	preg f: Jan; n1; Sri Lanka	8494, 8501
	birth: Apr; India	1340
Rhinolophus maclaudi	preg f: Aug; n1; central Africa	73
	birth: Sept-Oct; se Zaire, Rwanda	255
Rhinolophus macrotis	preg f: Feb-Mar; Malaysia	7161
Rhinolophus megaphyllus	testes: peak Apr-May; ne NSW, Australia	2830
	birth: mid Nov-Dec; ne NSW, Australia	2830
	wean: Jan; ne NSW, Australia	2830
Rhinolophus mehelyi	mate: Mar; Sardinia, Italy	3517
	preg f: end Mar-ear June; Algeria	5988
Rhinolophus rouxi	testes: peak Sept-Nov, regress Dec; Bangalore, India	8875
	mate: Nov-early Dec; Bangalore, India	8875
	mate: Nov-Dec; Bangalore, India	8882
	mate: Dec; w Ghats, India	8882
	preg f: Jan-Apr; India	10285
	preg f: Jan-May; India	3942
	birth: Apr; India	1340, 1343
	birth: Apr-May; Sri Lanka	8494, 8501
	birth: end Apr-early May; Bangalore, India	8875, 8882
	lact f: Apr-June; India	10285
Rhinolophus silvestris	birth: Oct; few data; central Africa	252
Rhinolophus simulator	mate: Aug; central Africa	252
	birth: est Oct-Nov; S Africa	10158
	birth: Nov; se Zaire	252, 255
	birth: est Nov; Cameroun	2955
Rhinolophus stheno	preg f: Feb-Apr; Pahang, Malaysia	7161
Rhinolophus swinnyi	mate: July; se Zaire	252
	birth: Oct; se Zaire	252, 255
Rhinolophus trifoliatus	preg f: May-June, Sept-Nov; Selangor, Malaysia	7161
	preg f: Mar; Perlis, Malaysia	7161
	preg f: Apr, Oct; n2; Malaysia	6446
Triaenops persicus	sperm: high June-Oct, low Dec-Jan; Tanzania	6735
	spermatogenesis: end Dec; central Africa	252
	preg f: Dec; Mhulumuzi, e Africa	5759
	birth: Jan; Tanzania	252

Order Chiroptera, Family Noctilionidae, Genus *Noctilio*

Bulldog bats (*Noctilio albiventris, N. leporinus*) are 30-80 g, piscivores and insectivores from Central and South America. Males are larger than females and may be differently colored (178a; cf 2808a). Ovulation may alternate between left and right ovaries and conception is reported to occur in the ipsilateral oviduct (8944, 8945). Blastocysts begin differentiation in the oviduct before the superficial, lateral to antimesometrial implantation of an endotheliochorial, discoid placenta near the cranial end of the endometrial ridge of the bipartite or bicornuate, symmetrical uterus (275, 4957, 6627, 9178, 11799). Neonates are naked, but have open eyes (1371). The clitoris is large (9178). Seminal vesicles are present (9178), but a baculum is not. Pockets of skin associated with the scrotum may be glandular (2808a, 3900a). Sperm have very large heads (3414). Details of anatomy (4957) and reproductive failure (8949) are available.

LITTER SIZE		
Noctilio albiventris	1 cl; 1 emb; no twins	275
	1 emb; n1 f	1006
	1 emb; n1 f	11471a
	1 emb ea; n2 f	10974
	1 emb ea; n9 f	3999a
	1 emb ea; n13 f	4966
	1 emb ea; n17 f	2345
	1 emb ea; n26 f	5425
	1.01 emb; mode 1; r1-2; n72 f	8944, 8949
Noctilio leporinus	1 emb; n1 f	5425
	1 emb; n1 f	10974
	1 emb ea; n2 f	3900
	1 emb ea; n2 f	2345
	1 emb ea; n3 f	4671
	1 emb ea; n3 f	4966
	1 emb ea; n3 f	3999a
	1 emb ea; n7 f	1641
	1 emb ea; n30 f	178a
LACTATION		
Noctilio albiventris	solid food: 44.7 d; sd4.3; r36-49; n9	1371
	nurse: 80.5 d; sd5.2; r75-90; n8	1371
SEASONALITY		
Noctilio albiventris	testes: small Apr; n1; Guatemala	2552
	mate: Mar; Cauca Valley, Colombia	8944
	mate: end Nov-Dec; Central America	275
	mate: end Nov-early Jan; Panama, Costa Rica	3322
	ovulate: Mar-Apr; Cauca Valley, Colombia	8944
	conceive: Mar; Cauca Valley, Colombia	8945
	preg f: Feb-mid Apr; Cauca Valley, Colombia	8944, 8949
	preg f: Apr; n1; Jalisco, Mexico	11471a
	preg f: Apr; Nicaragua	5425
	preg f: Apr; n1; Costa Rica	2345
	preg f: June-Aug; n9; Peru	3999a, 10974
	preg f: Dec; central America	275
	birth: Apr; central America	275
	birth: end Apr-May; Panama, Costa Rica	3322
	lact f: Jan; n1; Jalisco, Mexico	11471a
	lact f: Feb-Mar; neotropics	4966
	lact f: Mar; n1; Costa Rica	2345
	lact f: end May; Central America	275
Noctilio leporinus	preg f: Feb; Trinidad	3900a
	preg f: Feb; n3; Michoacan, Mexico	4671
	preg f: Feb; n4; Panama, Guatemala	1641
	preg f: Feb; n30; Panama	178a
	preg f: Mar; n1; Nicaragua	5425
	preg f: Apr; n1; Guatemala	2552
	preg f: July-Aug; n3; Peru	3999a, 10974
	preg f: Sept-Jan; Brazil	11736
	preg f: Oct; n3; Chiapas, Mexico	1641

	birth: Dec-Feb; Mexico, Cuba	1346
	lact f: Feb-Mar; Trinidad	3900a
	lact f: June; n3; Sinaloa, Mexico	5413a
	lact f: July; n7; Campeche, Mexico	5424f
	lact f: July; n2; Lesser Antilles	5411a
	lact f: Sept; n2; Guatemala	2552

Order Chiroptera, Family Mormoopidae

The insectivorous, ghost-faced bats (*Mormoops* 13-18 g) and moustached bats (*Pteronotus* 7.5-25 g) range from southwest United States to Brazil, Bolivia, and Peru. Ovulation occurs only from the right ovary and implantation is predominantly in the cranial portion of the right uterine horn (*Pteronotus* 3613, 4957, 11804). The endotheliochorial placenta implants in a bicornuate uterus (4957, 11799). Males have inguinal testes (*Pteronotus* 3613). Embryos can reach 30% of maternal mass. Details of anatomy (4957) are available.

NEONATAL MASS		
Pteronotus fuliginosus	1.8 g; large emb	9213
NEONATAL SIZE		
Pteronotus davyi	25 mm; term emb	62
LITTER SIZE		
Mormoops megalophylla	1 emb	577
	1 emb; n1 f	690a
	1 emb; n1 f	1475
	1 emb; n1 f	4671
	1 emb ea; n3 f	1937
	1 emb ea; n5 f	522
	1 emb ea; n15 f	950
Pteronotus davyi	1 emb; n1 f	1937
	1 emb; n1 f	5405e
	1 emb ea; n3 f	3205
	1 emb ea; n4 f	5425
	1 emb ea; n16 f	11471a
Pteronotus fuliginosus	1 emb; n1 f	5873
Pteronotus gymnonotus	1 emb; n1 f	3205
	1 emb ea; n3 f	3999a
	1 emb ea; n3 f	5425
	1 emb ea; n4 f	1475
Pteronotus parnellii	1; n1 lt	452
	1 emb; n1 f	10112a
	1 emb ea; n2 f	3705c
	1 emb ea; n2 f	5405b
	1 emb ea; n2 f	5425
	1 emb ea; n3 f	950
	1 emb ea; n5 f	3999a
	1 emb ea; n6 f	1937
	1 emb ea; n9 f	1475
	1 emb ea; n11 f	11471a
	1 emb ea; n21 f	5405e
	1 emb ea; n40 f	3205
Pteronotus personatus	1 emb; n1 f	1937
	1 emb; n1 f	5405e
	1 emb; n1 f	5413a
GESTATION		
Pteronotus parnellii	~3.5 mo	3613
LITTERS/YEAR		
Mormoops megalophylla	1	1474
SEASONALITY		
Mormoops megalophylla	preg f: Feb; n2; Campeche, Mexico	5424f
	preg f: Mar; n1; El Salvador	1475
	preg f: Mar, May; n5; Coahuila, Mexico	522
	preg f: May; n1; Michoacan, Mexico	4671
	preg f: June; n1; AZ	690a

	birth: June-July; sw US	1474
	lact f: Mar; Trinidad	3900a
Pteronotus davyi	preg f: Mar; n1; Guatemala	5405e
	preg f: Mar, May; n3; none Jan; El Salvador	3205
	preg f: Apr-May; n10; Yucatan, Mexico	950
	preg f: May; n1; N America	1937
	preg f: May; n4; Nicaragua	5425
	preg f: May-June; n16; Jalisco, Mexico	11471a
	preg f: June; n1; Sinaloa, Mexico	5413a
Pteronotus fuliginosus	preg f: May; n1; Haiti	5873
	preg f: Feb-June, peak May; Cuba	9213
	preg f: peak May; Puerto Rico	9213
Pteronotus gymnonotus	mate: Dec-Jan; few data; El Salvador	3205
	preg f: May; n1; none Aug-Sept, Dec; El Salvador	3205
	preg f: May; n4; El Salvador	1475
	preg f: May; n3; Nicaragua	5425
	preg f: Aug; n3; Peru	3999a
	lact f: Aug; n1; Costa Rica	2670a
Pteronotus parnellii	preg f: Jan, none Sept, Dec; El Salvador	3205
	preg f: Feb; n45; s Compeche, Mexico	5424f
	preg f: Feb-Mar; n2; Nicaragua	5425
	preg f: end Feb-early Mar; n1-3; Jamaica	7103a
	preg f: end Feb-May; Chinameca, Mexico	3613
	preg f: Mar; n1; El Salvador	1475
	preg f: Mar; n2; Michoacan, Mexico	4671
	preg f: Mar; n8; Panama	1475
	preg f: Mar; n21; Guatemala	5405e
	preg f: Apr; n1; Sonora, Mexico	1939
	preg f: Apr-May; n3; Yucatan, Mexico	950
	preg f: Apr-May; Costa Rica	6261
	preg f: May; n6; N America	1937
	preg f: May-June; n2; Sinaloa, Mexico	5413a
	preg f: May-June; n11; Jalisco, Mexico	11471a
	preg f: June; n1; Sinaloa, Mexico	523a
	preg f: June; n2; Durango, Mexico	5405b
	preg f: June, Aug; n5; Peru	3999a
	preg f: July; n1; Margarita I, Venezuela	10112a
	preg f: July; n2; Suriname	3705c
	preg/lact f: peak Mar-June, low July-Jan; Panama	1070
	birth: Mar-Apr; Panama, Costa Rica	3322
	lact f: Apr; Nicaragua	5425
	lact f: June; n1; Sinaloa, Mexico	523a
	lact f: July; n1; Guatemala	9113
Pteronotus personatus	mate: Jan; n1; El Salvador	3205
	preg f: Mar; n1; Guatemala	5405e
	preg f: May; n1; N America	1937
	preg f: June; n1; Sinaloa, Mexico	5413a

Order Chiroptera, Family Phyllostomidae

New-world, leaf-nosed bats (*Ametrida* 8-10 g; *Anoura* 4-20 g; *Ardops* 15-19 g; *Ariteus* 9-13 g; *Artibeus* 10-87 g; *Brachyphylla* 45-67 g; *Carollia* 10-25 g; *Centurio* 17-28 g; *Chiroderma* 13-29 g; *Choeroniscus* 5-9 g; *Choeronycteris* 10-22 g; *Chrotopterus* 73-90 g; *Desmodus* 15-50 g; *Diphylla* 24-34 g; *Ectophylla* 6-13 g; *Erophylla* 65-75 mm; *Glossophaga* 9-16 g; *Hylonycteris* 6-9 g; *Leptonycteris* 17-30 g; *Lichonycteris* 5-8 g; *Lionycteris* ~50 mm; *Lonchophylla* 7-16 g; *Lonchorhina* 11-17 g; *Macrophyllum* 6-10 g; *Macrotus* 12-20 g; *Micronycteris* 4-16 g; *Mimon* 9-23 g; *Monophyllus* 8-17 g; *Musonycteris* 70-79 mm; *Phylloderma* 57-71 g; *Phyllonycteris* 14-21 g; *Phyllops* ~48 mm; *Phyllostomus* 20-100 g; *Platalina* 72 mm; *Pygoderma* to 27 g; *Rhinophylla* 43-48 mm; *Scleronycteris* 57 mm; *Sphaeronycteris* ~53 mm; *Stenoderma* 25 g; *Sturnira* 15-67 g; *Tonatia* 7-36 g; *Trachops* 24-39 g; *Uroderma* 14-23 g; *Vampyressa* 7-10 g; *Vampyrodes* 33 g; *Vampyrops* 12-72 g; *Vampyrum* 120-190 g) have diverse diets and range throughout the tropics and subtropics of North, Central, and South America. Males of *Anoura* and *Phyllostomus* may be larger than females (1070, 7745), while female

Artibeus, *Choeroniscus*, *Desmodus*, *Phyllops*, *Pygoderma*, *Stenoderma*, and *Uroderma* are recorded as larger than males (3608, 3900a, 5423a, 5873, 7696, 10657). Sexual dichromatism may occur (6639; cf 279a).

Artibeus, *Carollia*, *Choeroniscus*, and *Pygoderma* males have an extra chromosome (3492, 7696). Ovulation and implantation in *Macrotus* occur in the right ovary and uterine horn, respectively (999, 1001, 1199, 1647, 11804). Unfertilized ova degenerate within the oviduct (*Glossophaga* 8946). The hemochorial or endotheliochorial, labyrinthine, discoid placenta has an interstitial implantation, usually into the intramural cornua of the simplex or bicornuate uterus (967, 1000, 1038, 1039, 3318, 3917, 4289, 4290, 4293, 4957, 6627, 7563, 8816, 8941, 8942, 8943, 8946, 9178, 10661, 11798, 11799, 11813, 11859) and occasionally is eaten after birth (5433, 9615, 11859).

Delayed development occurs in *Artibeus* (3318) and *Macrotus* (1198, 1199, 1460, 1461, 1462). *Macrotus*, *Leptonycteris*, and *Choeronycteris* are monoestrus (11754), *Glossophaga*, *Carollia*, *Sturnira*, *Uroderma*, *Vampyrops*, *Vampyrodes*, *Vampyresssa*, *Artibeus*, and *Desmodus* are continuously or more often bimodally polyestrous (2109, 3318, 3322, 9113, 10654, 10658, 10785, 11754, 11813; cf 8816). Concurrent gestation and lactation occur (*Artibeus* 4671). Neonates are 14-50% of adult mass (5857), have fur, and have open eyes (577, 2408, 3973, 4523a, 5432, 5433, 7353, 9615, 10661, 10663, 10664). Twinning is rare (595). *Diphylla* females have 2 mammae (4494). Males have a multipartite sex-gland complex whose components can be resolved histologically (6065, 6072, 9178).

Details of female anatomy (2374, 4957, 8941, 8942), endocrinology (6072), sperm (3413), mating (4048a), birth (902, 903, 7353, 10664), milk composition (5087, 5296, 10524), and female reproductive cycles (11754) are available.

NEONATAL MASS		
Anoura cultrata	3.35 g; r3.0-3.7; n2 aborted term emb	6372a
Anoura geoffroyi	5.0 g; r4.8-5.2; n2 largest emb	3900a
	5.2 g; r4.5-5.7; n3	3900a
Artibeus jamaicensis	5.9 g; n1; with placenta	903
Artibeus lituratus	7.8 g; n1 smallest young	10658
	8.9 g; n1	10658
	10.7 g; n1	10663
	12.2 g; n1; est r7.8-15.5 g; term emb, 1st observed yg	10661
Brachyphylla nana	9.9 g; largest emb	10594
	15.1 g; smallest yg	10594
Carollia perspicillata	5.0 g; r4.1-5.9; n13	5857
Choeronycteris mexicana	4.4 g; r4.3-4.5; n2 emb	7612
Diphylla ecaudata	4.4 g; with some membranes	902
Desmodus rotundus	5-7 g	4049
	6.9 g; r6.7-7.1; n2 smallest yg	3900a
	11.5 g; n1	11813
Leptonycteris sanborni	7.1 g; r6.2-7.9; n4	4523a
	8.3 g; n1 largest emb	4523a
Micronycteris minuta	1.8 g; n1 smallest yg	3900a
Phyllostomus discolor	7 g; n1; may not be neonate	3900a
Stenoderma rufum	7 g	10664
Sturnira aratathomasi	13.6 g; n1 term emb	10785a
Sturnira tildae	2.2 g; n1 large emb	3900a
Vampyressa pusilla	3.2 g; n1 large emb	8412
Vampyrops lineatus	8.3 g; n1	11737
NEONATAL SIZE		
Anoura cultrata	CR: 20.6 mm; r20.0-21.3; n2 aborted term emb	6372a
Artibeus jamaicensis	HB: 31 mm; n1; FA: 15 mm; n1	903
Artibeus lituratus	HB: 58 mm; n1	10661
Choeronycteris mexicana	CR: 29 mm; n2 emb	7612
Stenoderma rufum	HB: 45 mm; n1; FA: 29.4 mm; n1	10664
Sturnira aratathomasi	FA: 23 mm; n1 term emb	10785a
Vampyrops lineatus	HB: 46.5 mm; n1	11737
WEANING MASS		
Carollia perspicillata	~10 g; 4 wk from graph	5857
	~15 g; 8 wk from graph	5857
Leptonycteris sanborni	~15 g; 1st flight	4523a
Uroderma bilobatum	8.15 g; r7.7-8.6; n2 nursing juveniles	2552

LITTER SIZE

Species	Value	Reference
Ametrida centurio	1 emb ea; n17 f	1635
Anoura caudifer	1	10618
	1; n1	10663
	1; n1	452
	1 emb ea; n8 f	3999a
Anoura cultrata	1 emb; n1 f	3608
	1 emb ea; n3 f	3999a
	1 emb ea; n4 f	6372a
	2 yg; n1 lt	6372a, 7745
Anoura geoffroyi	1 emb ea; n2 f	10974
	1 emb ea; n22 f	3999a
Ardops nichollsi	1 emb ea; n5 f	5424d
Artibeus anderseni	1 emb ea; n4 f	3999a
Artibeus aztecus	1 emb; n1 f	523a
	1 emb ea; n>2 f	2342a
	1 emb ea; n17 f	5405d
Artibeus cinereus	1	1346
	1	10785
	1 emb; n1 f	5405e
	1 emb; n1 f	5425
	1 emb ea; n7 f	1475
	1 emb ea; n7 f	1635
	1 emb ea; n21 f	3999a
Artibeus fraterculus	1 emb ea; n46 f	3999a
Artibeus fuliginosus	1 emb ea; n23 f	3999a
Artibeus hartii	1	10785
	1 emb; n1 f	6258
	1 emb ea; n3 f	3608
	1 emb ea; n6 f	3999a
Artibeus hirsutus	1 emb; n1 f	280a
	1 emb; n1 f	3701
	1 emb ea; n4 f	1937
	1 emb ea; n8 f	11471a
	1 emb ea; n15 f	11938
Artibeus jamaicensis	1 emb	6307
	1; n1 lt	11859
	1 emb; n1 f	1162
	1 emb; n1 f	3701
	1 emb; n1 f	4671
	1 emb; n1 f	9113
	1 yg; n1 f	5411a
	1 emb ea; n3 f	1937
	1 emb ea; n5 lt	5405e
	1 yg; n5 lt	7978
	1 emb ea; n6 f	2390
	1 emb ea; n7 f	950
	1 emb ea; n8 f	1635
	1 emb ea; n9 f	11471a
	1 ea; n10	452
	1 emb ea; n13 f	5873
	1 emb ea; n16 f	1475
	1.02 emb; n195 f	595
Artibeus lituratus	1	10618
	1 emb; n1 f	1937
	1 emb; n1 f	4671
	1 emb ea; n2 f	1635
	1 emb ea; n2 f	2390
	1 yg ea; n2 f	8953
	1 emb ea; n3 f	3701
	1 emb ea; n5 f	5405e
	1 emb ea; n8 f	11471a
	1 emb ea; n10 f	217
	1 emb ea; n36 f	3999a

	1 emb ea; n40 f	10657, 10658, 10661, 10660, 10663
	1.4 cl or mature follicles; n49 f	10661
Artibeus phaeotis	1; n1 lt	452
	1 emb; n1 f	5405e
	1 emb; n1 f	7624
	1 emb ea; n3 f	9113
	1 emb ea; n18 f	11471a
Artibeus planirostris	1	10618
	1 emb ea; n4 f	10112a
	1 emb ea; n49 f	3999a
Artibeus toltecus	1 emb; n1 f	3701
	1 emb; n1 f	2390
	1 emb ea; n4 f	2345
	1 emb ea; n≥8 f	2342a
	1 emb ea; n33 f	11471a
Brachyphylla nana	1 emb; n1 f	10594
	1 emb ea; n12 f	1429c
Carollia brevicauda	1 emb; n1 f	11526
	1 emb ea; n28 f	3999a
Carollia castanea	1	10785
	1 emb ea; n2 f	5425
	1 emb ea; n3 f	11526
	1 emb ea; n24 f	3999a
Carollia perspicillata	1	3024
	1	6123
	1	10618
	1 emb; n1 f	1475
	1 yg ea; n2 lt	10660
	1 emb ea; n3 f	10974
	1 emb ea; n3 f	5405e
	1 ea; n13 lt	452
	1.02 emb; r1-2; n57 f	3999a
	1.05 emb; r1-2; n22 f	1635
	1.51 emb; r1-2; n11 f	181
Carollia subrufa	1 emb; n1 f	5405e
	1 emb; n1 f	6639
	1 emb ea; n49 f	3206
Centurio senex	1 yg; n1 f	3900a
	1 emb ea; n2 f	2345
	1 emb ea; n2 f	11471a
	1 emb ea; n3 f	5425
Chiroderma doriae	1	10618
Chiroderma salvini	1 emb; n1 f	1641
	1 emb ea; n2 f	6258
	1 emb ea; n7 f	11471a
Chiroderma trinitatum	1 emb; n1 f	3900a
	1 emb ea; n2 f	1635
	1 emb ea; n4 f	3999a
Chiroderma villosum	1; n1 lt	1635
	1 emb ea; n5 f	2345
	1 emb ea; n7 f	3999a
Choeroniscus godmani	1 emb; n1 f	5405d
	1 emb; n1 f	5425
Choeroniscus minor	1 emb ea; n2 f	3999a
Choeronycteris mexicana	1	1474
	1 yg	577
	1 emb; n1 f	522
	1 emb; n1 f	11471a
Chrotopterus auritus	1 emb; n1 f	4228
	1 yg; n1 f	10618
Desmodus rotundus	1	1346
	1 yg	2229
	1; n1 lt	9614

	1 emb; n1 f	2064
	1 yg; n1 lt	7353
	1 emb ea; n2 f	950
	1 emb ea; n4 f	280a
	1 emb ea; n4 f	5405e
	1 ea; n5 f	452
	1 emb ea; n10 f	3207
	1 emb ea; n16 f	1937
	1 emb ea; n31 lt	3999a
	mode 1 cl; sometimes 2	11813
	twins prob born premature	1463
Diphylla ecaudata	1	1346
	1 yg	2229
	1 emb; n1 f	3999a
	1 emb ea; n3 f	2228a
Ectophylla alba	1 emb; n1 f	3608
	1 emb ea; n2 f	1652
Ectophylla macconnelli	1 emb; n1 f	10974
	1 emb ea; n6 f	3999a
	1 emb ea; n7 f	1635
Erophylla sezekorni	1 emb; n1 f	11055a
	1 yg; n1 f	983a, 11754
	1.1 emb; r1-2; n10 f	595
Glossophaga commissarisi	1 emb; n1 f	5405e
	1 emb ea; n4 f	11471a
Glossophaga leachii	1 emb; n1 f	8285a
	1 emb ea; n5 f	4228
	1 emb ea; n≥5 f	11471a
	1 emb ea; n16 f	1937
Glossophaga longirostris	1 emb ea; n4 f	3900a
	1 emb ea; n8 f	10112a
Glossophaga mexicana	1 yg	11527
Glossophaga soricina	1	10618
	1	10785
	1; n1 f	452
	1 emb; n1 f	10974
	1 yg; n1 lt	1475
	1 emb ea; n3 f	5405e
	1 emb ea; n16 f	1937
	1 yg ea; n21 lt	10660
	1 emb ea; n38 f	3999a
	1.002; r1-2; n614 lt	595
Hylonycteris underwoodi	1 yg	5424
	1 emb ea; n3 f	8483a
	1 emb ea; n3 f	11471a
Leptonycteris curasoae	1 emb ea; n7 f	10112a
Leptonycteris nivalis	1-2	1474
	1-2 yg	577
	1 emb ea; n12 f	217
Leptonycteris sanborni	1 yg	577
	1 emb or yg; n97+ lt	4523a
Lichonycteris obscura	1 emb; n1 f	3608
	1 emb ea; n2 f	1641
Lionycteris spurrelli	1 emb; n1 f	10974
	1 emb ea; n2 f	3999a
Lonchophylla mordax	1 emb; n1 f	3608
Lonchophylla robusta	1 emb; n1 f	3999a
	1 emb; n1 f	10657
Lonchophylla thomasi	1 emb ea; n2 f	3999a
Lonchorhina aurita	1 emb ea; n2 f	3999a
Macrophyllum macrophyllum	1 yg	4401
Macrotus californicus	1	1474
	1	1199
	1 yg	1346

	1 yg	577
	1 emb ea; n4 f	1471a
	1 emb ea; n61 f	4079
	1.3 emb; r1-2; n3 f	1937
Macrotus waterhousii	1 emb ea; n5 f	1937
	1 emb ea; n5 f	5424e
	1 emb ea; n5 f	5873
	1 emb ea; n5 f	11471a
Micronycteris brachyotis	1 emb; n1 f	7154a
	1 emb; n1 f	9113
Micronycteris daviesi	1 emb; n1 f	10974
	1 emb; n1 f	3999a
Micronycteris hirsuta	1 emb; n1 f	10974
	1 emb; n1 f	3999a
	1 yg; n1 f	3900a
Micronycteris megalotis	1 emb; n1 f	4671
	1 emb; n1 f	1475
	1 emb ea; n2 f	10974
	1 emb ea; n2 f	5425
	1 emb ea; n8 f	3999a
Micronycteris minuta	1 emb; n1 f	3608
	1 ea; n5 lt	452
Micronycteris nicefori	1; n1 lt	452
Mimon cozumelae	1 emb; n1 f	9113
	1 emb; n1 f	1641
	1 emb ea; n2 f	4228
Mimon crenulatum	1; n1 lt	452
	1 emb; n1 f	3705c
	1 emb; n1 f	5405d
	1 emb ea; n2 f	3608
	1 emb ea; n2 f	10974
	1 emb ea; n7 f	3999a
Monophyllus redmani	est 1	3902
	1 emb ea; n2 f	5873
	1 emb ea; n4 f	4941
Monophyllus plethodon	1 emb	9685a
Phylloderma stenops	1 emb; n1 f	6260
Phyllonycteris aphylla	1 emb; n1 f	3902
Phyllonycteris poeyi	1 emb ea; n3 f	5873
	1 emb ea; n many	7328a
Phyllops haitiensis	1 emb ea; n19 f	5873
Phyllostomus discolor	1	10618
	1 emb; n1 f	10112a
	1 emb ea; n2 f	3999a
	1 emb ea; n2 f	5405c
	1 yg ea; n5 lt	10660
	1 yg ea; n6 f	8953
Phyllostomus elongatus	1 ea; n2 lt	452
	1 emb ea; n5 f	10974
	1 emb ea; n15 f	3999a
Phyllostomus hastatus	1	10785
	1 emb ea; n2 f	5425
	1 emb ea; n12 f	10974
	1 emb ea; n31 lt	3999a
Platalina genovensium	1 emb; n1 f	3999a
Pygoderma bilabiatum	1 emb ea; n>60 f	7696
Rhinophylla fischerae	1 emb ea; n3 f	3999a
Rhinophylla pumilio	1 emb ea; n9 f	3999a
Stenoderma rufum	1; n1 lt	10664
	1 emb ea; n6 f	5423a
Sturnira aratathomasi	1 emb ea; n2 f	10785a
Sturnira bidens	1 emb ea; n3 f	3608a
	1 emb ea; n15 f	3999a
Sturnira bogotensis	1 emb ea; n22 f	3999a

Sturnira erythromos	1 emb; n1 f	6372
	1 emb ea; n47 lt	3999a
Sturnira lilium	1	10618
	1 emb; n1 f	11938
	1 emb; n1 f	9113
	1 emb ea; n2 f	134
	1 emb ea; n2 f	950
	1 emb ea; n2 f	10974
	1 emb ea; n3 f	3701
	1 emb ea; n3 f	1635
	1 ea; n5 lt	452
	1 emb ea; n15 f	5405e
	1 emb ea; n32 f	3999a
Sturnira ludovici	1	10785
	1 emb ea; n5 f	5424c
	1 emb ea; n8 f	11471a
Sturnira magna	1 emb ea; n6 f	3999a
Sturnira mordax	1 emb ea; n2 f	3608
Sturnira nana	1 emb; n1 f	3999a
Sturnira tildae	1 emb; n1 f	3900a
	1 emb ea; n4 f	3999a
	1 emb ea; n5 f	1635
Tonatia bidens	1 emb; n1 f	3608
	1 emb ea; n2 f	3999a
Tonatia brasiliense	1 emb; n1 f	6258
	1 emb; n1 f	2345, 11055
Tonatia silvicola	1 emb ea; n4 f	10974
	1 emb ea; n10 f	3999a
Trachops cirrhosus	1 emb; n1 f	10974
	1 emb; n1 f	3999a
	1 emb ea; n3 f	1475
	1 emb ea; n5 f	1641
	1 emb ea; n10 f	5405e
Uroderma bilobatum	1 emb; n1 f	5405e
	1 emb; n1 f	10974
	1 emb ea; n4 f	1475
	1 emb ea; n4 f	3207
	1 emb ea; n4 f	5405c
	1 yg ea; n6 lt	10660
	1 emb ea; n8 f	2342
	1 emb ea; n28+ f	3999a
Uroderma magnirostrum	1 emb; n1 f	5425
Vampyressa bidens	1 emb; n1 f	3705a
	1 emb ea; n2 f	2343a
	1 emb ea; n27 f	3999a
Vampyressa melissa	1 emb ea; n4 f	3999a
Vampyressa nymphaea	1	10785
	1 emb ea; n2 f	3608
Vampyressa pusilla	1 emb; n1 f	8412
	1 emb; n1 f	2345
	1 emb; n1 f	9113
	1 emb; n1 f	11055
	1 emb; n1 f	5425
	1 emb ea; n15 f	3999a
Vampyrodes caraccioli	1 emb; n1 f	2345
	1 emb; n1 f	5405d
	1 emb; n1 f	10974
	1 emb ea; n3 f	5425
	1 emb ea; n4 f	3999a
Vampyrops brachycephalus	1 emb ea; n3 f	9318a
	1 emb ea; n12 f	3999a
Vampyrops dorsalis	1	10785
	1 emb ea; n33 f	3999a

Vampyrops helleri	1	10785
	1 emb; n1 f	2345
	1; n3 lt	452
	1 emb ea; n3 f	6258
	1 emb ea; n3 f	10974
	1 emb ea; n14 f	1635
	1 emb ea; n15 f	3999a
Vampyrops infuscus	1 emb ea; n6 f	3999a
Vampyrops lineatus	1	10618
	1 emb ea; n25 f	3999a
Vampyrops vittatus	1 emb; n1 f	3608
	1 emb ea; n2 f	2345
	1 emb ea; n4 f	3999a
Vampyrum spectrum	1; n1 lt	2109
SEXUAL MATURITY		
Artibeus lituratus	f: 46.3 g	10661
Carollia perspicillata	f: 1st birth: 13.5 mo; r11-16; n2	8683
	f/m: 1st mate: 8-9 mo	8683
Macrotus californicus	f: 1st mate: 3-4 mo	1199
	f: 1st conceive: 4 mo, no emb dev until 9 mo	1199
	m: 1st mate: 15-16 mo	1199
CYCLE LENGTH		
Desmodus rotundus	estrus: 2-3 d	9615
Glossophaga soricina	24 d; mode 24; r22-26; n43 cycles, 39 f	8941, 8942
GESTATION		
Anoura caudifer	est > 3.5 mo	10618
Artibeus jamaicensis	2.5 mo delay; ≤4 mo no delay, 7 mo with delay	3318
Artibeus lituratus	est > 3.5 mo	10618
Artibeus planirostris	est > 3.5 mo	10618
Carollia perspicillata	1.5-4 mo	6123
	est > 3.5 mo	10618
Chiroderma doriae	est > 3.5 mo	10618
Chiroderma villosum	~4 mo	10618
Chrotopterus auritus	99 d; n1; capture to birth	10618
Desmodus rotundus	>5 mo; n1; capture to birth	11813
	205 d; n1	9614, 9615
Glossophaga soricina	implant: 12-14 d	8941, 8943
	est > 3.5 mo	10618
Macrotus californicus	diapause: ~4.5 mo	999
	9 mo; with delayed development	1461, 1462
	~8 mo; with delayed development; field data	1199
	8.5 mo	999
	9 mo; with delayed development	1038, 1039
Micronycteris megalotis	~4 mo	10618
Phyllostomus discolor	est > 3.5 mo	10618
Phyllostomus hastatus	~4 mo	10618
Sturnira lilium	est > 3.5 mo	10618
Uroderma bilobatum	est 4-5 mo	527
Vampyrops lineatus	est > 3.5 mo	10618
LACTATION		
Brachyphylla cavernarum	1st fly: ~2 mo	7828a
Carollia perspicillata	1st fly: 2.5-4 wk	8683, 8684
	thick milk: 33 d; r21-49	5857
	thinner milk: until ~56 d; r42-72	5857
Desmodus rotundus	solid food: 2 mo	4049
	solid food: 6 mo	9615
	wean: 9 mo	9615
	nurse: 300 d	4049
Glossophaga soricina	1st fly: 25-28 d; n1	5857
	wean: ~2 mo; n1	5857
Leptonycteris sanborni?	wean: 4-8 wk	5296
Macrotus californicus	nurse: 1 mo	1199
Uroderma bilobatum	independent: 1 mo	527

INTERLITTER INTERVAL		
Artibeus lituratus	220 d; r200-240; n2 f	2552
Carollia perspicillata	159 d; r119-417	8683
Desmodus rotundus	9-10 mo	11703
SEASONALITY		
Ametrida centurio	preg f: July-Aug; Trinidad	1635
Anoura caudifer	epididymal sperm: yr rd, peak Apr-June; n57; São Paulo, Brazil	10618
	preg f: Jan; n2; Brazil	8368
	preg f: Jan-Feb; French Guiana	1350a
	preg f: May, July-Sept, Nov; n8; Peru	3999a
	preg f: May, Nov; n2; Colombia	10785
	preg f: June; n1; Colombia	10663
	preg f: est yr rd; São Paulo, Brazil	10618
	lact f: Feb; n1; Brazil	6084a
Anoura cultrata	preg f: Apr-June; n3; Peru	3999a
	preg f: July-Aug; n5; Colombia	6372a, 7745
	preg f: Aug; n1; Costa Rica	3608
	lact f: July; n1; Colombia	6372a, 7745
	lact f: Aug; n1; Peru	1639b
	lact f: Aug; few data; Peru	7745
Anoura geoffroyi	preg f: Mar; n2; Costa Rica	6809
	preg f: May-Aug; n8; Peru	3999a, 10974
	preg f: July; n1; Nicaragua	5425
	preg f: Nov; n56; Trinidad	3900a
	lact f: July; n3; Hidalgo, Mexico	1640a
	lact f: Nov; n5; Colima, Mexico	11754
Ardops nichollsi	preg f: Mar-Apr; n5; Dominica, Lesser Antilles	5424d
	preg f: July; n2; Guadeloupe, Lesser Antilles	527a
	lact f: Apr; n1; Dominica, Lesser Antilles	5424d
	lact f: July; n2; Guadeloupe, Lesser Antilles	527a
Artibeus anderseni	preg f: June, Aug, Nov; n4; Peru	3999a
Artibeus aztecus	preg f: Mar-Apr; middle America	2342a
	preg f: July; n1; Durango, Mexico	523a
	preg f: July; n18; Sinaloa, Mexico	5405d, 5413a
	preg f: July-Aug; n2; Tamaulipas, Mexico	217
	lact f: Sept; n1; Mexico	11754
Artibeus cinereus	preg f: Jan; n7; El Salvador	1475
	preg f: Feb-Apr, July-Aug, Nov; middle America	2342b
	preg f: Feb, Aug; n2; Nicaragua	5425
	preg f: Mar; n1; Guatemala	5405e
	preg f: Apr-May, July-Aug, Oct-Jan; n20; Colombia	10785
	preg f: June-July; n2; Brazil	11754
	preg f: June-Sept, Nov; n21; Peru	3999a
	preg f: June, Dec-Feb; n4; Panama	3322
	preg f: July; n1; Venezuela	11754
	preg f: Aug; n1; Colombia	352a
	lact f: Jan, Mar, May, Aug-Sept; n5; Colombia	10785
	lact f: Apr, Aug; n4; Panama	3322
	lact f: July-Aug; n4; Venezuela	11754
	lact f: Aug; n5; Colombia	352a
Artibeus concolor	preg f: Feb; n1; Colombia	10785
Artibeus fraterculus	preg f: June, Aug-Dec; n46; Peru	3999a
Artibeus fuliginosus	preg f: June-Dec; n24; Peru	3999a
Artibeus hartii	preg f: Jan, May-June; n3; Costa Rica	3608
	preg f: Apr-May; n19; Colombia	10785
	preg f: Aug-Sept, Nov-Dec; n6; Peru	3999a
	lact f: Apr; n1; Michoacan, Mexico	4671
	lact f: May-July; Costa Rica	3608
	lact f: Aug; Costa Rica	366b
	lact f: July, Sept; n2; Colombia	10785
Artibeus hirsutus	preg f: Feb, June, Aug; n8; Jalisco, Mexico	11741a
	preg f: Apr-May; n3; Sonora, Mexico	1137, 1139
	preg f: May; n8; Sonora, Mexico	279a

	preg f: May; n2; Guerrero, Mexico	279a
	preg f: July; n1; Chihuahua, Mexico	280a
	preg f: Aug; n5; Sinaloa, Mexico	5413a
	preg f: Sept; n15; Sonora, Mexico	3254b
	lact f: June; n1; Sinaloa, Mexico	5413a
	lact f: June, Aug; Jalisco, Mexico	11741a
	lact f: Sept; n6; Sonora, Mexico	3254b
Artibeus inopinatus	preg f: May; Honduras	11524
	lact f: July; n1; Honduras	2670a
Artibeus jamaicensis	preg f: Jan; n4; Providencia, Caribbean Sea, Colombia	10654
	preg f: Jan-Feb, Apr-July; Sinaloa, Mexico	5413a
	preg f: Jan-June; Jalisco, Mexico	11471a
	preg f: Feb; n1; Guerrero, Mexico	11754
	preg f: Feb; Veracruz, Mexico	4228
	preg f: Feb; n1; Tobago, West Indies	5433a
	preg f: Feb; n6; Tobago, West Indies	3900a
	preg f: Feb-Mar, Sept, Nov-Dec; n8; Michoacan, Mexico	4671
	preg f: Feb-Mar; n5; Guatemala	5405e
	preg f: Feb-Mar, June; St Croix, Virgin I	16052a
	preg f: Feb, Apr, July, peak July; Yucatan, Mexico	5424f
	preg f: end Feb-early Mar; n5; Jamaica	7103a
	preg f: Feb-May, July; Trinidad	3900a
	preg f: Mar; n6; Tamaulipas, Mexico	2390
	preg f: Mar-Apr; n7; Yucatan, Mexico	950
	preg f: Mar, May-Sept, Nov-Jan; n34; Colombia	10785
	preg f: Apr; n1; Yucatan, Mexico	1162
	preg f: Apr, July; Virgin I	595
	preg f: May; n1; Tamaulipas, Mexico	217
	preg f: May; n1; Guatemala	9113
	preg f: June; n3; San Luis Potosi, Mexico	1937
	preg f: June; n3; Sinaloa, Mexico	11754
	preg f: June; Puerto Rico	344a
	preg f: June; Trinidad	5433
	preg f: June; n1; Colombia	10657
	preg f: June, Aug, Dec-Jan; n13; Haiti	5873
	preg f: June or July; n3; Venezuela	8562a
	preg f: July; Hidalgo, Mexico	1640a
	preg f: July; n4; Morelos, Mexico	7978
	preg f: July; n1; Lesser Antilles	5411a
	preg f: July; n9; Guadeloupe, Lesser Antilles	527a
	preg f: July; n1; Colombia	352a
	preg f: Aug; n1; Veracruz, Mexico	595
	preg f: Nov; n1; Paraguay	7699a
	preg f: Nov-July, Sept; n42; Costa Rica	3322
	preg f: Dec; n16; El Salvador	1475
	preg f: Dec; n6; Jamaica	3902
	preg f: Dec-July; n138; Panama	3322
	preg f: end wet season; Brazil	11736
	preg f: yr rd, peak Jan-Mar, low Sept-Nov; Panama	1070
	preg/lact f: Apr-May 10% n289; July-Aug 9% n243; Sept 0% n152; Guatemala	2552
	birth: Mar, July; Guana, Virgin Islands	6307
	birth: peaks Mar-Apr, July-Aug; Panama	3318
	birth: July; n1; Lesser Antilles	5411a
	lact f: Jan-Apr, June-Aug, Oct; n29; Colombia	10785
	lact f: Feb-May, July, Dec; n48; Costa Rica	3322
	lact f: May, July-Aug; Yucatan, Mexico	5424f
	lact f: Mar-Apr; St Croix, Virgin I	16052a
	lact f: Mar-Oct; n134; Panama	3322
	lact f: Mar, Nov; Paraguay	7699a
	lact f: Apr-May, July-Oct; Jalisco, Mexico	11471a
	lact f: Apr, June-July, Sept; Trinidad	3900a
	lact f: June or July; n2; Venezuela	8562a
	lact f: July; n1; Guadeloupe, Lesser Antilles	527a

	lact f: July; n7; Lesser Antilles	5411a
	lact f: July-Aug; n4; Colombia	352a
	lact f: Sept, Nov; Sinaloa, Mexico	5413a
Artibeus lituratus	spermatogenesis: yr rd; Melgar, Colombia	10658, 10662
	preg f: Jan; n1; Querétaro, Mexico	10261a
	preg f: Jan-Feb, Apr, July; n6; Yucatan, Mexico	5424f
	preg f: Jan-Feb, May, July; n5; Costa Rica	3322
	preg f: Jan, Mar-May, Aug; n14; Panama	3322
	preg f: Jan, Mar-June, Aug, Oct-Nov; n31; Colombia	10661
	preg f: Feb; n1; Tobago	3900a
	preg f: Feb-Mar; n5; Guatemala	5405e
	preg f: Feb, Apr, June-July; Sinaloa, Mexico	5413a
	preg f: Feb-May, Nov; n9; Michoacan, Mexico	4671
	preg f: Mar; n2; Tamaulipas, Mexico	2390
	preg f: Mar; n3; Panama	1006
	preg f: Mar-Apr, June-July; n8; Jalisco, Mexico	11471a
	preg f: Mar-July; Trinidad	3900a
	preg f: Apr; n5; Oaxaca, Mexico	11754
	preg f: Apr, July-Aug, Oct-Dec; n16; Peru	3999a
	preg f: May; n10; Tamaulipas, Mexico	217
	preg f: May; n1; Morelos, Mexico	1937
	preg f: May; n2; Panama	4228a
	preg f: July; n5; Colombia	352a
	preg f: July; n1; Margarita I, Venezuela	10112a
	preg f: July; n2; Brazil	11754
	preg f: July-Aug; Brazil	8368
	preg f: Aug; n1; Venezuela	11754
	preg f: Aug-Sept; French Guiana	1350a
	preg f: few end wet season; Brazil	11736
	preg f: yr rd, low Aug-Sept; n183; Colombia	10785
	preg f: yr rd; Melgar, Colombia	10658
	preg f: est yr rd; São Paulo, Brazil	10618
	preg/lact f: Jan-June, Sept-Nov; Panama	1070
	preg/lact f: July-Aug; Hidalgo, Mexico	1640a
	preg/lact f: July-Aug 8.4% n107; Sept 8.5% n47; Apr-May 17.5% n143; Guatemala	2552
	lact f: Mar-Apr, Sept-Oct; n11; Panama	3322
	lact f: Mar, May-July, Oct-Nov; n11; Colombia	10661
	lact f: Apr-May, July; n5; Costa Rica	3322
	lact f: Apr-May, Oct; Trinidad	3900a
	lact f: May; n2; Guatemala	9113
	lact f: June; n2; Durango, Mexico	5405b
	lact f: July; n10; Colombia	352a
	lact f: July-Oct; Jalisco, Mexico	11471a
	lact f: July, Oct; Sinaloa, Mexico	5413a
	lact f: Sept; n1; Tobago	3900a
	lact f: yr rd; n100; Colombia	10785
Artibeus phaeotis	preg f: Jan-Mar; n10; Campeche, Mexico	5424e
	preg f: Jan-Apr, June, Aug; n15; Panama	3322
	preg f: Jan-Apr, June-Aug; middle America	2342b
	preg f: Jan-June, none Sept-Dec; Panama	1070
	preg f: Jan, Apr, June; n18; Jalisco, Mexico	11471a
	preg f: Mar; n1; Guatemala	5405e
	preg f: Mar-May, Dec; n13; Michoacan, Mexico	4671
	preg f: Apr; n1; Guatemala	7624
	preg f: Apr, Aug-Sept; s Mexico	11754
	preg f: Apr, Aug-Sept; La Selva, Costa Rica	6261
	preg f: May; n2; Guatemala	9113
	preg f: July; n4; Sinaloa, Mexico	5413a
	preg f: Aug; n2; Cozumel I, Mexico	5424f

	lact f: Mar; n1; Campeche, Mexico	5424f
	lact f: Apr; n1; Quintana Roo, Mexico	950
	lact f: peak May-Sept; Barro Colorado I, Panama	1070
	lact f: June, Oct; s Mexico	11754
	lact f: Aug; n1; Jalisco, Mexico	11471a
Artibeus planirostris	epididymal sperm: yr rd; n35; São Paulo, Brazil	10618
	preg f: Feb; n6; Tobago, W Indies	5433a
	preg f: Apr, June-Sept, Nov-Jan; n49; Peru	3999a
	preg f: July; n4; Margarita I, Venezuela	10112a
	preg f: est yr rd; São Paulo, Brazil	10618
	lact f: July; n2; Margarita I, Venezuela	10112a
Artibeus toltecus	preg f: Jan; n2; Puebla, Mexico	6258a
	preg f: Jan; n7; El Salvador	1475
	preg f: Jan-Apr, June-July; n33; Jalisco, Mexico	11471a
	preg f: Jan, May, Oct; Sinaloa, Mexico	5413a
	preg f: Jan-Aug; middle America	2342a
	preg f: Mar, Sept; n3; Michoacan, Mexico	4671
	preg f: Apr, June; n9; Nicaragua	5425
	preg f: Apr, Aug; Monte Verde, Costa Rica	6261
	preg f: end June-early July; n4; Chiapas, Mexico	2345
	preg f: July; n1; Tamaulipas, Mexico	2390
	preg f: July; n4; Hidalgo, Mexico	1640a
	lact f: Apr-Nov, peak Aug-Oct; Costa Rica	2587
	lact f: May; Sinaloa, Mexico	5413a
	lact f: May, Aug; n2; Chiapas, Mexico	2345
	lact f: June-Sept; Jalisco, Mexico	11471a
Brachyphylla cavernarum	preg f: Feb; Puerto Rico	7983, 11754
	preg f: Mar; St Croix, Virgin I	7983, 11754
	birth: end May-early June; Virgin I	7828a
	lact f: Apr; n1; Puerto Rico	7983, 11754
	lact f: Apr; n1; St Croix, Virgin I	7827a
	lact f: July; Puerto Rico	344a
	lact f: July; n3; Guadeloupe, Lesser Antilles	527a
	lact f: July; n4; Lesser Antilles	5411a
Brachyphylla nana	preg f: Mar; n12; Caicos I, Greater Antilles	1429c
	preg f: Dec-May; Cuba	10594
	lact f: Aug; n1; Haiti	5873
Carollia brevicauda	preg f: Jan-Mar; n20; Panama	8551
	preg f: Feb; n1; Guatemala	9113
	preg f: Feb-Mar; n5; Guatemala	8551
	preg f: Mar; Veracruz, Mexico	4228
	preg f: Mar; n1; Ecuador	8551
	preg f: Mar; n2; Oaxaca, Mexico	8551
	preg f: Mar; n18; Veracruz, Mexico	8551
	preg f: Mar-Apr; n4; Costa Rica	8551
	preg f: Mar-Apr, Aug; La Selva, Costa Rica	6261
	preg f: Apr; n1; Quintana Roo, Mexico	8551
	preg f: Apr; n1; Yucatan peninsula, Mexico	5424f
	preg f: Apr; n2; San Luis Potasi, Mexico	8551
	preg f: May-June; n6; Honduras	8551
	preg f: June-July; n7; Chiapas, Mexico	8551
	preg f: July; n1; Hidalgo, Mexcio	1640a
	preg f: July; n1; Nicaragua	8551
	preg f: Aug; n1; Guatemala	5405e
	preg f: Aug-Nov; n28; Peru	3999a
	preg f: Oct; n1; Peru	8551
	lact f: Feb; n1; Panama	8551
	lact f: Mar-Apr; n2; Costa Rica	8551
	lact f: Apr-May; n3; Tabasco, Mexico	8551
	lact f: Apr-June; n9; Honduras	8551
	lact f: July; n1; Nicaragua	8551
Carollia castanea	preg f: Jan-Mar; n15; Panama	5425
	preg f: Jan-Mar; French Guiana	1350a
	preg f: Jan, Mar-Apr, July-Aug; n10; Panama	3322

	preg f: Jan, Mar-Apr, Sept-Nov; n14; Colombia	10785
	preg f: end Jan-mid Mar; n15; Panama	8551
	preg f: Feb-Mar; n4; Costa Rica	8551
	preg f: Feb-Mar, Aug; n7; Costa Rica	5425
	preg f: Mar, July; Nicaragua	5425
	preg f: Apr, Aug; La Selva, Costa Rica	6261
	preg f: May, July; n4; Honduras	8551
	preg f: June-Sept, Nov-Dec; n24; Peru	3999a
	lact f: Jan, Apr-May; n5; Colombia	10785
	lact f: June; n1; Panama	3322
	lact f: Aug; n1: Costa Rica	8551
Carollia perspicillata	epididymal sperm: yr rd; n28; São Paulo, Brazil	10618
	m repro: May, Oct-Nov; Rio de Janeiro, Brazil	8368
	mate: Nov, Jan; El Salvador	3206
	preg f: Jan-Apr, June-Aug; n32; Costa Rica	3322
	preg f: Jan, Mar; n2; Colombia	10660
	preg f: Jan, Aug; Amazon Valley	181
	preg f: Jan, Aug; n7; Brazil	8551
	preg f: Jan, Sept-Oct; Rio de Janeiro, Brazil	8368
	preg f: Feb-Apr, June; n13; Panama	8551
	preg f: Feb-Apr, July-Aug; n14; Costa Rica	8551
	preg f: Feb-Aug, Oct; Trinidad	3900a
	preg f: Feb, Apr-Aug; n16; Nicaragua	8551
	preg f: few Feb-June; Brazil	11736
	preg f: Feb, Aug; La Selva, Costa Rica	6261
	preg f: Mar; n1; Ecuador	8551
	preg f: Mar; n1; El Salvador	1475
	preg f: Mar; n3; Guatemala	5405e
	preg f: Mar; n2; Guatemala, El Salvador	8551
	preg f: Mar; n1; Panama	3024
	preg f: Apr; Tuxtepec, Mexico	4201a
	preg f: May; n2; Campeche, Mexico	5424f
	preg f: May; n3; Veracruz, Mexico	11754
	preg f: May, July; n4; Honduras	8551
	preg f: June; Trinidad	4201a
	preg f: June-July; n2; Veracruz, Mexico	6148a
	preg f: June-July; n5; Colombia	8551
	preg f: June-July; n7; Venezuela	8562a
	preg f: July-Aug; n5; Quintana Roo, Mexico	8551
	preg f: July-Aug; n6; Quintana Roo, Mexico	5424f
	preg f: July-Sept; n55; Colombia	352a
	preg f: July-Nov; French Guiana	1350a
	preg f: July-Dec; n57; Peru	3999a, 10974
	preg f: Aug; n1; Peru	8551
	preg f: Aug-Sept; n2; Bolivia	8551
	preg f: Oct-Jan, Mar-Aug; n29; Colombia	10785
	preg f: Dec-Sept; n98; Panama	3322
	preg f: est yr rd; São Paulo, Brazil	10618
	preg/lact f: Jan-June, few Sept-Nov; Barro Colorado I, Panama	1070
	birth: Mar-Apr, July-Aug; captive	8684
	birth: Mar-Apr, Aug-Sept; Costa Rica	11713
	birth: mid Mar-mid May, mid Aug-mid Nov; zoo	6123
	birth: yr rd, peaks Mar-Apr, Aug-Sept; captive	8683
	lact f: Jan-June, Sept, Nov; n15; Colombia	10785
	lact f: Feb-Apr, Aug; n8; Costa Rica	8551
	lact f: Mar-Apr; n2; Panama	8551
	lact f: Mar-Oct; n47; Panama	3322
	lact f: Apr; n1; El Salvador	3206
	lact f: Apr-May, Aug; n26; Costa Rica	3322
	lact f: Apr, Oct; n2; Colombia	10660
	lact f: May; n1; Nicaragua	8551
	lact f: May, July; n7; Honduras	8551
	lact f: May-Oct; Trinidad	3900a
	lact f: June or July; n1; Venezuela	8562a

Species	Data	Reference
	lact f: Aug; n1; Chiapas, Mexico	8551
	lact f: Aug; n16; Colombia	
	repro: Jan-Aug, Oct-Nov; no data Dec; sw Colombia	10785
Carollia subrufa	preg f: Jan-Mar, Oct; n51; El Salvador	3206
	preg f: Feb; n6; Chiapas, Mexico	11754
	preg f: Feb; n6; Guatemala	8551
	preg f: May; n2; Oaxaca, Mexico	11754
	preg f: July-Aug; n7; Chiapas, Mexico	8551
	preg f: July-Aug; n2; Nicaragua	8551
	preg f: Aug; n1; Guatemala	5405e
	preg f: Dec; n1; Guerrero, Mexico	6639
	preg f: Dec-Mar; Panama	7893, 11754
	lact f: May, Oct; n2; Chiapas, Mexico	8551
	lact f: Nov; n2; Guatemala	8551
Centerio senex	preg f: Feb-Mar; n3; Nicaragua	5425
	preg f: Feb, July; n2; Yucatan, Mexico	5424f
	preg f: Mar; n2; Jalisco, Mexico	11471a
	preg f: Mar; n2; Oaxaca	11754
	preg f: Apr; n1; Michoacan, Mexico	4671
	preg f: Apr, July; n2; Chiapas, Mexico	2345
	preg f: June; n1; Tamaulipas, Mexico	217
	preg f: Aug; n2; Honduras	6258
	birth: Jan; n1; Trinidad	3900a
	lact f: Mar; n1; Oaxaca, Mexico	11754
	lact f: Mar, Aug; n2; Jalisco, Mexico	11471a
	lact f: Apr; n1; Michoacan, Mexico	4671
	lact f: Apr; n5; Veracruz, Mexico	11754
	lact f: July; n2; Hidalgo, Mexico	1640a
Chiroderma doriae	epididymal sperm: Aug; n9; São Paulo, Brazil	10618
Chiroderma villosum	preg f: Jan; n1; Colombia	10785
	preg f: Mar, July; n8; Nicaragua	5425
	preg f: Apr; n1; Panama	3322
	preg f: Apr-May, Aug; Guatemala	2552
	preg f: May; n1; Nicaragua	4048,
	preg f: July, Dec; n5; Chiapas, Mexico	2345
	preg f: Aug-Sept; n2; Trinidad	3900a
	lact f: Mar-Apr; n4; Panama	3322
	lact f: May; n1; Chiapas, Mexico	2345
Chiroderma salvini	preg f: Jan; n1; Sinaloa, Mexico	5413a
	preg f: Feb, June; n7; Jalisco, Mexico	11471a
	preg f: Mar-June, Oct, Dec-Jan; n11; Colombia	10785
	preg f: July; n1; Honduras	1641
	preg f: July-Aug; n2; Honduras	6258
	lact f: Mar-Apr, July; n6; Colombia	10785
	lact f: July-Aug; n2; Honduras	6258
Chiroderma trinitatum	preg f: Feb, May; n3; Panama	3322
	preg f: Mar; n1; Trinidad	3900a
	preg f: June-July; n2; Brazil	11754
	preg f: July; n1; Colombia	10785
	preg f: Oct-Nov; Peru	2345a
	preg f: Nov; n4; Peru	3999a
	lact f: May, Sept; n2; Panama	3322
Chiroderma villosum	preg f: June-Dec; n7; Peru	3999a
Choeroniscus godmani	preg f: Mar; n1; Nicaragua	5425
	preg f: Mar; n1; Costa Rica	6809
	preg f: July; n1; Sinaloa, Mexico	5405d
	lact f: Nov/Dec; n1; Oaxaca, Mexico	9542a
Choeroniscus intermedius	preg f: Aug; n1; Trinidad	3900a
Choeroniscus minor	preg f: Feb; n1; British Guiana	3900a
	preg f: June; n2; Peru	3999a
	lact f: Dec; n1; Colombia	10664a
Choeronycteris mexicana	preg f: Feb; n1; Sinaloa, Mexico	5413a
	preg f: Feb; n2; Guerrero, Mexico	11754
	preg f: Mar; n1; Coahuila, Mexico	522

	preg f: June; n2; NM	7612
	preg f: Sept; n1; Jalisco, Mexico	11471a
	lact f: June; n4; NM	7611, 7612
	lact f: June; n4; Coahuila, Mexico	522
	lact f: July-Aug; AZ	1568a
	lact f: Aug; Tamaulipas, Mexico	217
	repro: June-July; sw US	577, 1474
Chrotopterus auritus	preg f: Apr; n1; Veracruz, Mexico	4228
	preg f: Apr; n1; Yucatan, Mexico	5424f
	perg f: July; n1; Argentina	11754
	lact f: July; n1; Qunitana Roo, Mexico	5424f
Desmodus rotundus	sexually active: yr rd; Panama	11813
	mate: Mar, Aug-Dec; El Salvador	3207
	mate: mid Sept, mid Apr; lab	9614, 9615
	preg f: Jan; n1; Nayarit, Mexico	1937
	preg f: Jan; n1; Mexico, Mexico	1463
	preg f: Jan; n1; Morelos, Mexico	1463
	preg f: Jan; n2; Puebla, Mexico	6258a
	preg f: Jan; Brazil	8368
	preg f: Jan-Apr, June-July; n10; Yucatan, Mexico	5424f
	preg f: Jan-May, July, Nov; n17; Costa Rica	3322
	preg f: Jan-July, Sept, peak Jan-May; Jalisco, Mexico	11471a
	preg f: Feb; Veracruz, Mexico	4228
	preg f: Feb, Apr-May, July, Nov; n15; Panama	11813
	preg f: Mar, Aug; n4; Guatemala	5405e
	preg f: Mar, May, June; n4; Tamaulipas, Mexico	217
	preg f: Mar, May, July; n~90; Colima, Mexico	1465a
	preg f: Mar, May, July-Aug, Oct; n10; El Salvador	3207
	preg f: Mar, May, Dec-Jan; n4; Sinaloa, Mexico	5413a
	preg f: Apr; n2; Yucatan, Mexico	950
	preg f: Apr-May; n3; Panama	3322
	preg f: May; n4; Chihuahua, Mexico	280a
	preg f: May, Oct; n2; Colombia	10785
	preg f: June, Aug-Sept, Nov; n41; Guerrero, Mexico	3414a
	preg f: June-July; n5; Veracruz, Mexico	6148a
	preg f: June-Nov; n31; Peru	3999a
	preg f: July; n1; Hidalgo, Mexico	1640a
	preg f: July; n1; Michoacan, Mexico	4229a
	preg f: July; n3; Colombia	352a
	preg f: Oct; n1; Zacatecas, Mexico	1937
	preg f: Nov; Colombia	10657
	preg f: Dec; Querétaro, Mexico	9603a
	preg f: yr rd; Panama	11813
	preg f: yr rd; Trinidad	3900a
	preg f: yr rd; Brazil	11736
	preg/lact f: Feb-Sept; Veracruz, Mexico	4228
	birth: Dec; n1; captive	2109
	lact f: Jan; Brazil	8368
	lact f: Feb-Mar; n3; Costa Rica	3322
	lact f: Apr-May; n2; Panama	3322
	lact f: Apr, July-Aug; Yucatan, Mexico	5424f
	lact f: Apr, Oct; n2; Colombia	10785
	lact f: July; n1; Hidalgo, Mexico	1640a
	lact f: July; n3; Colombia	352a
	lact f: Aug; Jalisco, Mexico	11471a
	lact f: Nov; Panama	11813
	repro: yr rd: n Argentina	11304
Desmodus youngi	preg f: June; n1; Peru	3999a
	preg f: Oct; n4; Trinidad	3900a
	lact f: Aug, Oct; n3; Trinidad	3900a
Diphylla ecaudata	preg f: Mar; San Luis Potosí, Mexico	2228a
	preg f: Aug; n2; Honduras	11055
	preg f: Aug, Oct-Nov; n10; Mexico, Mexico	11754
	preg f: Oct; n1; Peru	3999a

	preg f: Nov; n2; Tamaulipas, Mexico	217
	preg f: Nov; n1; Yucatan, Mexico	4494
	lact f: May; n1; Yucatan, Mexico	950
	lact f: Aug; n1; Honduras	11055
	lact f: Nov; n1; Tamaulipas, Mexico	217
	lact f: Nov; n2; Yucatan, Mexico	4494
	lact f: Nov; n2; Mexico, Mexico	11754
Ectophylla alba	preg f: Feb, Aug; Costa Rica	6260
	preg f: Mar; n1; Costa Rica	3608
	lact f: Mar; n1; Costa Rica	6260
	lact f: Apr; n1; Costa Rica	3608
Ectophylla macconnelli	preg f: Jan; n1; Colombia	10785
	preg f: June-Aug, Oct; n6; Peru	3999a, 10974
	preg f: July; n1; Peru	11754
	lact f: May; n1; Peru	11754
Erophylla sezekorni	preg f: Feb-May; West Indies	1429b
	preg f: Mar-Apr; Puerto Rico	7983, 11754
	preg f: Apr; Puerto Rico	395
	preg f: June; n1; Puerto Rico	11055a
	birth: June; n1; Bahamas	983a, 11754
	lact f: May, July; Puerto Rico	7983, 11754
	lact f: June; n8; Bahamas	983a, 11754
	lact f: June; West Indies	1429b
Glossophaga commissarisi	preg f: Jan, July; n5; Sinaloa, Mexico	5413a
	preg f: Feb; n1; Guatemala	5405e
	preg f: Feb; n1; Nicaragua	5405c
	preg f: Feb, Apr, Sept; n4; Jalisco, Mexico	11471a
	preg f: Mar-Apr, Sept; La Selva, Costa Rica	6261
Glossophaga leachii	preg f: Jan-Mar, July-Sept, Nov-Dec; El Salvador	3206
	preg f: Feb-Apr, Sept-Oct; Jalisco, Mexico	11471a
	preg f: Aug; n1; Yucatan, Mexico	8285a
	preg f: Sept; n1; Veracruz, Mexico	4228
Glossophaga longirostris	preg f: Feb-Apr, Aug; n10; Trinidad	3900a
	preg f: Mar; n4; Patos I, Venezuela	3900a
	preg f: June or July; n3; Venezuela	8562a
	preg f: July; n8; Margarita I, Venezuela	10112a
	lact f: July; n2; Venezuela	11754
	lact f: Aug; n103; Bonaire, Dutch West Indies	3705b
Glossophaga mexicana	preg f: Mar; n1; Central America	11527
	lact f: May; n1; Central America	11527
Glossophaga soricina	epididymal sperm: yr rd; São Paulo, Brazil	10618
	preg f: Jan; n1; Puebla, Mexico	6258a
	preg f: Jan; n1; Jamaica	3902
	preg f: Jan-Feb, Aug; n7; Nayarit, Mexico	1937
	preg f: Jan-July; n40; Panama	3322
	preg f: Jan, Apr-June, Dec; Trinidad	3900a
	preg f: Feb; Chiapas, Mexico	11754
	preg f: Feb; n1; Panama	1006
	preg f: Feb-Mar; French Guiana	1350a
	preg f: Feb, Apr, July; Yucatan peninsula, Mexico	5424f
	preg f: Feb-Apr, Sept, Nov; n14; Michoacan, Mexico	4671
	preg f: Mar-Apr; n7; Oaxaca, Mexico	11754
	preg f: Mar-Apr, Aug, Oct-Jan; n23; Colombia	10660
	preg f: Mar-Apr, Sept; n5+; Veracruz, Mexico	4228
	preg f: Mar, May, Aug-Jan; Sinaloa, Mexico	5413a
	preg f: Mar, Aug; n3; Guatemala	5405e
	preg f: May; n3; Sonora, Mexico	1937
	preg f: May; Tres Marias I, Mexico	7235a
	preg f: May-July; Tabasco, Mexico	11754
	preg f: June-Nov; n38; Peru	10974
	preg f: June-July, Oct, Dec-Feb; n18; Colombia	10785
	preg f: June-Mar; São Paulo, Brazil	10618
	preg f: July; n1; Hidalgo, Mexico	1640a
	preg f: July; n7; El Salvador	10313a

	preg f: July-Aug; n2; Costa Rica	10313a
	preg f: July-Aug; n40; Colombia	352a
	preg f: July-Sept, Nov-Mar; n61; El Salvador	3205
	preg f: Aug; Chiapas, Mexico	595
	preg f: Aug; n1; Honduras	10313a
	preg f: Aug; n1; Venezuela	11754
	preg f: Sept; n1; Oaxaca, Mexico	1937
	preg f: Sept; n1; El Salvador	1475
	preg f: Nov-Dec; Colima, Mexico	11754
	preg f: Nov-Dec; Mato Grosso, Brazil	4291, 4293
	preg f: yr rd; Brazil	11736
	preg/lact f: Jan; Brazil	8368
	birth: Sept; n1; El Salvador	1475
	birth: yr rd; Mexico	1937
	lact f: Jan, Mar-Apr, Aug, Oct-Nov; n13; El Salvador	3205
	lact f: Jan-Mar, June; Trinidad	3900a
	lact f: Mar; n5; Michoacan, Mexico	4671
	lact f: Mar-Apr, Sept-Oct, Dec; n18; Colombia	10785
	lact f: Mar, Aug; n5; Panama	3322
	lact f: Apr; n1; Yucatan, Mexico	950
	lact f: June-July; Tabasco, Mexico	11754
	lact f: June-July, Sept, Dec; n5; Colombia	10660
	lact f: July; n1; Chihuahua, Mexico	280a
	lact f: Oct; French Guiana	1350a
	lact f: Oct; n1; Paraguay	11754
	lact f: Nov; Veracruz, Mexico	4228
	lact f: Nov; n1; El Salvador	1475
	lact f: Dec; n1; Sonora, Mexico	1939
Hylonycteris underwoodi	preg f: Jan-Apr, Oct-Nov; n7; Costa Rica	6260
	preg f: Sept; n3; Jalisco, Mexico	8483a
	lact f: Mar; n1; Guatemala	1641
	lact f: May; n1; Tabasco, Mexico	11754
	lact f: Nov; n1; Oaxaca, Mexico	11754
Leptonycteris curasoae	preg f: Nov; n7; Margarita I, Venezuela	10112a
Leptonycteris nivalis	preg f: Aug; n12; Tamaulipas, Mexico	217
	preg f: Sept; n6; Veracruz, Mexico	4228
	lact f: June; n3; TX	2865a
	repro: Apr-June; TX	2343
Leptonycteris sanborni	preg f: Feb, July, Nov; n3; Sinaloa, Mexico	5413a
	preg f: Mar-Apr; n12; Sonora, Mexico	1939
	preg f: Sept, Nov; n2; Mexico	11754
	birth: May; AZ	577
	birth: May; s AZ	4523a
	birth: Nov; Mexico	4523a
	lact f: Aug; AZ	4900a
	lact f: Dec; n1; Michoacan, Mexico	4671
Lichonycteris obscura	preg f: Feb; n2; Guatemala	1641
	preg f: Mar; n1; Costa Rica	3608
	lact f: Jan; n1; Costa Rica	3608
Lionycteris spurrelli	preg f: Aug; n2; Peru	3999a, 10974
Lonchophylla mordax	preg f: Aug; n1; Costa Rica	3608
	lact f: Mar; n1; Costa Rica	2345
Lonchophylla robusta	preg f: Nov; n1; Peru	3999a
Lonchophylla thomasi	preg f: July; n2; Peru	3999a
Lonchorhina aurita	preg f: Feb; Oaxaca, Mexico	7983, 11754
	preg f: Feb-Mar; n4; Panama	1006
	preg f: Feb-Mar; n4; Panama	3322
	preg f: Apr; n1; Trinidad	3900a
	preg f: Aug; n2; Peru	3999a
	lact f: July; n1; Peru	10974
	repro: Feb-Apr; Venezuela, Trinidad, Antilles	1346
Macrophyllum macrophyllum	mate: Dec; n3; El Salvador	3206
	preg f: Mar, May; Costa Rica	6260
	preg f: Oct; n2; El Salvador	4401

	preg f: Oct-Nov; French Guiana	1350a
Macrotus californicus	testes: enlarge July-Aug, regress end Nov; AZ	1199
	testes: enlarge June-Aug; AZ	6072
	mate: Aug-Nov; AZ	6072
	mate: Sept; s AZ	1462
	mate: Sept-Nov; AZ	1199
	mate: early Oct; s AZ	999
	preg f: Mar-Apr; n3; CA	1937
	preg f: Mar-Apr; n7; Sonora, Mexico	1939
	preg f: Apr; n9; CA	11754
	preg f: Apr; n60; CA	4079
	preg f: May; CA	5067a
	preg f: May; n4; Sonora, Mexico	1471a
	conceive: Sept-Nov; AZ	1199
	conceive: Sept; s AZ	1462
	conceive: Sept-Oct; AZ	1038, 1039
	implant: Sept-Oct; AZ	1000
	implant: Oct; placentation starts Feb; AZ	1038
	implant: end Oct; AZ	999
	birth: May-early July; s AZ	1198, 1199
	birth: end May-early June; s AZ	1462
	birth: June; s AZ	999
	birth: mid June; AZ	1038
	lact f: July; Sonora, Mexico	1471a
Macrotus waterhousii	preg f: Feb-Mar; n5; Jalisco, Mexico	11471a
	preg f: Feb, Apr; n4; West Indies	1429a
	preg f: Mar; n4; Cuba	280
	preg f: May; n5; Haiti	5873
	preg f: May; Tres Marias I, Mexico	7235a
	preg f: July; n5; Baja CA, Mexico	5424e
	lact f: May; Jalisco, Mexico	11471a
	lact f: July; Sinaloa, Mexico	5413a
	lact f: Dec; n4; Jamaica	3902
Micronycteris brachyotis	preg f: May, Aug; n2; Veracruz, Mexico	7154a
	preg f: July; n1; Guatemala	2520
	lact f: July; n6; Guatemala	2520
	lact f: Aug; n12; Veracruz, Mexico	7154a
Micronycteris daviesi	preg f: Aug; n1; Peru	10974
Micronycteris hirsuta	preg f: Mar, May; n3; Trinidad	3900a
	preg f: July; n1; Peru	10974
Micronycteris megalotis	preg f: Feb-Mar; n2; Trinidad	3900a
	preg f: Mar; n1; El Salvador	1475
	preg f: Mar, June; n3; Nicaragua	5425
	preg f: Apr; n1; Michoacan, Mexico	4671
	preg f: Apr; n1; Yucatan, Mexico	5424f
	preg f: May; n2; Oaxaca, Michoacan, Mexico	4671, 11754
	preg f: June, Aug; n3; Peru	3999a, 10974
	preg f: July; n1; Venezuela	11754
	lact f: May; n1; Panama	3024
	lact f: June; n2; Nicaragua	5425
	lact f: June; n1; Trinidad	3900a
	lact f: June; n1; Colombia	10785
	lact f: Aug; n3; Venezuela	11754
Micronycteris minuta	preg f: Mar; n1; Costa Rica	3608
	preg/lact f: May; n5; Trinidad	3900a
Micronycteris nicefori	lact f: July; n1; Nicaragua	527b
Micronycteris sylvestris	preg f: Feb-Mar; French Guiana	1350a
Mimon cozumelae	preg f: Mar; n1; Chiapas, Mexico	1641
	preg f: Mar; n1; Guatemala	9113
	preg f: Apr; n2; Veracruz, Mexico	4228
	preg f: Apr; n19; Yucatan, Mexico	5424f
	preg f: Apr; Costa Rica	6260
	lact f: May; Campeche, Mexico	5424f
	lact f: July; n1; Yucatan, Mexico	5424f

	lact f: July; n1; Honduras	11055
	lact f: Aug; n1; Guatemala	9113
	lact f: Aug; Costa Rica	6260
Mimon crenulatum	preg f: Feb; n1; Campeche, Mexico	5405d, 5424f
	preg f: Mar; n2; Trinidad	3900a
	preg f: Apr; n1; Costa Rica	6260
	preg f: Apr; n2; Costa Rica	3608
	preg f: June-July; n7; Peru	3999a, 10974
	preg f: July; n1; Suriname	3705c
Monophyllus redmani	preg f: Feb; n11; Jamaica	8134a
	preg f: end Feb-early Mar; n6; Jamaica	7103a
	preg f: May; n2; Haiti	5873
	preg f: Dec-Jan; n17; Jamaica	3902
	preg f: Dec-Feb; n5; Caribbean	4941
Monophyllus plethodon	preg f: Mar-Apr; Dominica, Lesser Antilles	9685a
	lact f: July; n1; Guadeloupe, Lesser Antilles	527a
Musonycteris harrisoni	preg f: Sept; n2; Colima, Mexico	11754
Phylloderma stenops	preg f: Feb; n1; Costa Rica	6260
Phyllonycteris aphylla	preg f: Jan; n1; Jamaica	3902
Phyllonycteris poeyi	preg f: June; Cuba	7328a
	preg f: Dec; n3; Haiti	5873
Phyllops haitiensis	preg f: Jan, May, Aug; n19; Haiti	5873
Phyllostomus discolor	preg f: Feb-Mar, June, Aug; Trinidad	3900a
	preg f: Feb, June, Aug-Sept, Nov; El Salvador	3206
	preg f: Feb, Oct; n5; Colombia	10660
	preg f: Mar; n2; Nicaragua	5405c
	preg f: Mar; n1; Costa Rica	3322
	preg f: July; n1; Margarita I, Venezuela	10112a
	preg f: Oct; n2; Colombia	11055b
	preg f: Nov; n2; Peru	3999a
	preg f: Dec; n1; El Salvador	1475
	preg/lact: Aug-Sept; Guatemala	2552
	lact f: Feb-Mar, May, Sept; n7; Colombia	10660
	lact f: July; n2; Margarita I, Venezuela	10112a
	lact f: July; n1; Brazil	7983, 11754
	lact f: Aug-Oct; Trinidad	3900a
	lact f: Dec; n11; Costa Rica	3322
Phyllostomus elongatus	preg f: June; n1; Colombia	10785
	preg f: July-Aug; n15; Peru	3999a, 10974
Phyllostomus hastatus	preg f: Mar; n2; Nicaragua	5425
	preg f: Mar, Sept-Oct; n4; Colombia	10785
	preg f: end Mar-Apr; Trinidad	3900a
	preg f: June-Oct; n31; Peru	3999a, 10974
	preg f: July; n1; Colombia	352a
	preg f: Aug; n1; Brazil	8368
	birth: Apr-Sept; Trinidad	1346
	birth: end Apr-early May; Trinidad	7036
	lact f: Apr-May; n2; Panama	3322
	lact f: Apr, June; Trinidad	3900a
	lact f: June-Aug; Nicaragua	5425
	lact f: Aug; Venezuela	11754
	preg f: Sept, Dec; n3; Colombia	10785
Platalina genovensium	preg f: Sept; n1; Peru	3999a
Pygoderma bilabiatum	preg f: Mar, July-Aug; Paraguay	7696
	preg f: Aug; n1; Brazil	8368
Rhinophylla fischerae	preg f: June-July; n3; Peru	3999a
	preg? f: Aug; n1; Peru	1639a
Rhinophylla pumilio	preg f: May, July; n2; Colombia	10785
	preg f: July-Sept, Nov; n9; Peru	3999a
	preg f: Dec; n1; Venezuela	7983, 11754
	lact f: Apr, June, Dec; n3; Colombia	19785
Stenoderma rufum	preg f: Feb-Mar, May, July-Aug; Puerto Rico	16052a
	preg f: July; n5; Puerto Rico	5423a
	preg f: July-Aug; n2; Puerto Rico	3699

	preg f: Aug; n1; Puerto Rico	10664
	birth: July; n1; Puerto Rico	5423a
	lact f: Feb, Nov; Puerto Rico	16052a
	lact f: July; n1; Puerto Rico	5423a
Sturnira aratathomasi	preg f: Feb, Aug; n2; Colombia	10785a
Sturnira bidens	preg f: June-Aug, Oct; n15; Peru	3999a
	preg f: Aug; n3; Peru	3608a
Sturnira bogotensis	preg f: June, Aug-Nov; n22; Peru	3999a
Sturnira erythromos	preg f: June-Oct, Dec; n47; Peru	3999a
	preg f: Aug; n10; Peru	3608a
	preg f: Dec; n2; Colombia	10785
Sturnira lilium	m repro: May, Aug, Nov; Rio de Janeiro, Brazil	8368
	preg f: Jan; n1; Querétaro, Mexico	10261a
	preg f: Jan, Mar-Apr, June-July, Sept; Jalisco, Mexico	11471a
	preg f: Jan, July; n6; Campeche, Mexico	5424f
	preg f: peak Feb-Mar, July-Sept, Nov-Dec; Mexico	4670, 4671
	preg f: Feb-Mar, June-Aug; Guatemala	5405e
	preg f: Mar-Apr, Aug; n24; Lesser Antilles	5424b
	preg f: Apr; n1; Sinaloa, Mexico	1939
	preg f: Apr; n2; Quintana Roo, Mexico	950
	preg f: Apr; n1; Oaxaca, Mexico	11754
	preg f: May; n1; Guatemala	9113
	preg f: May-June, Aug; n4; Sinaloa, Mexico	5413a
	preg f: June; n1; Nicaragua	5405c
	preg f: June or July; n3; Venezuela	8562a
	preg f: June-Aug; French Guiana	1350a
	preg f: June-Dec; n32; Peru	3999a
	preg f: July; n1; Durango, Mexico	523a
	preg f: July; n4; Veracruz, Mexico	6148a
	preg f: July; n1; Nicaragua	10313a
	preg f: July; n1; Brazil	11754
	preg f: July-Aug; n10; Colombia	352a
	preg f: July-Aug; n2; Peru	10974
	preg f: Aug; n1; Quintana Roo, Mexico	5424f
	preg f: Sept; n1; Sonora, Mexico	3254b
	preg f: Oct-Jan, Mar-Apr; n22; Colombia	10875
	preg f: Dec-Feb, May-June; n12; Costa Rica	3322
	preg/lact f: July; Hidalgo, Mexico	1640a
	lact f: Jan-May, July-Aug; n24; Costa Rica	3322
	lact f: Feb, Apr-Oct, Dec; n19; Colombia	10875
	lact f: Apr, June-Oct; Jalisco, Mexico	11471a
	lact f: May; n2; Guatemala	9113
	lact f: May-June; n5; Chiapas, Mexico	11754
	lact f: June; n3; Durango, Mexico	5405b
	lact f: July; n1; Durango, Mexico	523a
	lact f: July; n1; El Salvador	10313a
	lact f: July; n1; Nicaragua	10313a
	lact f: Aug; Rio de Janeiro, Brazil	8368
	lact f: Dec; n1; Oaxaca, Mexico	9542a
Sturnira ludovici	preg f: Feb-May, Aug, Oct-Dec; n38; Colombia	10785
	preg f: Apr, July; n8; Jalisco, Mexico	11471a
	preg f: Apr, Aug; Monte Verde, Costa Rica	6261
	preg f: Aug; n1; Chipas, Mexico	11754
	preg f: Nov; n5; Jalisco, Mexico	5424c
	lact f: Apr-Nov, peaks May, Sept-Oct; Monteverdi, Costa Rica	2587
	lact f: May-Sept; n13; Colombia	19785
	lact f: May, Aug; n3; Jalisco, Mexico	11471a
Sturnira magna	preg f: Nov; n3; Colombia	10653
	preg f: Aug, Oct, Dec; n6; Peru	3999a
	preg f: Dec; n1; Peru	10653
	lact f: May; n1; Peru	3607a
Sturnira mordax	preg f: Feb; n2; Costa Rica	3608
	preg f: Aug; n1; Costa Rica	366b
	lact f: May; n1; Costa Rica	3608

	lact f: May; Costa Rica	6260
Sturnira nana	preg f: Aug; n1; Peru	3999a
Sturnira thomasi	lact f: July; n2; Guadeloupe, Lesser Antilles	527a, 3705a
Sturnira tildae	preg f: Mar; n1; Trinidad	3900a
	preg f: June, Aug, Nov; Peru	3999a
	preg f: July; n1; Brazil	11754
	preg f: Aug; n5; Trinidad	1635
Tonatia bidens	preg f: Jan; n1; Costa Rica	3608
	preg f: Feb; n1; Guatemala	1641
	preg f: May; n2; Trinidad	3900a
	preg f: July; n2; Peru	3607a
	preg f: Aug; n2; Honduras	11055
Tonatia brasiliense	preg f: Feb; n1; Panama	2345
	preg f: Feb, Apr; n2; Costa Rica	6260
	preg f: July; n1; Nicaragua	11055
	lact f: Aug; n1; Honduras	11055
Tonatia evotis	preg f: Jan; Chiapas, Mexico	7154
	preg f: Mar; Belize	7154
	preg f: Apr; Quintana Roo, Mexico	7154
	lact f: Apr; n2; Quintana Roo, Mexico	9457a
Tonatia silvicola	preg f: Jan; n1; Colombia	10785
	preg f: Mar; n2; Panama	3322
	preg f: July; n1; Paraguay	7699a
	preg f: July-Oct; n10; Peru	3999a, 10974
	lact f: Mar; n1; Paraguay	7699a
Trachops cirrhosus	preg f: Feb; n3; El Salvador	1475
	preg f: Mar; n1; Oaxaca, Mexico	11754
	preg f: Mar; n1; Chiapas, Mexico	1641
	preg f: Mar; n2; Trinidad	3900a
	preg f: Mar-Apr; n10; Guatemala	5405e
	preg f: May; n4; Nicaragua	1641
	preg f: July; n1; Peru	10974
	preg f: Aug; n1; Panama	3322
	preg f: Dec; n1; Chiapas, Mexico	11754
	lact f: Mar; n1; Costa Rica	1641
	lact f: Aug; n1; Costa Rica	366b
	lact f: Aug; n1; Honduras	11055
Uroderma bilobatum	preg f: Jan; n4; El Salvador	3207
	preg f: Jan; n8; Panama	2342
	preg f: Jan-July; n54; Panama	3322
	preg f: Jan, May; n4; El Salvador	1475
	preg f: Jan, July, Sept, Nov; n6; Colombia	10660
	preg f: Feb; n1; Guatemala	5405e
	preg f: Feb; n4; Nicaraqua	5405c
	preg f: Feb-Apr; Panama	7983, 11754
	preg f: May; n3; Trinidad	3900a
	preg f: June-Jan; n28+; Peru	3999a
	preg f: July; n1; Veracruz, Mexico	6148a
	preg f: July; n72; El Salvador, Honduras, Nicaragua	526a
	preg f: July; n3; Brazil	11754
	preg f: Aug; n1; Peru	10974
	preg/lact f: Apr-May, 3.3% preg, 4.9% lact; n123; July-Aug, 1.8% preg, 14.0% lact; n171; Sept, 20% preg, 0% lact; n80; Guatemala	2552
	birth: yr rd; Costa Rica to Peru, s Brazil	1346
	lact f: Feb, May; n4; Trinidad	3900a
	lact f: Mar; Panama	1006
	lact f: Mar; n1; Colombia	10660
	lact f: Mar-May, July-Aug; n39; Panama	3322
	lact f: Apr; n1; Quintana Roo, Mexico	9457a
	lact f: May; Chiapas, Mexico	11754
	lact f: June or July; n1; Venezuela	8562a
	lact f: July; n1; El Salvador, Honduras, or Nicaragua	526a
	lact f: Aug; n2; Nicaragua	2345

Uroderma magnirostrum	preg f: Mar; n1; Nicaragua	5425
	preg f: June; n1; El Salvador	2342
	preg f: June; n1; Brazil	11754
	preg f: July; n1; Nicaragua	2342
	preg f: Sept; n10; Bolivia	2342
Vampyressa bidens	preg f: Aug; n1; Suriname	3705c
	preg f: Aug, Oct-Nov; n27; Peru	3999a
	preg f: Dec; n2; Peru	2343a
Vampyressa melissa	preg f: Aug, Nov; n4; Peru	3999a
Vampyressa nymphaea	preg f: Feb; n1; Nicaragua	5425
	preg f: Apr; n2; Costa Rica	3608
	preg f: Oct-Aug; n111; Colombia	10785
	lact f: Jan-Aug; n96; Colombia	10785
Vampyressa pusilla	preg f: Jan, Apr; n2; Panama	3322
	preg f: Feb; n1; Campeche, Mexico	5424f
	preg f: Feb; n2; Costa Rica	6809
	preg f: Mar; n1; Belize	8412
	preg f: Mar; n4; Nicaragua	5425
	preg f: Mar-May, Aug-Nov; n6; Colombia	10785
	preg f: June; n1; Paraguay	7700a
	preg f: July; n1; Chiapas, Mexico	2345
	preg f: July; n1; Guatemala	9113
	preg f: July; n1; Nicaragua	10313a
	preg f: Aug; n1; Honduras	11055
	preg f: Aug; n1; Colombia	352a
	preg f: Oct-Nov; Peru	2345a
	preg f: Nov-Dec; n15; Peru	3999a
	lact f: Mar-Apr; n3; Panama	3322
	lact f: Apr; n1; Panama	4228a
	lact f: July; n1; Colombia	10785
	lact f: July-Aug; n2; Costa Rica	366a
	lact f: Aug; n3; Colombia	352a
Vampyrodes caraccioli	preg f: Jan; n2; Panama	3322
	preg f: Jan-Apr, July, Oct-Nov; n14; Colombia	10785
	preg f: June; n1; Chiapas, Mexico	5405d
	preg f: July; n1; Chiapas, Mexico	2345
	preg f: July, Nov; n4; Peru	3999a, 10974
	preg f: July-Aug; n3; Nicaragua	5425
	preg f: end July-early Aug; n14; Honduras	11055
	preg f: Oct-Nov; Peru	2345a
	lact f: Jan-June, Aug; n14; Colombia	10785
	lact f: May; n1; Honduras	2345
Vampyrops brachycephalus	preg f: Feb; n1; Venezuela	9318a
	preg f: July-Nov; n12; Peru	3999a
	preg f: Aug; n2; Peru	9318a
	lact f: Oct; n1; Venezuela	9318a
Vampyrops dorsalis	preg f: June-Nov; n33; Peru	3999a
	preg f: Nov-July; n45; Colombia	10785
	lact f: Feb-Mar, May-July, Oct; n18; Colombia	10785
	lact f: Aug; n3; Columbia	352a
Vampyrops helleri	preg f: Mar; n1; Costa Rica	6809
	preg f: Mar-Apr, June-Aug; Nicaragua	5425
	preg f: Apr, July, Dec-Jan; n4; Panama	3322
	preg f: June; n2; El Salvador	6258
	preg f: June; n1; Nicaragua	5405c
	preg f: July; n1; Chiapas, Mexico	2345
	preg f: July-Dec; n15; Peru	3999a, 10974
	preg f: Aug; n1; Honduras	6258
	preg f: Aug; n14; Trinidad	1635
	preg f: Aug; n2; Colombia	352a
	preg f: Aug-Sept; French Guiana	1350a
	preg f: Oct-June; n35; Colombia	10785
	lact f: Apr, Sept; n2; Panama	3322
	lact f: May; n1; Tabasco, Mexico	11754

	lact f: May; n1; Chiapas, Mexico	11754
	lact f: May; n1; Guatemala	9113
	lact f: Aug; n2; Colombia	352a
	lact f: Dec-Jan, Apr-May; n29; Colombia	10785
Vampyrops infuscus	preg f: Mar; n1; Colombia	6814a
	preg f: July-Aug, Nov; n6; Peru	3999a
	lact f: Mar; n3; Colombia	6814a
Vampyrops lineatus	epididymal sperm: yr rd; n39; São Paulo, Brazil	10618
	preg f: Jan, Mar, Dec; Brazil	8368
	preg f: May-Mar; São Paulo, Brazil	11737
	preg f: July-Aug; n25; Peru	3999a
	preg f: July-Mar; Brazil	11736, 11737
	lact f: Nov-May; São Paulo, Brazil	11737
Vampyrops vittatus	preg f: Mar; Costa Rica	6260
	preg f: Apr; n2; Costa Rica	2345
	preg f: May, Oct; n2; Colombia	10785
	preg f: June; n1; Costa Rica	3608
	preg f: Aug; n4; Peru	3999a
	lact f: May-July; n6; Costa Rica	3608
Vampyrum spectrum	birth: June; n1; captive	4048b
	birth: July; n1; captive	2109
	lact f: May; n1; Trinidad	3900a

Order Chiroptera, Family Natalidae, Genus *Natalus*

Funnel-eared bats (*Natalus* 5-10 g) are insectivores from the neotropics of North and Central America. Females have a bicornuate uterus (3255) and are monoestrus (7375). Embryo development is slow and the left uterine cornu is used preferentially (7375). Males have seminal vesicles and a baculum (6065). Neonates are over 50% of adult mass (7375).

NEONATAL MASS		
Natalus stramineus	1.45-1.75 g; term emb	7375
	2.1 g; r1.7-2.6	7375
LITTER SIZE		
Natalus stramineus	1 emb	7375
	1 emb; n1 f	527a
	1 emb; n1 f	950
	1 emb; n1 f	1475
	1 emb ea; n8 f	11471a
	1 emb ea; n16 f	1937
	1 emb ea; n25 f	5424e
GESTATION		
Natalus stramineus	8-10 mo; slow growth	7375
SEASONALITY		
Natalus stramineus	sperm: Sept-Mar; n hemisphere	6065
	preg f: Jan; n2; none May, Aug, Dec; El Salvador	3208
	preg f: Feb-Mar; n10; Michoacan, Mexico	4671
	preg f: Apr; n1; El Salvador	1475
	preg f: May-June; n16; Mexico	1937
	preg f: May, July; n25; Baja CA, Mexico	5424e
	preg f: June; n1; Sinaloa, Mexico	5413a
	preg f: June; n8; Jalisco, Mexico	11471a
	preg f: July; n1; Guadeloupe, Lesser Antilles	527a
	implant: Dec-early Jan; Sonora, Mexico	7375
	birth: end July-Aug; Sonora, Mexico	7375

Order Chiroptera, Family Furipteridae

The Central and South American furipterid bats, *Amorphochilus schnablii* and *Furipterus horrens* (3.3 g), are insectivorous forest dwellers. Females are larger than males (5125, 11026). Two abdominal mammary glands are a familial distinction (5125, 11026). Bacula are present (5125).

LITTER SIZE		
Amorphochilus schnablii	1 emb ea; n8 f	5125
	1 emb ea; n16 f	3999a
Furipterus horrens	1 yg	11026
SEASONALITY		
Amorphochilus schnablii	preg f: June, Aug, Oct; n16; Peru	3999a

Order Chiroptera, Family Thyropteridae, Genus *Thyroptera*

Thyroptera tricolor and *T. discifera* are small (3-4 g) bats from tropical Central and South America (11753, 11757). The labyrinthine, hemodichorial placenta has an interstitial implantation into a bicornuate uterus (11805, 11806). A large corpus makes the uterus appear simplex (3255). Both sexes have a pair of axillary nipples (11753). The testes are partially inguinal and partially abdominal, and the penis lacks a baculum (11753).

NEONATAL MASS		
Thyroptera tricolor	0.98-1.2 g; near term	11806
NEONATAL SIZE		
Thyroptera tricolor	CR: 18 mm; near term	11806
LITTER SIZE		
Thyroptera tricolor	1 emb; n1 f	217a
	1 emb ea; n3 f	3255
	1 emb ea; n3 f	3999a
	1 cl, emb; n8 f	11805, 11806
GESTATION		
Thyroptera tricolor	~5 mo	11805, 11806
SEASONALITY		
Thyroptera tricolor	preg f: Jan-May; n8; Veracruz, Mexico	11805, 11806
	preg f: Feb-Mar; French Guiana	1350a
	preg f: May, June, Dec; Colombia	11806
	preg f: July; n3; Peru	3999a
	preg f: Aug; Costa Rica	3255
	lact f: May; Panama	4228a

Order Chiroptera, Family Myzopodidae, Genus *Myzopoda*

The rainforests of Madagascar have yielded only a few specimens of and no reproductive information on *Myzopoda aurita* (~57 mm). The penis is short and lacks a baculum (10784). A short review of its natural history is available (9599).

There are no tabular data for *Myzopoda*.

Order Chiroptera, Family Vespertilionidae

Vespertilionids (*Antrozous* 14-39 g; *Barbastella* 6-13 g; *Chalinolobus* 6-15 g; *Eptesicus* 4-40 g; *Euderma* 14-20 g; *Eudiscopus* 40-45 mm; *Glischropus* 3.5-4.5 g; *Harpiocephalus* 55-75 mm; *Hesperoptenus* 6.1-32 g; *Histiotus* 54-70 mm; *Ia* 89-104 mm; *Idionycteris* 8-20 g; *Kerivoula* 4-10 g; *Laephotis* 4-8 g; *Lasionycteris* 6-14 g; *Lasiurus* 6-35 g; *Mimetillus* 6-11 g; *Miniopterus* 4-20 g; *Murina* 3-12 g; *Myotis* 2-45 g; *Nyctalus* 11-55 g; *Nycticeius* 4-20 g; *Nyctophilus* 7-18 g; *Otonycteris* 18-20 g; *Pharotis* 50 mm; *Philetor* 9-13 g; *Phoniscus* 9-10 g; *Pipistrellus* 2-20 g; *Plecotus* 5-20 g; *Rhogeessa* 3-10 g; *Scotoecus* 4-5 g; *Scotomanes* 72-78 mm; *Scotophilus* 15-40 g; *Tomopeas* 2-4 g; *Tylonycteris* 3-11 g; *Vespertilio* 14-33 g) are insectivores with a worldwide distribution.

Females are often larger than males (217, 2045, 3216, 3944, 4231, 5405b$_1$, 5759, 6717, 8494, 8965, 9297, 9945, 9989, 11270, 11714; cf 191, 8127, 11759). *Lasiurus* exhibits sexual dichromatism (178a).

Male *Glischropus* have an extra chromosome (11339). *Miniopterus* exhibits extensive reproductive asymmetry: only one ovary, either left or right, is functional; one uterine horn is often much longer than the other; and implantation occurs only on the right side (820, 823, 2075, 3945, 6966, 7163, 7518, 9102, 11015, 11128, 11131, 11133, 11134, 11418). Other vespertilionids also exhibit uterine asymmetry (*Chalinolobus* 12042; *Eptesicus* 5834; *Myotis* 825, 1417, 1524, 2947, 3517, 4177a, 6070, 6106, 7695, 7950, 8843, 9645, 10070, 11318; *Nyctalis* 11318; *Pipistrellus* 1524, 6064; *Vespertilio* 2824, 9240).

The endotheliochorial or hemochorial, labyrinthine, discoid placenta has accessory placentae and an antimesometrial, superficial or interstitial implantation (1228, 1735, 2163, 3020, 3730, 3908, 3909, 3923, 3943, 4402, 5530, 5531, 5648, 5728, 6740, 7563, 8879, 9178, 11133, 11145a, 11318, 11793, 11796, 11799) into a bipartite or bicornuate uterus (820, 2824, 3020, 3730, 3907, 3909, 3917, 3939, 3943, 3944, 3945, 5530, 5531, 6064, 6966, 7759, 7950, 8695, 8839, 8843, 8879, 8883, 9002, 9102, 10070, 11094, 11131, 11133, 11134, 11418, 11793; cf 1735, 9178). Placentophagia occurs (1043, 2945, 2948, 5386, 5939, 7610, 8127, 9398, 9895, 11270, 12042; cf 9306). Breech births are common (1043, 4038, 12042).

Monoestry (6028, 6064, 9102, 11137), postpartum estrus (4038, 5830, 6026, 6716, 7695, 11759), vaginal plugs (5833, 6025; cf 2420), sperm storage (825, 1524, 2134, 2135, 2420, 3933, 3935, 4102, 4449, 4819, 4820, 5860, 6025, 6026, 6027, 6028, 6064, 6068, 6070, 7164, 7512, 7513, 7515, 7695, 8198, 8828, 8843, 8998, 9240, 9679, 11095, 11138, 11139, 11792, 11792a, 11811), delayed implantation (822, 5727, 8463, 8464, 9102, 11128, 11129, 11132, 11418), and delayed development (2947, 3564, 8826, 8840, 11017, 11145b) occur.

Most neonates are naked and blind (577, 1344, 1474, 2228, 2593, 2828, 2842, 2948, 3973, 5024, 5759, 6175, 6714, 6715, 6717, 7163, 7164, 8049, 8314, 8492, 8493, 8494, 10158, 11132, 12042; cf 3944, 6000, 6106, 6722, 8049, 8658, 11270). Neonates open their eyes at 2-10 d (*Antrozous* 7523, 2335, 8127; *Myotis* 1474, 2945, 3571, 6000, 7424, 7523, 8798, 9895; cf 3216, 3973; *Pipistrellus* 3932, 5845, 7523, 8798, 8865, 9175, 11622), 4-11 d (*Nyctalus* 2945, 3571, 5845, 6722, 7523, 8798; *Vespertilio* 3563, 9240), 1 wk (*Eptesicus* 5759, 5845, 8798; *Miniopterus* 3571), or 8-10 d (*Plecotus* 1474). *Lasiurus* neonates have fine hair at birth and open their eyes at 10-12 d (577, 1043, 5210, 7523, 7619, 8656), while *Nycticeius* neonates have a few hairs and open their eyes within hours of birth (577, 5386, 7523, 9398, 11470). Litter mass at birth may reach 40% of maternal mass (6107, 10295).

Females usually have 2 mammae (577, 1474, 3563, 3571, 3944, 6720, 8314, 9175), although *Lasiurus* has 4 (3900a, 8656) and *Otonycteris* may have multiple nipples per gland (6720). A baculum, ampullary and urethral glands are present in the males of most species (5704, 6064, 6068, 6957, 6966, 7439, 7440, 7441, 7664, 9178). *Miniopterus* and *Scotophilus* males have postanal scrota and *Miniopterus* has no baculum (191, 3906, 6966, 9102).

Details of female anatomy (347, 759, 820, 821, 823, 1416, 3609, 3907, 3945, 4176, 4177, 4177a, 4946, 7695, 8175, 9002, 10070, 11093, 11130, 11132, 11148, 11794, 11810, 11812), male anatomy (831a, 2069, 3906, 3907, 4161, 6069a, 6957, 8054, 8208, 8554, 11809), endocrinology (343, 344, 1417, 1418, 1419, 1555, 2074, 2185, 2237, 2238, 2239, 2240, 2241, 2242, 3372, 4163, 4164, 4167, 4177, 4662, 4666, 5729, 6138, 8176, 8465, 10051, 10107, 11143), ovulation (1414, 5860, 7517, 8831, 10050, 10069, 11014), sperm/spermatogenesis (1272, 2073, 3523, 4661, 4826a, 4881, 5729, 6062a, 8829), mating/conception (3940, 5210, 7514, 7516, 10523b, 11015), birth (3932, 6122a, 8127, 9895, 11470, 11795, 11801), milk composition (5296, 6114, 10296), reproductive energetics (340, 685, 2099, 6121, 6122, 6140, 6141, 8841, 10520, 10521, 10522, 10604), and light (683, 686, 688, 2237, 8833) are available.

NEONATAL MASS

Antrozous pallidus	3.1 g; r2.9-3.3; n2	8127
	m: 3.0 g; se0.13; r2.4-3.5; n9	2335
	f: 3.2 g; se0.19; r2.4-4.4; n11	2335
Chalinolobus morio	1.7 g; r1.5-2.0; twins: 1.5 g ea; n4	12042
Eptesicus capensis	2 g; n1 emb	11273
Eptesicus fuscus	2.5 g; r2.5-2.6; n5	1829
	2.85 g; r2.75-2.95; term emb	8485
	3.14 g; r2.7-3.6; n5	6107
	3.3 g; n16	1455
	4 g; r3.4-4.5; n10	2354

	f: 7.8 g	11627
Eptesicus serotinus	4.81 g; n1	2945
	5.8 g; r5.2-6.2; n4	5845
Eptesicus tenuipinnis	a little bigger than a bee	9297
Euderma maculatum	2.9 g; r2.6-3.25; n2	2866
	4 g; n1	2865
Lasionycteris noctivagans	m: 1.1 g; n1	7235
	f: 1.24 g; r1.10-1.38; n2	7235
	1.85 g; r1.8-1.9; n2	6105a, 6109
	2.5 g ea with placenta; n2	6122a
Lasiurus borealis	3.175 g; n4 prob not neonates	10298
	5.8 g; r5.3-6.8; n4 probably not neonates	10298
Lasiurus cinereus	nearly 5 g	5210
	6 g	1043
Lasiurus intermedius	3 g	577
Miniopterus schreibersi	2.416+ g; 237+ d emb	11132
	2.744 g; sd0.142; n2, 240+ d emb	11132
	est 2.5-3.0 g; smallest juv: 2.6-3.4 g	2828
	2.5 g; se0.2; r2.1-2.8; n18 neonates	11129
	2.7 g; se0.1; r2.4-3.1; n25 term emb	11129
	3 g; se0.2	3945
	3.4 g	6115
Myotis austroriparius	1.12 g; r1.10-1.15; n2	9895
Myotis daubentoni	f: 2.09 g; n1	2945
Myotis evotis	r1.08-1.36 g; n5	6780
Myotis grisescens	2.9 g; sd0.24; r2.4-3.4; n94	10975
Myotis lucifugus	1.36 g; n6 term emb	3254
	1.50 g; r1.45-1.55; n2	7424
	1.5-1.9 g	11795
	2.4 g; r2.0-2.8; n3	1526
	2.08 g; sd0.18; n19	10106
	2.17 g; n3 term emb	2842
	2.2 g; sd0.14; n36 with umbilicus	6111
	2.3 g; sd0.26; n121 neonates	1455, 6111
Myotis macrodactylus	0.6-2.3 g; unknown age emb	7522
	1.5 g	6115
Myotis myotis	5-6 g	9596
	6.1 g	6000
	6.2 g; r5.6-7.6; n28	2402, 2403
Myotis nigricans	1.0 g; n4	11752
Myotis thysanodes	22% of adult mass	8005
Myotis velifer	3.2 g; se0.03; n89, 1-2 d	6106
Nyctalus lasiopterus	r3.4-6.4 g; n8 twins	6722
	6.9 g; r5.5-8.3; n2 singletons	6722
Nyctalus noctula	4.0-5.5 g	2593
	m: 5.65 g; n1	2945
	5.7 g; r4.7-6.7; n10	5845, 5860
Nycticeius balstoni	f: 2.4 g; n1	9398
Nycticeius humeralis	1.4-1.6 g	3629
	2.0 g; n11	5386
Pipistrellus ceylonicus	1.2 g; term emb	3915, 3932, 6715
	1.27 g; r~1.2-1.4; neonate	3915, 3932, 6715
Pipistrellus javanicus	0.88 g; r0.9-1.1; n15 term emb, neonates	7521
	1.1 g	11016
Pipistrellus mimus	0.5 g	3944
Pipistrellus pipistrellus	1 g	2420
	~1 g	8827
	1.0 g; se0.16; r0.6-1.4; n18	8865
	1.027 g ea; twins	1179
	1.323 g; singletons	1179
	1.4 g; r1.3-1.5; n9	5845
Pipistrellus subflavus	0.94 g; r0.93-0.96; n2	6175
	52% of maternal postpartum mass	3549
Plecotus rafinesquii	2.3-2.6 g	5397

Plecotus townsendii	2.4 g; r2.1-2.7; n10	8314
Scotophilus heathi	5.0 g	6028
	5.5-6.0 g	6717
Tylonycteris pachypus	f: 0.6 g; n1	7164
	0.7 g; term emb	7164
Tylonycteris robustula	1.25 g; n2 term emb	7164
Vespertilio superans	2 g	3563
NEONATAL SIZE		
Antrozous pallidus	m: FA: 17.4 mm; se0.38; r16-20; n9	2335
	FA: 17.5 mm; r17-18; n2	8127
	45.5 mm; r45-46; n2	8127
Chalinolobus argentatus	15.1 mm; r11-20; n2	71
Chalinolobus gouldii	CR: 21.7 mm; term emb	5830
Chalinolobus morio	CR: ~24 mm	5833
Chalinolobus tuberculatus	m: CR: 13.2 mm; r11.9-14.6; n2 emb	2259
Eptesicus fuscus	FA: 16.1 mm; r15.1-17.0; n5	6107
	FA: 16.8 mm; n16	1455
Eptesicus regulus	f: CR: 18 mm; largest emb	5834
Euderma maculatum	HB: 39 mm; FA: 21 mm	2865
Lasionycteris noctivagans	TL: 45.8 mm; HB: 32.3; FA: 12.5; n1	6105a
	TL: 45.9 mm; r44.7-47.1; FA: 14.1 mm; r14.0-14.2; n2	6122a
Lasiurus borealis	TL: 43 mm; r41-44; n3 large emb	9961
Lasiurus cinereus	m: HB: 40.5 mm; n1; FA: 18.5 mm	7619
	46 mm ea; n2, 2 d	7847
	FA: 18-19 mm	577
Lasiurus intermedius	FA: ~16 mm	577
Myotis evotis	TL: r40-47 mm; n5	6780
Myotis keenii	CR: 50 mm	2864
Myotis lucifugus	HB: 30 mm	2842
	FA: 15.8 mm; sd0.78; n36 with umbilicus	6111
	FA: 16.1 mm; sd0.76; n121 neonates	1455, 6111
Myotis myotis	46.6 mm	6000
Nyctalus noctula	65 mm; n1	2593
Nyctophilus geoffroyi	35 mm; r34-36; n2	7074
Nycticeius balstoni	f: 31.3 mm; n1	9398
Pipistrellus javanicus	HB: 23.6 mm; r19.5-25.5; n15 neonates, term emb	7521
Pipistrellus pipistrellus	17.5 mm	8827
Pipistrellus stenopterus	37 mm; n1 3 d	4625
Plecotus townsendii	CR: ~24 mm	577
WEANING MASS		
Eptesicus fuscus	17.2 g; r11.4-22.6; n33	6298
	fledge: 75% adult size	1455
Kerivoula phalaena	f found carrying yg 3/4 her size	9297
Lasiurus borealis	m: 6.26 g; r5.6-7.0; n4 juveniles	4260
	f: 7.12 g; r6.4-8.1; n6 juveniles	4260
Lasiurus cinereus	9-10 g; 3 wk	5210
Miniopterus schreibersi	12.5 g; n28	2828
Myotis grisescens	lighter at weaning the farther they have to fly to water	10977
Myotis lucifugus	1st flight: 84% adult size; 22 d	1455
Myotis nigricans	adult mass before weaning	11752
Myotis thysanodes	adult mass at 3 wk	8005
Nyctalus lasiopterus	22-30 g; 30 d; fly 40-45 d	6722
Nycticeius balstoni	5.6 g; 29 d	9398
Nycticeius humeralis	2.6 g; 4 wk	5386
Pipistrellus ceylonicus	4.5-6.5 g; ~1 mo	6715
Pipistrellus mimus	~1.5 g; detach from mother; 2 g; end of lactation	3944
Pipistrellus subflavus	5.2 g; r4.1-7.5; n17	6298
WEANING SIZE		
Myotis nigricans	adult size at weaning	11752

LITTER SIZE

Antrozous dubiaquercus	1 emb; n1 f	8554
Antrozous pallidus	1-2 emb	502
	1-2 emb	4079
	1-2 yg	2335
	1.33 yg; r1-2; n6 f	10455
	1.5 emb; r1-2; n4 f	1471a
	1.5 emb; r1-2; n27 f	502
	1.75 emb; r1-2; n4 f	10455
	1.8 emb, yg; mode 2; r1-3; n28 lt	8127
	1.8; r1-3, 1/61 triplets; n61 lt	639
	1.82 yg; r1-2; n70 lt	640
	2 emb; n1 f	1196a
	2 emb ea; n2 f	522, 523a
	2 emb ea; n2 f	11228a
	3 emb; n1 f	4079
Barbastella barbastellus	1-2	3571
	2	4645
	2	8049
Chalinolobus dwyeri	1 ea; n30 lt	7075
	1.8 yg; n13 f	2829
Chalinolobus gouldii	1 ea; n29 lt	5340
	mode 2; 2-9 ova/ovary; 2-9 cl	5830
Chalinolobus morio	mode 1 yg; 1/32 f with twins	5833
	1.3 yg; r1-2; n3 lt	12042
	1.5 emb; n24 f	12042
Chalinolobus picatus	2; n1 lt	7075
Chalinolobus tuberculatus	1 emb ea; n2 f	2259
Chalinolobus variegatus	1.3 emb; r1-2; n3 f	191
Eptesicus brasiliensis	1 emb ea; n12 f	3999a
	2.2; r2-4; n9	452
Eptesicus capensis	1 emb; n1 f	11273
	1.6; mode 1-2; r1-3; 3 rare	10158
	2 emb; several f	9945
	2.25 emb; n4 f	8965
Eptesicus furinalis	1 implantation ea; n4 f	9113
	2 emb; n1 f	4671
	2 emb; n1 f	5425
	2 emb ea; n5 f	950
	2.0 emb; mode 2 1st lt, mode 1 2nd lt; r1-4; n24 f	7695
Eptesicus fuscus	1 yg	502
	1-2 yg	1279
	1-2 yg; 3-7 ova shed	11795
	mode 1; but twins observed	6059
	1 emb; n1 f	1649a
	1 emb; n1 f	11471a
	1 emb ea; n2 f	5424_2
	1 emb ea; n4 f	1119
	1 yg ea; n~36 f	5027
	1+; r1-2	10954
	1.13 emb; r1-2; n115 f	9643
	1.25 emb; mode 1; r1-2; n4 f	3703
	1.27 emb; mode 1; r1-2; n15 f	2197, $5405b_1$
	1.75 emb; n8 f	5405e
	1.75 yg; n51 f	2354
	1.89 emb; mode 2; r1-2; n54 f	11627
	1.9 emb; r1-2; n11 f	950a
	1.9 emb; r1-2; n11 late gestation	6107
	1.95 yg; r1-2; n21 f	1850
	mode 2 yg 83%	1829
	2-7 cl	11794
	2 emb; n1 f	5415
	2 emb ea; n2 f	10954a
	2 ps ea; n2 f	11956

	2.1 yg; n11 f	3628
	3.0 emb; r2-5; n10 early gestation	6107
	4 emb; n1 f	4387
Eptesicus melckorum	2 emb ea; n2 f	10158
Eptesicus nilssoni	1; r1-2	3571
	r1-2 yg; n35-40 f	5102
	2	8049
Eptesicus pumilus	1	4038
	1-2	7075
	1.17 yg; r1-2; n6 lt	6713
	1.21; r1-2; n28 f	6714
	2 yg	7074
Eptesicus regulus	1 yg	5834
Eptesicus serotinus	1 yg	8049
	1-2	3571
	2	4645
Eptesicus somalicus	2; n1 lt	9219
Euderma maculatum	1 yg	577
Hesperoptenus tickelli	mode 1	8493, 8494, 8501
Histiotus macrotus	1 emb ea; n3 f	3999a
Idionycteris phyllotis	1 yg	577
	1 emb; n1 f	969
	1 emb; n1 f	3607,
	1 emb; n1 f	5396
Kerivoula cuprosa	1 emb	252
Kerivoula lanosa	1 emb; n1 f	1347
Kerivoula papillosa	1 emb ea; n2 f	6447
Kerivoula picta	1	8494, 8501
Kerivoula smithi	1 emb; n1 f	252
Lasionycteris noctivagans	1 ps; n1 f	11956
	1+; r1-2; n11	10954
	1.78 emb; r1-2; n9 f	2868
	1.84 emb; n36 f	2733
	1.86 yg; r1-2; n7 f	8249
	2 emb; n1 f	261a
	2 emb; n1 f	4525
	2 emb; n1 f	10954a
	2 yg; n1 f	6122a
	2 emb ea; n4 f	495
	mode 2; r1-2	502
Lasiurus borealis	r1-4; 5 cases of 3; 3 cases of 4	181
	2-4 yg	4222
	2 emb; n1 f	1196a
	2 emb; n1 f	4525
	2 emb; n1 f	5396
	2 yg; n1 f	4054
	2 yg; n1 f	6105a
	2 ea; n2 lt	452
	2.67; mode 3; r2-4; n12 lt	11450
	2.78 yg; r1-4; n9 f	950a
	2.8 emb; mode 3; r2-3; n5 f	4079
	2.9 emb; mode 3; r2-3; n12 f	1290
	2.96 yg; r2-4; n27 f	5414a
	3 emb; n1 f	1973c
	3 emb; n1 f	4386
	3 yg; n1 lt	5405b,
	3.0 emb; r2-4; n2 f	589a
	3.0; r2-4; n4 lt	6677
	3.17 emb; mode 3; r2-5; n12 f	11627
	3.17 emb; r2-4; n18 f	1937
	3.24 emb; r2-4; n29 f	5414a
	3.35 emb; r2-5; n34 f	7608, 7610, 7611
	3.38 emb; r3-4; n13 f	6263
	3.45; r3-4; n11 f	6298

	3.5; r3-4; n4 f	7557
	3.8 yg; r3-4; n6 f	2197
	4 emb; n1 f	1871a
	4 emb; n1 f	2197, 5405b,
	4 yg; n1 f	1812a
	4 yg; n1 f	8187
	4 yg ea; n2 f	10298
Lasiurus cinereus	1+; r1-2	10954
	1.83; r1-2; n6 lt	1043
	1.9 yg; r1-2; n9 f	5414a
	2	181
	2-4 yg	502
	2 emb; n1 f	1196a
	2 emb; n1 f	10954a
	2 emb; n1 f	11244
	2 yg; n1 f	1812a
	2 yg; n1 f	7619
	2 yg; n1 f	7847
	2 yg; n1 f	8656, 8658
	2 ea; n2 lt	2197
	2 emb ea; n2 f	11627
	2 yg ea; n2 f	4222
	2 emb ea; n3 f	5414a
	2 emb ea; n4 f	1043
	2 yg ea; n4 f	9711
	2 emb ea; n8 f	8764
	max 4	2088
Lasiurus ega	1.5 emb; r1-2; n2 f	950
	2; n1 lt	190
	2 emb; n3 f	577
	2 emb ea; n3 f	5424e
	2.3; r1-3; n15	452
	2.9 emb; mode 3; r2-4; n17 f	7695
	3.0 emb; mode 3; r2-4; n7 f	528
Lasiurus intermedius	2 emb; n1 f	5413a
	2.0 emb; r1-3; n2 f	950
	3 emb; n1 f	6257
	3 yg ea; n2 f	9896a
	3.4 emb; r3-4; n13 f; est 2.8 born	577
Lasiurus seminolus	2-4	1474
	2 emb; n1 f	1973c
	2.5 emb; r2-3; n2 f	589a
	3.0 emb; r2-4; n2 f	7498a
	3.3 emb; mode 3-4; r1-4; n21 f	577
Mimetillus moloneyi	1 emb; n1 f	1346
Miniopterus australis	1	7163
	1	9102
	1; n many lt	7075
Miniopterus inflatus	1 emb	252
	1 emb ea; n3 f	10158
Miniopterus medius	1 yg	9457
Miniopterus mullatus?	1 yg	1351
Miniopterus schreibersi	1 emb	8965
	1 emb; n1 f	8049
	1 mode; 1/150 f with twins	10158
	1 emb ea; n11 f	4402
	1 emb ea; n40 f	8464
	1.0 emb; r1-2; n312 f; <1% twins	11128, 11131, 11134, 11135
	1 emb ea; n447 f	3927
Murina cyclotis	2 emb; n1 f	7161
Murina tubinaris	1.4 emb; r1-2; n5 f	4813b
Myotis adversus	1 ea; n22 lt	7075
Myotis albescens	1 emb; n10 f	10974
	1 emb ea; n11 f	3999a

Myotis auriculus	1 yg	577, 11465
Myotis austroriparius	1.18 wean; n1523 f; 1st flight	9089
	1.72; r1-2; n29 f	9895
	1.72 yg; n112 f	9089
	1.90 emb; r1-2; n1489 f	9089
	mode 2 yg	3428, 5096a
	3 emb; n1 f	3428
Myotis bechsteini	1	3571
Myotis blythii	1 emb ea; n5 f	5988
Myotis bocagei	1 emb; n1 f	8611
	1 ea; n2 lt	191
	1 emb ea; n3 f	1347
Myotis californicus	1	6060
	1 emb; n1 f	5413a
	1 emb; n1 f	11228a
	1 yg; n1 f	7611
	1 yg; n1 f	12104a
	1 emb ea; n3 f	3607_1
	1 emb ea; n3 f	7611
	1 emb ea; n6 f	1937
Myotis capaccinii	1 emb; 1 f	5988
Myotis chiloensis	1 emb ea; n3 f	3608
Myotis daubentoni	1	3571
	1 yg; n1 lt	11622
	1-2 emb	8049
Myotis elegans	1 emb; n1 f	5424f
Myotis emarginatus	1	3571
	1 emb; n1 f	5988
	1 emb; n1 f	8049
Myotis evotis	1 emb; n1 f	1940
	1 emb; n1 f	2415
	1 emb; n1 f	4079
	1 emb ea; n2 f	5415
	1 emb ea; n2 f	5424_2
	1 emb ea; n2 f	11228a
	1 emb ea; n2 f	11646a
	1 emb ea; n2 f	1937
	1 emb ea; n≥3 f	261a
Myotis fortidens	1 emb ea; n4 f	5413a
Myotis frater	1	6723
Myotis grisescens	1 emb; n1 f	6257
Myotis hasseltii	1 emb; n1 f	10872
	1; n1 lt	7161
Myotis keaysi	1 emb ea; n7 f	3999a
Myotis keenii	1 emb ea; n4 f	2197
	1 emb ea; n7 f	3607_1
	1 ea; n23 f	2043
Myotis leibii	1 emb	5415
	1 yg	5921
	1 emb; n1 f	502
	1 emb; n1 f	5424_2
	1 emb; n1 f	8807a
	1 emb ea; n2 f	3703
	1 emb ea; n4 f	10954, 10954a
Myotis lucifugus	1 ps ea; n2 f	10839
	1 emb; n1 f	7607
	1 emb; n1 f	11956
	1 emb; n>1 f	168
	1 emb ea; n5 f	2842
	1 emb ea; n5 f	5210
	1 emb ea; n14 f	261a
	1.01 emb; r1-2; n312 f	9645
	2	7651
	2 yg; n1 f	2352

Myotis macrodactylus	1	6115
Myotis muricola	1	179
Myotis myotis	1	8049
	1	9220
	1; always 1 emb	436
	1; no twins	3517
	1 emb ea; n74 f	11318
	1; r1-2	3571
	mode 1; r1-2; 2 rare	10045
	2	4645
Myotis mystacinus	1	3571
	1 emb ea; n8 f	10516
	1-2	4645
Myotis nattereri	1 yg	8049
	1	2045
	1 emb ea; n8 f	5988
Myotis nigricans	1	7695
	1 emb ea; n2 f	10974
	1 emb ea; n2 f	5425
	1 ea; n9 lt	452
	1 emb ea; n58 f	3999a
	1 emb; n~140 f	11752, 11755
Myotis oxyotus	1 emb; n1 f	3999a
Myotis riparius	1 emb; n1 f	3999a
Myotis siligorensis	1 emb; n1 f	10872
Myotis simus	1 emb; n1 f	7699a
	1 emb ea; n3 f	3999a
Myotis sodalis	1 emb; n1 f	4231
	1 emb; n1 f	7609
	1 emb; n7 f	2867
Myotis thysanodes	1 emb; n1 f	1641
	1 emb ea; n2 f	280a
	1 emb ea; n2 f	1937
	1 emb ea; n2 f	11471a
	1 ea; n3 lt	10954
Myotis tricolor	1 emb ea; n3 f	825
	1 emb ea; n30 f	4402
Myotis velifer	1 emb; n1 f	11471a
	1 emb; n1 f	1196a
	1 emb ea; n2 f	1937
	1 emb ea; n9 f	5405b, 5409
Myotis vivesi	1	1346
	1 emb ea; n2 f	1469a
	1 emb ea; n27 f	9002a
Myotis volans	1	1937
	1 emb; n1 f	2415
	1 emb; n1 f	5424e
	1 emb; n1 f	11228a
	1 emb ea; n2 f	261a
	1 emb ea; n2 f	1940
	1 emb ea; n2 f	5415
	1 emb ea; n2 f	5424_2
	1 emb ea; n2 f	11471a
	1 emb ea; n4 f	8807a
	1 emb ea; n20+	2231a
Myotis yumanensis	1 emb ea; n7 f	3803a
	1 emb ea; n8 f	7607
	1 emb ea; n62 f	2227
Nyctalus aviator	1.91; n11 lt	3033
Nyctalus lasiopterus	1.67 yg; r1-2; n9 lt	6722
Nyctalus leisleri	1-2	4645
	1-2	3571
	2; mode 2	8883

Nyctalus noctula	1-2	181
	1-2	11147
	1-2	4645
	1-2	8137
	1-3 yg; 3 rare	3577
	1 ea; n2 lt	11622
	1.15; r1-3; n181 lt	754
	1.2; 20% twins; r1-2	10071
	1.3 yg; r1-2; n7 f	5860, 8836
	1.4; r1-2; n7 lt	2593
	1.4 emb; r1-2; n12 f	11318
	1.5; r1-2	8049
	2.0; r1-3; n4 lt	9399
Nycticeius balstoni	1 yg; n1 lt	9398
	2 yg; n1 lt	7075
Nycticeius humeralis	1 emb; n1 f	5414a
	1.5 emb; r1-2; n2 f	6474
	1.60; r1-2; n5 f	1290
	1.75 emb; r1-2; n4 f	2868
	1.98 emb; mode 2; r1-4; n58 f	577
	2 emb; n1 f	950a
	2 emb; n1 f	5803
	2 emb ea; n2 f	2197, 5405b,
	2.0 emb; n3 f	11627
	2 yg ea; n several f	9709
	2 emb ea; n>9 f	3629
	2.0 yg; n14 lt	5386
	2 emb; r1-3; both 1 and 3 rare; n hundreds	11470
	2.3 yg; n115 lt	11472
	2.7 emb; r2-4; n3 f	589a
	3 emb; n1 f	4386
Nycticeius rueppellii	1; n1 lt	7075
Nycticeius schlieffeni	2.8 emb, morulae; mode 3; r1-5; n34 lt	11135, 11137, 11138
	2.8 cl; mode 3; r1-4; n32 f	11137, 11138
Nyctophilus geoffroyi	1-2	7075
	1.33; r1-2; n3 lt	9397
	mode 2; r1-2	4038, 4039
	2 yg ea; n4 f	7074
Nyctophilus gouldi	1.55; r1-2; n33 lt	8492
Nyctophilus timoriensis	2; n5 lt	7075
Otonycteris hemprichi	2 emb; n1 f	8049
Philetor brachypterus	1 emb; n1 f	6447
Pipistrellus spp	2 emb ea; n3 f	9136a
Pipistrellus ceylonicus	mode 2; r1-2	8493, 8494, 8501
	2 yg	1340, 1344
	2.01 emb; r1-3; n355 f	3915, 3927, 3932, 3933, 3939, 6715
Pipistrellus coromandra	1 yg ea; n2 f	8501
	2 emb; n1 f	1340, 1344
Pipistrellus dormeri	1+; r1-2; 1/12 twins	3943
	1.34 emb; r1-2; n271 f	3927
Pipistrellus hesperus	1.75; r1-2; n4 f	1937
	1.8 emb; mode 2; r1-2; n6 f	4079
	mode 2; r1-2	6064
	2 emb; n1 f	502
	2 emb; n1 f	11228a
	2 emb; n1 f	11471a
	2 emb ea; n2 f	4079
	2 emb ea; n4 f	7611
	2 emb ea; n6 f	280a
Pipistrellus javanicus	1-2	436
	2; r1-4	11016
	2.10 yg; mode 2; r1-3; n67 lt	11013, 11014
	2.50; r1-3; n32 f	10872

	2.94 emb; se0.13; mode 3; n17 f	12033
	3 emb; n1 f	179
	3 emb, yg; r2-4; n5 f	7521
Pipistrellus kuhlii	1-2	9175
	1 emb ea; n2 f	10158
	1.5; r1-2; n2 f	436
	1.5 emb; r1-2; n4 f	7720
Pipistrellus mimus	mode 1	9844
	1.93 emb; r1-3; n89 f	3927
	2	6026
	mode 2	5530
	mode 2; r1-2	8493, 8494, 8501
	mode 2 emb; r1-3	3944
	2 emb; n1 f	10872
	2.15 cl; 2.0 emb; r1.9-2.1; n115 lt	70
	3 emb; n1 f	5705
Pipistrellus nanulus	2 emb; n1 f	5387
Pipistrellus nanus	1+; r1-2; 2 rare	191
	1-2	6262
	1-2; occasionally 3	252
	1.5; r1-2; n6 f	1347
	1.9 emb; r1-2; n26 f	10158
	2 emb; n1 f	9945
	2 emb; n1 f	71
	2 emb, yg ea; n7 lt	6086
	2 ea; n16 f	8016
Pipistrellus nathusii	1-2	4645
	2	8049
	mode 2; r1-2	3571
Pipistrellus pipistrellus	1	8049
	1 ea; n3 lt	11622
	1 emb ea; n4 f	10872
	1 ea; n14 lt	8826
	1.05 emb; r1-2; n117 f	2420
	1.1 wean; r1-2; n35 lt	1179
	1.4; mode 1; r1-2; n12 lt	11318
	1.5 emb; r1-2; n64 f	8865
	mode 2	4645
Pipistrellus rusticus	mode 2 emb implant; r1-2	11139
	max 5 emb before implantation	11139
Pipistrellus savii	2	4645
	mode 2; r1-2	3571
Pipistrellus stenopterus	1 emb ea; n9 f	4625
Pipistrellus subflavus	1 emb ea ; n2 f	1937
	1.88 emb; r1-2; n24 f	1290
	2 emb; n1 f	4386
	2 emb; n1 f	4267
	2 emb; n1 f	8658
	2 yg; n1 f	2356
	2 emb ea; n2 f	6415
	2 emb ea; n7 f	6298
	2 emb ea; n17 f	2043
	2 emb ea; n19 f	6175
	2 yg; 2-4 ova	11795
	mode 2 yg; occasionally 3	2353
	3 emb; n1 f	7557
	3 emb; n1 f	4202
	4 emb; n1 f	6474
Pipistrellus tenuis	1; n1 lt	7075
	1 emb ea; n15 f	7073
Plecotus auritus	1	2045
	1-2	3571
	1-2	4645
Plecotus mexicanus	1 emb ea; n3 f	11471a

Plecotus rafinesquii	1 yg	587
	1 yg; n1 f	9578a
Plecotus townsendii	1	1937
	1	8314
	1 yg	5024
	mode 1	502
	1 emb; n1 f	10954a
Rhogeessa parvula	mode 2	11804
	2 yg ea; n2 lt	1937
	2 emb ea; n6 f	5413a
Rhogeessa tumida	1 emb ea; n2 f	5405e
	1.4; r1-2; n5 f	452
	2 emb ea; n2 f	950
	2 emb ea; n2 f	3608
Scotoecus hirundo	2 emb ea; n2 f	5759
Scotophilus dinganii	2 emb; n1 f	8084
	2 ea; n7 f	10158
	2.25 emb; r2-3; n4 f	10158
Scotophilus heathi	1.33 emb; r1-2; n3 f	10872
	1.93 emb; r1-2; n265	3927
	2 ova; 1 per ovary	6028
	mode 2 yg	1340, 1344
	mode 2	6717
Scotophilus kuhli	1-2	9175
	1.3 emb; r1-2; n3 f	8501
	2 emb	3905, 3915
	2 emb ea; n76 f	3927
Scotophilus leucogaster	1 emb; n1 f	8611
	2 emb	191
	2 emb; n4 f	5759
Scotophilus nigrita	1 emb ea; n2 f	191
	1-2 emb	9945
	1-2	5759
	1.57; r1-2; n7 f	252
	2; n1 f	5434
Scotophilus viridis	2 ea; 9 f	10158
Tylonycteris pachypus	1.92 emb; mode 2; n13 f	7161, 7164
Tylonycteris robustula	2.0 emb/yg; mode 2; n10 f	7161, 7164
Vespertilio murinus	1; r1-2	3571
	1-2	8049
	1 ea; n several hundred lt	2824
Vespertilio superans	2; r1-3	3563
SEXUAL MATURITY		
Eptesicus furinalis	f: preg: <1 yr	7695
	m: sperm: <1 yr	7695
Eptesicus fuscus	f: preg: 50% 1 yr	9643
	f/m: 2nd yr	8798
Lasionycteris noctivagans	f/m: 5 mo	2733
Lasiurus ega	f: preg: <1 yr	7695
	m: sperm: <1 yr	7695
Miniopterus australis	f: preg: 1st yr	7163
Miniopterus inflatus	f/m: 2 yr	1349
Miniopterus schreibersi	f: 19 mo	3945
	f: not 1st yr	2828
	m: 2 yr	3945
	f/m: 1.25-1.5 yr	3571
Myotis albescens	f: 1st mate: 2 mo	7695
	m: sperm: <1 yr	7695
Myotis austroriparius	f: 1st birth: 1 yr	9089
	m: 1st breeding season after birth	9089
Myotis blythii	f: may breed yr of birth	5988
Myotis daubentoni	f/m: 2nd yr	8798
Myotis emarginatus	f/m: ~15 mo	10070
Myotis frater	f: 1st birth: 2 yr	6723

Myotis grisescens	f: 1st birth: 2 yr	10976, 10978
	m: 1st mate: 15 mo	7330
Myotis lucifugus	f: 1st birth: 2 yr	9645
	f: 1st summer	4174
	m: 1st mate: 15 mo	7330
	m: spermatogenesis: 1 yr	4163
	f/m: 1st summer; 8 wk	4163
	f/m: 8 mo	5210
Myotis myotis	f: 1st birth: 2 yr	4663
	f/m: 1st mate: 1.25-1.5 yr	3571
	f/m: 2nd yr	10045
Myotis mystacinus	f: ~1 yr	10516
	f/m: 15 mo	10067
Myotis nigricans	m: spermatogenesis: 3.5-4 mo	11756
	f/m: 2-3 mo	11752
	f/m: 4 mo	7695
Myotis tricolor	f: 1st conceive: 4-16 mo	11818
Myotis velifer	f: 1st yr	6106
	f: 1 yr	6106
Myotis volans	f: 5 mo	2733
	m: yearlings	2733
Nyctalus noctula	f: 1st birth: 1 yr	5860, 8836
	f: 1st repro: 1 yr	3577
	f: 1st mate: 3 mo	5860
	m: 1st mate: 15 mo	5860
Nyctophilus gouldi	f: 7-9 mo	8492
	m: fertile: 12-15 mo	8492
Pipistrellus ceylonicus	f/m: 9-10 mo	6715
Pipistrellus mimus	f/m: 2 g	3944, 5530
Pipistrellus pipistrellus	m: spermatogenesis: 2 yr	2075
	f/m: 6 mo	6104
	f/m: ~2 yr	2420
Pipistrellus subflavus	f: 3-11 mo; n1	3549
Plecotus auritus	f: 1st birth: 75% 2nd yr, rest 3rd yr	2045
	m: 12 mo	2045
Scotophilus kuhli	m: epididymal sperm: 9 mo	3905, 3906
Tylonycteris pachypus	f/m: breed season following birth	7161
Tylonycteris robustula	m: 5 mo	7164
Vespertilio murinus	f/m: 2nd yr	9240
GESTATION		
Antrozous pallidus	~9 wk; r53-71 d	8127
Chalinolobus gouldii	~3 mo	5830
Eptesicus furinalis	3 mo; 1st lt; 2nd may be shorter	7695
Eptesicus fuscus	35 d	8798
	est 2 mo	1829
Eptesicus regulus	3 mo	5834
Eptesicus serotinus	2-2.5 mo	436
Lasionycteris noctivagans	implant: 10 d	2733
	50-60 d	2733
Lasiurus borealis	est 80-90 d growth	5210
	8 mo with delay	5210
Lasiurus cinereus	8 wk	2733
Lasiurus ega	3-3.5 mo	7695
Miniopterus australis	implant: <1 mo	9102
	4 mo without delay	2827, 2831, 2832
	~110 d	518
	4.5-5 mo; 134-149 d	7163
	6-6.5 mo with delay	2827, 2831, 2832
Miniopterus fraterculus	6.5 mo: ~4 mo growth + 2.5 mo delay	822
Miniopterus inflatus	3 mo	255
Miniopterus mullatus?	no delay	1351
Miniopterus schreibersi	implant: 4 mo	11128, 11129, 11131
	2 mo	5727
	~2.5 mo	11418

	120-125 d	3945
	~140 d	1340, 1344
	6 mo: 3 mo growth + 3 mo delay	2827, 2832
	7 mo	2828
	7.5 mo: 5 mo growth + 2.5 mo delay	9102
	8 mo: ~4 mo growth + 4 mo delay	821, 822
	8 mo: 4 mo growth + 4 mo delay	11128, 11129, 11131, 11132, 11135
	8 mo: 3 mo growth + 5 mo delay	8464
	260 d; with 2 mo delay	5727
	~9 mo: 2.5 mo growth + 6.5 mo delay	8463
	10-11 mo: 4 mo growth + 6-7 mo delay	8574
Myotis albescens	≤3 mo; later It may be shorter	7695
Myotis daubentoni	~54 d	3571, 8798
Myotis lucifugus	implant: 10 d post ovulation	1414
	implant: > 10 d post ovulation	11793
	50-60 d	11795
	2.5 mo	343
	~80 d	1474
Myotis myotis	49-59 d	10045
	~50 d; can be longer if weather unfavorable	4663
	60-70 d; delayed implantation	3571
	2.5-3 mo	436
Myotis nigricans	50-60 d; field data	11752, 11755, 11759
	≤3 mo	7695
Myotis tricolor	63 d; n20; fetal developemnt	825
	3 mo	255
Mytois thysanodes	50-60 d	8005
Myotis velifer	est 60-70 d	6106
Nyctalus noctula	70-73 d; delayed	3571
	~70-73 d	2945
	73 d	8798
Nycticeius balstoni	~7 mo	9398
Nycticeius schlieffeni	11 wk; field data	11135, 11138
Pipistrellus ceylonicus	50-55 d; does not include sperm storage	3932, 3933, 6715
Pipistrellus javanicus	~70 d; field data	11013
Pipistrellus mimus	~60 d	6026
Pipistrellus pipistrellus	35 d	8798
	41-49 d	8049
	~44 d; delayed	2420
	45 d; r41-51; arousal to birth; field data	8840
Pipistrellus subflavus	postimplantation: 44 d	3549
Pipistrellus rusticus	11-12 wk; field data	11139
Plecotus auritus	60-70 d; arousal to birth; field data	10248
Plecotus townsendii	56-100 d	8314
Scotophilus heathi	3.5 mo	9175
	116 d	6028
Scotophilus kuhli	est 105-115 d; field data	3905
	~4.5 mo	7415
Scotophilus leucogaster	delay: 6-8 wk as morula, 2 mo as blastocyst	11145b
	7.5 mo	11145b
Tylonycteris pachypus	~89 d	7164
Tylonycteris robustula	89 d	7164
Vespertilio murinus	~50 d; arousal +5-10 d until birth	9240
LACTATION		
Antrozous pallidus	solid food: 6 wk	8127
	den emergence: 7 wk	8127
Eptesicus fuscus	1st fly: 3 wk	6107
	forage: 4 wk	1455, 6107
	wean: est 32-40 d	6105a, 6107
Eptesicus pumilus	1st fly: 3-4 wk	6714
Eptesicus regulus	wean: ~2 mo	5834

Lasionycteris noctivagans	1st fly: 3-4 wk	1871a
	wean: 36 d	1871a
Lasiurus borealis	1st fly: 4-6 wk	1871a
	wean: est 38 d	6105a
Lasiurus cinereus	solid food: 22 d	1043
	1st fly: 3-4 wk	1871a
	1st fly: 4 wk	1474
	nurse: ~30 d	2733
	wean: 34 d	1043
Miniopterus schreibersi	wean: 6-7 wk	3562
	wean: 2 mo	8464
	wean: 3 mo	2828
Myotis adversus	nurse: 2 mo	2834
Myotis albescens	wean: 1 mo	7695
Myotis austroriparius	1st fly: 5-6 wk	9089
Myotis californicus	1st fly: 2 mo	1871a
Myotis lucifugus	wean: 3 wk	3216
	wean: 26 d	6122
	wean: 4 wk	5210
Myotis myotis	1st fly: 5-6 wk	3571
	wean: ~6 wk	4663
Myotis nattereri	wean: 2 mo	6248
Myotis nigricans	1st fly: 3 wk	11752
	nurse: 1 mo	7695
	wean: 5-6 wk	11752, 11755, 11759
Myotis tricolor	lact: 6 wk	825
Myotis thysanodes	1st fly: 17 d	1871a
	wean: est 3 wk	8005
Myotis velifer	1st forage: 4th wk	6106
	wean: est ~43 d	6106
Nyctalus noctula	solid food: shortly after birth	4645
	1st fly: 5 wk	3571
	wean: 2 mo	10071
Nycticeius humeralis	1st fly: 20 d	11470
	wean: 1 mo; nurse own yg 2 wk then nurse any yg	11472
	solid food: 4 wk	5386
Nyctophilus gouldi	wean: 6 wk	8492
Pipistrellus ceylonicus	wean: 25-30 d; max 7 wk	6715
Pipistrellus kuhlii	wean: est ~2 mo	6418
Pipistrellus mimus	wean: 2 wk	6026
	wean: 3 mo	9844
Pipistrellus pipistrellus	wean: ~6 wk	10603
	independent: 2 mo	3571
Pipistrellus subflavus	1st fly: 3 wk	3549
	1st fly: 4 wk	1474, 5210
Plecotus auritus	wean: 40-45 d	10248
	independent: 6 wk	3571
Plecotus rafinesquii	female/young association: 3 wk	5397
Plecotus townsendii	1st fly: 3 wk	1474
	wean: 6 wk	11870
Scotophilus heathi	nurse: 2-3 wk	6028
	nurse: 24-28 d	6717
Scotophilus kuhli	nurse: 2 mo	3907
Tylonycteris pachypus	roost emergence: 6 wk	7161, 7164
Vespertilio murinus	solid food: 50 d	9240
	wean: 3 mo	9240
Vespertilio superans	wean: 30-40 d	3563
LITTERS/YEAR		
Barbastella barbastellus	1	3571
	1	8049
Chalinolobus morio	1	5833
	1	12042
Eptesicus furinalis	2	7695
Eptesicus fuscus	1	1474

Eptesicus nilssoni	1	3571
Eptesicus serotinus	1	3571
Histiotus velatus	1	8368
Lasionycteris noctivagans	1	1474
	1	7235
Lasiurus borealis	1	1474
Lasiurus cinereus	1	1474
Lasiurus ega	1	7695
Lasiurus seminolus	1	1474
Miniopterus australis	1	7163
Miniopterus mullatus?	1	1351
Miniopterus schreibersi	1	1344
	1	3571
	1	11128, 11135
Myotis adversus	2; max 3	2833
Myotis albescens	1-3	7695
Myotis austroriparius	1	1474
Myotis bechsteini	1	3571
Myotis californicus	1	1474
Myotis daubentoni	1	3571
Myotis emarginatus	1	3571
	1	9240
Myotis evotis	1	1474
Myotis grisescens	1	10978
	1	1474
Myotis keenii	1	1474
Myotis leibii	1	1474
Myotis lucifugus	1	577
	1	4059
	1	1474
Myotis myotis	1	4663
	1	9220
	1	3571
Myotis mystacinus	1	3571
Myotis nattereri	1	2045
Myotis nigricans	3; r1-4	11755, 11759
Myotis thysanodes	1	577
	1	1474
Myotis velifer	1	1474
Myotis volans	1	1474
Myotis yumanensis	1	1474
Nyctalus leisleri	1	3571
Nyctalus noctula	1	3571
	1	8049
	1	10071
Nycticeius humeralis	1	1474
Pipistrellus hesperus	1	1474
Pipistrellus mimus	3	6025
Pipistrellus nathusii	1	3571
Pipistrellus pipistrellus	1	2420
	1	3571
	1	8049
Pipistrellus savii	1	3571
Pipistrellus subflavus	1	1474
Plecotus auritus	1	3571
	1	2045
Plecotus rafinesquii	1	1474
Plecotus townsendii	1	1474
Scotophilus heathi	1	6028
Scotophilus kuhli	1	9175
	1	3905, 3907
Tylonycteris pachypus	1	7161
Vespertilio murinus	1	9240
	1	8049

SEASONALITY

Antrozous dubiaquercus	preg f: Apr; n1; Honduras	8554
Antrozous pallidus	testes: enlarge end Aug, large Sept-Oct; CA	8127
	mate: Oct-Nov, Feb; CA	8127
	mate: Oct-Feb; WY	1871a
	mate: fall; sw US	8175
	ovulate: early Apr; n3/4 f; CA	8127
	conceive: spring; sw US	8175
	implant: mid Apr; CA	8127
	preg f: Apr; n2; CA	11228a
	preg f: May; n1; Coahuila, Mexico	522
	preg f: May; n1; Chihuahua, Mexico	1196a
	preg f: May-June; Chihuahua, Mexico	280a
	preg f: Aug-Oct; CA	8127
	birth: Apr-June; sw US	1474
	birth: May-June; CA	8127
	birth: est July; Sonora, Mexico	1471a
Barbastella barbastellus	birth: May-early Aug; Germany	3571
	birth: end July; Germany	7777
	birth: early in the yr; USSR	8049
Chalinolobus argentatus	birth: Mar; Cameroun, Kenya	252
	birth: Oct; Tanzania	252
	preg f: Jan; n2; Cameroun	71
Chalinolobus beatrix	birth: prob Mar; central Africa	252
Chalinolobus gouldii	mate: early winter; sperm stored; WA, Australia	5830
	conceive: end winter; WA, Australia	5830
	preg f: Dec; n22/25 f; VIC, Australia	9989
	birth: end Sept-early Oct; n Australia	5830
	birth: end Nov-early Dec; WA, Australia	5830
Chalinolobus morio	preg f: Aug-Oct; sw Australia	5833
Chalinolobus tuberculatus	birth: Jan-Feb; New Zealand	2259
Chalinolobus variegatus	sexually active m: July, Oct; Uganda	5759
	preg f: Mar; n3/10; Zaire	191
	birth: Mar; central Africa	252
Eptesicus brasiliensis	preg f: June, Aug; n12; Peru	3999a
	lact f: end July-early Aug; n1; Oaxaca, Mexico	523
Eptesicus capensis	preg f: Oct-early Nov; Transvaal, S Africa	8965
	preg f: Nov; Namibia	9945
	birth: end Feb-early Mar; Garamba Nat Park, Zaire	11270
	birth: mid Nov; Tanzania, Zambia	252
	birth: est early-mid summer; Transvaal, S Africa	8965
	birth: summer; S Africa	10158
Eptesicus furinalis	testes: large Apr-June, Sept-Dec; Paraguay	7695
	mate: May-June, Sept-Nov; Paraguay	7695
	preg f: May; n1; Michoacan, Mexico	4671
	preg f: July; n1; Nicaragua	5425
	preg f: end July-Dec; Paraguay	7695
	birth: Oct-Nov, Jan; Paraguay	7695
	lact f: July; n9; Nicaragua	2670a
Eptesicus fuscus	testes: large Aug-Sept, small Oct-June; IN	11627
	mate: fall; KS	8485
	mate: fall; IN	11627
	ovulate: early Apr; PA, MD	1829
	ovulate: spring; PA	11794
	preg f: Mar; n8; Guatemala	5405e
	preg f: mid Apr-May; IN	11627
	preg f: May; n1; NC	1290
	preg f: May; n1; Jalisco, Mexico	11471a
	preg f: June; n8; s central BC, Canada	3217
	preg f: June; n2; MT	5424_2
	preg f: June; n4/8; ND	3703
	preg f: June; n1; CO	1649a
	preg f: June; n11; KS	950a
	preg f: June-July; n2; w SD	10954a

	preg f: June-July; n13; NE	$5405b_1$
	birth: mid Mar-June; PA, MD	1829
	birth: May; IL	6298
	birth: May-mid June; SD	10954
	birth: end May-mid June, peak early June; KS	6107
	birth: end May-early June; CA	6059
	birth: early June; KY	2354
	birth: June-mid July; ALB, Canada	9643
	birth: June-mid July, peak end June; ALB, Canada	9643
	birth: July; BC, Canada	2088
	1st yg: June; IN	11627
	lact f: July; n21; MT	5424_2
	lact f: July; n3; ND	3703
	lact f: July; n1; SD	261a
	lact f: July; n8; w SD	10954a
	lact f: July; Baja CA, Mexico	5424e
	lact f: July; n1; Durango, Mexico	523a
	lact f: Sept; n1; NE	$5405b_1$
Eptesicus guineensis	birth: June; central Africa	252
Eptesicus melckorum	preg f: Oct; n2; Cape Provice, S Africa	10158
Eptesicus nilssoni	birth: June; Germany	3571
Eptesicus pumilus	testes: large Nov; TAS, Australia	4038
	birth: end Nov-mid Dec; central Australia	6713
	birth: end Nov-mid Dec; TAS, Australia	4038
Eptesicus regulus	mate: autumn; sw Australia	5834
	conceive: end winter; store sperm; sw Australia	5834
	birth: end Nov-Dec; sw Australia	5834
Eptesicus rendalli	birth: end Mar-early Apr; Garamba Nat Park, Zaire	252, 11270
Eptesicus serotinus	spermatogenesis: May-Sept; Europe	2075
	mate: Apr; n1; Pakistan	9175
	mate: mid Sept; Germany	2945
	mate: Sept, spring; Germany	3571
	mate: after mid Nov; Portugal	436
	mate: fall; Iraq	110
	ovulate: Mar; Portugal	436
	preg f: spring; Iraq	110
	birth: end May; Portugal	436
	birth: June; Germany	7777
	birth: June; Germany	3571
	birth: spring-early summer; Pakistan	9175
Eptesicus tenuipinnis	mate: Jan; central Africa	252
	birth: Mar; n1; Cameroun	2948
	birth: Mar; Cameroun	252
	birth: Oct; Albert Nat Park, Zaire	252
	lact f: July; n1; Liberia	184
Euderma maculatum	lact f: June; n3; NM	5396
Hesperoptenus tickelli	preg f: May (many), Dec (n1); Sri Lanka	8501
	birth: May; Sri Lanka	8493
	birth: early June; Sri Lanka	8494, 8501
Histiotus macrotus	preg f: Oct; n3; Peru	3999a
Histiotus montanus	lact f: Dec; n4; Chile	4054
Histiotus velatus	repro: mid Sept; Rio de Janeiro, Brazil	8368
Idionycteris phyllotis	preg f: June; n1; NM	5396
	preg f: June; AZ	2196
	preg f: June; Durango, Mexico	2196
	preg f: June; n1; Durango, Mexico	3607_1
	birth: June-July; US	577
	birth: est June-July; AZ, NV, CO	1474
	lact f: June; n4; Durango, Mexico	3607_1
	lact f: June-July; n9; NM	5396
	lact f: July; AZ	1939a
	lact f: Aug; n2; AZ	3254a
Kerivoula cuprosa	birth: est Jan; Cameroun, Kenya	252
Kerivoula lanosa	birth: Jan; central Africa	252

Kerivoula papillosa	preg f: June, Oct, Dec; n4; Malaysia	6446, 6447
Kerivoula smithi	birth: Sept; central Africa	252
Lasionycteris noctivagans	spermatogenesis: May-Oct, peak end Oct; NM	2733
	mate: July; BC, Canada	2088
	mate: end Sept-spring; NM	2733
	ovulate: Apr-May; NM	2733
	preg f: May-June; n4; MN	495
	preg f: June; n1; SD	261a
	preg f: June; n1; w SD	10954a
	preg f: June; n2; Long I, NY	7881a
	preg f: June-July; n9; IA	2868
	birth: end May-early July; SD	10954
	birth: end May-early July; NE	$5405b_1$
	birth: June; n1; MI	6122a
	birth: end June-early July; US	577
	birth: end June-early July; WI	5210
	birth: June-July; US	1474
	birth: early July; NY	7235
	birth: July; BC, Canada	2088
	lact f: June-July; n5; IA	2868
	lact f: end June-July; IA	6105a
	lact f: July; n10; MT	5424_2
	lact f: July; n1; MN	495
Lasiurus borealis	mate: Aug; NY	7651
	mate: Aug-Sept; WY	1871a
	preg f: Apr-May; n12; NC	1290
	preg f: May; n1; WY	1871a
	preg f: May; n1; NE	$5405b_1$
	preg f: May; n1; NM	5396
	preg f: May; n2; AL	589a
	preg f: May-early June; LA	6263
	preg f: end Apr-mid June; IN	7608, 11627
	preg f: May-June; n18; KS, IN	1937
	preg f: June; n1; SC	1973c
	preg f: June; n1; Chihuahua, Mexico	1196a
	birth: May-early June; IL	6298
	birth: end May-early June; KS	4222
	birth: end May-June; KS	5414a
	birth: June; US	1474
	lact f: June; n1; KS	1812a
	lact f: June; n3; NM	5396
	lact f: June; n1; Coahuila, Mexico	522
	lact f: June-mid July; LA	6263
	lact f: July; n<12; MT	5424_2
	lact f: July; n2; MN	495
	lact f: July; n2; SC	1973c
	lact f: Dec; n1; Chile	4054
Lasiurus cinereus	spermatogenesis: May-end Oct, peak early Sept; NM	2733
	mate: Aug; NY	7235
	mate: Sept-spring; NM	2733
	preg f: end May-early June; n1; Chihuahua, Mexico	1196a
	preg f: June; n1; w SD	10954a
	preg f: June; NE	$5405b_1$
	preg f: June; n1; IA	6105a
	birth: May-July; HI	10809
	birth: mid May-July; WY	1871a
	birth: end May-early June; IN	11627
	birth: peak end May-early June, May-July; SD	10954
	birth: June; N America	1474
	birth: June; n1; IL	7619
	birth: June; n6; captive, NM	1043
	birth: June; n1; SAS, Canada	7847
	birth: June-early July; BC, Canada	2088
	birth: mid June; KS	5414a

	birth: end spring-early summer?; NY	7235
	lact f: June; n1; w SD	10954a
	lact f: end June-July; IA	6105a
	lact f: July; n1; se MT	6166
Lasiurus ega	spermatogenesis: Dec-July; Paraguay	7695
	testes: large Mar-May; Paraguay	7695
	epididymal sperm: Mar-Sept; Paraguay	7695
	mate: May; Paraguay	7695
	ovulate: mid Aug; Asuncion, Paraguay	7695
	ovulate: mid Sept; Chaco, Paraguay	7695
	birth: June; NM	577
	birth: end Nov-early Dec; Paraguay	7695
	preg/lact f: July; n7; Baja CA, Mexico	5424e
	lact f: May; n1; Honduras	4048,
	lact f: May; n1; Costa Rica	3608
Lasiurus intermedius	preg f: June; n1; Sinaloa, Mexico	5413a
	birth: est end May-early June; FL	9896a
	lact f: July; n2; Hidalgo, Mexico	1640a
	lact f: July; n4; Veracruz, Mexico	522a
Lasiurus seminolus	preg f: May; n2; AL	589a
	preg f: May; n1; SC	1973c
	preg f: June; n2; FL	7498a
	birth: June; se US	1474
	lact f: June; n1; FL	7498a
	lact f: July; n1; NC	588a
	lact f: July; n1; SC	1973c
Mimetillus moloneyi	preg f: Feb; n1; tropical Africa	1346
	birth: Feb-Mar, Aug; Africa	5759
	birth: Mar; central Africa	252
	birth: bi-annual, end dry spells; e Africa	5759
Miniopterus australis	testes: large end May; NSW, Australia	2831
	testes: increase June-Aug; New Hebrides	518
	testes: large Nov; e Malaysia	7163
	mate: June-July; NSW, Australia	2827, 2832
	mate: Sept; New Hebrides	518
	mate: Nov-Dec; e Malaysia	7163
	conceive: mid Aug; Australia	9102
	implant: Sept; NSW, Australia	2827, 2832
	implant: mid Sept; Australia	9102
	preg f: Sept-Dec; New Hebrides	518
	birth: end Apr-early May; e Malaysia	7163
	birth: Dec; Australia	9102
	birth: Dec; NSW, Australia	2827, 2832
	birth: Dec; New Hebrides	518
Miniopterus fraterculus	mate: mid May; Natal, S Africa	822
	conceive: mid May; Natal, S Africa	822
Miniopterus inflatus	preg f: Oct; n3; Zimbabwe	10158
	birth: Oct; Cameroun, Gabon, se Zaire	252, 255
	birth: Oct; ne Gabon	1349
Miniopterus minor	birth: end Oct-early Nov; central Africa	252
	birth: Dec; e Africa	6966
	lact: ?-Dec; central Africa	252
Miniopterus mullatus?	birth: Oct; Gabon	1351
Miniopterus schreibersi	spermatogenesis: peak end Apr-mid Sept; Australia	9102
	spermatogenesis: Aug-Sept; s France	2075
	spermatogenesis: Sept; s France	8574
	testes: large Jan-Mar; S Africa	11134
	testes: large May-June, small Oct-early Nov; NSW, Australia	2828
	testes: small Dec-Jan; Germany	546
	follicular growth: Mar-Apr, Aug-Dec; Natal, S Africa	821
	mate: 1st 3 wk Feb; India	3945
	mate: mid Feb-early May, peak Mar-Apr; Transvaal, S Africa	11131, 11134, 11135
	mate: mid Apr; Natal, S Africa	822
	mate: May; 30°S; Aug; 15°S, Australia	2827, 2832

	mate: end May-early June; NSW, Australia	11418
	mate: end May-early June; Australia	9102
	mate: July?; few data; SW Africa	6966
	mate: mid Oct; Japan	11015
	mate: Sept; Pyrenees, France	8463
	mate: fall before hibernation; s France	2075
	mate: fall, spring; nw Sardinia, Italy	3517
	ovulate: Jan-Feb; Portugal	436
	ovulate: Feb-Mar; S Africa	11134
	ovulate: Mar-early Apr; s Transvaal, S Africa	11129
	ovulate: Mar-Apr; S Africa	11128
	ovulate: Mar-Apr, peak end Mar; s Transvaal, S Africa	11129
	ovulate: Apr; Natal, S Africa	821, 822
	ovulate: May; NSW, Australia	11418
	ovulate: Aug-Sept; Pyrenees, France	8464
	ovulate: mid Oct; diapause mid Oct-mid Dec; Japan	5727
	ovulate: mid Oct; Japan	11015
	ovulate: fall before hibernation; s France	2075
	implant: Feb; s France	8574
	implant: Mar; Pyrenees, France	8464
	implant: Mar-Apr; Pyrenees, France	8463
	implant: before mid Mar; Sardinia, Italy	3517
	implant: July-Aug, peak end July; Transvaal, S Africa	11129, 11131, 11133, 11135
	implant: Aug; Australia	9102
	implant: Aug-Sept; NSW, Australia	11418
	implant: mid Aug; Natal, S Africa	821
	implant: mid Sept; s Australia	2832
	implant: Aug-Sept; n Australia	2832
	implant: mid Dec-early Jan; Japan	5727, 5729
	preg f: Mar-Apr; Sardinia, Italy	3517
	preg f: May-Dec; NSW, Australia	2828
	preg f: Sept-Feb; s France	8574
	preg f: Oct, Nov; Transvaal, S Africa	8965
	preg f: mid Oct-early July; Japan	5729
	preg f: Nov; n11; Natal, S Africa	4402
	birth: mid Apr-end June; Algeria	5988
	birth: May; s France	8574
	birth: May-July; Germany	3571
	birth: June; Pyrenees, France	8463, 8464
	birth: end June-early July; Japan	3562, 5727
	birth: Oct-Nov; se Zaire	255
	birth: mid Oct; se Zaire	252
	birth: early Nov; Zambia	252
	birth: end Oct-mid Dec, peak mid-end Nov; s Transvaal, S Africa	11129, 11131, 11135
	birth: Dec; NSW, Australia	2827, 2828, 2832
	birth: Dec; Australia	9102
	birth: Dec; Natal, S Africa	821, 822
	birth: mid Oct-Jan; Australia	2840
	birth: early summer; Transvaal, S Africa	8965
	lact f: June-July; Pyrenees, Europe	8464
	lact f: July-Aug; Japan	3562
	lact f: Dec-Mar; NSW, Australia	2828
	wean: after July ?; s France	8574
Murina cyclotis	preg f: Feb; n1; Pahang, Malaysia	7161
Murina tubinaris	preg f: May; n5; Mishmi Hills, Assam, India	4813b
Myotis adversus	preg f: Mar; n2; Papua, New Guinea	7073
	birth: Jan, Oct; n19; QLD, Australia	2833
	lact f: Mar; n4; Papua, New Guinea	7073
Myotis albescens	spermatogenesis: Mar-Dec; Paraguay	7695
	testes: small Jan-Mar; Paraguay	7695
	mate: May; Oct; Paraguay	7695
	ovulate: July, Oct; Paraguay	7695
	preg f: Jan; n3; Costa Rica	6260

	preg f: June, Aug; n11; Peru	3999a
	preg f: July; n1; Honduras	2670a
	preg f: Aug; n10; Peru	10974
	birth: Oct, Jan; Paraguay	7695
Myotis auriculus	preg f: June; n3; Chihuahua, Mexico	280a
	birth: June; US	577
Myotis austroriparius	epididymis: large Nov-mid Apr; FL, captive	9089
	birth: end Apr-May; FL	577
	birth: end Apr-mid May; FL, captive	9089
	birth: May-June; US	1474
Myotis bechsteini	birth: June-July; Germany	3571
Myotis blythii	testes large: July; Lebanon	6418
	spermatogenesis: Aug-Sept, peak Sept; Yugoslavia	5902
	mate: July-Oct; Algeria, Kirghizia	4980
	mate: Aug-Sept; Czechoslovakia, Bulgaria	4980
	birth: Apr-May; Algeria	5988
Myotis bocagei	preg f: Mar; n1; Niger	8611
	birth: Mar; Gabon	252
	birth: Apr; Rwanda	252
	birth: Jan, June, Aug, Oct; Zaire	252
	birth: yr rd; central Africa	253
Myotis californicus	preg f: Apr; n1; Sinaloa, Mexico	5413a
	preg f: end Apr-mid June; CA	6060
	preg f: May-June; CA	11228a
	preg f: June; n3; Durango, Mexico	3607_1
	birth: May-June; US	1474
	birth: end May-mid June; US	577
	birth: July; BC, Canada	2088
	birth: July; n1; UT	12104a
Myotis capaccinii	spermatogenesis: none Nov-Apr; Europe	2075
Myotis chiloensis	preg f: Feb, July; n3; Costa Rica	3608
Myotis dasycneme	mate: 40% before mid Nov, rest mid Nov-mid Apr; St Petersburg, Russia	10495
	ovulate: Jan-mid Apr; Netherlands	11095
Myotis daubentoni	mate: Feb; Netherlands	11183
	mate: Sept-Jan; Poland	4074
	mate: 17% before mid Nov, rest mid Nov-mid Apr; St Petersburg, Russia	10495
	ovulate: Mar, Dec; Netherlands	11095
	birth: end June-early July; Germany	3571
	birth: early July; Germany	7777
Myotis elegans	preg f: Feb; n1; Campeche, Mexico	5424f
	preg f: Apr; Costa Rica	6260
	preg f: June; n2; Veracruz, Mexico	6148a
	lact f: June; n2; Caribbean, Guatemala	7024
Myotis emarginatus	mate: Aug; France	9240
	mate: Sept-Oct; Netherlands	10070
	birth: June; Netherlands	10070
	birth: end June-mid July; Germany	5175
	birth: end June-July; Germany	3571
Myotis evotis	preg f: May-June; n2; SD	6780
	preg f: May-June; n≥3; SD	261a
	preg f: May-June; n2; CA	11228a
	preg f: June; n2; ID	11646a
	preg f: June; n2; NV	6780
	preg f: mid June-mid July; WY	1871a
	preg f: July; n2; BC, Canada	2088
	preg f: July; n2; MT	5424_2
	birth: end June-July; w US	1474
	birth: July; n5; WA	6780
	lact f: July; n3; se MT	6166
	lact f: July; NM	6780
Myotis fortidens	preg f: June; n4; Sinaloa, Mexico	5413a
Myotis frater	birth: mid June-end July; Japan	6723

Myotis grisescens	spermatogenesis: June-Aug; MO	7330
	estrus: Oct-Nov; MO	7330, 10107
	mate: end fall; US	577
	birth: May-June; central US	1474
	birth: June; MO	4174
	birth: June; US	577
	lact f: June; MO	4173a
Myotis hasseltii	lact f: Jan; Malaysia	7161
Myotis keaysi	preg f: May-June, Aug, Nov; Costa Rica	6261
	preg f: July; n7; Peru	3999a
Myotis keenii	preg f: mid May-mid June; IA	6105a
	preg f: June; n7; Durango, Mexico	3607_1
	birth: early-mid June; IN	2043
	birth: June; n1; MO	2864
	birth: end June-July; e US	1474
	lact f: end June-July; IA	6105a
	lact f: June; n2; Durango, Mexico	3607_1
Myotis leibii	mate: Apr; n1; IN	4202
	preg f: May; n1; NM	502
	preg f: June; n1; MT	5424_2
	preg f: June; n3; ND	3703
	preg f: June; n1; w SD	10954a
	preg f: June-July; SD	10954
	preg f: July; n1; NE	8807a
	preg/lact f: June; s central BC, Canada	3217
	birth: May-July; US	1474
	lact f: July; n1; se MT	6166
Myotis lucifugus	spermatogenesis: Apr-Aug; ne US	343, 4161
	spermatogenesis: June-Aug; MO	7330
	uterus epithelium: high Oct-Feb; MO	9002
	mate: Aug-Sept; ONT, Canada	8843
	mate: end Aug-?; ONT, Canada	10779
	mate: Sept; AZ	6070
	mate: Sept-Oct, Apr; NY	2135
	mate: Sept-Mar; ne US	343
	mate: Oct; IN	4202
	mate: early fall; NY	7651
	mate: fall; MO	4174, 9002
	ovulate: Feb-Apr, peak Apr; MO	4174, 4175
	ovulate: end Apr-May; ONT, Canada	1414
	ovulate: spring; PA	11794
	preg f: May; n14; SD	261a
	preg f: May-June; MN	495
	preg f: June; IA	6105a
	preg f: mid June; 9/21; BC, Canada	3217
	preg f: end June-early July; n40; CO	2353a
	preg f: July; n10; NM	2353a
	birth: May-July; US	1474
	birth: May-July; NE	$5405b_1$
	birth: peak; end May; NY	784
	birth: mid May-mid July; IL	1526
	birth: end May-?; WI	5210
	birth: June-early July; Canada, NY	3214
	birth: June-early July; ONT, Canada	3214
	birth: June-mid July; ALB, Canada	9645
	birth: June-mid July; ONT, Canada	2842
	birth: June-July; ne US	2355
	birth: June-July; ONT, Canada	2842
	birth: mid June-mid July; ALB, Canada	9645
	birth: July; BC, Canada	2088
	birth: July; WY	3254
	birth: mid summer; NY	7651
	lact f: end June-July; n38; CO	2353a
	lact f: July; IA	6105a

	lact f: July; n9; NM	2353a
	lact: ends mid July; ne US	343
Myotis muricola	birth: May, June?; Fukien, Yunnan, China	179
Myotis myotis	mate: Aug; Terek Region, Russia	8049
	mate: Sept; AZ	6070
	mate: Sept-Oct; Netherlands	10070
	mate: mid Sept; Germany	2945
	mate: spring?; Germany	9679
	mate: spring, fall; Japan	4819
	mate: fall, also captive Jan; Germany	9679
	mate: fall; USSR	1524
	mate: est fall; Sardinia, Italy	3517
	ovulate: end Dec-mid Apr; Netherlands	11095
	ovulate: Jan; Netherlands	10069
	ovulate: Feb; Portugal	436
	ovulate: Mar-Apr; Germany	2947
	implant: before mid Mar; Sardinia, Italy	3517
	preg f: Mar-Apr; Sardinia, Italy	3517
	birth: May-June; Portugal	436
	birth: end May; Pyrenees, France	1657
	birth: end May-mid June; Eifel mts, Germany	9220
	birth: end May-mid June; Bohemia, Czechoslovakia	6001
	birth: end May-June/July; Germany	3571
	birth: end May-early July; Germany	5938
	birth: June; Belgium	4663
	birth: June; France	9596
	birth: June; Germany	2945
	birth: June; Netherlands	10070
	birth: June; Transcaucasus, Georgia	8049
	birth: end June; Netherlands	754
	birth: end June-early July; Guadalajara, Spain	2402
Myotis mystacinus	mate: Jan; Netherlands	11183
	mate: mid Nov-mid Apr; St Petersburg, Russia	10495
	ovulate: Dec-Apr; Netherlands	11095
	preg f: Feb, Apr, June, Aug, Oct; n7; Malaysia	6446
	preg f: yr rd, peak Apr-May; Malaysia	7161
	birth: June; Germany	7777
	birth: June-July; Germany	3571
Myotis nattereri	birth: end June-early July; Ireland, UK	2045
	lact f: mid June-mid Aug; Germany	6248
Myotis nigricans	spermatogenesis: Dec-Aug; Panama	11756
	spermatogenesis: yr rd; Paraguay	7695
	testes: small Feb; Banos, Ecuador	10659
	mate: Jan-mid Sept; Panama	11755
	mate: peak May-June; Paraguay	7695
	mate: est end Dec-early Jan; Banos, Ecuador	10659
	preg f: peak Jan/Feb, Apr/May, Aug; Panama	11755
	preg f: Jan-Nov; Costa Rica	3322
	preg f: Feb, Apr, June-Aug, Oct-Nov; n58; Peru	3999a
	preg f: June; n2; Peru	10974
	preg/lact f: 42% Aug, 84% Sept; Guatemala	2552
	preg/lact f: yr rd; Paraguay	7695
	birth: end Feb-Nov; Panama	11755
Myotis oxyotus	preg f: June; n1; Peru	3999a
Myotis riparius	preg f: Apr-June; Costa Rica	6261
	preg f: Aug; n1; Peru	3999a
Myotis simus	preg f: July-Aug; n3; Peru	3999a
	preg f: Oct; Paraguay	7699a
Myotis sodalis	mate: fall; MO	4174
	mate: Oct; midwest US	4231
	preg f: May-June; MO	2867
	preg f: June; n1; IN	7609
	birth: end June; midwest US	4231
	birth: end June-early July; IN	5096c

	lact f: June-Aug; n6; MI	6119a
Myotis thysanodes	preg f: Feb; n1; Chiapas, Mexico	1641
	preg f: May; n2; Jalisco, Mexico	11471a
	preg f: June; n2; Chihuahua, Mexico	280a
	preg f: June; n3; SD	10954
	birth: June-July; w US	1474
	lact f: June; n1; San Luis Potosí, Mexico	2228a
	lact f: Aug; n2; Hidalgo, Mexico	1640a
Myotis tricolor	mate: Apr?; Natal, S Africa	825
	ovulate: end Sept; Natal, S Africa	825
	conceive: Sept; Natal, S Africa	825
	implant: end Sept-Nov; Natal, S Africa	825
	birth: end Oct-mid Nov; Cape Province, S Africa	10158
	birth: mid Nov-mid Dec; Natal, S Africa	825
	birth: 1st 3 wk Dec; Natal, S Africa	11818
	birth: Dec-Jan; Zaire, S Africa	255
Myotis velifer	mate: Apr; n3; KS	3288
	mate: Sept-Oct; sw KS	6106
	mate: peak Sept-Oct, Dec-Mar; AZ	2135, 6070
	conceive: mid Apr; sw KS	6106
	preg f: Mar, Dec; few data; Veracruz, Mexico	4228
	preg f: May; n1; Jalisco, Mexico	11471a
	preg f: June; n1; Chihuahua, Mexico	1196a
	preg f: June; n9; Durango, Mexico	5405b
	preg f: June-July; n19+; Sinaloa, Mexico	5409, 5413a
	birth: June-July; sw US	1474
	birth: est mid June; sw KS	6106
	birth: end June-mid July; KS, OK	577
Myotis vivesi	preg f: May; n27; Pescadora, I, Mexico	9002a
	preg f: June; n2; Partida I, Mexico	1469a
	birth: May-June; CA to Sonora, Mexico	1346
	lact f: June; n25; Partida I, Mexico	1469a
Myotis volans	ovulate: Mar-May; NM	2733
	preg f: May; n1; CA	11228a
	preg f: May; n2; Jalisco, Mexico	11471a
	preg f: June; n2; SD	261a
	preg f: June; n100; CA	2231a
	preg f: June or July; n4+; NE	8807a
	preg f: mid June-mid July; NV	11466
	preg f: July; n1; se MT	6166
	preg f: July; n2; MT	5424_2
	preg f: July; NM	2353a
	preg f: July; NM	11466
	preg f: July; Baja CA, Mexico	5424e
	preg f: July-Aug; WY	3254
	birth: May-early Aug; NM	2733
	birth: June; n many; CA	2231a
	birth: June; w N America	1474
	birth: end June-mid Aug; WY	1871a
	lact f: May-Aug; NM	2733
	lact f: June; n1; Durango, Mexico	3607_1
	lact f: July; n1; se MT	6166
	lact f: July; n3; MT	5424_2
	lact f: Aug; n1; SD	261a
Myotis yumanensis	preg f: June; n7; OK	3803a
	preg f: June; n~34 f; NM	1996a
	preg f: June; n1; Sinaloa, Mexico	5413a
	birth: May-June; w US	1474
	birth: May-July; WY	1871a
	birth: May-July; US	577
	lact f: June; 26/59; sw BC, Canada	3217
	lact f: July; n1; CO	2353a
	lact f: Aug; n2; Jalisco, Mexico	11471a
Nyctalus aviator	birth: May-July; Japan	3033

Nyctalus lasiopterus	birth: mid June-mid July; Japan	6722
Nyctalus leisleri	preg f: June-early Aug, Jan; India	8883
	birth: May-July; Germany	3571
Nyctalus noctula	epididymal sperm: Nov-Mar; Germany	546
	testes: increase mid May-July; lab	5860
	testes: large July-Aug; Germany	4102
	testes: large Sept; n1; small Jan; n1; Germany	3631
	testosterone: peak Sept; lab	8829
	mate: July; Germany	4102
	mate: Aug; Voronesh, Russua	8137
	mate: end Aug; Germany	2945
	mate: Sept-?; Netherlands	10071
	mate: Sept-Oct; captive	5860
	mate: fall from end Aug, spring; Germany	3571
	mate: fall; Europe	11318
	mate: fall; Belgium	11147
	mate: fall; Germany	9679
	ovulate: May; Belgium	11148
	conceive: 5-10 d after hibernation; Germany	3571
	implant: Mar-Apr; Belgium	11147
	preg f: May-June; Belgium	11148
	birth: May-June; Russia	4645
	birth: May-June; Netherlands	754
	birth: end May-early June?; Russia	8049
	birth: end May-July; Germany	3571
	birth: early-mid June; Netherlands	10071
	birth: June; few data; UK	11622
	birth: June; Voronesh, Russia	8137
	birth: June or later; Belgium	11148
	birth: July; Germany	4102
	lact f: Aug; Netherlands	10071
Nycticeius humeralis	preg f: Apr-May; n5; NC	1290
	preg f: May; n1; KS	5414a
	preg f: May; n3; AL	589a
	preg f: June; n2; NE	5405b,
	preg f: June; n1; KS	950a
	preg f: June; n1; IA	6105a
	preg f: June-July; n4; IA	2868
	birth: mid May-mid June; se US	577
	birth: end May-early June; MS	5386
	birth: mid June; MO	11472
	lact f: May; FL	7498a
	lact f: June; n1; KS	5414a
	lact f: June-July; n9; IA	2868
Nycticeius schlieffeni	mate: June; Transvaal, S Africa	11138
	ovulate: Aug; Transvaal, S Africa	11138
	preg f: Sept-Nov; n11; Transvaal, S Africa	11137
	birth: Nov; central Africa	252
	birth: Nov-mid Dec; Transvaal, S Africa	11135, 11138
	birth: est early rainy season; few data; Burkina Faso, w Africa	5903
Nyctophilus geoffroyi	birth: Nov; TAS, Australia	4038, 4039
	lact f: Nov; n1; VIC, Australia	7074
Nyctophilus gouldi	spermatogenesis: summer; ACT, Australia	8492
	sperm storage: Jan-Sept; ACT, Australia	8492
	mate: early Apr; ACT, Australia	8492
	ovulate: Sept-early Oct; ACT, Australia	8492
	birth: Oct-Nov; ACT, Australia	8492
Philetor brachypterus	preg f: June; n1; Malaysia	6447
Pipistrellus spp	repro: yr rd; Papua, New Guinea	200
	preg f: May; n3; Honduras	9136a
Pipistrellus ceylonicus	mate: June; India	3933, 3940, 6715
	mate: yr rd; Sri Lanka	8493
	ovulate: mid July; India	3933, 3940, 6715
	preg f: July-Sept; India	3915

	preg f: Sept; Sri Lanka	8494, 8501
	birth: end Aug-mid Sept; India	6715
Pipistrellus coromandra	preg f: July; n1; Sri Lanka	8501
	lact f: July; n2; Sri Lanka	8501
	lact f: Aug; central India	5705
	juveniles: May; n2; sub-adult: Sept; n1; central India	5705
Pipistrellus dormeri	spermatogenesis: yr rd; India	6716
	preg f: Mar; n5; central India	5705
	preg f: yr rd; India	6716
	juveniles: Aug-Sept; central India	5705
Pipistrellus hesperus	spermatogenesis: June-Sept; CA	6064
	testes: large July-Oct, small Mar-May; CA	6064
	mate: Sept-Apr; CA	6064
	ovulate: mid Mar-end Apr; CA	6064
	preg f: Mar-Aug; CA	6064
	preg f: Apr; n1; Jalisco, Mexico	11471a
	preg f: May; n1; NM	502
	preg f: June; n1; CA	11228a
	birth: June-Aug; CA	6064
	lact f: June; Coahuila, Mexico	522
Pipistrellus javanicus	sperm: few Sept, many mid Oct; Japan	7759
	mate: Oct; Japan	7759
	mate: end Oct; Japan	11014
	ovulate: Apr; Japan	11013, 11014
	ovulate: May; Japan	7759
	birth: June; Japan	7521
	birth: early July; Japan	11013
Pipistrellus kuhlii	mate: fall; Iraq	110
	preg f: Oct; n2; Transvaal, S Africa	10158
	birth: Apr-May; Pakistan	9175
	birth: est end Apr-May; Lebanon	6418
	birth: est Nov; central Africa	252
	wean: end June; Lebanon	6418
Pipistrellus mimus	spermatogenesis: Nov, Feb-early June; India	6026
	spermatogenesis: yr rd; lab	3944
	testes: large Nov, Feb, small June-Sept; India	6026
	mate: end Nov-early Dec, Feb-June; India	6025, 6026
	mate: yr rd; Sri Lanka	8494, 8501
	conceive: end Dec, Feb-June; India	6025
	preg f: May; n1; India	5705
	preg f: peaks May-July, Sept-Oct; lab	3944
	preg f: Dec-Aug, peak Mar; India	6026
	preg f: yr rd, peaks July, Aug, Feb-Mar; India	70
	birth: Feb, May, Aug; India	6025, 6026
	lact f: Mar, May, Dec; Sri Lanka	8493
	yg observed: Mar-Apr; Thar Desert, India	9844
Pipistrellus nanulus	preg f: Mar; n1; Rio Muni, W Africa	5387
Pipistrellus nanus	mate: May-Sept; Kenya	8016
	preg f: Feb, May, July-Aug, Oct; Zaire	11272
	preg f: Aug, Dec-Jan; Zaire	191
	preg f: Aug-Sept; Zimbabwe	10158
	preg f: Sept-Oct; no data June-Sept; Africa	6262
	preg f: Oct; n1; Namibia	9945
	birth: Mar; Zaire	11270
	birth: Oct; no data other mo; Liberia	6086
	birth: Nov; Kenya	8016
	birth: yr rd; central Africa	252
	lact f: June, Nov; no data June-Sept; Africa	6262
	lact f: Aug; Rio Muni, W Africa	5387
Pipistrellus nathusii	birth: mid June; central Russia	8049
Pipistrellus pipistrellus	spermatogenesis: end May-Sept, none Oct-Apr; France	2074, 2075
	epididymal sperm: Nov-Mar; Germany	546
	testes: large Aug, small Nov; UK	8842
	testes: large Aug; n1; Lebanon	6418

	mate: Dec; n1; Spain	6104
	mate: fall; France	2075
	mate: fall; Europe	11318
	mate: fall; UK	2420
	mate: fall; USSR	8049
	mate: fall, spring; Germany	1524
	ovulate: May; Belgium	11148
	ovulate: May; UK	2420
	ovulate: spring; Germany	1524
	birth: May/June-July; Germany	3571
	birth: June; Germany	7777
	birth: June-early July; Azerbaijan	8865
	birth: June/July; Spain	6104
	birth: end June; UK	2420
	birth: end June; UK	1179
	birth: end June-?; Scotland, UK	10603
	birth: July; n3; UK	11622
	birth: early summer; Pakistan	9175
	wean: mid Aug; Scotland, UK	10603
Pipistrellus rueppelli	birth: Apr-May; Tanzania	252
	birth: Aug; Zambia	252
	juveniles: Sept; Zambia	331
Pipistrellus rusticus	birth: Nov; S Africa	11139
Pipistrellus savii	birth: June; Germany	3571
Pipistrellus stenopterus	preg f: Oct; n1; Singapore, Malaysia	7161
	lact f: Oct; n1; Singapore, Malaysia	7161
Pipistrellus subflavus	spermatogenesis: Aug-Feb; e US	6068
	testes: large Aug-Sept; e US	6068
	mate: Nov; IN	4202
	ovulate: spring; MO	4174
	preg f: Apr-May; n24; NC	1290
	preg f: May-mid June; MO	3549
	birth: May; se US	1474
	birth: end May-early June; n2; GA	4386
	birth: end May-mid June; FL	3549
	birth: June-July; ne US	1474
	birth: mid-end June; MA	3549
	birth: mid June-mid July; PA	6175
	birth: end June-early July; IN	3549
	lact f: June; FL	7498a
	lact f: July; n1; IA	6105a
Pipistrellus tenuis	preg f: Jan-Mar, May-June, Aug, Oct; Papua, New Guinea	7073
Plecotus auritus	epididymal sperm: Aug-Sept; Scotland, UK	10247
	mate: mid Oct-Nov; Ireland, UK	7413
	preg f: June-early July; ne Scotland, UK	10248
	birth: Apr?; May-July; Germany	3571
	birth: June; Germany	2945
	birth: June?; Portugal	436
	birth: July; ne Scotland, UK	10248
	birth: spring; UK	2045
	repro: Apr-mid May; Ireland, UK	7413
Plecotus mexicanus	preg f: Apr; n3; Jalisco, Mexico	11471a
	lact f: May; n1; Jalisco, Mexico	10942a
	lact f: July; San Luis Potosí, Mexico	10942a
Plecotus rafinesquii	preg f: July; WA	9578a
	birth: end May-early June; NC	587
Plecotus townsendii	mate: Oct-Apr; CA	8314
	mate: Oct-Feb; US	1474
	mate: est fall; CA	2228
	ovulate: Feb-Mar; CA	8314
	ovulate: Feb-Apr; US	1474
	ovulate: spring; CA	2228
	preg f: July; n1; SD	10954, 10954a
	birth: Apr-July; US	1474

	birth: June; TX	11870
	birth: mid July; BC, Canada	2088
	lact f: July; n19/22; SD	10954, 10954a
Rhogeessa parvula	preg f: Apr; n1; Michoacan, Mexico	4671
	preg f: May-June; n6; Sinaloa, Mexico	5413a
Rhogeessa tumida	preg f: Mar; n2; Costa Rica	3608
	preg f: Mar; n2; Guatemala	5405e
	preg f: Aug; n2; Guatemala	2552
	lact f: Mar; n2; Costa Rica	3608
Scotoecus hirundo	lact f: May; n1; e Africa	5759
Scotophilus heathi	spermatogenesis: Oct-Mar; 1st sperm: Jan; India	6028
	spermatogenesis: Oct-Nov; s India	3935
	testes: scrotal yr rd, small May-Oct; India	6028
	epididymal sperm: Sept-Oct; n2; central India	5705
	quiescent f: Aug-Nov; India	6028
	mate: Jan-early Feb; India	6027, 6028
	mate: Mar-Apr; Pakistan	9175
	mate: Nov-Dec; s India	3935, 6717
	ovulate: early Mar; India	6027, 6028
	ovulate: end Dec; s India	3935, 6717
	implant: mid Apr; India, Sri Lanka	6027
	preg f: Jan-Apr; India	6717
	birth: mid July; India	6027, 6028
	lact f: June-July; central India	5705
Scotophilus kuhli	epididymal sperm: Oct; n2; central India	5705
	spermatogenesis: Dec-?, peak Feb-Mar; India	7415
	testes: large Jan-Feb, small May-Oct; s India	3906
	testes: large Feb-Mar; s India	3907
	mate: mid Feb-mid Mar; s India	3905, 3906, 3937, 7415
	preg f: Mar-July; s India	3907, 3915
	preg f: Sept-Oct; n3; Sri Lanka	8501
	birth: Mar; India	9175
	birth: June-July; s India	3905, 3907, 7415
	lact f: June; Jaomam/Kimog, Asia	179
	juveniles: June-July; central India	5705
Scotophilus leucogaster	preg f: Feb-early Mar; n1; Africa	5759
	preg f: Mar; n1; Zaire	191
	preg f: Apr-Oct; S Africa	11145b
	preg f: May; n1; Niger	8611
	birth: prob Dec; Tanzania	252
Scotophilus nigrita	mate: end Dec-early Jan; central Africa	252
	preg f: Jan-Mar; boreal central Africa	252
	preg f: Aug-Oct; austral central Africa	252
	preg f: Mar; n2; Zaire	191
	birth: peak Jan-Mar, Aug-Sept; n12; e Africa	5759
	birth: Mar; Angola, Burundi	252
	birth: Oct-Nov; se Zaire	252
	lact f: ?-mid Apr/early May; ne Zaire	252
	lact f: end Dec-Jan; Rwanda, Uganda	252
Tylonycteris pachypus	testes: large Oct, small Jan-May; w Malaysia	7164
	mate: Nov; w Malaysia	7164
	mate: end Nov-mid Dec; Ula Gombak, w Malaysia	8839
	ovulate: Jan; w Malaysia	7164
	implant: Jan; w Malaysia	7164
	preg f: Mar-May; Selangor, Malaysia	7161
	preg f: Apr-May; Malaysia	7164
	birth: Apr-May; w Malaysia	7164
Tylonycteris robustula	testes: large Oct, small May; Malaysia	7164
	mate: mid Nov; Malaysia	7164
	ovulate: Jan; Malaysia	7164
	implant: end Jan; Malaysia	7164
	preg f: Apr; Malaysia	7164
	preg f: Apr-May; Malaysia	7164
	birth: Apr; Malaysia	7161, 7164

Vespertilio murinus	spermatogenesis: May-Sept; Europe	2075
	mate: Sept; s France	9240
	mate: end Sept; s France	9240
	ovulate: Mar-Apr; s France	9238, 9240
	birth: May-June; Russia	8049
	birth: end May-early June; s France	9240
	wean: Aug; s France	9240
Vespertilio superans	birth: summer; Ohzukue I, Jima, Fukuoka Prefecture, Japan	3563

Order Chiroptera, Family Mystacinidae, Genus *Mystacina*

The 12-35 g, New Zealand *Mystacina tuberculata* eats pollen, fruit, and insects (2257) in forested areas. Males form leks (2258). Monoestry is reported (2257). Young are born naked with their eyes open (2257).

NEONATAL MASS		
Mystacina tuberculata	3.2 g; n1 with placenta	2257
LITTER SIZE		
Mystacina tuberculata	1 emb ea; n2 f	2257
SEASONALITY		
Mystacina tuberculata	mate: Mar; New Zealand	2257, 2258
	birth: Dec; New Zealand	2257

Order Chiroptera, Family Molossidae

Free-tailed bats (*Chaerephon* 10-30 g; *Cheiromeles* 165-200 g; *Eumops* 30-70 g; *Molossops* 16-22 g; *Molossus* 8-35 g; *Mops* 7-65 g; *Mormopterus* 10-15 g; *Myopterus* 56-66 mm; *Nyctinomops* 10-30 g *Otomops* 25-50 g; *Promops* 15-20 g; *Tadarida* 10-40 g) are insectivores from the tropics and subtropics of the world. *Molossus* males may be slightly larger than females and possess a gular gland (4575, 10663).

Often only the right ovary is functional and the right uterine horn is used preferentially (191, 2133, 2337, 5305, 5306, 5548, 5603, 5835, 6067, 7120, 7682, 7683, 8872, 8951, 8952, 10375, 11136, 11140, 11804; cf 10154). The hemochorial or endotheliochorial, labyrinthine, diffuse or discoidal placenta has a superficial, mesometrial or antimesometrial implantation into a bicornuate or simplex uterus (181, 2133, 4289, 5306, 5835, 6067, 7563, 7683, 8952, 8952b, 9178, 10310, 10374, 10375, 10376, 10377, 11136, 11140, 11799). Placentophagia (4508, 11270; cf 9896), menstruation (8952a), monoestry (5546, 5835), postpartum estrus (252, 4400, 6067, 7120, 11136, 11735), delayed fertilization and sperm storage (2133, 6069, 8950), and communal nursing (7035) occur.

The naked neonates generally have open eyes except those of *Tadarida* who open their eyes at 1-2 wk (331, 577, 1938, 2337, 3973, 4508, 9933, 10158, 11270; cf 8197). Females have 2 mammae (2337, 6067, 6720). Males have seminal vesicles, ampullary glands, and a baculum (6065, 6069, 6071). *Mormopterus* has a horizontally bifid penis (6069) and *Mops* a penial scent gland (5759). Details of endocrinology (5306), female anatomy (5306, 5603, 9482, 11136), male anatomy (5604), sperm (1273), and milk composition (5087, 5296, 10524) are available.

NEONATAL MASS		
Chaerephon pumila	3.2 g	11140
	3.5 g; n1; term emb	7683
Molossus molossus	3.3 g ea; n2 small yg	3900a
	f: 3.3 g; se0.2; n12	4508
	m: 3.9 g; se0.3; n8	4508
Mops condylurus	10 g	7683
Mops midas	~9-10 g; term emb	10154, 10158
Tadarida brasiliensis	f: 2.40 g; n5	9933
	m: 2.66 g; n3	9933
	2.7 g; n1	4676
	3-4 g	502
	3-4 g; from graph	8197

NEONATAL SIZE		
Molossus molossus	CR: ~30 mm; term emb	6067
	57.5 mm; se3; n7	4508
Mops thersites	31.67 mm; r30-34; n3	71
Tadarida brasiliensis	f: FA: 17.6 mm; n5	9933
	m: FA: 18.6 mm; n3	9933
	CR: 23-27 mm	2337
	CR: 25.8 mm	11702
LITTER SIZE		
Chaerephon plicata	1 emb	5548
	1 emb; n1 f	10872
Chaerephon pumila	1	10158
	1	252
	1	331
	1 yg	7683
	1 emb; n1 f	8611
	1 emb ea; n5 f	187
	1 emb; n13 f	8965
	1 emb ea; n54 f	11140
	1.1 emb; mode 1; r1-2; n10 f	7120
Chaerephon russata	1	252
	1 emb ea; n20 f	1207
Eumops auripendulus	1 emb; n1 f	3999a
Eumops bonariensis	1 emb ea; n5 f	950, 1163
Eumops glaucinus	1 emb ea; n2 f	3608
	1.05 emb; r1-2; n18 f	950, 1163
Eumops perotis	1	6061
	1 yg	1938
	1 yg	577
	mode 1; occasionally 2	1474
	1 emb; n>1 f	5023
	1 emb ea; n3 f	4079
Eumops underwoodi	1 yg	577
	1	1474
Molossops abrasus	1 emb; n1 f	3999a
Molossops greenhalli	1 emb; n1 f	11055
	1 yg ea; n4 f	3900a
Molossops planirostris	1; n1	452
Molossops temminckii	1 emb ea; n2 f	10974
Molossus ater	1	1937
	1 emb; n1 f	217
	1 emb; n1 f	1937
	1 emb; n1 f	3208
	1 emb; n1 f	3999a
	1 emb; n1 f	5405e
	1 emb; n1 f	5409
	1 emb; n1 f	11507a
	1 emb ea; n2 f	950, 1162
	1 emb ea; n3 f	5413a
	1 emb ea; n7 f	1163
	1 emb ea; n8 f	5425
	1 emb ea; n30 f	5424f
Molossus bondae	1	1078
	1 emb; n1 f	3608
	1 emb ea; n8 f	2670a
Molossus molossus	1 emb; n1 f	6307
	1 cl; occasionally 2	6067
	1 emb ea; n6 f	1635
	1 emb ea; n7 f	5411a
	1 ea; n11 lt	452
	1 emb ea; n18 f	527a
	1 emb ea; n27 f	11471a
	1 emb ea; n34 f	3999a
Molossus pretiosus	1 emb ea; n2 f	2670a

Molossus sinaloae	1 emb ea; n2 f	9457a
	1 emb ea; n15 f	1162
	1.05 emb; r1-2; n20 f	950, 1163
Mops condylurus	1	11270
	1	5759
	1 emb; n1 f	8611
	1 yg	7683
	1 ea; n2 lt	331
	1 emb ea; n3 f	10154
	2 emb; n1 f	252
Mops congicus	1	252
	1 emb ea; n5 f	191
	1 emb ea; n5 f	1207
Mops demonstrator	1; n1 f	252
Mops leonis	1	252
	1 emb ea; n12 f	1207
Mops midas	1 ea; n14 lt	10154, 10158
	1-2	252
Mops nanulus	1	252
Mops niveiventer	1; a few 2	252
Mops thersites	1	252
Mops trevori	1 emb; n1 f	1207
	1 emb; n1 f	191
Mormopterus norfolkensis	1 ea; n2 lt	7075
Mormopterus planiceps	1	9100
	1 yg	2133
	1 ea; n18 lt	7075
Mormopterus setiger	1 yg ea; n3 f	252
Nyctinomops aurispinosus	est 1 yg	5125a, 5411
Nyctinomops femorosaccus	1 emb; n1 f	1937
	1 emb; n1 f	2865b
	1 emb ea; n2 f	781
	1 emb ea; n2 f	6099
Nyctinomops laticaudatus	1 emb; n1 f	1162
	1 emb ea; n9 f	1163
	1 emb ea; n16 f	5424f
	1 emb ea; n27 f	217
Nyctinomops macrotis	1 yg	577, 1474
	1 emb ea; n2 f	1937
	1 emb ea; n8 f	3607,
	1 yg ea; n36 f	7354
Otomops martiensseni	1 emb; n many	7682
Otomops wroughtoni	1 yg	1340, 1344
Promops centralis	1 emb ea; n2 f	3999a
	1 emb ea; n4 f	1163
Tadarida aegyptiaca	1 emb; n1 f	10158
	1 emb; n1 f	8501
	1 yg ea; n6 f	1340, 1344
Tadarida australis	1	5835
	1; n1 lt	7075
Tadarida brasiliensis	1	7035
	1 yg	10590
	1 yg	6062
	1 emb; 2 rare	577
	1; 2 occasional	4443
	1 emb; n1 f	527a
	1 yg; n1 lt	2197, 5405b,
	1 emb; n1 f	6639
	1 emb ea; n2 f	4079
	1 emb ea; n3 f	5873
	1 emb ea; n4 f	6259
	1 emb ea; n6 f	3999a
	1 yg ea; n6 f	485
	1 emb ea; n>50 f	178

	1 emb ea; n300 f	9896
	1 ea; n2000 f	2337
	1.25; r1-2; n4 f	7611
	1.58 emb; r1-3; n26 f	2539
Tadarida fulminans	1 ea; n6 f	10158
Tadarida teniotis	1 emb ea; n2 f	6418
SEXUAL MATURITY		
Chaerephon pumila	f: 1st preg: 3 mo	7120
	f: 1st preg: FA: >37.5 mm; shortly after weaning	6848
	m: 5 mo	7120
Molossus molossus	f: 1st mate: 3 mo?	6067
Mormopterus planiceps	f: 1st yr	2133
	m: est not 1st yr	6069
Otomops martiensseni	f/m: ~1 yr; ~25 g	7682
Tadarida aegyptiaca	f/m: 1st mate: 1 yr	5547
Tadarida brasiliensis	f: 9 mo	9896
	m: 18 mo	9932
GESTATION		
Chaerephon ansorgei	2-2.5 mo	255
Chaerephon pumila	60 d	11135, 11140
Molossus molossus	~3.5 mo	6067
Mops condylurus	est ~2 mo	7683
	3 mo	11270
Mormopterus planiceps	3-5 mo	2133
Otomops martiensseni	~3 mo	7682
Tadarida aegyptiaca	77-90 d	5546, 5547
Tadarida brasiliensis	77 d	8798
	11-12 wk	9896
	90+ d	2337
	< 100 d	6062
	~3 mo	4678
	~4 mo	10377
LACTATION		
Chaerephon ansorgei	nurse: est 2-3 mo; field data	252, 255
Chaerephon pumila	wean: < 21 d	11140
Molossus molossus	wean: ~6 wk	6067
	nurse: 65 d	4508
	wean: 4 mo	11735
Molossus sinaloae	nurse: 6-8 wk	4575
Tadarida aegyptiaca	wean: est 2 mo	5546
LITTERS/YEAR		
Chaerephon pumila	1-2	252
	max 3 lt	11135
	max 5	7120
Eumops perotis	1	1474
Eumops underwoodi	1	1474
Molossus molossus	1; sometimes 2	4508
	2	6067
Mops condylurus	1	11270
	2	7683
Mops congicus	1	252
Mops nanulus	1	252
Mormopterus planiceps	1	2133
Otomops martiensseni	1	7682
Tadarida aegyptiaca	1	5547
Tadarida brasiliensis	1	2337
	1	6259
SEASONALITY		
Chaerephon ansorgei	mate: Feb, Oct; Zaire	252, 255
	birth: Jan, May; Zaire	252, 255
	lact f: Jan-Mar; se Zaire	252
Chaerephon bivittata	preg f: Mar; n1; e Africa	5759
Chaerephon chapini	mate: est Oct-Jan; ne Zaire	252
	mate: Aug; Zambia	252

Chaerephon jobensis	yg found: Dec-Jan; Australia	10698
Chaerephon nigeriae	birth: Dec; se Zaire	252
	birth: early Jan; Zambia	252
Chaerephon pumila	spermatogenesis: Oct-Jan, Apr; no location given	252
	spermatogenesis: yr rd; Uganda	252
	epididymal sperm: yr rd; Victoria Nyanza, Uganda	6848
	mate: peaks Sept, Nov, Feb; S Africa	11135
	mate: yr rd; Victoria Nyanza, Uganda	6848
	conceive: Aug-Nov; Transvaal, S Africa	11140
	preg f: Jan; n1; Albert Park, Zaire	11272
	preg f: Apr-Oct; n Ghana	7120
	preg f: May; n1; Niger	8611
	preg f: Nov; n1; central Africa	187
	preg f: Nov; n13; Transvaal, S Africa	8965
	preg f: peak Dec; S Africa	10158
	preg f: Dec-Jan; few data; n Zimbabwe	331
	preg f: yr rd; Uganda	7683
	birth: early-mid Apr; Zaire	252
	birth: Aug; Uganda	252
	birth: Aug, Nov; few data; W Africa	5903
	birth: Aug; n Ghana	7120
	birth: Oct-Nov, Jan-Apr; Transvaal, S Africa	11135, 11140
	lact f: Dec; n9; Transvaal, S Africa	8965
Chaerephon russata	preg f: Sept; Africa	1207
	birth: Oct; central Africa	252
Eumops auripendulus	preg f: Feb; French Guiana	1350a
	preg f: Sept; n1; Peru	3999a
Eumops bonariensis	preg f: Mar, May-June; n33; Yucatan, Mexico	1163
	lact f: June-July; n76; Yucatan, Mexico	1163
Eumops glaucinus	preg f: mid Apr-June; n64; Yucatan, Mexico	1163
	preg f: May, Dec; n2; Costa Rica	3608
	birth: June-July; US	577
	lact f: Apr-May, Aug; n8; Costa Rica	3608
	lact f: mid June-July; n63; Yucatan, Mexico	1163
Eumops perotis	m repro: spring; se US, nw Mexico	1938
	preg f: Apr; n1; Coahuila, Mexico	522
	birth: May-June; CA, AZ, TX	1474
	birth: most by early July; CA	6061
Eumops underwoodi	birth: end June-July; US	577
	birth: July; AZ	1474
Molossops abrasus	preg f: July; n1; Peru	3999a
	preg f: Aug-Oct, Dec; Brazil	10618a
	preg f: Oct-Dec; se Brazil	7983
Molossops greenhalli	birth: June; n4; Trinidad	3900a
	lact f: May; n1; Costa Rica	3608
	lact f: July or Aug; n1; Honduras	11055
Molossops mattogrossensis	preg f; Aug-Dec; Brazil	11734
Molossops temminckii	preg f: Aug; n1; Peru	10974
Molossus ater	preg f: end Feb-early Apr; Veracruz, Mexico	4228
	preg f: Mar; n1; Guatemala	5405e
	preg f: Mar; n2; Trinidad	3900a
	preg f: Mar-June; n44; Yucatan, Mexico	1163
	preg f: 33% Mar; 0% July; n26; Nicaragua	5425
	preg f: Apr-May, July-Aug; Yucatan peninsula, Mexico	5424f
	preg f: June; n1; Durango, Mexico	5405b
	preg f: June-July; n≥3; Sinaloa, Mexico	5409, 5413a
	preg f: July; n1; Veracruz, Mexico	11507a
	preg f: Sept-Oct; French Guiana	1350a
	preg f: Nov; n1; El Salvador	3208
	preg f: Nov; n1; Peru	3999a
	birth: mid Sept; Trinidad, lab NY	8951
	lact f: June-mid Aug; n46; Yucatan, Mexico	1163
	lact f: July; n4; Hidalgo, Mexico	1640a
	lact f: July-Sept; Trinidad	3900a

Molossus bondae	preg f: Jan, Aug; n2; Costa Rica	6260
	preg f: Mar; n1; Costa Rica	3608
	preg f: July; n8; Nicaragua	2670a
	lact f: Sept-Dec, Mar-May; sw Colombia	10785
	lact f: July; n3; Nicaragua	2670a
	repro: Aug-Dec, Feb-June, none Jan, July; sw Colombia	10785
Molossus molossus	estrus: end Feb-Mar, May-June; Puerto Rico	6067
	preg f: Mar-Oct; Puerto Rico	6067
	preg f: Apr, July-Dec; n34; Peru	3999a
	preg f: May; Ecuador	1345a
	preg f: June; n1; Chihuahua, Mexico	1196a
	preg f: July; n3; El Salvador	2670a
	preg f: July; n7; Lesser Antilles	5411a
	preg f: July; n18; Guadeloupe, Lesser Antilles	527a
	preg f: mid July-mid Aug; Trinidad	1635
	preg f: Aug; n27; Jalisco, Mexico	11471a
	preg f: Sept-Feb; ne Brazil	11735
	preg f: Oct-Dec; French Guiana	1350a
	birth: Apr, end Sept-Oct; El Colegio, Colombia	4508
	birth: June, Sept; Puerto Rico	6067
	lact f: Jan-May; ne Brazil	11735
	lact f: July; n11; Lesser Antilles	5411a
	lact f: Sept; n10; Trinidad	3900a
	repro f: Mar-early July; Nicaragua	5425
Molossus pretiosus	preg f: Feb 0/18; Apr 3/4; July 1/1; Boaco, Nicaragua	5425
	preg f: Mar 2/3, mid Aug 3/18; nw Diriamba, Nicaragua	5425
	preg f: May, July, Oct; Costa Rica	6260
	preg f: Aug; n2; Costa Rica	2670a
Molossus sinaloae	preg f: Mar-June; n100; Yucatan, Mexico	1163
	preg f: min Mar-mid May; Yucatan peninsula, Mexico	5424f
	preg f: Apr; n2; Quintana Roo, Mexico	9457a
	preg f: July; n12; Honduras	2670a
	lact f: June-Aug; n518; Yucatan, Mexico	1163
	lact f: July; n12; Honduras	2670a
Mops condylurus	mate: Feb; se Africa	11270
	mate: Apr-May, Nov-Dec; Uganda	5759
	preg f: Jan; n2; n Zimbabwe	331
	preg f: Feb; n1; Niger	8611
	preg f: Dec-Mar, June-Sept; Uganda	7678, 7683
	preg f: yr rd, peak Aug-Oct, Nov-Feb; se Zaire	252
	birth: Feb-Mar, July-Aug; Uganda	5759
	birth: Mar, Sept; Uganda	7678
	birth: mid May; se Africa	11270
	juveniles seen: Feb; n Zimbabwe	331
Mops congicus	preg f: Aug-Sept: n5; Africa	1207
	preg f: Sept; n5; Zaire	191
	birth: Oct; ne Zaire	252
	lact: early Mar; Cameroun, e Zaire, w Uganda	5759
Mops demonstrator	preg f: Apr, June, Sept; ne Zaire, Uganda	252
	preg f: Apr, June, Sept; central Africa	11270
	birth: Apr, June; ne Zaire, Uganda	252
Mops leonis	preg f: Apr; Medge, Zaire	1207
	birth: Apr; ne Zaire	252
Mops midas	mate: Feb; central Africa	252
	preg f: Dec-Feb; Botswana	10154
	birth: est Feb-Mar; S Africa	10158
	birth: est Feb-Mar; Botswana	10154
	birth: May, Oct; central Africa	252
Mops mops	birth: Jan-Feb, July; Kuala Lumpur, Malaysia	3974
Mops nanulus	conceive: Jan; Uganda, w Kenya, Zaire	5759
	birth: end Mar-Apr, July-Sept; Uganda, w Kenya, Zaire	5759
	birth: end Aug-early Sept; central Africa	252
Mops niveiventer	birth: peak Nov-Feb/Mar; se Zaire, Zambia	252

Mops thersites	birth: end Apr; Cameroun	252
	birth: est end Apr; Cameroun	71
	birth: June; central Africa	252
	birth: Aug; ne Zaire	252
	no repro: Feb; Rwanda	252
Mops trevori	preg f: Sept; n1; Zaire	191
	preg f: Sept; n1; Africa	1207
Mormopterus planiceps	testes: large Feb-Mar; small June-Oct; SA, Australia	6069
	spermatogenesis: Sept-June; SA, Australia	6069
	epididymal sperm: Mar-Sept; SA, Australia	6069
	mate: end July-Aug; SA, Australia	2133
	ovulate: Aug; SA, Australia	2133
	implant: Sept; SA, Australia	2133
	birth: Dec-Jan; SA, Australia	2133
Mormopterus setiger	birth: Jan; se Kenya	252
	lact f: Feb; n2; central Africa	252
Nyctinomops aurispinosus	preg f: Sept; n3; Bolivia	5125a
Nyctinomops femorosaccus	preg f: July; n1; AZ	6099
	preg f: July; n1; NM	2865b
	preg f: June; n1; Sinaloa, Mexico	5413a
	preg f: June-July; TX	2864a
	lact f: June-July; TX	2864a
	lact f: July-early Aug; TX	6099
Nyctinomops laticaudatus	preg f: Apr-May; n16; Yucatan peninsula, Mexico	5424f
	preg f: June; n15; Yucatan, Mexico	1163
	preg f: June; n27; Tamaulipas, Mexico	217
	lact f: July; n1; Yucatan, Mexico	1163
	lact f: Aug; n many; Yucatan peninsula, Mexico	5424f
	no preg f: Mar-May; n11; Belize	7624
Nyctinomops macrotis	preg f: May-June; n4; Cuba	7354
	preg f: June; UT	8612
	preg f: June; n1; NM	7354
	preg f: June; n1; Chihuahua, Mexico	7354
	preg f: June; n8; Durango, Mexico	3607,
	preg f: June-July; n36; TX	7354
	birth: end May-early June; sw US	1474
	lact f: June-July; n10; Cuba	7354
	lact f: end June-Aug; TX	7354
	lact f: July-mid Sept; UT	8612
	lact f: July-Sept; NM	7354
	lact f: Aug-Sept; n27; NM	2026b
Otomops martiensseni	testes: large June-Oct, peak Aug; Kenya	5604, 7682
	preg f: Oct-Jan, occasionally Feb; Kenya	5603, 7682
	birth: Jan-Feb; se Zaire, Rwanda	255
	birth: Mar-Apr; Kenya	252
Promops centralis	preg f: Apr, June; n8; Yucatan, Mexico	1163
	preg f: Apr, Oct; n2; Peru	3999a
	preg f: May; n1; Michoacan, Mexico	4671
	lact f: Apr; n2; Trinidad	3900a
	lact f: mid June-mid Aug; n16; Yucatan, Mexico	1163
	lact f: Aug; n1; Suriname	3705c
Tadarida aegyptiaca	quiescence: July-Mar; India	5547
	mate: end May-June; India	5547
	ovulate: 2-3rd wk June; India	5547
	preg f: Mar; n1; Sri Lanka	8501
	preg f: end June-3rd wk Sept; India	5546
	preg f: Sept-Oct, Dec; Transvaal, S Africa	10158
	preg f: Nov; n1; Zimbabwe	10158
	birth: Jan; central Africa	252
	birth: 1-3rd wk Sept; India	5547
	birth: summer; S Africa	10158
	lact f: Dec; n Botswana	10158

Tadarida australis	conceive: end Aug; Australia	5835
	birth: Dec-Jan; Australia	5835
	wean: early May; Australia	5835
Tadarida brasiliensis	mate: July-Aug; NM	502
	mate: winter; TX	2337
	ovulate: early spring; sw US	5305
	conceive: end Feb; FL	10377
	preg f: May; n3; Querétaro, Mexico	9603a
	preg f: May-June; n3; Haiti	5873
	preg f: June; Puerto Rico	344a
	preg f: June; Chihuahua, Mexico	280a
	preg f: June; n1; Chihuahua, Mexico	1196a
	preg f: June; n25; Sinaloa, Mexico	5413a
	preg f: July; Baja CA, Mexico	5424e
	preg f: July; n1; Guadeloupe, Lesser Antilles	527a
	preg f: Aug, Oct; n6; Peru	3999a
	birth: end Jan-early July; CO	10590
	birth: May-June; NM	502
	birth: end May-June; se US	9896
	birth: early-mid June 86%; TX	9933
	birth: June; FL	10377
	birth: June; n1; NB, Canada	2197
	birth: mid June; TX	2337
	birth: end June; TX	2853
	birth: early July; sw US	5305
	birth: July; n1; NE	$5405b_1$
	birth: most by mid July; CA	6062
Tadarida teniotis	preg f: June; n2; Lebanon	6418
	lact f: Oct; n1; Lebanon	6418

Order Primates

A baculum is usually present except in Tarsiidae and Hominidae. Large primates produce only one young at a time but smaller species may produce two or more. Both lactation and gestation are long. Data on the reproductive physiology and endocrinology of primates are too extensive for thorough review in this context.

Reviews of cervix uteri (4189), anatomy (2625, 5842, 8616a), spermatogenesis (3993), reproductive cycles (6177, 10264a), copulation (2536, 2626), placentation (769, 6626), birth hour (5378), and litter size (6406) exist. Several journals are devoted solely to primates including: American Journal of Primatology, Contributions to Primatology, Folia Primatologica, International Journal of Primatology, Journal of Medical Primatology, Monographs in Primatology, and Primates.

Order Primates, Family Cheirogaleidae

Dwarf and mouse lemurs (*Allocebus* 100 g; *Cheirogaleus* 140-900 g; *Microcebus* 50-385 g; *Phaner* 350-500 g) are nocturnal, arboreal omnivores from Madagascar. The reproductive biology of *Allocebus* is unknown. The epitheliochorial and syndesmochorial, labyrinthine placenta has a central implantation into a bipartite or bicornuate uterus (4814, 8444, 9049, 10488, 10489) and is usually eaten after birth (3810, 8375, 8433, 8444). Estrus postpartum (3363) is reported. Neonates are furred and their eyes open at 1-4 d of age (3363, 3810, 3811, 4722, 6892, 8375, 8432, 8435, 8443, 8444). Sex of offspring may reflect maternal social or nutritive status (8379a). *Cheirogaleus* and *Microcebus* have 6 mammae (8443). Males have seminal vesicles, penial scales, and a baculum (5842, 8443, 8444). Details of female anatomy (4793, 8443, 8444), endocrinology (8376, 8377, 8378, 8379, 8382, 8454), and photoperiod (8446, 8447, 8449, 8450) are available.

NEONATAL MASS		
Cheirogaleus major	18.07 g; r17.2-19.5; n3 yg, n1 lt	8443, 8444
Cheirogaleus medius	18 g	4722
Microcebus coquereli	12 g	8435
Microcebus murinus	m: 2.7 g; n1	8443, 8444
	~4 g	4722
	f: 4.3 g; n1	8443, 8444
	f: 4.46 g; r3.5-7.0; n4	3810
	m: 6.46; r4-7.5; n7	3810
	6.8 g; sd1.3; n55	8375
Microcebus rufus	5-8 g	8435
NEONATAL SIZE		
Cheirogaleus major	HB: 76.7 mm; r73-82; n3	8443, 8444
Microcebus murinus	m: HB: 37 mm; n1	8443, 8444
	f: HB: 50 mm; n1	8443
LITTER SIZE		
Cheirogaleus major	mode 2; r2-3	8432, 8435
	3 yg; 1 lt	8443
Cheirogaleus medius	2.0 yg; r1-4; n32 lt, 9 f	3363
	2-3	4722
Microcebus coquereli	1.7; sd0.47; r1-2; n9	5192
	mode 2	8435
Microcebus murinus	1.83 yg; mode 2; r1-3; n12 lt	8443
	1.86 yg; mode 2; r1-3; n7 lt	1150, 8443, 8444
	2 yg	3810, 3811
	mode 2 emb, sometimes 3	8906
	mode 2; r2-3	8432, 8435
	2-3 yg	6892
	2-4; 1 rare	10282
	2.0; r1-3; n2 lt	4722
	2.17; mode 2; r1-3; n35 lt	8375
	3-4	8445
Microcebus rufus	2-3	8435
Phaner furcifer	1	1743, 8435

SEXUAL MATURITY		
Microcebus murinus	f: 1st mate: 1 yr	8435
	f: 1st birth: 11-12 mo	306
	f: 1st birth: 13 mo	3810
	f: 1st birth: 18 mo	8443, 8444
	m: testes large: 185.13 d; 95%ci 20.28	8450
	m: 1st sire: 11 mo	3810
	f/m: 1st mate: 7-8 mo	8432
	f/m: 1st mate: ~7-10 mo	8443, 8444
CYCLE LENGTH		
Cheirogaleus major	estrus: 2-3 d	8444
	~14 d	8445
	r23-31 d; n6 cycles, 3 f	8443
	25 d	8441
	~30 d	1150, 8443, 8444
Cheirogaleus medius	19.66 d; se1.58; r18-23; n9 cycles, 3 f	3363
Microcebus coquereli	16-46 d	5192
	~1 mo	8451
Microcebus murinus	estrus: 1 d	8435
	estrus: 2-3 d	8442, 8444
	estrus: 5.4 d; ±0.8; r3-13; n15	3810
	15-20 d	8442, 8445
	38.2 d; sd5.7; n4 isolated	8379
	40-55 d	8435
	mode 45-55 d 78%; r36-78; n23 cycles	1150, 8443, 8444
	50 d	6892
	mode 50-60 d; r35-108	306
	50.4 d; sd1.5; mode 39, 56; n34 cycles	3810
	53.7 d; sd5.9; n4 sensory cues (not tactile) from f	8379
	62.7 d; sd0.8; n4 tactile cues with other f	8379
Phaner furcifer	~15 d	8451
GESTATION		
Cheirogaleus major	70 d; n1	8432, 8443, 8444
Cheirogaleus medius	61.6 d; se0.96; mode 61 70%; r61-64; n10 lt, 6 f	3363
	70 d	4722
	70 d	8432
Microcebus coquereli	84-89 d	8435
	90.6 d; sd4.3; r85-100; n8	5192
	3 mo	8451
Microcebus murinus	3.5-4 wk	10282
	54-68 d	306
	58.3; sd0.4; n6	3810
	59-62 d; n7 lt	1150, 8443, 8444
	60.5 d; r59-62; n2	4722
	60.2 d; sd1.7; r56-64; n33	8374, 8375
LACTATION		
Cheirogaleus major	solid food: 25 d	8443, 8444
	nurse: ~1.5 mo	8443, 8444
Microcebus coquereli	wean: 134-138 d	5192
Microcebus murinus	emerge nest: 3 wk	6892
	solid food: 20-25 d	3810
	solid food: 3 wk	1150
	solid food: 1 mo	6892
	wean: 20-25 d	3810
	wean: 1 mo	8432
	wean: 40 d	8435
	wean: 1.5 mo	6892
	wean: 45 d	4722
	wean: 6 wk	1150, 8445
	independent: 60 d	4722
	wean: ~40 d	8443
Microcebus rufus	wean: 40 d	8435

INTERLITTER INTERVAL		
Microcebus murinus	71-75 d	8435
Microcebus rufus	71-75 d	8435
SEASONALITY		
Allocebus trichotis	birth: end Nov-Dec; native reports; Madagascar	7983
Cheirogaleus major	spermatogenesis: July-Oct; Madagascar	8444
	clitoris: open June, small Oct-June; n2; captive	8441
	mate: mid Nov; Madagascar	8435
	birth: Jan; Madagascar	8444
	birth: Jan; Madagascar	8432, 8435
	birth: Nov; captive, Madagascar	8445
Cheirogaleus medius	lethargy: Oct-Apr; captive	3363
	spermatogenesis: July-Oct; Madagascar	8444
	mate: Apr-June; captive	3363
	mate: Nov; Madagascar	4834
	preg f: Jan; n1; Madagascar	8445
	birth: Jan; Madagascar	4834, 8432
	birth: June-Aug; captive	3363
Microcebus coquereli	estrus: Apr, July; lab	8451
	mate: Apr; Madagascar	8435
	birth: July-Aug, Nov; lab	8451
Microcebus murinus	testes: increase Jan; n1; captive	8373
	testes: large Mar-June; captive	8445
	testes: large Oct, small July; s Madagascar	10282
	spermatogenesis: Oct, none July; few data; Madagascar	10282
	testosterone: peak Feb-Aug; captive	8377
	male repro: July-Jan; captive	306
	estrus: Mar-Aug; captive	8375, 8377
	estrus: Mar-Sept; captive, France	1150, 8442, 8446, 8448, 8451, 8453
	estrus: May-July; lab	6891
	estrus: Aug-Feb; captive	306
	estrus: Nov-Feb; captive, Madagascar	8442
	mate: Feb-Aug; captive, France	8379
	mate: Mar-July; captive	8378
	mate: Oct-Dec; Madagascar	8444
	mate: Sept-Jan; Madagascar	8432, 8435, 8445
	preg f: Oct-Nov; Madagascar	8906
	preg f: peak Oct-Nov; Madagascar	10282
	birth: Apr-May, July; captive, France	8444
	birth: May-June, Oct; captive	8375
	birth: May-Aug; captive, France	8443
	birth: May-Sept; zoo	4722
	birth: July; n1; lab	8451
	birth: Oct-Jan; Nosy Be I, Madagascar	8445
	birth: end Oct-early Feb; captive	306
	birth: Nov-Feb; Madagascar	8443, 8453
	birth: Nov-Mar; Madagascar	8432
	birth: end Nov-Feb; Madagascar	8444
	birth: Dec-Jan; ne Madagascar	8445
	repro: Oct-Feb; Madagascar	8446
Microcebus rufus	mate: Sept-Jan; Madagascar	8435
Phaner furcifer	testes: increase Jan-Mar; lab	8451
	estrus: mid Jan-June; lab	8451
	mate: mid Nov; Madagascar	8435
	birth: mid Nov; Madagascar	8440

Order Primates, Family Lemuridae

Lemurs (*Hapalemur* 0.8-2.5 kg; *Lemur* 1-3.7 kg; *Lepilemur* 500-900 g; *Varecia* 3-4 kg) are arboreal, often social, herbivores from Madagascar. The epitheliochorial, villous, diffuse placenta implants into a bipartite or bicornuate uterus (4770, 7563, 8443, 8444, 10310, 10962) and is sometimes eaten after birth (3362, 3538, 4718, 4722, 8433). The chorion produces a gonadotrophic complex (10636a). Neonates are furred with open eyes (3362, 3967, 4718, 4722, 8432, 8433, 8435, 8444). Females have 6 mammae (*Varecia* 8443). Nulliparous females may lactate (8369a). Males have a baculum and seminal vesicles (4793, 5842, 8435, 8443, 8444). Details of female anatomy (4793, 8443, 8444), endocrinology (305, 1045, 2636, 5525, 7957, 8371a, 8781, 9081, 9082, 9298, 9905, 9907, 11170, 11172), spermatogenesis (8959), birth (1440), and milk composition (1482) are available.

NEONATAL MASS

Hapalemur griseus	32 g; n1	4722
Lemur catta	50 g; n1	4718
	r50-70 g; n10	768
	60-80 g	4722
	m: 85 g; n1	4718
Lemur coronatus	70 g	7947
Lemur mongoz	60 g	4718, 4722
Varecia variegata	80-85 g	7175
	81.5 g; r75-88; n2; died within 24 h	1127
	m: 84 g; n1	7947
	93.7 g; r73-111; n7	768
	93.9 g; r70-140; n59	1300
	100 g	8435
	mode 100 g; r60-125; <75 g premature; n34	768
	104.63 g; sd19.10; r70-125; n8	9906

NEONATAL SIZE

Hapalemur griseus	HB: 60 mm	3967
	HB: 115 mm; n1	4722
Lemur coronatus	100 mm	7947
Lemur mongoz	138 mm; r136-140; n1	4718
Lepilemur mustelinus	HB: 100 mm	8443, 8444

LITTER SIZE

Hapalemur griseus	1 yg	8435, 8438, 8439
	1; n1 lt	4722
	1 ea; n2	379
	2; n1	3967
Lemur catta	1	8432, 8433
	mode 1	6476
	1 yg; 2 occasional	8443
	1 yg ea; n10 lt	768
	1 ea; n37 lt	3340
	1.12; r1-2; n17 lt	4722
	1.15; r1-3; n80 lt	12128
	1.17 yg; mode 1; r1-3; n547 lt	4736
	1.18; r1-2; n22 f, 60 lt	8253, 11171
	1.2; r1-2; n9 lt	4718
	1.25; sd0.48; n56 lt	8936
Lemur coronatus	1 ea; n25 lt	3340
	1.5; r1-2; n2 lt	7947
	1.5; r1-2; n6 lt, 1 f	5523
	1.60; sd0.55; n5 lt	8936
Lemur fulvus	1	5573
	mode 1	6476
	1 yg; occ 2	8443
	1 emb ea; n3 lt	5575
	1 ea; n7 lt	3340
	1 yg ea; n8 lt	1230
	1 yg ea; n18 lt	1440
	1 ea; n19 lt	3340

	1.03 yg; n170 lt	4736
	1.06; sd0.28; n31 lt	8936
	1.06; sd0.25; n99 lt	8936
	1.12; sd0.33; n17 lt	8936
	1.45; sd0.68; n33 lt	8936
	1.46; sd0.67; n13 lt	8936
	2 emb; n1 f	1440
Lemur macaco	1	8432, 8433
	1 yg; 2 occasional	8443
	1 ea; n2 lt	1230
	1.02 yg; n82 lt	4736
	1.04; mode 1; r1-2; n24 lt	3340
	1.39; sd0.53; n44 lt	8936
Lemur mongoz	1	8432, 8433
	1 ea; n3 lt	8936
	1 ea; n4 lt	4718
	1 ea; n5 lt	4722
	1.08 yg; n110 lt	4736
	1.5; r1-2; n2 lt	4722
	2; n1 lt	4718
Lemur rubriventer	1	8432, 8433
Lemur rufipes?	1 emb ea; n4 lt	10962
Lepilemur mustelinus	1	8432, 8433
	mode 1	8444
	1 emb ea; n9 lt	5573
	1 yg ea; n12 lt	8443
Varecia variegata	1.3; mode 1; r1-3; n7 lt	3340
	1.5 yg; r1-2; n2 lt, n1 f	7947
	1.89; sd0.74; n38 lt	8936
	1.92 yg; mode 2; r1-3; n12 lt	1127
	1.95 yg; r1-3; n22 lt	4736
	2-4	9905
	mode 2; max 6	8435
	2.0 yg; mode 2; r1-4; n38 lt	3362
	2.16; mode 2; r1-6; n326 lt	1300
	2.22; sd0.81; n18 lt	8936
	2.47 yg; r1-3; n15 lt	768
	2.63 yg; sd0.52; mode 3; r2-3; n8 lt	9906
	2.64 yg; mode 3; r1-6; n89 lt	1300
	3; n1 lt	1044
	max 5	6476
	6 yg; n1 lt	5525
SEXUAL MATURITY		
Lemur catta	f: 1st conceive: 19.56 mo; sd5.57; r10-31; n9 f	11171
	f: 1st birth: 2.5 yr	4718
	m: 1st sire: 3 yr	4718
Lemur coronatus	f: 1st conceive: ~20 mo; n1	5523
Lemur macaco	f: 1st birth: 3 yr	8435
	f/m: 1.5-2.5 yr	8444
Lepilemur mustelinus	f/m: est 1.5 yr	8443, 8444
Varecia variegata	f: 1st estrus: 9 mo	1300
	f: 1st estrus: 20-22 mo	3362
	f: 1st conceive: 20.5 mo; r20-21; n2	1127
	m: testes descend: 16-22 mo	3362
	f/m: 20 mo	6476
CYCLE LENGTH		
Lemur catta	estrus: <24 h	4722
	estrus: 24 h	6225
	vaginal estrus: 3.67 d; ±0.37; r2-5	11172
	luteal phase: 25 d; n4 f	1045
	follicular phase: 14 d; n4 f	1045
	39 d; n4 f	1045
	39.3 d; se0.6; 76% r37-42; r33-45; n17 cycles	3111
	39.87 d; sd2.66; r32-45; n40 cycles, 16 f	11171

	40 d; between ovulations	6225
	6 wk	4722
Lemur coronatus	estrus: 1 d	5523
	34.2 d; sd5.8; r26-46; n12 cycles, 4 f	5523
Lemur fulvus	estrus: 24-48 h	1056
	estrus: 4 d; metestrus: 8 d; diestrus: 14 d; preestrus 3 d	2114
	estrus: 6.7 d; r4-10; n10 cycles, 2 f	2094
	26.9; se0.2; n38 cycles, 4 f no visual contact	1128
	30.2; se0.3; n30 cycles, 4 f in visual contact	1128
	28.3 d; se2.3; r24-33; n22 cycles, 6 f	2114
	30 d; n2 f	1056
Lemur macaco	estrus: 25 h	6225
	folllicular phase: 5 d; n6 cycles, 2 f	1045
	luteal phase: 26 d	6225
	luteal phase: 28 d; n6 cycles, 2 f	1045
	33 d; n6 cycles, 2 f	1045
	33.3 d; r32-35; n3 cycles	1044
	40 d	6225
Lemur mongoz	37.3 d; r29-48; n9	9527
Varecia variegata	estrus: 12-24 h	6476
	estrus: 24 h	6225
	mate: 1 d; n8	9906
	genital changes: 9.85 d; sd1.67; n8	9906
	follicular phase: 14 d; n3 cycles, 1 f	1045
	luteal phase: 24 d; n3 cycles, 1 f	1045
	luteal phase: 26 d	6225
	14.8 d; r8-22; n85 cycles	1300
	30 d; n8 cycles, 4 f	1127
	38 d; n3 cycles, 1 f	1045
	40 d; between ovulations	6225
	40-42 d; n3 f	3362
	40-44 d	9907
GESTATION		
Hapalemur griseus	140 d; n1	8438
Lemur catta	4 mo; n1	8443
	4 mo	5377
	133 d; r132-134; n2	4718, 4722
	134.5 d; r131-136; n8	3111
	135 d	6476
	135 d; n54 lt, 16 f	8936
	135.64 d; sd1.63; r134-138; n14 f	11171
	4.5 mo	8432, 8433, 8445
	144 d	4582
Lemur coronatus	122 d; n4 lt, 1 f	8936
	123.5; r123-124; n2	7947
	124.6 d; sd2.9; n6 lt	5523
Lemur fulvus	~15 wk; field data	5573
	117 d; n33 lt, 9 f	8936
	117 d; n89 lt, 19 f	8936
	120 d; n25 lt, 7 f	8936
	4 mo	1056
	4 mo	2153
	128 d	6476
	4.5 mo; n1	8443
Lemur macaco	125 d; n41 lt, 11 f	8936
	127 d; n2	4722
	127-129 d	3538
	128-129 d; n3	1044
	4.5 mo	8432, 8433
	4 mo 3 wk; r120-135 d; field data	8444, 8445
	143 d	4582

Lemur mongoz	114 d; n1	4722
	126 d	4718, 4722
	4.5 mo	8432, 8433
	137 d; r126-159; n4; 159 d prob includes 1 cycle	9527
Lemur rubriventer	4.5 mo	8432, 8433
Lepilemur mustelinus	~14 wk; field data; not many	5573
	120-150 d; field data	8444
	>4 mo	8445
	4-5 mo; est 5 mo	8443
	4.5 mo	8432, 8433
Varecia variegata	90, 90-100, 95-100, 105 d; n4 lt	1127
	100.75 d; sd1.16; r100-103; n8 lt	9905, 9906
	101 d; n1; induced ovulation	5525
	102 d; n1	1044
	102 d; sd1; r99-106; n14	1300, 7175
	102 d; n18 lt, 7 f	8936
	102 d; n35 lt, 11 f	8936
	102.5 d; ±1.5; mode 102-103; r99-106; n21 lt	3362
	103 d	7947
	104 d	6476
	~104 d; n1 twins; vaginal swelling to birth	5880
LACTATION		
Hapalemur griseus	wean: 6 mo	379
Lemur catta	solid food: 1 mo	4718
	wean: 5 mo	8433
	independent: 6 mo	8433
Lemur macaco	solid food: 1 mo	8433
	wean: 5 mo; r4-6	8433
	independent: 6 mo	8433
Lepilemur mustelinus	solid food: 45 d nibbles	8444
	solid food: 1.5-2.5 mo	8443
	nurse: 2.5 mo	8443
	wean: ≤4 mo	8443
	wean: min 4 mo	8444
LITTERS/YEAR		
Lemur catta	1	768
	1	6476
	mode 1	8432
Lemur macaco	1	8444
Lemur mongoz	1	4722
Lepilemur mustelinus	1	8432
	1	8444
SEASONALITY		
Hapalemur aureus	birth: Dec; Hanomefana, Madagascar	326
Hapalemur griseus	mate: Jan; captive	8438
	birth: Jan-Feb; ne Madagascar	8439
	birth: Sept; zoo	8445
	birth: Dec-Jan; Madagascar	8435, 8438
	birth: June; captive	8438
Lemur catta	testosterone: high Sept, Dec-Jan, low Apr; captive	3111
	mate: Mar-June; zoo	12128
	mate: end Mar-early July; s Madagascar	5573
	mate: Apr; s Madagascar	8445
	mate: end Apr; Madagascar	5377
	mate: end Oct-Nov; zoo	4722
	birth: Mar-June, peak Mar; n21/37; zoo Egypt	3340
	birth: Mar-July; zoo	4722
	birth: Apr-June; Madagascar	8433
	birth: Aug-Nov; Madagascar	8432

	birth: end Aug; Madagascar	5377
	birth: end Aug; s Madagascar	8445
	birth: end Aug-Nov; Madagascar	1430
	birth: end Oct; s Madagascar	5573
	birth: Sept; Madagascar	5376
	birth: Sept-Oct; n30/31; zoo	1230
Lemur coronatus	estrus cycles: end Nov-Mar; captive	5523
	birth: Apr-July, peak Apr-May; zoo, Egypt	3340
	birth: May; n2; zoo	7947
Lemur fulvus	estrus: Feb-mid May; ne Madagascar	5573
	mate: Apr; Madagascar	8444
	mate: Apr-June; ne Madagascar	8435
	mate: Aug, Oct-Apr, peak Dec; captive	2153
	mate: Nov-Apr; captive, France	2114
	mate: Nov-Dec; captive	6476
	preg f: July-Aug; ne Madagascar	5573
	preg f: end July-mid Aug; n4; Madagascar	5575
	birth: mid Feb-mid July, peak Apr-May; captive	2153
	birth: Mar-June; n29; zoo	12120
	birth: Mar-July; n26; zoo, Egypt	3340
	birth: end Mar-May; captive	6476
	birth: Aug, Nov; ne Madagascar	8435
Lemur macaco	mate: Apr-June; ne Madagascar	8435
	mate: mid Apr-June; Madagascar	8444
	birth: Mar-Apr; zoo	4722
	birth: Mar-June; n14; zoo, England	12120
	birth: Mar-July, peak Apr; n24; zoo, Egypt	3340
	birth: Apr-June; Madagascar	8433
	birth: Aug-Nov; Madagascar	8432, 8435
	birth: Sept; n2; zoo	1230
	birth: Sept-Oct; Nosy Be I, Madagascar	8445
	birth: Sept-Nov; Madagascar	8444
Lemur mongoz	mate: Dec; zoo	4722
	birth: Apr; zoo	4722
	birth: Apr-June; Madagascar	8433
	birth: peak Apr-June, also Jan, Mar, July; captive	9527
Lemur rubriventer	birth: Apr-June; Madagascar	8433
Lepilemur mustelinus	spermatogenesis: May-Aug; Madagascar	8444
	estrus: Feb-May; Madagascar	5573
	mate: Apr-July; Madagascar	8444
	mate: May-June; Nosy Be I, Madagascar	8445
	mate: May-July; Nosy Be I, Madagascar	8432
	preg f: June-Sept; Madagascar	5573
	birth: May-July; Madagascar	8433
	birth: Sept-Nov; Nosy Be I, Madagascar	8432, 8443, 8445
	birth: mid Sept-Oct; Madagascar	8444
	birth: Oct; n1; Madagascar	5573
Lepilemur ruficaudatus	estrus: May; lab, Madagascar	8451
	estrus: Nov, Jan; captive	8451
	birth: June; n1; lab	8451
Varecia variegata	testes: low summer; captive	3362
	estrus: June-Mar; zoo	9906
	mate: Dec-Jan; captive	6476
	birth: Mar-July, peak Apr >50%; zoo	1300
	birth: peak Apr-May; zoo	9906
	birth: Apr-June; n7; zoo, Egypt	3340
	birth: June; captive	6476
	birth: Sept; zoo	8445
	birth: Oct-Nov; e Madagascar	8432, 8435

Order Primates, Family Indriidae

Indris (*Indri* 7-10 kg), avahis (*Lichanotus* 600-1000 g), and sifakas (*Propithecus* 3-9 kg) are social, arboreal herbivores from Madagascar's rainforests. The epitheliochorial, diffuse placenta implants in a bicornuate or bipartite uterus (7563, 8444, 10962) and is eaten after birth (2854, 8433, 9096). Neonates are furred and open-eyed (8432, 8443, 8444, 9096). Males have a baculum and seminal vesicles (8443, 8444). Details of female anatomy (4793, 8443, 8444) and birth (9096) are available.

NEONATAL MASS		
Propithecus verreauxi	35-40 g; n1	115
NEONATAL SIZE		
Lichanotus laniger	HB: 90 mm	8443, 8444
LITTER SIZE		
Lichanotus laniger	1	8444, 8445
	1	8432, 8435
Propithecus verreauxi	1	4374
	1	8435
	1 ea; n3 lt	8443
	1-2	5376
SEXUAL MATURITY		
Propithecus verreauxi	f/m: 1st mate: 2.5 yr	5376
	f/m: est > 2.5 yr	8444
GESTATION		
Indri indri	17-22 wk; field data	8648
Lichanotus laniger	4-5 mo	7983
Propithecus diadema	est ~5 mo	8435
Propithecus verreauxi	~15 wk; field data; not many	5573
	est 4-5 mo	5575
	est 5 mo; n1; swollen genitals to birth	2854
	~5 mo	8435
	~5 mo; field data	8443, 8444, 8445
	~5 mo; field data	9095
	156.75 d; r154-160; n4	4374
LACTATION		
Indri indri	wean: 8 mo-< 1 yr	8648
Propithecus spp	solid food: 53-57 d	8432
Propithecus verreauxi	solid food: 1-2 wk	9096
	solid food: 7 wk	4374
	solid food: 11.5 wk	2854
	wean: < 5 mo	9096
LITTERS/YEAR		
Lichanotus laniger	1	8444
	1	8432
Propithecus verreauxi	1	8445
SEASONALITY		
Indri indri	mate: Jan-Mar; e Madagascar	8648
	birth: end May; e Madagascar	8648
Lichanotus laniger	birth: Aug-Sept; ne, e Madagascar	8432, 8435
	birth: end Aug; ne Madagascar	8444, 8445
	birth: Oct-Nov; e Madagascar	8444, 8445
Propithecus diadema	birth: Sept-Oct; e Madagascar	8435
Propithecus verreauxi	estrus: mid Nov-Mar; nw Madagascar	5573
	mate: Jan-Mar; nw Madagascar Madagascar	8444, 8445
	mate: Jan-Mar; w,s,e coast Madagascar	8435
	mate: Feb; Madagascar	5376
	mate: Mar; nw, s Madagascar	9095
	preg f: end rains; none June-Dec; n Madagascar	5575
	birth: May-Aug; w,s,e coast Madagascar	8435
	birth: June-July; nw Madagascar	5573
	birth: end June-mid July; Madagascar	8444, 8445
	birth: early-mid July; Madagascar	5376, 5377
	birth: Aug; nw, s Madagascar	9095

Order Primates, Family Daubentoniidae, Genus *Daubentonia*

The rare, nocturnal aye-aye *Daubentonia madagascariensis* from Madagascar eats fruit, flowers, and insect larvae. Males of this 2.5-3 kg, solitary and arboreal omnivore have scrotal testes, extremely small seminal vesicles, large Cowper's glands, and a baculum (5572, 8163, 8444). Females have 2 inguinal teats (8435, 11821), a short clitoris, long perineum, and scrotal-like appendages lateral to the clitoris (4789). The villous, laminar, epitheliochorial, diffuse placenta (4765, 7563) implants into a bicornuate uterus (4765). Details of anatomy (8177, 8443, 8444) are available.

LITTER SIZE		
Daubentonia madagascariensis	1	8432, 8434, 8437
LACTATION		
Daubentonia madagascariensis	solid food: 3 mo; n1	11821
	wean: 6-7 mo; n1	11821
INTERBIRTH INTERVAL		
Daubentonia madagascariensis	2-3 yr	8437
SEASONALITY		
Daubentonia madagascariensis	birth: Feb-Mar; Madagascar	8432
	birth: Oct-Nov; Madagascar	8434, 8437

Order Primates, Family Lorisidae

Lorises (*Arctocebus* 150-450 g; *Loris* 100-350 g; *Nycticebus* 375-2000 g; *Perodicticus* 850-1600 g) are nocturnal, arboreal omnivores from India, Africa, and southeast Asia. The epitheliochorial, villous, diffuse placenta has a superficial circumferential implantation into a bicornuate uterus (4758, 4759, 4771, 4774, 7563, 7567, 10310) and is eaten after birth (5478). A vaginal plug is reported (8888). A postpartum estrus may occur if the neonate dies (1739, 5195, 5198, 5309, 6767, 8888). Neonates are furred and open-eyed (34, 974, 1736, 2092, 4004, 5758). Females have 2-6 mammae (1736, 4774, 5198, 7161) and pseudoscrota (5758). Males have abdominal testes, seminal vesicles, and a baculum (5842, 8443, 8880, 8881, 8886). Details of female anatomy/oogenesis (249, 1737, 2290, 2774, 2776, 3733, 5160, 8925), endocrinology (8371a), and birth (1736) are available.

NEONATAL MASS		
Arctocebus calabarensis	27 g; r24-30; n2	1736, 1742
	f: 37 g; n1	5309
Nycticebus coucang	33.0 g; n1 stillborn	5198
	50 g; n1	34
	f: 50.75 g; r50.5-51.0; n2	5198
	m: 51.57 g; r43.5-57.0; n3	5198
	60 g; n1, 3 d	4785
Perodicticus potto	28 g; n1	2109
	35.7 g; r30-42; n6	2092, 2095
	f: 52 g; n1	1738, 1742
NEONATAL SIZE		
Perodicticus potto	11.75 mm; n1	2109
LITTER SIZE		
Arctocebus calabarensis	1 yg ea; n3 lt, 2 f	1736
Loris tardigradus	1 yg ea; n9 lt	5195, 8937
	1 yg; 2 very occasional	8501
	1+; 2 occasional	8880
	1.5 emb; r1-2; n2 lt	4771
	1.6 emb; r1-2; n87 f	8888
Nycticebus coucang	1 yg	2313
	1	8619
	1 yg; n1 lt	34
	1 yg; n3 lt	4785
	1 yg ea; n19 lt	5198, 8937
	1.25; mode 1; r1-2; n8 lt	2109
	1.3; r1-2; n6 lt	851

Perodicticus potto	1; n1 lt	974
	1; n1 lt	11401
	1; r1-2; 2 rare	5758
	1 ea; n7 lt	2109
	1.2 yg; r1-2; n5 lt	2095
SEXUAL MATURITY		
Arctocebus calabarensis	f/m: ~9-10 mo	1742
Loris tardigradus	f: 1st estrus: 10 mo; n1	5195
	f: 1st estrus: 340 d; sd56.5; n4	8937
	f: 1st conceive: 364 d; n1	5195
	f: 1st repro: 2 yr	8881
	m: 1st spermatogenesis: 10-11 mo	8881
Nycticebus coucang	f: 1st estrus: 19 mo; r17-21; n2	5198
	f: 1st mate: 20.3 mo; r18-23; n3	5198
	f: 1st birth: 3 yr; n1	5198
	m: 1st sire: 17 mo; n1	5198
Perodicticus potto	f/m: 9 mo	5758
	f/m: ~18 mo	1742
CYCLE LENGTH		
Arctocebus calabarensis	39 d; r36-45; n4 cycles, 1 f	6767
Loris tardigradus	33.6 d; sd4.5; r29-40; n3 f	5195
Nycticebus coucang	36.4 d; sd4.5; r29-45; n28 cycles, 10 f; interestrus interval	5198
	42.3 d; mode 40; r37-54; n17 cycles, 2 f	6767
Perodicticus potto	37.9; se~1.0; n36 cycles, 2 f; vaginal smears	5159
	estrus: 1.8 d; metestrus: 4.6 d diestrus: 27.8 d; proestrus: 3.7 d	5159
	38.3 d; r21-48; n37 cycles, 3 f	6767
	50 d	8443
GESTATION		
Arctocebus calabarensis	134 d; r131-136; n3	6767, 6768
Loris tardigradus	163 d; r160-166; n2	6767, 6768
	167.2 d; sd1.5; r166-169; n4	5195, 8937
	174 d; n1	7881
Nycticebus coucang	174 d	11564
	192.2 d; sd3.6; r185-197; n9 lt, 5 f	5198, 8937
	193 d; n1	6767
Perodicticus potto	193 d; n1	1738, 1742
LACTATION		
Arctocebus calabarensis	solid food: 1.5 mo	1736
	wean: ~100-130 d	1742
Loris tardigradus	wean: 5 mo; from graph	8881
	milk present: 169 d; sd32.0; n5	5195, 8937
Nycticebus coucang	milk present: 180 d; sd11.3; n2	8937
	wean: ~6 mo; n5 lt, 4 f	5198
	wean: min 9 mo	2109
Perodicticus potto	solid food: 5 wk	5758
	wean: 120-180 d	1742
	wean: 2-3 mo	1740
INTERBIRTH INTERVAL		
Arctocebus calabarensis	131 d; n1	5309
	4.5-5.0 mo; n1	1739
Loris tardigradus	9.5 mo	5195
Nycticebus coucang	16.2 mo; r9-38; n7 f	5198
Perodicticus potto	min 6.5 mo	2092
SEASONALITY		
Arctocebus calabarensis	birth: Nov-May; Rio Muni, equatorial Guinea	8507
	birth: yr rd, low June-Aug; n33; Gabon	1742
Loris tardigradus	epididymal sperm: yr rd; Bangalore, India	8880, 8881
	accessory glands: peak May; Bangalore, India	8881
	estrus: June-July, Sept-Nov; Bangalore, s India	8886, 8888
	mate: June-July; Bangalore, India	8881
	term emb: Mar-May, Nov; Bangalore, s India	8888
	birth: Apr-May, Nov-Dec; Bangalore, s India	8880
Nycticebus coucang	preg f: yr rd, peak May; n146; captive	12120

	birth: Nov-Apr; n16/20; captive	851
	birth: Mar-May; n12/19; captive	5198
Perodicticus potto	testes: large Mar-May; no data after May; Africa	1739
	birth: Apr-May; n5; Nov-Feb; n9; e Africa	5758
	birth: June; n1; Zaire	8854
	birth: Aug-Jan; n17; Gabon	1739, 1742
	birth: not seasonal; n7; zoo	2109
	lact f: Jan; n1; s Cameroun, Benito, Equatorial Guinea	647
	lact f: Feb; n1; Kenya	4926

Order Primates, Family Galagidae

Bush babies (*Galago* 45-360 g; *Otolemur* 0.6-2 kg) are nocturnal, arboreal omnivores from African forests. Males are larger than females (4361, 4362, 5758, 7770). Oogenesis occurs in adults (1496, 1498, 3728, 3732, 3733, 4665, 5160). The endotheliochorial and epitheliochorial, diffuse placenta has a central, superficial, often antimesometrial implantation into a bicornuate uterus (1489, 1491, 1493, 1494, 1495, 1497, 2479, 3731, 3732, 4761, 4764, 5732, 7563, 8221, 8222, 10154, 10310) and is eaten after birth (2714). A postpartum estrus occurs in *Galago*, but not as regularly in *Otolemur* (1490, 2712, 2713, 2714, 4631, 5758, 5936, 9507, 11062, 11063, 11171). Neonates are furred and open eyed (973, 1491, 2711, 2788, 4122, 5758, 5936, 6444a, 6615, 8443, 9507, 10158). Females have 6 mammae (*Galago* 331, 1491, 1493; cf 10158). Males have seminal vesicles (1491) and androgen-dependent penial spines (2619, 2637). Details of anatomy (1491, 4211), gonadal development (12031), endocrinology (1444, 2268, 2636, 2872, 5194, 11170), birth (2714), milk composition (8541), and reproductive energetics (9511) are available.

NEONATAL MASS		
Galago alleni	24 g	1742
Galago demidovii	5-10 g	1742
	11.6 g; n1	2788
	25 g	5758
Galago senegalensis	9.3-9.6 g; near term	10154
	m: 11.45 g; n23	2712
	11.50 g; r8.5-15.5; n42; 12.11 g; n10 singletons; 11.31 g ea; n32 twins	2712, 2714
	f: 11.56 g; n19	2712
	f: 11.7 g; r8.5-14.8; n28	2711
	m: 11.9 g; r8.6-15.5; n28	2711
	15 g; n2	5936
	15.6-22.6 g; single births	2712, 2714
Galago zanzibaricus	14.1 g; r13-15; n3	4122
Otolemur crassicaudatus	40 g	5758
	47.4 g; sd6; twins each 45.9 g; n16	11062, 11063
	63 g; r46.4-72.7; n3	7870
	min 72.5 g; n5	11575
NEONATAL SIZE		
Galago demidovii	75 mm	5758
Galago senegalensis	58 mm; n2	5936
Otolemur crassicaudatus	101.0 mm; r81-108; n3	7870
	min 112 mm; n5	11575
WEANING MASS		
Galago senegalensis	90-110 g; mean +- 1sd; from graph; 70 d	2711
LITTER SIZE		
Galago demidovii	1.2 emb; r1-2; n18 f	1742
	1.3; r1-2; n3 lt	2788
	1.5; r1-2; n10 lt	5758
	2; n1 lt	973
	2; n1 lt	8443
	2.0 emb; n3 lt	8221
Galago elegantulus	1 emb ea; n9 f	1742
Galago inustus	1 emb; n1 f	5758
Galago senegalensis	1-2	9507
	1; north; 2; south	5758

	1 emb ea; n5 f	4926
	1 ea; n6 lt, 1 f	9137
	1.03; r1-2; n36 f	1488, 1491, 1493, 1495
	1.25 emb; r1-2; n4 lt	4665
	1.55 yg; sd0.50; n42	5196, 8937
	1.62; r1-3; n29 lt	2712
	1.875 yg; r1-3; n8 lt, 4 f	2714
	2	8443
	mode 2; n>11 lt	7769
	2; n1 lt	6615
Galago zanzibaricus	1 yg	7769
	1-2	5758
	1.29; r1-2, n7 lt	4122
Otolemur crassicaudatus	1-3	5196
	1; n1 lt	2109
	1 yg; n1	2334
	1 ea; n2 lt, 1 f	8443
	1.03 yg; r1-2; n86 lt	2855
	1.11 yg; mode 1; r1-2; n201 lt	4631
	1.12; r1-2; n69 lt	11062, 11063
	1.14; r1-2; n180 lt, 42 f	2637, 8253, 11171
	1.6 yg; se0.09; r1-3; n60 lt	2855
	1.64 yg; sd0.63; n77	5196, 8937
	2	11253
	2-3 88-100%; r1-3	7769
	2.0; mode 2; r1-3; n8 lt	681, 2713
SEXUAL MATURITY		
Galago alleni	f/m: 8-10 mo	1742
Galago demidovii	f/m: 8-10 mo	1742
Galago senegalensis	f: 1st estrus: 8 mo	5758
	f: 1st estrus: 265 d; sd95.6; n17	8937
	f: 1st conceive: mode 200 d; min 120	2711
	f: 1st mate: ~20 mo	1495
	f/m: 11 mo, 324 d; n1	2268
Galago zanzibaricus	f/m: 1 yr	5758
Otolemur crassicaudatus	f: 1st estrus: 445 d; sd89.8; n8	8937
	f: 1st mate: ~1 yr	1439
	f: 1st mate: min 16 mo; mode 19 mo	2855
	f: 1st mate: min 22 mo; mode 26.5 mo	2855
	f: 1st conceive: 1.89 yr; sd0.85; r0.99-4.58; n20	11171
	f: 1st conceive: 2.52 yr; sd0.97; r1.45-3.83; n5	11171
	m: 1st sire: min 10 mo; mode 18.5 mo	2855
	m: 1st sire: min 22 mo; mode 26 mo	2855
	f/m: 1.75 yr	1848
CYCLE LENGTH		
Galago senegalensis	estrus: 48 h	1495
	estrus: 3-6 d	6615
	estrus: 3-6 d	8442
	estrus: 4.8-6.7 d; n11 f	2268
	estrus: 7 d	9507
	14 d	8442
	28-30 d	2714
	4-6 wk	1490, 1491
	31.7 d; mode 31; r29-39; n31 cycles, 4 f	6767
	32.9 d; se0.6; r29.2-39.3 per f; n168 cycles, 11 f	2268
	39 d; r36-42; n3 cycles, 1 f	8443
	6 wk	9507
Otolemur crassicaudatus	estrus: 5.8 d	2870, 2872
	luteal phase: 24 d	2872
	39 d; se4.5; mode 34-38; n79 cycles	4631
	43.6; sd4.56; r39-49; n5 f	2637
	44 d; sd8.0; r35-59; n10 cycles	2870, 2872
	50.3 d; mode 17; r3-104; n54 cycles	11062, 11063

GESTATION		
Galago alleni	133 d	1742
Galago demidovii	1-3.7 mo; n1; pairing to birth	1589
	6 wk	5541
	3-4 mo	5758
	111-114 d	1742
Galago senegalensis	4 mo	8443
	4 mo; field data	9507
	~4 mo	6615
	~4 mo; field data	1488, 1490
	122-125 d; n4 lt, 2 f	2714
	123-124 d	5936
	123.5 d; r121-124 (± 1 d); n13	2712
	124.5 d; sd2.4; n28	5196, 8937
	139 d; ±3; n30	2268, 2714
	145 d; r144-146; n4	6767
Galago zanzibaricus	120 d; n4	4122
Otolemur crassicaudatus	125.2 d; se2.69; r121-133; n4	2855
	128 d; sd13; n69	11062, 11063
	18.5 wk	1439
	4 mo 10 d; n2	8443
	130.2 d; sd2.5; n12	5196
	131.9 d; sd3.4; r128-134; n20	4631
	132.2 d; sd4.76; r126-138	2637
	132.8 d; sd2.6; r130-136; n5	2872
	133 d	2713
	133 d; ±2	7870
	135.6 d; sd3.7; n21	5194, 5196, 8937
LACTATION		
Galago demidovii	solid food: 2 wk	1589
	suckle: min 5 wk	1589
	wean: 40-50 d	1742
	wean: ~6 wk; n5	5758
Galago senegalensis	solid food: 4 wk	10158
	solid food: 4 wk	9507
	solid food: 2 mo	5758
	nurse: 14 d; 1st decline; 45 d; 2nd decline	2711
	nurse: ~6 wk	5758
	nurse: 10-11 wk	2711, 2713
	nurse: 100-111 d	5192
	nurse: 100.3 d; sd9.9; n44	5193
	nurse: 101 d; sd11.1; n28	8937
	nurse: 3.5 mo	6615
Galago zanzibaricus	solid food: 1 mo	5758
	solid food: > 1 mo	4122
Otolemur crassicaudatus	solid food: mode 8 d	9507
	solid food: 1 mo	5758
	wean: begins 21 d	10158
	nurse: ~10 wk	10158
	nurse: ~3+ mo	5758
	nurse: 134 d; sd10.5; n9	8937
	nurse: 136.4 d; sd11.4; n19	5192, 5193
	nurse: 139.9 d; sd16.3; n17	5192, 5193
INTERLITTER INTERVAL		
Galago senegalensis	median 189 d; r124-325; n22 young die	5197
	7.5 mo	2712
	median 259.5 d; r167-454; n38 young live	5197
Otolemur crassicaudatus	min 119 d	2855
LITTERS/YEAR		
Galago demidovii	1	1742
	2?	11253
Galago zanzibaricus	mode 2	7769
	2	4122

SEASONALITY

Galago alleni	mate: yr rd; Cameroun	9468
	birth: Jan-Mar, May, July-Aug, Nov-Dec; n10; Gabon	1742
Galago demidovii	testes: mod-small Dec; Uganda	5758
	mate: end May; Katanga, Zaire	3732
	mate: yr rd, peak Aug; Uganda	5758
	birth: peak Jan-Apr, low June-July; Gabon	1742
	birth: Oct-Nov; Zaire	11253
	birth: yr rd, peaks Dec-Feb, June; Uganda	5758
Galago elegantulus	mate: low June-Oct; Cameroun	9468
	birth: Sept-Feb, Apr-May; n11; Gabon	1742
Galago inustus	testes: small July; n1; central Africa, s,w Uganda	5758
	mate: peak Nov-Dec; central Africa, s,w Uganda	5758
	preg f: Feb; n1; central Africa, s,w Uganda	5758
Galago senegalensis	anestrus: mid Feb; Tanzania	186
	estrus: Nov-Jan, Aug; 3.5 mo anestrus; Sudan	1491
	estrus: yr rd; Sudan	1490
	estrus: yr rd; captive	5197
	mate: Mar-Sept; captive	2714
	mate: end May; Katanga, Zaire	3732
	mate: end May-July, end Sept-Oct; S Africa	2713
	preg f: Dec-Jan, Mar-Apr, Oct; n24; Sudan	1488, 1491, 1495
	preg f: Feb, Nov-Dec; n5; Kenya	4926
	preg f: Sept; Rwanda	4665
	birth: Jan-Feb, mid Sept-Nov; Transvaal, S Africa	7769
	birth: Mar-Apr, June, Dec; n4; Sudan	1491
	birth: Apr; n1; captive	6615
	birth: July-Jan; captive	2714
	birth: Aug-Dec, peak Sept-Oct; Zambia, s Tanzania	5758
	birth: mid Sept-mid Nov; n5; zoo	1230
	birth: Oct, Feb-Mar; n Transvaal, S Africa	2713
	birth: Oct-Nov, Jan-Mar, peak Feb; S Africa	681
	birth: yr rd; Botswana	10154
	birth: yr rd; captive	5197
Galago zanzibaricus	preg f: Feb, June-Dec; Kenya	7769
	birth: Feb-Mar, Aug-Oct; Kenya	7769
Otolemur crassicaudatus	mate: end May; Katanga, Zaire	3732
	mate: June-July; Transvaal, S Africa	8965
	mate: est June-July; ne Transvaal, S Africa	2713
	birth: Mar; Somalia	5758
	birth: May; n1; zoo	2109
	birth: Aug-Sept; Zambia, Zimbabwe	10158
	birth: est Aug-Oct; few data; Kenya	7769
	birth: Aug-Nov; Tanzania, Zambia	5758
	birth: Oct-Nov; Zaire	11253
	birth: Oct-Nov; S Africa	1848
	birth: Oct-Nov; Transvaal, S Africa	7769
	birth: Nov, S Africa; ne Transvaal, S Africa	2713, 681
	birth: Nov; Transvaal, S Africa	8965
	birth: yr rd; captive	11063

Order Primates, Family Tarsiidae, Genus *Tarsius*

Tarsiers (*Tarsius bancanus*, *T. spectrum*, *T. syrichta*) are small (80-160 g), arboreal insectivores from tropical forests of southeast Asia. They are solitary except for reproduction (3364, 6310). Copulation lasts 60-90 sec with multiple thrusting and does not result in a copulatory plug (11943; cf 4788a). The 1.2-2.5 g, deciduate, hemochorial, labyrinthine, discoidal or diffuse placenta has a superficial, mesometrial implantation into the bicornuate or bipartite uterus (231, 1491, 4113, 4757, 4759, 5199, 5733, 6124, 7563, 11036, 11164). Neonates are mobile, furred, open-eyed, and up to 20% of maternal mass (557, 4374a, 6310, 11613). Females have 6 mammae (4788a). Male seminal vesicles are large (4788a). Details of anatomy (4788a, 11910a) are available.

NEONATAL MASS		
Tarsius bancanus	28.5 g; n1	5199
Tarsius spectrum	24 g	6310
	~28 g	557
Tarsius syrichta	23.2 g; r20-31.5; n9	4374a
	f: 25.3 g; n1	11036
	m: 27.2 g; n1	11036
NEONATAL SIZE		
Tarsius syrichta	m: 62 cm; n1	11036
	f: 66 cm; n1	11036
LITTER SIZE		
Tarsius bancanus	1	5192, 5199
	1	3364
	1 emb ea; n2 f	2313
	1 emb ea; n2 lt	2313
Tarsius spectrum	1	557
	1	6310
	1 emb ea; n5 lt	10482
Tarsius syrichta	1	11613
	1 yg; 1 lt	9647
CYCLE LENGTH		
Tarsius spp	23.5 d; ±0.7; r23-28; n1 f	1662
Tarsius bancanus	estrus: 2.2 d; r1-3	11942, 11943
	genitalia swollen: 6-9 d	11942, 11943
	22.5 d; ±3	5192
	24.0 d; sd3.2; r18-27; n6; smears	11942
	25.9 d; sd3.2; n9; visual swellings	11942
GESTATION		
Tarsius bancanus	178 d	5199
Tarsius spectrum	~6 mo	557
Tarsius syrichta	178 d	4374a
	est 6 mo	11036
LACTATION		
Tarsius spectrum	solid food: 3 wk	6310
INTERBIRTH INTERVAL		
Tarsius spectrum	262.4 d; r187-406; n5 intervals, 3 f	4374a
	1 yr	6310
SEASONALITY		
Tarsius bancanus	birth: Mar-July; Borneo, Indonesia	3364
	preg f: Aug; n2; N Borneo, Indonesia	2313
Tarsius spectrum	mate: yr rd; n88; captive	12120
	conceive: Oct-Dec; Borneo, Indonesia	6310
	preg f: yr rd, peak Oct-Nov; n1000+; Indonesia	11164
	preg/lact f: Oct-Feb; Borneo, Indonesia	6310
Tarsius syrichta	birth: Jan-Feb, Apr-May, Aug, Oct-Nov; n9; captive	4374a

Order Primates, Family Callithricidae

Marmosets and tamarins (*Callithrix* 190-600 g; *Cebuella* 100-150 g; *Leontopithecus* 360-800 g; *Saguinus* 225-900 g) are monogamous or social, arboreal omnivores from the forests of tropical Central and South America. Only dominant members of a group breed (*Callithrix* 3, 5, 8, 9314; *Saguinus* 2368, 3050, 3055, 3605, 4590, 9512, 11664; cf 10749) and a concealed estrus is reported (*Leontopithecus* 10498; *Saguinus* 1235). The 5-10 g, hemochorial, labyrinthine, bidiscoidal placenta (215, 770a, 1687, 4113, 4310, 4769, 4842, 5375, 7497, 11848, 11852, 11855, 11966) secretes gonadotrophins (10636a) and is shared by twins (767, 5854). Placentophagia occurs (1753, 6620, 6647, 6813, 8253, 9313, 9315, 10388). A postpartum estrus within 1-2 wk is usual (1686, 2366, 3049, 3120, 4540, 4544, 4545, 4549, 4550, 5850, 8487, 9313, 9314, 10184, 10185, 12098; cf 2153, 3503, 10199, 12098). Neonates are furred with open eyes (1753, 1821, 3049, 6222, 6621). Females have 2 mammae (*Callithrix* 5147). Males and siblings contribute to parental care by carrying young (*Callithrix* 4540, 4545, 6621, 6813, 10673, 11577; *Leontopithecus* 4838, 10184; *Saguinus* 3051, 5584, 6477, 7583, 10673, 11577, 11594, 11733).

Details of anatomy (10337, 11847, 11851), endocrinology (4306; *Callithrix* 2, 3, 5, 8, 1167, 1686, 2623, 2624, 2630, 2635, 2869, 3119a, 4307, 4308, 4372, 4378, 4379, 4380, 4540, 4542, 4545, 4549, 4550, 4566, 4567, 4568, 4844, 4858, 4859, 4862, 4864, 4865, 4865a, 4867, 4868, 5664, 5719, 5720, 5927, 6646, 7110, 8750, 9050, 9813, 10551, 11517; *Cebuella* 12100a; *Leontopithecus* 3507, 3508, 3509, 5859; *Saguinus* 1234, 1235, 3054, 3055, 3055a, 3504, 3505, 3694, 4590, 4591, 4864, 4866, 8748, 8752, 8767, 10671, 12098, 12099, 12100), sexual development (2, 6, 3052, 3053, 3397, 3695, 10671, 12099), spermatogenesis (4203), birth (10389), and reproductive energetics (5819) are available.

NEONATAL MASS

Callithrix jacchus	14.5 g; n34 twins, 17 lt	6647
	18.6 g; n36 triplets, 12 lt	6647
	22 g; n2	10388
	23.6 g; r18-27; n4 ea triplets	4306
	25 g; from graph	2
	25-30 g	4082
	25-32 g; varies inversely with litter size	1687
	28 g; n > 25	4544, 4545
	28.9 g; r27.5-30.0; n3 triplets, 1 d preterm	3049
	30 g; n1	6620, 6621
	30.3 g; r21-38; n9 ea twins	4306
	32.5 g; sd3.7; n3 singletons	6647
	34.7 g; n1	3049
Cebuella pygmaea	14-16 g	1821, 1822
	27 g; n1 stillborn	6222
Leontopithecus rosalia	49.6 g; r44.0-53.3	11749
	51.9 g; r50.2-53.65; n2, 1 d	2792
	m: 57 g; n1	215
	60.6 g; sd6.22; r52.1-74.6 g; n11; from graph	4838
Saguinus nigricollis	43.5 g; n10	11875
Saguinus oedipus	f: 30.5 g; r25.0-42.0; n2 triplets	11874
	34.3 g; r32-40; n3 ea triplets	4306
	36.1 g; r25-45; n20 ea twins	4306
	37.0 g; 34-40; n4 ea singletons	4306
	m: 38.2 g; r31.2-43.5; n6 triplets	11874
	41 g; n1	3049
	f: 42.1 g; r35.7-52.0; n12 twins	11874
	m: 44.0 g; n1 singleton	11874
	m: 44.0 g; r35.3-50.8; n20 twins	11874
	f: 46.4 g; r36.5-55.8; n1 singleton	11874

NEONATAL SIZE

Callithrix jacchus	CR: 68.6-91.8 mm; n14 neonates	8486
	CR: 70-72 mm; 146-149 d of gest	8486
	CR: 78.3 mm; r72-84; n6 ea triplets	4306
	80 mm	4545
	CR: 84.1 mm; r76-89; n11 ea twins	4306
Leontopithecus rosalia	HB: 100-114 mm	5854
	m: 122 mm; n1	215
Saguinus oedipus	CR: 89.3 mm; r87-91; n4 ea triplets	4306
	CR: 92.0 mm; r82-103; n20 ea twins	4306
	CR: 95.5 mm; r94-97; n4 ea singletons	4306

WEANING MASS

Callithrix jacchus	~60 g; 4 wk	5768
	~75 g; 6 wk	5768
	~125 g; 12 wk	5768
Saguinus oedipus	~70 g; 4 wk	5768
	~100-110 g; 7-8 wk	5768

LITTER SIZE

Callithrix spp	1.86; r1-3; n250 lt	4309
Callithrix argentata	1.5 yg; r1-2; n2 lt	4306
	1.83; mode 2; r1-2; n12 lt	4313
	2	10390
	2; n3 lt	5767
	2 yg ea; n7 lt	5768

Callithrix jacchus	1.33; r1-2; n3 lt, 1 f	1753
	1.54; r1-3; n41 lt	3697
	2	3
	mode 2; r1-3	1687, 4541, 4544, 4549, 7497
	2; n1 lt	6813
	2.0; r1-3; n2 lt	9313
	2.0 yg; r1-3; n14 lt	6620, 6621
	2.0 yg; mode 2; r1-3; n144 lt	4306, 4313
	2+; mode 2; r1-3	4545
	2.03 yg; mode 2; r1-4; n115 lt	5768
	2.05 yg; 1.88 alive; n66 lt, 30 f	10675
	2.05; r1-3; n87 lt	5767
	2.1; mode 2; r1-2	4551
	2.18; mode 2; r1-3; n17 lt	4550
	2.2 yg; mode 2; r1-3; n124 lt	1468
	2.22 yg; mode 2; r1-3; n159 lt	8487
	2.3; r2-3; n3 lt	12128
	2.33 yg; 1.7 alive; n6 lt, 2 pr	11874
	2.33 yg; 1.71 alive; mode 2; r1-3; n24 lt	5147
	2.42 yg; mode 2-3; r1-4; n50 lt	9315
	2.48 yg; mode 3; r1-4; n202 lt	9316
	2.5 yg; r2-3; n14 lt; triplets not raised	10388
	2.51 born; mode 3; r1-4; n87 lt	1468
	2.57; r2-3; n14 lt	3120
	2.61 yg; r1-4; n114 lt	9317
	2.68; mode 3; r1-4; n129 lt	8663
	2.69 yg; mode 3; r1-4; n202 lt; usually 2 nursed	1167
	2.82 yg; r2-4; n28 lt; no more than 2 yg reared	10391
Cebuella pygmaea	1.78; mode 2; r1-3; n9 lt	4306, 4313
	1.85 yg; mode 2; r1-3; n41 lt	12100a
	mode 2	10198
	2; n1 lt	5767
	2 yg; n1 lt	8891
	2 yg; n1 lt	5768
	2.02 yg, 1.74 alive; mode 2; r1-3; n39 lt	6222
	2.1 yg; r2-3; n8 lt, 3 f	1821, 1822
Leontopithecus rosalia	1.65 yg; n23 lt	2579a
	1.8; 2.1; n2 samples	5856
	1.9; mode 2; r1-3; n11 lt	4838
	1.94; mode 2; r1-3; n17 lt	5853
	2	1626
	2; n1 lt	5458
	2; n1 lt	767
	2 ea; n2 lt	3383
	2.0; n3 lt	2109
	2 ea; n3 lt, n1 f	1967
	2 ea; n4 lt, n1 f	8825
	2.1; r1-3; n11 lt, 1 f	4200, 11749
	2.17; r2-3; n6 lt	11035
Saguinus spp	1.63 yg; mode 2; r1-3; n295 lt	4309
Saguinus fuscicollis	1.65 yg; r1-2; n31 lt, 17 f	11874
	1.70 yg; mode 2; r1-2; n44 lt	4306
	1.72 yg; 1.44 alive; mode 2; r1-3; n43 lt, 17 f	10675
	1.82 yg; mode 2; r1-3; n148 lt	11874
	1.85 yg; mode 2; r1-3; n319 lt	3696, 3697
	1.92 yg; mode 2; r1-3; n184 lt	11874, 11875
	2.0; n5 lt, 1 f	1661
Saguinus labiatus	1.81 yg; 1.18 alive; r1-3; n27 lt	8043
	2 ea; n2 lt	3383
Saguinus midas	2 ea; n4 lt	1821, 1822
Saguinus mystax	1.5 yg; r1-2; n2 lt	4304
	1.93 yg; r1-3; n29 lt	8042
	mode 2	5584

	2; n1 lt	4306
Saguinus nigricollis	1.6; r1-2; n5 lt	4304
	1.7; r1-2; n9 lt	5767
	1.82; mode 2; r1-3; n88 lt	11874, 11875
	1.92 yg; r1-2; n13 lt	5768
Saguinus oedipus	1-2 yg; 3 rare	3503
	1.4 yg; r1-2; n5 lt	12099
	1.7 emb; r1-2; n10 f	2368
	1.72 yg; mode 2; r1-3; n138 lt	4304, 4305, 4306, 4311
	1.75 yg; mode 2; r1-2; n8 lt, 6 f	1233
	1.82 yg; mode 2; r1-3; n61 lt	7231
	1.84; mode 2; r1-3; n55 lt	10180
	1.88; mode 2; r1-3; n83 lt	3697
	1.9 emb; r1-2; n10 lt	2368
	1.91 yg; mode 2; r1-3; n68 lt	11874, 11875
	1.92 yg; r1-2; n12 lt	5768
	1.95 emb; mode 2 87.5%; r1-3; n40 lt	11852
	1.98 yg; 1.88 alive; n41 lt, 17 f	10672
	mode 2	7583
	2 ea; n2 lt	11510
	2.0; n9 lt	5767
	2.04; 1.62 alive; mode 2; r1-3; n106 lt, 47 f	10675
	2.05; mode 2; r1-3; n19 lt	3113
	2.09; mode 2; r1-3; n54 lt	5816, 5817
	2.15; r1-3	1628
	3.0; n2 lt	11733
SEXUAL MATURITY		
Callithrix jacchus	f: high progesterone: 300-400 d	2, 6, 4540, 4541, 4544, 4549
	f: 1st conceive: 335-903 d	2, 4544, 4551
	f: 1st conceive: ~14 mo; n1	6621
	f: 1st birth: 18.5 mo; r14-22; n11	8487
	f: 1st birth: 23 mo	10391
	f: 572 d; ±31.4	2, 4546, 4547
	m: high testosterone: 250-350 d	6, 4540, 4549
	m: proven fertile: ~15 mo; min 11 mo	5215
	m: testes large: min 250 d; mode 550-700 d	2, 4544, 4549, 4551
	f/m: 13 mo	8664
	f/m: 400-600 d	3
Cebuella pygmaea	f: 1st birth: 32.7 mo; median 32; sd7.6; r24-42; n7	12100a
	f/m: 2-3 yr	1822
Leontopithecus rosalia	f: 1st conceive: 29.3 mo	5856
	f: 1st birth: 2 yr	5850
	m: 1st mate: 28.3 mo; n16	5856
	f/m: 18 mo	5850
Saguinus fuscicollis	f: 1st conceive: 220 d: r211-229; n2	3053
	f: 1st conceive: 500 d; r398-631	3054
	f: 1st birth: 19-32 mo	11875
	m: 1st sire: 320 d; n2	3053
Saguinus mystax	f/m: glandular development: 18 mo	5584
Saguinus oedipus	f: 1st mate: 27.8 mo; se2.5	10672
	f: 1st birth: 24 mo; n1	7231
	f: 1st birth or abortion: 31 mo; sd8.5; n28 f	5817
	f: 1st birth: 32 mo	4306
	f: 1st birth: 34.0 mo	10672
	f: 10-11 mo; f with m; 15-17 mo; f with family	12099
	f: 436 d; ±30.2; housed with m	10671
	f: 590 d; ±54.6; housed in natal groups	10671
	m: 1st sire: 26 mo; n1	7231
	f/m: 1st mate: 15 mo	11733

CYCLE LENGTH		
Callithrix jacchus	estrus: 2-3 d	6620
	follicular phase: 8.25 d; ±0.30; n56	4379, 4545
	follicular phase: 8.8 d; ±3.7	4372
	follicular phase: 9.5 d; sd2.2; r6-12; n10	5719
	luteal phase: 16.6 d; sd1.1; r15-18; n9	5719
	luteal phase: 19.22 d; ±0.63	4379, 4545
	luteal phase: 21.5 d; ±2.2	4372
	mode 24-30 d; r9-32; n14 cycles, 8 f	3
	16.4 d; se1.7; n9 cycles	4549, 4550
	26.2 d; sd2.4; r24-29; n8 cycles, 5 f	5719
	27.5 d; n40 cycles, 14 f	4379, 4546, 4547
	28.54 d; se0.95; n18 cycles, 18 f with intact m	4378
	28.63 d; se1.01; n19 cycles, 14 f isolate or with castrate	4378, 4544, 4545
	30.1 d; sd3.8; median 29.5; r23-41; n30 cycles, 15 f	4372
Cebuella pygmaea	estrus: 6 d	10199
Leontopithecus rosalia	~15 d	3509
	2-3 wk	5850, 5851
	19.56 d; se1.41; n14 cycles, 4 f	3508
Saguinus fuscicollis	14.9 d; se0.7; r13.5-21.8; n18 cycles, 4 f with m?	3054
	17.3 d; se0.97; r13.5-21.8; n34 cycles, 6 f in family group	3054
	17.5 d; se1.0; r14.6-18.2 80%; r13.6-21.1; n10 f	3055
Saguinus oedipus	15.5 d; ±1.5; n15 cycles, 5 f; between progesterone peaks	8748
	21-32 d; between LH peaks	12099
	22.7 d; se1.7; r19-25; n10 cycles, 4 f	1234
	23.6 d; se1.2; n22 cycles, 5 f	3504
GESTATION		
Callithrix jacchus	implant: by 10 d	4542
	implant: 10-12 d	4378, 4545
	implant: 11-12.5 d	7497
	~132 d	7384
	133 d; r117-156; n3	6749, 6751
	138 d; r126-150; n2 lt, 1 f	1753
	est 140 d; r~138-170; n6	6620, 6621
	est 140-150 d	3049
	142 d	4567
	143.4 d; r141-146; n10	10388
	143.5; ±2.5	8253
	median est 144 d; n129	8663
	144 d; sd2; r141-146; n9	1686, 4544, 4545, 4546, 4547, 4549
	144-146 d	9314
	mode 144-185 d; r144-402; n28; pairing to birth	5952a
	145.3 d; r143-147; n3; pairing to birth	4307, 4308
	146 d; n1	9313
	min 146 d; 166 median; pairing to birth	1167
	21 wk	6647
	148 d; se4.3	4540, 4550
	151 d; r150-153; n4	4082, 8488
Cebuella pygmaea	107 d; min ibi	6222
	19.1 wk; r17-20; n5	1821, 1822
Leontopithecus rosalia	122 d	3508
	128.6 d; r125-132; n9	5850, 5851
	max 134 d	8825
	135 d	2109
	max 136.8 d; ibi	11749
Saguinus bicolor	~160 d	4591
Saguinus fuscicollis	145-152 d; n1	11874
Saguinus mystax	150 d	5584
Saguinus oedipus	>4.5 mo; n4; signs of pregnancy to birth	11594
	144-150 d	11510
	147 d; n1; pairing to birth	4304, 4305, 4306
	165-170 d	11733
	165.5 d	2153
	166-202 d; n1; pairing to birth	1233

	170-172 d; n1	1235
	183.7 d; se1.14; r179-187; n6 f; LH peak to birth	12098
	305 d; sd71.9; n5; pairing to birth	3118
LACTATION		
Callithrix humeralifer	nurse: 3-4 mo	9401
Callithrix jacchus	solid food: 23 d	10391
	solid food: 25 d	4550
	solid food: 4 wk	5147
	solid food: 4 wk	6813
	solid food: 4 wk; n1	6620
	solid food: 30.2 d; r26-34; n6	3049
	wean: 35 d; n1	6751
	nurse: 5-6 wk; n1	3289
	nurse: mode 40 d; max 80	4540, 4551, 4550
	wean: 40-100 d	4546
	wean: 6 wk; n1	6620
	wean: r52-85 d; n5; r52-60; n4	3049
	wean: 2 mo	1753
	wean: 60-81 d	1822
	wean: 80-100 d	6648
	nurse: 129 d; n1	1166a
Cebuella pygmaea	suckling decreases: 3 mo; n1	6222
	suckle: 3 mo	1821, 1822
	independent: 5 mo	1822
Leontopithecus rosalia	solid food: 4 wk	5854
	solid food: 5 wk	215
	nurse: 10 wk	215
	wean: 12 wk	5854
Saguinus oedipus	solid food: 25 d; n1 lt	11510
	solid food: 34.5 d; r29-40; n2	3049
	wean: 50 d	3503
INTERLITTER INTERVAL		
Callithrix argentata	median 6.5 mo; r5-13	10390
	median 7.5 mo; r5.5-10	10390
Callithrix jacchus	4.5-5.0 mo	4545
	most 147-167 d; r147-290; n166	5952a
	148 d; n1	4306
	5 mo	10391
	5-6 mo	4541
	153.6 d; r151-156; n9, excluding one at 275 d	10388
	mode 154 d; r145-190; n>60	8663
	~155 d	4540
	155.5 d; r155-156; n2	9313
	median 158 d; r145-382; n>130	1167
	159 d; n1	6749, 6751
	159.8 d; r150-267; n28 lt, 5 f	9314
	160-200 d	4550
	160-262 d; n3 f	3049
	171 d; r152-305; n28	8487
	179.1 d; sd28.1; median 175; r151-246; n22 after abortion	6648
	188 d; r126-335; n10	3697
	204.9 d; sd55.2; median 186.5; r147-441; n152 (n38 yg died)	6648
	212.6 d; sd57.7; median 206.0; r147-441; n114 yg lives	6648
Cebuella pygmaea	158.96 d; r148-164.8; n8 f; r107-191; n32 intervals	6222
	19-20 wk	1822
	5-7 mo	10198
	21-23 wk	1821
	185 d; n2 f	4309
	212.7 d; sd122.3; median 166; r149-746; n31 intervals	12100
Leontopithecus rosalia	132.5 d; r132-133; n2 minimum lengths	11035
	8.1 mo; r4-19; n11 lt, 10 intervals	4200
Saguinus fuscicollis	~6 mo	1661
	222 d; r162-367; n60	3697
Saguinus labiatus	224 d; r165-317	8043

Saguinus mystax	150 d	5584
	187 d; 1st to 2nd birth; 274 d; 2nd to 3rd; 201 d; 3rd to 4th	8042
Saguinus oedipus	172 d shortest	1628
	207 d; sd16.9; r195-238; n5 yg die	3113
	208.5 d; sd16.3; n35 intervals, 8 f	3503
	222.7 d; r154-265; n7	7231
	238 d; r187-302; n13 full term	4304
	245 d; r189-334; n8 aborted	4304
	259.1 d; r166-523; n17 hand reared	11874
	280.0 d; r191-578; n17 parent reared	11874
	7-8 mo	11594
	289 d; median 238; sd16.9; r182-374; n7 yg live	3113
	294 d; se18.2; r160-480; n24	5816, 5817
	295 d; r190-446; n27	3697
SEASONALITY		
Callithrix argentata	birth: yr rd; n41; zoo	1232
Callithrix humeralifer	mate: est Sept-Nov, Feb-Apr; Brazil	9401
Callithrix jacchus	mate: May, Feb, July; n1 pair; zoo	6749
	conceive: low July-Aug; captive	12102
	birth: Feb; few data; captive	9313
	birth: end Oct-Nov, end Mar-Apr; Brazil	5059
	birth: peak spring, fall; zoo	1230
	birth: yr rd; zoo	9314, 9315
	birth: yr rd; lab	4550
	birth: yr rd; n92; captive	8487
	birth: yr rd; n30; zoo	1232
	repro: yr rd; captive	1167
Cebuella pygmaea	birth: July; ne Peru	8891
	birth: peaks Nov-Jan, May-June; Peru	10198
Leontopithecus rosalia	birth: Feb, Apr, July; n1 pr; zoo	2109
	birth: peak Feb-May, July-Aug; captive	5850
	birth: Sept-Feb; Brazil	5854
	birth: Sept-Mar; Rio de Janeiro, São Paulo, Brazil	1972
	birth: yr rd; captive	10185
Saguinus fuscicollis	birth: peak Feb-Mar; lab	3696
	birth: yr rd; n40; zoo	1232
	birth: yr rd, peak Mar-June; lab	3697
Saguinus mystax	birth: middle & end yr; captive	5584
Saguinus oedipus	mate: peak Nov-Dec; captive	2153
	ovulate: Nov-Feb; Panama	2366
	preg f: Feb-July; Barro Colorado I, Panama	3024
	birth: Mar; only 3 mo data; Carribbean Atlantic Coast	6477
	birth: Mar-June; Panama	2366
	birth: peak Mar-Jun, low July-Sept; n44; zoo	1232
	birth: small peak Apr-May, none Sept; captive	2153
	birth: yr rd, peak Mar-May; lab	5816, 5817
	birth: yr rd, peak end Apr-early June; Panama	2368
	birth: yr rd, peak Nov-Mar 28/46, none Aug; captive	4304
	birth: yr rd; zoo	7231
	young seen: Feb-early June, peak May; n Colombia, central Panama	7583

Order Primates, Family Callimiconidae, Genus *Callimico*

Goeldi's marmoset (*Callimico goeldii* 450-850 g) is an arboreal omnivore from the Amazon basin. Placentophagia is reported (698, 8651). Estrus occurs 1-10 d postpartum (698, 2153, 4613, 6577, 6923, 12097). Neonates are furred with open eyes and suckle from axillary teats (698, 4580, 4613, 4792, 8651). Males and siblings carry preweaned young (4612, 4613, 8651, 8653). Details of endocrinology (1629, 1630, 1823, 12101) are available.

NEONATAL MASS		
Callimico goeldii	est 25-30 g	6579
	44 g	8651
	m: 52.3 g; r45-66; n3	698
	f: 54.7 g; r46-63; n3	698
	m: 59.8g; n1	4733
NEONATAL SIZE		
Callimico goeldii	f: 91.0 cm; n1; 4 d	4792
LITTER SIZE		
Callimico goeldii	1	698
	1	7584
	1 ea; n2 lt	8651
	1 yg ea; n7 lt, 3 f	4580
	1 yg ea; n7 lt	6579
	1 yg ea; n9 lt	8653
	1 yg ea; n11 lt	4306
	1 yg ea; n>20 lt	4613
	1 yg ea; n25 lt	4309, 4313
	1.02; r1-2; n188 lt	214
	2; n1 lt	4792
SEXUAL MATURITY		
Callimico goeldii	f: 1st conceive: min 12 mo	698
	f: 1st birth: 16.5 mo; n1	6577
	m: 1st mate: min 13 mo	698
CYCLE LENGTH		
Callimico goeldii	estrus: 7 d; r5-12	6577
	22 d; r21-24; n13 cycles, 3 f	6577
	24 d; few data	698
GESTATION		
Callimico goeldii	≤143, 144, 156 d; n3	8653
	143.75 d; r143.5-144; n2; pairing to birth	8651
	150-160 d; ibi	2153
	150-180 d	6923
	154 d; r151-159; n7 lt, 3 f	6577
	160 d; r154-167	698
LACTATION		
Callimico goeldii	solid food: 18-27 d; n3	6923
	independent: 33-40 d; n3	6923
	wean: max 10 wk	4580
	wean: 12 wk	4613
INTERBIRTH INTERVAL		
Callimico goeldii	165 d without extremes; r161-290	698
	167 d; n5 births, 1 f	6579
	187 d; n3 f	4309
	197.7 d; r165-245; n6, 3 f	8653
SEASONALITY		
Callimico goeldii	birth: Mar-Dec, none Jan-Feb; zoo	2153
	birth: Nov; n1; Bolivia	8654, 8655
	birth: yr rd; captive	698
	birth: yr rd; n37; zoo	1232

Order Primates, Family Cebidae

New World monkeys (*Alouatta* 4-10 kg; *Aotus* 600-1200 g; *Ateles* 3.5-8 kg; *Brachyteles* 9.5 kg; *Cacajao* 3.5-11 kg; *Callicebus* 500-1000 g; *Cebus* 1-4.5 kg; *Chiropotes* 2.7-3.2 kg; *Lagothrix* 4-11 kg; *Pithecia* 700-2500 g; *Saimiri* 500-1500 g) are arboreal, often social omnivores from the forests of Central and South America. Males are often larger than females (1409, 1609, 1974, 3382, 5110, 5394, 5395, 10807, 10808; cf 350, 11189, 11940).

A copulatory vaginal plug is reported (*Cebus* 4296). The hemochorial, labyrinthine, villous, discoid or bidiscoidal placenta, which secretes gonadotrophic hormones (10636a), has an antimesometrial, superficial implantation into the dorsal and ventral wall of the simplex uterus (4113, 4788, 4790, 7563, 10310, 10415,

10475, 11837, 11843, 11845, 11855, 11966). Placentophagia (1154, 1609, 1840, 4791, 4974, 5520, 9743, 10636, 11940), but not estrus (2622, 3800, 5100) occurs after the birth of an open-eyed and furred neonate (399, 2622, 3043, 3449, 4791, 7225, 7296, 9286). *Aotus* and *Callicebus* males contribute to parental care (3449, 5770, 6932, 11577, 11941). Seminal vesicles and a baculum (except in *Aotus*, *Ateles*, and *Lagothrix*) are present (4788, 5842).

Endocrinological data for squirrel monkeys (*Saimiri*) are not extensively cited here as these monkeys are common laboratory models for human endocrinology. Details of anatomy (1894, 1895, 2477, 4685, 4696, 4788, 5495, 11847, 11851, 11865), endocrinology (*Aotus* 1094, 1095, 1097, 2622, 2628, 2629, 4235, 5100, 8781, 9298, 9795, 10793, 11865; *Ateles* 4866; *Cebus* 1656, 2201, 4863, 4866, 6229, 7741, 7742, 7743; *Lagothrix* 1656), birth (1154), and milk composition (1481a, 1840) are available.

NEONATAL MASS		
Alouatta palliata	275-400 g	3800
Aotus trivirgatus	56 g; n1	5904
	90-105 g	2621
	91.92 g; sd21.04; n13	2982
	96.5 g; r95-105; n4	2622
	m: 98 g; n1	3043
Ateles fusciceps	<500 g	2931
Ateles geoffroyi	340 g; n1	7296
Ateles paniscus	66.7 g; r60-80; 3 stillborns, est premature	1558
	480 g, n1	3710
Callicebus moloch	74.4 g; n1	5390
	~100 g	3449
Cebus albifrons	f: 222.1 g; sd9.1; n14	3295
	226 g; se5.8; r184-244; n10	11694
	m: 250.7 g; sd18.9; n20	3295
Lagothrix lagotricha	f: 140.5 g; n1, 2 d	11717
Saimiri sciureus	62-247 g; high mortality <80 g	8938
	87 g; r68-100; n4	1154
	84.3 g; r78-90; n3 dead	4974
	80-140 g	536
	90-125 g	2790
	90-125 g	2042
	f: 93 g; n1	10889
	94.5 g; r84-113; n6	5113
	f: 100 g; min 80	10451
	f: 104.1 g; se1.3; n193	8938
	104.85 g; r73-115; n20	3963
	109.49 g; sd13.4	11406
	m: 111.4 g; se1.1; n254	8938
	120.9 g; n10?	5686
	129.1 g; sd7.14; n4	5475
NEONATAL SIZE		
Aotus trivirgatus	CR: 12.79 cm; sd1.99; n13	2982
Callicebus moloch	HB: 125 mm; n1	5390
Cebus apella, C albifrons	CR: 14.7 cm; sd0.5; n7	460
Saimiri sciureus	~12-12.5 cm	5475
	CR: 12.26 cm; sd0.54	11406
	12.6 cm; n1	4791
	CR: 12.9 cm; sd0.4; n8	460
	13.2 cm; n4	5475
WEANING MASS		
Saimiri sciureus	f: 403 g; n1, 177 d	10889
LITTER SIZE		
Alouatta caraya	1 yg ea; n7 lt	9921, 9922
Alouatta palliata	1	6387
	1; n1 lt	7624
Alouatta pigra	mode 1	3799
Alouatta seniculus	1	5110
	1	5070
	1 yg ea; n3 lt	3383

	2; n1 lt	2147
Aotus trivirgatus	1	11941
	1	2621
	1 yg; n1 lt	5904
	1; n1 lt	3043
	mode 1	3024
	1 yg; 0.5% twins; n179	3985a
	1.06; r1-2; n35 lt	1840
	2	11869
Ateles geoffroyi	1; n1 lt	7296
Ateles paniscus	1	11188
	1 emb; n1 f	5110, 5111
	1 yg ea; n5 lt	10021
Cacajao calvus	1 yg; n17	3380, 3382, 3383
Callicebus moloch	1	11941
Callicebus torquatus	1	5770
Cebus albifrons	1 yg	2429
Cebus apella	1 ea; n7 lt	1230
	1 yg; n19+ lt	10468
	2; n1 lt	2943
	2 yg; n1 lt	10468
Cebus capucinus	1; n1 lt	11086
Chiropotes albinasus	mode 1	10809
Chiropotes satanas	1	11189
	1 yg; n2 lt, 1 f	4722
Pithecia pithecia	1	9923
	1 ea; n2 lt	1893
	1 ea; n2 lt	4722
Saimiri sciureus	1	4791
	1 yg	7117
	1 ea; n21 lt	3650
	1.002; r1-2; n512 lt	8938
SEXUAL MATURITY		
Alouatta caraya	f: 1st conceive: 38.5 mo; r35-42; n2	9922
	m: 1st sire: 30.5 mo; r24-37; n2; perhaps earlier	9922
Alouatta palliata	f: 1st estrus: 4-5 yr	1609
	f: 1st birth: 3-5 yr	3531
	f: 1st birth: 43 mo	3800
	m: 30-48 mo	3800
	m: 6-8 yr	1609
Aotus trivirgatus	f: 1st birth: 40.56 mo; sd7.82; n9	3985a
	f: 1.5-2.0 yr	4547
	m: high testosterone: median 313 d; r211-337; n6	2622, 2629
	m: 1st sire: 42.17 mo; sd10.73; n12	3985a
	f/m: est 18-24 mo	2621
Ateles geoffroyi	f: 1st birth: 5.7, ~6.5, 7.6 yr; n3	7355
	f: 5 yr	2929
Cacajao calvus	f: 1st birth: 3.58 yr; n1	3382
	m: 1st mate: 17.5 mo; r17-18; n2	3382
	m: 1st sire: ~6 yr	3380, 3382
Cebus albifrons	f/m: 3.59 yr; se0.17; r2.83-4.4	11694
Cebus capucinus	f: physically mature: 4 yr	8117
	m: physically mature: 8 yr	8117
Lagothrix flavicauda	f/m: > 4 yr	6643
Lagothrix lagotricha	f: 1st birth: 8 yr; r6-10; n2	11717
	m: sexual interest: 3-4 yr	11717
Pithecia pithecia	m: > 4 yr	4722
Saimiri sciureus	f: 1st conceive: 35.03 mo; sd5.18; r32.11-42.78	9354
	f: 1st conceive: 58 mo; r44-67; n6	5520
	f: 1st preg: mode 3.5-4 yr n24/60; r2.5-8.5	10687
	f: 1st birth: 40.57 mo; sd5.18; r37.66-48.33	9354
	f: 3 yr	4547
	m: 1st sire: 42 mo	9354

CYCLE LENGTH

Alouatta palliata	estrus: several d	1606
	15.52 d; sd4.88; n25 cycles, 14 f	5395
	16.08 d; sd4.27; n12 cycles, 10 f	5395
	16.3 d; se0.7; r11-24; n23 cycles, 6 f	3800
Alouatta seniculus	estrus: 2.5, 3 d; n2	2149
	20.5 d; median 17; r16-27 d; n4 cycles, 4 f	2149
Aotus trivirgatus	follicular phase: ~6 d	2622
	luteal phase: ~10 d	2622
	10-15 d; n6 f	1840
	15.92 d; se0.26; n60 cycles, 55 f	1094, 1095, 1097, 2621, 2622
Ateles fusciceps	estrus: 2 d	2929
Ateles geoffroyi	menses: 3-4 d; n1 f	3895
	24-27 d	3895
Cacajao calvus	14-48 d	3381
Cebus spp	menses: 4.3 d; r1-7	4296
	18.3 d; mode 18; r16-23 d; n12 cycles, 2 f; vaginal smears	4296
Cebus albifrons	menses: 2-4 d	1656
	17-19 d	4866
	17.5 d; r15-21; n50 cycles	1656
Cebus apella	estrus: 5-6 d	5254
	menses: 2.8 d; ±0.4; r1-5	7741
	menses: 4.06 d; sd1.89; r0-8; n18	11937
	17.08 d between menses; ±4.27; r11-24; n13; 21.14 d with menses	11937
	20.8 d; mode 21; r13-28; n108 cycles	7741
Cebus griseus?	median 14-16 d; r10-19	3770
Lagothrix lagothricha	estrus: 3-4 d	11718
	menses: 3 d	11717
	r12-49 d; n25 cycles, 2 f	1656
	3 wk	11717
Saimiri boliviensis?	9.7 d; r6-12	148a
Saimiri sciureus	estrus: 12-36 h	9287
	6-12 d	2541
	7; mode 7; n15	9287
	mode 8; n42 cycles, 9 f	11776
	8.7 d; sd0.6; r8-10; between progesterone peaks	11868
	9.4 d; se0.6; r8-10; between estrogen peaks	11868
	9.44; r6-12; n9 cycles, 7 f	2541
	9.7 d; se0.4; n20 cycles, 10 f	12012
	10.9; se0.10; r9-13; n9 cycles, 6 f	10889
	11.3 d; n10 f; 8.2 d; 10 f smaller cage	5113
	12 d; sd2.7; median 10-11; n108 cycles, 24 f	1656
	12.54 d; ±2.01; mode 13; r6-18; n23 f	6176
	13.36 d; r8.3-25.3; n53 cycles, 5 f	5113
	18-20 d; n15 f	10287

GESTATION

Alouatta spp	139 d	11869
Alouatta caraya	187 d; n1	9921, 9922
Alouatta palliata	186 d; r180-194; n4; swelling to birth	3800
Alouatta pigra	180-194 d	2932
Alouatta seniculus	191 d; r186-194; n6	2149, 9743
Aotus trivirgatus	120-140 d; palpation, ibi	1840
	18 wk; n17 f	4235
	135.5 d; r133-138; n2	5100
	est 148-159 d	2982
	~5 mo; min ibi	7224
Ateles belzebuth	139 d; n1	1558
Ateles fusciceps	7-7.5 mo; max 226 d	2929
	226-232 d	2931
Ateles geoffroyi	~200 d; n1	7355
Callicebus spp	5-6 mo	7225
Cebus spp	162 d; r157-167	11869
Cebus apella	152-160 d	2943

	153 d; r149-158; n3	11937
Lagothrix lagothricha	7 mo	11718
	225 d; n1	6696
	7.5 mo	11717
Pithecia pithecia	est 101 d; n1	4316
	est ~150 d	9923
	163-176 d; n1	4722
	169.5; r163-176; n2	4722
Saimiri sciureus	146.9 d; sd3.28; median 146.5; mode 146; r141-154; n10	5686
	152.5 d; r141-157; n46	10451
	157 d; n1	10889
	159 d; se3	1974
	162.6 d; r152-168; n5 f	6578
	24-26 wk; n13	4974
	24-26 wk	3650
	170.8 d; r168-172; n19	3963
	177.75 d; sd7.2; n4	5475
	~6 mo	4791
	6 mo; field data	2791
LACTATION		
Alouatta caraya	nurse: min 8 wk	9921
Alouatta palliata	wean: min 6 mo	1609
Alouatta seniculus	wean: 10.5-14 mo	2147
	wean: >1 yr	6695
Aotus trivirgatus	solid food: 19 d	7224
	solid food: 2 mo	399
	wean: 8-10 wk	7224
	wean: 12-14 wk	7225
	wean: ~9 mo	11940
Ateles fusciceps	solid food: 8 mo	2931
	wean: min 16 mo	2931
	wean: ~10-12 mo; n6 yg	2929
Ateles geoffroyi	wean: 2+ yr; 26-28 mo	7355
Cacajao spp	wean: ~21 mo	3381
Callicebus spp	wean: 12-16 wk	7225
Callicebus torquatus	wean: > 4 mo	5770
Cebus albifrons	wean: 9 mo	11694
Lagothrix lagothricha	nurse: est 9-12 mo	11717
Pithecia pithecia	nurse: 4 mo	4316
Saimiri sciureus	solid food: 3 mo	9286
	wean: 177 d; n1	10889
INTERBIRTH INTERVAL		
Alouatta caraya	11.1 mo; r7-16; n16	9922
Alouatta palliata	22.5 mo; se0.6; n16	1886, 3800
Alouatta seniculus	10.5 mo; sd2.9; r7.75-15; n12 yg dies	2150
	17.4 mo; sd4.5; r10.5-35; n135 yg lives	2148
Aotus trivirgatus	204, 206, 274 d; n3	399
	213.5 d; r156-319; n9	7224
	253 d; r166-419	2621
	271 d; median 258; min 148; n51	5100
	1 yr	11940
	12.72 mo; sd5.72; r5-29; n75 intervals	3985a
Ateles fusciceps	22-36 mo	2931
Ateles geoffroyi	31.9 mo; ±3; r28-36; n7 birth, 3 f	7355
Ateles paniscus	34.5 mo; sd5.8; r25-42; n17 intervals	10609
	3 yr	11188
Brachyteles arachnoides	est >24 mo	7356
Cacajao calvus	23-28 mo; n5	3380, 3382
Callicebus moloch	1 yr	11941
Pithecia pithecia	10-11 mo	11739
Saimiri sciureus	1 yr	5520

SEASONALITY

Alouatta caraya	repro: not seasonal; zoo	9922
Alouatta palliata	sex behavior: not seasonal; Barro Colorado I Panama	1606
	all stages of embryonic development: July-Aug (2 yr); Panama	11845
	birth: Feb, Apr-June, Aug-Sept; n16; Guanacaste, Costa Rica	5394
	birth: yr rd; Guanacaste, Costa Rica	1886, 3800
Alouatta pigra	young of various age seen in most bands; Veracruz, Mexico	4228
Alouatta seniculus	birth: yr rd, peak end dry, low early wet season; Guarico, Venezuela	2147
Aotus trivirgatus	mate: Dec; n1; Barro Colorado I, Panama	3024
	birth: Jan, Mar, May, Aug, Oct-Nov; n9; zoo	7224
	birth: June; Barro Colorado I, Panama	3024
	birth: yr rd, peak Oct-Jan; n126; captive	3985a
	repro: yr rd; captive	2622, 5100
Ateles belzebuth/A geoffroyi	birth: yr rd; n15; zoo	5862
Ateles fusciceps	birth: Apr-June; n11/19; captive	2931
Ateles geoffroyi	all stages emb development: July-Aug; Panama	11845
	birth: Feb-Mar; Chiapas, Mexico	218
	birth: June-July; Nicaragua	11837
	birth: peak June-Dec; Barro Colorado I, Panama	7355
	birth: yr rd; Yucatan, Mexico	3633
	repro: yr rd; Panama	1607
Ateles paniscus	birth: peak Jan-Aug; Peru	10608
	young seen: yr rd; Suriname	5110
Cacajao calvus	mate: Oct-May; zoos, n hemisphere	3381, 3382
	birth: May-Oct; zoos, n hemisphere	3381, 3382
Callicebus moloch	birth: Nov-Mar; equatorial S America	5770, 6932
Callicebus torquatus	birth: Nov-Mar; Peru	5770
Cebus spp	ovulate: yr rd; s Goyaz, Brazil	4296
	birth: peaks May-June, Oct-Nov; s Goyaz, Brazil	4296
Cebus albifrons	conceive: Dec-Jan; Peru	1022
Cebus apella	estrus: yr rd, peak Apr-June; no data Jan-Feb; Peru	5254
	birth: mid-end Jan, early-mid May, Oct-mid Dec; zoo	1230
	birth: peak Oct-Dec; Peru	5254
	young seen: yr rd; Suriname	5110
Cebus capucinus	all stages of embryonic development: July-Aug; Panama	11845
	birth: 3/5 peak Jan, Mar-Apr; Barro Colorado I, Panama	8117
	birth: dry season, Dec-early Apr; Costa Rica	3501
Chiropotes satanas	birth: Dec-Jan; Guyana	11189
Lagothrix lagothricha	repro: not seasonal; captive	11718
Saimiri sciureus	testes: large Apr-Aug; captive, Argentina	1974
	vaginal cornification: high Jan-Aug, low Oct; lab	9287
	estrus: yr rd, peak June-Sept; lab	10889
	mate: June-Aug; captive, Argentina	1974
	mate: July-Sept; Amazon, S America	2790
	mate: Aug-Feb; lab	11776
	mate: Sept-Mar; zoo	2790
	mate: after Oct; lab	12012
	mate: Nov-Mar; captive	7202
	mate: Dec-Mar; lab	2542
	mate: mid Dec-Mar; captive	2791
	conceive: mid Dec-mid Mar; lab	11719
	preg f: Oct-Jan; captive	11869
	preg f: yr rd; captive	9288
	birth: Jan-Mar; Amazon, S America	2790
	birth: Feb-Sept; zoo	2790
	birth: May-July; captive	9354
	birth: June-Aug; captive	2791
	birth: June-Aug; lab	2542
	birth: 65% June-Aug, 87% May-Sept; lab	5520
	birth: peak Oct-Nov; captive, Argentina	1974
	birth: end July-Oct; Surinam and captive	3650
	birth: yr rd, 72% May-Aug; captive	8938

Order Primates, Family Cercopithecidae

Old World monkeys (*Allenopithecus* 3.5-6 kg; *Cercocebus* 3-18 kg; *Cercopithecus* 750 g-12 kg; *Colobus* 3-23 kg; *Erythrocebus* 4-25 kg; *Macaca* 1.5-20 kg; *Nasalis* 7-25 kg; *Papio* 6-54 kg; *Presbytis* 5-21 kg; *Pygathrix*, *Theropithecus* 10-20 kg) are primarily arboreal, social herbivores or omnivores in Africa and Asia. Males are often larger than females (92, 146, 790, 1015, 1116, 1297, 1781, 1952, 2501, 2955, 5692, 5703, 5758, 6970, 7270, 8496, 9667, 9970, 9982, 10078; cf 1115).

The 20-290 g, hemochorial, villous, bidiscoid placenta, which secretes gonadotrophins (10636a), has a superficial, antimesometrial implantation in the simplex uterus (679, 1787, 1903, 2580, 3748, 4113, 4721, 4723, 4786, 4796, 4842, 5009, 5010, 5011, 5228, 5316, 6624, 7563, 7946, 8172, 8174, 8222, 9707, 9728, 10209, 10310, 10311, 11199, 11843, 11852, 11861, 12129) and may be eaten after birth (16, 49, 1798, 1903, 2501, 3762, 3966, 4298, 4445, 4448, 5273, 5649, 5919, 6606, 7768, 7804, 7946, 9289, 9334, 9982, 10142, 10234, 10264, 10636, 10726, 10914, 11199; cf 1299, 4720, 4723, 7485). Estrus postpartum is uncommon, but has been observed in *Macaca fascicularis* (3385, 10844). Sex of offspring may be related to the social rank of the mother (8265, 8265a; cf 10162a). Neonates have fur and open eyes (184, 482, 1042, 1798, 2495, 3386, 3966, 4721, 5273, 5758, 6747, 8118, 8701, 8705, 9361, 9728, 10158, 11985). Females have 2 mammae (331, 1483, 10158). Males have seminal vesicles (2293).

Endocrinological data for macaques (*Macaca*) and baboons (*Papio*) are not cited as these data are summarized in reviews of human endocrinology (5891). Details of anatomy (3384, 5508, 12125), endocrinology (*Cercocebus* 91, 92, 1553a, 6770, 6771, 10291; *Cercopithecus* 256, 689, 1165, 1204, 2631, 2888, 2889, 2967, 2988, 2989, 4699, 4860, 5703, 6081, 6874, 7187, 9334, 9701, 9794, 10368a; *Erythrocebus* 92, 456, 11826; *Presbytis* 4786, 6547, 8884, 9832, 9833; *Pygathrix* 2201, 6229), copulation (3642), and birth (16, 1795a, 1798, 2800, 3892, 3966, 4298, 4616, 6606, 7083, 7507a, 8118, 8503, 8974) are available.

NEONATAL MASS

Cercopithecus aethiops	est 300-400 g	10154
	f: 310 g; r220-419; n26	11062
	m: 319 g; r219-411; n28	11062
	321.3 g; sd56.9; n38 (14 f, 24 m)	5364
	m: 352 g; r233-480; n26	9728
	f: 364 g; r247-480; n24	9728
	397 g; n1	6612
Cercopithecus ascanius	371 g; term	3748
Cercopithecus mitis	337 g	3748
	360 g; n1	9362
	402.2 g	8107
Cercopithecus talapoin	~170 g	3640
	230 g; n1	9338
Cercopithecus wolfi	435 g	3748
Colobus polykomos	m: 820 g; n1	2109
Macaca arctoides	450 g	11985
	495 g; n1	10914
	500 g; r380-650; n38	7908
Macaca cyclopis	402 g; sd61; n54	8347
	402 g; r300-530; n35	12000
Macaca fascicularis	272 g; sd50; r210-340; n≤11	11549
	m: 287 g; n1	10263
	~300 g	3550
	f: 302.2 g; sd49.5; r240-360; n9; gang cages	4954
	m: 302.7 g; sd52.6; r240-370; n5; gang cages	4954
	m: 309.7 g; sd38.0; r240-410; n41; individual cages	4954
	f: 320.0 g; sd48.6; r200-450; n46; individual cages	4954
	f: 335 g; r234-469; n83	11062, 11064
	338 g; sd80; r153-470; n24	814
	m: 359 g; r242-546; n81	11062, 11064
Macaca fuscata	475 g; r450-500; n2	4524
	f: 517 g; sd90; n4	2291
Macaca mulatta	412, 450 g; n2	4445
	421 g; sd78; n271 live, 390 dead; wild bred	11061

	f: 424 g; r385-475; n5	386
	f: 435 g; ±51; r351-570; n20	6880
	~442 g; r320-500; se49.3; n97	5228
	f: 450 g; sd12; n303	10077, 10078
	453 g; r340-660; n22	4448
	f: 456.2 g; sd69.9; n88	3545
	f: 457.7 g; r260-670; n161	11199
	m: 467.5 g; r395-540; n2	386
	f: 469 g; ±77; r247-752; n193	6880
	m: 470 g; sd70.5; n91	3545
	472 g; sd56; n27	5693
	472 g; sd65; n111, 160-169 d gestation	11201
	f: 474 g; sd74; n229 live, 23 dead	11061
	f: 477 g; sd77; n285	11064
	m: 479.1 g; r300-770; n150	11199
	m: 480 g; sd13; n262	10077, 10078
	480.5 g; sd78.2; n209	2580
	f: 482 g; sd52; r424-550; n4	4149
	f: 484 g; r340-678; n69	11062
	489 g; sd49; n469 live, 61 dead, 22 dead unsexed	11061
	m: 494 g; ±73; r377-756; n40	6880
	498 g; sd70.7; n100, 170-179 d gestation	11201
	m: 500 g; sd75; n276	11064
	m: 501 g; ±68; r320-760; n222	6880
	m: 504 g; r362-806; n77	11062
	m: 513.2 g; sd9.85; n67	1297
	m: 516 g; sd74; n5	4150
	536 g; sd85.5; n19, 180-189 d gestation	11201
	544.4 g; sd101.6; n8, 175 d gestation	5692a
Macaca nemestrina	400-600 g; 176 d gest	7865
	f: 451 g; sd70; n187	10009
	m: 493 g; sd70; n211	10009
	500-600 g; nll; from graph	10304
Macaca radiata	290 g; n1 stillborn	4450
	350 g; r330-370; n2	4450
	f: 388 g; sd34; n5	11062, 11064
	m: 411 g; sd50; n11	11062, 11064
Macaca silenus	m: 348.4 g; n1	2109
	f: 430 g; n1	5650
Macaca thibetana	~550 g	3386
Nasalis larvatus	454 g; n1	9667
Papio hamadryas	m: 600 g; n1 near term	4796
	600-800 g	3783
	f: 820.2 g; n15	7095
	m: 910.4 g; n9	7095
	f: 910 or 920 g; sd120; r0.70-1.40; n66	1952, 1953, 3807, 3808, 3809
	861 g; r590-1212; n17	3776
	m: 980 g; sd130; r0.67-1.22; n77	1952, 1953, 3807, 3808, 3809
Presbytis entellus	500 g; n1 term emb	4786
Presbytis obscura	113.3 g; n2	482
Presbytis senex	360 g; n1	4785
	490 g; r390-590; n2	4783
Pygathrix nemaeus	80 g; premature	4721
NEONATAL SIZE		
Macaca mulatta	CR: 16.6 cm; sd1.0; n5	460
	18.5 cm; se1.43; n12	5228
	30.5 cm; sd1.2; n8	5693
Macaca nemestrina	CR: r19.0-21.5 cm; 176 d gest	7865
Papio hamadryas	19 cm; near term	4796
	f: CR: 23.9 cm; sd1.1; r21.5-29.4; n59	1952, 3807
	m: CR: 24.1 cm; sd1.1; r21.0-27.0; n73	1952, 3807

WEANING MASS

Cercopithecus aethiops	m: 1015 g; r700-1300	9728
	f: 1058 g; r830-1450	9728
Macaca fascicularis	f: 686 g; sd206; r535-988	11595
	m: 844 g; sd76; r728-963	11595
	858 g; n28	11595
	m: 860 g; sd86; r780-930	11595
	f: 994 g; sd166; r759-1185	11595
Macaca fuscata	805 g; r780-830; n2, 4 mo, small	7758
Papio hamadryas	f: 3840 g; sd1060; r1000-8300; n241, 1 yr	1952
	m: 4060 g; sd1.15; r1600-8200; n247, 1 yr	1952

WEANING SIZE

Papio hamadryas	f: CR: 39.6 cm; sd3.3; r31.7-51.7; n242, 1 yr	1952
	m: CR: 40.0 cm; sd3.8; r28.4-51.6; n249, 1 yr	1952

LITTER SIZE

Allenopithecus nigroviridis	1 yg ea; n3 lt, 1 f	8705
Cercocebus albigena	1 ea; n5 lt	2109
Cercocebus galeritus	1 yg; n1 f	5768a
Cercocebus torquatus	1 ea; n5 lt	1230
Cercopithecus aethiops	1; 2 rare	9945
	mode 1; r1-2	2449
	1; n1 lt	187
	1 emb; n1 lt	4926
	1; n4 lt, 1+ wk	331
	1 ea; n5 lt	1230
	2; n1 lt; twins dead	10467
Cercopithecus ascanius	1	4182
	1 ea; n2 lt	10505
Cercopithecus cephus	mode 1	6742
Cercopithecus lhoesti	1 emb; n1 lt	2953, 2955
	1 emb; n1 lt	8856
Cercopithecus neglectus	1 yg; 0.4% twins; n272	1277a
Cercopithecus mitis	1 emb; n1 lt	8856
	1 emb; n1 lt	185
	1 emb ea; n2 lt	331
Cercopithecus mona	1 yg	4354a
	1 emb ea; n2 lt	2953, 2955
Cercopithecus nictitans	1 emb; n1 lt	2953
	1 emb or yg ea; n3 f	185
Colobus abyssinicus?	1 emb ea; n2 f	4926
Colobus angolensis	1 ea; n3 lt	2109
Colobus badius	1 emb ea; n2 lt	184
Colobus polykomos	1 yg	6747
	1 emb; n1 lt	8856
Colobus satanas	1 emb; n1 lt	8508
Erythrocebus patas	est 1	5522
	1.01 yg; r1-2; n501 lt	10072
Macaca arctoides	1 yg; n1 lt	11985
	1 yg ea; n26 lt	3107
Macaca fascicularis	1 emb; n1 lt	2313
	1 yg ea; n5 lt	5649
	1 ea; n65 lt	10844
Macaca fuscata	2 yg; n1 lt	7758
Macaca mulatta	1	4546
	1 yg ea; n17 lt	1230
	1 ea; n19 lt	6684
	1 ea; n132 lt, 45 f	264
	1 yg; 0.1% twins; r1-2; n1003 births	5920a
	1.01 yg; r1-2; n580 lt	11199
Macaca nigra	1 yg	4462
Macaca radiata	1	9982
	1 yg ea; n12 lt	1230
Macaca silenus	1; n1 lt	2109
Macaca sinica	mode 1; r1-2	8496, 8501

Macaca sylvanus	1.003 yg; r1-2; n572 lt	8265
	2 yg; n1 lt	1476
	2 yg; n1 lt	9556
Macaca thibetana	1 yg	12093b
Nasalis larvatus	mode 1; 2 occasional	11852
	1.5 emb; r1-2; n2 lt	9667
Papio hamadryas	1 emb ea; n5 lt	12118
	1 yg ea; n7 lt	1230
	1-2	1445
	2 rare	7494
Papio sphinx	1.20 yg; n5 lt	6508
Presbytis aygula	1	557
Presbytis cristata	mode 1	7161
	1 yg ea; n4 f	8215
Presbytis entellus	1	5273
	1	9175
	1 yg	8720
	1 yg ea; n19	1230
	1-2 yg; twins frequent	8020b
Presbytis melalophos	1	7161
Presbytis obscura	1 yg ea; n2 lt, 2 f	482
Presbytis rubicunda	1 yg	2313
Presbytis senex	mode 1; r1-2	8496, 8501
Pygathrix nemaeus	1 yg	4721
	1 yg ea; n4+ lt	1299
SEXUAL MATURITY		
Cercocebus albigena	f: 1st birth: median ~4 yr; from graph	9341
Cercocebus torquatus	f: 1st sex skin: median 32 mo; r30-39; n6	4183
	f: 1st swelling: 36.4 mo; se1.1; n14	4159a
	f: 1st menses: 37.6 mo; se1.3; n12	4159a
	f: 1st conceive: 51 mo; se1.8; n28	4159a
	f: 1st birth: median ~4.5 yr; from graph	9341
	f: 1st birth: 56.5 mo; se1.8; r40.8-81.6; n28	4159a
Cercopithecus aethiops	f: 1st conceive: 4.4-5.7 yr; n3 habitats	6350
	f: 1st birth: 3-4 yr	3152
	f: 1st birth: median ~4 yr; from graph	9341
	f: sexual maturity to 1st preg: 15 mo	401
	f: 34 mo	4995
	m: 60 mo	4995
	f/m: 1st repro: ~40 mo	1225
Cercopithecus mitis	f: 1st birth: median ~5.5 yr; from graph	9341
	f/m: 1st repro: 55 mo; n1	1225
Cercopithecus neglectus	f: 1st birth: mode 4-5 yr	1277a
	f: 1st birth: median 5 yr; from graph	9341
	m: 6-8 yr	1127a
Cercopithecus talapoin	f: menarche: 2.5-3.5 yr	9338
	f: menarche: 36.8 mo; r31-54; n6	9337
	f: 1st conceive: 51.5 mo; r42-58; n4	9337
	f: 1st conceive: 4.5 yr; n3	9338
	m: testes enlarge: 5 yr; n3	9337
Colobus guereza	f: 1st birth: median ~5.75 yr; from graph	9341
Colobus polykomos	f/m: ~2 yr	5758
Erythrocebus patas	f: menarche: 25.7 mo; r19-30.5; n5	9337
	f: 1st conceive: 157 wk; r134-184	10072
	f: 1st conceive: 29 mo; r19-40; n8	1797
	f: 1st birth: 35 mo; r25-46; n8	1797, 9341
	f: 3.5 yr	5758
	m: 1st sire: 3.67 yr	10072
	m: scrotum turns blue: 4 yr; n2	9337
	m: 5 yr	5758
Macaca arctoides	f: progesterone rise: 3.73 yr; r3.42-4.23; 6.2 kg; r5.7-7.0; n6	7908, 7909
	f: 1st conceive: 3.00 yr; n1	10913
	f: 1st conceive: 4.4 yr; n5	3107
	f: 1st birth: 4.90 yr; r4.02-6.55; n8	7909

	m: testes descend: 3.33 yr; r2.76-3.69; 5.7 kg; r5.2-6.3; n9	7908
	m: 1st sire: 3.25 yr; n1	10913
	f/m: est 4-5 yr	3386
Macaca cyclopis	f: menarche: 36.3 mo; sd6.0; r31.4-46.4; n7	8347
Macaca fascicularis	f: menarche: 2.79 yr; n12	10264
	f: 1st mate: 2.25-3.17 yr	4384
	f: conceive: 3.25-5.33 yr	4384
	f: 1st birth: 3.83 yr; r1.0-4.5	7954
	m: spermatogenesis: 44-53 mo	3434
	m: spermatogenesis: 3.9 yr; n6	10264
	m: sperm in ejaculate: 3.5-4.2 yr	4384
	m: testes: large 36-52 mo; 3-4 kg	10359
	m: mate: 3.83-4.5 yr	4384
Macaca fuscata	f: ovarian activity: 2 yr	5156
	f: menarche: 3-3.5 yr	7915
	f: 1st estrus: 3.5-6.5 yr	10633
	f: 1st conceive: 3.5 yr	4315
	f: 1st birth: 3.92 yr	7916
	f: 1st birth: 4.5 yr; r4-6.92	7954
	f: 1st birth: 5.15 yr; sd0.86; r4-7; n27	3187
	f: 1st birth: 5.5-6.4 yr	3980
	m: 1st mount: 2.5 yr	4315
	m: 1st spermatogenesis: 4 yr	7917
	m: 1st ejaculate: 4 yr	9304
	m: 1st ejaculate: 4.5 yr	4315
Macaca mulatta	f: 1st sex skin: 1.851 yr; se0.146; r1.159-2.178; 3316 g; se78; r3100-3500; n6	11693
	f: 1st sex skin: 2-4 yr	9333
	f: 1st sex skin: median 26.5 mo; r24-35; n14	4183
	f: menarche: 1.941 yr; se0.128; r1.480-2.440; 3354 g; se106; r2821-3600; n7	11693
	f: menarche: 722 d; r14-31 mo; ~3 kg	11198, 11199
	f: menarche: 27 mo; 3.7 kg	3427
	f: menarche: 127.3 wk; ±6.6; n22	9056
	f: menarche: 30.3 mo; se3.1; n22	10745
	f: menarche: 133 wk; ±3.0; 3.46 kg; ±0.08; n26	9056
	f: menarche: 30.7 mo; ±1.2; n30	10744
	f: menarche: 30.9 mo; ±1.1; n31	11771
	f: menarche: 31.1 mo; se2.6; n14	10742
	f: menarche: ~3 yr; 3.35 kg	4448, 4546
	f: menarche: 2.9-4.7 kg	173
	f: sexual maturity to 1st preg: 15 mo	401
	f: 1st ovulate: 3.5 yr	8143
	f: 1st ovulate: 31.5 mo; ±0.4; n31	11771
	f: 1st ovulate: 42.9 mo; ±0.2; n15 1st generation	11771
	f: 1st ovulate: 44 mo; ±2	8929
	f: 1st ovulate: 47 mo; se2.6; n14	10742
	f: 1st ovulate: 48.1 mo; ±2.2; n30	10744
	f: 1st ovulate: 51.2 mo; se3.3; n22	10745
	f: 1st conceive: 3.11, 3.13 yr; n2	6800
	f: 1st conceive: 3.5 yr, 75%; 2.5 yr, 14%	11216
	f: 1st conceive: ~5 yr; 5000 g	4448
	f: 1st birth: 32.9-59.4 mo; fall born	11768
	f: 1st birth: 3 yr 156 d-12 yr 74 d	264
	f: 1st birth: 36.6-59.4 mo; spring born	11768
	f: 1st birth: 4.0 yr; ±0.1; mode 4 59%; r3-7; n56	2721
	f: 1st birth: 4 yr	5312a
	f: 4 yr	5919
	m: testes descend: 3.32 yr; n25	11496
	m: testes large: 4.02 yr; n21	11496
	m: sperm: 2.83-3.42 yr	11200
	m: fully active: 4 yr	11200
	m: 1st sire: 3.30 yr; n2	6800
Macaca nemestrina	f: 1st sex skin: median 35 mo; r30-46; n16	4183
Macaca nigra	f: 1st sex skin: median 48.5 mo; r47-50; n4	4183
Macaca radiata	f: 1st birth: 4 yr	9973

	f: socially mature: ~4 yr	9982
	m: socially mature: ~6 yr	9982
	f/m: 1st mate: 2.5-3.5 yr	9982
Macaca silenus	f: 1st mate: 2.83 yr	6464
	f: 1st birth: 3-4.5 yr	6464
	f: 1st birth: 4.5 yr; r3-7.5	5650
	f: 1st birth: 5 yr	4044
	f: 1st birth: 8 yr	5650
	m: 1st mate: 5.5-6 yr	6464
Macaca sinica	f: puberty: 3 yr	8501
	f: 4-5 yr	8496
Macaca sylvanus	f: 1st sex skin: 3.25-3.5 yr; n11f	7199
	f: 1st conceive: 45.3 mo; r30-78; n8	9171
	f: 1st birth: 4.77 yr; ±0.66; r4-7; n62	8266
	f: 1st birth: 4.99 yr; sd1.13; r2.9-7.9	1477
	m: 1st mount: 4-5 yr	7199
Papio hamadryas	f: 1st estrus: 3.5-4 yr	2501
	f: 1st estrus: 4-4.75 yr	9699
	f: 1st estrus: 4-5.5 yr	10507
	f: menarche: ~3.5 yr	3783
	f: menarche: 183 wk; sd11.7; r168-202; n11	3808
	f: menarche: 189 wk; sd20.5; r141-214; n11	3808
	f: menarche: 200 wk	3809
	f: menarche: 4.3 yr; r4-5; n13 f	9970
	f: menarche: 4.5 yr	213
	f: menarche: 4-5.5 yr	212
	f: sex maturity to 1st preg: 10 mo	401
	f: 1st conceive: ~5.5 yr	213
	f: 1st birth: 4.42 yr; r3.75-4.92	7954
	f: 1st birth: 6.1 yr; r5.5-7; n8	9970
	f: 34.8 mo; r21-45; 7.6 kg	401
	f: 46.8 mo	7494
	m: testes descend: 4.8-6.8 yr	9970
	m: testes: large 3-4 yr	1658
	m: 1st mate: 5 yr	2501
	m: acquired 1st f: 8.5-11 yr	9970
	m: 51.6 mo	7494
Papio leucophaeus	f: 1st estrus: 3.5 yr	1042
Presbytis spp	f: menarche: 4 yr; n1	4784
Presbytis entellus	f: 1st estrus: 3.5-4 yr	5272, 5273
	f: menarche: 28.8 mo; r25.2-31.5; n3	3503
	f: menarche: 3 yr 4 mo 29 d; n1	8884
	f: 1st conceive: 34.2 mo; r31.7-36.9; n3	3503
	m: spermatogenesis; 4.5 yr	8884
	m: 1st mount: 4-6 yr	5272, 5273
Presbytis johnii/P senex	m: min 4 yr	4783
Presbytis senex	f/m: ~3 yr	4781
Theropithecus gelada	f: 1st birth: 4-4.5 yr	2796, 2797
	f: 5 yr	6970
	m: puberty: 3-4 yr, sexually mature: 8-9 yr	2797

CYCLE LENGTH

Cercocebus albigena	follicular phase: 16.5 d; se4; n4	1553a
	luteal phase: 14.8 d; se1; n4	1553a
	3.4 wk; r2.5-5; n4 cycles, 1 f; from graph	12117
	26.0 d; se0.8; r21-40; n26 intermenstrual intervals, 3 f	1553a
	median 29 d; mode 30; r21-36; n47 cycles, 3 f	9339
	29.75 d; r25-42; n44 cycles, 4 f	3643
	31.3 d; se5; n4 cycles; between LH peaks	1553a
	38.5 d; sd63; r29-?; n8 cycles; 45 d; sd19.6; r29-98; n9 cycles	11428
	47 d; r8-184; n26 cycles; between peak swellings	11428
Cercocebus galeritus	29.83 d; r22-39; n6 cycles, 2 f	3643

Cercocebus torquatus	30 d; r26-33; n5 cycles, 1 f	12117
	30.3 d; r30-31; n3 cycles, 1 f	12121
	32.5 d; r26-35; n8 cycles, 2 f (7 from 1 f)	3643
	33.0 d; n40 cycles, 1 f	7851
	33.4 d; se1.1; mode 30; r28-46; n22 cycles	12123
	median 34.5 d; r25-47 of medians; r10-177; n145 cycles, 14 f	4183
	45.5 d; r26-76	10291
Cercopithecus aethiops	follicular phase: 15.4 d; ±2.4	6081
	estrus: 0-66 d	10503
	luteal phase: 27.8 d; ±4.6	6081
	menses: 3.7 d	11062
	menses: 4.4 d; ±2.1; r1-12; n243	4699
	menses: 4.8 d	2989
	menses: 4.9 d; se0.2; r3-6; n24 cycles, ≤17 f	2967a
	27 d; r24-30	6750
	30.9 d; sd9.6; r16-50	5364
	30.9 d; n392 cycles; cycles >50 d excluded	11062
	31.4; sd5.3; r18-49; n183 cycles, 21 f	4699
	32 d; r18-50; n524 cycles, 28 f	2989
	32.4 d; se0.9; r29-45; n24 cycles, ≤17 f	2967a
	32.5 d; most r18-50; n692	2989
	median 33 d; r25-46; n32 cycles, 5 f	9334
	43.2 d; n16 f	6081
	72.3 d; ±36.23; r24-149; n29 cycles, 7 f	1492
Cercopithecus mitis	menses: 5.3 d; sd1.9; r1-10; n59 cycles, 5 f	2988
	median 30 d; r19-57; n72 cycles, 6 f	9334
	31.9 d; sd6.0; r24-48; n59 cycles, 5 f	2988
Cercopithecus nictitans	28 d; r27-54; n8 cycles, 1 f	3643
Cercopithecus talapoin	luteal phase: 14 d	9338
	menses: 1-4 d	10865
	mode 31 d; median 35; r27-83; n43 cycles, 10 f	9338
	32.9 d; se6.2; r28.0-37.7; n64 cycles, 14 f	9701
	33.0 d; se1.6; r29.1-36.8; n7 f	9701
	33.2 d; r27-43; n5 cycles	10865
	36 d; r29-53; n12 cycles, 2 f	3643
Colobus polykomos	estrus: 1-3 d	3532
Erythrocebus patas	estrus: 12 d; r1-66	6617
	estrus: 13.5 d; ±4.1; n80	6618
	menses: 3 d; r1-5; n16	10072
	30.6 d; sd2.6; r24-38; n64 cycles	10072
	31.9; r10-78	6617
	32.9 d; se1; n82 cycles	6618
Macaca arctoides	follicular phase: 11.9 d; se0.5; r10-14	11707
	luteal phase: 17.2 d; se0.4; r15-19	11707
	menses: 2.8 d	1385
	menses: 3.9 d; ±0.2; n167	6687
	r21-49 d; n10 cycles, 2 f	10912
	mode 28-34 d; r21-34; n1097 cycles, 58 f	10371
	29-30.5 d; across mo	2783
	29.1 d; se0.6; r27-32; n9 cycles, 5 f	11707
	29.4 d; n30 cycles, 6 f	10914
	29.9 d; sd4.4; r19-49; n1365 cycles, 95 f	1385, 2785
	30.2 d; mode 28; r21-42; n189 cycles	7909
	30.7 d; se0.4; n132 cycles, 11 f	6687
Macaca assamensis	follicular phase: 16 d; se1; r9-22; n14 f	11538
	luteal phase: 16 d; se1; most 15-19; n14 f	11538
	r25-35 d; n10 f	11537
	32 d; se2; n14 f	11538
Macaca cyclopis	menses: 4.2 d; sd3.2; mode 3; r1-16; n44 menses	8347
	29.4 d; sd15.5; mode 24; r6-109; n398 cycles, 54 f	8347
Macaca fascicularis	follicular phase: 14.1 d; sd3.11	9863
	follicular phase: 14.4 d; se0.8	2783, 8976, 8977
	luteal phase: 14.2 d; ±0.4; n8 f	7172
	luteal phase: 16.1 d; se0.50	2783, 8976, 8977
	luteal phase: 17.3 d; sd5.8	9863
	menses: 1-6 d	9863

	menses: ~2 d	9446
	menses: 2.8 d; sd1.6; mode 2; r1-9; n707 menses, 28 f	2781, 2782
	menses: 3-5 d 85%; r1-8 d	9827
	menses: 3-6 d	4384
	menses: 3.9 d; mode 4; r1-11; n509 menses, 87 f	11064
	menses: 4 d; ±1	11549
	menses: 4.2 d; se0.3; n35	6687
	menses: mode 6 d; r2-13	2056
	mode 27-28 d; r13-180; n272 cycles	3550
	25-45 d	5316
	28-30 d	11164
	28 d; mode 28-30 70%; r26-38; n20 f	7782
	28 d; median 28; r23-42	4384
	29-32 d	2783
	29.1 d; r26-31; n6 cycles, 2 f	10214
	29.2 d; r26-30; n18 cycles, 15 f	7511
	29.2 d; r22-37; n647 cycles, 93 f	9827
	29.4 d; sd4.3; mode 27, 38; r20-42; n240 cycles, 60 f; 28.8 d; sd2.9; without long cycles	12027
	30.3 d; sd4.1	1280
	30.9 d; sd4.8; median 33; mode 28; r19-43; n595 cycles, 28 f	2781, 2782, 2785, 2786, 5310, 8976, 8977
	31.0 d; sd1.7; n8 f	7172
	31 d; sd1; n25 f	11549
	31 d; median 31; mode 27; r23-41; n345 cycles, 19 f	10264
	31.3 d; se1.5; n21 cycles, 11 f	6687
	31.4 d; n17 cycles, (15/17 ovulatory)	9863
	31.5 d; se1.3; median 32, 36; n77 cycles, 30 f	9446
	32.5 d; se2.3; n6 cycles	7473
	mode 33 d; most 25-42; r15-50; n372 cycles	11062, 11064
	34.5 d; se1.3; r27-82; n275 cycles, 55 f	2247
	39.2 d; mode 32-34; r24-111; n62 cycles, 4 f	2056
	41 d; n4 cycles, 1 f	12117
	47.2 d; mode 26-28; r26-30 26%; median 32; r9-339; n326 cycles, 45 f	5687
Macaca fuscata	follicular phase: 12.2 d; ±1.4; r10-15; n21 cycles, 20 f	7918
	estrus: 8.0-8.9 d	11873
	luteal phase: 8.2 d; ±1.6; r6-10; n5 no progesterone rise	7918
	luteal phase: 14.0 d; ±1.1; r13-16; n13 progesterone rise	7918
	menses: 2-4 d	7954
	menses: 3.6 d	8139
	24-28 d	7954
	25 d; n2 cycles, 2 f	4516
	26.3 d; sd5.4	7915
	4 wk	429
	28.5 d; sd3.0; r23-36; n36 cycles, 7 f	8139
	29.39 d; sd16.9; n129 cycles, 70 f	11873
	30 d; n1 f	4315
	34.1 d; ±24.8; mode 26 d; r11-107; 30.6 d; ±22.8; excluding >60 d	10633
Macaca maura	20-40 d	8885
	median 32 d; r27-37 of medians; r14-85; n33 cycles, 5 f	4183
Macaca mulatta	follicular phase: 10.6 d; ±1.4	1131
	follicular phase: 12 d	12084
	follicular phase: 12.59 d; ±3.13	6732
	follicular phase: 14.1 d; ±0.6	455
	estrus: 9.4 d; ±5.1; r1-25	6466, 6467
	luteal phase: 14.85 d; ±1.23	6732
	luteal phase: 14.9 d; ±2.1	1131
	luteal phase: 15 d	12084
	luteal phase: 15.8 d; ±0.3	455

	menses: 2.7 d; ±1.0; n30 f	618
	menses: 2.8 d; ±0.2; n8 cycles	914
	menses: 3.3 d; ±0.1; n82	6687
	menses: 3.4 d	4534
	menses: 3.6 d; r1.7-6.6; n520 cycles, 30 f	4448
	menses: 3.9 d; mode 4; r1-11; r1-8 with conception	11064
	menses: mode 4 d; r2-8	173
	menses: mode 5 d; r2-11; n132, 10 f	2055
	20.4 d; sd9.2; r6-45; n204 cycles, 20 f	6616
	r24-30 d; n39 cycles, 13 f	7819
	r24.6-33.0 d per f; est r14-42; n481 cycles, 30 f	4448
	r25-27 d; n36 cycles, 12 f	7816
	r25-31 d; n23 f	8807, 10265
	r25-35 d; n9 f	11537
	r25-35 d; n25 f	1696
	mode 25-28 d; r10-70; n15 cycles	173
	25.5 d; sd2.3; r22-30; n8 cycles, 8 f	1131
	r26-30 d; n10 cycles, 10 f	12084
	26.27 d; r23-33; n35 cycles, 7 f	661
	26.5 d; n47 cycles, 15 f	4829
	26.7 d; sd2.1; r24-31; n12 cycles, 9 f	11423
	26.9 d; sd4.9; mode 27-28; r24-30 78%; r19-40; n18 cycles	1132
	mode 27 d; most 15-51; r15-190; n125 cycles, 10 f	2055
	27-28 d; mode 28; r14-42; n392 cycles, 22 f	4448
	27.25 d; se~1.5; n15 cycles, 5 f	2977
	27.36 d; r21-36; mode 27; sd3.18; n33 cycles, 13 ? f	6732
	27.5 d; se0.68; n19 cycles, 4 f	7818
	27.6 d; se1.0; n14 cycles, 7 f	7472
	mode 28 d; r21-37	8649
	mode 28, 30 d; most 23-35; r15-50; n2047 cycles	11062, 11064
	28.3 d; r23-41; n12 cycles, 1 f	12117
	28.4 d; sd2.56; mode 28; n101 cycles, 7 f	3065
	28.5 d; se2.18; r25-34; n2	4534
	28.5 d; se2.2; r22-28; n30 f	618
	29 d; sd3; n450 cycles, 30 f	8509
	29.2 d; se0.7; n74 cycles, 27 f	6687
	29.4 d; sd6.6; median 24; n30 cycles	6467
	29.5 d; sd5.01; r24-50; n39 cycles, 31 f	2195
	29.9 d; se0.5; n9 f	455
	median 30; spring; median 35; winter	9332
	30.6 d; r23-53; n44 cycles, 15 f	3064
	30.6 d; n183 cycles, 16 f; 35.4 d anovulatory cycles	11421
	30.7 d; n24 summer cycles; 26.2 d; n30 winter cycles, 9 f	9132
	30.8 d; mode 28; se0.4; r6-237; n1256 cycles, 50 f	2219
	31.5 d; mode 28; se1.18; n167 cycles	12123
	35.4 d; median 30d; mode 28; 35% r26-30; r14-192; n523 cycles, 65 f	5687
Macaca nemestrina	follicular phase: 13.9 d; r11-60; n23; perineal detumescence	11641
	follicular phase: 14.8 d; r11.5-21.5; n10; laparoscopy	11641
	follicular phase: 15.1 d; ±3.1	975
	follicular phase: 15.5 d; ±0.7; 18.6 d; ±4.5; 2 measures	2870
	follicular phase: 16.2 d; se0.47; r11-35; n25	12123
	follicular phase: 17.6-19.2 d; n21 cycles	986
	luteal phase: 13.6-15.2 d; n21 cycles	986
	luteal phase: 14.3 d; ±1.1	975
	luteal phase: 14.4 d; r4-19; n23; perineal detumescence	11641
	luteal phase: 14.6 d; r13-16.5; n10; laparoscopy	11641
	luteal phase: 15.1 d; se0.31; r11-18; n25	12123
	menses: 5 d; mode 5; r3-7; n22	11641
	29.2 d; n9 cycles, 5 f	975
	31.8 d; mode 28-29; r17-75; n28 cycles, 5 f	11641
	32.08 d; mode 30; sd0.8; r24-66	6177
	32.5 d; se1.5; mode 29; r27-68; n30 cycles, 5 f	12123
	32.8 d; n21 cycles	986
	34.2 d; sd4.5; n47 cycles, 5 f	2870

	median 35.25 d; r of medians 14-77; r6-396; n485 cycles, 32 f	4183
	38.2 d; r31-47; n5 cycles, 1 f	12117
	40.1 d; ±13.4; mode 31; r11-96; n69 cycles, 24 f	6083
Macaca nigra	menses: 4.6 d; n5	2620
	30.7 d; r28-35; n6 cycles, 1 f	10264
	~31 d	1016
	33.5 d; r23-45; n2 cycles, 3 f	2620
	median 35.5 d; r30-44.5 of medians; r10-115; n159 cycles, 12 f	4183
	median 40 d; n15 cycles, 7 f	838
Macaca radiata	menses: 3.8 d; mode 3; n101 menses	11064
	r25-27 d; n6 cycles, 4 f	7511
	r25-36 d; n4 cycles	4450
	26-27 d	10290
	27.06 d; r23-31; n15 cycles, 6 f	6228
	28 d	7009
	mode 28, 30 d; n76 cycles	11062, 11064
	28.0 d; se0.56; n6 f	8971
	29.5 d; mode 31; r25-38	6177
	29.6 d; r10-57; n62 cycles, 22 f	607
	33 d; 44 d with 2 long cycles	12117
Macaca silenus	follicular phase: 14 d	8889
	estrus: 9 d	2497
	luteal phase: 16 d	8889
	28 d	2497
	30 d; r30-36; n2 zoos	8889
	30 d; n8 cycles, 5 f	9908
	31 d; se0.63; n10 cycles, 4 f	9904
Macaca sinica	29.5 d; mode 31; r25-38	6177
Macaca sylvanus	menses: 3-4 d	8223
	27-33 d	8223
Papio hamadryas	follicular phase: 15.7 d; ±0.52; r6-25	9826
	follicular phase: 17.1 d; se0.35; n72	12123, 12129
	follicular phase: 21.8 d; se1.49; n55	12123
	follicular phase: 24.5 d	9699
	luteal phase: 14.4 d; se0.25; n55	12123
	luteal phase: 15 d	9699
	luteal phase: 15.1 d; se0.24; n72	12123, 12129
	luteal phase: 15.6 d; ±0.35; r10-20	9826
	luteal phase: 15.8 d; ±0.4	11599
	luteal phase: 18 d	128
	menses: 1.4 d	10147
	menses: 3 d	12129
	menses: 3 d; r1-6	4629
	menses: 3.7 d	10836
	menses: mode 4 d; median 4; r1-7	913
	13.78 d; r3-84; n29 perineal cycles, 19 f	7331
	23-37 d	916
	median 35 d; mode 36; r29-39; n32 cycles, 8 f	913
	29.3 d	690
	29 d; n10 f	5989
	30 d; n1 cycle, 1 f	12118
	30-33 d; n17 f	3884
	30.3 d; n100 cycles, 20 f	10147
	30.3 d; se0.22; n200 cycles	10385
	30.6 d; r23-40; n3 cycles, 8 f	12117
	30.9 d; se0.95; r25-37; n11 cycles, 1 f	12127
	31.4 d; se1.0; sd5.3; r23-47; n26 cycles, 5 f	4507
	31.4 d; se0.41; n72 cycles, 10 f	12129
	31.5 d; r20-38; n53 cycles, 53 f	9826
	32 d; r24-38; n300 f	3885
	32.98 d; se~0.9; n30 cycles	10836
	33.16 d; sd3.74; mode 31; r19-43; n96 cycles, 32 f	4628, 4629
	33.3 d; se0.7; mode 31-32; r25-41; n32 cycles, 2 f	12123
	34.5 d; n41 cycles, 11 f	11209

	34.6 d; r28-46; n5 f	3651
	34.7 d; se0.48; r30-41; n25 cycles, 1 f	12117, 12123, 12127
	35.61 d; sd3.206; se0.158; mode 35; r29-42; n404 cycles, 34 f	3783
	35.7 d; se0.66; n40 cycles, 11 f	11682
	36 d; r18-52; n150 cycles, 40 f	4627
	36.2 d; se1.5; mode 33; n55 cycles, 5 f	12117, 12123
	38.8 d; r38-42; n~144 cycles, 16 f	3782
	40.11 d; r33-65; n9 cycles, 1 f	2591
	40 d	9699
Papio leucophaeus	r25-42 d; n12 cycles, 3 f	1042
	30.8 d; n12 cycles, 1 f	7851
	32.6 d; se0.9; n18 cycles, 2 f	12123
	median 35 d; r30-46.5 of medians; r13-82; n88 cycles, 3 f	4183
Papio sphinx	32.0 d; n55 cycles, 3 f	7851
	median 33.5 d; r32-35 of medians; r20-244; n46 cycles, 2 f	4183
Presbytis entellus	estrus: 5-7 d	5272, 8119
	menses: 2.0 d; ±0.1; r1-3	6547
	menses: 2-4 d	2293
	menses: mode 2-4 d; r1-11; n86, 16 f	2292, 9833
	21 d; n4 cycles, 2 f	9833
	24.1 d; sd3.8; r17-36; n37 cycles, 9 f	3503
	26.3 d; sd3.6; median 22; r18-45; n125 cycles, 5 f	6547
	26.8 d; sd9.1; mode 22; r21-26 68%; r19-68; n76 cycles, 16 f	2292
	28 d; r24-34; n10 cycles	5044
	30 d	8119
Presbytis obscura	estrus: 3 d	6369
Pygathrix nemaeus	estrus: 2-3 d	4720
	3 wk	4720
	28-30 d	7983
Theropithecus gelada	estrus: 7-10 d	2799
	estrus: 8-10 d	1015
	menses: 3 d	1015
	29 d; r24-34; n2 f	2799
	34.6 d; sd2.3; r31-39; n22 cycles, 1 f?	10142, 10143
	~35 d	2796
	35.4 d	1015
GESTATION		
Cercocebus albigena	176 d; r174-179; last swelling to birth	9339
	177.3 d; r170-181; n3	3643
	184-189 d; n3	11428
Cercocebus galeritus	180 d; se4.49	5768a
Cercocebus torquatus	160-174 d	10291
	161.3 d; r160-162; n3	3643
	median 167 d; r155-199; n23 birth, 12 f	4183
	170.7 d; r164-175; n3 births, 2 f; sex skin swelling to birth	10387
Cercopithecus aethiops	implant: 15-21 d postcoitus	11062
	140-160 d	5758
	153 d	9341
	163 d; ±2	1225
	163.2 d; sd6.3; n38	5364
	165 d; r142-171; n11	9334
	175-178 d	6750
	~7.5 mo	7485
Cercopithecus ascanius	est 120-130 d	5758
	148.5 d; r122-175; n2	2047
	5-6 mo	3581
Cercopithecus mitis	120-130 d	5758
	r138-140 d; n4; (127 d; stillborn gestation not used)	9334
Cercopithecus neglectus	est 170 d	9341
	178 d; r126-229	1127a
Cercopithecus talapoin	156 d	9335
	158-166 d; n18	9338
	158 d; ±5; live birth	9701
	164 d; ±5; n2 stillborn	9701
	167 d; n1	3643

	est ~7 mo; field data	3640
Colobus badius	est 4.5-5.5 mo; field data	10503a
Colobus guereza	est 170 d	9341
Colobus polykomos	174.7 d; r169-178; n3	1956
	est 195-225 d	3532
Erythrocebus patas	167 d	9341
	167 d; se2.2; r159-174; n8	6618
	167.2 d; sd10.3; r133-190; n142	10072
	est 170 d; n1	3966
Macaca arctoides	implant: 16.8 d; se1.8; n5	6687
	172.4 d; n8	1385
	174.7 d; sd8.3; n9	2785
	175 d; r161-186; n12	10371
	176.6 d; r167-185; n25	7909
	177.5 d; se2.1; n10 lt	6687
	178 d; n5 timed	10912, 10914
	207 d; n1	11985
Macaca cyclopis	162 d; sd9.6; r142-175; n9	8347
Macaca fascicularis	implant: 17 d; ±4	11549
	implant: 19 d; r18-21; n3	5310
	implant: 19.2 d; se0.8; n14	6687
	160 d; sd3; n8	11549
	160 d; sd7; n109	11062, 11064
	160-162 or 224 d; est 160-162 is correct	10263
	mode 161-170 d; r151-196	3550
	162.7 d; se0.9; n10	6687
	164 d; r139-176; n101	4954
	165.4 d; sd2.2; n5	2785, 2786, 5310
	167 d; r153-179; n17 after last mating	10264
	169 d	8977
	180 d; r167-193; n31 after last menses	10264
Macaca fuscata	~5 mo; field data	5974
	150-165 d	7405
	171-180 d	5589
	173 d; sd6.9; r161-188; n17	7916
	~7 mo; field data	5139
Macaca mulatta	implant: 9 d; n1	9023
	implant: 9-10 d	11861
	implant: 9-11 d	4707
	implant: 9.5 d	3017
	implant: 10.5 d	1109
	implant: 20.2 d; se0.9; n13	6687
	implant: 21.4 d; ±7.29	6467
	135-171 d	3545
	139 d	6880
	~5 mo	10234
	150 d; n24	5228
	r152-173 d; n18; timed	8588
	r155-172 d; n10	4854
	r157-168 d; n6	7380
	r158-173 d; n16	6038
	159-174 d; n3	4445
	161.2 d; r145-169; n6	49
	164 d; ±4; r153-176	6880
	164 d; n38	4448
	164 d; sd6; n967	11061, 11062, 11064
	165 d; r162-169; n4	2557
	165.7 d viable birth	4448
	165.7 d; se1.5; n12	6687
	166.6 d	914
	167 d; ±7	8509
	167 d; ±6; r142-181; n441 wild	6880
	168.2 d; n5 primiparous	4448
	168.4 d; se1.3; n8	8589
	168.9 d; sd6.85; r148-194; n225	11201

	168.98 d; n57	11408
	169 d; 90% 156-179; r140-195; n268	5104, 11199
	170-175 d	1680
	~177 d	5312a
Macaca nemestrina	170 d; mode 168; r162-186; n21	6083
	median 171 d; r154-230; n56 birth, 23 f	4183
	171 d; n1	12119
	172.5 d; sd6.4; r154-189; n90	9373
Macaca nigra	170.9 d; mode 173; r165-190; n16; 142, 206 d excluded	2153
	median 176 d; r166-181; n15 birth, 8 f	4183
	≤ 7 mo	4462
Macaca radiata	153-169 d; n3	4450
	162 d; sd4; n16	11062, 11064
	169 d; r166.5-170.5; n4	10290
Macaca silenus	171.8 d; r166-176; n4	6464
	172 d; r154-186	8889
	173-188 d; n2	2495, 2497
Macaca sinica	~7 wk	8496
	~6 mo	8501
Macaca sylvanus	implant: 18.2 d	6127
	164.7 d; sd6.11; median 165; r145-177; n33	6127
	5-6 mo	6711
	163.2 d; sd5.38; r145-177; n56; m offspring	8265
	165.3 d; sd4.55; r145-178; n67; f offspring	8265
Papio spp	175 d; ±11; r164-186	5012, 6024
Papio hamadryas	implant: 9 d	5011
	154-185 d; sex skin detumescence to birth	12118
	5.5-7 mo	4582
	r164-179 d; n4	8357
	175 d; r164-186	7494
	25 wk; n23; sex skin detumescence to birth	213
	169-185 d; n3	7233
	175-185 d	4870
	180 d; se0.98; n13	10162a
	181 d; se4; n13	8358
	184 d; n1	3783
	187 d; r173-193; n14 lt, 11 f	3776
Papio leucophaeus	173 d; n1	4183
	181.3 d; r179-183; n3	1042
Papio sphinx	4.5 mo	4582
	median 175 d; r164-213; n4 birth, 1 f	4183
	183-225 d; n4 births, 2 f; sex skin swelling to birth	6508
Presbytis entellus	165-185 d	8119
	~180-196 d	9175
	6-7 mo	4377
	200.1 d; r196-202; n7; field data	11819
Presbytis obscura	140-150 d	6369
	5 mo	482
Presbytis senex	~6 mo	8501
	6.5-8.5 mo	9361
Pygathrix nemaeus	~165 d; n7	4723
	r180-190 d; n8	1299, 6503
Theropithecus gelada	~5.5 mo	2799
	6 mo	2796
	~200 d; n2	10142

LACTATION

Allenopithecus nigroviridis	solid food: 2.5 mo	8702
	wean: 2.5 mo	8705
Cercocebus albigena	wean: 140 d	9339
	wean: 33.33 mo; ±15.87; r17-48	11428
Cercopithecus aethiops	nurse: max 6 mo	3151
	nurse: 6 mo	9728
	wean: 6 mo	7485
	nurse: 206 d; n4	4995

	nurse: 263 d; ±3	10503
Cercopithecus ascanius	independent: ~6 mo	4182
	solid food: 2 mo	4182
Cercopithecus mitis	solid food: 1.5-2 mo	146
	independent: 2 mo	10158
Cercopithecus talapoin	solid food: ~1 mo	3642
	wean: 49-210 d	9338
Colobus polykomos	solid food: 4-5 wk	3532
	wean: 6 mo	3532
Erythrocebus patas	wean: 6 wk; n1	3966
Macaca arctoides	wean: 6-15 mo	10914
	wean: 7 mo	7909
	wean: 10-14 mo; n4	10912
Macaca fascicularis	nurse: 84 d; ±10.4; r52-127	2248
	wean: 100-120 d	4839
	wean: 207 d; sd71; r96-286; n12	11595
	wean: 245 d; sd30; r184-295; n16	11595
	wean: 9-18 mo	10264
Macaca fuscata	solid food: 3 mo	10667
	wean: est 6 mo	10667
Macaca mulatta	solid food: ~2 wk	6466
	solid food; 2 mo	4448
	wean: 7-14 mo	4448
	wean: 6-9 mo	8722
	wean: 8 mo	6800
	wean: ~10 mo	6466
	wean: ~1 yr	10234
Macaca nigra	solid food: 5 wk	4462
	wean: ~1 yr	4462
Macaca radiata	wean: 8-12 mo	9982
Macaca silenus	solid food: 20 d	2495, 2497
	solid food: 4 wk	2109
	wean: 1 yr	2495, 2497
Macaca thibetana	wean: ~6 mo	3386
Nasalis larvatus	solid food: 6 wk	8710
	wean: 7 mo	8710
Papio spp	wean: 14-18 mo	6342
Papio hamadryas	solid food: 33 d	3776
	wean: 6-8 mo	10158
	wean: ~6-~15 mo	5758
	wean: ~12 mo	1486
Papio leucophaeus	wean: 16 mo	1042
Presbytis entellus	solid food: 2-3 mo	5273
	wean: 8-10 mo	9175
	wean: 12-15 mo	5273
	wean: ~15-16 mo	4377
Presbytis senex	wean: 28 wk	9361
Theropithecus gelada	wean: 12-15 mo	10142
	wean: 12-18 mo	2799
INTERBIRTH INTERVAL		
Cercocebus albigena	median ~1.6 yr; n6; from graph	9341
Cercocebus torquatus	median ~1.3 yr; n9; from graph	9341
	median 390 d; r120-780; n19 intervals, 10 f	4183
	16.6 mo; se0.6; n110	4159a
Cercopithecus spp	3 yr	3748
Cercopithecus aethiops	10.7 mo; r5-24; n57	3151
	337.9 d; n29	1225
	11.7 mo; se0.4; with new m	3153
	353 d; yg survives	4995
	median ~1 yr; n42; from graph	9341
	12.15 mo; r8.75-19; n5	6750
	373.8 d; r223-512; n55 intervals	9728
	13 mo; r10-16; n3	7485
	13.6 mo; se0.8; with old m	3153

Cercopithecus mitis	413 d; n11	1225
	median ~1.5 yr; n30; from graph	9341
	19.2 mo; r14.4-25.8	2988
Cercopithecus neglectus	mode 12 mo	1127a
	median ~1.6 yr; n17; from graph	9341
Colobus badius	8-28 mo; field data	10503a
Colobus guereza	median ~1 yr; n16; from graph	9341
Colobus polykomos	14 mo; r12.5-15.5; n2	3532
	459 d; r283-731; n6 intervals, 4 f	1956
Erythrocebus patas	348.4 d; se16.4	6618
	11.8 mo; n31 intervals, 9 f	1797
	median ~1 yr; n11; from graph	9341
Macaca arctoides	523.1 d; r412-691; n17 intervals, 10 f; indoor	7909
	median 525 d; r390-1080; n10 intervals, 5 f	4183
	544.6 d; se36.5; r363-908; yg lives	4464
	619.4 d; r444-910; n21 intervals, 13 f; outdoor	7909
	19 mo 2 d; n9 f	3107
Macaca fascicularis	median 390 d; r330-720; n22 intervals, 11 f	4183
	448 d; r335-626	11595
	~15 mo; n832	7954
Macaca fuscata	274-448 d; n53; 564-796; n107; >796-1115; n8	10667
	median 386 d; mode 425.3; sd113.5; r325-833; 2 outliers 679, 833 d; n29	9702
	2 yr; n283	5974, 7954
Macaca mulatta	1 yr; n94; 2 yr; n45	10077
	median 360 d; r300-690; n29 intervals, 23 f	4183
	362 d; r294-441; n2	5312a
	406 d; se23; n19 high rank	3883
	15 mo	4546
	17.3 mo	401
	630 d; se94; n11 low rank	3883
Macaca nemestrina	median 405 d; r240-1080; n44 intervals, 13 f	4183
Macaca nigra	median 540 d; r300-720; n13 intervals, 7 f	4183
Macaca sylvanus	210.5 d; sd29.1; median 203; r183-310; n21	6127
	1 yr; 88.4%; 2 yr; 11.6%	8266
	431.716 d; sd165.941; median 373.75; mode 361; ~1 yr 85.2%; r231-1121; n155 ibi	1477
	31.5 mo; r22-49; n8; yg lives	9171
Macaca thibetana	2 yr	12093b
Papio hamadryas	11 mo; stillborn	213
	12-16 mo; forest; 18-24 mo; savanna	5758
	423.5 d; r382-465; n2 samples	6937
	~16.5 mo	401
	18 mo; yg lives	1486
	592-840 d; yg lives	10507
	?-22 mo; yg lives	213
	24 mo; n16 intervals, 12 f; yg lives	9970
	760 d; se34.68; n23	10162a
Papio leucophaeus	19 mo	1042
Presbytis entellus	14.1 mo; r8.2-20.2; n25 ibi, 13 f	3503
	15.3 mo; r11.2-20.2; n19 ibi, 11 f; yg lives	3503
	15.4 mo; yg lives	4377
	15.7 mo; n14 ibi, 9 f; yg lives	4377a
	2 yr	5273
Presbytis senex	16-24 mo	9361
Pygathrix nemaeus	1 yr	6502
Theropithecus gelada	median 525 d; r270-720; n11 intervals, 6 f	4183
	est 2-3 yr	2796
	2.14 yr; r1-6 yr; 1.64 yr if harem taken over	2797

SEASONALITY

Cercocebus albigena	birth: May, July, Sept, Dec; n2 f; zoo	2109
	birth: yr rd; W Africa	5758
Cercocebus galeritus	est Dec-Feb; ne Gabon	8820
Cercocebus torquatus	sex skin: Apr-early June; n1; zoo	12117
	birth: Apr-Feb; n32; captive	4183
	birth: May, July, Aug, Dec; few data; zoo	1230
	birth: yr rd, peak Oct-Mar; captive	4159a
Cercopithecus spp	preg f: July-Oct; Kisangani, Zaire	3748
	birth: Jan, Apr-Nov; n17; captive	537
	birth: July-Oct, peak Aug; Kisangani, Zaire	3748
Cercopithecus aethiops	testes: large June-Sept; captive	2967
	testosterone: high Dec-Mar, low Apr-Nov; captive	2967
	cycles: normal May-Nov; prolonged Jan-June; lab	2989
	mate: mid Mar; zoo	7485
	mate: May-mid Oct; Kenya	10503
	mate: June-July; Kenya	256
	mate: est yr rd; W Indies	2009
	mate: yr rd; e Africa	6612
	conceive: May-June; n Kenya	11654
	conceive: yr rd; captive	5364
	preg f: Feb-Mar; n7; Ethiopia	10311
	preg f: yr rd; Botswana	10154
	birth: Jan-Feb; Buffle Noir, Cameroun	5587
	birth: Jan-June; n11/12; July; n1/12; captive	9334
	birth: Feb, Sept-Nov; few data; zoo	1230
	birth: July-Aug; Kalamaloue, Cameroun	5587
	birth: peak Aug-Oct; Zambia	331
	birth: Sept-May; Zaire	11253
	birth: Oct-Nov; Bakosi, Cameroun	5587
	birth: Oct-Mar; Kenya	10503, 10504
	birth: Nov; zoo	7485
	birth: yr rd, peak Apr-July 73%; W Indies	4995
	birth: yr rd, peak May-Sept; captive	6126
	birth: yr rd; captive	3151
	birth: yr rd; S Africa	10158
Cercopithecus ascanius	estrus: yr rd, peak May-Sept; w Kenya	2047
	mate: Mar-Apr; Central Republic of Africa	3581
	mate: Dec-Apr; Uganda	5758
	mate: peak Dec-Apr; small emb Jan-May; Uganda	4182
	birth: Jan, Sept-Oct, Dec; Kenya	2047
	birth: Mar-Apr, June, Oct-Dec; few data; Uganda	10505
	birth: peaks May, Sept; Uganda	5758
	birth: Aug-Sept; Central Republic of Africa	3581
Cercopithecus campbelli	mate: est June-July; Ivory Coast	1148
	birth: end wet-early dry season; Ivory Coast	1148
Cercopithecus cephus	birth: Dec-Mar; n9; Gabon	3640
Cercopithecus lhoesti	preg f: Feb; few data; Cameroun	2955
	lact f: Feb, Oct, Nov; few data; Cameroun	2955
Cercopithecus mitis	mate: Jan, Mar, July-Dec; Uganda	146
	mate: Aug-Jan; Kenya	8106
	mate: yr rd; Rift Valley, Kenya	1116
	conceive: July-Nov; Kenya	8107
	conceive: peak June-Oct; Kenya	2048
	birth: Jan, May, July-Sept, Nov; n13; Uganda	146
	birth: June-July, Nov; n5; captive	9334
	birth: est Sept-Apr; S Africa	10158
	birth: Sept-Dec; Zimbabwe, Zambia, Malawi	10153
	birth: Nov-Mar; Kenya	8107
	birth: Dec-May; n14; Uganda	9362
	birth: est yr rd; captive	2988
	birth: yr rd; Kenya	2048
Cercopithecus mona	birth: Dec-Apr; n5; Gabon	3640

Cercopithecus neglectus	conceive: May-July, Oct-Nov; s Cameroun, Benito, Equatorial Guinea	647
	birth: Feb-Mar; n3; Gabon	3640
Cercopithecus nictitans	preg f: Mar; n1; e Africa	185
	repro: Mar; n1; Cameroun	2955
Cercopithecus talapoin	mate: Jan-Mar, peak Feb; Cameroun	9340
	mate: May-Aug, peak July; n7; Gabon	3640
	mate: Dec-Feb; Cameroun	9335
	conceive: yr rd; captive	9338
	preg f: Oct-Dec; Cameroun	9340
	birth: June-Aug; Cameroun	9340
	birth: Aug-Sept; Cameroun	9335
	birth: peak Nov-Mar; Gabon	3639
	birth: Nov-Apr; Gabon	3641
	birth: Dec-Apr; n18; Gabon	3640
Colobus abyssinicus?	preg f: June, Oct; n2; Kenya	4926
	birth: yr rd; Uganda	5758
Colobus angolensis	birth: Feb-Dec; few data; zoo	2109
Colobus badius	preg f: Sept; n2; Liberia	184
	birth: yr rd; n27; Kenya	6833
	lact f: Dec; n2; Tanzania	186
Colobus polykomos	mate: yr rd; Zaire	11271
	birth: Sept-Apr; Eq Guinea	8505
	birth: yr rd; e Africa	5758
Erythrocebus patas	mate: May-Dec; Kenya	1796
	mate: June-July; Uganda	4232
	conceive: June-Aug; Kenya	1796
	conceive: June-Aug; Kenya	1797
	birth: Jan-May; e Africa	5758
	birth: Dec-Jan; Kenya	1796
	birth: Dec-Feb; Kenya	1797
	birth: nearly yr rd; captive	5522
	birth: yr rd, peak Oct-Dec; n501, 169 f; lab	10072
Macaca arctoides	cycles: yr rd; captive	10371
	mate: yr rd; Veracruz, Mexico	3107
	birth: Mar, June-July, Nov; 63% wet season; Veracruz, Mexico	3107
	birth: Mar-June, Oct-Dec; n10; lab	6687
	birth: peak May/June; China	3386
	birth: Aug-June; n19; captive	4183
	birth: yr rd, peak July-Aug; captive	3106
	birth: yr rd, peak Aug 13/62; captive	10065
	birth: yr rd; captive	4464
	birth: yr rd; captive	7909
	birth: yr rd; captive	2783, 2785
Macaca assamensis	birth: peak Apr-June; few data; Thailand	3385
Macaca fascicularis	sperm: peak Apr-May, low Feb, July; lab	6729
	testes: peak Apr-Sept; lab	6729
	sex skin: peak Jan-Mar, low Apr-Aug, none Sept-Dec; Sumatra, Indonesia	11184
	conceive: Aug-Feb, peak Nov; n43; zoo	12120
	conceive: peak Sept-Jan, low June-Aug; lab	11064
	preg f: yr rd, peak May-June; Malaysia	5588
	preg f: Aug; n1 n Borneo, Indonesia	2313
	birth: peak Mar-May; Thailand	3385
	birth: peak Mar-June, low Sept; n111; semi wild, Japan	7954
	birth: Mar-Nov; n42; captive	4183
	birth: peak Aug-Oct; captive	3634
	birth: peak Aug-Dec; lab	10844
	birth: yr rd, peak Mar-Oct; captive	4839
	birth: yr rd, peak May-June; captive	10264
	birth: yr rd; lab	6687
	birth: yr rd; zoo	12117
	birth: yr rd; captive	2783

Macaca fuscata	spermatogenesis: Nov-Mar; Oita, Japan	10816
	spermatogenesis: June-Dec; Hiroshima, Japan	10816
	spermatogenesis: peak winter, low summer; lab	10815
	spermatogenesis: yr rd; Japan	5156
	cycles: Oct-Apr; semi-wild, Japan	7954
	cycles: Oct-May; captive, Japan	7915
	cycles: Dec-Apr; lab	429
	mate: end Jan-early Feb; Japan	7917
	mate: peak Sept-Jan; Japan	5589
	mate: Sept-Jan; captive, OR	147
	mate: Oct-Mar; Japan	10633
	mate: Oct-Apr, peak Jan-Feb; Japan	7405
	mate: Nov-Dec, none Feb-Sept; semi wild, Japan	5974
	mate: peak Dec-Jan, low Oct-Apr; Japan	5139
	mate: Dec-Feb; Oita, Japan	10816
	mate: Sept-Nov; Hiroshima, Japan	10816
	mate: fall-winter; captive	3981
	conceive: yr rd; Japan	5156
	birth: Feb-Aug, peak Mar-May; semi-wild, Japan	5974, 7954, 10667
	birth: Mar-Sept, peak May; n186; TX	3979
	birth: Mar-July; Japan	5590
	birth: peak Mar-July, peak Apr-June; Japan	7916
	birth: Apr-early June; captive	147
	birth: Apr-July, peak May; Japan	10633
	birth: May-June; Japan	5589
	birth: May-Sept; Japan	6168
	birth: June-Aug; Japan	7917
	birth: peak June-Aug, low May-Nov; Japan	5139
	repro: mid Oct-early Mar; Japan	11873
Macaca mulatta	sex maturity: peak Oct-Feb, low July-Aug; lab	401
	spermatogenesis: peak Oct, low Apr; Cayo Santiago, Puerto Rico	12060
	testes: large Aug-Dec; Puerto Rico	9417
	testes: large Sept-Nov, small Mar-May; Puerto Rico	2008
	testes: large Sept-Dec, small Jan-Aug; Cayo Santiago, Puerto Rico	9417
	testes: large Oct, small June-July; lab	11661
	testosterone: peak Mar; indoors; Sept; outdoors	4673
	ejaculation: yr rd, peak Dec, low Mar; lab	7265
	ejaculation: peak Dec-Jan, low June-July; lab	7270
	follicles: large Oct-Mar; lab	4448
	mate: Feb-May; India	8722
	mate: Apr-Sept; free ranging, Brazil	1970
	mate: none June-Sept; captive	3949
	mate: July-Jan, peak Sept-Oct; Cayo Santiago, Puerto Rico	2007
	mate: mid July-Aug, none Feb-June; Puerto Rico	5919
	mate: mid July-Dec; Cayo Santiago, Puerto Rico	6616
	mate: peak Aug-Sept; Nepal	7382
	mate: Sept-Dec; captive	3948
	mate: peak Sept-Jan; captive	11770
	mate: Sept-Mar; Puerto Rico	11219
	mate: end Sept-early Feb; India	6466
	mate: peak Oct, none June-Sept; captive	3949
	mate: Nov-Dec; India	6467
	mate: Nov-Dec; Pakistan	9175
	mate: peak Nov-Jan, none Mar; n India	10234
	ovulate: Feb-Mar, Oct-Dec; captive	8676
	ovulate: peak Sept-Nov, low May-July; captive	11772
	ovulate: Oct-Apr; captive	4451
	ovulate: peak Oct-May, low May-Sept; captive	11421
	ovulate: fall-mid winter; captive	8677
	conceive: peak Sept-Oct; Cayo Santiago, Puerto Rico	6616
	conceive: peak Sept-Dec, low Jan-Apr; lab	11064
	conceive: Oct-Nov; captive	11408

	conceive: Nov-Jan; captive	4448
	conceive: peak Nov-Feb; n15; lab	6880
	conceive: yr rd; captive	948
	conceive: yr rd; captive	8649
	birth: Jan-May, peak Mar-Apr; Puerto Rico	5919
	birth: Jan-June, peak Feb-Mar; Cayo Santiago, Puerto Rico	5920
	birth: Jan-June, peak Mar-Apr; Cayo Santiago, Puerto Rico	2007
	birth: Jan-mid July; Cayo Santiago, Puerto Rico	6616
	birth: Feb-Apr; Cayo Santiago, Puerto Rico	8975
	birth: Feb-Aug, peak Apr-May; captive	3427
	birth: Mar-May; Desechio I, Puerto Rico	7542
	birth: Mar-May, Sept-Oct; Rajasthan, India	8722
	birth: Mar-June, 71% of 31; captive, TN, NE	11216
	birth: Mar-June, 74% of 42; captive, GA	11216
	birth: Mar-Aug; captive	3950
	birth: Mar-Aug (Nov), 74% May-June; Puerto Rico	11219
	birth: Mar-Nov; n76; captive	4183
	birth: peak Mar-May; Nepal	7382
	birth: peak Mar-May, none Oct-Nov; lab	6687
	birth: mid Mar-early June; n11; captive	3948
	birth: end Mar-early Oct; Rajasthan, India	8719
	birth: end May-June; captive, Puerto Rico	11216
	birth: Apr-May, Sept-Oct; Pakistan	6766
	birth: Apr-May; India	6466
	birth: Apr-May; captive	11408, 11767
	birth: peak Apr-May, few Sept-Oct; n India	10234
	birth: Apr-Aug; Hainan I, China	5312a
	birth: May-June; free ranging, Puerto Rico	11218
	birth: peak May, July-Aug, few Nov-Mar; captive	9056
	birth: peak Oct-Jan; zoo	1230
	birth: Oct-Jan; zoos, s hemisphere	917
	birth: Oct-Mar or Oct-Dec; free ranging, Rio de Janeiro, Brazil	1970
	birth: end Dec-mid July, peak Feb-Mar; Cayo Santiago, Puerto Rico	5920
	birth: yr rd; Mar; zoo, India	4534
	birth: yr rd, peak Mar-June; n815; captive	10078
	birth: yr rd, peak Apr-May; lab	11062
	birth: yr rd, peak Apr-Sept; captive	1151
	birth: yr rd, peak late Apr-mid July; 60% May-June; captive	2721
	birth: yr rd; captive	5228
	birth: yr rd; lab	264
	birth: yr rd; zoo	12117
Macaca nemestrina	conceive: Jan-May, July-Aug, Oct-Nov; few data; Malaysia	1543
	preg f: Feb-Apr; n5; Thailand	3385
	birth: est Dec-Jan; Thailand	3385
	birth: yr rd; captive	6083
	birth: yr rd; n68; captive	4183
	lact f: Feb; n3; Thailand	3385
Macaca nigra	birth: Dec-Jan, Mar, May, June-Oct; n17; captive	4183
	birth: Dec-Oct; n41; captive	2620
Macaca radiata	mate: Sept-Oct; captive	8885
	birth: Feb-Mar; India	9982
	birth: peak Mar-May; no data Jan-Nov; captive	9973
	brith: Apr-Aug; n7; captive	4183
	birth: June-Aug; n3; zoo	44
	birth: peak spring-summer; zoo	1230
	birth: yr rd: peak Mar-July; captive	9974
	birth: yr rd; lab	11062
Macaca silenus	birth: Mar-Nov, peak Apr-May; zoo	2497
	birth: June; n1; zoo	2109
	genital swelling: peak June; India	6096
	mounting: peak June, low Mar-Apr; India	6096
	mate: yr rd; zoo	2495
	birth: Mar-Nov; zoo	2495

Macaca sinica	repro: not seasonal; Sri Lanka	8496, 8501
Macaca sylvanus	mate: mid Aug-Mar, peak Oct-Dec 80%; n1701; captive	6127, 8265
	mate: Oct-Nov, n95; Gibraltar	6711
	conceive: mid Nov-mid Dec 60%; captive	6127
	birth: Mar-Aug, Apr-May 80%; captive	8265, 8266
	birth: end Mar-July; captive	9556
	birth: peak Apr-June, none Jan-Feb; zoo	9171
	birth: Apr-July; few data; Algeria	7199
	birth: May-Sept; n145/147; Gibraltar	6711
Macaca thibetana	birth: Jan-May; Mt Emei, China	12093b
Nasalis concolor	neonates seen: June-July; Siberut I, Indonesia	10836a
Nasalis larvatus	repro: not seasonal; N Borneo, Indonesia	9667
Papio hamadryas	mate: yr rd; e Africa	6612
	mate: yr rd; few data; zoo	1230
	mate: yr rd; zoo	12120
	birth: peak Sept-Nov; captive	12118
	birth: peak Sept-Nov; Tanzania	9086
	birth: Oct-Dec; S Africa	2501
	birth: yr rd, peak mid Nov; n87; zoo	12120
	birth: yr rd, peak Dec, low Apr; n93; semi wild; Japan	7954
	birth: yr rd; forest; peak end wet season; savanna; e Africa	5758
	birth: yr rd, peak rainy season; Zaire	11253
	birth: yr rd; S Africa	10158
	birth: yr rd, except Oct-Nov; n36 over 20 mo; captive	6937
Papio leucophaeus	birth: yr rd; Cameroun	9468
	lact f: Aug; n1; s Cameroun, Benito, Equatorial Guinea	647
Papio sphinx	birth: Jan-Apr; Gabon	5457
Presbytis cristata	mate: yr rd; Malaya	7162
	preg f: Apr-May; n2; w Thailand	3385
	small yg: 1st half of the yr; Malaysia	7161
	lact f: Feb; n1; w Thailand	3385
Presbytis entellus	birth: Jan-May, Oct; Rajasthan, India	8722
	birth: Feb-Mar; nw India	4532
	birth: Mar, May-June; Himalayas, Pakistan	9175
	birth: Mar-Sept; n India	10539
	birth: Apr-May; n India	5273
	birth: yr rd, peak Dec-Apr; s India	10539
	birth: yr rd, peaks July-Aug, Dec-Mar; India	5044
	birth: yr rd; zoo	1230
Presbytis obscura	birth: Jan-Mar; Thailand	6369
Presbytis senex	mate: Oct-Jan; Sri Lanka	9361
	mate: yr rd; Sri Lanka	8619
	birth: peak Feb-Mar; sw Sri Lanka	8619
	birth: Feb-Mar; Sri Lanka	8496, 8501
	birth: May-Aug; Sri Lanka	9361
Pygathrix nemaeus	birth: peak Feb-June, low Sept-Dec; ne Vietnam	6502
Theropithecus gelada	birth: few Apr, Sept-Oct; Ethiopia	2796
	birth: Sept-Mar, July; n19; captive	4183
	birth: peak Nov-Mar, minor peak May-Aug; Ethiopia	2797

Order Primates, Family Hylobatidae, Genus *Hylobates*

Gibbons (*Hylobates* 4-13 kg) brachiate through tropical forests of southeast Asia feeding on leaves and fruit. Some species exhibit sexual dimorphism with respect to pelage color (7013a). The labyrinthine, hemochorial, discoidal placenta has an interstitial, antimesometrial implantation in the dorsal or ventral surface of the simplex placenta (2478, 4113, 7563, 10310, 10415, 11843) and may be eaten after birth (483, 1305). Neonates are furred (1305, 1810, 2109, 2489), cf 10212). Males have a baculum, seminal vesicles, penial spines (3738, 6967, 11851), and may carry yearlings (113, 1809, 2555a). Details of anatomy (2478, 5494, 6967, 11847) and birth (8629a) are available.

NEONATAL MASS		
Hylobates lar	m: 113-170 g; n1	483
	m: 380 g; r350-410; n2	9515
LITTER SIZE		
Hylobates hoolock	1 yg	988a
Hylobates lar	1 yg	988a
	1 yg ea; n2 lt, 1 f	1305
	1 yg ea; n3 lt	815
	1 yg ea; n7 lt	2109
	1 yg ea; n8 lt, 1 f	4198
	1 yg ea; n12 lt	1284
Hylobates moloch	1	557
	1 emb ea; n2 f	2313
Hylobates syndactylus	1	3665
	2 yg; n1 lt	2555a
SEXUAL MATURITY		
Hylobates concolor	f: 1st birth: 9 yr; n1	2610
Hylobates hoolock	f: 1st menses: ~7 yr	6967
Hylobates lar	f: 1st birth: 10 yr; n1	4198
CYCLE LENGTH		
Hylobates spp	follicular phase: 11-14 d	6230
	luteal phase: 10-11 d	6230
	r21-25 d; n10 cycles, 1 f	6230
Hylobates hoolock	menses: 2.6 d; sd0.66	6967
	27.83 d; sd4.07; r20-33; n6 cycles, 1 f	6967
Hylobates lar	menses: 2.38 d; sd0.54	1608
	3 wk-1 mo	483
	29.8 d; sd4.1; r21-43; n17 cycles, 2 f	1608
	30 d; r22-42; n20 f	1284
	31 d; n1 f	1305
GESTATION		
Hylobates concolor	200-214 d	8099
Hylobates lar	199 d; n1; pairing to birth	11371
	210 d; n1; menses to birth	1305
	7 mo	1284
	~7 mo; n4	9515
	215 d	2109
	7.5 mo	483
Hylobates moloch	7-9 mo	557
Hylobates syndactylus	232.0 d; r230-235; n3	4734
LACTATION		
Hylobates lar	solid food: 4 mo	2109
	dependent: 7 mo	988a
INTERBIRTH INTERVAL		
Hylobates lar	2-4 yr	4198
	~3 yr	2109
Hylobates moloch	2 yr	557
Hylobates syndactylus	289.75 d; r265-313; no lactation	4734
	619 d; yg lives	4734
	2-3 yr	1809
SEASONALITY		
Hylobates concolor	mate: end May-June; n1 pr; captive	8099
	birth: Dec; n1; captive	8099
Hylobates hoolock	birth: Nov-Feb; Naga Hills, India/Burma	7013a
	birth: Nov-Mar; s Asia	9265a
Hylobates lar	birth: early winter; s Asia	9265a
	lact f: Jan-early Apr; w Thailand	3385
Hylobates moloch	preg f: mid May-mid June; n2; N Borneo, Indonesia	2313

Order Primates, Family Pongidae

Great apes (gorillas, *Gorilla* 70-350 kg; chimpanzees, *Pan* 30-80 kg; orangutans, *Pongo* 30-140 kg) are diurnal, terrestrial to semiarboreal, often social herbivores from Africa and southeast Asia. Males are larger than females (2406, 3387, 3583, 4357, 5043, 5467, 5758, 5766, 8001, 8798, 9084, 9133, 9544). The 200-350 g, hemochorial, villous, discoid placenta, which secretes gonadotropins (10636a), implants into a simplex uterus (3737, 4000, 4113, 4842, 6183, 6633, 7554, 7563, 7714, 9293, 10310, 10415, 11843, 11847) and is usually eaten after birth (400, 697, 1068, 1889, 2407, 2962, 3584, 3888, 3890, 4385, 4583, 5786, 5787, 5788, 5793, 6583, 6758, 7938, 8798, 9084, 10399, 10400, 11087; cf 3440, 4000, 6183, 7721). Estrus does not occur (2153, 5766) after the birth of a furred, open-eyed neonate (1354, 5758, 5788, 6183, 6186, 7721, 8798, 10399; cf 6186). Males have seminal vesicles, penial spines, and a baculum (3738, 4787, 9472). The extensive endocrinological data on great apes are not reviewed here. Details of anatomy (3737, 5494, 9666a, 11847, 11851), birth (1068, 2407, 3278, 3584, 3890, 4000, 5786, 5788, 6583, 7721, 7938, 10399, 10400), mating (11495a), and milk composition (1152, 5471, 10695, 12015) are available.

NEONATAL MASS

Gorilla gorilla	1.276 kg; stillborn, premature	8914
	1.68 kg; se0.29; n4	2199
	est 1.75 kg; n1	3278
	1.8 kg; r1.7-1.9; n2	5793
	est 1.82 kg; n1	6183
	1.871 kg	3379
	f: 1.87 kg; n1	10786
	m: r1.986-2.270; n6	5791
	2.04 kg; 36 h	7721
	2.097 kg; 2 d	3536
	2.1 kg; n1	5792
	2.1 kg; n1, 21 h	6583
	f: 2.106 kg; r1.617-2.400; n9	5791
	2.14 kg; n1	9371
	2.2 kg; n1	1600
	2.2 kg; r1.6-2.6; n15	2076
	2.25 kg; n1	6583
	2.3 kg; r2.2-2.6; n3	1955
	2.375 kg; r2.20-2.55; n2	6758
	2.6 kg; n1 stillborn; ~2.6 kg; n1 live birth	4369
Pan paniscus	1.162 kg; 41 h; (infant had not eaten)	4735
Pan troglodytes	0.78 kg; r0.63-0.93; n2 stillborn, no brain, triplet birth	9293
	0.88 kg; r0.8-0.98; n3	11997
	0.89 kg; n1, 1 wk, from triplet birth	9293
	~1 kg; n5	5471
	1.2-1.5 kg; n~20	10679
	f: 1.69 kg; r1.20-2.44; n13	5614
	1.7785 kg; r1.502-2.055; n2	3379
	m: 1.81 kg; r1.35-2.20; n16	5614
	1.89 kg; r1.61-2.26; n17	12015
	m: 2 kg; n1	2962
Pongo pygmaeus	f: 0.85 kg; n1	4581
	1-2 kg; from graph	1073
	1.2 kg; small; 2.2 kg; normal	1072
	1.423 kg; r1.265-1.600; n3	9666a
	m: 1.47 kg; n1	4581
	f: 1.51 kg; r1.35, 1.67	6501
	f: 1.56 kg; sd0.39; r0.92-2.04; n14	3387
	~1.6 kg; r0.95-1.65	5878
	1.644 kg	3379
	f: 1.694 kg; r1.42-2.04; n5	9737
	1.71 kg; se0.24; n5	2199
	m: 1.74 kg; r1.59-2.015; n7	9737
	m: 1.93 kg; sd0.29; r1.59-2.26; n12	3387
	m: 2.16 kg; r2.06-2.26; n2	6501

NEONATAL SIZE		
Pan troglodytes	f: 39.1 cm; r34.3-45.7; n9	5614
	m: 40.3 cm; r37.0-44.5; n12	5614
Pongo pygmaeus	m: 32.5 cm; r32, 33; n2	11048
WEANING MASS		
Gorilla gorilla	est 4-7 kg; 6-8 mo; from graph	4816
LITTER SIZE		
Gorilla gorilla	1	9544
	1 yg	6583
	1 yg ea; n2 lt	7723
	1.04; r1-2; n56 lt	5791
	2; n1 lt	5793
	2 emb; n1 f	9283a
Pan paniscus	mode 1; r1-2	8798
	1 yg ea; n7 lt	5257
Pan troglodytes	1	10971
	1-2	11997
	mode 1; r1-2	12015
	1 yg; n1 f	1354
	1 ea; n4 lt	12128
	1.02; r1-2; n55 lt	3888, 3889
	1.04; r1-3; n291 lt	6876
	1.05; r1-2; n120 lt	8284
	1.06 yg; r1-2; n118 lt	8284
	3 yg; n1 lt	9293
Pongo pygmaeus	1	11034
	1 ea; n2 lt	3569
	1 yg; n2 lt	11048
	1 ea; n55 lt	5766
	1.1; r1-2; n28 lt	3387
	2 yg; n1 lt	320, 4581
	2 yg; n1 lt	9511a
SEXUAL MATURITY		
Gorilla gorilla	1st mate: 1st birth: 5.5 yr	5618
	f: genital swelling: 7.75 yr; r6.5-8.5; n4	4358, 4360
	f: menarche: 9 yr; n1	7945
	f: menarche: ~7 yr; r6.42-8.58; n13	3426
	f: 1st mate: median 6.33 yr; r5.75-7.1	11495b
	f: 1st mate: 6-7 yr	9544
	f: 1st mate: 7.5 yr; n5	4357
	f: 1st conceive: 7-9 yr	4360
	f: 1st preg: 9.83 yr; r6.33-15; n26	2076
	f: 1st birth: min 5.5 yr; mode 8.5	7650
	f: 1st birth: 6.75 yr	5791
	f: 1st birth: median 10.05 yr; r8.67-12.75	11495b
	f: 1st birth: 10.42 yr; r9.5-10.83	4357, 4358, 4360
	f: 2 yr	4359
	m: 1st mate: ~9 yr	9544
	m: 1st mate: 9-10 yr	11495b
	m: 1st mate: 12 yr	3426
	m: 1st mate: 15 yr	4358
	m: 1st sire: 7.67 yr	5791
	m: 1st sire: 10.08 yr; r6.33-17; n37	2076
	m: 8 yr	4357
Pan troglodytes	f: menarche: 7 yr	6171
	f: menarche: 7.7-9.3 yr	1946
	f: menarche: 8 yr	11964
	f: menarche: 8.92 yr; r7.33-10.17; n7	12047
	f: menarche: 12-15 yr	10972
	f: swelling, menarche: 6-8 yr	11177
	f: 1st cycle: 8-10 yr	12015
	f: 1st swelling: 10 yr	10972
	f: 1st mate: 11-13/14 yr	3889
	f: 1st mate: 13.25 yr; n1	10971

	f: 1st conceive: 8.2-10.7 yr; n6	1946
	f: 1st conceive: 10-10.5 yr	12047
	f: 1st preg: 13-15 yr	10970
	f: 1st birth: ~13-15 yr	10972
	f: 1st repro: >9 yr	11176
	f: 18 mo; precocious	11825
	m: sperm, testosterone: 11 yr	6879
	m: sperm: 8 yr	12015
	m: testes large, testosterone: 7-9 yr	5992
	m: testes large: 110 mo; r98-123; n3	8800
	m: 10-11 yr	2701
	m: 1st sire: 13-15 yr	3889
	m: 6-11 yr	3904
Pongo pygmaeus	f: menarche: 7-8 yr	9133
	f: menarche: 9-14 yr	2406
	f: 1st mate: 7-8 yr	4994
	f: 1st mate: 8 yr	11087
	f: 1st conceive: 8-9 yr	9133
	f: 1st birth: 8 yr	9736
	f: 1st birth: 10 yr; n1	4581
	f: 1st birth: 15 yr; n1	3583
	m: sperm: 6.4, 7.7 yr; n2	2634
	m: 1st mate: 6.6 yr	2634
	m: 1st mate: 10 yr	11087
	m: 1st sire: 7 yr	9736
CYCLE LENGTH		
Gorilla gorilla	follicular phase: 9-31 d; n4 cycles, 3 f	7731
	follicular phase: 17.1 d; ±2.0; r8-25; n5 f	2202
	follicular phase: 19.5 d; se1.0; r11-30; n21	7386
	estrus: median 1 d; r1-3; n3, 6 f	4358
	estrus: median 1 d; r1-4; n77	11495b
	estrus?: median 11 d; r1-5; n14, 5 f adolescents	4358
	luteal phase: 11-13 d; n4 cycles, 3 f	7731
	luteal phase: median 12 d; r9.5-14; n5 cycles, 5 f	7730
	luteal phase: 12.3 d; se0.3; r10-14; n21	7386
	luteal phase: 13 d; ±1.1; r8-17; n5 f	2202
	r23-48 d; n4 cycles, 3 f	7731
	r24-38 d; n1 f	7723
	median 26 d; r25-40; n7 nulliparous f, 24 cycles	3426
	r27-39 d; median 28; n5 cycles, 1 f; 38 d; 1 cycle; n1 f	4358
	28.8 d; median 28; r20-39 d; n25 cycles	11495b
	30 d; ±1.5; r25-36; n5 f	2202, 7385
	31.1 d; sd7.5; median 30.8; n56 cycles, 7 f; via tumescence	7722, 7724
	32 d; se1; r25-42; n22 cycles, 6 f	2203, 7386
	32 d; r24-44; n25 cycles, 7 f; via tumescence	7727
	32 d; r26-47; n53 cycles, 1 f	6758
	32.5 d; sd4.6; median 31.8; n25 cycles, 7 f; via urogenital cleft length	7722, 7724
	median 33 d; r21-49; n19 cycles, 5 f	7730
	33.24 d; sd4.74; se1.15; n17 cycles, 2 f	10832
	45 d; n3 cycles, 1 f	12123
	49 d; without 60, 72 d cycles; r36-72; n8 cycles, 1 f	7945
	1 mo	6183
Pan paniscus	menses 1 d	9513
	swelling phase: 22.4 d	10798
	35 d; r32-36; n3 cycles, 1 f	1068
	36.0 d; n12 cycles, 2 f	9513
	46.0 d	10798
	49 d; sd2.9; n23 intermenstrual intervals, 4 f	2213a, 2213b
Pan troglodytes	menses: 2-3 d	3889
	menses: 2-4 d	2960
	menses: 2-4 d	6171
	menses: 2-6 d	12047
	menses: 3 d	2961, 12016
	menses: 3 d	10864

	menses: 5-7 d	10850
	menses: 6 d	11177
	follicular phase: 10-32.5 d	12017
	follicular phase: 17-45 d	9072
	follicular phase: 20.4 d; se1.0; r15-25	7734
	estrus: 6 d	12016
	genital swelling: 6.5 d	11177
	luteal phase: 12-14 d	5033
	luteal phase: 12-16 d	9072
	luteal phase: 15.0 d; se0.5; r13-18	7734
	r22-43 d; n4 cycles, 4 f	5033
	r27-30 d; n6 cycles, 1 f	11964
	28 d; n2 cycles, 1 f	3066
	28 d; n2 cycles, 1 f	3997
	31 d; r27-37; n9 cycles, 3 f	7008
	32-36 d	10850
	34 d; mode 35; r32-36; n10 cycles, 1 f	4583
	median 34 d; r32-56; n6 cycles, 4 f	9072
	35 d	8443
	median 35 d; r28-53; n263 cycles, 13 f	2960
	35.4 d; se1.2; r29-43; n11 cycles, 7 f	7734
	35.5 d; r26-49; n21 cycles, 3 f	1896
	35.5 d; n50 cycles, 7 f	10864
	36 d; r31-41; n2 cycles, 1 f	2962
	36 d; sd4.82; median 35; mode 35; r29-53; n164 cycles, 11 f	2961, 12015, 12016, 12017
	median 36 d; r25-84; n46 cycles, 9 f	10970, 10972
	36.2 d; n2 f	9513
	37 d	3889
	37.28 d; se0.14; median 35; r22-187; n653 cycles, 22 f	12047
	37.6 d; n31 intermenstrual intervals	2213b
	39.7 d; r28-50; n7 cycles, 1 f	6171
	41 d	11177
Pongo pygmaeus	menses: median 2.8 d; r1-4	7726
	follicular phase 12.7 d; ±2.1	7729
	estrus: 7-10 d; n1	1679
	luteal phase: 13.5 d; ±1.3	7729
	~25 d	1679
	26.3 d; r24-28; sd1.7; n5 cycles, 5 f	7729
	26.8, 28.6 d; r24-32; n6 cycles, 2 f	1982, 3992
	28-29 d	4581
	30.1 d; median 31; r26-37; n11 cycles, 3 f	5766
	median 30.5 d; r26-32; n8 cycles, 3 f	7728
	31.3 d; r22-44; n3 cycles, 1 f; via consortship	3583
	32 d; n4 cycles, 1 f	457
GESTATION		
Gorilla gorilla	236-238 d; n1	4369
	245-247, 249-251 d; n2	7723
	251 d; n1	4979
	252 d; n1	6186
	252 or 267-268 d; n1	1955
	254 or 274 d	10832
	254.7 d; r238-274; n7	2076
	255 d	2199
	255 d; n1	10933
	256 d; n1	6887
	256-259 d; n1	9714
	257-259 d; n1	10786
	257 d; n1	697
	259.7 d; r245-280; n21; 2 outliers dropped	2153
	261 d; n1; hormonal data	6893
	8.5 mo	3426, 4357, 4358
	265-267 d; n1	6758

Pan troglodytes	~5 mo; n1	1889
	167 d; r153-179	4385
	180 d; r167-193; n31; last menses to birth	4385
	214 d; n1	4583
	226.8 d; sd13.3; r196-260; n118	8284
	227.5 d; sd12.6; median 228; r202-260; n48; detumescence to birth	7938
	228.7 d; sd12.7; median 228.5; r202-261; n48; last mating to birth	7938
	229 d; r203-244; n9	6878
	231 d; r202-261; n44	12015
	232 d; r202-261; n13	2962
	236 d; median 236; r232-261	12017
	8 lunar mo	3889
	246 d; n1	10850
	250 d	11964
	253 d; r232-278; n11; menses to birth	10864
	~9 mo	1355
Pongo pygmaeus	233 d	4106
	250 d; n1	3584
	240-259 d; n1	5766
	255-270 d; n1	11048
	255-275 d	8798
	261 d; n1	400
	263 d; n1	4000
	270.7 d; mode 270; n15	2153
	275 d	457
	294.5 d; r292-297; n2	6500
LACTATION		
Gorilla gorilla	solid food: 2.5 mo	9544
	solid food: 3 mo; regularly 5-6 mo	8798
	solid food: 12-14 wk; n1	6186
	wean: est 6-7 mo	4816
	wean: 8 mo	9544
	wean: 2 yr	8798
Pan paniscus	solid food: 3 mo; regularly 5-6 mo	8798
Pan troglodytes	solid food: 4-6 mo	3888
	solid food: 6 mo	9084
	solid food: 10 mo	10679
	wean: 15 d-10 mo	3904
	wean: 10-18 mo	6342
	wean: 18 mo	10679
	wean: 1-2 yr	9084
	wean: 1.5-2.5 yr	3888
	wean: 2-4 yr	11176
	wean: 64.6 mo; r50-86; n5 m	8800
	wean: 63.2 mo; r50-72; n6 f	8800
Pongo pygmaeus	solid food: 6-9 mo; regularly 1 yr	8798
	wean: 3 yr; n1	1679
	wean: 3-4 yr	8798
	wean: 1-2.5 yr	5766
INTERBIRTH INTERVAL		
Gorilla gorilla	3 yr	9544
	3-4.92 yr; shorter if yg die	4360
	3.5 yr; n9	4357
	median 3.83 yr; r3-4; n6 f; yg live	4358
	3.92 yg; r3.0-7.25; n26 ibi; yg live	9969a
	4.16 yr; median 4; r2.33-6.42; n16 ibi, 13 f; yg live	9969a
	median 4.42 yr; r3-7; n8 f; yg die	4358
Pan paniscus	4.5 yg; yg survives	5257
Pan troglodytes	14.5 mo	3904
	2.5-3 yr	3888
	min 3 yr	11176
	3-4 yr	9084

	4.38 yg; r3-7; n13 intervals, 6 f	10537
	4.5-7.5 yr	3889
	5-6 yr; n3 f; yg lives	7924
	5.67 yr; r4.58-7.58; n15 intervals, 12 f	10971
	5.83 yr; r4.33-7.5; n8 f	10970, 10972
	6.0 yr; r5.16-7.08; n7	11415
Pongo pygmaeus	2.5-3 yr	4994
	>5 (6-7)? yr	3583
SEASONALITY		
Gorilla gorilla	birth: yr rd; Virunga Volcanoes, Albert Nat Park, Zaire	9544
	birth: yr rd; zoo	2076
	repro: not seasonal; central Africa	8798
	repro: not seasonal; e Africa	5758
Pan paniscus	birth: not seasonal; Congo R, Zaire	8798
Pan troglodytes	birth: yr rd; captive	7938
	birth: yr rd; lab	6876
	repro: not seasonal; w, equatorial Africa	8798
Pongo pygmaeus	birth: Mar, May; n2; captive	1679
	birth: May; n1; captive	400
	repro: not seasonal; Indonesia	8798

Order Primates, Family Hominidae, Genus *Homo*

Humans, *Homo sapiens*, are bipedal, terrestrial omnivores with a global distribution. Males are larger than females. Females have 2 mammary glands and a simplex uterus. Menstruation and concealed ovulation are characteristic. Neonates are born with open eyes. Review of the extensive literature (5891, 12009a) on human reproduction is outside the scope of this compendium.

NEONATAL MASS		
Homo sapiens	<2500 g; low	2040
	2909-3249 g	8337
	3179 g; se105; n12	1078
	f: 3200 g; sd550; n6693	5541
	3250-3600 g	9285
	m: 3300 g; sd600; n7037	5541
	4070 g; se109; n4 large	1078
LITTER SIZE		
Homo sapiens	mode 1; 2 rare	8244
SEXUAL MATURITY		
Homo sapiens	f: menarche: 12-14 yr	11149
	f: menarche: 13.533 yr; r11.044-16.310; n250	3040
CYCLE LENGTH		
Homo sapiens	28.7 d; sd3.74; 26-28 45%; r of x 17-58; n770 f	4154
	30.4 d; 18-42 97%; r11-144; n747 cycles	3343a
	33.9 d; sd14.28; mode 28; 17-45 87%; r7-256; n3140 cycles, 250 f	3040
GESTATION		
Homo sapiens	259 d	2040
	279.66 d; sd15.41; n7037 f offspring	5541
	280.60 d; sd14.37; n6693 m offspring	5541
	43 wk	1078
LACTATION		
Homo sapiens	0.5-3 yr	8677a
INTERBIRTH INTERVAL		
Homo sapiens	15 mo; r8-38; n50 f no lactation	9519
	27 mo; r8-44; n209 f with lactation	9519
SEASONALITY		
Homo sapiens	birth: Mar-June >50%; Kenya	6400
	birth: yr rd; worldwide	2093

Order Carnivora

Perhaps artificially, carnivores are split into terrestrial forms (fissipeds: dogs, bears, raccoons, genets, mongooses, aardwolves, hyaenas, cats) and marine forms (pinnipeds: fur seals, walruses, seals). The aquatic pinnipeds generally give birth to a single, highly precocial offspring, while the terrestrial carnivores more often give birth to several somewhat altricial young. Gestational delays occur in both aquatic (all pinnipeds) and terrestrial (Ursidae, Mustelidae) forms.

Details of reproductive cyclicity (9350) and pinniped lactation (8037) are available. One journal, Carnivore, focuses on this order.

Order Carnivora, Family Canidae

Foxes, wolves, coyotes, jackels, and dogs (*Alopex* 2-9 kg; *Canis* 5-80 kg; *Chrysocyon* 20-32 kg; *Cuon* 10-20 kg; *Dusicyon* 4-13 kg; *Lycaon* 17-60 kg; *Nyctereutes* 4-11 kg; *Otocyon* 3-5 kg; *Speothos* 5-8 kg; *Urocyon* 2-7 kg; *Vulpes* 1-14 kg) are nocturnal carnivores with a worldwide distribution. Male canids often are larger than females (*Canis* 3571, 3649, 3768, 4645, 5210, 5632, 5761, 7979, 8495, 8501, 9175, 9327, 10499; cf 9945; *Cuon* 5330; *Nyctereutes* 11866; cf 4643; *Urocyon* 5210; *Vulpes* 502, 4253, 5210, 6420, 6520, 7979, 10462, 10499), although the reverse may be true for *Otocyon* (5761, 6743, 9945).

A copulatory tie is common (39, 209, 744, 2210, 2302, 3452, 3453, 3637, 4643, 5131, 5844, 6520, 7147, 10734, 11069, 11144, 11445, 11929, 11984; cf 2171, 5837, 9609). The endotheliochorial, zonary placenta implants into a bicornuate uterus (2479, 6520, 7563) and is usually eaten after birth (7147, 7158, 7710, 7712, 10462, 10499, 10734).

Neonates are furred, but blind (39, 333, 744, 1474, 2440, 3155, 3768, 4643, 5131, 5761, 6305, 9175, 10154, 10229, 10254, 10499, 11163, 11866). Their eyes open at 8-14 d (*Canis* 746, 1474, 3571, 3769, 4643, 4645, 5761, 7147, 7979, 9175, 9769, 10308, 10499, 11144, 11984; *Chrysocyon* 1206, 3005, 4297; *Nyctereutes* 562, 4643, 5130, 7979, 9611, 10499; *Vulpes* 2702, 3571, 3598, 4643, 5725, 6520, 7979, 9175, 10306, 10499; cf 3155, 3815, 5908), 8-17 d (*Speothos* 745, 909, 2734), 9 days (*Urocyon* 5210), 13-15 d (*Cuon* 745, 8798; *Dusicyon* 909, 1205; *Lycaon* 745, 2109, 11163), or 14-16 d (*Alopex* 745, 7979; cf 4643).

Females have 4 mammae (*Otocyon* 10158), 6 mammae (*Urocyon* 1474, 6305, 9885), 6-8 mammae (*Vulpes* 502, 1474, 3134, 3571, 4253, 6305, 9175, 9885, 10158), 8-10 mammae (*Canis* 744, 1474, 3571, 8613, 11144; cf 10158), or 12-16 mammae (*Cuon* 2303, 4643; *Lycaon* 331, 333, 918, 11163). Males have a baculum (1473, 9347).

Details of anatomy (261, 2123, 4295, 5317, 7449), endocrinology (*Alopex* 7447, 7450, 7452, 7453, 7462, 7463, 7981, 10083, 10084, 10085, 10086, 10087; *Canis* 395, 396, 3769, 8186, 9705a; *Lycaon* 11163; *Nyctereutes* 11068; *Urocyon* 6306; *Vulpes* 1100, 1104a, 1105, 1106, 1107, 5318, 5319, 5320, 5323, 6306, 6982, 6983, 6985, 6987, 7458, 7459, 7461, 8131, 8132), spermatogenesis (6982, 10082), conception (8311, 8313), birth (2017a), milk composition (4253, 5190, 6247), and light (3419, 3420, 3421) are available. Information on reproduction of domestic dogs is summarized in supplement 39 of the Journal of Reproduction and Fertility published in 1989.

Note that dingos, wolves, and domestic dogs are all considered *Canis lupus.*

NEONATAL MASS

Alopex lagopus	57 g	10254
	60-85 g	4643
	60-90 g	7979
Canis aureus	201-214 g; 2 d	4643
Canis latrans	154.9 g; sd10.81; n7 near term, 1 lt	5668
	177.8 g; sd6.78; n5 near term, 1 lt	5668
	200-250 g	3769
	275 g; r255-284	746
Canis lupus	300-500 g	4643
	mode 300-400 g; r300-675	3571
	350-400 g	3649
	f: 535 g	10308
	m: 547 g	10308
	m: ~1 kg	6137
	f: ~1 kg	6137
Canis mesomelas	159 g	745
Chrysocyon brachyurus	340-430 g	2578

	350 g; n1 lt	1206
	~500 g	8798
Cuon alpinus	200-350 g	10229
Dusicyon thous	120-160 g	1205
Lycaon pictus	f: 296.6 g; sd40.4; n3 dead, 1 d	11163
	m: 308 g; sd71.9; n5 dead, 1 d	11163
	350-380 g; n8 yg, 1 lt	2440
Nyctereutes procyonoides	60-80 g	10499
	60-90 g	7979
	60-110 g	4643
	60-110 g	562
	f: 90.4 g; r63-125; n11	9626
	m: 103.5 g; r50-150; n16	9626
	109.1 g; r105-115	5130
	122 g	582
Otocyon megalotis	r99.4-142.0; n5 yg, 1 lt	10158
	106.5-140.5 g; term emb	10158
Speothos venaticus	130-140 g; n1 lt, 6 yg	5256
	137.5 g; r126-154; n4	5837
	182.5 g; r175-190; n2, 1 d	5256
	~0.5 kg	2734
Urocyon cinereoargenteus	85 g; n5	11893
	~100 g	9885
	115 g	745
Vulpes bengalensis	52-65 g	39
Vulpes corsac	60-65 g	4643
Vulpes vulpes	50-150 g	4253
	70-100 g	10152
	71-120 g	10461
	80-98 g	10734
	86.9 g; r56-110	4643
	r95-106 g; n6 near term	7071
	99.91 g; sd13.28; n6 term lt, 21 emb	6305
	~100 g	9885
	100-130 g	6520
	f: 105 g; r100-110; n2	7710
	105.13 g; r101.5-110.6; n3 lt, 18 yg	10460
	114 g; n6	2702
	m: 140 g; r90-165; n5	7710
Vulpes zerda	23.75 g; r23.5-24; n2	11325
	m: 26.9 g; r23.1-29; n4	5908
	f: 28.2 g; r26.9-30.7; n3	5908
	r74-104 g; n4, 8 d	7183
NEONATAL SIZE		
Canis latrans	CR: 125 mm	522
	CR: 152.4-165.0 mm	746
Vulpes corsac	135-140 mm	4643
Vulpes velox	TL: 155 mm; r149-165; n8 term emb	2913
Vulpes vulpes	CR: 131 mm	6520
	145 mm	4643
	145 mm	7979
	est 150 mm	5941
Vulpes zerda	HB: 108.5 mm; r108-109; n2	11325
	HB: 115 mm; n1	8426
WEANING MASS		
Nyctereutes procyonoides	400-500 g	5130
	1100-1300 g; 2 mo	562
Vulpes vulpes	400-800 g	5941
	1057 g; r795-1107; n10, 28 d	2702
	~1600 g	9885
	f: 1731 g	6520
	m: 1997 g	6520
WEANING SIZE		
Vulpes vulpes	HB: 250-350 mm	5941

LITTER SIZE

Alopex lagopus	1-13	6546
	2.8 yg; r1-5; n16 dens	836
	3-12; max 18	10499
	mode 4-6; 7-8 occasional	3647
	4-12	1780
	4 yg; n1 lt	10217a
	5 yg; n1	5844
	5-6; r1-14	1474
	5.5; max 10; n4 lt	11041
	6.0; r3-12; n9 lt	5492
	6.4; r4-10; n17 lt	566
	7-10; max 21; 4-6 poor yr; max 26 emb	7979
	7.7 wean; n2030 lt	10052
	8-13 yg; max 22	4643
	8.7; r1-17; n1331 lt	5324
	8.75 yg; n2984 lt, 1401 f	10052
	9.8 yg; n2030 lt	10052
	10.0; n272 lt	5324
	11; max 22	10254
	14.0; n2 lt	10203
	20 emb; r17-23; n2 f	3647
	20; max 25	11042
Canis spp "gray jackals"	5 yg; n1 lt	5905
Canis adustus	2.0; r1-3; n2 lt	12128
	3-6	5761
	4; mode 5; r2-6	12128
	4-6	11144
	5; r3-7; n2 lt, 1 f	4180
	5.4 emb; r4-6; n5 f	10154
	7 emb	10158
	12 emb	9945
Canis aureus	1-5	5761
	2; r1-3	11965
	2.7; r2-4; n3 lt	12128
	3 emb; n1 f	8613
	3.5; r3-4; n2 lt	8495, 8501
	4-5	4645
	4-5	9175
	4-9	7979
	4.5 yg; r3-7; n6 lt	9066
	5; r3-8	4643
	6 emb ea; n2	10616
Canis latrans	1 yg; n1 lt	11645a
	2-15	3769
	3.3; r1-7; n19 lt	2109
	3.8 ps; 95%ci 1.7; n9 yearling f	10857
	3.8 emb; n26 yearling f	10856
	4.45; sd0.33	1560
	4.7 emb; n52 adult f	10856
	4.8 ps; 95%ci 0.9; n17 adult f	10857
	4.9 cl; 95%ci 1.5; n14 yearling f	10857
	mode 5-7 yg; usually 4-8; 9-12 not uncommon; 17, 19 recorded	5210
	5-10	1474
	5 emb; n1 f	522
	5.0; r3-7; n2 lt	5844
	5.0 yg; r3-6; n3 lt	3793
	5.0 yg; r2-8; n6 lt	9066
	5.2 cl; n41 yearling f	10856
	5.4 emb, ps, or yg; sd1.8; n20 lt	436a
	5.5; sd1.3; per pack; 5.3; sd1.5; per pair	747
	5.6 cl; n17, 2 yr f	5667
	5.70 yg; n1582 dens	4295

	6-7 yg; max 11	5146
	6.0; r5-7; n2 lt	5668
	6.13 emb, yg; r4-7; n13 lt, 73 emb, 7 yg	10954
	6.2 cl; n9, 3 yr f	5667
	6.23 emb; n1330 f	4295
	6.3 yg; n20 lt	2136
	6.6-8.0 cl; 7.1; r6.1-7.9	5276
	6.9 cl; 95%ci 1.3; n26 adult f	10857
	6.9 cl; n63 adult f	10856
	7 emb; n1 f	5405b,
	7.1 cl; n6, 4 yr f	5667
	8.4 cl; 6.15 ps; n23 f	3768
	9 yg; n1 den	495
Canis lupus	2.8-6.5; r1-11	7147
	max 3; n3 lt	7236
	3-5; young f; r6-8; adult f; occasionally 10-13	7979
	3-8	4643
	3-12	4645
	3.0; r3-5; n3 lt	11984
	3.4; r1-5; n5 lt	3714
	3.4 yg; r1-6; 7 lt	6239
	3.7; r1-9; 6 lt	2109
	3.8; r1-7; n15 lt	2109
	mode 4-6; r1-8	3571
	4.0 yg; se0.46; n15 lt	1663a
	4.1; mode 4; r1-8	12128
	4.3 ps; 95%ci 0.9; n7 f	3620
	4.33 yg; n316 lt	7149
	4.5; r2-10	6382
	4.5 yg; r2-7; n20 lt	3526
	4.6 emb; 95%ci 0.7; n5 f	3620
	4.7 yg; r1-8; n7 dens	502
	4.7 yg; se0.25; n48 lt	1663a
	4.75; r4-6; n4 lt	12001
	4.8; 3.1 weaned	3526
	4.8; r4-5; n6 lt	3649
	4.8; r1-8; n26 lt	1230
	4.9 yg; r3-7	8547a
	mode 5; r1-8	12128
	mode 5-8; r3-12	10499
	5 yg; n1 lt	2496
	5 ps; r3-7; n5 lt	8415
	5.0 yg; se0.41; n14 lt	1663a
	5.2 emb; r1-9; n30 f	6239
	5.4 cl; 95%ci 0.8; n9 f	3620
	5.41 emb; 6.08 cl; 2 yr old f	8964
	5.5 emb, ps; sd1.4; r2-9; n48 f	5402
	5.6; se0.67; r2-8	8695,
	5.7 emb; se0.22; n44 f	1663a
	mode 6-7; r3-14	1474
	6.4 ca; se0.6; 5.5 yg; r2-9; per pack	543
	6.4 yg; r4-9; n8 lt	10371a
	6.5 emb; 7.3 cl; adult f	8964
	6.6 ps, yg; r4-8; n9 lt	3556a
	6.6 emb; 6.8 cl	3620
	6.6 emb; n18 f	8964
	6.8 cl; n56 f	8964
	~7 yg	5146
	7 emb; n1 f	3526
	7.1 ps; n45 f	8964
Canis mesomelas	1-2	8514
	2-8	9327
	mode 3; r1-6	10158
	mode 3-4; r2-6	9945

	3-4; r2-7	11965
	3; r2-4; n4 lt	12128
	4-10	11144
	4; max 9	5761
	4.3; r1-8; n16	1230
Canis rufus	2-7	10809
	2.6; r1-4; n5 lt	12128
	4-7	1474
	6.2 emb; r2-10; n29 f	8227
Chrysocyon brachyurus	1.6; r1-2; n5 lt	1206
	1.67; r1-3; n3 lt	2210
	2; n1 lt	4297
	2.47; r1-5; n17 lt	2577, 2579
	3.0; r2-4; n3 lt	3180
	4 ea; n2 lt	6499
	4 ea; n2 lt	3003
	max 7	1648
Cuon alpinus	2 or 4; r2-6	6654
	2-6 yg	1478
	3 yg; n1 lt	4466
	3.5; r2-5	12128
	4-5	10229
	~8	5330
	8 emb; n1 f	1960, 1961
	8 emb; r7-9; n2 f	1478
	9 emb; n1 f	61
Dusicyon culpaeus	5.16 emb; n6 f	2129
Dusicyon gymnocercus	2.5; r1-4; n2 lt	12128
	3.35 emb; se0.40; r1-8; n20 f	2129
Dusicyon thous	3 yg; n1 lt	1968
	r3-6; n4 lt	1205
Dusicyon vetulus	4 yg; n1 lt	1968
Lycaon pictus	2-6	11145
	2-8	331, 333
	2-19	5761
	4 yg	1521
	4-6	5277
	6; n1 lt	11616
	6.0 yg; r1-9; n3 lt	2440
	6.5 yg; r6-7; n2 dens	334
	6.33; r4-11; n3 lt	2109
	mode 7; r2-12	2449
	mode 7-10; r2-19	10158
	7.0; r4-12; n3 lt	2171
	7.11 yg; r2-12; n18 lt	918, 1230, 1231
	7.6 yg; r6-9; n5 lt	9066
	8.1 yg; sd1.88; n20 lt	11163
	9.0; n2 lt	5221
	10.1 yg; r6-16; n15 lt not birth	3453
	11 yg; r4-16; n4 dens	9546
	16 emb; n1 f	2284
	max 16	743
Nyctereutes procyonoides	3.0 yg; r2-5; n7	8086
	mode 4-6; r4-12	5131
	4.0 yg; r2-5; n4 lt	5130
	r4.9-7.9; mode 6-7; max 15-16; n7 regions	562
	5-7	4643
	mode 5-7; max 14-19	7979
	5-8	9609
	5.0 yg; r2-8; n2 lt	2109
	5.0; sd2.3; n9 lt	11069
	5.49; farm	6739
	5.5-7.0 yg	11067
	6-8; max 19	10499

	6.33 yg; r1-10; n6 lt	9626
	7-8; r5-12	9611
	7; n1 lt	12128
	9.4; r5-14; n8 lt	11866
	10.2 emb; r7-15; n10 f	581, 582
	11 normal +2 resorbed emb	579
Otocyon megalotis	2-5	5761
	3.2 cubs; n5 lt	6743
	4 yg; n1 nest	9945
	5 yg; n1 lt	9284
	5.0 emb; r4-6; n4 f	10153, 10154, 10158
Speothos venaticus	3 yg; n1 lt	5837
	3.8 yg; r1-6; n32 lt	8689
	5.0; r1-9; n13 lt	2734, 2735
	6 yg; n1 lt	2109
	6 yg ea; n3 lt	5256
Urocyon cinereoargenteus	2 emb; n1 f	4249
	2 yg; n1 lt	9066
	3 emb; n1 f	11228a
	3 yg ea; n3 lt	7887
	3.49 ps; mode 4; r1-9; n85 f includes juveniles	9267
	3.59 ps; r1-9; n41 adult f	9267
	3.62 ps; se0.15; r2-5; n32 f	6299
	3.66 emb, ps; r1-7; n35 f	9885
	3.80 ps; n20 f	10542
	3.9 ps; se0.2; n44 f	9101
	3.96 emb; se0.24; r2-6; n24 f	6299
	4-6	190
	4-6	9164
	4.1 ps; n29 f	3767
	4.1 emb; n18 f	3767
	4.16 emb; n6 f	10542
	4.31 ps; sd1.01; n67 f	11893
	4.4 ps; r3-7; n32 f	6306
	4.42 ps; r3-5; n12	11669
	4.5 emb; r3.7; n10 f	6306
	4.90 emb; sd1.12; n40 f	11893
	5.2 cl; n30 f	3767
	5.2 cl; r2-9; n47 f	6306
	5.67 cl; r4-7; n3	11669
	5.67 cl; n3 f	11669
	4.42 ps; r3-5; n12 f	11669
Vulpes bengalensis	3 emb; n1 f	7382
	4; n1 lt	39
Vulpes chama	1 emb; n1 f	10154
	r1-4; n5 lt	1230
Vulpes corsac	est 2-6, occasionally larger	2280
	2-6	4645
	2-11	10499
	2-11; 16 rare	7979
	5.0; mode 3-6; r2-16; n100	4643
	6.1; r3-9; n28 lt	9966
	16 emb; n1 f	10499
Vulpes ferrilata	2-5 yg	7983
Vulpes macrotis	2-7	6387
	4 emb; n1 f	502
	4.0 yg; mode 4; r3-5; n5 lt	7525
	4.6 yg; r3-7; n6 f	8014
	4.55 yg; r4-5; n11 lt	2905, 2908
Vulpes rueppelli	2 yg, n1 lt	9175
Vulpes velox	3-6 yg	1871a
	3.5 emb; r3-4; n4 f	5725
	4 emb ea; n2 f	5405b,
	4.25 yg; r3-6; n4	5725

	4-7	1474
Vulpes vulpes	1-4	397
	2; n1 lt	12128
	2-5 yg	6420
	3 yg; n1 lt	8720
	mode 3-5; r3-12	4253
	r3.0-6.6; n345 f	3044
	r3.1-5.8 live emb; n136 f	197
	3.3; r2-5	12128
	3.5; r3-4; n2 lt	7710
	3.5 yg; r1-10; n384 lt	10461
	3.5-5	11358
	3.7 ps; sd1.5; mode 3; r1-9; n142 lt	9390
	3.8 emb; r1-6; n8 f	7071
	3.8 yg; r1-12; n175 lt	10461
	4-6; max 13	4643
	mode 4-8; occasionally >10	9164
	4 yg; r3-5; n2 dens	5405b,
	4.0 emb; sd1.6; mode 3; r1-8; n35 f	9390
	4.05 yg; n333 lt	4153
	4.16; se0.15; sd1.52; r1-8; 7-10 d; n100 yearling f	8310
	4.25 yg; r1-5; n6 lt in den	7071
	4.29 emb; n37 f	6518
	4.3 ps; r1-8; n38 f	7071
	4.3; n436 lt	391
	4.43 yg; r3-6; n7 dens	9640
	4.5 cl; sd1.5; r2-9; 4.2 ps; sd1.6; r1-9; n75 f	9390
	4.52 yg; sd1.35; mode 4-5; r1-8; n793 lt, 361 f	5322, 5323, 5324
	4.59; se0.10; sd1.56; r1-9; 7-10 d; n254 adult f	8310
	4.6 emb; r1-10; n35 f	6306
	4.64 ps; se0.13; r1-10; n146	4395
	4.71 emb; n14 f	6518
	4.72 ps; sd1.55; n252 f	4397
	4.76; mode 5; r1-10; n108 lt	10511, 10515, 10518
	4.76 ps; sd1.52; n192 f	4397
	4.8; r3-7	466
	4.9 yg; ±0.9; n19 lt	8131
	4.92 yg; n210 lt	10606
	5 yg; r3-7; n10 dens	2441
	5.0 cl; 3.9 ps; n27 f	7071
	5.0 emb; n38 f	5940
	5.0 ps; n41 f	7832
	5.1 yg; se0.046; n1809 lt	9633
	5.1 yg; se0.3; n25 lt	9101
	5.1 ps; se0.2; n103 f	9101
	5.36 emb; se0.18; max 8; n73 f	3155
	5.37 emb, ps; r1-9; n95 f	9885
	5.4 ps; r1-12; n541 f	6306
	5.41 ps; se0.15; n114 f	3155
	r5.48-7.50 cl; r3.13-6.50 emb; mode 5; n136 f	198
	5.7 cl; n6 f	9347
	5.9 ps; n≥38	5940
	5.9 cl; r2-12; n590 f	6306
	5.99 cl; se0.16; max 12; n76 f	3155
	6 ps; n24 f	3767
	6.0 cl; n2 f	7832
	6.1 yg; ±0.3; n23 lt	8131
	6.5 emb; n2 f	7832
	6.7 emb; n7 f	3767
	6.7 emb; r3-12; n48 f	10461
	6.8 emb; se0.338; r4-13; n30 f	4882
	6.8 emb; r2-9; n34 f	10461
	7; n1 lt	10761
	7.1 ps; sd1.9; r4-12; n29 f	10461
	7.2 cl; n15 f	3767

	11 emb; n1 f	10606
	max 12 ps; max 8 live emb	197
	17 yg; n1 den	4912
Vulpes zerda	1 yg; n1 lt	8426
	1.5; r1-2; n8 lt	5908
	1-2	3637
	2.0 yg; r1-3; n2 lt, 1 f	9439
	3.0 yg; n2 lt	3598
	3-4	8223
SEXUAL MATURITY		
Alopex lagopus	f: 1st conceive: 1 yr	1780
	m: 1st sperm: 9-10 mo	4643
	m: testes: large 9 mo	10083
	m: testes: large 10 mo	1780
	f/m: 1st mate: 9 mo	745
	f/m: 9-10 mo	7979
	f/m: end 1st yr	4643
	f/m: 2nd yr	10499
Canis adustus	f: 1st mate: 9 mo	745
	m: 1st mate: 9 mo	745
Canis aureus	f: 10-11 mo	4643
	f: 11 mo	5761
	f: 1st mate: 9 mo	745
	m: 1 yr	5761
	m: 21-22 mo	4643
	m: 1st mate: 9 mo	745
	f/m: 2nd yr	7979
Canis latrans	f: 1st mate: 9 mo	745
	f: 1st mate: 10 mo	3769
	f: 1st mate: 10 mo	5666
	f: 1st mate: 1st yr	9769
	f: 1st birth: ~1 yr	1474
	m: epididymal sperm; 43-45 wk	5666
	m: 1st mate: 9 mo	745
	m: 1st mate: 20 mo	3769
Canis lupus	f: 1st estrus: <2 yr	5402
	f: 1st estrus: 342.4 d; 95%ci 15.7; n118	10148
	f: 1st mate: 12 mo	745
	f: 1st mate: 20 mo	745
	f: 1st mate: 22 mo	8415
	f: 1st birth: 2nd yr	1474
	f: 1st birth: 2nd yr	4645
	f: 1st birth: 2 yr	7236
	f: 22 mo	6382
	f: end 2nd-early 3rd yr	4643
	f: 2 yr	7979
	f/m: 1st mate: 22 mo	8964
	f/m: 1st mate: 22 mo	7147
	f/m: 1st repro: ~12 mo	7158
	m: sperm: ≥10 kg	5402
	m: 1st mate: 12 mo; dingo	745
	m: 1st mate: 20 mo	745
	m: 1st mate: 2 yr	11929
	m: 22 mo	6382
	m: 2-3 yr	7979
	m: end 2nd-4th yr	4643
	f/m: 11-24 mo	3571
Canis mesomelas	f: 1st mate: 7 mo	745
	f: 1st mate: 12-14 mo	11144
	m: 1st mate: 6 mo	745

Cuon alpinus	f: 1st mate: 12 mo	745
	m: 1st mate: 12 mo	745
Lycaon pictus	f: 1st estrus: 23 mo	11163
	f: 1st birth: 22-39 mo	3452
	f: 2 yr	9008
	m: 1st mate: 1.75-5 yr	3453
	m: 1st mate: 11 mo; n1	11163
Nyctereutes procyonoides	f: 1st mate: 10 mo	745
	f: 1st birth: 2 yr	9626
	m: 1st mate: 10 mo	745
	f/m: 1st mate: ~1 yr	9611
	f/m: 8-10 mo	4643
	f/m: 8-10 mo	562
	f/m: 9-11 mo	7979
	f/m: 9-11 mo	10499
	f/m: 10-11 to 22-23 mo	8798
Speothos venaticus	f: 1st mate: 12 mo	745
	f: 1st conceive: 10 mo	8689
	m: 1st mate: 12 mo	745
Urocyon cinereoargenteus	f: 1st mate: 12 mo	745
	f: 10 mo	3529
	f: 1st yr	11893
	m: 1st mate: 12 mo	745
	f/m: ~1 yr	11669
Vulpes corsac	f: 1st repro: 3 yr; n1	2280
	f: 9-10 mo	4643
Vulpes vulpes	f: 1st mate: 10 mo	10462
	f: 1st mate: 10 mo	8310
	f: 9-10 mo	7071
	f: 9-10 mo; mode end 2nd yr	4643
	f: 10 mo	9347
	m: spermatogenesis: 9 mo	5318
	m: 1st mate: 10 mo	745
	m: ~9 mo	9347
	f/m: 1st mate: 10 mo	6520
	f/m: 1st mate: 1 yr	4153
	f/m: after 7 mo; 1st yr	10462
	f/m: 9-10 mo	4253
	f/m: 9-11 mo	7979
	f/m: 10-12 mo	3571
	f/m: 2nd yr	10499
Vulpes zerda	f: 1st mate: 9 mo	745
	m: 1st mate: 9 mo	745
	f/m: sexual activity: 1 yr; n1 pair	9439
CYCLE LENGTH		
Alopex lagopus	estrus: 3-5 d	7453, 7462
	estrus: 4-5 d	4643
	estrus: 12-14 d	7979
Canis aureus	estrus: 3-4 d	4643
Canis latrans	proestrus: 2-3 mo; n20 f	5667
	estrus: 10.1 d; r4-15	5666
	estrus: 10.2 d; n20 f	5667
Canis lupus	estrus: 7-9 d	4535
	estrus: ~2 wk	4643
	6 mo	12001
	218 d; 95%ci 4.0	10148
Canis mesomelas	estrus: ~7 d	11144
Chrysocyon brachyurus	estrus: 5 d	8798
	estrus: 7.8 d	2577
Lycaon pictus	proestrus: 14 d; n29 cycles	11163
	estrus: 20 d; n29 cycles	11163
Nyctereutes procyonoides	proestrus: 7.6 d; sd3.5; r2-14; n35 cycles	11067, 11069
	estrus: 3.9 d; sd1.2; r2-6; n35 cycles	11067, 11069
	estrus: max 6 d	4643

	20-24 d	4643
Speothos venaticus	estrus: 4.1 d; r1-12; n22	8689
Vulpes vulpes	proestrus: ~10 d	4253
	proestrus: 14.4 d; ±3.5	8131
	proestrus: 18.8 d; ±3.0	8131
	estrus: mode 1-4 d; sometimes 5-6 d	4153
	estrus: 1-5 d	5322
	estrus: 1-6 d	8310
	estrus: 1-6 d	10461
	estrus: 2-3 d	8131
	estrus: 2-3 d	4253
	estrus: 2-4 d	10312
	estrus: 3 d; r1-5	8311
	estrus: 3-8 d	7979
	estrus: 7-9 d	4535
	10-12 d	10312
Vulpes zerda	estrus: 1-1.5 d	5908
	estrus: 1-2 d	3637
GESTATION		
Alopex lagopus	46-58 d	6546
	50 d; n1	5844
	51-52 d; r49-57	7979
	51-54 d	1474
	51-57 d	10254
	52 d	1780
	52 d	566
	52 d	4643
	52-53 d	10499
	53.66 d; mode 54; r47-62; n2030	10052
	53-54 d; n9	7447
Canis adustus	57-70 d	5761
	60 d	4180
	60-70 d	10153
	70 d	11697
Canis aureus	~60 d	4645
	60-62 d	7979
	60-63 d	8495, 8501
	60-63 d	4643
	2 mo	5761
	63 d	7712
	63 d	9175
Canis latrans	50-65 d	2088
	59.5 d; r58-61; n2	5844
	~60 d	3768
	2 mo	257
	60-63 d	1474
	60-63 d	9769
	62 d; n2	11645a
	~63 d	747
	est 63 d; r60-65	5666, 5668
	63 d; r58-65	744
	63-65 d	6387
	64 d	1355
	65 d; n1	3594
Canis lupus	~2 mo	5632
	mode 60-65 d; r56-70	3571
	60-66 d; n1	7625
	2 mo; n1	7158, 8185
	62-65 d	7979
	62-65 d	10499
	62-75 d	4643
	62 d; se3-4	11929
	62.4 d; r56-70	10308
	63 d	5146

	63 d	7712
	9 wk	1355
	9 wk	1474
	63-65 d	4645
	65.5 d; r59-73; n4	5844
	68 d	9175
	68 d; n1	11984
	2.5 mo	10698
Canis mesomelas	60-65 d	11144
	~2 mo	5761
Canis rufus	60-62 d; n1	8227
Chrysocyon brachyurus	56-64 d	6499
	~2 mo	209
	60-62, 60-65, 70 d; n3	3180
	62 d	3003
	65.5 d; r65-66; n2	2210
	r64-66 d; n4	1206
	66-67 d	4297
Cuon alpinus	60-62 d	5330
	60-62 d	10229
	60-63 d	1961
Dusicyon culpaeus	est 55-60 d	2129
Dusicyon gymnocercus	58 d	745
Dusicyon thous	56 d; r52-59; n4	1205
Lycaon pictus	2 mo	5277
	60-80 d	5761
	69 d ea; n2	2440
	69.3 d; r69-70; n3	5221
	72-73 d	1521
	72.4 d; sd1.6; n7	11163
	80 d; mode 80; r78-86; n6	2171
Nyctereutes procyonoides	59 d; r51-70	562
	59 d; r58-61	8273
	59 d; r59-70	4643
	59-64 d	7979
	69 d; r59-79; n2	9626
	~60 d	11068
	~60 d; temperate zone, ~max 79 d; north	8798
	~60 d; field data	581, 582
	60-64 d	10499
	60-65 d	5131
	61.0 d; sd2.0; r58 or 59-64; n9	11067, 11069
	61 d	9609
	61-63 d	9611
	61.7 d; r60-63; n6	8086
	~64 d	4645
Otocyon megalotis	60-70 d	10153
	70 d	11697
	75 d; n1	9284
Speothos venaticus	65 d	2109
	67 d	8689
	67-70 d	2734
	76-83 d; n1; last vs 1st mating to birth	5837
Urocyon cinereoargenteus	~51 d	1474
	~60 d	9164
	63 d	6387
Vulpes bengalensis	50-51 d	39
Vulpes chama	51-52 d	9945
Vulpes corsac	49-50, 50, 51 d; n3	2280
	50-60 d	10499
	50-60 d	4645
	52 d	4643
	est 60 d	7979

Vulpes vulpes	implant: 8-10 d	4643
	implant: 10-14 d	6520
	30-50 d	5323
	mode 51-52 d; r49-55	4253
	51 d	9347
	51 d	9164
	51.99 d; sd0.88; se0.036; mode 52; r49-55; n587	5322
	52 d	3044
	52 d; n1	7710, 7712
	52 d; r49-58	4643
	mode 52 d; r51-56; n171 f	10152
	52-53 d	8131
	52-54 d; field data	10511
	55-60 d	397
	~2 mo	2702
Vulpes zerda	50 d; n1	11325
	50.5 d; r50-51; n2	5908
	50.5 d; r50-51; n2	7952
	51 d; n1	8426
	51.0 d; r50-52; n2	9439
	62.5 d; r62-63; n2	3598
LACTATION		
Alopex lagopus	den emergence: 3 wk	7979
	den emergence: 3 wk	10499
	den emergence: 3-4 wk	4643
	solid food: 4 wk	7979
	solid food: 6 wk	4643
	nurse: 1-2.5 mo	4643
	nurse: 6 wk	10499
	nurse: 1.5-2 mo	7979
Canis adustus	wean: 6 wk	745
Canis aureus	solid food: 15-20 d	4643
	den emergence: 2-2.5 mo	4643
	wean: 50-70 d or 90 d; n2 locations	4643
Canis latrans	den emergence: 2-3 wk	744
	den emergence: 3-4 wk	747
	solid food: 12-15 d	3768, 3769
	solid food: 3 wk	744
	solid food: 5 wk; n1 yg	11645a
	solid food: < 6 wk	4398
	wean: 5-7 wk	744
	wean: 6-7 wk	747
	wean: 2-4 mo	3768, 3769
Canis lupus	den emergence: ≤3 wk	1871a
	den emergence: 21 d	7147
	den emergence: 3 wk	4643
	den emergence: 3 wk	7979
	den emergence: 3 wk	9638
	solid food: 3 wk	6137
	solid food: ~23 d	7147
	solid food: 3-4 wk	4643
	nurse: 4-6 wk	7979
	wean: 1.5 mo	4643
	nurse: 34-51 d	6137
	wean: 5-6 wk	10499
	wean: 5-8 wk	7147, 7148
	wean: 38-39 d	9638
	nurse: 8-10 wk	3714
	wean: 4 mo	11984
	catch own food: 7-8 mo	8793
Canis mesomelas	den emergence: 14 d	11144
	den emergence: 3 wk	7412
	solid food: 2-3 wk	11144
	wean: 8-9 wk	7412

Chrysocyon brachyurus	wean: 17 wk	3003
	wean: 4-7 mo	1206
Cuon alpinus	solid food: <4 wk	5330
	solid food: 2 mo	10229
	wean: 58 d	5330
Dusicyon thous	solid food: 20 d	1205
	solid food: 30 d	848
	wean: 90 d	1205
Lycaon pictus	den emergence: 7 d	11253
	den emergence: 3.4 wk	11163
	solid food: 14 d	10158
	wean: 3 mo	10158
Nyctereutes procyonoides	den emergence: 25-30 d	4643
	solid food: 18-20 d or 25-30 d	7979
	solid food: 21-30 d	562, 4643
	wean: 30 d	5130, 5131
	wean: 30-40 d	11452
	wean: 45-60 d	562, 4643, 10499
	wean: 10 wk	9609
	nurse: 80 d	5130
Otocyon megalotis	solid food: 1 mo	5761
	wean: shortly after 1 mo	5761
	wean: 4 wk	10158
Speothos venaticus	wean: 4 wk	745
	nurse: 4-5 mo	2735
Urocyon cinereoargenteus	solid food: est 2-3 wk	5210
	solid food: ~5 wk	9164
	wean: 12 wk	745
Vulpes corsac	solid food: 28 d	4643
Vulpes velox	wean: 6-7 wk	5725
Vulpes vulpes	den emergence: 3-4 wk	4643
	den emergence: 24 d	4253
	den emergence: 6 wk	6520
	den emergence: ~6 wk	7979
	den emergence: 1.5-2 mo	397
	solid food: 20 d	4253
	solid food: 3-4 wk	2702
	solid food: 4 wk	6686
	solid food: 4 wk	3571
	solid food: 1 mo	10461
	solid food: 6 wk	9164
	solid food: 8 wk	6520
	wean: ~5 wk	6686
	wean: ~6 wk	3815
	wean: 1.5 mo	4643
	wean: 1.5 mo	7979
	wean: 6-8 wk	6520
	wean: 7-8 wk	3571
	wean: 8-10 wk	9885
Vulpes zerda	wean: 61-70 d	5908
INTERLITTER INTERVAL		
Alopex lagopus	1 yr	7979
Canis aureus	1 yr	5761
Canis latrans	1 yr	3769
	1 yr	744
Canis lupus	2 lt in 1 yr; n1	11083
	1 yr per pack	8415
Canis mesomelas	0.5-1 yr	11144
Cuon alpinus	1 yr	4643
Dusicyon culpaeus	1 yr	2129
Dusicyon thous	8 mo; n1	1205
Lycaon pictus	11 mo; r9-12 mo; n5	2171
	11.7; se0.6; n9	11163
	12-14 mo; min 6 mo if litter dies	3452

Nyctereutes procyonoides	1 yr	562
Otocyon megalotis	max 2 lt in 1 yr	5761
Speothos venaticus	249 d; r152-302; n7	8689
Vulpes bengalensis	1 yr	9175
Vulpes corsac	1 yr	10499
	1 yr	4643
Vulpes vulpes	1 yr	7071
	1 yr	10761
	1 yr	9347
	1 yr	4253
	1 yr	6520
Vulpes zerda	1 yr; yg live	5908
SEASONALITY		
Alopex lagopus	spermatogenesis: Jan-July; Canada, AK	1780
	testes: large Feb-Apr, small Aug-Oct; farm	10084, 10086
	estrus: Feb-Mar; USSR	7979
	mate: Feb-May; USSR	4643
	mate: Mar-Apr; n AK	1780
	mate: Mar-Apr; captive	7453, 7462, 10086, 10087
	mate: Apr-early May; w Siberia; later e Siberia	10499
	mate: Apr-May; Siberia	1780
	mate: end Apr-early May; Finnish Lapland, Finland	5492
	birth: Mar-May; USSR	7979
	birth: Apr-June; n Canada	1474
	birth: end Apr-early July; USSR	4643
	birth: May-June; Commander I, Aleutians	566
	birth: May-June; captive	10083
	birth: May-July; N America	1780
	birth: June; n1; Finnish Lapland, Finland	5492
	birth: peak June; AK	836
Canis adustus	mate: yr rd, peak Sept-Nov; Transvaal, S Africa	10153
	birth: end Mar-mid June; n11; zoo	12128
	birth: June-July, Sept; Uganda	5761
	birth: Aug-Jan; S Africa	10158
	birth: Dec-Apr; Angola	11697
	birth: June; S Africa	11144
Canis aureus	mate: mid Jan-Feb; USSR	7979
	mate: mid Jan-Apr; Europe, w USSR	4645
	mate: Feb-early Mar; USSR	4643
	mate: spring-summer; Pakistan	9175
	birth: end Mar-early May; USSR	4643
	birth: Apr-early May; Europe, w USSR	4645
	birth: Apr-mid May; USSR	7979
	birth: Dec-Apr; Tanzania	5761
	lact f: Mar-May; Nepal	7382
	yg found: yr rd; Bangladesh	8613
	wean: June; USSR	7979
	wean: mid July-early Aug; USSR	4643
Canis latrans	spermatogenesis: mid Oct-mid Mar; captive	5666
	mobile sperm: Dec-May; captive	4033
	testes: large Nov-Mar; AR	3793
	estrus: Jan-Mar; WY	1871a
	estrus: Mar; captive	5844
	estrus: Mar; captive	5666
	mate: Jan-Feb; WY	747
	mate: Jan-Feb; N America	1474
	mate: Jan-Feb; San Patricia Co, TX	257
	mate: Jan-early Mar; KS	3768
	mate: Jan-mid Mar; ID	6249
	mate: peak Feb; Canada	9769
	mate: Feb; Jasper ALB, BC, Canada	2088
	mate: Feb; WI	5210
	ovulate: Feb-Mar; KS	3768
	preg f: Jan-May; OR	4295

	preg f: Jan-June; AZ	4295
	preg f: Feb-Apr; WY	4295
	preg f: Feb-May; ID	4295
	preg f: Feb-May; NM	4295
	preg f: Feb-May; NV	4295
	preg f: Feb-May; WA	4295
	preg f: Feb-June; CA	4295
	preg f: Feb-June; n1; SD	4295
	preg f: Feb-June; n2; CO	4295
	preg f: Feb-July; n1; UT	4295
	preg f: Mar; n1; NE	$5405b_1$
	preg f: Mar-May; MT	4295
	preg f: Mar-May; TX	4295
	preg f: mid Mar-Apr; ID	6249
	preg f: Apr; n2; Coahuila, Mexico	522
	preg f: end Apr-May; Black Hills, SD, WY	10954
	birth: Mar-Apr; NM	502
	birth: Mar-July; Chiapas, Mexico	3633
	birth: end Mar-May; KS	3768
	birth: Apr; WI	5210
	birth: Apr; San Patricia Co, TX	257
	birth: Apr-May; N America	1474
	birth: end Apr; MAN, Canada	2136
	birth: May; WY	747
Canis lupus	testes: large Mar, small Jan; Russia	3649
	testosterone: low July-Jan; captive	1663
	estrus: Jan-Feb; captive	396
	estrus: Jan-Apr; WY	1871a
	mate: Jan; n Tibet	8740
	mate: Jan-Mar; Canada	1474
	mate: Feb-Mar; captive	12103
	mate: Feb-Mar; MN, AK	8185
	mate: Feb-Mar; WI	5210
	mate: Feb-Mar; w US	5146
	mate: Feb-Mar; central, n USSR	7979
	mate: Feb-Apr, peak Mar; AK	8964
	mate: end Feb-?; MI	7147
	mate: Mar; AK	7625
	mate: end Mar-early Apr; AK	5632
	mate: Apr-May; n central Australia	1663a
	mate: Apr-June; captive	1663
	mate: Dec; n1; zoo, India	2496
	mate: Dec-Mar, peak Feb; zoo	11929
	mate: Dec-Mar; south 1-1.5 mo before north; USSR	4643
	mate: end Dec-early Jan; USSR	4645
	mate: end Dec-Feb; s USSR	7979
	mate: end Dec-Mar; Germany	3571
	mate: end Dec-Apr; USSR	10499
	birth: Feb-Mar; USSR	4645
	birth: end Feb-mid Apr; s USSR	7979
	birth: Apr-May; central, n USSR	7979
	birth: Mar-May; Germany	3571
	birth: Mar-Sept; n25; zoo	1230
	birth: Mar-Sept, peak June-Aug; VIC, Australia	5402
	birth: Apr-May; Canada	1474
	birth: Apr-May; MN, AK	8185
	birth: Apr-May; WI	5210
	birth: Apr-early June; USSR	4643
	birth: early May; AK	7625
	birth: peak May; Russia	3649
	birth: May; AK	8415
	birth: May-June; NWT, Canada	6137
	birth: mid May; n1; zoo	9638
	birth: mid May-early June; captive	6382

	birth: end May-early June; AK	5632
	birth: June-July; n central Australia	1663a
	birth: June-Aug; captive	1663
Canis mesomelas	mate: Apr-Aug, peak May-July; S Africa	11144
	mate: mid-Aug-mid Nov; zoo	1230
	birth: June-Aug 79%; Natal, S Africa	9327
	birth: July-Nov, peak Oct; S Africa	11144
	birth: July-Sept; Kenya, Uganda, n Tanzania	5761
	birth: Aug-Oct; ne S Africa	3145
	yg seen: July-Oct; S Africa	8514
Canis rufus	mate: Jan-Feb; se TX, sw LA	9134
	preg f: Mar-mid May; AR	8227
	preg f: Mar-mid May; TX	8227
	preg f: mid Mar-mid Apr; OK	8227
	birth: Mar-Apr; se TX, sw LA	9134
	birth: Apr-May; s central US	1474
Chrysocyon brachyurus	estrus: Oct-Jan; captive	1206
	mate: Apr; Geraes State, Brazil	2579
	mate: end Oct-mid Nov; zoo	209
	birth: early Jan; zoo	209
	birth: Jan-Feb; n2; zoo	3003
	birth: Aug; Geraes State, Brazil	2579
Cuon alpinus	mate: Jan-Feb; zoo, China	4643
	mate: Feb; zoo	10229
	mate: Sept-Jan; India	1961
	birth: Jan-Feb; Nepal	7382
	birth: Jan-Mar; India	988a, 6654
	birth: Apr; zoo, China	4643
	birth: Nov-Dec; India	61
	birth: Nov-Mar; India	2303
	yg seen: Dec; Karnataka, India	5330
Dusicyon culpaeus	estrus: Aug-Oct; Argentina	2129
	birth: Oct-Dec; Argentina	2129
Dusicyon gymnocercus	testes: increase mid June-Sept; Argentina	2129
	preg f: end Aug-mid Dec; Argentina	2129
	birth: end Oct-mid Dec; Argentina	2129
Dusicyon thous	birth: peak Jan-Feb; Venezuela	7983
Lycaon pictus	estrus: end Jan-early May; S Africa	11163
	mate: Mar-May; Africa	5775
	mate: Mar-Sept, peak Apr-June; S Africa	11145
	birth: Jan, May, Nov; zoo	2171
	birth: Mar; Zaire	11253
	birth: Mar-July; zoo	1230
	birth: peak Mar-July; zoo	1230
	birth: Mar-Sept, peak Apr; Namibia	9945
	birth: Apr; n1; Natal	11616
	birth: Apr-July; S Africa	11163
	birth: Apr-Nov, none July, Sept; zoo	918
	birth: peak May-June; ne S Africa	3145
	birth: May-June; S Africa	9008
	birth: May-July; n Zimbabwe	331, 333, 334
	birth: yr rd, peak Mar-June, none Sept; Serengeti, Tanzania	3452
	yg seen: Mar, Sept; S Africa	8514
	yg seen: June-July, Sept; Africa	5775
Nyctereutes procyonoides	testes: large Dec-Feb, small May-Nov; se Europe	581, 582
	mate: early Feb-Apr; USSR	4643
	mate: Feb-?; captive	9611
	mate: Feb-Apr; USSR	562
	mate: Feb-Mar; captive	11069
	mate: mid Feb-mid Mar; se Europe	581, 582
	mate: end Feb; farm	9609
	mate: Mar; USSR	4645
	mate: peak Mar, some Feb; USSR	10499
	mate: Dec-Mar; zoo, Germany	9626

	ovulate: Mar; Japan	5131
	birth: Mar-Apr; Japan	11451a
	birth: Apr, June; captive	2109
	birth: Apr, peak May-June; USSR	4643
	birth: Apr-May; USSR	562
	birth: Apr-June; n10; captive, Japan	8086
	birth: mid Apr-early May; n6; zoo, Germany	9626
	birth: end Apr-mid May; se Europe	581, 582
	birth: May; USSR	4645
Otocyon megalotis	mate: Nov-Apr; w, sw Zimbabwe	10153
	birth: end Aug-Oct; sw Kenya	6743
	birth: Sept-Nov; S Africa	10158
	birth: Oct; Tanzania	5761
	birth: Mar; ne Uganda	5761
	birth: Nov; n1; sw Africa	9945
	birth: Dec-Apr; Angola	11697
Speothos venaticus	birth: Oct; n1; captive	2109
	birth: yr rd; zoo	8689
Urocyon cinereoargenteus	testes: large Feb; MD	11669
	testes: large Nov-Feb; e US	11669
	mate: Jan-Apr; WY	1871a
	mate: mid Jan-mid May; NY	9885
	mate: peak end Jan-early Feb; s GA, n FL	11893
	mate: end Jan-May, Mar 52.5%; NY	6306
	mate: peak Feb; AL	10542
	mate: Feb-Mar; US	1474
	mate: mid Feb-Mar; WI	5210, 9101
	mate: end winter; Mexico	6387
	preg f: Jan-Mar; s GA, n FL	11893
	preg f: Feb-Mar; IL	6299
	preg f: Mar; n1; CA	11228a
	birth: Mar-Apr; Mexico	6387
	birth: Apr-early May; PA	9164
	birth: Apr-May; US	1474
	birth: early spring; w US	5146
	lact f: May; n3; NM	502
Vulpes bengalensis	mate: Nov-Jan; India	988a
	birth: Feb-Apr; Pakistan	9175
	birth: Feb-Apr; India	988a
Vulpes chama	birth: mid Sept-mid Oct; Transvaal, S Africa	8965
	birth: mid Sept-mid Oct; n5; zoo	1230
Vulpes corsac	estrus: Jan-Feb; USSR	7979
	mate: Jan-Feb; USSR	4643
	mate: end Jan-Feb; n Tibet	8740
	mate: Feb; s USSR	4645
	mate: Feb; w Siberia, Transbaikal, Russia	10499
	mate: end Feb-Mar; Altai, Russia	10499
	birth: Mar; s USSR	4645
	birth: Mar-Apr; USSR	7979
	birth: mid Mar-mid May, peak Apr; USSR	4643
Vulpes ferrilata	birth: Apr-May; Nepal	7983
Vulpes macrotis	mate: Jan; UT	2905
	mate: Dec-Feb; CA	7525
	preg f: Feb; n1; AZ	502
	birth: Feb; Mexico	6387
	birth: Feb-Mar; CA	7525
	birth: Feb-Apr; sw US, nw Mexico	1474
	birth: Mar-Apr; UT	2905
	birth: mid Mar-mid Apr; n6; NE	8014
Vulpes rueppelli	lact f: Mar; n1; Pakistan	9175
Vulpes velox	preg f: Apr; n2; OK	$5405b_1$
	birth: Feb-Apr; central US	1474

Vulpes vulpes	sperm: yr rd, peak May-Aug; n434; NSW, Australia	9390
	spermatogenesis: Apr-Sept; ACT, Australia	7071
	spermatogenesis: Oct-Jan; captive	10312
	spermatogenesis: Nov-Feb, peak 3rd wk Jan; IL, IA	10461
	spermatogenesis: Nov-Mar, peak Jan; UK	6520
	spermatogenesis: end Nov-Mar; NY	954
	testes: large Jan-Feb, small Sept; UK	2123
	testes: large Oct-Mar; England, UK	2124
	testes: large Nov-Feb, small May-Aug; Ireland, UK	3155
	testes: large Dec-Jan, small July-Aug; England, UK	6521
	testes: large Dec-Feb, small July; Wales, England, UK	2124
	testes: large Dec-Feb; USSR	4643
	testosterone: high Jan-Mar; captive	6983
	epididymal sperm; Nov-Apr; Ireland, UK	3815
	estrogen: peaks Dec, Apr; lab	7461
	estrus: Jan; captive	7710
	estrus: Jan-Feb; NY	6306
	estrus: Jan-Feb; zoo	397
	estrus: Jan-May, peak Feb; captive	8310
	mate: Jan; captive	6686
	mate: Jan-Feb; w US	5146
	mate: Jan-Feb; Ireland, UK	3815
	mate: Jan-Feb; central Europe	4253
	mate: Jan-Feb; USSR	4643
	mate: Jan-mid Feb; Ireland, UK	3155
	mate: Jan-Mar; farm	5322
	mate: Jan-Mar; captive	8131
	mate: Jan-Apr; Scotland, UK	5940
	mate: Jan-Apr; USSR	4645
	mate: mid Jan-mid Feb; Ireland, UK	3156
	mate: mid Jan-mid Feb; Germany	10511
	mate: mid Jan-Feb; WI	9101
	mate: mid Jan-early Apr; farm	5323
	mate: mid Jan-3rd wk Apr, end Jan-early Feb 88.3%; NY	6306
	mate: end Jan; UK	9347
	mate: end Jan-Feb; s USSR	7979
	mate: end Jan-Mar; NY	954
	mate: Feb; farm	8310
	mate: Feb-Apr; USSR	10499
	mate: Feb-Apr; captive	10052
	mate: end Feb-early Apr; n Europe	4253
	mate: Mar; central USSR	7979
	mate: Oct-May; captive	10926
	mate: Dec-Jan; USSR	466
	mate: Dec-Jan; s Europe	4253
	mate: Dec-early Feb; PA	9164
	mate: Dec-Mar; WY	1871a
	mate: mid Dec-early Feb; IA, IL	10461
	mate: mid Dec-mid Feb; captive	5319
	mate: mid Dec-Mar, peak end Jan-Feb; NY	5323, 9885
	mate: end Dec-mid Mar; Germany	3571
	ovulate: Jan-Feb; England, UK	2123
	ovulate: Jan-Mar; Ireland, UK	3155
	ovulate: Feb-July, peak Mar; Sweden	3044
	CL present: Jan-Mar; Ireland, UK	3815
	conceive: Dec-Apr; IL, IA	10461
	preg f: peak Jan-Feb; se England, UK	6520
	preg f: Feb-Mar; Wales, UK	6520, 6521
	preg f: Feb-Mar; NY	6306
	preg f: peak Mar-July, low Nov-Feb; Luxemburg, Belgium	818
	preg f: peak Aug; NSW, Australia	9390
	birth: Feb; captive	7710
	birth: Feb-May; Germany	3571
	birth: mid Feb-May; Pyrenees, Spain	11255

	birth: Mar; captive	6686
	birth: Mar; Ireland, UK	3155
	birth: Mar; USSR	466
	birth: Mar-Apr; central US	1474
	birth: Mar-Apr; Europe, mid east	6420
	birth: Mar-Apr; Nepal	7382
	birth: Mar-Apr; USSR	4643, 4645
	birth: Mar-Apr; zoo	397
	birth: Mar-Apr; farm	8310
	birth: Mar-May; Scotland, UK	10761
	birth: Mar-May; farm	5322
	birth: mid Mar; SD	261a
	birth: mid Mar; lab	5318
	birth: mid-end Mar; France	391
	birth: mid Mar-mid Apr; Germany	10511
	birth: peak mid Mar-mid Apr; ND	9501
	birth: mid Mar-Apr; WI	5210
	birth: end Mar-early Apr; UK	9347
	birth: end Mar-early Apr; IL, WI	2060
	birth: Apr-early May; Scotland, UK	2702
	birth: ~Apr-May; USSR	7979
	birth: Apr-May; n9; zoo	12128
	lact f: Mar-mid May; Ireland, UK	3155
Vulpes zerda	mate: Jan-Feb; n African, Arabian desert	3637
	mate: Feb-Apr; captive	9439
	birth: mid Mar-Apr; n African, Arabian desert	3637
	birth: Apr-June; captive	9439

Order Carnivora, Family Ursidae

Bears (*Ailuropoda* 75-160 kg; *Helarctos* 30-65 kg; *Melursus* 55-145 kg; *Tremarctos* 60-175 kg; *Ursus* 50-800 kg) are often diurnal, solitary, terrestrial omnivores from Europe, Asia, and the New World. Males are often larger than females (3571, 4383, 4643, 6469, 8495, 8501, 8623, 8895, 8900, 9175, 9549, 10414, 10434, 10499). Copulation may trigger ovulation (*Ursus* 3060, 11802) and implantation can be delayed (8799, 11802, 11952). The endotheliochorial, labyrinthine, discoid or zonary placenta with villous, circumperipheral hematomae implants into a bicornuate or bipartite uterus (3060, 7278, 7563, 8895, 8968, 9707, 10532, 11802). Placentophagia is recorded (*Ursus* 2276).

Neonates are blind, but usually have some fur (1002, 2109, 2475, 4643, 4851, 6387, 8895, 9175, 9308, 10579, 11207). Eyes open at 3 wk (*Melurus* 2802, 8495, 8501), 3-6 wk (*Ursus* 507, 1505, 2416, 4375, 4704, 5970, 6469, 7234, 7277, 7979, 9600, 9624, 10308, 10357, 10499), or 40-48 d (*Ailuropoda* 5314, 9549). Females have 2-6 mammae (3058, 4745, 9549, 11207). Males have a baculum (1473, 2314, 3059) and prepenial testes (*Ailuropoda* 2314). Details of anatomy (3059, 4641, 5961), endocrinology (*Ailuropoda* 1098, 1763, 4861, 7471, 7622, 9903a; *Ursus* 3407, 4599, 7103, 8213, 10927a, 10928), and milk composition (512, 4852, 5292, 8896, 9233) are available.

NEONATAL MASS

Ailuropoda melanoleuca	90-130 g	5314
	100 g	325
	104 g; r60-130	9549
Helarctos malayanus	325 g; n1	5121
Ursus americanus	200-340 g	1474, 5210
	225-400 g	5435
	~227 g	5942
	~250 g	6387
	250-280 g; n1	1505
	250-400 g	4850
	252-336 g	507
	f: 292 g; n1	9600
	300 g	10308
	m: 301 g; n1	9600

	310.5 g; r265-362; n4	9069
	316 g; se9; n9, <3 d	4599
	<453.6 g	7234
	<454 g	6823
Ursus arctos	280-780 g	1474
	298 g; r286-310; n2 near term	8116
	300-625 g	10308
	~450 g	2060
	500 g	7979
	501-510 g	4643
	f: 629 g; n1	10579
Ursus maritimus	500 g	10434
	540-660 g	11328
	566 g; r520-610; n5 dead	9739
	~600 g; r400-700	8895
	603 g; r453-753; n2	2109
	666 g; r652-680; n2	4704
	675 g; 2 d	10357
	680 g; 1 d	7277
	697.5 g; r675-720	3179
	~700 g	4375
	755 g; r650-840; 3 d	4643
	~1 kg	10433
Ursus thibetanus	305 g; n1	9064
	300-450 g	4643, 7979, 10499
	710 g; n2; a few d after birth	10499
NEONATAL SIZE		
Ailuropoda melanoleuca	15-17 cm	5314, 9549
Helarctos malayanus	19 cm; n1	5121
Tremarctos ornatus	22.5 cm; n1	9308
Ursus americanus	~15-20 cm	5942
	<15.2 cm	7234
	20 cm	2088
Ursus arctos	20-25 cm	2060
	23-23.5 cm	4643
	23-28 cm	10499
	25 cm	3571
Ursus maritimus	~25 cm	4375
	25 cm; r24-26; n2	2109
	28-32 cm; 3 d	4643
	~30 cm	7979
	f: 34 cm; n1 nose to tail	1002
Ursus thibetanus	26.5 cm; n2; a few d after birth	10499
WEANING MASS		
Ailuropoda melanoleuca	22 kg; 6 mo; from graph	2314
Ursus arctos	26 kg; n1 7 mo cub self-sufficient	5350
Ursus maritimus	5.78 kg; r2.8-10.25	9739
	den emergence: 10-15 kg; 1 yr: 45-80 kg; 2 yr: 70-140 kg	2475
LITTER SIZE		
Ailuropoda melanoleuca	1.5	5855
	1.56 born; mode 2; r1-3; n55 lt, 24 f	11574
	1.7; r1-2; n37 lt	9549
	2; n1 lt	4861
Helarctos malayanus	1	9064
	1 ea; n2 lt	5121
	1 yg ea; n2 lt	9066
	1 yg ea; n3 lt, 1 f	7044
	1.1; r1-2; n18 lt	2277, 2279, 2281, 2283
Melursus ursinus	1 yg ea; n2 lt	9066
	1-2 yg; n5 lt	5220
	1.3; r1-2; n3 lt	9052
	1.6; r1-2; n24 lt	6250
	2	7652
	2; n1 lt	1230

	2; r1-2	8495, 8501
	2; sometimes 3	2802
	2; occasionally 3	6654
Tremarctos ornatus	1	9308
	1.4; r1-3	2286
	1.5 yg; r1-2; n2 lt	3711
	2 yg; 1 lt	1019, 1020
Ursus americanus	1-4	7652
	1.5; r1-2; n2 lt	9600
	1.7	5437
	1.73 yg; r1-3; n30 lt	4493
	1.77 ps; r1-2; n9	8623
	mode 2	502
	mode 2; r1-4	2088
	mode 2; r1-4	5942
	mode 2; r2-4	6387
	mode 2-3; r2-4	7234
	2-4 cl	11802
	2 yg; n1	4851
	2 yg ea; n3 lt	9066
	2.05 yg; r1-4; n219 lt	3060
	2.1 cl; r1-3; n11	8623
	2.2 yg; n5	4599
	2.2; r1-4; n10 lt	4383
	2.26; n15 lt, 2 f	508
	2.4 cl; r1-3; n12 f	3060
	2.4 yg; n38 lt	10260
	2.4 yg; r1-4; n47	5943
	2.4 yg; max 6; n284 lt	5210
	2.5 emb; r2-3; n2 f	2545
	2.5; r2-3; n4 lt	3407
	2.6 emb, yg; n19 lt	10414
	2.7, 2.8 cl; max 4; 2 samples	1986
	2.75 cl	5961
	2.9; r1-5; n68 lt	205
	mode 3	10839
	3.2 ps; r1-4; n13 f	3060
	3.5 yg; r3-4; n4 lt, 2 f	9319
	5 yg; n1 f; anecdote	6950
	6 yg; n1 f	9320
Ursus arctos	1-2, occasionally more	7979
	1-5	5824
	1.5 yg; r1-2; n6 f	10929
	1.6 yg; n22 f	8288, 8289
	1.64 new ca; n14 f	10929
	1.69; mode 2; r1-3; n13 lt	6469
	1.7 yg, 1.8 yearlings; n65 lt	6917
	1.76 ps; n16 f	10929
	1.8; r1-3	3986
	1.9 cl; n9 f	8289
	1.98 yg; r1-3; n168 f	7615
	2	4645
	2; r1-4	1474
	2; n1 lt	10350
	2; n3 lt	1230
	mode 2; 3 rare	6387
	mode 2; r1-3	11330
	mode 2; r1-4	2060
	mode 2; r1-5	4643
	mode 2-3; r2-5	3571
	rarely > 2	12049
	~2; r1-3; 4 rare; n46 lt	8791
	2.05; mode 2; r1-4; n213 lt	2611, 2612
	2.23 yg; se0.06; r1-4; n98 f	4641

	2.26 yg; mode 2; r1-3; n31 lt, ~145 d	7092
	2.3; mode 2; r1-4; n18 lt	6469
	2.4 yg; r2-3; n8 lt	2276
	2.53 ps; n15 tracts	4641
	2.55	2106
	2.8 born; r2-3; n4 lt	542
	3 emb	7980
	3 emb; r2-4; n2 f	8116
	3; n1 lt	10352
	5 yg; n1 f	10499
	6 emb; n1 f	10499
	6 yg; n1 f	11697a
Ursus maritimus	r1-3; n4 lt	76
	1 yg; yearling; 2 yg; adult; 3,4 rare	7979
	1; n1	5514
	1 ea; n2 lt	2109
	1.4; r1-2; n14 lt	9739
	1.43; mode 1; r1-2; n14 lt	11327
	1.46; n13 lt	11328
	1.5; r1-2; n4 lt	5970
	1.52; r1-3; n54 lt	8899
	est 1.6 yg	5436
	1.64 yg < 1 yr; n36 f	3568
	1.69; r1-3; n65 lt	11049
	1.7; n10 lt, 3 f	9233
	1.8; r1-3; n6 lt	2916
	1.8; r1-2; mode 2; n10 lt	12128
	1.8; r1-2; n13 lt	9733
	1.81 yg emerge; r1-4; n129 f	6220
	1.82; n34 lt	10433
	1.9 yg; r1-3; n118 lt	8898, 10434
	mode 2	12057
	mode 2; r1-4	4375
	mode 2; r1-4	4643
	2 yg; n1 lt	9066
Ursus thibetanus	mode 1-2; r1-3	7979
	1 yg; n1 f	7382
	1 ea; n2 lt	1230
	2; 1,3 rare	10499
	mode 2	6653, 6654
	mode 2; r1-2	9175
	mode 2; 3 rare	1310
	2 yg; n1 lt	9064
SEXUAL MATURITY		
Ailuropoda melanoleuca	f/m: 4-10 yr; native reports	9884a
	f/m: 5.5-6.5 yr	9549
	f/m: 6-7 yr	1816
Ursus americanus	f: 1st cl: 3.5 yr	3060
	f: 1st estrus: 4.5 yr	5437
	f: 1st mate: 38% 2.5 yr; 100% breed 3.5 yr	5961
	f: 1st mate: 3 yr	2088
	f: 1st mate: 3.4 yr	8623
	f: 1st conceive: 2.5 yr; n1	1986
	f: 1st conceive: 3.5 yr	6387
	f: 1st birth: 3.5-4.8 yr; n4 groups, 23 f	7091
	f: 1st birth: 5.7 yr; r5-8; n18	5943
	f: 1st lt raised: 6.5-7.51 yr	5435, 5437
	m: spermatids in testes: 3.5 yr	3060
	f/m: 1st mate: 3.5 yr	1474
	f/m: 3.5 yr	5942
Ursus arctos	f: ovarian histology: 1-4 yr	10927a
	f: 1st estrus: 4.5 yr; r3.5-5.5	542
	f: 1st conceive: 3 yr 4 mo	4641
	f: 1st conceive: 6.5-8.5 yr	8288, 8289

	f: 1st birth: 2 yr; n1	2611, 2612
	f: 1st birth: 3 yr	6509
	f: 1st birth: 5th yr	10499
	f: 1st birth: 6 yr; n5	7092
	f: 6-7 yr	12049
	m: testes histology: 2-4 yr	10927a
	f/m: 1st mate: 3 yr	1474
	f/m: 1st mate: 4-5 yr	6387
	f/m: 2.5-4 yr	3571
	f/m: 3rd yr	7979
Ursus maritimus	f: 1st mate: 4 yr	10433
	f: 1st mate: 50% 4 yr; rest 5-6	8895
	f: 1st birth: 4-5 yr	8898, 8899
	f: 1st birth: 5 yr	3568
	f: 1st birth: 6-7 yr	2475, 5435
	m: 1st sperm: 3 yr	2475
	f/m: 3-4 yr	10499
	f/m: 3-4 yr wild; 5 yr; zoo	7979
	f/m: 5 yr	11327
Ursus thibetanus	f: 1st birth: 3 yr	6509
	f/m: 3rd yr	7979
CYCLE LENGTH		
Ailuropoda melanoleuca	estrus: 1 d	4860
	estrus: mode 10 d; r4-14	1816
	estrus: 12-25 d; peak receptivity 2-7 d	9549
Helarctos malayanus	estrus: a few d	2279
Ursus americanus	estrus: a few d	9227
Ursus arctos	estrus: 27 d	7615
	4-18 d; interestrous interval	7615
	16 d	11330
	24 d; r18-30; n3 cycles, 1 f	10350
	max 27 d	2106
Ursus maritimus	estrus: 3 d	11327
GESTATION		
Ailuropoda melanoleuca	~135 d; r97-161; n19	9549
	16 wk	7471
	140 d; ±8; r122-163	1816, 11952
	159 d; n1	4861
	2-5 mo	5314
Helarctos malayanus	95.7 d; r95-96; n3	2277, 2279, 2283
	96-97 d	10352
	214 d; r174-240; n3	7044
Melursus ursinus	est 169-176 d	5220
	7 ? mo	7652
	7 mo	7974
	~7 mo	2802, 8501
Tremarctos ornatus	194-208 d	3711
	215 d; r175-240; n6	1019, 1020
	7 mo 3 wk-8.5 mo; n1	2286
	8 mo; n1	9308
Ursus americanus	implant: 6 mo	11802
	~6-7 mo	5435
	200-217 d	10308
	7 ? mo	7652
	~7 mo; field data	3060
	~7 mo	6387
	210-253	3407
	7 mo 1 wk	1355
	215 d; r205-226; n9	507, 4851
	~220 d; r210-234; n1	4851
	~225 d	5210
	8 mo with delay	11802

Ursus arctos	implant: 1-2 mo	3571
	implant: 62.5 d	6469
	implant: r18-33 wk; n20	2612
	postimplantation: 7 mo	3571
	min 50 d	2106
	115-136 d; n1; 1st, last mating to birth	10352
	180-240 d	10308
	6-9 mo	3571
	est 189-251 d	8289
	196 d; r179-213	10350
	r197-288 d; n20 with delay	2611, 2612
	~7 mo	4582
	~7 mo	4645, 7979, 10499
	7-8 mo	8741
	7-8 mo	7652
	227 d; r185-251; n8	4643
	236 d	1355
	237 d	6469
	est 8 mo	4641
	8 mo; r209-272 d	11330
	8 mo	3986
	9 mo	2276
Ursus maritimus	6-7 mo	5435
	201-303 d	11328
	207 d; r201-214; n3	3179
	226 d; r195-253; n4	10357
	230-250 d	4643
	mode 235-254 d; r228-303	11327
	~8 mo	10499
	~8 mo; n13	9733
	8 mo; r230-250 d	7979
	246 d; r245-248	5970
	246 d; r228-262; n4	76
	10-11 mo	12057
Ursus thibetanus	~6 mo	6654
LACTATION		
Ailuropoda melanoleuca	solid food: 5 mo	5314
	independent foraging: 3-4 mo	1816
	nurse: 6 mo	5314
	wean: 46 wk	9549
Helarctos malayanus	wean: ~3 mo	2281
Melursus ursinus	wean: 2-3 mo	2802
Tremarctos ornatus	solid food: 89 d; n1 lt	1020
Ursus americanus	independent: 5.5 mo	3057
	wean: 6-7 mo	1474
	mode 1st yr; max 18-22 mo	5435
Ursus arctos	den emergence: 3-4 mo	10499
	independent: 7 mo; n1	5350
	wean: 1.5 yr	2276
	wean: 1-2 yr	8288, 8289
	wean: 4 mo	10499
	wean: ~5 mo	7979
	wean: 6 mo	4643
Ursus maritimus	den emergence: 45 d	7979
	den emergence: 2 mo	4643
	solid food: 12 wk	2416
	solid food: 6 mo	10499
	wean: 52 d	10357
	wean: 73.4 d; n19	9739
	wean: 5-8 mo	6342
	wean: 6-8 mo	4643
	wean: 9-21 mo	4375
	wean: ~15 mo	7979
	nurse: max 15 mo	10499

	wean: 20+ mo	11327
	wean: 24-28 mo	2475, 8899
Ursus thibetanus	den emergence: ~2 mo	7979
	den emergence: ~2-2.5 mo	10499
	wean: 3-3.5 mo	7979
	wean: 3-3.5 mo	10499
INTERLITTER INTERVAL		
Ailuropoda melanoleuca	1 yr; yg die	9549
	2 yr; yg live	9549
Melursus ursinus	3 yr	2802
Ursus americanus	1 yr; yg die before June	205
	2 yr; yg live	205
	2 yr	6387
	2 yr	2088
	2-3 yr	5435
Ursus arctos	1 yr; yg die	10499
	2 yr; yg live	10499
	2 yr	2106
	2 yr	3571
	2 yr	4643
	2-3 yr	6387
	2-3 yr	1474
	2.67 yr; n9	7092
	3 yr	8289
	3 yr	12049
	mode 3 yr	542
Ursus maritimus	1 yr	5970
	1 yr	9733
	2 yr	1474
	> 2 yr	11327
	2-3 yr	5435
	2.9 yr; 40% 2 yr; r1-5; n66 intervals	8895, 8898, 8899, 10433
	3 yr	4643
Ursus thibetanus	1 yr	6654
SEASONALITY		
Ailuropoda melanoleuca	estrus: Mar-June; zoo	5855
	mate: mid Mar-mid May; captive	9549
	mate: end Apr-May; zoo	11952
	mate: May; zoo	1098
	birth: Jan; w China	9884a
	birth: Sept; zoo	11952
Helarctos malayanus	birth: Jan; n1; zoo	10352
	birth: yr rd; zoo	2281
Melursus ursinus	mate: June; India	7652
	mate: yr rd, peak 1st half of yr; Sri Lanka	8495, 8501
	birth: June; n1; captive	1230
	birth: Sept-Jan, one exception June-July; Nepal	6250
	birth: yr rd, more yg seen Aug-Sept; Sri Lanka	7974
	birth: cold weather; India	2802
Tremarctos ornatus	estrus: May-June; zoo	9308
	mate: Aug; zoo	2286
	birth: Jan, Mar; n2; zoo	1020
	birth: mid Jan; zoo	9308
	birth: end Dec-mid Mar, peak Jan; zoo	2286
Ursus americanus	spermatogenesis: Apr-mid Sept; MI	3060
	estrus: May-Aug; Rocky Mt, N America	5437
	mate: Jan-Feb; BC, Canada	2088
	mate: May-June, peak early June; N America	5435
	mate: May-June; zoo	3407
	mate: May-Aug, peak June-July; WA	8623
	mate: June-early July; WI	5210
	mate: June-July; US	11207
	mate: June-July; captive	507
	mate: June-July; PA	205

	mate: mid-end June; AK	4850
	mate: mid June-early July; MN	9227
	mate: mid June-mid July; MI	3060
	mate: end June-early Aug; N America	7652
	mate: early summer; Sierra Madre Occidental, Mexico	6387
	ovulate: May-mid June; Adirondacks, NY	11802
	implant: Nov; US	11207
	implant: mid Nov-early Dec; PA	5961
	implant: early Dec; Adirondacks, NY	11802
	implant: Dec-?; Augusta Co, VA	10414
	birth: early Jan; MI	3060
	birth: Jan; US	11207
	birth: Jan; Adirondacks, NY	7234
	birth: Jan; n32 lt; PA	205, 206
	birth: Jan; Sierra Madre Occidental, Mexico	6387
	birth: Jan-early Feb; WI	5210
	birth: Jan-Feb; BC, Canada	2088
	birth: Jan-Feb; WA	8623
	birth: Jan-Feb; WY	1871a
	birth: mid Jan; ME	10260
	birth: end Jan; n9; captive	507
	birth: end Jan-Feb; Adirondacks, NY	11802
	birth: end Nov-Feb, peak Jan-Feb; N America	5435
	birth: Dec-Mar, peak mid Jan; AK	4850
	yg appear: May; Coahuila, Mexico	522
	wean: Oct; AK	8623
Ursus arctos	sperm: end May-Aug; YUK, Canada	8289
	testes: large May, small Oct-Nov; AK	3059
	estrus: Apr-July; zoo	2612
	estrus: Apr-Aug; Europe	8741
	mate: Apr-May; zoo	2276
	mate: Apr-June; Alps, Europe	3571
	mate: May; Djunegarsky Altau, USSR	3986
	mate: May-early June; captive, Japan	10928
	mate: May-June; n, ne European USSR	4645
	mate: May-June; Canada	7615
	mate: May-June; ne Siberia	5824
	mate: peak May-June; AK	542
	mate: May-July; US	1474
	mate: May-July, some to Nov; AK	3059
	mate: May-Aug; Poland, Rumania	6469
	mate: mid May-mid July, peak early-mid June; WY	2106
	mate: mid May-mid July; WY	1871a
	mate: mid May-mid July; Tschita, USSR	4643
	mate: end May-June; s central AK	542
	mate: June-July; USSR	7979
	mate: mid June-mid Aug; Europe, central Asia	7652
	implant: Nov?; zoo	2612
	birth: early-mid Jan; USSR	4643
	birth: Jan; US	1474
	birth: Jan; St Petersburg, Russia	7980
	birth: Jan; zoo	2611
	birth: Jan; zoo	11330
	birth: Jan-Feb; Cerro Campana, Mexico	6387
	birth: Jan-Feb; zoo	2276
	birth: Jan-Mar; Altai Mt, USSR	10499
	birth: mid Jan; n, ne European USSR	4645
	birth: mid-end Jan; Djunegarsky Altau, USSR	3986
	birth: end Jan-early Feb; IL, WI	2060
	birth: June-July; upper Yang-tse-kiang, China	3034
	birth: July; few data; zoo	1230
	birth: Dec-Feb; Europe	8741
	birth: Dec/Jan-Feb; Alps, Europe	3571
	birth: mid Dec-early Mar, peak Jan; zoo	2611, 2612

Ursus maritimus	spermatogenesis: Feb-May poss June; Arctic	2475
	mate: mid Feb-mid May; zoo	11327
	mate: Mar; arctic	12057
	mate: Mar-Apr; zoo	5970
	mate: Mar-Apr; USSR	7979
	mate: Mar-Apr; zoo	9733
	mate: Apr; n Canada	4375
	mate: Apr-May; zoo	3179
	mate: Apr/May-?; N America	5435
	birth: ~Jan-Feb; N America	5435
	birth: Jan-Feb+; USSR	7979
	birth: Nov-Dec; zoo, USSR	7979
	birth: Jan-Apr; arctic	4643
	birth: Jan-May; arctic	12057
	birth: Feb-Mar; Finland, Greenland	2594
	birth: end Jan-May; Siberia	10499
	birth: Nov; few data; zoo	2109
	birth: Nov; zoo	11327, 11328
	birth: Nov-early Dec; n10; zoo	12128
	birth: Nov-Dec; zoo	9733
	birth: Nov-Jan; n Canada	11049
	birth: mid Nov-Dec; zoo	2594
	birth: end Nov-early Dec; n Canada	4375
	birth: end Nov-Jan; se Baffin I, Canada	10433, 10434
	birth: Dec; zoo	3179
	birth: Dec; zoo	5970
	birth: end Dec-Jan; MAN, Canada	8899
	wean: July; circumpolar	4375
Ursus thibetanus	den emergence: Apr-mid May; USSR	4643
	mate: June-July; Primocsky, USSR	1310
	mate: June-July; Siberia	10499
	mate: mid June-mid Aug; USSR	4643
	mate: Oct; Pakistan	9175
	mate: fall; Nepal, s Tibet	6654
	birth: mid Jan-mid Feb; USSR	4643
	birth: Feb; Pakistan	9175
	birth: May; few data; zoo	1230
	birth: end Dec-Mar, peak Jan-Feb; USSR	7979
	birth: spring; Nepal, s Tibet	6653, 6654
	birth: end Dec-Mar, peak mid-Feb; Siberia	10499

Order Carnivora, Family Procyonidae

Raccoons and their allies (*Ailurus* 3-6 kg; *Bassaricyon* 940-1700 g; *Bassariscus* 700-1400 g; *Nasua* 3-6 kg; *Nasuella*, *Potos* 1.4-4.5 kg; *Procyon* 2-12 kg) are semiarboreal to arboreal, generally nocturnal (except *Nasua*) carnivores or omnivores primarily from the New World, although red pandas, *Ailurus*, inhabit the Himalayas. Female *Bassaricyon* are larger than males, but the reverse is recorded for *Nasua* and *Procyon* (5580, 7979, 8628, 9155b).

Ovulation may be induced by copulation (*Potos* 8798) or by the presence of males (*Procyon* 7534; cf 9467). Long copulations (0.5-1 h) suggest a copulatory lock (8625, 9174, 9549, 10297; cf 3870, 5580, 8628). Transuterine migration occurs (6514a). Latitudinal variation in gestation length suggests delayed implantation in *Ailurus* (9174). The 9-15 g (for the smaller genera), deciduate, zonary or discoidal, labyrinthine, hemoendothelial or endotheliochorial placenta with hemophagous organs has an antimesometrial implantation into a bicornuate or bipartite uterus (932, 2125, 2126, 7563, 8628, 8632, 9467, 9707, 11483) and is eaten after birth (8624, 8628, 8632, 10877). A postpartum estrus is not common, but may occur at 10-16 d in *Procyon* (8626, 8628, 9467).

Neonates are furred, but blind at birth (2109, 7624, 8624, 8625, 9174). *Ailurus* neonates open their eyes either at 17-18 d (9170, 9549) or 36-41 d (3074, 9051). Eye-opening in *Bassaricyon* is at 21-25 d (8628), in *Bassariscus* is at 21-34 d (1474, 5146, 8633, 9105, 10877), in *Nasua* is at 4-11 d (5580), in *Potos* is at 1-3 wk (900, 8629, 8798), and in *Procyon* is at 14-23 d (1474, 3571, 4271b, 7979). Procyonids have 2 (*Potos* 8798), 4

(*Ailurus* 9549; *Bassariscus* 9105), or 6 (*Bassariscus* 1474; *Procyon* 3571) mammae. Males have a baculum (1473, 6558, 7464, 8624, 9467, 11483). Details of anatomy (9467), birth (9173), photoperiod (962, 963), and reproductive energetics (3796) are available.

NEONATAL MASS		
Ailurus fulgens	f: 73 g; n1	626
	m: 95 g; n1	626
	110-130 g	9173
	118 g	9174
Bassaricyon spp	m: 51 g; n1 some tail eaten	8625
	52.5 g; r49.6-60.2; n8	8628
Bassariscus astutus	14-40 g	8633
	25-30 g	1871a
	37.5 g; sd0.7; r37-38; n2, 2 d	10877
Nasua nasua	100-180 g	5580
Potos flavus	125-200 g	8798
	150 g; n1	900
	173.2 g; r148.7-191.4; n4	8624
	191.4 g; n1	1904
	203 g; r175.4-248; n8	8629
Procyon lotor	75 g	4271b
	80 g	7979
	85 g	1474, 5210
NEONATAL SIZE		
Ailurus fulgens	f: 190 mm; n1	626
	m: 220 mm; n1	626
Bassaricyon spp	HB: 119.3 mm; r112-127; n6	8628
Bassariscus astutus	110.5 g; sd0.7; r110-111; n2, 2 d	10877
Nasua nasua	155-165 mm	5580
Potos flavus	HB: 180.6 mm; r168-208; n4	8624
	HB: 185.5 mm; r165-197; n8	8629
Procyon lotor	CR: ~95 mm; term emb	6513
WEANING MASS		
Bassariscus astutus	258.7 g; sd78.3; r168.3-304.5; n3, 40 d	10877
Procyon lotor	600-900 g; 40-50 d	4271b
	1.4-1.8 kg; 10-12 wk	5210
WEANING SIZE		
Bassariscus astutus	220.7 mm; sd19.3; r198-234; n3, 40 d	10877
LITTER SIZE		
Ailurus fulgens	1.5 yg; r1-2; n2 lt	7618
	1.67; r1-2; n3 lt	12128
	1.7 yg; r1-4; n74	9170, 9174
	mode 2 yg	7577, 7578
	2 yg	2592
	2 yg; n1 lt	5176
	2 yg; n1 lt	11415
	2 yg; n1 lt	626
	2 ea; n3 lt	2285
	2.0 yg; mode 2; r1-4; n39 lt	9165
	2.5 yg; r2-3; n2 lt	9066
Bassaricyon spp	1 yg ea; n12 lt	8625, 8628
Bassariscus astutus	2-4	10721
	2.5 yg; r1-4; n2 f	9105
	3 yg; n1 lt	2109
	3.5 emb; r3-4; n2 f	502
Bassariscus sumichrasti	mode 2	3633
Nasua spp	4.0 yg; r3-5; n3 lt	9066
Nasua nasua	2 ea; n4 lt	12128
	3.5; r3-4; n2 lt	5580
	4 emb; 1 f	4228
	4 emb; n1 f	7624
	4-6	1474
	5 yg ea; n2 f	7118

Potos flavus	1 yg ea; n2 lt	900
	1 emb; n1 f	2348
	1 yg ea; n17 lt	8624, 8629
	1.5 emb; r1-2; n2 f	4228
Procyon cancrivorus	2.5; r2-3; n2 lt	12128
Procyon lotor	1.83 yg; n6 lt	9588
	2-6; max 8	7979
	2; r1-3	12128
	2 emb; n1 f	6298
	2.2 yg; r1-6; n55	2109
	2.3; n11 lt	9155a
	2.46 recent ps	5332a
	2.5 yg; r2-3; n2 lt	9066
	2.5; n101 lt	5332a
	2.6; n10 lt	9155a
	2.6 implantations; se0.1; r1-4; n95	3248
	2.62 emb; se0.11; n21 f	5332a
	2.7; n45 lt	9155a
	2.7; n88 lt	9155a
	2.7 ps; se0.1; r1-6; n115	3248
	2.75 cl; se0.35; n4 f	935a
	2.8 emb; r2-4; n5 f	11892
	2.8; n20 lt	9155a
	2.8; n31 lt	9155a
	2.8 ps; mode 3 75%; n20 f	7975a
	2.9; n12 lt	9155a
	2.9; n25 lt	9155a
	2.87 emb; se0.26; n9 lt	935a
	2.89 ps; se0.21; n26 f	935a
	2.9 ps; r2-5; n26 f	10138
	2.93 ps; r2-5; n14 f	12092
	3 ps; n1 f	7498a
	3 emb; n1 f	9588
	3.15 yg; r2-4; n13 lt	10523a
	3.2 emb; se0.18; r2-5; n17 f	7078
	3.23 ps; sd0.837; r1-5; n74 lt	2808
	3.30 emb; sd0.675; r2-4; n10 f	2808
	3.3; n11 lt	9155a
	3.33 cl; sd0.072; r2-4; n15 f	2808
	3.33; n24 lt	963
	3.4 ps; n69	5466a
	3.5 yg, 4.0 ps; n8 lt	9463
	3.5 yg; r3-4; n2 lt	11892
	3.56 ps; se0.03; n1421 f	9466
	3.58 ps; mode 3; r1-6; est n287	3530
	3.6; n217 lt	9155a
	3.63 ps; r2-4; n19 f	11892
	3.66 ps; mode 3; r1-10; est n200	3530
	3.7; n23 lt	9155a
	3.7; n212 lt	9155a
	3.8; n14 lt	9155a
	3.83 emb; r3-5; n6 f	1525
	4-5 cl	7534
	4 ps; n1 f	10523a
	4.0 ps; r3-5; n2 f	6298
	4 yg; r3-7; n10 dens	10523a
	4.25 emb; r4-5; n4 f	2125
	4.6 ps; r2-7; n5 f	7149a
	4.6 yg; n8 lt	10297
	4.67 yg; r3-6; n3 nests	3723
	4.8 emb, yg; mode 4; r3-7; n25	3528
	4.8; mode 4; r1-5	1230
	4.8; n52 lt	9155a
	mode 5-6; r2-8	3571

SEXUAL MATURITY

Species	Data	Reference
Ailurus fulgens	f: 1st conceive: 18 mo	9172
	f: 1st birth: 24-26 mo; n7	9174
	f/m: 18-20 mo; n7	9174
Bassaricyon spp	f: 2 yr	8628
	m: 1.5 yr	8628
Bassariscus astutus	f: 1st birth: ~1 yr	8632
	m: testes descend: 16 wk	8633
Potos flavus	f/m: 1st mate: 3 yr	8626
	f/m: 18-30 mo	8798
Procyon lotor	f: 1st mate: 1 yr 9.3%	2808
	f: 1st mate: 1 yr	9467
	f: 1st preg: yearling	3528
	m: sperm, testes wt: 1 yr	9467
	m: sperm: 1.5 yr	2808
	m: 1st mate: 2 yr	2088

CYCLE LENGTH

Species	Data	Reference
Ailurus fulgens	estrus: 18-24 h	9172
	26-44 d	9173
Bassaricyon spp	estrus: 1-3 d; possibly 4	8628
	23.9 d; r19-40; n22 cycles, 3 f	8628
Bassariscus astutus	estrus: 24-36 h	8633
Potos flavus	vulvar swelling: 2 d; r1-3; n4 f	8629
	estrus: 17.6 d; r12-24	8629
	62.8 d; r46-92; n43 cycles, 4 f	8629
Procyon lotor	cornified epithelium: 3.4 d; partially isolated f	7534
	cornified epithelium: 14.3 d; n6 isolated f	7534
	r80-140 d; n5 f	9467

GESTATION

Species	Data	Reference
Ailurus fulgens	112 d	2592
	125-132, 155? d; via f behavior	3074
	131 d; sd9.27; r114-145; n9	9172, 9174
	131.5 d; r131-132; n2	7618
	132 d; n1	9051
	134.2 d; sd14.7; r112-158; n17 [literature compilation]	9173
Bassaricyon spp	73.25 d; r72-75; n12	8625, 8628
Bassariscus astutus	51.92 d; r51.33-53.58; n9	8632
	~56 d	1871a
Nasua spp	71 d; n1	3594
Nasua nasua	67-73 d	4582
	10-11 wk; field data	5580
	77 d	1355
Potos flavus	115.5 d; r111.5-118; n2 f; 8 lt	8629
	119-120 d	8626
Procyon lotor	60-63 d	3571
	63 d	4271b
	63 d	9467
	~9 wk	1474, 2088, 5146, 6387, 7979
	65 d; n1	3594
	69 d	1355

LACTATION

Species	Data	Reference
Ailurus fulgens	den emergence: 3-4 mo	9174
	solid food: 90 d	9173
	solid food: 125-135 d	9170, 9174
	wean: 4-5 mo	9172, 9174
Bassariscus astutus	solid food: 3 wk	6387
	solid food: 30-40 d	10877
	wean: ~35-50 d	10877
Nasua nasua	den emergence: 3-5 wk	5582
	nurse: 4 mo	5580
Potos flavus	solid food: 35 d	8798
	solid food: 8 wk	8624
	nurse: 2.5-4 mo	8626, 8629

	wean: 16-18 wk	8798
Procyon lotor	den emergence: 6-8 wk	2109
	solid food: ~1 mo	7979
	solid food: 40 d	4271b
	solid food: 9 wk	7481
	wean: 10-12 wk	5210
	wean: 13-17 wk	7480
	nurse: 16 wk	7481
LITTERS/YEAR		
Ailurus fulgens	1	9174
Bassariscus astutus	1	1474
Potos flavus	1	8624
Procyon lotor	1	10297
	1	7234
	max 2/season; only 1 raised	9467
SEASONALITY		
Ailurus fulgens	mate: Jan; zoo	9051
	mate: Jan-Feb; zoo	3074
	mate: Jan-mid Mar; n hemisphere zoos	9172, 9173
	mate: July-Aug; s hemisphere zoos	9173
	birth: May-Aug, peak June 79%; n199; n hemisphere zoos	9173
	birth: June; Himalayas, Asia	9173
	birth: June; zoo	9051
	birth: June; zoo	7578
	birth: June; zoo	3074
	birth: June-Aug, peak June; n39; zoo	9165
	birth: mid June-mid July; zoo	9172
	birth: Dec-Mar; n11; s hemisphere zoos	9173
Bassaricyon spp	birth: Jan-Mar, May-June, Sept, Dec; n12; zoo	8628
Bassariscus astutus	preg f: mid Apr-May; TX	10721
	preg f: May; n1; NM	502
	birth: Apr-July; WY	1871a
	birth: May-June; sw US	1474
	birth: May-June; w US	5146
	birth: June; n1; zoo	2109
	lact f: July; n1; Durango, Mexico	523a
Bassariscus sumichrasti	birth: Mar; Yucatan, Mexico	3633
Nasua nasua	mate: end Jan-mid Mar; Barro Colorado I, Panama	5580
	birth: Mar-June; n1; Chiapas, Mexico	6387
	birth: Apr-June; Barro Colorado I, Panama	5580
	birth: June-July; n2; nw Chihuahua, Mexico	6387
	birth: est July; sw US	1474
	birth: July; n1; Sinaloa, Mexico	6387
	birth: July-Sept; n1; Guerrero, Mexico	6387
Potos flavus	preg f: Apr; Panama	3398
	preg f: Sept; Suriname	3398
	preg f: Dec-Jan; n2; Veracruz, Mexico	4228
	birth: Jan-Feb, Aug-Sept; captive (literature compilation)	3398
	birth: est yr rd; zoo	8624, 8629
Procyon lotor	no sperm: July-Oct; IL	9467
	sperm: Oct-Apr; MD	2808
	testes large: peak Feb-Mar; AL	5332a
	estrus: Feb; Europe	7979
	mate: Jan-early Apr, peak mid Feb; IL	9467
	mate: peak end Jan-early Feb; NY	4271b
	mate: peak end Jan-mid Mar; WI	5210
	mate: end Jan-Mar; lab	963
	mate: peak Feb; PA	9164
	mate: Feb-Mar; n US; earlier; s US	1474
	mate: Feb-Mar; Germany	3571
	mate: Feb-June, peak Feb; MD	2808
	mate: peak Mar; GA, FL	7078
	mate: Dec-June, peak Feb; KS	10297
	mate: mid Dec-early Apr; IL	9467

mate: yr rd; zoo	1230
conceive: Feb-mid June; n18; AL	5332a
preg f: Apr-Sept; n9; s FL	935a
birth: Jan-Apr; IL	9467
birth: Mar; MI	3723
birth: Mar-May; MI	6587
birth: Mar-July; zoo	2109
birth: Apr-early May; KS	10297
birth: peak Apr-May; WI	5210
birth: Apr-May; IL, WI	2060
birth: Apr-May; Europe	3571
birth: Apr-May; ND	3528
birth: Apr-June, Oct; n6; zoo	12128
birth: Apr-Aug, peak Apr; MD	2808
birth: peak mid Apr; IL	9467
birth: mid Apr; Adirondacks, NY	7234
birth: end Apr-early May; PA	9164
birth: end Apr-early May; Europe	7979
birth: peak May; GA, FL	7078
birth: May-Aug; AL	6587
birth: Aug; n1; WV	786
birth: est Sept; as 3 mo animals killed Dec; IN	6363
birth: Dec-Apr; w US	5146
lact f: June; n1; SD	261a
lact f: July; NE	5405b,

Order Carnivora, Family Mustelidae

Mustelids (*Aonyx* 1-34 kg; *Arctonyx* 7-14 kg; *Conepatus* 0.9-4.5 kg; *Eira* 4-6 kg; *Enhydra* 15-45 kg; *Galictis* 1 kg; *Gulo* 7-35 kg; *Ictonyx* 420-1400 g; *Lutra* 3-16 kg; *Lyncodon* 300-350 mm; *Martes* 0.6-5.5 kg; *Meles* 6.5-34 kg; *Mellivora* 7-14 kg; *Melogale* 1-3 kg; *Mephitis* 700-6300 g; *Mustela* 25-1600 g; *Mydaus* 1.3-3.6 kg; *Poecilictis* 200-250 g; *Poecilogale* 115-400 g; *Pteronura* 22-34 kg; *Spilogale* 200-1000 g; *Taxidea* 4-12 kg; *Vormela* 370-715 g) are a diverse group of terrestrial to semiaquatic, primarily carnivorous inhabitants of the New World, Eurasia, and Africa. Males are often larger than females (190, 1304, 2068, 2421, 3076, 3797, 3802, 4225, 4269, 4285, 4576, 5210, 5583, 5631, 5678, 5681, 5761, 6927, 7195, 7250, 7599, 8495, 8501, 8636, 8715, 9327, 10499; cf 8637).

Ovulation is triggered by lengthy coitus (ovulation: 834, 958, 1632, 1633, 2422, 2687, 2689, 3026, 3028, 4299, 4300, 4340, 6156, 7655, 7975, 8276, 9391, 10189, 11277, 11381, 11382; cf 7126; duration of coitus: 958, 2687, 4299, 4300, 4340, 4391, 4458, 4643, 5681, 6089, 6725, 8276, 8715, 9326, 9610 8627, 10499). In *Martes* development may be slow (5966). Implantation may also be delayed (1578, 11945; *Martes* 487, 1580, 1586, 3030, 5438, 6385; *Meles* 1102, 1103, 1104, 1572, 1573, 1574, 1575, 1576, 1577, 1579, 1581, 1584, 1585, 2822, 4416, 4429, 6160, 6167, 7457, 7460, 7975, 9502; *Mephitis* 11383; *Mustela* 489, 2255, 2421, 3028, 3406, 4340, 5325, 5326, 6906, 7132, 7134, 7135, 7637, 7638, 7640, 7643, 7643a, 7644, 8522, 9271, 9460, 9880, 9881, 10469, 11946; *Taxidea* 3245). Some *Spilogale* subspecies have delayed implantation, while others do not (840, 3019, 3185, 3407a, 3408, 3410, 7126, 7127, 7129, 7129a, 7130, 7132a, 7133, 7140, 7141, 8969, 8970, 10002, 10003). Precocious implantation in *Mustela vison* is induced experimentally by a) transfer from natural daylight to long photoperiods (6907); b) administration of prolaction which stimulates secretion of progesterone by the corpus luteum (6905, 7641, 7642, 8226a); c) administration of pimozide which stimulates prolactin secretion by the hypophysis (7639); and d) the administration of progestagens (2018, 5264; cf 4303, 7638).

The labyrinthine, endotheliochorial, zonary or bidiscoid placenta has a mesometrial or antimesometrial, superficial implantation into a bipartite or bicornuate uterus (1412, 2126, 2789, 3006, 3026, 4261, 4340, 4576, 7126, 7563, 7655, 8303, 8715, 10000, 10003, 10004, 10532, 11277, 11857, 11950) and may be eaten after birth (2421, 4269, 4689, 7655, 7709, 8627, 9822). A postpartum estrus occurs (1104, 1474, 1572, 1573, 1577, 1578, 2088, 4225, 4285, 4299, 4416, 4643, 6047, 6559, 7979; cf 1578, 1582, 6559, 7975, 9028).

Most mustelids are blind and naked or sparsely furred at birth (334, 576, 1240, 3416, 5146, 5210, 5761, 6385, 6387, 6559, 7234, 8276, 9325, 11277; cf 9610). Their eyes open at 20-30 d (*Conepatus* 8798), 3-5 wk (*Mephitis* 10330, 11277), ~4 wk (*Aonyx* 5761; *Mellivora* 296, 6559), 4-6 wk (*Meles* 3571, 4643, 6559, 7813,

7975, 7979, 10499; *Mustela* 2862, 3571, 4269, 4458, 4576, 4689, 7598, 7599, 8769, 9344, 10499, 10571, 11321; cf 1304, 4645), 4-7 wk (*Martes* 410, 1240, 2068, 3571, 4225, 7137, 10499), 41-48 d (*Arctonyx* 8232; *Ictonyx* 9326, 9327, 10158), 51-54 d (*Poecilogale* 9326, 9327, 10158), or 62 d (*Eira* 8631). Other neonates have fur, but do not open their eyes until 4 wk (*Pteronura* 465, 2812; *Spilogale* 2100, 10758) or 4-5 wk (*Gulo* 1474, 4643, 6046; *Lutra* 5242, 6927, 8318, 10499; cf 1941, 7979, 9175; *Taxidea* 5210, 6559, 7250). *Enhydra* neonates are fully developed with fur and open eyes (565, 5678, 5681, 8303, 9589, 10499).

Females have 4 mammae (*Aonyx* 10158; *Ictonyx* 10158; *Martes* 1474, 2068, 3571, 9175; cf 1474; *Mellivora* 6559, 9945, 10158), 5 mammae (*Lyncodon* 8618), 6 mammae (*Arctonyx* 6559; *Conepatus* 502; *Meles* 3571; *Mydaus* 6559), 4-6 mammae (*Lutra* 1474, 3571, 4643, 5277, 9175, 10158; *Poecilogale* 332, 4744, 9325, 10158), 6-10 mammae (*Mustela* 502, 1474, 3571, 4269, 4643, 5210, 7655), 8 mammae (*Spilogale* 502, 1474; *Taxidea* 502, 1119, 1474), or 10-15 mammae (*Mephitis* 502, 1474, 3134, 6387). Males have a baculum (636, 1473, 2068, 4576, 4643, 4776, 7128, 7979, 10499, 11277).

For *Mustela*, details of anatomy (1517, 2963, 3026, 5945, 6734, 6849, 11497), endocrinology (161, 664, 699, 995, 996, 1049, 1053, 1054, 1413, 1632, 1633, 2018, 2424, 2425, 2686, 2687, 2690, 2691, 2692, 3083, 3084, 3409, 3473, 4143, 4144, 4145, 4527, 4775, 5769, 5839, 6905, 6912, 7085, 7113, 7136, 7137, 7142, 7448, 7451, 7645, 7798, 7799, 8966, 9127, 9393, 9394, 9395, 9879, 10010, 10011, 10228, 10855, 11716a, 11961, 11962), spermatogenesis (1049, 1129, 5317, 5971, 7799, 8338, 9123, 10815a, 10815b, 11439), semen (9958), mating (6540, 7316a), conception (59, 1694, 4300), fertility (3777, 7635, 7636, 10553, 10556, 10557), sex differentiation (662, 3083, 3085, 6883, 10854), temperature (7654), light (454, 665, 959, 960, 961, 1050, 1051, 1646, 2688, 2762, 4302, 4651, 4652, 4653, 4777, 4911, 5046, 5977, 6312, 6851, 6906, 6907, 9391, 9392, 9393, 9394, 9396, 10800, 10812), milk composition (1999, 7580, 9635), and reproductive energetics (3172, 11448a) are available.

For other genera details of anatomy (*Enhydra* 10000; *Lutra* 9584; *Meles* 1108; *Taxidea* 11950), endocrinology (*Gulo* 7138; *Martes* 444, 445, 488, 834, 1101, 1586, 9955; *Meles* 100, 446, 447, 448, 1103, 1587, 3264, 6981, 6983, 6985, 6986, 6987, 7457, 7460; *Mephitis* 4194, 11277; *Spilogale* 7133, 8970; *Taxidea* 3245), oogenesis (1517), ovulation (4051), copulation (4391), milk composition (5297), light (*Martes* 756, 3030; *Meles* 1574, 1576, 6984), and reproductive energetics (4381, 8716) are available. One journal, Lutra, focuses on mustelids.

NEONATAL MASS

Aonyx cinerea	m: 57 g; n1	6393
Arctonyx collaris	58 g	8232
Eira barbara	m: 77.95 g; r74.4-81.5; n2	8631
	f: 85.05 g; r78.0-92.1; n2	8631
Enhydra lutris	1300 g	1474
	1500 g; r1400-1600	4643, 7979
	m: 1750 g; r1200-2500; n4	5677, 5678, 5681
	f: 1960 g; r1020-2830; n6	5677, 5678, 5681
	2000 g; n1	565
	2000 g; r1600-2500; n3	9589
	2000 g; r1700-2400; n3	7478
	3000 g; n1	3617
Gulo gulo	m: 73 g; near term	4643
	f: 83 g; near term	4643
	84 g; n3	7171
	88.1 g; r84.5-94.1; n3, 1 d	9916a
	107.7 g; n4; term emb	548a
Ictonyx striatus	10.0-15.0 g	10158
Lutra canadensis	132 g; n1, near term	4286
	r140-145 g; n4 near term	4739
Martes americana	28 g; n1	1240
	30-40 g	5210
Martes martes	30 g	9610
Martes pennanti	f: 28.9 g; r26.0-30.7; n3, n1 lt	6385
	m: 30.0 g; n1	6385
	m: 39.0 g; n1, 3 d	8715
	f: 41.2 g; n1, 3 d	8715

Martes zibellina	30 g; r25-34; (34 g singletons; 25 g ltsz >1)	4643
	30 g	9610
	30-35 g	7979, 8276
Meles meles	50-132 g	6559
	75 g	10499
	75 g; r63-84	4643, 7979
	89.5 g; r85-94; n2	11255
Mellivora capensis	210 g; n1	6559
Mephitis mephitis	32-35 g	11384
	33.4 g; r26.7-41.4; n12	11277
Mustela erminea	1.7 g; n6, 1 d	4269
	3 g; r2.7-4.2	5741
Mustela eversmanni	4-6 g	10499
	4.5 g	4643
	~8.4 g	7137
Mustela frenata	3.1 g; n6, 1 lt, 1 d	4269
Mustela lutreola	6.5 g	4643
	f: 7.6 g	12050
	m: 8.4 g	12050
Mustela nivalis	1.42 g; r1.09-1.70; n8	4576
	1.5 g	4458
	3.0 g; r2.0-4.0; n14, 1 d	2863
	4.5 g; n6 (1 m, 5 f)	2862
Mustela putorius	7 g	4643
	9-10 g	3571
	~10 g	9344
	10 g ea; n2	4689
Mustela vison	6.2 g; r5.9-6.5; n2	10571
	6-11 g	4643
	9-11 g	10499
Poecilogale albinucha	4 g	10158
Pteronura brasiliensis	m: 189 g; r161-213; n3	465
	f: 219.5 g; r209-230; n2	465
Spilogale putorius	9.5 g; n1 (alcohol)	2100
Spilogale pygmaea	6.9 g	10758
Taxidea taxus	93.5 g; r89.5-97.5; n2	6559
NEONATAL SIZE		
Eira barbara	f: 137 mm; n1	8631
	m: 137 mm; r134-140; n2	8631
Enhydra lutris	380 mm	4643
	500-550 mm	7979
	500-560 mm	4643
Gulo gulo	CR: 112.5 mm; n4 term emb	548a
	CR: 121 mm [text says cm]; n3	7171
	122-125 mm; a few days before birth	4643
	HB: 157.3 mm; r155-160; n3, 1 d	9916a
Lutra canadensis	CR: 136 mm; n4 near term	4739
	~270 mm	1871a
Martes zibellina	110-120 mm	8276
	110-120 mm	4643
	110-120 mm	7979
Meles meles	max 130 mm	10499
Mellivora capensis	100 mm	296
Mephitis mephitis	94.2 mm; n5, 1 lt	11277
Mustela eversmanni	65-70 mm	4643
	65-70 mm	10499
	80 mm	11321
Mustela frenata	52 mm; n1, 1 d	4269
	~56 mm	1871a
Mustela lutreola	TL: f: 71.9 mm	12050
	TL: m: 73.9 mm	12050
Mustela nivalis	42 mm; r38-43; n8	4576
	HB: 50 mm; n2	2863
	60 mm; n1, 4 d	2862

Mustela putorius	53 mm; r51-55	4689
	70 mm	4643
Mustela vison	50-90 mm	10499
Pteronura brasiliensis	m: 327 mm; r310-335; n3	465
	f: 330 mm; r325-335; n2	465
Spilogale pygmaea	HB: 63 mm	10758
WEANING MASS		
Eira barbara	1290 g; r1134-1146; n2, 60 d	8630
	2125 g; r2100-2150; n2, 90 d	8630
Enhydra lutris	est 6000-15000 g	3617
	10000-13600 g	5681
	f: est 15200 g	7478
	m: est 16200 g	7478
Martes americana	327 g; n3	1240
	435-480 g	6821
Martes martes	680 g	9610
Martes zibellina	600 g	9610
Mustela erminea	f: 28.2 g; n1, 35 d	4269
	m: 32.5 g; r30-34.7; n4, 35 d	4269
Mustela frenata	f: 30.5 g; 4 wk	4269
	m: 38.7 g; 4 wk	4269
Mustela nivalis	20 g; 3 wk; from graph	4576
	f: ~50 g; 8 wk; from graph	4576
	m: ~60 g; 8 wk; from graph	4576
Taxidea taxus	half grown	6559
WEANING SIZE		
Martes zibellina	390-400 mm; 2 mo	4643
LITTER SIZE		
Aonyx capensis	mode 2 yg	11507b
	mode 2-3	9327
	2-3	9945
	2-5	5761
	max 4	10153
Aonyx cinerea	1.25; r1-2; n4 lt, 1 f	6393
Arctonyx collaris	3 yg; r1-4; n3 lt	6559
	3 yg; r2-4; n2 lt	8232
Conepatus leuconotus	2-4	1474
Conepatus mesoleucus	mode 3-5; r2-10	8798
	3-6	190
Eira barbara	1.75; mode 2; r1-2; n4 lt	8627, 8631
Enhydra lutris	1 yg	7634
	1; est max twins	5246
	1 yg; 2 emb observed once	565
	1; 2 emb seen; never 2 yg	5681
	1 emb; n1 f	8303
	1 emb ea; n11 f	10004
	1 ea; n278 lt	5677, 5678, 10000
	2% twin emb, 4% twin ovulations	3098
Gulo gulo	mode 2-3; r1-5	10499
	2; r2-3; n7	4643
	2.00 blastocysts; se0.19; n18 f	6505
	2.2; r2-3; n13 lt	6046
	2.2; r2-3; n15 lt	8793a
	2.5; r1-4; n38 lt	8793a
	2.5 yg; r1-4; n58 lt	8793a
	2.6; r2-3	8792
	2.60 emb; se0.24; n5 f	6505
	2.6; r1-4; n45 lt	8793a
	2.78 cl; se0.18; n14 f	6505
	2.9; mode 2-3; r1-4; n33 lt	12091
	3 yg; n1 lt	7171
	3-4	11949
	3.2 emb; sd0.8d; n23 f	548a
	3.3 ps; sd1.2; n18 f	548a

	3.5 yg; r1-6	1871a
	3.9 cl; sd1.0; n23 f	548a
	4; n1 lt	4201
	5 emb; n1 f	8793a
Ictonyx striatus	1-3	9326, 9327
	1-3 wild; 3-5 captive	10153, 10158
	2-3 yg	4354a
Lutra canadensis	2; r1-5	1474
	2-4 yg; r1-6	1871a
	2.22 emb; r1-3; n9	4286
	2.68 emb; ±0.71	4739
	2.875 emb; r2-4; n8 f	11763
	2.92 cl; ±0.79; n60 f; 1.9 blastocysts; n7 f	4739
	mode 3; r1-5	2088
	3 yg; n2 nests	11763
	3.0 yg; r2-4; n3 lt	2109
	3.43 yg; r2-4; n7 lt	6442
	4 ?; young with f, but some mixing of pups	7195
Lutra felina	mode 2 or 4-5	10809
Lutra longicaudis	4 emb; n1 f	4228
Lutra lutra	r1-2 weaned; n13 lt	5286
	1.5; mode 1; r1-3; n4 lt; n6	8318
	1.78 yg; sd0.57; mode 2; r1-3	6075
	2; n2 lt	12128
	2.0 yg; sd0.6; r1-3; n19 f	4569
	2.1 emb, implantations; sd0.7; r1-3; n17 f	4569
	2.1; mode 2; r1-5; n160 lt	6927
	2.2 cl; sd0.8; r1-3; n14 f	4569
	2.3; mode 2; r1-4; n70 lt	10513, 10514
	2.8; r2-5; n48 lt	11206
Lutra maculicollis	1-2 yg; n2	9327
	mode 2; r1-3	10158
	2-3	5277
Lutra perspicillata	5; n1	9175
Martes americana	1-3 cl	6863
	2-3 or 4 yg; n3 lt	11402
	2.2 yg; r1-3; n5 lt	1240
	2.6 yg; r1-4; n20	6821
	2.85 yg; r1-5	1869
	3 yg ea; n3 lt	410, 411
	3.02 cl; r2-5	11949
Martes flavigula	2-3	4643
	2-3; sometimes 5	10499
	2-3	9175
	max 5	7979
Martes foina	2-3	9175
	2-6	4645
	3-4; r1-8	7979
	mode 3-4; max 8	10499
	3-5	7652
	mode 3-5; r2-7	3571
	3-7; 2 occasional	4643
Martes martes	2-5	3571
	2; n1 lt	1942
	3; r1-4	9610
	3; r2-4	4643
	3-4; max 8	7979
	mode 3-4; r2-7	4643
	3-5	7652
	3-5	4645
	mode 3-5; r2-8	10499
	3.8; r3-5	4643
	5; r3-7	4643

Martes pennanti	1-5	2060
	mode 2-3; r1-5	2088
	mode 3; most 2-5; r1-6	8714, 8715
	2 yg; max 5; r1-4	5146
	2.66; r1-4; n2 lt + blastocysts counts of 7 f	4285
	2.7 yg; mode 3; r1-4; n26 lt	4225
	2.72 cl; mode 3; r2-4; n23 f	2852
	3-4	11949
	3-5 cl	3031
	3.28 cl; n44 f	2068, 11953
	3.45 emb; mode 3; r3-4; n11 lt	2068, 11953
	3.5 cl; mode 3; n16 f	6385
	3.67 cl; ±0.26; n12 f	5631
	4; n1	8716
Martes zibellina	mode 3-4; r1-10	8276
	2.8-3.0; across years	9610
	3.2 emb; n153 f	7987
	3.2; r1-5; n116 lt	756
	3.4 emb; n14 f	675
	3.49 cl	834
Meles meles	1-5	1572
	2 yg; n1 lt	11255
	2; mode 2; r1-4	12128
	2.23 cl; r2-11; n43 lt	4429
	3-4 cl; r1-10	89
	3.6; r2-7; n16 lt	11205
	4.7 cl; 3.1 emb	11444
Mellivora capensis	1-4	5761
	2	9175
	2	10158
	2	9945
	2-4	11253
	2; n1 lt	1230
	mode 2; r1-4; n3 lt	6559
Melogale personata	2 emb; n1	5403
	3 yg	6369
Mephitis macroura	3 emb; n1 f	2348
	5 emb; n1 f	502
	5 emb; n1 f	1474
	5; r3-8	6387
Mephitis mephitis	2.5 yg; se0.4; n6 dens	966
	3 yg; n1 f	9822
	3.3 yg; r2-5; n3 lt	9066
	3.5; r3-4; n2 lt	9821
	4.28 yg; sd2.23; r1-10; n118 lt, 67 f	11382
	5-19 cl	11381
	5 yg; n1 lt	576
	5.2 yg; se0.8; n9 dens	966
	5.7 yg; r4-7; n3 lt	502
	6 yg ea; n2 f	10330
	6; r2-14	9164
	6.0 ps; se0.5; n15 f	966
	6.0 emb, ps; se0.33; n3-10; n25	3557
	6.3 cl; n116 f	9644
	6.45 cl, emb, yg; n11 f	498
	6.7 ps; n274 f	9644
	6.8 emb; n87 f	9644
	7.15 ps; se0.28; r4-11; n39 f	11277
	7.33 emb; r5-9; 6.33 live emb; r0-5; n21 f	11277
	8.0; r5-9; n3 lt	8245
Mustela altaica	2-8	10499
	5.5; adult f; 3.6; young f	4643
	7-8	7979

Mustela erminea	3-7	3571
	3-9 emb	2418
	4-8	1474
	mode 4-8; r2-13	4645
	mode 4-8; r2-18	4643
	4-13	7652
	4-13	5146
	4; n1 lt	8769
	4 emb; n1 f	502
	5; n1 lt	7598
	5-10 weaned	9175
	7 yg; n1 f	4269
	7 blastocysts; n1 lt	4283
	8-9; r3-14; max 18	7979
	9.7; r9-10; n3 lt	10512
	9.7 cl; 8.8 emb; r6-13	5744
	11 emb; n1 f	4073
	max 19 ova	5741
Mustela eversmanni	4-14; mode 8-9	10499
	6.8 yg; r4-10; primiparous f	7137
	8-11; max 18	4645
	9.5; r3-18	4643
	13.25; n4 lt	11321
	8-11; max 18	7979
Mustela frenata	3-9 cl	11946
	4 yg; n1 f	4224
	6-9 yg	11948
	mode 6; r3-8	6387
	6.5 yg; r6-7; n2 lt	4269
	9 yg; n1 nest	576
Mustela lutreola	mode 4-5; r2-7	3571, 4643, 4645, 7979, 10499
Mustela nigripes	2.4 wean; sd1.7; r1-6; n62	11716
	3.0 yg; sd1.4; r1-6; n62	11716
	3.3 yg; sd0.89; mode 3; r1-5; n68 lt; over 4 yr	3416
	3.33 yg; r1-5; n3 f	9876
	3.43 yg; n21 f	327, 328
	3.5; r1-5; n11 f	4802
	4; r3-5	1867
Mustela nivalis	4; n1 lt	1304
	4.4; r4-5; n8 lt	1010
	4.5; r3-6; n4 lt	4458
	4.7 yg; r2-7; n3 lt	2863
	4.7; mode 6; r1-6; n9 lt	4576
	5.0; mode 4; r3-10; n7 lt	4227
	5.2 emb; sd1.25; r4-8; 4.9 live; sd1.49; n14 f	5277a
	5.5 emb, 5.0 alive; n2 f	580
	5.6 emb; n12 f	5740
	6; n1 lt	2862
	6; n1 lt	4457
	6; r4-12	10512
	6.75 yg; r5-9; n4 lt	8636
	7.1; r4-11	2422
	9; n1 lt	8637
Mustela nudipes	4; n1 lt	7161
Mustela putorius	5-11 cl	1633
	5-13	4299
	5.7; r2-12; n13 lt	1230
	5.83; r2-8; n6 lt	7652, 7654
	7; n1 lt	4689
	8 implantations; se1.5; n4 lt, 20 d postcoitus	7637
	8-9	6156
	9	9028
	9.3; r7-12; n3 lt	7709

	9.86 cl; mode 7-8; r6-10	7655
	13.5; se1.3	7135
Mustela sibirica	2-12	7979, 10499
	7; r4-10	4643, 4645
Mustela vison	5 blastocysts; n1 f	161
	4.0 implantations; r1-7; n2 f	161
	1.20-4.32; 3 wk; across strains	2778
	2.29-4.70; across strains	4911
	3.04; n84 lt	3473
	3.8; r3-6; n10 lt	3735
	3.87 yg; se0.6; n15 young f	8522
	4.2; mated before Mar 21	5133
	4.25 yg; n697	1166
	4.31-5.38; n295 lt	5325
	4.40; se0.5; r3-6; n5 adult	2018
	4.48 yg; n>600 lt	3026
	4.62; n2032 lt	5954
	4.74; se0.17; n158	7636
	4.8 yg; ±2.6; n10 f, 10 lt	5264
	5.16 mated after Mar 21	5133
	5.20; r1-8; n5 young f	2018
	5.3 yg; se0.9; n10 adult f	8522
	5.34; se0.09	7638
	5.47; se0.09	7638
	5.5; se0.1	7639
	5.7 emb; r5-7; n3 f	161
	6 emb; n1 f	1290
	6.17; r4-12; n30 lt	57
	9.8 cl; n50	3026
	max 17 rare	7979, 10499
Poecilictis libyca	1-3 yg	4354a
	2 yg; n1 lt	8427
Poecilogale albinucha	1-3	9326, 9327
	2	11253
	2 yg	4744
	2-3 yg	10153, 10158
Pteronura brasiliensis	1 yg; n1 lt	10891
	r1-3 yg; n19 sightings	2812
	2.75 yg; r2-5; n4 lt	465
Spilogale putorius	1-4 blastocysts	10003
	5.3; r4-7; n3 lt	2100, 2101
	5.5	7126
Spilogale pygmaea	4.0; r2,6 yg; n2 lt	10758
Taxidea taxus	mode 2-3; r1-7	6559
	2; max 5	5146
	2-5	1474
	2 emb; n1 f	4249
	3 emb; n1 f	1119
	4-5 yg	502
Vormela peregusna	4-8	7979
	4-8	10499
	4-8	4645
	4.3; r3-8; n13	4643
SEXUAL MATURITY		
Aonyx congica	f/m: ~1 yr	7983
Conepatus mesoleucus	f/m: 1st mate: ~10-11 mo	8798
Eira barbara	f: 1st estrus: 23 mo; n1	8630
	m: testes: descend 6 mo	5583
Enhydra lutris	f: 1st mate: 2 yr	7979
	m: 5-6 yg	3098
	f/m: 2 yr	4643
Gulo gulo	f: 1st mate: 1 yr	11949
	f: 12-15 mo	1871a
	f: 2 yr	6505

	f: 4 yr	6046
	m: >3 yr	6505
	f/m: 1-2 yr	6048
	f/m: 2 yr	10499
Ictonyx striatus	f: 1st birth: 10 mo	9326
	m: testes: scrotal 16-20 wk	9326
	m: 1st mate: 22 mo	9326
Lutra canadensis	f: 1st birth: 2 yr; n1 f	6443
	f/m: 1st mate: 24 mo	7195
	f/m: 2nd yr	857
	f/m: ~2 yr; max 3 yr	8798
Lutra lutra	f: 1 yr	10499
	m: 1st sire; 18 mo	11502
	m: scrotal sac length: 20-22 mo; n1	5287
	m: 2 yr	10499
	f/m: 1st mate: 3 yr (some earlier)	6927
	f/m: 2 yr; wild; 1 yr	4643
	f/m: 2-3 yr, some 1 yr	3571
Martes americana	f: 1st mate: 2 yr	5146
	f: 1st mate: 2 yr	11949
	f: 1st mate: not as yearling	5438
	f: 1st birth: 1st yr	1474
	m: 1st mate: 1 yr	5438
	f/m: 1st mate: 15 mo	6821
Martes foina	f: 1st birth: 1 yr	4645
	f/m: ~2 yr	3571
Martes martes	f: 1st mate: 456 d; n1	6047
	m: 1st mate: 14 mo	6047
	f/m: 2 yr, 3 mo	7979
	f/m: 2-3 yr	3571
Martes pennanti	f: 1st mate: 1 yr	2068
	f: 1st mate: 1 yr; n1	4225
	f: 1st birth: 2 yr	2068
	f: 1st birth: 2 yr	2088
	m: sperm: 1 yr	2068
	m: 2 yr	8714
	f/m: 1st mate: 1 yr	11949
	f/m: 1st mate: 1 yr	6385
	f/m: 1 yr	4285
	f/m: 1-2 yr	1871a
Martes zibellina	f: 1st mate: 18 mo (20%)	9610
	f: 1st preg: 26-33% 1 yr	4643
	f/m: 1st mate: 15 mo	9955
	f/m: ~15 mo	8276
	f/m: 15-16 mo	7979
	f/m: 15-18 mo	488
Meles meles	f: 1st mate: 12-17 mo	4416
	f: 1st ovulate: 13 mo	89
	f: 1st birth: 1 yr	4645
	m: spermatogenesis: 1 yr	89
	m: spermatogenesis: 2 yr	10419
	m: spermatogenesis: > 2 yr	4416, 4429
	m: 1st mate: 1 yr	4645
	m: ~15 mo	3988
	f/m: 1 yr	7979
	f/m: 1 yr	10499
	f/m: ~1.5 yr	3571
	f/m: 2 yr	4643
Mephitis mephitis	f: 1st mate: 1 yr	11277
	f/m: 10 mo	1871a
Mustela erminea	f: 1st mate: 17-75 d; n29	10752
	f: 1st mate: 6-8 wk; before weaning	11564
	f: 1st mate: 3 mo; n1	7598
	f: conceive: 5-6 wk	7599

	f: 1st preg: 3 mo	4283
	f: 3 mo	10512
	f: 3-4 mo	1871a
	m: spermatogenesis: 1 yr	4144, 11564
	m: 1 yr	4283
Mustela eversmanni	f: 1st mate: 10 mo	7137
	f: 1st birth: 1 yr	7137
	f/m: ~9 mo	10499
	f/m: ~10 mo	7979
	f/m: 10 mo; completely developed 2 yr	4643
	f/m: 1 yr	4645
Mustela frenata	f: 1st mate: 3-4 mo	11945, 11948
	m: 1st mate: 1 yr	11947, 11948
Mustela lutreola	f: 9-10 mo	12050
	f/m: ~9 mo	3571
	f/m: 1 yr	7979
	f/m: next spring	4645
	f/m: yr after birth	4643
Mustela nigripes	f/m: 1 yr	3416
	f/m: 1 yr	11716
Mustela nivalis	f: 1st mate: 16 wk	4576
	m: testes: ~4 mo	4776
	m: 1st mate: 32 wk	4576
	m: 1st mate: 1 yr	11564
	f/m: 3-4 mo	1871a
	f/m: 3-9 mo	3571
Mustela putorius	f: vulvar edema: 30-50 wk	9394
	f: 1st mate: 6-9 mo	9344
	m: spermatogenesis: 8 mo	5167
	f/m: 1st mate: 1 yr	9028
	f/m: ~9 mo	3571
	f/m: ~1 yr	4643
	f/m: 1 yr	7979
Mustela vison	m: spermatogenesis: 8-9 mo	1129
	f/m: 10 mo	3026
Poecilogale albinucha	m: 1st mate: 33 mo	9326
Spilogale putorius	f: 10 mo	7126
	f/m: 5 mo	1871a
Taxidea taxus	f: 1st mate: 4-5 mo	7250
	f: 6-9 mo	6559
	m: sperm: 15 mo	11951
	m: 1st mate: 16-17 mo	7250
	m: >1 yr	6559
	f/m: 1st mate: ~1 yr	11950
CYCLE LENGTH		
Aonyx cinerea	24-30 d	2813
	28 d; n1 f	6393
Eira barbara	estrus: 12.1 d; r5-20; n20, 3 f	6395, 8627
Lutra lutra	estrus: 14 d	11502
	24-28 d	2813
	36 d; via estradiol peaks	10918
	40-45 d	11502
Lutra perspicillata	estrus: 14 d	9175
	~1 mo	2813
Martes martes	estrus: 3-4 d; n34, 24 f	9610
	estrus: ~10 d	6047
	3-7 d; r3-18; n34 cycles, 24 f	9610
Martes zibellina	estrus: 1-2 d; rarely 3	8276
	estrus: 1-3 d	4643
	estrus: 2 d; n116 f	9610
	7-20 d	4643
	mode 9-12 d; r7-17; n177 cycles, 116 f	9610
	10-12 d	8276
Mephitis mephitis	estrus: 1 wk	4194

Mustela erminea	4 wk	4643
Mustela eversmanni	estrus: 41 d; ± 27	11716
Mustela lutreola	estrus: 5 d	7559
	metestrus: 10 d; r1-32	12050
	diestrus: 17 d; r3-41; n4	12050
Mustela nigripes	estrus: 32-42 d; n1 f	11716
	proestrus: 2-3 wk; r8-31 d	11716
Mustela putorius	estrus: 3-5 d	7979
	estrus: 4-5 mo	6156
	estrus: max 17 wk	4299
Mustela vison	estrus: mode 9 d; r1-40	7979
	estrus: ~3 wk	3026
	estrus: 1 mo; n some f	4643
	7-10 d; interestrus interval	1871a
GESTATION		
Aonyx capensis	63 d	11697
	~63 d	5761
	~63 d	10153
Aonyx cinerea	60-64 d	2813
Arctonyx collaris	postimplantation: 6 wk	8232
	5-9.5 mo	8232
Conepatus leuconotus	~2 mo	1474
Conepatus mesoleucus	42 d	8798
Eira barbara	63-67 d; n≥7	8627, 8630, 8631
	~70 d; n1	11235a
Enhydra lutris	6.5-9 mo	3098
	est 8 mo with delay	5678, 5681
	est 8-9 mo	10499
	238 d; n1; r8-9 mo	4643
Gulo gulo	active development: 30-40 d	1871a
	4 mo	7652
	215 d; n1	7429a
	~250 d with delay	1871a
	272 d; n1	7171
Ictonyx striatus	36 d ea; n3	9326
Lutra canadensis	active development: ~50 d	1871a
	max 43 d; n1	2088
	245-365 d	2813
	9 mo 18 d; n1	6443
	9.5-10 mo	1474
	9 mo 18 d-12 mo 15 d	6442
	est 12 mo with 9 mo delay	4286
Lutra felina	60-120 d	10809
Lutra lutra	~59 d; r53-100; n5	8318
	60-62 d	2813
	61 d; n1	1941
	61-63 d	3571, 4643
	61-74 d	11502
	61-87 d; 1st vs last mating to birth	1355
	63 d	10513
	9 wk	7652
	9-10 wk	10499
Lutra maculicollis	2 mo	8760
	2 mo	5277
Lutra perspicillata	61-63 d	9175
	62 d	2813
Martes americana	implant: est 60-102 d	411
	implant: ≥6 mo	11402
	implant: 7-8.5 mo	5438
	implant: 248-250 d	11949
	postimplantation: ~27 d	5438
	postimplantation: 28-29 d	11949
	220-230 d	1240
	244 d	410

	8.5-9 mo	3030
	259-275 d	411
	259-276 d; with delay	11949
	268.5 d; r261-276; n2	6821
	9 mo	8312
	~9 mo with delay	11402
Martes foina	implant: 2-3 mo	3571
	implant: 240 d	1586
	postimplantation: 28-30 d	1586
	postimplantation: 5.5-7.5 mo	3571
	236-274 d with delay	7979, 10499
	236-275 d with delay	4643
	est 8-9 mo	7652
	8.5-9.5 mo with delay	3571
	268-270 d with delay	1586
Martes martes	implant: 5-7.5 mo	3571
	implant: ~240 d	1582
	postimplantation: 27-28 d	4643
	postimplantation: 30 d	1582
	postimplantation: 2-3 mo	3571
	postimplantation: est 2-3 mo	9610
	r94-106 d; n3	1942
	230-270 d	7979
	230-275 d with delay	10499
	236-274 d with delay	4643, 4645
	8-9 mo	7652
	8-9.5 mo with delay	3571
	264-274 d; n1	6047
	270 d with delay	1582
	267-291 d	9610
Martes pennanti	implant: 9 mo	2088
	implant: 10 mo	4285
	postimplantation: 1 mo	2088
	postimplantation: 2 mo	4285
	9 mo	3031
	327-358 d	11949
	11.5 mo	8715
	350 d; r211-370	2088
	352 d; r338-358; n15	4225
	~1 yr	4285
	1 yr	2068
Martes zibellina	implant: 5 mo	4643
	implant: 7-9 mo	7979
	postimplantation: 27-28 d	4643
	postimplantation: 25-40 d	7979
	postimplantation: est 2-3 mo	9610
	236-315 d; repeated matings	8276
	~8 mo	7652
	245-298 d with delay	4643
	250-300 d; 1 mating	8276
	253-297 d with delay	7979
	267-291 d	9610
	~11 mo	756
Meles meles	implant: 2-10 mo	6559
	implant: 3-9 mo	4416, 4429
	implant: 4 mo	3264
	implant: 4-5 mo	3571
	implant: 6-10 mo	11444
	implant: 300-320 d	1575, 1584
	implant: 10 mo	1577
	postimplantation: 45 d	1104, 1572, 1575
	postimplantation: 6 wk	6559
	postimplantation: 2 mo	1577
	postimplantation: 2-3 mo	3571

	postimplantation: 7-8 wk	3264
	5 mo--~1 yr	4416, 4429
	6 mo	3264
	6 mo	7652
	min 6 mo; 2 mo	7813
	6-8 mo	3571
	200-350 d	4645
	271-284 d; summer mating	4643
	271-284 d; mate mid-July	7979
	10-11 mo	11444
	12 mo	1577
	12 mo; with delay	1104, 1572, 1575
	~1 yr; spring mating	4643
	~1 yr; r339-376 d; non-July matings	7979
	14-15 mo	10419
	14-15 mo; winter mating	4643
Mellivora capensis	180 d	12059
	~180 d	9175
	6 mo	11697
	est 6 mo	6559
	~6 mo	5761
Mephitis macroura	~8 wk	6387
Mephitis mephitis	implant: ~19 d	11381
	59-77 d	11383
	59-77 d; 61.5 d 2nd preg of yr	11382
	61 d; r59-62 d; n4	8245
	61-67 d	9822
	62 d	11674
	63 d	1355
	63.2 d; sd2.2; r60-69; n26 mate on or after mid Mar	11382
	63.3 d; r62-66; n3	11277
	64, 66-75 d; n2	9821
	69.1 d; sd4.1; r63-77; n33 mate before mid Mar	11382
Mustela altaica	30-40 d	4643
	~40 d	7979
	~40 d	10499
Mustela erminea	implant: ~5 wk; longer 2nd mating	3571
	implant: est 5-6 mo	7979
	implant: 9-10 mo	2421
	active development: 21-28 d	1871a
	postimplantation: ~6 wk	3571
	postimplantation: est 2-3 mo	7979
	4 wk	9175
	42 d	11564
	6 wk or 9-10 mo	4643
	8-9 wk, mate Mar; 8-9 mo; other matings	11497
	77 d; n1 capture to birth	4269
	10-11 wk	7652
	10 wk, mate Feb-Apr; 9-10 mo other matings	10512
	10.5-11 wk or 9-10 mo 2nd mating	3571
	7 mo; adult f	7598
	8 mo; young f	7598
	r240-393 d; n35	10752
	8.5-10 mo	1474
	8-9 mo	5744
	est 9-10 mo	7979
	270 d with delay	1871a
	280 d; r>71-378 d; n16	7599
	~11 mo	4645
	1 yr; with delay	2421
Mustela eversmanni	implant: <13 d	7137
	implant: 14 d; with 7-8 d delay	4643
	postimplantation: 29 d	7137
	36-40 d	7979

	36-43 d with delay	4643
	38-41 d	10499
	40.8 d; ±1.2; r39-43; n9	7137
	40-42 d	4645
Mustela frenata	implant: 8 mo	11949
	active development: 21-28 d	1871a
	postimplantation: 23-24 d; r21-28	11944, 11948, 11949
	min 70 d; n1; capture to birth	4269
	≥131 d; n1	4224
	270 d with delay	1871a
	279 d; r205-337; n18	11944, 11945, 11946, 11948
Mustela lutreola	35-42 d	7979
	39-48 d; sometimes 71-76 d	3571
	40-43 d	12050
	42-46 d	4643
	6-8 wk	4645
	43-72 d	10499
Mustela nigripes	42.7 d; sd0.7; r42-45; n62	11716
Mustela nivalis	5 wk	1304
	5 wk	10512
	5 wk	3467
	35; mode 35; r34-36; n9	4576
	35.1 d; r33-36	1010
	35.7 d; r35-36; n3	2863
	36 d; r35-37; n4	4457, 4458
	5-6 wk; after main mating; 8-10 wk; rest of yr	3571
	~42 d	11564
	min 7 wk	4645
	~50 d	8223
Mustela putorius	implant: 11-13 d	7655
	implant: 12 d	7135, 7142
	implant: 13 d	11962
	implant: 12-13 d	3018
	40-43 d	3571
	~40 d	7979
	40 d 17 h; r39 d 5 h-42.0 d; n5	7654, 7655
	41 d	9028
	41 d	4689
	42 d; mode 42; r41-43; n21	4299, 4527
	42 d; sd3	6156
Mustela sibirica	28-30 d	10499
	28-30 d or 2 mo	7979
	~30 d	4645
	35-42 d	4643
Mustela vison	implant: 1-46 d	7979, 10499
	implant: 9-20 d	7448
	implant: 14-16 d	7641
	implant: 2-3 wk; max 1 mo	4643
	implant: 16-20 d	4340
	implant: 21-23 d	3028
	postimplantation: 25-26 d	4643
	postimplantation: 28-30 d	3028
	postimplantation: 29-35 d	7979
	postimplantation: ~31-33 d	8312
	postimplantation: 40 d	7448
	36-75 d with delay	7979, 10499
	38-85 d	5146
	42 d; r39-76	1474
	46.4-56.8 d	2779
	48.0-51.5; across treatments	4911
	7-8 wk	7652
	49.05 d; se0.19; n374	7638
	49.66 d; sd4.89; r39-76; n924	3026
	50.95 d; r39-76; n747	1166

	51 d	3028
	51 d; sd5.3; r40-76; n233	8312
	51.3 d; r40-65	4643
	51.7 d; sd5.2; r40-71; n459	345a
	52.2 d; ±5.1; n10 f	5264
	53.3 d; se0.3; n390; last mating to birth	7638, 7639
	56.2 d; se1.8; adult	8522
	60 d	7448
	2 mo; n2	161
	61.8 d; se1.5; young f	8522
	75 d; n1	10571
Poecilictis libyca	37 d; n1	8427
Poecilogale albinucha	31.75 d; mode 31; r31-33; n4	9326
Pteronura brasiliensis	66.7 d; r65-70; n3	465
	70 d; n1	10891
Spilogale putorius	delay: est 14-16 d	7126
	implant: 180-230 d	7130, 7133
	postimplantation: 28-31 d	7130, 7133
	33 d	10003
	est 50-60 d	7126
	120 d; capture to birth	2026a
	210-266 d	7130, 7133
	218 d; r203-234; n7	7129
Spilogale pygmaea	43-51 d; ~47 d after mating; n1	10758
Taxidea taxus	implant: 6-7 mo	7250
	postimplantation: 5 wk	6559
	postimplantation: 6 wk	4287
	postimplantation: 2-3 mo	7250
	6.5 mo	11950
	9-10 mo	7250
Vormela peregusna	est 8 wk	4643
	~8 wk	4645
	~2 mo	7979
LACTATION		
Arctonyx collaris	solid food: 85 d	8232
Conepatus mesoleucus	den emergence: ~35 d; r6-8 wk	8798
Eira barbara	solid food: 62-64 d	8631
	wean: 75-100 d	8631
Enhydra lutris	wean: ~2 mo	7979
	dependent: 169 d; r76-333; n23	7478
	dependent: 6-8 mo	3098
	independent: 7 mo	7979
Gulo gulo	solid food: the middle of lactation period	10499
	wean: 2.5-3 mo	10499
	wean: 3 mo	4643
Ictonyx striatus	solid food: ~33 d	9327
	solid food: 32 d	9326
	wean: ~8 wk	9326
Lutra canadensis	solid food: ~8 wk	8798
	solid food: 63-75 d	1871a
	wean: ~4 mo	8798
Lutra lutra	solid food: 2 mo	6927
	wean: 21 d	8318
	wean: ~3 mo	10499
Lutra maculicollis	den emergence: ~2 mo	5761
Lutra perspicillata	wean: 6 wk	9175
Martes americana	solid food: 6 wk	1869
	weight decrease: 6 wk	1240
Martes foina	solid food: 6 wk	3571
	wean: 40-45 d	4643
	wean: ~3 mo	3571
Martes martes	solid food: 6 wk	3571
	solid food: 6 wk	4643
	den emergence: 44 d	9610

	wean: 8 wk	9610
	wean: ~3 mo	3571
	wean: 6-7 wk	10499
	wean: 1.5 mo	7979
Martes pennanti	den emergence: 9-10 wk	2088
	wean: begins 8-10 wk	8714
	wean: ~10 wk	1871a
	nurse: 114 d	2068
	effective predation: 4 mo	8714
Martes zibellina	den emergence: 44 d	9610
	solid food: 50-52 d	4643
	wean: 8 wk	9610
Meles meles	den emergence: 6-9 wk	4643
	solid food: 2.5-3 mo	10499
	wean: 2.5 mo	7979
	wean: 3 mo	4643
	independent: 150 d	7975
	wean: 2-2.5 mo	3571
	wean: 75 d	7975
	wean: 12-18 wk	6559
	wean: 4 mo	4429
Mellivora capensis	solid food: 1 wk	11253
	den emergence: 6 wk	11253
Mephitis mephitis	solid food: 36 d	10330
	wean: ~2 mo; max 4	11277
Mustela altaica	independent: 2 mo	7979
	wean: 34-41 d	4643
	wean: ≤8 wk	11384
	wean: ~2 mo	10499
Mustela erminea	solid food: 3 wk	4269
	solid food: 4-5 wk	5741
	solid food: 35 d	8769
	wean: 4 wk	3571
	wean: 5 wk	4269
	wean: ~5 wk	9175
	wean: 7-12 wk	5741
	den emergence: 2 mo	7979
Mustela eversmanni	solid food: 20 d	11321
	solid food: 28-34 d	4643
	solid food: 6 wk	4645
	solid food: 2 mo; ~2.5 mo	10499
	nurse: >40 d	11321
	wean: ~6 wk	4645
	wean: 1.5-2 mo; 1.5 mo	7979
Mustela frenata	solid food: 3 wk	4269
	nurse: 5 wk	11948
	wean: 36 d	4269
Mustela lutreola	solid food: 20-25 d	4643
	wean: 4-5 wk	3571
	prey capture: 52-62 d	12050
	wean: ~2 mo	4645
	wean: 2-2.5 mo	4643
	wean: 10 wk	7979
Mustela nivalis	solid food: 2.5 wk	2863
	solid food: 18 d	4576
	den emergence: 28 d	2862
	wean: 2-9 wk	3467
	wean: 24 d	4457, 4458
	wean: 28-35 d	1871a
	wean: 42-56 d	4576
	wean: 6-8 wk	11564
	nurse: 45 d	1304
	wean: ~12 wk	2862

Mustela putorius	nurse: ~10 wk	4299
	wean: 6-8 wk	9344
	wean: < 3 mo	10512
	independent: 2 mo	7979
	wean: 4-5 wk	3571
Mustela sibirica	wean: 2 mo	4643
	wean: 2 mo	10499
Mustela vison	solid food: 20-25 d	4643
	wean: 2-2.5 mo	4643
Poecilogale albinucha	solid food: 35 d	9326, 9327
	wean: ~11 wk	9326
Pteronura brasiliensis	wean: ~4 mo	2812
Spilogale putorius	solid food ≤36 d	1871a
	solid food: ~40 d	2100
	wean: 50 d	1474
Taxidea taxus	wean: near 8 wk	5210
INTERLITTER INTERVAL		
Enhydra lutris	1 yr	10499
	2+ yr	5678
Gulo gulo	1 yr	6048
	2-3 yr	1474
Lutra canadensis	1 yr	7195
	1 yr	7234
Lutra felina	1 yr	10809
Lutra lutra	1 yr	9175
Lutra maculicollis	1 yr	5761
Martes americana	1 yr	1474
Martes foina	1 yr	3571, 4645, 7979, 10499
	1 yr	1586
Martes martes	1 yr	3571, 4645, 7979
Martes pennanti	1 yr	2068
	1 yr	1474, 2088
Meles meles	1 yr	4429
	1 yr	4645
	at times 2 yr?	3571
Mephitis mephitis	2.5 mo; n2 f	8245
	1 yr	1474
	1 yr	502
Mustela altaica	1 yr	10499
Mustela erminea	1 yr	4145
	1 yr	1474, 3571, 7979, 9175
Mustela eversmanni	1 yr	4643, 4645, 7979
	mode 1; 2 if 1st dies	10499
Mustela frenata	1 yr	11947, 11948
Mustela lutreola	1 yr	3571, 4645, 7979
Mustela nivalis	35.3 d; r34-38	3467
	1 yr	4645
Mustela mustela?	~1 yr	2421
Mustela putorius	20 wk	9344
Mustela vison	1 yr	4645, 5146
Vormela peregusna	1 yr	4645
LITTERS/YEAR		
Conepatus mesoleucus	sometimes 2	8798
Gulo gulo	0.5	4643
	1/2-3 yr	1474
Ictonyx striatus	mode 1/season	9327
Lutra lutra	1; r1-2 (?)	3571
Meles meles	0.5-1	3571
Mellivora capensis	max 2	6559
	2	11253
	2	9175
Mephitis mephitis	2	8245
Mustela nivalis	1	4645
	1; r1-2	3571

	> 1	1474
	1-2	2422
	3	3467
Mustela putorius	1	7979, 10499
	1; 2 if 1st lt removed	6156
	1; r1-2	3571
	2	9344
	2; r0-2	4299
Poecilogale albinucha	mode 1/season	9327
	2	11253
SEASONALITY		
Aonyx capensis	mate: July-Aug; Zambia	10153
	mate: Mar; Zimbabwe	10153
	birth: yr rd?; S Africa	10158
	yg emerge: Mar-Apr; central, s Africa	5761
Aonyx cinerea	repro: not seasonal; captive	2813
Conepatus leuconotus	mate: Feb; sw US, n Mexico	1474
	birth: Apr-May; sw US, n Mexico	1474
Eira barbara	mate: Jan-Apr, June, Oct; zoo	8627
Enhydra lutris	spermatogenesis: yr rd; w US	5678
	mate: peak Mar-Apr; USSR	7979
	mate: peak Mar-May; USSR	10499
	mate: Mar-May, July-Aug; USSR	4643
	mate: peak spring; Commander I, far e Russia	565
	mate: yr rd, peak fall; w US	5678
	unimplanted blastocysts: yr rd; AK	10000
	birth: Feb-May, Oct-Nov; w US	5678
	birth: est June; w US	1474
	birth: yr rd; CA	3273
	birth: yr rd, peak May; AK	7478
	birth: yr rd, peak spring; USSR	4643
	lact f: yr rd; USSR	10499
Gulo gulo	testes: regressed Aug-Jan; captive, WA	7138
	spermatogenesis: end Apr-May; USSR	4643
	spermatogenesis: end June; AK	4643
	mate: Apr-May; Finland, Sweden	6046, 6048
	mate: Apr-Aug; AK	1474
	mate: Apr-Oct; AK	11955
	mate: early June; n1; central AK	6049
	mate: June, Aug; n2; AK	6725
	mate: June-Sept; USSR	4643
	mate: Apr-Oct; AK	4643
	mate: Sept-Oct; w Siberia	10499
	mate: Dec-Jan; n Europe, n Asia, n N America	7652
	implant: Jan; AK	11955
	birth: end Jan-Mar; Finland	8793a
	birth: Feb-Mar; Finland, Sweden	6046, 6048
	birth: Feb-Mar; Sweden	4201
	birth: Feb-Apr; AK	1474
	birth: end Feb-early Mar; USSR	4643
	birth: Mar-Apr; AK	11955
	birth: end Mar; Sweden	12091
	birth: Apr-early May; zoo	4643
	birth: Apr-June; w US	5146
	birth: end winter-mid spring; w Siberia	10499
Ictonyx striatus	testes: large Sept-Apr, small May-Aug; captive	9326
	birth: Sept-Dec; captive	9326
	lact f: Feb, June, Sept, Oct; few data; Kenya, Tanzania	5761
Lutra canadensis	estrus: Oct-May; captive	2813
	mate: est Jan-Mar; NC	11763
	mate: Apr; ID	7195
	mate: winter-early spring; MN	6442
	birth: Jan-May; WI	5210
	birth: Mar; ID	7195

	birth: est Mar-Apr; NC	11763
	birth: Apr-May; US	1474
	birth: mid Apr; NY	7234
	birth: Nov-May, peak Mar-Apr; WY	1871a
Lutra lutra	mate: Jan-Mar; Iraq	10767
	mate: end Jan-Mar; Germany	10513
	mate: Feb-Apr; USSR	4643, 7979, 10499
	mate: July-Aug; captive	1941
	mate: peak winter; s Sweden	3075
	mate: yr rd, peaks Feb, July; Germany	3571
	birth: Mar-Oct; Shetland, UK	6075
	birth: Apr-May, sometimes June, rare Dec; USSR	7979
	birth: Apr-May; Germany	10513
	birth: Apr-May; USSR	10499
	birth: end Apr-early May; Petschora, USSR	4643
	birth: June-early July; Volga delta, USSR	4643
	birth: May; Eur USSR	4645
	birth: mid Oct; captive	1941
	birth: peak spring; lower Saxony, Germany	9063
	birth: spring-early summer; Pakistan	9175
	birth: yr rd, low Mar-Apr; nw Scotland, UK	6075
	birth: yr rd, peak Apr-June; Germany	3571
	birth: yr rd; Aberdeenshire, UK	5287
	birth: yr rd; Sri Lanka	8495, 8501
	birth: yr rd; n240; England, UK	6927
Lutra maculicollis	mate: July; Tanzania	8760
	birth: Sept; Tanzania	8760
Lutra perspicillata	birth: end Aug; n1; Pakistan	9175
Lutra provocax	mate: est yr rd; Chile	10809
Martes americana	mate: June-Aug, peak July; WI	5210
	mate: mid June-Aug; ALB, Canada	1869
	mate: end June-July; AK	1869
	mate: July-Aug; nw MT	5438
	mate: July-Aug; captive	8312
	mate: July-Aug; captive, ID	411
	mate: July-Aug; MT, ID	11949
	mate: mid July-early Sept; captive	6821
	mate: mid July-Sept; NY	3029
	mate: end July-early Aug; n N America	1474
	mate: end July-mid Aug; captive	411
	mate: Aug; captive	410
	mate: Aug; zoo	1240
	unimplanted blastocysts: Nov-Jan; ID	6863
	implant: Feb-Apr; nw MT	5438
	implant: after Feb 19; no location given	3030
	birth: Mar-Apr; MT, ID	11949
	birth: Mar-Apr; zoo	1240
	birth: end Mar-mid May, peak Apr; WI	5210
	birth: Apr; captive	8312
	birth: Apr; captive, ID	411
	birth: Apr; n N America	1474, 5146
	birth: mid Apr; captive	410
	birth: end Apr-early May; IL, WI	2060
Martes flavigula	mate: June-July, perhaps Mar; USSR	10499
	birth: spring; USSR	10499
	birth: spring-early summer; Pakistan	9175
Martes foina	testes: small Nov-Jan, large May; captive	444, 445
	mate: June-July; Belgorod, Ukraine	4643
	mate: June-Aug; s, central Europe	7652
	mate: July; USSR	7979
	mate: July; France	1586
	mate: July-Aug; Germany	3571
	mate: ~Aug; Eur USSR	4645
	mate: mid-summer; USSR	10499

	mate: end summer; Pakistan	9175
	diapause: until Mar; Crimea, USSR	5966
	birth: early May; Europe, w USSR	4645
	birth: Mar-May; Germany	3571
	birth: end Mar-Apr; USSR	4643
	birth: end Mar-early Apr; France	1586
	birth: early spring; Baluchistan, Pakistan	9175
Martes martes	sexual quiescence: Sept-Mar; Europe, w USSR	4643
	mate: Jan; n1 pr; captive	1942
	mate: June-Aug; n, central Europe	7652
	mate: end June-early Aug, peak July; captive	9610
	mate: end June-early Aug, peak end July-early Aug; Europe, w USSR	4643
	mate: end June-Aug; Germany	3571
	mate: July; Austria	6047
	mate: July; captive	1578, 1580
	mate: July-Aug; Europe, w USSR	4645
	mate: summer; USSR	7979
	mate: summer; USSR	10499
	ovulate: July; captive	1580
	implant: Mar; captive	1578
	implant: Mar; Europe, w USSR	4643
	birth: Mar-May; Germany	3571
	birth: mid Mar-Apr; captive	1578, 1580
	birth: Apr; Austria	6047
	birth: Apr; n1; captive	1942
	birth: Apr-early May; Europe, w USSR	4645
	birth: mid-end Apr; captive	9610
Martes pennanti	mate: Mar?; indirect evidence; Canada	2068
	mate: end Mar-early May; WI	5210
	mate: end Mar-mid May; BC, Canada	2088
	mate: early Apr; ONT, Canada	8715
	mate: Apr; captive, CA	4225
	mate: Apr-May; Canada	3031
	implant: Feb; NY	2852
	implant: end Feb; ONT, Canada	8715
	implant: after Feb; Canada	3031
	birth: Feb-May; WY	1871a
	birth: end Feb-early Apr; Canada	2068
	birth: Mar-Apr; captive, CA	4225
	birth: est mid-end Mar; NY	2852
	birth: mid Mar-mid Apr; WI	5210
	birth: mid Mar-Apr; BC, Canada	2088
	birth: end Mar; ONT, Canada	8715
	birth: end Mar-early Apr; MAN, Canada	6385
	birth: end Mar-early Apr; n N America	1474
	birth: end Mar-early Apr; w US	5146
	birth: end Apr-early May; IL, WI	2060
	birth: ~early May; NY	7234
Martes zibellina	spermatogenesis: mid May-July, USSR	4643
	mate: end June-mid Aug, peak mid July; captive	9610
	mate: June-July; USSR	488
	mate: June-July; 30-35 d; USSR	7979
	mate: June-July; captive	756
	mate: June-July; captive	834
	mate: mid June-mid Aug; Asia	8276
	mate: July-Aug; n, e Siberia	7652
	implant: end Feb-early Mar; USSR	4643
	implant: Mar; delayed; Asia	8276
	birth: end Mar-early May; USSR	4643
	birth: Apr; Asia	8276
	birth: Apr; captive	9955
	birth: Apr-May; USSR	7979
	birth: mid-end Apr; captive	9610

	birth: May; captive	756
Meles meles	epididymal sperm: yr rd, peak Jan-Feb; Switzerland	3988
	fertile m: yr rd; Sweden	89
	spermatogenesis: spring-summer; USSR	7979
	spermatogenesis: yr rd; France	448
	spermatogenesis: yr rd, peak June-Feb; France	449
	spermatogenesis: yr rd; France	447
	testes: large Jan-Feb; France	449
	testes: large Jan-Feb; Switzerland	3988
	testes: large Feb, low July; France	9027
	testes: large Feb-Mar, small Oct-Nov; Sweden	89
	testes: large early Aug; Germany	3522
	testes: regressed Oct-Dec; Germany, Austria	10419
	testosterone: high Feb-Apr; captive	6983
	mate: Jan-Feb; France	1572, 1577, 1578
	mate: Feb-Mar; adult; Apr-June; yearlings; Switzerland	11444
	mate: Feb-Apr; w France	6986
	mate: late Mar-May, July-Sept; Somerset, England, UK	4429
	mate: May; n1; Sweden	6564
	mate: early Aug-early Nov; Europe, w, n Asia	7652
	mate: Apr-June; s Sweden	7975
	mate: (late Apr) June-Aug; Germany	3571
	mate: June; zoo	7813
	mate: July-Aug; Germany	3522
	mate: July-Aug?; Germany	3263
	mate: summer, Oct; Germany	3264
	ovulate: Feb, June, Sept-Oct; sw England, UK	4416
	ovulate: July-Aug; Germany	3522
	conceive: Feb; lab	1104
	implant: Nov-Jan; sw England, UK	4416
	implant: mid Nov-Jan; Germany	3263, 3264
	implant: after Nov; Germany	3522
	implant: early-mid Dec; sw France	1572, 1573
	implant: Dec-Jan; Switzerland	11444
	implant: ~mid-end Dec; Bordeaux, France	1577, 1578, 1582
	implant: early Jan; s Sweden	7975
	implant: Jan; lab	1104
	birth: mid Jan-mid Mar; sw England, UK	4416
	birth: mid Jan-mid Apr; Germany	3264
	birth: end Jan-Mar; Switzerland	11444
	birth: end Jan-early Apr; Germany	3571
	birth: Feb; sw France	1572, 1573, 1577
	birth: Feb-mid Mar; Germany, Austria	10419
	birth: Feb-Mar; n11; zoo	12128
	birth: Feb-Mar; s Sweden	7975
	birth: Feb-Mar; Germany	3522
	birth: mid Feb-mid Mar; Sweden	6564
	birth: Mar; zoo	7813
	birth: Mar-Apr; Europe, w USSR	4645
	birth: Mar-Apr; Lebanon	6420
	birth: Mar-Apr; USSR	7979
	birth: est Mar-Apr; Siberia, USSR	10499
	birth: end Mar-Apr; n7; zoo	4643
Mellivora capensis	birth: Feb; n1; zoo	1230
	birth: twice/yr, rainy periods; Zaire	11253
	birth: yr rd?; S Africa	10158
Melogale personata	birth: just before rainy season; Thailand	6369
Mephitis macroura	mate: end winter; Mexico	6387
	birth: May-June; n Mexico	1474
Mephitis mephitis	spermatogenesis: end Dec-Aug; NY, NJ	4194
	testes: large Mar, small Aug; n IL	11277
	estrus: end Feb-early Mar; NY, NJ	4194
	mate: Feb-Mar; IL	11277
	mate: mid Feb-mid Apr; captive	11382, 11383

	mate: early Mar; OR	11674
	mate: Mar, May; captive	8245
	mate: Apr-May; TX	2192
	birth: early May; US	1474
	birth: May; ALB, Canada	966
	birth: May; OR	11674
	birth: May-early June; WY	1871a
	birth: May, July; captive	8245
	birth: May-mid June; WI	5210
	birth: May-June; ALB, SAS, Canada	9644
	birth: May-July; captive	11382
	birth: peak May; PA	9164
	birth: end May; IL	11277
Mustela altaica	spermatogenesis: end Dec-mid Jan; USSR	4643
	mate: mid Jan-early Apr; Kazakhstan	4643
	mate: Feb-Mar; Kazakhstan	7979
	mate: Feb-Mar; Kazakhstan, Altai, w Sayan, central Russia	10499
	birth: early Apr-mid June; USSR	4643
	yg found: early May; USSR	7979
Mustela erminea	spermatogenesis: Apr-Aug; lab	4144
	testes: large mid Feb-Aug, decline Sept; lab	4144
	testes: peak Apr; Netherlands	11190
	sperm in testes: Mar; Germany	11497
	ovarian wt: peak Nov; Netherlands	11190
	ovulate: Mar, July, Aug; Germany	11497
	mate: Feb-Mar; n, w Europe, w Asia	7652
	mate: Feb-Mar, June-July; Germany	3571
	mate: Feb-mid June; n Kazakhstan	4643
	mate: Feb-Oct, peak May-July; captive	7599
	mate: mid Feb-early Apr, mid June-early Aug; Germany	10512
	mate: May-July; lab	4144
	mate: June-July; n2; lab	7598
	mate: July-Aug; WI	5210
	mate: Nov-Dec; New Zealand	5744
	mate: spring-summer; Europe, w USSR	4645
	mate: summer; Pakistan	9175
	implant: after Feb; MT	11945
	implant: Mar; captive	4145
	implant: Aug-Oct; New Zealand	5744
	birth: mid Jan-mid Feb; n2; lab	7598
	birth: end Feb-mid May; Europe, w USSR	4645
	birth: Mar-Apr; Germany	10512
	birth: Mar-May; May-July; 2 regions; USSR	4643
	birth: Apr; captive	4145
	birth: Apr-May; ne, nw N America	1474
	birth: Apr-May; w US	5146
	birth: Apr-June; Germany	3571
	birth: mid Apr-early May; WI	5210
	birth: May; n1; captive	4073
	birth: Sept-Oct; New Zealand	5743, 5744
	birth: spring; Pakistan	9175
	birth: spring; USSR	7979
Mustela eversmanni	testes: large end Dec-mid May, peak end Feb; captive	7137
	estrus: Feb-mid Mar; Kazakhstan	9772
	estrus: Feb-Mar; USSR	7979
	mate: mid Feb-mid Mar; USSR	10499
	mate: end Feb-mid Mar; Kazakhstan	9772
	mate: Mar; captive	4643
	mate: Mar-?; Europe, w USSR	4645
	mate: end Mar-Apr; captive	7137
	birth: mid Apr-May; USSR	10499
	birth: end Apr-early May; USSR	7979
	birth: end Apr-May; Europe, w USSR	4645
	birth: mid May; captive	7137

Mustela frenata	testes: large Apr-July; n US	11947
	mate: Mar-Apr; Yucatan, Mexico	3633
	mate: June-Aug; captive	11946
	mate: July-Aug; MT	11945
	mate: mid summer; MN	11948
	implant: after Jan; MT	11945
	birth: Apr; BC, Canada	2088
	birth: peak Apr; MN	11948
	birth: Apr-May; MT	11945
	birth: mid Apr-mid May; WI	5210
	birth: May-June; Yucatan, Mexico	3633
Mustela lutreola	mate: Feb-?; Europe, w USSR	4645
	mate: Feb-Apr; USSR	7979
	mate: Feb-Aug; Germany	3571
	mate: Mar-Apr; USSR	10499
	mate: Apr; n1; zoo, Russia	4643
	mate: Apr-May; captive	7559
	birth: Apr-?; Europe, w USSR	4645
	birth: Apr-May; Germany	3571
	birth: Apr-May; USSR	7979
	birth: Apr-May; Europe, w USSR	10499
	birth: June; n1; zoo, Russia	4643
Mustela nigripes	mate: mid Mar-early Apr; WY	1871a
	mate: end Mar-early May; captive	11716
	birth: June; midcentral US	1474
	birth: spring; midcentral US	3416
Mustela nivalis	anestrus: Nov-Mar; England, Scotland, UK	5740
	estrus: Mar-Apr; USSR	7979
	mate: Feb-Mar; Germany	10512
	mate: Mar-summer; Europe, w USSR	4645
	mate: Apr; captive	2862
	mate: mid Apr; captive	1304
	mate: June-July; Sept-Oct; location not stated	11564
	mate: mid Oct; lab	4458
	mate: yr rd; Europe, n Africa, n Asia	7652
	mate: yr rd, peak Feb-Apr; Germany	3571
	mate: yr rd; Germany	8637
	mate: yr rd; USSR	4643
	preg f: Mar, Aug; n2; Romania	580
	preg f: Mar-Aug; England, UK	2422
	preg f: Apr-June; England, Scotland, UK	5740
	preg f: Apr-Sept; Poland	5277a
	birth: Apr-Aug; zoo	1010
	birth: Apr-Sept; England, Wales, UK	4776
	birth: early May; NY	7234
	birth: mid May; captive	1304
	birth: end June; captive	2862
	birth: July-Aug; Nov-Dec; location not stated	11564
	birth: mid Nov; lab	4458
	birth: yr rd; n N America	1474
	birth: est yr rd; America	4227
	lact f: Apr-July, Oct; England, Scotland, UK	5740
Mustela putorius	spermatogenesis: Mar-July; captive	5167
	testes: large Feb-Oct; lab	958
	testes: large Dec-Apr; captive	1049
	testes: large Mar-May; UK	11439
	testes: large Apr-May; captive	163
	anestrus: Aug-Apr; lab	4299
	estrus: Feb-May; Germany	7652, 7655
	estrus: Feb-Aug; captive England, UK	4299
	estrus: Mar-Apr; USSR	7979
	estrus: Mar-Aug; lab	9344
	estrus: mid Mar-end May; captive	10800
	estrus: June-Aug; captive	961

	mate: mid Feb-June, peak Apr-May; zoo	4643
	mate: end Feb-early June; Germany	3571
	mate: early Mar; lab	664
	mate: Mar-?; Europe, w USSR	4645
	mate: Mar-May; captive	9028
	mate: Mar-Aug; captive	6851
	mate: Mar-Aug; Sweden; captive	6156
	mate: end Mar-mid Apr; Germany	10512
	mate: end Mar-early May; USSR	10499
	mate: Apr; n1; lab	4689
	ovulate: Mar-Aug; lab	958
	birth: Jan, Mar-Apr, Aug, Nov-Dec; n13; zoo	1230
	birth: Apr; captive	961
	birth: Apr-July, rarely Aug; Germany	3571
	birth: May; captive	9028
	birth: May; Europe, w USSR	4645
	birth: May; n1; lab	4689
	birth: ~May-June; USSR	7979
	birth: May-June; Germany	10512
	birth: mid June-mid July; zoo	4643
Mustela sibirica	estrus: end winter-early summer; USSR	7979
	mate: end Feb-Mar; USSR	10499
	mate: spring; Europe, w USSR	4645
	birth: Apr; w Siberia, USSR	10499
	birth: end Apr-early May; sw Transbaikalia, Siberia, Russia	10499
	birth: May; Nepal	7382
	lact f: Aug; n1; Nepal	7382
Mustela vison	epididymal sperm: Feb-mid Apr; captive	5971
	seminiferous tubule lumen: present Dec-Mar, absent Apr-Nov; captive	8338
	spermatogenesis: Nov-Mar; captive	4911
	testes: large Dec-May; captive	1129
	testes: large mid Jan-mid Apr; captive	5971
	testes: large Dec-Mar, peak Feb; captive	1049
	testes: large Mar, small May; captive	636
	testes: large Mar; captive, NY	8522
	testes: recrudesce Oct-Dec; captive	1050
	testosterone: peak Jan; captive	10553
	estrus Feb-Mar; captive	5046
	estrus: end Feb-Mar; captive, NY	8522
	mate: Jan-Mar; N America	1474
	mate: Feb-Mar; n10; lab	161
	mate: Feb-Mar; USSR	7979
	mate: Feb-Apr peak mid Mar; ne US	3028
	mate: Feb-Apr; USSR	10499
	mate: end Feb-early Mar; N America	7652
	mate: end Feb-early Mar; PA	9164
	mate: end Feb-early Apr, peak mid Mar; USSR	4643
	mate: end Feb-early Apr; captive	3026
	mate: Mar; captive	1129
	mate: Mar; captive	5325
	mate: Mar; captive	2018
	mate: Mar; lab	2778, 2779
	mate: Mar; lab	6907
	mate: Mar-?; Europe, w USSR	4645
	mate: mid Mar; SAS, Canada	7644
	mate: end Mar-early Apr; captive	4340
	ovulate: Feb-Apr; ne US	3028
	preg f: Mar; n1; NC	1290
	preg f: Mar-Apr; captive	4911
	implant: after Mar 21; captive	9272
	implant: mid Apr; captive	3026
	birth: Apr; n4; lab	161
	birth: Apr-May; WI	5210

	birth: Apr-May; N America	1474
	birth: Apr-May; USSR	10499
	birth: mid Apr; PA	9164
	birth: end Apr-early May; captive	3026
	birth: end Apr-early May; captive	4340
	birth: end Apr-mid May; WY	1871a
	birth: early May; captive	8522
	birth: early May; NY	7234
	birth: May; captive	2778
	birth: May; captive	5046
	birth: May; lab	6907
	birth: May; USSR	4643
Poecilogale albinucha	testes: large Sept-Apr, small May-Aug; S Africa	9326
	mate: Oct-Dec; Zambia	10153
	mate: spring-summer; Natal, S Africa	9327
	birth: Sept-Apr; captive	9326
	birth: Nov-Mar; S Africa	10158
Pteronura brasiliensis	mate: end Dec; zoo	465
	birth: mid Mar; zoo	465
	birth: Apr-Oct; Surinam	2812
Spilogale putorius	testes: regress end Apr-May; e US	7126
	estrus cycles: Sept-Jan; captive	4051
	mate: Apr; e US	7126
	mate: end Sept-early Oct; w US	7130, 7133
	mate: ~early Oct; lab	7129
	implant: early Apr; OR	7140
	implant: Apr-early May; w US	7130, 7133
	implant: end Apr-May; lab	10003
	birth: May-mid June; WI	5210
	birth: May-June; WY	1871a
	birth: May-June; US	1474
	birth: mid May; lab	7133
	birth: end May-June; e US	7126
Spilogale pygmaea	testes: scrotal Dec-May; Pacific coast, Mexico	6570
	birth: July-Aug; Pacific coast, Mexico	6570
Taxidea taxus	testes: large May-end Aug; SD	11951
	mate: July-Aug; America	11951
	mate: end July-mid Aug; SD, MT	11950
	ovulate: Aug; SD, MT	11950
	conceive: July-Aug; ID	7250
	implant: Feb; ID	7250
	implant: Feb; SD, MT	11950
	implant: mid Feb; KS	4287
	birth: Feb-May; w US	1474
	birth: Feb-June; US	6559
	birth: Mar-early Apr; WY	1871a
	birth: Mar-Apr; ID	7250
	birth: ~early Apr; SD, MT	11951
	birth: end Apr-early June; WI	5210
	wean: May-June; SD, MT	11950
Vormela spp	birth: Mar-Apr; Pakistan	9175
Vormela peregusna	mate: Mar; Europe, w USSR	4645
	birth: Feb-early Mar; Kazakhstan; later; n Caucasus Mountains	4643
	birth: Feb-Mar; Kazakhstan, central Asia	10499
	birth: est end Feb; Balkhash, Kazakhstan	7979
	birth: May; Europe, w USSR	4645

Order Carnivora, Family Viverridae

Viverrids (*Arctictis* 9-22 kg; *Arctogalidia* 2-2.5 kg; *Chrotogale* 51-64 mm; *Civettictis* 7-20 kg; *Cryptoprocta* 7-12 kg; *Cynogale* 3-5 kg; *Eupleres* 2-4.6 kg; *Fossa* 1.4-2.6 kg; *Genetta* 1-3.5 kg; *Hemigalus* 745 g-3 kg; *Macrogalidia* 4-6 kg; *Nandinia* 1.5-5 kg; *Osbornictis* 1.5 kg; *Paguma* 3.5-5 kg; *Paradoxurus* 1.5-5.5 kg; *Poiana* 650 g; *Prionodon* 600-800 g; *Viverra* 2-11 kg; *Viverricula* 1.8-4 kg) are semiarboreal, solitary, nocturnal carnivores or omnivores from Eurasia, Africa, and Madagascar. Males are often larger than females (120, 122, 8495, 8501). The endothelial, zonary placenta has a mesometrial implantation into a duplex or bipartite uterus (1597, 6563, 7418, 7563) and is eaten after birth (120, 9216). The furred neonates open their eyes at 0-4 d *Arctictis, Civettictis, Eupleres, Fossa* (119, 120, 121, 349, 3135, 3713, 5761, 6125, 6753, 6754, 10154, 11589); 8-11 d *Arctogalidia, Genetta, Hemigalus, Paradoxurus, Viverra* (42, 47, 652, 2301, 5761, 6603, 8615, 9216, 11326); or 12-25 d *Cryptoprocta, Paguma, Prionodon* (116, 120, 122, 179, 5929, 6603). Males have a baculum (1596). Females have 2-6 mammae (119, 331, 6446a, 7161, 10158). *Cryptoprocta* females have a large, erectile clitoris with spines and an os clitoridis (1596, 6563). Details of anatomy (5274, 5601, 10459) and the hormonal control of male scent glands (10230) are available.

NEONATAL MASS		
Arctictis binturong	307.4 g; n9	11591
	310 g; r270-350; n2	6171a
	319 g; r283.7-340.5; n4	349
	333 g; n1	6125
Civettictis civetta	162 g; n1, 6 d	3135
	f: 319 g; n1 near term	10154
	m: 324 g; n3 near term	10154
	717 g; r680-792; n3, 4 d	6754
Cryptoprocta ferox	100 g	116, 120
Eupleres goudotii	150 g	120
Fossa fossa	65-70 g	119, 120
Genetta spp	61-82 g	11329
Genetta genetta	60-85 g	9216
	83 g; r77-88; n3	11326
Hemigalus derbyanus	125 g; n2	6603
Paradoxurus hermaphroditus	83 g; r69-102	47
	94.3 g; r92-98	42
Prionodon linsang	40 g; n2	6603
NEONATAL SIZE		
Arctictis binturong	HB: 21.5 cm; n1	6125
Civettictis civetta	f: 14 cm; n1	10154
	m: 14.8 cm; n3	10154
	m: HB: 20.35 cm; n1, 1.5 d	6746
Eupleres goudotii	15 cm	120
Genetta genetta	HB: 14.6 cm; r14-15; n3	11326
WEANING MASS		
Civettictis civetta	500 g; r440-530; n3, 17 d	3135
	763 g; r680-810; n3, 33 d	3135
	2.94-3.62 kg; n5, 73-80 d	6754
LITTER SIZE		
Arctictis binturong	1.98; se0.1; mode 2; r1-6; n148 lt	11591
	2 yg; n1 lt	3707, 3708, 3713
	2; r1-3; n3	6125
	2.3; r1-3; mode 2; n3 lt	8046
	2.5; r2-3; n2 lt	349
	3 yg; n1 lt	11974
	3 yg; n1 lt	2109
	3 yg; n2 lt	1443
	4; n1 lt	9634
Arctogalidia trivirgata	2 emb; n1 f	8619
	3 yg; n1 lt	652
Civettictis civetta	1 yg; n1 f	185
	2.0 yg; r1-3; n2 lt	12128
	2.2 yg; r1-4; n22 lt, 5 f	6745, 6746, 6753, 6754

	2.4 yg; r2-3; n5 lt, 1 f	3135
	2.5 yg; r2-3; n2 lt	6612
	2.5; r2-3; n2 lt	647
	3.0; n4 lt	1230
	4 emb; n1 f	10153, 10154
Cryptoprocta ferox	3.0; r2-4; n2 lt	116, 122
Cynogale bennettii	2	556
	2-3 emb	7983
Eupleres goudotii	1 yg	120, 121
Fossa fossa	1; n1 lt	11589
	1 ea; n5 lt	119, 120
Genetta spp	2-3 yg	11329
Genetta felina	1 yg; n1 lt	12128
	2.5; r2-3; n2 lt	9945
Genetta genetta	2 emb; n1 f	8965
	2 emb; n1 f	4924
	2 ea; n2 lt	10714
	2.14; r2-3; n7 lt	9216
	2.25 yg; r1-3; n4 lt	1149
	2.34; mode 2; r1-3; n35 lt	471
	2.6 emb; r2-4; n11 f	10158
	2.75; r1-3; n8 lt, 1 f	11326
	2.8; n5 lt	10154
	max 5	10153
Genetta maculata	2 emb; 1 f	4924
	2 yg; 1 f	185
	2-3 yg	4354a
Genetta servelina	1-2	5277
Genetta tigrina	2 yg ea; n4 lt	10714
	2 yg; r1-3; n2 lt	12128
	2.9; max 5; n10	10158
	3 yg; n1	9327
	3 emb; n2 f	8965
Hemigalus derbyanus	2 yg; n1 f	6603
Nandinia binotata	1 yg; n1 lt	12128
	mode 2-3 yg	4354a
	2; max 4	5761
	2 emb; n1 f	10153, 10158
Osbornictis piscivora	1 emb; n1 f	4434
Paguma larvata	1.71; r1-3; n7 lt, n1 f	2109
	2; n1 lt	179
	2 yg ea; n3 lt	12128
	3; n1 lt	8615
	3.0 yg; r1-4; n7 lt	10873
Paradoxurus hermaphroditus	2-3 yg	557
	3 yg; n1 lt	38, 42, 47
	mode 3	8495
	3-4 yg	8501
	5 yg; n1 f	2538
Paradoxurus zeylonensis	2-3 yg	8501
Poiana richardsoni	2	647
Prionodon linsang	2 yg; n1 f	6603
	2.5 emb; r2-3; n2 f	6446a
Prionodon pardicolor	2 yg	988a
	mode 2	8619
Viverra tangalunga	2 emb ea; n2 f	2313
Viverra zibetha	3	2301
Viverricula indica	2-4	120
	3-5 yg	8495, 8501
	3-5	9175
	4-5 yg	988a
	4 emb; n1 f	8906

SEXUAL MATURITY		
Arctictis binturong	f: 1st mate: 30.4 mo; se3.7; r12.9-47.9; n10	11591
	f: 1st conceive: 30 mo; se5.6; r13-47.9; n6	11591
	m: 1st mate: 27.7 mo; se3.2; r20.5-37.8; n5	11591
Civettictis civetta	f: 1st birth: 14 mo	3135
	f: 1st birth: 1 yr 308 d; r1 yr 210 d-1 yr 357 d	6753
	f: 1 yr	3135
	m: 1st mate: 7 mo	3135
Genetta spp	f: 4 yr; n1 f	11329
Nandinia binotata	f/m: 3rd yr	1742a
CYCLE LENGTH		
Arctictis binturong	82.5 d (or 81.8); se11.8; r18-187; n15 cycles	11591
Arctogalidia trivirgata	6 mo	652
Civettictis civetta	estrus: 1-6 d	3135
Prionodon linsang	estrus: 11 d	3597
GESTATION		
Arctictis binturong	88-89 d; n1	11974
	90-92 d	3707, 3708, 3713
	91.5 d text (or 91.1 abstract); se1; r84-99; n13	11591
	93 d; r88-98; n2	349
	96 d; n1	6125
Arctogalidia trivirgata	~45 d	652
Civettictis civetta	48.75 d; r43-54; n4 last mating to birth	6746, 6753, 6754
	63-68, 66, 79, 80, 81 d; n5	3135
	75.3 d; r72-82; n3; male removed to birth	6753, 6754
Cryptoprocta ferox	~3 mo; few data; field data	116, 120
Fossa fossa	82, 88-89 d; n2; [+ 3 others ~3 mo]	119, 120
Genetta genetta	10-11 wk	11326
	70 d	9216
	3 mo; field data	9215
Genetta tigrina	70 d	9216
Nandinia binotata	64 d	5761
Paradoxurus hermaphroditus	60 d; n1	47
LACTATION		
Arctictis binturong	den emergence: 1 mo	3708
	solid food: 2 wk	3708
	solid food: 8 wk	3707
	solid food: 65 d	11974
	nurse: 5 mo maybe longer	11974
Arctogalidia trivirgata	solid food: 3 wk	652
Civettictis civetta	leave nest: 8 d	11253
	solid food: 21 d	6753, 6754
	solid food: 6 wk	3135
	wean: 84 d	6753
	nurse: 14-20 wk	3135
Cryptoprocta ferox	solid food: 3 mo	116
	wean: 4-4.5 mo	116, 120
Eupleres goudotii	wean: ~9 wk	121
Fossa fossa	solid food: 1.5 mo	119, 120
	wean: 2.5 mo	119, 120
Genetta genetta	solid food: ~6 wk	5761
	wean: 2 mo	11326
	wean: 9 wk	10158
Nandinia binotata	wean: 64 d	5761
Paguma larvata	solid food: 8 wk	7161
Viverricula indica	wean: 4-4.5	120
INTERLITTER INTERVAL		
Arctictis binturong	271.9 d; ±47.4; r107-926; n17 young die	11591
	317.6 d; ±17.6; r111-446; n22 m present	11591
	334.0 d; ±27.2; r107-1168; n53 young live	11591
	468.3 d; ±94.8; r128-1168; n11 m absent	11591
Civettictis civetta	221 d; r193-258; n4 with lactation	6753, 6754

SEASONALITY

Arctictis binturong	mate: Oct-Nov; zoo	3708, 3713
	birth: Apr; n1; captive	2109
	birth: Nov-early Feb; zoo	3713
	birth: yr rd, peak Jan-Mar; Himalayan foothills	11591
Arctogalidia trivirgata	preg f: May; n1; India	8619
Civettictis civetta	mate: Aug-Oct; s Africa	10153
	birth: Feb-Mar, Nov, peak Mar; Nigeria	4354a
	birth: May-Oct; n19/22 lt; zoo	6753, 6754
	birth: end May-June, end Dec; Zaire	11253
	birth: July-Mar; few data; captive, Ghana	3135
	birth: Aug-Jan; S Africa	10158
	birth: Oct-Nov; n12; zoo	1230
	birth: Oct-Nov; n Zimbabwe	331
	lact f: Apr, Aug, Oct; n3; s Cameroun, Benito, Eq Guinea	647
	lact f: Jan, Feb; n2; e Africa	185
Cryptoprocta ferox	mate: Sept-Oct; Madagascar	120
	birth: Dec-Jan; Madagascar	120
Cynogale bennettii	young found: May; Borneo, Indonesia	556
Eupleres goudotii	spermatogenesis: July-Sept; captive	120
	spermatogenesis: Aug-Sept, end winter; Madagascar	121
	birth: Nov; Madagascar	121
Fossa fossa	mate: Aug-Sept; Madagascar	119, 120
	birth: Nov-Dec; Madagascar	119, 120
Genetta spp	lact f: Jan; n1; s Cameroun, Benito, Eq Guinea	647
Genetta felina	young seen: Oct; Angola	11697
Genetta genetta	mate: Jan; captive	9215
	mate: Jan-Sept, peak Feb-Mar; Spain	471
	mate: Oct-Feb; Zimbabwe, Zambia, Malawi	10153
	preg f: Jan; n1; Transvaal, S Africa	8965
	preg f: Oct-Feb; few data; Botswana	10154
	preg f: Nov; n1; Uganda, e Africa	4924
	birth: Mar-June Sept-Nov; captive	9216
	birth: Mar-Nov, peak Apr-June; Spain	471
	birth: Apr; captive	9215
	birth: Aug-Apr; S Africa	10158
	birth: Nov; n1; Tanzania	5761
	birth: spring, fall; zoo, Czechoslovakia	5761
	birth: summer; Transvaal, S Africa	8965
	lact f: Feb, Apr; Transvaal, S Africa	8965
Genetta maculata	preg f: Feb; n1; e Africa	4924
	lact f: Jan-Feb; n1; e Africa	185
Genetta tigrina	preg f: Nov; Transvaal, S Africa	8965
	birth: Aug-Mar; S Africa	10158
	birth: Aug-Nov, Mar; Natal, S Africa	9327
	birth: Feb-May; few data; Zaire	11271
Nandinia binotata	birth: peaks May, Oct; e Africa	5761
	juvenile: Oct; n1; Zambia	331
Osbornictis piscivora	preg f: Dec; n1; Zaire	4434
Paguma larvata	sperm: Mar-July; captive	10930
	testes: large Mar-Aug, small Sept-Feb; captive	10930
	mate: May-June; captive	10930
	lact f: June; n1; central Asia	179
Paradoxurus hermaphroditus	birth: yr rd, peak before monsoon; Sri Lanka	8495, 8501
Poiana richardsoni	lact f: Oct; n1; s Cameroun, Benito, Eq Guinea	647
Prionodon linsang	preg f: May; n2; w Malaysia	6446a
	lact f: May, Oct; n2; w Malaysia	6446a
Prionodon pardicolor	repro: Feb, Aug; Sri Lanka	988a, 8619
Viverra tangalunga	preg f: June; n2; n Borneo, Indonesia	2313
Viverricula indica	birth: Sept-Oct; Madagascar	120
	birth: yr rd; e of Indus, Pakistan	9175
	birth: yr rd; Sri Lanka	8501

Order Carnivora, Family Herpestidae

Mongooses (*Atilax* 2.5-5 kg; *Bdeogale* 1-3 kg; *Crossarchus* 450-2000 g; *Cynictis* 410-900 g; *Dologale* 300-400 g; *Galidia* 700-900 g; *Galidictis* 320-340 mm; *Helogale* 185-680 g; *Herpestes* 225 g-4 kg; *Ichneumia* 2-5 kg; *Liberiictis* 2.3 kg; *Mungos* 1-2 kg; *Mungotictis* 600-900 g; *Paracynictis* 1.0-2.5 kg; *Rhynchogale* 2-3 kg; *Salanoia* 780 g; *Suricata* 620-1000 g) are primarily terrestrial, often diurnal, sometimes colonial carnivores from Africa, Madagascar, central and southeast Asia. Females have an extra chromosome (3492). Male *Herpestes* are often larger than females (5761, 8495, 8501, 9263, 9945).

Induced ovulation may occur in *Herpestes* (4883). *Helogale* family groups are distinctive as usually only the dominant female produces young (2126a, 2126b, 2128, 8930, 9261, 9262), but lactation may occur spontaneously in other females (2127). Placentophagia is reported for *Mungotictis* (120). Females have a bipartite uterus (*Herpestes* 8308). Estrus occurs 1-2 wk (2763, 3129, 9175, 9259) after the birth of usually blind (not *Mungotictis* 120), but furred neonates (757, 4180, 5761, 6215, 9175, 11253). Eyes open at 6-8 d (*Galidia* 120), 8-12 d (*Atilax* 515, 3515; *Mungos* 9984, 10158, 11301), 12 d (*Crossarchus* 3866), 13-14 d (*Helogale* 5761; *Suricata* 3129), or 16-21 d (*Herpestes* 757, 5225, 9175). Females have 2-6 mammae (331, 3129, 3134, 5761, 8308, 9945, 10158). Males have a baculum (8308; cf 9175).

Details of anatomy (7441, 7442, 8308, 9965), endocrinology (4883, 4884, 7838, 10190, 10191, 10192), reproductive energetics (2126a), and light (10193, 10194) are available.

NEONATAL MASS		
Atilax paludinosus	~100 g; from graph	515
Galidia elegans	50 g	117, 120
Herpestes auropunctatus	r19.8-22.1 g; n3 term emb	10863
	r22.3-36.4 g; n18 term emb	7829
	29 g; n1	10863
Mungos mungo	20 g	10158
	24.2 g; n1, 3 d	9984
	73 g; n3, 1 lt [5.4% of 1.35 kg]	11301
Mungotictis decemlineata	50 g	120
Suricata suricatta	25-30 g	9476
	30.5 g; r25-36; n4	3129
NEONATAL SIZE		
Crossarchus obscurus	HB: 9-10 mm [*sic*, est units should be cm]	3866
Galidia elegans	110-130 mm	117, 120
Herpestes auropunctatus	CR: r51-58 mm; n3 term emb	10863
Mungos mungo	HB: 99 mm; n1, 3 d	9984
	HB: 100 mm	5761
	~180 mm	9985
WEANING MASS		
Atilax paludinosus	~550 g; from graph	515
Galidia elegans	183 g; r150-200; n3	117
Herpestes auropunctatus	250-300 g; 10-12 wk	7829
Herpestes ichneumon	800 g; 12 wk [1 mo after weaning]	757
Suricata suricatta	225.5 g; r172-275; n6, 5 wk	3129
LITTER SIZE		
Atilax paludinosus	1 emb; n1 f	331
	1.67; r1-2; n3 f	10158
	2-4 yg	11253
	2 yg; n1 lt	12128
	2.5 yg; sd0.54; r2-3; n6 lt	515
	3 yg ea; n2 lt	3515
Bdeogale crassicauda	1 emb; n1 f	4924
	1 yg; n1 f	4924
Crossarchus alexandri	max 4 yg	5761
Crossarchus obscurus	mode 4 yg; r2-4	3866
	5 emb; n1 f	2955
Cynictis penicillata	1.3 yg; r1-2; n3 lt	12128
	1.67; r1-2	1230
	2.33 emb; r2-3; n3 f	9327
	3.2 emb; r2-5; n6 f	10154, 10158

Galidia elegans	1 yg ea; n3 lt	6215
	1 ea; n11 lt	117, 120
Galidictis fasciata	est 1 yg	120
Helogale parvula	1-6	9260
	2-4; max 7	10153, 10158
	2 emb, n1 f	331
	3 yg, n1 f	331
	3; r2-4; n2 lt, 1 f	4180, 10154
	4 yg; r1-7	8930a
Herpestes auropunctatus	1.7 yg; r1-3; n3 lt	12128
	1.8 emb; se0.35; r1-4	3952
	2-3 yg; n9 f	538
	2.1 emb; r2-4; n52 f	8546
	2.12 ps; r1-3; n154 f	1932
	2.14 emb; r1-3; n69	7829
	2.19 emb; r1-5; n283	7829
	2.25 yg; n27	7829
	2.4 implantations; r2-4; n17 f	8308
	2.4 ps; n35 f	8308
	2.7 cl; r1-4; n16 f	8308
	2.7 emb; r2-4	538
	2.7 emb; r2-4; n9 f	8308
	2.9 emb; r1-3; n68 f	1932
Herpestes edwardsi	1 yg; n1 lt	12128
	mode 2	8501
	2 emb ea; n2 f	7382
	2 ea; n6 lt, 1 f	9175
	3-4	8495
Herpestes fuscus	mode 3-4	8495, 8501
Herpestes ichneumon	2-4	11253
	2-5	8223
	2; n1 lt	2763
	3 emb; n1 f	1149
	3.3 yg; sd0.67; r1-4; n10 lt	757
Herpestes pulverulentus	2-3; 4 not uncommon	9945
Herpestes sanguineus	1.45; mode 1; r1-3; n11 lt	9263
	1.8 emb; r1-2; n5 f	10154, 10158
	2 emb; n1 f	2096
	3 emb; n1 f	11271
	3 yg; r2-4; n3 lt	10714
Herpestes vitticollis	2-3	8501
Ichneumia albicauda	2-4	5277
	2 emb; n1	9327
	2.4 emb, yg; r1-3; n5 f	10153, 10158
Mungos mungo	1 emb; n1 f	6611
	1 yg ea; n2 lt	12128
	2-8	9327
	mode 2; max 3	10153
	2; n1 lt	9441
	2.4 yg	9259
	3-5 yg	9985
	3 yg; 1 lt	11301
Mungotictis decemlineata	mode 1 yg; r1-2	120
Paracynictis selousi	2 yg	4744
	max 2	10153
	2-4	10158
Rhynchogale melleri	3 emb; n1 f; 2 yg ea; n2 nests	10153, 10158
Suricata suricatta	2.9; r2-5; n34 f	10158
	3; n1 lt	4180
	3.75 yg; r3-4; n4 lt	9476
	mode 4; r2-5	3134
	4.0 yg ea; n2 lt	3129
	4.0 yg; r2-5; n4 lt	1230
	4.1 yg; r2-7?; n3 lt	12128

SEXUAL MATURITY		
Atilax paludinosus	m: testes descend: 255.5 d; r204-302	515
Crossarchus obscurus	f/m: ~9 mo	3866
Helogale parvula	f: 1st preg: 15 mo; n1	9260
Herpestes auropunctatus	f: 1st mate: 9 mo	8308
	f: 1st conceive: min 130 d; n1	7829
	m: 1st mate: <6 mo; 400-500 g	7829
Herpestes ichneumon	f/m: 1st mate: 2 yr	2763
Herpestes sanguineus	f: 1st estrus: 13.5 mo	5225
	f: 1st preg: early 2nd yr	9263
Mungos mungo	f: 1st estrus: 9-10 mo	9985
	f: 1st birth: ~1 yr	9985
	f: 1st birth: ~1 yr	9441
Suricata suricatta	f: 1st birth: 1 yr	3129
	m: 1st fertile mating: 1 yr	3129
CYCLE LENGTH		
Galidia elegans	mode 12 d	6215
Helogale parvula	estrus: 1-6 d; n1 f	8930a, 8931
Herpestes auropunctatus	estrus 3 d	7829
	3 wk	7829
Mungos mungo	estrus: 3-6 d	9985
GESTATION		
Crossarchus obscurus	~8 wk	3866
	~10 wk	4354a
Galidia elegans	60.67 d; r52-77; n3	6215
	82.6 d; r74-90; n11	117, 120
Helogale parvula	54-56 d	8930a
Herpestes auropunctatus	6 wk; n1	9175
	46-49 d	3952
	~7 wk	538, 8308
	49.3 d; r49-50; n3	7829
	49.6 d; r47-53; n8	10863
Herpestes ichneumon	60 d	757
	11 wk	5761
Herpestes sanguineus	58-62 d	5225
Mungos mungo	~60 d	9985
	60-62 d	2449
	~2 mo	9259
	~2 mo	9984
Mungotictis decemlineata	100 d; r90-105; n3	120
Suricata suricatta	≤77 d	3129, 3134
LACTATION		
Atilax paludinosus	solid food: 14 d; prechewed; 30 d; not prechewed	3515
	wean: 35.9 d; r30-46	515
Crossarchus obscurus	solid food: ≤3 wk	3866
Galidia elegans	solid food: 1 mo	117
	wean: 2 mo	117, 120
Helogale parvula	solid food: 10-12 d	8930a
	wean: after ~3 wk; den emergence: 13 d	5761
Herpestes auropunctatus	nurse: 3-5 wk	3952
	nurse: est 4 wk	8308
	wean: 4-8 wk	7829
	independent: 12 wk	10863
Herpestes ichneumon	in nest: 6 wk	757
	solid food: 32 d	757
	nurse: 2 mo	757
Herpestes sanguineus	wean: ~50-65 d	9263
	wean: 52 d	5225
Mungos mungo	den emergence: 3 wk	9258
Mungotictis decemlineata	solid food: 15 d	120
	wean: 2 mo	120
Suricata suricatta	solid food: 4-6 wk	3129
INTERLITTER INTERVAL		
Herpestes auropunctatus	4 mo	7829

LITTERS/YEAR		
Atilax paludinosus	2	11253
Crossarchus obscurus	2-3	3866
Galidia elegans	1	117
Helogale parvula	2	11253
	2-3	9260
Herpestes auropunctatus	est 2	538, 8308
	2	7829
	2	8546
	2	3952
Herpestes edwardsi	3.3; 5 lt/18 mo; n1 f	9175
Herpestes ichneumon	mode 1; 2 with good conditions	757
	2?	11253
Herpestes sanguineus	2	11253
Mungos mungo	1	11253
	max 4	9259
Suricata suricatta	1	3134
SEASONALITY		
Atilax paludinosus	preg f: Oct; n1; n Zimbabwe	331
	birth: July, & rainy season; Zaire	11253
	birth: Aug-Dec; S Africa	10158
Bdeogale crassicauda	preg f: Dec; n1; Mt Kilimanjaro, e Africa	4924
	lact f: Dec; n1; Mt Kilimanjaro, e Africa	4924
Crossarchus obscurus	preg f: Mar; n1; Cameroun	2955
	birth: yr rd; captive	3866
Cynictis penicillata	preg f: Sept; n3; Natal, S Africa	9327
	birth: peak Oct-Jan; S Africa	10158
	lact f: Nov; Natal, S Africa	9327
Galidia elegans	mate: Apr, July-Oct, Dec; captive	120
	mate: July-Nov; Madagascar	117, 120
	birth: Apr, Aug-Sept; n3; zoo	6215
Galidictis fasciata	birth: summer; Madagascar	120
Helogale parvula	estrus: end Mar-early Apr; zoo	8931
	mate: Oct-Dec; Zambia, w Zimbabwe	10153
	preg f: Feb; n2; Botswana	10154
	birth: Oct-Mar; S Africa	10158
	birth: Nov-May; Tanzania	9260
	birth: twice/yr; Zaire	11253
Herpestes auropunctatus	spermatogenesis: yr rd; Fiji I	3952
	testes: large Feb-Mar; HI	8308
	testes: large Sept-Dec, small Feb-July; Fiji I	3952
	mate: Jan-Oct; Puerto Rico	8546
	mate: Feb-Aug; Oahu, HI	10190, 10194
	mate: end Feb-July; HI	538
	mate: end summer, with monsoon; Pakistan	9175
	preg f: Feb-Aug; Virgin I	1932
	preg f: Feb, Aug-Sept; Oahu, HI	4884
	preg f: end Feb-July; HI	8308
	preg f: Aug-Feb, peak Oct-Dec; Fiji I	3952
	preg f: yr rd, low Nov-Dec; Grenada	7829
	birth: peaks Mar-Apr, July-Aug; Puerto Rico	8546
	lact f: Mar-Oct; Oahu, HI	4884
Herpestes edwardsi	preg f: Apr; n2; Nepal	7382
	birth: Apr, July; zoo	44
	not seasonal; Sri Lanka	8495, 8501
Herpestes fuscus	birth: yr rd, peak 1st few mo of yr; Sri Lanka	8495, 8501
Herpestes ichneumon	mate: end Feb-Mar; Israel	757
	mate: Mar-Apr; n1 pr; zoo	2763
	birth: Feb-Sept; s Uganda	5761
	birth: Apr-May; Israel	757
	birth: July; n1 pr; zoo	2763
	birth: July-Aug; s Spain	5761
	birth: twice/yr, rainy season; Zaire	11253

Herpestes sanguineus	mate: Aug-Sept, sometimes Apr; captive, S Africa	5225
	preg f: Feb-Apr, Oct-Nov; e Africa	10715
	preg f: Dec; n1; Botswana	10154
	birth: Oct-Mar, peak Oct-Dec; captive, S Africa	5225
	birth: mid dry season, rainy season; Zaire	11253
	lact f: Feb; n2; Botswana	10154
	lact f: June; n1; Kenya	4924
	repro: yr rd; Serengeti, Tanzania	9263
Herpestes vitticollis	lact f: May; n1; Sri Lanka	8501
Ichneumia albicauda	birth: Sept-Dec; S Africa	10158
	birth: Oct-Jan; s Africa	10153
Liberiictis kuhni	lact f: Aug; n1; Liberia	3867
Mungos mungo	mate: end Aug, Oct; Natal, S Africa	9327
	mate: Oct-Jan; Zambia	10153
	preg f: Oct-Jan; captive	9984
	birth: Oct-Feb; S Africa	10158
	birth: Oct, Dec, Feb; Natal, S Africa	9327
	birth: Nov-Dec; Zaire	11253
	birth: peak Dec; captive	9984
	repro: yr rd; Uganda	9259
Mungotictis decemlineata	mate: June-Dec; captive, Madagascar	120
	birth: Sept-Mar; captive, Madagascar	120
Paracynictis selousi	birth: Aug-Mar; S Africa	10158
	birth: Oct; s Zambia, Zimbabwe	10153
Rhynchogale melleri	birth: ~Nov; Zimbabwe	10153
Suricata suricatta	mate: Sept-Oct; captive	3129, 3134
	birth: Jan, Nov, Sept; n4; zoo	1230
	birth: Oct-Mar; S Africa	10158
	birth: Oct-Apr; via natives; sw Africa	4560
	birth: Nov-Dec; captive	3129, 3134

Order Carnivora, Family Protelidae, Genus *Proteles*

The 5-14 kg aardwolf, *Proteles cristatus*, from the arid plains and woodlands of Africa eats termites by night and rests in burrows by day. Monogamous pairs copulate 1-4.5 hr (9104). Neonates are born blind and helpless (5761), and reach adult size by 9 mo (5761). Females have 4 teats (9945). Details of anatomy (11584) are available.

LITTER SIZE		
Proteles cristatus	mode 2-3; r1-5	5761
	most 2-4 yg; max 5	5907
	2.36 yg; r1-4; n11 lt, 5 f	9103, 9104
	3 yg; n1 nest	9945
	3 yg; r2-4	10153, 10158
CYCLE LENGTH		
Proteles cristatus	estrus: 1-3 d	5907, 9104
GESTATION		
Proteles cristatus	55 d; r54-56; n2; field data	9103
	~90 d	5907
	~3 mo; r90-110 d	5761
LACTATION		
Proteles cristatus	den emergence: 1 mo	5907
	den emergence: 2 mo	9103
	solid food: 9-12 wk	5907
	solid food: 105-120 d	9103
	wean: 120-145 d	9103, 9104
SEASONALITY		
Proteles cristatus	mate: Oct-Nov; Zimbabwe	10153
	mate: mid May-mid Aug; S Africa	9103, 9104
	birth: May; n1; s Kenya	5761
	birth: yr rd, peak Oct-Feb; S Africa	10158
	lact f: Nov; n1; Namibia	9945

Order Carnivora, Family Hyaenidae

Hyaenas (*Crocuta* 35-85 kg; *Hyaena* 25-55 kg) are terrestrial, nocturnal carnivores from Africa and southwest Asia. *Crocuta* females are larger than males and have male-like genitalia (1716, 2315, 4075, 4265, 6074, 6963, 7802, 7858, 11173, 11481; cf 11615), while *Hyaena* males are larger than females (5761, 10033). The labyrinthine, hemochorial, villous, zonary placenta (232, 2479, 6963, 6969, 7555, 8031, 11967, 11968) implants in a bipartite uterus and is probably eaten after birth (3861). A postpartum estrus occurs 2-3 wk (*Hyaena* 9621) after the birth of furred neonates that either have their eyes open (*Crocuta* 2427, 5761, 6964, 8709, 9621, 10158) or whose eyes open 1-2 wk later (*Hyaena* 5761, 7979, 9128, 9670, 10158). Females have 4-6 mammae (9175, 9621, 9945, 11482). Males have penial spines (11584). Details of anatomy (6963, 11482, 11584) and endocrinology (*Crocuta* 3471, 3880, 6471, 6472, 8838; *Hyaena* 3046) are available.

NEONATAL MASS		
Crocuta crocuta	m: 1.1 kg; n1	9621
	1.46 kg; n3	8709
	1.5 kg; r1.35-1.6; n3	3861
Hyaena brunnea	f: 635 g; n1	6182
	m: 635 g; n1	6182
	f: 657.5 g; r645-680; n4	9670
	693.2 g; se17.6; n13	7349
	710.5 g; r635-812; n4	2879
	m: 727.5 g; r690-805; n4	9670
	760 g; n1	6182
Hyaena hyaena	650 g	9444
	f: 660 g; n1	10033
	m: 668 g; r661-675; n2	10033
	700 g	9130
WEANING MASS		
Crocuta crocuta	14.5 kg; n1, 100 d	8709
Hyaena brunnea	27.4 kg; r24.9-29.9; n2, ~10 mo	7349
LITTER SIZE		
Crocuta crocuta	1-2	6964
	1 emb; n1 f	6612
	1.5 yg; r1-2; n2 lt	3861
	1.5 yg; r1-2; n9 lt	12128
	1.83 yg; se0.56; mode 2; r1-3; n24 lt	3470
	2-4	8514
	mode 2; r1-3	2427
	2; r1-4	10158
	2 emb, 2 yg; n2 f	4924
	2 yg ea; n2 lt	9026
	2; r1-3; n9 lt, 2 f	7555
	2.2 yg; r1-3; n5	8709
Hyaena brunnea	1; n1 lt	1230
	1.8; mode 4; r1-5; n6 lt	11616
	2.0; r1-4; n6 lt	8170
	2.3 yg; r1-4; n9 lt	2879
	2.3; r1-5; n61 lt (literature compilation)	7348
	2.8 yg; r1-5; n5	10024
	3.0 yg; r1-4; n3 lt, 1 f	9670
	3.0 yg; r2-4; n5 lt, 2 f	7350
Hyaena hyaena	r1-4; n21 lt	9621
	1.33; n3 lt	1230
	mode 2-4; r1-6	5761
	2.3; r2-3; n3 lt	10033
	2.4 yg; r1-5; n43 lt	9128, 9129
	2.5 yg; r2-3; n2 lt	12128
	4 yg; n1 lt	9444

SEXUAL MATURITY		
Crocuta crocuta	f: 3 yr	9621
	m: 1-2 yr	5761, 10158
Hyaena brunnea	f: 1st birth: 3 yr; from graph	7350
Hyaena hyaena	f/m: 2-3 yr, rarely 1.5	9128
CYCLE LENGTH		
Crocuta crocuta	estrus: 1 d	9627
	14 d; +14 d anestrus	4075
Hyaena brunnea	estrus: 1-4 d	2879
	estrus: 1 wk	7352
	30 d; n3-4 cycles/f, 2 f	2879
Hyaena hyaena	estrus: mode 1 d; r1-2	9130
	40-50 d	9130
GESTATION		
Crocuta crocuta	110 d; r97-132	9621
Hyaena brunnea	74-93 d; n1	6182
	~3 mo	7345
	95.8 d; r93-100; n5	2879
	98-99 d; n1	4075
Hyaena hyaena	89.3 d; r86-92; n6 lt	9128, 9129
	90 d	7979
	90-91 d	4582
	90-92 d	8798
LACTATION		
Crocuta crocuta	den emergence: 5 wk	11253
	solid food: 2-4 wk	9026
	solid food: 6 mo	6074
	solid food: 7-8 mo	5761, 10158
	wean: 8 wk	9026
	wean: ~6 mo	6963
	wean: 12-14 mo	6074
	wean: ~1 yr-16 mo	5761
Hyaena brunnea	solid food: 3 mo	7352
	wean: 9 mo	7352
	min 10 mo	8169
	wean: 12-15 mo	10158
Hyaena hyaena	solid food: 30 d	9128
	solid food: >2 mo	7979
	nurse: 4-5 mo	9130
INTERLITTER INTERVAL		
Crocuta crocuta	~12 mo	6963
	1 yr	11253
	15.9-19.3 mo	10158
	17.4 mo; r9-26; n7 intervals	6074
Hyaena brunnea	1 yr	7352
	1.8 yr; r0.9-3.5; n5 intervals; from graph	7350
Hyaena hyaena	est 0.3-1 yr	7979
SEASONALITY		
Crocuta crocuta	mate: yr rd; central, s Africa	5761
	preg f: May; n1; e Africa, Eritrea, Somalia	4924
	preg f: yr rd; Serengeti, Tanzania	6963, 6964
	birth: Mar-Apr; Africa	5774
	birth: early rainy season; Zaire	11253
	birth: yr rd, peak June-Aug; Kruger Nat Pk, S Africa	3145
	birth: yr rd, peak end summer; S Africa	10158
	lact f: May; n1; e Africa, Eritrea, Somalia	4924
	repro: Oct-Apr; Elosha Game Reserve, Namibia	2427
	repro: yr rd, Jan-Mar, peak end summer; S Africa	6470
Hyaena brunnea	estrogen: high Sept; zoo	3046
	mate: end Nov; zoo	2879
	birth: May; n1; zoo	1230
Hyaena hyaena	birth: Jan-Feb; n3; zoo	1230
	birth: Apr-May; central Asia	7979
	birth: yr rd; captive	9129

Order Carnivora, Family Felidae

Cats (*Acinonyx* 22-72 kg; *Felis* 1-100 kg; *Lynx* 4-32 kg; *Neofelis* 16-37 kg; *Panthera* 17-320 kg) are terrestrial, primarily nocturnal, usually solitary carnivores ranging over most of the earth. Males are larger than females (833, 852, 3571, 4924, 5761, 6468, 7979, 8238, 9175, 9360, 9546, 9559, 9567, 9675, 9945, 10172, 10499, 10760).

Ovulation is probably triggered by coitus (7496, 9706, 9708, 11687; cf 9704). The 25 g (small *Felis* spp), labyrinthine, zonary or annular placenta (6367, 6965, 7563, 7839, 9563) implants into a bicornuate uterus and is eaten after birth (6268, 10158, 10788, 10870). Postpartum estrus may occur (2875, 2877, 2878, 5400, 5592, 5838, 9358, 10158, 12045; cf 1018, 1230). Cats are born with fur and closed eyes (2494, 3202, 3625, 5761, 7088, 7624, 9175, 9546, 9563, 10157). Their eyes open at 1-15 d (*Acinonyx* 2883, 3339, 5761, 10020, 10158, 10869, 11284; *Felis* 41, 300, 370, 2375, 3571, 4462, 5761, 6422, 6423, 6605, 7259, 7781, 7979, 8407, 8639, 8822, 9560, 9564, 9565, 9573, 9574, 9674, 9675, 10870, 10871, 11038, 11331, 11378, 11936, 12045; *Panthera* 29, 833, 2109, 2375, 3537, 5465, 5761, 5930, 6829, 7979, 9546, 10308, 10499) or 8-17 d (*Lynx* 1474, 3571, 3982, 4314, 5761, 5995, 7979, 9175, 10499). Females have 4-8 mammae (331, 1474, 3134, 3571, 6965, 9546, 10760, 11633). Males have a baculum (6926, 9559, 9568, 9569).

Details of anatomy (2770, 6428, 6965), endocrinology (*Acinonyx* 9796, 11684, 11686, 11688; *Felis* 1096; *Panthera* 968, 1357, 1358, 9604, 9605, 9704, 9706, 11687), semen (11680, 11684, 11686), capicitation (1515), and ovulation (*Felis* 7495, 7496, 7822; *Panthera* 2718, 9348) are available. Information on the reproduction of domestic cats is reviewed in supplement 39 of the Journal of Reproduction and Fertility published in 1989.

NEONATAL MASS

Acinonyx jubatus	250-280 g; n3	3339
	~300 g; n1	3338
	m: 520 g; n1	11333
	m: 666.7 g; r600-700; n3	11284
	f: 695.0 g; r690-700; n2	11284
Felis aurata	215.0 g; r195-235; n2	10870
Felis bengalensis	82.5 g; r75-95; n4	2282
	92.5 g; r90-95; n2	8639
	113.14 g; r93-120; n7, 4 lt	26
Felis chaus	106.21 g; r83-125; n24	35, 41
	156.25 g; r150-161; n4	9565
Felis concolor	225-450 g	12045
	390-450 g	5210
	450 g	7088
Felis geoffroyi	65 g ea; n2	9574
	91.7 g; n1	6268
	122.9 g; n1 with placenta	6268
Felis manul	89 g; n1	9563
Felis margarita	39 g	9573
	39 g; se2	4646
	105 g; n1	4644
Felis marmorata	70-100 g	9564
Felis nigripes	72.0 g; r60-84; n2, 1 d	6423
	m: 60 g; n1	9674
	85.2 g; r82.5-87.9; n2	370
Felis pardalis	200-276 g	9567
	206-340 g	8407
Felis serval	246.0 g; r230-260; n5	11378
	280 g; r270-290; n2	3168
Felis silvestris	~40 g	3571
	65 g; n1 stillborn	11331
	82-149 g; median 116; n71	813
	83-120 g	9560
	97.0991 g; 95%ci 23.1204; n78	7839
	m: 97.75 g; r82.7-107.4; n6	6240
	100 g	10871
	~100-163 g live; r65-163; n54, with stillborns	9561
	f: 103.92 g; r97.0-119.5; n6	6240

	106.4 g; sd13.5; se0.81; r70-144; n128, 33 lt	4237
	106.5 g; r103-110; n2 term emb	2490
	119 g; r97-141; n6; 5 stillborn, 1 alive	6437
	119.44 g; r80-135; n9	11331
	134.67 g; r100-163; n24	7259
Felis temmincki	110 g; n1 black fur, did not survive	24
	250 g; n1	6605
Felis viverrina	170 g; without tail	11038
Felis wiedii	m: 163 g; n1	8407
	f: 170 g; n1	8407
Lynx canadensis	204 g; r197-211; n2	9510a
Lynx caracal	165 g; n1	5592
Lynx rufus	112-226 g	1474
	~300 g	1871a
Lynx lynx	289 g; r250-360	10942
Neofelis nebulosa	142-170 g; n4	3378
	200 g; n1	7648
Panthera leo	1000-2000 g	5761
	f: 1152 g	10308
	1199 g; r1100-1350; n5	2278
	1200-2100 g	9546
	f: 1266 g; n571	9730
	1284.5 g; r1210-1359; n2	2375
	m: 1288 g	10308
	1306 g; r540-2190; n1173	9730
	m: 1345 g; n601	9730
	1466.3 g; r1243-1800; n3	2109
	~1500 g	10158
Panthera onca	680 g; n1, 2 d	5098
	800 g; r700-900	9811
	840 g; r510-1070; n29	9730
	m: 847 g; n1	2109
	m: 950 g; n1, 1 d	10332
Panthera pardus	43-60 g	5761
	50-60 g	10158
	710 g; r500-1000; n7	2494
	560 g; n1	2109
	498.5 g; r430-567; n2	2375
	553.4 g; r338-858; n174	9730
Panthera tigris	f: 678-1243 g	2109
	785-1043 g	75
	785-1500 g	10499
	1100-1430 g	29
	f: 1300.0 g; r950-1600; n6	11283
	f: 1341.4 g; r990-1575; n24	9729
	m: 1386-1700 g	2109
	m: 1387.5 g; r1100-1610; n4	11283
	1400 g	7979
	1496 g; r808-1893; n128	9730
	m: 1519.75 g; r1210-1760; n20	9729
Panthera uncia	300-380 g	6829
	f: 320 g; n1	5465
	r320-567 g; n7	4617
	m: 350 g; n1	5465
	540.9 g; r368-708; n10	3537
	581.1 g; r510-652; n10, 1 d	5838
	623 g; n1	3496
	666 g; r623-709; n2	3499
NEONATAL SIZE		
Acinonyx jubatus	HB: ~250 mm	3338, 3339
Felis bengalensis	178.5 mm; r175-182; n2	8639
	TL: 241.4 mm; r220-250; n7, 4 lt	26
Felis chaus	TL: 231.7 mm; r220-226; n24	35, 41

Felis concolor	200-300 mm	12045
	300 mm	7088
Felis geoffroyi	HB: 135 mm; n2	9574
Felis manul	HB: 152 mm; n1	9563
Felis margarita	143 mm	7979
Felis nigripes	150 mm	9675
Felis serval	f: 190 mm; n1	8856
	m: 195 mm; n1	8856
Felis silvestris	m: 152 mm; n1	185
	f: 155 mm; n1	185
	CR: 126.7 mm; 95%ci 9.2; n78	7839
Felis temmincki	HB: 200 mm; n1 black fur, did not survive	24
Lynx canadensis	CR: 160.5 mm; r158-163; n2	9510a
Panthera onca	400 mm	9811
Panthera pardus	150 mm	5761
Panthera tigris	315-400 mm	75
	364 mm	29
Panthera uncia	HB: 233.5 mm; r230-237; n2	4617
WEANING MASS		
Felis chaus	2212.5 g; r2100-2300; n4, 90 d	9565
Felis concolor	3-4 kg; 1-2 mo	2186
Felis silvestris	m: 362.73 g; r266.4-486.7; n6, 4 wk	6240
	398 g; sd70; r264-521; n70, 30 d	4237
	f: 402.42 g; r330.2-475.4; n6, 4 wk	6240
	604 g; sd102; r390-876; n70, 50 d	4237
	600 g	2020
	f: 683.8 g; r644.8-760.0; n6, 8 wk	6240
	m: 713.92 g; r558.7-820.4; n6, 8 wk	6240
	968.0 g; r890-1230; n5, 2 mo	7259
	1055 g; r1030-1080; n2, 62 d	11331
Neofelis nebulosa	f: 1055 g; r1050-1060; n2, 6 wk	3202
	7500 g ea; n2, 32 wk	3202
	m: 1155 g; n1, 6 wk	3202
	12000 g; n1, 32 wk	3202
Panthera tigris	11003 g; r10760-11850; n3, 88 d	11283
LITTER SIZE		
Acinonyx jubatus	1 yg; n1	3338
	1 yg; n1 lt	6794
	2 yg; n1 lt	9504
	2 yg; n1 lt	6793
	2.5 yg; r2-3; n2 lt	10788
	3; n1 lt	3339
	3; n1 lt	4659
	3 yg; n1 lt	11065
	3 yg; n1 lt	334
	3 yg; n2 lt, 1 f	6792
	3; r1-5; max 6	10158
	3.1 yg; n14 lt	6795
	3.48; n217 lt	7992
	3.7; r1-8; n15 lt	2873, 2880, 2883
	4 yg; n1 lt	10020
	4 yg; n1 lt	3001, 3002
	4; r3-5; n2 lt	11284
	5 yg; n1 lt	10869
	5 ea; n2 lt	11333
Felis aurata	2 yg ea; n2 lt	10870
Felis bengalensis	1.75 yg; r1-2; n4 lt	26
	2	6448a
	2-3	7781
	2.0; r1-3; n3 lt	3514
	2.4; n18 lt	8639
	3-4	8619
	3; n1 lt	179
	3.0; n5 lt, 1 f	9175

	3 yg ea; n5 lt	2282
Felis chaus	2.89; mode 3; r1-6	9565
	3; n1 lt	8822
	3 emb; n1 f	7382
	3-5; max 10	7979
	3.5; mode 3; r3-5; n8 lt	35, 41
Felis concolor	r1-4 yg; n14 f	9385
	2 emb; n1 f	7324
	2.4; r1-4	7088
	2.4; mode 3; r1-4; n35	1230
	2.5 yg; r2-3; n2 lt, 1 f	8822a
	2.7; n11 lt	2881
	2.7 yg; n11 lt, 9 f	6545
	2.76; mode 3; r1-4	11338
	2.9; r1-4; n7 lt, 1 f	7234
	2.9 yg; r1-5; n258 lt	9185
	3 emb; n1 f	6723a
	3 yg; n1 f	8468
	3.0 yg; r2-4; n4 lt	9066
	3.4 emb; r1-6; n66 f	9185
	4 yg; n1 lt	502
Felis geoffroyi	1	5476
	1.67 yg; r1-2; n3	6268
	2.0 yg; r1-3; n2	9574
	2.57 yg; r2-3; n7 lt	300
	3 emb; n1 f	11977
Felis manul	1; n1 lt	9563
	4-6 yg	9558
	5-6	10499
Felis margarita	mode 3; r2-5	9559
	4	4618
	6.0; r4-8; n2 lt	9573
	8; n1 lt	4646
Felis marmorata	2 yg	9564
Felis nigripes	1.6; r1-2; n5 lt	370
	1.7 yg; 1-2; n3	6423
	2 emb; n1 f	11311
	2; r1-3; n3 lt	9675
	2; r1-3; n5 lt	9674
	2.5 yg; r2-3; n2	11311
Felis pardalis	1-2 yg	9567
	1.3; r1-2; n3 lt	12128
	2 yg; n1 lt	5900
	2 emb; n1 f	8422
Felis rubiginosa	2 yg; n1 lt	8502
	2-3 yg	8501
Felis serval	1 yg; n1 lt	1018
	2.17 yg; n6 lt, 1 f	1014
	2.35 yg; r1-4; n20 lt, 3 f	11378
	2.4 yg; r1-4; n11 lt, 1 f	3168
	2.5 emb; r1-3; n7 f	10153, 10155, 10157
	3 yg; n1 lt	334
	4 cl; n1 f	9327
Felis silvestris	2.0; r1-4; n4 lt	10871
	2.4; r2-3; n5 lt	348
	3-5; max 10	7979
	mode 3-5; r3-10	4645
	3; r1-5; n16 lt	11331
	3.3 emb; r1-6; n39 f	2490
	3.4; r2-5; n7 lt	10154
	3.42 yg; r1-8; n61 lt	2019, 9560
	3.59 yg; sd1.55; mode 3; r1-8; n51 lt	6362
	3.6; mode 3; r2-6; n25 lt	2020
	3.63 yg; sd1.30; mode 3-4; r1-8; n119 lt	6362

	3.88 yg; mode 3,4; r1-8; n33 lt	4237
	4 yg; n1 lt	6395
	4; n6 lt	1973
	mode 4; r3-5	9843
	mode 4-5; r2-6	6468
	4; r1-8; n16 lt, 4 f	7259
	4.20 yg; se0.05; r1-10; n973 lt	9197
	4.35 emb; sd1.27; r2-7; 5.07 cl; sd1.28; r3-7; n17 f	5400
	4.59 emb; sd1.12; r2-7; 5.88 cl; sd1.12; r4-8; 2.66 weaned; sd0.76; n17	11072, 11074
	4.7 emb; r1-9; n14	2044
Felis temmincki	1 yg	24
	1 yg	6605
	2 yg	4813b
Felis tigrina	1-2	6422
Felis viverrina	1.5 yg; r1-2; n2	11038
	2; n1 lt	5275
	2; n1 lt	9368
Felis wiedii	1 emb; n1	7624
	2 yg; n1 lt	8207
Felis yagouaroundi	mode 2	5090
	2-3	1474
Lynx canadensis	1 ps, 4 ca; n1 f	10839
	2; r1-4	1474
	2 yg; n1 lt	9510a
	4.3-11.9 cl; across age	1229
	mode 4; r1-6	5620
	max 4 yg	5146
Lynx caracal	r1-4 emb; n7 f	8758
	1-6	5761
	max 2	10153
	2-4; max 5	10158
	2-4	9945
	est 2-4	8223
	2-5	10156
	2-6	5995
	2-6	9175
	2; r1-3	5592
	2; r1-2; n6	1230
	2 yg; n1 f	9327
	2.2 yg; r1-4; n37 lt	833
	2.3; r2-3; n3 lt	12128
	3-4	9496
Lynx lynx	1 yg; 1 lt	11501
	mode 2; max 4 rare	11226
	mode 2-3; r1-5	3571
	mode 2-3; r1-5	7979
	mode 2-3; r1-5	10499
	2 yg; n1 lt	4314
	2 yg; n1 lt	9066
	2.0; r1-3; n6 lt	8772
	2.2; r1-3; n5 lt, 4 f	6102
	2.26 yg; mode 2; r1-4; n19 lt	736
	2.5 emb; n4 f	3527
	2.50 emb; sd0.53; r2-3; 2.88 cl; sd0.64; r2-4; n8 f	6137b
	3.0; n3 lt	6103
	3.10 cl; se0.16; r1-10; n104 f	6137a, 6137b
	4.18 ps; n95 f	8813
	4.2 cl; r2-10; 2.5 ps; r1-4; n64 f	3527
Lynx pardinus	2-3 yg	10809
	2.5; mode 2; r1-4; n14 lt	11070
Lynx rufus	1 emb ea; n2 f	11228a
	2 emb; n1 f	8316
	2 yg; n1 f	4249

	2.02 ps; n664 f	5355
	2.22-2.74 ps; r1-6; n3 ages	8239
	2.3 yg; r1-3; n9 lt	5884a
	2.54 ps; ±1.09; n157	5885
	2.71 ps; ±0.79; n53	5885
	2.8 yg	1871a
	3.0 emb; r2-4; n3 f	502
	3.0 yg ea; n3 lt	9066
	3.0 emb; r2-4; n8 f	12044
	3.2 emb; mode 3; r1-8; n356 f	3625
	3.3; r2-5; n6 lt, n1 f	11815
	3.5 yg; mode 4; r1-6; n47 f	3625
	3.9 ps; mode 4; r3-5; n12 f	3625
	4.8 cl; mode 5; r2-9; n21 f	3625
Neofelis nebulosa	1.5 yg; r1-2; n2 lt	3660
	1.82; r1-2; n11 lt	7648
	3 yg; 1 lt	3202
	4 yg; 1 lt	3378
Panthera leo	1 yg; n1 lt	30
	1.7-2.3	9546
	mode 2-4; r1-9	5761
	2.1; r1-4; n7 lt	2278
	2.4; n41 lt	2875, 2882
	2.48 yg; r1-5; n64 lt	2039
	2.5; mode 3; r1-5	12128
	2.5 yg; r2-3; n59 lt, 4-6 wk	852
	2.57 yg; r1-6; n112 lt	9360
	2.6; r1-4; max 6; n19	10158
	2.75; r1-4; n4 lt	2583
	2.9 yg; r2-4; n15 lt	9066
	3; mode 3 52%; r1-4; n19 lt, 4 wk	9358
	3.01; r1-8; mode 3; n707 lt	9730
	3.02 yg; mode 3; r1-4; n47 lt	10170
	3.04; r1-6; mode 3; n323 lt	10412
	3.08 cl; mode 3; r1-5; n25 f	10170
	3.08 emb; mode 3; r1-5; n12 f	10170
	3.1; mode 3; r1-5; n84 lt	1230
	3.3; r3-4; n6 lt	9360
	7; n1 lt	11167
Panthera onca	1-4	10809
	1 emb; n1 f	7324
	1.67; r1-2; n3	10332
	1.93; mode 2; n15 lt	9730
	2-4	1474
	2; n5	1230
	3 yg; n1 lt	2109
Panthera pardus	1-2	9546
	1-4	2439
	1-4	8223
	1.5 yg; r1-2; n2 lt	2109
	1.6 yg; mode 1-2; r1-3; n39 lt	2494
	1.7 yg; r1-3; n60 lt, 8 f	2878
	1.70; se0.06; r1-3; mode 2; n159 lt	9196
	1.75; mode 1; r1-3	12128
	1.77 yg; n39 lt	2494
	1.78; r1-3	1230
	mode 2; r2-4	9175
	2-3; r2-5	10499
	2-3; max 6	331
	mode 2-3; max 6	10158
	2-4	6654
	2-4	9945
	2-5	7979
	2; n1 lt	6712

	2; r1-3; n17 lt	2936
	2.09; se0.06; mode 2, 4 rare; r1-4; n171 lt	9196
	2.2 yg; r1-3; n10 lt	2651
	2.32; mode 2; n75 lt	9730
	2.4 yg; r2-3	9066
	2.5 yg; r2-3; n2 lt	9726
	3 yg; n1 f	9725
	mode 3 born; mode 2 weaned	8501
	mode 3-4; r2-6	5761
	4	11253
Panthera tigris	1.53; mode 2; r1-4; n30 lt	1230
	mode 2-3; r1-5	19
	2.3; mode 3	12128
	2.3; n530 lt	9706
	2.33 yg; r1-3; n12 lt	29
	2.5 yg; r2-4; n15 lt	9066
	2.53 yg; r1-5; n652 lt	7407
	2.55 yg; r1-4; n27 lt	11089, 11090
	2.6; mode 2,4; r1-5; n22 lt	2272
	2.65; n148 lt	7004
	2.79; mode 2-3; r1-7; n107 lt	9474
	2.8 yg; r1-5; n15 lt	11283
	3 yg; n1 lt	2333
	3.63; n49 lt	9729, 9730
	5 emb; n2 f	3425
Panthera uncia	mode 2-3; r2-5	10499
	2; n1 lt	5465
	2 yg; n1 lt	1557
	mode 2; r1-4; n6 lt	6829
	2.04; r1-4; n47 lt	3499
	2.2; mode 2; r1-3; 19 lt	5930
SEXUAL MATURITY		
Acinonyx jubatus	f: 1st conceive: 21-22 mo	5761
	f: 1st birth: 3.67 yr	3338
	f: 21-24 mo	10158
	f/m: 1st birth: 2.5-3.5 yr	6795
	f/m: 14-16 mo	2873
	f/m: ~2 yr	8798
Felis bengalensis	f: 1st birth: 13-14, 31, 36 mo; n3	9566
Felis chaus	f: 11 mo	9565
Felis concolor	f: 1st mate: 2-3 yr	12045
	f: 1st birth: 20-21 mo	2881
	f: repro: 3-4 yr	6545
	m: 1st mate: 16-29 mo	2881
Felis nigripes	f: 1st estrus: 251 d	370
	m: spraying: 228 d	370
Felis pardalis	f/m: 1st mate: 20-24 mo	9567
	f/m: 16-18 mo	9567
Felis serval	f: 1st estrus: 26 mo	3168
	f: 1st mate: 28 mo	1018
	f: 1st birth: 28 mo	3168
	m: 1st mate: 17 mo	1018
Felis silvestris	f: 1st cl: ~8 mo	11072
	f: 1st birth: 11-13 mo; n5	2019, 9560
	f: 1st birth: 1 yr	7259
	f: 10 mo	11331
	f: 10 mo	3571, 9175
	f: 10-12 mo; 2500 g	5400
	f: < 12 mo	6965
	m: spermatogenesis: ~7 mo; 2.4 kg; ±0.17; min 2.2	11072
	m: spermatogenesis: 2600 g	5400
	m: 9-10 mo	6965
	m: 1 yr	2019, 9560
	m: 12-14 mo; 3200-3800 g	5400

	f/m: 1st season	10760
	f/m: ~1 yr	5761
Lynx canadensis	f: 1st conceive: 9 mo	8238
	f/m: 1st mate: 1 yr	5620
	f/m: 1st mate: 22 mo	8238
	f/m: 22-23 mo	1871a
Lynx caracal	f: 1st birth: 15 mo	5592
	f: 21 mo	9175
	m: 1st sire: 12.5-14 mo	833
Lynx lynx	f: large ovary: 7.5-11.5 mo 50%	6137a
	f: 1st repro: 1-2 yr	3527
	f: 1.75 yr	3571
	m: large testes: 19.5-23.5 mo 50%	6137b
	m: spermatogenesis: 1.0-1.5 yr	3527
	m: 2.75 yr	3571
	f/m: 1st mate: end 2nd yr	11226
	f/m: 2 yr	10499
Lynx rufus	f: 1st cl: >12 mo	8765
	f: 1st mate: 1-2 yr	1871a
	f: 1st birth: 1 yr	8239
	f: 1st birth: 2 yr; n1	11815
Neofelis nebulosa	f: 1st estrus: 21.8 mo; sd2.9; r17-28; n9	11991a
	f: 1st conceive: 26.3 mo; sd7.9; r13-38; n13	11991a
	m: 25.5 mo; sd6.6; r16-38; n16	11991a
Panthera leo	f: 1st estrus: 3.5-4.5 yr	9546
	f: 1st mate: 2-3 yr; n21	9730
	f: 1st conceive: 2.5 yr	9358
	f: 1st conceive: 48 mo; r43-55	10170
	f: 1st birth: 24 mo; n1	2874
	f: 1st birth: 2.87 yr	30
	f: 1st birth: 4.75 yr	852
	m: 1st sperm: median 30 mo; r26-34	10170
	m: 1st mate: 2.17-3.5 yr; n8	9730
	f/m: 1st mate: 20-24 mo	9704
	f/m: 1st mate: <2 yr	10412
Panthera pardus	f: 1st mate: 28 mo; r23-32; n8	2878
	f: 1st mate: <3 yr	9730
	f: 1st birth: 2.25-4.17 yr; n6	2494
	f: 1st birth: 36.9 mo; r27-52; n8	2878
	m: 1st sire: 1.5 yr; n1	9730
	f/m: 22 mo	9546
Panthera tigris	f: 1st estrus: 3.5-4.83 yr; n6	2272
	f: 1st mate: <3 yr	9730
	f: 1st mate: 3 yr	38
	f: 1st birth: 3.25 yg; n2	11090
	f: 1st birth: 40 mo; n1	10555
	f: 1st birth: 4.08-5.17 yr; n2	2272
	f: 4 yr	19
	m: 1st mate: <3 yr, 2 in 1 case	9730
	m: 1st mate: 4 yr	38
	m: 4-5 yr	7004
CYCLE LENGTH		
Acinonyx jubatus	27 d; n1 cycle, 1 f	4659
Felis bengalensis	4-7 d	7781
Felis concolor	estrus: 6.6 d; r4-9; n8, 3 f	9422
	estrus: 8-11 d; n9	8822a
	estrus: ~9 d	12045
	22.8 d; interestrous interval	8822a
Felis pardalis	estrus: 5.35 d; r1-22; n20 cycles, 8 f	2877
Felis silvestris	estrus: 2-3 d	11271
	estrus: 4.39 d; sd1.46; n18	7776
	estrus: 5-6 d; max 8	2019
	27.2 d; ±1.1; 12-21 49.8%; r6-120	11683
	32.9 d; r13-53	11271

Felis wiedii	estrus: 10.13 d; r7-13; n8	8406
	26.5 d	8406
	32 d; n1 f, 5 mo of obs	8207
Lynx caracal	estrus: 4.5 d; r3-6; n20	833
	14 d; n15 cycles	833
Lynx rufus	~44 d	1871a
Neofelis nebulosa	estrus: 6.2 d; sd4.1; r1-17; via behavior	11991a
	estrus: 6.1 d; sd4.2; r1-16; via mating	11991a
	29.9 d; sd13.8; median 27; r10-55; n72 cycles, 28 f	11991a
Panthera leo	estrus: 3 d	1017
	estrus: 4-8 d	9730
	estrus: 6.76 d	2875
	estrus: 7.2 d; r4-16; n14, 3 f	9422
	19-66 d	2882
	r3-8 wk; n8 cycles, 3 f	9605
	28 d	1017
	55.4 d; n9 cycles, 3 f	9422
	76.45 d; n192 cycles, 24 f, n3 prides	2875
Panthera onca	estrus: 7.4 d; r5-11; n13, 2 f	9811, 10322
	estrus: 12.0 d; ±1.0; n7	11687
	estrus: 12.9 d; r6-17; n10, 1 f	9422
	27.4 d; r13-58; n11, 2 f	9811, 10322
	42.6 d; r25-60; n8 cycles, 1 f	9422
	47.2 d; se5.4; n6 cycles, 1 f	11687
Panthera pardus	estrus: 6.7 d; r4-12; n34, 3 f	9422
	estrus: 7.5 d; r6-9	6712
	3.4 wk; n33 cycles no luteal phase, 3 f	9604
	45.8 d; r19-115; n27 cycles, 3 f	9422
	47 d; r27-58; n5 cycles, 1 f	6712
	7.3 wk; n31 cycles with luteal phase, 3 f	9604
	54.96 d; r6-210; n24 cycles, 7 f	2878
Panthera tigris	estrus: 1 wk	7979
	estrus: 3-13 d	75
	estrus: 5.3 d; se0.2; n12, 4 f	9706, 9708
	estrus: 7.1 d; r3-10; n14, 1 f	9422
	estrus: 7.2 d; r2-22	5848
	estrus: 7.6 d; r5-10	29
	estrus: 10-15 d	11283
	20 d	9474
	25.0 d; se1.3; n12 cycles, 4 f	9706, 9708
	25; r20-30; n5 cycles, 1 f	10555
	41 d; r33-53; n12 cycles, 2 f	29
	49.12 d; r12-86 d; n69 cycles, 2 f	5848
	r45-55 d; n3, 1 f, from graph	9422
	65.9 d; r25-138; n15 cycles, 2 f	43
Panthera uncia	estrus: 4-5 d	9131
	estrus: 6.2 d; r5-7; n5, 1 f	9422
	55-65 d; n3, 1 f, from graph	9422
GESTATION		
Acinonyx jubatus	90 d	11697
	~3 mo; n1	10788
	91-92 d; n1	3338, 3339
	91-95 d	6792
	94 d	10020
Felis bengalensis	~2 mo	7979
	~2 mo	10499
	63-69 d; 1st, last mating to birth	7781
Felis chaus	56 d	9175
	~2 mo	4645
	64-66 d; n1	9565
Felis concolor	82-98 d	1871a
	87 d	4582
	88-90 d; n2	8822a
	88-97 d	1474

	~90 d	11936
	3 mo	1355
	91 d	7234
	~92 d; r88-96	11338
	91-97 d	7088
	92-104 d	10809
	92.3 d; r84-106; n12	2881
	93.2 d; r86-98; n10	9185
	95 d; n1; artificial insemination	7495, 7496
Felis geoffroyi	65.3 d; r62-70; n9	300
	73.3 d; r72-74; n3	6268
	74-75 d	5476
	74, 75-76 d; n2	9574
Felis manul	74-75 d; n1	9563
Felis margarita	59-63 d	4646
	61 d; r59-63; n2	9573
	77-82 d	9564
Felis nigripes	~67 d	9675
	67 d; n1	9674
	67-68 d; n3	6423
Felis pardalis	79.7 d; r77-82; n23	2877
	82 d; r77-90; n32	9567
Felis serval	56 d	11697
	64-78 d	5761
	71-76 d; n2	1014
	74 d	3168
	74 d; r67-77; n3 f, 14 births	11378
Felis silvestris	56 d	8223
	~56 d	2439
	63-67 d	11331
	65 d	7839
	65-69 d	10871
	9 wk; median 66 d; r60-70; n75	813
	66 d; 63-69.5	2019
	~68 d	6965
	69 d; n1	7259
Felis tigrina	74-75, 76, 76 d; n3	6422
Felis viverrina	63 d; n1	11038
	80 d; n1	9368
Felis wiedii	81 d; n1	8207
Felis yagouaroundi	72-75 d; n3	5090
Lynx canadensis	min 60 d	5146
	60-65 d	5620
	~62 d	1474
	est 63-70 d	10942
Lynx caracal	66 d; r64-68; n2	10156, 10158
	69 d ea; n2	3982
	69-70 d	9175
	69-78 d	5761
	79 d; r78-81; n3	5592
	79 d; r78-81; n5	833
Lynx lynx	~9-10 wk	3571, 4645, 7979, 10499
	67, 68-70, 70, 70 d; n4	1451
	10-11 wk	11226
	73-74 d	6102
Lynx pardinus	63-73 d	10809
Lynx rufus	~50 d	2088
	~50 d	6387
	50-60 d	1474
	50-60 d; hunting records	12044
	50-70 d	1871a
Neofelis nebulosa	86-94 d	3203
	90-97, ~95 d	3660
	mode 91 d 56%; r88-109; n9	7648

	93.4 d; sd6.3; r85-121	11991a
Panthera leo	100-113 d	10308
	100-119 d	5761
	100.48 d; n42 lt, 3 prides	2875
	102.1 d; sd3.02; n7	2882
	105 d; r102-112	4582
	~3.5 mo	852
	106.2 d; n398; last mating to birth	9730
	est 108 d	10170
	~108 d	1230
	109.74 d; r100-114; n51	2039
	110 d; sd10	9704
	111 d	1355
	111.9; n419; 1st mating to birth	9730
	112-115 d	1017
	114.75 d; r111-119; n4 lt, 3 f	9422
	115 d; n1	2583
Panthera onca	93-105 d	10809
	95-108 d	1230
	98-109 d	2109
	3.33 mo	1355
	101, 103, 103-105 d; n3	10332
	3.5 mo	4582
	109.5 d; r108-111; n2, 1 f	9422
Panthera pardus	~69 d; stillborn after artificial insemination	2718
	90 d; n1 stillborn	9422
	90-100 d	11697
	3 mo	1355
	3 mo; r87-99 d	4582
	~3 mo	10499
	~3 mo	7979
	91.9 d, r84-98	2494
	92-95 d	2439
	92-95 d	8223
	92-105 d	5761
	93-103 d	1230
	96 d	6712
	98-105 d	9175
	100 d ea; n2	9422
	112 d	2651
Panthera tigris	95-107 d	19
	102 d; n1	5848
	102-106 d; last mating to birth; 108-111 d; 1st mating to birth	9730
	mode 103; r95-107	75
	103 d; r102-104; n5 f, n5 births	2272
	103-104 d; r98-110	4582
	104 d; n4	9708
	~15 wk	6654
	105 d	7979
	105.4 d; n7, 1 f	9422
	106.27 d; r93-114; n11 lt, 5 f	9474
	109 d	1355
	111.7 d; r93-105; n9 lt	11283
	~113 d	2333
Panthera uncia	54 d; n1 premature birth	9422
	~3 mo	10499
	90-100 d	7979
	98 d	5465
	98-103 d; n5	6829
	98-103 d	3498
	98-105 d	5838
	99 d; n1	1557
	99-100 d; n1	3496

LACTATION

Acinonyx jubatus	solid food, den emergence: ~4 wk	8798
	solid food: 1 mo; n1 lt	3339
	solid food: 5-6 wk	10158
	solid food: 47 d	6793
	wean: 3 wk	9546
	wean: ~3 mo	10158
	wean: 14-24 wk	5761
	wean: 5 mo	3339
	wean: 4-5 mo	8798
Felis bengalensis	solid food: 4th wk	3514
Felis chaus	solid food: 49 d	9565
	wean: 3 mo, 102 d	9565
Felis concolor	wean: 4-5 wk	11936
	wean: 4-5 wk	12045
Felis geoffroyi	wean: 8-10 wk	300
Felis nigripes	solid food: 1 mo	9675
	solid food: 34-35 d	6423
Felis pardalis	wean: 6 wk	9567
Felis serval	solid food: 39 d	1018
Felis silvestris	solid food: 25-30 d	4237
	solid food: 1 mo	7259
	solid food: 35-36 d	10871
	solid food: 1.5 mo	3571
	wean: 50-60 d	11331
	wean: 3 mo	2490
	wean: ~4 mo	4252
	driven off: 3 mo	11331
	wean: 6-8 wk	7259
	wean: ~4 mo	3571
Felis tigrina	solid food: 55 d	6422
Felis viverrina	solid food: 53 d	9175
	wean: 6 mo	9175
	independent: 4 wk	9368
Felis wiedii	solid food: 53.6 d; r52-57; n3	8407
Lynx caracal	solid food: 1.5 mo	3982
	solid food: 2 mo	5761
	wean: 6 mo	5995
	wean: 2.5 mo	3982
	wean: 15-20 wk	833
Lynx lynx	solid food: 6 wk; n1 yg, 1 lt	11501
	wean: ~4 mo	11226
	wean: 2 mo	3571
	wean: 2-3 mo	10499
	wean: 3 mo	4645
Lynx rufus	f took food to den: 7-8 wk	11815
	wean: 60-70 d	5210
	nurse: 16th wk	11815
Neofelis nebulosa	solid food: 6 wk	3202
	nurse: 93 d	3660
Panthera leo	solid food: 5 wk	9546
	solid food: ~2 mo	852
	solid food: ~2 mo	5761
	solid food: 2.5 mo	1017
	nurse: 5-8 mo	5761
	nurse: 6-7 mo	10158
	wean: 5.5 mo	1017
	wean: 7-8 mo	9546
	wean: max 15 mo	9360
	wean: 6-8 mo	9704
	wean: 6-8 mo	852
	wean: 8-9 mo	10170
Panthera onca	solid food: 10-11 wk	9811
	wean:5-6 mo	9811

Panthera pardus	solid food, den emergence: 42 d	5761
	wean: 3 mo	5761, 10158
	wean: max 129 d; r114-130	2494
Panthera tigris	solid food: ~1 mo	7979
	solid food: 2 mo	7004
	solid food: 43 d	75
	wean: 3 mo	11283
	nurse: 3 mo	9474
	nurse: 3-6 mo	7004
	independent: 18 mo	10555
	wean: ~6 mo	10499
Panthera uncia	wean: > 2 mo	5465
	wean: 48 d	3499
INTERLITTER INTERVAL		
Felis bengalensis	163.3 d; r81-305; n3 intervals, 1 f	26
Felis chaus	93-131 d	9565
Felis concolor	17.5 mo; r16-19; n2	6545
	2-3 yr	1474
	2-3 yr	12045
Felis pardalis	174.3; n15	2877
	2 yr; n2 f; circumstantial evidence	2999
Felis serval	5-6 mo	3168
	184 d w/normal lact	11378
Felis silvestris	114 d; r112-116; n2	11072
Panthera leo	1 yr 3 mo 23 d; n7	36
	19 mo	852
	24 mo	9546
	2 yr	10158
	2 yr	9358
Panthera pardus	8.44 mo; r 3.5-40; n51 intervals, 7 f	2878
	1 yr 2 mo 27 d; n10	36
	20-21 mo	9726
Panthera tigris	1 yr 6 mo 5 d; n5	36
	1.71-2.21 yr	29
	2-2.5 yr	10555
	2-3 yr	6654
LITTERS/YEAR		
Felis chaus	2	9565
Felis concolor	1/2.5 yr	7848
	est 0.5	9185
Felis geoffroyi	1-2	300
Felis nigripes	4	9675
Felis pardalis	1; 2 if yg die	2877
Felis serval	2	5761
	3?	11253
Felis silvestris	1	7259
	est 1	10760
	2	2044
	2	5400
	2, occasionally 3	6965
	3.11; mode 4; r1-6	2019
Felis yagouaroundi	est 2	1474
Lynx lynx	1	3571
Panthera leo	1	11253
	max 3; yg removed ~6 wk	1230
Panthera pardus	1	10499
	1	8223
	1; occasionally 2	9196
	2?	11253
Panthera tigris	1/2-3 yr	7979
Panthera uncia	1	10499

SEASONALITY

Acinonyx jubatus	mate: Jan; n1; captive	10788
	mate: peak July-Sept; captive	2880
	birth: Jan-Aug, peak Mar-June; e Africa	5761
	birth: Jan-Aug; Serengeti, Tanzania	9546
	birth: Apr; n1; captive	10788
	birth: Nov, Mar-Apr; few data; Mumbwa, n Zimbabwe	331
	birth: fall-winter; Kruger Nat Park, S Africa	8514
	birth: yr rd; n67; captive	11684
	yg seen: yr rd; Natal	11616
	repro: yr rd; n14; zoo	6795
Felis bengalensis	estrus: Mar; Amur Basin, Siberia, Russia	7979
	mate: Feb-Mar, Aug-Sept; Malaysia	6448a
	mate: Mar; e USSR	10499
	mate: irregular; captive	7781
	birth: Feb-Aug; zoo	8639
	birth: Mar; central Asia	179
	birth: May; Amur Basin, Siberia, Russia	7979
	birth: mid-end May; e USSR	10499
Felis chaus	estrus: Feb-Mar; USSR	7979
	mate: spring; Europe, USSR	4645
	preg f: Apr; n1; Nepal	7382
	birth: May; USSR	7979
	birth: Dec-May; captive	9565
	birth: winter-early spring; Pakistan	9175
	lact f: Dec; n2; Nepal	7382
Felis concolor	mate: Dec-Mar; WI	5210
	mate: yr rd; zoo	1230
	birth: May-June; n2; FL	6723a
	birth: yr rd; most Apr; w US	5146
	birth: yr rd, 60% June-Sept, peak July; UT, NV	9185
	birth: yr rd, except Nov, peak May-Oct; zoo	2881
	birth: yr rd, peak spring-summer; Canada	11936
	birth: yr rd; w US	12045
	repro: yr rd; MAN, Canada	7848
Felis geoffroyi	birth: Dec-May; Uruguay	11977
Felis manul	birth: Apr-May; Transbaikalia, Russia	9558
	birth: end Apr-May; Transbaikalia, Russia	10499
Felis margarita	mate: Jan-Feb; n Africa, Asia, Arabia	9559
	birth: early-mid Apr; USSR	7979
	birth: Apr, Sept; zoo	4618
	birth: Apr-May; n Africa, Asia, Arabia	9559
	birth: spring, end summer; Pakistan	9175
Felis marmorata	repro: not seasonal; Asia, Indonesia	9564
Felis nigripes	birth: Nov-Dec; S Africa	11311
Felis pardalis	birth: Nov, Apr; TX	9567
	birth: Dec-Feb; S America	9567
	birth: Jan; Yucatan, Mexico	9567
Felis serval	mate: Jan; n1; zoo	1018
	preg f: Nov, Jan-Mar; S Africa	10157
	birth: Jan-Apr; s Africa	10155
	birth: Jan-June: n part of range	10155
	birth: Mar-Apr, Sept-Nov; n8; Uganda, e Zaire	5761
	birth: Apr; n1; zoo	1018
	birth: Aug-Feb; Zimbabwe	10153
	birth: peak Sept-Apr; Zimbabwe	10157
	birth: Sept-Apr; S Africa	10158
	birth: twice/yr, start/end rainy season; Zaire	11253
	repro: est yr rd; zoo	11378
Felis silvestris	estrus: end Jan-Mar; USSR	7979
	estrus: Feb-Mar; central Europe	4252
	mate: Feb-Mar; Europe, USSR	4645
	mate: Feb-Mar, May-early June; Germany	3571
	mate: Feb-Apr, Oct; India	9843

	mate: end Feb-early Mar; Caucasus, Asia	9370
	mate: Mar; Zambia	10153
	mate: Apr, Aug, Oct; Zimbabwe	10153
	mate: Dec-June; captive	2019
	mate: spring, fall; Scotland, UK	10760
	preg f: Oct-Apr; Macquarie I, Australia	2044
	birth: Mar-June; zoo	11331
	birth: Mar-Aug; zoo	7259
	birth: Mar-Aug, peak May; n60; captive	2019
	birth: Mar-Oct, peak Apr; ne France	2020
	birth: end Mar-Apr; USSR	7979
	birth: Apr-May; central Europe	4252
	birth: peak Apr-May; Europe	2019, 9560
	birth: Apr-June, end July-Aug; Germany	3571
	birth: Apr-June; Europe, USSR	4645
	birth: early May, Aug, end Nov-early Dec; Scotland, UK	6965
	birth: May-June; Caucasus, Asia	7979
	birth: Aug-Apr, peaks Oct-Nov, Jan; S Africa	348
	birth: peak Sept-Mar; S Africa	10158
	birth: mid Sept-mid Mar; subantarctic	11074
	birth: peak Sept-Mar; se Australia	5400
	birth: Oct-Dec; Angola	11697
	birth: peak Oct-Jan, none May-July; Kerguelen, subantarctic	2490
	birth: Nov-Mar; Macquarie I, Australia	5399
	birth: yr rd; captive	9197
	wean: June; USSR	7979
	yg found: Mar-Oct; France	2020
	repro: Oct-May, peaks Nov, Apr; subantarctic	2491
Felis wiedii	estrus: Dec-Apr; captive	8406
Felis yagouaroundi	mate: Sept-Nov; Paraguay	10809
	mate: Nov-Dec; Mexico	10809
Lynx canadensis	mate: Jan-Feb; Canada	1474
	mate: Jan-Feb; w US	5146
	mate: Jan-Feb; WI	5210
	mate: Feb-Mar; Canada	5620
	mate: Mar-early Apr; NEW, Canada	10942
	mate: Mar-Apr; AK	10942
	mate: Apr-May; ALB, Canada	10942
	birth: Mar-Apr; Canada	1474
	birth: Mar-Apr; w US	5146
	birth: Mar-Apr; WI	5210
	birth: Apr-May; Canada	5620
	birth: end May-early June; NEW, Canada	10942
	birth: end May-early June; WY	1871a
	birth: spring; Cape Breton, NS, Canada	8238
Lynx caracal	mate: Sept-Dec; Zambia	10153
	preg f: Apr, Sept-Nov; n8; Bedford District, S Africa	8758
	birth: Apr; Karakum, USSR	9496
	birth: Sept; n1; Natal, S Africa	9327
	birth: peak Oct-Jan; S Africa	833
	lact f: Nov; few data; n Zimbabwe	331
	repro: yr rd; Morocco	8223
Lynx lynx	estrus: Feb-early Mar; Romania	11226
	mate: Jan-Mar; Europe	3571
	mate: end Jan-Apr; Europe, USSR	4645
	mate: Feb-Mar; zoo	4058
	mate: Feb-early Apr; Norway	6137b
	mate: mid Feb-Mar; USSR	10499
	mate: Mar; some regions earlier; USSR	7979
	birth: Apr-May; Europe	3571
	birth: May-June; USSR	10499
	birth: mid-end May; zoo	4058
	birth: end May; Russia	10942
	birth: end May-early June; Norway	6137b

Lynx pardinus	mate: Jan; Spain	11070
Lynx rufus	follicular size: large Jan-Mar; SC	8765
	CL present: early Mar; ne US	8647
	mate: mid Jan-July; WY	1871a
	mate: peak Feb-May; UT	12044
	mate: end Feb-Mar; WI	5210
	preg f: Jan-Sept, peak Mar-Apr; UT	3625
	preg f: Feb-June; N America	12044
	preg f: Mar; n2; CA	11228a
	birth: Apr-May; BC, Canada	2088
	birth: Apr-May for 5 yr; Apr, Oct 1 yr; FL	11815
	birth: mid Apr-early Sept; WY	1871a
	birth: May; ne US	8647
	birth: est Aug; n1 lt; MA	9886
	birth: spring; Cape Breton, NS, Canada	8239
	birth: spring-summer; w US	5146
	birth: yr rd; UT	12044
Neofelis nebulosa	estrus: peak Oct-Feb; captive	11991a
	birth: peak Mar; captive	11991a
Panthera leo	mate: yr rd; Kenya	9358
	birth: Feb-Mar, June; n3; zoo	44
	birth: Mar-Apr; S Africa	5780
	birth: peak Dec-Jan; Kenya	9358
	birth: dry season; Zaire	11253
	birth: yr rd; S Africa	10170
	birth: yr rd; S Africa	8514
	birth: yr rd; n Zimbabwe	331
	birth: yr rd; n38; zoo	12128
	birth: yr rd; S Africa	3145
	birth: yr rd; Serengeti, Tanzania	852
	birth: yr rd; zoo	1230
	birth: yr rd; zoo	10412
Panthera onca	mate: Jan; n Mexico	1474
	birth: Jan-Feb, June, Sept; few data; zoo	1230
	birth: peak Mar-July; Argentina	9811
	birth: Apr-May; n Mexico	1474
	birth: peak June-Aug; Belize	9811
	birth: peak: July-Sept; Mexico	9811
	birth: peak Nov-Dec; Paraguay	9811
	birth: peak Dec-May; Brazil	9811
Panthera pardus	estrus: Jan; e Asia	7979
	estrus: peak May, Oct; zoo	2878
	estrus: yr rd; captive	2494
	mate: Jan-Feb; Amur, Siberia, Russia	10499
	mate: Nov-Dec; Nepal	7382
	mate: Dec-Jan?; Wilpatti Nat Park, Sri Lanka	2936
	mate: est not seasonal; few data; n Zimbabwe	331
	birth: Feb-Mar; India	988a
	birth: Feb-Mar; India	6654
	birth: Feb-Mar; Nepal	7382
	birth: Apr-May; USSR	7979
	birth: Apr-May; Amur, USSR	10499
	birth: Apr-June, Dec; n5; zoo	44
	birth: peak Aug, low Oct; zoo	2878
	birth: spring-early summer; Pakistan	9175
	birth: rainy season; Angola	11697
	birth: start rainy & start dry seasons; Zaire	11253
	birth: not seasonal; Namibia	9945
	birth: yr rd; S Africa	10158
	birth: yr rd, peak May, low Dec; n146 lt; zoo	3497
	birth: yr rd; S Africa	3145
	birth: yr rd; n27; zoo	12128
	birth: yr rd; S Africa	3145
	repro: yr rd; zoos	1230, 9196

Panthera tigris	estrus: end Jan-early June; zoo	9706, 9708
	estrua: Dec-Jan; USSR	7979
	estrus: not seasonal; e Asia	7979
	mate: end Dec-early Jan; Khabaroush, Siberia, Russia	10499
	mate: Mar-Apr, July-Aug; n12; zoo	11283
	mate: not seasonal; zoo	1230
	birth: Mar-July, Nov-Dec; zoo	29
	birth: Apr; Khabaroush Terr, Siberia, Russia	10499
	birth: peak May-June; n1239; zoo	9706
	birth: July, Dec; n2; zoo	44
	birth: low Oct-Mar, peak Apr-June; zoo	9706
	birth: yr rd, peak May; zoo	9706
	birth: yr rd, peak Mar-June; Rajasthan, India	9474
	birth: yr rd; n17; zoo	12128
Panthera uncia	estrus: winter-early spring; USSR	7979
	mate: Feb; zoo	9131
	mate: Mar; n1; zoo	5465
	mate: end winter-early spring; USSR	10499
	birth: Apr; USSR	7979
	birth: Apr-May; USSR	10499
	birth: May 62%; zoo	3499
	birth: peak May; zoo	3497
	birth: end May-early June; n Chitral, Pakistan	9175
	birth: none Oct-Jan; zoo	3497

Order Carnivora, Family Otariidae

Eared or fur seals (*Arctocephalus* 20-700 kg; *Callorhinus* 30-280 kg; *Eumetopias* 270-1000 kg; *Neophoca* 60-300 kg; *Otaria* 140-340 kg; *Phocarctos* 230-400 kg; *Zalophus* 40-450 kg) are polygynous, semiaquatic, marine carnivores from the southern hemisphere and north Pacific. Males are larger than females (359, 1083, 1086, 1090, 1474, 2121, 3717, 3720, 4257, 6170, 6828, 6939, 6940, 6977, 8024, 8278, 8279, 8417, 8908, 8910, 8911, 9459, 9570, 9677, 11237, 11238, 11239, 11403, 11404).

The 500-1000 g, zonary, deciduous placenta implants into a bicorunate uterus (359, 2062, 2102, 2666, 3716, 3717, 4258, 4845) and is not usually eaten after birth (1083, 1605, 3716, 4257, 8024, 8908, 11404). Implantation can be delayed (*Arctocephalus* 9570, 11461; *Callorhinus* 2252, 2254, 2255, 3032, 10736). Estrus occurs 0-3 wk after birth (359, 621, 1083, 1090, 2102, 2299, 2667, 3716, 3717, 3719, 5683, 5752, 6170, 6287, 6485, 6827, 6940, 7318, 7319, 8024, 8417, 8568, 8908, 8911, 9458, 9459, 9658, 10158, 10736, 10809, 10898, 10899, 11461). The precocial neonates are furred with open eyes (1604, 1605, 3716, 7104, 8908, 8909, 11239). Females have 2-6 mammae (1604, 3716, 5755). Males have a baculum (1473, 9587, 11239).

Details of anatomy (236, 2102, 4258, 9586, 9969, 10728, 12029), endocrinology (4845), birth (11220), suckling (1604, 4730), milk composition (420, 1083, 2062, 3718, 8035, 8040, 8542, 8697, 8908, 9608, 10902, 10904), and light (10276) are available.

NEONATAL MASS

Arctocephalus australis	3-5 kg	5755
	3.5-5.5 kg	11238
Arctocephalus forsteri	3.5 kg	5755
	3.7 kg; r2.90-4.40; n5	6977
	4.3 kg	2121
Arctocephalus galapagoensis	3.0-4.5 kg	10898
	f: 3.1-3.4 kg; n36	9509
	m: 3.3-3.9 kg; n34	9509
Arctocephalus gazella	5-6 kg	1086
	f: ~5.6 kg; n12	8279
	6 kg	5755
	m: 6 kg; n9	8279
	7 kg	2666
Arctocephalus pusillus	3.6-5.4 kg	1605
	4.5-6.4 kg	8908, 8911
	4.5-7 kg	9570, 9864

	f: 7.1 kg; r4.5-10	11461
	m: 8.1 kg; r5-12.5	11461
Arctocephalus tropicalis	4-5 kg	5755
	~4.2 kg	10158
	4.9 kg	1090
Callorhinus ursinus	4-6 kg	359
	f: 4.8 kg; sd0.7; r3.3-6.0; n16	9591
	5 kg	5755
	f: 5.10 kg; se0.11; n11	2062, 3718
	m: 5.4 kg; sd0.9; r4.1-7.1; n23	9591
	5.5 kg	5683
	f: 5.55 kg; r5.3-5.8; n2	8419
	m: 6.23 kg; se0.27; n8	2062, 3718
Eumetopias jubatus	15.9 kg	5683
	16-23 kg	6940
	18 kg	5146
	18-22 kg	5755
	18.25 kg; r17-22; n4	9585
	19-22 kg	6944
Neophoca cinerea	6-8 kg	5755
	6.4-7.9 kg	6481a, 6483, 11403
Otaria byronia	f: 11.43 kg; r10.28-13.40; n6	11237, 11239
	f: 12.3 kg; sd1.4; n38	1591a
	m: 13.7 kg; sd1.6; n37	1591a
	m: 14.15 kg; r13.26-14.77; n4	11237, 11239
Zalophus californianus	5-6 kg	6939
	~6.0 kg	3718
	6.2 kg; n1	2109
	f: 7.7 kg; n28	8024
	m: 7.82 kg; n1	9623
	m: 9 kg; n33	8024
NEONATAL SIZE		
Arctocephalus forsteri	40-45 cm	2121
	55 cm	2122
Arctocephalus gazella	60-66 cm	1086
	65 cm	5755
Arctocephalus pusillus	60-70 cm	9570, 9864
	65 cm	5755
	m: 69.7 cm; r62-79	11461
	80 cm	8911
Arctocephalus tropicalis	60 cm	5755
	63 cm	1090
Callorhinus ursinus	f: 63.1 cm; sd4.45; r54.4-68.7; n16	9591
	m: 65.9 cm; sd4.58; r54.4-75.4; n23	9591
Eumetopias jubatus	100 cm	6940
Neophoca cinerea	62-68 cm	6481a, 11403
	70 cm	5755
	75 cm	5752
Otaria byronia	f: 78.5 cm; r73.0-82.0; n6	11237, 11239
	f: 79 cm; sd3; n29	1591a
	m: 81.7 cm; r79.0-85.0; n4	11237, 11239
	m: 82 cm; sd4; n29	1591a
	m: 83.8 cm; n1	4257
Phocarctos hookeri	~75 cm	6828
	75-80 cm	5755
Zalophus californianus	f: 70.5 cm	8024
	m: 74 cm; n33	8024
	75 cm	6939
	m: 78 cm; n1	9623
WEANING MASS		
Arctocephalus galapagoensis	f: 14 kg; 18-36 mo	3718
	m: 16 kg; 18-36 mo	3718
Arctocephalus gazella	f: 14.5 kg; 4 mo	3718
	m: 17.1 kg; 4 mo	3718

Arctocephalus pusillus	m: 25.0 kg; 9-11 mo	3718
	27.3 kg	8911
Callorhinus ursinus	f: 11.7 kg; 4 mo	3718
	f: 12.0 kg; sd2.2; r7.0-16.6; n91; 3 mo	9591
	m: 13.9 kg; sd2.2; r8.9-18.8; n82; 3 mo	9591
	m: 14.1 kg; 4 mo	3718
Zalophus californianus	m: ~25 kg; 10-12 mo	3718
WEANING SIZE		
Eumetopias jubatus	179 cm; yearlings	6599
LITTER SIZE		
Arctocephalus pusillus	1	1605
	1; 2 rare	8908, 8911
Arctocephalus tropicalis	1 ea; n2 lt	1428, 1429
	2 wean; n1 f	891
Callorhinus ursinus	1	3032
	1.04; r1-2; n84 lt	2102
	2 emb; n3 f	7914
Eumetopias jubatus	1	6944
	1 emb; n1 f	566
	1; 2 rare; 1 case in 3 yr	3716
Neophoca cinerea	1	10698
	2 aborted; n1 f	6481
Otaria byronia	1	4257
Zalophus californianus	1	2914
	1 ea; n12 lt	12128
	1 yg	5146
	2; n1 f	11020
SEXUAL MATURITY		
Arctocephalus australis	f: 3 yr	11238
	m: 7 yr	11238
Arctocephalus forsteri	f: 4-6 yr	2121
	m: 4-6 yr	2121
	m: socially mature: 10-12 yr	2121
Arctocephalus galapagoensis	f: 1st mate: 3 yr	10898
Arctocephalus gazella	f: 1st birth: 3-4 yr	1086
	f: 1st birth: 3-4 yr	8279
	m: 1st territory: ~8 yr	8279
	m: 3-4 yr	8279
	m: 6-7 yr	1086
Arctocephalus pusillus	f: 1st mate: 6 yr	9570
	f: 1st mate: after 2nd winter	8908
	f: 1st birth: 3rd yr	8911
	f: 1st birth: 4 yr	9864
	f: 3-6 yr	11461
	m: 4-5 yr	11461
	m: socially mature: 9-12+ yr	11461
Callorhinus ursinus	f: 1st estrus: mode 4 yr; min 2 yr	1705
	f: 1st mate: 3 yr; Asia; 4 yr; N America	6170
	f: 1st ovulate: 70% 4 yr	2102
	f: 1st cl: 3-7 yr	3717
	m: sperm: 3 yr; mode 4-5 yr	9591
	m: spermatogenesis: 3 yr	359
	m: 1st mate: 6 yr	359
	m: 1st mate: 7-9 yr	3717
	m: 4-5 yr	6170
	m: 5 yr	3717
Eumetopias jubatus	f: 1st ovulate: 4.6 yr; 95%ci 0.8; n ≥ 38	8568
	f: 3-8 yr	6599
	f: 4-5 yr	6940
	m: 1st mate: 9-13 yr	8568
	m: socially and physically competitive: 7-9 yr	6940
	m: 3-7 yr	6599
	m: 5-7 yr	6940
	m: est 5-7 yr	8568

	f/m: 1st repro: 4.9 yr; 95%ci 1.2	8568
Neophoca cinerea	f: 1st mate: 3 yr	5752
	m: 1st mate: 6+ yr	5752
Otaria byronia	f: 1st birth: 4 yr	4257, 4258
	m: 6 yr	4257
Phocarctos hookeri	m: 6 yr	6828
Zalophus californianus	f: 1st mate: 3 yr	1474
	m: 5 yr	1474
CYCLE LENGTH		
Arctocephalus forsteri	birth to estrus: 8 d	5755
Arctocephalus gazella	birth to estrus: 8 d	5755
Arctocephalus pusillus	birth to estrus: 6 d	5755
Arctocephalus tropicalis	birth to estrus: 8-12 d	5755
Callorhinus ursinus	birth to estrus: 7 d	5755
	behavioral estrus: 48 h	3717
	physiological estrus: ~1 mo	3717
Eumetopias jubatus	birth to estrus: 10-14 d	5755
	estrus: short	9459
Neophoca cinerea	birth to estrus: 4-9 d	5755
Zalophus californianus	birth to estrus: ~14 d	5755
GESTATION		
Arctocephalus galapagoensis	emb development: 7 mo	10898
Arctocephalus gazella	51 wk	1086
	4 mo; 1 yr; 8 mo	6287
	~1 yr	8278
Arctocephalus pusillus	implant: ~3 mo	11461
	implant: 4 mo	8908, 8910, 8911
	359 d with delay; primiparous longer up to 15 mo	8908, 8910, 8911, 8912
	51 wk	11461
	est 365-414 d; n10	1605
Callorhinus ursinus	implant: 3-5 mo	2254
	implant: 3.5-4 mo	6170
	implant: 4 mo	2102
	implant: min 4? mo	3032
	implant: 125 d	10736
	51 wk	6170
	360 d	2254
	1 yr	2102
	~1 yr	621
Eumetopias jubatus	implant: 3 mo	8568
	implant: 3 mo	6940
	11 mo	8128
	11.5 mo total	8568
	1 yr	1474, 5146
	~1 yr	3716
	~1 yr	5683
	~1 yr	6940
Neophoca cinerea	~12 mo	5752
	18 mo	11403
Otaria byronia	~1 yr; field data	11236
Phocarctos hookeri	~12 mo	6828
Zalophus californianus	11 mo; field data	8023, 8024
	~11 mo	4582
	50 wk	6939
	11.5-12 mo	1355
	~1 yr	8417
LACTATION		
Arctocephalus australis	wean: 6-12 mo	11238
Arctocephalus forsteri	solid food: 1 d (?)	7104
	wean: 10-11 mo	2121
	wean: ~1 yr; field data	7318
Arctocephalus galapagoensis	wean: 18-36 mo	3718
	wean: 2 yr	10898, 10899

Arctocephalus gazella	wean: 117 d; sd8.0; n8	2667
	nurse: 4 mo; r110-115 d	8278, 8279
Arctocephalus pusillus	solid food: 4 mo	8911
	solid food: 7 mo	9864
	wean: 8-9 mo	1605
	wean: most 9-11 mo; r8-18 mo	2299, 8908, 8911
	wean: 11-12 mo	11461
	wean: 12+ mo	9570
	wean: 1-2 yr	10903
Arctocephalus tropicalis	nurse: ~100 d	1083
	wean: 2 yr; but author states lactation 7 mo	1090
Callorhinus ursinus	wean: 2-3 mo	8416
	wean: 3 mo	5683, 9591
	wean: 4 mo	6170
	wean: 125 d	3719
	nurse: 3-4 mo	10276
Eumetopias jubatus	wean: min 3 mo	5683
	wean: 3 mo	5146
	wean: 8-11 mo	6940
	wean: 1 yr	9658
	wean: 12-15 mo	3716
	wean: 12-16 mo	9458
	nurse: est 1 yr	8568
Neophoca cinerea	wean: >1 yr	6827
	wean: 2 yr	10429
Otaria byronia	wean: ~1 yr	11239
Phocarctos hookeri	wean: est 6-8 mo	11404
Zalophus californianus	solid food: 4-5 mo	10900
	wean: 5-12 mo	6939
	wean: 10 mo-1 yr, exceptionally 2 yr	10900
	nurse: 1 yr	8417
INTERLITTER INTERVAL		
Arctocephalus galapagoensis	1 yr	10898
Arctocephalus gazella	1 yr	1086
Arctocephalus pusillus	1 yr	8911
	1 yr	9570
	1 yr	11461
Arctocephalus tropicalis	1 yr	1090
Eumetopias jubatus	1 yr	8568
	1 yr	3716
Neophoca cinerea	17-18 mo	6484, 6485, 11403
	est 18 mo	10698
Otaria byronia	1 yr	4257
Zalophus californianus	2 yr	2914
SEASONALITY		
Arctocephalus australis	mate/birth: end Nov-Dec; Falkland I	11238
	implant: Mar-Apr; Falkland I	11238
	birth: Nov-Dec; Uruguayan I	11236
	birth: Oct-early Dec; s hemisphere	5755
Arctocephalus forsteri	mate: mid Nov-mid Jan, peak mid Dec; South I, New Zealand	7318, 7319
	mate: end Nov-mid Jan, peak mid Dec; New Zealand	2122
	birth: Nov-mid Jan, peak mid Dec; New Zealand	2122
	birth: mid Nov-mid Dec; s hemisphere	5755
	birth: end Nov-mid Dec, peak early-mid Dec; South I, New Zealand	7318
Arctocephalus galapagoensis	birth: Aug-Nov, peak Oct; Galapagos, Ecuador	10898
	birth: mid Aug-mid Nov; Galapagos, Ecuador	10899
	repro: Aug-Nov; Galapagos, Ecuador	10809
Arctocephalus gazella	mate/birth: Oct-Nov; S Georgia I, Antarctic	8278
	mate/birth: Nov-Dec; S Georgia Sea, Antarctica	1086
	mate/birth: Nov-Dec; n7; S Georgia I, Antarctic	2667
	mate/birth: Dec; Antarctic	6287
	ovulate: before Jan; Marion I, Antarctic	8909
	birth: end Nov; S Africa	10158

	birth: early Dec; S Georgia I, Antarctic	8279
	birth: Dec; S Georgia I, Antarctic	2666
	wean: end Feb-Mar; Bird I, Antarctic	6287
	wean: Apr; S Georgia I, Antarctic	2667
Arctocephalus philippii	mate: Nov-Jan; Juan Fernandez Archipelago, Chile	10809
Arctocephalus pusillus	mate: Oct-Jan, median early Dec; Namibia, S Africa	9570
	mate: Nov-Dec; sw Africa	2299, 8908, 8912
	mate/birth: Nov-Dec; NSW, s TAS, Australia	11461
	birth: Mar-May; captive, ONT, Canada	1604
	birth: June; n hemisphere	5755
	birth: Oct-Dec; Punta San Juan, Peru	10903
	birth: Oct-Jan, median early Dec; Namibia, S Africa	9570
	birth: mid Oct-Dec; sw Africa	8908, 8911, 8912
	birth: Nov-Dec; s hemisphere	5755
	wean: Sept-Oct; S Africa	8908, 8911
Arctocephalus townsendi	mate/birth: May-July; Guadalupe, Mexico	8418
Arctocephalus tropicalis	mate/birth: Nov-Dec; Tristan da lunha Gough, Prince Edward, New Amsterdam, St Paul I, Indian Ocean	1090
	mate/birth: Dec; S Georgia, Antarctica	1083
	birth: est end Jan; Falkland I	8013
	birth: Dec; Gough I, s Atlantic Ocean	890
	birth: mid Dec; S Africa	10158
Callorhinus ursinus	mate: June, primiparous until July; Pribilofs, US	359
	mate: June-Aug; Pribilofs, US	3032
	mate/birth: June-July; Pribilofs, US	3719
	mate/birth: end June-July; Bering Sea, Okhotsk Sea	5683
	ovulate: mid July, primiparous Aug; e Pacific, Pribilofs, US	2102
	implant: early Nov; Bering Sea, n Pacific	6170
	implant: Nov; e Pacific, Pribilofs, US	2102
	birth: end May-July; n831; San Miguel I, CA	10736
	birth: June-July; St Paul I, Pribilofs, US	8416
	birth: June-early Aug; St Paul I, Pribilofs, US	621
	birth: mid June; Pribilof I, US	359
	birth: mid June-July, peak mid July; n433; St George I, Pribilofs, US	10736
	birth: end June-early Aug; Bering Sea, n Pacific	6170
	birth: mid July; e Pacific, Pribilofs, US	2102
	nurse: mid July-mid Oct; AK	9591
	wean: Nov; Pribilofs, US	3719
Eumetopias jubatus	mate: end May-mid July, peak June; n215; AK	8568
	mate: June; Año Nuevo I, CA	3716
	mate: June-July; Año Nuevo I, CA	9459
	mate: mid June-mid July; OR	6938
	mate: mid June-mid July; Año Nuevo I, CA	8128
	mate: mid June-mid July; Año Nuevo I, CA	9350b
	mate/birth: mid May-June; n Pacific	6940
	mate/birth: end May-early July; peak mid June; AK	9458
	mate/birth: June-July; CA, Bering Sea	5683
	implant: end Sept-Oct; AK	8568
	implant: Sept-Oct; n Pacific	6940
	birth: mid May-mid June; peak early June; AK	8568
	birth: mid May-mid July, peak June; Año Nuevo I, CA	3716
	birth: end May-early July; Año Nuevo I, CA	8128
	birth: end May-June; nw Pacific	6944
	birth: June; San Miguel, CA	9658
Neophoca cinerea	birth: est July; WA, Australia	6480
	birth: Oct; Dangerous Reef, Australia	6827
	birth: Oct-Dec; Australia	5752
	birth: Dec-Jan; s hemisphere	5755
	birth: yr rd; Australia	6484, 11403
Otaria byronia	mate: end Dec-Feb, Aug-Sept; Uruguayan I	11236
	mate/birth: Dec-Jan; Falkland I	4258
	birth: Dec-Jan; Falkland I	4257
	birth: mid Dec-early Feb, peak Jan; Uruguayan I	11236, 11237, 11239

	birth: mid Dec-mid Feb; Argentina	11239
Phocarctos hookeri	birth: Nov-Dec; s hemisphere	5755
	birth: Dec; Auckland I	3160
	birth: Dec-early Jan; Auckland I	6828
Zalophus californianus	m: territorial: May-Aug, peak June; San Nicolas I, CA	8417
	mate: May-June; CA, Mexico	5683
	mate: May-Aug; w US	5146
	mate: June-July; San Nicolas I, CA	8023, 8024
	mate/birth: June-July; San Nicolas I, CA	8417
	birth: Jan; Galapagos I, Ecuador	2924
	birth: mid May/June-July/Aug; Santiago I, Galapagos, Ecuador	10900
	birth: end May-early June; n12; zoo	12128
	birth: end May-June; CA, Mexico	6939
	birth: mid May-June; San Nicolas I, CA	8023, 8024
	birth: June 90%; n38; zoo	10635
	birth: June; n5 same f; zoo	2914
	birth: June-July; Pacific coast	1474
	birth: peak Aug-Oct, low Apr-May; Galapagos, Ecuador	2922
	birth: Oct-Dec; Galapagos, Ecuador	6939
	birth: mid? Dec-mid Jan; Galapagos, Ecuador	5755

Order Carnivora, Family Odobenidae, Genus *Odobenus*

Walruses (*Odobenus rosmarus* 600-1900 kg) are circumpolar, gregarious carnivores from arctic waters of the northern hemisphere. Males are larger than females and have larger tusks (1282, 1474, 3181, 3182, 6600). Implantation occurs 3-3.5 mo post coitus into a bicornuate uterus (1282, 3182). Gulls eat the placenta after birth (3182). Neonates can can swim at birth (1457). Males have inguinal testes and a baculum (1332, 1473, 3182, 5755). Details of anatomy (3182, 6786, 7627) and milk composition (3181, 3182) are available.

NEONATAL MASS		
Odobenus rosmarus	f: 33.2 kg; n1	1332
	41 kg	1457
	55 kg	6789
	f: 63.3 kg; r33.8-85; n15	3181, 3182
	m: 63.8 kg; r45-77; n5	3181
NEONATAL SIZE		
Odobenus rosmarus	f: 100 cm	1332
	105 cm	1457
	m: 112 cm; r104-122; n6	3182
	f: 113 cm; r95-123; n11	3182
	m: 119.4 cm; n1, term emb	6786
	f: 121.4 cm; r102-133; n30	3181
	m: 122.4 cm; r112-137; n18	3181
WEANING MASS		
Odobenus rosmarus	200 kg; 1 yr	6789
LITTER SIZE		
Odobenus rosmarus	1	1514
	1	6789
	1-2	5171
	1; 2 occasional	753
	1; 2 rare	3181, 3182
SEXUAL MATURITY		
Odobenus rosmarus	f: 1st cl: 5-6 yr	3181, 3182
	f: 1st cl: 6.25 yr; r5-9; n4	6786
	f: 1st cl: 6-7 yr	3183
	f: 1st mate: 10 mo	5171
	f: 1st mate: 4 yr	6789
	f: 1st birth: 5 yr; few data	1332
	f: repro: 5 yr 20%, 7 yr 50%	6077
	f: 3-4 yr	1514
	f: 4-5 yr	753
	f: 5-7 yr	1282

	m: spermatogenesis: 7-10 yr	3181, 3182
	m: testes large: 7 yr; few data	1332
	m: fertile: 8-10 yr	3183
	m: 5 yr	1514
	m: 5-6 yr	753
	m: 6 yr	6786, 6789
	m: 5-10 yr	1282
GESTATION		
Odobenus rosmarus	implant: 3-3.5 mo	1282
	implant: 4 mo	3181
	implant: 4-5 mo	5755
	implant: ~5 mo	3182
	postimplantation: 10 mo	3182
	postimplantation: 11.5-12 mo	1282
	est 11 mo	1984
	11 mo; field data	753
	~1 yr	1332
	~1 yr	5171
	15 mo with delay	1282
	~15 mo	6789
	15-16 mo with delay	3181, 3182, 3183
LACTATION		
Odobenus rosmarus	solid food: 5-6 mo	3182
	wean: 1-2 yr	1332
	wean: 1-2 yr; 2 yr more common	1282
	wean: >1 yr; perhaps 2 yr	753
	wean: >1 yr; probably 2 yr	6789
	wean: ≥ 16 mo	6600
	wean: 18-24 mo	1457
	wean: ≥2 yr	1514
	wean: ~2 yr; some 3 yr	3182, 3181
	wean: 2.5 yr	5171
INTERLITTER INTERVAL		
Odobenus rosmarus	2 yr	1514
	2 yr	6600
	~2 yr	3182
	≥2 yr	3181
	2-3 yr	1332
	2-3 yr	1474
	2-4 yr	6789
SEASONALITY		
Odobenus rosmarus	mate: Jan-Mar; circumpolar	3181
	mate: peak Feb-Mar; Greenland	6789
	mate: Feb-Mar; Greenland, Canadian arctic, Chukchi/Laptev seas	1282
	mate: early-mid June; Bering Sea, Pacific	753
	mate: Nov-May, peak Apr-May; e Canadian Arctic	6786
	mate: Dec-Mar; Bering Sea	3182
	ovulate: Jan-Feb; Bering Sea	3181, 3182
	implant: June; Bering Sea	3182
	birth/mate: Apr-June; Novaja Zeml'a, Russia	1514
	birth: Apr-early May; Bering Sea	1332
	birth: Apr-June; Eskimo account; Hudson Bay, Canada	6600
	birth: Apr-June; AK	1474
	birth: mid Apr-early June, peak May; Bering Sea	3181, 3182
	birth: early May; Bering Sea, Pacific	753
	birth: Apr-June, peak May; Greenland, w Canadian Arctic	6786, 6789

Order Carnivora, Family Phocidae

Earless seals (*Cystophora* 140-410 kg; *Erignathus* 110-370 kg; *Halichoerus* 100-310 kg; *Hydrurga* 115-590 kg; *Leptonychotes* 325-550 kg; *Lobodon* 200-240 kg; *Mirounga* 250-3700 kg; *Monachus* 200-400 kg; *Ommatophoca* 130-300 kg; *Phoca* 30-250 kg) are marine carnivores with a worldwide distribution. Males are larger than females in *Cystophora* (9005, 9783), *Halichoerus* (1979, 3571), *Mirounga* (2097, 6275, 6286, 6329, 6482, 7017, 7018, 7019), and *Phoca* (925, 927, 1088, 1142, 4511, 7087, 7753, 10145; cf 6655, 10430), while females are larger than males in *Erignathus* (1458; cf 772), *Hydrurga* (4904), *Leptonychotes* (854), *Monachus* (1281; cf 189, 5682), and *Ommatophoca* (4903, 6289).

The hemoendothelial, endotheliochorial, or epitheliochorial, 500 g-2 kg, labyrinthine, zonary placenta has an antimesometrial or central implantation into a bipartite or bicornuate uterus (928, 1456, 4427, 4431, 4708, 4845, 5470, 5755, 6277, 7563, 7864, 9965, 10001, 11099) and is not usually eaten after birth (854, 5166, 5875, 6276, 6333, 10426). Implantation can be delayed (*Halichoerus* 478, 1172, 4708; *Hydrurga* 6971; *Leptonychotes* 4427, 6971; *Lobodon* 4427, 6971; *Mirounga* 3763; *Phoca* 3275, 4515). Estrus usually does not occur immediately after birth, but at the end of the short lactation period (*Cystophora* 9005; *Erignathus* 1458; *Halichoerus* 1079, 1171, 4416; cf 1979, 2187; *Leptonychotes* 1906; cf 6287; *Mirounga* 1622, 1623, 2023, 5055, 6276, 6277, 6286, 6329, 7018, 7019, 7062; cf 622, 6328, 6482; *Phoca* 1088, 10014; cf 3188, 3275, 7087, 7786).

Neonates are furred with open eyes (234, 622, 854, 970, 1142, 1359, 1374, 1457, 1622, 1979, 2023, 2024, 4415, 5682, 6275, 6276, 6277, 6333, 6787, 7853, 8703, 9244, 10158, 10781, 11831). Females have 2-4 mammae (854, 5682, 5755, 6788, 9243). Males have inguinal testes and a baculum (1395, 1473, 4259, 5755).

Details of anatomy (233, 236, 926, 1146, 1171, 4062, 4415, 4427, 5754, 5755, 6277, 6971, 7087, 9586, 10727), endocrinology (1169, 1170, 1177, 4060, 4843, 4845, 8849, 9473), birth (1479, 5875, 9246), suckling (2142, 3365, 5552a, 5552b, 5552c, 8130), milk composition (234, 240, 1034, 2034, 5959, 6265, 6331, 7244, 8036, 8285, 9126, 9244, 10407, 10525, 10729, 11168, 11505, 11698), reproductive energetics (285, 2061, 3186, 10402, 10405), and light (929) are available.

NEONATAL MASS

Cystophora cristata	4-5.5 kg	625
	10-15 kg	9783
	22.0 kg; se0.61; n21	1156, 1158, 8038
	30 kg	9005
Erignathus barbatus	29-43 kg	8691
	m: 32 kg; r30-34; n2; 1 term emb, 1 pup	5353
	33.6 kg; n13	1456, 1457, 1458
	43 kg	10431
Halichoerus grypus	4.5-6 kg	3571
	9.1-17.2 kg	1092, 4725
	9.2 kg	2187
	10-15 kg; from graph	3186
	13.6 kg; r11.8-15	240
	f: 14 kg; r10-19; n34	2067
	m: 14.24 kg; r13.63-14.54; n3	233, 234
	14.5 kg	1087
	f: 14.6 kg	5980
	f: 14.8 kg; r10.9-19.5; n26	1180
	m: 15 kg; r9-20; n61	2067
	m: 15.8 kg; r12.3-19.9; n25	1180
	m: 16.3 kg	5980
Hydrurga leptonyx	29.0 kg; 1 stillborn	10158
	29.5 kg; n1	1359
	35 ? kg	4904
Leptonychotes weddelli	22-25 kg	2400
	24 kg; n9	10729
	27.24 kg	854
	f: 27.7 kg; n8	1399
	m: 29.0 kg; n8	1399
	29 kg; r23-35; n17	6479
	30-40 kg	5755

Lobodon carcinophagus	20 kg	6285
	22 kg	854
Mirounga angustirostris	30-34 kg	7062
	30 kg; r22-45.5	6333
	f: 31.5 kg	6329
	34 kg	6330, 6331
	m: 36.0 kg	6329
	40 kg	2061, 9024
Mirounga leonina	29.1 kg; r26.7-30.5; n3	7853
	30.6 kg; r30-31; n4	5881
	m: 31 kg; n1	5881
	f: 35.0 kg; n19	6507
	m: 37 kg; r30-50; n13, Sept birth	1624
	37.7 kg; n46	6507
	f: 38.1 kg; se0.8; n7	2024
	38.5 kg; ±7.0; n29	4137
	38.5 kg; ±6.5; n56	4137
	m: 38.9 kg; se2.1; n8	2024
	f: 39 kg; n3	1397
	m: 39.09 kg; n27	6507
	f: 39.3 kg; sd7.29; n3	6507
	f: 39.4 kg; sd4.9; n28	6507
	40.9 kg; sd5.1; n6	6507
	41 kg; coefficient of variation 36.6	1396
	m: 41 kg; r28-54; n14, Oct birth	1624
	41.02 kg; sd5.3; n61	6507
	41.8 kg; stillborn, 38.8 kg; dead within 24 hr	5879
	m: 42.4 kg; sd5.3; n33	6507
	m: 42.5 kg; sd2.14; n3	6507
	m: 43 kg; n3	1397
	45-50 kg	6482
	f: 45 kg; r39-51; n8	6275
	45.2 kg; sd5.7; n12	7019
	m: 49 kg; r43-56; n7	6275
Monachus albiventer	17 kg; term emb	10910
Monachus monachus	17-26 kg	1145
	20 kg	9788
Monachus schauinslandi	10.65 kg; n10	11831
	15.2 kg stillborn, 18.6, 19.3 kg	9090
	~16 kg	5679
	16-17 kg	1281
Ommatophoca rossi	16.8 kg; n1	10781
	est 27 kg	6289
Phoca caspica	5 kg	8681
Phoca fasciata	8.5 kg; near term	1457
	10 kg	10430
	10.5 kg	1459
Phoca groenlandica	2.7-3.6 kg	3571
	3-4 kg	625
	6.2-14 kg	10402
	9.30 kg; se0.62; n6	8038
	9.9 kg; sd1.7; n32	5978, 6264
	10-14 kg	4012
	10.7 kg; sd1.6; n48	5978, 6264
	10.8 kg; 95%ci 0.65; n40	5152, 10404
Phoca hispida	3.5-4 kg	5353
	4.0 kg	1457
	4-5 kg	8679
	7.5 kg	3188
	mode 8-12; r5.5-19.5 kg	3571
	8.1 kg	3534
Phoca largha	7.1 kg; r6.9-7.3	1089
Phoca sibirica	1.5-4.5 kg	10782
	3 kg	8680

Phoca vitulina	6.9-15 kg	3571
	mode 8-9 kg; r7-10.5	11829
	m: 8.9 kg; n1	233
	m: 9 kg; n1	9246
	9-11 kg	4415
	f: 9.49 kg	1143
	9.5-11.81 kg	1457
	m: 9.53 kg	1143
	10.2 kg; 95%ci 1.5; n9	925
	f: 10.4 kg	1143
	m: 10.59 kg	1143
	f: 10.71 kg; sd2.76; n13	7864
	f: 11 kg	1142
	m: 11 kg	1142
	m: 14.81 kg; sd2.74; n5	7864
	15 kg	4511
	est ~15-18 kg; n1	5166
NEONATAL SIZE		
Cystophora cristata	m: 87-115 cm	9005
	100 cm	9783
Erignathus barbatus	87 cm	10431
	120 cm	7086
	m: 129.25 cm; r127.5-131.0; n2; 1 term emb, 1 pup	5353
	130 cm	1457
	131.3 cm; n13	1456
Halichoerus grypus	60-110 cm	3571
	76 cm	5755
	83.5 cm; r87-100; n2	2187
	m: 87-105 cm	1087
	95-105 cm	1091
Hydrurga leptonyx	150-160 cm	4904
	est 150-160 cm	8121
	157.5 cm; n1	1359
Leptonychotes weddelli	120 cm	854
	120-130 cm	5755
Lobodon carcinophagus	nose-tail: 113.03 cm; r102-120; term emb	6278a
	114-127 cm; n19	6278a
	120 cm	6285
	130 cm; r115-145; n35 term emb	8121
	132 cm; n1	8121
	150 cm	5755
Mirounga angustirostris	120 cm	5755
	~127 cm	7062
	141.8 cm; se10	6333
	f: 147 cm	6329
	m: 153 cm	6329
Mirounga leonina	f: 116.0 cm; se2.6; n7	2024
	m: 116.0 cm; se9.0; n8	2024
	120-127 cm	6275, 6277, 6286
	130 cm	6482
	136.5 cm; r133-142; n4	5881
Monachus albiventer	120 cm; term emb	10910
Monachus monachus	80 cm	9788
	98.8 cm; r87-120	1145
Monachus schauinslandi	100 cm	1281
	100 cm	5679
Ommatophoca rossi	96 cm	5755
	est ~105 cm	8121
	est 105-120 cm	6289
Phoca caspica	65-79 cm	8681
Phoca fasciata	80 cm	10430
	85 cm; near term	1457, 1459
Phoca groenlandica	75-90 cm	10402
	90-105 cm	6264

	95 cm	3571
	105 cm	5152
Phoca hispida	50-60 cm	8679
	54 cm	3534
	58 cm	1457
	64-72 cm	3188
	65 cm; n5; n3 term emb, 2 new born	7087
Phoca largha	76-85 cm	1089
	~85 cm; r78.0-92.5; term emb	7752, 7753
Phoca sibirica	70 cm	8680
	226 mm; emb	10782
Phoca vitulina	m: 75 cm; n1	233
	f: 75.76-77.26 cm; n2 samples	1143
	m: 75.94-78.16 cm; n2 samples	1143
	m: HB: 76 cm; n1	9246
	76.4 cm; r75.9-77.7 for 3 yr data	1146
	mode 80 cm; r71-90	11829
	81.6 cm; 95% ci 6.2; n9	925
	84 cm	1457
	85 cm	5166
	95-105 cm	4511
	98 cm; n20	7752, 7753
	98.2 cm; se3.2	1088
WEANING MASS		
Cystophora cristata	42.6 kg; se1.37; n11, 4 d	1158
	43.6 kg; se1.78; n5	8038
	43.7 kg; se1.9; n12	1156
Erignathus barbatus	43-90 kg	8691
	85 kg; n11	1456, 1458
Halichoerus grypus	f: 35.8 kg	5980
	m: 40.0 kg	5980
	41-45 kg; 17 d	2067
	41.8 kg	240
	42.5 kg	2187
	45.1 kg; sd7.7; n17	11933
	50 kg; from graph	3186
Leptonychotes weddelli	m: 93.75 kg; n8, 5 wk, 2 sites	1399
	f: 101.67 kg; n8, 5 wk, 2 sites	1399
	110 kg	6787
	120-140 kg	10729
	125 kg	2400
	135 kg; n1, 52 d	6479
Mirounga angustirostris	f: 129.73 kg; se10.55; n4	9025
	136 kg; r113-238	6331, 9024, 9126
	m: 137.53 kg; se19.30; n5	9025
	158 kg	2061, 6329, 6330
Mirounga leonina	m: 70.5 kg; r63-78; n2	2024
	106.5 kg; ±26.5; n36, 3 wk	4137
	109.0 kg; ±21.0; n22, 3 wk	4137
	~110-160 kg	6482
	119.5 kg; ±13.0; n20, 3 wk	4137
	127.0 kg; ±22.5; n40, 3 wk	4137
	m: 129 kg; n3, 3 wk	1397
	f: 141 kg; n3, 3 wk	1397
	182 kg	6277
Monachus schauinslandi	~64 kg	5679
Phoca groenlandica	m: 20-55 kg	10402
	22-40 kg; 12-15 d	5978
	31.4 kg; sd4.8; n15	11933
	34.4 kg; 95%ci 2.6; n29	10404
	38.6 kg; se1.45; n5	8038
Phoca hispida	14 kg	3534
Phoca vitulina	~17 kg; from graph; 3 wk	11829
	m: 20.9 kg; r19.1-22.7; n2	925
	f: 25.3 kg; r20.9-28.6; n8	925

	f: 26.2 kg; n6	1146
	m: 27 kg; n1	9246
	m: 27.27 kg; n11	1146
WEANING SIZE		
Mirounga leonina	f: 123.0 cm; se5.3; n7	2024
	m: 125.4 cm; se4.8; n7	2024
Erignathus barbatus	147.3 cm; n48	1456
Phoca groenlandica	87-110 cm	10402
Phoca vitulina	f: 90.0 cm; n5, 30-36 d	1146
	m: 90.7 cm; n11, 30-36 d	1146
LITTER SIZE		
Cystophora cristata	1	1474
Erignathus barbatus	1	7086
Halichoerus grypus	1	1979
	1	1079
	1 ea; n3 lt	6396
	1 emb ea; n36 f	4843
Hydrrurga leptonyx	1 emb ea; n19 f	8121
Leptonychotes weddelli	1	10846
	1	1374
	1; n~81	6637
	1; 2 pr twins found dead	6479
	1; twins 1%	854
Lobodon carcinophagus	1 emb; 0.4% twins; r1-2; n226 f	8121
Mirounga angustirostris	1	2097, 6330
	1	5055
Mirounga leonina	1	4062
	1; 2 rare	6275, 6276, 6277
Monachus schauinslandi	1	9090
Ommatophoca rossi	1 emb ea; n7 f	8121
Phoca fasciata	1	1474
Phoca groenlandica	mode 1; occasionally 2	1474
	1; r1-2	3571
Phoca hispida	1; 2 uncommon	3534
Phoca sibirica	~1.02 yg; 2.2-3.9% twins; r1-2	8252
Phoca vitulina	1	925
	1 yg; 1 report of twin emb	9590
	1 yg; n1 lt	9246
	1 yg ea; n2 lt	7391
	1.01; 1.26% twins; r1-2;; n556 lt	11829
SEXUAL MATURITY		
Cystophora cristata	f: 1st ovulate: median 2.8 yr; r2-9; n114	8122
	f: 1st birth: median 3.8 yr; r3-11; n114	8122
	m: 4-6 yr	9005
Erignathus barbatus	f: 1st cl: 3 yr	1456
	f: 1st cl: 3 yr; 4 yr 20% have cl	8691
	f: 1st cl: 6 yr	7086
	f: 1st conceive: 5 yr	1456
	f: 1st birth: 6 yr	1456
	f: 3 yr (8%), 4 yr (70%)	10833
	f: 5-6 yr	10431
	m: increase testes: 4-5 yr	8691
	m: testes large: 7 yr	7086
	m: 5 yr (50%), 6 yr (66%), 7 yr (100%)	10833
	m: 6-7 yr	10431
Halichoerus grypus	f: 1st mate: 3-4 yr	1979
	f: 1st preg: 4 yr 50%	1174
	f: 1st birth: 5 yr 16%; 6 yr 45%; 7 yr 39%	4465
	f: 5 yr; few data	2187
	m: 1st mate: 3-4 yr	1979
	m: 6 yr; few data	2187
Hydrurga leptonyx	f: 1st ovulate: est 2 yr; n1	8121
	f: 1st cl: 2nd yr	4259
	f: 1st cl: 3-6 yr	10834

Species	Data	Reference
	f: 1st conceive: 3 to just over 4 yr	6278
	f: 1st conceive: 5 yr; n1	8121
	m: epididymal sperm: 3-6 yr	8121
	m: histology testes: 3rd yr	4259
	m: testes large: 4-5 yr	10834
Leptonychotes weddelli	f: 1st conceive: 26 mo	854
	f: 1st preg: 4 yr 50%; n8/16	10428
	f: 1st preg: 5 yr	6787
	f: 1st birth: 4-5 yr; r2-8	2161
	f: 1st birth: 4-6 yr 26%; n45/170	2886
	f: 1st birth: mode 6 yr; mean 7.1 yr	10759
	f: 2 yr	6479
	f: 3-6 yr	2400
	m: 3-6 yr; social maturity: 7-8 yr	2400
	f/m: 1st mate: 6-8 yr	10427
Lobodon carcinophagus	f: 1st ovulate: 3 yr; median 3.5 yr	8121
	f: 1st ovulate: 3.8 yr	761
	f: 1st cl: 3-5 yr	10834
	f: 1st birth: 5 yr	761
	f: 2-6 yr	6285
	m: epididymal sperm: 3-6 yr	8121
	m: testes large: 4-7 yr	10834
Mirounga angustirostris	f: 1st birth: 3-5 yr	6329, 9024
	f: 1st birth: 3-5 yr; mode 4 yr	5055
	f: 1st repro: 3.9-4.6 yr	5056
	m: 3 yr n1	10278
	m: 4-5 yr; social maturity: 9-10 yr	6329
Mirounga leonina	f: 1st mate: 2 yr	6276, 6277, 6286
	f: 1st mate: 3 yr; r2-5	6287
	f: 1st birth: 3 yr	6276, 6286
	f: 1st birth: 4-7 yr	1622
	f: 1st birth: mode 4 yr; r3-5	7017, 7019
	m: 1st sperm: 3-4 yr	6275, 6276, 6277, 6286
	m: 1st mate: 4-6 yr	6287
	m: 1st mate: 6-7 yr	6276, 6286
	m: 1st mate: 10 yr	1622
	m: 1st mate: 10 yr	5401
	m: 5-6 yr	1622
Monachus monachus	f: 1st mate: 4 yr	9243
	f/m: 4 yr	4417
	f/m: est 4 yr; few data	1145
Monachus schauinslandi	f: 5 yr	5332
Ommatophoca rossi	f: 1st ovulate: 3.6 yr; r2-7; n7	8121
	f: 1st cl: 3-5 yr	10834
	f: 3-4 yr; few data	6289
	m: testes large: 3-5 yr	10834
	m: 2-7 yr; few data	6289
Phoca caspica	f: 1st birth: 5-7 yr	8681
	m: 7 yr	8681
Phoca fasciata	f: 1st birth: 2 yr; 50% 3 yr	1459
	f: 2 yr 44%; 3 yr 86%	10833
	f: 2-4 yr	10430
	m: 2 yr 15%; 3 yr 69%	10833
	m: 3 yr 22%; 4 yr 75%	1459
	m: 3-5 yr	10430
Phoca groenlandica	f: 1st cl: 3 yr	7786
	f: 1st cl: 4.3-5.5 yr	9782
	f: 1st ovulate: 3-6 yr	9245
	f: 1st ovulate: 4-5 yr	9780
	f: median 5 yr	6264
	m: testes large: 50% 3-5 yr	9782
	m: median 6 yr	6264
	f/m: 3-5 yr	7786
	f/m: 4-5 yr	9244

Phoca hispida	f: 1st cl: 5 yr n1/12; 6 yr n10/10	5353
	f: 1st cl: 5-8 yr	7087
	f: 1st conceive 4-7 yr	3534
	f: 1st ovulate: 4 yr	3261
	f: 1st ovulate: 6 yr	10432
	f: 1st cl: 3 yr	3534
	f: 1st cl: 4 yr, 12.5% in Cumberland, 40.9% Home Bay	10145
	f: 1st preg: 7-8 yr 52%	4594
	f: 4 yr (18%), 5 yr (74%)	10833
	f: 5-7 yr	3188
	m: spermatogenesis: 7th yr	5353
	m: testes size, sperm in ejaculate: 7 yr	7087
	m: 5 yr (40%), 6 yr (43%), 7 yr (100%)	10833
	m: 5-7 yr	3534
	m: 6-8 yr	3188
	m: 7 yr	10145
	f/m: 3 yr 20%; 4 yr 60%; 5 yr 80%	6655
Phoca largha	f: 3-4 yr	1089
	m: mating attempts: 4 yr	739
	m: 4-5 yr	1089
Phoca sibirica	f: 1st mate: 5-6 yr	8680
	f: 1st mate: mode 6 yr; r4-6	10782
	m: 1st mate: ≤7 yr	10782
	m: 1st mate: 7-8 yr	8680
Phoca vitulina	f: 1st ovulate: 3.7 yr	4515
	f: 1st cl: 2 yr 20%; 3 yr 38%; 4 yr 34%; 5 yr 8%	925
	f: 1st cl: 3 yr 20%; 4 yr 100%	7752
	f: 1st conceive: end 3rd yr	4511
	f: 1st preg: 4.6 yr	4515
	f: 1st birth: 3 yr	9590
	f: 2-5 yr	1088
	f: 3 yr 10%; 4 yr 96%	10833
	f: 4 yr	11829
	f: 6 yr	9246
	m: large baculum: 3 yr	4511
	m: spermatogenesis 3 yr 20%, 4 yr 100%	7752
	m: end 3rd yr	11829
	m: 3 yr 33%; 4 yr 100%	10833
	m: 3-6 yr	925
	f/m: 3-7 yr	1146
CYCLE LENGTH		
Mirounga leonina	estrus: 4 d	7021
	estrus: ~5 d	6276
Phoca vitulina	estrus: 1-9 wk; proestrus: 3-4 wk	926
GESTATION		
Cystophora cristata	implant: 16 wk; field data	9005
	implant: 4 mo	9783
	postimplantation: 7.7 mo	9783
	postimplantation: 240-250 d; field data	9005
	11+ mo	1991
	11.5 mo	5755
	50-51 wk; field data	9005
	11.7 mo	9783
Erignathus barbatus	implant: 1.5-3.5 mo	8691
	implant: 2 mo	10833
	implant: 2 mo; field data	1458
	implant: 2.3-3 mo	10431
	implant: 2.5 mo	1456
	implant: 2.5 mo	7086
	postimplantation: 7 mo	7086
	postimplantation: ~8 mo	10431
	postimplantation: 8.5-10.5 mo	8691
	postimplantation: 9 mo	10833
	postimplantation: 9 mo; field data	1458

	9.5 mo	7086
	10.5-11 mo	10431
	11 mo	10833
	~11 mo; field data	1458
	12 mo	8691
Halichoerus grypus	implant: 3 mo	1087
	implant: 130 d	4708
	implant: 135.69 d; sd14	1171, 1172
	334 d; n1	5552b
	343 d; n1	5552c
	355 d	6396
	11.5 mo	1087
	1 yr	1171
	1 yr	4708
	1 yr	2187
	373 d; n1	5552a
Hydrurga leptonyx	implant: 1 mo	6287
	implant: 2.5-3.5 mo	5755
	postimplantation: 11 mo	6287
	~8 mo	4259
	est 8-9 mo	1374
	11 mo with 2.5-3 mo delay	10834
	~1 yr	6287
Leptonychotes weddelli	implant: 1 mo	6287
	implant: 1 mo	854
	implant: 2 mo	2400
	postimplantation: 9 mo	2400
	postimplantation: 11 mo	6287
	est > 9 mo and < 10 mo 10 d	6479
	est 9.5-10 mo	854
	~11 mo	2400
	11.5 mo	5755
	~1 yr	6287
Lobodon carcinophagus	implant: 2-3 mo	6285, 6287
	postimplantation: 8-9 mo	6285, 6287
	9 mo	854
	11 mo	6285, 6287
Mirounga angustirostris	implant: 3 mo	6329
	implant: 4 mo	5755
	postimplantation: 8.3 mo	6329
	11 mo; field data	7062
	11.3 mo	6329
Mirounga leonina	implant: 16 wk	6482
	implant: 4 mo	6275, 6276, 6277, 6286, 6287
	implant: ~4 mo	3763
	postimplantation: ~7 mo	6482
	postimplantation: 7.5 mo	6275, 6276, 6277, 6286, 6287
	postimplantation: ~8 mo	3763
	~11 mo	6482
	11.5 mo	6275, 6276, 6277, 6286, 6287
	1 yr	3763
	12 mo; n2	5879
Monachus monachus	10 mo	4417
	est 11 mo	1145
Ommatophoca rossi	implant: 3.5-4.5 mo	5755
	~11.5 mo	5755
	almost 1 yr	10698
Phoca caspica	11 mo	5755, 8681
Phoca fasciata	implant: 2 mo	10833
	postimplantation: 9 mo	10833
	9 mo	10062

	10.5-11 mo	10430
	11 mo	10833
Phoca groenlandica	implant: 2-3 mo	9244
	implant: 11 wk	9245
	implant: 4.5 mo	6264, 10406
	postimplantation: 7 mo	6264, 10406
	postimplantation: 7.5 mo	9245
	postimplantation: 8 mo	9244
	11 mo	9245
	11+ mo	1991
	11.5 mo	6264, 10406
	11.5 mo	9244
Phoca hispida	implant: 2 mo	10833
	implant: 3.5 mo	10432
	implant: 3.5 mo	7087
	postimplantation: 7-7.5 mo	10432
	postimplantation: 8.5 mo	7087
	postimplantation: 9 mo	10833
	8 mo	10062
	~9 mo	1474
	10.5-11 mo	10432
	11 mo	10833
	~11 mo	3571
	~1 yr	7087
Phoca largha	10.5 mo	1089
Phoca sibirica	~9 mo with delay	10782
Phoca vitulina	implant: 1-3 mo	926
	implant: 2 mo	10833
	implant: ~2 mo	566
	implant: est 2-3 mo	4415
	implant: 3 mo	1146
	implant: ~3 mo	4515
	postimplantation: 7-9 mo	926
	postimplantation: 8 mo	1146
	postimplantation: 9 mo	10833
	9 mo	10062
	9.5 mo; field data	11245
	~10 mo	4511
	10.5-11 mo	1088
	~11 mo	8680
	11 mo	7752
	11 mo	5470
	11 mo	10833
	11 mo	1146
	11 mo with 2-2.5 mo delay	925, 928, 3275
	341 d	4582
	341 d; r10-11 mo	3571
LACTATION		
Cystophora cristata	wean: 4 d; r3-5	1156, 1158, 8038
	wean: 7-12 d	9005
	wean: 10 d	9783
Erignathus barbatus	solid food: 13-14 d	1458
	wean: 12-18 d	1456, 1457, 1458
	wean: 3-4 wk	10833
	wean: 6-7 wk	5755
Halichoerus grypus	wean: 14 d	4708
	wean: 16-21 d	1084, 1085, 1092, 2067
	wean: 17 d	5980
	abandoned: 17-18 d	234
	wean: 18 d	3186
	wean: 20 d	2187
	wean: 20.3 d; 18-24 d; n3	5552a, 5552b, 5552c
	wean: 23 d	6396
Hydrurga leptonyx	wean: ~1 mo	10834

Leptonychotes weddelli	wean: 12-18 d	5755
	wean: 4-5 wk	8682
	wean: 5-6 wk	6637
	abandoned: 33-44 d	5578
	wean: 6 wk	2400
	wean: 6-8 wk	1906
	wean: 45 d	10729
	wean: 7 wk	854
	wean: 7 wk	6787
	permanent separation: 50.3 d; r40-55; n18	6479
Lobodon carcinophagus	wean: est 4 wk	10008
Mirounga angustirostris	wean: 22-28 d; varies with age of mother	2061, 9024
	wean: 22-32 d	5055
	wean f: 26.6 d; ±1.42	2097, 6331, 6332, 6333, 9025
	wean m: 27.8 d; ±1.54	2097, 6331, 6332, 6333, 9025
	independent: 3-4 mo	622
Mirounga leonina	wean: 14-22 d	10158
	wean: 20-34 d	4578
	wean: 3 wk	6482
	wean: 3 wk	1397
	wean: 3 wk	10698
	wean: 21.8 d; ±2.4; n62	4137
	wean: 22.5 d; ±3.5; n18	2023, 2024
	wean: 23 d	6275, 6276, 6277, 6286, 6287
	wean: 23 d	889
	wean: 23 d	7021
	wean: 23-25 d	7019
	wean: 24.2 d; ±2.3; n10	2024
Monachus monachus	wean: 5 wk	4417
	wean: 6-7 wk	9243
	wean: 6-7 wk	1145
Monachus schauinslandi	wean: ~1 mo	11831
	wean: 5-6 wk	5679, 5680, 5682
Phoca fasciata	wean: 3-4 wk	1457, 1459
	wean: 3-4 wk	10833
Phoca groenlandica	wean: 8-12 d	6264, 10404
	wean: 10 d	10402
	wean: 10-12 d	9245
	wean: 12-15 d	5978
	wean: ~13 d	8038
Phoca hispida	wean: ~3 wk	3571
	wean: 3-4 wk	10833
	wean: 4-6 wk	10432
	wean: ~4.5-7 wk	3534
	wean: 5-7 wk	1457
	wean: 2 mo	5755
	wean: 5 mo	8679
Phoca largha	wean: 2-3 wk	7752
Phoca sibirica	wean: 1.5-2.5 mo	10782
	wean: 2-2.5 mo	8680
Phoca vitulina	wean: 2-3 wk	11829
	wean: 2-6 wk	926
	wean: 3 wk	4415
	wean: 28 d	9246
	wean: 30 d	1146
	wean: 3-4 wk	4416
	wean: 4-5 wk	5470
	wean: 4-6 wk	1084
	wean: 4-6 wk	9590
	wean: 4-6 wk	7752, 7753
	wean: 4-6 wk	7864
	wean: 4-6 wk; then 8-14 d fast before solid food	3571

	wean: 5-6 wk	925, 928
	wean: 2 mo	10195
INTERLITTER INTERVAL		
Erignathus barbatus	1 yr	10833
	2 yr	7086
	2 yr	10062
Halichoerus grypus	1 yr	1079
	1 yr	1979
	1 yr	2187
Leptonychotes weddelli	1 yr	6787
	est 2 yr	10007
Lobodon carcinophagus	2 yr	854
Mirounga angustirostris	1 yr	2097, 6329
	1 yr	7062
Mirounga leonina	1 yr	1622
	1 yr	6277, 6286
Monachus monachus	2 yr	9243
Monachus schauinslandi	382 d; r361-393	11831
	1.5 yr	5332
Phoca fasciata	1 yr	10833
Phoca groenlandica	1 yr	3571
Phoca hispida	1 yr	3571
	1 yr	10062
	1 yr	10833
Phoca vitulina	1 yr	925
	1 yr	10062
	1 yr	10833
SEASONALITY		
Cystophora cristata	mate: mid-end Mar; N Atlantic	9005
	birth: end Feb; N Atlantic	1474
	birth: Mar-Apr; n hemisphere	5755
	birth: mid Mar; ne NEW, Canada	1156
	birth: mid-end Mar; Greenland	9783
	birth: mid-end Mar; N Atlantic	9005
Erignathus barbatus	spermatogenesis: Mar-Apr; Barents Sea	8691
	mate: end Mar-mid May; Barents Sea	8691
	mate: 1st 3 wk May; Bering Strait/Sea, Okhotsk Sea	1458
	mate: May; N Atlantic, Pacific	10431
	mate: June-July; Okhotsk Sea	10062
	ovulate; Mar-June; Barents Sea	8691
	ovulate; May; e Canadian Arctic	7086
	implant: mid July-early Aug; Bering Strait & Sea, Okhotsk Sea	1458
	implant: end July-Aug; e Canadian Arctic	7086
	birth: mid Mar-early May, peak Apr; Bering Strait/Sea, Okhotsk Sea	1456, 1458
	birth: end Mar-mid May; Barents Sea	8691
	birth: Apr; Bering, Chukchi Seas	1457
	birth: Apr-May; n hemisphere	5755
	birth: Apr-May; N America	1474
	birth: mid Apr; Okhotsk, Bering Seas	10833
	birth: mid-end Apr; AK	5353
	birth: mid Apr-May; N Atlantic & Pacific	10431
	birth: Apr-May; e Canadian Arctic	7086
Halichoerus grypus	mate: Feb-Mar; Baltic Sea	2187
	mate: Feb-Mar; Baltic Sea	3571
	mate: Feb-Mar; Baltic Sea	1085
	mate: Mar; Baltic Sea	4961
	mate: Mar-Apr; E Anglia, England, UK	4416
	mate: Sept-Oct; Pembrokeshire, UK	478
	mate: Oct; UK	233
	mate: peak end Oct; Orkney, Shetland I, UK	4708
	mate: end Oct; Farne I, UK	1084
	mate: end Oct-Nov; Farne I, Northumberland, UK	2066
	mate: end Dec-mid Feb; Sable I, Canada	1079

	mate: fall; Fro I, Norway	1979
	ovulate: Mar-Apr; E Anglia, England, UK	4416
	conceive: end Nov; Farne I, UK	1172
	implant: after early Jan; n1; Pembrokeshire, UK	478
	implant: end Jan; Farne I, UK	1084
	implant: Mar-May; Farne I, UK	1172
	birth: Jan; e Canada	4961
	birth: Jan; n1; UK	1288
	birth: Jan-Feb; w Atlantic	1087
	birth: Jan-Feb; Netherlands	1
	birth: Jan-Mar; Baltic Sea	3571
	birth: Feb-Mar; Baltic Sea	1087
	birth: Feb-Mar; Baltic Sea, St Lawrence	5755
	birth: end Feb-early Mar; Baltic Sea	2187
	birth: Mar; E Anglia, England, UK	4416
	birth: Mar; n2; Rumsey I, Pembrokeshire, UK	479
	birth: end Aug-early Dec; UK	5755
	birth: Sept-Nov; UK	4961
	birth: Sept-Oct; Norway, Sweden	4961
	birth: Sept-Jan; e Atlantic	1087
	birth: mid Sept-early Nov; Orkney, Shetland I, UK	4708
	birth: end Sept-mid Oct; Fro I, Norway	1979
	birth: Oct; Iceland	4961
	birth: Oct; n Rona, UK	285
	birth: mid Oct-Dec, peak Nov; Farne I, Northumberland, UK	1091, 1092, 1093, 2066, 2067, 4725, 4726, 4727
	birth: Nov-Dec; N Atlantic	1171
	birth: Dec; Kola Peninsula, Baltic Sea	4961
	birth: Dec-Feb, peak Jan; Netherlands	11240a
Hydrurga leptonyx	mate: Jan-Mar; Falkland I	4259
	mate: end Dec; Antarctic Peninsula	6287
	mate: autumn; Australia	10698
	birth: Oct-Dec; S Africa	10158
	birth: mid Oct-mid Nov; Pacific Antarctic	10834
	birth: est Nov; sw Atlantic	8121
	birth: Nov-Dec; Antarctica	4904
	birth: Nov-Dec; s hemisphere	5755
	birth: mid Nov; n1; Heard I, Southern Ocean	1359
	birth: end Nov; Antarctic Peninsula	6287
	wean: end Dec; Antarctic Peninsula	6287
Leptonychotes weddelli	spermatogenesis: Nov; Graham Land, Antarctica	854
	testes: large Sept-Nov, small June; Antarctic	4427
	mate: ear-mid Nov; King George I, Antarctica	8682
	mate: Nov-Dec; Ross I, Antarctica	10007
	mate: mid Nov-mid Dec; Graham Land, Antarctica	854
	mate: mid Dec; McMurdo Sound, Antarctica	6287
	mate: summer; Australia	10698
	ovulate: end Nov; Graham Land, Antarctica	854
	ovulate: mid Dec; McMurdo Sound, Antarctica	10428
	implant: mid Jan-mid Feb; McMurdo Sound, Antarctica	10428
	implant: mid Jan-mid Feb; Ross I, Antarctica	10007
	birth: Aug-Sept; Signy I, S Orkneys, UK	6787
	birth: end Aug-early Oct; King George I, Antarctica	8682
	birth: end Aug-Sept; Graham Land, Antarctica	1374
	birth: Sept-Oct; S Africa	10158
	birth: Sept-Oct; s hemisphere	5755
	birth: Sept-early Nov; Antarctica	2400
	birth: 3rd wk Sept; Antarctic	854
	birth: Oct; Antarctica	6637
	birth: mid Oct-Nov; Hutton Cliffs, Antarctica	5578
	birth: Nov; McMurdo Sound, Antarctica	6287
	birth: Nov-Dec; Ross I, Antarctica	6444
Lobodon carcinophagus	epididymal sperm: Oct; sw Atlantic	8121
	mate: Sept-Oct; Antarctica	10008

	mate: Oct-Dec; Graham Land, Antarctica	854
	mate: mid Nov; Antarctic Peninsula	6287
	mate: end spring; Australia	10698
	ovulate: Oct; sw Atlantic	8121
	birth: Sept; Graham Land, Antarctica	854
	birth: Sept-Oct; Antarctica	10008
	birth: Sept-Oct; Antarctica	6285
	birth: Sept-early Nov; s hemisphere	5755
	birth: mid Sept-early Nov; S Africa	10158
	birth: Oct; sw Atlantic	8121
	birth: Oct-early Nov; s hemisphere	5755
	birth: end Oct; Antarctic Peninsula	6287
Mirounga angustirostris	estrus: Feb; CA	622
	mate: Dec-Feb; Baja CA Mexico, n CA US	6329
	mate: Jan; Año Nuevo I, CA	9024
	mate: mid Jan; w US, Mexico	7062
	mate: mid Jan-mid Mar; Año Nuevo I, CA	6328
	birth: Dec-Feb; CA	622
	birth: Dec-Feb; Baja CA Mexico, n CA US	6329
	birth: mid Dec; w US, Mexico	7062
	birth: end Dec-early Feb; Farallon I, CA	5055
	birth: Jan; Año Nuevo I, CA	6332, 9024
	birth: peak Jan-Feb; CA	6330
Mirounga leonina	spermatogenesis: Aug-Nov; Macquarie I, Heard I, Southern Ocean	4060, 6482
	testes: large Aug-Nov; Falkland I	6277
	testes: large Nov, small May; Macquarie I, Southern Ocean	4060
	mate: Aug-Oct, peak Sept; S Georgia, s Atlantic Ocean	7021
	mate: Sept-Nov; Macquarie I, Southern Ocean	4060, 4061, 4062
	mate: Oct-mid Nov; Macquarie I, Heard I, Southern Ocean	6482
	mate: mid Oct-mid Nov; Heard I, Southern Ocean	3763
	mate: peak end Oct-early Nov; Marion I, S Africa	2023
	mate: Nov; Cumberland E Bay, S Georgia, s Atlantic Ocean	7018
	mate: Nov; S Georgia, s Atlantic Ocean	6287
	implant: end Feb; Macquarie I, Heard I, Southern Ocean	6482
	birth: Sept; Macquarie I, S Georgia, s Atlantic Ocean	7017
	birth: Sept-Oct; Australia	10698
	birth: Sept-Oct; S Georgia, s Atlantic Ocean	7021
	birth: Sept-Oct, few Nov; Signey I, S Orkney, S Georgia, s Atlantic Ocean	6276
	birth: Sept-Oct; S Georgia, s Atlantic Ocean, Macquarie I, Heard I, Southern Ocean	1622, 1623
	birth: Sept-Oct, peak mid Oct; Macquarie I, Heard I, Southern Ocean	6482
	birth: Sept-Nov; Gough I, s Atlantic Ocean	889
	birth: Sept-Nov; S Georgia, s Atlantic Ocean, Macquarie I, Southern Ocean	1622
	birth: Sept-Nov; S Orkney, S Georgia, s Atlantic Ocean, Macquarie I, Southern Ocean	6286
	birth: Sept-Nov; Marion I, S Africa	2023, 2024
	birth: mid Sept-Oct; Kergeulen I, Indian Ocean	8269
	birth: end Sept-Oct; s hemisphere	5755
	birth: end Sept-early Nov; S Georgia, s Atlantic Ocean	7020
	birth: Oct; Cumberland E Bay, S Georgia, s Atlantic Ocean	7018, 7019
	birth: peak Oct; S Africa	10158
	birth: Oct-Nov; Antarctica	4578
	birth: Oct-Nov; Heard I, Southern Ocean	3763
	birth: mid Oct; S Georgia, s Atlantic Ocean	6287
	wean: mid Nov; S Georgia, s Atlantic Ocean	6287
Monachus monachus	mate: Mar-July; Mediterranean	4417
	mate: Aug; n1; Mediterranean, n Atlantic	9788
	birth: Mar-May; Mediterranean	4417
	birth: May-Nov, peak Sept; Mediterranean, n Atlantic	9243, 9788
	birth: Sept-Oct; n hemisphere	5755

	yg found: yr rd; Mediterranean	1145
Monachus schauinslandi	birth: Dec-July, peak Apr-May; Fr Frigate Shoals, Lisianske, Laysen, Pearl, Hermes Rf, Kure Atoll, N Pacific	1281
	birth: Dec-Aug, peak Mar-May; Kure, Pearl, Hermes reef, Lisianski, Laysan, French Frigate shoals, N Pacific	5679
	birth: Jan-June; Leeward I, Pacific	5682
	birth: Jan-Aug, peak Mar-Apr; HI	5332
	birth: Feb-June, peak Mar-May; n47; HI	11831
	birth: end Mar-early June; n hemisphere	5755
Monachus tropicalis	preg f: Dec; n1; W Indies	189
	birth: Nov-Dec; n hemisphere	5755
	birth: Dec; Caribbean	5753
	lact f: Dec; n1; W Indies	189
Ommatophoca rossi	mate: est Nov; sw Atlantic	8121
	birth: early-mid Nov; Pacific Antarctic	10834
	birth: early-mid Nov ?; Antarctic	10781
	birth: Nov-Dec; s oceans	6289
	birth: Nov-Dec; Australia	10698
Phoca caspica	birth: Jan-Feb; n hemisphere	5755
	birth: end Jan-Feb; Caspian sea	8681
Phoca fasciata	mate: end Apr-early May; Bering, Okhotsk Seas	10430
	mate: end Apr-early May; Bering, Okhotsk, Chukchi Seas	1459
	mate: end July-Aug; Far East	10062
	ovulate: ?-early May; Bering, Okhotsk, Chukchi Seas	1459
	birth: Mar-Apr; n hemisphere	5755
	birth: end Mar-mid Apr; Bering, Okhotsk Seas	10430
	birth: Apr-early May; Bering, Okhotsk, Chukchi Seas	1459
	birth: mid Apr; Okhotsk, Bering Seas	10833
Phoca groenlandica	mate: Mar-Apr; Europe	3571
	mate: mid-end Mar; White Sea	7786
	mate: est mid-end Mar; Gulf of St Lawrence	9244, 9245
	conceive: mid Mar; nw Atlantic	10406
	delay: Mar-June; nw Atlantic	10406
	implant: Aug; nw Atlantic	10406
	birth: Jan-Apr, peak end Feb-early Mar; n Atlantic	9245
	birth: Feb-early Mar; NEW, Canada	1991
	birth: end Feb-early Mar; White Sea	6264
	birth: end Feb-early Apr; Europe	3571
	birth: Mar; NEW, Canada	6264, 10403, 10406
	birth: Mar; White Sea	10014
Phoca hispida	spermatogenesis: Mar-mid May; e Canadian Arctic	7087
	spermatogenesis: Mar-May; AK	5353
	mate: Mar-early May; arctic	10432
	mate: Mar-mid May; Okhotsk Sea	3188
	mate: Mar-mid May, some June; NWT, Canada	10145
	mate: Mar-May; e Canadian Arctic	7087
	mate: Apr-June; Europe	3571
	mate: end Apr-early May; n Atlantic	3534
	mate: Aug-mid Sept; June-Aug; Far East	10062
	ovulate: Apr; AK	5353
	birth: mid Feb-mid Mar; Lake Ladoga, Russia	8679
	birth: Mar-Apr; Europe	3571
	birth: end Mar?; NWT, Canada	10145
	birth: end Mar-Apr; arctic	10432
	birth: end Mar-May; Okhotsk Sea	3188
	birth: Apr; N Atlantic	3534
	birth: mid Apr; Okhotsk, Bering Seas	10833
Phoca largha	mate: Apr-May; captive	739
	birth: end Jan-early May, peak mid Mar-mid Apr; Bering, Okhotsk Seas	1089
Phoca sibirica	birth: Feb-Mar; Lake Baikal, Russia	5755
	birth: peak mid Mar; Lake Baikal, Russia	8680
Phoca vitulina	epididymal sperm: Mar-Nov; se Vancouver I, BC, Canada	925
	estrus Aug-Sept; se Vancouver I, BC, Canada	928

mate: Apr-July; NEW, NB, Canada	1146
mate: June; Baltic Sea	4961
mate: June; Hokkaido, Japan	4515
mate: June-July; Okhotsk Sea	10062
mate: end June; Hokkaido	7752
mate: end July-early Aug; E Anglia, England, UK	233, 4416
mate: Aug; zoo	5470
mate: Aug-early Sept; Europe	3571
mate: Aug-Sept; Netherlands	4511
mate: Aug-Sept; se Vancouver I, BC, Canada	925, 928
mate: Sept; e Pacific	5683
mate: Sept; w US	5146
mate: Sept-early Oct; Shetland I, UK	11245
mate: mid Sept-mid Oct; Europe	1084
ovulate: mid June; NEW, NB, Canada	1146
ovulate: end July-Aug, a few May; E Anglia, England, UK	4416
ovulate: Aug-Oct; se Vancouver I, BC, Canada	928
conceive: June; NEW, NB, Canada	1146
conceive: Aug-Sept; se Vancouver I, BC, Canada	928
implant: Sept; Nova Scotia, NB, Canada	3275
implant: mid Sept; NEW, NB, Canada	1146
implant: end Sept; Hokkaido, Japan	4515
implant: end Oct-mid Dec; Europe	1084
implant: Nov-early Dec; E Anglia, England, UK	4416
implant: mid Nov; se Vancouver I, BC, Canada	925, 928
birth: mid-end Mar; Hokkaido, Japan	7752
birth: end Mar-early May; CA	926
birth: Apr-July?; Kurile I, Russia	10195
birth: mid Apr; Okhotsk, Bering Seas	10833
birth: May; Hokkoido, Japan	7753
birth: May; NEW, NB, Canada	1146
birth: May-early June, peak end May; Sable I, NS, Canada	1143
birth: May-early June; Salile I, NS, Canada	1142
birth: May-June; Kurile I, Russia, AK	7753
birth: May-June; NS, NB, Canada	3275
birth: May-June; n hemisphere	5755
birth: May-June; Yaquina estuary, OR	674
birth: May-Aug; e Pacific	5683
birth: mid-end May; Hokkaido, Japan	7752
birth: mid-end May; Pacific	7864
birth: mid May-mid June; southern range	6788
birth: end May-early July; Hokkaido I, Japan	7753
birth: June; E Anglia, England, UK	4415, 4416
birth: June; Baltic Sea	4961
birth: June; n1; captive	9246
birth: June-mid July; Europe	1084
birth: June-July; Netherlands	4511
birth: June-early Sept; BC, Canada	926
birth: mid-end June; UK	233
birth: mid June-early July; Shetland I, UK	11245
birth: mid June-mid July; Europe	3571
birth: mid June-mid July; N Sea, Germany	11829
birth: end June-early July; Bering, Chukchi Seas	1457
birth: end June-early Sept, peak July; se Vancouver I, BC, Canada	925, 928
birth: July; zoo	5470
birth: mid Aug-mid Sept; Gertrude I, WA	7864
birth: spring; w US	5146
birth: early summer; w Atlantic	1474
lactation: ?-Sept; Kurile I, Russia	10195

Order Cetacea

The cetaceans are divided into two groups, Odontoceti (toothed cetaceans: dolphins, porpoises) and Mysticeti (baleen whales). The birth of a singleton, highly precocial offspring is normal for this order. Males have abdominal testes. Females have elongate mammary glands whose teats are enclosed within deep pockets on each side of the urogenital slit.

Several periodicals cover cetacean reproduction, among them are Aquatic Mammals, Cetology, Hvalradets Skrifter, Investigations on Cetacea, Marine Mammal Science, Norsk Hvalfangst-tidende (no longer in publication), Reports of the International Whaling Commission, and Scientific Reports of the Whales Research Institute.

Order Cetacea, Family Platanistidae

The 10-240 kg river dolphins (*Inia, Lipotes, Platanista, Pontoporia*) inhabit large river systems of Asia and South America, where they feed on a variety of vertebrates and invertebrates. Female *Pontoporia* are larger than males (5560) and ovulate primarily from their left ovary (4422). The villous, diffuse, epitheliochorial placenta implants into a bicornuate uterus which is asymmetrical in *Lipotes* (1778, 3739, 4418, 7563). In one observed captive birth, the young was born tail first (*Inia* 10949). Females have 2 U-shaped mammary glands (*Lipotes* 1778) and may have a postpartum estrus (*Pontoporia* 4423). Details of anatomy (1778, 4418, 4422, 4423), copulation (8529), and birth (5069, 10949) are available.

NEONATAL MASS		
Inia geoffrensis	6.8 kg; n1	10949
Platanista minor	7 kg	9175
NEONATAL SIZE		
Inia geoffrensis	76-80 cm	4421
	81 cm; n1	10949
	min 81 cm; n1	873
Lipotes vexillifer	est 80 cm	9049a
Platanista minor	45 cm	9175
Pontoporia blainvillei	est 59 cm; n1	4421
	TL: 63-70 cm; n5 term emb	1377
	70-85 cm	4423, 5560
LITTER SIZE		
Inia geoffrensis	1 yg	4421
	1 yg; n1 lt	10949
Lipotes vexillifer	1 yg	9049a
Platanista gangetica	1 emb; n1 f	8532
Platanista minor	1; sometimes 2	8530
Pontoporia blainvillei	1 yg	4421, 4423
SEXUAL MATURITY		
Inia geoffrensis	f: ovulate: 183 cm	873
	m: spermatogenesis: 198.5 cm	873
Lipotes vexillifer	f: ovulate: 8 yr, 184.5 cm	9049a
Platanista spp	~10 yr, 170 cm	5555
Platanista gangetica	f: 1st cl: 200 cm	4422
	m: testes: large 180-185 cm	4422
Pontoporia blainvillei	f: 1st cl: 2.7 yr; r2-4 yr, r137-146 cm	5560
	f: 1st cl: 151-170 cm	4422
	f: 137-145 cm	1377
	m: testes: large 2-3 yr, 131 cm, 25-29 kg	5560
GESTATION		
Inia geoffrensis	8.5 mo; n1	10949
Lipotes vexillifer	est 10-11 mo	9049a
Platanista minor	8-9 mo	9175
	12 mo; hunter reports	9175
	1 yr	8530
Pontoporia blainvillei	10.5 mo	5560
	est <11 mo; r10.5-12	4423
LACTATION		
Pontoporia blainvillei	wean: 9 mo	4423
	wean: ~1 yr; milk in stomach	5560

INTERLITTER INTERVAL		
Lipotes vexillifer	est 2 yr	9049a
Pontoporia blainvillei	2 yr	5560
	2 yr	4423
SEASONALITY		
Inia geoffrensis	birth: May-July; Amazon R, S America	873
	birth: July-Sept; upper Amazon R, S America	4421
Lipotes vexillifer	birth: Feb; Changjiang R, China	9049a
	birth: Feb-Mar; few data; Yangtze R, China	1778
	lact f: Dec; n1; China	1379
Platanista spp	birth: yr rd, small peaks Oct-Mar, June; India	5555
Platanista minor	mate: Mar; local fisherman; Pakistan	9175
	mate/birth: Mar; Indus R, Pakistan	8530
Pontoporia blainvillei	mate: Dec-Feb, peak early-mid Jan; La Plata, S America	4423
	birth: mid Nov-Dec; La Plata, S America	4423

Order Cetacea, Family Delphinidae

Marine dolphins (*Cephalorhynchus* 25-75 kg; *Delphinus* 40-160 kg; *Feresa* 160-170 kg; *Globicephala* 800-2900 kg; *Grampus* 230-450 kg; *Lagenodelphis* 130-210 kg; *Lagenorhynchus* 30-235 kg *Lissodelphis* 80-115 kg; *Orcaella* 125-200 kg; *Orcinus* 2200-9000 kg; *Peponocephala* 180-205 kg; *Pseudorca* 370-1700 kg; *Sotalia* 30-80 kg; *Sousa* 140-285 kg; *Stenella* 35-165 kg; *Steno* 100-145 kg; *Tursiops* 115-650 kg) are social, aquatic carnivores distributed throughout the world's oceans and some tropical rivers. Males are often larger than females (393, 1975, 7123, 7370, 8022, 9784, 9787), but not in *Cephalorhynchus* (6539a, 10065a).

Both induced and spontaneous ovulation are reported for *Delphinus* and *Tursiops* (induced: 4420, 5085, 9407, 11980; spontaneous: 766, 4413, 4420, 4422, 5784, 8390). The left ovary may be dominant to the right one (1975, 4419, 4421, 5086, 6841, 7123, 7396, 8387, 8389, 9300; cf 4430, 8064). Vaginal plugs are reported (3093, 11980). The 0.5-26 kg, epitheliochorial, villous, diffuse placenta often implants in the left horn of the bicornuate or bipartite uterus (192, 765, 766, 1678, 1998, 4146, 4413, 4845, 7563, 10061, 10956, 11858). A postpartum estrus is suggested for *Delphinus* and *Stenella* (192, 5564, 8387, 8389, 10158, 11980). Neonates are precocious (5146, 7012, 10691). Testes and epididymides can reach 3% of body mass (*Cephalorhynchus* 10065a).

Details of anatomy (1975, 1976a, 1998, 4036, 4413, 4420, 4422, 4430, 4823, 5085, 6841, 9653, 9777), intersex (7926), endocrinology (4845, 5459, 5784, 9108, 9109, 9629, 9653, 9654, 9655, 11405, 11586), sperm (3309), copulation (3093, 6704, 8780, 11320), birth (431, 3092a, 3094, 5385, 7012, 10691), milk composition (8319, 8398, 8543, 11980), fetal growth rate (8390a), and reproductive energetics (817, 1767a, 10277) are available.

NEONATAL MASS		
Cephalorhynchus commersonii	m: 4.5-5.5 kg; n1	5385
	m: 7.3 kg; n1 term emb	6539a
Cephalorhynchus heavisidii	m: 9.5 kg; n1 presumed newborn	869a
Delphinus delphis	6.8 kg; n1	6544
	7.35 kg; r6.8-7.9; n2	192
Globicephala melaena	180 kg; r165-190	9777
Lagenorhynchus acutus	24 kg	9789
Orcinus orca	est 140 kg; n1	431
	155 kg stillborn; 158 kg 11 d; est 160 kg; n3	765
	~180 kg	6945a
Stenella attenuata	>10 kg	7400
Stenella coeruleoalba	>10 kg	7400
Tursiops truncatus	11.3-15.9 kg	4022
	m: 12 kg; n1	9407
	f: 24.5 kg; r19-30; n2 stillborn	5567
	26.5 kg; r23-30; n2 3-4 d; 25 kg; 10 d; 30 kg; 32 d	5567
	m: 29 kg; n1 stillborn	5567

NEONATAL SIZE

Cephalorhynchus commersonii	m: 55-65 cm; n1	5385
	m: 73.4 cm; n1 term emb	6539a
Cephalorhynchus heavisidii	m: 84.7 cm; n1 presumed newborn	869a
Delphinus delphis	TL: 75-85 cm	10158
	75-90 cm	4420
	est 79.0-81.3 cm	8390a
	80-90 cm	8390a
	84 cm	8066
	86.4 cm; n1	6544
	90 cm	1975
	91.5 cm; r86-97; n2	192
	est 105 cm	9695a
Globicephala spp	160 cm; n1 smallest yg	4156a
Globicephala macrorhynchus	135-146 cm	12021a
	139.5 cm	5562
	140 cm	7370
Globicephala melaena	est 140 cm	8390a
	TL: ~160-170 cm	9776
	174 cm	8066
	f: 174 cm	9777
	175 cm	10313b
	m: 178 cm	9777
	m: 187 cm; n1	4135
Grampus griseus	est 110-120	8390a
	150 cm	7370
	150 cm	8390a
Lagenodelphis hosei	~100 cm	7983
Lagenorhynchus acutus	95 cm	7370
	~110-120 cm	9789
Lagenorhynchus albirostris	~122 cm	7983
	125 cm	8390a
	est 125 cm	9695a
Lagenorhynchus obliquidens	est 95-110 cm	4420
Orcinus orca	208-220 cm	8390a
	210 cm	5441
	210-250 cm	8390a
	TL: 210-270 cm	10158
	236.2 cm; n4	927a
	240 cm	7370
	246.4 cm; n1	927a
	270 cm	7929
Peponocephala electra	75-90 cm	2363
Pseudorca crassidens	est 150-210 cm	10292a
	est 160 cm	8390a
	193 cm	8797
Sotalia fluviatilis	71-83 cm	7983
	est ~75 cm	873
Sousa chinensis	est 100 cm	8390a
Stenella attenuata	est 82-85 cm; r71-92; n609 term emb, yg	4909
	82.5 cm; n86 emb, yg	8387
	89 cm	5564
Stenella coeruleoalba	est 99.8 cm	5554
	100 cm	7370
	100 cm; n68 emb, 20 yg	7396, 7399
	105 cm	8066
Stenella longirostris	est 76.5 cm	8388
	est 76.9 cm; from emb & sm yg	8389
	~80 cm	10158
Tursiops truncatus	81-122 cm	4022
	r83.8-112.0 cm; n8	9300
	95-115 cm	9653
	100 cm	9787
	110.3 cm; r98-118; n6	4420, 4422, 4430

	m: 112 cm; n1	9407
	115 cm	4908a
	r115-140 cm; n12	5567
	116 cm; r95-132; n9 f	4908a
	118 cm; r110-134; n9 m	4908a
	128 cm; n20, 0-11 d	5567
	est 130 cm	9695a
	135 cm	8066
WEANING MASS		
Delphinus delphis	40 kg	192
WEANING SIZE		
Delphinus delphis	150 cm	192
Orcinus orca	430 cm	4715
Stenella attenuata	f: r115-130 cm	8387
Stenella coeruleoalba	174 cm	7397
Tursiops truncatus	~150 cm; n1, 10 mo	9300
	195 cm; n1, 18 mo	9300
LITTER SIZE		
Cephalorhynchus commersonii	1; n1 lt	5385
	1 emb; n1 f	6539a
Cephalorhynchus eutropia	1 emb; n1 f	3890b
Cephalorhynchus heavisidii	1 emb; n1 f	869a
Cephalorhynchus hectori	1 yg	10065a
Delphinus delphis	1	192
	1 cl	10061
Feresa attenuata	1	8064
Globicephala macrorhynchus	1 emb ea; n141 f	5562
Globicephala melaena	1 emb ea; n2 f	7099
	1.7 ova/cycle	9777
Grampus griseus	1 emb; n1 f	9301a
Lagenodelphis hosei	1	7983
Lagenorhynchus acutus	1-2; 1 set of twins	5442
Lagenorhynchus albirostris	1 yg	8390a
Lagenorhynchus obliquidens	1 yg	4420
Orcaella brevirostris	1 yg ea; n2 lt	10678
Orcinus orca	1 yg	431
	1 yg	6945a
	1 yg; n5	765
Pseudorca crassidens	1 emb; n1 f	10158
	1 emb ea; n4 f	9696
Sotalia fluviatilis	1 emb; n1	5111
Sousa chinensis	1	8390a
Stenella attenuata	1.02 emb; r1-2; n59 f	5564
Stenella coeruleoalba	1	7396
	1.8 cl/cycle	7397
	2 emb; n1 f	10853
	2; n1 f	10853
Stenella longirostris	1 emb; 1 f	7401
Tursiops truncatus	1 cl	11858
	1 cl/yr	9787
	1 yg	7012
	1 yg ea; n3 f	244
	1 emb ea; n6 f	4156
	1 ea; n18 lt	12073
SEXUAL MATURITY		
Cephalorhynchus commersonii	f: ovaries mature: ~5 yr; ~165 cm	1976a
	f: ovaries mature: 5-6 yr; ~130 cm	3890a
	m: testicular sperm: 5-6 yr; 127-131 cm	3890a, 6539a
	m: testes mature: ~8 yr; ~165 cm	1976a
Cephalorhynchus hectori	f: 1st birth: est 7-9 yr	10065a
	m: testes large: 6-9 yr	10065a
Delphinus delphis	f: 1st cl: 7 yr	5085
	f: 1st cl: 165-182 cm	5086
	f: 1st cl: 195-196 cm; 6 layers dentine	1975, 1976

	f: 3 yr	4420
	f: 3 yr	10061
	f: 3.5 yr	6544
	f: TL 210-220 cm	10158
	m: spermatogenesis: 187-198 cm	4420
	m: testes large: 175-190 cm	5086
	m: testes large: >190 cm; 12 dentine layers	5085
	m: testes large: 200 cm; 5-7 layers dentine	1975, 1977
	m: 4 yr	10061
	m: 4.5 yr	6544
Globicephala macrorhynchus	f: 1st ovulate: 7-12 yr	6841
	f: 1st cl: 300-320 cm	7370
	f: sperm in cloaca: 8.25 yr	5562
	f: 9 yr	5563
	f: 380-390 cm	10158
	m: epididymal sperm: min 394 cm	5562
	m: 480 cm	7370
Globicephala melaena	f: 6-8 yr; 366 cm	9777, 9784
	m: 12 yr; 487 cm	9777, 9784
Grampus griseus	m: 300+ cm	7370
Lagenodelphis hosei	f: 1st preg: 236 cm; n1	10158
	f/m: ~7 yr; 230 cm	11096
Lagenorhynchus acutus	f: 1st cl: 201-222 cm	9789
	m: testes large: 230-240 cm; from graph	9789
Lagenorhynchus obliquidens	f: 1st cl: 173 cm; n1	4420, 4422
	f: 175-186 cm	7370
	m: testes large: >180 cm	4422
	m: sperm present: 200 cm; n1	4420
	m: 170-180 cm	7370
Lissodelphis borealis	f: 1st cl, 1st preg: 200-201 cm	6323
	m: spermatogenesis: 212-219.5 cm	6323
Orcinus orca	f: 1st cl: 480 cm	5441
	f: 1st preg: min 6.7 yr; 490 cm	927a
	f: 1st preg: 460 cm	1826
	f: 455-485 cm	6945a
	f: 460-540 cm	4715
	f: 490 cm	1375
	m: testes large: 570 cm	5441
	m: min 12 yr; 580-600 cm	927a
	m: 520-620 cm	4715
	m: 670 cm	1375
Peponocephala electra	f: <12 yr	7983
	f: >225 cm	1398
	m: <7 yr	7983
Pseudorca crassidens	f: 1st cl: est 358-440 cm	8022
	f: 1st cl: 395-427 cm	3482
	f: 1st cl: <427 cm	1998
	f: 1st preg: 360-420 cm	9784
	f: 8-14 yr; 366-427 cm	8797
	m: 8-14 yr; 366-427 cm	8797
	m: 511 cm	9784
	f/m: 8-10 yr	2045
Sotalia fluviatilis	f: 1st cl: 138.5 cm	873
	m: testes large: 139 cm	873
Stenella spp	f: 1st cl: 165-170 cm	4420
	m: sperm: 180-214 cm	4420
Stenella attenuata	f: 1st cl: >170 cm; n1	4422
	f: 1st conceive: 4.5 yr; r4.1-7.6; 181 cm; 50-60 kg	8387
	f: 1st cl: median est 9 yr	5556
	m: sperm: 180-214 cm; n6	4420, 4422
	m: testes large: 6 yr; 195 cm; 75 kg	8387
	m: testes large: median est 11.8 yr	5556
Stenella coeruleoalba	f: 5-6 yr; 210 cm	7370
	f: est 7.4-9.7 yr	5558

	f: 8.7-8.8 yr; 219 cm; 8.9 yr; 50% mature	7396, 7397, 7399
	f: 9 yr; 212 cm	5554, 5556
	f: est TL 210 cm	10158
	m: testes large: 8.8 yr, 50% mature	7399
	m: socially mature: 17 yr	7399
	m: 6.7-7.1 yr	7396
	m: 7-11 yr	5556
	m: 9.0 yr; 217 cm	7397
	m: 9 yr; 220 cm	5554
	f/m: 220 cm	4824
Stenella longirostris	f: 1st cl: 7-10 yr; 180 cm; 55 kg	7123
	f: est 3.7-5 yr; 165 cm	8389
	m: testes large: 192 cm; 65 kg	7123
	m: est 6-11.5 yr; 170 cm	8389
	f/m: ~5-6 yr; 164.1-167.2 cm	8388
	f/m: 170 cm	4422
Steno bredanensis	max 225 cm	7398
Tursiops truncatus	f: adult hormone levels: 5-7 yr	9653
	f: 1st cl: 12 yr; 235 cm	9787
	f: 1st cl: 202-231 cm; possibly 6-9 yr	9300
	f: 1st preg: 11-13 yr; 227-238 cm	1767b
	f: 1st birth: 5 yr; n1	7012
	f: ~5 yr; >220 cm	4420, 4430
	f: 6 yr	9175
	f: ~160-222.5 cm	4424
	m: spermatogenesis: ~9-11 yr	9300
	m: testes large: 13 yr; 245 cm	9787
	m: 1st sire: 10 or 11, 13 yr; n2	1767b
	m: 6-7 yr	4420
	m: ~10 yr; 260 cm	4430
	m: ~202 cm	4424
CYCLE LENGTH		
Stenella attenuata	1 mo	5564
Stenella coeruleoalba	4 mo	7396
GESTATION		
Delphinus delphis	5-6 mo	10061
	276 d	4422
	10-11 mo	192
	10-11 mo; n3 sources	8390a
	11 mo	1975
	11 mo	4420
	~11 mo	11980
Globicephala macrorhynchus	452 d; r402-512	5562
Globicephala melaena	15.5-16 mo	9777
	16 mo	3488
Grampus griseus	est ~1 yr	10158
Lagenorhynchus acutus	10 mo; n2 sources	8390a
	~10 mo	7370
	~11 mo	9789
Lagenorhynchus obliquidens	~10 mo	4146
Orcaella brevirostris	est 14 mo	10678
Orcinus orca	<1 yr	7929
	12 mo; n3 sources	8390a
	~15 mo; n1	2218
	est 15-16 mo	6945a
	~517 d; n1	431
Peponocephala electra	est ~1 yr	1398
	~1 yr	10158
Pseudorca crassidens	~13 mo	2045
	462 d; fetal growth data	8797
	15.5-15.7 mo	10292a
Sotalia fluviatilis	10.2 mo; fetal growth data	873
Stenella attenuata	11.2 mo; fetal growth data	5564
	11.5 mo; n6243 f; fetal growth data	8387, 8390

Stenella coeruleoalba	~12 mo; fetal growth data	5554, 7397
	13.4 mo; fetal growth data	7399
Stenella longirostris	est 10.6 mo; from brain size	8388, 8389
Tursiops truncatus	>11 <13 mo	7012
	11-12 mo	9175
	~1 yr	9300
	12 mo	244
	12 mo	9407
	12 mo	4430
	~12 mo	11980
	~12 mo	10691
	12 mo; r11.5-12.5	9653
	12-13 mo	5146
LACTATION		
Cephalorhynchus hectori	solid food: 6 mo	10065b
Delphinus delphis	solid food: 6.5 mo	192
	wean: 10 mo	192
	wean: 10 mo	1975
	wean: est 14 mo	8390a
	wean: est 19 mo; n4 sources	8390a
Globicephala macrorhynchus	wean: min 2 yr	5562
Globicephala melaena	wean: 20 mo	10158
	wean: 22 mo	9777
Lagenorhynchus acutus	wean: ~18 mo	9789
Orcinus orca	solid food: est <2 mo; 260 cm; n1	4714
	solid food: 5 mo; n1	4714
	solid food: 11 wk	431
	nurse: 12 mo	4715
	nurse: ~18 mo	431
	independent: 2 yr	4715
Pseudorca crassidens	wean: 6 mo	2045
	wean: est 18 mo	8390a
Sousa chinensis	wean: 4-6 mo	9175
	young accompany mother: min 6 mo	9175
Stenella attenuata	solid food: 0.25-0.5 yr	5564
	wean: est 11.2 mo; r9-12	8387, 8390
	wean: est 24.5 mo; demographic statistics	5564
Stenella coeruleoalba	solid food: 3 mo	7396
	solid food: 0.5 yr	7397
	wean: 6-12 mo	10158
	wean: 15 mo	2045
	wean: 1.4 yr	7399
	wean: 1.5 yr; demographic statistics	5554
	wean: min 18 mo	7370
	wean: 1.5 yr; r1-3	7396, 7397
Stenella longirostris	wean: 13.8 mo; r10.1-17.5; n2	8389
	wean: 15-19 mo; n3 sources	8390a
Tursiops truncatus	solid food: 6 mo	9300
	solid food: 6-7 mo	9653
	wean: 1 yr	10698
	wean: 18 mo	9653
	wean: ~18 mo	10691
	wean: max 1.5 yr	7012
	wean: ~2 yr	9175
	nurse: 29 mo	9300
INTERLITTER INTERVAL		
Cephalorhynchus hectori	min 2 yr	10065b
Delphinus delphis	1 yr	10061
	~2 yr	1975
Globicephala macrorhynchus	3 yr	10158
Globicephala melaena	3-4 yr	2045
	40 mo	9777
Orcinus orca	3 yr	927a
Pseudorca crassidens	3 yr	2045

Stenella attenuata	r25.4-26.2 mo, 3 different yr	8387, 8390
	38.6-57.5 mo	5564
Stenella coeruleoalba	3 yr	5554
	3 yr	7397
	4.3 yr	7399
Stenella longirostris	2.11-2.22 yr	8389
Tursiops truncatus	1 yr; n1	9300
	2 yr	7012
	~2 yr	6322
	2-4 yr	11889
	22.7-27.6 mo; 2 samples	2051
	4 yr; n1 f	9407
SEASONALITY		
Cephalorhynchus commersonii	birth: Jan-Mar; Argentina	3890a, 6539a
Cephalorhynchus eutropia	preg f: Mar; n1; Chile	3890b
Cephalorhynchus heavisidii	preg f: Apr; n1; near Walvis Bay, S Africa	869a
Cephalorhynchus hectori	birth: Nov-Feb; New Zealand	10065b
Delphinus delphis	spermatogenesis: Jan-July; English Channel	1975
	spermatogenesis: Feb-Mar, low Aug; Bay of Biscayne	1977
	spermatogenesis: Mar-Aug; Gulf of Gascogne	1975
	testes: active Dec-June/July; e N Atlantic	1977
	mate: Feb-May; captive, St Augustine, FL	3093
	mate: June-Oct, peak July-Aug; Black Sea	10061
	mate: not seasonal; San Diego, CA	5085
	birth: peak May/June-end summer; France	1975
	birth: Dec-Mar; CA	4420, 4422
	birth: spring-fall, peak June-July; Black Sea	10061
	birth: yr rd, peak summer; S Africa	10158
	birth: yr rd, peaks Feb-May, Sept-Nov; CA, FL	4430
	birth: mid-winter; s CA	7960
	young seen: July-Jan, peak Nov; Napier Hawk's Bay, New Zealand	192
Globicephala macrorhynchus	conceive: peak Apr-May; Japan	5562
	fetal length varied Oct; w North Atlantic	7370
	preg f: high June, low Dec; Japan	5562
	birth: peak July-Aug; Japan	5562
	birth: Aug-winter; CA, Mexico	7960
Globicephala melaena	mate: Apr-May; e NEW, Canada	9777
	birth: ?-Oct, peak Aug; e NEW, Canada	9777
Grampus griseus	birth: est Dec-Apr; S Africa	10158
Lagenorhynchus acutus	birth: May-Aug, peak June/July; ME, MA	9789
Lagenorhynchus obliquidens	birth: May-Sept; s CA	4420
	varied length of young: summer-fall; CA, Mexico	7960
Lagenorhynchus obscurus	mate: late summer; S Africa	10158
Orcinus orca	mate: May-July; Japan	7929
	mate: Nov-Jan; n Atlantic	10158
	mate: May-July; n Pacific	10158
	conceive: peak Sept-Jan; Norway	1826
Peponocephala electra	preg f: Aug; e Australia	2363
	birth: Aug-Dec, s hemisphere	1398
Sotalia fluviatilis	conceive: Dec-Jan; Amazon, S America	873
	birth: Oct-Nov; Amazon, S America	873
Sousa spp	birth: yr rd, 50/75 Dec-Feb; S Africa	9408
Stenella spp	birth: Apr-June; Mexico	4420
Stenella attenuata	mate: Feb-Mar, July, Nov; Japan	5564
	birth: spring, fall; Carribean, Atlantic	4422
	birth: peak spring, fall, some summer; tropical Pacific	8387
	birth: yr rd, peak Feb-Apr; north popn also peak Sept; Pacific	594
Stenella coeruleoalba	testes: peak Oct, min Feb; Japan	7399
	conceive: Jan, July; Japan	7399
	mate: Jan-Feb, May-June, Sept-Oct; Japan	7397
	mate: Feb-May, July-Sept, Dec; Japan	7396
	birth: yr rd, peaks Dec-Feb, July-Aug; Japan	7396

Stenella longirostris	testosterone: high Apr-July; captive	11586
	birth: Mar-Apr, Sept; Pacific	594
	birth: mid May-mid June; FL	7123
	birth: yr rd; Pacific	594
Tursiops truncatus	mate: yr rd, peak Sept-Dec; captive	9407, 10693
	birth: Feb-May, Aug-Dec; n33; captive	11889
	birth: Mar-May; Pakistan	9175
	birth: Feb-June; England, UK	4420
	birth: yr rd, peak Feb-May, Sept-Nov; captive	4420, 4430
	birth: yr rd, peak Mar-Apr, Oct; FL	2051
	birth: est May; CA; few data; captive	2051
	birth: yr rd, peak Oct-Mar; captive, S Africa	9407
	birth: yr rd; few data; S Africa	9300
	birth: yr rd; n Gulf of Mexico	6322
	birth: yr rd; n18; captive	12073

Order Cetacea, Family Phocoenidae

Porpoises (*Neophocaina* 25-40 kg; *Phocoena* 25-90 kg; *Phocoenoides* 80-150 kg) are piscivores in temperate and arctic waters of both hemispheres. Females are larger than males (*Phocoena* 11191). Ovulation is usually from the left ovary (1777, 3276, 7506, 10063; cf 8064). A 700 g (n1), epitheliochorial, diffuse or discoidal placenta implants into the left horn of the bicornuate or bipartite uterus (1777, 2287, 4146, 4428, 5238, 5843, 7506, 7563, 10063, 10737, 11849). Details of anatomy (1777, 4428, 7506, 7507) and milk composition (1034, 11698a, 11980) are available.

NEONATAL MASS		
Neophocaena phocaenoides	m: 6.6 kg; n1 stillborn	5567
	m: 7.2, 7.4 kg; n2; 4, 17 d	5567
Phocoena phocoena	m: 3.1 kg; n1 stillborn	5238
	5-8 kg	5517
	m: 9.1 kg; n1 near term	3276
	f: 9.5 kg; n1 near term	3276
NEONATAL SIZE		
Neophocaena phocaenoides	77 cm	2365
Phocoena dioptrica	188.5 cm; r186-191; n2 term emb	5278
Phocoena phocoena	m: 40.3 cm; n1 stillborn	5238
	60 cm; n1 near term	5843
	67-80 cm	11191
	70-99 cm	9983
	70.5 cm; r70-71; n2	3626
	75 cm	8066
	76.2-99.1 cm	3626
	m: 78.5 cm; r70-87 mm; n2 near term	3276
	est 80 cm	9695a
	80-86 cm	3626
	82-85 cm	12058
	f: 84 cm; n1 near term	3276
Phocoena sinus	m: 74 cm; n1 yg with fetal folds	1378
Phocoena spinipinnis	44 cm; n1 near term	1380
Phocoenoides dalli	95-100 cm	5279
	100 cm	5557
WEANING MASS		
Neophocaena phocaenoides	24.5 kg; n1 immature	4428
WEANING SIZE		
Neophocaena phocaenoides	130 cm; n1 immature	4428
LITTER SIZE		
Neophocaena phocaenoides	1 emb; n1 f	5567
Phocoena dioptrica	1 emb ea; n2 f	5278
Phocoena phocoena	1 emb; n1 f	5238
	1 emb; n1 f	5843
	1 emb ea; n2 f	3276

Phocoena sinus	1 yg; n1 f	1378
Phocoena spinipinnis	1 emb; n1 f	7370
Phocoenoides dalli	1 emb ea; n39 f	5557
SEXUAL MATURITY		
Neophocaena phocaenoides	f: 1st cl: 141 cm; n1	4428
	m: spermatogenesis: 140-150 cm	4428
Phocoena phocoena	f: 1st preg: 14 mo, ~50 kg	7422
	f: ~4 yr, 145 cm	3276
	m: ~3 yr, 133 cm	3276
	m: 6 yr	11191
	f/m: 3-4 yr	3626
Phocoenoides dalli	f: 3 yr, 170-171 cm	5279
	f: 6.8 yr, 186.5 cm	5557
	f: 7 yr	11191
	f: 170 cm	7370
	m: 5-6 yr, ~182 cm	5279
	m: 7.9 yr, 195.7 cm	5557
	m: 185 cm	7370
	f/m: min 2 yr	5566
GESTATION		
Neophocaena phocaenoides	10 mo	1777
	~11 mo	5561
Phocoena spp	10 mo	4146
Phocoena phocoena	10-11 mo	7422
	~11 mo	3276
	332 d	6279
Phocoenoides dalli	7-9 mo	5279
	<1 yr	7370
	~12 mo; field data	5557
LACTATION		
Neophocaena phocaenoides	wean: 6-15 mo	5561
Phocoena phocoena	wean: 8 mo	7422
Phocoenoides dalli	wean: 1.6-4 mo	5279
	wean: 3 mo	5279
	wean: 2 yr	5557
INTERLITTER INTERVAL		
Neophocaena phocaenoides	mode 2 yr; r1-2	5561
Phocoena phocoena	1 yr	12058
	1 yr	7422
Phocoenoides dalli	3 yr	5557
SEASONALITY		
Neophocaena phocaenoides	mate: est May-Sept; Japan	5561
	mate: est June-July; Yangtse R, China	1777
	birth: est Apr; Yangtse R, China	1777
	birth: Apr-Aug, est peak Apr-May; Japan	5561
	f with young: peak Oct-Nov; few data; Pakistan	9175
Phocoena dioptrica	preg f: July-Aug; n2; se S America	5278
Phocoena phocoena	testes: large May-June, decline mid-Aug; w Atlantic	3276
	spermatogenesis: early phases May-July, peak mid-end July; w Atlantic	3276
	mate: end June/July-early Aug; w Atlantic	3276
	mate: July-Oct; Russia	12058
	conception: peak Aug; Denmark, Netherlands	6279
	birth: Apr-mid June; Russia	12058
	birth: est May-June; n Pacific	9983
	birth: June-July; w Atlantic	3276
Phocoena spinipinnis	preg f: Feb; n1; Uruguay	1380
Phocoenoides dalli	mate: mid Aug-Oct; Japan	5557
	birth: end July-early Aug; CA	7370
	birth: Aug-Sept; Japan	5557
	birth: est yr rd; few data; Pacific	7506

Order Cetacea, Family Monodontidae

Belugas (*Delphinapterus leucas*) and narwhals (*Monodon monoceros*) are 1000-2000 kg, gregarious carnivores of arctic waters. The long, spiral tusk commonly protruding from the upper jaw of male, and occasionally female, narwhals may be used in male-male competition (9977). Males are usually larger than females (2643, 6790, 9786, 11312). Ovulation may be induced (11980). The left ovary is slightly dominant (8064). The epitheliochorial placenta of *Delphinapterus* has branched chorionic villi (11980) and usually implants into the left uterine horn (10393). The uterus is bicornuate (*Delphinapterus* 11980) or bipartite (*Monodon* 358). Concurrent gestation and lactation as well as accessory corpora lutea have been reported for *Delphinapterus* (1303, 9781, 11980). Details of anatomy (358) and milk composition (6245, 11980) are available.

NEONATAL MASS		
Delphinapterus leucas	35-85 kg	10393
	57 kg	2643
Monodon monoceros	80+ kg	6790
NEONATAL SIZE		
Delphinapterus leucas	150 cm	6279
	150 cm	8066
	150 cm; r120-183; term emb, small calves	9781
Monodon monoceros	153 cm; n1 smallest calf	9006
	~160 cm	6790
	165 cm; n1 term emb	9006
LITTER SIZE		
Delphinapterus leucas	1	9781
	1.02; r1-2; n49 lt	2643
	1.06; r1-2; n17 lt	11312
Monodon monoceros	1 yg	6790
	mode 1 yg; r1-2	9006
SEXUAL MATURITY		
Delphinapterus leucas	f: 1st cl: 5 yr	9781
	f: 1st cl: 3.45 m	11312
	f: 1st preg: 5 yr; r4-7	10393
	f: 1st preg: 3.15-3.20 m	11312
	f: 3 yr, 270 cm	2643
	f: 5 yr	7370
	m: testicular histology: 7-9 yr	9781
	m: 4 yr	11312
	m: 8-9 yr	7370
	f/m: 2-5 yr	11980
	f/m: 5 yr	1302
GESTATION		
Delphinapterus leucas	11.5 mo	11980
	12 mo?	11312
	est 14 mo	9781
	14 mo	7370
	14 mo; fetal growth data	9781
	14 mo	1474
	14.5 mo	1302
	14-15 mo	10393
Monodon monoceros	est ~14 mo; fetal growth data	874
	est 15 mo	6790
LACTATION		
Delphinapterus leucas	solid food: 1 yr	10393
	wean: 20-24 mo	10393
	wean: est 21 mo; field data	9781
	wean: 2 yr	1302
Monodon monoceros	wean: est 20 mo; field data	6790
INTERLITTER INTERVAL		
Delphinapterus leucas	2-3 yr	7370
	3 yr	9781
Monodon monoceros	2-3 yr	874

SEASONALITY

Delphinapterus leucas	mate: end Feb-June, est peak Mar; AK	10393
	mate: May; Cumberland Sound, Baffin I, NWT, Canada	1302
	mate: May-June; Gulf of St Lawrence	11312
	mate: June-Sept; Hudson Bay	10393
	preg f: July; Gulf of St Lawrence	11312
	birth: peak end Mar; w Greenland	10393
	birth: Mar-May; Arctic, n Atlantic, Hudson Bay	1474
	birth: May-June; Gulf of St Lawrence	6279, 11312
	birth: June-Aug, peak July; Gulf of St Lawrence	9785
	birth: peak end June-July, some Sept; w Hudson Bay	9781
	birth: July; Chukcki Sea, Mackensie, Bering Sea	10393
	birth: end July-early Aug; Cumberland Sound, Baffin I, NWT, Canada	1302
Monodon monoceros	mate: mid Apr; Canada, nw Greenland	6790
	birth: end June-Aug; arctic	874
	birth: mid July; Canada, nw Greenland	6790

Order Cetacea, Family Physeteridae

Sperm whales (*Kogia* 150-400 kg; *Physeter* 12,000-48,000 kg) are worldwide, oceanic predators primarily of squid. Pygmy sperm-whale males and females are of equal size, while males of *Physeter macrocephalus* are much larger than females (183, 393, 871, 2045). Mating occurs with ventral-ventral orientation (871). The placenta is epitheliochorial (11980). Females ovulate late in lactation (861, 3591, 11980). Details of anatomy (860, 861, 862, 863, 3590, 6961), sperm counts (1007), birth (8399), and milk composition (871, 5293, 11980, 12090) are available.

NEONATAL MASS

Kogia breviceps	82 kg	7498c
Physeter macrocephalus	774 kg	10158
	1054 kg	3590

NEONATAL SIZE

Kogia breviceps	113.5 cm; n1 large emb	9301
	116.5 cm; r110-123; n2 large emb	2766
	124.8 cm; r120-129.5; n2 small yg	1546, 1546a
	TL: 163 cm	7498c
Kogia simus	est 100 cm	9301
Physeter macrocephalus	360-430 cm	1474, 2045
	m: 382 cm; se12; n6	871
	390 cm	8066
	392 cm	1887
	f: 393 cm; se10; n9	871
	395-425 cm	5146
	~400 cm	6961
	TL: ~400 cm	10158
	404 cm	861
	est 405 cm	8065
	est 415 cm	9695a
	425-455 cm	7403

WEANING MASS

Kogia breviceps	62.6 kg; n1 small m with squid in stomach	10078a

WEANING SIZE

Kogia breviceps	TL: 155 cm; n1 small m with squid in stomach	10078a
Physeter macrocephalus	670 cm	8065
	< 760 cm	861

LITTER SIZE

Kogia breviceps	1 emb ea; n2 f	2766
	1 emb; n1 f	9301
Kogia simus	1 yg	9301

Physeter macrocephalus	1.00 emb; twins 0.40%; n503 f	871
	1.00 emb; twins 0.45%; n2664 f	8065
	1.01 cl; 0.67% 2, 0.22% 3; n446 f	8065
	1.02 emb; twins 1.3%, triplets 0.01%; n748 f	5498
	1.17 emb; r1-2; n6 f	5559
SEXUAL MATURITY		
Kogia breviceps	f: 1st preg: est 2.7-2.8 m	9301
	m: testes enlarge: est 2.7-3.0 m	9301
Kogia simus	f: 1st preg: est 2.1-2.2 m	9301
	m: 1st sperm: est 2.0-2.2 m	9301
Physeter macrocephalus	f: 1st cl: 15 mo	6961
	f: 1st cl: 50% 9 yr, 8.5 m	861, 864
	f: 1st cl: 9.2 yr	8065
	f: 1st cl: 8.4-8.6 m	1888
	f: 1st cl: 8.7 m	1887
	f: 50% mature: 12-13 yr; 8.8 m; n256	564
	f: 1st preg: 9.0-11.1 m	7403
	f: 1st preg/lact: 9.75-10.05 m; n5	5559
	f: 54 mo	393
	f: 8.8 m	1375
	f: 9.15 m	8110
	m: spermatogenesis: 2 yr	6961
	m: spermatogenesis: 9.6-9.8 m	1888
	m: sperm: est 10.5-11 m	7927, 7931, 7932, 7934
	m: sperm: 11.27 m; ±0.44; n162	6527
	m: sperm: 12.5 m	871
	m: testes large: 9.33 m; ±0.51; n162	6527
	m: testes large: 12.5 m	7374
	m: 50% mature: 11 yr; 8.5-8.8 m; n781	564
	m: puberty: 10.7-11.6 m; sexual maturity: 25 yr, 13.72 m	7371
	m: puberty: 11.9-12.2 m; sexual maturity: 13.7-14.0 m	862, 864
	m: 9.5-10 m	1007
	m: 11.9 m	1375
	m: 14.20 m	393
	f/m: 9-10 yr	2045
	f/m: <9.15 m	7403
GESTATION		
Kogia simus	questionable est ~9 mo	7744
Physeter macrocephalus	14-15 mo; field data	871
	14.6 mo; fetal growth data	861
	14.75 mo	3591
	15-16 mo; n1706; fetal growth data	3488
	473 d; n166; fetal growth data	3588
	479 d; n55; fetal growth data	564
	16 mo	393
	16 mo; fetal growth data	1887
	~16 mo	6961
	16.4 mo; field data	8065
	17 mo	7403
	518 d; n277; fetal growth data	1888
LACTATION		
Kogia breviceps	solid food: 1.6 m; n1	1546
Physeter macrocephalus	solid food: ~1 yr, 2 dental layers	871
	nurse: est 5-6 mo or 17-18 mo	1888
	wean: 6 mo	5146
	wean: ~6 mo	2045
	wean: > 6 mo	6961
	wean: >1 yr	10698
	wean: ~13 mo	1887
	wean: 1.6-2.4 yr; f <20 yr; 2.2-4.0 yr; f >20 yr	871
	wean: 24 mo	393
	wean: 24-25 mo	861
	wean: 24-25 mo	3590, 3591

INTERLITTER INTERVAL		
Physeter macrocephalus	~2 yr	6961
	3 yr	564
	3-4 yr	8065
	4 yr	3590
	4 yr	861
SEASONALITY		
Kogia breviceps	birth: Feb; n1; e US	183
Physeter macrocephalus	ovarian activity: low June, peak Aug east, Oct west; S Africa	871
	spermatogenesis: yr rd; w S Africa	862
	mate: Mar; Japan	7403
	mate: Mar-May; n Hemisphere	3488
	mate: Apr-May; WA, Australia	7373
	mate: peak Apr-May; Azores	7373
	mate: Apr-July; CA	7373
	mate: Apr-Aug; ne Pacific	9091
	mate: Apr-Feb, most June-Dec, peak Sept; se Pacific	1888
	mate: peak Aug-Dec, most Oct; S Africa	6961
	mate: Sept; s Pacific	7373
	mate: Sept-Dec; s Hemisphere	3488
	mate: est Oct-Dec; S Africa	871
	mate: Oct-Apr, peak Dec; S Africa	3588
	mate: peak Dec-Jan; WA, Australia	564
	mate: yr rd, peak Apr; Japan	8065
	ovulate: end Aug-Mar, peak Dec; w S Africa	861
	conceive: Feb-June, older f earlier; Japan	8065
	conceive: Sept-Mar, 80% Nov; S Africa	3590, 3591
	conceive: end Aug-Mar, peak early Dec; w S Africa	861
	birth: Jan-Apr, peak Feb-Mar; S Africa	871
	birth: end Jan-mid Aug, peak Apr; S Africa	3588
	birth: peak Feb; s Pacific	7373
	birth: Apr-May; WA, Australia	564
	birth: May-Dec, peak July-Dec; Azores	1887
	birth: end July-Oct; Japan	7403
	birth: Aug-Sept; Japan	3726
	birth: Aug-Dec; ne Pacific	9091
	birth: peak Aug-Dec; CA	7373
	birth: Spet-July, most Nov-May, peak Feb; se Pacific	1888
	birth: Nov-June, peak Feb-Mar; w S Africa	861
	birth: peak Dec-Jan; S Georgia, s Atlantic Ocean	7373
	birth: yr rd, peak Jan-Mar; S Africa	6961

Order Cetacea, Family Ziphiidae

The relatively unstudied, beaked whales (*Berardius* 9000 kg; *Hyperoodon* 3500-5000 kg; *Indopacetus* est 2200 kg; *Mesoplodon* 1300-1500 kg; *Tasmacetus* 2500 kg; *Ziphius* 2450-2950 kg) are carnivores from both the Atlantic and the Pacific in arctic, temperate, and tropical waters. *Indopacetus* is known only from two skulls and *Tasmacetus* from a dozen specimens. *Berardius* has an epitheliochorial placenta (11980). The presence of old and new corporea lutea in the same female (*Berardius*, *Ziphius*), as well as the capture of a concurrently lactating and pregnant female (*Hyperoodon*), suggests a postpartum estrus (8059, 11980). Details of anatomy (2438, 7370) and milk composition (11031, 11980) are available.

NEONATAL SIZE		
Berardius bairdii	460 cm	8066
Hyperoodon ampullatus	305 cm	771
	350 cm; n1 term emb	8059
Mesoplodon bidens	150-240 cm	5440
Mesoplodon carlhubbsi	250 cm	7124
Ziphius cavirostris	200-305 cm	7370
	TL: 228-240 cm	10158
WEANING SIZE		
Mesoplodon bidens	315 cm; n1; est recently weaned	5440

LITTER SIZE		
Berardius bairdii	1 emb; n1 f	9091
Hyperoodon ampullatus	1	8059
Mesoplodon bidens	1 yg	5440
Mesoplodon carlhubbsi	1 emb; n1 f	7124
Mesoplodon mirus	1 emb; n1 f	1291
Ziphius cavirostris	1	10158
SEXUAL MATURITY		
Berardius bairdii	f: 10 m	6953
Hyperoodon ampullatus	f: 1st cl: 8-13 yr	773
	f: 1st cl: ~9 yr; 7.2-7.3 m	1824
	f: 1st preg: 6.60-6.80 m	771
	f: 6.7-7.01 m	7370
	m: spermatogenesis: 7-9 yr	773
	m: testes histology: 9-11 yr, 7.20 m	771
	m: 7.3 m	7370
Mesoplodon densirostris	f: 9 yr	7983
Ziphius cavirostris	f: 1st cl?: 6.1 m	7370
	m: testes size?: 5.4 m	7370
GESTATION		
Berardius arnuxii	~10 mo; few data	7014
Berardius bairdii	10 mo; few data	7014
	est 17 mo; fetal growth	5556a
Hyperoodon ampullatus	est 12 mo	8059
	~12 mo	771
Mesoplodon bidens	est 1 yr	5440
Mesoplodon carlhubbsi	12 mo	7983
LACTATION		
Hyperoodon ampullatus	wean: 1 yr; n1; squid and milk in stomach	773
Mesoplodon bidens	wean: est 1 yr	5440
LITTERS/YEAR		
Hyperoodon ampullatus	1	8059
SEASONALITY		
Berardius arnuxii	birth: end spring-early summer; New Zealand	7014
Hyperoodon ampullatus	mate: peak Apr; n Atlantic	771
	mate: Apr-May; Arctic	8059
	birth: peak Apr; n Atlantic	771
	birth: end Apr-June; Arctic Sea	8059
Hyperoodon planifrons	birth: spring-early summer; S Africa	10158
Mesoplodon bidens	birth: end winter-spring; Norway	5440
Mesoplodon carlhubbsi	birth: est June-Aug; n4; CA	7124
Mesoplodon densirostris	birth: summer; S Africa	10158
Mesoplodon stejnegeri	lact f: Oct; n1; AK	11031
Ziphius cavirostris	birth: summer; S Africa	10158

Order Cetacea, Family Eschrichtidae, Genus *Eschrichtius*

The migratory, grey whale (*Eschrichtius robustus*) is a 20,000 to 37,000 kg, baleen whale that inhabits in temperate and arctic Pacific waters and feeds on krill. Females are reportedly larger than males (302). The non-deciduate, epitheliochorial, diffuse placenta implants into a bipartite uterus (9094). Conception may occur during lactation (9094, 11979). Details of anatomy (9094), copulation (532, 9506), and milk composition (11979, 11980, 12090) are available.

NEONATAL MASS		
Eschrichtius robustus	f: 409 kg; n1 term emb	9094
	500 kg	11880
	680 kg	3788
NEONATAL SIZE		
Eschrichtius robustus	360-570 cm	302
	366-518 cm	9154
	425-515 cm	5146
	446.7 cm; r354-540; n69	8755

	450 cm	8066
	m: 456 cm; n24 term emb	9094
	458 cm; sd49; r350-630; n133 term emb, yg	10548
	468.3 cm; r395.5-540.5; n6	2887
	f: 469 cm; n17 term emb	9094
	~560 cm	3788
WEANING SIZE		
Eschrichtius robustus	850 cm	9094
LITTER SIZE		
Eschrichtius robustus	1 yg	9094
SEXUAL MATURITY		
Eschrichtius robustus	f: 1st cl: 8 yr; r5-11, 11.7 m	9094, 11880
	f: 1st cl: 7-11 yr	9018
	f: 1st cl, emb: 8-12 yr, ~12 m	11979
	f: 1st cl: 11.7-11.8 m	1008, 1009
	f: 1st preg: 12.7 m; n1	9091
	m: spermatogenesis: median 8 yr; r5-11	9018
	m: testes large: 6-8 yr, 11 m	11979
	m: 5-11 yr, 11.1 m	9094
	m: ~11.5 m	1008
	m: 11.9 m	9091
	f/m: 4-5 yr	3788
CYCLE LENGTH		
Eschrichtius robustus	40 d	9094, 11880
GESTATION		
Eschrichtius robustus	11-12 mo	3789
	est ~1 yr	5054
	~1 yr	302
	~1 yr	11979
	400 d; field data	9094, 11880
	est 418 d	8755
LACTATION		
Eschrichtius robustus	wean: 137 d	11979
	wean: 5 mo	5146
	wean: 6 mo	10548
	wean: 7 mo	9018
	wean: 7 mo	9094, 11880
INTERLITTER INTERVAL		
Eschrichtius robustus	1-2 yr	11979
	2 yr?	3789
	2 yr	9094
SEASONALITY		
Eschrichtius robustus	spermatogenesis: peak Nov-Jan; CA	11880
	estrus: end Nov-early Dec; CA	9094, 11880
	mate: Jan-Apr; CA	532
	mate: June; AK, Siberia	9506
	mate: Dec; Korea	302
	mate: Dec-May; e Pacific	5054
	mate: Dec; e Pacific	8755
	preg f: Jan; n14 near term; Mexico	9154
	birth: Jan-Feb; CA, Mexico	3789
	birth: Jan-Feb; ne Pacific	8755
	birth: Nov-Dec; Korea	302
	birth: Dec-Feb; Mexico	5054
	birth: end Dec-early Feb, peak Jan; CA	9094, 11880
	birth: winter; AK	6726
	wean: July-Aug; CA, Chukchi Sea	1009

Order Cetacea, Family Balaenopteridae

Rorquals (*Balaenoptera* 8000-172000 kg; *Megaptera* 25000-48000 kg) are aquatic, baleen whales with a worldwide distribution. Females are larger than males (1801, 2045, 3593). The diffuse, villous placenta implants into a bicornuate or bipartite uterus (691, 6958, 7563, 8108, 10526, 10969, 11361). Ovulation may be spontaneous (6836). A postpartum estrus occurs except in *Balaenoptera acutorostrata* (869, 1802, 3591, 5157, 6252, 6281). Details of anatomy (224, 691, 1799, 1800, 2480, 3038, 3589, 6702, 6958, 6960, 9186, 11178), endocrinology (392, 8743), sperm counts (1007), intersex (563), and milk composition (859, 869, 1802, 2439, 4056, 5295, 6245, 6702, 8072, 8073, 8321, 10611, 11634, 11980, 12090) are available.

NEONATAL MASS		
Balaenoptera borealis	650 kg	6279
Balaenoptera musculus	2500 kg	6279
Balaenoptera physalus	1900 kg	6279
Megaptera novaeangliae	1300 kg	6279
NEONATAL SIZE		
Balaenoptera acutorostrata	224 cm; n1	7099
	240-270 cm	6952
	250 cm	8066
	250-300 cm	4147
	est 270 cm	9695a
	280 cm	5157
	280 cm	5182
Balaenoptera borealis	TL: 360 cm	10158
	est 400 cm	4147
	420-450 cm	7403
	440 cm	6279
	440 cm	8066
	450 cm	2045
	450 cm	6700
	est 450 cm	9695a
	800-900 cm	6960
Balaenoptera edeni	TL: 340 cm	10158
	396 cm	866
Balaenoptera musculus	700 cm	6279
	700 cm	6700, 6702
	700 cm	8066
	est 700 cm	9695a
	750-800 cm	4147
Balaenoptera physalus	630-660 cm	7403
	640 cm	6279
	640 cm; r625-670	8063, 8066, 8071
	640-670 cm	7403
	650 cm	6700, 6702
	est 650 cm	9695a
Megaptera novaeangliae	410 cm; mode 420; r240-540; n240 term emb	1802
	430 cm	8066
	450-500 cm	6700
	450-500 cm	6958
	456 cm	6279
	est 460 cm	9695a
WEANING MASS		
Balaenoptera musculus	17,000 kg; 6 mo; from graph	3592
WEANING SIZE		
Balaenoptera borealis	TL: 8.5 m	10158
Balaenoptera musculus	TL: 16 m; 6 mo; from graph	3592
	16 m	5630
Balaenoptera physalus	est 12 m	8063
	12 m	5630
	~12.6 m	6702
Megaptera novaeangliae	7.5-8.0 m	6958
	TL: 8-9 m	10158

LITTER SIZE

Balaenoptera acutorostrata	1 cl	10969
	1.01 yg; r1-2; 0.65% twins; n1998 lt	8069
	Siamese twin emb; n1 f	12109
Balaenoptera borealis	1.02 yg; r1-2; n1098 lt	5730
	1.07 cl; r1-2; n27 f	3589
Balaenoptera edeni	1 emb ea; n2 f	8101
Balaenoptera musculus	1 emb; n~50 f	6702
	1.01 emb; 0.67% twins; n593 f; 7 emb; n1 f	9154
	1.01 emb; 0.74% twins; n272 f	1294
	1.01 emb; 0.78% twins; n19057 f	5730
	1.01 yg; r1-2; 0.87% twins; n19005 lt	5730
	2; n1 lt	10158
Balaenoptera physalus	1 emb; n~80 f	6702
	mode 1 ovulation; 2-3 rare	11619
	1.0; 0.43% twins; n>900 lt	8071
	1.01; 0.65% twins; n462 lt	9154
	1.01 emb; r1-6; 0.82% twins; n39947 f	5730
	1.02 yg; r1-4; n525 lt	8061
	1; 2.38% twins; n569 lt	1294
	1.1 cl; n20 f	3591
	2 emb ea; n3 f	9091
	3 emb; n1	856
Megaptera novaeangliae	mode 1 ovulation; 2 rare	1803
	1.01 yg; r1-2; 0.57% twins; n2979 lt	5730
	1.03; 2.68% twins; n149 lt	1802

SEXUAL MATURITY

Balaenoptera acutorostrata	f: 1st cl: 7 yr; 1960's; 13-14 yr; 1940's	5568
	f: 1st cl: 7-8 yr	8070
	f: 1st cl: 7.3 m	7372
	f: 1st cl: 7.3 m	1825
	f: 1st cl: 7.6 m	6218, 6219
	f: 3 yr; 7.5 m	9778
	f: 6 yr; 50% 8.6 m; r7.9-9.4	8069
	f: 10.5 yr; 1945; 7.7 yr; 1973; 7.98 m	869
	f: 7.2 m	8114
	m: spermatogenesis: 7-8 yr	8070
	m: 6.6-6.9 m	8114
	m: 7.47 m	869
	m: 50% 7.7 m; r6.9-8.2	8069
	m: 8.1-8.9 m	1007
	f/m: 50% 6.2 yr	5157
	f/m: 7.5 m	1375
	f/m: 8 m	5181
Balaenoptera borealis	f: 1st cl: 12.2-13.4 m	5439
	f: 1st cl: 12.9-13.2 m	6539
	f: 1st cl: 18 mo	6960
	f: 1st mate: 2 yr	6960
	f: 10-11 yr; 13.9 m	3593
	f: 1212 cm; 5% fiducial limits 1197-1257	7933
	f: 12.5-13.7 m	8110
	f: 13.3 m	9092
	f: 13.7 m	9091
	m: testes large: 12.8-13.1 m	5439
	m: testes large: > 13.5 m	3589
	m: 10-11 yr; 13.5 m	3593
	m: 1212 cm; 5% fiducial limits 1194-1227	7933
	m: 12.2-13.1 m	8110
	m: 12.6-12.9 m	6539
	m: 12.8 m	9091, 9092
	f/m: 1st mate: 2.5 yr; 13.5 m	9388
	f/m: 8-9 yr	2045
	f/m: 11.6 m	7403
	f/m: 13.7 m	1375

Balaenoptera edeni	f: 1st cl: 11.7 m	859
	f: 1st cl: 12.2 m	8067
	f: 75% 1st cl: 12.5 m; 95%ci 12.0-12.8	7933
	f: ~5 yr	10158
	m: testes large: 11.9-12.2 m; inshore 12.2-13.1 m; offshore	859, 866
	m: testes large: <12.2 m	8067
	m: ~5 yr	10158
	m: 75% sperm: 12.2 m; 95%ci 12.1-12.5	7933
	m: 12.6-13.2 m	1007
Balaenoptera musculus	f: 1st cl: ~2 yr	6252
	f: 1st cl: 22.2 m	7930
	f: 1st cl: 22.8 m	7404
	f: 1st cl: 23.7 m	6700, 6702
	f: 1st mate: 3-7 yr; r19.5-25.2 m	9389
	f: 1st birth: 23.4 m	1294
	f: ~5 yr	10158
	f: 20.4 m	9091
	f: 22.5 m	5498
	f: 27 m; from graph	1375
	m: testes large: 2-7 yr, 22.3 m	9389
	m: testes large: 21.6 m	7404
	m: testes large: 22.6 m	6700, 6702
	m: testes increase: 23.4 m	7930
	m: ~5 yr	10158
	m: 22.35 m; r21.3-23.4; n2	5498
	m: 25.5 m; from graph	1375
	f/m: increase anterior pituitary size: 22.5 m	7937
	f/m: 2nd winter	5630
	f/m: 8-10 yr	2045
	f/m: 18.9 m	1375
Balaenoptera physalus	f: increase anterior pituitary wt: 20.1 m	7937
	f: 1st cl: 6 yr, 18.3 m	8071
	f: 1st cl: 6-7 yr	7402
	f: 1st cl: 6-7 yr, 20 m	6534
	f: 1st cl: 17.1-17.4 m	4498
	f: 1st cl: 19.5 m	7404
	f: 1st cl: 19.9 m, 3 yr?	6700
	f: 1st cl: 20.1 m	7930
	f: 1st preg: 18.3 m	4147
	f: 1st birth: 19.5-19.8 m	1294
	f: 6-8 yr	11192
	f: 17.1-19.5 m	8110
	f: 17.4 m; 12 yr; 1954; 6 yr; 1975	8068
	f: 18.0 m; r18.3-18.6	4498
	f: 18.6 m	5498
	f: 18.6 m	8071
	f: 18.6 m	9091
	f: mode 19.2 m; r16.5-21.3	6875
	m: increase anterior pituitary wt: 18.6 m	7937
	m: testes large: 5-13 yr varies	3591
	m: testes large: 6 yr; 17 m	8071
	m: testes large: 17.1-17.4 m	4498
	m: testes large: 18.3 m	7404
	m: testes large: 19.2 m	6700
	m: testes large: 19.2 m; 6-7 yr	6534
	m: testes increase: 18.9 m	7930
	m: 17.5 m	9091
	m: 17.7 m	8071
	m: 17.7-18.0 m	8110
	m: 18.0-20.1 m	5498
	m: 18.3 m; 11 yr; 1954; 4 yr; 1975	8068
	f/m: 1st mate: 3 yr, 18.0 m	9388
	f/m: min 4 yr; 6 yr	8071

	f/m: 50% ~7 yr; 6-7 growth layers ear plugs	6538
	f/m: 8 yr; 1955-1979	6537
	f/m: 10.5 yr; 1941-1950	6537
	f/m: 11.5-12 yr; 1900-1928	6537
	f/m: 15.2 m	7403
	f/m: 19.5 m	1375
	f/m: est 21 m	8063
Megaptera novaeangliae	f: 1st cl: 22 mo	6958
	f: 1st cl: 5-6 yr, 11.7-12.0 m	7928
	f: 1st cl: 11.4 m	7404
	f: 1st cl: 11.4 m (33% mature); 12.3 m (100%)	8111
	f: 1st ovulate: 11.55 m; ±0.05; n29	1801
	f: 1st preg: 13.7 m	4147
	f: 5.0 yg; r4-6; n2 f	1846
	f: 11.4 m	5498
	f: 12 m	9091
	m: spermatogenesis: est 3 yr	3038
	m: spermatogenesis: 11-12 m	6958
	m: testes large: 5 yr; 11.1-11.7 m	7928
	m: testes large: 11 m; r10-12.25	1800
	m: testes large: 11.4 m (75% mature); 12.0 m (100%)	8111
	m: 11 m	10158
	m: 11.7 m	9091
	f/m: 11.9 m	1375
GESTATION		
Balaenoptera acutorostrata	9.5-10 mo; from size of newborn	5182
	10 mo	4147
	10 mo	5157
	10 mo	8114
	~10 mo	10394
	10-11 mo	5181
	10-11 mo	6952
Balaenoptera borealis	324 d; fetal growth data	6539
	<11 mo	7403
	12 mo	3589, 3593
	~12 mo	6960
Balaenoptera edeni	12.7 mo; fetal growth data	9092
Balaenoptera musculus	9 mo; fetal growth data	3488
	10-12 mo	6700, 6702
	10.75 mo; fetal growth data	6279
	>1 yr	4147
Balaenoptera physalus	9-10 mo; fetal growth data	3488
	10-11 mo	6701
	11 mo; fetal growth data	7713
	11.25 mo	6281
	11.25 mo; fetal growth data	6279
	est 11.5 mo; fetal growth data	6702
	12 mo 10 d	7403
	~1 yr; field data	4498
Megaptera novaeangliae	~11 mo	6958
	~11 mo; fetal growth data	4147
	11.5 mo	10158
	12 mo; fetal growth data	3488
	~12 mo	1802
LACTATION		
Balaenoptera acutorostrata	wean: ~6 mo	5157
Balaenoptera borealis	wean: ~5 mo	2045
	wean: 6 mo	3589, 3593
	wean: est 6 mo	6539
	wean: 1.5 yr	6960
Balaenoptera edeni	wean: 9 mo	9092
Balaenoptera musculus	wean: 6 mo	6702
	wean: ~7 mo	5630

Balaenoptera physalus	wean: ~6 mo	6701, 6702
	wean: ~6 mo	5630
Megaptera novaeangliae	wean: ~5 mo	6958
	wean: 10-11 mo	10158
	nurse: 1 yr	10698
	wean: 2nd winter	1845

INTERLITTER INTERVAL

Balaenoptera acutorostrata	1.3 yr	10158
	14 mo	5157
	15.3 mo; min 14 mo	869
	18-24 mo	8114
Balaenoptera borealis	2 yr	6960
	2 yr	3589
Balaenoptera musculus	~2 yr	6702
	2-3 yr	6252
Balaenoptera physalus	~16 mo	6092
	2 yr	6701, 6702
Megaptera novaeangliae	~1 yr	6092
	2 yr; r1.5-2	6958
	2.4 yr; mode 2-3; r1-4; n28	1845

SEASONALITY

Balaenoptera acutorostrata	mate: Jan-Mar; n Atlantic	4147
	mate: Feb-Mar or Aug-Sept; 2 popn; Pacific	8114
	mate: Apr?; NEW, Canada	7372
	mate: est June-Sept; Durban, S Africa	869
	mate: Dec-Mar; sw seas of Japan	6952
	birth: Oct-Jan; sw seas of Japan	6952
	birth: Nov-Jan; n Atlantic	4147
	birth: Dec-Jan or June-July; 2 popn; Pacific	8114
Balaenoptera borealis	testes: same yr rd; s hemisphere	3589
	mate: Jan; Japan	7403
	mate: Mar-Oct, peak July; S Africa	6960
	mate: est winter; n Atlantic	4147
	mate: yr rd, peak May; Atlantic	303
	mate/birth: winter; Antarctica	3591, 3593
	conceive: Apr-Aug; s hemisphere	3589
	conceive: end Nov-mid Mar; Antarctica	9154
	birth: May-Oct, peak July; S Africa	6960
	birth: ~July; S Africa	10158
	birth: Sept-Mar, peak Nov; ne Pacific	9091, 9092
	birth: Nov; Japan	7403
Balaenoptera edeni	conceive: est yr rd; few data; S Africa	859
	conceive: yr rd; inshore; fall; offshore; w Cape Province, S Africa	866
	birth: Sept-Mar, peak end Nov; e N Pacific	9092
Balaenoptera musculus	testes: large Apr-June; S Georgia, Antarctica	6700
	estrus: peak winter; S Africa	10158
	ovulate: June-Aug; S Africa, Antarctica	6702
	mate: Apr; n hemisphere	3488
	mate: Apr-Aug, peak end May-early June; S Africa	3592
	mate: June-Aug; s hemisphere	3488
	mate: yr rd, peak June-Aug; Antarctica	9154
	mate/birth: winter; Antarctica	3591
	mate/birth: yr rd; n Atlantic	4147
	birth: Mar-June; S Africa, Antarctica	6702
	birth: peak mid Apr; S Africa	3592
Balaenoptera physalus	spermatogenesis: Apr-June; S Georgia, Antarctica	6700
	estrus: 21% summer; Iceland	6538
	mate: Jan; Japan	7403
	mate: Mar-Apr; n hemisphere	3488
	mate: Apr-Aug, peak June-July; s hemisphere	7713
	mate: May; Antarctica	6701
	mate: July-Sept; s hemisphere	3488
	mate: Dec; Iceland	6538

	mate: yr rd, most Nov-Jan, peak Dec; n Pacific	8071
	mate/birth: Jan; Japan	7403
	mate/birth: Jan-Mar; n Atlantic	4147
	ovulate: peak June-Sept; s hemisphere	11192
	ovulate/sperm in testes: May-Aug; S Africa, Antarctica	6702
	conceive: end Jan-early Feb; Norway	4498
	conceive: June; Antarctica	6701
	conceive: June-Oct 70%; Antarctica	9154
	conceive: mid Nov-early Apr 68%; Norwegian Sea	9154
	conceive: yr rd, peak Apr-Aug, low Feb; Antarctica	6281
	conceive/birth: Apr-Aug; Antarctica	3591
	birth: peak early Jan; Norway	4498
	birth: Apr-Aug, peak June-July; S Africa, Antarctica	6702
	repro: Sept-Mar, peak Nov; ne Pacific	9091
Megaptera novaeangliae	testes: large yr rd; w Australia	1800
	CL present: June-Nov, peak July; w Australia	1799, 1803
	mate: Apr-May; n Atlantic	4147
	mate: Apr-May; n hemisphere	3488
	mate: June-Nov, peak Aug; Australia	10526
	mate: July-Sept; s hemisphere	3488
	mate: Aug-Nov, peak Sept; S Africa	6958
	mate/birth: winter; Australia	10698
	birth: July-Oct, peak Aug; w Australia	1802
	repro: Oct-Apr, peak Dec; ne Pacific	9091

Order Cetacea, Family Balaenidae

Right whales (*Balaena glacialis, B. mysticetus, Caperea marginata*) inhabit temperate and arctic waters. These 23,000-67,000 kg whales subsist on krill filtered through baleen. Females have 2 teats (2166, 9695). Details of anatomy (6959, 9695), copulation (1454, 2683, 3125), and milk composition (11980) are available.

NEONATAL SIZE		
Balaena glacialis	500-600 cm	8066
	600 cm	6700
Balaena mysticetus	304-427 cm	9695
	~400-405 cm	7846
Caperea marginata	est 190 cm; r160-220	9302
WEANING SIZE		
Caperea marginata	est 3.5 m; r3.2-3.8	9302
LITTER SIZE		
Balaena glacialis	1	2684
Balaena mysticetus	1; 2 rare	9695
Caperea marginata	1 emb; n1 f	4131
	1 yg; n1 f	7099
	1 emb ea; n2 f	7613
	1 ea; n2 lt	7370
SEXUAL MATURITY		
Balaena glacialis	f: 1st ovulate: 14.5-15.5 m	8113
	f: 1st birth: est 11 yr	3625a
	f: ~10 yr, 15.5 m	2166
	m: spermatogenesis: 15-16 m	8113
	m: ~10 yr, 15 m	2166
Balaena mysticetus	m: 11.58 m	6830
Caperea marginata	f: smallest preg: 6 m	510
GESTATION		
Balaena glacialis	9-10 mo	2684
	est 11-12 mo	865
	est 13-14 mo	3625a
Balaena mysticetus	~12 mo	6830
	est 13-14 mo	7846

LACTATION		
Balaena mysticetus	wean: est 1 yr	7846
INTERLITTER INTERVAL		
Balaena glacialis	2 yr	2684
	min 2 yr	865
	est 3 yr; r2-5	3625a
SEASONALITY		
Balaena glacialis	mate: Aug-Sept; Algoa Bay, S Africa	2683, 2684
	mate: winter; n Pacific, n Atlantic	2166
	preg f: June-July; n Atlantic	1980
	birth: June-July; S Africa	2684
	birth: peak Aug; s,w S Africa	868
	birth: Dec-Mar; n Atlantic	3625a
	birth: spring; n Pacific, n Atlantic	2166
Balaena mysticetus	mate: May; n1; n AK	3125
	mate: end summer; Greenland	9695
	birth: Feb-Mar; Greenland	9695
	birth: Mar-Aug, peak May; AK	6830, 7846
Caperea marginata	preg f: June, Sept, Nov-Dec; n3; s hemisphere	510

Order Sirenia

Manatees and dugongs may be the marine equivalent of terrestrial elephants. These large, aquatic herbivores give birth to well developed, singleton offspring who suckle from a pair of axillary mammae.

Order Sirenia, Family Dugongidae, Genus *Dugong*

Dugongs (*Dugong dugon* 250-1000 kg) and the extirpated Stellar's sea cow (*Hydrodamalis gigas* 4000 kg) are tropical and subtropical, aquatic grazers. The non-deciduate, epitheliochorial or hemochorial, villous, diffuse or zonary placenta implants into a bicornuate uterus (231, 4435, 6835, 6838, 6842, 7563, 7936, 10966, 10967, 10968). Conception can occur during lactation (6840). Neonates are furred, although the hair falls out during development (5758, 10158). Males have tusks, inguinal testes, a diffuse prostate and bulbourethrals, and large seminal vesicles (4417, 6837, 6839, 7936). Details of anatomy (6835, 6838), birth (5106, 6707a), and suckling (276) are available.

NEONATAL MASS		
Dugong dugon	20 kg; n1 stillborn	193
	20-35 kg	6840, 6842, 7936
NEONATAL SIZE		
Dugong dugon	100-130 cm	6840, 6842, 7936
	109-150 cm; small calves with solid food in stomachs	5106
	128 cm; n1 stillborn	193
	140 cm; n1 emb	5106
LITTER SIZE		
Dugong dugon	mode 1	2449
	mode 1 yg; twins recorded	6840, 6842
	mode 1; r1-2	10809
	1; 2 rare	10158
	1; 2 rare	5758
Hydrodamalis gigas	1 yg	4217
SEXUAL MATURITY		
Dugong dugon	f: 1st ovulate: 9-10 yr	6840
	m: est 9-10 yr	6840
	f/m: ~2 yr; 2.4 m; 248 kg	4586, 6842, 10242
	f/m: 9-9+ yr	6834
	f/m: 9-10 yr, sometimes 15-17 yr	10809
	f/m: 9-14 yr	10698
GESTATION		
Dugong dugon	11-12 mo	5758
	333 d	2449
	est ~1 yr	6840, 6842
	13-14 mo	7936
Hydrodamalis gigas	>1 yr	4217
LACTATION		
Dugong dugon	solid food: 3 mo	5106
	wean: ≥1.5 yr	6840
	with f: 18+ mo	10698
	with f: 1-2 yr	10809
INTERLITTER INTERVAL		
Dugong dugon	≥3 yr	10698
	3-6 yr	10809
	3-7 yr	6840, 7936
SEASONALITY		
Dugong dugon	spermatogenesis: May-Oct; n QLD, Australia	6834
	preg f: Feb-Mar; India, Sri Lanka	8499
	birth: Mar-June; nw Australia	6707a
	birth: Aug-Sept; n QLD, Australia	4586
	birth: Aug-Dec; n QLD, Australia	6840
	birth: Sept-Nov; n QLD, Australia	6834
	birth: Sept-Dec; QLD, Australia	10809
	birth: Oct-Apr; New Guinea	10809
	birth: peak Sept-Dec; n QLD, Australia	7936
Hydrodamalis gigas	birth: fall; Commander I, Bering Sea	4217

Order Sirenia, Family Trichechidae, Genus *Trichechus*

Manatees (*Trichechus inunguis*, *T. manatus*, *T. senegalensis*) are 150-650 kg, aquatic grazers from tropical and subtropical west Africa, the Amazon basin, and the New World Atlantic. The hemochorial, labyrinthine, zonary placenta implants into a bicornuate or bipartite uterus and may have a partly endotheliochorial accessory placenta (231, 1545, 7563, 7628, 11374, 11850). Females have 2 axillary, mammae (1474, 1545, 7498b). Males have seminal vesicles and abdominal testes (1545, 11374). Manatees do not have a pineal (8869). Details of anatomy (7498b), birth (578, 7498b), and milk composition (475, 8398) are available.

NEONATAL MASS		
Trichechus manatus	11-27 kg	5108
	~27 kg	5111
	28.6 kg; n1	578
	m: 31.8 kg; n1	11636
	34.3 kg; r27.7-36.3; n7	11636
	42.0 kg; n1	4454
NEONATAL SIZE		
Trichechus inunguis	73.6 cm; n1 small calf	5105
Trichechus manatus	99 cm; n1	578
	120 cm	9079
	m: 120 cm; n1	11636
Trichechus senegalensis	~100 cm	5107
	104.5 cm; r104-105; n2 small calves	5107
WEANING MASS		
Trichechus inunguis	65-70 kg; hand reared 1 yr after capture	875
LITTER SIZE		
Trichechus manatus	1 yg; n1 f	7500
	1 yg; n1 f	578
	1-2 yg; 2 rare	2211
	1; 2 occasional	5111
Trichechus senegalensis	1 yg	5107
SEXUAL MATURITY		
Trichechus inunguis	f/m: 3 yr	2211
Trichechus manatus	f: 3-5 yr, 2.6 m	4454
	m: 3-5 yr, 2.5 m	4454
	f/m: 4-4.5 yr	2211
GESTATION		
Trichechus inunguis	180-270 d	2211
	12-14 mo	10809
Trichechus manatus	5-6 mo	5111
	≥152 d	7498b
	180-270 d	2211
	~11 mo	1474
	385-400 d	5108
	~13 mo	4454
LACTATION		
Trichechus inunguis	wean: 1 yr after capture; hand reared	875
	wean: ~2 yr	2211
Trichechus manatus	solid food: 38 d	4454
	solid food: 3 mo	1545
	solid food: 4-5 mo	578
	wean: 1 yr	2211
	wean: 1-2 yr	4454
	nurse: 18 mo	578

INTERLITTER INTERVAL		
Trichechus manatus	3-3.5 yr	4454
SEASONALITY		
Trichechus inunguis	conceive/birth: June; Amazon R, S America	10809
	birth: Dec-July, 63% Feb-May; Amazon R, S America	872
Trichechus manatus	m fertile: yr rd; Atlantic	1545
	mate: Dec; Crystal R, w FL	4454
	birth: May, Aug, Nov; captive	7499
	birth: yr rd; e US, Mexico	1474
	birth: yr rd; Suriname	5111
	young found: yr rd; FL	7499

Order Proboscidea, Family Elephantidae

Elephants (*Elephas maximus*, *Loxodonta africana*) are large (1000-7500 kg), diurnal, terrestrial herbivores in Africa and Asia. Males are larger than females (281a, 6292) and have a temporal gland which is active during musth, a period of male sexual activity (1239a, 1484, 1928, 2040a, 2058, 2856, 2936, 4245, 5235, 7262, 8660, 8661, 9659). Flehmen-like behavior occurs (8939).

The labyrinthine, deciduate, zonary placenta has been described as hemochorial, endotheliochorial, epitheliochorial, and vasochorial, and implants into a bicornuate or duplex uterus (241, 434, 706, 1040, 1715, 2041, 2391, 7262, 7563, 8395, 8396, 8397, 9972, 10310, 11485). Placentophagia is occasional (2596, 3063, 5762, 6412, 6681, 9972). Estrus does not occur postpartum (4320, 6119). A breech birth is reported (6733a). Neonates are precocial with open eyes and developed body hair (330, 2596, 5762, 6681, 7751, 9972, 10158). Females have 2 thoracic mammae located between the forelegs (331, 1715, 7262). Males have abdominal testes, seminal vesicles, and a sigmoid penis (3823, 4322, 5427, 9665, 9941, 11479).

Details of anatomy (540, 541, 4321, 4327, 4937, 6282, 6283, 8396, 9665, 9935a, 10135, 11484, 11485, 11487), endocrinology (1239, 1592, 1731, 2412, 4327, 4701, 4702, 4869, 5235, 5549, 6733a, 7111, 8048, 8602, 8607, 8894, 8940, 9935a, 9936, 10116, 10691a), reproductive cyclicity (9350), sperm maturation (2264, 4937, 5357, 5426, 5427, 5429, 5430), copulation (2936, 2941, 5233, 5762), birth (6187), suckling (6351), and milk composition (7039, 8401) are available.

NEONATAL MASS		
Elephas maximus	r79.4-96.8 kg; n3	3341
	90 kg, n1	308
	~90 kg	7751
	~90.7 kg; 2 d	8501
	91 kg	8498
	m: 91.6 kg; r50-150; n14	2597
	92.5 kg; r83-102; n2	6681
	f: 93 kg; n1	11047
	97 kg; n1	1715
	est 100 kg	3063
	~100 kg	10491
	101 kg; n1	1283
	f: 101.2 kg; r77-127	2597
	104.35 kg; r70-133; n10	9065
	107 kg; r50-150; n25	2596
	114.25 kg; r106.5-122; n2	2597
	138 kg; n1, 1.5 d	6733a
Loxodonta africana	73 kg; term emb	331
	78.5 kg; r78-79; n2	1485
	89.8 kg; r84.1-95.5; n2 term	9310
	f: 90-100 kg	9972
	94.98 kg; r80.0-113	9310
	113 kg	9065
	113 kg	6187
	~118 kg	4319
	120 kg	6290, 6292
	m: ~120 kg	
LITTER SIZE		
Elephas maximus	1	6119
	mode 1; r1-2	8501
	1 stillborn; n1 lt	706
	1 yg; n1 lt	6032
	1 yg; n1 lt	11047
	twins; n1 set: 1 m, 1 f	12018
Loxodonta africana	1	8514
	1	331
	1	11253
	1 cl	4327
	1 egg fertilized	8394
	1; 2 occasional	6292
	1; 2 rare	9972
	1; n1	6187
	1; r1-2; 0.02% twins; n6000 lt	7204

	<1% twins	10135
	1.01 emb; r1-2; 0.6% twins; n351 f	10163, 10165
	2 emb; n1 f	9310
	2 emb; n1 f	9801
SEXUAL MATURITY		
Elephas maximus	f: 1st conceive: 6-7 yr	2941
	f: 1st mate: ~8 yr	6119
	f: 1st birth: ~10 yr	6119
	f: 1st birth: 13, 14-16 yr; n2	8501
	f: 1st birth: 15-16 yr	3341
	m: 1st sire: 7 yr	2941
	m: 1st sire: 14-15 yr	3341
	f/m: min 9 yr	2597
Loxodonta africana	f: follicles large: 14.42 yr; ±1.17	6290, 6292
	f: 1st mate: 13-14 yr	6282
	f: 1st ovulate: 11 yr	11731
	f: 1st ovulate: 12-13 yr	5698
	f: 1st ovulate: 14 yr	4320
	f: 1st cl: 12-13 yr	9901
	f: 1st cl: 17.8 yr; r12-22	6292
	f: 1st conceive: 7 yr	5202
	f: 1st conceive: 12-14 yr	6349
	f: 1st preg: min 9 yr; 75% 12 yr; few data	10163
	f: 1st preg: 7-15 yr	10134
	f: preg/lact: ~15 yr	7469
	f: 1st birth: 8-12 yr	8395
	f: 1st birth: ~9 yr	1485
	f: ~11 yr	10158
	f: 13-15 yr	4319
	f: 15-20 yr	6057
	m: sperm: 7-8 yr; abundant 9 yr	5358
	m: sperm: 8-12 yr	8395
	m: sperm: 10 yr; r3-15	5356
	m: spermatogenesis: 13 yr	9901
	m: spermatogenesis: 14 yr	5698
	m: spermatogenesis: 14.43 yr; ±0.76	6290, 6292
	m: spermatogenesis: 15 yr	4320
	m: testes large: 6-8 yr	4322
	m: tubule diam increases: 15 yr	4322
	m: 1st repro: 17.2 yr; r10-17; sperm 2-3 yr earlier	6292
	m: ~10 yr	10158
	m: 15-20 yr	6057
	m: 1800 kg	6291
	f/m: 1st mate: 7 yr	1485
	f/m: 8-13 yr	9972
	f/m: 12-13 yr	8514
CYCLE LENGTH		
Elephas maximus	estrus: mode 4 d; r2-8	2936, 2941, 5233
	follicular phase: 4.2 wk; ±0.5	8607
	follicular phase: 6.8 wk; ±1.1	4702
	luteal phase: 10.5 wk; ±0.3	4701
	luteal phase: 10.6 wk; ±0.6	8607
	luteal phase: 10.7 wk; ±0.3	4702
	17.7 d; r18-26; n3 cycles, 2 f	8894
	22 d; r18-27; n6 cycles, 6 f	2936, 2941, 5233
	3.5 mo; behavior	8939
	14.7 wk; se0.5; n10 cycles, 2 f	8607
	16.1 wk; sd2.1; n5 cycles, 1 f, between progesterone peaks	6733a
	16.3 wk; n15 cycles, 6 f; behavior	4701
	16.6 wk; sd1.6; n5 cycles, 1 f, between estrogen peaks	6733a
	16.8 wk; se0.6; n14 cycles, 2 f, between progesterone peaks	10691a
	18 wk; n14 cycles, 7 f	4702

Loxodonta africana	estrus: 1-2 d	5762
	estrus: 2-10 d; n7 f	7560
	follicular phase: 5.9 wk; ±0.6	8607
	luteal phase: 10.0 wk; ±0.3	8607
	16 d; n1 f; smears	11487
	13.3 wk; ±1.3; n11 cycles, 3 f	1239
	15.9 wk; se0.6; n4 cycles, 25 f	8607
GESTATION		
Elephas maximus	17 mo 17 d-24 mo 13 d	1453
	18-22 mo	308
	18-23 mo	8501
	mode 19 mo; r18-22	8498
	20 mo 18 d	2058
	20 mo 20 d	1355
	20 mo 21 d	4582
	21 mo	6119
	21 mo 18 d; n1	102
	630-656 d; last vs first mating to birth	1715
	630.3 d; r624-637; n3	3063
	634.5 d; r634-635; n2	6681
	643.7 d; r614-668; n15	2597
	646 d	11047
	666 d; n1	6032
	644 d; r628-668; n3	2596
	~25 mo; n1	10491
Loxodonta africana	20-20.5 mo; stated; 20-24 mo; field data	11697
	21-22 mo	5762
	649-661 d; n1; last vs first mating to birth	6187
	~22 mo	4320
	22 mo	5202
LACTATION		
Elephas maximus	solid food: est 3 mo	7751
	solid food: 4-5 mo	6681
	wean: ≤18 mo	9946
Loxodonta africana	solid food: 3 mo	6349
	dependent on milk: 24 mo	6349, 6351
	nurse: 2 yr	9972
	nurse: 4-8 yr	6292
	wean: 6-18 mo	5762
	wean: 36 mo	6349
	wean: est 4.8 yr	10134
INTERLITTER INTERVAL		
Elephas maximus	min 2 yr	3658
	2-2.5 yr; n1	12018
	4 yr; r3-5 yr	6119
Loxodonta africana	2-2.5 yr	9972
	2.5-9 yr	5762
	3.3-3.7 yr	9901
	3.35 yr	4320
	3.5 yr	5698
	3.5 yr	6412
	3.5 yr; good habitat; 7.6 yr; poor habitat	7469
	3.9 yr; sd1.1; r2.2-5.3; n15	5202
	4 yr	11731
	4 yr	8514
	4 yr	4319
	~4 yr	8395
	4.5 yr	10165
	5 yr	7560
	5.6 yr; r4.8-6.8	6290, 6292
	6.7 yr	6282
	8 yr	10134
	9.1 yr; r2.6-13.5	6292

SEASONALITY

Elephas maximus	mate: yr rd; India	2058
	repro: yr rd; Sri Lanka	8501
	repro: yr rd, peak Mar-Oct, Dec, low Jan-Feb, Sept-Nov; zoo	2597
Loxodonta africana	hormones: high yr rd; Uganda	7111
	mate: Jan-Feb; Angola	11697
	conceive: Oct-July, peak Jan-Mar, 88% Nov-Apr; Zambia	4318, 4320
	conceive: Nov-Apr 70%; S Africa	10165
	conceive: Nov-Apr 88%; Zimbabwe	11731
	conceive: Nov-Apr; Zimbabwe	5698
	birth: Apr-June; Zaire	11253
	birth: est Oct-Dec; Okawango, Angola	11697
	birth: yr rd, peaks Feb, July, Nov; Bunyory, Uganda	6292
	birth: yr rd; Zimbabwe	331
	birth: yr rd; Uganda	8394
	birth: yr rd; S Africa	3145
	birth: yr rd, peak before peak rain; central, e, se Africa	5762
	repro: yr rd; S Africa	10158

Order Perissodactyla

Horses and their relatives, tapirs, and rhinos form a group colloquially referred to as odd-toed ungulates as they have one or three functional toes on each of their hindfeet. These large herbivores have precocial, singleton offspring which suckle from inguinal mammae. Males lack a baculum. Reviews of secondary sex ratios (7407a) and reproductive cyclicity (9350) are available.

Order Perissodactyla, Family Equidae, Genus *Equus*

Horses, asses, and zebra (*Equus* 175-800 kg) are social, terrestrial grazers predominantly in Africa and Asia. The 2.5-3.5 kg, epitheliochorial, diffuse placenta, which secretes gonadotrophic hormones (78a, 198a, 198b, 751, 7059b, 9296a, 11048a), implants in a bicornuate uterus (2673, 5598, 6635, 10168, 11963), and may be eaten after birth (5868, 10168; cf 5872, 6193, 11377). Estrus usually occurs within 10 d (789, 5762, 5868, 5869, 10168, 10984, 11377; cf 11336) of the birth of a precocial, furred, and open-eyed neonate (558, 4643, 5453, 5868, 6193, 11253, 11377, 11963). Females have 2 mammae (331, 5453, 5756).

Details of anatomy (1121, 10168), endocrinology (1761, 1973b, 2201a, 5749, 5804a, 6954, 9428a), embryo transfer (10550), spermatogenesis (8356), birth (5872, 11377), and milk composition (251, 5756, 8041, 9657, 10620) are available. Supplements 23, 27, 32, and 35 of the Journal of Reproduction and Fertility, published in 1975, 1979, 1982, and 1987 respectively, contain a wealth of information on virtually all aspects of reproduction in the domestic horse, thus such is not reviewed here.

NEONATAL MASS		
Equus asinus	27-31.5 kg; n3	6193
Equus burchelli	28.4 kg; n9	5169
	30-35 kg	5756
	f: 31.5 kg; r25-35; n18	11377
	m: 33.3 kg; r28-37.5; n14	11377
	33.7 kg; se1.39; n4	10166
	34.82 kg; r31.78-39.04; n13 term emb	10166
Equus caballus	79.2 kg; sd10.0; 1 wk	2694
Equus grevyi	40 kg	5756
Equus zebra	f: 35 kg; n1	5189
WEANING MASS		
Equus burchelli	205.0 kg; se10.75; n6, 1 yr	10166
LITTER SIZE		
Equus spp	1 emb; n1 f	5052
Equus asinus	1	5365
Equus burchelli	1	11253
	1	5869
	1 emb; n1 f	6612
	1 ea; n2 lt	2109
	1 ea; n111 lt	1230
	1; n151+ births	5868
	1; 2 very rare	331
Equus caballus	1 ea; 9 lt	2109
	1.0; r1-2; 0.18% twins; n567 lt	10253
Equus grevyi	1 yg	2109
	1 ea; n13 lt	12128
Equus hemionus	1	2300
	1	4643
	1	8972
	1	558
	1 emb; n1 f	182
	1 yg ea; n24 lt	5964
Equus zebra	1	5189
	1 ea; n10 lt	1230
SEXUAL MATURITY		
Equus asinus	f: 1st preg: 1-2 yr	9367
	f: 1st birth: <2 yr	8060
	f: 1st birth: 2 yr	11908
	f/m: 2.5-3 yr	6193

Equus burchelli	f: 1st estrus: 12-15 mo	5867, 5869
	f: 1st estrus: 15.5 mo	11377
	f: 1st estrus: 18 mo; min 13	5868
	f: 1st estrus: 1.83 yr; n1	5756
	f: 1st conceive: 28 mo; r22-33	5756
	f: aborted: 2 yr 19 d	11377
	f: 1st birth: 3 yr	5867
	f: 1st birth: 3.5 yr	5868, 5869
	f: 1st birth: 3 yr 7 mo 14 d	11377
	f: 2.5 yr	7204
	m: 1st mate: 1 yr 6 mo 19 d	11377
	m: 1st mate: 5-6 yr	5868, 5869
	m: epididymal sperm: mode 3.5 yr; min 2	10169
	m: testes: large 2.5-4 yr	5756
Equus caballus	f: 1st estrus: yearling	9450
	f: 1st estrus: ~2 yr	10984
	f: 1st mate: mode 3 yr; min 2 yr	3195
	f: 1st preg: 28.75 mo; r26-31.5; n2	11286
	f: 1st birth: 2 yr	9705
	f: 1st birth: mode 3 yr; min 2	9450
	f: 2 yr	11334
	m: 1st sire: 30.5 mo; r30-31; n2	11286
	m: 31-36 mo	11334
	m/f: 2 yr; n large	3616a, 3616b
Equus grevyi	f: 3-4 yr	5756
	m: 4 yr	5756
Equus hemionus	f: 1st birth: 2-3 yr	559
	f: 1st birth: 4 yr	10896
	m: 3-4 yr	559
	f/m: 3 yr	8322
Equus zebra	f: 1st estrus: 12-13 mo	5454
	f: 1st birth: 3 yr	5454
	f: 1st birth: 4-5 yr	10158
	f: 1st birth: 53.4 ml; r46-58; n5	8355
	f: 1st birth: 66.5 mo; r38-105; n29	8355
	f: 1st birth: median 67 mo	8354
	m: 14-16 mo expelled from group, but was sexually immature	5453
	m: testes large: 3.5 yr	5454
CYCLE LENGTH		
Equus burchelli	estrus: 6 d	10168
	17-24 d	11377
Equus caballus	estrus: ~1 wk	10984
	19.7 d; n15 cycles, 6 f	394
	21 d; se0.6; n5 f	8608
Equus hemionus	estrus: 3-7 d	8933
GESTATION		
Equus asinus	11 mo	152
	r377-389 d; n15	6193
Equus burchelli	11 mo 6 d; 11 mo 20 d; n2	1355
	359-379 d	11963
	360-390 d	10158
	~1 yr	11697
	371.2 d; r361-390; n28	11377
	372 d; n18	5169
	396 d; n1; last mating to birth	10168
	11 mo 20 d-11 mo 25 d	2109
Equus caballus	320 d; (11+ mo)	11286
	337 d; se8.3; n5 before Apr 1	789
	337.2-343.3 d; across breeds	1125
	min 339 d	10984
	340 d	9705
	346 d; se8.1; n33 after Apr 1	789
Equus grevyi	406 d	345

Equus hemionus	334-358 d	8933
	~11 mo	558, 559
	~11 mo	8322
	340 d; se5	10896
	~1 yr	10207
Equus kiang	7 mo; field data	8740
	10.5 mo; field data	3034
Equus zebra	362 d; n1	5453, 5454
	12 mo	4582
	~1 yr	10158
LACTATION		
Equus burchelli	solid food: 1 wk	5762
	wean: 11 mo; r9-16	10168
	wean: 15 mo; n1	11377
Equus caballus	wean: 6 mo	10253
	wean: ~1 yr	10984
Equus grevyi	wean: 32-42 wk	5756
Equus hemionus	solid food: 1 mo	558
	nurse: ~9-11 mo	558
	wean: 1-1.5 yr	10207
Equus zebra	solid food: 3 d	5452, 5453
	reduced suckling: 2 mo	5453
	reduced suckling: 6 mo	5452
	wean: 10 mo	5452, 5453
INTERLITTER INTERVAL		
Equus burchelli	381.5 d; r378-385; n2	5867, 5868, 5869
	2 yr	11253
Equus caballus	~1 yr	789
	mode 2 yr	11336
Equus hemionus	1-2 yr	8322
Equus zebra	median 25 mo; r1-6.5 yr; n100 birth, 49 f	8354
	18 mo	10158
SEASONALITY		
Equus asinus	mate: Aug-Oct; Kutch, India	152
	mate: yr rd; CA	11908
	birth: est Mar-July; AZ	9367
	birth: July-Sept; Kutch, India	152
	birth: yr rd; n15; zoo	6193
Equus burchelli	testes large: Nov-Mar, peak Jan; S Africa	10169
	birth: 61% Jan-Mar, 85.5% Oct-Mar; Tanzania	5868
	birth: Apr-June 96%; n294; Britain	10984
	birth: May, Aug; n2; captive	2109
	birth: peak July-Sept, some Feb-Apr; Zaire	11253
	birth: July-Sept, peak Aug-Sept; n Zimbabwe	331
	birth: peak Aug-Sept; e Africa	7854
	birth: Aug-Dec; Angola	11697
	birth: Sept-Feb, peak Nov-Feb; S Africa	8514
	birth: Sept-Mar, peak Nov-Feb; S Africa	3145
	birth: Oct-Apr, peak Dec-Jan; S Africa	10167, 10168
	birth: none Aug-Sept; s Zimbabwe	2275
	birth: yr rd, 85.5% Jan-Mar; n151; e Africa	5868
	birth: yr rd, peak Oct-Mar; S Africa	10158
	birth: yr rd, many Oct-Mar, 61% Jan; Tanzania	5867, 5869
	birth: yr rd; zoo	1230
	birth: yr rd; n25; zoo	5818
	birth: yr rd; Zambia	331
	repro: yr rd, peak Dec-Apr; Tsavo East, Kenya	6413
Equus caballus	birth: mid Apr-June, peak May-mid June; WY, MT	3194, 3195
	birth: yr rd, most Apr-June, peak May; captive	11334, 11335
Equus ferus?	birth: Feb-Oct; n28; zoo	5818
Equus grevyi	mate: July-Aug, Oct-Nov; n Kenya	5762
	birth: peak Aug-Sept; e Africa	7854
	birth: yr rd?; n Kenya	5762
	birth: yr rd; n13; zoo	12128

Equus hemionus	estrus: end May-mid June; Kazakhstan	8933
	estrus: June-Sept; Kazakhstan, Mongolia	558, 559
	mate: Feb-Mar; Badchys, Turkmenia	559
	mate: mid Aug; Tibet	8972
	mate: Sept; Koko Nor, Mongolia	182
	mate/birth: mid Apr-mid May; Turkmenia	10207
	preg f: June; n1; Mongolia	182
	birth: Apr-June; Kazakhstan, Mongolia	558, 559
	birth: Apr-June; USSR	4643
	birth: May-July; n229; zoo	8322
	birth: end May-early June; n25; Ukraine, captive	10896
	birth: June-Aug, n56; zoo	5818
Equus kiang	estrus: mid Aug-mid Sept; Asia	3034
	mate: Sept; according to Mongols; Mongolia	8740
	birth: May; according to Mongols; Mongolia	8740
	birth: mid June-July; n14; zoo	12128
	birth: mid July; Asia	3034
Equus zebra	testes: large Jan, small Sept; Namibia	5454
	mate: peak Feb; Namibia	5454
	birth: June-Aug; n11; zoo	5818
	birth: yr rd, peak Nov-Apr; Namibia	5454
	birth: yr rd, peak Dec; S Africa	8354
	birth: yr rd; zoo	1230
	repro: yr rd; S Africa	10158

Order Perissodactyla, Family Tapiridae, Genus *Tapirus*

Tapirs (*Tapirus bairdii*, *T. indicus*, *T. pinchaque*, *T. terrestris*) are large (100-300 kg) herbivores from tropical Central and South America, and southeast Asia. Tapirs have an epitheliochorial, villous, diffuse placenta, (7563) which is eaten after birth (5789, 8990, 9673, 11352). Estrus occurs 22-35 d (11352) after the birth of a furred (3237) neonate who can walk within hours (8990). Males are ascrotal (281a). Details of endocrinology (5550) and milk composition (8124) are available.

NEONATAL MASS		
Tapirus spp	m: 6 kg; n1	12032
	8.45 kg; n1	6098
Tapirus indicus	~3 kg	3237
	m: 3.1 kg; n1	8990
	6-7 kg	6369
	6.7 kg; n1	9625
	f: 7.3 kg; n1	8990
	f: 9.40 kg; n1	9738
	m: 10.20 kg; n1	9738
Tapirus pinchaque	4.72 kg; r4.05-5.4; n2	1099
Tapirus terrestris	4.3 kg; n1	12046
	f: 4.54 kg; n1	6748
	10 kg	281a
NEONATAL SIZE		
Tapirus indicus	72 cm; n1	9625
LITTER SIZE		
Tapirus spp	1; n1	12032
Tapirus pinchaque	1 yg	1099
Tapirus terrestris	1	12128
	1 yg ea; n1 lt	6748
	1 yg ea; n9 lt	509
SEXUAL MATURITY		
Tapirus indicus	f: 1st conceive: 36 mo	8990
	f: 1st conceive: 4-4.5 yr	9738
CYCLE LENGTH		
Tapirus indicus	29.4 d; r29-31; n1 f	8990
	30 d; n1 cycle, 1 f	11352
Tapirus terrestris	64 d; r28-101; n5 cycles, 3 f	11352

GESTATION

Tapirus indicus	9 mo; n1	3237
	13 mo	1355
	~13 mo; n1	9625
	390-395 d	6369
	401 d; r396-407; n4	8990
	403 d	9738
	405 d; n1	11352
Tapirus pinchaque	392-393 d; n1	1099
Tapirus terrestris	385-402 d	11352
	r392-405 d; n8 lt, 2 f	509
	~13 mo	1355

LACTATION

Tapirus spp	wean: 3 mo	12032
Tapirus indicus	wean: 153 d	8990
	wean: 6-8 mo	6369
Tapirus terrestris	milk present: 9 mo	6748

INTERLITTER INTERVAL

Tapirus indicus	2 yr	6369

SEASONALITY

Tapirus indicus	mate: Apr-May; Thailand	6369
Tapirus terrestris	birth: Feb, May-June, Aug-Oct; n7; zoo	12128

Order Perissodactyla, Family Rhinocerotidae

Rhinoceroses (*Ceratotherium* 1400-3600 kg; *Dicerorhinus* 900-2000 kg; *Diceros* 720-1800 kg; *Rhinoceros* 1600-4000 kg) are large, terrestrial herbivores in Africa, India and southeast Asia. The 4960 g (n1), epitheliochorial, diffuse placenta (232, 2674, 6635) implants into a bicornuate uterus (3844a, 4108) and may be eaten after birth (4024, 6185, 6255, 9571). Estrus occurs from 2 wk to 8 mo (1433, 2599, 5762, 6185, 6251, 8167, 8973) after the birth of an open-eyed, mobile neonate (3983, 4024, 5504, 5762, 8167, 8798, 10118, 10158, 10867). Females have 2 teats (9155, 10158). Males have seminal vesicles, but no scrotum (3395).

Details of anatomy (3844a), endocrinology (5549, 5552, 8893), copulation (3827), birth (6255, 6943), suckling (6703), and milk composition (407, 4025, 4057, 4117, 5551, 6185) are available.

NEONATAL MASS

Ceratotherium simum	m: est 32-36 kg; n1	6943
	~40 kg	10158
	~47.7 kg; n1, 10 d	920
	48.5 kg; n1	10118
	f: 55 kg; n1, 1 wk diseased	11785
	m: est 100 kg	2604
Dicerorhinus sumatrensis	23 kg; n1	4108
Diceros bicornis	est 22-34 kg	4521
	35 kg; r25-45	8798
	39.04 kg; n1 term emb	3830
	~40 kg	5762, 10158
Rhinoceros unicornis	33.75 kg; r33-34.5; n2	1820
	44.0 kg; r33.6-54.4; n2	155, 3656
	48.6 kg; n1, 2 d	7406
	56.8 kg; n1	6703
	m: 62.7 kg; r40-79; n9	153, 6181, 6185, 6194
	m: 63.5 kg; r59-68; n2	2674
	f: 68.6 kg; r55.7-81; n9	153, 6181, 6185, 6194
	70 kg	6190
	71.3 kg; r40-81; n20	6255

NEONATAL SIZE

Dicerorhinus sumatrensis	914 mm; n1	4108
Rhinoceros unicornis	HB: 96.5-122 cm	6255
	99.0 cm; r96.5-101.6; n2	155
	105 cm; n1	6181

LITTER SIZE		
Ceratotherium simum	1 yg	4642a
	1 yg	5762, 8798, 10158
	1 yg	8514
	1 yg; n5 lt	5973
Dicerorhinus sumatrensis	1 yg	4108
Diceros bicornis	1	5762, 8798, 10158
	mode 1	2449
	1 yg	2599
	1 yg	9155
	1; n1 lt	5504
Rhinoceros unicornis	1 yg	1820
	1; n1 lt	6133
	1 yg; n1 lt	4197
	1 yg; n1 lt	6033
	1 ea; n24 lt	994
	1 yg ea; n27 lt	6194
	2 yg; with 1 f	3657
SEXUAL MATURITY		
Ceratotherium simum	f: 1st estrus: 3 yr	4642a
	f: 1st birth: 6.5-7 yr	8167
	f: 3 yr	4107
	m: 12 yr	8167
	f/m: 1st mate: ~5 yr	8973
	f/m: 4-5 yr	10158
	f/m: 7-10 yr	8798
Diceros bicornis	f: 1st conceive: 4.5 yr; n1	3829
	f: 1st birth: 4.75-5.25 yr	9592
	f: 1st birth: 6.42 yr	4831
	m: spermatogenesis: 7-8 yr	4831
	m: 1st mate: 5.5-6 yr	9592
	m: hold territory: 9 yr	4831
	m: 1st sire: 4.3 yr; n1	3829
	m: 6-7 yr	8798
	f/m: ~5-6 yr	5762
	f/m: 5-6 yr	5762
Rhinoceros unicornis	f: 1st estrus: 4 yr	6255
	f: 1st birth: 7.1 yr; r6-8	6251
	f: 4 yr	6190
	m: 7 yr	6190
CYCLE LENGTH		
Ceratotherium simum	estrus: 30 d	8167
Diceros bicornis	estrus: 6-7 d	4831
	18 d	4251
	26-30 d	2599
	28-30 d	11998
	30-35 d	3983
	35 d; sd7.7; r26-46; n10 cycles, 4 f	4831
Rhinoceros unicornis	estrus: 24 h	6181
	24 d; r21-33; n1 f	6255
	27-42 d	6251
	36-58 d	6181, 6185
	40-50 d; n2 yr, 1 f	10868
	40-50 d; n10 cycles; 1 78 d cycle, 1 126 d cycle	10867
	43 d; sd2; r41-45; n4 cycles, 2 f	5552
	45 d; r24-57; n3	1433
GESTATION		
Ceratotherium simum	>15 mo 3 wk	10118
	484 d; n1	9571
	~16 mo	5762, 10158
	16 mo	8167
	est 17 mo; ibi minus ppe	8973
	547 d; n1	8798
	18 mo	8514

	584 d; n1	11785
	~20 mo; n2	3062
Dicerorhinus sumatrensis	8 mo; questionable	4108
Diceros bicornis	431.67 d; r419-438; n3	4024
	15-17 mo; n5	4251
	458 d; n1	3983
	458.0 d; r454-463; n3	4521
	462 d; r446-478; n2	3828
	463.5 d; r462-465; n2	11998
	469 d	2599
	16-18 mo	5776
	530-550 d	2449
	540 d	9155
Rhinoceros unicornis	463 d; r462-464; n2	994
	464 d; n1	4197
	474 d; r471-477; n2	994
	478 d; r465-489; n17	994
	478.5 d; r462-489; n27	153, 6181, 6185, 6188, 6194
	479 d; r470-490; n3	994
	479 d; r462-491; n31	6255
	484 d; n1	901
	484 d; n1	7406
	486 d; n1	6033
	16 mo	1433
	488 d	10867, 10868
	18 mo	1820
	est 18.5-19 mo	155, 3656
LACTATION		
Ceratotherium simum	solid food: 1 wk	8798
	solid food: 2-3 wk	3062
	solid food: 2 mo	8167
	wean: ~1 yr	8167
Diceros bicornis	solid food: 9 d	8798
	solid food: 30 d	5504
	solid food: few wk	10158
	nurse: ~1 yr	10158
	nurse: 2 yr	9155
	wean: 2 yr	11697
	wean: >17.5 mo; n1f	3828
	wean: 1.5-2 yr	8798
Rhinoceros unicornis	solid food: 2-3 mo	6251
	solid food: 7 wk	6703
	wean: 12-18 mo	6251
	wean: 15 mo	6703
INTERLITTER INTERVAL		
Ceratotherium simum	21-22 mo; est from 6-7 mo ppe & gest	7204
	2-3 yr	8167
	3-5 yr	4246a
Diceros bicornis	2-4 yr	5762
	27.3 mo; r25-29; n3	3828
	2.5-3.25 yr	9592
	30.6 mo; r22-55; n13; 28 mo if drop >40 mo	4831
	~3 yr	10158
	44.5 mo; r20-89; n32; 32 mo if drop >40 mo	4831
Rhinoceros unicornis	2 yr	6188
	25.8 mo	9364
	median 34 mo; max 50; n50	6251

SEASONALITY

Ceratotherium simum	mate: Feb-May; Uganda	4642a
	mate: peaks Feb, Nov, few July-Sept; Zululand, S Africa	8168
	mate: peak Oct-Dec; Zululand, S Africa	8167
	birth: peak Mar-May; Zululand, S Africa	8167
	birth: peaks Mar, July; Zululand, S Africa	8168
	birth: yr rd; S Africa	10158
Diceros bicornis	estrus: Jan-Feb; Okawango, Angola	11697
	mate: peak est Mar-Apr, Sept-Nov; Kenya	5762
	mate: Aug-Dec; Kaokoveld, Namibia	5455
	mate: Nov-Dec; Okawango, Angola	11697
	mate: yr rd; Kenya	9155
	mate: yr rd; Tanzania	3827
	birth: Jan-May; Okawango, Angola	11697
	birth: peak Jan, June, low Mar-May; n Natal, S Africa	4831
	birth: Feb, May, July, Sept-Nov; n32; zoo	5818
	birth: peak June-July; Kenya	9592
	birth: Dec-Aug; Kaokoveld, Namibia	7204
	birth: yr rd; est peaks Jan, June; n Natal, S Africa	4831
	birth: yr rd; S Africa	10158
	birth: yr rd; Tanzania	5762
Rhinoceros unicornis	estrus: Aug; n1; zoo	1433
	mate: est Feb-Apr; India	3656
	birth: Oct; n1; zoo	10867
	birth: yr rd; Chitwan valley, Nepal	6251
	birth: yr rd; n36; zoo	6194

Order Hyracoidea, Family Procaviidae

Hyraxes (*Dendrohyrax* 1.4-4.5 kg; *Heterohyrax* 475 g-4.5 kg; *Procavia* 2-4 kg) are nocturnal, terrestrial herbivores from Arabia and Africa. Females may be slightly larger than males (*Heterohyrax* 4875). The hemochorial, labyrinthine, zonary, deciduate placenta implants in a bicornuate uterus with a long corpus (231, 433, 2391, 2479, 4526, 5605, 5607, 7314, 7563, 8002, 8032, 10310, 10527, 10959, 11141, 11842, 11844, 11862). Gestation is long. The precocial neonates are open-eyed and furred (4874, 5758, 6420, 7201, 7314, 7445, 9945, 10158). Females have 2-6 mammae (331, 4078, 5277, 9200, 9945, 10158). Males do not have a baculum but do have seminal vesicles and abdominal testes (708, 3823, 3824, 5277, 7315). Details of anatomy (3879, 5605, 5607, 8002), endocrinology (4526, 7313), reproductive cyclicity (9350), suckling (4874, 9359), and light (7316) are available.

NEONATAL MASS		
Dendrohyras spp	380 g; n1	9200
Dendrohyrax dorsalis	200 g; r180-220; n3	5389
Heterohyrax brucei	220-230 g; term emb	10158
Procavia capensis	142.5 g; r140-145; n2	9200
	f: 165.0 g; se14.0	7314
	195 g; r110-310	7314
	200 g; r180-220; n3	7445
	200-280 g	10771
	202 g; r170-240	7201
	m: 231.2 g; se2.01	7314
	240 g; n6	4331
	240-405 g	5758
	m: 340 g; n2	9447
NEONATAL SIZE		
Dendrohyrax arboreus	35 mm; n1	8856
Procavia capensis	HB: 190-193 mm	9200
	204 mm; term emb	9945
WEANING MASS		
Procavia capensis	~500 g; 6 wk	4331
LITTER SIZE		
Dendrohyrax spp	mode 1 yg; 2 occasional	7445
	1.45 emb, yg; mode 1; r1-3; n11 lt	9200
Dendrohyrax arboreus	1.0 ea; n14 lt	9449
	1.46 ov; r1-2	8002
	1.75; mode 2; r1-2; n4 lt	9359
	3 emb, 1 yg; n1 lt	8856
Dendrohyrax dorsalis	1-2 yg	4354a
Heterohyrax spp	1.88 emb; mode 2; r1-3; n8 f	9200
Heterohyrax brucei	1.4 yg	4874
	1.6; r1-3; n55 lt	4875
	1.7; r1-2; n16 lt	9449
	1.8 emb; r1-2; n6 f	4526
	2 emb ea; n5 f	4926
	3; n1 lt	6203
	3 emb; n1 f	4926
Procavia capensis	1-3	10771
	1 ea; n2 lt	7658
	1.4 yg	4874
	1.5 emb; r1-2; n2 f	185
	1.5 emb; r1-2; n2 f	11844
	1.8-3.5 depending on area	7314
	1.9; r1-3; n45 lt	9449
	mode 2-3; r1-3	12128
	2; n1 lt	9447
	2 emb; n1 f	2096
	2.00 emb; r1-3; n8	185
	2.23; r1-5	1230
	2.36 emb, yg; mode 2; r1-4; n14 lt	9200
	2.39 emb; mode 2; r1-4; n114 f	11103
	2.4; n14 lt	4875
	2.4 emb; mode 2 51%; r1-5; n49 f	11141

	2.57 yg; max 5; n14 lt	4331
	2.65 emb; mode 2-3 76%; r1-5; n95 f	3435
	2.7 cl	11141
	3 yg; n1 f	7445
	3.2; mode 3-4; r1-5; n36 lt	7201
	3.3 yg; r1-5; n16	4331
	3.67 emb; r3-4; n3 f	6420
SEXUAL MATURITY		
Heterohyrax brucei	f/m: 16 mo	4875
Procavia capensis	f: 1st conceive: min 5 mo; max 16-17	7314
	f: 1st birth: 3 yr	9449
	f: 1st birth: 9 yr	7658
	f: min 4-5 mo; mode 16-17 mo	3435
	m: spermatogenesis: min 17 mo; mode 28-29 mo	3435
	f/m: ~15 mo	5758
	f/m: 16 mo	7201
CYCLE LENGTH		
Heterohyrax brucei	estrus: max 5 d	4876
Procavia capensis	13.42 d; ±1.31; r11-16; n12 cycles, 7 f	3879
GESTATION		
Dendrohyrax spp	est 8 mo	9200
Dendrohyrax arboreus	230 d	2449
	7-8 mo	5758
Dendrohyrax dorsalis	~7 mo	4354a
Heterohyrax brucei	7-8 mo	9449
	7.5 mo	4874
	230 d	2449
Procavia capensis	est 6-7 mo; from emb size	11103
	196 d	9200
	205-208 d; n3	7201
	214 d; n1	9447, 9448
	225 d	4331
	7.5 mo	4874
	7.5 mo; n1	7658
	226-237 d; n9	7201
	230 d	11141
	230 d	7314
	233-285 d; n6	7201
LACTATION		
Dendrohyrax arboreus	wean: 5.0; r3-7 mo; n4	9359
Heterohyrax brucei	wean: 3 mo	9449
Procavia capensis	solid food: 1 d	4331
	solid food: 2 d	5758
	solid food: nibble 3-4 d, eat 2 wk	7201
	solid food: few d	9449
	wean: 1-5 mo	7314
	wean: 5-6 wk	4331
	wean: mode 3 mo; r1-5	10158
	nurse: 3-5 mo	7201
INTERLITTER INTERVAL		
Procavia capensis	1 yg	10771
SEASONALITY		
Dendrohyrax arboreus	fertile m: yr rd; Ruwenzori, Uganda	8002
	mate: July, Oct, Jan; Malawi	10153
	preg f: May; n1; Lake Kivu, Zaire	8856
	birth: yr rd, peak est mid yr; Ruwenzori, Uganda	8002
	birth: yr rd, peak June-July; e Africa	5758
Dendrohyrax dorsalis	mate: June-Aug; Zaire	9098
	birth: Mar-Apr; Gabon	9098
	birth: May-Aug; Zaire	9098
	birth: Dec-Feb, Oct; Nigeria	4354a

Heterohyrax brucei	testes: large May; Kenya	5607
	testes: large July; Kenya	9449
	mate: July; Kenya	5607
	mate: Dec; Zambia	10153
	preg f: May, July; n4; Kenya	4926
	preg f: end July-early Mar; Kenya	5607
	birth: Feb-Mar; Kenya	9449
	birth: May-June, Dec-Jan; Tanzania	4875
	birth: prob yr rd; S Africa	10158
Procavia capensis	spermatogenesis: Feb-early Apr; Cape Province, S Africa	7315
	testes: large Feb-Mar, small May-Jan; Cape Province, S Africa	7313, 7314, 7315, 7316
	testes: large Feb-May, small June-Oct; S Africa	11141
	testes: large Feb-May; Cape Province, S Africa	3435
	testes: large Nov; Kenya	9449
	mate: Mar-May; Kenya	9447, 9449
	mate: Aug-Sept; Israel, Jordan	7201
	mate: Oct-Nov; S Africa	10771
	mate: peak Nov-Feb; S Africa	4331
	mate: peak mid-end Nov; zoo	1230
	ovulate: Mar-May; S Africa	11141
	preg f: Mar; n3; Lebanon	6420
	preg f: May-Dec; S Africa	11103
	preg f: Jan, Mar, Apr; n8; e Africa	185
	preg f/birth: Mar; n3; Kaokoveld, Namibia	9945
	birth: Feb; n S Africa	7314
	birth: Mar-Apr, 3 wk; Tanzania	4875
	birth: mid Mar-mid May; n40; captive	7201
	birth: June-July; Kenya	9449
	birth: June-Sept, Nov; n6; zoo	12128
	birth: est Aug-Nov; Mt Kenya	1949
	birth: Aug-Jan; Mt Kenya	5758
	birth: Sept; n1; nw Ovamboland, Namibia	9945
	birth: Sept-Oct s S Africa	7314
	birth: Oct-Dec; S Africa	11141
	birth: Nov; n1; captive	9447
	birth: Nov-Dec; S Africa	11103

Order Tubulidentata, Family Orycteropodidae, Genus *Orycteropus*

The solitary, semi-fossorial, nocturnal aardvarks (*Orycteropus afer* 50-80 kg) subsist on ants and termites in Africa south of the Sahara. Although not easily classified, the placenta (280 g, n1, 2787) is predominantly zonary and endotheliochorial (2391, 2396, 2397, 7563, 7566, 10310, 10690, 10961, 11109) and implants into a bicornuate (10961) or duplex (10690, 11109) uterus. Neonates are naked with open eyes (2787, 5219, 9947). Breeding may be initiated by rain because floods concentrate animals on high ground (5758). Males are ascrotal and have a short penis (281a). Details of anatomy (11109) and milk composition (5822, 11635) are available.

NEONATAL MASS		
Orycteropus afer	1-2 kg	2384
	1.8-2 kg	9947
	1.87 kg; n1	9454
NEONATAL SIZE		
Orycteropus afer	TL: ~550 mm	9947
WEANING MASS		
Orycteropus afer	3.5 kg, 1 mo	9947
	9 kg, 3 mo	9947
WEANING SIZE		
Orycteropus afer	TL: 680 mm, 1 mo	9947
	TL: 930 mm, 3 mo	9947
LITTER SIZE		
Orycteropus afer	1	3865
	1	5277
	1	11109
	1; 2 rare	2384, 11046, 11253
	1 emb; n1 f	10154
	1 ea; n4 lt	5219
	1.2 yg; r1-2; n5 lt, 1 f	2787
SEXUAL MATURITY		
Orycteropus afer	f: 1st birth: 3 yr; n1	3865
GESTATION		
Orycteropus afer	est 7 mo	2384, 7196, 7361
	252.5 d; r221-282; dropped 53, 110 d gests	5219
LACTATION		
Orycteropus afer	leave nest: ~2 wk	11046
	den emergence: 3rd wk	11253
	solid food: 3 wk	4354a
	solid food: 3 mo	7196
LITTERS/YEAR		
Orycteropus afer	1	11253
SEASONALITY		
Orycteropus afer	mate: Apr-May; native guides; Ruwenzori Mt, e Africa	7361
	mate: end Apr-May; Zaire	11253
	mate: May-?; Zaire	11046
	birth: May-June; Ethiopia	9947
	birth: May-July; S Africa	9947
	birth: May-Aug; s Africa	10153
	birth: Oct-Nov; Zaire	11253
	birth: Oct-Nov; native guides; Ruwenzori Mt, e Africa	7361
	birth: early Nov; Uganda	5758

Order Artiodactyla

Even-toed ungulates, so-named because they usually have two or four toes on each foot, are a diverse group of relatively large, often gregarious, and usually herbivous, mammals of great economic importance. The reproductive physiology of the domesticated species (pigs, cattle, sheep, and goats) is extensive and outside the realm of this compendium. A review of reproductive cyclicity (9350) is available.

Order Artiodactyla, Family Suidae

Pigs (*Babyrousa* ?-100 kg; *Hylochoerus* 130-275 kg; *Phacochoerus* 40-150 kg; *Potamochoerus* 40-130 kg; *Sus* 3-350 kg) are terrestrial, often diurnal, omnivores from Eurasia and Africa. Males are often larger than females (937, 3571, 4645, 5762, 6752, 8466, 8501; cf 10941). The epitheliochorial, diffuse placenta implants into a bicornuate uterus and is not eaten after birth (*Phacochoerus* 1911). Neonates are furred and open-eyed (4152, 4638, 4643, 5762, 10028, 11253). Females usually have 4 (*Phacochoerus*), 6 (*Potamochoerus*), or 8-12 (*Sus*) mammae (331, 333, 1474, 3571, 5762, 9175, 9945, 10028, 10158). Males have seminal vesicles, a spiral penis (1911), and Sertoli cells with crystalloid inclusions (*Sus* 8140a, 10883a). Details of anatomy (1911, 6930), endocrinology (1760, 3221, 9641), and milk composition (3178, 9309) are available. The reproductive biology of domestic pigs is reviewed in supplements 33 and 40 of the Journal of Reproduction and Fertility published in 1985 and 1990 respectively.

NEONATAL MASS		
Babyrousa babyrussa	380-1050 g	9607a
Phacochoerus aethiopicus	400-550 g; term emb; via graph	1793
	f: 437 g; r330-490; n5	7184
	480-850 g; varies with litter size	937
	m: 553 g; r510-600; n3	7184
	est 720 g	9212
	860 g; injured; 910 g; intact	3443
Potamochoerus porcus	700-800 g	10158
Sus scrofa	500 g	2973
	557.43 g; r325-665; n7	35
	637.2 g; r603.1-753.5; n4	2588
	681 g; r630-715; n7	27
	720 g	8466
	740 g; sd170; n580	8220
	750 g; r600-1000	4643
	870 g; sd240; n28	10941
	1156 g; sd272; n>5000	2040
	f: 1432 g; se18.0; n296	4216
	f: 1440 g; se13.9; n676	4216
	m: 1470 g; se20.1; n257	4216
	m: 1534 g; se14.6; n631	4216
NEONATAL SIZE		
Babyrousa babyrussa	15-20 cm	7434
Sus scrofa	HB: 24 cm; sd2; n28	10941
	CR: 24.08 cm; r21.9-25.5; n2 lt, n5, 110 d emb	4638
WEANING MASS		
Sus scrofa	3.86 kg; sd1.32; n28, 4 wk	10941
	6.47 kg; sd2.10; 60 d	8220
	6.5-6.8 kg	2973
WEANING SIZE		
Sus scrofa	HB: 42 cm; sd6; n28, 4 wk	10941
LITTER SIZE		
Babyrousa babyrussa	1; n1 lt	12128
	1-2	7434
	mode 1-2; r1-3	1161
Hylochoerus meinertzhageni	2 yg; n1 lt	2206
	2-6	2449
	2-8 yg	4354a
	2-11	5762
	4-6 yg	2207

Phacochoerus aethiopicus	r1-4; n23	1230
	2-3; 4 being a normal full lt	9945
	mode 2-3; max 8	5762
	2-6	8514
	2 emb; n1 f	10154
	2.6 emb; r1-4; n21 f	1126
	2.79; r1-4	9212
	2.9 emb; mode 3; r1-4; n15 f	1793
	3-4	2439
	mode 3-4; r1-8	2449
	mode 3-4; r2-7	331, 333
	3-4 emb; n5 f	2275
	3-8 yg	10772
	3.0 emb; mode 2-4 82%; r1-8; n97 f	1793
	3.1 emb; mode 3-4; n119 f	1793
	3.12; r1-5; n16 lt	4142
	3.2 yg; n20 f	2275
	3.26; r1-5; n61	6929
	3.4 emb; r2-4; n7 f	3661
	3.5; r2-5; n2 lt	7184
	3.5; mode 3-4; r2-5	937
	3.6 cl	1911
	3.6 yg; r1-9	7467
	mode 4	5053
	4; n1 lt	345
	4; r3-5	11253
	6.0 yg; r5-7; n2 lt	3443
Potamochoerus porcus	1-4	12128
	1.5; r1-2; n4 lt	5575
	2-6 yg	10772
	2; n1 lt	2109
	mode 3-4; r2-6	331, 333, 334
	mode 3-4; max 8	10158
	3-6; primiparous: 3-4; adult: 5-6	11253
	3 emb; n1 f	8906
	3; mode 3-4; r1-6; n14 lt	6680, 10028
	4; n1 lt	647
	4 yg; max 10	8490
	mode 5-6	5771
	8-12	5304
Sus barbatus	4; n1 lt	7603
	8.2 emb; r4-11; n5 f	2313
Sus salvanius	2.25; r1-4; n4 lt	6752, 6756
	3.45; mode 3; r2-6; n11 lt	8095
	3.9; r3-6; n9 lt	6756, 6757
Sus scrofa	2.82 emb; r1-8; n56 f	2588
	3.2; r1-6	1230
	3.3	11254
	3.3; r1-5; n117 lt	5349
	3.5 emb; r3-4; n4 f	11255
	4 emb; n1 f	10079
	4 yg; n1 f	5136
	4.3 emb; se0.2; mode 4-5; r1-6; n29 f	9428
	4.6; r4.4-4.8; n2 samples	458
	4.5; mode 5; r1-8	2588
	4.62 emb; n21 f	6979
	4.70 emb; se0.33; r1-10; n34 f	1931
	5; r2-7	4152
	5.0 yg; r4-6; n2 lt	35
	5.00 emb; se0.36; 6.67 cl; se0.30; n27 f	473
	5.0; r1-9; n81 lt	1287
	5.3	10519
	5.4; r4-12	8466
	5.56 yg; r5.36-5.76; n2 morphs, 1st lt	2973

	6-12 yg	5146
	6.12; sd1.78	8220
	7.35 yg; r6.76-7.94; n2 morphs, 2nd lt	2973
	7.4 emb; r5-12; n8 f	10600
	8 yg; n1 lt	8720
	r9.0-10.6; n7 samples	4216
SEXUAL MATURITY		
Babyrousa babyrussa	f/m: 1-2 yr	9607a
Hylochoerus meinertzhageni	f: 1st mate: est 1 yr	5762
	f: 1st birth: 17-18 mo; n1	2206
	m: 18-19 mo	2206
	f/m: 18 mo	2207
Phacochoerus aethiopicus	f: 1st cl: 17-18 mo	1911
	f: 1st mate: 19 mo	1793
	f: 1st mate: 20 mo	1126
	f: 1st birth: 2 yr	937
	f: 18-19 mo	10158
	m: epididymis large: 18-19 mo	1911
	m: 1st sperm: 17-18 mo	6930
	m: elongation of spermatids: 17-18 mo	6930
	m: epididymal sperm: 26 mo	1793
	m: 18-20 mo	2165
	m: 24 mo	9212
Potamochoerus porcus	f: 1st conceive: 3 yr	10028
	f/m: 1st mate: ~86 wk	10239
	f/m: min 3 yr	5575
Sus scrofa	f: 1st ovulate: 15-18 mo	8360
	f: 1st mate: ~1 yr	4643
	f: 1st mate: 21 mo	4181
	f: 1st conceive: min 5-7 mo; mode >1 yr	473
	f: 1st conceive: 6 mo	2588
	f: 1st preg: 8-10 mo	2588
	f: 1st birth: 37 wk	2973
	f: 1st birth: 12 mo	9175
	f: 4-7 mo	8220
	f: 20 wk	2973
	f: 6 mo; r5-8	5349
	f: ~10 mo	10600
	f: 18-20 mo	4643
	m: 1st mate: 2 yr	4643
	m: 12-15 wk	2973
	m: 5-7 mo	10600
	m: 7.5-12 mo	5349
	m: 4-5 yr	4643
	m: 4-6 yr	8466
	f/m: 0.75-1.75 yr	3571
	f/m: 18-20 mo	8466
CYCLE LENGTH		
Babyrousa babyrussa	30 d; n1 cycle	1161
Phacochoerus aethiopicus	estrus: 48 h	1911
	6 wk; n2 cycles, 1 f	1911
Sus scrofa	20 d	8220
	~21 d	2936
	21.8 d; 64% r21-23; n11 cycles; isolated	4639
	23.8 d; 53% r21-23; n34 cycles; crowded	4639
GESTATION		
Babyrousa babyrussa	150-157 d	8242
	157-166 d	1161
Hylochoerus meinertzhageni	125 d	2449
	125 d	5762
	151.5 d; r149-154; n2	2206, 2207
Phacochoerus aethiopicus	140-170 d; field data	1126
	5 mo	11697
	~5 mo; r125-175 d	2439

	150-170 d	1793
	167-175 d	10158
	170-175 d	5762
	171 d	6929
	171-175 d	2449
	173 d; r171-175; n2 lt	1355
Potamochoerus porcus	~10 wk; field data, not many	5573
Sus salvanius	100-121 d	8095
	110-120 d	6756, 6757
Sus scrofa	15.5 wk	9175
	110 d	2588
	110-115 d	8220
	~112 d	4795
	113.38 d; r105-120; n2023 lt	2098
	16-17 wk	1474
	114 d	2040
	115 d	2973
	115.2 d; ±2.3; mode 115; 80% r114-118; r108-120; n41 lt	4639
	115.6 d; sd1.9	6885
	4 mo	5146
	~4 mo	8498, 8501
	~4 mo	8223
	~4 mo	2936
	~4 mo	4645
	~4 mo; field data	8466
	18 wk	4181
	18 wk; r16-20	3571
	r133-140 d; adults; r114-130 d; primiparous	4643
LACTATION		
Babyrousa babyrussa	wean: 6-8 mo	8242
Hylochoerus meinertzhageni	wean: 8-10 wk	2206
Phacochoerus aethiopicus	solid food: grass almost at once	5762
	solid food: ~1 wk	10158
	solid food: 17 d; n1 lt	2165
	solid food: <2 mo	1793
	den emergence: 2 wk	11253
	den emergence: 2 wk	5762
	wean: 9 wk	10158
	nurse: ~5 mo	10158
	wean: 3-4 mo	1793
Sus scrofa	nest emergence: 1 wk	4643
	solid food: 2-3 wk	4643
	wean: 2.5-3.5 mo	4643
	wean: ~4 mo	2588
INTERLITTER INTERVAL		
Hylochoerus meinertzhageni	12 mo	2207
Phacochoerus aethiopicus	10 mo	2439
	1 yr	9945
	1 yr	11253
	1.5 yr	8490
Potamochoerus porcus	1 yr	11253
Sus scrofa	230 d	2588
	230 d	2588
LITTERS/YEAR		
Phacochoerus aethiopicus	1-2	937
Sus scrofa	1; r1-2	5349
	1; r1-2	3571
	2	10348
	2	8501
SEASONALITY		
Hylochoerus meinertzhageni	birth: peak Mar, Sept; Zaire	2206
	birth: yr rd, peak July-Aug; Africa	5762
Phacochoerus aethiopicus	testes: large Mar-May, small Aug-Dec; Zululand, S Africa	6930
	estrus: May-June?; Angola	11697

	mate: May; Zimbabwe	1793
	mate: July-Dec, peak July-Oct; zoo	1230
	mate: fall; Zululand, S Africa	6929
	birth: May-July, Oct-Jan; n7; W Africa	937
	birth: peak May-July, Nov-Dec; Nigeria	4354a
	birth: June-Oct, peak July-Aug; nw plateau, Zambia	331, 333
	birth: Aug-Jan, peak Oct-Nov; Tanzania	9212
	birth: Sept-Nov; Tanzania	3661
	birth: Sept-Nov, peak Oct; Tanzania	1126
	birth: mid Sept-mid Nov, peak Oct; Rwanda	7467
	birth: Sept-Dec, peak Oct-early Nov; Zimbabwe	1793
	birth: Sept-Dec, peak Oct-Nov; Luangwa Valley, Zambia	331, 333
	birth: Oct-Nov; Namibia	9945
	birth: Nov-Dec, May-June; Zaire	11253
	birth: peak Nov-Dec, some Jan-Apr; S Africa	8514
	birth: peak Nov-Dec; S Africa	3145
	birth: peak Nov; e Africa	5053
	birth: Dec; Zimbabwe	2275
	birth: early summer; Zululand, S Africa	6929
	birth: yr rd; n Africa	2439
	birth: yr rd; Zimbabwe, Uganda	5762
	repro: Nov-Jan; S Africa	10772
Potamochoerus porcus	mate: June-Aug; S Africa	10772
	mate: dry period; Zaire	11253
	preg f: Apr-mid June; Madagascar	5573
	preg f: Aug; n1; Madagascar	8906
	preg f: Sept-Nov; n4; nw Madagascar	5575
	birth: Jan, June-Nov; n6; zoo	12128
	birth: May; n1; zoo	2109
	birth: July; s Cameroun	647
	birth: Oct-Jan; Tanzania, s central Africa	5762
	birth: peak Oct-Mar; Zambia	331, 333
	birth: Nov-Jan; S Africa	10772
	birth: end Nov-Feb; captive	10239
	birth: Dec-Jan; se Africa	5771
	birth: peak rainy season, none June, 1 lt July; n S Africa	10028
	birth: 1st half of rainy period; Zaire	11253
Sus salvanius	mate: Feb; nw Assam, India	8095
	mate: Dec-Feb; nw Assam, India	8095
	birth: Apr-May; n9; Assam, India	6756
	birth: May-June; n4; zoo	6756
	birth: Apr-May; captive	8095
	birth: Apr-May; Assam, India	6757
	birth: peak Apr-June, 1/16 Aug, 1/16 Sept; nw Assam, India	8095
Sus scrofa	sperm: yr rd; Smokey Mt, TE	5349
	anestrus: July-Oct; Chizé, France	6980
	estrus: Dec-Jan; Pyrenees, Spain	11255
	estrus: Dec-Feb; Chizé, France	6980
	mate: peak end Oct/Nov-Jan; Germany	3571
	mate: Oct-May, peak Nov-Dec; E Germany	1287
	mate: Nov-Jan; Europe, w USSR	4643, 4645
	mate: Nov-Feb; Chizé, France	6980
	mate: Nov-Dec; central Asia	8466
	birth: Jan-Aug, peak Mar-May; E Germany	1287
	birth: Jan-Sept; sw France	8360
	birth: peak mid Feb-end Apr; central Spain	9428
	birth: peak end Feb-May; Germany	3571
	birth: Mar, end July-Sept; Chizé, France	6979
	birth: Mar; central Asia	8466
	birth: Mar-Apr, July-Sept; zoo	10348
	birth: Mar-May; Europe, USSR	4643
	birth: mid Mar-mid Apr; Poland	4181
	birth: peaks Apr, Aug; Europe	1892
	birth: Apr-May, Oct; n6; zoo	44

	birth: Apr-May; Chizé, France	6980
	birth: Apr-May; Europe, USSR	4645
	birth: peak Apr-July; Santa Catalina I, CA	473
	birth: peak July-Oct; Pakistan	9175
	birth: peak Oct-Feb 11/15, some July-Aug; s Pakistan	10079
	birth: Feb-Mar; Sri Lanka	9487
	birth: spring-summer, peak Sept; zoo	1230
	birth: winter-spring; Morocco	8223
	birth: yr rd, peak mid winter; SC	10600
	birth: yr rd; Galapagos, Ecuador	1931
	birth: yr rd; TN, NC	5349
	birth: yr rd; WV	5129

Order Artiodactyla, Family Tayassuidae

Peccaries (*Catagonus* 30-40 kg; *Tayassu* 14-30 kg) are terrestrial, social omnivores that range from the southern United States through South America. A copulatory plug is reported (10238). The epitheliochorial, villous, diffuse or zonary placenta has central implantation into a bicornuate uterus (231, 6613, 7563, 11846) and is eaten after birth (8184). A postpartum estrus occurs 5-12 d (6613, 10238) after the birth of a furred, open-eyed neonate who can walk immediately (1474, 6989, 7624, 8798, 10136). Females have 2 (1474) or 8 (6613) teats. Details of anatomy (6613, 11846), endocrinology (4597, 7050, 10240), and milk composition (1376, 6529, 6532, 9309, 10241, 10524) are available.

NEONATAL MASS		
Catagonus wagneri	500 g; n2, ~1 wk	1516a
	678-763 g; n3, 1 wk	6989, 6990
Tayassu tajacu	454 g; r320-680; n6	7787
	561 g; n7; term emb	10136
	f: 608 g; n18 m & f	10136
	618 g; ±134; n7	8183a
	m: 639 g; n18 m & f	10136
	693 g; se18; n38, 21 lt	6528
	711 g; se24; n8	6530, 6531
	715 g; r681-766; n7	6613
	726 g; ±92; n9	8183a
	~900 g	3167
WEANING MASS		
Tayassu tajacu	3862 g; se163; n4, 6 wk	6530
	5400-6400 g	3167
LITTER SIZE		
Catagonus wagneri	mode 2	764
	2.46 yg; r1-4; n10 f	6989
	2.72 cl	6989
	2.73 emb; mode 2; r2-4; n11 f	6989
Tayassu pecari	1.9 yg; r1-2; n10 lt	6991
	mode 2; 4 rare; r1-4	8798
	2 emb; n1 f	7624
	2 ea; n3 lt	9268
	2.3 emb; r2-3; n3 f	5111
Tayassu tajacu	1.29 emb; mode 1; r1-4; n7 f	5298
	1.3 yg seen/f; mode 2; n82	957
	1.4 yg; mode 1; r1-3; n43 lt	5889
	1.5 yg; r1-3; n18 lt	9066
	1.5 yg; r1-3; n26 lt	7787
	1.53 yg; r1-2; n17 lt	3167
	1.6 yg; mode 2; r1-2; n11 lt	12128
	1.8 yg; se0.1; mode 2; r1-3; n32 lt	4598, 6528, 8184
	1.81 emb; r1-4	6613
	2 emb; n1 f	3024
	2 emb; n1 f	7040
	2 emb; n1 f	2341
	mode 2; r1-3; n71 lt	1230

	2.0; se0.3; n15 lt	4597
	2.13 emb; mode 2; r1-4; n23 f	10136
	2.17; r1-4; n29 lt	10238
	5 emb; n1 f	4248
	5.5 emb; r5-6; n2 f	5889
SEXUAL MATURITY		
Catagonus wagneri	f: 3 yr	7983
Tayassu pecari	f: 18 mo	9268
	f/m: ~2 yr	8798
Tayassu tajacu	f: 1st mate: 10.2 mo	6613
	f: 1st conceive: 11.3 mo	6613
	f: 9 mo	8798
	m: 1 yr	8798
	f/m: 1st mate: 33-67 wk	10238
CYCLE LENGTH		
Tayassu tajacu	estrus: 2 d	2656
	estrus: 2.60 d; r1-15; mate: 2.71 d; r1-6	6528
	estrus: 3.6 d	10238
	estrus: 5.8 d; r3-9; n4	6613
	23.46 d; r17-30; n64 cycles, 7 f	10238
	26 d; r8-42; n5 cycles, 1 f	6613
GESTATION		
Tayassu pecari	158.3 d; r156-162; n3	9268
	~158 d	8798
Tayassu tajacu	144.17 d; r142-148; n6	10237
	144.3 d; se0.6; r140-146; n13	6613
	144.88 d; r142-149; n9	10238
	145.5 d; se~0.5; n15	4597
	145.6 d; r144-149; n32	6528
LACTATION		
Tayassu pecari	solid food: end 1st wk	8798
Tayassu tajacu	wean: 6 wk	6530
	wean: 6-8 wk	10238
	wean: 6-8 wk	3167
	wean: ≤8 wk	5889
LITTERS/YEAR		
Tayassu tajacu	2	6613
SEASONALITY		
Catagonus wagneri	preg f: July-Sept; n11; Paraguay	6989
	birth: Sept-early Dec; Paraguay	6990
Tayassu pecari	birth: warm season; subtropics/temperate, New World	8798
	birth: yr rd; tropics, New World	8798
Tayassu tajacu	estrus: yr rd; zoo	2656
	mate: Apr-July; TX	6613
	mate: yr rd; zoo	1230
	mate: yr rd; sw US, n Mexico	1474
	mate: yr rd; TX	957
	conceive: Dec-Mar; TX	6613
	conceive: yr rd, except Feb; few data; TX	5298
	preg f: Aug; n1; Mexico	2341
	birth: Jan, May-Aug, Oct-Dec; n13; zoo	12128
	birth: Mar-Aug, Oct-Dec; n17; captive	3167
	birth: peak May-Oct; TX	957
	birth: peak July-Sept; zoo	2656
	birth: yr rd, peak May-July; TX	6613
	birth: yr rd, 74% June-Aug; AZ	7787
	birth: yr rd, peak June-Aug; n146; AZ	10238
	young found: yr rd; AZ	5889

Order Artiodactyla, Family Hippopotamidae

Hippopotamuses (*Choeropsis liberiensis* 160-275 kg; *Hippopotamus amphibius* 655-2100 kg) are semiaquatic, nocturnal herbivores in Africa south of the Sahara. The 9 kg (n1), epitheliochorial, villous, diffuse placenta has a central implantation into the horn of the bicornuate uterus that is ipsilateral to the side of ovulation (220, 231, 237, 4323, 6288, 7563, 10173; cf 238) and is not eaten after birth (220, 8798). A postpartum estrus occurs 4 (*Choeropsis* 9307) or 50 d (*Hippopotamus* 6288) after the birth of a mobile neonate (1353, 2109, 8798, 10158, 11253). Females have 2 inguinal mammae (331, 333, 9945, 10158). Males have seminal vesicles and a sigmoid penis (6288). Details of anatomy (6288, 9520), endocrinology (10035), and milk composition (4023) are available.

NEONATAL MASS		
Choeropsis liberiensis	4.5-6.5 kg	8798
	5-7 kg	8525
	5.73 kg; r3.6-7.0	6405
	6.56 kg; r5.0-7.0; n5	10365
Hippopotamus amphibius	24.4 kg; n1 stillborn	2109
	~25 kg; n1	220
	25.4-55 kg	5762
	30 kg	4023
	30-40 kg	11153
	30-50 kg	8798
	36.5-40 kg	5689
	44 kg; n1; heaviest fetus	6862
	50 kg	6288
NEONATAL SIZE		
Hippopotamus amphibius	748 mm	238
	750 mm; n1	220
	800 mm	2439
	m: 940 mm; n1	2595
	1270 mm	6288
LITTER SIZE		
Choeropsis liberiensis	1; n1 lt	6196
	1; n1 lt	2109
Hippopotamus amphibius	1	9064
	1; 2 very rare	5762
	1; 2 very rare	331, 333
	1 yg ea; n4 lt, 1 f	1353
	1 ea; n8 lt	1230
	1; r1-2; 0.7% twins; n274 lt	6288
SEXUAL MATURITY		
Choeropsis liberiensis	f: 3-4 yr	8798
	m: 4-5 yr	8798
Hippopotamus amphibius	f: 1st cl: 50% 9.5 yr; r3-17	6288
	f: 1st mate: 4 yr	11153
	f: 1st conceive: 3.5 yr; n1 pr	11332
	f: 1st preg: min 5 yr; mode 9-10	10173
	f: 1st birth: ~4 yr	10158
	f: ~5-6 yr; n1 pr	2109
	f: 9 yr; r7-15	5762
	f: mode 11-13 yr; r7-20	6862, 9520
	m: sperm: few 2-5 yr; many 6 yr	10173
	m: 1st sperm: 7-8 yr	10035
	m: testes large: 7-8 yr; r6-13	6288
	m: 1st mate: 3 yr	11153
	m: 1st sire: 3.5 yr; n1 pr	11332
	m: histology: 6-8 yr	6862
	m: r4-11 yr	5762
	m: ~6 yr; n1 pr	2109
	m: min 6 yr; mode by 8 yr	9520
	f/m: 3 yr; n1 pair	318
	f/m: mode 3-4 yr; r2.5-8	2609

CYCLE LENGTH		
Choeropsis liberiensis	estrus: ~3 d	8798
	estrus: 3 d	10365
	28-30 d	10500
	1 mo	10365
Hippopotamus amphibius	estrus: 3-4 d	8798
	estrus: 4-10 d	5689
	3-4 wk	5689
	~35 d; n1 pr	2109
GESTATION		
Choeropsis liberiensis	188 d; r187-204	6405
	r192-196 d; n4	10500
	199 d; r190-210	8798
	200 d; n1	2109
	205.3 d; r201-210; n3	10365
	7 mo; n2	9307
	~9 mo	1355
Hippopotamus amphibius	201-212 d	1230
	>7 mo	11332
	227-240 d	2449
	233.2 d; mode 235; r225-238; n15	5689
	236.75 d; r227-242; n4	11290
	~237 d	2439
	238.25 d; r234-243; n4 lt, 1 f	1353, 1355
	240 d; r225-257	8798
	~240 d	6288
	est 8 mo	11697
	est ~8 mo	10173
LACTATION		
Choeropsis liberiensis	solid food: 15 d	2109
	wean: 3 mo	10500
	wean: 6 mo	6196
	wean: 9-12 mo	8798
Hippopotamus amphibius	solid food: 1 mo, graze 5 mo	5762
	wean: 9 mo; n1	1353
	wean: min 10 mo; n1	6862
	wean: 10-12 mo	6288
	wean: ~1 yr	5762
	wean: 14 mo	6288
INTERLITTER INTERVAL		
Hippopotamus amphibius	17 mo; r13-26; n1 pr	5790
	21.8 mo	10173
	2 yr	6288
SEASONALITY		
Choeropsis liberiensis	mate: Aug; n1; zoo	2109
	birth: Jan, Mar, Aug-Sept, Nov; n41; zoo	5818
	birth: Mar, May-June; n3; zoo	2109
Hippopotamus amphibius	m fertile: yr rd; w Uganda	6288
	mate: peak Feb, Aug; w Uganda	6288
	mate: June, Sept-Oct; Zambia	333
	mate: July-Aug; Zaire	11253
	mate: yr rd; Namibia	9945
	mate: yr rd, peak May; n23; zoo, Netherlands	8339
	conceive: yr rd, peaks Feb, Apr, low Oct-Nov; w Uganda	6288
	birth: Feb-Mar; Zaire	11253
	birth: peak Mar-June; ne S Africa	3145
	birth: Apr-Dec; Zambia	333
	birth: Apr, Oct; w Uganda	6288
	birth: Oct-Mar 70%; S Africa	10173
	birth: yr rd, peak Feb; S Africa	10173
	birth: yr rd; Zambia	331

Order Artiodactyla, Family Camelidae

Camels (*Camelus* 300-1019 kg; *Lama* 40-155 kg; *Vicugna* 35-65 kg) are terrestrial herbivores from Asia, Africa, and South America. Ovulation is triggered by mating (1772, 1774, 1775, 2991, 3035, 3036, 6811, 6812, 7667, 7783, 7785, 8217, 9828) and sperm storage occurs (10366). The 21 kg (*Camelus*), epitheliochorial, villous, crescent-shaped placenta usually implants into the left horn of a bicornuate or bipartite uterus (1772, 2991, 3230, 3233, 5173, 7556, 7563, 7668, 7785, 9824, 9828, 9846, 10546, 10692) and is not usually eaten after birth (1420, 5915, 5994). Right horn implantations survive less than 50 d (389, 390). Estrus occurs 4-6 wk (*Camelus* 13, 11987) after the birth of a mobile, furred, and open-eyed neonate (2971, 3636, 5762, 5915, 5917). Males have a sigmoid penis and ampullary glands (151, 2432, 2456, 2456a, 2457, 2992, 5174, 7408, 10546).

Details of anatomy (389, 842, 2991, 3518, 3754, 9829, 10692), endocrinology (*Camelus* 11, 12, 77, 78, 713, 2146, 2968, 2969, 3882, 4947, 6811, 6812, 7982, 8268a, 9290, 9828, 11978, 11986, 11987; *Lama* 3036, 3234, 3235, 6383), spermatogenesis (10, 2457a, 8140, 8141), copulation (1751, 5709, 9998), birth (2971), and milk composition (23, 476, 2027, 2767, 3164, 3826, 4115, 6224, 8035, 10620, 11988) are available.

NEONATAL MASS		
Camelus bactrianus	35 kg; n1	3004
Camelus dromedarius	f: 24.5 kg; se2.5; n2	2971
	25.82 kg; sd2.14; n33	1450
	26-51 kg	896
	m: 31.3 kg; se1.69; n6	2971
	f: 32.5 kg; n3	2431
	f: 37 kg	896
	37.3 kg; r26.4-52.3	5926
	m: 38 kg	896
	f: 38.25 kg; se1.0	8870
	m: 41.84 kg; se1.2	8870
	42; ±5.6; n63	9846
Lama glama	8-16 kg	3475
Lama guanicoe	8-15 kg	3475
Lama pacos	6-7 kg	3475
	7.92 kg; r6.5-9.5; n4	9455, 10547
Vicugna vicugna	4-6 kg	3475
	6 kg; r5-6.8; n3	1420
	f: 6.2 kg; r5.0-7.9; n6	9607
	m: 6.5 kg; r5.9-7.8; n4	9607
LITTER SIZE		
Camelus bactrianus	1	8798
	1; 2 rare	8739
	2 mummified emb; n1 f	6682
Camelus dromedarius	1	896
	1; 2 rare	5762
	1; twins rare, 2 ovulations not rare	10692
	mode 1 cl; 12.45% 2-3 cl; r1-3	9831
	mode 1 cl; 15% 2-3 cl; n491 f	7668
	mode 1 emb; 0.26% 2-3; r1-3; n942 f	7784, 9828, 9831
	1.004 emb; r1-2; n497 f	7668
	14% twin conceptions	389
Lama glama	mode 1	2109
	mode 1	12128
	1 yg ea; n19 lt	9066
	1 ea; n40 lt	1230
Lama pacos	1.04 emb; r1-2; n22 f	3233
Vicugna vicugna	1; r1	3474
SEXUAL MATURITY		
Camelus bactrianus	f: 1st estrus: 4 yr	1772
	m: 3 yr	1772
	f/m: ~3 yr	8798
Camelus dromedarius	f: 1st ovulate: 3 yr	6942
	f: 1st mate: 4-5 yr	5926
	f: 1st birth: 3-4 yr	13

	f: 1st birth: 4 yr	389
	f: 1st birth: 54.3 mo; se6.3; r45.6-71.3; n37	11779a
	f: 1st repro: ~5 yr	5762
	f: 8-12 mo	13
	m: 1st rut: 3 yr	5926
	m: mature testes: 9-10 yr	5926
	m: 6 yr	6942
Lama guanicoe	m: testes anatomy: ~2 yr	2456
Lama pacos	f: ~1 yr	10546
Vicugna vicugna	f: 1st birth: 2-4.25 yr	9607
	f: 2 yr	3474
	m: 1st mate: 2.33-2.5 yr	9607
CYCLE LENGTH		
Camelus bactrianus	estrus:: 2-4 d	8798
	follicular maturation: 6.8 d; with ovulation	855
	follicular maturation: 12.4 d; no ovulation	855
	5-8 to 31-40 d	842
	19.0 d; ±4.25; follicular development to regression	1773
Camelus dromedarius	estrus: 3-4 d	6942
	estrus: 4.64 d; ±2.29; r1-15	7785
	estrus: 5.00 d; se0.26; r3-6; n11	5446
	14 d	6942
	17.2 d; n4	2969
	23.4 d; r22-24; se0.22; n9 cycles, 6 f	5446
	28 d; n35 cycles, 5 f	7667
GESTATION		
Camelus bactrianus	384 d; r370-395	2843
	385-406 d	8798
	13 mo	8739
	398 d; r375-412; bactrianus/dromedarius cross	2843
	402.22 d; se11.5; n23	1772
Camelus dromedarius	r345-360 d; n24	11987
	12-12.5 mo	1355
	12-13 mo	5762
	370-375 d	390
	370-390 d	6942
	377 d; r360-411	5926
	377.1 d; se12.08; r360-411; n142	11779a
	386.52 d; se1.75; n60	584
	389.87 d; se2.1; n33	7173
	391.1 d; ±16.7; n296	9847
	~13 mo	14
	398 d; r375-412; bactrianus/dromedarius cross	2843
	404.32 d; se4.82; n56	8870
	412 d; r380-439	2843
Lama glama	245-260 d	1046
	11 mo	1355
	330 d; se9.5	3035
	348-368 d	3475
	~13 mo	5994
Lama guanicoe	11 mo	1355
	345-360 d	3475
Lama pacos	342-345 d	3475
	345 d	10547
Vicugna vicugna	330-350 d	3474, 3475
	346-356 d	9607

LACTATION

Camelus spp	wean: 6-10 mo	6342
Camelus bactrianus	solid food: ~14 d	8798
	wean: 10-11 mo	8798
Camelus dromedarius	solid food: <2 mo	5762
	wean: 47-67 wk	476
	wean: max 20 mo	5762
Lama glama	wean: 4-6 mo	6342
Vicugna vicugna	solid food: 1 wk	5915
	nurse: max 1 yr	5915

INTERLITTER INTERVAL

Camelus bactrianus	2 yr	8798
Camelus dromedarius	380 d; r365-395	11987
	15 mo without lactation	2431
	20.2 mo; se5.3; r8-46; n460	11779a
	mode 2 yr	6942
	2 yr	896
	2 yr	3636
Vicugna vicugna	18 mo; n6+ lt ea, 8 f	9607

SEASONALITY

Camelus bactrianus	mate: Feb; captive	8739
	mate: Mar; n17; zoo, Netherlands	8339
	mate: May-June; USSR	8798
	mate: Dec-mid Apr; captive	1772
	birth: Feb-Jan, Oct, n89; zoo	5818
	birth: Feb-May, occasionally June, July; USSR	8798
	birth: Mar-Apr; n6; zoo	12128
Camelus dromedarius	sperm: May-Oct; captive	1752
	spermatogenesis: peak Feb-Apr, low Aug-Oct; captive	11323
	spermatogenesis: peak Mar, decrease June, increase Sept; captive	10
	spermatogenesis: yr rd, peak Nov-Jan, low June; captive	10848
	spermatogenesis: yr rd; captive	3518
	testes: large Dec-Mar, small Apr-Nov; captive	9996, 9997
	testes: large Dec-Apr, small June-Aug; captive	1751
	rut: Mar-Apr; Egypt	5926
	rut: June-Sept; Australia	5926
	rut: Nov-Feb; India	5926
	rut: mid Dec-May; Morocco	5926
	ovarian activity: peak Mar-June; captive	9828
	follicles large: yr rd; captive	7783
	estrus: Jan-Feb; captive	5446
	estrus: Feb-?; captive	7667
	mate: Jan-Apr; Sinai desert	11986
	mate: mid Jan-May; captive	13
	mate: peak Feb-Mar; n Africa	5762
	mate: Apr-June, Sept-Nov; Somalia, n Kenya	5762
	mate: May-Feb; n10; captive	9
	mate: mid Oct-mid Mar; captive	5709
	mate: Nov-Mar; captive	6942
	mate: Dec-Mar; captive	713
	mate: Dec-Mar; captive	3518
	mate: Dec-Apr; captive	3636
	ovulate: peak Dec-May; captive	9831
	birth: Jan-Mar, rare Apr; captive	3636
	birth: Feb-mid May; captive	14
	birth: mid Dec-May; captive	896
	birth: yr rd, peak Dec-Feb; captive	389
	birth: yr rd; captive, Kenya	11779a
	birth: not seasonal; n9; zoo	1230
Lama glama	mate: peak Nov-Mar, low ?-Aug; captive	3035, 3036
	mate: sporadic; n40; zoo	1230
	ovulations after mating: rare June-Aug; captive	3035
	birth: Nov-Feb; captive	3035

	birth: yr rd; n38; zoo	12128
	birth: yr rd; n76; zoo	5818
Lama guanicoe	mate: Dec-Mar; s Chile	11779
	birth: yr rd; n141; zoo	5818
Vicugna vicugna	mate: Apr-June; Puna Zone, Peru	5917
	birth: Feb-Apr; Peru	3474
	birth: Feb-Apr; Peru	5915
	birth: Mar-May; Puna Zone, Peru	5917
	birth: Apr-early Dec; zoo	1420
	birth: July-Oct; zoos, n hemisphere	9607

Order Artiodactyla, Family Tragulidae

Chevrotains, *Hyemoschus* (10-15 kg) from central and west Africa and *Tragulus* (0.7-8 kg) from southeast Asia, are terrestrial, nocturnal herbivores of equatorial forests and savannas. Females are usually larger than males (5762, 8501). Copulatory bouts have multiple intromissions (1522, 2756). The villous, epitheliochorial placenta is transitional between cotyledonary and diffuse, and has a superficial implantation into a bicornuate uterus (3616, 7563, 9746, 11351). Placentophagia occurs (1522). A postpartum estrus (1522, 2109, 2332, 8868, 11614a) follows the birth of a furred and open-eyed neonate that can walk almost immediately (2332, 7161). Females have 4-6 mammae (5762, 7161, 8798). Details of birth (1522) and milk composition (9242) are available.

NEONATAL MASS		
Tragulus meminna	319 g; r288-382; n4	27
Tragulus napu	371 g; n1	2109
	375 g; r371-379; n2, 2 d	2332
NEONATAL SIZE		
Hyemoschus aquaticus	210 mm; n1	5762
LITTER SIZE		
Hyemoschus aquaticus	1	5762
	1-2	8798
Tragulus javanicus	1 yg	556, 557
	1 emb ea; n6 f	2313
	mode 1; r1-2	7161
	mode 1 yg; r1-2; n23 lt	613b
Tragulus meminna	1	8501
	1-2 yg	7382
	mode 2; r1-2	6653, 6654
Tragulus napu	1 yg	2332
	1 emb ea; n3 f	2313
	1 ea; n4 lt	2109
	1 yg; r1-2; n176 lt	11614a
SEXUAL MATURITY		
Hyemoschus aquaticus	f/m: 9 mo	5762
	f/m: 9-26 mo	2757
Tragulus napu	f: 1st conceive: 4.5 mo	2332
	f: 1st birth: 4.5 mo + 152 d	2332
	f: 1st birth: 9.5 mo	11614a
	f: 4.5 mo	11614a
	m: 1st mate: 135 d	11614a
CYCLE LENGTH		
Tragulus napu	14 d	8868
GESTATION		
Hyemoschus aquaticus	4-5? mo	8798
	5-6 mo	5762
	6-9 mo	2755, 2757
Tragulus javanicus	~5 mo; field data	2251
Tragulus meminna	150-160 d	7382
Tragulus napu	153.5 d; r152-155; n6; excludes 172 d outlier	2332

LACTATION		
Hyemoschus aquaticus	wean: 2-3 mo	2755
	wean: 8 mo	2756
	wean: max 9 mo	5762
	wean: 3-6 mo	2757
Tragulus javanicus	nurse: short time	7161
	wean: 21 d; n1; survived mother's death	9242
Tragulus napu	solid food: 2 wk	2332
	nurse: 3 mo	2332
INTERLITTER INTERVAL		
Tragulus napu	5 mo	2109
SEASONALITY		
Hyemoschus aquaticus	birth: Mar; n1; Lake Kivu, Zaire	8854
	birth: Apr, July; n2; captive, e Zaire	5762
	repro: yr rd; ne Gabon	2755
Tragulus javanicus	mate: Nov-Dec; Vietnam	2251
	preg f: end Apr-early Aug; n6; n Borneo, Indonesia	2313
	birth: end Apr-May; Vietnam	2251
	birth: Dec-Jan; Borneo, Indonesia	556
Tragulus meminna	mate: June-July; s India, Sri Lanka	6653, 6654
	mate: June-July; Nepal	7382
	preg f: latter part of yr; Sri Lanka	8501
	birth: end rainy season; India, Sri Lanka	6653, 6654
	birth: yr rd; Sri Lanka	8501
Tragulus napu	preg f: June-early Aug; n3; n Borneo, Indonesia	2313
	birth: Jan, Apr, June, Nov; n1 pr; zoo	2109

Order Artiodactyla, Family Cervidae

Deer (*Alces* 200-825 kg; *Blastocerus* 100-150 kg; *Capreolus* 15-50 kg; *Cervus* 18-475 kg; *Elaphodus* 17-50 kg; *Elaphurus* 150-200 kg; *Hippocamelus* 45-65 kg; *Hydropotes* 8-15 kg; *Mazama* 8-35 kg; *Moschus* 7-17 kg; *Muntiacus* 14-34 kg; *Odocoileus* 14-215 kg; *Ozotoceros* 30-41 kg; *Pudu* 6-11 kg; *Rangifer*; 44-318 kg) are terrestrial grazers and browsers from the New World, Eurasia, and Africa. Males are usually larger than females and have antlers (52, 806, 1474, 1523, 1924, 2109, 2288, 2618, 3571, 4390, 4643, 4645, 5331, 5637, 6387, 8235, 8501, 8711, 9019, 9175, 10730, 11084; cf 412, 6387). *Rangifer* females have antlers.

Implantation is preferentially in the right horn in *Muntiacus* (1700a, 1712, 1713) and is delayed in *Capreolus* (96, 97, 98, 99, 101, 2259a, 5616, 6459, 9061, 9939). The epitheliochorial, syndesmochorial, or endotheliochorial, villous, cotyledonary or diffuse placenta implants in a bicornuate uterus (297, 371, 1700a, 1708, 1712, 1713, 3127, 4262, 4263, 4414, 4425, 4426, 6116, 6817, 7055, 7563, 9158, 9707, 9713, 10005, 10477, 10510, 10789, 10964, 10965, 11057, 11157, 11630, 11860) and is eaten after birth (1708, 1924, 4505, 5897, 7046, 7174, 7320, 7323, 9545, 9809, 10881, 11157, 11500). Estrus may occur postpartum (2204, 3447a, 4836, 9175; cf 1924, 10730) or be synchronous (1924, 5124, 9978). Ovulation may occur without estrus (480, 804, 7058, 10776, 11267).

Neonates are precocial with fur and open eyes (322, 360, 361, 806, 1474, 1708, 1713, 1924, 3092, 3477, 3571, 4426, 4505, 4643, 4645, 4717, 4719, 5066, 5232, 5338, 6491, 7174, 7263, 7320, 7543, 7623, 7624, 9545, 9688, 9809, 9931, 10221, 11156, 11500, 11638). The sex of the offspring may vary with the social rank of the mother (*Cervus* 509, 609, 1921, 1923; *Odocoileus* 2378a). Females have 4 mammae (182, 1474, 3571, 4494, 6491, 7921). Males have ampullae and seminal vesicles (1730).

Details of anatomy (*Capreolus* 6738; *Cervus* 2703, 3367; *Odocoileus* 3877, 11856; *Rangifer* 6320), intersex (2144, 6804), endocrinology (1705a, 6596; *Alces* 10409; *Capreolus* 98, 620, 1385, 4880, 6738, 9552, 9764, 9766, 9767, 9768, 9940, 9953, 11341; *Cervus* 51, 371, 413, 414, 415, 416, 417, 418, 613, 619, 1385, 1924, 2175, 2178, 2921, 4141, 4207, 4208, 4214, 5229a, 5634, 5636, 6450, 6462, 6463, 6595, 6596, 7475, 7592, 10562a, 10563, 11182, 11509; *Elaphurus* 2175, 6594, 6596; *Odocoileus* 18, 1372, 1401, 1401a, 1402, 1403, 1404, 1405, 1697, 2965, 4367, 7102, 7359, 8596, 8601, 8603, 8605, 8606, 9661, 9662, 9663, 11884, 11909; *Rangifer* 2696, 7240, 9400a, 10449, 11644), spermatogenesis (*Capreolus* 7920; *Cervus* 4207, 4208, 6451, 6591; *Odocoileus* 6164; *Rangifer* 6319), birth (360, 6208, 10881), suckling (3635, 7322), milk composition (4643, 5179; *Alces* 2036, 3479, 9031; *Cervus* 362, 6078; *Odocoileus* 5841, 7588; *Pudu* 2204; *Rangifer* 406, 4489, 8744), reproductive energetics (124, 126, 1920, 1921, 1922, 1923, 1926, 5932, 6595, 11266), and light (*Cervus* 50, 52, 53, 2178, 2682, 6458, 6850, 10564a, 11520, 11520a; *Odocoileus* 1373, 1400, 1406, 8597, 8598, 8599, 8600, 9664) are available.

NEONATAL MASS

Alces alces	f: 6-14 kg	4643
	m: 8-16 kg	4643
	f: 9.35 kg; n1 stillborn	6243
	11-16 kg	3477
	11.2 kg; n6	11262
	~12 kg	3571
	17 kg; n1 stillborn	9031
	17.7 kg; n1, 2 wk	2483
Capreolus capreolus	0.93 kg; r0.5-1.26	11442
	1.09 kg; r0.95-1.23; n2 term emb	4643
	~1 kg	3571
	1.34 kg; r1.28-1.39; n2	6459
Cervus axis	f: 2.77 kg; r1.53-3.45; n3 stillborn	6243
	2.991 kg; r2.2-4.0; n34	27
	m: 3.45 kg; n1	6243
	4.06 kg; r3.63-4.5; n2	9475
Cervus dama	m: 3.2 kg; n1	6243
	f: 4.4 kg; n93 m & f	1708
	4.5 kg; n1, 230 d emb	10380
	f: r4.55-6.4 kg; n5	1244
	m: 4.6 kg; n93 m & f	1708
	f: 4.7 kg; n7	6079
	f: 4.9 kg; sd0.8; n28, <3 d	1245
	m: 5.3 kg; n3	6079
	m: 5.7 kg; sd0.7; n29, <3 d	1245
Cervus elaphus	5.8-7.2 kg	1924
	f: 6.00 kg; se0.22; r3.4-7.6; n25 1st calves	361, 362
	f: 6.23 kg	1920
	6.35 kg; r3.63-9.99; n204	7364
	f: 6.44 kg	4138
	m: 6.59 kg; se0.21; r4.0-7.7; n21 1st calves	361, 362
	6.71 kg; r3.17-10.43; n244	7364
	m: 6.72 kg	1920
	m: 6.90 kg	4138
	7.5 kg; r5.5-9.1; n13	2200
	f: 7.6 kg; n1 not first calf	361, 362
	m: 8.12 kg; se0.41; r7.9-9.2; n4 not 1st calf	361, 362
	f: 8.3 kg; r7.77-9.1; n3	6243
	m: 8.95 kg; r8.7-9.2; n2	6243
	9 kg	7042
	11-22 kg; 1st wk	4643
	<11.4 kg; <50% survive	10811
	11.5 kg; se2.4; moribund	4840
	13 kg	1182
	mode 13-16 kg	9385
	15 kg; r8.6-20; n23	5338
	>16 kg; 90% survive	10811
	m: 17.4 kg; n1 term	7543
	18.5 kg; se1.8	4840
Cervus eldi	4.54-6.1 kg; n8	9475
	f: 4.7 kg; se1.1; n18	11590
	m: 4.8 kg; se1.4; n16	11590
Cervus nippon	2.25 kg; r1.5-3.0; n2	9475
	f: 2.4 kg; n1	6243
	m: 2.52 kg; n1	6243
	f: r4.2-6.2 kg; n9	4643
	4.5-7.0 kg	3198
	m: r4.75-7.32 kg; n12	4643

Cervus porcinus	2.365 kg; r2.0-2.74; n6	27
	m: 2.70 kg; n1 stillborn	6243
Cervus unicolor	10.270 kg; r7.0-12.3; n20	27
Elaphurus davidianus	3.1 kg; n1 emb, prob not near term	4262
	9.73 kg; n11 stillborn or dead within 24 h	11590a
	f: 11.49 kg; sd1.707; n49	11590a
	m: 11.66 kg; sd1.434; n43	11590a
Hydropotes inermis	m: 0.91 kg; n1	6243
	1.175 kg	2054
Mazama gouazoubira	f: 1.3 kg; r1.2, 1.4; n2	3444
	0.510-0.567 kg	10788a
Moschus berezovskii	0.350-0.558 kg; n15	4032a
	0.455-0.609 kg	4032a
Moschus chrysogaster	0.6 kg; n2	4032a
Moschus moschiferus	0.3-0.5 kg	4643
	0.400-0.635 kg	4032a
Moschus sifanicus	1 kg; 2 d	5232
Muntiacus muntjak	f: 1.32 kg; n1	6243
	m: 1.42 kg; n1	6243
	1.554 kg; r1.2-2.01; n22	27
Muntiacus reevesi	1.05 kg; n1	1713
Odocoileus hemionus	2.5 kg; n1	7320, 7321
	f: 2.5 kg; n53; 2.7 kg; n11 singletons; 2.5 kg; n42 twins	7590
	2.7 kg; sd0.4; r1.7-3.8; n101	7590
	f: 2.8 kg; n1, 174 d fetus	5066
	m: 2.8 kg; n48; 3.3 kg; n9 singletons; 2.7 kg; n39 twins	7590
	f: 2.89 kg; r2.8-2.95; n2	2618
	2.97 kg; r2.0-3.6; n11	2089
	~3 kg; n3	9376
	3.03 kg; n4	3876
	3.2 kg; r2.1-4.5; n11	3876
	3.3 kg; r2.3-4.5	9183
	m: 3.30 kg; r3.01-3.46; n3	2618
	3.4 kg; r3.2-3.7; n11	7878
Odocoileus virginianus	1.8-3.6 kg; n4 emb	10909
	2.1 kg; r1.9-2.5; n3	7878
	2.3-2.7 kg; r1.13-3.97; 1-2 d	11638
	f: 2.5 kg; n8	4500
	2.54 kg; r2.49-2.58; n2	7263
	f: 2.6 kg; r1.47-3.74; n12	4503
	m: 2.61 kg; n1	6243
	2.75 kg; r2.27-3.24; n2 samples	11263
	2.9 kg; sd0.3; r2.5-3.3; n8, <1 d	7845
	2.95 kg; r0.9-4.58; n201	11259
	m: ~3 kg; n10	4500
	3.2-3.6 kg	4250
	m: 3.37 kg; r2.04-6.58; n17	4503
	3.4 kg; ±0.59; n large	11261
	f: 3.5 kg; se0.1; n74 twin	11266
	m: 3.7 kg; se0.1; n78 twin	11266
	f: 3.9 kg; se0.1; n30 singleton	11266
	3.9 kg; n8?, 3 d	10790
	m: 4.1 kg; se0.1; n28 singleton	11266
Ozotoceros bezoarticus	2.1 kg; r1.2-2.5	3446
Pudu mephistophiles	0.4 kg; n1	3445
Pudu pudu	0.78 kg; n1	2204
	1.0 kg	4719
Rangifer tarandus	2.5-3.5 kg; n3	9022
	2.75 kg; r2.5-3.0; n2 twins	7057
	2.98 kg; se0.47; n7 dead	10047
	3.2-5.9 kg	8210
	3.72 kg; se0.27; n11	10047
	f: 4.5 kg	5637
	4.7 kg; se0.2; n14	10221

	4.8 kg; r2.7-6.16; n5	55
	5-9 kg	806
	5 kg ea; singletons	7057
	f: 5.0 kg; se0.1; n397	2987
	5.1 kg; se0.04; r1.8-8.5; n823	2987
	m: 5.3 kg; se0.1; n423	2987, 10837
	f: 5.52 kg; r4.63-6.36; n3	6243
	5.6 kg; se0.1; n37	10221
	f: 5.7 kg; twin	7985
	6-7 kg	4643
	m: 6.14 kg; r3.81-7.42; n5	6243
	m: 6.2 kg; twin	7985
	6.4 kg	2288
	6.76-9.07 kg; across n8 yr	801
	f: 6.8 kg; singleton	7985
	m: 7.2 kg; singleton	7985
	f: 7.2 kg; n153	804
	f: 7.3 kg	2288
	m: 7.8 kg; n121	804
NEONATAL SIZE		
Alces alces	80 cm	3571
Cervus dama	HB: 66.5-73.8 cm	3199
Cervus elaphus	92-101 cm	4643
	m: CR 92.5 cm; n1 term	7543
Cervus nippon	TL: 570 mm	3198
Hydropotes inermis	45 cm; n3 stillborn	2054
WEANING MASS		
Cervus elaphus	56.5 kg; se4.9; text; 59.9 kg; 92 d table	4840
Odocoileus hemionus	21.6 kg; ±1.80; n16, 105 d	9731
	21.9 kg; sd4.71; n7, 105 d	8240
	23.9 kg; ±1.44; n16, 105 d	2089, 8240
	~25 kg; r20-30; 111-120 d	7590
Odocoileus virginianus	f: 35 kg; 6 mo	9975
	m: 37 kg; 6 mo	9975
Rangifer tarandus	~20 kg	5637
LITTER SIZE		
Alces alces	1	9542
	1-2	9427
	1-2	2483
	1-2; twins not rare	11156
	1; r1-3	3477
	1 yg ea; n6 lt	11262
	mode 1 yg; 4.5% twins; r1-2	1871a
	mode 1; <5% twins; r1-2	7623
	5-33% twins; n13 studies	3478
	1.06; r1-2; n160 lt	5993
	1.1 emb; r1-2; n64 f	8862
	1.11-1.16/100 f; r1-2	3620
	1.12 cl; 1 emb; yg f; 1.25 cl; 1.16 emb; adult; n424 lt	9597
	1.12-1.50 cl; across regions, yr	6818
	1.135; 13.5% twins; 3 rare; r1-3; n1217 f	8411
	1.14 emb; r1-2; n330 f	1012
	1.15; 15% twins; r1-2; n1158	8547
	1.15; r1-2; n2047 lt	8411
	1.22; r1-2; n129 lt	2903
	1.28 cl; r1.06-1.65; n8 regions; 11-29% twins 5 regions	9597
	1.33 emb; se0.05; mode 1 55%; r1-3	5990
	1.38 yg; 38% twins; n238 f	543a
	1.47 cl; r1-3; n121 f	9979
	1.543 yg; r1-2; n151 lt	3480
	1.59 emb; se0.06; mode 2 53%; r1-3	5990
	1.89; r1-2; n19 lt, 2 f	10056
	3 yg; n1 lt; anecdote	5004

Blastocerus dichotomus	mode 1	10809
	1 emb; n1 f	9551
Capreolus capreolus	1-2; r1-3	3571
	mode 1 73%; r1-2	8759
	1+; mode 2	3388
	1.25; r1-2; n4 lt	953
	1.3; r1-2; n3 lt	6243
	1.54-1.86	6118
	1.76 emb; mode 2; r1-3; n17 f	1699
	1.8	266
	1.83; r1-3; n12 lt	3092
	1.85 emb; mode 2; r1-3; n149 f	10510
	1.91 emb; mode 2; r1-4; n108 f	1707
	1.92; mode 2; r1-3; n38 lt	8759
	2 emb; mode 2; r1-4; n73 f	11443
	2 yg each of 2 lt	11500
Cervus axis	1	9486
	1	2936
	1; n1 lt	6654
	1 emb ea; n4 f	7382
	1 ea; n9 lt	6243
	1; 2 sets twins	9545
	1 yg; 1 set twins; n225 lt	2109
	1.01; mode 1; 1.2% twins; r1-2; n80 lt	12128
	1.03 yg; r1-2; n38 lt	9066
	1.25 yg; r1-2; n4 lt	5062
Cervus dama	1-2	7652
	1 ea; n17 lt	6243
	1 emb ea; n52 f	10380
	1 ea; n93 lt	1230
	1; 0.4% twins; r1-2; n250 lt	4426
	1 yg; 0.5% twins; r1-2; n200 lt	2109
	mode 1; 11% twins	11023
Cervus duvauceli	1	9545
	mode 1	12128
Cervus elaphus	1; mode 1	4214
	1	5635
	1	1924
	1	1230
	1	7652
	1; twins rare	1182
	1; twins extremely rare	5636
	1; r1-3	5146
	1 ea; n40 lt	6243
	1 emb ea; n211 f	10884
	1 emb; 0.08% twins; r1-2; n1186 uteri	3336
	1 emb; 0.1% twins; r1-2; n896 f	5840
	1 conceptions; 0.2% twins; r1-2; n600 lt	7365
	1 yg; 0.26% twins; r1-2; n380 lt	2109
	1 emb; 0.7% twins; r1-2; n149 f	12002
	1; 0.8% twins; r1-2; n117	12128
	1.09; n11 lt	2200
	1.4 cl; n10 lt	7647
	1.85; r1-2; n26 lt	4263
Cervus eldi	1	6654
	1 ea; n3 lt	12128
	1 yg ea; n8 lt	9475
	1; 97.5% single; n158 lt; twins died	11590
	1.04 yg; 4% twins; r1-2; n24 lt	2109
Cervus nippon	1 ea; n9 lt	6243
	1 emb ea; n9 f	7594
	1 yg ea; n187 lt	2109
	1.05; 4.6% twins; r1-2; n108 lt	12128

Cervus porcinus	1 yg ea; n32 lt	2109
	1.01; r1-2; n124 lt	1230
Cervus timorensis	1	11180
	1 ea; n21 lt	12128
Cervus unicolor	1	1230
	mode 1	12128
	1 ea; n4 lt	12128
	1 emb ea; n5 f	5710
	1 ea; n6 lt	6243
	1.02 yg; 1.5% twins; r1-2; n64 lt	2109
Elaphodus cephalophus	1; n>1	4592
Elaphurus davidianus	mode 1	11592
	1-2	4425
	1 yg; n1 lt	10536
	1 ea; n6 lt	1230
	1 ea; n6 lt	6243
	1 yg ea; n16 lt	2109
	1 yg ea; n143 lt	11590a
Hydropotes inermis	mode 2; max 3	2109
	mode 2-3 yg; 1,4 rare	2653
	2.3 yg; mode 2; r2-4; n12 lt	7289a
	3; 1 dead; n1 lt	2054
	3 yg; 1 lt	11999
	3.5; r3 yg, 4 emb; n2 lt	3127
	4 cl	3127
	5-6	182
Mazama americana	1-2	3633
	1-2	11642
	1	11387
	1 emb; n1 f	7624
	2; mode 2	4263
Mazama gouazoubira	1 emb ea; n12 f	10302
	1 yg; n17 lt	3444
Mazama rufina	1	7652
Moschus chrysogaster	mode 1 yg; r1-2; n7	4032a
Moschus moschiferus	mode 1; 2 rare	9836
	mode 1; 2 occasional	6654
	mode 1; r1-2	9175
	1-2	4263
	1-2	7652
	1-2	11642
	1.82; n119	613b
	mode 2	3034
	2	4643
Muntiacus muntjak	1	12128
	mode 1; 2 occasional	8498, 8501
	1-2	6653, 6654
	1-2	7652
	1-2	4263
	1 emb; n1 f	7382
	1 emb ea; n2 f	2313
	1 ea; n3 lt	6243
	1 ea; n38 lt	1230
Muntiacus reevesi	1	12128
	1 yg ea; n5 lt	9066
	1 emb ea; n11 f	1700a
	1 emb ea; n21 f	1712, 1713
	1 yg ea; n100 lt	2109
Odocoileus hemionus	1-2	2086, 10775, 10776
	1-3; mode 1 yearling, mode 2 adult	5146
	1.2 emb; r1-2; n9 yearling	7156
	1.375; r1-2; n8 lt	3876
	1.38 emb; n26 f	10613
	1.4; mode 1; r1-2	9183

	1.44; r1-2; n9 lt	11369
	1.44; n152 lt	6232
	1.45 yg; mode 1; 3 rare; r1-3; n64 lt	6491
	1.47; r1-3; n127 lt	6244
	1.5; r1-2; n2 lt	6243
	1.5; r1-2; n1 f, 2 lt	4185
	1.5 emb; r1-2; n16 f	10595
	1.54 emb; r1-4; n50 f	9713
	1.58 emb; r1-3; n246 f	9182
	1.6 yg; r1-2; n9 lt	9066
	1.60; n47 lt; r1.38-1.91 by age	10773
	1.63; r1-2; n73 lt	6244
	1.67 emb; r1-2; n3 f	9000
	1.67; mode 2; r1-2; n61 lt	7590
	1.67; mode 2; r1-3; n113 lt; r1.27-1.91; n6 herds	952
	1.67; r1-2; mode 2; n162 lt, 5 herds	952
	1.67 emb; mode 2; r1-3; n562 f	9184
	1.7 emb; mode 2 58%; r1-3; n50 f	5065
	1.75; mode 2; r1-3	1758
	1.80 emb	1466
	1.8 emb; r1-3; n41 adult	7156
	1.81 cl; r1-2; n79 f; 1.67 emb; r1-2; n85 f	7825
	1.83 yg; r1-2; n6 lt	7878
	1.9 emb; mode 2; r1-3; n8 f	10595
	1.92 emb; mode 2 67%; r1-3; n24 f	5303
	1.98 cl; n40 lt	7156, 7157
	2 repeatedly; n1 f	9303
	2.06 cl; 1.73 emb; r1-3; n59 f	12072
	5 emb; n1 f	7827
Odocoileus virginianus	mode 1; 2 occasional	3024
	mode 1; 2 occasional	9385
	mode 1; r1-2	7145
	1-1.6 cl	11637
	1-3 cl	10730
	1-3 yg	8182
	1-4	1770, 9809
	r1.0-2.267 emb; n~100+?	2263
	1 yg; n1 lt	6142
	1.0 emb ea; n6 fawns	5874
	1.09 cl; fawn; 2.27 cl; adult	4506
	1.1 cl; r1-4	5464
	1.2 emb; n5 f	8605
	1.2 yg; r1-2 n6 lt	1013a
	1.21 cl; n19 f	1309
	r1.22-2.18 cl; r1.00-1.92 emb; n8 samples	8917
	1.25 cl; fawns; 2.04 cl yearling; 2.22 cl adult	7942
	1.25; r1-2; n16 lt	12128
	1.25 emb; se0.07; n91 f	8180
	1.28 emb; n32 f	5134
	1.28 emb; mode 1; r1-2; n186 f; incl 4 fawns	9112
	1.3 fawns; minimal production 1974-1977	3620, 7837
	1.3; r1-2; n7 lt	6243
	1.3; r1-2; n10 lt	4503
	1.4 emb; n12-15 f	5861
	1.45; mode 1; r1-2; n42 lt	10005
	1.5 yg; r1-2; n2 lt	7878
	1.5 emb; n100 f	7614
	~1.5 cl	6783
	1.52 yg; r1-3; n132 lt	11259
	1.59 emb; 1.85 cl; n331 f	615
	1.61 emb; 1.75 cl; mode 2; r1-3; n130 f	5208
	1.61 emb; n318 f	5329
	1.62 emb; n441 f	5723
	1.63 emb; n109 f	4502

	1.63 emb; n1103 f	9088
	1.66 yg; 1.77 cl; r1-3	11780
	1.66 emb; n108 f	11780
	1.67; r1-2; n12 lt	4500
	1.67 emb; n936 f	4706
	1.7 emb; n12 yearling	5874
	1.74; r1-3; 3 rare; n27 lt	11260
	1.79 emb; se0.09; n389 f	8180
	1.8 emb	5464
	1.8 emb; n65 f	548
	1.86 emb; mode 2 70%; r1-3; n361 f	9282
	1.86-2.25	7941
	2	7652
	2 emb	4504
	mode 2	9164
	mode 2; r1-2	7235
	2; r1-3	4250
	2 cl; 3 emb	7043
	2 emb; n1 f	10965
	2 cl; n1 f	4705
	twins n10 birth	10881
	2 yg; n10 lt, 1 f	11258
	2.0 emb; n11 adult f	5874
	2.03 cl; n460 f	9282
	2.27; r1-3; n44 lt	6244
	2.33 yg; r1-3; n12 lt, 1 f	8211
	3 emb; n1 f	4705
	4 emb; n1 f	10909
Ozotoceros bezoarticus	1 ea; n20 lt	3446
Pudu mephistophiles	mode 1; r1-2	10809
Pudu pudu	1 yg; n1 lt	4719
	1-2	11642
Rangifer tarandus	1	7652
	1 yg ea; n9 lt	2109
	1 ea; n15 lt	6243
	1; 2 rare	806
	1 set of twins, died w/in 12 h	7057
	2 rare; 1/10,000 lt	7055
	2; n1 lt	7046
	2.9 cl; 11% 5-7; r1-7; n73	2288
SEXUAL MATURITY		
Alces alces	f: 1st cl: 24/68 yearling; 84/99 adult	9978
	f: 1st cl: 50% yearling	6816
	f: 1st estrus: 2nd autumn	4643
	f: 1st mate: 1.5 yr	1871a
	f: 1st mate: 2-3 yr	1474
	f: 1st mate: 3-4 yr	4643
	f: 1st preg: yearling	1012
	f: 1st preg: 80% yearling; n87	8862
	f: 1st preg: 3rd yr	4643
	f/m: 1st mate: yearling	8547
	f/m: 1st mate: 16-17 mo	11156
	f/m: 1st repro: 1 yr; n1 ea; high nutritional plane	9686
	f/m: 16-18 mo	3477
	f/m: 16.5-30 mo	3571
Capreolus capreolus	f: 1st cl: 4-12 mo	11443
	f: 1st mate: 14 mo	8759
	f: 1st mate: 5-6 mo	10925
	f: 1st birth: 2 yr	3388
	m: 1st mate: 3 yr	3388
	m: >1 yr	10925
	f/m: ~14 mo	3571
	f/m: 2nd yr	4645

Cervus axis	f: 14-17 mo	9545
	f/m: 1st mate: 3-4 mo; n1	3989
	f/m: 1st mate: 12-24 mo	2936
Cervus dama	f: 1st birth: 3 yr	11023
	f: 15-18 mo	4426
	f: ~16 mo	371
	f: 16 mo	1708
	f: 16-17 mo	10380
	m: testes: large 13 mo	3267
	m: spermatogenesis: 7 mo	3199
	m: spermatogenesis: yearling	1701
	m: sperm: 14 mo	1706, 1708
	m: sperm: 16 mo	1709
	m: epididymal sperm: 13 mo	3267
	m: eversion of prepuce transition zone: stops 9 mo, recommences 17 mo	1711
	m: 2nd yr	10380
	f/m: 1st mate: 17-20 mo	3266
	f/m: 1st mate: active 4-5 yr	4643
	f/m: 2.25 yr	3571
Cervus duvauceli	f: 1st birth: ~3 yr	9545
Cervus elaphus	f: 1st mate: 1-3 yr	7365, 7367
	f: 1st mate: 2 yr 4 mo	644
	f: 1st mate: 2-4 yr	7366
	f: 1st mate: 3.5-3.9 yr	1921
	f: 1st conceive: 1.5 yr	11532
	f: 1st conceive: 20 mo; n1	643
	f: 1st conceive; 2-3 yr	6463
	f: 1st preg: 1 yr	2377
	f: 1st preg: 18 mo; n1	9510
	f: 1st preg: 20 mo; n1	1957
	f: 1st preg: 0-78% yearling	3336
	f: 1st preg: 8-17% yearling	3368
	f: 1st preg: yearling; but not many	5840
	f: 1st birth: 14-16 mo; 65 kg	5636
	f: 1st birth: 2 yr	88
	f: 1st birth: 2 yr	5146
	f: 1st birth: 3.25-3.64 yr	1923
	f: 1st birth: 3-4 yr	4139
	f: 1st birth: 3-4 yr	6213
	f: 1st lact: 2 yr; n4	11531
	f: 1 yr	1871a
	m: mature sperm: yearling	2000
	m: motile sperm: 14-16 mo	5636
	m: epididymal sperm: 2nd yr	7365
	m: secondary sex characters: 9-15 mo	6450
	m: testes large: ~4 yr	5634
	m: 1st sire: yearling	7505
	f/m: mature gonads: 18 mo	12002
	f/m: 1st repro: ~3 yr	12002
	f/m: active 5-6 yr	4643
Cervus eldi	f: 1st conceive: 468.9 d; r340-583; n16 f	11590
	f: 1st birth: 31.5 mo; r21-42; n2	8640
Cervus nippon	f: 1st preg: 8 mo; n1	1714
	f: 1st preg: yearling	2306
	f: 1st birth: 3 yr	4643
	f: min 6 mo	7594
	f/m: 1st mate: 3-4 yr	4643
	f/m: 1.25-1.5 yr	3571
	f/m: 16-18 mo	3198
	f/m: 1.5 yr	4643
Cervus timorensis	f: 1st mate: 11-12 mo	11060
	f: 1st conceive: 2 yr 3 mo	11084
	f: 18-20 mo	11180

	f: 54-59 kg	6473
Cervus unicolor	f: 1st birth: 2 yr 1 mo 28 d	30
	f: 1st birth: ~3 yr	9545
Elaphodus cephalophus	f: 18 mo	4592
Hydropotes inermis	f: 1st preg: 6 mo	7289a
Mazama gouazoubira	f: 20.5 mo	3444
Moschus moschiferus	f: 1st mate: < 1 yr	6653, 6654
	f: 18 mo	9175
	m: 1st mate: 21 mo	3034
	m: 1st sire: 1 yr	3444
	m: 3 yr	9175
	f/m: 1st mate: 15-17 mo	4643
Moschus sifanicus	f: 1st birth: 2 yr; n1	5232
	f/m: 1st mate: 1.5 yr; n1	5232
Muntiacus crinifrons	f/m: 1st mate: subadult	4836
Muntiacus muntjac	f: 8 mo	613a
	m: 12 mo	613a
Muntiacus reevesi	f: 1st ovulate: <12 mo	9423a
	f: ~6 mo	1709a
Odocoileus hemionus	f: 1st cl: 6 mo	5444
	f: 1st conceive; 6-8 mo; but few do	9182
	f: 1st mate: 16 mo	2086
	f: 1st mate: 17 mo	10614
	f: 1st mate: 18 mo	10615
	f: 1st mate: 1.5 yr	12072
	f: 1st mate: 2.5 yr; some yearling	10595
	f: 1st mate: 85% yearling	10776
	f: 1st birth: 50% 1 yr; 22.5-33.5 kg	7587, 7589
	f: 1st birth: 2 yr	7156
	f: 1st lact: 1.5 yr; n1	10777
	f: ~1 yr; n1	2110
	f: 2nd yr	9184
	f/m: 1st mate: 18 mo	7878
	f/m: 1st mate: few as fawns, yearling	10595
	f/m: 1st repro: 1 yr	10775
	f/m: 195-212 d	5146
	f/m: some 16 mo	2618
	f/m: some 1.5 yr	952
Odocoileus virginianus	f: 1st estrus: 11 mo	1227a
	f: 1st ovulate: 1/25 yearling; est 2 yr	5464
	f: 1st cl: 4-7 mo	4506
	f: 1st cl: 6-8 mo 5/21; 8-10 mo 17/17	1308, 1309
	f: 1st conceive: 0.5 yr; n3/73; mode 1.5 yr	5723
	f: 1st conceive: 6 mo; 5/160	10730
	f: 1st conceive; 6 mo 24%; 18 mo 21%	548
	f: 1st conceive: 214 d; n1	9976
	f: 1st mate: 5-8 mo	7942
	f: 1st mate: 0.5 yr	5329
	f: 1st mate: 1.5 yr	1474
	f: 1st preg: 6 mo	9088
	f: 1st preg: 9 mo; mode 10-15	1308
	f: 1st birth: 1 yr	7145
	f: 1st birth: 463 d; n1	9976
	f: 6-9 mo	11265
	m: epididymal sperm: 48% fawns	3369
	m: 1st sire: 202 d; n1	9976
	m: 1st sire: <1.5 yr; n5	9667a
	m: 16 mo	1871a
	f/m: 1st mate: 4-7 mo; 83.6% of fawns	4506
	f/m: 1st mate: 6-18 mo	1769, 9809
	f/m: 1st mate: no fawns	1651
	f/m: 1st birth: 1.5 yr	4250
Ozotoceros bezoarticus	f: 1 yr	3446

Pudu pudu	f: 1st mate: 0.5 yr	11221
	m: 1st sire: 9 mo	2204
Rangifer tarandus	f: 1st mate: 18 mo; some 4-5 mo	8744
	f: 1st ovulate: 1.5 yr; 2/100 yearling, 46/100 2 yr	2288
	f: 1st conceive: 6-18 mo	9021
	f: 1st conceive: 16-17 mo	6320
	f: 1st conceive: 3-4 yr	7055
	f: 1st preg: 8 mo; occasionally 5 mo	7921
	f: 1st preg: 17 mo	7055
	f: 1st preg: 21 mo if >46 kg	10774
	f: 1st preg: 22 mo; 9/21	8235
	f: 1st preg: 29-30 mo	801
	f: 1st preg: 2.5 yr	2987
	f: 29-41 mo	806
	m: sperm: 17-18 mo	7055
	m: testes large: 7 yr	6319
	m: 1st mate: 18 mo	7921
	m: 16 mo; n1	2109
	f/m: 1st mate: 3rd yr	4643
	f/m: 2 yr	4643
CYCLE LENGTH		
Alces alces	estrus: <24 yr	1871a
	estrus: mode 4-5 d; r2-5	4643
	estrus: several d	11156
	receptivity: 7-12 d	1871a
	20-22 d	1871a
	25 d	6817
Capreolus capreolus	estrus: 3-4 d	3388
	estrus: 4-5 d	4643
Cervus dama	22.4 d; r20-27; n142 cycles	413
	24-26 d	3199
Cervus elaphus	18.2 d; se0.4; n17 cycles	5636
	18.3 d; se1.7; r5-22; n78 cycles, 11 f	2200, 6463
	21.2 d; r19-25; n9 cycles	7539
Cervus eldi	mode 17 d; n3 f	11590
Cervus nippon	estrus: 1 wk	4643
Cervus timorensis	estrus: 2 d	11060
	~13 d	11060
Elaphurus davidianus	18.464 d; sd2.017; r17-25; n33 cycles, 4 f	11590a
	19.5 d; se0.6; n5 f	2175
	~31 d	11592
Muntiacus muntjac	estrus: <48 h	613a
	14-21 d	613a
Odocoileus hemionus	estrus: 24-36 h	10773
	estrus: 24-36 h	11598
	8+ d; r6.5-9.0	10773, 10776
	3-4 wk	6491
	21-23 d	7878
	22-28 d	11598
	23.7 d; sd2.9; r18-30; n20 cycles, 4 f	11882
Odocoileus virginianus	estrus: ~24 h	4250
	estrus: 24-36 h	1769, 9809
	26.2 d; sd2.1; r21-30; n8 f	5899
	28 d	4250
	28-29 d	1769, 9809
Rangifer tarandus	estrus: 1-3 d	804
	10-12 d	2561
	10-12 d	806
	~2 wk	4643
	2-3 wk	7921
	17 d; r11-26; n3	7058
	r18-25; n12 cycles, 2 f	804
	23.5 d; n14 f	2696

GESTATION

Alces alces	217 d; r216-218; n2	10401
	7 mo; field data	6817
	225 d; n2	10056
	227-235 d	11156
	234 d; r226-244	3478
	~236 d	480
	~240 d	2483
	240-246 d	3478
	~8 mo	7623
	~8 mo; field data	5993
	~8 mo	7652
	242.3 d; n3	11262
	35-38 wk	3571
Blastocerus dichotomus	271 d; n1	3447a
Capreolus capreolus	implant: 2-5 mo	4643
	implant: 3 mo	12096
	implant: 4-5 mo	10417, 10418
	implant: 4-5 mo	3571
	implant: 150 d	8759
	implant: 5.5 mo	11443
	implant: 5.5 mo	9939
	postimplantation: 4 mo	9939
	postimplantation: 4.5 mo	11443
	postimplantation: 144 d	8759
	postimplantation: 5 mo	10417, 10418
	postimplantation: 5-6 mo	3571
	postimplantation: 28 wk	12096
	4 mo	3388
	~22 wk; at 2nd mating	3571
	~9 mo	4645
	~9 mo; r6-10	4643
	9-9.5 mo	7652
	9-10 mo	10417, 10418
	9-10 mo	3571
	40 wk	953
	40 wk	12096
	9.5 mo	9939
	294 d	8759
	10 mo	11443
Cervus albirostris	10 ? mo; field data	3034
Cervus axis	210-238 d	4263
	7 mo 8 d	1355
	~229 d	3989
	8-8.5 mo	9545
	~8 mo; some observers say ~6 mo	8498, 8501
	8.25 mo	7652
Cervus dama	7 mo	1523
	~7 mo; field data	4426
	7 mo	6243
	229 d; r225-238; n8	7425
	230 d; r225-234; n11	8742
	230-240 d	4263
	33-35 wk	3199
	7.5 mo	11023
	7.5 mo; field data	4414
	8 mo	1355
	8 mo; field data	8344
	8 mo; field data	2682
Cervus duvauceli	8 mo	7652
	256 d; n1	7425
Cervus elaphus	implant: 6 wk	10789
	~7 mo; field data	9222
	~7 mo; field data	644

	7.5 mo; ~8.5 mo	7652
	225-246 d	4263
	~226 d	1230
	228 d; r227-229; n2	6463
	231-238 d	3571
	230.2 d; sd4.4; n11 1st preg; 234.3; sd0.95; n6 later preg	6460
	231.0 d; sd4.5; n11	2200
	231.491 d; captive; 236.587 d; r220-251; wild	4139
	~233 d	7367
	233.1 d; se0.6; n38	5636
	234 d; r230-239; n9	8742
	234 d; se0.5; n32	51
	234.2 d; ±5.04; f emb; 236.1 d; ±4.75; m emb; n70	1920, 1924
	234.25 d; r231-238; n4	5635
	235 d	2151
	237 d; field data	3873
	237 d; r230-244; n4	7425
	34 wk	7256
	8 mo	2682
	8-8.5 mo	1355
	249-262 d	9385
	~250 d	7042
	255 d	5146
	~8.5 mo; r249-262 d	6213
	265 d; n1	4390
	8.5-9 mo; field data	7366
	9 mo; field data	11642
	~9 mo	7652
	10-11 mo	4263
Cervus eldi	> 190 d	8640
	~7 mo	6654
	240.4 d; r236-244; n5	9475
	241.7 d; r240-244; n3	985
Cervus nippon	~30 wk	3198
	7.25-7.5 mo	7652
	222 d; r218-229; n11	8742
	223 d; r218-230; n8	7425
	225-246 d	4263
	7.5; field data	11642
	8 mo	1355
Cervus porcinus	est 7 mo	1013a
	7.5 mo	7652
	est max 8 mo	9175
	8 mo	1355
	~8 mo	8498, 8501
	~8 mo	6654
Cervus timorensis	est 7.5 mo	6698
	240-264 d	11060
	248.6 d; sd13.4; r217-277; n19	11180
	267-284 d	11084
Cervus unicolor	~240 d	1230
	8 mo; field data	2682
	8 mo	8498, 11642
	~8 mo	8501
	246 d; n1	7425
	8-9 mo	7652
Elaphodus cephalophus	~6 mo	3034
Elaphurus davidianus	8.5 mo; field data	4262
	283.38 d; sd6.11; mode 280; n21	11590a
	9.5 mo	4582
	285-300 d; field data	9548
	288 d	11592
	11 mo	9939
Hippocamelus antisensis	240-270 d	7233a

Mazama gouazoubira	206 d; courtship to birth	3444
	~225 d	10788a
Mazama rufina	7.25 mo	7652
Moschus berezovskii	178-192 d	4032a
Moschus chrysogaster	170-198 d; n4	4032a
Moschus moschiferus	3-4 mo; field data	3034
	~5 mo	6653
	~160 d	9175
	5.5 mo	7652
	185-195 d	4643
	7.5 mo; field data	9836
Muntiacus muntjak	5-6 mo	6653, 6654
	~180 d	9175
	~6 mo	8501
	~6 mo	7652
	210 d; r205-215; n40	3447a
Muntiacus reevesi	6 mo	7652
	209-220 d	1700
Odocoileus hemionus	199.7 d; r194-205; n9 lt, 4 f	7878
	r200-203 d; n207	10773
	202-212 d	6387
	203 d; r199-207; n5	3876
	~205 d	6491
	207 d; r205-209; n4	2618
	7 mo; field data	6386
	7 mo; r183-212 d; n10	2086
	~7 mo; r205-212 d; field data	2618
Odocoileus virginianus	implant: 5-6 wk	9809
	6 mo; field data	11637
	193.8 d; r187-198; n12	4500
	196 d	1769
	mode 196 d; r189-210; 7 lunar mo	372
	199.2 d; se0.4; n119	8605, 11266
	199.4 d; n21	9809
	~200 d; n2 lt, 2 f	7878
	200.8 d; se0.4; n174	11263
	202 d	5213
	202 d; r195-212	4250
	202.1 d; ±4.1; r196-213; n55	11261
	204 d; 90% 197-208; r197-222; n10	4503
	7 mo; field data	11642
	~7 mo	627
	~7 mo	1355
	~7 mo	7652
	~7 mo; r205-212 d	6213
	270 d	10529
Ozotoceros bezoarticus	est 7 mo	3446
	7+ mo	5212
Pudu pudu	5.5 mo	4717
	est 202-203 d	2204
	202-223 d	6293
	207 d; n1	11221
Rangifer tarandus	implant: 27-29 d postovulation	7055
	~6 mo	4389
	200-210 d; n1	3714a
	207 d; r204-213; n3	7425
	208.0 d; se2.9; n5	7058
	7 mo; field data	6319
	216.4; se1.7; n15	7058
	217 d; n2; artificial insemination; 209, 214 d; n2 f	2696

	220-225 d	7921
	~223 d; gestation of m emb 2.7-5.4 d > for f emb	4643
	~227 d	7055
	229 d; r227-230; n4	800, 804, 806
	240 d	8210
	8 mo	7652
LACTATION		
Alces alces	solid food: 14 d	11156
	solid food: 2 wk	4643
	solid food: 2 wk	3477
	solid food: 1.5 mo	4645
	solid food: 1.5-2 mo	3571
	wean: 3-3.5 mo	4645
	wean: max 92 d	4840
	wean: 3.5-4 mo	4643
	wean: ~4 mo	3571
Capreolus capreolus	solid food: 6-8 d	3092
	solid food: 1 mo	4643
	nurse: max > 15 mo	8759
	wean: ~2 mo	4645
	wean: 2-3 mo	4643
	wean: 2-3.5 mo	6342
	wean: 4.5 mo	3092
Cervus axis	wean: 4 mo	3989
Cervus dama	solid food: 4 wk	4643
	wean: min 20 d	3635
	wean: 3.5-4 mo	3571
	nurse: 4 mo	1708
	nurse: min 6 mo	1701
	milk present: ~7 mo; field data	5211
Cervus elaphus	solid food: 3-4 wk	5338
	solid food: 3-4 wk	9688
	solid food: 1 mo	4643
	solid food: 6 wk	9385
	wean: 4 mo	9688
	wean: 4-5 mo	7042
	wean: 5-7 mo	1924
Cervus eldi	wean: 35 wk	11590
Cervus nippon	solid food: 10-12 d	4643
	wean: 3-4 mo	3571
	wean: 4-5 mo	4643
	wean: 8-10 mo	3198
Cervus timorensis	wean: 7-8 mo	11180
Elaphurus davidianus	nurse: 4-7 mo	11592
Odocoileus hemionus	solid food: few d after birth	6491
	solid food: 2-3 wk	7878
	solid food: 2-3 wk	2926
	solid food: 1 mo	9376
	wean: 5-16 wk; some 32-56	263
	1st rumination: 50.2 d; r36-58; n≤7	8240
	wean: 60 d	7878
	wean: 60-75 d	2618
	nurse: 3 mo	9376
	wean: 3-4 mo	9183
Odocoileus virginianus	solid food: 2 wk	11259
	solid food: 2-3 wk	9809
	wean: min 20 d	3635
	wean: 4 mo	9809
	wean: 10 wk	10150
	wean: 5 mo	1308
	wean: 5-7 mo	9975
	wean: min 4 mo; max 6-7	9525

Pudu pudu	solid food: 12 d	4717
	lact: 60 d; n1	4719
Rangifer tarandus	wean: 2-2.5 mo	5637
	wean: 20 wk	11640
	wean: 5+ mo	5932
INTERLITTER INTERVAL		
Alces alces	1 yr	2483
Capreolus capreolus	1 yr	8759
Cervus axis	~14 mo	3989
Cervus dama	1 yr	11023
	1 yr	1708
Cervus eldi	11 mo; r9-13; n5	8640
	380.7 d; se105.0; n60 yg live	11590
	384.7 d; se93.3; n28 yg die	11590
	387 d; r298-635; n13 intervals, 4 f	985
Cervus timorensis	366 d; sd6.8; n34	11180
Elaphurus davidianus	359.5; sd17.7; n102	11590a
Muntiacus muntjac	241 d; n60	3447a
Muntiacus reevesi	209-220 d	1700
	216 d; r214-219; n4	613a
Odocoileus hemionus	1 yr	7156
	1 yr; n1	4185
Odocoileus virginianus	~10 mo	1309
Ozotoceros bezoarticus	252-408 d	3446
Rangifer tarandus	1 yr	7055
SEASONALITY		
Alces alces	spermatogenesis: Aug-Oct; Europe, w USSR	4645
	estrus: Aug-Jan; zoo	6243
	estrus: mid Sept-Oct; zoo	11156
	estrus: end Sept-Oct; n Sweden	6817
	mate: ?-Sept 15; Mongolia	9542
	mate: mid Aug-Oct; USSR	4643
	mate: end Aug-Oct; Europe, w USSR	4645
	mate: end Aug-Oct; n Poland	3571
	mate: end Aug-Oct; Trondelag, Norway	7652
	mate: early Sept-?; n Sweden	6817
	mate: Sept; BC, Canada	3666
	mate: Sept; MI	11262
	mate: Sept-mid Oct; MI	7623
	mate: Sept-Oct; Canada, N US	1474
	mate: mid Sept-mid Oct; n USSR	4643
	mate: mid Sept-mid Oct; NEW, ONT, Canada, AK	8411
	mate: mid Sept-Oct; Sweden	480
	mate: mid Sept-early Nov; WY	2483
	mate: end Sept-early Oct; Canada	6379
	mate: end Sept-1st 3 wk Oct; se Norway	5993
	ovulate: Sept-early Oct; Sweden	480
	ovulate: Oct; Sweden	6817
	ovulate: Oct; Trondelag, Norway	4179
	ovulate: Oct-Nov; no data other mo; Sweden	6816
	conceive: yearling before Oct 16; Finland	8862
	conceive: end Sept-early Oct; nw ONT, Canada	9978
	conceive: est mid-Dec; Fairbanks, AK	1929
	birth: mid Mar-Oct; Trondelag, Norway	4179
	birth: Apr-May; WY	2483
	birth: Apr-June; n33; zoo	5818
	birth: end Apr-early May; n Poland	3571
	birth: early May; Europe, w USSR	4645
	birth: May; n1; zoo	6243
	birth: May-June; Canada, n US	1474
	birth: peak mid-end May 90%; AK	3620
	birth: mid-end May; Sweden	480
	birth: mid May-early June; NEW, ONT, Canada, AK	8411
	birth: mid May-mid June; MI	7623

	birth: mid May-mid June; se Norway	5993
	birth: mid May-mid June; USSR	4643
	birth: end May-early June; WY	1871a
	birth: end May-early June; Sweden	6817
	birth: end May-June; N Am	3477
	birth: Aug-Sept; Mongolia	4645, 9542
	birth: mid Aug, rare; QUE, Canada	7437
Blastocerus dichotomus	birth: May-Sept; Pantanal, Brazil	10809
Capreolus capreolus	sperm: epididymis end May-Dec; Germany	10418
	sperm: epididymis ?-Dec; Bern, Switzerland	11443
	spermatogenesis: Apr-?; Germany	12096
	spermatogenesis: May-Sept; USSR	10925
	spermatogenesis: mid May-mid Aug; Germany	10418
	spermatogenesis: June-Aug; Germany	10417
	spermatogenesis: peak June-Aug; Hungary	7920
	spermatogenesis: end July/early Aug-Sept; Europe, w USSR	4645
	testes: large Apr-July, small Sept-Mar; Bern, Switzerland	11443
	testes: large June-Aug; Bavaria, Germany	1384
	testes: peak June-Aug; Germany	10417
	testes: peak mid July-mid Aug; England, UK	9940
	testes: increase Aug; Germany	12096
	estrus: June-July; zoo	6243
	estrus: July-Aug, Oct-Nov; Germany	10418
	mate: July; w USSR; a few wk later; e USSR	4643
	mate: July-Aug; UK	9939
	mate: July-Aug; Bavaria, Germany	1384
	mate: July-Aug; central Germany	10510
	mate: July-Aug; Switzerland	11443
	mate: July-Aug; captive	9553
	mate: mid July-mid Aug; Europe	7652
	mate: mid July-mid Aug; Dorset, England, UK	1226, 8759
	mate: mid July-mid Aug; England, UK	9940
	mate: mid July-mid Aug; UK	3388
	mate: mid July-mid Aug; captive	620, 9554
	mate: mid July-Aug, some Oct-Nov; Germany	10417, 10418
	mate: peak mid July-early Sept, minor peak end Nov-early Dec; Germany	3571
	mate: end July-Aug; Germany	953
	mate: end July-Aug; Dorset, England, UK	8759
	mate: end July-Sept; Europe, w USSR	4645
	mate: Aug; Germany	12096
	mate: mid Aug-Sept; USSR	10925
	ovulate: end Aug; Germany	953
	cl: well formed Nov, regressed Dec; Germany	12096
	preg f: Nov-Jan; n4; Germany	953
	implant: Dec; UK	3388
	implant: Dec; Germany	10417, 10418
	implant: ~Dec; Germany	953
	implant: Dec-Jan; central Germany	10510
	implant: end Dec; Dorset, England, UK	8759
	implant: end Dec-mid Jan; USSR	4643
	implant: Jan-Feb; Bern, Switzerland	11443
	birth: Apr-?, peak May; Europe, w USSR	4645
	birth: Apr-May; Germany	10417, 10418
	birth: end Apr-June, peak May; England, UK	8759
	birth: end Apr-early July; Germany	953
	birth: May, sometimes Apr; Europe, w USSR	4645
	birth: May; UK	9939
	birth: peak May; w USSR; 2-3 wk later; e USSR	4643
	birth: May-mid June; Germany	4880
	birth: May-June; Dorset, England, UK	1226
	birth: May-June; Germany	3571
	birth: May-June; Switzerland	9430
	birth: May-early July; zoo	6243

	birth: end May-early June; UK	3388
	birth: end May-early July; captive, Sweden	3092
	birth: peak end May-early June; Switzerland	6118
	birth: June; Germany	12096
Cervus albirostris	mate: Sept-Oct; native accounts; e, central Tibet	3034
	birth: July; native accounts; e, central Tibet	3034
Cervus axis	spermatogenesis: increase Mar-June, peak May; India	9545
	mate: Apr-May; Nepal	7382
	mate: Apr-Aug; HI	3989
	mate: June-Aug, Jan-Feb; Sri Lanka	2936
	mate: yr rd, peak May-July; India	9545
	mate: yr rd, except Jan-Feb; zoo	6243
	conceive: peak May; India	9545
	birth: Jan-Feb; Nepal	7382
	birth: yr rd, peak early in year; Sri Lanka	8498, 8501
	birth: yr rd, peak Jan-early Mar; India	9545
	birth: yr rd, peak Feb-Mar; n110; zoo	44
	birth: yr rd, peak May-July; captive	2109
	birth: yr rd, peak Sept-Oct, Dec-Jan; Sri Lanka	2936
	birth: yr rd, peak Nov-Apr; HI	3989
	birth: yr rd, peak cold season; India	6654
	birth: yr rd, low May-Aug; n576; Chitwan Nat Park, Nepal	7360a
	birth: yr rd; n215; zoo	5818
	birth: yr rd; n80; zoo	12128
Cervus dama	ovarian activity: ceased Aug-Sept; New Zealand	414
	sperm: Mar, June, Sept, none Nov-Jan; captive, New Zealand	415
	sperm: Aug-Jan; England, UK	1706
	testes: peak Oct; s England, s Scotland, UK	1701
	testes: peak Oct-Nov, small Jan-Feb; Essex, England, UK	1706, 1708
	testosterone: high Apr; captive, New Zealand	415
	estrus: Apr, Oct-Dec; zoo	6243
	estrus: end Apr-Aug; captive	413
	estrus: Sept-Mar, peak Oct; zoo	4005
	mate: Feb-Apr; TAS, Australia	11642
	mate: Apr-May; New Zealand	2682
	mate: Apr-May; captive, New Zealand	415, 418
	mate: mid Apr-mid May; New Zealand	11642
	mate: end Aug-early Sept; captive	4486
	mate: end Aug-Sept; Essex, England, UK	1708
	mate: end Sept-Oct; USSR	4643
	mate: Sept/Oct-?; captive	2109
	mate: Oct; se Europe, Asia	7652
	mate: peak Oct; Spain	1243
	mate: peak Oct; Alsace, France	9528
	mate: Oct-early Nov; England, UK	1706
	mate: Oct-early Nov; UK	1523
	mate: Oct-early Nov; UK	4263, 4414, 4426
	mate: Oct-Nov; captive	8344
	mate: Oct-Dec; zoo	6243
	mate: mid-end Oct; W Germany	11023
	mate: mid Oct-Nov; 2 wk; Germany	3571
	mate: Nov-Dec; zoo	1230
	conceive: end Sept-early Dec; Essex, England, UK	1708
	conceive: Oct-Feb; captive	3266
	conceive: mid-end Oct; e Europe	10380
	birth: May-July; captive, s England, UK	4263, 4414, 4426
	birth: May-Sept; n8; zoo	7425
	birth: mid May-early June; W Germany	11023
	birth: end May-June; UK	1523
	birth: peak end May-early June; Spain	1244, 1245
	birth: June; Essex, England, UK	1708
	birth: June; s England, s Scotland, UK	1701
	birth: June; USSR	4643
	birth: June; captive	8344

	birth: June; zoo	7174
	birth: June, Sept; n3; W Germany	3267
	birth: June, Oct; captive	3266
	birth: mid-end June; Hampshire, England, UK	5211
	birth: June-early July; Germany	3571
	birth: June-Aug; zoo	6243
	birth: June-Dec, peak June; n202; captive	2109
	birth: June-Dec, peak June; n51; zoo	5818
	birth: Dec; Australia, TAS, Australia	11642
	birth: Dec-Jan; New Zealand	2682
	birth: Dec-Jan; New Zealand	11642
Cervus duvauceli	mate: peak Sept; Assam, India	11033
	mate: Oct; India	7652
	mate: mid-end Oct; India	6653
	mate: mid-end Oct; captive	4488
	mate: mid Dec-mid Jan; India, Nepal	11642
	mate: Aug-early Oct; England, UK	11642
	birth: May-Nov; n54; captive	5818
	birth: Apr-Dec, peak Apr-July; zoo, Germany	3447a
	birth: June-July; England, UK	9545
	birth: June-Aug; n8; captive	12128
	birth: July-Aug; n India	9545
	birth: Sept-Oct; central India	9545
Cervus elaphus	sperm: Sept-Mar; captive	4208
	spermatogenesis: mid Aug-Oct; Europe, w USSR	4645
	testes: peak Apr-May, low Aug; New Zealand	613
	testes: peak Aug; captive	4208
	testes: peak Sept; Canada	3336
	testes: large Sept-Oct; Bavaria, Germany	1384
	testosterine: peak Aug; captive	4208
	testosterone: peak Sept; captive	10563
	estrus: Apr; Invernary, New Zealand	5634, 5636
	estrus: end Aug-Oct, peak mid Sept-mid Oct; Germany	10789
	estrus: Oct-May/June, some Oct-Nov; Rhum, Inner Hebrides, UK	2200
	estrus: yr rd; zoo	4535
	mate: Mar-Apr; New Zealand	2682
	mate: Mar-Apr; QLD, Australia	9222
	mate: Mar-July; CA	4390
	mate: mid Mar-mid Apr; New Zealand	11642
	mate: mid Apr-mid May; New Zealand	5633, 5635
	mate: July; location not specified	4263
	mate: Aug-Sept; MT	5888
	mate: Aug-Nov, peak mid Sept-mid Oct; USSR	4643
	mate: mid Aug-Oct; Europe, w USSR	4645
	mate: end Aug-Sept; s Sweden	88
	mate: end Aug-Sept; Algeria, Tunisia	7256
	mate: end Aug-early Oct; N America	7652
	mate: Sept-Oct; Rhum, Inner Hebrides, UK	6463
	mate: Sept-Oct; Bavaria, Germany	1384
	mate: Sept-Nov; Rhum, Inner Hebrides, UK	4139, 5124
	mate: mid Sept-Oct; OR, WA	1182
	mate: mid Sept-Oct; Rhum, Inner Hebrides, UK	6450
	mate: mid Sept-mid Oct, & 1 Mar; Apache Co, AZ	11420
	mate: mid Sept-mid Oct; ID	9385
	mate: mid Sept-mid Nov; Amu-Daria R, USSR	6765
	mate: mid Sept-mid Nov; captive	3873
	mate: mid Sept-Nov; Rhum, Inner Hebrides, UK	7365
	mate: end Sept-early Oct; WI	5210
	mate: end Sept-mid Oct; Afognak I, AK	644
	mate: end Sept-Oct; Asia	6654
	mate: end Sept-mid Nov, peak Oct; Scotland, UK	7366
	mate: early Oct-Feb; Rhum, Inner Hebrides, UK	2200
	mate: Oct; according to natives; w Tatsienlic, Tibet, China	3034

mate: Oct; Rhum, Inner Hebrides, UK	1920
mate: Oct; Rhum, Inner Hebrides, UK	2151
mate: Oct-Feb; semiwild, central Bohemia, Czechoslovakia	631
mate: Oct-Nov; captive	50
mate: Oct-Nov; zoo	6243
mate: Nov; Germany	12096
ovulate: mid Mar-May; New Zealand	5633
ovulate: Sept-Oct; Rhum, Inner Hebrides, UK	6460
ovulate: Sept-Nov, peak Oct; WA, MT, OK	4214
ovulate: peak mid Oct; captive	52
ovulate: Oct-Jan; Rhum, Inner Hebrides, UK	6463
ovulate: Nov; Germany	12096
cl: mid Mar-June; New Zealand	5633
conceive: Sept-Oct; Germany	10789
conceive: Sept-Oct; Canada	3336
conceive: Sept-Nov, peak Oct; Rhum, Glen Feskie, Inner Hebrides, UK	7367
conceive: mid Sept-Oct, +1 mid Dec; Rhum, Inner Hebrides, UK	6463
conceive: mid Sept-Oct; MT	7543
conceive: Oct; Rhum, Inner Hebrides, UK	1924, 2151
preg f: Oct-May; Crimea, Russia	12002
preg f: end Nov-Mar; Germany	10789
birth: Jan; New Zealand	5635
birth: Apr; Asia	6654
birth: Apr-May; Owens Valley, CA	7042
birth: Apr-Nov; n168; zoo	2109
birth: peak end Apr; Algeria, Tunisia	7256
birth: May; Crimea, USSR	12002
birth: May-early June; WY	1871a
birth: May-June; location not specified	4263
birth: May-June; w US	6213
birth: May-Aug; Rhum, Glen Feskie, Inner Hebrides, UK	7366, 7367
birth: May-Aug; Rhum, Inner Hebrides, UK	4139
birth: May-Aug, peak June; n64; zoo	5818
birth: peak May-June, some Oct-Nov; Rhum, Inner Hebrides, UK	1920, 1924, 2151, 2200, 6460, 6463
birth: May-Oct; n218; zoo	2109
birth: May-Sept, peak June-July; n50; zoo	12128
birth: mid May-early July; Europe, w USSR	4645
birth: mid May-June; Germany	3571
birth: mid May-mid June; ID	9385
birth: mid May-mid June; OR, WA	1182
birth: mid May-mid June, & 1 Nov; Apache Co, AZ	11420
birth: mid May-mid June; USSR	4643
birth: end May; Afognak I, AK	644
birth: end May-early Nov; n67; zoo	12128
birth: peak early June; WA	9688
birth: June; WY	2105
birth: June-July; n5; zoo	2109
birth: June-Oct; n4; zoo	7425
birth: Aug-Sept; CA	4390
birth: Nov-Dec; QLD, Australia	9222
birth: Nov-Dec, late births Jan, Mar; zoo	1230
birth: mid Nov-Dec; New Zealand	2682
birth: end Nov-early Dec; Invernary, New Zealand	5636
birth: Dec; n3; New Zealand	11520
birth: end Dec; New Zealand	11642
wean: Jan; Canada	3336
wean: Sept; Crimea, USSR	12002
wean: Sept; CA	7042
wean: Dec-Feb; Rhum, Inner Hebrides, UK	2151

Cervus eldi	mate: mid Mar-mid May; Burma	6654
	mate: Nov-June; zoo	985
	mate: winter; zoo	4488
	birth: May-Nov, peak Oct; zoo Germany	3447a
	birth: Aug-Jan; zoo	985
	birth: Sept-Mar; n25; zoo	2109
	birth: 80% mid Sept-Nov; captive	11590
	birth: Oct-Nov; Burma	6654
Cervus nippon	mate: May; New Zealand	11642
	mate: Sept-?; captive	3964
	mate: Sept-Jan, peak Oct-Nov; Japan	7388
	mate: Sept-mid Nov; Europe, w USSR	4645
	mate: mid Sept-mid Nov; Dnjepr, USSR	3126
	mate: end Sept-Nov; 25-35 d; USSR	4643
	mate: Oct-Nov; Germany	3571
	mate: Oct-Nov; s Ukraine, USSR	10368
	birth: Jan; New Zealand	11642
	birth: end Mar-mid Oct, Nov-Dec, peak May-June; n115; zoo	12128
	birth: Apr-Oct; n8; zoo	7425
	birth: May-June; Dnjepr, USSR	3126
	birth: May-June; s Ukraine, USSR	10368
	birth: May-June; USSR	4643
	birth: May-Oct; n187; captive	2109
	birth: May-Dec, peak July; n95; zoo	5818
	birth: end May; Europe, w USSR	4645
	birth: end May-June; Dnjepr, USSR	3126
	birth: June; Germany	3571
	birth: peak mid Dec; New Zealand	2306
Cervus porcinus	antlers cast: Nov-Feb; Orissa, India	45
	estrus: Aug, Nov-Dec; zoo	6243
	mate: Apr-May; Sri Lanka	8501
	mate: Sept-Oct; Asia	6654
	mate: Sept-Dec; zoo	6243
	mate: Nov-Dec; Sri Lanka, India	11642
	birth: Feb-May, peak May; Chitwan Nat Park, Nepal	7360a
	birth: Feb-June; Baewan I, Indonesia	1013a
	birth: Mar-July; Pakistan	9175
	birth: Mar-Aug, Nov-Dec; zoo	6243
	birth: May-June; Asia	6654
	birth: Nov-Dec; Sri Lanka	8498, 8501
	birth: yr rd; captive	2109
	birth: yr rd; zoo	1230
	birth: yr rd; zoo	5818
	lact f: Mar; n1; Nepal	7382
Cervus timorensis	mate: May, Aug, Oct, Dec; captive	11060
	mate: July-Aug; New Zealand, New Caledonia	11642
	birth: Jan, May, July-Sept; captive	11060
	birth: Feb, Apr-June, Sept-Dec; n21; zoo	12128
	birth: Mar-July; VIC, Australia	11180
	birth: Mar-Sept; Java, Indonesia	11642
	birth: Apr-May; New Caledonia	11642
	birth: Oct-Nov; Flores, Indonesia	11642
	birth: peak Apr-June; NSW, Australia	4255
	birth: Apr-July; Pulan Rusa, Java, Indonesia	11084
	birth: Aug-Sept; Indo-Australian archipelago	11084
	birth: almost yr rd; Celebes, Indonesia	11642
	birth: yr rd; n18; zoo	5818
	lact f: 67% summer, 44% winter; captive	6698
	wean: Nov; VIC, Australia	11180
Cervus unicolor	mate: Feb-June, Aug, Nov-Dec; zoo	6243
	mate: Mar-Apr; New Zealand	2682
	mate: Mar-Apr; North I, New Zealand	11642
	mate: July-Nov; Borneo, Indonesia	11642
	mate: Oct-Nov; Nepal	7382

	mate: Oct-Nov; India, China	7652
	mate: Oct-Nov; plains India	6654
	mate: Oct-Nov; Sri Lanka	8498
	mate: Nov-Apr; s India	5331
	mate: peak Nov-Dec; India	9545
	mate: yr rd, peak May; n21; zoo	8339
	birth: May; Nepal	7382
	birth: May-Sept; Borneo, Indonesia	557
	birth: June-July; India	8498
	birth: June-July; India	6653, 6654
	birth: Nov-Dec; New Zealand	2682
	birth: Dec; North I, New Zealand	11642
	birth: yr rd, peak May-Aug; n89; Chitwan Nat Park, Nepal	7360a
	birth: yr rd, low Apr-May; India	9545
	birth: yr rd, except June-July; zoo	6243
	birth: yr rd; England, UK	6653
	birth: yr rd; n35; zoo	44
	birth: yr rd; n58; zoo	12128
	birth: yr rd; zoo	1230
	birth: yr rd; captive	2109
Elaphodus cephalophus	birth: end spring-early summer; China	4592
Elaphurus davidianus	estrus: almost yr rd; zoo	6243
	mate: mid May-Aug; zoo	9548
	mate: June-July; captive	4425
	mate: June-Oct; zoo	6243
	mate: end June-early Sept; Bedford, England, UK	9939
	mate: July; captive, England, UK	4262
	mate: mid Dec-early Jan; captive, Bedford, England, UK	2175
	birth: Mar-May; zoo	6243
	birth: Mar-July; n16; captive	2109
	birth: Mar-Aug, peak end Mar-May; n144	11590a
	birth: Mar-Sept, peak Apr; n17; zoo	5818
	birth: Mar-Nov; zoo	11592
	birth: Apr-May; zoo, Netherlands	11154
	birth: Apr-June, peak end Apr-mid May; zoo	9548
	birth: peak mid Apr-May; captive, England, UK	4262, 4425
	birth: May-Aug; Bedford, England, UK	9939
Hippocamelus antisensis	birth: Jan-Mar; s Peru	7233a
Hippocamelus bisulcus	courtship: mid Feb-mid Apr; 15 d; few data; Rio Claro, Chile	8711
Hydropotes inermis	estrus: Oct-early Dec; zoo	6243
	birth: May, July; zoo	6243
	birth: May; Hunan, Hupeh, China	182
	birth: end May-June; captive	2109
Mazama americana	birth: June; Yucatan, Mexico	3633
	birth: Sept-Apr, peak Nov-Feb; Suriname	1227a
	birth: yr rd; zoo, NY	6709a
Mazama gouazoubira	repro: not seasonal; zoo	3444, 3447a
Mazama rufina	mate: June; Guyana to Paraguay	7652
Moschus chrysogaster	birth: June-mid July; Gharwal, Himalaya, India	4032a
Moschus moschiferus	mate: Jan; Himalayas, India	6653, 6654
	mate: Jan; s Asia	7652
	mate: Jan; China	3034
	mate: est Jan; China	11642
	mate: Sept-Feb, peak Nov-Dec; USSR	4643
	mate: Nov-Dec; Pakistan	9175
	mate: end Nov; Altai, USSR	9836
	birth: Apr-May; China	3034
	birth: May-July, peak June; USSR	4643
	birth: end May-early June; Pakistan	9175
	birth: June; Nepal	7382
	birth: June; Himalayas, India	6653, 6654
	birth: est June; China	11642
	birth: June-early July; Altai, USSR	9836
Muntiacus crinifrons	preg f: yr rd; e China	4836

Muntiacus muntjak	antlers cast: Apr-May; zoo, Orissa, India	25
	estrus: Jan, Mar, May, Aug, Nov-Dec; zoo	6243
	mate: Jan-Feb; India	7652
	mate: Jan-Feb; n India	6653, 6654
	mate: Jan, Mar, May; zoo	6243
	mate: Aug-Sept; Sri Lanka	8501
	mate: yr rd; s India	6653, 6654
	preg f: Mar; n1; Nepal	7382
	birth: Feb-Mar; Pakistan	9175
	birth: Feb-Mar, July, Nov-Dec; zoo	6243
	birth: end Feb-Nov; zoo	12128
	birth: May-May, Aug-Sept, Nov-Dec; Chitwan Nat Park, Nepal	7360a
	birth: July-Aug; n India; yr rd; s India	6653, 6654
	birth: nearly yr rd; zoo	1230
	birth: yr rd; Java, Indonesia	11085
	repro: yr rd; zoo	44
Muntiacus reevesi	mate: end Jan-Feb; Kwantung, China	182
	birth: Apr-Sept, Nov-Dec; n17; zoo	12128
	birth: early June; Kwantung, China	182
	birth: yr rd, peak Apr-July; captive	2109
	birth: yr rd, peak Apr-July; captive, Japan	12030
	birth: yr rd; England, UK	1710
Odocoileus hemionus	antlers shed: Apr; few data; zoo	6243
	spermatogenesis: June-Dec; CA	6491
	estrus: Mar; few data; zoo	6243
	mate: Sept; BC, Canada	2086
	mate: end Sept-mid Nov; 10-17 d; CA	6491
	mate: Oct, fawns; captive	7587
	mate: Oct-Dec; CA	10615
	mate: mid Oct-Jan; w coast N America	2926
	mate: mid Oct-Nov; ID	9385
	mate: end Oct-Dec peak mid Nov; CA	10613
	mate: Nov usually; WY	9376
	mate: Nov-Dec; captive, Vancouver I, Canada	7589
	mate: Nov-Dec; CO	7157
	mate: Nov-Dec, rarely to mid Jan; CO	7156
	mate: Nov-Dec; WA	12072
	mate: Nov-mid Dec; CA	6386
	mate: peak mid-end Nov; WI	5210
	mate: mid Nov-Dec; CA, NV	6232
	mate: mid Nov-mid Dec; central UT	9182, 9183
	mate: peak mid Nov-early Dec, some to Jan; OR, CA	1758
	mate: early Dec; captive, AZ	7878
	mate: Dec; UT	9183
	mate: Dec-Feb; AZ	10595
	mate: mid Dec-Jan; CA	2618
	mate: mid Dec-Feb, peak mid-Dec; captive, AZ	7878
	mate: end fall-early winter; Baja CA, Mexico	6387
	ovulate: Oct-Nov; OR	5444
	ovulate: Nov; Vancouver I, Canada	10773, 10776
	conceive: Jan-Feb; AZ	10595
	conceive: end Sept-Oct; CA	6491
	conceive: mid Nov-early Feb, peak Nov-Dec; CO	7156, 7157
	birth: Mar-Nov; BC, Canada	2086
	birth: Apr; Baja CA, Mexico	10809
	birth: mid Apr-July, peak early-mid May; CA	6491
	birth: May-Aug; n11; captive	2109
	birth: last 3 wk May-early June; CA	10614, 10615
	birth: peak end May-early June; WA	10334
	birth: end May-early June; WI	5210
	birth: end May-mid July; WA	12072
	birth: early June-Aug, peak July; CA	6386
	birth: ?-June; w US	5146
	birth: June; n207; Vancouver I, Canada	10773

	birth: June; captive, Vancouver I, Canada	7587, 7589
	birth: June; WY	1871a
	birth: June; few data; zoo	6243
	birth: June; zoo	11369
	birth: peak June; ID	9385
	birth: June-mid July; central UT	9182, 9183
	birth: June-Aug; Baja CA, Mexico	6387
	birth: June-Aug; w coast N America	2926
	birth: June-early Oct, peak June-July; n22; zoo	12128
	birth: end June-early July; WY	9376
	birth: July; CA	2618
	birth: mid July; captive, AZ	7878
	birth: mid July-Sept, peak end July-early Aug; AZ	10595
	birth: end July; n1; NM	3761
	birth: end spring-early summer; AZ	7145
	wean: Aug-Sept; CA	6491
	wean: end Aug; CA	10615
	wean: Oct; UT	9183
Odocoileus virginianus	antlers shed: Feb-Apr; zoo	6243
	epididymal sperm: mid Aug-Feb, SC	8282
	sperm: peak Nov; PA	6164
	sperm: peak Nov-Dec; TX	9198
	spermatogenesis: July-Dec-Jan; MA	11856
	testes: large Aug-Jan; AR	11637
	testes: peak Oct-Nov; captive	8599
	testes: large Sept-Jan; TX	5134
	testosterone: peak Aug-Nov; captive	7102
	testosterone: peak Nov-Dec; captive	1373
	estrus: Oct-early Mar; zoo	6243
	estrus: mid-end Nov; Finland	5464
	mate: mid Jan-Mar; captive, AZ	7878
	mate: Apr-May; New Zealand	11642
	mate: Apr-July; Virgin I	11507
	mate: Apr-Oct; Suriname	1227a
	mate: June-Nov; Honduras	5861
	mate: Oct-Nov; ID	9385
	mate: Sept-Oct; TX	10730
	mate: Sept-Nov; central N America	4250
	mate: Sept-Dec; e N America	7652
	mate: end Sept-Feb, peak mid Nov; NY	9809
	mate: Oct; PA	4368
	mate: Oct-Dec; Canada	8916
	mate: Oct-Dec; NY	7235
	mate: Oct-Feb, peak early Nov; OH	7942
	mate: mid Oct-mid Jan; s IL	9282
	mate: end Oct-mid Dec; MI	8181
	mate: end Oct-Dec; ME	548
	mate: end Oct-Dec; NY	5213
	mate: Nov; zoo, Germany	627
	mate: Nov-Jan; zoo	6243
	mate: mid Nov-mid Dec; SC	8282
	mate: mid Nov-mid Mar; n387; MS	5208
	mate: peak end Nov-early Dec; NY	5214
	mate: end Dec-Feb; fawns; mid Sept-Nov; adult; SC	5329
	mate: yr rd; Venezuela	1309
	ovulate: Nov; OH	4367, 11909
	cl: none Aug-Nov; MS	5208
	conceive: end Sept-mid Dec; AR	11637
	conceive: Sept-mid Feb, peak mid Nov-early Dec; AR	11780
	conceive: end Oct-mid Feb; TX	5723
	conceive: Nov-Dec; Finland	5464
	conceive: Nov-mid Mar; n AL	6636
	birth: Mar-July; Guarica, Venezuela	10529
	birth: Mar-Sept, peak June; NY	9809

	birth: Apr-June; AR	11637
	birth: Apr-June; NY	7235
	birth: May-June; PA	4368
	birth: May-Aug, peak May-June; n16; zoo	12128
	birth: May-Sept; n177; captive	2109
	birth: mid May-early June; n10; ONT, Canada	10881
	birth: mid May-mid June; ID	9385
	birth: mid May-mid June; NY	5213
	birth: mid May-Aug, peak end May-early June; OH	7942
	birth: peak end May-early June; PA	9164
	birth: peak end May-early June; VA	7061
	birth: end May-June; central N America	4250
	birth: end May-mid Oct, peak June; TX	1506
	birth: June; ME	548
	birth: peak June; AR	11780
	birth: June; zoo, Germany	627
	birth: June-Apr; n13; captive	2109
	birth: June-mid July; e MT	2821a
	birth: June-Aug; zoo	6243
	birth: peak mid June; NY	5214
	birth: July; sw US, Sonora, Mexico	7145
	birth: end Aug; captive, AZ	7878
	birth: Sept; n1; MI	3945a
	birth: Sept; n1; MI	11257
	birth: Sept-Mar; Colombia	1013a
	birth: Sept-Apr; Suriname	1227a
	birth: mid Oct-mid Nov; New Zealand	11642
	birth: yr rd; Venezuela	1309
	preg f: end Nov-mid May, SC	8282
	lact: none Apr-June; MS	5208
	wean: July; ID	9385
Ozotoceros bezoarticus	territorial m: Apr-Oct; zoo, Germany	3446
	birth: July-Dec, peak Oct-Nov; central Brazil	8999a
	birth: Oct-Nov; Buenos Aires District, Argentina	10809
	birth: yr rd, peak Sept-Nov; Buenos Aires District, Argentina	5212a
	birth: yr rd; zoo, Germany	3447a
Pudu pudu	estrus: May; Chile	11221
	mate: Apr, July; n3; zoo	4717
	birth: Jan; n1; zoo	4717
	birth: end Apr-July; zoo	3445
	birth: Oct-Feb; zoo, NY	6709a
	birth: est Nov-Jan; w S America	11642
	birth: Dec; Chile	11221
Rangifer tarandus	tubluar sperm: Sept-early Dec; NWT, Canada	7055
	spermatogenesis: mid Sept-mid Nov; Sweden	7240
	testosterone: high Sept-Nov; captive	11644
	estrus: early Oct; Finland	5934
	mate: Mar-Apr; S Georgia, Antarctica	6319, 6320
	mate: mid Mar; S Georgia, Antarctica	1082
	mate: end Aug-Sept; captive	2109
	mate: end Aug-mid Oct; AK	8210
	mate: end Aug-Oct; AK	4184
	mate: Sept-Oct; captive, USSR	8744
	mate: Sept-Oct; AK	2561, 11640
	mate: Sept-Oct; Canada	1474
	mate: Sept-Oct; zoo	6243
	mate: Sept-Nov; USSR	4643
	mate: mid Sept-Oct; AK	7625
	mate: mid Sept-Oct; captive	8210
	mate: end Sept; Swedish Lapland	3091
	mate: end Sept-mid Oct; QUE, Canada	802
	mate: end Sept-Oct; n N America, n Europe, n Asia	7652
	mate: end Sept-Oct; WI	5210
	mate: Oct; Canada	2288, 2289

mate: Oct; NEW, Canada	803, 805
mate: Oct; nw AK	6377
mate: Oct-mid Nov; captive	7240
mate: Oct-Nov; Mackenzie, BC, Canada	5637
mate: Oct-Nov; NWT, Canada	7055
mate: Oct-Nov; captive	804, 806
mate: mid Oct-Nov?; Canada	4389
conceive: Oct; Canada	2288
birth: Apr-June; AK	8210
birth: Apr-June; zoo	6243
birth: Apr-June; n24; zoo	12128
birth: Apr-Aug; n3; zoo	7425
birth: mid Apr; AK	11640
birth: mid Apr-mid May; captive	2109
birth: May; n1; Scotland, UK	2696
birth: May; s Norway	9021
birth: May-June; Canada	4389
birth: May-June; Canada	1474
birth: May-June; n N America	804, 805, 806
birth: May-June; USSR	4643
birth: mid May; AK	7625
birth: end May-early June; captive	2499
birth: end May-June; Mackenzie, BC, Canada	5637
birth: end May-mid June; nw AK	6377, 6380
birth: end May-early July; NWT, Canada	7055
birth: June; AK	1568
birth: June-July; Canada	2288
birth: Nov; S Georgia, Antarctica	6319, 6320
wean: Aug-Sept; Mackenzie, BC, Canada	5637
wean: Sept; Canada	4389

Order Artiodactyla, Family Giraffidae

Giraffes (*Giraffa camelopardalis* 500-1930 kg) and okapis (*Okapia johnstoni* 200-250 kg) are terrestrial herbivores in Africa south of the Sahara. Male giraffe are larger than females, while female okapis are reported slightly larger than males (2109, 5762). The epitheliochorial, cotyledonary or diffuse placenta implants into a bipartite or bicornuate uterus (297, 4241, 6634, 7563, 7711, 8160, 8161) and is occasionally eaten (2109, 3441, 6034, 8290, 8798, 8823). Neonates are mobile, furred, and have open eyes (2109, 5153, 5762, 6034, 8290, 8798, 8823, 9191). Fetal ovaries have corpora lutea with progesterone concentrations similar to those of preimplantation gestation (3881, 5600). These fetal corpora lutea do not disappear until puberty (5600).

One captive giraffe returned to estrus 49 d after birth (5935). Giraffes have 2-4 mammae (331, 8798). Males have seminal vesicles and a sigmoid penis (11243). Details of anatomy (3823, 4240, 4242, 5600, 5627, 8160), endocrinology (5549, 6582; *Giraffa* 3881, 5627, 11704; *Okapia* 6580, 6581), birth (8823), and milk composition (408, 4243, 9770) are available.

NEONATAL MASS

Giraffa camelopardalis	f: 30 kg; n1	6178
	f: 34-45 kg; n1	6034
	36-45 kg	3441
	m: 39 kg; n1	2109
	r48.5-63.5 kg; n4	9514
	m: 50-55 kg; n1	6178
	~54 kg	5153
	54.5 kg	3247
	54.5 kg; n16	2212
	67.5 kg; n1 stillborn	9068
	77 kg; n1	11785
	f: 89.0 kg	4239
	m: 95.1 kg	4239
	102.0 kg	10030

Okapia johnstoni	f: r14-22.5 kg	6191
	17 kg; r14.5-19.5; n4	8823
	18 kg; n1	8798
	18.5 kg	6191
NEONATAL SIZE		
Giraffa camelopardalis	164 no units; n1; stretched out, tip nose to base of tail	11785
LITTER SIZE		
Giraffa camelopardalis	1	9068
	1	9064
	1; 2 occasional	5153
	1-2	4241
	1 yg ea; n4 lt	319
	1; n40 lt	3431
	1 ea; n77 lt	8699
	1.12 yg; r1-2; n8 lt	9066
	2 stillborn; n1 lt	2212
Okapia johnstoni	1; n1	6184
	1; n1 lt	2109
	1 emb; n1 f	7711
SEXUAL MATURITY		
Giraffa camelopardalis	f: 1st mate: 56 mo	4239
	f: 1st conceive: 3.83 yr; ± 3 mo; captive	4241
	f: 1st conceive: 4.67 yr, ± 3 mo; wild	4241
	f: 1st preg: 2nd half 4th yr	8798
	f: 1st preg: 6 yr; ±3 mo	4241
	f: < 4 yr	5762
	f: 4.67 yr	10158
	m: spermatogenesis, testes large: 3-4 yr	4242
	m: min 2.67 yr; 3-4 yr	10158
	m: 7 yr	5762
	f/m: 3 yr; n1 pr	318
	f/m: 3-4 yr	8798
Okapia johnstoni	f: 1st mate: 2.83 yr; n1	8798
	f: 1st conceive: 2 yr	6191
CYCLE LENGTH		
Giraffa camelopardalis	estrus: max 24 h	2109
	r14-19 d; n6 cycles, 1 f	2109
	15 d	2439
	15.5 d; n12 cycles, 4 f	6582
Okapia johnstoni	estrus: 24 h	8798
	estrus: several d	6179
	~14.5 d	6581
	15.2 d; n12 cycles, 4 f	6582
	r20-21 d; n3 cycles, 1 f	4860
	3 wk	8798
GESTATION		
Giraffa camelopardalis	400-480 d	2449
	13.75 mo; r13 mo 4 d, 14.5 mo; n2	6178
	~14 mo	4582
	~14 mo; n1	10612
	427-488 d	5762
	434 d; n1	11785
	14.5 mo	3441
	~14.5 mo	9068
	444 d	8161
	r445-446 d; n10	2439
	450 d	8421
	450 d	3247
	15 mo; n5	3441
	15 mo; n35	2212
	455.25 d; r450-461; n4	12088
	457 d	2109
	457 d	4239, 4240
	462 d; r459-465; n3	9514
	468 d; n1	11704

	488 d; n1	9191
Okapia johnstoni	435-449 d; n1; 421 d; n1	8290
	441 d; n1	2109
	427-457 d	5762
	439.6 d; r427-452; n6	6191
	444 d; r435-453; n5	8823
	446 d; n1	6184
	450 d	8421
	15 mo	6581
LACTATION		
Giraffa camelopardalis	solid food: 13 d; n1	2109
	solid food: 2-3 wk	8798
	solid food: 2 wk	5762
	wean: min 1 mo	3431
	wean: min 1 mo; wild; to 1 yr; captive	5762
	ruminate: 6-8 wk	8823
	wean: 6 mo	8823
	wean: 6 mo	2439
	wean: 6-8 mo	6206
	wean: 7-11 mo	6342
Okapia johnstoni	solid food: 6 wk	2109
	ruminate: 6-7 wk	8823
	wean: 6 mo	8823
	wean: 6-9 mo	8798
INTERLITTER INTERVAL		
Giraffa camelopardalis	14 mo	10158
	19.9 mo; ±0.6; wild	4241
	21-22 mo; n4; yg live	3431
	21.5 mo; ±0.7; captive	4241
	645 d	4239
	2 yr; yg live	5762
SEASONALITY		
Giraffa camelopardalis	mate: yr rd; Africa, south of Sahara	8798
	mate: yr rd; S Africa	4239
	conceive: yr rd, 60% Dec-Mar; n143; Transvaal, S Africa	4244
	mate: est June-Sept; Okawango, Angola	11697
	birth: Jan-Mar; ne Uganda	5762
	birth: May-Aug; Serengeti, Tanzania	5762
	birth: not after Apr; n Uganda	5762
	birth: Feb-Mar, Aug-Oct; S Africa	3145
	birth: Mar-Sept; n32; Africa, south of Sahara	8798
	birth: Mar, Oct; n2; zoo	2109
	birth: May-Sept; Zimbabwe	2275
	birth: Mar-?; Okawango south bank, Angola	11697
	birth: yr rd, except Dec; captive	2109
	birth: yr rd, peaks Jan-Apr, Sept-Oct; S Africa	8514
	birth: yr rd, peaks Feb-Mar, Aug-Oct; S Africa	3145
	birth: yr rd, peak near end of rains; Luangwa valley, Zambia	843
	birth: yr rd; Kenya, Uganda, Tanzania	5762
	birth: yr rd, may peak dur dry season; Kenya	3431
	birth: yr rd; Transvaal, S Africa	5153
	birth: yr rd; Zambia	331
Okapia johnstoni	birth: Oct; n1; zoo	2109
	birth: yr rd; n125; zoo	8823

Order Artiodactyla, Family Bovidae

Bovids (*Addax* 60-125 kg; *Aepyceros* 34-83 kg; *Alcelaphus* 100-218 kg; *Ammodorcas* 23-34 kg; *Ammotragus* 40-145 kg; *Antidorcas* 26-48 kg; *Antilocapra* 28-70 kg; *Antilope* 19-57 kg; *Bison* 350-1020 kg; *Bos* 272-1000 kg; *Boselaphus* 108-270 kg; *Bubalus* 228-1200 kg; *Budorcas* 196-350 kg; *Capra* 16-150 kg; *Capricornis* 30-140 kg; *Cephalophus* 4-80 kg; *Connochaetes* 118-295 kg; *Damaliscus* 56-172 kg; *Dorcatragus* 9-27 kg; *Gazella* 2.5-98 kg; *Hemitragus* 39-100 kg; *Hippotragus* 150-300 kg; *Kobus* 50-309 kg; *Litocranius* 28-52 kg; *Madoqua* 3-7 kg; *Nemorhaedus* 22-35 kg; *Neotragus* 2-9 kg; *Oreamnos* 45-154 kg; *Oreotragus* 8-16 kg; *Oryx* 91-238 kg; *Ourebia* 8-21 kg; *Ovibos* 143-650 kg; *Ovis* 19-200 kg; *Pantholops* 25-55 kg; *Pelea* 20-30 kg; *Procapra* 20-40 kg; *Pseudois* 25-80 kg; *Raphicerus* 6-16 kg; *Redunca* 14-95 kg; *Rupicapra* 16-50 kg; *Saiga* 21-69 kg; *Sigmoceros* 125-204 kg; *Sylvicapra* 12-25 kg; *Syncerus* 265-900 kg; *Tetracerus* 17-21 kg; *Tragelaphus* 24-1000 kg) are large, often diurnal, terrestrial herbivores from North America, Eurasia, and Africa. Males are usually larger than females (269, 491, 546b, 1237, 1432, 1649, 1890, 2079, 2439, 3571, 4643, 4645, 4656, 4832, 4950, 5260, 5300, 5763, 5764, 6189, 6192, 6241, 6387, 6407, 6409, 6525, 6584, 6922, 7144, 7855, 7884, 8477, 8593, 8740, 8972, 9121, 9122, 9175, 9540, 9541, 9545, 9547, 9636, 10027, 10637, 10740, 11697, 11784, 11787, 12019) except in *Capricornis* (7389; cf 5947), *Madoqua* (5625, 9944, 9945), *Neotragus* (2716), *Ourebia* (5763, 10631), *Raphicerus* (9945; cf 11783), and *Sylvicapra* (4926, 5763, 11789) where females are reported larger. For *Nemorhaedus* (9547) the sexes are reported equal in size.

Female *Tragelaphus* have an extra chromosome (3492). The 1-3 kg, epitheliochorial or syndesmochorial, villous, cotyledonary placenta implants into a bicornuate or duplex (*Addax* 2487; *Connochaetes* 10472) uterus (22, 239, 269, 271, 297, 1424, 1435, 1649, 2296, 2308, 2746, 4190, 4907, 5597, 5625, 5626, 5699, 6043, 6044, 6525, 7563, 7616, 7617, 8008, 9330, 9415, 9593, 9715, 10271, 10310, 11790, 11797, 11854, 11860). Only the right uterine horn is used in *Aepyceros* (4907, 5597, 6352, 7562, 10158), *Antidorcas* (10038, 11211), *Antilope* (22), *Gazella* (5122, 9180; cf 6044), *Kobus* (1434, 6044, 9181, 10158, 10450, 11731a; cf 1431), *Madoqua* (5625), *Neotragus* (6581a), *Oreotragus* (10158), *Oryx* (10154), and *Sylvicapra* (1791, 6352, 10610). The right ovary is preferentially used in *Kobus* (11731a).

In utero siblicide occurs (*Antilocapra* 8008). Placentophagia is common (208, 269, 785, 1047, 2081, 3099, 3100, 3667, 3871, 3955, 4086, 4196, 5018, 5115, 5229, 5248, 5249, 5258, 5763, 5789, 5828, 5876, 6801, 9311, 9593, 9617, 9691, 10158, 10269, 10638, 10939, 11044, 11506, 11758, 11789, 11854; cf 2296, 2746, 3105, 6657, 8641, 12107). Estrus may return within a month of birth (269, 270, 493, 922, 1434, 2608, 2815, 4972, 5229, 5763, 5764, 5896, 5999, 7541, 7616, 9691, 10158, 10269, 10628, 11697, 11758; cf 1432, 2605, 11253), except in *Bos* (10905, 11088). *Connochaetes* births are synchronous (3099, 3102, 3105, 9993; *Ovis* 8844).

Neonates are mobile, furred and have open eyes (207, 334, 459, 560, 561, 1047, 1474, 1649, 1684, 2079, 2296, 2598, 2602, 2746, 3099, 3102, 3105, 3146, 3377, 3571, 3667, 3871, 4254, 4643, 4645, 4950, 4972, 5229, 5249, 5258, 5763, 5764, 5828, 5876, 5996, 6586, 6657, 6801, 7051, 7395, 7625, 7910, 8458, 8739, 8798, 9121, 9545, 9945, 9994, 10043, 10158, 10216, 10535, 10638, 10946, 11021, 11044, 11253, 11433, 11578, 11789, 12019; cf 7910).

Females have 2 mammae (*Alcelaphus* 9945; *Capra* 2081, 3571; *Connochaetes* 324, 331, 9945, 10158; *Damaliscus* 324, 331, 9945, 10158; *Ovis* 1474, 3571; *Sigmoceros* 331, 333, 9030), 2-4 mammae (*Antidorcas* 324, 921, 9945, 10158; *Ovibos* 1474, 2364), or 4 mammae (*Aepyceros* 324, 331, 333, 10158; *Antilocapra* 1474, 8009; *Bison* 1474, 3571; *Budorcas* 3034, 8798; *Capricornis* 3034; *Cephalophus* 324, 331, 10158; *Hippotragus* 324, 10158; *Kobus* 331, 333, 374, 9945, 10158; *Madoqua* 324, 5625, 10158; *Nemorhaedus* 9175; *Neotragus* 324, 10158; *Oreamnos* 1474, 9122; *Oreotragus* 324, 331, 10158; *Oryx* 324; *Ourebia* 324, 331, 10158; *Pelea* 324, 10158; *Raphicerus* 324, 9945, 10158, 10392; *Redunca* 324, 331, 9945, 10158; *Rupicapra* 2079, 3571; *Sylvicapra* 331, 333, 10158; *Syncerus* 331; cf 8515; *Tragelaphus* 324, 331, 10158).

Males have seminal vesicles (1424, 10602). *Damaliscus* and *Kobus* may mate on leks (546a, 546b).

Details of anatomy (3715, 3823; *Aepyceros* 3048, 5596; *Antilocapra* 8010; *Capricornis* 5825; *Gazella* 6922, 10271; *Madoqua* 5606, 5625; *Ovibos* 9330; *Kobus* 7540; *Rupicapra* 3715; *Tragelaphus* 1381), intersex (10031), endocrinology (9428a, *Addax* 2486; *Aepyceros* 1227, 3148, 5132; *Alcelaphus* 5608; *Ammotragus* 3332; *Antilope* 4860, 4938; *Bubalus* 380, 381, 467, 649, 650, 4191, 4615, 5496, 5509, 8206, 8926, 8927; *Damaliscus* 10029; *Oryx* 2818; *Ovibos* 2498, 9331; *Ovis* 8897, 11645; *Redunca* 10025; *Rupicapra* 3663), spermatogenesis (938, 10817, 10818), birth (2296, 3105, 3955, 5018, 6586, 7068, 9971), milk composition (408a, 561, 1776, 1927, 2035, 2079, 2581, 3417, 4116, 4118, 4643, 4950, 4972, 5595, 6058, 6246, 6522, 8690, 8892, 9594, 10299,

10300, 10620, 10739, 10740, 11214, 11291, 11350), reproductive energetics (789a, 4047, 4048), and light (270, 9318) are available. Supplements 30 and 43 of the Journal of Reproduction and Fertility, published in 1981 and 1991, review the reproduction of domestic ruminants (cattle, sheep, and goats).

NEONATAL MASS

Addax nasomaculatus	f: 4.8 kg	2603
	m: 5.6 kg; r4.5-6.4; n4	6802
	f: 5.9 kg; r4.5-7.3; n4	6802
	6.1 kg; n1	6819
Aepyceros melampus	~5 kg	3146
	6.1 kg; n1	633
Alcelaphus buselaphus	m: ~5-6 kg; n1	9741
	12.5 kg	3247
Ammotragus lervia	4.5 kg	4020
Antidorcas marsupialis	3.82 kg; se0.12; n13	11213
	m: 4 kg	10040
Antilocapra americana	1.04-5.78 kg	11208
	2.3-3.2 kg	12019
	3.0 kg; r1.5-4.16	2607
	~3.2 kg	4647
	3.5 kg	8009
	f: 3.96 kg; n46, 1-6 d	1515a
	m: 4.13 kg; n44, 1-6 d	1515a
Antilope cervicapra	3-4 kg	7616
	3.276 kg; r2.6-4.2; n33	27
	4.5-5.0 kg	10628
Bison bison	est 15-25 kg	9386
Bison bonasus	22-23 kg	4643
	24.25 kg; r24.0-24.5; n2	9505
Bos frontalis	m: 13 kg; n1	2109
	f: 23.1 kg; r19.5-26.5; n4	9593
	m: 33.0 kg; n1	9593
Boselaphus tragocamelus	5.875 kg; r5.5-6.25; n2	9475
Capra caucasica	3.5-4.2 kg	4643
Capra hircus	m: 1.55 kg; n1 stillborn	7051
	f: 1.7 kg; n15	9223
	m: 1.9 kg; n11	9223
	f: 2.5 kg; se0.23; r2.0-3.0; n4	11054
	m: 2.9 kg; se0.18; r2.5-3.3; n4	11054
	<3 kg	11389
Capra ibex	m: 2.21 kg; n1 term emb	2081
	m: 2.55 kg; se2.43; r1.7-3.15; n5	8594
	f: 2.75 kg; se3.21; r1.97-3.45; n5	8594
	2.89 kg; n20	8594
	2.95 kg; r2.9-3; n2	8178
	3.5-4 kg	4643
Capricornis crispus	3.313-3.708 kg; extrapolated	10535
	3.5 kg; n1	5177
Cephalophus dorsalis	m: 1.60 kg; n1	2605
	f: 1.62 kg; n1	2605
Cephalophus maxwelli	710-954 g	8866
Cephalophus monticola	10% of mother's weight	5763
Cephalophus rufilatus	m: 0.79 kg; n1	2605
	f: 1.14 kg; n1	2605
Cephalophus sylvicultor	6 kg; n1, 4 d	6642
Cephalophus zebra	1.615 kg; n1	9691
	~1.8 kg est n1	11021
Connochaetes gnou	~8 kg	5901
Connochaetes taurinus	f: ~15 kg	10638
	16.5 kg; n1 emb	10272
	m: ~19 kg	10638
	~22 kg	10158

Damaliscus dorcas	6.36 kg	5702
	6.4-7.3 kg	2746
	7 kg	10023, 10027
Damaliscus hunteri	8.35 kg; n1	5249
Damaliscus lunatus	10-12 kg; term emb	1792
Gazella dorcas	1.65 kg; sd0.23; n33	6764
Gazella granti	5.2 kg	3247
	m: 5.4 kg; n1	2603
	5.63 kg; r5.2-6.2; n3 near term	10271, 10272
Gazella leptoceros	f: 4.64 kg; n1	2603
Gazella spekei	f: 1.150 kg; r0.7-1.7	8991
	1.25 kg; n26	8991
	m: 1.36 kg; r0.7-1.8	8991
Gazella subgutturosa	2.4 kg; r2-3.2; n10	4643
	r2.6-3.1 kg; n5	4643
Gazella thomsoni	2.315 kg; r2.0-3.5; n64	1196
	~2.5 kg	9180
	2.7 kg	3247
Hippotragus equinus	f: 11 kg; n1	334
	f: 16.5 kg; n1	2602, 2603
	m: 18 kg; n1	2603
Hippotragus niger	m: 13.6 kg; n1	4085
	m: 13.75 kg; r12.3-14.5; n4	6802
	f: 14.0 kg; r11.4-16.1; n4	6802
	15-20 kg	5972
	f: 16.5 kg; n1	2603
	m: 18.95 kg; se0.76; r13.50-19.80; n26	4086
	f: 20.33 kg; se1.05; r13.50-22.50; n22	4086
Kobus kob	f: 5.405 kg	1434
Kobus leche	5.1 kg; n5	9181
Madoqua kirki	f: 0.6 kg; n1	2603
	0.605-0.795 kg; n5	2598
Madoqua saltiana	f: r0.56-0.68 kg	5763
	m: r0.725-0.795 kg	5763
Nemorhaedus goral	~2 kg; ~10-15 d before birth	4643
Neotragus batesi	max 0.5 kg; term emb	2716
Oreamnos americanus	f: 2.95 kg; n1	1237
	m: 3.19 kg; sd0.43; n8	5112
	3.32 kg; sd0.50; n17	5112
	f: 3.45 kg; sd0.50; n9	5112
Oreotragus oreotragus	1.130 kg; n1 term emb	11788
Oryx dammah	m: 10.05 kg; n1	2603
Oryx gazella	m: 11.14 kg; r9.5-12.3; n7	6802
	f: 12.12 kg; r11.4-13.6; n3	6802
	12.8 kg	11212
Ourebia ourebi	m: 2.2 kg; n1	2603
	2.27 kg; n1	6315
Ovibos moschatus	7.3 kg; n1	4985
	r9-ll kg; n14	11639
	f: 10.5 kg; n7; 2 stillborn twins: 7, 8 kg	11705
	f: 14.0 kg; r9.1-18.1; n29	6241
	m: 15.5 kg; r10.4-20.9; n32	6241
Ovis ammon	2.25-2.5 kg	3571
	2.8-6 kg; to 6 d; includes subspecies variation	4643
Ovis aries	m: 1.8 kg; r1.4-2.4; singletons; 1.5 kg; 1.1-1.8; twins	5307
	f: 1.8 kg; r1.4-2.5; singletons; 1.5 kg; 1.2-1.8; twins	5307
	1.9 or 2.0 kg; r0.85-2.9	1449
	2.8 kg; se0.21; r2.0-4.3; n10	11054
	3.5 kg; se0.18; r2.5-4.3; n11	11054
Ovis canadensis	3.3 kg; r2.8-3.7; n9	3667
	~3.6 kg	4333
	4.2 kg; se0.27; n16	4482
	4.4 kg; se0.37; n4	4482
	4.5 kg; r4-5; n4	11910

	5.4 kg; near term	3417
	f: 5.4 kg; r5.3-5.5; n2	1021
Ovis dalli	3.2-4.1 kg	7884
	3.3 kg; sd0.4	1446, 9732
	3.3 kg; r2.9-3.6; n5	3667
	4.025 kg; n1 fetus	7883
Ovis musimon	2 kg; r1.6-2.6	8467
	2.25 kg; r2.0-3.4	1449
	f: 2.523 kg; r2.3-3.07; n5	9415
	2.7 kg; n1	10946
	m: 2.852 kg; r2.1-3.39; n5	9415
Procapra gutturosa	2.8-3 kg	4643
Raphicerus campestris	0.84 kg; term emb; 0.936 kg newborn	11790
	1 kg; n1	5763
Redunca arundinum	4.5 kg	5763
Redunca fulvorufula	2.5 kg? term emb, from graph	5165
	>3 kg	5763
Rupicapra rupicapra	f: 1.89 kg; n1 near term	2079
	2-2.5 kg	2079
	2-3 kg	8477
	m: 2.38 kg; n1 near term	2079
	2.8-3.2 kg; 3-5 d	4643
Saiga tatarica	m: 3.135 kg; n4	8641
	3.5 kg; r2-4.4	560
	3.5 kg; m ~200 g heavier; r2-4.4 kg; n45 f, n45 m	4643
	f: 3.657 kg; n6	8641
Sigmoceros lichtensteini	~15 kg	10158
Sylvicapra grimmia	f: 1.35-1.8 kg; n3	11789
	1.61 kg	334
	m: 1.65 kg; n1	11789
Syncerus caffer	f: 34 kg; r32-36; n2	8515
	m: 37.2 kg; r26-42; n6	8515
	40 kg; r38-45	9994
	47 kg; r37-54; n3	11292
	f: 50.0 kg; n1	11292
	m: 53.6 kg; n1	11292
Tetracerus quadricornis	0.942 kg; r0.740-1.065; n7	27
	1.15 kg; r1.0-1.3; n2	9956, 9957
	1.04 kg; r0.75-1.2; 0.75 kg yg died 13 d	37
Tragelaphus angasi	4.15-5.5 kg	6525
	5.6 kg; n10	270
	m: 5.68 kg; n12	269
Tragelaphus eurycerus	16-18 kg	11022
	19.5 kg; n3	8867
	20.4 kg; n1	2720
	22.3 kg; n1	11973
Tragelaphus imberbis	m: 5.2 kg; n1	6411
	m: 6.1 kg; r5.3-7.6; n8	6189
	f: 6.3 kg; n1 term emb	6411
Tragelaphus oryx	22-36 kg	5763
	f: 28.4 kg; n21	10895
	f: 28.6 kg; r24.5-32.7; n4	8690
	f: 29.1 kg	11784
	m: 31.3 kg; n25	10895
	31.5 kg	10894
	f: 32 kg	10158
	m: 36 kg	10158
	m: 36.4 kg; r32.7-40.9; n3	8690
	m: 36.8 kg	11784
Tragelaphus scriptus	m: 2.948 kg; r2.778-3.118; n2, 1-7 d	11787
	f: 3.044 kg; r2.664-3.203; n4, 1-7 d	11787
	f: 3.375 kg; n1	201
	3.5-4.5 kg	10158
	m: 4.175 kg; n1	201

Tragelaphus spekei	~4 kg; n4	6192
Tragelaphus strepsiceros	~15 kg; n1	11291
	m: 16.3 kg; n1, 3 d	11781
NEONATAL SIZE		
Antilocapra americana	58.4 cm	11208
Aepyceros melampus	~50 cm	5597
Bison bonasus	88.5 cm; r87-90; n2	9505
Capra hircus	m: HB: 37 cm; n1 stillborn	7051
Capricornis crispus	48-50 cm; extrapolated	10535
Cephalophus monticola	m: 31 cm; n1	185
Gazella granti	f: max 62.5 cm	10271
	m: max 71.0 cm	10271
Gazella subgutturosa	54.3 cm; r49.0-59.0; n10	4643
	r55-58 cm; n5	4643
Hippotragus equinus	33 cm	5764
Kobus leche	70.3 cm; n1 term emb	9945
Oreamnos americanus	TL: 55.9 cm; n1	1237
Oreotragus oreotragus	f: 39 cm; emb	9945
Ovis dalli	63.7 cm; n1 fetus	7883
Procapra gutturosa	51.35 cm; r51-56; n2	4643
Redunca fulvorufula	HB: 57.5 cm; term fetus; from graph	5165
Saiga tatarica	60 cm; r53-67	560
Tragelaphus angasi	81-83 cm	6525
Tragelaphus oryx	body: 52.0 cm	10894
Tragelaphus scriptus	m: 56.1 cm; n1	11787
	f: 58.2 cm; r54-61; n4	11787
WEANING MASS		
Antilocapra americana	8.9 kg; 2 mo	11596
Bison bison	135-180 kg; 8-9 mo	7143
Ovis canadensis	f: ~27 kg; 6 mo	4333
	m: ~29 kg; 6 mo	4333
Ovis musimon	17.50 kg; 180 d	8467
LITTER SIZE		
Addax nasomaculatus	1 yg ea; n5 lt	2109
	1.002 yg; 0.2% twins; n1012 lt	2487
Aepyceros melampus	1; twins 1 ignored & died	10158
	1 yg	5259
	1 emb ea; 2 f	6612
	1 emb; n7 f	4907
	1 yg ea; n58 f	2274, 2275, 7562
	1 yg; n144 lt	1230
	1; sometimes 2	331, 333
Alcelaphus buselaphus	1	3956
	1	5764
	1 yg; n1 lt	2603
	1 yg; n1	9741
	1 yg; n8 lt	1230
Ammodorcas clarkei	1	7167
Ammotragus lervia	1.01 yg; 1% twins; r1-2; n96 lt	1230
	1.07; n190 lt	12128
	1.2 yg; r1-2; n162 lt	2109
	1.3 emb; mode 1; r1-3; n44 f	4020
	1.35 emb; n20 f	4020
	1.5; r1-2; n4 lt, 1 f	9734
	1.54 yg; r1-2; n13 lt	9066
Antidorcas marsupialis	1	10038
	1 yg; 1 lt	2109
	mode 1; 2 rare; r1-2	8351
	1.01 yg; 1.4% twins; r1-2; n146 lt	1230
Antilocapra americana	1-2	10809
	1.5; r1-2; n4	11370
	1.78; r1-2; n14 lt	2607
	1.80; mode 2; r1-3	11208
	1.94 emb; r1-2; n18 f	1759

	2	5018
	mode 2	8644
	mode 2; 1,3 rare	4647
	2; r1-2; at first birth 1 fawn, thereafter 2	12019
	2; n1 lt	7376
	2 yg; n1 lt	2109
	2 emb; n5 f	7145
	2 emb; n6 f	1424
	2 emb; n6 f	10924a
	2 yg; n17 lt	5828
	2.1 emb; mode 2 92%; r1-3; n37 f	6217
	4-6 cl	11860
	5.1 cl; r2-7; n66	8008
Antilope cervicapra	1-2	6653, 6654
	1 ea; n4 lt	6541
	1 yg ea; n38 lt	9066
	1 yg ea; n46 lt	1230
	1 yg ea; n97 lt	2109
	1 yg ea; n338 lt	7616
Bison bison	1	7144
	mode 1 yg; r1-2	7068
	mode 1; 2 occasional	10216
	1; 2 occasional	7068
	1 yg ea; n9 lt	1230
	1 yg ea; n19 lt	9066
	1 yg ea; n91 lt	2109
	1 yg ea; n111 lt	4047
	1 emb ea; n481 f	3559, 3560
	2 very rare	4501
Bison bonasus	1	3571
	1	4643
	1 yg	3822a
	1 yg ea; n2 lt	2109
	1 yg ea; n2	9505
	1 ea; min n132 lt	5999
	2 rare	5229
Bos frontalis	1	9545
	mode 1	7161
	1-2	2802
	1 yg; n16 lt	2109
	1 ea; n16 lt	12128
Bos grunniens	mode 1	12128
	1 yg ea; n2 lt	9066
Bos javanicus	1	6652
Boselaphus tragocamelus	1-2	6654
	mode 1; r1-2	12128
	1 yg ea; n8 lt	2109
	1.3 yg; r1-2; n17 lt	9066
	1.3 yg; r1-2; n~60 lt	1230
	1.5; r1-2	9175
	2 yg; n1 lt	3871
	2 yg; n1 lt	6145
	2 yg; n1 lt	30
Bubalus bubalis	1-2	6654
	1 yg; r1-2	8501
	1 yg ea; n27 lt	1230
	2 yg	7382
Bubalus depressicornis	1 ea; n7 lt	12128
	1 yg ea; n8 lt	1230
Budorcas taxicolor	1	11419
	1 yg	9547
	1; seldom 2	8798
	maybe 1-2/lt; anecdotal	154
	1 yg; n1	459

Capra caucasica	1	4643
	mode 1	8458
	1.08; r1-2; n13 lt	12128
Capra cylindricornis	1.2; r1-2; n9 lt	8328
Capra falconeri	2 common; r1-2	9175
	2 yg; n1 lt	2109
	3 nursing; 1 lt	8694
Capra hircus	1-2	4643
	1-2 yg	9223
	1-2 yg	9175
	mode 1; 2 rare; r1-2	2080
	1-3 yg	11389
	1 emb; n1 f	6293
	1.22 yg; r1-2; n9 lt	2109
	1.25; r1-2; n8 lt	12128
	1.25 yg; r1-2; n60 f	9357
	1.26 emb; r1-2; n27 f	9357
	1.5 emb; r1-2	12020
	1.63 cl; n27 f	9357
	1.93 yg; n15 lt	4934
	2.05 yg; n18 lt	4934
Capra ibex	1; seldom 2	3571
	1; 2 rare	2081
	1-2	6652
	mode 1; r1-2	4643
	1 stillborn	8707
	1.1; mode 1; r1-2	6764
	1.11; r1-2; n213 lt	7910, 7911
	1.20 yg; mode 1; r1-2; n365 lt	10530
Capricornis crispus	1 yg; 1 lt	5177
	1 yg; 1 lt	5948
	1.01; r1-3; n209 f; 1 set each twins, triplets	7390, 10535
Capricornis sumatraensis	1 yg; n1 lt	496
Cephalophus dorsalis	1 yg ea; n2 lt	2603, 2605
Cephalophus leucogaster	1 yg ea; n3 lt, 1 pair	2109
Cephalophus maxwelli	1	74
	1 yg	4354a
Cephalophus monticola	1 emb; n1 f	11342
	1 emb; n1 f	331
	1 yg ea; n12 lt	1230
	1 yg ea; n16 lt	2109
	1 ea; n39 lt	11349
Cephalophus natalensis	1 yg ea; n7 lt	1230
Cephalophus ogilbyi	1	5277
Cephalophus rufilatus	1 yg ea; n3 lt	2603, 2605
Cephalophus sylvicultor	1	5277
	1	254
	1	5575
	1 yg	5996
Cephalophus zebra	1 yg	11021
Connochaetes gnou	1 yg	5901
	1 yg ea; n5 lt	2109
	1 yg ea; n66 lt	1230
	1.02; r1-2; n50 lt	4557
Connochaetes taurinus	1	10158
	1	10638
	1 emb	2275
	1 yg; n1 lt	2109
	1 yg ea; n3 lt	9066
	1 yg ea; n43 lt	1230
	mode 1; r1-2	2449
	1; 2 rare	331
	2 rare	8514

Damaliscus dorcas	1 yg	2746
	1 yg; n1 lt	2603
	1 yg ea; n74 lt	1230
	2 yg; n1 lt	7774
Damaliscus hunteri	1; n4 lt	5249
Damaliscus lunatus	1	331
	1 emb; n1 f	1792
	1? yg ea; n2 lt	1230
Gazella cuvieri	1.41 yg; r1-2; n195 lt	8100
	1.5 yg; n18 lt	3567
Gazella dama	1	5575
	1	254
	1 yg	3567
Gazella dorcas	1	493
	1	8223
	1; 2 rare	5997
	1-2	5575
	1-2	254
	1 yg; 1 lt	2109
	1 yg ea; n4 lt	2600, 2603
	1; n7	12128
	1; mode 1; n33 lt	6764
Gazella gazella	1-2	9175
	1-3 yg	2109
	1 yg; 2 rare	492, 493
	1 emb; n1 f	6293
Gazella granti	1 yg; 1 lt	9294
	1 yg ea; n3 lt	2600, 2603
Gazella leptoceros	1 yg	3567
	1 yg ea; n2 lt	2600, 2603
Gazella rufifrons	1 yg	3567
	1 yg ea; n2 lt	2600, 2603
Gazella spekei	1 yg	3567
Gazella subgutturosa	1-2	4643
	1-2; one third of births are twins	9175
	est 1.5	3567
	1.3; r1-2	12128
	1.3; r1-2; n3 lt	5876
	2 yg; n1	2273
Gazella thomsoni	1	6922
	1	4972
	mode 1	2449
	1; n1 lt	9180
	1 emb ea; n2 f	6612
	1.02 yg; r1-2; n42 f	5122, 5123
Hemitragus hylocrius	1 yg ea; n7 lt, 2 f	8693
	1 yg ea; n22 lt	11750
	2	6654
Hemitragus jemlahicus	mode 1	6652, 6654
	1 yg ea; n105 lt	1230
	1.01 emb; n157 births	1669
	1.01 yg; r1-2; n180 lt	2109
	1.09; r1-2; n115 lt	12128
Hippotragus equinus	1	5764
	1	11253
	1	11758
	1	2449
	1 emb; n1 f	6612
	1 yg; n1 lt	5248
	1 emb ea; n2 f	5827
	1 yg ea; n2 lt	2602, 2603
Hippotragus niger	1	11758
	1 ea; n2 lt	5972
	1 yg ea; n3 lt	2603

	1 ea; n4 lt	1437
	1 yg ea; n18 lt	1230
Kobus ellipsiprymnus	1	8350
	1	11253
	1	5763
	1 yg	2109
	1; r1-2	333
	1; perhaps 2	331
	mode 1; r1-2	2449
	mode 1; r1-3	10158
	1 emb; n1 f	186
	1 emb; n1 f	6612
	1 emb ea; n3 f	5052
	1 yg ea; n62 lt	1230
	2; n1 lt	9248
Kobus kob	1	1432
	1 yg ea; n11 lt	2109
	1 ea; n17 f	3543
Kobus leche	1	10158
	1	12128
	1	9945
	1	331, 333
	mode 1	2449
	1; no twins observed	7368
	1 ea; n78	9181
	1 yg ea; n149 lt	1230
Kobus vardoni	1	331, 333
	1	5763
	1	11253
Litocranius walleri	1	6410
	1 yg; n1 lt	2109
	1	5789
Madoqua guentheri	1 emb ea; n2 f	5606
Madoqua kirki	1	10158
	1	5763
	1 emb; n1 f	6612
	1 emb; n1 f	185
	1 yg ea; n3 lt	2603
Madoqua saltiana	1	5763
Nemorhaedus goral	1	6654
	1	4643
	mode 1; 2 rare; r1-2	9175
Neotragus moschatus	1	10158
	1	5763
	est 1 yg; n8 lt	5191
Oreamnos americanus	1; 2 reported	9121
	mode 1; r1-2	4933
	mode 1; r1-2	1237
	1-2; 3 occasional	1474
	most 1-2 yg; r1-3	6381a
	1.4; r1-2; n15 lt	5112
Oreotragus oreotragus	1	10158
	1	9945
	1	5763
	1	11253
	1 emb; n1 f	331
	1 yg ea; n2 lt	1230
	1 yg ea; n3 lt	2109
Oryx dammah	1 yg; n1 lt	2603
	1 yg ea; n2 lt	2109
	1 ea; n12 lt	12128
Oryx gazella	1	5764
	1 yg; n1 lt	2603
	1 yg; n1 lt	10787

	1 yg ea; n33 lt	1230
Oryx leucoryx	1 yg	6619
	1 yg	10395
	1 yg ea; n4 lt	2109
Ourebia ourebi	1	10158
	1	2439
	mode 1	2449
	1; seldom 2	11253
	1 emb ea; n2 f	331
Ovibos moschatus	1 yg	4221
	1; <1% twins	10259
	1; 2 rare	10740
	mode 1; 2 rare	10259
	1 yg; n8 lt	8033
	1.01; 1.3% twins; r1-2; n75 lt; twins stillborn	11705
	1.04; 3.9% twins; n125 f	6381
Ovis ammon	mode 1 especially first birth; r1-2	4643
	mode 1; 2 occasional; r1-2	9175
	mode 1; 2 rare	3571
	1-2	3034
Ovis aries	1-2 yg	1181
	1.1; r1-2; n76 lt	9547
	1.3; n120 lt	11053
	1.5 yg; r1-2; n2 lt	2109
	2.5/lt; r1-7; highest ovulation rate 4.2 r1-10	939
Ovis canadensis	1	4643
	1	3667
	1; no 2	11578
	1; no 2	10257
	1; occasionally 2	1474
	1+; few 2	4950
	1.36; r1-2; n11 f	10243
	11 records of twinning	2890
Ovis dalli	1	1447
	1	3667
	mode 1; 2 occasional	7884
	1-2; one set twins	7625
	1 emb ea; n21 f	7883
	1 yg ea; n73 lt	9980
	1 record of twins	2890
	2 rare	4878
Ovis musimon	1 yg	8467
	mode 1; r1-2	10946
	1.07 yg; r1-3; n159 lt	1230
	1.09 yg; 9.3% twins; r1-2; n86 lt	2109
	1.10; r1-2; n115 lt	12128
	mode 2	10945
Ovis nivicola	1	10722
Ovis vignei	1-2	6652, 6654
	1-2	12128
Pantholops hodgsoni	1	6654
	1 yg	9547
Pelea capreolus	1	10158
	1 yg ea; n14 lt	1230
Procapra gutturosa	1	4643
	1 yg	7395
	2	11419
Pseudois nayaur	1-2	9547
	1-2 yg	11449
Raphicerus campestris	1	11697
	1	5763
	1; 2 rare	10158
	1 yg ea; n2 lt	1230
	1 yg ea; n6 lt	1684

	1 emb ea; n7 f	4907
Redunca arundinum	1	10158
	1	5763
	1 emb	11697
	mode 1; r1-2	2449
	1; 2 occasional	11253
	1; 2 rare	331
	1 emb; n1 f	186
	1 yg ea; n31 lt	1230
Redunca fulvorufula	1	10158
	1	5763
	1 emb; n1 f	4926
	1 yg; n1 lt	2603
	1 yg ea; n6 lt	2109
	1 yg ea; n30 lt	1230
Redunca redunca	1	5763
	1 yg	4354a
Rupicapra rupicapra	1; yg animals; 2; older animals; r1-3	3571
	1; <1% twins; r1-2	2079, 2080
	1; 2 rare	10888
	1+; 2 rare; r1-2	8477
	1 emb; n1 f	11255
	1 emb ea; n38 f	11056
Saiga tatarica	1 ea; n4 lt	11370
	1.3; r1-2; n18 lt	8641
	1.5 yg; r1-2; n2 lt	9066
	1.75 yg; r1-3; n4 lt	3377
	2; 1,3 rare	4645
	mode 2	2215
	mode 2 >70%; r1-2	560, 561
	mode 2 70-90%	4643
Sigmoceros lichtensteini	1	10158
	1	2439
	1	7369
	mode 1	333
	mode 1; r1-2	2449
	mode 1; 2 rare	331
	mode 1; 2 sometimes	11253
	1 emb; n1 f	5052
Sylvicapra grimmia	1	11697
	1	5763
	1	11253
	1	333
	1; 2 rare	10158
	mode 1; r1-2	2449
	mode 1; 2 occasional	331
	1 emb; n1 f	185
	1 emb ea; n48 f	11789
	1 yg ea; n75 lt	1230
Syncerus caffer	1	9994
	1	331
	1	11253
	1	2449
	1 yg ea; n8 lt	1230
	1.12 emb; r1-2; n17 f	8515
	2 stillborn; n1 lt	7411
Tetracerus quadricornis	1-2	6654
	1 yg ea; n2 lt	9957
	1.83 yg; n6 lt	37
	2 yg	7382
Tragelaphus angasi	1	10158
	1 ea; n4 lt	11506
	1 yg ea; n43 lt	1230
	1.005; r1-2; n217 lt	269, 271

Tragelaphus eurycerus	1 yg ea; n11 lt	8867
Tragelaphus imberbis	1 yg; n1 lt	2603
Tragelaphus oryx	1	5277
	1	5763
	1	11253
	1; 2 rare	10158
	1; 2 sometimes	331
	1 emb ea; n3 f	5052
	1 yg ea; n4 lt	9066
	1 yg ea; n63 lt	1230
	1.02; r1-2; n94 lt	12128
	1 yg; 1 case stillborn twins n272	10895
Tragelaphus scriptus	1	10158
	1	2439
	1	5763
	1	331
	1	11253
	1 yg	201
	mode 1	2449
	1 emb; n1 f	185
	1 emb ea; n2 f	334
	1 yg ea; n3 lt	2603
	1 ea; n6 lt	8712
	1 yg ea; n32 lt	1230
	2 emb; n1 f	2603
Tragelaphus spekei	1	5763
	1	331
	1; n1 lt	5688
	1.0 yg; 0.2% twins; r1-2; n1085 lt	2485
Tragelaphus strepsiceros	1	2275
	1	331
	1	11781
	1 yg; n1 lt	11291
	1 yg ea; n2 lt	2109
	1 yg ea; n11 lt	1230
SEXUAL MATURITY		
Addax nasomaculatus	f: 1st birth: 36.15 mo; r33.333-39; n2	2603
	f: 1st birth: 34.0 mo; sd6.6	2487
Aepyceros melampus	f: 1st mate: ~12-18 mo	5597
	f: 1st conceive: ~18 mo	5764
	f: 1st birth: 2 yr	3144
	m: 1st sire: ~13 mo	5764
	m: testis wt, spermatogenesis: 13 mo	5697
	m: 1st spermatogenesis: 18 mo	1334
	m: 1st territorial: 5+ yr	1334
Alcelaphus buselaphus	f: 1st mate: <2 yr	3247
	f: 1st birth: 25 mo; well fed; n3 of 5 f	7774
	f: 1st birth: ~3 yr	10158
Ammotragus lervia	f: 1st birth: 13.4 mo; n1	4020
	f: sperm: 11 mo; n1	4020
Antidorcas marsupialis	f: 1st ovulation: 28 wk	10023, 10038
	f: 1st conceive: 43-44 wk; n1	6510
	m: 1st sperm: 48 wk; n2	10040
	m: 1st mate: 52 wk	10023
	m: 1st mate: 24 mo; n1	6510
Antilocapra americana	f: 1st mate: 16 mo	11208
	f: 1st preg: ~8 mo; n1	7376
	f: 1st birth: 1 yr; n1	11954
	f: 5 mo min	8009
	f: 1.33 yr	2607
	f: 18 mo	1871a
	m: 1st mate: 16-17 mo	11208
	m: 2.33 yr	2607
	f/m: 1st mate: 15-16 mo	12019

	f/m: 1st mate: 1.5 yr	1474
Antilope cervicapra	f: 1st estrus: 7 mo	10628
	f: 1st birth: 14 mo	10628
	f: 1st birth: 22 mo	7617
	f: 1st birth: 26.5 mo; r24-29; n2	32
Bison bison	f: 1st mate: 1 yr rare	7068
	f: 1st mate: 15 mo 5%; 27 mo 40%; 39 mo 50%	3559
	f: 1st mate: 2-3 yr	6584
	f: 1st conceive: 1 yr rare; 2 yr 38%; 3 yr 52%	3560
	f: 1st conceive: 2 yr	4501
	f: 1st preg: 3.5 yr	1871a
	f: 1st birth: 3 yr	7068
	f: 1st birth: 3 yr	7805
	f: 1st birth: 3 yr; r2-4; n84	4047
	f: 2 yr	7068
	m: testicular sperm: yearling	4501
	m: epididymal sperm: 2-3 yr; 1 yr rare	3560
	m: 1st mate: 1 yr 16%, 2 yr 33%, 3 yr 100%	3559
	m: 1st breed: 6 yr	6584
	f/m: 1st mate: 2-3 yr	1474
Bison bonasus	f: 1st birth: 32-36.5 mo	5229
	f: 1st birth: 36+ mo	3822a
	f: 1st birth: 47.8 mo; mode 35-38; min 33 mo; max 6th yr; n52	5999
	f: 2 yr; active 3 yr	4643
	m: fertile insemination: 15-29 mo	5229
	m: active 6-7 yr	4643
	f/m: 2-3 yr	3571
Bos frontalis	f: 1st estrus: 451 d; n1	9593
	m: 1st mate: 1 yr; n1	9593
	f/m: 1st mate: ~2 yr	9545
Bos javanicus	f: 1st birth: 2.5 yr	6524
	f: 2 yr	11088
Boselaphus tragocamelus	f: 1st birth: 2 yr 9 mo 19 d; n1	30
	f: 1st birth: 3 yr 20 d; n1	32
Bubalus bubalis	f: 1st conceive: 28.5 mo; puberty 14-19 mo (cl present)	10940
Budorcas taxicolor	f/m: ~2.5 yr	8798
Capra caucasica	f: 1st estrus: 2 yr; often do not mate then	4643
	m: 1st mate: 4-5 yr	4643
Capra cylindricornis	m: < 3 yr	8328
Capra falconeri	f: 1st estrus: 30 mo	9175
	f: 1st birth: 2 yr	9547
	f: during 2nd yr	4643
	m: after 3rd yr	4643
Capra hircus	f: 1st mate: 6 mo	9357
	f: 1st mate: 1.5 yr; captive; 2.5 yr; wild	2080
	f: 1st conceive: 3 mo	9223
	f: 1st preg: 5 mo	12020
	f: 1st birth: 8 mo	9223
	f: 1st birth: 3 yr	2080
	f: 1st birth: 3 yr	9175
	f: 2nd yr; 1st birth: 3rd yr	4643
	m: 1st sire: 6 mo	9357
	m: 1st mate: 1.5 yr	2080
	m: 1st mate: 3-4 yr	4643
	f/m: 2 yr	9547
Capra ibex	f: prob repro: 212-242 d	8707
	f: 1st birth: 2-3 yr	4643
	f: 1 1/2 yr	2081
	m: 1st mate: active 5-6 yr	4643
	m: 1.5 yr	4643
	f/m: 1st mate: 3-4 yr; rarely 2 yr	8594
	f/m: 4th yr	3571

Capricornis crispus	f: 1st repro: 2-5 yr	7390
	f: ovarian histology: 2.5 yr	5825, 10535
	m: spermatogenesis: 6 mo	10817
Cephalophus dorsalis	f: 1st conceive: 14 mo	2605
	f: 1st birth: 24 mo	2603
Cephalophus maxwelli	f: 1st birth: 3 yr	74
Cephalophus monticola	f: 1st mate: 6-17 mo	1047
	m: 1st mate: 9 mo	1047
	f/m: ~3 yr	5763
Cephalophus rufilatus	f: 1st birth: 26.25 mo; r16.75-35.75; n2	2603
Connochaetes gnou	f: 1st conceive: 28 mo	11354
	f: 1st birth: most 3 yr; min 2 yr	11353, 11356
	f: 16-18 mo	11356
	m: 1st sire: 16 mo; n1	11356
	f/m: 1st mate: 3 yr	10042
Connochaetes taurinus	f: 1st mate: 1 yr	10637
	f: 1st mate: 15 mo	3099
	f: 1st conceive: 22% of yearlings, all by 2-3 yr	441
	f: 1st birth: 2 yr	3105
	f: 1st birth: 2 yr	10638
	f: 1.5-2.5 yr	5764
	m: 1st mate: 28 mo; mode 40 mo	3099
	m: over 5 yr	5764
Damaliscus dorcas	f: 1st mate: 18 mo	5702
	f: 1st mate: mode 28 mo	2746
	f: 1st birth: 2-3 yr	10158
	f: 1st birth: 36 mo	9328
	m: 1st mate: 18 mo some; most later	5702
	f/m: 1st mate: 18 mo	10023, 10027
	f/m: 28 mo	2746
Damaliscus lunatus	f: 1st mate: yearling	5308
	f: 1st birth: 2 yr 7/10, 3 yr 6/6	5101
	f: 1st birth: ~36 mo	1790, 1792
	m: testis wt: 32-34 mo	1792
	m: spermatogenesis: 40-42 mo	4084, 4085
	m: epididymal sperm: 18, 30 mo; n2	5308
	m: 1st mate: 2 yr 1/8, 3 yr 3/5, 4 yr 5/5	5101
	m: 1st repro: 40-42 mo	1792
Gazella cuvieri	f: 1st birth: min 344 d	8100
	m: 1st sire: 180 d	3567
Gazella dama	m: 1st sire: 120 d	3567
Gazella dorcas	f: 1st conceive: 18 mo	493
	f: 1st conceive: 1 yr 9 mo 23 d	2600
	f: 1st conceive: 716 d; n14	5997
	f: 1st birth: 27.5 mo	2603
	m: 1st sire: 589 d	3567
Gazella gazella	f: 1st conceive: 6-18 mo	493
	f: 1st birth: 12 mo	491
	f: 1st birth: 2 yr	492
	f: 18 mo	492
Gazella granti	f: 1st mate: 7-14 mo	3247
	f: 1st conceive: 152 d	2600
Gazella leptoceros	f: 1st birth: 10.5 mo	2603
Gazella soemmerringi	f: 1st conceive: 1 yr 7 mo	2600
	f: 1st birth: 26.375 mo; r25-27.75; n1	2603
Gazella spekei	f: 1st conceive: 16.8 mo; n6	8991
Gazella subgutturosa	f: 1st mate: 18-19 mo	4643
	m: 1st rut: 19 mo	9175
	m: 1st mate: active 2.5 yr	4643
Gazella thomsoni	f: 1st mate: 10-12 mo	1196
	f: 1st mate: ~13 mo	5122
	f: 1st mate: 1 yr	4972
	f: 1st mate: 1st yr	1328
	f: 1st conceive: 9-11 mo est	9180

	f: 1st conceive: 1 yr 4 d	2600
	f: 1st birth: ~18 mo	5122
	f: 1st birth: 18.33 mo	2603
	m: spermatogenesis: ~12 mo	5122
	m: 1st fertile: 480 d	3567
	m: 1st mate: 2 yr	4972
	m: 1st breed: 3.5 yr	5122
Hemitragus hylocrius	f: 1st birth: earliest 22 mo	11750
Hemitragus jemlahicus	f: 1st birth: 2 yr	9547
	m: 5-6 yr	9540
Hippotragus equinus	f: 1st conceive: 23-25 mo	10158
	f: 1st birth: 32-34 mo	10158
	f: 1st birth: 3 yr	11758
	f: 1st birth: 3 yr; n1	5101
	f: ~2 yr	5764
	m: electroejaculation: 16-19 mo	11758
	m: 1st mate: 3 yr; n1	5101
Hippotragus niger	f: 1st estrus: 27 mo; n2	4085, 4086
	f: 1st birth: 28.66 mo	2603
	f: 1st birth: 3 yr	11758
	f: 1st birth: 3 yr, 1/3; 4+ yr, 18/25	5101
	m: electroejaculation: 16-19 mo	4086
	m: sexual activity: 3 yr	4086
	m: 1st mate: 5 yr 1/2	5101
	m: 3 yr	11758
Kobus ellipsiprymnus	m: own territory: 6 yr	10269
Kobus kob	f: 1st ovulation: 13-14 mo	1434
	f: 1st birth: 13 mo; n1	1432
	m: fertile: 1 yr	1434
Kobus leche	f: 1st mate: 2-3 yr	9181
	f: 1st preg: 3rd yr	11731a
	f: 1st birth: yearling	10158
	f: 1st birth: 2-3 yr	7368
	m: testes size: 4-5 yr	9181
	m: ~15 mo	10158
Litocranius walleri	f: 1st birth: 18-19 mo; n2	6410
	f: end second yr	5789
Madoqua kirki	f: 1st birth: 15-18 mo	4622
	f/m: ≤8 mo	5625
Madoqua saltiana	f: ~10 mo	5763
	m: ~9 mo	5763
Nemorhaedus goral	f/m: 1st mate: 3 yr	1310a
	f/m: some 2nd yr	4643
	f/m: 2-3 yr	7125
Neotragus batesi	f: 1st birth: 8-16 mo	3191, 3192
	m: 8-18 mo	3192
Neotragus moschatus	f/m: 6 mo	5191
	f: ~6 mo	5763
Oreamnos americanus	f: ovarian histology: 2.5 yr	1237
	f: 1st mate: 18 mo; captive; later in wild	4997
	f: 1st mate: 2.5 yr	1871a
	f: 1st cl: 27.5 mo	6381a
	f: 1st birth: 3 yr; n2	5112
	f: preg: 34 mo; n1	6381a
	f: lact: 38 mo; n1	6381a
	m: sperm present: 1-2 yr	4620
	m: 1st mate: yearling	4997
	m: rut: 18 mo	4339
	m: large testes: 39.5 mo; n1	6381a
	f/m: 1st mate: 2.5 yr	1474
	f/m: >2 yr	5146
Oreotragus oreotragus	f/m: ~1 yr	5763
Oryx dammah	f: 1st birth: 32.92 mo; r30.333-35.5; n2	2603

Oryx gazella	f: 1st birth: 26 mo	11212
	f: 1st birth: 29-33 mo	10158
	f: ~24 mo	10158
Ourebia ourebi	f: 1st birth: 22.17mo; r19.333- 25; n2	2603
	f: ~10 mo	5763
	m: 14 mo; n1	1520
Ovibos moschatus	f: 1st estrus: 14-15 mo	8798
	f: 1st mate: 3 yr	5300
	f: 1st birth: min 3 yr; mode ~4	10259
	f: 1st birth: 4 yr; few data	10740
	f: 3 yr	8090
	m: 6 yr	5300
	m: 6 yr; few data	10740
	f/m: 3-4 yr	1474
Ovis ammon	f: 8/9 mo-1.5 yr	3571
	m: horn growth: ~4 mo	3571
	m: 1.5-2.5 yr	3571
	m: 2.5 yr	4643
Ovis aries	f: 1st mate: 18 mo	9175
	f: 1st birth: 2 yr	9547
	m: 1st mate: 2.5 yr; in herds only 4.5+ yr mate	9175
	m: spermatogenesis: ~6 mo	4109
Ovis canadensis	f: 1st mate: 1.5 yr few; mode 2.5 yr	10105
	f: 1st mate: 2 yr	4950
	f: 1st mate: 2.5 yr	10257
	f: 1st mate: ~2.5 yr	4643
	f: 1st birth: 290-313 d	7414
	f: 1-2 yr	1871a
	m: 1st mate: >3 yr	4643
	m: 1.5-2.5 yr	3667
	m: 18-36 mo	1871a
	f/m: 1-2 yr	788
	f/m: 18 mo	9815
Ovis dalli	f: 1st mate: 18-30 mo	7883
	f: 1st birth: some 2 yr; mode 3 yr	4878
	f: 1st repro: 2 yr captive; 50% 4 yr; adult wt 5 yr	1447
	m: 1.5-2.5 yr	3667
	f/m: ~18 mo	7884
Ovis musimon	f: 1st mate: 18 mo	10946
	m: spermatogenesis: ~6 mo	7579
	m: 1st mate: 30 mo	10946
Ovis nivicola	f: 2 yr	10722
Pseudois nayaur	f: 1st birth: 2 yr	9547
	f: yearling	11449
Raphicerus campestris	f: 1st conceive: 9.5 mo; n1	1684
	f: 1st preg: when the second molars erupt	11790
	f: 6-7 mo	10158
	m: 1st sire: 8.5 mo; n1	1684
	f/m: 6-9 mo	5763
Raphicerus sharpei	f: 1st cl; when the second molars erupt	5699
	m: spermatogenesis: when the third molars erupt	5699
Redunca fulvorufula	f: 9-24 mo	5165
	f: min 9 mo; mode 12-14 mo	10158
	m: 1st sperm: 8-12 mo	5165
	m: 1 yr	10158
Rupicapra rupicapra	f: corpora albicantia: 2.5 yr; no data <2.5 yr	3663
	f: 1st mate: 1.5 yr	2080
	f: 1st mate: 96% 3 yr	8477
	f: 1st preg: 7 mo; n1	658
	f: 1st preg: 2 yr, n1 of 3; 3 yr, n3 of 7	11056
	f: 1st birth: 4 yr	9650
	f: 1st birth: yearling	657
	f: 1st birth: 3-4 yr	4643
	f: min 2 yr	10888

	f: 2nd yr	4643
	m: spermatogenesis: 18 mo	8477
	m: 1st sire: 2+ yr	8477
	m: yearling	658
	f/m: 18-20 mo some; mode 3-4 yr	3571
Saiga tatarica	f: 1st mate: 7-8 mo	4643
	f: 1st mate: 7-8 mo	560, 561
	f: 1st mate: 1.75 yr	12107
	f: 7-8 mo	2215
	m: 1st mate: 20 mo	560, 561
	m: 1st mate: 2nd yr	4643
Sigmoceros lichtensteini	f: 1st birth: 2 yr	10158
	f: 16-18 mo	1118
Sylvicapra grimmia	f: 1st mate: 8 mo; n1	334
	f: 1st mate: ~16 mo; n1	11789
	f: 1st conceive: 8-9 mo	10158
Syncerus caffer	f: 50% cl present: 5 yr	4077
	f: 50% ovulate: 3.5 yr	9991, 9994
	f: 50% preg: 4.8 yr	1599
	f: 50% preg: 5 yr	9991, 9994
	f: 1st birth: 4 yr	8515
	m: testes wt, sperm: 2.5-3 yr	4077
	m: 1st mate: 7-8 yr	8515
	m: 2.5-3 yr	8515
	f/m: ~2 yr	5763
Tetracerus quadricornis	f: 1st birth: 1 yr 9 mo; n2	37
Tragelaphus angasi	f: 1st ovulate: 14-18 mo; 50% 20 mo	269, 271
	f: 1st birth: 21.75 mo	2603
	f: 1st birth: 2 yr	7774
	m: spermatogenesis: 12 mo-2 yr	269, 271
	m: 1st mate: > 5 yr	269, 271
Tragelaphus eurycerus	f: 1st conceive: 29 mo; r27-31; n2	8867
	f: ~2 yr	11022
	m: ~2.5 yr	11022
Tragelaphus imberbis	f: 1st birth: 36.66 mo	2603
Tragelaphus oryx	f: 1st estrus: 15-24 mo	10367
	f: 1st estrus: 20.2 mo; r17-37; n19	9311
	f: 1st mate: 28.7 mo; ±2.3	10023
	f: 1st conceive: 13 mo	10895
	f: 1st birth: 2 yr min; wild 4 yr	10158
	f: 1st birth: 26-37 mo	10300
	f: 1st birth: 26-46 mo	8690, 9311
	m: 15-24 mo	10367
	m: 18 mo min	10158
Tragelaphus scriptus	f: 1st conceive: 18 mo	8712
	f: 1st ovulation: 11 mo	7537
	f: 1st birth: 535.5 d; r509-562; n2	11347
	f: 1st birth: 24.75 mo	2603
	f: 14 mo	10158
	m: spermatogenesis: 11 mo	7537
	m: 10.5 mo	10158
Tragelaphus spekei	f: 1st birth: 16-23 mo	2485
	f: 1st birth: 20 mo; n1	9693
	f: 1st birth: 22-28 mo	6192
	f/m: full maturity: ~4 yr	5763
Tragelaphus strepsiceros	f: 1st mate: 17 mo	10023, 10027
	f: 1st preg: 17 mo	9986
	f: 1st birth: 2 yr	2275
	f: 1st birth: ~3 yr	11786

CYCLE LENGTH

Addax nasomaculatus	estrus: 1.5-2 d	5115
Aepyceros melampus	estrus: 24-48 h	3147
	12-29 d	3147
Ammotragus lervia	estrus: few h	4535

Antidorcas marsupialis	16 d; mode 16; r14-17; n18 cycles, 1 f	6510
Antilocapra americana	27.4 d; sd6.6; r12-40; n11 cycles, 4 f	8644
Antilope cervicapra	estrus: ~24 h	1649
	estrus: 24 h	10628
	5-6 d; not cycle length	10628
	20 d	4860
Bison bison	estrus: 9-28 h	7143
	~3 wk	7143
Bison bonasus	estrus: lasts ~6-8 d	4643
	20.6 d; 80% 18-22; r12-28	5999
	mode 20-30 d; r10-70; interestrous intervals	5229
Bos frontalis	estrus: 1-4 d	9593
	3 wk	9545
	26.3 d; r7-37; n15 cycles, 1 f	9593
Bos javanicus	3-4 wk	11088
Bos taurus	29.3 d; mode 21-22; 60% 18-25; n500 cycles, 200 f	10905
Boselaphus tragocamelus	estrus: 4 d	9175
Bubalus bubalis	estrus: 19.6-19.8 h	467
	estrus: 21.69 h; r16-27	9445
	estrus: 2-3 d	2936
	r20-22 d; n14 f	5496
	21.39 d; r19-25; n46 cycles, 7 f	9445
	22-28 d	2936
	22 d; se0.6; 30% 31-60; 49% 11-30; r10-88; n235 cycles, 87 f	4187
	25.05 d; se0.41	8927
	32-50 d	9994, 10938
Capra hircus	18.4 d; ±5.6; 70% 18-24; r5-27; n20 cycles, 6 f	8230
	19-21 d	9223
Capra ibex	estrus: <24 h - several d	2081
	20 d; sd1; n13 cycles, 10 f	10530
Capricornis crispus	20-21 d	5178
Cephalophus rufilatus	estrus: 0.5-1 d	2605
Cephalophus zebra	estrus: 0.5-1 d; n1	9691
Damaliscus dorcas	estrus: <24 h	2296
Gazella gazella	28 d; n2 cycle, 1 f	9175
Gazella subgutturosa	estrus: 12 h	9175
Gazella thomsoni	estrus 1-3 ?d	4972
	1 ?wk	4972
Kobus kob	estrus: <24 h	1434
	6-13 d	7540, 7541
	20-27 d	1434
Madoqua kirki	estrus: ~2 d	5763
Madoqua saltiana	estrus: ~2 d	5763
Oryx dammah	luteal phase: 16-18 d	2815
	21-22 d	2815
	21.7 d; r20-23; n3 cycles, 1 f; from graph	6581
Ovibos moschatus	19.6 d; sd0.96; n19 cycles, 6 f	9331
Ovis ammon	estrus: 3 wk to 1-1.5 mo	4643
Ovis aries	estrus: ~1 d; lambs; ~1-3 d; ewes	4109
	estrus: 45.3 h; n49	12011
	15 d; r10-18 d; 1st ovulation no estrus	4109, 4111
	16.7; n43 cycles, 5 f	12011
	24.9 d; se3.3; n8 cycles, 3 f	12010
Ovis canadensis	estrus: 1-3 d	4908
	estrus: 2 d	10950
	~28 d	10950
Raphicerus campestris	estrus: ~4 d	1684
	3 mo	1684
Syncerus caffer	23 d	9994
Tragelaphus angasi	19 d	270
	24 d; r10-34; n5 cycles, 1 f	269, 271
Tragelaphus eurycerus	estrus: 1-2 d	1381
	estrus: ~3 d	8867
	21-22 d	8867

	22 d; n1 f	1381, 2720
Tragelaphus oryx	estrus: 1-2 d	10367
	estrus: 3 d	8690
	22.16 d; r22-23; se0.04	10367
	21-26 d	8690
GESTATION		
Addax nasomaculatus	260.5 d; r257-264; n2	2603
	265 d; r292-315	8421
	10-11 mo; according to natives	1352
Aepyceros melampus	5-6 mo	11697
	6 mo; field data	10041
	6 mo	3145
	195-200 d	3147, 10158
	195-204 d	8965
	~6.5 mo	5764
	196 d	3146
	196 d	10023
	196 d	1230
	204 d ea; n2	633
Alcelaphus buselaphus	238 d; n1	2603
	8 mo	4582
	~8 mo	10042
	~8 mo	3956
	~242 d	2382
	~9 mo	5764
Ammotragus lervia	12-15 wk	4535
	5 mo	1352
	154-161 d	8223
	22-23 wk	1355
	155 d; sd5	6523
	160 d	6652
	165 d	9734
Antidorcas marsupialis	167 d	1230
	168 d	10023, 11213
	~171 d	923
	177 d; r174-180; n1 f	6510
Antilocapra americana	217-252 d	11208
	7.5 mo	6387
	~246 d; field data	6217
	250 d	12019
	252 d; r249-255	4647
	~9 mo; field data	2607
Antilope cervicapra	5 mo	7616
	5 mo; field data	6541
	6 mo	1355
	6 mo	10628
Bison bison	~9 mo; field data	3559, 3560
	~9 mo	188
	9 mo	1355
	9-9.5 mo; not reliable	5229
	~285 d	4501
	~9.5 mo	7144
Bison bonasus	261-283 d	4643
	262 d	2109
	264.29 d; sd4.20; r254-272; n132	5999
	265 d; 56% 260-269; r256-283; n52	5229
	40-41 wk	3571
	9 mo	3822a
Bos frontalis	~8-9 mo	2802
	9 mo	5047
	9 mo	4582
	298.2 d; r293-301; n5	9593
Bos grunniens	8.5-9 mo	1355
	9 mo	4582

	~9 mo; field data	8740
	280 d	1969
Bos javanicus	285-300 d	11088
	297 d; se5	6524
	300 d; r329-338	8421
Boselaphus tragocamelus	8 mo	4582
	<8 mo 7 d	1355
	247 d	9175
	250 d; n1	6145
	280-300 d	1046
Bubalus bubalis	~10 mo	1355
	~10 mo	6654
	317 d	4187
	320 d; r312-334	10939, 10940
Bubalus depressicornis	9.5-10 mo	4582
Budorcas taxicolor	200-220 d; n1	459
	7-8 or 10-11 mo	7800
	8 mo; field data	11419
Capra caucasica	5 mo? (no data provided)	8458
	150-160 d	4643
Capra falconeri	155 d	9547
	162-170 d	9175
	~6 mo	4643
Capra hircus	<5 mo	11389
	5 mo	9223
	5 mo; field data	9668
	150-155 d	9175
	<156 d	9357
	165-170 d; field data	2080
Capra ibex	22-23 wk	3571
	155-170 d	9175
	165-170 d	2081
	167 d; sd3; n63	10530
	170-180 d	4643
	~6 mo; field data	6652
Capricornis crispus	200-230 d	11996
	210-220 d	5947
	212 d; r211-213	5178
Cephalophus dorsalis	7.5-8 mo	2605
	256 d; n1	2603
Cephalophus maxwelli	120 d	74
Cephalophus monticola	167 d	1230
	205 d	7204
	207 d; se6; mode 206-210; r196-220; n16	1047
Cephalophus niger	126 d; n1	3175
Cephalophus ogilbyi	4 mo	5277
Cephalophus rufilatus	223 d daughters, 241 d sons	2605
	236.3 d; r223-245; n3	2603
Cephalophus sylvicultor	4 mo	5277
	151 d; n1	3175
Cephalophus zebra	222.8 d; r190-241; n5	9691
Connochaetes spp	8-8.5 mo	1355
Connochaetes gnou	8.5 mo	4582
	8.5 mo	10042
	250-260 d	7204
	9 mo; field data	10041
Connochaetes taurinus	239-243 d	2449
	~8 mo; field data	441
	8-8.5 mo	3099
	251 d; n1	10349
	252 d; r249-255; n2	10351
	8.5 mo	11697
	~255 d; r8-9 mo	9993
	~9 mo	10637

Damaliscus dorcas	223 d; n1	2603
	225 d	10023, 10027
	7.5 mo; field data	5702
	238-254 d	10158
	8 mo	9328
	~8 mo; field data	6657
	~8 mo; field data	6204
	~8 mo; field data; 237/8, 254 d; n2; 1st mating to birth	2295, 2296
	288-308 d	1230
Damaliscus hunteri	279 d; n1	5249
Damaliscus lunatus	7-8 mo; field data	1790, 1792
	235-241 d	5764
	8 mo; field data	11697
	8 mo; field data	7470
Gazella cuvieri	~160 d	8100
	165-180 d	3567
Gazella dama	200 d	3567
	~7 mo	345
Gazella dorcas	3 mo	8223
	3 mo	254
	3 mo	5575
	169-174 d; n4; 1 stillborn 181 d	10059
	172.5 d; r171-174; n4	2600, 2603
	6 mo	493
Gazella gazella	6 mo	493
Gazella granti	~6 mo	5764
	195 d	3247
	198.7 d; r198-199; n3	2600, 2603, 2606
Gazella leptoceros	162.5 d; r156-169; n2	2600, 2603
Gazella rufifrons	5 mo	5277
	exactly 6 mo	345
	187.5 d; r186-189; n2	2600, 2603
Gazella soemmerringi	6-7 mo	5764
Gazella spekei	178.6 d; r169-190; n15 lt, 8 f	8991
	~7 mo	4462
Gazella subgutturosa	5-5.5 mo	9175
	5.5-6 mo	4643
Gazella thomsoni	implant: 60 d	4972
	115 d; r114-116; n2; mating to birth	4972
	3.5-3.75 mo; n2; presumed mating to birth	4972
	150 d	3247
	est 150-170 d	5122
	5-6 mo	5764
	165 d	2449
	168 d; r152-186; from fetal wt	1196
	191 d; r185-195; n8	9180
	199 d; r189-214; n19 f	4972
	223 d; r222-224; n2	9180
Hemitragus hylocrius	180-190 d; postpartum estrus to birth	11750
Hemitragus jemlahicus	6 mo	9540
	~6 mo	1355
	~6 mo; field data	6652
Hippotragus equinus	274 d; r268-280; n2	2603
	275 d	11758
	~275 d	2449
	~275 d	5764
	276-287 d	10158
	10 mo; field data	254
	10 mo; field data	5575
Hippotragus niger	240-248 d	11758
	265 d; r261-271; n3	2603
	266 d; r259-272; n4; 1st mount to birth	4085, 4086
	9 mo	4582
	9 mo	11697

	272 d	6801
	275, 266 d, n2	5972
	~10 mo; field data	3104
Kobus ellipsiprymnus	4+ mo	5052
	est 8 mo	4326
	243 d	2439
	9 mo	4582
	272-287 d	2449
	280 d	10158
Kobus kob	~6 mo	345
	235-240 d	7540
	236-252 d	1432, 1434
	261-271, 266-267 d, 8 mo 26 d	11037
	~9 mo	5763
	10 mo	7591
Kobus leche	7 mo	11697
	215-248 d	2449
	7-8 mo	9181
	8 mo; field data	333
Kobus vardoni	~8 mo	10158
	~9 mo	5763
Litocranius walleri	6.5-7 mo	6410
	203 d; n1	9636
Madoqua kirki	170-174 d	5763
	170-174 d; n1	5625
	171 d; r169-174; n3	2598, 2603
	172 d	4977
	25 wk	4622
Nemorhaedus goral	~6 mo	6654
	6-8 mo	9175
	~7 mo; field data	3034
	~8-9.5 mo; n4	2652
	250-260 d	4643
Neotragus batesi	~6 mo	5763
Neotragus moschatus	180 d	5191
Oreamnos americanus	147 d	5299
	6 mo; field data	1237
	185.8 d; sd1.34; n9	5112
	~191 d; field data	4933
Oreotragus oreotragus	5 mo	10158
	7-7.5 mo	2170
	214 d	7531
Oryx dammah	242-256 d	5896
	246 d; n1	2603
	9 mo; field data	3781
Oryx gazella	8.5-10 mo	4582
	264 d; n1	2603
	~264 d	1230
	265 d	8421
	265-300 d	5764
	~9 mo	11212
Oryx leucoryx	240 d	7531
Ourebia ourebi	192 d	6315
	210 d	7531
	210-211 d	10158
	7 mo	11697
	7 mo	5763
	217-220 d	2608
Ovibos moschatus	~8-8.5 mo; field data and 1 captive birth	10738, 10740
	244-252 d; n4	6381
	250 d	9330
	250-275 d	8798
	8.5-9 mo	5300
	9 mo	2364

Ovis ammon	~5 mo	6654
	~5 mo	4643
	21-22 wk	3571
	5.5 mo	9175
	~6 mo; field data	6652
	~7 mo; field data	8740
Ovis aries	134 d	9175
	150 d	9547
	150 d	7735
	151 d; se1.34; n162	1181, 4111
Ovis canadensis	174.2 d; sd1.70; r171-178; n20	9816
	175 d	3667
	176 d; n1	1021
	179 d; sd6	10950
	180 d (maybe slightly more); field data	3535
	~180 d	10257
	6 mo; field data	10105
	~6 mo	1355
	~6-7 mo; field data	6652
Ovis dalli	5.5-6 mo; field data	7625
	171 d; se2.21; field data	7883, 7884
	174 d; r172-176; n2	3667
Ovis musimon	148-159 d	8467
	153 d; n2	9318
	~5.5 mo	10946
Ovis vignei	~6 mo	1355
	~6 mo	6654
	~6 mo; field data	6652
Pantholops hodgsoni	~7 mo; field data	8972
	~8 mo; field data	8740
Pelea capreolus	~261 d	1230
Procapra gutturosa	~6 mo	4643
	186, 189-192 d; n2	7395
Pseudois nayaur	est 4-5 mo; few field data	11419
	est 159-160 d; field data	11778
	160 d	9547
	~6 mo; field data	9541
Raphicerus campestris	168 d; min interbirth interval	922
	210 d	7531
	7 mo	11697
Raphicerus sharpei	7 mo	10158
Redunca arundinum	7-7.5 mo	5763
	~215 d	2449
	7.75 mo	11697
Redunca fulvorufula	~6 mo; field data	10025
	223 d; n1	2603
	236-251 d; n1	5165
Redunca redunca	~7.5 mo	5763
Rupicapra rupicapra	160-170 d	2079, 2080
	165-185 d	8477
	25-27 wk	3571
Saiga tatarica	5 mo	4643
	~5 mo	4645
	~5 mo; r139-152 d	560, 561
	153 d; n2	3377
Sigmoceros lichtensteini	214 d; ~8 mo	2439
	237 d	2449
	240 d	2710
Sylvicapra grimmia	3 mo	10158
	est 4.5-7 mo	7204
	123 d	2449
	170 d; n1	9971
Syncerus caffer	300 d	5575
	300 d	254

	11 mo	11697
	11 mo	3145
	~11 mo	5763
	~335 d	1230
	340 d	9994
	343-346 d	2449
	345 d; r343--~346; n2	11292
Tetracerus quadricornis	~6 mo	6654
	~180 d	7382
	235.5 d; r228-243; n2	9957
	240-255 d	37
Tragelaphus angasi	~6 mo	1230
	220 d; 7 mo	269, 270, 271
	252 d	7774
Tragelaphus eurycerus	282 d; n1; [reported as 294-296 d]	8867, 11972
	285 d; n1	2720
	298 d; embryo transferred 8 d after ovulation & insemination	2719
Tragelaphus imberbis	~7 mo	5763
	222 d; n1 stillborn	2603
	8 mo 10 d; r8 mo 1 d-8 mo 20 d; n3	6189
	8-8.5 mo; r8-11	6411
Tragelaphus oryx	~4.5 mo; field data	5052
	255 d	10894
	8.5 mo	8690
	8.5 mo	11697
	8.5-9 mo	4582
	260 d	2439
	260-284 d	5763
	9 mo exactly	1355
	9 mo	5277
	9 mo; field data	6058
	271 d; se2.9	10023
	273 d; r264-278; n12	10300
	276 d; ±4; n40 lt 1st calves	10895
	280.4 d; n5	9311
Tragelaphus scriptus	~5.5 or ~6 mo	5763
	179.7 d; r178-187; n3	2603, 2608
	6 mo	201
	~6 mo	7537
	est r6-7 mo	11697
	214 d	2439
Tragelaphus spekei	~7.5 mo	5763
	7 mo 24 d; n1	9693
	239 d; n1	5688
	8 mo + few d; r8 mo 4 d-8 mo + 15 d; n7	6192
	247 d; r225-258	2485
	251-260 d	8421
Tragelaphus strepsiceros	~4 mo; field data	5052
	210 d	5671
	210-214 d	10023, 10027
	7-8 mo	11697
	214 d	5299
	~8 mo	11786
	261 d	8421
	~9 mo	5763

LACTATION

Aepyceros melampus	solid food: 2.4 d; sd0.6; r2-3	7503a
	wean: 4-7 mo	5259
Alcelaphus buselaphus	solid food: 2 wk	10158
	wean: 4 mo	3956
	wean: 7 mo	10158
	wean: 8 mo	10042
Ammotragus lervia	wean: 4 mo	1352

Antidorcas marsupialis	solid food: 2 wk	10158
	wean: 120-122 d; n1 f	6510
Antilocapra americana	solid food: 3 wk	12019
	wean: 3.5-4 mo	5828
	wean: 12 wk	1515a
	wean: 4 mo	8009
Antilope cervicapra	solid food: 5 d nibbled on grass	9545
	wean: ~8-9 wk; nursing infrequent; max 12	785
Bison bison	solid food: 5 d	7143
	solid food: ~1 wk	4048
	can survive: 7-8 wk	7143
	wean: 5-9 mo	6342
	wean: 9-12 mo; n8 f with subsequent birth	4047
	wean: 17-21 mo; n5 f without subsequent birth	4047
Bison bonasus	wean: 5-9 mo	6342
	wean: 5 mo-2 yr	5229
	wean: 6-8 mo	3571
Bos frontalis	wean: 4.5 mo	3447
Bos grunniens	wean: 5-9 mo	6342
	wean: 8 mo	1969
Bos javanicus	wean: 10 mo	11088
Budorcas taxicolor	solid food: 2 wk	8798
	solid fodd: 2nd mo	7800
Capra caucasica	solid food: 1 mo	4643
Capra falconeri	wean: 5-6 mo; into winter	4643
Capra hircus	wean: 4-5 mo; rejection by mother	9175
	wean: 6 mo	11389
Capra ibex	solid food: < 1 mo	4643
	wean: 3 mo	6764
	wean: 1 yr	8594
	wean: ~1 yr	3571
Capricornis crispus	solid food: 1 mo	11996
Cephalophus monticola	solid food: browse at 2 wk	5763
Cephalophus sylvicultor	wean: 4-6 wk	6642
Cephalophus zebra	solid food: within 1 wk	11021
Connochaetes gnou	solid food: 4 wk	11356
	wean: 6-9 mo	11356
Connochaetes taurinus	solid food: 10 d	10637
	solid food: 10-14 d	10158
	wean: 7-8 mo	11697
	wean: 8 mo	10158
	wean: ~1 yr with new calf, 16 mo no new calf	10637, 10638
Damaliscus dorcas	solid food: 7 wk grazes 40% of the day	2296
	wean: 4 mo	2746
	wean: 6 mo	6204
Gazella dorcas	can wean: 3 wk	5997
	wean: 75-90 d	493
Gazella gazella	wean: 75-100 d	493
Gazella subgutturosa	solid food: 8-10 d	4643
	wean: 1-6 mo	4643
Gazella thomsoni	wean: ~2 mo	5764
	wean: 4.7 mo; r1.5-7	4972
Hemitragus hylocrius	solid food: 14 d; nibbled	11750
Hippotragus equinus	solid food: 40 d	5248
	wean: 6 mo	10158
Hippotragus niger	wean: 6 mo	11758
	wean: 8 mo	4086
Kobus ellipsiprymnus	wean: 6-8 mo	10269
	wean: ~6-8 mo	5763
	wean: 276 d; lact stops 180-210 d	10158
Kobus kob	wean: ~180 d	7541
	wean: 6-7 mo	1432
Kobus leche	wean: 3-4 mo	7368
	wean: ~5-6 mo	10158

Madoqua kirki	nurse: short period	5625
	wean: ~6 wk	5763
	leave parents: ~6 mo	4977
Madoqua saltiana	wean: ~6 wk	5763
Nemorhaedus goral	solid food: ~1 mo	4643
Neotragus moschatus	solid food: few d	5763
Oreamnos americanus	solid food: 1st d	1237
	solid food: 2-3 d	4933
	nurse: est 3 mo; stay with f 1 yr	9122
	wean: ~4 mo	1237
Oreotragus oreotragus	wean: 4-5 mo	10158
Oryx leucoryx	solid food: 30 d	4943
Ourebia ourebi	wean: 4-5 mo	2439
Ovibos moschatus	solid food: 2-3 wk	8798
	wean: 5-9 mo	10770
	nurse: est until next birth; n1	8033
Ovis ammon	solid food: ~2 wk	3571
	solid food: 1 mo	4643
	wean: 4-5 mo	3571
Ovis canadensis	solid food: 1 wk	3667
	solid food: 10 d	4643
	solid food: 14 d	9815
	solid food: ~2 wk?	1011
	solid food: 3 wk-1 mo	2476
	wean: 4-6 mo	3667
	wean: 5-6 mo	1011
Ovis dalli	solid food: 1 wk	3667
	wean: 3-5 mo	1166a
	wean: 4-5 mo	3667
Ovis musimon	wean: 6 mo	10946
Procapra gutturosa	solid food: 10 d	4643
	solid food: 40 d	7395
Raphicerus campestris	solid food: 2 wk	1684
	wean: 3 mo	1684
	wean: 3 mo	5763
Rupicapra rupicapra	solid food: 1 mo	4643
	solid food: 2 mo	2079
	wean: 2-3 mo	2079
Saiga tatarica	solid food: 2-2.5 mo	4643
	solid food: 3-4 d	561
	solid food: 4-8 d	4643
	solid food: 8 d	560
	solid food: 3 wk	3377
	solid food: 1 mo	4645
	wean: 6 wk	3377
	wean: 2.5-3 mo	560, 561
	wean: 2.5-4 mo	4643
	wean: 4 mo	10200
Syncerus caffer	suckle: 4-5 mo	8515
	wean: 5-6 mo	8515
	wean: 10-12 mo	1599
	wean: 12 mo	4077
Tragelaphus angasi	wean: 7 mo	269, 271
Tragelaphus eurycerus	solid food: 1-2 wk	11022
Tragelaphus oryx	solid food: min 1 d	11044
	dependent: 3 mo	4801
Tragelaphus strepsiceros	wean: 6 mo	9986
INTERLITTER INTERVAL		
Addax nasomaculatus	355 d; 33.7% 275; r227-741	2487
	1 yr	6741
Aepyceros melampus	365 d	10023
Alcelaphus buselaphus	326 d; r271-366	3955
	1 yr	3956

Ammotragus lervia	182 d; r178-184; n3	9734
	0.5-1 yr	1352
	<7 mo; n2	4020
Antidorcas marsupialis	1 yr	10038
Antilocapra americana	1 yr	12019
Antilope cervicapra	6 mo; r5-7; n2	6541
	6 mo	7616
	6-9 mo	10628
	197 d; n6	36
	1 yr	9545
Bison bison	44-76 wk	4048
	1 yr	5229
	54 wk; se5; n37	4047
	1.5 yr	3559
	1-2 yr	6585
Bison bonasus	1-2 yr	3571
	2 yr	4643
Bos frontalis	1 yr	9545
	14-15 mo; r394-906 d; n1 f	3447
	not until 3rd yr	2802
Boselaphus tragocamelus	1 yr	6654
Bubalus bubalis	1 yr	6654
Budorcas taxicolor	1 yr	8798
Boselaphus tragocamelus	436 d; n8	36
Capra caucasica	1 yr	4643
Capra hircus	7-10 mo	9357
	1 yr	4643
	2 yr	2080
Capra ibex	1 yr	10530
	1 yr	4643
	2 yr	2081
Capricornis crispus	18 mo; ovarian histology	5826
Cephalophus monticola	265 d; se21.4; r208-446; n39 lt, 9 f	11349
Cephalophus sylvicultor	399 d median	5996
Connochaetes taurinus	1 yr	10637, 10638
Damaliscus dorcas	1 yr	10023, 10027
Gazella cuvieri	0.5-1 yr	8100
Gazella dama	0.5 yr	254
	0.5 yr	5575
Gazella dorcas	310.7 d; median 271; r177-965; n95 intervals, 13 f	5997
	1 yr	8223
Gazella gazella	7-8 mo or 1 yr; 2 locations	492, 493
Gazella spekei	201-212 d	8991
Gazella thomsoni	0.5 yr	5122, 5123
	0.5 yr	1328
	6-7.5 mo	4972
Hemitragus hylocrius	7.8, 8.4 mo; r7-10; n14 intervals, 2 f	11750
Hemitragus jemlahicus	1 yr	6654
Hippotragus equinus	10-10.5 mo	11758
	317 d	10158
	1 yr	11253
Hippotragus niger	1 yr	11253
	1 yr	11758
Kobus ellipsiprymnus	1 yr	11253
Kobus vardoni	1 yr	11253
Litocranius walleri	min 223 d	6410
	8-9 mo; n5, 1 f	6410
Madoqua kirki	~6 mo	5625
	0.5 yr	5625
	1 yr	4622
Nemorhaedus goral	1 yr	6654
	1 yr	4643
Oreamnos americanus	1 yr	9121
Ourebia ourebi	0.5 yr	11253

Oreotragus oreotragus	1 yr	11253
Ovibos moschatus	1 yr	1474
	1 yr	8033
	2 yr	10770
	2 yr	10740
	2 yr	10770
	2 yr	1474
Ovis ammon	1 yr	3571
	1 yr	4643
Ovis canadensis	1 yr	4643
Ovis dalli	~1 yr	1447
Ovis musimon	1 yr	10946
Ovis vignei	1 yr	6654
Pantholops hodgsoni	1 yr	6654
Procapra gutturosa	1 yr	4643
Raphicerus campestris	r168-270 d; n5; 4 intervals 168-177, 1 was 270	922
	6 mo?	11697
	~8 mo; n3 intervals	1684
Redunca arundinum	0.5-1 yr	11253
Redunca fulvorufula	10.2 mo; r8-13; n6	5165
	12-14 mo	10158
Rupicapra rupicapra	1 yr	3571
	1 yr	2079, 2080
	1 yr	4643
Saiga tatarica	1 yr	4645
	1 yr	561
Sigmoceros lichtensteini	1 yr	11253
Sylvicapra grimmia	4 mo	11253
	259.5 d; 2se11.6; r232-298	11348
Syncerus caffer	15 mo; 95%ci 13.3-17.5	9991, 9994
	18 mo; 95%ci 15.6-21.3	9994
	18-24 mo; ovaries	5763
	18 or 24 mo; 2 locations	4077
	20-26 mo	1599
	2 yr	11253
	2 yr	10158
Tetracerus quadricornis	316 d; r285-347; n2	37
Tragelaphus angasi	231-303 d	270
	r257-293 d; n6 birth, 1 f	7774
	297 d	269, 271
Tragelaphus imberbis	min 8 mo; n300	6408
	8-10 mo	6411
Tragelaphus oryx	~10 mo	5763
	331.7 d; r277-553; n11	8690
	337 d; r284-553; n51	9311
	354.5 d; se19.9	10023
	370.8 d; r289-602; n6	10300
Tragelaphus scriptus	0.5 yr	11253
	7.5 mo; n1	7537
	8 mo; n1 f	8712
	249.1 d; se8.6; n7	11347
Tragelaphus spekei	min 11.6 mo, 64% < 12 mo	2485
Tragelaphus strepsiceros	1 yr	10023, 10027
SEASONALITY		
Addax nasomaculatus	birth: Jan-mid Apr; Chad	2897
	birth: peak Sept-Oct; Circle d'Agadez, Niger	1352
	birth: yr rd; captive	2487
	birth: yr rd; n19; zoo	2603
	birth: yr rd; Chad	6741
Aepyceros melampus	testes size: peak Jan-Mar, low Oct-Dec; Zimbabwe	5697
	testes size: peak Apr, low June; S Africa	1334
	testes physiology: peak May, low Sept; Transvaal, S Africa	10022, 10041
	testes size: peak Nov-June, low July-Oct; Zululand, S Africa	267
	mate: Mar; Zululand, S Africa	267

	mate: Mar?; Angola	11697
	mate: Mar-May; Zululand, S Africa	7204
	mate: Apr-May; S Africa	1334
	mate: Apr-May; S Africa	6928
	mate: Apr-mid July; Kruger Nat Park, S Africa	3145, 3148
	mate: peak mid May; n Botswana	9157
	mate: end Apr-May; 10-20 d; nw Zimbabwe	7660
	mate: peak major May, minor Oct; Zululand, S Africa	268
	mate: May-Aug, peak May; Zambia	11463
	mate: mid May-mid June; Zimbabwe	2275
	mate: peak May-June; Transvaal, S Africa	10022, 10023, 10041
	mate: yr rd, peaks Dec, May; Rwanda	7470
	conceive: Mar-Apr; Serengeti, e Africa	5260
	conceive: peak May; S Africa	3149
	ovulation: May; Zululand, S Africa	267
	birth: Sept-Oct; Angola	11697
	birth: Sept-Dec; S Africa	331, 333
	birth: end Oct-early Nov; Botswana	9157
	birth: Nov-Dec; ne S Africa	3145, 3148
	birth: Nov-Dec, if dry Jan-Feb; Kruger Nat Park, S Africa	8514
	birth: Nov-Dec; S Africa	5778
	birth: Nov-mid Jan, peak Dec, low July-Oct; zoo	1230
	birth: Nov-Jan; Zululand, S Africa	7204
	birth: Nov-Jan; S Africa	10158
	birth: Nov-Feb, peak Nov; Matopos Nat Park, Zimbabwe	11785
	birth: peak Nov-Jan; S Africa	10041
	birth: Dec-Jan; Zimbabwe	2274, 2275
	birth: Dec-Jan, 2 wk; Transvaal, S Africa	8965
	birth: Dec (60%)-Jan (20%); n547; S Africa	10041
	birth: end Dec-Feb, peak Jan; Namibia	5451
	birth: yr rd, peaks Feb-May, Oct-Dec; Kenya	5597
	repro: yr rd; zoo	2603
	repro: yr rd; Kenya	6413
Alcelaphus buselaphus	mate: Feb-Mar; Transvaal, S Africa	10041
	mate: Apr-May; Zaire	477
	mate: end Apr-June; Uganda	5764
	mate: peak mid Oct-mid Nov; S Africa	1153
	mate: yr rd, peaks June-July, Nov; Kenya	3956
	conceive: yr rd; n41, 2 yr; Kenya	3247
	conceive: yr rd, peaks June-July, minor Nov; Kenya	3956
	conceive: peak Feb-Apr; S Africa	1153
	birth: peaks Feb-Mar, minor July; Kenya	3956
	birth: Aug-Oct, peak Sept; Cape Province, S Africa	7204
	birth: peak Sept-Oct; S Africa	2382
	birth: peak Oct; zoo	1230
	birth: Oct-Dec, 82% Oct-Nov; n278; Transvaal, S Africa	10041, 10042
	birth: mid Dec-mid Feb; Ethiopia	6416
	birth: end Dec-mid Feb; W Africa	375
	birth: yr rd, peaks before rain; Kenya	3955
	repro: after Oct; s of Botswana	7204
	repro: yr rd, peaks Feb-Mar, July-Aug; Kenya	6413
Ammodorcas clarkei	mate: yr rd, peak Mar-May; Mudugh Province, Somalia	7167
	mate: 30% Oct; Somalia	1890
	birth: peak Oct-Nov; Somalia	1890
	birth: yr rd; Mudugh Province, Somalia	7167
Ammotragus lervia	mate: mid July-mid Aug; Niger	1352
	mate: Oct-Nov; zoo, Ukraine	6523
	mate: yr rd, peak spring; TX	4021
	birth: start spr, sometimes Jan; Morocco	8223
	birth: 70% Mar-Apr; n26 1 yr; panhandle, TX	4021
	birth: peak Mar-Apr, low Jan, July; n176; zoo	5818
	birth: Mar-May; zoo, Ukraine	6523
	birth: May, June, Oct; zoo	9734
	birth: Dec-Jan; Niger	1352

	birth: yr rd, peak Mar; n162 lt	2109
	birth: yr rd, peak Mar-May; n190; zoo	12128
	birth: yr rd, peak spring; zoo	1230
Antidorcas marsupialis	testes size: peak Apr-May, low Jan; S Africa	10039, 10041
	mate: Feb-Apr, some June, Sept, Nov; S Africa	2297
	mate: June-Aug, peak end July; Namibia	923
	mate: yr rd, peak Mar-July; S Africa	10023, 10039, 10041
	mate: yr rd, peak May; n23; zoo	8339
	ovulation: end May-June; Transvaal, S Africa	10038
	birth: Mar-June, Aug-Nov, peak Oct; Cape, S Africa	923
	birth: peak Apr-Sept; s Africa	9703
	birth: Aug; Namaqualand, Namibia	9945
	birth: Aug-Sept; S Africa	2297
	birth: Sept-Dec; Botswana	1393
	birth: peak Sept-Jan, few Apr-July; n1655; S Africa	10023, 10036, 10038, 10041
	birth: Dec-Jan; W Africa	8351
	birth: yr rd, peak spring-early summer; zoo	1230
Antilocapra americana	testes large: July-Oct; WY, MT, CO	8010
	mate: Aug-Sept; Mexico	6387
	mate: Aug-Oct; w US	1474
	mate: mid Aug-Oct; zoo	2607
	mate: peak Sept; NM	6217
	mate: Sept-Oct; MT	1424
	mate: mid Sept-early Oct; WY	1871a
	mate: mid Sept-mid Oct; WY	4647
	mate: Oct; w US	5146
	conceive: Sept-Oct; MT	5828
	birth: Feb; Baja CA, Mexico	10809
	birth: Feb; s TX	11208
	birth: Apr-May; sw US	1474
	birth: May; NM	502
	birth: May-June; nw US	1474
	birth: May-June; w N America	12019
	birth: mid May-mid June; zoo	2607
	birth: end May-June; WY	4647
	birth: June; Mexico	7145
	birth: July; ID	11208
Antilope cervicapra	estrus: yr rd; zoo	785
	spermatogenesis: yr rd; TX	1649
	spermatogenesis: yr rd; captive	3109
	spermatogenesis: yr rd; captive, Italy	10628
	spermatogenesis: yr rd, peak Apr, mid Aug-mid Oct; India	9545
	mate: Feb-Mar; India	6653, 6654
	mate: Feb-Apr; Nepal	6365
	mate: May-June; Nepal	7382
	mate: yr rd; zoo, S Africa	1230
	mate: yr rd, peak May; n82; zoo	8339
	birth: Jan-Mar, May, Sept; n11 lt; zoo	44
	birth: Feb-May, Sept-Nov; few data; zoo, Italy	10628
	birth: Mar, June-July, Sept, Nov; n9; captive	785
	birth: Apr-May, Oct-Nov; captive	6541
	birth: May-July, peak June; n143; zoo	5818
	birth: Nov-Dec; Nepal	7382
	birth: yr rd; India	6653, 6654
	birth: yr rd; captive	2109
	birth: yr rd; captive, TX	7617
Bison bison	mate: mid June-Sept; WY	7068
	mate: peak Aug; US	188
	mate: July-Oct; NWT, Canada, WY	1474
	mate: July-Oct; WY	1871a
	mate: mid July-early Sept; NWT, Canada, WY	7144
	mate: peak July 21-mid Aug; SD, NE	4501
	mate: end July-mid Aug, 46% July 30-Aug 2; MT	6584
	mate: Aug; NWT, Canada	10216

	mate: Aug; Canada	3559, 3560
	birth: Mar-June; US	188
	birth: Mar-Oct; n36; zoo	5818
	birth: peak mid Apr-May; Yellowstone Park, WY	7068
	birth: Apr-Nov; n91; captive	2109
	birth: mid Apr-May; WY	7068
	birth: 80% end Apr-mid May; MT	9386
	birth: end Apr-Oct; SD, NV	4501
	birth: May; Canada	3559, 3560
	birth: May-early June; NWT, Canada	10216
	birth: peak May; WY, NWT, Canada	7144
	birth: peak May; NE, SD	4501
	birth: May-July; n Canada	3559
Bison bonasus	mate: Aug-mid Sept; USSR	4643
	mate: Aug-Sept; Germany	3571
	mate: Aug-Dec, peak Aug-Oct; Poland	5999
	mate: end summer; Poland	5229
	birth: low Jan, Feb n29; zoo	5818
	birth: Apr-June; WY	1871a
	birth: Apr-Dec, peak May-July; Poland	5999, 8846, 8954
	birth: May-early June; Poland	6652
	birth: May-mid June; USSR	4643
	birth: July; n2; zoo	2109
	birth: yr rd, peak May; n187; captive	5229
	birth: yr rd, peak May-July; Germany	3571
	birth: yr rd, peak May-July; zoo	11879
Bos frontalis	mate: Dec-Jan; India, se Asia	6652
	mate: peak Dec-Jan; India	2802
	mate: yr rd; India	9545
	birth: Jan-May; Aug, Sept; n16; zoo	12128
	birth: May-Oct; India	2802
	birth: peak June; n48; captive	2109
	birth: Aug-Sept, some Apr-June; India, se Asia	6652
	birth: peak Aug-Sept; India	2802
	birth: Sept-mid Mar, peak Dec-Jan; India	9545
	birth: yr rd; Malaysia	5047
	birth: yr rd; captive	3447
Bos grunniens	mate: Sept; Tibet	8740
	birth: June; according to natives; Tibet	8740
	birth: yr rd, low Feb-Mar; n31; zoo	5818
	birth: yr rd; n4; zoo	12128
Bos javanicus	birth: 14/25 Feb-Mar; captive; USSR	6524
	birth: rainy season; se Asia	6652
Boselaphus tragocamelus	estrus: May-July; zoo	4535
	mate: peak Nov; n23; zoo	8339
	mate: possible peak Nov-Dec; e Rajasthan, India	9545
	birth: Jan-Mar; n5; zoo	44
	birth: Apr-Oct, peak July; n126; zoo	5818
	birth: June-Oct; India	9545
	birth: yr rd; n61; zoo	12128
	birth: yr rd; zoo	1230
	birth: yr rd; India	9545
Bubalus bubalis	m fertile 8 mo/yr; Australia	10940
	mate: Sept; Nepal	7382
	mate: fall; India	6654
	mate: yr rd; zoo	1230
	birth: June-July; Nepal	7382
	birth: 90% Dec-Apr; Australia	10939, 10940
	birth: yr rd, peak July-Sept, low Dec-Apr; Pakistan	6737
Bubalus depressicornis	mate: yr rd; zoo	1230
	birth: Feb-July; n14; captive	8708
	birth: Feb-Mar, May-June, Oct; few data; zoo	2669
	birth: Mar, May, Sept-Nov; n7; zoo	12128
	birth: yr rd, Feb-July, Oct; zoo	2669

Budorcas taxicolor	mate: end July-early Sept; China	8798
	mate: end July-Sept; China	9547
	birth: May-June; Szetchwan, China	3034
Capra caucasica	mate: Jan?; zoo	8458
	mate: end Nov-early Jan; USSR	4643
	birth: mid May-3rd wk June; captive to mid July; USSR	4643
	birth: end May-June; n24; zoo	12128
	birth: June; zoo	8458
Capra falconeri	estrus: Dec; Kashmir	4535
	mate: Oct; s Pakistan	9175
	mate: Dec; n Pakistan	9175
	mate: early Nov-3rd wk Dec; USSR	4643
	mate: mid Dec-early Jan; Pakistan, Afghanistan	9547
	birth: end Apr-early May; captive to early June, peak May; USSR	4643
	birth: peak May, few Apr, June; n48; zoo	5818
Capra hircus	estrus: June-early July; Giura I, Greece	9668
	estrus: end Oct-Nov; Crete, Greece	9668
	estrus: Aug-early Sept; Antimilos, Greece	9668
	mate: Sept-Oct; Pakistan	9175
	mate: mid Sept-mid Oct; Pakistan	9547
	mate: Oct; Scotland, UK	7051
	mate: mid Nov-mid Dec; 2 wk; USSR	4643
	mate: Dec-mid Jan; Swiss Alps	2080
	mate: Dec-Feb; Taurus Mt, Turkey	11389
	mate: yr rd; New Zealand	9357
	birth: Apr-May; Taurus Mt, Turkey	11389
	birth: end Apr-mid June; n22; zoo	12128
	birth: end May-early June; USSR	4643
	birth: end May-mid June; Swiss Alps	2080
	birth: Jan; Antimilos, Greece	9668
	birth: end Feb-Mar; Samotha Lake, Greece	9668
	birth: end Jan-Mar; Pakistan	9175
	birth: end Feb-Mar 20; Pakistan	9547
	birth: Mar; Scotland, UK	7051
	birth: Mar; Tavolara I, ne of Sardinia, Italy	9668
	birth: Apr-May; Crete, Greece	9668
	birth: Nov; Giura I, Greece	9668
	birth: every 2-4 mo; Santa Catalina I, CA	1930
Capra ibex	estrus: Nov-Dec; Himalayas, India, Tibet	4535
	estrus: Nov-Jan; semiwild; Germany	10363
	mate: Nov-early Jan; USSR	4643
	mate: Nov-Mar; Switzerland	7912
	mate: Dec-Jan; Switzerland	4210
	mate: Dec-mid Jan; Alps, Europe	2081
	mate: Dec-early Feb; Switzerland	10530
	mate: mid Dec-early Jan; Germany	3571
	mate: Jan; Swiss Alps	6652
	mate: yr rd; Ethiopia	7912
	birth: Apr-May, later higher elevations; n Tian Shan, China	3189
	birth: end Apr-3rd wk June, peak May; USSR	4643
	birth: May; Pakistan	9175
	birth: May-June; central Asia	6652
	birth: May-June; Switzerland	7911
	birth: May-July; zoo	8595
	birth: May-Aug; Switzerland	10530
	birth: (end May) June-early July; Germany	3571
	birth: peak early-mid June; Alps	2081
	birth: end June-early July; Swiss Alps	6652
	birth: spring; Israel	6764
Capra pyrenaica	birth: mid Apr-early May; Pyrenees, France/Spain	6652
	birth: May; Spain	112

Capricornis crispus	mate: Oct-Dec; n2 yr; captive	5178
	mate: Oct-Nov; Japan	10817
	birth: end May-early June; Kasabori, Japan	104
Cephalophus adersi	mate: yr rd; captive	5763
Cephalophus dorsalis	m fertile: yr rd; zoo	2605
Cephalophus maxwelli	birth: Jan-Mar, Aug-Sept; captive	74
	birth: peak Jan-Mar, Aug-Sept; Nigeria	4354a
Cephalophus melanorheus	lact f: Apr; n1; Tanzania	186
Cephalophus monticola	birth: Jan-Feb; Uganda, w Kenya	5763
	birth: Feb; n2; e Africa	185
	birth: peak Sept-Nov, low June; n49; zoo	11346
	birth: yr rd; captive	1230
	birth: yr rd; n38; captive	1047
	birth: yr rd; Cape Province, S Africa	5777
	repro: yr rd; S Africa	10158
Cephalophus natalensis	birth: Oct-Nov; S Africa	5781
Connochaetes gnou	testes size: peak Mar-May; S Africa	10041, 10042
	mate: Jan, May-June; n26; zoo	8339
	mate: mid Mar-mid May; Transvaal, S Africa	10041, 10042
	birth: Nov-Mar, slight peak Dec; captive	1230
	birth: mid-end Nov; Pretoria Game Reserve, S Africa	11355
	birth: Dec (54%)-Jan (38%); n408; S Africa	10041, 10042
	birth: mid-end Dec; Lombard Nature Reserve, S Africa	11355
Connochaetes taurinus	mate: end Mar-Aug, peak Apr-mid June; e Africa	10638
	mate: mid Apr-mid May; Tanzania	9993
	mate: June-?; Angola	11697
	conceive: Mar-Apr, peak Apr, none Sept-Oct; Zululand, S Africa	440, 441
	conceive: peak Mar-mid June; Tanzania	9994
	birth: Jan-Feb, 80% 3 wk period; Tanzania	3102, 3105
	birth: prob Jan-Feb; Zimbabwe	2275
	birth: end May-early July; zoo	6522
	birth: Aug-Nov; w Zambia	331
	birth: Sept; Angola	11697
	birth: Sept-early Nov; S Africa	1392
	birth: Oct-Jan; Namibia	9945
	birth: Oct-Nov; Tanzania	3099
	birth: peak Nov-Dec; Luangwa Valley, Zambia	331
	birth: Nov-Dec, peak Dec; Zululand, S Africa	441
	birth: Nov-Jan, peak Dec; Matopos Nat Park, Zimbabwe	11785
	birth: end Nov-Jan; Kruger Nat Park, S Africa	8514
	birth: mid Nov-Dec; ne S Africa	3145
	birth: mid Nov-mid May, peak Jan-Mar; Kenya	10637, 10638
Damaliscus dorcas	testosterone: peak Feb; S Africa	10029
	testes size: peak Mar-Apr, low Oct; Transvaal, S Africa	10032, 10041
	mate: peak Jan; Cape, S Africa	6204
	mate: peak Jan-mid Mar; S Africa	2295, 2296
	mate: Mar-Apr; Transvaal, S Africa	6657
	mate: Mar-Apr; Transvaal, S Africa	2746
	mate: peak Mar-May; S Africa	10032, 10041
	mate: Apr-June, peak mid Apr; S Africa	5702
	mate: Nov-mid Feb; zoo	1230
	conceive: Mar-May; Transvaal, S Africa	7976
	birth: June, Aug; n2; zoo	208
	birth: Sept-Oct; S Africa	1391
	birth: peak Sept-Oct; Cape Province, S Africa	6204
	birth: Sept-Nov; S Africa	2295, 2296
	birth: Nov, sometimes Feb; S Africa	5702
	birth: Nov-Dec; S Africa	9328
	birth: 88% Nov-Dec; n364; S Africa	10032, 10041
	birth: mid Nov-Dec; Transvaal, S Africa	2746
	birth: mid Nov-Dec; Transvaal, S Africa	6657

Damaliscus lunatus	mate: Dec-Feb; Rwanda	7470
	mate: peak mid Dec-Jan; S Africa	5456
	mate: Feb-Mar; Uganda	5308
	mate: Mar-Apr; Botswana	1792
	mate: May-July; Angola	11697
	birth: June-Sept, peak July-Aug; Zambia	331
	birth: Aug-Nov; Rwanda	7470, 7476
	birth: Sept-Nov; Kruger Nat Park, S Africa	3145
	birth: Sept-Dec; Angola	11697
	birth: Sept-Dec, peak Oct; S Africa	5456
	birth: Oct; S Africa	3618
	birth: Nov; Botswana	1792
Gazella cuvieri	birth: peak spring, smaller peak fall; captive	8100
Gazella dama	birth: Sept-Oct; Circle d'Agadez, Niger	1352
Gazella dorcas	mate: Sept; Israel	493
	birth: Mar; Israel	493
	birth: Mar-May, July-Aug, Oct; n7; zoo	12128
	birth: Aug-Oct; Circle d'Agadez, Niger	1352
	repro: yr rd; Morocco	8223
Gazella gazella	mate: peak Dec-Jan; Upper Galilee, Israel	492
	conceive: Dec; Upper Galilee, Israel	493
	preg f: Jan; n1; nw Iran	6293
	birth: peak May-July; Upper Galilee, Israel	492
	birth: June; Upper Galilee; yr rd; lower Galilee, Israel	491, 493
	repro: end monsoon-early Oct, Mar-Apr; Pakistan	9175
Gazella granti	conceive: peak Dec; Tanzania	10271
	conceive: yr rd; n44 2 yr; Kenya	3247
	birth: Jan-Feb, Apr, Sept; e Africa	7856
	birth: Jan-Feb, June-Aug, Nov-Dec; e Africa	5764
	birth: peak June; Tanzania	10271
	birth: yr rd; n24; zoo	2603
	repro: yr rd; Kenya	6413
Gazella soemmerringi	mate: Oct; Somalia	5764
	birth: Apr; Somalia	5764
	birth: est Apr; Somalia	2978
Gazella subgutturosa	mate: Nov-Dec; USSR	4643
	mate: Dec; Pakistan	9175
	birth: Apr-May; USSR	4643
	birth: Apr-mid July, Sept-Oct, peak Apr-May; n22; zoo	12128
	birth: peak end Apr-early June; captive	9175
	wean: Oct; USSR	4643
Gazella thomsoni	sperm present: yr rd; s Kenya, n Tanzania	5122
	mate: dry season, Jan-Feb, July-Aug; captive; Kenya	4972
	mate: yr rd, peak Aug-Oct; Kenya	6922
	mate: yr rd; Tanzania	1328
	mate: 1 mo after birth: e Africa	5764
	conceive: yr rd; n44, 2 yr; Kenya	3247
	preg f; yr rd; Tanzania	1196
	birth: Jan-Feb, July; e Africa	5764
	birth: peaks Nov-Jan, July-Aug; captive, Kenya	4972
	birth: yr rd, peaks Jan-Mar, June-July; Tanzania	1328
	birth: yr rd, peak Apr, Nov; s Kenya, n Tanzania	5122
	birth: yr rd, peak May, Nov; n Tanzania	9180
	birth: yr rd; n59; zoo	5818
Hemitragus hylocrius	birth: yr rd; Asia	6654
Hemitragus jemlahicus	estrus: Dec; Kashmir, Himalayas	4535
	estrus: winter; zoo	4487
	mate: Oct-mid Jan; Nepal	9547
	mate: mid Nov-Dec; Tibet	9540
	birth: May-Sept, peak June; n181; captive	2109
	birth: May-early Oct, peak May-July; n115; zoo	12128
	birth: June; zoo	4487
	birth: June-July; Himalayas	6652, 6654

Hippotragus equinus	mate: Jan; Zaire	254
	mate: Jan; Zaire	5575
	mate: peak May-Aug; Zaire	11253
	mate: Oct-Dec (calculated); Angola	11697
	mate: yr rd; e Africa	5764
	birth: Jan-Mar, May, Nov; Zimbabwe	1794
	birth: peak Feb-mid May; Zaire	11253
	birth: Apr-Aug?; few data; Kruger Nat Park, S Africa	3145
	birth: Aug-?; Angola	11697
	birth: Aug-Oct; s, s central Africa	9752
	birth: Sept-Apr; s, e Africa	5764
	birth: Nov; Zaire	254
	birth: Nov; Zaire	5575
	birth: yr rd, peak Mar-July; Percy Fyfe Nature Reserve, S Africa	10027
	birth: yr rd; few data; Zimbabwe	1794
	repro: yr rd; S Africa	10158
Hippotragus niger	mate: Mar-May; Angola	11697
	mate: Aug-Sept; Angola	3104
	mate: yr rd; zoo	1230
	conceive: May-July; Transvaal, S Africa, Zimbabwe	11758
	birth: Dec-Feb; Angola	11697
	birth: Dec-June, peak early Feb; Matopos Nat Park, Zimbabwe	4085
	birth: Jan-early Feb; ne Botswana; S Africa	10158
	birth: Jan-Mar, peak Feb; n Transvaal; S Africa	10158
	birth: Jan-Mar, peak Feb; Transvaal, S Africa, Zimbabwe	11758
	birth: end Jan-mid Mar; Kruger Nat Park, S Africa	8514
	birth: Feb-Mar; S Africa	3145
	birth: Feb-Apr, peak Mar; Matopos Nat Park, Zimbabwe	11785
	birth: Mar, Aug, Oct-Nov; Zimbabwe	1794
	birth: Mar-Aug, peak Feb-Mar; Percy Fyfe Nature Reserve, S Africa	10027
	birth: Apr-May, Oct-Nov; Zaire	11253
	birth: Apr-June; Kenya	9742
	birth: Apr-Aug; zoo	1230
	birth: May-July; Angola	3103, 3104
	birth: yr rd, July, Nov-Dec; few data; Zimbabwe	4084, 4086
	birth: Sept-Oct; s, s central Africa	9751
Kobus ellipsiprymnus	estrus: May-July; 3 wk; zoo	4535
	mate: June-Aug; Zambia	333
	mate: yr rd; zoo	1230
	preg f: Mar; n1; Tanzania	186
	birth: Jan-Feb, July-Aug; n27; zoo	5818
	birth: peaks Jan-Mar, Sept-Oct; Kruger Nat Park, S Africa	3145
	birth: Feb-Nov; Zambia	333
	birth: end Feb-Apr; Zimbabwe	2275
	birth: Aug; W Africa	8350
	birth: Oct-May, peaks Oct, Feb-Mar; Kruger Nat Park, S Africa	8514
	birth: peak Oct-Nov; Zaire	11253
	birth: peak Dec-Mar, low Nov; zoo	1230
	birth: Dec-July, peak Feb-Mar; n Natal, S Africa	7197
	birth: mid Dec-Feb; British E Africa	5206
	birth: end Dec; Niger	374
	birth: yr rd; Zambia	331
	birth: yr rd, poss peak dry season; Zaire	11271
	birth: yr rd, peak Jan-Feb; Transvaal, S Africa	4655, 4656
	repro: yr rd; Kenya	6413
Kobus kob	mate: Mar-Apr; Pibor, Akobo Valleys, Uganda	5763
	mate: May-Aug; Ivory Coast	7591
	mate: yr rd; w Rift Valley, Uganda	1434, 1436
	mate: yr rd; Uganda	6407
	conceive; Jan-Apr; n16/17; Uganda, Sudan	3543
	birth: est peak Jan-Feb; Zaire	11271
	birth: ~Nov-Dec; Senegal, s Sudan	5763

	birth: Dec-Mar; Ivory Coast	7591
	birth: yr rd; Ivory Coast	3659
	birth: yr rd; Uganda	5763
	birth: yr rd, slight peak Jan-Feb; Uganda	1432
Kobus leche	mate: Oct; Angola	11697
	mate: peak end Oct-Dec; Zambia	333
	mate: end Oct-Dec; Zambia	7368
	mate: mid Nov-mid Feb; Zambia	9676
	mate: yr rd, peak May; nw Botswana	6378
	mate: yr rd, peak mid Nov-Jan; Zambia	9181
	preg f: Oct; n1; Namibia	9945
	birth: Jan, Mar-May, July-Aug, Dec; n17; zoo	12128
	birth: Apr-Sept, peak mid July-mid Aug; Zambia	331, 333
	birth: May-Sept, some Oct, Nov; Zambia	194
	birth: May-Dec, peak mid July-mid Aug; Zambia	7368
	birth: Aug-Sept; Angola	11697
	birth: Oct-Nov; upper Zambesi, Zambia	9750
	birth: peak Oct-Dec; Botswana	10158
	birth: yr rd, peak mid Aug; Zambia	9181
Kobus vardoni	birth: Jan-Feb, June-July; Zaire	11253
	birth: peak Jan-Apr; Luangwa Valley, Zambia	
	birth: peak May-Sept, minor July-Aug; Zambia	331, 333
	birth: Apr-Dec; S Africa	2413
	birth: probably Nov-Dec; Zambia	9749
	birth: yr rd, poss peak during 1st heavy rains; central Africa	5763
Litocranius walleri	birth: during rains; ne Africa	5764
	birth: prob yr rd; n12; Kenya	6410
	repro: yr rd; Kenya	6413
Madoqua kirki	testes size: peak May; Tanzania	5625
	spermatogenesis: Nov, low Jan-Feb, July-Aug; Tanzania	5625
	mate: yr rd: peak June-July, Nov-Dec; Tanzania	5625
	birth: Dec-Apr; S Africa	10158
Madoqua saltiana	preg f: yr rd; Ethiopia, Horn of Africa	5763
Nemorhaedus goral	mate: Sept-Nov; Korea	1310a
	mate: Sept-Jan; captive	2652
	mate: end Sept-Nov; USSR	4643
	mate: early-mid Nov; China	3034
	birth: Apr-May; hunters report; Pakistan	9175
	birth: May; China	3034
	birth: May-June: India	6654
	birth: end May-mid June, few July-Aug; USSR	4643
	lact f: May, Aug; n2; Nepal	7382
Neotragus batesi	mate: yr rd, peaks between wet-dry seasons; Gabon	5763
	birth: yr rd, peaks end each rainy season; Gabon	3191, 3192
Neotragus moschatus	mate: yr rd; e Africa	5763
	birth: Aug-Feb; S Africa	10158
	birth: peak Nov-Feb; e Africa	5763
	birth: mid Nov-mid Dec; Zululand, S Africa	5772
	birth: prob yr rd; captive	5191
Neotragus pygmaeus	birth: Sept; n1; Liberia	184
Oreamnos americanus	mate: Oct-Dec; Rocky Mt, w N America	1474
	mate: end Oct-early Dec; ID	1237
	mate: Nov; BC, Canada	4933
	mate: Nov-Dec; ID, MT	1237
	mate: Nov-mid Dec; MT, SD	9121
	mate: mid Nov-early Dec; WY	1871a
	birth: May-June; Rocky Mt, N America	1474
	birth: mid May-mid June; ID, MT	1237
	birth: mid May-mid June; WY	1871a
	birth: end May-early June; MT, SD	9121
	birth: end May-early June; BC, Canada	4933
	birth: end May-mid June; MT	6381a
	wean: Sept; MT, SD	9121

Oreotragus oreotragus	preg f: Feb-Nov; Zambia	11788
	preg f: Sept; n1; Namibia	9945
	preg f: yr rd; S Africa	10158
	birth: July-Aug; Zaire	11253
	birth: Nov; Ethiopia	2798
	birth: Nov-Dec?; Kruger Nat Park, S Africa	8514
Oryx dammah	mate: Oct-Nov; Chad	3781
	birth: Jan, Mar-June, Sept-Dec; n12; zoo	12128
	birth: July-early Aug; Chad	3781
	birth: yr rd; n41; zoo	5818
Oryx gazella	mate: peak Nov-Dec; S Africa	10041
	birth: Jan-Mar; e Africa	5207
	birth: prob yr rd, peak Aug-Sept; n22; S Africa	10041, 11212
	birth: yr rd, peak Sept; S Africa	2985
	birth: yr rd; zoo	1230
	birth: yr rd; Circle d'Agadez, Niger	1352
	birth: yr rd; e Africa, Namibia	5764
	birth: yr rd; S Africa	10158
	repro: yr rd, peak June-Aug; Kenya	6413
Oryx leucoryx	birth: May-Sept; Saudi Arabia	10395
Ourebia ourebi	birth: July; Angola	11697
	birth: Sept-Nov; Kruger Nat Park, S Africa	8514
	birth: Oct-Dec; Zimbabwe	10803
	birth: Dec-Jan, May-June; Zaire	11253
Ovibos moschatus	estrus: Aug; e Greenland	5300
	mate: July-Aug; n N America	1474
	mate: July-Aug; Norway	8593
	mate: Aug; NWT, Canada	10738, 10740
	mate: Aug-Sept; Jameson Land, ne Greenland	10770
	mate: Aug-early Oct; arctic	8798
	mate: Aug-Oct; Arctic, Canada	9330
	birth: June-Feb; captive, ALB, Canada	8033
	lact: May-June; captive, ALB, Canada	8033
	birth: Apr-early May; e Greenland	5300
	birth: Apr-May; NWT, Canada	6241
	birth: Apr-May; Arctic, Canada	9330
	birth: Apr-mid June; Nunivak I, AK	6381
	birth: Apr-July, peak May; n23; zoo	5818
	birth: mid Apr-mid June; NWT, Canada	10738, 10740
	birth: end Apr-mid May, peak early May; Jameson Land, ne Greenland	10770
	birth: May-early June; Arctic	8798
Ovis ammon	mate: Aug; Mongolia	8739, 8740
	mate: Oct; Nepal	7382
	mate: Oct-Nov; according to natives; Tibet	3034
	mate: Oct-Dec, rare Aug-Jan; Germany	3571
	mate: mid Oct-mid Jan; USSR	4643
	mate: Nov?; Chinese Turkestan, Pakistan	9175
	mate: Dec-Jan; Tibet	6652, 6654
	birth: end Feb-June, peak end Mar-early May; Germany	3571
	birth: Mar, June; Mongolia	8740
	birth: Apr-June; USSR	4643
	birth: May-June; Tibet	3034
	birth: May-June; Tibet	6652, 6654
	birth: May-June; Nepal	7382
Ovis aries	estrus: yr rd, peak June-Nov; captive, Niger	12010
	estrus: mid Oct-Nov; Outer Hebrides, UK	4109, 4111
	estrus: July-Jan; captive, Niger	12011
	mate: Oct-mid Dec; Scotland, UK	1181
	ovulation: yr rd, peak June-Nov; captive, Niger	12010
	birth: Mar-Apr; Scotland, UK	1181, 4111, 5307
	birth: Apr-June; India	9547
	birth: Apr-May; Corsica, France, Sardinia, Italy	6652
	birth: mid Apr-early May; Pakistan	9175

	birth: Nov-June; captive, Niger	12011
Ovis canadensis	estrus: mid Nov-mid Dec; n21; Rocky Mt, N America	3667
	mate: end June-Nov, peak Sept-Oct, some Jan-Mar; CA	11578
	mate: July-Sept; AZ	10950
	mate: Nov-Dec; ID	10950
	mate: July-mid Oct; NV	6392
	mate: end July-early Dec; NV	144
	mate: peak Aug; AZ	9384
	mate: Oct-Nov; Rocky Mt, N America	6652
	mate: mid Oct-early Mar; WY	7346
	mate: Nov-Dec; Rocky Mt, N America	1474
	mate: Nov-Dec; ALB, Canada	3667
	mate: mid-end Nov; BC, Canada	1011
	mate: mid Nov-mid Dec; USSR	4643
	mate: mid Nov-mid Dec; CO	10257
	mate: mid Nov-Dec; ID	10105
	mate: mid Nov-Dec; w US	5146
	mate: end Nov-mid Jan, peak Dec; WY	7016
	birth: Jan; CA	11578
	birth: Jan-Mar; AZ	10950
	birth: May-June; ID	10950
	birth: Jan-Mar, peak Feb-Mar; AZ	9384
	birth: Jan-mid Apr; NV	6392
	birth: Feb-July; NV	144
	birth: Mar-Apr; TX	2351a
	birth: end Apr-mid June, peak end May; BC, Canada	1011
	birth: May; Rocky Mt, N America	6652
	birth: May-June; Rocky Mt, N America	1474
	birth: May-July, Oct; n15; zoo	5818
	birth: mid May-?, peak June 1; ID	10105
	birth: mid May-mid June; CO	10257
	birth: peak end May; ALB, Canada	3243
	birth: end May-mid June; USSR	4643
	birth: end May-mid July, peak mid June; WY	4950
	birth: end June-early July; MT	10411
	birth: yr rd; NV	4333
Ovis dalli	estrus: end Nov-mid Dec; n16; BC, Canada	3667
	mate: Oct-Nov; AK	7625
	mate: Nov-Dec; AK	7883
	mate: mid Nov-early Dec; n N America	7884
	birth: May; YUK, Canada	1448
	birth: May; BC, Canada	3667
	birth: May; AK	7883
	birth: May; AK	8844
	birth: May-June, peak May; AK	7625
	birth: end May-early June; n N America	7884
Ovis musimon	spermatogenesis: peak Oct-Dec; Czechoslovakia	7579
	estrus: mid Mar-early May; Czechoslovakia	10945
	mate: mid Oct-Dec; Corsica, France	8467
	mate: Oct-Dec; Germany	10946
	mate: Nov-Jan; zoo	207
	mate: Nov-Dec; Corsica, France	9318
	birth: Jan-Oct, peak Mar-May; n115; zoo	12128
	birth: Feb-Dec, peak Mar-Apr; n>100; zoo	5818
	birth: mid Mar-early May; Czechoslovakia	10945
	birth: Mar-May; Germany	10946
	birth: Apr, Oct; n2; E Germany	1048
	birth: mid-end Apr; Corsica, France	8467
	birth: mid Apr-early May; zoo	207
	birth: spring-early summer; zoo	1230
Ovis nivicola	mate: Dec; Koryak Mt, Siberia, Russia	10722
	birth: mid-end June; Koryak Mt, Siberia, Russia	10722

Ovis vignei	mate: Sept; India	6652, 6654
	birth: Mar-Apr; India	6654
	birth: Mar-July, peak May-June; n14; zoo	12128
	birth: June; Astor	6652
Pantholops hodgsoni	mate: Nov-Dec; Russia	9547
	mate: mid Nov-Dec; Tibet	3034
	mate: end Nov; Tibet	8972
	mate: end Nov-Dec; Tibet	8740
	mate: winter; according to natives; Tibet	6653, 6654
	birth: end June-early July; Tibet	8972
	birth: July; Tibet	8740
	birth: summer; according to natives; Tibet	6653, 6654
Pelea capreolus	mate: Apr; S Africa	10158
	birth: Nov-Dec; e Transvaal highveld, S Africa	3095
	birth: mid Nov-Dec; S Africa	5783
	birth: yr rd, peak Sept-Dec; zoo	1230
Procapra gutturosa	mate: end Nov-early Jan; USSR	4643
	mate: end fall; Mongolia	182
	birth: May; Mongolia	8739
Procapra picticaudata	mate: Oct-Jan, peak Dec-Jan; Tibet	3034
	mate: end Dec-Jan; Tibet	8740
	birth: June-Aug; Tibet	3034
Pseudois nayaur	estrus: Nov; Tibet	9541
	estrus: Nov-Jan; Nepal	11449
	mate: mid Dec-end Jan, peak early Jan; Nepal	11778
	birth: mid May-mid June; Tibet	9541
	birth: end May-June; Nepal	11778
	birth: end May-early Sept; n43; zoo	12128
	birth; June; zoo	9175
	birth: June-Aug; Pakistan	9175
Raphicerus campestris	mate: spring; S Africa	5782
	mate: yr rd; s Africa	5763
	preg f: yr rd; n32; Botswana	10154
	birth: Apr-May, Sept-Dec; Namibia	9945
	birth: Apr-June, Oct, Jan; Kruger Nat Park, S Africa	8514
	birth: May, Sept; Angola	11697
	birth: Dec; S Africa	5782
	birth: yr rd; S Africa	10158
Raphicerus melanotis	conceive: peak Mar-Apr; S Africa	7977
	conceive: peaks Apr-May, Sept-Nov; S Africa	269
	birth: peak Oct-Dec; n18; S Africa	7977
Raphicerus sharpei	testes size; no change; Zimbabwe	5699
	birth: peak early summer; Nyassaland to ne Transvaal, S Africa	10392
	repro: est yr rd; S Africa	10158
Redunca arundinum	mate: yr rd, peak May-June; S Africa	5469
	preg f: Mar; n1; Tanzania	186
	birth: May, Dec; Angola	11697
	birth: Dec-Mar; S Africa	5779
	birth: yr rd; zoo	1230
	birth: yr rd; Zambia	331
	birth: yr rd, peak Dec, May; Zululand, se Africa	5763
	birth: yr rd, peak Dec-May; S Africa	5468, 5469
Redunca fulvorufula	mate: Mar-May; Transvaal, S Africa	5164
	mate: peak Apr-May; S Africa	10025
	preg f: May; n1; Kenya	4926
	birth: Sept-Feb; Kruger Nat Park, S Africa	8514
	birth: peak Oct-Jan; S Africa	10158
	birth: Dec-Feb, wet season, some Nov; Transvaal, S Africa	5164
	birth: peak summer; zoo	1230
	birth: yr rd, peak 78% Oct-Dec; S Africa	10025
Redunca redunca	mate: end June-early July; e Africa	5203

Rupicapra rupicapra	mate: Oct-early Dec; French Alps	8477
	mate: Oct-Dec; Europe	2079, 2080
	mate: end Oct-early Dec; USSR	4643
	mate: end Oct-mid Dec; Germany	3571
	mate: mid-end Nov; Austria	10888
	mate: Dec-Feb; Austria	7177
	conceive: Nov 12%, Dec 70%; Alps, Europe	11056
	ovulation: mid Nov; Switzerland	3663
	birth: end Apr-mid May; USSR	4643
	birth: end Apr-June; Germany	3571
	birth: May; French Alps	8477
	birth: mid May-mid June; Europe	2079, 2080
	birth: est June-July; Austria	7177
Saiga tatarica	mate: Dec-Jan; USSR	560, 561
	mate: Dec-Jan; Europe, w USSR	4645
	mate: end Nov-Dec; Europe	560
	mate: last 1/3 Dec; zoo	8641
	birth: Apr; zoo	12107
	birth: Apr; Europe	560
	birth: Apr-May, peak early May; USSR	2215
	birth: Apr-May; zoo	2668
	birth: end Apr-May, peak early May; USSR	560, 561
	birth: May; Russia	3162
	birth: May; USSR	4643
	birth: May-early June; Germany	8641
	birth: May-June; Europe, w USSR	4645
Sigmoceros lichtensteini	mate: mid Dec; Zambia	2710
	birth: May-Sept, peak July: nw plateau, Zambia	331, 333
	birth: June-Oct, peak July-Aug; Zambia	7369
	birth: peak July-Sept; Zaire	11253
	birth: Sept; e Africa	5052
	birth: Sept; S Africa	10158
	birth: Oct-Nov; S Africa	9030
	birth: Oct-Nov: Luangwa Valley, Zambia	331, 333
Sylvicapra grimmia	preg f: yr rd; n39; Botswana	10154
	birth: end May-July; Angola	11697
	birth: yr rd; Zimbabwe	11789
	birth: yr rd; Zimbabwe	331, 333
	birth: yr rd; S Africa	10158
	birth: yr rd, peaks Jan-Feb, June-Aug; Zimbabwe	9135
	birth: yr rd, peak summer; zoo	1230
	lact f: Jan; n1; Tanzania	186
	repro: yr rd, peak dry season; Zimbabwe	2275
Syncerus caffer	estrus: Jan?; Angola	11697
	mate: Mar-May; Kruger Nat Park, S Africa	3145
	mate: Mar-May; S Africa	8515
	mate: peak June-July, low Aug-Sept, some Oct-May; Tanzania	9994
	conceive: peak Apr-July; Tanzania	9994
	birth: Jan-Mar; S Africa	9748
	birth: Jan-Apr, peak Jan-Feb; S Africa	8515
	birth: Jan-Apr, peak Jan-Feb; Kruger Nat Park, S Africa	3145
	birth: perhaps Jan-Apr; Zambia	2488
	birth: peak Mar-May; Tanzania	9994
	birth: Apr, Oct, Dec; Zaire	11271
	birth: Apr-June; Zaire	11253
	birth: Oct-Jan; Angola	11697
	birth: Nov-Apr, peak Jan-Feb; Botswana	1599
	birth: Nov-July, peak May-Apr, low Aug-Oct; e Africa	9991
	birth: Dec-Oct, peak July; Tanzania	9994
	birth: Dec-Feb; British E Africa	5205
	birth: yr rd; Zambia	331
	birth: yr rd; Kruger Nat Park, S Africa	8514
	birth: yr rd; n8; zoo	1230
	birth: yr rd, peak Apr-June; Zaire	11253

	birth: yr rd, peaks May, Dec; Uganda	4077
	repro: yr rd; Zimbabwe	2275
Tetracerus quadricornis	mate: July-Sept; Nepal	7382
	mate: during rainy season; India	37
	birth: Jan-Feb; Nepal	7382
	birth: Jan-Feb; India	6654
Tragelaphus angasi	spermatogenesis: yr rd; ne Natal S Africa	269
	conceive: yr rd, peaks Feb-May, Oct; ne Natal, S Africa	269, 270
	mate: peak Apr-May, Oct; ne Natal, S Africa	269
	mate: July, Oct; but not sampled all yr; Mozambique, Malagasy	6525
	preg f; June-Nov; se Zimbabwe	2357
	birth: Feb, Apr, Sept, Oct; few data; Mozambique, Malagasy	6525
	birth: Apr-May, July-Nov peak Aug-Oct; Kruger Nat Park, S Africa	8514
	birth: July-Aug; Zululand, S Africa	7857
	birth: Aug-Nov; few data; Kruger Nat Park, S Africa	3145
	birth: Oct-Jan; ne Natal, S Africa	269
	birth: Nov; se Zimbabwe	2357
	birth: yr rd, bimodal; S Africa	10158
	birth: yr rd; zoo	1230
Tragelaphus eurycerus	mate: peak Apr, Nov; e Africa	5763
	birth: peak July-Sept; e Africa	5763
Tragelaphus imberbis	mate: probably yr rd; e Africa	5763, 6413
	birth: yr rd; n40; Kenya	6411
Tragelaphus oryx	testes size: peak Mar; S Africa	10041
	estrus: May-July; 3 wk; zoo	4535
	estrus: Dec-Jan; Angola	11697
	estrus: yr rd, peak May-July; captive	10367
	mate: May-June; zoo	6058
	mate: Dec-Jan; Angola	11697
	mate: yr rd, peak probably Feb; S Africa	10023, 10041
	mate: yr rd, peak May; n16; zoo	8339
	birth: Feb, May, Sept-Dec; few data; captive	10300
	birth: Feb-Mar; captive	6058
	birth: Feb-Mar; zoo	6058
	birth: Mar; e Africa	5052
	birth: peak Apr-Sept; Zambia	2488
	birth: May-Nov, peak July-Sept; e Africa	5763
	birth: June-July; Zaire	11253
	birth: June-Aug, peak Aug; se Africa	9753
	birth: July-Aug; Zimbabwe	11784
	birth: Aug-Sept calculated; Angola	11697
	birth: Aug-Sept; Kruger Nat Park, S Africa	3145
	birth: Aug-Oct; captive	8690
	birth: Aug-Nov; Matopos Nat Park, Zimbabwe	11785
	birth: Dec-Aug, 71% Feb-Apr; zoo	10895
	birth: yr rd; zoo	1230
	birth: yr rd; n103; zoo	12128
	birth: yr rd; n408; zoo	10894
	birth: yr rd; e Africa	4801
	birth: yr rd, peak Aug-Sept; Kruger Nat Park, S Africa	8514
	birth: yr rd, peak Aug-Oct; Zimbabwe	9311
	birth: yr rd, 30% Nov; S Africa	10023, 10041
Tragelaphus scriptus	conceive: Apr-May; Zambia	7537
	birth: Feb-Mar; e Africa	5204
	birth: bimodal Apr-May, Oct-Nov; S Africa	10158
	birth: June; Angola	11697
	birth: July-Nov, peak Oct; few data; Kruger Nat Park, S Africa	3145
	birth: Oct-Nov; Zambia	7537
	birth: mid Oct-mid Dec, sometimes Feb; Cape Province, S Africa	5773
	birth: peak rainy season; Zaire	11253
	birth: prob yr rd; n24; Kenya	201

	birth: yr rd; zoo	1230
	birth: yr rd; n7; captive	11347
	birth: yr rd; n20; zoo	2603
Tragelaphus spekei	mate: Mar; n1; zoo, Netherlands	5688
	mate: yr rd, peak May; n21; zoo	8339
	birth: peak Apr; Uganda	5763
	birth: Nov; n1; zoo, Netherlands	5688
	birth: probably yr rd; Kenya	8165
	birth: probably yr rd; Zambia	331
	birth: yr rd; zoo	6192
	birth: yr rd; n61; zoo	5818
	birth: yr rd; S Africa	10158
	repro: yr rd; captive	2485
Tragelaphus strepsiceros	testes size: peak June, low Nov; Transvaal, S Africa	10031
	mate: May-July calculated; Angola	11697
	mate: June; Transvaal, S Africa	10023, 10027, 10031
	mate: June-July; S Africa	11043
	mate: July; Zimbabwe	9986
	mate: peak Sept; Tanzania	5763
	mate: yr rd, peak May; n21; zoo	8339
	preg f: July-Feb; few data; Zambia	11781
	birth: Jan-Feb; Kruger Nat Park, S Africa	8514
	birth: peak Jan-Apr; Zambia	11781
	birth: Jan-Apr, peak Feb; Zimbabwe	11786
	birth: mid Jan-Feb; ne S Africa	3145
	birth: Feb-Mar; Zimbabwe	9986
	birth: Feb-Mar; Zimbabwe	2275
	birth: peak Mar; S Africa	11043
	birth: July-Jan; n21; zoo	5818
	birth: Oct; Angola	11697
	birth: during rains; Tanzania	5763
	birth: end summer; S Africa	10023, 10027
	birth: yr rd; Zambia	331
	birth: yr rd, peak Jan-Mar; zoo	1230

Order Pholidota, Family Manidae, Genus *Manis*

Throughout Africa and across southern Asia, the semifossorial, solitary, armoured pangolins (*Manis* 2-33 kg) dig for ants and termites. Males may be larger than females (5226a, 8501, 9175). The 13 g (n1), non-deciduate, epitheliochorial or endotheliochorial, villous, diffuse placenta with interanastomosing placental strips has a superficial, antimesometrial implantation into a bicornuate uterus with endometrial glands (2391, 2396, 2397, 4552, 6936, 7563, 10310, 10958, 11185). Placentophagia has not been observed (8202). A postpartum estrus occurs 9-16 d (8200, 8202) after the birth of a well-developed, scaled neonate whose eyes open either at birth or within 9 d (31, 5758, 6936, 7206, 7421, 8202, 9175, 11162). The 2 pectoral mammae are well developed in both sexes (9175, 10158). Males are ascrotal with inguinal testes and lack bulbourethral glands (281a). Details of endocrinology (4664) are available.

NEONATAL MASS		
Manis crassicaudata	235 g	31
	242 g	7421
	400 g	8047
	450 g	9175
Manis gigantea	500 g; n2	5758
Manis pentadactyla	92.5 g; r92-93; n2	4556
Manis temminckii	120 g; n1 est premature	5226a
	256 g; n1	6180
	336-420 g	11162
	340 g	5758
Manis tricuspis	r90-150 g; n10	8200, 8202
	100 g; n1	7206
NEONATAL SIZE		
Manis crassicaudata	310 mm; 3 d	9175
Manis gigantea	450 mm; n2	5758
Manis temminckii	150 mm	5758
	TL: 225 mm; n1 est premature	5226a
	290 mm	6180
Manis tricuspis	r300-350 mm; n10	8200, 8202
WEANING MASS		
Manis gigantea	6 kg	8202
Manis tetradactyla	400 g	8202
Manis tricuspis	600 g	8202
	750 g; n1	7206
LITTER SIZE		
Manis crassicaudata	1 yg; n1 lt	8720
	mode 1; r1-2	8497, 8501
Manis javanica	1	11185
	1 yg	556, 557
Manis pentadactyla	1 yg ea; n2 f	4556
Manis temminckii	1	10158
	mode 1 yg; r1-2	5226a
Manis tetradactyla	1.05; r1-2; n20 lt	8199
Manis tricuspis	1	8202
	1	5277
	1 emb; n1 f	2953
	1 emb; n1 f	8854
	1; n1	7206
SEXUAL MATURITY		
Manis tricuspis	2 yr	8200
CYCLE LENGTH		
Manis tricuspis	9-14 d	8200
	9.3 d; mode 4-8 62%; r3-29; n36 cycles, 4 f	10621
GESTATION		
Manis crassicaudata	65-70 d	9175
Manis temminckii	139 d	11162
	139 d	8202
Manis tricuspis	est ~4 mo	5758
	>5 mo; n1; capture to birth	8200, 8202

LACTATION		
Manis pentadactyla	milk production: 86 d	6936
Manis temminckii	solid food: 1 mo	5758
Manis tetradactyla	nurse: until next gestation	8199
Manis tricuspis	solid food: 3 mo	8202
	wean: 3-4 mo	8202
	wean: 7.5 mo; n1	7206
INTERLITTER INTERVAL		
Manis tricuspis	0.5 yr	8202
SEASONALITY		
Manis crassicaudata	preg f: July; n1; Sri Lanka	8501
Manis gigantea	birth: end Sept-Oct; Uganda	5758
Manis javanica	birth: Feb-Mar, Aug, Oct; n5; Borneo, Indonesia	556
Manis temminckii	preg f: July; S Africa	10158
Manis tetradactyla	lact f: Jan; n1; Nigeria	4354a
	repro: est yr rd; few data; Gabon	8199
Manis tricuspis	testes: large yr rd; Gabon	8202
	mate: est Dec-Feb; British Camerouns	9468
	preg f: Mar; n1; Lake Kivu, Zaire	8854
	preg f: yr rd, not equally distributed; Gabon	8200
	birth: peak est Nov-Feb; few data; Uganda	5758
	repro: est yr rd; few data; Gabon	8199

Order Rodentia

As rodents are generally small bodied and often herbivorous, they are of tremendous ecological importance in nearly all terrestrial habitats. They also comprise about 40% of all mammalian species. Unfortunately, except for domesticated lab species, knowledge of their reproductive biology is not in keeping with either their importance or their diversity.

Details of anatomy (11372), sperm (4826a), developmental time (1449a), reproductive cyclicity (9350), copulation (2513, 6207), and copulatory plugs (6235) are available.

Order Rodentia, Family Aplodontidae, Genus *Aplodontia*

The 800-1800 g mountain beavers (*Aplodontia rufa*) are semi-fossorial herbivores of temperate forests in western North America. Mountain beavers exhibit little sexual dimorphism (9578, 10720). The hemochorial, labyrinthine, discoidal placenta implants in a duplex uterus (8470, 8474). Naked neonates develop eye slits at 10 d but may not open their eyes completely until 45-54 d (3200, 4817, 6609), and are suckled from 6 teats (8470, 8474). Reproductive females have dense patches of black hair around their nipples which can be induced by estrogen (8471). Males have semi-scrotal testes, a forked baculum, and seminal vesicles (1473, 6609, 8470, 8473). Details of birth (2107) are available.

NEONATAL MASS		
Aplodontia rufa	19.8 g; r18.0-21.6; n2, 2 d	6609
	27 g; r25.0-29.5; n3, 1 d	2107
NEONATAL SIZE		
Aplodontia rufa	66 mm; n2	6609
	87.3 mm; r85-90; n3, 1 d	2107
WEANING MASS		
Aplodontia rufa	347 g; n2, 8 wk	6608
	m: 357.44 g; se19.69; r280-440; n9; 1st trapped	6608
	f: 473.13 g; se35.62; r330-590; n8; 1st trapped	6608
LITTER SIZE		
Aplodontia rufa	2-4 yg	5048
	2.18 emb, yg; r2-3; n16 f	9578
	2.4 emb; n12 f	8470, 8474
	2.5 ps; n25 f	8474
	2.8-3.1 cl	8470, 8474
	3 yg; n1 lt	2107
	3 yg; n1 nest	4817
	3 yg; r2-4; n2 f	6608, 6609
	3-4 yg	5146
SEXUAL MATURITY		
Aplodontia rufa	f: 1st ovulate: yearlings, but none preg	8470
	f: 1st mate: 2 yr	1474
	f: 1st birth: 2 yr	3200
	m: hypertrophy of genital tract: some yearlings	8470
	m: 1st breed: 2 yr	5048
GESTATION		
Aplodontia rufa	est 28-30 d	8474
	~30 d	9578
	~1 mo; field data	8470
LACTATION		
Aplodontia rufa	solid food: 7 wk	6609
	wean: 6-8 wk	6608
	den emergence: ~10 wk	3200
LITTERS/YEAR		
Aplodontia rufa	1	1474
	1	3200

SEASONALITY

Aplodontia rufa	large testes: Jan-Mar; WA	8473
	large testes: Jan-Mar; WA	5048
	mate: Feb-Mar; WA	9578
	estrus: mid-Feb; w N America	3200
	ovulate: Jan-Mar; CA	8470
	birth: Feb-Apr; w US	8474
	birth: end Mar-early Apr; few data; captive, OR	6609
	birth: Apr; WA, BC	2088
	birth: July; WA	8470
	lact f: Apr-May; OR	6608

Order Rodentia, Family Sciuridae

Squirrels (*Aeretes* ~305 mm; *Aeromys* 1128-1250 g; *Ammospermophilus* 80-186 g; *Atlantoxerus* 300-350 g; *Belomys* 180-260 mm; *Callosciurus* 45-500 g; *Cynomys* 465-1700 g; *Dremomys* 140-270 g; *Epixerus* 388-604 g; *Eupetaurus* 515-610 mm; *Exilisciurus* 15 g; *Funambulus* 70-168 g; *Funisciurus* 200-350 g; *Glaucomys* 32-185 g; *Glyphotes* ~128 mm; *Heliosciurus* 200-480 g; *Hylopetes* 30-570 g; *Hyosciurus* 200-240 mm; *Iomys* 119-191 g; *Lariscus* 120-230 g; *Marmota* 1.6-13 kg; *Menetes* 184 g; *Microsciurus* 120-160 mm; *Myosciurus* 16 g; *Nannosciurus* 75-115 mm; *Paraxerus* 37-700 g; *Petaurillus* 25 g; *Petaurista* 1-2.5 kg; *Petinomys* 22-110 g; *Prosciurillus* 100-135 mm; *Protoxerus* 388-1000 g; *Pteromys* 140 g; *Pteromyscus* 134-315 g; *Ratufa* 0.9-3 kg; *Rheithrosciurus* 1-2 kg; *Rhinosciurus* 187-255 g; *Rubrisciurus* 150-500 g; *Sciurillus* 90-110 mm; *Sciurotamias* 195-250 mm; *Sciurus* 60-1080 g; *Spermophilopsis* 520-620 g; *Spermophilus* 85-1260 g; *Sundasciurus* 55-420 g; *Syntheosciurus* 150-170 mm; *Tamias* 25-142 g; *Tamiasciurus* 127-312 g; *Tamiops* 27-85 g; *Trogopterus* 270-305 mm; *Xerus* 260-1022 g) are primarily diurnal, arboreal to terrestrial herbivores from the New World, Eurasia, and Africa.

Males are reported larger in *Cynomys* (5916, 8570, 8572), *Funambulus* (8500; cf 8726), and *Spermophilus* (5658, 9890, 10220, 10862; cf 1552, 7287), while females are larger in *Funisciurus* (9945) and *Tamias* (2805, 5658, 10315). Each sex has been reported larger for *Ratufa* (2313, 8501) and *Sciurus* (1361, 4230, 8188).

Ovulation is induced by coitus in *Spermophilus* (2729, 3432, 4850, 10862). *Cynomys*, *Spermophilus*, and *Sciurus* have copulatory plugs (589, 7286, 7633, 10437). The hemochorial, labyrinthine, discoid, deciduate placenta has a superficial, eccentric or mesometrial implantation into a bicornuate or duplex uterus (2015, 3011, 4561, 6353, 6354, 7152, 7563, 7567, 7575, 7944, 8050, 8053, 9707, 10187, 10310, 10532, 11242; cf 231). A postpartum estrus is recorded for *Funisciurus* (2997), *Paraxerus* (11298), and *Tamias* (12014).

Sciurids are usually blind and naked at birth (40, 139, 502, 554, 1474, 1552, 2060, 4083, 4274, 5803, 6504, 6798, 7161, 7235, 8050, 8053, 8573, 8798, 9175, 10445, 10584; cf 2901). Eyes open at 1-3 wk (*Funambulus* 105, 8500, 8501, 9175), 1-5 wk (*Spermophilus* 978, 1286, 1864, 2896, 3238, 3281, 3595, 5020, 5210, 5663, 6992, 7080, 7082, 7550, 7788, 8050, 8348, 9873, 10235, 10862, 10948, 11224, 11322, 11379; cf 9872; *Xerus* 10158, 12134), 20-28 d (*Marmota* 3571, 4270, 8770, 8771, 8798), 22 d (*Callosciurus* 11975; *Ratufa* 33), 20-30 d (*Trogopterus* 11448), 25-33 d (*Tamias* 175, 1295, 3390, 4825, 9416, 11385), 26-36 d (*Tamiasciurus* 3238, 6297, 8745, 10585), 29-36 d (*Ammospermophilus* 7788, 8348), 30 d (*Petaurista* 9175), 4-6 wk (*Glaucomys* 1117, 1672, 3238, 3239, 6497a, 7685, 10206, 10211; *Sciurus* 589, 1361, 2923, 3571, 3749, 5002, 7433, 8053, 8798, 9214, 9943, 10293, 10809, 11024, 12139; cf 1361), or 33-42 d (*Cynomys* 339, 1474, 5342, 8798, 10139). *Paraxerus* is finely furred at birth and opens its eyes within 1-2 wk (2997, 2998, 10158, 11297, 11302). *Funisciurus* also has fine hair at birth, but opens its eyes at 16-20 d (184, 5760, 11302).

Females have 2-4 mammae (*Myosciurus* 2997, 5393), 4 mammae (*Exilisciurus* 7502; *Funambulus* 7502, 7383, 8500, 105; *Nannosciurus* 7502; *Rhinosciurus* 7502; *Rubrisciurus* 7502), 4-6 mammae (*Callosciurus* 182, 7161, 7383, 7502, 10826, with lactation hair 6172; *Funisciurus* 2955, 2997, 5760, 7502, 9945, 10158; *Xerus* 5760, 7502, 9945, 10158), 4-10 mammae (*Tamiasciurus* 1474, 4496, 7502, 8798), 6 mammae (*Belomys* 7383; *Dremomys* 7502; *Heliosciurus* 331, 2955, 2997, 7502; *Lariscus* 7502; *Menetes* 7502; *Microsciurus* 7502; *Paraxerus* 2955, 2997, 7502, 9945, 10158; *Petaurista* 9175; *Prosciurillus* 7502; *Sundasciurus* 2313, 7502; *Tamiops* 7161, 7502), 6-8 mammae (*Atlantoxerus* 7502; *Sciurillus* 7502; *Sciurotamias* 182, 7502; *Syntheosciurus* 7502), 6-10 mammae (*Ratufa* 7161, 7502, 8798; *Sciurus* 1361, 1474, 3022, 3024, 3571, 4494, 4745, 7433, 7502, 8798, 10516), 6-14 (*Spermophilus* 145, 182, 1474, 3571, 4081, 6489, 7433, 7502, 8050, 8798), 8 mammae

(*Epixerus* 2997, 7502; *Glaucomys* 1474; *Protoxerus* 2955, 2997, 7502; *Rheithrosciurus* 7502; *Spermophilopsis* 7502, 8053; *Tamias* 1295, 1474, 4745, 7502, 8053, 9612, 11385), 8-10 mammae (*Cynomys* 1474, 4745, 6561, 7502, 8572), 8-12 mammae (*Marmota* 182, 1474, 3571, 7433, 7502, 8798; cf 8771), 10-12 mammae (*Ammospermophilus* 1474, 4081, 7502). Males have a baculum and seminal vesicles (164, 182, 338, 1473, 2435, 4105, 4782, 7570, 7572, 8053, 8734, 8735, 9964, 10862, 11296; cf 3402). *Glaucomys* sperm form rouleaux in the epididymis (6868).

Details of anatomy (*Cynomys* 3403, 10438, 10439; *Funambulus* 11242; *Marmota* 1828a, 8934; *Sciurus* 2426, 4510, 8777, 8779; *Spermophilus* 3432), endocrinology (*Cynomys* 3401, 3404, 3405; *Funambulus* 5516, 8995, 9798, 10013; *Marmota* 533, 2015, 2016, 2017, 2469, 3337, 8934; *Sciurus* 8779, 10015, 10630, 11513, 11514, 11515, 11516; *Spermophilus* 598, 599, 600, 601, 602, 603, 1480, 1993, 2729, 2983, 2984, 3256, 3502, 4917, 4918, 4919, 5344, 5346, 6429, 7079, 10806, 11580, 11582, 12054, 12055, 12115, 12116), spermatogenesis (6869, 8997, 10629, 11581, 11582, 12056), reproductive energetics (4373a, 4509, 5662, 5663, 7182, 10617), and light (*Ammospermophilus* 5654, 5656, 5659; *Cynomys* 3400; *Funambulus* 4215; *Glaucomys* 7686; *Marmota* 3337; *Spermophilus* 2983, 5655, 6429, 7493, 8349, 10307, 11579, 11583) are available.

NEONATAL MASS

Ammospermophilus harrisii	3.6 g; r3.0-4.1; n12 yg, 2 lt	7788
Ammospermophilus leucurus	2.92 g; sd1.02; r2.0-3.6; n17	6988
	3-4 g; n4 lt	8348
Ammospermophilus nelsoni	4.88 g; n9 yg, 1 lt	4512a
Callosciurus prevosti	16.35 g; r15.9-16.8; n2	11975
Cynomys ludovicianus	15.5 g; n1	5342
	16 g	10139
Funisciurus congicus	9.91 g; ±0.64; n4	11302
Glaucomys sabrinus	5.01 g; r4.8-5.4; n14, 1 d	3239
	5.8 g; n4	7685
	6.0 g; n1	1117
Glaucomys volans	3.0 g; n3, 1 lt	1672
	3.4 g; n1	10584
	3.63 g; r3-5.3; n10	10206
	4.2 g; se0.12; n17	6497a
Marmota camtschatica	33 g	5518
Marmota flaviventris	33.8 g; n3	3481
Marmota marmota	29-30 g	3571
	30 g	8798
Marmota monax	23.7 g; n22	3240
	26.5 g; r22-39; n20	4270
	27.2 g; n11	4083
	32.3 g; r30.2-34.0; n3	10188
Paraxerus cepapi	~10.0 g	10158
	12.75 g; ±2.14; n6	11302
	13 g	11297
Paraxerus palliatus	12.75 g; ±1.06; n2	11302
	17.67 g; ±4.05; n3	11302
Paraxerus poensis	9.91 g; r9.52-10.3; n2	2997, 2998
Petaurista leucogenys	5 g; n1	295
Ratufa bicolor	77 g; n1	33
Sciurus aberti	12 g; n7	5618a
Sciurus carolinensis	13-17 g	9943
	f: 13.9 g; r13.3-14.5; n2	9943
	14.5-16.0 g	10158
	15-18 g; 1 d	11024
Sciurus granatensis	9.5 g; r9-10; n2, 1 lt	7940
Sciurus niger	14-17 g	170
	14.65 g; r14.6-14.7; n2	7943a
Sciurus vulgaris	7-8.5 g	8053
	8-12 g	3571
	8-12 g	8798
Spermophilus beecheyi	8.9 g; n42 yg, 6 lt; r7.2-11.7, r by lt	3281
	9.3 g; r7.7-12.6; n34	10862
	11.8 g; n5	2896

Spermophilus beldingi	6.87 g; sd0.62; r5.24-7.82; n33	7550
Spermophilus columbianus	8.6 g; n23 yg, 5 lt	9873
	~9 g	6798
	10.1 g; n46 yg, 7 lt	5909
	11.3 g; n22	11959
Spermophilus elegans	5.9 g; sd0.45; r5.5-6.6; n6 yg, 2 d	1864
	5.96 g; r5-7; n6 lt	2484
	6.2 g	5909
Spermophilus lateralis	3.6 g; n4	7080
	6-8 g; n14 lt	8348
	6.1 g; 1 d	1871
	6.6 g; n3	7080
Spermophilus mexicanus	4.31 g; r3.85-5.05; n4	2901
Spermophilus mohavensis	4-5 g; n4 lt	8348
Spermophilus richardsoni	6.8 g; r4.4-9.6; n147	7289
Spermophilus saturatus	5.97 g; sd0.39; r5.22-6.55; n18 lt	5658, 5663
Spermophilus tereticaudus	3-5 g; n10 lt	8348
	3.7 g; r2.7-4.7; n26 yg, 4 lt	7788
Spermophilus townsendii	3.26 g; n19 yg, 2 lt	9873
	3.7 g; se0.7; r2.2-4.9; n33	9116
	3.87 g; sd0.62; r2.72-5.15; n21	9115
Spermophilus tridecemlineatus	2.6 g; n10	11379
	3 g; n2	11379
	3.2 g; min 2; n67 yg, 9 lt	5343
Spermophilus variegatus	7.8 g; n4	8019
Spermophilus undulatus	9.7 g; r8.1-11.5; n8	6992
	r9.7-13.2 g; n6 lt	6992
	10.8 g; r9-12.1; n9	6992
	12 g; r11.1-12.8; n5	6992
Tamias amoenus	2.65 g; r2.6-2.8; n10	1295
Tamias minimus	~2.3 g	5210
Tamias quadrivittatus	2-3 g; from graph?	11385
Tamias striatus	2.5-5 g	10182
	3.0 g; n7	175
Tamias townsendii	r3.2-3.9 g; n8	3391
Tamiasciurus douglasii	est 5 g	5922
Tamiasciurus hudsonicus	6 g; n1	10585
	6.7 g; n1	6297
	6.8 g	8745
	7.5 g	4274
	7.5 g; r7.0-8.3; n4	10585
Xerus inauris	20 g	10158
NEONATAL SIZE		
Ammospermophilus nelsoni	58 mm; n2	4512a
Cynomys gunnisoni	CR: 38 mm; near term emb	8572
Cynomys ludovicianus	65-70 mm	10139
Glaucomys sabrinus	HB: 48.7 mm; r47-50; n14, 1 d	3239
	TL: ~71 mm	5210
Glaucomys volans	10 mm; n3 yg, 1 lt	1672
Marmota camtschatica	107 mm	5518
Marmota flaviventris	111.0 mm; n3	3481
Marmota monax	85 mm; n22; from graph	3240
	89 mm; n8	4270
Sciurus aberti	TL: 60 mm	5618a
Sciurus carolinensis	78-80 mm; 1 d	11024
	109-119 mm	589
Sciurus niger	50-60 mm	1871a
Sciurus vulgaris	51-56 mm	8053
Spermophilus beldingi	5.07 cm; sd0.15; r4.91-5.28	7550
Spermophilus citellus	25-30 mm	1552
Spermophilus elegans	CR: 46 mm; n6 lt	2484
Spermophilus fulvus	125-155 mm	8050
Spermophilus lateralis	~61 mm	1871a
Spermophilus mexicanus	TL: 66.5 mm; r63-69; n4	2901

Spermophilus richardsoni	HB: 55.7 mm	7289
Spermophilus townsendii	52.6 mm; se0.4; r42-56; n33	9116
Spermophilus undulatus	55 mm	8050
Tamias amoenus	35-38 mm	1295
Tamias striatus	TL: 66 mm; n1	10182
Tamias townsendii	55-60 mm; n8	3391
WEANING MASS		
Ammospermophilus leucurus	35 g; 6 wk	5661
Ammospermophilus nelsoni	41 g; r40-42; n2, ~8 d after emergence	4512a
Cynomys ludovicianus	f: 145.8 g; se1.6; n287, emergence	3371
	m: 150.9 g; se1.5; n328, emergence	3371
Glaucomys sabrinus	39.74 g; r25.2-53.7; n14, 55 d	3239
Glaucomys volans	46.2 g; se1.11; n26, 8 wk	6497a
Marmota flaviventris	f: 419 g; n13, 1st capture	259
	446 g; 33 d	259
	~454 g; ~30 d	2708
	m: 464 g; n20, 1st capture	259
	f: 500 g	3481
	m: 590 g	3481
Marmota monax	247 g	4270
	249 g; n22, 42 d	3240
Sciurus aberti	355 g; 9 wk	5618a
Sciurus carolinensis	200 g	1361
Spermophilus armatus	130 g; 30 d	5909
Spermophilus beecheyi	50-70 g; emergence	10920
	73 g; ~8 wk	2896
Spermophilus beldingi	36.4 g; r33.5-41.2; emergence	7548
	59.5 g; se2.4	1480
	69.00 g; sd13.27; r47.70-92.13; n33, 25 d	7550
Spermophilus columbianus	70.6-75.7 g; 28 d	5909
	85 g; 30 d	5909
	138.7 g; se10.5; n15 captive; 143.4 g; se7.99; n22 wild; 30 d	11959
	151.2 g; se8.03; n15 captive; 130.6 g; se5.79; n16 wild; 30 d	11959
Spermophilus elegans	80 g; 30 d	5909
Spermophilus lateralis	72.3 g; 35 d partially weaned	1871
	100.8 g; 42 d	1871
Spermophilus major	200 g; 1 mo	8050
Spermophilus richardsoni	105 g; 30 d	5909
Spermophilus saturatus	85 g; 1st capture	5658
Spermophilus spilosoma	40-50 g; emergence	10496
Spermophilus townsendii	42.16 g; sd9.40; r17.88-55.78; n41, 34 d	9115
Spermophilus undulatus	109.9 g; 20 d	6992
Spermophilus washingtoni	22-44 g; emergence	9119
Tamias palmeri	21.22 g; se0.63; n12	4825
Tamias panamintinus	17.68 g; se1.20; n10	4825
Tamias striatus	50.5 g; r27.9-70.9; emergence; regression estimate	8511
Tamias townsendii	35 g; 1st capture	5658
Tamiasciurus hudsonicus	~70-80 g; 7-8 wk	6297
	125 g; r120-142; juveniles	2671
WEANING SIZE		
Glaucomys sabrinus	HB: 10.77 cm; r9.2-12.4; n14, 55 d	3239
Marmota monax	HB: 21 cm; n22, 42 d; from graph	3240
Spermophilus beldingi	13.78 cm; sd1.23; r11.26-15.21; n33	7550
Spermophilus xanthoprymnus	HB: 13-15 cm	8050
LITTER SIZE		
Ammospermophilus harrisii	6 emb; n1 f	7145
	6.4 ps; r4-8; n11 f	7789
	6.5 yg; r6-7; n2 lt	7788, 7789
	6.5; n34 records	7788, 7789
	7.3 emb; r6-10; n7 f	7789
Ammospermophilus interpres	5-14 yg	10110a
Ammospermophilus leucurus	5-11 yg	10110a
	6-8 emb	502
	6 emb; n1 f	6560

	7.40 emb, ps, yg; sd1.73; r5-11; n43 f	5661
	8.0 yg; r7-9	6988
	8-11; n4 lt	8348
	~9 emb; r5-14; n9 f	4081
Ammospermophilus nelsoni	8.9 emb; r6-11	4512a
	12 ps; n1 f	4512a
Callosciurus adamsi	2 emb; n1 f	2313
Callosciurus caniceps	2.2; mode 2; r1-5; n14 lt	4406
Callosciurus erythraeus	2 emb; n1 f	10826, 10827
	2 emb; n1 f	182
Callosciurus nigrovittatus	2.2; mode 2; r1-4; n23 lt	4406
Callosciurus notatus	2.2; mode 2; r1-4; n25 lt	4406
	3 emb ea; n3 f	2313
	3-4 yg	557
Callosciurus prevosti	1 yg; n1 lt	2904
	1.2; mode 1; r1-2; n5 lt	11975
	2-3 yg	557
	2.33 emb; r2-3; n3 f	2313
Cynomys gunnisoni	3-4	9583
	3.5 ps; r3-4; n2 f	145
	mode 4	8572
	4.1 ps; n12 young f	6561
	4.17 emerge; n12 lt	3291
	4.69 emb; mode 4; r1-7; n29 f	145
	4.9 ps; n34 f	6561
	mode 5; r4-6	502
	5.6 emerge; r3-10; n39 lt	8989
Cynomys leucurus	4.90 ps; sd0.77; n20 f	531
	5 yg; r1-8	1871a
	5.40 cl; sd0.72; n48 f	531
	5.48 emb; r2-10; n67 f	10436, 10437
	5.64 emb; sd0.74; r3-8; n25 f	531
	5.65 ps; r2-8; n40 f	10835
Cynomys ludovicianus	2.7 yg; sd1.2; n28	3612
	3.11 emerge; se0.08; r1-6; n198 lt, 124 f	3371
	3.3 emb; yearling f	5916
	3.3 yg; sd1.0; n17	3612
	r4-7 emb; n>12 f	11380
	4 yg; n1 f	5342
	4.00 ps; sd0.30; n10 f yearlings	10441
	4.24 ps; sd0.49; n21 f	10441
	4.3 ps; yearling f	5916
	4.4; sd1.2; n153 lt	5895
	4.56 ps; r2-7; n25 f	10835
	4.6 emb; r1-8; n13 f	339
	4.61 ps; sd0.24; n38 f	10441
	5.0 emb; adult f	5916
	5.02 emb; mode 5; r2-10; n50 f	11380
	5.4 ps; adult f	5916
	5.5 yg; n8 lt	1649a
	5.61 ps; sd0.35; n36	10441
	8 ps, 8 ca; n1 f	6166
Cynomys mexicanus	3 emb; n1 f	522
	6 yg; n1 lt, 12 d	8573
Cynomys parvidens	2-8	10809
	4.8 yg; r3-6; n4 lt	8571
Dremomys lokriah	2-5	7383
	4 emb; n1 f	7383
Epixerus ebii	2; n1	2997, 2998
Exilisciurus exilis	2 emb ea; n2 f	2313
	2.5 yg; r2-3; n2 f	2006
	3 emb ea; n2 f	2006
Funambulus layardi	2.5 emb; r2-3; n2 f	8500, 8501

Funambulus palmarum	2-3	8501
	2.75; r2-3; n4 lt	6163, 8500
Funambulus pennanti	2.9 emb; r1-4; n20 f	6588
	3 yg; 2,4 rare	554
	3 emb ea; n10 f	8720
	3.23 emb; ±0.119; mode 3; r2-5; n30 f	1764
	5 yg ea; n2	8723
Funambulus sublineatus	2 yg; n1 nest	8501
Funisciurus anerythrus	1.08 emb; mode 1; r1-2; n37 f	2446a
	1.2 emb; r1-2; n77 f	8855
	1.5; r1-2; n4	2997, 2998
	2 emb ea; n2 f	8610
Funisciurus congicus	2 yg ea; n3 nests	11300
	2; n8 lt	10158
	2.0 yg; n8 lt	11302
Funisciurus isabella	1 emb ea; n6 f	2997, 2998
Funisciurus lemniscatus	1.8; r1-3; n4	2997, 2998
Funisciurus leucogenys	1 emb; n1 f	2953, 2955
Funisciurus pyrrhopus	1 emb; n1 f	2998
	1 emb; n1 f	187
	mode 1 emb; r1-2	8855
	1.6 yg; n10 lt	2821, 9522
	1.6; r1-3; n31 lt	6755
	2; n1 lt	184
	2 yg	5304
Glaucomys sabrinus	2 emb; n1 f	2085, 2088
	mode 2-3 yg	9385
	3 yg ea; n3 nest	2085, 2088
	4 yg; r2-6	1871a
	4; n1 lt	7685
	5 ps; n1 f	7809
Glaucomys volans	2 emb; n1 f	11896
	r2-4 yg; n9 nests	4716
	2.1 yg; n41 lt	3853
	2.5 emb; r2-3; n2 f	495
	2.54; se0.18; r2-4; n13 lt	3790
	2.6 yg; r2-4; n5 nests	1290
	2.8 old yg; n60 nests	8978a
	3 yg; n1 lt	1672
	3 yg; n1 nest	5210
	3 yg; n1 nest	5747
	3 yg; n1 nest	5803
	3 yg; r2-5; n6 f	10206
	3.0 emb, yg; r2-4; n9 f	4716
	3.1 yg; r1-4; n24 f	10206
	3.17 yg; sd0.36; r2-6; n24 lt	6497a
	3.40 emb, yg; r1-6; n63 f	11025
	3.5 yg; n33 nests	8978a
	3.67; r3-4; n3 lt	12128
	4 yg; n1 f	10584
	4 emb; n1 f	576
	4 emb; n1 f	4716
	4 yg; n1 f	10205
	4.5 yg; n2 nests	495
Heliosciurus gambianus	3.0 emb; n2 f	331
Heliosciurus rufobrachium	1-5	5760
	1-5	10153
	1.5 emb; r1-2; n2 f	185
	mode 2 yg	4354a
	4 emb; n1 f	10158
Heliosciurus ruwenzorii	3 emb; n1 f	8856
Hylopetes alboniger	2 yg; n1 nest	7382
	mode 2-3 yg	7383

Hylopetes fimbriatus	mode 3-4	9175
	4; n1	105
Hylopetes lepidus	2.0 emb; r1-4; n39 f	7687
Iomys horsfieldii	2.5 yg; n2 lt	7161
	3 emb; n1 f	7161
Lariscus hosei	2 emb; n1 f	2313
Marmota baibacina	6; r3-8	8050
Marmota bobak	4-5	5172
	4-5; 6 occasional	8050
	5-6	10516
	max 7	4645
	max 12	8798
Marmota caligata	4-5	1474
	mode 4-5	2088
	5 emb; n1 f	5029
Marmota camtschatica	mode 5; r3-11	5518
Marmota caudata	3; r2-4; 5 occasional	8050
	3-6 forest; 4-9; subalpine	2361
	3.4 emerge; r2-4; n5 lt	9175
Marmota flaviventris	3.64-5.13 yg; n7 locations	11194a
	est 4-5; n1 lt	502
	4.07 weaned; se0.45; n15 lt	5328
	4.14; se0.857; r2-6; n7 lt satellite	2708
	4.15; se0.19; r1-8; n65 lt; colonial	366
	4.2 cl; n4 subadult	7803
	4.46; se0.39; r1-8; n13 lt; satellite	366
	4.50; se0.187; r2-8; n52 lt; colonial	2708
	4.7 cl; n7 adult	7803
	4.8 emb; r3-6	7803
	5 ps; r2-8; n2 f	10954
	mode 5-6 emb; r3-6; n5 f	5029
	5.26; 19 lt	366
	5.3 yg; r5-6; n3 f	7803
	5.5 emb; r5-6; n2 f	10954
	8 emb; n1 f	5029
Marmota marmota	2-4	7433
	2-7	8798
	mode 3-5; r2-8	3571
	7; n1 lt	8770
	7; n1 lt	8771
Marmota menzbieri	1-5	6924
Marmota monax	2.5 cl/ovary; se0.3; n5 postparturient f	2016
	mode 3-4; r1-9	12043
	3.7 ps; se0.4; n8 lt	2015
	3.75 emerge; r2-5; n8 lt	2414
	4 emb; n1 f	9639a
	4 yg; n1	5038
	4-6 yg; max 9	5029
	4.07 emb, ps; n49 f	4270
	4.1 yg; se0.2; r2-6; n17 lt	2015
	4.32 emb; n122 f	10187
	4.58 ps; n169 f	10187
	4.6 emb, ps; max 8; n29 f	4083
	4.7 implantations; se0.6; n6 f	2015
	4.7; r4-5; n3 lt	570
	5 emb; n1 f	10217
	5 yg; n1 f	11956
	5.19 cl; n238 f	10187
	6 ps; 1 f	10839
	9 yg; n1 lt	5210
	13.6 cl/ovary; se1.5; n7 nonparturient f	2016
	15.3 cl/ovary; se2.2; n3 nonparturient yearling f	2016
Marmota sibirica	3-6	8050

Marmota vancouverensis	2.7 emerge; n6 lt	7746
	3.0 emerge; n5 lt	7746
	4.6 emerge; n5 lt	7746
Myosciurus pumilio	2	2997
Paraxerus alexandri	mode 1 emb; r1-2	8855
Paraxerus boehmi	1 emb; n1 f	8856
	1.1 emb; r1-2; n61 f	8855
	1.12 emb; mode 1; r1-2; n34 f	2446a
Paraxerus cepapi	1.93 yg; ±0.59; n29	11302
	mode 2	11295, 11299
	2 emb; n1 f	6272
	2-4	11253
	2-4	9945
	2.1 emb; r1-3; n39 f	10154, 10158
	3 emb; n1 f	334
Paraxerus flavivittis	1; n1 lt	187
	1-2 yg	5760
Paraxerus ochraceus	2 emb ea; n2 f	4925
Paraxerus palliatus	1-2	10158
	1.60 yg; ±0.52; n10	11302
	1.71 yg; ±0.49; n7	11302
Paraxerus poensis	1.3; r1-2; n6	2997, 2998
	2 emb; n1 f	1948
	2 emb; n1 f	2955
Petaurista elegans	mode 1; r1-2	7382, 7383
Petaurista magnificus	1 emb ea; n2 f	7382, 7383
Petaurista petaurista	1 yg	8142
	1 emb; n1 f	8501
	1 emb ea; n2 f	7161
	mode 1; r1-2	7383
	mode 2	9175
Petinomys fuscocapillus	2 yg	8501
Protoxerus stangeri	mode 1 emb; r1-2	8855
	1.5 emb; r1-2; n2 f	2997, 2998
	2 emb; n1 f	187
Pteromys volans	2 yg; n1	11441
	mode 2-3; r1-4	8053
	2-4	4645
Pteromyscus pulverulentus	1.3 emb; r1-2; n14 f	7687
Ratufa bicolor	1; n1	8270
	1 yg ea; 3 lt	11741
	1	40
	1-2 yg	7383
	2-3 yg	10828
Ratufa indica	1 yg ea; n2 lt	10559
Ratufa macroura	mode 1; r1-2	8501
	mode est 3-4	8500
	3-5	8798
	4 yg; n1 f	11507b
Rhinosciurus laticaudatus	1.3; r1-2; n3 lt	4406
Sciurillus pusillus	2 emb; n1 f	8087
Sciurus aberti	2.9 emb, ps	10378
	3-5 yg	1871a
	3 emb; n1 f	6387
	3.20; ±0.65; r2-4; n5 lt	3165
	3.3 emb; r3-4; n3 f	7145
	3.4 yg; ±0.9; r2-5; n8 nests	5618a
Sciurus alleni	3; n1 lt	522
Sciurus aureogaster	2; n1 lt	4228
	2 emb ea; n2 f	2341
Sciurus carolinensis	1.45 wean; r1-2; n11 lt, 10 f	9190
	1.6 wean/season	2088
	2 emb; n1 f	5803
	2.25; n8 lt	8188
	2.5 emb; se0.74; r1-4; n26 f	7297

	2.56 yg; r1-4; n27 nests	4716
	2.56 ps; n43 f	5808
	2.60 yg; n93 lt	5808
	2.68 yg; r1-4; n19 nests	6562
	2.7; n16 lt	3898
	2.7; mode 2-4; r1-7; 2.50; n148 spring; 3.23; n55 fall	9943
	2.74; n27 lt	1361
	2.75 yg; n122 lt	5808
	3 emb	2088
	3.0 ps; r1-4; n16 lt	9214
	3.0; r2-4; n2 lt	1241
	3 ps ea; n3 f	9190
	3.02; n447 lt	589
	3.05 lg emb; r1-5; n21 f	6562
	3.1 emerge; se0.4; n58 f	10791
	3.14 ps; r2-5; n37 f	6562
	3.19 emb; n16 f	5808
	3.43 ps; se0.09; r1-6; n98 f; 2.7; se0.14; winter; 3.49; se0.11; spring	7944
	3.6	2426
	3.8 emb; n5 f	9214
	3.96 sm emb; r1-6; n24 f	6562
	4 emb; n1 f	7498a
	4 yg; n1 nest	6197
	5 yg; n1 nest	7498a
	7 emb; n1 f	9943
	8 yg; 1 nest, <3 d	588
	8.5 yg; r8-9; n2	588
Sciurus colliaei	2.5 emb; r2-3; n2 f	5409
	3 emb; n1 f	5405b
Sciurus deppei	2 emb; n1 f	5423b
	3 emb; n1 f	7624
	5; n1 lt; hybrid with *S. yucatanensis*	3633
Sciurus granatensis	1.5 emb; r1-2; n2 f	8562a
	1.8; n4 lt	2998
	1.9 yg; n12 lt	3801
	mode 2 emb; max 3	3024
	2.6 emb; r2-3; n5 f	7940
Sciurus griseus	2 emb; n1 f	11228a
	3-5	1474
	3-5 yg	5146
Sciurus nayaritensis	2.5 emb; r2-3; n2 f	6387
Sciurus niger	2.19 ps; se0.32; n16 yearlings winter	7943
	2.27; r1-4; n11 lt	7501
	2.58; n110 lt	1361
	2.6; r1?-4; n14 lt	8188
	2.77 yg; n13 lt	170
	2.8 yg	1871a
	2.80 ps; se0.33; n10 yearlings summer	7943
	2.85 ps; se0.11; n52 f summer	7943
	2.97 emb; n38 f	170
	3.0; r2-4; n2 lt	522
	3.01 ps; se0.10; n80 f winter	7943
	3.11 emb; n18 f	6562
	3.23 ps; n22 f	6562
	3.5 yg; r3-4; n4 nests	10445
	5.5 ps; r5-6; n2 f	6298
	7 emb; n1 f	4971
Sciurus richmondi	2.67 emb; r2-3; n6 f	5417
Sciurus vulgaris	2; n1 lt	12139
	3-4	2468
	3-5	8798
	3-6	7433
	3-6; yg f	4645, 8053

	mode 3-7; r3-10	3571
	3.0 emb; r1-4; n5 f	11255
	3.4 emb	2469
	4; n1 lt	3749
	4 yg; n1 lt	11255
	5-10; adult f	4645, 8053
	6; n1 lt	2923
	9; hunted; 4; unhunted	8053
Sciurus yucatanensis	3 emb; n1 f	5423b
	3 yg; n1 f	950
	5; n1 lt; hybrid with *S. deppei*	3633
Spermophilopsis leptodactylus	3-6	8053
	3-6	6273
	4.1; r1-9	10568
	5.2 emb; mode 5; r2-8; n5 f	9774
Spermophilus armatus	4-6 yg	1474
	5.8 emb; r4-8; n22 f	11646a
	6-8 yg	1871a
Spermophilus beecheyi	3 emb; n1 f	11228a
	3 yg; n1 f	5020
	4.14 yg; r3-8; n7 lt	6843
	5 yg; n1 f	2896
	5.14 emb; r3-7; n7 f	2896
	5.46 implantation sites; n13 lt	1725
	5.5-8.1; r4-15	10454
	5.86; n22	1726
	6.06 emb, yg; n34 lt	10862
	6.11 ps; est r1-9; n114 f	10862
	6.25 emerge; n8 nests	3114
	6.86 yg; mode 7; r6-8; n7 lt	3281
	7; r4-15	1474
	7.2 emb; r4-11	4081
	7.5; r4-11	5030
	7.5 emb; r5-12	6489
	9.84 emb; n86 f	5227
Spermophilus beldingi	1-9	4916
	4.0 yg; sd1.0; n5 yearling f	7547
	4.3 emerge; sd2.1; n115 lt	12064
	4.42 emerge; sd2.06; r3-6; n219 lt	9900
	4.57 emb; sd1.07; n20 yearling f	7547
	5.0 ps; n2 yearling f	7547
	5.71 yg; sd1.45; n17 adult f	7547
	5.8 yg; r5-7; n9 lt	6988
	6.33 emb; sd2.01; n21 adult f	7547
	6.88 ps; sd1.22; n17 adult f	7547
	7.1; r3-10; n37 lt	7082
	7.4 emb; ±0.8; n175 f	10541
	8; r4-12	5030
Spermophilus citellus	mode 6-8; r3-11	3571, 7433
	6; n2 lt	1552
Spermophilus columbianus	r1-7; r2.57-5.07; r of 6 populations	12061
	2-7	1474
	2.3-4.6; n12 populations	2661
	2.3 emerge; r1-3; n55 lt	7631
	2.3; se0.09; n61 lt; r1.8-2.5, over 5 yr	7630, 7632
	2.69; n29 lt	2658
	2.7 ps; r1-4; n22 lt	7631
	2.7 emerge; r1-5; n31 lt	7631
	2.78 emerge; se0.83; n23 lt [but n72 yg/23 lt = 3.13]	11469
	2.8 cl; r2-4; n17 lt	7631
	2.9 yg; r1-4; n14	11959
	2.9; se0.12; n62 lt; r2.4-3.3, over 5 yr	7630, 7632
	2.94-4.22 emb, ps; sd0.73-1.45; n6 samples	12063
	3.04 emerge; se0.96; n28 lt [but n86 yg/28 lt = 3.07]	11469

	3.2 emb; r1-5; n28 f	7631
	3.3 cl; r2-4; n29 lt	7631
	3.3 ps; r1-5; n54 lt	7631
	3.36; n121 lt with food added	2658
	3.4 yg; r1-5; n15 f	3242
	3.5 juv; r3-4; n4 lt	7631
	3.5; r2-7	2088
	3.57 emb; r3-4; n7 f	2546
	3.6 yg; r1-6; n11	11959
	3.65 emerge; r1-6; n92 lt	3244
	3.70 yg; mode 3-4; n174	12062
	3.9 cl; r1-7; n12 lt	7631
	~4; r2-7	6798
	4.04; r2.9-6.7 across 5 populations; n37 lt	11959
	4.2 emb; r4-5; n5 f	7631
	4.6 yg; n5 lt	9873
	4.6 ps; r3-7; n8 lt	7631
	5.4 emb; r2-7; n56 f	9872
	6.6 yg; r5-8; n7 lt	5909
Spermophilus dauricus	5-8	8050
	5 emb; n1 f	182
Spermophilus elegans	2.7 emerge; se0.23; n15 lt	8469
	3.6 emerge; se0.75; n12 yearling f	8469
	4.8 emerge; se0.50; n19 adult f	8469
	5.2 emerge; se0.40; n21 lt	8469
	5.3 emb; r4-9; n16 f	1865
	6.0 yg; r5-8; n5 f	1865
	6.1 ps; r4-9; n40 f	1865
	7.0 yg; r5-9; n6 lt	2484
	7.5 yg; r5-11; n8 lt	5909
Spermophilus erythrogenys	mode 7-9 emb; n426 f	8050
Spermophilus franklinii	mode 6-7 yg; r5-8	5210
	7.5 emb; r5-11; n26 f	10235
	9.4 ps, emb, yg; r2-13	5185
	9.7; se0.6; r7-12; n10 lt	10948
	10 emb; n1 f	495
Spermophilus fulvus	6; r4-13; more accurate account: 5-9	8050
	6; mode 5-9; r4-13	4645
	6; r5-9	10845
Spermophilus lateralis	2-8	1474
	3-7; n14 lt	8348
	4.3 yg; ±0.6; r3-9; n3	10055
	4.4 emb; r3-6; n10 f	2415
	4.5 yg; n2 lt	1871
	4.5; r4-5; n13 lt	8489
	5 emb; n1 f	145
	5.0 emb; r2-6; n9 f	4081
	5.0 emb; r3-8; n36 f	7080, 7082
	mode 5 ps, yg, emb; n138 f	7080, 7082
	5.1; ±1.32; r3-9; n35	10055
	5.1; r3-8; n43 lt	10762
	5.2 ps; r2-7; n88 f	7080, 7082
	5.28 emb; 2se0.19; n179 f	1323, 1324
	5.38 emb, ps; 2se0.12; n284 f	1323, 1324
	5.4 emb; ±0.72; r4-6; n8 f	10055
	5.9 ps; ±1.41; r3-9; n24 f	10055
	6.0 ps; r5-7; n3 f	7809
	6.0; r4-9; n25 lt	8489
	6.5 emb; r6-7; n2 f	7593a
	6.5 emb; r6-7; n4 f	6560
Spermophilus madrensis	4 emb; r2-5; n4 f	886
	5 emb; n1 f	886

Spermophilus mexicanus	4-10	1474
	5 yg; r1-10	12035
	5 emb; n1 f	522
	5 yg; n1 lt	2901
	10 emb; n1 f	2901
Spermophilus mohavensis	5-9; n4 lt	8348
	6 emb; n1 f	1471
Spermophilus parryii	6 yg; n1 nest	6993
	6.1 emerge; sd1.00; r4-8; n14 lt	7089
	6.1 yg; sd0.99; n19 lt	7089
	6.5 ps; r6-7; n2 f	6993
	7 emb; n1 f	5030
Spermophilus perotensis	6.5 emb; r6-7; n2 f	2341
Spermophilus pygmaeus	5.43, 6.13, 7.6 emb; n3 authors	8050
	6.7; mode 7; r2-12	9453
	7.7 emb; mode 7; r4-13; n252 f	8050
	7.75 emb; mode 8; r4-13; n302 f	11322
Spermophilus richardsonii	2-10	1474
	3.4; se0.05; n7 yearling f	7284
	3.8 emerge; sd1.3; r1-6; n24 f	7283
	4.3; se0.4; n8 lt, 2 f; 3.5; se0.4; n15 lt, 11 f; 4.9 yg; se0.3; n15 f	7284
	4.4 emerge; se0.5; 11 lt	7284
	4.8; se0.03; n10 adult f	7284
	6.3 emerge; sd2.1; n26 adult f	7288
	6.4 emerge; sd2.1; n55 yearling f	7288
	7.4 emerge; sd2.2; n32 yearling f	7288
	7.4 emerge; sd1.4; n53 yearling f	7288
	7.5; r6-11	5030
	7.9 emerge; sd2.7; n19 adult f	7288
	7.93 emb; n15 f near term	7826
	8.1 emerge; sd1.8; n18 adult f	7288
	8.19 cl; 6.93 live emb; se0.18; n126 f	9890
	8.25 yg; r6-11; n4 lt	5909
	8.75 emb; r4-13; n84 f	7826
	9.85 ps; r6-15; n13 f	7826
Spermophilus saturatus	4.14; sd0.86; r1-5; n37	5658
	4.17 yg; sd0.62; r3-5; n18 f	5663
	7 emb; n1 f	2088
Spermophilus spilosoma	4.57; r3-7; n7 lt	522
	5-12	10496
	5 emb; n1 f	502
	6 yg; n1 f	978
	6 emb; r5-7; n3 f	502
	6.6; r4-11	10496
	6.6 yg; r4-12	1871a
	7 emb; n1 f	145
	8; n1 f	2349
	8 emb ea; n2 f	5405b$_1$
Spermophilus suslicus	mode 4-7; r4-12	4645
	mode 7; r5-11; n120 f	8050
	5.5 emb; r3-9	8424
Spermophilus tereticaudus	1.86 yg; n14 lt	7788
	3.5 emb; r3-4; n2 f	4081
	3-8; n10 lt	8348
	6-12 emb	1474
	6.1 emb; r5-10; n11 f	7789
	6.43; n92	7789
	6.5; r4-9; n26 lt	7789
	6.5 emb; n222 f; r3.3-9.0; different sites	9078
	6.7 ps; r2-9; n35 f	7789
Spermophilus townsendii	4.8 yg; r1-10; n12 lt	9116
	5-10, rarely 15	1474
	7.7 emb, ps; se0.2; n51 yearling f	10110

	7.99 emb, ps; r7.15-9.11 over 6 yr; n257 lt	10110
	8.3 emb; n22 f	9582
	8.3 yg; sd1.7; n135 yearling f	9117
	8.5 yg; r6-13; n20 lt	5343
	8.5 emb; n129 f	6354
	8.6 yg; r6-12; n4 f	9115
	8.6 emb; r4-16; n52	9582
	8.75 emb; r8-10; n4 f	5030
	8.9 emb; r7-10	5030
	9.0 emb; r4-12; n21 f	5343
	9.2 emb, ps; se0.3; n21 adult f	10110
	9.3 emb; n163 f	139
	9.4 yg; sd2.2; n60 adult f	9117
	15 emb; n1 f	5312
Spermophilus tridecemlineatus	5-13	5030
	4.2 emerge; se0.8; yearling f	7023
	4.9 emerge; se0.6; yearling f 1st lt	7023
	5.7 emerge; se1.3; adult f 2nd lt	7023
	mode 6-7; r5-10	2060
	6-10 emb	502
	6.0 yg; r2-10; n2 lt	11379
	6.1 emerge; se0.8; yearling f	7023
	6.9 emerge; se0.8; adult f	7023
	7-10; rarely 14	1474
	7 ps, 8 ca; n1 f	6166
	7.0 emerge; se0.3; adult f 1st lt	7023
	7.4 emerge; se0.4; adult f 1st lt	7023
	7.5 yg; r1-10	1871a
	8.1 emb; n269 f	2140
	8.4; r6-11; n12 lt	1286
	8.57 yg; r5-13; n28 lt	12108
	8.67 emb; sd1.14; r7-11; n18 f	3294
	9 yg	5343
	9 ps; n1 f	3703
	9.1 emb; r6-13; n9 f	495
	9.3 emb; r6-12; n6 f	5405b,
	10.2; r10-11; n5 lt; prob birth	7023
	11 emb; n1 f	1812a
Spermophilus undulatus	1-4 yg	7656
	5-11	10516
	r7-9 emb; n160 f	8050
	7.8 yg; r5-10; n6 f, 6 lt	6992
Spermophilus variegatus	3 emb; n1 f	1649a
	3.8 yg; n6 lt	5347
	4 emerge; r1-7	8019
	4 emb; n1 f	4962
	5-7	1474
	5 emb; r3-9	8019
	5; n1 lt	522
	5 emb; n1 f	7145
	5 emb; n1 f	4914
	6 emb; r5-7; n2 f	502
	7 emb; n1 f	2341
Spermophilus washingtoni	5-11	1474
	8 emb, ps; n11	9582
	8 emb; r5-11; n26	9582
Spermophilus xanthoprymnus	4-6	8050
Sundasciurus lowii	2.5 emb; r2-3; n2 f	2313
Sundasciurus tenuis	3.0; mode 3; r2-4; n4 lt	4406
Tamias amoenus	4-8	2088
	4.9	5662
	5 yg; r2-9	1871a
	r5-7 emb; n5 f	1295
	5.0 emb; n1 f	11646a

	5.0 yg; r4-6; n2 f	1295
	5.14 cl; 4.6 emb; se0.14; n42 f	9889
	5.8 emb; n7 f	1295
	6.1; r4-8; n17 lt	10762
Tamias bulleri	2.5 emb; r2-3; n2 f	523a
Tamias dorsalis	4-6 yg	1871a
	5.0 emb; r2-8; n4 f	4433a
	5.3 emerge; r5-6; n3 lt	4433a
Tamias minimus	2-6	1474
	4 yg; n1 lt	10053
	4 ps; n1 f	3703
	4 emb; r3-5; n2 f	261a
	4.2 ps; r3-5; n5	10839
	4.5 emb; r3-6; n4 f	10954
	4.54 cl; 4.03 emb; se0.18; n59 f	9889
	5; r2-6	2088
	5 ps; n1 f	261a
	5 yg; n1 lt	3390
	5 ca; r3-7; n2	10839
	5.0 emb; mode 5; r3-7; n9 f	279
	5.1 ps; r3-8; n17 lt	10954
	5.4 ps; se0.25; r3-8; n24 f	10053
	5.5 yg; r3-9	1871a
	5.7 yg; se0.52; r5-8; n7	10053
	5.7 emb; se0.15; r3-8; n52 f	11230
	5.7 ps; se0.15; r2-10; n97 f	11230
	6.4 emb; se0.31; r5-9; n11 f	10053
	7 emb; n1 f	10218, 10219
Tamias palmeri	3.75; n4 lt	4825
	4.20 emb; mode 4; r3-6; n15 f	2415
Tamias panamintinus	4.0; n4 lt	4825
	5.75 emb; r4-9; n4 f	2415
Tamias quadrimaculatus	4.4; r2-6; n10 lt	10762
	8 yg; n1 lt	9302a
Tamias quadrivittatus	3.25 emb; r2-4; n4 f	4962
	4-6 emb	502
	5.0 ps; r4-6; n2 f	4914
Tamias ruficaudus	4.65; se0.13; n24 lt	726
	5.16; se0.20; n24 lt	726
Tamias sibiricus	r3.2-6.8; n~12 colonies	982
	mode 4-5; r3-12; rarely 1-2	4645
	4-6	10177
	4.4 emerge; sd1.6; 66% 3-5; r1-8; n176 lt, 56 f	983
	4.8-6.9 emb, yg; varies w/region & yr	9953
	5 yg; n1 lt	9416
	5.57 emb; r3-10	8053
Tamias speciosus	4.3; n3 lt	10762
Tamias striatus	1.8 emerge; r1-4; n44	11989
	2.8-5.3 emerge; over 3 yr	8511
	3 emb; n1 f	1472
	4-5	9164
	4 emb, ps; r1-6; n13 f	12014
	4.67 yg; r4-5; n3 f	8225
	4.7 yg; r4-6; n11 lt	11956
	5 emb; n1 f	10839
	5 emb; n1 f	175
	5; r3-7; n2 lt	175
	5.02 emb; mode 5; r2-7; n41 f	2022
	5.2 ps; r2-7; n5 f	10839
	6 emb; n1 f	9612
	6 emb; n1 f	9639a
	7.0 emb; r6-8; n2 f	10220
Tamias townsendii	3-6	1474
	3-6	2088

	3.75; sd0.45; r3-4; n12	5658
	4.5 yg; r4-5; n2 lt	3391
	4.5; r3-5; n6 lt	10762
Tamias umbrinus	max 7 emb	1871a
Tamiasciurus douglasii	4-8	1474
	5.0; r4-8; n5 lt	10098
	5 yg	5146
	mode 5-6; r3-7	6387
Tamiasciurus hudsonicus	3 emb; n1 f	495
	3 emb; n1 f	6560
	3.0 yg; r2-4; n2 nest	4496
	3 ps ea; n2 f	6166
	3.2; 4.1 cl	2324
	3.2 yg; n9 f	2324
	3.3 emb; mode 3; r2-5; n16 f	2671
	3.4 ps; mode 4; 95% ci 3.1-3.7; r1-5; n43	5641
	3.4; r2-5; n61	10097, 10098
	3.5 yg; r1-10	1871a
	3.7 ps; n34 f	7299
	4.0 ps; n20 f	11894
	4-5 yg; n9 lt	4274
	max 4 emb; 5 ps; n1 f	12049
	4 ps; n4	10839
	4; r1-6	2088
	4.1 emb, ps; n164 f	9374
	4.2 yg; r2-7; n11 nests	6297
	4.2; n12 lt	6158
	4.2 emb; r1-6; n17 f	6297
	4.2 emb; r3-5; n38 f	4274
	4.3 ps; mode 4; 95%ci 3.9-4.7; r2-8; n39	5641
	4.37 cl; r2-7; n83 lt	7299
	4.5 yg; r4-5; n2 f	10585
	4.7 ps; r3-8; n46 f	6297
	4.7 cl; n123 f	9374
	5 ps; r4-6; n3 f	11956
	5.0 emb; r4-6; n4 f	10954
	5.4 yg; r4-8; n7 f	3241
	5.4 yg; r4-8; n8 f	3241, 8745
Tamiops macclellandi	2 emb; n1 f	4813b
	3 emb; n1 f	7161
Tamiops swinhoei	2-5 emb	10825
Trogopterus xanthipes	1.64 yg; mode 1-2; r1-4; n25 lt	11447, 11448
Xerus erythropus	2-4 yg	4354a
	3-4	5277
	3-4	2439
	3.5 emb; r3-4; n2 f	1149
	4; r2-6	5760
Xerus inauris	1-4	12134
	1.9; r1-3	4697
	2.2 emb; r1-3; n27 f	10154, 10158
Xerus rutilus	2 emb; n1 f	8017
	2 emb; n1 f	8017
SEXUAL MATURITY		
Ammospermophilus harrisii	f: vag perf: 10-11 mo	7788
	m: scrotal testes: 14-19 wk	7788
Ammospermophilus leucurus	f: ~1 yr	5661
	m: testes large: ~1 yr	5661
Callosciurus caniceps	f: 1st preg: 230 g	4406
	m: fertile: 50% 210 g; 95% 259 g	4406
Callosciurus nigrovittatus	f: 1st preg: 130 g	4406
	m: fertile: 50% 149 g; 95% 269 g	4406
Callosciurus notatus	f: 1st preg: 150 g	4406
	m: fertile: 50% 113 g; 95% 257 g	4406
Cynomys gunnisoni	f: 1st preg: 1 yr	6561

Cynomys leucurus	f: 1st mate: most 1 yr	531
Cynomys ludovicianus	f: 1st mate: 2 yr; some 1 yr	4958
	f: 1st conceive: yearling	5916
	f: preg: 40% yearlings; 88% 2 yr	5895
	m: 1st mate: 2 yr	4958
	f/m: 1st mate: 2 yr	1474
	f/m: 2 yr	10139
Cynomys parvidens	f/m: 1 yr	10809
Funambulus pennanti	f: 1st mate: 7-10 mo	8735, 9799
	f: 6-8 mo	554
	f: 6-8 mo	9175
	m: spermatogenesis: 7-9 mo	8996
	f/m: 7-11 mo	1764
Lariscus insignis	f: 1st preg: 190 g	4406
Marmota bobak	f/m: ~2 yr	8798
	f/m: 3rd summer	8050
	f/m: 1st mate: 4 yr	5172
Marmota caligata	f: 1st mate: est 2 yr	2088
Marmota caudata	f/m: 3rd yr	8050
	f/m: 4th yr	2361
Marmota flaviventris	f: 1st mate: 2 yr	2708
	f/m: 1st birth: 2 yr; median 3 yr	5328
Marmota marmota	f/m: ~2 yr	8798
	f/m: 3-4 yr; full grown: 2 yr	3571
Marmota menzbieri	f/m: 3rd summer; rarely 2nd yr	6924
Marmota monax	f: 1st mate: 1 yr; n6/21	4270
	m: 1st mate: 1 yr; n15/45	4270
	f/m: 1st mate: mode 2 yr	4083
Marmota sibirica	f/m: 3rd summer	8050
Paraxerus poensis	f: 1st birth: ~126 d; n1	2998
Ratufa macroura	f/m: ~2 yr	8798
Rhinosciurus laticaudatus	f: 1st preg: 170 g	4406
Sciurus carolinensis	f: 1st mate: 124 d	589
	f: 1st birth: yearling	2088
	f: 1st birth: mode 1 yr; 1/99 f <10 mo	10133
	f: fertile: ~1 yr	9214
	f: 1st repro: few 4-6 mo; most 10-14 mo	7297
	f: > 500 g	9943
	m: fertile: ~10 mo	9214
	m: testes: large 9-11 mo	589
	m: testes: large 10-13 mo; spring born	7297
	m: testes: large 14-16 mo; summer born	7297
	f/m: 1st mate: ~10-11 mo	2426
	f/m: 1st mate: 2nd yr	9943
Sciurus niger	f: 1 yr	1474
	f/m: 1st mate: ~1 yr	170
	f/m: 10-11 mo	1871a
	f/m: ~1 yr	8798
Sciurus vulgaris	f: ovulation: ~1 yr; sometimes 7-8 mo	6526
	f: 1st birth: 8-10 mo	4645
	f: 8-12 mo	8053
	f/m: 8-10 mo	3571
	f/m: ~1 yr	8798
Spermophilus beecheyi	f/m: 1st mate: ~1 yr	10862
Spermophilus beldingi	f: 1st repro: 1 yr	9900
	m: mate: 1 yr	4931
	m: 1st repro: 2 yr	9900
Spermophilus citellus	f/m: 9-10 mo	3571
Spermophilus columbianus	f/m: 1st mate: r1-2.4 yr; n82	2658, 2659
	f/m: 1 yr; n4	3242
Spermophilus franklinii	f/m: 1st repro: 10-11 mo	5185
Spermophilus lateralis	f: vag perf: 11 mo	12115
	f: 1-3 yr	1323
	m: ~1 yr	601

	m: 2-3 yr	1323
Spermophilus major	m: 1 yr	8050
Spermophilus richardsonii	f: 1st birth: yearlings	9890
	f/m: 11 mo	7289
Spermophilus spilosoma	f/m: ~1 yr	9723
Spermophilus tereticaudus	f/m: 1st mate: 10-11 mo	7788
Spermophilus townsendii	f: 1st preg: yearlings	9117
Spermophilus tridecemlineatus	f: 1st mate: 1 yr	5345
	m: testes descent: 72 d; n2	5343
	m: sperm: 7.5 mo	11581, 11582
Sundasciurus? lowii	f: 1st preg: 62 g	4406
Sundasciurus tenuis	f: 1st preg: 66 g	4406
	m: fertile: 50% 50.2 g; 95% 88.3 g	4406
Tamias amoenus	f: 1st birth: ~1 yr	1295
	m: 1st mount: 51 d	1295
	f/m: ~1 yr	1295
	f/m: spring after birth	1474
	f/m: yearlings	9889
Tamias minimus	f: 1st mate: yearlings	11230
	f/m: 1st repro: min 8 mo, but not likely in wild	10053
	f/m: 1st mate: yearlings	9889
Tamias quadrivittatus	f: 1st birth: ~12 mo	11385
	f/m: 10-11 mo	11385
Tamias ruficaudus	f: 1st birth: 18-29% 1 yr	726
Tamias sibiricus	f/m: 1st mate: yearlings	983
	f/m: 11 mo	8053
	f/m: ~1 yr	10422
Tamias striatus	f: 1st preg: 3.5 mo; most 1 yr	12014
	f: ~3 mo	1472
	m: testes: large 7-8 mo	12014
Tamiasciurus douglasii	f: 1st repro: 4 mo; n1	5922
Tamiasciurus hudsonicus	f: 10 mo	6158
	f: season of birth	6297
	f/m: 1st mate: 1 yr	2324
	f/m: ~1 yr	8798
Trogopterus xanthipes	f/m: 1st mate: 22 mo; n1 pr	11447, 11448
Xerus inauris	f/m: ~1 yr	4697
CYCLE LENGTH		
Cynomys ludovicianus	estrus: 1 d	4958
Funambulus pennanti	estrus: ~16 h	9175
Sciurus vulgaris	estrus: 2 wk; standing estrus: 3-4 d	8053
Spermophilus beldingi	estrus: 4.7 h; ±0.04; n54	4317
Spermophilus columbianus	14-15 d; interestrous interval	9873
Spermophilus parryii	estrus: 1 d	7090
Spermophilus richardsonii	5.1 d; sd1.5; r3-7; n10	7285
Tamias sibiricus	estrus: 12 h	6736
	12 d ea; n7 cycles, 1 f	6736
	13.6 d; sd1.6; mode 13 37%; 80% 12-14; r11-21; n82 cycles, 3 f	983
Tamias striatus	estrus: 3-10 d	10102
GESTATION		
Ammospermophilus harrisii	min 29 d; n1	7788
Ammospermophilus leucurus	4 wk	5661
	min 29 d; n4 lt	8348
Ammospermophilus nelsoni	est 26 d; field data	4512a
Callosciurus prevosti	47.2 d; r46-48; n5	11975
Cynomys gunnisoni	30 d	8572
Cynomys leucurus	est 27-33 d	10436, 10437
	est 30 d	531
Cynomys ludovicianus	28-32 d	10139
	30-35 d	10954
	32 d	5916
	34.8 d; se0.1; r34-37; n32 lt	3371
	35 d; r33-37	339, 3404

Funambulus palmarum	within 28-40 d; n1 ?	6163
Funambulus pennanti	est ~30 d	9799
	40-42 d	554
	41 d; r40-42	105
	42 d	8720
	<46 d	6588
Funisciurus congicus	~52 d	11302
Glaucomys sabrinus	37 d; n1	7685
	~40 d	5146
	42 d	11588
Glaucomys volans	40 d	10206
Marmota baibacina	40 d	8050
Marmota bobak	40 d	10516
	~40 d	4645
	40-42 d	8798
	40-42 d	8050
Marmota caligata	~30 d	571
Marmota caudata	~30 d	9175
Marmota flaviventris	4 wk	5328
	30 d	3481
	~1 mo	5146
Marmota marmota	33-34 d; n1	8770, 8771
	33-35; 42?; d	3571
	35-38 d	8798
	35-42 d	7433
Marmota monax	~4 wk; field observation	4270
	30 d	2015
	31-32 d; n3	5037, 5038
	31-33 d; n3	4083
	35-40 d	2088
Marmota sibirica	40-42 d	8050
Paraxerus cepapi	56.5 d; r56-58; n4	11295, 11297, 11298
Paraxerus palliatus	59 d; n1	11302
	60-65 d	10158
Pteromys volans	4 wk	8053
Ratufa bicolor	28-35 d	7383
Ratufa indica	2.5-3 mo	10559
Ratufa macroura	28 d	8798
Sciurus aberti	≥38-≤46 d; n1	5618a
	40 d; field data	5618a
	46 d; n3	3165
Sciurus carolinensis	~44 d	11514
	~44 d	9943
	44 d; r43-46	4274
	44-45 d; n1	1241
Sciurus griseus	> 43 d	5143
	est > 43 d	1474
	44 d	5146
Sciurus niger	44 d	1474
	45 d	8798
	~45 d	10809
Sciurus vulgaris	32-40 d	6526
	32-40 d	7433
	35-40 d	8053
	35-40 d	10293
	35-40 d	4645
	38 d	8798
	38 d	2923
	~38 d; r32-40	3571
	38-40 d	12139
Spermophilopsis leptodactylus	1.5 mo	9774
Spermophilus armatus	24 d	2984
Spermophilus beecheyi	implant: 8 d	10862
	25-30 d	1474

	4 wk	10806
	~30 d	10862
Spermophilus beldingi	23-26 d; mating to decrease in f mass	9897
	27-31 d	10953
Spermophilus citellus	25-30 d	1552
	28 d	7433
	28 d; r25-30 d	3571
Spermophilus columbianus	23.75 d; r23-24; n4 lt, 4 f	9872, 9873
	24.2 d; se0.21; r23.5-25.0; n7; sperm in smear to decrease in f mass	7632
Spermophilus dauricus	30 d	8050
Spermophilus elegans	22-23 d; n1	3143
Spermophilus franklinii	~28 d	5210
Spermophilus fulvus	~1 mo	4645
	~1 mo	8050
Spermophilus lateralis	27 d; n1 f; capture to birth	7080, 7082
	est 27-28 d; n1; pairing to birth	1562
	min 33 d; n14 lt	8348
	~35 d	8349
Spermophilus mohavensis	min 24 d; n4 lt	8348
Spermophilus parryii	~25 d	6993
Spermophilus pygmaeus	25-26 d; field data?	11322
	25-30 d	8050
Spermophilus richardsonii	22-23 d; r21 d 15.5 h-23 d 14 h; n28; sperm in vagina to birth	7281, 7282, 7285
	22.08 d; sd0.43; r21.18-23.26; n46	7288
	22.3 d; sd0.5; n22; adult	7288
	22.4 d; sd0.5; n32; yearlings	7288
	22.5 d; sd0.5; n13; adult	7288
	22.7 d; sd0.6; n52; yearlings	7288
	22.8 d; sd0.7; n14; adult	7288
	22.8 d; sd0.6; n29; yearlings	7288
	28-32 d	5030
Spermophilus suslicus	min 3 wk	8050
	22-26 d	4645
Spermophilus tereticaudus	≥26 d; n10 lt	8348
	~27 d	9078
	est 28-29 d	7788, 7789
Spermophilus townsendii	>20 d; est 24 d	139
	est 24 d	9116
Spermophilus tridecemlineatus	27 d; n1	1286
	27 d; n7	3432
	27-28 d	12108
	27-28 d; n1	5343
	27-28 d; n2	11379
	28 d; mode 28; n7	5345, 5346
	4 wk	7493
Spermophilus undulatus	25 d	6992
	~1 mo; from graph	4850
Spermophilus variegatus	est ~30 d	1474
Tamias amoenus	28-30 d	1871a
Tamias cinereicollis	min 30 d	8549
Tamias dorsalis	28-31 d	1871a
Tamias palmeri	min 33 d; capture to birth	4825
Tamias panamintinus	min 36 d; capture to birth	4825
Tamias quadrimaculatus	31 d	9302a
Tamias quadrivittatus	est 30-33 d	11385
Tamias sibiricus	28 d	9416
	31.3 d; sd1.1; mode 31 57%; r28-35; n56 lt, 17 f	983
	35-40 d	4645
	35-40 d	8053
Tamias striatus	~30 d	2981
	31 d; n1 f	1472
	31 d	12014
	31 d	10102

	32 d	175
Tamiasciurus douglasii	36-40 d	5146
	est 40 d	5922
Tamiasciurus hudsonicus	~1 mo; field data	2088
	33 d; r31-35; n4	6159
	35 d; n1	3241, 8745
	~36 d	1871a
	38 d	1474
	40 d	8798
	est 40 d	4274
Trogopterus xanthipes	80.3 d; r74-89 d	11447, 11448
Xerus inauris	est 42-49 d	12134
LACTATION		
Ammospermophilus harrisii	wean: 7 wk	7788
Ammospermophilus leucurus	lact: 2 mo; field data	5661
	wean: min 65 d; 100% survival 67 d; n4 lt	8348
	wean: 2+ mo	5661
Ammospermophilus nelsoni	emergence: ~30 d	4512a
Cynomys spp	emergence: 33-37 d	8798
	wean: 7-8 wk	8798
Cynomys gunnisoni	lact: 1+ mo	8572
Cynomys ludovicianus	solid food: ~40 d	1871a
	solid food: 6th wk	339
	emergence: 6 wk	1474
	emergence: 43.4 d; sd3.5; r38-50; n17 lt	4959
	emergence: 44 d	5895
	wean: 6-7 wk text; 7-8 wk summary	339
	survive without milk: 7 wk	5342
	wean: est 7-8 wk	10139
	wean: 3 mo	4958
Cynomys mexicanus	wean: 41-50 d	1673
Funambulus pennanti	lact: 6 wk	9799
	lact: 2 mo	554
	lact: 2-2.5 mo	8735
	wean: 8 wk	9175
Funisciurus congicus	wean: 50 d	11302
Glaucomys sabrinus	solid food: 47 d	1117
	wean: 55 d	3239
	wean: 65 d	1117
	wean: ~90 d	1871a
Glaucomys volans	wean: 44 d	10211
	wean: 55 d	1672
	wean: ~2 mo	10584
Hylopetes fimbriatus	independent: 2.5 mo	9175
Marmota bobak	wean: 35-40 d	5172
	emergence: soon after birth	5172
Marmota caligata	lact: 3-4 wk; & 1st emergence	571
Marmota flaviventris	emergence: 20-30 d	3481
	wean: 25 d	1871a
	emergence: 30 d	1474
	wean: 3-4 wk	5328
Marmota marmota	emergence: 20-40 d; 2 mo	3571
	emergence: 39-40 d	8798
	emergence: 40 d	8771
	solid food: 2 mo	8771
	wean: 2 mo or longer	8771
	wean: 8-10 wk	8798
Marmota monax	solid food: 32.1 d; sd2.0; r29-38; n22	3240
	solid food: 33 d	570
	emergence: 39-40 d	8798
	emergence: 6 wk	1830
	wean: 4-5 wk	1830
	wean: ~38 d	570
	wean: ~44 d	10187

	wean: 45 d	2015
	wean: 8-10 wk	8798
Paraxerus cepapi	solid food: 18-29 d	11297, 11302
	solid food: 3 wk	11299
	wean: ~5 wk	10158
	wean: 5-6 wk	11297, 11302
Paraxerus ochraceus	nest emergence: 3-4 wk	5760
Paraxerus palliatus	solid food: 28-57 d	11302
	wean: 39-57 d	11302
Petaurista petaurista	wean: 2-2.5 mo	9175
Ratufa bicolor	solid food: 5 wk	33
Ratufa macroura	wean: 8-10 wk	8798
Sciurus spp	solid food: 45-47 d	8798
	emergence: 45-47 d	8798
	wean: 8-10 wk	8798
Sciurus aberti	wean: 10 wk	5618a
Sciurus carolinensis	wean: 7-10 wk	9943
	wean: 8 wk	1361
	wean: 2 mo	1474
	wean: 10 wk	10158
	wean: 82 d	7297
Sciurus niger	solid food: 50-54 d	8798
	emergence: 50-54 d	8798
	wean: 8 wk	1361
	wean: 8-10 wk	8798
	wean: 2-3 mo	1474
	wean: 9-12 wk	10809
	wean: 11-12 wk	7943a
Sciurus vulgaris	solid food: 40 d	8053
	solid food: 40-42 d	8798
	emergence: 40-42 d	8798
	emergence: 42 d	2923
	wean: 52 d	12139
	wean: 56-68 d	3571
	wean: 8-10 wk	8798
	wean: 70-75	8053
	lact: 10-12? wk	2468
Spermophilopsis leptodactylus	emergence: 6 wk	9774
Spermophilus armatus	emergence: 21 d	2984
Spermophilus beecheyi	solid food: 34 d	10862
	solid food: 8 wk	2896
	emergence: 36-50 d; ~6 wk in nest	6489
	emergence: 6 wk	1474
	wean: 5-6 wk	10806
	nurse: max 74 d	10862
Spermophilus beldingi	emergence: 3-4 wk	7548
	emergence: 25-28 d; nurse 1-3 d after emergence	9897
	wean: 24 d	7550
Spermophilus columbianus	solid food: 25-26 d	11959
	emergence: ~30 d	11959
	wean: 25-32 d	11959
	nurse: 30 d	9872, 9873
Spermophilus elegans	wean: 28-35 d	1864
	emergence: 29-33 d	3143
	wean: 5 wk	1871a
Spermophilus franklinii	wean: 28 d	10948
Spermophilus lateralis	solid food: 6 wk	7082
	wean: 28-34 d	8489
	nurse: 7-8 wk	7082
	wean: min 29 d; 100% survival 32 d; n14 lt	8348, 8349
Spermophilus major	solid food: 2 wk	8050
	emergence: 2 mo	8050
Spermophilus mohavensis	wean: min 32 d; 100% survival 40 d; n4 lt	8348

Spermophilus pygmaeus	solid food: 8 d	8050
	emergence: 15 d; 20-22 d	8050
	emergence: 82 d	11322
Spermophilus richardsonii	emergence: 29.0 d; sd1.0; n35; adult f	7282, 7288
	emergence: 29.0 d; sd1.1; n41; yearling f	7282, 7288
	emergence: 29.4 d; sd1.3; n11; adult f	7282, 7288
	emergence: 29.4 d; sd1.6; n51; yearling f	7282, 7288
	emergence: 29.5 d; sd1.3; n25; yearling f	7282, 7288
Spermophilus saturatus	1st food handling: 34.0 d; sd1.8	5663
	1st capture: 6 wk	5658
	lact: 7 wk	5658
Spermophilus suslicus	solid food: ~20 d	4645
	emergence: ~20 d	4645
	wean: 4-5 wk	8050
Spermophilus tereticaudus	wean: 20 d	9078
	wean: 5th wk	7788
	wean: min 35 d; 100% survival 50 d; n10 lt	8348
Spermophilus townsendii	wean: 19-22 d	9116
	solid food: 26-28 d	9116
	nurse: 34-36 d	9115
Spermophilus tridecemlineatus	emergence: ~28 d	1286
	wean: start 18-28 d	1286
	wean: ~26-27 d	9249
	wean: 28 d; n1 lt	11379
	solid food: 30 d	5343
	wean: 5 wk	3432
Spermophilus undulatus	solid food: 22 d	6992
	solid food: 35 d	8050
	emergence: 35 d	8050
	care of young: 11 wk	4850
Spermophilus variegatus	wean: 6-8 wk	5347
Tamias amoenus	solid food: 40 d	5662
	lact: 35-40 d	9889
	lact: 7-8 wk	5662
	wean: 6 wk	1474
Tamias cinereicollis	solid food: 36-40 d	8549
	nurse: 41-45 d	8549
Tamias dorsalis	wean: 30 d	1871a
Tamias minimus	lact: 35-40 d	9889
	wean: 40-60 d	1871a
	lact: 49 d; n1; one other shorter	10053
Tamias palmeri	solid food: by 6 wk	4825
Tamias panamintinus	solid food: by 6 wk	4825
Tamias quadrivittatus	wean: 6-7 wk	11385
Tamias sibiricus	emergence: 33.9 d; sd3.9; mode 30-38; r24-40; n23	982
	wean: 28-30 d	8053
	wean: 1 mo	4645
Tamias striatus	emergence: 5-7 wk; n22	8511
	emergence: 40 d	2981
	emergence: when 2/3 grown	1474
Tamias townsendii	1st capture: 6 wk	5658
	lact: 7-8 wk	5658
Tamiasciurus hudsonicus	solid food: nibbled 34 d, eaten 37-41 d	8745
	solid food: 5 wk	4274
	solid food: ~37 d	7877a
	wean: 69-72 d	8745
	wean: 7-8 wk	6297
	wean: 8-10 wk	8798
	wean: 62-78 d; n3 lt	10097
	wean: 9-11 wk	1871a
Trogopterus xanthipes	wean: 90-120 d	11448
Xerus inauris	wean: begins 50 d	10158

INTERLITTER INTERVAL		
Funambulus pennanti	~5 mo	9799
Funisciurus congicus	60 d; n3	11302
Glaucomys volans	~69 d; n1	8978a
	100 d; n1	10307a
	3.5-12 mo	6497a
	124-143 d	6359a
Marmota bobak	2 yr	5172
Marmota flaviventris	1-1.5 yr	2708, 5328
Marmota marmota	mode 2-3 yr; max 4	3571
Paraxerus palliatus	r61-89 d; n6	11302
Paraxerus poensis	40-65 d	2997
	63 d; r61-65; n2	2998
Ratufa bicolor	149 d; r94-235; n4 yg died 1-2 mo	40
	407 d; n1 yg lived	40
Tamias amoenus	1 yr	9889
Tamias minimus	1 yr	9889
Tamias sibiricus	103.5 d; sd20.0; r77-165; n26 intervals, 15 f	983
LITTERS/YEAR		
Ammospermophilus interpres	1-2	10110a
Ammospermophilus leucurus	1	5661
	1-2	10110a
	max 2	1474
Cynomys gunnisoni	1	8572
Cynomys ludovicianus	1	10954
Funambulus pennanti	2	9799
	3	554
	3	8735
Glaucomys sabrinus	1; sometimes 2	2088
Glaucomys volans	2	8798
	2	7686
	2	10206
	max 2	9164
Hylopetes fimbriatus	2	9175
Marmota baibacina	1	8050
Marmota bobak	0.5	5172
	1	4645
	1	8050
Marmota caligata	0.5	571
Marmota caudata	1	8050
Marmota flaviventris	1	2708
Marmota marmota	1	3571
	1	7433
Marmota monax	1	4083
	1	10187
	1	2015
Marmota sibirica	1	8050
Paraxerus cepapi	2	2411
	min 3	11253
Paraxerus palliatus	1 wild; several captive	10158
Pteromys volans	1-2	8053
Ratufa bicolor	>1	7383
	2	10828
Ratufa macroura	1-2	8798
Sciurus aberti	1-2	7145
	est 2	502
Sciurus carolinensis	1-2	2426
	1 yearling; 2 adult	9943
	2	9164
	2	11956
	2	8188
	2	1474
	2	9190
	max 2	7944

	2; adult f	6562
	mode 2	9214
	mode 2	7297
	2; r1-3	589
Sciurus griseus	1	5146
	1	1474
Sciurus niger	0-2 lt/10 mo season	7943
	1-2	10809
	1-2	8798
	1-2	170
	2	8188
	2 adult; 1 yearlings	1474
Sciurus vulgaris	1-2	8798
	1-3	8053
	est 2	2435
	2-3; r2-5 ?	3571
	4	6526
Spermophilopsis leptodactylus	1	10568
Spermophilus armatus	1	1474
Spermophilus beecheyi	1	6489
	1	3114
	1	4081
	1	10862
	1	10806
Spermophilus beldingi	1	9897, 9900
	1	4081
Spermophilus citellus	1-2	7433
	1	3571
Spermophilus columbianus	1	9872
Spermophilus dauricus	1	8050
Spermophilus erythrogenys	1	8050
Spermophilus franklinii	1	5185
	1	10235
Spermophilus fulvus	1	10845
	1	4645
	1	8050
Spermophilus lateralis	1	10762
	1	4081
Spermophilus major	1	8050
Spermophilus pygmaeus	1	8050
Spermophilus richardsonii	1	5030
Spermophilus spilosoma	1	9723
	est 2	2349
	est 2	1474
	est 2	502
Spermophilus suslicus	1	4645
	1	8050
Spermophilus tereticaudus	1	9078
	rarely 2	7789
Spermophilus townsendii	1	10110
Spermophilus tridecemlineatus	1	11379
	1 est	502
	1; possibly 2nd lt late in summer	1474
	2	7023
Spermophilus undulatus	1	10516
Spermophilus variegatus	1-2	7145
Tamias spp	1	6918
Tamias amoenus	1	1295
	1	9889
	1	10762
	1	1295, 1296
Tamias dorsalis	1	2805
	2	502

Tamias minimus	1	9889
	1	2088
	1	10053
	max 2	1474
Tamias quadrimaculatus	1	10762
Tamias ruficaudus	1	726
Tamias sibiricus	1	10177
	1	10422
	1	4645
	1	8053
	mode 1; 3 rare	982, 983
	1-2	9953
Tamias speciosus	1	10762
Tamias striatus	1	9164
	1	10839
	1-2	1472
	1-2	8511
	2	12014
	2	175
	2 ?	11989
Tamias townsendii	1	10762
Tamiasciurus douglasii	1 mode; 1-2	5922
	2	10098
	2?	5146
Tamiasciurus hudsonicus	0/4 f with 2 lt	10839
	1	2671
	1	10098
	1	3241
	1-2	8798
	1-5; r1-2; n10	6158
	mode 2	4274
	2	1474
	max 2	6297
	2?	10954
Xerus inauris	1	10158
	1	4697

SEASONALITY

Ammospermophilus harrisii	mate: mid Jan-mid Mar; Mexico, sw AZ	7145
	preg f: Feb-May; s AZ	7789
Ammospermophilus interpres	lact f: Mar; n2; Coahuila, Mexico	522
	repro: Feb; TX	10110a
Ammospermophilus leucurus	testes: peak Nov-Dec; captive	5654, 5657
	testes: peak Feb; CA	5654
	mate: early-mid Mar; CA	5661
	birth: mid Mar-Apr; n4 lt; CA	8348
	birth: early-mid Apr; CA	5661
	birth: mid-end Apr; CA	5661
	birth: est May; NM	502
	birth: peak May; CA	4081
	repro: Feb-July; TX	10110a
Ammospermophilus nelsoni	preg f: Feb-Mar; CA	4512a
Callosciurus erythraeus	preg f: Aug; n1; China	182
	preg/lact f: July; n4; n Vietnam	10826
	repro: Mar-May, Oct-Dec; according to hunters; n Vietnam	10827
Callosciurus nigrovittatus	m without sperm: peak Jan-Mar, July-Sept, low Apr-June, Oct-Dec; Malaysia	4406
	preg f: yr rd; Malaysia	4406
Callosciurus notatus	preg f: mid Apr-July; n3/15; n Borneo, Indonesia	2313
	preg f: yr rd, peak Apr-June; Malaysia	4405, 4406
	preg f: yr rd, peak Apr-June; Selangor, Malaysia	7161
Callosciurus prevosti	preg f: end Apr-early Aug; n3/17; n Borneo, Indonesia	2313
Cynomys gunnisoni	mate: Mar-Apr; w central US	1474
	mate: mid Apr-early May; CO	3291
	preg f: Apr-May; n29; CO	145

	preg f: June; n1; CO	8572
	birth: end Apr-early May; AZ	9583
	birth: early May; w central US	1474
	lact f: Apr; n13; CO	145
	lact f: May; n2; AZ	145
Cynomys leucurus	conceive: Mar-Apr; WY	531
	mate: end Mar; CO	10835
	mate: end Mar-early Apr; WY	1871a
	birth: end Apr-early May; WY	10437
	birth: early May; WY	531
	lact f: none after 15 June; WY	531
Cynomys ludovicianus	testes: large Mar, small Apr; MT	5895
	testes: large Dec-Feb, small Mar-Oct; OK	338
	mate: end Jan-mid Feb; KS	10139
	mate: end Feb; CO	10835
	mate: Feb; zoo	10149
	mate: Feb-Mar; SD	10954
	mate: Feb-Mar; SD	4958
	mate: Mar-early Apr; WY	1871a
	mate: Mar-Apr; Great Plains, N America	5916
	preg f: end Mar-early May; Great Plains, N America	5916
	birth: Mar; captive	339
	birth: Mar-Apr; KS	10139
	birth: early Apr; SD, WY	10954
	birth: mid-end Apr; MT	5895
	birth: Apr-May; Great Plains, N America	5916
	birth: May; NM	502
	lact f: June; n2; ND	3703
	pups emerge: May; CO	10835
	wean: May-June; SD	4958
Cynomys mexicanus	preg f: Mar; n1; Coahuila, Mexico	522
	birth: May; n1; captive	8573
Cynomys parvidens	birth: Apr; n1; UT	10809
	birth: Apr; n4; captive	8571
Dremomys lokriah	birth: May-Aug; Nepal	7383
Exilisciurus exilis	preg f: end May-mid Aug; n2/7; n Borneo, Indonesia	2313
Funambulus palmarum	spermatogenesis: yr rd; India	8732
	preg f: Mar-Apr, July-Aug; India	11242
	birth: May, Sept, Jan; India	8732
	birth: yr rd, peak Nov-Apr; ne monsoon; Sri Lanka	8501
Funambulus pennanti	spermatogenesis: Jan-July; n226; India	8996
	testes: large Jan-July, regress Aug-Sept, quiescent Oct-Nov; Delhi, India	8995, 10013
	testes: large Jan-Sept, small Oct-Dec; India	5516
	testes: large Mar, small Sept; India	6588
	mate: Jan-Aug; Delhi, India	9799
	mate: yr rd; Pakistan	105
	preg f: Feb-Sept, peaks Mar, July; n Bangalore, India	8735, 9799
	preg f: Apr-Sept; Pakistan	1764
	preg f: May 40%, Nov 0%; India	6588
	preg f: May-June; n10; Rajahstan, India	8720
	birth: Feb-Sept; w Rajahstan, India	8723
	birth: Mar-Aug; India	8996
Funisciurus anerythrus	preg f: yr rd, peak Aug-Oct; Zaire	8855
	repro: small peak Apr, large peak Aug-Oct; Gabon	2997
Funisciurus carruthersi	lact f: May; n1; e Africa	5760
Funisciurus congicus	birth: Oct, Mar; Namibia	11300
Funisciurus lemniscatus	repro: July-Jan; Gabon	2997
Funisciurus leucogenys	preg f: Feb; n1; Cameroun	2955
Funisciurus pyrrhopus	preg f: end Jan; n1; Budorigo, Uganda	5760
	preg f: Jan-Mar, May; n9; Zaire	8855
	birth: yr rd; captive	6755
	birth: est yr rd; n10; zoo	9522
	repro: July-Jan; Gabon	2997

Glaucomys sabrinus	conceive: est Oct; n1; CA	8928
	birth: Mar-June; w US	5146
	birth: May, occasionally Oct; BC, Canada	2088
	birth: May-June; US	1474
	birth: early June; ID	9385
Glaucomys volans	testes: scrotal Feb, Apr, Aug, Oct; few data; Mexico	1672
	mate: Feb-Mar, July; e US	10205, 10206
	mate: mid Feb-Mar, June-July; WI	5210
	preg f: Mar; n2; MN	495
	preg f: July; n1; KY	11588a
	birth: Feb, Sept; n12; captive	6497a
	birth: end Mar-early May; PA	9164
	birth: early Apr-mid Sept, peak Apr, Aug; MI, MA	7686
	birth: Apr; IL, WI	2979
	birth: Apr-May, Aug-Sept; e US	1474
	birth: end Apr; NY	7235
	birth: May, Aug; MI	1472
	birth: end Aug-mid Jan; FL	8978a
	birth: Sept-Feb; n16; AL	6497a
	birth: Oct; n1; LA	3848
	birth: yr rd, peak Jan-Mar; LA	3853
	lact f: Mar; n1; Mexico	1672
	repro: peaks end summer-early fall, spring; VA	10211
Heliosciurus gambianus	young found: Feb; Karamoja, e Africa	5760
	lact f: Nov; Cameroun	2955
Heliosciurus rufobrachium	testes: large Aug, Feb; n2; Africa	5760
	mate: June-Oct; Malawi, Zambia, e Zimbabwe	10153
	preg f: Dec-Jan; n2; e Africa	185
Heliosciurus ruwenzorii	preg f: Mar; n1; Zaire	8856
Hylopetes alboniger	mate: end Feb-early Mar; Nepal	7383
	birth: mid May-June; Nepal	7383
	repro: Apr-mid June; Nepal	7382
Hylopetes lepidus	preg f: erratic; Malaysia	7687
Iomys horsfieldii	preg f: Mar; n1; Malaysia	7161
Marmota baibacina	mate: Apr; USSR	8050
	mate: spring; USSR	9837
	birth: May; USSR	8050
Marmota bobak	mate: Mar-Apr; Mongolia	10516
	mate: ~Apr; USSR	8050
	mate: spring; USSR	4645
	birth: mid June; USSR	5172
	emergence: mid-end May; USSR	8050
Marmota caligata	mate: Apr-May; nw BC, Canada	2088
	mate: mid May; WA	571
	birth: early-mid June; WA	571
	birth: end spring-early summer; w N America	1474
	juveniles: end July; Skeena R, BC, Canada	10597
Marmota camtschatica	birth: 2 wk after hibernation; USSR	5518
Marmota caudata	mate: immediately after hibernation; Pakistan?	9175
	birth: Apr-early July; USSR	8050
Marmota flaviventris	spermatogenesis: Mar-May; Europe	2075
	mate: Mar; w US	5146
	mate: Apr; 2 wk after thaw; CA	7803
	mate: after hibernation; 2 wk; WY	364
	preg f: Mar-Apr; CA	7803
	preg f: Apr; N America	5029
	preg f: Apr; SD	10954
	birth: Mar-Apr; w US	1474
	birth: early May; BC, Canada	2088
	birth: early June; CO	2708
	birth: June; CO	5328
	lact f: May-June; CA	7803
	lact f: June-July; SD	10954

Marmota marmota	mate: Apr; captive	8770
	mate: Apr-early May; Alps, Europe	12083
	mate: Apr-May; n Austria, n Switzerland	7433
	mate: end Apr; Germany	7605
	birth: Apr-May; Germany	3571
	birth: Apr-May; n Austria, n Switzerland	7433
	birth: May; captive	8770
	birth: May-early June; Alps, Europe	12083
Marmota monax	seminal vesicles: large Feb-Mar, low end Apr; PA	1830
	testes: small Aug-Oct; NY	8934
	mate: early Feb; MO	10981
	mate: mid Feb-mid Mar; PA	1830
	mate: end Feb-Mar; MD	4083
	mate: 1st 3 wk Mar; NY	533
	mate: est Mar-Apr; captive	4270
	mate: Mar-mid Apr; WI	5210
	mate: Mar-Apr; e US	1474
	preg f: mid Feb-Apr; PA	10187
	preg f: Apr; n1; NY	9639a
	birth: Feb; lab	10186
	birth: Mar; lab	2016
	birth: Mar-Apr; lab	12043
	birth: end Mar-Apr; lab	2015, 2016
	birth: Apr; lab	570
	birth: Apr; PA	1830
	birth: Apr-mid May; MD	4083
	birth: Apr-mid May; NY	4270
	birth: end Apr; NY	5029
	birth: end Apr-early May; NY	7235
	lact f: Apr-early June; PA	10187
	emergence: mid May; PA	1830
Marmota sibirica	mate: Apr; USSR	8050
	birth: end May-?; USSR	8050
Paraxerus alexandri	mate: not seasonal; central Africa	187
	preg f: Mar-Apr, July, Sept; n5; Zaire	8855
Paraxerus boehmi	preg f: end Dec; n1; Lake Kivu, Zaire	8856
	preg f: yr rd; Zaire	8855
Paraxerus cepapi	testes: large July, small Apr; Transvaal, S Africa	11296
	preg f: peak Oct-Apr, none May, Sept; Botswana	10154
	birth: June, Nov; few data; Transvaal, S Africa	11298
	birth: Sept-Jan; Transvaal, S Africa	11299
	birth: yr rd; Botswana	11299
	birth: yr rd, peak Oct-Apr; S Africa	10158
	lact f: May; n1; Lake Tanganyika, e Africa	186
Paraxerus lucifer	testes: large Aug-Sept; e Africa	5760
	mate: may peak drier half of yr; e Africa	5760
	preg f: Mar-Apr; n1; Tanzania	186
Paraxerus ochraceus	preg f: July, Nov; e Africa	5760
	preg f: Nov-Dec; n2; Kenya	4925
	yg found: Mar, June-July, Sept; e Africa	5760
Paraxerus palliatus	birth: Aug-Mar; S Africa	10158
	lact f: Mar; n1; s Kenya	5760
	lact f: Oct; n1; Mafia I, Tanzania	5760
	juveniles: Apr; central Africa	187
Paraxerus poensis	lact f: Sept; Gabon	2997
Petaurista magnificus	preg f: Nov; n1; Nepal	7382
	birth: Feb-Mar; Nepal	7383
Petaurista petaurista	preg f: Feb; n2; Malaysia	7161
	preg f: Apr; n1; Sri Lanka	8501
	birth: June; n1; Sri Lanka	8501
Protoxerus stangeri	preg f: none dry season; Cameroun	2955
	lact f: Jan-Feb, Nov; Gabon	2997
Pteromys volans	mate: summer; Eur USSR	4645
	birth: Feb-June; USSR	8053

Pteromyscus pulverulentus	birth: nearly yr rd; Malaysia	7687
Ratufa affinis	preg f: June; n2; n Borneo, Indonesia	2313
	lact f: Aug; n2; n Borneo, Indonesia	2313
Ratufa bicolor	birth: Apr-May, Aug-Sept; Vietnam	10828
Sciurus aberti	mate: Mar-Apr; sw US	1474
	mate: Apr-May; CO	3166
	birth: Apr-May; WY	1871a
	birth: Apr-May; sw US	1474
	birth: May-Aug; sw US	7145
	birth: mid June-mid July; AZ	5618a
	lact f: July; n1; Durango, Mexico	523a
	repro: mid Apr-June; AZ	10378
Sciurus alleni	preg f: Mar; n1; Coahuila, Mexico	522
	lact f: Apr; n2; Coahuila, Mexico	522
Sciurus anomalus	juveniles: end Apr-early May; Lebanon	6419
Sciurus aureogaster	preg f: July-Aug; n2; Mexico	2341
Sciurus colliaei	preg f: June; n2; Sinaloa, Mexico	5409
	preg f: June; n1; Durango, Mexico	5405b
	lact f: June; n1; Durango, Mexico	5405b
Sciurus carolinensis	anestrus: Aug-Dec; England, UK	2426
	spermatogenesis: Oct-July; ONT, Canada	10015
	testes: scrotal Dec-Feb, July; KE	1241
	testes: large Dec-May, small June-Mar; s England, UK	11516
	testes: large Jan-Aug, small Sept-Nov; BC, Canada	9190
	testes: functional Feb-Mar, June-July; England, UK	164
	testes: large May-Jan, small Feb-Apr; S Africa	7297
	testes: regressed June-Mar; s England, UK	10629
	testes: regressed July-Aug; NC	589
	testes: regressed July-Aug; IL	1361
	testes: inguinal ~50% peak Mar, July, Dec; KY	1241
	mating condition: Aug; n4/15; NY	9214
	mate: Jan-Feb, July; ne US	1474
	mate: peaks Jan-Mar, June-July; UK	9943
	mate: Feb, May-June; KS	8188
	mate: Jan-mid July; IL	1361
	mate: peaks Feb, June, none Aug-Nov; WI, IN	7570
	mate: Mar-June; BC, Canada	2088
	mate: June, Dec; se US	1474
	mate: peak July, Dec; S Africa	10158
	mate: Nov-Feb; S Africa	7297
	mate: mid Dec-mid Jan; NC	11478
	mate: end Dec-Sept, peaks Jan, end May-early June; OH	7944
	mate: est peaks Mar-Apr, June-July; BC, Canada	9190
	preg f: yr rd, peak Jan-Feb, Aug; e TX	3898
	preg f: Jan-July; England, UK	2426
	preg f: Mar 46%, Apr 54%, Aug 75%; KY	1241
	preg f: Mar-Apr, June; n4; KS	8188
	preg f: Apr-June, Sept-Oct; IL	1361
	preg f: June; n1; FL	7498a
	birth: peaks Feb-Mar, July; OH	7944
	birth: peaks Feb-Mar, July-Aug; VA	5808
	birth: mid Feb-Mar; IL	1361
	birth: Mar, July-Aug; NC	11478
	birth: end Mar-early Apr, end July-early Aug; WI	5210
	birth: May, July; BC, Canada	2088
	birth: May, Aug; QUE, Canada	11956
	birth: June, Oct; NY	7235
	birth: June-Jan, peak Oct-Jan; S Africa	2323
	birth: Aug-Oct, Dec-Mar; S Africa	7297
	lact f: 67% Sept; KY	1241
	lact f: Oct; n1; FL	7498a
Sciurus deppei	preg f: Dec; n1; Campeche, Mexcio	5423b
	birth: Feb; n1 hybrid; Yucatan, Mexico	3633
	birth: July; San Luis Potosí, Mexico	2228a

Sciurus granatensis	mate: June; Panama	3024
	mate: Dec-July; Panama	3801
	preg f: Mar, Sept-Oct; Venezuela	7940
	preg f: June; Panama	3024
Sciurus griseus	mate: Jan-May; CA	5143
	preg f: Mar; n1; CA	11228a
	birth: Feb-June; w US	1474
Sciurus niger	testes: functional yr rd; IN	5801
	testes: regress July-Aug; IL	1361
	estrus: Jan, June; WY	1871a
	mate: Jan-Feb, end May-June; WI	5210
	mate: Jan-Feb, June-July; n N America; 1 mo earlier s N America	1474
	mate: peaks Feb-Mar, July-Aug; DE, MD	10809
	mate: peaks Feb, June, none Aug-Nov; WI	7570
	mate: Dec-June; KS	8188
	mate: mid Dec-June; IL	1361
	preg f: Jan-Mar, July; MI	170
	preg f: Feb-Mar, June; n5; KS	8188
	preg f: Feb-June, Aug-Oct; IL	1361
	birth: peak mid Mar, July; WY	1871a
	preg f: May-Aug, Dec-Feb; FL	7501
	preg f: mid June; Coahuila, Mexico	522
	birth: Jan-Feb, June-Aug; FL	7501
	birth: Feb-Apr, Aug-Sept; n 1 mo earlier s N America	1474
	birth: mid Feb-Mar; IL	1361
	birth: Dec-Sept, peaks Dec-Jan, May-June; IL	7943
	lact f: end Mar-early Apr; Coahuila, Mexico	522
Sciurus nayaritensis	lact f: July; n2; Durango, Mexico	523a
Sciurus richmondi	mate: low Feb-Sept; Nicaragua	5417
	preg f: Feb, Apr, June; Nicaragua	5421
	lact f: June, Sept; Nicaragua	5421
Sciurus vulgaris	spermatogenesis: Dec-July; quiescent Aug-Nov; s France	2469
	testes: large Dec-Jan, small Aug; England, UK	9343
	testes: large Jan; Switzerland	4561
	mate: Jan-Feb, ~2 wk; Eur USSR	4645
	mate: Feb-Mar; USSR	8053
	mate: Mar; n1; W Germany	2923
	mate: Dec-July; s France	2468
	mate: Dec-July; Germany	3571
	mate: spring-summer; Germany	7433
	ovulation: peaks Jan-Mar, mid May-June, low mid Mar-Apr, mid Aug-mid Sept; USSR	6526
	preg f: Feb-June; s France	2469, 2470
	preg f: Dec-July; s France	2468
	preg f: Dec-July; n15; England, UK	9343
	birth: ~Jan; n1; zoo	3271
	birth: Jan-Aug; Germany	3571
	birth: Feb-Aug; Denmark	2435
	birth: Feb-June/July; s France	2468
	birth: May; n1; captive	12139
	lact f: Mar-mid Sept, peak July 38.4%; s France	2469
Sciurus yucatanensis	preg f: July; n1; Quintana Roo, Mexico	5423b
Spermophilopsis leptodactylus	testes: large Mar; n1; USSR	8053
	mate: Feb-Mar; Turkmenia	6273
	mate: mid Feb-; several wk; USSR	8053
	mate: mid Apr, June; Turkmenia	9774
	birth: Apr & later; Turkmenia	6273
	birth: end May; Turkmenia	9774
	repro: Feb-June; Karakumy Desert, Turkmenia	9498, 9499
	repro: Apr; Turkmenia	10568
Spermophilus armatus	mate: Mar-Apr; UT	2983, 5893
	preg f: mid Mar-early May, peak early Apr; ID	11646a
	birth: Apr; w US	1474

	birth: May-?; UT	2983
Spermophilus beecheyi	spermatogenesis: Nov-Apr; CA	6489
	testes: large Feb-Apr; CA	10862
	mate: Jan-Mar; CA	3281
	mate: end Jan-mid Mar; CA	6489
	mate: Feb-Apr; CA	3114
	mate: Feb-mid Apr; low elevations, CA	5030
	mate: Feb-June; high elevations, CA	5030
	mate: end Feb-early Mar; CA	10806
	preg f: Apr-May; OR	2896
	preg f: Mar; n1; CA	11228a
	birth: ?-Mar; CA	6489
	birth: Mar-June; CA	10862
	birth: peak Mar-Apr; w US	1474
Spermophilus beldingi	testes: large Mar-May; CA	7079, 7082
	testes: large Apr; Big Bend, CA	7547
	testes: large May; Tioga Pass, CA	7547
	mate: Apr-May; Big Bend, CA	7547
	mate: May-June; Tioga Pass, CA	7547
	mate: May; MI	1480
	mate: end May-early June; CA	4317, 9900
	birth: Apr; CA	5030
Spermophilus citellus	birth: Apr-May; Europe	7433
	birth: Apr-May; se Europe	3571
	birth: May; Russia	1552
Spermophilus columbianus	mate: Apr-May; MT	6798
	mate: f: 3.8 d after emergence; se0.20; r2-8	7632
	preg f: Mar-Apr; w Canada	9872
	preg/lact f: mid Apr-June; sw ALB, Canada	539
	birth: end Mar-early Apr; w N America	1474
	birth: May-June; ALB, Canada	11959
	wean: mid May; w Canada	9872, 9875
	yg emerge: June; 2 wk; ALB, Canada	11469
Spermophilus dauricus	preg f: May; n1; Mongolia	182
	birth: May; USSR	8050
	lact f: end May; USSR	8050
	yg found: July; Mongolia	182
Spermophilus elegans	mate: Apr; CO	3143
	preg f: Apr-early May; WY	1865
Spermophilus erythrogenys	mate: mid Apr-early May; USSR	8050
	birth: end May; USSR	8050
Spermophilus franklinii	testes: scrotal May-July; MAN, Canada	5185
	mate: Feb-mid Apr; low elevations, MAN Canada, ND, MN	5030
	mate: Feb-June; high elevations, MAN Canada, ND, MN	5030
	preg f: May; n1; MN	495
	preg f: mid May-June; MAN, Canada	5185
	birth: May-June; MAN Canada, ND, MN	5030
	birth: end May-mid June; WI	5210
Spermophilus fulvus	mate: Feb; Turkmenia	5545
	mate: before Mar; Siberia, e Europe	9773
	mate: end Mar-early Apr; Europe, USSR	4645
	mate: after hibernation; USSR	8050
	birth: Apr-May; Russia	10845
	birth: Apr-May; USSR	8050
	birth: end Apr-early May; Eur USSR	4645
	yg emerge: Apr; Turkmenia, USSR	5545
Spermophilus lateralis	spermatogenesis: Apr-May; Lassen Co, CA	601
	testes: increase Dec-Apr; Lassen Co, CA	601
	testes: large Mar, small Aug; WY	10055
	testes: large Mar-Apr; CA	7080
	testes: large Mar-June; Lassen Co, CA	7082
	mate: end Apr-early May; WY	10055
	CL present: Apr; Lassen Co, CA	7082
	preg f: Apr; n2; CA	7593a

	preg f: May; n1; AZ	145
	preg f: May-Aug; n9; CA	4081
	birth: May; BC, Canada	2088
	birth: May; captive	602
	birth: May-early June; n14; CA	8348
	birth: end May-early June; WY	10055
	birth: early-mid June; lab	12115
	birth: June-July; AZ	7145
	birth: before July; CO	279
	birth: early spring; w US	1474
	yg emerge: end June; WY	10055
Spermophilus madrensis	preg f: May; n1; Chihuahua, Mexico	866
	preg f: June; n4; Chihuahua, Mexico	866
	lact f: July; n7; Chihuahua, Mexico	866
Spermophilus major	mate: mid May?; USSR	8050
	emergence: mid-end June; USSR	8050
Spermophilus mexicanus	mate: Apr; TX, Mexico	1474
	preg f: Apr; n1; TX	2901
	preg f: June; n1; Coahuila, Mexico	522
	birth: May; n1; TX, captive	2901
	birth: May; TX, Mexico	1474
Spermophilus mohavensis	birth: Mar; CA	1474
	birth: Apr-mid May; n4; CA	8348
Spermophilus parryii	mate: Apr, 3 wk; YUK, Canada	7090
	birth: mid May-mid June; nw BC, Canada	2088
	birth: June-July; n N America	1474
Spermophilus perotensis	preg f: July; n2; Mexico	2341
Spermophilus pygmaeus	mate: mid Mar; Russia	11224
	mate: mid Mar; USSR	9453
	birth: early-mid May; USSR	8050
Spermophilus richardsonii	testes: large Mar, small May; WY	1865
	conceive: mean date: Apr 12; SAS, Canada	9890
	conceive: Apr-early May; ALB, Canada	7826
	birth: Apr-May; SAS, Canada	2872
	birth: May; nw US, sw Canada	1474
	birth: 27 d after hibernation; ALB, Canada	7285
	lact f: May-early June; SAS, Canada	7279
	emergence: June; WY	1865
Spermophilus saturatus	testes: large Mar; captive	600
	mate: end Apr; WA	5658
	birth: end May; n22 lt; WA	5658
Spermophilus spilosoma	testes: scrotal Mar-Aug; TX	9723
	preg f: Apr; n1; NM	145
	preg f: May-July; WY	1871a
	preg f: June; n2; NE	5405b,
	preg f: June; n2; NM	502
	preg f: June-July; few data; Coahuila, Mexico	522
	preg/lact f: end Apr-Sept; TX	9723
	birth: July; San Luis Potosí, Mexico	2228a
	lact f: June; n1; NM	502
Spermophilus suslicus	mate: after hibernation; Eur USSR	4645
	birth: end May-early June; USSR	8050
Spermophilus tereticaudus	fertile m: Jan-Apr; s AZ	7789
	birth: peak mid Mar-mid Apr; s AZ	7789
	birth: end Mar-mid Apr; n10 lt; CA	8348
	birth: Apr, possibly other mo; sw US	1474
Spermophilus townsendii	mate: Jan; WA	9871
	preg f: Mar; NV	139
	birth: Apr; NV	139
	birth: Mar; w central US	1474
	yg emerge: Mar; WA	9871

Spermophilus tridecemlineatus	sperm: Dec-June; IL	7493, 11580, 11582, 11583
	testes: scrotal Jan-July; lab	5345
	testes: large: Dec-May, small July; IL	11579, 11580, 11583
	vag perf: Jan-July, peak Mar-Apr; lab	5345
	ovulate: Mar-Apr; MN	2729
	mate: Mar-Apr; MN?	2729
	mate: Mar-June; TX	7023
	mate: Apr; 2 wk; MI	9681
	mate: Apr; lab	1286
	mate: Apr; n1; lab	5346
	mate: Apr-May; MAN, Canada	2140
	mate: Apr-May; WI?	3432
	mate: peak end Apr; NB	11379
	preg f: Apr-May; n9; MN	495
	preg f: Apr-May; n6; NE	5405b,
	preg f: May; n1; KS	1812a
	birth: est Apr-May; WI	9249
	birth: May; n1; lab	5346
	birth: May; lab	1286
	birth: May; KS	5343
	birth: May-early June; MAN, Canada	2140
	birth: May-June; N America	5030
	birth: May-June; IL	7493, 11581
	birth: May-June; IL, WI	2060
	birth: mid-end May; KS	1812a
	birth: end May; WI?	3432
	lact f: June; n1; ND	3703
	lact f: June; NE	5405b,
	emergence: mid June; MAN, Canada	2140
	repro: May; ne CO	3294
Spermophilus undulatus	sperm: May; AK	7381
	mate: early-mid May; AK	4850
	mate: May; AK	6992
	birth: early May-early June; USSR	8050
	birth: early-mid June; AK	4850
	birth: June; AK	6992
	wean: end June-early July; USSR	8050
Spermophilus variegatus	mate: Mar-Apr; TX	5347
	mate: Mar-July; N America	1474
	preg f: Apr; n1; Coahuila, Mexico	522
	preg f: May; n1; CO	1649a
	preg f: June; n2; NM	502
	birth: Apr-Aug; N America	1474
	birth: peak May-June; NM	502
Spermophilus washingtoni	birth: Feb, Mar; Columbia R Basin, WA	5030
	birth: Mar; WA, OR	1474
	emergence: Apr; WA, OR	1474
Spermophilus xanthoprymnus	lact f: June; USSR	8050
	emergence: end June; USSR	8050
Sundasciurus hippurus	preg f: none end Apr-Aug; n15; n Borneo, Indonesia	2313
Sundasciurus lowii	preg f: end Apr-Aug; n3/16; n Borneo, Indonesia	2313
	lact f: Aug; n1; n Borneo, Indonesia	2313
Syntheosciurus brochus	lact f: Apr; n1; Panama	11585
Tamias spp	birth: early June; Yellowstone, MT	6918
Tamias amoenus	testes: large Apr, small June; SAS, Canada	9889
	mate: Apr; WA	1295
	mate: Apr-May; SAS, Canada	9889
	preg f: end Apr-mid May; CA	10762
	preg f: May; n1; ID	11646a
	preg f: May-June; SAS, Canada	9889
	birth: ~early May; WA?	5662
	birth: May; nw US sw Canada	1474
	birth: May; WA	1295
	birth: May or later; sw BC, Canada	2088

	birth: May-June; SAS, Canada	9889
	birth: May-June; WY	1871a
	lact f: June; n1; ID	11646a
	weanling emergence: June; nw US, sw Canada	1474
Tamias bulleri	preg f: June-July; n2; Durango, Mexico	523a
	lact f: June-July; n≥6; Durango, Mexico	523a
	lact f: July-Aug; n2; Coahuila, Mexico	522
Tamias cinereicollis	birth: peak June; AZ	8549
Tamias dorsalis	m fertile: Jan-June; AZ	2805
	mate: May-early June; AZ	2805
	preg f: none July; n23; UT	2340
	birth: May-Sept; NM	502
	birth: end June-early July; AZ	2805
	lact f: July; Sonora, Mexico	1471a
	wean: Aug; AZ	2805
Tamias minimus	testes: large Mar, small mid summer, scrotal Mar-June; WY	10053
	testes: large May, small mid June; no Apr data; Canada	9889
	mate: Apr; WI	5210
	mate: Apr-Aug; MI	6796
	mate: May; no Apr data; Canada	9889
	mate: after hibernation; CO	11230
	preg f: Apr-June; WY	10053
	preg f: May; n2; SD	261a
	preg f: May-June; no Apr data; Canada	9889
	preg f; May-mid July; CO	11230
	preg f: mid June-early July; n9/20; CO	279
	preg/lact f: June; SD	10954
	birth: May; BC, Canada	2088
	birth: May-early June; WY	1871a
	birth: May-June; no Apr data; Canada	9889
	lact f: Apr; NV	6488
	lact f: May; n1; SD	261a
	lact f: May-Aug; WY	10053
	lact f: June-July; n2; se MT	6166
	lact f: June-July; n5; ND	3703
	lact f: ?-Aug; MI	6796
Tamias quadrimaculatus	preg f: end May-mid June; CA	10762
Tamias quadrivittatus	mate: est end Feb-Mar; UT	11385
	mate: early-mid June; Mt Taylor Valencia Co, NM	4962
	birth: May; n1; NM	502
Tamias ruficaudus	mate: end Apr-May; MT	726
	birth: end May-June; MT	726
Tamias sibiricus	estrus Apr-May; Amur-Zeya R, Siberia, Russia	10177
	mate: Apr; USSR	10422
	mate: Apr-early May; USSR	8053
	mate: Apr-May; zoo	2654
	mate: peak Apr-May; Germany	3662
	mate: end Apr-early May; Amur-Zeya R, Siberia, Russia	10177
	mate: early spring; Europe, w USSR	4645
	birth: end May; Amur-Zeya R, Siberia, Russia	10177
	birth: end May-early June; 20-30 d later mts; Siberia, Russia	9953
	birth: May-early June; USSR	10422
	birth: May-early June; USSR	8053
	emergence: early July; USSR	8053
Tamias speciosus	repro: seasonal; CA	10762
Tamias striatus	spermatogenesis: peak Mar; WI	8224
	testes: large mid Feb-early Oct; NY	12014
	conceives: mid Mar-early Apr, mid June-mid July; NY	12014
	mate: Feb-Apr, June-Aug; NY	2981
	mate: mid Feb-early Mar, July; OH	11989
	mate: mid Mar-?; ONT, Canada	10102
	mate: mid Mar-mid Apr, July; WI	5210
	mate: Apr, July-Aug; e US	1474
	mate: Apr-Sept; MI	6796

	preg f: Mar; n1; MN	495
	preg f: Apr-Sept; MI	1472
	preg f: Mar, June; NY	175
	preg f: Mar-Oct; OH	2022
	preg f: July; n1; NY	9639a
	birth: Mar-May, July-Sept; NY	2981
	birth: Apr; July; NY	175
	birth: mid Apr-mid May, mid July-mid Aug; NY	12014
	birth: end Apr-early May, Aug; s QUE, Canada	11956
	birth: peaks May, Aug; OH	2022
	birth: May, Aug-Sept; e US	1474
	lact f: end May-Aug; few data; MI	6796
Tamias townsendii	mate: Apr; BC, Canada	2088
	mate: Apr; w US	1474
	mate: end Apr-early May; WA	5658
	birth: May; BC, Canada	2088
	birth: end May-early June; n11 lt; WA	5658
Tamiasciurus douglasii	testes: scrotal Mar-July; n50%; BC, Canada	10543
	mate: early spring; Mexico	6387
	mate: Mar-July; CA	5922
	birth: June, Oct; w US	1474
	lact f: May-Sept; n50%; BC, Canada	10543
Tamiasciurus hudsonicus	epididymal sperm: end Jan-Sept; NY	4274
	testes: large Apr-early Sept; QUE, Canada	11956
	testes: inactive Sept-Nov; NY	6297
	mate: mid Jan-Sept; WI	5210
	mate: Feb-Mar, June-July; n N America	1474
	mate: mid Feb-Sept; NY	4274
	mate: peaks Mar-Apr, June-July; QUE, Canada	6158
	mate: Mar-May; n SAS, Canada	2324
	mate: Apr-June; CO	2671
	mate: end Apr; BC, Canada	2088
	preg f: Feb-Sept; NY	6297
	preg f: Apr; n1; MN	495
	birth: Mar-May; WY	1871a
	birth: Mar-July; SD	10954
	birth: ~Apr 1; NY	7235
	birth: Apr-May, Aug-Sept; n N America	1474
	birth: early May; Canada	5882
	birth: May-early June; BC, Canada	2088
	lact f: May-July; n5/7; se MT	6166
Tamiops macclellandi	preg f: Apr; n1; Malaysia	7161
Tamiops swinhoei	repro: spring-fall; n Vietnam	10825
Trogopterus xanthipes	birth: Mar-mid Apr; captive	11448
Xerus erythropus	yg found: peak Aug-Oct; s Uganda	5760
	repro: yr rd; Senegal	1149
Xerus inauris	preg f: July-Nov; Transvaal, S Africa	12134
	preg f: Aug, Oct; n2; sw Transvaal, S Africa	8965
	preg f: yr rd; Transvaal, S Africa	4697
	preg f: yr rd, none Apr, Sept; few data; Botswana	10154
	birth: Aug-Nov; Transvaal, S Africa	12134
	repro: yr rd; S Africa	10158
Xerus rutilus	preg f: Aug; n1; Ethiopia	8017
	preg f: Sept; n1; Kenya	8017
	preg f: Sept, Nov; n2; e Africa	5760
	juveniles: yr rd; e Africa	5760

Order Rodentia, Family Geomyidae

Pocket gophers (*Geomys* 100-450 g; *Orthogeomys* 500-800 g; *Pappogeomys* 110-900 g; *Thomomys* 45-545 g; *Zygogeomys* 450 g) are solitary, subterranean herbivores from the United States south to northern South America. We found no reproductive information on *Zygogeomys*. Males are larger than females (*Geomys* 5670, 11728, 11729; *Thomomys* 1474, 3740, 5021, 10100). The hemochorial, discoid placenta has a partially interstitial, antimesometrial implantation into a bipartite uterus (7563, 7569, 7574, 10310, 10924). A postpartum estrus may occur (7488, cf 1367) after the birth of naked and blind neonates whose eyes open at 22-23 d (*Geomys* 10534), or 26 d-5 wk (*Thomomys* 258, 2138, 4743, 8798, 9575, 9646, 11671). Females have 4-14 mammae (499, 502, 1474, 1475, 2344, 4745). Males have a baculum, seminal vesicles, coagulating glands, inguinal testes, and scrotal epididymides (1473, 4157a, 5669, 10924). Details of anatomy (3128, 5739), endocrinology (4157), and conception (7574) are available.

NEONATAL MASS		
Geomys bursarius	4.3 g; n1	11891
	5.1 g; r4.9-5.4; n4	11229
	5.125 g; n4	10534
	6-7 g	5210
Pappogeomys castanops	6.07 g; r3.8-7.9; n3	4565
Thomomys bottae	2-3 g	9646
	4.1 g; r4-4.2; n2	4743
Thomomys talpoides	2.77 g; se0.14; n6 yg, 1 lt	258
	3.58 g; se0.09; n20 yg, 4 lt	258
Thomomys bulbivorus	6.12 g	11671
NEONATAL SIZE		
Geomys bursarius	CR: 38 mm; n1	11891
	CR: ~40 mm; n4	11229
	TL: 47 mm; n4	10534
Pappogeomys castanops	40.97 mm; r38-43; n3	4565
Thomomys bulbivorus	50 mm	11671
WEANING MASS		
Geomys arenarius	CR: max 30 mm	11727
Geomys bursarius	22.5 g; nibble food	11891
Pappogeomys spp	m: 90 g; lowest wgt 1st capture	10160
	f: 110 g; lowest wgt 1st capture	10160
Thomomys bulbivorus	86.0 g	11671
WEANING SIZE		
Thomomys bulbivorus	16.4 cm	11671
LITTER SIZE		
Geomys arenarius	4.7 emb; r4-6; n3 f	5424a
Geomys bursarius	1.7 ps; se0.23; n34 f	11719a
	mode 2-3, r1-5	495
	2.0 yg; r1-3; n2 f	10534
	2.3 emb; se0.35; n16 f	11719a
	2.36 emb, yg; r1-3; n25 f	3042
	2.5 emb; r2-4; n10 f	5670
	2.5 cl; se0.30; n61 f	11719a
	2.52 ps; r1-4; n23 f	11890
	2.66 yg; r1-4; n6 nests	11890
	2.70 emb; r1-5; n31 f	11890
	3-4; r2-6	5210
	3-5; r1-8	1474
	3 yg ea; n3 nests	11891
	3.40 emb; ±0.13; mode 3-4 68%; r1-8; n73 f	11229
	3.46 ps; ±0.03; mode 3-4 67%; n155	11229
	3.6 emb; r2-6; n13 f	5405b,
	3.9 emb; r3-6; n20 f	11957a
	mode 4-5; r3-6	2060
	4 yg; n1 lt	11229
	4.0 ps; r2-8; n19 f	11957a
	4.15 emb; mode 4; r1-6; n34 f	9574a
Geomys personatus	3.18 emb; r2-4; n11 f	5670

Geomys pinetis	1-3	1474
	1.52 emb; ±0.11; r1-2; n19 f	11816
	1.73 ps; ±0.09; r1-5; n60 f	11816
	1.74 emb; sd0.51; mode 2; r1-3; n64 f	1367
	2.10 cl; sd0.54; n64 f	1367
Orthogeomys hispidus	2; n1 lt	4228
	2 emb; n1 f	2341
Pappogeomys bulleri	1 emb; n1 f	3702
Pappogeomys castanops	1-3	1474
	1.8 emb; r1-3	522
	2.08 emb; r1-4; n25 f	2305
	3.8 emb, ps, yg; r2-5; n22 f	4565
Pappogeomys merriami	2 emb; n1 f	2341
Pappogeomys tylorhinus	2.0 emb; r1-3; n4 lt	2050
Thomomys bottae	2.5; r1-4; n2	9646
	3; r2-4; n4 lt	2700
	3.2 ps; 6 lt	552
	3.33 emb; r2-4; n3 f	522
	3.8 emb; r1-7	145
	3.91 emb; n>50 f	9580
	4.33 lg viable emb; se0.47; n21 f	6542
	4.5 ps; r1-9; n59 f	6294
	4.6 yg; se0.11; n36? lt	5021
	4.75 med viable emb; se0.34; n40 f	6542
	4.92 emb; se0.15; r1-9; n110 f	7329
	5.53 sm viable emb; se0.24; n70 f	6542
	5.74 emb; r1-10; n74 f	6294
	5.85 emb; n>50 f	9580
	6 emb; n1 f	502
	6.223 emb; mode 6; r3-12 (16?); n220 f	1074
	6.3 emb; n10 f	9580
	8 emb; n1 f	7488
Thomomys bulbivorus	3-9	9575
	4-9 yg	11671, 11672
	4.2 emb; n8 f	9580
	4.68 ps; se0.17; mode 5; r2-8; n59 f, 10 locations	11280
	5 yg ea; n3 lt	11671
	8.33 emb; r8-9; n3 f	11671
Thomomys clusius	mode 6-7	499
Thomomys monticola	mode 3-4	5145
Thomomys talpoides	2 emb; n1 f	261a
	2.89; r1-5; n9 f	2138
	3-9	9575
	3.3 emb, ps; n400? f	4336
	3.65 emb, ps; n670 f	4336
	3.67 emb; r2-5; n3 f	279
	4 emb; n1 f	4962
	4 emb; n1 f	4745
	4 emb; n1 f	6488
	4 yg; r3-10	1871a
	4.0 emb; n2 f	10954
	4.0 emb; se0.53; r2-7; n12 f	11230
	r of x 4.3-5.2 emb, ps; r of se 0.2-0.4	4337
	4.4 emb, ps; se0.13	10924
	4.4 emb, ps; se0.21; n33	4336
	4.7 emb, ps; se0.24; n33	4336
	4.8 emb, ps	11830
	4.9 emb, ps; se0.14; n12	4336
	4.9 large ps; 5.6 all ps; r3-8; n12 f	260
	5 emb; n>50 f	9580
	5.1 ps; se0.16; r2-8; n66 f	11230
	5.2; r5-6; n5 lt	258
	5.5 emb; r5-6; n2 f	10219
	6 ps; r5-7; n3 f	3703

	6.32 emb; r4-9; n25 f	11673
	6.4 emb, ps; se0.18; n86 lt	4336
	mode 6-7	499
	9.7 ps; r6-13; n3 f	6166
Thomomys townsendii	5 emb ea; n2 f	11646a
	5.7 emb; r5-6; n3 f	11646a
	6.83; mode 7; r3-10; n12 f	4983
	7.1 emb; r5-9; n7 f	11646a
Thomomys umbrinus	2 emb; n1 f	522
	2 emb; n1 f	2341
	2-3	499
SEXUAL MATURITY		
Geomys bursarius	f: loss of pubic symphysis: 100-127 g	11706
	m: testes scrotal: 140 g	11706
	f/m: 3 mo	11890
Geomys pinetis	f/m: 6 mo	11816
Pappogeomys spp	f: 1st birth: min 1st yr	10160
Thomomys bottae	f: 1st birth: 3? mo	5021
	f: season of birth	2232
	m: yr after birth	2232
	f: 1st mate: min 1st yr	2700
Thomomys monticola	f: 1st mate: yr after birth	5145
	f/m: yr after birth	5144
Thomomys talpoides	f: 1st mate: yearling	11230
CYCLE LENGTH		
Geomys bursarius	estrus: 29.5 h	2500
	8 d; r7-9; n3	2500
GESTATION		
Geomys bursarius	18 d; n1; capture to birth	5334
	18-19 d	1474
	min 51 d; n1; capture to birth	10534
Thomomys bottae	19 d; n2	9646
	est 30 d	5021
Thomomys talpoides	~28 d	2088
LACTATION		
Geomys bursarius	food in cheek pouches: 32 d	10534
	leave den: 5-6 wk	11706
Thomomys talpoides	solid food: 17 d	258
Thomomys bulbivorus	solid food: 3 wk	
	wean: 6 wk	11671
INTERLITTER INTERVAL		
Thomomys talpoides	1 yr	11230
LITTERS/YEAR		
Geomys bursarius	1	3042
	1	11229
	1; north; 2+; south	1474
	est 2; 1.31; est 1.31-1.70	11890
	perhaps 2	11706
Geomys pinetis	min 2	1474
	min 2	11816
Pappogeomys castanops	max 4	2305
Thomomys bottae	est 1	502
	1; 2 occasional	5021
	2.0	7329
Thomomys bulbivorus	mode 1	9575
Thomomys monticola	1	5144, 5145
Thomomys talpoides	mode 1	9575
	1	502
	1	4336
	1	11230
	1	11673
	1-2	1474
Thomomys townsendii	2	1474

SEASONALITY

Geomys arenarius	preg f: June, Aug; n3; TX	5424a
Geomys bursarius	fertile m: Jan-June; CO	11229
	mate: Jan/Feb-July; TX	3042
	mate: end Mar-May; WI	5210
	preg f: Jan-Mar, June, none Apr-May; Bee Co, TX	5670
	preg f: Jan-May; KS	9574a
	preg f: Feb-Aug, TX	11890
	preg f: peak end Mar-Apr; TX	11229
	preg f: Apr-May; MN	495
	preg f: Apr-July; n13; NE	$5405b_1$
	preg f: end Apr-July; MAN, Canada	11957a
	birth: Feb-Aug; s US	1474
	birth: Apr; IL, WI	2060
	birth: peak Apr-May; CO	11229
	birth: Apr-July; n US	1474
	lact f: Jan, June-July; few data; TX	11706
	lact f: Apr, June; NE	$5405b_1$
	lact f: Oct; n2; KS	1812a
	repro f: peak Jan-Apr; TX	11719a
Geomys personatus	preg f: Dec-Mar, May; Bee C, TX	5670
Geomys pinetis	epididymal sperm: low Oct-Jan; FL	11816
	epididymal sperm: yr rd; FL	1367
	mate: est yr rd; FL	1367
	repro: yr rd; FL	11816
Orthogeomys hispidus	preg f: July; n1; Mexico	2341
	lact f: May; n1; San Luis Potosí, Mexico	2228a
	repro: est seasonal; Veracruz, Mexico	4228
Pappogeomys castanops	mate: June-Aug, Dec-Mar; Coahuila, Mexico	522
	preg f: June-Aug; TX	2305
	preg f: yr rd; TX	2305
Pappogeomys merriami	preg f: Aug; n1; Mexico	2341
Pappogeomys tylorhinus	sperm: peak Sept-Oct; Mexico	2050
Thomomys spp	preg f: June; n high altitude; spring-summer, s N Am	499
Thomomys bottae	uteri: large Apr, Aug-Sept; Mesa Verde, CO	2700
	mate: Mar-Sept; Mesa Verde, CO	2700
	mate: yr rd, low fall; CA	7329
	preg f: Jan-mid Apr; Sierra Nevada, San Joaquin Valley, CA	5021
	preg f: Feb-Mar; NM	145
	preg f: peaks Mar-Apr, Oct-Nov; CA	1074
	preg f: end Apr-May; Mesa Verde, CO	2700
	preg f May; n1; NM	502
	preg f: June-July, Dec; n3; Coahuila, Mexico	522
	birth: Feb-mid Apr; Sierra Nevada, San Joaquin Valley, CA	5021
	birth: Oct-June; w US	1474
	lact f: May, not Aug-Sept; Mesa Verde, CO	2700
	juvenile caught: May-Aug; n AZ	552
	repro m: Feb-July; n AZ	552
	repro f: Mar-June; n AZ	552
Thomomys bulbivorus	mate: Apr-June; Willamette Valley, OR	9575
	birth: Apr-July; Willamette Valley, OR	11671, 11672
Thomomys monticola	birth: July-Aug; CA	5145
Thomomys talpoides	testes: large Mar-May; MT	10924
	fertile m: July-Sept; Grand Co, CO	11230
	mate: Mar-July; OR	11673
	mate: Apr-June; sw WA	9575
	mate: May-June; CO	11230
	preg f: end Mar-mid June; CO	4336
	preg f: Apr-June; MT	10924
	preg f: mid May-June; WY	11830
	preg f: May-July; WY	1871a
	preg f: June; n1; SD	261a
	preg f: June; n1; Mirabel Spring, Valencia Co, NM	4962
	preg f: June-July; Grand Co, CO	11230

	preg f: July; n2; Black Hills, SD, WY	10954
	birth: Apr; BC, Canada	2088
	birth: mid May-mid June; MT	10924
Thomomys townsendii	preg f: Apr-June; n12; ID	11646a
	birth: Mar-Apr; s ID, n NV	1474
Thomomys umbrinus	preg f: Jan; n1; Coahuila, Mexico	522
	preg f: July; n1; Mexico	2341

Order Rodentia, Family Heteromyidae

Pocket mice and kangaroo rats (*Dipodomys* 30-180 g; *Heteromys* 35-85 g; *Liomys* 30-90 g; *Microdipodops* 10-25 g; *Perognathus* 5-50 g) are nocturnal, terrestrial to semifossorial herbivores in the New World. Males are often larger than females (4913, 5651, 6488, 6978). The hemochorial, endotheliochorial, or hemoendothelial, labyrinthine, discoid, deciduous placenta has a mesometrial or antimesometrial, interstitial implantation into a duplex uterus (5737, 6355, 7563, 7896, 10820). As simultaneously lactating and pregnant females are observed, a postpartum estrus is possible (1120, 2928, 4518, 4913, 5661, 6978, 8007; cf 2236). Neonates are blind and naked (1785, 4913, 8798, 9221, 10197, 11364). Eyes open at 9-17 d (*Dipodomys* 1474, 1503, 1782, 2164, 2693, 2927, 2935, 3282, 6148, 9075, 9076, 9221a, 10398a, 10670; *Perognathus* 2927, 2935, 4517) or 17-21 d (*Heteromys* 3321, 9252; *Liomys* 2927, 2935; cf 3321). Females have 6 mammae (502, 1120, 1474, 1475, 2236, 3024, 4080, 4494, 4745, 11364). Males have a baculum (884a, 1473, 6431, 9224). Details of anatomy (2768, 8472, 8475), intersex (671), copulation (162a, 737, 5652), milk composition (5958, 10524), and light or green vegetation (882, 1197, 5657, 5660, 6978, 8145, 9010a, 9353, 10197a, 11097) are available.

NEONATAL MASS		
Dipodomys deserti	3.04 g; r2.2-4.6	881
	5-6 g; from graph	1503
Dipodomys heermanni	2.8 g; r2.3-3.2; n4 premature	10670
	3.65 g; r3.4-3.9; n4	10670
	4.3 g; sd0.46; r3.8-4.8; n6	9221, 9221a
	max 5 g	3282
Dipodomys merriami	3.0 g	5661
	~3 g; n6 lt	10197
	3.04 g; r2.2-4.6; n9 lt	1782
	3.5 g; r3.2-3.8; n3	2693
	4.0 g; n6	9076
Dipodomys microps	4.0 g	5661
Dipodomys nitratoides	2.2 g; r1.5-3.5; n7 yg, 4 lt	2164
	4.0 g; n8	2927
Dipodomys ordii	est 5 g; 1 d; from graph	10111
Dipodomys panamintinus	4.5 g; n3	2927
Dipodomys spectabilis	7.7 g; r6.0-8.7; n7	4913
	7.8 g; r5.3-9.3; n3	502
Dipodomys stephensi	4.4 g; r3.8-4.7	6148
Heteromys anomalus	3.3 g	9252
Heteromys desmarestianus	3.0 g	3321
Liomys pictus	2.5 g; n5	2927
Liomys salvini	1.9 g	3321
Microdipodops pallidus	1.0 g; n5	2927
Perognathus californicus	1.5 g; n1, 2 d	2927
Perognathus formosus	1.0 g	5661
Perognathus longimembris	0.5 g	5661
	~1.3 g; n26	4517
NEONATAL SIZE		
Dipodomys deserti	TL: ~52 mm	881
Dipodomys merriami	42.5 mm; n6	9076
Dipodomys ordii	TL: est 65 mm; 1 d; from graph	10111
Dipodomys spectabilis	HB: 40 mm; n1	502
	CR: 60 mm; largest emb	877
Dipodomys stephensi	HB: 40.0 mm; r37-44	6148

WEANING MASS		
Dipodomys heermanni	36.2 g; sd1.02; r29.3-47.5; n24, 25-32 d	9221a
Dipodomys merriami	10 g; 2 wk	9076
	emergence: 17 g; 4 wk	5661
	1st capture: 20-23 g	1784
Dipodomys microps	emergence: 21 g; 4 wk	5661
Dipodomys nitratoides	20 g; 20 d; from graph	2927
Dipodomys stephensi	25-28 g; 18-20 d	6148
Liomys pictus	20-25 g; 25-30 d; from graph	2927
Perognathus californicus	10 g; 20 d; from graph	2927
Perognathus formosus	emergence: 8 g; 4 wk	5661
Perognathus longimembris	emergence: 4 g; 4 wk	5661
LITTER SIZE		
Dipodomys agilis	2.6; ±0.7; r2-4; n15 f	6708
Dipodomys deserti	3.29; mode 3; r2-5; n7 lt	1503, 1504
	3.43 emb; mode 3; r1-6	881
	3.5 emb; r3-5; n4 f	11044a
Dipodomys elator	2-4 emb	1642
Dipodomys californicus	2.65 emb; r2-4; n20 f	2224
Dipodomys compactus	2 emb; n1 f	666
	2 ps; n1 f	666
Dipodomys heermanni	2-5	5146
	2-5	1474
	2.64 yg; r2-3; n14 lt	3282
	2.65; n20 f	3282
	2.76 yg; n25 lt	9221a
	3 emb; n1 f	11044a
	3.0 emb; r3-4; n4 f	10398a
	3.5 yg; r3-4; n2 lt	10398a
	3.5 emb; r3-4; n2 f	10670
Dipodomys ingens	3-4; max 6	1474
	4.6 emb, yg, ps; n7	11715
	5.25 emb; r5-6; n4 f	4080
Dipodomys merriami	1.67; n3 lt; n5	522
	2 emb; n1 f	502
	2.02 emb; r1-3; n133 f	9075, 9076
	2.11; r2-3; n9 lt	2693
	2.25; r2-3; n4 lt	2927, 2935
	2.46 emb, ps, yg; sd0.72; r2-5; n70 f	5661
	2.5; r1-6; n129 lt, 51 f	2236
	2.6 ps; n157 f	1197, 6978
	2.6 emb; r1-5; n163 f	1197, 6978
	2.67 emb; r2-3; n6 f	4897
	2.67 emb; r1-5; n127 f	6430
	2.83 yg; n12 lt	1782
	3 yg; n1 f	1786
	3.1 emb; r2-4; n32 f	140
Dipodomys microps	2.4 yg; r1-4; n49 lt, 43 f	2236
	2.42 emb, ps, yg; sd0.53; r1-3; n65 f	5661
	2.5 emb; r2-4; n6 f	6488
	3 emb; n1 f	7489
	4 emb ea; n2 f	6560
Dipodomys nelsoni	2 emb; n1 f	523a
	2 emb; n1 f	878
	2 emb; n2 f	878
Dipodomys nitratoides	est 2; r1-3	2164
	2.3 yg; mode 2; r1-3; n12 lt	2927, 2928, 2935
Dipodomys ordii	1-6 yg	1871a
	1.67 emb; r1-2; n3 f	145
	2.2 emb; n33 non adult	882
	2.35 emb; n40 f	7041
	2.37 emb; r2-3	5372
	2.5 cl; max 5	3611
	2.5 emb; r1-5	3611

	2.5 emb; r2-3; n2 f	1812a
	2.5 ps; n38 f	4873
	2.5 emb; n133 adult	882
	2.8 emb; n18 f	4873
	2.87 emb; sd0.56; r2-4; n31 f	3294
	3 emb; n1 f	3703
	3.0 emb; r2-4; n2 f	4745
	3.0; r2-4; n2 lt	522
	3.0 yg; r2-4; n10 lt, 6 f	2370
	3.1 emb; r2-5; n13 f	$5405b_1$
	3.1 cl; n49 f	4873
	3.25 emb; n4 f	502
	3.25 emb; n73 f	7041
	3.6 emb; r2-5; n50 f	140
	3.8 emb	9353
	4 emb; n1 f	502
	4 emb; n1 f	1470
	4 emb; n1 f	1119
	5 emb; n1 f	4962
Dipodomys panamintinus	3.5; r3-4; n2 lt	2927, 2935
	4.2 emb; r4-5; n5 f	5158
Dipodomys phillipsii	2.67 emb; r2-3; n3 lt	3702a
Dipodomys spectabilis	1 emb ea; n2 f	2228a
	1.81 emb, ps, yg; mode 2; r1-3; n69 f	1120
	2.75 emb, ps, yg; r2-4; n12 f	4913
	3-4 yg	502
Dipodomys stephensi	2.67 yg; n6 lt	6148
Dipodomys venustus	1 yg; n1 nest	4512
Heteromys anomalus	2; r1-3; n2 lt	9252
	4.14 emb; r3-5; n7 f	11044a
Heteromys australis	2.83 emb; r2-4; n6 f	11044a
Heteromys desmarestianus	2.57; r1-4; n14 f	11044a
	2.7 weaned; mode 3; r3-5; n6 lt	3321
	3.1 emb	3321
	3.3 yg; n6 lt	3321
	3.67 emb; r3-4; n3 f	7624
	4.0 emb; r3-6; n4 f	2590
Heteromys gaumeri	2 emb; n1 f	4494
	3 ps; r1-5; n2 f	950
	3.0 emb; r2-4; n3 f	950
	4 emb; r3-5; n2 f	5423b
Heteromys oresterus	3 emb; n1 f	9223a
	4 yg; r3-5	3990_1
Liomys adspersus	3.2 emb; r2-4; n5 f	3319
	4.8 emb; r4-6; n9 f	11044a
Liomys irroratus	3.5 emb; r3-4; n4 f	4228
	4.3 emb; r1-6; n6 f	217
	4.39 emb; mode 4; r2-8; n52	2707
	5 emb; n1 f	2707
Liomys pictus	3.0	9176
	3.5; r2-5; n6 lt	2927, 2928, 2935
	3.80 emb; mode 3; r2-6; n49 f	7060
	4.29 emb; r3-6; n17 f	11388
	4.5 emb; r4-5; n2 f	4228
Liomys salvini	2.5; r2-3; n2 lt	3209
	3 emb ea; n2 f	1475
	3.55 emb; mode 3; r2-6	1634
	3.8 emb	3321
	4 yg; n1 lt	3321
	5 emb; n1 f	11044a
Microdipodops megacephalus	1-7 emb	1474
	3.9 emb; r1-7; n54 f	4228b
	4.0 emb; r1-6; n37 f	4229
	5.0 emb; r4-6; n2 f	11044a

Microdipodops pallidus	3.5 emb; n6 f	11044a
	4.0 emb; r1-6; n56 f	4228b
Perognathus anthonyi	4.2 emb; r3-5; n5 f	11044a
Perognathus arenarius	2 emb; n1 f	6151
	4.5 emb; r2-6; n6 f	11044a
Perognathus baileyi	3-4 emb	1474
	3.5 emb	8271
	3.6 ps	8271
	mode 4 emb	9077
Perognathus californicus	4.0; r2-5; n3 lt	2927, 2928, 2935
Perognathus fallax	3 emb; n1 f	11044a
Perognathus fasciatus	4 cl; n1 f	6166
	5.1 emb; r2-7; n9 f	3703
	5.42 emb; r2-9	6770
	5.64 ps; r2-12	6770
	5.7 emb; r3-9; n7 f	6166
	6 ps; n1 f	3703
	6 ps; n1 f	5405b$_1$
	6 ps; r3-9; n6 f	6166
	6.0 cl; r4-9	6770
	6.0 ps; r3-11; n14 f	11957a
	6.1 emb; r4-10; n8 f	11957a
Perognathus flavescens	3.67 emb; r2-5; n3 f	5405b$_1$
	4 emb ea; n3 f	495
	5 yg; n1 f	495
	6 emb; n1 f	495
Perognathus flavus	2-6	1474, 1871a
	2.75 emb; r2-3; n4 f	5405b$_1$
	3 emb; n1 lt	4962
	3 emb; r2-4; n2 f	502
	3-6	1474
	4; n1 lt	2935
	4 emb; n1 f	4745
	4 emb; n1 f	522
	4.5 emb; r4-5; n2 f	11044a
	6 yg; n1 lt	502
Perognathus formosus	5.34 emb, ps, yg; sd0.94; r4-7; n35 f	5661
	5.38; mode 5	2773
Perognathus hispidus	2-6	1474
	2-9	8272
	5 emb; n1 f	2547
	5 ps; n1 f	5405b$_1$
	5.5 emb; r5-6; n2 f	5405b$_1$
	6 emb; n1 f	10954
Perognathus inornatus	~4	2928
Perognathus intermedius	3-6	1474
	7 emb; n1 f	11044a
Perognathus longimembris	3-7	1474
	4; r1-6; n57 lt	4518
	4.25; r3-6; n4 lt	1785
	4.75 emb, ps, yg; sd0.77; r4-6; n16 f	5661
	5.38; mode 5	2773
Perognathus nelsoni	3-5 yg	2228a
	3.0; r2-5; n5 lt	522
Perognathus parvus	3.7; n3 lt	8007
	4 emb; n1 f	2544
	4.67; r4-6; n3 lt juv f	10262
	4.77 ps; r4-6; n16 f	5184
	4.91 emb; r4-6; n23 f	5184
	4.93 emb; n11 f	11281

	5.16 emb; mode 4; r2-8; n132 f	9579, 9581
	5.38; mode 5	2773
	5.5 emb; r5-6; n4 f	1119
	5.7 emb; mode 6; r4-7; n6 f	6488
	8 emb; n1 f	2544
Perognathus penicillatus	r3-4 emb; n3 f	523a
	3.4 emb; r2-4	9077
	4.38 emb; r3-6; n16 f	377
	4.5 emb; r3-6; n2 f	522
Perognathus spinatus	3.25 emb; r3-4; n4 f	11044a
	4 emb ea; n≥2 f	4079a
SEXUAL MATURITY		
Dipodomys deserti	f: vag perf: 24-33 d	881
	m: scrotal testes: 85 d	881
Dipodomys heermanni	f: 1st estrus: 40-43 d	9221a
	f: 1st mate: 1st yr	1474
	m: scrotal testes: 35-38 d	9221a
	f/m: 1st yr	5146
Dipodomys merriami	f: vag perf: 24-33 d	1782
	f: vag perf: 24-33 d	1782
	f: 1st conceive: 64 d	2236
	f: 1st birth: 97 d	2236
	f: 1st yr	5661
	m: scrotal testes: 84 d	1782
	m: scrotal testes: 85 d	1782
	m: yearling	5661
Dipodomys microps	f: 1st estrus: 34.2 d; ±2.9; r29-39	2236
	f: 1st yr	5661
	m: testes large; est ~1 yr	5661
Dipodomys nitratoides	f: 1st conceive: 84 d	2928, 2935
Dipodomys ordii	f: 1st estrus: 41 d; n1	3611
	f: 1st mate: 3 mo	3611
	f: 1st yr	7041
Liomys adspersus	f: 1st conceive: min 3-4 mo; most 1 yr	3319
	m: fertile: 1 yr	3319
Liomys pictus	f: 1st conceive: 98 d	2928, 2935
Perognathus fallax	f/m: 166 d	4518
Perognathus formosus	f: 1st conceive: ~1 yr	5661
	m: testes large: ~1 yr	5661
Perognathus longimembris	f: 1st conceive: 41 d	4518
	f: 1st conceive: ~1 yr	5661
	m: testes large: ~1 yr	5661
Perognathus parvus	f: 1st birth: 1st yr	10262
CYCLE LENGTH		
Dipodomys heermanni	mode 17-18 d; r6-43; n~100 cycles	9221a
Dipodomys merriami	mode 13 d; median 13.4; n106 cycles, 13 f	2236, 11764
Dipodomys microps	mode 12 d; median 12.5; n422 cycles, 27 f	2236, 11764
Dipodomys ordii	4-5 d	5374
	5-6 d	2928
Perognathus inornatus	5-6 d	2928
Perognathus longimembris	~10 d	4518
GESTATION		
Dipodomys deserti	29-30 d; n4	1504
	29-32 d	8798
Dipodomys heermanni	30 d	10809
	31.3 d; r30-33; n19	9221, 9221a
Dipodomys merriami	17-23 d; n1	1786
	4 wk	5661
	mode 33 d; r32-35; n65	2236
	33 d; ±2	1782
Dipodomys microps	4 wk	5661
	mode 31 d; r30-34; n33	2236
Dipodomys nitratoides	32 d; n2	2935

Dipodomys ordii	est 29-30 d	2370
	29-30 d	8798
Dipodomys panamintinus	29-30 d; n1	2927, 2935
Dipodomys spectabilis	17 d; n1	502
	22-27 d	877
	~27 d	1474
Heteromys desmarestianus	est 27-28 d; r23-30; n4	3321
Liomys pictus	25 d; r24-26; n2	2927, 2928, 2935
Liomys salvini	27-28 d; n1	3321
Perognathus californicus	25 d; n2	2927, 2928, 2935
Perognathus fallax	r24-26 d; n4	4518
Perognathus fasciatus	~4 wk	1474
Perognathus flavus	≤26 d; n1	2935
Perognathus formosus	4 wk	5661
Perognathus longimembris	3 wk	5661
	mode 22-23 d 74%; r21-31; n31; 31 d not successsful	4518
Perognathus parvus	21-25 d	5184
	est 21-28 d	9581
LACTATION		
Dipodomys deserti	solid food: 15 d	1503
	wean: 15-25 d	1503
Dipodomys heermanni	solid food: 18-24 d	9221a
	solid food: 23 d	3282
	wean: 25 d	3282
	wean: 25-32 d	9221a
	wean: 26 d	10398a
	wean: 6 wk	10809
Dipodomys merriami	solid food: 9 d	1782
	solid food: 10 d	9076
	solid food: 11 d	9075
	solid food: 13 d	1503
	nurse: 3 wk	2236
	nurse: 3 wk	9075
	wean: 17-22 d	1503, 1782
	wean: 19-22 d	2935
	wean: 20-25 d	1784
Dipodomys microps	wean: 3 wk	5661
Dipodomys nitratoides	solid food: 9 d	2927
	solid food: 20 d; n1	2164
	wean: 21-24 d	2927, 2935
Dipodomys panamintinus	wean: 27-29 d	2927, 2935
Dipodomys spectabilis	wean: 20-25 d	877
	wean: 3-4 wk	4913
Dipodomys stephensi	solid food: 15 d	6148
	nurse: 18 d; n14; 22 d; n2	6148
Liomys pictus	solid food: 13-14 d	2927
	wean: 24-28 d	2927, 2935
Perognathus californicus	solid food: 13 d	2927
	wean: 22-24 d	2927, 2935
Perognathus formosus	wean: 3 wk	5661
Perognathus hispidus	independent: 1 mo	1871a
Perognathus longimembris	solid food: 10 d	4517
	f died, young survived: 14 d	4517
	self-sufficient: 18 d	4517
	wean: 30 d	4518
	wean: 3 wk	5661
Perognathus parvus	wean: est 3 wk	5184
INTERLITTER INTERVAL		
Dipodomys merriami	min 63 d	2236
Dipodomys ordii	4.7 mo; r3-8	3611
Dipodomys spectabilis	5-7 wk	4913

LITTERS/YEAR

Dipodomys agilis	2 rare	6708
Dipodomys deserti	2-5	1474
Dipodomys californicus	2/season	2224
Dipodomys heermanni	1-3	5146
	1-3	1474
Dipodomys merriami	>1 in good yr	5661
	1-2	1474
Dipodomys microps	>1 in good yr	5661
Dipodomys ordii	est 1	502
	est 2	5372
	2	7041
	2	3611
Dipodomys spectabilis	1-3	4913
Heteromys goldmani	3-4	11388
Liomys adspersus	1.44; n19 f	3319
Perognathus fallax	1	6708
Perognathus fasciatus	1	1474
Perognathus flavescens	est 2	1474
Perognathus flavus	2	1474
	2+; irregular breeding	502
Perognathus formosus	1	5661
Perognathus hispidus	est 2; in north	1474
Perognathus longimembris	1	5661
	1-2	1474
Perognathus parvus	1.1; r1-3; n50	8007
	1-2	10262
	2	9581

SEASONALITY

Dipodomys agilis	testes: large Dec-July, small Aug-Nov; San Gabriel, CA	6708
	spermatogenesis: Jan-July; San Gabriel, CA	6708
	preg/lact f: Jan-July; CA	6708
	repro f: peak Mar-Aug; CA	7028
Dipodomys californicus	preg f: Feb-Sept; CA	2224
Dipodomys compactus	preg f: Aug; n1; TX	666
Dipodomys deserti	mate: Feb-June; sw US	1474
	preg f: Feb-Apr; n4; AZ	11044a
Dipodomys elator	preg f: Feb, June-July, Sept; TX	1642
	lact f: July; n2; TX	11044a
Dipodomys heermanni	mate: Feb-Aug; CA	10670
	mate: Feb-Oct, peak Apr; CA, OR	1474
	mate: Mar-Aug; w US	5146
	mate: low Nov-Jan; CA	3282
	preg f: Mar-Aug; CA	10670
	preg f: June; n1; CA	11044a
	preg f: July; n6; CA	10398a
	birth: yr rd, peak Feb-Aug; Morro Bay, CA	10809
Dipodomys ingens	preg f: Feb, May; n4; CA	4080
	repro: est Jan-May; CA	4080
Dipodomys merriami	testes: large yr rd; NV	6978
	mate: Feb-Oct; sw US	1474
	mate: Mar-May or later; Alamogordo, NM	979
	mate: peak Mar-June; se AZ	1784
	preg f: Jan-July; Coahuila, Mexico	522
	preg f: Feb-Aug; NV	1197
	preg f: peak Mar-May; CA	1783
	preg f: Mar-Oct, not Nov-Feb; s AZ	9076
	preg f: Mar-Oct, peaks May, Sept; AZ	9075
	preg f: Aug; n1; NM	502
	preg f: Dec-May; CA	1783
	birth: mid Feb-mid June, peak Mar ; CA	5661
	lact f: Feb-Aug, except July; small July sample; NV	6978
	repro f: peak Mar-May; CA	7028
	repro f: June-Oct, Feb; AZ	9010

	repro: yr rd, peak m June-Sept, peak f Sept-Oct; AZ	12089
Dipodomys microps	conceive: Feb-mid Mar; Owens Valley, CA	5661
	preg f: Apr-June; no data other mo; NV	6488
	birth: Mar-mid Apr; CA	5661
	birth: May-June; w US	1474
Dipodomys nelsoni	preg f: July; n2; Coahuila, Mexico	878
	preg f: July; n1; Durango, Mexico	523a
	preg f: July; n1; Zacatecas, Mexico	878
	lact f: June, Sept; San Luis Potosí, Mexico	2228a
Dipodomys nitratoides	mate: Dec-Aug; captive	2927
	mate: yr rd; CA	2164
Dipodomys ordii	sperm: yr rd, peak June-Feb, low Mar-June; TX	3611
	sperm: yr rd; central OK	4873
	testes: large yr rd; NM	882
	estrus: Feb-early July; lab	5374
	mate: peaks Jan-Mar, end Aug-Oct; UT	2769
	mate: Feb-Sept; w US	5146
	mate: Aug-Mar; OK, TX	7041
	mate: Dec-Oct; captive	2370
	mate: spring, fall; few data; Alamogordo, NM	979
	CL present: Aug-May, peaks Sept-Oct, Jan-Feb; TX	3611
	preg f: Jan; n1; AZ	145
	preg f: Feb; n2; Coahuila, Mexico	522
	preg f: May-Sept; n13; NE	5405b,
	preg f: June, Oct; n2; NM	502
	preg f: July; n1; ND	3703
	preg f: Aug-Apr; TX	3611
	preg f: Sept-Oct; n2; KS	1812a
	preg f: Oct; n2; NM	145
	preg f: yr rd, low Mar-Apr; NM	882
	birth: mid-Feb-mid-July; NM	5372
	birth: May-June; w US	1474
	lact f: May; n1; SD	261a
	lact f: Nov; n4; NM	502
	repro: Feb-Aug; ne CO	3294
Dipodomys panamintinus	preg f: apr-May; n5; CA	5158
	repro: peak Feb-Mar; CA	5158
Dipodomys phillipsii	preg f: June; n1; Durango, Mexico	3702a
	preg f: Oct; n2; Jalisco, Mexico	3702a
Dipodomys spectabilis	mate: Jan-Aug; AZ	1120
	preg f: Aug; n2; San Luis Potosí, Mexico	2228a
	birth: Jan-Aug; sw US	1474
	lact f: Apr, Oct; n4; Socorro, NM	9648
	repro: Dec-Aug; NM	4913
Dipdomys stephensi	preg f: June-July; CA	6148
Heteromys anomalus	scrotal testes: yr rd; Venezuela	451
	preg f: Jan; n2; Colombia	11044a
	preg f: Mar-Apr; n4; Tobago	11044a
	preg f: Apr; n1; Venezuela	11044a
	preg f: May-June, Sept, Dec; 45 km s of Calabazo, Venezuela	451
	lact f: Feb, Oct; 45 km s of Calabazo, Venezuela	451
	lact f: Apr, June-July; n5; Venezuela	11044a
	repro: yr rd, peaks at seasonal transitions; Venezuela	7996
Heteromys australis	preg f: Jan-Mar; n6; Panama	11044a
	lact f: Jan-Feb; n3; Panama	11044a
Heteromys desmarestianus	preg f: Apr; n1; Mexico	11044a
	preg f: Mar-Apr; n7; Panama	11044a
	preg f: Mar, May-June; n4; Guatemala	11044a
	lact f: Mar, June; n5; Panama	11044a
Heteromys gaumeri	preg f: end Apr-early May; n3; Yucatan peninsula, Mexico	950
	preg f: July; n1; Quintana Roo, Mexico	5423b
	preg f: Dec; n1; Campeche, Mexico	5423b

Heteromys goldmani	testes: large yr rd; s Mexico	11388
	ovaries: regress dry season; s Mexico	11388
	birth: May-Dec; Mexico	11388
Heteromys oresterus	preg f: July or Aug; n1; Costa Rica	9223a
Liomys adspersus	testes: large Oct-Apr, small June-July; Panama	3319
	mate: peak Mar-Apr; Panama	3319
	preg f: Feb; n9; Panama	11044a
	preg f: Dec-Apr; Panama	3319
	lact f: Mar-Apr; n2; Panama	11044a
	yg found: Dec-May; Panama	3317
Liomys irroratus	preg f: Aug; n1; Mexico	2341
	preg f: July-Jan, Mar, May, peak Aug-Nov; n52; Mexico	2707
Liomys pictus	preg f: Feb,-Sept, Nov-Dec; n49; Mexico	7060
	preg f: Mar; Villa Flores, s Mexico	11388
	preg f: dry season; n8/10; Oaxaca, Mexico	9176
	repro: Sept-May; Veracruz, Mexico	4228
Liomys salvini	preg f: July; n1; Guatemala	11044a
Microdipodops megacephalus	preg f: May; n2; NV	11044a
	birth: May-early July; w US	1474
	birth: peak May-June; NV	4229
	lact f: May; n1; NV	11044a
Microdipodops pallidus	preg f: end Mar-Sept; NV	8004
	preg f: May; n6; NV	11044a
	lact f: May; n1; NV	11044a
Perognathus amplus	repro f: June-Aug; AZ	9010
Perognathus anthonyi	preg f: May-June, Oct; n5; Baja CA, Mexico	11044a
	lact f: May; n2; Baja CA, Mexico	11044a
Perognathus arenarius	preg f: Mar; n1; Cerralvo I, Baja CA, Mexico	6151
	preg f: Mar-Apr; n6; Baja CA, Mexico	11044a
	lact f: Apr; n1; Baja CA, Mexico	11044a
Perognathus baileyi	mate: June-Oct, peak June, Sept; AZ	9010
	preg f: Apr-Aug; AZ	9077
	birth: Apr-May; sw US	1474
Perognathus fallax	mate: spring; few data; CA	6708
	preg/lact f: Oct; n3; CA	11228a
Perognathus fasciatus	preg f: June-Aug; n9; ND	3703
	preg f: July; n7; se MT	6166
	preg f: July-mid Aug; MAN, Canada	11957a
	birth: Apr-Aug; WY	1871a
	lact f: July; n1; se MT	6166
Perognathus flavescens	preg f: May; n1; MN	495
	preg f: May, July; n3; NE	$5405b_1$
	birth: May; n1; MN	495
	lact f: July; n1; NE	$5405b_1$
	repro: Apr-July; central US	1474
Perognathus flavus	mate: Apr-Nov; w central US	1474
	preg f: Mar-Oct; NM	1871a
	preg f: May; n1; Valencia Co, NM	4962
	preg f: May-June; n2; NM	502
	preg f: June; n1, none Mar-Apr; Coahuila, Mexico	522
	preg f: July; n1; NM	11044a
	preg f: July-Aug; n2; TX	11044a
	birth: June; n1; NM	502
	lact f: July; n1; TX	11044a
	lact f: Aug; n1; San Luis Potosí, Mexico	2228a
	lact f: Dec; NM	502
Perognathus formosus	mate: early Apr; CA	5661
	preg f: peak June; UT	2773
	birth: 1st 3 wk May; Owens Valley, CA	5661
	birth: May-July; CA, NV, UT	1474
Perognathus hispidus	preg f: Aug; n1; SD, WY	10954
	preg f: Aug-Sept; n2; NE	$5405b_1$
	birth: yr rd; s US; seasonal; n, mid US	1474

Perognathus intermedius	preg f: Apr; n1; az	11044a
	birth: May-July; sw US	1474
	repro f: June-Sept, peak June; AZ	9010
Perognathus longimembris	mate: end Apr-early May; CA	5661
	preg f: peak June; UT	2773
	birth: Apr-July; sw US	1474
Perognathus nelsoni	preg f: end Mar-July; Coahuila, Mexico	522
	birth: Aug; San Luis Potosí, Mexico	2228a
Perognathus parvus	testes: large spring; e WA	5657
	estrus: Apr-June; WA	8007
	preg f: May-early June; n6; central NE	6488
	preg f: May-Aug, peak June-July; WA	8007
	preg f: peak June; UT	2773
	preg/lact f: May-Sept; no data other mo; ID	10262
	birth: May-June; WY	1871a
	birth: June-Aug; n39; BC, Canada	5184
	birth: end spring-early summer; w US	1474
	lact f: May-Sept peak June-July; WA	8007
	repro m: Mar/Apr-June/July; WA	8007
Perognathus penicillatus	mate: end Apr-May; no data June-Feb; Alamogordo, NM	979
	preg f: Feb, June; few data; Coahuila, Mexico	522
	preg f: May-Sept; AZ	9077
	preg f: June-July; n3; Durango, Mexico	523a
	birth: May-Sept; sw US, n Mexico	1474
Perognathus spinatus	preg f: Apr; n≥2; lower Colorado R, CA	4079a
	preg f: Apr; n4; Baja CA, Mexico	11044a

Order Rodentia, Family Castoridae, Genus *Castor*

Beavers (*Castor canadensis*, *C. fiber*) are 5-45 kg, semiaquatic herbivores of temperate North America and Eurasia. Males have been reported to be larger than, smaller than and equal to females (4570, 4815, 8050). Urogenital and digestive products exit via a cloaca (8050, 11791, 12065). A vaginal plug occurs (6056). The hemodichorial, labyrinthine, reniform placenta with a subplacenta has a superficial, mesometrial implantation into a Y-shaped, duplex uterus (2003, 2649, 3268, 3766, 7563, 7565) and is eaten after birth (4564, 8255, 9817). A postpartum estrus (2109) follows the birth of furred and open-eyed neonates (323, 1474, 3571, 4124, 4564, 5146, 7433, 8050, 8133, 8255, 8798, 9817, 10922; cf 11045). Females have 4 teats (1474, 3571, 4032, 7433, 8050, 9387). Males are ascrotal (testes descend into a subpelvic diverticulum during the breeding season) and have a baculum and seminal vesicles (1473, 2648, 2649, 3524, 8050, 11791). Details of anatomy (2645, 2646, 3766, 8133, 8763), birth (9817), and milk composition (12137, 12138) are available.

NEONATAL MASS		
Castor canadensis	361.7 g; r265-490; n5	9817
	~380 g; r365-390; n2	4124
	387.9-614.5 g; term emb	8133
	490 g; r390-700; n10, 3 lt	1203
Castor fiber	300-700 g	11791
	380-620 g	8050
	475 g; n1 stillborn	4564
	f: 548.5 g; r527-570; n2, 1 d	2645
	555 g; n9, 3 lt	12138
	700 g; n1	11791
NEONATAL SIZE		
Castor fiber	245 mm; n1 stillborn	4564
WEANING MASS		
Castor fiber	4.2 kg; n8, n2 lt, 90 d	12138
LITTER SIZE		
Castor canadensis	1-6 yg	4011
	1.5 emb; r1-2; n2 f	7325a
	1.7; r0.9-2.5; f <2 yr	8248
	2 yg; n1	4818
	2.6 emb; se0.16; mode 2; n40 f	11675, 11676

	2.8 ps; se0.18; mode 3; n26 f	11675, 11676
	2.8 cl; sd1.3; n89 f	8280
	2.9 emb; sd1.1; n60 f	8280
	2.9 ps; sd1.3; n243 f	8280
	2.91 emb; mode 2; r1-5; n22 f	8133
	3 emb ea; n2 f	3037
	mode 3 ps, emb; 76% 2-4; r1-7; n504 f	6360
	3.0 cl; se0.16; mode 2; n48 f	11675, 11676
	3.02 emb; n176 f	9387
	3.1 emb	1168
	3.1 ps; n232 f	6360
	3.2 ps; r1-8; n43 lt	6679
	3.37 ps; r1-6; n19 f	8133
	3.4 ps, emb; n22 f	8408
	3.4 emb; n272 f	6360
	3.5; r3-4; n6 f	9577
	3.6 cl; n27 f	8408
	3.66; mode 3-4; r2-5; n9 f	8133
	3.72 yg; mode 4; r1-8; n65	1202
	3.8 emb	4011
	3.9 ps; n28 f	4634
	4.0; n2 lt	8254
	4.0; r3.3-4.7; f 2.5 yr	8248
	4.08 emb; r2-7; n11 f	10922
	4.4 cl; n48 f	4634
	4.5 emb; r3-6; n4 f	5405b,
	4.81 emb; 5.19 cl; n16 f	4634
	5.3 ps; r4-7; n18 f	11956
	5.50 emb; 5.04 without resorption; r1-9; n24 f	1278
	5.6 emb; r3-7; n17 f	11956
	6 yg; n1 f	9817
	7 emb; n1 f	11658
	8 emb, 9 cl; n1 f	4514
	10 emb; n1 f	5210
Castor fiber	1.7 yg; sd0.67; mode 2; r1-4; n120 yg, 71 sites	12136
	mode 2-5; r2-7	3571
	2.57 emb; r1-4; n18 f	4570
	2.7; r1-6; n117 lt	2649
	3.0; r1-4; n3 lt	12138
	3.0 ps; r1-5; n5	4570
	3.0; r2-5; n10 lt	2109
	4.7; r4-5; n3 lt	4564
	4.7 emb; max 8; n9 f	6157
SEXUAL MATURITY		
Castor canadensis	f: 1st cl, ps: yearling	6679
	f: 1st mate: 2nd season after birth	8133
	f: 1st mate: 2.5 yr	1474
	f: 1st preg: 1-2 yr	500
	f: 1st preg: 1.5-2 yr	8248
	f: 1st preg: est ~2 yr	1278
	f: parous: by 2.5-3 yr thus mature 1.5-2 yr	6221
	f: 1st repro: 1.5 yr	11675
	m: sperm: 1.5-2 yr	6221
	m: epididymal sperm: yearling	6679
	m: 1st repro: est ~2 yr	1278
	f/m: ~21 mo	5210
Castor fiber	f: 1st cl: 6 mo	2645
	f: 1st estrus: 2.5 yr	8050
	f: 1st mate: 1.5-2 yr	11791
	f: 1st birth: 2 yr 29%	2649
	f: 1st birth: 3 yr	4645
	f/m: 1st mate: 3 yr	6056
	f/m: mode 3-4 yr	3571

CYCLE LENGTH

Castor fiber	estrus: 3 h	6056
	estrus: 10-12 h	4815
	estrus: 12-24 h	2649
	estrus: 1-2 d	2647
	r7-9 d; n19 f	2647
	7-12 d	2649
	14 d	4815
	14-15 d	8050

GESTATION

Castor canadensis	~90 d	2088
	~120 d	5210
	est 128 d; n1	1203
Castor fiber	90 d	2109
	~90 d; max 107	3571
	3.5 mo	6056
	105-107 d	4645
	105-107 d	8050
	105-107 d	4815
	106.3 d; r104-111; n3	4564
	107 d; sd2.5; n39	2649, 12135
	128 d	7433

LACTATION

Castor canadensis	den emergence: 4 wk	4032
	nurse: 50 d	4818
	wean: ≤2 mo	1871a
Castor fiber	solid food: ~4 wk	3571
	solid food: after 30 d	12138
	nurse: 8 wk	3571
	wean: 2 mo	11791
	nurse: 90 d	12135, 12138

LITTERS/YEAR

Castor canadensis	1	1202
	1	11675
	1	8254
Castor fiber	1	2109
	1	3571, 4645
	1	7433
	1	8050

SEASONALITY

Castor canadensis	spermatogenesis: peak Jan-May; SAS, Canada	9201
	mate: Jan-Apr; BC, Canada	2088
	mate: end Jan-Feb; WI	5210
	mate: end Jan-Feb; PA	1278
	preg f: Feb; n2; WA	7325a
	preg f: end Feb-?; WY	4011
	preg f: Mar-Apr; n4; NE	5405b,
	birth: Feb-July; MS	11675
	birth: Apr-Mar; w US	5146
	birth: Apr-July; N America	1474
	birth: end Apr-early June; WY	8133
	birth: end Apr-early July; BC, Canada	2088
	birth: May; IL, WI	2060
	birth: May; n2; QUE, Canada	8255
	birth: May-June; WY	1871a
	birth: May-Aug; n25; end Sept; n1; zoo	12128
	birth: mid May-June, sometimes July; CO	9387
	birth: end May; MAN, Canada	4032
	birth: June-?; WY	4011

Castor fiber	estrus: Jan-Mar; Europe	7433
	mate: Jan-Mar; Europe, USSR	4645
	mate: Jan-May; Russia	4815
	mate: Feb-Mar; zoo	4564
	mate: Feb-Mar; Europe	3571
	mate: Dec-May, peak mid Jan; Poland	2649
	birth: Apr-May; Europe	7433
	birth: Apr-May; Europe, USSR	4645
	birth: Apr-Aug, peak May; Poland	2649
	birth: end Apr-early May; Europe	3571
	birth: May; according to local farmers; Ukraine	9555
	birth: May; Lapland, Finland	6056
	birth: May-June; Finland	6157
	birth: end May; Mongolia	8512
	birth: end May-early June; n Europe	11791

Order Rodentia, Family Anomaluridae

The west and central African, scaly-tailed, flying-squirrels (*Anomalurus, Idiurus, Zenkerella*) are arboreal herbivores that range from 15 g (*Idiurus*) to 1050 g (*Anomalurus*) in tropical and subtropical forests. The natural history of the flightless *Zenkerella* has not been recorded and that of *Idiurus* is poorly known. *Anomalurus* is reported to be either social (5760) or solitary (331). A copulatory plug is present (2439, 5760). The hemomonochorial, villous, discoid placenta has a superficial, mesometrial attachment (6625, 7563). Neonates are active with fur and open eyes (5760). Females produce copious milk from 2 (5760) or 4 (331) mammae. Both sexes bring food to young (5760, 9468).

NEONATAL MASS		
Anomalurus derbianus	~160 mm	5760
LITTER SIZE		
Anomalurus beecrofti	1 emb; n1 f	2439
	1 yg ea; n2 f	4354a
Anomalurus derbianus	1	9468
	1	331
	1 yg	2446a
	1 emb; n1 f	2446a
	mode 1; r1-3	5760
	1-2	11253
Anomalurus peli	mode 2-3; max 4; hunter's reports	65
Idiurus macrotis	1 emb ea; n3 f	8854
Idiurus zenkeri	mode 1	5760
LITTERS/YEAR		
Anomalurus derbianus	2	11253
Anomalurus peli	2	65
SEASONALITY		
Anomalurus beecrofti	lact f: Jan, Oct; n2; Nigeria	4354a
Anomalurus derbianus	large testes: Feb, July, Dec; Uganda, w Kenya	5760
	birth: Jan; n4; Uganda, w Kenya	5760
	birth: Sept-Nov; British Cameroun	9468
	birth: dry season; n1; wet season; n1; Zaire	11253
	lact f: Feb, May; Rio Muni	5388a
	f with juvenile: Oct; Zambia	331
	juvenile alone Oct-Nov; Zambia	331
Anomalurus peli	birth: Sept & other mo; Gold Coast	65
Idiurus macrotis	preg f: Aug; n3; Lake Kivu, Zaire	8854
Idiurus zenkeri	testes: small Nov-Dec; n2; Camerouns, central Forest Refuge	5760
	no preg f: Nov-Dec; n5; Bwamba, Camerouns, central Forest Refuge	5760

Order Rodentia, Family Pedetidae, Genus *Pedetes*

Springhares (*Pedetes capensis* 2-4 kg) are nocturnal, semi-fossorial, saltatorial, solitary herbivores in arid and semi-arid regions of central and southern Africa. A vaginal plug is present after mating (1508, 1509). The labyrinthine, endotheliochorial, discoid placenta implants into a bicornuate uterus (1508, 1509, 7565, 7568, 8173, 8247, 10154) and is eaten after birth (9295). The precocious neonate is furred with eyes which open 0-3 d after birth (1507, 1508, 1950, 4563, 5760, 9295, 9945, 11102). Females have a postpartum estrus, an erectile clitoris, and 2-4 teats (1509, 1951, 5760, 8247). Males have visible teats, semi-external testes, seminal vesicles, penial spines, and a baculum (1508, 1509, 1951, 3823, 9945, 10096). Details of endocrinology (11142) are available.

NEONATAL MASS		
Pedetes capensis	145 g; n1; est premature	1950
	240-280 g	5760
	250 g; n1	4563
	252 g; ±19; r222-319; n12	1508, 1509, 1510, 1512
	255 g; n1	1950
	289 g; r278-300; n2	9295
	~300 g	1507
WEANING MASS		
Pedetes capensis	1.3 kg; den emergence, independent	1507, 1509
LITTER SIZE		
Pedetes capensis	1	11102
	1 emb	4925
	1; n1 lt	4563
	1 emb; n1 f	8247
	1 yg ea; n2 lt	9295
	1 yg ea; n2 lt	1950
	1.01 emb; r1-2; n152 f	1507, 1508, 1509
SEXUAL MATURITY		
Pedetes capensis	f: 1st conceive: ~2.7 kg	1508, 1509
	m: spermatogenesis: ~1 yr	10096
	m: sperm: 2.3-2.8 kg	1508
	m: 2.9 kg	1509
	f/m: 2.6-2.8 kg	1507
GESTATION		
Pedetes capensis	est ~2 mo	5760
	~77 d	1509
	~79 d; high progesterone	5923
	est 80-82 d	9295
LACTATION		
Pedetes capensis	den emergence: 6-7 wk	1509, 1510
	wean: min 46 d	1508, 1509
INTERLITTER INTERVAL		
Pedetes capensis	est 101 d	1509
SEASONALITY		
Pedetes capensis	testes: active yr rd; Botswana	1508, 1512
	mate: yr rd; Botswana	1507, 1512
	preg f: Mar, Oct-Nov; Transvaal, S Africa	8965
	preg f: Dec-Feb; Zimbabwe	5923
	preg f: yr rd; Botswana	10154
	preg f: yr rd, except Jan, Mar; S Africa	11142
	preg/lact f: yr rd; Botswana	1508
	lact f: Mar, Nov; Transvaal, S Africa	8965
	lact f: yr rd; Botswana	1509
	juveniles appear: May; Zimbabwe	5923

Order Rodentia, Family Cricetidae

Cricetids (*Abrawayaomys* ~200 mm; *Aepeomys* 100-120 mm; *Akodon* 10-60 g; *Ammodillus* 85-105 mm; *Andalgalomys* 45 g; *Andinomys* 80-91 g; *Anotomys* 100-120 mm; *Auliscomys* 45-105 g; *Baiomys* 4-13 g; *Beamys* 50-150 g; *Bibimys* 23-34 g; *Blarinomys* 130-160 mm; *Bolomys* 25-63 g; *Brachiones* 80-95 mm; *Brachytarsomys* 200-250 mm; *Brachyuromys* 85-105 g; *Calomys* 17-59 g; *Calomyscus* 15-30 g; *Chilomys* 85-100mm; *Chinchillula* 120-180 mm; *Cricetomys* 0.6-1.6 kg; *Cricetulus* 20-40 g; *Cricetus* 112-908 g; *Daptomys* 47 g; *Delanymys* 5-6 g; *Dendromus* 5-27 g; *Dendroprionomys*; *Deomys* 40-70 g; *Desmodilliscus* 11 g; *Desmodillus* 39-70 g; *Dipodillus* 16-58 g; *Eligmodontia* 10-32 g; *Eliurus* 35-103 g; *Euneomys* 78-86 g; *Galenomys* 105-120 mm; *Gerbillurus* 11-50 g; *Gerbillus* 18-81 g; *Graomys* 105-165 mm; *Gymnuromys* 125-160 mm; *Habromys* 19-48 g; *Hodomys* 290-450 g; *Holochilus* 280 g; *Hypogeomys* 300-350 mm; *Ichthyomys* max 120 g; *Irenomys* 30-65 g; *Isthmomys*; *Juscelinomys*; *Kunsia* 160-185 mm; *Leimacomys* ~120 mm; *Lenoxus* 150-170 mm; *Lophiomys* 590-920 g; *Macrotarsomys* 80-100 mm; *Malacothrix* 7-22 g; *Megadendromys* 49-66 g; *Megadontomys* 77-79 g; *Meriones* 30-275 g; *Mesocricetus* 90-258 g; *Microdillus* 60-80 mm; *Microxus* ~90 mm; *Myospalax* 150-288 g; *Mystromys* 75-185 g; *Neacomys* 18-20 g; *Nectomys* 160-420 g; *Nelsonia* 120-130 mm; *Neotoma* 94-585 g; *Neotomodon* 37-66 g; *Neotomys* 63-69 g; *Nesomys* 185-230 mm; *Nesoryzomys* 100-200 mm; *Neusticomys* 40 g; *Notiomys* 25-95 g; *Nyctomys* 39-61 g; *Ochrotomys* 15-30 g; *Onychomys* 9-60 g; *Oryzomys* 10-120 g; *Osgoodomys* 48-68 g; *Otomys* 47-255 g; *Otonyctomys* 29-36 g; *Ototylomys* 53-120 g; *Oxymycterus* 46-125 g; *Pachyuromys* 40 g; *Parotomys* 89-155 g; *Peromyscus* 11-110 g; *Petromyscus* 11-28 g; *Phaenomys* ~150 mm; *Phodopus* 20-53 g; *Phyllotis* 20-100 g; *Platacanthomys* 75 g; *Podomys* 27-44 g; *Podoxymys* ~100 mm; *Prionomys* 60 mm; *Psammomys* 88-190 g; *Pseudoryzomys* 30-56 g; *Punomys* 130-155 mm; *Reithrodon* 77-95 g; *Reithrodontomys* 5-28 g; *Rhagomys*; *Rheomys* 90-140 mm; *Rhipidomys* 68-100 g; *Rhombomys* 140-430 g; *Saccostomus* 30-96 g; *Scapteromys* 150-200 mm; *Scolomys* ~90 mm; *Scotinomys* 12-15 g; *Sekeetamys* 41-61 g; *Sigmodon* 57-275 g; *Steatomys* 5-70 g; *Tatera* 30-227 g; *Taterillus* 42-62 g; *Thomasomys* 35-61 g; *Tylomys* 63-370 g; *Typhlomys* 90-100 mm; *Wiedomys* 100-130 mm; *Xenomys* 155-165 mm; *Zygodontomys* 50-104 g) are primarily terrestrial herbivores from the New World, Asia, and Africa.

Males are larger than females in *Akodon* (2221, 2223, 7213), *Calomys* (667), *Cricetulus* (7503, 8457), *Gerbillus* (4713, 8134), *Meriones* (8961, 9175, 11905), *Neotoma* (217, 3087, 8962), *Oryzomys* (2221), *Phodopus* (9516), *Sigmodon* (1795, 7254; cf 3850), *Tatera* (9175; cf 4925, 8501), while females are larger than males in *Eligmodontia* (8307). Male *Gerbillus* have an additional chromosome (3492, 6972).

Copulatory ties occur (*Baiomys* 3096; *Calomys* rarely, 667); *Neotoma* 2507; cf 3087, 3096; *Ochrotomys* 2507; *Onychomys* 2533, 6209) as do vaginal plugs (*Calomys* 5474; *Dipodillus* 4675, 6234, 6237; *Meriones* 4459, 4460, 7972, 9451; cf 6237, 9468; *Mesocricetus* 1382, 2419, 4460; *Neotoma* 11696; *Pachyuromys* 4675, 6234, 6237; *Peromyscus* 4459, 4460, 12005; *Tatera* 7146; *Zygodontomys* 11373).

The hemochorial or hemoendothelial, labyrinthine, discoid placenta has an antimesometrial, interstitial implantation into the duplex or bicornuate uterus (987, 993, 1138, 2419, 3011, 3269, 3270, 3752, 5735, 7563, 9451, 9707) and is eaten after birth (1853, 4991, 6237, 6302, 6803, 6807, 10573, 11319, 11696). Implantation be delayed by lactation (6238, 8805, 9114). Estrus occurs postpartum (*Akodon* 2409; *Baiomys* 5064; *Calomys* 2410, 4853, 8430; *Cricetomys* 3130; *Cricetus* 11319; *Deomys* perhaps 5760; *Desmodilliscus* 8696; *Desmodillus* 5684; *Dipodillus* 6235, 6237, 6238; *Gerbillus* 4713; *Malacothrix* 10154; *Meriones* 56, 1035, 4645, 6236, 6237, 6803, 6867, 7150, 7971, 7972, 8263, 9175, 9451, 9692; *Mystromys* 4220; *Neotoma* 2415, 2909, 3087; cf 3262; *Onychomys* 2907, 10697; *Oryzomys* 10572, 11388; *Ototylomys* 4606, 4607; *Peromyscus* 1236, 1363, 2527, 2529, 2806, 3897, 4119, 5797, 6149, 6150, 6493, 6708, 7072, 10573, 10574; *Psammomys* 2234; *Reithrodontomys* 9107; cf 10587a; *Rhombomys* 7778; *Scotinomys* 4967; *Sigmodon* 4898, 7253; *Tatera* 987, 1815; cf 7146; *Tylomys* 4606, 10755; *Zygodontomys* 87, 11373) except in *Beamys* (2912), *Mesocricetus* (2419), *Phyllotis* (7241), and *Saccostomus* (11082).

Some cricetids are furred at birth (*Lophiomys* 5760) and open their eyes at 0-2 d (*Sigmodon* 4898, 7030, 7253, 9572, 10570), 0-4 d (*Otomys* 331, 2338, 2339, 2571, 5760, 10158), 6 d (*Ototylomys* 4606, 7624), 14 d (*Akodon* 7212), or 18-21 d (*Saccostomus* 2861, 4329, 10158). Others are nearly naked (few hairs or sparse coat) and open their eyes at 6-9 d (*Calomys* 5474, 7189, 8430), 6-10 d (*Oryzomys* 1474, 7190, 10572, 11305, 11931), 11-16 d (*Eligmodontia* 6807), 12-20 d (*Podomys* 6301), 2 wk (*Beamys* 298, 2912), or 15-18 d (*Nyctomys* 946). *Neotoma* and *Tylomys* are born either furred or naked and open their eyes in 2-15 d (*Tylomys* 4606, 4607, 4608, 10755) or 11-21 d (*Neotoma* 1474, 2679, 2906, 2909, 3041, 3087, 3201, 4281, 6490, 7030, 8315, 8657, 8659,

8859, 8860, 9106, 9689, 10581, 11888, 11930). Other cricetids are naked and blind at birth (*Bolomys* 2371; *Calomys* 1502; *Calomyscus* 9175; *Delanymys* 5760; *Dendromus* 187; *Gerbillus* 8727; *Holochilus* 7192; *Malacothrix* 10158; *Meriones* 7098; *Peromyscus* 1474, 5886, 7885; *Reithrodontomys* 1474, 10587; *Tatera* 8500, 8501). The eyes of these altricial cricetids open at 6-8 d (*Zygodontomys* 87, 11373, 11931), 7-9 d (*Calomys* 4853), 8-13 d (*Reithrodontomys* 4915, 6300, 6391, 8189, 9107, 10099, 10587a; cf 5609), 8-23 d (*Desmodillus* 5684, 7824, 10158), 10-14 d (*Otomys* 11710a), 10-15 d (*Cricetulus* 2731, 4695, 6602, 7503, 9175), 10-20 d (*Onychomys* 504, 1474, 4986, 4991, 9366, 10588), 10-28 (*Peromyscus* 1236, 2727, 2910, 3897, 4125, 4609, 4780, 5748, 6149, 6150, 6302, 6493, 7303, 7304, 7311, 8698, 8907, 9114, 9603, 9802, 10573, 10574, 10575, 10576, 10577, 11288), 11-12 d (*Phodopus* 9516, 11285, 11548), 11-14 d (*Ochrotomys* 6497), 12-14 d (*Petromyscus* 2481), 12-15 d (*Baiomys* 977), 12-16 d (*Mesocricetus* 1071, 1382, 3571, 4098, 6656), 12-22 d (*Meriones* 1035, 2233, 2917, 6237, 6803, 7096, 7754, 7778, 8428, 8961, 9175, 9692, 11252, 11905; *Scotinomys* 4967), 13 d (*Calomyscus* 3587), 13-15 d (*Psammomys* 2233, 2234, 8731), 14 d (*Cricetus* 3571, 7433, 8457), 14-22 d (*Gerbillus* 109, 4348, 4351, 4713, 5760, 5785, 9175, 11397), 14-24 d (*Gerbillurus* 403, 2481), 14-25 d (*Tatera* 987, 5051, 7168, 8725, 9175), 14-30 d (*Cricetomys* 103, 1147, 3130, 5051, 5760, 7528, 10158), 16-20 d (*Mystromys* 4246, 7168, 10158), 22 d (*Malacothrix* 5887), 22-24 d (*Dendromus* 2572), or 28 d (*Desmodilliscus* 8696).

Females have 4 mammae (*Deomys* 7983; *Eliurus* 7983; *Mystromys* 4220, 10158; *Neotoma* 502, 1474, 1563, 2679, 3041, 4281, 4745, 8659, 10581, 11888; *Otomys* 331, 4495, 9945, 10158; *Otonyctomys* 4494; *Ototylomys* 4494, 4606; *Parotomys*, 9945, 10158), 4-6 mammae (*Deomys* 4495, 5760; *Peromyscus* 502, 1474, 1475, 1475, 2910, 3323, 4494, 9612; *Petromyscus* 9945; *Rheomys* 1475, 4965), 6 mammae (*Brachyuromys* 7983; *Calomyscus* 9175; *Gerbillus* 4497; *Onychomys* 502, 1474, 4745; *Podomys* 1474; *Reithrodontomys* 502, 1474, 4745, 4963; *Rhipidomys* 5111), 6-8 mammae (*Cricetomys* 331, 3130, 4495, 7528; *Gerbillurus* 2381, 9945), 8 mammae (*Cricetulus* 7814, 9175, 10516; *Cricetus* 7433, 7814, 7815, 8457; *Dendromus* 331, 2439, 4495, 5760, 9945, 10158; *Holochilus* 5111; *Malacothrix* 5887, 9945; *Meriones* 1138, 4497, 8961; *Myospalax* 8050; *Nectomys* 5111; *Oryzomys* 1474, 1475, 3024, 5111, 10572; *Oxymycterus* 596; *Pachyuromys* 6237; *Phodopus* 3334; *Scapteromys* 596; *Tatera* 331, 4495, 9945, 10158; *Zygodontomys* 5111), 8-12 mammae (*Sigmodon* 502, 1470, 1474, 7253, 10570), 8-16 mammae (*Steatomys* 331, 5760, 9945, 10158), 10-12 mammae (*Saccostomus* 331, 5760, 9945, 10158, 11082, 11607), or 12-16 mammae (*Mesocricetus* 278, 3571, 7814, 7815). Males have seminal vesicles, preputial glands, ampullary glands, penial spines, and a baculum (352, 1473, 1818, 2731, 3752, 4105, 4964, 6495, 8457, 9603).

Details of anatomy (1138, 2419, 2732, 3331, 7877, 7965, 8243, 9878, 10244), endocrinology (4166; *Gerbillus* 5707; *Mesocricetus* 642, 2270, 6324, 8085; *Neotoma* 1884; *Peromyscus* 114, 751a, 3253, 4120, 5800, 10795; *Phodopus* 10768; *Psammomys* 5706, 5708), conception (3433, 8510), spermatogenesis (94, 989a, 7797c, 7843), copulation (2507, 2540, 3096, 5676; *Calomys* 667, 2188; *Cricetomys* 3130; *Meriones* 84, 1452a, 2528, 4928, 4929, 8858; *Neotoma* 2504, 2508, 3310; *Onychomys* 2519, 2533, 4991, 6209, 9366; *Oryzomys* 2502; *Peromyscus* 1898, 1899, 2503, 2509, 2510, 2511, 2512, 2515, 2516, 2517, 2526, 2534, 2535, 6607, 10119; *Phodopus* 9516; *Psammomys* 2234; *Reithrodontomys* 2331; *Tatera* 8728; *Tylomys* 4460), birth (998a, 2481), pregnancy failure (83, 1318, 2520, 2521, 2523, 2966, 4205, 4610, 5673, 7962, 7967, 9229, 10202, 11608, 11969, 11970), birth (1853, 5001, 8096, 8698, 10573), geographic variation (1563a), reproductive energetics (3812, 3814, 6975, 6976, 7030, 7303, 7305, 8147, 8915, 8915, 9424, 9619, 10322, 10843, 11548), and light (629, 10947, 12114; *Cricetulus* 6951; *Cricetus* 1571; *Meriones* 2614, 9451; *Mesocricetus* 1645, 4948, 4951, 4952, 7486, 8516, 8635, 10818a; *Onychomys* 3533; *Peromyscus* 684, 2265, 2706, 3412, 3804, 3805, 4002, 4554, 5019a, 5367, 5369, 5370, 5371, 5894, 6659, 6661, 6810, 8455, 8455a, 8756, 10245, 10751, 10751a, 11547, 11648, 11649, 11650, 11651, 11652; *Phodopus* 1183, 1643, 1644, 2803, 3864, 4885, 4890, 10252, 10382, 12008, 12009; *Sigmodon* 1837, 5368, 7254; *Zygodontomys* 4573a) are available.

NEONATAL MASS

Akodon dolores	f: 3 g; from graph	7213
	m: 3 g; from graph	7213
Akodon longipilis	3.3 g	
Akodon molinae	m: 3 g; from graph	7213
	f: 3 g; from graph	7213
Baiomys musculus	1.4 g; n3 emb, 1 lt	2341

Baiomys taylori	1.1 g; n21; <12 hr old	977
	1.20 g; sd0.08; n47	5064
	f: 1.2 g; n94	7545
	m: 1.2 g; n100	7545
Beamys hindei	3.2 g; r2.1-4.3; n13	2912
Bolomys lasiurus	3 g	2371
	f: 3.65 g; ±1.1; n10, 1 d	7193
	m: 3.70 g; ±1.1; n10, 1 d	7193
Calomys callosus	1.3-2.5 g	8430
	f: 2.2 g; n21, 1 d	7189
	m: 2.4 g; n21, 1 d	7189
	2.5 g	5474
	m: 2.7 g; n39	7545
	f: 2.8 g; n31	7545
Calomys lepidus	f: 2.0 g; n85	7545
	m: 2.1 g; n90	7545
Cricetomys gambianus	r14-36.4 g; n33	103
	19.5 g; r19-20; n2	4329
	f: 22.5 g; r22-23; n2	3130
Cricetulus barabensis	2 g	4695
Cricetulus griseus	1.5-2.5 g	7503
Cricetulus triton	≤5.25 g	6602
Cricetus cricetus	3.18 g; r2.4-3.8; n9	9440
	~7 g	3571
Desmodilliscus braueri	0.9 g	8696
Desmodillus auricularis	1.84 g; n5	7824
	4.4 g; n19	5684
Eligmodontia typus	1.725 g; r1.5-2.8; n16 yg, 4 groups	6807
Gerbillurus paeba	1.9 g; n23	2481
	2.3 g; n≥5	403
Gerbillurus tytonis	1.9 g; n22	2481
Gerbillus gleadowi	1.9 g; r1.8-2.0; n4	8727
Gerbillus perpallidus	2.24 g; sd0.29; n19 lt	11397
Gerbillus pyramidum	2-3 g; from graph	4348
	2.5 g	4713
Holochilus brasiliensis	4-6 g	7192
Malacothrix typica	1.1 g; sd0.32	5887
	1.3 g; r1.1-1.5; n2	10154
Megadontomys thomasi	4.5 g	6302
Meriones crassus	2.5-4.5 g	8428
	3.2 g; sd0.13	6803
Meriones libycus	4-7 g	8428
	5 g	11905
Meriones persicus	5 g	8428
Meriones shawi	3-5 g	9451
	3.5-6.0 g	8428
Meriones tamariscinus	~2.5 g; n6; from graph	8961
	3.55 g; mode 4.0-4.5; r2.5-6.5; n68	1035
Meriones tristrami	3 g	8428
Meriones unguiculatus	r2.5-3.3 g; means by lt sz & sex; n1220, 357 lt	7964
	2.8 g; sd0.28; n22 lt	7096, 7098
	3 g	7778
	3.0 g	6867
Meriones vinogradovi	3 g	8428
Mesocricetus auratus	1.9 g	4830
	m: 2.179 g; sd0.5074; n39	1071
	f: 2.225 g; sd0.4797; n40	1071
	2.5 g	4830
	m: 2.92 g; r2.0-4.0; n54, 1 d	8643
	f: 2.96 g; r2.0-3.8; n56, 1 d	8643
Mystromys albicaudatus	6.5 g; r5.0-7.8; n39	4246, 7168
Neotoma albigula	10.9 g; r9.3-12.5; n>9	9106
	f: 11.6 g; sd1.3; n17	9689
	m: 12.0 g; sd1.3; n10	9689
Neotoma cinerea	13.5 g; r12.0-17.3	2909

Neotoma floridana	10.4 g; r9.8-11.0; n5	3285, 8859
	r10.5-14 g; n6	10581
	12.25 g; r12-2-12.3; n2	7649
	14 g; r11.8-14.8; n21	4281
	14.8 g; ±0.3	7030
	~15 g	8659
	15.3 g; n16	5892
Neotoma fuscipes	10 g; n5	2679
	r12.5-13.8 g; n6 yg, 3 lt	11888
	~14 g	6490
Neotoma lepida	4.6 g; r4.4-4.8; n2; est premature	8860
	6.1 g; se0.05; litter size 5	1563
	f: 9.5 g; sd1.5; n10	9689
	m: 9.6 g; sd0.9; n8	9689
	10.4 g	2906
	10.5 g; se0.24; litter size 1	1563
Neotoma mexicana	f: 10.0 g; r8.9-11.9; n6	8105
	m: 10.8 g; r10.0-11.7; n6	8105
Neotoma micropus	10-13.4 g	1242
Neotoma stephensi	10.6 g; se1.1; n14	11232
Neotomodon alstoni	3.4 g; sd0.67	8096
Nyctomys sumichrasti	4.7 g ea; n2	946
Ochrotomys nuttalli	2.7 g; r1.8-3.6; n47	6497
Onychomys leucogaster	2.2 g; r1.7-2.7; n22 yg, 7 lt	10588
	2.8 g	4986
Onychomys torridus	2.34 g; r2.1-2.6; n7	4991
Oryzomys capito	3.7-4.0 g	11931
Oryzomys palustris	3.14 g; r2.35-4; n13	10572
Oryzomys subflavus	5.89 g; r4.42-7.51	11305
Otomys denti	~20 g	2571
Otomys irroratus	12.5 g; r9.6-15.5; n28	2338, 2339
	14-17 g	2446a
	f: 14.5 g; r13-16; n2	2571
Otomys sloggetti	10.6-12.2 g; n12	11710a
Otomys typus	~12.5 g	5760
Otomys unisulcatus	12.3 g; n1 emb possible near term	11267a
Ototylomys phyllotis	10.2 g; se0.3; n86	4606, 4607
Peromyscus boylii	2.2 g; n1	502
Peromyscus californicus	4.31 g	7013
	4.92 g; ±0.78; r3.6-5.8; n19	10573
Peromyscus crinitus	2.2 g; r1.8-2.6	2910
Peromyscus eremicus	2.23 g; r1.85-2.60	1236
	2.54 g; ±0.65; r2.1-2.9; n9	10573
Peromyscus gossypinus	2.19 g; ±0.33; n45	8698
Peromyscus hooperi	2.62 g; n16	9603
Peromyscus interparietalis	2.74 g; r2.55-3.00	1236
Peromyscus leucopus	1.5-2 g	5210
	1.73 g; se0.03; n66	3323
	1.79 g; se0.02; n232 yg, 52 lt	7304
	1.80 g; sd0.25; n101	4125
	1.8 g; n312	7303
	1.87 g; ±0.03; r1.4-2.4; n114	10573
	2.5 g; sd0.38; 2 d	4780
Peromyscus maniculatus	r0.9-1.6 g; n16	10574
	1.295 g; se0.027; n37 near term	9225
	1.46-1.91 g; n1835, 389 lt; r across litter size	7697
	1.48 g; n11	7013
	1.58 g; n73	7698
	1.60 g; n296	7699
	1.62 g; ±0.13; r1.3-1.9	10576
	1.63 g; ±0.12; r1.1-2.3	10577
	1.65 g; se0.1	7305
	1.65 g; sd0.178; n14	4218
	1.65 g; n103	7698
	1.7 g	5658

	1.70 g; sd0.178; n18	4218
	1.70 g; se0.16; n165	7308
	1.73 g sib pairs; se0.04; 1.85 g non sib pairs; se0.03	4749
	1.73 g; r0.8-2.4; n248	10573
	1.74 g; n57	7698
	1.77 g; sd0.87; n24	4218
	1.80 g; ±0.68; r1.3-2.2; n≥13	10575
	1.8 g; se0.02; n80	6493
	1.86 g; se0.039; n18	5751
	1.95 g; se0.16; n257	7308
	1.98 g; se0.082; n18	5751
	1.98 g; se0.05; n35, lt size 5; varies w/lt size	7312
	f: 2.0 g; se0.1; n37, 1 d	2727
	m: 2.1 g; se0.1; n55, 1 d	2727
	f: 2.6 g; n77	7545
	m: 2.6 g; n117	7545
	m: 2.7 g; n104	7545
	f: 3.1 g; n102	7545
Peromyscus megalops	3.9 g; r2.9-5.0; n3	6302
Peromyscus melanocarpus	4.5 g; se0.1; n41	9114
Peromyscus melanotis	1.67 g; ±0.20; r1.3-2.0; n44	10573
Peromyscus mexicanus	4.4 g; se0.2; n17	9114
Peromyscus perfulvus	3 g	4609
Peromyscus polionotus	1.2 g	6302
	1.36 g; se0.029; n37	9225
	1.41-1.70 g; across lt sz	5576
	1.59 g; se0.04; n44	5577
	1.70 g; sd0.158	11725
	1.74 g; sd0.268	11725
	m: 2.10 g; sd0.30; n50, 1 d	1603
	f: 2.19 g; sd0.28; n60, 1 d	1603
Peromyscus truei	2.31 g; ±0.42; r1.7-3.0; n21	10573
	2.32 g	7013
Peromyscus yucatanicus	2.5 g; se0.0; n92	6149
Petromyscus collinus	2.1 g; n5	2481
Phodopus sungorus	1.67 g; ±0.18; n many	11548
	2.5 g; n15	11285
Podomys floridanus	2.4 g; r1.9-2.9; n15	6301
Psammomys obesus	5.5 g; sd0.6; n9	8731
	~7 g	3513
Reithrodontomys fulvescens	1.11 g; r1.07-1.15; n2	10587
Reithrodontomys humulis	r0.82-1.12 g; n8	2793
	1.2 g; r0.8-1.6	6300
	1.2 g; n27	5609
Reithrodontomys megalotis	1-1.5 g	5210
	1.5 g	10587a
Reithrodontomys montanus	0.91 g; r0.88-0.94; n2 dwarfs	6391
	1.1 g; r1.0-1.3	6391
Rhombomys opimus	4.5-5 g	7778
Saccostomus campestris	2.4 g; sd0.47; n173	11607
	2.8 g	2861
Scotinomys teguina	~0.5 g	4967
Scotinomys xerampelinus	~0.5 g	4967
Sigmodon hispidus	6.19 g; sd0.58; n10 1 d	9572
	6.3 g; sd0.82; n10 1 d	9572
	6.6 g; ±0.4	7030
	6.6 g; se0.1	8915
	6.7 g; se0.12	6975
	6.8 g; se0.07	6975
	7.23 g; r6.5-8; n19	10570
Sigmodon ochrognathus	4.5-6.6 g	4898
Steatomys pratensis	1.55 g; r1.5-1.6; n2	3693
Tatera afra	~4.5 g	7146
Tatera brantsii	4.4 g; n1	7168

	~5 g	2401
	6 g; n1 fullterm	7146
Tatera indica	3 g	8725
Tatera leucogaster	2.8 g; r2.6-3.0; n7	7797a, 7797b
Tylomys nudicaudus	10.2 g; ±0.3; n86	4607
	19.9 g; r15-24; n47	10755
	20.2 g; se0.3; n95	4606, 4608
Zygodontomys brevicauda	3.1-4.4 g	11931
	f: 3.5 g; ±0.15; n10	87
	m: 3.6 g; ±0.12; n15	87
	3.6 g; r2.8-4.8; n76 lt	11373
NEONATAL SIZE		
Beamys hindei	35.5 mm	298
Bolomys lasiurus	42 mm	2371
Cricetomys gambianus	HB: 69-87 mm	5051
	f: 89 mm; r89-89; n2	3130
Cricetulus griseus	30 mm	2731
Desmodillus auricularis	HB: 28.2 mm; n5	7824
Gerbillurus paeba	HB: 29.6 mm	403
	HB: 31.8 mm; n10	2481
Gerbillurus tytonis	HB: 34.2 mm; n17	2481
Gerbillus pyramidum	HB: 33 mm	4348
Malacothrix typica	HB: ~26 mm	5887
Meriones meridianus	30.5 mm; se0.5	11252
Meriones unguiculatus	37.1 mm; n22 lt	7096
Myospalax myospalax	80-100 mm	8050
Mystromys albicaudatus	HB: 53 mm	4246
Neotoma floridana	87.4 mm; n5	8859
Neotoma lepida	36.4 mm; r34-38.9; n2 est premature	8860
Neotomodon alstoni	CR: 29 mm; largest emb	11730
	40.5 mm; sd5.3	8096
Nyctomys sumichrasti	f: HB: 52.2 mm; r50-54.4; n2	946
Ochrotomys nuttalli	TL: 50.8 mm; r38-58; n47	6498
Otomys irroratus	HB: 71.1 mm; r64-76	2339
Ototylomys phyllotis	CR: 41-45 mm	4607
	HB: 67 mm; ±3.2; n86	4606
	TL: 119.5; ±0.9; n86	4606
Peromyscus gossypinus	TL: 47.16 mm; ±0.22; n45	8698
Peromyscus leucopus	32.1 mm; sd2.7; n101	4125
Peromyscus maniculatus	23 mm; se0.2; n37	9225
Peromyscus megalops	39.5 mm; r38-41.5; n2	6302
Peromyscus melanocarpus	44.8 mm; se0.2; n41	9114
Peromyscus mexicanus	43.9 mm; se0.7; n17	9114
Peromyscus polionotus	23.4; se00.2; n37	9225
Petromyscus collinus	HB: 33.7 mm; n3	2481
Reithrodontomys megalotis	7-8 mm	5210
Saccostomus campestris	47 mm w/tail	2861
Scotinomys teguina	~32 mm	4967
Scotinomys xerampelinus	~32 mm	4967
Steatomys pratensis	r73-75 mm; n3	4329
Tatera brantsii	33.8 mm; n1 fullterm	7146
Tatera inclusa	HB: 45 mm; n1	5050
Tylomys nudicaudus	CR: 50-60 mm	4607, 4608
	HB: 84.3 mm; se0.7	4606, 4608
WEANING MASS		
Beamys hindei	59.2 g; n10	2912
Calomys callosus	f: 9.4 g; n21, 16 d	7189
	m: 9.6 g; n21, 16 d	7189
	18 g	5474
	~20 g; 3 wk	5474
Cricetulus barabensis	14 g	4695

Cricetulus griseus	10.2 g; r7.9-15.4; 21 d	8685
	f: 15.1-16.8 g	468
	m: 15.9-17 g	468
Cricetus cricetus	f: 101.2 g; ±13; n63	9085
	m: 108.9 g; ±11.97; n60	9085
Gerbillurus paeba	12-15 g; ~20 d; from graph	403
Gerbillus perpallidus	17.23 g; n16	11397
Gerbillus pyramidum	11 g; 19-23 d	4351
	~20 g; from graph	4348
Meriones libycus	45 g; 30 d	11905
Meriones shawi	20-30 g; depends on lt sz	9451
Meriones tamariscinus	11.5 g; r10-13; n2	1035
	20 g; 21 d	8961
Meriones unguiculatus	12-15 g; 20 d	9692
	12.9 g; sd2.29	7098
	13 g; 21 d	6867
	f: 17.8 g; se0.3; r12-23; n82	7969
Mesocricetus auratus	22.93 g; sd6.9; n43, 7 lt, 21 d	4098
Mystromys albicaudatus	40 g	4220
Neotoma albigula	103.3 g; r86.1-113.9; n4, 66 d	9106
Neotoma cinerea	75 g	3087
	f: 85-132.3 g	2909
	m: 117.7-150.2 g	2909
Neotoma floridana	45.9 g; ±2.3	7030
Neotoma lepida	34.4 g; se0.16; lt sz 5	1563
	35.5 g; se1.64; lt sz 1	1563
Neotoma stephensi	35 g	11231
Ochrotomys nuttalli	12.5 g; r6.8-13.1; n40, 3 wk	6497
Onychomys leucogaster	13.1 g; n3, 23 d	4986
Onychomys torridus	9.2 g; n1, 20 d	4991
Oryzomys palustris	22 g; from graph	8231
Oryzomys subflavus	~35 g	11305
Otomys sloggetti	25-30 g; n10, 2 wk; from graph	11710a
Otomys irroratus	~30 g; n18; from graph	2338
Peromyscus californicus	f: 20.5 g	4119
Peromyscus crinitus	13.2-15.0 g; 4 wk	2910
Peromyscus hooperi	10.07 g	9603
Peromyscus leucopus	7.9 g; n100, 18 d	7303
	8.9 g; sd1.61; 20 d	4780
	8.9 g; se0.12; r6.2-13.1; n102	2492
	9.0 g; se0.4; n8; leave nest	11288
	m: 11.5 g; n45	8455
Peromyscus maniculatus	f: r7.10-8.80; n3 samples	4218
	7.3 g; se0.3; n8; leave nest	11288
	8 g	5658
	8.00 g; se0.492; n16,; 20 d	5751
	8.08 g; se0.310; n17, 20 d	5751
	m: 8.47 g; se0.25; n26, 20 d	10795
	f: 8.6 g; se0.1; n18, 21 d	2727
	8.65-11.76 g; n433 lt, 23 d; by litter size	7697
	f: 8.95 g; r8.8-9.1; n2, 21 d	10723
	m: 9.3 g; se0.2; n33, 21 d	2727
	m: 11.1 g; se0.3; n22, 21 d	2727
	f: 11.2 g; se0.4; n19, 21 d	2727
	11.6 g; se0.25; n64, 4 wk	6493
Peromyscus polionotus	5.29-6.57 g; by litter size	5576
	m: 5.56-8.41 g; 17-25 d	1603
	f: 5.71-8.76 g; 17-25 d	1603
Peromyscus yucatanicus	13.1 g; n50, 28 d	6149
Phodopus sungorus	f: ~10 g	3249
	m: ~17 g; 21 d	3249
	17.4 g; r11.7-21.5	5443
Podomys floridanus	~10-15 g; 3-4 wk	6301
Psammomys obesus	~40 g; 28 d	8731

Reithrodontomys fulvescens	3.0-3.5 g; n4 lt	8189
Reithrodontomys montanus	5.4 g; r4.3-7.5; 15 d	6391
Sigmodon hispidus	10-20 g	3850
	17.1 g; ±0.6	7030
	18.2 g; se0.46	6975
	21.7 g; se0.32	6975
	~27 g; 18 d; from graph	9572
Tatera leucogaster	16-22 g; 18-28 d	7797a, 7797b
Tatera valida	30 g	7794
Tylomys nudicaudus	20-60 g	10755
Zygodontomys brevicauda	~22 g; r15.9-27.6	11373
WEANING SIZE		
Beamys hindei	HB: 105-122 mm	4330
Gerbillurus paeba	HB: 48-58 mm; 20 d; from graph	403
Peromyscus leucopus	144.4 mm; n45	8455
LITTER SIZE		
Akodon azarae	3.5 yg; ±0.10; mode 3; r1-7; n222	2409
	4.1 emb; r3-7; n13 f	596
	4.6 yg	2221
	5.0 emb; r3-7; n12 f	596
Akodon boliviensis	5 emb; n1 f	8302
Akodon dolores	3.9 yg; r2-6	7213
Akodon longipilis	3.7; r2-5; n9 lt	4054
	3.78 emb; se0.11; r2-5; n51 f	8305a
	5 emb; r4-6; n2 f	7241
Akodon molinae	3.65 yg; mode 4-5; n275 lt	7212
	3.9 yg; r2-6	7213
Akodon olivaceus	5.08 emb; se0.38; r4-8; n12 f	8305a
	5.5 emb; se0.27; n9 f	3551
	5.5; r3-8; n18 lt	4054
	5.6 emb; sd1.1; n11 f	7241
Akodon urichi	5.0; r4-6; n9	7995, 7998
Andinomys edax	3 emb; n1 f	8302
Anotomys leander	1 emb; n1 f	11372a
Anotomys trichotis	1 emb; n1 f	11372a
Auliscomys boliviensis	3-4 emb	8302
Auliscomys micropus	4.07 emb; se0.26; r1-7; n27 f	8305a
	4.5 emb; r4-5; n2 f	4054
Baiomys musculus	1-4 emb	4228
	2.0	9176
	2 emb ea; n4 f	1475
	2.92 emb; n26 f	8188a, 8190
	3 emb; n1 f	2341
	3.9 emb; n9 f	11388
Baiomys taylori	1-5	1474
	2.2 weaned	7544
	2.49 yg; sd0.072; r1-5; n58 lt	8805
	2.5; r2-3; n4 lt	1785
	2.6 emb; r2-4; n8 f	8963
	2.72 yg; se0.13; r1-5; n39 lt	977
	2.8 yg	7544
	2.8; r1-5	217
Beamys hindei	2.8 yg; mode 3; r1-5; n39 lt	2912
	3.5; r3-4; n2 lt	298
	mode 4 yg; 4-7	4330
Blarinomys breviceps	1.25 emb; r1-2; n4 f	21
Bolomys lasiurus	3.3; sd0.5; r3-4; n7 lt	10494
	3.62 yg; mode 4; r1-6; n69 lt, 20 f	2371
	5.6; r2-9	7193
	mode 6.0 emb; max 13; many f	5540
Bolomys obscurus	3.5 emb; r3-4; n2 f	596
	5 ps; n1 f	596
	6.2 emb; se0.3; n32 f	7347

Calomys callosus	4.3; n324 lt	7191
	4.4; sd2.3; r2-8; n5 lt	10494
	4.5 yg; ±1.8; r2-9; n27 lt, 13 f	7188, 7189
	5.0 weaned	7544
	5.5 yg	7544
	5.5; r1-10; n65 lt	8430
	6.21 yg; n112 lt	5474
Calomys laucha	3.75 emb; r3-4; n4 f	1502, 4682
	4.0 yg; r3-5; n2 f	4682
	5.3 yg	7347
	5.7 or 6.0 emb; r4-8; n7 f	596
	6.2 emb; se0.1; n179 f	7347
Calomys lepidus	3.9 weaned	7544
	4.4 yg	7544
Calomys musculinus	5.4 yg; ±0.11; r1-15	2410
	7.5 emb; se0.2; n114 f	7347
Calomyscus bailwardi	2; n1 lt	9175
	3-5	3587
	4.29 ps; r2-7; n7 f	6293
Calomyscus mystax	3-7	9488
Cricetomys emini	1 emb; n1 f	2446a
Cricetomys gambianus	1-5	228
	1 emb; n1 f	2452
	2-4 juveniles	8078
	mode 3-4 yg	7528
	3 yg; n1 f	337
	3 emb, yg; n2 f	4329
	3.0; r2-4; n2 lt	1147
	3 yg; r2-4; n2 lt	2446a
	4; n1 lt	2439
	4 emb; n1 f	331
	4; r1-5	103
Cricetulus barabensis	5.33; r1-9; n12 lt, n5 f	4695
	7.1 emb; n106 f	3334
Cricetulus curtatus	5.0 emb; r4-6; n2 f	10516
	5.7 emb; n10 f	3334
Cricetulus eversmanni	4-6	4645
Cricetulus griseus	4.8 yg; 4.7 born live; mode 5; r1-8; 4.2 wean; n500 lt	468
	4.8 born alive, 4.35 wean; n31 lt, 10 pr	8685
	5.0	1693
	5.52 yg; mode 6; r1-9; n190 lt	7503
	5.53 yg; mode 6; r2-10; n39 lt	2730
	6.1 yg; r1-11; n>500	1554
	7.6 ovulations; ±0.2; r5-10	8510
	7.92 cl; ±0.41	2731
Cricetulus longicaudatus	5-6	3335
	6.1 emb; n36 f	3334
Cricetulus migratorius	2.3; r1-4; n3 lt	12128
	4-6	9175
	5.57 ps, emb; r4-7; n7 lt	6293
	6; r3-10	4645
Cricetulus triton	3; r1-5	12128
	mode 6; most 4-6	6602
Cricetus cricetus	3.4 emb; sd0.3; n25 f	3951
	5.83; r3-11; n6	9440
	mode 6-12; r4-21	8457
	6.75 yg; n6 lt	7436, 9085
	6.8 yg; sd1.6; n5 lt	3951
	mode 7-8; r5-12	7815
	7.6; r4-10; n27 lt	11319
	8.79 yg; ±1.35; n14 lt	7436, 9085
Delanymys brooksi	3 emb ea; n2 f	2446a
	3-4	5760
Dendromus kahuziensis	3.5; r3-4; n2 lt	187
Dendromus melanotis	mode 2-4; max 7	10158

	3-6	2449
	3-8 yg	5760
	3.5 yg; n4 lt	2572
	3.67 emb; r3-5; n9 f	2444, 2446a, 2452
	3.8 emb; n5 f	2572
	4.4 emb; r4-6; n5 f	331, 10154
	8 yg; n1 f	186
Dendromus mesomelas	2.75; max 4; n14 lt	8855
	3-8 yg	5760
	4 yg; n1 f	9945
	4.0 emb; n3 f	2572
	4.0 emb; r2-6; n3 f	4329
	5 emb; n1 f	2955
Dendromus messorius	3; n1 lt	187
	3 emb; n1 f	4495
Dendromus mystacalis	2.6 emb; mode 2-3; r2-4; n13 f	2754
	mode 3; max 4	2452
	mode 3-4; r3-7	5760
	mode 3-4; max 8	10158
	3 emb; n1 f	4925
	3.5 emb; r3-4; n2 lt	2572
	4 emb; r4-6	331
	6 yg; n1 lt	2572
	7 yg; n1 lt	187
Dendromus nyikae	3-8	5760
	4 emb; n1 f	4329
Deomys ferrugineus	1.59 emb; mode 2; r1-2; n32 f	2446a
	1.69 emb; mode 2; r1-2; n67 f	2573, 2574
	1.75; r1-2; n24 lt	8855
	mode 2 yg; r1-3	5760
Desmodilliscus braueri	2.6 yg; r2-3; n8	8696
Desmodillus auricularis	2-6	9945
	2 emb; n1 f	7823
	2 yg; r1-3; n5 lt	7822, 7824
	2.3 yg; r1-3; n3 lt, 1 f	5684
	3.9 emb; r1-7; n13 f	10154, 10158
Dipodillus simoni	4.68 yg; r1-7; n25 lt	6235, 6237, 6238
Eligmodontia typus	4.0; r2-7; n4	6807
	5.90 emb; se0.39; r3-9; n20 f	8307
	7.0 emb; r6-8; n2 f	4054
Gerbillurus henleyi	4.0; n2 lt	8134
Gerbillurus paeba	3-4	2381
	3.0 yg	7822
	3.5 yg; sd1.4; n11 lt	403
	3.7 emb; r2-5; n39 f	10154, 10158
	4 emb; n1 f	9945
	4.1 emb; n20 f	7823
	4.6 yg; n7	2481
Gerbilurus tytonis	4.4 yg; n7	2481
Gerbillurus vallinus	5 yg; n1 nest	10158
Gerbillus spp	3 emb; n1 f	299
Gerbillus andersoni	3.9; r3-7; n10 lt	8134
Gerbillus campestris	4.0 emb; mode 4; n2 f	7906
	5.0 emb; r3-6; n3 f	8134
Gerbillus cheesmani	8 ea; n2 lt	9175
Gerbillus dasyurus	3 ps; n1	10616
	4 emb; n1 f	4497
	4.4	109
Gerbillus gerbillus	3-5	12128
	4.3; r3-6; n7 lt	8134

Gerbillus gleadowi	2 yg ea; n2 lt	8727
	3; r2-5	9175
	3.5 emb; r3-4; n2 f	8727
	5.5; r5-6	8723
Gerbillus latastei	5 emb; n1 f	7906
Gerbillus nanus	3-7	5785
	3.25 ps, emb; r2-4; n4 f	6293
Gerbillus perpallidus	4.46; sd1.25; r1-8; n109 lt	11397
	5 emb; n1 f	8134
Gerbillus pyramidum	2 yg ea; n2 lt	4344
	3 ps; 5 emb	8134
	3.12; r1-6; n498 lt	4713
	3.5; r2-5; n2 lt	12128
	4.75 yg; r4-7; n4 lt	4348
Graomys griseoflavus	6.0; r3-8; n237 lt	2388
	8.1; r7-10; n7 lt	2387
Gymnuromys roberti	2 emb	7983
Habromys lepturus	1.9	9176
Hodomys alleni	1 emb; n1 f	3700
	2 yg; n1 lt	3700
Holochilus brasiliensis	1-5	7192
	4 emb; n1 f	596
	5 emb; n1 f	5111
Holochilus magnus	3 ps; n1 f	596
Holochilus sciureus	3.12 emb; r1-8; n354 f	10983
	3.47; r2-5; n30	10982
Hypogeomys antimena	1 yg	7983
Ichthyomys tweedi?	2 emb; n1 f	11372a
Irenomys tarsalis	3.5 emb; r3-4; n2 f	8305a
Isthmomys pirrensis	2.5 emb; r2-3; n4 f	11044a
Lophiomys imhausi	1 yg ea; n2 lt	2446a
	mode 2-3 yg	5760
Macrotarsomys bastardi	2.45 yg; r1-3; n11 lt, 5 f	6402
Malacothrix typica	3.9; r2-6; n10 lt	10154, 10158
	4.2 yg; r2-8; n9 lt	5887
Megadontomys thomasi	2.0	9176
	2.6; se0.19; max 4; n21 lt	2728
	3.5	6302
Meriones crassus	1 yg; n1 lt	2109
	3.3 emb, ps, nestlings; n10 lt	8134
	3.5 yg; r3-4; n4 lt	2233
	3.7; mode 4; r1-8	12128
	4-5	9175
	4.4 yg; sd0.23; 3.6 wean; sd0.33; n15 lt	6803
	4.56 yg; r3-6; n9 lt	6236, 6237, 6238
	max 8	8428
Meriones hurrianae	3-5 emb; n13 f	8720
	4.0 emb; r3-5; n2 f	10616
	4.0 emb; r3-5; n4 f	79
Meriones libycus	3-4	9175
	3.7 yg; r1-7; n3 lt	2233
	4 emb; n1 f	6293
	4 emb; n1 f	7906
	mode 5-6	7778
	5.3 yg; r2-8	7778
	8.25; r8-9; n4 lt	11905
Meriones meridianus	4-7	4645
	4.6 yg; r1-9	7778
	5.8 yg; r1-9	7778
	6.4 yg; r4-12	7778
Meriones persicus	2 ea; n2 lt	9175
	6 emb; n1 f	2404a
	max 7 yg	7778
Meriones sacramenti	max 6	8428

Meriones shawi	4.91 yg; r2-7; n11 lt	6236, 6237, 6238
	5.79 yg; n159 lt	9451
	5.85; r1-10; n159 lt	9451
	6.5; mode 6; r1-16	12128
	max 8	8428
Meriones tamariscinus	3.64; mode 4; r1-7; n66 lt	1035
	mode 4-5; r1-8	7778
Meriones tristrami	2-7 yg	7778
	5.8; se0.3; n9 f	2528
	max 6	8428
Meriones unguiculatus	2-5	10596
	3.2 wean; late maturing f	1863
	3.2; r1-7; n48 lt	4588
	3.42; mode 3-7; r1-9; n357 lt	7964
	4.0 emb; 47 d gestation	7971
	4.1-5.2; varies with f age	56
	4.1; r1-11	9692
	4.5 born, 4.4 wean; r1-9; n187 lt	6867
	4.7 wean; early maturing f	1863
	4.83 yg; n272 f	7973
	4.89 yg; r2-9; n27 lt	7754
	4.9 yg; r2-8; n35 lt	7969
	5.0 yg; r1-9; n284 lt	382
	5.04 yg; r1-9; n757 lt, 103 f	1768
	5.12 yg; sd1.62; r1-9; n198	84
	5.2 emb; 26 d gestation	7971
	5.3 yg; se0.58; late maturing f	1863
	6.0 yg; se0.31; early maturing f	1863
	6.3 yg; r1-9; n40 lt; 5.35 wean	2917
	6.4; r2-11	7778
	6.6 oocytes	6866
	7-9 cl	1138
Meriones vinogradovi	mode 6-8	7778
	max 8	8428
Mesocricetus auratus	1-12	4830
	5.8 emb; se0.8; left; 9.2 emb; se1.9; right; n15 f	7989
	mode 6-12; r3-15	3571
	6.1 wean; sd3.8; 21 d	4098
	6.41 yg; max 12; n34 lt	1382
	6.5 cl; se0.3; left; 9.3 cl; se0.1; right; n70 f	7989
	6.93 yg; se0.23; sd2.414; r2-15; n107 lt	1071
	8.6; se0.63; n24	6210
	9.26; sd2.86; mode 8-10; r2-16; n128 lt	278
	r10.3-14.3 cl; n5 wgt gps; 2-16 implantations	9160
	11.1 yg; se0.3; n25 multiparous	5061
	11.6 yg; se0.8; n25 primiparous	5061
	11.7 yg; sd2.1; n7 lt	4098
	12.5 emb; se0.6; 15.7 cl; se0.7; n11 f	6210
	13-15 emb	2419
Mesocricetus brandti	5.97; r1-13; n171 lt	6656
	14.0 emb; r13-15; n2 f	7815
Mesocricetus raddei	8-10	4645
Myospalax myospalax	mode 1-2 yg; r1-4	8050
	4-5 yg; hunters reports	8050
	6 emb ea; n4 f	8050
Myospalax psilurus	2 emb; n1 f	12105
	4-6	1817
Mystromys albicaudatus	2-5	10158
	2.9; n51 lt	7168
	3 yg	4220
Neacomys tenuipes	3.8; r2-5; n12	7995, 7998
Nectomys squamipes	2.5 yg; r2-3; n2 f	3077
	3 emb; n1 f	3077
	4-5 yg; n3 f	3077

	5 yg	3077
	6 emb; r5-7; n2 f	3077
Neotoma albigula	1.5 emb; r1-3; n14 f	522
	2 yg; n1 lt	6173
	2.1 emb; r1-3; n13 f	3262
	2.1 yg; mode 2; r1-3; n29 lt	3201
	2.25 emb; r1-3; n8 f	502
	2.5 yg; r2-3; n12 f	9689
	2.7 emb; r2-3; n3 f	11462
Neotoma alleni	1 emb; n1 f	3700
Neotoma cinerea	2.75 emb; r1-4; n4 f	3262
	3 emb; n1 f	4745
	3.2 emb; r3-4; n5 f	2415
	3.4 yg; mode 3; r1-6; n34 lt	2909
	3.6 emb; r3-4; n8 f	10954
	3.75 emb; mode 4; r1-5; n8 f	2615
	4; n1 lt	5001
	4 yg; n1 lt	$5405b_1$
	4.7 emb; r4-6; n3 f	11462
Neotoma chrysomelas	1 emb ea; n2 f	11044a
Neotoma floridana	1 emb; n1 f	6298
	mode 2-3; r1-4	11930
	2-4	1474
	2.0; r1-3; n10 lt	8659
	2.3 yg; mode 2; r1-5; n9 lt	8859
	2.6; ±0.2	7030
	2.8; mode 3; n5 lt	949
	3 yg; n1 f	8657
	3; r2-4; n7 lt	8315
	mode 3-4; r1-4; n5 lt	10581
	3.2; r2-5; n8 lt	5892
	3.2 emb, ps; r2-7; n50 f	3851
	3.41 yg; r2-4; n12 f	949
	3.5 yg; r2-5; n4 lt	8859
	3.67 emb; r3-4; n6 f	949
	4 emb; n1 f	1812a
	4 emb; n1 f	3262
	4 emb; n1 f	$5405b_1$
	4 yg; n2 lt	$5405b_1$
	4.25 ps; r3-6; n4 f	6298
Neotoma fuscipes	2 yg ea; n3 f	11888
	2.6 yg; r1-4; n18 nests	11289
	2.667 emb; r2-3; n3 f	3041
	2.8 yg; r2-3; n7 f	3041
	3 emb; n1 f	3593a
	3.0 yg; mode 2; r1-3; n10 lt	6490
	3.5 emb; r3-4; n2 f	11289
Neotoma lepida	2 emb; n1 f	3262
	2.29 yg; mode 2; r1-5; n100 lt	2906
	2.6 emb; r1-5	1563
	2.64 yg; r2-4; n14 f	9689
	2.7; ±0.7; r1-4	6708
	2.8 yg; r1-5 yg; n211 lt	2907
	2.86 emb; r2-3; n7 f	6560
	3 emb ea; n2 lt	6488
	3.11 emb; r2-4; n9 f	2415
	4 yg; n1 lt	10453
	4 yg; n1 lt	8860
	5 emb; n1 f	11462
Neotoma mexicana	1-2 emb	4745
	r1-4 emb; n3 f	523a
	1.33 emb; r1-2; n3 f	522
	2 emb ea; n2 lt	4962
	2.1 yg; r1-3 yg; n8 lt	8105

	2.44 emb; sd0.27; n14 yg f	1366
	2.52 cl; sd0.31; n19 yg f	1366
	2.57 emb; r2-4; n7 f	3262
	2.67 emb; r2-4; n3 f	11462
	3 emb; n1 f	502
	3.43 emb; sd0.35; r2-5; n19 f	1366
	3.47 ps; sd0.39; r2-5; n22 f	1366
	3.66 cl; sd0.31; r2-5; n45 f	1366
Neotoma micropus	2 emb; n1 f	522
	2 yg ea; n4 f	8962
	2 emb; mode 2; n5 f	4962
	2 yg ea; n11 lt	3201
	2.12 emb; r1-3; n8 f	949
	2.24 emb; n42 f	8962
	2.4 emb; r2-3; n5 f	3262
	2.63 emb; se0.77; mode 2; r1-4; n35 f	11696
	2.74 yg; se0.83; mode 2; r1-4; n38 lt	11696
	2.81 ps; se0.81; mode 3; r1-4; n63 f	11696
	2.85 yg; r1-4; n27 f	949
	3 emb; n1 f	502
	3 emb ea; n2 f	11462
Neotoma phenax	2 emb; n1 f	1471a
	mode 2 yg; r1-2	5423
	2.7 emb; r2-3; n3 f	5423
Neotoma stephensi	1.08; se0.07; n159	11232
	2 emb ea; n4 f	4900
Neotomodon alstoni	2 emb; n1 f	4228
	3.1 yg; mode 3; r1-6; n137 lt	8096
	3.4 emb; r2-5; n12 f	2341, 2346, 2351
Neusticomys monticolus	2 emb; n1 f	11372a
Notiomys macronyx	4.5 emb; r4-5; n2 f	8305a
Notiomys valdivianus	3.5 emb; r3-4; n2 f	8305a
Nyctomys sumichrasti	2.0	9176
	2 emb; n1 f	11388
	2.0 yg; r1-3; n11 lt	946
	2.5 emb; r1-4; n2 f	3704
	2.5 emb; r2-4; n4 f	2590
Ochrotomys nuttalli	2 yg; n1 lt	2863a
	2.63 emb, yg; ±0.10; r1-5; n71 lt, several sources 2.7 emb; n10 f; 2.54 yg; n61 lt	1027
	2.65 yg; mode 2-3; r1-4; n85 lt	6497
	2.7 yg	6498
	3.5 emb, yg; r3 yg, 4 emb; n1 f	5187
Onychomys leucogaster	r2-5; n10 lt	10588
	3 ps; n1 f	4525
	3.54 or 3.6 yg; r1-6 yg; n205 lt	2907
	3.7; r1-6; n153 lt	8559
	4 emb ea; n1 f	1812a
	4 yg ea; n3 lt	4986
	4 emb; r2-5; n10 f	11957a
	4.25 emb; r4-5; n4 f	502
	4.58 emb; sd1.02; r3-7; n36 f	3294
	4.7 ps; r4-6; n3 f	11957a
	4.7 emb; r3-8; n9 f	5405b,
	5.0 emb; n2 f	522
Onychomys torridus	2.4; r1-5; n213 lt	8559
	2.6 emb, yg; r1-5; n105	10697
	3.0 emb; r2-4; n2 f	5409
	3.2 emb; r3-4; n5 f	522
	3.95 emb; r2-7; n21 f	10697
	4 emb; n1 f	502
	4 emb; n1 f	6560
	4 emb; r3-5; n4 f	1470
	7; n1 lt	2109

Oryzomys albigularis	3-4 emb	3024a
Oryzomys alfaroi	3 emb; n1 f	2590
	3.2-4.0; n2 locations	9176
	3.5 emb; r3-4; n4 f	11388
	4 emb; n1 f	5414
	4.0 emb; n2 f	3210
	4.0 emb ea; n2 f	4228
	4.7 emb; r4-6; n3 f	2228a
	7 emb; n1 f	7624
Oryzomys argentatus	1-5	10809
Oryzomys bicolor	3 emb; n1 f	3024
	3.5 emb; r3-4; n2 f	5111
Oryzomys bombycinus	3.33 emb; r2-4; n3 f	8550
Oryzomys caliginosus	1.5 emb; r1-2; n2 f	3317
Oryzomys capito	2 emb; n1 f	5111
	2.8 yg; max 5; n8 lt	3122
	2.9 yg; max 4; n11 lt	3122
	3.5 wean; max 6; n33 lt	11931
	3.68 yg; sd0.17; n35 f	3316
	3.75 emb; r3-4; n4 f	3024
	3.8; r2-6; n10	7995, 7998
	3.92 emb; 95%ci 0.56; r2-5; n14 f	3319
Oryzomys concolor	2.5 emb; r1-4; n2 f	5111
	3.6; r2-5; n7	7995, 7998
Oryzomys couesi	3.0 emb; r2-4; n3 f	3210
	3.3 emb; r1-6; n3	2590
	3.5 emb; r2-5; n2 f	11388
	4 emb; n1 f	2228a
	4 emb; n1 f	9542a
	4.125 emb; r1-6; n8 f	7624
	5 emb; n1 f	2351
Oryzomys delicatus	3.0 emb; r2-4; n2 f	5111
Oryzomys delticola	3.4 emb; r2-4; n5 f	596
Oryzomys flavescens	3.4 yg	7347
	5.1 emb; r3-7; n8 f	596
	5.2 emb; se0.3; n22 f	7347
Oryzomys fulvescens	3 emb ea; n2 f	1475
	3.75 emb; r3-4; n4 f	4228
	4 emb; n1 f	217
	4 emb; n1 f	5423b
	4 emb; n1 f	5423b
	4.0 emb; r3-6; n4 f	11388
	4.0 emb; r2-6; n8 f	5414
Oryzomys longicaudatus	4.5 emb; r3-6; n4 f	3551
	4.9 emb; r3-9; n44 f	4054
	4.93 emb; se0.23; r2-11; n44 f	8305a
	7.29 emb; r6-9; n7 f	8556
Oryzomys melanotis	3.3 emb; r3-4; n3 f	4228
	4 emb; n1 f	5423b
	4.6 emb; r3-6; n3 f	2590
Oryzomys nigripes	2-8 emb	5540
	3.0; ±1.3; r1-5; n18 lt	7190
	3.6 yg	2221
Oryzomys palustris	2-4 yg	11930
	3.0-3.8; n2 locations	9176
	3 emb; n1 f	5423b
	3.6 wean; mode 1-2; n149 lt	8231
	3.7 emb; r3-4; n3 f	1475
	4 emb; n1 f	5188
	4.2 emb; r3-7; n6 f	4228
	4.8 emb; r2-7; n20 f	7810
	5 emb; n1	8907
	5 emb; n1 f	950
	5 emb; n1 f	217

	5 emb; r3-7	1290
	5 yg; ±0.2; r4-6; n10 lt	2002
	5.9 cl; r3-12; n20 f	7810
Oryzomys subflavus	1-8 emb	5540
	4.1 yg; r3-6; n14 lt	11305
	5.5; n2 lt	10494
Osgoodomys banderanus	3 emb; n1 f	2341
Otomys angoniensis	2 emb; n1 f	337
	2.0 emb; r1-3; n3 f	4329
	2.5; r2-4; n10 f	10158
	2.8 emb; r1-5; n109 f	10712
	2.9 emb; r2-5; n14 f	8965
	3 emb; n1 f	4925
	3 emb; n1 f	334
	3.5 emb; r3-4; n2 f	2454
	3.7 emb; r3-5; n3 f	10154
	4 yg; n1 f	5051
Otomys denti	1.07; mode 1; r1-2; n14 lt	2571
	2 emb; n1 f	5051
	2.5; r1-4; n2 lt	4329
Otomys irroratus	1 emb ea; n2 f	4925
	1.25; max 3; n23 lt	8855
	1.33 emb; r1-2; n3 f	4925
	1.37 emb; mode 1; r1-3; n73 f	2571
	1.48; ±0.61; n81	1312
	1.59 yg; n44 lt	8384
	1.7 emb; r1-2; n3 f	2444, 2452
	2 emb; n1 f	4495
	2; n1 lt	185
	2.33 emb, yg; mode 2; r1-4; n39 f	2338, 2339
	3.2 emb; r1-7; n9 f	8965
Otomys maximus	3.7; r3-5; n3 f	10158
Otomys sloggetti	1.44 yg; se0.17; r1-2; n9 lt, 5 f	11710a
Otomys typus	1-2; r1-4	5760
	2 emb; n1 f	337
Otomys unisulcatus	1 emb; n1 f	11267a
Ototylomys phyllotis	1.75; r1-2; n8 f	1475
	2 emb; n1 f	6105
	2 ps; n1 f	950
	2 emb ea; n2 f	4494
	2 emb ea; n4 f	11044a
	2.0; r1-3; n5 f	7624
	2.0 emb; r1-3; n5 f	7624
	2.15; r1-5; n40	6270
	2.25 emb; r1-3; n16 f	5423b
	2.29 emb; mode 2; r1-4; n75 f	2590
	2.4 yg; r1-4; n36 lt	4606, 4607
	2.43; r1-4; n7 f	950
Oxymycterus rutilans	2.1 emb; r1-4; n7 f	596
	3.1 yg	2221
Pachyuromys duprasi	3.36; r1-6; n11 lt	6237
	3.5; r3-4; n2 lt, n1 f	4674
	5; r1-9	12128
Parotomys brantsi	2.1 emb; r1-3; n9 lt	10154
Parotomys littledalei	3 emb; n1 f	9945
Peromyscus attwateri	3 emb; n1 f	9601
	3.3 emb; r3-4; n3 f	11044a
	4.0 emb; r3-5; n3 f	9601
Peromyscus aztecus	3.0 emb; r1-4; n4 f	11044a
	3.5	9176
	4 ps; n1 f	11044a
Peromyscus boylii	2-6	1474
	2 emb; n1 f	217
	2 emb; n1 f	11228a

	2-3 emb; n4 f	502
	2.2-2.3; n2 locations	9176
	2.80 yg; sd0.41; mode 3; r1-4; n12	1363
	3-5 emb	2228a
	3 emb; n1 f	4962
	3 emb; n1 f	4228
	3 emb; r1-4; n8 f	523a
	3.1 emb; se0.12; n42 f	5243
	3.28 ps; sd0.35; n99	1363
	3.39 emb; sd0.81; n23 f	1363
	3.40 cl; sd0.42; n179	1363
	3.5 emb; r3-4; n4 f	6560
	4 emb; n1 f	1470
	4 yg; n1 lt	502
	5 emb; n1 f	11388
Peromyscus californicus	1.79; r1-2; n34 lt	7013
	1.87 yg; se0.09; r1-3; n15 lt	10573
	1.99 yg; r1-3; n190	9255
	1.9; se0.04; max 4; n493 lt	2728
	2.11; r1-3; n19 lt	7013
	2.5; ±0.8; r2-4; n6 f	6708
Peromyscus crinitus	2.9; se0.07; max 6; n206 lt	2728
	2.96 yg; r1-6; n122	9255
	3 emb; n1 f	2415
	3.07 yg; mode 4; r1-5; n130 lt	2910
	3.6 emb; r3-5	4226
	4.0 emb ea; n3 f	6488
Peromyscus difficilis	2.5 emb	2341
	2.8; se0.11; max 5; n88 lt	2728
	3 emb ea; n2 f	523a
	3 emb ea; n2 f	2341
	3.3 emb; r2-4; n3 f	2228a
	3.5 emb; r3-4; n2 f	522
	4 emb; r3-5	4962
	5.0; r4-6; n2 lt	502
	4.19 cl; sd0.92; r2-7; n57	1841
	4.24 ps; sd0.84; r2-6; n24	1841
	4.25 emb; sd1.38; r2-6; n31 f	1841
Peromyscus eremicus	1 emb; n1 f	5405b
	2-3 emb	2228a
	2.22 yg; r1-4; n14 lt	1236
	2.42 emb; se0.15; n12 f	3812, 3813
	2.42 yg; sd0.95; n404 lt	2319
	2.53 emb; r1-4; n13 f	6414
	2.55 yg; r1-5; n153	9255
	2.60 yg; ±0.24; r2-4; n5 lt	10573
	2.7 emb; r1-5; n37 f	522
	2.8; se0.03; max 6; n372 lt	2728
	2.9; ±1.2; r2-6; n21 f	6708
	3 emb; n1 f	1470
	3 emb; r1-4; n7 f	502
Peromyscus furvus	2 emb ea; n2 f	4228
Peromyscus gossypinus	3.5; r3-4; n4 lt	8907
	3.7 yg; ±0.14; mode 4; r1-7; n72 lt	8698
	3.9 emb; ±0.16; mode 4; r2-6; n32 f	8698
	5.67; r4-6 emb, 7 yg; n3 f	5187
Peromyscus grandis	2.5 ps; r2-3; n2 f	11044a
Peromyscus guatemalensis	2 emb ea; n2 f	11044a
Peromyscus hooperi	2.9 emb; n10 f	9603
	3.5 emb; n4 f	9603
Peromyscus interparietalis	2.4 yg; mode 3; r1-4; n40 lt	1236
Peromyscus leucopus	2.75 emb; r2-3; n4 f	950
	3 emb; n1 f	2341
	3 emb; r2-4; n2 f	5423b
	3.33 yg; r3-4; n3 f	9802

	3.33 emb; r3-4; n3 f	2025
	3.43 yg; ±0.13; r1-5; n21 lt	10573
	3.48 emb; r2-5; n83 f	10564
	3.5 emb, yg; se0.21; r1-6; n35 lt	11877
	3.5 yg; mode 3; r1-6; n60 lt	4125
	3.58 emb, ps; sd1.31; n12 f	67
	3.7; r1-7	9164
	3.74 emb; sd0.51; r1-6; n42 f	1363
	3.8 emb; r2-6; n24 f	6298
	3.86 emb; r3-5; n7 f	2548
	3.9 emb; r1-6; n10 f	522
	3.91 emb; se0.34; n11 f	3812, 3813
	4 emb; n1 f	575
	4 emb ea; n3 f	2547
	4 ea; n6 f	6487
	4 emb; r3-5; n2 f	1812a
	4.0 yg; mode 4; r2-7; n91 lt	5460
	mode 4-5; r1-8	6321
	4.12 ps; r2-6; n8 f	950
	4.2 yg; se0.22; n39 lt	3978
	4.2; mode 4; r1-7; n137 lt	7533
	4.25 emb; r4-5; n4 f	495
	4.26; r2-6; n39 f	1472
	4.29; se0.10; mode 4; r1-8; n128 lt	6150
	4.3 ps; r1-7; n9 f	6298
	4.32 ps; sd0.97; n85	1363
	4.33 emb, ps; sd1.35; n21 f	67
	4.36 yg; ±0.10; r1-6; n53 lt	10573
	4.4 emb; r2-6; n43 f	5405b,
	4.46 cl; sd1.24; r2-8; n177	1363
	4.5 emb; r4-5; n2 f	576
	4.5 emb; se0.8; n12 f	654
	4.59 emb?; sd0.98; n116 f	2318
	4.60 yg; se0.13; mode 4.5; r2-7; n107 lt	7304
	4.67 yg; r1-9; n229	9255
	4.7; r3-6; n9 lt	11956
	4.7; r1-8; n400 lt	4779
	4.7; se0.06; max 9; n871 lt	2728
	4.76 yg; se0.21; mode 5; r2-8; n59 lt	3323
	4.76; se0.10; mode 5; r2-8; n110 lt	6150
	4.77 emb; ±0.18; mode 5; r3-7; n31 f	6555
	4.8 yg; se0.6; n12	2529
	4.8 emb; r4-6; n5 f	4228
	4.8-4.9 emb	4205
	4.89; mode 5; r4-5; n9 lt	7169
	4.94 emb; se0.34; mode 5; r2-7; n90 f	7303, 7304
	5 yg; n1 nest	4267
	5.0 emb; r4-6; n2 f	6797
	5.04 yg; ±0.08; r3-7; n50 lt	2082
	5.1 emb; r4-9; n40 f	10954
	5.20 emb; se0.36	758
	5.3 emb; r4-7; n3 f	502
	5.3; se0.2; n7 lt	684
	5.52 emb; se0.30	758
	6 emb; n1 f	261a
	6 emb; n1 f	3703
	6 emb; n1 f	9612
Peromyscus madrensis	3 emb; n1 f	11044a
Peromyscus maniculatus	r2-7; n19 f	7311
	3.00-4.80 ps; n26 lt, 4 samples	2806
	3 emb; n1 f	4228
	3.28 weaned; n72 lt, 21+ d	5019
	3.29 emb; r2-4; n7 f	575
	3.4 emb, yg; se0.07; r3-4; n52 lt	11877
	3.5 wean	7544

3.5 emb; r2-4; n14 f	6493
3.7; ±0.3; n25 lt, 7 f	2727
3.85 yg; n27 lt	9225
3.91 lg viable emb; se0.19; n55 f	6542
4.00-4.84 emb; n40 f, 4 samples	2806
4 emb; n1 f	11228a
4.0 emb; n2 f	502
4.0 emb; r3-5; n2 f	2548
4.0 emb, yg; r2-5 emb, 5 yg; n4 f	2544
4.08 yg; r1-9; n152 lt	10573
4.1 emb; r3-8; n7 f	522
4.1 emb; ±0.21; n14 f	6814
4.1 yg; n98 lt	4987
4.15 yg	7544
4.16 yg; r1-7; n32 lt	6493
4.25 yg; n25 lt, 0-2 d	5019
4.29 emb; r1-7; 4.38 ps; n405 f	9235
4.30 emb; se0.42; n10 f	3812, 3813
4.3; ±1.3; r2-7; n11 f	6708
4.3; n568 lt	3418
4.33 yg; se0.14; n80 lt	3150
4.4 emb; n71 lt	7081
4.4 ps; se0.06; mode 4; r1-9; n252	3624
4.4 emb; se0.02; mode 4; r1-9; n427 f	3624
4.42 med viable emb; se0.07; n325 f	6542
4.5; mode 4; r2-8	2088
4.5; ±0.31; n21 lt	2727
4.5 emb; sd1.12; n57 f	9423
4.5 emb; mode 5; r2-9; n62 f	5405b,
4.5 yg; r1-9; n186 lt	2907
4.5 emb; n217 f	722
4.55; sd1.10; r1-7; n65 lt	5658
4.575; mode 5; r2-7	7305
4.58 emb; mode 4; r1-9; n140 f	10256
4.6 emb; se0.11; n96 f	5243
4.6; se0.01; max 10; n711 lt	2728
4.6 cl; se.05; mode 4; r1-11; n747	3624
4.65 emb; se1.67; r1-8	2699
4.67; r4-5; n3 lt	1785
4.7; se0.4; n11 lt	5673
4.70 emb; sd1.14; r2-8; n71 f	3294
4.7; se0.04; max 11; n2285 lt	2728
4.71; mode 5; r1-9; n69 lt	4218
4.72 yg; r2-9; n389 lt	7697
4.73 sm viable emb; se0.04; n808 f	6542
4.75 cl; ±1.12; n57 f	9423
4.75 yg; r1-11; n610	9255
4.8 yg; se0.2	4749
4.8 emb; r3-6; n5 f	2415
4.8; r4-7; n6 lt	10574
4.8 yg; r1-7; n64 lt	5797
4.84 yg; r1-9; n70	9255
4.89 emb; r4-6; n9 f	6560
4.89 yg; mode 5; r2-7; n37 lt	7013
4.9; r1-7; n22 lt	9722
4.9 ps; n82 f	7081
4.96 emb; r1-7; n59 f	3703
4.98 yg; n804 lt	7312
mode 5; r3-6	2060
5.0; r3-7	5748
5 emb; n1 f	738
5 emb; n1 f	7487
5 emb ea; n2 f	1812a
5 emb ea; n2 f	6796
5.0 emb; r4-6; n3 f	4962

Species	Litter size	Reference
	5.0 ps; r3-6; n3 f	261a
	5.0 emb; r4-6; n7 f	261a
	5.0 emb; r4-6; n7 f	11646a
	5.0 emb; r3-7; n22 f	6166
	5 emb; mode 5; r3-9; n48 f	9576
	5.03; se0.18; r1-9; n98	7306
	5.06 yg; mode 5; r3-7; n52 lt	7013
	5.1 emb; ±0.34; n11 f	6814
	5.14; se0.14; n104	7308
	5.15 live emb; r2-9; n285 f	723
	5.2 emb; se0.54; r3-7; n6 [5 in table]	10839
	5.2 yg; n465 lt	7698
	5.21 emb; mode 5; r4-6; n14 f	5798
	5.25; se0.09; n102	7308
	5.25; se0.07; n132 lt	7310
	5.27 emb; r4-7; n11 f	1119
	5.28 ps; r2-9; n288 f	723
	5.3 ps; r4-7; n18 f	11956
	5.3 yg; n332 lt	7699
	5.31 emb; ±0.49; n13 f	1364
	5.33 ps; ±0.41	1364
	5.33 yg; r3-7; n3 f	1853
	5.35 emb	551
	5.38 yg; ±0.13; r2-8; n50 lt	2082
	5.4 ps; r4-9; n14 f	6166
	5.5; mode 5-6; r2-9; n53 lt	9891
	5.57 cl; ±0.46; n25	1364
	5.6 emb; r3-7; n17 f	11956
	5.6 emb; se0.12; r2-9; n111 f	11230
	5.71 emb; n62 f	6015
	5.74 cl; r1-12; n288 f	723
	5.9 ps; se0.31; r2-10; n34 f	11230
	6 emb; n1 f	10219
	6.0; se0.7; n12	2517
	6.1; r5-9; n21 lt	9891
	6.2; se0.6; n13 lactating f	2517
	6.5 emb; r6-7; n2 f	10218, 10220
	8.3 ps; 5.2 emb; ±0.81; n10	4610
Peromyscus megalops	2.0	9176
	3 emb; n1 f	2341
Peromyscus melanocarpus	1.8 yg; mode 2; r1-3; n28 lt	9114
	2.3 emb, ps; mode 3; r1-3; n30 f	9114, 9176
Peromyscus melanophrys	2 emb ea; n2 f	523a
	2.6 yg; se0.3; r1-4; n24 f, 24 lt	3228
	3 ps; n1 f	2351
	3 emb ea; n2 f	2228a
	3.3; se0.16; max 4; n33 lt	2728
	3.5 emb; r3-4; n2 f	217
Peromyscus melanotis	3.7 emb; r1-5	2341
	3.8 emb; r1-5; n13 f	522
	4.26 yg; se0.18; r1-8; n27 lt	10573
Peromyscus melanurus	2.7 emb; r2-3; n3 f	6149
	2.9	9176
Peromyscus merriami	3.5 emb; r3-4; n2 lt	5409
Peromyscus mexicanus	1 emb; n1 f	2341
	mode 2	6149
	2.1 yg; r1-3; mode 2; n9 lt	9114
	2.4 emb; r2-3; n10 f	4228
	2.5 emb; r1-4; mode 3; n17 f	11388
	2.6 emb; r2-3; mode 3; n20 f	9114, 9176
	2.7 emb; r2-3; n6 f	3211
	2.7 yg; mode 2-3; r1-4; n20	6149
	3.0 emb; r2-4; n2 f	2228a
	3 emb ea; n4 f	1475
Peromyscus ochraventer	5 ps ea; n2 lt	217

Species	Litter size	Reference
Peromyscus pectoralis	2.9 emb; r2-5; n20 f	522
	4 emb; r3-5; n3 f	3108
	4.6 ps; r3-9; n8 f	3108
	5.0 emb; r3-7; n5 f	2343
Peromyscus perfulvus	2.6 yg; r1-3; n11 lt, 4 f	4609
Peromyscus polionotus	2.67; n27 lt	2262
	3.1 emb; n63 f	1547
	3.135 emb; sd0.86; 52% mode 3; r1-5; n172 f	1548
	3.35 yg; n27 lt	9225
	3.59 yg; sd1.10; mode 3-4; r1-7; n290 lt	11725
	3.64 yg; se0.39; n11 lt	3812, 3813
	3.8; se0.09; max 9; n636 lt	2728
	3.89 yg; r1-9; n217	9255
	3.97 yg; n32 lt	8907
	4.0 emb; n13 f	8907
	4.68; mode 4; r1-8; n179 lt	5576
	5.5 emb, yg; r5 emb, 6 yg; n2 f	5187
Peromyscus sitkensis	6	1474, 2088
Peromyscus stirtoni	2.0 emb; r1-3; n8 f	11044a
	3 emb; n1 f	5408
	3.0 ps; r2-5; n3 f	11044a
Peromyscus truei	2.67 emb; r2-5; n6 f	522
	2.84 yg; ±0.14; r2-5; n19 lt	10573
	3 emb ea; n2 f	6560
	3 emb ea; n3 f	4962
	3.32; r2-5; n22 lt	7013
	3.5 emb; r3-4; n2 f	11228a
	3.52; r1-5; n27 lt	7013
	3.6 yg; r1-6; n110 lt	2907
	4.29 emb; r2-6; n7 f	2415
	4.65; se0.912; r1-8	2699
Peromyscus yucatanicus	2 emb; n1 f	4494
	2.8 emb; mode 3; n9	6149
	3.0 emb ea; n5 f	6269
	3.33 emb; r3-4; n3 f	950
	3.5 yg; mode 3-4; r1-5; n63	6149
	5 ps; r3-8; n5 f	950
Peromyscus zarhynchus	2.0 emb; r1-3; n14 f	6149
Petromyscus collinus	2 emb; n1 f	9945
	2-3	11864
	2-3 emb ea; n12 f	9945
	2-3; n13 f	10158
	2.8 yg; n7	2481
Phodopus campbelli	9 emb; n1 f	12105
Phodopus roborovskii	6.3 emb; r3-9; n13 f	3334
Phodopus sungorus	2.6 wean	5443
	3.6 yg	5443
	4.99 yg; r1-9; mode 5; n281 lt	3249
	5.6 yg; ±1.7; r2-10; n173 lt, 30 pairs	11548
	8.0 emb; r5-12; n25 f	3334
Phyllotis darwini	3 emb; n1 f	8556
	3.94 emb; r2-7; n16 f	4682
	4.0 emb; n6 f	8302
	4.14 emb; r2-7; n14 f	4682
	4.17; n70 f	11407
	5.1 emb; sd1.0; n11 f	7241
	5.2 emb; se0.58; n5 f	3551
Phyllotis osilae	4.38 emb; r2-6; n8 f	4682
Podomys floridanus	1.72 yg; r1-3; n25	9255
	2.1 yg; r1-3; n10 lt	2551
	2.25 emb; se0.48; n4 f	3813
	2.4; se0.05; max 5; n239 lt	2728
	2.6 yg; r1-4; n≤19	6301
	3.1 yg; r2-4; n8 lt	6301
	3.2 emb; r2-5; n57 f	6301

Psammomys obesus	mode 3-4; max 5	10481
	3.2; r1-8; n10 lt	8134
	3.3; r2-5; mode 3; n6 lt	2233, 2234
	3.6 yg; sd1.3; r1-6; n56 lt	3513, 9825
	4.1; sd1.4; n91 lt	8731
Punomys lemminus	2 emb; n1 f	8302
Reithrodon physodes	mode 2; r1-4	12128
	3.5 emb; r3-4; n2 f	596
	4.53 emb; r1-8; n17 f	8306
Reithrodomtomys brevirostris	3.5 emb; r3-4; n2 f	5413
Reithrodontomys chrysopsis	3.5 emb; r3-4; n2 f	4963
Reithrodontomys creper	2 emb; n1 f	4963
	2 emb ea; n2 f	3024a
	2.6 emb; r2-4; n10 f	11044a
Reithrodontomys darienensis	4 emb; n1 f	11044a
Reithrodontomys fulvescens	2 yg; n1 f	10587
	3 emb; n1 f	523a
	3.0 yg; r2-4; n9 lt	1563a
	3.5 emb; r3-4; n4 f	4228
	4-5 emb	2228a
	4.0; r3-5; n2 f	4963
	5 emb; n1 f	1475
	6 emb; n1 f	10261
Reithrodontomys gracilis	3.0 emb; r2-4; n2 f	12037
	3.2 emb; r3-4; n5 f	5423b
	3.3 emb; r3-4; n3 f	12037
Reithrodontomys hirsutus	4 emb; n1 f	4963
Reithrodontomys humulis	1.88; n8 lt	6300
	2 yg; n1 f	4915
	3.2; mode 3; r2-4; n9 lt, 9 f	5609
	3.4 yg; mode 3; r1-8; n25 lt	2794
Reithrodontomys megalotis	2.6 yg; r1-7; n25 lt	10587a
	3 yg; n1 f	10099
	3 emb ea; n2 f	2341
	3.5 emb; r3-4; n2 f	6560
	3.5 emb; r2-4; n4 f	10954
	3.83; r1-7; n198 lt	9107
	3.84 yg; n114 lt, 32 f	550
	4 emb; n1 f	4228
	4 emb ea; n4 f	1812a
	4.3 emb; r2-8; n75 f	5405b,
	4.3 emb; r1-7; n7 f	502
	4.4 emb; r2-6; n39 f	3399
	4.67 emb, ps; r5,5 emb, 4 ps; n3 f	11628
	5 emb; n1 f	6487
	5.17 emb; r5-6; n6 f	3703
	5.27 emb; r3-8; n10 f	551
	6 emb; n1 f	9642
	6 emb; r5-7; n4 f	261a
	7 emb; n1 f	4745
	8.5 emb; r8-9; n2 f	6552
Reithrodontomys mexicanus	2.5 emb; r2-3; n2 f	4963
	3.0	9176
	3 emb; n1 f	2228a
	3 emb ea; n2 f	11388
Reithrodontomys montanus	2.86 yg; r2-4; n14 lt	6391
	3 emb; n1 f	261a
	3-7 yg	1871a
	~4; r1-9	11700
	4 emb; r3-5; n3 f	5405b,
	5 emb; n1 f	1812a
	5 yg; n1 f	5405b,
Reithrodontomys raviventris	3.7-4.0	9887
	~4	10809

Reithrodontomys sumichrasti	2.7 emb; r2-4; n10 f	4963
	3.5	9176
	4 emb; n1 f	11044a
Rheomys raptor	1 emb; n1 f	11372a
Rheomys thomasi	1 emb; n1 f	1475
	1 emb; n1 f	11372a
Rhipidomys mastacalis	3.33 emb, yg; r3-5 emb, 2 yg; n3 f	5111
	3.8; r3-5; n5	7995, 7996, 7998
Rhombomys opimus	3.84; n32 lt	10202
	5-6	10516
	mode 5-6; most 4-7; r1-14	7778
Saccostomus campestris	4; n1 lt	11935
	4.8; r2-8	2860, 2861
	5 emb; n1 f	9893
	5.1 emb, yg; r2-9; n10 f	4329
	5.1 ps; r3-9	4329
	6.5 emb; n2 f	331
	6.7; r1-10; n7	10158
	7 emb; n1 f	2452
	7.1 yg; sd3.99; r3-13; n18 primiparous	11082
	7.4; r5-10; n8 lt	10154
	7.9 yg; sd2.34; r3-11; n27 multiparous	11607
	8 emb; n1 f	9945
	10 emb; n1 f	9892
	11 yg, 4 wean; n1 lt	8385
Saccostomus mearnsi	5.0 emb; r4-6; n2 f	4925
	7 emb; n1 f	2443
Scapteromys tumidus	3.4; r2-4; n7 lt	4683
	4.0 emb; r3-5; n11 f	596
Scotinomys teguina	2 emb ea; n4 f	3024a
	2.28 emb; sd0.643; r1-3; n21 f	4967
	2.73 yg; sd0.845; mode 2-3; r1-5; n156 lt	4967
	3 emb; n1 f	1475
Scotinomys xerampelinus	2.67 emb; sd0.707; r2-4; n9 f	4967
	3.12 yg; sd0.669; mode 3; r2-4; n83 lt	4967
Sekeetamys calurus	2.9 yg; max 6; n47 lt	7983
Sigmodon alstoni	5.0; r2-6; n6	7995, 7996, 7998
Sigmodon fulviventer	mode 7-8	1527
Sigmodon hispidus	r2-8 emb; n10 f	1475
	2.79 emb, yg; n14 lt	1155
	3 emb; n1 f	11507a
	3 yg; n2 lt	6303
	3.3 emb; r2-7; n14 f	4228
	3.5 emb, yg; n2 lt	1155
	3.5 emb; r2-5; n2 f	8907
	3.5 emb; r3-5; n4 f	3317
	3.5 emb; r2-4; n4 f	11388
	3.52 emb; mode 3; r2-6; n27 f	7624
	3.65 emb; r2-6; n11 f	8316
	3.85 emb; r1-6; n25 f	6303
	3.86 emb; r3-7; n7 f	950
	3.9 emb; r2-6; n27 f	7624
	4.0 emb; r2-6; n4 f	6303
	4.0 emb, yg; n24 lt	1155
	4 cl; n1	1852
	4.2 yg; se0.1; n63 lt	8146
	4.5 emb; n2 f	2228a
	4.6 ps; r3-7; n5 f	6303
	4.6 emb, yg; n14 lt	1155
	4.7 yg; se0.52; n7	6975
	4.7 yg; se0.28; 4.4 at 12 d; se0.27; n16 lt	6976
	4.7 emb; r2-10; n56 f	2590
	4.75 yg; r3-6	10570
	4.8; ±0.3	7030

	5 emb; n1 f	523
	5 emb; n1 f	8196
	5 emb; n1 f	9199
	5 emb; r3-9	522, 523
	5.0 emb; r3-10; n7 f	5423b
	5.0 yg; se0.4; r2-7; n16 lt	8915
	5.0; r2-8; n41 lt	1563b
	5.1; r1-8; n25 lt	1563b
	5.12 emb; r3-8; n26 f	8026a
	5.3 yg; se0.2; n52 lt	8146
	5.3; r1-13; n56 lt	1563b
	5.39; se0.10; r1-15; n626 table, n627 text	1563b
	5.47 yg; r2-11; n60 lt	5806
	5.5 yg; r1-8; n10 lt	1563a
	5.6 emb, yg; n30 lt	1155
	5.6 yg; r2-10; n44 lt	7253
	5.7 emb, yg; n3 lt	1155
	5.81 emb; r1-11; n75 f	3849
	5.9-6.3 emb; 8-9 cl	4213, 4213a
	6 emb; r3-9	1290
	6 ps; r5-7; n2 f	950
	6.10 yg; r2-10	9572
	6.1; r4-12; n11 lt	1563b
	6.1 yg; se0.2; n73 lt	8146
	6.2 emb; r1-15; n67 f	4212
	6.3 emb, yg; n6 lt	1155
	6.5 yg; r6-7; n2 nests	10492
	6.53 yg; r1-12	9572
	6.6 emb; r4-10; n16 f	9710
	6.6 yg; r2-15; n28 lt	5726
	6.7	3308a
	6.9 yg; se0.41; 6.6 at 12 d; se0.35; n17 lt	6976
	7 emb; n1 f	1812a
	7.0 yg; se0.39; n15	6975
	7.4; n27	1163a
	7.5 emb; r4-11; n2 f	502
	7.5 emb; r5-10; n6 f	3211
	7.6 emb; r6-9; n9 f	551
	8.3 ps; r3-30; n35 f	3849
	8.5 emb; r6-11; n2 f	2351
	9.04 emb; ±0.31; r3-14; n70 f	7029, 1563b for range
	10 emb; n1 f	3850
	12 emb; r8-17; n6 f	1470
	14.3 emb; r14-15; n3 f	5806
Sigmodon ochrognathus	3.0 emb; r2-4; n4 f	4898
	3.0 yg; r2-6; n8 lt	4898
	5.67 emb; r2-9; n6 f	522
Steatomys krebsii	4.4 yg; n5 lt; 2-8 d	331
	5 emb; n1 f	331
Steatomys minutus	5; n1 lt; ~12 d	331
Steatomys pratensis	3 emb; n1 f	5760
	3.0 emb; n2 f	331
	3.2 emb; r1-9; n5 f	10158
	3.5; r3-4; n2 lt	4329
	4.5; r2-7; n2 lt	10154
	5 emb; n1 f	4495
	6.0; n2 lt	9893
Tatera afra	3.98 emb; mode 4; r2-6; n62 f	7146

Tatera boehmi	3.5 ps; r2-5; n2 lt	4329
	5 emb; n1 f	2446a
Tatera brantsii	2.0 yg	7822
	2; n3 lt	7168
	2.64 emb; mode 3; r1-4; n42 f	7146
	2.94 cl; mode 3; r1-4; n49 f	7146
	3.2 emb; n6 f	7823
	3.3 emb; r1-5; n12 f	10154, 10158
	4.0; n2 lt	12128
Tatera indica	2-4 yg	8500, 8501
	4; r1-9; n13 lt	8725
	4.76 yg [text], 5.11 yg [tables]; r2-7; n27 lt	987
	4.78 emb; mode 5; r1-8; n74 f	5230, 5231
	5-8 yg; n>9	8720
	5 emb; n1 f	6293
	5.25 ps/emb; r3-4 ps, 5-7 emb; n4 lt	10616
	6 ps; n1 f	7382
	6.33 emb; ±0.318; mode 5-7; r2-10; n36 f	729
	6.38 ps; ±0.304; mode 5-8; r2-13; n50 f	729
Tatera leucogaster	mode 4-5; r2-8	9945
	4.2 emb; n13 f	9893
	4.38 emb; ±1.2; r2-6; n69 f	8384, 8386
	4.44 ps; ±1.63; r2-6; n148 lt	8386
	4.49 emb; r2-9; n53 f	10154, 10158
	4.6; r3-7; n7	4329
	4.6; r3-7; n21 lt	8965
	4.8 emb; ±0.36; r3-6; n10 f	7797a, 7797b
	5.6 emb; ±0.29; r3-8; n25 f	7797a, 7797b
	5.67 emb; r4-7; n3	9892
	5.67 emb; n6 f	331
Tatera nigricauda	4 emb; n1 f	2443
	5.22 live emb; se0.24; r1-8; n37 f	7794
Tatera robusta	4.6; r2-7; n5 lt	7794
	5 emb ea; 2 f	4925
	5.7 yg; r5-6; n3 f	5051
	8 emb; n1 f	5051
Tatera valida	2-6 yg	5058
	3.7 ps, emb; r2-5; n9 lt	10799
	3.8 emb; r3-4; n5 f	7794
	4 emb; n1 f	2443
	5 emb; n1 f	331
	5.0 nestlings; r4-6; n2 lt	4495
Taterillus emini	3.5 nestlings; r3-4; n2 lt	4495
Taterillus gracilis	2-6 yg	4354a
	3 emb; n1 f	10799
	4-6 yg	5058
Taterillus pygargus	3-5	5058
	4 yg	8695a
Tylomys nudicaudus	2.0 yg; n3 lt	9176
	2.3 yg; mode 2; r1-4; n47 lt	4606, 4608
	2.67 yg; r1-4; n24 lt	10755
	4.0 emb; n1 f	2590
Tylomys watsoni	3 emb; n1 f	3024a
Wiedomys pyrrhorhinos	3.8; sd1.8; r1-6; n5 lt	10494
Zygodontomys brevicauda	2 emb; n1 f	5111
	3.0 emb, yg; r2 yg, 3-4 emb; n4 f	3024
	3.7 wean; max 7; n33 lt	11931
	3.7 emb; r3-5; n7 f	3317
	4.1; r3-7; n44	7995, 7996, 7998
	4.5 emb, yg; r4 emb, 5 yg; n2 lt	451
	4.64 yg; ±0.27; n91	87
	4.85 yg; sd1.72; r1-11; n283 lt	11373

SEXUAL MATURITY

Akodon azarae	f: 75.3 d	2409
	m: 83.7 d	2409
	f/m: 60 d	2221
	f/m: min 2 mo	8305
Akodon dolores	f: 1st birth: 33.3% < 3 mo	7213
Akodon longipilis	f: 1st preg: 1-2 mo	8305a
	f: 1st birth: 112.3 d; sd3.3; n6 f	7241
	m: 47.5 g	7241
Akodon molinae	f: vag perf: 30 d: r25-30	7212
	f: 1st birth: < 3 mo	7212, 7213
	m: testes descend: 25 d; r21-30	7212
Akodon olivaceus	f: 1st preg: 2 mo	8305a
	f: 1 yr	7241
	f/m: 22-24 g	7241
Akodon urichi	f/m: 2.7 mo; n9	7995, 7998
Auliscomys micropus	f: 1st preg: ~2 mo	8305a
Baiomys taylori	f: 1st conceive: 28 d	5064
	f: 1st conceive: 60-70 d	8805
	f: 1st birth: 81.5 d; r64-101; n8 f	977
	f: 40 d	10110a
	m: testes descend: 70-80 d	8805
Beamys hindei	f: 1st repro: 5 mo; majority 6.5-8 mo	2912
	f/m: 136 mm	5760
Bolomys lasiurus	f: 1st birth: 4-5 mo	2371
	f: 40.9 d	7193
	m: 34.9 d	7193
Calomys callosus	f: vag perf: 20 d	8430
	f: vag perf: 40.1 d; ±7.6	7188, 7189
	f: vag perf: 42-46 d	5474
	f: most mate: 2 mo	8430
	f: 1st preg: 50 d	5474
	f: 1st birth: 6 wk	8430
	m: testes descend: 30 d	5474
	m: fertile: 42-45 d	5474
	m: 19.6 d; ±6.6	7188
Calomys callidus?	m: testes descend: 14.9 d; r14-16; n43	4853
Calomys musculinus	f: 72.5 d, 16.88 g	2410
	m: testes descend: 14.6 d; r14-19; n77	4853
	m: 50% 82.03 d, 19.8 g	2410
Cricetomys gambianus	f: vag perf; 21.63 wk; sd1.38; n4	103
	f: 1st estrus: 22.63 wk; sd1.75; n4	103
	f: 1st birth: 7 mo	5760
Cricetulus griseus	f: 1st mate: 48 d	1693
	m: testes descend: 1 mo	7503
	f/m: 8-12 wk	1693
Cricetus cricetus	f: vag perf: 60-90 d	11319
	f: 1st birth: 3 mo	11319
	f: 1st birth: 298.17 d; ±92.59; n23	7436, 9085
	f: 12 wk	7436, 9085
	m: 8 wk	7436, 9085
	f/m: 2nd yr	4645
Dendromus mesomelas	f: 67 mm	4329
	m: 74 mm	4329
Dendromus mystacalis	m: fecund: 60-63 mm	4329
Dendromus nyikae	f: 1st preg: 71 mm; n1	4329
	m: fecund: > 73 mm	4329
Dipodillus simoni	f: 1st mate: 2 mo	6237
	f: 1st birth: 5 mo	6235
	m: 1st mate: 3.5 mo	6237
	m: 1st sire: 9.5 mo	6235
Eligmodontia typus	f: 1st conceive: 6-8 wk	8307
Gerbillus pyramidum	f: vag perf: 70 d	4713
	m: 1st mate: 45 d	4713
	m: 1st mate: 75 d; n1	4348

	m: 1st sire: 93 d; n1	4348
	f/m: 75-80 d	4351
Gerbillurus paeba	f: 1st mate: 84 d; n1	403
	m: 1st sire: 63 d; n1	403
Holochilus brasiliensis	f: 2-4.5 mo; 107-165 g	7192
	m: 2-3 mo; 115-205 g	7192
Holochilus sciureus	f: 1st preg: 20-40 g	10983
	m: 1st sperm: most 30-50 g	10983
Malacothrix typica	f: 1st birth: 111 d; r102-120; n2	5887
	f/m: 1st mate: 51-72 d	10154
	f/m: 1st mate: >70 d	5887
Megadontomys thomasi	m: spermatogenesis: 75 d	6302
Meriones crassus	f: 1st mate: end of 2 mo	6237
	m: 1st sire: 82 d	6237
Meriones libycus	f/m: 3 mo	11905
	f/m: est ~3 mo	2235
Meriones shawi	f: vag perf: 59.6 d; mode 54-63; r44-83	9451
	f: 1st birth: 2.5-6 mo	9451
	f/m: > 4 mo	6237
Meriones unguiculatus	f: vag perf: 26.8 d; r13-43	1862, 1863
	f: vag perf: 40-60 d	6867
	f: vag perf: 41.3 d; ±4; r33-53; 27.7 g; ±0.4; r21-36; n82	7969
	f: vag perf: ~44-54 d	7964
	f: vag perf: 45 d	9692
	f: vag perf: 48.7 d; r40-76; n29	7754
	f: ovary/uterus wt: 90 d	7965
	f: 1st mate: 127 d; ±7; r60-272	56
	f: 1st birth: 72 d	1768
	f: 1st birth: 62 d	1862
	f: 1st birth: 110 d; ±9; n40	7969
	f: 1st birth: 118.7-161 d	1863
	f: 63-84 d	6867
	m: testes descend: 35 d	9692
	m: testes descend: 36.3 d; r28-45; n492	7754
	f/m: 1st mate: 10-12 wk	9692
Mesocricetus auratus	f: vag perf: 8-14 d; 8 g	1071
	f: 1st birth: 73.6 d; se1.50; sd8.99; r59-90	1071
	m: ~7-8 wk; 60-70 g	2419
	f/m: 36-40 d	3571
	f/m: 7-8 wk	1382
Mesocricetus brandti	f: 1st estrus: 46 d; se1; n60	4952
	f: 1st birth: 50 d	6656
	m: testes enlarge: 5-6 wk	4952
Mystromys albicaudatus	f: 1st birth: 146 d	7168
	f: 1st birth: 4.8 mo	4220
Neacomys tenuipes	f/m: 1 mo; n12	7995, 7998
Neotoma albigula	f: vag perf: 83.5 d; r80-87; n2	9689
	m: testes descend: 101 d; n4	9689
Neotoma cinerea	f/m: ~1 yr	3087
	f/m: 2 yr	1871a
Neotoma floridana	f: 5-6 wk	3285
	f: ~5 mo, 160 g	4680
	m: testes descend: ~200 g	4680
Neotoma lepida	f: 2-3 mo	2906
	f/m: 60 d	1474
	f/m: 1 yr	6708
Neotoma mexicana	f: ~2 mo	1366
Neotoma micropus	f: 5 mo	8962
	f: yr of birth	3262
Neotoma stephensi	f: 9-10 mo	11232
Neotomodon alstoni	f: vag perf: 49.0 d; sd10.1; r35-76; n78	8096
	f: 1st repro: 174.5 d; r65-241	8096

Onychomys leucogaster	f: 1st preg: 95 d	10588
	f: 1st birth: min 4 mo; most 5-6	8559
	f: 1st birth: 191 d; n1	2907
	m: 1st sire: 4 mo	8559
Onychomys torridus	f: 1st birth: 4 mo	8559
	m: 1st sire: 3-4 mo	8559
	f/m: 6 wk	4991
	f/m: 7 wk; field bred; 4-5 mo; lab	10697
Oryzomys bicolor	f/m: 3? mo	7995, 7996
Oryzomys capito	f: 1st ovulate: 50 d	11931
	f: 1st preg: 45-50 d	3316
	m: 1st sire: 46 d	11931
	f/m: <2 mo	3319
	f/m: 2 mo; n10	7995, 7998
Oryzomys concolor	f/m: 3 mo; n7	7995, 7998
Oryzomys longicaudatus	m: 1st mate: 1 mo	8305a
Oryzomys nigripes	f: vag perf: 57.6 d; ±8.2; r46-66	7190
	m: testes descend: 37.1 d; ±7.8; r32-43	7190
	f/m: next breeding season	2221
Oryzomys palustris	f: vag perf: 40-45 d	7810
	f: 1st mate: 7 wk	4280
	f/m: 50 d	1474
Otomys angoniensis	f: 139 mm	4329
	m: 158 mm	4329
Otomys denti	f: fecund: 142 mm	4329
	f/m: 90 g	2571
Otomys irroratus	f: 9-10 wk; 76 g	10158
	f: >5 mo; 90 g	2571
	m: 13 wk; 96.0 g	10158
	m: >5 mo; 80-100 g	2571
	f/m: 1st mate: 3-4 mo	2338, 2339
Otomys sloggetti	f: 11 wk, 90 g	11710a
	m: 16 wk, 119 g	11710a
Otomys typus	f/m: 1st mate: < 3 mo	5760
Ototylomys phyllotis	f: vag perf: 21-70 d; n40	4606, 4607
	f: 1st mate: 29 d	4606, 4607
	m: testes descend: 15-38 d	4606, 4607
	m: 1st mate: 175 d	4606, 4607
Oxymycterus rutilans	f/m: 90 d; based on wt	2221
Pachyuromys duprasi	m: 1st sire: 2 mo	6237
Peromyscus boylii	f: 1st estrus: 50.93 d; ±1.92; min 37; n33	1854
	m: 1st sperm: 60 d	1363
	m: 1st epididymal sperm: 65 d	1363
Peromyscus californicus	f: vag perf: 38.7 g; se0.7; n30	4119
	f: 1st estrus: 44.1 d; se1.0; n30	4119
Peromyscus crinitus	f/m: 1st mate: 70 d min, most 4-6 mo	2910
Peromyscus eremicus	f: 1st estrus: 39.21 d; ±1.54; min 28; n61	1854
	f: 1st birth: 10 mo; min 3-4	2319
	f: 1st repro: 103.75 d; sd23.61; r71-168; n87 f excludes 13 f >180 d	10054
Peromyscus gossypinus	f: vag perf: 43 d; n27	8698
	f: 1st conceive: 73 d	8698
	m: epididymal sperm: ~45 d	8698
Peromyscus hooperi	f: 1st mate: 69 d; sd6.99; r67-76; n3	9603
Peromyscus leucopus	f: vag perf: 38-40 d	4205
	f: vag perf: 38.6 d; se2.6; r27-75; n20	9226
	f: 1st estrus: 42-44 d	4205
	f: 1st estrus: 46.22 d; ±3.18; min 28; n32	1854
	f: 1st estrus: 51.0 d; se3.6; r30-101; n20	9226
	f: 1st mate: 10-11 wk	1474
	f: 1st birth: 3.5 mo	1472
	f: median 13.15 g; 95%ci 9.11-18.99; island	67
	f: median 17.20 g; 95%ci 13.98-21.16; mainland	67
	m: median 14.06 g; 95%ci 8.90-22.23; island	67

	m: median 22.85 g; 95%ci 17.21-30.33; mainland	67
Peromyscus maniculatus	f: vag perf: 32 d 50%	6550
	f: 1st estrus: 48.72 d; ±1.23; min 28; n186	1854
	f: 1st conceive: min 31-42 d; across generations	7312
	f: 1st birth: 60-74 d; n4 groups	4749
	f: 1st birth: 3-4 mo	11956
	f: yearlings	11230
	m: testes enlarge: 40-60 d	10795
	f/m: 1st mate: ~5-6 wk	1474
	f/m: 1st repro: 84.1 d; sd20.38	4204
	f/m: breed: most 1 yr	3774
	f/m: breed yr of birth rare	7310
	f/m: 4.5-9 wk (see section f)	5019
	f/m: 7 wk	3150
	f/m: ~50 d	2806
	f/m: 7-11 mo	6796
	f/m: yr of birth	2699
Peromyscus megalops	f: 46-48 d	6302
Peromyscus melanocarpus	f: vag perf: 79.3 d; r70-98; n6	9114
	m: scrotal swell; 124.4 d; r86-154; n9	9114
Peromyscus mexicanus	f: vag perf: 46 d; n1	9114
	m: scrotal swell; 66.5 d; r49-77; n4	9114
Peromyscus polionotus	f: vag perf: 26 d	8907
	f: 1st estrus: 29.64 d; ±0.47; min 23; n121	1854
Peromyscus truei	f: 1st estrus: 50.09 d; ±2.44; min 28; n34	1854
	f/m: yr of birth	2699
Peromyscus yucatanicus	f: vag perf: 51.5 d	6149
Phodopus roborovskii	f/m: 3 wk; r14-24 d	3334
Phodopus sungorus	f: 1st birth: 139 d; r109-233	5443
Phyllotis darwini	f: vag perf: 60.0 d; sd10.2; 40.8 g; sd6.5; n3	7241
	m: scrotal testes: 50.3 d; sd7.2; 46.3 g; sd6.4; n6	7241
Podomys floridanus	f: vag perf: 35 d; r18-48	6301
Psammomys obesus	f: vag perf: 53 d	2234
	f: vag perf: 57 d	8731
	f: 1st conceive: 92-106 d	2234
	f: 1st birth: 8-11.5 mo	9825
	m: testes descend: 54 d	8731
Reithrodon physodes	f: 1st preg: est 2 mo	8306
Reithrodontomys megalotis	f: 1st mate: 4.5 mo	1474
	f: 1st birth: 62 d; n1	9107
	m: 1st mate: 59 d	9107
	f/m: 1st repro: 4 mo 8 d	10587a
	f/m: 2 mo	1871a
Reithrodontomys montanus	f: 1st birth: 12 wk; n1	6391
	f/m: 2 mo	1871a
Rhipidomys mastacalis	f/m: 3? mo; n5	7995, 7996, 7998
Rhombomys opimus	f/m: 3-4 mo	7778
Saccostomus campestris	f: vag perf: 34.6; sd2.27; n62	11082, 11608
	f: 1st estrus: 44.6 d; sd3.17; n19	11082, 11608
	f: 1st conceive: 56.4 d; sd8.12; r44-71; n21	11082
	f: 1st birth: 96 d; min 67 d	2861
	f: >110 mm; wild; ~102 mm	4329
	m: 100-109 mm	4329
Scotinomys teguina	f: vag perf: 33.8 d; r28-39; n41	4967
	m: epididymal sperm; 6-8 wk	4967
Scotinomys xerampelinus	f: vag perf: 51.8 d; r44-60; n15	4967
Sigmodon alleni	f: 1st birth: 87 d	9959
Sigmodon alstoni	f/m: 2 mo; n6	7995, 7998
Sigmodon fulviventer	f: 1st birth: 77 d	524
Sigmodon hispidus	f: vag perf: 30-40 d	7254
	f: vag perf: 40-47 d	3110
	f: 1st estrus: r43-56 d; n9 (assorted treatments)	3110
	f: 1st preg: 57 g	522
	f: 1st birth: 65 d; spring; 84 d; Sept	1795

	f: 32 g (Mar-June), 60 g (July-Oct), 105 g (Nov-Feb)	7029
	m: testes descend: 20-30 d	7254
	m: epididymal sperm: 2 mo 2/5; 3 mo 100%	1795
	m: 40 g (Mar-June), 60 g (July-Oct), 100 g (Nov-Feb)	7029
	f/m: 40 d	1474
Sigmodon ochrognathus	f: 1st breed: 45 d	4898
	f: 1st birth: 71 d	525
Tatera afra	f: 3 mo; born early in season; 5-7 mo; born later	7146
	m: 6-8 wk born spring-early summer; 6-7 mo born later	166
Tatera brantsii	f: 3 mo; born spring; 8 mo; born summer-fall	7146
	m: 122 mm; testes wt 1.59 g; full sperm activity; n1	166
Tatera inclusa	f: 1st birth: 125 d; n1	5050
Tatera indica	f: 10-12 wk	5230
	m: 12-14 wk	5230
Tatera leucogaster	f: preg f: 120-129 mm 12%; 130-139 mm 75%	4329
	m: fecund 137-138 mm	4329
Tatera valida	f/m: 11-15 wk	5058
Taterillus gracilis	f/m: 12 wk	5058
Taterillus pygargus	f/m: 12 wk	5058
Tylomys nudicaudus	f: 1st mate: 90 d	4606, 4608
	m: 1st mate: 144 d	4606, 4608
	f/m: 10-11 wk	10755
Zygodontomys brevicauda	f: 1st ovulate: 50 d	11931
	f: 1st preg: 21 d	11373
	f: 25.6 d; ±1.21; n7	87
	m: abundant sperm: 40 d	11373
	m: 1st sire: 51 d	11931
	m: 42.3 d; ±1.94; n6	87
	f/m: 2 mo; n44	7995, 7996, 7998

CYCLE LENGTH

Baiomys taylori	4.9 d; r3-6; n33 cycles, 30 f	5064
	7.54 d; se0.23; n19 f	8805
Bolomys lasiurus	4-8 d; n9 f	7193
Brachytarsomys albicauda	8-9 d; n1 f	7477
Calomys callosus	proestrus: 1.3 d; ±0.4; estrus: 1.5 d; ±0.8	7189
	metestrus: 2.0 d; ±1.7; diestrus: 2.1 d; ±0.82	7189
	6 d; n6 f	5474
	6.6 d; ±1.4; n17 f	7189
Calomys musculinus	5.7 d	2410
Cricetomys gambianus	4.4 d; ±1.9	312
	5-6 d	103
	6 d; n1 f	7477
Cricetulus griseus	proestrus: 0.5, estrus: 1.5, metestrus: 2.5, diestrus: 2.5 d	8243
	4 d	468
	4-5 d	1693
	4.46 d; r3-7; n69 cycles, 5 f	8243
Cricetus cricetus	estrus: several hr-2 d; diestrus: 6 d	9085
	7.5 d	9085
Dipodillus simoni	estrus: ~1 d	6237
	10 d; r9-14; n4 cycles, 2 f	6237
Gerbillurus paeba	6.2 d; sd1.2; n8 cycles	2482
	7.2 d; sd1.5; n9 cycles	2482
Gerbillurus tytonis	6.2 d; sd1.2; n5 cycles	2482
Gerbillurus vallinus	11.3 d; sd1.2; n3 cycles	2482
Holochilus brasiliensis	6-8 d	7192
Meriones crassus	estrus: ~1 d	6237
Meriones libycus	5 d; smears	11905
Meriones shawi	estrus: ~1 d	6237
	4.74 d; r2-35; mode 4 (603/671); n671 cycles, 39 f	9451
Meriones tristrami	4.2 d; mode 4; median 4.0; se0.1; n775 cycles, 41 f	2527
Meriones unguiculatus	estrus: 5-23+ hr; n36, 9 f	583
	4.6 d; r4-6; n14 cycles	583
	14.6 d; r10-18; n10 cycles; pseudopregancy?	583
Mesocricetus auratus	4 d	9160
Mesocricetus brandti	4 d	6656

Neotoma floridana	mode 4-6 d; r3-8; n13 f; no true cornified cells	1703
Neotomodon alstoni	4.5 d; sd0.4; mode 4; r3-7	8096
Onychomys leucogaster	4.9; se0.3; n73 cycles	3533
	6.3; mode 4; median 4.4; se0.6; n402 cycles, 32 f	2527
Oryzomys capito	5-6 d	11931
Oryzomys nigripes	proestrus: 1.6 d; ±0.9; estrus: 2.6 d; ±1.1	7190
	metestrus: 2.3 d; ±1.3; diestrus: 2.9 d; ±1.1	7190
	10 d; ±1.5; n14 f	7190
Oryzomys palustris	7.62 d; ±0.19; n21 cycles, 7 f	2002
Ototylomys phyllotis	8.8 d; sd0.5; n14 f	4606, 4607
Pachyuromys duprasi	7.11 d; r3.5-11.5; n15 cycles, 1 f	6237
Peromyscus californicus	6.4 d; se0.2; mode 6; median 6.2; n161 cycles, 12 f	2527
	9 d; se0.4; mode 6,10; median 9; r5-20; n59 cycles, 18 f	4119
Peromyscus eremicus	5.3 d; se0.3; mode 4; median 4.7; n939 cycles, 58 f	2527
Peromyscus crinitus	6.1 d; se1.3; mode 5; median 4.9; n30 cycles, 7 f	2527
Peromyscus gossypinus	4.7 d; se0.2; mode 4; median 4.2; n733 cycles, 22 f	2527
	5.26 d; r3.5-10; n13 cycles, 6 f	8698
Peromyscus leucopus	6.0 d; se0.2; mode 4; median 4.9; n780 cycles, 59 f	2527
Peromyscus maniculatus	4-6 d	2806
	4.8 d; r4-5.8 d; n18 cycles, 39 f	1851
Peromyscus melanophrys	4.6 d; se0.3; r4.4-4.9; n40 cycles, 10 f	3229
Phodopus campbelli	estrus: ~10 h	11971
	4 d; mode 4; r4-5 (15% of f)	11971
Psammomys obesus	4-5 d	3513
Saccostomus campestris	4.1 d; sd1.35; n38 cycles, 8 f	11607
	4.1 d; sd7.5; n71 cycles, 19 yg f	11607
Scotinomys teguina	8.2 d; mode 6; median 6.4; se0.4; n13 cycles, 4 f	2527
Sigmodon hispidus	proestrus: 14.5 h; r12-21; estrus: 46.6 h; r21-123	1852
	metestrus: 145 h; r9-21; diestrus: 116 h; r42-156	1852
	estrus: 2.6 d; r1-12; n106 cycles, 21 f	7253, 7254
	6.2 d; ±0.45; r4-8; n9 cycles, 6 f	10656
	8 d; r5-9	1852
	8 d; r4-20; n106 cycles, 21 f	7253, 7254
Tatera brantsii	proestrus: 16-18 h; estrus: 20-28 h	7146
	metestrus: ~24 h; diestrus: 2 d	7146
	r4-6; one case 3 d	7146
Tatera indica	r4-5 d; n250 cycles	987
Tatera valida	7 d	1997
Tylomys nudicaudus	6.8 d; se0.4; n15 f	4606, 4608
Wiedomys pyrrhorhinos	1-6	10493
Zygodontomys brevicauda	r4-7 d; n11 f	11373
	r4-14 d; n10 f	11373
	r5-6 d; n16 cycles	11931

GESTATION

Akodon azarae	22.7 d; may delay implantation	2221
	24.5 d; sd0.86	2409
Akodon molinae	23 d; r21-25	7212
Baiomys taylori	≤20 d; min ibi	977
	20 d	8805
	20 d	10110a
	22 d with lactation	8805
	23 d; r21-26; n7; ppe to birth	5064
	25.5 d; r22-29; n31; pairing to birth	5064
Beamys hindei	22-23 d; n7 lt; pairing to birth	2912
	30 d	5049
Bolomys lasiurus	21 d; min ibi	2371
	~23 d	7193
Calomys callosus	19-22 d	8430
	21.8 d; ±1.04; r20-23; n14	7188, 7189
	24.7 d; mode 25; r21-29; n70; pairing to birth	5474
Calomys laucha	25 d; n1	4682
Calomys musculinus	20.96 d; sd1.07; n15	2410
	21 d; ±1	4853
Cricetomys gambianus	27-36 d	103
	31.7 d; r31-32; n3; pairing to birth	3130

	32.3 d; se2.8; r30-38; n7	227, 312
	42 d; n2	2439, 2446a, 2449, 7528
Cricetulus griseus	implant: 5-6 d	8510
	implant: 6 d	993
	19.5 d; r19-21	468
	20 d	7503
	20-21 d	1693
Cricetulus migratorius	11-13 d	4645
Cricetus cricetus	15.5-21 d	7436, 9085
	19-20 d	3571
	19.32 d; mode 17; r17-37; n25	11319
	20 d	8457
	20 d	7433
	4-5 wk	4645
Dendromus melanotis	23-27 d	2572
Desmodilliscus braueri	26 d	8696
	35 d with lactation	8696
Desmodillus auricularis	21 d	5684
	35 d with lactation	5684
Dipodillus simoni	20.4 d; r20-21; n11 lt, 6 f	6235, 6237, 6238
	32 d; r29-36; n6 with lactation	6238
Eligmodontia typus	18 d; n1	6807
Gerbillurus paeba	26 d; n1	10528
Gerbillus dasyurus	r24-26 d; n4	109
Gerbillus gleadowi	~20 d	9175
Gerbillus nanus	~20 d	9175
	21-23 d	5785
Gerbillus perpallidus	20.22 d; sd0.55; r19-21; n18 lt	11397
Gerbillus pyramidum	21 d	4713
	~22 d	4344, 4348, 4351
	37-42 d; with lactation	4713
Holochilus brasiliensis	28.4 d; r26-30	7192
Macrotarsomys bastardi	>24 d; separate m from f	6402
Malacothrix typica	23.11 d; sd1.59; n19 lt; without lactation	5887
	~26 d	10158
	26.75 d; r25-35; n6	10154
	27.45 d; sd0.69; n11 lt; with lactation	5887
Meriones crassus	21 d	8428
	21 d; n3	6236, 6238
	21.6 d; r21-24	6803
	23.5 d; mode 21; r21-31; n4	6237
	31 d; n1 nursing 3 yg	6236, 6238
	32.6 d; nursing ≥ 2 yg	6803
Meriones hurrianae	28-30 d	9175
Meriones libycus	22.3 d; r22-23; n3	11905
	29 d; r23-35	7778
Meriones meridianus	20-21 d	11252
	28 d; r22-30	7778
Meriones persicus	~28 d	9175
Meriones shawi	implant: 4 d	9451
	20 d	9451
	20.5 d	6237, 6238
	24.8 d; nursing 2 yg	9451
	31.85 d; r30-32; n4 nursing 4-7 yg	6237, 6238
	36.3 d; nursing 8 yg	9451
Meriones tristrami	24 d	8428
Meriones unguiculatus	implant: 6 d	11960
	implant: 8 d	3269
	24-26 d	7754
	24-26 d	6867
	r24-26 d; n86	56
	24 d + 8-30 h	9692
	25 d	82
	25.2 d; ±0.1; r24-26; n67 lt	7970

	min 25-26 d; lactation prolongs gestation	1768
	29-30 d	7778
Meriones vinogradovi	21.5 d	8428
Mesocricetus auratus	15 d 21 h; n28 f	1071
	16 d	3571
	16 d; r15-17; n9	1382
Mesocricetus brandti	15 d	6656
Mystromys albicaudatus	37 d	10158
	38 d; r36-39	4220
Neotoma albigula	<30 d	3201
Neotoma cinerea	25-35 d; ibi	3087
	r27-31 d; n16	2909
Neotoma floridana	30-36 d	8659
	33-39 d; n1 f	8315
	33-41 d; recaptured f	8859
	34 d; r33.3-35	5892
	34.7 d; r32-37	7030
	est 6 wk	4281
Neotoma fuscipes	est 23 d; ≥16 to ≤29; n1	2679
	33 d; n2	11888
Neotoma lepida	30-36 d	2906
Neotoma mexicana	32.7 d; r31-34; n8, 5 f	8105
Neotoma micropus	30-39 d; pairing to birth	11696
	<33 d	3201
	mode 33-35 d	1242
Neotoma stephensi	31 d	11232
Neotomodon alstoni	27.3 d; sd1.1; mode 27; r26-30; n50 primiparous f	8096
Nyctomys sumichrasti	30 or 31 d; 1 f	946
Ochrotomys nuttalli	29-30 d	3897
Onychomys leucogaster	26 d; 30 d with lactation	8006
	27 d	4986
	27-32 d	1871a
	29-32 d	2907
	32 d; n1; pairing to birth	10588
	32-38 d; with lactation	2907
	32-47 d; with lactation	1871a
	39.7 d; r33-47; n3 with lactation	10588
Onychomys torridus	27-29 d	10697
Oryzomys argentatus	25 d	10809
Oryzomys capito	27.6 d; r25-31; n20; pairing to birth	11931
Oryzomys palustris	mode 21-24 d; r21-28; pairing to birth	8231
	25 d; n5	10572
Otomys irroratus	est 35-40 d	2339
	max 42 d	1312
Otomys sloggetti	38 d; min ibi	11710a
Ototylomys phyllotis	51 d; sd0.6; r50-52; n6	4606, 4607
	53.6 d; sd7.6; r49-69; n9 with lactation	4606, 4607
Pachyuromys duprasi	20 d; n2 lt, 1 f	4674, 4675
	20.4 d; mode 20; r19-22; n7	6234, 6237, 6238
	24 d	4582
Peromyscus boylii	23 d; n1	1363
	29.3 d; r26-32; n6 with lactation	1363
Peromyscus californicus	23.6 d; ±0.36; r21-25; n5	10573
	31.6 d; se0.2; r31-33; n18	4119
	34.5 d; se0.6; n15 with lactation	4119
Peromyscus crinitus	24-25 d	1871a
Peromyscus eremicus	21 d; n1	10573
	28-32 d	2526
Peromyscus gossypinus	22.86-23.34 d; n7 non lact	8698
Peromyscus hooperi	33.50 d; sd0.71; r33-34; n2	9603

Peromyscus leucopus	23.05 d; r22-25; n19	10573
	25.95 d; se0.5; n62	684
	29.56 d; r23-37; n25 with lactation	10573
Peromyscus maniculatus	3-4 wk	5019
	21-40 d	2806
	~23 d	10750
	23-24 d	2517, 2523
	23.2 d; se0.2; n11	5673
	23.51 d; r22-27 d; n43	10573, 10576, 10577
	mode 24 d (10/19); r23-26 d; n19 lt	7311
	25-26 d; with lactation	2517, 2523
	26-35 d; n1	10574
	27.14 d; r22-35; n21 with lactation	10573, 10576, 10577
	28.54 d; se0.81	6493
Peromyscus melanocarpus	30 d; n1	9114
	37.0 d; r31-41; n9 with lactation	9114
Peromyscus melanophrys	24.7 d; se0.1; r23-26; n24 lt, 24 f	3228
Peromyscus melanotis	23.55 d; ±0.11; r23-24; n9	10573
	26.57 d; ±0.73; r23-32; n7 with lactation	10573
Peromyscus mexicanus	30 d; n1 lact f	9114
	35 d; r31-39; n2; min ibi	9114
Peromyscus perfulvus	43.25 d; r39-46; n4	4609
Peromyscus polionotus	implant: 5 d	6302
	23.8 d; r23-24	6302
Peromyscus truei	26.20 d; ±0.26; r25-27; n5	10573
	40 d; n1 with lactation	10573
Peromyscus yucatanicus	30 d; pairing to birth	6149
	31-33 d; n5 with lactation	6149
Phodopus sungorus	18 d; min ibi	5443
	19 d; mode 19; r18-19; n29; pairing to birth	3249
	20-22 d	3334
	21-22 d	11285
Phyllotis darwini	33.6 d; n59; pairing to birth	11407
Psammomys obesus	23 d; r23-31; pairing to birth	8731
	24 d; min pairing to birth	9825
	24.8 d; n1; ibi	2234
	36 d; n1 lactating f	2234
Reithrodontomys humulis	21-22 d; n1; pairing to birth	5609
	≤25 d	6300
Reithrodontomys megalotis	22 d; min ibi	9107
	23-24 d; n4 lt, 2 f	10587a
	23-24 d	1474
Reithrodontomys montanus	21 d; min ibi	6391
Rhombomys opimus	23-32 d	7778
Saccostomus campestris	implant: ext 4-5 d	11607
	20-21 d	2861
	21.2 d; sd0.45; n36	11082
Sigmodon leucotis	~35 d	9960
Sigmodon hispidus	26-27 d	6975, 7030, 8915
	~27 d	7253
Sigmodon ochrognathus	34.5 d; r33-36; n2	4898
Tatera brantsii	22.5 d	7146
Tatera inclusa	23 d; n2	5050
Tatera indica	22.3 d; r18-30; mate at ppe: 36 d; r34-40; delayed with lactation of 3+ yg	987
	28.22; r27-30 d; ibi	8725
Tatera valida	~25 d	5058
Taterillus gracilis	26 d	5058
Taterillus pygargus	3 wk	8695a
	30 d	5058
Tylomys nudicaudus	implant: 4 d	4606
	min 39 d	10755
	40.6 d; se0.5; r35-51; n47 lt	4606, 4608

Zygodontomys brevicauda	25.0 d; ±0.19; n29	87
	25.1 d; sd1.0; r24-27; n10 f	11373
	26.0 d; r25-32; n16; pairing to birth	11931
	27.9 d; sd2.4; n78; pairing to birth	11373
LACTATION		
Akodon azarae	wean: 14-15 d	2221
Akodon molinae	wean: 26 d; r21-30	7212
Baiomys taylori	teat attachment: 18-22 d	977
Beamys hindei	solid food: ~3 wk	2912
	wean: 5-6 wk	2912
Calomys callosus	wean: 15-17 d	7189
	wean: 21 d	5474
Cricetomys gambianus	solid food: 17 d	3130
	solid food: 17-18 d	10158
	solid food: 26-28 d	103
	wean: 26-28 d	103
	wean: 5 wk	7528
	den emergence: 43-52 d	3130
Cricetulus barabensis	wean: 20 d	4695
Cricetulus griseus	wean: 21 d	8685
	wean: 25 d	468
	wean: < 4 wk	2730
	den emergence: 4 wk	2730
Cricetulus migratorius	wean: 3rd-4th wk	9175
Cricetulus triton	solid food: 15 d	6602
Cricetus cricetus	solid food: 5 d	8457
	solid food: 8 d-3 wk	3571
	solid food: 10 d	9440
	wean: 25 d	7436
	den emergence: 25 d	8457
Desmodillus auricularis	solid food: 18 d	5684
	solid food: 21 d	7824
	wean: 33 d	7824
	nest emergence: 10 d	5684
Gerbillurus paeba	wean: 19 d	10158
	wean: 20-30 d	403
	wean: 28-30 d	2481
Gerbillurus tytonis	wean: 28-30 d	2481
Gerbillus nanus	wean: end 3 wk	9175
	mother leaves yg: 4 wk	5785
Gerbillus pyramidum	wean: 25-30 d	4348
	wean: ~4 wk; occasional 2 mo	4713
Holochilus brasiliensis	wean: 3 wk	7192
Lophiomys imhausi	independent: 40 d	2446a
Malacothrix typica	lact: max 32 d	5887
Megadontomys thomasi	wean: 30 d	2728
Meriones crassus	solid food: 16 d	2233
	wean: 20 d	6236
Meriones hurrianae	wean: 21 d	9175
Meriones libycus	solid food: 17 d	2233
	wean: 30 d	11905
	independent: 1 mo	7778
Meriones meridianus	independent: 20-22 d	11252
Meriones persicus	wean: 18-20 d	9175
Meriones shawi	solid food: 3 wk	9451
	wean: 3 wk	9451
Meriones tamariscinus	wean: 14-15 d	1035
	wean: 21 d or more?	8961
Meriones unguiculatus	solid food: 17-20 d	7098
	wean: 20-25 d	7778
	wean: 21 d	7098
	wean: 23-28 d	1768
	wean: 32 d	7964

Mesocricetus auratus	solid food: 8 d	1071
	independent: 20-25 d	3571
Mesocricetus brandti	wean: <20 d	6656
Mystromys albicaudatus	solid food: 20 d	4246
	teat attachment: 15-20 d	4246, 7168
	teat attachment: to 3rd wk	4220
	wean: 21-28 d	4220
	wean: 32 d	7168
	wean: 38 d	10158
Neotoma albigula	lact: 20-25 d	3201
	1st trapped: 28-35 d	9689
	wean: 27-40 d	9689
	wean: min 37 d; max 62-72	9106
Neotoma cinerea	1st trapped: 3 wk	6895
	wean: 26-30 d	2909
	wean: 5 wk	3087
Neotoma floridana	solid food: 16-20 d	8859
	solid food: 19-22 d	7030
	solid food: 24 d	4281
Neotoma fuscipes	wean: ~3 wk	3041
	wean: 36 d; n1	2679
Neotoma lepida	solid food: 3 wk	2906
	wean: 27-40 d	9689
	wean: 4 wk	2906
	1st trapped: 35-41 d	9689
Neotoma stephensi	solid food: 2 wk	11231
	wean: 5-6 wk	11232
Nyctomys sumichrasti	teat attachment: 2 wk	946
Ochrotomys nuttalli	wean: 17-18 d	6497
Onychomys leucogaster	nest emergence: 5 d	9366
	solid food: 10 d	9366
	solid food: 23 d	4986
	practically wean: 19-20 d	10588
	wean: 24 d	9366
	wean: ≤1 mo	1871a
	wean: 31 d; n1 lt of 4	8006
Onychomys torridus	wean: 20 d	4991
Oryzomys nigripes	wean: 14-15 d	2221
Oryzomys palustris	wean: 11 d	10572
	wean: 20 d	8231
Oryzomys subflavus	wean: 4 wk	11305
Otomys irroratus	wean: 12 d	2446a
Otomys sloggetti	solid food: 12 d	11710a
	wean: 16 d	11710a
Otomys irroratus	solid food: 2 d	2338
	teat attachment: 7-14 d	10158
	wean: 13 d	2338
	wean: 4-5 wk	2571
Ototylomys phyllotis	teat attachment: 30 d	4606, 4607
Oxymycterus rutilans	wean: 14 d	2221
Pachyuromys duprasi	solid food: 20 d	6237
	wean: 29 d	6237
Peromyscus californicus	wean: 30 d	2728
Peromyscus crinitus	wean: 23 d	2728
	wean: 4 wk	2910
Peromyscus difficilis	wean: 26 d	2728
Peromyscus eremicus	wean: 20-22 d	1236
	wean: 21 d	2728
Peromyscus hooperi	wean: 22.67 d; sd2.89, n3	9603
Peromyscus interparietalis	wean: 16-21 d	1236
Peromyscus leucopus	leave nest: 14-16 d	11288
	solid food: 13-16 d	7304
	solid food: 18 d	7303
	independent: 20 d	7303

	wean: 21 d	2728
	wean: 22 d	7304
	wean: 22 d	4780
	wean: 24-30 d	1871a
Peromyscus maniculatus	leave nest: 14-16 d	11288
	1st emergence: 16.0 d; se0.4; r15-17; n4	7308
	1st emergence: 16.6 d; se0.3; r16-18; n8	7308
	wean: r18.01-18.93 d; n3 areas	4218
	wean: 3 wk	5658
	wean: 21-23 d	2728
	wean: 21-25 d	7843
	wean: 22-37 d	2806
	wean: 23-25 d; begins at eye opening	10575, 10576, 10577
	wean: 30-40 d	2699
	no wgt loss after 24 hr: 18-24 d; n2 subspecies	5750
	zero growth: 19.2 d	7308
	zero growth: 25.3 d	7308
	1st trapped: 4 wk	5658
Peromyscus megalops	wean: 21-23 d	6302
Peromyscus melanocarpus	no wt loss after 24 hr: 4-5 d after eye opening	9114
Peromyscus melanophrys	wean: 26 d	2728
	wean: 33-35 d	3228
Peromyscus mexicanus	no wt loss after 24 h: 4-5 d after eye opening	9114
Peromyscus polionotus	teat attachment: 2 wk	8907
	wean: 21 d	2728
	wean: 28 d; r25-31	6302
Peromyscus truei	wean: ≤30 d	1871a
	wean: 30 d	2699
Petromyscus collinus	wean: 30-33 d	2481
Phodopus sungorus	solid food: 10 d	11548
	solid food: 16 d	11285
	den emergence: 16 d	11285
	wean: 20 d	5443
	lact: 20-21 d	11285
Podomys floridanus	wean: 3-4 wk	6301
	wean: 26 d	2728
Psammomys obesus	solid food: 11 d	2233, 2234
	solid food: 14 d	8731
	den emergence: 11-12 d	2234
	wean: 28 d	8731
	wean: 4 wk	10481
Reithrodontomys fulvescens	wean: 13-16 d	8189
Reithrodontomys humulis	solid food: ~9 d	5609
	wean: 3 wk	5609
Reithrodontomys megalotis	solid food: 11-12 d	10587a
	wean: 16 d max	9107
	wean: ≤20 d	1871a
	wean: 21 d	10797
	wean: ~24 d	5210
Reithrodontomys montanus	wean: 14 d	6391
	leave nest: 3-4 wk	1871a
Rhombomys opimus	solid food: 12-13 d	7778
	wean: 22 d	7778
Saccostomus campestris	coprophagy: 5 d	2861
	solid food: 13 d	2861
	solid food: 17 d; sd1.22; n71	11082
	solid food: 25 d	4329
	leave nest: 2.5-3 wk	2861
	1st use of cheek pouch: 3 wk	2861
	wean: 25 d; n181	11082
	lact: 35 d	4329
Scotinomys teguina	solid food: 16-18 d	4967
	wean: 3 wk	4967

Scotinomys xerampelinus	solid food: 21-24 d	4967
	wean: 3.5 wk	4967
Sigmodon hispidus	solid food: 10 d	9572
	wean: 10 d	10570
	wean: 10-12 d	6975, 7030
	wean: 12 d	6975, 8915
	wean: 18-21 d	9572
	wean: 20 d	7253
Sigmodon ochrognathus	solid food: 8 d	4898
	wean: 15 d	4898
Tatera brantsii	teat attachment: 17 d	7168
	wean: 22-24 d	7146
	wean: 29 d	7168
Tatera inclusa	wean: 22 d; n1 lt	5050
Tatera indica	wean: 20 d	987
Tatera leucogaster	wean: 28-28 d	7797a
Taterillus pygargus	wean: 3 wk	8695a
Tylomys nudicaudus	teat attachment: 2-3 wk	10755
	wean: 2-3 wk	10755
Zygodontomys brevicauda	solid food: 9-10 d	11373
	can survive without mom: 10 d	87
	wean: 16-21 d	11373
INTERLITTER INTERVAL		
Akodon longipilis	min 16-18 d	7241
Akodon molinae	30 d	7212
Baiomys taylori	20.5-21 d	7544
	23 d; r21-26; n7	5064
	27.6 d; se1.8; r20-64; n26 intervals, 6 f	977
Beamys hindei	min 62 d	2912
Calomys callosus	20.5-21 d	7544
	mode 22 d; r20-23; n28	8430
	min 21 d	5474
Calomys lepidus	20.5-21 d	7544
Cricetulus griseus	min 19 d	8685
Cricetus cricetus	131.40 d; ±67.77; n11	9085
Dipodillus simoni	46.3 d; r30-57; n10	6237
Meriones crassus	59-75 d; n4	6237
Meriones shawi	31.25 d; r30-32; n4	6237
Meriones tamariscinus	25-29 d; r28-84	1035
Meriones unguiculatus	38.6 d; se0.46; r25-123	1768
	54.7 d; mode 30-39; n192 lt	382
	40-52 d; depending on age	56
Mystromys albicaudatus	min 36 d	7168
Neotoma albigula	49 d; n1	9106
Neotoma cinerea	25-35 d	3087
Neotoma floridana	83 d; n1	8659
Neotoma lepida	60 d	2906
Neotoma stephensi	130 d est	11232
Nyctomys sumichrasti	min 38 d	946
Ochrotomys nuttalli	min 25 d; n3; r25-28; n31 young live	6497
	min 25 d; n1; r25-28; n5 no lactation	6497
Onychomys leucogaster	26-37 d no lactation	8559
Onychomys torridus	27-30 d	10697
	28-30 d; r26-35	8559
	1 mo	979
Otomys sloggetti	49 d; r38-74; n4 intervals, 2 f	11710a
Peromyscus californicus	24 d	9255
Peromyscus crinitus	25 d	9255
	min 27 d; most 29-31; 24-25 no lactation	2910
Peromyscus eremicus	27 d	9255
	50 d; min 28	2319
Peromyscus interparietalis	31.9 d; r27-37; n25; est lactational gestation	1236

Peromyscus leucopus	20 d	9255
	mode 24 d; r22-24; n182 lts	6150
	29.2 d; r23-42; n151 intervals, 27 f	4779
	29.56 d; r23-37; n25	10573
Peromyscus maniculatus	20.5-21 d	7544
	21-23 d	9255
	27.14 d; r22-35; n21	10573, 10576, 10577
	1 mo; young removed 2 wk	5797
	35.2 d; min 18; n489	7312
Peromyscus melanocarpus	30 d; n1 young die	9114
	37.1 d; n8 young live	9114
Peromyscus melanotis	26.57 d; se0.73; r23-32; n7	10573
Peromyscus polionotus	21 d	9255
	23-24 d; no lactation	11725
	24-26 d	8907
	29.43 d; ±3.81; r23-41; with lactation	11725
Peromyscus truei	40 d; n1	10573
Peromyscus yucatanicus	27 d; n1	6149
	28 d; n4; young die	6149
Phodopus sungorus	min 18 d; n4	5443
	mode 19-23 d 35%; r18-87	11548
	32.0 d; r18-80	7255
	35.6 d; r16-120	7255
	93 d; r22-360	7255
Podomys floridanus	33 d	9255
Psammomys obesus	38.3 d; r35-44 d; n6, 1 f	2234
Reithrodontomys humulis	31.6 d; r24-49 d; n5	6300
Reithrodontomys megalotis	27.6 d; r22-38; n26, 2 f	550
Saccostomus campestris	51.8 d; sd9.06; r46-74; n14	11082
Scotinomys teguina	30.8 d; r27-36; n16 young die	4967
	31.1 d; r29-36; n32 young live	4967
Scotinomys xerampelinus	31.9 d; r30-38; n10 young die	4967
	34.3 d; r29-38; n17 young live	4967
Tatera brantsii	131 d; n1	7168
Tatera inclusa	2 mo; n1	5050
Tatera indica	28.22 d; r27-30	8725
	67 d	1689
Tatera nigricauda	40-44 d; wet season	7794
	97-107; dry season	7794
Tatera valida	109 d	7794
Tylomys nudicaudus	min 39 d	10755
Zygodontomys brevicauda	~25 d	87
	mode 25 d 75%; r24-28; n71	11373
LITTERS/YEAR		
Akodon azarae	mode 1; r1-4	2221
Akodon urichi	3.6	7995, 7998
Baiomys taylori	10	10110a
Beamys hindei	5 max	2912
Calomys callosus	3 max	5474
Calomyscus bailwardi	0-2	3587
Calomyscus mystax	2	9488
Cricetomys gambianus	2	11253
	5-6	103
Cricetulus eversmanni	2	4645
Cricetulus migratorius	2; in the south perhaps more	4645
Cricetus cricetus	1-2	9440
	1-2	4645
	1-3	11319
	2; r2-3	3571
	2-3	7433
	2-3	9085
	3 max	8457
Gerbillus nanus	3	5785
Gerbillus pyramidum	5	4713

Malacothrix typica	>1	10158
Meriones crassus	2	9175
Meriones meridianus	2-3	7778
	3-4	4645
Meriones tamariscinus	2; r2-3 ?	4645
Meriones unguiculatus	mode 5; r1-10; n657; 13 dead/2160 born	56
	3; r2-3	10516
	3 max	7778
	3.0; se0.30; late maturing	1863
	5.2; se0.84; early maturing	1863
Mesocricetus auratus	3 lt/6 mo; n1 f	1382
	3-4	2419
	3-5; r3-8	3571
Mesocricetus brandti	3 max	6656
Mesocricetus raddei	2	4645
Myospalax myospalax	1	8050
Neacomys tenuipes	6.1	7998, 7995
Neotoma cinerea	1	2615
	1	1474
	1	2088
	2 lt	6895
	2-3	3087
	3.0; max 7	2909
Neotoma floridana	2-3	8659
	2-3	3285
Neotoma fuscipes	1	11289
	1-5	6490
Neotoma lepida	1-4 lt/season	6708
	4	2906
Neotoma mexicana	2	1366
Neotoma micropus	1	3201
	2+; r1-5	8962
	2+; min 2-3	11696
Neotoma stephensi	1.9; sd0.86; r0-5; n51 [note includes 0]	11232
Nyctomys sumichrasti	5 lt/7 mo; n1 f	946
Onychomys leucogaster	est >1	502
	2-3	1474
Onychomys torridus	2-3	1474
	min 2	2109
	juv: 3.8 lt/yr; 9.8 yg/yr; adult: 1.6 lt/yr; 4.0 yg/yr	10697
Oryzomys capito	5.8	7995, 7998
	6.06; n34 f	3316, 3319
Oryzomys concolor	4.2	7995, 7998
Otomys irroratus	3-5	2338
	mode 4-5; 7 poss	10158
Otomys typus	max 5	5760
Peromyscus californicus	5.5; r5-6; n2	10573
Peromyscus crinitus	2.05; max 8	2910
Peromyscus eremicus	max 3	6708
	est 3-4	1474
Peromyscus gossypinus	est 4+	1474
Peromyscus leucopus	1-4/5 mo season	4376
	2-4	1474
	2.0	11877
	4-5 lt/season	1472
	5; r4-6; n2	10573
	5.3; r5-6; n3	10573
Peromyscus maniculatus	1.5-5.2; field	4218
	1.9-2.3	5151
	2 max	6708
	2	11230
	2 lt/season not uncommon; no f w/3 n35f	10839
	2 or more	11956
	2-4	1474

	2-4 or more	2088
	2.1-2.9; lab	4218
	2.5-2.9; poor vs good seed crop	3624
	5.5; r4.7; n2	10573
	5; r4-6; n2	10573
	6; r5-7; n2	10573
	max 8 lt/8 mo; yg removed 2 wk	5797
	1: yg not removed 13 lt/370 d & 12 lt/512 d	5797
	13; n1 pr	6493
Peromyscus melanotis	7.5; r7, 8	10573
Peromyscus pectoralis	est > 1	1474
Peromyscus polionotus	est 2+	1474
Peromyscus sitkensis	2	2088
	est 2	1474
Peromyscus truei	est > 1	1474
	5; r3-7; n4	10573
Petromyscus collinus	1	11864
Phodopus sungorus	mode 3,4	11285
Reithrodontomys megalotis	4-5 lt/season max	9107
Reithrodontomys raviventris	1	10809
Rhipidomys mastacalis	3.5	7995, 7998
Rhombomys opimus	2-3	10516
	max 4; most 2-3	7778
Sigmodon alstoni	5.0	7995, 7998
Sigmodon hispidus	2	6303
	max 9	1474
	9 lt/10 mo	5146
Tatera afra	6-7	7146
Tatera brantsii	4; r4-5 or 5-6*	7146
Tatera indica	5.5	1689
Tatera nigricauda	5; 4 wet season, 1 dry season; r4.9-5.4	7794
Tatera valida	2	7794
Tylomys nudicaudus	6 lt/13 mo; n1 f	10755
Zygodontomys brevicauda	5.5	7995, 7998
SEASONALITY		
Akodon azarae	mate: June-Aug or yr rd; few data; Uruguay	596
	preg f: peak Nov-Apr; central Argentina	7347
	birth: Oct-Apr; Buenos Aires, Argentina	8305
	birth: Nov-Apr; Argentina	2221
Akodon boliviensis	testes: large Dec; few data; Peru	8302
	preg f: Dec; n1; Peru	8302
Akodon dolores	birth: Jan-Mar, July-Aug, Nov-Dec; captive	7213
Akodon longipilis	testes descend: July-Sept; n Chile	7241
	preg f: June-Sept, none Nov-June; n Chile	7241
	preg f: end Oct-Dec; Patagonia, Argentina	8305a
	preg/lact f: Feb-Mar; n12/113 f; Chile	4054
	repro: Oct-May; s Chile	7242
Akodon molinae	birth: Jan-May, July-Sept, Dec, peaks Mar-Apr, July-Aug, Dec; lab	7213
Akodon olivaceus	testes descend: June-Jan; n Chile	7241
	mate: Sept-Mar; Santiago, Chile	3551
	mate: Nov-Dec; Fray Jorge, Chile	3551
	preg f: Sept-Nov; Chile	7241
	preg f: Oct-Dec; Patagonia, Argentina	8305a
	preg/lact f: Nov-Dec, Feb; n27/147; Malleco, Chile	4054
Akodon sanborni	repro: Oct-May; Chile	7242
Akodon urichi	preg f: est yr rd; n Venezuela	7995, 7998
Auliscomys boliviensis	preg f: Oct; n1; Caccacharo, Peru	8302
Auliscomys micropus	preg f: Mar; n2/15; Malleco, Chile	4054
	preg f: Nov-Dec; Patagonia Argentina	8305a
	lact f: Feb-Mar; n2/15; Malleco, Chile	4054
Baiomys musculus	preg f: July; n1; Mexico	2341
	preg f: yr rd; Veracruz, Mexico	4228
	preg f: wet season; s Mexico	11388

Baiomys taylori	mate: Jan-Oct; s TX, n Mexico	1474
	preg f: yr rd, peak Oct-Nov; n54; s TX	8963
Beamys hindei	repro: wet season; Nyasaland	4330
	mate: coincides with rains; Malawi	5760
	preg f: Nov-May; Nyasaland	4330
	birth: yr rd; captive	2912
Bolomys lasiurus	preg f: peak Apr-June; ne Brazil	5540
Bolomys obscurus	preg f: Oct-Feb; Uruguay	596
	preg f: est yr rd; central Argentina	7347
Calomys callosus	preg/lact f: Feb, Apr, Sept, Nov; ne Brazil	10494
	repro: yr rd; captive	8430
Calomys laucha	preg f: summer; n4; grasslands, S America	4682
	preg f: yr rd, low July-Aug; central Argentina	7347
Calomys musculinus	preg f: est yr rd, low July-Sept; central Argentina	7347
Calomyscus bailwardi	mate: est spring-fall; Pakistan	9175
	birth: Mar-June; Armenia USSR	3587
	lact f: June; n1; Iran	2404a
	lact f: Aug, Oct, Dec; Iran	6293
Calomyscus mystax	mate: mid Mar-mid June; USSR	9488
Cricetomys gambianus	epidydimal sperm: peak Oct-Jan; Kenya	312
	mate: est rainy season; Nyasaland	7528
	conceive: Apr; captive, Malawi	4329
	preg f: Jan, Mar-Apr, Aug; n4; Kenya	312
	preg f: Sept; Malawi	4329
	preg/lact f: Feb; n2; e Africa	185
	birth: Dec-Mar; s Africa	10153
	lact f: Jan; n2; Zambia	331
	lact f: Apr; n1; Tanzania	186
	repro: twice/yr; Zaire	11253
Cricetulus barabensis	repro: Apr-Oct; est Transbaikal, Siberia, Russia	3334
Cricetulus curtatus	birth: Apr-Sept; Mongolia	3334
Cricetulus eversmanni	birth: summer; EurUSSR	4645
Cricetulus griseus	preg f: Feb-Nov; no data Dec-Jan; Peking, China	2730
Cricetulus longicaudatus	mate: early Mar; USSR	3335
	repro: end Mar-Aug/Sept; Transbaikal, Russia	3334
Cricetulus migratorius	testes: large Dec; Lebanon	6419
	testes: large spring-mid May; Pakistan	9175
	preg f: Mar; n1; Lebanon	6419
	preg f: Aug, Nov-Jan; no data other mo; Iran	6293
	birth: spring-summer; Pakistan	9175
	lact f: May; n1; Iran	2404a
	lact f: Oct; n1; Lebanon	6419
Cricetulus triton	mate: Jan-May; Manchuria, China	6602
	preg/lact f: June, Aug; Korea	5424
Cricetus cricetus	birth: May-Aug; Germany	3571
	mate: mid Apr-July; lab	11319
	mate: end Apr; Germany	8457
	mate: end Apr-July; Germany	3571
	mate: May-July; Europe, USSR	7433
	mate: yr rd; lab	9085
	birth: May-July; Eur, USSR	7433
	repro: see DC; Yugoslavia	6050
Dendromus kahuziensis	birth: Feb; n2; central Africa	187
Dendromus melanotis	birth: Jan-Feb, Oct; Zambia	331
	lact f: May; n1; Tanzania	186
Dendromus mesomelas	preg f: Mar, May-June; n3; Malawi	4329
	preg f: Nov; n1; Cameroun	2955
	juveniles: end Feb; Tanzania	186
Dendromus messorius	preg f: Apr; n1; central Africa	4495
	juveniles: Aug; central Africa	4495
Dendromus mystacalis	preg f: Jan; n1; Kenya	4925
	birth: peak Nov-Jan; e Zaire, Uganda	5760
	lact f: Apr; n1; central Africa	187

Dendromus nyikae	testes large: Jan, Aug; n2; Tanzania, Angola	5760
	preg f: Nov; n1; Malawi	4329
	lact f: May; n1; Malawi	4329
	lact f: mid Aug; n1; Tanzania, Angola	5760
Deomys ferrugineus	estrus: yr rd; Zaire	2573
	mate: peak Oct-Mar, low June-Aug; e Zaire	2574
	preg f: Apr, June, Dec; Cameroun, Uganda	5760
	preg f: est yr rd, peak May-Nov; Zaire	8855
Desmodillus auricularis	preg f: Jan; n1; Kalahari Gemsbok Nat Park, S Africa	7823
	birth: Mar, Sept; n3; lab	5684
	juveniles: Jan, May, July-Aug, Oct-Nov; Botswana	10154
Dipodillus simoni	birth: yr rd; lab	6237
Eligmodontia typus	preg f: Oct-Apr; Patagonia, Argentina	8307
Gerbillurus paeba	preg f: Jan-Feb, Nov; n44; Kalahari Gemsbok Nat Park, S Africa	7823
	preg f: Oct; n1; Namibia	9945
	preg f: Dec-Mar, June-Aug, Oct; Botswana	10154
	birth: Oct-June; Cape Province, S Africa	404
	birth: yr rd; S Africa	2381
Gerbillus andersoni	mate: Feb-May; Israel	20
	mate: Sept-June; Egypt	8134
Gerbillus campestris	mating condition: Nov; n1 m; Sudan	4345
	preg f: Mar, Dec; n3; Egypt	8134
	lact f: May, Dec; n2; Egypt	8134
Gerbillus cheesmani	mate: est yr rd; Pakistan	9175
	lact f: Feb, Nov; Pakistan	9175
Gerbillus dasyurus	testes: large Mar-Apr; lab	712
	mate: Feb-Apr; Ngev Desert, Israel	715
	mate: Feb-July; Saudi Arabia	109
Gerbillus gerbillus	testes: large Jan-May; Egypt	8134
	testes: large Mar-Apr, small Sept-Oct; Sahara	5707
	preg/lact f: Jan-May; Egypt	8134
Gerbillus gleadowi	preg f: May-June, Nov-Jan; w Rajasthan, India	8723
	birth: peak post monsoon & winter; Pakistan	9175
Gerbillus nanus	mate: summer, winter; w Rajasthan, India	8723
	mate: yr rd, peak summer, winter; Pakistan	9175
	preg f: Nov; n1; no data other mo; se Iran	6293
	lact f: Dec; no data other mo; se Iran	6293
	juveniles: Jan; no data other mo; se Iran	6293
Gerbillus pyramidum	mate: June-Mar; Sudan	4348
	birth: June-Feb; Sudan	4344, 4348, 4351
	repro: yr rd; captive	4713
Graomys griseoflavus	preg f: peak Dec, Feb-Mar, none June-Aug; n Mendozo, Argentina	2388
Gymnuromys roberti	preg f: June-July; Madagascar	7983
Hodomys alleni	preg f: Feb; n1; Nayarit, Mexico	3700
	lact f: Sept; n1; Sinaloa, Mexico	3700
Holochilus brasiliensis	mate: yr rd; Brazil	4681
Holochilus magnus	mate: yr rd; e Uruguay	4681
Holochilus sciureus	preg f: yr rd; British Guyana	10983
Irenomys tarsalis	preg f: Nov; n2; Patagonia, Argentina	8305a
Isthmomys flavidus	lact f: Feb, June; n2; Panama	11044a
Isthmomys pirrensis	preg f: Feb; n4; Panama	11044a
	lact f: Feb-Mar; n6; Panama	11044a
Malacothrix typica	birth: Aug-Mar; Botswana	10154
Meriones crassus	mate: Nov-June; few data; Egypt	8134
	birth: yr rd; n35; zoo	12128
	birth: peak spring-summer; Pakistan	9175
Meriones hurrianae	scent gland: large July-Sept, small Oct-Nov; India	6097
	mate: Aug-Oct; Rajasthan, India	8721
	mate: end Oct-Dec; India	79
	preg f: Sept-Nov; Rajasthan, India	8720
	preg f: yr rd, peak Feb, low Dec-Jan; India	6097, 8723
Meriones libycus	mate: yr rd; USSR	7778
	mate: spring, fall; Pakistan	9175

	preg f: Oct; n1; Iran	6293
	juveniles: end Oct; Iran	6293
Meriones meridianus	repro: Feb-Nov; USSR	8275
	mate: Mar-Sept; EurUSSR	4645
	preg f: peaks Sept, Apr; USSR	8275
	preg f: yr rd; USSR	7778
	repro: mid Feb-mid June; USSR	9499
Meriones persicus	large follicles: Sept; n1; Iran	6293
	preg f: May; n1; Iran	2404a
	juveniles: Sept-Oct; Iran	6293
Meriones shawi	birth: yr rd; n51; zoo	12128
Meriones tamariscinus	repro: Feb-Nov; USSR	8275
	birth: yr rd, peak Apr-Sept, 46/67; lab	1035
	birth: Apr-Oct, peak May, Aug; USSR	7778
Meriones tristrami	juveniles: Sept; n1 lt; Iran	6293
Meriones unguiculatus	preg f: mid May-Sept; Mongolia	10516
	preg f: June-July; Mongolia	10516
Mesocricetus auratus	mate: Mar-Oct; Germany; yr rd; captive	3571
	birth: yr rd, peak May-Oct; captive	1382
	birth: yr rd; lab	1382
	repro: yr rd; captive	1071
Mesocricetus brandti	mate: none Nov-Mar; lab	6656
Mesocricetus raddei	birth: summer; Europe, USSR	4645
Myospalax myospalax	birth: end Apr-early June; USSR	8050
Myospalax psilurus	birth: Mar-Apr; Mongolia	1817
Neacomys tenuipes	preg f: est yr rd; n Venezuela	7995, 7998
Nectomys squamipes	preg f: Aug, Oct-Nov; Brazil	3078
	preg f: Oct-Nov; Argentina	3077
Neotoma albigula	mate: Jan-Aug; sw US	1474
	mate: yr rd; Coahuila, Mexico	522
	preg f: Mar, Sept-Oct; n8; NM	502
	preg f: Apr-June; no data other mo; CO	3262
	preg f: Mar-May, July; n14; Coahuila, Mexico	522
	preg f: June-July; n3; Durango, Mexico	523a
Neotoma bryanti	lact f: May; n1; Baja California	11044a
Neotoma cinerea	mate: Feb-July; captive	3087
	mate: May-Sept; w US, BC, Canada	1474
	preg f: May-Sept; LA	2615
	preg f: end May-July; n1; CO	3262
	preg f: June; few data; SD	10954
	birth: mid Mar-July; lab	3087
	birth: mid Apr-mid Aug; NV	3087
	birth: early May-June; BC, Canada	2088
	lact f: end May-July; n6; CO	3262
	lact f: June; n2; CO	1649a
	lact f: July; n1; NE	$5405b_1$
Neotoma chrysomelas	preg f: Feb, Sept; n2; Honduras	11044a
Neotoma floridana	testes: peak Jan, low Sept-Oct; KS	8859
	estrus: yr rd; lab	1703
	mate: end Jan-fall; KS	3285
	mate: Feb-Aug, peaks early Feb, end Mar, early Apr, May; KS	8859
	mate: yr rd, peak summer; FL	4680
	mate: yr rd, low winter; OK	3851
	preg f: June; n1; KS	1812a
	preg f: July; n1; NE	$5405b_1$
	birth: Mar-Apr; central US	949
	birth: Mar-Sept; PA	9164
	birth: mid Mar-early Sept; PA	8659
	birth: yr rd; s US	1474
Neotoma fuscipes	mate: Mar-Aug; CA	11289
	mate: end Mar-?; CA	2679
	conception: Jan-Sept, peak Feb-Apr; CA	6490
	preg f: Dec; n1; CA	3593a
	birth: Feb-May; OR	3041

	birth: Apr-early May; n6; CA	2679
	birth: peak May-June, few Jan-Oct; OR, CA	1474
	lact f: Mar, Oct; CA	11228a
Neotoma goldmani	lact f: July-Aug; San Luis Potosí, Mexico	2228a
Neotoma lepida	testes: peak Aug-May; CA	6708
	preg/lact f: Nov-May; CA	6708
	lact f: Sept; n1; CA	11228a
Neotoma mexicana	testes: small Aug-Nov; CO	1366
	ovulation: Mar-July, peak Mar-May; CO	1366
	preg f: Apr, June; few data; CO	3262
	preg f: Apr, Aug; n3; Coahuila, Mexico	522
	preg f: May; n1; NM	502
	preg f: July; n3; Durango, Mexico	523a
	lact f: end May-early June; n2; CO	3262
Neotoma micropus	epididymal sperm: yr rd; TX	11696
	mate: Mar-May; no data other mo; NM	979
	preg f: Mar-Sept; TX	11696
	preg f: June; n1; NM	502
	preg f: Dec; n1; Coahuila, Mexico	522
	preg/lact f: May-June; no data other mo; CO	3262
	birth: Feb-June; central N America	949
	birth: yr rd; s TX	8962
Neotoma phenax	preg f: Feb; n1; Sonora, Mexico	1471a
	preg f: Apr; n1; Mexico	5423
	preg f: Dec-Jan; Sinaloa, Mexico	5423
Neotoma stephensi	birth: most Feb-July, peak Mar-May; AZ	11232
Neotomodon alstoni	repro: mid June-Sept; Mexico	2346
	repro: Mar-Sept; Cerrodil, Ajusco, Mexico	9232
Notiomys macronyx	preg f: Nov; n1; Patagonia, Argentina	8305a
	lact f: Feb; n1; Malleco, Chile	4054
Notiomys valdivianus	preg f: Nov; Patagonia Argentina	8305a
Nyctomys sumichrasti	preg/lact f: July-Aug; n3/25; nw Nicaragua	3704
Ochrotomys nuttalli	mate: mid Mar-early Oct, peaks end spring, early fall; TN	6496
	mate: Sept-Apr; e TX	7022
	mate: yr rd, peak spring-summer; TN	6497
	preg f: Mar; n1; TX	8189a
	preg f: Apr; n1; MO	2863a
	preg f: Apr-June, Aug-Sept; NC	1290
	birth: est Mar-Oct; KE	3897
	lact: Apr; n1; KY	11588a
	lact f: Dec; n1; TX	8189a
Onychomys leucogaster	repro: Mar-Aug; ne CO	3294
	mate: Jan-Feb, Apr; few data, none after May; NM	979
	preg f: Mar-Apr; n2; Coahuila, Mexico	522
	preg f: Apr, June, Aug; MAN, Canada	11957a
	preg f: Apr, June-Aug; n9; NE	$5405b_1$
	preg f: Apr, Aug; n2; NM	502
	preg f: Apr-Dec; TX, AZ	504
	preg f: May-Aug; n US	504
	preg f: June; n2; KS	1812a
	birth: yr rd, peak May-Sept; lab	8559
	birth: peak Feb-Aug; UT	2907
	birth: Feb-Sept; sw US	1474
Onychomys torridus	mate: Mar-May; NM	979
	preg f: May-Aug; n US	504
	preg f: Apr-Dec; TX, AZ	504
	preg f: June-July; n5; Coahuila, Mexico	522
	preg f: June-July; n3; Durango, Mexico	523a
	preg f: Nov; n2; Sinaloa, Mexico	5409
	birth: Feb-Sept; sw US	1474
	birth: July, Sept; n2; captive	2109
	birth: yr rd, peak Jan-July; captive	8559
	repro: yr rd; lab	10697

Species	Data	Ref
Oryzomys alfaroi	preg f: Nov; n3; San Luis Potosí, Mexico	2228a
	preg f: yr rd; Oaxaca, Mexico	9176
Oryzomys argentatus	mate: Feb-Nov; LA	10809
Oryzomys capito	mate: yr rd; Panama	3316, 3317, 3319
	preg f: Jan, July; n4; Panama	3024
	preg f: peak Aug-Jan, Oct-Nov; Panama	3319
	birth: yr rd; Trinidad	11931
Oryzomys concolor	preg f: est yr rd; Venezuela	7995, 7996
Oryzomys couesi	preg f: Feb, Sept, Nov; n3; El Salvador	3210
	preg f: Nov; n1; San Luis Potosí, Mexico	2228a
Oryzomys delticola	mate: end summer-early fall; few data; Uruguay	596
Oryzomys flavescens	mate: Jan-May; n12/14; Uruguay	596
	preg f: Sept-May; central Argentina	7347
Oryzomys fulvescens	preg f: Dec; n1; Campeche, Mexcio	5423b
Oryzomys longicaudatus	preg f: Mar; n5/7; Santiago, Chile	3551
	preg f: Apr, Nov-Dec; Patagonia, Argentina	8305a
	preg/lact f: peak Dec-Feb; Malleco, Chile	4054
	repro: Oct-May; Chile	7242, 7666
Oryzomys melanotis	preg f: July; n1; Yucatan, Mexico	5423b
Oryzomys nigripes	preg f: Jan-May; Balcarce, Argentina	2221
Oryzomys palustris	mate: Feb-Oct; captive	10572
	mate: yr rd; se US	1474
	preg f: Mar-Apr, June-Sept, Nov; NC	1290
	preg f: Mar-Sept; Veracruz, Mexico	4228
	preg f: May; n1; Qunitana Roo, Mexico	950
	preg f: Aug; n1; Cozumel, Quintana Roo, Mexico	5423b
	preg f: yr rd, peak end spring-fall; MS	11871
	repro: yr rd; LA	7810
Oryzomys subflavus	preg f: peak Apr-May; lab	11305
Osgoodomys banderanus	lact f: Mar; n1; Mexico	11044a
Otomys angoniensis	mate: yr rd; Transvaal, S Africa	7200
	preg f: May, July, Nov; Malawi	4329
	preg f: June; n1; Kenya	4925
	birth: perhaps Oct-Mar; S Africa	10158
	birth: summer; Transvaal, S Africa	8965
	birth: yr rd, peak Apr-Sept; nw Kenya	10712
Otomys denti	preg f: May, June, Nov; Malawi	4329
	preg f: yr rd; Zaire	2571
Otomys irroratus	mate: yr rd, except June-July; captive	2338, 2339
	mate: yr rd; Zambia	331
	preg f: Jan, Oct; few data; e Africa	4925
	preg f: yr rd; Zaire	2571
	birth: Mar-Apr; no data other mo; central Africa	4495
	birth: Aug-Apr; Transvaal, S Africa	8965
	birth: yr rd except June-July; captive	2338
	lact f: Dec, Feb; few data; e Africa	185
	repro: June-Aug; Namibia	9945
Otomys maximus	birth: est Aug-Mar; S Africa	10158
Otomys typus	preg f: May-Aug; few data; Ethiopia	7600
	birth: yr rd; Kenya	5760
Ototylomys phyllotis	preg f: peak Jan-June; Belize	2590
	preg f: end Apr-early May; n7; Yucatan peninsula, Mexico	950
	preg f: May-June; n5; Guatemala	11044a
	preg f: Sept; n1; Honduras	11044a
	lact f: Jan; n1; Guatemala	11044a
	repro: Dec-Apr, July-Aug; Yucatan peninsula, Mexico	5423b
	repro: yr rd; Central America	6270
Oxymycterus rutilans	preg f: yr rd; Balcarce, Argentina	2221
Pachyuromys duprasi	birth: summer; captive	6237
	birth: yr rd; n23; captive	12128
Parotomys littledalei	preg f: Aug; n1; Namibia	9945
	juveniles: Dec; Namibia	9945

Peromyscus attwateri	preg f: Mar; n3; OK	11044a
	preg f: Dec; n1; KS	9601
	lact f: Apr, Oct; AR	9601
Peromyscus aztecus	preg f: Feb, May, July; n4; Guatemala	11044a
	preg f: dry season; n2/7; Oaxaca, Mexico	9176
Peromyscus boylii	ova: yr rd; CA	5243
	preg f: Apr-Oct; CA	5243
	mate: yr rd, peak spring-summer; sw US	1474
	preg f: May, Aug, Oct, Dec; n4; NM	502
	preg f: June-July; n8; Durango, Mexico	523a
	preg f: Nov; n1; CA	11229a
	preg f: Nov; San Luis Potosí, Mexico	2228a
Peromyscus californicus	testes: peak Apr-mid Oct; CA	7013
	preg f: Feb; n2; CA	11228a
	preg f: mid Apr-mid Oct; CA	7013
	preg f: yr rd; captive	9255
	preg/lact f: Feb-Oct; n20, 4 f; CA	6708
	birth: Apr-mid Oct; CA	7013
	lact f: Feb, Oct; CA	11228a
Peromyscus crinitus	mate: est Mar-Oct; CA	2807
	preg f: May; n3; NV	6488
	preg f: yr rd; captive	9255
	birth: Jan-Aug; 109/135 lt; lab	2910
Peromyscus difficilis	testes: peak June, low Jan; n CO	1841
	mate: Apr-Aug, peak May-June; n CO	1841
	preg f: June; n2; Durango, Mexico	523a
	preg f: July-Aug; n2; NM	502
	preg f: July, Dec; n2; Coahuila, Mexico	522
	preg f: Aug, Oct; n3; San Luis Potosí, Mexico	2228a
	birth: Spring-Oct; sw US, n Mexico	1474
	lact f: Dec; n3; Coahuila, Mexico	522
Peromyscus eremicus	testes: peak May-Sept; CA	6708
	nonfecund m: Oct-Nov; CA	6708
	preg f: Jan-Apr; n7; NM	502
	preg f: July-Aug, Oct, Dec; San Luis Potosí, Mexico	2228a
	preg f: Dec; n2; CA	11228a
	preg f: yr rd; Coahuila, Mexico	522
	preg f: yr rd; captive	9255
	preg/lact f: Jan, Feb, June, Sept; est yr rd; AZ	6414
	preg/lact f: yr rd; CA	6708
	birth: yr rd; captive	1236
	lact f: Sept; n1; CA	11228a
	repro f: Jan, June-July; n9; Durango, Mexico	523a
	repro f: June-Nov; AZ	9010
Peromyscus gossypinus	mate: Aug-May; se US	1474
	mate: yr rd; FL	10091
	preg f: Jan, July; n2; FL	6303
	preg f: Aug-May; FL	8698
Peromyscus guatemalensis	preg f: July; n2; Guatemala	11044a
Peromyscus hooperi	birth: June-Dec, none Jan-May; n14; Coahuila, Mexico	9603
Peromyscus interparietalis	birth: yr rd; captive	1236
Peromyscus leucopus	fertile m: Mar-Nov; MA	7715
	testes: peak end May-mid Sept; ONT, Canada	4376
	sperm present: yr rd, low Dec-Jan; TX	5460
	mate: Mar-June, Sept-Nov; n US	1474
	mate: yr rd; s US	1474
	mate: Mar-Oct; MI	7885
	mate: Mar-Dec; MA	67
	mate: end Mar-mid Nov, peaks Apr-June, Sept-Oct; sw VA	11877
	mate: Apr-May, Nov-Dec; MD	7169
	mate: Apr-May; data only for Mar-May; NM	979
	mate: May-Nov; MA	7715
	conceive: mean 1st Mar 6-Apr 15 across yr; ONT, Canada	7309
	preg f: Jan, Mar-Apr, July, Nov; NC	1290

	preg f: May; n1; Yucatan, Mexico	5423b
	preg f: Mar-Sept; WI	6555
	preg f: Mar-Oct, peaks May-June, Aug-Sept; MA	2724
	preg f: Mar-Dec; Coahuila, Mexico	522
	preg f: mid Mar-mid Nov; MI	1472
	preg f: end Mar-early Oct; PA	9164
	preg f: Apr-May; n4; MN	495
	preg f: Apr-Aug; no data Sept-Nov, Jan; QUE, Canada	758
	preg f: Apr-Sept; ONT, Canada	2082
	preg f: Apr-Oct; n43; NE	5405b$_1$
	preg f: end Apr-early May; n4; Yucatan peninsula, Mexico	950
	preg f: May-Sept; QUE, Canada	11956
	preg f: June; n1; SD	261a
	preg f: June; n1; SD	10954
	preg f: July; n1; ND	3703
	preg f: July; n1; Campeche, Mexico	5423b
	preg f: Oct; n3; NM	502
	preg f: Oct; MI	6797
	preg f: Sept, Dec; n2; KS	1812a
	preg f: yr rd, peak July-Aug, low Dec-Mar; TX	5460
	preg f: yr rd; captive	9255
	birth: Jan-Oct; TX	4125
	birth: Apr-Nov; NY	7235
	lact f: Mar-Oct; WI	6555
	lact f: June; n1; SD	10954
	lact f: June, Aug; n2; SD	261a
	lact f: Aug; n1; Quintana Roo, Mexico	5423b
Peromyscus madrensis	preg f: Mar; n1; Mexico	11044a
	lact f: Mar; n1; Mexico	11044a
Peromyscus maniculatus	testes: peak Mar-June; BC, Canada	9423
	testes: peak mid Mar-mid Oct; CA	7013
	testes: large mid Apr-early Nov; plains, CO	4218
	testes: large end Apr-early Sept; mountains, CO	4218
	testes: large May-mid Aug; tundra, CO	4218
	testes: scrotal ?-Sept; CO	2699
	repro m: end Apr-mid July; YUK, Canada	6015
	ova: yr rd; CA	5243
	mate: mid Mar-Aug; BC, Canada, WA	9891
	mate: end Mar-mid Nov, peaks Apr-June, Sept-Oct; sw VA	11877
	mate: Apr-Aug; NY	1364
	mate: Apr-early Oct; CO	2699
	mate: Apr-Oct; MI	6796
	mate: mid Apr-mid Sept; s QUE, Canada	11956
	mate: May-Aug; CO	11230
	mate: May-Sept; ALB, Canada	7310
	mate: May-Sept; YUK, Canada	3774
	mate: spring; only studied Mar-May; NM	979
	preg f: Jan; n2; CO	9001
	preg f: Jan-May; ID	11646a
	preg f: Feb-Dec; CA	5243
	preg f: Mar-Oct; WA	5658
	preg f: Mar-Oct; MN	722
	preg f: Mar-Nov; n62; NE	5405b$_1$
	preg f: Mar-Nov; Coahuila, Mexico	522
	preg f: Mar-Dec; CA	7081
	preg f: mid Mar-mid Nov, Jan (n1); CA	7013
	preg f: Apr-Aug, peak May-July, none Mar, Sept, Oct; MA	2724
	preg f: mid Apr-Aug; QUE, Canada	11956
	preg f: May; n7; SD	261a
	preg f: May, Sept; n3; NM	502
	preg f: May-June; n2; KS	1812a
	preg f: May-mid July; YUK, Canada	6015
	preg f: peak May-Aug; BC, Canada	9423
	preg f: May-Aug; CO	11230

Species	Data	Ref
	preg f: June-July; n6; Durango, Mexico	523a
	preg f: June-Aug; n59; ND	3703
	preg f: July; n22; se MT	6166
	preg f: peak Sept-Nov, Mar-Apr; KS	551
	preg f: Oct, Dec; CA	11228a
	preg f: Oct; n1; CO	738
	preg f: yr rd; WA	9576
	preg f: yr rd; captive	9255
	birth: Mar-Nov; WI, IL	2060
	birth: Apr-Oct, peak Sept-Oct; MI	5019
	birth: Apr-Oct; CA	7013
	birth: peak mid June-mid July; ALB, Canada	7310
	birth: yr rd; captive	6493
	lact f: May, July-Aug; SD	261a
	lact f: July; n2; se MT	6166
	lact f: Nov; n1; CA	11228a
	repro: Feb-Sept; BC, Canada	4667
	repro: Feb-Nov; ne CO	3294
Peromyscus megalops	preg f: Aug; n1; Mexico	2341
Peromyscus melanocarpus	birth: Dec, none Apr-May; captive, Oaxaca, Mexico	9114
Peromyscus melanophrys	preg f: June; n2; Durango, Mexico	523a
	preg f: Dec; n2; San Luis Potosí, Mexico	2228a
	lact f: Mar; n1; Coahuila, Mexico	522
Peromyscus melanotis	preg f: Jan, Apr, June-Aug; n13; Coahuila, Mexico	522
	preg f: June-Aug; n6; Mexico	2341
	repro f: June-July; n5; Durango, Mexico	523a
Peromyscus melanurus	preg f: seasonal; Oaxaca, Mexico	9176
Peromyscus merriami	preg f: June; n1; AZ	11044a
	preg f: Nov; n2; Sinaloa, Mexico	5409
Peromyscus mexicanus	m fertile: Jan-Mar, Aug; few data; El Salvador	3211
	preg f: Mar-Apr; few data; El Salvador	3211
	preg f: July; n1; Mexico	2341
	preg f: July; n2; San Luis Potosí, Mexico	2228a
	preg f: Aug; n1; Oaxaca, Mexico	523
	preg f: yr rd; Oaxaca, Mexico	9176
	preg f: yr rd; Oaxaca, Mexico	11388
	preg f: yr rd; Veracruz, Mexico	4228
	lact f: Feb, Sept; few data; El Salvador	3211
	lact f: May, Sept; Costa Rica	4574
	juveniles: Jan-Feb, May, Nov-Dec; El Salvador	1475
Peromyscus pectoralis	mate: Apr-Oct; w TX, n Mexico	1474
	preg f: Mar-Dec, none Jan; Coahuila, Mexico	522
	preg f: June-Aug; n6; Durango, Mexico	523a
	repro: est yr rd; TX	3108
Peromyscus perfulvus	preg f: Feb, Aug-Nov; w, central Mexico	4609
Peromyscus polionotus	mate: yr rd, peak winter; se US	1474
	preg f: yr rd; captive	9255
	repro: yr rd; FL	8907
Peromyscus simulus	lact f: Oct; n1; Mexico	11044a
Peromyscus stirtoni	preg f: Feb-Mar, June, Sept; n8; Honduras	11044a
	preg f: Apr; n1; Nicaragua	5408
	lact f: Sept; n1; Honduras	11044a
Peromyscus truei	testes: peak Mar-Sept; CA	7013
	mate: June-Sept; captive	2699
	mate: est not winter; NV	6512
	birth: May-Oct; CA	7013
	preg f: Mar, June-July, Sept; n6; Coahuila, Mexico	522
	preg f: May-June; NM	502
	preg f: June-July; n9; Durango, Mexico	523a
	preg f: Oct; n2; CA	11228a
	repro: Feb-?; CO	2699
Peromyscus yucatanicus	preg f: Apr, July-Aug, Dec; n5; Mexico	6269
	preg f: end Apr-early May; n3; Yucatan peninsula, Mexico	950
	preg/lact f: Apr, July-Aug, Dec; Yucatan peninsula, Mexico	5423b

	lact f: July-Aug; n2; Mexico	6269
	repro: yr rd; captive	6149
Petromyscus collinus	preg f: Feb; Namib Desert, Namibia	11863
	preg f: Sept; n1; Namibia	9945
	birth: perhaps summer; S Africa	10158
Phodopus campbelli	preg f: Aug; n1; China	12105
Phodopus sungorus	testes: peak May-Aug, low Nov-Feb; captive	4885
	testes: peak summer, low winter; lab	3249
	mate: end Apr-Sept; Mongolia	3334
	birth: peak Mar-Aug; none Dec-Jan, inside; none Nov-Feb; outside	3249
	birth: yr rd, peak spring-summer; Siberia, Russia	11285
Phyllotis darwini	testes descend: July-Nov; n Chile	7241
	testes scrotal: Aug; Fray Jorge Nat Pk, Chile	3551
	perforate vaginae: 27% of f Aug; Fray Jorge Nat Pk, Chile	3551
	preg f: Mar; n1; Malleco, Chile	4054
	preg f: May, Aug, Oct-Feb; n14; Peru	4682
	preg f: end July-early Aug; Santa Rose, Peru	8302
	preg f: 30-80% Sept-Nov, <10% Jan-July; n Chile	7241
	lact f: Mar; n1; Malleco, Chile	4054
Phyllotis osilae	preg f: Feb, Apr-June; n8; Peru	4682
Podomys floridanus	mate: yr rd; captive, FL	1474
	preg f: June-Mar, peak Sept-Oct; FL	6301
	preg f: yr rd, except Apr; captive	9255
Psammomys obesus	testes: peak Dec-Mar, low June; Algeria	5706
	mate: Sept-May; few data; Sinai Penninsula	8134
	mate: Dec-July, peak Mar; captive	9825
	mate: yr rd; Algeria	2234
	birth: yr rd, peak Mar; captive	3513
Reithrodon physodes	preg f: Jan, May; n2; Uruguay	596
	preg f: peak Nov; Patagonia	8306
Reithrodontomys brevirostris	preg f: June; n2; Nicaragua	5413
Reithrodontomys creper	preg f: Mar, June-July; n10; Panama	11044a
Reithrodontomys darienensis	preg f: Feb; n1; Panama	11044a
	lact f: Feb; n2; Panama	11044a
Reithrodontomys fulvescens	gonads: regressed Mar-Apr; south Mexico	11388
	mate: est Feb-Oct; sw US, n Mexico	1474
	mate: peaks Mar, July; TX	8189
	preg f: July; n1; Durango, Mexico	523a
Reithrodontomys gracilis	preg f: Mar; n1; Guatemala	12037
	preg f: July; n1; El Salvador	12037
	preg f: July; n2; Nicaragua	12037
	preg f: July-Aug; n5; Yucatan peninsula, Mexico	5423b
Reithrodontomys humulis	testes: large Feb, May, June, Aug, Dec; FL	6303
	preg f: Feb; n1; FL	6303
	preg f: peak May-Nov, some winter; TN	2794
	preg f: May, Nov; NC	1290
	birth: May-Nov; se US	1474
	birth: July; n1; FL	6303
	repro: yr rd, except Feb, May, Nov; lab	6300
Reithrodontomys megalotis	mate: begins Mar; BC, Canada	2088
	mate: almost yr rd, except Jan-Mar; CA; w US	1474
	preg f: none Jan, Oct; CA	10099
	preg f: none Feb, Nov; Latin America	4963
	preg f: peak Mar, Sept; KS	551
	preg f: Apr-Nov; n75; NE	5405b,
	preg f: May-June, Sept-Oct, Dec-Jan; n7; NM	502
	preg f: June; n4; SD	261a
	preg f: June-July; n6; ND	3703
	preg f: June-July; n4; KS	1812a
	preg f: July-Aug, Oct-Nov; CA	997
	preg f: July-Aug; n3; Mexico	2341
	preg f: yr rd; CA	3279
	birth: nearly yr rd; El Salvador	1475

Reithrodontomys mexicanus	preg/lact f: Nov; n2; San Luis Potosí, Mexico	2228a
Reithrodontomys montanus	preg f: Mar, June-July; n4; NE	5405b,
	preg f: June; n1; SD	261a
	preg f: Oct; n1; KS	1812a
	birth: yr rd; s US	1474
Reithrodontomys raviventris	fertile m: yr rd; CA	3279
	mate: Mar-Nov; n CA	10809
	preg f: yr rd; CA	3279
	repro f: Mar-Nov; n CA	9887
Reithrodontomys sumichrasti	preg f: Jan; n1; Guatemala	11044a
	preg f: Mar; n1; Panama	11044a
Rhipidomys mastacalis	preg f: est yr rd; n Venezuela	7996, 7998
Rhombomys opimus	birth: end Apr-early Sept; Mongolia	10516
	repro: Jan-mid June; USSR	9499
	repro: peak Mar, none Nov; lab	10202
Saccostomus campestris	mate: summer; S Africa	2861
	conceive: Apr-May, Aug-Sept, Dec-Jan; Malawi	4329
	preg f: Jan-Apr; Botswana	10154, 10158
	preg f: Feb, Apr, July, Oct; captive; Malawi	4329
	preg f: Dec; n1; Namibia	9945
	birth: Oct-Apr; Transvaal, S Africa	8965
	juvenile: yr rd; e Africa	5760
	juvenile: yr rd; Zambia	331
Saccostomus mearnsi	preg f: July; n2; e Africa	4925
Scapteromys tumidus	sperm present: yr rd; Uruguay	596
	preg f: Apr-May, Oct-Dec; Uruguay	596
	preg f: peak Dec; few data; Uruguay	4683
Scotinomys teguina	birth: yr rd; captive	4967
Scotinomys xerampelinus	birth: yr rd; captive	4967
Sekeetamys calurus	birth: yr rd; captive	7983
Sigmodon alstoni	preg f: est yr rd; n Venezuela	7995, 7996, 7998
Sigmodon fulviventer	birth: yr rd; captive	1527
Sigmodon hispidus	testes: peak Feb-Aug, low small Nov-Dec; n175 TX	4212
	uteri: large Feb-Oct, small Nov-Jan; TX	4212
	ovaries: peak Mar, low Dec; TX	4212
	ovulation: yr rd, few Oct-Jan; TX	4212
	mate: peaks Feb-July, Sept-Nov; AR	9710
	mate: Apr-Nov, peak May-Sept; ne KS	7029
	mate: yr rd; captive	7253
	preg f: Jan-Mar, July; n7; Yucatan peninsula, Mexico	5423b
	preg f: Jan, May, July, Oct; NC	1290
	preg f: Apr, Aug; Coahuila, Mexico	522
	preg f: Apr-Oct; KS	7029
	preg f: Apr-Nov; FL	6303
	preg f: end Apr-early May; n7; Yucatan peninsula, Mexico	950
	preg f: July; n1; Veracruz, Mexico	11507a
	preg f: July; n1; Oaxaca, Mexico	523
	preg f: July-Aug, none Jan-Feb, June, Oct, Dec; few data; El Salvador	3211
	preg f: Sept; n1; KS	1812a
	preg f: Sept; n2; NM	502
	preg f: yr rd; s Prusia Valley, Mexico	11388
	preg f: yr rd; Veracruz, Mexico	4228
	preg f: yr rd; Belize	2590
	birth: yr rd; Panama	3317
	birth: yr rd, except May-June; El Salvador	1475
	lact f: peaks spring, fall, winter; few data; FL	6303
Sigmodon ochrognathus	preg f: Apr, Oct; n6; Coahuila, Mexico	522
Steatomys krebsii	preg/lact f: Oct; Zambia	331
Steatomys parvus	birth: est summer; S Africa	10158
Steatomys pratensis	preg f: Nov; n1; subSaharan Africa	5760
	birth: Feb; n1; Zambia	331
	birth: July; n2; Malawi	4329
	birth: Oct-May; S Africa	10158

Tatera afra	testes: large June-Dec; S Africa	166
	preg f: Aug-Mar; S Africa	7146
Tatera brantsii	preg f: Jan, July, Nov; n6; Kalahari Gemsbok Nat Park, S Africa	7823
	preg f: yr rd; n63; S Africa	7146
	birth: Apr-June 40%, July-Sept 4%, Oct-Mar 13-17%; Zaire	8564
	birth: yr rd, peak summer; Namibia	9945
	repro: yr rd; S Africa	10158
Tatera indica	fertile m: low Nov-Jan; Pakistan	729
	scent gland: peak Dec-June, low July-Nov; India	6097
	sperm present: July-Apr; India	8733
	mate: Sept-Oct, Jan-Feb; Rajasthan, India	8721
	mate: yr rd; Pakistan	729
	preg f: Jan; n1; Iran	6293
	preg f: Mar-Nov, peak Mar-May, July-Sept; Pakistan	729
	preg f: yr rd; India	6097
	preg f: yr rd, peak monsoon; Rajasthan, India	5230
	birth: yr rd, peak Oct-Apr; Sri Lanka	8500, 8501
	lact f: Apr; n2; Nepal	7382
	lact f: May-Dec; n161; Rajasthan, India	6775
	repro: yr rd; India	8723
	repro: yr rd; Mysore, India	1689
Tatera leucogaster	mate: Sept-Apr; Transvaal, S Africa	8386
	mate: peak rainy season; Zambia	9893
	mate: yr rd; Zambia	331
	preg f: peak Feb-Mar; S Africa	10158
	preg f: 80-100% Oct-May, <20% July-Sept; Transvaal, S Africa	8384
	preg/lact f: Mar, May-June, Oct; Malawi	4329
	birth: summer; Zimbabwe	1815
	birth: yr rd; Botswana	10154
	birth: yr rd, peak summer; Namibia	9945
Tatera nigricauda	testes: large yr rd, peak rains; Kenya	7794
	mate: low end dry season; Kenya	7796, 7797
	preg f: peaks Nov-Feb, Apr-May; Kenya	7794
Tatera robusta	preg f: July-Oct; Senegal	5057
	preg/lact f: peak July-Aug, Nov-Jan; Kenya	7794
Tatera valida	testes: large yr rd; Uganda	7794
	mate: yr rd; Zambia	331
	preg f: 50% June, Nov, near 0% other mo; Uganda	2453
	preg f: July-Oct; Senegal	5057
	preg f: yr rd, peaks Apr-May, Nov-Dec; Uganda	7794
	young found: Mar, Nov-Dec, none Feb, May; no data Jan, Apr, June-Oct; central Africa	4495
Tatera vicina?	preg f: Dec; n2; Kenya	4925
Taterillus gracilis	preg f: July-Oct; Senegal	5057
Taterillus pygargus	birth: Mar, Nov; Senegal	8695a
Tylomys nudicaudus	gonads: regressed Mar-Apr; s Mexico	11388
	preg f: Mar; n1; Guatemala	11044a
Wiedomys pyrrhorhinos	lact f: May, Aug-Nov; Caatinga, Brazil	10494
Zygodontomys brevicauda	preg f: Jan, Mar-Apr, Nov; Panama	3024
	preg f: June, Sept-Nov, est yr rd; Venezuela	451
	preg f: est yr rd; n Venezuela	7996, 7998
	birth: est yr rd; Panama	3317

Order Rodentia, Family Spalacidae, Genus *Spalax*

Spalax giganteus, *S. leucodon*, and *S. microphthalmus* are 150-550 g, subterranean, solitary herbivores that tunnel under grasslands from the eastern Mediterranean to Mongolia. Neonates are naked (7860) and suckle from 6 mammae (8050). Males have a baculum and seminal vesicles (6002, 8050).

NEONATAL MASS		
Spalax leucodon	~5 g	7860
LITTER SIZE		
Spalax leucodon	2-3	6419
	mode 3-4; r1-6	8050, 10479
	mode 3-4; r1-6	7860
	4-5; 3 rare	9452
Spalax microphthalmus	2-4	4645
	2.83; r1-5	8154
GESTATION		
Spalax leucodon	>28 d; n1	7860
LACTATION		
Spalax leucodon	wean: est 1-2 mo	10479
LITTERS/YEAR		
Spalax leucodon	1	7860
	1	9452
	1	10479
Spalax microphthalmus	1	8050
	1; r1-2?	4645
SEASONALITY		
Spalax leucodon	birth: peak mid Jan-mid Feb 67%; Israel	7860, 7863
	birth: early Feb-?, peak early Apr; Lebanon	6419
	birth: end Feb-Mar; Odessa, Crimea	9452
	birth: Mar-Apr; Sofia, Bulgaria	10479
Spalax microphthalmus	mate: est Mar; USSR	8050
	mate: spring; Europe, w Ukraine, w Kazakstan	4645
	preg f: ?-Apr; USSR	8050
	birth: end Feb-mid May; Voronesh, Russia	8154

Order Rodentia, Family Rhizomyidae

Bamboo (*Cannomys badius* 500-800 g; *Rhizomys* 1-4 kg) and mole (*Tachyoryctes macrocephalus* 330-930 g; *T. splendens* 160-280 g) rats are fossorial herbivores from tropical and subtropical regions of east Africa (*Tachyoryctes*) and Nepal through southeast Asia (*Cannomys, Rhizomys).* Males are larger than females (*Tachyoryctes* 4925, 5760). *Tachyoryctes* are solitary (5268). A copulatory plug is present (*Cannomys* 2938). *Tachyoryctes* probably has a postpartum estrus (5267). Neonates are naked and open their eyes at 3-4 wk (*Cannomys* 2938; *Rhizomys* 182, 7161, 8045). *Rhizomys* have 8-10 teats (182, 7161).

NEONATAL MASS		
Tachyoryctes splendens	~15 g	2446a
	15-25 g; n10	8857
NEONATAL SIZE		
Tachyoryctes splendens	HB: 55-60 mm; n10	8857
LITTER SIZE		
Cannomys badius	mode 2; n5 lt	2938
Rhizomys pruinosus	2.0 emb; r1-3; n3 f	10830
	3 yg; n1 f	10830
	5-6 yg; hunter reports	10830
Rhizomys sinensis	3 yg; n1 nest	182
Rhizomys sumatrensis	2-4 yg	10830
	3-5 yg	6369
	5; n1 lt	7161
	5 yg; n1 lt, 1 wk	8045

Tachyoryctes splendens	1 emb ea; n2 f	4925
	1.20 emb; n38 f	2446a, 5267
	1.25; max 4; n322 lt	8855
	1.3; r1-2	5268
	1.46 emb; mode 1; r1-4; n356 f	2446a, 2449
	1.51 emb; mode 1; r1-4; n82 f	8857
	1.64 emb; n37 f	2446a, 5267
	2 yg; n1 f	4925
SEXUAL MATURITY		
Tachyoryctes splendens	f: 120 d	5268
GESTATION		
Cannomys badius	r40-43 d; n3	2938
Rhizomys sumatrensis	≥22 d; n1	7161
Tachyoryctes splendens	37-40 d	2449
	38.5 d; r37-40	5268
	44-53 d; n6	8857
LACTATION		
Cannomys badius	solid food: 23 d	2938
	wean: 8 wk	2938
Rhizomys pruinosus	independent: 4 mo	10830
Rhizomys sumatrensis	solid food: 29 d	7161
	nurse: 3 mo	7161
Tachyoryctes splendens	wean: 4-6 wk	5267
	wean: 43 d; r35-50	5268
INTERLITTER INTERVAL		
Tachyoryctes splendens	173 d	5268
SEASONALITY		
Rhizomys pruinosus	birth: Feb-Apr; Vietnam	10830
Rhizomys sinensis	birth: Apr-May; central Asia	182
	lact f: May; n1; China	182
	young in nest: June-July, Nov; China	182
Rhizomys sumatrensis	birth: Feb-Apr, June-Aug; Vietnam	10830
Tachyoryctes splendens	preg f: Oct; n1; e Africa	4925
	preg f: yr rd, except May; Kenya	5267
	preg/lact f: Feb; n3; e Africa	4925

Order Rodentia, Family Arvicolidae

Voles and lemmings (*Alticola* 22-69 g; *Arborimus* 28 g; *Arvicola* 70-310 g; *Clethrionomys* 10-59 g; *Dicrostonyx* 20-112 g; *Dinaromys* 30-82 g; *Ellobius* 30-88 g; *Eolagurus*; *Eothenomys* 27 g; *Hyperacrius* 22-60 g; *Lagurus* 11-51 g; *Lemmus* 21-113 g; *Microtus* 10-170 g; *Neofiber* 156-357 g; *Ondatra* 330-2400 g; *Phenacomys* 16-56 g; *Pitymys* 13-58 g; *Proedromys*; *Prometheomys* 60-88 g; *Synaptomys* 15-50 g) are terrestrial to fossorial herbivores from North America and north Eurasia. Females are reported larger than males in *Lemmus* (9762), while males are reported larger in *Microtus* (1933, 2466, 3112, 5511, 5794, 6632; cf 5486, 9894, 10954, 11878), *Neofiber* (944), and *Ondatra* (9503). Each sex has been reported larger in *Clethrionomys* even within the same species (716, 2524, 5501, 5798, 7238, 11679). The sexes are of equal size in *Pitymys* (6199, 6881, 9273, 10954; cf 7904), *Prometheomys* (10275), and *Synaptomys* (2026).

Ovulation is induced (289, 462, 481, 794, 1248, 1249, 1250, 1251, 1267, 1268, 1269, 1270, 1271, 1638, 1749, 1750, 1807, 1875, 1876, 1880, 1881, 1918, 1919, 1964, 2330, 3220, 3610, 3968, 4014, 4015, 4052, 4053, 4169, 4264, 4475, 5000, 5281, 5283, 5284, 5672, 5674, 5675, 5676, 5809, 6348, 6389, 6692, 7291, 7334, 7335, 7336, 7339, 7341, 7342, 7343, 7595, 8021, 8251, 8958, 9009, 9110, 9111, 9478, 9517, 9530, 10421, 11603, 11604, 11605).

The usual sex determination does not obtain in *Dicrostonyx torquatus* (3780), *Ellobius lutescens* (2389, 6651, 6973, 7736), *Lemmus schisticolor* (3493, 3494, 5502, 10372, 11375), and *Microtus oregoni* (1370, 3492, 6973, 8062, 8557, 8998a). Phenolic plant extracts may alter reproduction (793, 795, 796, 797, 1501, 1837, 2113, 4813, 5963, 7812, 8558, 8562, 9462, 9538). A copulatory plug is present (*Clethrionomys* 4172, 5594, 8425; *Lemmus* 3466; *Microtus* 1218, 4278, 4459, 4460; *Pitymys* 783, 9111).

The 5 g (*Ondatra*), hemochorial, labyrinthine, discoid placenta has an interstitial, mesometrial implantation into a bicornuate, bipartite, or duplex uterus (1222, 2464, 2467, 3081, 3082, 5735, 7271, 7563, 7567, 9479,

10421) and is eaten after birth (4278, 4893, 5240, 9503). Estrus occurs postpartum (501, 660, 944, 981, 1218, 1222, 1250, 1407, 1806, 1875, 2026, 2045, 2077, 2087, 2529, 2562, 2563, 3061, 3457, 3459, 3462, 3465, 3466, 3468, 3571, 3622, 3652, 3745, 4014, 4053, 4172, 4264, 4275, 4278, 4469, 4645, 4660, 5039, 5135, 5240, 5593, 6041, 6347, 6348, 6397, 6693, 6782, 6806, 7271, 7904, 8051, 8052, 8920, 8921, 9013, 9110, 9111, 9536, 9656, 9762, 10383, 10446, 10569, 10589, 10681, 11225, 11227, 11362, 11647, 12049, 12070; cf 3558). Lactation may delay implantation (*Clethrionomys* 7271).

Arvicolids are born blind and naked (1474, 2243, 4219, 5311, 8179, 9656, 11467, 12007). A few papers report *Ondatra* (3081, 8051, 10580) and *Pitymys* (783, 3283, 6051) as furred at birth. Eyes open at 8-11 d (*Arvicola* 2045, 3571, 7433, 8052; cf 4660), 8-13 d (*Microtus* 501, 1965, 2087, 3457, 3465, 3468, 3571, 4275, 4278, 4490, 4970, 4999, 5148, 5313, 6128, 6347, 7433, 8052, 9009, 9013, 9269, 9355, 9724, 9747, 9967, 11225, 12049, 12105), 9-13 d (*Clethrionomys* 3571, 5148, 7002, 7238, 7433, 7546, 8052, 8425; cf 10569; *Lagurus* 1474, 3516, 5240; cf 8634; *Lemmus* 2386, 3462, 5135, 6806, 8051, 10210; cf 716; *Pitymys* 783, 3283, 3571, 4273, 6199, 7433, 7904, 9111; cf 6051), 10-12 d (*Synaptomys* 2026, 10329), 10-14 d (*Dicrostonyx* 4335, 8051, 8810), 11-20 d (*Ondatra* 3081, 3082, 3571, 7433, 8051, 9503; cf 7291), 14 d (*Neofiber* 944), 14-16 d (*Phenacomys* 3429, 5150), 18-23 d (*Arborimus* 4264, 5026), or 21 d (*Ellobius* 6403).

Females have 4 mammae (*Arborimus* 1474; *Eothenomys* 182), 4-6 mammae (*Pitymys* 1474, 2376, 3571, 4645, 6199, 8052, 9175, 9530, 9536), 4-8 mammae (*Synaptomys* 1474, 2026, 6486, 10329), 6 mammae (*Hyperacrius* 9175; *Neofiber* 944, 1474), 6-10 mammae (*Ondatra* 1474, 3571, 7433, 8051, 10580, 11934), 8 mammae (*Alticola* 8052; *Arvicola* 3571, 7433, 8052; *Clethrionomys* 502, 1474, 2548, 2564, 3547, 3571, 4035, 7433, 8052, 9612, 10516, 10569, 11362; *Ellobius* 182; *Lagurus* 1474, 8052; *Lemmus* 655, 8051; *Microtus* 502, 1218, 1474, 1965, 2376, 2562, 3571, 4278, 4745, 7433, 8052, 8986, 9612, 9724, 9967; *Phenacomys* 1474, 5551; *Prometheomys* 8051). Males have ampullary glands, preputial glands, seminal vesicles, penial scales, and a baculum (352, 2460, 2464, 2467, 4891, 5091, 6805, 6897, 8051, 8052, 8984, 12081; cf 10323, 10327).

Details of anatomy (668, 1878, 2463, 4468, 8402, 9479, 11606), endocrinology (7707; *Clethrionomys* 3424, 6055a, 7849, 10625, 11602; *Dicrostonyx* 1441, 1779, 9521; *Ellobius* 10328, 10335, 10336; *Lemmus* 304; *Microtus* 64, 724, 1746, 1748, 1749, 1882, 2266, 2657, 3424, 4015, 7048, 7339, 7341, 7344, 8409, 8410, 9152, 11274, 11932; *Ondatra* 719, 723a; *Pitymys* 9537), oogenesis (8402), parthenogenesis (9478), spermatogenesis (4091, 4092, 8517), copulation (1637, 1639, 2385, 2506, 2514, 2531, 4016, 4017, 4018, 4019, 5675), implantation (7850), pregnancy failure (1336, 1916, 1917, 5250, 5594, 5673, 6760, 6760a, 6822, 7337, 7338, 7340, 8958, 9531, 9533, 9539, 10073, 10332a, 10333, 10457, 10458, 11486), milk (7065a), reproductive energetics (5149; *Clethrionomys* 5477; *Microtus* 7292, 10911; *Pitymys* 6533), and light (*Clethrionomys* 3193, 6052, 10622, 10624, 10748; *Dicrostonyx* 4471, 4472, 8809; *Microtus* 519, 1747, 1873, 1882, 1883, 2087, 2115, 2116, 2220, 4087, 4088, 4089, 4090, 4998, 4999, 5139a, 5731, 6339, 6340, 6341, 6356, 6357, 6358, 6359, 6900, 6901, 6910, 6911, 7842, 9087, 9352, 10249, 10250, 10251, 11275; *Pitymys* 2493, 7840, 7841, 7844, 10074) are available. A monograph on New World *Microtus* (including some *Pitymys*) is available (10648a).

NEONATAL MASS

Arborimus longicaudus	2.33 g; n35	4264
	2.58 g; r2.56-2.60; n2	1362
	3.1 g	5026
Arvicola terrestris	m: r3.19-7.84 g; n55	4660
	3.2-7.8 g	3571
	f: r3.28-7.32 g; n55	4660
	5.0 g	2045
	5-6 g	8052
Clethrionomys gapperi	1.18 g	5148
	1.75; se0.03; n229	5149
	1.9 g; r1.7-2.3	10569
Clethrionomys glareolus	f: 1.55 g; se0.032; r1.3-1.8; n15	12080
	m: 1.59 g; se0.051; r1.2-1.9; n17	12080
	1.6-1.7 g	1032
	1.77 g; spring-summer; 2.03 g; fall	7002
	1.8 g; n4, 18 d emb	8179
	1.815 g; 18 d emb	8425
	1.91 g; r1.67-2.50; n80	11362
	2.7 g	1222, 2077
Clethrionomys rufocanus	~2.0-2.5 g; n13, 1 d	8287

Clethrionomys rutilus	f: 2.9 g; n98	7545
	m: 3.0 g; n100	7545
Dicrostonyx groenlandicus	3.8 g; r2.7-4.8; n34	4335
	4.4 g	4469
	f: 4.6 g; n66	7545
	m: 4.6 g; n39	7545
Dicrostonyx torquatus	2.5-3 g	8810
	3.4 g; r2.0-4.2; n23	8051
Dicrostonyx unalascensis	5 g; from graph	2560
	m: 5.2 g; n112	7545
	f: 5.4 g; n122	7545
Eothenomys smithi	2.13 g; se0.06; r1.98-2.26; n16	294a
Lagurus curtatus	0.85 g	6925
	1.47 g; n8	5240
Lagurus lagurus	1.2 g; n1	11647
	1.5 g	3516
Lemmus lemmus	2 g	11679
	f: 3.6 g; r2.2-4.3; n41	3462
	3.67 g; r2.74-4.59; n45; from graph	6806
	3.9 g	4830
	m: 3.9 g; r2.6-4.6; n32	3462
	4.5-5 g	8051
	f: 4.6 g; n55	7545
	m: 4.9 g; n74	7545
Lemmus schisticolor	3.2 g; r2-4	5135
Lemmus sibiricus	3 g	716
	3.6-4.0 g	8051
	f: 4.0 g; n47	7545
	m: 4.3 g; n59	7545
Microtus breweri	3.5 g	7714a
Microtus abbreviatus	f: 3.3 g; n70	7545
	m: 3.6 g; n91	7545
Microtus arvalis	f: 1.09-1.29 g	2462
	1.1-3.2 g	3571
	1.22-2.77 g	9269
	m: 1.35-1.51 g	2462
	1.75 g	7292
	1.79 g; r1.26-2.64	11867
	1.9 g	8153
	1.94 g; n29	6361
	f: 1.95; se0.0015; n1418	9013
	1.99 g; se0.0006; n2738	9013
	m: 2.03; se0.0015; n1320	9013
	2.53 g; r1.5-3.4; n348	3459, 3460, 3465
Microtus brandti	1.9 g; r1.4-2.0; n140	12105
Microtus californicus	2.8 g; n19 yg, 5 lt	4490
	3.18 g; n10, 2 lt	9747
	3.4 g; r2.7-4.8; n35; from graph	6041
Microtus canicaudus	2.4-2.5 g; se0.03-0.06; n4 groups	4196a
Microtus guentheri	~3 g	8115
Microtus miurus	2.29 g; r1.65-3.0; n17	12049
	f: 2.8 g; n88	7545
	m: 3.0 g; n120	7545
Microtus montanus	2.2 g; n18	9724
	3.9 g	7714a
Microtus montebelli	2.5 g; sm lt: 2.7; lg lt: 2.4 g; n95 babies	7762
Microtus nivalis	3.7 g	3571
	3.7 g; r3.6-3.8; n2	5486
Microtus oeconomus	1.86 g; ±0.13; 1 d	9355
	1.9-3.1 g	3571
	f: 2.06 g; se0.03; n32	5313
	m: 2.10 g; se0.03; n36	5313
	2.3 g; r1.9-2.6; n27	3468
	2.4 g; se0.1; n46	3655

	f: 2.95 g; n124	2563, 7545
	m: 2.95 g; n143	2563, 7545
Microtus oregoni	1.56 g	4970
	1.7 g; r1.59-2.2; n41	2087
Microtus pennsylvanicus	2-3 g	6347
	2.07 g; r1.6-3.0; n24 lt	4272, 4278
	2.26 g; se0.05; n197	5149
	2.35 g	5148
	2.4 g; r1.4-3.5; n776	2660
	f: 2.4 g; n368	2660
	m: 2.5 g; n408	2660
	2.92 g; sd0.35; n59 yg, 15 lt	9283
	~3 g	501
	3.2 g; n212	2562, 7545
Microtus richardsoni	5.05 g; n6 emb, 1 f	4969
Microtus socialis	2.1 g	8052
Microtus transcaspicus	2.8 g	8153
Microtus xanthognathus	3.5 g; r2.7-4.2; n28	12049
Neofiber alleni	7.2 g; n3	4282
	12.1 g; r10.0-14.9; n7	944
Ondatra zibethicus	16-20 g	6267
	21 g	11467
	21.3 g; median 21; r16-28; n41	3081, 3082
	23.0 g; sd2.56; n70, 1 d	2695
	25 g	3061
	28 g	9503
Phenacomys intermedius	1.89 g; se0.07	5150
	2.05 g; n14, 3 lt	4219
	2.36 g; n18, 4 lt; 1 d	4219
	2.4 g; r2.0-2.7	3429
Pitymys duodecimcostatus	f: 2.2 g; ±0.03; n50	4123
	m: 2.2 g; ±0.02; n39	4123
	2.24 g; r2-2.8; n21	6361
Pitymys lusitanicus	1.7 g; r1.5-1.9; n3	11820
Pitymys ochrogaster	2.8 g; se0.3; n16	6881
	2.84 g; n264	9111
	2.9 g; ±0.05; r2.0-3.8; n67 yg, 27 lt	3283
	3.1 g; sd0.27; se0.05; r2.4-3.5; n34	6051
Pitymys pinetorum	2.0 g; se0.1; n19 lt	6533
	2.2 g	4273
	2.3 g; n15, n4 lt	3852
Pitymys subterraneus	f: 1.42 g; n1	6199
	m: 1.43 g; r1.41-1.45; n2	6199
	1.58 g; n59	3654
	1.95 g; se0.26; n28	9656
Synaptomys cooperi	2.55 g; r2.4-2.7; n2	10329
	3.9 g; r3.1-4.3; n11	2026
NEONATAL SIZE		
Clethrionomys gapperi	CR: 30 mm	7238
Clethrionomys glareolus	22.0 mm; se0.53; r18-26; n32	12080
	25-35 mm	8052
	CR: 26.5 mm; 18 d emb	8425
	38.5 mm; n32	7002
Dicrostonyx groenlandicus	CR: 20-27 mm	4335
	HB: 35 mm; n1, 1 d	2434
Dicrostonyx torquatus	33-38 mm	8051
Lagurus curtatus	HB: 27 mm; n8	5240
Lagurus lagurus	32 mm; n1	11647
Lemmus lemmus	15-18 mm	8051
Lemmus sibiricus	32-40 mm	8051
Microtus montanus	33 mm; n13	9724
Neofiber alleni	TL: 82 mm; n3	4282

Ondatra zibethicus	28.5 mm; sd2.15; n70	2695
	96 mm	9503
	100.4 mm; median 102; r85-115; n41	3082
Phenacomys intermedius	CR: 30 mm	3429
Pitymys ochrogaster	CR: 30-35 mm	6051
	40-50 mm	3283
Pitymys pinetorum	CR: 27-29 mm	8268
	TL: 39-48 mm	10159
Pitymys sikimensis	27.5 mm; largest emb	7382
Pitymys subterraneus	28.25 mm; se4.33	9656
Prometheomys schaposchnikowi	36 mm	12007
WEANING MASS		
Arborimus longicaudus	17-25 g	4264
Clethrionomys gapperi	9.16 g; 14 d	5148
	11.5-12 g; 16-18 d	7238
Clethrionomys glareolus	8.43 g; n563 lt	4173
	9.1; sd1.9; n219	4343
	9.44 g; n134 lt	4173
	10.0; sd1.5; n77	4343
	12 g; 20 d	1835
Clethrionomys rufocanus	10 g	5501
Dicrostonyx groenlandicus	20 g	4469
	26.0 g; r16.5-34.6; 22 d	4335
Dicrostonyx torquatus	r23-28 g; n> 30, 19 d	5466
Lagurus lagurus	5.5-6.0 g	11647
Lemmus schisticolor	m: 15.6 g; r11.2-21.2; n60, 20-22 d	5135
	f: 15.7 g; r9.6-23.8; n60, 20-22 d	5135
Lemmus sibiricus	r24-30 g; n> 30, 19 d	5466
Microtus agrestis	7.64 g; se0.13; n37 lt, 170 yg	1805
	10.23 g; se0.18; n19 lt, 66 yg	1805
Microtus arvalis	6.2-8.2 g	6910
	8.90 g; 3 wk	9013
	9.18 g; 17 d	11867
	9.69 g; 3 wk	9013
	9.87 g	7292
Microtus californicus	17 g; r12.5-23; n29, 16 d	6041
	21 g; n5, 20 d	9747
Microtus canicaudus	m: 10.7-16.8 g; se0.4-1.0; n4 groups, 18 d	4196a
	f: 12.0-16.4 g; se0.2-0.8; n4 groups, 18 d	4196a
Microtus guentheri	~15 g; from graph	1965
Microtus montanus	9.33 g; n18, 4 lt, 15-16 d	9967
Microtus montebelli	17.6 g	7762
Microtus oeconomus	12 g; 11-12 d; first outside nest	10681
	15.6 g; r14.7-16.4	3468
	r17-19 g; n> 30, 19 d	5466
Microtus pennsylvanicus	11.52 g; 14 d	5148
	14-16 g	4278
Ondatra zibethicus	109.3 g; 4 wk	8051
	median 150 g	3082
	~150-165 g	2695
	197 g; n9, 31 d	3081
Pitymys ochrogaster	19.0 g; sd2.61; se0.51; r13.9-24.1; 21 d	6051
Pitymys pinetorum	7.5-9.5 g; 3 wk	8268, 10159
	8.7 g; se0.3; n19 lt, 18 d	6533
	10-11 g; 16 d	4273
	11.1 g; se0.4; n19 lt, 21 d	6533
Pitymys subterraneus	9.44 g; se1.23; n37	9656
Synaptomys cooperi	~20 g; 3 wk; from graph	2026
WEANING SIZE		
Lagurus lagurus	40 mm	11647
Pitymys pinetorum	TL: 69-98 mm; 3 wk	8268, 10159

LITTER SIZE

Alticola roylei	4.3; r4-5; n3 lt	9175
	5.6 emb; n64 f	10516
	6 ps; n1 f	8052
Alticola stoliczikanus	8 emb; n1 f	8052
Alticola strelzowi	3.6 emb; r2-6 (2-3 early season, 2-6 late season)	8020a
	7-11 emb	8052
	7.1; r3-13	3088
Arborimus albipes	2-3 yg	5245
Arborimus longicaudus	2; r1-3	5026
	2.5 yg; r1-4; n14 lt	4264
	3 yg; n1 nest	5311
	3 emb; n1 f	1362
	3 emb; n1 f	782
	4 yg; n1 nest	11672a
Arvicola sapidus	3.31 emb; r2-5; n19 f	11248
Arvicola terrestris	2-7	7433
	mode 2-8; r2-14	3571
	3.3; r1-8; n107 lt	4660
	3.42; se0.098; sd1.294; n173 f	11668
	3.82 yg; mode 4; r2-5; n17 lt	11204
	4.2; r2-7	981
	4.48 emb; sd1.23; se0.13; mode 4; r2-9; n80 f	11249, 11250
	4.5 emb; r4-5; n2 f	1590
	4.52; se0.120; n119 lt	11668
	4.87 emb; mode 5; r1-9; n116 f	11204
	5.25; se0.396; n20 lt	11668
	5.5; se0.2; r1-10	7772
	5.7 emb; n14 f	8392
	5.7 emb; n122 f	6007
	mode 6-8; r6-14	4645
	6.4 cl	8392
	6.4 emb; by palpation	10447
	6.9; r3-14	8052
Clethrionomys spp	3.12-4.91; r for 7 yr	8256
Clethrionomys californicus	4 emb; n1 f	6708a
Clethrionomys centralis	1-7	634
	4-8 emb	8052
Clethrionomys gapperi	2.3 ova; r1-3; n6 f	3220
	3.33 emb; r3-4; n6 f	575
	3.6 cl; r1-7; n9 f	3220
	4 emb; n1 f	502
	4 emb; n1 f	4035
	4 yg; n1 nest	495
	4 emb ea; n2 f	2548
	4.17 yg; n6 lt, 5 f	10569
	4.33 emb, yg; r3 yg, 5,5 emb; n3 f	6796, 6799
	4.5 emb; r4-5; n2 f	4280
	4.6 yg; r3-7; n20 lt	11956
	mode 5 yg; r3-8	5210
	5 yg; n1 f	9882
	5 emb; n1 f	11504
	5; r2-6	9164
	5.2 emb; se0.22; r3-7; n20 f	10839
	5.29 emb; r3-7; n7 f	279
	5.3 emb; r3-7; n6 f	2025
	5.3 emb, ps; ±1.6; n22 f	8713
	5.38 emb, ps; r4-8; n21 f	10101
	5.47 yg; ±0.13; r3-8; n49 lt	2082
	5.58; se0.19; mode 6; r2-8; n41 lt	5149
	5.7 emb; r5-6; n3 f	279
	5.9 emb, ps; ±0.6; n15 f	8713
	6 emb; n1 f	9612
	6 emb; n1 f	10218, 10220
	6.0 emb; r5-7; n6 f	10144

	6.0 yg; ±1.68; n10 lt	8383
	6 ps; r5-7; n3 f	3703
	6.01 emb, ps; n115 f	7895
	6.1 emb; se0.50; r4-8; n10 f	11230
	6.33 emb; r6-7; n3 f	7809
	6.4 emb stated, [5.86 from data]; r4-7; n14 f	5798
	6.65 cl; se0.127; r4-13; 6.19 ps; se0.118; r3-10; 6.07 emb; se0.123; r3-10; n107 f	723
	7 emb; n1 f	4388
	7.0 emb; r6-8; n4 f	3703
	8 emb; n1 f	5333
Clethrionomys glareolus	2-10	10589
	2.5 emb; r2-3; n2 f	10274
	3 emb; n1 f	7888
	3 emb; n1 f	6548
	3.0 yg; ±0.1; n62 non lactating f	1881
	3.2; ±0.1; n189 primiparous	1874
	3.3 emb; se0.4; 6.7 cl; se0.7; n12 f	1877
	3.4-6.2; n3 locations	8052
	3.42 yg live; ±1.51; r1-7; n83 lt	8021
	3.44; se0.04; mode 4; r1-7; n140 lt	10383
	3.52 yg; n1712 lt	1874
	3.6; n26 lt	7002
	3.62 emb; r1-6; n21 f	12076
	3.65; ±0.1; n299 multiparous	1874
	3.66 yg; mode 4; r1-10; n782 lt, 264 f	1408
	3.71; n14 lt	5307
	3.77 emb; mode 4; r1-6; n90 f	2077
	3.795; r1-8; n83 lt	8021
	3.97 yg; se0.11; n65 lt, 13 f	157
	4 emb; mode 3-5	516
	4.0 wean; r3-6	5477
	4 ea; n2 lt	3555
	4.00 yg; mode 4; r2-6; n20 lt	11362
	4.0 emb; r2-7; n58 f	10175
	4.04 cl; r2-7; n23 f	12076
	4.06 emb; r2-6; n49 f	1844
	4.09 emb; se0.104; mode 4; r2-8; n94 f	12079
	4.11; mode 4; r1-6; n70 lt	1222
	4.2 yg; ±0.2; n43 lactating f	1881
	4.3-5.5	453
	4.3; se0.1; n104 lt	1874
	4.31 ps; se0.155; mode 4; r1-11; n94 f	12079
	4.45 emb; n11 f	955
	4.5 yg; sd1.4; n85 lt	4343
	4.6-6.1 cl	7271
	4.73 emb; mode 5; r3-7; n44 f	10339, 10340, 10344
	4.84 yg; mode 5; r1-10; n799 lt	4172, 4173
	4.85; r1-9; n54 lt	5103
	4.88 emb; se0.055; n546 f	12082
	4.9 yg; sd1.2; n141 lt	4343
	4.902 emb; se0.065; mode 5; r1-10; n388 f	12077, 12079
	4.980 ps; se0.085; mode 4-5; r1-11; n347 f	12077, 12079
	5.0 yg; r3-7	5477
	5.01 cl; ±0.2; n65	1874
	5.03; n726 lt	1874
	5.2 emb; n110 f	9404
	5.22 emb; mode 6; r3-9; n45 f	11670
	5.24 emb; mode 5-6; r3-7; n29 f	11679
	5.32 emb; n368 f	10373
	5.82 yg; mode 6; r1-13; n342 lt	4173
	5.96 emb; n82 f	1076
	6.0-6.8 emb	223
	6; r1-10	4645
Clethrionomys rufocanus	3	5493

	3 emb; mode 5; r2-9	7760
	3.50 yg; se1.5; max 8	7760
	4 emb; n1 f	10516
	5.0 emb; r4-6; n4 f	11679
	5.16 emb, ps; mode 5; r3-8; n125 f	5140
	5.3 emb; mode 5,6; 88% 4-7; r1-10; n747 f	3547
	5.9; r4-11; n11 lt	12105
	6.0 emb; r5-7; n6 f	1913
	6 emb; mode 6; r2-10; n285 f	5501
	6.1 cl; r2-14; n439 f	3547
	6.2 emb; r3-13	8052
Clethrionomys rutilus	3-4	5493
	3.6 wean; ±3.7	2564, 7544
	4 ps; n1 lt	10516
	4.9 yg; ±2.1	2564, 7544
	5 emb; n1 f	10598
	5.0 emb; r4-6; n2 f	2545
	5.2 emb; r4-6; n5 f	12105
	5.3 ps; r4-6; n3 f	6782
	5.50 emb; n78 primiparous	6015
	6.05 emb; r4-9; n18 f	6782
	6.1 yg; r5-8; n8 lt	6782
	6.25 emb; n94 multiparous	6015
	7 emb; n1 f	8808
	7.0 emb; r3-9; n4 f	1913
	7.6 emb; r5-9	8052
	8.3 emb; r6-10; 6 f	5039
Dicrostonyx groenlandicus	1.50-3.33; r1-6; varies w/age, parity	4469
	2.35 wean; n52 lt	4724
	2.52 yg; se0.09; n52 lt	4724
	3-4; max 7	6327
	3.1 yg; se0.13; r1-6; n116 lt	6760, 8958
	3.1; sd1; n8 lt	9697
	3.3; r1-8	4467a
	3.41; se0.21; r1-7; n56 lt	6781
	4-8.5 cl	6013
	4.1 yg; r2-7; n8	2434
	4.78 emb, yg, ps; r3-9; n9 f	10101
	5.47 emb, ps; se0.33; r2-11; n30 f	6781
	5.5; r3-8; n17 lt	6710
	5.7 emb, ps; n47 f	3561
Dicrostonyx hudsonius	3-4; r1-7	1474
Dicrostonyx rubricatus	3.0 emb; n3 f	177
	7.2 emb; r5-11; n5 f	177
Dicrostonyx torquatus	2.63; r1-4; n8 lt	8810
	3.7-6.1; r1-11; varies age & parity	8051
	3.83; r2-6; n9 lt	8810
Dicrostonyx unalascensis	2.8 yg; sd1.2; 2.1 yg survive; sd1.0; survive; n188 lt	2560, 7544
Dinaromys bogdanovi	2-3 yg	7983
Ellobius fuscocapillus	3 yg; n1 lt	8052
	3-5	9175
Ellobius lutescens	4.8 emb; ±0.38; n24 f	6651, 11367
	12.9 ovulations; ±1.7; n17 f	11367
Ellobius talpinus	2-4	8052
	3.0	12111
	mode 4-5; r1-7	4645
Eolagurus luteus	mode 6-9 yg; r4-10	9951
Eothenomys inez	2 emb; n1 f	182
Eothenomys melanogaster	1.5 emb; r1-2; n2 f	182
	2.67 emb; r2-3; n3 f	182
Eothenomys smithi	2.56 emb; mode 2-3; r1-5; n120 f	10665
	4.1; r2-7; n13 lt	12026
	4.45 yg; se0.12; r1-9; n188 lt	294b
Hyperacrius fertilis	2-3	9175
Hyperacrius wynnei	2.5; r2 emb, 3 ps; n2 f	9175

Lagurus curtatus	3 emb; n1 f	4223
	4.0 emb; r3-5; n2 f	1119
	4.4 emb; r2-13; n42 f	6925
	5 ps; n1 f	3703
	5.26 emb; mode 6; r1-11; n281 f	5240
	5.6 emb; se1.2; n11 f	7606
	6 emb; n1 f	7325
	6.11; r4-10; n66 f	7491
	7.0 emb; r6-8; n2 f	10218, 10219
Lagurus lagurus	1.2 weaned	11647
	3.11 yg; 73% 2-5; r1-8; n448 lt	3652
	3.3 yg	11647
	3.3; r1-5; n14 lt	11647
	4.0; r2-7; n88 lt	3516
	4.5-8.1; r1-12; n>100	9950
	most 5-6; r3-7	8052
	mode 5-6; r4-10	8634
	6.17 emb; r1-12; n80 f	9950
Lemmus lemmus	3.5 weaned	7544
	3.75-6.62	5967
	4.0 yg; sd1.87; r1-16; n630 lt	9762
	4.1; r1-7; n77 lt	3462
	4.5; r1-7	4830
	4.8 yg	7544
	5	5493
	5.29 yg; mode 5-6; r1-9; n80 lt	6806
	6.4; mode 5; r2-13; n50 lt	11679
	7.5; r6-9; n2 lt	10210
	max 9 yg	2386
Lemmus schisticolor	4.1 emb; se0.36; r2-6; n35 f	10016
	5.3 emb; r3-7	8051
	5.5 emb, ps; sd1.25; n95 f	1077
	7.0 emb, ps; sd1.6; n30 f	1077
Lemmus sibiricus	1.7 weaned	7544
	3.1 yg	7544
	3.8-8.1 cl	6013
	4.7-8.1; varies with season	716
	5.5; r3-9	9071
	6-13 emb	8051
	6.2 yg; r2-9; n25 lt	7595
	6.9 emb; r3-10; n14 f	6710
	~7 emb; r1-13; n286 f	7595
	7.89; r6-12; n9 lt; 2 d	655
Microtus spp	4.6 cl; ±0.48	5674
Microtus abbreviatus	2.2 weaned	7544
	3.0 yg	7544
Microtus agrestis	2.5-4.6	1872
	r3.4-5.9 yg; n133 lt; varies with parity	1877
	3.5 weaned; se0.30; n19 lt	1805
	3.50 yg; n490 lt, 140 f	6397
	3.5-3.6	519
	3.7 emb; r3-4; n3 f	8986
	3.73; r1-7; n389 lt	8921
	3.9 cl	1248
	mode 4-7; r3-11	3571
	4-7 emb; n6 f	1590
	4.0 yg; r3-6; n9 nests	3835
	4.0-7.5 emb; n2 locations	7703
	4.1; r2-7; n60 lt	6008
	4.3 emb; se0.4; 5.8 cl; se0.6; n11 f	1877
	4.6 weaned; se0.25; n37 lt	1805
	4.66; mode 5; r1-7; n32 lt	12070
	4.7 cl; ±0.50	7339
	4.7; r4-6; n7 lt	10340

	5.0 emb; 5.5 ova ovulated	1218
	5.07 emb; n329 f	10373
	5.1; se0.3; n23 lt	1250
	5.61 emb, ps; se0.35; r4-8; n13 f	8336
	6.1; n80 lt; small islands; 5.3; n140 lt; large islands; r3-10	8645
	max 8	10345
	max 8	8920
Microtus arvalis	mode 1-12; r1-13	3571
	2.1; r1-3; n6 lt	8153
	2.67; r1-8; n508 lt	6398
	2.7 wean; n12 lt	11427
	3-7	2466
	3-8	6902
	3.25 emb; n8 f	955
	3.73; median 5; r1-11; n2518 lt	3459
	3.79 yg; mode 3-4; r1-9; n1016 lt	9013
	mode 4-9 emb; r2-15; mode 4-7 yg	8052
	4-5 emb; n4 f	1590
	4 ps; n1 f	6293
	4 ps; n1 f	6293
	4.0 emb; r3-5; n2 f	6293
	4.1 emb; r3-7; n18 f	2465
	4.2; r1-8; n19 lt	9269
	4.25; r3-5; n4 lt	6237
	4.25; n17 lt	7292
	4.31 emb; r1-8; n16 f	11593
	4.5 emb; r4-5; n2 f	11255
	4.8; n111 lt	9269
	4.92; n67 light f	11867
	5-10	11225
	mode 5-6; most 4-7; r4-15	4645
	5.17 emb; r3-9; n30 f	10344
	5.24 emb; r3-9; n105 f	741
	5.39; mode 5; r2-9 n44 lt	6361
	5.41 emb; se0.039; n1567 f	12082
	5.47; r1-9; n75 lt	3465
	5.49; r2-9; n84 f	10343
	5.86; r4-10; n7 lt	7899
	5.96; n32 heavy f	11867
	6.0 emb; r5-7; n5 f	7905
	7; max 12	3460
	8.0 emb; r5-11; n2 f	10516
Microtus brandti	5.0; max 8; n22 lt	9012
	5.0; r2-8; n28 lt	12105
	r5-9; r6-12; n2 samples	8052
Microtus breweri	3.4	10648
	mode 4-5 yg	7326
Microtus californicus	2-7 yg	4490
	4.17 old ps; n35 f	4893
	4.17; se0.20; r1-10; n220	6040
	4.20 emb; se0.11; sd1.41; mode 4; r1-9; n154 f	4052, 4053
	4.24; se0.11; sd1.8; mode 4; n295	6041
	4.25; se0.28; r1-10; n65	6040
	4.29 new ps; n42 f	4893
	4.69 reared; r1-9; n42 lt	1995
	4.8 yg; r4-5; n5 lt	9747
	4.94 emb; r1-10; n73 f	4893
	5 emb; n1	9747
	5.05 emb; 4.82 healthy emb; 280 f	6433
	5.57 ps; 108 f	6433
	9.29 cl; max 29; n82 f	4053
	11.6 cl; se2.4; 5.0 live emb; se0.3; n10 f	5675
	11.6-12.2 cl	5675
	13.12 cl; r2-50; n526 episodes	6433

Microtus canicaudus	3.23; r2-10; n44 lt	11011
	4.1-5.2 yg; se0.2-0.4; n4 groups	4196a
Microtus chrotorrhinus	2.88 emb; mode 3; r2-5; n8 f	5794
	3.5 emb, cl, ps; n13 f	10838, 10842
	3.56 yg; ±0.19; r2-5; n9 lt	2082
	3.71 emb; se0.19; mode 4; r1-7	6896
	3.91 emb; mode 4; r2-5; n11 museums	5794
	5.32 emb; r3-8; n16 f	5796
Microtus fortis	3-7 yg	5424,
	5.5 emb; r5-6	182
	8.33 emb; r7-9; n3 f	8052
Microtus gregalis	5-13; r8-9; n2 locations	8052
	mode 7-9; r7-15	4645
	7.5 emb; r7-8; n2 f	8052
	11 emb; n1 f	10516
Microtus gud	6 emb; n1 f	8052
Microtus guentheri	mode 4-8; r1-21	1037
	5.5; r4.5-6.3; n~365 lt	1036
	5.6 emb; median 6; r4-8; n21 f	8115
	6.1; mode 7; r1-10	1965
	6.2; max 15	3610
	8 emb; max 13	6419
	8.82 emb; sd1.29; r6-11; n17 f	1962, 1966
	9.3 emb; r9-11; n3 f	7905
	10.73 ps; sd5.35; r6-19; n11	1966
Microtus longicaudus	2 emb; n1 f	2544
	3.5 emb; r3-4; n2 f	10598
	3.7 emb; r2-5; n9 f	12049
	3.9; r2-6	6632
	3.95 reared; r1-7; n37 lt	1995
	4-7 emb	502
	4-8	1474
	4 ps; n1 f	7809
	4.5 emb; r4-5; n2 f	11646a
	4.76 emb; r4-6; n17 f	10954
	4.8 emb; r4-6; n5 f	1119
	4.9 emb; r4-6; n10 f	4897
	5; r3-7	2088
	5.0; n26	10161
	5.2; mode 6; r3-6; n5 lt	6488
	5.5 yg; r5-6; n2 nests	10598
	5.6; r2-8; n39	4226
	6.0; r5-7; n3	10161
	7 emb; n1 f	6560
	7 emb; n1 f	4745
	7 ps; n1 f	12049
Microtus maximowiczii	5-9 emb	8052
Microtus mexicanus	1.8	9176
	r2-3 emb; n3 f	523a
	3.5 emb; r2-5; n2 f	502
	2.23 emb; ±0.32; mode 2; r1-4; n22 f	1365
	2.25 emb; ±0.22; mode 2; r1-4; n8 f	1365
	2.3 emb; mode 3; r1-4; n59 f	1813
	2.6 emb; r1-3; n7 f	4228
	2.6; r1-4; n15 lt	522
	3 emb	2341
	mode 3-4; r2-5	1474
Microtus middendorffi	mode 5-6; r3-8	8052
Microtus miurus	2.8 wean	7544
	3.9 birth	7544
	4-12 emb	1474
Microtus montanus	r4-7 emb; n10 f	2226
	4-8 emb	7809
	4-10 emb	502
	4.16 yg; r1-10; n254 lt, 31 f	7811

	4.3 cl; ±0.6	5672
	4.5 yg; r3-6; n4 lt	9967
	4.65 yg; r1-10; n60 lt, 12 f	7811
	4.7 emb; se0.5; 5.6 cl; se0.4; n12 f	2330
	r4.8-6.6 emb, ps; n17 seasons	8561
	5 emb; r4-6; n8 f	11646a
	5.1 cl; se0.93; n12	4014
	5.5 emb; r5-6; n2 f	6488
	5.8 emb; se0.25; r2-10; n46 f	11230
	5.98 reared; r3-6; n42 lt	1995
	6.02 ps; n75 f	4893
	6.47 emb; r2-10; n109 f	4893
	6.75 cl; r1-13; n194 f	4893
	7 emb; n1 f	10305
	7.0 emb; r6-8; n2 f	2544
	r7-8 ps	7809
Microtus montebelli	~3	3969
	3.5 emb	7392
	3.75 emb; se0.28; r2-5; n12 f	5511
	4.11 yg; ±1.49; mode 4; r1-8; n168	7762
	4.3; n88 lt	15
	4.5	9443
	4.9 emb; r2-8; n8 f	9919
	5.1 emb; se0.3; mode 5; r2-9; n40 f	5510, 5512
	6.00 emb; se0.29; r5-7; n9 f	5511
Microtus nivalis	2-4 emb	5984
	2.67; r1-4; n3 lt	3457
	3.0 emb; r2-4; n23 f	6313
	3.1 emb; mode 4; r2-4; n12 f	5486
	mode 4-8 emb; r1-9	6919
	4 emb; n1 f	6293
	5 emb; n1 f	6293
Microtus oaxacensis	1.2	9176
Microtus oeconomus	2-12	7433
	2.2-2.3 weaned	2563, 7544
	3-11	8845
	3.9-4.0 yg	2563, 7544
	mode 4-7; r2-12	3571
	4.04 emb; mode 7; r4-8; n25 f	3468
	4.41 emb; n80 f	3468
	4.6; r4-6	3655
	5.0-7.8	716
	5.19 emb; n31 f	6015
	5.3 emb; r4-6; n3 f	10516
	5.33 yg; n6 lt	9355
	5.48 emb; mode 5; r2-8; n25 f	11679
	5.67 emb; mode 6; r3-8; n18 f	11679
	5.7 emb; r4-7; n3 f	10516
	5.7 yg; mode 5; r3-9; n4 lt	659
	5.7; mode 5; r3-9; n25 lt	10340
	mode 6-9; r3-12	8052
	6.0 emb; r5-9; n8 f	10339
	6.1 emb; n16 f	12049
	6.4; r3-8; n10 lt	10344
	6.5 emb; r6-7; n2 f	8808
	6.74; se0.27; n41 f	5969
	6.94 emb; mode 7; r1-12; n373 f	5493
	7.15; mode 7; n119 f	10681
	7.7 emb; r6-9; n6 f	1913

Microtus oregoni	2.79 yg; mode 3; r1-5; n28 lt	2087
	3.11 emb; mode 3; r1-5; n26 lt	2087
	3.2 emb, ps (3.1 emb, 3.5 ps); mode 3; r1-8; n29 lt	3622
	3.3 emb; n8 f	1615
	3.41 cl; n34 lt	3622
	3.5; se0.21; n12 lt	1370
	3.8 reared; r1-6; n15 lt	1995
	4.0 emb; r2-6; n21 f	4970
Microtus pennsylvanicus	2-9	501
	2.6 weaned; ±1.6	2562, 7544
	3.7; se0.6; n12 lt	5673
	3.72 yg; sd1.88; r1-9; n416 lt	8642
	3.78 yg; se0.23; n37	7116
	4-8 emb	502
	4 yg; r1-9	6347, 6348
	4 emb; n1 f	495
	4 yg; n1 nest	9612
	4 yg; r3-5; n3 nests	2902
	4.0 yg; mode 4; r1-6; n23	5976
	4.1 yg; ±1.7	2562, 7544
	4.12 emb; r3-6; n8 f	2548
	4.18 yg; n51 lt	2059
	4.25 emb; r4-5; n4 f	2549
	4.37 emb; sd1.43; n51 f	1832
	4.46 emb; se0.13; mode 4; r1-9; 4.80 cl; n153 f	2059
	4.48 emb, ps; r1-8; n35 f	2376
	4.5	10648
	4.5 emb; r3-6; n2 f	261a
	4.54 emb; sd1.51; n152 f	5629
	4.72 yg; se0.26; n25	7116
	4.8 yg; mode 5; r2-7; n73	5976
	5 emb; n1 f	6796
	5 emb ea; n2 f	2902
	5.0; r4-6; n2 f	7809
	5 yg; r2-9; n17 lt, 1 f	501
	5.0 emb; mode 5; r3-8; n24	5976
	5.00 yg; ±0.17; r3-9; n41 lt	2082
	5.05; mode 5; r3-7; n39 lt	5149
	5.09 emb; n87 f	6015
	5.1 yg; 4.8 yg alive; r1-11; n106 lt, 27 f	10463
	5.24 cl; sd1.10; n151 f	5629
	5.25 emb; r4-7; n16 text; 5.20 emb; n15 table	10839
	5.3 emb; r5-6; n4 f	5798
	5.3 emb; r2-10; n67 f	12049
	5.32 uterine swellings; sd1.61; n25 f	1832
	5.49 yg; se0.27; n47	7116
	5.52 yg; se0.31; n23	7116
	5.52 reared; r2-8; n31 lt	1995
	5.5 emb; n4 f	11646a
	5.5 emb; mode 5; r1-11; n440 f	4278
	5.6-5.9 cl	1919
	5.67 ps; r5-6; n3 f	261a
	5.8 yg; r4-9; n11 lt	11956
	5.86; r2-10; n144 f	2660
	5.9 yg; mode 6; r4-8; n28	5976
	6 yg; n1 f	6769
	6 emb ea; n2 f	10218, 10219
	6 emb; r4-8; n2 f	10220
	6.0 emb; r5-7; n2 f	4745
	6.0 yg; n14 lt	11518
	6.00 cl; sd1.27; 4.25 emb; sd0.83; n16	4018
	6.09 cl; se0.115; r3-12; 5.59 emb; se0.110; r1-11; n251 f	721, 723
	6.2 emb; r4-9; n12 f	5405b,
	6.2 emb; mode 7; r3-8; n13 f	5976
	6.3 emb; r5-8; n15 f	3703

	6.3 yg; n24 lt	4272
	6.6 emb; r2-9; n36 f	10954
	6.7 emb; r5-9; n10 f	6166
	6.9 emb; r4-12; n17 f	10101
	7 emb; n1 f	2545
	7 emb; n1 f	10652
	7 ps; n1 f	3703
	7.8 ps; r6-9; n5 f	10101
	8 emb; n1 f	10218
	8 yg; n1 lt	5210
	8.14 emb; r6-9; n7 f	11504
Microtus richardsoni	4-7 yg	503
	5 yg; r2-8	2088
	5.17 emb; se0.44; n19 f	6630
	5.42 ps; se0.58; n12 f	6630
	5.59 emb; se0.28; n37 f	6630
	5.60 emb; n16 f	6630
	5.6 emb; n20 f	6630
	5.62 emb; se0.18; mode 5; r2-9; n86 f	6631, 6632
	5.91 cl; se0.24; n46 f	6630
	6 emb; n1 f	4969
	6 emb; r5-9; n10 f	7809
	6.11 ps; se0.43; mode 8; r2-10; n26 f	6631, 6632
	6.35 ps; se0.32; n23 f	6630
	7.0 emb; n3 f	6630
	7.10 ps; se0.34; n7	6630
	7.33 emb; n3 f	6630
	7.85 emb; ±0.51; mode 8; r6-10; n26 f	1369
Microtus roberti	2.0 emb; r2-4; n11 f	10274
Microtus sachalinensis	7.9	11366
Microtus socialis	2.00-6.88	2995
	4 emb; n1 f	6293
	5.67 emb; r5-7; n3 f	5985
	mode 6; r4-11	8052
	max 9	4645
Microtus townsendii	2-10	2053
	4; r1-9	2088
	4-5; r1-9	1474
	4.92-5.42 emb; se1.00-1.19; n3 seasons	2053
	4.8 yg; r2-6; n13	6693
	5.4 yg; r2-7; n12	6693
	6.0 emb; r5-7; n20 f	6693
	7; r5-8	2053
Microtus transcaspicus	2.1; r1-4; n17 lt	8153
Microtus xanthognathus	8.0 yg; r7-10; n11 lt	12049
	8.0 emb; r6-10; n7 f	11878
	8.33 emb; r7-10; n3 f	6376
	8.8 emb; sd1.62; r6-13; n71 f	11878
Neofiber alleni	1.5 yg; n12 lt	11519
	2.2 emb, ps; mode 2; r1-4; n104 f	944
	2.7 cl; r1-5; mode 3; n46 f	944
	3 yg ea; n2 lt	4282
Ondatra zibethicus	1.5 emb; r1-2; n2 f	1290
	2.65 yg; r1-5; n17 nests	10580
	3-9; (7-9 start of season, 3-5 end of season)	7291
	3.3 yg; mode 3; r1-5; n108 lt	11762
	3.47 emb; r1-5; n103 f	8015
	3.62 emb; mode 4; r1-6; n66 f	10580
	3.7 emb; mode 4; r1-6; n206 f	11762
	3.9 yg; r1-8; n23 lt	4832a
	4.5; 95%ci 4.0-4.9; r1-10	2243
	4.8; 95%ci 4.3-5.3; r1-10; lt <9 d	2243
	5.0; 1st lt; 5.6; 2nd lt; r1-12	7093
	5.2 yg; n55 families	8762

	5.38 emb; mode 5; r3-7; n13 f	9529
	5.4 ps; n50 lt, 36 f	1915
	5.4 yg; r2-9; n62 nests	3621
	r5.5-8.5; n25 locations	5884
	5.8	11174
	5.8 yg; r4-8; n5 lt	6805
	6-8 yg	90
	6 emb; r3-9; n23 f	2616
	6.0 emb; r5-7; n2 f	4491
	6.0 ps; n44 sets, 38 f	3061
	6.1 yg; mode 6; r2-13; n15 lt	11934
	6.1; mode 6-7; r1-11; n418 lt	2697
	6.15 yg; n27 lt, 0-5 d	3061
	6.27; se0.18; r1-13; n132 lt	705
	6.28 yg; n60 lt	9503
	6.33 emb; n27 f	2685
	6.4 yg; n26 f	10112
	6.5 yg; r1-11; n158 lt <1 wk old	3081
	6.6 emb; se0.3; r5-8; n16 f	10410
	6.6 ps; mode 8; r2-11; n25 f	9003
	6.7 emb; mode 7; r1-16; n28330 f	4892
	6.78 yg; n188 lt	3082
	6.79; se0.16; r1-13; n115 lt	705
	6.79 emb; mode 7; r1-12; n1294 lt	4891
	6.87 ps; n510 sets, 197 f	9503
	6.9 ps; r3-15; n24 f; prob 2 lt/f	351
	6.91 emb; n65 f	703
	7 emb; n1 f	3558
	7 emb; n1 f	5333
	7.0 emb; r1-10	11467
	7.0-9.0; varies with parity	10066
	7.0 yg; mode 6; r2-9; n35 lt	9003
	7.03 emb; n39 f	2685
	7.1 yg; r5-11; n22 lt	0000
	7.1 emb; r2-10; n45 f	3621
	7.2 emb; r5-10; n11 f	6805
	7.3 yg; r5-10; n31 lt; 8.0; ≤1 wk	9003
	7.3 yg; modes 6,8; r4-10; n80 lt	8104
	7.4; n76 lt	719
	7.5 emb; n8 yg f	8237
	7.5 yg; n364 lt; r6.35-8.41; n22 yr	3082
	7.7 ps; r7-9; n3 yg f	8761
	7.7 emb; r4-11; n33 f	9690
	8.2 emb; r5-11; n6 f	3081
	8.4 emb; sd1.2; n9 f	8237
	8.7 ps; n7 f	3558
	9 emb; n1 f	8234
	9	6267
	9.1 emb; r2-10; n16 f	9690
	10.93 ps; n16, 2 sites	2685
	12.06 ps; n103	2685
	12.7 ps; n48 f	8762
	17 ps; r12-24	8236
	18.31 ps; ±1.01	10112
	19.8 ps; n36 f	8237
Phenacomys intermedius	mode 4-5; max 8	2088
	4.24 yg; se0.12; mode 4; r2-7; n59 lt	5150
	4.4 yg; n8 lt	4219
	4.8 yg; r2-8; juveniles: 3.8, adults: 5.9	555a
	4.8 emb; se0.75; r3-6; n4 f	11230
	4.8 emb, yg; r2-8; n43 lt	3429
	4.9 yg, emb, ps; r3-8; n37 f	10101
	5.3 ps; se0.52; r2-9; n11 f	11230
	7 emb; n1 f	10220

	7 emb ea; n2 f	279
Pitymys afghanus	3.5 emb; r3-4; n2 f	8052
Pitymys duodecimcostatus	2.0 emb; sd0.8; mode 2; r1-4; n37 f	8216
	2.2 emb; mode 2	11255
	2.4 yg; r1-3; n5	11820
	2.66; mode 3; r1-4; n32 lt	6361
	2.7 emb; r1-5	1843a
	3.0 emb; r1-4; n9 f	11820
Pitymys juldaschi	3.43; r2-6; n60 lt	1057
	3.63; r2-7; n70 lt	1057
Pitymys leucurus	2 emb; n1 f	7382
Pitymys lusitanicus	2.0 yg; r1-3; n56 lt	11820
	2.26 emb; r1-4; n57 f	11820
Pitymys majori	2.75 live emb; r1-4; n4 f	10274
	3-7 emb	8052
Pitymys multiplex	3	6370
Pitymys ochrogaster	1-8	2137
	2.3 yg; se0.5; n6	2529
	3.18; se0.24; median 3; r1-6; n65 lt	6881
	3.27 emb	5629
	3.27 emb; sd1.28; n160 f	5629
	3.3; se0.2; n22 lt	10333
	3.37 yg; ±0.075; n82 lt	3283
	3.38 emb; 5.16 cl; n20 f	4019
	3.40 emb, ps; r1-5; n35 f	2376
	3.4 emb; r1-7; n58 f	5241
	3.42 emb; 3.79 cl; n182 f	9273
	3.5 emb; r2-5; n11 f	6298
	3.6; r1-6; n28 lt	10954
	3.6 emb; mode 3; r1-6; n110 f	5405b$_1$
	3.7 emb; r1-6; n3 f	3703
	3.73 cl; sd1.06; n158 f	5629
	3.77; r1-8; n280 lt	9111
	3.87 yg; n54 lt	2059
	3.89 emb; se0.10; mode 4; r2-7; 4.35 cl; n134 f	2059
	3.93 reared; r1-7; n28 lt	1995
	4.0 emb; r3-5; n2 f	495
	4.19 emb; r1-9; 4.17 ps; n198 f	9235
	4.25; n8 lt	6051
	4.4 emb; r3-6; n9 f	261a
	4.5; se0.6; n13 lt	5673
	4.7 ps; r3-7; n7 f	6298
	5.0; se0.4	3648
	5.0 emb; r4-6; n2 f	1812a
	5.0 emb; r3-7; n2 f	2547
	5.8 emb; r5-7; n5 f	11957a
	6 ps ea; n2 f	261a
	7.2 ps; r6-10; n15 f	11957a
	10 ps; n1 f	4525
Pitymys pinetorum	1.9 emb	11059
	mode 2-4	783
	2 emb; n1 f	2547
	2 emb; n1 f	8988
	2 emb ea; n2 f	6475
	2 emb ea; n3 f	6298
	2.2 yg; se0.1; r1-3; n19 lt	6533
	2.25 emb; r1-5; n138 f	8268
	2.33; r2, 2 emb, 3 yg; n2 f	6487
	2.33 emb; r2-3; n3 f	5405b$_1$
	2.44 ps; r2-5; n9 lt	3852
	2.5 emb; r2-3; n2 f	3747
	2.5 emb; r2-3; n2 f	3803
	2.54 emb; r2-5; n13 f	3852
	2.59 wean; r1-6; n150 lt	9530

	2.75 wean; se0.09	9536
	2.79; r2-4; n14 lt	4273
	3-4; r2-7	1474
	3.0 ps; r2-4; n3 f	3747
	3 yg ea; n4 lt	3852
	3.2 yg; se0.1; n50 lt	2493
	3.31 yg; se0.09; mode 3; r1-6; n150 lt	9530, 9536
	3.4 yg; se0.3; n25 lt	2493
	4 ps; n1 f	9164
	8 emb; n1 f	9164
Pitymys quasiater	1 emb; n1 f	2341
	2.1 emb; r1-4; n12 lt	4228
Pitymys sikimensis	2 emb; n1 f	2256
	2.32 emb; r1-4; n22 f	7382
	3 emb; n1 f	182
Pitymys subterraneus	mode 2-5; r1-7	3571
	2.0; r1-4; n6 lt	7904
	2.17-2.45 ovulations	5281
	2.2	10379
	2.2 emb; r1-3; n5 f	4114
	2.25 emb; se0.077; r1-4; n71 f	6004
	2.4; r1-4	9656
	2.4 yg; se0.2; n28 primiparous	5282
	2.5 emb; r2-3; n2 f	5986
	2.5; r2-3; n22 lt	6199
	2.65 yg; 2.36 wean; se0.1; mode 3; r1-5; n317 lt	5282
	2.66 emb; se0.095; r1-4; n38 f	8335
	2.74 yg; mode 3; r1-5; 2.28 wean; n143 lt	1407
	2.875 ps; n8 lt	4114
	2.88; r2-4; n32 lt	7903
Pitymys tatricus	3 emb; n1 f	5986
	3.0 emb; se0.081; r2-4; n61 f	6004
Pitymys thomasi	4.45 emb; sd1.58; r1-8; n20 f	7905
Prometheomys schaposchnikowi	2.7 emb, ps; r1-4; n6 lt	10275
	3.5 emb; r3-4; subadult f	12007
	mode 4-5 emb; r1-5; adult f	12007
Synaptomys borealis	4 yg; r4-8	555a
	mode 4-5; r2-8	2088
	4-5; r2-8	1474
	4.4 emb; r3-6; n10 f	12049
Synaptomys cooperi	1 emb; n1 f	$5405b_1$
	1-7 yg	1469
	2-4 emb	10329
	2 emb; n1 f	2025
	2.5 yg; r2-3; n2 f	10329
	2.8 emb; r2-4; n5 f	6475
	mode 3 yg; r2-5	5210
	3 yg; r1-7	555a
	3 emb; n1 f	6298
	3 emb; n1 f	11956
	3.20 yg; ±0.19; r2-5; n9 lt	2082
	3.33 emb; r3-4; n3 f	4525
	3.7; n6 lt, 1 f	2026
	4.0 emb; r3-5; n4 f	5798
	4.5 emb; r4-5; n2 f	6486
	4.7 emb; r4-6; n6 f	$5405b_1$
SEXUAL MATURITY		
Arvicola terrestris	f: 1st conceive: 15 cm	11204
	f: 1st preg: 76-96 g	11250
	f: 1st birth: 1st summer	4645
	f: ~6 wk; 120 g	8392
	f: 35-40 d	10944
	f: 60-65 g	8052
	f: 116 g	8392

	m: spermatogenesis: 15 cm	11204
	m: testes mature: 6-10 wk; 65-95 g	11250
	m: testes mature: 85-90 g	11668
	m: 1st mate: 43-47 d	11668
	m: 110 g	8392
	f/m: mate: not yr of birth	10447
	f/m: 2-2.5 mo	3571
Clethrionomys gapperi	f: 1st birth: 1st summer	11230
	f: ≤4 mo	10569
	m: sperm: 1st summer	11230
	f/m: mate: 7 mo	7238
	f/m: mate: yr of birth	6796
	f/m: 2-4 mo	7238
Clethrionomys glareolus	f: vag perf: 26.9 d; se3.1; n14	1075
	f: vag perf: 29.3 d; se4.6; n17	1075
	f: mate: yr of birth	2077
	f: 1st cl: 16 g	12076
	f: 1st conceive: 27 d; n1	4172, 12081
	f: 1st preg: ~5 wk	12077
	f: 1st preg: 22 g	2447
	f: 1st birth: 60 d	4645
	f: 1st birth: 3.5 mo	11362
	f: 1 mo	10589
	f: 1-1.5 mo	1408
	f: est 13 g	516
	m: 2 mo	8425
	m: <3 mo	12077
	m: est 14 g	516
	m: testes mature: 18 g	12076
	m: 1st sire: ~67 d	1408
	f/m: 1st mate: ~1 yr	11679
	f/m: 1st mate: ~1 yr	1076
	f/m: mate: 1 mo	11670
	f/m: mate: yr of birth	1222, 9342
	f/m: mate: yr of birth	7988
	f/m: 1st repro: yr of birth	4341
	f/m: 8-9 wk	3571
	f/m: ~3 mo	5307
Clethrionomys rufocanus	f: 1st birth: ~110 d	6134
Clethrionomys rutilus	f: preg: 1st yr	6782
	f: yr of birth	3775
Dicrostonyx groenlandicus	f: vag perf: 25-27 d	4335
	f: 1st birth: 49 d; n1	4335
	f: 1st birth: 58-90 d	4469
	m: 1st sire: 85 d	4469
	m: ~1 yr	6013
Ellobius talpinus	f: 1st birth: ~5 mo	4645
	f/m: 1st repro: 90 d	12111
Eolagurus luteus	f/m: 3-4 wk	9951
Eothenomys smithi	f: 1st birth: 120 d	294b
	f/m: 1st repro: >25 g	12026
Lagurus curtatus	f: 1st mate: min 47 d; mode 60 d	5240
	f: 1st mate: 2 mo	5146
	f: preg f: 103 mm	6925
Lagurus lagurus	f: 1st mate: 35 d-1.5 mo	8052
	f: 1st birth: 49-54 d	3516
	f: 1st birth: 50-65 d	4645
	f: 30 d; r19-35	3652
	m: 1st sire: 51-56 d	3516
	m: 4-6 wk	8052
	m: 45-46 d	3652
	f/m: 1st mate: 40-45 d	8634

Lemmus lemmus	f: vag perf: 10 d	6806
	f: 1st mate: 20 d	6806
	f: 1st birth: 39 d; mode 2nd mo	3462
	f: 16-49 d	9762
	f/m: 1-1.5 mo	5967
	f/m: ~80 d, 45-60 g	8051
Lemmus schisticolor	f: vag perf: 22 d; 24 g	5135
	f: 1st mate: 3 wk	3466
	f: 1st birth: 2 mo; n15/73	5135
	f: 16 g; 82 mm	10016
	m: testes descend: 44 d; 28.2 g	5135
	m: 22 g; 88 mm	10016
	f/m: ≥20 g	5135
Lemmus sibiricus	f: ~2 mo	9071
	f: 110 d	716
	f: 3-4 wk	7595
	m: 3-4 wk	7595
	f/m: summer of birth	6013
Microtus agrestis	f: vag perf: 19.8 d; se0.2; n21 f with m present	1875
	f: vag perf: 20 d	8920
	f: vag perf: 22.0 d; se0.5; n21 f no male present	1875
	f: fecund: 3 wk	6397
	f: 1st conceive: 42 d	6397
	f: 1st birth: 7-8 wk	9011
	f: 2-3 wk	3571
	f: 3 wk	1872
	f: 6 wk	11294
	f: >12 g; 5-6 wk; n1	1218
	m: testes large: 6 wk	521, 6397, 8920
	m: fertile: ~20 g	1218
	m: 6 wk	1872
	f/m: 50-60 d	12070
Microtus arvalis	f: vag perf: 11 d	3460
	f: 1st mate: 21 d	6902
	f: 1st birth: 33-34 d	3460
	f: 1st birth: 40 d; mode 60-65	4645
	f: 1st birth: 40 d; mode 80-89	3459
	f: 1st birth: 41-57 d	9013
	f: 1st birth: 47-144 d	9269
	f: 1st birth: 50 d	3465
	f: 20-25 d	2465
	f: 47-90 d or 59-90 d	8052
	m: epididymal sperm: 35 d	6902
	m: 2nd mo	8052
	f/m: 1-2 mo to >4 mo	6903
	f/m: 9-12 wk; full grown 6-10 wk	3571
	f/m: 68 d	6237
Microtus brandti	f: vag perf: 3 wk	9012
	f: 1st birth: 8 wk	9012
Microtus californicus	f: 1st mate: 20 d	4490
	f: 1st mate: 25-30 g	4893
	f: 1st preg: 3 wk-9 mo	4053
	f: 10-15 g	4052
	m: sperm: 40 g	4053
	m: 1st mate: 6 wk	4490
	m: 1st mate: 35-40 g	4893
	f/m: 25-30 g	6434
Microtus canicaudus	f: 1st conceive: 18-22 d	4196a
	f: 1st conceive: 30-50 d; n1/20; 90-120 d; n4/8	8410
Microtus chrotorrhinus	f: preg f: 21 g	5796
	f: preg f: TL: 131 mm; n1	5796
	f: HB: >140 mm; 30 g	6896
	m: HB: >150 mm; 30 g	6896
Microtus fortis	f: 3.5-4 mo; 106 mm	8052

Microtus gregalis	m: ~2 mo; 20 g	8052
	f/m: repro: yr of birth	10516
Microtus guentheri	f: vag perf: 30 d	1965, 1966
	f: 1st mate: 50 d	3610
	f: 1st preg: 60 d	1965
	m: sperm: 30 d; n1	1966
	m: sperm: 85 d; mode >100 d	1965
	m: 1st mate: 62 d; mode 95 d	3610
	m: 1st mate: 11 mo	1966
	f/m: mate: most 9-12 mo; few season of birth	1962
Microtus longicaudus	f/m: repro: yr of birth	502
Microtus middendorffi	f/m: ~2 mo	8052
Microtus montanus	f: vag perf: 18-42 d; n42	9518
	f: 1st mate: 19-29 g	4893
	m: 35 g	4893
	f/m: 3-4 wk	7808
	f/m: summer of birth	11230
Microtus montebelli	f/m: 1st repro; 60 d; 32.2 g	7762
	f: 25 g	5510
Microtus nivalis	f/m: repro: yr of birth	6919
Microtus oeconomus	f: vag perf: 51.38 d	5313
	f: 1st birth: 6 wk	3571
	f: 3 wk	3571
	f: summer of birth	10681
	m: testes large: 48.4 d	5313
	f/m: mate: yr of birth	10516
Microtus oregoni	f: 1st mate: 22-24 d	2087, 2088
	f: 1st conceive: 27-36 d	2087
	m: 1st mate: 34-38 d	2087
	m: 42 d	2088
Microtus pennsylvanicus	f: 1st mate: 25 d	501
	f: 1st conceive: 4 wk; 24 g	4272, 4278
	f: 1st birth: 45 d	501
	f: 3-4 wk	2850
	f: median 18.8 g; 95%ci 17.7-20.1	7293
	f: yr of birth; 45-49 g	12007
	m: testes: 1 mo; 25-30 g	4278
	m: 1st mate: 45 d	501
	m: 5-6 wk	2850
	m: median 23.6 g; 95%ci 20.2-27.4	7293
	f/m: mate: 35 g	6347
	f/m: 30-40 d	2088
Microtus richardsoni	f: mate: season of birth	6632
Microtus socialis	f: 1st birth: 45-60 d	4645
Microtus townsendii	f: vag perf: 2.5-7.5 wk; n16/20; 15-19 g	6691
	m: >30 g	6693
Neofiber alleni	f/m: 90-100 d; 230-280 g	944
Ondatra zibethicus	f: vag perf: 4 mo	10066
	f: preg: 3-4 mo	4891
	f: preg: 1 yr; some yr of birth	702, 705
	f: 1st birth: 1st summer	4645
	m: 4-5 mo; mode 7-8 mo	10490
	m: 7-8 mo	11027
	m: ~1 yr	704
	f/m: 4-8 mo	3571
	f/m: 1+ yr	11467
	f/m: yr of birth; n3/594	8761
Phenacomys intermedius	f: 1st mate: 1st yr	1474
	f: 1st mate: 4-6 wk	3429
	f: 1st birth: 7 wk	555a
	f: 1st birth: yr of birth	5150
	f: 4-6 wk	1474
	f/m: 1st mate: 1st summer	11230

Pitymys lusitanicus	f: 1st preg: 5 wk; n1	11820
	m: 1st sperm: 33 d	11820
	m: 1st sire: 7 wk; n1	11820
Pitymys ochrogaster	f: 1st conceive: ~1 mo; mode 7-11 wk	3283
	f: 1st mate: 30 d	1474
	f: 1st mate: 33-34 d not fertile; >60 d fertile	9111
	m: 1st sire: 49 d	9111
	f/m: ≤30 d	1871a
Pitymys pinetorum	f: 1st cl: 10-12 wk	9536
	f: 1st conceive: 62 d; median 50; r32-115; n13	6388, 6389
	f: 1st conceive: 11-24 wk; n21	9530
	f: 1st conceive: 105 d	9536
	f: 1st preg: TL: 108 mm; n2	8268
	m: 1st sperm 6-8 wk or 60 d	9536
	m: 1st sire: 51-77 d; n12	9530
	m: 1st sire: 69 d; median 57; r44-99; n12	6388, 6389
Pitymys subterraneus	f/m: by 100 d	1407
	f: 16.3 g	8335
	f: 1st cl: 13.5 g	6004
	f: 1st conceive: 90 d	9656
	f: 1st preg: 13-14 g	10379
	m: 18.6 g	8335
	m: fertile: 13-17 g	10379
	m: testes large: yearlings	6005
Pitymys tatricus	f: 1st cl: 23 g; sometimes 18 g	6004
Prometheomys schaposchnikowi	f: 45-49 g; yr of birth	12007
CYCLE LENGTH		
Clethrionomys glareolus	anestrus: 0.5 d; r0-5; proestrus: 1.5 d; r0-5	10204
	estrus: 1.8 d; 0-7; postestrus 2.7 d; 0-7	10204
	mode 3 d; most 2-8; n300 cycles, 30 f	6824
	4 d; smears	1408
	4-8 d	8425
	6.5 d; r3-12; n29 f	10204
Dicrostonyx groenlandicus	9.66 d; sd2.2; n41 cycles, 18 f	9521
Lagurus lagurus	7 d; vaginal smears	11647
Microtus agrestis	3-4 d; during lactation	1806
	5.18 d; r2.5-11.5; n28 cycles, 7 f; diestrous intervals	7334
Microtus arvalis	mode 4 d; most 2-9; n70 cycles, 10 f	6824
Microtus guentheri	6-10 d	3610
Microtus oeconomus	8.5 d; r6-11; n6 cycles	5039
Microtus townsendii	estrus: 4.5 d; r1-19; diestrus: 3.2 d; r1-18; n12 f	6692
	8 d; mode 4; most 2-8 70%; 5 cycles >14; r2-28; n40 cycles, 12 f	6692
Neofiber alleni	≤15 d	7983
Ondatra zibethicus	estrus: 1 d	719
	6.1 d; mode 4; r2-22; n136 cycles, 10 f	7094
	28.7 d; r24-34; n11 cycles, 7 f	719, 720
Pitymys pinetorum	estrus: 1-22 d; diestrus: 1-9 d; n10 f, no cycles	9530
GESTATION		
Arborimus longicaudus	r27.0-28.5 d; n4 f	4264
	33-42 d; n2 f	4264
Arvicola terrestris	2.5-3 wk	8052
	21 d	7772
	21 d	7433
	mode 21 d; r21-26	981
	21-22 d	3571
	22-24	4660
Clethrionomys gapperi	mode 17 d; r17-19; n6 cases, 5 f	10569
Clethrionomys glareolus	16-35 d; lactation delay possible	8425
	17.5 d	7002
	17.5-18 d	8052
	18 d	10589
	18 d	5477
	18 d; r18-23; n25; primiparous	4173

	18.3 d; se0.07; n12; primiparous	4172
	18.32 d; r17.5-21.5; n7	11362
	19 d	8179
	19-20 d; 3 d longer with lactation	2077
	19.5 d; se0.01; n74	1875, 1881
	est 20 d; min ibi	1408
	mode 20-21 d; r19-34; pairing to birth	8021
	20 d	10383
	20 d; n1	11679
	21 d	7433
	21-21.5 d; with lactation	7002
	25 d; r20-30	4645
	max 37 d; with lactation	11362
Dicrostonyx groenlandicus	implant: 5 d	4468
	19-21 d	6781
	19.5-20.5 d	8958
	20 d	4335
	20.0 d; se0.1; r19.5-21.5; n38 no lactation	4468, 4469
	20.7 d; se0.2; r19.5-23.0; n25 with lactation	4468, 4469
	21.0 d; se0.6; r20-22; n3 no lactation	6759
	22.4 d; se0.7; r20-27; n9 f with lactation	6759
	28.0 d; se1.0; r24-31; n8 f with lactation	6759
Dicrostonyx hudsonius	19-21 d	1474
Dicrostonyx torquatus	15-21 d	8810
	18 d; r18-19	8051
Dicrostonyx unalascensis	mode 21 d	2560
Dinaromys bogdanovi	1 mo	7983
Ellobius talpinus	26 d	12111
Eothenomys smithi	19 d; min ibi	294b
Lagurus curtatus	25 d; r24-26	5240
Lagurus lagurus	implant: 6-7 d	142
	0.5 mo	8052
	19-20 d; min ibi	11647
	20 d; r15-23	4645
	20+ d; mode 20; r20-21	3516
	20-21 d; vaginal smears	3652
	20-22 d	8634
Lemmus lemmus	~16 d	8051
	20.2 d; r19.54-21.29; n67; pairing to birth, min ibi	6806
	20 d; r20-21; mating to birth, min ibi	3462
	3 wk	2386
	mode 3 wk	9762
Lemmus schisticolor	14 d	7705
	est 22-25 d	5135
	24 d	3466
Lemmus sibiricus	18 d	8051
	20-20.5 d; n12	7595
	21-21.5 d; n2	6013
	23 d	716
Microtus agrestis	implant: 3-5 d	1249
	18-21 d	3571
	18-22 d	1875
	20-23 d	1250
	20.5 d; nursing	12070
	20.7 d; n32	12070
	20.76 d; r20.5-21.5; n13	1806
	21 d	7433
	21 d	6397, 8920
Microtus arvalis	18 d	7292
	18 d	10911
	19 d; r16-23	4645
	mode 19-21 d; r16-24	3571
	20 d	2466
	20-28 d	6339

	20 d; r16-36; ibi	3459, 3460
	20.56 d; se0.08; mode 20; r18-24; n273	9013
	21 d	11225
	21 d	2465
	21 d; r18-32	3459, 3460
	21-23 d; r20-31; n21; pairing to birth	9269
	22.8 d; r20-27; ibi	3465
	< 22 d	2462
Microtus breweri	est 21 d	10648
Microtus californicus	21 d	4052
	~21 d	4490
	21.1 d; r20.5-22; n4	9747
Microtus canicaudus	≤23 d	4196a
Microtus guentheri	21 d	8115
	21 d; r21-25; min ibi	1965, 1966
Microtus miurus	21 d	12049
Microtus montanus	20.5-21 d	9967
	21 d; n314 lt, 43 f	7811
	mode 21 d; r21-26; ibi	4014
Microtus montebelli	29 d; ibi	3969
Microtus nivalis	20 d; estrus to birth; 22 d; ibi	3457
	21 d	7433
	21 d	5486
Microtus oeconomus	20-23 d; mating to birth, ibi	3468
	21 d	7433
Microtus oregoni	23.8 d; r23.33-24.5; n4 lt	2087
Microtus pennsylvanicus	20.0 d; sd0.0; n6 without lactation	9898
	21 d	2562
	21 d	10648
	21 d	501
	21 d	4278
	21 d; r19-22	6347, 6348
	21.0 d; se0.2; n12	5673
	21.5 d; sd1.4; n11 with lactation	9898
Microtus richardsoni	22 d; n2; pairing to birth	5252
	22 d; n3 min ibi	5252
	42 d; with lactation	1369
Microtus socialis	19-20 d	4645
Microtus townsendii	~21 d	1474
	21-24 d; n3	6693
	24 d; n1 nursing 5 babies	6693
Neofiber alleni	26 d; min ibi; 29 d; death of m to birth; n2	944
	1 mo; n1; removal of m to birth	4282
Ondatra zibethicus	21 d	1355
	21-23 d	8051
	25 d	6267
	28 d	11934
	4 wk	11174
	29 d; r28-30; n2	3061
	30 d	7291
	est ~30 d	11934
	~30 d; ibi	7093
	~1 mo	2243
	~1 mo; n76	3080, 3082
Phenacomys intermedius	19 d; capture to birth	5150
	21-24, 19-24 d; n2	3429
Pitymys lusitanicus	20 d; n2; min ibi	11820
Pitymys ochrogaster	~21 d	1474
	~21 d	3648
	21.5 d; r20.5-23; n18	9110, 9111
	22.8 d; se0.2; n13	5673
	24.8 d; se0.3; n22; pairing to birth	10333
Pitymys pinetorum	~21 d	1474
	24 d; mode 24; r24-25; n8	5809

	24 d; n several hundred	6822, 9530, 9531
Pitymys subterraneus	20 d; n2; min ibi	7904
	21 d	9656
	21 d; mode 21; r20-21; n4	6199
	22.9 d; se0.3; r21-25; n28 primiparous	5282
Synaptomys cooperi	23 d; nursing	2026
LACTATION		
Arborimus longicaudus	den emergence: 29 d	5026
	solid food: 30-40 d	4264
	wean: 21 d	4264
Arvicola terrestris	solid food: 12 d	4660
	wean: 11 d	4660
	wean: ~14 d	2045
Clethrionomys gapperi	solid food: 14 d; 8.1 g	7238
	solid food: at eye opening	10569
	wean: 12-17 d	7238
	wean: 14.7 d	5148
	lact: 17-21 d	10569
	independent: 1 mo	1871a
Clethrionomys glareolus	solid food: >14 d	3571
	solid food: 2 wk	8052
	wean: ~3 wk	3571
	leave nest: 3 wk	8425
	wean: 18 d	5477
	wean: 20-21 d	8052
Dicrostonyx groenlandicus	solid food: 14 d	4335
	wean: 14 d	6781
	wean: 16-18 d	4335
Dicrostonyx torquatus	wean: 1 mo	8051
Ellobius talpinus	wean: 60 d	4645
Eothenomys smithi	solid food: 15 d	294a
	nurse: 25-30 d	294a
Lagurus curtatus	leave nest: 12 d; r10-15; n7 lt	4905
	solid food: 14 d; r12-16	4905
	may nurse: 25+ d	4905
	wean: 21 d	5240
Lemmus lemmus	solid food: 10 d	6806
	solid food: 11 d	8051
	den emergence: 13-30 d	8051
	independent: 15 d	2186a
	den emergence: 2-3 wk	11679
Lemmus schisticolor	solid food: 11 d	5135
	leave nest: 12-14 d	5135
	lact: ~20 d	5135
Microtus agrestis	wean: 12-14 d	8920
	lact: 14 d; 13-16	8921
	independent: 14 d	1805
Microtus arvalis	solid food: 14 d	3571
	solid food: 14 d	8052
	independent: 14 d	9269
	wean: 15-20 d	2465
	wean: 16 d	7292
	wean: 16 d	10911
	wean: 17-20 d	3460
Microtus brandti	den emergence: 2-3 wk	9012
Microtus californicus	solid food: ~10 d	9747
	solid food: 2 wk	4490
	nurse: 14 d	4052, 4053
	wean: 16 d	6041
	nurse: min 23 d	9747
	nurse: 3 wk	4490
Microtus canicaudus	wean: ≤18 d	4196a
Microtus gregalis	solid food: 10 d	4645
	solid food: 10 d	8052

Microtus guentheri	solid food: 10 d	1965
	wean: 2 wk	3610
	wean: 15-21 d	1965, 1966
	wean: 18-20 d	12003
Microtus montanus	solid food: 11 d	9967
	wean: 15 d	9967
	abandon young: 13-16 d	5250
	independent: 2-3 wk	1871a
Microtus montebelli	wean: 20 d	7762
Microtus nivalis	wean: 3 wk	3571
Microtus oeconomus	leave nest: 11-12 d	10681
	solid food: 2 wk	3571
	wean: 20 d	3468
Microtus oregoni	wean: ~15 d	4970
Microtus pennsylvanicus	solid food: 9-12 d	4278
	solid food: 13-14 d	7066
	wean: 11-12 d	4272, 4275, 4278
	wean: 12 d	501
	nurse: 12-14 d	7066
	wean: 14 d	6347
	wean: 14.4 d	5148
	wean: 18-21 d	10463
Microtus socialis	solid food: 20 d	4645
	solid food: 20 d	8052
	den emergence: 15 d	4645
Microtus townsendii	wean: 15 d; n15 lt	6693
Microtus xanthognathus	wean: ~1 mo	12049
Neofiber alleni	wean: 3 wk	944
Ondatra zibethicus	den emergence: 3 wk	10364
	den emergence: 3 wk	3571
	wean: ~24 d	2695
	wean: 4th wk	3081
	independent: 4-6 wk	11934
	wean: 30 d	8051
	independent: 2 mo	1871a
Phenacomys intermedius	solid food: 14 d	3429
	solid food: <18 d	5150
	nurse: 19 d	3429
	independent: ~2 mo	1871a
Pitymys lusitanicus	independent: 20 d	11820
Pitymys ochrogaster	solid food: 10-14 d	9111
	solid food: 12-16 d	7066
	solid food: 17 d	3283
	nurse: 18-21 d	7066
	wean: 21 d	9111
Pitymys pinetorum	solid food: 15-16 d	6533
	solid food: 15-18 d	7066
	wean: 17 d	4273
	wean: 21 d	6533
	wean: 3rd wk	783
	nurse: 20-21 d	7066
Pitymys subterraneus	wean: ~20 d	9656
	wean: 27 d	6199
Prometheomys schaposchnikowi	wean: 3 wk	12007
Synaptomys cooperi	wean: ~3 wk	2026
INTERLITTER INTERVAL		
Arvicola terrestris	3-12 wk	10944
	mode 21-22 d; r21-25	11668
	22 d	4660
Clethrionomys gapperi	17 d min	7238
Clethrionomys glareolus	mode 19 d; n large; no lactation	4172, 4173
	20.16 d; most 16-35; r20-100; n212	1408
	20-21 d	10383
	21 d	157

	mode 21; r17-24, 78%; r37-42, 7.5%; n1164 lt	1875, 1881
	mode 22-23 d; n large; with lactation	4172, 4173
	max 5.5 wk; with lactation	11362
	45-55 d	8052
	50-60 d	4645
Clethrionomys rufocanus	20.2 d; sd1.4; n12 intervals	5593
	40-50 d	6134
Clethrionomys rutilus	17.5 d; r14-22; n36 intervals	7544
	min 20.5 d	2564
	r28-25 d; n12 intervals	7544
Dicrostonyx groenlandicus	19-37 d	6781
	mode 21 d; r19-45	4469
Dicrostonyx torquatus	min 23 d	8051
Dicrostonyx unalascensis	20 d; r16-27	7544
Ellobius talpinus	34-36 d	12111
Eothenomys smithi	mode 22 d	294a
Lagurus lagurus	76% 19-23 d; r18-25	3652
	19-91 d	11647
	25-30 d	4645
Lemmus lemmus	20 d; r18-22	7544
	21 d	6806
	21 d	3462
	40 d	8051
Lemmus schisticolor	mode 25 d; 60% 23-28; r22-40; n170 intervals	5135
Lemmus sibiricus	21 d; r18-27	7544
Microtus abbreviatus	18-26 d	7544
Microtus agrestis	20.26 d; r19-24; n42	1806
Microtus arvalis	20 d; r16-36	3459
	mode 20 d; r16-36; n417	9013
	18-132 d	9269
	22.8 d; r19-49	3460, 3465
	25-30 d	4645
Microtus fortis	40-45 d	8052
Microtus guentheri	25 d	1966
	33.2 d	3610
Microtus miurus	21 d; r18-30	7544
Microtus montanus	min 21-26 d; n10	4014
	21 d; n314 lt, 43 f	7811
Microtus montebelli	min 18 d; mode 20-21	7762
Microtus nivalis	33; r22-44; n2	3457
Microtus oeconomus	20 d; r16-25	2563, 7544
	~3 wk	10681
	25.3 d; n17 lt, 1 f	3468
	7 wk	659
Microtus oregoni	23.5-54 d	2087
Microtus pennsylvanicus	mode 20 d; 70% 18-24; r18-48; n71 lt, 19 f	10463
	21 d; r19-23	7544
	95% ~21 d	7116
	21 d	501
	21 d	4278
	21-22 d	6347, 6348
Microtus socialis	30-35 d	4645
	30-35 d	8052
Ondatra zibethicus	27.9 d; r18-35; n10 f	8104
	30 d	7093
	33 d; r29-35; n11 lt, 5f	2695
Pitymys juldaschi	25.6 d; r20-41	1057
	27.6 d; r27-66	1057
Pitymys lusitanicus	23.4 d; mode 24; r20-27; n19 intervals	11820
Pitymys ochrogaster	mode 21 d; r19-63; n431 intervals	9110, 9111
	r23-27 d; n5 intervals, 5 f	3283
	25-50 d	1871a
	~27 d	9273

Pitymys pinetorum	24-25 d	9530
	27.6 d; se1.8	2493
Pitymys subterraneus	66.1%; 18-25 d; 13.5%; 26-36 d; 20.4%; 37-88 d	5282
	22-29 d	1407
	23.8 d; mode 24; r20-26; n78	7904
	24.58 d	9656
LITTERS/YEAR		
Alticola roylei	10	8646
Alticola strelzowi	1-2	3088
	2-3	8646
	3 summer	8052
Arvicola terrestris	1.5-2.3 lt	10447
	2-4/season; south, more north	8052
	3-4	7433
	3-4; r2-4	3571
	4; mode 2 54%; r1-4	10944
	4-5	11668
	13-14	4660
Clethrionomys centralis	10	8646
Clethrionomys gapperi	>1	11230
	2	2088
	2-3	10839
	2.2-3.3	5151
	2.4; r1-6	6632
Clethrionomys glareolus	mode 1; r1-3	1844
	mode 2; r1-3	12077
	2; r1-3	7271
	2-4	4645
	3-4	3571
	3-4	7433
	3-4	8052
	max 4	11362
	4-5; wild; 4-7; lab	10589
	est 4-5	1222
Clethrionomys rufocanus	1-3	5140
	2-3	8646
	3-4	5501
	max 7	6134
Clethrionomys rutilus	3.8; ±3.3	2564
Dicrostonyx groenlandicus	0-17	4469
Dicrostonyx hudsonius	1-2	1474
Dicrostonyx torquatus	2; occasionally 3 in summer	8051
Ellobius talpinus	6-7	12111
	4	8052
	2; adult f	4645
Eolagurus luteus	min 3-4	9951
Hyperacrius fertilis	2-3	9175
Lagurus lagurus	mode 4-5; r4-7	4645
	6	8052
	10-12	8634
	11/8 mo; n1 f	3516
Lemmus lemmus	2/winter; r1-3	5967
	3; favorable yr	8051
	4?	6806
	10-12	3462
Lemmus schisticolor	2	8051
	max 2	10016
Lemmus sibiricus	min 2	1474
	3-4	8051
	min 12	2088
Microtus agrestis	2-5	1872
	3-4	3571
	3-4	7433
	4	12070

	4	8052
Microtus arvalis	1-11	9269
	2; r1-5	9013
	3-10; r3-12	3571
	5-7	8052
	16	11225
Microtus brandti	min 2-3	8052
	3; r1-5; n5	9012
Microtus californicus	4.4-5.6	4053
	9.30-11.22; over 3 yr	4893
Microtus chrotorrhinus	min 2	6632
Microtus fortis	6; good yr	8052
Microtus gregalis	2	8052
	3-4	10516
Microtus gud	2-3	8052
Microtus guentheri	mode 1-3; r1-5	1965
	6-7; 7 is theoretical	1962, 1966
	10	3610
Microtus longicaudus	1.9; r1-4	6632
Microtus montanus	4.38-5.15; over 3 yr	4893
Microtus nivalis	1-2; r1-3	3571
	2	7433
	2	6313
	2-3	6919
Microtus oeconomus	2.9	2563
	min 3	8052
	3-4	10681
	3-4	4645
	3-4	7433
	3-4	3571
	3.8	2563
Microtus oregoni	4.8	3622
	max 5	2088
Microtus pennsylvanicus	1.6-2.3	5151
	2-3	10839
	4.0; ±3.9	2562
	max 5 lt/season	7116
	5-10	4278
	est 5-6; theoretical max 17	4278
	max 17	501
Microtus richardsoni	2	1369
	max 2/season	6632
Microtus sachalinensis	0-2	11366
Microtus socialis	3-5	4645
	4-8	8052
Microtus xanthognathus	2 lt/season	11878
Neofiber alleni	4-5	944
Ondatra zibethicus	>1	9003
	2	3558
	1-2	11934
	1-2; north; 3; south	6266
	mode 1; r1-3	2243
	1-4	8104
	mode 5-6; r1-8	8015
	1; r1-2; 2 rare, 4/133 f	7093
	2	719
	2	10410
	2	6267
	mode 2	8762
	2	10410
	2; 3 rare	11467
	2	7015
	2	8762
	mode 2	3080

	est 2-3 lt/season	12112
	2-3	1474
	2-3	720
	2-3	90
	2-3; north; 3-4 south	4645
	2.0; mode 2; r1-4; n76 f	3082
	2.3	9529
	2.36	8237
	2.7; r1-4; 2.5; 2 yr	9503
	2.8	10112
	3	11027
	3-4; r3-5	3571
	3-4	9164
	3-5	6214
	3.28	2697
	4-5	7433
	mode 4	10421
	mode 4; max 6	7291
	3-4	8051
Phenacomys intermedius	min 2	1474
	max 3	5150
Pitymys ochrogaster	3-4/season	1474
	min 5 lt/summer	2137
Pitymys pinetorum	1-4	3852
	3-4	1474
Pitymys subterraneus	min 2	8052
	2-3; r2-6	3571
	5-6	7433
	8.7/9 mo	9656
	max 10	7904
Prometheomys schaposchnikowi	2	12007
	2?	8051
Synaptomys cooperi	2-3	1474
SEASONALITY		
Alticola strelzowi	repro: Mar-Oct, peak Apr-June; Kazakhstan	8020a
	repro: end Apr-mid July; USSR	3088
Arborimus longicaudus	mate: yr rd; CA	782
	mate: yr rd; n Canada, nw US	5026
	birth: yr rd; OR, CA	1474
Arvicola sapidus	preg f: Feb-Oct; Spain	11248
Arvicola terrestris	fertile m: Feb-Nov; n Austria	11668
	testes: large Feb-Aug; Netherlands	11204
	testes: large Mar-May, regressed Dec; s France	2471
	mate: Mar-Apr; GB	2045
	mate: end Mar-Sept; lab	8392
	mate: end Mar-early Oct; Germany	3571
	mate: Apr-Sept; USSR	10944
	mate: Apr-Oct; Europe	7433
	mate: peak July-Aug; lab	7772
	preg f: Jan-Mar; Switzerland	7261
	preg f: Mar-Sept; s France	2470
	preg f: Mar-Nov; Spain	11250, 11251
	preg f: Apr-Oct; Czechoslovakia	6007
	preg f: Apr-mid Oct; n Austria	11668
	birth: Feb-Oct; n Austria	11668
	birth: Mar-Oct; Netherlands	11204
	birth: begins Apr-May; USSR	8052
	birth: Apr-Sept; s France	2470
	birth: Apr-Oct; Europe	7433
	birth: Apr-Oct; Germany	3571
	birth: May-Aug; Scotland, UK	10447
	lact f: Sept; n1; Iran	6293
Clethrionomys californicus	preg f: Aug; n1; OR	6708a

Clethrionomys centralis	testes large: June-mid Aug; USSR	8052
	birth: June-mid Aug; USSR	8052
	birth: yr rd; lab	634
Clethrionomys gapperi	mate: end Mar-Nov; PA	9164
	mate: May-Sept; MI	6796
	mate: May-Nov; CO	11230
	preg f: end Apr-early Sept; s QUE, Canada	11956
	preg f: May, July; n3; NS, Canada	10144
	preg f: mid June-early Sept; MAN, Canada	10101
	preg f: July; n4; ND	3703
	preg f: Sept; n1; NM	502
	birth: Mar-Oct; boreal N America	1474
	birth: Apr-Aug; MAN, Canada	8383
	birth: June-Sept; BC, Canada	2088
	repro: Mar-Oct; WI	5210
Clethrionomys glareolus	fertile: yr rd, peak summer-fall; lab	8021
	sperm: Feb-Oct, peaks Apr-May, low Aug; England, UK	9342
	spermatogenesis: Mar-Oct?; Czechoslovakia	5041
	spermatogenesis: peak June, low Jan; UK	516
	spermatogenesis: peak July-Aug; UK	2077
	testes: large mid May-mid Sept; Norway	11679
	testes: large June-Aug, small Aug-Sept; AK	716
	epididymal sperm: Apr-Aug; Wales, UK	5307
	ovulation: early Apr; UK	2077
	ovulation: Apr-Aug; s Sweden	7988
	mate: Jan-mid Apr; lab	6052
	mate: Mar-Sept; Alps, Europe	3571
	mate: Mar-Oct; Poland	6053
	mate: Apr-Aug; USSR	12095
	mate: Apr-Sept; Europe, UK	7433
	mate: Apr-Sept, some Dec; France	943
	mate: Apr-Sept, some winter; Czechoslovakia	12076
	mate: Apr-Oct, peak May, some winter; England, UK	7873
	mate: mid Apr-early Oct; Germany	7271
	mate: May-Aug; USSR	8052
	mate: May-summer, peak June; Channel I, UK	955
	mate: peak May, low Oct-Jan; Czechoslovakia	12078
	mate: June-Aug; AK	716
	mate: June-Aug; Smolensk, Russia	8052
	mate: ends Aug-Sept; n Sweden; 2 mo earlier s Sweden	7988
	mate: Dec; England, UK	10175
	mate: Dec-Mar; s Sweden	4342
	mate: Dec-Mar; s Sweden	7251
	mate: low winter; s Sweden	3073
	preg or vag perf: Mar-Oct; England, UK	3342
	preg f: Feb-July, peak May; Czechoslovakia	12077
	preg f: Feb-Oct; Germany	6548
	preg f: Mar-Nov, peak July; Pyrenees, Spain	11255
	preg f: Apr-?; Germany	10340
	preg f: Apr-June; Germany	10344
	preg f: Apr-Sept, peak 76.9% Aug; Germany	7271
	preg f: Apr-mid Sept; Germany	10339
	preg f: Apr-Oct, peak June; England, UK	1222
	preg f: peak Apr-Sept, low Nov-Mar; Czechoslovakia	12079
	preg f: May-Aug; Wales, UK	5307
	preg f: May-Oct; s Sweden	7988
	preg f: mid May-Sept; Czechoslovakia	5103
	preg f: mid May-Sept; Norway	11679
	preg f: peak June-Aug; UK	2077
	preg f: 6% Oct; USSR	12095
	preg/lact f: May-Sept; Switzerland	1844
	birth: Mar-Sept, peak June-July; lab	11362
	birth: Mar-Oct, peak July-Sept; France	2470
	birth: Apr-Oct; Alps, Europe	3571

	birth: Apr-Oct; Europe, UK	7433
	birth: begins May; ne Turkey	10274
	birth: May-Aug; Poland	6055
	birth: June-Sept; Norway	11679
	birth: Nov, Jan; Germany	6091
	birth: peak May-Aug; lab	10383
	repro: end Mar-Sept; s Sweden	810
	repro: Apr-Sept; Denmark	5302
	repro: Apr-Sept; Germany	10339
	repro: yr rd, low winter; Germany	6091
	repro: yr rd, low Jan-May; se France	6370
Clethrionomys rufocanus	testes: large Apr-Aug; USSR	8052
	mate: Apr-early Oct; Mongolia	10516
	mate: mid May-mid July; n Norway	5140
	mate: June-Aug; n Sweden	11603
	preg f: ?-Aug; few data; Manchuria	12105
	preg f: May-Sept; Norway	11679
	preg f: May-Sept; Finland	5501
	preg f: mid May; USSR	8052
	birth: May-early June; Finland	5493
	birth: synchronized early June; Japan	5593
	lact f: ?-Sept; few data; Manchuria	12105
Clethrionomys rutilus	spermatogenesis: peak mid May; USSR	8052
	mate: ?-Aug; n Norway	5039
	mate: Apr-Aug; Altai, Mongolia	8052
	mate: mid June-?; w Siberia	8052
	mate: May-Sept; sw YUK, Canada	3774
	mate: June-?; n Canada	6782
	mate: Jan; n3; n Norway	6645
	preg f: ?-mid Aug; n Canada	6782
	preg f: low Mar; Finland	5493
	preg f: July; Manchuria	12105
	birth: May-early June; Finland	5493
	birth: May-Sept; nw Canada	1474
	birth: June-Aug; USSR	8052
	repro: f: May-Aug; m: end Apr-mid Sept; YUK, Canada	6015
Dicrostonyx groenlandicus	mate: June-July; NWT, Canada	6013
	preg f: June-July; MAN, Canada	10101
	preg f: end June-early Aug; NWT, Canada	6710
	repro f: July-Sept; NWT, Canada	9211
Dicrostonyx rubricatus	mate: ?-mid Aug; Arctic	177
	birth: end June-early July; Arctic	177
Dicrostonyx torquatus	mate: Mar-July, Oct-Nov; Russia	1779
	preg f: peak end June; USSR	8051
	birth: end winter-early July; USSR	8051
Dinaromys bogdanovi	repro: Mar, June; Yugoslavia, nw Greece	7983
Ellobius lutescens	spermatogenesis: yr rd; captive	10323
Ellobius talpinus	mate: yr rd; USSR	12111
	birth: Apr-Oct; USSR	8052
	repro: Apr-Oct; Jarantaj, Mongolia	10516
Eolagurus luteus	repro: summer; USSR; yr rd; lab	9951
Eothenomys andersoni	yg appear: July-Oct; Japan	7393
Eothenomys melanogaster	mate: most of yr; China	182
	conceive: Oct; China	182
Eothenomys smithi	preg f: yr rd, peaks end spring, Oct-Nov, low July-Aug; Japan	10665
	repro: m: Nov-Mar, f: Nov-Apr; Japan	12026
Hyperacrius fertilis	birth: spring, end summer; Pakistan	9175
Lagurus curtatus	sterile: July-Nov; WA	5240
	mate: yr rd; OR	6925
	mate: yr rd; OR	7491
	mate: yr rd, peak summer; ID	7606
	mate: yr rd; w US	1474
	birth: yr rd, low summer; WY	1871a
	lact f: June; n1; ND	3703

	repro: est yr rd; CA	2807
Lagurus lagurus	mate: Apr-Oct; Balkash, Kazakhstan	9950
	mate: Apr-Oct; USSR	8052
	birth: end Apr-early Oct; USSR	8052
	birth: peak warm season, low winter; USSR	4645
	repro: yr rd, peak fall; USSR	9903
	repro: yr rd; lab	3652
Lemmus lemmus	testes: large May-Oct; Norway	11679
	testes: large July; Lapland	811
	testes: small Oct; Jantland, n Sweden	811
	vaginal plugs: mid Aug-?; Norway	5301
	mate: May-Aug; Sweden	6806
	mate: spring-fall; USSR	5967
	preg f: May-Sept; Norway	11679
	preg f: July; Lapland, USSR	811
	birth: May-early June; Finland	5493
	birth: June-Aug; Lapland, USSR	8051
	birth: July; n1; Telemarkin, Norway	10210
	repro: est yr rd; lab	2386
Lemmus schisticolor	preg f: May-Sept, peak July; Finland	10016
	preg f: July-Sept, maybe earlier; USSR	8051
	birth: June-Aug; USSR	8051
	juveniles: winter; Norway	7706
	repro: Mar-Dec; lab	3466
Lemmus sibiricus	mate: June-Aug; NWT, Canada	6013
	preg f: Mar-Apr, July 57%, Dec-Feb; Yamaf Peninsula, n Siberia	653
	preg f: end June-early Aug; NWT, Canada	6710
	preg f: peak July; AK	716
	preg f: peak June-Aug; AK	7595
	repro: May-Sept; USSR	9071
	birth: peak mid June-July; USSR	8051
	repro f: mid July-Aug; n11; NWT, Canada	9211
Microtus agrestis	anestrus f: winter; UK	1218
	testes: peak Apr-Oct; England, UK	11932
	testes large: June-mid Oct; Czechoslovakia	6008
	mate: Jan-Mar; Sweden	4342
	mate: Feb-?; UK	521
	mate: Mar-Oct; Germany	3571
	mate: mid-Mar-Oct; UK	1218
	mate: end June-?; s Sweden	11603
	mate: spring-fall; Europe, UK	7433
	mate: winter if acorn crop good; England, UK	10175
	preg f: ?-Dec; UK	521
	preg f: Mar-Dec; France	2470
	preg f: Mar-Oct; Finland	8645
	preg f: Apr; England, UK	1804
	preg f: May-Sept; Brandenburg, Germany	10340
	preg f: end May-Aug; s Ural, USSR	8052
	preg f: June-Oct; Czechoslovakia	6008
	preg f: Nov-Dec; Berlin, Germany	6091
	preg/lact f: June, none winter; England, UK	1878
	birth: Mar-Sept; captive	519
	birth: Mar-Nov; Germany	3571
	birth: Mar-Dec; France	2470
	birth: May-?; USSR	8052
	birth: May-early June; Finalnd	5493
	birth: Apr-Dec; captive	1872
	birth: June; England, UK	1878
	birth: spring-fall; Europe, UK	7433
	lact f: mid Mar; Sweden	6223
Microtus arvalis	scrotal testes: Jan; Belgium	819
	testes: large Mar-Oct; Poland	6337
	testes: large Apr-July, small winter; France	2460

	testes: regress Oct; Poland	5091
	testes: regress Oct; Poland	6336
	mate: mid Jan-Oct; France	6902
	mate: Feb-Sept; Channel I, UK	955
	mate: Feb-Oct, low winter; w Europe	3571
	mate: Apr-Sept; Czechoslovakia	741
	mate: Apr-Oct; USSR	2650
	preg f: Jan; Belgium	819
	preg f: Jan-Feb; Germany	11593
	preg f: Jan-May; Germany	10344
	preg f: Feb-Oct; Mecklenburg, Germany	9013
	preg f: Feb-Oct; Poland	10343
	preg f: Mar, June, Aug, Oct; n8; Channel I, UK	955
	preg f: Mar-Sept; France	2466
	preg f: Mar-Sept; France	2465
	birth: Feb-Oct; Poland	10343
	birth: Mar-Sept, peak July-Sept; France	2470
	birth: Mar-early Oct; Germany	10343
	birth: peak Mar-Oct; w Europe	3571
	birth: May-Oct; captive	9013
	birth: yr rd; captive, Russia	8153
	lact f: Oct; n2; Channel I, UK	955
	lact f: Nov; Belgium	819
	repro: mid Jan-Nov; France	6902
	repro: low Dec; Germany	3463
	repro: yr rd, low Mar-May; Alps, France	6370
Microtus brandti	birth: mid Mar-?; lab	9012
	birth: May, Aug/Sept; n2; Mongolia	182
	birth: mid May-?; USSR	8052
	repro: mid Feb-Aug; lab	9012
Microtus breweri	testes: quiescent Oct-Dec; MA	64
	preg f: Apr 90%-Oct 30%; MA	9351, 10648
	lact f: Mar-Oct; MA	10647
Microtus californicus	mate: yr rd, peak spring; CA	4052, 4053
	mate: yr rd; CA, OR	1474
	preg f: yr rd, peaks spring, fall; CA	4893
	preg f: low July-Oct, Jan; CA	4053
	birth: yr rd, peak Jan-May; CA	6040, 6041
Microtus chrotorrhinus	mate: end Mar-mid Oct; Canada	6896
	preg f: July n6/12, Oct n1/13; no data other mo; WV	5794
Microtus fortis	mate: Apr-Oct; USSR	8052
	conceive: Mar; China	182
	repro: end Mar-mid Nov; Korea	5424,
Microtus gregalis	mate: summer; USSR	8052
	birth: May-Sept; Mongolia	10516
	birth: warm season; Europe, USSR	4645
	birth: est yr rd as yg found Nov-Dec; Siberia, Russia	7919
Microtus gud	birth: July; USSR	8052
Microtus guentheri	mate: Feb-Oct; lab	1965
	preg f: Dec-Feb; n21; no data other mo; Greece	8115
	birth: peak Mar; lab	1965
	birth: yr rd; Lebanon	6419
	birth: yr rd; captive	1036
	birth: yr rd; lab	3610
	repro: Oct-Apr; captive, Israel	1966
	repro: Nov-Apr; Israel	1962
Microtus longicaudus	testes: large May-June; CO	10255
	mate: Apr-Sept; BC, Canada	2088
	preg f: May, Sept; n2; ID	11646a
	preg f: end May-June; no data other mo; NV	6488
	preg f: June-July; no data other mo; SD	10954
	preg f: June-Aug; n9; YUK, Canada	12049
	preg f: Aug; n10; AZ	4897
	birth: May-Sept; CA	1474

Microtus mexicanus	mate: yr rd?; NM	502
	preg f: Jan, July-Aug; Coahuila, Mexico	522
	preg f: July; n3; Durango, Mexico	523a
	preg f: Aug, Oct; NM	502
	preg f: yr rd; Mexico	9176
	preg f: est yr rd; Mexico	1813
Microtus middendorffi	birth: early Apr-mid Oct; USSR	8052
Microtus miurus	mate: June-Aug; AK, nw Canada	1474
Microtus montanus	mate: Mar-mid Nov, none mid Nov-early Mar; WA	9724
	mate: Apr-Aug; no winter data; CO	11230
	preg f: Jan, Mar, Apr; n8; ID	11646a
	preg f: Feb-Sept; UT	7808
	preg f: Apr; n1; UT	10305
	preg f: Apr-Oct; CA	4893
	preg f: yr rd, low winter; WY	5251
	birth: yr rd; WY	1871a
Microtus montebelli	testes: regress Sept; Japan	7392
	mate: Apr-Dec, peaks early summer, end fall; Japan	15
	mate: May-June, Sept-Oct; low winter; Japan	5510
	mate: yr rd, peaks spring, fall; Japan	9443
	preg f: Apr 70%, Aug 60%; Japan	9443
	preg f: May-Nov, peak June; Japan	7392
	preg f: none July; Japan	5510
Microtus nivalis	testes: large June-Oct, small Nov-Jan, no data Jan-June; Alps, Germany	5486
	estrus: Sept; Germany	6128
	estrus: Apr; captive	6128
	mate: Apr; n1; captive	3457
	mate: May-Aug; France	6313
	mate: June-Aug; USSR	7433
	mate: June-early Aug; Germany	3571
	preg f: July-Sept; no data Jan-June; Bavarian Alps, Germany	5486
	preg f: Aug; n1; Iran	6293
	birth: Apr; n1; captive	3457
	birth: May; captive	6128
	birth: June-July; Germany	3571
	birth: June-Aug; USSR	7433
	repro: mid Mar-early Sept; USSR	6919
Microtus oaxacensis	preg f: Apr-June, Nov; n11; Oaxaca, Mexico	9176
Microtus oeconomus	testes: large May-Sept; Norway	11679
	testes: regress Aug; n1; n Norway	5039
	testes: small Sept; Germany	10340
	mate: ?-Sept; Mongolia	10516
	mate: Apr-Sept; Scandinavia, Germany	7433
	mate: mid Apr-mid Sept, peak May-June; USSR	8052
	mate: prob mid Apr-Oct; Finland	5969
	mate: May-Sept; Finland	10681
	mate: end July-early Sept; s Norway	5301
	mate: winter; s Norway	4342
	preg f: Apr; n1; USSR	8052
	preg f: Apr-early Aug; Germany	10339, 10340
	preg f: May-Sept; Norway	11679
	birth: Feb-Apr, Aug-Nov; lab, Berlin, Germany	3468
	birth: yr rd; lab, Oldenburg, Germany	3468
	birth: Apr-Sept; Scandinavia, Germany	7433
	birth: May-early June, low winter; Finland	5493
	birth: yr rd, peak Mar-Oct; Germany	3571
	repro: f; mid May-Aug; m; May-Aug; YUK, Canada	6015
Microtus oregoni	sperm: Feb-Sept; OR	3622
	mate: mid Mar-mid Sept; BC, Canada	2087, 2088
	mate: May-Aug+; w US, s BC, Canada	1474
	preg/lact f: Mar-Sept; OR	3622
	yg trapped: yr rd, peak June-July; OR	4970

Microtus pennsylvanicus	mate: end Mar-Nov; MAN, Canada	7293
	mate: Apr-June, Oct-Nov; MI	3741, 3742
	mate: end Apr-Sept; s QUE, Canada	11956
	mate: yr rd, low winter; NY	4278
	mate: none winter; MN	10839
	preg f: Jan, Mar-Apr, Oct-Nov; NC	1290
	preg f: Feb-Nov; MN	495
	preg f: Mar-Nov; NY	4278
	preg f: Mar-Dec; MI	3878
	preg f: May-June; n2; SD	261a
	preg f: Apr-May, July; n12; NE	5405b,
	preg f: peak Apr-Sept; OH	2376
	preg f: Apr-Oct; ONT, Canada	2082
	preg f: est May-Aug; MAN, Canada	10101
	preg f: June; ID	11646a
	preg f: June-Aug; n15 f; ND	3703
	preg f: mid June-Aug; no data other mo; SD	10954
	preg f: July; n10; se MT	6166
	preg f: peak mid July-mid Sept; YUK, Canada	12049
	preg f: Nov-Feb, prob also other mo; MA	10648
	preg f: yr rd; MN	721
	birth: yr rd; WY	1871a
	birth: yr rd; IL	10463
	lact f: Mar-Oct; MA	10647
	lact f: June; n1; SD	261a
	lact f: July; n4; se MT	6166
	sex active f: few Oct-Mar; OH	2376
	repro f: mid May-Aug; m: May-mid Sept; YUK, Canada	6015
	repro: yr rd, peak Mar-June; IN	2059
	repro: yr rd, most Mar-Nov; WI	5210
Microtus richardsoni	conceive: end May-early Sept; ALB, Canada	6631, 6632
	preg f: July-Aug; WY	1871a
	preg/lact f: July; n4; OR	4969
	birth: end June-Sept; BC, Canada	2088
Microtus roberti	yg trapped: early June-?; Turkey	10274
Microtus sachalinensis	mate: May-Sept; ne Sachalin, Russia	11366
Microtus socialis	mate: mid Jan-Apr, end Aug-mid Sept; USSR	8052
	preg f: Nov; Iran	6293
	birth: spring-fall; Europe, USSR	4645
	juveniles: Aug-Sept; Iran	6293
Microtus townsendii	scrotal testes: yr rd, peak 100% Apr-July, low 50% Nov-Dec; BC, Canada	3792
	mate: Mar-Sept; BC, Canada	2088
	mate: Apr-Aug, no data after Aug; BC, Canada	3792
	birth: Mar-Sept; w US, s BC, Canada	1474
Microtus transcaspicus	birth: yr rd; captive, Turkmenia	8153
Microtus xanthognathus	mate: est May-Oct; NWT, Canada	2704
	preg f: May-mid July; AK	11876, 11878
Neofiber alleni	mate: est Jan-fall; few data; GA	4386
	preg f: yr rd, not Mar; FL	944
	birth: mid June; few data; FL	8267
Ondatra zibethicus	gonads: increase Mar-July, regress Aug-Nov; WI	719
	sperm: Dec-Oct; MD	3392
	sperm: yr rd; TN	9529
	spermatogenesis: Feb-Aug; Germany	7291
	spermatogenesis: Mar-July; WI	719
	testes: enlarge Dec, large Mar-Aug; Germany	704
	testes: enlarge Mar; Canada	3558
	testes: large Feb-Sept; Netherlands	11174
	testes: large Apr-May; Kazakhstan	10490
	testes: max May, prob sperm yr rd; MD	12112
	testes: large May-Aug, small Sept-Dec; Germany	7291
	testes: large June-Sept; USSR	9690
	m fertile: Mar-Sept; Netherlands	2697

cl present: Jan-Feb; MD	3392
cl present: Jan-Aug; OH	2685
uterus: large Feb-June; Berlin, Germany	705
mate: Jan-Oct; Germany	3571
mate: Feb-Mar; Berlin, Germany	705
mate: mid Feb-mid Aug, peak Apr-June; WI	720
mate: end Feb-Mar; UK	11467
mate: end Feb-mid Nov; IA	3082
mate: Mar-?; Kazakhstan	10066
mate: Mar-Apr; Netherlands	11174
mate: Mar-Aug; N Am	6214
mate: mid Mar-mid Apr; NY	3061
mate: Apr-May; WY	1871a
mate: Apr-Aug; n N America	1474
mate: Apr-early Oct; Mongolia	2362
mate: Apr-Oct; Europe, UK	7433
mate: end Apr-early May; Canada	3558
mate: May-Sept; s BC, Canada	2088
mate: July; MAN, Canada	7093
mate: June-Aug; e Siberia	5946
mate: yr rd, peaks Nov, Mar, low July-Aug; LA	8015
mate: yr rd; CA	2616
ovulation: Mar-June; WI	719
preg f: Feb-Aug; OH	2685
preg f: Mar, July; n2; NC	1290
preg f: Mar-early Aug; s Bavaria	705
preg f: Mar-Sept, peak June; Netherlands	2697
preg f: Mar-Sept; Germany	10421
preg f: Apr-Aug; TN	9529
preg f: Apr-Sept; Netherlands	11174
preg f: end May-Aug, sometimes Sept; ne Sweden	6805
preg f: June-Dec; East Germany	4891
preg f: low Oct-Mar; Germany	7291
preg f: Nov; n1; NB, Canada	8234
preg f: yr rd, peak Nov-Apr; LA	10580
birth: Feb-Nov; Germany	3571
birth: peak mid Feb-Oct; CA	2616
birth: Mar-early Sept; WI	2695
birth: Mar-Oct, peak Apr-July, rare Oct; IA	3082
birth: Apr-?, peak end June; USSR	9690
birth: Apr, June-Sept; N Am	6214
birth: Apr-July; ONT, Canada	11934
birth: Apr-Aug, peak Apr-June; NE	9503
birth: Apr-Aug; e Canada	8237
birth: Apr-Sept; N America	6214
birth: Apr-Oct; Europe, UK	7433
birth: end Apr-early July, peak after mid May; WI	719
birth: end Apr-July; s ONT, Canada	11934
birth: end Apr-early Oct; UK	11467
birth: end May-July; Canada	3558
birth: end May-June; MAN, Canada	7093
birth: end May-mid Aug	9003
birth: May-Aug; NB, Canada	8236
birth: May-mid Aug; USSR	8051
birth: May-Sept; Sweden	2243
birth: May-Sept; captive	6267
birth: May-Sept, peaks monthly; WI	2695
birth: mid May-Aug; MAN, Canada	8104
birth: mid May-Aug, peak early June; captive	7094
birth: peak May-June; QUE, Canada	10410
birth: peak May-July; IA	3080
birth: June-July; Netherlands	11174
birth: yr rd, few Nov-Feb; w US	5146
lact f: July; n1; ND	3703

	repro: mid Feb-mid Aug; WI	5210
	repro: Mar-Aug; Germany	702, 704
	repro: Apr-Sept; Germany	11027
	repro: ends Sept; Dnepr, Ukraine	787
Phenacomys intermedius	preg f: end May-early Sept; ALB, Canada	5150
	preg f: June-Sept; CO	11230
	preg f: mid June-early Sept; MAN, Canada	3429, 10101
	birth: May-Sept; ALB, Canada	5150
	birth: June-Sept; N America	1474
Pitymys duodecimcostatus	preg f: peak Oct-June; s Spain	8216
	preg f: Dec-Apr; Pyrenees, Spain	11255
Pitymys majori	mate: peak June; USSR	8052
Pitymys ochrogaster	m fertile: yr rd; KS	5241
	spermatogenesis: yr rd; KS	5241
	mate: yr rd; captive	9111
	mate: yr rd, low winter, July; KS	9273
	preg f: peak Mar-June, none Jan; KS	5241
	preg f: Mar-Sept; SD	10954
	preg f: Mar, Oct; n2; KS	1812a
	preg f: Apr; n2; MN	495
	preg f: Apr-Nov; n110; NE	$5405b_1$
	preg f: peak Apr-Sept, few Oct-Mar; OH	2376
	preg f: May, July-Aug; n5; MAN, Canada	11957a
	preg f: May-June; n9; SD	261a
	preg f: June-July; n3; ND	3703
	preg f: Dec-Jan; n1; KS	5241
	preg f: yr rd, low Dec-Mar; TX	3284
	preg f: yr rd, low Dec-Jan; KS	3283
	birth: peak Mar-Sept; central, nw US	1474
	birth: mid May-Sept; MAN, Canada	2137
	lact f: May; n2; SD	261a
	repro: f: yr rd, peak Apr-July; IL	682
	repro: yr rd, peak Mar-June; IN	2059
	repro: yr rd, peak July-Sept, low Dec-Feb; WI	5210
	scrotal m: yr rd; OK	3852
Pitymys pinetorum	mate: Jan-Oct; n US	1474
	mate: Jan-Oct; NY	783
	mate: mid Feb-mid Nov; CN	3747
	mate: Apr-Sept; captive; NY	4273
	preg f: Feb-Nov, peak July-Aug, low Feb-Mar; no data Dec-Jan; PA	9987
	preg f: Apr, Aug, Dec; n3; NE	$5405b_1$
	preg f: Apr-Oct; VA	11059
	preg f: yr rd, low Dec-Jan; NC	8268
	repro: yr rd, most Mar-Sept; WI	5210
	repro: yr rd, peak Oct-May; OK	3852
Pitymys sikimensis	preg f: Mar, Aug-Nov; n22; Nepal	7382
Pitymys subterraneus	spermatogenesis: Mar-Nov; France	2470
	testes large: June-Oct; Germany	4114
	m fertile: Mar-Nov; no data Nov-Feb; Germany	6199
	m fertile: Apr-Sept; no data before Apr; Czechoslovakia	6005
	mate: ?-Oct; Europe	7433
	mate: Feb, Apr-early Oct; Germany	3571
	mate: Apr-?; Tatry Mt, Czeckoslovakia	10379
	preg f: mate: mid Feb-Sept, peak May; Czechoslovakia	8335
	preg f: Apr-Sept, peak June; Czechoslovakia	6004
	preg f: Aug-Oct; n5; Germany	4114
	preg f: yr rd; Germany	6199
	birth: ?-Sept; Tatry Mt, Czechoslovakia	10379
	birth: ?-Nov; Europe	7433
	birth: Apr-Sept; Europe, USSR	4645
	birth: May-Nov; Germany	3571
	birth: June-Nov; France	2470
	repro: peak Mar-Sept; lab	9656

Pitymys tatricus	m fertile: Apr-Sept; no data before Apr; Czechoslovakia	6005
	preg f: Apr-Aug; Czechoslovakia	6004
Prometheomys schaposchnikowi	mate: Apr; USSR	12007
	birth: May; USSR	12007
	birth: end May-mid July?; USSR	8051
	repro: spring-fall; Turkey	10275
Synaptomys borealis	birth: May-Aug; n N Am	1474
	birth: May-Aug; BC, Canada	2088
Synaptomys cooperi	mate: Apr-Nov, peak May; no data Nov-Apr; NJ	2026
	mate: yr rd; mid e N America	1474
	preg f: May, Oct; n7; NE	$5405b_1$
	preg f: yr rd; KS	1469
	birth: May-Sept; e central N America	555a
	repro: Mar-Nov; WI	5210
	repro f: yr rd; IL	682

Order Rodentia, Family Muridae

Rats and mice (*Acomys* 11-90 g; *Aethomys* 25-150 g; *Anisomys* 500-600 g; *Anonymomys* ~125 mm; *Apodemus* 13-64 g; *Apomys* 19-50 g; *Arvicanthis* 35-183 g; *Bandicota* 100-1500 g; *Batomys* ~205 mm; *Berylmys* 143-400 g; *Bullimus* 240-275 mm; *Bunomys* 60-150 g; *Carpomys* ~200 mm; *Celaenomys* ~195 mm; *Chiromyscus* >47 g; *Chiropodomys* 15-43 g; *Chiruromys* 115-175 mm; *Chrotomys* 115-160 g; *Colomys* 50-75 g; *Conilurus* 100-200 g; *Crateromys* ~255 mm; *Cremnomys* 30-60 g; *Crossomys* ~205 mm; *Crunomys* 55 g; *Dacnomys* 230-290 mm; *Dasymys* 47-164 g; *Diomys* 40-70 g; *Diplothrix* ~230 mm; *Echiothrix* 200-250 mm; *Eropeplus* 195-240 mm; *Gatamiya*, *Golunda* 50-80 g; *Grammomys* 24-80 g; *Hadromys* 66 g; *Haeromys* 55-75 mm; *Hapalomys* 120-165 mm; *Hybomys* 30-70 g; *Hydromys* 400-1300 g; *Hylomyscus* 8-42 g; *Hyomys* 1 kg; *Kadarsanomys* 176-230 g; *Komodomys* 125-200 mm; *Leggadina* 15-25 g; *Lemniscomys* 18-72 g; *Lenomys* 235-290 mm; *Lenothrix* 81-273 g; *Leopoldamys* 200-495 g; *Leporillus* 140-200 mm; *Leptomys* 145-160 mm; *Limnomys* ~125 mm; *Lophuromys* 23-111 g; *Lorentzimys* 12-13 g; *Macruromys* 150-250 mm; *Malacomys* 48-145 g; *Mallomys* 980 g-2 kg; *Margaretamys* 115-150 mm; *Mastacomys* 100-200 g; *Maxomys* 29-284 g; *Mayermys* 17-21 g; *Melasmothrix* 40-58 g; *Melomys* 30-200 g; *Mesembriomys* <1 kg; *Microhydromys* ~80 mm; *Micromys* 3-12 g; *Millardia* 47-100 g; *Muriculus* 70-95 mm; *Mus* 2-50 g; *Mylomys* 46-190 g; *Neohydromys* ~92 mm; *Nesokia* 112-185 g; *Niviventer* 26-240 g; *Notomys* 20-60 g; *Oenomys* 15-121 g; *Papagomys* 130-175 mm; *Parahydromys* ~240 mm; *Paraleptomys* 120-140 mm; *Paruromys* 350-500 g; *Pelomys* 46-170 g; *Phloeomys* 1.5-2 kg; *Pithecheir* 58-146 g; *Pogonomelomys* 51-65 g; *Pogonomys* 35-75 g; *Praomys* 8-100 g; *Pseudohydromys* 20 g; *Pseudomys* 12-105 g; *Rattus* 30-550 g; *Rhabdomys* 25-75 g; *Thynchomys*; *Solomys* 230-330 mm; *Srilankamys* ~145 mm; *Stenocephalemys* 120-195 mm; *Stochomys* 27-104 g; *Taeromys* 185-250 mm; *Tarsomys* ~135 mm; *Tateomys* 35-98 g; *Thallomys* 37-128 g; *Thamnomys* 30-100 g; *Tokudaia* 125-175 mm; *Tryphomys* 130-175 mm; *Uranomys* 31-53 g; *Uromys* 600-1000 g; *Vandeleuria* 10 g; *Vernaya* ~90 mm; *Xenuromys* ~1000 g; *Xeromys* 40-60 g; *Zelotomys* 38-85 g; *Zyzomys* 35-143 g) are primarily terrestrial herbivores from Europe, Asia, Africa, and Australia.

Males are larger than females in *Acomys* (2532, 4329), *Aethomys* (4329; cf 9893), *Apodemus* (2193), *Bandicota* (8500; cf 728), *Golunda* (8500, 8501), *Millardia* (730), *Nesokia* (728), *Oenomys* (5760), *Praomys* (4329; cf 4350, 4925), *Pseudomys* (10117), *Rattus* (899, 2841, 4841, 10501, 10696, 11475, 11832, 11833; cf 2954, 4925, 8500), and *Uranomys* (4350), while females are larger than males in *Notomys* (1258, 1261, 10117; cf 2532). In *Grammomys* either males or females may be larger (4329, 4350). A copulatory plug is present in *Acomys* (2532), *Hydromys* (8103), *Notomys* (1258, 1266; cf 2532; *Pseudomys* 1263), *Praomys* (1215, 5366), *Pseudomys* (5643), and *Rattus* (6162, 6237, 8393). Sex chromosome abnormalities occur (3492; *Acomys* 6974).

The 50 g (*Rattus* 6366), hemochorial, hemoendothelial, or epitheliochorial, labrinthine, discoid or diffuse placenta has a mesometrial or antimesometrial, interstitial implantation into a bicornuate uterus (231, 432, 1215, 1256, 3011, 5735, 7563, 8326, 8794, 9356) and is eaten after birth (4992, 9175, 9213, 10696). Estrus occurs postpartum (427, 845, 947, 1215, 1255, 1335, 1814, 1815, 2045, 2131, 2132, 2443, 2445, 2446, 2450, 2518, 2530, 2566, 2568, 2836, 2911, 3133, 3461, 3756, 4353, 4355, 5170, 5366, 5643, 5760, 6237, 7688, 7791, 7792, 8034, 8103, 8393, 8904, 8965, 9429, 10117, 10158, 10481, 10696, 10698, 10704, 10707, 11434, 11614, 11711; cf 1771, 8103). Lactation may delay gestation (1256, 3047, 3798).

Acomys are relatively precocious at birth often with fur and open eyes (1035, 2532, 2566, 2567, 5051, 8686, 10481, 11397; cf 187, 299, 1035, 1080, 2568, 4329, 5760). Other murids have hair at birth (*Mastacomys* 1542, 10698; *Pithecheir* 6448) and open their eyes at 4-7 d (*Lophuromys* 299, 1771, 2445, 5760; cf 5051), 7-8 d (*Lemniscomys* 299, 5051, 5760, 8429; cf 11614), 7-10 d (*Rhabdomys* 1335, 1814, 10158; cf 5051), 1-2 wk (*Aethomys* 331, 1333, 1771, 1815, 4329, 5051, 7168), 9-21 d (*Pseudomys* 670, 3259, 3438, 5642, 10117; cf 4042, 4355), 10-12 d (*Grammomys* 299, 334, 5051, 5760, 8218, 10158), 11 d (*Mesembriomys* 2131), 11-15 d (*Thallomys* 7168), or 17-28 d (2132, 2532, 4355, 10117). *Praomys* neonates are either naked or furred and open their eyes after 10-19 d (247, 299, 514, 1815, 2445, 4353, 5051, 5760, 7166, 7168, 8034, 10158). Many murids are born naked and blind (187, 334, 2091, 2452, 4329, 4410, 5760, 6317, 8500, 8501, 8861, 9175, 9265, 10158, 10709, 11710). Eyes open at 5-22 d (*Rattus* 1330, 2954, 3133, 3260, 3437, 3571, 7433, 9175, 10696; cf 7165, 11833), 7 d (*Arvicanthis* 5051; cf 2450), 7-10 d (*Micromys* 2045, 2565, 3461, 3571, 5553, 7433), 10 d (*Melomys* 7209), 10-16 d (*Hydromys* 8103), 11-16 d (*Apodemus* 2045, 2569, 3571, 7433), 12 d (*Millardia* 12034), 12-15 d (*Mus* 245, 329, 670, 845, 5685, 7433, 8501, 9175, 10158, 11711), 13-17 d (*Bandicota* 108, 9175, 9429, 10268), 15-21 d (*Hylomyscus* 2445), 16-17 d (*Zelotomys* 947, 10158), or 17-18 d (*Niviventer* 12105).

Females have 2 mammae (*Phloeomys* 9562), 2-4 mammae (*Melomys* 7209), 4 mammae (*Colomys* 4495; *Conilurus* 7209, 10698; *Chrotomys* 7983; *Hydromys* 7108, 8103, 10698; *Hyomys* 7209; *Leporillus* 7983; *Macruromys* 7209; *Mallomys* 7209; *Pithecheir* 7983; *Pseudohydromys* 7983; *Pseudomys* 5619, 10704; *Tateomys* 7675; *Thamnomys* 4495; *Uromys* 7209; *Xeromys* 10698), 4-6 mammae (*Acomys* 331, 5760; *Aethomys* 331, 4495, 9945, 10158; *Grammomys* 331, 988, 4495, 8218; *Pogonomelomys* 7209), 4-8 mammae (*Oenomys* 4495, 5760), 6 mammae (*Anisomys* 7209; *Chiropodomys* 4408; *Chiruromys* 7209; *Dasymys* 331, 4495, 5760, 9945, 10158; *Lophuromys* 331; *Lorentzimys* 7209; *Malacomys* 331, 4495; *Melasmothrix* 7675; *Pogonomys* 7209; *Thallomys* 9945), 6-8 mammae (*Apodemus* 182, 3571, 7433; *Hylomyscus* 5760; *Maxomys* 573, 11186; *Millardia* 9175), 6-14 mammae (*Rattus* 573, 610, 1254, 2321, 2555, 2911, 3260, 3571, 4041, 5111, 5277, 7052, 7209, 7433, 9175, 10696, 10917), 6-24 mammae (*Praomys* 184, 186, 331, 1215, 4495, 5760, 9468, 9945, 10158), 8 mammae (*Kadarsanomys* 7672; *Lemniscomys* 331, 4495, 9945; *Leopoldamys* 182; *Micromys* 182, 3571, 7433; *Mylomys* 4495; *Nesokia* 728; *Pelomys* 331, 9945; *Rhabdomys* 9945), 10 mammae (*Komodomys* 7983; *Lenothrix* 7674; *Tryphomys* 573; *Zelotomys* 5760, 10158), 10-12 mammae (*Mus* 331, 573, 4495, 5111, 6202, 7209, 7433, 9175, 9945, 11711), 10-22 mammae (*Bandicota* 108, 728, 7907, 9175, 11436). Commonly, males have a baculum, penial spines, seminal vesicles, and coagulating glands (1257, 1259, 1264, 1265, 1276, 4683, 7546a, 7791, 8034, 8983, 11883). *Mus* exhibits no photoperiodic regulation (1314).

Details of anatomy (988, 1258, 1263, 1266a, 1275, 1326, 3068, 5513, 8979, 8980, 9431, 9432, 9878, 10565), endocrinology (*Acomys* 714, 715; *Notomys* 1256, 1261, 1275; *Praomys* 7737; *Rattus* 5163, 5695), spermatogenesis (1260, 1262, 1274a, 1276, 1277, 3292, 3293, 7205, 8324, 10927; *Acomys* 8327; *Apodemus* 5093, 5094, 5479; *Rattus* 5695), copulation (1266, 2525, 3097, 3133), birth (*Apodemus* 9213), reproductive energetics (669, 670), and light (*Bandicota* 9433, 9434; *Notomys* 1252, 1253) are available.

NEONATAL MASS

Acomys cahirinus	4-5 g; term emb	7795
	f: 4.9 g; n67	7545
	5-6 g; r3.0-8.5	1035
	5.49 g; sd0.64; n20 lt	11397
	m: 5.6 g; n81	7545
	5.78 g; r4-8; n74	2566, 2567
	6.37 g	2568
Acomys wilsoni	3-4 g; term emb	7795
Aethomys chrysophilus	4.1 g	1333
	4.8 g	7797a
Aethomys kaiseri	6.1 g; r5.2-7.4; n42	1771
Aethomys namaquensis	2.5 g	7797a
Apodemus agrarius	1.8 g	3571
Apodemus flavicollis	2.3 g	1032
Apodemus mystacinus	3 g	2569
Apodemus peninsulae	2.0 g; r1.6-2.5; n38	12105
Apodemus sylvaticus	1-2 g	2045
Arvicanthis niloticus	2.5-3.5 g	7793
	~3 g	2446a

	4-6 g	8811
	4.25 g; n6	2450
	4.65 g; ±0.18; n19	2450
Bandicota bengalensis	3-5 g	9175
	4.85 g; r4.0-5.7; n26	10268
	5.8 g; n25, 3 lt, youngest litters seen	4409
Bandicota indica	f: 9.43 g; se0.14; n65	10370
	m: 10.54 g; se0.15; n77	10370
Grammomys dolichurus	4.2 g; n8	8218
Hybomys univittatus	4-4.8 g	3692a
Hydromys chrysogaster	f: 22.23 g; sd2.72; n17	8103
	m: 22.89 g; sd4.38; n19	8103
	f: 25.49 g; sd0.27; n161	8103
	m: 26.93 g; sd0.20; n209	8103
Hylomyscus alleni	1.2-1.8 g	2446a
Hylomyscus stella	1.47 g; r1.3-1.7; n5	2445
Lemniscomys striatus	1.6-1.8 g	7792
	~3 g	2446a
	~3 g	8429
Lophuromys flavopunctatus	4-5 g; near term	2452
	5-8 g	5760
Lophuromys sikapusi	6.5-7.5 g	2572a
	6.6-9 g	3692a
	7.89 g; r6.6-8.9; n8	1771
Lophuromys woosnami	9-10 g; 2 d	2572a
Mastacomys fuscus	7.25 g; n1	1542
Mesembriomys gouldii	34.70 g; r33.00-37.50; n6	2131
Micromys minutus	0.7 g; r0.65-0.8	2045
	0.8-0.9 g	3571
	1.1 g; n41 ?	5553
Mus caroli	1.2 g	10370
Mus cervicolor	1-2 g	10370
Mus domesticus	1.70 g; se0.05	6864
Mus minutoides	0.8 g	245
	0.8 g	11711
	~1.3 g	2446a
Mus musculus	0.8 g; n6, 1 lt	4409
	~1 g	845
	1.14 g	7704
Mus triton	1.3 g; n1	4329
Niviventer niniventer	3-4.5 g	12105
Notomys alexis	2.5 g	10117
Praomys daltoni	1.5 g	247
Praomys morio	2.435 g; n8	2954
	2.97 g; r2.57-3.09; n1 lt, 4 yg, 2 wk	2445
Praomys natalensis	1.8 g; late term	7791
	2.2 g; r1.9-3.0; n137, 15 lt, 3 f	7166, 7168
Praomys tullbergi	2.7 g; r2.0-3.0	4353
Pseudomys australis	~4 g; from graph	10117
Pseudomys gracilicaudatus	f: 4.2 g; 95%ci 3.6-4.8; n23	3438
	m: 4.5 g; 95%ci 3.6-5.4; n28	3438
Pseudomys higginsi	5 g; from graph	4042
Pseudomys novaehollandiae	1.0-2.4 g	5642
	2 g ea; n2	8801
Rattus exulans	1.6-3.5 g	1330
	2.0 g; n11 lt	4409
	2.2 g	10370
	f: 2.2 g; ±0.4; n44	1330
	m: 2.4 g; ±0.4; n60	1330
	f: 2.8 g; n7	11833
	m: 3.1 g; n3	11833
	3.4 g; r3.2-3.6; n3 f	10578
Rattus fuscipes	4.5 g; r2.5-6.0; n156	10696

Rattus lutreolus	f: 4.14 g; n1	4041
	f: 5.0 g; 95%ci 4.9-5.2; n69 (m&f)	3437
	m: 5.12 g; r4.89-5.34; n2	4041
	m: 5.2 g; 95%ci 5.1-5.2; n69 (m&f)	3437
Rattus norvegicus	5.27 g; r2.800-6.380; n204 yg, 18 lt	7708
	5.6 g; n11, 2 lt	4409
	5.72-5.99 g	409
	5.8 g; n333, 32 lt	5503
	7.35 g	1843
Rattus rattus	4.0 g; n21, 5 lt	4409
	4.15 g; r2.6-5.1; n16	10268
	f: 4.6 g; ±0.1; d2; n92 m&f	2091
	m: 4.9 g; ±0.1; d2; n92 m&f	2091
	5.1 g; median 5.4 g; sd1.46; n136	8861
Rattus sordidus	~3 g; from graph	7053
Rattus tiomanicus	4.1 g	10370
Rhabdomys pumilio	~2 g	1814
	2.5 g; se0.05	1335
	2.5-3.0 g	4640
Thallomys paedulcus	2.5-2.8 g	7168
Uromys caudimaculatus	20 g	10698
NEONATAL SIZE		
Acomys cahirinus	HB: 40-46 mm	5051
	51 mm; n1, 12 h	1080
	52 mm; r44-57	2566, 2568
Acomys wilsoni	HB: 40 mm; n2	5051
Aethomys chrysophilus	m: 42 mm	1333
Aethomys kaiseri	42-45 mm; n6	5051
Bandicota bengalensis	35 mm	9175
	HB: 47 mm; n25, 3 lt, youngest litters seen	4409
Bandicota indica	f: 52.0 mm; se0.4; n65	10370
	m: 55.0 mm; se0.3; n77	10370
Grammomys dolichurus	HB: 35 mm; n5	299
	HB: 40.0 mm	8218
Hydromys chrysogaster	CR: 52.4 mm; n3, 33 d emb	8103
	f: HB: 83.06 mm; sd1.39; n17	8103
	m: HB: 83.84 mm; sd1.83; n19	8103
	f: HB: 87.14 mm; sd0.34; n161	8103
	m: HB: 89.22 mm; sd0.28; n209	8103
Lemniscomys griselda	<80 mm; n1	9179
Lophuromys flavopunctatus	HB: 48 mm; n3	299
Mastacomys fuscus	63 mm; n1	1542
Mesembriomys gouldii	f: 92 mm	2131
Mus minutoides	HB: 25 mm	11711
	HB: 26 mm	11711
	HB: 26 mm	245
Mus musculus	HB: 22 mm; n6, 1 lt	4409
Mus triton	26 mm; n1	4329
Praomys natalensis	HB: 25-33 mm; n12	299
	HB: ~30-35 mm	7166
Pseudomys gracilicaudatus	f: 43.6 mm; 95%ci 40.0-47.2; n25	3438
	m: 44.8 mm; 95%ci 40.9-48.6; n29	3438
Pseudomys higginsi	48 m	4042
Rattus exulans	HB: 37 mm; n11 lt	4409
	f: HB: 42.4 mm; n7	11833
	m: HB: 42.6 mm; n3	11833
	56 mm; r55-57; n3	10578
Rattus fuscipes	46 mm; r38.9-51.4; n110	10696
Rattus lutreolus	f: 43.8 mm; 95%ci 42.9-44.6; n69 m & f	3437
	m: 45.6 mm; 95%ci 44.6-46.6; n69 m & f	3437
	f: 50 mm; n1	4041
	m: 50 mm; r49-51; n2	4041
Rattus norvegicus	HB: 52 mm; n11, 2 lt	4409

Rattus rattus	HB: 40 mm	9265
	HB: 45 mm; n21, 5 lt	4409
Rhabdomys pumilio	HB: 35 mm; se0.0	1335
	HB: 45 mm; n4	299
WEANING MASS		
Acomys cahirinus	~10 g; 1st trapped	7795
	15.16 g; n14	11397
Acomys wilsoni	~10 g; 1st trapped	7795
Aethomys chrysophilus	30-32 g; 1st trapped	7797a
Aethomys kaiseri	f: 46.0 g	1771
	47.8 g; n33	1771
	m: 48.9 g	1771
Aethomys namaquensis	10-15 g; 1st trapped	7797a
Arvicanthis niloticus	15-25 g	7793
	~25 g	2450
Bandicota bengalensis	22.9 g; r13.1-41.6; n12, 30 d	10268
Bandicota indica	41.0 g; r31.4-46.4; n9, 30 d	10268
Dasymys incomtus	30 g	2570
Hydromys chrysogaster	~150 g; 30 d; from graph	8103
Lemniscomys striatus	10-15 g	7792
	20 g	2570
Lophuromys flavopunctatus	6.5-9.5 g; from graph	2445
	30 g	2570
Lophuromys sikapusi	13.61 g; n8 yg, n2 lt	1771
Melomys burtoni	11 g; lightest trapped	10109
	50 g; juvenile	735
Melomys cervinipes	17-18 g; lightest trapped	10109
Mesembriomys gouldii	481 g; n1, 42 d; from graph	2131
Mus minutoides	3.18 g; 24 d	245
	~3.5 g; 3 wk; from graph	11711
Mus musculus	6.07 g; n36 yg, 3 lt of 12, 21 d	10095
	~10 g	845
	10.33 g; n15 yg, 5 lt of 3, 21 d	10095
	f: 10.7 g; se0.3	2723
	f: 12.8 g; se0.1	2723
Mus triton	4-5 g	2570
Notomys alexis	10-15 g; 25 d	10117
Oenomys hypoxanthus	30 g	2570
Praomys daltoni	~8 g	247
Praomys erythroleucus	12 g; lightest trapped; est 20-25 g at weaning	1215
Praomys jacksoni	20 g	2570
Praomys natalensis	10-15 g; lightest trapped	7791
	10-20 g	2570
	11.7 g; n125, 14 lt, 4 f, 21 d	7166
	f: 13 g; ±0.72; r10-16.5 d; n12, 21 d	5366
Praomys tullbergi	8-13 g; 28 d	4353
Pseudomys apodemoides	est 9 g; sd0.8; n6	1934
	13 g; 6 wk	3259
Pseudomys australis	10-15 g; 25 d; from graph	10117
Pseudomys gracilicaudatus	~24 g; from graph	3438
Rattus exulans	18-23 g; solid food	11833
	m: 26.6 g; sd2.8; n9	4821
	f: 27.2 g; sd1.8; n9	4821
Rattus fuscipes	m: 12.3 g; r11-13; n3, 5 wk	3260
	f: 13 g; n1, 5 wk	3260
	17.6 g; 3 wk	10696
	29.9 g; 28 d	10696
	40.9 g; r36.9-42.1; n2	10696
Rattus lutreolus	25 g	4041
	< 40 g; young juveniles	614
Rattus norvegicus	f: 61.2 g; sd8.7; n10	4821
	m: 66.3 g; sd9.9; n10	4821
Rattus rattus	28.6 g; r25.3-40.1; n7, 30 d	10268
	30-50 g; est 6-7 wk	8861

	m: 42.4 g; sd7.7; n10	4821
	f: 47.4 g; sd11.7; n10	4821
Rhabdomys pumilio	8.0 g; n9	4640
	8.9 g; 14 d	1335
Tateomys rhinogradoides	82.8 g; sd12.7; n4	7675
Zyzomys argurus	14 g m; lightest trapped	733
WEANING SIZE		
Mus minutoides	HB: ~50 mm; 3 wk; from graph	11711
Praomys tullbergi	f: HB: 70-80 mm, 28 d	4353
LITTER SIZE		
Acomys spp	1.88 emb; r1-4; n25 f	299
Acomys cahirinus	mode 1-2; r1-4	2449, 2452, 4344
	1-2 emb; n9+ f	4925
	1-4	10481
	1.7 wean	7544
	1.94 emb; ±0.05; 1.85 live emb; ±0.05; r1-5; n179 f	7795
	2.0 yg	7544
	2 emb; n1 f	331
	2.0; n26	2532
	2.08; sd0.85; r1-4; n87 lt	11397
	2.13 emb; se0.09; mode 2; r1-4; n55 f	11475a
	2.37; mode 2; most 1-4; r1-7; n408 lt	12038
	2.38 yg; n75 lt	2568
	2.44; mode 2; r1-5; n123 lt	2566, 2568
	2.5 emb	11475a
	2.68 emb; mode 3; r1-4; n25 f	2567
	2.75; n4 lt	12128
	2.76 yg; r1-5; n84 lt	1035
	2.8 ps; n21	4329
	3; r2-4	1080
	3 emb, ps; r1-6; n13 f	8134
	3.2 emb; n17 f	4329
	4.5 yg; r2-11; n17 lt	12128
	4.5 yg; 5.51 emb; r3-7; n49 lt	109
Acomys russatus	2.5; r2 yg, 3 emb; n2 lt	8134
Acomys spinosissimus	2.5 emb, ps; sd1.3; mode 2; r1-5; n6 f	3821
	2.6 emb; r2-3; n5 f	9893
	3.0 emb; r2-5; n12 f	10154, 10158
	3.3 emb; r2-6; n9 f	8965
Acomys wilsoni	2.23 emb; 2.21 live emb; ±0.09; r1-5; n79 f	2443, 2452, 7795
	2.5 emb; r1-4; n2 f	4925
Aethomys chrysophilus	2-4 emb	9945
	2 emb; n1 f	337
	2.5 emb; r2-3; n2 f	9893
	2.6 emb; r1-4; n9 f	299
	2.8 emb; n5 f	331
	3 emb; n1 f	2096
	3.1; r1-5; n37 lt	1333
	3.2 emb; r2-5; n15 f	4329
	3.3 ps; r1-7; n12 f	4329
	3.3 emb; r1-5; n41 f	8965
	3.9 emb; r3-6; n10 f	10154, 10158
	4.1 yg; ±0.27; n14 lt	7797a
Aethomys hindei	2.05; r1-4; n18 lt	8083
Aethomys kaiseri	2.6; r1-4	1771
	3 emb; n1 f	4495
	3 yg; n1 lt	187
	3.0 emb ea; n4 f	2443, 2446a, 2452
	3.4 emb, yg; r1-5; n5 f	299
	3.5 emb; r3-4; n2 f	4925
Aethomys namaquensis	2-5	11864
	2-5	9945
	2.6 ps/emb; sd1.1; mode 3; r1-4; n15 f	3821
	3.1 emb; r2-7; n42 f	10154, 10158

	3.3 emb; r1-5; n43 f	8965
	3.4 yg; ±0.16; n53 lt	7797a
	3.6; r1-5; n11	10158
	4; n3 lt	7168
	5 yg; n1 f	7165
	7.0 ps; r3-11; n2 f	4329
Aethomys nyikae	3.67 emb; n3 lt	331
Aethomys walambae	3 emb; n1 f	337
Apodemus spp	6.8 emb; r6-7; n5 f	2442
Apodemus agrarius	2-8	7433
	mode 4-9; r2-12	3571
	5-6	4645
	mode 5-6 yg; r2-9	5424_1
	5.0; r4-6; n4 lt	11710
	5.8 emb; mode 6; r2-8; n15 f	10340
	6.64; r4-9; n79 lt	8330, 8331, 8332
Apodemus argenteus	4.0 emb; n34 f	3546
	4.1; n14 lt	7925
	4.13; mode 5; r3-6; n23 lt	12025
Apodemus flavicollis	3-8	7433
	mode 3-8; r3-12	3571
	3.9; n104 lt	5170
	4.42 emb; mode 5; n91 f	11255
	4.76 emb; r3-7; n13 f	10340
	5.04 emb	8331
	5.12 emb; sd1.28; mode 5; r1-8; n31 lt, n159 emb	10358
	6	6370
	6; n1 lt	9213
	6; r1-12	4645
	6.8 emb; mode 7; 66% 5-7; r3-11; n61 f	5461
Apodemus microps	4.83 emb; sd0.76; mode 5; r4-6; n6 f	10358
	5.89 emb; r3-8; n73 lt	10480
	6.40; se0.159; mode 6; r3-10; n80 lt	8329, 8331
Apodemus mystacinus	1.67 yg; r1-2; n3 lt, 1 f	2569
	4 emb; r2-9	8317
	5 emb; n1 f	6419
Apodemus peninsulae	4; r1-7; n20 lt	12105
	2-8 emb, yg	5424_1
Apodemus speciosus	4.3 emb; r2-7; n18 f	12024
	4.6 emb; 3.2 spring, 5.6 fall	10632
	4.7 emb	7394
Apodemus sylvaticus	mode 2-8; r2-9	3571
	2-9	7433
	4 emb; n1 f	4403
	4 emb; n1 f	182
	4 emb; n1 f	6419
	4 emb; n1 f	2448
	4 emb ea; n2 f	7185
	4.1 emb; se0.3; n9 f	1877
	4.1; mode 4; r2-9; n10 lt	9438
	4.23; se0.10; n100 lt	3067
	4.3 emb; r2-6; n3 f	10340
	4.3 yg; se0.2; n51 primiparous	1874
	4.5-5.7; across breeding season	2990
	4.8 emb; r3-7; n25 f	9477
	4.85 emb; r2-6; n33 f	741
	mode 5-6; r1-12	4645
	5 emb; r3-7	516
	5.05 yg; se0.2; n405 lt	1874
	5.2 cl; ±0.3; n9 f	1877
	5.25; r4-7; n8 lt	955
	5.37 emb; sd1.31; r4-7; n16 f	10358
	5.4 emb	5307
	5.4 emb; r2-10; n6 f	6293

	5.4 emb; ±0.2; n18 f	3154
	5.5 emb; r5-6; n2 f	7382
	5.57 emb	8331
	5.6 emb; r5-7; n5 f	846
	5.67 emb; r2-9; n3 f	1590
	5.84; se0.147; mode 5-6; r3-10; n143 lt	8329
	6.5 emb; mode 7; 78% 5-7; r4-11; n69 f	5461
	6.6 emb; r5-7; n6 f	7779
	6.70; r4-9; n10 lt	955
Arvicanthis niloticus	2-10	4344
	mode 3; r3-10	2449
	3.0; r2-4; n2 lt	12128
	3.60 emb; se0.27; n15 f	7793
	3.7 yg; ±1.0; r1-6; n12 lt	2450
	3.7; r1-7; den emergence	8193
	3.93 emb; se0.29; n46 f	7793
	4-6 yg	5760
	4 emb ea; n2 f	2443, 2452
	4.5 emb; r3-10; n44 lt	7790
	4.64 yg; mode 6; r2-7; n14 lt	8431
	4.66 emb; mode 4; r3-10; n41 f	2446a
	4.875 emb; r3-6; n8 f	4925
	5-6	4484
	5-10	8811
	5 emb; n1 f	10799
	5 emb, yg; r3-7; n3 f	299
	5.38 emb; se0.23; n53 f	7793
	5.4; mode 4-8; max 12	3756
	6 emb; n1 f	186
	6.00 emb; r2-12; n130 f	10712
	7.125 emb; r5-9; n8 f	2454
	8 ps; r7-9; n2 f	10799
	max 8 yg	4347
Bandicota bengalensis	2 yg	11507b
	2-14	8724
	2-13 emb; n33 f	4409
	3.87-4.56 cl	9432
	4.0 emb; n3 f	8905
	mode 5-10; most of yr; mode 14-18; Sept, Nov	9175
	5.4; r1-10; n29 lt	9429
	6.10 emb; n345 f	10267
	6.2 emb; se0.67; n14 f	728
	7.4 emb; sd2.4; mode 7; r1-14; n227 f	11436
	8.88 emb (or ps?); modes 7-11; r5-13; 2 yr	5586
	8.89; n137 lt	3554
	mode 10-12	8500, 8501
	10.3; se0.67; n65 lt	5718
	14-18; Oct-Nov: 5-10; rest of yr	108
Bandicota indica	5.3 emb; r4-7; n16 f	8794
	5.49 yg; n43 lt	10821
	6.84; se0.52; r2-13; n25 lt	10370
	10 yg; n1 lt	10268
Berylmys bowersi	4.0 emb; r2-5; n3 lt	4406
Chiropodomys gliroides	1.25; n4 lt, 1 f	7159
	2.2 yg; mode 2; r1-3; n18 lt	4406
Chiruromys forbesi	3 emb; n1 f	7209
Chiruromys vates	1 yg; n1 f	7209
Colomys goslingi	2; n1 lt	4495
	2 emb; max 3; n6 f	8855
	2.0 yg; n2 lt	2574
Conilurus penicillatus	1-3	10698
	2 emb ea; 3 lt	10705
Cremnomys blanfordi	2.5; r2-3; n4 lt	1339

Cremnomys cutchicus	4.03 emb; 64.6% 3-4; r2-8	8723, 8729
	7.2 cl; mode 6-9; r3-11	8723, 8729
Dasymys incomtus	1; n1 lt	12128
	1.5 emb; r1-2; n2 f	7362
	mode 2-3; r1-4	5760
	2-4	9945
	2; n1 lt	187
	2.3 emb; r2-3; n6 f	10799
	2.5 emb; r2-3; n2 f	4329
	2.5 emb; r2-3; n2 f	4495
	2.5 emb; r2-3; n6 f	2452
	2.75 emb; max 4; n32 f	8855
	3 yg; n1 f	186
	3 emb or yg; n2 f	4925
	4 emb; n1 f	334
	4 emb; n1 f	9893
	4.5 emb; n2 f	331
	5.2 ps; r3-9; n5 lt	10799
	5.3 emb; r2-9; n4 f	10154, 10158
Golunda ellioti	mode 3	8500
	mode 3-4 yg	8501
	4 yg; n1 lt	9175
	6 emb; n1 f	8501
	6.6 emb; r5-10	8723, 8724
Grammomys cometes	2.9 emb; r2-5; n6 f	4329
	2.9 ps; r1-4; n6 f	4329
	3 emb ea; n3 f	4925
Grammomys dolichurus	70% 2-4; r1-7; n634 lt	988
	2.25 emb; max 4; n20 f	8855
	2.7 yg; r1-4; n3 lt	8218
	3 emb; n1 f	4925
	3 emb ea; n2 f	4925
	3 emb ea; n3 f	2452
	3 nestlings; n1 lt	6272
	3.6 emb, yg; r3-5; n5 lt	299
	3.8 ps; r1-8	4329
	4.0 emb; n3 f	331
	4.6 emb, yg; r4-6; n5 lt	4329
	4.67 emb; r4-6; n3 f	334
Grammomys rutilans	1 yg; n1 lt	2574
	1.3 emb; mode 1; r1-3; n12 f	2754
	2.0 emb; r1-3; n3 f	2446a
	2.1 emb; mode 2; r2-3; n7 f	4350
Grammomys surdaster	2 yg; n1 f	186
Graomys spp	6 yg; r3-10	4682
Hybomys fumosus?	4 emb; n1 f	2754
Hybomys trivirgatus	2 emb; n1 f	10799
	2 emb ea; n3 f	7362
	2.11 emb; r1-3; n9 f	2154
	2.8 emb; mode 3; r2-3; n4 f	4350
Hybomys univittatus	2 emb; max 4; n50 f	8855
	2.2 emb; r1-3; n5 f	8856
	2.24 emb [2.35]; mode 2; r1-5; n126 f, 282 emb	2573, 2574
	2.5 emb; mode 2; r1-5; n38 f	2754
	2.7 emb; r2-3; n3 f	2443
	2.7 emb; r2-3; n3 f	8854
	3 yg	3692a
	3 emb ea; n2 f	2452
Hydromys chrysogaster	3; n1 lt	2214
	3.29; ±1.26; median 3; r1-7; n131 lt, 40 f	8103
	4 yg	7209
	mode 4-5 yg; r1-6	7108
Hylomyscus aeta	4.3 emb; r3-6; n3 f	2955

Hylomyscus alleni	2.5 emb; max 3; n6 f	8855
	mode 3; r1-5	2449
	3 emb; n1 f	4495
	3 emb ea; n3 f	7362
	3.0 emb; mode 2; r2-6; n11 f	4406
	3.21 emb; mode 3; r2-4; n19 f	2446a
	3.38 emb; mode 3; r2-5; n26 f	2573, 2574
Hylomyscus denniae	4.75 emb; max 5; n6 f	8855
Hylomyscus stella	2.9 emb; mode 3; r1-4; n14 f	4350
	3 emb; n1 f	10799
	3.2 ps; n19 lt	2445
	3.4; mode 4; r1-6; n35 lt	2754
	3.57 ps; n5 lt	2445
Hyomys goliath	1 yg	7209
Kadarsanomys sodyi	max 4 yg	7672
Lemniscomys barbarus	4; mode 4; r3-6	12128
	4.5 emb; mode 3; r4-5; n4 f	10799
	5 emb; n1 f	2443, 2452
	5.5 ps; r4-7; n2 f	10799
Lemniscomys griselda	2-5	10158
	3 yg; n1 lt, 12 d	331
	3; n1 lt	10154
	4.0 emb, yg; r2-6; n3 lt	299
	5.7 emb; r1-3; n9 f	8965
Lemniscomys striatus	2.7-7.0 emb; varies with season	3645
	3.0 emb; r2-4; n5 f	4925
	3.91 yg; r2-7; n11 lt	8429
	4; mode 5; max 12	5760
	4.25 emb; max 5; n43 f	8855
	4.4 emb; r3-7; n9 f	10799
	4.7 emb, yg; n3 f	187
	4.7 emb; r1-5; n3 f	8856
	4.78 live emb; se0.17; r2-8 1st; n58 f	7790, 7792
	4.8 yg; n5 lt	2574
	mode 5; r1-8	2449
	5 emb; n1 f	2443
	5.14 emb; r3-6; n7 f	8429
	6 ps ea; n2 f	10799
Lenothrix canus	3.0 emb; mode 2; r2-6; n11 f	4406
Leopoldamys edwardsi	2 emb; n1 f	182
	14 ps; n1 lt	10823
Leopoldamys sabanus	3.1 emb; mode 3; r1-7; n54 f	4406
	5 emb; n1 f	5403
	5.0 ps; r4-6; n2 lt	10823
Leporillus conditor	3.5; r3-4; n2	6317
Lophuromys flavopunctatus	r1-16 ps; n43 f	4328
	1.67 emb; r1-2; n3 f	4925
	1.83 emb; mode 2; r1-2; n70 f	2572a, 2573, 2574
	1.99 emb; mode 2; r1-4; n107 f	2572a
	2.0 emb ea; n6 f	2443
	2.0 emb; r1-3; n18 f	8856
	2 emb; max 4; n105 f	8855
	2.16 emb; mode 2; r1-4; n38 f	299
	2.16 emb; mode 2; r1-4; n275 f	2572a
	2.17 emb; mode 2; r1-3; n53 f	2445
	2.17 emb; n276 lt; r2.00-2.47; by month	2570
	2.20 yg; mode 2; r2-3; n5 lt	2445
	2.3; r2-3; n6 lt	2452
	2.33; r2-3; n3 f	337
	2.4 emb; mode 3 51%; r1-4; n43 f	4328
	2.5 emb; r2-3; n2 f	186
	2.5 emb; r2-3; n2 f	8854
	2.5 emb; r1-3; n4 f	10231
	3 emb; n1 f	187

	3.0 emb; n2 f	331
	3.0 yg ea; n2 lt	299
	3.7 emb; n6 f	4495
	4 emb; n1 f	187
	7.0 emb; n2 f	10231
Lophuromys luteogaster	1 emb; max 2; n5 f	8855
	2.0 yg; n10 lt	2572a, 2574
Lophuromys nudicaudus	1.6 emb; r1-2; n5 f	2572a
Lophuromys rahmi	2.0 emb; n2 f	2572a
Lophuromys sikapusi	2 emb; n1 f	2443, 2452
	2 emb; n1 f	2953
	2.13 emb; mode 2; r2-3; n8 f	7362
	2.5 emb; r2-3; n6 f	2154
	3 yg	3692a
	3 emb ea; n2 lt	10799
	3.0 emb; mode 2; r2-5; n13 f	4350
	3.24 emb; mode 3; r1-5; n46 f	2446a, 7790
	4.0; r3-5; n2 lt	1771
	4.7 ps; r4-5; n3 lt	10799
Lophuromys woosnami	2 emb ea; n3 f	2443, 2452
	2 emb; n20 f	8855
Lorentzimys nouhuysi	2 emb; n1 f	7209
Malacomys edwardsi	2.0 emb ea; n4 f	7362
	2.5 emb; r2-4; n4 f	2154
	3 emb; n1 f	10799
Malacomys longipes	2.5 emb; mode 2-3; r1-4; n12 f	2754
	3 emb; n1 f	331
	3 emb; n1 f	2452
	3 emb; r1-5; n27 f	8855, 8856
	3.00 emb; mode 2; r1-5; n41 f	2573, 2574
Mallomys rothschildi	1 emb ea; n2 f	3294a
Mastacomys fuscus	1 yg; n1	9194
	2	9716
	2.0; r1-3; n3 lt, 1 f	1542
	2.0 ea; n7 lt	4042
Maxomys inas	3 emb; n1 f	7161
Maxomys rajah	3.3 emb; mode 3; r2-5; n19 f	4406
Maxomys whiteheadi	3.0 emb; mode 3; r1-6; n57 f	4406
Melomys burtoni	2	10917
	2.5; max 4; n105 lt	3606
	2.85 emb; r1-4; n20 f	7053
Melomys cervinipes	2.7 yg; n3 lt	1255
	2.8; r2-3; n11 lt	2309
Melomys leucogaster	2 yg	7209
Melomys levipes	1-2 yg	7209
	1.45 emb; 1.09 live emb; mode 1; r1-2; n11 f	2836
Melomys lorentzi	1	7209
Melomys moncktoni	2 yg; n1 f	7209
Melomys platyops	2 emb; n1 f	6436a
Melomys rubex	mode 2 yg	7209
	2.00 emb; 1.60 live emb; mode 2; n5 f	2836
Melomys rufescens	mode 2 yg	7209
	2.14 emb; r2-3; n7 f	6436a
	2.27 emb; 1.91 live emb; mode 2; r1-4; n11 f	2836
Mesembriomys gouldii	1.75; r1-3; n4 lt	2131
	2	2214
	2	10917
	2 yg; n1	9194
Mesembriomys macrurus	2	2214
Micromys minutus	2.68; n16 lt	5553
	mode 3-6; r3-12	3571
	3.77 yg; r2-6; n35 lt	3461
	4; n1 lt	2565
	5-9 yg	5424,

	5.40; ±0.16; n62 lt	4396
	6.0 emb; r5-7; n2 f	5968
	max 12	7433
Millardia gleadowi	2.3; r2-3	8723, 8724
Millardia meltada	3.44 yg; r1-8; n27 lt	940
	4.1; r1-10; n21 lt	12034
	5.03; r3-7	8904
	5.53 emb; r3-9; n32 f	730
	5.62; n101 lt	3554
	5.9 yg; r4-10	8723, 8724
	6-8 yg	8500, 8501
	6 ps; n1 f	7382
	6; n1 lt	9175
	6 yg; r5-8; n6 lt	940
	6.15; r4-9	8904
	6.78 ps; r3-11; n23 f	730
	8	107
Mus spp	2-4.5 emb	3645
Mus booduga	4 emb; n1 f	7382
	4-8	9175
	5.58 yg; mode 3-4; r2-11; n96 f	1688
	6-10	8501
	6.1 emb; mode 6; r2-12; n199 f	1688
Mus bufo/minitoides/triton	4 emb; max 7; n40 f	8855
Mus cervicolor	4.4; r2-6	8723
Mus domesticus	1-8	6774
	4.4; ±0.09; n180 lt	4031
	5.25 emb; r3-6; n4 f	10616
	5.33 ps; r4-6; n3 lt	10616
	10.9 yg; se0.7	6864
Mus gratus	3 emb; n1 f	185
	4; n1 lt	187
	4 emb; n1 f	4495
	5 yg; n1 nest	4495
Mus indutus	4.9 emb; r2-8; n17 f	10158
Mus mayori	3 emb; n1 f	8500, 8501
	3 yg; n1 nest	8501
Mus minutoides	2.0 emb; r1-3; n2 f	7362
	3 yg; n1 f	334
	3 yg; n1 f	186
	3.0; r2-4; n4 lt	5685
	3.00 yg; mode 3; r1-5; n18 lt, 7 f	245
	3.38 emb; mode 2; r2-6; n13 f	245
	3.5 emb; r3-5; n4 f	4350
	3.6 yg; n5 lt	2574
	3.85 ps; mode 3; r2-6; n33 f	245
	4 yg; n1 lt	329
	4.0 emb; r2-7; n6 f	9893
	4.0 yg; r1-7; n27 lt, 12 f	11711
	4.3 emb; r4-5; n3 f	2443, 2452
	4.3 ps; r3-6; n12 f	10799
	4.5 emb; n2 f	331
	4.5 emb; mode 4; r2-10; n13 f	10799
	4.67 emb, ps; n3 lt	4329
	4.9; r2-8; n17 lt	10154
	5 emb; n1 f	184
	5.67 emb; r4-8; n3 f	10158
	6.0 yg; r3-8; n4 lt	187
	6.75 emb; r5-10; n4 f	9892
	7 emb; n1 f	187
Mus musculus	1.63 ps; n63 f	1246
	2.7 emb; 1.7 healthy emb; n3 f	6432
	3-8	4344
	3.9 emb	6855

	4.29 yg; n55 lt, 21 f	2725
	4.3 emb; mode 5; r1-7; n82 f	4406
	4.5 emb; sd1.3; n150 f	11437
	4.7-6.3; n244 lt; monthly	844, 845
	4.8 emb; se0.09; n292 f	10151
	4.9 ps; n24 f	6432
	5 emb; n1 f	8907
	5 emb; n1 f	2341
	5 yg; n1 f	11956
	5 emb ea; n2 f	4409
	mode 5-6; r4-8	8500, 8501
	5.0-10.7 ps; n58 f; monthly	844, 845
	5.2 yg; n13 lt, 6 f	8102
	5.3 yg; 95%ci 5.1-5.5; 5.2 wean; 95%ci 5.0-5.4; n205 lt	2459
	5.37 emb; se0.10	9323
	5.4; ±0.04; n1400 lt	4031
	5.7 emb; n4 f	1547
	5.7 emb; n~550 f	10233
	5.76 emb; n582 f	6254
	5.9 emb; mode 6; r1-10; n48 f	1702
	mode 6 emb; r3-9; n10 f	1702
	6.0	10632
	6.0 emb; r5-7; n2 f	6298
	6.1 emb; ±1.5; r4-9	8804
	6.12 yg; sd2.32; mode 6-7; r1-12; n431 lt	11490
	6.17; r5-9; n6 f	2547
	6.3 ps; r5-8; n6 f	6298
	6.5; n41 lt	8768
	6.9 cl; n8 f	6432
	7.5; r5-9; n4 lt	1785
	7.5 emb; n8 f	847
	max 8 yg	4347
	9.1; se0.7; r1-9; n16 lt	1315
	9.78	7704
	r12.3-14.3; n2624 lt; across months	2723
Mus platythrix	3-10 emb	8723, 8724
	3.8 yg; r2-7; n15 lt	8924
	4.56 emb; mode 4,6; r2-8; n22 f	8924
	7.6 yg; mode 7; r2-12; n87 f	1688
	8.3 emb; mode 6,8,12; r4-14; n83 f	1688
Mus setulosus	3.25 emb; mode 3; r2-5; n8 f	2754
	5 emb ea; n3 f	2955
Mus spretus	5.3 emb; sd1.37; mode 5; r2-10; n193 f	11222a
Mus tenellus	2 emb; 1 f	4344, 4345
Mus triton	4.5 emb; mode 5; r2-6; n14 f	7790
	6; r5-7	4329
Mylomys dybowskii	4.0 emb; r3-5; n3 f	2443, 2444
	4.32 emb; r2-6; n28 f	2446a, 7790
Nesokia indica	r2-8 emb; n8 f	6293
	4.0; r1-7; n2 lt	12128
	4.07 emb; se0.54; n13 f	728
	4.18; n101 lt	3554
	7 emb; n1 f	10616
Niviventer cremoriventer	3.7 emb; mode 3; r2-5; n10 f	4406
Niviventer eha	5 emb; r3-7; n2 f	7382
Niviventer fulvescens	2.5 emb; r2-3; n2 f	182
	5 emb; r3-6; n3 f	7382
Niviventer niviventer	mode 1-7	12105
	5.0 recent ps; r4-6; n2 lt	10824
	6 emb; n1 f	7382
Notomys alexis	mode 3 emb; r2-5	3258
	3.0; sd1.0; mode 3; n23 lt	4355
	4.0; sd0.96; mode 4-5; n14 lt	4355
	4.0; mode 4; r1-9; n176 lt	10117

	4.04 yg; mode 4; r2-8; n3 f	1255
	4.4; n8	2532
	4.6 emb; se0.3; n150 f	1255
Notomys cervinus	1-4 yg; 50% of 3; n8 lt	2132
	2.5	670
	3.5; sd0.84; mode 4; r2-4; n6 lt	4355
Notomys fuscus	2.7; mode 3; r1-5; n74 lt	427
Notomys mitchellii	2-3	12128
	3	670
	3.6; sd0.85; mode 4; r2-5; 9 lt	4355
	4	10917
	4; r4-6	10917
Oenomys hypoxanthus	2.25 emb; max 4; n54 f	8855
	2.3 emb; r1.85-2.71; n130 f; across months	2570
	2.5 emb; mode 3; r1-5; n33 f	2754
	2.61 emb; mode 2; r1-5; n13 f	2573
	2.71 emb; mode 3; n14 f	4495
	3-4; max 6	5760
	3; n1 lt	187
	3.17 yg; mode 2-3; r1-5; n18 lt	2573, 2574
	3.4 emb; r3-4; n9 f	2452
	5-8	9468
Pachyuromys dupresi	5; n1 lt	8614
Pelomys fallax	2-3 yg	5760
	3 yg; n1 f	5051
	4 emb; max 8; n21 f	8855
	5 emb; n1 f	334
	7.9 emb; r6-9; n12 f	9893
	8.0 emb; n2 f	331
Pelomys minor	2-3 yg	5760
Phloeomys cumingi	1	9562
Pithecheir parvus	mode 2 yg	6448
Pogonomelomys sevia	mode 1 yg	7209
	1.00 emb, 0.75 live emb; mode 1; r1; n4 f	2836
Pogonomys macrourus	2.45 emb, 2.36 live emb; mode 2; r2-3; n11 f	2836
	4.00 emb/ps; sd2.00; n3 f	7114
Pogonomys mollipilosus	2.45 emb, 2.36 live emb; mode 2-3; r2-3; n11 f	2836
Pogonomys sylvestris	2.64 emb, 2.45 live emb; mode 3; r2-3; n11 f	2836
Praomys daltoni	3.3-6.5 emb; varies with season	3645
	3.93 yg; 4 mode; r1-7	246
	5.55 emb; mode 5; r3-10; n18 f	245, 246, 247
	6.0 emb; r5-7; n4 f	10799
	7 ps ea; n2 f	10799
Praomys delectorum	3.67 emb; r2-6; n12 f	5051
	4 emb ea; n3 f	337
	6 yg; n1 lt	5051
Praomys derooi	3.75; r2-5	4354a
Praomys erythroleucus	4.0; n2 lt	12128
	7-15	5058
	11.8 emb; r7-17; n17 f	1215
	12.1 cl; r5-19; n51 f	1215
Praomys fumatus	3.0 emb; r2-4; n2 f	4925
Praomys huberti/erythroleucus	14.7 emb; max 28; n21 f	3646
Praomys jacksoni	78% 2-4 ps; r1-9; n18 f	4329
	2-6	5760
	2 yg; n1 lt	187
	mode 3; r1-8	2449
	3.0 emb; r2-4; n6 f	8856
	3.19 emb; mode 3; r1-5; n99 f	2573, 2574
	3.75 emb; max 8; n65 f	8855
	3.75 emb; n102 f	2570
	3.8 emb; mode 4; r2-5; n11 f	2754
	mode 4 emb	4329
	4 emb; n1 f	334

	4.0 emb; n3 f	4495
	5 emb; n1 f	331
Praomys morio	mode 3-4 yg; r2-6	2443, 2444, 2452
	3 emb; n1 f	2953
	3.0 yg; mode 4; r1-4; n9 lt	2445
	3.25 emb; mode 3; r1-5; n44 f	2754
	3.33 emb; mode 3; r2-5; n67 f	2445
	4.02; mode 4-5; r2-6; n61 lt	2954
	4.4 emb; r3-5; n5 f	2443, 2444, 2452
Praomys natalensis	2-4	9945
	4.25; r1-8; n4 lt	12128
	4.7 emb; max 12; n33 f	8855
	5 emb; n1 f	10231
	5.1; r1-13; n28 lt	4588
	5.4; sd2.9; n5 lt	8385
	mode 6-12; max 22	10158
	6.5 yg; r6-7; n2 nests	334
	7 ps; n1 f	10799
	7.32	8091
	7.5 ps, emb; sd0.9; mode 7; r7-9; n6 f	3821
	8 emb; n1 f	334
	8; max 16	8034
	8.0 emb; r6-11; n4 f	6612
	8.1 emb; r2-16; n30 f	299
	8.2 emb; r7-9; n5 f	2454
	8.4 weaned; r2-14; n16 lt	7166
	8.53 yg; mode 7; r1-17; n19 lt, 4 f	7166
	9.0 emb ea; n2 f	2154
	9.2 emb; r3-16; n5 f	4495
	9.4 emb; r7-13; small f	4329
	9.5; ±0.57; r6-15; n19 f	9892
	9.5 emb	1954
	9.97; r7-22; n34 lt	10154
	10.0 emb; r9-11; n2 f	7362
	10.3 yg; r4-21; n57 lt	4579
	10.4 emb; r7-13; n5 f	4925
	11.0 emb; se0.46; r2-20; n72 f	9893
	11.2 emb; r3-16	1704
	11.27 yg; sd2.28; r8-17; n11 lt	514
	11.63 emb; r5-17; n41 f	8965
	11.99 emb; r5-24; n76 f	10712
	mode 12; r1-19	2449, 2452
	12.0 emb; r9-14; n3 f	10799
	12.1 emb; se0.26; r6-19; n41 f	7790, 7791
	12.4 emb; r1-20; n95 f	1789
	12.8 emb; r8-21; n43 f	1788
	13.4 emb; r7-19; n34 f	9014
	13.6 emb; r11-17; large f	4329
	14.0 emb; r13-15; n2 f	184
	15 emb; n1 f	187
Praomys tullbergi	2.71 ps; r1-4; n7 f	10799
	3.0 yg; mode 3; r1-6; n103 lt	4353
	3.4 emb; mode 4; r2-5; n16 f	10799
	3.45 emb; r2-4; n15 f	2154
	3.46 emb; mode 3; r3-4; n11 f	7362
	3.6 emb; r2-6; n36 f	4350, 4353
	3.67 emb; r3-4; n3 f	4925
	4 yg; n1 f	186
Pseudohydromys murinus	1 emb; n1 f	6436a
Pseudomys albocinereus	3.8; sd0.96; mode 4; r2-5; n21 lt	4355
	4 yg; n1	9194
Pseudomys apodemoides	3-6 yg	3259
	4-7 emb	3259

Pseudomys australis	2.5 yg; r2-3; n2 f	11495
	3.0; sd0.52; mode 3; r2-4; n10 lt	4355
	3.6; mode 4; r1-7; n140 lt	10117
Pseudomys delicatulus	3 yg; n1	9194
	3.25 emb; r3-4; n8 f	10704
Pseudomys desertor	3.0; sd0.66; mode 3; r1-4; n24 lt	4355
Pseudomys gracilicaudatus	2-4 cl	10708
Pseudomys hermannsburgensis	2 yg; n1 f	11495
Pseudomys higginsi	2.3; r2-3; n3 lt	4042
Pseudomys nanus	3 yg; n1	9194
Pseudomys novaehollandiae	2.5; r2-3; n2 lt, 1 f	4355
	4.0; sd1.03; r1-6; n95 lt	5643
	4.56 emb; r2-6	5645
	5 emb; n1 f	9194
Pseudomys praeconis	3.5; r3-4; n2 lt	9189
Pseudomys shortridgei	3 yg ea; n6 lt	4355
Punomys lemminus	2 emb; n1 f	7983
Rattus argentiventer	6.0 emb; r5-7; n9 f	4404
	6.94 yg; se0.19; mode 7; r1-18; n395 lt	6162
	10.81 ps; se0.29; 10.63 emb; se0.29; r1-18; n118 f	6162
	11.06 ps; se0.31; 10.87 emb; se0.31; r2-18; n103 f	6162
	11.50 ps; se0.25; 11.30 emb; se0.25; r4-18; n129 f	6162
	12.70 ps; se0.55; 12.50 emb; se0.56; r4-18; n40 f	6162
Rattus exulans	2.5 live emb; n201 f	5216
	2.5-4.5 yg	5217
	3; r1-5	7209
	3.0 emb; r1-5; n9 f	7886
	3.1; r1-6; n14 f	6436a
	3.12 emb; sd0.70; 2.88 live emb; sd1.05; n17 f	2839
	3.2 ps; r3-4; n5	10702
	3.35 emb; 3.09 live emb; mode 3; r1-5; n75 f	2836
	3.43 emb; sd0.65; 3.14 live emb; sd0.94; n51 f	2839
	3.6 yg; sd0.21; r2-6; n30 lt	1330, 11434
	3.66 emb; sd1.06; 3.50 live emb; sd1.30; n50 f	2839
	3.69 live emb; sd1.11; n23 f	7114
	3.8 emb	10578
	3.8 yg; mode 4; r1-10; n221 lt, 37 f	2911
	3.81 emb; mode 3-4; r1-8; n87 f	11832
	3.9; r3-6; n7	4821
	3.96 emb; sd1.24; mode 4; r1-9; n152 f	11434
	3.97, 4.07; n59 lt, 2 groups	11833
	3.97 yg; mode 4; r2-7; n212 f	11832
	3.98 emb; ±1.05; mode 4; r2-6; n62 f	4408, 4409
	4.0 emb; 3.7 live emb; n7 f	10650
	4.0 emb; r1-7; n25 f	10702
	4.07 emb; r2-19	10545
	4.07 ps/emb; mode 3-4; r1-11; n59 lt	11832
	4.3 emb; mode 4; r1-8; n98 f	4406
	4.48 ps; mode 3-5; r2-11; n125 f	11832
	4.5 emb; mode 4; r1-7; n38 f	4404
	4.61 emb, ps; sd1.37; n13 f	7114
	4.7 emb; mode 5; r2-6; n10 f	11477
	4.7 emb; n~140 f	10955
Rattus fuscipes	3 yg; n1	9194
	3.81 yg; mode 4; r1-7; n79 lt	10696
	4 yg; n1 lt	3260
	4.2; r3-7	9188
	4.21 yg; n24 lt	10696
	4.7 emb, yg; r2-7; n45 f	4988, 4990, 10707
	5.0 emb, ps; r3-7; n11 f	10709
	6 emb ea; n2 f	3260
	6.0 cl; r2-13; n16	10709
	6.1 new cl; n16	1254
	6.2 cl; r1-13; n68	4990, 10707

	7 cl; r3-11	10696
Rattus hoffmanni	4-5 emb	7675a
Rattus hoxaensis	5 ps; n1 lt	10824
Rattus leucopus	2-5	7209
	3 cl; n1	10709
	3.35 emb; r2-5; n22 f	10702
	3.5 new cl; se0.4; n5	1254
	5 emb; n1 f	9194
Rattus lutreolus	3.85; r3-5	4041
	3.9 emb; sd1.0; r1-6; n16 f	1208
	4.1 emb; sd0.9; r3-6; n11 f	4041
	5.0 emb; sd2.0; r1-11; n35 f	1208
	6 yg; n1	9194
	7.0 new cl; se1.0; n5	1254
	12.0 ps; sd4.2; r7-24; n21	1208
	13.4 ps; sd8.2; r2-31; n68	1208
Rattus mindoriensis	4-9 emb	3025
Rattus muelleri	3 emb; n1 f	2313
	3 emb; n1 f	2313
	3.8 emb; mode 3; r1-9; n28 f	4406
Rattus niobe	2; r1-3	7209
	2.0 emb; r1-3; n9 f	10702
	2.00 emb; 1.91 live emb; mode 2; r1-3; n11 f	2836
	2.7 ps; r2-4; n3	10702
Rattus norvegicus	2 emb; n1 f	4483
	3-9 emb; n9 f	4409
	5-14	2555
	5-20	7433
	5.2; r3-8; n6	4821
	5.8	2455
	6.4; se1.0; n20 lt	1849
	6.84 emb; n270 medium f	10456
	7 emb; n1 f	1590
	7-8	8393
	mode 7-8; r5-12	4645
	7.0-9.1; r of x	7033
	7.9; n83	2317b
	8.09 yg; sd2.81; n11 f	956
	8.2 emb; se0.309; n59 f	2317a
	8.4 emb; n402 f	2317b
	8.7 emb; sd2.1; r5-12; n21 f	11435
	8.83 emb; ±2.62; r5-17; n18 f	6242
	8.9 emb; n288 f	2317b
	mode 9-10	9175
	9.0; n216 lt	4485
	9.0; r8-12; n7	8802
	9.00 emb; n2768 large f	10456
	9.036; r1-17; n775 lt	6399
	9.28 emb; se0.28; n105 f	2321
	9.3 emb; r3-16; n22 f	7886
	9.35 emb, ps; 61% 7-9; r2-19; 10.39 cl; r7-17; n17 f	6242
	9.5 emb; r7-12; n4 f	2547
	9.6 emb; r6-13; n95 f	4234
	9.69 emb; sd2.66; n29 f	956
	9.9 emb; 9.0 late gest emb; n212 f	2996
	9.9 emb; n345	2317b
	10 emb; n1 f	5111
	10.0 ps; r2-19; n12 lt	6242
	10.1 emb; se0.188; n213 lt	2317a
	10.4 yg; n32 f	5503
	10.5; n8 lt	6238
	11.2 cl; n95 f	4234
	max 12	7209

Rattus praetor	3 emb; r2-5	6436a
	3 emb; r1-6	7209
	3.10 emb; 2.60 live emb; mode 1-4; r1-6; n10 f	2836
	4.5 emb; r2-7; n4 f	10702
	5 yg; n1 lt	10702
Rattus rattus	1-9	8723
	1-9	8723
	2.65; mode 3-4; r1-5; n46 lt	8861
	3-4; r3-8	8501
	3-8 emb; n15 f	4409
	3.6 live emb; n97 f	5216
	4 emb; n1 f	1590
	mode 4-9; r3-15	3571
	4.17 yg; r1-9; n6 lt	9265
	4.49 emb; se0.10; r1-8; n176 f	899
	4.83 emb; r2-9; n6 f	9265
	4.9; r3-7; n8	4821
	5 emb; n1 f	5403
	5-10	2555
	5 emb; r3-6; n6 f	7886
	5.2 emb; n19 f	10650
	5.25 emb; r1-11; n357 f	4404
	5.4	2455
	5.6 emb; n118 medium f	10456
	5.67 emb; r4-9; n3 f	10616
	5.8 emb	8423
	5.88 yg; se0.59; n99 lt	1765
	5.89 emb (6-8 cl); se0.31; n194 f	1765
	5.89 emb; n3015 f	11632
	5.9; n21 lt	858
	6	7209
	mode 6-7; max 17	5760
	6-8 emb	10823
	6 emb; n54 f	2452
	6.05 emb; se0.23; mode 6; r1-11; n74 f	11475a
	6.1; ±0.233; r1-15	11475
	6.4; r4-9; n9 lt	5315
	6.44 wean; r1-9	10545
	6.60; se0.18; mode 5-7; r1-13	8902
	6.7 emb; r4-8; n6 f	10702
	6.77 emb; r3-11; n13 f	2590
	6.8 emb; n99 f	2316
	6.9; ±0.147; r1-16	11475
	7 emb; n1 f	182
	r7-8 emb; n3 f	10824
	7.2 yg; ±0.5; r4-10; n13 lt, 9 f	2091
	7.3 emb; n978 large f	10456
	7.5 emb; ±0.181	11475
	8-20	7433
	mode 8; r3-10; n13 lt	3133
	8 ps; n1 f	10616
	9 emb; n1 f	185
	9 emb; n1 f	7498a
	9.23 emb; r2-18	10545
	9.5 emb; r9-10; n2 f	1590
	10.5 yg; r9-12; n8 lt	6237
Rattus sordidus	1.5 emb; r1-2 n4 f	7209
	4.5 ps; r4-5; n2	10702
	5.8 emb; r5-9; n14 f	10702
	6 yg; n1 f	11495
	6 emb; max 13	7053
	7.4 new cl; n19	1254
Rattus steini	2.7 emb; r1-4; n7 f	10702
	3.4 ps; r2-5; n8	10702
	4 yg; n1 lt	10702

	4.67 emb, ps; sd1.37; n6 f	7114
Rattus taerae	4.5 emb; r4-5; n2 f	4925
Rattus tiomanicus	3.0 yg; n8	11885
	3.7 yg; n24 nests	11885
	3.98 emb; r1-10; n82 f	4404
	4.39 emb; mode 4; r2-7; n442 f	11885
Rattus tunneyi	5.5 new cl; se0.6; n5	1254
	7.5; r4-11; n263 lt	3606
Rattus turkestanicus	3.8 emb; r2-6; n5 f	7382
	4-6 emb	9175
Rattus verecundus	2.13 emb; mode 2; r1-3; n15 f	2836
	2.3 emb; r2-3; n3 f	10702
Rhabdomys pumilio	r2-31 cl; n20 f	2530
	3-10	1814
	4-12	5760
	4.0 emb; r3-5; n2 f	4925
	4.3 emb; r3-5; n3 f	2452
	4.36 emb; ±0.24; n42 primiparous f	2298
	4.5; ps: 5.0; r2-7	4329
	4.75 emb; r3-6; n4 f	299
	4.90 healthy emb	2298
	5.0; r3-9; n11 lt	10154
	5.00 emb; ±0.12; n85 parous f	2298
	5.43 emb; r2-10; n80 f	10712
	5.75 emb; r4-7; n4 f	337
	5.9; r2-9	10158
	6-7	9945
	6 emb; n1 f	334
	6.19 yg; n20 lt	8384
	6.5 yg; r3-10; n42 lt	1335
	7 emb; n1 f	7822, 7823
Stochomys defua	2.75; mode 3; r2-3; n4 lt	7362
Stochomys longicaudatus	2 emb; r1-4; n18 f	8855
	2.4 emb; mode 2; r1-4; n30 f	2573, 2574
	2.5 emb; r1-4; mode 2; n42 f	2754
	3 yg ea; n2 lt	4354a
Thallomys paedulcus	2 yg; n1 f	4329
	2.5 emb; r1-4; n2 f	9945
	2.7 yg	7822
	2.7; n23 lt	7168
	3 yg; n1 f	9945
	3.6; r2-5; n5 lt	10154, 10158
	4 emb; n1 f	4925
	4 emb; n1 f	4329
	4 yg; n1 f	1836
	7 emb; n1 f	9893
Thamnomys venustus	1.42 emb; r1-2; n12 f	2446a
Uranomys ruddi	2.6-5.7 yg	7983
	2.9; r1-4	4350
	3.96 emb; n112 f	752a
	4 ea; n3 lt	4350
	5 emb; r4-6; n4 f	10799
	8 ps; n1	4329
Uromys caudimaculatus	1 yg; n1	9194
	1-3	7209
	1-3	10698
Vandeleuria nolthenii	4 yg ea; n2 nests	8501
Vandeleuria oleracea	3; 4 occasional	8501
	mode 3-4	8500
Xeromys myoides	2	10698
Zelotomys hildegardeae	5 emb; n1 f	2444
	5 emb; n1 f	2452
	7 emb; n1 f	4925
Zelotomys woosnami	4.7; r4-5; n3 lt	947

	5.0	10154
	7.0 emb; r5-11; n3 f	10158
Zyzomys argurus	2.7 emb; sd1.2; r1-6; n17 f	1192
	4.3 ps; sd2.8; r1-11; n24	1192
SEXUAL MATURITY		
Acomys cahirinus	f: vag perf: 45 d; ±2.7; n26 f; 36.2 g; ±1.66	8326
	f: 1st mate: 39-50 d	2568
	f: 1st birth: 103 d; ±4.04	8326
	f: 80-84 mm	4329
	m: sperm present: 56 d; n1; 90-94 mm	4329
	m: 46 d	1080
	m: 2-3 mo	2568
	f/m: 2-3 mo	5760
Aethomys chrysophilus	f: vag perf: 56-70 d	1333
	f: 1st birth: 138 d; r108-187	1333
	f: 110-119 mm	4329
	m: 120-129 mm	4329
Aethomys hindei	f: vag perf: 70 d; ±3; r54-84; r101.9-108.2 g	8083
	m: scrotal testes: 44-51 d; n5; r54.9-109 g	8083
Apodemus agrarius	f: 1st cl: 15 g; 50% 18.16	8330, 8333
	m: testes large: 17 g; 50% 17.79	8330, 8333
	m: sire yg: 30% 16.1-18.0 g; 100% 22.1	8332
	f/m: 1st mate: 2.5 mo	4645
	f/m: > 22 g	8343
Apodemus argenteus	f/m: 2+ mo	3546
Apodemus flavicollis	f: 1st cl: 10 g	54
	f: 1st cl: 50% 23.81 g	8333
	f: ~1 mo; 14-16 g	10358
	m: testes large: 50% 26.96 g	8333
	m: testes large: ~1 mo	10358
	m: sperm present: 15-20 g	54
	m: 1 yr	6548
	f/m: 7-8 wk	3571
Apodemus microps	f: 1st cl: 50% 12.24 g	8333
	f: 1st preg: 16 g; rarely 10.1-14	10480
	f: 18 g; active	10358
	m: testes large: 50% 14.73 g	8333
Apodemus speciosus	f/m: season of birth	12024
Apodemus sylvaticus	f: 1st cl: 50% 16.05 g	8333
	f: 1st birth: 80-90 d	4645
	f: 16 g	10358
	m: est 14 g	516
	m: testes large: 60-65 d	3070
	m: testes large: 21% 13.0 g; 91% 21	8333
	m: 1 yr	6548
Arvicanthis niloticus	f: 1st repro: 40-50 d	8811
	f: 1st birth: 60 d	2450
	f: 3 mo	4484
	f: 20-35 g	7793
	f: 45 g	10712
	f: 55-65 g	7793
	m: scrotal testes: 40-50 d	2450
	m: testes large: ~6 mo	7600
	m: 33-40 g	7793
	m: 60-80 g	7793
Bandicota bengalensis	f: 1st birth: 3 mo	9175
	f: 60 d	9429
	f: vag perf: 50% 40-79 g; 100% > 90	3554
	f: vag perf: 90-100 d	10268
	m: scrotal testes: 51 d	9429
	m: testes descend: 58-70 d; n3	10268
	m: visible epididymal tubules: 70-159 g	3554

Bandicota indica	f: ovarian mass: 350 g	8794
	f: vag perf: 190-200 d, ~300 g	10268
	f: 1st birth: 130-415 d	10370
	m: epididymal sperm: 50% 150 d	10370
	m: testes descend: 160 d; n1	10268
Berylmys bowersi	f: 1st preg: 290 g	4406
	m: fertile: 50% 166.5 g; 95% 338.1	4406
Chiropodomys gliroides	f: 1st preg: 11 g	4406
	m: fertile: 50% 7.2 g; 95% 165	4406
Dasymys incomtus	f: fecund: 127 mm	4329
	f/m: 60-80 d	2570
Grammomys cometes	f: 1st preg f: 105-109 mm	4329
	f: 1st multiparous: 115-119 mm	4329
Grammomys dolichurus	f: 1st estrus: 58-69 d	988
	f: 1st mate: 75-94 d	988
	f: ps present: min 95-99 mm; mode 105	4329
Grammomys rutilans	f: 1st preg: 40 g	4350
	m: scrotal testes: 40-45 g	4350
Hybomys trivirgatus	f: 1st preg: 40 g	2154
Hybomys univittatus	f: 1st mate: 36-90 d; most 4.5 mo	2446a
	f/m: 45 g	2573
Hydromys chrysogaster	f: vag perf: ~95 d	8103
	f: 1st estrus: 124-130 d; 507.00 g; se46.59; n8	8103
	f: 1st birth: 158 d	8103
	f: 425 g	7108
	m: scrotal testes: 90-120 d; 518.67 g; se48.05; n12	8103
	m: sperm present: 130-140 d	8103
	m: 400-600 g	7108
Hylomyscus alleni	f: 1st preg: 110 g	4406
	m: fertile: 50% 67.4 g; 95% 145.4	4406
	m: scrotal testes: 12-14 g	2573
Hylomyscus stella	f: 1st mate: 38 d; n1	2445
	f: 1st preg: 15 g	4350
	m: scrotal testes: 15 g	4350
Leggadina forresti	f/m: 90 d	11493
Leggadina lakedownensis	f/m: 90 d	11493
Lemniscomys striatus	f: fertile: 3 mo	11614
	f: vag perf: 7 wk	7477, 8429
	f: 1st mate: ~1 yr	5760
	f: 1st preg: 5-9 mo	11614
	f: 1st birth: 32 g	4350
	m: fertile: 2 mo+	11614
	m: scrotal testes: 30 g	4350
	f/m: 70-90 d	2570
Leopoldamys sabanus	f: 1st preg: 190 g	4406
Leporillus conditor	f/m: 240 d	11493
Lophuromys flavopunctatus	f: 1st preg: 45 g	2572a
	m: scrotal testes: 45 g	2573
	m: testes large: 50 g	2572a
	f/m: 50-70 d	2570
Lophuromys sikapusi	f: 1st preg: 40-50 g	4350
Lophuromys woosnami	f: 1st preg: 39 g	2572a
	m: testes large: 40-44 g	2572a
Malacomys edwardsi	f: 1st preg: 61 g	2154
	m: testes large: ~60 g	2154
Malacomys longipes	m: scrotal testes: 90 g	2573
Maxomys burtoni?	f: 1st repro: 50% 52-60 g	5647
	m: spermatogenesis: 50% 49-81 g	5647
Maxomys rajah	f: 1st preg: 130 g	4406
	m: fertile: 50% 67.5 g; 95% 145.1	4406
Maxomys whiteheadi	f: 1st preg: 30 g	4406
	m: fertile: 50% 35.6 g; 95% 44.7	4406
Melomys cervinipes	f: vag perf: 80 d; se5	1255
	f: 1st mate: 7 mo	2309

Micromys minutus	f: 1st birth: 52 d; n1	3461
	f/m: 4.5-6 wk	3571
Millardia meltada	f: vag perf: 50 d; r35-66	12034
	f: 1st birth: 3-4.5 mo; r108-335 d; n6 f	940
	f: 30-49 g	3554
	f: 40 g	730
	m: epididymal sperm: 45 g	8904
	m: testes large: 50 g	730
	m: 30-59 g	3554
Mus domesticus	f/m: 4-8 wk	11217
	f: 1st birth: 177 d; ±6.2; n57	4031
Mus minutoides	f: 1st birth: 62 d; n1	11711
	f/m: 1st mate: ~42 d	10158
	f/m: 70-84 d	11711
Mus musculus	f: vag perf: 42.2 d; se1.4; n18 with males	2725
	f: vag perf: 57.2 d; se1.7; n20	2725
	f: vag perf: 68.4 d; se2.2; n17 with females	2725
	f: vag perf: 8-12 g; n258/546	9323
	f: 1st cl: 8-12 g; n323/546	9323
	f: 1st estrus: 27-47 d	2726
	f: 1st estrus: 38.1-39.9 d; n98, 5 samples	6935
	f: 1st estrus: < 8 wk	8501
	f: 1st preg: 5 g	4406
	f: 1st birth: 102 d; ±1.1; n317	4031
	f: <2 mo	7209
	f: 2.5 mo	10151
	f: 12 wk	9175
	m: scrotal testes: 8-12 g; n87/151	9323
	m: fertile: 50% 9.0 g; 95% 12.1	4406
	f/m: 1st mate: 2 mo	4645
Mus spretus	f: 1st cl 1/18 3-5 wk; 100% 5 wk)	2814
	f: 1st preg: 28-35 d	2814
	f: 6-7 wk	11222a
	m: sperm present: 3-5 wk	2814
	m: 8 wk	11222a
Mus triton	f: 70-74 mm	4329
	m: > 74 mm	4329
	f/m: 50 d	2570
Nesokia indica	f: 63-89 g	3554
	m: 110 g	3554
Niviventer cremoriventer	f: 1st preg: 40 g	4406
	m: fertile: 50% 19.0 g	4406
Notomys alexis	f: vag perf: 40-55 d	1255, 10792
	f: 1st estrus: 53.0-82.3 d; n42, 5 gps	10792
	f: 1st cl: min 44 d; mode >54 d	1255
	m: sperm present: 60 d; 22.5 g min	1255
	f/m: 1st repro: 12 wk	10117
	f/m: 2-3 mo	4355
	f/m: 80 d	11493
Notomys cervinus	f/m: 180 d	11493
Notomys fuscus	f: 1st conceive: 70 d	427
	f: 1st repro: 105 d	427
	f/m: 80 d	11493
Notomys mitchellii	f/m: 90 d	11493
Oenomys hypoxanthus	f/m: 100-130 d	2570
Praomys daltoni	f/m: 1st birth: 4.5-5.5 mo	247
Praomys erythroleucus	f: 40 g	1215
	m: 45 g	1215
	f/m: 1st breed: 4-5 mo	5057, 5058
Praomys jacksoni	f: 1st cl or ps: 90-94 mm	4329
	m: 18% 80-89 mm; 100% >109 mm	4329
	m: testes large: 30 g	2573
	f/m: 50-90 d	2570
	f/m: ~2.5 mo	5760

Praomys morio	f: 1st mate: 2.5 mo	2445
	f: 1st birth: 72-211 d	2954
Praomys natalensis	f: vag perf: 38-47 d	8034
	f: vag perf: 76 d; ±5.88; r64-145 d; n12	5366
	f: 1st birth: 77 d; n1	7166
	f: 1st birth: 99.2 d; n231	8034
	f: 35 g	7791
	m: epididymal sperm: > 40 d	514
	m: 35 g	7791
	f/m: 104 d; ~39 g	8091
Praomys tullbergi	f: 1st conceive: 62 d; 27 g	4350, 4353
	f: 1st preg: 34 g	2154
	f: 1st birth: 108 d; r86-139; n13	4353
	f: 1st birth: 4 mo	4354
	m: scrotal testes: 25 g, sometimes 15 g	4350
	m: scrotal testes: 40-60 d	4353
	m: testes large: 40 g	2154
	m: ~100 d	4353
Pseudomys albocinereus	f/m: 2-3 mo	4355
Pseudomys australis	f/m: 180 d	11493
Pseudomys desertor	f/m: 2-3 mo	4355
	f/m: 80 d	11493
Pseudomys hermannsburgensis	f/m: 90 d	11493
Pseudomys higginsi	f/m: 270 d	11493
Pseudomys nanus	f/m: 100 d	11493
Pseudomys novaehollandiae	f: vag perf: 13.1 wk; sd4.59; r7-23; n30	5643, 5644
	f: 10.5 g; 7 wk; lab; yr of birth; wild	5645
	f: 14.6 g; sd2.06; r10.5-19.0; n29	5643
	m: 1st mate: 20.2 wk; sd9.03; r11-36; n11	5643, 5644
	m: 14.275 g; r13.6-14.8; n4	5643
	f/m: 120 d	11493
Pseudomys shortridgei	f/m: 10-11 mo	4355
Rattus argentiventer	f: vag perf: 48.6 d; ±0.7; r43-53	6162
	f: vag perf: 32.9 d; ±0.4; 50% 31; r26-46; n130	6162
	f: 1st mate: r43-53 d	6162
	f: 1st preg: ~85 g	4404
	m: scrotal testes: 55% by 30 d; r26-82; n98	6162
	m: sperm present: 89.6 d; ±2.5; r85-105	6162
Rattus exulans	f: vag perf: min 48 d; mode 8 wk	11833
	f: vag perf: 40 g	11833
	f: 1st preg: 22.9 g; se2.1	4404
	f: 1st birth: mode 30+ wk; min 17	11833
	f: 1st birth: 137.1 d; r72-239	2911
	f: median 39.2 g; 95%ci 33.2-46.2	10650
	m: scrotal testes: 6-7 wk	11833
	m: fertile: 12 wk	11833
	m: 1st repro: 63 d	11833
	m: median 36.6 g; 95%ci 30.2-44.3	10650
Rattus fuscipes	f: vag perf: 21-56 d	10706, 10707
	f: vag perf: 46 d; r35-57	10696
	f: 1st ovulation: 56-66 d	10706
	f: 1st conceive: 16-17 wk	10696
	f: 1st birth: 4.5 mo	10696
	m: mature sperm: 8-9 wk	10706
	m: epididymal sperm: 10-11 wk	10706
	m: 1st sire: 2.5 mo	10696
	m: 1st sire: 17 wk	11494
Rattus leucopus	f: 1st birth: 23 wk	11494
	m: 1st sire: 16 wk	11494
Rattus lutreolus	f: vag perf: 34 d; r26-45	3437
	f: 1st conceive: 73 d	3437
	f: 1st birth: 16 wk	11494
	f/m: >100 g; immature <75 g	614
Rattus mindoriensis	f: 120 g	3025
Rattus muelleri	f: 1st preg: 220 g	4406

	m: fertile: 50% 172 g; 95% 387.7 g	4406
Rattus norvegicus	f: vag perf: 42.5 d; ±5.7; 81.1 g; ±9.2	1849
	f: 1st estrus: 45.7 d; ±6.9; 110.7 g; ±16.5	1849
	f: 1st conceive: 55.7 d; ±22.6; 103.2 g; ±23.3	1849
	f: 1st preg: 105 g	8393
	f: 1st birth: 3 mo	4645
	m: 95-104 g	6399
	m: sperm present: 42.1 d; ±1.9; 111.5 g; ±17.3	1849
	m: 1st mate: 64.6 d; ±20.5; 185.5 g; ±62.8	1849
	f/m: 2 mo	2555
Rattus rattus	f: vag pref: 50-70 d; n3	10268
	f: 1st mate: 12 wk	9175
	f: 1st conceive: 50% 80 g	3250
	f: 1st preg: 62-115 g	4404
	f: 1st birth: 95-143 d	3133
	f: 1st birth: 5-6 mo	8500, 8501
	f: 3-4 mo	11475a
	f: 1.5-6 mo	10545
	f: 7 mo, 165 g	5315
	f: median 76.7 g; 95%ci 70.9-82.9	10650
	f: 78-97 g; n3 locations	11475
	m: testes descend: 40-50 d; n3	10268
	m: sperm present: 24% 80-100 g; 74% 120-140	3250
	m: median 107.9 g; 95%ci 101.2-115.0	10650
	m: 80 g	8902
	m: 6 mo; 165 g	5315
	f/m: 3 mo	2555
	f/m: 3-4 mo	5760
	f/m: 3-6 mo	3571
Rattus sordidus	f: vag perf: 37 d; 35 g	7053
	f: 1st estrus: 63-70 d; 50-55 g	7053
	f: 1st birth: 11 wk	11494
	m: 6 wk	11494
	m: 9-10 wk	7053
Rattus tiomanicus	f: vag perf: 21-30 g	11885
	m: scrotal testes: 50% 46-59 g; r30-100	11885
Rattus tunneyi	f: 1st birth: 8 wk	11494
	f/m: 1st mate: ≤ 2 mo	3606
Rhabdomys pumilio	f: 1st mate: 5-6 wk	2298
	f: 1st conceive: 34 d	1335
	f: 1st birth: 3 mo	5760
	f: 1st birth: 90 d; r57-107; n11	1335
	f: 1st birth: 100% >119 mm; min 100-104	4329
	f: > 20 g	10712
	m: sperm: 11 wk	2298
	m: 52-58 d	1814
	m: 85 mm	4329
	f/m: 2 mo	10158
Stochomys longicaudatus	m: testes large: 75 g	2573
Thallomys paedulcus	f: 1st birth: min 107 d	7168
Uranomys ruddi	f: HB: 90 mm	752a
	m: HB: 95 mm	752a
Uromyx caudimaculatus	f/m: 180 d	11493
Zyzomys argurus	m: scrotal testes: 5-6 mo	733
	f/m: > 29 g	733
Zyzomys woodwardi	m: scrotal testes: 5-6 mo	733
	f/m: > 70 g	733

CYCLE LENGTH

Acomys cahirinus	8-14 d	2568
	11.1 d; sd1.9; mode 11; r6-18; n116 cycles, 88 f	8325, 8326
Arvicanthis niloticus	6 d; n1 f	1997
Bandicota bengalensis	proestrus: 9.5 h; se0.75; estrus: 43.52 h; se4.33	9436
	metestrus: 24.02 h; se1.5; diestrus: 47.46 h; se5.7	9436
	4.19 d; se0.30; n24 f	9435
	5.19 d; n20 cycles	9436

Chiropodomys gliroides	10 d; sd3.7; r7-13; n6 cycles, 1 f	7159
Dasymys incomtus	7 d; n1 f	7477
Grammomys dolichurus	4-5 d	988
Grammomys rutilans	7 d; n1 f	1997
Grammomys surdaster	5 d; n1 f	1997
Hybomys univittatus	4-5 d; n4 f	1997
Hydromys chrysogaster	proestrus 1 d, estrus 2, metestrus 2, anestrus 5	8103
	10.79 d; se1.85; median 10; r7-17; n169 cycles, 38 f	8103
Leggadina forresti	7 d	7983
Lemniscomys striatus	6 d; r5-7	7477, 8429
Lophuromys sikapusi	11 d; n1 f	1997
Malacomys longipes	4-6 d	1997
Mesembriomys gouldii	26.4 d; ± 3.5; mode 27; r21-35; n21 cycles, 7 f	2131
Mus minutoides	4 d	7477
Notomys alexis	6.2 d; ±2.94; r4-8; mode 6; n42 cycles, 5 f	10117
	6.5 d; r6-7; n6 cycles, 1 f	4355
	6.8 d; se0.2; n29 cycles, 10 f	10732
	8.0; mode 7; r5-10; n30 f	2132
	8.1 d; se0.9; n29 cycles, 6 f	1253, 1275
	8.3 d; se0.2; n21 cycles, 9 f	10732
Notomys cervinus	median 20 d; r8-38; n86 cycles, 18 f	2132
Notomys fuscus	7.4 d; mode 7; r5-10; n68 cycles, 13 f	427
	8.5 d; modes 7,9; r5-13; n18 cycles, 4 f	2132
Notomys mitchellii	8.3 d; mode 7; r6-17; n69 cycles, 8 f	2132
Pelomys campanae	6.5 d; n1 f	7477
Praomys jacksoni	4-5 d	1997
Praomys morio	6 d	1997
Praomys natalensis	7-8 d	8034
	8.8 d; ±0.40; mode 6; r4-20; n83 cycles	5366
Praomys tullbergi	6-7 d	4353
Pseudomys albocinereus	prob 7-10 d; modes 6-9, 14; r6-19; n19 cycles, 6 f	4355
Pseudomys australis	8.5 d; ±1.12; mode 7; r5-21; n52 cycles, 9 f	10117
Pseudomys desertor	7.75 d; r7-9; n4 cycles, 4 f	4355
Pseudomys novaehollandiae	6.01 d; sd1.26; r4-10; n73 cycles, 25 f	5643
Rattus argentiventer	4.3 d; ±0.1; r3-6	6162
Rattus fuscipes	4.5 d; mode 4-5 99%; r3-6; n1603 cycles, 37 f	10696
	5.3; se0.2; n34 cycles, 5 f	1254
Rattus leucopus	4.8; se0.5; n33 cycles, 6 f	1254
Rattus lutreolus	5.2 d; se0.2; n40 cycles, 6 f	1254
Rattus norvegicus	4.5; se1.0; n11 cycles, 11 f; 1st cycle	1849
Rattus rattus	4.5; se0.1; mode 4; median 4.1; n1077 cycles, 68 f	2527
	10 d	5760
Rattus sordidus	4.3 d; r3-6; n132 f	7053
	5.5 d; n88 cycles, 14 f	1254
Rattus tunneyi	4.8 d; se0.1; n35 cycles, 6 f	1254
Rhabdomys pumilio	proestrus 1.1 d; estrus 4.6 d; nonestrus 6.2 d	2530
	11.0 d; mode 7; median 7.4; sd8.9; se0.4; n475 cycles, 37 f	2530
Stochomys longicaudatus	5-6 d; n2 f	1997
Uromys caudimaculatus	7 d	10698
GESTATION		
Acomys cahirinus	11 d 14 h; n1	1080
	35.875 d; mode 36; r35-37; n32	2568
	36-40 d	2449
	37-45 d; n27	1035
	37.6 d; mode 38; r36-40; n87; ibi	2566
	38 d	8326
	38 d	9356
	38.1 d; n26	2532
	39 d	12038
	39.9 d; sd1.19; r38-44; n46 lt	11397
	44.2 d; r42-46; n5	109

Aethomys chrysophilus	21-23 d	5760
	~26 d; min ibi	1333
	31 d; n1 ibi	4329
Aethomys hindei	r24-25 d; n18	8083
Aethomys kaiseri	27 d; ±1	1771
Aethomys namaquensis	> 22 d	7165
Apodemus agrarius	18 d; r18-21; min ibi	11710
	mode 21-23 d; r20-23	3571
	21-23 d	7433
Apodemus flavicollis	23-26 d	3571
	23-26 d	7433
	24-25 d; mode 28; r21-35; min ibi	5170
Apodemus mystacinus	23 d; n3	2569
Apodemus sylvaticus	21 d; n6; 2-3 d delay if nursing	3067
	mode 21-23 d; r21-25	3571
	25-26 d	2045
Arvicanthis niloticus	~18 d; via 1205 Davis 1963, ill 24 se 91	2446, 2449
	20 d	4484
	prob 21-23 d	3756
	22-24 d; min ibi	2450
	24-26 d	8811
Bandicota bengalensis	17 d	3554
	21 d	9429
Bandicota indica	20-21 d; pairing to birth	10370
Chiropodomys gliroides	19-21 d; n1	7159
Conilurus penicillatus	35 d	10698
	36 d	11493
Grammomys dolichurus	24 d	988
Grammomys rutilans	25 d	2446a, 2449
	~25 d	5760
Hybomys univittatus	29-31 d	2446a, 2449
Hydromys chrysogaster	34.87 d; se1.82; mode 34; r33-41; n38	8103
	35 d	11493
	36.64 d; se2.20; r35-41; n11 f with lactation	8103
Hylomyscus alleni	25-33 d	2449
Hylomyscus stella	29.5 d; r25-33; n6	2445
Leggadina lakedownensis	33 d	11493
Lemniscomys striatus	21 d	11614
	22.6 d; ±1.4; n29	3644
	28 d	2446a
	28 d; min ibi	8429
Leporillus conditor	44 d	11493
Lophuromys flavopunctatus	30.5 d; r30-31; n2	2445
Lophuromys woosnami	≥32 d	2572a
Mastacomys fuscus	5 wk	9716
	38-40 d	11493
Melomys cervinipes	38-40 d	11493
Mesembriomys gouldii	43.75 d; mode 44; r43-44; n4	2131
Micromys minutus	18-21 d	3571
	21 d	7433
	21 d; ibi	3461
Millardia meltada	16 d	3554
	20 d	12034
	20 d; min ibi	940
Mus caroli	18.1 d; lactating 23.1 d	10370
Mus cervicolor	17.6 d; lactating 21.0 d	10370
Mus domesticus	20.0 d; se0.5	6864
Mus minutoides	est 18-19 d	11711
	23.2 d; r22-24; n6	245
Mus musculus	18-20 d	7433
	18-21 d	9175
	19 d	8501
	19-21 d	845
	19.6 d	7704

	~20 d	4645
	31 d; n1 f nursing 8 yg	6238
Nesokia indica	17 d	3554
Notomys alexis	implant: 7-8 d	1255
	implant: 11-13 d with 3-4 yg suckling	1255
	implant: 14-17 d with 5 yg suckling	1255
	32 d	2132
	32-47 d	11493
	32-47 d; mode 35, 41; n13 sperm in urine to birth	10117
	32.5 d; se1.5; n4 f; postpartum estrus to birth	1255, 1258
	32.6 d; n8	2532
	32.67 d; r32-34; n3	4355
	34 d with 1-2 yg; 39.8 d with 4-6 yg	1255
Notomys cervinus	40 d; mode 38; r38-43; n11 f	2132
	47 d; r39-51; n8 f with lactation	2132
Notomys fuscus	32-35 d	11493
	34.3 d; r32-38; n6; vaginal sperm-birth usually > 61 d	427
Notomys mitchellii	35.5 d; r34-37; n4	2132
	34-42 d	11493
Praomys daltoni	29 d; r21-37; sd1.4; n40	247
Praomys erythroleucus	21 d	5058
Praomys jacksoni	34-37 d	2445, 2446, 2449
Praomys morio	26.5 d; r26-27; n2	2954
	36.0 d; r34-37; n4	2445
Praomys natalensis	23 d	2446a
	23 d	8034
	~23 d	5366
Praomys tullbergi	23-25 d; min ibi	4353
Pseudomys albocinereus	>23 d	9189
	38.5 d; r38-39; n2; ibi	4355
Pseudomys apodemoides	34-55 d	11493
Pseudomys australis	30-31 d; mode 31; n5; sperm in urine to birth	10117
Pseudomys delicatulus	29-30 d	11493
Pseudomys desertor	27-34 d	11493
	27.5 d; n3	4355
Pseudomys gracilicaudatus	27 d	3438
Pseudomys hermannsburgensis	30-33 d	11493
Pseudomys higginsi	31-32 d	11493
Pseudomys nanus	23-25 d	11493
Pseudomys novaehollandiae	30 d	11493
	31.5 d; sd0.96; r29-33; n25	5643
	33.2 d; sd1.75; r32-37; n8 with lactation	5643
Pseudomys praeconis	30 d	11493
Rattus argentiventer	21.4 d; ±0.2 d; r20-23; n18	6162
Rattus exulans	19 d	10545
	19-21 d	11833
	≥ 21 d	10578
	21.7 d; lactating 23.6 d	10370
	23 d; pairing to birth; 3-7 d longer with lactation	2911
Rattus fuscipes	22.8 d; mode 23 60%; r22-24; n20 cases, 15 f	10696, 10707
	28-32 d; with lactation	10696, 10707
Rattus norvegicus	20-24 d	7433
	21-22 d	2555
	22 d; n8	6238
Rattus rattus	15-20 d	8500, 8501
	mode 20-24 d; r20-26	3571
	21 d	10545
	21 d	2452
	21-22 d	3133
	21-30 d	5760
	22 d	5216
	22 d ea; n7	6237
	22 d	2555
	25 d	8861

	26 d; n2	5315
Rattus sordidus	r21-22 d; n3	7053
Rattus tiomanicus	22.0 d; lactating rats 24.3 d	10370
Rhabdomys pumilio	22 d; ibi	1814
	25.4 d; min 23	10158
	25.5 d; n6	2298
	26.4 d; r26-28; n7	2530
Uromys caudimaculatus	36 d	7983
	41 d	10698
Zelotomys woosnami	~31 d; n2 cases, 1 f; min ibi	947
Zyzomys argurus	25 d	11493
LACTATION		
Acomys cahirinus	solid food: 1 d	10481
	solid food: 5 d	8686
	den emergence: 6-7 d	2566
	independent: 14 d	5051
	wean: 2 wk	5760
	wean: 28 d; can wean: 17 d	12038
	wean: ~1 mo	8686
Aethomys chrysophilus	wean: 24 d	4329
	wean: 31-33 d	7797a
Aethomys kaiseri	teat attachment: 12 d	1771
	solid food: 12 d	1771
	wean: 26 d	1771
Aethomys namaquensis	teat attachment: 18 d	7168
	teat attachment: 3 wk	1815
	wean: 18 d	7165
	wean: 21-26 d	7797a
	wean: 27-28 d	7168
Apodemus sylvaticus	wean: ~18 d	2045
Arvicanthis niloticus	wean: 3 wk	2450
Bandicota bengalensis	solid food: ~20 d	10268
	wean: 28 d	9175
	no milk: ~30 d	10268
	leave nest: ~1 mo	108
Bandicota indica	wean: 25 d	10821
	no milk: ~30 d	10268
Conilurus penicillatus	independent: 20 d	10698
	wean: 30 d	11493
Cremnomys blanfordi	teat attachment: 1 mo	1339
Dasymys incomtus	wean: 30 d	2570
Grammomys cometes	wean: 19-26 d	10158
Grammomys dolichurus	teat attachment: 1st few d	8218
	wean: 19 d	8218
	wean: 30 d; n3	988
Hydromys chrysogaster	solid food: 3 wk	8103
	wean: 29.45; sd2.68; median 29; r17-35; n85 lt	8103
	wean: 30 d	11493
	independent: 35 d	10698
Leggadina forresti	wean: 30 d	11493
Leggadina lakedownensis	wean: 30 d	11493
Lemniscomys striatus	solid food: 14 d	11614
	wean: 19-21 d	11614
	wean: 30-50 d	2570
Leporillus conditor	wean: 40 d	11493
Lophuromys flavopunctatus	wean: ~1 mo	2445
	wean: 50-70 d	2570
Lophuromys sikapusi	solid food: 8 d	1771
	wean: 12 d	1771
Mastacomys fuscus	teat attachment: 3 wk	10698
	wean: 40 d	11493
Melomys cervinipes	wean: 37 d; se3	1255
	wean: 40 d	11493

Mesembriomys gouldii	teat attachment: 4 wk	2131
	solid food: 43 d	2131
	wean: 40 d	11493
	wean: 6-8 wk	2131
Micromys minutus	den emergence: 11-12 d	5553
	den emergence: 12-13 d	3461
	solid food: 12-13 d	3461
	solid food: 12-13 d	3571
	wean: 15 d	5553
	wean: 16 d; r15-16	3571
Mus minutoides	wean: 17 d	11711
	wean: 24 d	245
Mus musculus	wean: 21 d	9175
	wean: 3 wk	845
	wean: 3 wk	10095
	wean: 21-22 d	8501
	wean: 26 d	7704
Mus triton	wean: 25-30 d	2570
Notomys alexis	wean: 24-30 d	10117, 11493
	wean: 30 d	1252, 1255
	wean: 1 mo	2132
	wean: 33 d	4355
Notomys cervinus	wean: 1 mo	2132
	wean: 35 d	11493
Notomys fuscus	wean: 30 d	427, 11493
	wean: 1 mo	2132
Notomys mitchellii	wean: 30 d	11493
	wean: 1 mo	2132
Oenomys hypoxanthus	wean: 30 d	2570
Praomys jacksoni	wean: 30 d	2570
Praomys morio	solid food: 17-19 d	2954
	wean: 27 d	2954
Praomys natalensis	solid food: 17 d	514
	independent: ~19 d	11241
	wean: 19-21 d	7166, 7168
	wean: 21 d	2446a
	wean: 21 d	8034
Praomys tullbergi	wean: 24-25 d	4353
Pseudomys albocinereus	wean: 25 d	4355
	wean: 35 d	11493
Pseudomys apodemoides	wean: 35 d	11493
Pseudomys australis	wean: 22-30 d	10117, 11493
Pseudomys delicatulus	wean: 30 d	11493
Pseudomys desertor	wean: 20 d	4355
	wean: 30 d	11493
Pseudomys gracilicaudatus	solid food: 12-14 d	3438
	wean: 26 d; r20-35	3438
Pseudomys hermannsburgensis	wean: 30 d	11493
Pseudomys higginsi	wean: 30 d	11493
Pseudomys nanus	wean: 25 d	11493
Pseudomys novaehollandiae	solid food: 4th wk	5642
	wean: 25.2 d; r23-28; n14 lt	5642, 5643, 5644
	wean: 31.5 d; r29-33	5645
Pseudomys praeconis	wean: 30 d	11493
Rattus exulans	solid food: end 3rd wk	11833
	wean: 28 d	11833
	wean: 4 wk	4821
Rattus fuscipes	wean: 21-28 d	10696
Rattus lutreolus	solid food: 19-21 d	3437
	wean: 25 d; r23-30	3437
	wean: 25 d	4041
Rattus norvegicus	wean: est 21-25 d	2321
	wean: 4 wk	4821

Rattus rattus	leave nest: 17-23 d	3133
	solid food: 15-18 d	3133
	wean: ~25 d	9265
	wean: 28 d	2091
	wean: 4 wk	4821
	wean: mode 6-7 wk; r4-10	8861
Rhabdomys pumilio	leave nest: 14 d	1335
	wean: 14 d	7168
	wean: 14 d	4640
	wean: 16 d	1335
Thallomys paedulcus	teat attachment: 15 d	7168
	wean: 28-31 d	7168
Uromys caudimaculatus	wean: 40 d	11493
Zelotomys woosnami	leave nest: 16 d	947
Zyzomys argurus	wean: 35 d	11493
Zyzomys woodwardi	wean: 35 d	11493
INTERLITTER INTERVAL		
Acomys cahirinus	34 d	7544
Apodemus agrarius	18-21	11710
Apodemus flavicollis	mode 20-35 d; r20-126; n90	5170
	40-60 d	4645
Apodemus sylvaticus	27.56 d; mode 23-24; r18-54; n86	3067
	50-60 d	4645
Arvicanthis niloticus	20.4 d; wet season; 49 d; dry season; Rojewero Plains	7793
	22-24 d	2450
	30 d; wet season; 70 d; dry season; Crater Track	7793
	42.4 d; wet season; 68.4 d; dry season; Mweya Pen	7793
Bandicota bengalensis	61.9 d	11436
Bandicota indica	71.25 d; r40-105; n12	10370
Grammomys dolichurus	mode 28 d; r20-80; n500	988
	48 d	8218
Lemniscomys striatus	21-23 d	11614
Micromys minutus	min 21 d	3461
Mus minutoides	r22-58 d; n5	245
	22.4 d; r19-26; n11	11711
Notomys alexis	32.75 d; r32-34; n4	4355
Notomys fuscus	min 41-42 d; mode 80-89	427
Notomys mitchellii	36.5 d; r36-37; n2	4355
Praomys natalensis	25 d	8034
	26 d	8091
	4 wk	4588
	min 29 d	514
	33.1 d; ±2.5; r24-58; n15 intervals, 4 f	7166
Praomys tullbergi	23-25 d; 29 lt; modes 24,29 d	4353
Pseudomys albocinereus	38.5 d; r38-39; n2	4355
Pseudomys desertor	29.75 d; r28-34; n4	4355
Pseudomys novaehollandiae	33.5 d; sd1.29; r32-39; n18	5643
Rattus argentiventer	mode 20-25 d; r20-60+; n196	6162
Rattus exulans	74 d	5216
Rattus sordidus	22-35 d	7053
Rattus rattus	< 1 mo	3133
	31 d; r22-57; n7	6237
	31.5 d; ±0.1; r27-38; n10	2091
	50 d	5216
Rhabdomys pumilio	min 23 d	1335
	26.8-43.5 d; n39 intervals, 5 pr	2530
Thallomys paedulcus	26 d min	7168
Zelotomys woosnami	31 d min	947
Zyzomys argurus	7.2 mo; sd3.0; r3-13	733
Zyzomys woodwardi	7.6 mo; sd3.1; r2-14; prob overestimate	733

LITTERS/YEAR

Acomys cahirinus	potential 7.6; mode 3-4	7795
Acomys wilsoni	potential 7.9; mode 3-4	7795
Aethomys namaquensis	2	11864
Apodemus agrarius	1-3	10340
	3-4	7433
	mode 3-4; r3-5	3571
	mode 3-4; r3-5	4645
Apodemus argenteus	max 4	7925
Apodemus flavicollis	2-3	7433
	2-4	4645
	mode 2-4; r2-5	3571
	max 11	5170
Apodemus mystacinus	min 2	6419
Apodemus sylvaticus	2-4	4645
	3	9438
	3; r2-4	3571
	4-5	7433
	4-5	9175
	max 6	107
Arvicanthis niloticus	3-4	4484
Bandicota bengalensis	>1	8501
	5.9	11436
	max 9	9429
Grammomys cometes	max 3	4329
Hydromys chrysogaster	1	7108
	2-3	7209
	2.55/season; ±0.97; r1-5	8103
	2.64; ±1.10; r1-5	8103
Lemniscomys striatus	1 each rainy season	2446a
	2; mode 2; r1-3	7792
Melomys cervinipes	1.3/season; ±0.77; r1-2; n18	10109
Mesembriomys gouldii	4	2131
Micromys minutus	2-3	7433
	2-3; r2-7	3571
	3	2045
Millardia meltada	9.28	8904
Mus minutoides	2-3 each rainy season	2446a
	3-8	5760
Mus musculus	3-5	8500, 8501
	4-5	4645
	4-6	7433
Mus triton	3?	10158
	max 3	4329
Praomys natalensis	2	7791
Pseudomys apodemoides	max 3	1934
Pseudomys shortridgei	1	4355
Rattus exulans	1-2; 2 rare	11832, 11833
	3.9	5216
	5.2; r1-13	2911
	max 17	10578
Rattus fuscipes	3.6; n34 f	10696
Rattus norvegicus	1	8802
	2-5	4645
	2-7	2555
	2-7	7433
	2.65	6242
	max 3	2455
Rattus rattus	2+	7433
	2-5	2555
	2-6	3571
	max 3	2455
	3	5315
	3.2	5216

	4	10544
	4+	7185
	4-5	8501
	4.2; calculated	858
	6-7	9175
Rhabdomys pumilio	4	5760
	max 10; 1 record 21 lt/27 mo	1814
Zyzomys woodwardi	86% 1 lt/2+yr; 14% 2 lt/2+yr	733
SEASONALITY		
Acomys cahirinus	fertility: Feb-July, minor peak Nov; sw Saudi Arabia	109
	fertility: 80+% Dec-Feb, Apr-May, July, 40% Oct; Kenya	7797
	conception: peak Dec-May, low June-Sept; Malawi	4329
	mate: Feb-Apr; Negev desert, Israel	715
	mate: 1-3 mo after rains; Sudan	4344
	mate: yr rd; Kenya	7795, 7796
	mate: yr rd; captive	715
	preg f: Jan, Apr, July, Oct-Sept; Kenya	4925
	preg f: Feb-Oct; few data; Egypt	8134
	preg f: peak Apr-Sept; Cyprus	11475a
	preg f: May, July-Aug; few data; Tanzania	299
	preg f: Nov-Dec; Sudan	4344
	birth: low Nov, high Apr-Oct; lab	2568
	birth: yr rd, peak Apr-Sept, n58/85; captive	1035
	birth: yr rd, low Oct-Jan; captive	12038
	birth: yr rd; n17; zoo	12128
	placental scars: Sept-Jan; Sudan	4344
Acomys spinosissimus	breed: Nov-Mar; Zimbabwe	1815
	preg f: Dec-Jan, Mar-Apr; n12; Botswana	10154
	birth: Oct-Apr; Transvaal, S Africa	8965
	birth: Nov-Apr; S Africa	10158
	repro: Oct-Apr; Mozambique	3821
Acomys wilsoni	fertility: low Mar; Kenya	7797
	mate: yr rd; Kenya	7795, 7796
	preg f: July; n2; Kenya	4925
	preg f: Nov-Feb; Tanzania	299
Aethomys chrysophilus	spermatogenesis: peak Feb-May; Malawi	4329
	mate: yr rd; lab	1333
	preg f: peaks Feb-May, Oct-Nov, none Aug-Sept; Malawi	4329
	preg f: Oct-Nov; Namibia	9945
	birth: Oct-Apr; n10; Botswana	10154
	birth: yr rd, peak summer; Zimbabwe	1815
	birth: yr rd; Transvaal S Africa	8965
	juveniles: yr rd; Zambia	331
	repro: Oct-Apr; Mozambique	3821
	repro: yr rd; S Africa	10158
Aethomys hindei	preg f: peaks Oct-Dec, Mar-May; s Uganda	8083
	birth: peaks Dec-Feb, June-Aug; s Uganda	8083
Aethomys kaiseri	mate: Oct-Jan; Tanzania	299
	preg f: Jan, Aug; n2; e Africa	4925
Aethomys namaquensis	preg f: Jan; n2; Malawi, Nyasaland	4329
	preg f: Jan-May, Oct, peak Mar-Apr; Botswana	10154
	preg f: Feb; Namib Desert, Namibia	11863
	preg f: Aug-Sept; Namibia	9945
	birth: Mar; Namibia	9945
	birth: yr rd, peak summer; Zimbabwe	1815
	lact f: Nov; n1; Malawi	4329
	repro: Oct-Apr; Mozambique	3821
Aethomys nyikae	juveniles: most mo; Zambia	331
Apodemus agrarius	m sexually active: Mar-Dec; Poland, Czechoslovakia	8330
	testes: peak Mar-Sept; Poland	5094
	testes: regress Sept; Poland	5093
	mate: Mar-Sept; Germany	3571
	mate: peak Mar-Apr; Korea	12048
	mate: spring-early fall; Finland	7433

	preg f: Mar-Nov; east Hessen, Germany	8343
	preg f: Apr-Sept; Poland, Czechoslovakia	8330, 8331, 8332
	preg f: end Apr-Aug; Germany	10340
	preg/lact f: May-Oct; Poland	5093
	birth: Apr-Oct; Germany	3571
	birth: spring-early fall; Finland	7433
	repro: end Mar-mid Oct; Korea	5424_1
Apodemus argenteus	testes: large Mar-Sept; Japan	3546
	mate: Feb-Apr, Sept-Nov; Japan	7925
	mate: Oct-Apr; Japan	12025
	birth: May-Aug, Oct-Nov; Japan	7761
	yg appear: June-Sept; Japan	7393
Apodemus flavicollis	f without cl: Feb, May-Nov; s Sweden	3072
	fertile m: Apr-Oct; Austria	8848
	fertile m: yr rd; Denmark	5302
	spermatogenesis: Feb-Oct; s Sweden	3071
	testes: increase Jan, May, regress Nov; Germany	5461
	testes: large Feb-Aug, small Sept-Dec; Austria	10358
	testes: large Apr, small Aug-Jan; s Sweden	3071
	mate: Feb-Mar; s Sweden	4342
	mate: Feb-Sept; Europe	7433
	mate: Feb-Sept; Germany	6548
	mate: Feb-Oct; Germany	3571
	mate: early Apr-Oct; s Sweden	810
	mate: low winter, 37% f, 42% m fertile; s Sweden	3073
	preg f: Feb-June, Aug-Oct; s Sweden	3072
	preg f: Feb-Sept; Austria	10358
	preg f: Feb-Oct, 293 d; s Moravia, Czechoslovakia	8331
	preg f: Feb-Oct; Austria	10358
	preg f: mid Feb-mid Oct; Brandenburg, Germany	10340
	preg f: Mar-Sept, none July; n Germany	5461
	preg f: June-July, Oct; Pyrenees, Spain	11255
	birth: Mar-Oct; Germany	3571
	birth: yr rd; lab	5170
	lact f: Jan-Feb; Berlin	6091
	yg appear: Apr-Aug or Aug-Dec; 2 generations; Poland	54
	repro f: peak Mar-Oct, rare Dec-Jan; Denmark	5302
	repro: Mar-Sept	5461
	repro: end Mar-Oct; Sweden	810
Apodemus microps	sperm: Feb; Bulgaria	10480
	testes: large Feb-Sept, small Oct-Dec; Austria	10358
	preg f: Feb-Mar, Aug-Sept; Austria	10358
	preg f: Mar-Oct; Czechoslovakia	8331
	preg f: Mar-Oct, peak June; Bulgaria	10480
	preg/lact f: Mar-Oct; May 62%, none Nov-Feb; s Moravia, Czechoslovakia	8329
Apodemus mystacinus	testes; large Apr-May, July-Sept; Lebanon	6419
	birth: Feb, Apr; captive	2569
	birth: warm mo; Lebanon	6419
	repro: Mar-Oct; Bulgaria	8317
Apodemus peninsulae	preg/lact f: Apr-Oct; Korea	5424_1
Apodemus speciosus	testes: peaks Feb-Mar, Aug-Oct; Japan	7620
	testes: increase Feb, peak Mar-Oct, small Nov; Japan	7394
	testes: large Sept-Mar, small Apr; Japan	12024
	estrus f: Oct-Feb; Japan	12024
	mate: mid Apr-early May, end Aug-mid Sept; Japan	5951
	preg f: Mar-May, Sept-Dec; Japan	10632
	preg f: Apr-Sept, peak Aug; Japan	7394
	preg f: May, Sept-Oct, not June-Aug; Japan	5951
	preg f: Oct-Apr; Japan	12024
	birth: May, mid Sept-mid Oct; Japan	5951
	lact f: Oct-May; Japan	12024
	repro f: Mar-Apr, Oct; Japan	7620

Apodemus sylvaticus	f without cl: Jan-June, Aug-Dec; s Sweden	3072
	testes: large Feb-early Sept, small Sept-Dec; Austria	10358
	testes: large Feb-Oct; Germany	5461
	testes: enlarge Mar; sw France	8982
	testes: large Apr, small June; Sweden	3071
	testes: large Sept-Feb; Algeria	5987
	testes: large summer-Oct, regress: Dec-Jan; s France	8985
	testes: large summer, small winter, increase Mar; UK	516
	testes: small Oct-mid Jan; France	9438
	fertile m: 42-100% Feb-Apr, June-Sept, none May, Oct-Jan; Austria	10358
	fertile m: Mar-Oct; s France	5247
	fertile m: Apr-Sept, none Oct-Nov, Jan; few data, Mellum I, Germany	7779
	fertile m: few end fall-early winter; UK	516
	conceives: end Apr; St Kilda, Scotland, UK	5307
	mate: Feb; Germany	11530
	mate: Feb-Sept; Germany	6548
	mate: Feb-mid Oct; Scillly, Cornwall, England, UK	9254
	mate: end Feb-Sept, peak June-Aug; Fair Isle, Scotland, UK	2448
	mate: Mar-Sept; Germany	3571
	mate: Mar-Nov, peak June-July; England, UK	7332
	mate: Apr-Sept; nw Bohemia, Czechoslovakia	741
	mate: Apr-mid Oct; s Sweden	810
	mate: May-fall; Europe	7433
	mate: June-Aug; Ireland, UK	3154
	mate: Dec; England, UK	10175
	mate: Dec-Mar; Sweden	4342
	mate: summer-fall; Channel I, UK	955
	mate: low winter, 26% f, 21% m fertile: s Sweden	3073
	mate: winter if acorn crop good; England, UK	10175
	mate: yr rd; France	943
	preg f: Jan-Nov; 296 d; s Moravia, Czechoslovakia	8331
	preg f: Feb-Apr, Aug-Oct; s Sweden	3072
	preg f: Feb-Aug; Austria	10358
	preg f: Mar-Sept; Germany	5461
	preg f: Mar-Sept; France	9438
	preg f: Mar-Oct; England, UK	3342
	preg f: Apr-Aug; St Kilda, Scotland, UK	5307
	preg f: July; few data; Mellum I, Germany	7779
	preg f: peak July, low Oct-Mar; UK	516
	preg f: Aug; n1; China	182
	preg f: Oct; n1; Lebanon	6419
	preg f: Oct-Dec; Algeria	5987
	preg/lact f: 72.6% May, none Dec-Jan; no data Nov-Jan; s Moravia, Czechoslovakia	8329
	birth: Mar; Scilly, Cornwall, UK	9254
	birth: Apr-Oct; Germany	3571
	birth: May-Aug; St Kilda, Scotland, UK	5307
	birth: May-fall; Europe	7433
	birth: May-Oct, winter; France	10897
	birth: yr rd; lab	3067
	lact f: July-Aug; n3; Fair Isle, Scotland, UK	2448
	lact f: Aug-early Nov; Iran	6293
	lact f: Dec-Feb; Algeria	5987
	repro: Feb-Oct; s Moravia, Czechoslovakia	8329
	repro: Mar-Sept; Germany	5461
	repro: Apr-; Germany	10340
	repro: end Mar-early Oct; s Sweden	810
	repro: summer-Oct, none winter; sw France	8982, 8983
	repro: yr rd; lab	8985
Arvicanthis niloticus	fertile f/m: yr rd; Uganda	7793
	testes: large Aug-Dec; Ethiopia	7600
	testes: regress Feb; Kenya	10712

	mate: June-Nov; Egypt	4484
	mate: 1st half dry season; Ethiopia	7600
	mate: rainy-early dry season, peak end rains-early dry; Kenya	10712
	mate: wet season; Kenya	2454
	mate: most mo; Khartoum, Sudan	4347
	mate: yr rd	3756
	preg f: Jan, May-June; few data; e Africa	4925
	preg f: June; n1; Tanzania, Uganda	186
	preg f: July, Sept, Feb; few data; Sudan	4344
	preg f: peak Sept-Oct; Ethiopia	7600
	preg f: Nov-Feb, May-Sept; Uganda	7793
	preg f: yr rd, low dry season; n179; Uganda	2446a, 2453
	birth: Mar-May; lab	8193
	birth: shortly after rain period; Ethiopia	7600
	juveniles: yr rd; Sudan	4344
Bandicota bengalensis	mate: mid Feb-mid Oct, peak Apr, Aug-Sept; Punjab, India	5586
	mate: low winter; n42; Ludhiana, India	6775
	mate: yr rd; Pakistan	108
	preg f: yr rd; Calcutta, India	10267
	birth: peak Oct-Nov; Pakistan	108
Bandicota indica	preg f: 50% Mar, none Feb, 10-30% other mo; n199 f; India	8794
	birth: yr rd; Sri Lanka	8500, 8501
Chiropodomys gliroides	preg f: 18% Jan-Mar, 6% Apr-Sept; Malaysia	4406
Colomys goslingi	fertile m: Mar; w Uganda, e Zaire	5760
	fertile f: July; few data; Zaire	8855
	birth: June-July; n4; w Uganda, e Zaire	5760
	preg f: Apr; n1; w Uganda, e Zaire	5760
	lact f: June-July; n1; w Uganda, e Zaire	5760
Conilurus penicillatus	repro: through winter; Australia	10698
Cremnomys cutchicus	epididymal sperm: yr rd, low Oct-Dec; India	8729
	mate: Mar-Oct; w Rajahstan, India	8723
	birth: Mar-Oct; India	8729
Dasymys incomtus	preg f: June; Malawi	4329
	preg f: Aug; n1; Kenya	4925
	preg/lact f: 60-80% Jan-June, 45% July, 0-30% Aug-Nov; Zaire	2570
	lact f: Jan; n1; Kenya	4925
	lact f: Apr; n1; Tanzania	186
	lact f: Aug; few data; Malawi	4329
	birth: June-Oct; Namibia	9945
	birth: Aug-Jan; S Africa	10158
Golunda ellioti	testes: regressed Dec-Jan, no mating until Apr; Pakistan	9175
	mate: Mar-Sept; n33; Ludhiana, India	6775
	preg f: Mar-Aug; Rajahstan, India	8724
	birth: early part of the yr; Sri Lanka	8501
	birth: May; n1; Pakistan	9175
Grammomys cometes	preg f: Aug-Sept; n2; Kenya	4925
	preg f: peak wet season; Malawi	4329
Grammomys dolichurus	testes: peak Feb-Mar n8; Malawi	4329
	mate: extended season; S Africa	10158
	preg f: Feb; n1; e Africa	4925
	preg f: Nov; n2; Kenya	4925
	birth: Dec-June, peak Feb-May; Malawi	4329
	yg found: almost yr rd; Zambia	331
	repro: peak Feb-May; S Africa	8218
Grammomys surdaster	lact f: Feb; n1; Tanzania	186
Graomys spp	preg f: Dec-Mar; neotropics, S America	4682
Hybomys univittatus	repro f: yr rd, peak Dec-Jan, low Apr-Aug; Zaire	2573
	mate: peak Oct-Mar, low June-Aug; e Zaire	2574
	preg f: Jan; n3; Zaire	8854
	preg f: Apr-Aug, Oct, Dec-Jan; Zaire	8855
	yg found: Feb-Mar, June; Zaire	8856
Hybomys trivirgatus	preg f: Mar, July, Sept; Nigeria	4354a

Hydromys chrysogaster	sperm: yr rd, some regression fall-winter; captive	8103
	preg f: July-Feb; VIC, Australia	7108
	birth: yr rd, peak Sept-Mar 88%, low June; captive	8103
Hylomyscus aeta	preg/lact f: Nov-Dec, Feb; Cameroun	2955
Hylomyscus alleni	mate: peak Oct-Mar, low June-Aug; e Zaire	2574
	preg f: Jan, July; Zaire	8855
	repro f: peak Dec-Mar, low June-Aug, Oct; Zaire	2573
Hylomyscus stella	mate: peak May-June; Uganda	2445
	mate: peak June, Dec; w Uganda	5760
	preg f: Jan-Feb, Nov; Nigeria	4352
	preg f: Mar-Oct; Nigeria	4350
	lact f: Feb-Apr; Uganda	2445
	juveniles: Oct; July-Aug, Dec; e Africa	5760
Lemniscomys griselda	birth: Dec; Malawi	4329
	birth: perhaps summer; S Africa	10158
	birth: summer; Transvaal S Africa	8965
Lemniscomys striatus	fertile: Mar-May, Aug-Dec; Uganda	7797
	preg f: Jan-Feb, June-July; n5; e Africa	4925
	preg f: Jan, Nov; Zaire	8855
	preg f: Mar-May, Sept-Dec; Uganda	7792
	preg f: Apr-June, Sept-Dec, peaks May, Oct-Nov; n213; Uganda	2446a, 2453
	preg f: Oct-Nov; Nigeria	5190a
	preg f: Oct-Dec; Cameroun	9468
	preg f: Dec-Feb; n4; Zaire	8856
	preg/lact f: 50-70% Dec-Mar, June-July; 10-30% Apr-May, Aug-Nov; Zaire	2570
	birth: Apr-July, Sept-Nov; Ivory Coast	3645
Leopoldamys sabanus	m without sperm: 24% Jan-Mar, 4-5% Apr-Sept, 49% Oct-Dec; Malaysia	4406
	preg f: 8% Jan-Mar, 11% Apr-June, 22% July-Sept, 14% Oct-Dec; Malaysia	4406
	preg f: yr rd; Malaysia	4405
Lophuromys flavopunctatus	sexually active: yr rd; Zaire	2573
	testes non-spermatic: July-Sept 3/7, Nov-Jan 5/31; Malawi	4329
	mate: peak Sept-Nov, Mar-June; Uganda	2445
	mate: peak with rainfall; s Tanzania	5760
	preg f: Jan; n2; Tanzania	186
	preg f: Jan-Feb; Zaire	8856
	preg f: Jan-May, Oct-Dec, none June-Sept; n171; Tanzania	299
	preg f: Jan, Sept; n2n; Zaire	8854
	preg f: peak Mar, low May-Sept, increase Oct-Dec; Zaire	2570
	preg f: May-Aug; few data; Ethiopia	7600
	preg f: Sept-Oct; n2; e Africa	4925
	preg f: peak Sept-Dec; Zaire	2572a
	preg f: Oct-May; Nyasaland	4328
	preg f: yr rd, not Jan, low Dec, July-Aug; Uganda	2445
	preg/lact f: 80% Feb-Mar, 10% July-Aug, 60% Dec; Zaire	2570
	yg present: few May-June, abundant July-Aug; no data other mo; Uganda, Kenya	10231
	yg: peak end rainy-early dry season; Zaire	8706
	repro: peak Nov; Malawi, Nyasaland	4329
Lophuromys sikapusi	preg f: Mar-July, Sept-Oct; Nigeria	4350
	preg f: May, Aug, Oct-Nov; Nigeria	5190a
	preg f: yr rd, peak wet season; Uganda	2446a, 2453
	preg f: end dry season, few data probably; Nimba, Liberia	7362
	fertile: 100% Mar-June, Aug-Dec; < 20% July; Uganda	7797
Lophuromys woosnami	preg f: Sept-May; Zaire	2572a
Malacomys edwardsi	preg f: Nov-July; Nigeria	4352
Malacomys longipes	sexual activity f: peak Dec-Feb; none May, Aug; Zaire	2573
	mate: Mar-June; British Cameroun	9468
	mate: peak Oct-Mar, low June-Aug; e Zaire	2574
	preg f: Mar, May, July, Sept, Dec-Jan; Zaire	8855

Mastacomys fuscus	preg f: Oct-Feb; TAS, Australia	4042
	repro: spring-summer; Australia	10698
Maxomys rajah	m without sperm: 35% Jan-Mar, 4% Apr-June, 14% July-Sept, 21% Oct-Dec; Malaysia	4406
Maxomys whiteheadi	m without sperm: 30% Jan-Mar, 6% Apr-June, 4% July-Sept, 26% Oct-Dec; Malaysia	4406
	preg f: yr rd; Malaysia	4405, 4406
Melomys burtoni	mate: none dry periods; QLD, Australia	3606
	mate: rainy season; Australia	10917
	mate: yr rd, higher wet season; NT, Australia	735
	preg f: 1/32 June, 1/9 Aug; none July, Sept-Nov; QLD, Australia	7053
	preg/lact f: yr rd; NT, Australia	735
	birth: Mar-June, Aug; QLD, Australia	7053
	repro: May-Aug; 26o14'S 153o02'E	10109
Melomys cervinipes	preg f: Sept; n1; QLD, Australia	2841
	repro: May-July; 26o14'S 153o02'E	10109
Melomys levipes	mate: low June-Sept; New Guinea	2836
	mate: end yr; New Guinea	7209
Melomys lorentzi	preg f: Jan, May; New Guinea	7209
Melomys rubex	mate: peak end yr; New Guinea	7209
Melomys rufescens	mate: low July-Sept; New Guinea	2836
	repro: yr rd; New Guinea	7209
Mesembriomys macrurus	mate: est yr rd, n Australia	2214
Micromys minutus	mate: peak Apr-Aug; Germany	3571
	mate: May-Sept; Europe	7433
	mate: end May-Dec; Great Britain	2045
	birth: Apr-Sept; Germany	3571
	birth: May-Sept; Europe	7433
	birth: May-Dec, 74% Aug-Sept; Britain	4396
	repro: peak Sept-Oct; Korea	5424,
Millardia gleadowi	mate: Aug-Oct; India	8723
Millardia meltada	scent gland: large Oct-Nov, Mar-June; India	6097
	scrotal m: Feb-Oct; large sample; Ludhiana, India	6775
	testes: large Oct-Dec, quiescent Jan-May; India	2428
	fertile m: yr rd; Pakistan	730
	preg f: Jan-Oct; large sample; Ludhiana, India	6775
	preg f: none winter; Pakistan	730
	preg f: yr rd, peak June-Aug; India	8904
	preg f: yr rd, peak Oct; w Rajahstan, India	8723
	preg/lact f: peaks Mar-May, Aug-Oct; lab, India	940
Mus spp	repro: Apr-May, Oct; Ivory Coast	3645
Mus booduga	scrotal m: Mar-Oct; Ludhiana, India	6774
	preg f: Apr; n1; Nepal	7382
	preg f: July-Mar; India	1688
	birth: yr rd, peak Mar-Apr; Sri Lanka	8501
Mus bufo/minitoides/triton	preg f: Aug, Nov; Zaire	8855
Mus cervicolor	preg f: Dec, July; India	8723
Mus domesticus	sub adult: yr rd; Ludhiana, India	6774
Mus flavipectus?	birth: Apr, June, Oct, Dec; Canton, China	7185
Mus gratus	preg f: Jan; n1; e Africa	185
	juveniles: yr rd; Katanga, Zaire	8565
Mus indutus	repro: yr rd; S Africa	10158
Mus mayori	preg f: Feb; n1; Sri Lanka	8501
	lact f: Dec; n1; Sri Lanka	8501
Mus minutoides	mate: yr rd; e Africa	5760
	preg f: May, Dec; Nigeria	5190a
	preg/lact f: Feb, Apr, July; Malawi	4329
	birth: Dec, Feb, June; n3; Zambia	331
	birth: wet season; S Africa	10158
	birth: yr rd; n17; Botswana	10154
	lact f: May; n1; Tanzania	186
	lact f: July; n1; Kalahari Gemsbok Nat Park, S Africa	7823
	juveniles: peak July-Aug, low Nov-May; Katanga, Zaire	8565

	repro: Apr-June, Sept-Dec; Uganda	2446a
	repro: Aug-May; sw Nigeria	245
Mus minutoides/triton	% preg/lact f: 80% Jan, 20% Apr-May; 50% June, 30% July, 0% Aug-Sept; 14-31% Oct-Dec; Zaire	2570
	preg f: low Mar-Dec; Zaire	2570
Mus musculus	repro active f: end Apr-July; CA	6432
	m with sperm: yr rd; CA	1246
	mate: Oct-May, est yr rd; Japan	10632
	mate: spring-fall; Europe	7433
	mate: est yr rd; Sudan?	4344
	preg f: Mar-Sept; UK	844
	preg f: Apr-Sept; PA	9164
	preg f: mid Apr-mid Dec; CA	1246
	preg f: May-June, Aug, Oct-Nov; n41; Nepal	7382
	preg f: May-Oct; MD	10413
	preg f: June-Oct, Dec; NC	1290
	preg f: yr rd, low mid June-mid July, peak mid July-Aug; England, UK	9323
	birth: spring-fall; Europe	7433
	birth: yr rd, peak spring-summer; lab	2723
	birth: yr rd; Sri Lanka	8501
	repro: yr rd, peaks Apr-May, Aug-Sept; WI	5210
	repro: est yr rd; Sudan	4344
	repro: est yr rd; Khartoum, Sudan	4347
	repro: yr rd; MS	10151
	repro: yr rd; New Guinea	7209
Mus platythrix	preg f: Aug, Oct; few data; India	8723
	preg f: Aug-Mar; India	1688
	preg f: Sept-May, peak Oct-Jan; India	8924
	lact f: Jan-Oct; Ludhiana, India	6774
Mus setulosus	preg/lact f: Jan; n5; Cameroun	2955
Mus spretus	scrotal testes: 50% Mar-Oct; s Spain	11222
Mus triton	repro m: peak Apr-May; Malawi	4329
	preg f: Apr-July, none Aug-Nov; Malawi	4329
	preg f: peak May-Dec; Uganda	2453
	repro: low Aug-Nov; S Africa	10158
	repro: Mar-May, Aug-Jan; Uganda	7797
Mylomys dybowskii	fertile: Mar-May, Aug-Dec; no data Jan-Feb; Uganda	7797
	preg f: July; n2; Uganda	2444
	preg f: yr rd, none July-Aug; n47; Uganda	2453
Nesokia indica	mate: yr rd; lab	6293
Niviventer eha	preg f: Sept; n2; Nepal	7382
	lact f: Oct; n5; Nepal	7382
Niviventer fulvescens	preg f: May, Aug; n3; Nepal	7382
	preg f: July-Aug; n2; China	182
	lact f: May, Nov; n7; Nepal	7382
Niviventer niviventer	lact f: June, Aug-Sept; Nepal	7382
Notomys alexis	mate: yr rd, peak after heavy rain; Australia	4355
	preg f: Dec-Jan; n5; NT, Australia	1266a
Notomys cervinus	mate: winter; ne SA, sw QLD, Australia	11495
	mate: yr rd; captive	4355
Notomys fuscus	birth: yr rd; captive	427
Notomys mitchellii	repro: yr rd; lab	4355
Oenomys spp	juv found: yr rd, peak Nov-Dec, Feb; Kantanga, Zaire	8565
Oenomys hypoxanthus	mate: peak Oct-Mar, low June-Aug; e Zaire	2574
	preg f: Feb, Apr, June, Aug, Oct; Zaire	8855
	preg/lact f: 50-60% Nov-Mar, 10% July-Aug; Zaire	2570
	birth: dry season; British Cameroun	9468
Pelomys fallax	preg f: Mar, June, Aug, Nov; Zambia	9893
	preg f: May-June; n2; Malawi	4329
	birth: est Aug-Apr; S Africa	10158
	juveniles: yr rd; Zambia	331
	sub adults: June-July; Malawi	4329
Pithecheir parvus	mate: est yr rd; Malaysia	6448

Pogonomelomys sevia	mate: end yr; New Guinea	7209
Pogonomys macrourus	mate: low July-Sept; New Guinea	2836
Pogonomys sylvestris	mate: low July-Sept; New Guinea	2836
Praomys daltoni	mate: yr rd, peaks start end rainy season; Nigeria	246, 247
	repro: Feb-May, Nov-Dec; Ivory Coast	3645
	repro: yr rd, peaks Feb-Apr, Oct-Nov; Nigeria	4354a
Praomys erythroleucus	mate: yr rd, peak end wet-early dry season; Sierra Leone	1215
	preg f: Sept-Oct; Senegal	5057
	preg f: peak Oct-Nov; Sierra Leone	1215
Praomys fumatus	preg f: Sept, Nov; n2; Kenya	4925
Praomys jacksoni	fecundity: 100% Dec-Jan, 16% June-July; Malawi	4329
	sexually active: yr rd, peak Oct-Apr, low June, Aug; Zaire	2573
	testes regress: dry season; Malawi	4329
	mate: peak Oct-Mar, low June-Aug; e Zaire	2574
	preg f: Feb-June, Aug-Sept, Dec; Zaire	8855
	preg f: Nov-Dec; n6; Zaire	8856
	preg f: Sept-Nov, Apr; Malawi	4329
	preg/lact f: 50-70% Nov-July; 20-30% Aug-Oct; Zaire	2570
	yg: increase Feb, abundant Apr-June, rare Sept; Zaire	8564
Praomys morio	mate: peak Apr-May; Kenya, Uganda	10231
	preg f: Aug-Sept; n3/17; Uganda	2444
	birth: yr rd; s Uganda	8080
	lact f: May-Aug; n6; Uganda, Kenya	10231
	repro f: yr rd, low 24% Aug, peak 77% June; Uganda	2445
Praomys natalensis	fertile: Apr-May, Sept-Dec; Uganda	7797
	fertile m: Oct-June peak Nov-Apr; Zambia	1789
	fertile m: Nov; Sudan	4345
	scrotal testes: Sept; Zimbabwe	10591
	testes: large Mar-Apr, Oct-Dec; Uganda	7791
	testes: large Nov-June, peak Jan-Mar; Zambia	1788
	testes: small June-Sept; Zambia	9893
	preg f: May-mid June, Oct-Dec; Uganda	7791
	mate: none Mar-May; Kenya	10712
	mate: Aug-May; Transvaal, S Africa	1954
	mate: yr rd, low Oct-Apr; Zambia	331
	mate: yr rd; Botswana	9892
	preg f: Jan-Mar, July, Sept, Nov; n30/225; Tanzania	299
	preg f: Jan, May; n5; e Africa	4925
	preg f: Feb-May; Zambia	1788
	preg f: Feb, Apr-May, Oct; Nigeria	5190a
	preg f: Apr-May, Oct-Dec; Uganda	7791
	preg f: May-June, Oct-Dec, peak May, Oct-Nov; n119; Uganda	2446a, 2453
	preg f: none June-Jan; Malawi	4329
	preg f: Sept; n2; Liberia	184
	preg f: Sept-Feb; Natal, S Africa	10592
	preg f: Oct-Dec; British Cameroun	9468
	preg f: Dec-June; Zambia	1789
	preg f: yr rd, peak Apr; Botswana	10154
	preg/lact f: June, Sept; no data Dec; Zimbabwe	10591
	birth: peak Feb-Apr; Transvaal, S Africa	8965
	birth: peak Feb-May; Malawi	4329
	birth: peak end rainy season; Zambia	331
	birth: yr rd, low winter; Zimbabwe	1815
	lact f: mid May-mid Aug; n2; Uganda	10231
	juveniles: peak spring, autumn; Transvaal, S Africa	7200
	yg: abundant June-Oct, Jan; Zaire	8564
	repro: Oct-Apr; Mozambique	3821
	repro: yr rd, low June-July; S Africa	10158
	repro: yr rd; Namibia	9945
Praomys tullbergi	scrotal testes: yr rd; Nigeria	4353
	mate: yr rd, peak Dec-Apr; Sierre Leone	2154
	mate: yr rd, peak Dec-Apr; Nigeria	4354
	preg f: Jan-Feb; n3; e Africa	4925

	preg f: end dry period; Mt Nimba, Liberia	7362
	preg f: yr rd; Nigeria	4350
	preg f: yr rd; W Africa	4353
	lact f: Feb; n1; Uganda, Tanzania	186
	repro: peaks July-Aug, Nov-Mar; Nigeria	4352
Pseudomys albocinereus	mate: yr rd; lab	4355
Pseudomys apodemoides	testes: enlarge Mar-Apr; SA, Australia	3259
	mate: fall-winter; SA, Australia	3259
	lact f: peak Aug-Feb; n2 pop; VIC, Australia	1934
Pseudomys australis	fertile: May; QLD, Australia	10708
	mate: yr rd; captive	4355
Pseudomys desertor	mate: yr rd; lab	4355
Pseudomys gracilicaudatus	mate: June; NT, Australia	10708
Pseudomys higginsi	mate: summer; TAS, Australia	4042
	birth: mid Nov-mid Mar; TAS, Australia	4042
Pseudomys novaehollandiae	preg f: Jan-Mar; TAS, Australia	8801
	preg f: Aug-Mar, peak Aug-Dec; NSW, Australia	5645
Pseudomys shortridgei	preg f: Oct-Dec; VIC, Australia	4355
Punomys lemminus	birth: Nov-Apr; Peru	7983
Rattus argentiventer	bimodal, peaks harvesting rice; Malaysia	6162
Rattus exulans	testes: large yr rd; Ponape I, Micronesia	5216
	mate: July-Sept; Ponape I, Micronesia	5216
	mate: Dec-Aug, peak Dec-Mar; New Guinea	2839
	mate: yr rd, low Aug-Sept; Marshall I, n Pacific Ocean	2911
	mate: yr rd, low June-Oct; New Guinea	2836
	preg f: est yr rd, peak Feb-July; New Guinea	10702
	preg f: yr rd; Burma	11434
	preg f: yr rd; Malaysia	4405, 4406
	preg f: yr rd, peak Feb-Apr, Sept-Nov; Java, Indonesia	10955
	birth: Jan-Sept, peak Mar-Aug; HI	11833
	birth: peak Mar-Aug; HI	11832
	repro: peaks Nov, Apr; New Calidonia	7886
	repro: yr rd; New Guinea	7209
Rattus fuscipes	m fertile: yr rd; NSW, Australia	10696
	scrotal m: July-Feb; VIC, Australia	3260
	mate: Nov-Dec; Bass Strait, Australia	4841
	mate: peak Dec-Jan; s Australia	6344
	mate: spring-autumn; QLD, Australia	2841
	birth: Jan-Mar; VIC, Australia	9188
	birth: yr rd; NSW, Australia	10696
	birth/lact: Sept-Feb; VIC, Australia	3260
Rattus hoffmanni	lact f: Mar-May, Aug-Dec; Sulawesi, Indonesia	7675a
Rattus leucopus	preg f: Jan-Feb, Apr, Oct, Dec; New Guinea	10702
Rattus lutreolus	mate: none Apr-July; SA, Australia	614
	mate: Oct-Feb; TAS, Australia	4041
Rattus muelleri	m without sperm: Oct-Dec 27%, Jan-Mar 22%, Apr-June 0%, July-Sept 10%; Malaysia	4406
	preg f: 3% Jan-Mar, 13% Apr-June, 26% July-Sept, 19% Oct-Dec; Malaysia	4406
Rattus niobe	mate: Sept-June; New Guinea	2836
	preg f: est yr rd; New Guinea	10702
	repro: yr rd; New Guinea	7209
Rattus norvegicus	m fertile: yr rd; MD	6242
	testes: large Dec, small July; s GA	8802
	testes: large yr rd; MD	2320
	mate: peak Mar-June; Wales, UK	8393
	mate: Dec-Feb; s GA	8802
	mate: spring-fall; Europe	7433
	mate: yr rd, low Jan, peak Mar; n620 f; VA	4485
	preg f: Feb-Oct, peak June-July; MD	6242
	preg f: yr rd; Wales, UK	8393
	preg f: yr rd; India	5141
	birth: Jan-Feb; s GA	8802
	birth: spring-fall; Europe	7433

	repro: low Jan; New Caledonia	7886
	repro: yr rd; New Guinea	7209
Rattus praetor	preg f: May-June, Oct; n5; New Guinea	10702
	lact f: July; n1; New Guinea	10702
	repro: yr rd, low June-Oct; New Guinea	2836
Rattus rattus	testes: large yr rd; Ponape I, n Pacific Ocean	5216
	testes: peak Jan-Feb, June-July; India	8902
	mate: peaks Jan-Feb, Sept-Oct; Kenya	10607
	mate: peaks Mar-Oct, Nov-Feb; Germany	3571
	mate: Apr-Sept; w Rajahstan, India	8723
	mate: peaks Apr, Sept, low Jan-Feb; Cyprus	11475
	mate: spring-fall; Europe	7433
	mate: yr rd, peak Jan-Mar; Ponape I, n Pacific Ocean	5216
	mate: yr rd, peak Feb-Mar, May-June, low July-Aug; TX	2316
	mate: 7 mo; Philippines	10544
	preg f: peaks Jan-Mar, Sept, or Feb-Mar, Nov; n2 populations; New Zealand	858
	preg f: Jan, Mar, Sept-Oct; n8; New Guinea	10702
	preg f: Jan 0/98; Feb 4.1% n103; Mar 7% n67; Apr-Dec 11-35.3%; India	1765
	preg f: 38% end Jan-early Mar; Vietnam	10823
	preg f: Feb-Dec, peak Mar-Apr, Sept; Cyprus	11475a
	preg f: peak Feb-May, low Nov-Jan; n2041; India	315
	preg f: peaks Apr-July, Oct-Nov, low Mar; n11,096; Sri Lanka	4826
	preg f: Apr-May 9-10%, Oct-Feb 20%; Perth, Australia	1897
	preg f: peak June, Dec; Jodhpur, India	8902
	preg f: Aug 25%, Nov-Jan 4%; HI	3090
	preg f: Oct; n1; FL	7498a
	preg f: peak summer, low Nov-Jan; n1877 f; Egypt	8423
	preg f: yr rd, 22% Apr, 80% July; few data; Czechoslovakia	3250
	preg f: yr rd, low May; India	899
	preg f: yr rd; Nigeria	1513
	preg f: yr rd; Malaysia	4405, 4406
	preg f: yr rd; India	8902
	birth: May-June; Rajasthan, India	8720
	birth: spring-fall; Europe	7433
	repro: low Mar, Aug; no data other mo; Zaire	2570
	repro: low May-June; New Calidonia	7886
	repro: mid Nov-mid Mar; England, UK	11475
	repro: yr rd; Bohemia, Czechoslovakia	5315
	repro: yr rd; New Guinea	7209
Rattus richardsoni	birth: May-June; few data; New Guinea	10702
Rattus steini	lact f: est Apr-Dec; New Guinea	10702
Rattus sordidus	mate: yr rd; QLD, Australia	7053
Rattus taerae	preg f: Nov; n2; Kenya	4925
Rattus tiomanicus	preg f: yr rd, peak Dec-Jan, low Apr-June; Malaysia	11885
Rattus tunneyi	birth: none dry period; QLD, Australia	3606
Rattus turkestanicus	preg f: May, Aug, Oct; n5; Nepal	7382
Rattus verecundus	mate: end Oct-May; New Guinea	2836
	preg f: Mar; n3; New Guinea	10702
	lact f: Sept; n1; New Guinea	10702
	repro: Nov-June; New Guinea	7209
Rhabdomys pumilio	sperm: 50% m Aug-Feb; dry summer; S Africa	2298
	testes: large Aug, declines Dec; Kenya	10712
	mate: Oct-Apr; S Africa	1335
	mate: peak wet season; Malawi	4329
	preg f: Jan; n1; Kalahari Gemsbok Nat Park, S Africa	7823
	preg f: Jan-Apr, July, Nov-Dec; n45; Transvaal, S Africa	8965
	preg f: Sept; Kenya	5760
	preg f: Sept; n3; Kenya	4925
	preg f: Sept-Apr; S Africa	2298
	birth: Mar, May, Nov; Malawi	4329
	birth: few May-July; captive	1814
	birth: Sept-Mar; Cape Flats, S Africa	4640

	juveniles: Oct-May; Cape Flats, S Africa	2298
	juveniles: peak fall, spring; Transvaal, S Africa	7200
Stochomys longicaudatus	sexually active f: Feb-Apr 100%, Aug-Nov 80%; Zaire	2573
	mate: peak Oct-Mar, low June-Aug; e Zaire	2574
	mate: est yr rd; Nigeria	5760
	preg f: Jan, Apr-Sept; Zaire	8855
	preg f: Mar, June; Nigeria	4354a
Thallomys paedulcus	preg f: Apr; n1; Kenya	4925
	preg f: Oct; n2; Namibia	9945
	preg/lact f: Jan, Apr, May; Malawi	4329
	birth: Aug-Apr; S Africa	10158
	lact f: Sept; n1; Namibia	9945
	juveniles: July-Aug; e Africa	5760
Uranomys ruddi	conceive: est Mar-July; Malawi	4329
	preg f: May-July, Nov; Ivory Coast	3645
	preg f: yr rd, peak Oct-Dec; Ivory Coast	752a
Uromys caudimaculatus	birth: Nov-Dec; New Guinea	7209
Zelotomys hildegardeae	preg f: July; n1; w Uganda	5760
	preg f: Nov; n1; s Kenya	5760
	preg f: Nov; n1; Kenya	4925
Zelotomys woosnami	mate: Dec-Mar; Botswana	947
	preg f: Feb; n3; Botswana	10154
	birth: Dec-Apr; S Africa	10158
	repro: Dec-Mar; S Africa	947
Zyzomys argurus	mate: yr rd; NT, Australia	733
	preg f: yr rd, peak Apr; nw Australia	1192
	preg/lact: yr rd; nw Australia	733
	birth: peak Mar-May; nw Australia	733
Zyzomys woodwardi	mate: yr rd, low Nov-Dec; NT, Australia	733
	preg/lact: yr rd; nw Australia	733
	birth: peak Mar-May; nw Australia	733
	repro: May-Sept; nw Australia	733

Order Rodentia, Family Gliridae

Dormice (*Dryomys* 17-44 g; *Eliomys* 45-170 g; *Glirulus* 16-17 g; *Graphiurus* 15-85 g; *Muscardinus* 15-40 g; *Myomimus* 60-110 mm; *Myoxus* 70-180 g) are primarily granivorous, terrestrial inhabitants of Africa south of the Sahara (*Graphiurus*) or the Palaearctic. *Eliomys* has a vaginal plug after mating (6237). Neonates are blind and naked (4180, 8050). Their eyes open at 16-26 d, (*Muscardinus* 6237, 8050, 9672, 11376), 18-21 d (*Eliomys* 3571, 7433), 21 d (*Dryomys* 3571, 7433), or 20-22 d (*Myoxus* 8050). Females have 6-12 mammae (331, 3571, 4495, 5760, 7433, 8050). Males have seminal vesicles and a baculum (4105, 5040, 8050). Details of anatomy (3570), endocrinology (5236, 5236a, 5237), and spermatogenesis (5317) are available.

NEONATAL MASS		
Dryomys nitedula	1.3 g	7862
	1.9 g	8050
Graphiurus murinus	3.5 g; n2	5760
Muscardinus avellanarius	0.80 g	9672
NEONATAL SIZE		
Graphiurus murinus	40 mm; n2	5760
WEANING MASS		
Graphiurus ocularis	25.6 g; lightest trapped	1695
LITTER SIZE		
Dryomys laniger	4.0 emb; r3-5; n5 f	10273
Dryomys nitedula	2 ps ea; n2 f	3578
	mode 2-4; r1-7	3571
	2-4	7433
	2.9 yg; r1-5; n7 lt	3578
	3	7862
	mode 3-4; r2-5	8050
	mode 3-4; r2-6	4645
	4 emb; n1 f	3578

Eliomys quercinus	2-8	7433
	2.86; r2-5; n7 lt	5484
	mode 3-6; r2-8	3571
	3-6	4645
	3-8	5490
	4.12; r1-7; n64 lt	5488
	5 emb; n1 f	7900
	5 emb; n1 f	5489
	5.1 emb; mode 5; r3-7; n9 f	5487
	5.75 yg; r5-7; n4 lt	5489
	6.6 emb; n9 f	6314
	7 yg; 1 nest	8050
Glirulus japonicus	4 yg; r3-7	7983
Graphiurus crassicaudatus	2 yg; n1 lt	184
Graphiurus hueti	2 emb; n1 f	7362
Graphiurus murinus	1 emb; n1 f	7362
	r2-3 nestlings	4495
	mode 3 yg	4354a
	3; n1 lt	4180
	3 emb; n1 f	4495
	3 emb; n1 f	331
	3 yg; n1 f	334
	3.0 emb; r2-4; n3 f	4925
	4 emb; n1 f	9892
	4 yg; n1 nest	4925
	max 5	5760
Graphiurus ocularis	4-6 yg	1695
Graphiurus parvus	4 emb ea; n2 f	4925
Graphiurus platyops	5.5 emb; r5-6; n2 f	331
Muscardinus avellanarius	2-4	11376
	3-5	4645
	3-5	8050
	mode 3-7; r1-9	3571
	3-9	7433
	3.65; r3-5; n17 lt	5491
	3.87; mode 3,4; r2-8; n78 lt	9672
	4.3; r4-5; n3 lt	641
	4.5 yg; r4-5; n4 lt	3578
	4.9 ps; r3-6; n7 f	3578
Myoxus glis	2.0-6.4; r2-9	11359
	mode 2-9; r2-11	3571
	2.7 yg; r2-3; n3 lt	3578
	3-7	7433
	3-10 emb	4645, 8050
	4.9 ps; r3-8; n10 f	3578
	5.5 emb; r5-6; n2 f	3578
	5.7; r1-9; n3 lt, 1 f	11360
	10 emb ea; n2 f	6293
SEXUAL MATURITY		
Eliomys quercinus	f/m: 1 yr	3571
	f: 1st mate: mode 2nd yr	5488
Graphiurus spp	m: testes scrotal: 14-15 g	4350
	f: 1st preg: 15 g	4350
Graphiurus murinus	m: spermatogenesis: 14-15 g	4350
	f: 1st mate: 15 g	4350
Muscardinus avellanarius	f: 1st birth: 3 yr; n1	641
	f/m: 10-12 mo	3571
Myoxus glis	f: 1st birth: ~10-11 mo	4645
	f: 2nd yr	8050
	m: 10-11 mo	8050
	m: sperm: 1 yr	5040
	f/m: 10-12 mo	3571
CYCLE LENGTH		
Eliomys quercinus	mode 10 d; r6-10; interestrous interval	5488

GESTATION		
Dryomys nitedula	21-28 d	7433
	mode 23-25 d; r21-28	3571
	min 1 mo	8050
Eliomys quercinus	21-22, 23 d	5488
	22 d; n1; vaginal plug to birth	6237
	23 d	7433
	23 d; r21-28	3571
Glirulus japonicus	~1 mo	7983
Graphiurus murinus	~24 d	5760
Muscardinus avellanarius	20 (?)-28 d	3571
	20 d	7433
	~26 d	5491
	30 d; r20-43; pairing to birth	9672
Myoxus glis	20-25 d	4645, 8050
	est ~4 wk	11359
	28-30 d	3578
	30-32 d	3571, 7433
LACTATION		
Eliomys quercinus	wean: 4 wk	3571
Muscardinus avellanarius	solid food: 19 d	6237
	wean: 1 mo	8050
	independent: 5-6 wk	4645, 8050
	independent: 40 d	11376
	wean: ~1 mo	3571, 7433
Myoxus glis	solid food: 24-25 d	8050
	wean: ~3 wk	4645
	wean: 25-30 d	8050
	wean: >5 wk; r4-5 wk	3571
INTERLITTER INTERVAL		
Graphiurus ocularis	8 wk	1695
Muscardinus avellanarius	50 d; n1	3578
LITTERS/YEAR		
Dryomys laniger	1	10273
Dryomys nitedula	1	3578
	1	4645
	1; perhaps 2	8050
	1; r1-2 (?)	3571
	1-2	7433
Eliomys quercinus	1	3570
	1	5488
	1; r1-2	3571
	1; r1-2	4645
	1-2	7433
	2?	8050
	est 2	5487
Glirulus japonicus	1-2	7983
Graphiurus ocularis	2	1695
Muscardinus avellanarius	1	11376
	1+; max 2	3578
	1; r1-2 (rare)	9672
	1.5; r1, 2	641
	1-2 (3)	3571
	2	4645
	2	7433
	2/summer	8050
Myoxus glis	1	3578
	1	4645
	1	11360
	1	8050
	1; r1-2 (?)	3571
	1-2	7433

SEASONALITY

Dryomys laniger	testes: large June, small Aug; s Turkey	10273
	preg f: June; few data; s Turkey	10273
	lact f: end July-early Aug; few data; s Turkey	10273
Dryomys nitedula	testes: active June-July; no May data; n Moravia, Czechoslovakia	3578
	testes: decrease mid June-Aug; USSR	8050
	mate: Apr-May; central Europe	3571
	mate: Apr-May; Eurasia	7433
	mate: May-June; n Moravia, Czechoslovakia	3578
	ovulate: May-June; n Moravia, Czechoslovakia	3578
	CL: present May-June, absent July-Aug; n Moravia, Czechoslovakia	3578
	birth: peaks Mar-Apr, July, Sept-Oct; Israel	7862
	birth: May-June; Eurasia	7433
	birth: May-Aug, peak May-June; central Europe	3571
	birth: June; USSR	8050
	birth: June-July; n Moravia, Czechoslovakia	3578
Eliomys quercinus	spermatogenesis: Feb-Aug; peak May-June; lab	3570
	testes: large Feb-July; Germany	5488
	testes: large Mar; n15; Germany	5490
	testes: large Apr; Formentera I, Spain	5487
	m active: Apr-Aug; Alps, Europe	6314
	estrus: Apr-May; Germany	7433
	mate: Feb-?; few data; Tunisia	5490
	mate: Feb-July, peak May-June; Germany	5488
	mate: Apr-May, Aug; Germany	3571
	mate: mid June-mid Aug; captive outside	1140
	ovulate: May; lab	3570
	preg f: Apr, Aug; no data other mo; Formentera I, Spain	5487
	preg f: Apr-June, Aug, Oct; Germany	5488
	preg f: Oct, few data; Mallorca, Baleares, Spain	5489
	preg/lact f: May-Aug, peak end June-early July; Alps, Europe	6314
	birth: Mar; Germany	5488
	birth: May-June, Sept; Germany	3571
	birth: May-June; Germany	7433
	birth: mid June-July; France	6314
	nest: June; n1; USSR	8050
Graphiurus murinus	preg f: Jan; n1; Zambia	331
	preg f: Feb-Mar, May, Aug; Nigeria	4354a
	preg f: June, Nov; central Ivory Coast	3645
	preg f: Nov; n3; Kenya	4925
	lact f: Jan; e Africa	185
	lact f: May-June, Dec; central Ivory Coast	3645
	juveniles: Nov; Zambia	331
	yg found: Feb-Mar; S Africa	10158
Graphiurus ocularis	birth: spring-summer; S Africa	1695
Graphiurus parvus	preg f: Nov; n2; Kenya	4925
Graphiurus platyops	birth: Feb; n1; Zambia	331
	juveniles: Nov-Dec; Zambia	331
Muscardinus avellanarius	testes: large May-Sept; n Moravia, Czechoslovakia	3578
	estrus: May-July; Europe	7433
	mate: May-Aug; central, s Germany	3571
	mate: mid May-early July; n Moravia, Czechoslovakia	3578
	mate: spring; Europe, USSR	4645
	conception: ~end May; near Munich, Germany	11376
	cl present: end Apr-Aug; no Sept data; n Moravia, Czechoslovakia	3578
	birth: May-mid Sept; s Harz Mtns, Germany	9672
	birth: June-Aug; Europe	7433
	birth: June-Sept; central, s Germany	3571
	birth: June-mid Oct; USSR	8050
	birth: mid June-July; n Moravia, Czechoslovakia	3578
	birth: peak end June-early July; near Munich, Germany	11376

	birth: July-Aug; German & Austrian Alps	5491
	birth: spring-early summer; Europe, USSR	4645
Myoxus glis	sperm: July, none May-June, Aug-Sept; s Slovakia, Czechoslovakia	5040
	testes: active June-Aug; Moravia, Czechoslovakia	3578
	testes: large end May-mid July; captive, Germany	11360
	estrus: May-July; Europe	7433
	mate: end May/early June-mid Aug; Germany	3571
	mate: end June; ~3 wk after emergence; USSR	8050
	mate: end June-early Aug; Germany	11359
	mate: July; n Moravia, Czechoslovakia	3578
	mate: 3 wk after hibernation; Europe, USSR	4645
	preg f: June-Aug; n Moravia, Czechoslovakia	3578
	preg f: Aug-Sept; n2; Iran	6293
	birth: June-Aug; Europe	7433
	birth: June-Sept; Germany	3571
	birth: July, perhaps Aug; USSR	8050
	birth: Aug-early Sept; n Moravia, Czechoslovakia	3578
	birth: Aug-Sept; lower Saxony, Germany	11360
	birth: spring; Europe, USSR	4645

Order Rodentia, Family Seleviniidae, Genus *Selevinia*

Selevinia betpakdalaensis is a 20-25 g inhabitant of the Bet-Pak-Dala Desert in Kazakhstan. This crepuscular rodent preys on terrestrial invertebrates. *Selevinia*'s molt includes epidermal tissues as well as fur.

LITTER SIZE		
Selevinia betpakdalaensis	6-8 emb	7983
SEASONALITY		
Selevinia betpakdalaensis	mate: May, July; Kazakhstan	7983

Order Rodentia, Family Zapodidae

Jumping mice (*Eozapus, Napaeozapus, Sicista, Zapus*) are 5-35 g, nocturnal, terrestrial omnivores from North America and northern Eurasia. We found no reproductive information on *Eozapus*. Female *Napaeozapus* and *Zapus* are larger than males (5034, 11624, 11957). Neonates are naked and open their eyes at 18-21 d (*Sicista* 3571, 7433), 22-25 d (*Zapus* 2112, 4271, 8812, 10582, 11624), or 26-31 d (*Napaeozapus* 4271, 6304, 9883, 9884, 10183). Females have 6-8 mammae (502, 1474, 2548, 3571, 6304, 7433, 9612, 11957). Males have penial spines, a baculum, and seminal vesicles (1473, 5983, 8051, 11957).

NEONATAL MASS		
Napaeozapus insignis	0.87 g; n1	6304
	1 g	9883, 9884
	1.0 g; n9 yg, 2 lt	4271
Zapus hudsonius	0.78 g; r0.65-0.9; n6	8812
	0.8 g; r0.7-1.0; n14	11624
Zapus trinotatus	0.82 g; r0.7-0.9; n6	10582
NEONATAL SIZE		
Napaeozapus insignis	HB: 24.2 mm; n1	6304
	HB: 29.5-31 mm; n3	9883
	HB: 30 mm; n9	4271
Zapus hudsonius	TL: 34.4 mm; r30-39; n19	11625
	TL: 38.3 mm; r37-39; n7 stillborn	10954
	52 mm; n2	4271
WEANING MASS		
Napaeozapus insignis	8.5-9.3 g	6304

LITTER SIZE		
Napaeozapus insignis	3-5; r1-6	1474
	3.75 emb; r2-6; n4 f	9883
	r4-6 emb; n14 f	8738
	4.0 yg; r2-5; n3 f	9883
	4.2 emb; r4-5; n10 f	4271
	4.3 ps; r4-5; n6 f	4271
	4.3 ps; r2-7; n45 f	11957
	4.6 emb; ±0.11; r2-7; n80 f	11957
	mode 5 yr; r2-6(8?)	5210
	5 yg; n1 nest	10183
	5.3 emb; r4-7; n6 f	5798
	7 ps; n1 f	10839
Sicista betulina	2-6	3571
	2-6	7433
Sicista napaea	1-6	9954
Sicista tianschanica	3-6 emb	8051
Zapus hudsonius	3.5; r1-6; n2 f	9883
	3.5 yg; r2-5; n2 lt	4271
	3.5 emb; r2-4; n4 f	4271
	4 ps ea; n2 f	4271
	4.5; r3-7; n17 lt	11625
	4.5 emb; r2-8; n62 f	11625
	4.6 ps; r2-7; n12 f	10954
	5 emb; n1 f	11504
	5.3 yg; n8 lt	8812
	5.4 emb; r2-8; n62 f	5405b,
	5.53 emb; r2-9; n78 f	11624, 11625
	5.8 emb; r4-7; n14 f	8812
	5.9 emb; r4-8; n25 f	10954
	6 emb; n1 f	495
	6 emb; n1 f	2025
	6 ps; 6 emb; n2	10101
	6 yg; r5-7; n4 lt	11956
	6.0 emb; n6 f	6166
	7 emb; n1 f	261a
	7 emb; n1 f	1290
	7 emb; n1 f	9164
	7.8 emb; r6-9; n5 f	3703
	8 yg; n1 f	1290
Zapus princeps	2-7	1474
	4 emb; n1 lt	6488
	mode 4-6	502
	4.46 emb; se0.23; n22 f	3159
	4.7 emb, yg; r2-7; n139 lt	2112
	5 ps; n1 f	11957a
	5.4 emb; r4-7; n5 f	1119
	5.5 emb; r5-6; n2 f	10597
	5.9 emb; r4-8; n18 f	11957a
	6.0 emb; r5-7; n3 f	7809
Zapus trinotatus	4-6 yg	5146
	6 yg; n1 f	10582
SEXUAL MATURITY		
Napaeozapus insignis	f: not 1st yr	8152
Zapus hudsonius	f: 1st birth: 2 mo	8812
	f: median 14.5 g; 95%ci 12.1-17.4	68
	m: median 12.5 g; 95%ci 6.4-24.6	68
	f/m: 2 mo	7882
Zapus trinotatus	f/m: yr after birth	3600
GESTATION		
Napaeozapus insignis	19, 22, 23, ~29+ d; n4; capture to birth	9883
	20-23 d	5210
	23 d; n1; pairing to birth	9884
	25 d; n1; capture to birth	9884

Sicista betulina	28-35 d	3571, 7433
Sicista napaea	~20 d	9954
Zapus hudsonius	17-22 d	1871a
	17.5, 18, 20/21 d; n3 f with lactation	8812
	18-21 d	1474
Zapus princeps	18 d	1871a
LACTATION		
Napaeozapus insignis	solid food: 34 d	6304
	wean: ~5 wk	6304
Sicista betulina	wean: >5 wk	3571
Zapus hudsonius	solid food: at eye opening	8812
	nurse: 3-4 wk	8812
	wean: 4 wk	1871a
Zapus princeps	den emergence: ~30 d	2112
	wean; >28 d	2112
	nurse: 1 mo	1871a
	nurse: 33 d; ±2.5	2112
	independent: 2-3 mo	1871a
Zapus trinotatus	independent: 1 mo	3600
LITTERS/YEAR		
Napaeozapus insignis	1	11956
	1	5186
	mode 1	4271
	1-2	8152
	2/season	11957
	possibly 2	1474
Sicista betulina	1	7433
	1; r1-2	3571
Sicista napaea	est 1	8051
Sicista subtilis	1	4645
Sicista tianschanica	1	8051
Zapus hudsonius	1	2088
	1; possibly 2	11956
	max 2	10839
	2-3	11624
	2-3	8812
	2-3/season	1474
	2-3/season	7882
Zapus princeps	1	1474
SEASONALITY		
Napaeozapus insignis	epididymal sperm: end May-early June; MAN, Canada	5186
	mate: May-Sept; ne US, se Canada	11957
	preg f: end May-Aug; NH	8738
	preg f: June; s QUE, Canada	11956
	preg f: peak June; NS, Canada	8152
	birth: early June-mid July; NY	4271
	birth: June-Sept; ne Canada, ne US	1474
Sicista betulina	estrus: May-June; Scandinavia, s Norway, Germany, Russia	7433
	birth: June-Aug; Scandinavia, s Norway, Germany, Russia	7433
	birth: June-Aug; Germany	3571
	mate: mid May-June; Germany	3571
Sicista napaea	repro: end June-mid Aug; USSR	9954
Sicista tianschanica	testes: large June, decrease end summer-fall; USSR	8051
	preg f: mid June; USSR	8051
	lact f: July; n1; USSR	8051
Zapus hudsonius	testes: scrotal May-Sept; MI	7882
	mate: May-Oct; NY	11624
	mate: June-Aug; n N America	1474
	preg f: May; n1; MN	495
	preg f: June; n1; SD	261a
	preg f: June; few data; SD, WY	10954
	preg f: June-Aug; n5; ND	3703
	preg f: June-early Sept; ONT, Canada	5034
	preg f: July; n6; se MT	6166

	preg f: Sept; n1; NC	1290
	preg/lact f: June-Aug; MI	7882
	preg/lact f; June-Sept; MI	7585
	birth: peak end May-early June; WI	5210
	birth: June; NY	4271
	birth: June; BC, Canada	2088
	birth: mid June-Aug; MN	8812
	birth: peak spring; s QUE, Canada	11956
	lact f: June; n4; ND	3703
	lact f: June; n1; NC	1290
	lact f: Aug-Sept; n2; NE	5405b,
Zapus princeps	testes: large 12 d after emergence; UT	2112
	mate: end May-mid July; UT	2112
	conceive: peak mid June-early July; ALB, Canada	3159
	preg f: June-July; BC, Canada	10597
	preg f: June, Aug; MAN, Canada	11957a
	birth: June-July; w Canada, w US	1474
	birth: mid July-?; BC, Canada	10597
Zapus trinotatus	birth: July-Aug; OR	3600

Order Rodentia, Family Dipodidae

Jerboas (*Alactagulus* 45-60 g; *Allactaga* 45-420 g; *Cardiocranius* 50-75 mm; *Dipus* 70-95 g; *Euchoreutes* 70-90 mm; *Jaculus* 50-135 g; *Paradipus* 110-155 mm; *Pygeretmus* 75-120 mm; *Salpingotus* 40-55 mm; *Salpingotulus* 35-45 mm; *Stylodipus* 100-130 mm) are nocturnal, saltatorial, often semifossorial herbivores from arid regions from northern Africa through central Asia. Male *Jaculus* are slightly larger than females (8134). Corpora lutea are found only in pregnant females suggesting that ovulation is induced (*Jaculus* 3755). A postpartum estrus has been observed (8051). Neonates are naked and blind (4346, 4349, 4351, 7778, 8051). *Jaculus* neonates open their eyes at 38 d (4349, 4351). Females have 8 mammae (8051). Males have seminal vesicles and penial spines, but usually no baculum (4105, 8051, 11822).

NEONATAL MASS		
Jaculus jaculus	2 g; n6	4346, 4349
NEONATAL SIZE		
Jaculus jaculus	25 mm; n6	4346, 4349
Salpingotulus michaelis	HB: 17 mm	9175
LITTER SIZE		
Alactagulus pumilio	1-6 emb	10516
	2 yg	10516
	3 emb ea; n2 f	11307
	3.0 emb; ±0.08; mode 3; r1-4; n71 f	7778
	3.3-4.2 emb, ps; r1-6 emb	9888
	3.3 emb; ±0.06; mode 3; r1-6; n231 f	7778
	3.4 emb; ±0.17; mode 3; r2-6; n43 f	7778
	3.4 emb; ±0.11; mode 3; r1-6; n104 f	7778
	3.9 emb; ±0.22; mode 4; r2-7; n28 f	7778
	4; r3-6	8051
	5.0 uterine blood clots; r3-7; n4 f	6293
Allactaga bobrinskii	5.3 emb; ±0.22; r3-7; n32 f	7778
Allactaga bullata	2.5 emb; r2-3; n2 f	8051
	2.7 emb; n3 f	10516
Allactaga elater	2-8 emb	10516
	4 emb ea; n2 f	6293
	4.1; r2-6	7005
	4.1 emb; ±0.21; mode 5; r1-6; n34 f	7778
	4.4 emb; ±0.14; mode 4; r2-8; n88 f	7778
	4.5 emb; 1st lt; 3.7 emb; 2nd lt	742
	4.5 emb; ±0.09; mode 4; r2-8; n147 f	7778
	4.7 emb; ±0.10; mode 5; r2-8; n148 f	7778
	4.8 emb; ±0.21; mode 5; r2-8; n33 f	7778
	4.9 emb; ±0.20; mode 4,6; r2-8; n48 f	7778
	5; r2-6	8051
Allactaga euphratica	2-8	7778
Allactaga jaculus	1-6	8051

	3.4; r2-5	7005
Allactaga major	3.0 emb; ±0.16; r2-4; n8 f	7778
	4.1 emb; ±0.49; r4-6; n9 f	7778
	4.6 emb; ±0.38; r2-7; n3 f	7778
Allactaga severtzovi	3.5 emb; r1-8	7778
	4.0 emb; r3-6	7778
	4.1 emb; ±0.14; r2-7; n>48 f	7778
	5.0 emb; ±0.38; r2-7; n14 f	7778
Allactaga sibirica	2.8 emb; ±0.19; mode 3; r1-4; n18 f	7778
	~3; 5-6 rare	8051
	3.4 emb; ±0.29; mode 3; r2-5; n9 f	7778
	3.7; r2-6	7005
	3.7 emb; ±0.26; mode 3; r1-7; n40 f	7778
	3.75 emb; r1-7; n4 f	10516
Dipus sagitta	3; r2-5	8051
	2.9 emb; mode 3; r2-5; n80 f	3213
	3.0 emb; ±0.08; mode 3; r2-5; n80 f	7778
	3.01 emb; mode 3; r2-5; n85 f	10516
	3.5 yg; r3-4; n2 nests	11307
	3.6 emb; ±0.08; mode 4; r1-5; n20 f	7778
	4 emb; n1 f	11307
	4.2 emb; ±0.09; mode 4; r1-8; n201 f	7778
	4.4 emb; ±0.08; mode 4; r2-8; n127 f	7778
Jaculus blanfordi	3 emb; n1 f	6293
Jaculus jaculus	1.8, 3.8, 4.8 emb; r1-5; n3 samples	3755
	mode 3 emb; r2-5	4344, 4346, 4347, 4349, 4351
	3-7 cl; r2-7	5480
Jaculus lichtensteini	mode 4-6 emb; r2-8	7778
	4.7, 6.5 emb; n2 samples	7778
	5 emb; n1 f	8051
Jaculus orientalis	2 emb; n2 f	8134
	2 ps; n1 lt	8134
	mode 3 yg; r2-4	4960
	3 ea; n2 lt	12128
Jaculus turcmenicus	4.0 emb; ±0.27; r3-6; n13	7778
Paradipus ctenodactylus	3 emb; n1 f	8051
	3.5 emb; r2-6; n8 f	7778
Pygeretmus platyurus	4 ps; n1 f	7778
	5 emb; n1 f	7778
	6 yg; n1 lt	7778
Salpingotus crassicauda	2.3-2.7; mode 2-3	9952
	2.7 yg; ±0.07	7778
Salpingotulus michaelis	2-4	9175
Stylodipus telum	2 yg	10516
	3-4	4645, 8051
	3-4 emb	10516
	4.1 emb	7778
	5.2 emb; ±0.19; mode 5; r2-8; n164 f	7778

SEXUAL MATURITY

Allactaga elater	f/m: 3-3.5 mo	742
Dipus sagitta	f/m: yr following birth	8051
Jaculus jaculus	f: 1st birth: 8-12 mo	4346
	f: 12-14 mo	8051
	m: sperm: 4-5 mo; 42-49 g	3755

GESTATION

Alactagulus pumilio	25-30 d	10516
Dipus sagitta	est 25-30 d	3213
	25-30 d	10516
Jaculus jaculus	44-46 d	3755

LACTATION		
Jaculus jaculus	solid food: 38 d	4349
	independent: ~2 mo	8051
LITTERS/YEAR		
Alactagulus pumilio	2	8051
Allactaga elater	1	8051
	2 lt/season	742
	2	7005
	est 2	9175
Allactaga sibirica	1	7821
	1	8051
Dipus sagitta	2	10516
	2	8051
Jaculus jaculus	1-2	8051
Jaculus lichtensteini	1	7778
Paradipus ctenodactylus	several, few data	9774
Salpingotus crassicauda	2	9952
Stylodipus telum	1	8051
SEASONALITY		
Alactagulus pumilio	testes: large end July, small Aug; USSR	8051
	mate: spring, fall; no data summer, winter; USSR	9888
	preg f: est Oct; few data; ne Iran	6293
Allactaga bobrinskii	mate: Apr-June, Sept-Oct; w Kizil-Kum, n Turkmenia	7778
Allactaga elater	testes: large end Mar; USSR	8051
	testosterone: high Feb, Aug, low June; Kyzylum, Ustyurt, Uzbekistan	742
	mate: Apr-?; USSR	8051
	mate: est spring-fall; sw Pakistan	9175
	preg f: peak Apr-Aug; Kazakhstan	7005
	preg f: Oct; n2; Iran	6293
	birth: May-?; USSR	8051
	lact f: Oct; n1; Iran	6293
	repro: Mar-Aug; n Kyzylum, Ustyurt, Uzbekistan	742
Allactaga euphratica	testes: large Feb-Apr; Iraq	5481
	preg f: May-Oct; Armenia	7778
Allactaga major	preg f: Apr-June; Aral Sea, Kara-Kum, Turkmenia	7778
Allactaga sibirica	testes: peak May-June, decrease July; Transbaikal, Russia	8051
	preg f: May-June; Kazakhstan	7005
	birth: end June-July; USSR	8051
	repro: ?-mid July; se Transbaikal, Russia	7821
Cardiocranius paradoxus	repro: est spring; USSR	7778
Dipus sagitta	mate: mid Mar; USSR	8051
	preg f: Apr-Aug; w Kazakhstan	3213
	preg f: May; n1; Turkestan	11307
	lact f: June; n2; Turkestan	11307
	repro: Feb-mid June; Karakumy desert, Turkmenia	9499
	repro: Mar-Aug, peak June-July; USSR	8051
	repro: peak spring-summer; nw Kyzylkumy, Uzbekistan	9409
Jaculus blanfordi	preg f: Jan; n1; Iran	6293
Jaculus jaculus	spermatogenesis: Feb-Mar, May, Aug, Oct-Nov; n Iraq	5480
	spermatogenesis: yr rd; Khartoum, Sudan	4346
	testes: large Feb, May, Oct; n Iraq	5480
	testes: large yr rd; Sudan	3755
	mate: June-July, Oct-Dec; Khartoum, Sudan	4346, 4347
	preg f: Feb, June, Nov; n Iraq	5480
	preg f: Apr-Jan; Kazakhstan	7005
	preg f: peak Oct-Nov; USSR	4351
	preg f: peak after rain ~80%, none June; Sudan	3755
	preg f/birth: peak Oct-Nov; Sudan	4344
	birth: end Apr-June; USSR	8051
	birth: Oct-Feb; n Sudan	4344

Jaculus lichtensteini	repro: Feb-mid June; Karakumy desert, Turkmenia	9499
Paradipus ctenodactylus	repro: Feb-mid June; Karakumy desert, Turkmenia	9499
Salpingotus crassicauda	birth: May-June; Zaisan Hollow, Kazakhstan	9952
Stylodipus telum	birth: May-June; Europe, USSR	4645
	birth: est May-mid July; USSR	8051
	repro: est end Mar/early Apr-mid June; USSR	8051

Order Rodentia, Family Hystricidae

Old world porcupines (*Atherurus* 1.5-4 kg; *Hystrix* 8-30 kg; *Thecurus* 4-5.5 kg; *Trichys* 1.5-2.5 kg) are terrestrial, nocturnal herbivores from Africa and southern Asia. Usually, males and females are of equal size, however, female *Aethurus* are larger than males, while male *Hystrix* have been reported to be larger than females (8500, 8501, 8853, 10158, 11202, 11203). The placenta is hemochorial with a distinct subplacenta and interstitial implantation (6628). A postpartum estrus is recorded at 10 d (10831; cf 11076, 11078). The well-developed and mobile neonates have quills and open eyes at birth (2384, 5760, 8501, 8853, 9175, 10158). Females have 4-6 teats (331, 4495, 5760, 9945, 10158). Males have seminal vesicles, a baculum, and a sigmoid penis (8246, 11081). Details of anatomy (9161, 11075, 11080, 11565), endocrinology (11079, 11081a, 11208a), and copulation (7532) are available.

NEONATAL MASS		
Atherurus africanus	146.25 g; r100-175; n12	8853
Hystrix africaeaustralis	r107-294 g; n1 lt; term emb	10158
	~300 g	2384
	311.3 g; r300.0-330.0	10158
	351 g; sd47.4; r300-400; n19	11076, 11077
Hystrix brachyura	350-540 g	10831
NEONATAL SIZE		
Atherurus africanus	HB: 180 mm	8853
Hystrix indica	200-225 mm	8501
LITTER SIZE		
Atherurus africanus	1	8853
	1	12128
	1-4	5760
	mode 1; r1-3	2449
	mode 1-2; r1-4	8798
	1 emb ea; n2 f	4495
Atherurus macrourus	1 ea; n2 lt	12128
	1-3 yg	10829
Hystrix spp	mode 1-2; r1-4	8798
Hystrix africaeaustralis	1-3	2384
	1-4	331
	1-4	2439
	1.5 yg; sd0.66; mode 1 59%; r1-3; n165 lt	11076, 11078
	1.51; r1-4; n151	10158
	1.81; mode 2; r1-3; n77 lt	1230
	2	11253
	2	4180
Hystrix cristata	mode 1; r1-2	12128
	2-4	5277
	2-4 yg	4354a
Hystrix indica	1	8500, 8501
	1-2	12128
	2-4	9175
	2; n1	10616
	2; r2-4	106
SEXUAL MATURITY		
Hystrix africaeaustralis	f: 1st conceive: 619 d; ±172	11076
	m: testosterone: 413 d; ±118; r273-532	11076
	m: 8-18 mo	11081

CYCLE LENGTH		
Hystrix africaeaustralis	follicular phase: 1.9 d	11075
	follicular phase: 6.0 d; r4-8; n2	7532
	luteal phase: 19.5 d; sd7.2; r11-26; n4	7532
	luteal phase: 29.3 d; sd4.7	11075
	28.4 d; sd3.86; r25-35; n5 cycles	7532
	31.2 d; se6.5; most 28-36; r17-42; n43 cycles, 12 f	11076, 11078
	33 d; sd11.6; r18-51; n16 cycles	11081a
	34 d; sd6.6; r23-42; n12	11075
	36.9 d; sd11.5; r16-62; n34 cycles	11081a
Hystrix cristata	35 d	11551
GESTATION		
Atherurus africanus	100-110 d	8853
Hystrix spp	6-8 wk	8798
Hystrix africaeaustralis	42-56 d	2384
	est~8 wk	5760
	93.5 d; r93-94; se0.6; n4	10036, 11076, 11078, 11080
	112 d	2439
Hystrix brachyura	~4 mo	10831
Hystrix cristata	est ~8 wk	5760
Hystrix indica	~112 d	9175
LACTATION		
Hystrix africaeaustralis	den emergence: 1+ wk	5760
	wean: start 4 wk	10158
	wean: 120 d; r90-148	10036
	wean: ~7 wk	10158
Hystrix cristata	den emergence: 1+ wk	5760
INTERLITTER INTERVAL		
Hystrix africaeaustralis	345 d; sd66; r200-500; n22 young live	11078
	385 d; se60.4; r296-500; n8	11076
	405.8 d; se86.2; n5	10036
Hystrix indica	109 d; n1	8724
LITTERS/YEAR		
Atherurus africanus	2-3	8798
	max 3	5760
Hystrix africaeaustralis	1	10036, 11078
	2	11253
	2	4180
Hystrix cristata	2 possible	5760
SEASONALITY		
Atherurus africanus	birth: yr rd, peak end spring-early fall; location not given	8798
	repro: not seasonal; w Africa, w,s Uganda, s Kenya	5760
Atherurus macrourus	preg f: Apr, July-Aug; n3; n Vietnam	10829
Hystrix africaeaustralis	estrus: yr rd; S Africa	11076
	mate: May-Dec; Orange R, S Africa	10036
	mate: summer; s Africa	5760
	mate: yr rd; zoo	1230
	birth: July-Dec; S Africa	2384
	birth: Aug-Mar; S Africa	10158
	birth: Aug-Mar; Orange R, S Africa	10036, 11076
	birth: rainy season; Zaire	11253
	birth: yr rd; S Africa	11078
Hystrix cristata	birth: yr rd; n32; zoo	12128
Hystrix indica	birth: peak June-July, none Oct, Jan, Mar; w Rajahstan, India	8723
	birth: peak Oct-Apr; Sri Lanka	8501

Order Rodentia, Family Erethizontidae

The New World porcupines (*Coendou* 1-5 kg; *Echinoprocta* TL 500 mm; *Erethizon* 3-18 kg; *Sphiggurus*) from North, Central, and South America are nocturnal herbivores. *Erethizon* males are larger than females (2665). Ovulation is usually from the right ovary and may be induced (*Erethizon* 2665). A copulatory plug is present after mating (*Erethizon* 11906). The discoidal, labyrinthine, hemochorial placenta has a partially interstitial, mesometrial attachment usually into the right horn of the uterus which is transitional between duplex and bicornuate (*Erethizon* 2665, 7571, 8391). A postpartum estrus occurs after the birth of precocious, furred or quilled, open-eyed neonates (502, 2665, 5146, 9166, 9167, 9823, 10258, 10508, 10923, 11906). Females have 4-12 teats (219, 502, 1474, 5111, 11906). Males have a baculum (1473). Details of anatomy (2665, 7571) are available.

NEONATAL MASS		
Coendou prehensilis	390 g	5858
	f: 413.0 g; r398-432; n4	9167
	m: 413.5 g; r410-422; n4	9167
	415.3 g; sd26.1; r364-447; n4	9166
Erethizon dorsatum	340-640 g	11906
	~450 g	2665
	491.2 g; r402.7-604.5; n7	9818, 9819, 9823
	510 g; n1	10508
	588 g; n1	10923
LITTER SIZE		
Coendou mexicanus	1 emb; n1 f	2228a
	1 emb; n1 f	4228
	1 emb; n1 f	5423b
	1 yg ea; n3 lt	219
Coendou prehensilis	1.0	5858
	1 emb; n1 f	8562a
	1 ea; n23 lt, 4 f	9166, 9167
Erethizon dorsatum	mode 1	7571
	1	2665
	1	502
	1; 2 possible	5146
	1 emb; n1 f	10839
	1 emb; n1 f	10954
	1 yg; n1 lt	10923
	1 emb ea; n2 f	9834
	1 ps ea; n2 f	10954
	1 ea; n4 f, 4 lt	3171
	1 yg ea; n7 lt	9818, 9819
	1 emb ea; n20 f	6016
	1.0196; r1-2; n52 f	10508
Sphiggurus villosus	1; n1 lt	12128
SEXUAL MATURITY		
Coendou prehensilis	f: 19 mo; n1	9166, 9167
Erethizon dorsatum	f: 1st ovulate: 1.5 yr	2665
	f: 18 mo	11906
	m: 1st sire: 16 mo; n1	9820
	f/m: 2 yr	1871a
	f/m: 2.5 yr	8798
	f/m: 3 yr	1474
CYCLE LENGTH		
Erethizon dorsatum	estrus: 8-12 hr	11906
	25-30 d	2665
GESTATION		
Coendou prehensilis	192 d; min ibi	9167
	est 195-210 d	9166
Erethizon dorsatum	16 wk; n1	10508
	6.5-7 mo	2665
	29-31 wk	1871a
	7 mo; n2 f	5146

	~7 mo	1474
	210.25 d; r205-217; n4	9818, 9819
	213 d	11564
LACTATION		
Coendou prehensilis	solid food: 4 wk	9166, 9167
	independence: 10 wk	9167
	independence: 15 wk	9166
Erethizon dorsatum	solid food: a few hours after birth	1474
	nurse: 2-3 wk	1871a
	wean: min 2 mo	11906
INTERLITTER INTERVAL		
Coendou prehensilis	192-674 d	9167
SEASONALITY		
Coendou mexicanus	preg f: Jan; n1; Veracruz, Mexico	4228
	preg f: Feb; n1; Campeche, Mexico	5423b
	preg f: May; n1; San Luis Potosí, Mexico	2228a
	lact f: Apr; n2; Yucatan, Mexico	5423b
Erethizon dorsatum	testes: scrotal June-Jan; MA	2665
	mate: Aug/Sept-Nov; MA	2665
	mate: Sept-Oct; US	1474
	mate: Nov; BC, Canada	2088
	mate: Nov-early Dec; NH	10508
	mate: fall; UT	6560
	preg f: Mar; n1; SD	10954
	birth: Mar; BC, Canada	2088
	birth: Apr-May; US	1474
	birth: end Apr-early May; MA	2665
	birth: ~early May; IL, WI	2060
	birth: May-June; w US	5146
	birth: spring; WI	7571
	lact f: July; n1; ND	3703

Order Rodentia, Family Caviidae

Cavies and guinea pigs (*Cavia* 200-500 g; *Dolichotis* 1.5-16 kg; *Galea* 300-700 g; *Kerodon* 450-1000 g; *Microcavia* 200-500 g) are herbivores from South America. *Galea* males may be larger than females (5953). *Cavia* sperm form rouleaux in the epididymis (3519). Coitus triggers ovulation in *Galea* (9264, 11559, 11561, 11565) and a copulatory plug follows mating in *Cavia* (6237, 11553). Four-cell embryos and morulas are found in the uterus of *Galea* at 4.5 d (9161). The 9.5 g (n1, *Galea*) labyrinthine, hemoendothelial, discoidal placenta with subplacenta has a mesometrial, interstitial implantation into a bicornuate uterus (3011, 6237, 7563, 9161, 9162, 9484) and is eaten after birth (2759, 9256). A postpartum estrus follows (2759, 5953, 6146, 6147, 6237, 9161, 9168, 9256, 9257, 9264, 10494, 10680, 11553) the birth of open-eyed, furry, precocial neonates (2759, 6147, 6601, 8798, 9168, 9256, 9257, 10493, 10680, 11553). Females have 2 (*Cavia*), 4 (*Galea, Dolichotis, Microcavia*), or 8 mammae (*Dolichotis*) (5111, 6237, 9257, 11563). Males have penial spines and a baculum (6237). Details of anatomy (9161, 9162, 11565), endocrinology (4528, 6601, 10645), spermatogenesis (4936), copulation (2759, 10811a), zygote transport (7968), birth (2759), and milk composition (7211) are available.

NEONATAL MASS		
Cavia aperea	m: 56.6 g; se4.7; r39-86; n9	9257
	f: 58.7 g; se3.1; r40-91; n17	9257
	m: 61.2 g; se1.5; n27	11553
	f: 61.8 g; se1.7; n38	11553
Cavia porcellus	f: 75.3 g; se4.7; r59-108; n9	9257
	m: 78.45 g; r65-114; n50 1 d	8643
	f: 80.34 g; r61-125; n50 1 d	8643
	m: 91.6 g; se5.2; r55-120; n11	9257
Dolichotis patagonum	524 g; se55; n5	992, 6601
	639.7 g; r481-733; n6	9291
Dolichotis salinicola	199.0 g	5858
Galea musteloides	f: 32.0 g; sd50; n9	5953
	m: 36.4 g; se1.0; r24-50; n42	9257

	f: 37.6 g; se1.2; r23-54; n46	9257
	f: 39.4 g; se0.9; n81	11553
	m: 40.0 g; sd8.6; n17	5953
	m: 40.5 g; se0.7; n85	11553
Kerodon rupestris	76.0 g; sd11.99; n10	6147
	80.0 g	5858
	85 g; singleton lt heavier	9168
Microcavia australis	m: 29.9 g; se1.9; r17-44; n17	9257
	f: 30.9 g; se1.2; r24-39; n16	9257
NEONATAL SIZE		
Cavia aperea	m: TL: 122.0 mm; se3.3; r108-136; n9	9257
	f: TL: 123.9 mm; se1.9; n17	9257
Cavia porcellus	f: TL: 128.0 mm; se2.7; r112-138; n9	9257
	m: TL: 138.0 mm; se3.1; r121-156; n11	9257
Dolichotis patagonum	CR: 249 mm; se9; n5	992, 6601
Galea musteloides	m: TL: 99.8 mm; se1.0; r88-116; n42	9257
	f: TL: 99.9 mm; se1.0; r87-112; n46	9257
Kerodon rupestris	145.2 mm; sd10.50; n10	6147
Microcavia australis	f: TL: 97.1 mm; se1.2; r89-104; n16	9257
	m: TL: 99.4 mm; se1.8; r81-104; n17	9257
WEANING MASS		
Cavia aperea	f: 169.6 g; se8.0; n37	11553
	m: 172.5 g; se8.5; n22	11553
Cavia porcellus	228 g	6610
Galea musteloides	f: 127.5 g; se5.5; n58	11553
	m: 146.0 g; se5.7; n52	11553
LITTER SIZE		
Cavia aperea	1-2 emb	596
	2.1 yg; se0.2; r1-5; n28	9257, 9264, 11553
	2.5 yg; r1-5; n31 lt	9264
Cavia porcellus	2.1 yg; r1-4; n25 lt	9257, 9264
	2.68 yg; r1-8; n7285 lt, 39 lineages	11958
	3.7 wean; n297 lt	6610
	4.3-4.5 yg; n16 yr (thousands of litters)	5033
	4.4 yg; n297 lt	6610
	7; n1 lt	6237
Dolichotis patagonum	mode 1; r1-3	12128
	1.33 yg; r1-2; n6 lt	9291
	1.54; r1-3; n39 lt	2109
	1.67; r1-2; n3 f	6601
	1.75 yg; n8 lt	6709
	1.79 yg; mode 2 70%; r1-3; n63 lt	2759
	2; r1-3	7425
	2.03; r1-3; n73 lt	1230
Dolichotis salinicola	1.5; r1-3	5858
	2 emb; n1 f	2944
	2; r1-3	7432
Galea musteloides	mode 2-3; r1-5	9264, 11553
	2 emb; r1-4; n12 f	8302
	2.1; sd1.2; r1-5; n10 lt	5953
	2.7; se0.1; n69 lt	9257
	4.47 cl	7968
Galea spixii	1-5	10493
	2.2 yg; sd0.87; n10 lt	6147
	3.0; sd1.0; r1-5; n24 lt	10494
Kerodon rupestris	1.38 or 1.41 yg; sd0.5; r1-2; n17 lt	6146, 6147
	1.53; mode 1; r1-3; n75	9168
	1.58; se0.73; r1-3; n26 lt	10680
	1.6; r1-2	5858
Microcavia australis	2.21; mode 2 64%; r1-3; n14 lt	9256
	2.8 yg; se0.3; r1-5; n16 lt	9257, 9264

SEXUAL MATURITY

Cavia aperea	f: 1st conceive: 30-89 d	9257, 9264
	f: vag perf: 58.6 d; se3.0	11553
	f: 1st conceive: 84.4 d; se9.9; min 27	11553
	f/m: 10-12 wk	8798
Cavia porcellus	f/m: 1st mate: >2.5 mo	6237
	f/m: 2 mo	8798
Dolichotis patagonum	f: 1st birth: 8 mo	2759
	m: 6 mo	2759
Galea musteloides	f: vag perf: 48.5 d; se1.2	11553
	f: 1st conceive: 50 d	11553
	f: 1st conceive: 58 d; r28-93; n9 spring-born	9257, 9264
	f: 1st conceive: 60.6 d; r17-152; n342	9264
	f: 1st conceive: 167 d; r113-255; n6 fall-born	9257, 9264
	m: 1st spermatids: 21-25 d	4936
	m: 1st sperm: 31-35 d	4936
	m: epididymal sperm: 60 d	4936
Galea spixii	f: vag perf: 80 d	6147
	m: testes descend: 135 d; n2	6147
Kerodon rupestris	f: vag perf: 60 d	6147
	f: 1st conceive: 55 d prob unusual	9168
	f: 1st conceive: 82 d; n1	6146
	f: 1st birth: 156 d	6147
	f: 133 d	9168
	m: testes descend: 115 d; ±2.7; n4	6147
Microcavia australis	f: 1st mate: 45 d; r40-50; n2	9256, 9257
	f: 1st conceive: 82-85 d	9256, 9257, 9264

CYCLE LENGTH

Cavia aperea	20.6 d; se0.8; n67 cycles, 21 f	9264, 11553
Cavia porcellus	estrus: ~24 h	6237
	estrus: ~24 h	10440
	16 d; r13-21; 2o sources 2 strains	2715
	50 d; n1 cycle, 1 f	6237
Dolichotis patagonum	31.4 d; se1.3; r24-39; n6 f	7669
	35.3 d; se5.0; r28-42; n7 f	991
Galea musteloides	17.3 d; r17-18; n3 cycles, 3 f	9257, 9264
	19.0 d; se1.2; n4 f with vasectomized m	11553
	19.1 d; r12-36; n21 cycles with vasectomized m	9264
	19.5 d; r12-27; n2 cycles, 2 f	11559
Microcavia australis	15 d; n1 cycle, 1 f; post partum	9257

GESTATION

Cavia aperea	60.9 d; se0.3 or 0.4; mode 60; r59-63; n9	9257, 9264, 11553, 11564
	63.9 d; se0.9; mode 60; r59-74; n26	9264
Cavia porcellus	>57 d	6237
	65-68 d	11551
	67.0 d; se0.8; n6	9257
	68 d; r58-75	2715
Dolichotis patagonum	~58 d	8798
	3 mo; min ibi	2759
	96 d; se0.7; n6	991, 992, 6601
Dolichotis salinicola	~77 d; n4	5858, 11563
Galea musteloides	implant: 5 d	7968
	implant: 6 d	9161
	mode 54 d; r48-61; n90; ppe to birth?	9264
	52 d	4528
	52-53 d	11553, 11564
	53 d; n1	10646
	59 d; sd5; r65-71	5953
Galea spixii	49-52 d	6147
	~7-8 wk	10494
Kerodon rupestris	75.0 d; sd1.42; r72-77	6146
	75.87 d; sd1.57; n24 lt	9168
	75.9 d; se1.14; r74-78; n17	10680
	77 d	5858
Microcavia australis	54 d	11564

LACTATION		
Cavia aperea	wean: 1 wk; most 2-3 wk	11553
	wean: 3 wk	8798
Cavia porcellus	solid food: 2-3 d	8798
	wean: 2 wk	8798
Dolichotis patagonum	solid food: 2 d	6709
	wean: 6-11 wk	2759
Galea musteloides	wean: 1 wk	11553
	wean: 3 mo	5953
	nurse: ~50 d	9257
Galea spixii	wean: 42.25 d; ±5.7; n4	6147
Kerodon rupestris	wean: 35.17 d; ±10.1; n6	6147
INTERLITTER INTERVAL		
Cavia porcellus	96.3 d	6610
Dolichotis patagonum	3-3.6 mo	2759
Galea musteloides	53.2 d; se0.2; r50-58; n41	9257
Microcavia australis	54.2 d; sd0.4; r53-55; n6	9256, 9257, 9264
LITTERS/YEAR		
Cavia aperea	max 5	9257, 9264
Dolichotis patagonum	3-4	2759
Galea musteloides	max 7	9257
Galea spixii	est max 6-7	10494
Microcavia australis	max 4	9257
	5 possible; 4 observed	9256
SEASONALITY		
Cavia aperea	mate: Sept-May; few data; Uruguay	596
	birth: yr rd, peak Nov; captive	9257, 9264, 11553
Dolichotis patagonum	mate: yr rd; zoo	1230
	birth: Feb-Nov; captive	2109
	birth: Mar-Oct; n19; zoo	12128
	birth: yr rd, peak May-Sept; captive	8798
	repro: yr rd; free ranging	2759
Dolichotis salinicola	preg f: Aug; n1; Bolivian Chaco	2944
Galea musteloides	mate: yr rd; captive	11553
	preg f: Dec; n12; Lake Titicaca, Peru	8302
	birth: yr rd; Argentina	9257
	birth: yr rd; captive	9257, 9264
Galea spixii	preg/lact f: yr rd; Caatinga, ne Brazil	10493, 10494
	repro: yr rd; Brazil	6808
Kerodon rupestris	birth: yr rd, peak July-Mar; captive	9168
	birth: yr rd; captive	6146
Microcavia australis	preg/lact f: Sept-Apr; no Aug data; Argentina	9256, 9257
	preg f: Oct-Feb; captive	9257, 9264

Order Rodentia, Family Hydrochaeridae, Genus *Hydrochaeris*

At 40-60 kg, capybaras (*Hydrochaeris hydrochaeris*) are the largest extant rodents. These social, semi-aquatic, crepuscular and nocturnal herbivores inhabit marshy areas of South America. They mate in the water (10892) and subsequently give birth to well-furred and open-eyed neonates that can walk within 15 minutes (8798, 10892, 12066). Females have 14 teats (5111). Details of anatomy (11565) and copulation (8075) are available.

NEONATAL MASS		
Hydrochaeris hydrochaeris	1.4 kg	8076
	1.5 kg	7466
	1.75 kg; r136-226	12066
LITTER SIZE		
Hydrochaeris hydrochaeris	2; n1 lt	12128
	2 yg ea; n2 f	5111
	3 yg; n1 f	10885
	3.3 yg; r2-5; n13 lt	12066
	3.5; r1-6	8076

	4; r1-8	10892
	4 emb; n1 f	10885
	4.8 emb; r1-8; n68 f	7466
	7	4558
SEXUAL MATURITY		
Hydrochaeris hydrochaeris	f/m: 15 mo	12066
	f/m: 15-18 mo	10892
	f/m: 1.5 yr, 30-40 kg	7466
GESTATION		
Hydrochaeris hydrochaeris	104-111 d; n1	10885
	119-126 d	8798
	120 d	8076
	148 d; r144-154; n3	4558
	152.5 d; r149-156; n2	12066
LACTATION		
Hydrochaeris hydrochaeris	solid food: 2-3 d	12066
	nurse: 3-4 mo	7466
	wean: 4 mo	12066
LITTERS/YEAR		
Hydrochaeris hydrochaeris	1	8798
	1.5	10892
SEASONALITY		
Hydrochaeris hydrochaeris	mate: peak Apr-May; Venezuela	7466
	mate: Oct-Nov; Mato Grosso, Brazil	7466
	birth: peak Sept-Nov; Venezuela	7466
	birth: peak Sept-Dec; e Argentina to Panama	10892

Order Rodentia, Family Dinomyidae, Genus *Dinomys*

The herbivorous pacaranas (*Dinomys branickii* 10-15 kg) inhabit the Andes Mountains of western South America. Term embryos have fur and open eyes (1988). Females have 8 teats (11563). Details of mating (1988) are available.

NEONATAL MASS		
Dinomys branickii	622.7 g; r570-660; n3	7228
	950.75 g; r825-1076.5; n2 term emb, dead with placenta	1988
NEONATAL SIZE		
Dinomys branickii	HB: 275 mm; r265-285; n2 term emb, dead	1988
LITTER SIZE		
Dinomys branickii	1-2 yg	11563
	2 emb; n1 f	1988
	2 emb; n1 f	3847
	2 yg; anecdote	10682
	2.4 yg; r1-4; n5 lt	7228
GESTATION		
Dinomys branickii	223-283 d; n1	1988
	< 252 d; n1	7228
LACTATION		
Dinomys branickii	solid food: 2 d	7228

Order Rodentia, Family Agoutidae, Genus *Agouti*

The 4-12 kg paca (*Agouti paca*) from lowland habitats of Central and northern South America is a solitary, nocturnal herbivore. Pacas have a hemichorial, discoidal placenta which implants into a bicornuate uterus (4728). A postpartum estrus occurs (1981). Neonates are furred with open eyes and can walk as well as swim shortly after birth (4728). Females have 2 (5111) or 4 (4728) teats. Males are ascrotal with seminal vesicles and coagulating glands (4728). Details of anatomy (1981, 4728) are available.

NEONATAL MASS		
Agouti paca	500-800 g	1981
	650 g; r600-800	4728
NEONATAL SIZE		
Agouti paca	24-30 cm	4728
LITTER SIZE		
Agouti paca	1 emb; n1 f	3209
	1 emb; n1 f	8562a
	1 ea; n2 lt	12128
	1 emb ea; n2 f	2228a
	1 ea; n55 lt	1981
	1.03; r1-2; n31 lt	4728
	2	3633
	max 18 cl	4728
SEXUAL MATURITY		
Agouti paca	f: 1 yr	1981
	m: testes size, spermatogenesis: 1 yr	1981
	m: 3 mo	4728
GESTATION		
Agouti paca	96.75 d; sd3.44; r93-101; n3	4728
	est 116 d	1981
LACTATION		
Agouti paca	wean: 82 d	4728
INTERLITTER INTERVAL		
Agouti paca	191 d; assuming 116 d gestation	1981
SEASONALITY		
Agouti paca	mate: early winter; Yucatan, Mexico	3633
	preg f: Jan; n1; El Salvador	3209
	preg f: June; n2; San Luis Potosí, Mexico	2228a
	preg f: yr rd; n50; ne Colombia	1981
	birth: Sept-Mar; lab, Costa Rica	4728
	birth: winter-early spring; Yucatan, Mexico	3633

Order Rodentia, Family Dasyproctidae

Agoutis (*Dasyprocta* 2-4 kg) and acouchis (*Myoprocta* 1 kg) are herbivores in tropical regions of Central and South America. A copulatory plug is formed (11551). The hemochorial, labyrinthine, zonary placenta, associated with a mesoplacenta, implants into a bicornuate uterus (692, 693, 3023, 7563, 9707, 10476), and is eaten after birth (3023, 5846). A postpartum estrus may occur midway through lactation, earlier if the young die, or not at all (5846, 5847, 11555). Neonates are furred with open eyes (3023, 3024, 5846, 10176) and suckle from 8 teats (3024, 5111, 11563). Details of anatomy (11555, 11556, 11565), endocrinology (9349, 10646), and copulation (10176) are available.

NEONATAL MASS		
Myoprocta acouchy	f: 74.3 g; se1.9; n24	5846
	m: 79.3 g; se2.5; n23	5846
WEANING MASS		
Myoprocta acouchy	300-400 g	5846
LITTER SIZE		
Dasyprocta cristata	2; n4, n2 lt	12128
Dasyprocta leporina	mode 1; r1-2	12128
	1.4 yg; r1-3; n53 lt	1230
	2-4 yg	5111
Dasyprocta prymnolopha	1-2	12128
Dasyprocta punctata	1; n1 lt	12128
	1 emb; n1 f	1475
	1 ea; n2 lt	12128
	1.25 emb; r2-4; n4 f	3023
	1.25; r1-2; n12 lt	10176
	1.9 yg; mode 2; r1-3; n20 lt	7227
	2	3317
	mode 2; max 4	3024

	2 emb; n1 f	1475
	2 yg; n1 f	3023
	2 yg ea; n7 lt	2109
Myoprocta acouchy	1.5; r1-2; n6 lt	5847
	mode 2	5846
	2-5 cl	9349
	2 emb; n6 f	9349
SEXUAL MATURITY		
Myoprocta acouchy	f: vag perf: 147 d; ±14	5846
	f: 1st conceive: 249 d; ±24	5846
CYCLE LENGTH		
Dasyprocta leporina	vag perf: 1-10 d	11556
	34.2 d; se2.1; n29 cycles, 3 f; fall to spring	11556
	87.2 d; ±8.9; n6; summer	11556
Myoprocta acouchy	vag perf: 10-22 d; r3-22; n62	5846
	40 d; se2; r16-68; n22 cycles	5846
	42.7 d; se2.5; n21 cycles	11555
GESTATION		
Dasyprocta leporina	104 d	1355
Dasyprocta punctata	35 d; n1	3024
	~3 mo	10176
	<127 d; min ibi	7227
Myoprocta acouchy	99 d; n5	11555
	99 d; se1; r94-115; n21; vag perf or mating to birth	5846, 9349
LACTATION		
Dasyprocta punctata	solid food: d of birth	3024
Myoprocta acouchy	solid food: middle 1st wk	5846
	wean: 6-8 wk	5846
INTERLITTER INTERVAL		
Dasyprocta punctata	min 127 d	7227
SEASONALITY		
Dasyprocta leporina	birth: yr rd; n27; zoo	12128
	birth: yr rd; zoo	1230
Dasyprocta prymnolopha	birth: yr rd; n17: zoo	12128
Dasyprocta punctata	preg f: July-Aug, Nov-Dec; few data; Panama	3024
	preg f: Dec; n2; El Salvador	1475
	birth: yr rd; zoo	7227
	birth: yr rd, peak Mar-July; Panama	10176
	young found: Jan, Apr, Aug, Dec; El Salvador	1475
	young found: yr rd; Panama	3317
Myoprocta acouchy	estrus: fall, spring, long cycles summer; zoo	11555
	estrus: Dec-May; zoo	5847
	mate: yr rd; zoo	5846
	birth: peak Aug, none Oct; zoo	5846

Order Rodentia, Family Chinchillidae

Chinchillas (*Chinchilla* 350-800 g) and viscachas (*Lagidium* 1-3 kg; *Lagostomus* 2-8 kg) are terrestrial to semifossorial, colonial herbivores in southern South America. *Chinchilla* and *Lagostomus* are primarily nocturnal, while *Lagidium* is diurnal (8301, 8798, 9345). *Chinchilla* females are larger than males (9266), while the reverse is true for *Lagostomus* (11557, 11562). *Lagidium* exhibits reproductive asymmetry with the right ovary and uterus used preferentially (8300, 8301, 11565). *Lagostomus* ovulates about 800 oocytes (9161, 9163, 11553, 11558, 11565), but only 2 embryos implant (3891, 6511, 6815, 11553, 11557). A copulatory plug is present after mating (8301, 11551, 11553, 11557).

The 9-11 g, hemochorial, discoidal, labyrinthine placenta with a subplacenta has a mesometrial, interstitial implantation into a duplex or bicornuate uterus (4800, 5738, 8301, 9161, 9162, 9163, 9266, 9707, 10819, 11558). Estrus may occur postpartum, but is more common after lactation (1499, 3891, 5263, 8301, 9161, 11551, 11553, 11557). The precocious neonates are furred with open eyes and can walk (2109, 3891, 6095, 8300, 11551, 11553, 11557, 11564). Females have 4 (*Lagidium, Lagostomus)*, or 6 (*Chinchilla*) teats (3891, 8301, 11557, 11563). Males have seminal vesicles, coagulating glands, and a baculum (4800, 8301, 9266).

Details of anatomy (4799, 4800, 5263, 8301, 9162, 9266, 11550, 11554, 11558, 11565), endocrinology (1559, 4095, 4096, 4528, 10646), copulation (936), and milk composition (3891, 11324) are available.

NEONATAL MASS

Species	Data	Reference
Chinchilla lanigera	r37.4-43.8 g; n4	6095
	f: 38.7 g; n10	9266
	39.7 g; se1.39; n19	9266
	m: 41.0 g; n9	9266
Lagidium peruanum	~180 g	8300, 8301
Lagidium viscacia	f: 260 g; n1	11554
Lagostomus maximus	f: 185.0 g; se4.6; n49	11553, 11557
	m: 207.0 g; se5.2; n38 or se6.1; n28	11553, 11557

WEANING MASS

Species	Data	Reference
Lagostomus maximus	m: 645.5 g; se35.7; n29	11553, 11557
	f: 659.0 g; se18.3; n42	11553, 11557

LITTER SIZE

Species	Data	Reference
Chinchilla brevicaudata	1.45	10809
Chinchilla lanigera	1.0; n68275 f	4800
	mode 1-2; r1-6	11550, 11551
	1.05 yg; n44454 f	4800
	1.45; mode 1; r1-3; n20 lt	2109
	1.87 yg; n648 lt	4800
	1.90 yg; ±0.76; mode 2 46%; r1-4; n123 lt	6117
	2.3; r2-3; n3 lt; n7 live fet, 8 total	10643
Lagidium peruanum	1 yg	8300, 8301
Lagidium viscacia	1 ps; n1 f	8556
	2	2944
	mode 2; r1-2	12128
Lagostomus maximus	1.89 emb; mode 2; r1-3; n18 f	6511
	1.94 yg; mode 2 90%; r1-3; n62 lt	3891, 11553, 11557
	2.2 yg; r1-5; n5 lt, 3 f	6815
	43-255 oocytes; max 544	9161
	200-800 oocytes	9163, 11553, 11565
	max 845	11558

SEXUAL MATURITY

Species	Data	Reference
Chinchilla lanigera	f: vag perf: 173.2 d; ±57.6; r<71-308; n45 f	6117
	f/m: 8+ mo	11551
Lagidium peruanum	f: 1st mate: ~6 mo	8301
	f: ~900 g	8300
	m: sperm: ~7 mo	8300, 8301
Lagostomus maximus	f: vag perf: 153.7 d; se3.4; n31	11557
	f: 1st conceive: 253.8 d; se20.2; min 83; n15	11557

CYCLE LENGTH

Species	Data	Reference
Chinchilla lanigera	estrus: 2 d	1499
	estrus: 2-3 d	11551
	vag perf: 7.4 d; r3-18	5263
	vag perf: 8 d	1499
	vag perf: 8.0 d; ±1.11; n25 f	1327
	33 d; r11-49; n14 cycles	1327
	35.7 d; ±7.9; r15-62; n100 cycles, 24 f	6117
	35 d	11564
	40 d	1499
	40.91 d; r21-54; n118 cycles, 23 f	5263
	mode 41; most 30-50; n70 cycles	11550, 11551
	41.05 d; r37.4-43.4; n154 cycles, 23 f	5263
	42 d	11552
	68.17 d; r36-130; n36 cycles, 23 f	5263
Lagostomus maximus	vag perf: 3-15 d	11553
	vag perf: 7.3 d; se0.2; n213	11557
	41 d; r25-55 75%; r15-75	11553
	45.0 d; se1.2; 30-35 68%; n152 cycles, 46 f	11557

GESTATION		
Chinchilla brevicaudata	128 d	10809
Chinchilla lanigera	implant: 5.25 d	9161
	110.4 d; r108-112; n5 gest, 5 f	6117
	111 d; r105-115	11550, 11551, 11564
	111 d	10646
Lagidium spp	est 3 mo	8798
Lagidium peruanum	~100 d	8301
	104 d	9161
Lagidium viscacia	4-4.5 mo	11554
Lagostomus maximus	implant: 18.5 d	9161, 9163
	153.7 d; se0.5; r145-166; n23	3891, 9163, 11557, 11564
	155 d; r149-162	11553
LACTATION		
Chinchilla lanigera	nurse: min 3 wk	11564
	wean: 6-8 wk	11551
	wean: 60 d	2109
Lagidium peruanum	solid food: day of birth	8300
	wean: >1 mo	8300, 8301
Lagostomus maximus	wean: 29-70 d; m earlier	11557
	wean: 54 d	11553
LITTERS/YEAR		
Chinchilla brevicaudata	1-2	10809
Chinchilla lanigera	max 2	10809
	2	4800
Lagidium peruanum	est 2; r1-3; field data	8301
SEASONALITY		
Chinchilla brevicaudata	mate: biannual; n Chile	10809
Chinchilla lanigera	m fertile: yr rd; lab	11551
	mate: Nov-May; lab	11552
	mate: Dec-May; captive	1499
	birth: Mar-Aug 81%; captive	6117
	birth: Apr-May, Aug-Oct, Dec; n19; zoo	12128
	birth: spring; coastal Cordillera, Chile	10809
	birth: yr rd, peak Apr-May; captive	8798
	repro: Nov-May; lab	11551
Lagidium peruanum	sperm: est yr rd; Peru	8301
	anestrus: Sept-early Oct; Peru	8301
	mate: yr rd, peak mid Oct-Nov; Peru	8301
	birth: mid Jan-Feb; Peru	8301
Lagidium viscacia	mate: end May-early June; n1; Argentina	11554
	birth: yr rd; n18; zoo	12128
Lagostomus maximus	mate: peak est Mar-Apr; Argentina	11557
	birth: yr rd, peak Apr-July; location not given	8798
	birth: yr rd; lab	11553, 11557

Order Rodentia, Family Capromyidae

Hutias (*Capromys*, *Geocapromys*, *Isolobodon* (possibly extinct), *Plagiodontia*) are 700 g to 7 kg, crepuscular or nocturnal herbivores, primarily distributed in the West Indies. *Capromys pilorides* is usually solitary, but is sometimes found in pairs (8798). *Capromys melanurus* enters estrus 28 d postpartum (1421). Neonates are furred, have open eyes and prehensile tails, and can walk (*Capromys* 1421, 8798; *Geocapromys* 5022, 8092). *Capromys* have 4 teats, while *Plagiodontia* have 6 (7430, 8798). Males have inguinal testes (*Capromys* 311).

NEONATAL MASS		
Capromys pilorides	199 g; r150-244	5354
Geocapromys ingrahami	70 g; n1	5354
	84 g; n1; 70-90 g	5022
Plagiodontia aedium	101.7 g; n1	5354
	110.0 g	5858

WEANING MASS		
Geocapromys ingrahami	200-300 g; 1-2.5 mo	1914
LITTER SIZE		
Capromys melanurus	1.75 yg; r1-2; n4 lt	1421
Capromys pilorides	1-3 yg	9139
	1.8; r1-3	5858
	2; r1-4	5354
	2 yg; n2 lt	1421
	2.0; r1-3; n4 lt	12128
	2.8 yg; 90% 1-4; r1-6; n38	1570a
Capromys prehensilis	2 yg; n1 lt	1421
Geocapromys brownii	1-2 yg	7509
	1-3	5354
	1.49 yg; 6% triplets; r1-3; n47 lt	282
	2.3; r2-3; n3 lt, 3 f	8092
Geocapromys ingrahami	1 yg ea; n8 lt	1914, 5022
Plagiodontia aedium	1	10809
	1 ea; n4 lt	5354
	1.3; r1-2	5858
SEXUAL MATURITY		
Capromys pilorides	f: 10 mo	5354
	f: HB: 416 mm	1570a
	m: 7-10 mo	5354
	m: HB: 440 mm	1570a
CYCLE LENGTH		
Capromys pilorides	16.3 d; r13-19	5354
	~1 mo	9139
Geocapromys ingrahami	10.08 d; r9-11	5354
Plagiodontia aedium	10 d	5354
GESTATION		
Capromys melanurus	199 d; r195-203; n2	1421
Capromys pilorides	17-18 wk	9139
	120-130 d	8798
	123 d; ±2	5354
	123 d	1421
	125 d	5858
	4.7 mo	1570a
Geocapromys brownii	est 6 wk	5354
	123 d; n1	8092, 8094
	~123 d	282
Geocapromys ingrahami	88 d; n1	5022
	108 d; n1; mating to death carrying a near term emb)	5022
	126-156 d; n1	5354
Plagiodontia aedium	≤119 d; n1	5354
	125-150 d	10809
LACTATION		
Capromys melanurus	leave nest: 2 d	1421
	solid food: 10 d	1421
Capromys pilorides	solid food: 10 d	8798
	solid food: 16 d	1421
	wean: 6 wk	8798
	nurse: 153 d	5354
Geocapromys brownii	solid food: ~30 h	282
	solid food: 34 h	8092
Geocapromys ingrahami	solid food: ≤ 3 d	5022
INTERLITTER INTERVAL		
Capromys melanurus	222 d; r221-223; n2	1421
Geocapromys brownii	168 d; n27	282
Geocapromys ingrahami	min 325 d	5022
SEASONALITY		
Capromys pilorides	birth: yr rd, peak June; zoo	5354
	birth: Oct; n1; captive	1421
Geocapromys brownii	birth: Aug-Oct; zoo	8092
Geocapromys ingrahami	birth: yr rd; Bahamas	1914

Order Rodentia, Family Myocastoridae, Genus *Myocastor*

Nutrias (*Myocastor coypus* 3-14 kg) are semiaquatic herbivores from temperate wetlands of central and southern South America. Males are larger than females (2045, 3571, 11743). The 12.2-14.5 g (n2), labyrinthine, hemochorial, discoidal placenta has a prominent subplacenta and implants mesometrially into a bicorunate or duplex uterus (4797, 4798, 9161, 9162). A postpartum estrus occurs (2336, 3957, 7875) after the birth of furred, open-eyed neonates that can swim within a day of birth (443, 2045, 2336, 2680, 3960, 6253, 7433, 8798). Ovulation may be induced (3957). Females have 6-10 thoracic, lateral teats (2336, 3571, 3957, 3960, 7433). Males have seminal vesicles (4798, 6776). Details of anatomy (3825, 4798, 8518, 9161, 9162, 10049, 11565), endocrinology (4528, 5280, 9346, 9552), milk composition (2918), and reproductive energetics (3960) are available.

NEONATAL MASS		
Myocastor coypus	100-160 g	2919
	150-250 g	2336
	150-250 g	159
	f: 201 g; sd46; n59	3960
	m: 213 g; sd53; n58	3960
	225 g; r175-332	7875
	234 g; n15	443
WEANING MASS		
Myocastor coypus	1.5-2 kg; 13 wk	3960
LITTER SIZE		
Myocastor coypus	1-9 yg; 12-14 rare	2919
	3.02; rofx 3.6-4.8; across months	11742
	4.1; mode 4; r1-12; n3279 lt	158
	4.23 emb; r1-9; n26 f	443
	~4.5 yg; r1-9	3115
	5; max 11	2336
	5.03 emb; mode 5; r1-11; n233 f	66
	5.06 live emb; se0.13; r1-9; n136 f; 1st lt	7875
	5.5 born; 4.64 born alive; n123 lt	9838
	5.63 emb; r2-11; n35 f	443
	5.63 live emb; se0.11; r1-13; n287 f; 2+ lt	7875
	5.75 emb; ±1.51; mode 5; r3-12; n148 f	1368
	6 emb; n582 f	3962
	6.05	5724
	7.3 emb; r5-9; n3 f	6253
	mode 16-18 emb; r12-20	2919
SEXUAL MATURITY		
Myocastor coypus	f: 1st cl: 10-12 mo	5952
	f: 1st mate: 4-8 mo	2336
	f: 1st preg: 2-4 mo; n1	5724
	f: 1st birth: 6-15 mo	11907a
	f: min 4 mo	2919
	f: < 6 mo	11742
	f: 8-9 mo	6253
	m: epididymal sperm: 5.5-6.5 mo; n1	5724
	m: testes large: 6 mo	11742
	m: spermatogenesis: 4-5 mo	8518
	m: 12-15 mo	6253
	f/m: 1.5-2.5 kg	7875
CYCLE LENGTH		
Myocastor coypus	estrus: 1 d; nulliparous; 2-4 d; multiparous	10049
	estrus: 1-4 d	3115
	estrus: 2-8 d; metestrus: 2 d; proestrus: 1 d	7295
	estrus: ~4 d	2336
	r5-11 d; n30 f	7295
	r5-28 d; n10 cycles	11760
	17 d; nulliparous; 19 d; multiparous; vaginal smears	10049

	24-26 d	3115
	26 d; r1-6 wk	7875
	28 d	2336
GESTATION		
Myocastor coypus	implant: 10 d	3957
	120-130 d	6253
	128 d	2336
	mode 128-130 d; r120-132	2919
	mode 128-135 d; r110-150	3571
	130 d	1046
	~130-134 d	443
	131 d; 128-133 72%; r126-141; n446	158
	134 d; r129-139; n16	10049
	135 d	4528
	135-150 d	7433
	19 wk	3961
LACTATION		
Myocastor coypus	solid food: at once	7875
	solid food: 24 hr	6253
	solid food: 1 d	443
	wean: 5 wk	3115
	wean: 6-8 wk	8798
	wean: 6-10 wk	7875
	nurse: 7.7 wk; 95%ci 1.0	3957, 3960
	nurse: 10.9 wk; n5	3957, 3960
INTERLITTER INTERVAL		
Myocastor coypus	128 d	2336
	~132 d	3957
LITTERS/YEAR		
Myocastor coypus	2	5724
	2-2.5	2336
	2.76	11742
	3/13-14 mo	1046
SEASONALITY		
Myocastor coypus	mate: yr rd; e Anglia, UK	3962
	preg f: yr rd, except Apr; MD	11742
	birth: yr rd, peak Feb-May; captive	8798
	birth: yr rd; Hessen, Germany	3571
	birth: yr rd, England, UK	7875
	birth: yr rd; N America	1474

Order Rodentia, Family Octodontidae

Southern South America hosts the 60-300 g, herbivorous, octodontid rodents (*Aconaemys, Octodon, Octodontomys, Octomys, Spalacopus).* While all genera burrow, only *Spalacopus* is considered fossorial (9016). Only the diurnal *Octodon degus* has received significant study. Morulae enter the bicornuate uterus at 3.5 d. At 5.5 d the mesometrial, interstitial attachment of a 200-300 g, discoidal, hemochorial placenta and prominent subplacenta begins (9161, 9162). A postpartum estrus occurs, but few captive females mate (9083, 11553). The eyes of the sparsely furred neonates open after 1 wk, although neonates are able to stand and right themselves at birth (9083, 11553, 11564). Females have 6-8 teats (11563). Males have a baculum, penial spines, seminal vesicles, and abdominal testes (2027a). Details of anatomy (9161, 9162, 11565) are available.

NEONATAL MASS		
Octodon degus	13.52 g; sd1.72; n89 yg, 16 lt	9283
	14.0 g; se0.1; n114	11553
	14.4 g	5858
	14.6 g; se0.06; n35	9083
Octodontomys gliroides	20 g	5858
WEANING MASS		
Octodon degus	68.0 g table, 69.8 g text; 28 d	9083
	74.4 g; r40-122; n106	11553

LITTER SIZE		
Aconaemys fuscus	3 emb; n1 f	8305a
	5 emb; n1 f	4054
Octodon degus	4.9; r1-8	5858
	5 yg; r1-8	11553, 11564
	5 yg; r1-10	11563
	5; n1	9161
	5.0 emb; r3-7; n2 f	8556
	5.3; mode 3; r3-8; n9	3552
	5.7 emb; sd1.3; n9 f	7241
	5.75 yg; se0.48; n4 lt	9083
	6.8 yg	11907
Octodontomys gliroides	2 yg; r1-3	11563
	2.1; r1-4	5858
Spalacopus cyanus	3 emb ea; n2 f	9016
SEXUAL MATURITY		
Octodon degus	f/m: 6 mo	9083
	f: vag perf: 127.5 g	7241
CYCLE LENGTH		
Octodon degus	vag perf: 3-21 d	11553
GESTATION		
Octodon degus	implant: 6.5-7 d	9230
	implant: 6.5-7 d	9161
	84-90 d	9230
	90 d	9083
	90 d	5858
	90 d; r87-93	11553
	>90 d	11564
Octodontomys gliroides	104 d	5858
LACTATION		
Octodon degus	solid food: 6-7 d	9083
	wean: 4-6 wk	9083
	wean: 4-6 wk	11553
SEASONALITY		
Aconaemys fuscus	preg f: Oct; n1; Neuquen, Argentina	8305a
	preg f: Nov; n1; Malleco, Chile	4054
Octodon degus	mate: June-July; lab, Chile	9230
	preg f: June-July; Chile	7241
	preg f: July-Sept; Chile	7241
	birth: peak July-Aug, Dec; captive, VT	11907
	repro: peak Sept; Chile	3552

Order Rodentia, Family Ctenomyidae, Genus *Ctenomys*

The 33 species of tuco-tucos (*Ctenomys* 100-700 g) are fossorial herbivores distributed from Peru to Patagonia. Males may be larger than females and solitary, while females are colonial (596, 8304). The 100-200 g, hemochorial, discoidal placenta has a prominent subplacenta (9161) and implants in either horn mesometrially (8302, 8304, 9161). Neonates are naked and blind (11564), but their eyes open at 2-3 d (1449c). Females have 6 teats (8304, 11563). Males have a baculum and penial spines (6401). Details of anatomy (11565), endocrinology (4528), and sperm (3196) are available.

NEONATAL MASS		
Ctenomys peruanus	34.7; n3 term emb	8304
Ctenomys talarum	8 g	1449c
LITTER SIZE		
Ctenomys opimus	1.8 emb; r1-3; n11 f	8304
	2-3 cl	8302
	2 emb; n1 f	8556
	2 emb ea; n2 f	8302
	3 ps; n1 f	8556

Ctenomys peruanus	3.1 viable emb; r1-5; n19 f	8304
Ctenomys talarum	1-6	11564
	5 yg; r1-7	11563
Ctenomys torquatus	2-3	596
SEXUAL MATURITY		
Ctenomys opimus	f/m: 1st mate: <1 yr	8304
Ctenomys peruanus	f/m: 1st mate: <1 yr	8304
GESTATION		
Ctenomys opimus	est ~2 mo; indirect est	8304
Ctenomys peruanus	est 4 mo	8304
Ctenomys talarum	>100 d	11564
	102 d; r93-120	1449c
Ctenomys torquatus	107 d	10639
LACTATION		
Ctenomys talarum	35 d	1449c
LITTERS/YEAR		
Ctenomys opimus	1	8304
Ctenomys peruanus	1?	8304
SEASONALITY		
Ctenomys opimus	mate: est Aug-Feb; Peru	8304
	preg f: Sept-Apr; Peru	8304
	birth: Oct-Mar; Peru	8304
Ctenomys peruanus	mate: end Nov-Dec; s Peru	8304
	birth: Mar-Apr; s Peru	8304
Ctenomys torquatus	mate?: Jan-Oct; Uruguay	10639
	preg f: Sept-Jan; Uruguay	596

Order Rodentia, Family Abrocomidae, Genus *Abrocoma*

South American chinchilla rats, *Abrocoma bennetti* and *A. cinerea*, range from 150 to 300 g. Females have 4 mammae.

NEONATAL MASS		
Abrocoma cinerea	8.0 g	5858
LITTER SIZE		
Abrocoma bennetti	5 emb; r4-6; n3 f	3552
Abrocoma cinerea	2.2; r1-3	5858
GESTATION		
Abrocoma cinerea	102 d	5858

Order Rodentia, Family Echimyidae

Spiny rats (*Chaetomys* 430-455 mm; *Carterodon* 155-200 mm; *Clyomys* 21-39 g; *Dactylomys* 600-700 g; *Diplomys* 250-480 mm; *Echimys* ~640 g; *Euryzygomatomys* 170-270 mm; *Hoplomys* ~450 g; *Isothrix* 220-410 mm; *Kannabateomys* ~250 mm; *Lonchothrix* ~200 mm; *Makalata* 150-215 g; *Mesomys* 150-200 mm; *Proechimys* 300-500 g; *Thrichomys* 300-450 g; *Thrinacodus* 180-240 mm) are a diverse assemblage of herbivores from Central and South American forests. Ovulation in *Proechimys* is induced (6650) and occurs 24-48 h before birth (11565). The morula reaches the uterus at 3.5 d (*Proechimys* 9161). *Proechimys* has a hemochorial, discoid, 8.5 g placenta with prominent subplacenta that begins as an antimesometrial attachment 5.5 d postcoitus, but becomes mesometrially attached (9161, 9162). Females have a bicornuate uterus with an extremely muscular cervix (*Proechimys* 9161). Dystocia is common in captive *Proechimys* (5398). A postpartum estrus occurs (*Proechimys* 9161, 11560). Neonates have open eyes at birth (*Hoplomys* 10756) or after 1-2 d *Proechimys* 11560). They are furred (*Diplomys* 10757; *Hoplomys* 10756; *Proechimys* 11560) and ambulatory (*Hoplomys* 10756). *Makalata* has 2-6 teats, one pair of which may be nonfunctional (5111, 7983). *Proechimys* (5111, 11560) has 4-6 mammae. *Euryzygomatomys* has 6 mammae (7983), while *Echimys* has 8 mammae (5111). Details of anatomy (6650, 9162, 11565) and endocrinology (4528) are available.

NEONATAL MASS		
Hoplomys gymnurus	24.3 g; r20-28.5; n6	10756
Proechimys guairae	20.9 g; se0.3; n190	11560
	26 g; n1 emb	5398
Proechimys guyannensis	24.7 g; from growth curve	3122
Trichomys apereoides	21.1 g; sd4.7; n127	9174a
LITTER SIZE		
Diplomys labilis	1.2 emb; r1-2; n13 f	10757
Echimys chrysurus	2 emb; n1 f	5111
Euryzygomatomys spinosus	1 emb ea; n2 f	7983
Hoplomys gymnurus	2.1 yg, emb; r1-3; n8 f	10756
Isothrix bistriatus	1 emb; n1 f	7983
Kannabateomys amblyonyx	mode 1 yg	7983
Makalata armata	1 yg; n1 f	5111
Mesomys hispidus	1 emb; n1 f	7983
Proechimys guairae	r1-3 emb; n8 f	5398
	2.7; r1-6; n123 lt	11560
	2.8; r2-5; n10	7995, 7998
	4 emb; n1 f	8562a
Proechimys guyannensis	2 emb; n1 f	5111
	2.4 yg; max 4	3122
Proechimys semispinosus	2.4 yg; se0.02; r1-2; n51 lt	3319
Trichomys apereoides	3.2 yg; sd1.43; mode 3; r1-7; n174 lt	9174a
SEXUAL MATURITY		
Proechimys guairae	f: vag perf: 55.9 d; sd12.8; n15 f	6650
	f: vag perf: 65.6 d; se2.4; n78 f	11560
	f: 1st conceive: 62.5 d; se3.6; n29	11560
	m: 1st mate: 3 mo	11560
	f/m: 5 mo; n10	7995, 7998
Proechimys guyannensis	f: 1st birth: 208 d	3122
Proechimys semispinosus	f: 1st preg: 9.2 mo; r6-12	3319
Trichomys apereoides	f: 1st conceive: ~73 d	9174a
	f: 1st preg: 170 d; sd13.5; n5	9174a
CYCLE LENGTH		
Proechimys guairae	14.9 d; r2-51; n54 cycles, 6 f	6650
	20.7 d; se3.4; n41 cycles, 27 f	11560
GESTATION		
Proechimys guairae	implant: 6-6.5 d	9161
	62.6 d; se0.2; n106	11560
Proechimys guyannensis	62-64 d; r55-71	3122
Trichomys apereoides	97.0 d; sd1.2; r95-98	9174a
LACTATION		
Proechimys guairae	wean: min 5 d	11560
Proechimys guyannensis	wean: 21-35 d	3122
Trichomys apereoides	wean: 6 wk	9174a
INTERLITTER INTERVAL		
Proechimys guyannensis	66 d; n1	3122
LITTERS/YEAR		
Proechimys guairae	4.4	7995, 7998
Proechimys semispinosus	4.68	3319
SEASONALITY		
Diplomys labilis	preg f: est yr rd; Panama	10757
Proechimys guairae	estrus: yr rd; lab	11560
Proechimys semispinosus	preg f: yr rd, peak Jan-June; Panama	3319
Trichomys apereoides	preg f: yr rd; captive	9174a
	repro: yr rd, low Dec-Jan; Brazil	7983

Order Rodentia, Family Thryonomyidae, Genus *Thryonomys*

The 2-9 kg cane rats (*Thryonomys gregorianus*, *T. swinderianus*) are semi-fossorial or semi-aquatic, nocturnal herbivores of marshes and savannas in Africa south of the Sahara. The hemochorial placenta with a well-developed subplacenta has an interstitial, mesometrial implantation (8029, 8029a) and is eaten after birth (421, 10158). The precocial neonates are furred with open eyes and can walk or run (331, 2439, 5760, 9945, 10158). Females have 2-3 pairs of lateral mammae placed high up on the sides of the abdomen (331, 421, 3132, 5760, 9945, 10158). The testes are semi-internal (9945).

NEONATAL MASS		
Thryonomys swinderianus	70-84 g; n4	3132
	110.0 g; sd34.0; n66	422
	131.0 g; sd28.0; r72-190; n46	421, 422
LITTER SIZE		
Thryonomys gregorianus	2-6	5760
	2.5 emb; r2-3; n2 f	10158
	3 emb ea; n2 f	4925
Thryonomys swinderianus	2-3 emb	8027
	3.6; sd1.2; n66	422
	3.9; sd1.3; 2-6 96%; r1-8; n46	421, 422
	4 emb; n1 f	331
	4-6	3132
	r4-11 cl; n4 lt	8027
SEXUAL MATURITY		
Thryonomys swinderianus	f: vag perf: 5 mo	421
	f: 1st birth: 12-18 mo	421
	f/m: 1st mate: ~1 yr	5760
CYCLE LENGTH		
Thryonomys swinderianus	6.62 d; sd3.1; n35 cycles	9652
GESTATION		
Thryonomys gregorianus	~3 mo	5760
Thryonomys swinderianus	est 10 wk	3132
	~3 mo	5760
	155 d; sd9; r137-172; fetal growth data	421
LACTATION		
Thryonomys swinderianus	solid food: 1 d?	11253
	wean: ~1 mo	10158
LITTERS/YEAR		
Thryonomys gregorianus	2 possible	5760
Thryonomys swinderianus	2	3132
	2	5760
	3	11253
SEASONALITY		
Thryonomys gregorianus	preg f: Nov-Dec; n2; Kenya	4925
	birth: Feb-Mar; n Uganda	5760
Thryonomys swinderianus	mate: Mar-Nov; Africa	5760
	preg f: 26% Mar, 95% June; n Ghana	421
	preg f: 80%/mo except 64% Feb, Aug; s Ghana	421
	birth: Mar; n1; Zambia	331
	birth: June-Aug; sw Africa	9945
	birth: June-Dec; Zimbabwe, Zambia, Malawi	10153
	birth: Aug-Dec; S Africa	10158
	birth: once dry, twice rainy season; Zaire	11253
	young found: mid-Oct-Jan; w Africa	5760

Order Rodentia, Family Petromyidae, Genus *Petromus*

The 100-300 g, terrestrial *Petromus typicus* from arid regions of southwest Africa are solitary or found in pairs. These diurnal herbivores are furred at birth (2380, 9945). Females have 6 lateral teats at the scapular level (2380, 7983, 9945, 10158), while males have semi-internal, inconspicuous testes and a short baculum (2380, 9945).

LITTER SIZE		
Petromus typicus	1-2	2380
	1-2	9945
	1-2 table or 2-4 text	11864
LITTERS/YEAR		
Petromus typicus	1	11864
SEASONALITY		
Petromus typicus	mate: Nov-Mar; Namibia	9945
	preg f: Jan; Namib desert, Namibia	11863
	preg f: Sept, Nov; S Africa	2380

Order Rodentia, Family Bathyergidae

African mole-rats (*Bathyergus* 350-1500 g; *Cryptomys* 50-200+ g; *Georychus* 125-360 g; *Heliophobius* 140-170 g; *Heterocephalus* 25-80 g) are fossorial herbivores from Africa south of the Sahara. Colonial life is highly developed (*Cryptomys* 780; *Heterocephalus* 5270), usually only one female and one or two males are reproductively active in a colony (1449c, 3176, 6144a). Males have internal testes (*Cryptomys* 9945). The hemochorial placenta has a subplacenta (*Bathyergus* 6628) and implants into a bipartite uterus (4794). Neonates are naked and open their eyes by 9 d (*Georychus* 779), 13-50 d (*Cryptomys* 780, 1449b), 3-4 wk (*Heterocephalus* 5267, 5269). Females have 6 teats (331, 2379, 3692, 4495, 5760, 9945, 10158). Males have seminal vesicles, penial spines, and abdominal testes (4794, 11127). Details of endocrinology (779, 3175a, 3176, 3177, 8116a), copulation (779, 780, 4727a), and birth (779) are available.

NEONATAL MASS		
Bathyergus janetta	15.4 g	5270b
Bathyergus suillus	34 g	5270b
Cryptomys hottentotus	7.9 g; sd0.5; r7.2-8.8; n18	1449b
	8-10 g	5270b
Georychus capensis	5.37 g; sd0.28; n6	779
	7.55 g; sd1.74; n4	779
	12.27 g; sd0.76; n10	779
Heliophobius argenteocinereus	7 g	5270b
Heterocephalus glaber	1.86 g; r1-2	5270a
	1.9 g	5267, 5269
NEONATAL SIZE		
Cryptomys hottentotus	2-2.5 cm	778a
	TL: 59.9 mm; sd2.4	1449b
Georychus capensis	3-4 cm	779
WEANING MASS		
Cryptomys hottentotus	18 g	1449c
	34.1 g; sd0.3; r33.8-34.5	1449b
Georychus capensis	36 g	1449c
Heterocephalus glaber	11 g	1449c
SEXUAL MATURITY		
Cryptomys hottentotus	f: 1st conceive: 340 d; n1 [aborted]	1449b
	f: 1st conceive: 73 wk	1449c
LITTER SIZE		
Bathyergus janetta	3.5; r1-7	5270b
Bathyergus suillus	2.4; r1-4; n34 lt	5267
	3-4 yg	10158
Cryptomys hottentotus	1-5	5760
	1.75 emb; r1-2; n8 f	8965
	2-4	11253

	2 emb; r1-3; n4 f	331
	2.2 yg; sd0.7; r1-3; n9+ lt	1449b, 1449c
	2.8; r2-4; n10 lt	780
	3 yg; n1 lt	646
	4.0 emb; n2 f	10154
	5 emb ea; n2 f	2379
	5 emb ea; n2 f	9945
Cryptomys ochraceocinereus	2-4	11253
Georychus capensis	5.9; r3-10; n19	779
	6.0 yg; mode 6; r1-10; n9 lt	10718
Heliophobius argenteocinereus	1-4	10158
	2; n1 lt	5267
	2-4	11253
	2.5 emb; r2-3; n2 f	4925
Heterocephalus glaber	3-11 yg	5267, 5269
	7.7; se2.3; r4-13; n26 lt	6144a
	9.7 weaned; se2.81	1283a
	max 12	5270
	12.3; se5.7; r1-27; n84 lt	5270a
CYCLE LENGTH		
Heterocephalus glaber	estrus: ≥24 h	6144a
GESTATION		
Bathyergus suillus	est 2-2.5 mo	5267
Cryptomys hottentotus	55-66 d; n2	778a
	78-92 d	780
	85-87 d; progesterone levels	780
	98 d; sd9.2; r84-112; n9	1449b, 1449c
Georychus capensis	44 d	779
Heliophobius argenteocinereus	87 d; n1	5267
Heterocephalus glaber	44 d	9899
	66-76 d; n121 lt, 16 f	5270a
	72-77 d	6144a
LACTATION		
Bathyergus janetta	wean: 28 d	5270b
Bathyergus suillus	wean: 21 d	5270b
Cryptomys hottentotus	independent: 18 d	780
	solid food: 19 d	1449b
	wean: 3-4 wk	778a
	wean: 82.5 d; sd15.4; r72-105	1449b, 1449c
Georychus capensis	independent: 17 d	779
Heterocephalus glaber	wean: 3-4 wk	5270a
	wean: 4-5 wk	6144a
	wean: ~1-2 mo	5270
LITTERS/YEAR		
Bathyergus janetta	1-2	5270b
Bathyergus suillus	1-2	5270b
Cryptomys hottentotus	1-2	780
	min 2	11253
Cryptomys ochraceocinereus	2	11253
Georychus capensis	2/season	779, 10718
Heliophobius argenteocinereus	1	10158
	2	11253
Heterocephalus glaber	1-4	5270
	4-5	5270a
	max 5	6144a
SEASONALITY		
Bathyergus janetta	preg f: Aug-Dec; S Africa	5270b
Bathyergus suillus	testes: large June-Sept; S Africa	11127
	mate: Apr-early Oct; S Africa	11127
	preg f: Aug-Sept; S Africa	10158
	repro: mid June-mid Oct; S Africa	5267, 5270b
Cryptomys hottentotus	preg f: Feb-Mar, Aug; Transvaal, S Africa	8965
	preg f: Apr; n2; Namibia	2379
	birth: end rainy season; Zaire	11253

	lact f: Jan, Apr; Transvaal, S Africa	8965
	juveniles/subadult found: yr rd; Zambia	331
	repro: yr rd; S Africa	5270b
Cryptomys hottentotus	preg f: Dec; n1; Karamoja, Uganda	5760
	preg f: Apr; n2; Namibia	9945
	est yr rd; S Africa	10158
Cryptomys ochraceocinereus	mate: Apr-June; Zaire	11253
	birth: Apr-June; Zaire	11253
Georychus capensis	inguinal testes: May-Sept; S Africa	779
	preg f: Sept-Dec; S Africa	779
Heliophobius argenteocinereus	preg f: May; n1; Kenya	4925
Heterocephalus glaber	yg found: Feb-Apr; Kenya	5267
	repro: yr rd; S Africa	1283a
	repro: yr rd; captive	5270a, 5270b

Order Rodentia, Family Ctenodactylidae

Gundis (*Ctenodactylus, Felovia, Massoutiera, Pectinator*) are 150-350 g, diurnal, terrestrial herbivores from the Sahara. A copulatory plug is formed and placentophagia occurs as does a prepartum estrus (3971). Neonates are furred, homeothermic, and have their eyes open (3727, 3970, 6237). Females have vaginal closure membranes, a bicornuate uterus, and 2-4 mammae (3727, 8404). Males have a baculum and seminal vesicles (8404). Details of placentation (2392, 2399) are available.

NEONATAL MASS		
Ctenodactylus gundi	29.9 g; se11.9; r18-40; n15	3971
Ctenodactylus vali	17 g; n1	3972
	f: 18 g; n1	3727
	19.9 g; n14	3727
Massoutiera mzabi	20.5 g; r20-21; n2	3727
LITTER SIZE		
Ctenodactylus gundi	1.7; mode 2; r1-3; n17 lt	3971
	2	3727
	~5 emb; 2.45; mode 2; r1-5; n148 f	2398
Ctenodactylus vali	2; r2-3	3727
Massoutiera mzabi	2	3727
Pectinator spekei	1	3727
	2-3	8404
SEXUAL MATURITY		
Ctenodactylus gundi	f: vag perf: 75 d	3971
	9-12 mo	3727
Ctenodactylus vali	f: vag perf: 10-12 mo; n2	3727
Massoutiera mzabi	f: vag perf: 10-12 mo; n1	3727
	f: vag perf: 6 mo; n1; wild-caught	3727
Pectinator spekei	f: vag perf: 10-12 mo; n1	3727
CYCLE LENGTH		
Ctenodactylus gundi	vagina open: 4.7 d; closed 23.6 d	3727
	vagina open: 7.4 d	3971
	28.7 d; mode 31; r21-34; n15 cycles, 2 f	3727
Ctenodactylus vali	vagina open: 3.8 d; closed 19.6 d	3727
	23.4 d; mode 22; r14-33; n34 cycles, 5 f	3727
Massoutiera mzabi	vagina open: 4.9 d; closed 19.9 d	3727
	24.9 d; mode 26; r14-36; n57 cycles, 7 f	3727
Pectinator spekei	vagina open: 4.9 d; closed 17.6 d	3727
	22.7 d; mode 25; r14-29; n31 cycles	3727
GESTATION		
Ctenodactylus gundi	est >57 d	6237
	73.2 d; se6.6; r70-79; n5: vaginal plug-birth	3971
Ctenodactylus vali	est 56-67 d	3727
	68 d; n1 capture to birth	3972
INTERLITTER INTERVAL		
Ctenodactylus gundi	69.7 d; se0.9; r69-70; n3	3971
Ctenodactylus vali	64 d; n1 after stillborn	3972

SEASONALITY

Ctenodactylus gundi	vagina open: yr rd; lab	3971
	diestrus: May-Dec; Tunis	2398
	estrus: Jan-Mar; Tunis	2398
	estrus: Feb-Aug; Algeria	3727
	preg f: Feb-Mar; Tunis	2398
	birth: est Jan-Mar; Algeria	3727
	birth: Feb-Mar, May, Nov; lab	3971
	yg found: Feb-June; Algeria	3971
Ctenodactylus vali	birth: Mar, July, Nov; n3; captive	3727
	birth: Mar-Apr; semi-captive, Algeria	3727
	repro: Feb-June, peaks Mar-Apr, end June; Algeria	3727
Massoutiera mzabi	estrus: Apr-Aug; Algeria	3727
	yg found: Mar-June; Algeria	3727
Pectinator spekei	preg f: Sept-Oct; few data; ne Africa	8404

Order Lagomorpha

Pikas, hares, and rabbits range from 100-3000 g, yet their reproductive biology is remarkably constant. The reported gestation and lactation lengths are unusually homogeneous. Ovulation is induced by copulation and a postpartum estrus is present. Testes are abdominal outside of the reproductive season. All species are polytoccous but some variation in litter size and the developmental state at birth exists. Delineating the differences between hares and rabbits should be especially useful as the two groups form a natural control for the effects of diet, habitat, and phylogeny on the production of altricial versus precocial young because both are polytocous, both have similar diets and similar phylogenetic ancestry, and they can be found sympatrically.

Order Lagomorpha, Family Ochotonidae, Genus *Ochotona*

Pikas (*Ochotona* 100-400 g) are terrestrial, often gregarious herbivores in Asia and North America. Ovulation is triggered by coitus (8787, 12007b). The deciduate, hemochorial, discoid placenta has a mesometrial, superficial implantation into a duplex uterus (4463, 8053) and is probably eaten after birth (11655). Estrus occurs (6820, 9808, 10089) after the birth of blind and naked or slightly furred (4193, 5146, 8783, 8784, 8786, 9175, 11657) neonates whose eyes open in 8-10 d (8053, 8783, 8798, 9808, 9949, 11039, 11655). Females have 6-12 mammae (1474, 2550, 8784, 9175) and a cloaca (2771). Testes descend during the breeding season (8053, 9808). Details of anatomy (2772, 5804, 9808) and milk composition (8786) are available.

NEONATAL MASS		
Ochotona alpina	6-8 g	8798
	9.6 g; n1	4193
Ochotona curzoniae	11.2 g; sd1.2; n36	12007a
Ochotona pallasi	6-8 g	8798
Ochotona princeps	8.5 g; r4.1-12.5; n34	9808
	9.0 g; r8.0-10.23; n5	6884
	m: 11.5 g	11657
	12.4 g; n71	3875, 11655, 11656
	f: 12.7 g	11657
Ochotona pusilla	6-8 g	8798
Ochotona roylei	6-8 g	8798
Ochotona rufescens	11.4 g; sd1.9; r8.5-12.25; n51	8783, 8784, 8786, 8788
NEONATAL SIZE		
Ochotona curzoniae	HB: 57.6 mm; sd4.3; n36	12007a
WEANING MASS		
Ochotona princeps	52.24 g; n28, 28 d	9808
	75 g; sd13.2; r50.9-103.2; n37,; 6th wk	11655, 11656
	f: 121.5 g; 59 d	11657
	m: 132 g; 59 d	11657
Ochotona rufescens	74.9 g; sd15.5; n51	8788
	82 g; 21 d	8783
LITTER SIZE		
Ochotona alpina	~1 wean	1238
	2.2 yg	9070
	2.4 emb, yg; r2-3; n5 f	4193
	3-8	1033
	mode 3; r1-5	5717
	3 yg; n1 lt	8053
	3 emb; n1 f	10516
	4-6	10517
	4.8 ps; r2-6	5519
Ochotona collaris	2-6	2088
	3 emb; n1 f	6689
	4 emb; n1 f	6689
	4 emb; n1 f	12049
Ochotona curzoniae	2-5 yg	7382
	4.8 yg; ±1.8; r1-8; n17 lt	12007b
Ochotona daurica	3-5	8053
	5.6 emb; r2-7; n5 f	10516
	6.1-7.7; r1-11; n >1 population	10090

Ochotona macrotis	1-4 yg	7382
	4.08; r2-6	10090
	5 yg; r3-7	12104
Ochotona pallasi	1 yg; n1 f	8053
	6.5-7.5; r1-12; n >1 population	10090
	8 emb; n1 f	10516
Ochotona princeps	1.62 yg; n26 lt	11657
	1.75; r1-3	10092
	1.83 wean; ±0.08; mode 2; n35 lt	7301, 7302
	2-3	2359
	2.17; r1-3; n6 lt	10089
	2.33 emb; ±0.07; mode 2-3; r1-4; n80 f	7301, 7302
	2.64 cl, ca; mode 2,3; r2-4; n175 lt	7301, 7302
	2.67 emb; r2-3; n3 f	279
	2.7 emb; r2-3; n3 f	279
	2.74 yg; mode 3; r1-4; n26 f	3875, 11655, 11656
	2.78; mode 2-3; r1-5; n18 lt	10089
	2.9 yg; se0.2; r1-6; n21	6820
	mode 3-4	2360, 6884
	mode 3-5 yg	9807
	3 emb; n1 f	5025
	3 emb; n1 f	6560
	3 yg; n1 lt	2360
	3 yg; n1 lt	2550a
	3 yg; n1	11039
	3.0 yg; r2-4; n2 lt, 1 f	10089
	3.0; r2-4; n10 lt	5804, 6820, 10089
	3.13 emb; n53 f	10088a, 10089
	3.4; r2-5; n50 lt	6820
	3.6 ps; se0.2; n33 f	12049
	4 emb; n1 f	6553
	4 emb; n1 f	2550
	5 emb; n1 f	9949
Ochotona pusilla	r3-13 yg; 8.1-8.8; 1st lt; 9.1-10.6; 2nd, 3rd lt; 7.3-7.4; last lt	1033
	4-12	8053
	6 yg n1 f	8798
	7-13	8053
	9.7 emb; r6-12; n3 f	4645
	max 12	10090
Ochotona roylei	3 emb; r1-5	7382
	3.0; r1-5	8784
Ochotona rufescens	4.82 wean; n85 lt	3553
	5.2 emb; se0.36; r4-10; n23 f	9175
	6; r4-9	8783, 8784, 8786, 8788
	6.12 yg; r4-10; n85 lt, 24 f	8785
	6.59; n182 lt	10090
Ochotona rutila	4.2	8798
SEXUAL MATURITY		
Ochotona alpina	f/m: 25-30 d	5717
	f/m: not 1st summer	10516
Ochotona collaris	f: 1st birth: ~1 yr	6689
Ochotona daurica	f: 1st birth: 1st yr	10090
Ochotona macrotis	f/m: 1st mate: 1 yr	12104
	f/m: 1st repro: 1 yr	10516
Ochotona pallasi	f/m: 1st mate: 1st yr	8798
	f/m: 25-30 d	7300
Ochotona princeps	m: sperm: yearling	9808
	f/m: spring of yr after birth	9949
Ochotona pusilla	f: ~4-5 wk	9949
	m: yearling	8798
	f/m: 25-30 d	8798

Ochotona roylei	f/m: 21-23 d	3553
Ochotona rufescens	f: 1st preg: 5-8 wk	8784
	f/m: 65 d	8785
	f/m: 60 d	10090
Ochotona rutila	f/m: 1st mate: yearling	10517
GESTATION		
Ochotona alpina	est ~4 wk	8798
	30 d	8798
Ochotona curzoniae	22.5 d; ±0.9; r21-24; n17 lt	12007b
Ochotona pallasi	25 d	7300, 7302
Ochotona princeps	30 d	11656
	30 d	10092
	30 d; n2	2359, 9807, 9808
	30 d 19 h; n8	8798
Ochotona pusilla	20-24 d	8783, 8784, 8785, 8786, 8788
Ochotona rufescens	25-26.5 d	8798
LACTATION		
Ochotona alpina	den emergence: 15 d	8798
	solid food: 18-20 d	8798
	wean: 20-22 d	8798
Ochotona pallasi	den emergence: 15 d	8798
	solid food: 18-20 d	3875, 11655, 11656
	wean: 20-22 d	11039
Ochotona princeps	solid food: 7 d	11656
	solid food: 12 d	9808
	wean: mode 35 d; min 18	2360
	wean: ~3-4 wk	11655
	wean: <4 wk	7302
	wean: end 6th wk	2359
	wean: 3-4 wk	11657
	wean: 21-29 d	8798
	wean: 3-5 wk	8798
Ochotona pusilla	den emergence: 15 d	8798
	solid food: 18-20 d	9949
	wean: 20-22 d	8798
	wean: max 3 wk	8798
Ochotona roylei	den emergence: 15 d	8798
	solid food: 18-20 d	8783, 8784
	wean: 20-22 d	9175
Ochotona rufescens	solid food: 8 d	8783, 8784, 8788
	wean: 2-3 wk	9808
	wean: 21 d; min 9-10	10092
INTERLITTER INTERVAL		
Ochotona princeps	~1 mo	5519
	~1 mo	1033
LITTERS/YEAR		
Ochotona alpina	est 1	9070
	2	2088
	2	5717
	2	6601a
	2/season	8798
	2	10090
	2-3/summer	8053
	2-3	10517
Ochotona collaris	mode 1; r1-2	10516
Ochotona daurica	1-3	8798
	2	10516
Ochotona pallasi	2-3	2088
	3	7300, 7301, 7302
Ochotona princeps	mode 1; r1-2	6820
	2	2359
	2	2772
	2	10089, 10092

	max 2	9808
	3	1033
	3.0; r2-5	8053
Ochotona pusilla	2	8798
	2	9949
	2+; r2-3	4645
	3-4	10090
	3-5 adult; 1-3 juveniles	9175
Ochotona roylei	1-2	8785
Ochotona rufescens	est 2	8786
	2.8-5.2	8053, 10090
	max 4	4193
Ochotona rutila	2-3	9070
SEASONALITY		
Ochotona alpina	estrus: end May-early June; Sojr-Charchan, Mongolia	10516
	preg f: June-early July; Japan	5717
	repro: mid Apr-?; Yakutia, USSR	2088
	repro: May-mid Aug; Mongolia	10516
	repro: ?-July; USSR	12104
Ochotona collaris	preg f: May-June; n2; AK	6689
	preg f: June; n1; AK	6689
	preg f: June; n1; YUK, Camada	12049
	birth: May-June; n BC, Canada, AK	179
	lact f: June; n1; AK	6689
	lact f: June-July; n2; YUK, Canada	12049
Ochotona curzoniae	birth: est May-Aug; Nepal	7382
Ochotona daurica	preg f: May-July; Mongolia	7300
Ochotona macrotis	repro: Apr-Aug; Tersky-Alatau, USSR	9808
Ochotona pallasi	birth: end May-July; Coney, Gobi, Mongolia	9808
Ochotona princeps	m fertile: May-June; Rocky Mt, ALB, Canada	2359, 2360
	testes: large Mar-May, small Oct, Dec; CA	7302
	estrus: early Mar-?; CA	10092
	mate: mid Apr-June; CO	2772
	mate: May-July; sw ALB, Canada	7300, 7302
	mate: end May-June; lab?	5804, 6820
	ovulate: June-July; UT, CO	12049
	conceive: May-early June; sw ALB, Canada	9808
	preg f: end Apr-July; CO	2088
	preg f: June; n1; YUK, Canada	5146
	preg f: ?-July; CA	1474
	birth: May-June; BC, Canada	10092
	birth: May-June; w US	7302
	birth: May-Aug; w US	9949
	birth: mid June-early Aug; lab?	9175
	birth: Aug; n1; captive	2550a
	wean: end June-early Aug; sw ALB, Canada	8786
Ochotona pusilla	repro: Apr-June/Aug; Irtysh R, USSR	8053
Ochotona roylei	preg f: May, Aug; Nepal	7382
Ochotona rufescens	birth: 2nd lt June-July; Pakistan	9175
	repro: Apr-Sept; Afghanistan	8786
Ochotona rutila	birth: early June; USSR	8053

Order Lagomorpha, Family Leporidae

Rabbits and hares (*Bunolagus* 1-1.5 kg; *Caprolagus*, *Lepus* 1.3-7 kg; *Nesolagus* 365-390 mm; *Oryctolagus* 0.8-3 kg; *Pentalagus* 430-510 mm; *Poelagus* 2-3 kg; *Pronolagus* 1-3 kg; *Romerolagus* 340-530 g; *Sylvilagus* 250-3280 g) are terrestrial herbivores from the New World, Eurasia, and Africa. *Lepus* and *Sylvilagus* females are often larger than males (*Lepus* 1518, 3344, 3346, 3348, 4710, 5210, 6419, 6452, 8500, 8501, 8730, 11398; cf 265, 9080; *Sylvilagus* 1720, 5210, 6614, 10236), while each sex has been reported larger in *Oryctolagus* (1213, 4054, 8491, 10386).

Ovulation is induced by coitus (617, 1213, 1307, 1530, 1531, 1654, 4536, 5944, 6904, 6914, 8521). Superfetation may be caused by sperm storage (6913, 6914). The discoid, deciduate, labyrinthine, hemochorial or hemoendothelial placenta has a mesometrial, superficial implantation into a duplex uterus (231, 1005, 1214, 2959, 3011, 6934, 7563, 7567, 8053, 10310, 10487, 11966) and may be eaten after birth (4099, 8053). A postpartum estrus is common (1004, 1213, 1214, 1307, 1653, 1654, 1722, 2326, 3344, 3396, 4099, 4499, 5239, 5624, 5760, 5865, 6832, 7561, 7692, 8053, 9805, 11398), but in hares a prepartum estrus also occurs which may lead to superfetation (1003, 5883, 6909).

Leporids exhibit a range of neonatal development. The extremes are the furred *Lepus* neonates with open eyes (143, 502, 680, 1110, 1474, 2088, 2139, 2439, 2458, 3571, 4006, 4467, 4645, 5146, 5760, 6441, 7433, 8053, 8501, 8798, 9080, 9175, 9687, 10154, 10158, 11058, 11253, 11363, 11365) and the naked *Oryctolagus* neonates whose eyes open at 10 d (3571, 4645, 7433, 8053, 8798). Other genera are blind at birth, but furred (*Poelagus* 5760; *Pronolagus* 10158; *Romerolagus* eyes open 5-7 d, 1676, 2404, 2819, 2820; *Sylvilagus* eyes open 2-10 d, 893, 1290, 1474, 2891, 5146, 7561, 8798, 10222, 10583, 10731, 10908; cf 5099).

Females have 6 mammae (*Caprolagus* 3759; *Pronolagus* 9945), 6-10 mammae (*Lepus* 331, 1474, 2548, 2550, 3571, 4745, 6335, 7433, 8053, 9175, 9945; *Oryctolagus* 3571, 7433, 8053), or 8-10 mammae (*Sylvilagus* 1474, 1720, 2548, 2550, 4745). Males have seminal vesicles, a proprostate, paraprostate, prostate, and bulbourethral glands (*Oryctolagus* 1611, 4940) or inguinal and coagulating glands (*Sylvilagus* 2959).

Details of anatomy (1005, 1723, 4981), endocrinology (*Lepus* 1528, 1529, 1530, 1532, 1533, 1534, 1535, 1535a, 1535b, 2326, 2327, 2328, 2329, 5621, 6452, 6453, 10316, 10317, 10318; *Sylvilagus* 1535, 2964, 9524, 10384), copulation (9350a), reproductive cyclicity (9350), birth (1653), milk composition (277, 511, 6424), suckling (1306, 6908), and light (*Lepus* 2326; *Sylvilagus* 964) are available.

NEONATAL MASS

Bunolagus monticularis	40-50 g	2821c
Lepus americanus	52 g; r48-58; n3	8005a
	65.9 g; n1 emb	1110
	67 g; sd13.2; n128	9805
	68.7 g; n3	3899
	81.7 g; r64-96; n22	2458
Lepus californicus	96-114 g	4099
	110 g	4467
Lepus capensis	80-130 g	3350
	97 g; n21	3349
	100-130 g	6101
	100-165 g	3344, 3346
	107 g; r65-115; n54, 1-2 d	8513
	113.2 g; ±15.6; r60-180; n283	6909
	130.2 g; r100-170; n11	4562
	150 g	11058
Lepus saxatilis	98 g; n8	3349
Lepus timidus	70-130 g	8798
	80-140 g	4645
	80.7-132.7 g; across litter size	8323
	84-140 g	8053
	87.3 g; r71-106; n9 term emb	3348
	104.8 g; n1, 2 d	11414
Lepus townsendii	87 g; r84-90; n2 yg	680
	89.25 g; r84-93; n4 term emb	680
	105.33 g; r101-110; n3, 1 d	680
Oryctolagus cuniculus	30-40 g	8798
	f: 35.23 g; se1.140; n40	1175
	m: 36.99 g; se0.833; n55	1175
	40-45 g	3571
	42.0-44.5 g	1213
	45.8 g; sd1.62; n32 emb, 7 lt, 31 d gestation	9270
	46.95 g; sd8.42; n126 emb, 20 lt, 30 d gestation	10763
Romerolagus diazi	24.1 g; r22.6-25.0; n5	1677
	f: 25-27 g	6953b
	m: 32 g	6953b

Sylvilagus aquaticus	53.5 g; r48.5-56.9; n6, 1 lt	4920
	55.7 g; n1	10583
	61.4 g; n18	10222
Sylvilagus audubonii	32.8 g; n9 stillborn	5142
	36 g; r33-40; n3	2550b
Sylvilagus bachmani	24.3-30.2 g; term emb	7561
Sylvilagus floridanus	24.8-31.2 g; 25.4 d gestation	9251
	29.54 g; r23.3-33.0; n5	893
	30 g	8798
	30.3-38.0 g; n6 term emb	2891
	37.8 g; n22 yg, 1 d	1137
	42.2 g; n6	4742
NEONATAL SIZE		
Lepus americanus	105.3 mm; r97.0-113.4; n8	2458
	CR: 110.7 mm	1110
	HB: 153.7 mm; n3	3899
Lepus capensis	130-170 mm	11058
Oryctolagus cuniculus	f: 117.70 mm; se1.181; n40; nose-tail	1175
	m: 119.07 mm; se0.964; n55; nose-tail	1175
Romerolagus diazi	TL: 78.3 mm; r74-83; n3 term emb	1677
	TL: 93.8 mm; r83-106; n5	1677
Sylvilagus aquaticus	TL: 130 mm; n4	4920
Sylvilagus audubonii	TL: 90 mm; n1 "very young"	5142
	117 mm; n9 stillborn	5142
Sylvilagus floridanus	CR: 87 mm	1137
	90-98 mm; n6 term emb	2891
	HB: 121-134 mm; 25.4 d gestation	9251
Sylvilagus transitionalis	CR?: 73.8 mm	10731
WEANING MASS		
Lepus alleni	3.0 kg; sd0.34; n13	9616
Lepus americanus	~400 g; 30 d	8005a
	486 g; sd84.4; n46; 28 d	9805
Lepus californicus	1.47 kg; 12-13 wk	10246
	2.3 kg; sd0.64; n9	9616
Lepus capensis	200 g; solid food	3346
	mode ~1 kg weaned	3346
	808-1022 g	6909
Lepus townsendii	737 g; ~4 wk	10246
Oryctolagus cuniculus	198 g; ±23	11887
	200-250 g; 1st capture	11887
Romerolagus diazi	~80 g; est 20 d	5007
	f: 80-130 g; 3 wk	6953b
	m: 95-128 g; 3 wk	6953b
Sylvilagus floridanus	~112 g; leave nest	893
LITTER SIZE		
Bunolagus monticularis	1 yg	2821c
Caprolagus hispidus	2-5	3759
Lepus alleni	1.94 emb; mode 1,2; r1-5; n124 f	11365
	3 emb; n1 f	11363
	6 emb; n1 f	5410
Lepus americanus	1 emb; n1 f	2548
	r1.64-3.05; across yr & season	6132
	2 emb; n1 f	9968
	2 ps; n2	10839
	2-4 cl	7872
	r2.2-4.9; 1st lt smaller	5624
	2.27 emb; 1st lt: r2.96-3.26; later lt	11895
	2.4 yg; mode 2; r1-4; n36	143
	2.5 emb; r1-4; n2 f	10839
	2.68 yg; mode 2; r1-5; n71 f	1111
	2.7 emb; mode 4; r1-4; n7 f	2663
	2.73 emb, yg; r1-5; n26 lt	6706
	2.82 emb; mode 3; r1-7; n266 f	143
	2.88; r1-5; n81 lt	9805

	2.90 emb; r1-5; n140 f; r2.54-3.58; n7 yr	4037
	2.90-5.67 across 4 mo; 2.6-6.5 by mo; n83 lt, 5 yr	2326, 2328
	2.96 yg; mode 3; r1-6; n80 lt	9806
	3.00; r2-4; 1st lt; 4.75; r2-7; later lt	2672
	3.00; 1st lt; 4.84; later lt	7243
	3 emb; n1 f	4745
	3.0 yg; r1-4; n6 lt	4006
	3.0; r1-4; n7	2663
	3.0 emb; r2-4; n11 f	12049
	3.1 yg; n16	2664
	3.35; se0.12; n245 lt	3510
	3.4 emb; 95%ci 0.3; n100 f	1229
	3.44 emb; r1-4; n149 f	2139
	3.56; r2-5; 1st lt; 5.91; r1-9; later lt	2672
	3.7 cl; n146 tracts	2664
	3.8 yg; n5 lt	9579a
	3.82 emb; mode 4; r1-7; n82 f	9321
	3.83 yg; r2-6; n6	9579a
	3.9 emb; n137 f	2664
	3.9 emb; 95%ci 0.2; n129 yearlings	1229
	4 emb; n1 f	4006
	4.5 emb; r3-6; n2 f	2545
	4.6 emb; 95%ci 0.4; n78 f	1229
	4.7 emb; r1-8; n3 f	10597
	5 emb; n1 f	7809
	5 emb; n1 f	8125
	5 emb; n1 f	10598
	5 emb ea; n2 f	4081a
	5 emb ea; n3 f	9579a
	5.4 emb; r3-8; n5 f	2550
	6 emb ea; n2 f	5254a
	7 emb; n1 f	6553
	7 emb; n1 f	10218
Lepus californicus	1.5-2.6 emb; across mo	4332
	1.79 yg	4467
	mode 2-4; max 6	502
	2.24 emb; mode 1-2; r1-6; n70 f	11365
	2.26 emb; mode 1-2; r1-8; n114 f	5254a
	2.5 yg	6335
	2.5 emb; r2-3; n2 f	522
	2.6 emb; se0.19; r1-5; n32 f	1322
	2.8 emb; r2-4; n6 f	2550
	3 emb; n1 f	523a
	3 emb; n1 f	11228a
	3.8 cl; r1-6	4099
	4 yg; r1-8	1871a
	4.9 emb; ±1.4; r2-7; n22 f	3197
	5.23 emb; r3-8; n39 f	8126
Lepus capensis	1.0-2.8; across mo; r1-5	3344
	1-3	6452
	1-5	9125
	1-5	8798
	1-6	7433
	1 emb; n1 f	2439
	1 emb; n1 f	4925
	1 emb ea; n4 f	6293
	1.17 emb	6517
	1.2 emb; r1-2; n4 f	331
	1.3	10318
	1.5	3349
	1.5-3.3 yg; across seasons	5944
	1.6-1.8; 1st lt; 4.23-4.27; 2nd, 3rd lt	8370
	1.65; se0.14; post partum mating; 2.24; se0.06; n>4000; prepartum mating	1530, 1531

	1.69; r1-3; n13 lt	1452
	1.7-3.7 emb; across seasons	4712, 5944
	1.7; r1-2; n3 lt	8965
	1.77; mode 2; r1-4; n31 lt	10487
	1.8; young f	226
	1.9 emb; r1-3; n13 f	10154
	1.9 emb; r1-6; n44 f	11058
	1.9 emb; mode 1 50%; r1-4; n211 f	9080
	2-4	5277
	mode 2-4; r1-7	3571
	2-5; across seasons	5944
	2 emb; n1 f	6612
	2 emb; r1-4	8359
	2.1	2576
	2.1; mode 2; r1-3; n8 lt	4562
	2.21 emb without resorption	4712
	2.21 large emb; mode 1-3; r1-5; r1.0-2.8; across mo; n311 lt	3346
	2.27; adult f	226
	2.3; n155 lt	7446
	2.3	8814
	2.3 emb; r2-3; n3 f	8134
	2.34 yg; r1-5; n154 lt	6909
	2.46	225
	2.5	8476
	2.55 ps; mode 3; r1-7; r1.0-3.6; across mo; n320 lt	3346
	2.6; r1-4; n10 lt, 2 f	8782
	2.64	6131
	2.66-3.40 yg; n4 groups	3572
	2.68; r1.4-3.3; across mo	1307
	2.69 emb; r1-4; n144 lt	4709
	2.69; mode 2-3; r1-5; n960 lt	4712
	2.76	4712
	2.8 cl; mode 3; r1-9; r1.0-3.8; across mo; n407 lt	3346
	2.84; sd1.23; n248 lt	3540
	3-5; rare 7	8053
	mode 3-5; r3-7	4645
	3-6	3778
	3.25; r1-6; n63 lt	9497
	3.75 emb; r3-4; n4 f	10516
	4 yg; n1 lt	1004
	5 emb ea; n4 f	9774
	max 6-7	6909
Lepus nigricollis	1.33; r1-2; n3	10616
	1.5 emb; r1-2; n2 f	7382
	1.84 emb; mode 1; r1-4; n63 f	8730
	2	9175
	mode 2	8500
	mode 2; r1-2	8501
	2.0 cl; r1-7; n60 f	8730
Lepus oiostolus	2 emb; n1 f	179
Lepus saxatilis	1-2	11253
	1-2 yg	4354a
	1; r1-2	8965
	1 emb ea; n2 f	4925
	1.6	3349
	1.6 emb; r1-3; n86 f	10154, 10158
	2	9945
	mode 2 yg; r1-2	1149
Lepus timidus	1-7	7433
	1.59 emb; r1-3; n278 f	4709
	1.74 emb; se0.05	4710a
	1.88 emb; mode 1; r1-5; n72 f	3348
	1.98 emb; se0.069	4710a
	2.02 emb; se0.045	4710a

	2.06 yg; n16 fall lt	7214
	2.10 implantation sites; mode 2; n78	3348
	2.3 emb; r1-5; n14 late-term f	3348
	2.32 emb; se0.063	4710a
	2.39 cl; mode 2; r1-5; n75	3348
	3	6441
	3; 1st lt; 5; r3-5; 2nd lt	4645
	3; 1st lt; 5; 2nd lt; 3; 3rd lt	8053
	3.11 yg; n84 spring lt	7214
	3.3 viable emb; 95%ci 0.5; r2.5-4.5; across yr	4548
	3.55; se0.2; r2-7; n40 lt	8323
	3.72; n29 1st lt	10876
	3.8; r3-5; n9	8233
	4-8	1474
	4.0 emb; n2 f	6710
	4.33; n88 lt (without 1st lt)	10876
	4.63; n8 lt	10886
	mode 6-8 emb	5029a
	6.3 emb, ps, yg; sd1.34; r3-8; n10	265
	6.5 cl; sd1.7; max 11; n15 f	8233
	7 yg; n1 nest	11414
	max 8 yg	5146
Lepus townsendii	3-6	1474
	3 emb; n1 f	8126
	3 emb; n1 f	6166
	3.3 emb; r1-5; n3 f	5405b,
	3.56 emb, 5.9 cl; n9 f	5865
	3.6 yg; ±0.6; r1-5; n15 f	5865
	4.5; r4-5	502
	4.5 emb; r4-5; n2 f	1649a
	4.6 yg	5239
	4.9 emb	5239
	5 emb, ps; r1-11; n25 f	680
	5.55 emb; r4-8; n9 f	5254a
	5.67 emb; r5-6; n3 f	261a
	5.75 cl; ±0.94; r3-8; n12 f	5865
	5.8 cl	5239
	7 emb; n1 f	3703
Oryctolagus cuniculus	1-14	8798
	2.0-5.2; across season and yr	8228
	mode 3-6 yg; n33 f; 4.5; n>100 lt	7691
	3.3-6.1 emb; n13 sites	5081
	4-7	11620
	4.16 yg; se0.19; r1-6; n56 lt	1175
	4.3 emb; n233 f	11887
	4.36 emb; r3.39-5.70; across months	8491
	4.7 emb; se0.14; n101 f	291, 293
	4.72; mode 5; r1-9; n191 lt	7260
	4.8 yg; sd1.5; n22 lt	7689
	r4.85-5.22 emb; n4 locations	8229a
	4.88 emb; 5.81 cl; n1819 f	1214
	4.9-5.6 cl	1213
	4.923 ps; ±0.067	196
	5.0 yg; sd1.4; n61 lt	7689
	5.0-5.33 emb; r3-7; n3 locations	2898
	5.47 all emb; se0.28; 5.38 live emb; se0.10; n17 f; r3.0-5.49; across age	8229
	5.54	752
	5.5 yg; n9 lt, 3 f	7693
	5.676 cl; ±0.062	196
	5.77 cl; sd1.57; mode 5; r3-15; n656 f	7070
	5.83 cl; sd1.76; mode 5; r3-15; n745 f	11478
	5.95 emb; mode 5,7; r3-10; n57 f	6516
	6 cl; 4-6 emb	1178
	6-7.8; across 5 mo	9918

	6.32 yg; n22	7701
	6.69 emb; mode 4; r1-7; n71 f	9270
	6.80 yg; se0.53; n20 lt	60
	6.92 live emb; sd2.81; se0.17; r1-14; n270 f	10763
	7.07 cl; ±1.49; 6.50 emb	10044
	7.29; r5-9	10386
	9.7 cl; se0.2	10875
	11.6 cl; se0.2	10875
Poelagus marjorita	1 emb; n1 f	4495
	1-2	5760
Pronolagus crassicaudatus	1-2	4180
Pronolagus randensis	1.1 emb; r1-2; n8 f	10158
Pronolagus rupestris	2 yg; n1 nest	10158
Romerolagus diazi	1-3 emb	9231
	1.5; r1-2; n2 lt	6478
	2.0 yg; n2 lt, n1 f	2819, 2820
	2.1; r1-5	6569
	2.3 yg; n18 lt, 4 f	5007
	2.33 yg; r2-3; n3 lt	1676
	2.33 emb, ps; n12 f	1677
	2.5 yg; n4 lt	6953b
	~3 yg; r1-4	6387
Sylvilagus aquaticus	1-6 yg	10583
	2.0 yg; r1-3; n5	8214
	2-3; r1-5	1474
	2.6 emb, yg; r1-4; n5 lt	6614
	2.8 large emb; mode 3; r1-4; n14 f	10861
	2.83 emb; mode 2-3; r1-5; n29 f	5099
	2.91; r1-6; n224 lt	4740
	3.0 all emb; n24 f	10861
	3.08; mode 3; r1-6; n38 lt	10222
	3.3 ps; r3-4; n3 f	6298
	3.4 ps; r3-4; n7 f	10861
	3.7 emb; r3-5	10583
	3.7 cl; r2-6; n46 f	10861
	4.8 yg; r4-6; n4 lt, 3-6 d	4920
	6.7 cl; n5 f	10753
Sylvilagus audubonii	1 emb; 1 f	2550
	1-5 yg	5146
	2 emb ea; n2 f	502
	2.5; r1-4; n2 lt	3705
	2.6 emb; r1-3; n10 f; 3.30 cl	1727
	2.7 ps; n8 f	10470
	2.9 emb; r2-4; mode 3; n56 f	10236
	3 emb; n1 f	5409, 5410
	3.0 yg; r1-5; n3 lt	2550b
	3 emb, yg; r1-5; n6 f	5142
	r3-6 emb; n≥5 f	523a
	3.3 emb; r2-6; n10 f	10470
	3.6 emb; r2-6; n19 f	8126
	4 emb; r3-5; n3 f	5254a
	4 ps; n1 f	3703
	4.3 cl; r2-7; n18 f	10470
	5 emb; n1 f	5405b,
	6 emb; n1 f	6166
	6 emb; n1 f	10954
	6.4 emb; r5-7; n5 f	261a
Sylvilagus bachmani	2.67-3.27 emb, ps; n2 yr	1722
	3.5 emb; r2-5; n11 f	8126
	3.69 emb; r2-6; n26 f	7561
	4 ps; r3-6; n14 f	7561
	5 yg; n1 lt	7561

Species	Value	Reference
Sylvilagus brasiliensis	1.18 nestlings; mode 1; r1-2; n11 lt	2817
	1.9	2816
	3-6	8798
Sylvilagus cunicularius	3 emb; n1 f	5410
	5 emb; n1 f	2341
Sylvilagus floridanus	2.6-6.6; 1st lt; r3.4-7.0 2nd lt across locations	2010
	2.8-5.0 viable emb	5807
	2.8 emb; n404 f	10822
	2.9 emb, yg; r2-5; n9 lt	6614
	2.9 emb; r2-4; n56 f	10236
	3.0 emb; r1-4; n3 f	3212
	3.0 emb; r2-3; n5 f	522
	3.02 emb; mode 3; r1-6; n280 f	1136
	3.1; r1-7; n36	10236
	3.13 cl; n107 lt	8077
	3.14; n209 lt	8340, 8341
	3.41 emb; r1-6; n29 f	586
	3.5-5.6 yg; n3 samples	3117
	3.5 emb; max 7; se0.0416; n611 lt	4741, 4742
	3.52 emb; r2-5; n34 f	11712
	3.57 emb; ±1.32; r1-6; n21 f	965
	r3.64-4.78 cl; r3.63-4.86 implantations; n3 yr	2012
	3.64; sd0.23; n12 1st lt	2014
	4 emb; n1 f	4897
	4.0 emb; r3-5; n3 f; 4.66 cl	1727
	4.0 emb, ps; n4 subadult	7806
	4.06 cl; ±0.83; r3-5; n17 f	965
	4.1-5.3 cl	5807
	4.3 yg; n17 nests	4276
	4.39 emb; ±0.28	10384
	4.4 emb; r1-8; n40 f	9687
	4.4; mode 4; r3-9; n109 lt	1386a
	4.5 emb; r4-5; n2 f	1290
	4.5 emb; r4-5; n2 f	3703
	4.5 emb; r4-5; n2 f	6298
	4.5; r4-5; n2 lt	3703
	4.5 emb; r2-7; n22 f	4276
	4.64 yg; r3-9; n11 lt	586
	4.7 ps; r3-6; n5 f	2891
	4.7 emb; r2-6; n21 f	6515
	4.797 cl; n142 lt	9382
	4.8 emb; 1st lt; 7.1 emb; 2nd lt	5864
	4.8 yg; r1-7; mode 1-2; n35 lt, 7 f	9877
	4.9 yg; n13 nests	2891
	4.92; se0.13; mode 5; r2-9; n140 lt; 4.95 emb; se0.29; n20 f	9250
	4.98 emb, yg; mode 4/5; r2-7; n44 f	893
	5 emb; n1 f	575
	5 yg ea; n2 nests	169
	5.04; n27 lt	10908
	5.06 emb; ±0.47	10384
	5.2; mode 5; r3-7; n38 lt	1386a
	5.22 emb; r4-7; n9 f	169
	5.25 emb; ±0.42	10384
	5.31 emb; ±0.68; n~281 f	6576
	5.34 emb; ±0.53	10384
	5.4 emb, yg; r4-7; n12 lt	4499
	5.42 yg; r3-8; n26 lt	892
	5.5 emb, ps; n124 f adult	7806
	5.58 swellings; se0.14; 5.10 emb; se0.15; r2-8; n106 f	10893
	5.86; n85 lt	2014
	6 emb; n1 f	2548
	6 emb; n1 f	6166
	6 emb; r2-10; n82 f	6371
	6.4 emb; r3-10; n8 f	5405b,
	6.5 emb; r3-8; n13 f	2891

	6.8 emb; n24 lt	5003a
	7 emb; n1 f	2228a
	8 emb; n1 f	261a
	9 yg; n1 nest	585
	10 emb; n1 f	6371
	12 emb; n1 f	5003a
	12 yg; n1 nest	5802
Sylvilagus idahoensis	5.89 ps, emb; r4-8; n9 f	3277
	5.93 emb; r4-8; n14 f	5254a
	max 6	5146
	6 emb ea; n3 f	8126
	6.2-6.3 ova; r5-8; n10 f	3277
Sylvilagus nuttallii	2 yg	2088
	3 emb; n1 f	502
	3.5; r3-4; n2 lt	10954
	4.6 emb; se0.2; r2-6; 4.3 live emb; se0.1; r1-6; 31 f	8717
	4.7 emb; r4-5; n3 f	2550
	5.0 emb; r4-6; n2 f	1119
	5.0 cl; se0.2; r3-8; n31 f	8717
	5.9 emb; r4-8; n14 f	4034
	6 emb ea; n2 f	4034
	6.1 emb; r4-8; n8 f	8126
	6.62 emb; r5-8; n15 f	5254a
	7.7 cl; r7-8; n3 f	4034
	max 8	5146
Sylvilagus palustris	2.82 ps; se0.093; n121 f	4921
	mode 3-5	10866
	3.09 cl; se0.054; n132 f	4921
	4.4 emb; n4 f	976
Sylvilagus transitionalis	5.0; n3 lt, 1 f	10731
	5.2 emb; n5 lt	5003a
	5.2 emb; r3-8; n19 f	1719
SEXUAL MATURITY		
Lepus americanus	f: 1st mate: yr of their birth	11228
	f: est 90-100 d; 1060 g probably conceived 2 mo; 400 other juv f did not breed	5622
	f: 7-8 mo	8798
	m: 5-7 mo	8798
	f/m: 1.2-2.3 yr	5624
Lepus californicus	f: 7-8 mo	8798
	m: 5-7 mo	8798
Lepus capensis	f: rapid growth of ovaries: 3-3.5 kg	4981
	f: 1st cl: 4 mo; mode 6-7 mo	1307
	f: 1st conceive: 5 mo	6452
	f: 7-8 mo	8798
	m: epididymal sperm: 3.2 mo	6452
	m: 4 mo; @2.0-2.5 kg	3346
	m: 5-7 mo	8798
	f/m: 1st repro: 5-6 mo	8359
	f/m: 1st repro: 6 mo; born June; 9-12 mo; born Jan-Apr	6909
	f/m: 6-9 mo; fully grown: 1.25-1.5 yr	3571
	f/m: 6.5 mo	1452
	f/m: 8 mo	5760
Lepus nigricollis	f: 6 mo	8501
Lepus timidus	f: 7-8 mo	8798
	m: 5-7	8798
	f/m: 8-9 mo	3571
Lepus townsendii	f/m: 1st mate: ~1 yr	5239
Oryctolagus cuniculus	f: 1st mate: 6 mo	11887
	f: 1st birth: 4 mo; n1	7701
	f: 120-150 d	8798
	f: 5 mo	11478
	m: spermatogenesis: 7 mo	293
	m: 9 mo	11478

	f/m: 1st mate: 4 mo	1213
	f/m: 5-8 mo	8053
	f/m: 6 mo	4645
	f/m: 6-8 mo	3571
Romerolagus diazi	f: 1st mate: 5-8 mo; 410-650 g	5007
	f: 403 g	6569
	m: testes descend: ~5 mo; n5	5007
	m: 1st mate: 6.5 mo; 440 g	5007
	m: 543 g	6569
Sylvilagus aquaticus	m: sperm: 3-6 mo	2013
	f/m: 23-30 wk	10222
Sylvilagus audubonii	f: 1st mate: summer of birth	10236
	f: 1st conceive: est ~80 d	10470
	m: epididymal sperm: 107 d	10470
Sylvilagus brasiliensis	f/m: 125 d	8798
Sylvilagus floridanus	f: 1st mate: 2-5 mo	8077
	f: 1st mate: 1st summer; n1	4624
	f: 1st mate: summer of birth	2038
	f: 1st conceive: 85, 90 d	1655
	f: 1st conceive: ~100 d	4742
	f: 1st preg: 1st summer	2891
	f: 2.5 mo	7806
	juv f, but not juv m, breed	5722
	m: epididymal sperm: 80 d	8077
	f/m: 125 d	8798
Sylvilagus idahoensis	m: spermatogenesis: 6 mo	3277
CYCLE LENGTH		
Lepus capensis	7 d	1530
Lepus timidus	estrus: 10-24 h	8053
Oryctolagus cuniculus	7 d; mode 4.5-8; r4.5-41.5; n67 cycle, 34 f	7260, 7690, 7692
Romerolagus diazi	13 d	2404
Sylvilagus aquaticus	12 d	6832
GESTATION		
Lepus americanus	34 d	1110
	35 d	5624
	35.5 d; ibi	11895
	36 d	4006
	36.56 d; r35.85-37.23; 18 lt	9805
	36.6 d; r34-38; n8 f	2664
	37 d	2458
	37 d	2326
	37.2 d; r36-40; n37 f	9806
	37.9 d; 19 lt	9805
	39.375 d; n8	9805
Lepus californicus	40 d; field data	4099
	42 d	8798
	43 d; r41-47; n8 f	4467
Lepus capensis	39.5 d; r38-41	10487
	40.2 d; se0.1; n66	6909
	41.1 d; se0.2; n43	6909
	41-42 d; n1	1005
	41.5 d; r41-42	1530
	42 d	9125
	42 d	3344
	~1.5 mo	4645
	~6 wk; r40-44 d	1307
	mode 42,43 d; r42-69; n6	4562
	44 d; artificial insemination	10318
	44-46 d	10316
Lepus nigricollis	~1 mo	8501
	42-44 d	9175
Lepus saxatilis	5-6 wk	4354a

Lepus timidus	44-52 d	3571
	44-53 d	7433
	~45 d	10294
	est 46 d	265
	50 d; ppe to birth	11398
	50 d; r48-51	8053
	~50 d	3348
	~50 d; 1st cl to 1st yg seen	8233
	51 d; se~0.2; n40	8323
	60 d	6441
Lepus townsendii	1 mo	6445
	36-43 d	6445
	42-43 d; field data	5865
Oryctolagus cuniculus	implant: 7 d	2162
	implant: 7 d	1214
	28-30 d	1214
	28-31 d; max 37	3571
	min 29 d	1175
	30 d	10763
	30-31 d	7510
	30-32 d	7690
	30-37 d	7433
	30.27 d; r28-33; n191 lt, 91 f	7260
	33 d	196
Poelagus marjorita	~5 wk	5760
Romerolagus diazi	1 mo	9231
	38 d; n1	1676
	38-40 d; n1	2819, 2820
	39 d ea; n3	1677
	39.8 d; r39-41; n20	6953a, 6953b
	40 d ea; n3	2404
Sylvilagus aquaticus	35-39 d; n18 lt, 9 f; assuming immediate ppe	10222
	mode 37 d 75%; r36-38; n8 lt	4920
	39-40 d; n3	5099
Sylvilagus audubonii	26-30 d	5146
	≤28 d; n3	2550b
Sylvilagus bachmani	~27 d; n1	7561
	27 d; ±3	1722
Sylvilagus brasiliensis	26-30 d	8798
Sylvilagus floridanus	25-28 d	9251
	26-28 d; interval between peaks of mating	3116
	26.5 d; n1	4623
	26.5 d; r25-28; n2	5003a
	26.67; mode 26; r26-28; n6	6831
	29.3 d	4499
	>29-<36 d	1386a
Sylvilagus idahoensis	39 d; field data	3277
Sylvilagus nuttallii	28-30 d	2088
Sylvilagus transitionalis	28 d; n1	5003a
LACTATION		
Lepus americanus	solid food: immediately	2088
	solid food: 7-9 d	8798
	solid food: 10 d	5210
	solid food: 11 d	4006
	solid food: 10-12 d	9805
	wean: 14-21 d	8798
	wean: max 3 wk	2139
	wean: 25-28 d	9805
	wean: 30 d	8005a
	independent: ~6 wk	1871a
Lepus californicus	solid food: 7-9 d	8798
	wean: 14-21 d	8798
	wean: 2-4 wk	1871a

Lepus capensis	solid food: 7 d	6424
	solid food: 7-9 d	8798
	solid food: 2 wk	4645
	wean: 14-21 d	8798
	wean: 1 mo	5760
	independent: ~1 mo	10158
	wean: 3 wk	6908
	wean: 3 wk; max 4 wk	3571
	wean: mode ~4 wk	1306
	wean: 8 wk	8053
	wean: ~2 mo	4645
Lepus timidus	den emergence: 2-4 d	8053
	solid food: 7-9 d	8798
	solid food: 8-9 d	4645
	wean: 14-21 d	8798
	wean: 3 wk	3571
Lepus townsendii	solid food: 15 d	1871a
Oryctolagus cuniculus	den emergence: 21 d	8798
	solid food: 21 d	8798
	wean: 22-25 d	8798
	independent: 25-28 d	8798
	wean: 4 wk	3571
	wean: 31 d	7894
Romerolagus diazi	solid food: 15-16 d	2819, 2820
	wean: 3 wk	6953b
	independent: 25-30 d	2819, 2820
Sylvilagus audubonii	leave nest: ≤2 wk	1871a
	wean: 3-4 wk	1871a
	wean: est 4 wk	2550b
Sylvilagus aquaticus	leave nest: 10-17 d; still nursed	10222
Sylvilagus bachmani	nurse: 2-3 wk	7561
Sylvilagus brasiliensis	den emergence: 14 d	8798
	solid food: 14-18 d	8798
	independent: 25-28 d	8798
Sylvilagus floridanus	den emergence: 14 d	8798
	solid food: by 2 wk	10908
	solid food: 14-16 d	2891
	solid food: 14-18 d	8798
	wean: 14.7 d; r14-15; n3	5003a
	wean: ~15 d	443a
	nurse: 20-25 d	2891
	independent: 25-28 d	8798
Sylvilagus nuttallii	independent: 4-5 wk	1871a
Sylvilagus transitionalis	wean: 16 d; n1	5003a
	leave nest: 16 d	10731
INTERLITTER INTERVAL		
Lepus americanus	34.4 d	7243
	35.5 d	11895
	37 d	6132
	39 d; r38-40; n2	3899
	38.4 d	2672
Lepus capensis	25-42 d	11058
	28-56 d	9125
	30.25 d; r12-38; n4	1452
	38 d; r35-39	1530
	40.2 d; mode 38 (24%); r26-59; n108	6909
	~48 d	1307
Lepus timidus	50.1 d; sd3.5; r46-54; n8	307
Oryctolagus cuniculus	29 d; min ibi	1175
Sylvilagus bachmani	29.5 d; n2	1722
Sylvilagus floridanus	1 mo	4742
	30.5 d	1386a
Sylvilagus idahoensis	39 d	3277

LITTERS/YEAR

Caprolagus hispidus	2-3	3759
Lepus americanus	1-4	2326
	1-5	5623
	min 2	2550
	2; CO; >2; UT	2672
	2.3	4037
	2.60; r1-3	11895
	2.75	9321
	2.75-3	9806
	2.9-3.5	2664
	3	7872
	3.27	7243
	3.75, 3.92; n2 yr; ~4 central WI; ~3 n WI	6132
	max 4	9805
	~4	5624
Lepus californicus	3.6; r3-5	4099
Lepus capensis	1.77	225
	2/summer	8053
	mode 3,4; r1-6	6909
	2; central Russia; 4 Caucasus	5944
	2-3	6452
	2-4	3572
	2.2; 3.0; 2.95; r1-4; 3 locations	3540
	2.46/season	226
	3	9497
	3	9125
	3	8370
	3-4	5944
	3-4	8476
	3-4; max 5	3571
	4.29	2576
	4.8; field data	3344
	7.73	3349
	max 5	11058
Lepus nigricollis	8-9	8730
Lepus saxatilis	min 2	11253
	8.7	3349
Lepus timidus	1	265
	1	4548
	1	8233
	est 1	5146
	2-3	6441
	2-3	307
	2.6	3348
	3	4711
Lepus townsendii	2; max 3-4	5865
	est 3.29; max 4	5239
Oryctolagus cuniculus	1-3	10386
	2.8/7 mo	11887
	3-5; r3-6	3571
	4	1213
	~4	1178
	4-5	4645
	4-5; max 7	8053
	5-6	7433
	5-7	8798
Sylvilagus aquaticus	3.09; r2-5; n34 lt, 11 f	10222
Sylvilagus bachmani	3-4	5146
Sylvilagus brasiliensis	4-5	8798
	4.7; enclosure	2816
Sylvilagus floridanus	1-4/season; n11 f	4499
	mode 2; max 4	10908
	2.9	9250

	3-4	1386a
	3-4	2891
	4.7	10822
	4.81	965
	7-8	2014
	max 7	3117
	max 7	4742
	7; n1	6832
Sylvilagus idahoensis	2	3277
Sylvilagus nuttallii	2-3	2088
	min 3/season	2550
	4-5	8717
SEASONALITY		
Caprolagus hispidus	mate: Jan-Mar, perhaps to June; Assam, India	3759
	juveniles found: June-July; Assam to Uttar Pradesh, India	3759
Lepus alleni	mate: Dec-Sept; NM, n Mexico	1474
	preg f: Jan-Oct; AZ	11365
Lepus americanus	spermatogenesis: Feb-Aug; MT	1111
	testes: large Apr-July, small Aug; Rochester, ALB, Canada	2326
	testes: large May-July, small Nov-early Feb; Mautoulin I, ONT, Canada	7872
	mate: Mar-?; ME	9805
	mate: Mar-Aug; NEW, Canada	2664
	mate: mid Mar-?; Canada	2139
	mate: Apr-?; UT	2672
	mate: Apr/May-?; CO	2672
	mate: Dec; NEW, Canada	2663
	conceive: Mar-July; central ALB, Canada	5624
	conceive: Apr-?; ALB, Canada	7243
	conceive: Apr-July; MI	1111
	conceive: Apr-July; Mautoulin I, Lake Huron, ONT, Canada	7872
	preg f: Feb-July; ID	3510
	preg f: Mar-Aug; MN	143
	preg f: Apr; Rochester, ALB, Canada	2326
	preg f: Apr-June; NS, Canada	11895
	preg f: Apr-July; ALB, Canada	9321
	preg f: Apr-Sept; Canada	2139
	preg f: May-June; YUK, Canada	12049
	preg f: June-July; n BC, Canada	10597
	preg f: Nov; n1; NH	9968
	preg f: end Nov-Jan; NEW, Canada	2663
	birth: Apr-Aug, peak June-July; WY	1871a
	birth: Apr-Aug; WI	5210
	birth: Apr-Aug; ME	9805, 9806
	birth: Apr-Oct; ONT, Canada	6706
	birth: May-June; NM	502
	birth: May-July; AK	8005a
	birth: May-Sept; MI	1111
	repro: summer; 9-21 wk; YUK, Canada	6014
Lepus californicus	mate: Dec-Sept; s N Am	1474
	preg f: Jan-Apr; UT	5254a
	preg f: Jan-Aug, peak Apr; n94; KS	1322
	preg f: Jan-Aug, peak Feb-Apr; UT	4099
	preg f: Jan-Sept; s AZ	11365
	preg f: Feb-early Mar; se CO	3086
	preg f: Feb-Mar; n2; CA	11228a
	preg f: Apr, June; n2; Coahuila, Mexico	522
	preg f: Apr, June, Dec-Feb; CA	8126
	preg f: June; n1; NE	5405b,
	preg f: July; n1; Durango, Mexico	523a
	preg f: yr rd, except Oct; sw TX	4332
	preg f: yr rd, peak Jan-Apr; CA	6335
	birth: yr rd; w US	5146
	lact f: Jan, Mar-Apr; few data; Coahuila, Mexico	522

	lact f: Apr-Sept; NM	502
	lact f: July; n2; Durango, Mexico	523a
	repro: mid Feb-mid June; ID	3197
Lepus capensis	spermatogenesis: reactivated end Nov; Netherlands	1307
	spermatogenesis: Dec-Aug; s Kazakhstan	8370
	testes: large Feb-Aug, small Sept-Oct; Germany	7446
	testes: small Sept-Dec; Norfolk, UK	6453
	testosterone: low July-Feb; Norfolk, UK	6453
	estrus: Jan-Aug; Germany	7433
	mate: Jan-June; Biserni Otak I, Yugoslavia	11058
	mate: Jan-Aug; Leeds, UK	4922
	mate: Jan-Aug; Germany	3571
	mate: Jan-Sept; Norfolk, UK	6452
	mate: Feb & earlier; Moscow, Russia	8053
	mate: Mar/Apr-?; n Eur USSR	4645
	mate: Jan-?; s Eur USSR	4645
	mate: July/Aug-Apr/May; Argentina	2576
	mate: Aug-Jan; Patagonia, Argentina	226
	mate: peak Oct, none Feb-July; Snowy Plain, Australia	3778
	mate: Dec-Sept; captive	1534
	mate: end Dec-?; Netherlands	1307
	mate: est yr rd; NEW, Canada	3346
	ovulation: Jan-Aug; captive	1533
	conceive: Mar-mid Sept; Germany	9125
	preg f: Jan; n4; Iran	6293
	preg f: Jan-Aug; ONT, Canada	9080
	preg f: Jan-Sept; Scotland, UK	4709
	preg f: 95% Feb-June, 10% Aug, none Dec; s ONT, Canada	9080
	preg f: >50% Feb-July, <40% Aug-Oct; no data Nov-Jan; Germany	7446
	preg f: Feb, June, Nov; few data; Botswana	10154
	preg f: Feb-Aug/Sept; Germany	10420
	preg f: mid Mar-Aug; Biserni Otak I, Yugoslavia	11058
	preg f: May; n1; Kenya	4925
	preg f: Aug; n1; El Vergel, Malleco Province, Chile	4054
	preg f: yr rd, 90% Aug-Feb; New Zealand	3344, 3345, 3346
	preg f: yr rd, peak Aug-Oct; Riverine District, Australia	3778
	preg f: yr rd; n12 yr; Scotland, UK	4712
	preg f: Jan-Sept; n1 yr; Poland	4712
	birth: Jan-June; Karakun, USSR	9497
	birth: end Jan-Nov; captive	6909
	birth: Feb-Aug; captive	10487
	birth: Feb-Sept/Oct; s Sweden	3540
	birth: Feb-Oct; Germany	3571
	birth: Mar-July; s Moravia, Czechoslovakia	4981
	birth: Mar-Aug; Germany	7433
	birth: Mar-Aug; zoo	1452
	birth: Mar-Sept; Moscow, Russia	8053
	birth: Apr-May, summer; no data fall, winter; Turkestan	9774
	birth: Oct-Apr; Angola	11697
	birth: yr rd, peak Oct-Feb; S Africa	10158
	yg found: Apr-early May; Baluchistan	9175
	lact f: Feb; n1; El Vergel, Malleco Province, Chile	4054
	lact f: Feb, July, Nov-Dec; n8; Transvaal, S Africa	8965
	lact f: Aug-Sept; n2; Iran	6293
	lact f: < 10% Dec-Jan; Germany	7446
	repro: Jan-mid June; Karakumy desert, Turkmenia	9499
	repro: Feb-Sept; Byelorussia	3572
	repro: yr rd, peak spring-summer; Caucasus, Russia	5944
	repro: mid winter-fall; USSR	5944
	repro: Jan-Sept; captive	6909
Lepus nigricollis	testes: large Mar-Sept; India	8730
	preg f: Apr; n2; Nepal	7382
	preg f: yr rd, peak July-Sept; India	8730

	birth: spring-summer; Pakistan	9175
	birth: yr rd; Sri Lanka	8501
	lact f: Sept; n1; Nepal	7382
	repro: not seasonal; Sri Lanka	8500
Lepus saxatilis	preg f: Jan, Mar-Apr, Aug, Nov-Dec; n13; Transvaal, S Africa	8965
	preg f: July; n2; Kenya	4925
	preg f: yr rd; Botswana	10154
	birth: Apr, Oct-Nov; Namibia	9945
	birth: almost yr rd; Transvaal, S Africa	8965
	birth: yr rd, peak Sept-Feb; S Africa	10158
	lact f: Feb, May, Oct; n8; Transvaal, S Africa	8965
	repro: yr rd	1149
Lepus timidus	estrus: mid Feb-July; Europe	8053
	estrus: Mar-Aug; Alps, Europe	7433
	mate: Jan-June; Scotland, UK	4710
	mate: early-mid Feb; USSR	4645
	mate: Mar-June; Germany	3571
	conceive: Feb-Aug; e Scotland, UK	4709
	conceive: Mar-?; Sweden	307
	conceive: Apr; w AK	265
	conceive: peak Apr; sw NEW, Canada	4548
	preg f: Jan-Sept, peak Mar-June; ne Scotland, UK	4710a
	preg f: Mar-June; ne Scotland, UK	3348
	birth: Mar-Aug; ne Scotland, UK	4710, 4711
	birth: Mar-Sept; Europe	8053
	birth: Apr/May-July/Aug; Germany	3571
	birth: Apr-Aug; Alps, Europe	7433
	birth: May-early July; captive	10294
	birth: May-end Aug/Sept; Norway	11398
	birth: end May-early June; w AK	265
	birth: peak end May-mid June; sw NEW, Canada	4548
	birth: June; NWT, Canada	8233
	birth: June-July; N America	1474
Lepus townsendii	testes: large Feb-June; ND	5239
	mate: Feb-Mar; IA	5865
	preg f: end Feb-July; WY	9228a
	preg f: Mar-July; UT	5254a
	preg f: May; n3; SD	261a
	preg f: May-mid July; CO	680
	preg f: June; n1; ND	3703
	preg f: July; n1; se MT	6166
	preg f: July-Aug; n3; NE	$5405b_1$
	preg f: Aug; n2; CO	1649a
	birth: May-Aug; NM	502
	lact f: July; n1; se MT	6166
	repro: Feb-early July perhaps longer; IA	5865
Oryctolagus cuniculus	sperm: yr rd; sw Australia	11620
	spermatogenesis: yr rd; n Cambridgeshire, England, UK	1178
	testes: large Mar-Aug; s Sweden	291
	testes: large Apr-May, low Oct-Nov; n Cambridgeshire, England, UK	1178
	testes: large May, small Oct; s Sweden	293
	testes: large Sept, small Jan; New Zealand	11478
	testes: large spring, small summer; France	9228
	m accessory glands: peak May-June; Sweden	2216
	ovaries: large spring, small summer-fall; France	9228
	uterine mass: peak May-June, low Dec; s Sweden	293
	estrus: Feb-July; Wales, UK	1213
	mate: Jan-June; w Wales, UK	8491
	mate: Feb-June; Netherlands	11422
	mate: Feb-Aug; Germany	7433
	mate: Feb/Mar-Sept; Germany	3571
	mate: early Mar-Apr; sw Australia	5746
	mate: Mar/Apr-Oct; WA, Australia	5745

	mate: mid Apr-mid Aug; s Sweden	2217
	mate: June-Nov; New Zealand	11478
	mate: July-May; sub Antarctic	9918
	mate: Aug-Mar; Macquarie I, sub Antarctic	10044
	mate: Nov-May, peak Mar-Apr; Sierra Morena, Spain	10226
	mate: yr rd; NSW, Australia	11887
	conceive: mid Mar/June-?; San Juan I, WA	10386
	preg f: Mar-Aug; captive	1173
	preg f: Mar/Apr-Aug; s Sweden	291
	preg f: mid June-early Nov; New Zealand	7070
	preg f: none Oct-Dec; n Cambridgeshire, England, UK	1178
	preg f: peak spring; France	9228
	preg f: yr rd, peak June-Oct; New Zealand	11476, 11478
	preg f: yr rd, high Aug-Nov, peak Oct; New Zealand	7070
	birth: Feb-July, peak Apr-June; Oxfordshire, s England, UK	2084
	birth: Mar-July, peak May; Netherlands	11422
	birth: Mar-July; s Sweden	293
	birth: Mar-Oct; Germany	3571
	birth: Mar-Oct; Germany	7433
	birth: peak May; Scotland, UK	4619
	birth: mid June-?; NSW, Australia	8228
	repro: June-Dec; captive, Australia	7701
Poelagus marjorita	mate: est yr rd; Angola, Uganda	5760
Pronolagus randensis	birth: est yr rd; S Africa	10158
Romerolagus diazi	mate: Mar-May; zoo	2404
	mate: Mar-July; Mexico	9231
	mate: Dec-July, peak Jan-Apr; Mexico	6569
	mate: yr rd; captive	5007
	preg f: May-June; Mexico	6387
	birth: May; n2; zoo	6478
	birth: peak spring; captive	5007
	lact f: Feb-Dec; Mexico	1677
	repro: Mar-June; Mexico	9231
Sylvilagus aquaticus	anestrus: Nov-Jan; MD	10861
	mate: Jan-Sept; mid-s US	1474
	mate: Feb; n AL	4740
	preg f: Feb-Sept; captive	10222
	preg f: Feb-?; MS	8214
Sylvilagus audubonii	sperm: yr rd, low Sept-Nov; AZ	10470
	testes: low Sept-Oct; AZ	10236
	mate: fall-spring; central CA	3280
	preg f: Jan-Aug; AZ	10236
	preg f: Feb-June, Dec; CA	8126
	preg f: Apr, July, Dec; few data; Mexico	522, 523a
	preg f: May-July; n5; SD	261a
	preg f: June; n1; Sinaloa, Mexico	5409
	preg f: July; n1; se MT	6166
	preg f: July; n2; NM	502
	preg f: July; n1; NE	$5405b_1$
	birth: yr rd; sw US	1474
	lact f: Apr, June-July; few data; Mexico	522, 523a
	lact f: May; n1; NM	502
Sylvilagus bachmani	sperm: Oct-July, peak Jan-May; CA	7561
	mate: Jan-June; w US	5146
	mate: Jan-June; w US	1474
	preg f: Jan, Mar-June; CA	8126
	preg f: mid Feb-mid Aug; OR	1722
	preg f: Dec-June; CA.	7561
	birth: Jan-May; w US	1474
	birth: est Aug; n1; OR	11276
Sylvilagus brasiliensis	repro: yr rd, low Mar-Apr; Venezuela	2816
Sylvilagus floridanus	sperm: yr rd, except Oct; OK	935
	spermatogenesis: Jan-July; VA	9524
	testes: large Feb-June/July, small July-Jan; piedmont GA	8340

	testes: large Feb-July; IL	9687
	testes: large Feb-Aug; OK	935
	testes: large Mar-Aug, small Sept-Jan; coastal GA	8340
	testes: large Apr; VA	9524
	testes: large May, quiescent Sept-Dec; WI	2964
	testes: large end Dec-Oct; NY	4276
	epididymal sperm: Feb-Aug; IL	2891
	mate: Jan-Aug; OR	10893
	mate: Jan-Aug; MI	10908
	mate: end Jan-Aug; WI	5210
	mate: Feb-Aug, peak mid Feb, early July; se MO	3116
	mate: mid Feb-mid July, peak Apr-May; St Clements I, MD	965
	mate: mid Feb-Aug, if warm winter Dec-Oct; AL	4742
	mate: mid Feb-mid Sept; OK	935
	mate: Mar-?; IL	9687
	mate: Apr-Aug; OH	10384
	mate: yr rd; Venezuela	8077
	conceive: Jan-Sept; GA	8342
	conceive: Jan-Sept; GA	8342
	preg f: Jan; n1; IL	6575
	preg f: Jan-Feb; n1; MI	169
	preg f: Jan-Sept, peak June; KY	1386a
	preg f: Feb-Aug, MO	2014
	preg f: Feb-Sept; NY	4276
	preg f: Feb-Sept; piedmont, GA	8340
	preg f: Feb-Oct; coastal, GA	8340
	preg f: Feb-Oct, peak Mar-June; GA	8340
	preg f: end Feb; captive, IA	5864
	preg f: end Feb-Aug; IL	2891
	preg f: Mar-Apr; FL	7498a
	preg f: Mar-June; mountain, GA	8340
	preg f: Mar, Sept; OK	935
	preg f: peak Apr-May; VA	6515
	preg f: Mar-June; few data; Coahuila, Mexico	522
	preg f: Mar-Aug; MI	4499
	preg f: Apr-Sept; n24; CT	5003a
	preg f: May-July; n8; NE	5405b,
	preg f: June; n1; se MT	6166
	preg f: June; n1; SD	261a
	preg f: June-July; n2; ND	3703
	preg f: Nov; n1; WI	6371
	preg f: Dec-Jan; n2; PA	1160
	preg f: yr rd, low June-July; TX	1136
	birth: ?-early Sept; IL	7290
	birth: Feb-July; MD	965
	birth: end Feb-Sept, peak Apr-June; OH	443a
	birth: Mar-Sept; PA	893
	birth: Mar-?; captive	9250
	lact f: Mar-June; n4; FL	7498a
	lact f: Apr; few data; Coahuila, Mexico	522
	lact f: Apr-Sept; PA	892
	lact f: May, July-Aug; NE	5405b,
	yg found: ?-July; MI	169
	repro: Jan-Sept; AL	586
	repro: end Jan-early Oct; NC	586
Sylvilagus idahoensis	spermatogenesis: Jan-July, none Aug-Dec; no data Apr-Aug; ID	3277
	mate: spring-early summer; w US	5146
	conceive: end Feb-Apr; ID	3277
	preg f: end Feb-Mar; UT	5254a
	preg f: end Mar-May; ID	4034
	birth: June-July; nw US	1474

Sylvilagus nuttallii	testes: large Jan-July; OR	8717
	mate: end Feb-early July; OR	8717
	preg f: end Feb-July; OR	8717
	preg f: Apr-July; ne CA	8126
	preg f: Aug; n1; NM	502
	birth: Apr-July; BC, Canada	2088
Sylvilagus palustris	mate: Feb-Sept; se US	1474
	preg f: Feb; GA	10866
	preg f: yr rd; s FL	4921
	lact f: Nov; GA	10866
Sylvilagus transitionalis	preg f: Apr-June; n5; CT	5003a

Order Macroscelidae, Family Macroscelididae

Elephant shrews (*Elephantulus* 25-100 g; *Macroscelides* 30-50 g; *Petrodromus* 155-280 g; *Rhynchocyon* 320-550 g) are terrestrial insectivores from Africa. *Elephantulus* has a copulatory plug (6640) and produces an exceptionally large number of oocytes, up to 89 in one ovary, although usually no more than 2 embryos implant (10906, 11115, 11118). The hemochorial and endotheliochorial, labyrinthine, discoidal placenta and subplacenta have a mesometrial implantation 7-9 d postcoitus (*Elephantulus*) at the caudal end of the bipartite or bicorunate uterus near a blood vessel and is resorbed (2393, 3729, 8028, 8030, 10310, 10906, 11104, 11107, 11110, 11112, 11113, 11117, 11120, 11121, 11123, 11124). Estrus occurs (*Elephantulus* 5759, 6641, 8957, 10906, 10907, 11113, 11114) after the birth of open-eyed, furred neonates (331, 334, 1356, 4879, 5759, 8798, 9945, 10158, 11253). Females have 2-6 mammae (331, 8798, 8965, 9945) and menstruate (11114, 11117, 11119, 11123). Males have multiple prostate glands and abdominal testes (10435, 11863). Details of anatomy (7084, 11105, 11106, 11106a, 11110, 11114, 11118, 11119, 11122, 11125, 11126), and ovulation (10906, 11115, 11116) are available.

NEONATAL MASS		
Elephantulus intufi	10.0 g; r8.5-10.5; n6	10907
Elephantulus myurus	8.1 g; sd1.8; n9	10907
Elephantulus rufescens	10.6 g; r9.3-13.0; n11, 1 d	8957
Petrodromus tetradactylus	m: 31.5 g; n1	10907
Rhynchocyon chrysopygus	80 g	8955
NEONATAL SIZE		
Elephantulus myurus	5 cm	10158
Petrodromus tetradactylus	~76 mm; n2	5759
	HB: 93 mm	334
WEANING MASS		
Rhynchocyon chrysopygus	180 g	8956
LITTER SIZE		
Elephantulus brachyrhynchus	1 emb; n1 f	1836
	1.33; r1-2; n6 lt	10907
	1.67 yg; r1-2; n3 lt	4924
	1.8 emb; r1-2; n5 f	10906
	1.8 emb; mode 2; r1-2; n5 f	8965
	2	3729
	2 emb; n1 f	331, 334
	mode 2 emb; r1-2; n9	1356
	8.5 oocytes/ovary; sd6.6; r0-23; n16 ovaries	10906
Elephantulus edwardi	2 emb; n1 f	10906
	44 oocytes/ovary; r31-55; n3 ovaries	10906
Elephantulus fuscipes	1 emb; n1 f	10906
Elephantulus fuscus	1 yg; n1 lt	6272
	2.0 emb ea; n2 f	6272
Elephantulus intufi	1.29 yg; n7 lt	10907
	1.3 emb; r1-2; n6 f	10906
	1.5; r1-2; n2 lt	10907
	1.6 emb; r1-2; n6 f	10158
	1.9 emb; r1-3; n9 f	10154
	2 emb ea; n4 f	9945
	2.5 oocytes/ovary; sd2.5; r0-8; n29 ovaries	10906
Elephantulus myurus	1.7; r1-2; n6 lt	10154
	1.89; mode 2; r1-2; n18	8965
	1.89 yg; mode 2; r1-2; n19 f	10907
	2	7146
	2 usual; r1-2	10158
	2.0 emb ea; n2 f	10906
	2 yg ea; 2 lt	10907
	49.0 oocytes/ovary; sd14.2; r25-89; n50 ovaries	10906
	60 oocytes/ovary	11118
Elephantulus rozeti	1.2 oocytes/ovary; sd0.85; r0-3; n23 ovaries	10906
	1.5 emb; r1-2; n2 f	10907
	2 emb ea; n4 f	7906
	2.4 emb; mode 2; r1-4; n36 f	10906

Elephantulus rufescens	0.7 oocytes/ovary; sd0.6; r0-2; n13 ovaries	10906
	1 yg; n1 lt	10907
	1 emb ea; n5 f	4924
	r1-2; n77 lt	8957
	r1-2 emb, yg; n8 f	1356
	r1-2; n15 lt	8030
	1.4 emb; r1-2; n7 f	10906
	1.4 emb; r1-2; n7 f	2443
Elephantulus rupestris	0.5 oocytes/ovary; r0-1; n2 ovaries	10906
	1.75 emb; r1-2; n4 f	9945
	2	8798
Macroscelides proboscideus	1-2	9508
	2 emb ea; n8 f	9945
	21 ooctyes/ovary; n2 ovaries	10906
Petrodromus tetradactylus	0.6 oocytes/ovary; sd0.69; r0-2; n18 ovaries	10906
	1 yg	6611
	1 yg; n1 f	334
	1 yg; n1 lt	10907
	1 emb ea; n3 f	4924
	1 emb ea; n6 f	10906
	1 emb ea; n17 f	1356
	1-2	11253
	2 emb; n1 f	9893
Rhynchocyon chrysopygus	1 yg	8955
Rhynchocyon cirnei	1	5759
	2 emb ea; n2 f	1356
	2 emb ea; n3 f	337
Rhynchocyon petersi	1 oocyte/ovary; r0-2; n2 ovaries	10906
	2 emb; n1 f	10906
SEXUAL MATURITY		
Elephantulus intufi	f: 1st mate: 11 mo; n1	10907
	m: 1st mate: 6 mo; n1	10907
Elephantulus myurus	f: 1st mate: 5 wk	11114
	f: 5-6 wk	10158
Elephantulus rufescens	f: 1st birth: 144 d; r104-184; n2	8957
	f/m: 2 mo	5759
CYCLE LENGTH		
Elephantulus myurus	~14 d	11114
Elephantulus rufescens	7-17 d; serial laporotomies	6641
	11.9 d; se1.48; mode 12.5; r6-19	6641
	median 13 d; r12-49; n22 cycles, 4 f	6640
GESTATION		
Elephantulus intufi	est 50-52 d; ibi + assumed 14-15 d infertile cycle	10907
Elephantulus myurus	49 d; capture to birth	10907
	~8 wk	11107
Elephantulus rufescens	50 d	5759
	est 57 d; min ibi 58 d	8957
	61-65 d	5955
Elephantulus rupestris	56 d	8798
Rhynchocyon chrysopygus	~42 d	8956
LACTATION		
Elephantulus spp	solid food: 8 d	4879
Elephantulus intufi	solid food: 3 wk	10907
	nurse: 5 wk	10907
Elephantulus rufescens	wean: 15 d	10797
	wean: 1 mo	5759
Macroscelides proboscideus	wean: 5 d	9508
Rhynchocyon chrysopygus	wean: 2 wk	8955
INTERLITTER INTERVAL		
Elephantulus intufi	64-70 d	10907
Elephantulus myurus	49 d; interval between lt up to 49 d	10907
Elephantulus rufescens	84.7 d; r58-145; n50, 18 f	8957
Rhynchocyon chrysopygus	81 d	8956

LITTERS/YEAR		
Elephantulus myurus	3	11114
Petrodromus tetradactylus	2	11253
Rhynchocyon cirnei	4-5	5759
SEASONALITY		
Elephantulus spp	repro: Oct-Apr; sw Africa	4879
Elephantulus brachyrhynchus	preg f: Jan, Mar, Oct, Dec; n5; Transvaal, S Africa	8965
	preg f: Sept; n1; Bulawayo Matabele, S Africa	1836
	birth: Oct-Feb; Zimbabwe	331, 334
	birth: peak May; Zimbabwe	5759
	birth: summer; S Africa	10158
Elephantulus intufi	preg f: Mar, Aug; Transvaal, S Africa	8965
	birth: Feb, Aug, Oct-Nov; few data; Botswana	10154
Elephantulus myurus	accessory glands: active July-Jan; S Africa	10435
	spermatogenesis: yr rd; S Africa	10435
	estrus: July-Aug; S Africa	11117
	mate: July-Dec; Transvaal, S Africa	11112
	mate: July-Jan; ne Transvaal, S Africa	10906
	mate: end July-Jan; Transvaal, S Africa	7146
	birth: Sept-Mar; S Africa	10158
	preg f: end July-early Mar; S Africa	11114
Elephantulus rufescens	preg f: Jan; n1; ne Zaire	2372
	suckling yg: Dec; n1; Tanzania	186
	preg f: Apr-May, none Dec-Jan; Kenya, Tanzania	5759
	repro: yr rd, 78% Oct-Dec, Apr, June; Kenya	7796, 7797
Elephantulus rupestris	breed: yr rd; Namib Desert, sw Africa	11863
Macroscelides proboscideus	birth: Sept-Feb; S Africa	10158
Petrodromus tetradactylus	preg f: Jan, Apr, July; Zambia	5759
	birth: Aug-Oct; S Africa	10158
	birth: Oct; e Africa	6611
Rhynchocyon cirnei	birth: seasonal; Uganda, Tanzania, Kenya	5759

Literature Cited

1. 't Hart, L, Moesker, A, Vedder, L, et al. 1988. On the pupping period of grey seals, *Halichoerus grypus* (Fabricius, 1791), reproducing on a shoal near the island of Terschelling, the Netherlands. Z Säugetierk. 53:59-60.

2. Abbott, DH. 1978. The physical, hormonal, and behavioural development of the common marmoset, *Callithrix jacchus*. Biology and Behaviour of Marmosets. 99-106, Rothe, H, Wolters, HJ, & Hearn, JP, eds, Eigenverlag Hartmut Rothe, Göttingen.

3. Abbott, DH. 1984. Behavioral and physiological suppression of fertility in subordinate marmoset monkeys. Am J Primatol. 6:169-186.

4. Abbott, DH. 1984. Differentiation of sexual behaviour in female marmoset monkeys: Effect of neonatal testosterone or a male co-twin. Prog Brain Res. 61:349-352.

5. Abbott, DH. 1987. Behaviourally mediated suppression of reproduction in female primates. J Zool. 213:455-470.

6. Abbott, DH & Hearn, JP. 1978. Physical, hormonal and behavioural aspects of sexual development in the marmoset monkey, *Callithrix jacchus*. J Reprod Fert. 53:155-166.

7. Abbott, DH, Batty, KA, Dubey, AK, et al. 1985. The passage of 5α-dihydrotestosterone from serum into cerebrospinal fluid and LH negative feedback in castrated rhesus monkeys. J Endocrinol. 104:325-330.

8. Abbott, DH, McNeilly, AS, Lunn, SF, et al. 1981. Inhibition of ovarian function in subordinate female marmoset monkeys (*Callithrix jacchus jacchus*). J Reprod Fert. 63:335-345.

9. Abdel-Raouf, M & El-Naggar, MA. 1964. Studies on reproduction in camels *Camelus dromedarius*: I Mating techinque and collection of semen. J Vet Sci UAR. 1:113-119.

10. Abdel-Raouf, M, Fatch El-Bab, MR, & Owaida, MM. 1975. Studies on the reproduction of the camel (*Camelus dromedarius*) V Morphology of the testis in relation to age and season. J Reprod Fert. 43:109-116.

11. Abdo, MS & El Mougy, SA. 1976. Hormonal variations in the blood of the one humped camel (*Camelus dromedarius*) during the various reproductive stages. Vet Med J (Cairo). 24:71-76.

12. Abdou, MSS, ElWishy, AB, Abdo, MS, et al. 1971. Hormonal activities of the placenta of the one-humped camel, *Camelus dromedarius* Part I Gonadotrophic, adrenocorticotrophic and thyrotrophic hormones. J Anim Morph Physiol. 18:11-16.

13. Abdunazarov, NH. 1970. Biological characteristics of reproduction in the one-humped camel. Tr Turkm S Kh Inst. 15:134-141.

14. Abdunazarov, NH. 1971. Calving season and parturition in Turkmen dromedaries. Tr Turkm S Kh Inst. 16:29-37.

15. Abe, T. 1974. An analysis of age structure and reproductive activity of *Microtus montebellii* population based on yearly trapping data. Jap App Ent Zool. 18:21-27.

16. Abegglen, H & Abegglen, JJ. 1976. Field observations of a birth in Hamadryas baboons. Folia Primatol. 26:54-56.

17. Abensperg-Traun, M. 1989. Some observations on the duration of lactation and movements of a *Tachyglossus aculeatus acanthion* (Monotremata: Tachyglossidae) from Western Australia. Aust Mamm. 12:33-34.

18. Abler, WA, Buckland, DE, Kirkpatrick, RL, et al. 1976. Plasma progestins and puberty in fawns as influenced by energy and protein. J Wildl Manage. 40(3):442-446.

19. Abramov, VK. 1962. A contribution to the biology of the Amur tiger *Panthera tigris longipilis*. Věstn Česk Spol Zool. 26:189-202.

20. Abramsky, Z. 1984. Population biology of *Gerbillus allenbyi* in northern Israel. Mammalia. 48:197-206.

21. Abravaya, JP & Matson, JO. 1975. Notes on a Brazilian mouse, *Blarinomys breviceps* (Winge). Los Angeles Co Mus Nat Hist Contrib Sci. 270:1-8.

22. Abromavich, CE, Jr. 1930. Uterus and fetal membranes of the Indian antelope (*Antelope cervicapra*). Anat Rec. 46:105-124.

23. Abu-Lehia, IH. 1987. Composition of camel milk. Milchwissenschaft. 42:368-371.

24. Acharjyo, LN. 1971. A note on the birth of a golden cat (*Felis temmincki*) in captivity. J Bombay Nat Hist Soc. 68:241.

25. Acharjyo, LN. 1984. A note on antler casting of barking deer (*Muntiacus muntjak*) in captivity. J Bombay Nat Hist Soc. 81:690.

26. Acharjyo, LN & Mishra, CG. 1980. A note on the breeding of the leopard-cat (*Felis bengalensis*) in captivity. J Bombay Nat Hist Soc. 77:127-128.

27. Acharjyo, LN & Mishra, CG. 1981. Notes on weight and size at birth of Indian wild ungulates in captivity. J Bombay Nat Hist Soc. 78:373-375.

28. Acharjyo, LN & Mishra, CG. 1983. A note on the longevity of two species of Indian otters in captivity. J Bombay Nat Hist Soc. 80:636.

29. Acharjyo, LN & Mishra, CG. 1985. On some aspects of reproduction among the tigers (*Panthera tigris*) of Nandankanan biological park (Orissa). J Bombay Nat Hist Soc. 82:628-632.

30. Acharjyo, LN & Misra, R. 1971. Age of sexual maturity of three species of wild mammals in captivity. J Bombay Nat Hist Soc. 68:446.

31. Acharjyo, LN & Misra, R. 1972. Birth of an Indian pangolin (*Manis crassicaudata*) in captivity. J Bombay Nat Hist Soc. 69:174-175.

32. Acharjyo, LN & Misra, R. 1973. A note on age of sexual maturity of two species of antelopes in captivity. J Bombay Nat Hist Soc. 70:378.

33. Acharjyo, LN & Misra, R. 1973. A note on the birth of a Malayan giant squirrel (*Ratufa bicolor*) in captivity. J Bombay Nat Hist Soc. 70:375.

34. Acharjyo, LN & Misra, R. 1973. Notes on the birth and growth of a slow loris (*Nycticebus coucang*) in captivity. J Bombay Nat Hist Soc. 70:193-195.

35. Acharjyo, LN & Misra, R. 1974. Weight and size at birth of two species of wild mammals in captivity. J Bombay Nat Hist Soc. 71:137-138.

36. Acharjyo, LN & Misra, R. 1975. A note on inter-parturition interval of some captive wild mammals. J Bombay Nat Hist Soc. 72:841-845.

37. Acharjyo, LN & Misra, R. 1975. A note on the breeding habits of four-horned antelope (*Tetracerus quadricornis*) in captivity. J Bombay Nat Hist Soc. 72:529-530.

38. Acharjyo, LN & Misra, R. 1975. Age of sexual maturity of two species of wild carnivores in captivity. J Bombay Nat Hist Soc. 72:196-197.

39. Acharjyo, LN & Misra, R. 1976. A note on the breeding of the Indian fox (*Vulpes bengalensis*) in captivity. J Bombay Nat Hist Soc. 73:208.

40. Acharjyo, LN & Misra, R. 1976. Some notes on the breeding habits and growth of the Malayan giant squirrel (*Ratufa bicolor*) in captivity. J Bombay Nat Hist Soc. 73:380-382.

41. Acharjyo, LN & Mohapatra, S. 1977. Some observations on the breeding habits and growth of junglecat (*Felis chaus*) in captivity. J Bombay Nat Hist Soc. 74:158-159.

42. Acharjyo, LN & Mohapatra, S. 1978. Birth and growth of common palm-civet (*Paradoxurus hermaphroditus*) in captivity. J Bombay Nat Hist Soc. 75:204-206.
43. Acharjyo, LN & Mohapatra, S. 1979. Some observations on inter-oestrus interval in captive tigresses (*Panthera tigris* (Linneaus)). J Bombay Nat Hist Soc. 76:495-497.
44. Acharjyo, LN & Padhi, GS. 1972. Some observations on distribution of zoo births among wild common mammals. J Bombay Nat Hist Soc. 69:175-178.
45. Acharjyo, LN & Patnaik, SK. 1983. Some observations on the antler cycle of hog-deer (*Axis porcinus*) in captivity. J Bombay Nat Hist Soc. 80:631-632.
46. Acharjyo, LN & Patnaik, SK. 1984. A note on the longevity of two species of wild carnivores in captivity. J Bombay Nat Hist Soc. 81:461-462.
47. Acharjyo, LN & Tripathy, AP. 1974. A note on body colour and breeding habits in captivity of common palm civet (*Paradoxurus hermaphroditus*) of Orissa. J Bombay Nat Hist Soc. 71:601-603.
47a. Acharya, L. 1992. *Epomophorus wahlbergi*. Mamm Species. 394:1-4.
48. Acher, R, Chauvet, J, & Chauvet, MT. 1973. Neurohypophysial hormones and evolution of tetrapods. Nature New Biol. 244:124-126.
49. Adachi, M, Saito, R, & Tanioka, Y. 1982. Observation of delivery behavior in the rhesus monkey. Primates. 23:583-586.
50. Adam, CL & Atkinson, T. 1984. Effect of feeding melatonin to red deer (*Cervus elaphus*) on the onset of the breeding season. J Reprod Fert. 72:463-466.
51. Adam, CL, Moir, CE, & Atkinson, T. 1985. Plasma concentrations of progesterone in female red deer (*Cervus elaphus*) during the breeding season, pregnancy and anoestrus. J Reprod Fert. 74:631-636.
52. Adam, CL, Moir, CE, & Atkinson, T. 1986. Induction of early breeding in red deer (*Cervus elaphus*) by melatonin. J Reprod Fert. 76:569-573.
53. Adam, CL, Moir, CE, & Shiach, P. 1989. Melatonin can induce year-round ovarian cyclicity in red deer (*Cervus elaphus*). J Reprod Fert. 87:401-408.
54. Adamczewska, KA. 1961. Intensity of reproduction of the *Apodemus flavicollis* (Melchior 1834) during the period 1954-1959. Acta Theriol. 5:1-21.
55. Adamczewski, JZ, Gates, CC, Hudson, RJ, et al. 1987. Seasonal changes in body composition of mature female caribou and calves (*Rangifer tarandus groenlandicus*) on an arctic island with limited winter resources. Can J Zool. 65:1149-1157.
56. Adams, CE & Norris, ML. 1973. Observations on reproduction in the Mongolian gerbil, *Meriones unguiculatus*. J Reprod Fert. 33:185-188.
57. Adams, CE. 1973. The reproductive status of female mink, *Mustela vison*, recorded as failed to mate. J Reprod Fert. 33:527-529.
58. Adams, CE. 1975. Stimulation of reproduction in captivity of the wild rabbit, *Oryctolagus cuniculus*. J Reprod Fert. 43:97-102.
59. Adams, CE. 1981. Observations on the induction of ovulation and expulsion of uterine eggs in the mink, *Mustela vison*. J Reprod Fert. 63:241-248.
60. Adams, CE. 1983. Reproductive performance of rabbits on a low protein diet. Lab Anim. 17:340-345.
61. Adams, EGP. 1949. Jungle memories Part IV: Wilddogs and wolves etc. J Bombay Nat Hist Soc. 48:645-655.
62. Adams, JK. 1989. *Pteronotus davyi*. Mamm Species. 346:1-5.
63. Adams, LE. 1903. A contribution to our knowledge of the mole (*Talpa europaea*). Mem Manchr Lit Phil Soc Old Ser. 47(4):1-39.
64. Adams, MR, Tamarin, RH, & Callard, IP. 1980. Seasonal changes in plasma androgen levels and the gonads of the beach vole, *Microtus breweri*. Gen Comp Endocrinol. 41:31-40.
65. Adams, WH. 1894. On the habits of the flying-squirrels of the genus *Anomalurus*. Proc Zool Soc London. 1894:243-249.
66. Adams, WH, Jr. 1956. The nutria in coastal Lousiana. LA Acad Sci. 19:28-41.
67. Adler, GH & Tamarin, RH. 1984. Demography and reproduction in island and mainland white-footed mice (*Peromyscus leucopus*) in southeastern Massachusetts. Can J Zool. 62:58-64.
68. Adler, GH, Reich, LM, & Tamarin, RH. 1984. Demography of the meadow jumping mouse (*Zapus hudsonius*) in eastern Massachusetts. Am Mid Nat. 112:387-391.
69. Adolf, TA. 1966. On the reproductive biology of *Erinaceus europeus* [*sic*] L. Zool Zhurn. 45:1108-1111.
70. Advani, R. 1983. Reproductive biology of *Pipistrellus mimus mimus* (Wroughton) in the Indian desert. Z Säugetierk. 48:211-217.
71. Aellen, V. 1952. Contribution a l'étude des Chiroptères du Cameroun. Mém Soc Neuchatel Sci Nat. 7(1):1-121.
72. Aellen, V. 1955. *Rhinolophys blasii* Peters (1866), chauve-souris nouvelle pour l'Afrique du Nord. Mammalia. 19:361-366.
73. Aellen, V. 1973. Un Rhinolophus nouveau d'Afrique centrale. Period Biol. 75:101-105.
74. Aeschlimann, A. 1963. Observations sur *Philamtomba maxwelli* (Hamilton-Smith), une antilope de la forêt eburnéene. Acta Trop. 20:341-368.
75. Afonskaja, RI & Krumina, MK. 1956. Zur Biologie des ussurischen Tiger. Mosk Zoopark Sbornik Trudov. 1:50-60.
76. Afonskaja, RI & Krumina, MK. 1958. Beobachtungen an Eisbären. Mosk Zoopark Sbornik Statej. 2:56-63.
77. Agarwal, SP, Agarwal, VK, Khanna, ND, et al. 1989. Serum estrogen and progesterone levels in camel (*Camelus dromedarius*) during estrous cycle. Indian Vet J. 66:605-608.
78. Agarwal, SP, Rai, AK, & Khanna, ND. 1991. Serum progesterone levels in female camels during oestrous cycle. Ind J Anim Sci. 61:37-39.
78a. Aggarwal, BB, Farmer, SW, Papkoff, H, et al. 1980. Purification and characterization of donkey chorionic gonadotrophins and pregnant mare serum gonadotrophin. J Endocrin. 85:449-455.
79. Agrawal, VC. 1965. Field observations on the biology and ecology of the desert gerbil, *Meriones hurrianae* (Rodentia, Muridae). J Zool Soc India. 17:125-134.
80. Agrawal, VC & Chakroborty, S. 1971. Notes on a collection of small mammals from Nepal, with the description of a new mouse-hare (Lagomorpha: Ochotonidae). Proc Zool Soc (Calcutta). 24:41-46.
81. Agrawal, VC & Ghosal, KK. 1969. A new field-rat (Mammalia: Rodentia: Muridae) from Kerala, India. Proc Zool Soc (Calcutta). 22:41-45.
82. Ågren, G. 1981. Two laboratory experiments on inbreeding avoidance in the Mongolian gerbil. Behav Process. 6:291-297.
83. Ågren, G. 1984. Alternative mating strategies in the Mongolian gerbil. Behaviour. 91:229-244.
84. Ågren, G. 1990. Sperm competition, pregnancy initiation and litter size: Influence of the amount of copulatory behaviour in Mongolian gerbils, *Meriones unguiculatus*. Anim Behav. 40:417-427.

85. Aguayo L, A. 1971. The present status of the Juan Fernandez fur seal. K Nozski Vidensk Selsk Skr. 1:1-4.
86. Aguayo L, A. 1979. Juan Fernandez fur seal. Mammals in the Seas, FAO Fish Ser 5. 2:28-33.
87. Aguilera M, M. 1985. Growth and reproduction in *Zygodontomys microtinus*. Mammalia. 49:75-83.
88. Ahlén, I. 1965. Studies on the red deer, *Cervus elaphus* L, in Scandinavia. Viltrevy. 3:177-376.
89. Ahnlund, H. 1980. Sexual maturity and breeding season of the badger, *Meles meles* in Sweden. J Zool. 190:77-95.
90. Ahrens, TG. 1921. Muskrats in central Europe. J Mamm. 2:236-237.
91. Aidara, D, Badawi, M, Tahiri-Zagret, C, et al. 1981. Changes in concentrations of serum prolactin, FSH, oestradiol and progesterone and of sex skin during the menstrual cycle in the mangabey monkey (*Cercocebus atys lunulatus*). J Reprod Fert. 62:475-481.
92. Aidara, D, Tahiri-Zagret, C, & Robyn, C. 1981. Serum prolactin concentrations in mangabey (*Cercocebus atys lunulatus*) and patas (*Erythrocebus patas*) monkeys in response to stress, ketamine, TRH, sulpirude and levo dopa. J Reprod Fert. 62:165-172.
93. Ainsworth, L & Ryan, KJ. 1969. Steroid hormone transformation by endocrine organs from pregnant mammals V The biosynthesis and metabolism of progesterone and estrogens by orangutan placental tissue *in vitro*. Steroids. 14:301-314.
94. Aire, TA. 1980. The cycle of the seminiferous epithelium of the domesticated giant rat (*Cricetomys gambianus* Waterhouse). Acta Anat. 108:160-168.
95. Aitken, PF. 1972. *Planigale gilesi* (Marsupialia: Dasyuridae): A new species from the interior of eastern Australia. Rec S Aust Mus. 16:1-14.
96. Aitken, RJ, Burton, J, Hawkins, J, et al. 1973. Histological and ultrastructural changes in the blastocyst and reproductive tract of the roe deer, *Capreolus capreolus* during delayed implantation. J Reprod Fert. 34:481-493.
97. Aitken, RJ. 1974. Calcium and zinc in the endometrium and uterine flushings of the roe deer (*Capreolus capreolus*) during delayed implantation. J Reprod Fert. 40:333-340.
98. Aitken, RJ. 1974. Delayed implantation in roe deer (*Capreolus capreolus*). J Reprod Fert. 39:225-233.
99. Aitken, RJ. 1974. Sex chromatin formation in the blastocyst of the roe deer (*Capreolus capreolus*) during delayed implantation. J Reprod Fert. 40:235-239.
100. Aitken, RJ. 1979. The hormonal control of implantation. CIBA Found Symp 64 (Maternal Recognition of Pregnancy). 53-74, Exerpta Medica, Amsterdam.
101. Aitken, RJ. 1981. Aspects of delayed implantation in the roe deer (*Capreolus capreolus*). J Reprod Fert Suppl. 29:83-95.
102. Aiyappan, A. 1946. Period of gestation of the Indian elephant (*Elephas maximus*). J Bombay Nat Hist Soc. 46:182.
103. Ajayi, SS. 1975. Observations of the biology, domestication and reproductive performance of the African giant rat *Cricetomys gambianus* Waterhouse in Nigeria. Mammalia. 39:343-364.
104. Akasaka, T & Maruyama, N. 1977. Social organization and habitat use of Japanese serow in Kasabori. J Mamm Soc Japan. 7:87-102.
105. Akhtar, SA. 1958. The rodents of West Pakistan, Part 1. Pak J Sci. 10:5-18.
106. Akhtar, SA. 1958. The rodents of West Pakistan, Part 2. Pak J Sci. 10:79-90.
107. Akhtar, SA. 1958. The rodents of West Pakistan, Part 3. Pak J Sci. 10:269-292.
108. Akhtar, SA. 1960. The rodents of West Pakistan, Part 4. Pak J Sci. 12:17-37.
109. Al-Khalib, AD & Delany, MJ. 1986. The postembryonic development and reproductive strategies of two species of rodents in south-west Saudi Arabia. Cimbebasia. 8:175-185.
110. Al-Robaae, K. 1966. Untersuchungen der Lebensweise irakischer Fledermäuse. Säugetierk Mitt. 14:177-211.
111. Al-Robaae, K. 1968. Notes on the biology of the tomb bat, *Taphozous midiventrus magnus* v Wettstein 1913, in Iraq. Säugetierk Mitt. 16:21-26.
112. Alados, CL & Escos, J. 1988. Parturition dates and mother-kid behavior in Spanish ibex (*Capra pyrenaica*) in Spain. J Mamm. 69:172-175.
113. Alberts, S. 1987. Parental care in captive siamangs (*Hylobates syndactylus*). Zoo Biol. 6:401-406.
114. Albertson, BD, Bradley, EL, & Terman, CR. 1975. Plasma progesterone concentrations in prairie deermice (*Peromyscus maniculatus bairdii*) from experimental laboratory populations. J Reprod Fert. 42:407-414.
115. Albignac, R. 1969. Élevage d'un jeune propithèque, Lémurien folivore de Madagascar. Mammalia. 33:341-343.
116. Albignac, R. 1969. Naissance et élevage en captivité de jeunes *Cryptoprocta ferox*, Viverridés malgaches. Mammalia. 33:93-97.
117. Albignac, R. 1969. Notes éthologiques sur quelques Carnivores malgaches: Le *Galidia elegans*. Terre Vie. 22:202-215.
118. Albignac, R. 1970. Notes éthologiques sur quelques Carnivores malgaches: Le *Cryptoprocta ferox* (Bennett). Terre Vie. 117:395-402.
119. Albignac, R. 1970. Notes éthologiques sur quelques Carnivores malgaches: Le *Fossa fossa* (Schreber). Terre Vie. 117:383-394.
119a. Albignac, R. 1972. The Carnivora of Madagascar. Monogr Biol. 21:667-682.
120. Albignac, R. 1973. Mammiferès carnivores. Faune de Madagascar. 36:5-206.
121. Albignac, R. 1974. Observations éco-éthologiques sur le genre *Eupleres*, Vivérride de Madagascar. Terre Vie. 28:321-351.
122. Albignac, R. 1975. Breeding the fossa *Cryptoprocta ferox* at Montpellier zoo. Int Zoo Yearb. 15:147-150.
123. Albignac, R. 1976. L'écologie de *Mungotictis decemlineata* dans les forêts decidués de l'ouest de Madagascar. Terre Vie. 30:347-376.
124. Albon, SD. 1983. Fertility and body weight in female red deer: A density-dependent relationship. J Anim Ecol. 52:969-980.
125. Albon, SD, Guinness, FE, & Clutton-Brock, TH. 1983. The influence of climatic variation on the birth weights of red deer (*Cervus elaphus*). J Zool. 200:295-298.
126. Albon, SD, Mitchell, B, Huby, BJ, et al. 1986. Fertility in female red deer (*Cervus elaphus*): The effects of body composition, age and reproductive status. J Zool. 209A:447-460.
127. Albrecht, ED. 1980. A role for estrogen in progesterone production during baboon pregnancy. Am J Obstet Gynecol. 136:569-574.
128. Albrecht, ED, Haskins, AL, Hodgen, GD, et al. 1981. Luteal function in baboons with administration of the antiestrogen ethamoxytriphetol (MER-25) throughout the luteal phase of the menstrual cycle. Biol Reprod. 25:451-457.
129. Albrecht, ED, Nightingale, MS, & Townsley, JD. 1977. Stress decreases serum progesterone concentrations in pregnant baboons (*Papio papio*). Fed Proc. 36:390.

130. Albrecht, ED, Nightingale, MS, & Townsley, JD. 1978. Stress-induced decreases in the serum concentration of progesterone in the pregnant baboon. J Endocrinol. 77:425-426.
131. Albrecht, ED & Pepe, GJ. 1984. Source and regulation of 17α-hydroxyprogesterone during baboon pregnancy. Biol Reprod. 31:471-479.
132. Albrecht, ED & Townsley, JD. 1976. Metabolic clearance and production rates of progesterone in non-pregnant baboons (*Papio papio*). Endocrinology. 99:1291-1294.
133. Albrecht, ED & Townsley, JD. 1976. Serum progesterone in the pregnant baboon (*Papio papio*). Biol Reprod. 14:610-612.
134. Albuja, L. 1979. On a collection of bats from the caves of Jumandi and the surrounding area. Southampton University Expedition to Ecuador Report. 1981:179-184, Brown, P, ed, Univ Southampton, Southampton.
135. Alcala, AC. 1976. Philippine land vertebrates: Field biology. New Day, Philippines.
136. Alcala, JR & Conaway, CH. 1968. The gross and microscopic anatomy of the uterus masculinus of tree shrews. Folia Primatol. 9:216-245.
137. Alcober, RC. 1971. The last Almiquis (*Solenodon cubanus*) in captivity. Zool Garten NF. 40:1-3.
138. Alcorn, GT & Robinson, ES. 1983. Germ cell development in female pouch young of the tammar wallaby (*Macropus eugenii*). J Reprod Fert. 67:319-325.
139. Alcorn, JR. 1940. Life history notes on the Piute ground squirrel. J Mamm. 21:160-170.
140. Alcorn, JR. 1941. Counts of embryos in Nevadan kangaroo rats (genus *Dipodomys*). J Mamm. 22:88-89.
141. not used
142. Aldeen, SM & Finn, CA. 1970. The implantation of blastocysts in the Russian steppe lemming (*Lagurus lagurus*). J Exp Zool. 173:63-78.
143. Aldous, CM. 1937. Notes on the life history of the snowshoe hare. J Mamm. 18:46-57.
144. Aldous, MC, Craighead, FC, Jr, & Devan, GA. 1958. Some weights and measurements of desert bighorn sheep (*Ovis canadensis nelsoni*). J Wildl Manage. 22:444-445.
145. Aldous, SE. 1935. Some breeding notes on rodents. J Mamm. 16:129-131.
146. Aldrich-Blake, FPG. 1970. The ecology and behaviour of the blue monkey *Cercopithecus mitis stuhlmanni*. Ph D Diss. Univ Bristol, Bristol.
147. Alexander, BK. 1970. Parental behavior of adult male Japanese monkeys. Behaviour. 36:270-285.
148. Alexander, PS, Lin, L-K, & Huang, B-M. 1987. Ecological notes on two sympatric mountain shrews (*Anourosorex squamipes* and *Soriculus fumidus*) in Taiwan. J Taiwan Mus. 40:1-7.
148a. Alexander, SE, Yeoman, RR, Williams, LE, et al. 1991. Confirmation of ovulation and characterization of luteinizing hormone and progesterone secretory patterns in cycling, isosexually housed Bolivian squirrel monkeys (*Saimiri boliviensis boliviensis*). Am J Primatol. 23:55-60.
149. Alho, CJR, Campos, ZMS, & Conçalves, HC. 1987. Ecologia de capivara (*Hydrochaeris hydrochaeris* Rodentia) do Pantanal: I Habitats, densidades e tamanho de grupo. Rev Bras Biol. 47:87-97.
150. Alho, CJR, Campos, ZMS, & Conçalves, HC. 1987. Ecologia de capivara (*Hydrochaeris hydrochaeris* Rodentia) do Pantanal: II Atividade, sazonalidade, uso do espaço e manejo. Rev Brasil Biol. 47:99-110.
151. Ali, HA, Tingari, MD, & Moniem, K. 1978. On the morphology of the accessory male glands and histochemistry of the ampulla ductus deferentis of the camel (*Camelus dromedarius*). J Anat. 125:277-292.
152. Ali, S. 1946. The wild ass of Kutch. J Bombay Nat Hist Soc. 46:472-277.
153. Ali, S & Santapay, H. 1958. Birth of a great Indian rhinoceros in captivity. J Bombay Nat Hist Soc. 55:157-158.
154. Ali, S & Santapay, H. 1959. Does the takin produce twin calves?. J Bombay Nat Hist Soc. 56:130-131.
155. Ali, SA. 1927. The breeding of the Indian rhinoceros (*Rhinoceros unicornis*) in captivity. J Bombay Nat Hist Soc. 31:1031.
156. Alibhai, SK. 1982. Persistence of placental scars in the bank vole, *Clethrionomys glareolus*. J Zool. 197:300-303.
157. Alibhai, SK. 1985. Effects of diet on reproductive performance of the bank vole (*Clethrionomys glareolus*). J Zool. 205A:445-452.
158. Aliev, FF. 1956. Theoretical and practical foundations of coypus (*Myocastor coypus* Molina) raising in Azerbaijan. Trans Inst Zool Acad Sci Azerbaijan SSR. 19:5-96.
159. Aliev, FF. 1965. Growth and development of nutrias' functional features. Fur Trade J Can. 1965(7):2-3.
160. Alikina, EV. 1959. Influence of the metabolism on the ovo and spermatogenesis in the field vole (*Microtus arvalis* Pall and *Microtus socialis* Pall). Zool Zhurn. 38:610-625.
161. Allais, C & Martinet, L. 1978. Relation between daylight ratio, plasma progesterone levels and timing of nidation in mink (*Mustela vison*). J Reprod Fert. 54:133-136.
162. Allan, CJ, Holst, PJ, & Hinch, GN. 1991. Behaviour of parturient Australian bush goats I Doe behaviour and kid vigour. Appl Anim Beh Sci. 32:55-64.
162a. Allan, PF. 1944. Mating behavior of *Dipodomys ordii richardsoni*. J Mamm. 25:403-404.
163. Allanson, M. 1932. The reproductive processes of certain mammals III The reproductive cycle of the male ferret. Proc R Soc London Biol Sci. 110B:295-312.
164. Allanson, M. 1933. The reproductive processes of certain mammals V Changes in the reproductive organs of the male grey squirrel (*Sciurus carolinensis*). Phil Trans R Soc London Biol Sci. 222B:79-96.
165. Allanson, M. 1934. The reproductive processes of certain mammals VII Seasonal variation in the reproductive organs of the male hedgehog. Phil Trans R Soc London Biol Sci. 223B:277-303.
166. Allanson, M. 1958. Growth and reproduction in the males of two species of gerbil, *Tatera brantsii* (A Smith) and *Tatera afra* (Gray). Proc Zool Soc London. 130:373-396.
167. Allanson, M, Rowlands, IW, & Parkes, AS. 1934. Induction of fertility in the anestrous ferret. Proc R Soc London Biol Sci. 115B:410-421.
168. Allen, AA. 1921. Banding bats. J Mamm. 2:53-57.
169. Allen, DL. 1938. Breeding of the cottontail rabbit in southern Michigan. Am Mid Nat. 20:464-469.
170. Allen, DL. 1942. Populations and habits of the fox squirrel in Allegan County, Michigan. Am Mid Nat. 27:338-379.
171. Allen, E. 1926. The menstrual cycle in the monkey: Effect of double ovariectomy and injury to large follicles. Proc Soc Exp Biol Med. 23:434-436.
172. Allen, E. 1926. The time of ovulation in the menstrual cycle of the monkey, *Macacus rhesus*. Proc Soc Exp Biol Med. 23:381-383.
173. Allen, E. 1927. The menstrual cycle of the monkey, *Macacus rhesus*: Observations on normal animals, the effects of removal of the ovaries and the effects of injections of ovarian

and placental extracts into the spayed animals. Contrib Embryol Carnegie Inst. 19(98):1-44.
174. Allen, E, Diddle, AW, & Elder, JH. 1935. Theelin content of pregnancy urine and placenta of the chimpanzee. Am J Physiol. 110:593-596.
175. Allen, EG. 1938. The habits and life history of the eastern chipmunk, *Tamias striatus lysteri*. NY State Mus Bull. 314:1-122.
176. Allen, GM. 1910. *Solenodon paradoxus*. Mem Mus Comp Zool. 40:1-54.
177. Allen, GM. 1919. The American collared lemmings (*Dicrostonyx*). Bull Mus Comp Zool. 82:509-540.
178. Allen, GM. 1922. Bats from New Mexico and Arizona. J Mamm. 3:156-162.
178a. Allen, GM. 1927. Sex dichromatism in *Noctilio*. J Mamm. 18:514.
179. Allen, GM. 1938. The mammals of China and Mongolia. Natural History of Central Asia. II Part 1:1-620, Granger, W, ed, Am Mus Nat Hist, New York.
180. Allen, GM. 1938. Zoological results of the second Dolan expedition to western China and eastern Tibet 1934-1936 Part 3 Mammals. Proc Acad Nat Sci Philadelphia. 90:261-294.
181. Allen, GM. 1939. Bats. Harvard Univ Press, Cambridge, MA.
182. Allen, GM. 1940. The mammals of China and Mongolia. Natural History of Central Asia. II Part 2:621-1350, Granger, W, ed, Am Mus Nat Hist, New York.
183. Allen, GM. 1941. Pygmy sperm whale in the Atlantic. Publ Field Mus Nat Hist Zool Ser. 27:17-36.
184. Allen, GM & Coolidge, HJ, Jr. 1930. Mammals of Liberia. The African Republic of Liberia and the Belgian Congo. 2:569-622, Strong, RP, ed, Harvard Univ Press, Cambridge.
185. Allen, GM & Lawrence, B. 1936. Scientific results of an expedition to rain forest regions in eastern Africa III Mammals. Bull Mus Comp Zool. 79:31-126.
186. Allen, GM & Loveridge, A. 1933. Reports on the scientific results of an expedition to the southwestern highlands of Tanganyika Territory II Mammals. Bull Mus Comp Zool. 75:47-140.
187. Allen, GM & Loveridge, A. 1942. Scientific results of a fourth expedition to forested areas in east and central Africa 1 Mammals. Bull Mus Comp Zool. 89:145-214.
188. Allen, JA. 1876. The American bisons, living and extinct. Mem Mus Comp Zool. 4(10):1-246.
189. Allen, JA. 1890. The West Indian seal (*Monachus tropicalis* Gray). Bull Am Mus Nat Hist. 2:1-34.
190. Allen, JA. 1906. Mammals from the states of Sinaloa and Jalisco, Mexico, collected by JH Batty during 1904 and 1906. Bull Am Mus Nat Hist. 22:191-262.
191. Allen, JA, Lang, H, & Chapin, JP. 1917. The American Museum Congo Expedition collection of bats. Bull Am Mus Nat Hist. 37:405-478.
192. Allen, JF. 1977. Dolphin reproduction in oceanaria in Australasia and Indonesia. Breeding Dolphins Present Status, Suggestions for the Future. 85-108, Ridgway, SH & Benirschke, K, eds, US Mar Mamm Comm, Washington, DC.
193. Allen, JF, Lépes, M, Budiarso, IT, et al. 1976. Some observations on the biology of the dugong (*Dugong dugong*) from the waters of south Sulawesi. Aquatic Mamm. 4:33-48.
194. Allen, LDC. 1963. The lechwe (*Kobus leche smithemani*) of the Bangweulu swamps. Puku. 1:1-8.
195. Allen, NT & Bradshaw, SD. 1980. Diurnal variation in plasma concentration of testosterone, 5α-dihydrotestosterone, and corticosteroids in the Australian brush-tailed possum, *Trichosurus vulpecula* (Kerr). Gen Comp Endocrinol. 40:455-458.
196. Allen, P, Brambell, FWR, & Mills, IH. 1947. Studies on sterility and prenatal mortality in wild rabbits I The reliability of estimates of prenatal mortality based on counts of corpora lutea, implantation sites and embryos. J Exp Biol. 23:312-331.
197. Allen, SH. 1983. Comparison of red fox litter sizes determined from counts of embryos and placental scars. J Wildl Manage. 47:860-863.
198. Allen, SH. 1984. Some aspects of reproductive performance in female red fox in North Dakota. J Mamm. 65:246-255.
198a. Allen, WR. 1975. The influence of fetal genotype upon endometrial cup development and PMSG and progestagen production in equids. J Reprod Fert Suppl. 23:405-413.
199. Allendorf, FW, Christiansen, FB, Dobson, T, et al. 1979. Electrophoretic variation in large mammals I The polar bear, *Thalarctos maritimus*. Hereditas. 91:19-22.
200. Allison, A & Woolley, PA. 1984. Reproduction in a New Guinea pipistrelle. Bull Aust Mamm Soc. 8:93.
201. Allsopp, R. 1971. Seasonal breeding in bushbuck (*Tragelaphus scriptus* Pallus, 1776). E Afr Wildl J. 9:146-148.
202. Almatov, LA. 1968. Early implantation stages of the embryo of the bat *Rhinolophus ferrumequinum*. Dokl Akad Nauk SSSR Translation. 182:524-526.
203. Alonso-Mejía & Medellín, RA. 1991. *Micronycteris megalotis*. Mamm Species. 376:1-6.
204. Alsina, G & Brandani, A. 1981. Population dynamics of the European hare in Patagonia, Argentina. Proc World Lagomorph Conf, Univ Guelph, Ontario. 1979:486-492.
205. Alt, G. 1982. Reproductive biology of Pennsylvania black bear. Game News PA. Feb:9-15.
206. Alt, GL. 1983. Timing of parturition of black bears (*Ursus americanus*) in northeastern Pennsylvania. J Mamm. 64:305-307.
207. Altmann, D. 1970. Ethologische Studie an Mufflons, *Ovis ammon musimon* (Pallas). Zool Garten NF. 39:297-303.
208. Altmann, D. 1971. Zur Geburt beim Buntbock, *Damaliscus dorcas dorcas*. Zool Garten NF. 40:80-96.
209. Altmann, D. 1972. Verhaltensstudien an Mähnenwölfen, *Chrysocon brachyurus*. Zool Garten NF. 41:278-298.
210. Altmann, F. 1924. Beiträge zur Anatomie des weiblichen Genitales der Dasypodiden. Z Ges Anat. 72:390-400.
211. Altmann, F. 1927. Untersuchungen über das Ovarium von *Talpa europaea* mit besonderer Berücksichtigung seiner cyclischen Veränderungen. Z Anat Entwicklungsgesch. 82:482-569.
212. Altmann, J, Altmann, S, & Hausfater, G. 1981. Physical maturation and age estimates of yellow baboons, *Papio cynocephalus*, in Amboseli National Park, Kenya. Am J Primatol. 1:389-399.
213. Altmann, J, Altmann, SA, Hausfater, G, et al. 1977. Life history of yellow baboons: Infant mortality, physical development and reproductive parameters. Primates. 18:315-330.
214. Altmann, J, Warneke, M, & Ramer, J. 1988. Twinning among *Callimico goeldii*. Int J Primatol. 9:165-168.
215. Altmann-Schönberner, D. 1965. Beobachtungen über die Aufzucht und Entwicklung des Verhaltens beim groszen Löwenäffchen, *Leontocebus rosalia*. Zool Gärten NF. 31:227-239.
216. Alvarez, J, Willig, MR, Jones, JK, Jr, et al. 1991. *Glossophaga soricina*. Mamm Species. 379:1-7.
217. Alvarez, T. 1963. The recent mamals of Tamaulipas, Mexico. Univ KS Publ Mus Nat Hist. 14:363-473.

217a. Alvarez, T & Ramirez-Pulido, J. 1972. Notas acerca de murciélagos Mexicanos. An Esc Nat Cienc Biol Méx. 19:167-178.
218. Alvarez del Toro, M. 1952. Los animales silvestres de Chiapas. Edicienes del Gobierno del Estado Tuxtla Gutierrez, Chiapas.
219. Alvarez del Toro, M. 1967. A note on the breeding of the Mexican tree porcupine *Coendou mexicanus* at Tuxtla Zoo. Int Zoo Yearb. 7:118.
220. Alving, T. 1932. Fluszpferdgeburt und andere Ereignisse im Zoo Kopenhagen. Zool Gärten NF. 5:34-37.
221. Alzina, V, Puig, M, de Echaniz, L, et al. 1986. Prostaglandins in human milk. Biol Neonate. 50:200-204.
222. Amano, J, Messer, M, & Kobata, A. 1985. Structures of the oligosaccharides isolated from milk of the platypus. Glycoconjugate J. 2:121-135.
223. Amantaeva, RA. 1974. Age changes of fecundity and embryonic mortality in the common redbacked vole (*Clethrionomys glareolus*). Zool Zhurn. 53:1865-1868.
224. Amasaki, H, Ishikawa, H, & Daigo, M. 1989. Development of the external genitalia in fetuses of the southern minke whale, *Balaenoptera acutorostrata.* Acta Anat. 135:142-148.
225. Amaya, JN, Alsimo, ML, Brandani, AA. 1979. Ecología de la liebre europea II Reproductíon y peso corporal de una poblacion del área USC de Bariloche. Informe Técnico SINTA Bariloche/Argentina. 9:36.
226. Amaya, JW. 1981. The European hare in Argentina. Proc World Lagomorph Conf, Univ Guelph, ONT. 1979:493-494.
227. Amizoba, MA. 1986. Detecting pregnancy in the African giant rat, *Cricetomys gambianus* Waterhouse (Mammalia Rodentia), through vaginal smearing. Rev Zool Afr. 99:419-422.
228. Amizoba, MA. 1986. Observations on the variation of body weight and maturational moult with age in laboratory-raised African giant rat (*Cricetomys gambianus* Waterhouse). Rev Zool Afr. 99:243-247.
229. Amori, G, & Contoli, L. 1986. Variazoni allometrische e isometrische in *Apodemus sylvaticus* e *Apodemus flavicollis* italiani, rispetto alle condizioni allopatra e simpatra. Hystrix. 1:161-188.
230. Amoroso, EC. 1951. The interaction of the trophoblast and endometrium at the time of implantation in the Ungulata. Proc Zool Soc London. 121:202.
231. Amoroso, EC. 1952. Placentation. Marshall's Physiology of Reproduction, 3rd ed. 2:127-311, Parkes, AS, ed, Longmans, Green & Co Ltd, London.
232. Amoroso, EC. 1959. Comparative anatomy of the placenta. Ann NY Acad Sci. 75:855-872.
233. Amoroso, EC, Bourne, GH, Harrison, RJ, et al. 1965. Reproductive and endocrine organs of foetal, newborn and adult seals. J Zool. 147:430-486.
234. Amoroso, EC, Goffin, A, Halley, G, et al. 1951. Lactation in the grey seal. J Physiol London. 113:4P-5P.
235. Amoroso, EC, Hancock, JL, & Rowlands, IW. 1948. Ovarian activity in the pregnant mare. Nature. 161:355-356.
236. Amoroso, EC, Hancock, JL, Matthews, LH, et al. 1951. Reproductive organs of near term and new-born seals. Nature. 168:771.
237. Amoroso, EC & Hancock, NA. 1956. The placenta and foetal membranes of the hippopotamus (*Hippopotamus amphibius* L). Proc Zool Soc London. 126:486-487.
238. Amoroso, EC, Hancock, NA, & Kellas, L. 1958. The foetal membranes and placenta of the hippopotamus (*Hippopotamus amphibius* Linnaeus). Proc Zool Soc London. 130:437-447.
239. Amoroso, EC, Kellas, LM, & Harrison, ML. 1953. The foetal membranes of an African waterbuck, *Kobus defassa.* Proc Zool Soc London. 123:477.
240. Amoroso, EC & Matthews, JH. 1951. The growth of the grey seal *Halichoerus grypus* (Fabricius) from birth to weaning. J Anat. 85:427-428.
241. Amoroso, EC & Perry, JS. 1965. The foetal membranes and placenta of the African elephant (*Loxodonta africana*). Phil Trans R Soc London Biol Sci. 248B:1-34.
242. Allen, P, Brambell, FWR, & Mills, IH. 1947. Studies on the sterility and prenatal mortality in wild rabbits I The reliability of estimates of prenatal mortality based on counts of corpora lutea, implantation sites and embryos. J Exp Biol 23: 312-331.
243. Amschler, JW. 1931. Die Yakzucht im sibirischen Altai. Deutsch Landwirtsch Tierz. 35:662-665.
244. Amundin, M. 1986. Breeding the bottle-nosed dolphin *Tursiops truncatus* at the Kolmarden Dolphinarium. Int Zoo Yearb. 24/25:263-271.
245. Anadu, PA. 1976. Observations on reproduction and development in *Mus musculoides* (Rodentia, Muridae). Mammalia. 40:175-186.
246. Anadu, PA. 1978. Laboratory and field studies of *Myomys daltoni* reproductive parameters. Acta Theriol. 23:519-526.
247. Anadu, PA. 1979. Gestation period and early development in *Myomys daltoni* (Rodentia, Muridae). Terre Vie. 33:59-69.
248. Anand Kumar, TC. 1965. Reproduction in the rat-tailed bat *Rhinopoma kinneari.* J Zool. 147:147-155.
249. Anand Kumar, TC. 1968. Oogenesis in lorises; *Loris tardigradus leydekkerianus* and *Nycticebus coucang.* Proc R Soc London Biol Sci. 169B:167-176.
250. Anand Kumar, TC, David, GFX, & Puri, V. 1980. Levels of estradiol and progesterone in the cerebrospinal fluid of rhesus monkey during the menstrual cycle. J Med Primatol. 9:222-232.
251. Anantakrishoran, CP. 1941. Studies on ass's milk: Composition. J Dairy Res. 12:119-130.
252. Anciaux de Faveaux, M. 1972. Répartition Biogéographique et Cycles Annuels des Chiroptères d'Afrique. Centrale Thèse Docteur en Sciences Zoologiques. Univ Paris, Paris.
253. Anciaux de Faveaux, M. 1973. Essai de synthèse sur la reproduction de Chiroptères d'Afrique (Région faunistique Ethiopienne). Period Biol. 75:195-199.
254. Anciaux de Faveaux, M. 1977. Définition de l'équateur biologique en fonction de la reproduction de Chiroptères d'Afrique central. Ann Soc R Zool Belg. 107:79-89.
255. Anciaux de Faveaux, M. 1978. Les cycles annuels de reproduction chez les Chiroptères cavernicoles du Shaba (S-E Zaire) et du Rwanda. Mammalia. 42:453-490.
256. Andelman, SJ, Else, JG, Hearn, JP, et al. 1985. The non-invasive monitoring of reproductive events in wild vervet monkeys (*Cercopithecus aethiops*) using urinary pregnanediol-3 -glucuronide and its correlation with behavioural observations. J Zool. 205A:467-477.
257. Andelt, WF. 1985. Behavioral ecology of coyotes in south Texas. Wildl Monogr. 94:1-45.
258. Andersen, DC. 1978. Observations on reproduction, growth, and behavior of the northern pocket gopher (*Thomomys talpoides*). J Mamm. 59:418-422.
259. Andersen, DC, Armitage, KB, & Hoffmann, RS. 1976. Socioecology of marmots: Female reproductive strategies. Ecology. 57:552-560.
260. Andersen, DC & MacMahon, JA. 1981. Population dynamics and bioenergetics of a fossorial herbivore, *Thomomys talpoides* (Rodentia: Geomyidae) in a spruce-fir sere. Ecol Monogr. 51:179-202.

261. Andersen, K, Sundby, A, & Hansson, V. 1981. Fine structure and FSH binding of Sertoli cells in the blue fox (*Alopex lagopus*) in different stages of reproductive activity. Int J Androl. 4:570-581.
261a. Andersen, KW & Jones, JW, Jr. 1971. Mammals of northwestern South Dakota. Univ KS Publ Mus Nat Hist. 19:361-393.
262. Anderson, AE. 1981. Morphological and physiological characteristics. Mule and Black-tailed Deer of North America. 27-97, Wallmo, OC, ed, Univ NE Press, Lincoln.
263. Anderson, AE & Wallmo. OC. 1984. *Odocoileus hemionus*. Mamm Species. 219:1-9.
264. Anderson, DM & Simpson, MJA. 1979. Breeding performance of a captive colony of rhesus macaques (*Macaca mulatta*). Lab Anim. 13:275-281.
265. Anderson, HL & Lent, PC. 1977. Reproduction and growth of the tundra hare. (*Lepus othus*). J Mamm. 58:53-57.
266. Anderson, J. 1961. Biology and management of roe deer in Denmark. Terre Vie. 108:41-53.
267. Anderson, JL. 1965. Annual changes in testis and kidney fat weight of impala (*Aepyceros melampus melampus* Lichtenstein). Lammergeyer. 3(2):57-59.
268. Anderson, JL. 1975. The occurence of a secondary breeding peak in the southern impala. E Afr Wildl J. 13:149-151.
269. Anderson, JL. 1978. Aspects of the ecology of the nyala (*Tragelaphus angasi* Gray, 1849) in Zululand. Ph D Diss. London Hospital Med Coll, Univ London.
270. Anderson, JL. 1979. Reproductive seasonality of the nyala *Tragelaphus angasi*: The interaction of light, vegetation phenology, feeding style and reproductive physiology. Mammal Rev. 9:33-46.
271. Anderson, JL. 1984. Reproduction in the nyala (*Tragelaphus angasi*) (Mammalia: Ungulata). J Zool. 204:129-142.
272. Anderson, JM & Benirschke, K. 1962. Tissue transplantation in the nine-banded armadillo, *Dasypus novemcinctus*. Ann NY Acad Sci. 99:399-414.
273. Anderson, JM & Benirschke, K. 1963. Fetal circulations in the placenta of *Dasypus novemcinctus* Linn and their significance in tissue transplantation. Transplantation. 1:306-310.
274. Anderson, JW & Wimsatt, WA. 1953. The fetal membranes and placentation of the tropical American noctilionid bat *Dirias albiventer minor*. Anat Rec. 117:573-574.
275. Anderson, JW & Wimsatt, WA. 1963. Placentation and fetal membranes of the Central American noctilionid bat, *Noctilio labialis minor*. Am J Anat. 112:181-201.
276. Anderson, PK. 1984. Suckling in (*Dugong dugon*). J Mamm. 65:510-511.
277. Anderson, RR, Sadler, KC, Knauer, MW, et al. 1975. Composition of cottontail rabbit milk from stomachs of young and directly from gland. J Dairy Sci. 58:1449-1452.
278. Anderson, RR & Sinha, KN. 1972. Number of mammary glands and litter size in the golden hamster. J Mamm. 53:382-384.
279. Anderson, S. 1959. Mammals of the Grand Mesa Colorado. Univ KS Publ Mus Nat Hist. 9:405-414.
279a. Anderson, S. 1960. Neotropical bats from western México. Univ KS Publ Mus Nat Hist. 14:1-8.
280. Anderson, S. 1969. *Macrotus waterhousii*. Mamm Species. 1:1-4.
280a. Anderson, S. 1972. Mammals of Chihuahua: Taxonomy and distribution. Bull Am Mus Nat Hist. 148:149-410
281. Anderson, S. 1982. *Monodelphis kunsi*. Mamm Species. 190:1-3.
281a. Anderson, S & Jones, JK, Jr. 1984. Orders and Families of Recent Mammals of the World. J Wiley & Sons. New York.
282. Anderson, S, Woods, CA, Morgan, GS, et al. 1983. *Geocapromys brownii*. Mamm Species. 201:1-5.
283. Anderson, SA, Burton, RW, & Summers, CF. 1975. Behaviour of grey seals (*Halichoerus grypus*) during a breeding season at North Rona. J Zool. 177:179-195.
284. Anderson, SS, Baker, JR, Prime, JH, et al. 1979. Mortality in grey seal pups: Incidence and causes. J Zool. 189:407-417.
285. Anderson, SS & Fedak, MA. 1987. Grey seal, *Halichoerus grypus*, energetics: Females invest more in male offspring. J Zool. 211:667-669.
286. Anderson, TJC, Berry, AJ, & Amos, JN. 1988. Spool-and-line tracking of the New Guinea spiny bandicoot, *Echymipera kalubu* (Marsupialia, Peramelidae). J Mamm. 69:114-120.
287. Andersson, CB & Gustafsson, TO. 1979. Delayed implantation in lactating bank voles, *Clethrionomys glareolus*. J Reprod Fert. 57:349-352.
288. Andersson, CB & Gustafsson, TO. 1981. Relation between fertility and adrenal growth after mating in the bank vole, *Clethrionomys glareolus*. Can J Zool. 59:329-331.
289. Andersson, CB & Gustafsson, TO. 1982. Effect of limited and completed matings on ovaries and adrenals in bank voles, *Clethrionomys glareolus*. J Reprod Fert. 64:431-435.
290. Andersson, M, Borg, B, & Meurling, P. 1982. Biology of the wild rabbit, *Oryctolagus cuniculus*, in southern Sweden II Modifications in the onset of breeding, in relation to weather conditions. Viltrevy. 11:129-137.
291. Andersson, M, Dahlbäck, M, & Meurling, P. 1982. Biology of the wild rabbit, *Oryctolagus cuniculus*, in southern Sweden I Breeding season. Viltrevy. 11:103-128.
292. Andersson, M & Meurling, P. 1977. The maturation of the ovary in wild rabbits, *Oryctolagus cuniculus*, in south Sweden. Acta Zool (Stockholm). 58:95-101.
293. Andersson, M, Meurling, P, Dahlbäck, M, et al. 1981. Reproductive biology of the wild rabbit in southern Sweden, an area close to the northern limit of its distribution. Proc World Lagomorph Conf, Univ Guelph, ONT, Canada. 1979:175-181.
294. Andelman, SJ, Else, JG, Hearn, JP, et al. 1985. The non-invasive monitoring of reproductive events in wild vervet monkeys (*Cercopithecus aethiops*) using urinary pregnanediol-3α-glucuronide and its correlation with behavioural observations. J Zool 205A:467-477.
294a. Ando, A, Shirashi, S, & Uchida, TA. 1987. Growth and development of the Smith's red-backed vole, *Eothenomys smithi*. J Fac Agr Kyushu Univ. 31:309-320.
294b. Ando, A, Shirashi, S, & Uchida, TA. 1988. Reproduction in a laboratory colony of the Smith's red-backed vole, *Eothenomys smithi*. J Mamm Soc Japan. 13:11-20.
295. Ando, M & Shiraishi, S. 1984. Relative growth and gliding adaptations in the Japanese giant squirrel, *Petaurista leucogenys*. Sci Bull Fac Agr Kyushu Univ. 39:49-57.
296. Ando, Y. 1979. Breeding of the ratel. Animals & Zoos. 31(4):6-7.
297. Andresen, A. 1927. Die Placenta der Wiederkäuer. Morphol Jb. 57:410-485.
298. Andresen Hubbard, C. 1970. A first record of *Beamys* from Tanzania with observations on its breeding and habits in captivity. Zool Afr. 5:229-236.
299. Andresen Hubbard, C. 1972. Observations on the life histories and behaviour of some small rodents from Tanzania. Zool Afr. 7:419-429.

300. Andreson, D. 1977. Gestation period of Geoffroy's cat *Leopardus geoffroyi* bred at Memphis Zoo. Int Zoo Yearb. 17:164-166.
301. Andrewartha, HG & Barker, S. 1969. Introduction to a study of the ecology of the Kangaroo Island wallaby *Protemnodon eugenii* (Desmarest) within Flinders Chase, Kangaroo Island, South Australia. Trans R Soc S Aust. 93:127-133.
302. Andrews, RC. 1914. The California gray whale (*Rhachianectes glaucus* Cope). Mem Am Mus Nat Hist NS. 1(5):229-290.
303. Andrews, RC. 1916. The sei whale (*Balaenoptera borealis* Lesson). Mem Am Mus Nat Hist. NS. 1(5):293-388.
304. Andrews, RV, Ryan, K, Strohbehn, R, et al. 1975. Physiological and demographic profiles of brown lemmings during thêir cycle of abundance. Physiol Zool. 48:64-83.
305. Andriamiandra, A & Rumpler, Y. 1968. Rôle de la testostérone sur le déterminisme des glandes brachiales et antébrachieles chez le *Lemur catta*. CR Soc Biol. 162:1651-1655.
306. Andriantsiferana, R, Rarijaona, Y, & Randrianaivo, A. 1974. Observations sur la reproduction du microcèbe (*Microcebus murinus*, Miller 1777) en captivité à Tananarive. Mammalia. 38:234-243.
307. Angersbjörn, A. 1986. Reproduction of mountain hares (*Lepus timidus*) in relation to density and physical condition. J Zool. 208A:559-568.
308. Anghi, CG. 1962. Breeding Indian elephants *Elephas maximus* at the Budapest zoo. Int Zoo Yearb. 4:83-86.
309. Anghi, CG, Lehoczky, Z, & Orbanyi, I. 1978. Superfoetation by Panthern. Zool Garten NF. 48:189-191.
310. Angulo, JJ. 1947. Teat location in the Cuban solenodon. J Mamm. 28:298-299.
311. Angulo, JJ & Alvarez, MT. 1948. The genital tract of the male conga hutia, *Capromys pilorides* (Say). J Mamm. 29:277-285.
312. Anizoba, MA. 1982. Reproductive cycles of the African giant rat (*Cricetomys gambianus* Waterhouse) in the wild. Rev Zool Afr. 96:833-840.
313. Anizoba, MA. 1986. Observations on the variation of body weight and maturational moult with age in laboratory-reared African giant rat (*Cricetomys gambianus* Waterhouse). Rev Zool Afr. 99:243-247.
314. Anizoba, MA. 1986. Detecting pregnancy in the African giant rat, (*Cricetomys gambianus* Waterhouse) (Mammalia Rodentia), through vaginal smearing. Rev Zool Afr. 99:419-422.
315. Anonymous. 1912. Report on plague investigations in India: Observations on plague in eastern Bengal and Assam. J Hygiene Plague Suppl. 1:157-192.
316. Anonymous. 1929. The Chinese porcupine. China J. 10:147.
317. Anonymous. 1960. Annual report of the biologist 1958/1959. Koedoe. 3:1-205.
318. Anonymous. 1960. Breeding notes on the hippopotamus and giraffe at Cleveland Zoo. Int Zoo Yearb. 2:90.
319. Anonymous. 1960. Breeding notes on the reticulated giraffe at Cheyenne Mountain Zoo. Int Zoo Yearb. 2:90.
320. Anonymous. 1968. Twin orangs born at Woodland Park Zoo. Animal Kingdom. 61:30.
321. Anonymous. 1972-74. Australian *Tachyglossus*. CSIRO Div Wildl Res Report. 72-74:32-41.
322. Anonymous. 1973. Caribou. Canadian Wildlife Service.
323. Anonymous. 1978. Beaver. Canadian Wildlife Service.
324. Anonymous. 1981. Two or four teats?. Custos. 10:29-31.
325. Anonymous. 1986. Panda's birth in China hailed. New York Times. 21 Aug:A17.
326. Anonymous. 1988. Discovery of a new species of lemur. Nature. 333:206.
327. Anonymous. 1989. Captive population doubles in 1989. Black-footed Ferret Newsl. 6(2):1,4.
328. Anonymous. 1989. Ferret focus. The Drumming Post. 2:8.
328a. Ansell, WFH. 1957. Some mammals from Northern Rhodesia. Ann Mag Nat Hist Bot Geol. 12:529-551.
329. Ansell, WFH. 1960. *Mus minutoides* born in captivity. J Mamm. 41:405.
330. Ansell, WFH. 1960. Contributions to the mammalogy of northern Rhodesia. S Rhodesia Nat Mus Occas Pap. 24B:351-398.
331. Ansell, WFH. 1960. Mammals of Northern Rhodesia. The Government Printer, Lusaka, Northern Rhodesia.
332. Ansell, WFH. 1960. The African striped weasel, *Poecilogale albinucha*. Proc Zool Soc London. 134:59-64.
333. Ansell, WFH. 1960. The breeding of some larger mammals in northern Rhodesia. Proc Zool Soc London. 134:251-274.
334. Ansell, WFH. 1963. Additional breeding data on northern Rhodesian mammals. Puku. 1:9-28.
335. Ansell, WFH. 1964. Captivity behaviour and post-natal development of the shrew *Crocidura bicolor*. Proc Zool Soc London. 142:123-127.
336. Ansell, WFH. 1986. Records of bats in Zambia carrying non-volant young in flight. Arnoldia (Zimbabwe). 9:315-318.
337. Ansell, WFH & Ansell, PDH. 1973. Mammals of the north-eastern montane areas of Zambia. Puku. 7:21-69.
338. Anthony, A. 1953. Seasonal reproductive cycle in the normal and experimentally treated prairie dog; *Cynomys ludovicianus*. J Morphol. 93:331-369.
339. Anthony, A & Foreman, D. 1951. Observations on the reproductive cycle of the black-tailed prairie dog (*Cynomys ludovicianus*). Physiol Zool. 24:242-248.
340. Anthony, ELP. 1981. Night roosting and the nocturnal time budget of the little brown bat, *Myotis lucifugus*: Effects of reproductive status, prey density, and environmental conditions. Oecologica. 51:151-156.
341. Anthony, ELP & Gustafson, AW. 1984. A quantitative study of pituitary colloid in the bat *Myotis lucifugus lucifugus* in relation to age, sex and season. Am J Anat. 169:89-100.
342. Anthony, ELP & Gustafson, AW. 1984. Seasonal variation in pituitary thyrotropes of the hibernating bat *Myotis lucifugus lucifugus*: An immunohistochemical study. Anat Rec. 209:363-372.
343. Anthony, ELP & Gustafson, AW. 1984. Seasonal variations in pituitary LH gonadotropes of the hibernating bat *Myotis lucifugus lucifugus*: An immunohistochemical study. Am J Anat. 170:101-115.
344. Anthony, ELP, Weston, PJ, Montvilo, JA, et al. 1989. Dynamic aspects of the LHRH system associated with ovulation in the little brown bat (*Myotis lucifugus*). J Reprod Fert. 87:671-686.
344a. Anthony, HE. 1925/26. Mammals of Porto [*sic*] Rico, living and extinct. NY Acad Sci. 9:1-238.
345. Antonius, O. 1932. Einige bemerkenswerte Zuchterfolge in Schönbrunn im Jahre 1931. Zool Gärten. 5:91-97.
345a. Apelgren, R. 1941. Kullstorleken och draktighetstidens langd hos mink. Vara Palsdjur. 12:349-351.
346. Appleby, MC. 1982. The consequences and causes of high social rank in red deer stags. Behaviour. 80:259-273.
347. Appley, MB & Richter, KM. 1970. Ciliated granulosa cells of the bat *Myotis grisescens*. Ann Proc Electron Microsc Soc Am. 28:102-103.

348. Apps, PJ. 1983. Aspects of the ecology of feral cats on Dassen Island, South Africa. S Afr J Zool. 18:393-399.
349. Aquilina, GD & Beyer, RH. 1979. The exhibition and breeding of binturongs (*Arctictis binturong*) as a family group at Buffalo Zoo. Int Zoo Yearb. 19:185-188.
350. Aquino, R & Encarnación, F. 1986. Population structure of *Aotus nancymai* (Cebidae: Primates) in Peruvian Amazon lowland forest. Am J Primatol. 11:1-7.
351. Arata, AA. 1959. Ecology of muskrats in strip-mine ponds in southern Illinois. J Wildl Manage. 23:177-186.
352. Arata, AA. 1964. The anatomy and taxonomic significance of the male accessory reproductive glands of muroid rodents. Bull FL State Mus Biol Sci. 9:1-42.
352a. Arata, AA & Vaughn, JB. 1970. Analyses of the relative abundance and reproductive activity of bats in southwestern Colombia. Caldasia. 10:517-528.
353. Archer, M. 1974. Some aspects of reproductive behaviour and the male erectile organs of *Dasyurus geoffroii* and *D hallucatus* (Dasyuridae; Marsupialia). Mem QLD Mus. 17:63-67.
354. Archer, M. 1975. *Ningaui*, a new genus of tiny dasyurids (Marsupialia), and two new species, *N timealeyi* and *N ridei*, from arid Western Australia. Mem QLD Mus. 17:237-249.
355. Archer, M. 1976. Revision of the marsupial genus *Planigale* Troughton (Dasyuridae). Mem QLD Mus. 17:341-365.
356. Archer, M. 1982. Carnivorous Marsupials. R Zool Soc NSW. Mossman, NSW, Australia.
357. Archer, M. 1987. Possums and Opossums: Studies in Evolution. R Zool Soc NSW, Surrey Beatty & Sons, NSW, Australia.
358. Arendsen Hein, SA. 1914. Contributions to the anatomy of *Monodon monoceros*. Verh Kon Akad Wetensch Amsterdam Tweede Sectie. 18(3):1-35.
359. Aretas, R. 1952. L'Otarie à fourrure de l'Alaska *Callorhinus ursinus* (Linné). Terre Vie. 7:25-34.
359a. Arlton, AV. 1936. An ecological study of the mole. J Mamm. 17:349-371.
360. Arman, P. 1974. A note on parturition and maternal behaviour in captive red deer (*Cervus elaphus*). J Reprod Fert. 37:87-90.
361. Arman, P, Hamilton, WJ, & Sharman, GAM. 1978. Observations on the calving of free-ranging tame red deer (*Cervus elaphus*). J Reprod Fert. 54:279-283.
362. Arman, P, Kay, RNB, Goodall, ED, et al. 1974. The composition and yield of milk from captive red deer (*Cervus elaphus* L). J Reprod Fert. 37:67-84.
363. Armati-Gulson, P & Lowe, J. 1984. The development of the reproductive system of the common ringtail possum, (*Pseudocheirus peregrinus*) (Marsupialia: Petauridae). Aust Mamm. 7:75-87.
364. Armitage, KB. 1965. Vernal behaviour of the yellow-bellied marmot (*Marmota flavieventris*). Anim Behav. 13:59-68.
365. Armitage, KB. 1974. Male behaviour and territoriality in the yellow-bellied marmot. J Zool. 172:233-265.
366. Armitage, KB & Downhower, JF. 1974. Demography of yellow-bellied marmot populations. Ecology. 55:1233-1245.
366a. Armitage, KB, Downhower, JF, & Svendsen, GE. 1976. Seasonal changes in weights of marmots. Am Mid Nat. 96:36-51.
366b. Armstrong, DM. 1969. Noteworthy records of bats from Costa Rica. J Mamm. 50:808-810.
367. Armstrong, DM & Jones, JK, Jr. 1971. *Sorex merriami*. Mamm Species. 2:1-2.
368. Armstrong, DM & Jones, JK, Jr. 1972. *Megasorex gigas*. Mamm Species. 16:1-2.
369. Armstrong, DM & Jones, JK, Jr. 1972. *Notiosorex crawfordi*. Mamm Species. 17:1-5.
370. Armstrong, J. 1975. Hand-rearing black-footed cats (*Felis nigripes*) at the National Zoological Park, Washington. Int Zoo Yearb. 15:245-249.
371. Armstrong, N, Chaplin, RE, Chapman, DI, et al. 1969. Observations on the reproduction of female wild and park fallow deer (*Dama dama*) in southern England. J Zool. 158:27-37.
372. Armstrong, RA. 1950. Fetal development of the northern white-tailed deer (*Odocoileus virginiana borealis* Miller). Am Mid Nat. 43:650-666.
373. Årnbåck-Christie-Linde, A. 1907. Der Bau der Soriciden und ihre Beziehungen zu andern Säugetieren. Morphol Jb. 36:463-514.
374. Arnold, AJ. 1899. The western sing-sing (*Cobus defassa unctuosus*). Great and Small Game of Africa. 276-280, Bryden, HA, ed, Rowland Ward, London.
375. Arnold, AJ. 1899. West African hartebeest (*Bubalis major*). Great and Small Game of Africa. 135-140, Bryden, HA, ed, Rowland Ward, London.
376. Arnold, J & Shield, J. 1970. Growth and development of the chuditch (*Dasyurus geoffroii*). Bull Aust Mamm Soc. 2:198.
377. Arnold, LW. 1942. Notes on the life history of the sand pocket mouse. J Mamm. 23:339-341.
378. Arnold, R & Shorey, CD. 1985. Structure of the uterine luminal epithelium of the brush-tailed possum (*Trichosurus vulpecula*). J Reprod Fert. 74:565-573.
379. Arnoult, J. 1954. Naissances d'*Halalemur griseus* au vivarium de l'IRSM. Le Nat Malgache. 6:131.
380. Arora, RC & Pandey, RS. 1982. Changes in peripheral plasma concentrations of progesterone, estradiol-17β, and luteinizing hormone during pregnancy and around parturition in the buffalo (*Bubalus bubalis*). Gen Comp Endocrinol. 48:403-410.
381. Arora, RC & Pandey, RS. 1982. Pattern of plasma progesterone, oestradiol-17β, luteinzing hormone and androgen in non-pregnant buffalo *Bubalus bubalis*. Acta Endocrinol. 100:279-284.
382. Arrington, LR, Beaty, TC, Jr, & Kelley, KC. 1973. Growth, longevity, and reproductive life of the Mongolian gerbil. Lab Anim Sci. 23:262-265.
383. Arroyo-Cabrales, J & Jones, JK, Jr. 1988. *Balantiopteryx plicata*. Mamm Species. 301:1-4.
384. Arroyo-Cabrales, J & Jones, JK, Jr. 1988. *Balantiopteryx io* and *Balantiopteryx infusca*. Mamm Species. 313:1-3.
385. Arroyo-Cabrales, J, Hollander, RR, & Jones, JK, Jr. 1987. *Choeronycteris mexicana*. Mamm Species. 291:1-5.
386. Arslan, M. 1969. Hormonal requirements during early pregnancy and effects of gonadotrophins in the rhesus monkey (*Macaca mulatta*). Ph D Diss. Univ WI, Madison.
387. Arslan, M, Mahmood, S, Khurshid, S, et al. 1986. Changes in circulating levels of immunoreactive follicle stimulating hormone, luteinizing hormone, and testosterone during sexual development in the rhesus monkey, *Macaca mulatta*. J Med Primatol. 13:351-359.
388. Arslan, M, Meyer, RK, & Wolf, RC. 1967. Chorionic gonadotropin in the blood and urine of pregnant rhesus monkeys (*Macaca mulatta*). Proc Soc Exp Biol Med. 125:349-352.
389. Arthur, GH & Al-Rahim, AT. 1982. Aspects of reproduction in the female camel (*C dromedarius*) in Saudi Arabia. Vet Med Rev. 1:83-88.
390. Arthur, GH, Al-Rahim, AT, & Al-Hindi, AS. 1985. The camel in health and disease 7 Reproduction and genital diseases of the camel. Br Vet J. 141:650-659.

391. Artois, M, Aubert, MFA, & Gérard, Y. 1982. Reproduction du Renard roux (*Vulpes vulpes*) en France: Rhythme saisonnier et fecondité des femelles. Acta Aecol, Aecol Applic. 3:205-216.
392. Arvy, L. 1971. Endocrine glands and hormonal secretion in cetaceans. Invest Cetacea. 3:229-300.
393. Arvy, L & Pilleri, G. 1983. Four centuries of observations on the bottlenosed whale, the sperm whale and the killer whale. Invest Cetacea. 15 Suppl:1-265.
394. Asa, CS, Goldfoot, DA, & Ginther, OJ. 1979. Sociosexual behavior and the ovulatory cycle of ponies (*Equus caballus*) observed in harem groups. Horm Behav. 13:49-65.
395. Asa, CS, Seal, US, Letellier, M, et al. 1987. Pinealectomy or superior cervical ganglionectomy do not alter reproduction in the wolf (*Canis lupus*). Biol Reprod. 37:14-21.
396. Asa, CS, Seal, US, Plotka, ED, et al. 1986. Effects of anosmia on reproduction in male and female wolves (*Canis lupus*). Behav Neur Biol. 46:272-284.
397. Asakura, S. 1968. A note on the breeding records of Japanese foxes (*Vulpes vulpes japonica*) at Tokyo Tama Zoo. Int Zoo Yearb. 8:20-21.
398. Asakura, S. 1970. Breeding common tupaias *Tupaia glis* at Tama Zoo, Tokyo. Int Zoo Yearb. 10:70.
399. Asakura, S & Okada, S. 1972. Breeding douroucoulis *Aotus trivirgatus* at Tama Zoo, Tokyo. Int Zoo Yearb. 12:47-48.
400. Asano, M. 1967. A note on the birth and rearing of an orang-utan *Pongo pygmaeus* at Tama Zoo, Tokyo. Int Zoo Yearb. 7:95-96.
401. Asanov, SS. 1971. Comparative features of the reproductive biology of Hamadryas baboons (*Papio hamadryas*), grivet monkeys (*Cercopithecus aethiops*) and rhesus monkeys (*Macaca mulatta*). Acta Endocrinol Suppl. 166:458-471.
402. Asawa, SC & Mathur, RS. 1981. Quantitative evaluation of the interrelationship of certain endocrine glands of *Hemiechinus auritus collaris* (Gray) during the reproductive cycle. Acta Anat. 111:259-267.
403. Ascaray, CM & McLachlan, A. 1991. Postnatal growth and development of the hairy-footed gerbil, *Gerbillus paeba exilis*. S Afr J Zool. 26:70-77.
404. Ascaray, CM, Perrin, MR, McLachlan, A, et al. 1991. Population ecology of the hairy-footed gerbil, *Gerbillurus paeba*, in a coastal dunefield of South Africa. Z Säugetierk. 56:296-305.
405. Asch, RH, Eddy, CA, & Schally, AV. 1981. Lack of luteolytic effect of D-TRP-6 LH-RH in hypophysectomized rhesus monkey (*Macaca mulatta*). Biol Reprod. 25:963-968.
406. Aschaffenburg, R, Gregory, ME, Kon, SK, et al. 1962. The composition of the milk of the reindeer. J Dairy Res. 29:325-328.
407. Aschaffenburg, R, Gregory, ME, Rowland, SJ, et al. 1961. The composition of the milk of the Afican black rhinoceros (*Diceros bicornis* Linn). Proc Zool Soc London. 137:475-479.
408. Aschaffenburg, R, Gregory, ME, Rowland, SJ, et al. 1962. The composition of the milk of the giraffe (*Giraffa camelopardalis reticulata*). Proc Zool Soc London. 139:359-363.
408a. Aschaffenburg, R, Sen, A, & Thompson, MP. 1968. The caseins of buffalo milk. Comp Biochem Physiol. 27:621-623.
409. Asdell, SA, Bogart, R, & Sperling, G. 1941. The influence of age and rate of breeding upon the ability of the female rat to reproduce and raise young. Cornell Univ Agric Exp Sta Mem. 238:1-26.
410. Ashbrook, FG & Hanson, KB. 1927. Breeding martens in captivity. J Hered. 18:498-503.
411. Ashbrook, FG & Hanson, KB. 1930. The normal breeding season and gestation period of martens. USDA Circ. 107:1-6.
412. Ashby, KR & Henry, BAM. 1979. Age criteria and life expectancy of roe deer (*Capreolus capreolus*) in coniferous forest in north-eastern England. J Zool. 189:207-220.
413. Asher, GW. 1985. Oestrous cycle and breeding season of farmed fallow deer, *Dama dama*. J Reprod Fert. 75:521-529.
414. Asher, GW, Barrell, GK, & Peterson, AJ. 1986. Hormonal changes around oestrus of farmed fallow deer, *Dama dama*. J Reprod Fert. 78:487-496.
415. Asher, GW, Day, AM, & Barrell, GK. 1987. Annual cycle of liveweight and reproductive changes of farmed male fallow deer (*Dama dama*) and the effect of daily oral administration in summer on the attainment of seasonal fertility. J Reprod Fert. 79:353-362.
416. Asher, GW, Fisher, MW, Smith, JF, et al. 1990. Temporal relationship between the onset of oestrus, the preovulatory LH surge and ovulation in farmed fallow deer, *Dama dama*. J Reprod Fert. 89:761-767.
417. Asher, GW & Macmillan, KL. 1986. Induction of oestrus and ovulation in anoestrous fallow deer (*Dama dama*) by using progesterone and GnRH treatment. J Reprod Fert. 78:693-697.
418. Asher, GW & Smith, JF. 1987. Induction of oestrus and ovulation in farmed fallow deer (*Dama dama*) by using progesterone and PMSG treatment. J Reprod Fert. 81:113-118.
419. Ashfaque, M & Tungare, SM. 1960. Observations on the structure of the female reproductive organs in some Indian bats. Bull Zool Soc Coll Sci, Nagpur. 3:1-7.
420. Ashworth, US, Ramaiah, GD, & Keyes, MC. 1966. Species difference in the composition of milk with special reference in the northern fur seal. J Dairy Sci. 49:1206-1211.
421. Asibey, EOA. 1974. Reproduction in the grass cutter (*Thryonomys swinderianus* Temminck) in Ghana. Symp Zool Soc London. 34:251-263.
422. Asibey, EOA. 1981. Maternal and neonatal weight in the grasscutter, (*Thyronomys* [sic] *swinderianus*) (Temminck) in Ghana. Afr J Ecol. 19:335-360.
423. Aslin, H. 1974. The behaviour of *Dasyuroides byrnei* (Marsupialia) in captivity. Z Tierpsychol. 35:187-208.
424. Aslin, HJ. 1975. Reproduction in *Antechinus maculatus* Gould (Dasyuridae). Aust Wildl Res. 2:77-80.
425. Aslin, HJ. 1980. Biology of a laboratory reared colony of *Dasyuroides byrnei* (Marsupialia: Dasyuridae). Aust Zool. 20:457-471.
426. Aslin, HJ. 1983. Reproduction in *Sminthopsis ooldea* (Marsupialia: Dasyuridae). Aust Mamm. 6:93-95.
427. Aslin, HJ & Watts, CHS. 1980. Breeding of a captive colony of *Notomys fuscus* Wood Jones (Rodentia: Muridae). Aust Wildl Res. 7:379-383.
428. Aso, T, Goncharov, N, Cekan, Z, et al. 1976. Plasma levels of unconjugated steroids in male baboons (*Papio hamadryas*) and rhesus monkeys (*Macaca mulatta*). Acta Endocrinol. 82:644-651.
429. Aso, T, Tominaga, T, Oshima, K, et al. 1977. Season changes of plasma estradiol and progesterone in the Japanese monkey (*Macaca fuscata fuscata*). Endocrinology. 100:745-750.
430. Aso, T & Williams, RF. 1985. Lactational amenorrhea in monkeys: Effects of suckling on prolactin secretion. Endocrinology. 117:1727-1734.
431. Asper, ED, Young, WG, & Walsh, MT. 1988. Observations on the birth and development of a captive born killer whale. Int Zoo Yearb. 27:295-304.
432. Assheton, R. 1905. On the foetus and placenta of the spiny mouse (*Acomys cahirinus*). Proc Zool Soc London. 1905:280-288.

433. Assheton, R. 1906. The morphology of the ungulate placenta and notes upon the placenta of the elephant and hyrax. Phil Trans R Soc London Biol Sci. 198B:143-220.
434. Assheton, R & Stevens, TG. 1905. Notes on the structure and development of the elephant's placenta. Q J Microsc Sci. 49NS:1-37.
435. Atherton, RG & Haffenden, AT. 1982. Observations on the reproduction and growth of long-tailed pygmy possum, *Cercartetus caudatus* (Marsupialia: Burramyidae) in captivity. Aust Mamm. 5:253-259.
436. Athias, M. 1920. Recherches sur les cellules interstitielles de l'ovaire des Cheiroptères. Arch Biol. 30:89-212.
436a. Atkinson, KT & Shackleton, DM. 1991. Coyote, *Canis latrans*, ecology in a rural-urban environment. Can Field-Nat. 105:49-54.
437. Atkinson, LE, Hotchkiss, J, Fritz, GR, et al. 1975. Circulating levels of steroids and chorionic gonadotropin during pregnancy in the rhesus monkey, with special attention to the rescue of the corpus lutuem in early pregnancy. Biol Reprod. 12:335-345.
438. Atramentowicz, M. 1982. Influence du milieu sur l'activité locomotrice et la reproduction de *Caluromys philander* (L). Terre Vie. 36:373-395.
439. Atramentowicz, M. 1986. Dynamique de population chez trois Marsupiaux Didelphidés de Guyane. Biotropica. 18:136-149.
440. Attwell, CAM. 1977. Reproduction and population ecology of the blue wildebeest *Connochaetes t taurinus* in Zululand. Ph D Diss. Univ Natal, Natal, S Africa.
441. Attwell, CAM & Hanks, J. 1980. Reproduction of the blue wildebeest *Connochaetes taurinus taurinus* in Zululand, South Africa. Säugetierk Mitt. 28:264-281.
442. Attwell, RIG. 1963. Surveying Luangwa hippo. Puku. 1:29-49.
443. Atwood, EL. 1950. Life history studies of nutria in Louisiana. J Wildl Manage. 14:249-265.
443a. Atzenhofer, DR & Leedy, DL. 1947. The cottontail rabbit and its management in Ohio. Wildl Cons Bull OH. 2:1-15.
444. Audy, MC. 1976. Influence du photopériodisme sur la physiologie testiculaire de la Fouine (*Martes foina* Erx). CR Hebd Séances Acad Sci Ser D Sci Nat. 283:805-808.
445. Audy, MC. 1978. Variations saisonnières de la testostérone chez la Fouine (*Martes foina* Erx). CR Hebd Séances Acad Sci Sér D Sci Nat. 287:721-724.
446. Audy, MC. 1980. Variations saisonnières du rythme nycothéméral de la testostérone plasmatique chez le Blaireau européen *Meles meles* L. CR Hebd Séances Acad Sci Sér D Sci Nat. 291:291-294.
447. Audy, MC, Bonnin, M, Souloumiac, J, et al. 1985. Seasonal variations in plasma luteinizing hormone and testosterone levels in the European badger *Meles meles* L. Gen Comp Endocrinol. 57:445-453.
448. Audy, MC & Bonnin-Laffargue, M. 1975. L'activité endocrine du testicule chez le Blaireau européen *Meles meles*. Arch Biol. 86:223-232.
449. Audy-Relexans, MC. 1972. Le cycle sexuel du Blaireau mâle (*Meles meles* L). Ann Biol Anim Bioch Biophys. 12:355-366.
449a. Augee, ML, ed. 1978. Monotreme biology. Aust Zool. 20:1-257.
449b. Augee, ML, ed. 1992. Platypus and echidnas. R Zool Soc NSW, Sydney, Australia.
450. August, PV. 1981. Population and community ecology of small mammals in northern Venezuela. Diss Abstr Int. 42:2202B.
451. August, PV. 1984. Population ecology of small mammals in the llanos of Venezuela. Sp Publ Mus TX Tech. 22:71-104.
452. August, PV & Baker, RJ. 1982. Observations on the reproductive ecology of some neotropical bats. Mammalia. 46:177-181.
453. Aulak, W. 1973. Production and energy requirements in a population of the bank vole, in a deciduous forest of *Circaeo-Alnetum* type. Acta Theriol. 18:167-189.
454. Aulerich, RJ, Holcomb, LC, Ringer, RK, et al. 1963. Influence of photoperiod on reproduction in mink. Q Bull MI St Univ Agr Exp Sta. 46:132-138.
455. Auletta, FJ, Paradis, DK, Wesley, M, et al. 1984. Oxytocin is luteolytic in the rhesus monkey (*Macaca mulatta*). J Reprod Fert. 72:401-406.
456. Auletta, FJ, Speroff, L, & Caldwell, BV. 1973. Prostaglandin F2α induced steroids genesis and luteolysis in the primate corpus luteum. J Clin Endocrinol Metab. 36:405-407.
457. Aulmann, G. 1932. Geglückte Nachzucht eines Orang-Utan im Düsseldorfer Zoo. Zool Gärten NF. 5:81-90.
458. Aumaitre, A, Quere, JP, & Peiniau, J. 1984. Influence du milieu sur la production hivernale et la prolificité de la laie. Colloques Internationals. 22:71-78.
459. Aung, H. 1968. A note on the birth of a Mishmi takin *Budorcas t taxicolor* at Rangoon Zoo. Int Zoo Yearb. 8:145.
460. Ausman, LM, Hayes, KC, Lage, A, et al. 1970. Nursery care and growth of old and new world monkeys. Lab Anim Care. 20:907-913.
461. Austad, SN & Sunquist, ME. 1986. Sex-ratio manipulation in the common opossum. Nature. 324:58-60.
462. Austin, CR. 1957. Oestrus and ovulation in the field vole (*Microtus agrestis*). J Endocrinol. 15:4.
463. Austin, CR & Bishop, MWH. 1959. Presence of spermatozoa in the uterine-tube mucosa of bats. J Endocrinol. 18:viii-ix.
464. Austin, CR & Short, RH, eds. 1972-1982. Reproduction in Mammals, 2 editions, 8 volumes. Cambridge Univ Press. NY.
465. Autuori, MP & Deutsch, LA. 1977. Contribution to the knowledge of the giant Brazilian otter, *Pteronura brasiliensis* (Gmelin 1788), Carnivora, Mustelidae. Zool Gärten NF. 47:1-8.
466. Avaliani, RS. 1965. The ecology of the Transcaucasian steppe fox *Vulpes vulpes alpherakyi* in the Gruzinian SSR. Biol Abstr. 46:Ref 96995.
467. Avenell, JA, Saepudin, Y, & Fletcher, IC. 1985. Concentrations of LH, estradiol-17β and progesterone in the peripheral plasma of swamp buffalo cows, (*Bubalus bubalis*) around the time of oestrus. J Reprod Fert. 74:419-424.
468. Avery, TL. 1968. Observations on the propagation of Chinese hamsters. Lab Anim Care. 18:151-159.
469. Axelrod, LR. 1967. The synthesis of testosterone from precursors by the baboon testes. The Baboon in Medical Research. 2:633-636, Vagtborg, H, ed, Univ TX Press, Austin.
470. Axelson, M, Graham, CE, & Sjövall, J. 1984. Identification and quantitation of steroids in sulfate fractions from plasma of pregnant chimpanzee, orangutan, and rhesus monkeys. Endocrinology. 114:337-344.
471. Aymerich, M. 1982. Contribution à l'étude de la biologie de la Genette (*Genetta genetta* L) en Espagne. Mammalia. 46:389-393.
472. Babb, TE & Terman, CR. 1982. The influence of social environment and urine exposure on sexual maturation of male prairie deer mice (*Peromyscus maniculatus bairdi*). Res Pop Ecol. 24:318-328.

473. Baber, DW & Coblentz, BE. 1986. Density, home range, habitat use, and reproduction in feral pigs on Santa Catalina Island. J Mamm. 67:512-525.
474. Babero, BB, Yousef, MK, & Wawerna, JC. 1971. Histo-pathological changes in cold-exposed kangaroo rats, *Dipodomys merriami*. Comp Biochem Physiol. 39A:361-366.
475. Bachman, KC & Irvine, AB. 1979. Composition of milk from the Florida manatee, *Trichechus manatus latirostris*. Comp Biochem Physiol. 62A:873-878.
476. Bachmann, MR & Schulthess, W. 1987. Lactation of camels and composition of camel milk in Kenya. Milchwissenschaft. 42:766-768.
477. Backhaus, D. 1959. Beobachtungen über das Freileben von Lelwel-Kuhantilopen (*Alcelaphus buselaphus lelwel* Heuglin 1877) und Gelegenheitsbeobachtungen an Sennar-Pferdeantifopen (*Hippotragus equinus bakeri*, Heuglin 1863) 3 Mitteilung. Z Säugetierk. 24:1-34.
478. Backhouse, KM & Hewer, HR. 1956. Delayed implantation in the grey seal, *Halichoerus grypus* (Fabi). Nature. 178:550.
479. Backhouse, KM & Hewer, HR. 1957. A note on spring pupping in the grey seal. Proc Zool Soc London. 128:593-594.
480. Backstrom, AvK. 1952. Avslutade studier över älgbrunsten. Svenk Jakt. 90:70-71.
481. Baddaloo, EGY & Clulow, FV. 1981. Effects of the male on growth, sexual maturation, and ovulation of young female meadow voles, *Microtus pennsylvanicus*. Can J Zool. 59:415-421.
482. Badham, M. 1967. A note on breeding the spectacled leaf monkey *Presbytis obscura* at Twycross Zoo. Int Zoo Yearb. 7:89.
483. Badham, M. 1967. A note on breeding the pileated gibbon *Hylobates lar pileatus* at Twycross Zoo. Int Zoo Yearb. 7:92-93.
484. Baecker, R. 1930. Zur Histologie des Urogenitalsystems der Didelphiden (*Metachirus crassicaudatus*). Z Mikr-Anat Forsch. 21:614-641.
485. Baer, GM & Holquim, GM. 1971. Breeding Mexican freetail bats in captivity. Am Mid Nat. 85:515-517.
486. Baevskii, YB. 1967. Cytometric and karyometric investigations on the blastocyst of the mole (*Talpa altaica*) during the period of delayed implantation. Dokl Akad Nauk SSSR Biol Sci Transl. 176:570-572.
487. Baevskii, YB. 1971. Levels of subcellular differentiation corresponding to diapause and activation of the sable (*Martes zebellina*) embryo. Dokl Akad Nauk SSSR Transl. 197:179-181.
488. Baevskii, YB & Shvarts, TV. 1976. Growth and hormonal conditions of maturation in sables. Dokl Akad Nauk SSSR Transl. 226:78-80.
489. Baevsky, UB. 1963. The effect of embryonic diapause on the nuclei and mitotic activity of mink and rat blastocysts. Delayed Implantation. 141-153, Enders, AC, ed, Univ Chicago Press, Chicago.
490. Baggott, LM, Davis-Butler, S, & Moore, HDM. 1987. Characterization of oestrus and timed collection of oocytes in the grey short-tailed opossum, *Monodelphis domestica*. J Reprod Fert. 79:105-114.
491. Baharav, D. 1974. Notes on the population structure and biomass of the mountain gazelle, *Gazella gazella gazella*. Israel J Zool. 23:39-44.
492. Baharav, D. 1983. Observation on the ecology of the mountain gazelle in the upper Galilee, Israel. Mammalia. 47:59-69.
493. Baharav, D. 1983. Reproductive strategies in female mountain and dorcas gazelles (*Gazella gazella gazella* and *Gazella dorcas*). J Zool. 200:445-453.
494. Baig, KJ, Mahmood, A, & Arslan, M. 1986. Seasonal changes in reproductive organs and androgen levels of the musk shrew, *Suncus murinus*. Pak J Zool. 18:229-237.
495. Bailey, B. 1929. Mammals of Sherburne county, Minnesota. J Mamm. 10:153-164.
496. Bailey, FM. 1910. Note on the serow (*Nemorhoedus bubalinus*) from the Chambi valley. J Bombay Nat Hist Soc. 19:822.
497. Bailey, LF & Lemon, M. 1966. Specific milk proteins associated with resumption of development by the quiescent blastocyst of the lactating red kangaroo. J Reprod Fert. 11:473-475.
498. Bailey, TN. 1971. Biology of striped skunks on a southwestern Lake Erie marsh. Am Mid Nat. 85:196-207.
499. Bailey, V. 1915. Revision of the pocket gophers of the genus *Thomomys*. N Am Fauna. 39:1-136.
500. Bailey, V. 1922. Beaver habits, beaver control, and possibilities in beaver farming. USDA Bulletin. 1078:1-29.
501. Bailey, V. 1924. Breeding, feeding, and other life habits of meadow mice (*Microtus*). J Agric Res. 27:523-535.
502. Bailey, V. 1931. Mammals of New Mexico. N Am Fauna. 53:1-412.
503. Bailey, V. 1936. The mammals and life zones of Oregon. N Am Fauna. 55:1-416.
504. Bailey, V & Sperry, CC. 1929. Life history and habits of grasshopper mice, genus *Onychomys*. USDA Tech Bull. 145:1-19.
505. Baird, DD & Birney, EC. 1985. Bilateral distribution of implantation sites in small mammals of 22 North American species. J Reprod Fert. 75:381-392.
506. Baird, DD, Timm, RM, & Nordquist, GE. 1983. Reproduction in the arctic shrew, *Sorex arcticus*. J Mamm. 64:298-301.
507. Baker, AB. 1903. A notable success in the breeding of black bears. Smithsonian Misc Coll. 45:175-179.
508. Baker, AB. 1912. Further notes on the breeding of the American black bear in captivity. Smithsonian Misc Coll. 59(10):1-4.
509. Baker, AB. 1920. Breeding of the Brazilian tapir. J Mamm. 1:143-144.
510. Baker, AN. 1985. Pygmy right whale *Caperea marginata* (Gray, 1846). Handbook of Marine Mammals. 3:345-354.
511. Baker, BE, Cook, HW, Bider, JR, et al. 1970. Snowshoe hare (*Lepus americanus*) milk I Gross composition, fatty acid and mineral composition. Can J Zool. 48:1349-1352.
512. Baker, BE, Harrington, CR, & Symes, AL. 1963. Polar bear milk I Gross composition and fat constitution. Can J Zool. 41:1035-1039.
513. Baker, BL, Karsch, FJ, Hoffman, DL, et al. 1977. The presence of gonadotropic and thyrotropic cells in the pituitary pars tuberalis of the monkey (*Macaca mulatta*). Biol Reprod. 17:232-240.
514. Baker, CM & Meester, J. 1977. Postnatal physical and behavioural development of *Praomys* (*Mastomys*) *natalensis* (A Smith, 1834). Z Säugetierk. 42:295-306.
515. Baker, CM & Meester, J. 1986. Postnatal physical development of the water mongoose *Atilax paludinosus*. Z Säugetierk. 51:236-243.
516. Baker, JR. 1930. The breeding-season in British wild mice. Proc Zool Soc London. 1930:113-126.
517. Baker, JR & Baker, Z. 1936. The seasons in a tropical rain forest (New Hebrides) Part 3 Pteropidae. J Linn Soc Lond Zool. 40:123-141.

518. Baker, JR & Bird, TF. 1936. The seasons in a tropical rain forest (New Hebrides) Part 4 Insectivourous bats (Vespertilionidae and Rhinolophidae). J Linn Soc Lond Zool. 40:143-161.
519. Baker, JR & Ranson, RM. 1932. Factors affecting the breeding of the field mouse (*Microtus agrestis*) Part I Light. Proc R Soc London Biol Sci. 110B:313-322.
520. Baker, JR & Ranson, RM. 1932. Factors affecting the breeding of the field mouse (*Microtus agrestis*) Part II Temperature and food. Proc R Soc London Biol Sci. 112B:39-46.
521. Baker, JR & Ranson, RM. 1933. Factors affecting the breeding of the field mouse (*Microtus agrestis*) Part III Locality. Proc R Soc London Biol Sci. 113B:486-495.
522. Baker, RH. 1956. Mammals of Coahuila, Mexico. Univ KS Publ Mus Nat Hist. 9:125-335.
522a. Baker, RH & Dickerman, RW. 1956. Daytime roost of the yellow bat in Veracruz. J Mamm. 37:443.
523. Baker, RH & Greer, JK. 1960. Notes on Oaxacan mammals. J Mamm. 41:413-415.
523a. Baker, RH & Greer, JK. 1962. Mammals of the Mexican state of Durango. Publ Mus MI State Univ Biol Ser. 2:25-154.
524. Baker, RH & Shump, KA, Jr. 1978. *Sigmodon fulviventer.* Mamm Species. 94:1-4.
525. Baker, RH & Shump, KA, Jr. 1978. *Sigmodon ochrognathus.* Mamm Species. 97:1-2.
526. Baker, RJ, August, PV, & Steuter, AA. 1978. *Erophylla sezekorni.* Mamm Species. 115:1-5.
526a. Baker, RJ, Bleier, WJ, & Atchley, WR. 1975. A contact zone between karyotypically characterized taxa of *Uroderma bilobatum* (Mammalia: Chiroptera). Syst Zool. 24:133-142.
527. Baker, RJ & Clark, CL. 1987. *Uroderma bilobatum.* Mamm Species. 279:1-4.
527a. Baker, RJ, Genoways, HH, & Patton, JC. 1978. Bats of Guadeloupe. Occas Pap Mus TX Tech Univ. 50:1-16.
527b. Baker, RJ & Jones, JK, Jr. 1975. Additional records of bats from Nicaragua, with a revised checklist of Chiroptera. Occas Pap Mus TX Tech Univ. 32:1-13.
528. Baker, RJ, Mollhagen, T, & Lopez, G. 1971. Notes on *Lasiurus ega.* J Mamm. 52:849-852.
529. Baker, RJ & Spencer, DL. 1965. Late fall reproduction in the desert shrew. J Mamm. 46:330.
530. Baker, RJ & Williams, SL. 1974. *Geomys tropicalis.* Mamm Species. 35:1-4.
531. Bakko, EB & Brown, LN. 1967. Breeding biology of the white-tailed prairie dog, *Cynomys leucurus* in Wyoming. J Mamm. 48:100-112.
532. Baldridge, A. 1974. Migrant gray whales with calves and sexual behavior of gray whales in the Monterey area of central California, 1967-73. Fish Bull. 72:615-618.
533. Baldwin, BH, Tenant, BC, Reimers, TJ, et al. 1985. Circannual changes in serum testosterone concentrations of adult and yearling woodchucks (*Marmota monax*). Biol Reprod. 32:804-812.
534. Baldwin, J, Temple-Smith, P, & Tidemann, C. 1974. Changes in testis specific lactate dehydrogenase isoenzymes during the seasonal spermatogenic cycle of the marsupial *Schoinobates volans* (Petauridae). Biol Reprod. 11:377-384.
535. Baldwin, JD. 1970. Reproductive synchronization in squirrel monkeys (*Saimiri*). Primates. 11:317-326.
536. Baldwin, JD & Baldwin, JI. 1981. The squirrel monkeys, genus *Saimiri.* Ecology and Behavior of Neotropical Primates, 1:277-330, Coimbra-Filho, AF & Mittelmeier, RA, eds. Academia Brasileira de Ciêcias Rio de Janeiro.
537. Baldwin, LA & Teleki, G. 1974. Field research on gibbons, siamangs, and orang-utans: An historical, geographical, and bibliographical listing. Primates. 15:365-376.
538. Baldwin, PH, Schwartz, CW, & Schwartz, ER. 1952. Life history and economic status of the mongoose in Hawaii. J Mamm. 33:335-356.
539. Balfour, D. 1983. Infanticide in the Columbian ground squirrel, *Spermophilus columbianus.* Anim Behav. 31:949-950.
540. Balke, JME, Barker, IK, Hackenberger, MK, et al. 1988. Reproductive anatomy of three nulliparous female Asian elephants: The development for artificial breeding techniques. Zoo Biol. 7:99-113.
541. Balke, JME, Boever, WJ, Ellersieck, MR, et al. 1988. Anatomy of the reproductive tract of the female African elephant (*Loxodonta africana*) with reference to development of techniques for artificial breeding. J Reprod Fert. 84:485-492.
542. Ballard, WB, Miller, SD, & Spraker, TH. 1982. Home range, daily movements, and reproductive biology of brown bear in southcentral Alaska. Can Field-Nat. 96:1-5.
543. Ballard, WB, Whitman, JS, & Gardner, CL. 1987. Ecology of an exploited wolf population in south-central Alaska. Wildl Monogr. 98:1-54.
543a. Ballard, WB, Whitman, JS, & Reed, DJ. 1991. Population dynamics of moose in south-central Alaska. Wildl Monogr. 114: 1-49.
544. Ballenberghe, VV, Erickson, AW, & Byman, D. 1975. Ecology of the timber wolf in north eastern Minnesota. Wildl Monogr. 43:1-43.
545. Balli, A. 1940. Observazioni biologiche su *Talpa europaea L.* Riv Biol. 29:35-54.
546. Ballowitz, E. 1890. Über das Vorkommen des *Miniopterus schreibersii natterer* in Deutschland nebst einige Bemerkungen über die Fortpflanzung deutscher Chiropteren. Zool Anz. 13:531-536.
546a. Balmford, A, Albon, S, & Blakeman, S. 1992. Correlates of male mating success and female choice in a lek-breeding antelope. Behav Ecol. 3:112-123.
546b. Balmford, A & Blakeman, S. 1991. Horn and body measurements of topi in relation to a variable mating system. Afr J Ecol. 29:37-42.
547. Bamberg, F. 1983. Zu der Wiedereinbürgerung des Gamswildes (*Rupicapra rupicapra*) im Schwarzwald und der Einbürgerung in den Vogesen. Z Jagdwiss. 29:23-30.
548. Banasiak, CF. 1961. Deer in Maine. ME Game Div Bull. 6:1-159.
548a. Banci, V & Harestad, A. 1988. Reproduction and natality of wolverine (*Gulo gulo*) in Yukon. Ann Zool Fennici. 25:265-270.
549. Bancroft, BJ. 1973. Embryology of *Schinobates volans* Kerr (Marsupialia: Petauridae). Aust J Zool. 21:33-52.
550. Bancroft, WL. 1967. Record fecundity of *Reithrodontomys megalotis.* J Mamm. 48:306-308.
551. Bancroft, WL. 1969. Notes on reproduction of three rodents of Douglass County, Kansas. Trans KS Acad Sci. 72:67-69.
552. Bandoli, JH. 1981. Factors influencing seasonal burrowing activity in the pocket gopher, *Thomomys bottae.* J Mamm. 62:293-303.
553. Banerjee, S & Karim, KB. 1982. Female reproductive cycle in the mouse-tailed bat, *Rhinopoma hardwickei hardwickei.* Bat Res News. 23:58.
554. Banerji, A. 1957. Further observations on the family life of the five-striped squirrel, *Funambulus pennanti* WR. J Bombay Nat Hist Soc. 54:335-343.

555. Banfield, AWF. 1955. A provisional life table for the barren ground caribou. Can J Zool. 33:143-147.
555a. Banfield, AWF. 1974. The Mammals of Canada. Univ Toronto Press, Toronto.
556. Banks, E. 1931. A popular account of the mammals of Borneo. J Malay Branch R Asiat Soc. 9(III):1-139.
557. Banks, E. 1978. Mammals from Borneo. Brunei Mus J. 4:165-227.
558. Bannikov, AG. 1958. Zur Biologie des Kulans *Equus hemionus* Pallas. Z Säugetierk. 23:157-168.
559. Bannikov, AG. 1961. Ecologie et distribution d'*Equus hemionus* Pallas: Les variations de sa limite de distribution septentrionale. Terre Vie. 108:86-100.
560. Bannikov, AG. 1961. L'écologie de *Saiga tatarica* L en Eurasie, sa distribution et son exploitation rationelle. Terre Vie. 108:77-85.
561. Bannikov, AG. 1963. Die Saiga-Antilope (*Saiga tatarica* L). A Ziemsen Verlag, Wittenberg Lutherstadt.
562. Bannikov, AG. 1964. Biologie du Chien viverrin en URSS. Mammalia. 28:1-39.
563. Bannister, JL. 1963. An intersexual fin whale *Balaenoptera physalus* L from South Georgia. Proc Zool Soc London. 141:811-822.
564. Bannister, JL. 1969. The biology and status of the sperm whale off Western Australia: An extended summary of results of recent work. Rep Int Whal Comm. 19:70-76.
565. Barabash-Nikiforov, I. 1935. The sea otters of the Commander Islands. J Mamm. 16:255-261.
566. Barabash-Nikiforov, I. 1938. Mammals of the Commander Islands and the surrounding sea. J Mamm. 19:423-429.
567. Baranga, J. 1980. The adrenal weight changes of a tropical fruit bat, *Rousettus aegyptiacus* E Geoffroy. Z Säugetierk. 45:321-336.
568. Baranga, J, Schliemann, H, Kanjanja, FIB, et al. 1984. The epididymus of the fruit bat, *Epomophorus anurus*. Afr J Ecol. 22:55-61.
569. Barash, DP. 1973. Social variety in the yellow bellied marmot *Marmota flaviventris*. Anim Behav. 21:579-584.
570. Barash, DP. 1974. Mother-infant relations in captive woodchucks (*Marmota monax*). Anim Behav. 22:446-448.
571. Barash, DP. 1980. The influence of reproductive status on foraging by hoary marmots (*Marmota caligata*). Behav Ecol Sociobiol. 7:201-205.
572. Barbehenn, KR. 1962. The house shrew on Guam. Bull Bernice P Bishop Mus 225:247-256.
573. Barbehenn, KR, Sumangil, JP, & Libay, JL. 1972-73. Rodents of the Philippine croplands. Philippine Agriculturalist. 56:217-242.
574. Barbour, RW. 1941. Three new mammal records from Kentucky. J Mamm. 22:195-196.
575. Barbour, RW. 1951. The mammals of Big Black Mountain, Harlan County, Kentucky. J Mamm. 32:100-110.
576. Barbour, RW. 1951. Notes on mammals from West Virginia. J Mamm. 32:368-371.
577. Barbour, RW & Davis, WH. 1969. Bats of America. Univ Press KY. Lexington, KY.
578. Barbour, T. 1937. Birth of a manatee. J Mamm. 18:106-107.
579. Barbu, P. 1967. Contributii la cunoaşterea ecologiei cîineluienot (*Nyctereutes procyonoides* Gray) din Delta Dunarii. Ocrot Nat. 11:75-83.
580. Barbu, P. 1968. Systématique et écologie de la Belette *Mustela nivalis* L provenant de quelques forêts des districts d'Ilfov et de Prahova-Roamanie. Trav Mus Hist Nat "Grigore Antipa". 8:991-1002.
581. Barbu, P. 1970. Sur la reproduction du Nyctéreute (*Nyctereutes procyonoides ussuriensis*, Matschie, 1907) dans le delta du Danube. Trav Mus Hist Nat "Grigore Antipa". 10:331-345.
582. Barbu, P. 1972. Beiträge zum Studium des Marderhundes, *Nyctereutes procyonoides ussuriensis* Matschie, 1907, aus den Donaudelta. Säugetierk Mitt. 20:375-405.
583. Barfield, MA & Beeman, EA. 1968. The oestrous cycle in the Mongolian gerbil *Meriones unguiculatus*. J Reprod Fert. 17:247-251.
584. Barhat, NK, Chowdhary, MS, & Gupta, AK. 1979. Note on relationship among gestation length birth weight, placental weight, and intra uterine development index in Bikaneri camel. Indian J Anim Res. 13:115-117.
585. Barkalow, FS. 1961. A large cottontail litter. J Mamm. 42:254.
586. Barkalow, FS, Jr. 1962. Latitude related to reproduction in the cottontail rabbit. J Wildl Manage. 26:32-37.
587)8. Barkalow, FS, Jr. 1966. Rafinesque's big-eared bat. Wildl NC. Jan:14-15.
588. Barkalow, FS, Jr. 1967. A record grey squirrel litter. J Mamm. 48:141.
588a. Barkalow, FS, Jr & Funderburg, JB, Jr. 1960. Probable breeding and additional records of the seminole bat in North Carolina. J Mamm. 41:394-395.
589. Barkalow, FS, Jr & Shorten, M. 1973. The World of the Gray Squirrel. JB Lippincott & Co, Philadelphia.
589a. Barkalow, FS, Jr. 1948. The status of the seminole bat, *Lasiurus seminolus* (Rhoads). J Mamm. 29:415-416.
589b. Barker IK, Beveridge, I, Bradley, AJ, et al. 1978. Observations on spontaneous stress-related mortality among males of the dasyurid marsupial, *Antechinus stuartii* Macleay. Aust J Zool. 26:435-447.
590. Barker, S. 1971. The dama wallaby *Protemnodon eugenii* in captivity. Int Zoo Yearb. 11:17-20.
591. Barker, S & Barker, J. 1959. Physiology of the quokka. J R Soc W Aust. 42:72-76.
592. Barlow, GW. 1972. A paternal role for bulls of the Galapagos Islands sea lion. Evolution. 26:307-310.
593. Barlow, GW. 1974. Galapagos sea lions are paternal. Evolution. 28:476-478.
594. Barlow, J. 1984. Reproductive seasonality in pelagic dolphins (*Stenella* spp): Implications for measuring rates. Rep Int Whal Comm Sp Issue. 6:191-198.
595. g. Barlow, JC & Tamsitt, JR. 1968. Twinning in American leaf-nosed bats (Chiroptera: Phyllostomatidae). Can J Zool. 46:290-292.
596. Barlow, JC. 1969. Observations on the biology of rodents in Uruguay. R Ont Mus Life Sci Contr. 75:1-59.
597. Barnes, A & Gemmell, RT. 1984. Correlations between breeding activity in the marsupial bandicoots and some environmental variables. Aust J Zool. 32:219-226.
598. Barnes, BM. 1982. Influence of energy stores on activation of reproductive function in male golden-mantled squirrels. Am Zool. 22:981.
599. Barnes, BM. 1984. Influence of energy stores on activation of reproductive function in male golden-mantled ground squirrels. J Comp Physiol. 154B:421-425.
600. Barnes, BM. 1986. Annual cycles of gonadotropins and androgens in the hibernating golden-mantled ground squirrel. Gen Comp Endocrinol. 62:13-22.

601. Barnes, BM, Kretzmann, M, Licht, P, et al. 1986. The influence of hibernation on testis growth and spermatogenesis in the golden-mantled ground squirrel, *Spermophilus lateralis*. Biol Reprod. 35:1289-1297.
602. Barnes, BM, Kretzmann, M, Zucker, I, et al. 1988. Plasma androgen and gonadotropin levels during hibernation and testicular maturation in golden-mantled ground squirrels. Biol Reprod. 38:616-622.
603. Barnes, BM, Licht, P, & Zucker, I. 1987. Temperature dependence of *in vitro* androgen production in testes from hibernating ground squirrels, *Spermophilus lateralis*. Can J Zool. 65:3020-3023.
604. Barnes, RD. 1968. *Marmosa mitis*, a small marsupial for experimental biology. Animal Models for Biomedical Research. 1594:88-98, Nat Acad Sci, Washington, DC.
605. Barnes, RD. 1968. Small marsupials as experimental animals. Lab Anim Care. 18:251-257.
606. Barnes, RD & Barthold, SW. 1969. Reproduction and breeding behaviour in an experimental colony of *Marmosa mitis* Bangs (Didelphidae). J Reprod Fert Suppl. 6:477-482.
607. Barnes, RD, Lasley, BL, & Hendrickx, AG. 1978. Midcycle ovarian histology of the bonnet monkey, *Macaca radiata*. Biol Reprod. 18:537-553.
608. Barnes, RD & Wolf, HG. 1971. The husbandry of *Marmosa mitis* as a laboratory animal. Int Zoo Yearb. 11:50-54.
609. Barnett, AP. 1991. Records of the grey-bellied shrew opossum, *Caenolestes caniventer* and Tate's shrew opossum, *Caenolestes tatei* (Caenolestidae, Marsupialia) from Ecuadorian montane forests. Mammalia. 55:443-446.
609a. Barnett, JL. 1973. A stress response in *Antechinus stuartii* (Macleay). Aust J Zool. 21:501-513.
610. Barnett, JL, How, RA, & Humphreys, WF. 1977. Small mammal populations in pine and native forests in north-eastern New South Wales. Aust Wildl Res. 4:233-240.
611. Barnett, SA, Munro, KMH, Smart, JR, et al. 1975. House mice bred for many generations in two environments. J Zool. 177:153-169.
612. Barr, AB. 1973. Timing of spermatogensis in four nonhuman primate species. Fert Steril. 24:381-389.
613. Barrell, GK, Muir, PD, & Sykes, AR. 1985. Seasonal profiles of plasma testosterone, prolactin, and growth hormone in red deer stags. R Soc NZ Bull. 22:185-190.
613a. Barrette, C. 1977. Some aspects of the behaviour of muntjacs in Wilpattu National Park. Mammalia. 41:1-34.
613b. Barrette, C. 1987. The comparative behavior and ecology of chevrotains, musk deer and morphologically conservative deer. Biology and Management of the Cervidae. 200-213, Wemmer, CM, ed, Smithsonian Inst Press, Washington, DC.
614. Barritt, MK. 1976. Breeding of *Rattus lutreolus* (Gray, 1841) in South Australia. S Aust Nat. 51:14-15.
615. Barron, JC & Harwell, WF. 1973. Fertilization rates of south Texas deer. J Wildl Manage. 37:179-182.
616. Barry, J. 1979. Modifications périodiques du tractus préoptica terminal à LRH chez le singe Écureuil femelle au cours du cycle oestral. Ann d'Endocrinol (Paris). 40:191-202.
617. Barry, M. 1839. Researches in Embryology-Second series. Phil Trans R Soc London. 1839:307-380.
618. Barsotti, DA, Abrahamson, LJ, Marlar, RJ, et al. 1980. Effects of climate-controlled housing on reproductive potential of rhesus monkeys. J Reprod Fert. 59:15-20.
619. Bartecki, R & Jaczewski, Z. 1983. Seasonal variation in the plasma androgens concentration of red deer. Acta Theriol. 28:333-336.
620. Barth, D, Gimenéz, T, Hoffmann, B, et al. 1976. Testosteronkonzentrationen im peripheren Blut beim Rehbock (*Capreolus capreolus*). Z Jagdwiss. 22:134-148.
621. Bartholomew, GA, Jr & Hoel, PG. 1953. Reproductive behavior of the Alaska fur seal, *Callorhinus ursinus*. J Mamm. 34:417-436.
622. Bartholomew, GA. 1952. Reproductive and social behavior of the northern elephant seal. Univ CA Publ Zool. 47:369-472.
623. Bartholomew, GA, Leitner, P, & Nelson, JE. 1964. Body temperature, oxygen consumption and heart rate in three species of Australian flying foxes. Physiol Zool. 37:179-198.
624. Bartke, A & Parkening, TA. 1981. Effects of short photoperiod on pituitary and testicular function in the Chinese hamster, *Cricetulus griseus*. Biol Reprod. 25:958-962.
625. Bartlett, RA. 1927. Newfoundland seals. J Mamm. 8:207-212.
626. Bartmann, H & Bartmann, W. 1977. Gelungene künstliche Aufzucht eines Kleinen Pandas (*Ailurus fulgens*) im Zoologischen Garten Dortmund. Fr Kölner Zoo. 20:107-112.
627. Bartmann, W. 1971. Superfetation beim Virginia-Hirsch (*Odocoileus virginianus* Zimmermann 1780)?. Z Säugetierk. 36:200-201.
628. Bartmann, W. 1983. Haltung und Zucht von groszen Ameisenbären, *Myrmecophaga tridactyla* Linné, 1758, im Dordtmunder Tierpark. Zool Gärten NF. 53:1-31.
629. Bartness, TJ, Wake, GN, & Goldman, BD. 1987. Are the short-photoperiod-induced decreases in serum prolactin responsible for the seasonal changes in energy balance in Syrian and Siberian hamsters. J Exp Zool. 244:437-454.
630. Barton, AS. 1823. Facts, observations and conjectures, relative to the generation of the opossum of North America In a letter from Prof Barton to Mons Roume, of Paris. Ann Philos (new ser 6). 22:349-354.
631. Bartoš, L. 1982. Reproductive and social aspects of the behaviour of "white" red deer. Säugetierk Mitt. 30:89-117.
632. Baruś, Š & Zejda, J. 1981. The European otter (*Lutra lutra* in the Czech Socialist Republic. Acta Sci Nat Brno. 15:1-41.
633. Basch, WSB. 1964. Gestation period of impala. Afr Wildl. 18:162.
634. Bashenina, NV. 1969. Investigation of *Clethrionomys frater* Thomas, 1908 under laboratory conditions. Vestn Zool (Kiev). 5:11-21.
635. Bashkirov, IS & Zharkov, IV. 1934. On the biology and trapping of the mole in Tartary. Uchenye Zapiski Kazan Univ. 94:1-66.
636. Basrur, PK & Ramos, AS. 1973. Seasonal changes in the accessory sex glands and gonaducts of male mink. Can J Zool. 51:1125-1132.
637. Bassett, CF & Llewellyn, LM. 1946. Timing fox matings for maximum production. Am Fur Breeder. 19(6):46,48,50,52.
638. Bassett, CF & Llewellyn, LM. 1947. The effect of duration and interruption of coupling on the production of young by silver fox vixen. Am Fur Breeder. 19(8):38,40.
639. Bassett, J, Schultz, M, Stamps, L, et al. 1983. Birth of triplets in the pallid bat, *Antrozous pallidus*. Bat Res News. 24:2-4.
640. Bassett, JE. 1984. Litter size and postnatal growth rate in the pallid bat, *Antrozous pallidus*. J Mamm. 65:317-319.
641. Bast, H. 1931. Einige Beobachtungen an Haselmäusen. Z Säugetierk. 6:239-240.
642. Bast, JD & Greenwald, GS. 1974. Daily concentrations of gonadotrophins and prolactin in the serum of pregnant or lactating hamsters. J Endocrinol. 63:527-532.

643. Batchelor, R. 1963. Evidence of yearling pregnancies in the Roosevelt elk. J Mamm. 44:111-112.
644. Batchelor, RF. 1965. The Roosevelt elk in Alaska: Its Ecology and Management. AK Dept Fish Game, Juneau.
645. Bate, DMA. 1905. On the mammals of Crete. Proc Zool Soc London. 1905:315-323.
646. Bateman, JA. 1960. Observations on young mole rats. Afr Wildl. 14:227-234.
647. Bates, GL. 1905. Notes on the mammals of southern Cameroons and the Benito. Proc Zool Soc London. 1:65-85.
648. Bates, PC, Bigger, TRL, Hulse, EV, et al. 1972. The management of a small colony of the marsupial *Potorous tridactylus*, and a record of its breeding in captivity. Lab Anim. 6:301-313.
649. Batra, SK, Pahwa, GS, & Pandey, RS. 1982. Hormonal milieu around parturition in buffalos (*Bubalus bubalis*). Biol Reprod. 27:1055-1061.
650. Batra, SK & Pandey, RS. 1983. Prostaglandin $F_{2\alpha}$ in blood and milk on non-pregnant and pregnant buffalos (*Bubalus bubalis*). Acta Endocrinol. 102:314-320.
651. Batta, SK, Stark, RA, & Brackett, BG. 1978. Ovulation induction by gonadotropin and prostaglandin treatments of rhesus monkeys and observations of the ova. Biol Reprod. 18:264-278.
652. Batten, P & Batten, A. 1966. Notes on breeding the small-toothed palm civet *Arctogalida trivirgata* at Santa Cruz Zoo. Int Zoo Yearb. 6:172-173.
653. Batzli, GO. 1975. The role of small mammals in arctic ecosystems. Small Mammals their Productivity and Population Dynamics. 243-268, Golley, FB, Petrusewicz, K, & Ryszkowski, L, eds, Cambridge Univ Press, Cambridge.
654. Batzli, GO. 1977. Population dynamics of the white-footed mouse in floodplain and upland forests. Am Mid Nat. 97:18-32.
655. Batzli, GO, Stenseth, NC, & Fitzgerald, BM. 1974. Growth and survival of suckling brown lemmings, *Lemmus trimucronatus*. J Mamm. 55:828-831.
656. Bauchot, R & Stephan, H. 1964. Le poids encéphalique chez les Insectivores Malgaches. Acta Zool. (Stockholm) 45:63-76.
657. Bauer, JJ. 1985. Fecundity patterns of stable and colonising populations of New Zealand and Europe. The Biology and Management of Mountain Ungulates. 155-165, Lovari, S, ed, Croom Helm, Dover, NH.
658. Bauer, JJ. 1987. Factors determining the onset of sexual maturity in New Zealand chamois (*Rupicapra rupicapra* L). Z Säugetierk. 52:116-125.
659. Bauer, K. 1953. Zur Kenntnis von *Microtus oeconomus mehelyi* EHIK. Zool Jb. 82:70-94.
660. Bauer, K. 1954. Die Streifenmaus (*Sicista subtilis trizona* Petenyi) in Österreich. Zool Anz. 152:206-213.
661. Baulu, J. 1976. Seasonal sex skin coloration and hormonal fluctuations in free-ranging and captive monkeys. Horm Behav. 7:481-494.
662. Baum, MJ, Gallagher, CA, Martin, JT, et al. 1982. Effects of testosterone, dihydrotestosterone, or estradiol administered neonatally on sexual behavior of female ferrets. Endocrinology. 111:773-780.
663. Baum, MJ, Keverne, EB, Everitt, BJ, et al. 1977. Effects of progesterone and estradiol on sexual attractivity of female rhesus monkeys. Physiol Behav. 18:659-670.
664. Baum, MJ, Lynch, HJ, Gallagher, CA, et al. 1986. Plasma and pineal melatonin levels in female ferrets housed under long or short photoperiods. Biol Reprod. 34:96-100.
665. Baum, MJ & Schretlen, PJM. 1978. Oestrogenic induction of sexual behaviour in ovariectomized ferrets housed under short or long photoperiods. J Endocrinol. 79:295-296.
666. Baumgardner, GD. 1991. *Dipodomys compactus*. Mamm Species. 369:1-4.
667. Baumgardner, DJ & Dewsbury, DA. 1979. Copulatory behavior of *Calomys callosus*. Bull Psychon Soc. 14:127-128.
668. Baumgartner, LL & Bellrose, FC, Jr. 1943. Determination of sex and age in muskrats. J Wildl Manage. 7:77-81.
669. Baverstock, PR, Spencer, L, & Pollard, C. 1976. Water balance of small lactating rodents - II concentration and composition of milk of females on *ad libitum* and restricted water intakes. Comp Biochem Physiol. 53A:47-52.
670. Baverstock, PR & Watts, CHS. 1975. Water balance of small lactating rodents I *Ad libitum* water intakes and effects of water restriction on growth of young. Comp Biochem Physiol. 50A:819-825.
671. Bawdon, ED. 1915. A hermaphroditic kangaroo rat. J Mamm. 46:684.
672. Baxter, JS. 1935. On the female genital tract in the Caenolestids (Marsupialia). Proc Zool Soc London. 1935:157-162.
673. Baxter, RM & Lloyd, CNV. 1980. Notes on the reproduction and postnatal development of the forest shrew. Acta Theriol. 25:31-38.
674. Bayer, RD. 1985. Six years of harbor seal censusing at Yaquina Estuary, Oregon. Murrelet. 66:44-49.
675. Bayevsky, YB. 1956. Changes in fertility of the Barguzin sable. Bull Soc Nat Moscow, Sec Biol. 61(6):15-26.
676. Beach, FA. 1939. Maternal behavior of the pouchless marsupial *Marmosa cinerea*. J Mamm. 20:315-322.
677. Beacham, TD. 1980. Survival of cohorts in a fluctuating population of the vole *Microtus townsendii*. J Zool. 191:49-60.
678. Beacham, TD. 1981. Some demographic aspects of dispensers in fluctuating populations of the vole *Microtus townsendii*. Oikos. 36:273-280.
679. Beamer, N, Hagemenas, F, & Kittinger, GW. 1972. Protein binding of cortisol in the rhesus monkey (*Macaca mulatta*). Endocrinology. 90:325-327.
680. Bear, GD & Hansen, RM. 1966. Food habits, growth and reproduction of white-tailed jackrabbits in southern Colorado. CO Agric Exp Stn Bull. 90:1-59.
680a. Beard, LA, Grigg, GC, & Augee, ML. 1992. Reproduction by echidnas in a cold climate. Platypus and Echidnas. 93-100, Augee, ML, ed, R Zool Soc NSW, Sydney.
681. Bearder, SK & Doyle, GA. 1974. Ecology of bushbabies *Galago senegalensis* and *Galago crassicaudatus*, with some notes on their behaviour in the field. Prosimian Biology. 109-130, Martin, RD, Doyle, GA, & Walker, AC, Pittsburgh Univ Press, Pittsburgh.
682. Beasley, LE & Getz, LL. 1986. Comparison of demography of sympatric populations of *Microtus ochrogaster* and *Synaptomys cooperi*. Acta Theriol. 31:385-400.
683. Beasley, LJ. 1986. Seasonal cycles of pallid bats (*Antrozous pallidus*): Proximate factors. Myotis. 23-24:115-123.
684. Beasley, LJ, Johnston, PG, & Zucker, I. 1981. Photoperiodic regulation of reproduction in post partum *Peromyscus leucopus*. Biol Reprod. 24:962-966.
685. Beasley, LJ & Leon, M. 1986. Metabolic strategies of pallid bats (*Antrozous pallidus*) during reproduction. Physiol Beh. 36:159-166.
686. Beasley, LJ, Smale, L, & Smith, ER. 1984. Melatonin influences the reproductive physiology of male pallid bats. Biol Reprod. 30:300-305.

687. Beasley, LJ & Smith, ER. 1983. "Circannual" cycles of testosterone in *Antrozous pallidus*. Bat Res News. 24:49-50.
688. Beasley, LJ & Zucker, I. 1984. Photoperiod influences the annual reproductive cycle of the male pallid bat (*Antrozous pallidus*). J Reprod Fert. 70:567-573.
689. Beattie, CW & Bullock, BC. 1978. Diurnal variation of serum androgen and estradiol-17β in the adult male green monkey (*Cercopithecus* sp). Biol Reprod. 19:36-39.
690. Beattie, IA. 1972. Some problems of breeding baboons under laboratory conditions. Breeding Primates. 48-54, Beveridge, WIB, ed, Karger, Basel.
690a. Beatty, LD. 1955. The leafchin bat in Arizona. J Mamm. 36:290.
691. Beauregard, H & Boulart, ni. 1882. Appareils génito-urinaires des Balaenidés. J Anat Physiol. 18:158-201.
692. Becher, H. 1921. Der feinere Bau der reifen Placenta von Aguti (*Dasyprocta azarae* Schl). Z Anat Entw-Gesch. 61:439-454.
693. Becher, H. 1921. Die Entwicklung des Mesoplacentariums und die Placenta bei Aguti (*Dasyprocta azarae* Schl). Z Anat Entw-Gesch. 61:337-364.
694. Becher, H. 1921. Zur Kenntnis der Placenta von *Bradypus tridactylus*. Z Anat Entw-Gesch. 61:114-136.
695. Becher, H. 1931. Placenta und Uterusschleimhaut von *Tamandua tetradactyla* (Myrmecophaga). Morphol Jb. 67:381-485.
696. Beck, AJ & Lim, BL. 1973. Reproductive biology of *Eonycteris spelaea*, Dobson (Megachiroptera) in west Malaysia. Acta Trop. 30:251-260.
697. Beck, BB. 1984. The birth of a lowland gorilla in captivity. Primates. 25:378-383.
698. Beck, BB, Anderson, D, Ogden, J, et al. 1982. Breeding the Goeldi's monkey *Callimico goeldii* at Brookfield Zoo, Chicago. Int Zoo Yearb. 22:106-114.
699. Beck, F. 1974. The development of a maternal pregnancy reaction in the ferret. J Reprod Fert. 40:61-69.
700. Beck, U. 1972. Über die künstliche Aufzucht von Borstengürteltieren (*Euphractus villosus*). Zool Gärten NF. 41:215-222.
701. Beck, W & Wuttke, W. 1979. Annual rhythm of luteinizing hormone, follicle stimulating hormone, prolactin and testosterone in the serum of male rhesus monkeys. J Endocrinol. 83:131-139.
702. Becker, K. 1967. Populationsstudien an Bisamratten (*Ondatra zibethicus* L) I Altersaufbau von Populationen der Bisamratte. Zool Beitr. 13:369-396.
703. Becker, K. 1969. Populationsstudien an Bisamratten (*Ondatra zibethicus* L) II Geschlechterverhältnis. Zool Beitr. 15:363-373.
704. Becker, K. 1970. Populationsstudien an Bisamratten (*Ondatra zibethicus* L) III Fortpflanzungszyklus der Männchen. Z Angew Zool. 57:211-227.
705. Becker, K. 1973. Populationsstudien an Bisamratten (*Ondatra zibethicus* L) IV Fortpflanzungsbiologie der Weibchen. Z Angew Zool. 60:343-363.
706. Beddard, FE. 1902. Report on the birth of an Indian elephant in the society's menagerie. Proc Zool Soc London. 1902:320-323.
707. Beddard, FE & Treves, F. 1890. On the anatomy of the Sondaic rhinoceros. Trans Zool Soc London. 12:183-198.
708. Bedford, JM & Millar, RP. 1978. The character of sperm maturation in the epididymis of the ascrotal hyrax, *Procavia capensis* and armadillo, *Dasypus novemcinctus*. Biol Reprod. 19:396-406.
709. Bedford, JM & Rifkin, JM. 1979. An evolutionary view of male reproductive tract and sperm maturation in a monotreme mammal: The echidna *Tachyglossus aculeatus*. Am J Anat. 156:207-230.
710. Bedford, JM, Rodger, JC, & Breed, WG. 1984. Why so many spermatozoa - A clue from marsupials?. Proc R Soc London Biol Sci. 221B:221-233.
711. Bediz, GM & Whitsett, JM. 1979. Social inhibition of sexual maturation in male prairie deer mice. J Comp Physiol Physch. 93:493-500.
712. Bedrak, E. 1972. Enzymes of androgen biosynthesis in the desert mouse *Gerbillus dasyurus*. Gen Comp Endocrinol. 18:524-533.
713. Bedrak, E, Rosenstrauch, A, Kafka, M, et al. 1983. Testicular steroidogenesis in the camel (*Camelus dromedarius*) during the mating and the nonmating seasons. Gen Comp Endocrinol. 52:255-264.
714. Bedrak, E, Samoiloff, V, & Finkelstein, Z. 1971. Testosterone biosynthesis in the desert mouse, *Acomys cahirinus*. J Endocrinol. 51:7-16.
715. Bedrak, E, Samoiloff, V, & Shachak, M. 1973. Androgen biosynthesis by testicular tissue of the desert rodents, *Acomys caharinus* and *Gerbillus dasyurus*. J Reprod Fert. 34:93-104.
716. Bee, JW & Hall, ER. 1956. Mammals of northern Alaska. Univ KS Mus Nat Hist Misc Publ. 8:1-309.
717. Beebe, W. 1926. The three-toed sloth *Bradypus cuculliger cuculliger*. Zoologica. 7:1-67.
718. Beeck, DM. 1955. Observations on the birth of the grey kangaroo (*Macropus ocydromus*). W Aust Nat. 5:9.
719. Beer, JR. 1949. Studies on reproduction and survival in Wisconsin muskrats. Ph D Diss. Univ WI, Madison.
720. Beer, JR. 1950. The reproductive cycle of the muskrat in Wisconsin. J Wildl Manage. 14:151-156.
721. Beer, JR & MacLeod, CF. 1961. Seasonal reproduction in the meadow vole. J Mamm. 42:483-489.
722. Beer, JR & MacLeod, CF. 1966. Seasonal population changes in the prairie deer mouse. Am Mid Nat. 76:277-289.
723. Beer, JR, MacLeod, CF, & Frenzel, LD. 1957. Prenatal survival and loss in some cricetid rodents. J Mamm. 38:392-402.
723a. Beer, JR & Meyer, RK. 1951. Seasonal changes in the endocrine organs and behavior patterns of the muskrat. J Mamm. 32:173-191.
724. Beers, PC & Wittliff, JL. 1973. Identification and partial characterization of specific estrogen receptors in the vole (*Microtus montanus*). Comp Biochem Physiol. 46B:647-652.
725. Beg, MA. 1971. Population dynamics of the red-tailed chipmunk (*Eutamias ruficaudus*), in western Montana. Pak J Zool. 3:133-145.
726. Beg, MA. 1971. Reproductive cycle and reproduction in the red-tailed chipmunk *Eutamias ruficaudus*. Pak J Zool. 3:1-13.
727. Beg, MA. 1972. Seasonal changes in body weight of the red-tailed chipmunk, *Eutamias ruficaudus*. Pak J Zool. 4:13-16.
728. Beg, MA, Adeeb, N, & Rana, SA. 1981. Observations on reproduction in *Bandicota bengalensis* and *Nesokia indica*. Biologia. 27:45-50.
729. Beg, MA & Ajmal, M. 1977. Reproduction in the Indian gerbil *Tatera indica indica* (Hardwicke). Mammalia. 41:213-220.
730. Beg, MA & Rana, SA. 1978. Ecology of the field rat, *Rattus meltada pallidior*, in central Punjab, Pakistan. Pak J Zool. 10:163-168.
731. Begg, RJ. 1981. The small mammals of Little Nourlangie Rock, NT II Ecology of *Antechinus bilarni*, the sandstone antechinus (Marsupialia: Dasyuridae). Aust Wildl Res. 8:57-72.

732. Begg, RJ. 1981. The small mammals of Little Nourlangie Rock, NT III Ecology of *Dasyurus hallucatus*, the northern quoll (Marsupialia: Dasyuridae). Aust Wildl Res. 8:73-85.
733. Begg, RJ. 1981. The small mammals of Little Nourlangie Rock, NT IV Ecology of *Zyzomys woodwardi*, the large rock-rat, and *Z argurus*, the common rock-rat, (Rodentia: Muridae). Aust Wildl Res. 8:307-320.
734. Begg, RJ, Martin, KC, & Price, NF. 1981. The small mammals of Little Nourlangie Rock, NT: V The effects of fire. Aust Wildl Res. 8:515-527.
735. Begg, RJ, Walsh, B, Woerle, F, et al. 1983. Ecology of *Melomys burtoni*, the grassland melomys (Rodentia: Muridae) at Cobourg Peninsula, NT. Aust Wildl Res. 10:259-267.
736. Behm, A. 1933. Zucht von Luchsen in Gefangenschaft. Zool Garten NF. 6:196.
737. Behrends, PR. 1981. Copulatory behavior of *Dipodomys microps* (Heteromyidae). Southwest Nat. 25:262-263.
738. Beidleman, RG. 1954. October breeding of *Peromyscus* in north central Colorado. J Mamm. 35:118.
739. Beier, JC & Wartzok, D. 1979. Mating behaviour of captive spotted seals (*Phoca largha*). Anim Behav. 27:772-781.
740. Beischer, DE & Furry, DE. 1964. *Saimiri sciureus* as an experimental animal. Anat Rec. 148:615-624.
741. Bejček, V. 1979. Notes on population dynamics of the common vole, *Micrtous arvalis* (Pall, 1778) and the wood mouse, *Apodemus sylvaticus* (L, 1758) on spoil banks after surface mining of lignite in the most basin (northwestern Bohemia). Lynx. 20:3-24.
742. Bekenov, A & Mirzabekov, J. 1977. Reproduction of *Allactaga elater* in north Kyzylkum and Ustyurt. Zool Zhurn. 56:769-778.
743. Bekoff, M. 1975. Social behavior and ecology of the African Canidae: A review. The Wild Canids. 120-142, Fox, MW, ed, Van Nostrand Reinhold Co, New York.
744. Bekoff, M. 1977. *Canis latrans*. Mamm Species. 79:1-9.
745. Bekoff, M, Diamond, J, & Mitton, JB. 1981. Life history patterns and sociality in canids: Body size, reproduction, and behavior. Oecologia. 50:386-390.
746. Bekoff, M & Jamieson, R. 1975. Physical development in coyotes (*Canis latrans*), with a comparison to other canids. J Mamm. 56:685-692.
747. Bekoff, M & Wells, MC. 1982. Behavioral ecology of coyotes: Social organization, rearing patterns, space use, and resource defense. Z Tierpsychol. 60:281-305.
748. Belanger, C, Shome, B, Friesen, H, et al. 1971. Studies of the secretion of monkey placental lactogen. J Clin Invest. 50:2660-2667.
749. Belk, MC & Smith, HD. 1991. *Ammospermophilus leucurus*. Mamm Species. 368:1-8.
750. Bell, BD. 1981. Breeding and condition of possums *Trichosurus vulpecula* in Orongorongo Valley, near Wellington, New Zealand, 1966-1975. Zool Publ Victoria Univ (Wellington). 74:87-139.
751. Bell, ET, Loraine, JA, Jennings, S, et al. 1967. Serum and urinary gonadotrophin levels in pregnant ponies and donkeys. Q J Exp Physiol. 52:68-75.
751a. Bell, FE & Dawson, WD. 1983. Comparative progesterone concentrations in two *Peromyscus* species. Comp Biochem Physiol. 74:703-708.
752. Bell, J. 1977. Breeding season and fertility of the wild rabbit, *Oryctolagus cuniculus* (L) in North Canterbury, New Zealand. Proc NZ Ecol Soc. 24:79-83.
752a. Bellier, L. 1968. Contribution a l'étude d'*Uranomys ruddi* Dollman. Mammalia. 32:419-446.
753. Belopolsky, LO. 1939. Some materials on the propagation of the Pacific walrus *Odobenus rosmarus divergens* (Ill). Zool Zhurn. 18:762-778.
754. Bels, L. 1952. Fifteen years of bat banding in the Netherlands. Publ Natuurhist Genotsch Limburg. 5:1-99.
755. Belyaev, DK & Evsikov, VI. 1967. Genetics and the fertility of animals I Effects of the fur color genes in the fertility of mink (*Lutreola vison* Bresson). Genetika. 2:21-33. (Biol Abs 47 ref 11103, 1968).
756. Belyaev, DK, Pereldik, NS, & Portnova, NT. 1951. Experimental reduction of the period of embryonal development in sables (*Martes zibellina*). Z Obsc Biol. 12:260-265.
757. Ben-Yaacov, R & Yom-Tov, Y. 1983. On the biology of the Egyptian mongoose, *Herpestes ichneumon*, in Israel. Z Säugetierk. 48:34-45.
758. Bendell, JF. 1959. Food as a control of a population of white-footed mice, *Peromyscus leucopus noveboracensis* (Fischer). Can J Zool. 37:173-209.
759. Benecke, B. 1879. Über Reifung und Befruchtung des Eies bei den Fledermäusen. Zool Anz. 2:304-305.
760. Beneski, JT, Jr & Stinson, DW. 1987. *Sorex palustris*. Mamm Species. 296:1-6.
761. Bengtson, JL & Siniff, DB. 1981. Reproductive aspects of female crabeater seals (*Lobodon carcinophagus*) along the Antarctic peninsula. Can J Zool. 59:92-102.
762. Benirschke, K. 1968. Why armadillos?. Animal Models for Biomedical Research. 45-54, Nat Acad Sci, Washington, DC.
763. Benirschke, K, Anderson, JM, & Brownhill, LE. 1962. Marrow chimerism in the marmoset. Science. 138:513-515.
764. Benirschke, K, Byrd, ML, & Meritt, D. 1990. New observations on the Chacoan peccary, *Catagonus wagneri*. Int Symp Erkrankungen Zoo-und Wildtiere. 32:341-347.
765. Benirschke, K & Cornell, LH. 1987. The placenta of the killer whale, *Orcinus orca*. Mar Mamm Sci. 3:82-86.
766. Benirschke, K, Johnson, ML, & Benirschke, RJ. 1980. Is ovulation in dolphins, *Stenella longirostris* and *Stenella attenuata*, always copulation-induced?. Fish Bull USA. 78:507-522.
767. Benirschke, K & Layton, W. 1969. An early twin blastocyst of the golden lion marmoset, *Leontocebus rosalia*. Folia Primatol. 10:131-138.
768. Benirschke, K & Miller, CJ. 1981. Weights and neonatal growth of ring-tailed lemurs (*Lemur catta*) and ruffed-lemurs (*Lemur variegatus*). J Zoo Anim Med. 12:107-111.
769. Benirschke, K & Miller, CJ. 1982. Anatomical and functional differences in the placenta of primates. Biol Reprod. 26:29-53.
770. Benirschke, K & Powell, HC. 1985. On the placentation of sloths. The Evolution and Ecology of Armadillos, Sloths, and Vermilinguas. 237-241, Montgomery, GG, ed, Smithsonian Inst Press, Washington, DC.
770a. Benirschke, K & Richart, R. 1963. The establishment of a marmoset breeding colony and its four pregnancies. Lab Anim Care. 13:70-83.
771. Benjaminsen, T. 1972. On the biology of the bottlenose whale, *Hyperoodon ampullatus* (Foster). Norw J Zool. 20:233-241.
772. Benjaminsen, T. 1973. Age determination and the growth and age distribution from cementum layers of bearded seals at Svalbard. Fiskerdir Skr Ser Havunders. 16:159-170.
773. Benjaminsen, T & Christensen, I. 1979. The natural history of the bottlenose whale, *Hyperoodon ampullatus* (Forster). Behaviour of Marine Animals. 3:143-164, Winn, HE & Olla, BL, eds, Plenum Press, New York.

774. Bennett, EL, Davison, GWH, & Kavanagh, M. 1983. Social change in a family of siamang (*Hylobates syndactylus*). Malay Nat J. 36:187-196.
775. Bennett, G. 1835. Notes on the natural history and habits of the *Ornithorhynchus paradoxus*, Blum. Trans Zool Soc London. 1:229-258.
776. Bennett, GF. 1877. Notes on *Ornithorhynchus paradoxus*. Proc Zool Soc London. 1877:161-166.
777. Bennett, GJ. 1881. Observations on the habits of the *Echidna hystrix* of Australia. Proc Zool Soc London. 1881:737-739.
778. Bennett, JH, Smith, MJ, Hope, RM, et al. 1982. Fat-tailed dunnart *Sminthopsis crassicaudata*: Establishment and maintenance of a laboratory colony. The Management of Australian Mammals in Captivity. 38-44, Evans, DD, ed, Zool Bd Victoria, Melbourne.
778a. Bennett, NC. 1989. The social structure and reproductive biology of the common mole-rat *Cryptomys h hottentotus* and remarks on the trends in reproduction and sociality in the family Bathyergidae. J Zool. 219:45-49.
778b. Bennett, NC. 1990. Behavior and social organization in a colony of the Damaraland mole-rat *Cryptomys damarensis*. J Zool. 220:225-248.
779. Bennett, NC & Jarvis, JUM. 1988. The reproductive biology of the Cape mole-rat, *Georychus capensis* (Rodentia, Bathyergidae). J Zool. 214:95-106.
780. Bennett, NC & Jarvis, JUM. 1988. The social structure and reproductive biology of colonies of the mole-rat, *Cryptomys damarensis* (Rodentia, Bathyergidae). J Mamm. 69:293-302.
781a. Benson, BN, Binz, H, & Zimmermann, E. 1992. Vocalizations of infant and developing tree shrews (*Tupaia belangeri*). J Mamm. 73:106-119.
781. Benson, SB. 1940. Notes on the pocketed free-tailed bat. J Mamm. 21:26-29.
782. Benson, SB & Borell, AE. 1931. Notes on the life history of the red tree mouse, *Phenacomys longicaudus*. J Mamm. 12:226-233.
783. Benton, AH. 1955. Observations on the life history of the northern pine mouse. J Mamm. 36:52-62.
784. Benton, AH & Scharoun, J. 1958. Notes on a breeding colony of *Myotis*. J Mamm. 39:293-295.
785. Benz, M. 1973. Zum Sozialverhalten der Sasin (Hirschziegenantilope, *Antilope cervicapra* L. 1758). Zool Beitr NF. 19:403-466.
786. Berard, EV. 1952. Evidence of a late birth for the raccoon. J Mamm. 33:247-248.
787. Berestennikov, DS. 1971. On some peculiarities of *Ondatra zibethicus* reproduction in the low Dnieper area. Vestn Zool (Kiev). 2:43-47.
788. Berger, J. 1982. Female breeding age and lamb survival in desert bighorn sheep (*Ovis canadensis*). Mammalia. 46:183-190.
789. Berger, J. 1983. Induced abortion and social factors in wild horses. Nature. 303:59-61.
789a. Berger, J. 1991. Pregnancy incentives, predation constraints and habitat shifts: Experimental and field evidence for wild bighorn sheep. Anim Beh. 41:61-77.
790. Berger, ME. 1972. Live weights and body measurements of olive baboons (*Papis anubis*) in Laikipia district of Kenya. J Mamm. 53:404-406.
791. Berger, PJ. 1966. Eleven-months embryonic diapause in a marsupial. Nature. 211:435-436.
792. Berger, PJ. 1970. The reproductive biology of the tammar wallaby, *Macropus eugenii* (Demarest (Marsupialia). Diss Abstr. 31:3760b-3761b.
793. Berger, PJ & Negus, NC. 1974. Influence of dietary supplements of fresh lettuce on ovariectomied *Microtus montanus*. J Mamm. 55:747-750.
794. Berger, PJ & Negus, NC. 1982. Stud male maintenance of pregnancy in *Microtus montanus*. J Mamm. 63:148-151.
795. Berger, PJ, Negus, NC, & Rowsemitt, CN. 1987. Effect of 6-methoxybenzoxazolinone on sex ratio and breeding performance in *Microtus montanus*. Biol Reprod. 36:255-260.
796. Berger, PJ, Negus, NC, Sanders, EH, et al. 1981. Chemical triggering of reproduction in *Microtus montanus*. Science. 214:69-70.
797. Berger, PJ, Sanders, EH, Gardner, PD, et al. 1977. Phenolic plant compounds functioning as reproductive inhibitors in *Microtus montanus*. Science. 195:575-577.
798. Berger, PJ & Sharman, GB. 1969. Embryonic diapause initiated without the suckling stimulus in the wallaby, *Macropus eugenii*. J Mamm. 50:630-632.
799. Berger, PJ & Sharman, GB. 1969. Progesterone-induced development of dormant blastocysts in the tammer wallaby, *Macropus eugenii* Desmarest; Marsupialia. J Reprod Fert. 20:201-210.
800. Bergerud, AT. 1961. The reproductive season of Newfoundland caribou. NE Wildl Conf. 1961:1-31.
801. Bergerud, AT. 1971. The population dynamics of Newfoundland caribou. Wildl Monogr. 25:1-55.
802. Bergerud, AT. 1973. Movement and rutting behavior of caribou (*Rangifer tarandus*) at Mount Albert, Quebec. Can Field-Nat. 87:357-369.
803. Bergerud, AT. 1974. Rutting behaviour of Newfoundland caribou. The Behaviour of Ungulates and its Relation to Management. 395-435, Geist, V & Walther, F, eds, IUCN, Morges, Switzerland.
804. Bergerud, AT. 1975. The reproductive season of Newfoundland caribou. Can J Zool. 53:1213-1221.
805. Bergerud, AT. 1976. The annual antler cycle in Newfoundland caribou. Can Field-Nat. 90:449-463.
806. Bergerud, AT. 1978. Caribou. Big Game of North America. 83-101, Schmidt, JL & Gilbert, DL, eds, Stackpole Books, Harrisburg, PA.
807. Bergin, TJ. The Koala. Zool Parks Bd NSW. NSW, Australia.
807a. Bergman, HC & Housley, C. 1968. Chemical analysis of American opossum (*Didelphys virginiana*) milk. Comp Biochem Physiol. 25:213-218.
807b. Bergmans, W. 1975. A new species of *Dobsonia* Palmer, 1898 (Mammalia, Megachiroptera) from Waigeo, with notes on other members of the genus. Beaufortia. 23:1-13.
807c. Bergmans, W. 1976. A revision of the African genus *Myonycteris* Matschie, 1899 (Mammalia, Megachiroptera). Beaufortia. 24:189-216.
808. Bergmans, W. 1979. Taxonomy and zoogeography of the fruit bats of the People's Republic of Congo, with notes on their reproductive biology (Mammalia, Megachiroptera). Bijdr Dierk. 48:161-186.
809. Bergmans, W & Rozendaal, FG. 1988. Notes on collections of fruit bats from Sulawesi and some off-lying islands (Mammalia, Megachiroptera). Zool Verh. 248:1-74.
810. Bergstedt, B. 1965. Distribution, reproduction, growth and dynamics of the rodent species *Clethrionomys glareolus* (Schreber), *Apodemus flavicollis* (Melchior) and *Apodemus sylvaticus* (Linné) in southern Sweden. Oikos. 16:132-160.
811. Bergström, U. 1968. Observations on Norwegian lemmings (*Lemmus lemmus*) in the autumn of 1963 and the spring of 1964. Ark Zool. 20:321-363.

812. Berkelbach van den Sprenkel, H. 1932. Persistenz der Dottergefäsze in den Embryonen der Fledermäuse und ihre Ursache. Z Mikr Anat Forsch. 28:185-268.
813. Berkson, G. 1967. Producing and hand-rearing kittens. Lab Anim Care. 17:365-378.
814. Berkson, G. 1968. Weight and tooth development during the first year in *Macaca irus*. Lab Anim Care. 18:352-355.
815. Berkson, G & Chaicumpa, V. 1969. Breeding gibbons (*Hylobates lar entelloides*) in the laboratory. Lab Anim Care. 19:808-811.
816. Berliner, AF & Jones-Witters, P. 1975. Early effects of a lethal cadmium dose on gerbil testis. Biol Reprod. 13:240-247.
817. Bernard, HJ & Hohn, AA. 1989. Differences in feeding habits between pregnant and lactating spotted dolphins (*Stenella attenuata*). J Mamm. 70:211-215.
818. Bernard, J. 1959. Notes sur la période de reproduction du Renard (*Vulpes vulpes* Linne, 1758) dans le Luxembourg belge. Säugetierk Mitt. 7:111-113.
819. Bernard, J. 1960. Note sur la reproduction en hiver du Campagnol des champs, *Microtus arvalis* (Pall). Z Säugetierk. 25:91-94.
820. Bernard, RTF. 1980. Female reproductive anatomy and development of ovarian follicles in *Miniopterus fraterculus*. S Afr J Zool. 15:111-116.
821. Bernard, RTF. 1980. Monthly changes in the reproductive organs of femle *Miniopterus schreibersi natalensis* (A Smith, 1834). Z Säugetierk. 45:217-224.
822. Bernard, RTF. 1980. Reproductive cycles of *Minioterus* [sic] *schreibersi natalensis* (Kuhl, 1819) and *Miniopterus fraterculus* Thomas and Schwann, 1906. Ann Transvaal Mus. 32:55-64.
823. Bernard, RTF. 1981. Changes in the oviduct of *Miniopterus fraterlucus* Thomas and Schwann, 1906, associated with reproduction. Säugetierk Mitt. 29:16-20.
824. Bernard, RTF. 1982. Female reproductive cycle of *Nycteris thebaica* (Microchiroptera) from Natal, South Africa. Z Säugetierk. 47:12-18.
825. Bernard, RTF. 1982. Monthly changes in the female reproductive organs and the reproductive cycle of *Myotis tricolor* (Vespertilionidae: Chiroptera). S Afr J Zool. 17:79-84.
826. Bernard, RTF. 1983. Reproduction of *Rhinolophus clivosus* (Microchiroptera) in Natal, South Africa. Z Säugetierk. 48:321-329.
827. Bernard, RTF. 1984. The occurrence of spermiophagy under natural conditions in the cauda epidiymidis of the Cape horseshoe bat (*Rhinolophus capensis*). J Reprod Fert. 71:539-543.
828. Bernard, RTF. 1985. Reproduction in the Cape horseshoe bat (*Rhinolophus capensis*) from South Africa. S Afr J Zool. 20:129-135.
829. Bernard, RTF. 1985. The occurence of abnormal sperm in the cauda epididymis of *Rhinolophus capensis* (Mammalia: Chiroptera). J Morphol. 183:177-183.
830. Bernard, RTF. 1986. Seasonal changes in plasma testosterone concentrations and Leydig cell and accessory gland activity in the Cape horseshoe bat (*Rhinolophus capensis*). J Reprod Fert. 78:413-422.
831. Bernard, RTF. 1988. Prolonged sperm storage in male Cape horseshoe bats: An alternative solution to the reproductive limitations of winter hibernation. Naturwissenschaften. 75:213-214.
831a. Bernard, RTF & Hodgson, AN. 1989. Ultrastructural changes in the seminiferous epithelium of two seasonally reproducing bats (Mammalia: Chiroptera). J Morphol. 199:249-258.
832. Bernard, RTF & Meester, JAJ. 1982. Female reproduction and the female reproductive cycle of *Hipposideros caffer caffer* (Sundevall, 1846) in Natal, South Africa. Ann Transvaal Mus. 33:131-144.
833. Bernard, RTF & Stuart, CT. 1987. Reproduction of the caracal *Felis caracal* from the Cape Province of South Africa. S Afr J Zool. 22:177-182.
834. Bernatskii, VG, Snytko, ÉG, & Nosova, HG. 1977. Natural and induced ovulation in the sable (*Martes zibellina* L). Dokl Akad Nauk SSSR Transl. 230:458-459.
835. Bernhardt, FW. 1961. Correlation between growth-rate of the suckling and percentage of total calories from protein in the milk. Nature. 191:358-360.
836. Berns, VD. 1969. Notes on the blue fox of Rat Island, Alaska. Can Field-Nat. 83:404-405.
837. Bernstein, IS. 1968. The lutong of Kuala Selangor. Behaviour. 32:1-16.
838. Bernstein, IS, Bruce, K, & Williams, L. 1982. The influence of male presence or absence on the reproductive cycle of Celebes black ape females (*Macaca nigra*). Primates. 23:587-591.
839. Bernstein, IS, Rose, RM, & Gordon, TP. 1977. Behavioural and hormonal responses of male rhesus monkeys introduced to females in the breeding and non-breeding seasons. Anim Behav. 25:609-614.
840. Berria, M, Joseph, MM, & Mead, RA. 1989. Role of prolactin and luteinizing hormone in regulating timing of implantation in the spotted skunk. Biol Reprod. 40:232-238.
841. Berrie, PM. 1978. Home range of a young female Geoffroy's cat in Paraguay. Carnivore. 1(1):132-133.
842. Berrukov, NI. 1968. On the morphology of ovary of the bactrian camel. Dokl Akad Nauk SSSR Biol Sci Transl. 179:202-205.
843. Berry, PSM. 1973. The Luangwa valley giraffe. Puku. 7:71-92.
844. Berry, RJ. 1968. The ecology of an island population of the house mouse. J Anim Ecol. 37:445-470.
845. Berry, RJ. 1970. The natural history of the house mouse. Field Studies. 3:219-262.
846. Berry, RJ, Evans, IM, & Sennitt, BFC. 1967. The relationships and ecology of *Apodemus sylvaticus* from the Small Isles of the Inner Hebrides, Scotland. J Zool. 152:333-346.
847. Berry, RJ, Peters, J, & van Aarde, RJ. 1978. Sub-antarctic house mice: Colonization, survival and selection. J Zool. 184:127-141.
848. Berta, A. 1982. *Cerdocyon thous*. Mamm Species. 186:1-4.
849. Berta, A. 1986. *Atelocynus microtis*. Mamm Species. 256:1-3.
850. Berthoud, G. 1980. Le Hérisson (*Erinaceus europaeus* L) et la route. Terre Vie. 34:361-372.
851. Bertram, B. 1983. Lorisids in captivity. Management of Prosimians and New World Primates, Proc Symp Assoc Br Wild Animal Keepers, 8. 35-38.
852. Bertram, BCR. 1975. Social factors influencing reproduction in wild lions. J Zool. 177:463-482.
853. Bertram, BCR. 1976. Kin selection in lions and in evolution. Growing points in Ethology. 281-301, Bateson, PPG & Hinde, RA, eds, Cambridge Univ Press, Cambridge.
854. Bertram, GCL. 1940. The biology of the Weddell and crabeater seals. British Graham Land Expedition, 1934-37, Sci Rep. 1:1-139.

855. Besrukov, NI. 1972. Hypophysis gonad correlation characteristics in the oestrous cycle process of Bactrian female camels. Int Congr Anim Reprod Artif Insem. 8:478.
856. Besson, J, Duguy, R, & Tardy, G. 1982. Note sur un cas de multiparité chez un rorqual commun (*Baleanoptera physalus*). Mammalia. 46:408.
857. Best, A. 1962. The Canadian otter *Lutra canadensis* in captivity. Int Zoo Yearb. 4:42-44.
858. Best, LW. 1973. Breeding season and fertility of the roof rat, *Rattus rattus rattus*, in two forest areas of New Zealand. NZ J Sci. 16:161-170.
859. Best, PB. 1960. Further information on Bryde's whale (*Balaenoptera edeni* Anderson) from Saldanha Bay, South Africa. Norsk Hvalfangst Tid. 49:201-215.
860. Best, PB. 1967. The sperm whale (*Physeter catodon*) off the west coast of South Africa 1 Ovarian changes and their significance. Div Sea Fish S Afr Invest Rep. 61:1-27.
861. Best, PB. 1968. The sperm whale (*Physeter catodon*) off the west coast of South Africa 2 Reproduction in the female. Div Sea Fish S Afr Invest Rep. 66:1-32.
862. Best, PB. 1969. The sperm whale (*Physeter catodon*) off the west coast of South Africa 3 Reproduction in the male. Div Sea Fish S Afr Invest Rep. 72:1-20.
863. Best, PB. 1969. The sperm whale (*Physeter catodon*) off the west coast of South Africa 4 Distribution and movements. Div Sea Fish S Afr Invest Rep. 78:1-12.
864. Best, PB. 1970. The sperm whale (*Physeter catodon*) off the west coast of South Africa 5 Age, growth and mortality. Div Sea Fish S Afr Invest Rep. 79:1-27.
865. Best, PB. 1970. Exploitation and recovery of right whales *Eubalaena australis* off the Cape Province. Div Sea Fish S Afr Invest Rep. 80:1-15.
866. Best, PB. 1977. Two allopatric forms of Bryde's whale off South Africa. Rep Int Whal Commn Sp Issue. 1:10-35.
867. Best, PB. 1980. Pregnancy rates in sperm whales off Durbau. Rep Int Whal Comm Sp Issue. 2:93-97.
868. Best, PB. 1981. The status of right whales (*Eubalaena glacialis*) off the coast of South Africa, 1969-1970. Div Sea Fish S Afr Invest Rep. 123:1-44.
869. Best, PB. 1982. Seasonal abundance, feeding, reproduction, age and growth in minke whales off Durban (with incidental observation from the Antarctic). Rep Int Whal Comm. 32:759-786.
869a. Best, PB. 1988. The external appearance of Heaviside's dolphin, *Cephalorhynchus heavisidii* (Gray, 1828). Rep Int Whal Comm Sp Issue. 9:279-299.
870. Best, PB & Bannister, JL. 1963. Functional polyovuly in the sei whale *Balaenoptea borealis* Lesson. Nature. 199:89.
871. Best, PB, Canham, PAS, & MacLeod, N. 1984. Patterns of reproduction in sperm whales *Physeter macrocephalus*. Rep Int Whal Commn Sp Issue. 6:51-79.
872. Best, RC. 1982. Seasonal breeding in the Amazonian manatee, *Trichechus inuguis* (Mammalia: Sirenia). Biotropica. 14:76-78.
873. Best, RC & Da Silva, VMF. 1984. Preliminary analysis of reproductive parameters of the boutu, *Inia geoffrensis*, and the tucuxi, *Sotalia fluviatilis*, in the Amazon river system. Rep Int Whal Comm Sp Issue. 6:361-369.
874. Best, RC & Fisher, HD. 1974. Seasonal breeding of the narwhal (*Monodon monoceros* L). Can J Zool. 52:429-431.
875. Best, RC, Ribeiro, GA, Yamakoshi, M, et al. 1982. Artificial feeding for unweaned Amazonian manatees *Trichechus inunguis*. Int Zoo Yearb. 22:263-267.
876. Best, TL. 1986. *Dipodomys elephantinus*. Mamm Species. 255:1-4.
877. Best, TL. 1988. *Dipodomys spectabilis*. Mamm Species. 311:1-10.
878. Best, TL. 1988. *Dipodomys nelsoni*. Mamm Species. 326:1-4.
879. Best, TL. 1991. *Dipodomys nitratoides*. Mamm Species. 381:1-7.
879a. Best, TL. 1992. *Dipodomys margaritae*. Mamm Species. 400:1-3.
880. Best, TL, Caesar, K, Titus, AS, et al. 1990. *Ammospermophilus insularis*. Mamm Species. 364:1-4.
881. Best, TL, Hildreth, NJ, & Jones, C. 1989. *Dipodomys deserti*. Mamm Species. 339:1-8.
882. Best, TL & Hoditschek, B. 1986. Relationships between environmental variation and the reproductive biology of Ord's kangaroo rat (*Dipodomys ordii*). Mammalia. 50:173-183.
883. Best, TL & Lackey, JA. 1985. *Dipodomys gravipes*. Mamm Species. 236:1-4.
884. Best, TL, Lewis, CL, Caesar, K, et al. 1990. *Ammospermophilus interpres*. Mamm Species. 365:1-6.
884a. Best, TL & Schnell, GD. 1974. Bacular variation in kangaroo rats (genus *Dipodomys*). Am Mid Nat. 91:257-270.
885. Best, TL & Thomas, HH. 1991. *Dipodomys insularis*. Mamm Species. 374:1-3.
886. Best, TL & Thomas, HH. 1991. *Spermophilus madrensis*. Mamm Species. 378:1-2.
887. Best, TL, Titus, AS, Caesar, K, et al. 1990. *Ammospermophilus harrisii*. Mamm Species. 366:1-7.
888. Best, TL, Titus, AS, Lewis, CL, et al. 1990. *Ammospermophilus nelsoni*. Mamm Species. 367:1-7.
889. Bester, MN. 1980. The southern elephant seal *Mirounga leonina* at Gough Island. S Afr J Zool. 15:235-239.
890. Bester, MN. 1982. Distribution, habitat selection and colony types of the Amsterdam Island fur seal *Arctocephalus tropicalis* at Gough Island. J Zool. 196:217-231.
891. Bester, MN & Kerley, GIH. 1983. Rearing of twin pups to weaning by subantarctic fur seal *Arctocephalus tropicalis* female. S Afr J Wildl Res. 13(3):86-87.
892. Beule, JD. 1940. Cottontail nesting-study in Pennsylvania. Trans N Am Wildl Conf. 5:320-328.
893. Beule, JD & Studholm, AT. 1942. Cottontail rabbit nests and nestlings. J Wildl Manag. 6:133-140.
894. Bever, K & Sprankel, H. 1986. A contribution to the longevity of *Tupaia glis* Diard, 1820 in captivity. Z Versuchstierk. 28:3-5.
895. Bhargava, KK, Sharma, VD, & Singh, M. 1963. A study of mortality rate, sex ratio and abortions in camel (*Camelus dromedarius*). Indian J Vet Sci Anim Husb. 33:187-188.
896. Bhargava, KK, Sharma, VD, & Singh, M. 1965. A study of birth-weight and body measurements of camel (*Camelus dromedarius*). Indian J Vet Sci Anim Husb. 35:358-362.
897. Bhat, HR & Sreenivasan, MA. 1981. Observations on the biology of *Hipposideros lankadiva* Kelaart, 1850 (Chiroptera, Rhinolophidae). J Bombay Nat Hist Soc. 78:436-442.
898. Bhat, HR, Sreenivasan, MA, & Jacob, PG. 1980. Breeding cycle of *Eonycteris spelaea* (Dobson, 1871) (Chiroptera, Pteropidae, Macroglossinae) in India. Mammalia. 44:343-347.
899. Bhat, SK, Sujatha, A, Advani, R, et al. 1987. Population structure and breeding season in *Rattus rattus wroughtoni* Hinton. Proc Indian Acad Sci (Anim Sci). 96:657-665.
900. Bhatia, CL & Desai, JH. 1972. Growth and development of a hand-reared kinkajou *Potos flavus* at Delhi Zoo. Int Zoo Yearb. 12:176-177.

901. Bhatia, CL & Desai, JH. 1975. Breeding the Indian rhinoceros in Delhi Zoological Park. Breeding Endangered Species in Captivity. 303-307, Martin, RD, ed, Academic Press, New York.
902. Bhatnagar, KP. 1978. Breech presentation in the hairy-legged vampire, *Diphylla ecaudata*. J Mamm. 59:864-866.
903. Bhatnagar, KP. 1978. Head presentation in *Artibeus jamaicensis* with some notes on parturition. Mammalia. 42:359-363.
904. Bhattacharya, AN, Dierschke, DJ, Yamaji, T, et al. 1972. The pharmacological blockade of the circhoral mode of LH secretion in the ovariectomized rhesus monkey. Endocrinology. 90:778-786.
905. Bhiwgade, DA. 1973. Development of the foetal membranes and placentation in the Indian horse-shoe bat, *Rhinolophus rouxi* Temminck. Ph D Diss. Fac Sci, Nagpur Univ, Nagpur.
906. Bhiwgade, DA. 1976. Observations on some early stages of development and implantation of the blastocyst of *Rhinolophus rouxi* (Temminck). Proc Indian Acad Sci. 84B:201-209.
907. Bhiwgade, DA. 1977. Development of the foetal membranes in the Indian horse-shoe bat, *Rhinolophus rouxi* (Temminck). Proc Indian Acad Sci. 86B:61-72.
908. Bhiwgade, DA. 1979. An analysis of implantation in Indian hipposiderid bats. J Anat. 128:349-364.
909. Biben, M. 1982. Ontogeny of social behaviour to feeding in the crab-eating fox (*Cerdocyon thous*) and the bush dog (*Speothos venaticus*). J Zool. 196:207-216.
910. Bibikov, DI. 1956. Biological observations on the hedgehog *Erinaceus auritus* GM as reservoirs of the leptospirae. Zool Zhurn. 35:1059-1063.
911. Bideau, E, Vincent, JP, & Maire, F. 1983. Evolution saisonniere de la taille des groupes chez le Chevreuil en milieu forestier. Terre Vie. 37:161-169.
912. Bidwai, P & Bawa, SR. 1971. Existence of degenerating cells in the seminiferous tubules of the normal adult and retrogressed testes of the Indian hedgehog, *Paraechinus micropus*. J Reprod Fert. 26:359-360.
913. Bielert, C. 1986. Sexual interactions between captive adult male and female chacma baboons (*Papio ursinus*) as related to the female's menstrual cycle. J Zool. 209A:521-536.
914. Bielert, C, Czaja, JA, Eisele, S, et al. 1976. Mating in the rhesus monkey (*Macaca mulatta*) after conception and its relationship to oestradiol and progesterone levels throughout pregnancy. J Reprod Fert. 46:179-187.
915. Bielert, C, Howard-Tripp, ME, & van der Walt, LA. 1981. Diurnal variations in serum testosterone concentrations of captive adult male chacma baboons (*Papio ursinus*). Am J Primatol. 1:421-425.
916. Bielert, C & van der Walt, LA. 1982. Male chacma baboon (*Papio ursinus*) sexual arousal: Mediation by visual cues from female conspecifics. Psychoneuroendocrinology. 7:31-48.
917. Bielert, C & Vandenbergh, JG. 1981. Seasonal influences on births and male sex skin coloration in rhesus monkeys (*Macaca mulatta*) in the southern hemisphere. J Reprod Fert. 62:229-233.
918. Bigalke, R. 1961. The size of the litter of the wild dog *Lycaon pictus* (Temminck). Fauna Flora (Pretoria). 12:9-16.
919. Bigalke, RC. Personal communication. [Cited by Mentis, 1972]
920. Bigalke, RC. 1947. Pretoria Zoo has a baby white rhinoceros. Animal Kingdom. 50:48-55.
921. Bigalke, RC. 1958. On the mamillae of the springbok, *Antidorcas marsupialis* Zimmermann. S Afr J Sci. 54:291.
922. Bigalke, RC. 1963. A note on reproduction in the steenbok *Raphicerus campestris* Thunberg. Ann Cape Prov Mus. 3:64-67.
923. Bigalke, RC. 1970. Observations on springbok populations. Zool Afr. 5:59-70.
924. Bigg, MA. 1969. Clines in the pupping season of the harbor seal *Phoca vitulina*. J Fish Res Bd Can. 26:449-455.
925. Bigg, MA. 1969. The harbour seal in British Columbia. Bull Fish Res Bd Can. 172:1-33.
926. Bigg, MA. 1973. Adaptations in the breeding of the harbour seal, *Phoca vitulina*. J Reprod Fert Suppl. 19:131-142.
927. Bigg, MA. 1981. Harbour seal *Phoca vitulina* Linnaeus, 1758 and *Phoca largha* Pallas, 1811. Handbook of Marine Mammals. 2:1-27, Ridgeway, SH & Harrison, RJ, eds. Academic Press, New York.
927a. Bigg, MA. 1982. An assessment of killer whale (*Orcinus orca*) stocks off Vancouver Island, British Columbia. Rep Int Whal Comm. 32:655-666.
928. Bigg, MA & Fisher, HD. 1974. The reproductive cycle of the female harbour seal of southeastern Vancouver Island. Functional Anatomy of Marine Mammals. 2:329-347, Harrison, RJ, ed, Academic Press, New York.
929. Bigg, MA & Fisher, HD. 1975. Effect of photoperiod on annual reproduction in female harbour seals. Rapp P-V Réun Cons Int Explor Mer. 169:141-144.
930. Biggers, JD. 1966. Reproduction in male marsupials. Symp Zool Soc London. 15:251-280.
931. Biggers, JD. 1967. Notes on reproduction of the woolly opossum (*Caluromys derbianus*) in Nicaragua. J Mamm. 48:678-680.
932. Biggers, JD & Creed, RFS. 1962. Two morphological types of placentae in the raccoon. Nature. 194:103-105.
933. Biggers, JD & Creed, RFS. 1962. Conjugate spermatozoa of the North American opossum. Nature. 196:1112-1113.
934. Biggers, JD, Creed, RFS, & DeLamater, ED. 1963. Conjugated spermatozoa in American marsupials. J Reprod Fert. 6:324.
934a. Biggers, JD & DeLamater, ED. 1965. Marsupial spermatozoa pairing in the epididymis of American forms. Nature. 208:402-404.
935. Bigham, SR. 1966. Breeding season of the cottontail rabbit in north-central Oklahoma. Proc OK Acad Sci. 46:217-219.
935a. Bigler, WJ, Hoff, GL, & Johnson, AS. 1981. Population characteristics of *Procyon lotor murinus* in estuarine mangrove swamps of southern Florida. FL Sci. 44:151-157.
936. Bignami, G & Beach, FA. 1968. Mating behaviour in the chinchilla. Anim Behav. 16:45-53.
937. Bigourdan, J. 1948. Le Phacocherè et les Suidés dans l'Ouest Africain. Bull Inst Fr Afr Noire. 10:285-360.
938. Bilaspuri, GS & Guraya, SS. 1980. Quantitative studies on spermatogenesis in buffalo (*Bubalus bubalis*). Reprod Nutr Dév. 20:975-982.
939. Bindon, BM. 1984. Reproductive biology of the Booroola merino. Aust J Biol Sci. 37:163-189.
940. Bindra, O & Sagar, P. 1968. Breeding habits of the field rat *Millardia meltada*. J Bombay Nat Hist Soc. 65:477-481.
941. Binkerd, PE, Hendrickx, AG, Rice, JM, et al. 1984. Embryonic development in *Erythrocebus patas*. Am J Primatol. 6:15-29.
942. Birdsall, DA & Nash, D. 1973. Occurrence of successful multiple insemination of females in natural populations of deer mice (*Peromyscus maniculatus*). Evolution. 27:106-110.
943. Birkan, M. 1968. Repartition écologique et dynamique des populations d'*Apodemus sylvaticus* et *Clethrionomys glareolus* en pinède à Rambouillet. Terre Vie. 22:231-273.

944. Birkenholz, DE. 1963. A study of the life history and ecology of the round-tailed muskrat (*Neofiber alleni* True) in north-central Florida. Ecol Monogr. 33:255-280.
945. Birkenholz, DE. 1972. *Neofiber alleni*. Mamm Species. 15:1-4.
946. Birkenholz, DE & Wirtz, WO, II. 1965. Laboratory observations on the vespar rat. J Mamm. 46:181-189.
947. Birkenstock, PJ & Nel, JAJ. 1977. Laboratory and field observations on *Zelotomys woosnami* (Rodentia: Muridae). Zool Afr. 12:429-433.
948. Birkner, FE. 1970. Photic influences on primate (*Macaca mulatta*) reproduction. Lab Anim Care. 20:181-185.
949. Birney, EC. 1973. Systematics of three species of woodrats (genus *Neotoma*) in central North America. Univ KS Mus Nat Hist Misc Publ. 58:1-173.
950. Birney, EC, Bowles, JB, Timm, RM, et al. 1974. Mammalian distributional records in Yucatan and Quintana Roo, with comments of reproduction, structure, and status of peninsular populations. Bell Mus Nat Hist Univ MN Occas Pap. 13:1-25.
950a. Birney, EC & Rising, JD. 1967. Notes on distribution and reproduction of some bats from Kansas, with remarks on incidence of rabies. Trans KS Acad Sci. 70:519-524.
951. Bischoff, AI. 1957. The breeding season of some California deer herds. CA Fish Game. 43:91-96.
952. Bischoff, AI. 1958. Productivity in some California deer herds. CA Fish Game. 44:253-259.
953. Bischoff, TLW. 1854. Entwicklungsgeschichte des Rehes. J Ricker'sche Buchhandlung, Giessen.
954. Bishop, DW. 1942. Germ cell studies in the male fox (*Vulpes fulva*). Anat Rec. 84:99-115.
955. Bishop, IR & Delany, MJ. 1963. Life histories of small mammals in the Channel Islands in 1960-61. Proc Zool Soc London. 141:515-526.
956. Bishop, JA & Hartley, DJ. 1976. The size and age structure of rural populations of *Rattus norvegicus* containing individuals resistant to the anticoagulant poison warfarin. J Anim Ecol. 45:623-646.
957. Bissonette, JA. 1982. Ecology and Social Behavior of the Collered Peccary in Big Bend National Park. US Dept Int Nat Park Serv Sci Monogr Ser. 16:1-75.
958. Bissonnette, TH. 1932. Modification of mammalian sexual cycles: Reactions of ferrets (*Putorius vulgaris*) of both sexes to electric light added after dark in November and December. Proc R Soc London. 110:322-336.
959. Bissonnette, TH. 1935. Modification of mammalian sexual cycles III Reversal of the cycle in male ferrets (*Putorius vulgaris*) by increasing periods of exposure to light between October second and March thirtieth. J Exp Zool. 71:341-367.
960. Bissonnette, TH. 1935. Modification of mammalian sexual cycles IV Delay of oestrus and induction of anoestrus in female ferrets by reduction of intensity and duration of daily light periods in the normal oestrous season. J Exp Biol. 12:315-320.
961. Bissonnette, TH. 1938. Influence of light on the hypophysis: Effects of long-continued "night lighting" on hypophysectomized female ferrets and those with optic nerves cut. Endocrinology. 22:92-103.
962. Bissonnette, TH & Csech, AG. 1937. Modifications of mammalian sexual cycles VII Fertile matings of raccoons in December instead of February induced by increasing daily period of light. Proc R Soc London. 122B:246-254.
963. Bissonnette, TH & Csech, AG. 1939. A third year of modified breeding behavior with raccoons. Ecology. 20:156-162.
964. Bissonnette, TH & Csech, AG. 1939. Modified sexual photoperiodicity in cottontail rabbits. Biol Bull. 77:364-367.
965. Bittner, SL & Chapman, JA. 1981. Reproductive and physiological cycles in an island population of *Sylvilagus floridanus*. Proc World Lagomorph Conf, Univ Guelph, ONT, Canada. 1979:182-203.
966. Bjorge, RR, Gunson, JR, & Samuel, WM. 1981. Population characteristics and movements of striped skunks (*Mephitis mephitis*) in central Alberta. Can Field-Nat. 95:149-155.
967. Björkman, NH & Wimsatt, WA. 1968. The allantoic placenta of the vampire bat (*Desmodus rotundus murinus*): A reinterpretation of its structure based on electron microscopic observations. Anat Rec. 162:83-97.
968. Black, D, Seal, US, Plotka, ED, et al. 1979. Uterine biopsy of a lioness and a tigress after melengestrol implant. J Zoo An Med. 10:53-58.
969. Black, HL. 1970. Occurence of the Mexican big-eared bat in Utah. J Mamm. 51:190.
970. Black, JD. 1936. Mammals of northwestern Arkansas. J Mamm. 17:29-35.
971. Blackburn, DG, Hayssen, V, & Murphy, CJ. 1989. The origins of lactation and the evolution of milk: A review with new hypotheses. Mamm Rev. 19:1-26.
972. Blackwell, K. 1967. Breeding and hand-rearing of fruit bats *Epomops frangueti* and *Eidolon helvum* at the University Zoo. Int Zoo Yearb. 7:79-80.
973. Blackwell, K. 1969. Rearing and breeding Demidoff's galago *Galago demidovii* in captivity. Int Zoo Yearb. 9:74-76.
974. Blackwell, KF & Menzies, JI. 1968. Observations on the biology of the potto (*Perodicticus potto* Miller). Mammalia. 32:447-451.
975. Blaine, CR, White, RJ, Blakeley, GA, et al. 1975. Plasma progesterone during the normal menstrual cycle of the pig tailed monkey (*Macaca nemestrina*). Int Congr Primatol. 5:141-151.
976. Blair, WF. 1936. The Florida marsh rabbit. J Mamm. 17:197-207.
977. Blair, WF. 1941. Observations on the life history of *Baiomys taylori subater*. J Mamm. 22:378-383.
978. Blair, WF. 1942. Rate of development of young spotted ground squirrels. J Mamm. 23:342-343.
979. Blair, WF. 1943. Populations of the deer-mouse and associated small mammals in the mesquite association in southern New Mexico. Contrib Lab Vert Biol Univ MI. 21:1-40.
980. Blair, WF. 1951. Population structure, social behavior, and environmental relations in a natural population of the beach mouse (*Peromyscus polionotus leucocephalus*). Contrib Lab Vert Biol Univ MI. 48:1-45.
981. Blake, BH. 1982. Reproduction in captive water voles, *Arvicola terrestris*. J Zool. 198:524-529.
982. Blake, BH & Gillett, KE. 1984. Reproduction of Asian chipmunks (*Tamias sibiricus*) in captivity. Zoo Biol. 3:47-63.
983. Blake, BH & Gillett, KE. 1988. Estrous cycle and related aspects of reproduction in captive Asian chipmunks, *Tamias sibiricus*. J Mamm. 69:598-603.
983a. Blake, HA. 1885. Note on the parturition of a West Indian bat. Sci Proc R Dublin Soc. 4:449-450.
984. Blakely, GA. 1969. Vaginal cytology and luteinizing hormone levels in *Pongo pygmaeus* (orangutan). Recent Adv Primatol. 22-30, Hofer, HO, ed, Karger, Basel.
985. Blakeslee, CK, Rice, CG, & Ralls, K. 1979. Behavior and reproduction of captive brow-antlered deer, *Cervus eldi thamin* (Thomas, 1918). Säugetierk Mitt. 27:114-127.

986. Blakley, GB, Beamer, TW, & Dukelow, WR. 1981. Characterisitcs of the menstrual cycle in nonhuman primates: IV Timed mating in *Macaca nemestrina*. Lab Anim Sci. 15:351-353.
987. Bland, KP. 1969. Reproduction in the female Indian gerbil (*Tatera indica*). J Zool. 157:47-61.
988. Bland, KP. 1973. Reproduction in the female African tree rat (*Grammomys surdaster*). J Zool. 171:167-175.
988a. Blanford, WT. 1891. The Fauna of British India, including Ceylon and Burma. Taylor & Francis, London.
989. Blank, JL, Nelson, RJ, Vaughan, MK, et al. 1988. Pineal melatonin content in photoperiodically responsive and non-responsive phenotypes of deer mice. Comp Biochem Physiol. 91A:535-537.
989a. Blank, JL & Desjardins, C. 1984. Spermatogenesis is modified by food intake in mice. Biol Reprod. 30:410-415.
990. Blank, MS, Gordon, TP, & Wilson, ME. 1983. Effects of capture and venipuncture on serum levels of prolactin, growth hormone and cortisol in outdoor compound-housed female rhesus monkeys (*Macaca mulatta*). Acta Endocrinol. 102:190-195.
991. Blankenship, LT, Muse, PD, & Louis, TM. 1981. Plasma progesterone and progesterone binding protein during the estrous cycle and pregnancy in *Dolichotis patagona*. Fed Proc. 40:471 (ref 1373).
992. Blankenship, LT, Muse, PD, Pryor, WH, Jr, et al. 1982. Maternal plasma estrogen and progesterone profiles during pregnancy and neonatal physical measurements in *Dolichotis patagona* (Mara). Fed Proc. 41:1490 (ref 7132).
993. Blankenship, TN, Given, RL, & Parkening, TA. 1990. Blastocyst implantation in the Chinese hamster (*Cricetulus griseus*). Am J Anat. 187:137-157.
994. Blaskiewitz, B. 1980. Gedanken zur Haltung des Panzernashorns (*Rhinoceros unicornis* Linné, 1758). Zool Beitr. 26:69-108.
995. Blatchley, FR & Donovan, BT. 1972. Peripheral plasma progestin levels during anoestrus, oestrus and pseudopregnancy and following hypophysectomy in ferrets. J Reprod Fert. 31:331-333.
996. Blatchley, FR & Donovan, BT. 1976. Progesterone secretion during pregnancy and pseudopregnancy in the ferret. J Reprod Fert. 46:455-456.
997. Blaustein, AR. 1981. Population fluctuations and extinctions of small rodents in coastal southern California. Oecologia. 48:71-78.
998. Bleich, VC. 1977. *Dipodomys stephensi*. Mamm Species. 73:1-3.
998a. Bleich, VC & Schwartz, OA. 1975. Parturition in the white-throated woodrat. Southwest Nat. 20:271-272.
999. Bleier, WJ. 1975. Early embryology and implantation in the California leaf-nosed bat, *Macrotus californicus*. Anat Rec. 182:237-254.
1000. Bleier, WJ. 1979. Embryology. Sp Publ Mus TX Tech Univ. 16(III):379-386.
1001. Bleier, WJ & Ehteshami, M. 1981. Ovulation following unilateral ovariectomy in the California leaf-nosed bat (*Macrotus californicus*). J Reprod Fert. 63:181-183.
1002. Blix, AS & Lentfer, JW. 1979. Modes of thermal protection in polar bear cubs - at birth and on emergence from the den. Am J Physiol. 236:R67-R74.
1003. Bloch, S, Hediger, H, Lloyd, HG, et al. 1967. Beobachtungen zur Superfetation beim Feldhasen (*Lepus europaeus*). Z Jagdwiss. 13:49-52.
1004. Bloch, S, Hediger, H, Miller, C, et al. 1954. Probleme der Fortpflanzung des Feldhasen. Rev Suisse Zool. 61:485-490.
1005. Bloch, S & Strauss, F. 1958. Die weibliche Genitalorgane von *Lepus europaeus* Pallas. Z Säugetierk. 23:66-80.
1006. Bloedel, P. 1955. Observations on the life histories of Panama bats. J Mamm. 36:232-235.
1007. Blokhin, SA. 1981. On the functional activity of the sexual system of sperm whale males. Rep Int Whal Comm. 31:719-721.
1008. Blokhin, SA. 1984. Investigations of gray whales taken in the Chukchi Coastal waters USSR. The Gray Whale, *Eschrichtius robustus*. 487-509, Jones, ML, Swartz, SL, & Leaterwood, S, eds, Academic Press, New York.
1009. Blokhin, SA. 1984. Some aspects of reproduction in the California-Chukchi Sea stock of gray whales. Rep Int Whal Commn. 34:457-460.
1010. Blomquist, L, Muuronen, P, & Rantanen, V. 1981. Breeding the least weasel (*Mustela rixosa*) in Helsinki Zoo. Zool Gärten NF. 51:363-368.
1011. Blood, DA. 1963. Some aspects of behavior of a bighorn herd. Can Field-Nat. 77:77-94.
1012. Blood, DA. 1973. Variation in reproduction and productivity of an enclosed herd of moose (*Alces alces*). Int Congr Game Biol. 11:59-66.
1013. Blossom, PM. 1932. A pair of long-tailed shrews (*Sorex cinereus cinereus*) in captivity. J Mamm. 13:136-143.
1013a. Blough, RA. 1987. Reproductive seasonality of the white-tailed deer on the Colombian llanos. Biology and Management of the Cervidae. 340-343, Wemmer, CM, ed, Smithsonian Inst Press, Washington, DC.
1013b. Blough, RA & Atmosoedirdjo, S. 1987. Biology of Bawean deer and prospects for its management. Biology and Management of the Cervidae. 320-327, Wemmer, CM, ed, Smithsonian Inst Press, Washington, DC.
1014. Bloxam, QMC. 1968. Serval (*Felis leptailurus serval*). Ann Rep Jersey Wildl Preserv Trust. 5:44-45.
1015. Bloxam, QMC. 1969. Common gelada *Theropithecus gelada gelada*. Ann Rep Jersey Wildl Preserv Trust. 6:19-22.
1016. Bloxam, QMC. 1970. Celebesian black ape *Cynopithecus niger niger*. Ann Rep Jersey Wildl Preserv Trust. 7:6-9.
1017. Bloxam, QMC. 1972. Indian lion *Panthera leo persica*. Ann Rep Jersey Wildl Preserv Trust. 9:39-42.
1018. Bloxam, QMC. 1973. The breeding of second generation (F_2) Serval cat *Felis leptailurus serval*. Ann Rep Jersey Wildl Preserv Trust. 10:41-43.
1019. Bloxam, QMC. 1975. Breeding of the spectacled bear *Tremarctos ornatus* at the Jersey Zoological Park. Ann Rep Jersey Wildl Preserv Trust. 12:42-47.
1020. Bloxam, QMC. 1977. Breeding the spectacled bear *Tremarctos ornatus* at Jersey Zoo. Int Zoo Yearb. 17:158-161.
1021. Blunt, FM, Dawson, HA, & Thorne, ET. 1977. Birth weights and gestation in a captive Rocky Mountain bighorn sheep. J Mamm. 58:106.
1022. Bluntschli, H. 1913. Entwicklungsgeschichte platyrrhiner Affen, von *Didelphys marsupialis*, *Tamandua bivittata* und *Bradypus marmoratus*. Verh Anat Ges. 27:196-202.
1023. Bluntschli, H. 1937. Die Frühentwicklung eines Centetinen (*Hemicentetes semispinosus* Cuv). Rev Suisse Zool. 44:271-282.
1024. Bluntschli, H. 1937. Die Frühentwicklung eines tief stehenden Placentaliers und ihre Bedeutung fur die Auffasung der Säugetierontogenese überhaupt. Mitt naturf Ges Bern. 1937:VII-XII.
1025. Bluntschli, H. 1938. Die Einbettung des Eies beim *Hemicentetes*. Biomorphosis. 1:332-333.
1026. Bluntschli, H. 1938. Le développement primaire et l'implantation chez un Centenité (*Hemicentetes*). Bull Assoc Anat. 44:39-46.

1027. Blus, LJ. 1966. Relationship between litter size and latitude in the golden mouse. J Mamm. 47:546-547.
1028. Blus, LJ. 1971. Reproduction and survival of short-tailed shrews (*Blarina brevicauda*) in captivity. Lab Anim Sci. 21:884-891.
1029. Boag, DA & Murie, JO. 1981. Population ecology of Columbian ground squirrels in southwestern Alberta. Can J Zool. 59:2230-2240.
1030. Boag, DA & Murie, JO. 1981. Weight in relation to sex, age, and season in Columbian ground squirrels (Sciuridae: Rodentia). Can J Zool. 59:999-1004.
1031. Boas, JEV. 1891. Zur Morphologie der Begattungsorgane der amnioten Wirbelthiere. Gegenb Morphol Jb. 17:271-287.
1032. Bobek, B. 1969. Survival, turnover and production of small rodents in a beech forest. Acta Theriol. 14:191-210.
1033. Bobrinoskoy, N, Kuznetzow, B, & Kuzyakin, A. 1944. Mammals of the USSR. USSR Gov Publ Office, Moscow.
1034. Bock, HD, Wünsche, J, Hoffman, B, et al. 1968. Über die Aminosäuren-Zusamensetzung der Milch der Kegelrobbe, *Halichoerus grypus* Fab, und des Schweinswals, *Phocoena phocoena*. Zool Garten NF. 35:69-73.
1035. Bodenheimer, FS. 1949. Ecological and physiological studies on some rodents. Physiol Comp Oecol. 1:376-389.
1036. Bodenheimer, FS & Dvoretzky, A. 1951. A dynamic model for the fluctuation of populations of the Levante vole (*Microtus guentheri* D A). Bull Res Council Israel. 1(4):62-80.
1037. Bodenheimer, FS & Sulman, F. 1946. The estrous cycle of *Microtus guentheri* D and A and its ecological implications. Ecology. 27:255-256.
1038. Bodley, HD. 1974. The development of the chorio allantoic placenta in the bat *Macrotus waterhousii*. Anat Rec. 178:313.
1039. Bodley, HD. 1974. Ultrastructural development of the chorioallantoic placental barrier in the bat *Macrotus waterhousii*. Anat Rec. 180:351-368.
1040. Boecker, E. 1907. Zur Kenntnis des Baues der Placenta von *Elephas indicus* L. Arch f Mikr Anat. 71:297-323.
1041. Böer, M. 1983. Several examinations on the reproductive status of lowland gorillas (*Gorilla g gorilla*) at Hanover Zoo. Zoo Biol. 2:267-280.
1042. Böer, M. 1987. Beobachtungen zur Fortpflanzung und zum Verhalten des Drill (*Mandrillus leucophaeus* Ritgen, 1824) im Zoo Hannover. Z Säugetierk. 52:265-281.
1043. Bogan, MA. 1972. Observations on parturition and development in the hoary bat, *Lasiurus cinereus*. J Mamm. 53:611-614.
1044. Bogart, MH, Cooper, RW, & Binirschke, K. 1977. Reproductive studies of black and ruffed lemurs *Lemur macaco macaco* and *L variegatus* ssp. Int Zoo Yearb. 17:177-182.
1045. Bogart, MH, Kumamoto, AT, & Lasley, BL. 1977. A comparison of the reproductive cycle of three species of *Lemur*. Folia Primatol. 28:134-143.
1046. Bognar, L & Bourdelle, E. 1956. Les Mammiferès du "Jardin Animé" du Cap Ferrat. Terre Vie. 11:22-27.
1047. Böhner, J, Vogler, K, & Hendrichs, H. 1984. Zur Fortpflanzungsbiologie des Blauduckers (*Cephalophus monticola*). Z Säugetierk. 49:306-314.
1048. Bohr, G. 1981. Nachweis doppelter Empfängnis in einem Jahr bei einem Muffelschaf (*Ovis ammon musimon*). Zool Gärten NF. 51:271-272.
1049. Boissin-Agasse, L & Boissin, J. 1979. Variations saisonnières du volume testiculaire et de la testostéronemie chez deux Mustelidés: Le Furet (*Mustela furo* L) et le Vison (*Mustela vison* S). J Physiol (Paris). 75:227-232.
1050. Boissin-Agasse, L & Boissin, J. 1985. Incidence of a circadian cycle of photosensitivity in the regulation of the annual testis cycle in the mink: A short-day mammal. Gen Comp Endocrinol. 60:109-115.
1051. Boissin-Agasse, L, Boissin, J, & Ortavant, R. 1982. Circadian photosensitive phase and photoperiodic control of testis activity in the mink (*Mustela vison* Peale and Beauvais), a short-day mammal. Biol Reprod. 26:110-119.
1052. Boissin-Agasse, L, Maurel, D, & Boissin, J. 1981. Seasonal variations in thyroxine and testosterone levels in relation to the moult in the adult male mink *Mustela vison* (Peale and Beauvais). Can J Zool. 59:1062-1066.
1053. Boissin-Agasse, L & Ortavant, R. 1978. Mise en évidence d'une séquence circadienne de photo gonado sensibilité chez le Furet (*Mustela furo* L). CR Hebd Scéances Acad Sci Ser D Sci Nat. 287:1313-1316.
1054. Boissin-Agasse, L, Ravault, JP, & Boissin, J. 1983. Photo sensibilité circadienne et contrôle photopériodique du cycle annuel de la prolactinemie chez le Vison. CR Hebd Séances Acad Sci Sér III Sci Vie. 296:707-710.
1055. Boisvert, M & Grisham, J. 1988. Reproduction of the short-nosed echidna *Tachyglossus aculeatus* at the Oklahoma City Zoo. Int Zoo Yearb. 27:103-108.
1056. Bokoff, KJ. 1978. Behavioural fluctuations in *Lemur fulvus*: Within and without the breeding season. Rec Adv Primatol. 1:503-506.
1057. Bol'shakov, VN & Pokrovskii, AV. 1969. The extent of reproductive isolation between Pamir vole (*Microtus juldaschi severtzoo*) and Carruther's vole *M carruthersi* Thomas). Dokl Akad Nauk SSSR Transl. 188:776-777.
1058. Bolliger, A. 1940. *Trichosurus vulpecula* as an experimental animal. Aust J Sci. 3:59-61.
1059. Bolliger, A. 1943. Functional relations between scrotum and pouch and the experimental production of a pouch-like structure in the male of *Trichosurus vulpecula*. J Proc R Soc NSW. 76:283-293.
1060. Bolliger, A. 1944. An experiment on the complete transformation of the scrotum into a marsupial pouch in *Trichosurus vulpecula*. Med J Aust. 11:56-58.
1060a. Bolliger, A. 1944. The distinctive brown patch of sternal fur of *Trichosurus vulpecula* and its response to sex hormones. Aust J Sci. 6:181.
1061. Bolliger, A. 1954. Organ transformation induced by oestrogen in an adolescent marsupial (*Trichosurus vulpecula*). J Proc R Soc NSW. 88:33-39.
1062. Bolliger, A & Carrodus, A. 1938. Changes in and around the pouch in *Trichosurus vulpecula*, as occurring naturally and as the result of the administration of oestrone. J Proc R Soc NSW. 71:615-622.
1063. Bolliger, A & Carrodus, A. 1939. The effect of oestrogens on the pouch of the marsupial *Trichosurus vulpecula*. J Proc R Soc NSW. 73:218-227.
1064. Bolliger, A & Carrodus, A. 1940. The effect of testosterone propionate on pouch, scrotum, clitoris, and penis of *Trichosurus vulpecula*. Med J Aust. 2:368-373.
1065. Bolliger, A & Carrodus, A. 1940. The effect of testosterone propionate on pouch and scrotum of *Trichosurus vulpecula*. Aust J Sci. 2:120.
1066. Bolliger, A & Pascoe, JV. 1953. The composition of kangaroo milk (wallaroo, *Macropus robustus*). Aust J Sci. 15:215-217.
1067. Bolliger, A & Tow, AJ. 1946. Late effects of castration and administration of sex hormones on the male *Trichosurus vulpecula*. J Endocrinol. 5:32-41.

1068. Bolster, L & Savage-Rumbaugh, S. 1989. Periparturitional behavior of a bonobo (*Pan paniscus*). Am J Primatol. 17:93-103.
1069. Bolwig, N. 1959. A study of the behaviour of the chacma baboon, *Papio ursinus*. Behaviour. 14:136-163.
1070. Bonaccorso, FJ. 1979. Foraging and reproductive ecology in a Panamanian bat community. Bull FL State Mus Biol Ser 24:359-408.
1071. Bond, CR. 1945. The golden hamster (*Cricetus auratus*): Care, breeding and growth. Physiol Zool. 18:52-59.
1072. Bond, MR. 1979. Second-generation captive birth of an orang-utan *Pongo pygmaeus*. Int Zoo Yearb. 19:165-167.
1073. Bond, MR & Block, JA. 1982. Growth and development of twin orang-utans *Pongo pygmaeus*. Int Zoo Yearb. 22:256-261.
1074. Bond, RM. 1946. The breeding habits of *Thomomys bottae* in Orange County, California. J Mamm. 27:172-174.
1075. Bondrup-Nielsen, S & Ims, RA. 1986. Comparison of maturation of female *Clethrionomys glareolus* from cyclic and noncyclic populations. Can J Zool. 64:2099-2102.
1076. Bondrup-Nielsen, S & Ims, RA. 1986. Reproduction and spacing behaviour of females in a peak density population of *Clethrionomys glareolus*. Holarctic Ecol. 9:109-112.
1077. Bondrup-Nielsen, S & Ims, RA. 1988. Demography during population crash of the wood lemming, *Myopus schisticolor*. Can J Zool. 66:2442-2448.
1078. Bonds, DR, Mwape, B, Kumar, S, et al. 1984. Human fetal weight and placental weight growth curves. Biol Neonate. 45:261-274.
1079. Boness, DJ & James, H. 1979. Reproductive behaviour of the grey seal (*Halichoerus grypus*) on Sable Island, Nova Scotia. J Zool. 188:477-500.
1080. Bonhote, JL. 1911. Exhibition of, and remarks upon, a young Cairo spiny mouse (*Acomys cahirinus*) about twelve hours old. Proc Zool Soc London. 1911(I):5-6.
1081. Bonnell, ML & Selander, RK. 1974. Elephant seals: Genetic variation and near extinction. Science. 184:908-909.
1082. Bonner, WN. 1958. The introduced reindeer of South Georgia. Falkl Is Depend Surv Sci Rep. 22:1-8.
1083. Bonner, WN. 1968. The fur seal of South Georgia. Br Antarctic Surv Sci Rep. 56:1-81.
1084. Bonner, WN. 1972. The grey seal and common seal in European waters. Oceanogr Marine Bio. 10:461-507.
1085. Bonner, WN. 1973. Grey seals in the Baltic. IUCN Suppl Pap. 39:164-174.
1086. Bonner, WN. 1979. Antarctic (Kerguelen) fur seal. Mammals in the Seas, FAO Fish Ser 5. 2:49-51.
1087. Bonner, WN. 1979. Grey seal. Mammals in the Seas, FAO Fish Ser 5. 2:90-94.
1088. Bonner, WN. 1979. Harbour (common) seal. Mammals in the Seas, FAO Fish Ser 5. 2:58-62.
1089. Bonner, WN. 1979. Largha seal. Mammals in the Seas, FAO Fish Ser 5. 2:63-65.
1090. Bonner, WN. 1979. Subarctic fur seal. Mammals in the Seas, FAO Fish Ser 5. 2:52-54.
1091. Bonner, WN. 1981. Grey seal *Halichoerus grypus* Fabricius, 1791. Handbook of Marine Mammals 2:111-144, Ridgway, SM & Harrison, RJ, eds. Academic Press, New York.
1092. Bonner, WN & Hickling, G. 1971. The grey seals of the Farne Islands: Report for the period October 1969-July 1971. Trans Nat Hist Soc Northumb. 17:141-162.
1093. Bonner, WN & Hickling, G. 1974. The grey seals of the Farne Islands: 1971-1973. Trans Nat Hist Northumb. 42:65-84.
1094. Bonney, RC, Dixson, AF, & Flemming, D. 1979. Cyclic changes in the circulating and urinary levels of ovarian steriods in the adult female owl monkey (*Aotus trivirgatus*). J Reprod Fert. 56:271-280.
1095. Bonney, RC, Dixson, AF, & Flemming, D. 1980. Plasma concentrations of oestradiol-17β, oestrone, progesterone and testosterone during the ovarian cycle of the owl monkey (*Aotus trivirgatus*). J Reprod Fert. 60:101-107.
1096. Bonney, RC, Moore, HDM, & Jones, DM. 1981. Plasma concentrations of oestradiol-17β and progesterone, and laporoscopic observations of the ovary in the puma (*Felis concolor*) during oestrus, pseudopregnancy and pregnancy. J Reprod Fert. 63:523-531.
1097. Bonney, RC & Setchell, KDR. 1980. The excretion of gonadal steroids during the reproductive cycle of the owl monkey (*Aotus trivirgatus*). J Steroid Biochem. 12:417-421.
1098. Bonney, RC, Wood, DJ, & Kleiman, DG. 1982. Endocrine correlates of behavioural oestrus in the female giant panda (*Ailuropoda melanoleuca*) and associated hormonal changes in the male. J Reprod Fert. 64:209-215.
1099. Bonney, S & Crotty, MJ. 1979. Breeding the mountain tapir *Tapirus pinchaque* at the Los Angeles Zoo. Int Zoo Yearb. 19:198-200.
1100. Bonnin, M, Audy, MC, Mondain-Monval, M, et al. 1986. LHRH et liberation de LH chez le renard roux (*Vulpes vulpes* L). CR Acad Sci Ser 3. 303:569-574.
1101. Bonnin, M, Canivenc, R, & Aitken, J. 1977. Variations saisonnières du taux de la progestérone plasmatique chez la Fouine (*Martes foina*), espèce à ovo-implantation déférée. CR Hebd Séances Acad-Sci Ser D Sci Nat. 285:1479-1481.
1102. Bonnin, M, Canivenc, R, & Charron, G. 1981. Variations du taux de la progestérone plasmatique chez la femelle de Blaireau après ablation de l'utérus gravide. CR Hebd Séances Acad Sci Sér III Sci Vie. 293:143-145.
1103. Bonnin, M, Canivenc, R, & Ribos, CL. 1978. Plasma progesterone levels during delayed implantation in the European badger (*Meles meles*). J Reprod Fert. 52:55-58.
1104. Bonnin, M, Martin, B, Charron, G, et al. 1984. C21 Steroids and transcortin-type protein during delayed implantation in the European badger *Meles meles* L. J Steroid Biochem. 20:575-580.
1104a. Bonnin, M, Mondain-Monval, M, Audy, MC, et al. 1989. Basal and gonadotropin releasing hormone stimulated gonadotropin levels in the female red fox (*Vulpes vulpes* L): Negative feedback of ovarian hormones during anoestrus. Can J Zool. 67:759-765.
1105. Bonnin, M, Mondain-Monval, M, & Dutourne, B. 1978. Oestrogen and progesterone concentrations in peripheral blood in pregnant red foxes (*Vulpes vulpes*). J Reprod Fert. 54:37-41.
1106. Bonnin-Laffargue, M & Canivenc, R. 1970. Biologie lutéale chez le Renard (*Vulpes vulpes* L): Persistance du corps jaune après la mise-bas. CR Hebd Séances Acad Sci Sér D Sci Nat. 271:1402-1405.
1107. Bonnin-Laffargue, M, Canivenc, R, & Lafus-Boul, M. 1972. Biologie lutéale chez le Renard (*Vulpes vulpes* L): L'hypophyse intervient-elle dans la persistance du corps jaune apres la mise-bas?. CR Hebd Séances Acad Sci Sér D Sci Nat. 274:85-88.
1108. Bonnin-Laffargue, M, Lajus, M, Relexans, MC, et al. 1962. Les circatrices placentaires chez le Blaireau européen *Meles meles* L. CR Soc Biol. 156:2035-2038.
1109. Booher, CB, Prahalada, S, & Hendrickx, AG. 1983. Use of a radioreceptorassay RRA for human luteinizing hormone/chorionic gonadotropin (hLH/CG) for detection of early pregnancy and estimation of time of ovulation in macaques. Am J Primatol. 4:45-53.

1110. Bookhout, TA. 1964. Prenatal developemnt of snowshoe hares. J Wildl Manage. 28:338-345.
1111. Bookhout, TA. 1965. Breeding biology of snowshoe hares in Michigan's upper peninsula. J Wildl Manage. 29:296-303.
1112. Boonstra, R & Sinclair, ARE. 1984. Distribution and habitat use of caribou, *Rangifer tarandus caribou*, and moose, *Alces alces andersoni*, in the Spatsizi Plateau Wilderness Area, British Columbia. Can Field-Nat. 98:12-21.
1113. Boorman, GA, Niswender, GD, Gay, VL, et al. 1973. Radioimmunoassay for follicle-stimulating hormone in the rhesus monkey using an anti-human FSH serum and rat FSH ^{131}I. Endocrinology. 92:618-623.
1114. Boorman, GA, Speltie, TM, & Fitzgerald, GH. 1974. Urinary chorionic gonadotropin excretion during pregnancy in the chimpanzee. J Med Primatol. 3:269-275.
1115. Booth, AH. 1957. Observations on the natural history of the olive colobus monkey, *Procolobus verus* (van Beneden). Proc Zool Soc London. 129:421-430.
1116. Booth, C. 1962. Some observations on behaviour of cercopithecus monkeys. Ann NY Acad Sci. 102(II):477-487.
1117. Booth, ES. 1946. Notes on the life history of the flying squirrel. J Mamm. 27:28-30.
1118. Booth, VR. 1985. Some notes on Lichtenstein's hartebeest, *Alcelaphus lichtensteini* (Peters). S Afr J Zool. 20:157-160.
1119. Borell, AE & Ellis, R. 1934. Mammals of the Ruby mountains region of north-eastern Nevada. J Mamm. 15:12-44.
1120. Borhies, CT & Taylor, WP. 1922. Life history of the kangaroo rat, *Dipodomys spectabilis spectabilis* Merriam. USDA Tech Bull. 1091:1-40.
1121. Born, L. 1874. Über die Entwicklung des Eierstockes des Pferdes. Archf Anat Physiol Wissensch Med. 1874:118-151.
1122. Borodulina, TL. 1951. Latent period of embryonic development in the Siberian mole. Dokl Akad Nauk SSSR. 80:689-692.
1123. Borowski, S & Dehnel, A. 1952. Materialy do biologii Soricidae. Ann Univ Mariae Curie-Sklodowska. 7:305-448.
1124. Borut, A, Dmi'el, R, & Shkolnik, A. 1979. Heat balance of resting and walking goats: Comparison of climatic chamber and exposure in the desert. Physiol Zool. 52:105-112.
1125. Bos, H & Vandermey, GJW. 1980. Length of gestation periods of horses and ponies belonging to different breeds. Livest Prod Sci. 7:181-187.
1126. Boshe, JI. 1981. Reproductive ecology of the warthog *Phacochoerus aethiopicus* and its significance for management in the eastern Selous Game Reserve, Tanzania. Biol Conser. 20:37-44.
1127. Boskoff, KJ. 1977. Aspects of reproduction in ruffed lemurs (*Lemur variegatus*). Folia Primatol. 28:241-250.
1128. Boskoff, KJ. 1978. The oestrous cycle of the brown lemur, *Lemur fulvus*. J Reprod Fert. 54:313-318.
1129. Bostrom, RE, Aulerich, RJ, Ringer, RK, et al. 1968. Seasonal changes in the testes and epididymides of the ranch mink (*Mustela vison*). MI Agric Exp Stn Q Bull. 50:538-558.
1130. Bosu, WTK. 1975. Implantation and maintenance of pregnancy in mated rhesus monkeys following bilateral oophorectomy or luteectomy with and without progesterone replacement. Acta Endocrinol. 79:598-609.
1131. Bosu, WTK, Holmdahl, TH, Johansson, EDB, et al. 1972. Peripheral plasma levels of oestrogens, progesterone and 17-α hydroxyprogesterone during the menstrual cycle of the rhesus monkey. Acta Endocrinol. 71:775-764.
1132. Bosu, WTK, Johansson, EDB, & Gemzell, C. 1973. Ovarian steroid patterns in peripheral plasma during the menstural cycle in the rhesus monkey. Folia Primatol. 19:218-234.
1133. Bosu, WTK, Johansson, EDB, & Gemzell, C. 1973. Peripheral plasma levels of oestrone, oestradiol-17β and progesterone during ovulatory menstrual cycles in the rhesus monkey with special reference to the onset of menstruation. Acta Endocrinol. 74:732-742.
1134. Bosu, WTK, Johansson, EDB, & Gemzell, C. 1973. Patterns of circulationg oestrone, oestradiol-17β and progesterone during pregnancy in the rhesus monkey. Acta Endocrinol. 74:743-755.
1135. Bosu, WTK, Johansson, EDB, & Gemzell, C. 1974. Influence of oophorectomy, luteectomy, foetal death and dexamethasone on the peripheral plasma levels of oestrogens, and progesterone in the pregnant *Macaca mulatta*. Acta Endocrinol. 75:601-616.
1136. Bothma, JP & Teer, JG. 1977. Reproduction and productivity in south Texas cottontail rabbits. Mammalia. 41:253-281.
1137. Bothma, JP, Teer, JG, & Gates, CE. 1972. Growth and age determination of the cottontail in south Texas. J Wildl Manage. 36:1209-1221.
1138. Böttger, I, von Benten, Ch, Wissdorf, H, et al. 1970. Untersuchungen zu Antomie, Topographie, Grösze und Gewicht der Weiblichen Geschlechtsorgane von *Meriones unguiculatus*. Z Versuchstierk. 18:263-284.
1139. Boulay, MC & Robbins, CB. 1989. *Epomophorus gambianus*. Mamm Species. 344:1-5.
1140. Boulouard, R. 1971. Etude de la fonction corticosurrénalienne au cours du cycle annuel chez un hibernant, le Lérot (*Eliomys quercinus* L) I Activité durant l'été: Période de reproduction. Gen Comp Endocrinol. 16:465-477.
1141. Boulouard, R. 1971. Etude de la fonction corticosurrénalienne au cours du cycle annuel chez un hibernant, le Lérot (*Eliomys quercinus* L) II Activité durant l'automme: Période de préparation à l'hibernation. J Physiol (Paris). 63:77-86.
1142. Boulva, J. 1971. Observations on a colony of whelping harbour seals, *Phoca vitulina concolor*, on Sable Island, Nova Scotia. J Fish Res Bd Can. 28:755-759.
1143. Boulva, J. 1975. Temporal variations in birth period and characteristics of newborn harbour seals. Rapp P-V Réun Cons Explor Mer. 169:405-408.
1144. Boulva, J. 1979. Carribean monk seal. Mammals in the Seas, FAO Fish Ser 5. 2:101-103.
1145. Boulva, J. 1979. Mediterranean monk seal. Mammals in the Seas, FAO Fish Ser 5. 2:95-100.
1146. Boulva, J & McLaren, IA. 1979. Biology of the harbor seal, *Phoca vitulina*, in eastern Canada. Bull Fish Res Bd Can. 200:1-24.
1147. Bourlière, F. 1948. Sur la reproduction et la croissance du *Cricetomys gambianus*. Terre Vie. 3:45-48.
1148. Bourlière, F, Bertrand, M, & Hunkeler, C. 1969. L'écologie de la Mone de Lowe (*Cercopithecus campbelli lowei*) en Côte d'Ivoire. Terre Vie. 116:135-163.
1149. Bourlière, F, Morel, G, & Galat, G. 1976. Les grands Mammifères de la basse, valleé du Sénégal et leurs saisons de reproduction. Mammalia. 40:401-412.
1150. Bourlière, F, Petter-Rousseaux, A, & Petter, JJ. 1961. Regular breeding in captivity of the lesser mouse lemur (*Microcebus murinus*). Int Zoo Yearb. 3:24-25.
1151. Bourne, GH, Keeling, ME, & Golarz de Bourne, MN. 1975. Breeding anthropoid apes and rhesus monkeys at the Yerkes Primate Center. Lab Anim Handb. 6:63-76.

1152. Bourne, GH. 1972. Breeding chimpanzees and other apes. Breeding Primates. 24-33, Beveridge, WIB, ed, Karger, Basel.
1153. Bourquin, O. 1975. Bestuursplan Willem Pretorius Wildtuin. Orange Free State Conservation.
1154. Bowden, D, Winter, P, & Ploog, D. 1967. Pregnancy and delivery behavior in the squirrel monkey (*Saimiri sciureus*) and other primates. Folia Primatol. 5:1-42.
1155. Bowdre, LP. 1971. Litter size in *Sigmodon hispidus*. Southwest Nat. 16:126-128.
1156. Bowen, WD, Boness, DJ, & Oftedal, OT. 1987. Mass transfer from mother to pup and subsequent mass loss by the weaned pup in the hooded seal, *Cystophora cristata*. Can J Zool. 65:1-8.
1157. Bowen, WD, Capstick, DK, & Sergeant, DE. 1981. Temporal changes in the reproductive potential of female harp seals (*Pagophilus groenlandicus*). Can J Fish Aquat Sci. 38:495-503.
1158. Bowen, WD, Oftedal, OT, & Boness, DJ. 1985. Birth to weaning in four days: Remarkable growth in the hooded seal, *Cystophora cristata*. Can J Zool. 63:2841-2846.
1159. Bowers, CL & Elton, RH. 1982. Synchronization of menstrual cycles in pigtailed macaques using photoperiod. J Med Primatol. 11:252-256.
1160. Bowers, GL. 1955. Unusual breeding of cottontails in Pennsylvania. J Mamm. 36:303.
1161. Bowles, D. 1986. Social behaviour and breeding of babirusa *Babyrousa babyrussa* at the Jersey Wildlife Preservation Trust. Dodo. 23:86-94.
1162. Bowles, JB. 1972. Notes on reproduction in four species of bats from Yucatan, Mexico. Trans KS Acad Sci. 75:271-272.
1163. Bowles, JB, Heideman, PD, & Erickson, KR. 1990. Observations on six species of free-tailed bats (Molossidae) from Yucatan, Mexico. Southwest Nat. 35:151-157.
1164. Bowley, EA. 1939. Delayed parturition in *Dromicia*. J Mamm. 20:499.
1165. Bowman, LA, Dilley, SR, & Keverne, EB. 1978. Suppression of oestrogen-induced LH surges by social subordination in talapoin monkeys. Nature. 275:56-58.
1166. Bowness, ER. 1942. Duration of pregnancy in minks. Can Silver Fox Fur. 8(2):12-15.
1166a. Bowyer, RT & Leslie, DM, Jr. 1992. *Ovis dalli*. Mamm Species. 393:1-7.
1166b. Box, HO. 1975. A social developmental study of young monkeys (*Callithrix jacchus*) within a captive family group. Primates. 16:419-435.
1167. Box, HO & Hubrecht, RC. 1987. Long-term data on the reproduction and maintenance of a colony of common marmosets (*Callithrix jacchus jacchus*) 1972-1983. Lab Anim. 21:249-260.
1168. Boyce, MS. 1981. Beaver life-history resonses to exploitation. J Appl Ecol. 18:749-753.
1169. Boyd, IL. 1983. Luteal regression, follicle growth and the concentration of some plasma steroids during lactation in grey seals *Halichoerus grypus*. J Reprod Fert. 69:157-164.
1170. Boyd, IL. 1984. Development and regression of the corpus luteum in grey seal (*Halichoerus grypus*) ovaries and its use in determining fertility rates. Can J Zool. 62:1095-1100.
1171. Boyd, IL. 1984. The occurrence of hilar rete glands in the ovaries of grey seals (*Halichoerus grypus*). J Zool. 204:585-588.
1172. Boyd, IL. 1984. The relationship between body condition and the timing of implantation in pregnant grey seals (*Halichoerus grypus*). J Zool. 203:113-123.
1173. Boyd, IL. 1985. Effect of photoperiod and melatonin on testes development and regression in wild European rabbits (*Oryctolagus cuniculus*). Biol Reprod. 33:21-29.
1174. Boyd, IL. 1985. Pregnancy and ovulation rates in grey seals (*Halichoerus grypus*) on the British coast. J Zool. 205A:265-272.
1175. Boyd, IL. 1985. The investment in growth by pregnant wild rabbits in relation to litter size and sex of the offspring. J Anim Ecol. 54:137-148.
1176. Boyd, IL. 1986. Photoperiodic regulation of seasonal testicular regression in the wild European rabbit (*Oryctolagus cuniculus*). J Reprod Fert. 77:463-470.
1177. Boyd, IL. 1990. Mass and hormone content of gray seal placentae related to fetal sex. J Mamm. 71:101-103.
1178. Boyd, IL & Myhill, DG. 1987. Seasonal changes in condition, reproduction and fecundity in the wild European rabbit (*Oryctolagus cuniculus*). J Zool. 212:223-233.
1179. Boyd, IL & Myhill, DG. 1987. Variations in the post-natal growth of pipistrelle bats (*Pipistrellus pipistrellus*). J Zool. 213:750-755.
1180. Boyd, JM & Campbell, RN. 1971. The grey seal (*Halichoerus grypus*) at North Rona, 1959 to 1968. J Zool. 164:469-512.
1181. Boyd, JM, Doney, JM, Gunn, RG, et al. 1964. The Soay sheep of the Island of Hirta, St Kilda: A study of a feral population. Proc Zool Soc London. 142:129-163.
1182. Boyd, RJ. 1978. American Elk. Big Game of North America. 11-29, Schmidt, JL & Gilbert, DL, eds, Stackpole Books, Harrisburg, PA.
1183. Brackmann, M & Hoffmann, K. 1977. Pinealectomy and photoperiod influence testicular development in the Djungarian hamster. Naturwissensch. 64:341-342.
1184. Bradbury, JW. 1976. Personal communication. [to Lopez-Forment, W, 1976]
1185. Bradbury, JW. 1977. Lek mating behavior in the hammer-headed bat. Z Tierpsychol. 45:225-255.
1186. Bradbury, JW & Emmons, LH. 1974. Social organization of some Trinidad bats I Emballonuridae. Z Tierpsychol. 36:137-183.
1187. Bradbury, JW & Vehrencamp, SL. 1976. Social organization and foraging in emballonurid bats. Behav Ecol Sociobiol. 1:337-381.
1188. Bradbury, JW & Vehrencamp, SL. 1977. Social organization and foraging in emballonurid bats III Mating systems. Behav Ecol Sociobiol. 2:1-17.
1189. Bradbury, JW & Vehrencamp, SL. 1977. Social organization and foraging in emballonurid bats IV Parental investment patterns. Behav Ecol Sociobiol. 2:19-29.
1190. Bradley, AJ. 1987. Stress and mortality in the red-tailed phascogale, *Phascogale calura* (Marsupialia: Dasyuridae). Gen Comp Endocrinol. 67:85-100.
1191. Bradley, AJ. 1990. Seasonal effects on the haematology and blood chemistry in the red-tailed phascogale, *Phascogale calura* (Marsupialia: Dasyuridae). Aust J Zool. 37:533-543.
1192. Bradley, AJ, Kemper, CM, Kitchener, DJ, et al. 1988. Population ecology and physiology of the common rock rat, *Zyzomys argurus* (Rodentia: Muridae) in tropical northwestern Australia. J Mamm. 69:749-764.
1192a. Bradley, AJ, McDonald, IR, Lee, AK. 1975. Effect of exogenous cortisol on mortality of a dasyurid marsupial. J Endocrinol. 66:281-282.
1193. Bradley, AJ, McDonald, IR, & Lee, AK. 1976. Cortisosteriod-binding globulin and mortality in a dasyarid marsupial. J Endocrinol. 70:323-324.
1194. Bradley, AJ, McDonald, IR, & Lee, AK. 1980. Stress and mortality in a small marsupial (*Antechinus stuartii*, Macleay). Gen Comp Endocrinol. 40:188-200.

1195. Bradley, AJ & Monamy, V. 1991. A physiological profile of male dusky antchinuses, (*Antechinus swainsonii*) (Marsupialia: Dasyuridae) surviving post-mating mortality. Aust Mamm. 14:25-27.
1196. Bradley, RM. 1977. Aspects of the ecology of the Thomson's gazelle in the Serengeti National Park Tanzania. Ph D Diss. TX A & M Univ, College Station.
1196a. Bradley, WG & Mauer, RA. 1965. A collection of bats from Chihuahua, Mexico. Southwest Nat. 10:74-75.
1197. Bradley, WG & Mauer, RA. 1971. Reproduction and food habits of Merriam's kangaroo rat, *Dipodomys merriami*. J Mamm. 52:497-507.
1198. Bradshaw, GVR. 1961. Le cycle de reproduction de *Macrotus californicus* (Chiroptera: Phyllostomatidae). Mammalia. 25:117-119.
1199. Bradshaw, GVR. 1962. Reproductive cycle of the California leaf-nosed bat, *Macrotus californicus*. Science. 136:645-646.
1200. Bradshaw, SD. 1983. Recent endocrinological research on the Rottnest Island quokka (*Setonix brachyurus*). J R Soc W Aust. 66:55-61.
1200a. Bradshaw, SD, Hahnel, R, & Heller, H. 1975. A possible endocrine role for the marsupial placenta. Aust Mamm. 1:407-408.
1201. Bradshaw, SD, McDonald, IR, Hahnel, R, et al. 1975. Synthesis of progesterone by the placenta of a marsupial. J Endocrinology. 65:451-452.
1202. Bradt, GW. 1938. A study of beaver colonies in Michigan. J Mamm. 19:139-162.
1203. Bradt, GW. 1939. Breeding habits of beaver. J Mamm. 20:486-489.
1204. Brady, AG & Koritnik, D. 1983. Testicular volumes, responses to administered gonadotropin releasing hormone, and ejaculate characteristics in peripubertal and adult male African green monkeys. Biol Reprod, 28(Suppl 1):95.
1205. Brady, CA. 1978. Reproduction, growth and parental care in crab-eating foxes *Cerdocyon thous* at the National Zoological Park, Washington. Int Zoo Yearb. 18:130-134.
1206. Brady, CA & Ditton, MK. 1979. Management and breeding of maned wolves *Chrysocyon brachyurus* at the National Zoological Park, Washington. Int Zoo Yearb. 19:171-176.
1207. Braestrup, FW. 1933. On the taxonomic value of the subgenus *Lophomops* (Nyctinomine bats) with remarks on the breeding time of African bats. An Mag Nat Hist Ser 10. 11:269-274.
1208. Braithwaite, RW. 1980. The ecology of *Rattus lutreolus* II Reproductive tactics. Aust Wildl Res. 7:53-62.
1209. Braithwaite, RW & Lee, AK. 1979. A mammalian example of semelparity. Am Nat. 113:151-155.
1210. Brambell, FWR. 1935. Reproduction in the common shrew (*Sorex araneus* Linnaeus) I The oestrous cycle of the female. Phil Trans R Soc London Biol Sci. 225B:1-50.
1211. Brambell, FWR. 1935. Reproduction in the common shrew (*Sorex araneus* Linnaeus) II Seasonal changes in the reproductive organs of the male. Phil Trans R Soc London Biol Sci. 225B:51-62.
1212. Brambell, FWR. 1942. Intra-uterine mortality of the wild rabbit, *Oryctolagus cuniculus* (L). Proc R Soc London Biol Sci. 130B:462-479.
1213. Brambell, FWR. 1944. The reproduction of the wild rabbit, *Oryctolagus cuniculus* (L). Proc Zool Soc London. 114:1-45.
1214. Brambell, FWR. 1948. Prenatal mortality in mammals. Biol Revs. 23:370-407.
1215. Brambell, FWR & Davis, DHS. 1941. Reproduction of the multimammate mouse (*Mastomys erythroleucus* Temm) of Sierra Leone. Proc Zool Soc London. 111B:1-11.
1216. Brambell, FWR & Hall, K. 1936. Anatomical and histological studies of an intersexual lesser shrew (*Sorex minutus* L) with special reference to the effects of the male hormones on the uterus and the vagina. J Anat. 70:339-348.
1217. Brambell, FWR & Hall, K. 1936. Reproduction of the lesser shrew (*Sorex minutus* Linnaeus). Proc Zool Soc London. 1936:957-969.
1218. Brambell, FWR & Hall, K. 1939. Reproduction of the field vole, *Microtus agrestis hirtus* Bellamy. Proc Zool Soc London. 109:133-138.
1219. Brambell, FWR & Mills, IH. 1947. Studies on the sterility and prenatal mortality in wild rabbits Part III The loss of ova before implantation. J Exp Biol. 24:192-210.
1220. Brambell, FWR & Mills, IH. 1948. Studies on sterility and prenatal mortality in wild rabbits Part IV The loss of embryos after implantation. J Exp Biol. 25:241-269.
1221. Brambell, FWR & Perry, JL. 1945. The development of the embryonic membranes of the shrews, *Sorex araneus* Linn and *Sorex minutus* Linn. Proc Zool Soc London. 115:251-278.
1222. Brambell, FWR & Rowlands, IW. 1936. III Reproduction of the bank vole (*Evotomys glareolus*, Schreber) I The oestrous cycle of the female. Phil Trans R Soc London Biol Sci. 226B:71-97.
1223. Brambell, MR. 1974. Breeding fennec foxes *Fennecus zerda* at London Zoo. Int Zoo Yearb. 14:117-118.
1224. Brambell, MR & Matthews, SJ. 1977. Criteria for the management of pudu in captivity. Proc Symp Assoc Brit Wild Animal Keepers. 1:25-28.
1225. Bramblett, CA, Pejaver, LD, & Drickman, J. 1975. Reproduction in captive vervet and sykes monkeys. J Mamm. 56:940-946.
1226. Bramley, PS. 1970. Territorality and reproductive behaviour of roe deer. J Reprod Fert Suppl. 11:43-70.
1227. Bramley, PS & Neaves, WB. 1972. The relationship between social status and reproductive activity in male impala, *Aepyceros melampus*. J Reprod Fert. 31:77-81.
1227a. Brannan, WV, & Marchinton, RL. 1987. Reproductive ecology of white-tailed and red brocket deer in Suriname. Biology and Management of the Cervidae. 344-351, Wemmer, CM, ed, Smithsonian Inst Press, Washington, DC.
1228. Branca, A. 1927. Recherches sur la placentation des Chéiroptères. Arch Zool Exp Gen. 66:291-450.
1229. Brand, CJ & Keith, LB. 1979. Lynx demography during a snowshoe hare decline in Alberta. J Wildl Manage. 43:827-849.
1230. Brand, DJ. 1963. Records of mammals bred in the National Zoological Gardens of South Africa during the period 1908 to 1960. J Zool. 140:617-659.
1231. Brand, DJ & Cullen, L. 1967. Breeding the Cape hunting dog *Lycaon pictus* at Pretoria Zoo. Int Zoo Yearb. 7:124-126.
1232. Brand, HM. 1980. Influence of season on birth distribution in marmosets and tamarins. Lab Anim. 14:301-302.
1233. Brand, HM. 1981. Husbandry and breeding of a newly-established colony of cotton-topped tamarins (*Saguinus oedipus oedipus*). Lab Anim. 15:7-11.
1234. Brand, HM. 1981. Urinary oestrogen excretion in the female cotton-topped tamarin (*Saguinus oedipus oedipus*). J Reprod Fert. 62:467-473.
1235. Brand, HM & Martin, RD. 1983. The relationship between female urinary estrogen excretion and mating behavior in cotton-topped tamarins, *Saguinus oedipus oedipus*. Int J Primatol. 4:275-290.

1236. Brand, LR & Ryckman, RE. 1968. Laboratory life histories of *Peromyscus eremicus* and *Peromyscus interparietalis*. J Mamm. 49:495-501.
1237. Brandborg, SM. 1955. Life history and management of the mountain goat in Idaho. ID Dept Fish Game Wildl Bull. 2:1-142.
1238. Brandt, CA. 1983. Den location by and reproductive success of pikas: A resource defence mating system. Am Zool. 23:932.
1239. Brannian, JD, Griffin, F, Papkoff, H, et al. 1988. Short and long phases of progesterone secretion during the oestrous cycle of the African elephant (*Loxodonta africana*). J Reprod Fert. 84:357-365.
1239a. Brannian, JD, Griffin, F, & Terranova, PF. 1989. Urinary androstenedione and luteinizing hormone concentrations during musth in a mature African elephant. Zoo Biol. 8:165-170.
1240. Brassard, JA & Bernard, R. 1939. Observations on breeding and development of marten, *Martes a americana* (Kerr). Can Field-Nat. 53:15-21.
1241. Brauer, A & Dusing, A. 1961. Sexual cycles and breeding seasons of the gray squirrel, *Sciurus carolinensis* Gmelin. Trans KY Acad Sci. 22:16-27.
1242. Braun, JK & Mares, MA. 1989. *Neotoma micropus*. Mamm Species. 330:1-9.
1243. Braza, F, Garcia, JE, & Alvarez, F. 1986. Rutting behavior of fallow deer. Acta Theriol. 31:467-478.
1244. Braza, F & San Jose, C. 1988. An analysis of mother-young behaviour of fallow deer during lactation period. Behav Processes. 17:93-106.
1245. Braza, F, San Jose, C, & Blom, A. 1988. Birth measurements, parturition dates, and progeny sex ratio of *Dama dama* in Donana, Spain. J Mamm. 69:607-610.
1246. Breakey, DR. 1963. The breeding season and age structure of feral house mouse populations near San Francisco Bay, California. J Mamm. 44:153-168.
1247. Breckon, G & Hulse, EV. 1972. Difficulties in the management of *Sminthopsis crassicaudata* due to iodine deficiency and thyroid diseases. Lab Anim. 6:109-118.
1248. Breed, WG. 1967. Ovulation in the genus *Microtus*. Nature. 214:826.
1249. Breed, WG. 1968. Reproductive physiology of the female short-tailed field vole (*Microtus agrestis*) with special reference to ovarian function. D Phil Diss. Univ Oxford, Oxford.
1250. Breed, WG. 1969. Oestrus and ovarian histology in the lactating vole (*Microtus agrestis*). J Reprod Fert. 18:33-42.
1251. Breed, WG. 1972. The question of induced ovulation in wild voles. J Mamm. 53:185-187.
1252. Breed, WG. 1975. Environmental factors and reproduction in the female hopping mouse *Notomys alexis*. J Reprod Fert. 45:273-281.
1253. Breed, WG. 1976. Effect of environment on ovarian activity of wild hopping mice (*Notomys alexis*). J Reprod Fert. 47:395-397.
1254. Breed, WG. 1978. Ovulation rates and oestrous cycle lengths in several species of Australian native rats (*Rattus* spp) from various habitats. Aust J Zool. 26:475-480.
1255. Breed, WG. 1979. The reproductive rate of the hopping-mouse *Notomys alexis* and its ecological significance. Aust J Zool. 27:177-194.
1256. Breed, WG. 1981. Early embryonic development and ovarian development during concurrent pregnancy and lactation in the hopping-mouse *Notomys alexis*. Aust J Zool. 29:589-604.
1257. Breed, WG. 1981. Histology of accessory sex organs and extra-gonadal sperm reserves in the male hopping mouse *Notomys alexis*. Archs Androl. 7:357-360.
1258. Breed, WG. 1981. Unusual anatomy of the male reproductive tract in *Notomys alexis* (Muridae). J Mamm. 62:373-375.
1259. Breed, WG. 1982. Morphological variation in the testes and accessory sex organs of Australian rodents in the genera *Pseudomys* and *Notomys*. J Reprod Fert. 66:607-613.
1260. Breed, WG. 1983. Variation in sperm morphology in the Australian rodent genus, *Pseudomys* (Muridae). Cell Tiss Res. 229:611-625.
1261. Breed, WG. 1983. Sexual dimorphism in the Australian hopping-mouse, *Notomys alexis*. J Mamm. 64:536-539.
1262. Breed, WG. 1984. Sperm head structure in the Hydromyinae (Rodentia: Muridae): A further evolutionary development of the subacrosomal space in mammals. Gamete Res. 10:31-44.
1263. Breed, WG. 1985. Morphological variation in the female reproductive tract of Australian rodents in the genera *Pseudomys* and *Notomys*. J Reprod Fert. 73:379-384.
1264. Breed, WG. 1986. Comparative morphology and evolution of the male reproductive tract in the Australian hydromyine rodents (Muridae). J Zool. 209A:607-629.
1265. Breed, WG. 1990. Reproductive anatomy and sperm morphology of the long-tailed hopping-mouse, *Notomys longicaudatus* (Rodentia: Muridae). Aust Mamm. 13:201-204.
1266. Breed, WG. 1990. Copulatory behaviour and coagulum formation in the female reproductive tract of the Australian hopping mouse, *Notomys alexis*. J Reprod Fert. 88:17-24.
1266a. Breed, WG. 1992. Reproduction of the spinifex hopping mouse (*Notomys alexis*) in the natural environment. Aust J Zool. 40:57-71.
1267. Breed, WG & Clarke, JR. 1970. Ovulation and associated histological changes in the ovary following coitus in the vole (*Microtus agrestis*). J Reprod Fert. 22:173-176.
1268. Breed, WG & Clarke, JR. 1970. Effect of photoperiod on ovarian function in the vole, *Microtus agrestis*. J Reprod Fert. 23:189-192.
1269. Breed, WG & Clarke, JR. 1970. Ovarian changes during pregnancy and pseudopregnancy in the vole, *Microtus agrestis*. J Reprod Fert. 23:447-456.
1270. Breed, WG & Charlton, HM. 1968. The control of ovulation in the vole *Microtus agrestis*. J Physiol. 198:2-4.
1271. Breed, WG & Charlton, HM. 1971. Hypothalamo-hypophysical control of the ovulation in the vole (*Microtus agrestis*). J Reprod Fert. 25:225-229.
1272. Breed, WG & Inns, RW. 1985. Variation in sperm morphology of Australian Vespertilionidae and its possible phylogenetic significance. Mammalia. 49:105-108.
1273. Breed, WG & Leigh, C. 1985. Sperm head morphology of Australian molossid bats with special reference to the acrosomal structure. Mammalia. 49:403-406.
1274. Breed, WG & Leigh, C. 1990. Morphological changes in the oocyte and its surrounding vestments during in vivo fertilization in the dasyurid marsupial *Sminthopsis crassicaudata*. J Morphol. 204:177-196.
1274a. Breed, WG & Musser, GG. 1991. Sulawesi and Philippine rodents (Muridae): A survey of spermatozoal morphology and its significance for phylogenetic inference. Am Mus Nov. 3003:1-15.
1275. Breed, WG & Papps, M. 1976. Corpus luteum activity during the oestrous cycle of the hopping mouse *Notomys alexis*. Theriogenology. 6:600.
1276. Breed, WG & Sarafis, V. 1978. On the phylogenetic significance of spermatozoal morphology and male reproductive

tract anatomy in Australian rodents. Trans R Soc S Aust. 103:127-135.

1277. Breed, WG & Sarafis, V. 1983. Variation in sperm head morphology in the Australian rodent *Notomys alexis*. Aust J Zool. 31:313-316.

1277a. Brennan, EJ. 1989. Demographics of captive De Brazza's guenons. Zoo Biol. 8:37-47.

1278. Brenner, FJ. 1964. Reproduction of the beaver in Crawford county, Pennsylvania. J Wildl Manage. 28:743-747.

1279. Brenner, FJ. 1968. A three-year study of two breeding colonies of the big brown bat, *Eptesicus fuscus*. J Mamm. 49:775-778.

1280. Brenner, RM, Carlisle, KS, Hess, DL, et al. 1983. Morphology of the oviducts and endometria of cynomolgus macaques during the menstrual cycle. Biol Reprod. 29:1289-1302.

1281. Brenton, C. 1979. Hawaiian monk seal. Mammals in the Seas, FAO Fish Ser 5. 2:104-105.

1282. Brenton, C. 1979. Walrus. Mammals in the Seas, FAO Fish Ser 5. 2:55-57.

1283. Bressou, C & Vandel, G. 1939. Mensurations d'un Élephant d'asie à la naissance. Mammalia. 3:49-52.

1283a. Brett, RA. 1991. The population structure of naked mole-rat colonies. The Biology of the Naked Mole-Rat. 97-136, Sherman, PW, Jarvis, JUM, & Alexander, RD, ed, Princeton Univ Press, NJ.

1284. Breznock, AW, Harrold, JB, & Kawakami, TG. 1977. Successful breeding of the laboratory-housed gibbon (*Hylobates lar*). Lab Anim Sci. 27:222-228.

1285. Bridges, W. 1936. The Haitian *Solenodon*. Bull NY Zool Soc. 39:13-18.

1286. Bridgwater, DD. 1966. Laboratory breeding, early growth, development and behavior of *Citellus tridecemlineatus* (Rodentia). Southwest Nat. 11:325-337.

1287. Briedermann, L. 1971. Zur Reproduktion des Schwarzwildes in der Deutschen Demokratischen Republik. Tag Ber Dt Akad Landwirtsch-Wiss Berlin. 113:169-186.

1288. Brien, Y. 1974. La reproduction du Phoque gris, *Halichoeurus grypus* Fabricius, en Bretagne. Mammalia. 38:346-347.

1289. Briese, D. 1970. The oestrous cycle of the koala (*Phascolarctos cinereus*). B Sc (Honors) Thesis. Dept Zool, Univ Adelaide.

1290. Brimley, CS. 1923. Breeding dates of small mammals at Raleigh, North Carolina. J Mamm. 4:263-264.

1291. Brimley, HH. 1943. A second specimen of True's beaked whale, *Mesoplodon mirus* True, from North Carolina. J Mamm. 24:199-203.

1292. Brinck-Johnsen, T. 1970. Hormonal steroids in the armadillo, *Dasypus novemcinctus* 2 Oestrone and 17β-oestradiol in pregnancy and their in vitro formation by preparations of placentae, early and late in development. Acta Endocrinol. 63:696-704.

1293. Brinck-Johnsen, T, Benirschke, K, & Brinck-Johnsen, K. 1967. Hormonal steroids in the armadillo, *Dasypus novemcinctus*. Acta Endocrinol. 56:675-690.

1294. Brinkmann, AJ. 1948. Studies on female fin and blue whales. Hvalradets Skrifter. 31:1-38.

1924a. Broadbooks, HE. 1952. Nest and behavior of a short-tailed shrew, *Cryptotis parva*. J Mamm. 33:241-243.

1295. Broadbooks, HE. 1958. Life history and ecology of the chipmunk, *Eutamias amoenus*, in eastern Washington. Misc Publ Mus Zool Univ MI. 108:1-42.

1296. Broadbooks, HE. 1970. Populations of the yellow-pine chipmunk, *Eutamias amoenus*. Am Mid Nat. 83:472-488.

1297. Broadhurst, PL & Jinks, JL. 1965. Parity as a determinant of birth weight in the rhesus monkey. Folia Primatol. 3:201-210.

1298. Brockie, RE, Bell, BD, & White, AJ. 1981. Age structure and mortality of possum *Trichosurus vulpecula* populations from New Zealand. Proc First Symp on Marsupials in New Zealand. Victoria Univ Wellington Zool Publ. 74:63-83

1299. Brockman, DK & Lippold, LK. 1975. Gestation and birth of a douc langur *Pygathrix n nemaeus* at San Diego Zoo. Int Zoo Yearb. 15:126-129.

1300. Brockman, DK, Willis, MS, & Karesh, WB. 1987. Management and husbandry of ruffed lemurs, *Varecia variegata*, at San Diego Zoo II Reproduction, pregnancy, parturition, litter size, infant care, and reintrodution of hand-raised infants. Zoo Biol. 6:349-363.

1301. Brodie, ED, III, Brodie, ED, Jr, & Johnson, JA. 1982. Breeding the African hedgehog *Atelerix pruneri* in captivity. Int Zoo Yearb. 22:195-197.

1302. Brodie, PF. 1971. A reconsideration of aspects of growth, reproduction, and behavior of the white whale (*Delphinapterus leucas*), with reference to the Cumberland Sound, Baffin Island, population. J Fish Res Bd Can. 28:1309-1318.

1303. Brodie, PF. 1972. Significance of accessory corpora lutea in odontocetes with reference to *Delphinapterus leucas*. J Mamm. 53:614-616.

1304. Brodmann, K. 1952. Mauswiesel frei im Haus. Bonn Zool Beitr. 3:210.

1305. Brody, EJ & Brody, AE. 1974. Breeding Müller's Bornean gibbon *Hylobates lar muelleri*. Int Zoo Yearb. 14:110-115.

1306. Broekhuizen, S & Maaskamp, F. 1980. Behaviour of does and leverets of the European hare (*Lepus europaeus*) whilst nursing. J Zool. 191:487-501.

1307. Broekhuizen, S & Maaskamp, F. 1981. Annual production of young in European hares (*Lepus europaeus*) in the Netherlands. J Zool. 193:499-516.

1308. Brokx, PA. 1972. Ovarian composition and aspects of the reproductive physiology of Venezuelan white-talied deer (*Odocoileus virginiana gymnotis*). J Mamm. 53:760-773.

1309. Brokx, PAJ. 1972. A Study of the Biology of Venuzuelan White-tailed Deer (*Odocoileus virginianus gymnotis* Wiegmann, 1833), with a Hypothesis on the Origin of South American Cervids. Ph D Diss. Univ Waterloo, Waterloo, Canada.

1310. Bromlej, GF. 1956. Himalayan bear (*Selenarctos tibetanus ussuricus* Heude, 1901). Zool Zhurn. 35:111-129.

1310a. Bromlej, GF. 1956. Goral (*Nemorhaedus caudatus raddeanus* Heude, 1894). Zool Zhurn. 35:1395-1405.

1311. Bromley, PT. 1969. Territoriality in pronghorn bucks on the National Bison Range, Moiese, Montana. J Mamm. 50:81-89.

1312. Bronner, G, Gordon, S, & Meester, J. 1988. *Otomys irroratus*. Mamm Species. 308:1-6.

1313. Bronner, GN & Meester, JAJ. 1988. *Otomys angoniensis*. Mamm Species. 306:1-6.

1314. Bronson, FH. 1979. The reproductive ecology of the house mouse. Q Rev Biol. 54:265-299.

1315. Bronson, FH. 1979. Light intensity and reproduction in wild and domestic house mice. Biol Reprod. 21:235-239.

1316. Bronson, FH. 1985. Mammalian reproduction: An ecological perspective. Biol Reprod. 32:1-26.

1317. Bronson, FH. 1989. Mammalian Reproductive Biology. Univ Chicago Press. IL.

1318. Bronson, FH & Eleftheriou, BE. 1963. Influence of strange males on implantation in the deer mouse. Gen Comp Endocrinol. 3:515-518.

1319. Bronson, FH, Eleftheriou, BE, & Dezell, HE. 1969. Strange male pregnancy block in deer mice: Prolactin and adrenocortical hormones. Biol Reprod. 1:302-306.
1320. Bronson, FH & Marsden, MM. 1964. Male-induced synchrony of estrus in deer mice. Gen Comp Endocrinol. 4:634-637.
1321. Bronson, FH & Pryor, S. 1983. Ambient temperature and reproductive success in rodents living at different latitudes. Biol Reprod. 29:72-80.
1322. Bronson, FH & Tiemeier, OW. 1958. Reproduction and age distribution of black-tailed jack rabbits in Kansas. J Wildl Manage. 22:409-414.
1323. Bronson, MT. 1977. Altitudinal variation in the annual cycle and life history of the golden-mantled ground squirrel *Spermophilus lateralis*. Ph D Diss. Univ CA, Berkeley.
1324. Bronson, MT. 1979. Altitudinal variation in the life history of the golden-mantled ground squirrel (*Spermophitus lateralis*). Ecology. 60:272-279.
1325. Brook, FA. 1984. Studies of ovarian structure and function in the woodmouse, *Apodemus sylvaticus*. D Phil Diss. Univ Oxford, Oxford, UK.
1326. Brook, FA & Clarke, JR. 1989. Overian interstitial tissue of the wood mouse, *Apodemus sylvaticus*. J Reprod Fert. 85:251-260.
1327. Brookhyser, KM & Aulerich, RJ. 1980. Consumption of food, body weight, perineal colour and levels of progesterone in the serum of cyclic female chinchillas. J Endocrinol. 87:213-219.
1328. Brooks, AC. 1961. A Study of the Thomson's Gazelle (*Gazella thomsoni* Gunther) in Tanganyika. Col Res Publ 25. Her Majesty's Stationery office, London.
1329. Brooks, DE, Gaughwin, M, & Mann, T. 1978. Structural and biochemical characteristics of the male accessory organs of reproduction in the hairy-nosed wombat (*Lasiorhinus latifrons*). Proc R Soc London Biol Sci. 201B:191-207.
1330. Brooks, JE & Htun, PT. 1980. Early post-natal growth and behavioural development in the Burmese house rat, *Rattus exulans*. J Zool. 190:125-136.
1331. Brooks, JE, Htun, BT, Walton, DW, et al. 1980. The reproductive biology of *Suncus murinus* L in Rangoon, Burma. Z Säugetierk. 45:12-22.
1332. Brooks, JW. 1954. A contribution to the life history and ecology of the Pacific walrus. AK Coop Wildl Res Unit Report. 1:1-103.
1333. Brooks, PM. 1972. Post-natal development of the African bush rat *Aethomys chrysophilus*. Zool Afr. 7:85-102.
1334. Brooks, PM. 1978. Relationships between body condition and age, growth, reproduction and social status in impala and its application to management. S Afr J Wildl Res. 8:151-157.
1335. Brooks, PM. 1982. Aspects of the reproduction, growth and development of the four-striped field mouse, *Rhabdomys pumilio* (Sparrman, 1784). Mammalia. 46:53-63.
1336. Brooks, RJ, Donald, MA, & Schwarzkopf, L. 1985. Failure of strange females to cause pregnancy block in collared lemmings, *Dicrostonyx groenlandicus*. Behav Neural Biol. 44:485-491.
1337. Brooks, RJ & Webster, AB. 1984. Relationship of seasonal change to changes in age structure and body size in *Microtus pennsylvanicus*. Sp Publ Carnegie Mus Nat Hist. 10:275-284.
1338. Broom, R. 1895. Note on the period of gestation in *Echidna*. Proc Linn Soc NSW. 10:576-577.
1339. Brosset, A. 1961. Some notes on Blanford's or the whitetailed wood rat (*Rattus blanfordi* (Thomas)) in western India. J Bombay Nat Hist Soc. 58:241-248.
1340. Brosset, A. 1962. La reproduction des Chiroptères de l'ouest et du centre de l'Inde. Mammalia. 26:176-213.
1341. Brosset, A. 1962-63. The bats of central and western India. J Bombay Nat Hist Soc. 59:1-57, 583-624, 707-746; 60:337-355.
1342. Brosset, A. 1962. The bats of central and western India. J Bombay Nat Hist Soc. 59:1-57.
1343. Brosset, A. 1962. The bats of central and western India. J Bombay Nat Hist Soc. 59:583-624.
1344. Brosset, A. 1962. The bats of central and western India. J Bombay Nat Hist Soc. 59:707-746.
1345. Brosset, A. 1963. The bats of central and western India. J Bombay Nat Hist Soc. 60:337-355.
1345a. Brosset, A. 1965. Contribution a l'étude des chiroptères de l'ouest de l'Ecuador. Mammalia. 29:211-227.
1346. Brosset, A. 1966. La Biologie des Chiroptères. Masson et Cie, Paris.
1347. Brosset, A. 1966. Les Chiroptères du Haut-Ivindo (Gabon). Biol Gabon. 2:47-86.
1348. Brosset, A. 1968. La permutation du cycle sexual saisonnier chez le Chiroptère *Hipposideros caffer*, au voisinage de l'équateur. Biol Gabon. 4:325-341.
1349. Brosset, A. 1969. Recherches sur la biologie des Chiroptères troglophiles dans le nord-est du Gabon. Biol Gabon. 5:93-116.
1350. Brosset, A. 1976. Social organization in the African bat, *Myotis boccagei*. Z Tierpsychol. 42:50-56.
1350a. Brosset, A & Dubost, G. 1967. Chiroptères de la Guyane Française. Mammalia. 31:583-594
1351. Brosset, A & St Girons, H. 1980. Cycles de reproduction des Microchiroptères troglophiles du nord-est du Gabon. Mammalia. 44:225-232.
1352. Brouin, G. 1950. Note sur les Ongulés du Cercle d Agadez et leur chasse. Mém Inst Francais Afr Noire. 10:425-455.
1353. Brown, CE. 1924. Rearing hippopotamuses in captivity. J Mamm. 5:243-246.
1354. Brown, CE. 1930. Birth of second chimpanzee in the Philadelphia Zoological Garden. J Mamm. 11:303-305.
1355. Brown, CE. 1936. Rearing wild animals in captivity and gestation periods. J Mamm. 17:10-13.
1356. Brown, JC. 1964. Observations on the elephant shrews (Macroscelididae) of equatorial Africa. Proc Zool Soc London. 143:103-119.
1357. Brown, JL, Wildt, DE, Phillips, LG, et al. 1989. Adrenal-pituitary-gonadal relationships and ejaculate characteristics in captive leopards (*Panthera pardus kotiya*) isolated on the island of Sri Lanka. J Reprod Fert. 85:605-613.
1358. Brown, JL, Goodrowe, KL, Simmons, LG, et al. 1988. Evaluation of the pituitary-gonadal response to GnRH, and adrenal status, in the leopard (*Panthera pardus japonensis*) and tiger (*Panthera tigris*). J Reprod Fert. 82:227-236.
1359. Brown, KG. 1952. Observations on the newly born leopard seal. Nature. 170:982-983.
1360. Brown, KR. 1924. Hermaphroditism in a mole with male external genitalia. J Anat. 58:355-358.
1361. Brown, LC & Yeager, LE. 1945. Fox squirrels and gray squirrels in Illinois. Bull IL Nat Hist Surv. 23:449-536.
1362. Brown, LN. 1964. Breeding records and notes on *Phenacomys silvicola* in Oregon. J Mamm. 45:647-648.
1363. Brown, LN. 1964. Reproduction of the brush mouse and white-footed mouse in the central United States. Am Mid Nat. 72:226-240.
1364. Brown, LN. 1966. Reproduction of *Peromyscus maniculatus* in the Laramie Basin of Wyoming. Am Mid Nat. 76:183-189.

1365. Brown, LN. 1968. Smallness of mean litter size in the Mexican vole. J Mamm. 49:159.
1366. Brown, LN. 1969. Reproductive characteristics of the Mexican woodrat at the northern limit of its range in Colorado. J Mamm. 50:536-541.
1367. Brown, LN. 1971. Breeding biology of the pocket gopher, *Geomys pinetis* in southern Florida. Am Mid Nat. 85:45-53.
1368. Brown, LN. 1975. Ecological relationships and breeding biology of the nutria (*Myocaster coypus*) in the Tampa, Florida, area. J Mamm. 56:928-930.
1369. Brown, LN. 1977. Litter size and notes on reproduction in the giant water vole (*Arvicola richardsoni*). Southwest Nat. 22:281-282.
1370. Brown, LN & Conaway, CH. 1964. A balanced lethal sex-chromosome system. J Hered. 55:7-8.
1371. Brown, PE, Brown, TW, & Grinnell, AD. 1983. Echolocation, development, and vocal communication in the lesser bulldog bat, *Noctilio albiventris*. Behav Ecol Sociobiol. 13:287-298.
1372. Brown, RD, Chao, CC, & Faulkner, LW. 1983. The endocrine control of the initiation and growth of antlers in white-tailed deer. Acta Endocrinol. 103:138-144.
1373. Brown, RD, Cowan, RL, & Kavanaugh, JF. 1978. Effect of pinealectomy on seasonal androgen titers, antler growth and feed intake in white-tailed deer. J Anim Sci. 47:435-440.
1374. Brown, RNR. 1915. The seals of the Weddell Sea: Notes on their habits and distribution. Scott Nat Antarctic Exp Rep. 4:185-198.
1375. Brown, SG & Lockyer, CH. 1984. Whales. Antarctic Ecology. 2:717-781, Laws, RM, ed, Academic Press, New York.
1376. Brown, WH, Stull, JW, & Sowls, LK. 1963. Chemical comparison of the milk fat of the collared peccary. J Mamm. 44:112-113.
1377. Brownell, RL. 1975. Progress report on the biology of the Franciscana dolphin, *Pontoporia blainvillei*, in Uruguayan waters. J Fish Res Bd Can. 32:1073-1078.
1378. Brownell, RL, Jr. 1983. *Phocoena sinus*. Mamm Species. 198:1-3.
1379. Brownell, RL, Jr & Herald, ES. 1972. *Lipotes vexillifer*. Mamm Species. 10:1-4.
1380. Brownell, RL, Jr & Praderi, R. 1984. *Phocoena spinipinnis*. Mamm Species. 217:1-4.
1381. Brownscheidle, CM, Dresser, BL, & Russell, PT. 1979. The estrous cycle of an African bongo-breeding behavior and cytology of vaginal smears. J Zoo Anim Med. 10:41-49.
1382. Bruce, HM & Hindle, E. 1934. The golden hamster, *Cricetus* (*Mesocricetus*) *auratus* Waterhouse: Notes on its breeding and growth. Proc Zool Soc London. 1934:361-366.
1383. Bruce, WS. 1913. Measurements and weights of Antarctic seals. Trans R Soc Edinb. 49:567-577.
1384. Brüggemann, J, Adam, A, & Karg, H. 1965. ICSH-Bestimmungen in Hypophyse von Rehböcken (*Capreolus capreolus*) und Hirschen (*Cervus elaphus*) unter Berücksichtigung des Saisoneinflusses. Acta Endocrinol. 48:569-580.
1385. Brüggemann, S & Dukelow, WR. 1980. Characteristics of the menstrual cycle in nonhuman primates III Timed mating in *Macaca arctoides*. J Med Primatol. 9:213-221.
1386. Bruhin, H. 1979. *Callithrix jacchus*, Marmoset ein Neues Versuchsmodell. Z Versuchstierk. 21:209-221.
1386a. Bruna, JF. 1952. Kentucky Rabbit Investigations. Dept Fish Wildl Res. Frankfort, KE.
1387. Bryant, S. 1988. Maintenance and captive breeding of the eastern quoll *Dasyurus viverrinus*. Int Zoo Yearb. 27:119-124.
1388. Bryant, SL. 1986. Seasonal variation of plasma testosterone in a wild population of male eastern quoll *Dasyurus viverrinus* (Marsupialia: Dasyuridae) from Tasmania. Gen Comp Endocrinol. 64:75-79.
1389. Bryant, SL & Rose, RW. 1985. Effect of cadmium on the reproductive organs of the male potoroo *Potorous tridactylus* (Macropodidae). Aust J Biol Sci. 38:305-311.
1390. Bryant, SL & Rose, RW. 1986. Growth and role of the corpus luteum throughout delayed gestation in the potoroo, *Potorous tridactylus*. J Reprod Fert. 76:409-414.
1391. Bryden, HA. 1899. The blesbok (*Damaliscus albifrons*). Great and Small Game of Africa. 183-190, Bryden, HA, ed, Rowland Ward, London.
1392. Bryden, HA. 1899. The brindled gnu or blue wildebeest (*Connochaetes taurinus typicus*). Great and Small Game of Africa. 194-199, Bryden, HA, ed, Rowland Ward, London.
1393. Bryden, HA. 1899. The springbuck (*Antidorcas euchore*). Great and Small Game of Africa. 332-340, Bryden, HA, ed, Rowland Ward, London.
1394. Bryden, MM. 1966. Twin foetuses in the southern elephant seal, *Mirounga leonina* (L). Pap Proc R Soc Tas. 100:89-90.
1395. Bryden, MM. 1967. Testicular temperature in the southern elephant seal, *Mirounga leonina* (Linn). J Reprod Fert. 13:583-584.
1396. Bryden, MM. 1969. Growth of the southern elephant seal, *Mirounga leonina* (Linn). Growth. 33:69-82.
1397. Bryden, MM. 1972. Body size and composition of elephant seals (*Mirounga leonina*): Absolute measurements and estimates from bone dimensions. J Zool. 167:265-276.
1398. Bryden, MM, Harrison, RJ, & Lear, RJ. 1977. Some aspects of the biology of *Peponocephala electra* (Cetacea: Delphinidae) I General and reproductive biology. Aust J Mar Freshwater Res. 28:703-715.
1399. Bryden, MM, Smith, MSR, Tedman, RA, et al. 1984. Growth of the Weddell seal, *Leptonychotes weddelli* (Pinnipedia). Aust J Zool. 32:33-41.
1400. Bubenik, GA. 1983. Shift of seasonal cycle in white-tailed deer by oral administration of melatonin. J Exp Zool. 225:155-156.
1401. Bubenik, GA, Bubenik, AB, Brown, GM, et al. 1975. Growth hormone and cortisol levels in the annual cycle of white-tailed deer (*Odocoileus virginianus*). Can J Physiol Pharmacol. 53:787-792.
1401a. Bubenik, GA, Bubenik, AB, Schams, D, et al. 1983. Circadian and circannual rhythms of LH, FSH, testosterone (T), prolactin, cortisol, T_3 and T_4 in plasma of mature, male white-tailed deer. Comp Biochem Physiol. 76A:37-45.
1402. Bubenik, GA, Bubenik, AB, & Zamecnik, J. 1979. The development of circannual rhythm of estradiol in plasma of white-tailed deer (*Odocoileus virginianus*). Comp Biochem Physiol. 62A:869-872.
1403. Bubenik, GA & Leatherland, JF. 1984. Seasonal levels of cortisol and thyroid hormones in intact and castrated mature male white-tailed deer. Can J Zool. 62:783-787.
1404. Bubenik, GA & Schams, D. 1986. Relationship of age to seasonal levels of LH, FSH, prolactin and testosterone in male, white-tailed deer. Comp Biochem Physiol. 83A:179-183.
1405. Bubenik, GA, Schams, D, & Sempere, AJ. 1987. Assessment of the sexual and antler potential of the male white-tailed deer (*Odocoileus virginianus*) by Gn-RH stimulation test. Comp Biochem Physiol. 86A:767-771.
1406. Bubenik, GA & Smith, PS. 1987. Circadian and circannual rhythms of melatonin in plasma of male white-tailed deer and the effect of oral administration of melatonin. J Exp Zool. 241:81-89.

1407. Buchalczyk, A. 1961. *Pitymys subterraneus* (de Selys-Longchamps 1835) under laboratory conditions. Acta Theriol. 4:282-284.
1408. Buchalczyk, A. 1970. Reproduction, mortality and longevity of the bank vole under laboratory conditions. Acta Theriol. 15:153-176.
1409. Buchanan, DB, Mittermeier, RA, & van Roosmalen, MGM. 1981. The saki monkeys, genus *Pithecia*. Ecology and Behavior of Neotropical Primates. 1:391-417. Coimbra-Filho, AF & Mittermeier, RA, eds, Academia Brasileira de Ciêcias, Rio de Janeiro.
1410. Buchanan, DG. 1963. Probable delayed implantation in Bennett's wallaby. J Mamm. 44:430-431.
1411. Buchanan, GD. 1957. Variation in litter size of nine-banded armadillos. J Mamm. 38:529.
1412. Buchanan, GD. 1966. Reproduction in the ferret (*Mustela furo*) I Uterine histology and histochemistry during pregnancy and pseudopregnancy. Am J Anat. 118:195-216.
1413. Buchanan, GD. 1969. Reproduction in the ferret (*Mustela furo*) II Changes following ovariectomy during early pregnancy. J Reprod Fert. 18:305-316.
1414. Buchanan, GD. 1987. Timing of ovulation and early embryonic development in *Myotis lucifugus* (Chiroptera: Vespertilionidae) from northern central Ontario. Am J Anat. 178:335-340.
1415. Buchanan, GD, Enders, AC, & Talmage, RV. 1956. Implantation in armadillos ovariectomized during the period of delayed implantation. J Endocrinol. 14:121-128.
1416. Buchanan, GD, Garfield, RE, & YoungLai, EV. 1988. Innervation and gap junction formation in the myometrium of pregnant little brown bats, *Myotis lucifugus*. Anat Rec. 221:611-618.
1417. Buchanan, GD & Ryan, ED. 1983. Steroid receptors in the uteri of hibernating little brown bats, *Myotis lucifugus*. Biol Reprod. 28(Suppl):126.
1418. Buchanan, GD & YoungLai, EV. 1986. Plasma progesterone levels during pregnancy in the little brown bat *Myotis lucifugus* (Vespertilionidae). Biol Reprod. 34:878-884.
1419. Buchanan, GD & YoungLai, EV. 1988. Plasma progesterone concentrations in female little brown bats (*Myotis lucifugus*) during hibernation. J Reprod Fert. 83:59-65.
1420. Bucher, F. 1968/1969. Haltung und Zucht von Vikunjas (*Vicugna vicugna*) in Zürcher Zoo. Zool Garten NF. 36:153-159.
1421. Bucher, GC. 1937. Notes on life history and habits of *Capromys*. Mem Soc Cuban Hist Nat. 11(2):93-107.
1422. Bucher, JE & Fritz, HI. 1977. Behavior and maintenance of the woolly opossum (*Caluromys*) in captivity. Lab Anim Sci. 27:1007-1012.
1423. Bucher, JE & Hoffmann, RS. 1980. *Caluromys derbianus*. Mamm Species. 140:1-4.
1424. Buck, PD. 1947. The biology of the antelope (*Antilocapra americana*) in Montana. MS Thesis. MT State Coll, Bozeman.
1425. Buckley, P & Caine, A. 1979. A high incidence of abdominal pregnancy in the Djungarian hamster (*Phodopus sungorus*). J Reprod Fert. 56:679-682.
1426. Buckner, CH. 1966. Populations and ecological relationships of shrews in tamarack bogs of south-eastern Manitoba. J Mamm. 47:181-194.
1427. Buckner, CH. 1969. Some aspects of the population ecology of the common shrew, *Sorex araneus*, near Oxford, England. J Mamm. 50:326-332.
1428. Budd, GM. 1972. Breeding of the fur seal at McDonald Islands, and further population growth at Heard Island. Mammalia. 36:423-427.
1429. Budd, GM & Downes, MC. 1969. Population increase and breeding in the Kerguelen fur seal, *Arctocephalus tropicalis gazella*, at Heard Island. Mammalia. 33:58-67.
1429a. Buden, DW. 1975. A taxonomic and zoogeographic appraisal of the big-eared bat (*Macrotus waterhousii* Gray) in the West Indies. J Mamm. 56:758-769.
1429b. Buden, DW. 1976. A review of the bats of the endemic West Indian genus *Erophylla*. Proc Biol Soc Washington. 89(1):1-16.
1429c. Buden, DW. 1977. First records of bats of the genus *Brachyphylla* from the Caicos Islands, with notes on geographic variation. J Mamm. 58:221-225.
1430. Budnitz, N & Dainis, K. 1975. *Lemur catta*: Ecology and behavior. Lemur Biology. 219-235, Tattersall, I & Sussman, RW, eds, Plenum Press, New York.
1431. Buechner, HK. 1961. Unilateral implantation in the Uganda kob, *Adenota kob thomasi* (PL Slater). Nature. 190:738.
1432. Buechner, HK. 1974. Implications of social behavior in the management of Uganda kob. The Behaviour of Ungulates and its Relation to Management. Geist, V & Walther, F, eds, IUCN, Morges, Switzerland.
1433. Buechner, HK & Mackler, SF. 1978. Breeding behavior in captive Indian rhinoceros. Zool Gärten NF. 48:305-322.
1434. Buechner, HK, Morrison, JA, & Leuthold, W. 1966. Reproduction in Uganda kob with special reference to behaviour. Symp Zool Soc London. 15:69-88.
1435. Buechner, HK & Mossman, HW. 1969. The opening between the allantoic vesicle and the uterine cavity in the kob conceptus. J Reprod Fert Suppl. 6:185-187.
1436. Buechner, HK & Schloeth, R. 1965. Ceremonial mating behavior in Uganda kob (*Adenota kob thomasi* Neumann). Z Tierpsychol. 22:209-225.
1437. Buechner, HK, Stroman, HR, & Xanten, WA. 1974. Breeding behavior of sable antelope *Hippotragus niger* in captivity. Int Zoo Yearb. 14:133-136.
1438. Buechner, HK & Swanson, CW. 1955. Increased natality resulting from lowered population density among elk in southwestern Washington. Trans N Am Wildl Conf. 20:560-567.
1439. Buettner-Janusch, J. 1964. The breeding of galagos in captivity and some notes on their behavior. Folia Primatol. 2:93-110.
1440. Buettner-Janusch, J. 1967. A leumr research colony. Int Zoo Yearb. 7:197-200.
1441. Buhl, AE, Hasler, JF, Tyler, MC, et al. 1978. The effects of social rank on reproductive indices of groups of collared lemmings (*Dicrostonyx groenlandicus*). Biol Reprod. 18:317-324.
1442. Bujalska, G. 1973. The role of spacing behaviour among females in the regulation of reproduction in the bank vole. J Reprod Fert Suppl. 19:465-474.
1443. Bulir, L. 1972. Breeding binturongs *Arctictis binturong* at Liberec Zoo. Int Zoo Yearb. 12:117-118.
1444. Bullard, SC. 1984. Effects of testosterone upon the chest-rubbing behavior of *Galago crassicaudatus umbrosus*. Folia Primatol. 42:70-75.
1445. Bungartz, MAH. 1949. Mantelpavian-Zwillinge im Zoo Hannover. Zool Garten NF. 16:133.
1446. Bunnell, FL & Olsen, NA. 1976. Weights and growth of Dall's sheep in Kluane park reserve, Yukon Territory. Can Field-Nat. 90:157-162.
1447. Bunnell, FL & Olsen, NA. 1981. Age-specific natality in Dall's sheep. J Mamm. 62:379-380.
1448. Bunnell, FL. 1980. Factors controlling lambing period of Dall's sheep. Can J Zool. 58:1027-1031.

1449. Bunnell, FL. 1982. The lambing period of mountain sheep: Synthesis, hypotheses, and tests. Can J Zool. 60:1-14.
1449a. Burda, H. 1989. Relationships among rodent taxa as indicated by reproductive biology. Z Zool Syst Evolut-forsch. 27:49-57.
1449b. Burda, H. 1989. Reproductive biology (behaviour, breeding, and postnatal development) in subterranean mole-rats, *Cryptomys hottentotus* (Bathyergidae). Z Säugetierk. 54:360-376.
1449c. Burda, H. 1990. Constraints of pregnancy and evolution of sociality in mole-rats. Z Zool Syst Evolut-Forsch. 28:26-39.
1450. Burgemeister, R. 1974. Probleme der Dromedarhaltung und-Zucht in Südtunesien. Thesis Justus Liebig Univ, Giessen, Germany. (abstract in Farid, MFA. 1981. Camelids Bibliography. page 83, Arab Centre Stud Arid Zones Dry Lands)
1451. Burger, M. 1966. Breeding of the european lynx *Felis lynx* at Magdeburg Zoo. Int Zoo Yearb. 6:182-183.
1452. Bürger, M. 1973. Weitere Beobachtungen zur Zucht des Europäischen Feldhasen, *Lepus europaeus* Pallas, in Gefangenschaft. Zool Garten NF. 43:275-277.
1452a. Burley, RA, Holman, SD, & Hutchinson, JB. 1983. The regulation of precopulatory behavior by ovarian hormones in the female Mongolian gerbil. Horm Behav. 17:374-387.
1453. Burne, EC. 1943. A record of gestation periods and growth of trained Indian elephant calves in the southern Shan states, Burma. Proc Zool Soc London. 113A:27.
1454. Burnell, AF, Burnell, SR, & Tagg, M. 1991. Observations on an apparent mating sequence in three southern right whales, *Eubalaena australia* (Cetacea: Balaenidae). Aust Mamm. 14:33-34.
1455. Burnett, CD & Kunz, TH. 1982. Growth rates and age estimation in *Eptesicus fuscus* in comparison with *Myotis lucifugus*. J Mamm. 63:33-41.
1456. Burns, JJ. 1967. The Pacific bearded seal. Fed Aid Wildl Restor Proj Rep, AK Dept Fish Game. 8:1-66.
1457. Burns, JJ. 1970. Remarks on the distribution and natural history of pagophilic pinnipeds in the Bering and Chukchi seas. J Mamm. 51:445-454.
1458. Burns, JJ. 1981. Bearded seal *Erignathus barbatus* Erxleben, 1777. Handbook of Marine Mammals. 2:145-170. Ridgway, SH & Harrison, RJ, eds. Academia Press, New York.
1459. Burns, JJ. 1981. Ribbon seal *Phoca fasciata* Zimmermann, 1783. Handbook of Marine Mammals. 2:89-109, Ridgway, SH & Harrison, RJ, eds. Academia Press, New York.
1460. Burns, JM. 1981. Aspects of endocrine control of delay phenomena in bats with special emphasis on delayed development. J Reprod Fert Suppl. 29:61-66.
1461. Burns, JM, Baker, RJ, & Bleier, WJ. 1972. Hormonal control of delayed development in *Macrotus waterhousii* I Changes in plasma thyroxine during pregnancy and lactation. Gen Comp Endocrinol. 18:54-58.
1462. Burns, JM & Wallace, WE. 1975. Hormonal control of delayed development in *Macrotus waterhousii* II Radioimmunoassay of plasma estrone and estradiol-17β during pregnancy. Gen Comp Endocrinol. 25:529-533.
1463. Burns, RJ. 1970. Twin vampire bats born in captivity. J Mamm. 51:391-392.
1463a. Burns, RK. 1939. The differentiation of sex in the opossum (*Didelphys virginiana*) and its modification by the male hormone testosterone propionate. J Morphol. 65:79-119.
1463b. Burns, RK. 1939. Sex differentiation during the early pouch stages of the opossum (*Didelphys virginiana*) and a comparison of the anatomical changes induced by male and female sex hormones. J Morphol. 65:497-547.
1463c. Burns, RK. 1942. Hormones and experimental modification of sex in the opossum. Biol Symp. 9:125-146.
1463d. Burns, RK. 1955. Experimental reversal of sex in the gonads in the opossum *Didelphis virginiana*. Proc Nat Acad Sci. 41:669-676.
1463e. Burns, RK. 1956. Hormones versus constitutional factors in the growth of embryonic sex primordia in the opossum. Am J Anat. 98:35-67.
1463f. Burns, RK. 1956. Transformation of the embryonic testicle of the opossum into an ovatestis or into an 'ovary' under the action of the female hormone estrogen dipropionate. Arch Microsc Anat Exp Morphol. 45:174-200.
1464. Burns, RK & Burns, LM. 1956. Vie et reproduction de l'Opossum américain *Didelphis marsupialis virginiana* Kerr. Bull Soc Zool France. 81:230-246.
1465. Burns, RK & Burns, LM. 1957. Observations on the breeding of the American opossum in Florida. Rev Suisse Zool. 64:595-605.
1465a. Burns, RK & Crespo, RF. 1975. Notes on local movement and reproduction of vampire bats in Colima, Mexico. Southwest Nat. 19:446-449.
1466. Burrell, GC. 1982. Age-related fecundity of mule deer (*Odocoileus hemionus hemionus*) on Entiat winter range. Murrelet. 63:26-27.
1467. Burrell, H. 1927. The Platypus. Angus and Robertson, Sydney.
1468. Burt, DA & Plant, M. 1983. Observations on marmoset breeding at Fisons. Anim Technol. 34:29-36.
1469. Burt, WH. 1928. Additional notes on the life history of the Goss lemming mouse. J Mamm. 9:212-216.
1469a. Burt, WH. 1932. The fish-eating habits of *Pizonyx vivesi* (Menegaux). J Mamm. 13:363-365.
1470. Burt, WH. 1933. Additional notes on the mammals of southern Arizona. J Mamm. 14:114-122.
1471. Burt, WH. 1936. Notes on the habits of the Mohave ground squirrel. J Mamm. 17:221-224.
1471a. Burt, WH. 1938. Faunal relationships and geographic distribution of mammals in Sonora, Mexico. Misc Publ Mus Zool Univ MI. 39:1-77.
1472. Burt, WH. 1940. Territorial behavior and populations of some small mammals in southern Michigan. Misc Publ Mus Zool Univ MI. 45:5-58.
1473. Burt, WH. 1960. Bacula of North American mammals. Misc Publ Mus Zool Univ MI. 113:5-70.
1474. Burt, WH & Grossenheider, RP. 1964. A Field Guide to the Mammals. Houghton Miffin, Boston.
1475. Burt, WH & Stirton, RA. 1961. The mammals of El Salvador. Misc Publ Mus Zool Univ MI. 117:1-69.
1476. Burton, FD & dePelham, A. 1979. A twinning event in *Macaca sylvanus* of Gibraltar. J Med Primatol. 8:105-112.
1477. Burton, FD & Sawchuk, LA. 1982. Birth intervals in *Macaca sylvanus* of Gibraltar. Primates. 23:140-144.
1478. Burton, RW. 1940. The Indian wild dog. J Bombay Nat Hist Soc. 41:692-715.
1479. Burton, RW, Anderson, SA, & Summers, CF. 1975. Perinatal activities in the grey seal (*Halichoerus grypus*). J Zool. 177:197-201.
1480. Bushberg, DM & Holmes, WG. 1985. Sexual maturation in male Belding's ground squirrels: Influence of body weight. Biol Reprod. 33:302-308.
1481. Buss, DH. 1968. Gross composition and variation of the components of baboon milk during natural lactation. J Nutr. 96:421-426.

1481a. Buss, DH & Cooper, RW. 1972. Composition of squirrel monkey milk. Folia Primatol. 17:285-291.
1482. Buss, DH, Cooper, RW, & Wallen, K. 1976. Composition of lemur milk. Folia Primatol. 26:301-305.
1483. Buss, DH & Hamner, JE, III. 1971. Supernumerary nipples in the baboon (*Papio cynocephalus*). Folia Primatol. 16:153-158.
1484. Buss, IO & Johnson, OW. 1967. Relationships of Leydig cells characteristics and intratesticular testosterone levels to sexual activity in the African elephant. Anat Rec. 157:191-196.
1485. Buss, IO & Smith, NS. 1966. Observations on reproduction and breeding behavior of the African elephant. J Wildl Manage. 30:375-388.
1486. Busse, C & Hamilton, WJ, III. 1981. Infant carrying by male chacma baboons. Science. 212:1281-1283.
1487. Busse, CD & Gordon, TP. 1983. Attacks on neonates by a male mangabey (*Cercocebus atys*). Am J Primatol. 5:345-356.
1488. Butler, H. 1957. The breeding cycle of the Senegal galago *Galago senegalensis*. Proc Zool Soc London. 129:147-149.
1489. Butler, H. 1959. An early blastocyst of the lesser bush baby (*Galago senegalensis senegalensis*): A preliminary account. J Anat. 93:257-261.
1490. Butler, H. 1960. Some notes on the breeding cycle of the Senegal galago *Galago senegalensis senegalensis* in the Sudan. Proc Zool Soc London. 135:423-430.
1491. Butler, H. 1964. The reproductive biology of a strepsirhine (*Galago senegalensis senegalensis*). Int Rev Gen Exp Zool. 1:241-296.
1492. Butler, H. 1966. Obervations on the menstrual cycle of the grivet monkey (*Cercopithecis aethiops aethiops*) in the Sudan. Folia Primatol. 4:194-205.
1493. Butler, H. 1967. Seasonal breeding of the Senegal galago (*Galago senegalensis senegalensis*) in the Nuba Mountains, Republic of the Sudan. Folia Primatol. 5:165-175.
1494. Butler, H. 1967. The giant cell trophoblast of the Senegal galago (*Galago senegalensis senegalensis*) and its bearing on the evolution of the primate placenta. J Zool. 152:195-207.
1495. Butler, H. 1967. The oestrus cycle of the Senegal galago (*Galago senegalensis senegalensis*) in the Sudan. J Zool. 151:143-162.
1496. Butler, H. 1969. Post pubertal oogenesis in prosimiae. Int Congr Primat. 2:215-221.
1497. Butler, H & Adam, KR. 1964. The structure of the allantoic placenta of the Senegal bush baby (*Galago senegalensis senegalensis*). Folia Primat. 2:22-49.
1498. Butler, H & Juma, MB. 1970. Oogenesis in an adult prosimian. Nature. 226:552-553.
1499. Butler, J. 1985. The estrous cycle of the chinchilla. Empress Chinchilla Breeder. 41(3):8-9.
1500. Butler, WR, Krey, LC, Lu, KH, et al. 1975. Surgical disconnection of the medial basal hypothalamus and pituitary function in the rhesus monkey IV Prolactin secretion. Endocrinology. 96:1099-1105.
1501. Butterstein, GM & Schadler, MH. 1988. The plant metabolite 6-methoxybenoxazoline interacts with follicle-stimulating hormone to enhance ovarian growth. Biol Reprod. 39:465-471.
1502. Butterworth, BB. 1960. The cricetid mouse, *Calomys*, from Venezuela. J Mamm. 41:517-518.
1503. Butterworth, BB. 1961. A comparative study of growth and development of the kangaroo rats, *Dipodomys deserti* Stephens and *Dipodomys merriami* Mearns. Growth. 25:127-139.
1504. Butterworth, BB. 1961. The breeding of *Dipodomys deserti* in laboratory. J Mamm. 42:413-414.
1505. Butterworth, BB. 1969. Postnatal growth and development of *Ursus americanus*. J Mamm. 50:615-616.
1506. Butts, GL, Harmel, DE, Cook, RL, et al. 1978. Fawning dates of known-age white-tailed deer and their management implications. Proc Ann Southeastern Conf Fish Wildl Agencies. 32:335-338.
1507. Butynski, TM. 1973. Life history and economic value of the springhare (*Pedetes capensis forster*) in Botswana. Botswana Notes and Records. 5:209-213.
1508. Butynski, TM. 1978. Ecological studies on the springhare *Pedetes capensis* in Botswana. Ph D Diss. MI State Univ, E Lansing, MI.
1509. Butynski, TM. 1979. Reproductive ecology of the springhaas *Pedetes capensis* in Botswana. J Zool. 189:221-232.
1510. Butynski, TM. 1980. Growth and development of the foetal spring hare *Pedetes capensis* in Botswana. Mammalia. 44:361-369.
1511. Butynski, TM. 1982. Harem-male replacement and infanticide in the blue monkey (*Cercopithecus mitis stuhlmanni*) in the Kibale Forest, Uganda. Am J Primatol. 3:1-22.
1512. Butynski, TM & Hanks, J. 1979. Reproductive activity in the male springhare *Pedetes capensis* in Batswana. S Afr J Wildl Res. 9:13-17.
1513. Buxton, PA. 1936. Breeding rates of domestic rats trapped in Lagos, Nigeria and certain other countries. J Anim Ecol. 5:53-66.
1514. Bychov, VA. 1973. Atlantic walrus, *Odobenus rosmarus rosmarus* L, 1758 Novaya Zembla population. IUCN Suppl Pap. 39:56-58.
1515. Byers, AP, Hunter, AG, Seal, US, et al. 1989. In-vitro induction of capacitation of fresh and frozen spermatozoa of the Siberian tiger (*Panthera tigris*). J Reprod Fert. 86:599-607.
1515a. Byers, JA & Moodie, JD. 1990. Sex-specific maternal investment in pronghorn, and the question of a limit on differential provisioning in ungulates. Behav Ecol Sociobiol. 26:157-164.
1516. Bygott, JD, Bertram, BCR, & Hanby, JP. 1979. Male lions in large coalitions gain reproductive advantages. Nature. 282:839-841.
1516a. Byrd, ML, Benirschke, K, & Gould, GC. 1988. Establishment of the first captive group of the Chaco peccary, *Catagonus wagneri*. Zool Garten. 58:265-274.
1517. Byskov, AG. 1975. The role of the rete ovarii in meiosis and follicle formation in the cat, mink and ferret. J Reprod Fert. 45:201-209.
1518. Cabon-Raczynska, K. 1974. Variability of the body weight of European hares. Acta Theriol. 19:69-79.
1519. Cabrera, A, & Yepes, J. 1940. Historia Natural Ediar Mamiferos Sub-Americanos (Vida, Costumbres y Descripcion). Compania Argentina de Editores, Buenos Aires, Argentina.
1520. Cade, CE. 1966. A note on the mating behaviour of the Kenya oribi *Ourebia ourebi* in captivity. Int Zoo Yearb. 6:205.
1521. Cade, CE. 1967. Notes on breeding the Cape hunting dog *Lycaon pictus* at Nairobi Zoo. Int Zoo Yearb. 7:122-123.
1522. Cadigan, FC, Jr. 1972. A brief report on copulatory and perinatal behaviour of the lesser Malayan mouse deer (*Tragulus javanicus*). Malay Nat J. 25:112-116.
1523. Cadman, WA. 1966. The fallow deer. Forestry Comm Leafl. 52:1-39.
1524. Caffier, P & Kolbow, H. 1934. Anatomisch-physiologische Genitalstudien an Fledermäusen zur Klärung der therapeutischen Sexualhormonwirkung. Z Geburtsh. 108:185-232.
1525. Cagle, FR. 1949. Notes on the raccoon, *Procyon lotor megalodous* Lowery. J Mamm. 30:45-47.

1526. Cagle, FR & Cockrum, L. 1943. Notes on a summer colony of *Myotis lucifugus lucifugus*. J Mamm. 24:474-492.
1527. Cahalane, VH. 1939. Mammals of the Chiricahua Mountains, Cochise county, Arizona. J Mamm. 20:418-440.
1528. Caillol, M & Castanier, M. 1980. Évolution des oestrogènes circulants au course de la gestation et de la superfoetation chez la Hase. Path Biol. 28:375-376.
1529. Caillol, M & Martinet, L. 1976. Preliminary results on plasma progesterone levels during pregnancy and superfetation in the hare, *Lepus europaeus*. J Reprod Fert. 46:61-64.
1530. Caillol, M & Martinet, L. 1981. Estrous behaviour, follicular growth and pattern of circulating sex steroids during pregnancy and pseudopregnancy in the captive brown hare. Proc World Lagomorph Conf, Univ Guelph, ONT. 1979:142-154.
1531. Caillol, M & Martinet, L. 1983. Mating periods and fertility in the doe hare (*Lepus europaeus*) bred in captivity. Acta Zool Fennica. 174:65-68.
1532. Caillol, M, Martinet, L, & Lacroix, MC. 1989. Relative roles of oestradiol and of the uterus in the maintenance of the corpus luteum in the pseudopregnant brown hare (*Lepus europaeus*). J Reprod Fert. 87:603-612.
1533. Caillol, M, Meunier, M, Mondain-Monval, M, et al. 1986. Seasonal variations in the pituitary response to LHRH in the brown hare (*Lepus europaeus*). J Reprod Fert. 78:479-486.
1534. Caillol, M, Meunier, M, Mondain-Monval, M, et al. 1989. Seasonal variations in testis size, testosterone and LH basal levels, and pituitary response to luteinizing hormone releasing hormone in the brown hare, *Lepus europaeus*. Can J Zool. 67:1626-1630.
1535. Caillol, M, Mondain-Monval, M, Meunier, M, et al. 1990. Effect of ovariectomy at two periods of the year on LH and FSH basal concentrations and pituitary response to LHRH in the brown hare (*Lepus europaeus*). J Reprod Fert. 88:533-542.
1535a. Caillol, M, Mondain-Monval, M, Meunier, M, et al. 1991. Pituitary and ovarian responses to luteinizing-hormone-releasing hormone during pregnancy and after parturition in brown hares (*Lepus europaeus*). J Reprod Fert. 92:89-97.
1535b. Caillol, M, Mondain-Monval, M, & Rossano, B. 1991. Gonadotrophins and sex steroids during pregnancy and natural superfoetation in captive brown hares (*Lepus europaeus*). J Reprod Fert. 92:299-306.
1536. Caire, W, LaVal, RK, LaVal, ML, et al. 1979. Notes on the ecology of *Myotis keenii* (Chiroptera, Vespertilionidae) in eastern Missouri. Am Mid Nat. 102:404-407.
1537. Cake, MH, Owen, FJ, & Bradshaw, SD. 1980. Difference in concentration of progesterone in plasma between pregnant and non-pregnant quokkas (*Setonix brachyurus*). J Endocrinol. 84:153-158.
1538. Calaby, JH. 1960. The numbat of south-western Australia. Aust Mus Mag. 13:143-146.
1539. Calaby, JH. 1960. Observations on the banded ant-eater *Myrmecobius f fasciatus* Waterhouse (Marsupiala), with particular reference to its food habits. Proc Zool Soc London. 135:183-207.
1540. Calaby, JH & Taylor, JM. 1981. Reproduction in two marsupial-mice, *Antechinus bellus* and *A bilarni* (Dasyuridae), of tropical Australia. J Mamm. 62:329-341.
1541. Calaby, JH & Sharman, GB. Personal communication. [Cited by Waring et al, 1963]
1542. Calaby, JH & Wimbush, DJ. 1964. Observations on the broad-toothed rat, *Mastacomys fuscus* Thomas. CSIRO Wildl Res. 9:123-133.
1543. Caldecot, JO. 1986. An ecological behavioural study of the pig-tailed macaque. Contrib Primatol. 21:1-259.
1544. Caldwell, DK & Caldwell, MC. 1975. Pygmy killer whales and short-snouted spinner dolphins in Florida. Cetology. 18:1-5.
1545. Caldwell, DK & Caldwell, MC. 1985. Manatees *Trichechus manatus* Linnaeus, 1758; *Trichechus inunguis* (Ntterer, 1883). Handbook of Marine Mammals. 3:33-66, Ridgway, SH & Harrison, RJ, eds, Academic Press, New York.
1546. Caldwell, DK & Golley, FB. 1965. Marine mammals from the coast of Georgia to Cape Hatteras. J Elisha Mitchell Sci Soc. 81:24-32.
1546a. Caldwell, DK, Neuhauser, H, Caldwell, MC, et al. 1971. Recent records of marine mammals from the coasts of Georgia and South Carolina. Cetology. 5:1-12.
1547. Caldwell, LD. 1964. An investigation of competition in natural populations in mice. J Mamm. 45:12-30.
1548. Caldwell, LD & Gentry, JB. 1965. Natality in *Peromyscus polionotus* populations. Am Mid Nat. 74:168-175.
1549. Caldwell, WH. 1884. On the arrangement of embryonic membranes in marsupial animals. Q J Microsc Sci. 24:655-658.
1550. Caldwell, WH. 1887. The embryology of Monotremata and Marsupialia Part 1. Phil Trans R Soc London. 178B:463-486.
1551. Caley, MJ. 1987. Dispersal and inbreeding avoidance in muskrats. Anim Behav. 35:1225-1233.
1552. Calinescu, RJ. 1934. Taxonomische, biologische und biogeographische Forschungen über die Gatlung *Citellus* OKEN in Rumänien. Z Säugetierk. 9:87-141.
1553. Call, RN & Janssens, PA. 1984. Hypertrophied adrenocortical tissue of the Australian brush-tailed possum (*Trichosurus vulpecula*): Uniformity during reproduction. J Endocrinol. 101:263-267.
1553a. Calle, PP, Chaudhuri, M, & Bowen, R. 1990. Menstrual cycle characterization and artificial insemination in the black mangabey (*Cercocebus aterrimus*). Zoo Biol. 9:11-24.
1554. Calland, CJ, Wightman, SR, & Neal, SB. 1986. Establishment of a Chinese hamster breeding colony. Lab Anim Sci. 36:183-185.
1555. Callard, GV, Kunz, TH, & Petro, Z. 1983. Identification of androgen metabolic pathways in the brain of little brown bats (*Myotis lucifugus*): Sex and seasonal differences. Biol Reprod. 28:1155-1163.
1556. Callard, GV, Petro, Z, & Tyndale-Biscoe, CH. 1982. Aromatase activity in marsupial brain, ovaries, and adrenals. Gen Comp Endocrinol. 46:541-546.
1557. Calvin, LO. 1969. A brief note on the birth of snow leopards *Panthera uncia* at Dallas Zoo. Int Zoo Yearb. 9:96.
1558. Calvin, VP. 1978. Reproduction d'*Ateles* à Monaco. Mammalia. 42:260-261.
1559. Calvo, JC, Sagripanti, JL, Peltzer, LE, et al. 1986. Photoperiod, follicle-stimulating hormone receptors, and testicular function in vizcacha (*Lagostomus maximus maximus*). Biol Reprod. 35:822-827.
1560. Camenzind, FJ. 1978. Behavioral ecology of coyotes on the National Elk Refuge, Jackson Wyoming. Coyotes, Biology, Behavior and Management. 267-294, Bekoff, M, ed, Academic Press, New York.
1561. Cameron, DG & Vyse, ER. 1978. Heterozygosity in Yellowstone Park elk, *Cervus canadensis*. Biochem Genetics. 16:651-657.
1562. Cameron, DM. 1967. Gestation period of the golden-mantled ground squirrel (*Citellus lateralis*). J Mamm. 48:492-493.
1563. Cameron, GN. 1973. Effects of litter size on postnatal growth and survival in the desert woodrat. J Mamm. 54:489-493.

1563a. Cameron, GN. 1977. Experimental species removal: Demographic responses by *Sigmodon hispidus* and *Reithrodontomys fulvescens*. J Mamm. 58:488-506.
1563b. Cameron, GN & McClure, PA. 1988. Geographic variation in life history traits of the hispid cotton rat (*Sigmodon hispidus*). Evolution of Life Histories of Mammals: Theory and Pattern. 33-64, Boyce, MS, ed, Yale Univ Press, New Haven, CT.
1564. Cameron, GN & Spencer, SR. 1981. *Sigmodon hispidus*. Mamm Species. 158:1-9.
1565. Cameron, JL, McNeill, TH, Fraser, HM, et al. 1985. The role of endogenous gonadotropin-releasing hormone in the control of luteinizing hormone and testosterone secretion in the juvenile male monkey, *Macaca fascicularis*. Biol Reprod. 33:147-156.
1566. Cameron, JL & Stouffer, RL. 1981. Comparison of the species specificity of gonadotropin binding to primate and non primate corpora lutea. Biol Reprod. 25:568-572.
1567. Cameron, JL & Stouffer, RL. 1982. Gonadotropin receptors of the primate corpus luteum II Changes in available luteinizing hormone- and chorionic gonadotropin-binding sites in macaque luteal membranes during the nonfertile menstrual cycle. Endocrinology. 110:2068-2073.
1568. Cameron, RD & Whitten, KR. 1976. Seasonal movements and sexual segregation of caribou determined by aerial survey. J Wildl Manage. 43:626-633.
1568a. Campbell, B. 1934. Notes on bats collected in Arizona during the summer of 1933. J Mamm. 15:241-242.
1569. Campbell, TM, III, Clark, TW, & Grover, CR. 1982. First record of pygmy rabbits (*Brachylagus idahoensis*) in Wyoming. Great Basin Nat. 42:100.
1570. Camus, L & Gley, E. 1899. Action coagulante du liquide de la prostate externe du Hérisson sur le continu des vesicules séminales. CR Séances Hebd Acad Sci. 128:1417-1419.
1570a. Canet, RS & Alvarez, VB. 1984. Reproduccion y ecologia de la jutia conga (*Capromys pilorides* Say). Poeyana. 280:1-20.
1571. Canguilhem, B, Vaultier, JP, Pevet, P, et al. 1988. Photoperiodic regulation of body mass, food intake, hibernation, and reproduction in intact and castrated male European hamsters, *Cricetus cricetus*. J Comp Phys A. 163:549-557.
1572. Canivenc, R. 1957. Étude de la nidation différée du Blaireau européen *Meles meles* L. Ann Endocrinol. 18:716-736.
1573. Canivenc, R. 1966. A study of progestation in the European badger (*Meles meles* L). Symp Zool Soc London. 15:15-26.
1574. Canivenc, R & Bonnin, M. 1979. Delayed implantation is under environmental control in the badger (*Meles meles* L). Nature. 278:849-850.
1575. Canivenc, R & Bonnin, M. 1981. Environmental control of delayed implantation in the European badger (*Meles meles*). J Reprod Fert. 29:25-33.
1576. Canivenc, R, Bonnin, M, & Ribes, C. 1981. Déclenchement de l'ovo-implantation par allongement de la phase sombre de la photopériode chez le Blaireau européen *Meles meles* L. CR Hebd Séances Acad Sci Sér III Sci Vie. 292:1009-1012.
1577. Canivenc, R & Bonnin-Laffargue, M. 1963. Inventory of problems raised by the delayed ova implantation in the European badger (*Meles meles L*). Delayed Implantation. 115-125, Enders, AC, ed, Univ Chicago Press, Chicago.
1578. Canivenc, R & Bonnin-Laffargue, M. 1975. Les facteurs écophisologiques de régulation de la fonction lutéale chez les Mammifères à ovo-implantation différée. J Physiol (Paris). 70:533-538.
1579. Canivenc, R, Bonnin-Laffargue, M, & Lejus, M. 1967. Action locale de la progestérone sur l'utérus de Blaireau pendant la phase de diapause blastocytaire. CR Hebd Séances Acad Sci Ser D Biol. 264:1308-1310.
1580. Canivenc, R, Bonnin-Laffargue, M, & Lajus-Boué, M. 1969. Induction de nouvelles générations lutéales pendant la progestation chez la Martre européene (*Martes martes* L). CR Hebd Séances Acad Sci Sér D Sci Nat. 269:1437-1440.
1581. Canivenc, R, Bonnin-Laffargue, M, & Lajus-Boué, M. 1971. Réalisation accents expérimentale précoce de l'ovo-implantation chez le Blaireau européen (*Meles meles* L) pendant la période de latence blastocytaire. CR Hebd Séances Acad Sci Sér Sci Nat. 273:1855-1857.
1582. Canivenc, R, Bonnin-Laffargue, M, & Relexans, MC. 1968. Cycles génitaux de quelques Mustelidés européens. Cycles Génitaux Saisonniers de Mammifères Sauvages. 85-104, Canivenc, R, ed, Masson et Cie, Paris.
1583. Canivenc, R & Lachaud, JP. 1967. Modification expérimentale du rythme circadien chez le Blaireau européen *Meles meles* L. CR Hebd Séances Acad Sci Sér D Sci Nat. 264:1088-1091.
1584. Canivenc, R & Laffargue, M. 1956. Présence de blastocystes libres intra-utérins au cours de la gestation chez le Blaireau européen (*Meles meles*). CR Soc Biol. 150:1193-1196.
1585. Canivenc, R & Laffargue, M. 1958. Action de différentes équilibres hormonaux sur la phase de vie libre de l'oeuf fecondé chez le Blaireau européen (*Meles meles* L). CR Soc Biol. 152:58-61.
1586. Canivenc, R, Mauget, C, Bonnin, M, et al. 1981. Delayed implantation in the beech marten (*Martes foina*). J Zool. 193:325-332.
1587. Canivenc, R, Short, RV, & Bonnin-Laffargue, M. 1966. Étude histologique et biochemique du corps jaune du Blaireau européen (*Meles meles* L). Ann Endocrinol. 27:401-413.
1588. Cannon, JR, Bakker, HR, Bradshaw, SD, et al. 1976. Gravity as the sole navigational aid to the newborn quokka. Nature. 259:42.
1589. Cansdale, GS. 1944. The lesser bush baby *Galago demidovii demidovii* G Fisch. J Soc Pres Fauna Empire. 50:7-12.
1590. Cantuel, P. 1946. Période de reproduction et nombre de foetus de quelques Micromammifères de la faune de France. Mammalia. 10:140-144.
1591. Cao, X. 1989. Seasonal changes in spermatogenesis of tree shrew (*Tupaia belangeri chinensis*). Zool Res. 10:15-21.
1591a. Cappozzo, HL. 1991. Sexual dimorphism in newborn southern sea lions. Mar Mamm Sci. 7:385-394.
1592. Car, WR. 1969. Studies on the pituitary gonadotrophins on the African elephant. J Reprod Fert Suppl. 6:219-223.
1593. Cardoso, FM, Figueiredo, EL, Godinho, HP, et al. 1985. Variação sazonal da atividade secretória das glāndulas genitais acessórias masculinas de tatus *Dasypus novemcinctus* Linnaeus, 1758. Rev Bras Biol. 45:507-514.
1594. Carlisle, KS, Brenner, TM, & Montagna, W. 1981. Hormonal regulation of sex skin in *Macaca nemestrina*. Biol Reprod. 25:1053-1063.
1595. Carlson, JC, Wong, AP, & Perrin, DG. 1977. Luteininzing hormone secretion in the rhesus monkey and a possible role for prostaglandins. Biol Reprod. 16:622-626.
1596. Carlsson, A. 1911. Über *Cryptoprocta ferox*. Zool Jb Syst. 30:419-470.
1597. Carlsson, A. 1920. Über *Arctictis binturong*. Acta Zool (Stockholm). 1:337-380.

1598. Carmel, PW, Araki, S, & Ferin, M. 1976. Pituitary stalk portal blood collection in rhesus monkeys: Evidence for pulsatile release of gonadotrophin-releasing hormone (GnRH). Endocrinology. 99:243-248.
1599. Carmichael, IH, Patterson, L, Dräger, N, et al. 1977. Studies on reproduction in the African buffalo (*Syncerus caffer*) in Botswana. S Afr J Wildl Res. 7:45-52.
1600. Carmichael, L, Kraus, MB, & Reed, T. 1961. The Washington National Zoological Park gorilla infant, Tomoko. Int Zoo Yearb. 3:88-93.
1601. Carmichael, M & MacLean, PD. 1961. Use of squirrel monkey for brain research, with description of restraining chair. Clin Neurophysiol. 13:128-129.
1602. Carmichael, SW, Spagnoli, DB, Frederickson, RG, et al. 1987. Opossum adrenal medulla: I Postnatal development and normal anatomy. Am J Anat. 179:211-219.
1603. Carmon, JL, Golley, FB, & Williams, RG. 1963. An analysis of the growth and variability in *Peromyscus polionotus*. Growth. 27:247-254.
1604. Carnio, J. 1982. Observations on the mother and young interactions in captive Cape fur seals (*Arctocephalus pusillus pusillus*). Aquatic Mamm. 9:50-56.
1605. Carnio, J, Chupa, B, & Stoner, J. 1982. Management and breeding of the South African fur seal *Arctocephalus pusillus* at Metro Toronto Zoo. Int Zoo Yearb. 22:207-213.
1606. Carpenter, CR. 1934. A field study of the behavior and social relations of howling monkeys. Comp Psychol Monogr. 10:1-168.
1607. Carpenter, CR. 1935. Behavior of red spider monkeys in Panama. J Mamm. 16:171-180.
1608. Carpenter, CR. 1941. The menstrual cycle and body temperature in two gibbons (*Hylobates lar*). Anat Rec. 79:291-296.
1609. Carpenter, CR. 1965. The howlers of Barro Colorado Island. Primate Behavior Field Studies of Monkeys and Apes. 250-291, DeVore, I, ed, Holt, Rinehart & Winston, New York.
1610. Carpenter, GP. 1970. Some observations on the rusty spotted genet (*Genetta rubiginosa zuluensis*). Lammergeyer. 11:60-63.
1611. Carr, EB. 1954. The vesicula seminalis of the rabbit. Proc Zool Soc London. 124:675-683.
1612. Carr, WR. 1972. Radioimmunoassay of lutenizing hormone in the blood of Zebu cattle. J Reprod Fert. 29:11-18.
1613. Carraway, LN. 1985. *Sorex pacificus*. Mamm Species. 231:1-5.
1614. Carraway, LN. 1988. Records of reproduction in *Sorex pacificus*. Southwest Nat. 33:479-480.
1615. Carraway, LN & Verts, BJ. 1985. *Microtus oregoni*. Mamm Species. 233:1-6.
1616. Carraway, LN & Verts, BJ. 1991. *Neotoma fuscipes*. Mamm Species. 386:1-10.
1617. Carraway, LN & Verts, BJ. 1991. *Neurotrichus gibbsii*. Mamm Species. 387:1-7.
1618. Carrick, FN & Cox, RI. 1973. Testosterone concentrations in the spermatic vein plasma of marsupials. J Reprod Fert. 32:338-339.
1619. Carrick, FN & Cox, RI. 1977. Testicular endocrinology of marsupials and monotremes. Proc Int Symp Comp Biol Reprod. 4:137-141.
1620. Carrick, FN, Drinan, JP, & Cox, RI. 1975. Progestagens and estrogens in peripheral plasma of the platypus *Ornithorhynchus anatinus*. J Reprod Fert. 43:375-376.
1621. Carrick, FN & Hughes, RL. 1978. Reproduction in male monotremes. Aust Zool. 20:211-231.
1622. Carrick, R, Csordas, SE, & Ingham, SE. 1962. Studies on the southern elephant seal, *Mirounga leonina* (L) IV Breeding and development. CSIRO Wildl Res. 7:161-197.
1623. Carrick, R, Csordas, SE, Ingham, SE, et al. 1962. Studies on the southern elephant seal, *Mirounga leonina* (L) III The annual cycle in relation to age and sex. CSIRO Wildl Res. 7:119-160.
1624. Carrick, R & Ingham, SE. 1960. Ecological studies of the southern elephant seal, *Mirounga leonina* (L), at Macquarie Island and Heard Island. Mammalia. 24:325-342.
1625. Carroll, JB. 1981. The wild status and behaviour of the Rodrigues fruit bat, *Pteropus rodricensis*: A report of the 1981 field study. Dodo, J Jersey Wildl Preserv Trust. 18:20-29.
1626. Carroll, JB. 1982. Breeding of the golden lion tamarin *Leontopithecus rosalia rosalia* at the Jersey Wildlife Preservation Trust. Dodo, J Jersey Wildl Preserv Trust. 19:42-46.
1627. Carroll, JB. 1982. Maintenance of the Goeldi's monkey *Callimico goeldii* at Jersey Wildlife Preservation Trust. Int Zoo Yearb. 22:101-105.
1628. Carroll, JB. 1983. Breeding the cotton-topped tamarin, *Saguinus oedipus oedipus* at Jersey Wildlife Preservation Trust. Dodo, J Jersey Wildl Preserv Trust. 20:48-52.
1629. Carroll, JB, Abbott, DH, George, LM, et al. 1989. Aspects of urinary oestrogen excretion during the ovarian cycle and pregnancy in Goeldi's monkey, *Callimico goeldii*. Folia Primatol. 52:201-205.
1630. Carroll, JB, Abbott, DH, George, LM, et al. 1990. Urinary endocrine monitoring of the ovarian cyucle and pregnancy in Goeldi's monkey (*Callimico goeldii*). J Reprod Fert. 89:149-161.
1631. Carroll, LE & Genoways, HH. 1980. *Lagurus curtatus*. Mamm Species. 124:1-6.
1632. Carroll, RS, Erskine, MS, & Baum, MJ. 1987. Sex difference in the effect of mating on the pulsatile secretion of luteinizing hormone in a reflex ovulator. Endocrinology. 121:1349-1359.
1633. Carroll, RS, Erskine, MS, Doherty, PC, et al. 1985. Coital stimuli controlling luteinizing hormone sectetion and ovulation in the female ferret. Biol Reprod. 32:925-933.
1634. Carter, CH & Genoways, HH. 1978. *Liomys salvini*. Mamm Species. 84:1-5.
1635. Carter, CH, Genoways, HH, Loregnard, RS, et al. 1981. Observations on bats from Trinidad, with a checklist of species occurring on the island. Occas Pap Mus TX Tech Univ. 72:1-27.
1636. Carter, CS, Getz, LL, Gavish, L, et al. 1980. Male-related pheromones and the activation of female reproduction in the prairie vole (*Microtus ochrogaster*). Biol Reprod. 23:1038-1045.
1637. Carter, CS, Witt, DM, Auksi, T, et al. 1987. Estrogen and the induction of lordosis in female and male prairie voles (*Microtus ochrogaster*). Horm Behav. 21:65-73.
1638. Carter, CS, Witt, DM, Manock, SR, et al. 1989. Hormonal correlates of sexual behavior and ovulation in male- induced and postpartum estrus in female prairie voles. Physiol Behav. 46:941-948.
1639. Carter, CS, Witt, DM, Thompson, EG, et al. 1988. Effects of hormonal, sexual, and social history on mating and pair bonding in prairie voles. Physiol Behav. 44:691-697.
1639a. Carter, DC. 1966. A new species of *Rhinophylia* [*sic*] (Mammalia: Chiroptera: Phyllostomidae) from South America. Proc Biol Soc Washington. 79:235-238.
1639b. Carter, DC. 1968. A new species of *Anoura* (Mammalia: Chiroptera: Phyllostomidae) from South America. Proc Biol Soc Washington. 81:427-430.

1640. Carter, DC. 1970. Chiropteran reproduction. About Bats. 233-246, Slaughter, BH & Walton, DW, eds, S Methodist Univ Press, Dallas.
1640a. Carter, DC & Jones, JK, Jr. 1978. Bats from the Mexican state of Hidalgo. Occas Pap Mus TX Tech Univ. 54:1-12.
1641. Carter, DC, Pine, RH, & Davis, WB. 1966. Notes on middle American bats. Southwest Nat. 11:488-499.
1642. Carter, DC, Webster, WD, Jones, JK, Jr, et al. 1985. *Dipodomys elator.* Mamm Species. 232:1-3.
1643. Carter, DS & Goldman, BD. 1983. Antigonadal effects of timed melatonin infusion in pinealectomized male Djungarian hamsters (*Phodopus sungorus sungorus*): Duration is the critical parameter. Endocrinology. 113:1261-1267.
1644. Carter, DS & Goldman, BD. 1983. Prolonged role of the pineal in the Djungarian hamster (*Phodopus sungorus sungorus*): Mediation by melatonin. Endocrinology. 113:1268-1273.
1645. Carter, DS, Hall, VD, Tamarkin, L, et al. 1982. Pineal is required for testicular maintenance in the Turkish hamster (*Mesocricetus brandti*). Endocrinology. 111:863-871.
1646. Carter, DS, Herbert, J, & Stacey, PM. 1982. Modulation of gonadal activity by timed injections of melatonin in pinealectomized or intact ferrets kept under two photoperiods. J Endocrinol. 93:211-222.
1647. Carter, FD & Bleier, WJ. 1988. Sequential multiple ovulations in *Macrotus californicus.* J Mamm. 69:386-387.
1648. Carvalho, CT. 1976. Aspectos faunisticos de cerrado-oloboguará (Mammalia, Canidae). Bol Técnico Inst Florestal São Paulo. 21:1-16.
1649. Cary, ER. 1976. Territorial and reproductive behavior of the black buck antelope (*Antilope cervicapra*). Ph D Diss, TX A & M Univ, College Station.
1649a. Cary, M. 1911. A biological survey of Colorado. N Am Fauna. 33:1-256.
1650. Casanova, J. 1958. The dormouse or pygmy possum. Walkabout. 24:30-31.
1651. Case, DJ & McCullough, DR. 1987. The white-tailed deer of north Manitou island. Hilgardia. 66:1-57.
1652. Casebeer, RS, Linsky, RB, & Nelson, CE. 1963. The phyllostomid bats *Ectophylla alba* and *Vampyrum spectrum* in Costa Rica. J Mamm. 44:186-190.
1653. Casteel, DA. 1966. Nest building, parturition, and copulation in the cottontail rabbit. Am Mid Nat. 75:160-167.
1654. Casteel, DA. 1967. Timing of ovulation and implantation in the cottontail rabbit. J Wildl Manage. 31:194-197.
1655. Casteel, DA & Edwards, WR. 1964. Two instances of multiparous juvenile cottontails. J Wildl Manage. 28:858-859.
1656. Castellanos, H & McCombs, HL. 1968. The reproductive cycle of the New World monkey: Gynecologic problems in a breeding colony. Fert Steril. 19:213-227.
1657. Casteret, N. 1939. La colonie de Murins de la grotte des Tignahustes. Mammalia. 3:1-9.
1658. Castracane, VD, Copeland, KC, Reyes, P, et al. 1986. Pubertal endocrinology of yellow baboon (*Papio cynocephalus*): Plasma testosterone, testis size, body weight, and crown-rump length in males. Am J Primatol. 11:263-270.
1659. Castracane, VD, D'Eletto, R, & Weiss, G. 1983. Relaxin secretion in the baboon (*Papio cynocephalus*). Factors Regulating Ovarian Function. 415-419, Greenwald, GS & Terranova, PF, eds, Raven Press, New York.
1660. Castracane, VD & Goldzieher, JW. 1986. Timing of the luteal-placental shift in the baboon (*Papio cynocephalus*). Endocrinology. 118:506-512.
1661. Castro, R & Soini, P. 1977. Field studies on *Saguinus mystax* and other callitrichids in Amazonian Peru. The Biology and Conservation of the Callitrichidae. 73-89, Kleiman, DG, ed, Smithsonian Inst Press, Washington, DC.
1662. Catchpole, HR & Fulton, JF. 1943. The oestrus cycle in *Tarsius*: Observation on a captive pair. J Mamm. 24:90-93.
1663. Catling, PC. 1979. Seasonal variation in plasma testosterone and the testis in captive male dingoes, *Canis familiaris dingo.* Aust J Zool. 27:939-944.
1663a. Catling, PC, Corbett, LK, & Newsome, AE. 1992. Reproduction in captive and wild dingoes (*Canis familiaris dingo*) in temperate and arid environments of Australia. Wildl Res. 19:195-209.
1664. Catling, PC & Sutherland, RL. 1980. Effect of gonadectomy, season and the presence of female tammar wallabies (*Macropus eugenii*) on concentrations of testosterone, luteinizing hormone and follicle-stimulating hormone in the plasma of male tammar wallabies. J Endocrinol. 86:25-33.
1665. Catt, DC. 1977. The breeding biology of Bennett's wallaby (*Macropus rufogriseus fruticus*) in South Canterbury, New Zealand. NZ J Zool. 4:401-411.
1666. Catt, DC. 1981. Growth and condition of Bennett's wallaby (*Macropus rufogriseus fruticus*) in South Canterbury, New Zealand. NZ J Zool. 8:295-300.
1667. Caughley, G. 1967. Calculation of population mortality rate and life expectancy for thar and kangaroos from the ratio of juveniles to adults. NZ J Sci. 10:578-584.
1668. Caughley, G. 1970. Fat reserves of Himalayan thar in New Zealand by season, sex, area and age. NZ J Sci. 13:209-219.
1669. Caughley, G. 1971. The season of births for northern-hemisphere ungulates in New Zealand. Mammalia. 35:204-219.
1670. Cazanove, JL. 1932. Le problème du Rat dans le territoire de Dakar et Dépendances. Conférence Internationale du Rat, Paris, Oct 1931. 2:95-146.
1671. Ceballos, G & Medellín, RA. 1988. *Diclidurus albus.* Mamm Species. 316:1-4.
1672. Ceballos, G & Miranda, A. 1988. Notes on the biology of Mexican flying squirrels (*Glaucomys volans*) (Rodentia: Sciuridae). Southwest Nat. 25:157-172.
1673. Ceballos-G, G & Wilson, DE. 1985. *Cynomys mexicanus.* Mamm Species. 248:1-3.
1674. Cederlund, G & Lindström, E. 1983. Effects of severe winters and fox predation on roe deer mortality. Acta Theriol. 28:129-145.
1675. Cerquiera, R. 1984. Reproduction de *Didelphis albiventris* dans le nord-est du Brésil (Polyprotodontia, Didelphidae). Mammalia. 48:95-104.
1676. Cervantes, F & Lopez-Forment, W. 1981. Observations on the sexual behavior, gestation period, and young of captive Mexican volcano rabbits, *Romerolagus diazi.* J Mamm. 62:634-635.
1677. Cervantes, FA, Lorenzo, C, & Hoffmann, RS. 1990. *Romerolagus diazi.* Mamm Species. 360:1-7.
1678. Chabry, L & Boulart, R. 1883. Note sur un foetus de Dauphin et ses membranes. J de l'Anat et Physiol. 19:572-575.
1679. Chaffee, PS. 1967. A note on the breeding of orang-utans, *Pongo pygmaeus*, at Fresno Zoo. Int Zoo Yearb. 7:94-95.
1680. Challis, JRG, Davies, IJ, Benirschke, K, et al. 1974. The concentrations of progesterone, estrone and estradiol-17β in the peripheral plasma of the rhesus monkey during the final third of gestation, and after the induction of abortion with $PGF_{2\alpha}$. Endocrinology. 95:547-553.
1681. Challis, JRG, Davies, IJ, & Hendrickx, AG. 1974. Prostaglandin F in the peripheral plasma of the rhesus monkey in

normal pregnancy and after the administration of dexamethasone and $PGF_{2\alpha}$. Prostaglandins. 6:389-396.
1682. Challis, JRG, Hartley, P, Johnson, P, et al. 1977. Steroids in the amniotic fluid of the rhesus monkey (*Macaca mulatta*). J Endocrinol. 73:355-363.
1683. Challis, JRG, Socol, M, Murata, Y, et al. 1980. Diurnal variations in maternal and fetal steroids in pregnant rhesus monkeys. Endocrinology. 106:1283-1287.
1684. Chalmers, G. 1963. Breeding data: Steinbok, *Raphicerus campestris*. E Afr Wildl J. 1:121-122.
1685. Chambers, KC & Phoenix, CH. 1982. Decrease in sexual initiative and responsiveness in female rhesus macaques (*Macaca mulatta*) during pregnancy. Am J Primatol. 2:301-306.
1686. Chambers, PL & Hearn, JP. 1979. Peripheral plasma levels of progesterone, oestradiol-17β, oestrone, testosterone, androstenedione and chorionic gonadotropin during pregnancy in the marmoset monkey, *Callithrix jacchus*. J Reprod Fert. 56:23-32.
1687. Chambers, PL & Hearn, JP. 1985. Embryonic, foetal and placental development in the common marmoset monkey (*Callithrix jacchus*). J Zool. 207A:545-561.
1688. Chandrahas, RK. 1974. Ecology of the brown spiny mouse *Mus p platythrix* (Bennett) and the Indian field mouse *Mus b booduga* (Gray). Indian J Med Res. 62:264-280.
1689. Chandrahas, RK & Krishnaswami, AK. 1974. Studies on the ecology of the Indian gerbil *Tatera indica hardwickei* (Gray), in Kolar (Mysore State). Indian J Med Res. 62:971-978.
1690. Chandrashekar, V, Dierschke, DJ, & Wolf, RC. 1987. Excessive ovarian follicular development in pregnant pigtailed macaques (*Macaca nemestrina*). Am J Primatol. 13:145-153.
1691. Chandrashekar, V, Meyer, RK, Bridson, WE, et al. 1979. Circulating levels of chorionic gonadotropin and progesterone in the rhesus monkey treated with LH antiserum during early gestation. Biol Reprod. 20:889-895.
1692. Chandrashekar, V, Wolf, RC, Dierschke, DJ, et al. 1980. Serum progesterone and corpus luteum function in pregnant pigtailed monkeys (*Macaca nemestrina*). Steroids. 36:483-495.
1693. Chang, CY & Wu, H. 1938. Growth and reproduction of laboratory-bred hamsters (*Cricetulus griseus*). Chin J Physiol. 13:109-118.
1694. Chang, MC. 1965. Fertilizing life of ferret sperm in the female tract. J Exp Zool. 158:87-100.
1695. Channing, A. 1984. Ecology of the namtap *Graphiurus ocularis* (Rodentia: Gliridae) in Cedarberg, South Africa. S Afr J Zool. 19:144-149.
1696. Channing, CP, Anderson, LD, Hoover, DJ, et al. 1981. Inhibitory effects of porcine follicular fluid on monkey serum FSH levels and follicular maturation. Biol Reprod. 25:885-903.
1697. Chao, CC & Brown, RD. 1984. Seasonal relationships of thryoid, sexual and adrenocortical hormones to nutritional parameters and climatic factors in white- tailed deer (*Odocoileus virginianus*) of south Texas. Comp Biochem Physiol. 77A:299-306.
1698. Chapais, B. 1986. Why do adult male and female rhesus monkeys affiliate during the birth season?. The Cayo Santiago Macaques History, Behavior and Biology. 173-200, Rawlins, RG & Kessler, MJ, eds, State Univ NY Press, Albany.
1699. Chaplin, RE, Chapman, DI, & Prior, R. 1966. An examination of the uterus and ovaries of some female roe deer (*Capreolus capreolus* L) from Wiltshire and Dorset, England. J Zool. 148:570-574.
1700. Chaplin, RE & Dangerfield, G. 1973. Breeding records of muntjac deer (*Muntiacus reevesi*) in captivity. J Zool. 170:150-151.
1700a. Chaplin, RE & Harrison, RJ. 1971. The uterus, ovaries and placenta of the Chinese muntjac deer (*Muntiacus reevesi*). J Anat. 110:147.
1701. Chaplin, RE & White, RWG. 1972. The influence of age and season on the activity of the testes and epididymides of the fallow deer, *Dama dama*. J Reprod Fert. 30:361-369.
1702. Chapman, A. 1981. Habitat preference and reproduction of feral house mice, *Mus musculus*, during plague and non-plague situations in Western Australia. Aust Wildl Res. 8:567-579.
1703. Chapman, AO. 1951. The estrous cycle in the woodrat, *Neotoma floridana*. Univ KS Sci Bull. 34:267-299.
1704. Chapman, BM, Chapman, RF, & Robertson, IAD. 1959. The growth and breeding of the multimammate rat, *Rattus* (*Mastomys*) *natalensis* (Smith) in Tanganyika territory. Proc Zool Soc London. 133:1-9.
1705. Chapman, DG. 1961. Population dynamics of the Alaska fur seal herd. Trans N Am Wildl Conf. 26:356-369.
1705a. Chapman, DI. 1974. Reproductive physiology in relation to deer management. Mamm Rev. 4:61-74.
1706. Chapman, DI & Chapman, NG. 1970. Preliminary observations on the reproductive cycle of male fallow deer (*Dama dama* L). J Reprod Fert. 21:1-8.
1707. Chapman, DI & Chapman, NG. 1971. Further observations on the incidence of twins in roe deer, *Capreolus capreolus*. J Zool. 165:505-509.
1708. Chapman, DI & Chapman, NG. 1975. Fallow Deer. Terrence Dalton Ltd, Lavenham-Suffolk.
1709. Chapman, DI & Chapman, NG. 1980. Morphology of the male accessory organs of reproduction of immature fallow deer (*Dama dama* L) with particular reference to puberty and antler development. Acta Anat. 108:51-59.
1709a. Chapman, DI & Chapman, NG. 1982. the taxonomic status of feral muntjac deer (*Muntiacus* sp) in England. J Nat Hist. 16:381-387.
1710. Chapman, DI, Chapman, NG, & Dansie, O. 1984. The periods of conception and parturition in feral Reeves muntjac (*Muntiacus reevesi*) in southern England, based upon age of juvenile animals. J Zool. 204:575-578.
1711. Chapman, DI, Chapman, NG, & Kennaugh, JH. 1981. Development of the preputial gland of immature fallow deer (*Dama dama* Linnaeus) with particular reference to puberty. Z Säugetierk. 46:322-330.
1712. Chapman, DI & Dansie, O. 1969. Unilateral implantation in muntjac deer. J Zool. 159:534-536.
1713. Chapman, DI & Dansie, O. 1970. Reproduction and foetal development in female muntjac deer (*Muntiacus reevesi* Ogilby). Mammalia. 34:303-319.
1714. Chapman, DI & Horwood, MT. 1968. Pregnancy in a sika deer calf, *Cervus nippon*. J Zool. 155:227-228.
1715. Chapman, HC. 1880. The placenta and generative apparatus of the elephant. J Acad Nat Sci Philadelphia. 8:413-422.
1716. Chapman, HC. 1888. Observations on the female generative apparatus of *Hyaena crocuta*. Proc Acad Nat Sci Philadelphia. 1888:189-191.
1717. Chapman, HC. 1901. Observations upon the placenta and young of *Dasypus sexcinctus*. Proc Acad Nat Sci Philadelphia. 53:366-369.
1718. Chapman, JA. 1974. *Sylvilagus bachmani*. Mamm Species. 34:1-4.
1719. Chapman, JA. 1975. *Sylvilagus transitionalis*. Mamm Species. 55:1-4.
1720. Chapman, JA. 1975. *Sylvilagus nuttallii*. Mamm Species. 56:1-3.

1721. Chapman, JA & Feldhamer, GA. 1981. *Sylvilagus aquaticus*. Mamm Species. 151:1-4.
1722. Chapman, JA & Harman, AL. 1972. The breeding biology of a brush rabbit population. J Wildl Manage. 36:816-823.
1723. Chapman, JA, Harman, AL, & Samuel, DE. 1977. Reproductive and physiological cycles in the cottontail complex in western Maryland and nearby West Virginia. Wildl Monogr. 56:1-73.
1724. Chapman, JA, Hockman, JG, & Ojeda C, MM. 1980. *Sylvilagus floridanus*. Mamm Species. 136:1-8.
1725. Chapman, JA & Lind, GS. 1973. Latitude and litter size of the California ground squirrel, *Spermophilus beecheyi*. Bull S CA Acad Sci. 72:101-105.
1726. Chapman, JA & Lind, GS. 1973. Latitude and litter size of the California ground squirrel, *Spermophilus beecheyi*. Bull S CA Acad Sci. 72:101-105.
1727. Chapman, JA & Morgan, RP, II. 1974. Onset of the breeding season and size of first litter in two species of cottontails from southwestern Texas. Southwest Nat. 19:277-280.
1728. Chapman, JA & Willner, GR. 1978. *Sylvilagus audubonii*. Mamm Species. 106:1-4.
1729. Chapman, JA & Willner, GR. 1981. *Sylvilagus palustris*. Mamm Species. 153:1-3.
1730. Chapman, NG & Chapman, DI. 1979. Seasonal changes in the male accessory glands of reproduction in adult fallow deer (*Dama dama*). J Zool. 189:259-273.
1731. Chappel, SC & Schmidt, M. 1979. Cyclic release of luteinizing hormone and the effects of luteinizing hormone-releasing hormone injection in Asiatic elephants. Am J Vet Res. 40:451-453.
1732. Chappel, SC, Bethea, CL, & Spies, HG. 1984. Existence of multiple forms of follicle stimulating hormone within the anterior pituitaries of cynomolgus monkeys. Endocrinology. 115:452-461.
1733. Chappell, RW & Hudson, RJ. 1978. Winter bioenergetics of Rocky Mountain bighorn sheep. Can J Zool. 56:2388-2393.
1734. Chapskii, KK. 1938. The bearded seal (*Erignathus barbatus*, Fabr) of the Kara and Barents seas. Trans Arctic Inst Leningrad. 123:7-70.
1735. Chari, GC & Gopalakrishna, A. 1984. Morphogenesis of the foetal membranes and placentation in the bat, *Miniopterus schreibersii fuliginosus* (Hodgson). Proc Indian Acad Sci (Anim Sci). 93:463-483.
1736. Charles-Dominique, P. 1966. Naissance et croissance d'*Arctocebus calabarensis* en captivité. Biol Gabon. 2:331-345.
1737. Charles-Dominique, P. 1966. Glandes préclitoridiennes de *Perodicticus potto*. Biol Gabon. 2:355-359.
1738. Charles-Dominique, P. 1968. La durée de gestation chez *Perodicticus potto edwarsi* (A Bouvier). Biol Gabon. 4:3.
1739. Charles-Dominique, P. 1968. Reproduction des Lorisidés Africains. Cycles Génitaux Saisonniers de Mammifères Sauvages. 2-5, Canivenc, R, ed, Masson et Cie, Paris.
1740. Charles-Dominique, P. 1974. Vie social de *Perodicticus potto* (Primates, Lorisides): Etude de terrain en fôret équatoriale de l'oest africain au Gabon. Mammalia. 38:355-379.
1741. Charles-Dominique, P. 1971. Eco-éthologie des Prosimiens du Gabon. Biol Gabon. 7:121-228.
1742. Charles-Dominique, P. 1977. Ecology and Behaviour of Nocturnal Primates. Columbia U Press. New York.
1742a. Charles-Dominique, P. 1978. Écologie et vie sociale de *Nandinia binotata* (Carnivores, Viverrides): Comparaison avec les prosimiens sympatriques du Gabon. Terre Vie. 32:477-528.
1743. Charles-Dominique, P & Petter, JJ. 1980. Ecology and social life of *Phaner furcifer*. Nocturnal Malagasy Primates. 75-96, Charles-Dominique, P, Cooper, HM, Hladik, A, et al, eds, Academic Press, New York.
1745. Charlick, J, Manessis, C, Stanley, N, et al. 1981. Quantitative alterations of the aerobic bacterial flora of the pouch of *Setonix brachyurus* (quokka) during oestrus, anoestrus, pregnancy and lactating anoestrus (pouch young). Aust J Exp Biol Med Sci. 59:743-751.
1746. Charlton, HM, Chiappa, SA, Fink, G, et al. 1983. Hypothalamic gonadotropin-releasing hormone and pituitary luteinizing hormone contents in the male vole (*Microtus agrestis*) in the field during the breeding and nonbreeding seasons. Can J Zool. 61:2405-2410.
1747. Charlton, HM, Grocock, CA, & Ostberg, A. 1976. The effects of pinealectomy and superior cervical gonglionectomy on the testis of the vole, *Microtus agrestis*. J Reprod Fert. 48:377-379.
1748. Charlton, HM, Milligan, SR, & Versi, E. 1978. Studies on the control of the corpus luteum in the vole, *Microtus agrestis*. J Reprod Fert. 52:283-288.
1749. Charlton, HM, Naftolin, F, Sood, MC, et al. 1975. The effect of mating upon LH release in male and female voles of the species *Microtus agrestis*. J Reprod Fert. 42:167-170.
1750. Charlton, HM, Naftolin, F, Sood, MC, et al. 1975. Electrical stimulation of the hypothalamus and LH release in the vole *Microtus agrestis*. J Reprod Fert. 42:171-174.
1751. Charnot, Y. 1963. Synchronisme de croissance de l'expansion palatale et du testicule au cours du cycle sexuel chez le Dromadaire. Bull Soc Sci Nat Phys Maroc. 43:49-54.
1752. Charnot, Y. 1964. Le cycle testiculaire du Dromaidaire. Bull Soc Sci Nat Phys Maroc. 44:37-45.
1753. Chartin, J & Petter, F. 1960. Reproduction et élevage en captivité du Ouistiti. Mammalia. 24:153-155.
1754. Chasen, FN. 1935. On a collection of mammals from the Natuna Islands, South China Sea. Bull Raffles Mus. 10:5-42.
1755. Chasen, FN. 1940. A handlist of Malaysian mammals. Bull Raffles Mus. 15:1-209.
1756. Chasen, FN & Kloss, CB. 1927. Spolia mentawiensia: Mammals. Proc Zool Soc London. 1927:797-840.
1757. Chasen, FN & Kloss, CB. 1931. On a collection of mammals from the lowlands and islands of north Borneo. Bull Raffles Mus. 6:1-82.
1758. Chattin, JE. 1948. Breeding season and productivity in the interstate deer herd. CA Fish Game. 34:25-31.
1759. Chattin, JE & Lassen, R. 1950. California antelope reproductive potentials. CA Fish Game. 36:328-329.
1760. Chaudhuri, M, Carrasco, E, Kalk P, et al. 1990. Urinary estrogen excretion during estrus and pregnancy in the babirusa *Babirousa babyrussa*. Int Zoo Yearb. 29:188-192.
1761. Chaudhuri, M & Ginsberg, JR. 1990. Urinary androgen concentrations and social status in two species of free ranging zebra (*Equus burchelli* and *E grevyi*). J Reprod Fert. 88:127-133.
1762. Chaudhuri, M, Kleiman, DG, Wildt, DE, et al. 1986. Urinary steroid levels during natural and induced estrus, pregnancy and pseudopregnancy in the giant panda. Biol Reprod. 34(Suppl):146.
1763. Chaudhuri, M, Kleiman, DG, Wildt, DE, et al. 1988. Urinary steroid concentrations during natural and gonadotrophin-induced oestrus and pregnancy in the giant panda (*Ailuropoda melanoleuca*). J Reprod Fert. 84:23-28.
1764. Chaudhuri, MA & Beg, MA. 1977. Reproductive cycle and population structure of the northern palm squirrel, *Funambulus pennanti*. Pak J Zool. 9:183-189.

1765. Chaunan, NS & Saxena, RN. 1985. Reproductive cycle of the female *Rattus rattus brunneusculus* (Hodgson), a common field rat of Mizoram. Proc Indian Sci Acad. 51B:560-565.
1766. Chauvet, MT, Hurpet, D, Chauvet, J, et al. 1980. Évolution des vasopressines chez les Marsupiaux: Une nouvelle hormone, la phénylpressine (Phe^2-Arg^8-vasopressine présente chez les Macropedidés. CR Hebd Séances Acad Sér D Sci Nat. 291:541-543.
1767. Chauvet, J, Hurpet, D, Michel, G, et al. 1985. The neurohypophysial hormones of the egg-laying mammals: Identification of arginine vasopressin in the platypus (*Ornithorhynchus anatinus*). Biochem Biophys Res Comm. 127:277-282.
1767a. Chael, AJ & Gales, NJ. 1991. Body mass and food intake in captive, breeding bottlenose dolphins, *Tursiops truncatus*. Zoo Biol. 10:451-456.
1767b. Chael, AJ & Gales, NJ. 1992. Growth, sexual maturity and food intake of Australian Indian Ocean bottlenose dolphins, *Tursiops truncatus*, in captivity. Aust J Zool. 40:215-223.
1768. Cheal, M. 1983. Lifespan, ontogeny of breeding and reproductive success in Mongolian gerbils. Lab Anim. 17:240-245.
1768a. Cheal, PD, Lee, AK, & Barnett, BD. 1976. Changes in the haemotology of *Antechinus stuartii* (Marsupialia), and their association with male mortality. Aust J Zool. 24:299-311.
1769. Cheatum, EL. 1949. The use of corpora lutea for determining ovulation incidence and variations in the fertility of white-tailed deer. Cornell Vet. 39(3):282-291.
1770. Cheatum, EL & Sevringhaus, CW. 1950. Variations in fertility of white-tailed deer related to range conditions. Trans N Am Wildl Conf. 15:170-190.
1771. Cheeseman, CL. 1981. Observations in the reproductive biology and early post-natal development of two species of African rodents. Mammalia. 45:483-491.
1772. Chen, BX & Yuen, ZX. 1980. Reproductive pattern of the bactrian camel. Animal Breeding Abstracts. 48: ref 6267.
1773. Chen, BX, Yuen, ZX, Kang, CL, et al. 1980. Reproductive pattern of the bactrian camel: II The sexual activities of the camel. Acta Vet Zootech Sin. 11:65-76.
1774. Chen, BX, Yuen, ZX, & Pan, GW. 1985. Semen-induced ovulation in the bactrian camel (*Camelus bactrianus*). J Reprod Fert. 73:335-339.
1775. Chen, BX, Yuen, ZX, Pan, GW, et al. 1983. Studies on the ovulation mechanism in the bactrian camel: 2 The role of semen in induction of ovulation. Acta Vet Zootech Sin. 14:161-166.
1776. Chen, ECH, Blood, DA, & Baker, BE. 1965. Rocky mountain big horn sheep (*Ovis canadensis canadensis*) milk I Gross composition and fat constitution. Can J Zool. 43:885-888.
1777. Chen, PX, Liu, R, & Harrison, AJ. 1982. Reproduction and reproductive organs in *Neophocaena asiaeorientalis* from the Yangtse River. Aquatic Mamm. 9:9-16.
1778. Chen, PX, Ren-Jim, L, & Ke-Jie, L. 1984. Reproduction and the reproductive system in the beiji, *Lipotes vexillifer*. Rep Int Whal Comm Sp Issue. 6:445-450.
1779. Chernyavskii, FB, Tkachev, AV, & Ardashev, AA. 1979. Regulation of numbers of lemmings in the arctic. Dokl Akad Nauk SSSR Translation. 242:425-428.
1780. Chesemore, DL. 1975. Ecology of the arctic fox (*Alopex lagopus*) in North America: A review. The Wild Canids. 143-163, Fox, MW, ed, Van Nostrand Reinhold Co, New York.
1781. Chevalier-Skolnikoff, S. 1975. Heterosexual copulatory patterns in stumptail macaques (*Macaca arctiodes*) and in other macaque species. Arch Sex Behav. 4:199-220.
1782. Chew, RM & Butterworth, BB. 1959. Growth and development of Merriam's kangaroo rat, *Dipodomys merriami*. Growth. 23:75-95.
1783. Chew, RM & Butterworth, BB. 1964. Ecology of rodents in Indian cove (Mojave Desert) Joshua Tree National Monument, California. J Mamm. 45:203-225.
1784. Chew, RM & Chew, AE. 1970. Energy relationships of the mammals of a desert shrub (*Larrea tridentata*) community. Ecol Monogr. 40:1-21.
1785. Chew, RM & Spencer, E. 1967. Development of metabolic response to cold in young mice of four species. Comp Biochem Physiol. 22:873-888.
1786. Chew, RM. 1958. Reproduction by *Dipodomys merriami* in captivity. J Mamm. 39:597-598.
1787. Chez, RA, Schlesselman, JJ, Salazar, H, et al. 1972. Single placentas in the rhesus monkey. J Med Primatol. 1:230-240.
1788. Chidumayo, EN. 1980. Ecology of rodents at an old quarry in Zambia. S Afr J Zool. 15:44-49.
1789. Chidumayo, EN. 1984. Observations on populations of multimammate mice at Livingstone, Zambia. Mammalia. 48:363-376.
1790. Child, G. Personal communication. [Cited by Mentis, MT. 1972]
1791. Child, G & Mossman, AS. 1965. Right horn implantation in the common duiker. Science. 149:1265-1266.
1792. Child, G, Robbel, H, & Hepburn, CP. 1972. Observations on the biology of tsessebe, *Damaliscus lunatus lunatus*, in northern Botswana. Mammalia. 36:342-388.
1793. Child, G, Roth, HH, & Kerr, M. 1968. Reproduction and recruitment patterns in warthog (*Phacochoerus aethiopicus*) populations. Mammalia. 32:6-29.
1794. Child, G & Wilson, VJ. 1964. Observations on ecology and behaviour of roan and sable in three tsetse control areas. Arnoldia (Rhodesia). 1(16):1-8.
1795. Chipman, RK. 1965. Age determination of the cotton rat (*Sigmodon hispidus*). Tulane Stud Zool. 12:19-38.
1795a. Chism, JB, Olson, DK, Rowell, TE. 1983. Diurnal births and perinatal behavior among wild patas monkeys: Evidence of an adaptive pattern. Int J Primat. 4:167-184.
1796. Chism, JB & Rowell, TE. 1986. Mating and residence patterns in male patas monkeys. Ethology. 72:31-39.
1797. Chism, JB, Rowell, TE, & Olson, D. 1984. Life history patterns of female patas monkeys. Female Primates: Studies by Female Primatologists. 175-190, Small, MF, ed, AR Liss, Inc, New York.
1798. Chism, JB, Rowell, TE, & Richards, SM. 1978. Daytime births in captive patas monkeys. Primates. 19:765-768.
1799. Chittleborough, RG. 1954. Studies on the ovaries of the humpback whale *Megaptera nodosa* (Bonnaterre) on the western Australian coast. Aust J Mar Freshwater Res. 5:35-63.
1800. Chittleborough, RG. 1955. Aspects of reproduction in the male humpback whale, *Megaptera nodosa* (Bonnaterre). Aust J Mar Freshwater Res. 6:1-29.
1801. Chittleborough, RG. 1955. Puberty, physical maturity, and relative growth of the female humpback whale, *Megaptera nodosa* (Bonnaterre), on the western Australian coast. Aust J Mar Freshwater Res. 6:315-327.
1802. Chittleborough, RG. 1958. The breeding cycle of the female humpback whale *Megaptera nodosa* (Bonnaterre). Aust J Mar Freshwater Res. 9:1-18.
1803. Chittleborough, RG. 1965. Dynamics of two populations of the humpback whale, *Megaptera novaeangliae* (Borowski). Aust J Mar Freshwater Res. 16:33-128.

1804. Chitty, D. 1952. Mortality among voles *Microtus agrestis* at Lake Vyrnwy, Montgomeryshire 1936-1939. Phil Trans R Soc London. 236B:505-552.
1805. Chitty, D & Phipps, E. 1966. Seasonal changes in survival in mixed populations of two species of vole. J Anim Ecol. 35:313-331.
1806. Chitty, H. 1957. The oestrous cycle and gestation period in the lactating field vole, *Microtus agrestis.* J Endocrinol. 15:279-283.
1807. Chitty, H & Austin, CR. 1957. Experimental modification of oestrus in the vole. Nature. 179:592-593.
1808. Chitty, H & Clarke, JR. 1963. The growth of the adrenal gland of laboratory and field voles and changes in it during pregnancy. Can J Zool. 41:1025-1034.
1809. Chivers, D. 1971. The Malayan siamang. Malay Nat J. 24:78-86.
1810. Chivers, DJ & Chivers, ST. 1975. Events preceding and following the birth of a wild siamang. Primates. 16:227-230.
1811. Choate, JR. 1973. *Cryptotis mexicana.* Mamm Species. 28:1-3.
1812. Choate, JR & Fleharty, ED. 1974. *Cryptotis goodwini.* Mamm Species. 44:1-3.
1812a. Choate, JR & Fleharty, ED. 1975. Synopsis of native, recent mammals of Ellis County, Kansas. Occas Pap Mus Tx Tech Univ. 37:1-80.
1813. Choate, JR & Jones, JK, Jr. 1970. Additional notes on reproduction in the Mexican vole, *Microtus mexicanus.* Southwest Nat. 14:356-358.
1814. Choate, TS. 1971. Research on captive wild mammals: With special reference to *Rhabdomys pumilio.* Rhodesia Sci News. 5:47-51.
1815. Choate, TS. 1972. Behavioural studies on some Rhodesian rodents. Zool Afr. 7:103-118.
1816. Chorn, J & Hoffmann, RS. 1978. *Ailuropoda melanoleuca.* Mamm Species. 110:1-6.
1817. Chou, F. Personal communication. [Cited by Zimmerman 1964]
1818. Chow, PH. 1988. Scanning-electron-microscopical study of the seminal vesicle, coagulating gland, ampullary gland and ventral prostate in the golden hamster. Acta Anat. 133:269-273.
1819. Chowdhury, AK & Steinberger, E. 1976. A study of germ cell morphology and duration of spermatogenic cycle in the baboon, *Papio anubis.* Anat Rec. 185:155-170.
1820. Chowdhury, T. 1966. A note on breeding Indian rhinoceroses *Rhinoceros unicornis* at Gauhati Zoo. Int Zoo Yearb. 6:197.
1821. Christen, A. 1968. Haltung und Brutbiologie von *Cebuella.* Folia Primatol. 8:41-49.
1822. Christen, A. 1974. Fortpflanzungbiologie und Verhalten bei *Cebuella pygmaea* und *Tamarin tamarin* (Primates, Platyrrhina, Callithricidae). Adv Ethology. 14:1-79.
1823. Christen, A, Dobeli, M, Kempken, B, et al. 1989. Urinary excretion of oestradiol-17β in the female cycle of Goeldi's monkeys (*Callimico goeldii*). Folia Primatol. 52:191-200.
1824. Christensen, I. 1973. Age determination, age distribution and growth of bottlenose whales, *Hyperoodon ampullatus* (Forster), in the Labrador Sea. Norw J Zool. 21:331-340.
1825. Christensen, I. 1975. Preliminary report on the Norwegian fishery for small whales: Expansion of Norwegian whaling to Arctic and northwest Atlantic waters, and Norwegian investigation of the biology of small whales. J Fish Res Bd Can. 32:1083-1094.
1826. Christensen, I. 1984. Growth and reproduction of killer whales, *Orcinus orca,* in Norwegian coastal waters. Rep Int Whal Comm Sp Issue. 6:253-258.
1827. Christensen, P. 1975. The breeding burrow of the banded ant-eater or numbat (*Myrmecobius fasciatus*). W Aust Nat. 13:32-34.
1828. Christenson, TE & Le Boeuf, BJ. 1978. Aggression in the female northern elephant seal, *Mirounga angustirostris.* Behaviour. 64:158-172.
1828a. Christian, JJ. 1950. Polyovular follicles in the woodchuck (*Marmota monax*). J Mamm. 31:196.
1829. Christian, JJ. 1956. The natural history of a summer aggregation of the big brown bat *Eptesicus fuscus fuscus.* Am Mid Nat. 55:66-89.
1830. Christian, JJ. 1962. Seasonal changes in the adrenal glands of woodchucks (*Marmota monax*). Endocrinology. 71:431-447.
1831. Christian, JJ. 1969. Maturation and breeding of *Blarina brevicauda* in winter. J Mamm. 50:272-276.
1832. Christian, JJ & Davis, DE. 1966. Adrenal glands in female voles (*Microtus pennsylvanicus*) as related to reproduction and population size. J Mamm. 47:1-18.
1833. Christian, JJ & Lloyd, JA. 1969. Reproductive activity of individual females in three experimental freely growing populations of house mice (*Mus musculus*). J Mamm. 50:49-59.
1834. Christiansen, E, Wiger, R, & Eilertsen, E. 1978. Morphological variations in the preputial gland of wild bank voles, *Clethrionomys glareolus.* Holarctic Ecol. 1:321-325.
1835. Christov, L & Markov, G. 1972. A population of *Clethrionomys glareolus pirinus* on the Vitosha Mountain, Bulgaria III Individual growth curve. Acta Theriol. 17:343-346.
1836. Chubb, EC. 1909. The mammals of Matabeleland. Proc Zool Soc London. 1909:113-125.
1837. Chupasko, JM. 1986. Interaction between photoperiod and ferulic acid (a plant phenolic compound) on the reproductive indices of three rodents. MS Thesis. LA State Univ, Baton Rouge.
1838. Churchfield, JS. 1979. Studies on the ecology and behaviour of British shrews. Ph D Diss. Westfield Coll, Univ London, London.
1839. Ciani, P. 1961. Il testiculo durante l'ibernazione e nelle prime fasi risveglio (ricerche in *Erinaceus europaeus*). Arch Ital Anat Embriol. 66:340-363.
1840. Cicmanee, JC & Campbell, AK. 1977. Breeding the owl monkey (*Aotus trivirgatus*) in a laboratory environment. Lab Anim Sci. 27:512-517.
1841. Cinq-Mars, RJ & Brown, LN. 1969. Reproduction and ecological distribution of the rockmouse, *Peromyscus difficilis,* in northern Colorado. Am Mid Nat. 81:205-217.
1842. Cisar, CF. 1969. The rat kangroo (*Potorous tridactylus*): Handling and husbandry practices in a research facility. Lab Anim Care. 19:55-59.
1843. Cisar, CF & Jayson, G. 1967. Effects of frequency of cage cleaning on rat litters prior to weaning. Lab Anim Care. 17:215-217.
1843a. Claramunt, T. 1976. Sobre la actividad sexual de *Pitymys duodecimcostatus* de Sélys-Longchamps, 1839, en Cataluña. P Dept Zool. 1:47-54. [not seen]
1844. Claude, C. 1970. Biometrie and Fortpflanzungsbiologie der Rötelmaus *Clethrionomys glareolus* (Schreber, 1780) an versohiedenen Höhenstufen der Schweiz. Rev Suisse Zool. 77:435-480.
1845. Clapham, PJ & Mayo, CA. 1987. Reproduction and recruitment of individually identified humpback whales,

Megaptera novaeangliae, observed in Massachusetts Bay, 1979-1985. Can J Zool. 65:2853-2863.
1846. Clapham, PJ & Mayo, CA. 1987. The attainment of sexual maturity in two female humpback whales. Mar Mamm Sci. 3:279-283.
1847. Clapperton, K, Fordham, RA, & Sparksman, RI. 1987. Preputial glands of the ferret *Mustela furo* (Carnivora: Mustelidae). J Zool. 212:356-361.
1848. Clark, AB. 1978. Sex ratio and local resource competition in a prosimian primate. Science. 201:163-165.
1849. Clark, BR & Price, EO. 1981. Sexual maturation and fecundity of wild and domestic Norway rats (*Rattus norvegicus*). J Reprod Fert. 63:215-220.
1850. Clark, DR & Lamont, TG. 1976. Organochlorine residues and reproduction in the big brown bat. J Wildl Manage. 40:249-254.
1851. Clark, FH. 1936. The estrous cycle of the deermouse, *Peromyscus maniculatus*. Contrib Lab Vert Genetics. 1:1-8.
1852. Clark, FH. 1936. The estrous cycle of the cotton-rat, *Sigmodon hispidus*. Contrib Lab Vert Genetics. 2:1-2.
1853. Clark, FH. 1937. Parturition in the deer mouse. J Mamm. 18:85-87.
1854. Clark, FH. 1938. Age of sexual maturity in mice of the genus *Peromyscus*. J Mamm. 19:230-234.
1855. Clark, G. 1949. The failure of estrogen to induce changes in the sex skin of the male chimpanzee. Yale J Biol Med. 21:245-247.
1856. Clark, MJ. 1967. Pregnancy in the lactating pygmy possum, *Cercartetus concinnus*. Aust J Zool. 15:673-683.
1857. Clark, MJ. 1968. Termination of embryonic diapause in the red kangaroo, *Megalia rufa* by injection of progesterone or oestrogen. J Reprod Fert. 15:347-356.
1858. Clark, MJ, & Sharman, GB. 1965. Failure of hysterectomy to affect the ovarian cycle of the marsupial *Trichosurus vulpecula*. J Reprod Fert. 10:459-461.
1859. Clark, JR, Dierschke, DJ, Meller, PH, et al. 1979. Hormonal regulation of ovarian folliculogenesis in rhesus monkeys: Serum concentrations of estradiol-17β and follicle stimulating hormone associated with growth and identification of the preovulatory follicle. Biol Reprod. 21:497-503.
1860. Clark, JR, Dierschke, DJ, & Wolf, RC. 1978. Hormonal regulation of ovarian folliculogenesis in rhesus monkeys: I Concentrations of serum luteinizing hormone and progesterone during laparoscopy and patterns of follicular development during successive menstrual cycles. Biol Reprod. 17:779-783.
1861. Clark, MJ. 1966. The blastocyst of the red kangaroo, *Megaleia rufa* (Desm), during diapause. Aust J Zool. 14:19-25.
1862. Clark, MM, Spencer, CA, & Galef, BG, Jr. 1986. Improving the productivity of breeding colonies of Mongolian gerbils (*Meriones unguiculatus*). Lab Anim. 20:313-315.
1863. Clark, MM, Spencer, CA, & Galef, BG, Jr. 1986. Reproductive life history correlates of early and late sexual maturation in female Mongolian gerbils (*Meriones unguiculatus*). Anim Behav. 34:551-560.
1864. Clark, TW. 1970. Early growth, development, and behavior of the Richardson ground squirrel (*Spermophilus richardsoni elegans*). Am Mid Nat. 83:197-205.
1865. Clark, TW. 1970. Richardson's ground squirrel (*Spermophilus richardsonii*) in the Laramie Basin, Wyoming. Great Basin Nat. 30:55-70.
1866. Clark, TW. 1975. *Arctocephalus galapagoensis*. Mamm Species. 64:1-2.
1867. Clark, TW. 1976. The black-footed ferret. Oryx. 13:275-280.
1868. Clark, TW. 1977. Ecology and ethology of the white-tailed prairie dog (*Cynomys leucurus*). Milw Public Mus Publ Biol Geol. 3:1-97.
1869. Clark, TW, Anderson, E, Douglas, C, et al. 1987. *Martes americana*. Mamm Species. 289:1-8.
1870. Clark, TW, Hoffmann, RS, & Nadler, CF. 1971. *Cynomys leucurus*. Mamm Species. 7:1-4.
1871. Clark, TW & Skryja, DD. 1969. Postnatal development and growth of the golden-mantled ground squirrel, *Spermophilus lateralis lateralis*. J Mamm. 50:627-629.
1871a. Clark, TW & Stromberg, MR. 1987. Mammals in Wyoming. Univ KS Mus Nat Hist Public Ed Ser. 10:1-314.
1871b. Clark, WF. 1953. Gray shrew, *Notiosorex*, from eastern Oklahoma. J Mamm. 34:117-118.
1872. Clarke, JR. 1955. Influence of numbers on reproduction and survival of two experimental vole population. Proc R Soc London. 144:68-85.
1873. Clarke, JR. 1957. Light-induced changes in some endocrine organs of the vole (*Microtus agrestis*). J Endocrinol. 15:iv.
1874. Clarke, JR. 1985. The reproductive biology of the bank vole (*Clethrionomys glareolus*) and the wood mouse (*Apodemus sylvaticus*). Symp Zool Soc London. 55:33-59.
1875. Clarke, JR & Clulow, FV. 1973. The effect of successive matings upon bank vole (*Clethrionomys glareolus*) and vole (*Microtus agrestis*) ovaries. The Development and Maturation of the Ovary and its Functions. 160-170, Peters, H, ed, Excerpta Medica, Amsterdam.
1876. Clarke, JR, Clulow, FV, & Greig, F. 1970. Ovulation in the bank vole, *Clethrionomys glareolus*. J Reprod Fert. 23:531.
1877. Clarke, JR & Egan, EA. 1984. Wastage of ova: Experimental studies in the field vole (*Microtus agrestis*), the bank vole (*Clethrionomys glareolus*) and the wood mouse (*Apodemus sylvaticus*). Acta Zool Fennici. 171:141-144.
1878. Clarke, JR & Forsyth, IA. 1964. Seasonal changes in the gonads and accessory reproductive organs of the vole (*Microtus agrestis*). Gen Comp Endocrinol. 4:233-242.
1879. Clarke, JR & Frearson, S. 1972. Sebaceous glands on the hindquarters of the vole, *Microtus agrestis*. J Reprod Fert. 31:477-481.
1880. Clarke, JR & Hellwing, S. 1977. Remote control by males of ovulation in bank voles (*Clethrionomys glareolus*). J Reprod Fert. 50:155-158.
1881. Clarke, JR & Hellwing, S. 1983. Fertility of the post-partum bank vole (*Clethrionomys glareolus*). J Reprod Fert. 68:241-246.
1882. Clarke, JR & Kennedy, JP. 1967. Changes in the hypothalamo-hypophysial neurosecretory system and the gonads of the vole (*Microtus agrestis*). Gen Comp Endocrinol. 8:455-473.
1883. Clarke, JR & Kennedy, JP. 1967. Effect of light and temperature upon gonad activity in the vole (*Microtus agrestis*). Gen Comp Endocrinol. 8:479-488.
1884. Clarke, JW. 1975. Androgen control of the ventral scent gland in *Neotoma floridana*. J Endocrinol. 64:393-394.
1885. Clarke, MR, 1983. Infant-killing and infant disappearance following male takeovers in a group of free-ranging howling monkeys (*Alouatta palliatai*) in Costa Rica. Am J Primatol 5:241-247.
1886. Clarke, MR and Glander, KE 1984. Female reproductive *Alouatta palliatai* in Costa Rica. Female Primates: Studies by Women Primatologists, 111-126, Small, MFS, ed, Alan R Liss, New York.

1887. Clarke, R. 1956. Sperm whales of the Azores. Discovery Rep. 28:237-298.
1888. Clarke, R, Aguayo L, A, & Paliza G, O. 1964. Progress report on sperm whale research in the southeast Pacific ocean. Norsk Hvalfangst-Tid. 53:297-302.
1889. Clarke, RC. 1934. Notes on birth of a chimpanzee in the Clifton Zoological Gardens, Bristol. Proc Zool Soc London. 1934:731-732.
1890. Clarke, TWH. 1899. Dibatag or Clarke's gazelle (*Ammadorcas clarkei*). Great and Small Game of Africa, 368-370, Bryden, HA, ed, Rowland Ward, London.
1891. Claude, C. 1970. Biometrie und Fortpflanzungsbiologie der Rötelmaus *Clethrionomys glareolus* (Schreber, 1780) auf verschiedenen Hohenstufen der Schweiz. Rev Suisse Zool. 77:435-480.
1892. Claus, R & Weiler, U. 1985. Influence of light and photoperiodicity on pig prolificacy. J Repro Fert Suppl. 33:185-197.
1893. Claussen, R. 1982. Beobachtungen zur Aufzucht eines männlichen Weiszgesichtsakis *(Pithecia pithecia)* im Familienverband. Zool Garten NF. 52:188-194
1894. Claver, JA, Colillas, OJ, & Travi, BL. 1984. Histological and microbiological aspects of the vagina in captive howler monkeys (*Alouatta caraya*). Primates. 25:110-116.
1895. Claver, JA, von Lawzewitsch, I, & Colillas, OJ. 1984. Microscopic anatomy of the ovary of *Alouatta caraya*. Primates. 25:362-371.
1896. Clegg, MT & Weaver, M. 1972. Chorionic gonodotropin secretion during pregnancy in the chimpanzee (*Pan troglodytes*). Proc Soc Exp Biol Med. 39:1170-1174.
1897. Cleland, JB. 1918. Presidential address. J Proc R Soc NSW. 52:1-5.
1898. Clemens, LG. 1969. Experimental analysis of sexual behavior of the deermouse *Peromyscus maniculatus gambelli* Behaviour. 34:267-285.
1899. Clemens, LG & Pomerantz, SM. 1982. Testosterone acts as a prohormone to stimulate male copulatory behavior in male deer mice (*Peromyscus maniculatus*). J Comp Physiol Psychol. 96:114-122.
1900. Clendenon, AL & Rissman, EF. 1990. Prolonged copulatory behavior facilitates pregnancy success in the musk shrew. Physiol Beh. 47:831-835.
1901. Clermont, Y. 1972. Kinetics of spermatogenesis mammals: Seminiferous epithelium cycle and spermatogonial renewal. Physiol Rev. 52:198-236.
1902. Clermont, Y & Antar, M. 1968. Duration of the seminiferous epithelium and spermatogonial renewal in the monkey *Macaca arctoides*. Am J Anat. 136:153-164.
1903. Clewe, TH. 1969. Observations on reproduction of squirrel monkeys in captivity. J Reprod Fert Suppl. 6:151-156.
1904. Clift, CE. 1967. Notes on breeding and rearing a kinkajou, *Potos flavus*, at Syracuse Zoo. Int Zoo Yearb. 7:126-127.
1905. Clifton, DK, Steiner, RA, Resko, JA, et al. 1975. Estrogen-induced gonadotropin release in ovariectomized rhesus monkeys and its advancement by progesterone. Biol Reprod. 13:190-194.
1906. Cline, DR, Siniff, DB, & Erickson, AW. 1971. Underwater copulation of the Weddell seal. J Mamm. 52:216-218.
1906a. Close, RL. 1976. The disappearance of sex-chromosome from somatic tissues of marsupials *Perameles* and *Isoodon*. Bull Aust Mamm Soc. 3:36-37.
1907. Close, RL. 1977. Recurrence of breeding after cessation of suckling in the marsupial *Perameles nasuta*. Aust J Zool. 25:641-645.
1908. Close, RL. 1979. Sex chromosome mosaicism in liver, thymus, spleen and regenerating liver of *Perameles nasuta* and *Isoodon macrourus*. Aust J Biol Sci. 32:615-624.
1909. Close, RL & Bell, JN. 1990. Age estimation of pouch young of the allied rock-wallaby (*Petrogale assimilis*) in captivity. Aust Wild Res. 17:359-367.
1910. Clothier, RR. 1955. Contribution to the life history of *Sorex vagrans* in Montana. J Mamm. 36:214-221.
1911. Clough, G. 1969. Some prelilminary observatons on reproduction in the warthog, *Phacochoerus aethiopicus* Pallas. J Reprod Fert Suppl. 6:323-337.
1912. Clough, GC. 1963. Biology of the arctic shrew, *Sorex arcticus*. Am Mid Nat. 69:69-81.
1913. Clough, GC. 1968. Small mammals of Pasvikdal, Finnmark, Norway. Nyt Mag Zool. 15:68-80.
1914. Clough, GC. 1972. Biology of the Bahamian hutia, *Geocapromys ingrahami*. J Mamm. 53:807-823.
1915. Clough, GC. 1987. Ecology of island muskrats, *Ondatra zibethicus*, adapted to upland habitat. Can Field-Nat. 101:63-69.
1916. Clulow, FV & Clarke, JR. 1968. Pregnancy block in *Microtus agrestis*, an induced ovulator. Nature. 219:511.
1917. Clulow, FV & Langford, PE. 1971. Pregnancy-block in the meadow vole, *Microtus pennsylvanicus*. J Reprod Fert. 24:275-277.
1918. Clulow, FV & Mallory, FF. 1970. Oestrus and induced ovulation in the meadow vole, *Microtus pennsylvanicus*. J Reprod Fert. 23:341-343.
1919. Clulow, FV & Mallory, FF. 1974. Ovaries of meadow voles, *Microtus pennsylvanicus* after copulation with a series of males. Can J Zool. 52:265-267.
1920. Clutton-Brock, TH, Albon, SD, & Guinness, FE. 1981. Parental investment in male and female offspring in polygynous mammals. Nature. 289:487-489.
1921. Clutton-Brock, TH, Albon, SD, & Guinness, FE. 1984. Maternal dominance, breeding success and birth sex ratios in red deer. Nature. 308:358-360.
1922. Clutton-Brock, TH, Albon, SD, & Guinness, FE. 1985. Parental investment and sex differences in juvenile mortality in birds and mammals. Nature. 313:131-133.
1923. Clutton-Brock, TH, Albon, SD, & Guinness, FE. 1986. Great expectations: Dominance, breeding success and offspring sex ratios in red deer. Anim Behav. 34:460-471.
1924. Clutton-Brock, TH, Guinness, FE, & Albon, SD. 1982. Red Deer: Behavior and Ecology of Two Sexes. Univ Chicago Press, Chicago.
1925. Clutton-Brock, TH & Harvey, PH. 1977. Primate ecology and social organization. J Zool. 183:1-39.
1926. Clutton-Brock, TH, Iason, GR, Albon, SD, et al. 1982. Effects of lactation on feeding behavior and habitat use in wild red deer hinds. J Zool. 198:227-236.
1927. Cmelilk, SHW. 1962. Fatty acid composition of the milk fat of the eland antelope (*Taurotragus oryx*). J Sci Food Agric. 13:662-665.
1928. Cnyrim, E. 1914. Zur Schläfendrüse und zum Lidapparate des Elefanten. Anat Anz. 46:273-279.
1929. Coady, JW. 1974. Late pregnancy of a moose in Alaska. J Wildl Manage. 38:571-572.
1930. Coblentz, BE. 1980. A unique ungulate breeding pattern. J Wildl Manage. 44:929-933.
1931. Coblentz, BE & Baber, DW. 1987. Biology and control of feral pigs on Isla Santiago, Galapagos, Ecuador. J Appl Ecol. 24:403-418.

1932. Coblentz, BE & Coblentz, BA. 1985. Reproduction and the annual fat cycle of the mongoose of St John, US Virgin Islands. J Mamm. 66:560-563.
1933. Cockburn, A & Lidicker, WZ, Jr. 1983. Microhabitat heterogeneity and population ecology of an herbivorous rodent, *Microtus californicus*. Oecologia. 59:167-177.
1934. Cockburn, A. 1981. Population processes of the silky desert mouse, *Pseudomys apodemoides* (Rodentia), in mature heathlands. Aust Wildl Res. 8:499-514.
1935. Cockburn, A, Lee, AK, & Martin, RW. 1983. Macrogeographic variation in litter size in *Antechinus* (Marsupialia: Dasyuridae). Evolution. 37:86-95.
1936. Cockburn, A, Scott, MP, & Scotts, DJ. 1985. Inbreeding avoidance and male-biased natal dispersal in *Antechinus* spp (Marsupialia: Dasyuridae). Anim Behav. 33:908-915.
1937. Cockrum, EL. 1955. Reproduction in North American bats. Trans KS Acad Sci. 58:487-511.
1938. Cockrum, EL. 1960. Distribution, habitat and habits of the mastiff bat, *Eumops perotis*, in North America. J AZ Acad Sci. 1:79-84.
1939. Cockrum, EL & Bradshaw, GVR. 1963. Notes on the mammals from Sonora, Mexico. Am Mus Novit. 2138:1-9.
1939a. Cockrum, EL & Musgrove, BF. 1964. Additional records of the Mexican big-eared bat, *Plectous phyllotis* (Allen), from Arizona. J Mamm. 45:472-474.
1940. Cockrum, EL & Ordway, E. 1959. Bats of the Chiricahua Mountains, Cochise County, Arizona. Am Mus Novit. 1938:1-35.
1941. Cocks, AH. 1881. Note on the breeding of the otter. Proc Zool Soc London. 1881:249-250.
1942. Cocks, AH. 1900. Note on the gestation of the pine-marten. Proc Zool Soc London. 1900:836-837.
1943. Cocks, AH. 1903. On the gestation of the badger. Zoologist Fourth Ser. 7:441-443.
1944. Cocks, AH. 1904. The gestation of the badger. Zoologist Fourth Ser. 8:108-114.
1945. Cockson, A & McNeice, R. 1980. Survival in the pouch: The role of macrophages and maternal milk cells. Comp Biochem Physiol. 66A:221-225.
1946. Coe, CL, Connolly, AC, Kraemer, HC, et al. 1979. Reproductive development and behavior of captive female chimpanzees. Primates. 20:571-582.
1947. Coe, CL, Murai, JT, Wiener, SG, et al. 1986. Rapid cortisol and corticosteroid-binding globulin responses during pregnancy and after estrogen administration in the squirrel monkey. Endocrinology. 118:435-440.
1948. Coe, M. 1975. Mammalian ecological studies on Mount Nimba, Liberia. Mammalia. 39:523-584.
1949. Coe, MJ. 1962. Notes on the habits of the Mount Kenya hyrax (*Procavia johnstoni mackinderi* Thomas). Proc Zool Soc London. 138:639-644.
1950. Coe, MJ. 1967. Preliminary notes on the eastern Kenya springhare, *Pedetes surdaster larvalis* Hollister. E Afr Wildl J. 5:174-177.
1951. Coe, MJ. 1969. The anatomy of the reproductive tract and breeding in the spring haas, *Pedetes surdaster larvalis* Hollister. J Reprod Fert Suppl. 6:159-174.
1952. Coelho, AM, Jr. 1985. Baboon dimorphism: Growth in weight, length and adiposity from birth to 8 years of age. Monogr Primatol 6:125-159.
1953. Coelho, AM, Jr, Glassman, DM, & Bramblett, CA. 1984. The relation of adiposity and body size to chronological age in olive baboons. Growth. 48:445-454.
1954. Coetzee, CG. 1965. The breeding season of the multimammate mouse *Praomys* (*Mastomys*) *natalensis* (A Smith) in the Transvaal highveld. Zool Afr. 1:29-39.
1955. Coffey, P & Pook, J. 1974. Breeding, hand rearing and development of the third lowland gorilla *Gorilla g gorilla* at the Jersey Zoological Park. Ann Rep Jersey Wildl Pres Trust. 11:45-52.
1956. Coffey, PF. 1970. A breeding analysis of a group of captive black and white colobus (*Colobus polykomos* Zimmerman, 1780 at the Jersey Wildlife Preservation Trust. Ann Rep Jersey Wildl Pres Trust. 7:10-15.
1957. Coffin, AL & Remington, JD. 1953. Pregnant yearling cow elk. J Wildl Manage. 17:223.
1958. Cogen, PH, Antunes, JL, Louis, KM, et al. 1980. The effects of anterior hypothalamic disconnection on gonadotropin secretion in the female rhesus monkey. Endocrinology. 107:677-683.
1959. Cognié, Y, Mariana, JC, & Thimonier, J. 1970. Étude du moment d'ovulation chez la brebis normale ou traitée par un progestagène associé ou non à une injection de PMSG. Ann Biol Anim Bioch Biophys. 10:15-24.
1960. Cohen, JA. 1977. A review of the biology of the dhole or asiatic wild dog (*Cuon alpinus* Pallas). Anim Regul Stud. 1:141-158.
1961. Cohen, JA. 1978. *Cuon alpinus*. Mamm Species. 100:1-3.
1962. Cohen, L. 1981. Aspects of the biology of the vole, *Microtus guentheri* in Israel. Israel J Zool. 30:104.
1963. Cohen, S. 1955. Variations in plasma protein mass during the menstrual cycle of the chacma baboon (*Papio ursinus*). J Endocrinol. 12:196-204.
1964. Cohen-Parsons, M & Carter, CS. 1987. Males increase serum estrogen and estrogen receptor binding in brain of female voles. Physiol Behav. 39:309-314.
1965. Cohen-Shlagman, L, Hellwing, S, & Yom-Tov, Y. 1984. The biology of the Levant vole, *Microtus guentheri* in Israel II The reproduction and growth in captivity. Z Säugetierk. 49:148-156.
1966. Cohen-Shlagman, L, Yom-Tov, Y, & Hellwing, S. 1984. The biology of the Levant vole, *Microtus guentheri* in Israel I Population dynamics in the field. Z Säugetierk. 49:135-147.
1967. Coimbra Filho, AF. 1965. Breeding lion marmosets *Leontideus rosalia* at Rio de Janeiro Zoo. Int Zoo Yearb. 5:109-110.
1968. Coimbra Filho, AF. 1966. Notes on the reproduction and diet of the Azara's fox *Cerdocyon thous azarae* and the hoary fox *Dusicyon vetulus* at Rio de Janeiro Zoo. Int Zoo Yearb. 6:168-169.
1969. Coimbra Filho, AF. 1967. Notes on the yak *Bos grunniens* in the humid tropical climate of Rio de Janeiro. Int Zoo Yearb. 7:224.
1970. Coimbra-Filho, AF & de A Maia, A. 1977. As fases do processo reproductivo de *Macaca mulatta* Zimmermann, 1780, na Ilha do Pinheiro, Rio de Janeiro, Brasil (Cercopithecidae, Primates). Rev Bras Biol. 37:71-78.
1971. Coimbra-Filho, AF & Mittermeier, RA. 1972. Taxonomy of the genus *Leontopithecus* Lesson, 1840. Saving the Lion Marmoset. 7-22, Bridgwater, DD, ed, Wild Animal Propag Trust, Wheeling, WV.
1972. Coimbra-Filho, AF & Mittermeier, RA. 1977. Conservation of the Brazilian lion tamarins (*Leontopithecus rosalia*). Primate Conservation. 59-94, Prince Rainier III of Monaco, & Bourne, GH, eds, Academic Press, New York.
1973. Colby, ED. 1970. Induced estrus and timed pregnancies in cats. Lab Anim Care. 20:1075-1080.

1973a. Cole, HH & Hart, GH. 1930. The potency of blood serum of mares in progressive stages of pregnancy in effecting sexual maturity of the immature rat. Am J Physiol. 93:57-68.
1973b. Cole, HH & Saunders, PJ. 1935. The concentrations of gonad-stimulating hormone in blood serum and of estrin in the urine throughout pregnancy in the mare. Endocrinology. 19:199-208.
1973c. Coleman, RH. 1950. The status of *Lasiurus borealis seminolus* (Rhoads). J Mamm. 31:190-192.
1974. Colillas, OJ, Ruiz, JC, & Patino, EM. 1982. Modificaciones estancionales y ciclo reproductivo del mono ardilla (*Saimiri sciureus*). Acta Physiol Lat Am. 32:11-20.
1975. Collet, A. 1981. Biologie du Dauphin commun *Delphinus delphis* L en Atlantique nord-est. Thèse du troisième cycle. Université de Poitiers.
1976. Collet, A & Harrison, RJ. 1981. Ovarian characteristics, corpora lutea and corpora albicantia in *Delphinus delphis* stranded on the Atlantic coast of France. Aquatic Mamm. 8:69-76.
1976a. Collet, A & Robineau, D. 1988. Data on the genital tract and reproduction in Commerson's dolphin, *Cephalorhynchus commersonii* (Lacépède, 1804), from the Kerguelen Islands. Rep Int Whal Comm Sp Issue. 9:119-141.
1977. Collet, A & St Girons, H. 1984. Preliminary study of the male reproductive cycle in common dolphins, *Delphinus delphis*, in the eastern north Atlantic. Rep Int Whal Comm Sp Issue. 6:355-360.
1978. Collett, F. 1981. Population characteristics of *Agouti paca* (Rodentia) in Columbia. Publ Mus MI State Univ. 5:489-601.
1979. Collett, R. 1881. On *Halichoerus grypus* and its breeding on the Fro Islands off Tronhjems-fjord in Norway. Proc Zool Soc London. 1881:380-387.
1980. Collett, R. 1909. A few notes on the whale *Balaena glacialis* and its capture in recent years in the north Atlantic by Norwegian whalers. Proc Zool Soc London. 1909:91-98.
1981. Collett, SF. 1981. Population characteristics of *Agouti paca* (Rodentia) in Columbia. Publ MI State Univ Mus Biol Ser. 5:489-601.
1982. Collins, DC, Graham, CE, & Preedy, JKR. 1975. Identification and measurement of urinary estrone, estradiol-17β, estriol, pregnanediol and androsterone during the menstrual cycle of the orangutan. Endocrinology. 96:93-101.
1983. Collins, DC & Kent, HA, Jr. 1964. Polynuclear ova and polyovular follicles in the ovaries of young guinea pigs. Anat Rec. 148:115-119.
1984. Collins, G. 1940. Habits of the Pacific walrus (*Odobenus divergens*). J Mamm. 21:138-144.
1985. Collins, JG & Bradbury, HJ. 1981. Structures of four new oligosaccharides from marsupial milk, determined mainly by C-nmr spectroscopy. Carbohydrate Res. 92:136-140.
1986. Collins, JM. 1974. Some aspects of reproduction and age structures in the black bear in North Carolina. Proc Ann Conf SE Assoc Game Fish Comm. 27:163-170.
1987. Collins, LC. 1973. Monotremes and marsupials: A reference for zoological institutions. book. Smithsonian Inst Press, Washington, DC.
1988. Collins, LR & Eisenberg, JR. 1972. Notes on the behavior and breeding of pacaranas, *Dinomys branickii*, in captivity. Int Zoo Yearb. 12:108-114.
1989. Collins, PM & Tsang, WN. 1987. Growth and reproductive development in the male tree shrew (*Tupaia belangeri*) from birth to sexual maturity. Biol Reprod. 37:261-267.
1990. Collins, PM, Tsang, WN, & Lofts, B. 1982. Anatomy and function of the reproductive tract in the captive male tree shrew (*Tupaia belangeri*). Biol Reprod. 26:169-182.
1991. Colman, JS. 1937. The present state of the Newfoundland seal fisheries. J Anim Ecol. 6:145-159.
1992. Colmenares, F. 1991. Greeting behaviour between male baboons: Oestrous females, rivalry and negotiation. Anim Behav. 41:49-60.
1993. Colosi, P, Holekamp, KE, Thordarson, G, et al. 1987. Purification and partial characterization of prolactin from the California ground squirrel (*Spermophilus beecheyi*). Biol Reprod. 36:1017-1023.
1994. Colvin, JD. 1986. Proximate causes of male emigration at puberty in rhesus monkeys. The Cayo Santiago Macaques: History, Behavior and Biology. 131-157, Rawlins, RG & Kessler, MJ, eds, State Univ NY, Albany.
1995. Colvin, MA & Colvin, DV. 1970. Breeding and fecundity of six species of voles (*Microtus*). J Mamm. 51:417-419.
1996. Comfort, A. 1957. Survival curves of mammals in captivity. Proc Zool Soc London. 128:349-364.
1996a. Commissaris, LR. 1959. Notes on the yuma myotis in New Mexico. J Mamm. 40:441-442.
1997. Compoint-Monmignault, C. 1968. Cycle oestral d'un Gerbillidé et de quelques Muridés d'Afrique centrale. Mammalia. 32:524-525.
1998. Comrie, LC & Adam, AB. 1938. The female reproductive system and corpora lutea of the false killer whale *Pseudorca crassidens* Owen. Trans R Soc Edinb. 59:521-531.
1999. Conant, RA. 1962. A milking technique and the composition of mink milk. Am J Vet Res. 23:1104-1106.
2000. Conaway, C. 1952. Age at sexual maturity of male elk. J Wildl Manage. 16:313-315.
2001. Conaway, CH. 1952. Life history of the water shrew (*Sorex palustris navigator*). Am Mid Nat. 48:219-248.
2002. Conaway, CH. 1954. The reproductive cycle of rice rats (*Oryzomys palustris palustris*) in captivity. J Mamm. 35:263-266.
2003. Conaway, CH. 1958. The uterus masculinus of *Castor canadensis*. J Mamm. 39:97-108.
2004. Conaway, CH. 1958. Maintenance, reproduction and growth of the least shrew in captivity. J Mamm. 39:507-512.
2005. Conaway, CH. 1959. The reproductive cycle of the eastern mole. J Mamm. 40:180-194.
2006. Conaway, CH. 1968. Post partum estrus in a sciurid. J Mamm. 49:158-159.
2007. Conaway, CH & Koford, CB. 1964. Estrous cycles and mating behavior in a free-ranging band of rhesus monkeys. J Mamm. 45:577-588.
2008. Conaway, CH & Sade, DS. 1965. The seasonal spermatogenic cycle in free ranging rhesus monkeys. Folia Primatol. 3:1-12.
2009. Conaway, CH & Sade, DS. 1969. Annual testis cycle of the green monkey (*Cercopithecus aethiops*) on St Kitts, West Indies. J Mamm. 50:833-835.
2010. Conaway, CH, Sadler, KC, & Hazelwood, DH. 1974. Geographic variation in litter size and onset of breeding in cottontails. J Wildl Manage. 38:473-481.
2011. Conaway, CH & Sorenson, MW. 1966. Reproduction in tree shrews. Symp Zool Soc London. 15:471-492.
2012. Conaway, CH & Wight, HM. 1962. Onset of reproductive season and first pregnancy of the season in cottontails. J Wildl Manage. 26:278-290.
2013. Conaway, CH & Wight, HM. 1963. Age at sexual maturity of young male cottontails. J Mamm. 44:426-427.

2014. Conaway, CH, Wight, HM, & Sadler, KC. 1963. Annual production by a cottontail population. J Wildl Manage. 27:171-175.
2015. Concannon, P, Baldwin, B, Lawless, J, et al. 1983. Corpora lutea of pregnancy and elevated serum progesterone during pregnancy and postpartum anestrus in woodchucks (*Marmota monax*). Biol Reprod. 29:1128-1134.
2016. Concannon, P, Baldwin, B, & Tennant, B. 1984. Serum progesterone profiles and corpora lutea of pregnant, postpartum, barren and isolated females in a laboratory colony of woodchucks (*Marmota monax*). Biol Reprod. 30:945-951.
2017. Concannon, P, Fullam, L, Baldwin, B, et al. 1989. Effects of induction versus prevention of hibernation on reproduction in captive male and female woodchucks (*Marmota monax*). Biol Reprod. 41:255-261.
2017a. Concannon, P, McCann, JP, & Temple, M. 1989. Biology and endocrinology of ovulation, pregnancy and parturition in the dog. J Reprod Fert Suppl 39:3-25.
2018. Concannon, P, Pilbeam, T, & Travis, H. 1980. Advanced implantation in mink (*Mustela vison*) treated with medroxyprogesterone acetate during early embryonic diapause. J Reprod Fert. 58:1-6.
2019. Condé, B & Schauenberg, P. 1969. Reproduction du Chat forestier d'Europe (*Felis silvestris* Schreber) en captivité. Rev Suisse Zool. 76:183-210.
2020. Condé, B & Schauenberg, P. 1974. Reproduction du Chat forestier (*F silvestris* Schreber) dans le nord-est de la France. Rev Suisse Zool. 81:45-52.
2021. Conder, P. 1970. Breeding of common wombat in captivity. Vict Nat. 87:322.
2022. Condrin, JM. 1936. Observations on the seasonal and reproductive activities of the eastern chipmunk. J Mamm. 17:231-234.
2023. Condy, PR. 1979. Annual cycle of the southern elephant seal *Mirounga leonina* (Linn) at Marion Island. S Afr J Zool. 14:95-102.
2024. Condy, PR. 1980. Postnatal development and growth in southern elephant seals (*Mirounga leonina*) at Marion Island. S Afr J Wildl Res. 10:118-122.
2025. Connor, PF. 1953. Notes on the mammals of a New Jersey pine barrens area. J Mamm. 34:227-235.
2026. Connor, PF. 1959. The bog lemming *Synaptomys cooperi* in southern New Jersey. Publ MI State Univ Mus Biol Ser. 1:161-248.
2026a. Constantine, DG. 1961. Gestation period in the spotted skunk. J Mamm. 42:421-422.
2026b. Constantine, DG. 1961. Spotted bat and big free-tailed bat in northern New Mexico. Southwest Nat. 6:92-97.
2027. Conti, A, Godovac-Zimmermann, J, Napolitano, L, et al. 1985. Identification and characterization of two α-lactalbumins from Somali camel milk (*Camelus dromedarius*). Milchwissenschaft. 40:673-675.
2027a. Contreras, L & Bustos-Obregòn, E. 1980. Anatomy of reproductive tract in *Octodon degus* Molina: A nonscrotal rodent. Arch Androl. 4:115-124.
2028. Conway, K. 1988. Captive management and breeding of the tiger quoll *Dasyurus maculatus*. Int Zoo Yearb. 27:108-119.
2029. Cook, B, Karsch, FJ, Graber, JW, et al. 1977. Luteolysis in the common opossum, *Didelphis marsupialis*. J Reprod Fert. 49:399-400.
2030. Cook, B, McDonald, IR, & Gibson, WR. 1978. Prostatic function in the brush-tailed possum, *Trichosurus vulpecula*. J Reprod Fert. 53:369-375.
2031. Cook, B & Nalbandov, AV. 1968. The effect of some pituitary hormones on progesterone synthesis *in vitro* by the luteinized ovary of the common opossum (*Didelphis marsupialis virginiana*). J Reprod Fert. 15:267-275.
2032. Cook, B, Sutterlin, NS, Graber, JW, et al. 1977. Synthesis and action of gonadal steroids in the American opossum, *Didelphis marsupialis*. Proc Symp Comp Biol Reprod, 1976. 4:253-254.
2033. Cook, C. 1904. Gestation of badgers. Zoologist, Fourth Series. 8:30.
2034. Cook, HW & Baker, BE. 1969. Seal milk I Harp seal (*Pagophilus groenlandicus*) milk: Composition and pesticide residue content. Can J Zool. 47:1129-1132.
2035. Cook, HW, Pearson, AM, Simmons, NM, et al. 1970. Dall sheep (*Ovis dalli dalli*) milk I Effects of stage of lactation on the composition of milk. Can J Zool. 48:629-633.
2036. Cook, HW, Rausch, RA, & Baker, BE. 1970. Moose (*Alces alces*) milk: Gross composition, fatty acid, and mineral composition. Can J Zool. 48:213-215.
2037. Cooke, JG. 1985. Has the age at sexual maturity of southern hemisphere minke whales declined?. Rep Int Whal Commn. 35:335-340.
2038. Cooley, ME. 1946. Cottontails breeding in their first summer. J Mamm. 27:273-274.
2039. Cooper, JB. 1942. An exploratory study on African lions. Comp Psy Mono. 17:1-48.
2040. Cooper, JE. 1975. The use of the pig as an animal model to study problems associated with low birthweight. Lab Anim. 9:329-336.
2040a. Cooper, KA, Harder, JD, Clawson, DH, et al. 1990. Serum testosterone and musth in captive male African and Asian elephants. Zoo Biol. 9:297-306.
2041. Cooper, RA, Connell, RS, & Wellings, SR. 1964. Placenta of the Indian elephant *Elephas indicus*. Science. 146:410-412.
2042. Cooper, RW. 1968. Small species of primates in biomedical research. Lab Anim Care. 18:267-279.
2043. Cope, JB & Humphrey, SR. 1972. Reproduction of the bats *Myotis keenii* and *Pipistrellus subflavus* in Indiana. Bat Res News. 13:9-10.
2044. Copson, GR, Brothers, NP, & Skira, IJ. 1985. Biology of the feral cat, *Felis catus*, on Macquarie Island. Aust Wildl Res. 12:428-436.
2045. Corbet, GB & Southern, HN. 1977. The Handbook of British Mammals. Blackwell, Oxford.
2046. Cordonnier, JL & Roux, M. 1978. Étude en histofluorescence des monoamines épiphysaires chez le Lérot (*Eliomys quercinus* L) dans diverses conditions expérimentales au cours de la période hivernale. Ann Endocrinol. 39:403-410.
2047. Cords, M. 1984. Mating patterns and social structure in redtail monkeys (*Cercopithecus ascianus*). Z Tierpsychol. 64:313-329.
2048. Cords, M, Mitchell, BJ, Tsingalia, HM, et al. 1986. Promiscuous mating among blue monkeys in the Kakamega Forest, Kenya. Ethology. 72:214-226.
2049. Cork, SJ & Dove, H. 1989. Lactation in the tammar wallaby *Macropus eughenii* II Intake of milk components and maternal allocation of energy. J Zool. 219:399-410.
2050. Cornejo, BV & Fernandez, VS. 1984. Algunos aspectos reproductivos de a tuza *Pappogeomys tylorhinus tylorhinus* (Rodentia: Geomyidae) en el norte de la ciudad de México. An Inst Biol Univ Nac Auton México. 54:199-209.
2051. Cornell, LH, Asper, ED, Antrim, JE, et al. 1987. Progress report: Results of a long-range captive breeding program for the

bottlenose dolphin, *Tursiops truncatus* and *Tursiops truncatus gilli*. Zoo Biol. 6:41-53.
2052. Cornely, JE & Baker, RJ. 1986. *Neotoma mexicana*. Mamm Species. 262:1-7.
2053. Cornely, JE & Verts, BJ. 1988. *Microtus townsendii*. Mamm Species. 325:1-9.
2054. Cornély, JM. 1877. Letter to Zoological Society of London. Proc Zool Soc London. 1877:533.
2055. Corner, GW. 1923. Ovulation and menstruation in *Macacus rhesus*. Contrib Embryol Carnegie Inst Wash. 15(75):73-102.
2056. Corner, GW. 1932. The menstrual cycle of the Malayan monkey, *Macaca irus*. Anat Rec. 52:401-410.
2057. Corner, GW, Hartman, CG, & Bartelmez, GW. 1945. Development, organization, and breakdown of the corpus luteum in the rhesus monkey. Contrib Embryol Carnegie Inst Wash. 31(204):117-146.
2058. Corse, J. 1799. Observations on the manners habits, and natural history of the elephant. Phil Trans R Soc London. 89:31-55.
2059. Corthum, KW, Jr. 1967. Reproduction and duration of placental scars in prairie vole and the eastern vole. J Mamm. 48:287-292.
2060. Cory, CB. 1912. The mammals of Illinois and Wisconsin. Field Mus Nat Hist Zool Ser. 11:1-505.
2061. Costa, DP, LeBoeuf, BJ, Huntley, AC, et al. 1986. The energetics of lactation in the northern elephant seal, *Mirounga angustirostris*. J Zool. 209A:21-33.
2062. Costa, DP & Gentry, RL. 1986. Free-ranging energetics of northern fur seals. Fur Seals Maternal Strategies on Land and at Sea. 79-101, Gentry, RL & Kooyman, GL, eds, Princeton Univ Press, Princeton.
2063. Cothran, EG, Aivaliotis, MJ, & Vandeberg, JL. 1985. The effects of diet on growth and reproduction in gray short-tailed opossums (*Monodelphis domestica*). J Exp Zool. 236:103-114.
2064. Cott, HB. 1926. Observations on the life-habits of some batrachians and reptiles from the lower Amazon, and a note on some mammals from Marajo Island. Proc Zool Soc London. 1926:1159-1178.
2065. Cotton, MJ. 1974. Observations on a population of the Greenland lemming *Dicrostonyx groenlandicus* Traill. J Zool. 174:531-534.
2066. Coulson, JC & Hickling, G. 1961. Variation in the secondary sex-ratio of the grey seal, *Halichoerus grypus* (Fabricius) during the breeding season. Nature. 190:281.
2067. Coulson, JC & Hickling, G. 1964. The breeding biology of the grey seal, *Halichoerus grypus* (Fab), on the Farne Islands, Northumberland. J Anim Ecol. 33:485-512.
2068. Coulter, MW. 1966. Ecology and management of fishers in Maine. Ph D Diss. Syracuse Univ, Syracuse, NY.
2069. Courrier, R. 1920. Sur l'éxistence d'une sécrétion épididymaire chez la Chauve-souris hibernante et sa signification. CR Soc Biol. 83:67-69.
2070. Courrier, R. 1921. Sur le rôle physiologique des sécrétions utérine et tubaire chez le Chauvre-souris hibernante. CR Soc Biol. 84:572-574.
2071. Courrier, R. 1923. Cycle annuel de la glande interstielle du testicule chez les Chéiroptères: Coexistence du repos seminal et de l'activité génitale. CR Soc Biol. 88:1163-1166.
2072. Courrier, R. 1923. Sur le cycle de la glande interstitielle et l'évolution des charactères sexuels secondaires chez les mammifères a spermatogénèse périodique. CR Soc Biol. 89:1311-1313.
2073. Courrier, R. 1926. Les effets de la castration chez les Chéiroptères. CR Soc Biol. 94:1386-1388.
2074. Courrier, R. 1926. Un cas d'eunuchodisme avec spermatogenèse normale chez un Pipistrelle. CR Ass Anat. 21:176-182.
2075. Courrier, R. 1927. Etude sur le déterminisme de caracterès sexuels secondaires chez quelques Mammifères à activité testiculaire périodique. Arch Biol. 37:173-334.
2076. Cousins, D. 1976. The breeding of gorillas, *Gorilla gorilla* in zoological collections. Zool Garten NF. 46:215-236.
2077. Coutts, RR & Rowlands, IW. 1969. The reproductive cycle of the Skomer vole (*Clethrionomys glareolus skomerensis*). J Zool. 158:1-25.
2078. Couture, RJ. 1980. The development of thermoregulation in *Rhabdomys pumilio nambiensis*. S Afr J Zool. 15:201-202.
2079. Couturier, MAJ. 1938. Le Chamois, *Rupicapra rupicapra* L. B Arthaud, Grenoble.
2080. Couturier, MAJ. 1961. Ecologie et protection du Bouquetin (*Capra aegagrus ibex ibex* L) et du Chamois (*Rupicapra rupicapra rupicapra* (L) dans les Alpes. Terre Vie. 16:54-73.
2081. Couturier, MAJ. 1962. Le Bouquetin des Alpes. Grenoble.
2082. Coventry, AF. 1937. Notes on the breeding of some Cricetidae in Ontario. J Mamm. 18:489-496.
2082a. Coventry, AJ & Dixon, JM. 1984. Small native mammals from the Chinaman Well area, northwestern Victoria. Aust Mamm. 7:111-115.
2083. Cowan, DP. 1987. Aspects of the social organisation of the European wild rabbit (*Oryctolagus cuniculus*). Ethology. 75:197-210.
2084. Cowan, DP. 1987. Patterns of mortality in a free-living rabbit (*Oryctalagus cuniculus*) population. Symp Zool Soc London. 58:59-77.
2085. Cowan, IM. 1936. Nesting habits of the flying squirrel (*Glaucomys sabrinus*). J Mamm. 17:58-60.
2086. Cowan, IM. 1956. Life and times of the coast blacktailed deer. The Deer of North America. 523-617, Taylor, WP, ed, Stockpole Co, Harrisburg & Wildl Manage Inst, Washington, DC.
2087. Cowan, IM & Arsenault, M. 1954. Reproduction and growth in the creeping vole, *Microtus oregoni serpens* Merriam. Can J Zool. 32:198-208.
2088. Cowan, IM & Guiguet, CJ. 1956. The Mammals of British Columbia. Handbook No 11. BC Prov Mus.
2089. Cowan, IM & Wood, AJ. 1955. The growth rate of the black-tailed deer. J Wildl Manage. 19:331-336.
2090. Cowan, JM. 1947. The timber wolf in the Rocky Mountain parks of Canada. Can J Res. 25:139-174.
2091. Cowan, PE. 1981. Early growth and development of roof rats, *Rattus rattus* L. Mammalia. 45:239-250.
2092. Cowgill, UM. 1969. Some observations on the prosimian *Perodicticus potto*. Folia Primatol. 11:144-150.
2093. Cowgill, UM. 1969. The season of birth and its biological implications. J Reprod Fert Suppl. 6:89-103.
2094. Cowgill, UM, Bishop, A, Andrew, RJ, et al. 1962. An apparent lunar periodicity in the sexual cycle of certain prosimians. Proc Nat Acad Sci. 48:238-241.
2095. Cowgill, UM & Zeman, LB. 1980. Life span in captive nocturnal prosimians (*Perodicticus potto*) with reproductive and mortality records. Primates. 21:437-439.
2096. Cowles, RB. 1936. Notes on the mammalian fauna of Umzumbe Valley, Natal, South Africa. J Mamm. 17:121-130.
2097. Cox, C & LeBoeuf, BJ. 1977. Female incitation of male competition: A mechanism in sexual selection. Am Nat. 111:317-335.

2098. Cox, DF. 1967. Survival and gestation in pigs. Evolution. 21:195-196.
2099. Cox, TJ. 1965. Seasonal change in the behavior of the western pipistrelle because of lactation. J Mamm. 46:703.
2100. Crabb, WD. 1944. Growth, development and seasonal weights of spotted skunks. J Mamm. 25:213-221.
2101. Crabb, WD. 1948. The ecology and management of the prairie spotted skunk in Iowa. Ecol Monogr. 18:201-232.
2102. Craig, AM. 1964. Histology of reproduction and the estrus [*sic*] cycle in the female fur seal, *Callorhinus ursinus*. J Fish Res Bd Can. 21:773-814.
2103. Craig, SA. 1985. Social organization, reproduction and feeding behavior of a population of yellow-bellied gliders, *Petaurus australis* (Marsupialia: Petauridae). Aust Wildl Res. 12:1-18.
2104. Craig, SA. 1986. A record of twins in the yellow-bellied glider (*Petaurus australis* Shaw) (Marsupialia: Petauridae) with notes on the litter size and reproductive strategy of the species. Vict Nat. 103:72-75.
2105. Craighead, JJ, Craighead, FC, Ruff, RL, et al. 1973. Home ranges and activity patterns of nonmigratory elk of the Madison Drainage herd as determined by biotelemetry. Wildl Monogr. 33:1-50.
2106. Craighead, JJ, Hornocker, MG, & Craighead, FC. 1969. Reproductive biology of young female grizzly bears. J Reprod Fert Suppl. 6:447-475.
2107. Cramblet, HM & Ridenhour, RL. 1956. Parturition in *Aplodontia*. J Mamm. 37:87-90.
2108. Cranbrook, Earl of & Medway, Lord. 1962. The Malayan mole. Malay Nat J. 16:205-208.
2109. Crandall, LS. 1964. The Management of Wild Mammals in Captivity. Univ Chicago, Chicago.
2110. Crane, HS & Jones, DA. 1953. Initial proof of mule deer fawns breeding is found in Utah. J Wildl Manage. 17:225.
2111. Cranford, JA. 1982. Effect of green vegetation and cotton nest material on reproduction and survival of pine voles (*Microtus pinetorum*). Proc E Pine Meadow Vole Symp. 6:124-129.
2112. Cranford, JA. 1983. Ecological strategies of a small hibernator, the western jumping mouse *Zapus princeps*. Can J Zoo). 61:232-240.
2113. Cranford, JA. 1983. Effect of 6-MBOA on *Microtus pinetorum* and *Microtus pennsylvanicus* of different ages. Proc E Pine Meadow Vole Symp. 7:111-116.
2114. Cranz, C, Ishak, B, Brun, B, et al. 1986. Study of morphological and cytological parameters indicating oestrus in *Lemur fulvus mayottensis*. Zoo Biol. 5:379-386.
2115. Craven, RP & Clarke, JR. 1982. Gonadotrophin levels in male voles (*Microtus agrestis*) reared in long and short photoperiods. J Reprod Fert. 66:709-714.
2116. Craven, RP & Clarke, JR. 1986. Gonadal development and gonadotrophin secretion in the male vole (*Microtus agrestis*) after an abrupt change in photoperiod. J Reprod Fert. 76:513-518.
2117. Crawford, RJM. 1984. Activity, group structure and lambing of blue duikers *Cephalophus monticola* in the Tsitsikamma National Parks, South Africa. S Afr J Wildl Res. 14:65-68.
2118. Crawley, MC. 1969. Movement and home range of *Clethrionomys glareolus* Schreber and *Apodemus sylvaticus* L in northeast England. Oikos. 20:310-319.
2119. Crawley, MC. 1970. Longevity of Australian brush-tailed opossums (*Trichosurus vulpecula*) in indigenous forest in New Zealand. NZ J Sci. 13:348-351.
2120. Crawley, MC. 1973. A live-trapping study of Australian brush-tailed possums, *Trichosurus vulpecula* (Kerr) in the Orongorongo Valley, Wellington, New Zealand. Aust J Zool. 21:75-90.
2121. Crawley, MC & Warneke, R. 1979. New Zealand fur seal. Mammals in the Seas, FAO Fish Ser 5. 2:45-48.
2122. Crawley, MC & Wilson, GJ. 1976. The natural history and behaviour of the New Zealand fur seal (*Arctocephalus forsteri*). Tuatara. 22:1-29.
2123. Creed, RFS. 1960. Gonad changes in the wild red fox (*Vulpes vulpes crucigera*). J Physiol London. 151:19P-20P.
2124. Creed, RFS. 1960. Observations on reproduction in the wild red fox (*Vulpes vulpes*): An account with special reference to the occurrence of fox-dog crosses. Brit Vet J. 116:419-426.
2125. Creed, RFS & Biggers, JD. 1963. Development of the raccoon placenta. Am J Anat. 113:417-431.
2126. Creed, RFS & Biggers, JD. 1964. Placental haemophagous organs in the Procyonidae and Mustelidae. J Reprod Fert. 8:133-137.
2126a. Creel, SR & Creel, NM. 1991. Energetics, reproductive suppression and obligate communal breeding in carnivores. Behav Ecol Sociobiol. 28:263-270.
2126b. Creel, SR, Creel, NM, Wildt, DE, et al. 1992. Behavioural and endocrine mechanisms of reproductive suppression in Serengeti dwarf mongooses. Anim Behav. 43:231-245.
2127. Creel, SR, Monfort, SL, Wildt, DE, et al. 1991. Spontaneous lactation is an adaptive result of pseudopregnancy. Nature. 351:660-662.
2128. Creel, SR & Waser, PM. 1990. Failures of reproductive suppression in dwarf mongosses (*Helogale parvula*): Accident or adaptation. Behav Ecol. 2:7-15.
2129. Crespo, JA. 1975. Ecology of the pampas gray fox and the large fox (Culpeo). The Wild Canids. 179-191, Fox, MW, ed, Van Nostrand Reinhold Co, New York.
2130. Creutz, G. 1967. Wiederfang einer Wasserfledermaus (*Myotis daubentoni*) nach 9 Jahren. Säugetierk Mitt. 15:69.
2131. Crichton, EG. 1969. Reproduction in the pseudomyine rodent, *Mesembriomys gouldii*. Aust J Zool. 17:785-797.
2132. Crichton, EG. 1974. Aspects of reproduction in the genus *Notomys* (Muridae). Aust J Zool. 22:439-447.
2133. Crichton, EG & Krutzsch, PH. 1987. Reproductive biology of the female little mastiff bat, *Mormopterus planiceps* (Chiroptera: Molossidae) in southeast Australia. Am J Anat. 178:369-386.
2134. Crichton, EG, Krutzsch, PH, & Chvapil, M. 1982. Studies on prolonged spermatozoa survival in Chiroptera II The role of zinc in the spermatozoa storage phenomenon. Comp Biochem Physiol. 71A:71-77.
2135. Crichton, EG, Krutzsch, PH, & Wimsatt, WA. 1981. Studies on prolonged spermatozoa survival in Chiroptera I The role of uterine free fructose in the spermatozoa storage phenomenon. Comp Biochem Physiol. 70A:387-395.
2136. Criddle, NE & Criddle, S. 1923. The coyote in Manitoba. Can Field-Nat. 37:41-45.
2137. Criddle, S. 1926. The habits of *Microtus minor* in Manitoba. J Mamm. 7:193-200.
2138. Criddle, S. 1930. The prairie pocket gopher, *Thomomys talpoides rufescens*. J Mamm. 11:265-280.
2139. Criddle, S. 1938. A study of the snowshoe rabbit. Can Field-Nat. 52:31-40.
2140. Criddle, S. 1939. The thirteen-striped ground squirrel in Manitoba. Can Field-Nat. 53:1-6.

2141. Crisp, EA, Cowan, PE, & Messer, M. 1989. Changes in milk carbohydrates during lactation in the common brushtail possum, *Trichosurus vulpecula* (Marsupialia: Phalangeridae). Reprod Fert Dév. 1:309-314.
2142. Crisp, EA, Messer, M, & Shaughnessy, PD. 1988. Intestinal lactase and other disaccharidase activities of a suckling crabeater seal (*Lobodon carcinophagus*). Comp Biochem Physiol. 90B:371-374.
2143. Crisp, EA, Messer, M, & VandeBerg, JL. 1989. Changes in milk carbohydrates during lactation in a didelphid marsupial, *Monodelphis domestica*. Physiol Zool. 62:1117-1125.
2144. Crispens, CG, Jr & Doutt, JK. 1973. Sex chromatin in antlered female deer. J Wildl Manage. 37:422-423.
2145. Cristoffer, C. 1991. Scaling of body weight in bats and rodents. J Mamm. 74:731-733.
2146. Cristofori, P, Aria, G, Seren, E, et al. 1986. Endocrinological aspects of reproduction in the female camel. World Anim Rev. 57:22-25.
2147. Crockett, CM & Rudran, R. 1987. Red howler monkey birth data I Seasonal variation. Am J Primatol. 13:347-368.
2148. Crockett, CM & Rudran, R. 1987. Red howler monkey birth data II Interannual, habitat, and sex comparisons. Am J Primatol. 13:369-384.
2149. Crockett, CM & Sekulic, R. 1982. Gestation length in red howler monkeys. Am J Primatol. 3:291-294.
2150. Crockett, CM & Sekulic, R. 1984. Infanticide in red howler monkeys (*Alouatta seniculus*). pp 173-191 in Infanticide: Comparative and Evolutionary Perspectives, Hausfater, G & Hrdy, SB, eds. Aldine Press, New York.
2151. Croiness, FE, Hall, MJ, & Cockerill, RA. 1979. Mother-offspring association in red deer (*Cervus elaphus* L) on Rhum. Anim Behav. 27:536-544.
2152. Crook, GA & Skipper, G. 1987. Husbandry and breeding of Matschie's tree kangaroo *Dendrolagus m matschiei* at Adelaide Zoological Gardens. Int Zoo Yearb. 26:212-216.
2153. Cross, JF & Martin, RD. 1981. Calculation of gestation period and other reproductive parameters for primates. Dodo, J Jersey Wildl Pres Trust. 18:30-43.
2154. Cross, RM. 1977. Population studies on *Praomys tullbergi* (Thomas) and other rats of forest regions of Sierra Leone. Rev Zool Afr. 91:345-367.
2155. Crowcroft, P. 1955. Notes on the behaviour of shrews. Behaviour. 8:63-80.
2156. Crowcroft, P. 1957. The Life of the Shrew. Max Reinhardt, London.
2157. Crowcroft, P. 1964. Note on the sexual maturation of shrews (*Sorex araneus* Linnaeus 1758) in captivity. Acta Theriol. 8:89-93.
2158. Crowcroft, P & Soderlund, R. 1977. Breeding wombats (*Lasiorhinus latifrons*) in captivity. Zool Garten NF. 47:313-322.
2159. Crowe, TM & Liversidge, R. 1977. Disproportionate mortality of males in a population of springbok (Artiodactyla: Bovidae). Zool Afr. 12:469-473.
2160. Crowley, HM, Woodward, DR, & Rose, RW. 1988. Changes in milk composition during lactation in the potoroo, *Potorous tridactylus* (Marsupialia: Potoroinae). Aust J Biol Sci. 41:289-296.
2161. Croxall, JP & Hiby, L. 1983. Fecundity, survival and site fidelity in Weddell seals, *Leptonychotes weddelli*. J Appl Ecol. 20:19-32.
2162. Cruikshawk, W. 1797. Experiments in which, on the third day after impregnantion, the ova of the rabbits were found in the Fallopian tubes; and on the fourth day after impregnantion in the uterus itself; with the first appearances of the foetus. Phil Trans R Soc London. 87:197-214. [abridged 1809 18:129-137]
2163. Cukierski, MA. 1987. Synthesis and transport studies of the intrasyncytial lamina: An unusual placental basement membrane in the little brown bat, *Myotis lucifugus*. Am J Anat. 178:387-409.
2164. Culbertson, AE. 1946. Observations on the natural history of the Fresno kangaroo rat. J Mamm. 27:189-203.
2165. Cumming, DHC. 1975. A field study of the ecology and behaviour of warthog. Nat Mus Monuments Rhodesia Mus Mem. 7:1-179.
2166. Cummings, WC. 1985. Right whales *Eubalaena glacialis* (Muller, 1776) and *Eubalaena australis* (Desmoulins, 1822). Handbook of Marine Mammals. 3:275-304. Ridgway, SH & Harrison, RJ, eds, Academic Press, New York.
2167. Cummins, JM. 1981. Sperm maturation in the possum *Trichosurus vulpecula*: A model for comparison with eutherian mammals. Zool Publ Victoria Univ (Wellington). 74:23-37.
2168. Cummins, JM, Temple-Smith, PD, & Renfree, MB. 1986. Reproduction in the male honey possum (*Tarsipes rostratus*: Marsupialia): The epididymis. Am J Anat. 177:385-401.
2169. Cummins, JM & Woodall, PF. 1985. On mammalian sperm dimensions. J Reprod Fert. 75:153-175.
2170. Cuneo, F. 1965. Observations on the breeding of the klipspringer antelope, *Oreotragus oreotragus*, and the behavior of their young born at the Naples Zoo. Int Zoo Yearb. 5:45-48.
2171. Cunningham, DJ. 1905. Cape hunting dogs (*Lycaon pictus*) in the gardens of the Royal Zoological Society of Ireland. Proc R Soc Edinb. 25:843-848.
2172. Cupp, CJ & Uemura, E. 1981. Body and organ weights in relation to age and sex in *Macaca mulatta*. J Med Primatol. 10:110-123.
2173. Curie-Cohen, M, Yoshihara, D, Luttrell, L, et al. 1983. The effect of dominance on mating behavior and paternity in a captive troup of rhesus monkeys (*Macaca mulatta*). Am J Primat. 5:127-138.
2174. Curlewis, JD, Axelson, M, & Stone, GM. 1985. Identification of the major steroids in ovarian and adrenal venous plasma of the brush-tail possum (*Trichosurus vulpecula*) and changes in the peripheral plasma levels of oestradiol and progesterone during the reproductive cycle. J Endocrinol. 105:53-62.
2175. Curlewis, JD, Loudon, ASI, & Coleman, APM. 1988. Oestrous cycles and the breeding season of the Père David's deer hind (*Elaphurus davidianus*). J Reprod Fert. 82:119-126.
2176. Curlewis, JD & Loudon, ASI. 1988. Experimental manipulations of prolactin following removal of pouch young or bromocriptine treatment during lactational quiescence in the Bennet's wallaby. J Endocrinol. 119:405-411.
2177. Curlewis, JD & Loudon, ASI. 1989. The role of refractoriness to long daylength in the annual reproductive cycle of the female Bennett's wallaby (*Macropus rufogriseus rufogriseus*). J Exp Zool. 252:200-206.
2178. Curlewis, JD, Loudon, ASI, Milne, JA, et al. 1988. Effects of chronic long-acting bromocriptine treatment on liveweight, voluntary food intake, coat growth and breeding season in non-pregnant red deer hinds. J Endocrinol. 119:413-420.
2179. Curlewis, JD & Stone, GM. 1985. Peripheral androgen levels in the male brush-tail possum (*Trichosurus vulpecula*). J Endocrinol. 105:63-70.
2180. Curlewis, JD & Stone, GM. 1986. Effects of oestradiol, the oestrous cycle and pregnancy on weight, metabolism and cytosol receptors in the uterus of the brush-tail possum (*Trichosurus vulpecula*). J Endocrinol. 108:201-210.

2181. Curlewis, JD & Stone, GM. 1986. Reproduction in captive female brushtail possums, *Trichosurus vulpecula*. Aust J Zool. 34:47-52.
2182. Curlewis, JD & Stone, GM. 1987. Effects of oestradiol, the oestrous cycle and pregnancy on weight, metabolism and cytosol receptors in the oviduct and vaginal complex of the brushtail possum (*Trichosurus vulpecula*). Aust J Biol Sci. 40:315-322.
2183. Curlewis, JD, White, AS, & Loudon, ASI. 1987. The onset of seasonal quiescence in the female Bennett's wallaby (*Macropus rufogriseus rufogriseus*). J Reprod Fert. 80:119-124.
2184. Curlewis, JD, White, AS, Loudon, ASI, et al. 1986. Effects of lactation and season on plasma prolactin concentrations and response to bromocriptine during lactation in the Bennett's wallaby (*Macropus rufogriseus*). J Endocrinol. 110:59-66.
2185. Currie, WB, Blake, M, & Wimsatt, WA. 1988. Fetal development, and placental and maternal plasma concentrations of progesterone in the little brown bat (*Myotis lucifugus*). J Reprod Fert. 82:401-407.
2186. Currier, MJP. 1983. *Felis concolor*. Mamm Species. 200:1-7.
2186a. Curry-Lindahl, K. 1962. The irruption of the Norway lemming in Sweden during 1960. J Mamm. 43:171-184.
2187. Curry-Lindahl, K. 1970. Breeding biology of the Baltic grey seal (*Halichoerus grypus*). Zool Garten NF. 38:16-29.
2188. Cutrera, RA, Yunes, RMF, & Castro-Vazquez, A. 1988. Postpartum sexual behavior of the corn mouse (*Calomys musculinus*): Repertoire, measurements, and effects of removal of pups. J Comp Psychol. 102:83-89.
2189. Cuttle, P. 1982. Life history strategy of the dasyurid marsupial *Phascogale tapoatufa*. Carnivorous Marsupials. 1:13-22.
2190. Cutts, JH, Krause, WJ, & Leeson, CR. 1978. General observations on the growth and development of the young pouch opossum *Didelphis virginiana*. Biol Neonate. 33:264-272.
2191. Cuyler, WK. 1924. Cinnamon and albino opossums found at Austin, Texas. J Mamm. 5:130.
2192. Cuyler, WK. 1924. Observations on the habits of the striped skunk (*Mephitis mesomelas varians*). J Mamm. 5:180-189.
2193. Cygan, T. 1985. Seasonal changes in thermoregulation and maximum metabolism in the yellow-necked field mouse. Acta Theriol. 30:115-130.
2194. Czaja, JA & Bielert, C. 1975. Female rhesus sexual behavior and distance to a male partner: Relation to stage of the menstrual cycle. Arch Sex Behav. 4:583-598.
2195. Czaja, JA, Robinson, JA, Eisele, SG, et al. 1977. Relationship between sexual skin colour of female rhesus monkeys and midcycle plasma levels of oestradiol and progesterone. J Reprod Fert. 49:147-150.
2196. Czaplewski, NJ. 1983. *Idionycteris phyllotis*. Mamm Species. 208:1-4.
2197. Czaplewski, NJ, Farney, JP, Jones, JK, Jr, et al. 1979. Synopsis of bats of Nebraska. Occas Pap Mus TX Tech Univ. 61:1-24.
2198. Czekala, NM & Benirschke, K. 1974. Observations on a twin pregnancy in the African long-tongued fruit bat (*Megaglossus woermanni*). Bonn Zool Beitr. 25:220-230.
2199. Czekala, NM, Benirschke, K, McClure, H, et al. 1983. Urinary estrogen excretion during pregnancy in the gorilla (*Gorilla gorilla*), orangutan (*Pongo pygmaeus*) and the human. Biol Reprod. 28:289-294.
2200. Czekala, NM, Hodges, JK, Gause, GE, et al. 1980. Annual circulating testosterone levels in captive and free-ranging male armadillos (*Dasypus novemcinctus*). J Reprod Fert. 59:199-204.
2201. Czekala, NM, Hodges, JK, & Lasley, BL. 1981. Pregnancy monitoring in diverse primate species by estrogen and bioactive luteinizing hormone determination in small volumes of urine. J Med Primatol. 10:1-16.
2201a. Czekala, NM, Kasman, LH, Allen, J, et al. 1990. Urinary steroid evaluations to monitor ovarian function in exotic ungulates: VI Pregnancy detection in exotic Equidae. Zoo Biol. 9:43-48.
2202. Czekala, NM, Mitchell, WR, & Lasley, BL. 1987. Direct measurements of urinary estrone conjugates during the normal menstrual cycle of the gorilla (*Gorilla gorilla*). Am J Primatol. 12:223-229.
2203. Czekala, NM, Roser, JF, Mortensen, RB, et al. 1988. Urinary hormone analysis as a diagnostic tool to evaluate the ovarian function of female gorillas (*Gorilla gorilla*). J Reprod Fert. 82:255-261.
2204. Czernay, S. 1977. Zur Haltung und Zucht von Südpudus (*Pudu pudu* Molina 1782) in Thüringer Zoopark Erfurt. Zool Garten NF. 47:226-240.
2205. D'Addamio, GH, Roussel, JD, & Storrs, EE. 1977. Response of the nine-banded armadillo (*Dasypus novemcinctus*) to gonadotropins and steroids. Lab Ani Sci. 27:482-489.
2206. d'Huart, JP. 1978. Ecologie de l'Hylochere (*Hylochoerus meinertzhageni* Thomas) au Parc National des Virunga. Explor Parc Nat Virunga. 25:1-152.
2207. d'Huart, JP. 1991. Monographie des Riesenwaldschweines (*Hylochoerus meinertzhageni*). Sitzungsberichte der Tagung über Wildschweine und Pekaris im Zoo Berlin. 103-118.
2208. D'Sousa, F. 1974. A preliminary field report on the lesser tree shrew *Tupaia minor*. Prosimian Biology. 167-182, Martin, RD, Doyle, GA, & Walker, AC, eds, Univ Pittsburgh Press, Pittsburgh.
2209. D'Souza, F & Martin, RD. 1974. Maternal behaviour and the effects of stress on tree shrews. Nature. 251:309-311.
2210. Da Silveira, EKP. 1968. Notes on the care and breeding of the maned wolf *Chrysocyon brachyurus* at Brasilia Zoo. Int Zoo Yearb. 8:21-23.
2211. Da Silveira, EKP. 1975. The management of Caribbean and Amazonian manatees *Trichechus m manatus* and *T inunguis* in captivity. Int Zoo Yearb. 15:223-226.
2212. Dagg, AI. 1971. *Giraffa camelopardalis*. Mamm Species. 5:1-8.
2213. Dahl, JF. 1985. The external genitalia of female pygmy chimpanzees. Anat Rec. 211:24-28.
2213a. Dahl, JF. 1986. Cyclic perineal swelling during the intermenstrual intervals of captive female pygmy chimpanzees (*Pan paniscus*). J Hum Evol. 15:369-385.
2213b. Dahl, JF, Nadler, RD, & Collins, DC. 1991. Monitoring the ovarian cycles of *Pan troglodytes* and *P paniscus*: A comparative approach. Am J Primatol. 24:195-209.
2214. Dahl, K. 1897. Biological notes on north-Australian mammalia. Zoologist Fourth Ser. 1:189-216.
2215. Dahl, SK, Gussev, VM, & Bedny, SN. 1958. On the bionomics and reproduction of *Saiga tatarica* L. Zool Zhurn. 37:447-456.
2216. Dahlbäck, M & Andersson, M. 1981. Biology of the wild rabbit, *Oryctolagus cuniculus*, in southern Sweden IV Leydig cell activity and seasonal development of two male accessory organs of reproduction. Acta Zool (Stockholm). 62:113-120.
2217. Dahlbäck, M & Andersson, M. 1981. Biology of the wild rabbit, *Oryctolagus cuniculus*) in southern Sweden V Seasonal

variation in weight of the anal and inguinal glands. Z Säugetierk. 46:280-283.

2218. Dahlheim, no initials. 1982. Personal communication. [Cited by Christensen, 1984]

2219. Dailey, RA & Neill, JD. 1981. Seasonal variation in reproductive hormones of rhesus monkeys: Anovulatory and short luteal phase menstrual cycles. Biol Reprod. 25:560-567.

2220. Daketse, M-J & Martinet, L. 1977. Effect of temperature on the growth and fertility of the field-vole, *Microtus arvalis*, raised in different day length and feeding conditions. Ann Biol Anim Bioch Biophys. 17:713-721.

2221. Dalby, PL. 1975. Biology of pampa rodents, Balcarce area, Argentina. Publ Mus MI State Univ Biol Ser. 5:149-271.

2222. Dalby, PL. 1976. Some general behavioral and ecological observations on the South American short bare-tailed opossum, *Monodelphis dimidiata*. SE Reg Meeting, Anim Beh Soc. Charlottesville, Va.

2223. Dalby, PL & Heath, AG. 1976. Oxygen consumption and body temperature of the Argentine field mouse, *Akodon azarae*, in relation to ambient temperature. J Therm Biol. 1:177-179.

2224. Dale, FH. 1939. Variability and environmental responses of the kangaroo rat, *Dipodomys heermanni saxatilis*. Am Mid Nat. 22:703-731.

2225. Dalke, PD. 1942. The cottontail rabbits in Connecticut. Bull CT Geol Nat Hist Surv. 65:1-97.

2226. Dalquest, WW. 1941. An isolated race of *Microtus montanus* from eastern Washington. Proc Biol Soc Wash. 54:145-148.

2227. Dalquest, WW. 1947. Notes on the natural history of the bat, (*Myotis yumanensis*), in California, with a description of a new race. Am Mid Nat. 38:224-247.

2228. Dalquest, WW. 1947. Notes on the natural history of the *Corynorhinus rafinesquii* in California. J Mamm. 28:17-30.

2228a. Dalquest, WW. 1953. Mammals of the Mexican state of San Luis Potosí. LA State Univ Press, Baton Rouge.

2229. Dalquest, WW. 1955. Natural history of the vampire bats of eastern Mexico. Am Mid Nat. 53:79-87.

2230. Dalquest, WW. 1957. Observations on the sharp-nosed bat, *Rhynchiscus nasio* [*sic*] (Maximilian). TX J Sci. 9:219-226.

2230a. Dalquest, WW & Hall, ER. 1949. Five bats new to the known fauna of Mexico. J Mamm. 30:424-427.

2231. Dalquest, WW & Orcutt, DR. 1942. The biology of the least shrew-mole, *Neurotrichus gibbsii minor*. Am Mid Nat. 27:387-401.

2231a. Dalquest, WW & Ramage, MC. 1946. Notes on the long-legged bat (*Myotis volans*) at Old Fort Tejon and vicinity, California. J Mamm. 27:60-63.

2232. Daly, JC & Patton, JL. 1986. Growth, reproduction and sexual dimorphism in *Thomomys bottae* pocket gophers. J Mamm. 67:256-265.

2233. Daly, M. 1975. Early use of solid food by a leaf-eating gerbil (*Psammomys obesus*). J Mamm. 56:509-511.

2234. Daly, M & Daly, S. 1975. Behavior of *Psammomys obesus* (Rodentia: Gerbilinae) in the Algerian Sahara. Z Tierpsychol. 37:298-321.

2235. Daly, M & Daly, S. 1975. Socio-ecology of Saharan gerbils, especially *Meriones libycus*. Mammalia. 39:289-311.

2236. Daly, M, Wilson, M, & Behrends, P. 1984. Breeding of captive kangaroo rats, *Dipodomys merriami* and *D microps*. J Mamm. 65:338-341.

2237. Damassa, DA & Gustafson, AW. 1984. Control of plasma sex steroid-binding protein (SBP) in the little brown bat: Effects of photoperiod and orchiectomy on the induction of SBP in immature males. Endocrinology. 115:2355-2361.

2238. Damassa, DA & Gustafson, AW. 1985. Relationship of food intake to the induction of plasma steroid-binding protein and testicular activity in immature male little brown bat (*Myotis lucifugus lucifugus*). J Reprod Fert. 74:701-708.

2239. Damassa, DA, Gustafson, AW, & Chari, GC. 1983. Control of plasma sex-steroid-binding protein in the bat *Myotis lucifugus lucifugus*: Induction of steroid-binding activity in immature males. J Endocrinol. 97:57-64.

2240. Damassa, DA, Gustafson, AW, & King, JC. 1982. Identification of a specific binding protein for sex steroids in the plasma of the male little brown bat, *Myotis lucifugus lucifugus*. Gen Comp Endocrinol. 47:288-294.

2241. Damassa, DA, Gustafson, AW, Kwiecinski, GG, et al. 1985. Control of sex steroid-binding protein (SBP) in the little brown bat: Effects of thyroidectomy and treatment with L- and D-thyroxine on the induction of SBP in adult males. Biol Reprod. 33:1138-1146.

2242. Damassa, DM, Gustafson, AW, & Lang, JC. 1981. Sex steroid binding protein in the plasma of male *Myotis lucifugus lucifugus*: Identification and evidence for seasonal variation. N Am Symp Bat Res. 12:13.

2243. Danell, K. 1978. Population dynamics of the muskrat in a shallow Swedish lake. J Anim Ecol. 47:697-709.

2244. Dang, DC. 1970. Le Cycle de l'Épithelium Seminifère du Singe Crabier (*Macaca fascicularis*). Thèse Fac Sci Univ Paris.

2245. Dang, DC. 1971. Durée du cycle de l'épithelium séminifère du Singe crabier, *Macaca fascicularis*. Ann Biol Anim Bioch Biophys. 11:373-377.

2246. Dang, DC. 1971. Studes du cycle de l'épithelium séminifère du Singe crabier, *Macaca fascicularis* (=*irus* ou *cynomolgus*). Ann Biol Anim Bioch Biophys. 11:363-371.

2247. Dang, DC. 1977. Absence of seasonal variation in the length of the menstrual cycle and the fertility of the crab-eating macaque (*Macaca fascicularis*) raised under natural daylength ratio. Ann Biol Anim Bioch Biophys. 17:1-7.

2248. Dang, DC. 1979. Return of postpartum menstruation and fertility in laboratory *Macaca fascicularis*. Ann Biol Anim Bioch Biophys. 19:375-383.

2249. Dang, DC. 1979. Testosterone levels in umbilical cord blood, maternal peripheral plasma and amniotic fluid of the crab-eating monkey (*Macaca fascicularis*). Ann Biol Anim Bioch Biophys. 19:1307-1316.

2250. Dang, DC & Meusy-Dessolle, N. 1981. Annual plasma testosterone cycle and ejaculatory ability in the laboratory-housed crab-eating macaque (*Macaca fascicularis*). Reprod Nutr Dév. 21:57-68.

2251. Dang, ZH or SH. 1968. Some data concerning the biology of *Tragulus javanicus affinis* Gr (Artiodactyla, Tragulidae) in the Democratic Republic of Vietnam. Zool Zh. 47:432-437.

2252. Daniel, JC, Jr. 1971. Growth of the preimplantation embryo of the northern fur seal and its correlation with changes in uterine protein. Dev Biol. 26:316-328.

2253. Daniel, JC, Jr. 1975. Concentration of circulating progesterone during early pregnancy in northern fur seal, *Callorhinus ursinus*. J Fish Res Bd Canada. 32:65-66.

2254. Daniel, JC, Jr. 1981. Delayed implantation in the northern fur seal (*Callorhinus ursinus*) and other pinnipeds. J Reprod Fert Suppl. 29:35-50.

2255. Daniel, JC, Jr & Krishnan, RS. 1969. Studies on the relationship between uterine fluid components and the diapausing state of blastocysts from mammals having delayed implantation. J Exp Zool. 172:267-282.

2256. Daniel, M & Hanzak, J. 1985. Small mammals in eastern part of Nepal Himalaya. Rozpravy Ceskoslovenske Akademie Ved Rada Matematickych A Prirodnich Ved. 95:1-59.
2257. Daniel, MJ. 1979. The New Zealand short-tailed bat *Mystacina tuberculata*: A review of present knowledge. NZ J Zool. 6:357-370.
2258. Daniel, MJ & Pierson, ED. 1987. A lek mating system in New Zealand's short-tailed bat, *Mystacina tuberculata*. Bat Res News. 28:33.
2259. Daniel, MJ & Williams, GR. 1983. Observations of a cave colony of the long-tailed bat (*Chalinolobus tuberculatus*) in North Island, New Zealand. Mammalia. 47:71-80.
2259a. Daniklin, AA. 1990. Embryonal diapause in roe deer (*Capreolus*, Cervidae) and the ethological hypothesis of its origin. Zool Zhur. 69:120-124.
2260. Danilov, PI & Tumanov, IL. 1975. The reproductive cycles of some Mustelidae species. Byull Mosk O-Va Ispyt Oto Biol. 80:35-47.
2261. Dapson, RW. 1968. Reproduction and age structure in a population of short-tailed shrews, *Blarina brevicauda*. J Mamm. 49:205-214.
2262. Dapson, RW. 1979. Phenologic influences on cohort-specific reproductive strategies in mice (*Peromyscus polionotus*). Ecology. 60:1125-1131.
2263. Dapson, RW, Ramsey, PR, Smith, MH, et al. 1979. Demographic differences in contiguous populations of white-tailed deer. J Wildl Manage. 43:889-898.
2264. Darin-Bennett, A, Morris, S, Jones, RC, et al. 1976. The glyceryl phosphorylcholine and phospholipid pattern of the genital duct and spermatogenesis of the Africal elephant, *Loxodonta africana*. J Reprod Fert. 46:506-507.
2265. Dark, J, Johnston, PG, Healy, M, et al. 1983. Latitude of origin influences photoperiodic control of reproduction of deer mice (*Peromyscus maniculatus*. Biol Reprod. 28:213-220.
2266. Dark, J, Whaling, CS, & Zucker, I. 1987. Androgens exert opposite effects on body mass of heavy and light meadow voles. Horm Behav. 21:471-477.
2267. Dark, J & Zucker, I. 1983. Short photoperiods reduce winter energy requirements of the meadow vole, *Microtus pennsylvanicus*. Physiol Behav. 31:699-702.
2268. Darney, KJ, Jr & Franklin, LE. 1982. Analysis of the estrous cycle of the laboratory-housed Senegal galago (*Galago senegalensis senegalensis*): Natural and induced cycles. Folia Primatol. 37:106-126.
2269. Darrow, JM, Tamarkin, L, Duncan, MJ, et al. 1986. Pineal melatonin rhythms in female Turkish hamsters: Effects of photoperiods and hibernation. Biol Reprod. 35:74-83.
2270. Darrow, JM, Yogev, L, & Goldman, BD. 1987. Patterns of reproductive hormone secretion in hibernating Turkish hamsters. Am J Physiol. 253:R329-R336.
2271. Darwood, MY, Fuchs, F. 1980. Estradiol and protesterone in the maternal and fetal circulation in the baboon. Biol Reprod. 22:179-184.
2272. Das, AK. 1980. Observations on the reproductive behavior of the tiger, *Panthera tigris tigris* Linn in captivity. J Bombay Nat Hist Soc. 77:253-260.
2273. Dash, Y, Szaniawski, A, Child, GS, et al. 1977. Observations on some large mammals of the Transaltai, Djungarian and Shargin Gobi, Mongolia. Terre Vie. 31:587-597.
2274. Dasmann, RF & Mossman, AS. 1962. Population studies of impala in southern Rhodesia. J Mamm. 43:375-394.
2275. Dasmann, RF & Mossman, AS. 1962. Reproduction in some ungulates in southern Rhodesia. J Mamm. 43:533-537.
2276. Dathe, H. 1961. Beobachtungen zur Fortpflanzungsbiologie des Braunbären, *Ursus arctos* L. Zool Garten NF. 25:235-249.
2277. Dathe, H. 1961. Breeding the Malayan bear (*Helarctos malayanus*). Int Zoo Yearb. 3:94.
2278. Dathe, H. 1961. Superfoetation beim Löwen (*Panthera leo*). Zool Garten NF. 25:410-411.
2279. Dathe, H. 1963. Beitrag zir Fortpflanzungsbiologie des Malaienbären, *Helarctos m malayanus* (Raffl). Z Säugetierk. 28:155-162.
2280. Dathe, H. 1966. Breeding the corsac fox *Vulpes corsac* at East Berlin Zoo. Int Zoo Yearb. 6:166-167.
2281. Dathe, H. 1966. Einige Bemerkungen zur Zucht des Malaienbären, *Helarctos malayanus* (Raffl). Zool Garten NF. 32:193-198.
2282. Dathe, H. 1968. Breeding the Indian leopard cat *Felis bengalensis* at East Berlin Zoo. Int Zoo Yearb. 8:42-44.
2283. Dathe, H. 1970. A second generation birth of captive sun bears *Helarctos malayanus* at East Berlin Zoo. Int Zoo Yearb. 10:79.
2284. Dathe, H. 1973. Hohe Jungenzahl bei einer Hyänenhundin, *Lycaon pictus* (Temm). Zool Garten NF. 43:47.
2285. Dathe, H. 1974. Hohes Lebensalter eines Katzenbären, *Ailurus f fulgens* Cuv. Zool Garten NF. 44:311.
2286. Dathe, M. 1967. Bemerkungen zur Augzucht von Brillenbären, *Tremarctos ornatus* (Cuv), im Tierpark Berlin. Zool Garten NF. 34:105-133.
2287. Daudt, W. 1898. Beiträge zur Kenntnis des Urogenital-apparates der Cetaceen. Jena Z Naturwiss. 32:231-312.
2288. Dauphiné, TC, Jr. 1976. Biology of the Kaminuriak population of barren-ground caribou Part 4 Growth, reproduction and energy reserves. Can Wildl Serv Rep Ser. 38:1-69.
2289. Dauphiné, TC, Jr & McClure, RL. 1974. Synchronous mating in Canadian barren-ground caribou. J Wildl Manage. 38:54-66.
2290. David, GFX, Anand Kumar, TC, & Baker, TG. 1974. Uptake of tritiated thymidine by primordial germ cells in the ovaries of the adult slender loris. J Reprod Fert. 41:447-451.
2291. David, GFX, Hafez, ESE, & Kamash, MA. 1975. Histomorphology of cervical and uterine epithelia of crab-eating macaque, *Macaca fascicularis*. Folia Primatol. 23:124-134.
2292. David, GFX & Ramaswami, LS. 1969. Studies on menstrual cycles and other related phenomena in the langur, (*Presbytis entellus entellus*). Folia Primatol. 11:300-316.
2293. David, GFX & Ramaswami, LS. 1971. Reproductive systems of the north Indian langur (*Presbytis entellus entellus* Dufresne). J Morphol. 135:99-130.
2294. David, JHM. ms. The ecology and behaviour of the bontebok, *Damaliscus dorcas dorcas*. 2nd & 3rd Progress Reports. [Cited by Mentis 1972]
2295. David, JHM. 1973. The behaviour of the bontebok, *Damaliscus dorcas dorcas*, (Pallas 1766), with special reference to territorial behaviour. Z Tierpsychol. 33:38-107.
2296. David, JHM. 1975. Observations on mating behaviour, parturition, suckling and the mother-young bond in the bontebok (*Damaliscus dorcas dorcas*). J Zool. 177:203-223.
2297. David, JHM. 1978. Observations on social organization of springbok *Antidorcas marsupialis*, in the Bontebok National Park, Swellendam. Zool Afr. 13:115-122.
2298. David, JHM & Jarvis, JUM. 1985. Population fluctuations, reproduction and survival in the striped fieldmouse *Rhabdomys pumilio* on the Cape Flats, South Africa. J Zool. 207A:251-276.
2299. David, JHM & Rand, RW. 1986. Attendance behavior of South African fur seals. Maternal Strategies on Land and at Sea.

126-141, Gentry, RL & Kooyman, GL, eds, Princeton Univ Press, Princeton.

2300. David, R. 1966. Breeding the Indian wild ass, *Equus hemionus khur*, at Ahmedabad Zoo. Int Zoo Yearbk. 6:197-198.

2301. David, R. 1967. A note on the breeding of the large Indian civet, *Viverra zibetha*, at Ahmedabad Zoo. Int Zoo Yearbk. 7:131.

2302. Davidar, ERC. 1973. Dhole or Indian wild dog (*Cuon alpinus*) mating. J Bombay Nat Hist Soc. 70:373-374.

2303. Davidar, ERC. 1975. Ecology and behavior of the dhole or Indian wild dog *Cuon alpinus* (Pallas). The Wild Canids. 109-119, Fox, MW, ed, Van Nostrand Reinhold Co, New York.

2304. Davidge, C. 1978. Ecology of baboons (*Papio ursinus*) at Cape Point Zool Afr 13:329-350.

2305. Davidow-Henry, BR, Jones, JK, Jr, & Hollander, RR. 1989. *Cratogeomys castanops*. Mamm Species. 338:1-6.

2306. Davidson, MM. 1976. Season of parturition and fawning percentages of sika deer (*Cervus nippon*) in New Zealand. NZ J For Sci. 5:355-357.

2307. Davies, DV. 1950. The foetal membranes of the Weddell seal (*Leptonychotes weddelli*). J Anat. 84:408.

2308. Davies, J & Wimsatt, WA. 1966. Observations on the fine structure of the sheep placenta. Acta Anat. 65:182-223.

2309. Davies, SJJF. 1960. A note on two small mammals of the Darwin area. J Proc R Soc West Aust. 43:63-66.

2310. Davis, DD. 1938. Notes on the anatomy of the treeshrew *Dendrogale*. Field Mus Nat Hist Publ Zool Ser. 20:383-404.

2311. Davis, DD. 1940. Notes on the anatomy of the babirusa. Field Mus Nat Hist Publ Zool Ser. 22:363-430.

2312. Davis, DD. 1958. Mammals of the Kelabit Plateau, northern Serawak. Fieldiana Zool. 39:119-147.

2313. Davis, DD. 1962. Mammals of the lowland rain-forest of North Borneo. Bull Singapore Nat Nus. 31:1-129.

2314. Davis, DD. 1964. The giant panda. Fieldiana Zool Mem. 3:1-339.

2315. Davis, DD & Story, HE. 1949. The female external genitalia of the spotted hyaena. Fieldiana Zool. 31:277-283.

2316. Davis, DE. 1947. Notes on commensal rats in Lavaca County, Texas. J Mamm. 28:241-244.

2317. Davis, DE. 1949. The weight of wild brown rats at sexual maturity. J Mamm. 30:125-130.

2317a. Davis, DE. 1951. A comparison of reproductive potential of two rat populations. Ecology. 32:469-475.

2317b. Davis, DE. 1953. The characteristics of rat populations. Q Rev Biol. 28:373-401.

2318. Davis, DE. 1956. A comparison of natality rates in white-footed mice for four years. J Mamm. 37:513-516.

2319. Davis, DE & Davis, DJ. 1947. Notes on reproduction of *Peromyscus eremicus* in a laboratory colony. J Mamm. 28:181-183.

2320. Davis, DE & Hall, O. 1948. The seasonal reproductive condition of male brown rats in Baltimore, Maryland. Physiol Zool. 21:272-282.

2321. Davis, DE & Hall, O. 1951. The seasonal reproductive condition of female Norway (brown) rats in Baltimore, Maryland. Physiol Zool. 24:9-10.

2322. Davis, DE & Peek, F. 1970. Litter size of the star-nosed mole (*Condylura cristata*). J Mamm. 51:156.

2323. Davis, DHS. 1950. Notes on the status of the American grey squirrel (*Sciurus carolinensis*) in the south-western Cape (South Africa). Proc Zool Soc London. 120:265-268.

2324. Davis, DW. 1969. The behavior and population dynamics of the red squirrel, *Tamiasciurus hudsonicus*, in Saskatchewan. Ph D Diss. Univ AR, Fayetteville.

2325. Davis, DW & Sealander, JA. 1971. Sex ratio and age structure in two red squirrel populations in northern Saskatchewan. Can Field-Nat. 85:303-308.

2326. Davis, GJ. 1971. The role of photoperiod and gonadotropins in reproductive rate fluctuations of snowshoe hares. Ph D Diss. Univ WI, Madison, WI. [Diss Abstr Int. 32B:4825-4826.]

2327. Davis, GJ & Meyer, RK. 1973. FSH and LH in the pituitary gland of gonadectomized snowshoe hares. Endocrinology. 92:340-344.

2328. Davis, GJ & Meyer, RK. 1973. FSH and LH in the snowshoe hare during the increasing phase of the 10-year cycle. Gen Comp Endocrinol. 20:53-60.

2329. Davis, GJ & Meyer, RK. 1973. Seasonal variation in LH and FSH of bilaterally castrated snowshoe hares. Gen Comp Endocrinol. 20:61-68.

2330. Davis, HN, Gray, GD, Zerylnick, M, et al. 1974. Ovulation and implantation in montane voles (*Microtus montanus*) as a function of varying amounts of copulatory stimulation. Horm Behav. 5:383-388.

2331. Davis, HN, Jr & Dewsbury, DA. 1974. A quantitative description of the copulatory behavior of western harvest mice (*Reithrodontomys megalotis*). Z Tierpsychol. 35:437-444.

2332. Davis, JA, Jr. 1965. A preliminary report of the reproductive behavior of the small Malayan chevrotain *Tragulus napu* at New York Zoo. Int Zoo Yearb. 5:42-44.

2333. Davis, M. 1946. Parturition in a Bengal tiger. J Mamm. 27:393.

2334. Davis, M. 1960. Galago born in captivity. J Mamm. 41:401-402.

2335. Davis, R. 1969. Growth and development of young pallid bats, *Antrozous pallidus*. J Mamm. 50:729-736.

2336. Davis, RA & Shillito, E. 1967. The coypu or nutria (*Myocastor coypus* Molina). UFAW Handbook Care and Management of Laboratory Animals, 3rd ed. 457-467, E & S Livingston, Edinburgh.

2337. Davis, RB, Herreid, CF, & Short, HL. 1962. Mexican free-tailed bats in Texas. Ecol Monogr. 32:311-346.

2338. Davis, RM. 1972. Behaviour of the vlei rat, *Otomys irroratus* (Brants, 1827). Zool Afr. 7:119-140.

2339. Davis, RM & Meester, J. 1981. Reproduction and postnatal development in the vlei rat, *Otomys irroratus*, on the van Riebeeck Nature Reserve, Pretoria. Mammalia. 45:99-116.

2340. Davis, WB. 1934. Notes on the Utah chipmunk. Murrelet. 15:20-22.

2341. Davis, WB. 1944. Notes on Mexican mammals. J Mamm. 25:370-403.

2342. Davis, WB. 1968. Review of genus *Uroderma* (Chiroptera). J Mamm. 49:676-698.

2342a. Davis, WB. 1969. A review of the small fruit bats (genus *Artibeus*) of middle America. Southwest Nat. 14:15-29.

2342b. Davis, WB. 1970. A review of the small fruit bats (genus *Artibeus*) of middle America Part II. Southwest Nat. 14:389-402.

2343. Davis, WB. 1974. The mammals of Texas. TX Parks Wildl Dept Bull. 41:1-294.

2343a. Davis, WB. 1975. Individual and sexual variation in *Vampyressa bidens*. J Mamm. 56:262-265.

2344. Davis, WB & Buechner, HK. 1946. Pocket gophers (*Thomomys*) of the Davis Mountains, Texas. J Mamm. 27:265-271.

2345. Davis, WB, Carter, DC, & Pine, RH. 1964. Noteworthy records of Mexican and central American bats. J Mamm. 45:375-387.

2345a. Davis, WB & Dixon, JR. 1976. Activity of bats in a small village clearing near Iquitos, Peru. J Mamm. 57:747-749.
2346. Davis, WB & Follansbee, LA. 1945. The Mexican volcano mouse, *Neotomodon*. J Mamm. 26:401-411.
2347. Davis, WB & Joeris, L. 1945. Notes on the life history of the little short-tailed shrew. J Mamm. 26:136-138.
2348. Davis, WB & Lukens, PW, Jr. 1958. Mammals of the Mexican state of Guerrero, exclusive of Chiroptera and Rodentia. J Mamm. 39:347-367.
2349. Davis, WB & Robertson, JL, Jr. 1944. The mammals of Culberson County Texas. J Mamm. 25:254-273.
2350. Davis, WB & Russell, RJ, Jr. 1952. Bats of the Mexican state of Morelos. J Mamm. 33:234-239.
2351. Davis, WB & Russell, RJ, Jr. 1954. Mammals of the Mexican state of Morelos. J Mamm. 35:63-80.
2351a. Davis, WB & Taylor, WP. 1939. The bighorn sheep of Texas. J Mamm. 20:440-455.
2352. Davis, WH. 1967. A *Myotis lucifugus* with two young. Bat Res News. 8:3.
2353. Davis, WH. 1966. Population dynamics of the bat *Pipistrellus subflavus*. J Mamm. 47:383-396.
2353a. Davis, WH & Barbour, RW. 1970. Life history data on some southwestern *Myotis*. Southwest Nat. 15:261-263.
2354. Davis, WH, Barbour, RW, & Hassell, MD. 1968. Colonial behavior of *Eptesicus fuscus*. J Mamm. 49:44-50.
2355. Davis, WH & Hitchcock, HB. 1965. Biology and migration of the bat, *Myotis lucifugus*, in New England. J Mamm. 46:296-314.
2356. Davis, WH & Mumford, RE. 1962. Ecological notes on the bat *Pipistrellus subflavus*. Am Mid Nat. 68:394-398.
2357. Davison, G. 1971. Some observations on a population of nyala, *Tragelaphus angasi* (Gray) in the se lowveld of Rhodesia. Arnoldia (Rhodesia). 5(17):1-8.
2358. Davison, GWH. 1979. Some notes on Savi's pygmy shrew. Malay Nat J. 32:227-231.
2359. Davison, R. 1973. Progressions in adapting the Colorado pika to captivity. Proc Am Assoc Zoo Vet. 1973:303-307.
2360. Davison, R. 1975. Habituation and breeding of captive pikas *Ochotona princeps saxatillis*. Int Zoo Yearb. 15:137-140.
2361. Davydov, GS. 1973. Patterns of reproduction of the red-tailed marmot (*Marmota caudata*) in the forest and subalpine zones of Tadjiksan. Zool Zhurn. 52:589-595.
2362. Dawaa, N, Stubbe, M, & Dorzvaa, O. 1983. Die Bisamratte *Ondatra zibethica* (L, 1758) in der Mongolischen Volksrepublik. Beitr Z Jagd-und Wildforsch. 10:342-352.
2363. Dawbin, WA, Noble, BA, & Fraser, FC. 1970. Observations on the electra dolphin, *Peponocephala electra*. Bull Brit Mus Nat Hist. 20:173-201.
2364. Dawkins, WB. 1867. *Ovibos moschatus* (Blainville). Proc R Soc London. 15:516-517.
2365. Dawson, E. 1959. On a large catch of the finless black porpoise *Neomeris phocaenoides* Cuvier. J Mar Biol Assoc India. 1:259-260.
2366. Dawson, G. 1976. Behavioral ecology of the Panamanian tamarin, *Saguinus oedipus* (Callitrichidae). Ph D Diss. MI State Univ, Lansing, MI.
2367. Dawson, GA. 1977. Composition and stability of social groups of the tamarin, *Saguinus oedipus geoffroyi*, in Panama: Ecological and behavioral implications. The Biology and Conservation of the Callitrichidae. 23-37, Kleiman, DG, ed, Smithsonian Inst Press, Washington, DC.
2368. Dawson, GA & Dukelow, WR. 1976. Reproductive characteristics of free-ranging Panamanian tamarins (*Saguinus oedipus geofforyi*). J Med Primatol. 5:266-275.
2369. Dawson, TJ & Dawson, WR. 1982. Metabolic scope in response to cold of some dasyrurid marsupials and Australian rodents. Carnivorous Marsupials. 1:255-260.
2370. Day, BN, Egoscue, HJ, & Woodbury, AM. 1956. Ord kangaroo rat in captivity. Science. 124:485-486.
2371. de Almeida, CR, de Almeida, AMP, & Brasil, DP. 1981. Observations sur le comportement de fouissement de *Zygodontomys lasiurus pixuna* Moojen, 1943 Reproduction au laboratoire (Rongeurs, Cricétidés). Mammalia. 45:415-421.
2372. de Balsac, HH & Verschuren, J. 1968. Insectivores. Exploration du Parc National de la Garamba. 54:1-50.
2373. de Bavay, JM. 1951. Notes on the female urogenital system of *Tarsipes spenserae* (Marsupialia). Pap Proc R Soc Tas. 85:143-150.
2374. de Bonilla, D & Rasweiler, JJ, IV. 1974. Breeding activity, preimplantation development, and oviduct histology of the short-tailed fruit bat, *Carollia*, in capitivy. Anat Rec. 179:385-404.
2375. de Carvalho, CT. 1968. Comparative growth rates of hand-reared big cats. Int Zoo Yearb. 8:56-59.
2376. de Coursey, GE, Jr. 1957. Identification, ecology and reproduction of *Microtus* in Ohio. J Mamm. 38:44-52.
2377. de Crombrugghe, SA. 1964. Untersuchungen über die Reproduktion des Rotwildes in den Niederlanden. Z Jagdwiss. 10:91-101.
2378. de Faucaux, MA. 1978. Les cycles annuels de reproduction chez les Chiroptères cavernicoles du Thaba (S-E Zaire) et du Rwanda. Mammalia. 42:454-490.
2378a. de Gayner, EJ & Jordan, PA. 1987. Skewed fetal sex ratios in white-tailed deer: Evidence and evolutionary speculations. Biology and Management of the Cervidae. 178-188, Wemmer, CM, ed, Smithsonian Inst Press, Washington, DC.
2379. de Graaff, G. 1972. On the mole-rat (*Cryptomys hottentotus damarensis*) (Rodentia) in the Kalahari Gemsbok National Park. Koedoe. 15:25-35.
2380. de Graaff, G. 1973. The rodents of South Africa 4 Dassie rats (*Petromus typicus*). Custos. 2:23-26.
2381. de Graaff, G. 1974. The rodents of South Africa 11 The South African pygmy gerbil *Gerbillurus paeba*. Custos. 3:17-21.
2382. de Graaff, G. 1975. The red hartebeest. Custos. 4:37-41.
2383. de Graaff, G. 1978. Notes on the southern black-tailed tree rat *Thallomys paedulus* (Sundevall, 1846) and its occurrence in the Kalahari Gemsbok National Park. Koedoe. 21:181-190.
2384. de Graaff, G. 1979. Aardvarks, hedgehogs and porcupines. Fauna Flora. 35:8-11.
2385. de Jonge, G & Ketel, NAJ. 1981. An analysis of copulatory behaviour of *Microtus agrestis* and *M arvalis* in relation to reproductive isolation. Behaviour. 78:227-259.
2386. de Kock, LL. 1966. Breeding lemmings *Lemmus lemmus* for exhibition. Int Zoo Yearb. 6:164-165.
2387. de la Barrera, JM. 1939. Contribución al conociemento de la peste selvática en la Argentina-caracteres del brote de Mendoza en 1937. Rev Inst Bacteriologico, Buenos Aires. 8:423-454.
2388. de la Barrera, JM. 1940. Estudios sobre peste selvática en Mendoza. Rev Inst Bacteriologico, Buenos Aires. 9:565-586.
2389. de la Maza, LM & Sawyer, JR. 1976. The G and Q banding pattern of *Ellobius lutescens*, a unique case of sex determination in mammals. Can J Genet Cytol. 18:497-502.
2390. de la Torre, L. 1954. Bats from southern Tamaulipas, Mexico. J Mamm. 35:113-116.
2391. de Lange, D. 1933. Plazentarbildung. Handbuch der Vergleichenden Anatomie der Wirbeltiere. 6:155-234.

2392. de Lange, D. 1937/38. Some remarks on the early development of *Ctenodactylus gundi* Pall. Arch Neerland Zool. 3(Suppl):131-147.
2393. de Lange, D. 1949. Communication on the attachment and the early development of *Macroscelides (=Elephantulus) rozeti*, Duv, the north African jumping shrew. Bijdr Dierkd. 28:255-285.
2394. de Lange, D, Jr. 1919. Contribution to the knowledge of the placentation of the Cape gold mole (*Chrysochloris*). Bijdr Dierkd. 21:161-173.
2395. de Lange, D, Jr. 1926. Quelques remargues sur la placentation de "*Bradypus*". CR Assoc Anat. 21:321-333.
2396. de Lange, D, Jr. 1930. Le placenta des Édentés. CR Ass Anat. 25:189-199.
2397. de Lange, D, Jr. 1931. Die Placenta der Edentaten. Verh Anat Ges Anat Anz. 71(Suppl):234-236.
2398. de Lange, D, Jr. 1934. Beobachtungen an puerperalen and schwangeren Uteri von *Ctenodactylus gundi*. Z Mikr Anat Forsch. 36:488-496.
2399. de Lange, D, Jr. 1938. Early developmental stages of *Ctenodactylus*. Biomorphosis. 1:320-321.
2400. de Master, DP. 1979. Weddell seal. Mammals in the Seas, FAO Fish Ser 5. 2:130-134.
2401. de Moor, PP. 1969. Seasonal variation in local distribution, age classes and population density of the gerbil *Tatera brantsi* on the South African highveld. J Zool. 157:399-411.
2402. de Paz, O. 1985. Contribución al estudio eco-ethologico de los quirópteros cavernicolas de "La Canleja", Abanades, Guadalajara. Bol Estac Cent Ecol. 14:77-87.
2403. de Paz, O. 1986. Age estimation and postnatal growth of the greater mouse bat *Myotis myotis* (Borkhausen, 1797) in Guadalajara, Spain. Mammalia. 50:243-251.
2404. de Poorter, M & van der Loe, W. 1981. Report on the breeding and behavior of the volcano rabbit at the Antwerp Zoo. Proc World Lagomorph Conf, Univ Guelph, ONT, Canada. 1979:956-972.
2404a. de Roguin, L. 1988. Notes sur quelques mammifères du Baluchistan iranien. Rev Suisse Zool. 95:595-606.
2405. de Rooy, DG, van Alphen, MMA, & van de Kant, JHG. 1986. Duration of the cycle of the seminiferous epithelium and its stages in the rhesus monkey (*Macaca mulatta*). Biol Reprod. 35:387-391.
2406. de Silva, GS. 1971. Notes on the orang-utan rehabilitation project in Sabah. Malay Nat J. 24:50-77.
2407. de Silva, GS. 1972. The birth of an orang-utan *Pongo pygmaeus* at Sepilok Game Reserve. Int Zoo Yearb. 12:104-105.
2408. de Verteuil, RL & Urich, FW. 1936. The study and control of paralytic rabies transmitted by bats in Trinidad, British West Indies. Trans R Soc Trop Med Hyg. 29:317-347.
2409. de Villafane, G. 1981. Reproduccion y crecimiento de *Akodon azarae azarae* (Fischer, 1829). Historia Natural. 1:193-204.
2410. de Villafane, G. 1981. Reproduccion y crecimiento de *Calomys musculinus murillus* (Thomas, 1916). Historia Natural. 1:237-256.
2411. de Villiers, DJ. 1986. Infanticide in the tree squirrel, *Paraxerus cepapi*. S Afr J Zool. 21:183-184.
2412. de Villiers, DJ, Skinner, JD, & Hall-Martin, AJ. 1989. Circulating progesterone concentrations and ovarian functional anatomy in the African elephant (*Loxodonta africana*). J Reprod Fert. 86:195-201.
2413. De Vos, A & Dowsett, RJ. 1966. The bahaviour and population structure of three species of the genus *Kobus*. Mammalia. 30:30-55.
2414. de Vos, A & Gillespie, DI. 1960. A study of woodchucks on an Ontario farm. Can Field-Nat. 74:130-145.
2415. Deacon, JE, Bradley, WG, & Larsen, KM. 1964. Ecological distribution of mammals of Clanc Canyon, Chaneston Mountains, Nevada. J Mamm. 45:397-409.
2416. Dean, G. 1960. Birth of a polar bear cub at Auckland Zoo. Int Zoo Yearb. 2:95.
2417. Deanesley, R. 1934. The reproductive processes of certain mammals VI The reproductive cycle of the female hedgehog. Phil Trans R Soc London Biol Sci. 223:239-276.
2418. Deanesley, R. 1935. The reproductive processes of certain mammals IX Growth and reproduction in the stoat (*Mustela erminea*). Phil Trans R Soc London Biol Sci. 225:459-492.
2419. Deanesley, R. 1938. The reproductive cycle of the golden hamster (*Cricetus auratus*). Proc Zool Soc London. 108A:31-37.
2420. Deanesley, R. 1939. Observations on pregnancy in the common bat (*Pipistrellus pipistrellus*). Proc Zool Soc London. 109A:57-60.
2421. Deanesley, R. 1943. Delayed implantation in the stoat (Mustela mustela). Nature. 151:365-366.
2422. Deanesley, R. 1944. The reproductive cycle of the female weasel (*Mustela nivalis*). Proc Zool Soc London. 114:339-349.
2423. Deanesley, R. 1966. Observations on reproduction in the mole *Talpa europaea*. Symp Zool Soc London. 15:387-402.
2424. Deanesley, R. 1967. Experimental observations on the ferret corpus luteum of pregnancy. J Reprod Fert. 13:183-185.
2425. Deanesley, R & Parkes, AS. 1933. The effect of hysterectomy on the oestrous cycle of the ferret. J Physiol London. 78:80-84.
2426. Deanesley, R & Parkes, AS. 1933. The reproductive processes of certain mammals IV The oestrous cycle of the grey squirrel *Sciurus carolinensis*. Phil Trans R Soc London Biol Sci. 222:47-78.
2427. Deanne, NN. 1962. The spotted hyaena, *Crocuta crocuta crocuta*. Lammergeyer. 2(2):26-44.
2428. Dechamma, PA & Gopal-Dutt, NH. 1980. The testicular and epididymal changes in the soft-furred field rat, *Millardia meltada* (Gray) in relation to some environmental factors. Proc Indian Acad Sci (Anim Sci). 89:303-310.
2429. Defler, TR. 1979. On the ecology and behavior of *Cebus albifrons* in eastern Columbia I Ecology. Primates. 20:475-490.
2430. DeFrees, SL & Wilson, DE. 1988. *Eidolon helvum*. Mamm Species. 312:1-5.
2431. Degen, AA, Elias, E, & Kam, M. 1987. A preliminary report on the energy intake and growth rate of early-weaned camel (*Camelus dromedarius*) calves. Anim Prod. 45:301-306.
2432. Degen, AA & Lee, DG. 1982. The male gential tract of the dromedary (one-humped) camel (*Camelus dromedarius*): Gross and microscopic anatomy. Zb Veterinärmed. 11:267-282.
2433. Degerbøl, M. 1943. Pairing and pairing fights of the hedgehog (*Erinaceus europaeus* L). Vid Med Dansk Naturhist For. 106:427-430.
2434. Degerbøl, M & Mohl-Hansen, U. 1943. Remarks on the breeding conditions and moulting of the collared lemming (*Dicrostonyx*). Meddelelser om Grönland. 131:1-84.
2435. Degn, HJ. 1973. Systematic position, age criteria and reproduction of Danish red squirrels (*Sciurus vulgaris* L). Danish Rev Game Biol. 8 (2):1-24.
2436. Dehnel, A. 1949. Studies on the genus *Sorex* L. Ann Univ Mariae Curie-Sklodowska Sect C. 4:17-102.
2437. Dehnel, A. 1952. The biology of breeding of the common shrew, *S araneus* L in laboratory conditions. Ann Univ Mariae Curie-Sklodowska Sect C. 6:359-377.

2438. Dehon, G, Crespo, EA, & Pagnoni, G. 1987. Stranding of a specimen of Gray's beaked whale at Puerto Piramides (Chubut, Argentina) and its gonadal appraisal. Sci Rep Whales Res Inst. 38:107-115.
2439. DeKeyser, PL. 1955. Les Mammifères de l'Afrique Noire Francaise. IFAN, Dakar.
2440. Dekker, D. 1968. Breeding the Cape hunting dog *Lycaon pictus* at Amsterdam Zoo. Int Zoo Yearb. 8:27-30.
2441. Dekker, D. 1983. Denning and foraging habits of red foxes, *Vulpes vulpes*, and their interaction with coyotes, *Canis latrans*, in central Alberta, 1972-1981. Can Field-Nat. 97:303-306.
2442. Delany, MJ. 1963. A collection of *Apodemus* from the Island of Foula, Shetland. Proc Zool Soc, London. 140:319-320.
2443. Delany, MJ. 1964. A study of the ecology and breeding of small mammals in Uganda. J Zool. 142:347-370.
2444. Delany, MJ. 1964. An ecological study of the small mammals in the Queen Elizabeth Park, Uganda. Rev Zool Bot Afr. 70:129-147.
2445. Delany, MJ. 1971. The biology of small rodents in Mayanja Forest, Uganda. J Zool. 165:85-129.
2446. Delany, MJ. 1972. The ecology of small rodents in tropical Africa. Mammal Rev. 2:1-42.
2446a. Delany, MJ. 1975. The rodents of Uganda. Brit Mus Nat Hist, London.
2447. Delany, MJ & Bishop, IR. 1960. The systematics, life history and evolution of the bank-vole *Clethrionomys* Tilesius in north-west Scotland. Proc Zool Soc London. 135:409-422.
2448. Delany, MJ & Davis, PE. 1961. Observations on the ecology and life history of the Fair Isle fieldmouse *Apodemus sylvaticus fridariensis* (Kinnear). Proc Zool Soc London. 136:439-452.
2449. Delany, MJ & Happold, DCD. 1979. Ecology of African Mammals. Longman, London.
2450. Delany, MJ & Monro, RH. 1985. Growth and development of wild and captive Nile rats, *Arvicanthis niloticus* (Rodentia: Muridae). Afr J Ecol. 23:121-131.
2451. Delany, MJ & Monro, RH. 1986. Population dynamics of *Arvicanthis niloticus* (Rodentia: Muridae) in Kenya. J Zool. 209A:85-103.
2452. Delany, MJ & Neal, BR. 1966. A review of the Muridae (Order Rodentia) of Uganda. Bull Br Mus Nat Hist. 13:297-355.
2453. Delany, MJ & Neal, BR. 1969. Breeding seasons in rodents in Uganda. J Reprod Fert Suppl. 6:229-235.
2454. Delany, MJ & Roberts, CS. 1978. Seasonal population changes in rodents in the Kenya Rift Valley. Bull Carnegie Mus Nat Hist. 6:97-108.
2455. Delattre, P & LeLouarn, H. 1980. Cycle de reproduction du Rat noir (*Rattus rattus*) et du Surmulot (*Rattus norvegicus*) dans différents milieux de la Guadeloupe (Antilles francaise). Mammalia. 44:233-243.
2456. Delhon, G, Zuckerberg, C, von Lawzewitsch, I, et al. 1983. Estudio citologico de las gonadas de guanaco (*Lama guanicoe*) macho, en los estadios prepuberales, sexualmente maduros y seniles. Rev Fac Cienc Vet Univ Buenos Aires Rep Argentina. 1:47-60.
2456a. Delhon, G, Zuckerberg, C, von Lawzewitsch, I, et al. 1984. Estudio histologico de las vias seminales en el testiculo del guanaco (*Lama guanaco*). Rev Mus Arg Cienc Nat <<Bernardino Rivadavia>> Inst Nac Invest Cienc Nat. 13:369-373.
2457. Delhon, GA & von Lawzewitsch, I. 1986. Light and scanning electron microscopy on the male accessory sexual glands of the llama (*Lama glama*). Comunicaciones Biologicas. 5:209-224.
2457a. Delhon, GA & von Lawzewitsch, I. 1987. Reproduction in the male llama (*Lama glama*), a South American camelid 1 Spermatogenesis and organization of the intertubular space of the mature testis. Acta Anat. 129:59-66.
2458. Dell, J & Schierbaum, DL. 1974. Prenatal development of snowshoe hares. NY Fish Game J. 21:89-104.
2459. DeLong, KT. 1978. The effects of the manipulation of social structure on reproduction in house mice. Ecology. 59:922-933.
2460. Delost, P. 1951. État de l'appareil génital du Campagnol des champs (*Microtus arvalis*) de sexe mâle en hiver. Bull Soc Hist Nat Toulouse. 86:133-150.
2461. Delost, P. 1951. Variations saisonnières de l'activité thyroïdienne du Campagnol des champs (*Microtus arvalis* P). CR Soc Biol. 145:377-380.
2462. Delost, P. 1952. Croissance pondérale et staturale du Campagnol des champs (*Microtus arvalis* P). Bull Soc Zool France. 77:88-98.
2463. Delost, P. 1953. Existence de glandes prostatiques chez les Campagnoles des champs (*Microtus arvalis* P). CR Soc Biol. 147:758-760.
2464. Delost, P. 1955. Anatomie et structure histologique de l'appareil génital du Campagnol des champs (*Microtus arvalis* Pallas) adulte en activité sexuelle. Bull Soc Zool France. 80:207-222.
2465. Delost, P. 1955. Étude de la biologie sexuelle du Campagnol des champs (*Microtus arvalis* P). Arch Anat Microsc Morphol Exp. 44:150-190.
2466. Delost, P. 1955. Recherches sur la biologie générale du Campagnol des champs (*Microtus arvalis* Pallas). Bull Soc Zool France. 80:149-162.
2467. Delost, P. 1960. Développement sexuel normal du Campagnol des champs (*Microtus arvalis* P) de la naissance à l'age adulte. Arch Anat Microsc Morphol Exp. 45:11-47.
2468. Delost, P. 1966. Le cycle saisonnier de l'Écureuil. CR Soc Biol. 159:1141-1145.
2469. Delost, P. 1966. Reproduction et cycles endocriniens de l'Écureuil. Arch Sci Physiol. 20:425-456.
2470. Delost, P. 1968. Étude comparative de la reproduction chez quelques rongeurs sauvages non hibernants dans différentes régions de France. Cycles Génitaux Saisoniers de Mammifères Sauvages. 23-39, Canivenc, R, ed, Masson et Cie, Paris.
2471. Delost, P. 1969. Les variations sexuelles saisonnières du Campagnol amphibie (*Arvicola terrestris amphibus*). CR Soc Biol. 163:1742-1747.
2472. Delost, P. 1971. Étude expérimentale des causes du repos hivernal chez les Mammefères à cycle sexuel saisonnier. CR Soc Biol. 164:2475-2479.
2473. Delost, P & Delost, H. 1975. Déterminisme du cycle sexuel circannuel du Campagnol des champs de sexe mâle. J Physiol (Paris). 70:521-532.
2474. Delost, P & Guérin, M. 1962. Variations pondérales saisonnières des glandes surrénales de l'Écureuil (*Sciurus vulgaris*) dans le Tarn. CR Soc Biol. 156:1305-1308.
2475. DeMaster, DP & Stirling, I. 1981. *Ursus maritimus*. Mamm Species. 145:1-7.
2476. Deming, OV. 1955. Rearing bighorn lambs in captivity. CA Fish Game. 44:131-143.
2477. Dempsey, EW. 1939. The reproductive cycle of new world monkeys. Am J Anat. 64:381-405.
2478. Dempsey, EW. 1940. The structure of the reproductive tract in the female gibbon. Am J Anat. 67:229-253.
2479. Dempsey, EW. 1969. Comparative aspects of the placenta of certain African mammals. J Reprod Fert Suppl. 6:189-192.

2480. Dempsey, EW & Wislocki, GB. 1941. The structure of the ovary of the humpback whale (*Megaptera nodosa*). Anat Rec. 80:243-251.
2481. Dempster, ER & Perrin, MR. 1989. Maternal behaviour and neonatal development in three species of Namib Desert rodents. J Zool. 218:407-419.
2482. Dempster, ER & Perrin, MR. 1989. The estrous cycle and induction of estrous behavior in four species of hairy-footed gerbils (genus *Gerbillurus*). J Mamm. 70:809-811.
2483. Denniston, RH, II. 1956. Ecology, behavior and population dynamics of the Wyoming or Rocky Mountain moose, *Alces alces shirasi*. Zoologica. 41:105-118.
2484. Denniston, RH, II. 1957. Notes on breeding and size of young in the Richardson ground squirrel. J Mamm. 38:414-416.
2485. Densmore, MA. 1980. Reproduction of sitatunga *Tragelaphus spekei* in captivity. Int Zoo Yearb. 20:227-229.
2486. Densmore, MA, Bowen, MJ, Magyar, SJ, et al. 1987. Artificial insemination with frozen thawed semen and pregnancy diagnosis in addax (*Addax nasomaculatus*). Zoo Biol. 6:21-29.
2487. Densmore, MA & Kraemer, DC. 1986. Analysis of reproductive data on the addax *Addax nasomaculatus* in captivity. Int Zoo Yearb. 24/25:303-306.
2488. Department of Game and Tsetse Control, N Rhodesia. Personal communication. [Cited by Mentis 1972]
2489. Deputte, BL & Leclerc-Cassan, M. 1981. Sex determination and age estimation in the white-cheeked gibbon *Hylobates concolor leucogenys*: Anatomical and behavioural features. Int Zoo Yearb. 21:187-193.
2490. Derenne, P. 1976. Notes sur biologie du Chat haret de Kerguelen. Mammalia. 40:531-595.
2491. Derenne, P & Mousin, JL. 1976. Données écologiques sur les Mammifères introduits de l'Ile aux Cochons, Archipel Crozet (46°06'S, 50°14'E). Mammalia. 40:49-53.
2492. Derrickson, EM. 1988. Patterns of postnatal growth in a laboratory colony of *Peromyscus leucopus*. J Mamm. 69:57-66.
2493. Derting, TL & Cranford, JA. 1989. Influence of photoperiod on postnatal growth, sexual development, and reproduction in a semifossorial microtine, *Microtus pinetorium*. Can J Zool. 67:937-941.
2494. Desai, JH. 1975. Observations on the reproductive biology and early postnatal development of the panther, *Panthera pardus* L in captivity. J Bombay Nat Hist Soc. 72:293-304.
2495. Desai, JH. 1985. The lion-tailed macaque: Captive propagation and management at Delhi Zoological Park. The Lion Tailed Macaque: Status and Conservation. 293-295, Hiltne, PG, ed, Alan R Liss, Inc, New York.
2496. Desai, JH, Koshy, M, & Nainan, T. 1986. Note on the breeding of the Indian wolf (*Canis lupus pallipes* at the National Zoological Park, New Delhi. J Bombay Nat Hist Soc. 83:193-194.
2497. Desai, JH & Malhotra, AK. 1976. A note on the captive status of the lion-tailed macaque *Macaca silenus* in India and its breeding at Delhi Zoo. Int Zoo Yearb. 16:116-117.
2498. Desaulniers, DM, Goff, AK, Betteridge, KJ, et al. 1989. Reproductive hormone concentrations in faeces during the oestrus cycle and pregnancy in cattle (*Bos taurus*) and muskoxen (*Ovibos moschatus*). Can J Zool. 67:1148-1154.
2499. DesMeules, P & Simard, B. 1970. Dates of calving in northern Quebec Caribou (*Rangifer tarandus*). Naturaliste Can. 97:61-66.
2500. Desy, EA & Druecker, JD. 1979. The estrous cycle of the plains pocket gopher, *Geomys bursarius*, in the laboratory. J Mamm. 60:235-236.
2501. DeVore, I & Hall, KRL. 1965. Baboon ecology. Primate Behavior Field Studies of Monkeys and Apes. 20-53, DeVore, I, ed, Holt, Rinehart & Winston, New York.
2502. Dewsbury, DA. 1970. Copulatory behaviour of rice rats (*Oryzomys palustris*). Anim Behav. 18:266-275.
2503. Dewsbury, DA. 1971. Copulatory behaviour of old-field mice (*Peromyscus polionotus subgriseus*). Anim Behav. 19:192-204.
2504. Dewsbury, DA. 1971. Copulatory behavior of *Neotoma albigula* in relation to that of other muroid rodents (Motion picture). Am Zool. 11:.
2505. Dewsbury, DA. 1972. Copulatory behavior of cotton rats (*Sigmodon hispidus*). Z Tierpsychol. 30:477-487.
2506. Dewsbury, DA. 1973. Copulatory behavior of montane voles (*Microtus montanus*). Behaviour. 44:186-202.
2507. Dewsbury, DA. 1974. Copulatory behaviour of white-throated wood rats (*Neotoma albigula*) and golden mice (*Ochrotomys nuttalli*). Anim Behav. 22:601-610.
2508. Dewsbury, DD. 1974. Copulatory behavior of *Neotoma floridana*. J Mamm. 55:864-866.
2509. Dewsbury, DA. 1974. Copulatory behavior of wild-trapped and laboratory-reared cactus mice (*Peromyscus eremicus*) from two natural populations. Behav Biol. 11:315-326.
2510. Dewsbury, DA. 1974. Copulatory behavior of California mice (*Peromyscus californicus*). Brain Behav Evol. 9:95-106.
2511. Dewsbury, DA. 1975. Copulatory behavior of white-footed mice (*Peromyscus leucopus*). J Mamm. 56:420-428.
2512. Dewsbury, DA. 1975. Copulatory behavior of *Peromyscus crinitus* and *Peromyscus floridanus*. Am Mid Nat. 93:468-471.
2513. Dewsbury, DA. 1975. Diversity and adaptation in rodent copulatory behavior. Science. 190:947-954.
2514. Dewsbury, DA. 1976. Copulatory behavior of pine voles (*Microtus pinetorum*). Percept Motor Skills. 43:91-94.
2515. Dewsbury, DA. 1979. Copulatory behavior of four Mexican species of *Peromyscus*. J Mamm. 60:844-846.
2516. Dewsbury, DA. 1979. Copulatory behavior of deer mice (*Peromyscus maniculatus*): I Normative data, subspecific differences, and effects of cross-fostering. J Comp Physiol Psych. 93:151-160.
2517. Dewsbury, DA. 1979. Copulatory behavior of deer mice (*Peromyscus maniculatus*): III Effects on pregnancy initiation. J Comp Physiol Psychol. 93:178-188.
2518. Dewsbury, DA. 1979. Pregnancy and copulatory behavior in random-bred house mice mated in postpartum estrus. Bull Psychon Soc. 13:320-322.
2519. Dewsbury, DA. 1981. The Coolidge effect in northern grasshopper mice (*Onychomys leucogaster*). Southwest Nat. 26:193-197.
2520. Dewsbury, DA. 1982. Pregnancy blockage following multiple-male copulation or exposure at the time of mating in deer mice, *Peromyscus maniculatus*. Behav Ecol Sociobiol. 11:37-42.
2521. Dewsbury, DA. 1983. A pregnancy block resulting from multiple-male copulation or exposure at the time of mating in deer mice (*Peromyscus maniculatus*). Chemical Signals in Vertebrates. 317-319, Silverstein, RM & Müller-Schwarze, D, eds, Plenum, New York.
2522. Dewsbury, DA. 1985. Interactions between males and their sperm during multi-male copulatory episodes of deer mice (*Peromyscus maniculatus*). Anim Behav. 33:1266-1274.
2523. Dewsbury, DA. 1985. Studies of pericopulatory pregnancy blockage and the gestation period in deer mice (*Peromyscus maniculatus*). Horm Behav. 19:164-173.

2524. Dewsbury, DA, Baumgardner, DJ, Sawrey, DK, et al. 1982. The adaptive profile: Comparative psychology of red-backed voles (*Clethrionomys gapperi*). J Comp Physiol Psychol. 96:649-660.
2525. Dewsbury, DA & Dawson, WW. 1979. African four-striped grass mice (*Rhabdomys pumilio*), a diurnal-crepuscular muroid rodent, in the behavioral laboratory. Behav Res Methods Instrumentation. 11:329-333.
2526. Dewsbury, DA & Estep, DQ. 1975. Pregnancy in cactus mice: Effects of prolonged copulation. Science. 187:552-553.
2527. Dewsbury, DA, Estep, DQ, & Lanier, DL. 1977. Estrous cycles of nine species of muroid rodents. J Mamm. 58:89-92.
2528. Dewsbury, DA, Estep, DQ, & Oglesby, JM. 1978. Copulatory behaviour and the initiation of pregnancy in Israeli gerbils (*Meriones tristrami*). Biol Behav. 3:243-257.
2529. Dewsbury, DA, Evans, RL, & Webster, DG. 1979. Pregnancy initiation in postpartum estrus in three species of muroid rodents. Horm Behav. 13:1-8.
2530. Dewsbury, DA, Ferguson, B, & Webster, DG. 1984. Aspects of reproduction, ovulation, and the estrous cycle in African four-striped grass mice (*Rhabdomys pumilio*). Mammalia. 48:417-424.
2531. Dewsbury, DA & Hartung, TG. 1982. Copulatory behavior of three species of *Microtus*. J Mamm. 63:306-309.
2532. Dewsbury, DA & Hodges, AW. 1987. Copulatory behavior and related phenomena in spiny mice (*Acomys cahirinus*) and hopping mice (*Notomys alexis*). J Mamm. 68:49-57.
2533. Dewsbury, DA & Jansen, PE. 1972. Copulatory behavior of southern grasshopper mice (*Onychomys torridus*). J Mamm. 53:267-278.
2534. Dewsbury, DA & Lanier, DL. 1976. Effects of variations in copulatory behavior on pregnancy in two species of *Peromyscus*. Physiol Behav. 17:921-924.
2535. Dewsbury, DA & Lovecky, DV. 1974. Copulatory behavior of old-field mice (*Peromyscus polionotus*) from different natural populations. Behav Genetics. 4:347-355.
2536. Dewsbury, DA & Pierce, JD, Jr. 1989. Copulatory patterns of primates as viewed in broad mammalian perspective. Am J Primatol. 17:51-72.
2537. Dhiman, RC & Rama-Rao, KV. 1984. Longevity of some microchiropteran bats in captivity. Indian J Zool. 12(2):44-46.
2538. Dhungel, SK & Edge, WD. 1985. Notes on the natural history of *Paradoxurus hermaphroditus*. Mammalia. 49:302-303.
2539. di Salvo, AP, Palmer, J, & Ajello, L. 1969. Multiple pregnancy in *Tadarida brasilienses cynocephala*. J Mamm. 50:152.
2540. Diakow, C & Dewsbury, DA. 1978. A comparative description of the mating behaviour of female rodents. Anim Behav. 26:1091-1097.
2541. Diamond, EJ, Aksel, S, Hazelton, JM, et al. 1984. Seasonal changes of serum concentrations of estradiol and progesterone in Bolivian squirrel monkeys (*Saimiri sciureus*). Am J Primatol. 6:103-113.
2542. Diamond, EJ, Aksel, S, Hazelton, JM, et al. 1987. Serum oestradiol, progesterone, chorionic gonadotrophin and prolactin concentrations during pregnancy in the Bolivian squirrel monkey (*Saimiri sciureus*). J Reprod Fert. 80:373-381.
2543. Diamond, JM. 1982. Big-bang reproduction and aging in male marsupial mice Florida. Nature. 298:115-116.
2544. Dice, CR. 1919. The mammals of southeastern Washington. J Mamm. 1:10-22.
2545. Dice, LR. 1921. Notes on the mammals of interior Alaska. J Mamm. 2:20-28.
2546. Dice, LR. 1922. Notes on a few mammals at Missoula, Montana, 1917-1918. J Mamm. 3:262-263.
2547. Dice, LR. 1923. Notes on some mammals of Riley county, Kansas. J Mamm. 4:107-112.
2548. Dice, LR. 1925. A survey of the mammals of Charlevoix County, Michigan, and vicinity. Occas Pap Mus Zool Univ MI. 159:1-33.
2549. Dice, LR. 1925. The mammals of Marion Island, Grand Transverse County, Michigan. Occas Pap Mus Zool Univ MI. 160:1-8.
2550. Dice, LR. 1926. Notes on pacific coast rabbits and pikas. Occas Pap Mus Zool Univ MI. 166:1-28.
2550a. Dice, LR. 1927. The Colorado pika in captivity. J Mamm. 8:228-231.
2550b. Dice, LR. 1929. An attempt to breed cottontail rabbits in captivity. J Mamm. 10:225-229.
2551. Dice, LR. 1954. Breeding of *Peromyscus floridanus* in captivity. J Mamm. 35:260.
2552. Dickerman, RW, Koopman, KF, & Seymour, C. 1981. Notes on bats from the pacific lowlands of Guatemala. J Mamm. 62:406-411.
2552a. Dickman, CR. 1980. Ecological studies of *Antechinus stuartii* and *Antechinus flavipes* (Marsupialia: Dasyuridae) in open-forest and woodland habitats. Aust Zool. 20:433-445.
2553. Dickman, CR. 1982. Some ecological aspects of seasonal breeding in *Antechinus* (Dasyuridae, Marsupialia). Carnivorous Marsupials. 1:139-150.
2553a. Dickman, CR & Braithwaite, RW. 1992. Postmating mortality of males in the dasyurid marsupials, *Dasyurus* and *Parantechinus*. J Mamm. 73:143-147.
2553b. Dickman, CR, Hayssen, V, & King, DH. 1987. Effects of seasonal reversal of photoperiod on the reproductive rhythm of a small marsupial. J Zool. 213:766-768.
2554. Dickman, CR & McKechnie, CA. 1985. A survey of the mammals of Mount Royal and Barrington Tops, New South Wales. Aust Zool. 21:531-543.
2555. Didier, R & Rode, P. 1942. Les Micromammifères de la faune francaise II Rats, Souris, Mulots. Mammalia. 6:36-45,120.
2555a. Dielentheis, TF, Zaiss, E, & Geissmann, T. 1991. Infant care in a family of siamangs (*Hylobates syndactylus*) with twin offspring at Berlin Zoo. Zoo Biol. 10:309-317.
2556. Dierschke, DJ, Bhattacharya, AN, Atkinson, LE, et al. 1970. Circhoral oscillations of plasma LH levels in the ovariectomized rhesus monkeys. Endocrinology. 87:850-853.
2557. Dierschke, DJ, Wehrenberg, WB, Wolf, RC, et al. 1978. A reversal in the ratio of estrone and estrdiol during late pregnancy in the uterine vein of ovariectomized rhesus monkeys. Endocrinology. 103:486-491.
2558. Dierschke, DJ, Weiss, G, & Knobil, E. 1974. Sexual maturation in the female rhesus monkey and the development of estrogen-induced gonadotropic hormone release. Endocrinology. 94:198-206.
2559. Dierschke, DJ, Yamaji, T, Karsch, FJ, et al. 1973. Blockade by progesterone of estrogen-induced LH and FSH release in the rhesus monkey. Endocrinology. 92:1496-1501.
2560. Dieterich, RA. 1975. The collared lemming (*Dicrostonyx stevensoni nelson*) in biomedical research. Lab Anim Sci. 25:48-54.
2561. Dieterich, RA & Luick, JR. 1971. Reindeer in biomedical research. Lab Anim Sci. 21:817-824.
2562. Dieterich, RA & Preston, DJ. 1977. The meadow vole (*Microtus pennsylvanicus*) as a laboratory animal. Lab Anim Sci. 27:494-499.

2563. Dieterich, RA & Preston, DJ. 1977. The tundra vole (*Microtus oeconomus*) as a laboratory animal. Lab Anim Sci. 27:500-506.
2564. Dieterich, RA & Preston, DJ. 1977. The red-backed vole (*Clethrionomys rutilus*) as a laboratory animal. Lab Anim Sci. 27:507-511.
2565. Dieterlen, F. 1960. Bemerkungen zu Zucht und Verhalten der Zwergmaus (*Micromys minutus sorcinus* Hermann). Z Tierpsychol. 17:552-554.
2566. Dieterlen, F. 1961. Beiträge zur Biologie der Stachelmaus, *Acomys cahirinus dimidiatus* Cretzschmar. Z Säugetierk. 26:1-13.
2567. Dieterlen, F. 1963. Vergleichende Untersuchungen zur Ontogenese von Stachelmaus (*Acomys*) und Wanderratte (*Rattus norvegicus*) Beiträge zum Nesthocker-Nestflüchter Problem bei Nagetieren. Z Säugetierk. 28:193-227.
2568. Dieterlen, F. 1963. Zur Kenntnis der Kreta-stachelmaus, *Acomys* (*cahirinus*) *minnous* Bate. Z Säugetierk. 28:47-57.
2569. Dieterlen, F. 1965. Von der Lebensweise und dem Verhalten der Felsenmaus, *Apodemus mystacinus* (Danford und Alston, 1877) nebst Beiträgen zur vergleichenden Ethologie der Gattung *Apodemus*. Säugetierk Mitt. 13:152-161.
2570. Dieterlen, F. 1967. Jahreszeiten und Fortpflanzangsperioden bei den Muriden des Kivusee-Gebietes (Congo). Z Säugetierk. 32:1-44.
2571. Dieterlen, F. 1968. Zur Kenntnis der Gattung *Otomys* (Otomyine, Muridae, Rodentia) Beiträge zur Systematik, Ökologie, und Biologie zentral afrikanischer Formen. Z Säugetierk. 33:321-352.
2572. Dieterlen, F. 1971. Beiträge zur Systematik, Ökologie and Biologie der Gattung *Dendromus* (Dendromurinae, Cricetidae, Rodentia) insbesondere ihrer zentralafrikanischen Formen. Säugetierk Mitt. 19:97-132.
2572a. Dieterlen, F. 1976. Die afrikanische Muriden gattung *Lophuromys* Peters 1874. Stuttgarter Beitr Naturk A. 285:1-96.
2573. Dieterlen, F. 1985. Daten zur Fortpflanzung und Populationsstruktur der myomorphen Nager eines afrikanischen Tieflandregenwaldes. Z Säugetierk. 50:68-88.
2574. Dieterlen, F. 1986. Seasonal reproduction and population dynamics in rodents of an African lowland rainforest. Cimbebasia. 8:1-7.
2575. Dieterlen, F & de Balsac, HH. 1979. Zur Ökologie und Taxonomie der Spitzmäuse (Soricidae) des Kivu Gebietes. Säugetierk Mitt. 27:241-287.
2576. Dietrich, U. 1985. Populationsokölogie des in Argentinien eingebürgerten europaischen Feldhasen (*Lepus europaeus*). Z Jagdwiss. 31:92-102.
2577. Dietz, JM. 1984. Ecology and social organization of the maned wolf (*Chrysocyon brachyurus*). Smithson Contrib Zool. 392:1-51.
2578. Dietz, JM. 1985. *Chrysocyon brachyurus*. Mamm Species. 234:1-4.
2579. Dietz, JM. 1987. Grass roots of the maned wolf. Nat Hist. 96(3):52-59.
2579a. Dietz, JM & Kleiman, DG. 1986. Reproductive parameters in groups of free-living golden lion tamarins. Primate Rep. 14:77.
2580. Digiacomo, RF, Shaughnessy, PW, & Tomlin, ST. 1978. Fetal-placental weight relationships in the rhesus (*Macaca mulatta*). Biol Reprod. 17:749-753.
2581. Dill, CW, Tylor, PT, McGill, R, et al. 1972. Gross composition and fatty acid constitution of blackbuck antelope (*Antilope cervicapra*) milk. Can J Zool. 50:1127-1129.
2582. Dimpel, H & Calaby, JH. 1972. Further observations on the mountain pygmy possum (*Burramys parvus*). Vict Nat. 89:101-106.
2583. Din, NA. 1978. Notes on two lion prides near Mweya in the Rwenzori National Park, Uganda. Pak J Zool. 10:133-138.
2584. Din, NU. 1983. Bonnet monkey cervical mucus glycoproteins: Study of the minor glycoprotein components of periovulatory phase mucus. Biol Reprod. 28:1189-1199.
2585. Dinale, G. 1965. Alcuni risultati dell'inanellamento di *Rhinolophus ferrum equinum* Schreber et di *Rhinolophus euryale* Blasius in Liguria (1957-1964) e nel Lazio (1962-1965). Bol Zool. 32:815-822.
2586. Dinale, G. 1968. Studi sui Chirotteri italianiv Sul raggiungimento della maturità sessuale nei Chirotteri europei ed in particulare nei Rhinolophidae. Arch Zool Ital. 53:51-71.
2587. Dinerstein, E. 1986. Reproductive ecology of fruit bats and the seasonality of fruit production in a Costa Rican cloud forest. Biotropica. 18:307-318.
2588. Diong, CH. 1973. Studies of the Malayan wild pig in Perak and Johore. Malay Nat J. 26:120-151.
2589. Dippenaar, NJ. 1979. Notes on the early post-natal development and behavior of the tiny musk shrew, *Crocidura bicolor* Bocage, 1889 (Insectivora: Soricidae). Mammalia. 43:83-91.
2590. Disney, RHL. 1968. Observations on a zoonosis: Leishmaniasis in British Honduras. J Appl Ecology. 5:1-59.
2591. Distant, WL. 1897. Notes on the chacma baboon. Zoologist Fourth Ser. 1:29-32.
2592. Dittoe, G. 1944. Lesser pandas. Zoonooz. 17(Dec):4-5.
2593. Dittrich, L. 1958. Haltung und Aufzucht von *Nyctalus noctula* Schreb. Z Säugetierk. 23:99-107.
2594. Dittrich, L. 1961. Zur Werfzeit des Eisbären (*Ursus maritimus*). Säugetierk Mitt. 9:13-15.
2595. Dittrich, L. 1962. Versuchte künstliche Aufzucht eines Fluszpferdes (*Hippopotamus amphibius* L). Zool Garten NF. 26:175-190.
2596. Dittrich, L. 1966. Breeding Indian elephants *Elephas maximus* at Hanover Zoo. Int Zoo Yearb. 6:193-196.
2597. Dittrich, L. 1967. Beitrag zur Fortpflanzung und Jugendentwicklung des Indischen Elefanten, *Elephas maximus* im Gefangenschaft mit einer Übersicht über die Elefantengeburten in europaischen Zoos und Zirkussen. Zool Garten NF. 34:56-92.
2598. Dittrich, L. 1967. Breeding Kirk's dik dik *Madoqua kirki thomasi* at Hanover Zoo. Int Zoo Yearb. 7:171-173.
2599. Dittrich, L. 1967. Breeding the black rhinoceros *Diceros bicornis* at Hanover Zoo. Int Zoo Yearb. 7:161-162.
2600. Dittrich, L. 1968. Keeping and breeding gazelles at Hanover Zoo. Int Zoo Yearb. 8:139-143.
2601. Dittrich, L. 1969. Birth weights and weight increases of African antelopes born at Hanover Zoo. Int Zoo Yearb. 9:118-120.
2602. Dittrich, L. 1969. Breeding the roan antelope *Hippotragus equinus* at Hanover Zoo. Int Zoo Yearb. 9:116.
2603. Dittrich, L. 1970. Beitrag zur Fortpflanzungsbiologie afrikanischer Antilopen in Zoologischen Garten. Zool Garten NF. 39:16-40.
2604. Dittrich, L. 1971. Beobachtungen zur Jugendentwicklung eines Breitmaulnashorns (*Ceratotherium s simum*) im Zoo Hannover. Fr Kölner Zoo. 14:73-81.
2605. Dittrich, L. 1972. Beobachtungen bei der Haltung von *Cephalophus*-Arten sowie zur Fortpflanzung und Jugendentwicklung von *C dorsalis* und *C rufilatus* in Gefangenschaft. Zool Garten NF. 42:1-16.

2606. Dittrich, L. 1972. Gestation periods and age at sexual maturity of some Africa antelopes. Int Zoo Yearb. 12:184-187.
2607. Dittrich, L. 1974. Erfahrungen bie der Haltung und Zucht des Gabelbockes (*Antilocapra americana*). Zool Garten NF. 44:216-234.
2608. Dittrich, L. 1974. Postpartum conception in African antelope. Int Zoo Yearb. 14:181-182.
2609. Dittrich, L. 1976. Age of sexual maturity in the hippopotamus *Hippopotamus amphibius*. Int Zoo Yearb. 16:171-173.
2610. Dittrich, L. 1979. Jugendentwicklung, Geschlechtsreife und Wechsel der Farbung des Haarkleides beim Schopfgibbon (*Hylobates concolor leucogenys*). Bijdr Dierkd. 49:247-254.
2611. Dittrich, L & Einsiedel, IV. 1961. Bemerkungen zur Fortpflanzung und Jugendentwicklung des Braunbären (*Ursus arctos* L) im Liepziger Zoo. Zool Garten NF. 25:250-269.
2612. Dittrich, L & Kronberger, H. 1963. Biologisch-anatomische Untersuchungen über die Fortpflanzungsbiologie des Braunbären (*Ursus arctos* L) und anderer Ursiden in Gefangenschaft. Z Säugetierk. 28:130-155.
2613. Dixit, VP & Jain, HC. 1978. Effects of chronically administered quinacrine hydrochloride on the histology and biochemistry of intact, ovariectomized and estrogen treated female genital tract of gerbil *Meriones hurrianae* Jerdon. Indian J Exp Biol. 16:984-986.
2614. Dixit, VP, Sharma, OP, & Agrawal, M. 1977. The effects of light deprivation/blindness on testicular function of gerbil (*Meriones hurricnae* [*sic*] Jerdon). Endokrinologie. 70:13-18.
2615. Dixon, J. 1919. Notes on the natural history of the bushy-tailed wood rats in California. Univ CA Publ Zool. 21:49-74.
2616. Dixon, J. 1922. Rodents and reclamation in the Imperial Valley. J Mamm. 3:136-146.
2617. Dixon, J. 1924. Notes on the life history of the gray shrew. J Mamm. 5:1-6.
2618. Dixon, JS. 1934. A study on the life history and food habits of mule deer in California. CA Fish Game. 20:181-282; 315-345.
2619. Dixson, AF. 1976. Effects of testosterone on the sternal cutaneous glands and genitalia of the male greater galago (*Galago crassicaudatus crassicaudatus*). Folia Primatol. 26:207-213.
2620. Dixson, AF. 1977. Observations on the displays, menstrual cycles and sexual behaviour of the "Black ape" of Celebes (*Macaca nigra*). J Zool. 182:63-84.
2621. Dixson, AF. 1982. Some observations on the reproductive physiology and behaviour of the owl monkey *Aotus trivirgatus* in captivity. Int Zoo Yearb. 22:115-118.
2622. Dixson, AF. 1983. The owl monkey (*Aotus trivirgatus*). Reproduction in New World Primates. 69-113, Hearn, J, ed, MTP Press Ltd, Lancaster, England.
2623. Dixson, AF. 1986. Proceptive displays of the female common marmoset (*Callithrix jacchus*): Effects of ovariectomy and oestradiol 17β. Physiol Behav. 36:971-973.
2624. Dixson, AF. 1986. Plasma testosterone concentrations during postnatal development in the male common marmoset. Folia Primatol. 47:166-170.
2625. Dixson, AF. 1987. Observations on the evolution of genitalia and copulatory behavior in primates. J Zool. 213:423-443.
2626. Dixson, AF. 1987. Baculum length and copulatory behavior in Primates. Am J Primatol. 13:51-60.
2627. Dixson, AF, Everitt, BJ, Herbert, J, et al. 1973. Hormonal and other determinants of sexual attractiveness and receptivity in rhesus and talapoin monkeys. Symp 4th Int Cong Primatol 1972. 2:36-63.
2628. Dixson, AF & Gardner, JS. 1981. Diurnal variations in plasma testosterone in a male nocturnal primate, the owl monkey (*Aotus trivirgatus*). J Reprod Fert. 62:83-86.
2629. Dixson, AF, Gardner, JS, & Bonney, RC. 1980. Puberty in the male owl monkey (*Aotus trivirgatus griseimembra*): A study of physical and hormonal development. Int J Primatol. 1:129-139.
2630. Dixson, AF & George, L. 1982. Prolactin and parental behaviour in a male New World primate. Nature. 299:551-553.
2631. Dixson, AF & Herbert, J. 1974. The effects of testosterone on the sexual skin and genitalia of the male talapoin monkey. J Reprod Fert. 38:217-219.
2632. Dixson, AF & Herbert, J. 1977. Testosterone, aggressive behavior and dominance rank in captive adult male talapoin monkeys (*Miopithecus talapoin*. Physiol Behav. 18:539-543.
2633. Dixson, AF, Herbert, J & Rudd, BT. 1973. Gonadal hormones and behaviour in captive groups of talapoin monkeys (*Miopithecus talapoin*). J Endocrinol. 57:xli.
2634. Dixson, AF, Knight, J, Moore, HDM, et al. 1982. Observations on sexual development in male orang-utans *Pongo pygmaeus*. Int Zoo Yearb. 22:222-227.
2635. Dixson, AF & Lunn, SF. 1987. Post-partum changes in hormones and sexual behaviour in captive groups of marmosets (*Callithrix jacchus*). Physiol Behav. 41:577-583.
2636. Dixson, AF & van Horn, RN. 1976. Diurnal variations in plasma testosterone in male bushbabies (*Galago c crassicaudatus*) and ringtailed lemurs (*Lemur catta*). J Endocrinol. 71:99P-100P.
2637. Dixson, AF & van Horn, RN. 1977. Comparative studies of morphology and reproduction in two subspecies of the greater bushbaby, *Galago crassicaudatus crassicaudatus* and *G c argentatus*. J Zool. 183:517-526.
2638. Djakiew, D & Jones, RC. 1981. Structural differentiation of the male genital ducts of the echidna (*Tachyglossus aculeatus*). J Anat. 132:187-202.
2639. Djakiew, D & Jones, RC. 1982. Stereological analysis of the epididymis of the echidna, *Tachyglossus aculeatus*, and wistar rat. Aust J Zool. 30:865-875.
2640. Djakiew, D & Jones, RC. 1982. Ultrastructure of the ductus epididymidis of the echidna, *Tachyglossus aculeatus*. J Anat. 135:625-634.
2641. Djakiew, D & Jones, RC. 1983. Sperm maturation, fluid transport, and secretion, and absorption of protein in the epididymis of the echidna *Tachyglossus aculeatus*. J Reprod Fert. 68:445-456.
2642. Dluzen, DE, Ramirez, VD, Carter, CS, et al. 1981. Male vole urine changes luteinizing hormone-releasing hormone and norepinephrine in female olfactory bulb. Science. 212:573-575.
2643. Doan, KH & Douglas, CW. 1953. Beluga of the Churchill region of Hudson Bay. Bull Fish Res Bd Can. 98:1-27.
2644. Dobbelaar, MJ & Arts, THM. 1972. Estimation of the foetal age in pregnant rhesus monkeys (*Macaca mulatta*) by radiography. Lab Anim Sci. 6:235-240.
2645. Doboszyńska, T. 1977. A macrometric and micrometric description of the ovary in the european beaver. Acta Theriol. 22:261-270.
2646. Doboszyńska, T. 1978. Histomorphology of the female reproductive system of the european beaver. Acta Theriol. 23:99-125.
2647. Doboszyńska, T & Żurowski, W. 1977. Vaginal smears during a sexual cycle of the beaver. Acta Theriol. 22:153-155.

2648. Doboszyńska, T & Żurowski, W. 1981. Anatomical studies of the male genital organs of the european beaver. Acta Theriol. 26:331-340.
2649. Doboszyńska, T & Żurowski, W. 1983. Reproduction of the European beaver. Acta Zool Fenn. 174:123-126.
2650. Dobrokhotov, BP, Baranovskii, PM, & Demidova, TN. 1985. Peculiarities of habitat distribution of the sibling species, *Microtus arvalis* and *M rossiaemeridionalis* (Rodentia, Microtinae) and their role in natural tularemia foci of field-meadow type. Zool Zhurn. 64:269-275.
2651. Dobroruka, LJ. 1968. A note on the gestation period and rearing of young in the leopard *Panthera pardus* at Prague Zoo. Int Zoo Yearb. 8:65.
2652. Dobroruka, LJ. 1968. Breeding group of gorals *Nemorhaedus goral* at Prague Zoo. Int Zoo Yearb. 8:143-145.
2653. Dobroruka, LJ. 1970. Fecundity of the Chinese water deer, *Hydropotes inermis* Swinhoe, 1870. Mammalia. 34:161-164.
2654. Dobroruka, LJ. 1972. Scent marking and courtship in the Siberian chipmunk, *Tamias sibiricus lineatus* (Siebold 1924), with notes on the taxonomic relations of chipmunks (Mammalia). Vestn Čcsl Spol Zool. 36:12-16.
2655. Dobroruka, LJ. 1977. Ein hohes Alter beim Ameisenbär, *Myrmecophaga tridactyla* Linnaeus 1758. Zool Garten NF. 47:312.
2656. Dobroruka, LJ & Horbowyjová, R. 1972. Notes on the ethology of the collared peccary *Dicotyles tajacu* (Linnaeus, 1766) in the Prague Zoological Garden. Lynx. 13:85-94.
2657. Dobrowolska, A & Gromadska, J. 1978. Relationship between haematological parameters and progesterone blood concentration in different stages of estrous cycle in common vole, *Microtus arvalis*. Comp Biochem Physiol.61A:483-485.
2658. Dobson, FS & Kjelgaard, JD. 1985. The influence of food resources on population dynamics in Columbian ground squirrels. Can J Zool. 63:2095-2104.
2659. Dobson, FS & Kjelgaard, JD. 1985. The influence of food resources on life history of Columbian ground squirrels. Can J Zool. 63:2105-2109.
2660. Dobson, FS & Myers, P. 1989. The seasonal decline in the litter size of meadow voles. J Mamm. 70:142-152.
2661. Dobson, FS, Zammuto, RM, & Murie, JO. 1986. A comparison of methods for studying life history in Columbian ground squirrels. J Mamm. 67:154-158.
2662. Dobson, GE. 1882-1890. A Monograph of the Insectivora. Van Voorst, Gurney & Jackson, London.
2663. Dodds, DG. 1962. Late breeding in Newfoundland snowshoe hare. Can Field-Nat. 76:60-61.
2664. Dodds, DG. 1965. Reproduction and productivity of snowshoe hares in Newfoundland. J Wildl Manage. 29:303-315.
2665. Dodge, WE. 1967. The biology and life history of the porcupine (*Erithizon dorsatum*) in western Massachusetts. Ph D Diss. Univ MA, Amherst.
2666. Doidge, DW, Croxall, JP, & Ricketts, C. 1984. Growth rates of Arctic fur seal *Arctocephalus gazella* pups at South Georgia. J Zool. 203:87-93.
2667. Doidge, DW, McCann, TS, & Croxall, JP. 1986. Attendance behavior of Antarctic fur seals. Fur Seals Maternal Strategies on Land and at Sea. 102-114, Gentry, RL & Kooyman, GL, eds, Princeton Univ Press, Princeton.
2668. Dolan, J. 1977. The saiga: A problematic ungulate under captive conditions. Breeding Dolphins Present Status, Suggestions for the Future, 288-293, Ridgway, SH & Benirschke, K, eds, US Mar Mamm Comm Rep #MMC-76/07 NTIS No PB273673, US Dept Comm Nat Tech Info Serv, Arlington, VA.
2669. Dolan, JM. 1965. Breeding of the lowland anoa, *Bubalus (Anoa) d depressicornis* (H Smith, 1827) in the San Diego Zoological Garden. Z Säugetierk. 30:241-245.
2670. Dolan, PG & Carter, DC. 1977. *Glaucomys volans*. Mamm Species. 78:1-6.
2670a. Dolan, PG & Carter, DC. 1979. Distributional notes and records from middle American Chiroptera. J Mamm. 60:644-649.
2671. Dolbeer, RA. 1973. Reproduction in the red squirrel (*Tamiasciurus hudsonicus*) in Colorado. J Mamm. 54:536-540.
2672. Dolbeer, RA & Clark, WR. 1975. Population ecology of snowshoe hares in the central Rocky Mountains. J Wildl Manage. 39:535-549.
2673. Dolinar, ZJ, Ludwig, KS, & Müller, E. 1963. Ein weiterer Beitrag zur Kenntnis der Placenten der Ordnung Perissodactyla: Eine Geburtsplacenta von *Equus asinus* L. Acta Anat. 53:81-96.
2674. Dolinar, ZJ, Ludwig, KS, & Müller, E. 1965. Ein weiterer Beitrag zur Kenntnis der Placenten der Ordnung Perissodactyla: Zwei Geburtsplacenten des Indischen Panzernashorns (*Rhinoceros unicornis*). Acta Anat. 61:331-354.
2675. Dollar, JR, Graham, CE, & Reyes, FI. 1982. Postovulatory predecidual development in the baboon, chimpanzee, and human. Am J Primatol. 3:307-313.
2676. Dollman, G. 1914. Notes on a collection of East African mammals presented to the British Museum by Mr GP Cosens. Proc Zool Soc London. 1914:307-318.
2677. Dollman, G. 1930. On mammals obtained by Mr Shaw Moyer in New Guinea, and presented to the British Museum by Mr J Spedan Lewis, FZS. Proc Zool Soc London. 1930:429-435.
2678. Dominic, CJ & Pandey, SD. 1978. Pheromonal influences on estrous cycle of the Indian field mouse *Mus booduga* Gray. Indian J Exp Biol. 16:1134-1136.
2679. Donat, F. 1933. Notes on the life history and behavior of *Neotoma fuscipes*. J Mamm. 14:19-26.
2680. Doncaster, P & Jouventin, P. 1989. The nutria. La Recherche. 20:754-761.
2681. Dönhoff, C. 1942. Zur Kenntnis des afrikanischen Waldschweines *Hylochoerus meinertzhageni* Thos. Zool Garten NF. 14:192-200.
2682. Donne, TE. 1924. The Game Animals of New Zealand. John Murray, London.
2683. Donnelly, BG. 1967. Observations on the mating behaviour of the southern right whale *Eubalaena australis*. S Afr J Sci. 63:176-181.
2684. Donnelly, BG. 1969. Further observations on the southern right whale, *Eubalaena australis*, in South African waters. J Reprod Fert Suppl. 6:347-352.
2685. Donohue, RW. 1966. Muskrat reproduction in areas of controlled and uncontrolled water-level units. J Wildl Manage. 30:320-326.
2686. Donovan, BT. 1963. The effect of pituitary stalk section on luteal function in the ferret. J Endocrinol. 27:201-212.
2687. Donovan, BT. 1967. The control of corpus luteum function in the ferret. Arch Anat Microsc Morphol Exp. 56 Suppl to 3-4:315-325.
2688. Donovan, BT. 1986. Is there a critical weight for oestrus in the ferret?. J Reprod Fert. 76:491-497.
2689. Donovan, BT & Gledhill, B. 1981. Changes in FSH and LH secretions in the ferret associated with the induction of ovulation by copper acetate. Biol Reprod. 25:72-76.
2690. Donovan, BT & Gledhill, B. 1984. Half-life of FSH and LH in the ferret. Acta Endocrinol. 105:14-18.
2691. Donovan, BT, Matson, C & Kilpatrick, MJ. 1983. Effect of exposure to long days on the secretion of estradiol, oestrone,

progesterone, testosterone, androstenedione, cortisol and FSH in intact and spayed ferrets. J Endocrinol. 99:361-368.
2692. Donovan, BT & ter Haar, MB. 1977. Effects of luteinizing hormone releasing hormone on plasma follicle-stimulating hormone and luteinizing hormone levels in the ferret. J Endocrinol. 73:37-52.
2693. Doran, DJ. 1952. Observations on the young of the Merriam kangaroo rat. J Mamm. 33:494-495.
2694. Doreau, M, Boulot, S, Martin-Rosset, W, et al. 1986. Relationship between nutrient intake, growth and body composition of the nursing foal. Reprod Nutr Dév. 26:683-690.
2695. Dorney, RS & Rusch, AJ. 1953. Muskrat growth and litter production. WI Cons Dept Techn Wildl Bull. 8:1-32.
2696. Dott, HM & Utsi, MNP. 1973. Artificial insemination of reindeer (*Rangifer tarandus*). J Zool. 170:505-508.
2697. Doude van Troostwyk, WJ. 1976. The musk-rat (*Ondatra zibethicus* L) in the Netherlands, its ecological aspects and their consequences for man. Ph D Diss. Univ Leiden, Leiden.
2697a. Douglas, AM. 1962. *Macroderma gigas saturata* (Chiroptera, Megadermatidae) a new subspecies from the Kimberley division of Western Australia. W Aust Nat. 8:59-61.
2698. Douglas, AM. 1967. The natural history of the ghost bat, *Macroderma gigas* (Microchiroptera, Megdermatidae), in Western Australia. W Aust Nat. 10:125-137.
2699. Douglas, CL. 1969. Comparative ecology of pinyon mice and deer mice in Mesa Verde National Park, Colorado. Univ KS Publs Mus Nat Hist. 18:421-504.
2700. Douglas, CL. 1969. Ecology of pocket gophers of Mesa Verde, Colorado. Univ KS Mus Nat Hist Misc Publ. 51:147-175.
2701. Douglas, JD & Butler, TM. 1970. Chimpanzee breeding at the 6571st aeromedical research laboratory. Lab Anim Care. 20:477-482.
2702. Douglas, MJW. 1965. Notes on the red fox (*Vulpes vulpes*) near Braemar, Scotland. J Zool. 147:228-233.
2703. Douglas, MJW. 1966. Occurrence of accessory corpora lutea in red deer, *Cervus elaphus*. J Mamm. 47:152-153.
2704. Douglass, RJ. 1977. Population dynamics, home ranges, and habitat associations of the yellow-cheeked vole, *Microtus xanthognathus* in the Northwest Territories. Can Field-Nat. 91:237-247.
2705. Dove, H & Cork, HJ. 1989. Lactation in the tammar wallaby *Macropus eughenii* I Milk consumption and the algebraic description of the lactation curve. J Zool. 219:385-398.
2706. Dowell, SF & Lynch, GR. 1987. Duration of the melatonin pulse in the hypothalamus controls testicular function in pinealectomized mice (*Peromyscus leucopus*). Biol Reprod. 36:1095-1101.
2707. Dowler, RC & Genoways, HH. 1978. *Liomys irroratus*. Mamm Species. 82:1-6.
2708. Downhower, JF & Armitage, KB. 1971. The yellow-bellied marmot and the evolution of polygamy. Am Nat. 105:355-370.
2709. Downhower, JF & Armitage, KB. 1981. Dispersal of yearling yellow-bellied marmots (*Marmota flavieventris*). Anim Behav. 29:1064-1069.
2710. Dowsett, RJ. 1966. Behaviour and population structure of hartebeest in the Kafue National Park. Puku. 4:147-154.
2711. Doyle, GA. 1979. Development of behavior in prosimians with special reference to the lesser bushbaby, *Galago senegalensis moholi*. The Study of Prosimian Behavior. 157-206, Doyle, GA & Martin, RD, eds, Academic Press, New York.
2712. Doyle, GA, Andersson, A, & Bearder, SK. 1971. Reproduction in the lesser bushbaby (*Galago senegalensis moholi*) under semi-natural conditions. Folia Primatol. 14:15-22.
2713. Doyle, GA & Bearder, SK. 1977. The galagines of South Africa. Primate Conservation. 1-35, Prince Rainier III of Monaco & Bourne, GH, eds, Academic Press, New York.
2714. Doyle, GA, Pelletier, A, Bekker, T, et al. 1967. Courtship, mating and parturition in the lesser bushbaby (*Galago senegalensis moholi*) under semi-natural conditions. Folia Primatol. 7:169-197.
2715. Doyle, RE, Sharp, GC, Irvin, WS, et al. 1976. Reproductive performance and fertility testing in strain 13 and Hartley guinea pigs. Lab Anim Sci. 26:573-580.
2716. Dragesco, J, Feer, F, & Genermont, J. 1979. Contribution à la connaissance de *Neotragus batesi* de Winton, 1903 (position systématique, données biométriques). Mammalia. 43:71-81.
2717. Dräseke, J. 1931. Zur Kenntnis des Gehirns von *Echidna aculeata lawesi* (Rams). Deutsch Z f Nervenheilk 117/119:103-112.
2718. Dresser, BL, Kramer, L, Reece, B, et al. 1982. Induction of ovulation and successful artificial insemination in Persian leopard (*Panthera pardus saxicolor*). Zoo Biol. 1:55-57.
2719. Dresser, BL, Pope, CE, Kramer, L, et al. 1984. Nonsurgical embryo recovery and successful interspecies embryo transfer from bongo (*Tragelaphus euryceros*) to eland (*Tragelaphus oryx*). Am Assoc Zoo Vet Abstracts Ann Meeting. 1984:180.
2720. Dresser, BL, Romo, JS, Brownscheidle, CM, et al. 1980. Reproductive behaviour and birth of a bongo *Boocercus eurycerus* at the Cincinnati Zoo. Int Zoo Yearb. 20:229-234.
2721. Drickamer, LC. 1974. A ten-year summary of reproductive data for free-ranging *Macaca mulatta*. Folia Primatol. 21:61-80.
2722. Drickamer, LC. 1974. Social rank, observability, and sexual behaviour of rhesus monkeys (*Macaca mulatta*. J Reprod Fert. 37:117-120.
2723. Drickamer, LC. 1977. Seasonal variation in litter sizs, body weight and sexual maturation in juvenile female house mice (*Mus musculus*). Lab Anim Sci. 11:159-162.
2724. Drickamer, LC. 1978. Annual reproduction patterns in populations of two sympatric species of *Peromyscus*. Behav Biol. 23:405-408.
2725. Drickamer, LC. 1979. Acceleration and delay of first estrus in wild *Mus musculus*. J Mamm. 60:215-216.
2726. Drickamer, LC. 1981. Selection for age of sexual maturation in mice and the consequence for population regulation. Behav Neural Biol. 31:82-89.
2727. Drickamer, LC & Bernstein, J. 1972. Growth in two subspecies of *Peromyscus maniculatus*. J Mamm. 53:228-231.
2728. Drickamer, LC & Vestal, BM. 1973. Patterns of reproduction in a laboratory colony of *Peromyscus*. J Mamm. 54:523-528.
2729. Drips, D. 1919. Studies on the ovary of the spermophile (*Spermophilus citellus tridecemlineatus*) with special reference to the corpus luteum. Am J Anat. 25:117-184.
2730. Droogleever Fortuyn, AB. 1927. Notes on the striped hamster (*Cricetulus griseus*, Thomas). China Med J. 41:859-863.
2731. Droogleever Fortuyn, AB. 1928. Further notes on the striped hamster ((*Cricetulus griseus*). China Med J. 42:524-525.
2732. Droogleever Fortuyn, AB. 1929. Prenatal death in the striped hamster (Cricetulus griseus, M Edw). Arch Biol. 39:583-606.
2733. Druecker, JD. 1973. Aspects of reproduction in *Myotis volans*, *Lasionycteris noctivagans*, and *Lasiurus cinereus*. Diss Abstr Int. 33B:5065.
2734. Drüwa, P. 1977. Beobachtungen zur Geburt und natürlichen Aufzucht von Waldhunden (*Speothos venaticus*) in der Gefangenschaft. Zool Garten NF. 47:109-137.

2735. Drüwa, P. 1982. Perro de grulleiro, der Südamerikanische Walahund: Ein Rätsel fur die Hundeforschung. Z Kölner Zoo. 25:71-90.
2736. Dryden, GL. 1968. Growth and development of *Suncus murinus* in captivity on Guam. J Mamm. 49:51-62.
2737. Dryden, GL. 1969. Reproduction in *Suncus murinus*. J Reprod Fert Suppl. 6:377-396.
2738. Dryden, GL. 1985. Development and regression of the corpus luteum of pseudopregnancy in *Suncus murinus*. Acta Zool Fenn. 173:263-264.
2739. Dryden, GL & Anderson, JN. 1977. Ovarian hormone: Lack of effect on reproductive structures of female Asian musk shrews. Science. 197:782-784.
2740. Dryden, GL & Anderson, RR. 1978. Milk composition and its relation to growth rate in the musk shrew, *Suncus murinus*. Comp Biochem Physiol. 60A:213-216.
2741. Dryden, GL, Gebczynski, M, & Douglas, EL. 1974. Oxygen consumption by nursling and adult musk shrews. Acta Theriol. 19:453-461.
2742. Dryden, GL & McAllister, HY. 1970. Sustained fertility after $CdCl_2$ injection by a non-scotal mammal, *Suncus murinus* (Insectivora, Soricidae). Biol Reprod. 3:23-30.
2743. Dryden, GL & Pucek, Z. 1976. Insectivores in reproduction studies, with emphasis on ovulation in American, Asian, and European shrews. The Laboratory Animal in the Study of Reproduction 6th Symposium ICLA, Thessaloniki 9-11 July, 1975. 39-50, Antikatzides, T, Ericksen, S, & Spiegel, A, eds, Fischer, New York.
2744. Dryden, GL & Ross, JM. 1971. Enhanced growth and development of captive musk shrews, *Suncus murinus*, on an improved diet. Growth. 35:311-325.
2745. du Plessis, SS. 1968. Ecology of blesbok (*Damaliscus dorcas phillipsi* Burchell) on the van Riebeek Nature Reserve, Pretoria, with special reference to productivity. D Sc Diss. Pretoria Univ, Pretoria.
2746. du Plessis, SS. 1972. Ecology of blesbok with special reference to productivity. Wildl Monogr. 30:1-70.
2747. Duangkhae, S. 1986. Ecology and behavior of Kitti's hog-nosed bat (*Craseonycteris thonglongyai*). Masters Thesis. Mahidol Univ, Bangkok, Thailand.
2748. Dubey, AK, Camerone, JL, Steiner, RA, et al. 1986. Inhibition of gonadotropin secretion in castrated male rhesus monkeys (*Macaca mulatta*) induced by dietary restriction: Analogy with the prepubertal hiatus of gonadotropin release. Endocrinology. 118:518-525.
2749. Dubey, AK & Plant, TM. 1985. Testosterone administration to ovariectomized female rhesus monkeys (*Macaca mulatta*) reduces the frequency of pulsatile luteinizing hormone secretion. Biol Reprod. 32:1109-1115.
2750. Dubock, AC. 1979. Male grey squirrel (*Sciurus carolensis*) reproductive cycles in Britain. J Zool. 188:41-51.
2751. Dubois, E. 1924. Over de hersenhoeveelheid van gespecialiseerde zoogdiergeslachten. Versl Gewone Verg Wis Natuurk Afd Kon Akad Wetensch. 33:319-326.
2752. Dubost, G. 1964. Un Muridé arboricole du Gabon, *Dendromus pumilio* Wagner, possesseur d'un cinquième orteil opposable. Biol Gabon. 1:187-190.
2753. Dubost, G. 1965. Quelques renseignements biologiques sur *Potamogale velox*. Biol Gabon. 1:257-272.
2754. Dubost, G. 1968. Aperçu sur le rhythme annuel de reproduction des Muridés du Nord-Est du Gabon. Biol Gabon. 4:227-239.
2755. Dubost, G. 1968. Le rhythme annuel de reproduction du Chevrotain aquatique *Hyemoschus aquaticus* Ogilby, dans le secteur forestier du Nord-Est du Gabon. Cycles Génitaux Saisonniers de Mammifères Sauvages. 51-61, Canivenc, R, ed, Masson et Cie, Paris.
2756. Dubost, G. 1975. Le comportement du Chevrotain africain *Hyemoschus aquaticus* Oglby (Artiodactyla, Ruminantia). Z Tierpsychol. 37:403-501.
2757. Dubost, G. 1978. Un aperçu sur l'ecologie du Chevrotain africain *Hyemoschus aquaticus*, Artiodactyle Tragulid. Mammalia. 42:2-62.
2758. Dubost, G. 1980. L'écologie et la vie sociale du Cephalophe blue *Cephalophus monticola* (Thunberg) petit ruminant forestier africain. Z Tierpsychol. 54:205-266.
2759. Dubost, G & Genest, H. 1974. Le comportement social d'une colonie de Maras *Dolichotis patagonum* Z dans le Parc de Branfère. Z Tierpsychol. 35:225-302.
2760. Dubrovskaya, RM. 1965. Characteristics of the milk of Arctic foxes. Krolikov Zverov. 8(8):21.
2761. Dubrovskaya, RM. 1969. Milk production in female blue foxes and its effect on the development of the young. Uch Zap Petrozavodsk Gos Universiteta. 17(4):80-91.
2762. Duby, RT & Travis, HF. 1972. Photoperiodic control of fur growth and reproduction in the mink (*Mustela vision*). J Exp Zool. 182:217-226.
2763. Dücker, G. 1959. Beobachtungen über das Paarungsverhalten des Ichneumons (*Herpestes ichneumon* L). Z Säugetierk. 25:47-51.
2764. Ducsay, CA, Stanczyk, FZ, & Novy, MJ. 1985. Maternal and fetal production rates of progesterone in rhesus macaques: Placental transfer and conversion to cortisol. Endocrinology. 117:1253-1258.
2765. Dufau, ML, Hodgen, GD, Goodman, AL, et al. 1977. Bioassay of circulating luteinizing hormone in the rhesus monkey: Comparison with radioimmunoassay during physiological changes. Endocrinology. 100:1557-1565.
2766. Duguy, R. 1966. Quelques données nouvelles sur un Cétacé rare sur les Côtes d'Europe: Le Cachalot a tête courte, *Kogia breviceps* (Blainville, 1838). Mammalia. 30:259-269.
2767. Duhaiman, AS. 1988. Purification of camel milk lysozyme and its lytic effect on *Escherichia coli* and *Micrococcus lysodeikticus*. Comp Biochem Physiol. 91B:793-796.
2768. Duke, KL. 1940. A preliminary histological study of the ovary of the kanagroo rat, *Dipodomys ordii columbianus*. Great Basin Nat. 1:63-73.
2769. Duke, KL. 1944. The breeding season in two species of *Dipodomys*. J Mamm. 25:155-160.
2770. Duke, KL. 1949. Some notes on the histology of the ovary of the bobcat (*Lynx*) with special reference to the corpora lutea. Anat Rec. 102:111-131.
2771. Duke, KL. 1951. The external genitalia of the pika, *Ochotona princeps*. J Mamm. 32:169-173.
2772. Duke, KL. 1952. Ovarian histology of *Ochotona princeps*, the Rocky Mountain pika. Anat Rec. 112:737-759.
2773. Duke, KL. 1957. Reproduction in *Perognathus*. J Mamm. 38:207-210.
2774. Duke, KL. 1964. Histological observations on the ovary of the slow loris, *Nycticebus coucang*. Anat Rec. 148:414.
2775. Duke, KL. 1966. Histological observations on the ovary of the white-tailed mole, *Parascaptor leucurus*. Anat Rec. 154:527-532.
2776. Duke, KL. 1967. Ovogenetic activity of the fetal-type in the ovary of the adult slow loris, *Nycticebus coucang*. Folia Primatol. 7:150-154.

2777. Duke, KL & Luckett, WP. 1965. Histological observations on the ovary of several species of tree shrews (Tupaiidae). Anat Rec. 151:450.
2778. Dukelow, WR. 1966. Effects of age and strain of female, and of ethylene dichloride extracted wheat germ oil on reproduction in mink (*Mustela vison*). J Reprod Fert. 11:181-184.
2779. Dukelow, WR. 1966. Variation in gestation length of mink (*Mustela vison*). Nature. 211:211.
2780. Dukelow, WR. 1970. Induction and timing of single and multiple ovulations in the squirrel monkey (*Saimiri sciureus*). J Reprod Fert. 22:303-309.
2781. Dukelow, WR. 1975. The morphology of follicular development and ovulation in non-human primates. J Reprod Fert Suppl. 22:23-51.
2782. Dukelow, WR. 1977. Ovulatory cycle characteristics in *Macaca fascicularis*. J Med Primatol. 6:33-42.
2783. Dukelow, WR. 1978. Ovulation detection and control relative to optimal time of mating in non-human primates. Symp Zool Soc London. 43:195-206.
2784. Dukelow, WR. 1983. The squirrel monkey. Reproduction in New World Primates. 151-179, Hearn, J, ed, MTP Press, Lancaster, England.
2785. Dukelow, WR, Grauwiler, J, & Brüggemann, S. 1979. Characteristics of the menstrual cycle of nonhuman primates I Similarities and dissimilarities between *Macaca facicularis* and *Macaca arctoides*. J Med Primatol. 8:39-47.
2786. Dukelow, WR, Jewett, DA, & Rawson, JMR. 1973. Follicular development and ovulation in *Macaca fascicularis*, *Saimiri sciureus* and *Galago senegalensis*. Am J Phys Anthropol. 38:207-210.
2787. Dulaney, MW. 1987. A mother-reared second-captive-generation aardvark *Orycteropus afer* at the Cincinnati Zoo. Int Zoo Yearb. 26:281-285.
2788. Dulaney, MW. 1987. Successful breeding of Demidoff's galago *Galago d demidovi* at the Cincinnati Zoo. Int Zoo Yearb. 26:229-231.
2789. Dumartin, B & Canivenc, R. 1984. Étude ultrastructurale de l'érythrophagocytose dans la poche choriale paraplacentaire chez deux Mustelidés: Le Blaireau *Meles meles* L et le Furet *Mustela putorius furo*. Arch Biol. 95:445-460.
2790. DuMond, FV. 1968. The squirrel monkey in a seminatural environment. The Squirrel Monkey. 87-145, Rosenblum, LA & Cooper, RW, eds, Academic Press, New York.
2791. DuMond, FV & Hutchinson, TC. 1967. Squirrel monkey reproduction: The "fatted" male phenomenon and seasonal spermatogenesis. Science. 158:1067-1070.
2792. DuMond, FV, Hoover, BL, & Norconk, MA. 1979. Hand-feeding parent-reared golden lion tamarins *Leontopithecus rosalia rosalia* at Monkey Jungle. Int Zoo Yearb. 19:155-158.
2793. Dunaway, PB. 1962. Litter-size record for eastern harvest mouse. J Mamm. 43:428-429.
2794. Dunaway, PB. 1968. Life history and population aspects of the eastern harvest mouse. Am Mid Nat. 79:48-67.
2795. Dunbar, RIM. 1978. Sexual behaviour and social relationships among gelada baboons. Anim Behav. 26:167-178.
2796. Dunbar, RIM. 1980. Demographic and life history variables of a population of gelada baboons (*Theropithecus gelada*). J Anim Ecol. 49:485-506.
2797. Dunbar, RIM. 1984. Reproductive Decisions: An Economic Analysis of Gelada Baboon Social Strategies. Princeton Univ Press, Princeton, NJ.
2798. Dunbar, RIM & Dunbar, EP. 1974. Social organization and ecology of the klipspringer (*Oreotragus oreotragus*) in Ethiopia. Z Tierpsychol. 35:481-493.
2799. Dunbar, RIM & Dunbar, EP. 1974. The reproductive cycle of the gelada baboon. Anim Behav. 22:203-210.
2800. Dunbar, RIM & Dunbar, EP. 1974. Behaviour related to birth in wild gelada baboons (*Theropithecus gelada*). Behaviour. 50:185-191.
2801. Dunbar, RIM & Dunbar, EP. 1977. Dominance and reproductive success among female gelada baboons. Nature. 266:351-353.
2802. Dunbar-Brander, AA. 1931. Wild Animals in Central India. Edward Arnold, London.
2803. Duncan, MJ, Goldman, BD, DiPinto, MN, et al. 1985. Testicular function and pelage color have different critical day lengths in the Djungarian hamster, *Phodopus sungorus sungorus*. Endocrinology. 116:424-430.
2804. Duncan, PE. 1979. The biology of the honey possum (*Tarsipes spenserae*) (Gray 1842). B Sc Thesis, Murdoch Univ.
2805. Dunford, C. 1974. Annual cycle of different chipmunks in the Santa Catalina Mountains, Arizona. J Mamm. 55:401-416.
2806. Dunmire, WW. 1960. An altitudinal survey of reproduction in *Peromyscus maniculatus*. Ecology. 41:174-182.
2807. Dunmire, WW. 1961. Breeding season of three rodents on White Mountain, California. J Mamm. 42:489-493.
2808. Dunn, JP & Chapman, JA. 1983. Reproduction, physiological responses, age structure, and food habits of raccoon in Maryland, USA. Z Säugetierk. 48:161-175.
2808a. Dunn, LH. 1934. Notes on the little bulldog bat, *Dirias albiventer minor* (Osgood), in Panama. J Mamm. 15:89-99.
2809. Dunnet, GM. 1956. A live trapping study of the brush-tailed possum *Trichosurus vulpecula* Kerr (Marsupialia). CSIRO Wildl Res. 1:1-18.
2810. Dunnet, GM. 1964. A field study of local populations of the brush-tailed possum *Trichosurus vulpecula* in eastern Australia. Proc Zool Soc London. 142:665-695.
2811. Dunnett, GM. 1962. A population study of the quokka *Setonix brachyurus* (Quoy and Gaimand) (Marsupialia) II Habitat, movements, breeding and growth. CSIRO Wildl Res. 7:13-32.
2812. Duplaix, N. 1980. Observations on the ecology and behavior of the giant river otter *Pteronura brasiliensis* in Suriname. Terre Vie. 34:495-620.
2813. Duplaix-Hall, N. 1975. River otter in captivity: A review. Breeding Endangered Species in Captivity. 315-327, Martin, RD, ed, Academic Press, New York.
2814. Durán, AC & Sans-Coma, V. 1986. Geschlechtsreife bei *Mus spretus* Lataste, 1883. Z Säugetierk. 51:345-349.
2815. Durant, B. 1983. Reproductive studies of oryx. Zoo Biol. 2:191-197.
2816. Durant, P. 1981. Ecological study of *Sylvilagus biasibensis meridensis* in an Andean Venezuelan Paramo. Proc World Lagomorph Conf, Univ Guelph, Ontario. 1979:549-558.
2817. Durant, P. 1983. Estudio écologico del Conejo silvestre, *Sylvilagus brasiliensis meridensis* (Lagomorpha: Leporidae), en Los Páramos de los Andes Venezolanos. Carib J Sci. 19:21-29.
2818. Durrant, BS. 1983. Reproductive studies of the oryx. Zoo Biol. 2:191-197.
2819. Durrell, G & Mallinson, JJC. 1968. The volcano rabbit or teporingo. Ann Rep Jersey Wildl Preserv Trust. 5:29-36.
2820. Durrell, G & Mallinson, JJC. 1970. The volcano rabbit *Romerolagus diazi*. Int Zoo Yearb. 10:118-122.
2821. Durrell, J. 1969. Keeping tropical squirrels in captivity. Ann Rep Jersey Wildl Preserv Trust. 6:53-55.
2821a. Dusek, GL, Mackie, RJ, Herriges, JD, Jr, et al. 1989. Population ecology of white-tailed deer along the lower Yellowstone River. Wildl Monogr. 104:1-68.

2821b. Dusi, JL. 1959. *Sorex longirostris* in eastern Alabama. J Mamm. 40:438-439.
2821c. Duthie, A. 1987. The endangered riverine rabbit: A research update. Afr Wildl. 41:168-171.
2822. Dutourné, B & Canivenc, R. 1971. Dynamique de la migration blastocytaire chez le Blaireau européen (*Meles meles* L). CR Hebd Séances Acad Sci Sér D Sci Nat. 273:2579-2582.
2823. Dutourné, B & Saboureau, M. 1983. An endocrine and histophysiological study of the testicular annual cycle in the hedgehog (*Erinaceus europaeus* L). Gen Comp Endocrinol. 50:324-332.
2824. Duval, M. 1895. Etudes sur l'embryologie des Chéiroptères. J Anat Physiol Paris. 31:93-160.
2825. Duval, M. 1895. Sur l'accouplement des Chauves-souris. CR Soc Biol. 2:135-136.
2826. Duvall, SW, Bernstein, IS, & Gordon, TP. 1976. Paternity and status in a rhesus monkey group. J Reprod Fert. 47:25-31.
2827. Dwyer, PD. 1963. Reproduction and distribution in *Miniopterus* (Chiroptera). Aust J Sci. 25:435-436.
2828. Dwyer, PD. 1963. The breeding biology of *Miniopterus schreibeisi blepotis* (Temminck) (Chiroptera) in northeastern New South Wales. Aust J Zool. 11:219-240.
2829. Dwyer, PD. 1966. Observations on *Chalinolobus dwyeri* (Chiroptera: Vespertilionidae) in Australia. J Mamm. 47:716-718.
2830. Dwyer, PD. 1966. Observations on the eastern horse-shoe bat in northeastern New South Wales. Helictite. 4:73-82.
2831. Dwyer, PD. 1968. The biology, origin, and adaptation of *Miniopterus australis* (Chiroptera) in New South Wales. Aust J Zool. 16:49-68.
2832. Dwyer, PD. 1968. The little bent-winged bat - Evolution in progess. Aust Nat Hist. 16:55-58.
2833. Dwyer, PD. 1970. Latitude and breeding season in a polyestrus [*sic*] species of *Myotis*. J Mamm. 51:405-410.
2834. Dwyer, PD. 1970. Social organization in the bat *Myotis adversus*. Science. 168:1006-1008.
2835. Dwyer, PD. 1975. Notes of *Dobsonia moluccensis* (Chiroptera) in the New Guinea highlands. Mammalia. 39:113-118.
2836. Dwyer, PD. 1975. Observations on the breeding biology of some New Guinea murid rodents. Aust Wildl Res. 2:33-45.
2837. Dwyer, PD. 1977. Notes on *Antechinus* and *Cercartetus* (Marsupialia) in the New Guinea highlands. Proc R Soc Queensl. 88:69-73.
2838. Dwyer, PD. 1977. Notes on *Antechinus* and *Cercartetus* (Marsupialia) in the New Guinea highlands. Proc R Soc Queensl. 88:69-73.
2839. Dwyer, PD. 1978. A study of *Rattus exulans* (Peale) (Rodentia: Muridae) in the New Guinea Highlands. Aust Wildl Res. 5:221-248.
2840. Dwyer, PD & Hamilton-Smith, E. 1965. Breeding caves and maternity colonies of the bent-winged bat in southeastern Australia. Helictite. 4:3-21.
2841. Dwyer, PD, Hockings, M, & Willmer, J. 1979. Mammals of Cooloola and Beerwah. Proc R Soc QLD. 90:65-84.
2842. Dymond, JR. 1936. Life history notes and growth studies on the little brown bat, *Myotis lucifugus lucifugus*. Can Field-Nat. 50:114-116.
2843. Dzhumagulov, I. 1981. Camelids Bibliography: Pregnancy duration in camels. Konevodstvo plodonocheniya Konnyi Sport No 10, 29. Arab Ctr Stud Arid Zones Dry Land, Damascus.
2844. Eadie, WR. 1939. A contribution to the biology of *Parascalops breweri*. J Mamm. 20:150-173.
2845. Eadie, WR. 1947. Homologies of the male accessory reproductive glands in *Sorex* and *Blarina*. Anat Rec. 98:347-359.
2846. Eadie, WR. 1947. The accessory reproductive glands of *Parascalops* with notes on homologies. Anat Rec. 97:239-251.
2847. Eadie, WR. 1948. Corpora amylacea in the prostatic secretion and experiments on the formation of a copulatory plug in some insectivores. Anat Rec. 102:259-271.
2848. Eadie, WR. 1948. The male accessory reproductive glands of *Condylura* with notes on a unique prostatic secretion. Anat Rec. 101:59-79.
2849. Eadie, WR. 1951. A comparative study of the male accessory genital glands of *Neurotrichus*. J Mamm. 32:36-43.
2850. Eadie, WR. 1962-63. Grassland studies of small mammals. The Cornell Plantations. 18:65-66.
2851. Eadie, WR & Hamilton, WJ, Jr. 1956. Notes on reproduction in the star-nosed mole. J Mamm. 37:223-231.
2852. Eadie, WR & Hamilton, WJ, Jr. 1958. Reproduction in the fisher in New York. NY Fish Game J. 5:77-83.
2853. Eads, RB, Wiseman, JS, & Menzies, GC. 1957. Observations concerning the Mexican free-tailed bat, *Tadarida mexicana*, in Texas. TX J Sci. 9:227-242.
2854. Eaglen, RH & Boskoff, KJ. 1978. The birth and early development of a captive sifaka, *Propithecus verreauxi coquereli*. Folia Primatol. 30:206-219.
2855. Eaglen, RH & Simons, EL. 1980. Notes on the breeding biology of thick-tailed and silvery galagos in captivity. J Mamm. 61:534-537.
2856. Eales, NB. 1925. External characters, skin and temporal gland of a foetal African elephant. Proc Zool Soc London. 1925:445-456.
2857. Ealey, EHM. 1963. The ecological significance of delayed implantation in a population of the hill kangaroo (*Macropus robustus*). Delayed Implantation. 33-47, Enders, AC, ed, Univ Chicago Press, Chicago.
2858. Ealey, EHM. 1964. L'Euro, ou Kangourou des Collines, *Macropus robustus*, dans le Nord-Ouest de l'Australie. Terre Vie. 111:3-19.
2859. Ealey, EHM. 1967. Ecology of the euro (*Macropus robustus* Gould) in northwest Australia: IV Age and growth. CSIRO Wildl Res. 12:67-80.
2860. Earl, Z. 1977. Female *Saccostomus campestris* (Rodentia: Muridae) carrying young in cheek pouches. J Mamm. 58:242-243.
2861. Earl, Z. 1978. Postnatal development of *Saccostomus campestris*. African Small Mammal Newsletter. 2:10-12.
2862. East, K & Lockie, JD. 1964. Observations on a family of weasels (*Mustela nivalis*) bred in captivity. Proc Zool Soc London. 143:359-363.
2863. East, K & Lockie, JD. 1965. Further observations on weasels (*Mustela nivalis*) and stoats (*Mustela erminea*) born in captivity. J Zool. 147:234-238.
2863a. Easterla, DA. 1968. Terrestrial home site of golden mouse. Am Mid Nat. 79:246-247.
2864. Easterla, DA. 1968. Paturition of Keen's myotis in southwestern Missouri. J Mamm. 49:770.
2864a. Easterla, DA. 1970. First record of the pocketed free-tailed bat for Coahuila Mexico and additional Texas records. TX J Sci. 22:92-93.
2865. Easterla, DA. 1971. Notes on young and adults of the spotted bat *Euderma maculatum*. J Mamm. 52:475.
2865a. Easterla, DA. 1972. Status of *Leptonycteris nivalis*, (Phyllostomatidae) in Big Bend National Park, Texas. Southwest Nat. 17:287-292.
2865b. Easterla, DA. 1973. Additional record of the pocketed free-tailed bat for New Mexico. TX J Sci. 24:543.

2866. Easterla, DA. 1976. Notes on the second and third newborn of the spotted bat, *Euderma maculatum* and comments on the species in Texas. Am Mid Nat. 96:499-501.
2867. Easterla, DA & Watkins, LC. 1969. Pregnant *Myotis sodalis* in northwestern Missouri. J Mamm. 50:372-373.
2868. Easterla, DA & Watkins, LG. 1970. Breeding of *Lasionycteris noctivagans* and *Nycticeius humeralis* in southwestern Iowa. Am Mid Nat. 84:254-255.
2869. Eastman, SAK, Makawiti, DW, Collins, WP, et al. 1984. Pattern of excretion of urinary steroid metabolites during the ovarian cycle and pregnancy in the marmoset monkey. J Endocrinol. 102:19-26.
2870. Eaton, GG. 1973. Social and endocrine determinants of sexual behavior in simian and prosimian females. Symp 4th Int Cong Primatol. 2:20-35.
2871. Eaton, GG & Resko, JA. 1974. Ovarian hormones and sexual behavior in *Macaca nemestrina*. J Comp Physiol Psychol. 86:919-925.
2872. Eaton, GG, Slob, A, & Resko, JA. 1973. Cycles of mating behaviour, oestrogen and progesterone in the thick-tailed bushbaby (*Galago crassicaudatus crassicaudatus*) under laboratory conditions. Anim Behav. 21:309-315.
2873. Eaton, RL. 1970. Notes on the reproductive biology of the cheetah *Acinonyx jubatus*. Int Zoo Yearb. 10:86-89.
2874. Eaton, RL. 1972. Reproductive age in lions (*Panthera leo*): A record. Mammalia. 36:165-166.
2875. Eaton, RL. 1974. The biology and social behavior of reproduction in the lion. The World's Cats. 3-55, Eaton, RL, ed, Feline Res Gp Woodland Pk Zoo, Seattle.
2876. Eaton, RL. 1976. The brown hyena: Review of biology, status and conservation. Mammalia. 40:377-399.
2877. Eaton, RL. 1977. Breeding biology and propagation of the ocelot (*Leopardus* [*Felis*] *pardalis*). Zool Garten NF. 47:9-23.
2878. Eaton, RL. 1977. Reproductive biology of the leopard. Zool Garten NF. 47:329-351.
2879. Eaton, RL. 1981. The ethology, progagation and husbandry of the brown hyena (*Hyaena brunnea*). Zool Garten NF. 51:123-149.
2880. Eaton, RL & Craig, SJ. 1973. Captive management and mating behavior of the cheetah. The World's Cats. 1:217-254. Eaton, RL, ed, Carn Res Inst Burke Mus, Univ Wash, Seattle.
2881. Eaton, RL & Velander, KA. 1977. Reproduction in the puma: Biology, behavior and ontogeny. The World's Cats. 3:45-70, Eaton, RL, ed, Carn Res Inst, Burke Mus, Univ WA, Seattle.
2882. Eaton, RL & York, W. 1971. Reproductive biology, and preliminary observations on mating preferences, in a captive lion *Panthera leo* population. Int Zoo Yearb. 11:198-202.
2883. Eaton, RL & Yost, R. 1978. The birth and development of cheetahs in a wild animal park. Zool Garten NF. 48:81-93.
2884. Eberhard, IH. 1978. Ecology of the koala, *Phascolarctos cinereus* (Goldfuss) Marsupialia: Phascolarctidae, in Australia. The Ecology of Arboreal Folivores. 315-327, Montgomery, GG, ed, Smithsonian Inst Press, Washington, DC.
2885. Eberhardt, L, Peterle, TJ, & Schofield, R. 1963. Problems in a rabbit population study. Wildl Monogr. 10:1-51.
2886. Eberhardt, LL & Siniff, DB. 1977. Population dynamics and marine mammal management policies. J Fish Res Bd Can. 34:183-190.
2887. Eberhardt, RL & Norris, KS. 1964. Observations of newborn Pacific gray whales on Mexican calving grounds. J Mamm. 45:88-95.
2888. Eberhart, JA, Herbert, J, Keverne, EB, et al. 1980. Some hormonal aspects of primate social behavior. Endocrinology 1980. 622-625, Cumming, IA, Funder, JW, & Mendelsohn, FAO, eds, Aust Acad Sci, Canberra.
2889. Eberhart, JA, Keverne, EB, & Miller, RE. 1980. Social influences on plasma testosterone levels in male talapoin monkeys. Horm Behav. 14:247-266.
2890. Eccles, TR & Shackleton, DM. 1979. Recent records of twinning in North American mountain sheep. J Wildl Manage. 43:974-976.
2891. Ecke, DH. 1955. The reproductive cycle of the Mearn's cottontail in Illinois. Am Mid Nat. 53:294-311.
2892. Eckhardt, RB. 1975. The relative body weights of Bornean and Sumatran orangutans. Am J Phys Anthrop. 42:349-350.
2893. Eckstein, P. 1950. The induction of progesterone withdrawal bleeding in spayed rhesus monkeys. J Endocrinol. 6:405-411.
2894. Eddy, CA, Garcia, RG, Kraemer, DC, et al. 1975. Detailed time course of ovum transport in the rhesus monkey (*Macaca mulatta*). Biol Reprod. 13:363-369.
2895. Eddy, CA, Shenken, RS, & Pauerstein, CJ. 1986. Pregnancy rate in time-mated, spontaneously cyclic rhesus monkeys, *Macaca mulatta*. J Reprod Fert. 78:705-710.
2896. Edge, ER. 1931. Seasonal activity and growth in the Douglas ground squirrel. J Mamm. 12:194-200.
2897. Edmond-Blanc, F. 1955. Note sur l'époque des mises-bas des Addax et des Oryx. Mammalia. 19:427-428.
2898. Edmonds, JW, Backholer, JR, & Shepherd, RCH. 1981. Some biological characteristics of a feral rabbit, *Oryctolagus cuniculus* (L) population of wild and domestic origin. Aust Wildl Res. 8:589-596.
2899. Edmunds, RM, Goertz, JW, & Linscombe, G. 1978. Age ratios, weights, and reproduction of the Virginia opossum in north Louisiana. J Mamm. 59:884-885.
2900. Edwards, RG. 1970. Are oocytes formed and used sequentially in the mammalian ovary?. Phil Trans R Soc London Biol Sci. 259B:103-105.
2901. Edwards, RL. 1946. Some notes on the life history of the Mexican ground squirrel in Texas. J Mamm. 27:105-115.
2902. Edwards, RL. 1963. Observations on the small mammals of the southeastern shore of Hudson Bay. Can Field-Nat. 77:1-12.
2903. Edwards, RY & Ritcey, RW. 1958. Reproduction in a moose population. J Wildl Manage. 22:261-268.
2904. Edwards, TJ. 1978. Breeding Prevost's squirrel *Callosciurus prevostii* in captivity. Int Zoo Yearb. 18:124.
2905. Egoscue, HJ. 1956. Preliminary studies of the kit fox in Utah. J Mamm. 37:351-357.
2906. Egoscue, HJ. 1957. The desert woodrat: A laboratory colony. J Mamm. 38:472-481.
2907. Egoscue, HJ. 1960. Laboratory and field studies of the northern grasshopper mouse. J Mamm. 41:99-110.
2908. Egoscue, HJ. 1962. Ecology and life history of the kit fox in Toole County, Utah. Ecology. 43:481-497.
2909. Egoscue, HJ. 1962. The bushy-tailed wood rat: A laboratory colony. J Mamm. 43:328-336.
2910. Egoscue, HJ. 1964. Ecological notes and laboratory life history of the canyon mouse. J Mamm. 45:387-396.
2911. Egoscue, HJ. 1970. A laboratory colony of the Polynesian rat, *Rattus exulans*. J Mamm. 51:261-266.
2912. Egoscue, HJ. 1972. Breeding the long-tailed pouched rat, *Beamys hindei*, in captivity. J Mamm. 53:296-302.
2913. Egoscue, HJ. 1979. *Vulpes velox*. Mamm Species. 122:1-5.
2914. Ehlers, K. 1957. Über die Seelöwin (*Eumetopias californianus*) "Inge" der Tiergrotten Bremerhaven. Zool Garten NF. 23:189-194.

2915. Ehlers, K. 1966. Über zwei weitere Klappmutzen (*Cystophora cristata* Erxl) in den "Tiergrotten Bremerhaven". Zool Garten NF. 32:1-19.
2916. Ehlers, K. 1973. Über die unbefriedigende Zucht von Eisbären (*Thalarctos maritimus* Phipps 1774) in Zoologischen Garten. Zool Garten NF. 43:48-58.
2917. Ehrat, H, Wissdorf, H, & Isenbügel, E. 1974. Postnatal Entwicklung von *Meriones unguiculatus* (Milne Edwards, 1867) vom Zeitpunkt der Geburt bis zum Absetzen der Jungtiere im Alter von 30 Tagen. Z Säugetierk. 39:41-50.
2918. Ehrlich, S. 1958. The biology of the nutria. Bamidgeh Bull Fish Cult Israel. 10(2):36-43.
2919. Ehrlich, S. 1966. Ecological aspects of reproduction in nutria *Myocastor coypus* Mol. Mammalia. 30:142-152.
2920. Eibatov, JM. 1976. Natural life span in *Phoca caspica*. Zool Zhurn. 55:1893-1896.
2921. Eiben, B, Scharla, S, Fischer, K, et al. 1984. Seasonal variation of serum 1,25-dihydorxy vitamin D_3 and alkaline phosphatase in relation to antler formation in the fallow deer (*Dama dama*). Acta Endocrinol. 107:141-144.
2922. Eibl-Eilesfeldt, I. 1984. The Galapagos seals Part 1 Natural history of the Galapagos sea lion *Zalophus californianus wollebaeki*, Silvertsen. Key Evironments: Galapagos. 207-214, Perry, R, ed, Pergamon, Oxford, UK.
2923. Eibl-Eibesfeldt, I. 1951. Beobachtungen zur Fortpflanzungsbiologie und Jugendentwicklung des Eichhörnchens (*Sciurus vulgaris* L). Z Tierpsychol. 8:370-400.
2924. Eibl-Eibesfeldt, I. 1955. Einige Bemerkungen über den Galapagos-Seelöwen, *Zalophus wollebaeki*, Sivertsen, 1953. Säugetierk Mitt. 3:102-103.
2925. Eimer, no initial. 1879. Über die Fortpflanzung der Fledermäuse. Zool Anz. 2:425-426.
2926. Einarsen, AS. 1956. Life of the mule deer. The Deer of North America. 363-390, Taylor, WP, ed, Wildl Management Inst, Washington, DC.
2927. Eisenberg, JF. 1963. The behavior of heteromyid rodents. Univ CA Publ Zool. 69:1-100.
2928. Eisenberg, JF. 1967. The heteromyid rodents. UFAW Handbook on the Care and Management of Laboratory Animals, 3rd ed. 390-395, E & S Livingston, Edinburgh.
2929. Eisenberg, JF. 1973. Reproduction in two species of spider monkeys, *Ateles fusciceps* and *Ateles geoffroyi*. J Mamm. 54:955-957.
2930. Eisenberg, JF. 1975. Tenrecs and solenodons in captivity. Int Zoo Yearb. 15:6-12.
2931. Eisenberg, JF. 1976. Communication mechanisms and social integration in the black spider monkey *Ateles fusciceps robustus* and related species. Smithsonian Contr Zool. 213:1-108.
2932. Eisenberg, JF. 1977. Comparative ecology and reproduction of New World monkeys. The Biology and Conservation of the Callitrichidae. 13-22, Kleiman, DG, ed, Smithsonian Inst Press, Washington, DC.
2933. Eisenberg, JF & Gould, E. 1966. The behavior of *Solenodon paradoxus* in captivity with comments on the behavior of other Insectivora. Zoologica. 51:49-58.
2934. Eisenberg, JF & Gould, E. 1970. The tenrecs: A study in mammalian behavior and evolution. Smithsonian Contrib Zool. 27:1-138.
2935. Eisenberg, JF & Isaac, DE. 1963. The reproduction of heteromyid rodents in captivity. J Mamm. 44:61-66.
2936. Eisenberg, JF & Lockhart, M. 1972. An ecological reconnaissance of Wilpattu National Park, Sri Lanka. Smithsonian Contrib Zool. 101:1-118.
2937. Eisenberg, JF & Maliniak, E. 1967. Breeding the murine opossum *Marmosa* sp in captivity. Int Zoo Yearb. 7:78-79.
2938. Eisenberg, JF & Maliniak, E. 1973. Breeding and captive maintenance of the lesser bamboo rat *Cannomys badius*. Int Zoo Yearb. 13:204-207.
2939. Eisenberg, JF & Maliniak, E. 1974. The reproduction of the genus *Microgale* in captivity. Int Zoo Yearb. 14:108-110.
2940. Eisenberg, JF & Maliniak, E. 1985. Maintenance and reproduction of the two-toed sloth *Choloepus didactylus* in captivity. The Evolution and Ecology of Armadillos, Sloths, and Vermilinguas. 327-331, Montgomery, GG, ed, Smithsonian Inst Press, Washington, DC.
2941. Eisenberg, JF, McKay, GM, & Jainudeen, MR. 1971. Reproductive behavior of the Asiatic elephant (*Elephas maximus maximus* L). Behaviour. 38:193-225.
2942. Eisenberg, JF & Muckenhirn, N. 1968. The reproduction and rearing of tenrecoid insectivores in captivity. Int Zoo Yearb. 8:106-110.
2943. Eisenstein, N & D'Amato, MR. 1972. Twinning in the New World monkey, *Cebus apella*. J Mamm. 53:406-407.
2944. Eisentraut, M. 1933. Biologische Studien im bolivianischen Chaco III Beitrag zur Biologie der Säugetierfauna. Z Säugetierk. 8:47-69.
2945. Eisentraut, M. 1936. Zur Fortpflanzungsbiologie der Fledermäuse. Z Morphol Ökol Tiere. 31:27-63.
2946. Eisentraut, M. 1937. Die Deutschen Fledermäuse; Eine Biologische Studie. Monographien der Wildsäugetiere. 1-184, Hilzheimer, M, ed, Verlag, Dr Paul Schops, Leipzig.
2947. Eisentraut, M. 1937. Die Wirkung niedriger Temperaturen auf die Embryonentwicklung bei Fledermäusen. Biol Zbl. 57:59-74.
2948. Eisentraut, M. 1941. Beitrag zur Oekologie Kameruner Chiropteren. Mitt Zool Mus Berlin. 25:245-273.
2949. Eisentraut, M. 1950. Beobachtung über Begattung bei Fledermäusen im Winterquartier. Zool Jb Syst. 78:297-300.
2950. Eisentraut, M. 1950. Beobachtungen über Lebensdauer und jährliche Verlustziffern bie Fledermäusen insbesondere bei *Myotis myotis*. Zool Jahrb Syst. 78:193-216.
2951. Eisentraut, M. 1952. Contribution a l'étude biologique de *Paraechinus aethiopicus* Ehrenb. Mammalia. 16:232-252.
2952. Eisentraut, M. 1952. Vom Kugel-Gürteltier (*Tolypeutes conurus*). Natur Volk. 82:43-48.
2953. Eisentraut, M. 1957. Beitrag zur Säugetierfauna des Kamerungebirges und Verbreitung der Arten in den verschiedenen Hohenstufen. Zool Jahrb Syst. 83:619-672.
2954. Eisentraut, M. 1961. Gefangenschaftsbeobachtungen an *Rattus* (*Praomys*) *morio* (Trouessart). Bonn Zool Beitr. 12:1-21.
2955. Eisentraut, M. 1963. Die Wirbeltiere des Kamerungebirges. Paul Parey, Berlin.
2956. Eisentraut, M. 1970. Beitrag zur Fortpflanzungsbiologie der Zwergbeutelratte *Marmosa murina* (Didelphidae, Marsupialia). Z Säugetierk. 35:159-172.
2957. El Omari, B, Lacroix, A, & Saboureau, M. 1989. Daily and seasonal variations in plasma LH and testosterone concentrations in the adult male hedgehog (*Erinaceus europaeus*). J Reprod Fert. 86:145-155.
2958. Elapata, SAI. 1969. The sexual behavior of wild elephants in Ceylon. Loris. 11:246-247.
2959. Elchlepp, JG. 1952. The urogenital organs of the cottontail rabbit (*Sylvilagus floridanus*). J Morphol. 91:169-198.
2960. Elder, JH. 1938. The time of ovulation in chimpanzees. Yale J Biol Med. 10:347-364.
2961. Elder, JH & Yerkes, RM. 1936. The sexual cycle of the chimpanzee. Anat Rec. 67:119-143.

2962. Elder, JH & Yerkes, RM. 1936. Chimpanzee births in captivity: A typical case history and report of sixteen births. Proc R Soc London. 120B:409-421.
2963. Elder, WH. 1952. Failure of placental scars to reveal breeding history in mink. J Wildl Manage. 16:107.
2964. Elder, WH & Finerty, JC. 1943. Gonadotropic activity of the pituitary gland in relation to the seasonal sexual cycle of the cottontail rabbit *Sylvilagus floridanus* Mearnsi. Anat Rec. 85:1-16.
2965. Eleftheriou, BE, Boehlke, KW, Zolovich, A, et al. 1966. Free plasma estrogens in the deer. Proc Soc Exp Biol Med. 121:88-90.
2966. Eleftheriou, BE, Bronson, FH, & Zarrow, MX. 1962. Interaction of olfactory and other environmental stimuli on implantation in the deer mouse. Science. 137:764.
2967. Eley, RM, Else, JG, Gulamhusein, N, et al. 1986. Reproduction in the vervet monkey (*Cercopithecus aethiops*): I Testicular volume, testosterone, and seasonality. Am J Primatol. 10:229-235.
2967a. Eley, RM, Tarara, RP, Worthman, CM, et al. 1989. Reproduction in the vervet monkey (*Cercopithecus aethiops*): III The menstrual cycle. Am J Primatol. 17:1-10.
2968. Elias, E, Bedrak, E, & Cohen, D. 1985. Induction of oestrus in the camel (*Camelus dromedarius*) during seasonal anoestrus. J Reprod Fert. 74:519-525.
2969. Elias, E, Bedrak, E, & Yagil, R. 1984. Estradiol concentration in the serum of the one-humped camel (*Camelus dromedarius*) during the various reproductive stages. Gen Comp Endocrinol. 56:258-264.
2970. Elias, E, Bedrak, E, & Yagil, R. 1984. Peripheral blood levels of progesterone in female camels during various reproductive stages. Gen Comp Endocrinol. 53:235-240.
2971. Elias, E & Cohen, D. 1986. Parturition in the camel (*Camelus dromedarius*) and some behavioral aspects of their newborn. Comp Biochem Physiol. 84A:413-419.
2972. Ellefson, JO. 1974. A natural history of white-handed gibbons in the Malayan peninsula. Gibbon & Siamang. 3:1-136.
2973. Ellendorff, F, Parvizi, N, & Folkmar, E. 1977. The miniature pig as an animal model in endocrine and neuroendocrine studies of reproduction. Lab Anim Sci. 27:822-830.
2974. Ellinwood, WE, Baughman, WL, & Resko, JA. 1982. The effects of gonadectomy and testosterone treatment on luteinizing hormone secretion in fetal rhesus monkeys. Endocrinology. 110:183-189.
2975. Ellinwood, WE, Norman, RL, & Spies, HG. 1984. Changing frequency of pulsatile luteinizing hormone and progesterone secretion during the luteal phase of the menstrual cycle of rhesus monkeys. Biol Reprod. 31:714-722.
2976. Ellinwood, WE & Resko, JA. 1980. Sex differences in biologically active and immunoreactive gonadotropins in the fetal circulation of rhesus monkeys. Endocrinology. 107:902-907.
2977. Ellinwood, WE & Resko, JA. 1983. Effect of inhibition of estrogen synthesis during the luteal phase on function of the corpus luteum in rhesus monkeys. Biol Reprod. 28:636-644.
2978. Elliot, DG. 1899. Soemmerring's gazelle (*Gazella soemmerringi*). Great and Small Game of Africa. 361-363, Bryden, HA, ed, Rowland Ward, London.
2979. Elliot, DG. 1901. A list of mammals obtained by Thaddeus Surber in North and South Carolina, Georgia and Florida. Field Columbian Mus Zool Ser. 3:31-57.
2980. Elliott, CL & Flinder, JT. 1991. *Spermophilus colombianus*. Mamm Species. 372:1-9.
2981. Elliott, L. 1978. Social behavior and foraging ecology of the eastern chipmunk (*Tamias striatus*) in the Adirondack Mountains. Smithsonian Contrib Zool. 265:1-107.
2982. Elliott, MW, Sehgal, PK, & Chalifoux, LV. 1976. Management and breeding of *Aotus trivirgatus*. Lab Anim Sci. 26:1037-1040.
2983. Ellis, LC & Balph, DF. 1976. Age and seasonal differences in the synthesis and metabolism of testosterone by testicular tissue and pineal HIOMT activity of Uinta ground squirrels (*Spermophilus armatus*). Gen Comp Endocrinol. 28:42-51.
2984. Ellis, LC, Palmer, RA, & Balph, DF. 1983. The reproductive cycle of male Uinta ground squirrels: Some anatomical and biochemical correlates. Comp Biochem Physiol. 74A:239-245.
2985. Eloff, FC. 1961. Observations on the migration and habits of the antelopes of the Kalahari Gemsbok Park, Part III. Koedoe. 4:18-30.
2986. Eloff, FC. 1973. Ecology and behavior of the Kalahari lion. The World's Cats. 90-126, Eaton, RL, ed, World Wildlife Safari, Winston, OR.
2987. Eloranta, E & Nieminen, M. 1986. Calving of the experimental reindeer herd in Kaamanen during 1970-85. Rangifer Sp Issue. 1:115-121.
2988. Else, JG, Eley, RM, Suleman, MA, et al. 1985. Reproductive biology of sykes and blue monkeys (*Cercopithecus mitis*). Am J Primatol. 9:189-196.
2989. Else, JG, Eley, RM, Wangula, C, et al. 1986. Reproduction in the vervet monkey (*Cercopithecus aethiops*): II Annual menstrual patterns and seasonality. Am J Primatol. 11:333-342.
2990. Elton, C, Ford, EB, Baker, JR, et al. 1931. The health and parasites of a wild mouse population. Proc Zool Soc London. 1931:657-721.
2991. ElWishy, AB. 1988. A study of the genital organs of the female dromedary (*Camelus dromedarius*). J Reprod Fert. 82:587-593.
2992. ElWishy, AB, Mobarak, AM, & Fouad, SM. 1972. The accessory genital organs of the one-humped male camel (*Camelus dromedarius*). Anat Anz. 131:1-12.
2993. ElWishy, AB & Omar, AM. 1975. On the relation between testes size and sperm reserves in the one-humped camel (*Camelus dromedarius*). Beitr Trop Subtrop Landwirts Veterinärmed. 13:391-398.
2994. Elwood, RW. 1975. Paternal and maternal behaviour in Mongolian gerbils. Anim Behav. 23:766-772.
2995. Emelyanov, IG. 1979. Ecological-and-morphological characteristics and peculiarities in the number dynamics of *Microtus socialis* Pall (Mammalia, Cricetidae) in the steppe zone of the Ukraine. Vestnik Zool (Kiev). 4:56-61.
2996. Emlen, JT, Jr & Davis, DE. 1948. Determination of reproductive rates in rat populations by examination of carcasses. Physiol Zool. 21:59-65.
2997. Emmons, LH. 1975. Ecology and behavior of African rainforest squirrels. Ph D Diss. Cornell Univ, Ithaca, NY.
2998. Emmons, LH. 1979. Observations on litter size and development of some African rainforest squirrels. Biotropica. 11:207-213.
2999. Emmons, LH. 1988. A field study of ocelots (*Felis pardalis*) in Peru. Terre Vie. 43:133-157.
3000. Encke, B. 1978. Sieben Jahre Tamanduas (*Tamandua tetradactyla*) in Krefelder Zoo. Zool Garten NF. 48:19-30.
3001. Encke, W. 1960. Birth and rearing of cheetahs at Krefeld Zoo. Int Zoo Yearb. 2:85-86.

3002. Encke, W. 1963. Bericht über Geburt und Aufzucht von Geparden, *Acinonyx jubatus* (Schreb), im Krefelder Zoo. Zool Garten NF. 27:177-181.
3003. Encke, W. 1970. Beobachtungen und Erfahrungen bei der Haltung und Zucht von Mähnenwölfen im Krefelder Tierpark. Fr Kölner Zoo. 13:69-75.
3004. Encke, W. 1970. Birth and hand-rearing of a bactrian camel *Camelus bactrianus* at Krefeld Zoo. Int Zoo Yearb. 10:90.
3005. Encke, W, Gandras, R, & Bieniek, HJ. 1970. Beobachtungen am Mähnenwolf (*Chrysocyon brachyurus*). Zool Garten NF. 38:47-67.
3006. Enders, AC. 1957. Histological observations on the chorio-allantoic placenta of the mink. Anat Rec. 127:231-245.
3007. Enders, AC. 1960. Development and structure of the villous haemochorial placenta of the nine-banded armadillo (*Dasypus novemcinctus*). J Anat. 94:34-45.
3008. Enders, AC. 1960. Electron microscopic observations on the villous haemochorial placenta of the nine-banded armadillo (*Dasypus novemcinctus*). J Anat. 94:205-215.
3009. Enders, AC. 1962. The structure of the armadillo blastocyst. J Anat. 96:39-48.
3010. Enders, AC, ed. 1963. Delayed Implantation. Univ Chicago. IL.
3011. Enders, AC. 1965. Comparative study of the fine structure of the trophoblast in several hemochorial placentas. Am J Anat. 116:29-67.
3012. Enders, AC. 1966. The reproductive cycle of the nine-banded armadillo (*Dasypus novemcinctus*). Symp Zool Soc London. 15:295-310.
3013. Enders, AC & Buchanan, GD. 1959. Some effects of ovariectomy and injection of ovarian hormones in the armadillo. J Endocrinol. 19:251-258.
3014. Enders, AC & Buchanan, GD. 1959. The reproductive tract of the female nine-banded armadillo. TX Rep Biol Med. 17:323-340.
3015. Enders, AC, Buchanan, GD, & Talmage, RV. 1958. Histological and histochemical observations on the armadillo uterus during the delayed and post-implantation periods. Anat Rec. 130:639-658.
3016. Enders, AC & Enders, RK. 1969. The placenta of the four-eyed opossum (*Philander opossum*). Anat Rec. 165:431-450.
3017. Enders, AC, Hendrickx, AG, & Schlafke, S. 1983. Implantation in the rhesus monkey: Initial penetration of the endometrium. Am J Anat. 167:275-298.
3018. Enders, AC & Schlafke, S. 1972. Implantation in the ferret: Epithelial penetration. Am J Anat. 133:291-316.
3019. Enders, AC, Schlafke, S, Hubbard, NE, et al. 1986. Morphological changes in the blastocyst of the western spotted skunk during activation from delayed implantation. Biol Reprod. 34:423-437.
3020. Enders, AC & Wimsatt, WA. 1968. Formation and structure of the hemodichorial chorio-allantoic placenta of the bat (*Myotis lucifugus lucifugus*). Am J Anat. 122:453-490.
3021. Enders, RK. 1930. Banana stowaways again. Science. 71:438-439.
3022. Enders, RK. 1930. Notes on some mammals from Barro Colorado Island, Canal Zone. J Mamm. 11:280-292.
3023. Enders, RK. 1931. Parturition in the agouti, with notes on several pregnant uteri. J Mamm. 12:390-396.
3024. Enders, RK. 1935. Mammalian life histories from Barro Colorado Island, Panama. Bull Mus Comp Zool. 78:386-502.
3024a. Enders, RK. 1945. Personal communication to Asdell, SA.
3025. Enders, RK. 1949. Field study of rats: Marianas and Palaus. Sci Invest Micronesia. 3:1-24.
3026. Enders, RK. 1952. Reproduction in the mink (*Mustela vison*). Proc Am Philos Soc. 96:691-755.
3027. Enders, RK. 1966. Attachment, nursing, and survival of young in some didelphids. Symp Zool Soc London. 15:195-203.
3028. Enders, RK & Enders, AC. 1963. Morphology of the female reproductive tract during delayed implantation in the mink. Delayed Implantation. 129-139, Enders, AC, ed, Univ Chicago Press, Chicago.
3029. Enders, RK & Leekley, JR. 1941. Cyclic changes in the vulva of the marten (*Martes americana*). Anat Rec. 79:1-5.
3030. Enders, RK & Pearson, OP. 1943. Shortening gestation by inducing early implantation with increased light in the marten. Am Fur Breeder. 15(7):18.
3031. Enders, RK & Pearson, OP. 1943. The blastocyst of the fisher. Anat Rec. 85:285-287.
3032. Enders, RK, Pearson, OP, & Pearson, AK. 1946. On certain aspects of reproduction in the fur seal. Anat Rec. 94:213-226.
3033. Endo, K. 1963. Notes on pregnant females of *Nyctalus lasiopterus aviator* Thomas. J Mamm Soc Japan. 2:61-62.
3034. Engelmann, C. 1938. Über die Groszsäuger Szetchwans, Sikongs und Osttibets. Z Säugetierk Sp Issue. 13:1-76.
3035. England, BG, Foote, WC, Cardozo, AG, et al. 1971. Estrus and mating behavior in the llama (*Llama glama*). Anim Behav. 19:722-726.
3036. England, BG, Foote, WC, Matthews, DH, et al. 1969. Ovulation and corpus luteum function in the llama (*Lama glama*). J Endocrinol. 45:505-513.
3037. Engle, ET. 1924. Breeding record of beaver. J Mamm. 5:202.
3038. Engle, ET. 1927. Notes on the sexual cycle of the Pacific Cetacea of the genera *Megaptera* and *Balaenoptera*. J Mamm. 8:48-51
3039. Engle, ET. 1946. No seasonal breeding cycle in dogs. J Mamm. 27:79-81.
3040. Engle, ET & Shelesnyak, MC. 1934. First menstruation and subsequent menstrual cycles of pubertal girls. Hum Biol. 6:431-453.
3041. English, PF. 1923. The dusky-footed wood rat (*Neotoma fuscipes*). J Mamm. 4:1-9.
3042. English, PF. 1932. Some habits of the pocket gopher, *Geomys breviceps breviceps*. J Mamm. 13:126-132.
3043. English, WL. 1934. Notes on the breeding of a douroucouli (*Aotus trivirgatus*) in captivity. Proc Zool Soc London. 1934:143-144.
3044. Englund, J. 1970. Some aspects of reproduction and mortality rates in Swedish foxes (*Vulpes vulpes*), 1961-63 and 1966-69. Viltrevy. 8:1-82.
3045. Engstrom, MD, Lee, TE, & Wilson, DE. 1987. *Bauerus dubiaquercus*. Mamm Species. 282:1-3.
3046. Ensley, PK, Wing, AE, Gosink, BB, et al. 1982. Application of noninvasive techniques to monitor reproductive function in a brown hyena (*Hyaena brunnea*). Zoo Biol. 1:333-343.
3047. Enzmann, EV, Saphir, NR, & Pincus, G. 1932. Delayed pregnancy in mice. Anat Rec. 54:325-341.
3048. Epelu-Opio, J, Kayanja, FIB, Muwazi, RT, et al. 1987. Oviduct of the impala, *Aepyceros melampus* (Lichtenstein, 1812). African J Ecology. 25:173-183.
3049. Epple, G. 1970. Maintenance, breeding, and development of marmoset monkeys (Callithricidae) in captivity. Folia Primatol. 12:56-76.

3050. Epple, G. 1972. Social behavior of laboratory groups of *Saguinus fuscicollis*. Saving the Lion Marmoset. 50-58, Bridgwater, DD, ed, The Wild Animal Propagation Trust, Wheeling, WV.
3051. Epple, G. 1975. Parental behaviour in *Saguinus fuscicollis* ssp (Callithricidae). Folia Primatol. 24:221-238.
3052. Epple, G, Alvaerio, MC, & St Andre, E. 1987. Sexual and social behavior of adult saddle-back tamarins (*Saguinus fuscicollis*), castrated as neonates. Am J Primatol. 13:37-49.
3053. Epple, G & Katz, Y. 1980. Social influences on first reproductive success and related behaviors in the saddle-back tamarin (*Sanguinus fuscicollis*, Callithrichidae). Int J Primatol. 1:171-183.
3054. Epple, G & Katz, Y. 1983. The saddle back tamarin and other tamarins. Reproduction in New World Primates. 117-148, Hearn, J, ed, MTP Press, Lancaster, England.
3055. Epple, G & Katz, Y. 1984. Social influences on estrogen excretion and ovarian cyclicity in saddle back tamarins (*Saguinus fuscicollis*). Am J Primatol. 6:215-227.
3055a. Epple, G, Küderling, I, Belcher, AM, et al. 1991. Estimation of immunoreactive testicular androgen metabolites in the urine of saddle-back tamarins. Am J Primatol. 23:87-98.
3056. Erb, L, Lasley, BL, Czekala, NM, et al. 1982. A dual radioimmunoassay and cytosol receptor binding assay for the measurement of estrogenic compounds allied to urine, fecal and plasma samples. Steroids. 39:33-46
3057. Erickson, AW. 1959. The age of self-sufficiency in the black bear. J Wildl Manage. 23:401-405.
3058. Erickson, AW. 1960. Supernumerary mammae in the black bear. J Mamm. 41:409.
3059. Erickson, AW, Mossman, HW, Hensel, RJ, et al. 1968. The breeding biology of the male brown bear (*Ursus arctos*). Zoologica. 53:85-105.
3060. Erickson, AW, Nellor, JE, & Petrides, G. 1964. The black bear in Michigan. MI Sta Univ Agr Exp Sta E Lansing Res Bull. 4:1-102.
3061. Erickson, HR. 1963. Reproduction, growth, and movement of muskrats inhabiting small water areas in New York State. NY Fish Game J. 10:90-117.
3062. Eriksen, E. 1977. Birth of two white rhinoceroses (*Ceratotherium simum simum*) at the Copenhagen Zoo. Zool Garten NF. 47:33-44.
3063. Eriksen, E. 1978. Geburten von asiatischen Elefanten (*Elephas maximus*) im Zoologischen Garten von Kopenhagen. Zool Garten NF. 48:421-432.
3064. Erikson, LB. 1961. Sampling of vaginal and cervical Muellerian duct derivatives in the adult rhesus monkey. Acta Anat. 47:233-260.
3065. Erikson, LB. 1964. Light-dark periodicity and the rhesus monkey menstrual cycle. Fertil Steril. 15:352-366.
3066. Erikson, LB. 1966. Determination of the period of ovulation in chimpanzee (*Pongidae pan* [*sic*]). Anat Rec. 154:500-501.
3067. Eriksson, M. 1980. Breeding in a laboratory colony of wood mice, *Apodemus sylvaticus* (Linné, 1758). Säugetierk Mitt. 28:29-80.
3068. Eriksson, M. 1980. The ovary of the wood mouse, *Apodemus sylvaticus* (Linné, 1758) at inhibition and regression. Säugetierk Mitt. 28:251-254.
3069. Eriksson, M. 1980. The adrenals of the wood mouse, *Apodemus sylvaticus* (Linné, 1758), in breeding and non-breeding conditions. Säugetierk Mitt. 28:255-256.
3070. Eriksson, M. 1981. The bulbourethral gland of the wood mouse, *Apodemus sylvaticus*. J Mamm. 62:375-378.
3071. Eriksson, M. 1983. Variation in size of testes and in spermatogenic activity in two species of wild mice, *Apodemus flavicollis* (Melchior) and *Apodemus sylvaticus* (L). Säugetierk Mitt. 31:55-60.
3072. Eriksson, M. 1983. The follicle population in the ovaries of two wild mice, *Apodemus flavicollis* and *Apodemus sylvaticus*. Can J Zool. 61:195-201.
3073. Eriksson, M. 1984. Winter breeding in three rodent species, the bank vole *Clethrionomys glareolus*, the yellow-necked mouse *Apodemus flavicollis* and the wood mouse *Apodemus sylvaticus* in southern Sweden. Holarctic Ecol. 7:428-429.
3074. Erken, AHM & Jacobi, EF. 1972. Successful breeding of lesser panda (*Ailurus fulgens* F Cuvier, 1825) and loss through inoculation. Bijdr Dierkd. 42:92-96.
3075. Erlinge, S. 1967. Home range of the otter *Lutra lutra* in southern Sweden. Oikos. 18:186-209.
3076. Erlinge, S. 1979. Adaptive significance of sexual dimorphism in weasels. Oikos. 33:233-245.
3077. Ernest, KA. 1986. *Nectomys squamipes*. Mamm Species. 265:1-5.
3078. Ernest, KA & Mares, MA. 1986. Ecology of *Nectomys squamipes*, the neotropical water rat, in central Brazil: Home range, habitat selection, reproduction and behaviour. J Zool. 210:599-612.
3079. Ernest, KA & Mares, MA. 1987. *Spermophilus tereticaudus*. Mamm Species. 274:1-9.
3080. Errington, PL. 1937. The breeding season of the muskrat in northwest Iowa. J Mamm. 18:333-337.
3081. Errington, PL. 1939. Observations on young muskrats in Iowa. J Mamm. 20:465-478.
3082. Errington, PL. 1963. Muskrat Populations. IA State Univ Press, Ames.
3083. Erskine, MS & Baum, MJ. 1982. Plasma concentrations of testosterone and dihydrotestosterone during perinatal development in male and female ferrets. Endocrinology. 111:767-772.
3084. Erskine, MS & Baum, MJ. 1984. Plasma concentrations of oestradiol and oestrone during perinatal development in male and female ferrets. J Endocrinol. 100:161-166.
3085. Erskine, MS, Tobet, SA, & Baum, MJ. 1988. Effect of birth on plasma testosterone, brain aromatase activity, and hypothalamic estradiol in male and female ferrets. Endocrinology. 122:524-530.
3086. Esch, GW, Beidleman, RC, & Long, LE. 1959. Early breeding of the black-tailed jack rabbit in southeastern Colorado. J Mamm. 40:442-443.
3087. Escherich, PC. 1981. Social biology of the bushy-tailed woodrat, *Neotoma cineria*. Univ CA Publs Zool. 110:1-132.
3088. Eshelkin, II. 1976. On the reproduction of *Alticola strelzovi* in the southeast Altai. Zool Zhurn. 55:437-442.
3089. Eshelman, BD & Cameron, GN. 1987. *Baiomys taylori*. Mamm Species. 285:1-7.
3090. Eskey, CR. 1934. Epidemiological study of plague in the Hawaiian Islands. US Publ Health Serv Bull #213. [Cited by Buxton 1936]
3091. Espmark, Y. 1964. Rutting behaviour in reindeer (*Rangifer tarandus* L). Anim Behav. 12:159-163.
3092. Espmark, Y. 1969. Mother-young relations and development of behaviour in roe deer *Capreolus capreolus*. Viltrevy. 6:461-540.
3092a. Essapian, FS. 1953. The birth and growth of a porpoise. Nat Hist. 62:392-399.

3093. Essapian, FS. 1962. Courtship in captive saddle-backed porpoises, *Delphinus delphis* L 1758. Z Säugetierk. 27:211-217.

3094. Essapian, FS. 1963. Observations on abnormalities of parturition in captive bottle-nosed dolphins *Tursiops truncatus* and concurrent behavior of other porpoises. J Mamm. 44:405-414.

3095. Esser, J. 1971. The biology and behaviour of the vaal rhebuck (*Pelea capreolus*). Roneoed Progress Reports. [Cited by Mentis, 1972]

3096. Estep, DQ & Dewsbury, DA. 1976. Copulatory behavior of *Neotoma lepida* and *Baiomys taylori*: Relationships between penile morphology and behavior. J Mamm. 57:570-573.

3097. Estep, DQ, Kenney, AM, & Dewsbury, DA. 1978. Copulatory behavior of roof rats (*Rattus rattus*). J Comp Physiol Psych. 92:322-334.

3098. Estes, JA. 1980. *Enhydra lutris*. Mamm Species. 133:1-8.

3099. Estes, RD. 1966. Behaviour and life history of the wildebeest *Connochaetes taurinus* (Burchell). Nature. 212:999-1000.

3100. Estes, RD. 1967. The comparative behavior of Grant's and Thompson's gazelles. J Mamm. 48:189-209.

3101. Estes, RD. 1974. Social organization of the African bovidae. The Behaviour of Ungulates and its Relation to Management. 166-205, Geist, V & Walther, F, eds, IUCN, Morges, Switzerland.

3102. Estes, RD. 1976. The significance of breeding synchrony in the wildebeest. E Afr Wildl J. 14:135-152.

3103. Estes, RD & Estes, RK. 1970. Preliminary report on the giant sable. Progress Report. 3:1-22. [Cited by Mentis 1972]

3104. Estes, RD & Estes, RK. 1974. The biology and conservation of the giant sable antelope, *Hippotragus niger variani* Thomas, 1916. Proc Acad Nat Sci Philadelphia. 126:73-104.

3105. Estes, RD & Estes, RK. 1979. The birth and survival of wildebeest calves. Z Tierpsychol. 50:45-95.

3106. Estrada, A & Estrada, R. 1976. Birth and breeding cyclicity in an outdoor living stumptail macaque (*Macaca arctoides*) group. Primates. 17:225-231.

3107. Estrada, A & Estrada, R. 1981. Reproductive seasonality in a free ranging troop of stumptail macaques (*Macaca arctoides*): A five-year report. Primates. 22:503-511.

3108. Etheredge, DR & Engstrom, MD. 1991. Notes on reproduction of the white-ankled mouse, *Peromyscus pectoralis*, in west-central Texas. TX J Sci. 43:205-207.

3108a. Etgen, AM & Fadem, BH. 1987. Estrogen binding macromolecules in hypothalamus-preoptic area of male and female gray short-tailed opossums (*Monodelphis domestica*). Gen Comp Endocrinol. 66:441-446.

3109. Etkin, W. 1954. Social behavior of the male black buck under zoo conditions. Anat Rec. 120:736.

3110. Evans, AM & McClure, PA. 1986. Effects of social environment on sexual maturation in female cotton rats (*Sigmodon hispidus*). Biol Reprod. 35:1081-1087.

3111. Evans, CS & Goy, RW. 1968. Social behaviour and reproductive cycles in captive ring-tailed lemurs (*Lemur catta*). J Zool. 156:181-197.

3112. Evans, DM. 1973. Seasonal variations in the body composition and nutrition of the vole *Microtus agrestis*. J Anim Ecol. 42:1-18.

3113. Evans, E. 1983. Breeding of the cotton-top tamarin *Saguinus oedipus oediupus*: A comparison with the common marmoset. Zoo Biol. 2:47-54.

3114. Evans, FC & Holdenried, R. 1943. A population study of the Beechey ground squirrel in central California. J Mamm. 24:231-260.

3115. Evans, J. 1970. About nutria and their control. US Dept Int Res Pub. 86:1-65.

3116. Evans, RD. 1962. Breeding characteristics of southeastern Missouri cottontails. Proc Ann Conf SE Assoc Game Fish Comm. 16:140-142.

3117. Evans, RD, Sadler, KC, Conaway, CH, et al. 1965. Regional comparisons of cottontail reproduction in Missouri. Am Mid Nat. 74:176-184.

3118. Evans, S. 1983. Breeding of the cotton-top tamarin *Saguinus oedipus oedipus*: A comparison with the common marmoset. Zoo Biol. 2:47-54.

3119. Evans, S. 1983. The pair-bond of the common marmoset, *Callithrix jacchus jacchus*: An experimental investigation. Anim Behav. 31:651-658.

3119a. Evans, S & Hodges, JK. 1984. Reproductive status of adult daughters in family groups of common marmosets (*Callithrix jacchus jacchus*). Folia Primat. 42:127-133.

3120. Evans, S & Poole, TB. 1983. Pair-bond formation and breeding success in the common marmoset *Callithrix jacchus jacchus*. Int J Primatol. 4:83-96.

3121. Evans, SM, Tyndale-Biscoe, CH, & Sutherland, RL. 1980. Control of gonadotrophin secretion in the female tammar wallaby (*Macropus eugenii*). J Endocrinol. 14:13-23.

3122. Everard, COR & Tikasingh, ES. 1973. Ecology of the rodents, *Proechimys guyannensis trinitatis* and *Oryzomys capito velutinus* on Trinidad. J Mamm. 54:875-886.

3123. Everett, NB. 1942. The origin of ova in the adult opossum. Anat Rec. 82:77-91.

3124. Everitt, BJ & Herbert, J. 1971. The effects of dexamethasone and androgens on sexual receptivity of female rhesus monkeys. J Endocrinol. 51:575-588.

3125. Everitt, RD & Krogman, BD. 1979. Sexual behavior of bowhead whales observed of the north coast of Alaska. Arctic. 32:277-280.

3126. Evtushevsky, NN. 1974. Reproduction of *Cervus nippon hortularum* SW under conditions of the middle Dnieper Area. Vestn Zool. 7(4):23-28.

3127. Ewart, JC. 1878. The fecundity and placentation of the Shanghai river deer. J Anat. 12:225-228.

3128. Ewel, KC. 1972. Patterns of change in the reproductive organs of the male pocket gopher, *Geomys pinetis*. J Reprod Fert. 30:1-6.

3129. Ewer, RF. 1963. The behaviour of the meerkat *Suricata suricata* (Schreber). Z Tierpsychol. 20:570-607.

3130. Ewer, RF. 1967. The behaviour of the African giant rat (*Cricetomys gambianus* Waterhouse). Z Tierpsychol. 24:6-79.

3131. Ewer, RF. 1968. A preliminary survey of the behaviour in captivity of the dasyurid marsupial, *Sminthopsis crassicaudata* (Gould). Z Tierpsychol. 25:319-365.

3132. Ewer, RF. 1969. Form and function in the grass cutter, *Thryonomys swinderianus* Temm (Rodentia, Thryonomyidae). Ghana J Sci. 9:131-141.

3133. Ewer, RF. 1971. The biology and behaviour of a free living population of black rats (*Rattus rattus*). Anim Behav Monogr. 4:127-174.

3134. Ewer, RF. 1973. The Carnivores. Cornell Univ Press, Ithaca.

3135. Ewer, RF & Wemmer, C. 1974. The behaviour in captivity of the African civet, *Civettictis civetta* (Schreber). Z Tierpsychol. 34:359-394.

3136. Fadem, BH. 1985. Evidence for the activation of female reproduction by males in a marsupial, the gray short-tailed opossum (*Monodelphis domestica*). Biol Reprod. 33:112-116.
3137. Fadem, BH. 1987. Activation of estrus by pheromones in a marsupial: Stimulus control and endocrine factors. Biol Reprod. 36:328-332.
3138. Fadem, BH. 1989. The effects of pheromonal stimuli on estrus and peripheral plasma estradiol in female gray short-tailed opossums (*Monodelphis domestica*). Biol Reprod. 41:213-217.
3139. Fadem, BH & Rayve, RS. 1985. Characteristics of the oestrous cycle and influence of social factors in grey short-tailed opossums (*Monodelphis domestica*). J Reprod Fert. 73:337-342.
3140. Fadem, BH, Rayve, RS, Trupin, GL, et al. 1982. The estrous cycle of the gray short-tailed opossum (*Monodelphis domestica*). unpublished ms.
3141. Fadem, BH & Tesoriero, JV. 1986. Inhibition of testicular development and feminization of the male genitalia by neonatal estrogen treatment in a marsupial. Biol Reprod. 34:771-776.
3142. Fadem, BH, Trupin, GL, Maliniak, E, et al. 1982. Care and breeding of the gray, short-tailed opossum (*Monodelphis domestica*). Lab Anim Sci. 32:405-409.
3143. Fagerstone, K. 1988. The annual cycle of Wyoming ground squirrels in Colorado. J Mamm. 69:678-687.
3144. Fairall, N. Personal communication. [cited by Mentis, MT. 1972]
3145. Fairall, N. 1968. The reproductive seasons of some mammals in the Kruger National Park. Zool Afr. 3:189-210.
3146. Fairall, N. 1969. Prenatal development of the impala, *Aepyceros melampus* Licht. Koedoe. 12:97-103.
3147. Fairall, N. 1970. Research on the reproduction of wild ungulates. Proc S Afr Soc Anim Prod. 9:57-61.
3148. Fairall, N. 1972. Behavioural aspects of the reproductive physiology of the impala, *Aepyceros melampus* (Licht). Zool Afr. 7:167-174.
3149. Fairall, N. 1983. Production parameters of the impala, *Aepyceros melampus*. S Afr J Anim Sci. 13:176-179.
3150. Fairbairn, DJ. 1977. Why breed early? A study of reproductive tactics in *Peromyscus*. Can J Zool. 55:862-871.
3151. Fairbanks, LA & McGuire, MT. 1984. Determinants of fecundity and reproductive success in captive vervet monkeys. Am J Primatol. 7:27-38.
3152. Fairbanks, LA & McGuire, MT. 1986. Age, reproductive value, and dominance related behaviour in vervet monkey females: Cross-generational influences on social relationships and reproduction. Anim Behav. 34:1710-1721.
3153. Fairbanks, LA & McGuire, MT. 1987. Mother-infant relationships in vervet monkeys: Response to new adult males. Int J Primatol. 8:351-366.
3154. Fairley, JS. 1970. Foetal number and resorption in wood mice from Ireland. J Zool. 161:276-277.
3155. Fairley, JS. 1970. The food reproduction, form, growth, and development of the fox *Vulpes vulpes* (L) in northeast Ireland. Proc R Irish Acad. 69B(5):103-137.
3156. Fairley, JS. 1971. Notes on the breeding of the fox (*Vulpes vulpes*) in County Galway, Ireland. J Zool. 164:262-263.
3157. Fajer, AB & Bechini, D. 1971. Pregnenolone and progesterone concentrations in ovarian venous blood of the artificially ovulated squirrel monkey (*Saimiri sciureus*). J Reprod Fert. 27:193-199.
3158. Falconer, J, Mitchell, MD, Mountford, LA, et al. 1980. Plasma oxytocin concentrations during the menstrual cycle in the rhesus monkey, *Macaca mulatta*. J Reprod Fert. 59:69-72.
3159. Falk, JW & Millar, JS. 1987. Reproduction by female *Zapus princeps* in relation to age, size, and body fat. Can J Zool. 65:568-571.
3160. Falla, RA. 1965. Birds and mammals of the subantarctic islands. Proc NZ Ecol Soc. 12:63-68.
3161. Falvo, RE & Nalbandov, AV. 1974. Radioimmunoassay of peripheral testosterone in males from eight species using a specific antibody without chromatography. Endocrinology. 95:1466-1468.
3162. Fandeyev, AA. 1960. Saiga lambing in the right bank area of the Volga. Zool Zhurn. 39:906-911.
3163. Fanning, FD. 1982. Reproduction, growth and development in *Ningaui* sp (Dasyuridae, Marsupialia) from the Northern Territory. Carnivorous Marsupials. 1:23-37, Archer, M, ed, R Zool Soc NSW. Mossman, NSW.
3164. Farah, Z & Farah-Riesen, M. 1985. Separation and characterization of major components of camel *Camelus dromedarius* milk casein. Milchwissenschaft. 40:669-671.
3165. Farentinos, RC. 1972. Observations on the ecology of the tassel-eared squirrel. J Wildl Manage. 36:1234-1239.
3166. Farentinos, RC. 1972. Social dominance and mating activity in the tassel-eared squirrel (*Sciurus alberti ferreus*). Anim Behav. 20:316-326.
3167. Farmer, C. 1970. Collared peccary *Tayassu tajacu*. Ann Rep Jersey Wildl Preserv Trust. 7:16-19.
3168. Farmer, C. 1971. Notes on the breeding and hand-rearing of the serval cat *Felis leptailurus serval*. Ann Rep Jersey Wildl Preserv Trust. 8:37-40.
3169. Farmer, SW, Licht, P, Bona Gallo, A, et al. 1981. Studies on several marsupial anterior pituitary hormones. Gen Comp Endocrinol. 43:336-345.
3170. Farmer, SW & Papkoff, H. 1974. Studies on the anterior pituitary hormones of the kangaroo. Proc Soc Exp Biol Med. 145:1031-1036.
3171. Farrell, BC & Christian, DP. 1987. Energy and water requirements of lactation in the North American porcupine, *Erethizon dorsatum*. Comp Biochem Physiol. 88A:695-700.
3172. Farrell, DJ & Wood, AJ. 1968. The nutrition of the female mink (*Mustela vison*) I The metabolic rate of the mink. Can J Zool. 46:41-45.
3173. Farris, EJ. 1946. The time of ovulation in the monkey. Anat Rec. 95:337-345.
3174. Farris, EJ. 1950. The opossum. The Care and Breeding of Lab Anim. 256-267, Farris, EJ, ed, John Wiley & Sons, New York.
3175. Farst, DD, Thompson, DP, Stones, GA, et al. 1980. Maintenance and breeding of duikers *Cephalophus* spp at Gladys Porter Zoo, Brownsville. Int Zoo Yearb. 20:93-99.
3175a. Faulkes, CG & Abbott, DH. 1991. Social control of reproduction in breeding and non-breeding male naked mole-rats (*Heterocephalus glaber*). J Reprod Fert. 93:427-435.
3176. Faulkes, CG, Abbott, DH, & Jarvis, JUM. 1990. Social suppression of ovarian cyclicity in captive and wild colonies of naked mole-rats, *Heterocephalus glaber*. J Reprod Fert. 88:559-568.
3177. Faulkes, CG, Abbott, DH, Jarvis, JUM, et al. 1990. LH responses of female naked mole-rats, *Heterocephalus glaber*, to single and multiple doses of exogenous GnRH. J Reprod Fert. 89:317-323.
3178. Faust, R. 1961. Milchanalyse vom Warzenschwein (*Phacochoerus aethiopicus* L). Zool Garten NF. 26:121.
3179. Faust, R & Faust, I. 1959. Bericht über Aufzucht und Entwicklung eines isolierten Eisbären, *Thalarctos maritimus* (Phipps). Zool Garten NF. 25:143-165.

3180. Faust, R & Scherpner, C. 1967. A note on the breeding of the maned wolf *Chrysocyon brachyurus* at Frankfurt Zoo. Int Zoo Yearb. 7:119.
3181. Fay, FH. 1981. Walrus *Odobenus rosmarus* (Linnaeus, 1758). Handbook of Marine Mammals. 1:1-23, Ridgway, SH & Harrison, RJ, eds, Academic Press, New York.
3182. Fay, FH. 1982. Ecology and biology of the Pacific walrus, *Odobenus rosmarus divergens* Illiger. N Am Fauna. 74:1-279.
3183. Fay, FH. 1985. *Odobenus rosmarus*. Mamm Species. 238:1-7.
3184. Fayenuwo, JO & Halstead, LB. 1974. Breeding cycle of the straw-colored fruit bat, *Eidolon helvum*, at Ile-Ife, Nigeria. J Mamm. 55:453.
3185. Fazleabas, AT, Mead, RA, Rourke, AW, et al. 1984. Presence of an inhibitor of plasminogen activator in uterine fluid of the western spotted skunk during delayed implantation. Biol Reprod. 30:311-322.
3186. Fedak, MA & Anderson, SS. 1982. The energetics of lactation: Accurate measurements from a large wild mammal, the grey seal *Halichoerus grypus*. J Zool. 198:473-479.
3187. Fedigan, LM, Fedigan, L, Gouzoules, S, et al. 1986. Lifetime reproductive success in female Japanese macaques. Folia Primatol. 47:143-157.
3188. Fedoseev, GA. 1964. On the embryonic and postembryonic growth and sexual maturation in *Phoca hispida ochotensis* Pall. Zool Zhurn. 43:1228-1235.
3189. Fedosenko, AK & Blank, DA. 1983. On behaviour of the Siberian ibex (*Capra sibirica*) during the period of reproduction in north Tian-Shan. Zool Zhurn. 61:428-435.
3190. Fedyk, S. 1980. Chromosone polymorphism in a population of *Sorex aranaeus* at Bialowieza. Folia Biol (Kraków). 28:83-120.
3191. Feer, F. 1979. Observations écologiques sur le Néotrague de Bates (*Neotragus batesi* de Winton 1903, Artiodactyle, ruminant, bovidé) du nord-est du Gabon. Terre Vie. 33:159-239.
3192. Feer, F. 1982. Maturité sexuelle et cycle annuel de reproduction de *Neotragus batesi* de Winton 1903 (Bovidé forestier africain). Mammalia. 46:65-74.
3193. Feist, DD & Feist, CF. 1986. Effects of cold and melatonin on thermogenesis, body weight and reproductive organs in Alaskan red-backed voles. J Comp Physiol B. 156:741-746.
3194. Feist, JD & McCullough, DR. 1975. Reproduction in feral horses. J Reprod Fert Suppl. 23:13-18.
3195. Feist, JD & McCullough, DR. 1976. Behavior patterns and communication in feral horses. Z Tierpsychol. 41:337-371.
3196. Feito, R & Gallardo, M. 1982. Sperm morphology of the Chilian species of *Ctenomys* (Octodontidae). J Mamm. 63:658-661.
3197. Feldhamer, GA. 1979. Age, sex ratios and reproductive potential in black-tailed jackrabbits. Mammalia. 43:473-478.
3198. Feldhamer, GA. 1980. *Cervus nippon*. Mamm Species. 128:1-7.
3199. Feldhamer, GA, Farris-Renner, KC, & Barker, CM. 1988. *Dama dama*. Mamm Species. 317:1-8.
3200. Feldhamer, GA & Rochelle, JA. 1982. Mountain beaver. Wild Mammals of North America Biology, Management and Economics. 167-175, Chapman, JA & Feldhamer, GA, eds, Johns Hopkins Univ Press, Baltimore.
3201. Feldman, HW. 1935. Notes on two species of wood rats in captivity. J Mamm. 16:300-303.
3202. Fellner, K. 1965. Natural rearing of clouded leopards *Neofelis nebulosa* at Frankfurt Zoo. Int Zoo Yearb. 5:111-113.
3203. Fellner, K. 1970. Einige Beobachtungen zur zweiten natürlichen Nebelparder-Aufzucht im Vergleich zur ersten. Zool Garten NF. 38:68-72.
3204. Felten, H. 1955. Fledermäuse (Mammalia, Chiroptera) aus El Salvador Teil 1. Senck Biol. 36:271-285.
3205. Felten, H. 1956. Fledermäuse (Mammalia, Chiroptera) aus El Salvador Teil 2. Senck Biol. 37:69-86.
3206. Felten, H. 1956. Fledermäuse (Mammalia, Chiroptera) aus El Salvador Teil 3. Senck Biol. 37:179-212.
3207. Felten, H. 1956. Fledermäuse (Mammalia, Chiroptera) aus El Salvador Teil 4. Senck Biol. 37:341-367.
3208. Felten, H. 1957. Fledermäuse (Mammalia, Chiroptera) aus El Salvador Teil 5. Senck Biol. 38:1-22.
3209. Felten, H. 1957. Nagetiere (Mammalia, Rodentia) aus El Salvador Teil 1. Senck Biol. 38:145-155.
3210. Felten, H. 1958. Nagetiere (Mammalia, Rodentia) aus El Salvador Teil 2. Senck Biol. 39:1-10.
3211. Felten, H. 1958. Nagetiere (Mammalia, Rodentia) aus El Salvador Teil 3. Senck Biol. 39:133-144.
3212. Felten, H. 1958. Weitere Säugetiere aus El Salvador (Mammalia: Marsupialia, Insectivora, Primates, Edentata, Lagomorpha Carnivora und Artiodactyla). Senck Biol. 39:213-326.
3213. Feniuk, BK & Kazantzeva, JM. 1937. The ecology of *Dipus sagitta*. J Mamm. 18:409-426.
3214. Fenton, MB. 1970. Population studies of *Myotis lucifugus* (Chiroptera: Vespertilionidae) in Ontario. R Ont Mus Life Sci Contr. 77:1-34.
3215. Fenton, MB. 1975. Observations on the biology of some Rhodesian bats, including a key to the Chiroptera of Rhodesia. R Ont Mus Life Sci Contr. 104:1-27.
3216. Fenton, MB & Barclay, RMR. 1980. *Myotis lucifugus*. Mamm Species. 142:1-8.
3217. Fenton, MB, van Zyll de Jong, CG, Bell, GP, et al. 1980. Distribution, parturition dates, and feeding of bats in south-central British Columbia. Can Field-Nat. 94:416-420.
3218. Feremutsch, K & Strauss, F. 1949. Beitrag zum weiblichen Genitalzyklus der madagassischen Centetinen. Rev Suisse Zool. 56 Suppl 1:1-110.
3219. Feremutsch, K. 1948. Der praegravide Genitaltrakt und die Praeimplantation. Rev Suisse Zool. 55:567-622.
3220. Ferguson, B, Webster, DG, & Dewsbury, DA. 1984. Stimulus control of ovulation in red-backed voles (*Clethrionomys gapperi*). Bull Psychon Soc. 22:365-367.
3221. Ferguson, DR. 1969. The genetic ditribution of vasopressins in the peccary (*Tayassu angulatus*) and wart hog (*Phacochoerus aethiopicus*). Gen Comp Endocrinol. 12:609-613.
3222. Ferguson, DR & Pickering, BT. 1969. Arginine and lysine vasopressins in the *Hippopotamus* neurohypophysis. Gen Comp Endocrinol. 13:425-429.
3223. Ferguson, JWH, Nel, JA, & DeWet, MJ. 1983. Social organization and movement patterns of black-backed jackals *Canis mesomelas* in South Africa. J Zool. 199:487-502.
3224. Ferin, M, Carmel, PW, Zimmerman, EA, et al. 1974. Location of intrahypothalamic estrogen-responsive sites influencing LH secretion in the female rhesus monkey. Endocrinology. 95:1059-1068.
3225. Ferin, M, Dyrenfurth, I, Cowchock, S, et al. 1974. Active immunization to 17β-estradiol and its effects upon the reproductive cycle of the rhesus monkey. Endocrinology. 94:765-776.
3226. Ferin, M, Rosenblatt, H, Carmel, PW, et al. 1979. Estrogen-induced gonadotropin surges in female rhesus monkeys after pituitary stalk section. Endocrinology. 104:50-52.

3227. Ferin, M, van Vugt, D, & Chernick, A. 1983. Central nervous system peptides and reproductive fuction in primates. Neuroendocrine Aspects of Reproduction. 69-91, Norman, RL, ed, Academic Press, New York.
3228. Ferkin, MH. 1987. Parental care and social interactions of captive plateau mice, *Peromyscus melanophrys*. J Mamm. 68:266-274.
3229. Ferkin, MH. 1987. Reproductive correlates of aggressive behavior in female *Peromyscus melanophrys*. J Mamm. 68:698-701.
3230. Fernandez Baca, S, Sumar, J, Novoa, C, et al. 1973. Relacíon entre la ubicación del cuerpo lúteo y la localización del embrión de la alpaca. Rev Invest Pecuarias. 2:131-135.
3231. Fernandez, M. 1909. Beiträge zur Embryologie der Gürteltiere I Zur Keimblatterinversion und spezifischen Polyembryonie der Mulita (*Tatusia hybrida* Desm). Morphol Jb. 39:302-333.
3232. Fernandez, M. 1914. Zur Anordnung der Embryonen und Form der Placenta bei *Tatusia novemcincta*. Anat Anz. 46:253-258.
3233. Fernandez-Baca, S, Hansel, W, & Novoa, C. 1970. Embryonic mortality in the alpaca. Biol Reprod. 3:243-251.
3234. Fernandez-Baca, S, Hansel, W, & Novoa, C. 1970. Corpus luteum function in the alpaca. Biol Reprod. 3:252-261.
3235. Fernandez-Baca, S, Hansel, W, Saatman, R, et al. 1979. Differential luteolytic effects of right and left uterine horns in the alpaca. Biol Reprod. 20:586-595.
3236. Ferrell, CS & Wilson, DE. 1991. *Platyrrhinus helleri*. Mamm Species. 373:1-5.
3237. Ferris, WB. 1906. Note on a Malay tapir (*Tapirus indicus*) in captivity. J Bombay Nat Hist Soc. 17:242-243.
3238. Ferron, J. 1981. Comparative ontogeny of behaviour in four species of squirrels (Sciuridae). Z Tierpsychol. 55:193-216.
3239. Ferron, J & Ouellet, JP. 1985. Developpement physique post-natal chez le grand Polatouche (*Glaucomys sabrinus*). Can J Zool. 63:2548-2552.
3240. Ferron, J & Ouellet, JP. 1991. Physical and behavioral postnatal development of woodchucks (*Marmota monax*). Can J Zool. 69:1040-1047.
3241. Ferron, J & Prescott, J. 1977. Gestation, litter size, and number of litters of the red squirrel (*Tamiasciurus hudsonicus*) in Quebec. Can Field-Nat. 91:83-84.
3242. Festa-Bianchet, F. 1981. Reproduction in yearling female Columbian ground squirrels (*Spermophilus columbianus*). Can J Zool. 59:1032-1035.
3243. Festa-Bianchet, M. 1988. Age-specific reproduction of bighorn ewes in Alberta, Canada. J Mamm. 69:157-160.
3244. Festa-Bianchet, M & King, WJ. 1991. Effects of litter size and population dynamics on juvenile and maternal survival in Columbian ground squirrels. J Anim Ecol. 60:1077-1090.
3245. Fevold, HR & Wright, PL. 1969. Steriod metabolism by badger (*Taxidea taxus*) ovarian tissue homogenates. Gen Comp Endocrinol. 13:60-67.
3246. Fèvre, J. 1967. Oestrogènes urinaires chez la brebis gestante ovariectomisée. Ann Biol Anim Bioch Biophys. 7:29-32.
3247. Field, CR & Blankenship, LH. 1973. Nutrition and reproduction of Grant's and Thomson's gazelles, Coke's hartebeest and giraffe in Kenya. J Reprod Fert Suppl. 19:287-301.
3248. Fiero, BC & Verts, BJ. 1986. Age-specific reproduction in raccoons in northwestern Oregon. J Mamm. 67:169-171.
3249. Figala, J, Hoffmann, K, & Goldau, G. 1973. Zur Jahresperiodik beim Dsungarischen Zwerghamster *Phodopus sungorus* Pallas. Oecologia. 12:89-118.
3250. Figala, J. 1964. The reproduction and population structure of the black rat, *Rattus rattus* (L), in the Czechoslovak habitats. Vestn Čcsl Spol Zool. 28:48-67.
3251. Findlay, L. 1982. The mammary glands of the tammar wallaby (*Macropus eugenii*) during pregnancy and lactation. J Reprod Fert. 65:59-66.
3252. Findlay, L, Ward, KL, & Renfree, MB. 1983. Mammary gland lactose, plasma progesterone and lactogenesis in the marsupial *Macropus eugenii*. J Endocrinol. 97:425-436.
3253. Finlay, M. 1974. Comparative responses of *Peromyscus* (white-footed mice) to estradiol benzoate. Comp Biochem Physiol. 48A:229-236.
3254. Findley, JS. 1954. Reproduction in two species of *Myotis* in Jackson Hole, Wyoming. J Mamm. 35:434.
3254a. Findley, JS & Jones, C. 1961. New United States record of the Mexican big-eared bat. J Mamm. 42:97.
3254b. Findley, JS & Jones, C. 1965. Northernmost records of some neotropical bat genera. J Mamm. 46:330-331.
3255. Findley, JS & Wilson, DE. 1974. Observations on the neotropical disk-winged bat, *Thyroptera tricolor* Spix. J Mamm. 55:562-571.
3256. Finkel, MP. 1945. The relation of sex hormones to pigmentation and to testis descent in the opossum and ground squirrel. Am J Anat. 76:93-151.
3257. Finlayson, HH. 1932. *Caloprymnus campestris* its recurrence and characters. Trans Proc R Soc S Aust. 56:146-167.
3258. Finlayson, HH. 1940. On central Australian mammals Part 1 The Muridae. Trans R Soc S Aust. 64:125-136.
3259. Finlayson, HH. 1944. A further account of the murid, *Pseudomys* (*Gyomys*) *apodemoides*. Trans R Soc S Aust. 68:210-224.
3260. Finlayson, HH. 1960. On *Rattus greyi* Gray and its derivatives. Trans R Soc S Aust. 83:123-148.
3261. Finley, KJ, Miller, GW, Davis, RA, et al. 1983. A distinctive large breeding population of ringed seals (*Phoca hispida*) inhabiting the Baffin Bay pack ice. Arctic. 36:162-173.
3262. Finley, RB, Jr. 1958. The wood rats of Colorado: Distribution and ecology. Univ KS Publs Mus Nat Hist. 10:213-552.
3263. Fischer, E. 1900. Zur Entwicklungsgeschichte des Dachses. Mitt Bad Zool Vereins Karlsruhe. 6:105-111.
3264. Fischer, E. 1931. Die Entwicklungsgeschichte des Dachses mid die Frage der Zwillingsbildung. Verhandl Anat Anz. 40:22-34.
3265. Fischer, GM. 1978. Los pequenos mamiferos de Chile: marsupiales, quiropteros, edentados y roedores. Gayana Zoologia. 40:1-342.
3266. Fischer, K. 1983. Untersuchungen zur Fortpflanzungsfähigkeit von jungen weiblichem und männlichem Damwild (*Dama dama* L). Z Jagdwiss. 29:137-142.
3267. Fischer, K, Gosch, B, & Mennerich, B. 1985. Individueller Pubertätsverlauf bei in Juni und verspätet im September gestezten Damhirschen (*Dama dama*). Z Jagdwiss. 31:211-221.
3268. Fischer, TV. 1971. Placentation in the American beaver (*Castor canadensis*). Am J Anat. 131:159-183.
3269. Fischer, TV & Floyd, AD. 1972. Placental development in the Mongolian gerbil (*Meriones unguiculatus*) I Early development to the time of chorio-allantoic contact. Am J Anat. 134:309-319.
3270. Fischer, TV & Floyd, AD. 1972. Placental development in the Mongolian gerbil (*Meriones unguiculatus*) II From the establishment of the labyrinth to term. Am J Anat. 134:321-335.
3271. Fischer, W. 1957. Winterwurf des Eichhörnchen. Z Säugetierk. 22:105-106.

3272. Fish, WR, Young, WC, & Dorfman, RI. 1941. Excretion of estrogenic and androgenic substances by female and male chimpanzees with known behavior records. Endocrinology. 28:585-592.
3273. Fisher, EM. 1940. Early life history of a sea otter pup. J Mamm. 21:132-137.
3274. Fisher, HD. 1952. The status of the harbour seal in British Columbia, with particular reference to the Skeena River. Bull Fish Res Bd Canada. 93:1-58.
3275. Fisher, HD. 1954. Delayed implantation in the harbour seal, *Phoca vitulina* L. Nature. 173:879-880.
3276. Fisher, HD & Harrison, RJ. 1970. Reproduction in the common porpoise (*Phocoena phocoena*) of the North Atlantic. J Zool. 161:471-486.
3277. Fisher, JS. 1979. Reproduction in the pygmy rabbit in southeastern Idaho. MS thesis. ID State Univ, Pocatello.
3278. Fisher, LE. 1972. The birth of a lowland gorilla, *Gorilla g gorilla*, at Lincoln Park Zoo, Chicago. Int Zoo Yearb. 12:106-108.
3279. Fisler, GF. 1971. Age structure and sex ratio in populations of *Reithrodontomys*. J Mamm. 52:653-662.
3280. Fitch, HS. 1947. Ecology of a cottontail rabbit population in central California. CA Fish Game. 33:159-184.
3281. Fitch, HS. 1948. Ecology of the California ground squirrel on grazing lands. Am Mid Nat. 39:513-596.
3282. Fitch, HS. 1948. Habits and economic relationships of the Tulare kangaroo rat. J Mamm. 29:5-35.
3283. Fitch, HS. 1957. Aspects of reproduction and development in the prairie vole (*Microtus ochrogaster*). Univ KS Publ Mus Nat Hist. 10:129-161.
3284. Fitch, HS, Fitch, VR, & Kettle, WD. 1984. Reproduction, population changes and interactions of small mammals on a natural area in northeastern Kansas. Occas Pap Mus Nat Hist Univ KS. 109:1-37.
3285. Fitch, HS & Rainey, DG. 1956. Ecological observations on the woodrat, *Neotoma floridiana*. Univ KS Publ Mus Nat Hist. 8:499-533.
3286. Fitch, HS & Sandidge, LL. 1953. Ecology of the opossum on a natural area in northeastern Kansas. Univ KS Publ Mus Nat Hist. 7:305-338.
3287. Fitch, JH & Shump, KA, Jr. 1979. *Myotis keenii*. Mamm Species. 121:1-3.
3288. Fitch, JH, Shump, KA, Jr, & Shump, AU. 1981. *Myotis velifer*. Mamm Species. 149:1-5.
3289. Fitzgerald, A. 1935. Rearing marmosets in captivity. J Mamm. 16:181-188.
3290. Fitzgerald, AE, Clark, RJJ, Reid, CSW, et al. 1981. Physical and nutritional characteristics of the possum (*Trichosurus vulpecula*) in captivity. NZ J Zool. 8:551-562.
3291. Fitzgerald, JP & Lechleitner, RR. 1974. Observations on the biology of Gunnison's prairie dog in central Colorado. Am Mid Nat. 92:146-164.
3292. Flaherty, SP & Breed, WG. 1987. Formation of the ventral hooks on the sperm head of the plains mouse, *Pseudomys australis*. Gamete Res. 17:115-129.
3293. Flaherty, SP, Breed, WG, & Sarahs, V. 1983. Localization of actin in the sperm head of the plains mouse, *Pseudomys australis*. J Exp Zool. 225:497-500.
3294. Flake, LD. 1974. Reproduction of four rodent species in a short-grass prairie of Colorado. J Mamm. 55:213-216.
3294a. Flannery, TF, Alpin, K, Groves, CP, et al. 1989. Revision of the New Guinean genus *Mallomys* (Muridae: Rodentia) with descriptions of two new species from subalpine habitats. Rec Aust Mus. 41:83-105.
3295. Fleagle, JG & Samonds, KW. 1975. Physical growth of cebus monkeys (*Cebus albifrons*) during the first year of life. Growth. 39:35-52.
3296. Fleay, D. 1929. The fat-tailed pouched mouse. Vict Nat. 45:278-280.
3297. Fleay, D. 1935. Breeding of *Dasyurus viverrinus* and general observations on the species. J Mamm. 16:10-16.
3298. Fleay, D. 1935. Notes on the breeding of Tasmanian devils. Vict Nat. 52:100-105.
3299. Fleay, D. 1936. Observations on the birth of a wallaby. Aust Zool. 8:153-156.
3300. Fleay, D. 1937. Observations on the koala in captivity: Successful breeding in the Melbourne Zoo. Aust Zool. 9:68-80.
3301. Fleay, D. 1940. Breeding of the tiger-cat. Vict Nat. 56:159-163.
3302. Fleay, D. 1944. Observations on the breeding of the platypus in captivity. Vict Nat. 61:8-14.
3303. Fleay, D. 1950. New facts about the family life of the Australian *Platypus*. Zoo Life. 5(2):48-51.
3304. Fleay, D. 1952. The Tasmanian or marsupial devil: Its habits and family life. Aust Mus Mag. 10:275-280.
3305. Fleay, D. 1961. Breeding the mulgara. Vict Nat. 78:160-167.
3306. Fleay, D. 1962. The northern quoll, *Satanellus hallucatus*. Vict Nat. 78:288-293.
3307. Fleay, D. 1965. Australia's "needle-in-a-haystack" marsupial. Vict Nat. 82:195-204.
3308. Fleay, D. 1967. Planigale holds record family number. Vict Nat. 84:202.
3308a. Fleharty, ED & Choate, JR. 1973. Bioenergetic strategies of the cotton rat, *Sigmodon hispidus*. J Mamm. 54:680-692.
3309. Fleming, AD, Yanagimachi, R, & Yanagimachi, H. 1981. Spermatozoa of the Atlantic bottlenosed dolphin, *Tursiops truncatus*. J Reprod Fert. 63:509-514.
3310. Fleming, AS, Chee, P, & Vaccarino, F. 1981. Sexual behaviour and its olfactory control in the desert woodrat (*Neotoma lepida lepida*). Anim Behav. 29:727-745.
3311. Fleming, D, Cinderey, RN, & Hearn, JP. 1983. The reproductive biology of Bennett's wallaby (*Macropus rufogriseus rufogriseus*) ranging free at Whipsnade's Park. J Zool. 201:283-291.
3312. Fleming, MR. 1985. The thermal physiology of the mountain pygmy-possum *Burramys parvus* (Marsupialia: Burramyidae). Aust Mamm. 8:79-90.
3313. Fleming, MR & Cockburn, A. 1979. *Ningaui*: A new genus of dasyurid for Victoria. Vic Nat. 96:142-146.
3314. Fleming, MR & Frey, H. 1984. Aspects of the natural history of feathertail gliders (*Acrobates pygmaeus*) in Victoria. Possums and Gliders. 403-408, Smith, A & Hume, I, eds, Surrey Beatty & Sons Pty Ltd, Sydney, Australia.
3315. Fleming, MW & Harder, JD. 1981. Uterine histology and reproductive cycles in pregnant and non-pregnant opossums, *Didelphis virginiana*. J Reprod Fert. 63:21-24.
3316. Fleming, TH. 1970. Comparative biology of two temperate tropical rodent counterparts. Am Mid Nat. 83:462-471.
3317. Fleming, TH. 1970. Notes on the rodent faunas of the two Panamanian forests. J Mamm. 57:473-490.
3318. Fleming, TH. 1971. *Artibeus jamaicensis*: Delayed embryonic development in a neotropical bat. Science. 171:402-404.
3319. Fleming, TH. 1971. Population ecology of three species of neotropical rodents. Misc Publ Mus Zool Univ MI. 143:1-77.

3320. Fleming, TH. 1973. The reproductive cycle of three species of opossums and other mammals in the Panama Canal Zone. J Mamm. 54:439-455.
3321. Fleming, TH. 1977. Growth and development of two species of tropical heteromyid rodents. Am Mid Nat. 98:109-123.
3322. Fleming, TH, Hooper, ET, & Wilson, DE. 1972. Three Central American bat communities: Structure, reproductive cycles and movement patterns. Ecology. 53:555-569.
3323. Fleming, TH & Rauscher, RJ. 1978. On the evolution of litter size in *Peromyscus leucopus*. Evolution. 32:45-55.
3324. Fletcher, TJ. 1974. The timing of reproduction in red deer (*Cervus elaphus*) in relation to latitude. J Zool. 172:363-367.
3325. Fletcher, TP. 1985. Aspects of reproduction in the male eastern quoll, *Dasyurus viverrinus* (Shaw) (Marsupial: Dasyuridae), with notes on polyoestry in the female. Aust J Zool. 33:101-110.
3326. Fletcher, TP. 1989. Plasma progesterone and body weight in the pregnant and non-pregnant kowari, *Dasyuroides byrnei* (Marsupialia: Dasyuridae). Reprod Fert Dév. 1:65-74.
3327. Fletcher, TP. 1989. Luteinizing hormone in the kowari, *Dasyuroides byrnei* (Marsupialia: Dasyuridae) during the oestrous cycle and pregnancy. Reprod Fert Dév. 1:55-64.
3328. Fletcher, TP, Jetton, AE, & Renfree, MB. 1988. Influence of progesterone and oestradiol-17β on blastocysts of the tammar wallaby (*Macropus eugenii*) during seasonal diapause. J Reprod Fert. 83:193-200.
3329. Fletcher, TP & Renfree, MB. 1988. Effects of corpus luteum removal on progesterone, oestrabiol-17β and LH in early pregnancy of the tammar wallaby, *Macropus eugenii*. J Reprod Fert. 83:185-191.
3330. Fletcher, TP, Shaw, G, & Renfree, MB. 1990. Effects of bromocriptine at parturition in the tammar wallaby *Macropus eugenii*. Reprod Fert Dév. 2:79-88.
3331. Flickinger, CJ, Howards, SS, & English, HF. 1978. Ultrastructural differences in efferent ducts and several regions of the epididymis of the hamster. Am J Anat. 152:557-586.
3332. Flint, APF, Burton, RD, & Heap, RB. 1983. Sources of progesterone during gestation in Barbary sheep (*Ammotragus lervia*). J Endocrinol. 98:283-288.
3333. Flint, APF & Renfree, MB. 1982. Oestradiol-17β in the blood during seasonal reactivation of the diapausing blastocyst in a wild population of tammar wallabies. J Endocrinol. 95:293-300.
3334. Flint, VE. 1966. Die Zwerghamster der paläarktischen Fauna Nr 366. Neue Brem Bücherei, Wittenberg.
3335. Flint, VE. 1966. On the biology of *Cricetulus longicaudatus* Milne-Edwards, 1886. Zool Zhurn. 45:471-474.
3336. Flook, DR. 1970. Causes and implications of an observed sex differential in the survival of wapiti. Can Wildl Serv Rep Ser. 11:1-71.
3337. Florant, GL & Tamarkin, L. 1984. Plasma melatonin rhythms in euthermic marmots (*Marmota flaviventris*). Biol Reprod. 30:332-337.
3338. Florio, PL & Spenelli, L. 1967. Successful breeding of a cheetah *Acinonyx jubatus* in a private zoo. Int Zoo Yearb. 7:150-152.
3339. Florio, PL & Spenelli, L. 1968. Second successful breeding of cheetahs *Acinonyx jubatus* in a private zoo. Int Zoo Yearb. 8:76-78.
3340. Flower, SS. 1933. Breeding season of lemurs. Proc Zool Soc London. 1933:317.
3341. Flower, SS. 1943. Notes on age at sexual maturity, gestation period and growth of the Indian elephant, *Elephas maximus*. Proc Zool Soc London. 113A:21-26.
3342. Flowerdew, JR. 1973. The effect of natural and artificial changes in food supply on breeding in woodland mice and voles. J Reprod Fert Suppl. 19:259-269.
3343. Flowerdew, JR. 1985. The population dynamics of wood mice and yellow-necked mice. Symp Zool Soc London. 55:315-338.
3343a. Fluhmann, CF. 1934. The length of the human menstrual cycle. Am J Obs Gyn. 27:73-78.
3344. Flux, JEC. 1964. Hare reproduction in New Zealand. NZ J Agric. 109:483-486.
3345. Flux, JEC. 1965. Timing of the breeding season in the hare, *Lepus europaeus* Pallas, and rabbit, *Oryctolagus cuniculus* (L). Mammalia. 29:557-562.
3346. Flux, JEC. 1967. Reproduction and body weights of the hare *Lepus europaeus* Pallas, in New Zealand. NZ J Sci. 10:357-401.
3347. Flux, JEC. 1969. Current work on the reproduction of the African hare, *Lepus capensis* L, in Kenya. J Reprod Fert Suppl. 6:225-227.
3348. Flux, JEC. 1970. Life history of the mountain hare (*Lepus timidus scoticus*) in north-east Scotland. J Zool. 161:75-123.
3349. Flux, JEC. 1981. Reproductive strategies in the genus *Lepus*. Proc World Lagomorph Conf, Univ Guelph, ONT, Canada. 1979:155-174.
3350. Flux, JEC & Jarvis, JUM. 1970. Growth rates of two African hares, *Lepus capensis*. J Mamm. 51:798-799.
3351. Flynn, TT. 1910. Contributions to a knowledge of the anatomy and development of the marsupialia 1 The genitalia of *Sarcophilus satanicus* (female). Proc Linn Soc NSW. 35:873-886.
3352. Flynn, TT. 1911. Notes on marsupial anatomy II On the female genital organs of a virgin *Sarcophilus satanicus*. Pap Proc R Soc Tas. 1911:144-156.
3353. Flynn, TT. 1922. Notes on certain reproductive phenomena in some Tasmanian marsupials. Ann Mag Nat Hist Ser 9. 10:225-231.
3354. Flynn, TT. 1922. The phylogenetic significance of the marsupial alloplacenta. Proc Linn Soc NSW. 47:541-544.
3355. Flynn, TT. 1923. Photograph illustrating method of parturition in *Potorous tridactylus* exhibited at meeting. Proc Linn Soc NSW. 47:XXVIII.
3356. Flynn, TT. 1923. The yolk-sac and allantoic placenta in *Perameles*. Q J Microsc Sci. 67:123-182.
3357. Flynn, TT. 1928. The corpus luteum and the cause of birth. Nature. 121:1020-1021.
3358. Flynn, TT. 1930. On the unsegmented ovum of echidna (*Tachyglossus*). Q J Microsc Sci. 74:119-131.
3359. Flynn, TT. 1930. The uterine cycle of pregnancy and pseudopregancy as it is in the diprotodont marsupial *Bettongia cunniculus* with notes on other reproductive phenomena in this marsupial. Proc Linn Soc NSW. 55:506-531.
3360. Flynn, TT & Hill, JP. 1939. The development of the Monotremata Part IV Growth of the ovarian ovum, maturation, fertilization, and early cleavage. Trans Zool Soc London. 24:445-622.
3361. Flynn, TT & Hill, JP. 1947. The development of the Monotremata Part VI The later stages of cleavage and the formation of the primary germ-layers. Trans Zool Soc London. 26:1-151.
3362. Foerg, R. 1982. Reproductive behavior in *Varecia variegata*. Folia Primatol. 38:108-121.
3363. Foery, R. 1982. Reproduction in *Cheirogaleus medius*. Folia Primatol. 39:49-62.

3364. Fogden, MPL. 1974. A preliminary field study of the western tarsier, *Tarsianus bancanus* Horsefeld. Prosimian Biology. 151-165, Martin, RD, Doyle, GA, & Walker, AC, eds, Univ Pittsburgh Press, Pittsburgh.
3365. Fogden, SCL. 1968. Suckling behaviour in the grey seal (*Halichoerus grypus*) and the northern elephant seal (*Mirounga angustirostris*). J Zool. 154:415-420.
3366. Folk, GE, Jr. 1940. The longevity of sperm in the female bat. Anat Rec. 76:103-109.
3367. Follis, TB, Foote, WC, & Spillett, JJ. 1972. Observation of genitalia in elk by laparotomy. J Wildl Manage. 36:171-173.
3368. Follis, TB & Spillett, IJ. 1974. Winter pregnancy rates and subsequent fall cow/calf ratios in elk. J Wildl Manage. 58:789-791.
3369. Follmann, EH & Klimstra, WD. 1969. Fertility in male white-tailed deer fawns. J Wildl Manage. 33:708-711.
3370. Fölsch, D. 1967. Vor-und Nachgeburtphasen bei drei Flughundengeburten *Pteropus giaganteus*. Z Säugetierk. 32:375-377.
3371. Foltz, DW, Hoogland, JL, & Koscielny, GM. 1988. Effects of sex, litter size, and heterozygosity on juvenile weight in black-tailed prairie dogs (*Cynomys ludovicianus*). J Mamm. 69:611-614.
3372. Fonda, E & Peyre, A. 1965. Localization de la 5δ,3β-hydroxysteroïdo-déshydrogénase dans le placenta de Minioptère (Chiroptère) au cours de la gestation. CR Hebd Séances Acad Sci. 261:2963-2969.
3373. Fons, R. 1973. Modalités de la reproduction et développement postnatal en captivité chez *Suncus etruscus* (Savi, 1822). Mammalia. 37:288-324.
3374. Fons, R. 1974. La répertoire comportementale de la Pachyure étrusque, *Suncus etruscus* (Savi, 1822). Terre Vie. 28:131-157.
3375. Fons, R. 1974. Méthodes de capture et d'élevage de la Pachyure étrusque *Suncus etruscus* (Savi, 1822) (Insectivora, Soricidae). Z Säugetierk. 39:204-210.
3376. Fons, R. 1979. Durée de la vie chez la Pachyure étrusque, *Suncus etruscus* (Savi, 1822) en captivité (Insectivora, Soricidae). Z Säugetierk. 44:241-248.
3377. Fontaine, PA. 1965. Breeding saiga antelope, *Saiga tatarica*, at Dallas Zoo. Int Zoo Yearb. 5:57-58.
3378. Fontaine, PA. 1965. Breeding clouded leopards *Neofelis nebulosa* at Dallas Zoo. Int Zoo Yearb. 5:113-114.
3379. Fontaine, PA. 1968. Birth of four species of apes at Dallas Zoo. Int Zoo Yearb. 8:115-118.
3380. Fontaine, R. 1973. The individual nonsocial behavior of *Cacajao rubicundus* in a nonsocial environment. MS thesis. Bucknell Univ, Lewistown, PA.
3381. Fontaine, R. 1981. The uakaris, genus *Cacajao*. Ecology and Behavior of Neotropical Primates. 1:443-493. Coimbra Filho, AF & Mittermeier, RA, eds, Academic Brasileira de Ciências, Rio de Janeiro.
3382. Fontaine, R & Du Mond, FV. 1977. The red ouakari in a seminatural environment: Potentials for propagation and study. Primate Conservation. 167-236, Prince Rainier III of Monaca & Bourne, GH, eds, Academic Press, New York.
3383. Fontaine, R & Hench, M. 1982. Breeding New World monkeys at Miami's Monkey Jungle. Int Zoo Yearb. 22:77-84.
3384. Fooden, J. 1967. Complementary specialization of male and female reproductive structures in the bear macaque, *Macaca arctoides*. Nature. 214:939-941.
3385. Fooden, J. 1971. Reports on primates collected in western Thailand Jan-April, 1967. Fieldiana Zool. 59:1-62.
3386. Fooden, J, Guoquang, Q, Zongren, W, et al. 1985. The stumptail macaques of China. Am J Primatol. 8:11-30.
3387. Fooden, J & Izor, RJ. 1983. Growth curves, dental emergence norms, and supplementary morphological observations in known-age captive orangutans. Am J Primatol. 5:285-301.
3388. Fooks, HA. 1966. The roe deer. Forestry Com Leafl; HMSO. 45:1-16.
3389. Forbes, RB. 1964. Some aspects of the life history of the silky pocket mouse, *Perognathus flavus*. Am Mid Nat. 72:438-443.
3390. Forbes, RB. 1966. Notes on a litter of least chipmunks. J Mamm. 47:159-161.
3391. Forbes, RB & Turner, LW. 1972. Notes on two litters of Townsend's chipmunks. J Mamm. 53:355-359.
3392. Forbes, TR. 1942. The period of gonadal activity in the Maryland muskrats. Science. 95:382-383.
3393. Forbes, TR. 1953. Progesterone in the systemic blood of women and monkeys during the puerperium. Endocrinology. 52:236-237.
3394. Forbes, WA. 1882. On some points in the anatomy of the great anteater (*Myrmecophaga jubata*). Proc Zool Soc London. 1882:287-302.
3395. Forbes, WA. 1885. On the male generative organs of the Sumatran rhinoceros (*Ceratorhinus sumatrensis*). Trans Zool Soc London. 11:107-109.
3396. Forcum, DL. 1966. Postpartum behavior and vocalizations of snowshoe hares. J Mamm. 47(1):543.
3397. Ford, CE & Evans, EP. 1977. Cytogentic observations on XX/XY chimaeras and a reassessment of the evidence for germ cell chimaerism in heterosexual twin cattle and marmosets. J Reprod Fert. 49:25-33.
3398. Ford, LS & Hoffmann, RS. 1988. *Potos flavus*. Mamm Species. 321:1-9.
3399. Ford, SD. 1977. Range, distribution and habitat of the western harvest mouse, *Reithrodontomys megalotis*, in Indiana. Am Mid Nat. 98:422-431.
3400. Foreman, D. 1962. The normal reproductive cycle of the female prairie dog and the effects of light. Anat Rec. 142:391-405.
3401. Foreman, D. 1967. The effects of gonadotrophins on the metabolism of ovaries from prairie dogs (*Cynomys ludovicianus*). Gen Comp Endocrinol. 8:66-71.
3402. Foreman, D. 1974. Structural and functional homologies of the accessory reproductive glands of two species of sciurids, *Cynomys ludovicianus* and *Citellus tridecemlineatus*. Anat Rec. 180:331-340.
3403. Foreman, D. 1981. Follicular dynamics in a monestrous annually breeding mammal: Prairie dog (*Cynomys ludovicianus*). Dynamics of Ovarian Function. 245-251, Schwartz, NB & Hunzicker-Dunn, M, eds, Raven Press, New York.
3404. Foreman, D & Garris, D. 1984. Plasma progesterone levels and corpus luteum morphology in the female prairie dog (*Cynomys ludovicianus*). Gen Comp Endocrinol. 55:315-322.
3405. Foreman, D & Williams, D. 1967. The effects of gonadotrophic hormones on the oxygen consumption of testes from the prairie dog (*Cynomys ludovicianus*). Gen Comp Endocrinol. 9:287-294.
3406. Foresman, KR. 1987. Uterine luminal proteins during the preimplantation period in the ferret. J Exp Zool. 243:103-109.
3407. Foresman, KR & Daniel, JC, Jr. 1983. Plasma progesterone concentrations in pregnant and non-pregnant black bears (*Ursus americanus*). J Reprod Fert. 68:235-239.

3407a. Foresman, KR & Mead, RA. 1973. Duration of post-implantation in a western subspecies of the spotted skunk (*Spilogale putorius*). J Mamm. 54:521-523.
3408. Foresman, KR & Mead, RA. 1974. Pattern of luteinizing hormone secretion during delayed implantation in the spotted skunk (*Spilogale putorius latifrons*). Biol Reprod. 11:475-480.
3409. Foresman, KR & Mead, RA. 1978. Luteal control of nidation in the ferret (*Mustela putorius*). Biol Reprod. 18:490-496.
3410. Foresman, KR, Reeves, JJ, & Mead, RA. 1974. Pituitary responsiveness to luteinizing-hormone releasing hormone during delayed implantation in the spotted skunk (*Spilogale putorius latifrons*): Validation of LH radioimmunoassay. Biol Reprod. 11:102-107.
3411. Forger, NG, Dark, J, Barnes, BM, et al. 1986. Fat ablation and food restriction influence reproductive development and hibernation in ground squirrels. Biol Reprod. 34:831-840.
3412. Forger, NG & Zucker, I. 1985. Photoperiodic regulation of reproductive development in male white-footed mice (*Peromyscus leucopus*) born at different phases of the breeding season. J Reprod Fert. 73:271-178.
3413. Forman, GL & Gneoways, HH. 1979. Sperm morphology. Sp Publ Mus TX Tech Univ. 16(3):177-204.
3414. Forman, GL, Smith, JD, & Hood, CS. 1989. Exceptional seze and ususual morphology of spermatozoa in *Noctilio albiventris* (Noctilionidae). J Mamm. 70:179-184.
3414a. Forment, WL, Schmidt, U, & Greenhall, AM. 1971. Movement and population studies of the vampire bat (*Desmodus rotundus*) in Mexico. J Mamm. 52:227-228.
3415. Formozov, AN & Kodachova, KS. 1961. Les Rongeurs vivants en colonies dans la steppe eurasienne et leur influence sur les sols et la vegetation. Terre Vie. 108:116-129.
3416. Forrest, SC, Biggins, DE, Richardson, L, et al. 1988. Population attributes for the black-footed ferret (*Mustela nigripes*) at Meeteetse, Wyoming, 1981-1985. J Mamm. 69:261-273.
3417. Forrester, DJ. 1965. Fetal measurement and milk characteristics of bighorn sheep. J Mamm. 46:524-525.
3418. Forrester, DJ. 1975. Reproductive performance of a laboratory colony of the deermouse, *Peromyscus maniculatus gambelii*. Lab Anim Sci. 25:446-449.
3419. Forsberg, M, Fougner, JA, Hofmo, PO, et al. 1989. Photoperiodic regulation of reproduction in the male silver fox (*Vulpes vulpes*). J Reprod Fert. 87:115-123.
3420. Forsberg, M, Fougner, JA, Hofmo, PO, et al. 1990. Effect of melatonin implants on reproduction in the male silver fox (*Vulpes vulpes*). J Reprod Fert. 88:383-388.
3421. Forsberg, M & Madej, A. 1990. Effects of melatonin implants on plasma concentrations of testosterone, thyroxine and prolactin in the male silver fox (*Vulpes vulpes*). J Reprod Fert. 89:351-358.
3422. Forsten A & Youngman, PM. 1982. *Hydrodamalis gigas*. Mamm Species. 165:1-3.
3423. Forsyth, DJ. 1976. A field study of growth and development of nestling masked shrews (*Sorex cinereus*). J Mamm. 57:708-721.
3424. Forsyth, IA & Blake, LA. 1976. Placental lactogen (chorionic mammotrophin) in the field vole, *Microtus agrestis*, and the bank vole, *Clethrionomys glareolus*. J Endocrinol. 70:19-23.
3425. Forsyth, W. 1911. The number of cubs in a tiger litter. J Bombay Nat Hist Soc. 20:1148.
3426. Fosey, D. 1982. Reproduction among free-living mountain gorillas. Am J Primatol Suppl. 1:97-104.
3427. Foster, DL. 1977. Luteinizing hormone and progesterone secretion during sexual maturation of the rhesus monkey: Short luteal phases during the initial menstrual cycles. Biol Reprod. 17:584-590.
3428. Foster, GW & Humphrey, SR. 1978. Survival rate of young southern brown bats, *Myotis austroriparius*, in Florida. J Mamm. 59:299-304.
3429. Foster, JB. 1961. Life history of the *Phenacomys* vole. J Mamm. 42:181-198.
3430. Foster, JB. 1966. The giraffe of Nairobi National Park: Home range, sex ratios, the herd, and food. E Afr Wildl J. 4:139-148.
3431. Foster, JB & Dagg, AI. 1972. Notes on the biology of the giraffe. E Afr Wildl J. 10:1-16.
3432. Foster, MA. 1934. The reproductive cycle in the female ground squirrel *Citellus tridecemlineatus* (Mitchill). Am J Anat. 54:487-511.
3433. Fourquet, JP, Fraile, B, & Kann, ML. 1991. Sperm actin and calmodulin during fertilization in the hamster: An immune electron microscopic study. Anat Rec. 231:316-323.
3434. Fouquet, JP, Meusy-Dessolle, N, & Dang, DC. 1984. Relationships between Leydig cell morphometry and plasma testosterone during post-natal development of the monkey, *Macaca fascicularis*. Reprod Nutr Dév. 24:281-296.
3435. Fourie, LJ & Perrin, MR. 1987. Some new data on the reproductive biology of the rock hyrax. S Afr J Wildl Res. 17:118-122.
3436. Fowler, PA. 1988. Seasonal endocrine cycles in the European hedgehog, *Erinaceus europaeus*. J Reprod Fert. 84:259-272.
3437. Fox, BJ. 1979. Growth and development of *Rattus lutreolus* (Rodentia: Muridae) in the laboratory. Aust J Zool. 27:945-957.
3438. Fox, BJ & Kemper, CM. 1982. Growth and development of *Pseudomys gracilicaudatus* (Rodentia: Muridae) in the laboratory. Aust J Zool. 30:175-185.
3439. Fox, BJ & Whitford, D. 1982. Polyoestry in a predictable coastal environment: Reproduction, growth and development in *Sminthopsis murina* (Dasyuridae, Marsupialia). Carnivorous Marsupials. 1:39-48, Archer, M, ed, R Zool Soc NSW. Mossman, NSW.
3440. Fox, H. 1929. The birth of two anthropoid apes. J Mamm. 10:37-51.
3441. Fox, H. 1938. The Giraffe: Some notes upon the natural characters of this animal and upon its care and its misfortunes. Zool Soc Philadelphia Penrose Res Lab Rep. 1938:35-67.
3442. Fox, MW & Johnsingh, AJT. 1975. Hunting and feeding in wilddog. J Bombay Nat Hist Soc. 72:321-326.
3443. Frädrich, H. 1966. Breeding and hand-rearing warthogs *Phacochoerus aethiopicus* at Frankfurt Zoo. Int Zoo Yearb. 6:200-202.
3444. Frädrich, H. 1974. Notizen uber seltener gehaltene Cerviden Teil I. Zool Garten NF. 44:189-200.
3445. Frädrich, H. 1975. Notizen uber seltener gehaltene Cerviden Teil II. Zool Garten NF. 45:67-77.
3446. Frädrich, H. 1981. Beobachtungen am Pampahirsch, *Blastoceros bezoarcticus* (L, 1758). Zool Garten NF. 51:7-32.
3447. Frädrich, H & Klös, H-G. 1976. Zur Haltung und Zucht des Gaurs, *Bos gaurus*. Zool Garten NF. 46:417-425.
3447a. Frädrich, H. 1987. The husbandry of tropical adn temperate cervids in the west Berlin Zoo. Biology and Management of the Cervidae. 422-428, Wemmer, CM, ed, Smithsonian Inst Press, Washington, DC.

3448. Fränkel, L & Cohn, F. 1901. Experimentelle Untersuchungen ueber den Einfluss des Corpus luteum auf die Insertion des Eies. Anat Anz. 20:294-300.
3449. Fragaszy, DM, Schwarz, S, & Shimosaka, D. 1982. Longitudinal observations of care and developement of infant titi monkeys (*Callicebus moloch*). Am J Primatol. 2:191-200.
3450. Frahm, HD. 1980. Zuchterfolg und Hirngewicht bei *Tupaia glis*. Z Säugetierk. 45:129-133.
3451. Frame, LH & Frame, GW. 1976. Female African wild dogs emigrate. Nature. 263:227-229.
3452. Frame, LH, Malcolm, JR, Frame, GW, et al. 1979. Social organization of African wild dogs (*Lycaon pictus*) on the Serengeti Plains, Tanzania (1967-1978). Z Tierpsychol. 50:225-249.
3453. Frame, LH, Malcolm, JR, Frame, GW, et al. 1979. Social organization of African wild dogs (*Lycaon pictus*) on the Serengeti Plains, Tanzania (1967-1978). Z Tierpsychol. 50:225-249.
3454. Francke, H. 1961. Gefangenschaftsbeobachtungen an *Hemicentetes semispinosus*. Sitzungsber Ges Naturf Fr Berlin NF. 1:118-123.
3455. Frank, F. 1953. Beitrag zur Biologie, insbesondere Paarungsbiologie der Feldspitzmaus (*Crocidura leucodon*). Bonner Zool Beitr. 4:187-194.
3456. Frank, F. 1953. Untersuchungen über den Zusammeubruch von Feldmausplagen (*Microtus arvalis* Pallas). Zool Jb Syst. 82:95-136.
3457. Frank, F. 1954. Beitrag zur Biologie, insbesondere Jugendentwicklung der Schneemaus (*Chionomys nivalis* Mart). Z Tierpsychol. 11:1-9.
3458. Frank, F. 1954. Zur Jugendentwicklung der Feldspitzmaus (*Crocidura leucodon* Herm). Bonner Zool Beitr. 5:173-178.
3459. Frank, F. 1956. Beiträge zur Biologie der Feldmaus, *Microtus arvalis* (Pall) Teil II: Laboratoriums ergebnisse. Zool Jb Syst. 84:32-74.
3460. Frank, F. 1956. Das Fortpflanzungspotential der Feldmaus, *Microtus arvalis* (Pallas)-eine Spitzenleistung unter den Säugetieren. Z Säugetierk. 21:176-181.
3461. Frank, F. 1957. Zucht und Gefangenschafts-Biologie de Zwergmaus (*Micromys minutus subobscurus* Fritsche). Z Säugetierk. 22:1-44.
3462. Frank, F. 1962. Zur Biologie des Berglemmings, *Lemmus lemmus* (L) Ein Beitrag zum Lemming-Problem. Z Morph Ökol Tiere. 51:87-164.
3463. Frank, F. 1964. Die Feldmaus, *Microtus arvalis* (Pallas), im nordwestdeutschlanden Rekordwinter 1962/63. Z Säugetierk. 29:146-152.
3464. Frank, F. 1967. Die Wurfzeit des Waldlemmings, *Myopus schisticolor* (Liljeborg, 1844). Z Säugetierk. 32:172-173.
3465. Frank, F. 1968. Zur Kenntnis der spanischen Feldmaus (*Microtus arvalis asturianus* Miller, 1908). Bonn Zool Beitr. 19:189-197.
3466. Frank, F. 1974. Sexualzyklys, Vaginal-pH und Geschlecntsverhältnis der Nachkommen beim Waldlemming, *Myopus schisticolor* (Liljeborg). Z Säugetierk. 39:269-276.
3467. Frank, F. 1974. Wurfzahl und Wurffolge beim nordischen Wiesel (*Mustela nivalis rixosa* Bangs, 1896). Z Säugetierk. 39:248-250.
3468. Frank, F & Zimmermann, K. 1956. Zur Biologie der nordischen Wuhlmaus (*Microtus oeconomus stimmingi* Nehring). Z Säugetierk. 21:58-83.
3469. Frank, LG. 1986. Social organization of the spotted hyaena *Crocuta crocuta* I Demography. Anim Behav. 34:1500-1509.
3470. Frank, LG. 1986. Social organization of the spotted hyaena *Crocuta crocuta* II Dominance and reproduction. Anim Behav. 34:1510-1527.
3471. Frank, LG, Smith, ER, & Davidson, JM. 1985. Testicular origin of circulating androgen in the spotted hyaena *Crocuta crocuta*. J Zool. 207A:613-615.
3472. Frankiewicz, J & Marchlewska-Koj, A. 1985. Effect of conspecifics on sexual maturation in female European pine voles *Pitymys subterraneus*. J Reprod Fert. 74:153-156.
3473. Franklin, BC. 1958. Studies on the effects of progesterone on the physiology of reproduction in the mink, *Mustela vison*. OH J Sci. 58:163-170.
3474. Franklin, WL. 1974. The social behavior of the vicuna. The Behaviour of Ungulates and its Relation to Management. 477-487, Geist, V & Walther, F, eds, IUCN, Morges, Switzerland.
3475. Franklin, WL. 1982. Biology, ecology, and relationship to man of the South American Camelids. Pymatuning Lab Ecol Spec Publ Ser. 6:457-489.
3476. Franz, C & Longcope, C. 1979. Androgen and estrogen metabolism in male rhesus monkeys. Endocrinology. 105:869-874.
3477. Franzmann, AW. 1978. Moose. Big Game of North America. 67-81, Schmidt, JL & Gilbert, DR, eds, Stackpole Books, Harrisburg, PA.
3478. Franzmann, AW. 1981. *Alces alces*. Mamm Species. 154:1-7.
3479. Franzmann, AW, Ameson, PD, & Ullrey, DE. 1975. Composition of milk from Alaskan moose in relation to other North American wild ruminants. J Zoo An Med. 6:12-14.
3480. Franzmann, AW & Schwartz, CC. 1985. Moose twinning rates: A possible population condition assessment. J Wildl Manage. 49:394-396.
3481. Frase, BA & Hoffmann, RS. 1980. *Marmota flaviventris*. Mamm Species. 135:1-8.
3482. Fraser, FC. 1936. Recent strandings of the false killer whale, *Pseudorca crassidens*, with special reference to those found at Donna Nook, Lincolnshire. Scott Nat. 1936:105-114.
3483. Fraser, HM, Abbott, M, Laird, NC, et al. 1986. Effects of an LH-releasing hormone antagonist on the secretion of LH, FSH, prolactin and ovarian steroids at different stages of the luteal phase in the stumptailed macaque (*Macaca arctoides*). J Endocrinol. 111:83-90.
3484. Fraser, HM, McNeilly, AS, Abbott, M, et al. 1986. Effect of LHRH immunoneutralization on follicular development, the LH surge and luteal function in the stump tailed macaque monkey (*Macaca arctoides*). J Reprod Fert. 76:299-309.
3485. Fraser-Smith, AC. 1975. Male-induced pregnancy termination in the prairie vole, *Microtus ochrogaster*. Science. 187:1211-1213.
3486. Frawley, LS, Mulchahey, JJ, & Neill, JD. 1983. Nursing induces a biphasic release of prolactin in rhesus monkeys. Endocrinology. 112:558-561.
3487. Frawley, LS & Neill, JD. 1979. Age related changes in serum levels of gonadotropins and testosterone in infantile male rhesus monkeys. Biol Reprod. 20:1147-1151.
3488. Frazer, JFD & Huggett, A St G. 1958. The breeding season and length of pregnancy in four species of large whales. Int Congr Zool. 15:311-313.
3489. Frazer, JFD & Huggett, A St G. 1959. The growth rate of foetal whales. J Physiol. 146:21P-22P.
3489a. Frazer, JFD & Hugget, A St G. 1973. Specific foetal growth rates of cetaceans. J Zool. 169:111-126.

3490. Frazer, JFD & Huggett, A St G. 1974. Species variation in the fetal growth rates of eutherian mammals. J Zool. 174:481-509.
3491. Frechkop, S. 1927. Remarques sur les poids du cerveau chez les Mammifères. Ann Soc Zool Belg. 58:109-116.
3492. Fredga, K. 1970. Unusual sex chromosome inheritance in mammals. Phil Trans R Soc London. 259B:15-36.
3493. Fredga, K, Gropp, A, Winking, H, et al. 1976. Fertile XX- and XY-type females in the wood lemming *Myopus shisticolor*. Nature. 261:225-227.
3494. Fredga, K, Gropp, A, Winking. H, et al. 1977. A hypothesis explaining the exceptional sex ratio in the wood lemming (*Myopus schisticolor*). Hereditas. 85:101-104.
3495. Free, SL & McCaffrey, E. 1972. Reproductive synchrony in the female black bear. Bears: Their Biology and Management. 2:199-206.
3496. Freeman, H. 1975. A preliminary study of the behavior of captive snow leopards *Panthera uncia*. Int Zoo Yearb. 15:217-222.
3497. Freeman, H. 1977. Breeding and behavior of the snow leopard. The World's Cats. 36-44, Eaton, RL, ed, Carn Res Inst Burke Mus, U Wash, Seattle.
3498. Freeman, H & Braden, K. 1977. Zoo location as a factor in the reproductive behavior of captive snow leopards, *Uncia uncia*. Zool Garten NF. 47:280-288.
3499. Freeman, H & Hutchins, M. 1980. Captive management of snow leopard cubs: An overview. Zool Garten NF. 50:377-392.
3500. Freeman, MMR. 1968. Winter observations on beluga (*Delphinapterus leucas*) in Jones sound NWT. Can Field-Nat. 182:276-286.
3501. Freese, CH & Oppenheimer, JR. 1981. The capuchin monkeys, genus *Cebus*. Ecology and Behavior of Neotropical Primates. 1:331-390. Coimbra Filho, AF & Mittermeier, RA, eds, Academic Brasileira de Ciências, Rio de Janeiro.
3502. Frehn, JL, Urry, RL, Balph, DF, et al. 1973. Photoperiod and crowding effects on testicular serotonin metabolism and lack of effect on melatonin synthesis in Unita ground squirrels (*Spermophilus armatus*). J Exp Zool. 183:139-144.
3503. French, JA. 1983. Lactation and fertility: An examination of nursing and inter-birth intervals in cotton-top tamarins (*Saguinus o oedipus*). Folia Primatol. 40:276-282.
3504. French, JA, Abbott, DH, Scheffler, G, et al. 1983. Cyclic excretion of urinary oestrogens in female tamarins (*Saguinus oedipus*). J Reprod Fert. 68:177-184.
3505. French, JA, Abbott, DH, & Snowdon, CT. 1984. The effect of social environment on estrogen excretion, scent marking, and sociosexual behavior in tamarins (*Saguinus oedipus*). Am J Primatol. 6:155-167.
3506. French, JA & Cleveland, J. 1984. Scent-marking in the tamarin, *Saguinus oedipus*: Sex differences and ontogeny. Anim Behav. 32:615-623.
3507. French, JA, Inglett, BJ, & Dethlefs, TM. 1989. The reproductive status of nonbreeding group members in captive golden lion tamarin social groups. Am J Primatol. 18:73-86.
3508. French, JA & Stribley, JA. 1985. Patterns of urinary oestrogen excretion in female golden lion tamarins (*Leontopithecus rosalia*). J Reprod Fert. 75:537-546.
3509. French, JA & Stribley, JA. 1987. Synchronization of ovarian cycles within and between social groups of golden lion tamarins (*Leontopithecus rosalia*). Am J Prim. 12:469-478.
3510. French, NR, McBride, R, & Detmer, J. 1965. Fertility and population density of the black-tailed rabbit. J Wildl Manage. 29:14-26.
3511. French, TW. 1980. *Sorex longirostris*. Mamm Species. 143:1-3
3512. French, TW. 1980. Natural history of the southeastern shrew, *Sorex longirostris* Bachman. Am Mid Nat. 104:13-31.
3513. Frenkel, G, Shaham, Y, & Kraicher, PF. 1972. Establishment of conditions for colony-breeding of the sand-rat *Psammomys obesus*. Lab Anim Sci. 22:40-47.
3514. Frese, R. 1980. Some notes on breeding the leopard cat *Felis bengalensis* at West Berlin Zoo. Int Zoo Yearb. 20:220-223.
3515. Frese, R. 1981. Notes on breeding the marsh mongoose *Atilax paludinosus* at Berlin Zoo. Int Zoo Yearb. 21:147-151.
3516. Freye, HA. 1961. Zur Gefangenschaftsbiologie des Steppenlemmings (*Lagurus lagurus* Pallas 1773). Z Versuchstierk. 1:49-65.
3517. Frick, H & Felten, H. 1952. Ökologische Beobachtungen an sardischen Fledermäusen. Zool Jb Syst Ökol Geogr Tiere. 81:175-189.
3518. Friedländer, M, Rosenstrauch, A, & Bedrak, E. 1984. Leydig cell differentiation during the reproductive cycle of the seasonal breeder *Camelus dromedarius*: An ultra structural analysis. Gen Comp Endocrinol. 55:1-11.
3519. Friend, DS & Fawcett, DW. 1974. Membrane differentiations in freeze-fractured mammalian sperm. J Cell Biol. 63:641-664.
3520. Friend, GR. 1985. Ecological studies of a population of *Antechinus bellus* (Marsupialia: Dasyuridae) in tropical northern Australia. Aust Wildl Res. 12:151-162.
3521. Friend, JA & Barrows, RG. 1983. Bringing up young numbats. Swans. 13:3-9.
3522. Fries, S. 1880. Über die Fortpflanzung von *Meles taxus*. Zool Anz. 3:486-492.
3523. Fries, S. 1895. Über die Fortpflanzung der einheimischen Chiropteren. Zool Anz. 2:355-357.
3524. Friley, CE, Jr. 1949. Use of the baculum in age determination of Michigan beaver. J Mamm. 30:261-267.
3525. Frith, JJ & Sharman, GB. 1964. Breeding in wild populations of the red kangaroo, *Megaleia rufa*. CSIRO Wildl Res. 9:86-114.
3526. Fritto, SH & Mech, LD. 1981. Dynamics, movement and feeding ecology of a newly protected wolf population in northwestern Minnesota. Wildl Monogr. 80:1-79.
3527. Fritts, SH & Sealander, JA. 1978. Reproductive biology and population characteristics of bobcat (*Lynx rufus*) in Arkansas. J Mamm. 59:347-353.
3528. Fritzell, EK. 1978. Reproduction of raccoons (*Procyon lotor*) in North Dakota. Am Mid Nat. 100:253-256.
3529. Fritzell, EK & Haroldson, KJ. 1982. *Urocyon cinereoargenteus*. Mamm Species. 189:1-8.
3530. Fritzell, EK, Hubert, GF, Jr, Meyen, BE, et al. 1985. Age-specific reproduction in Illinois and Missouri raccoons. J Wildl Manage. 49:901-905.
3531. Froehlich, JW, Thorington, RW, Jr, & Otis, JS. 1981. The demography of howler monkeys (*Alouatta palliata*) on Barro Colorado Island, Panama. Int J Primatol. 2:207-236.
3532. Frolka, J. 1981. Die Aufzucht der Guerezas (*Colobus polykomos occidentalis*) im Zoo Lesna. Zool Garten NF. 51:377-381.
3533. Frost, D & Zucker, I. 1983. Photoperiod and melatonin influence seasonal gonadal cycles in the grasshopper mouse (*Onychomys leucogaster*). J Reprod Fert. 69:237-244.
3534. Frost, KJ & Lowry, LF. 1981. Ringed, Baikal and Caspian seals *Phoca hispida* Schreber, 1775; *Phoca sibirica* Gmelin, 1788 and *Phoca caspica* Gmelin, 1788. Handbook of Marine

Mammals. 2:29-53. Ridgway, SH & Harrison, RJ, eds, Academic Press, New York.
3535. Frost, NM. 1942. Gestation period of bighorn sheep, *Ovis canadensis*. J Mamm. 23:215-216.
3536. Frueh, RJ. 1968. A captive-born gorilla *Gorilla gorilla* at St Louis Zoo. Int Zoo Yearb. 8:128-131.
3537. Frueh, RJ. 1968. A note on breeding snow leopards *Panthera unica* at St Louis Zoo. Int Zoo Yearb. 8:74-76.
3538. Frueh, RJ. 1979. The breeding and management of black lemurs *Lemur macaco macaco* at St Louis Zoo. Int Zoo Yearb. 19:214-217.
3539. Frylestam, B. 1979. Population ecology of the European hare in Sweden. Ph D Diss. Univ Lund.
3540. Frylestam, B. 1980. Reproduction in the European hare in southern Sweden. Holarctic Ecol. 3:74-80.
3541. Frylestam, B. 1980. Growth and body weight of European hares in southern Sweden. Holarctic Ecol. 3:81-86.
3542. Fryxell, JM. 1987. Lek breeding and territorial aggression in white-eared kob. Ethology. 75:211-220.
3543. Fryxell, JM. 1987. Seasonal reproduction of white-eared kob in Boma National Park, Sudan. Afr J Ecol. 25:117-124.
3544. Fuchs, W. 1980. Hohes Alter eimes Rothirsch (*Cervus elaphus*). Säugetierk Mitt. 28:160.
3545. Fujikura, T & Niemann, WH. 1967. Birth weight, gestational age, and type of delivery in rhesus monkeys. Am J Obst Gynec. 97:76-78.
3546. Fujimaki, Y. 1969. Reproductive activity in *Apodemus argenteus* Temminck. J Mamm Soc Japan. 4:74-80.
3547. Fujimaki, Y. 1981. Reproductive activity in *Clethrionomys rufocanus bedfordiae* 4 Number of embryos and prenatal mortality. Jap J Ecol. 31:247-256.
3548. Fujino, K. 1963. Intra-uterine selection due to maternal-fetal incompatibility of blood types in the whales. Sci Rep Whales Res Inst. 17:53-65.
3549. Fujita, MS & Kunz, TH. 1984. *Pipistrellus subflavus*. Mamm Species. 228:1-6.
3550. Fujiwara, T, Honjo, S, & Imaizumi, K. 1969. Practice of the breeding of cynomolgus monkeys (*Macaca irus*) under laboratory conditions. Jikken Dobutsu (Experimental Animals). 18:29-39.
3551. Fulk, GW. 1975. Population ecology of rodents in the semiarid shrublands of Chile. Occas Pap Mus TX Tech Univ. 33:1-40.
3552. Fulk, GW. 1976. Notes on the activity, reproduction, and social behavior of *Octodon degus*. J Mamm. 57:495-505.
3553. Fulk, GW & Khokhar, AR. 1980. Observations on the natural history of a pika (*Ochotona rufescens*) from Pakistan. Mammalia. 44:51-58.
3554. Fulk, GW, Lathiya, SB, & Khokhar, AR. 1981. Rice-field rats of lower Sind: Abundance, reproduction and diet. J Zool. 193:371-390.
3555. Fullagar, PJ, Jewell, PA, Lockley, RM, et al. 1963. The Skomer vole (*Clethrionomys glareolus skomerensis*) and long-tailed field mouse (*Apodemus sylvaticus*) on Skomer Island, Pembrokeshire in 1960. Proc Zool Soc London. 140:295-314.
3556. Fuller, GB, Yates, DE, Helton, ED, et al. 1981. Diethylstilbestrol reversal of gonadotropin patterns in infant rhesus monkeys. J Steroid Biochem. 15:497-500.
3556a. Fuller, TK. 1989. Population dynamics of wolves in north-central Minnesota. Wildl Monogr. 105:1-41.
3557. Fuller, TK & Kuehn, DW. 1985. Population characteristics of striped skunks in north central Minnesota. J Mamm. 66:813-815.
3558. Fuller, WA. 1951. Natural history and economic importance of the muskrat in the Athabasca-Peace Delta, Wood Buffalo Park. Can Wildl Serv Wildl Manage Bull Ser 1. 2:1-82.
3559. Fuller, WA. 1961. The ecology and management of the American bison. Terre Vie. 108:286-304.
3560. Fuller, WA. 1962. The biology and management of the bison of Wood Buffalo National Park. Can Wildl Serv Wildl Manage Bull Series. 1(16):1-52.
3561. Fuller, WA, Martell, AM, Smith, RFC, et al. 1975. High-arctic lemmings, *Dicrostonyx groenlandicus* II Demography. Can J Zool. 53:867-878.
3562. Funakoshi, K. 1986. Maternal care and postnatal development in the Japanese long-fingered bat, *Miniopterus schreibersi fuliginosus*. J Mamm Soc Japan. 11:15-26.
3563. Funakoshi, K & Uchida, TA. 1981. Feeding activity during the breeding season and postnatal growth in the Namie's frosted bat, *Vespertilio superans superans*. Jap J Ecol. 31:66-77.
3564. Funakoshi, K & Uchida, TA. 1982. Annual cycles of body weight in the Namie's frosted bat, *Vespertilio superans superans*. J Zool. 196:417-430.
3565. Funkenstien, B & Hellwing, S. 1977. The effect of PMSG on ovulation in the shrew, *Crocidura russula monocha*. Gen Comp Endocrinol. 32:446-453.
3566. Funmilayo, O. 1979. Ecology of the straw-coloured fruit bat in Nigeria. Rev Zool Afr. 93:589-600.
3567. Furley, CW. 1986. Reproductive parameters of African gazelles: Gestation, first fertile matings first parturition and twinning. Afr J Ecol. 24:121-128.
3568. Furnell, DJ & Schweinsburg, RE. 1984. Population dynamics of central Canadian Arctic Island polar bears. J Wildl Manage. 48:722-728.
3569. Furuse, T & Nakamura, T. 1970. Birth of orang-utans. Animals & Zoos. 22(11):6-9.
3570. Gabe, M, Agid, R, Martoja, M, et al. 1964. Données histophysiologiques et biochimiques sur l'hibernation et le cycle annuel chez, *Eliomys quercinus*. Arch Biol. 75:1-87.
3571. Gaffrey, G. 1961. Merkmale der Wildlebenden Säugetiere Mitteleuropas. Akademische Verlagsgesellschafts, Leipzig.
3572. Gaiduk, VE. 1973. Reproduction of *Lepus europaeus* Pall in Byelorussia. Vestnik Zool (Kiev). 3:51-54.
3573. Gaisler, J. 1965. The female sexual cycle and reproduction in the lesser horseshoe bat (*Rhinolophus hipposideros hipposideros* Bechstein, 1800). Vestn Čcsl Spol Zool. 29:336-352.
3574. Gaisler, J. 1966. Reproduction in the lesser horseshoe bat (*Rhinolophus hipposideros hipposideros*). Bijdr Dierkd. 36:45-64.
3575. Gaisler, J. 1970. The bats (Chiroptera) collected in Afghanistan by the Czechoslovak Expeditions of 1965-1967. Acta Sc Nat Acad Sci Bohemoslovacae Brno. 4(6):1-56.
3576. Gaisler, J. 1986. The smallest mammal?. Ziva. 34:197.
3577. Gaisler, J, Hanak, V, & Dungel, J. 1979. A contribution to the population ecology of *Nyctalus noctula* (Mammalia; Chiroptera). Acta Sc Nat Acad Sci Bohemoslovacae Brno. 13:1-38.
3578. Gaisler, J, Holas, V, & Homolka, M. 1977. Ecology and reproduction of Gliridae (Mammalia) in northern Moravia. Folia Zool. 26:213-228.
3579. Gaisler, J & Klíma, M. 1968. Das Geschlechterverhältnis bei Feten und Jungen einiger Fledermausarten. Z Säugetierk. 33:352-357.
3580. Gaisler, J & Titlbach, M. 1964. The male sexual cycle in the lesser horseshoe bat (*Rhinolophus hipposideros hipposideros* Bechstein, 1800). Vestn Čcsl Spol Zool. 28:268-277.

3581. Galat-Luong, A. 1975. Notes préliminaires sur l'écologie de *Cercopithecus ascanius schmidti* dans les environs de Bangui (RCA). Terre Vie. 29:288-297.

3582. Galbreath, GJ. 1985. The evolution of monozygotic polyembryony in *Dasypus*. The Evolution and Ecology of Armadillos, Sloths, and Vermilinguas. 243-246, Montgomery, GG, ed, Smithsonian Inst Press, Washington, DC.

3583. Galdikas, BMF. 1981. Orangutan reproduction in the wild. Reproductive Biology of the Great Apes. 281-300, Graham, CE, ed, Academic Press, New York.

3584. Galdikas, BMF. 1982. Wild orangutan birth at Tanjung Puting reserve. Primates. 23:500-510.

3585. Galla, RR, van de Walle, C, Hoffman, WH, et al. 1977. Lack of a circannual cycle of day time serum prolactin in man and monkey. Acta Endocrinol. 86:257-262.

3586. Galli, FE, Irusta, O, Wasserman, GF, et al. 1972. Biosynthesis of steroid hormones by testes and ovaries of the peludo, *Chaetopharactus villosus* (Edentata, Mammalia). Acta Physiol Lab Am. 22:22-31.

3587. Gambarian, PP & Martirosyan, BA. 1960. On the ecology of *Calomyscus bailwardi* Thomas. Zool Zhurn. 39:1408-1413.

3588. Gambell, R. 1966. Foetal growth and the breeding season of sperm whales. Norsk Hvalfangst-Tidende. 55:113-118.

3589. Gambell, R. 1968. Seasonal cycles and reproduction in sei whales of the southern hemisphere. Discovery Rep. 35:35-131.

3590. Gambell, R. 1972. Sperm whales off Durban. Discovery Rep. 35:199-357.

3591. Gambell, R. 1973. Some effects of exploitation on reproduction in whales. J Reprod Fert Suppl. 19:533-553.

3592. Gambell, R. 1979. The blue whale. Biologist. 26:209-215.

3593. Gambell, R. 1985. Sei whale *Balaenoptera borealis* Lesson, 1828. Handbook of Marine Mammals. 3:155-170. Ridgeway, SH & Harrison, RJ, eds Academic Press, New York.

3593a. Gander, FF. 1927. Experiences with wood rats, *Neotoma fuscipes macrotis*. J Mamm. 10:52-58.

3594. Gander, FF. 1928. Period of gestation in some American mammals. J Mamm. 9:75.

3595. Gander, FF. 1930. Development of captive squirrels. J Mamm. 11:315-317.

3596. Ganfini, C. 1903. Les cellules interstielles du testicule chez les animanx hibernants. Arch Ital Biol. 40:323-324.

3597. Gangloff, B. 1975. Beitrag zur Ethologie der Schleichkatzen (Banderlinsang, *Prionodon linsang* [Hardw], und Banderpalmenroller *Hemigalus derbyansu* [Gray]. Zool Garten NF. 45:329-376.

3598. Gangloff, L. 1972. Breeding fennec foxes *Fennecus zerda* at Strasbourg Zoo. Int Zoo Yearb. 12:115-116.

3599. Gannon, MR, Willig, MR, & Jones, JK, Jr. 1989. *Sturnira lilium*. Mamm Species. 333:1-5.

3600. Gannon, WL. 1988. *Zapus trinotatus*. Mamm Species. 315:1-5.

3601. Ganslosser, U. 1981. Erfolgreiche Aufzucht von Kowaris (*Dasyuroides byrnei*, Marsupalia: Dasyuridae) in der Familiengruppe. Z Kölner Zoo. 24:127-132.

3602. Ganslosser, U & Meissner, K. 1984. Behavioural signs of oestrus in *Dasyuroides byrnei* (Marsupialia: Dasyuridae). Aust Mamm. 7:223-224.

3603. Garagna, S, Zuccotti, M, Searle, JB, et al. 1989. Histological description of the seminiferous epithelium cycle in the common shrew (*Sorex araneus* L). Boll Zool. 56:299-304.

3604. Garagna, S, Zuccotti, M, Searle, JB, et al. 1989. Spermatogenesis in heterozygotes for Robertsonian chromosomal rearrangements from natural populations of the common shrew, *Sorex araneus*. J Reprod Fert. 87:431-438.

3605. Garber, PA, Moya, L, & Malaga, C. 1984. A preliminary field study of the moustached tamarin monkey (*Saguinus mustax*) in northeastern Peru: Questions concerned with the evolution of a communal breeding system. Folia Primatol. 42:17-32.

3606. Gard, KR. 1935. The rat pest in cane areas. Proc Int Soc Sugar Cane Techn 5th Congr, Brisbane, QLD. 5:594-603.

3607. Garde, ML. 1930. The ovary of *Ornithorhynchus* with special reference to follicular atresia. J Anat. 64:422-453.

3607_1. Gardner, AL. 1965. New bat records from the Mexican state of Durango. Proc W Found Vert Zool. 1:101-106.

3607a. Gardner, AL. 1976. The distributional status of some Peruvian mammals. Occas Pap Mus Zool LA State Univ. 48:1-18.

3608. Gardner, AL, LaVal, RK, & Wilson, DE. 1970. The distributional status of some Costa Rican bats. J Mamm. 51:712-729.

3608a. Gardner, AL & O'Neill, JP. 1969. The taxonomic status of *Sturnira bidens* (Chiroptera: Phyllostomatidae) with notes on its karyotype and life history. Occas Pap Mus Zool LA State Univ. 38:1-18.

3609. Garfield, RE & Buchanan, GD. 1985. Myometrial innervation during pregnancy in the bat, *Myotis lucifugus*. Anat Rec. 211:65A.

3610. Garman, A. 1981. Reproduction of levant vole, *Microtus guentheri*, in the laboratory. Israel J Zool. 30:103-104.

3611. Garner, HW. 1974. Population dynamics, reproduction, and activities of the kangaroo rat, *Dipodomys ordii*, in western Texas. Grad Studies, TX Tech Univ. 7:1-28.

3612. Garrett, MG, Hoogland, JL, & Franklin, WL. 1982. Demographic differences between an old and a new colony of prairie dogs (*Cynomys ludovicianus*). Am Mid Nat. 108:51-59.

3613. Garrido-Rodriguez, D & Lopez-Forment, W. 1981. Preliminary data on the reproductive pattern of *Pteronotus parnelli mexicanus* Miller. Bat Res News. 22:39.

3614. Garrison, TE & Best, TL. 1990. *Dipodomys ordii*. Mamm Species. 353:1-10.

3615. Garrod, AH. 1879. Notes on the manatee (*Manatus americanus*) recently living in the Society's Gardens. Trans Zool Soc London. 10:137-146.

3616. Garrod, AH & Turner, W. 1878. On the gravid uterus and placenta of *Hyemoschus aquaticus*. Proc Zool Soc London. 1878:682-686.

3616a. Garrott, RA. 1991. Sex ratios and differential survival of feral horses. J Anim Ecol. 60:929-937.

3616b. Garrott, RA, Eagle, TC, & Plotka, ED. 1991. Age-specific reproduction in feral horses. Can J Zool. 69:738-743.

3617. Garshelis, DL & Garshelis, JA. 1987. Atypical pup rearing strategies by sea otters. Mar Mamm Sci. 3:263-270.

3618. Garstang, R. 1982. An analysis of home range utilization by the tsessebe *Damaliscus lunatus lunatus* (Burchell) in PW Willis Private Nature Reserve. MS Thesis. Univ Pretoria, Pretoria. [Cited by Skinner 1985]

3619. Gartlan, JS. 1969. Sexual and maternal behaviour of the vervet monkey, *Cercopithecus aethiops*. J Reprod Fert Suppl. 6:137-150.

3620. Gasaway, WC, Stephenson, RO, Davis, JL, et al. 1983. Interrelationships of wolves, prey and man in interior Alaska. Wildl Monogr. 84:1-50.

3621. Gashwiler, JS. 1950. A study of the reproductive capacity of Maine muskrats. J Mamm. 31:180-185.

3622. Gashwiler, JS. 1972. Life history notes on the Oregon vole, *Microtus oregoni*. J Mamm. 53:558-569.

3623. Gashwiler, JS. 1976. Notes on the reproduction of trowbridge shrews in western Oregon. Murrelet. 57:58-62.

3624. Gashwiler, JS. 1979. Deer mouse reproduction and its relationship to the tree seed crop. Am Mid Nat. 102:95-104.
3625. Gashwiler, JS, Robinette, W, & Morris, OW. 1961. Breeding habits of bobcats in Utah. J Mamm. 42:76-84.
3625a. Gaskin, DE. 1991. An update on the status of the right whale, *Eubalaena glacialis*, in Canada. Can Field-Nat. 105:198-205.
3626. Gaskin, DE, Arnold, PW, & Blair, BA. 1974. *Phocoena phocoena*. Mamm Species. 42:1-8.
3627. Gatenby, JB. 1922. Some notes on the gametogenesis of *Ornithorhinychus paradoxus*. Q J Microsc Sci. 66:475-499.
3628. Gates, WH. 1937. Notes on the big brown bat. J Mamm. 18:97-98.
3629. Gates, WH. 1941. A few notes on the evening bat, *Nycticeius humeralis* (Rafinesque). J Mamm. 22:53-56.
3630. Gauckler, A. 1962. Beitrag sur Fortpflanzangsbiologie der Wasserspitzmaus (*Neomys fodiens*). Bonn Zool Beitr. 13:321-323.
3631. Gauckler, A & Kraus, M. 1966. Winterbeobachtungen am Abendsegler (*Nyctalus noctula* Schreber, 1774). Säugetierk Mitt. 14:22-27.
3632. Gaughwin, MD. 1982. Southern hairy-nosed wombat *Lasiorhinus latifrons*: Its maintenance, behaviour and reproduction in captivity. The Management of Australian Mammals in Captivity. 144-155, Evans, DD, ed, Zool Bd VIC, Melbourne.
3633. Gaumer, GF. 1917. Mamiferos de Yucatán. Dept Talleres Graficos Secretaria, Mexico.
3634. Gauquelin, MF. 1968. Le cycle annuel de reproduction du Macaque, *Macaca irus*. Bull Biol Fr Belg. 102:261-270.
3635. Gauthier, D & Barrette, C. 1985. Suckling and weaning in captive white-tailed and fallow deer. Behaviour. 94:128-149.
3636. Gauthier-Pilters, H. 1959. Einige Beobachtungen zum Droh-Angriffs-und Kampfverhalten des Dromedarhengstes, so wie über die Geburt und Verhaltensentwicklung des Jungtieres, in der nordwestlechen Sahara. Z Tierpsychol. 16:593-604.
3637. Gauthier-Pilters, H. 1967. The fennec. Afr Wildl. 21:117-125.
3638. Gauthier-Wright, F, Baudot, N, & Mauvais-Jarvis, P. 1973. Testosterone-binding globulin in stump-tailed macaque monkeys (*Macaca speciosa*). Endocrinology. 93:1277-1286.
3639. Gautier-Hion, A. 1966. L'écologie et l'éthologie du Talapoin *Miopithecus talapoin talapoin*. Biol Gabon. 2:311-329.
3640. Gautier-Hion, A. 1968. Etude du cycle annuel de reproduction du Talpoin (*Miopithecus talapoin*) vivant dans son milieu naturel. Biol Gabon. 4:163-173.
3641. Gautier-Hion, A. 1971. L'écologie du Talapoin du Gabon. Terre Vie. 118:427-490.
3642. Gautier-Hion, A. 1971. Répertoire comportemental du Talapoin (*Miopithecus talapoin*). Biol Gabon. 7:295-391.
3643. Gautier-Hion, A & Gautier, J-P. 1976. Growth, sexual maturity, social maturity and reproduction in African forest cercopithecines. Folia Primatol. 26:165-184.
3644. Gautun, JC. 1972. Note sur la dureé de gestation de *Lemniscomys striatus* en Cote d'Ivoire. Mammalia. 36:309-310.
3645. Gautun, JC. 1975. Periodicité de la reproduction de quelques Rongeurs d'une savanne préforestiere du centre de la Cote-d'Ivoire. Terre Vie. 29:265-287.
3646. Gautun, JC & Sicard, B. 1985. Records de fertilité de *Mastomys huberii* au Burkina Faso (ex Haute-Volta). Mammalia. 49:579.
3647. Gavin, A. 1945. Notes on mammals observed in the Perry River district of Queen Maud Sea. J Mamm. 26:226-230.
3648. Gavish, L, Carter, CS, & Getz, LL. 1981. Further evidences for monogamy in the prairie vole. Anim Behav. 29:955-957.
3649. Gavrin, WF & Donaurov, SS. 1954. Der Wolf in Walde von Bielowieza. Zool Zhurn. 33:904-924.
3650. Gawe, L, Goldau, A, Künne, WD, et al. 1974. Zur Haltung und Zucht von Totenkopfaffen *Saimiri sciureus* Paramaribo (Surinam). Fr Kölner Zoo. 17:111-118.
3651. Gear, HS. 1926. The oestrous cycle of the baboon. S Afr J Sci. 23:706-712.
3652. Gębczynska, Z. 1967. Reproduction of *Lagurus lagurus* (Pallas, 1773) in the laboratory. Acta Theriol. 12:521-531.
3653. Gębczynski, M. 1966. Altersvariabilität des Metabolismus und der Aktivität bei *Sorex araneus* L. Lynx. 6:41-44.
3654. Gębczynski, M & Gębczynska, Z. 1984. The energy cost of nesting growth in the European pine vole. Acta Theriol. 29:231-241.
3655. Gębczynski, M & Gębczynska, Z. 1985. Parental energy investment into offspring growth in the root vole. Zesz Nauk Filii UW. 10:99-104.
3656. Gee, EP. 1953. The life history of the great Indian one-horned rhinoceros (*R unicornis* Linn). J Bombay Nat Hist Soc. 51:341-348.
3657. Gee, EP. 1955. Great Indian one-horned rhinoceros (*R unicornis* Linn) cow with (presumptive) twin calves. J Bombay Nat Hist Soc. 53:256-257.
3658. Gee, EP. 1955. The Indian elephant (*E maximus*): Early growth gradient and intervals between calving. J Bombay Nat Hist Soc. 53:125-128.
3659. Geerling, C & Bokdam, J. 1971. The Senegal kob, *Adenota kob kob* (Erxleben), in the Comoe National Park, Ivory Coast. Mammalia. 35:17-24.
3660. Geidel, B & Gensch, W. 1976. The rearing of clouded leopards *Neofelis nebulosa* in the presence of the male. Int Zoo Yearb. 16:124-126.
3661. Geigy, R. 1955. Observations sur les Phacochéres du Tanganyika. Rev Suisse Zool. 62(Suppl):139-163.
3662. Geinitz, Ch. 1980. Beiträge zur Biologie des Streifenhörnchens (*Eutamias sibericus* Laxmann, 1769), auf einem Friedhof in Freiburg (Süd deutschland). Z Säugetierk. 45:279-287.
3663. Geiser, F, Huber, W, & Wandeler, A. 1976. Zum Geschlechtszyklus der Gemsgeiss (*Rupicapra rupicapra*). Rev Suisse Zool. 83:948-951.
3664. Geiser, F, Matwiejczyk, L, & Baudinette, RV. 1986. From ectothermy to heterothermy: The energetics of the kowari, *Dasyuroides byrnei* (Marsupialia: Dasyuridae). Physiol Zool. 59:220-229.
3665. Geissmann, T. 1986. Mate change enhances duetting activity in the siamang gibbon (*Hylobates syndactylus*). Behaviour. 96:17-27.
3666. Geist, V. 1963. On the behaviour of the North American moose (*Alces alces andersoni* Peterson 1950) in British Columbia. Behaviour. 20:377-416.
3667. Geist, V. 1971. Mountain Sheep. Univ Chicago Press, Chicago.
3668. Geist, V. 1981. Behavior: Adaptive strategies in mule deer. Mule and Black-tailed Deer of North America. 157-223, Walmo, CC, ed, Univ NE, Lincoln, NE.
3669. Gemmell, RT. 1979. The fine structure of the luteal cells in relation to the concentration of progesterone in the plasma of the lactating bandicoot, *Isoodon macrourus* (Marsupialia: Peramelidae). Aust J Zool. 27:501-510.

3670. Gemmell, RT. 1981. The role of the corpus luteum of lactation in the bandicoot *Isoodon macrourus* (Marsupialia: Peramelidae). Gen Comp Endocrinol. 44:13-19.
3671. Gemmell, RT. 1982. Breeding bandicoots in Brisbane (*Isoodon macrourus*; Marsupialia, Peramelidae). Aust Mamm. 5:187-193.
3672. Gemmell, RT. 1984. Plasma concentrations of progesterone and 13,14-dihydro-15-keto-prostaglandin $F_{2\alpha}$ during regression of the corpora lutea of lactation in the bandicoot (*Isoodon macrourus*). J Reprod Fert. 72:295-299.
3673. Gemmell, RT. 1985. The effect of prostaglandin $F_{2\alpha}$ analog on the plasma concentration of progesterone in the bandicoot, *Isoodon macrourus* (Marsupialia: Peramelidae), during lactation. Gen Comp Endocrinol. 57:405-410.
3674. Gemmell, RT. 1986. Sexual maturity in the female bandicoot *Isoodon macrourus* (Gould) in captivity. Aust J Zool. 34:199-204.
3675. Gemmell, RT. 1987. Effect of melatonin and removal of young on the seasonality and births in the marsupial possum. J Reprod Fert. 80:301-307.
3676. Gemmell, RT. 1987. Influence of melatonin on the initiation of the breeding season of the marsupial bandicoot, *Isoodon macrourus*. J Reprod Fert. 79:261-265.
3677. Gemmell, RT. 1987. Sexual maturity in the captive male bandicoot, *Isoodon macrourus*. Aust J Zool. 35:433-441.
3678. Gemmell, RT. 1988. The oestrous cycle length of the banicoot *Isoodon macrourus*. Aust Wildl Res. 15:633-635.
3679. Gemmell, RT. 1989. The persistence of the corpus luteum of pregnancy into lactation in the marsupial bandicooot, *Isoodon macrourus*. Gen Comp Endocrinol. 75:355-362.
3680. Gemmell, RT. 1989. Survival of pouch young and juvenile bandicoots, *Isoodon macrourus* (Marsupialia: Peramelidae), in captivity. Aust Mamm. 12:73-76.
3681. Gemmell, RT. 1989. Breeding season and litter size of the bandicoot, *Isoodon macrourus* (Marsupialia: Peramelidae), in captivity. Aust Mamm. 12:77-79.
3682. Gemmell, RT. 1990. Influence of daylength on the initiation of the breeding season of the marsupial possum, *Trichosurus vulpecula*. J Reprod Fert. 88:605-609.
3683. Gemmell, RT, Cepon, G, & Barnes, A. 1987. The development of thermoregulation in the rat kangaroo, *Potorous tridactylus*. Comp Biochem Physiol. 88A:257-261.
3684. Gemmell, RT, Jenkin, G, & Thorburn, GD. 1980. Plasma concentrations of progesterone and 13,14-dihydro-15-keto-prostaglandin $F_{2\alpha}$ at parturition in the bandicoot, *Isoodon macrourus*. J Reprod Fert. 60:253-256.
3685. Gemmell, RT & Johnston, G. 1985. The development of thermoregulation and the emergence from the pouch of the marsupial bandicoot *Isoodon macrourus*. Physiol Zool. 58:299-302.
3686. Gemmell, RT, Johnston, G, & Barnes, A. 1984. The uniformity of growth within the litter of the marsupial, *Isoodon macrourus*. Growth. 48:221-233.
3687. Gemmell, RT, Johnston, G, & Barnes, A. 1985. Seasonal variations in plasma testosterone concentrations in the male marsupial bandicoot *Isoodon macrourus* in captivity. Gen Comp Endocrinol. 59:184-191.
3688. Gemmell, RT & Sernia, C. 1989. Immunocytochemical location of oxytocin and mesotocin within the hypothalamus of two Australian marsupials, the bandicoot *Isoodon macrourus*and the brushtail possum *Trichosurus vulpecula*. Gen Comp Endocrinol. 75:96-102.
3689. Gemmell, RT & Sernia, C. 1989. The localization of oxytocin and mesotocin in the reproductive tract of the male marsupial bandicoot *Isoodon macrourus*. Gen Comp Endocrinol. 75:103-109.
3690. Gemmell, RT, Singh-Asa, P, Jenkin, G, et al. 1982. Ultrastructural evidence for steroid hormone production in the adrenal of the marsupial, *Isoodon macrourus*, at birth. Anat Rec. 203:505-512.
3691. Gemmell, RT, Walker, MT, Johnston, G, et al. 1984. The number of young present in sequential litters of the marsupial bandicoot, *Isoodon macrourus*, in captivity. Aust J Zool. 32:623-629.
3692. Genelly, RE. 1965. Ecology of the common mole-rat (*Cryptomys hottentotus*) in Rhodesia. J Mamm. 46:647-665.
3692a. Genest-Villard, H. 1968. L'estomac de *Lophuromys sikapusi* (Temminck) (Rongeurs, Muridé). Mammalia. 32:639-656.
3693. Genest-Villard, H. 1979. Ecologie de *Steatomys opimus* Pousargues, 1894 (Rongeurs, Dendromurides) en Afrique centrale. Mammalia. 43:275-294.
3694. Gengozian, N & Meritt, CB. 1970. Effect of unilateral ovariectomy on twinning frequency in the marmoset. J Reprod Fert. 23:509-512.
3695. Gengozian, N, Batson, JS, & Eide, P. 1964. Hematologic and cytogenetic evidence for chimerism in the marmoset, *Tamarinus nigricollis*. Cytogenetics. 3:384-393.
3696. Gengozian, N, Batson, JS, & Smith, TA. 1977. Breeding of tamarins (*Saguinus* spp) in the laboratory. The Biology and Conservation of the Callitrichidae. 207-213, Kleiman, DG, ed, Smithsonian Inst Press, Washington, DC.
3697. Gengozian, N, Batson, JS, & Smith, TA. 1978. Breeding of marmosets in a colony environment. Marmosets in Experimental Medicine. 71-78, Gengozian, N & Deinhardt, F, eds, S Karger, New York.
3698. Genoud, M & Vogel, P. 1990. Energy requirements during reproduction and reproductive effort in shrews (Soricidae). J Zool. 220:41-60.
3699. Genoways, HH & Baker, RJ. 1972. *Stenoderma rufum*. Mamm Species. 18:1-4.
3700. Genoways, HH & Birney, EC. 1974. *Neotoma alleni*. Mamm Species. 41:1-4.
3701. Genoways, HH & Jones, JK, Jr. 1968. Notes on bats from the Mexican state of Zacatecas. J Mamm. 49:743-745.
3702. Genoways, HH & Jones, JK, Jr. 1969. Notes on pocket gophers from Jalisco, Mexico, with descriptions of two new subspecies. J Mamm. 50:748-755.
3702a. Genoways, HH & Jones, JK, Jr. 1971. Systematics of southern banner-tailed kangaroo rats of the *Dipodomys phillipsii* group. J Mamm. 52:265-287.
3703. Genoways, HH & Jones, JK, Jr. 1972. Mammals from southwestern North Dakota. Occas Pap Mus TX Tech Univ. 6:1-36.
3704. Genoways, HH & Jones, JK, Jr. 1972. Variation and ecology in a local population of the vesper mouse (*Nyctomys sumichrasti*). Occas Pap Mus TX Tech Univ. 3:1-22.
3705. Genoways, HH & Jones, JK, Jr. 1973. Notes on some mammals from Jalisco, Mexico. Occas Pap Mus TX Tech Univ. 9:1-22.
3705a. Genoways, HH & Jones, JK, Jr. 1975. Additional records of the stenodermine bat, *Sturnira thomasi*, from the Lesser Antillean island of Guadelopue. J Mamm. 56:924-925.
3705b. Genoways, HH & Williams, SL. 1979. Notes on bats (Mammalia: Chiroptera) from Bonaire and Curaçao, Dutch West Indies. Ann Carnegie Mus. 48:311-321.

3705c. Genoways, HH & Williams, SL. 1979. Records of bats (Mammalia: Chiroptera) from Suriname. Ann Carnegie Mus. 48:323-335.
3706. Gensch, W. 1962. Breeding tupaias *Tupaia glis* Diard. Int Zoo Yearb. 4:75-76.
3707. Gensch, W. 1962. Successful rearing of the binturong *Arctitis binturong*. Int Zoo Yearb. 4:79-80.
3708. Gensch, W. 1963. Erste erfolgreiche Nachzucht des Binturong (*Arctictis binturong* Raffl) im Zoo Dresden. Zool Garten NF. 28:89-92.
3709. Gensch, W. 1964. Hysterektomie bei einem Spitzhörnchen (*Tupaia glis* Diard). Zool Garten, NF. 29:194-195.
3710. Gensch, W. 1965. Birth and rearing of a black-faced spider monkey *Ateles paniscus chamek* Humboldt at Dresden Zoo. Int Zoo Yearb. 5:110.
3711. Gensch, W. 1965. Birth and rearing of a spectacled bear *Tremarctos ornatus* at Dresden Zoo. Int Zoo Yearb. 5:111.
3712. Gensch, W. 1966. Kurze Mitteilung uber die Zucht des Borstengürteltieres (*Euphractes villosus villosus* Fisch) im Zoologischen Garten Dresden. Zool Garten NF. 32:216-218.
3713. Gensch, W. 1966. Nochmals zur Nachzucht des Binturong (*Arctitis binturong* Raffl) im Zoologischen Garten Dresden. Zool Garten NF. 33:126-128.
3714. Gensch, W. 1968. Notes on breeding timber wolves at Dresden Zoo. Int Zoo Yearb. 8:15-16.
3714a. Gensch, W. 1969. Bemerkenswert kurze Tragzeit beim Hausrentier (*Rangifer tarandus* L). Zool Garten. 37:150-151.
3715. Gentile, R, Sciscioli, V, Petrosino, G, et al. 1989. The presence of epithelial crypts in the epididymis of chamois *Rupicapra rupicapra* L Preliminary report. Boll Soc Ital Biol Sper. 65:549-554.
3716. Gentry, RL. 1970. Social behavior of the Steller sea lion. Ph D Diss. Univ CA, Santa Cruz.
3717. Gentry, RL. 1981. Northern fur seal *Callorhinus ursinus* (Linnaeus, 1758). Handbook Marine Mammals. 1:143-160. Ridgway, SH & Harrison, RJ, eds, Academic Press, New York.
3718. Gentry, RL, Costa, DP, Croxall, JP, et al. 1986. Synthesis and conclusions. Fur Seals, Maternal Strategies on Land and at Sea. 220-264, Gentry, RL & Kooyman, GL, eds, Princeton Univ Press, Princeton, NJ.
3719. Gentry, RL & Holt, JR. 1986. Attendance behavior of northern fur seals. Fur Seals, Maternal Strategies on Land and at Sea. 41-60, Gentry, RL & Kooyman, GL, eds, Princeton Univ Press, Princeton.
3720. Gentry, RL & Kooyman, GL. 1986. Introduction. Fur Seals, Maternal Strategies on Land and at Sea. 3-27, Gentry, RL & Kooyman, GL, eds, Princeton Univ Press, Princeton.
3721. George, GG. 1990. Monotreme and marsupial breeding programs in Australian zoos. Aust J Zool. 37:181-205.
3722. George, FW, Hodgins, MB, & Wilson, JD. 1985. The synthesis and metabolism of gonadal steroids in pouch young of the opossum, *Didelphis virginiana*. Endocrinology. 116:1145-1150.
3723. George, JL & Stitt, M. 1951. March litters of raccoons (*Procyon lotor*) in Michigan. J Mamm. 32:218.
3724. George, SB. 1989. *Sorex trowbridgii*. Mamm Species. 337:1-5.
3725. George, SB, Choate, JR, & Genoways, HH. 1986. *Blarina brevicauda*. Mamm Species. 261:1-9.
3726. George, W. 1974. Notes on the ecology of gundis (F Ctenodactylidae). Symp Zool Soc London. 34:143-160.
3727. George, W. 1978. Reproduction in female gundis (Rodentia: Ctenodactylidae). J Zool. 185:57-71.
3727a. Geraci, JR & St Aubin, DJ. 1977. Mass stranding of the long-finned pilot whale, *Globicephala melaena*, on Sable Island, Nova Scotia. J Fish Res Bd Can. 34:2196-2199.
3728. Gérard, P. 1920. Contribution à l'étude de l'ovaire des Mammifères l'ovaire de *Galago mossambicus* (Young). Arch Biol. 30:357-391.
3729. Gérard, P. 1923. Etude sur les modifications de l'uterus pendant la gestation chez *Nasilio brachyrhynchus* (Smith). Arch Biol. 33:197-227.
3730. Gérard, P. 1928. Recherches histophysiologiques sur les annexes foetales des Chéiroptères (*Vesperugo noctula* Schreb). Arch Biol. 38:327-354.
3731. Gérard, P. 1929. Contribution à l'étude de la placentation chez les Lemuriens: A propos d'une anomalie de la placentation chez *Galago demidoffi* (Fisch). Arch Anat Micr. 25:56-68.
3732. Gérard, P. 1932. Etudes sur l'ovogenèse et l'ontogenése chez les Lemuriens de genre Galago. Arch Biol. 43:93-151.
3733. Gérard, P & Herlant, M. 1953. Sur la persistance de phénomènes d'oogenèse chez les Lémuriens adultes[s]. Arch Biol. 64:97-111.
3734. Gerell, R. 1970. Home ranges and movements of the mink *Mustela vison* Schreber in southern Sweden. Oikos. 21:160-173.
3735. Gerell, R. 1970. Population studies on mink, *Mustela vison* Schreber, in southern Sweden. Viltrevy. 8:83-109.
3736. Gerell, R, Gambell, R, & Brown, SG. 1969. Activity patterns of the mink *Mustela vison* Schreber in southern Sweden. Oikos. 20:451-460.
3737. Gerhardt, U. 1905. Bemerkungen über das Urogenital system des weiblechen Gorilla. Verh Deutschen Zool Gesellsch. 15:135-140.
3738. Gerhardt, U. 1909. Über das Vorkommen eines Penis-und Clitorisknochen bei Hylobatiden. Anat Anz. 35:353-358.
3739. Gervais, HP. 1883. Sur un utérus gravide de Pontoporia Blainvillei. CR Hebd Séances Acad Sci (Paris). 97:760-762.
3740. Gettinger, RD. 1984. Energy and water metabolism of free-ranging pocket gophers, *Thomomys bottae*. Ecology. 65:740-751.
3741. Getz, LL. 1960. A population study of the vole, *Microtus pennsylvanicus*. Am Mid Nat. 64:392-405.
3742. Getz, LL. 1961. Home ranges, territoriality, and movement of the meadow vole. J Mamm. 42:24-36.
3743. Getz, LL. 1989. A 14-year study of *Blarina brevicauda* populations in east-central Illinois. J Mamm. 70:58-66.
3744. Getz, LL & Carter, CS. 1980. Social organization in *Microtus ochrogaster* populations. Biologist. 62:56-69.
3745. Getz, LL, Carter, CS, & Gavish, L. 1981. The mating system of the prairie vole, *Microtus ochrogaster*: Field and laboratory evidence for pair-bonding. Behav Ecol Sociobiol. 8:189-194.
3746. Getz, LL, Dluzen, D, & McDermott, JL. 1983. Suppression of female reproductive maturation in *Microtus* by a female urine pheromone. Behav Proc. 8:59-64.
3747. Getz, LL & Miller, DH. 1969. Life-history notes on *Microtus pinetorum* in central Connecticut. J Mamm. 50:777-784.
3748. Gevaerts, H & Upoki, A. 1987. Le cycle de reproduction observé chez les Cercopithèques de la forêt ombrophile dans la région de Kisangani Pres de l'équateur. Rev Zool Afr. 101:284-287.
3749. Gewalt, W. 1952. Beobachtungen über die Aufzucht von Eichhörnchen (*Sciurus vulgaris*) in der Gefangenschaft. Zool Garten NF. 19:26-33.

3750. Gewalt, W. 1968. Kleine Beobachtungen an seltenere Beuteltieren im Berliner Zoo V Zwergbeutelratte (*Marmosa mexicana* Merriam 1897). Zool Garten NF. 35:288-303.
3751. Gewalt, W. 1980. Über einige seltenere Nachzuchten im Zoo Duisburg 2 Beutelteufel (*Sarcophillus harrisi*) (Boitard). Zool Garten NF. 50:138-154.
3752. Geyer, H. 1972. Anatomische Untersuchungen am Harn und Geschlechtsapparat des Chinesisches Zwerghamsters (*Cricetulus griseus*). Z Versuchstierk. 14:107-123.
3753. Geyer, H, Bertschinger, H, Strittmatter, J, et al. 1975. Erbliche Unfruchtbarkeit bei männlichen weiszen Chinesenhamstern (*Cricetulus griseus* Milne Edwards 1867). Z Versuchstierk. 17:78-90.
3754. Ghazi, R. 1981. Angioarchitectural studies of the utero-ovarian component in the camel (*Camelus dromedarius*). J Reprod Fert. 61:43-46.
3755. Ghobrial, LI & Hodieb, ASK. 1973. Climate and seasonal variations in the breeding of the desert jerboa, *Jaculus jaculus*, in the Sudan. J Reprod Fert Suppl. 19:221-233.
3756. Ghobrial, LI & Hodieb, ASK. 1982. Seasonal variations in the breeding of the Nile rat (*Arvicanthis niloticus*). Mammalia. 46:319-333.
3757. Ghose, RK. 1964. A new rat of the genus *Rattus* Fischer, 1803 (Mammalia: Rodentia), from Darjeeling district, West Bengal, India. Proc Zool Soc (Calcutta). 17:193-197.
3758. Ghose, RK. 1965. A new species of mongoose [Mammalia: Carnivora: Viverridae] from West Bengal, India. Proc Zool Soc (Calcutta). 18:173-178.
3759. Ghose, RK. 1981. On the ecology and status of the hispid hare. Proc World Lagomorph Conf, Univ Guelph, ONT, Canada. 1979:918-924.
3760. Ghosh, M, Hutz, RJ, & Dukelow, WR. 1982. Serum estradiol-17β, progesterone and relative luteinizing hormone levels in *Saimiri sciurus*: Cyclic variations and the effect of laparoscopy and follicular aspiration. J Med Primatol. 11:312-318.
3761. Gianini, CA. 1932. Birth of mule deer fawns. J Mamm. 13:80.
3762. Gibber, JR. 1986. Infant-directed behavior of rhesus monkeys during their first pregnancy and parturition. Folia Primatol. 46:118-124.
3763. Gibbney, L. 1953. Delayed implantation in the elephant seal. Nature. 172:590-591.
3764. Gibson, RA, Neumann, M, Grant, TR, et al. 1988. Fatty acids of the milk and food of the platypus (*Ornithorhynchus anatinus*). Lipids. 23:377-379.
3765. Gibson, RM & Guiness, FE. 1980. Behavioural factors affecting male reproductive success in red deer (*Cervus elaphus*). Anim Behav. 28:1163-1174.
3766. Gienc, J & Doboszyńska, T. 1972. Macromorphological description of the genital organs of the female beaver. Acta Theriol. 17:399-406.
3767. Gier, HT. 1947. Populations and reproductive potentials of foxes in Ohio. Midwest Wildl Conf. 9:1-6.
3768. Gier, HT. 1968. Coyotes in Kansas. KS Agr Exp Sta Bull. 393:1-118.
3769. Gier, HT. 1975. Ecology and behavior of the coyote (*Canis latrans*). The Wild Canids. 247-262, Fox, MW, ed, Van Nostrand Reinhold Co, New York.
3770. Gieszczykiewicz, J. 1979. Observations on periodic changes in the sexual behaviour of a grey capuchin monkey *Cebus griseus* Cuvier. Folia Biol (Kraków). 27:235-238.
3771. Gihr, M & Pilleri, G. 1969. On the anatomy and biometry of *Stenella styx* Gray and *Delphinus delphis* L (Cetacea, Delphinidae) of the western Mediterranean. Invest Cetacea. 1:15-65.
3772. Gihr, M & Pilleri, G. 1979. Interspecific body length-body weight ratio and body weight-brain weight ratio in Cetacea. Invest Cetacea. 10:245-253.
3773. Gihr, M, Pilleri, G, & Zhou, K. 1979. Cephalization of the Chinese river dolphin *Lipotes vexillifer* (Platanistoidea, Lipotidae). Invest Cetacea. 10:257-274.
3774. Gilbert, BS & Krebs, CJ. 1981. Effects of extra food on *Peromyscus* and *Clethrionomys* populations in the southern Yukon. Oecologica. 51:326-331.
3775. Gilbert, BS, Krebs, CJ, Talarico, D, et al. 1986. Do *Clethrionomys rulilus* females suppress maturation of juvenile females?. J Anim Ecol. 55:543-552.
3776. Gilbert, C & Gillman, J. 1951. Pregnancy in the baboon (*Papio ursinus*). S Afr J Med Sci. 16:115-124.
3777. Gilbert, FF & Bailey, ED. 1967. The effect of visual isolation on reproduction in the female ranch mink. J Mamm. 48:113-118.
3778. Gilbert, N & Myers, K. 1981. Comparative dynamics of the Australian rabbit. Proc World Lagomorph Conf, Univ Guelph, ONT, Canada. 1979:648-653.
3779. Gilbert, O & Stebbings, R. 1958. Winter roosts of bats in West Suffolk. Proc Zool Soc London. 131:329-333.
3780. Gileva, EA & Chebotar, NA. 1979. Fertile XO males and females in the varying lemming, *Dicrostonyx torquatus* Pall: A unique genetic system of sex determination. Heredity. 42:67-77.
3781. Gillet, H. 1965. L'Oryx algazelle et l'Addax au Tchad. Terre Vie. 19:257-272.
3782. Gillman, J. 1942. Effects on the perineal swelling and on the menstrual cycle of single injections of combinations of estradiol benzoate and progesterone given to baboons in the first part of the cycle. Endocrinology. 30:54-60.
3783. Gillman, J & Gilbert, C. 1946. The reproductive cycle of the chacma baboon (*Papio ursinus*) with special reference to the problems of menstrual irregularities as assessed by the behaviour of the skin. S Afr J Med Sci Biol Suppl. 11:1-54.
3784. Gilmore, DP. 1969. Seasonal reproductive periodicity in the male Australian brush-tailed possum (*Trichosurus vulpecula*). J Zool. 157:75-98.
3785. Gilmore, DP. 1984. Organ-body weight relationships in the common brushtail posssom *Trichosurus vulpecula* (Marsupialia: Phalangeridae). Aust Mamm. 7:131-138.
3786. Gilmore, DP, Pere's da Costa, C, Valenca, M, et al. 1991. Effects of exogenous LHRH on plasma LH and sex steroid levels in the three-toed sloth *Bradypus tridactylus*. Med Sci Res. 19:333-336.
3787. Gilmore, DP, Sirett, NE, & Purves, HP. 1968. Gonadotrophins in the pituitary of the male possum *Trichosurus vulpecula* during the seasonal breeding cycle. J Endocrinol. 42:155-156.
3788. Gilmore, RM. 1950. The California gray whale. Zoonooz. Feb:4-5.
3789. Gilmore, RM. 1971. A census of the California gray whale. US Fish Wildl Serv Sci Rep Fish. 342:1-30.
3790. Gilmore, RM & Gates, JE. 1985. Habitat use by the southern flying squirrel at a hemlock-northern hardwood ecotone. J Wildl Manage. 49:703-710.
3791. Ginther, OJ, Dierschke, DJ, Walsh, SW, et al. 1974. Anatomy of arteries and veins of uterus and ovaries in rhesus monkeys. Biol Reprod. 11:205-219.
3792. Gipps, JHW, Taitt, MJ, Krebs, CJ, et al. 1981. Male aggression and the population dynamics of the vole, *Microtus townsendii*. Can J Zool. 59:147-157.

3793. Gipson, PS, Gipson, IK, & Sealander, JA. 1975. Reproductive biology of wild *Canis* (Canidae) in Arkansas. J Mamm. 56:605-612.
3794. Girod, C & Curé, M. 1965. Etude des corrélations hypophyso-testiculaire au cours du cycle annuel chez le Herisson (*Erinaceus europaeus* L). CR Hebd Séances Acad Sci. 261:257-260.
3795. Girod, C, Dubois, P, & Curé, M. 1967. Recherches sur les corrélations hypophyso-génitales chez la femelle de Hérisson (*Erinaceus europaeus* L). Ann Endocrinol. 28:281-610.
3796. Gittelman, JL. 1988. Behavioral energetics of the lactation in a herbivorous carnivore, the red panda (*Ailurus fulgens*). Ethology. 79:13-24.
3797. Giuliano, WM, Litvaitis, JA, & Stevens, CL. 1989. Prey selection in relation to to sexual dimorphism of fishers (*Martes pennanti*). J Mamm. 70:639-640.
3798. Given, RL & Enders, AC. 1978. Mouse uterine glands during the delayed and induced implantation periods. Anat Rec. 190:271-284.
3799. Glander, KE. 1975. Habitat and resource utilization: An ecological view of social organization in mantled howling monkeys. Ph D Diss. Univ Chicago, Chicago.
3800. Glander, KE. 1980. Reproduction and population growth in free-ranging mantled howling monkeys. Am J Phys Anthrop. 53:25-36.
3801. Glanz, WE, Thorington, RW, JR, Giacalone-Madden, J, et al. 1982. Seasonal food use and demographic trends in *Sciurus*. The Ecology of a Tropical Forest. 239-252, Leith, EG, Jr, Rand, AS, & Windsor, DM, eds, Smithsonian Inst Press, Washington, DC.
3802. Glas, GH. 1974. Over lichaamsmaten en gewichten van de bunzing *Mustela putorius* Linnaeus, 1758, in Nederland. Lutra. 16:13-19.
3803. Glass, BP. 1949. Reproduction in the pine vole, *Pitymys nemoralis*. J Mamm. 30:72-73.
3803a. Glass, BP & Ward, CM. 1959. Bats of the genus *Myotis* from Oklahoma. J Mamm. 40:194-201.
3804. Glass, JD. 1986. Short photoperiod-induced gonadal regression: Effects on the gonadotropin-releasing hormone (GnRH) neuronal system of the white-footed mouse, *Peromyscus leucopus*. Biol Reprod. 35:733-743.
3805. Glass, JD & Lynch, GR. 1981. The effect of superficial pinealectomy on reproduction and brown fat in the adult white-footed mouse, *Peromyscus leucopus*. J Comp Physiol. 144:145-152.
3806. Glasser, SR, Wright, C, & Heyssel, RM. 1968. Transfer of iron across the placenta and fetal membranes in the rat. Am J Physiol. 215:205-210.
3807. Glassman, DM & Coelho, AM, Jr. 1987. Principal components analysis of physical growth in savannah baboons. Am J Phys Anthrop. 72:59-66.
3808. Glassman, DM & Coelho, AM, Jr. 1988. Formula-fed and breast-fed baboons: Weight growth from birth to adulthood. Am J Primatol. 16:131-142.
3809. Glassman, DM, Coelho, AM, Jr, Carey, KD, et al. 1984. Weight growth in savannah baboons: A longitudinal study from birth to adulthood. Growth. 48:425-433.
3810. Glatston, ARH. 1979. Reproduction and behaviour of the lesser mouse lemur (*Microcebus murinus*, Miller 1777) in captivity. Ph D Diss. Univ London, Univ College.
3811. Glatston, ARH. 1981. The husbandry, breeding and hand-rearing of the lesser mouse lemur *Microcebus murinus* at Rotterdam Zoo. Int Zoo Yearb. 21:131-137.
3812. Glazier, DS. 1979. An energetic and ecological basis for different reproductive rates in species of *Peromycus* (mice). Ph D Diss. Cornell Univ.
3813. Glazier, DS. 1985. Energetics of litter size in five species of *Peromyscus* with generalizations for other mammals. J Mamm. 66:629-642.
3814. Glazier, DS. 1985. Relationship between metabolic rate and energy expenditure for lactation in *Peromyscus*. Comp Biochem Physiol. 80A:587-590.
3815. Gleeson, SK. 1987. Fitness, reproductive value, and Cole's result. Oikos. 48:116-119.
3816. Glendenning, R. 1959. Biology and control of the coast mole, *Scapanus orarius orarius* True in British Columbia. Can J Anim Sci. 39:34-44.
3817. Glick, BB. 1984. Male endocrine responses to females: Effects of social cues in cynomolgus macaques. Am J Primatol. 6:229-239.
3818. Glitsch, C. 1865. Beiträge zur Naturgeschichte der *Antelope saiga* Pallas. Bull Soc Imp Natur Moscow. 36:207-245.
3819. Gliwicz, J. 1973. A short characteristic of a population of *Proechimys semispinosus* (Tomes, 1860): A rodent species of the tropical rain forest. Bull Acad Polon Sci Biol Ser. 21:413-418.
3820. Gliwicz, J. 1984. Population dynamics of the spiny rat *Proechimys semispinosus* on Orchid Island (Panama). Biotropica. 16:73-78.
3821. Gliwicz, J. 1985. Rodent community of dry African savanna: Population study. Mammalia. 49:509-516.
3822. Glockner-Ferrari, DA & Ferrari, MJ. 1984. Reproduction in humpback whales, *Megaptera novaeangliae* in Hawaiian waters. Rep Int Whal Comm Sp Issue. 6:237-242.
3822a. Glover, R. 1947. The wisent or European bison. J Mamm. 28:333-342.
3823. Glover, TD. 1973. Aspects of sperm production in some east African mammals. J Reprod Fert. 35:45-53.
3824. Glover, TD & Sale, JB. 1968. The reproductive system of the male rock hyrax (*Procavia* and *Heterohyrax*). J Zool. 156:351-362.
3825. Gluchowski, W & Maciejowski, J. 1958. Investigation on factors controlling fertility in the coypu II: Attempts at determining potential fertility based on histological studies of the ovary. Ann Univ Mariae Curie-Sklodowska. 13E:345-361.
3826. Gnan, SO & Sheriha, AM. 1986. Composition of Libyan camel's milk. Aust J Dairy Tech. 41:33-35.
3827. Goddard, J. 1966. Mating and courtship of the black rhinoceros (*Diceros bicornis*). E Afr Wildl J. 4:69-75.
3828. Goddard, J. 1967. Home range, behaviour, and recruitment rates of two black rhinoceros populations. E Afr Wildl J. 5:133-150.
3829. Goddard, J. 1970. A note on age at sexual maturity in wild black rhinoceros. E Afr Wildl J. 8:205.
3830. Goddard, J. 1970. Age criteria and vital statistics of a black rhinoceros population. E Afr Wildl J. 8:105-122.
3831. Godet, R. 1947. Action des gonadotrophines sériques sur l'ovaire de la Taupe. CR Soc Biol. 141:1100-1102.
3832. Godet, R. 1947. Variations périodiques de l'interstielle medullaire dans l'ovaire de la Taupe. CR Soc Biol. 141:1102-1103.
3833. Godet, R. 1949. Recherches d'anatomie, d'embryologie normale et expérimentale sur l'appariel génital de la Taupe (*Talpa europoea* L) [*sic*]. Bull Biol Fr Belg. 83:25-111.
3834. Godet, R. 1951. Contribution à l'éthologie de la Taupe (*Talpa europaea* L). Bull Soc Zool France. 76:107-128.
3835. Godfrey, GK. 1953. A technique for finding *Microtus* nests. J Mamm. 34:503-505.

3836. Godfrey, GK. 1956. Reproduction of *Talpa europaea* in Suffolk. J Mamm. 37:438-440.
3837. Godfrey, GK. 1957. Observations of the movements of males (*Talpa eruopaea* L) after weaning. Proc Zool Soc London. 128:287-295.
3838. Godfrey, GK. 1969. Reproduction in a laboratory colony of the marsupial mouse *Sminthopsis larapinta* (Marsupialia: Dasyuridae). Aust J Zool. 17:637-654.
3839. Godfrey, GK. 1969. The influence of increased photoperiod on reproduction in the dasyurid marsupial, *Sminthopsis crassicaudata*. J Mamm. 50:132-133.
3840. Godfrey, GR. 1975. A study of oestrus and fecundity in a laboratory colony of mouse opossums (*Marmosa robinsoni*). J Zool. 175:541-555.
3841. Godfrey, GR. 1979. Gestation period in the common shrew, *Sorex coronatus* (*araneus*) *fretalis*. J Zool. 189:548-551.
3842. Godfrey, GK & Crowcroft, P. 1960. The Life of the Mole. Museum Press, London.
3843. Godfrey, GK & Crowcroft, P. 1971. Breeding the fat-tailed marsupial mouse *Sminthopsis crassicaudata* in captivity. Int Zoo Yearb. 11:33-38.
3844. Godfrey, GK & Oliver, WLR. 1978. The reproduction and development of the pigmy hedgehog tenrec *Echinops telfairi*. Dodo, J Jersey Wildl Preserv Trust. 15:38-51.
3844a. Godfrey, RW, Pope, CE, Dresser, BL, et al. 1991. Gross anatomy of the reproductive tract of female black (*Diceros bicornis michaeli*) and white rhinoceros (*Ceratotherium simum simum*). Zoo Biol. 10:165-175.
3845. Godinho, HP, Cardoso, FM, & Nogueira, JC. 1977. Blood supply to the testis of a Brazilian marsupial (*Didelphis azarae*) and its abdominotesticular temperature gradient. Acta Anat. 99:204-208.
3846. Godsell, J. 1982. The population ecology of the eastern quoll *Dasyurus viverrinus* (Dasyuridae, Marsupialia) in southern Tasmania. Carnivorous Marsupials. 1:199-207. Archer, M, ed, R Soc NSW, Mossman, NSW.
3847. Goeldi, EA. 1904. On the rare rodent *Dinomys branickii*. Proc Zool Soc London. 1904:158-162.
3848. Goertz, JW. 1965. Late summer breeding of flying squirrels. J Mamm. 46:510.
3849. Goertz, JW. 1965. Reproductive variation in cotton rats. Am Mid Nat. 74:329-340.
3850. Goertz, JW. 1965. Sex, age and weight variation in cotton rats. J Mamm. 46:471-477.
3851. Goertz, JW. 1970. An ecological study of *Neotoma floridana* in Oklahoma. J Mamm. 51:94-104.
3852. Goertz, JW. 1971. An ecological study of *Microtus pinetorum* in Oklahoma. Am Mid Nat. 86:1-13.
3853. Goertz, JW, Dawson, RM, & Mowbray, EE. 1975. Response to nest boxes and reproduction by *Glaucomys volans* in northern Louisiana. J Mamm. 56:933-939.
3854. Goetz, RH. 1936. Studien zur Placentation der Centetiden I Eine Neu-Untersuchung der Centetesplacenta. Z Anat Entwickl Gesch. 106:315-342.
3855. Goetz, RH. 1937. Studien zur Placentation der Centetiden II Die Implantation und Frühentwicklung von *Hemicentetes semispinosus* (Cuvier). Z Anat Entwickl Gesch. 107:274-318.
3856. Goetz, RH. 1937. Studien zur Placentation der Centetiden III Die Entwicklung der Fruchthüllen und der Placenta bei *Hemicentetes semispinosus* (Cuvier). Z Anat Entwickl Gesch. 108:161-200.
3857. Goetz, RH. 1938. On the early development of the Tenrecoidea (*Hemicentetes semispinosus*). Biomorphosis. 1:67-79.
3858. Goldfoot, DA. 1981. Olfaction, sexual behavior, and the pheromone hypothesis in rhesus monkeys: A critique. Am Zool. 21:153-164.
3859. Goldfoot, DA, Slob, AK, Scheffer, G, et al. 1975. Multiple ejaculations during prolonged sexual tests and lack of resultant serum testosterone increases in male stump tail macaques (*M arctoides*). Arch Sex Behav. 4:547-560.
3860. Goldfoot, DA, Wallen, K, Neff, DA, et al. 1984. Social influences on the display of sexually dimorphic behavior in rhesus monkeys: Isosexual rearing. Arch Sex Behav. 13:395-412.
3861. Golding, RR. 1969. Birth and development of spotted hyaenas *Crocuta crocuta* at the University of Ibadan Zoo, Nigeria. Int Zoo Yearb. 9:93-95.
3862. Goldingay, RL & Kavanagh, RP. 1990. Socioecology of the yellow-bellied glider, *Petaurus australis*, at Waratah Creek, NSW. Aust J Zool. 38:327-341.
3863. Goldman, B, Hall, V, Hollister, C, et al. 1981. Diurnal changes in pineal melatonin content in four rodent species: Relationship to photoperiodism. Biol Reprod. 24:778-783.
3864. Goldman, BD, Darrow, JM, & Yogev, L. 1984. Effects of timed melatonin infusions on reproductive development in the Djungarian hamster (*Phodopus sungorus*). Endocrinology. 114:2074-2083.
3865. Goldman, CA. 1986. A review of the management of the aardvark *Orycteropus afer* in captivity. Int Zoo Yearb. 24/25:286-294.
3866. Goldman, CA. 1987. *Crossarchus obscurus*. Mamm Species. 290:1-5.
3867. Goldman, CA & Taylor, ME. 1990. *Liberiictis kuhni*. Mamm Species. 348:1-3.
3868. Goldman, EA. 1911. Revision of the spiny pocket mice (genera *Heteromys and Liomys*). N Am Fauna. 34:1-70.
3869. Goldman, EA. 1921. Two new rodents from Oregon and Nevada. J Mamm. 2:232-233.
3870. Goldman, EA. 1950. Raccoons of North and Middle America. N Am Fauna. 60:1-153.
3871. Goldman, JE & Stevens, VJ. 1980. The birth and development of twin nilgai *Boselaphus tragocamelus* at Washington Park Zoo, Portland. Int Zoo Yearb. 20:234-240.
3872. Goldsmith, PC, Lamberts, R, & Brezina, LR. 1983. Gonadotropin-releasing hormone neurons and pathways in the primate hypothalamus and forebrain. Neuroendocrine Aspects of Reproduction. 7-43, Norman, RL, ed, Academic Press, New York.
3873. Goldspink, CR. 1987. The growth, reproduction and mortality of an enclosed population of red deer (*Cervus elaphus*) in north-west England. J Zool. 213:23-44.
3874. Goldstone, AD & Nelson, JE. 1986. Aggressive behaviour in two female *Peradorcas concinna* (Macropodidae) and its relation to oestrus. Aust Wildl Res. 13:375-385.
3875. Golian, SC & Whitworth, MR. 1985. Growth of pikas (*Ochotona princeps*) in Colorado. J Mamm. 66:367-371.
3876. Golley, FB. 1957. Gestation period, breeding and fawning behavior of Columbian black-tailed deer. J Mamm. 38:116-120.
3877. Golley, FB. 1957. An appraisal of ovarian analyses in determining reproductive performance of black-tailed deer. J Wildl Manage. 21:62-?.
3878. Golley, FB. 1961. Interaction of natality, mortality and movement during one annual cycle in a *Microtus* population. Am Mid Nat. 66:152-159.
3879. Gombe, S. 1983. Reproductive cycle of the rock hyrax (*Procavia capensis*). Afr J Ecol. 21:129-133.

3880. Gombe, S. 1985. Short term fluctuation in progesterone, oestradiol and testosterone in pregnant and non-pregnant hyaena (*Crocuta crocuta* Erxleben). Afr J Ecol. 23:269-271.
3881. Gombe, S & Kayanja, FIB. 1974. Ovarian progestins in Masai giraffe (*Giraffa camelopardalis*). J Reprod Fert. 40:45-50.
3882. Gombe, S & Oduor-Okelo, D. 1977. Effect of temperature and relative humidity on plasma and gonadal testosterone concentrations in camels. J Reprod Fert. 50:107-108.
3883. Gomendio, M. 1990. The influence of maternal rank and infant sex on maternal investment trends in rhesus macaques: Birth sex ratios, inter-birth intervals and suckling patterns. Behav Ecol Sociobiol. 27:365-375.
3884. Goncharov, N, Antonichev, AV, Gorluschkin, VM, et al. 1979. Luteinizing hormone levels during the menstrual cycle of the baboon (*Papio hamadryas*). Acta Endocrinol. 91:49-58.
3885. Goncharov, N, Aso, T, Cekan, Z, et al. 1976. Hormonal changes during the menstrual cycle of the baboon (*Papio hamadryas*). Acta Endocrinol. 82:396-412.
3886. Gontscharow, NP, Woronzow, WI, Rutschinzkaja, SJ, et al. 1974. Steroidstoffwechsel bei Primaten XV Ausscheidung von C_{19}- and C_{21}-Steroiden im Harn beim Pavian (*Papio hamadryas*) nach Langzeitbehandlung mit einer Kombination von Mestranol Chlormadinonacetat. Endokrinologie. 64:74-80.
3887. Gonzalez, CA, Hennessy, MB, & Levine, S. 1981. Subspecies differences in hormonal and behavioral responses after group formation in squirrel monkeys. Am J Primatol. 1:439-452.
3888. Goodall, J. 1965. Chimpanzees of the Gombe Stream Reserve. Primate Behavior Field Studies of Monkeys and Apes. 425-473, DeVore, I, ed, Holt, Rinehart & Winston, New York.
3889. Goodall, J. 1983. Population dynamics during a 15 year period in one community of free-living chimpanzees in the Gombe National Park, Tanzania. Z Tierpsychol. 61:1-60.
3890. Goodall, J & Athumani, J. 1980. An observed birth in a free-living chimpanzee (*Pan troglodytes schweinfurthii*) in Gombe National Park, Tangania. Primates. 21:545-549.
3890a. Goodall, RNP, Galeazzi, AR, Leatherwood, S, et al. 1988. Studies of Commerson's dolphins, *Cephalorhynchus commersonii*, off Tierra del Fuego, 1976-1984, with a review of information on the species in the south Atlantic. Rep Int Whal Comm Sp Issue. 9:3-70.
3890b. Goodall, RNP, Norris, KS, Galeazzi, AR, et al. 1988. On the Chilean dolphin, *Cephalorhynchus eutropia* (Gray, 1846). Rep Int Whal Comm Sp Issue. 9:197-257.
3891. Goode, JA, Peaker, M, & Weir, BJ. 1981. Milk composition in the plains viscacha (*Lagostomus maximus*). J Reprod Fert. 62:563-566.
3892. Goodlin, BL & Sackett, GP. 1983. Parturition in *Macaca nemestrina*. Am J Primatol. 4:283-307.
3893. Goodman, AL, Descalzi, CD, Johnson, DK, et al. 1977. Composite pattern of circulating LH, FSH, estradiol and progesterone during the menstrual cycle of cynomolgus monkeys. Proc Soc Exp Biol Med. 155:479-481.
3894. Goodman, AL & Hodgen, GD. 1978. Post partum patterns of circulating FSH, LH, prolactin, estradiol, and progesterone in nonsuckling cynomolgus monkeys. Steroids. 31:731-744.
3895. Goodman, L & Wislocki, GB. 1935. Cyclical uterine bleeding in a New World monkey (*Ateles geofforyi*). Anat Rec. 61:379-388.
3896. Goodman, RL, Hotchkiss, J, Karsch, FJ, et al. 1974. Diurnal variations in serum testosterone concentrations in the adult male rhesus monkey. Biol Reprod. 11:624-630.
3897. Goodpaster, WW & Hoffmeister, DF. 1954. Life history of the golden mouse, *Peromyscus nuttalli*, in Kentucky. J Mamm. 35:16-27.
3898. Goodrum, PD. 1940. A population study of the gray squirrel in eastern Texas. TX Agric Exp Sta Bull. 591:1-34.
3899. Goodwin, DL & Currie, PO. 1965. Growth and development of black-tailed jack rabbits. J Mamm. 46:96-98.
3900. Goodwin, GG. 1928. Observations on *Noctilio*. J Mamm. 9:104-113.
3990$_1$. Goodwin, GG. 1946. Mammals of Costa Rica. Bull Am Mus Nat Hist. 87:271-474.
3900a. Goodwin, GG & Greenhall, AM. 1961. A review of the bats of Trinidad and Tobago. Bull Am Mus Nat Hist. 122:187-302.
3901. Goodwin, GG & Greenhall, AM. 1964. New records of bats from Trinidad and comments in the status of *Molossus trinitatus* Goodwin. Am Mus Novit. 2195:1-23.
3902. Goodwin, RE. 1970. The ecology of Jamaican bats. J Mamm. 51:571-580.
3903. Goodwin, RE. 1979. The bats of Timor: Systematics and ecology. Bull Am Mus Nat Hist. 163:75-122.
3904. Goosen, C, Schrama, A, Brinkhof, H, et al. 1983. Housing conditions and breeding success of chimpanzees at the primate center TNO. Zool Biol. 2:295-302.
3905. Gopalakrishna, A. 1947. Studies on the embryology of Microchiroptera Part I Reproduction and breeding season in the south Indian vespertilionid bat, *Scotophilus wroughtoni* (Thomas). Proc Indian Acad Sci. 26B:219-232.
3906. Gopalakrishna, A. 1948. Studies on the embryology of Microchiroptera Part II Reproduction of the male vespertilionid bat *Scotophilus wroughtoni* (Thomas). Proc Indian Acad Sci. 27B:137-151.
3907. Gopalakrishna, A. 1949. Studies on the embryology of Microchiroptera Part III The histological changes in the genital organs and accessory reproductive structures during the sex-cycle of the Vespertilionid bat *Scotophilus wroughtoni* (Thomas). Proc Indian Acad Sci. 30B:17-46.
3908. Gopalakrishna, A. 1949. Studies on the embryology of Microchiroptera Part IV An analysis of implantation and early development in *Scotophilus wroughtoni* (Thomas). Proc Indian Acad Sci. 30B:226-242.
3909. Gopalakrishna, A. 1950. Studies on the embryology of Microchiroptera Part V Placentation in the vespertilionid bat *Scotophilus wroughtoni* (Thomas). Proc Indian Acad Sci. 31B:235-251.
3910. Gopalakrishna, A. 1950. Studies on the embryology of Microchiroptera Part VI Structures of the placenta of the Indian vampire bat, *Lyroderma lyra* (Geoffroy), Megadermatidae. Proc Nat Inst Sci India. 16:93-98.
3911. Gopalakrishna, A. 1954. Breeding habits of the Indian sheath tailed bat *Taphozous longimanus* (Hardwicke). Curr Sci. 23:60-61.
3912. Gopalakrishna, A. 1955. Observations on the breeding habits and ovarian cycle in the Indian sheath-tailed bat, *Taphozous longimanus* (Hardwicke). Proc Nat Inst Sci India. 21B:29-41.
3913. Gopalakrishna, A. 1958. Foetal membranes of some Indian Microchiroptera. J Morphol. 102:157-197.
3914. Gopalakrishna, A. 1964. Post-partum pregnancy in the Indian fruit-bat, *Rousettus leschenaulti* (Desm). Curr Sci. 33:558-559.
3915. Gopalakrishna, A. 1969. Gestation period in some Indian bats. J Bombay Nat Hist Soc. 66:317-322.

3916. Gopalakrishna, A. 1969. Unusual persistence of the corpus luteum in the Indian fruit-bat, *Rousettus leschenaulti* (Desmaret). Curr Sci. 38:388-389.
3917. Gopalakrishna, A. 1971. Uterus-blastocyst relationship in Chiroptera: Part I Topographical relationship between the uterus and the blastocyst in bats. J Zool Soc India. 23:55-61.
3918. Gopalakrishna, A & Bhatia, D. 1982. Breeding habits and associated phenomenon in some Indian bats: VII *Hipposideros speoris* (Schaeider) (Hipposideridae) from Chandrapus, Maharashta. J Bombay Nat Hist Soc. 79:549-556.
3919. Gopalakrishna, A & Bhiwgade, DA. 1974. Foetal membranes in the Indian horseshoe bat *Rhinolophus rouxi* (Temminck). Curr Sci. 43:516-517.
3920. Gopalakrishna, A & Chari, GC. 1983. A review of the taxonomic position of *Miniopterus* based on embryological characters. Curr Sci. 52:1176-1180.
3921. Gopalakrishna, A & Choudhari, PN. 1977. Breeding habits and associated phenomena in some Indian bats: I *Rousettus leschenaulti* (Desmarest), Megachiroptera. J Bombay Nat Hist Soc. 74:1-16.
3922. Gopalakrishna, A & Karim, KB. 1971. Localized progestational endometrial reaction in the uterus of the Indian fruit-bat, *Rousettus leschenaulti* (Desmaret). Curr Sci. 40:490-491.
3923. Gopalakrishna, A & Karim, KB. 1972. Arrangement of the foetal membranes and the occurrence of a haemodichorial placenta in the vespertilionid bat *Pipistrellus mimus mimus*. Curr Sci. 41:144-146.
3924. Gopalakrishna, A & Karim, KB. 1972. The yolk-sac gland in the Indian fruit bat *Rousettus leschenaulti* Desmarest. Curr Sci. 41:639-641.
3925. Gopalakrishna, A & Karim, KB. 1975. Development of the foetal membranes in the Indian leaf-nosed bat *Hipposideros fulvus fulvus* Graff Part II Placentation. Rev Roum Biol Ser Zool. 20:257-267.
3926. Gopalakrishna, A & Karim, KB. 1979. Fetal membranes and placentation in Chiroptera. J Reprod Fert. 56:417-429.
3927. Gopalakrishna, A, Karim, KB, & Chari, G. 1979. Pregnancy record of some Indian bats. Curr Sci. 48:716-718.
3928. Gopalakrishna, A & Khaparde, MS. 1972. Variable orientation of the embryonic mass during the implantation of the blastocyst in the Indian false vampire bat, *Megaderma lyra lyra* (Geoffroy). Curr Sci. 41:738-739.
3929. Gopalakrishna, A & Khaparde, MS. 1978. Early development and amniogenesis in the India vampire bat, *Megaderma lyra lyra* (Geoffroy). Proc Indian Acad Sci. 87B(Anim Sci 2):91-104.
3930. Gopalakrishna, A, Khaparde, MS, & Sapkal, VM. 1976. Parturition in the Indian false vampire bat, *Megaderma lyra lyra* Geoffroy. J Bombay Nat Hist Soc. 73:464-467.
3931. Gopalakrishna, A & Madhavan, A. 1970. Sex ratio in some Indian bats. J Bombay Nat Hist Soc. 67:171-175.
3932. Gopalakrishna, A & Madhavan, A. 1971. Parturition in the Indian vespertilionid bat *Pipistrellus ceylonicus chrysothrix* (Wroughton). J Bombay Nat Hist Soc. 68:666-670.
3933. Gopalakrishna, A & Madhavan, A. 1971. Survival of spermatozoa in the female genital tract of the Indian vespertilionid bat *Pipistrellus ceylonicus chrysothrix* (Wroughton). Proc Indian Acad Sci. 73:43-49.
3934. Gopalakrishna, A & Madhavan, A. 1977. Breeding habits and associated phenomena in some Indian bats: III *Hipposideros ater ater* (Templeton), Hipposideridae. J Bombay Nat Hist Soc. 74:511-517.
3935. Gopalakrishna, A & Madhavan, A. 1978. Viability of inseminated spermatozoa in the Indian vespertilionid bat *Scotophilus heathi* (Horsefield). Indian J Exp Biol. 16:852-854.
3936. Gopalakrishna, A & Moghe, MA. 1960. Development of the foetal membranes in the Indian leaf-nosed bat, *Hipposideros bicolor pallidus*. Z Anat Entw Gesch. 122:137-149.
3937. Gopalakrishna, A & Moghe, MA. 1960. Observations on the ovaries of some Indian bats. Proc Nat Inst Sci India. 26(Silver Jub Number):11-19.
3938. Gopalakrishna, A & Murthy, KVR. 1976. Studies on the male genitalia of Indian bats Part I Male genitalia of the Indian fruit bat *Rousettus leschenaulti* (Desmaret). J Zool Soc India. 28:13-24.
3939. Gopalakrishna, A & Phansalkar, RB. 1970. A rare case of monozygotic synchorial twins in the vespertilionid bat, *Pipistrellus ceylonicus chrysothrix*. Curr Sci. 39:309-310.
3940. Gopalakrishna, A, Phansalkar, RB, & Sahasrabudhe, JD. 1970. Degeneration of the inseminated spermatozoa after ovulation in two species of Indian bats. Curr Sci. 39:489-490.
3941. Gopalakrishna, A & Ramakrishna, PA. 1977. Some reproductive anomalies in the Indian rufus horse shoe bat, *Rhinolophus rouxi* (Temminck). Curr Sci. 46:767-770.
3942. Gopalakrishna, A & Rao, KVB. 1977. Breeding habits and associated phenomena in some Indian bats: II *Rhinolophus rouxi* (Temminck), Rhinolophidae. J Bombay Nat Hist Soc. 74:213-219.
3943. Gopalakrishna, A & Sapkal, VM. 1974. Foetal membranes of the Indian pipistrelle, *Pipistrellus dormeri* (Dobson) [Mammalia: Chiroptera]. J Zool Soc India. 26:1-9.
3944. Gopalakrishna, A, Thakur, RS, & Madhavan, A. 1975. Breeding biology of the southern dwarf pipistrelle, *Pipistrellus mimus mimus*, (Wroughton) from Maharashtra, India. Zool Soc India. Dr BS Chauhan Commemoration Vol:225-240.
3945. Gopalakrishna, A, Varute, AT, Sapkal, VM, et al. 1985. Breeding habits and associated phenomena in some Indian bats Part XI *Miniopterus schreibersii fuliginosus* (Hodgson) Vespertilionidae. J Bombay Nat Hist Soc. 82:594-601.
3945a. Gordinier, EJ. 1948. September birth of a Michigan white-tailed deer. J Mamm. 29:184-185.
3946. Gordon, G. 1974. Movements and activity of the short-nosed *Isoodon macrourus*, Gould (Marsupialia). Mammalia. 38:405-431.
3947. Gordon, K, Fletcher, TP, & Renfree, MB. 1988. Reactivation of the quiescent corpus luteum and diapausing embryo after temporary removal of the sucking stimulus in the tammar wallaby (*Macropus eugenii*). J Reprod Fert. 83:401-406.
3948. Gordon, TP. 1981. Reproductive behavior in the rhesus monkey: Social and endocrine variables. Amer Zool. 21:185-195.
3949. Gordon, TP & Bernstein, IS. 1973. Seasonal variation in sexual behavior of all-male rhesus troops. Am J Phys Anthrop. 38:221-226.
3950. Gordon, TP, Rose, RM, & Bernstein, IS. 1976. Seasonal rhythm in plasma testosterone levels in the rhesus monkey (*Macaca mulatta*): A three year study. Horm Behav. 7:229-243.
3951. Gorecki, A. 1977. Energy flow through the common hamster population. Acta Theriol. 22:25-66.
3952. Gorman, ML. 1976. Seasonal changes in the reproductive pattern of feral *Herpestes auropunctatus* (Carnivora: Viverridae), in the Fijian Islands. J Zool. 178:237-246.
3953. Gorman, ML. 1979. Dispersion and foraging of the small Indian mongoose, *Herpestes auropunctatus* (Carnivora: Viverridae) relative to the evolution of social viverrids. J Zool. 187:65-73.

3954. Gorwill, RH, Snyder, DL, Lindholm, UB, et al. 1971. Metabolism of pregnenolone-4-^{14}C and pregnenolone-7α^{3}H sulfate by the *Macaca mulatta* fetal adrenal in vitro. Gen Comp Endocrinol. 16:21-29.
3955. Gosling, LM. 1969. Parturition and related behaviour in Coke's hartebeest, *Alcephalus buselaphus cokei* Gunther. J Reprod Fert Suppl. 6:265-286.
3956. Gosling, LM. 1974. The social behaviour of Coke's hartebeest (*Alcelaphus buselaphus cohei*). The Behaviour of Ungulates and its Relation to Management. 488-511, Geist, V & Walther, F, eds, IUCN, Morges, Switzerland.
3957. Gosling, LM. 1980. The duration of lactation in feral coypus (*Myocastor coypus*). J Zool. 191:461-474.
3958. Gosling, LM. 1981. Climatic determinants of spring littering by feral coypus, *Myocastor coypus*. J Zool. 195:281-288.
3959. Gosling, LM & Baker, SJ. 1982. Coypu (*Myocastor coypus*) potential longevity. J Zool. 197:285-288.
3960. Gosling, LM, Baker, SJ, & Wright, KMH. 1984. Differential investment by female coypus (*Myocastor coypus*) during lactation. Symp Zool Soc London. 51:273-300.
3961. Gosling, LM & Petrie, M. 1981. The economics of social organization. Physiological Ecology. 315-345, Townsend, CR & Calow, P, eds, Blackwell Sci Publ, Oxford.
3962. Gosling, LM, Watt, AD, & Baker, SJ. 1981. Continuous retrospective census of the east Anglian coypu population between 1970 and 1979. J Anim Ecol. 50:885-901.
3963. Goss, CM, Popejoy, LT, II, Fusiler, JL, et al. 1968. Observations on the relationship between embryological development, time of conception, and gestation. The Squirrel Monkey. 171-191, Rosenblum, L & Cooper, RW, eds, Academic Press, New York.
3964. Goss, RJ. 1968. Inhibition of growth and shedding of antlers by sex hormones. Nature. 220:83-85.
3965. Goswami, SB & Nair, AP. 1964. Effect of some climatological factors on reproduction of buffaloes (*Bubalus bubalis*). Indian J Vet Sci Anim Husb. 34:127-134.
3966. Goswell, MJ & Gartlan, JS. 1965. Pregnancy, birth and early infant behavior in the captive patas monkey *Erythrocebus patas*. Folia Primatol. 3:189-200.
3967. Goto, A & Miyake, T. 1975. Breeding of grey gentle lemur. Animals & Zoos. 27(4):6-9.
3968. Goto, N & Hashizume, R. 1978. Pattern of ovulation in *Microtus montebelli*. J Mamm Soc Japan. 7:181-188.
3969. Goto, N, Hashizume, R, & Sai, I. 1977. Litter size and vaginal smear in *Microtus montebelli*. J Mamm Soc Japan. 7:75-85.
3970. Gouat, J & Gouat, P. 1987. The behavioural repertoire of the gundi *Ctenodactylus gundi* (Rodents, Ctenodactylidae) II Ontogenesis. Mammalia. 51:173-193.
3971. Gouat, J. 1985. Notes sur la reproduction de *Ctenodactylus gundi* Rongeur Ctenodactylidae. Z Säugetierk. 50:285-293.
3972. Gouat, J. 1986. A propos d'une gestation énigmatique chez *Ctenodactylus vali* (Rodentia). CR Hebd Séances Acad Sc Paris. Sér III 303:665-667.
3973. Gould, E. 1975. Neonatal vocalizations in bats of eight genera. J Mamm. 56:15-29.
3974. Gould, E. 1978. Rediscovery of *Hipposideros ridleyi* and seasonal reproduction in Malaysian bats. Biotropica. 10:30-32.
3975. Gould, E & Eisenberg, JF. 1966. Notes on the biology of the Tenrecidae. J Mamm. 47:660-686.
3976. Gould, KG & Martin, DE. 1981. The female ape genital tract and its secretions. Reproductive Biology in the Great Apes. 105-125, Graham, CE, ed, Academic Press, New York.
3977. Goulden, EA & Meester, J. 1978. Notes on the behaviour of *Crocidura* and *Myosorex* (Mammalia: Soricidae) in captivity. Mammalia. 42:197-202.
3978. Goundie, TR & Vessey, SM. 1986. Survival and dispersal of young white-footed mice born in nest boxes. J Mamm. 67:53-60.
3979. Gouzoules, H, Gouzoules, S, & Fedigan, L. 1981. Japanese monkey group translocation: Effects on seasonal breeding. Int J Primatol. 2:323-334.
3980. Gouzoules, H, Gouzoules, S, & Fedigan, L. 1982. Behavioural dominance and reproductive success in female Japanese monkeys (*Macaca fuscata*). Anim Behav. 30:1138-1150.
3981. Gouzoules, H & Goy, RW. 1983. Physiological and social influences on mounting behavior of troop-living female monkey (*Macaca fuscata*). Am J Primatol. 5:39-49.
3982. Gowda, CDK. 1967. A note on the birth of caracal lynx *Felis caracal* at Mypore Zoo. Int Zoo Yearb. 7:133.
3983. Gowda, CDK. 1967. Breeding the black rhinoceros *Diceros bicornis* at Mypore Zoo. Int Zoo Yearb. 7:163-164.
3984. Goyal, RP & Mathur, RS. 1974. Certain biochemical observations on the testes and male accessory glands of two insectivores, *Suncus murinus sindensis* Anderson and *Hemiechinus auritus collaris* Gray. Acta Anat. 90:462-466.
3985. Goyal, RR & Mathur, RS. 1974. Anatomic, histologic and certain enzymatic studies on the male genital organs of *Hemiechinus auritus collaris* Gray, the Indian long eared hedgehog. Acta Zool (Stockholm). 55:47-58.
3985a. Gozalo, A & Montoya, E. 1990. Reproduction of the owl monkey (*Aotus nancymai*) (Primates: Cebidae) in captivity. Am J Primatol. 21:61-68.
3986. Grachev, YA & Fedosenko, AK. 1977. *Ursus arctos* in Djungarsky Alatun. Zool Zhurn. 56:120-129.
3987. Grady, RM & Hoogland, JL. 1986. Why do black-tailed prairie dogs (*Cynomys ludovicianus*) give a mating call?. Anim Behav. 34:108-112.
3988. Graf, M & Wandeler, AI. 1982. Der Geschlechtszyklus mannlicher Dachse (*Meles meles* L) in der Schweiz. Rev Suisse Zool. 89:1005-1008.
3989. Graf, W & Nichols, L, Jr. 1967. The axis deer in Hawaii. J Bombay Nat Hist Soc. 63:629-734.
3990. Graham, CE. 1970. Reproductive physiology of the chimpanzee Vol 3 The Chimpanzee. 183-220, Bourne, GH, ed, Karger, New York.
3991. Graham, CE. 1976. The chimpanzee: A unique model for human reproduction. The Laboratory Animal in the Study of Reproduction 6th Symp ICLA, Tessaloniki, 9-11 Fischer, New York.
3992. Graham, CE. 1981. Menstrual cycle of the great apes. Reproductive Biology of the Great Apes. 1-43, Graham, CE, ed, Academic Press, New York.
3993. Graham, CE. 1981. Endocrine control of spermatogenesis in primates. Am J Primatol. 1:157-165.
3994. Graham, CE & Bradley, CF. 1971. Polyovular follicles in squirrel monkeys after prolonged diethylstilboestrol treatment. J Reprod Fert. 27:181-185.
3995. Graham, CE, Collins, DC, Robinson, H, et al. 1972. Urinary levels of estrogens and pregnanediol and plasma levels of progesterone during the menstrual cycle of the chimpanzee: Relationship to sexual swelling. Endocrinology. 91:13-24.
3996. Graham, CE, Gould, KG, Collins, DC, et al. 1979. Regulation of gonadotropin release by luteinizing hormone-releasing hormone and estrogen in chimpanzees. Endocrinology. 105:269-275.

3997. Graham, CE, Guilloud, N, & McArthur, JW. 1969. Reproductive endocrinology of the chimpanzee. Proc Int Congr Primat. 2:66-72.
3998. Graham, CE, Keeling, M, Chapman, C, et al. 1973. Method of endoscopy in the chimpanzee: Relations of ovarian anatomy, endometrial histology and sexual swelling. Am J Phys Anthrop. 38:211-215.
3999. Graham, CE, Warner, H, Misener, J, et al. 1977. The association between basal body temperature, sexual swelling and urinary gonadal hormone levels in the menstrual cycle of the chimpanzee. J Reprod Fert. 50:23-28.
3999a. Graham, GL. 1987. Seasonality of reproduction in Peruvian bats. Fieldiana Zool NS. 39:173-186.
4000. Graham-Jones, O & Hill, WCO. 1962. Pregnancy and parturition in a Bornean orang. Proc Zool Soc London. 139:503-510.
4001. Grainger, JP & Fairley, JS. 1978. Studies on the biology of the pygmy shrew *Sorex minutus* in the west of Ireland. J Zool. 186:109-141.
4002. Gram, WD, Heath, HW, Wichman, HA, et al. 1982. Geographic variation in *Peromyscus leucopus*: Short-day induced reproductive regression and spontaneous recrudescence. Biol Reprod. 27:369-373.
4003. Granados, H. 1981. Basic information on the volcano rabbit. Proc World Lagomorph Conf, Univ Guelph, ONT, Canada. 1979:940-947.
4004. Grand, T, Duro, E, & Montagna, W. 1964. Potto born in captivity. Science. 145:663.
4005. Grandl-Grams, M. 1977. Verhaltensstudien an Damwild (*Cervus dama* L 1758) in Gefangenschaft. Zool Garten NF. 47:81-108.
4006. Grange, WB. 1932. Observations on the snowshoe hare, *Lepus americanus phaeonotus* Allen. J Mamm. 13:1-19.
4007. Grant, PR. 1974. Reproductive compatibility of voles from separate continents (Mammalia: *Clethrionomys*). J Zool. 174:245-254.
4008. Grant, TR & Anink, PJ. 1976. Maintenance of body temperature in the platypus, *Ornithorhynchus anatinus*. Bull Aust Mamm Soc. 3:19.
4009. Grant, TR & Griffiths, M. 1984. Aspects of lactation in the platypus, *Ornithorhynchus anatinus*. Bull Aust Mamm Soc. 8:120.
4009a. Grant, TR & Griffiths, M. 1992. Aspects of lactation and determination of sex ratios and longevity in a free-ranging population of platypuses, *Ornithorhynchus anatinus*, in the Shoalhaven River, NSW. Platypus and Echidnas. 80-89, Augee, ML, ed, R Zool Soc NSW, Sydney.
4010. Grant, TR, Griffiths, M, & Leckie, RMC. 1983. Aspects of lactation in the platypus, *Ornithorhynchus anatinus* (Monotremata) in waters of eastern New South Wales. Aust J Zool. 31:881-889.
4011. Grasse, JE & Putnam, EF. 1950. Beaver management and ecology in Wyoming. Bull WY Game Fish Comm. 6:1-52.
4012. Grav, JH, Blix, AS, & Påsche, A. 1974. How do seal pups survive birth in arctic winter?. Acta Physiol Scand. 92:427-429.
4013. Graves, JAM, Hope, RM, & Cooper, DW, eds. 1990. Mammals from pouches and eggs: Genetics, breeding and evolution of marsupials and monotremes. Aust J Zool. 37:143-479.
4014. Gray, GD, Davis, HH, Zerylnick, M, et al. 1974. Oestrus and induced ovulation in montane voles. J Reprod Fert. 38:193-196.
4015. Gray, GD, Davis, HN, Kenney, AM, et al. 1976. Effect of mating on plasma levels of LH and progesterone in montane voles (*Microtus montanus*). J Reprod Fert. 47:89-91.
4016. Gray, GD & Dewsbury, DA. 1973. A quantitative description of copulatory behavior in prairie voles (*Microtus ochrogaster*). Brain Behav Evol. 8:437-452.
4017. Gray, GD & Dewsbury, DA. 1975. A quantitative description of the copulatory behaviour of meadow voles (*Microtus pennsylvanicus*). Anim Behav. 23:261-267.
4018. Gray, GD, Kenney, AM, & Dewsbury, DA. 1977. Adaptive significance of the copulatory behavior pattern of male meadow voles (*Microtus pennsylvanicus*) in relation to induction of ovulation and implantation in females. J Comp Physiol Psy. 91:1308-1319.
4019. Gray, GD, Zerylnick, M, Davis, HN, et al. 1974. Effects of variations in male copulatory behavior on ovulation and implantation in prairie voles, *Microtus ochrogaster*. Horm Behav. 5:389-396.
4020. Gray, GG & Simpson, CD. 1980. *Ammotragus lervia*. Mamm Species. 144:1-7.
4021. Gray, GG & Simpson, CD. 1983. Population characteristics of free-ranging barbary sheep in Texas. J Wildl Manage. 47:954-962.
4022. Gray, KH & Chapman, CG. 1974. Pregnancy diagnosis in bottle nosed dolphin. Proc Am Assn Zoo Vet. 1974:157-160.
4023. Gray, RF. 1959. Hippopotamus milk analysis. Int Zoo Yearb. 1:46.
4024. Greed, GR. 1967. Notes on the breeding of the black rhinoceros *Diceros bicornis* at Bristol Zoo. Int Zoo Yearb. 7:158-161.
4025. Greed, RE. 1960. The composition of the milk of the black rhinoceros. Int Zoo Yearb. 2:106.
4026. Green, B. 1984. Composition of milk and energetics of growth in marsupials. Symp Zool Soc London. 51:369-387.
4027. Green, B & Griffiths, M. 1984. Energetics of growth in the echidna (*Tachyglossus aculeatus*). Bull Aust Mamm Soc. 8:121.
4028. Green, B, Griffiths, M, & Newgrain, K. 1985. Intake of milk by suckling echidnas (*Tachyglossus aculeatus*). Comp Biochem Physiol. 81A(2):441-444.
4028a. Green, B, Griffiths, M, & Newgrain, K. 1992. Energy costs of lactation in free-living echidnas. Platypus and Echidnas. 90-92, Augee, ML, ed, R Zool Soc NSW, Sydney.
4029. Green, B, Merchant, J, & Newgrain, K. 1987. Milk composition in the eastern quoll, *Dasyurus viverrinus* (Marsupialia: Dasyuridae). Aust J Biol Sci. 40:379-387.
4029a. Green B, Newgrain, K, Catling, P, et al. 1991. Patterns of prey consumption and energy use in a small carnivorous marsupial, *Antechinus stuartii*. Aust J Zool. 39:539-547.
4029b. Green B, Newgrain, K, & Merchant, J. 1980. Changes in milk composition during lactation in the tammar wallaby (*Macropus eugenii*). Aust J Biol Sci. 33:35-42.
4030. Green, B & Rowe-Rowe, DT. 1987. Water and energy metabolism in free-living multi-mammate mice, *Praomys natalensis* during summer. S Afr J Zool. 22:14-17.
4031. Green, CV. 1932. Breeding habits in captivity of *Mus bactrianus* Blyth. J Mamm. 13:45-47.
4032. Green, HU. 1936. The beaver of the Riding Mountain, Manitoba. Can Field-Nat. 50:1-8, 21-23, 36-50, 61-67, 85-92.
4032a. Green MJB. 1987. Some ecological aspects of a Himalayan population of musk deer. Biology and Management of the Cervidae. 308-319, Wemmer, CM, ed, Smithsonian Inst Press, Washington, DC.

4033. Green, JS, Adair, RA, Woodruff, RA, et al. 1984. Seasonal variation in semen production by captive coyotes. J Mamm. 65:506-509.
4034. Green, JS & Flinders, JT. 1980. *Brachylagus idahoensis*. Mamm Species. 125:1-4.
4035. Green, MM. 1925. Notes on some mammals of Montmorency county, Michigan. J Mamm. 6:173-178.
4036. Green, RF. 1977. Gross anatomy of the reproductive organs in dolphins. Breeding Dolphins Present Status, Suggestions for the Future, US Mar Mamm Comm Rep No MMC-76/07 NTIS PB 273673. 185-194, Ridgway, SH & Benirschke, K, eds, US Comm Nat Tech Info Serv, Arlington, Va.
4037. Green, RG & Evans, CA. 1940. Studies on a population cycle of snowshoe hares on the Lake Alexander area III Effect of reproduction and mortality of young hares on the cycle. J Wildl Manage. 4:347-358.
4038. Green, RH. 1965. Observations on the little brown bat *Eptesicus pumilus* Gray in Tasmania. Rec Queen Vict Mus. 20:1-16.
4039. Green, RH. 1966. Notes on lesser long-eared bat *Nyctophilus geoffroyi* in northern Tasmania. Rec Queen Vict Mus. 22:1-4.
4040. Green, RH. 1967. Notes on the devil (*Sarcophellus harrisii*) and the quoll (*Dasyurus ververrinus*) in northeastern Tasmania. Rec Queen Vic Mus. 27:1-12.
4041. Green, RH. 1967. The murids and small dasyurids in Tasmania Parts 1 and 2. Rec Queen Vict Mus. 28:1-19.
4042. Green, RH. 1968. The murids and small dasyurids in Tasmania Parts 3 and 4. Rec Queen Vict Mus. 32:1-19.
4043. Green, RH. 1972. The murids and small dasyurids in Tasmania Parts 5, 6 and 7. Rec Queen Vict Mus. 46:1-34.
4043a. Green, RH. 1979. The little pigmy possum *Cercartetus lepidus* in Tasmania. Rec Queen Vict Mus. 68:1-12.
4044. Green, S & Minkowski, K. 1977. The lion-tailed monkey and its south Indian rain forest habitat. Primate Conservation. 289-337, Prince Rainier III of Monaco & Bourne, GH, eds, Academic Press, New York.
4045. Green, SH & Zuckerman, S. 1948. A comparison of the growth of the ovum and follicle in normal rhesus monkeys, and in monkeys treated with oestrogens and androgens. J Endocrinol. 5:207-219.
4046. Green, SH & Zuckerman, S. 1954. Further observations on oocyte numbers in mature rhesus monkeys (*Macaca mulatta*). J Endocrinol. 10:284-290.
4047. Green, WCH. 1990. Reproductive effort and associated costs in bison (*Bison bison*): Do older mothers try harder?. Beh Ecol. 1:148-160.
4048. Green, WCH & Rothstein, A. 1991. Sex bias or equal opportunity? Patterns of maternal investment in bison. Behav Evol Sociobiol. 29:373-384.
4048₁. Greenbaum, IF & Jones, JK, Jr. 1978. Noteworthy records of bats from El Salvador, Honduras, and Nicaragua. Occas Pap Mus TX Tech Univ. 55:1-7.
4048a. Greenhall, AM. 1965. Notes on behavior of captive vampire bats. Mammalia. 29:441-451.
4048b. Greenhall, AM. 1968. Notes on the behavior of the false vampire bat. J Mamm. 49:337-340.
4049. Greenhall, AM, Joermann, G, & Schmidt, U. 1983. *Desmodus rotundus*. Mamm Species. 202:1-6.
4050. Greenhall, AM, Schmidt, U, & Joermann, G. 1984. *Diphylla ecaudata*. Mamm Species. 227:1-3.
4051. Greensides, RD & Mead, RA. 1973. Ovulation in the spotted skunk (*Spilogale putorius latifrons*). Biol Reprod. 8:576-584.
4052. Greenwald, GS. 1956. The reproductive cycle of the field mouse, *Microtus californicus*. J Mamm. 37:213-222.
4053. Greenwald, GS. 1957. Reproduction in a coastal California population of the field mouse, *Microtus californicus*. Univ CA Publ Zool. 54:421-446.
4054. Greer, JK. 1965. Mammals of Malleco Province Chile. Publ MI State Univ Mus Biol Ser. 3:49-152.
4055. Gregor, DH, Jr. 1985. Ecology of the little hairy armadillo *Chaetophractus vellerosus*. The Evolution and Ecology of Armadillos, Sloths, and Vermilinguas. 397-403, Montgomery, GG, ed, Smithsonian Inst Press, Washington, DC.
4056. Gregory, ME, Kon, LK, Rowland, SJ, et al. 1955. The composition of the milk of the blue whale. J Dairy Res. 22:108-112.
4057. Gregory, ME, Rowland, SJ, Thompson, SY, et al. 1965. Changes during lactation in the composition of the milk of the African black rhinoceros (*Diceros bicornis*). Proc Zool Soc London. 145:327-333.
4058. Gressner, JL. 1964. Die Geburten der Luchse (*Lynx lynx L*) im Zoologischen Garten Bojnice. Lynx. 3:40-41.
4059. Griffin, DR. 1940. Notes on the life histories of New England cave bats. J Mamm. 21:181-187.
4060. Griffiths, DJ. 1984. The annual cycle of the testis of the elephant seal (*Mirounga leonina*) at Maquarie Island. J Zool. 203:193-204.
4061. Griffiths, DJ & Bryden, MM. 1981. The annual cycle of the pineal gland of the elephant seal (*Mirounga leonina*). Pineal Function. 57-66, Matthews, CD & Seamark, RF, eds, Elsevier/North Holland, Biomedical, Amsterdam.
4062. Griffiths, DJ & Bryden, MM. 1986. Adenohypophysis of the elephant seal (*Mirounga leonina*): Morphology and seasonal histological changes. Am J Anat. 176:483-495.
4063. Griffiths, M. 1965. Rate of growth and intake of milk in a suckling echidna. Comp Biochem Physiol. 16:383-392.
4064. Griffiths, M. 1978. The Biology of the Monotremes. Academic Press, New York.
4064a. Griffiths, M. 1984. Mammals: Monotremes. Marshall's Physiology of Reproduction. 351-385, Lamming, GE, ed, Churchill Livingstone, London.
4065. Griffiths, M, Elliott, MA, Leckie, RMC, et al. 1973. Observations of the comparative anatomy and ultrastructure of mammary glands and on the fatty acids of the triglycerides in platypus and echidna milk fats. J Zool. 169:255-279.
4066. Griffiths, M, Friend, JA, Whitford, D, et al. 1988. Composition of the milk of the numbat, *Myrmecobius fasciatus* (Marsupialia: Myrmecobiidae), with particular reference to the fatty acids of the lipids. Aust Mamm. 11:59-62.
4067. Griffiths, M, Green, B, Leckie, RCM, et al. 1984. Constituents of platypus and echidna milk, with particular reference to the fatty acid complement of the triglycerides. Aust J Biol Sci. 37:323-329.
4068. Griffiths, M, Kristo, F, Green, B, et al. 1988. Observations on free-living, lactating echidnas, *Tachyglossus aculeatus* (Monotremata: Tachyglossidae), and sucklings. Aust Mamm. 11:135-143.
4069. Griffiths, M, McIntosh, DL, & Coles, REA. 1969. The mammary gland of the echidna, *Tachyglossus aculeatus* with observations on the incubation of the egg and on the newly-hatched young. J Zool. 158:371-386.
4070. Griffiths, M, McIntosh, DL, & Leckie, RMC. 1972. The mammary glands of the red kangaroo with observations of the

fatty acid components of the milk triglycerides. J Zool. 166:265-275.
4071. Grigg, G, Jarman, P, & Hume, I. 1989. Kangaroos, Wallabies and Rat-Kangaroos. Surrey Beatty & Sons, NSW, Australia.
4072. Grigor, MR. 1980. Structure of milk triacylglycerols of five marsupials and one monotreme: Evidence for an unusual pattern common to marsupials and eutherians but not found in the echidna, a monotreme. Comp Biochem Physiol. 65B:427-430.
4073. Grigoriev, ND. 1938. On the reproduction of the stoat (*Mustela erminea*). Zool Zhurn. 17:811-814.
4074. Grimmberger, E, Hackethal, H, & Urbanczyk, Z. 1987. Beitrag zum Paarungsverhalten der Wasserfledermaus, *Myotis daubentoni* (Kuhl, 1819), im Winterquartier. Z Säugetierk. 52:133-140.
4075. Grimpe, G. 1916. Hyanologische studien. Zool Anz. 48:49-61.
4076. Grimpe, G. 1923. Neues über die Geschlechtsverhältnisse der Gefleckten Hyäne (*Crocutta crocutta* Erxl). Verh Deutsch Zool Ges. 1923:77-78.
4077. Grimsdell, JJR. 1973. Reproduction in the African buffalo, *Syncerus caffer*, in western Uganda. J Reprod Fert Suppl. 19:303-318.
4078. Griner, LA. 1968. The rock hyrax (*Procavia capensis*): A potential laboratory animal. Lab Anim Care. 18:144-150.
4079. Grinnell, HW. 1918. A synopsis of the bats of California. Univ CA Publ Zool. 17:223-404.
4079a. Grinnell, J. 1914. An account of the mammals and birds of the lower Colorado Valley with especial reference to the distributional problems presented. Univ CA Publ Zool. 12:51-294
4080. Grinnell, J. 1932. Habitat relations of the giant kangaroo rat. J Mamm. 13:305-320.
4081. Grinnell, J & Dixon, J. 1918. Natural history of the ground squirrels of California. Monthly Bull State Comm Horticulture. 7:597-708.
4081a. Grinnell, J, Dixon, J, & Linsdale, JM. 1930. Vertebrate natural history of a section of northern California through the Lassen Peak region. Univ CA Publ Zool. 35:1-594.
4082. Grist, SM. 1976. The common marmoset (*Callithrix jacchus*): A valuable experimental animal. J Inst Animal Techn. 27:1-7.
4083. Grizzell, RA, Jr. 1955. A study of the southern woodchuck, *Marmota monax monax*. Am Mid Nat. 53:257-293.
4084. Grobler, JH. 1973. Biological data on tsessebe, *Damaliscus lunatus* (Mammalia: Alcephalinae), in Rhodesia. Arnoldia (Rhodesia). 6(12):1-16.
4085. Grobler, JH. 1974. Aspects of the biology, population ecology and behaviour of the sable *Hippotragus niger niger*, (Harris 1838) in the Rhodes Mafopos National Park, Rhodesia. Arnoldia (Rhodesia). 7(6):1-36.
4086. Grobler, JH. 1980. Breeding biology and aspects of social behaviour of sable *Hippotragus niger niger* (Harris, 1838) in the Rhodes Matopos National Park, Zimbabwe. S Afr J Wildl Res. 10:150-152.
4087. Grocock, CA. 1979. Testis development in the vole, *Microtus agrestis*, subjected to long or short photoperiods from birth. J Reprod Fert. 55:423-427.
4088. Grocock, CA. 1980. Effects of age on photo-induced testicular regression, recrudescence, and refractoriness in the short-tailed field vole *Microtus agrestis*. Biol Reprod. 23:15-20.
4089. Grocock, CA. 1981. Effect of different photoperiods on testicular weight changes in the vole, *Microtus agrestis*. J Reprod Fert. 62:25-32.
4090. Grocock, CA & Clarke, JR. 1974. Photoperiodic control of testis activity in the vole, *Microtus agrestis*. J Reprod Fert. 39:337-348.
4091. Grocock, CA & Clarke, JR. 1975. Spermatogenesis in mature and regressed testes of the vole (*Microtus agrestis*). J Reprod Fert. 43:461-470.
4092. Grocock, CA & Clarke, JR. 1976. Duration of spermatogenesis in the vole (*Microtus agrestis*) and bank vole (*Clethrionomys glareolus*). J Reprod Fert. 47:133-135.
4093. Grodziński, W. 1966. Bioenergetics of small mammals from Alaskan taiga forest. Lynx. 6:51-55.
4094. Grodziński, W. 1985. Ecological energetics of bank voles and wood mice. Symp Zool Soc London. 55:169-192.
4095. Gromadzka-Ostrowska, J & Szylarska-Gozdz, E. 1984. Progesterone concentration and their seasonal changes during the estrus [*sic*] cycle of the chinchilla. Acta Theriol. 29:251-258.
4096. Gromadzka-Ostrowska, J, Zalewska, B, & Szylarska-Gozdz, E. 1985. Peripheral plasma progesterone concentrationand haematological indices during normal pregnancy of chinchillas (*Chinchilla laniger*, M). Comp Biochem Physiol. 82A:661-665.
4097. Groome, JR. 1940. The seasonal modification of the interstitial tissue of the testes in the fruit bat (*Pteropus*). Proc Zool Soc London. 110(A):37-42.
4098. Gross, GH. 1977. Effects of normoxic prenatal low pressure on gestation and development in *Mesocricetus auratus*. Physiol Zool. 50:223-230.
4099. Gross, JE, Stoddart, LC, & Wagner, FH. 1974. Demographic analysis of a northern Utah jackrabbit population. Wildl Monogr. 40:1-68.
4100. Gross, R & Bolliger, A. 1958. The occurrence of carbohydrates other than lactose in the milk of a marsupial (*Trichosurus vulpecula*). Aust J Sci. 20:184-185.
4101. Gross, R & Bolliger, A. 1959. Composition of milk of the marsupial *Trichosurus vulpecula*. Am J Dis Child. 98:768-775.
4102. Grosser, O. 1903. Die physiologische bindegewebige Atresie des Genitalkanales von *Vesperugo noctula* nach erfolgter Kohabitation. Anat Anz Erganzhft 17. 23:129-132.
4103. Grosser, O. 1909. Vergleichende Anatomie und Entwicklung der Eihäute und der Placenta. Wilhelm Brawmüller, Vienna.
4104. Grosser, O. 1928. Die Placenta von *Centetes* und ihre Lehren betreffs der Stoffaufnahme in den Placenten. Z Anat Entw Gesch. 88:509-521.
4105. Grosz, S. 1905. Beiträge zur Anatomie der accessorischen Geschechtsdrusen der Insectivoren und Näger. Arch Mibrosk Anat. 66:567-608.
4106. Groves, CP. 1971. *Pongo pygmaeus*. Mamm Species. 4:1-6.
4107. Groves, CP. 1972. *Ceratotherium simum*. Mamm Species. 8:1-6.
4108. Groves, CP & Kurt, F. 1972. *Dicerorhinus sumatrensis*. Mamm Species. 21:1-6.
4109. Grubb, P. 1974. Mating activity and the social significance of rams in feral sheep communities. The Behaviour of Ungulates and its Relation to Management. 457-487, Geist, V & Walther, F, eds, IUCN, Morges, Switzerland.
4110. Grubb, P. 1981. *Equus burchelli*. Mamm Species. 157:1-9.
4111. Grubb, P & Jewell, PA. 1973. The rut and the occurrence of oestrus in the Soay sheep on St Kilda. J Reprod Fert Suppl. 19:491-502.
4112. Gruenberger, HB. 1970. On the cerebral anatomy of the Amazon dolphin *Inia geoffrensis*. Invest Cetacea. 2:129-144.

4113. Gruenwald, P. 1973. Lobular structure of hemochorial primate placentas and its relation to maternal vessels. Am J Anat. 136:133-151.
4114. Grummt, W. 1960. Zur Biologie und Ökologie der Kleinäugigen Wuhlmaus *Pitymys subterraneus* de Selys-Longchamps. Zool Anz. 165:129-144.
4115. Grzimek, B. 1957. Zusammensetzung von Kamelmilch. Zool Garten NF. 23:247.
4116. Grzimek, B. 1958. Die Milch der Giraffengazellen, *Litocranius walleri* (Brooke). Zool Garten NF. 24:283-284.
4117. Grzimek, B. 1960. Die Zusammensetzung der Nashornmilch. Zool Garten NF. 25:202-204.
4118. Grzimek, B. 1967. Zusammensetzung der Milch eines Bleszbocks (*Damaliscus dorcas philippsi* Harper) in Frankfurter Zoo. Zool Garten NF. 33:261.
4118a. Grzimek, B. 1990. Encyclopedia of Mammals. McGraw-Hill, New York.
4119. Gubernick, DJ. 1988. Reproduction in the California mouse, *Peromyscus californicus.* J Mamm. 69:857-860.
4120. Gubernick, DJ & Nelson, RJ. 1989. Prolactin and paternal behavior in the biparental California mouse, *Peromyscus californicus.* Horm Beh. 23:203-210.
4121. Gucwinska, H. 1971. Development of six-banded armadillos *Euphractus sexcinctus* at Wroclaw Zoo. Int Zoo Yearb. 11:88-89.
4122. Gucwinska, H & Gucwinski, A. 1968. Breeding the Zanzibar galago *Galago senegalensis zanaibaricus* at Wroclaw Zoo. Int Zoo Yearb. 8:111-114.
4123. Guedon, G, Pascal, M, & Maxouin, F. 1991. Le Campagnol provençal en captivité (*Pitymys duodecimcostatus*de Sely-Longchamps, 1839) (Rongeurs, Microtides) II La croissance. Mammalia. 55:397-406.
4124. Guenther, SE. 1948. Young beavers. J Mamm. 29:419-420.
4125. Guetzow, DD & Judd, FW. 1981. Postnatal growth and development in a subtropical population of *Peromyscus leucopus texanus.* Southwest Nat. 26:183-191.
4126. Guiler, ER. 1957. Longevity in the wild potoroo, *Potorous tridactylus.* Aust J Sci. 20:26.
4127. Guiler, ER. 1958. Observations on a population of small marsupials in Tasmania. J Mamm. 39:44-58.
4128. Guiler, ER. 1960. The breeding season of *Potorous tridactylus* (Kerr). Aust J Sci. 23:126-127.
4129. Guiler, ER. 1960. The pouch young of the potoroo. J Mamm. 41:441-451.
4130. Guiler, ER. 1961. Breeding season of the thylacine. J Mamm. 42:396-397.
4131. Guiler, ER. 1961. A pregnant pygmy right whale. Aust J Sci. 24:297-298.
4132. Guiler, ER. 1970. Observations on the Tasmanian devil, *Sarcophilus harrisii* (Marsupialia: Dasyuridae) II Reproduction, breeding and growth of pouch young. Aust J Zool. 18:63-70.
4133. Guiler, ER. 1971. The husbandry of the potoroo *Potorous tridactylus.* Int Zoo Yearb. 11:21-22.
4134. Guiler, ER. 1971. The Tasmanian devil *Sarcophilus harrisii* in captivity. Int Zoo Yearb. 11:32-33.
4135. Guiler, ER. 1978. Whale strandings in Tasmania since 1945 with notes on some seal reports. Pap Proc R Soc Tasmania. 112:189-213.
4136. Guiler, ER & Heddle, RWL. 1970. Testicular and body temperatures in the Tasmanian devil and three other species of marsupial. Comp Biochem Physiol. 33:881-891.
4137. Guinet, C. 1991. Growth from birth to weaning in the southern elephant seal (*Mirounga leonina*). J Mamm. 72:617-620.
4138. Guinness, FE, Albon, SD, & Clutton-Brock, TH. 1978. Factors affecting reproduction in red deer (*Cervus elaphus*) hinds on Rhum. J Reprod Fert. 54:325-334.
4139. Guinness, FE, Gibson, RM, & Clutton-Brock, TH. 1978. Calving times of red deer (*Cervus elaphus*) on Rhum. J Zool. 185:105-114.
4140. Guinness, FE, Hall, MJ, & Cockerill, RA. 1979. Mother-offspring association in red deer (*Cervus elaphus* L) on Rhum. Anim Behav. 27:536-544.
4141. Guinness, FE, Lincoln, GA, & Short, RV. 1971. The reproductive cycle of the female red deer, *Cervus elaphus.* J Reprod Fert. 27:427-438.
4142. Guiraud, M. 1948. Contribution à l'étude de *Phacochoerus aethiopicus.* Mammalia. 12:54-66.
4143. Gulamhusein, AP & Beck, F. 1977. Determination of a sensitive period for the induction of the maternal pregnancy reaction in the ferret. J Reprod Fert. 49:127-128.
4144. Gulamhusein, AP & Tam, WH. 1974. Reproduction in the male stoat, *Mustela erminea.* J Reprod Fert. 41:303-312.
4145. Gulamhusein, AP & Thawley, AR. 1972. Ovarian cycle and plasma progesterone levels in the female stoat, *Mustela erminea.* J Reprod Fert. 31:492-493.
4146. Guldberg, GA. 1894. On the duration of gravidity in certain Odontoceti. Bergen Museum. 4:61-65.
4147. Guldberg, GV. 1887. Zur Biologie der nordatlantischen Finwalarten. Zool Jb Syst. 2:127-174.
4148. Gulotta, EF. 1971. *Meriones unguiculatus.* Mamm Species. 3:1-5.
4149. Gulyas, BJ, Hodgen, GD, Tullner, WW, et al. 1977. Effects of fetal or maternal hypophsectomy on endocrine organs and body weight in infant rhesus monkeys (*Macaca mulatta*): With particular emphasis on oogenesis. Biol Reprod. 16:216-227.
4150. Gulyas, BJ, Tullner, WW, & Hodgen, GD. 1977. Fetal or material [*sic*] hypophysectomy in rhesus monkeys (*Macaca mulatta*): Effects on the development and testes and other endocrine organs. Biol Reprod. 17:650-660.
4151. Gulyas, BJ, Yuan, L, Tullner, WW, et al. 1976. The fine structure of corpus luteum from intact, hypophysectomized and fetectomized pregnant monkeys (*Macaca mulatta*) at term. Biol Reprod. 14:613-626.
4152. Gundlach, H. 1968. Brutfürsorge, Brutpflege, Verhaltungsontogenese und Tagesperiodik beim europaischen Wildschwein (*Sus scrofa* L). Z Tierpsychol. 25:955-995.
4153. Gunn, CK. 1945. The effect of mating time upon reproduction in foxes. Emp J Exp Agric. 13:193-198.
4154. Gunn, DL, Jenkin, PM, & Gunn, AL. 1937. Menstrual periodicity; Statistical observations on a large sample of normal cases. J Obst Gyn. 51:839-879.
4155. Gunson, JR & Bjorge, RR. 1979. Winter devining of the striped skunk in Alberta. Can Field-Nat. 93:252-259.
4156. Gunter, G. 1942. Contributions to the natural history of the bottle-nose dolphin, *Tursiops truncatus* (Montague), on the Texas coast, with particular reference to food habits. J Mamm. 23:267-276.
4156a. Gunter, G. 1946. Records of the blackfish or pilot whale from the Texas coast. J Mamm. 27:374-377.
4157. Gunther, WC. 1956. Studies on the male reproductive system of the California pocket gopher (*Thomomys bottae navus* Merriam). Am Mid Nat. 55:1-40.
4157a. Gunther, WC. 1960. The origin and homology of the male reproductive system in *Thomomys bottae.* J Mamm. 41:243-250.
4158. Gupta, BB. 1961. Birth and early development of Indian hedgehogs. J Mamm. 42:398-399.

4159. Guraya, SS & Bilaspuri, GS. 1976. Stages of seminiferous epithelial cycle in the buffalo (*Bos bubalis*). Ann Biol Anim Bioch Biophys. 16:137-144.
4159a. Gust, DA, Busse, CD, & Gordon, TP. 1990. Reproductive parameters in the sooty mangabey (*Cercocebus torquatus atys*). Am J Primatol. 22:241-250.
4160. Gustafson, AW. 1979. Male reproductive patterns in hibernating bats. J Reprod Fert. 56:317-331.
4161. Gustafson, AW. 1987. Changes in Leydig cell activity during the annual testicular cycle of the bat *Myotis lucifugus lucifugus*: Histology and lipid biochemistry. Am J Anat. 178:312-325.
4162. Gustafson, AW & Belt, WD. 1981. The adrenal cortex during activity and hibernation in the male little brown bat, *Myotis lucifugus lucifugus*: Annual rhythm of plasma cortisol levels. Gen Comp Endocrinol. 44:269-278.
4163. Gustafson, AW & Damassa, DA. 1984. Perinatal and postnatal patterns of plasma sex steroid-binding protein and testosterone in relation to puberty in the male little brown bat. Endocrinology. 115:2347-2354.
4164. Gustafson, AW & Damassa, DA. 1985. Annual variations in plasma sex steroid-binding protein and testosterone concentrations in the adult male little brown bat: Relation to the asychronous recrudescence of the testis and accessory reproductive organs. Biol Repro. 33:1126-1137.
4165. Gustafson, AW, Damassa, DA, & Pratt, RD. 1986. Plasma sex-steroid-binding protein (SBP) in the Djungarian hamster (*Phodopus sungorus*): Steroid-binding properties and postnatal variations in females. Biol Reprod. 34(Suppl):234.
4166. Gustafson, AW, Damassa, DA, Pratt, RD, et al. 1989. Post-natal patterns of plasma androgen-binding activity in Djungarian (*Phodopus sungorus*) and golden (*Mesocricetus auratus*) hamsters. J Reprod Fert. 86:91-104.
4167. Gustafson, AW & Shemesh, M. 1976. Changes in plasma testosterone levels during the annual reproductive cycle of the hibernating bat, *Myotis lucifugus lucifugus* with a survey of plasma testosterone levels in adult male vertebrates. Biol Reprod. 15:9-24.
4168. Gustafsson, TO. 1984. Strange animals suppress sexual maturation in male bank voles, *Clethrionomys glareolus*. Acta Zool Fennica. 171:149-150.
4169. Gustafsson, TO & Andersson, CB. 1980. Adrenal growth during pregnancy in the bank vole, *Clethrionomys glareolus*: initiation by mating. Can J Zool. 58:1458-1461.
4170. Gustafsson, TO & Andersson, CB. 1980. Social environment and sexual maturation in male bank voles, *Clethrionomys glareolus* (Schreber, 1780). Säugetierk Mitt. 28:310-312.
4171. Gustafsson, TO, Andersson, CB, & Nyholm, NEI. 1983. Comparison of sensitivity to social suppression of sexual maturation in captive male bank voles, *Clethrionomys glareolus*, originating from populations with different degrees of cyclicity. Oikos. 41:250-254.
4172. Gustafsson, TO, Andersson, CB, & Westlin, LM. 1980. Reproduction in a laboratory colony of bank vole, *Clethrionomys glareolus*. Can J Zool. 58:1016-1021.
4173. Gustafsson, TO, Andersson, CB, & Westlin, LM. 1983. Reproduction in laboratory colonies of bank vole, *Clethrionomys glareolus*, originating from populations with different degrees of cyclicity. Oikos. 40:182-188.
4173a. Guthrie, MJ. 1933. Notes on the seasonal movements and habits of some cave bats. J Mamm. 14:1-19.
4174. Guthrie, MJ. 1933. The reproductive cycles of some cave bats. J Mamm. 14:199-216.
4175. Guthrie, MJ & Jeffers, KR. 1935. The ovary of the bat during hibernation. Anat Rec. 61:21.
4176. Guthrie, MJ & Jeffers, KR. 1938. Growth of follicles in the ovaries of the bat *Myotis lucifugus lucifugus*. Anat Rec. 71:477-496.
4177. Guthrie, MJ & Jeffers, KR. 1938. The ovaries of the bat *Myotis lucifugus lucifugus* after injection of hypophyseal extract. Anat Rec. 72:11-36.
4177a. Guthrie, MJ, Jeffers, KR, & Smith, EW. 1951. Growth of follicles in the ovaries of the bat *Myotis grisescens*. J Morph. 88:127-144.
4178. Haacke, W. 1885. On the marsupial ovum, the mammary pouch, and the male milk glands of *Echidna hystrix*. Proc R Soc London. 38:72-74.
4179. Haagenrud, H & Markgren, G. 1973. The timing of estrus in the moose (*Alces alces* L) in a district of Norway. Int Congr Game Biol. 11:71-78.
4180. Haagner, A. 1920. South African Mammals. Witherby, London.
4181. Haber, A. 1961. Le Sanglier en Pologne. Terre Vie. 16:74-76.
4182. Haddow, AJ. 1952. Field and laboratory studies on an African monkey, *Cercopithecus ascanius schmidti* Matschie. Proc Zool Soc London. 122:297-394.
4183. Hadidan, J & Bernstein, IS. 1979. Female reproductive cycles and birth data from an old world monkey colony. Primates. 20:429-442.
4184. Hadwen, S & Palmer, LJ. 1922. Reindeer in Alaska. USDA Tech Bull. 1089:1-73.
4185. Haensel, J. 1966. Zwillingsgeburt des Groszohr-Hirsches, *Odocoileus hemionus* Raf im Tierpark Berlin. Z Säugetierk. 31:410-411.
4186. Haensel, J. 1968. Neues Hochalter für das Mausohr, *Myotis myotis* (Borkhausen, 1797). Säugetierk Mitt. 16:53.
4187. Hafez, ESE. 1953. Conception-rate and periodicity in the buffalo. Emp J Exp Agric. 21:15-21.
4188. Hafez, ESE. 1971. Reproductive cycles. Comparative Reproduction of Nonhuman Primates. 160-204, Hafez, ESE, ed, CC Thomas, Springfield, IL.
4189. Hafez, ESE & Jaszczak, S. 1972. Comparative anatomy and histology of the cervix uteri in non-human primates. Primates. 13:297-314.
4190. Hafez, S. 1954. The placentome in the buffalo. Acta Zool. 35:177-192.
4191. Hafez, S & Attar, T. 1956. Excretion of urinary estrogens in the pregnant buffalo. Acta Physiol Lat Am. 6:27-32.
4192. Haffenden, A. 1984. Breeding, growth and development in the Herbert river ringtail possum, *Pseudocheirus herbertensis herbertensis* (Marsupialia: Petauridae). Possums and Gliders. 277-281, Smith, AP & Hume, ID, eds, Aust Mamm Soc, Sydney.
4193. Haga, R. 1960. Observations on the ecology of the Japanese pika. J Mamm. 41:200-212.
4194. Hagedoorn, JP. 1966. Hypothalamic neurosecretory activity in relation to reproductive cycle in the common striped skunk (*Mephitis mephitis nigra*: order Carnivora). Z Zellforsch Mikrosk Anat. 75:1-10.
4195. Hagemenas, FC & Kittinger, GW. 1974. The effect of fetal sex on placental biosynthesis of progesterone. Endocrinology. 94:922-924.
4196. Hagen, H & Hagen, U. 1980. Freilandbeobachtung der Geburt einer Thomsongazelle, *Gazella thomsoni* Gunther, 1884, im Amboseli-National park, Kenia. Säugetierk Mitt. 28:256-263.

4196a. Hagen, JB & Forslund, LG. 1979. Comparative fertility of four age classes of female gray-tailed voles, *Microtus canicaudus*, in the laboratory. J Mamm. 60:834-837.
4197. Hagenbeck, CH. 1969. Notes on the artificial rearing of a great Indian rhinoceros *Rhinoceros unicornis* at Hamburg Zoo. Int Zoo Yearb. 9:99-101.
4198. Haggard, VD. 1965. Lar gibbon *Hylobates lar* breeding records. Int Zoo Yearb. 5:110-111.
4199. Hagino, N & Goldzieher, JW. 1970. Regulation of gonadotrophin release by the corpus luteum in the baboon. Endocrinology. 87:413-418.
4200. Hagler, E. 1975. Einige Notizen zur Zucht und Haltung von zwei Paar Löwenäffchen (*Leonpithecus rosalia rosalia*) im Oklahoma City Zoo. Fr Kölner Zoo. 18:126-127.
4201. Haglund, B. 1966. De stora rovdjurens vintervanor I. Viltrevy. 4:75-310.
4201a. Hahn, WL. 1907. A review of the bats of the genus *Hemiderma*. Proc US Nat Mus. 32:103-118.
4202. Hahn, WL. 1908. Some habits and sensory adaptations of cave-inhabiting bats. Biol Bull. 15:135-193.
4203. Haider, SG, Passia, D, Treiber, A, et al. 1989. Description of eight phases of spermiogenesis in the marmoset testis. Acta Anat. 135:180-184.
4204. Haigh, GR. 1983. Effects of inbreeding and social factors on the reproduction of young female *Peromyscus maniculatus bairdii*. J Mamm. 64:48-54.
4205. Haigh, GR, Cushing, BS, & Bronson, FH. 1988. A novel postcopulatory block of reproduction in white-footed mice. Biol Reprod. 38:623-626.
4206. Haigh, GR, Lounsbury, DM, & Gordon, TA. 1985. Pheromone-induced inhibition in young female *Peromyscus leucopus*. Biol Reprod. 33:271-276.
4207. Haigh, JC, Cates, WF, & Glover, GJ. 1982. Seasonal changes in serum testosterone, scrotal circumference and sperm morphology of male wapiti. Ann Proc Am Assoc Zoo Vet. 1982:121-126.
4208. Haigh, JC, Cates, WF, Glover, GJ, et al. 1984. Relationships between seasonal changes in serum testosterone concentrations, scrotal circumference and sperm morphology of male wapiti (*Cervus elaphus*). J Reprod Fert. 70:413-418.
4209. Haim, A, Skinner, JD, & Robinson, TJ. 1987. Bioenergetics, thermoregulation and urine analysis of squirrels of the genus *Xerus* from an arid environment. S Afr J Zool. 22:45-49.
4210. Hainard, R. 1953. Notes sur le rut du Bouquetin. Säugetierk Mitt. 1:26-28.
4211. Haines, DE, Holmes, KR, & Carmichael, SW. 1976. Sex determination on the lesser bushbaby (*Galago senegalensis*). Lab Anim Sci. 26:430-435.
4212. Haines, H. 1961. Seasonal changes in the reproductive organs of the cotton rat, *Sigmodon hispidus*. TX J Sci. 13:219-230.
4213. Haines, HB. 1961. Characteristics of a cotton rat (*Sigmodon hispidus*) population cycle. Ph D Diss. Univ TX, Austin.
4213a. Haines, HB. 1971. Characteristics of a cotton rat (*Sigmodon hispidus*) population cycle. TX J Sci. 23:3-27.
4214. Halazon, GC & Buechner, HK. 1956. Postconception ovulation in elk. Trans N Am Wildl Conf. 21:545-554.
4215. Haldar-Misra, C & Srivastava, M. 1987. Effect of pinealectomy on photoperiodic gonadal response of Indian palm squirrel, *Funambulus pennanti*. Can J Zool. 65:833-836.
4216. Hale, OM & Bondari, K. 1986. Effect of crossbreeding on genetic improvement of growth and reproduction in pigs. Growth. 50:526-536.
4217. Haley, D. 1980. The great northern sea cow. Oceans. 13:7-9.
4218. Halfpenny, JC. 1980. Reproductive strategies: Intra- and inter-specific comparison within the genus *Peromyscus*. Ph D Diss. Univ CO, Boulder.
4219. Halfpenny, JC & Ingraham, KP. 1983. Growth and development of heather voles. Growth. 47:437-445.
4220. Hall, A, III, Persing, RL, White, DC, et al. 1967. *Mystromys albicaudatus* (the African white-tailed rat) as a laboratory species. Lab Anim Care. 17:180-188.
4221. Hall, AB. 1964. Musk-oxen in Jameson Land and Scoresby Land, Greenland. J Mamm. 45:1-11.
4222. Hall, ER. 1923. Occurrence of the hoary bat at Lawrence Kansas. J Mamm. 4:192-193.
4223. Hall, ER. 1928. Notes on the life history of the sage-brush meadow mouse (*Lagurus*). J Mamm. 9:201-204.
4224. Hall, ER. 1938. Gestation period in the long-tailed weasel. J Mamm. 19:249-250.
4225. Hall, ER. 1942. Gestation period in the fisher with recommendations for the animal's protection in California. CA Fish Game. 28:143-147.
4226. Hall, ER. 1946. Mammals of Nevada. Univ CA Press, Berkeley, CA.
4227. Hall, ER. 1951. American weasels. Univ KS Publ Mus Nat Hist. 4:1-466.
4228. Hall, ER & Dalquest, WW. 1963. The mammals of Veracruz. Univ KS Publ Mus Nat Hist. 14:165-362.
4228a. Hall, ER & Jackson, WB. 1953. Seventeen species of bats recorded from Barro Colorado Island, Panama Canal Zone. Univ KS Publ Mus Nat Hist. 5:641-646.
4228b. Hall, ER & Kelson, KR. 1959. The Mammals of North America. Ronald Press, New York.
4229. Hall, ER & Linsdale, JM. 1929. Notes on the life history of the kangaroo mouse (*Microdipodops*). J Mamm. 10:298-305.
4229a. Hall, ER & Villa R, B. 1949. An annotated checklist of the mammals of Michoacan, Mexico. Univ KS Publ Mus Nat Hist. 1:431-472.
4230. Hall, JG. 1981. A field study of the Kaibab squirrel in Grand Canyon Park. Wildl Monogr. 75:1-54.
4231. Hall, JS. 1962. A life history and taxonomic study of the Indiana bat, *Myotis sodalis*. Reading Public Mus Art Gallery Sci Publ. 12:1-68.
4232. Hall, KRL. 1965. Ecology and behavior of baboon, patas, and vervet monkeys in Uganda. The Baboon in Medical Research. 43-61, Vagtborg, H, ed, Univ TX Press, Austin.
4233. Hall, LS. 1983. Observations on body weights and breeding of the northern brown bandicoot, *Isoodon macrourus*, trapped in south-east Queensland. Aust Wildl Res. 10:467-476.
4234. Hall, O & Davis, DE. 1950. Corpora lutea counts and their relation to the numbers of embryos in the wild Norway rats. TX Rep Bio Med. 8:564-582.
4235. Hall, RD & Hodgen, GD. 1979. Pregnancy diagnosis in owl monkeys (*Aotus trivirgatus*) evaluation of hemagglutination inhibition test for urinary choronic gonadotropin. Lab Anim Sci. 29:345-348.
4236. Hall, VD, Bartke, A, & Goldman, BD. 1982. Role of the testis in regulating the duration of hibernation in the Turkish hamster, *Mesocricetus brandti*. Biol Reprod. 27:802-810.
4237. Hall, VE & Pierce, GN, Jr. 1934. Litter size, birth weight and growth to weaning in the cat. Anat Rec. 60:111-124.

4238. Hall-Craggs, ECB. 1962. The testis of *Gorilla gorilla beringei*. Proc Zool Soc London. 139:511-514.
4239. Hall-Martin, AJ. 1975. Studies on the biology and productivity of the giraffe. D Sc Thesis. Univ Pretoria, Pretoria. [Cited by Skinner, 1984]
4240. Hall-Martin, AJ & Rowlands, IW. 1980. Observations on ovarian structure and development of the southern giraffe, *Giraffa camelopardales giraffa*. S Afr J Zool. 15:217-221.
4241. Hall-Martin, AJ & Skinner, JD. 1978. Observations on puberty and pregnancy in female giraffe (*Giraffa camelopardalis*). S Afr J Wildl Res. 8:91-94.
4242. Hall-Martin, AJ, Skinner, JD, & Hopkins, BJ. 1978. The development of the reproductive organs of the male giraffe, *Giraffa camelopardalis*. J Reprod Fert. 52:1-7.
4243. Hall-Martin, AJ, Skinner, JD, & Smith, A. 1977. Observations on lactation and milk composition of the giraffe *Giraffa camelopardalis*. S Afr J Wildl Res. 7:67-71.
4244. Hall-Martin, AJ, Skinner, JD, & van Dyk, JM. 1975. Reproduction in the giraffe in relation to several environmental factors. E Afr Wildl J. 13:237-248.
4245. Hall-Martin, AJ & van der Walt, LA. 1984. Plasma testosterone levels in relation to musth in the male African elephant. Koedoe. 27:147-149.
4246. Hallett, AF & Meester, J. 1971. Early postnatal development of the South African hamster *Mystromys albicaudatus*. Zool Afr. 6:221-228.
4247. Hallett, JG. 1978. *Parascalops breweri*. Mamm Species. 98:1-4.
4248. Halloran, AF. 1945. Five fetuses reported for *Pecari angulatus* from Arizona. J Mamm. 26:434.
4249. Halloran, AF & Blanchard, WE. 1954. Carnivores of Yuma county, Arizona. Am Mid Nat. 51:481-487.
4250. Halls, LK. 1978. White-tailed deer. Big Game of North America. 43-65, Schmidt, JL & Gilbert, DL, eds, Stackpole Books, Harrisburg, PA.
4251. Hallstrom, E. 1967. Notes on breeding the black rhinoceros *Diceros bicornis* at Sydney Zoo. Int Zoo Yearb. 7:165.
4252. Haltenorth, T. 1957. Die Wildkatze. Neue Brehm-Bücherei, Vol 189, A Ziomsen, A Wittenberg-Lutherstadt, Germany.
4253. Haltenorth, T & Roth, HH. 1968. Short review of the biology and ecology of the red fox *Canis vulpes vulpes* Linnaeus 1758. Säugetierk Mitt. 16:339-352.
4254. Hamann, U. 1979. Beobachtungen zum Verhalten von Bongoantilopen (*Tragelaphus euryceros*, Ogilby, 1836). Zool Garten NF. 49:319-375.
4255. Hamilton, CA. 1979. Aspects of the ecology of feral rusa deer in New South Wales: Deer refresher course. Proc Univ Sydney Postgraduate Comm Vet Sci. 49:85-92. [Cited by Mackenzie, 1985]
4256. Hamilton, GD & Bronson, FH. 1985. Food restriction and reproducive development in wild house mice. Biol Reprod. 32:773-778.
4257. Hamilton, JE. 1934. The southern sea lion, *Otaria byronia* (De Blainville). Discovery Rep. 8:269-318.
4258. Hamilton, JE. 1939. A second report on the southern sea lion *Otaria byronia* (De Blainville). Discovery Rep. 19:121-164.
4259. Hamilton, JE. 1939. The leopard seal *Hydrurga leptonyx* (De Blainville). Discovery Rep. 18:239-264.
4260. Hamilton, RB & Stalling, DT. 1972. *Lasiurus borealis* with live young. J Mamm. 53:190.
4261. Hamilton, WJ & Gould, JH. 1940. The normal oestrous cycle of the ferret: The correlation of the vaginal smear and the histology of the genital tract, with notes on the distribution of glycogen, the incidence of growth and the reaction to intravitum staining by trypan blue. Trans R Soc Edinb. 60:87-106.
4262. Hamilton, WJ & Harrison, RJ. 1951. Placentation in Père David's deer, *Elaphurus davidianus*. Nature. 167:443-444.
4263. Hamilton, WJ, Harrison, RJ, & Young, BA. 1960. Aspects of placentation in certain Cervidae. J Anat. 94:1-33.
4264. Hamilton, WJ, III. 1962. Reproductive adaptations of the red tree mouse. J Mamm. 43:486-504.
4265. Hamilton, WJ, III, Tilson, RL, & Frank, LG. 1986. Sexual monomorphism in spotted hyenas, *Crocuta crocuta*. Ethology. 71:63-73.
4266. Hamilton, WJ, Jr. 1929. Breeding habits of the short-tailed shrew, *Blarina brevicaudata*. J Mamm. 10:125-134.
4267. Hamilton, WJ, Jr. 1930. Notes on the mammals of Breathitt county, Kentucky. J Mamm. 11:306-311.
4268. Hamilton, WJ, Jr. 1931. Habits of the star-nosed mole, *Condylura cristata*. J Mamm. 12:345-355.
4269. Hamilton, WJ, Jr. 1933. The weasels of New York. Am Mid Nat. 14:289-344.
4270. Hamilton, WJ, Jr. 1934. The life history of the rufescent woodchuck, *Marmota monax rufescens* Howell. Ann Carnegie Mus. 23:85-178.
4271. Hamilton, WJ, Jr. 1935. Habits of jumping mice. Am Mid Nat. 16:187-200.
4271a. Hamilton, WJ, Jr. 1936. The food and breeding habits of the raccoon. OH J Sci. 36:131-140.
4272. Hamilton, WJ, Jr. 1937. Growth and life span of the field mouse. Am Nat. 71:500-507.
4273. Hamilton, WJ, Jr. 1938. Life history notes on the northern pine mouse. J Mamm. 19:163-170.
4274. Hamilton, WJ, Jr. 1939. Observations on the life hisotry of the red squirrel in New York. Am Mid Nat. 22:732-745.
4275. Hamilton, WJ, Jr. 1940. Life and habits of field mice. Sci Monthly. 50:425-434.
4276. Hamilton, WJ. Jr. 1940. Breeding habits of the cottontail rabbit in New York state. J Mamm. 21:8-11.
4277. Hamilton, WJ, Jr. 1940. The biology of the smoky shrew (*Sorex fumeus fumeus* Miller). Zoologica. 25:473-492.
4278. Hamilton, WJ, Jr. 1941. Reproduction in the field mouse, *Microtus pennsylvanicus* (Ord). Mem Cornell Univ Agr Exp Sta. 237:1-23.
4279. Hamilton, WJ, Jr. 1944. The biology of the little short-tailed shrew, *Cryptotis parva*. J Mamm. 25:1-7.
4280. Hamilton, WJ, Jr. 1949. The reproductive rates of some small mammals. J Mamm. 30:257-260.
4281. Hamilton, WJ, Jr. 1953. Reproduction and young of the Florida wood rat, *Neotoma floridana floridana*. J Mamm. 34:180-189.
4282. Hamilton, WJ, Jr. 1956. The young of *Neofiber alleni*. J Mamm. 37:448-449.
4283. Hamilton, WJ, Jr. 1958. Early sexual maturity in the female short-tailed weasel. Science. 127:1057.
4284. Hamilton, WJ, Jr. 1958. Life history and economic relations of the opossum (*Didelphis marsupialis virginiana*) in New York State. Mem Cornell Univ Agr Exp Sta. 354:1-48.
4285. Hamilton, WJ, Jr & Cook, AH. 1955. The biology and management of the fisher in New York. NY Fish Game J. 2:13-35.
4286. Hamilton, WJ, Jr & Eadie, WR. 1964. Reproduction in the otter, *Lutra canadensis*. J Mamm. 45:242-252.
4287. Hamlett, GWD. 1932. Observations on the embryology of the badger. Anat Rec. 53:283-303.

4288. Hamlett, GWD. 1932. The reproductive cycle in the armadillo. Z Wiss Zool. 141:143-157.
4289. Hamlett, GWD. 1934. Implantation und Embryonalhüllen bei zwei südamerikanischen Fledermäusen. Anat Anz. 79:146-149.
4290. Hamlett, GWD. 1934. Uterine bleeding in a bat, *Glossophaga soricina*. Anat Rec. 60:9-17.
4291. Hamlett, GWD. 1935. Breeding habits of the phyllostomid bats. J Mamm. 16:146-147.
4292. Hamlett, GWD. 1935. Delayed implantation and discontinuous development in the mammals. Q Rev Biol. 10:432-447.
4293. Hamlett, GWD. 1935. Notes on the embryology of a phyllostomid bat. Am J Anat. 56:327-354.
4294. Hamlett, GWD. 1935. The effects of antuitrin S and pituitary extract upon the armadillo ovary. Anat Rec. 62:201-267.
4295. Hamlett, GWD. 1938. The reproductive cycle of the coyote. USDA Techn Bull. 616:1-12.
4296. Hamlett, GWD. 1939. Reproduction in American monkeys I Estrous cycle, ovulation and menstruation in *Cebus*. Anat Rec. 73:171-188.
4297. Hammerling, F & Lippert, W. 1975. Beobachtung des Geburts-und Mutter-Kind-Verhaltens beim Mähnenwolf (*Chrysocyon brachyurus*) über ein Nachtsichtgerät (Infrarot-Beobachtungsanlage). Zool Garten NF. 45:393-415.
4298. Hammerschmidt, K & Ansorge, V. 1989. Birth of a barbary macaque (*Macaca sylvanus*): Acoustic and behavioural features. Folia Primatol. 52:78-87.
4299. Hammond, J & Marshall, FHA. 1930. Oestrus and pseudo-pregnancy in the ferret. Proc R Soc London Biol Sci. 105:607-630.
4300. Hammond, J & Walton, A. 1934. Notes on ovulation and fertilization in the ferret. J Exp Biol. 11:307-319.
4301. Hammond, J & Walton, A. 1934. Pregnancy during the anoestrous season in the ferret. J Exp Biol. 11:320-325.
4302. Hammond, J, Jr. 1951. Control by light of reproduction in ferrets and mink. Nature. 167:150-151.
4303. Hammond, J, Jr. 1951. Failure of progesterone treatment to affect delayed implantation im mink. J Endocrinol. 7:330-334.
4304. Hampton, JK, Jr & Hampton, SH. 1965. Marmosets (Hapalidae): Breeding seasons, twinning, and sex of offspring. Science. 150:915-917.
4305. Hampton, JK, Jr, Hampton, SH, & Landwehr, BT. 1966. Observations on a successful breeding colony of the marmoset, *Oedipomidas oedipus*. Folia Primatol. 4:265-287.
4306. Hampton, JK, Jr, Hampton, SH, & Levy, BM. 1971. Reproductive physiology and pregnancy in marmosets. Med Primatol. 1970:527-535. Goldsmith, EI & Moor-Jankowski, J, eds, Karger, Basel.
4307. Hampton, JK, Jr, Levy, BM, & Sweet, PM. 1969. Chorionic gonadotropin excretion during pregnancy in the marmoset, *Callithris jacchus*. Endocrinology. 85:171-174.
4308. Hampton, JK, Jr, Levy, BM, & Sweet, PM. 1969. Chorionic gonadotropin excretion during pregnancy of the marmoset, *Callithrix jacchus*. Fed Proc. 28:367 (ref 613).
4309. Hampton, SH, Gross, MJ, & Hampton, JK, Jr. 1978. A comparison of captive breeding performance and offspring survival in the family *Callitricidae*. Primates in Medicine. 10:88-95.
4310. Hampton, SH. 1975. Placental development in the marmoset. Int Cong Primat. 5:106-114.
4311. Hampton, SH & Hampton, JK, Jr. 1967. Rearing marmosets from birth by artificial laboratory techniques. Lab Anim Care. 17:1-10.
4312. Hampton, SH & Hampton, JK, Jr. 1977. Detection of reproductive cycles and pregnancy in tamarins (*Saguinus* spp). The Biology and Conservation of the Callitrichidae. 173-179, Kleiman, DG, ed, Smithsonian Inst Press, Washington, DC.
4313. Hampton, SH, Hampton, JK, Jr, & Levy, BM. 1972. Husbandry of rare marmoset species. Saving the Lion Marmoset. 70-85, Bridgwater, DD, ed, The Wild Animal Propagation Trust, Wheeling, WV.
4314. Hamura, N. 1980. Reproduction of the Siberian lynx. Animals & Zoos. 32(4):12-13.
4315. Hanby, JP, Robertson, LT, & Phoenix, CH. 1971. The sexual behavior of a confined troop of Japanese macaques. Folia Primatol. 16:123-143.
4315a. Handasyde, KA, McDonald, IR, & Evans, BK. 1992. Seasonal changes in plasma concentrations of progesterone in free-ranging platypus (*Ornithorhynchus anatinus*). Platypus and Echidnas. 75-79, Augee, ML, ed, Roy Zool Soc NSW, Sydney.
4316. Hanif, M. 1967. Notes on breeding the white-headed saki monkey *Pithecia pithecia* at Georgetown Zoo. Int Zoo Yearb. 7:81-82.
4317. Hanken, J & Sherman, PW. 1981. Multiple paternity in Belding's ground squirrel litters. Science. 212:351-353.
4318. Hanks, J. 1969. Seasonal breeding of the African elephant in Zambia. E Afr Wildl J. 7:167.
4319. Hanks, J. 1970. Elephant research project. Black Lechwe. 8(1):14-18.
4320. Hanks, J. 1972. Reproduction of elephant, *Loxodonta africana*, in the Luangwa valley, Zambia. J Reprod Fert. 30:13-26.
4321. Hanks, J. 1973. Growth and development of the ovary of the African elephant, *Loxodonta africana* Blumenbach. Puku. 7:126-131.
4322. Hanks, J. 1973. Reproduction in the male African elephant in the Luangwa valley, Zambia. J S Afr Wildl Mgt Assn. 3:31-39.
4323. Hanks, J & Albl, P. 1973. Observations on the amniotic nodules of the hippopotamus (*Hippopotamus amphibius* Linn). Puku. 7:133-136.
4324. Hanks, J, Cumming, DHM, Orpen, JL, et al. 1976. Growth, condition and reproduction in the impala ram (*Aepyceros melampus*). J Zool. 179:421-435.
4325. Hanks, J & McIntosh, JEA. 1973. Population dynamics of the African elephant (*Loxodonta africana*). J Zool. 169:29-38.
4326. Hanks, J, Price, MS, & Wrangham, RW. 1969. Some aspects of the ecology and behaviour of the defassa waterbuck (*Kobus defassa*) in Zambia. Mammalia. 33:471-494.
4327. Hanks, J & Short, RV. 1972. The formation and function of the corpus luteum in the African elephant, *Loxodonta africana*. J Reprod Fert. 29:79-89.
4328. Hanney, P. 1964. The harsh-furred rat in Nyasaland. J Mamm. 45:345-358.
4329. Hanney, P. 1965. The Muridae of Malawi (Africa: Nyasaland). J Zool. 146:577-633.
4330. Hanney, P & Morris, B. 1962. Some observations of the pouched rat in Nyasaland. J Mamm. 43:238-248.
4331. Hanse, WA. 1962. Preliminary studies on the dassie. Rep Dept Nat Cons Cape Good Hope. 19:108-121.
4332. Hanselka, CW, Inglis, JM, & Applegate, HG. 1971. Reproduction in the black-tailed jackrabbit in south-western Texas. Southwest Nat. 16:214-217.
4333. Hansen, CG. 1965. Growth and development of desert bighorm sheep. J Wildl Manage. 29:387-391.
4334. Hansen, RM. 1957. Communal litters of *Peromyscus maniculatus*. J Mamm. 38:523.

4335. Hansen, RM. 1957. Development of young varying lemmings, (*Dicrostonyx*). Arctic. 10:105-117.
4336. Hansen, RM. 1960. Age and reproductive characteristics of mountain pocket gophers in Colorado. J Mamm. 41:323-335.
4337. Hansen, RM & Bear, GD. 1964. Comparison of pocket gophers from alpine, subalpine and shrub-grassland habitats. J Mamm. 45:638-640.
4338. Hansen, RS. 1972. Observations on the Sumatran rhino (*Dicerorhinus sumatrensis*) at Copenhagen Zoo. Zool Garten NF. 42:296-303.
4339. Hanson, WM. 1950. The mountain goat in South Dakota. Ph D Diss. Univ MI, Ann Arbor.
4340. Hansson, A. 1947. The physiology of reproduction in mink (*Mustela vison*, Schreb) with special reference to delayed implantation. Acta Zool (Stockholm). 28:1-136.
4341. Hansson, L. 1983. Reproductive development related to age indicators in microtine rodents. J Wildl Manage. 47:1170-1172.
4342. Hansson, L. 1984. Winter reproduction of small mammals in relation to food conditions and population dynamics. Sp Publ Carnegie Mus Nat Hist. 10:225-234.
4343. Hansson, L. 1988. Parent-offspring correlations for growth and reproduction in the vole (*Clethrionomus* [*sic*] *glareolus*) in relation to the Chitty hypothesis. Z Säugtierk. 53:7-10.
4344. Happold, DCD. 1966. Breeding periods of rodents in the northern Sudan. Rev Zool Bot Afr. 74:357-363.
4345. Happold, DCD. 1966. The mammals of Jebel Marra, Sudan. J Zool. 149:126-136.
4346. Happold, DCD. 1967. Biology of the jerboa, *Jaculus jaculus butleri* (Rodentia, Dipodidae), in the Sudan. J Zool. 151:257-275.
4347. Happold, DCD. 1967. The natural history of Khartoum province: Mammals. Sudan Notes & Records. 8:111-132.
4348. Happold, DCD. 1968. Observations on *Gerbillus pyramidum* (Gerbillinae, Rodentia) at Khartoum, Sudan. Mammalia. 32:44-53.
4349. Happold, DCD. 1970. Reproduction and development of the Sudanese jerboa, *Jaculus jaculus butleri* (Rodentia, Dipodidae). J Zool. 162:505-515.
4350. Happold, DCD. 1974. The small rodents of the forest-savanna-farmland association near Ibadan, Nigeria, with observations on reproductive biology. Rev Zool Afr. 88:814-836.
4351. Happold, DCD. 1975. The ecology of rodents in the northern Sudan. Rodents in Desert Environments. 15-46, Prakash, I & Ghosh, PK, eds, Dr W Junk, The Hague.
4352. Happold, DCD. 1977. A population study of small rodents in the tropical rain forest of Nigeria. Terre Vie. 31:385-457.
4353. Happold, DCD. 1978. Reproduction, growth and development of a West African forest mouse, *Praomys tullbergi* (Thomas). Mammalia. 42:73-95.
4354. Happold, DCD. 1979. Age structure of a population of *Praomys tullbergi* (Muridae, Rodentia) in Nigerian rain forest. Terre Vie. 33:253-274.
4354a. Happold, DCD. 1987. The Mammals of Nigeria. Clarendon Press, Oxford, UK.
4355. Happold, M. 1976. Reproductive biology and developments in the conilurine rodents (Muridae) of Australia. Aust J Zool. 24:19-26.
4356. Haque, AKMA, Nishiwaki, M, Kasya, T, et al. 1977. Observations on the behaviour and other biological aspects of the Ganges susu, *Platanista gangetica*. Sci Rep Whales Res Inst. 29:87-94.
4357. Harcourt, AH, Fossey, D, & Sabater-Pi, J. 1981. Demography of *Gorilla gorilla*. J Zool. 195:215-233.
4358. Harcourt, AH, Fossey, D, Stewart, KJ, et al. 1980. Reproduction in wild gorillas and some comparisons with wild chimpanzees. J Reprod Fert Suppl. 28:59-70.
4359. Harcourt, AH, Stewart, KJ, & Fossey, D. 1976. Male emigration and female transfer in wild mountain gorilla. Nature. 263:226-227.
4360. Harcourt, AH, Stewart, KJ, & Fossey, D. 1981. Gorilla reproduction in the wild. Reproductive Biology of the Great Apes. 265-279, Graham, CE, ed, Academic Press, New York.
4361. Harcourt, CS & Nash, LT. 1986. Social organization of galagos in Kenyan coastal forests I *Galago zanzibaricus*. Am J Primatol. 10:339-355.
4362. Harcourt, CS & Nash, LT. 1986. Species differences in substrate use and diet between sympatric galagos in two Kenyan coastal forests. Primates. 27:41-52.
4363. Harder, JD & Fleming, MW. 1981. Estradiol and progesterone profiles indicate a lack of endocrine recognition of pregnancy in the opossum. Science. 212:1400-1402.
4364. Harder, JD & Fleming, MW. 1986. Annual pattern of circulating testosterone relative to the breeding season in male opossums (*Didelphis virginiana*). Biol Reprod. 34(Suppl):203.
4365. Harder, JD, Hinds, LA, Horn, CA, et al. 1984. Oestradiol in follicular fluid and in utero-ovarian venous and peripheral plasma during parturition and post-partum oestrus in the tammar, *Macropus eugenii*. J Reprod Fert. 72:551-558.
4366. Harder, JD, Hinds, LA, Horn, CA, et al. 1985. Effects of removal in late pregnancy of the corpus luteum, Graafian follicle or ovaries on plasma progesterone, oestradiol, LH, parturition, and post-partum oestrus in the tammar wallaby, *Macropus eugenii*. J Reprod Fert. 75:449-459.
4367. Harder, JD & Moorhead, DL. 1980. Development of corpora lutea and plasma progesterone levels associated with the onset of the breeding season in white-tailed deer (*Odocoileus virginianus*). Biol Reprod. 22:185-191.
4368. Harder, JD & Woolf, A. 1976. Changes in plasma levels of oestrone and oestradiol during pregnancy. J Reprod Fert. 47:161-163.
4369. Hardin, CJ, Danford, D, & Skeldon, PC. 1969. Notes on the successful breeding by incompatible gorillas *Gorilla gorilla* at Toledo Zoo. Int Zoo Yearb. 9:84-88.
4369a. Harding, HR. 1976. Comparative spermatology of some Australian marsupials. Bull Aust Mamm Soc. 3:59-60.
4370. Harding, HR, Carrick, FN & Shorey, CD. 1981. Marsupial phylogeny: New indications from sperm ultrastructure and development in *Tarsipes spenserae*?. Search. 12:45-47.
4370a. Harding, HR, Carrick, FN, & Shorey, CD. 1982. Crystalloid inclusions in the sertoli cells of the koala, *Phascolarctos cinereus*, (Marsupialia). Cell Tiss Res. 221:633-642.
4371. Harding, HR, Carrick, FN, & Shorey, CD. 1984. Sperm ultrastructure and development in the honey possum, *Tarsipes rostratus*. Possums and Gliders. 451-461, Smith, A & Hume, I, eds, Surrey Beatty & Sons Pty Ltd, Sydney, Australia.
4372. Harding, RD, Hulme, MJ, Lunn, SF, et al. 1982. Plasma progesterone levels throughout the ovarian cycle of the common marmoset (*Callithrix jacchus*). J Med Primatol. 11:43-51.
4373. Harding, RSO & Olson, DK. 1986. Pattern of mating among male patas monkeys (*Erythrocebus patas*) in Kenya. Am J Primatol. 11:343-358.
4373a. Hare, JF & Murie, JO. 1992. Manipulation of litter size reveals no cost of reproduction in Columbian ground squirrels. J Mamm. 73:449-454.

4374. Haring, DM. 1988. Natural history and captive management of Verreaux's sifaka *Propithecus verreauxi.* Int Zoo Yearb. 27:125-134.
4374a. Haring, DM & Wright, PC. 1989. Hand-raising a Philippine tarsier, *Tarsius syrichta.* Zoo Biol. 8:265-274.
4375. Harington, CR. 1964. Polar bears and their present status. Can Audubon. 26(1):4-12.
4376. Harland, RM, Blancher, PJ, & Millar, JS. 1979. Demography of a population of *Peromyscus leucopus.* Can J Zool. 57:323-328.
4377. Harley, D. 1985. Birth spacing in langur monkeys (*Presbytis entellus*). Int J Primatol. 6:227-242.
4377a. Harley, D. 1988. Patterns of reproduction and mortality in two captive colonies of hanuman langur monkeys (*Presbytis entellus*). Am J Primatol. 15:103-114.
4378. Harlow, CR. 1984. Endocrine and morphological aspects of pre-implantation development in the marmoset monkey (*Callithrix jacchus jacchus*). Ph D Diss. Univ London, London.
4379. Harlow, CR, Gems, S, Hodges, JK, et al. 1983. The relationship between plasma progesterone and timing of ovulation and early embryonic development in the marmoset monkey *Callithrix jacchus*). J Zool. 201:273-282.
4380. Harlow, CR, Hearn, JP, & Hodges, JK. 1984. Ovulation in the marmoset monkey: Endocrinology, prediction and detection. J Endocrinol. 103:17-24.
4381. Harlow, HJ, Miller, B, Ryder, T, et al. 1985. Energy requirements for gestation and lactation in a delayed implanter, the American badger. Comp Biochem Physiol. 82A:885-889.
4382. Harlow, HJ, Phillips, JA, & Ralph, CL. 1981. Day-night rhythm in plasma melatonin in a mammal lacking a distinct pineal gland, the nine-baded armadillo. Gen Comp Endocrinol. 45:212-218.
4383. Harlow, RF. 1961. Characteristics and status of Florida black bear. Trans N Am Wildl Nat Res Conf. 26:481-495.
4384. Harms, JW. 1956. Fortpflanzungsbiologie. Primatologia. 562-660, Hofer, H, Schultz, AH, & Starck, D, eds, S Karger, Basel.
4385. Harms, JW. 1956. Schwangerschaft und Geburt. Primatologia. 661-722, Hofer, H, Schultz, AH, & Starck, D, eds, S Karger, Basel.
4386. Harper, F. 1927. The mammals of the Okefinokee swamp region of Georgia. Proc Boston Soc Nat Hist. 1927:491-506.
4387. Harper, F. 1929. Mammal notes from Randolph county, Georgia. J Mamm. 10:84-85.
4388. Harper, F. 1932. Mammals of the Athabaska and Great Slave Lakes Region. J Mamm. 13:19-36.
4389. Harper, F. 1955. The barren ground caribou of Keewater. Univ KS Mus Nat Hist Misc Publ. 6:1-164.
4390. Harper, JA, Harn, JH, Bentley, WW, et al. 1967. The status and ecology of the Roosevelt elk in California. Wildl Monogr. 16:1-49.
4391. Harper, RJ & Jenkins, D. 1981. Mating behaviour in the European otter (*Lutra lutra*). J Zool. 195:556-558.
4392. Harrington, FH, Paquet, PC, Ryon, J, et al. 1982. Monogamy in wolves: A review of the evidence. Wolves of the World. 209-222, Harrington, FH & Paquet, PC, eds, Noyes Publ, Park Ridge, NJ.
4393. Harrington, JE. 1975. Field observations of social behavior of *Lemur fulvus fulvus* E Geoffroy 1812. Lemur Biology. 259-279, Tattersall, I & Sussman, RW, eds, Plenum Press, New York.
4394. Harris, MA & Murie, JO. 1986. Discrimination of oestrous status by scent in Columbian ground squirrels. Anim Behav. 32:939-940.
4395. Harris, S. 1979. Age-related fertility and productivity in red foxes, *Vulpes vulpes*, in suburban London. J Zool. 187:195-199.
4396. Harris, S. 1979. Breeding season, litter size and nestling mortality of the harvest mouse, *Micromys minutus* (Rodentia: Muridae), in Britain. J Zool. 188:437-442.
4397. Harris, S & Smith, GC. 1987. Demography of two urban fox (*Vulpes vulpes*) populations. J Appl Ecol. 24:75-86.
4398. Harrison, DJ & Harrison, JA. 1984. Foods of adult Maine coyotes and their known-aged pups. J Wildl Manage. 48:922-926.
4399. Harrison, DL. 1958. A new race of tomb bat *Taphozous perforatus* E Geoffroy, 1818, from northern Nigeria, with some observations on its breeding biology. Durban Mus Novit. 5:143-149.
4400. Harrison, DL. 1958. A note on successive pregnancies in an African bat (*Tadarida pumila websteri* Dollman). Mammalia. 22:592-595.
4401. Harrison, DL. 1975. *Macrophyllum macrophyllum.* Mamm Species. 62:1-3.
4402. Harrison, DL & Clancey, PA. 1952. Notes on the bats (Microchiroptera) from a cave in the Pietermaritzburg district of Natal. Ann Natal Mus. 12:177-182.
4403. Harrison, DR. 1947. A new central Mediterranean subspecies of field mouse (*Apodemus sylvaticus* Linn), and notes on surrounding forms. Proc Zool Soc London. 117B:650-652.
4404. Harrison, JL. 1951. Reproduction in rats of the subgenus *Rattus.* Proc Zool Soc, London. 121:673-694.
4405. Harrison, JL. 1952. Breeding rhythms of Selangor rodents. Bull Raffles Mus. 24:109-131.
4406. Harrison, JL. 1955. Data on the reproduction of some Malayan mammals. Proc Zool Soc London. 125:445-460.
4407. Harrison, JL. 1956. Survival rate of Malayan rats. Bull Raffles Mus. 27:5-26.
4408. Harrison, JL & Traub, R. 1950. Rodents and insectivores from Selangor Malaya. J Mamm. 31:337-346.
4409. Harrison, JL & Woodville, HC. 1949. Variation in size and weight in five species of house-rats (Rodentia: Muridae), in Rangoon, Burma. Rec Indian Mus. 47:65-71.
4410. Harrison, JL & Woodville, HC. 1950. The growth of a tame specimen of the Indian mole rat *Bandicota bengalensis* (Rodentia, Muridae), and an attempt to estimate the age-structure of a wild population. J Zool Soc India. 2:14-17.
4411. Harrison, LH. 1937. The female sexual cycle in the British horse-shoe bats, *Rhinolophus ferrumequinum insulanus* Barrett-Hamilton and *Rhinolophus hipposideros minutus* Montagu. Trans Zool Soc London. 323:224-255.
4411a. Harrison, RG. 1948. Vascular patterns in the testis, with particular reference to *Macropus.* Nature. 161:399-400.
4412. Harrison, RJ. 1949. Multiovular follicles in the ovaries of lower primates. Nature. 164:409-410.
4413. Harrison, RJ. 1949. Observations on the female reproductive organs of the ca'aing whale *Globiocephala melaena* Traill. J Anat. 83:218-253.
4414. Harrison, RJ. 1953. Some consideration of the growth of placenta in fallow deer, *Dama dama.* Proc Zool Soc London. 123:476.
4415. Harrison, RJ. 1960. Reproduction and reproductive organs in common seals (*Phoca vitulina*), in the Wash, East Anglia. Mammalia. 24:372-385.
4416. Harrison, RJ. 1963. A comparison of factors involved in delayed implantation in badgers and seals in Great Britain. Delayed Implantation. 99-114, Enders, AC, ed, Univ Chicago Press, Chicago.

4417. Harrison, RJ. 1969. Reproduction and reproductive organs. The Biology of Marine Mammals. 253-348, Andersen, HT, ed, Academic Press, New York.
4418. Harrison, RJ. 1972. Reproduction and reproductive organs in *Platanista indi* and *Platanista gangetica*. Invest Cetacea. 4:71-82.
4419. Harrison, RJ. 1977. Ovarian appearances and histology in *Tursiops truncatus*. Breeding Dolphins: Present Status, Suggestions For the Future. 195-204, Ridgway, SH & Benirschke, K, eds, US Mar Mamm Comm, Washington, DC.
4420. Harrison, RJ, Boice, RC, & Brownell, RL, Jr. 1969. Reproduction in wild and captive dolphins. Nature. 222:1143-1147.
4421. Harrison, RJ & Brownell, RL, Jr. 1971. The gonads of the South American dolphins, *Inia geoffrensis*, *Pontoporia blainvillei*, and *Sotalia fluviatilis*. J Mamm. 52:413-419.
4422. Harrison, RJ, Brownell, RL, Jr, & Boice, AC. 1972. Reproduction and gonadal appearances in some Odontocetes. Funct Anat Mar Mamm. 1:361-429, Harrison, RJ, ed, Academic Press, New York.
4423. Harrison, RJ, Bryden, MM, & McBrearty, DA. 1981. The ovaries and reproduction in *Pontoporia blainvillei* (Cetacea: Platanistidae). J Zool. 193:563-580.
4424. Harrison, RJ & Fanning, JC. 1973/74. Anatomical observations in the South Australian bottlenosed dolphin (*Tursiops truncatus*). Invest Cetacea. 6:203-217.
4425. Harrison, RJ & Hamilton, WJ. 1952. The reproductive tract and the placenta and membranes of Père David's deer (*Elaphurus davidianus* Milne Edwards). J Anat. 86:203-225.
4426. Harrison, RJ & Hyett, AR. 1954. The development and growth of the placentomes in the fallow deer (*Dama dama*). J Anat. 88:338-355.
4427. Harrison, RJ, Matthews, LH, & Roberts, JM. 1952. Reproduction in some Pinnipedia. Trans Zool Soc London. 27:437-540.
4428. Harrison, RJ & McBrearty, DA. 1973/74. Reproduction and gonads of the black finless porpoise, *Neophocaena phocaenoides*. Invest Cetacea. 6:225-230.
4429. Harrison, RJ & Neal, EE. 1956. Ovulation during delayed implantation and other reproductive phenomena in the badger (*Meles meles* L). Nature. 177:977-979.
4430. Harrison, RJ & Ridgway, SH. 1971. Gonadal activity in some bottlenose dolphins (*Tursiops truncatus*). J Zool. 165:355-366.
4431. Harrison, RJ & Young, BA. 1966. Functional characteristics of the pinniped placenta. Symp Zool Soc London. 15:47-67.
4432. Harrison, RM & Dukelow, WR. 1973. Seasonal adaption of laboratory-maintained squirrel monkey. J Med Primatol. 2:277-283.
4433. Hart, EB. 1982. The raccoon, *Procyon lotor*, in Wyoming. Great Basin Nat. 42:599-600.
4433a. Hart, EB. 1992. *Tamias dorsalis*. Mamm Species. 399:1-6.
4434. Hart, JA & Timm, RM. 1978. Observations on the aquatic genet in Zaire. Carnivore. 1:130-132.
4435. Harting, P. 1879. A thesis on the ovum and placenta of the Halicore dugong. Ph D Diss. Univ Utrecht, 1878. [abstract: Turner, no initial, J Anat, 13:116-117]
4436. Hartman, CG. 1920. The freemartin and its reciprocal; opossum, man, dog. Science. 52:469-471.
4437. Hartman, CG. 1920. Studies in the development of the opossum *Didelphys virginiana* L: V The phenomena of parturition. Anat Rec. 19:251-261.
4438. Hartman, CG. 1922. Breeding habits, development, and birth of the opossum. Smithsonian Inst Ann Rep. 1921:347-364.
4439. Hartman, CG. 1923. The oestrous cycle in the opossum. Am J Anat. 32:353-421.
4440. Hartman, CG. 1925. Hysterectomy and the oestrous cycle in the opossum. Am J Anat. 35:25-29.
4441. Hartman, CG. 1925. Observations on the functional compensatory hypertrophy of the opossum ovary. Am J Anat. 35:1-24.
4442. Hartman, CG. 1925. The interruption of pregnancy by ovariectomy in the aplacental opossum: A study in the physiology of implantation. Am J Physiol. 71:436-454.
4443. Hartman, CG. 1926. Polynuclear ova and polyovular follicles in the opossum and other mammals, with special reference to the problem of fecundity. Am J Anat. 37:1-51.
4444. Hartman, CG. 1928. The breeding season of the opossum (*Didelphis virginiana*) and the rate of intrauterine and postnatal development. J Morphol. 46:143-200.
4445. Hartman, CG. 1928. The period of gestation in the monkey, *Macacus rhesus*, first description of parturition in monkeys, size and behavior of the young. J Mamm. 9:181-194.
4446. Hartman, CG. 1929. Some excessively large litters of eggs liberated at a single ovulation in mammals. J Mamm. 10:197-202.
4447. Hartman, CG. 1931. The breeding season in monkeys with special reference to *Pithecus* (*Macacus*) *rhesus*. J Mamm. 12:129-142.
4448. Hartman, CG. 1932. Studies in the reproduction of the monkey *Macacus* (*Pithecus*) *rhesus*, with special reference to menstruation and pregnancy. Contrib Embryol Carnegie Inst. 23(134):1-161.
4449. Hartman, CG. 1933. On the survival of spermatozoa in the female genital tract of the bat. Q Rev Biol. 8:185-193.
4450. Hartman, CG. 1938. Some observations on the bonnet macaque. J Mamm. 19:468-474.
4451. Hartman, CG. 1938. Menstruation without ovulation ("Pseudomenstruation") incidence and treatment, with special reference to the rhesus monkey. Les Hormones Sexuelles, Comptes Rendus Collogue International, Collège de France 110-19, Juin, 1937. 103-131, Brouka, L, ed, Hermann et Cie, Paris.
4452. Hartman, CG. 1952. Possums. Univ TX Press, Austin.
4453. Hartman, CG & Corner, GW. 1947. Removal of the corpus luteum and of the ovaries of the rhesus monkey during pregnancy: Observations and cautions. Anat Rec. 98:539-546.
4454. Hartman, DS. 1979. Ecology and behavior of the manatee (*Trichechus manatus*) in Florida. Sp Publ Am Soc Mamm. 5:1-153.
4455. Hartman, GD & Gottschang, JC. 1983. Notes on sex determination, neonates, and behavior of the eastern mole, *Scalopus aquaticus*. J Mamm. 64:539-540.
4456. Hartman, GD & Yates, TL. 1985. *Scapanus orarius*. Mamm Species. 253:1-5.
4457. Hartman, L. 1964. Reproduction of the weasel, *Mustela nivalis* L, in captivity. Säugetierk Mitt. 12:90-91.
4458. Hartman, L. 1964. The behaviour and breeding of captive weasels (*Mustela nivalis* L). NZ J Sci. 7:147-156.
4459. Hartung, TG. 1976. A comparative analysis of copulatory plugs in rodents and their relationship to copulatory behavior. MS Thesis. Univ FL. [Cited by Voss, 1979]
4460. Hartung, TG & Dewsbury, DA. 1978. A comparative analysis of copulatory plugs in muroid rodents and their relationship to copulatory behavior. J Mamm. 59:717-723.

4461. Hartung, TG & Dewsbury, DA. 1979. Paternal behavior in six species of muroid rodents. Behav Neural Biol. 26:466-478.
4462. Hartz, G. 1966. Notes on some recent mammal breeding successes at Lincoln Park Zoo, Chicago. Int Zoo Yearb. 6:206-207.
4463. Harvey, EB. 1959. Placentation in Ochotonidae. Am J Anat. 104:61-85.
4464. Harvey, NC & Rhine, RJ. 1983. Some reproductive parameters of stumptailed macaques (*Macaca arctoides*). Primates. 24:530-536.
4465. Harwood, J & Prime, JH. 1978. Some factors affecting the size of British grey seal populations. J Appl Ecol. 15:401-411.
4466. Hashikawa, T, Inoue, K, & Iwasaki, T. 1981. Growth of the dhole. Animals & Zoos. 33(8):8-11.
4467. Haskell, HS & Reynolds, HG. 1947. Growth, developmental food requirements, and breeding activity of the California jack rabbit. J Mamm. 28:129-136.
4467a. Hasler, JF. 1975. A review of reproduction and sexual maturation in the microtine rodent. Biologist. 57:52-86.
4468. Hasler, JF & Banks, EM. 1974. Morphological changes in the ovaries and other organs of the collared lemming (*Dicrostonyx groenlandicus*) during pregnancy and pseudopregnancy. Can J Zool. 53:12-18.
4469. Hasler, JF & Banks, EM. 1975. Reproductive performance and growth in captive collared lemmings (*Dicrostonyx groenlandicus*). Can J Zool. 53:777-787.
4470. Hasler, JF & Banks, EM. 1975. The influence of mature males on sexual maturation in female collared lemmings (*Dicrostonyx groenlandicus*). J Reprod Fert. 42:583-586.
4471. Hasler, JF & Banks, EM. 1975. The influence of exteroceptive factors on the estrous cycle of the collared lemming (*Dicrostonyx groenlandicus*). Biol Reprod. 12:647-656.
4472. Hasler, JF, Buhl, AE, & Banks, EM. 1976. The influence of photoperiod on growth and sexual function in male and female collared lemmings (*Dicrostonyx groenlandicus*). J Reprod Fert. 46:323-329.
4473. Hasler, JF, Dziuk, PJ, & Banks, EM. 1974. Ovulation and related phenomena in the collared lemming (*Dicrostonyx groenlandicus*). J Reprod Fert. 38:21-28.
4474. Hasler, JF & Sorensen, MW. 1974. Behavior of the tree shrew, *Tupaia chinensis*, in captivity. Am Mid Nat. 91:294-314.
4475. Hasler, MJ & Conaway, CH. 1973. The effect of males on the reproductive state of female *Microtus ochrogaster*. Biol Reprod. 9:426-436.
4476. Hasler, MJ, Falvo, RE, & Nalbandov, AV. 1975. Testicular development and testosterone concentrations in the testis and plasma of young male shrews (*Suncus murinus*). Gen Comp Endocrinol. 25:36-41.
4477. Hasler, MJ, Hasler, JF, & Nalbandov, AV. 1977. Comparative breeding biology of musk shrews (*Suncus murinus*) from Guam and Madagascar. J Mamm. 58:285-290.
4478. Hasler, MJ & Nalbandov, AV. 1974. Body and peritesticular temperatures of musk shrews (*Suncus murinus*). J Reprod Fert. 36:397-399.
4479. Hasler, MJ & Nalbandov, AV. 1974. The effect of weanling and adult males on sexual maturation in female voles (*Microtus ochrogaster*). Gen Comp Endocrinol. 23:237-238.
4480. Hasler, MJ & Nalbandov, AV. 1978. Pregnancy maintenance and progesterone concentrations in the musk shrew, *Suncus murinus* (order: Insectivora). Biol Reprod. 19:407-413.
4481. Hasler, MJ & Nalbandov, AV. 1980. Ovulation, ovum maturation and changes in plasma and adrenal progesterone concentrations in the musk shrew (*Suncus murinus*). Biol Reprod. 22:377-381.
4482. Hass, CC. 1990. Alternative maternal-care patterns in two herds of bighorn sheep. J Mamm. 71:24-35.
4483. Hass, GE, Wilson, N, & Brighton, HD. 1986. Observations on Norway rats, *Rattus norvegicus*, in Kodiak, Alaska. Can Field-Nat. 100:383-386.
4484. Hassan, MS & Hegazy, A. 1968. Rats injurious to agriculture and their control. Tech Bull UAR Min Agric Cairo. [Cited by Osborn & Helmy, 1980]
4485. Hasseltine, HE. 1929. Rat-flea survey of Port Norfolk, VA. Public Health Rep Wash. 44:579-589.
4486. Hassenberg, L. 1977. Zum Fortpflanzungsverhalten des mesopotamischen Damhirsches *Cervus dama mesopotamica* Brooke, 1875, in Gefangenschaft. Säugetierk Mitt. 25:161-194.
4487. Hassenberg, L. 1981. Verhaltungsbeobachtungen am Himalaja-Tahr, *Hemitragus jemlahicus* (H Smith, 1826) in Gefangenschaft. Säugetierk Mitt. 29(2):1-45.
4488. Hassenberg, L & Klös, H-G. 1975. Zur Fortpflanzung und Jugendaufzucht des Baraschinga, *Cervus duvauceli* G Cuvier, 1823 in Gefangenschaft. Säugetierk Mitt. 23:64-73.
4489. Hatcher, VB, McEwan, EH, & Baker, BE. 1967. Caribou milk I Barren-ground caribou (*Rangifer tarandus groenlandicus*): Gross composition, fat and protein constitution. Can J Zool. 45:1101-1106.
4490. Hatfield, DM. 1935. A natural history study of *Microtus californicus*. J Mamm. 16:261-271.
4491. Hatfield, DM. 1939. Notes on sex ratio in Minnesota muskrats. J Mamm. 20:258.
4492. Hatfield, DM. 1939. Rate of metabolism in *Microtus* and *Peromyscus*. Murrelet. 20:54-56.
4493. Hatler, DF. 1967. Some aspects in the ecology of the black bear (*Ursus americanus*) in interior Alaska. MS Thesis. Univ AK. [Cited by Poelker & Hartwell, 1973]
4494. Hatt, RT. 1938. Notes concerning mammals collected in Yucatan. J Mamm. 19:333-337.
4495. Hatt, RT. 1940. Lagomorpha and Rodentia other than Sciuridae, Anomaluridae and Iduridae collected by the American Museum Congo expedition. Bull Am Mus Nat Hist. 76:457-604.
4496. Hatt, RT. 1943. The pine squirrel in Colorado. J Mamm. 24:311-345.
4497. Hatt, RT. 1959. The mammals of Iraq. Misc Publ Mus Zool Univ MI. 106:1-113.
4498. Haug, T. 1981. On some reproduction parameters in fin whales *Balaenoptera physalus* (L) caught off Norway. Rep Int Whal Comm. 31:373-378.
4499. Haugen, AO. 1942. Life history studies of the cottontail rabbit in southwestern Michigan. Am Mid Nat. 28:204-244.
4500. Haugen, AO. 1959. Breeding records of captive white-tailed deer in Alabama. J Mamm. 40:108-113.
4501. Haugen, AO. 1974. Reproduction in the plains bison. IA State J Res. 49:1-8.
4502. Haugen, AO. 1975. Reproductive performance of white-tailed deer in Iowa. J Mamm. 56:151-159.
4503. Haugen, AO & Davenport, LA. 1950. Breeding records of white-tailed deer in the upper peninsula of Michigan. J Wildl Manage. 14:290-295.
4504. Haugen, AO & Mustard, EW, Jr. 1960. Velvet-antlered pregnant white-tailed doe. J Mamm. 41:521-523.
4505. Haugen, AO & Speake, DW. 1957. Parturition and early reactions of white-tailed deer fawns. J Mamm. 38:420-421.
4506. Haugen, AO & Trauger, DL. 1962. Ovarian analysis for data on corpora lutea changes in white-tailed deer. IA Acad Sci. 69:231-238.

4507. Hausfater, G. 1975. Dominance and reproduction in baboons (*Papio cynocephalus*): A quantitative analysis. Contrib Primatol. 7:1-150.
4508. Häussler, U, Möller, E, & Schmidt, U. 1981. Zur Haltung und Jugendentwicklung von *Molossus molossus* (Chiroptera). Z Säugetierk. 46:337-351.
4509. Havera, SP. 1979. Energy and nutrient cost of lactation in fox squirrels. J Wildl Manage. 43:958-965.
4510. Havera, SP, Nixon, CM, & Belcher, HK. 1985. Ovarian characteristics of fox squirrels. Am Mid Nat. 114:396-399.
4511. Havinga, B. 1933. Der Seehund (*Phoca vitulina* Linn) in den holländischen Gewässern. Tijdschr Ned Dierk Ver. 3:79-111.
4512. Hawbecker, AC. 1940. The burrowing and feeding habits of *Dipodomys venustrus*. J Mamm. 21:388-396.
4512a. Hawbecker, AC. 1975. The biology of some desert-dwelling ground squirrels. Monographiae Biologicae. 28:277-303.
4512b. Hawkins, M & Fanning D. 1992. Courtship and mating behaviour of captive platypuses at Taronga Zoo. Platypus and Echidnas. 106-114, Augee, ML, ed, R Zool Soc NSW, Sydney.
4513. Hawkins, RW. 1946. Mating behaviour of the porcupine, *Erethizon dorsatum*. Can Field-Nat. 60:109.
4514. Hay, KG. 1957. Record beaver litter for Colorado. J Mamm. 38:268-269.
4515. Hayama, S, Suzuki, M, Uno, H, et al. 1986. Female sexual maturity and delayed implantation period of the Kuril seal. Sci Rep Whales Res Inst. 37:173-178.
4516. Hayashi, M, Oshima, K, Yamaji, T, et al. 1975. LH levels during various reproductive states in the Japanese monkey (*Macaca fuscata fuscata*). Int Congr Primatol. 5152-157.
4517. Hayden, PH & Gambino, JJ. 1966. Growth and development of the little pocket mouse, *Perognathus longimembris*. Growth. 30:187-198.
4518. Hayden, PH, Gambino, JJ, & Lindberg, RG. 1966. Laboratory breeding of the little pocket mouse, *Perognathus longimembris*. J Mamm. 47:412-423.
4519. Hayman, DL & Martin, PG. 1965. Sex chromosome mosaicism in the marsupial genera *Isoodon* and *Perameles*. Genetics. 52:1201-1206.
4520. Hayman, DL, Martin, PG, & Waller, PF. 1969. Parallel mosaicism of supernumerary chromosomes and sex chromosomes in *Echymipera kalabu* (Marsupialia). Chromosoma. 27:371-380.
4521. Hays, HR. 1967. Notes on breeding black rhinoceroses *Diceros bicornis* at Pittsburgh Zoo. Int Zoo Yearb. 7:164-165.
4522. Hayssen, V. 1980. Observations of the behavior of the Brazilian, short, bare-tailed opossum (*Monodelphis domestica* Wagner) in captivity. Bull Aust Mamm Soc. 6:37.
4523. Hayssen, V. 1991. *Dipodomys microps*. Mamm Species. 389:1-9.
4523a. Hayssen, V. in review. Empirical and theoretical constraints on the evolution of lactation. J Dairy Sci.
4523b. Hayward, BJ & Cockrum, EL. 1971. The natural history of the western long-nosed bat *Leptonycteris sanborni*. W NM Univ Publ. 1:71-123.
4524. Hazama, N. 1964. Weighing wild Japanese monkeys in Arashiyama. Primates. 5:81-104.
4525. Heaney, LR & Birney, EC. 1975. Comments on the distribution and natural history of some mammals in Minnesota. Can Field-Nat. 89:29-34.
4526. Heap, RB, Gombe, S, & Sale, JB. 1975. Pregnancy in the hyrax and erythrocyte metabolism of progesterone. Nature. 257:809-811.
4527. Heap, RB & Hammond, J, Jr. 1974. Plasma progesterone levels in pregnant and pseudopregnant ferrets. J Reprod Fert. 34:149-152.
4528. Heap, RB & Illingworth, DV. 1974. The maintenance of gestation in the guinea-pig and other hystricomorph rodents: Changes in the dynamics of progesterone metabolism and the occurrence of progesterone-binding globulin (PBG). Symp Zool Soc London. 34:385-415.
4529. Heap, RB, Renfree, MB, & Burton, RD. 1980. Steroid metabolism in the yolk sac placenta and endometrium of the tammar wallaby, *Macropus eugenii*. J Endocrinol. 87:339-349.
4530. Heape, W. 1886. The development of the mole (*Talpa europaea*) the ovarian ovum, and segmentation of the ovum. Q J Microsc Sci. 26:157-174.
4531. Heape, W. 1893. The menstruation of *Semnopithecus entellus*. Proc R Soc London. 54:169-172.
4532. Heape, W. 1894. The menstruation of *Semnopithecus entellus*. Phil Trans R Soc London. 185:411-471.
4533. Heape, W. 1896. The menstruation and ovulation of *Macacus rhesus*. Proc R Soc London. 60:202-205.
4534. Heape, W. 1897. The menstruation and ovulation of *Macacus rhesus* with observations undergone by the discharged follicle Part II. Phil Trans R Soc London. 188:134-166.
4535. Heape, W. 1901. The "sexual seasons" of mammals and the relation of the "pro-oestrum" to menstruation. Q J Microsc Sci. 44:1-70.
4536. Heape, W. 1905. Ovulation and degeneration of ova in the rabbit. Proc R Soc London Biol Sci. 76B:260-268.
4537. Hearn, JP. 1973. Pituitary inhibition of pregnancy. Nature. 241:207-208.
4538. Hearn, JP. 1973. The pituitary gland and pregnancy in the marsupial, *Macropus eugenii*, employing hypophysectomy and radioimmunoassay for gonadotrophin. J Reprod Fert. 32:328-329.
4539. Hearn, JP. 1975. The rôle of the pituitary in the reproduction of the male tammar wallaby, *Macropus eugenii*. J Reprod Fert. 42:399-402.
4540. Hearn, JP. 1977. The endocrinology of reproduction in the common marmoset, *Callithrix jacchus*. The Biology and Conservation of the Callitrichidae. 163-171, Kleiman, DG, ed, Smithsonian Inst Press, Washington, DC.
4541. Hearn, JP. 1978. Fertility and infertility in the marmoset monkey, *Callithrix jacchus*. Biology and Behaviour of Marmosets. 59-64, Rothe, H, Wolters, HJ, & Hearn, JP, eds, Eigenverlag Rothe, Göttingen.
4542. Hearn, JP. 1980. Endocrinology and timing of implantation in the marmoset monkey, *Callithrix jacchus*. Prog Reprod Biol. 7:262-269.
4543. Hearn, JP. 1982. Research Progress Report 1979-1981. J Zool. 197:12-123.
4544. Hearn, JP. 1982. The reproductive physiology of the common marmoset *Callithrix jacchus* in captivity. Int Zoo Yearb. 22:138-143.
4545. Hearn, JP. 1983. The common marmoset (*Callithrix jacchus*). Reproduction in New World Primates. 181-215, Hearn, J, ed, MTP Press, Lancaster, England.
4546. Hearn, JP. 1984. Lactation and reproduction in non-human primates. Symp Zool Soc London. 51:327-335.
4547. Hearn, JP. 1986. The embryo-maternal dialogue during early pregnancy in primates. J Reprod Fert. 76:809-819.
4548. Hearn, BJ, Keith, LB, & Rongstad, OJ. 1987. Demography and ecology of the arctic hare (*Lepus arcticus*) in southwestern Newfoundland. Can J Zool. 65:852-861.

4549. Hearn, JP, Abbott, DH, Chambers, PC, et al. 1978. Use of the common marmoset, *Callithrix jacchus*, in reproductive research. Primates in Medicine. 10:40-49.
4550. Hearn, JP & Lunn, SF. 1975. The reproductive biology of the marmoset monkey, *Callithrix jacchus*. Lab Anim Handb. 6:191-202.
4551. Hearn, JP, Lunn, SF, Burden, FJ, et al. 1975. Managment of marmosets for biomedical research. Lab Anim. 9:125-134.
4552. Heath, E & Amachree, M. 1967. Endotheliochorial placentation in a pangolin, *Manis tetradactyla*. Acta Zool (Stockholm). 48:309-315.
4553. Heath, E, Schaeffer, N, Merritt, DA, Jr, et al. 1987. Rouleaux formation by spermatozoa in the naked-tail armadillo, *Cabassous unicinctus*. J Reprod Fert. 79:153-158.
4554. Heath, HW & Lynch, GR. 1982. Intraspecific differences for melatonin-induced reproductive regression and the seasonal molt in *Peromyscus leucopus*. Gen Comp Endocrinol. 48:289-295.
4555. Heath, ME. 1987. Twenty-four-hour variations in activity, core temperature, metabolic rate, and respiratory quotient in captive Chinese pangolins. Zoo Biol. 6:1-10.
4556. Heath, ME & Hammel, HT. 1986. Body temperature and rate of oxygen consumption in Chinese pangolins. Am J Physiol. 250:R377-R382.
4557. Heck, H. 1971. Eine Zwillingsgeburt beim Weiszschwanzgnu, *Connochaetes gnou* (Zimmermann, 1780), der Catskill Game Farm. Zool Garten NF. 40:328.
4558. Heck, H. 1979. Notes on breeding capybara (*Hydrochoerus hydrochaeris*). Zool Garten NF. 49:80.
4559. Heck, L. 1968. Lange Lebensdauer einer Leopardin. Z Säugetierk. 33:382.
4560. Heck, L, Jr. 1956. Beobachtungen an südwestafrikanischen Scharrtieren, *Suricata suricata hahni* Thomas, 1927. Säugetierk Mitt. 4:33-34.
4561. Hediger, H. 1945. Zur Biologie des Eichörnchens (*Sciurus vulgaris* L). Rev Suisse Zool. 52:361-370.
4562. Hediger, H. 1948. Die Zucht des Feldhasen (*Lepus europaeus* Pallas) in Gefangenschaft. Physiol Comp Oecol. 1:46-62.
4563. Hediger, H. 1950. Gefangenschaftgeburt eines afrikanischen Springhasen, *Pedestes caffer*. Zool Garten NF. 17:166-169.
4564. Hediger, H. 1970. Zur Fortpflanzangsverhalten des Kanadischen Bibers (*Castor fiber canadensis*). Forma et Functio. 2:336-351.
4565. Hegdal, PL, Ward, AL, Johnson, AM, et al. 1965. Notes on the life history of the Mexican pocket gopher (*Cratogeomys castanops*). J Mamm. 46:334-335.
4566. Heger, HW & Neubert, D. 1983. Timing of ovulation and implantation in the common marmoset, *Callithrix jacchus*, by monitoring of estrogens and 6β-hydroxypregnanolone in urine. Arch Toxicol. 54:41-52.
4567. Heger, W & Neubert, D. 1987. Determination of ovulation and pregnancy in the marmoset (*Callithrix jacchus*) by monitoring of urinary hydroxypregnanolone excretion. J Med Primatol. 16:151-164.
4568. Heger, W, Hoyer, G-A, & Neubert, D. 1988. Identification of the main gestagen metabolite in marmoset (*Callithrix jacchus*) urine by NMR and MS spectroscopy. J Med Primatol. 17:19-29.
4569. Heggberget, TM. 1988. Reproduction in the female European otter in central and northern Norway. J Mamm. 69:164-167.
4570. Heidecke, D. 1984. Untersuchungen zur Ökologie und Populationsentwicklung des Elbebibers, *Castor fiber albicus* Matschie, 1907 Teil 1 Biologische und populations ökologische Ergebnisse. Zool Jb Syst. 111:1-41.
4571. Heideman, PD. 1986. Delayed development in a paleotropical fruit bat: Possible initiation and termination cues. Biol Reprod. 34(Suppl 1):239.
4572. Heideman, PD. 1988. The timing of reproduction in the fruit bat *Haplonycteris fischeri* Pteropodidae: Geographic variation and delayed development. J Zool. 215:577-595.
4573. Heideman, PD. 1989. Delayed development in Fischer's pygmy fruit bat, *Haplonycteris fischeri*. J Reprod Fert. 85:363-382.
4573a. Heideman, PD & Bronson, FH. 1990. Photoperiod, melatonin secretion, and sexual matruation in a tropical rodent. Biol Reprod. 43:745-750.
4574. Heideman, PD & Bronson, FH. 1992. A pseudo-seasonal reproductive strategy in a tropical rodent, *Peromyscus nudipes*: Correlates and causes. J Reprod Fert. 95:57-67.
4575. Heideman, PD, Erickson, KR, & Bowles, JB. 1990. Notes on the breeding biology, gular gland and roost habits of *Molossus sinaloae* (Chiroptera, Molossidae). Z Säugetierk. 55:303-307.
4576. Heidt, GA. 1970. The least weasel *Mustela nivalis* Linnaeus: Developmental biology in comparison with other North American *Mustela*. Publ Mus MI State Univ Biol Ser. 4:227-282.
4577. Heikura, K. 1984. The population dynamics and the influence of winter on the common shrew (*Sorex araneus* L). Sp Publ Carnegie Mus Nat Hist. 10:343-361.
4578. Heimark, RJ & Heimark, GM. 1986. Southern elephant seal pupping at Palmer Station, Antarctica. J Mamm. 67:189-190.
4579. Hein, H, Kebeci, E, & Neiss, A. 1971. Vergleich des Einflusses von Körnermischfutter und Preszfutter auf die Fortpflanzung und das Wachstum von *Mastomys natalensis* (Smith, 1834). Z Versuchstierk. 13:200-209.
4580. Heinemann, H. 1970. The breeding and maintenance of captive Goeldi's monkey *Callimico goeldii*. Int Zoo Yearb. 10:72-78.
4581. Heinrichs, WL & Dillingham, LA. 1970. Bornean orang-utan twins born in captivity. Folia Primatol. 13:150-154.
4582. Heinroth, O. 1908. Trachtigkeits-und Brutdauern. Zool Beob. 49:14-25.
4583. Heinroth-Berger, K. 1965. Über Geburt und Aufzucht eines männlichen Schimpansen im Zoologischen Garten Berlin. Z Tierpsychol. 22:15-35.
4584. Heinsohn, GE. 1966. Ecology and reproduction of the Tasmanian bandicoots (*Perameles gunni* and *Isoodon obesulus*). Univ CA Publ Zool. 80:1-107.
4585. Heinsohn, GE. 1968. Habitat requirements and reproductive potential of the macropod marsupial *Potorous tridactylus* in Tasmania. Mammalia. 32:30-43.
4586. Heinsohn, GE. 1972. A study of dugongs (*Dugong dugon*) caught in nets off Townsville, North Queensland. Aust Mamm. 1:71.
4587. Heinsohn, GF. 1970. World's smallest marsupial. Animals. 13:220-222.
4588. Heisler, C. 1979. "K-bzw r-Selektion" in Verbindung mit "Matriarchat" bzw "Patriarchat" bei der mongolischen Rennmaus, *Meriones unguiculatus* (Milne-Edwards, 1867), und der Vielzitzmaus, *Mastomys coucha* (A Smith, 1836). Säugetierk Mitt. 27:76-80.
4589. Heisler, C. 1980. Die Soziale Organisation bei der mongolischen Rennmaus (*Meriones unguiculatus*) und der Vielzitsenmaus (*Mastomys coucha*). Biol Beitr NF. 26:17-37.

4590. Heistermann, M, Kleis, E, Prove, E, et al. 1989. Fertility status, dominance, and scent marking behavior of family-housed female cotton-top tamarins (*Saguinus oedipus*) in absence of their mothers. Am J Primatol. 18:177-189.
4591. Heistermann, M, Pröve, E, Wolters, HJ, et al. 1987. Urinary oestrogen and progesterone excretion before and during pregnancy in a pied bare-face tamarin. J Reprod Fert. 80:635-640.
4592. Helin, S & Houji, L. 1982. Distribution, habits and resource status of the tufted deer (*Elaphodus cepholophus*). Acta Zool Sinica. 28:307-311.
4593. Helle, E. 1980. Age structure and sex ratio of the ringed seal *Phoca* (*Pusa*) *hispida* Schreber population in the Bothnian Bay, northern Baltic Sea. Z Säugetierk. 45:310-317.
4594. Helle, E. 1980. Lowered reproductive capacity in female ringed seals (*Pusa hispida*) in the Bothnian Bay, northern Baltic Sea, with special reference to uterine occlusions. Ann Zool Fennici. 17:147-158.
4595. Helle, T. 1984. Effects of insect harassment on weight gain and survival in reindeer calves. Rangifer. 4:24-27.
4596. Heller, H. 1973. The effects of oxytocin and vasopressin during the oestrous cycle and pregnancy on the uterus of a marsupial species, the quokka (*Setonix brachyurus*). J Endocrinol. 56:657-671.
4597. Hellgren, EC, Lockmiller, RL, Amoss, MS, Jr, et al. 1985. Serum progesterone, estradiol-17β, and glucocorticoids in the collared peccary during gestation and lactation as influenced by dietary protein and energy. Gen Comp Endocrinol. 59:358-368.
4598. Hellgren, EC, Lochmiller, RL, & Grant, WE. 1984. Demographic, morphologic and reproductive status of a herd of collared peccaries (*Tayassu tayacu*) in south Texas. Am Mid Nat. 112:402-407.
4599. Hellgren, EC, Vaughan, MR, Gwazdauskas, FC, et al. 1990. Endocrine and electrophoretic profiles during pregnancy and nonpregnancy in captive female black bears. Can J Zool. 69:892-898.
4600. Hellwing, S. 1970. Reproduction in the white-toothed shrew *Crocidura russula monacha* Thomas in captivity. Israel J Zool. 19:177-178.
4601. Hellwing, S. 1971. Maintenance and reproduction in the white toothed shrew *Crocidura russula monacha* Thomas, in captivity. Z Säugetierk. 36:103-113.
4602. Hellwing, S. 1973. Husbandry and breeding of white-toothed shrews (Crocidurinae) in the research zoo of the Tel-Aviv University. Int Zoo Yearb. 13:127-134.
4603. Hellwing, S. 1973. The postnatal development of the white-toothed shrew *Crocidura russula monacha* in captivity. Z Säugetierk. 38:258-270.
4604. Hellwing, S. 1975. Sexual receptivity and oestrus in the white-toothed shrew, *Crocidura russula monacha*. J Reprod Fert. 45:469-477.
4605. Hellwing, S & Funkenstein, B. 1977. Ovarian asymmetry in the shrew, *Crocidura russula monacha*. J Reprod Fert. 49:163-165.
4606. Helm, JD. III. 1973. Reproductive Biology of *Tylomys* and *Ototylomys*. Ph D Diss. MI State Univ, E Lansing.
4607. Helm, JD, III. 1975. Reproductive biology of *Ototylomys* (Cricetidae). J Mamm. 56:575-590.
4608. Helm, JD, III & Dalby, PL. 1975. Reproductive biology and postnatal development of the neotropical climbing rat, *Tylomys*. Lab Anim Sci. 25:741-747.
4609. Helm, JD, III, Sánchez Hernández, C, & Baker, RH. 1974. Observaciones sobre los ratones de las marismas, *Peromyscus perfulvus* Osgood (Rodentia Cricetidae). An Inst Biol Univ Nal Auton, México Ser Zool. 45:141-146.
4610. Helmreich, RL. 1960. Regulation of reproductive rate by intra-uterine mortality in the deer mouse. Science. 132:417-418.
4611. Heltne, PG. 1977. Census of *Aotus* in the north of Columbia. Rep PAHO project AMRO-3171, Pan Am Health Org, Washington, DC. [Cited by PC Wright, 1981]
4612. Heltne, PG, Turner, DC, & Wolhandler, J. 1973. Maternal and paternal periods in the development of infant *Callimico goeldii*. Am J Phys Anthrop. 38:555-560.
4613. Heltne, PG, Wojcik, JF, & Pook, AG. 1981. Goeldi's monkey, genus *Callimico*. Ecology and Behavior of Neotropical Primates. 1:169-209. Coimbra Filho, AF & Mittermeier, RA, eds, Academic Brasileira de Ciências, Rio de Janeiro.
4614. Hembeck, H. 1958. Zum Paarung der Murmeltiere. Z Jagdwiss. 4:40-41.
4615. Hemeida, NA, El-Baghdady, YR, & El-Fadaly, MA. 1985. Serum profiles of androstenedione, testosterone and LH from birth through puberty in buffalo bull calves. J Reprod Fert. 74:311-316.
4616. Hemmalin, H & Loy, J. 1989. Observations of parturition among captive patas monkeys (*Erythrocebus patas*). Am J Primatol. 19:183-188.
4617. Hemmer, H. 1972. *Uncia uncia*. Mamm Species. 20:1-5.
4618. Hemmer, H. 1977. Biology and breeding of the sand cat. The World's Cats. Carnivore Res Inst Burke Mus, 13-21, Eaton, RL, ed, Univ WA, Seattle.
4619. Henderson, BA. 1979. Regulation of the size of breeding population of the European rabbit, *Oryctolagus cuniculus*, by social behaviour. J Appl Ecol. 16:383-392.
4620. Henderson, RE & O'Gara, BW. 1978. Testicular development of the mountain goat. J Wildl Manage. 42:921-922.
4621. Henderson, SA & Edwards, RG. 1968. Chiasma frequency and maternal age in mammals. Nature. 218:22-28.
4622. Hendrichs, H. 1975. Changes in a population of dikdik, *Madoqua* (*Rhynotragus*) *kirki* (Günther 1880). Z Tierpsychol. 38:55-69.
4623. Hendrickson, GO. 1943. Gestation period in Mearns cottontail. J Mamm. 24:273.
4624. Hendrickson, GO. 1947. Cottontail breeding in its first summer. J Mamm. 28:63.
4625. Hendrickson, J. 1956. A maternity colony of bats. Malay Nat J. 11:121-125.
4626. Hendrickson, JR. 1954. Breeding of the tree shrew. Nature. 174:794-795.
4627. Hendrickx, AG. 1967. The menstrual cycle of the baboon as determined by the vaginal smear, vaginal biopsy, and perineal swelling. The Baboon in Medical Research. 2:437-459, Vagtborg, H, ed, Univ TX Press, Austin.
4628. Hendrickx, AG & Kraemer, DC. 1969. Observations on the menstrual cycle, optimal mating time and pre-implantation embryos of the baboon, *Papio anubis* and *Papio cynocephalus*. J Reprod Fert Suppl. 6:119-128.
4629. Hendrickx, AG & Kraemer, DC. 1971. Reproduction. Embryology of the Baboon. 3-30, Hendrickx, AG, ed, Univ Chicago Press, Chicago.
4630. Hendrickx, AG & Kriewaldt, FH. 1967. Observations on a controlled breeding colony of baboons. The Baboon in Medical Research. 2:69-83, Vagtborg, H, ed, Univ TX Press, Austin.
4631. Hendrickx, AG & Newman, LM. 1978. Reproduction of the greater bushbaby (*Galago crassicaudatus panganiensis*) under laboratory conditions. J Med Primatol. 7:26-43.

4632. Hendrickx, AG, Sawyer, RH, Lasley, BL, et al. 1975. Comparison of developmental stages in primates with a note on the detection of ovulation. Lab Anim Handb. 6:305-315.
4633. Hendrickx, AG, Thompson, RS, Hess, DL, et al. 1978. Artificial insemination and a note on pregnancy detection in the non-human primate. Symp Zool Soc London. 43:219-240.
4634. Henry, DB & Bookhout, TA. 1969. Productivity of beavers in northeastern Ohio. J Wildl Manage. 33:927-932.
4635. Henry, S. 1984. Social organisation of the greater glider in Victoria. Possums and Gliders. 221-228, Smith, A & Hume, I, eds, Surrey Beatty & Sons Pty Ltd, Sydney, Australia.
4636. Henry, SR & Craig, SA. 1984. Diet, ranging behaviour and social organization of the yellow-bellied glider in Victoria. Possums and Gliders. 331-341, Smith, A & Hume, I, eds, Surrey Beatty & Sons Pty Ltd, Sydney, Australia.
4637. Henry, SR & Suckling, G. 1984. A review of the ecology of the sugar glider (*Petaurus breviceps*). Possums and Gliders. 355-358, Smith, A & Hume, I, eds, Surrey Beatty & Sons Pty Ltd, Sydney, Australia.
4638. Henry, VG. 1968. Fetal development in European wild hogs. J Wildl Manage. 32:966-970.
4639. Henry, VG. 1968. Length of estrous cycle and gestation in European wild hogs. J Wildl Manage. 32:406-408.
4640. Henschel, JR, David, JHM, & Jarvis, JUM. 1982. Age determination and age structure of a striped field mouse, *Rhabdomys pumilio*, population from the Cape Flats. S Afr J Zool. 17:136-141.
4641. Hensel, RJ, Troyer, WH, & Erickson, AW. 1969. Reproduction in the female brown bear. J Wildl Manage. 33:357-365.
4642. Hensley, AP & Wilkins, KT. 1988. *Leptonycteris nivalis*. Mamm Species. 307:1-4.
4642a. Heppes, JB. 1959. The white rhinoceros in Uganda. Afr Wild Life. 12:272-290.
4643. Heptner, VG & Naumov, NP. 1966. Die Säugetiere der Sowjetunion Vol I. VEB Gustav Fischer Verlag, Jena.
4644. Heptner, W. 1970. Die turkestanische Sicheldünenkatze (Barchaukatze), *Felis margarita thinobia* Ogn, 1926. Zool Garten NF. 39:116-128.
4645. Heptner, WG, Morosowa-Turowa, LG, & Zalkin, WI. 1956. Die Säugetiere in der Schutzwaldzone. Deutscher Verlag der Wissenschafte, Berlin.
4646. Heptner, WG & Sludskij, AA. 1972. Mlekopita juscie Sovetskogo Sojuza T22 Chiscnje (Gieny i Koski). Vyssaja Skola, Moskva. [Cited by Hemer, 1977]
4647. Hepworth, W & Blunt, F. 1966. Research findings on Wyoming Antelope. WY Wildl. 30:24-29.
4648. Herbert, DC. 1978. Stimulation of prolactin secretion by testosterone in juvenile male rhesus monkeys. Biol Reprod. 18:448-453.
4649. Herbert, DC & Sheridan, PJ. 1983. Uptake and retention of sex steroids by the baboon pituitary gland: Evidence of sexual dimorphism with respect to dihydrotestosterone. Biol Reprod. 28:377-383.
4650. Herbert, DC & Sheridan, PJ. 1984. Identification of the progesterone target cells in the female baboon pituitary gland. Biol Reprod. 30:479-483.
4651. Herbert, H. 1972. Initial observations on pinealectomized ferrets kept for long periods in either daylight or artificial illumination. J Endocrinol. 55:591-597.
4652. Herbert, J. 1969. The pineal gland and light-induced oestrus in ferrets. J Endocrinol. 43:625-636.
4653. Herbert, J, Stacey, PM, & Thorpe, DH. 1978. Recurrent breeding seasons in pinealectomized or optic-nerve-sectioned ferrets. J Endocrinol. 78:389-397.
4654. Herbert, J & Trimble, MR. 1967. Effect of oestradiol and testosterone on the sexual receptivity and attractiveness of the female rhesus monkey. Nature. 216:165-166.
4655. Herbert, HJ. 1970. The population dynamics of the waterbuck, *Kobus ellipsiprymnus* (Ogilby, 1833) in the Sabi-Sand Wildtuin. M Sc Thesis. Univ Pretoria, Pretoria. [Cited by Mentis, 1972]
4656. Herbert, HJ. 1972. The Population Dynamics of the Waterbuck, *Kobus ellepsiprymnus* (Ogilby, 1833) in the Sabi-Sand Wildtuin. Paul Parey, Hamburg.
4657. Herbig-Sandreuter, A. 1964. Neue Beobachtungen am venezolanischen Dreizehenfaultier, *Bradypus infuscatus flaccidus* Gray 1849. Acta Tropica. 21:97-113.
4658. Herd, RM. 1983. *Pteronotus parnellii*. Mamm Species. 209:1-5.
4659. Herdman, R. 1973. Cheetah breeding program. The World's Cats Vol I. 225-262, Eaton, RL, ed, World Wildlife Safari, Winston, OR.
4660. Herfs, A. 1939. Über die Fortpflanzung und Vermehrung der "Gröszen Wühlmaus" (*Arvicola terrestris* L). Nachr Schädl Bekampf. 14:92-193.
4661. Herlant, M. 1934. Recherches sur les potentialités de developpement des oeufs obtenus par ovulation provoqueé chez la Chauve-souris en hibernation. Bull Acad R Belg Cl Sci Bruxeles (5e series). 20:359-366.
4662. Herlant, M. 1954. Influence des oestrogènes chez le Muron (*Myotis myotis*) hibernant. Bull Acad Belge Cl Sci. 40:408-415.
4663. Herlant, M. 1956. Corrélations hypophyso-génitales chez la femelle de la Chauve-souris, *Myotis myotis* (Borkhausen). Arch Biol. 67:89-180.
4664. Herlant, M. 1958. L'hypophyse et le système hypothalamo-hypophysaire du Pangolin (*Manis tricuspis* Raf et *Manis tetradactyla* L). Arch Anat Microsc Morphol Exp. 47:1-23.
4665. Herlant, M. 1960. L'activité génitale chez la femelle de *Galago senegalensis moholi* (Geoffr) et ses rapports avec la persistance de phenomènes d'ovogenèse chez l'adulte. Ann Soc R Zool Belg. 91:1-15.
4666. Herlant, M. 1967. Action de la gonadotropine FSH sur le tube seminifère de la Chauve-Souris hibernante. CR Hebd Séances Acad Sci Sér D. 264:2483-2486.
4667. Herman, TB. 1984. Dispersion of insular *Peromyscus maniculatus* in coastal coniferous forest, British Columbia. Sp Publ Carnegie Mus Nat Hist. 10:333-342.
4668. Hermanson, JW & O'Shea, TJ. 1983. *Antrozous pallidus*. Mamm Species. 213:1-8.
4669. Hermanson, JW, Woods, CA, & Howell, KM. 1982. Dental ontogeny in the old world leaf-nosed bats (Rhinolophidae, Hipposiderinae). J Mamm. 63:527-529.
4670. Hernandez, CS, Osorio, MTC, & Tapia, CBC. 1986. Patron reproductivo de *Sturnira lilium parvidens* (Chiroptera: Phyllostomidae) en la costa central del pacifico de México. Southwest Nat. 31:331-340.
4671. Hernandez, CS, Tapia, CBC, Garduno, AN, et al. 1985. Notes on distribution and reproduction of bats from coastal regions of Michoacan, Mexico. J Mamm. 66:549-553.
4672. Herndon, JG, Blank, MS, Mann, DR, et al. 1985. Negative feedback effects of oestradiol-17β on luteinizing hormone in female rhesus monkeys under different seasonal conditions. Acta Endocrinol. 108:31-35.

4673. Herndon, JG, Ruiz de Elvira, MC, Turner, JJ, et al. 1985. Resumption of seasonal breeding patterns in male and female rhesus monkeys transferred from an indoor to and outdoor environment. Biol Reprod. 32:733-744.
4674. Heron-Royer. 1881. Concrétions vagino-utérines observés chez le *Pachyuromys duprasi* (Lataste). Zool Anz. 4:623-628.
4675. Heron-Royer. 1882. A propos des bouchons vagino-utérines des Rongeurs. Zool Anz. 5:453-459.
4676. Herreid, CF, II. 1959. Notes on a baby free-tailed bat. J Mamm. 40:609-610.
4677. Herreid, CF, II. 1964. Bat longevity and metabolic rate. Exp Geront. 1:1-9.
4678. Herreid, CF, II. 1967. Temperature regulation, temperature preference and tolerance, and metabolism of young and adult free-tailed bats. Physiol Zool. 40:1-22.
4679. Herrick, CJ. 1921. The brain of *Caenolestes obscurus*. Publ Field Mus Nat Hist Zool Ser. 14:157-162.
4680. Hersh, SL. 1981. Ecology of the Key Largo woodrat (*Neotoma floridana smalli*). J Mamm. 62:201-206.
4681. Hershkovitz, P. 1955. South American marsh rats, genus *Holochilus*, with a summary of sigmodont rodents. Fieldiana Zool. 37:639-687.
4682. Hershkovitz, P. 1962. Evolution of neotropical cricetine rodents (Muridae) with special reference to the phyllotine group. Fieldiana Zool. 46:1-524.
4683. Hershkovitz, P. 1966. South American swamp and fossorial rats of the scapteromyine group (Cricetinae, Muridae) with comments on the glans penis in murid taxonomy. Z Säugetierk. 31:81-149.
4684. Hershkovitz, P. 1982. Neotropical deer (Cervidae) Part I Pudus, genus *Pudu* Gray. Fieldiana Zool NS. 11:1-86.
4685. Hertig, AT, King, NW, Jr, Barton, BR, et al. 1971. Observations on the ovary of the squirrel monkey, *Saimiri sciureus*, using the light and electron microscope. Med Primatol. 1970:472-503.
4686. Herter, K. 1932. Zur Fortpflanzungsbiologie des Igels. Z Säugetierk. 7:251-253.
4687. Herter, K. 1933. Gefangenschaftsbeobachtungen an europaischen Igeln II. Z Säugetierk. 8:195-218.
4688. Herter, K. 1938. Die Biologie der europaischen Igel. P Schöps, Leipzig.
4689. Herter, K. 1953. Über das Verhalten von Iltissen. Z Tierpsychol. 10:56-71.
4690. Herter, K. 1958. Die Säugetierkundlichen Arbeiten aus dem Zoologischen Institut der Freihen Universität Berlin. Z Säugetierk. 23:1-32.
4691. Herter, K. 1962. Uber die Borstenigel von Madagaskar (Tenrecinae). Sitzungsber Ges Naturf Fr Berlin NF. 2:5-37.
4692. Herter, K. 1964. Gefangenschaftsbeobachtungen an einem algerischen Igel (*Aethechinus algirus* (Duvernoy, Lereboullet). Zool Beitr.
4693. Herter, K. 1965. Über das Paarungsverhalten der Igel. Sitzungsber Geselsch Naturforsch Freunde Berlin NF. 5:57-76.
4694. Herter, K. 1971. Gefangenscchaftsbeobachtungen an mittelafrikanischen Igeln (*Atlerix pruneri* [Wagner]). Zool Beitr. 17:337-370.
4695. Herter, K & Rauch, H-G. 1956. Haltung un Aufzucht chinesischer Zwerghamster (*Cricetulus barabensis griseus* A Milne-Edwards 1867). Z Säugetierk. 21:161-171.
4696. Hertig, AT, Barton, BR, & MacKey, JJ. 1976. The female genital tract of the owl monkey (*Aotus trivirgatus*) with special reference to the ovary. Lab Anim Sci. 26:1041-1064.
4697. Herzig-Straschil, B. 1978. On the biology of *Xerus inauris* (Zimmermann, 1780) (Rodentia, Sciuridae). Z Säugetierk. 43:262-278.
4698. Herzig-Straschil, B & Robinson, GA. 1978. On the ecology of the fruit bat, *Rousettus aegyptiacus leachi* (A Smith, 1829) in the Tsitsikama Coastal National Park. Koedoe. 21:101-110.
4699. Hess, DL, Hendrickx, AG, & Stabenfeldt, GH. 1979. Reproductive and hormonal patterns in the African green monkey (*Cercopithecus aethiops*). J Med Primatol. 8:273-281.
4700. Hess, DL & Resko, JA. 1973. The effects of progesterone on the patterns of testosterone and estradiol concentrations in the systemic plasma of the female rhesus monkey during the intermenstrual period. Endocrinology. 92:446-453.
4701. Hess, DL, Schmidt, AM, & Schmidt, MJ. 1983. Reproductive cycle of the Asian elephant (*Elephas maximus*) in captivity. Biol Reprod. 28:767-773.
4702. Hess, DL, Schmidt, MJ, & Schmidt, AM. 1981. Endocrine and behavioral comparisons during the reproductive cycle in the Asian elephant *Elephas maximus*. Biol Reprod. 24(Suppl 1):93A.
4703. Hess, DL, Spies, HG, & Hendrickx, AG. 1981. Diurnal steroid patterns during gestation in the rhesus macaque: Onset, daily variation, and the effects of dexamethasone treatment. Biol Reprod. 24:609-616.
4704. Hess, JK. 1971. Hand rearing polar bear cubs *Thalarctos maritimus* at St Paul Zoo. Int Zoo Yearb. 11:102-107.
4705. Hesselton, WT. 1967. Two further incidents of polyovulation in white tailed deer in New York. J Mamm. 48:321.
4706. Hesselton, WT & Jackson, LW. 1974. Reproductive rates of white-tailed deer in New York State. NY Fish Game J. 21:135-152.
4707. Heuser, CH & Streeter, GL. 1941. Development of the macaque embryo. Contrib Embryol Carnegie Inst. 29(181):15-55.
4708. Hewer, HR & Backhouse, KM. 1968. Embryology and foetal growth of the grey seal *Halichoerus grypus*. J Zool. 155:507-533.
4709. Hewson, R. 1964. Reproduction in the brown hare and the mountain hare in north-east Scotland. Scott Nat. 71:81-89.
4710. Hewson, R. 1968. Weights and growth rates in the mountain hare *Lepus timidus scoticus*. J Zool. 154:249-262.
4710a. Hewson, R. 1970. Variation in reproduction and shooting bags of mountain hares on two moors in north-east Scotland. J Appl Ecol. 7:243-252.
4711. Hewson, R. 1976. A population study of mountain hares (*Lepus timidus*) in north-east Scotland from 1956-1969. J Anim Ecol. 45:395-414.
4712. Hewson, R & Taylor, M. 1975. Embryo counts and length of the breeding season in European hares in Scotland from 1960-1972. Acta Theriol. 20:247-254.
4713. Heyder, G. 1968. Zucht und Gefangenschaftbiologie der Wüstenrennmaus *Gerbillus pgramidum* Geoffroy, 1825. Z Versuchstierk. 10:298-313.
4714. Heyning, JE. 1987. Presence of solid food in a young calf killer whale (*Orcinus orca*). Mar Mamm Sci. 3:68-70.
4715. Heyning, JE. 1988. *Orcinus orca*. Mamm Species. 304:1-9.
4716. Hibbard, CW. 1935. Breeding seasons of gray squirrel and flying squirrel. J Mamm. 16:325-326.
4717. Hick, U. 1967/68. Geglückte Aufzucht eines Pudus (*Pudu pudu* Mol) im Kölner Zoo. Fr Kölner Zoo. 10:111-117.
4718. Hick, U. 1969. Einige Beobachtungen zur Fortpflanzung und Jugendentwicklung von Katta (*Lemur catta*) Monogozmaki

(*Lemur mongoz mongoz*), Kronen maki (*Lemur mongoz coronatus*) und Weiszkopfmaki (*Lemur fulvus albifrons*) im Kölner Zoo. Fr Kölner Zoo. 12:75-84.
4719. Hick, U. 1969. Successful raising of a pudu *Pudu pudu* at Cologne Zoo. Int Zoo Yearb. 9:110-112.
4720. Hick, U. 1970. Zucht und Haltung von Kleideraffen *Pygathrix nemaeus nemaeus* Linnaeus, 1771) im Zoologischen Garten Köln. Fr Kölner Zoo. 13:83-91.
4721. Hick, U. 1972. Breeding and maintenance of douc langurs *Presbytis nemaeus nemaeus* at Cologne Zoo. Int Zoo Yearb. 12:98-103.
4722. Hick, U. 1973. Wir sind umgezogen. Z Kölner Zoo. 16:127-145.
4723. Hick, U. 1975. Breeding and maintenance of douc langurs at Cologne Zoo. Breeding Endangered Species in Captivity. 223-233, Martin, RD, ed, Academic Press, New York.
4724. Hickie, JP, Lavigne, DM, & Woodward, WD. 1982. Vitamin D and winter reproduction in the collared lemming, *Dicrostonyx groenlandicus*. Oikos. 39:71-76.
4725. Hickling, G. 1959. The grey seals of the Farne Islands. Oryx. 5:7-11.
4726. Hickling, G & Hawkey, P. 1979. The grey seals of the Farne Islands: The 1978 breeding season. Trans Nat Hist Soc Northumbria. 43:35-44.
4727. Hickling, G, Hawkey, P, & Harwood, LH. 1977. The grey seals of the Farne Islands: The 1976 breeding season. Trans Nat Hist Soc Northumbria. 42:119-126.
4727a. Hickman, GC. 1982. Copulation of *Cryptomys hottentotus* (Bathyergidae) a fossorial rodent. Mammalia. 46:293-298.
4728. Hidalgo, YM. 1980. Contribución al Estudio de la Biologia del Tepezcuintle (*Cuniculus paca*) en Cantiverio. Thesis for Licensia en Biologia. Univ Costa Rica, Costa Rica.
4729. Hiddleston, WA. Manuscript. [Cited by Brahin, 1979]
4730. Higgins, LV, Costa, DP, Huntley, AC, et al. 1987. Behavioral and physiological measurements of maternal investment in the Steller sea lion, *Eumetopias jubatus*. Mar Mamm Sci. 3:44-58.
4731. Hikim, APS. 1985. Spermatogenesis in the bandicoot rat II Quantitation of the cells of the seminiferous epithelium. Arch Biol. 96:441-452.
4732. Hikim, APS & Maiti, BR. 1982. Quantitative studies of the accessory reproductive organs of the male bandicoot rat: Common rodent pest. Anat Anz. 151:483-495.
4733. Hill, CA. 1966. A callimico is born. Zoonooz. 39:14-16.
4734. Hill, CA. 1967. A note on the gestation period of the siamang *Hylobates syndactylus*. Int Zoo Yearb. 7:93-94.
4735. Hill, CA. 1968. Observations on the birth of a pigmy chimpanzee *Pan paniscus* at San Diego Zoo. Int Zoo Yearb. 8:119-120.
4736. Hill, CA. 1973. The frequency of multiple births in the genus *Lemur*. Mammalia. 37:101-104.
4737. Hill, CJ. 1933. The development of the Monotremata Part I The histology of the oviduct during gestation. Trans Zool Soc London. 21:413-443.
4738. Hill, CJ. 1938. The secretory activities of the oviduct in monotremes. Biomorphosis. 1:329-331.
4739. Hill, EP, & Lauhachinda, V. 1981. Reproduction in river otters from Alabama and Georgia. Proc Worldwide Furbearer Conf. 1:478-486.
4740. Hill, EP, III. 1967. Notes on the life history of the swamp rabbit in Alabama. Proc SE Assoc Game Fish Comm. 21:117-123.
4741. Hill, EP, III. 1972. Litter size in Alabama cottontails as influenced by soil fertility. J Wildl Manage. 36:1199-1209.
4742. Hill, EP, III. 1972. The cotton tail rabbit in Alabama. Auburn Univ Agric Exp Sta Bull. 440:1-103.
4743. Hill, JE. 1934. External characters of newborn pocket gophers. J Mamm. 15:244-245.
4744. Hill, JE. 1941. A collection of mammals from Dondi, Angola. J Mamm. 22:81-85.
4745. Hill, JE. 1942. Notes on mammals of northeastern New Mexico. J Mamm. 23:75-86.
4746. Hill, JE. 1974. A new family, genus and species of bat (Mammalia; Chiroptera) from Thailand. Bull Brit Mus (Nat Hist) Zool Ser. 27:301-336.
4747. Hill, JE & Morris, P. 1970. Bats from Ethiopia collected by the Great Abbai Expedition. Bull Brit Mus (Nat Hist). 21:25-50.
4748. Hill, JE & Smith, SE. 1981. *Craseonycteris thonglongyai*. Mamm Species. 160:1-4.
4749. Hill, JL. 1974. *Peromyscus*: Effect of early pairing on reproduction. Science. 186:1042-1044.
4750. Hill, JP. 1895. Preliminary note on the occurrence of a placental connection in *Perameles obesula* and on the foetal membranes of certain macropods. Proc Linn Soc NSW. 10:578-581.
4751. Hill, JP. 1897. The placentation of *Perameles*. Q J Microsc Sci. 40:385-468.
4752. Hill, JP. 1897. Contributions to the morphology and development of the female urogenital organs in the Marsupialia: I On the female urogenital organs of *Perameles*, with an account of the phenomena of parturition. Proc Linn Soc NSW. 24:42-82.
4753. Hill, JP. 1900. Contributions to the embryology of the Marsupialia II On a further stage in the placentation of *Perameles*. Q J Microsc Sci. 43:1-22.
4754. Hill, JP. 1900. On the foetal membranes, placentation and parturition of the native cat (*Dasyurus viverrinus*). Anat Anz. 18:364-373.
4755. Hill, JP. 1910. The early development of the Marsupialia, with special reference to the native cat (*Dasyurus viverrinus*). Q J Microsc Sci. 56:1-134.
4756. Hill, JP. 1918. The early development of *Didelphys aurita*. Q J Microsc Sci. 63:91-140.
4757. Hill, JP. 1919. The affinities of *Tarsius* from the embryological aspect. Proc Zool Soc London. 1919:476-494.
4758. Hill, JP. 1925. On a collection of pregnant uteri of the slender loris (*Loris gracilis*). Proc Zool Soc. 1925:1239.
4759. Hill, JP. 1932. The developmental history of primates. Phil Trans R Soc London Biol Sci. 221B:45-178.
4760. Hill, JP. 1933. The development of the Monotremata Part II The structure of the egg-shell. Trans Zool Soc London. 21:443-472.
4761. Hill, JP. 1938. Implantation of the blastocyst in *Galago demidoffi*. Biomorphosis. 1:333.
4762. Hill, JP. 1938. The macroscopic features of the placentation of the water shrew (*Potomogale velox*). Biomorphosis. 1:331-332.
4763. Hill, JP. 1949. The allantoic placenta of *Perameles*. Proc Linn Soc London. 161:3-7.
4764. Hill, JP. 1965. On the placentation of *Tupaia*. J Zool. 146:278-304.
4765. Hill, JP & Burne, RH. 1922. The foetal membranes and placentation of *Chiromys madagascariensis*. Proc Zool Soc London. 1922:1145-1165.
4766. Hill, JP & Fraser, EA. 1925. Some observations on the female urogenital organs of the Didelphydae. Proc Zool Soc London. 1925:189-219.

4767. Hill, JP & Gatenby, JB. 1923. The corpus luteum of *Ornithorhynchus*. J Anat. 59:92P-93P.
4768. Hill, JP & Gatenby, JB. 1926. The corpus luteum of the Monotremata. Proc Zool Soc London. 1926:715-763.
4769. Hill, JP & Hill, CJ. 1927. Demonstration. J Anat. 61:514.
4770. Hill, JP, Ina, FE, & Rau, S. 1928. The mode of vascularization of the chorion in the Lemuroidea and its physiological significance. CR Assoc Anat. 23:196-200.
4771. Hill, JP & Inge, FE. 1928. The development of the foetal membranes in *Loris*, with special reference to the mode of vascularization of the chorion in Lemuroidea and its phylogenetic significance. Proc Zool Soc London. 1928:699-716.
4772. Hill, JP & Martin, CJ. 1895. On a platypus embryo from the intrauterine egg. Proc Linn Soc NSW. 10:43-74.
4773. Hill, JP & O'Donoghue, CH. 1913. The reproductive cycle in the marsupial *Dasyurus viverrinus*. Q J Microsc Sci. 59:133-174.
4774. Hill, JP & Subba Rau, A. 1925. On a collection of pregnant uteri of the slender loris (*Loris gracilis*). Proc Zool Soc London. 1925:1239.
4775. Hill, M. 1932. Studies on the hypophysectomy of the ferret III Effect of post-coitus hypophysectomy on ovulation and the development of the corpus luteum. Proc R Soc London Biol Sci. 112B:153-158.
4776. Hill, M. 1939. The reproductive cycle of the male weasel (*Mustela nivalis*). Proc Zool Soc London. 109B:481-512.
4777. Hill, M & Parkes, AS. 1934. Effect of absence of light on the breeding season of the ferret. Proc R Soc London Biol Sci. 115B:14-17.
4778. Hill, P. 1822. Extract from the Minute-Book of the Linnean Society Letter. Trans Linn Soc London. 13:621-624.
4779. Hill, RW. 1972. The amount of maternal care in *Peromyscus leucopus* and its thermal significance for the young. J Mamm. 53:774-790.
4780. Hill, RW. 1976. The ontogeny of homeothermy in neonatal *Peromyscus leucopus*. Physiol Zool. 49:292-306.
4781. Hill, WCO. 1934. A monograph on the purple-faced leaf monkey (*Pithecus vetulus*). Ceylon J Sci. 19:23-88.
4782. Hill, WCO. 1936. The penis and its bone in Ceylonese squirrels; with special reference to its taxonomic importance. Ceylon J Sci. 20:99-113.
4783. Hill, WCO. 1936. Supplementary observations on the purple-faced leaf monkey (genus *Kasi*). Ceylon J Sci. 20B:115-133.
4784. Hill, WCO. 1936. On a hybrid leaf-monkey; With special remarks on the breeding of leaf-monkeys in general. Ceylon J Sci. 20B:135-148.
4785. Hill, WCO. 1937. On the breeding and rearing of certain species of primates in captivity. Ceylon J Sci. 20B:369-389.
4786. Hill, WCO. 1937. The pre-natal development of the grey langur *Semnopithecus priam thersites*. Ceylon J Sci. 20B:211-251.
4787. Hill, WCO. 1946. Note on the male external genitalia of the chimpanzee. Proc Zool Soc London. 116:129-132.
4788. Hill, WCO. 1952. Observations on the genitalia of the woolly monkey (*Lagothrix*). Proc Zool Soc London. 122:973-984.
4788a. Hill, WCO. 1952/53. The natural history, endoparasites and pseudo-parasites of the tarsiers (*Tarsius carbonarius*) recently living in the Society's menageris. Proc Zool Soc London. 122:79-119.
4789. Hill, WCO. 1953. Primates Comparative Anatomy and Taxonomy I Strepsirhini. Univ Press, Edinburgh.
4790. Hill, WCO. 1959. Aborted mid-term foetus of *Aotus trivirgatus infulatus* (Feline Douroucouli). Proc Zool Soc London. 132:148.
4791. Hill, WCO. 1962. Reproduction in the squirrel monkey, *Saimiri sciurea*. Proc Zool Soc London. 138:670-672.
4792. Hill, WCO. 1966. On the neonatus of *Callimico goeldii* (Thomas). Proc R Soc Edinb Biol. 69B:321-333.
4793. Hill, WCO & Davies, DV. 1954. The reproductive organs in *Hapalemur* and *Lepilemur*. Proc R Soc Edinb Biol. 65B:251-270.
4794. Hill, WCO, Porter, A, Bloom, RT, et al. 1957. Field and laboratory studies on the naked mole rat, *Heterocephalus glaber*. Proc Zool Soc London. 128:455-514.
4795. Hillaby, J. 1968. Wild porkers. New Scientist. 39:542.
4796. Hillemann, HH. 1955. Organization, histology, and circulatory pattern of the near-term placenta of the Guinea baboon, *Papio cynocephalus*, Demarest. OR State Monogr. 9:1-19.
4797. Hillemann, HH & Gaynor, AI. 1961. The definitive architecture of the placentae of nutria, *Myocastor coypus* (Molina). Am J Anat. 109:99-317.
4798. Hillemann, HH, Gaynor, AI, & Stanley, HP. 1958. The genital system of nutria (*Myocastor coypus*). Anat Rec. 130:515-531.
4799. Hilleman, HH & Tibbits, FD. 1956. Ovarian growth and development in *Chinchilla*. Northwest Sci. 30:115-126.
4800. Hillemann, HH, Tibbits, FD, & Gaynor, AI. 1959. Reproductive biology in *Chinchilla*. Nat Chinchilla Breeders Am, Middletown, New York.
4801. Hillman, C. 1974. Ecology and behavior of the wild eland. Afr Wild Life. 9:6-9.
4802. Hillman, CN & Clark, TW. 1980. *Mustela nigripes*. Mamm Species. 126:1-3.
4803. Hinds, LA. 1989. Plasma progesterone through pregnancy and the estrous cycle in the eastern quoll, *Dasyurus viverrinus*. Gen Comp Endocrinol. 75:110-117.
4804. Hinds, LA. 1991. Prostaglandin alone does not cause luteolysis in the non-pregnant tammar wallaby, *Macropus eugenii*. Repro Fert Dév. 3:17-24.
4805. Hinds, LA & den Ottolander, RC. 1983. Effect of changing photoperiod on peripheral plasma prolactin and progesterone concentrations in the tammar wallaby (*Macropus eugenii*). J Reprod Fert. 69:631-639.
4806. Hinds, LA, Evans, SM, & Tyndale-Biscoe, CH. 1983. In-vitro secretion of progesterone by the corpus luteum of the tammar wallaby, *Macropus eugenii*. J Reprod Fert. 67:57-63.
4807. Hinds, LA & Janssens, PA. 1986. Changes in prolactin in peripheral plasma during lactation in the brushtail possum *Trichosurus vulpecula*. Aust J Biol Sci. 39:171-178.
4808. Hinds, LA & Merchant, JC. 1986. Plasma prolactin concentrations throughout lactation in the eastern quoll, *Dasyurus viverrinus* (Marsupialia: Dasyuridae). Aust J Biol Sci. 39:179-186.
4809. Hinds, LA & Tyndale-Biscoe, CH. 1982. Plasma progesterone levels in the pregnant and non-pregnant tammar, *Macropus eugenii*. J Endocrinol. 93:99-107.
4810. Hinds, LA & Tyndale-Biscoe, CH. 1982. Prolactin in the marsupial *Macropus eugenii*, during the estrous cycle, pregnancy and lactation. Biol Reprod. 26:391-398.
4811. Hinds, LA & Tyndale-Biscoe, CH. 1985. Seasonal and circadian patterns of circulating prolactin during lactation and seasonal quiescence in the tammar, *Macropus eugenii*. J Reprod Fert. 74:173-183.

4812. Hinds, LA, Tyndale-Biscoe, CH, Shaw, G, et al. 1990. Effects of prostaglandin and prolactin on luteolysis and parturient behaviour in the non-pregnant tammar, *Macropus eugenii*. J Reprod Fert. 88:323-333.
4813. Hinkley, R, Jr. 1966. Effects of plant extracts in the diet of male *Microtus montanus* on cell types of the anterior pituitary. J Mamm. 47:396-400.
4813a. Hinton, HE & Dunn, AWS. 1967. Mongooses: Their Natural History and Behavior. Univ CA Press. Berkeley.
4813b. Hinton, MAC & Lindsay, HM. 1926. Report #41: Assam and Mishmi Hills. J Bombay Nat Hist Soc. 31:383-403.
4814. Hintzsche, E. 1940. Über Beiziehungen zwischen Placentarbau, Urniere und Allantois. Z Mikr Anat Forsch. 48:54-107.
4815. Hinze, G. 1957. Methodik der Biberzucht. Zool Garten NF. 23:177-184.
4816. Hion, A. 1965. Notes sur l'elevage de Gorilles de Belinga. Biol Gabon. 1:361-374.
4817. Hipsley, WB. 1924. Notes on the young of *Aplodontia*. Murrelet. 5:4-5.
4818. Hirai, N. 1975. Beavers breeding in Tokyo. Animals & Zoos. 27(7):6-9.
4819. Hiraiwa, YK & Uchida, T. 1955. Fertilization in the bat *Pipistrellus abramus abramus* (Temminck) II On the properties of semen stored in the uterus. Sci Bull Fac Agric Kyushu Univ. 15:255-266.
4820. Hiraiwa, YK & Uchida, T. 1956. Fertilization capacity of spermatozoa stored in the uterus after copulation in the fall. Sci Bull Fac Agric Kyushu Univ. 15:565-574.
4820a. Hiraiwa-Hasegawa, M, Hasegawa, T, & Nishida, T. 1984. Demographic study of a large-sized unit-group of chimpanzees in the Mahale Mountains, Tanzania: A preliminary report. Primates. 25:401-413.
4821. Hirata, DN & Nass, RD. 1974. Growth and sexual maturation of laboratory-reared, wild *Rattus norvegicus*, *R rattus*, and *R exulans* in Hawaii. J Mamm. 55:472-474.
4822. Hird, DG. 1976. Some aspects of ringtail possum ecology. Bull Aust Mamm Soc. 3:55.
4823. Hirose, K, Kasya, T, Kazihara, T, et al. 1970. Biological study of the corpus luteum and the corpus albicans of the blue white dolphin (*Stenella caerulo-alba*) [sic]. J Mamm Soc Japan. 5:33-39.
4824. Hirose, K & Nishiwaki, M. 1971. Biological study on the testis of the blue white dolphin, *Stenella caeruleoalba*. J Mamm Soc Japan. 5:91-98.
4825. Hirshfeld, JR & Bradley, WG. 1977. Growth and development of two species of chipmunks: *Eutamias panamintinus* and *E palmeri*. J Mamm. 58:44-52.
4826. Hirst, LF & Vadivelu, K. 1929. The rat flea survey of Kandy. Ceylon, Sessional Paper 13. [Cited by Buxton 1936]
4826a. Hirth, HF. 1960. The spermatozoa of some North American bats and rodents. J Morphol. 106:77-83.
4827. Hirunagi, K, Fujioka, T, Furumura, K, et al. 1984. Fine structure of the lutein cell in the house musk shrew, *Suncus murinus*. Cell Tiss Res. 238:447-452.
4828. Hirunagi, K, Fujioka, T, Furumura, K, et al. 1991. Scanning electron microscopy of the ovary in the house musk shrew, *Suncus murinus*. Anat Anz. 172:241-246.
4829. Hisaw, FL. 1944. The placental gonadotrophin and luteal function in monkeys (*Macaca mulatta*). Yale J Biol Med. 17:119-137.
4830. Hissa, R. 1968. Postnatal development of thermoregulation in the Norwegian lemming and the golden hamster. Ann Zool Fenn. 5:345-383.
4831. Hitchins, PM & Anderson, JL. 1983. Reproduction, population characteristics and management of the black rhinoceros *Diceros bicornis minor* in the Hluhluwe Corridor Umfolozi Game Reserve Complex. S Afr J Wildl Res. 13:78-85.
4832. Hitchins, PM. 1968. Live weights of some mammals from Hluhluwe Game Reserve, Zululand. Lammergeyer. 9:42.
4832a. Hjältén, J. 1991. Muskrat (*Ondatra zibethica*) territoriality, and the impact of territorial choice on reproduction and predation risk. Ann Zool Fennici. 28:15-21.
4833. Hladik, CM & Charles-Dominique, P. 1974. The behavior and ecology of the sportive lemur (*Lepilemur mustelinus*) in relation to its dietary peculiarities. Prosimian Biology. 23-37, Martin, RD, Doyle, GD, & Walker, AC, eds, Univ Pittsburgh Press, Pittsburgh.
4834. Hladik, CM, Charles-Dominique, P, & Petter, JJ. 1980. Feeding strategies of five nocturnal prosimians in the dry forest of the West Coast of Madagascar. Nocturnal Malagasy Primates. 41-73, Charles-Dominique, P, Cooper, HM, Hladik, A, et al, eds, Academic Press, NY.
4835. Hladik, CM & Hladik, A. 1972. Disponibilités alimentaires et domaines vitaux des Primates à Ceylan. Terre Vie. 26:149-215.
4836. Ho-gee, L & He-lin, S. 1984. Status of the black muntjac, *Muntiacus crinifrons*, in eastern China. Mammal Rev. 14:29-36.
4837. Hoage, RJ. 1977. Parental care in *Leontopithecus rosalia rosalia*: Sex and age differences in carrying behavior and the role of prior experience. The Biology and Conservation of the Callitrichidae. 293-305, Kleiman, DG, ed, Smithsonian Inst Press, Washington, DC.
4838. Hoage, RJ. 1982. Social and physical maturation in captive lion tamarins, *Leontopithecus rosalia rosalia* (Primates: Callitrichidae). Smithson Contrib Zool. 354:1-56.
4839. Hobbs, KR, Welshman, MD, Hazareno, JB, et al. 1987. Conditioning and breeding facilities for the cynomolgus monkey (*Macaca fascicularis*) in the Philippines: A progress report on the SICONBREC project. Lab Anim. 21:131-137.
4840. Hobbs, NT & Baker, DL. 1979. Rearing and training elk calves for use in food habits studies. J Wildl Manage. 43:568-570.
4841. Hobbs, RP. 1971. Studies of an island population of *Rattus fuscipes*. Vict Nat. 88:32-38.
4842. Hobson, BM. 1975. Chorionic gonadotropin in the placenta of a chimpanzee (*Pan troglodytes*). Folia Primatol. 23:135-139.
4843. Hobson, BM & Boyd, IL. 1984. Gonadotrophin and progesterone concentrations in placentae of grey seals (*Halichoerus grypus*). J Reprod Fert. 72:521-528.
4844. Hobson, BM, Hearn, JP, Lunn, SF, et al. 1977. Urinary excretion of biologically active chorionic gonadotropin by the pregnant marmoset (*Callithrix jacchus jacchus*). Folia Primatol. 28:251-258.
4845. Hobson, BM & Wide, L. 1986. Gonadotrophin in the term placenta of the dolphin (*Tursiops truncatus*), the Californian sea lion (*Zalophus californianus*), the grey seal (*Halichoerus grypus*) and man. J Reprod Fert. 76:637-644.
4846. Hobson, W, Faiman, C, Dougherty, WJ, et al. 1975. Radioimmunoassay of rhesus monkey chorionic gonadotropin. Fertil Steril. 26:93-97.
4847. Hobson, W & Fuller, GB. 1977. LH-RH induced gonadotropin release in chimpanzees. Biol Reprod. 17:294-297.
4848. Hobson, WC, Fuller, GB, Winter, JSD, et al. 1981. Reproductive and endocrine development in the great apes. Reproductive Biology of the Great Apes. 83-103, Graham, CE, ed, Academic Press, New York.

4849. Hochereau-de Reviers, MT & Lincoln, GA. 1978. Seasonal variation in the histology of the testis of the red deer, *Cervus elaphus*. J Reprod Fert. 54:209-213.
4850. Hock, RJ. 1960. Seasonal variations in physiological functions of arctic ground squirrels and black bears. Bull Mus Comp Zool. 124:155-159.
4851. Hock, RJ. 1966. Growth rate in newborn black bear cubs. Growth. 30:339-348.
4852. Hock, RJ & Larson, AM. 1966. Composition of black bear milk. J Mamm. 47:539-540.
4853. Hodara, VL, Kajon, AE, Quintans, C, et al. 1984. Parametros metricos y reproductivos de *Calomys musculinus* (Thomas, 1913) y *Calomys callidus* (Thomas, 1916) (Rodentia, Cricetidae). Rev Mus Argent Cienc Nat 'Bernardino Rivadavia' Inst Mac Invest Cienc Nat. 13:453-459.
4854. Hodgen, GD, Dufau, ML, Catt, KJ, et al. 1972. Estrogens, progesterone and chorionic gonadotropin in pregnant rhesus monkeys. Endocrinology. 91:896-900.
4855. Hodgen, GD, Stouffer, RL, Barber, DL, et al. 1977. Serum estradiol and progesterone during pregnancy and the status of the corpus luteum at delivery in cynolmogus monkeys (*Macaca fasciata*). Steroids. 30:295-301.
4856. Hodgen, GD & Tullner, WW. 1975. Plasma estrogens, progesterone and chorionic gonadotropin in pregnant rhesus monkeys (*Macaca mulatta*) after ovariectomy. Steroids. 25:275-282.
4857. Hodgen, GD, Wilks, JW, Vaitukaitis, JL, et al. 1976. A new radioimmunoassay for follicle-stimulating hormone in macaques: Ovulatory menstrual cycle. Endocrinology. 99:137-145.
4858. Hodges, JK. 1978. Effects of gonadectomy and oestradiol treatment on plasma luteinizing hormone concentrations in the marmoset monkey *Callithrix jacchus*. J Endocrinol. 76:271-281.
4859. Hodges, JK. 1980. Regulation of oestrogen-induced LH release in male and female marmoset monkeys (*Callithrix jacchus*). J Reprod Fert. 60:389-398.
4860. Hodges, JK. 1985. The endocrine control of reproduction. Symp Zool Soc London. 54:149-168.
4861. Hodges, JK, Bevan, DJ, Celma, M, et al. 1984. Aspects of reproductive endocrinology of the female giant panda (*Ailuropoda melonaleuca*) in captivity with special reference to the detection of ovulation and pregnancy. J Zool. 203:253-267.
4862. Hodges, JK, Brand, H, Henderson, C, et al. 1983. Levels of circulating and urinary oestrogens during pregnancy in the marmoset monkey (*Callithrix jacchus*). J Reprod Fert. 67:73-82.
4863. Hodges, JK, Czekala, NM, & Lasley, BL. 1979. Estrogen and luteinizing hormone secretion in diverse primate species from simplified urine analysis. J Med Primatol. 8:349-364.
4864. Hodges, JK & Eastman, SAK. 1984. Monitoring ovarian function in marmosets and tamarins by the measurement of urinary estrogen metabolites. Am J Primatol. 6:187-197.
4865. Hodges, JK, Eastman, SAK, & Jenkins, N. 1983. Sex steroids and their relationship to binding proteins in the serum of the marmoset monkey (*Callithrix jacchus*). J Endocrinol. 96:443-450.
4865a. Hodges, JK, Green, DI, Cottingham, PG, et al. 1988. Induction of luteal regression in the marmoset monkey (*Callithrix jacchus*) by a gonadotrophin-realeasing hormone antagonist and the effects of subsequent follicular development. J Reprod Fert. 82:743-752.
4866. Hodges, JK, Gulick, BA, Czekala, NM, et al. 1981. Comparison of uinary oestrogen excretion in South American primates. J Reprod Fert. 61:83-90.
4867. Hodges, JK & Hearn, JP. 1978. A positive feedback effect of oestradiol on LH release in the male marmoset monkey, *Callithrix jacchus*. J Reprod Fert. 52:83-86.
4868. Hodges, JK, Henderson, C, & Hearn, JP. 1983. Relationship between ovarian and placental steroid production during early pregnancy in the marmoset monkey (*Callithrix jacchus*). J Reprod Fert. 69:613-621.
4869. Hodges, JK, McNeilly, AS, & Hess, DL. 1987. Circulating hormones during pregnancy in the Asian and African elephants, *Elephas maximus* and *Loxodonta africana*: A diagnostic test based on the measurement of prolactin. Int Zoo Yearb. 26:285-289.
4870. Hodges, JK, Tarara, R, & Wangula, C. 1984. Circulating steroids and the relationship between ovarian and placental secretion during early and mid pregnancy in the baboon. Am J Primatol. 7:357-366.
4871. Hodges, JK, Tarara, R, Hearn, JP, et al. 1986. The detection of ovulation and early pregnancy in the baboon by direct measurement of conjugated steroids in urine. Am J Primatol. 10:329-338.
4872. Hodgson, BH. 1847. On a new form of hog kind or Suidae. J Asiatic Soc Bengal. 14:423-428.
4873. Hoditschek, B & Best, TL. 1983. Reproductive biology of Ord's kangaroo rat (*Dipodomys ordii*) in Oklahoma. J Mamm. 64:121-127.
4874. Hoeck, HN. 1977. "Teat order" in hyrax (*Procavia johnstoni* and *Heterohyrax brucei*). Z Säugertierk. 42:112-115.
4875. Hoeck, HN. 1982. Population dynamics, dispersal and genetic isolation in two species of hyrax (*Heterohyrax brucei* and *Procavia johnstoni*) on habitat islands in the Serengeti. Z Tierpsychol. 59:177-210.
4876. Hoeck, HN, Klein, H, & Hoeck, P. 1982. Flexible social organization in hyrax. Z Tierpsychol. 59:265-298.
4877. Hoefs, M. 1978. Twinning in dall sheep. Can Field-Nat. 92:292-293.
4878. Hoefs, M & Cowan, IM. 1979. Ecological investigation of a population of Dall sheep (*Ovis dalli dalli* Nelson). Syesis Suppl 1. 12:1-81.
4879. Hoesch, W. 1959. Zur Jugenentwicklung der Macroscelididae. Bonn Zool Beitr. 10:263-265.
4880. Hoffman, B, Barth, D, & Karg, H. 1978. Progesterone and estrogen levels in peripheral plasma of the pregnant and nonpregnant roe deer (*Capreolus capreolus*). Biol Reprod. 19:931-935.
4881. Hoffman, LH, Wimsatt, WA, & Olson, GE. 1987. Plasma membrane structure of bat spermatozoa: Observations on epididymal and uterine spermatozoa in *Myotis lucifugus*. Am J Anat. 178:326-334.
4882. Hoffman, RA & Kirkpatrick, CM. 1954. Red fox weights and reproduction in Tippecanoe County, Indiana. J Mamm. 35:504-509.
4883. Hoffmann, JC & Sehgal, A. 1976. Effects of exogenous administration of hormones on reproductive tract of the female Hawaiian mongoose, *Herpestes auropunctatus auropunctatus* (Hodgson). Indian J Exp Biol. 14:480-482.
4884. Hoffmann, JC, Soares, MJ, Nelson, ML, et al. 1984. Seasonal reproduction in the mongoose, *Herpestes auropunctatus* IV Organ weight and hormone changes in the female. Gen Comp Endocrinol. 55:306-311.
4885. Hoffmann, K. 1973. The influence of photoperiod and melatonin on testis size, body weight, and pelage colour in the Djungarian hamster (*Phodopus sungorus*). J Comp Physiol. 85:267-282.

4886. Hoffmann, K. 1975. Photoperiod influences age of puberty in the Djungarian hamster *Phodopus sungorus*. Pflügers Archiv. 359 Suppl R:144.
4887. Hoffmann, K. 1978. Effects of short photoperiods on puberty, growth and moult in the Djungarian hamster (*Phodopus sungorus*). J Reprod Fert. 54:29-35.
4888. Hoffmann, K, Illnerová, HI, & Vaněček, J. 1981. Effect of photoperiod and of one minute light at night-time on the pineal rhythm on N-acetyltransferase activity in the Djungarian hamster *Phodopus sungorus*. Biol Reprod. 24:551-556.
4889. Hoffmann, K & Küderling, I. 1975. Pinealectomy inhibits stimulation of testicular development by long photoperiod in a hamster (*Phodopus sungorus*). Experientia. 31:122-123.
4890. Hoffmann, K & Küderling, I. 1977. Antigonadal effects of melatonin in pinealectomized Djungarian hamsters. Naturwissensch. 64:339-340.
4891. Hoffmann, M. 1958. Die Bisamratte. Akademische Verlagsgesellschaft Geest und Portig, KG, Leipzig.
4892. Hoffmann, M. 1974. Ein Beitrag zur Populationsdynamik der Bisamratte (*Ondatra zibethicus*). Z Angew Zool. 61:385-394.
4893. Hoffmann, RS. 1958. The role of reproduction and mortality in population fluctuations of voles (*Microtus*). Ecol Monogr. 28:79-109.
4894. Hoffmann, RS & Owen, JG. 1980. *Sorex tenellus* and *Sorex nanus*. Mamm Species. 131:1-4.
4895. Hoffmann, RS, Newby, FE, & Wright, PL. 1969. The distribution of some mammals in Montana: I Mammals other than bats. J Mamm. 50:579-604.
4896. Hoffmann, RS & Taber, RD. 1960. Notes on *Sorex* in the northern rocky mountain alpine zone. J Mamm. 41:230-234.
4897. Hoffmeister, DF. 1956. Mammals of the Graham (Pinaleno) Mountains, Arizona. Am Mid Nat. 55:257-288.
4898. Hoffmeister, DF. 1963. The yellow-nosed cotton rat, *Sigmodon ochrognathus*, in Arizona. Am Mid Nat. 70:429-441.
4899. Hoffmeister, DF. 1981. *Peromyscus truei*. Mamm Species. 161:1-5.
4900. Hoffmeister, DF & de la Torre, L. 1960. A revision of the wood rat *Neotoma stephensi*. J Mamm. 41:476-491.
4900a. Hoffmeister, DR & Goodpaster, WW. 1954. The mammals of the Huachuca Mountains, southeastern Arizona. IL Biol Monogr. 24:1-152
4901. Hoffmeister, DF & Goodpaster, WW. 1962. Life history of the desert shrew *Notiosorex crawfordi*. Southwest Nat. 7:236-252.
4902. Hofman, R, Erickson, A, & Siniff, D. 1973. The Ross seal (*Ommatophoca rossi*). IUCN Suppl Pap. 39:129-139.
4903. Hofman, RJ. 1975. Distribution patterns and population structure of Antarctic seals. Ph D Diss. Univ MN.
4904. Hofman, RJ. 1979. Leopard seal. Mammals in the Seas, FAO Fish Ser 5. 2:125-129.
4905. Hofmann, JE, McGuire, B, & Pizzuto, TM. 1989. Parental care in the sagebrush vole (*Lemmiscus curtatus*). J Mamm. 70:162-165.
4906. Hofmann, L. 1923. Zur Anatomie des männlichen Elefanten-, Tapir- und Hippopotamus-Genitale. Zool Jb Abt Anat Ont. 45:161-212.
4907. Hofmeyr, JM & Skinner, JD. 1969. A note on ovulation and implantation in the steenbok and the impala. Proc S Afr Anim Prod. 8:175.
4908. Hogg, JT. 1984. Mating in bighorn sheep: Multiple creative male strategies. Science. 225:526-529.
4908a. Hohn, A. 1980. Age determination and age related factors in the teeth of western North Atlantic bottlenose dolphins. Sci Rep Whales Res Inst. 32:39-66.
4909. Hohn, AA & Hammond, PS. 1985. Early postnatal growth of the spotted dolphin, *Stenella attenuata*, in the offshore tropical Pacific. Fish Bull. 83:553-566.
4910. Hoke, J. 1987. Oh, it's so nice to have a sloth around the house. Smithsonian. April:88-98.
4911. Holcomb, LC. 1963. Reproductive physiology in mink (*Mustela vison*). Ph D Diss. MI State Univ, E Lansing.
4912. Holcomb, LC. 1965. Large litter size of red fox. J Mamm. 46:530.
4913. Holdenried, R. 1957. Natural history of the bannertail kangaroo rat in New Mexico. J Mamm. 38:330-350.
4914. Holdenried, R & Morlan, HB. 1956. A field study of wild mammals and fleas of Sante Fe county, New Mexico. Am Mid Nat. 55:369-381.
4915. Holding, BF & Royal, OL. 1952. The development of a young harvest mouse (*Reithrodomtomys*). J Mamm. 33:388.
4916. Holekamp, KE. 1986. Proximal causes of natal dispersal in Belding's ground squirrels (*Spermophilus beldingi*). Ecol Monogr. 56:365-391.
4917. Holekamp, KE, Nunes, S, & Talamantes, F. 1988. Patterns of progesterone secretion in free-living California ground squirrels (*Spermophilus beecheyi*). Biol Reprod. 39:1051-1059.
4918. Holekamp, KE, Nunes, S, & Talamantes, F. 1988. Circulating prolactin in free-living California ground squirrels (*Spermophilus beecheyi*). Gen Comp Endocrinol. 71:484-492.
4919. Holekamp, KE, Smale, L, Simpson, HB, et al. 1984. Hormonal influences on natal dispersal in free-living Belding's ground squirrels (*Spermophilus beldingi*). Horm Behav. 18:465-483.
4920. Holler, NR, Baskett, TS, & Rogers, JP. 1963. Reproduction in confined swamp rabbits. J Wildl Manage. 27:179-183.
4921. Holler, NR & Conaway, CH. 1979. Reproduction of the marsh rabbit (*Sylvilagus palustris*) in south Florida. J Mamm. 60:769-777.
4922. Holley, AJF & Greenwood, PJ. 1984. The myth of the mad March hare. Nature. 309:549-550.
4923. Hollis, DE & Lyne, AG. 1980. Ultrastructure of luteal cells in fully formed and regressing corpora lutea during pregnancy and lactation in the marsupials *Isoodon macrourus* and *Perameles nasuta*. Aust J Zool. 28:195-211.
4924. Hollister, N. 1918. East African mammals in the United States National Museum Part I Insectivora, Chiroptera, and Carnivora. Bull US Nat Mus. 99:1-185.
4925. Hollister, N. 1919. East African mammals in the United States National Museum Part II Rodentia, Lagomorpha, and Tubulidentata. Bull US Nat Mus. 99:1-184.
4926. Hollister, N. 1924. East African mammals in the United States National Museum Part III Primates, Artiodactyla, Perissodactyla, Proboscidea, and Hyracoidea. Bull US Nat Mus. 99:1-164.
4927. Holm, E. 1969. Contribution to the knowledge of the biology of the Namib Desert golden mole *Eremitalpa granti namibensis* Bauer & Niethammer. Sci Pap Namib Desert Res Stn. 41:37-42.
4928. Holman, SD & Hutchison, JB. 1982. Pre-copulatory behaviour in the male Mongolian gerbil: I Differences in dependence on androgen of component patterns. Anim Behav. 30:221-230.
4929. Holman, SD, Hutchison, JB, & Snelson, D. 1982. Pre-copulatory behavior in the male mongolian gerbil: II effects of post-castration sexual and aggressive interactions on responsiveness to androgen. Anim Behav. 30:231-239.

4930. Holmes, ACV & Sanderson, GC. 1965. Populations and movements of opossums in east-central Illinois. J Wildl Manage. 29:287-295.
4931. Holmes, WG. 1988. Body fat influences sexual maturation in captive male Belding's ground squirrel. Can J Zool. 66:1620-1625.
4932. Holmes, WG & Sherman, PW. 1982. The ontogeny of kin recognition in two species of ground squirrel. Amer Zool. 22:491-517.
4933. Holroyd, JC. 1967. Observations of Rocky Mountain goats on Mount Wardle, Kootenay National Park, British Columbia. Can Field-Nat. 81:1-22.
4934. Holst, PJ. 1981. Age, hair colour, live weight, and fertility of two samples of Australian feral goat, *Capra hircus*. Aust Wildl Res. 8:549-553.
4935. Holt, C & Jenness, R. 1984. Interrelationships of constituents and partition of salts in milk samples from eight species. Comp Biochem Physiol. 77A:275-282.
4936. Holt, WV. 1977. Postnatal development of the testes in the cuis *Galea musteloides*. Lab Anim. 11:87-91.
4937. Holt, WV, Jones, RC, & Skinner, JD. 1980. Studies of the deferent ducts from the testis of the African elephant, *Loxodonta africana* II Histochemistry of the epididymis. J Anat. 130:367-379.
4938. Holt, WV, Moore, HDM, North, RD, et al. 1988. Hormonal and behavioural detection of oestrus in blackbuck, *Antilope cervicapra*, and successful artificial insemination with fresh and frozen semen. J Reprod Fert. 82:717-725.
4939. Holt, WV & Tam, WH. 1973. Steroid metabolism by the chin gland of the male cuis, *Galea musteloides*. J Reprod Fert. 33:53-59.
4940. Holtz, W & Foote, RH. 1978. The anatomy of the reproductive system in male Dutch rabbits (*Oryctolagus cuniculus*) with special emphasis on the accessory sex glands. J Morphol. 158:1-20.
4941. Homan, JA & Jones, JK, Jr. 1975. *Monophyllus redmani*. Mamm Species. 57:1-3.
4942. Homan, JA & Jones, JK, Jr. 1975. *Monophyllus plethodon*. Mamm Species. 58:1-2.
4943. Homan, WG. 1975. Breeding the international herd of Arabian oryx at Phoenix Zoo. Breeding Endangered Species in Captivity. 285-292, Martin, RD, ed, Academic Press, New York.
4944. Home, E. 1802. A description of the anatomy of the *Ornithorhynchus paradoxus*. Phil Trans R Soc London. 1802:67-84.
4945. Home, E. 1802. Description of the anatomy of the *Ornithorhynchus hystrix*. Phil Trans R Soc London. 1802 Part 2:348-364.
4946. Homei, V & Barbu, Pr. 1962. Contributie la studiul gestatiei la *Miniopterus schreibersi*. Acad Repub Pop Rom Stud Cercet Biol Ser Zool. 14:57-64.
4947. Homeida, AM, Khalil, MGR, & Taha, AAM. 1988. Plasma concentrations of progesterone, oestrogens, testosterone and LH-like activity during the oestrous cycle of the camel (*Camelus dromedarius*). J Reprod Fert. 83:593-598.
4948. Hona, SM & Stetson, MH. 1983. Effects of day-length on testicular function in Turkish hamsters. Am Zool. 23:993.
4949. Honegger, RE & Noth, W. 1966. Beobachtungen bei der Aufzucht von Igeltanreks *Echinops telfairi* Martin. Zool Beitr NF. 12:191-218.
4950. Honess, RF & Frost, NM. 1942. A Wyoming big-horn sheep study. Bull WY Game Fish Comm. 1:1-126.
4951. Hong, SM & Stetson, MH. 1983. Effects of day length on testicular function in turkish hamsters. Am Zool. 23:993.
4952. Hong, SM & Stetson, MH. 1986. Functional maturation of the gonads of Turkish hamsters under various photoperiods. Biol Reprod. 35:858-862.
4953. Honigmann, H. 1935. Beobachtungen am groszen Ameisenbären. Z Säugetierk. 10:78-104.
4954. Honjo, S, Fujiwara, T, & Cho, F. 1975. A comparison of breeding performance of individual cage and indoor gang cage systems in cynomolgus monkeys. Int Congr Primat. 5:98-105.
4955. Hood, CS & Jones, JK, Jr. 1984. *Noctilio leporinus*. Mamm Species. 216:1-7.
4956. Hood, CS & Pitocchelli, J. 1983. *Noctilio albiventris*. Mamm Species. 197:1-5.
4957. Hood, CS & Smith, JD. 1983. Histomorphology of the female reproductive tract in phyllostomoid bats. Occas Pap Mus TX Tech Univ. 86:1-38.
4958. Hoogland, JL. 1982. Prairie dogs avoid extreme inbreeding. Science. 215:1639-1641.
4959. Hoogland, JL. 1985. Infanticide in prairie dogs: Lactating females kill offspring of close kin. Science 230:1037-1040.
4960. Hoogstraal, H. 1963. A brief review of the contemporary land mammals of Egypt (including Sinai) 2 Lagomorpha and Rodentia. J Egypt Public Health Assoc. 38:1-35.
4961. Hook, O & Johnels, AG. 1972. The breeding and distribution of the grey seal (*Halichoerus grypus*, Fab) in the Baltic Sea, with observations on other seals of the area. Proc R Soc London Biol Sci. 182B:37-58.
4962. Hooper, ET. 1941. Mammals of the lava fields and adjoining areas in Valencia county, New Mexico. Misc Publ Mus Zool Univ MI. 51:1-47.
4963. Hooper, ET. 1952. A systematic review of the harvest mice (genus *Reithrodontomys*) of Latin America. Misc Publ Mus Zool Univ MI. 151:1-52.
4964. Hooper, ET. 1959. The glans penis in five genera of cricetid rodents. Occas Pap Mus Zool Univ MI. 613:1-11.
4965. Hooper, ET. 1968. Habits and food of amphibious mice of the genus *Rheomys*. J Mamm. 49:550-553.
4966. Hooper, ET & Brown, JH. 1968. Foraging and breeding in two sympatric species of neotropical bats, genus *Noctilio*. J Mamm. 49:331-312.
4967. Hooper, ET & Carleton, MD. 1976. Reproduction, growth and development in two contiguously allopatric rodent species, genus *Scotinomys*. Misc Publ Mus Zool Univ MI. 151:1-52.
4968. Hooper, JHD & Hooper, WM. 1956. Habits and movements of cave-dwelling bats in Devonshire. Proc Zool Soc London. 127:1-26.
4969. Hooven, EF. 1973. Notes on the water vole in Oregon. J Mamm. 54:751-753.
4970. Hooven, EF. 1973. Response of the Oregon creeping vole to the clearcutting of a douglass-fir forest. Northwest Sci. 47:256-264.
4971. Hoover, RL. 1954. Seven fetuses in western fox squirrel (*Sciurius niger rufiventer*). J Mamm. 35:447-448.
4972. Hopcraft, D. 1976. Productivity comparison between Thomson's gazelle and cattle and their relation to the ecosystem in Kenya. Ph D Diss. Cornell Univ, Ithaca, NY.
4973. Hope, RM. 1971. The maintenance of the brush-tailed possum *Trichosurus vulpecula* in captivity. Int Zoo Yearb. 11:24-25.
4974. Hopf, S. 1967. Notes on pregnancy, delivery, and infant survival in captive squirrel monkeys. Primates. 8:323-332.
4975. Hopf, S & Ploog, D. 1979. Life span in captive squirrel monkeys (*Saimiri*) with pathological and reproductive records. Primates. 20:313-316.

4976. Hopkins, DD & Forbes, RB. 1979. Size and reproductive patterns of the Virginia opossum in northwestern Oregon. Murrelet. 60:95-98.
4977. Hoppe, PP. 1977. How to survive heat and aridity: Ecophysiology of the dikdik antelope. Vet Med Rev. 1:77-86.
4978. Hopper, BR & Tullner, WW. 1967. Urinary estrogen excretion patterns in pregnant rhesus monkeys. Steroids. 9:517-527.
4979. Hopper, BR, Tullner, WW, & Gray, CW. 1968. Urinary estrogen excretion during pregnancy in a gorilla (*Gorilla gorilla*). Proc Soc Exp Biol Med. 129:213-214.
4980. Horáček, I & Gaisler, J. 1986. The mating system of *Myotis blythi*. Myotis. 23-24:125-130.
4981. Hořácek, V & Uher, J. 1965. Zur Kenntnis des Geschlechtszyklus und der Superfötation bei dem Feldhasen *Lepus europaeus* Pall. Zool Listy. 14:107-116.
4982. Horn, CA, Fletcher, TP, & Carpenter, S. 1985. Effects of oestradiol-17β on peripheral plasma concentrations of LH and FSH in ovariectomized tammars (*Macropus eugenii*). J Reprod Fert. 73:585-592.
4983. Horn, EE. 1923. Some notes concerning the breeding habits of *Thomomys townsendi*, observed near Vale Malheur county, Oregon, during the spring of 1921. J Mamm. 4:37-39.
4984. Hornaday, WT. 1912. Our pygmy hippopotamus. Zool Soc Bull (NY). 16(52):877-879.
4985. Hornaday, WT. 1926. Two musk-ox calves in New York. J Mamm. 7:61.
4986. Horner, BE. 1968. Gestation period and early development in *Onychomys leucogaster brevicaudus*. J Mamm. 49:513-515.
4987. Horner, BE, Potter, GL, & van Ooteghem, S. 1980. A new black coat-color mutation in *Peromyscus*. J Hered. 71:49-51.
4988. Horner, BE & Taylor, JM. 1958. Breeding of *Rattus assimilis* in captivity. J Mamm. 39:301-302.
4989. Horner, BE & Taylor, JM. 1959. Results of the Archbold expeditions No 80 Observations on the biology of the yellow-footed marsupial mouse, *Antechinus flavipes flavipes*. Am Mus Novit. 1972:1-24.
4990. Horner, BE & Taylor, JM. 1965. Systematic relationships among *Rattus* in southern Australia: Evidence from cross-breeding experiments. CSIRO Wildl Res. 10:101-109.
4991. Horner, BE & Taylor, JM. 1968. Growth and reproductive behavior in the southern grasshopper mouse. J Mamm. 49:644-660.
4992. Horner, BE & Taylor, JM. 1969. Paternal behavior in *Rattus fuscipes*. J Mamm. 50:803-805.
4993. Hornsby, PE. 1978. A note on the pouch life of rock wallabies. Vict Nat. 95:108-111.
4994. Horr, DA. 1975. The Borneo orang-utan: Population structure and dynamics in relationship to ecology and reproductive strategy. Primate Behavior. 4:307-323.
4995. Horrocks, JA. 1986. Life-history characteristics of a wild population of vervets (*Cercopithecus aethiops aethiops*) in Barbados, West Indies. Int J Primatol. 7:31-48.
4996. Horst, HJ. 1979. Photoperiodic control of androgen metabolism and binding in androgen target organs of hamsters (*Phodopus sungorus*). J Steroid Biochem. 11:945-960.
4997. Horston, DB, Robbins, CT, & Stevens, V. 1989. Growth in wild and captive mountain goats. J Mamm. 70:412-416.
4998. Horton, TH. 1984. Growth and maturation in *Microtus montanus*: Effects of photoperiods before and after weaning. Can J Zool. 62:1741-1746.
4999. Horton, TH. 1984. Growth and reproductive development of male *Microtus montanus* is affected by the prenatal photoperiod. Biol Reprod. 31:499-504.
5000. Horton, TH & Schwartz, NB. 1986. Exposure to males required to induce compensatory ovarian hypertrophy in montane voles (*Microtus montanus*). Biol Reprod. 34(Suppl):63.
5001. Horvath, O. 1966. Observation of parturition and maternal care of the bushy-tailed wood rat (*Neotoma cinerea occidentalis* Baird). Murrelet. 47:6-8.
5002. Horwich, RH. 1972. The ontogeny of social behavior in the gray squirrel (*Sciurus carolinensis*). Advances in Ethology: Suppl to J Comp Ethology. 8:1-103.
5003. Horwood, JW, Donovan, GP, & Gambell, R. 1980. Pregnancy rates of the southern hemisphere sei whale (*Balaenoptera borealis*). Rep Int Whal Comm. 30:531-535.
5003a. Hosley, NW (Dalke, PD). 1942. The cottontail rabbits in Connecticut. Bull CT Wildl Res Unit. 65:1-97.
5004. Hosley, NW & Glaser, FS. 1952. Triplet Alaskan moose calves. J Mamm. 33:247.
5005. Hotchkiss, J, Atkinson, LE, & Knobil, E. 1971. Time course of serum estrogen and luteinizing hormone (LH) concentrations during the menstrual cycle of the rhesus monkey. Endocrinology. 89:177-183.
5006. Hotchkiss, J, Dierschke, DJ, Butler, WR, et al. 1982. Relation between levels of circulating ovarian steroids and pituitary gonadotropin content during the menstrual cycle of the rhesus monkey. Biol Reprod. 26:241-248.
5007. Hoth, J & Granados, H. 1987. A preliminary report on the breeding of the volcano rabbit, *Romerolagus diazi* at the Chapultepec Zoo Mexico City. Int Zoo Yearb. 26:261-265.
5008. Houlihan, RT. 1963. The relationship of population density to endocrine and metabolic changes in the California vole *Microtus californicus*. Univ CA Publ Zool. 65:327-362.
5009. Houston, ML. 1969. The villous period of placentogenesis in the baboon (*Papio* sp). Am J Anat. 126:1-15.
5010. Houston, ML. 1969. The development of the baboon (*Papio* sp) placenta during the fetal period of gestation. Am J Anat. 126:17-29.
5011. Houston, ML. 1971. Placenta. Embryology of the Baboon. 153-172, Hendrickx, AG, ed, Univ Chicago Press, Chicago.
5012. Houston, ML & Hendrickx, AG. 1968. Observations on the vasculature of the baboon placenta (*Papio* sp) with special reference to the transverse communicating artery. Folia Primatol. 9:68-77.
5013. How, RA. 1972. The ecology and management of *Trichosurus* species (Marsupialia) in New South Wales. Ph D Diss. Univ New England, Armidale, NSW Australia.
5014. How, RA. 1976. Reproduction, growth and survival of young in the mountain possum, *Trichosurus caninus* (Marsupialia). Aust J Zool. 24:189-199.
5015. How, RA. 1978. Population strategies in four species of Australian possums. The Ecology of Arboreal Folivores. 305-314, Montgomery, GG, ed, Smithsonian Inst Press, Washington, DC.
5016. How, RA. 1981. Population parameters of two congeneric possums, *Trichosurus* spp, in north-eastern New South Wales. Aust J Zool. 29:205-215.
5017. How, RA, Barnett, JL, Bradley, AJ, et al. 1984. The population biology of *Pseudocheirus peregrinus* in a *Leptospermum laevigatum* thicket. Possums and Gliders. 261-268, Smith, AP & Hume, ID, eds, Aust Mamm Soc, Sydney.
5018. Howard, VW, Jr. 1966. An observation of parturition in the pronghorn antelope. J Mamm. 47:708-709.
5019. Howard, WE. 1949. Dispersal, amount of inbreeding, and longevity in local populations of prairie deermice on the George Reserve, southern Michigan. Contrib Lab Vert Biol Univ MI. 43:1-50.

5019a. Howard, WE. 1950. Winter fecundity of caged male white-footed mice in Michigan. J Mamm. 31:319-321.
5020. Howard, WE. 1959. California ground squirrel breeding in captivity. J Mamm. 40:445-446.
5021. Howard, WE & Childs, HE, Jr. 1959. Ecology of pocket gophers with emphasis on *Thomomys bottae mewa*. Hilgardia. 29:277-358.
5022. Howe, R & Clough, GC. 1971. The Bahaman hutia *Geocapromys ingrahami* in captivity. Int Zoo Yearb. 11:89-93.
5023. Howell, AB. 1920. Contribution to the life history of the California mastiff bat. J Mamm. 1:111-117.
5024. Howell, AB. 1920. Some Californian experiences with bat roosts. J Mamm. 1:169-177.
5025. Howell, AB. 1924. The mammals of Mammoth, Mono county, California. J Mamm. 5:25-36.
5026. Howell, AB. 1926. Voles of the genus *Phenacomys*. N Am Fauna. 48:1-66.
5027. Howell, AB & Little, L. 1924. Additional observations on California bats; with observations upon the young of *Eumops*. J Mamm. 5:261-263.
5028. Howell, AH. 1914. Revision of the American harvest mice (genus *Reithrodontomys*). N Am Fauna. 36:1-97.
5029. Howell, AH. 1915. Revision of the American marmots. N Am Fauna. 37:1-77.
5029a. Howell, AH. 1936. A revision of the American arctic hares. J Mamm. 17:315-337.
5030. Howell, AH. 1938. Revision of the North American ground squirrels, with a classification of the North American Sciuridae. N Am Fauna. 56:1-256.
5031. Howell, DJ. 1979. Flock foraging in nectar-feeding bats: Advantages to the bats and to the host plants. Am Nat. 114:23-49.
5032. Howell, KM & Jenkins, PD. 1984. Record of shrews (Insectivora, Soricidae) from Tanzania. Afr J Ecol. 22:67-68.
5033. Howland, BE, Faiman, C, & Butler, TM. 1971. Serum levels of FSH and LH during the menstrual cycle of the chimpanzee. Biol Reprod. 4:101-105.
5033a. Hoyland, F. 1972. Guinea-pig production at the Animal Virus Research Institute, Pirbright, Woking, Surrey. Guinea-Pig Newsl. 5:13-17.
5034. Hoyle, JA & Boonstra, R. 1986. Life history traits of the meadow jumping mouse, *Zapus hudsonius*, in southern Ontario. Can Field-Nat. 100:537-544.
5035. Hoyt, RA & Altenbach, JS. 1981. Observations on *Diphylla ecaudata* in captivity. J Mamm. 62:215-216.
5036. Hoyt, RA & Baker, RJ. 1980. *Natalus major*. Mamm Species. 130:1-3.
5037. Hoyt, SF. 1952. Additional notes on the gestation period of the woodchuck. J Mamm. 33:388-389.
5038. Hoyt, SY & Hoyt, SF. 1950. Gestation period of the woodchuck, *Marmota monax*. J Mamm. 31:454.
5039. Hoyte, HMD. 1955. Observations on reproduction in some small mammals of arctic Norway. J Anim Ecol. 24:412-425.
5040. Hrabě, V. 1970. Seasonal changes in microscopical structure of the vesicular gland and the prostate in *Glis glis* (Gliridae, Rodentia). Zool Listy. 19:249-260.
5041. Hrabě, V. 1972. Contribution to the analysis of the sexual cycle of male *Clethrionomys glareolus* Schr (Rodentia). Zool Listy. 21:309-318.
5042. Hrabě, V & Zejda, J. 1981. Age determination and mean length of life in *Citellus citellus*. Folia Zool. 30:117-123.
5043. Hrdlička, A. 1925. Weight of the brain and of the internal organs in American monkeys. Am J Phys Anthrop. 8:201-211.
5044. Hrdy, SB. 1977. The Langurs of Abu. Harvard Univ Press, Cambridge, MA.
5045. Hrdy, SB & Hrdy, DB. 1976. Hierarchiral relations among female Hanuman langurs (Primates: Colobinae: *Presbytis entellus*). Science. 193:913-915.
5046. Hronopulo, NP. 1935. Increasing multifoetation in the mink by increasing daylight. Karackulevodstvo i Zverodstvo. 8(4):32-33. (Animal Breeding Abstr 24:ref 279, 1956).
5047. Hubback, TR. 1937. The Malayan guar or seladang. J Mamm. 18:267-279.
5048. Hubbard, CA. 1922. Some data upon the rodent *Aplodontia*. Murrelet. 3:14-18.
5049. Hubbard, CA. 1970. A first record of *Beamys* from Tanzania with observations on its breeding and habits in captivity. Zool Afr. 5:229-236.
5050. Hubbard, CA. 1970. A new species of *Tatera* from Tanzania with a description of its life history and habits studied in captivity. Zool Afr. 5:237-247.
5051. Hubbard, CA. 1972. Observations on the life histories and behaviour of some small rodents from Tanzania. Zool Afr. 7:419-449.
5052. Hubbard, WD. 1926. Notes on the antelopes and zebra of northern Rhodesia and Portuguese east Africa. J Mamm. 7:184-193.
5053. Hubbard, WD. 1929. Further notes on the mammals of northern Rhodesia and Portuguese east Africa. J Mamm. 10:294-297.
5054. Hubbs, CL. 1958. Natural history of the grey whale. Int Congr Zool. 15:313-316.
5055. Huber, HR. 1987. Natality and weaning success in relation to age of first reproduction in northern elephant seals. Can J Zool. 65:1311-1316.
5056. Huber, HR, Rovetta, AC, Fry, LA, et al. 1991. Age-specific natality of northern elephant seals at the south Farallon Islands, California. J Mamm. 72:525-534.
5057. Hubert, B. 1977. Ecologie des populations de Rongeurs de Bandia (Senegal), en zone Sahelo-Soudanienne. Terre Vie. 31:33-100.
5058. Hubert, B & Adam, F. 1975. Reproduciton et croissance en élevage de quatre espèces de Rongeurs Sénégalais. Mammalia. 39:57-73.
5059. Hubrecht, RC. 1984. Field observations on group size and composition of the common marmoset (*Callithrix jacchus jacchus*), at Tapacura, Brazil. Primates. 25:13-21.
5060. Huck, UW & Banks, EM. 1982. Male dominance status, female choice and mating success in the brown lemming, *Lemmus trimucronatus*. Anim Behav. 30:665-675.
5061. Huck, UW & Lisk, RD. 1985. Determinants of mating success in the golden hamster (*Mesocricetus auratus*): II Pregnancy initiation. J Comp Psychol. 99:231-239.
5062. Hudson, C. 1907. Breeding seasons of deer. J Bombay Nat Hist Soc. 18:187.
5063. Hudson, GE. 1932. On the food habits of *Marmosa*. J Mamm. 13:159.
5064. Hudson, JW. 1974. The estrous cycle, reproduction, growth, and development of temperature regulation in the pygmy mouse, *Baiomys taylori*. J Mamm. 55:572-588.
5065. Hudson, P. 1959. Fetal recoveries in mule deer. J Wildl Manage. 23:234-235.
5066. Hudson, P & Browman, LG. 1959. Embryonic and fetal development of the mule deer. J Wildl Manage. 23:295-304.
5067. Hudson, WS & Wilson, DE. 1986. *Macroderma gigas*. Mamm Species. 260:1-4.

5067a. Huey, LM. 1925. Food of the California leaf-nosed bat. J Mamm. 6:196-197.
5068. Huffman, MA. 1987. Consort intrusion and female mate choice in Japanese macaques (*Macaca fuscata*). Ethology. 75:221-234.
5069. Huffman, WE. 1970. Notes on the first captive conception and live birth of an Amazon dolphin in North America. Underwater Nat. 6:9-11.
5070. Hugget, AStG & Widdas, WF. 1951. The relationship between mammalian foetal weight and conception age. J Physiol London. 114:306-317.
5071. Hughes, RD. 1965. On the age composition of a small sample of individuals from a population of the banded hare wallaby *Lagostrophus fasciatus* Peron and Leseur. Aust J Zool. 13:75-95.
5072. Hughes, RL. 1962. Reproduction in the macropod marsupial *Potorous tridactylus* (Kerr). Aust J Zool. 10:193-224.
5073. Hughes, RL. 1962. The role of the corpus luteum in marsupial reproduction. Nature. 194:890.
5074. Hughes, RL. 1964. Sexual development and spermatozoon morphology in the male macropod marsupial *Potorous tridactylus* (Kerr). Aust J Zool. 12:42-51.
5075. Hughes, RL. 1974. Morphological studies on implantation in marsupials. J Reprod Fert. 39:173-186.
5076. Hughes, RL. 1976. Preliminary observations on reproduction and embryonic development in the marsupial devil, *Sarcophilus harrisii*. J Reprod Fert. 46:504-505.
5076a. Hughes, RL. 1976. The foetal membranes of the marsupial mouse *Sminthopsis crassicaudata* with particular reference to implantation. Bull Aust Mamm Soc. 3:24.
5077. Hughes, RL. 1982. Reproduction in the Tasmanian devil *Sarcophilus harrisii* (Dasyuridae, Marsupialia). Carnivorous Marsupials. 1:49-63.
5078. Hughes, RL & Carrick, FN. 1978. Reproduction in female monotremes. Aust Zool. 20:233-253.
5079. Hughes, RL, Carrick, FN, & Shorey, CD. 1975. Reproduction in the platypus *Ornithorhynchus anatinus* with particular reference to the evolution of viviparity. J Reprod Fert. 43:374-375.
5080. Hughes, RL & Hall, LS. 1984. Embryonic development in the common brushtail possum *Trichosurus vulpecula*. Possums and Gliders. 197-212, Smith, AP & Hume, ID, eds, Aust Mamm Soc, Sydney.
5081. Hughes, RL & Rowley, I. 1966. Breeding season of female wild rabbits in natural populations in the riverina and southern tablelands districts of New South Wales. CSIRO Wildl Res. 11:1-10.
5082. Hughes, RL & Shorey, CD. 1973. Observations on the permeability properties of the egg membranes of the marsupial, *Trichosurus vulpecula*. J Reprod Fert. 32:25-32.
5083. Hughes, RL, Thomson, JA, & Owen, WH. 1965. Reproduction in natural populations of the Australian ringtail possum, *Pseudocheirus peregrinus* (Marsupialia: Phalangeridae), in Victoria. Aust J Zool. 13:383-406.
5084. Huhtaniemi, IT, Koritnik, DR, Korenbrot, CC, et al. 1979. Stimulation of pituitary-testicular function with gonadotropin-releasing hormone in fetal and infant monkeys. Endocrinology. 105:109-114.
5085. Hui, CA. 1977. Growth and physical indices of maturity in the common dolphin, *Dolphinus delphis*. Breeding Dolphins:Present Status, Suggestions for the Future. US Nat Tech Inf Serv, Arlington, VA.
5086. Hui, CA. 1979. Correlates of maturity in the common dolphin, *Delphinus delphis*. Fish Bull (USA). 77:295-300.
5087. Huibregtse, H. 1966. Some chemical and physical properties of bat milk. J Mamm. 47:551-555.
5088. Hulbert, AJ. 1972. Growth and development of pouch young in the rabbit-eared bandicoot, *Macrotis lagotis* (Peramelidae). Aust Mamm. 1:38-39.
5089. Hulbert, AJ. 1982. Notes on the management of a captive breeding colony of the greater bilby *Macrotis lagotis*. The Management of Australian Mammals in Captivity, Evans, DD, ed. Zool Bd Victoria, Melborne.
5090. Hulley, JT. 1976. Maintenance and breeding of captive jaguarundis *Felis yagouaroundi* at Chester Zoo and Toronto. Int Zoo Yearb. 16:120-122.
5091. Humiński, S. 1958. The autumnal involution of the male sexual apparatus in the field vole (*Microtus arvalis* Pall). Zool Pol. 9:197-214.
5092. Humiński, S. 1963. Winter breeding in the field vole, *Microtus arvalis* (Pall) in the light of an analysis of the effect of environmental factors on the condition of the male sexual apparatus. Zool Pol. 14:157-203.
5093. Humiński, S. 1968. Maturation and seasonal variations of the testes and the male accessory glands in the field mouse, *Apodemus agrarius* (Pallas, 1771). Zool Pol. 18:69-78.
5094. Humiński, S & Kowalczyk, D. 1975. Seasonal variations in the epididymides in the striped field mouse, *Apodemus agrarius* (Pallas, 1771). Zool Pol. 24:189-199.
5095. Humiński, S & Lecyk, M. 1960. Influence of captivity on the involution of the male sexual apparatus in the field vole (*Microtus arvalis* Pall). Zool Pol. 10:75-83.
5096. Humiński, S & Wojcik-Migala, I. 1967. Note on *Crocidura suaveolens* (Pallas, 1811) from Poland. Acta Theriol. 12:168-171.
5096a. Humphrey, SR. 1975. Nursery roosts and community diversity of neoarctic bats. J Mamm. 56:321-346.
5096b. Humphrey, SR & Cope, RB. 1977. Survival rates of the endangered Indiana bat, *Myotis sodalis*. J Mamm. 58:32-36.
5096c. Humphrey, SR, Richter, AR, & Cope, JB. 1977. Summer habitat and ecology of the endangered Indiana bat, *Myotis sodalis*. J Mamm. 58:334-346.
5097. Humphreys, WF, How, RA, Bradley, AJ, et al. 1984. The biology of *Wyulda squamicaudata* Alexander 1919. Possums and Gliders. 162-169, Smith, AP & Hume, ID, eds, Surrey Beatty & Sons Pty Ltd, Sydney, Australia.
5098. Hunt, H. 1967. Growth rate of a new-born, hand-reared jaguar *Panthera onca* at Topeka Zoo. Int Zoo Yearb. 7:147-148.
5099. Hunt, TP. 1959. Breeding habits of the swamp rabbit with notes on its life history. J Mamm. 40:82-91.
5100. Hunter, J, Martin, RD, Dixson, AF, et al. 1979. Gestation and inter-birth intervals in the owl monkey (*Aotus trivirgatus griseimembra*). Folia Primatologica. 31:165-175.
5101. Huntley, BJ. 1971. Reproduction in roan, sable and tsessebe. S Afr J Sci. 67:454.
5102. Hůrka, L. 1967. Ökologische Beobachtungen in der Wochenstube von *Eptesicus nilssoni* Keyserling et Blassius, 1839 in der Tsechoslovakei. Zool Listy. 16:193-197.
5103. Hůrka, L. 1981. Analyse der Population von *Clethrionomys glareolus* (Mammalia: Rodentia) im Otava-Tal in Sumava (Böhmerwald). Vestn Čcsl Spol Zool. 45:47-55.
5104. Hurme, VO & van Wagenen, G. 1956. Emergence of permanent first molars in the monkey (*Macaca mulatta*): Association with other growth phenomena. Yale J Biol Med. 28:538-567.
5105. Husar, SL. 1977. *Trichechus inunguis*. Mamm Species. 72:1-4.
5106. Husar, SL. 1978. *Dugong dugon*. Mamm Species. 88:1-7.

5107. Husar, SL. 1978. *Trichechus senegalensis*. Mamm Species. 89:1-3.
5108. Husar, SL. 1978. *Trichechus manatus*. Mamm Species. 93:1-5.
5109. Husband, TP. 1976. Energy metabolism and body composition of the fox squirrel. J Wildl Manage. 40:255-263.
5110. Husson, AM. 1957. Notes on the primates of Suriname. Stud Fauna Suriname & other Guyanas. 1(2):14-48.
5111. Husson, AM. 1978. The Mammals of Suriname. Zoologische Monographieën van het Rijksmuseum van Natuurlijke Historie. EJ Brill, Leiden.
5111a. Hutchins, M, Smith, GM, Mead, DC, et al. 1991. Social behavior of Matschie's tree kangaroos (*Dendrolagus matschiei*) and its implications for captive management. Zoo Biol. 10:147-164.
5112. Hutchins, M, Thompson, G, Sleeper, B, et al. 1987. Management and breeding of the Rocky Mountain goat *Oreamnos americanus* at Woodland Park Zoo. Int Zoo Yearb. 26:297-308.
5113. Hutchinson, TC. 1970. Vaginal cytology and reproduction in the squirrel monkey (*Saimiri sciureus*). Folia Primatol. 12:212-223.
5114. Hutchison, SS & Zeleznik, A. 1983. The rhesus monkey corpus luteum is dependent upon gonadotropin secretion throughout the luteal phase of the menstrual cycle. Biol Reprod. 28(Suppl 1):79.
5115. Huth, H-H. 1980. Verhaltensstudien an Pferdeböcken (Hippotraginae) unter Berücksichtigung stammesgeschichtlicher und systematischer Fragen. Säugetierk Mitt. 28:161-245.
5116. Hutson, GD. 1976. Grooming behavior and birth in the dasyurid marsupial *Dasyuroides byrnei*. Aust J Zool. 24:277-282.
5117. Hutson, JM, Shaw, G, O, WS, et al. 1988. Müllerian inhibiting substance production and testicular migration and descent in the pouch young of a marsupial. Development. 104:549-556.
5118. Hutterer, R. 1976. Beobachtungen zur Geburt and Jugendentwicklung der Zwergspitzmaus, *Sorex minutus* L (Soricidae, Insectivora). Z Säugetierk. 41:1-22.
5119. Hutterer, R. 1977. Haltung und Lebensdauer von Spitzmäusen der Gattung *Sorex* (Mammalia, Insectivora). Z Angew Zool. 64:353-367.
5120. Hutz, RJ, Dierschke, DJ, & Wolf, RC. 1985. Seasonal effects on ovarian folliculogenesis in rhesus monkeys. Biol Reprod. 33:653-659.
5121. Hutzelsider, HB. 1940. Eine Malayenbärengeburt im Zoo Aarhus. Zool Garten NF. 12:157-161.
5122. Hvidberg-Hansen, H. 1970. Contribution to the knowledge of the reproductive physiology of the Thomson's gazelle (*Gazella thomsonii* Gunther). Mammalia. 34:551-563.
5123. Hvidberg-Hansen, H & de Vos, A. 1971. Reproduction, population and herd structure of two Thomson's gazelle (*Gazella thomsonii* Gunther) populations. Mammalia. 34:1-16.
5124. Iason, GR & Guinness, FE. 1985. Synchrony of oestrus and conception in red deer (*Cervus elaphus* L). Anim Behav. 33:1169-1174.
5125. Ibáñez, C. 1985. Notes on *Amorphochilus schnablii* Peters (Chiroptera, Furipteridae). Mammalia. 49:584-587.
5125a. Ibáñez, C & Ochoa G, J. 1989. New records of bats from Bolivia. J Mamm. 70:216-219.
5126. Ichihara, T. 1962. Prenatal dead foetus of baleen whales. Sci Rep Whales Res Inst. 16:47-60.
5127. Ifuta, NB & Gevaerts, H. 1987. Le cycle de reproduction d'*Epomops franqueti* (Tomes 1860) (Chiroptera, Pteropodidae) de la région de Kisangani. Rev Zool Afr. 101:288-289.
5128. Ifuta, NB, Gevaerts, H, & Kühn, ER. 1988. Thyroid hormones, testosterone, and estradiol-17β in plasma of *Epomops franqueti* (Tomes, 1860) (Chiroptera) in the rain forest of the equator. Gen Comp Endocrinol. 69:378-380.
5129. Igo, WK, Allen, TJ, & Michael, ED. 1979. Observations on European wild boars released in West Virginia. Proc Ann Conf SE Fish Wildl Ag. 33:313-317.
5130. Ikeda, H. 1983. Development of young and parental care of the raccoon dog *Nyctereutes procyonoides viverrinus* Temmick [*sic*] in captivity. J Mamm Soc Japan. 9:229-236.
5131. Ikeda, H. 1986. Old dogs, new treks. Nat Hist. 95(8):35-45.
5132. Ilius, AW, Haynes, NB, Lamming, GE, et al. 1983. Evaluation of LH-RH stimulation of testosterone as an index of reproductive status in rams and its application in wild antelope. J Reprod Fert. 68:105-112.
5133. Iljina, ED. 1950. Some data on oestrus in the mink. Karakulevodstvo i Zverovodstvo. 3(3):50-54. (An Breeding Abstr 19:ref 1333, 1951).
5134. Illige, D. 1951. An analysis of the reproductive pattern of white-tailed deer in south Texas. J Mamm. 32:411-421.
5135. Ilmen, M & Lahti, S. 1968. Reproduction, growth and behavior in the captive wood lemming, *Myopus schisticolor* (Lilljeb). Ann Zool Fennici. 5:207-219.
5136. Imaizumi, T & Chabata, T. 1977. Notes on the nest for reproduction of Ryukyu wild boar, *Sus riukiuanus*. J Mamm Soc Japan. 7:111-113.
5137. Imaizumi, Y. 1969. Reproduction in the Japanese shrew mole in Niigata Pref, Honshu. J Mamm Soc Japan. 4:81-86.
5138. Imaizumi, Y. 1972. Notes on the oestrous cycle of *Tamias sibiricus*. J Mamm Soc Japan. 5:186.
5139. Imanishi, K. 1960. Social organization of subhuman primates in their natural habitat. Current Anthopol. 1:393-407.
5139a. Imel, KJ & Amann, RP. 1979. Effects of duration of daily illumination on reproductive organs and fertility of the meadow vole (*Microtus pennsylvanicus*). Lab Ani Sci. 29(2):182-185.
5140. Ims, RA. 1987. Differential reproductive success in a peak population of the grey-sided vole *Clethrionomys rufocanus*. Oikos. 50:103-113.
5141. Indian Plague Commission. 1907. The epidemiological observations made by the commission in Bombay City. J Hygiene. 7:724-798.
5142. Ingles, LG. 1941. Natural history observations on the Audubon cottontail. J Mamm. 22:227-250.
5143. Ingles, LG. 1947. Ecology and life history of the California gray squirrel. CA Game Fish. 33:139-158.
5144. Ingles, LG. 1949. Field observations on the growth of young mountain pocket gophers. Murrelet. 30:35-36.
5145. Ingles, LG. 1952. The ecology of the mountain pocket gopher, *Thomomys monticola*. Ecology. 33:87-95.
5146. Ingles, LG. 1965. Mammals of the Pacific States. Stanford Univ Press, Stanford, CA.
5147. Ingram, JC. 1975. Husbandry and observation methods of a breeding colony of marmosets (*Callithrix jacchus*) for behavioral research. Lab Anim. 9:249-259.
5147a. Innes, DGL, Bendell, JF, Naylor, BJ, et al. 1990. High densities of the masked shrew, *Sorex cinereus*, in jack pine plantations in northern Ontario. Am Mid Nat. 124:330-341.
5148. Innes, DGL & Millar, JS. 1979. Growth of *Clethrionomys gapperi* and *Microtus pennsylvanicus* in captivity. Growth. 43:208-217.

5149. Innes, DGL & Millar, JS. 1981. Body weight, litter size, and energetics of reproduction in *Clethrionomys gapperi* and *Microtus pennsylvanicus*. Can J Zool. 59:785-789.
5150. Innes, DGL & Millar, JS. 1982. Life-history notes on the heather vole, *Phenacomys intermedius levis*, in the Canadian Rocky Mountains. Can Field-Nat. 96:307-311.
5151. Innes, DGL & Millar, JS. 1987. The mean number of litters per breeding season in small mammal populations: A comparison of methods. J Mamm. 68:675-678.
5152. Innes, S, Stewart, REA, & Lavigne, DM. 1981. Growth in northwest Atlantic harp seals *Phoca groenlandica*. J Zool. 194:11-24.
5153. Innis, AC. 1951. The behaviour of the giraffe, *Giraffa camelopardalis*, in the eastern Transvaal. Proc Zool Soc London. 131:245-278.
5154. Inns, RW. 1980. Occurrence of twins in macropod marsupials. Search. 11:118-119.
5155. Inns, RW. 1982. Seasonal changes in the accessory reproductive system and plasma testosterone levels of the male tammar wallaby, *Macropus eugenii*, in the wild. J Reprod Fert. 66:675-680.
5156. Inoue, M & Hayama, S. 1961. Histological studies of sexual cycles in the Japanese monkey (*Macaca fuscata*). Primates. 3:76.
5157. International Whaling Comm. 1980. Report of the special meeting on southern hemisphere minke whales, Seattle, May 1978. Rep Int Whal Comm. 29:349-358.
5158. Intress, C & Best, TL. 1990. *Dipodomys panamintinus*. Mamm Species. 354:1-7.
5159. Ioannou, JM. 1966. The oestrous cycle of the potto. J Reprod Fert. 11:455-457.
5160. Ioannou, JM. 1967. Oogenesis in adult prosimians. J Embryol Exp Morph. 17:139-145.
5161. not used.
5162. not used.
5163. Irby, DC, Kerr, JB, Risbridger, GP, et al. 1984. Seasonally and experimentally induced changes in testicular function of the Australian bush rat (*Rattus fuscipes*). J Reprod Fert. 70:657-666.
5164. Irby, LR. 1973. A preliminary report on the mountain reedbuck (*Redunca fulvowfula*) in the Loskop Dam Nature Reserve. J S Afr Wildl Manage Assn. 3:53-58.
5165. Irby, LR. 1979. Reproduction in mountain reedbuck (*Redunca fulvorufula*). Mammalia. 43:191-213.
5166. Isenbugel, E. 1968. Seehundgeburt im Aquarium des Botanischen Gartens der Stadt Essen. Fr Kölner Zoo. 11:25-26.
5167. Ishida, K. 1968. Age and seasonal changes in the testis of the ferret. Arch Histol Japan. 29:193-205.
5168. Ishii, N. 1982. Reproductive activity of the Japanese shrew-mole *Urotrichus talpoides* Pemminck. J Mamm Soc Japan. 9:25-36.
5169. Ishimaru, A. 1978. Breeding Grant's zebra. Animals & Zoos. 30(1):6-9.
5170. Ising, E & Niethammer, J. 1979. Zur Fortpflanzung der Gelbhalsmaus (*Apodemus flavicollis*) im Laboratorium. Z Säugetierk. 44:25-30.
5171. Isjumov, GJ. 1961. Einige Angaben über Wachstum und Enwicklung des Walrosses im Leningrader Zoologischen Garten. Zool Garten NF. 26:82-95.
5172. Ismaghlov, MI. 1956. Contribution to the ecology of *Marmota bobac centralis* Thom. Zool Zhurn. 35:908-915.
5173. Ismail, ST. 1987. A review of reproduction in the female camel (*Camelus dromedarius*). Theriogenology. 28:363-371.
5174. Ismail, ST. 1988. Reproduction in the male dromedary (*Camelus dromedarius*). Theriogenology. 29:1407-1418.
5175. Issel, B & Issel, W. 1953. Zur Verbreitung und Lebensweise der gewimperten Fledermaus, *Myotis emarginatus* (Geoffroy, 1806). Säugetierk Mitt. 1:145-148.
5176. Itayama, S & Shimura, R. 1978. Lesser pandas born. Animals & Zoos. 30(10):6-9.
5177. Ito, T. 1966. Breeding the Japanese serow. Animals & Zoos. 18(4):92-93.
5178. Ito, T. 1971. On the oestrous cycle and gestation period of the Japanese serow, *Capricornis crispus*. J Mamm Soc Japan. 5:104-108.
5179. Ivanova, GM. 1964. Chemical composition and nutritive value of elk's milk. Trudy Pechoro-Ilychsk gos Zapovedn. 11:55-60. (Abstract in Dairy Sci Abst 27:55,1965).
5180. Ivashin, MV. 1963. An instance of extra-uterine pregnancy in *Balaenoptera borealis* Lesson. Zool Zhurn. 42:1275-1277.
5181. Ivashin, MV. 1976. On the reproduction of *Balaenoptera acutorosrata* in the Indian sector of Antarctic. Zool Zhurn. 55:893-903.
5182. Ivashni, MV & Mikhalev, YUA. 1978. To the problem of the prenatal growth of minke whales *Balaenoptera acutorostrata* of the southern hemisphere and of the biology of their reproduction. Rep Int Whal Comm. 28:201-205.
5183. Iversen, JA. 1972. Basal metabolic rate of wolverines during growth. Norw J Zool. 20:317-322.
5184. Iverson, SL. 1967. Adaptations to arid environments in *Perognathus parvus* (Peale). Ph D Diss. Univ BC, Vancouver.
5185. Iverson, SL & Turner, BN. 1972. Natural history of a Manitoba population of Franklin's ground squirrels. Can Field-Nat. 86:145-149.
5186. Iverson, SL & Turner, BN. 1973. Ecological notes on Manitoba *Napaeozapus insignis*. Can Field-Nat. 87:15-19.
5187. Ivey, RD. 1949. Life history notes on three mice from the Florida east coast. J Mamm. 30:157-162.
5188. Ivey, RD. 1959. The mammals of Palm Valley, Florida. J Mamm. 40:585-591.
5189. Iwamoto, Y & Kobayashi, N. 1976. Breeding of Hartmann's mountain zebra. Animals & Zoos. 28(2):6-9.
5190. Iwata, H & Ishii, T. 1946. Studies on fur animals: On the milk of the raccoon dog. Japan J Zootech Sci. 17:99-101.
5190a. Iyawe, JG. 1988. Distribution of small rodents and shrews in a lowland rain forest zone of Nigeria, with observations on their reproductive biology. Afr J Ecol. 26:189-195.
5191. Izard, J & Umfleet, K. 1971. Notes on the care and breeding of the suni *Nesotragus moschatus* at Dallas Zoo. Int Zoo Yearb. 11:129.
5192. Izard, MK. Personal communication to V Hayssen, 1985.
5193. Izard, MK. 1987. Lactation length in three species of *Galago*. Am J Primatol. 13:73-76.
5194. Izard, MK & Fail, PA. 1988. Progesterone levels during pregnancy in the greater thick-tailed galago (*Galago crassicaudatus*). J Med Primatol. 17:125-133.
5195. Izard, MK & Rasmussen, DT. 1985. Reproduction in the slender loris (*Loris tardigradus malabaricus*). Am J Primatol. 8:153-165.
5196. Izard, MK & Simons, EL. 1986. Infant survival and litter size in primigravid and multigravid *Galago*. J Med Primatol. 15:27-35.
5197. Izard, MK & Simons, EL. 1987. Lactation and interbirth interval in the Senegal galago (*Galago senegalensis moholi*). J Med Primatol. 16:323-332.
5198. Izard, MK, Weisenseel, KA, & Ange, RL. 1988. Reproduction in the slow loris (*Nycticebus coucang*). Am J Primatol. 16:331-339.

5199. Izard, MK, Wright, PC, & Simons, EL. 1985. Gestation length in *Tarsius bancanus*. Am J Primatol. 9:327-331.
5200. Izawa, K. 1979. Studies on peculiar distribution pattern of *Callimico*. Kyoto Univ Overseas Res Rep New World Monkeys. 1979:1-19.
5201. Izor, RJ & Pine, RH. 1987. Notes on the black-shouldered opossum, *Caluromysiops irrupta*. Fieldiana Zool NS. 39:117-124.
5202. Jachmann, H. 1980. Population dynamics of the elephants in Kasungu National Park, Malawi. Neth J Zool. 30:622-634.
5203. Jackson, FJ. 1899. Bohor reedbuck (*Cervicapra bohor*) in east Africa. Great and Small Game of Africa. 310-312, Bryden, HA, ed, Rowland Ward, London.
5204. Jackson, FJ. 1899. Eastern race (*Tragelaphus scriptus roualeyni*) in east Africa. Great and Small Game of Africa. 481-483, Bryden, HA, ed, Rowland Ward, London.
5205. Jackson, FJ. 1899. The Cape buffalo (*Bos caffer typicus*) in British east Africa. Great and Small Game of Africa. 112-115, Bryden, HA, ed, Rowland Ward, London.
5206. Jackson, FJ. 1899. The defassa sing-sing (*Cobus defassa typicus*) in British east Africa. Great and Small Game of Africa. 282-285, Bryden, HA, ed, Rowland Ward, London.
5207. Jackson, FJ. 1899. The fringe-eared beisa (*Oryx callotis*). Great and Small Game of Africa. 391-393, Bryden, HA, ed, Rowland Ward, London.
5208. Jackson, HA, Guynn, DC, Jr, Griffin, RN, et al. 1979. Fecundity of white-tailed deer in Mississippi and periodicity of corpora lutea and lactation. Proc Ann Conf SE Fish Wildl Ag. 33:30-35.
5209. Jackson, HHT. 1915. A review of the American moles. N Am Fauna. 38:1-100.
5210. Jackson, HHT. 1961. Mammals of Wisconsin. Univ WI Press. Madison, WI.
5211. Jackson, J. 1977. The duration of lactation in New Forest fallow deer (*Dama dama*). J Zool. 183:542-543.
5212. Jackson, JE. 1987. *Ozotoceros bezoarticus*. Mamm Species. 295:1-5.
5212a. Jackson, JE & Langguth, A. 1987. Ecology and status of the pampas deer *Ozotoceros bezoarticus*) in the Argentinian pampas and Uruguay. Biology and Management of the Cervidae. 402-409, Wemmer, CM, ed, Smithsonian Inst Press, Washington, DC.
5213. Jackson, LW & Hesselton, WT. 1972. Breeding and parturition dates for the various age classes of female white-tailed deer (*O virginianus borealis*) in New York. Proc Ann NE Fish Wild Conf. 28:21-35.
5214. Jackson, LW & Hesselton, WT. 1973. Breeding and parturition dates of white-tailed deer in New York. NY Game Fish J. 20:40-47.
5215. Jackson, MR & Edmunds, JG. 1984. Morphological assesssment of testicular maturity in marmosets (*Callithrix jacchus*). Lab Anim. 18:173-178.
5216. Jackson, WB. 1962. Population studies: D Reproduction. Pacific Island Rat Ecology. 92-107, Storer, TI, ed, Bull BP Bishop Mus, Honolulu.
5217. Jackson, WB. 1965. Litter size in relation to latitude in two murid rodents. Am Mid Nat. 73:245-247.
5218. Jacob, S & Podar, S. 1986. Morphology and histochemistry of the ferret prostate. Acta Anat. 125:268-273.
5219. Jacobi, EF. 1972. Raising aardvark (*Orycteropus afer* Pallas) in Amsterdam Zoo. Zool Garten NF. 41:209-214.
5220. Jacobi, EF. 1975. Breeding sloth bears in Amsterdam zoo. Breeding Endangered Species in Captivity. 351-360, Martin, RD, ed, Academic Press, New York.
5221. Jacobi, EF & van Bemmel, AC. 1968/69. Breeding of cape hunting dogs (*Lycaon pictus* [Temminck]) at Amsterdam and Rotterdam Zoo. Zool Garten NF. 36:90-94.
5222. Jacobs, W. 1966. Nachgewiesene Superfötation beim Feldhasen (*Lepus europaeus*). Z Jagdwiss. 12:138.
5223. Jacobsen, NHG. 1974. Distribution, homerange and behaviour patterns of bushbuck in the Lutope and Sangwa Valleys, Rhodesia. J S Afr Wildl Mgt Assoc. 4:75-93.
5224. Jacobsen, NHG. 1982. A note of mating and neonate development of the South African hedgehog (*Erinaceus frontalis*). Säugetierk Mitt. 30:199-200.
5225. Jacobsen, NHG. 1982. Observations on the behaviour of slender mongooses, *Herpestes sanguineus*, in captivity. Säugetierk Mitt. 30:168-183.
5226. Jacobsen, NHG & du Plessis, E. 1976. Observations on the ecology and biology of the Cape fruit bat *Rousettus aegyptiacus lechi* in the eastern Transvaal. S Afr J Sci. 72:270-273.
5226a. Jacobsen, NHG, Newbery, RE, de Wet, MJ, et al. 1991. A contribution of the ecology of the steppe pangolin *Manis temminckii* in the Transvaal. Z Säugetierkd. 56:94-100.
5227. Jacobsen, WC. 1923. Rate of reproduction in *Citellus beecheyi*. J Mamm. 4:58.
5228. Jacobson, HN & Windle, WF. 1960. Observations on mating, gestation, birth and postnatal development of *Macaca mulatta*. Biol Neonat. 2:105-120.
5229. Jaczewski, Z. 1958. Reproduction of the European bison, *Bison bonasus* (L) in reserves. Acta Theriol. 1:333-376.
5229a. Jaczewski, Z & Galka, B. 1970. Effects of administration of testosteronum propionicum on the antler cycle in red deer. Finnish Game Res. 30:303-308.
5230. Jain, AP. 1970. Body weights, sex ratio, age structure and some aspects of reproduction in the Indian gerbil, *Tatera indica indica* Hardwicke, in the Rajasthan desert, India. Mammalia. 34:415-432.
5231. Jain, AP. 1984. Relationship between body weight of mother and size of litter in Antelope rat, *Tatera i indica* Hardwicke. Mammalia. 48:143.
5232. Jain, MS. 1980. Observations on birth of a musk deer fawn. J Bombay Nat Hist Soc. 77:497-498.
5233. Jainudeen, MR, Eisenberg, JF, & Tilakeratne, N. 1971. Oestrous cycle of the Asiatic elephant, *Elephas maximus*, in captivity. J Reprod Fert. 27:321-328.
5234. Jainudeen, MR & Hafez, ESE. 1973. Egg transplant in the macaque (*Macaca fascicularis*). Biol Reprod. 9:305-308.
5235. Jainudeen, MR, Katongole, CB, & Short, RV. 1972. Plasma testosterone levels in relation to musth and sexual activity in the male Asiatic elephant, *Elephas maximus*. J Reprod Fert. 29:99-103.
5236. Jallageas, M & Assenmacher, I. 1983. Annual plasma testosterone and thyroxine cycles in relation to hibernation in the edible dormouse *Glis glis*. Gen Comp Endocrinol. 50:452-462.
5236a. Jallageas, M & Assenmacher, I. 1984. External factors controlling annual testosterone and thyroxine cycles in the edible dormouse *Glis glis*. Comp Biochem Physiol. 77A:161-167.
5237. Jallageas, M & Assenmacher, I. 1986. Effects of castration and thyroidectomy on the annual biological cycles of the edible dormouse *Glis glis*. Gen Comp Endocrinol. 63:301-308.
5238. James, LH. 1914. Birth of a porpoise at the Brighton Aquarium. Proc Zool Soc London. 1914:1061-1062.
5239. James, TR & Seabloom, RW. 1969. Reproductive biology of the white-tailed jack rabbit in North Dakota. J Wildl Manage. 33:558-568.

5240. James, WB & Booth, ES. 1954. Biology and life history of the sagebrush vole. Walla Walla Coll Publ Dept Biol Sci Biol Sta. 4:1-21.
5241. Jameson, EW, Jr. 1947. Natural history of the prairie vole (mammalian genus *Microtus*). Univ KS Publ Mus Nat Hist. 1:125-151.
5242. Jameson, EW, Jr. 1950. Determining fecundity in male small mammals. J Mamm. 31:433-436.
5243. Jameson, EW, Jr. 1953. Reproduction of deer mice (*Peromyscus maniculatus* and *P boylei*) in the Sierra Nevada, California. J Mamm. 34:44-58.
5244. Jameson, EW, Jr. 1955. Observations on the biology of *Sorex trowbridgei* in the Sierra Nevada, California. J Mamm. 36:339-345.
5245. Jameson, EW, Jr & Peeters, HJ. 1988. California Mammals. Univ CA Press. Berkeley, CA.
5246. Jameson, RJ & Bodkin, JL. 1986. An incidence of twinning in the sea otter (*Enhydra lutris*). Mar Mamm Sci. 2:305-309.
5247. Jamon, M. 1986. The dynamics of wood mouse (*Apodemus sylvaticus*) populations in the Camargue. J Zool. 208A:569-582.
5248. Janecek, J. 1971. Acclimatisation and breeding of roan antelopes *Hippotragus equinus* at Dvur Kralove zoo. Int Zoo Yearb. 11:127-128.
5249. Janecek, J. 1975. Some notes on the breeding of hunter's antelope *Damaliscus hunteri* in captivity. Int Zoo Yearb. 15:166-169.
5250. Jannett, FJ, Jr. 1980. Social dynamics of the Montane vole, *Microtus montanus*, as a paradigm. Biologist. 62:3-19.
5251. Jannett, FJ, Jr. 1984. Reproduction of the Montane vole, *Microtus montanus* in subnivean populations. Sp Publ Carnegie Mus Nat Hist. 10:215-224.
5252. Jannett, FJ, Jr, Jannett, JA, & Richmond, ME. Notes on reproduction in captive *Arvicola richardsoni*. J Mamm. 60:834-837.
5253. Jansen, J. 1953. Studies on the Cetacean brain. Hvålrad Skr. 37:1-35.
5254. Janson, CH. 1984. Female choice and mating system of the brown capuchin monkey *Cebus apella* (Primates: Cebidae). Z Tierpsychol. 65:177-200.
5254a. Janson, RG. 1946. A survey of the native rabbits of Utah with references to their classification, distribution, life histories and ecology. Masters Thesis, UT State Agric College, UT.
5255. Janssens, PA, Jenkinson, LA, Paton, BC, et al. 1977. The regulation of gluconeogenesis in pouch young of the tammar wallaby, *Macropus eugenii* (Desmarest). Aust J Biol Sci. 30:183-195.
5256. Jantschke, F. 1973. On the breeding and rearing of bush dogs *Speothos venaticus* at Frankfurt zoo. Int Zoo Yearb. 13:141-143.
5257. Jantschke, F. 1975. The maintenance and breeding of pygmy chimpanzees. Breeding Endangered Species in Captivity. 245-251, Martin, RD, ed, Academic Press, New York.
5258. Jarman, MV. 1976. Impala social behavior: Birth behaviour. E Afr Wildl J. 14:153-167.
5259. Jarman, PJ & Jarman, MV. 1973. Social behaviour, population structure and reproductive potential in impala. E Afr Wildl J. 11:329-338.
5260. Jarman, PJ & Jarman, MV. 1974. Impala behaviour and its relevance to management. The Behaviour of Ungulates and its Relation to Management. 871-881, Geist, V & Walther, F, eds, IUCN, Morges, Switzerland.
5261. Jaroli, DP & Lall, SB. 1981. A comparison of ovarian follicular types and their enzymological characteristic in a mulliparous and pregnant *Taphozous melanopogon melanopogon* Temmnick [*sic*] (Microchiroptera: Mammalia). Bat Res News. 22:51.
5262. Jaroli, DP & Lall, SB. 1987. Comparison of histoenzymological alterations in the contralateral ovary of nulliparous and parous females of the bat *Taphozous melanopogon melanopogon* Temmnick [*sic*] (Microchiroptera) displaying absolute dextral dominance of the genital tract. Acta Anat. 129:140-148.
5263. Jarosz, S. 1973. The sexual cycle in chinchilla (*Chinchilla velligera*). Zool Pol. 23:119-128.
5264. Jarosz, SJ & Dukelow, WR. 1985. Effect of progesterone and medroxyprogesterone acetate on pregnancy length and litter size in mink. Lab Anim Sci. 35:156-161.
5265. Jarosz, SJ, Kuehl, TJ, & Dukelow, WR. 1977. Vaginal cytology, induced ovulation and gestation in the squirrel monkey (*Saimiri sciureus*). Biol Reprod. 16:97-103.
5266. Jarvis, JUM. 1969. Some aspects of the biology of east African mole rats. Ph D Diss. Univ East Africa, Nairobi.
5267. Jarvis, JUM. 1969. The breeding season and litter size of African mole-rats. J Reprod Fert Suppl. 6:237-248.
5268. Jarvis, JUM. 1973. The structure of a population of mole rats, *Tachyoryctes splendens*, (Rodentia: Rhizomyidae). J Zool. 171:1-14.
5269. Jarvis, JUM. 1978. Energetics of survival in *Heterocephalus glaber* (Ruppell), the naked mole-rat (Rodentia: Bathyergidae). Bull Carnegie Mus Nat Hist. 6:81-87.
5270. Jarvis, JUM. 1981. Eusociality in a mammal: Cooperative breeding in naked mole-rat colonies. Science. 212:571-573.
5270a. Jarvis, JUM. 1991. Reproduction in naked mole rats. The Biology of the Naked Mole-Rat. 384-425, Sherman, PW, Jarvis, JUM, & Alexander, RD, eds Princeton Univ Press, Princeton, NJ.
5270b. Jarvis, JUM & Bennett, NC. 1991. Ecology and behavior of the family Bathyergidae. The Biology of the Naked Mole-Rat. 66-96, Sherman, PW, Jarvis, JUM, & Alexander, RD, eds, Princeton Univ Press, Princeton, NJ.
5271. Jay, P. 1962. Aspects of maternal behaviour among langurs. Ann NY Acad Sci. 102(2):468-476.
5272. Jay, P. 1963. Ecologie et comportement social du Langur commun des Indes, *Presbytis entellus*. Terre Vie. 17:50-65.
5273. Jay, P. 1965. The common langur of North India. Primate Behavior Field Studies of Monkeys and Apes. 197-249, DeVore, I, ed, Holt, Rinehart & Winston, New York.
5274. Jayakar, SD & Spurway, H. 1968. Notes on the common palm civet or toddy cat *Paradoxurus hermaphroditus* (Pallus), with special reference to the age at shedding of the milk teeth. J Bombay Nat Hist Soc. 65:211-214.
5275. Jayewardene, EDW. 1975. Breeding the fishing cat *Felis viverrina* in captivity. Int Zoo Yearb. 15:150-152.
5276. Jean, Y & Bergeron, J-M. 1984. Productivity of coyotes (*Canis latrans*) from southern Quebec. Can J Zool. 62:2240-2243.
5276a. Jeanmaire-Besançon, F. 1988. Sexual maturity of *Crocidura russula* (Insectivora: Soricidae). Acta Theriol. 33:477-485.
5277. Jeannin, A. 1936. Les Mammifères Sauvages du Cameroun. Encycl Biol (Paris) vol 16. 1-252, Paul Lechevalier, Paris.
5277a. Jędrzejewska, B. Reproduction in weasels *Mustela nivalis* in Poland. Acta Theriol. 32:493-496.

5278. Jefferson, TA. 1975. *Phocoena dioptrica.* Mamm Species. 66:1-3.
5279. Jefferson, TA. 1988. *Phocoenoides dalli.* Mamm Species. 319:1-7.
5280. Jelińek, P, Pichová, D, & Pícha, J. 1981. The levels of plasma testosterone in coypus during postnatal development. Zivocisna Vyrobá. 26:225-234. [Abstr in Scientifur, 6:26-27, 1981]
5281. Jemiolo, B. 1983. Ovulation and fertilization in the vole *Pitymys subterraneus.* Biol Reprod. 28:523-527.
5282. Jemiolo, B. 1983. Reproduction in a laboratory colony of the female pine voles, *Pitymys subterraneus.* Acta Theriol. 28:197-207.
5283. Jemiolo, B. 1987. Male-related activation of reproduction in adult female European pine voles (*Pitymys subterraneus*). J Mamm. 68:714-717.
5284. Jemiolo, B, Marchlewski-Koj, A, & Buchalczyk, A. 1980. Acceleration of ovarian follicle maturation of female caused by male in *Microtus agrestis* and *Clethrionomys glareolus.* Folia Biol (Krakow). 28:270-272.
5285. Jenkin, G, Mitchell, MD, Hopkins, P, et al. 1980. Concentrations of melatonin in the plasma of the rhesus monkey (*Macaca mulatta*). J Endocrinol. 84:489-494.
5286. Jenkins, D. 1980. Ecology of otters in northern Scotland I Otter (*Lutra lutra*) breeding and dispersion in mid-Deeside, Aberdeenshire in 1974-79. J Anim Ecol. 49:713-735.
5287. Jenkins, D & Harper, RJ. 1982. Fertility in European otters *Lutra lutra.* J Zool. 197:299-300.
5288. Jenkins, GA, Bradbury, JH, Messer, M, et al. 1984. Determination of the structures of fucosyl-lactose and difucosyl-lactose from the milk of monotremes, using ^{13}C-nmr spectroscopy. Carbohydrate Res. 126:157-161.
5289. Jenkins, SH & Busher, PE. 1979. *Castor canadensis.* Mamm Species. 120:1-8.
5290. Jenkins, SH & Eshelman, BD. 1984. *Spermophilus beldingi.* Mamm Species. 221:1-8.
5291. Jenness, R. 1986. Lactational preformance of various mammalian species. J Dairy Sci. 69:869-885.
5292. Jenness, R, Erickson, AW, & Craighead, JJ. 1972. Some comparative aspects of milk from four species of bears. J Mamm. 53:34-47.
5293. Jenness, R & Odell, DK. 1978. Composition of milk of the pygmy sperm whale (*Kogia breviceps*). Comp Biochem Physiol. 61A:383-386.
5294. Jenness, R, Regehr, EA, & Sloan, RE. 1964. Comparative biochemical studies of milks II Dialyzable carbohydrates. Comp Biochem Physiol. 13:339-352.
5295. Jenness, R & Sloan, RE. 1970. The composition of milks of various species: A review. Dairy Sci Abstr. 32:599-612.
5296. Jenness, R & Studier, EH. 1976. Lactation and milk. Sp Publ Mus TX Tech Univ. 10:201-218.
5297. Jenness, R, Williams, TD, & Mullin, RJ. 1981. Composition of milk of the sea otter (*Enhydra lutris*). Comp Biochem Physiol. 70A:375-379.
5298. Jennings, HD & Harris, JT. 1953. The collared peccary in Texas. TX Game Fish Comm FA Rep Ser. 12:1-31.
5299. Jennison, G. 1927. Table of Gestation Periods and Number of Young. A & C Black, Ltd, London.
5300. Jennov, JG. 1933. Der Moschusochse in Ost-Grönland. Z Säugetierk. 8:40-46.
5301. Jensen, PM & Gustafsson, TO. 1984. Evidence for pregnancy failure in young female *Lemmus lemmus* and *Microtus oeconomus.* Can J Zool. 62:2568-2570.
5302. Jensen, TS. 1975. Population estimations and population dynamics of two Danish forest rodent species. Vidensk Meddr Dansk Naturh Foren. 138:65-86.
5303. Jensen, W & Robinette, WL. 1955. A high reproductive rate for Rocky Mountain mule deer (*Odocoileus h hemionus*). J Wildl Manage. 19:503.
5304. Jentink, FA. 1888. Zoological researches in Liberia: A list of mammals collected by J Buttikofer, CF Sala and FX Stampfli, with biological observations. Notes Leyden Mus. 10:1-58.
5305. Jerrett, DP. 1977. Structural-functional differences in the right and left ovaries in the Mexican free-tailed bat, *Tadarida brasiliensis mexicana.* Anat Rec. 187:615.
5306. Jerrett, DP. 1979. Female reproductive patterns in non hibernating bats. J Reprod Fert. 56:369-378.
5307. Jewell, PA. 1966. Breeding season and recruitment in some British mammals confined on small islands. Symp Zool Soc London. 15:89-116.
5308. Jewell, PA. 1972. Social organization and movements of topi (*Damaliscus korrigum*) during the rut, at Ishasha, Queen Elizabeth Park, Uganda. Zool Afr. 7:233-235.
5309. Jewell, PA & Oates, JF. 1969. Breeding activity in prosimians and small rodents in West Africa. J Reprod Fert Suppl. 6:23-38.
5310. Jewett, DA & Dukelow, WR. 1972. Cyclicity and gestation length of *Macaca fascicularis.* Primates. 13:327-332.
5311. Jewett, SG. 1923. A breeding record of *Phenacomys longicaudus.* J Mamm. 4:125.
5312. Jewett, SG. 1923. A breeding record of *Citellus mollis.* J Mamm. 4:191.
5312a. Jiang, H, LLiu, Z, Yuan, X, et al. 1988. Study on the reproduction of rhesus monkey in nature at Nanwan Peninsula, Hainan Island. Acta Theriol Sinica. 8:105-112.
5313. Jierong, L, Jinxiang, Z, Zuwang, W, et al. 1982. Studies on growth and development in the root vole (*Microtus oeconomus*). Acta Biol Plateau Sinica. 1:195-207.
5314. Jing, Z & Yangwen, L. 1980. The Giant Panda. Science Press, Beijing, China.
5315. Jirsik, J. 1955. Die Hausratte, *Rattus rattus* (Linné, 1758), wieder in der Tschechoslowakei. Säugetierk Mitt. 3:21-28.
5316. Joachimovits, R. 1928. Studien zu Menstruation, Ovulation, Aufbau und Pathologie des weiblichen Genitales bei Mensch und Affe (*Pithecus fasciculris mordax*). Biol Gen. 4:447-540.
5317. Joffre, J & Joffre, M. 1973. Seasonal changes in the testicular blood flow of seasonally breeding mammals: Dormouse, *Glis glis*, ferret, *Mustella furo*, and fox, *Vulpes vulpes.* J Reprod Fert. 34:227-233.
5318. Joffre, M. 1976. Puberté et cycle génital saisonnier du Renard mâle (*Vulpes vulpes*). Ann Biol Anim Bioch Biophys. 16:503-520.
5319. Joffre, M. 1977. La capsule testiculaire du Renard roux (*Vulpes vulpes* L): Relation avec l'activité testiculaire pendant la période prepubère et au cours du cycle saisonnier. Ann Biol Anim Bioch Biophys. 17:695-712.
5320. Joffre, M. 1977. Relationship between testicular blood flow, testosterone secretion and spermatogenic activity in young and adult wild red foxes (*Vulpes vulpes*). J Reprod Fert. 51:35-40.
5321. Johansson, EDB, Neill, JD, & Knobil, E. 1968. Periovulatory progesterone concentration in the peripheral plasma of the rhesus monkey with a methodologic note on the detection of ovulation. Endocrinology. 82:143-148.
5322. Johansson, I. 1938. Reproduction in the silver fox. Lantbruks högskolans Ann. 5:179-200.

5323. Johansson, I. 1941. Oestrus and mating in the silver fox. Lantbruks högskolans Ann. 9:239-263.
5324. Johansson, I. 1947. The inheritance of the platinum and whiteface characters in the fox. Hereditas. 33:152-174.
5325. Johansson, I. 1953. The influence of the mating method on the breeding results of standard mink. Våra Pälsdjur. 24:48-54.
5326. Johansson, I & Venge, O. 1951. Relation of the mating interval to the occurrence of superfetation in mink. Acta Zool (Stockholm). 32:255-258.
5327. Johns, A & Coons, LW. 1981. Physiological and pharmacological characteristics of the baboon (*Papio anubis*) oviduct. Biol Reprod. 25:120-127.
5328. Johns, DW & Armitage, KB. 1979. Behavioral ecology of alpine yellow-bellied marmots. Behav Ecol Sociobiol. 5:133-157.
5329. Johns, PE, Baccus, R, Manlove, MN, et al. 1977. Reproductive patterns, productivity and genetic variability in adjacent white-tailed deer populations. Proc Ann SE Conf Fish Wildl Ag. 32:167-172.
5330. Johnsingh, AJT. 1982. Reproductive and social behaviour of the dhole *Cuon alpinus* (Canidae). J Zool. 198:443-463.
5331. Johnsingh, AJT. 1983. Large mammalian prey-predators in Bandipur. J Bombay Nat Hist Soc. 80:1-57.
5332. Johnson, AM, Delong, RL, Fiscus, CH, et al. 1982. Population status of the Hawaiian monk seal (*Monachus schauinslandi*), 1978. J Mamm. 63:415-421.
5332a. Johnson, AS. 1970. Biology of the raccoon (*Procyon lotor varius* Nelson and Goldman) in Alabama. Bull Agric Exp Sta Auburn Univ. 402:1-148.
5333. Johnson, CE. 1922. Notes on the mammals of northern Lake county, Minnesota. J Mamm. 3:33-39.
5334. Johnson, CE. 1926. Notes on a pocket gopher in captivity. J Mamm. 7:35-37.
5335. Johnson, CN. 1987. Relationships between mother and infant red-necked wallabies (*Macropus rufogriseus banksianus*). Ethology. 74:1-20.
5336. Johnson, CN & Crossman, DG. 1991. Sexual dimorphism in the northern hairy-nosed wombat, *Lasiorhinus krefftii* (Marsupialia: Vombatidae). Aust Mamm. 14:145-146.
5337. Johnson, CN & Johnson, KA. 1983. Behavior of the bilby, *Macrotis lagotis* (Reid), (Marsupialia: Thylacomyidae) in captivity. Aust Wildl Res. 10:77-87.
5338. Johnson, DE. 1951. Biology of the elk calf, *Cervis canadensis nelsoni*. J Wildl Manage. 15:396-410.
5339. Johnson, DF, Modahl, KB, & Eaton, GG. 1982. Dominance status of adult male Japanese macaques: Relationship to female dominance status, male mating behaviour, seasonal changes, and developmental changes. Anim Behav. 30:383-392.
5340. Johnson, DH. 1964. Mammals of the Arnhem Land Expedition. Records of the American-Australian Expedition to Arnhem Land. 427-515, Specht, RL, ed, Melbourne Univ Press, Melbourne.
5341. Johnson, DW & Armstrong, DM. 1987. *Peromyscus crinitus*. Mamm Species. 287:1-8.
5342. Johnson, GE. 1927. Observations on young prairie-dogs (*Cynomys ludovicianus*) born in the laboratory. J Mamm. 8:110-115.
5343. Johnson, GE. 1931. Early life of the thirteen-lined ground squirrel. Trans KS Acad Sci. 34:282-291.
5344. Johnson, GE & Challans, JS. 1932. Ovariectomy and corpus luteum extract studies on rats and ground squirrels. Endocrinology. 16:278-284.
5345. Johnson, GE, Foster, MA, & Coco, RM. 1933. The sexual cycle of the thirteen-lined ground squirrel in the laboratory. Trans KS Acad Sci. 36:250-269.
5346. Johnson, GE & Wade, NJ. 1931. Laboratory reproductive studies on the ground squirrel (*Citellus tridecemlineatus pallidus*, Allen). Biol Bull. 61:101-114.
5347. Johnson, K. 1981. Social organization in a colony of rock squirrels (*Spermophilus variegatus*, Sciuridae). Southwest Nat. 26:237-242.
5348. Johnson, KA & Roff, AD. 1980. Discovery of ningauis (*Ningaui* sp Dasyruidae: Marsupialia) in the Northern Territory, Australia. Aust Mamm. 3:127-129.
5349. Johnson, KG, Duncan, RW, & Pelton, MB. 1982. Reproductive biology of European wild hogs in the Great Smoky Mountains National Park. Proc SE Assoc Fish Wildl Ag. 36:552-564.
5350. Johnson, LJ & LeRoux, P. 1973. Age of self-sufficiency in brown/grizzly bear in Alaska. J Wildl Manage. 37:122-123.
5351. Johnson, ML. 1989. Exploratory behavior and dispersal: A graphical model. Can J Zool. 67:2325-2328.
5352. Johnson, ML & Clanton, CW. 1954. Natural history of *Sorex merriami* in Washington state. Murrelet. 35:1-4.
5353. Johnson, ML, Fiscus, CH, Ostenson, BT, et al. 1966. Marine mammals. Environment of the Cape Thompson Region, Alaska. 877-924, Wilimovsky, NJ & Wolfe, JN, eds, US Atomic Energy Comm, Oak Ridge.
5354. Johnson, ML, Taylor, RH, & Winnich, NW. 1975. The breeding and exhibition of capromyid rodents at Tacoma Zoo. Int Zoo Yearb. 15:53-56.
5355. Johnson, NF & Holloran, DJ. 1985. Reproductive activity of Kansas bobcats. J Wildl Manage. 49:42-46.
5356. Johnson, OW. 1963. Histological and quantitative characteristics of the testes, observations on the teeth and pituitary gland, and the possibility of reproductive cyclicity in the African elephant (*Loxodonta africana*). Ph D Diss, WA State Univ.
5357. Johnson, OW & Buss, IO. 1967. The testis of the African elephant (*Loxodonta africana*) I Histological features. J Reprod Fert. 13:11-21.
5358. Johnson, OW & Buss, IO. 1967. The testis of the African elephant (*Loxodonta africana*) II Development, puberty and weight. J Reprod Fert. 13:23-30.
5359. Johnson, PM. 1978. Studies of Macropodidae in Queensland 9 Reproduction of the rufous rat-kangaroo (*Aepyprymnus rufescens* (Gray)) in captivity with age estimation of pouch young. Queensl J Agric Anim Sci. 35:69-72.
5360. Johnson, PM. 1979. Reproduction in the plain rock-wallaby, *Petrogale penicillata inornata* Gould, in captivity, with age estimation of the pouch young. Aust Wildl Res. 6:1-4.
5361. Johnson, PM. 1980. Observations of the behavior of the rufous rat-kangaroo, *Aepyprymnus rufescens* (Gray), in captivity. Aust Wildl Res. 7:347-357.
5362. Johnson, PM & Lyon, BJ. 1985. The spectacled hare-wallaby in Queensland. QLD Agric J. 111:116.
5363. Johnson, PM & Strahan, R. 1982. A further description of the musky rat-kangaroo, *Hypsiprymnodon moschatus* Ramsay, 1876 (Marsupialia, Potoroidae) with notes on its biology. Aust Zool. 21:27-46.
5364. Johnson, PT, Valerio, DA, & Thompson, GE. 1973. Breeding the African green monkey, *Cercopithecus aethiops*, in a laboratory environment. Lab Anim Sci. 23:355-359.
5365. Johnson, RA, Carothers, SW, & McGill, TJ. 1987. Demography of feral burros in the Mohave Desert. J Wildl Manage. 51:916-920.
5366. Johnston, HL & Oliff, WD. 1954. The oestrous cycle of female *Rattus* (*Mastomys*) *natalensis* (Smith) as observed in the laboratory. Proc Zool Soc London. 124:605-613.

5367. Johnston, PG, Boshes, M, & Zucker, I. 1982. Photoperiodic inhibition of testicular development is mediated by the pineal gland in white-footed mice. Biol Reprod. 26:597-602.
5368. Johnston, PG & Zucker, I. 1979. Photoperiodic influences on gonadal development and maintenance in the cotton rat, *Sigmodon hispidus*. Biol Reprod. 21:1-8.
5369. Johnston, PG & Zucker, I. 1980. Photoperiodic regulation of reproductive development in white-footed mice (*Peromyscus leucopus*). Biol Reprod. 22:983-989.
5370. Johnston, PG & Zucker, I. 1980. Photoperiodic regulation of the testes of adult white-footed mice (*Peromyscus leucopus*). Biol Reprod. 23:859-866.
5371. Johnston, PG & Zucker, I. 1980. Antigonadal effects of melatonin in white-footed mice (Peromyscus leucopus). Biol Reprod. 23:1069-1074.
5372. Johnston, RF. 1956. Breeding of the Ord kangaroo rat (*Dipodomys ordii*) in southern New Mexico. Southwest Nat. 1:190-193.
5373. Johnston, RF & Rudd, RL. 1957. Breeding of the salt marsh shrew. J Mamm. 38:157-163.
5374. Jollie, WP. 1956. Rearing the pallid Ord kangaroo rat in the laboratory. Symposium on Ecology of Disease Transmission in Native Animals. 54-56, Army Chemical Corps, Dugway, UT.
5375. Jollie, WP. 1973. Fine structural changes in the placenta membrane of the marmoset with increasing gestational age. Anat Rec. 176:307-320.
5376. Jolly, A. 1966. Lemur social behavior and primate intelligence. Science. 153:501-506.
5377. Jolly, A. 1967. Breeding synchrony in wild *Lemur catta*. Social Communication among Primates. 3-14, Altmann, SA, ed, Univ Chicago Press, Chicago.
5378. Jolly, A. 1973. Primate birth hour. Int Zoo Yearb. 13:391-397.
5379. Jolly, S. 1988. Vaginal smears and the reproductive cycle of the common sheath-tail bat, *Taphozous georgianus* (Chiroptera: Emballonuridae). Aust Mamm. 11:75-76.
5380. Jolly, S. 1990. The biology of the common sheath-tail bat, *Taphozous georgianus* (Chiroptera: Emballonuridae), in central Queensland. Aust J Zool. 38:65-77.
5381. Jolly, SE & Blackshaw, AW. 1987. Prolonged epididymal sperm storage, and the temporal dissociation of testicular and accessory gland activity in the common sheath-tail bat, *Taphozous georgianus*, of tropical Australia. J Reprod Fert. 81:205-211.
5382. Jolly, SE & Blackshaw, AW. 1988. Testicular migration, spermatogenesis, temperature regulation and environment of the sheath-tail bat, *Taphozous georgianus*. J Reprod Fert. 84:447-455.
5383. Jolly, SE & Blackshaw, AW. 1989. Sex steroid levels and leydig cell ultrastructure of the male common sheath-tail bat, *Taphozous georgianus*. Reprod Fert Dév. 1:47-53.
5384. Joly, N. 1878. Études sur le placenta de l'Ai (*Bradypus tridactylus*), Linné: Place que cet animal doit occuper dans la série des Mammifères. CR Hebd Séances Acad Sci. 87:283-287.
5385. Jones, BE, Antrim, JE, & Cornell, LH. 1987. Commerson's dolphin (*Cephalorhynchus commersoni*): A discussion of the first live birth within a marine zoological park. Zoo Biol. 6:69-77.
5386. Jones, C. 1967. Growth, development, and wing loading in the evening bat, *Nycticeus humeralis* (Rafinesque). J Mamm. 48:1-19.
5387. Jones, C. 1971. The bats of Rio Muni, West Africa. J Mamm. 52:121-140.
5388. Jones, C. 1972. Comparative ecology of three pteropid bats in Rio Muni, West Africa. J Zool. 167:353-370.
5388a. Jones, C. 1972. Notes on the anomalurids of Rio Muni and adjacent areas. J Mamm. 52:568-572.
5389. Jones, C. 1978. *Dendrohyrax dorsalis*. Mamm Species. 113:1-4.
5390. Jones, C & Anderson, S. 1978. *Callicebus moloch*. Mamm Species. 112:1-5.
5391. Jones, C & Hildreth, NJ. 1989. *Neotoma stephensi*. Mamm Species. 328:1-3.
5392. Jones, C & Manning, RW. 1989. *Myotis austroriparius*. Mamm Species. 332:1-3.
5393. Jones, C & Setzer, HW. 1970. Comments on *Myosciurus pumilio*. J Mamm. 51:813-814.
5394. Jones, CB. 1978. Aspects of reproductive behavior in the mantled howler monkey (*Alouatta palliata* Gray). Ph D Diss. Cornell Univ, Ithaca, New York.
5395. Jones, CB. 1985. Reproductive patterns in mantled howler monkeys: Estrus, mate choice and copulation. Primates. 26:130-142.
5396. Jones, CJ. 1961. Additional records of bats in New Mexico. J Mamm. 42:538-539.
5397. Jones, CJ. 1977. *Plecotus rafinesquii*. Mamm Species. 69:1-4.
5398. Jones, DM & Pugsley, SL. 1980. High incidence of dystokia in the casiragua (*Proechimys guairae*). Lab Anim. 14:291.
5399. Jones, E. 1977. Ecology of the feral cat, *Felis catus* (L.), (Carnivora: Felidae) on Macquarie Island. Aust Wildl Res. 4:249-262.
5400. Jones, E & Coman, BJ. 1982. Ecology of the feral cat, *Felis catus* (L), in southeastern Australia II Reproduction. Aust Wildl Res. 9:111-119.
5401. Jones, E. 1981. Age in relation to breeding status of the male southern elephant seal, *Mirounga leonina* (L), at Macquarie Island. Aust Wildl Res. 8:327-334.
5402. Jones, E & Stevens, PL. 1988. Reproduction in wild canids, *Canis familiaris*, from the eastern highlands of Victoria. Aust Wildl Res. 15:385-394.
For Jones, F Wood *see* Wood Jones, F
5403. Jones, GS. 1983. Ecological and distributional notes on mammals from Vietnam, including the first record of *Nyctalus*. Mammalia. 47:339-344.
5404. Jones, IC & Henderson, IW. 1963. The ovary of the 13-lined ground squirrel (*Citellus tridecemlineatus* Mitchell) after adrenalectomy. J Endocrinol. 26:265-272.
5405. Jones, IC, Vinson, GP, Jarrett, IG, et al. 1964. Steroid components in the adrenal venous blood of *Trichosurus vulpecula* (Kerr). J Endocrinol. 30:149-150.
5405a. Jones, JK, Jr. 1959. Comments on the biology of the Quelpart Island shrew, *Crocidura dsinezumi quelpartis* Kuroda. J Mamm Soc Japan. 1:105-114.
5405b. Jones, JK, Jr. 1963. Additional records of mammals from Durango, Mexico. Trans KS Acad Sci. 66:750-753.
5405b$_1$**.** Jones, JK, Jr. 1964. Distribution and taxonomy of mammals of Nebraska. Univ KS Publ Mus Nat Hist. 16:1-356.
5405c. Jones, JK, Jr. 1964. Bats new to the fauna of Nicaragua. Trans KS Acad Sci. 67:506-508.
5405d. Jones, JK, Jr. 1964. Bats from western and southern Mexico. Trans KS Acad Sci. 67:509-516.
5405e. Jones, JK, Jr. 1966. Bats from Guatemala. Univ KS Publ Mus Nat Hist. 16:439-472.
5406. Jones, JK, Jr. 1977. *Rhogeessa gracilis*. Mamm Species. 76:1-2.

5407. Jones, JK, Jr. 1982. *Reithrodontomys spectabilis*. Mamm Species. 193:1.

5408. Jones, JK, Jr. 1990. *Peromyscus stirtoni*. Mamm Species. 361:1-2.

5409. Jones, JK, Jr, Alvarez, T, & Lee, MR. 1962. Noteworthy mammals from Sinaloa, Mexico. Univ KS Publ Mus Nat Hist. 14:149-159.

5410. Jones, JK, Jr & Armstrong, DM. 1971. Mammals from the Mexican state of Sinaloa I Marsupialia, Insectivora, Edentata, Lagomorpha. J Mamm. 52:747-757.

5411. Jones, JK, Jr & Arroyo-Cabrales, J. 1990. *Nyctinomops aurispinosus*. Mamm Species. 350:1-3.

5411a. Jones, JK, Jr & Baker, RJ. 1979. Notes on a collection of bats from Montserrat, Lesser Antilles. Occas Pap Mus TX Tech Univ. 60:1-6.

5412. Jones, JK, Jr & Baker, RJ. 1980. *Chiroderma improvisum*. Mamm Species. 134:1-2.

5413. Jones, JK, Jr & Baldassarre, GA. 1982. *Reithrodontomys brevirostris* and *Reithrodontomys paradoxus*. Mamm Species. 192:1-3.

5413a. Jones, JK, Jr, Choate, JR, & Cadena, A. 1972. Mammals from the Mexican state of Sinaloa II Chiroptera. Occas Pap Mus Nat Hist Univ KS. 6:1-29.

5414. Jones, JK, Jr & Engstrom, MD. 1986. Synopsis of the rice rats (genus *Oryzomys*) of Nicaragua. Occas Pap Mus TX Tech Univ. 103:1-23.

5414a. Jones, JK, Jr, Fleharty, ED, & Dunnigan, PB. 1967. The distributional status of bats in Kansas. Misc Publ Univ KS Mus Nat Hist. 46:1-33.

5415. Jones, JK, Jr & Genoways, HH. 1967. Annotated checklist of bats from South Dakota. Trans KS Acad Sci. 70:184-196.

5416. not used.

5417. Jones, JK, Jr & Genoways, HH. 1971. Notes on the biology of the Central American squirrel, *Sciurus richmondi*. Am Mid Nat. 86:242-246.

5419. Jones, JK, Jr & Genoways, HH. 1973. *Ardops nichollsi*. Mamm Species. 24:1-2.

5420. Jones, JK, Jr & Genoways, HH. 1975. *Dipodomys phillipsii*. Mamm Species. 51:1-3.

5421. Jones, JK, Jr & Genoways, HH. 1975. *Sciurus richmondi*. Mamm Species. 53:1-2.

5422. Jones, JK, Jr & Genoways, HH. 1975. *Sturnira thomasi*. Mamm Species. 68:1-2.

5423. Jones, JK, Jr & Genoways, HH. 1978. *Neotoma phenax*. Mamm Species. 108:1-3.

5423a. Jones, JK, Jr, Genoways, HH, & Baker, RJ. 1971. Morphological variation in *Stenoderma rufum*. J Mamm. 52:244-247.

5423b. Jones, JK, Jr, Genoways, HH, & Lawlor, TE. 1974. Annotated checklist of mammals of the Yucatan peninsula, Mexico II Rodents. Occas Pap Mus TX Tech Univ. 22:1-24.

5423c. Jones, JK, Jr, Genoways, HH, & Smith, JD. 1974. Annotated checklist of mammals of the Yucatan peninsula, Mexico III Marsupialia, Insectivora, Primates, Edentata, Lagomorpha. Occas Pap Mus TX Tech Univ. 23:1-12.

5424. Jones, JK, Jr & Homan, JA. 1974. *Hylonycteris underwoodi*. Mamm Species. 32:1-2.

5424_1. Jones, JK, Jr & Johnson, DH. 1965. Synopsis of the Lagomorphs and rodents of Korea. Univ KS Publ Mus Nat Hist. 16:357-407.

5424_2. Jones, JK, Jr, Lampe, RP, Spenrath, CA, et al. 1973. Notes on the distribution and natural history of bats in southeastern Montana. Occas Pap Mus TX Tech Univ. 15:1-12.

5424a. Jones, JK, Jr & Lee, MR. 1962. Three species of mammals from western Texas. Southwest Nat. 7:77-78.

5424b. Jones, JK, Jr & Phillips, CJ. 1976. Bats of the genus *Sturnira* in the Lesser Antilles. Occas Pap Mus TX Tech Univ. 40:1-16.

5424c. Jones, JK, Jr & Phillips, GL. 1964. A new subspecies of the fruit-eating bat, *Sturnira ludovici*, from western Mexico. Univ KS Publ Mus Nat Hist. 14:475-481.

5424d. Jones, JK, Jr & Schwartz, A. 1967. Bredin-Archbold-Smithsonian biological survey of Dominica 6 Synopsis of bats of the Antillieean genus *Ardops*. Proc US Nat Mus. 124(3634):1-13.

5424e. Jones, JK, Jr, Smith, JD, & Alvarez, T. 1965. Notes on bats from the cape region of Baja California. Trans San Diego Soc Nat Hist. 14:53-56.

5424f. Jones, JK, Jr, Smith, JD, & Genoways, HH. 1973. Annotated checklist of mammals of the Yucatan Peninsula, Mexico I Chiroptera. Occas Pap Mus TX Tech Univ. 13:1-31.

5425. Jones, JK, Jr, Smith, JD, & Turner, RW. 1971. Noteworthy records of bats from Nicaragua, with a checklist of the chiropteran fauna of the country. Occas Pap Mus Nat Hist Univ KS. 2:1-35.

5426. Jones, RC. 1980. Luminal composition and maturation of spermatozoa in the genital ducts of the African elephant (*Loxodonta africana*). J Reprod Fert. 60:87-93.

5427. Jones, RC & Brosnan, MF. 1981. Studies on the deferent ducts from the testis of the African elephant, *Loxodonta africana*: I Structure differentiation. J Anat. 132:371-386.

5428. Jones, RC & Chulow, J. 1987. Regulation of the elemental composition of the epididymal fluids in the tammar, *Macropus eugenii*. J Reprod Fert. 81:583-590.

5429. Jones, RC & Holt, WV. 1981. Studies of the deferent ducts from the testis of the African elephant, *Loxodanta africana* III Ultrastucture and cytochemistry of the ductuli efferents. J Anat. 132:247-255.

5430. Jones, RC, Rowlands, IW, & Skinner, JD. 1974. Spermatozoa in the genital ducts of the African elephant, *Loxodonta africana*. J Reprod Fert. 41:189-192.

5431. Jones, RC, Stone, GM, Hinds, LA, et al. 1988. Distribution of 5α-reductase in the epididymis of the tammar wallaby (*Macropus eugenii*) and dependence of the epididymmis on systemic testosterone and luminal fluids from the testis. J Reprod Fert. 83:779-783.

5431a. Jones, RC, Stone, GM, & Zupp, J. 1992. Reproduction in the male echidna. Platypus and Echidnas, 115-126, Augee, ML, ed, R Zool Soc NSW, Sydney.

5432. Jones, TS. 1945. Unusual state of birth of a bat. Nature. 156:365.

5433. Jones, TS. 1946. Parturition in a west Indian fruit bat (Phyllostomidae). J Mamm. 27:327-330.

5433a. Jones, TS. 1951. Bat records from the islands of Grenada and Tobago, British West Indies. J Mamm. 32:223-224.

5434. Jones, TS. 1962. Twins in an African bat (*Scotophilus nigrita*). Mammalia. 26:136.

5435. Jonkel, C. 1978. Black, brown (grizzly), and polar bears. Big Game of North America. 227-248, Schmidt, JL & Gilbert, DL, eds, Stackpole Books, Harrisburg, PA.

5436. Jonkel, C, Land, E, & Redhead, R. 1978. The productivity of polar bears (*Ursus maritimus*) in the northeastern Baffin Island area, Northwest Territories. Can Wildl Serv. 91:1-7.

5437. Jonkel, CJ & Cowan, IM. 1971. The black bear in the spruce-fir forest. Wildl Monogr. 27:1-57.

5438. Jonkel, CJ & Weckwerth, RP. 1963. Sexual maturity and implantation of blastocysts in the wild pine marten. J Wildl Manage. 27:93-98.

5439. Jonsgård, Å & Darling, K. 1977. On the biology of the eastern north Atlantic sei whale, *Balaenoptera borealis* Lesson. Rep Int Whal Comm Sp Issue. 1:124-129.
5440. Jonsgård, Å & Høidal, P. 1957. Strandings of Sowerby's whale (*Mesplodon bidens*) on the west coast of Norway. Norsk Hvalf Tid. 46:507-512.
5441. Jonsgård, Å & Lyshoel, PB. 1970. A contribution to the knowledge of the biology of the killer whale *Orcinus orca* (L). Nytt Mag Zool. 18:41-48.
5442. Jonsgård, Å & Nordli, O. 1952. Concerning a catch of white-sided dolphins (*Lagenorhynchus acutus*) on the west coast of Norway, winter 1952. Norsk Hvalf Tid. 41:229-232.
5443. Jordan, J. 1971. The establishment of a colony of Djungarian hamsters (*Phodopus sungorus*) in the United Kingdom. J Inst Anim Tech. 22:56-61.
5444. Jordan, JW & Vohs, PA, Jr. 1976. Natality of black-tailed deer in McDonald State Forests, Oregon. Northwest Sci. 50:108-113.
5445. Jordan, SM & Morgan, EH. 1968. The changes in the serum and milk whey proteins during lactation and suckling in the quokka (*Setonix brachyurus*). Comp Biochem Physiol. 25:271-283.
5445a. Jordan, SM & Morgan, EH. 1969. The serum and milk whey proteins of the echidna. Comp Biochem Physiol. 29:383-391.
5446. Joshi, CK, Vyas, KK, & Pareek, PK. 1978. Studies on the oestrus [*sic*] cycle in Bikaneri she-camel (*Camelus dromedarius*). Indian J Anim Sci. 48:141-145.
5447. Joshi, SG, Szarowski, DH, & Bank, JF. 1981. Decidua-associated antigens in the baboon. Biol Reprod. 25:591-598.
5448. Josimovich, JB, Levitt, MJ, & Stevens, VC. 1973. Comparison of baboon and human placental lactogens. Endocrinology. 93:242-244.
5449. Josimovich, JB, Weiss, G, & Hutchison, DL. 1974. Sources and disposition of pituitary prolactin in maternal circulation, amniotic fluid, fetus and placenta in the pregnant rhesus monkey. Endocrinology. 94:1364-1371.
5450. Joubert, E. 1969. An ecological study of the black rhinoceros, *Diceros bicornis* Linn 1758 in South West Africa. M Sc Thesis. Univ Pretoria, Pretoria. [Cited by Mentis, 1972]
5451. Joubert, E. 1971. Observations on the habitat preferences and population dynamics of the black-faced impala, *Aepyceros petersi* Bocage, 1875, in South West Africa. Madoqua. 1(3):55-65.
5452. Joubert, E. 1972. Activity patterns shown by mountain zebra *Equus zebra hartmannae* in south west Africa with reference to climatic factors. Zool Afr. 7:309-331.
5453. Joubert, E. 1972. The social organization and associated behaviour in the Hartmann zebra *Equus zebra hartmannae*. Madoqua. 6:17-56.
5454. Joubert, E. 1974. Notes on the reproduction in Hartmamn zebra *Equus zebra hartmannae* in South West Africa. Madoqua Series 1. 8:31-35.
5455. Joubert, E & Eloff, FC. 1971. Notes on the ecology and behaviour of the black rhinoceros *Diceros bicornis* Linn 1758 in South West Africa. Madoqua Series 1. 3:5-53.
5456. Joubert, SCJ. 1975. The mating behaviour of the tsessebe (*Damaliscus lunatus lunatus*) in the Kruger National Park. Z Tierpsychol. 37:182-191.
5457. Jouventin, P. 1975. Observations sur la socio-écologie du Mandrill. Terre Vie. 29:493-532.
5458. Joy, JE. 1969. Successful breeding of golden lion marmosets *Leontideus rosalia* at Dallas Zoo. Int Zoo Yearb. 9:77-78.
5459. Judd, HL & Ridgway, SH. 1977. Twenty-four hour patterns of circulating androgens and cortisol in male dolphins. Breeding Dolphins Present Status, Suggestions for the Future. 269-277, Ridgway, SH & Benirschke, K, eds, US Tech Inf Serv. Arlington, VA.
5460. Judd, FW, Carpenter, G, & Wagner, M. 1984. Variation in reproduction of a subtropical population of *Peromyscus leucopus*. Sp Publ Mus TX Tech Univ. 22:125-135.
5461. Jüdes, U. 1979. Untersuchungen zur Ökologie der Waldmaus (*Apodemus sylvaticus* Linné, 1758) und der Gelbhalsmaus (*Apodemus flavicollis* Melchior, 1834) im Raum Kiel (Schleswig-Holstein) I Populationsdichte, Gewichtsveränderungen, Fortpflanzungs-Jahreszyklus, Populationbiologische Parameter. Z Säugetierk. 44:81-95.
5462. Jüdes, U. 1979. Untersuchungen zur Ökologie der Waldmaus (*Apodemus sylvaticus* Linné, 1758) und der Gelbhalsmaus (*Apodemus flavicollis* Melchior, 1834) im Raum Kiel (Schleswig-Holstein) II Pränatale Mortalität. Z Säugetierk. 44:185-195.
5463. Judin, BS. 1974. Reproduction of *Asioscalops altaica* Nikolsky, 1883. Acta Theriol. 19:355-365.
5464. Jukkara, A, Koivisto, I, & Rajakoski, E. 1966. Results of a study of the sexual organs of white-tailed deer (*Odocoileus virginianus*) in Finland. Suomen Riista. 19:72-78.
5465. Juncys, V. 1964. Zur Fortpflanzung des Schneeleoparden (*Uncia uncia*) im Zoologischen Garten. Zool Garten NF. 29:303-306.
5466. Jung, HJG & Batzli, GO. 1981. Nutritional ecology of microtine rodents: effects of plant extracts on the growth of arctic microtines. J Mamm. 62:286-292.
5466a. Junge, RE & Sanderson, GC. 1982. Age related reproductive success of female raccoons. J Wildl Manage. 46:527-529.
5467. Jungers, WL & Susman, RL. 1984. Body size and skeletal allometry in African apes. The Pygmy Chimpanzee. 131-177, Sussman, RL, ed, Plenum Press, New York.
5468. Jungius, H. 1970. Studies on the breeding biology of the reedbuck (*Redunca arundinum* Boddaert, 1785) in the Krüger National Park. Z Säugetierk. 35:129-146.
5469. Jungius, H. 1971. The Biology and Behaviour of the Reedbuck (*Redunca arundinum* Boddaert 1785). Paul Parey, Hamburg.
5470. Junker, H. 1940. Die Aufzucht der Seehunde in den Tiergrotten der Stadt Wesermunde. Zool Garten NF. 12:306-315.
5471. Juntke, C. 1980. Über die künstliche Aufzucht und Entwicklung von 5 Schimpansengeschwistern. Zool Garten NF. 50:227-252.
5472. Jurewicz, RL, Cary, JR, & Rongstad, OJ. 1981. Spatial relationships of breeding female cottontail rabbits in southwestern Wisconsin. Proc World Lagomorph Conf, Univ Guelph, ONT, Canada. 1979:295-309.
5473. Jurgelski, W, Jr & Porter, ME. 1974. The opossum (*Didelphis virginiana* Kerr) as a biomedical model III Breeding the opossum in captivity: Methods. Lab Anim Sci. 24:412-425.
5474. Justines, G & Johnson, KM. 1970. Observations on the laboratory breeding of the cricetine rodent *Calomys callosus*. Lab Anim Care. 20:57-60.
5475. Kaack, B, Walker, L, & Brizzee, KR. 1979. The growth and development of the squirrel monkey (*Saimiri sciureus*). Growth. 43:116-135.

5476. Kachuba, M. 1977. Sexual behavior and reproduction in captive Geoffrey's cats (*Leopardus geoffroyi* d'Orbigny and Gervais 1844). Zool Garten NF. 47:54-56.
5477. Kaczmarski, F. 1966. Bioenergetics of pregnancy and lactation in the bank vole. Acta Theriol. 11:409-417.
5478. Kadam, KM & Swayamprabha, MS. 1980. Parturition in the slender loris (*Loris tardigradus lydekkerianus*). Primates. 21:567-571.
5479. Kadim, AH, Stefanov, SK, & Penkov, VT. 1972. Polynucleate spermatogonial cells in the seminiferous epithelium of the field mouse (*Apodemus sylvaticus* L). CR Akad Bulgare Sci. 25:373-376.
5480. Kadhim, AH. 1979. Notes on the food, predators and reproduction of the lesser jerboa *Jaculus jaculus* (Linné, 1758) (Dipopidae: Rodentia) from Iraq. Säugetierk Mitt. 27:312-314.
5481. Kadhim, AH & Wahid, IN. 1986. Reproduction of male Euphrates jerboa *Allactaga euphratica* Thomas (Dipodidae: Rodentia) from Iraq. Mammalia. 50:107-111.
5482. Kahmann, H. 1951. Das Zwergwiesel (*Mustela minuta*) in Bayern. Zool Jb Syst. 80:171-188.
5483. Kahmann, H. 1970. Der Gartenschläfer *Eliomys quercinus ophiosae* Thomas 1925 von der Pityusen Insel Formentera. Veröff Zool Staatssamml (München). 14:75-90.
5484. Kahmann, H. 1981. Zur Naturgeschichte des Löffelbilches *Eliomys melanurus* Wagner, 1840. Spixiana. 4:1-37.
5485. Kahmann, H & Einlechner, J. 1959. Bionomische Untersuchung an der Spitzmaus (*Crocidura*) der Insel Sardinien. Zool Anz. 162:63-83.
5486. Kahmann, H & Halbgewachs, J. 1962. Beobachtungen an der Schneemaus, *Microtus nivalis* Martin, 1842, in den bayerischen Alpen. Säugetierk Mitt. 10:64-82.
5487. Kahmann, H & Lau, G. 1972. Der Gartenschläfer *Eliomys quercinus ophiusal* Thomas, 1925 von der Pityusen Insel Formentera (Lebensfuhrung). Veröff Zool Staatssaml München. 16:29-49.
5488. Kahmann, H & Staudenmayer, T. 1970. Über das Fortpflanzengsgeschehen bei dem Gartenschläfer *Eliomys quercinus* (Linnaeus, 1766). Säugetierk Mitt. 18:97-114.
5489. Kahmann, H & Thoms, G. 1974. Über den Gartenschläfer, *Eliomys quercinus* (Linne, 1766) auf der Insel Mallorca, Balearen. Säugetierk Mitt. 22:122-130.
5490. Kahmann, H & Thoms, G. 1987. Zur Biometrie und Bionomie des tunesischen Gartenschläfers *Eliomys quercinus tunetal* Thomas, 1903. Spixiana. 10:97-114.
5491. Kahmann, H & von Frisch, O. 1950. Zur Ökologie der Haselmaus (*Muscardinus avellanarius*) in den Alpen. Zool Jb Syst. 78:531-546.
5492. Kaikusalo, A. 1973. On the breeding of the arctic fox (*Alopex lagopus*) in NW Enontekio, Finnish Lapland. Suomen Riista. 23:7-16.
5493. Kaikusalo, A & Tast, J. 1984. Winter breeding of microtine rodents at Kilpisjärvi, Finnish Lapland. Sp Publ Carnegie Mus Nat Hist. 10:243-252.
5494. Kaiser, IH. 1947. Histological appearance of coiled arterioles in the endometrium of rhesus monkey, baboon, chimpanzee and gibbon. Anat Rec. 99:199-225.
5495. Kaiser, IH. 1947. Absence of coiled arterioles in the endometrium of menstruating New World monkeys. Anat Rec. 99:353-367.
5496. Kaker, ML, Razdan, MN, & Galhotra, MM. 1980. Serum LH concentrations in cyclic buffalo (*Bubalus bubalis*). J Reprod Fert. 60:419-424.
5497. Kakuskina, EA. 1937. The normal reproductive cycle of silver foxes. Bull Biol Med Exp URSS 4:26-28. (An Breeding Abstr. 7:141, 1989).
5498. Kakuwa, Z, Kawakami, T, & Iguchi, K. 1953. Biological investigation on the whales caught by the Japanese Antarctic whaling fleets in the 1951-52 season. Sci Rep Whales Res Inst. 8:147-213.
5499. Kaldor, I & Ezekiel, E. 1964. Milk iron and plasma iron transport in a lactating marsupial during short-term intravenous infusion of iron. Aust J Exp Biol Med Sci. 42:54-61.
5500. Kaldor, I & Morgan, EH. 1986. Iron metabolism during lactation and suckling in a marsupial, the quokka (*Setonix brachyurus*). Comp Biochem Physiol. 84A:691-694.
5501. Kalela, O. 1957. Regulation of reproduction rate in subarctic populations of the vole *Clethrionomys rufocanus* (Sund). Ann Acad Sci Fenn IV Biol. 34:7-58.
5502. Kalela, O & Oksala, T. 1956. Sex ratio in the wood lemming, *Myopus schisticolor* (Lilljeg), in nature and in captivity. Ann Univ Turku Ser A II Biol Geogr. 37:5-24.
5503. Kali, J & Amir, S. 1973. The effect of mating female rats at 35 or 80 days of age on pup production and maternal weight. Lab Anim. 7:61-64.
5504. Kamei, I. 1970. Birth of a black rhinoceros. Animals & Zoos. 22(3):6-9.
5505. Kamerling, JP, Dorland, L, van Halbeek, H, et al. 1982. Structural studies of 4-0-acetyl-α N acetylneuraminyl (2-3)-lactose, in the main oligosaccharide in echidina milk. Carbohydrate Res. 100:331-340.
5506. Kamiyama, N. 1978. The growth and development of the gonad in Japanese house shrew (*Suncus murinus riukiuanus*). J Mamm Soc Japan. 7:274-279.
5507. Kanagawa, H & Hafez, ESE. 1973. Copulatory behavior in relation to anatomical characteristics of three macaques. Am J Phys Anthrop. 38:233-240.
5508. Kanagawa, H, Hafez, ESE, Nawar, MM, et al. 1972. Patterns of sexual behavior and anatomy of copulatory organs in the macaques. Z Tierpsychol. 31:449-460.
5509. Kanai, Y & Shimizu, H. 1984. Plasma concentrations of LH, progesterone and oestradiol during the oestrous cycle in swamp buffaloes (*Bubalus bubalis*). J Reprod Fert. 70:507-510.
5510. Kaneko, Y. 1976. Reproduction of Japanese field voles, *Microtus montebelli* (Milne-Edwards), at Iwakura, Kyoto, Japan. Jap J Ecol. 26:107-114.
5511. Kaneko, Y. 1978. A comparison of number of embryos and measurements of *Microtus montebelli* in two types of habitats. Acta Theriol. 23:140-143.
5512. Kaneko, Y. 1978. Variations in the number of embryos in *Microtus montebelli* (Milne-Edwards). J Mamm Soc Japan. 7:121-129.
5513. Kang, YH & Anderson, WA. 1975. Ultrastructure of the oocytes of the Egyptian spiny mouse (*Acomys cahirinus*). Anat Rec. 182:175-200.
5514. Kanno, H. 1975. The birth of a polar bear. Animals & Zoos. 27(11):6-9.
5515. Kano, T. 1982. The social group of pygmy chimpanzees (*Pan paniscus*) of Wamba. Primates. 23:171-188.
5516. Kanwar, KC & Chaudhry, V. 1977. Adrenal testis inter-relationship in male Indian palm squirrel *Funambulus pennanti* Wroughton. Endokrinologie. 69:287-292.
5517. Kanwisher, J & Sundnes, G. 1965. Physiology of a small cetacean. Hvalrjadets Skrifter. 48:45-53.
5518. Kapitonov, VI. 1960. An essay on the biology of the marmot *Marmota camtschatica* Pall. Zool Zhurn. 39:448-457.

5519. Kapitonov, VI. 1961. Ecological observations on *Ochotona hyperborea* Pall in the lower part of the Lena River. Zool Zhurn. 40:922-933.
5520. Kaplan, JN. 1977. Breeding and rearing squirrel monkeys (*Saimiri sciureus*) in captivity. Lab Anim Sci. 27:557-567.
5521. Kaplan, JR, Adams, MR, Koritnik, DR, et al. 1986. Adrenal responsiveness and social status in intact and ovariectomized *Macaca fascicularis*. Am J Primatol. 11:181-193.
5522. Kaplan, JR, Anthony, M, & Wood, L. 1981. Domestic breeding of patas monkeys (*Erythrocebus patas*). Lab Anim Sci. 31:409-412.
5523. Kappeler, PM. 1987. Reproduction in the crowned lemur (*Lemur coronatus*) in captivity. Am J Primatol. 12:497-503.
5524. Kar, AB & Chandra, H. 1965. Response of the ovary of prepubertal langurs (*Presbytis entellus*) to heterologous mammalian gonadotropins. Acta Anat. 60:608-615.
5525. Karesh, WB, Willis, MS, Czekala, NM, et al. 1985. Induction of fertile mating in a red ruffed (*Varecia variegata rubra*) using pregnant mare serum gonadotropin. Zoo Biol. 4:147-152
5526. Karim, KB. 1972. Development of the yolk sac in the Indian fruit bat, *Rousettus leschenaulti* (Desmarest). J Zool Soc India. 24:135-147.
5527. Karim, KB. 1972. Foetal membranes and placentation in the Indian leaf-nosed bat, *Hipposideros fulvus fulvus* (Gray). Proc Indian Acad Sci. 76B:71-78.
5528. Karim, KB. 1973. A rare case of twinning in the bat *Hipposideros fulvus fulvus* (Gray). Curr Sci. 42:823-824.
5529. Karim, KB. 1974. Development of the foetal membranes in the Indian leaf-nosed bat *Hipposideros fulvus fulvus* (Gray) I Early development. Rev Roum Biol. 19:251-256.
5530. Karim, KB. 1975. Early development of the embryo and implantation in the Indian vespertilionid bat, *Pipistrellus mimus mimus* Wroughton. J Zool Soc India. 27:119-136.
5531. Karim, KB. 1977. Foetal membranes and placentation in the Indian vesperitilionid bat *Pipistrellus mimus mimus* (Wroughton). J Zool Soc India. 29:45-55.
5532. Karim, KB. 1987. Histogenesis of the chorioallantoic placenta in the Indian rhinopomatid bat, *Rhinopoma hardwickei hardwickei* Gray. Anat Rec. 218:70A.
5533. Karim, KB & Banerjee, S. 1985. Storage of spermatozoa in the epididymis of the tropical bat, *Rhinopoma hardwickei hardwickei* (Gray). Anat Rec. 211:95A.
5534. Karim, KB & Banerjee, S. 1989. Reproduction in the Indian mouse-tailed bat, *Rhinopoma hardwickei hardwickei* (Chiroptera, Rhinopomatidae). Reprod Fert Dév. 1:255-264.
5535. Karim, KB & Fazil, M. 1986. Post-implantation development of the Indian rhinopomatid, *Rhinopoma hardwickei hardwickei* (Gray). Myotis. 23-24:63-69.
5536. Karim, KB & Fazil, M. 1987. Early embryonic development and preimplantation changes in the uterus of the bat *Rhinopoma hardwickei hardwickei* (Gray) (Rhinopomatidae). Am J Anat. 178:341-351.
5537. Karim, KB & Gupta, N. 1986. A case of superfetation in the Indian fruit bat *Rousettus leschenaulti* (Desmarest). Bat Res News. 27:13-14.
5538. Karim, KB, Willis, MS, Czekala, NM, et al. 1973a. Occurrence of a bicornuate vagina in the Indian leaf-nosed bat *Hipposideros fulvus fulvus* (Gray). Curr Sci. 42:62-63.
5539. Karim, KB, Wimsatt, WA, Enders, AC, et al. 1979. Electron microscopic observations on the yolk sac of the Indian fruit bat, *Rousettus leschenaulti* (Desmarest) (Pteropidae). Anat Rec. 195:493-510.
5540. Karimi, Y, Rodrigues de Almeida, C, & Petter, F. 1976. Note sur les Rongeurs du nord-est du Bresil. Mammalia. 40:257-266.
5541. Karn, MN & Penrose, LS. 1951. Birth weight and gestation time in relation to maternal age, parity and infant survival. Ann Eugenics. 16:147-164.
5542. Karsch, FJ, Krey, LC, Weick, RF, et al. 1973. Functional luteolysis in the rhesus monkey: The role of estrogen. Endocrinology. 92:1148-1152.
5543. Karsch, FJ, Weick, RF, Butler, WR, et al. 1973. Induced LH surges in the rhesus monkey: Strength-duration characteristics of the estrogen stimulus. Endocrinology. 92:1740-1747.
5544. Karsch, FJ, Weick, RF, Hotchkiss, J, et al. 1973. An analysis of the negative feedback control of gonadotropin secretion utilizing chronic implantation of ovarian steroids in ovariectomized rhesus monkeys. Endocrinology. 93:478-486.
5545. Kashkarov, D & Lein, L. 1927. The yellow ground squirrel of Turkestan, *Cynomys fulvus oxianus* Thomas. Ecology. 8:63-73.
5546. Kashyap, SK. 1980. Reproductive cycle of the Indian molossid bat, *Tadarida aegyptiaca*. Curr Sci. 49:252-253.
5547. Kashyap, SK. 1981. Reproduction in the wrinkle-lipped bat *Tadarida aegyptiaca* breeding habits in colonies of east Nimar. Bat Res News. 22:44.
5548. Kashyap, SK. 1982. Observations on a roost of free-tailed bat *Tadarida plicata plicata* (Buchanan) in east-Nimar. J Bombay Nat Hist Soc. 79:182-183.
5549. Kasman, LH, Loskutoff, NM, Evans, KL, et al. 1983. Application of direct steroid conjugate measurements in the urine of domestic and exotic hoofstock. Biol Reprod. 28(Suppl):109.
5550. Kasman, LH, McCowan, B, & Lasley, BL. 1985. Pregnancy detection in tapirs by direct urinary estrone sulfate analysis. Zoo Biol. 4:301-306.
5551. Kasman, LH, Ramsay, EC, & Lasley, BL. 1986. Urinary steroid evaluations to monitor ovarian function in exotic ungulates: III Estrone sulfate and pregnanediol-3-glucuronide excretion in the Indian rhinoceros (*Rhinoceros unicornis*). Zoo Biol. 5:355-361.
5552. Kassam, AAH & Lasley, BL. 1981. Estrogen excretory patterns in the Indian rhinoceros (*Rhinoceros unicornis*), determined by simplified urinary analysis. Am J Vet Res. 42:251-255.
5552a. Kastelein, RA & Wiepkema, PR. 1988. Case study of the neonatal period of a grey seal pup (*Halichoerus grypus*) in captivity. Aqua Mamm. 14:33-38.
5552b. Kastelein, RA & Wiepkema, PR. 1990. The suckling period of a grey seal (*Halichoerus grypus*) while confined to an outdoor land area. Aqua Mamm. 16:120-128.
5552c. Kastelein, RA & Wiepkema, PR. 1991. The suckling period of a grey seal (*Halichoerus grypus*) while the mother had acces to a pool. Aqua Mamm. 17:42-51.
5553. Kästle, W. 1953. Die Jugendentwicklung der Zwergmaus, *Micromys minutus soricinus*. Säugetierk Mitt. 1:49-59.
5554. Kasuya, T. 1972. Growth and reproduction of *Stenella caeruleoalba* based on the age determination by means of dentinal growth layers. Sci Rep Whales Res Inst. 24:57-80.
5555. Kasuya, T. 1972. Some information on the growth of the Ganges dolphin with a comment on the Indian dolphin. Sci Rep Whales Res Inst. 24:87-108.
5556. Kasuya, T. 1976. Reconsideration of life history parameters of the spotted and striped dolphins based on cemental layers. Sci Rep Whales Res Inst. 28:73-106.

5556a. Kasuya, T. 1977. Age determination and growth of the Baird's beaked whale with a comment of the fetal growth rate. Sci Rep Whales Res Inst. 29:1-20.
5557. Kasuya, T. 1978. The life history of Dall's porpoise with special reference to the stock off the Pacific coast of Japan. Sci Rep Whales Res Inst. 30:1-64.
5558. Kasuya, T. 1985. Effect of exploitation on reproductive parameters of the spotted and striped dolphins off the Pacific coast of Japan. Sci Rep Whales Res Inst. 36:107-138.
5559. Kasuya, T. 1986. A note on the reproductive status of female sperm whales taken by Japanese coastal whaling, 1983/84. Rep Int Whal Comm. 36:185-186.
5560. Kasuya, T & Brownell, RL, Jr. 1979. Age determination, reproduction, and growth of Franciscana dolphin, *Pontoporia blainvillei*. Sci Rep Whales Res Inst. 31:45-67.
5561. Kasuya, T & Kureha, K. 1979. The population of finless porpoise in the Inland Sea of Japan. Sci Rep Whales Res Inst. 31:1-44.
5562. Kasuya, T & Marsh, H. 1984. Life history and reproductive biology of the short-finned pilot whale, *Globicephala macrorhynchus* off the Pacific coast of Japan. Rep Int Whal Comm Sp Issue. 6:259-310.
5563. Kasuya, T & Matsui, S. 1984. Age determination and growth of the short-finned pilot whale off the Pacific coast of Japan. Sci Rep Whales Res Inst. 35:57-91.
5564. Kasuya, T, Miyazaki, N, & Dawbin, WH. 1974. Growth and reproduction of *Stenella attenuata* in the Pacific coast of Japan. Sci Rep Whales Res Inst. 1974:157-226.
5565. Kasuyu, T & Ohsumi, S. 1966. A secondary sexual character of the sperm whale. Sci Rep Whales Res Inst. 20:89-93.
5566. Kasuya, T & Shiraga, S. 1985. Growth of Dall's porpoise in the western north Pacific and suggested geographical growth differentiation. Sci Rep Whales Res Inst. 36:139-152.
5567. Kasuya, T, Tobayama, T, Saiga, T, et al. 1986. Perinatal growth of delphinoids: information from aquarium reared bottlenose dolphins and finless porpoises. Sci Rep Whales Res Inst. 37:85-97.
5568. Kato, H. 1985. Further examination of the age at sexual maturity of the Antarctic minke whale as determined from earplug studies. Rep Int Whal Comm. 35:273-277.
5569. Kato, H & Shimadzu, Y. 1983. The foetal sex ratio of the Antarctic minke whale. Rep Int Whal Comm. 33:357-359.
5570. Kato, J, Onouchi, T, & Oshima, K. 1980. The presence of progesterone receptors in the sexual skin of the monkey. Steroids. 36:743-749.
5571. Kato, M. 1983. Some considerations on the decline in age at sexual maturity of the Antarctic minke whale. Rep Int Whal Comm. 33:393-399.
5572. Kaudern, W. 1914. Über die glandulae vesiculares bei *Chiromys madagascariensis*. Arkiv Zool. 9:1-5.
5573. Kaudern, W. 1914. Einige Beobachtungen über die Zeit der Fortpflanzung der madagassisschen Säugetiere. Arkiv Zool. (Stockholm): 9(1):1-22.
5574. Kaudern, W. 1915. Studien über die männlichen Geschlechtsorgane von Edentaten I Xenarthra. Arkiv Zool (Stockholm). 9(1):1-52.
5575. Kaudern, W. 1915. Säugetiere aus Madagaskar. Arkiv Zool (Stokholm). 9(18):1-101.
5576. Kaufman, DW & Kaufman, GA. 1987. Reproduction by *Peromyscus polionotus*: Number, size, and survival of offspring. J Mamm. 68:275-280.
5577. Kaufman, GA & Kaufman, DW. 1977. Body composition of the old-field mouse (*Peromyscus polionotus*). J Mamm. 58:429-434.
5578. Kaufman, GW, Siniff, DB, & Reichle, R. 1975. Colony behavior of Weddell Seals, *Leptonychotes weddelli*, at Hutton Cliffs, Antarctica. Rapp P-V Réun Int Explor Mer. 169:228-246.
5579. Kaufman, J-M, Kesner, JS, Wilson, RC, et al. 1985. Electrophysiological manifestation of luteinizing hormone-releasing hormone pulse generator activity in the rhesus monkey: Influence of α-adrenergic and dopaminergic blocking agents. Endocrinology. 116:1327-1333.
5580. Kaufmann, JH. 1962. Ecology and social behavior of the coati, *Nasua narica* on Barro Colorado Island Panama. Univ CA Publ Zool. 60:95-222.
5581. Kaufmann, JH. 1965. Studies on the behavior of captive tree shrews. Folia Primatol. 3:50-74.
5581a. Kaufmann, JH. 1974. Social ethology of the whiptail wallaby, *Macropus parryi*, in northeastern New South Wales. Anim Behav. 22:281-369.
5582. Kaufmann, JH & Kaufmann, A. 1963. Some comments on the relationship between field and laboratory studies of behaviour, with special reference to coatis. Anim Behav. 11:464-469.
5583. Kaufmann, JH & Kaufmann, A. 1965. Observations on the behavior of tayras and grisons. Z Säugetierk. 30:146-155.
5584. Kaumanns, W. 1982. Verhaltungsbeobachtungen an Schnurrbarttamarinen (*Saguinus m mystax*). Z Kölner Zoo. 45:107-117.
5585. Kaunas, VJ. 1964. Zur Fortpflanzung des Schnuleoparden (*Uncia uncia*) im Zologische Garten. Zool Garten NF. 29:202-206.
5586. Kaur, P & Guraya, SS. 1983. Body weight, sex ratio and seasonal reproductive changes in the Indian mole rat, *Bandicota bengalensis*, in the Punjab. Aust J Zool. 31:123-130.
5587. Kavanagh, M. 1983. Birth seasonality in *Cercopithecus aethiops* a social advantage from synchrony?. Perspectives in Primate Biology. 89-98, Seth, PK, ed, Today & Tomorrow's Prters & Pub, New Delhi.
5588. Kavanagh, M & Laursen, E. 1984. Breeding seasonality among long-tailed macaques, *Macaca fascicularis*, in peninsular Malaysia. Int J Primatol. 5:17-29.
5589. Kawai, M. 1966. A case of unseasonable birth in Japanese monkeys. Primates. 7:391-392.
5590. Kawai, M, Azuma, S, & Yoshiba, K. 1967. Ecological studies of reproduction in Japanese monkeys (*Macaca fuscata*) I Problems of the birth season. Primates. 8:35-74.
5591. Kawamichi, T & Kawimichi, M. 1982. Social system and independence of offspring in tree shrews. Primates. 23:189-205.
5592. Kawata, K, Bailey, B, & Siminski, P. 1975. Notes on feline reproduction at the Indianapolis zoo. Zool Garten NF. 45:436-440.
5593. Kawata, M. 1985. Mating system and reproductive success in a spring population of the red-backed vole, *Clethrionomys rufocanus bedfordiae*. Oikos. 45:181-190.
5594. Kawata, M. 1987. Pregnancy failure and suppression by female-female interaction in enclosed populations of the red-backed vole, *Clethrionomys rufocanus bedfordiae*. Behav Ecol Sociobiol. 20:89-97.
5595. Kay, DH. 1974. Milk and milk production. The Husbandry and Health of the Domestic Buffalo. 329-376, Cockrill, WR, ed, FAO, Rome.
5596. Kayanja, FIB. 1969. The ovary of the impala, *Aepyceros melampus* (Lichtenstein 1812). J Reprod Fert Suppl. 6:311-317.

5597. Kayanja, FIB. 1972. Reproduction in Antelopes. E Afr Literature Bureau, Nairobi.
5598. Kayanja, FIB. 1979. The fine structure of the placenta of the zebra *Equus burchellii*, Gray. Afr J Ecol. 17:105-113.
5599. Kayanja, FIB, Baranga, J, & Schliemann, H. 1984. Leydig cells of the fruit bat testis. Afr J Ecol. 22:223-227.
5600. Kayanja, FIB & Blankenship, LH. 1973. The ovary of the giraffe, *Giraffa camelopardalis*. J Reprod Fert. 34:305-313.
5601. Kayanja, FIB & Epelu-Opio, JE. 1983. The oviduct epithelium of *Genetta tigrina*. Afr J Ecol. 21:205-207.
5602. Kayanja, FIB & Jarvis, J. 1971. Histological observations on the ovary, oviduct and uterus of the naked-mole rat. Z Säugetierk. 36:114-121.
5603. Kayanja, FIB & Mutere, FA. 1975. The ovary of the insectivorous bat (*Otomops martiensseni*). Anat Anz. 137:166-175.
5604. Kayanja, FIB & Mutere, FA. 1978. The fine structure of the testis of the insectivorous bat, *Otomops martiensseni*. Proc Int Bat Res Conf. 4:245-254.
5605. Kayanja, FIB & Sale, JB. 1973. The ovary of rock hyrax of the genus *Procavia*. J Reprod Fert. 33:223-230.
5606. Kayanja, FIB & Sale, JB. 1975. The ovary of Gunther's dik-dik, *Madaqua guentheri* Thomas. E Afr Wildl J. 13:1-7.
5607. Kayanja, FIB & Sale, JB. 1977. The ovary of rock hyrax of the genus *Heterohyrax*. Anat Anz. 141:206-219.
5608. Kayanja, FIB & Stanley Price, MR. 1973. The ovary of the hartebeest *Alcelaphus buselaphus cokii*, Günther. Z Säugetierk. 38:373-378.
5609. Kaye, SV. 1961. Laboratory life history of the eastern harvest mouse. Am Mid Nat. 66:439-451.
5610. Kean, RI. 1959. Bionomics of the brush-tailed opossum (*Trichosurus vulpecula*) in New Zealand. Nature. 184:1388-1389.
5611. Kean, RI. 1971. Selection for melanism and for low reproductive rate in *Trichosurus vulpecula* (Marsupialia). Proc NZ Ecol Soc. 18:42-47.
5612. Kean, RI. 1975. Growth of possums (*Trichosurus vulpecula*) in the Orongorongo Valley, Wellington, New Zealand 1953-1961. NZ J Zool. 2:435-444.
5613. Kean, RI, Marryatt, RG, & Carroll, ALK. 1964. The female urogenital system of *Trichosurus vulpecula* (Marsupialia). Aust J Zool. 12:18-41.
5614. Keeling, ME & Riddle, KE. 1975. Reproductive, gestational, and newborn physiology of the chimpanzee. Lab Anim Sci. 25:822-828.
5615. Keep, ME. 1973. Factors contributing to a population crash of nyala in Ndumu game reserve. Lammergeyer. 19:16-23.
5616. Keibel, F. 1902. Die Entwicklung des Rehes bis zur Anlage des Mesoblast. Arch Anat Phys Anat Abt. 1902:292-314.
5617. Keibel, F. 1922. Zur Entwicklungsgeschichte einer Groszfledermaus (*Cynopterus marginatus*). Archif Mikr Anat. 96:528-554.
5618. Keiter, MD & Pichette, LP. 1979. Reproductive behavior of captive subadult lowland gorillas (*Gorilla g gorilla*). Zool Garten NF. 49:215-237.
5618a. Keith, JO. 1965. The Abert squirrel and its dependence on ponderosa pine. Ecology. 46:150-163.
5619. Keith, K & Calaby, JH. 1968. The New Holland mouse, *Pseudomys novaehollandiae* (Waterhouse), in the Port Stephens District, New South Wales. CSIRO Wildl Res. 13:45-58.
5620. Keith, LB. 1977. Canada lynx. Can Wildl Serv.
5621. Keith, LB. 1981. Population dynamics of hares. Proc World Lagomorph Conf, Univ Guelph, ONT. 1979:395-440.
5622. Keith, LB & Meslow, EC. 1967. Juvenile breeding in the showshoe hare. J Mamm. 48:327.
5623. Keith, LB, Rongstad, OJ, & Meslow, EC. 1966. Regional differences in reproductive traits of the snowshoe hare. Can J Zool. 44:953-961.
5624. Keith, LB & Windberg, LA. 1978. A demographic analysis of the snowshoe hare cycle. Wildl Monogr. 58:1-70.
5625. Kellas, LM. 1955. Observations on the reproductive activities, measurements, and growth rate of the dikdik (*Rhynchotragus kirkii thomasi* Neumann). Proc Zool Soc London. 124:751-784.
5626. Kellas, LM. 1966. The placenta amd foetal membranes of the antelope *Ourebia ourebi* (Zimmerman). Acta Anat. 64:390-445.
5627. Kellas, LM, vanLennep, EW, & Amoroso, EC. 1958. Ovaries of some fetal and prepubertal giraffes. Nature. 181:487.
5628. Keller, BL. 1985. Reproductive patterns. Biology of New World *Microtus*, Sp Publ Am Soc Mamm. 8:725-778.
5629. Keller, BL & Krebs, CJ. 1970. *Microtus* population biology III Reproductive changes in fluctuating populations of *M ochrogaster* and *M pennsylvanicus* in southern Indiana, 1965-67. Ecol Monogr. 40:263-294.
5630. Kellogg, R. 1931. Whaling statistics for the pacific coast of North America. J Mamm. 12:73-77.
5631. Kelly, GM. 1977. Fisher (*Martes pennanti*) biology in the White Mountain National Forest and adjacent areas. Ph D Diss. Univ MA, Amherest.
5632. Kelly, MW. 1954. Observations afield on Alaskan wolves. Proc AK Sci Conf. 5:35.
5633. Kelly, RW & Challies, CN. 1978. Incidence of ovulation before the onset of rut and during pregnancy in red deer hinds. NZ J Zool. 5:817-819.
5634. Kelly, RW, McNatty, KP, & Moore, GH. 1985. Hormonal changes about oestrus in female red deer. R Soc NZ Bull. 22:181-184.
5635. Kelly, RW, McNatty, KP, Moore, GH, et al. 1982. Plasma concentrations of LH, prolactin, estradiol, and progesterone in female red deer (*Cervus elaphus*) during pregnancy. J Reprod Fert. 64:475-483.
5636. Kelly, RW & Moore, GH. 1977. Reproductive performance in farmed red deer. NZ Agric Sci. 11:179-181.
5637. Kelsall, J. 1968. The Caribou. Can Wildl Serv Monogr. 3:1-340.
5638. Kelt, DA. 1988. *Dipodomys heermanni*. Mamm Species. 323:1-7.
5639. Kelt, DA. 1988. *Dipodomys californicus*. Mamm Species. 324:1-4.
5640. Kemp, GA. 1972. Black bear population dynamics at Cold Lake, Alberta 1968-70. Bears: Their Biology and Management. 2:26-31.
5641. Kemp, GA & Keith, LB. 1970. Dynamics and regulation of red squirrel (*Tamiasciurus hudsonicus*) populations. Ecology. 51:763-779.
5642. Kemper, CM. 1976. Growth and development of the Australian murid *Pseudomys novaehollandiae*. Aust J Zool. 24:27-38.
5643. Kemper, CM. 1976. Reproduction of *Pseudomys novaehollandiae* (Muridae) in the laboratory. Aust J Zool. 24:159-167.
5644. Kemper, CM. 1979. Growth of an Australian murid (*Pseudomys novaehollandiae*) in the wild. Acta Theriol. 24:257-266.
5645. Kemper, CM. 1980. Reproduction of *Pseudomys novaehollandiae* (Muridae) in the wild. Aust Wildl Res. 7:385-402.

5646. Kemper, CM, Kitchener, DJ, & Humphreys, WF. 1990. The biology of the northern brown bandicoot, *Isoodon macrourus* (Marsupialia: Peramelidae) at Mitchell Plateau, Western Australia. Aust J Zool. 37:627-644.
5647. Kemper, CM, Kitchener, DJ, Humphreys, WF, et al. 1987. The demography and physiology of *Melamys* sp (Rodentia: Muridae) in the Mitchell Plateau area, Kimberley, Western Australia. J Zool. 212:533-562.
5648. Kempermann, CT. 1929. Die Placenta der Fledermaus (*Miniopterus schreibersii*) und ihre funktionelle Bedeutung. Z Anat Entw Gesch. 91:292-303.
5649. Kemps, A & Timmermans, P. 1982. Parturition behaviour in pluriparous Java-macaques (*Macaca fascicularis*). Primates. 23:75-88.
5650. Kempske, S. 1985. *Macaca silenus*: Survey of North American and European zoo practices. Monogr Primatol 7:221-235.
5651. Kenagy, GJ. 1973. Daily and seasonal patterns of activity and energetics in a heteromyid rodent community. Ecology. 54:1201-1219.
5652. Kenagy, GJ. 1976. Field observations of male fighting, drumming, and copulation in the Great Basin kangaroo rat, *Dipodomys microps*. J Mamm. 57:781-785.
5653. Kenagy, GJ. 1980. Interrelation of endogenous annual rhythms of reproduction and hibernation in the golden-mantled ground squirrel. J Comp Physiol. 135:333-339.
5654. Kenagy, GJ. 1981. Endogenous annual rhythm of reproductive function in the non-hibernating desert ground squirrel *Ammospermophilus leucurus*. J Comp Physiol A. 142:251-258.
5655. Kenagy, GJ. 1981. Effects of day length, temperature, and endogenous control on annual rhythms of reproduction and hibernation in chipmunks (*Eutamias* spp). J Comp Physiol. 141:369-378.
5656. Kenagy, GJ. 1981. Endogenous annual rhythm of reproductive function in the non-hibernating desert ground squirrel *Ammospermophilus leucurus*. J Comp Physiol. 142:251-258.
5657. Kenagy, GJ & Barnes, BM. 1984. Environmental and endogenous control of reproductive function in the Great Basin pocket mouse *Perognathus parvus*. Biol Reprod. 31:637-645.
5658. Kenagy, GJ & Barnes, BM. 1988. Seasonal reproductive patterns in four coexisting rodent species from the Cascade Mountains, Washington. J Mamm. 69:274-292.
5659. Kenagy, GJ & Bartholomew, GA. 1979. Effects of day length and endogenous control on the annual reproductive cycle of the antelope ground squirrel, *Ammospermophilus leucurus*. J Comp Physiol. 130:131-136.
5660. Kenagy, GJ & Bartholomew, GA. 1981. Effects of day length, temperature, and green food on testicular development in a desert pocket mouse *Perognathus formosus*. Physiol Zool. 54:62-73.
5661. Kenagy, GJ & Bartholomew, GA. 1985. Seasonal reproductive patterns in five coexisting California desert rodent species. Ecol Monogr. 55:371-397.
5662. Kenagy, GJ, Biebach, H, Stevenson, RD, et al. 1981. Timing and energetics of reproduction and development in yellow pine chipmunks *Eutamias amoenus*. Am Zool. 21:965.
5663. Kenagy, GJ, Stevenson, RD, & Masman, D. 1989. Energy requirements for lactation and postnatal growth in captive golden-mantled ground squirrels. Physiol Zool. 62:470-487.
5664. Kendrick, KM & Dixson, AF. 1984. Ovariectomy does not abolish proceptive behaviour cyclicity in the common marmoset (*Callithrix jacchus*). J Endocrinol. 101:155-162.
5665. Kennard, MA & Willner, MD. 1941. Weights of brains and organs of 132 new and old world monkeys. Endocrinology. 28:977-984.
5666. Kennelly, JJ. 1978. Coyote reproduction. Coyotes, Biology, Behavior and Management. 73-93, Bekoff, M, ed, Academic Press, New York.
5667. Kennelly, JJ & Johns, BE. 1976. The estrous cycle of coyotes. J Wildl Manage. 40:272-277.
5668. Kennelly, JJ, Johns, BE, Breidenstein, CP, et al. 1977. Predicting female coyote breeding dates from fetal measurements. J Wildl Manage. 41:746-750.
5669. Kennerly, TE, Jr. 1958. The baculum in the pocket gopher. J Mamm. 39:445-446.
5670. Kennerly, TE. 1958. Comparison of morphology and life history of two species of pocket gophers. TX J Sci. 10:133-146.
5671. Kenneth, JH & Ritchie, GR. 1953. Gestation Periods: A Table and Bibliography; Tech Comm #5. Comm Bur Anim Breed & Genetics, Edinburgh.
5672. Kenney, AM & Dewsbury, DA. 1977. Effect of limited mating on the corpora lutea in montane voles, *Microtus montanus*. J Reprod Fert. 49:363-364.
5673. Kenney, AM, Evans, RL, & Dewsbury, DA. 1977. Postimplantation pregnancy disruption in *Microtus ochrogaster*, *M pennsylvanicus*, and *Peromyscus maniculatus*. J Reprod Fert. 49:365-367.
5674. Kenney, AM, Hartung, TG, Davis, HN, Jr, et al. 1978. Male copulatory behavior and the induction of ovulation in female voles: A quest for species specificity. Horm Behav. 11:123-130.
5675. Kenney, AM, Hartung, TG, & Dewsbury, DA. 1979. Copulatory behavior and the initiation of pregnancy in California voles (*Microtus californicus*). Brain Behav Evol. 16:176-191.
5676. Kenney, AM, Lanier, DL, & Dewsbury, DA. 1977. Effects of vaginal-cervical stimulation in seven species of muroid rodents. J Reprod Fert. 49:305-309.
5677. Kenyon, KW. 1959. The sea otter. Ann Rep Bd Regents Smithsonian Inst. 1958:399-407.
5678. Kenyon, KW. 1969. The sea otter in the eastern Pacific Ocean. N Am Fauna. 68:1-352.
5679. Kenyon, KW. 1973. Hawaiian monk seals (*Monachus schauinslandi*). IUCN Suppl Pap. 39:88-97.
5680. Kenyon, KW. 1981. Monk seals *Monachus* Fleming, 1822. Handbook of Marine Mammals. 2:195-220. Ridgway, SH & Harrison RJ, eds, Academic Press, New York.
5681. Kenyon, KW. 1981. Sea otter *Enhydra lutris* (Linnaeus, 1758). Handbook of Marine Mammals. 1:209-223, Ridgway SH & Harrison, PJ, eds, Academic Press, New York.
5682. Kenyon, KW & Rice, DW. 1959. Life history of the Hawaiian monk seal. Pacific Sci. 13:215-252.
5683. Kenyon, KW & Scheffer, VB. 1985. The seals, sea-lions, and sea otter of the Pacific coast. US Fish Wildl Serv Wildl Leaflet. 334:1-28.
5684. Keogh, HJ. 1973. Behaviour and breeding in captivity of the Namaqua gerbil *Desmodillus auricularis* (Cricetidae: Gerbillinae). Zool Afr. 8:231-240.
5685. Keogh, HJ & Cronjé, PA. 1975. A note on the birth of litters and post-natal development in captive *Leggada minutoides*. Zool Afr. 10:107-108.
5686. Kerber, WT & Conaway, DM. 1977. The duration of gestation in the squirrel monkey (*Saimiri sciureus*). Lab Anim Sci. 27:700-702.
5687. Kerber, WT & Reese, WH. 1969. Comparison of the menstrual cycle of cynomolgus and rhesus monkeys. Fertil Steril. 20:975-979.

5688. Kerbert, C. 1891. Gestation of *Tragelaphus gratus*. Proc Zool Soc London. 1891:213.
5689. Kerbert, C. 1922. Over dracht, geboorte, puberteit en levensduur van *Hippopotamus amphibius* L. Bijdr Dierkd. 22:185-191.
5690. Kerle, JA. 1984. Variation in the ecology of *Trichosurus*: its adaptive significance. Possums and Gliders. 115-128, Smith, A & Hume, I, eds, Surrey Beatty & Sons Pty Ltd, Sydney, Australia.
5691. Kerle, JA. 1984. Growth and development of *Burramys parvus* in captivity. Possums and Gliders. 409-412, Smith, AP & Hume, ID, eds, Surrey Beatty & Sons Pty Ltd, Sydney, Australia.
5692. Kern, JA. 1964. Observations on the habits of the proboscis monkey *Nasalis larvatus* (Wurmb), made in the Brunei Bay area, Borneo. Zoologica. 49:183-191.
5692a. Kerr, GR, Kennan, AL, Waisman, HA, et al. 1969. Growth and development of the fetal rhesus monkey I Physical growth. Growth. 33:201-213.
5693. Kerr, GR, Scheffler, G, & Waisman, HA. 1969. Growth and development of infant *M mulatta* fed a standardized diet. Growth. 33:185-200.
5694. Kerr, JB & Hedger, MP. 1983. Spontaneous spermatogenic failure in the marsupial mouse *Antechinus stuartii* Macleay (Dasyuridae: Marsupialia). Aust J Zool. 31:445-466.
5695. Kerr, JB, Keogh, EJ, Hudson, B, et al. 1980. Alterations in spermatogenic activity and hormonal status in a seasonally breeding rat, *Rattus fuscipes*. Gen Comp Endocrinol. 40:78-88.
5696. Kerr, JB, Knell, CM, & Irby, DC. 1987. Ultrastructure and possible function of giant crystalloids in the Sertoli cell of the juvenile and adult koala (*Phascolarctos cinereus*). Anat Embryol. 176:213-224.
5696a. Kerr, JB & Weiss, M. 1991. Spontaneous or experimentally induced formation of a special zone in the adrenal cortex of the adult brush-tailed possum (*Trichosurus vulpecula*). Am J Anat. 190:101-117.
5697. Kerr, MA. 1965. The age at sexual maturity in male impala. Arnoldia (Rhodesia). 1(24):1-6.
5698. Kerr, MA. 1978. Reproduction of elephant in the Mana Pools National Park, Rhodesia. Arnoldia (Rhodesia). 8(29):1-11.
5699. Kerr, MA & Wilson, VJ. 1967. Notes on reproduction in Sharpe's grysbok. Arnoldia (Rhodesia). 3(17):1-4.
5700. Kerridge, DC & Baker, RJ. 1978. *Natalus micropus*. Mamm Species. 114:1-3.
5701. Kershaw, JA. 1912. Notes on the breeding habits and young of the platypus *Ornithorhynchus anatinus*. Vict Nat. 29:102-106.
5702. Kettlitz, WK. 1965. The blesbok (*Damaliscus dorcas phillipsi*) with special reference to the herd in the Percy Fyfe Nature Reserve. Fauna Flora Pretoria. 18:36-47.
5703. Keverne, EB, Eberhart, JA, Yodyingyad, M, et al. 1984. Social influences on sex differences in the behavior and endocrine state of talapoin monkeys. Prog Brain Res. 61:331-347.
5704. Khajuria, H. 1982. External genitalia and bacula of some central Indian Microchiroptera. Säugetierk Mitt. 30:287-295.
5705. Khajuria, H. 1984. Breeding and feeding habits of some central Indian Microchiroptera. Säugetierk Mitt. 31:127-134.
5706. Khammar, F & Brudieux, R. 1984. Seasonal changes in testicular contents of testosterone and androstenedione and in metabolic clearance rate of testosterone in the sand rat (*Psammomys obesus*). J Reprod Fert. 71:235-241.
5707. Khammar, F & Brudieux, R. 1987. Seasonal changes in testicular contents and plasma concentrations of androgens in the desert gerbil (*Gerbillus gerbillus*). J Reprod Fert. 80:589-594.
5708. Khammar, F & Brudieux, R. 1989. Seasonal changes in plasma testosterone concentrations in response to administration of hCG in a desert rodent, the sand rat *Psammomys obesus*. J Reprod Fert. 85:171-175.
5709. Khan, AA & Kohli, IS. 1972. A study on sexual behaviour of male camel (*Camelus dromedarius*) Part I. Indian Vet J. 49:1007-1012.
5710. Khan, MKBM. 1968. Deer biological data, 1967. Malay Nat J. 21:159-164.
5711. Khan-Dawood, FS. 1986. Localization of oxytocin and neurophysin in baboon (*Papio anubis*) corpus luteum by immunocytochemistry. Acta Endocrinol. 113:570-575.
5712. Khan-Dawood, FS. 1987. Oxytocin in baboon (*Papio anubis*) corpus luteum. Biol Reprod. 37:659-664.
5713. Khan-Dawood, FS, Marut, EL, & Dawood, MY. 1984. Oxytocin in the corpus luteum of the cynomolgus monkey (*Macaca fascicularis*). Endocrinology. 115:570-574.
5714. Khaparde, MS. 1976. Behavior of the female *Taphozous melanopogon* after parturition. J Bombay Nat Hist Soc. 73:207-8.
5715. Khaparde, MS. 1976. Foetal membranes and placentation in the Indian sheath-tailed bat, *Taphozous melanopogon* (Temminck). J Zool Soc India. 28:25-34.
5716. Khaparde, MS. 1976. Notes on the breeding habits of the Indian sheath-tailed bat, *Taphozous melanopogon* (Temminck). J Bombay Nat Hist Soc. 73:321-324.
5717. Khmelevskaya, NV. 1961. On the biology of *Ochotona alpina* Pallas (in Russian). Zool Zhurn. 40:1583-1585.
5718. Khokhar, AR. 1986. Field ecology of *Bandicota bengalensis*. Pak J Zool. 18:61-65.
5719. Kholkute, SD. 1984. Plasma progesterone levels throughout the ovarian cycle of the common marmoset (*Callithrix jacchus*). Primates. 25:123-126.
5720. Kholkute, SD. 1984. Plasma estradiol and progesterone levels during post partum ovulation and early pregnancy in the common marmoset *Callithrix jacchus*. Primates. 25:538-543.
5721. Kholkute, SD, Aitken, RJ, & Lunn, SF. 1983. Plasma testosterone response to hCG stimulation in the male marmoset monkey (*Callithrix jacchus jacchus)*. J Reprod Fert. 67:457-463.
5722. Kibbe, DP & Kirkpatrick, RL. 1971. Systemic evolution of late summer breeding in juvenile cottontails, *Sylvilagus floridanus*. J Mamm. 52:465-466.
5723. Kie, JG & White, M. 1985. Population dynamics of white-tailed deer (*Odocoileus virginianus*) on the Welder Wildlife Refuge, Texas. Southwest Nat. 30:105-118.
5724. Kik, P. 1980. De beverrat, *Myocastor coypus* (Molina), in Nederland. Lutra. 23:55-64.
5725. Kilgore, DL, Jr. 1969. An ecological study of the swift fox (*Vulpes velox*) in the Oklahoma panhandle. Am Mid Nat. 81:512-534.
5726. Kilgore, DL, Jr. 1970. The effects of northward dispersal on growth rate of young, size of young at birth and litter size in *Sigmodon hispidus*. Am Mid Nat. 84:510-520.
5726a. Kiltie, RA. 1982. Intraspecific variation in the mammalian gestation period. J Mamm. 63:646-652.
5727. Kimura, K & Uchida, TA. 1983. Ultrastructural observations of delayed implantation in the Japanese long-fingered bat, *Miniopterus schreibersii fuliginosus*. J Reprod Fert. 69:187-193.
5728. Kimura, K & Uchida, TA. 1984. Development of the main and accessory placentae in the Japanese long-fingered bat, *Miniopterus schreibersii fuliginosus*. J Reprod Fert. 71:119-126.

5729. Kimura, K, Takeda, A, & Uchida, TA. 1987. Changes in progesterone concentrations in the Japanese long-fingered bat, *Miniopterus schreibersii fuliginosus*. J Reprod Fert. 80:59-63.
5730. Kimura, S. 1957. The twinning in southern fin whales. Sci Rep Whales Res Inst. 12:103-126.
5731. Kinch, RFT, Craven, RP, & Follett, BK. 1986. Effects of age and photoperiod on the responsiveness of the pituitary gland of the vole (*Microtus agrestis*) to stimulation by GnRH. J Reprod Fert. 76:75-82.
5732. King, BF. 1984. The fine structure of the placenta and chorionic vesicles of the bush baby, *Galago crassicandata*. Am J Anat. 169:101-116.
5733. King, BF. 1986. Morphology of the placenta and fetal membranes. Comparative Biology of Primates, Reproduction and Development. 311-331, Dukelow, WR, Erwin, J, eds, Alan R Liss, Inc, New York.
5734. King, BF, Enders, AC, & Wimsatt, WA. 1978. The annular hematoma of the shrew yolk-sac placenta. Am J Anat. 152:45-58.
5735. King, BF & Hastings, RA, II. 1977. The comparative fine structure of the interhemal membrane of chorioallantoic placentas from six genera of myomorph rodents. Am J Anat. 149:165-180.
5736. King, BF, Pinheiro, PBN, & Hunter, RL. 1982. The fine structure of the placental labyrinth in the sloth, *Bradypus tridactylus*. Anat Rec. 202:15-22.
5737. King, BF & Tibbitts, FD. 1969. The ultrastructure of the placental labyrinth in the kangaroo rat, *Dipodomys*. Anat Rec. 163:543-554.
5738. King, BF & Tibbitts, FD. 1976. The fine structure of the chinchilla placenta. Am J Anat. 145:33-56.
5739. King, BF & Tibbitts, FD. 1977. Ultrastructural observations on cytoplasmic lamellar inclusions in oocytes of the rodent, *Thomomys*. Anat Rec. 189:263-271.
5740. King, CM. 1980. Population biology of the weasel *Mustela nivalis* on British game estates. Holarctic Ecol. 3:160-168.
5741. King, CM. 1983. *Mustela erminea*. Mamm Species. 195:1-8.
5742. King, CM. 1984. The origin and adaptive advantages of delayed implantation in *Mustela erminea*. Oikos. 42:126-128.
5743. King, CM & Moody, JE. 1982. The biology of the stoat (*Mustela erminea*) in the National Parks of New Zealand III Morphometric variation in relation to growth, geographical distribution, and colonization. NZ J Zool. 9:81-102.
5744. King, CM & Moody, JE. 1982. The biology of the stoat (*Mustela erminea*) in the National Parks of New Zealand IV Reproduction. NZ J Zool. 9:103-118.
5745. King, DR & Wheeler, SH. 1981. Ecology of the rabbit in Western Australia. Proc World Lagomorph Conf, Univ Guelph, ONT, Canada. 1979:858-869.
5746. King, DR & Wheeler, SH. 1985. The European rabbit in south-western Australia I Study sites and population dynamics. Aust Wildl Res. 12:183-196.
5747. King, FH. 1883. Instinct and memory exhibited by the flying squirrel in confinement, with a thought on the origin of wings in bats. Am Nat. 17:36-42.
5748. King, JA. 1958. Maternal behavior and behavioral development in two subspecies of *Peromyscus maniculatus*. J Mamm. 39:177-190.
5749. not used.
5750. King, JA, Deshaies, JC, & Webster, R. 1963. Age at weaning in two subspecies of deer mice. Science. 139:483-484.
5751. King, JA & Efeftheriou, BE. 1960. Differential growth in the skulls of two subspecies of deermice. Growth. 24:179-192.
5752. King, JB & Marlow, BJ. 1979. Australian sea lion. Mammals in the Sea, FAO Fish Ser 5. 2:13-15.
5753. King, JE. 1956. The monk seals. Bull Brit Mus (Nat Hist) Zool. 3(5):201-256.
5754. King, JE. 1969. Some aspects of the anatomy of the Ross seal, *Ommatophoca rossi* (Pinnipedia: Phocidae). Brit Antarctic Surv Sci Rep. 63:1-54.
5755. King, JE. 1983. Seals of the World. Brit Mus (Nat Hist), London.
5756. King, JM. 1965. A field guide to the reproduction of the Grant's zebra and Grey's zebra. E Afr Wildl J. 3:99-117.
5757. King, JM. 1965. Comparative aspects of reproduction in Equidae. Ph D Diss. Univ Cambridge.
5758. Kingdon, J. 1971. East African Mammals, Vol I. Academic Press, New York.
5759. Kingdon, J. 1974. East African Mammals, Vol IIA. Academic Press, New York.
5760. Kingdon, J. 1974. East African Mammals, Vol IIB. Academic Press, New York.
5761. Kingdon, J. 1977. East African Mammals, Vol IIIA. Academic Press, New York.
5762. Kingdon, J. 1979. East African Mammals, Vol IIIB. Academic Press, New York.
5763. Kingdon, J. 1982. East African Mammals, Vol IIIC. Academic Press, New York.
5764. Kingdon, J. 1982. East African Mammals, Vol IIID. Academic Press, New York.
5765. Kingsley, S. 1982. Causes of non-breeding and the development of the secondary sexual characteristics in the male orang utan: A hormonal study. The Orang Utan its Biology and Conservation. 215-229, DeBoer, LEM, ed, Dr W Junk, The Hague.
5766. Kingsley, SR. 1981. Reproductive physiology and behaviour of captive orangutans (*Pongo pygmaeus*). Ph D Diss. Univ Coll London.
5767. Kingston, WR. 1972. The breeding of endangered species of marmosets and tamarins. Saving the Lion Marmoset. 86-91, Bridgwater, DD, ed, The Wild Animal Propagation Trust, Wheeling, WV.
5768. Kingston, WR. 1975. The breeding of endangered species of marmosets and tamarins. Breeding Endangered Species in Captivity. 213-222, Martin, RD, ed, Academic Press, New York.
5768a. Kinnaird, MF. 1990. Pregnancy, gestation and parturition in free-ranging Tana River crested mangabeys (*Cercocebus galeritus galeritus*). Am J Primatol. 22:285-289.
5769. Kintner, PJ & Mead, RA. 1983. Steroid metabolism in the corpus luteum of the ferret. Biol Reprod. 29:1121-1127.
5770. Kinzey, WG. 1981. The titi monkeys, genus *Callicebus*. Ecology and Behavior of Neotropical Primates. 1:241-276. Coimbra Filho, AF & Mittermeier, RA, eds, Academio Brasileira de Ciências, Rio de Janeiro.
5771. Kirby, FV. 1899. Common bush-pig (*Sus choeropotamus typicus*. Great and Small Game of Africa. 526-530, Bryden, HA, ed, Rowland Ward, London.
5772. Kirby, FV. 1899. Livingstone's suni (*Nesotragus livingstonianus typicus*). Great and Small Game of Africa. 256-259, Bryden, HA, ed, Rowland Ward, London.
5773. Kirby, FV. 1899. Southern race (*Tragelaphus scriptus sylvaticus*). Great and Small Game of Africa. 484-487, Bryden, HA, ed, Rowland Ward, London.
5774. Kirby, FV. 1899. Spotted hyaena (*Hyaena crocuta*). Great and Small Game of Africa. 592-595, Bryden, HA, ed, Rowland Ward, London.

5775. Kirby, FV. 1899. The African hunting-dog (*Lycaon pictus*). Great and Small Game of Africa. 602-606, Bryden, HA, ed, Rowland Ward, London.
5776. Kirby, FV. 1899. The black rhinoceros (*Rhinoceros bicornis*). Great and Small Game of Africa. 35-42, Bryden, HA, ed, Rowland Ward, London.
5777. Kirby, FV. 1899. The blue duiker or blue-buck (*Cephalophus monticola*). Great and Small Game of Africa. 229-231, Bryden, HA, ed, Rowland Ward, London.
5778. Kirby, FV. 1899. The common impala or palla (*Aepyceros melampus*). Great and Small Game of Africa. 323-327, Bryden, HA, ed, Rowland Ward, London.
5779. Kirby, FV. 1899. The common reedbuck (*Cervicapra arundinum*). Great and Small Game of Africa. 305-308, Bryden, HA, ed, Rowland Ward, London.
5780. Kirby, FV. 1899. The lion (*Felis leo*). Great and Small Game of Africa. 545-559, Bryden, HA, ed, Rowland Ward, London.
5781. Kirby, FV. 1899. The Natal or red duiker (*Cephalophus natalensis*). Great and Small Game of Africa. 218-219, Bryden, HA, ed, Rowland Ward, London.
5782. Kirby, FV. 1899. The steinbuck (*Raphiceros campestris typicus*). Great and Small Game of Africa. 251-253, Bryden, HA, ed, Rowland Ward, London.
5783. Kirby, FV. 1899. The vaal rhebuck (*Pelea capreolus*). Great and Small Game of Africa. 319-322, Bryden, HA, ed, Rowland Ward, London.
5784. Kirby, VL & Ridgway, SH. 1984. Hormonal evidence of spontaneous ovulation in captive dolphins, *Tursiops truncatus* and *Delphinus delphis*. Rep Int Whal Comm Sp Issue. 6:459-464.
5785. Kirchshofer, R. 1958. Freiland-und Gefangenschaftsbeobachtungen an der nordafrikanischen Renmaus. Z Säugetierk. 23:33-49.
5786. Kirchshofer, R. 1962. Beobachtungen bei der Geburt eines Zwergchimpansen (*Pan paniscus* Schwarz 1929) und einige Bemerkungen zum Paarungsverhalten. Z Tierpsychol. 19:597-606.
5787. Kirchshofer, R. 1962. Die erste Geburt eines Zwergschimpansen in einen Zoo. Umschau. 62:537-538.
5788. Kirchshofer, R. 1962. The birth of a dwarf chimpanzee *Pan paniscus* Schwarz 1929 at Frankfurt zoo. Int Zoo Yearb. 4:76-78.
5789. Kirchshofer, R. 1963. Das Verhalten der Giraffengazelle, Elenantilope, und des Flachland-Tapirs bei der Geburt; einige Bemerkungen zur Vermehrungsrate und Generationanfolge dieser Arten im Frankfurter Zoo. Z Tierpsychol. 20:143-159.
5790. Kirchshofer, R. 1970. Notiz über eine Steiszgeburt beim Fluszpferd (*Hippopotamus amphibius* Linne) Einnige Bemerkungen zur Vermehrungsrate der Frankfurter Fluszpferdepaares. Z Säugetierk. 35:27-34.
5791. Kirchshofer, R. 1970. Gorillazucht im Zoologischen Gärten und Forschungsstationen. Zool Garten NF. 38:73-96.
5792. Kirchshofer, R, Frädrich, H, Podolczak, D, et al. 1967. An account of the physical and behaviorial development of the hand-reared gorilla infant *Gorilla g gorilla* born at Frankfurt Zoo. Int Zoo Yearb. 7:108-113.
5793. Kirchshofer, R, Weisse, K, Berenz, K, et al. 1968. A preliminary account of the physical and behavioral development during the frist 10 weeks of the hand-reared gorilla (*Gorilla g gorilla*) twins born at Frankfurt Zoo. Int Zoo Yearb. 8:121-128.
5794. Kirkland, GL, Jr. 1977. The rock vole, *Microtus chrotorrhinus* (Miller) (Mammalia: Rodentia) in West Virginia. Ann Carnegie Mus. 46:45-53.
5795. Kirkland, GL, Jr. 1981. *Sorex dispar* and *Sorex gaspensis*. Mamm Species. 155:1-4.
5796. Kirkland, GL, Jr & Jannett, FJ, Jr. 1982. *Microtus chrotorrhinus*. Mamm Species. 180:1-5.
5797. Kirkland, GL, Jr & Linzey, AU. 1973. Observations of the breeding success of the deer mouse, *Peromyscus maniculatus nubiterrae*. J Mamm. 54:254-255.
5798. Kirkland, GL, Jr & Schmidt, DF. 1982. Abundance, habitat, reproduction and morphology of forest-dwelling small mammals of Nova Scotia and southeastern New Brunswick. Can Field-Nat. 96:156-162.
5799. Kirkland, GL, Jr & van Deusen, HM. 1979. The shrews of *Sorex dispar* group: *Sorex dispar* Batchelder and *Sorex gaspensis* Anthony and Goodwin. Am Mus Novit. 2675:1-21.
5800. Kirkland, LE & Bradley, EL. 1980. Reproductive inhibition and serum prolactin concentrations in laboratory populations of the prairie deermouse. Biol Reprod. 35:579-586.
5801. Kirkpatrick, CM. 1955. The testis of the fox squirrel in relation to age and season. Am J Anat. 97:229-255.
5802. Kirkpatrick, CM. 1960. Unusual cottontail litter. J Mamm. 41:119-120.
5803. Kirkpatrick, CM & Conaway, CH. 1948. Some notes on Indiana mammals. Am Mid Nat. 39:128-136.
5804. Kirkpatrick, JF & Satterfield, V. 1973. Histology and morphology of the female reproductive tract of *Ochotona princeps*. J Mamm. 54:855-861.
5804a. Kirkpatrick, JF, Lasley, BL, & Shideler, SE. 1990. Urinary steroid evaluations to monitor ovarian function in exotic ungulates: VII Urinary progesterone metabolites in the Equidae assessed by immunoassay. Zoo Biol. 9:341-348.
5805. Kirkpatrick, JF, Vail, R, Devous, S, et al. 1976. Diurnal variation of plasma testosterone in wild stallions. Biol Reprod. 15:98-101.
5806. Kirkpatrick, RD. 1965. Litter size and fetus number in the cotton rat. J Mamm. 46:514.
5807. Kirkpatrick, RL & Baldwin, DM. 1974. Population density and reproduction in penned cottontail rabbits. J Wildl Manage. 38:482-487.
5808. Kirkpatrick, RL, Coggin, JL, Mosby, HS, et al. 1976. Parturition times and litter sizes of gray squirrels in Virginia. Proc Ann SE Conf Fish Wildl Ag. 30:541-545.
5809. Kirkpatrick, RL & Valentine, GL. 1970. Reproduction in captive pine voles, *Microtus pinetorum*. J Mamm. 51:779-885.
5810. Kirkpatrick, TH. 1965. Studies of macropodidae in Queensland 3 Reproduction in the grey kangaroo (*Macropus major*) in southern Queensland. QLD J Agric Anim Sci. 22:319-328.
5811. Kirkpatrick, TH. 1966. Studies of Macropodidae in Queensland 4 Social organization of the grey kangaroo (*Macropus giganteus*). QLD J Agric Anim Sci. 23:317-322.
5812. Kirkpatrick, TH. 1968. Studies of the wallaroo. QLD Agr J. 94:362-365.
5813. Kirkpatrick, TH. 1970. The swamp wallaby in Queensland. QLD Agr J. 96:335-336.
5814. Kirkpatrick, TH & Johnson, PM. 1969. Studies of Macropodidae in Queensland 7 Age estimation and reproduction in the agile wallaby (*Wallabia agilis* (Gould)). QLD J Agric Anim Sci. 26:691-698.
5815. Kirkpatrick, TH & McEvoy, JS. 1966. Studies of Macropodidae in Queensland 5 The effects of drought on reproduction in the grey kangaroo (*Macropus giganteus*). QLD J Agric Anim Sci. 23:439-440.

5816. Kirkwood, JK, Epstein, MA, & Terlecki, AJ. 1983. Factors influencing population growth of a colony of cotton-top tamarins. Lab Anim. 17:35-41.
5817. Kirkwood, JK, Epstein, MA, Terlecki, AJ, et al. 1985. Rearing a second generation of cotton-top tamarins (*Saguinus oedipus oedipus*) in captivity. Lab Anim. 19:269-272.
5818. Kirkwood, JK, Gaskin, CD, & Markham, J. 1987. Perinatal mortality and season of birth in the captive wild ungulates. Vet Rec. 120:386-390.
5819. Kirkwood, JK & Underwood, SJ. 1984. Energy requirements of captive cotton-top tamarins (*Saguinus oedipus oedipus*). Folia Primatol. 42:180-187.
5820. Kirsch, JAW & Poole, WE. 1972. Taxonomy and distribution of the grey kangaroos *Macropus giganteus* Shaw and *Macropus fuliginosus* (Desmarest) and their subspecies (Marsupialia: Macropodidae). Aust J Zool. 20:315-339.
5821. Kirsch, JAW & Walker, PF. 1979. Notes on the trapping and behavior of the Caenolestidae (Marsupialia). J Mamm. 60:390-395.
5822. Kisling, VN, Jr & Sampsell, RN. 1976. Aardvark *Orycteropus afer* diets and milk composition. Int Zoo Yearb. 16:164-165.
5823. Kiso, Y, Yasufuku, K, Matsuda, H, et al. 1990. Existence of an endothelio-endothelial placenta in the insectivore *Suncus murinus*. Cell Tiss Res. 262:195-197.
5824. Kistchinski, AA. 1972. Life history of the brown bear (*Ursus arctos* L) in north-east Siberia. Bears: Their Biology and Management, Int Conf Bear Res Manage, 2:67-73, Univ ALB, Canada, 1970, IUCN, Morges, Switzerland.
5825. Kita, I, Sugimura, M, Suzuki, Y, et al. 1983. Reproduction of female Japanese serows, *Capricornis crispus*, based on pregnancy and macroscopical ovarian findings. Res Bull Fac Agric Gifu Univ. 48:137-146.
5826. Kita, I, Tiba, T, Sugimara, M, et al. 1987. Frequency of past parturition estimated by retrograde corpora lutea of pregnancy, elastoid bodies in Japanese serow ovary. Zool Anz. 219:40-49.
5827. Kita, K. 1974. Breeding of roan antelope. Animals & Zoos. 26(10):330-333.
5828. Kitchen, DW. 1974. Social behavior and ecology of the pronghorn. Wildl Monogr. 38:1-96.
5829. Kitchener, DJ. 1973. Reproduction in the common sheath-tailed bat, *Taphozous georgianus* (Thomas) (Microchiroptera: Emballonuridae) in western Australia. Aust J Zool. 21:375-389.
5830. Kitchener, DJ. 1975. Reproduction in female Gould's wattled bat, *Chalinolobus gouldii* (Gray) (Vespertilionidae), in western Australia. Aust J Zool. 23:29-42.
5831. Kitchener, DJ. 1976. Further observations on reproduction in the common sheath-tailed bat, *Taphozous georgianus* Thomas, 1915 in Western Australia, with notes on the gular pouch. Rec W Aust Mus. 4:335-347.
5832. Kitchener, DJ, Cooper, N, & Bradley, A. 1986. Reproduction in male *Ningaui* (Marsupialia: Dasyuridae). Aust Wildl Res. 13:13-25.
5833. Kitchener, DJ & Coster, P. 1981. Reproduction in female *Chalinolobus morio* (Gray) (Vespertilionidae) in south-western Australia. Aust J Zool. 29:305-320.
5834. Kitchener, DJ & Halse, SA. 1978. Reproduction in female *Eptesicus regulus* (Thomas) (Vespertilionidae), in south-western Australia. Aust J Zool. 26:257-267.
5835. Kitchener, DJ & Hudson, CJ. 1982. Reproduction in the female white-striped mastiff bat, *Tadarida australis* (Gray) (Molossidae). Aust J Zool. 30:1-14.
5836. Kitchener, DJ, Stoddart, J, & Henry, J. 1984. A taxonomic revision of the *Sminthopsis murina* complex (Marsupialia, Dasyuridae) in Australia, including description of four new species. Rec W Aust Mus. 11:201-248.
5837. Kitchener, SL. 1971. Observations on the breeding of the bush dog *Speothos venaticus* at Lincoln Park Zoo, Chicago. Int Zoo Yearb. 11:99-101.
5838. Kitchener, SL, Meritt, DA, & Rosenthal, MA. 1975. Observations on the breeding and husbandry of snow leopards *Panthera uncia* at Lincoln Park Zoo, Chicago. Int Zoo Yearb. 15:212-217.
5839. Kitner, PJ & Mead, RA. 1983. Steroid metabolism in the corpus luteum of the ferret. Biol Reprod. 29:1121-1127.
5840. Kittams, WH. 1953. Reproduction of Yellowstone elk. J Wildl Manage. 17:177-184.
5841. Kitts, WD, Cowan, IM, Bandy, J, et al. 1956. The immediate post-natal growth in the columbian black-tailed deer in relation to the composition of the milk of the doe. J Wildl Manage. 20:198-.
5842. Klaar, J & Krasa, FC. 1921. Zur Anatomie der akzessorischen Geschlechtsdrüsen der Prosimier und Primaten I Teil Vescicula seminalis und Prostata. Z Anat Entw Gesch. 61:41-75.
5843. Klaatsch, H. 1886. Die Eihüllen von *Phocaena cummunis* Cuv. Arch f mikr Anat. 26:1-50.
5844. Kleiman, DG. 1968. Reproduction in the Canidae. Int Zoo Yearb. 8:3-8.
5845. Kleiman, DG. 1969. Maternal care, growth rate, and development in the noctule (*Nyctalus noctula*), pipistrelle (*Pipistrellus pipistrellus*), and serotine (*Eptesicus serotinus*) bats. J Zool. 157:187-211.
5846. Kleiman, DG. 1970. Reproduction in the female green acouchi, *Myoprocta pratti* Pocock. J Reprod Fert. 23:55-65.
5847. Kleiman, DG. 1971. The courtship and copulatory behaviour of the green acouchi, *Myoprocta pratti*. Z Tierpsychol. 29:259-278.
5848. Kleiman, DG. 1974. The estrous cycle in the tiger (*Panthera tigris*). The World's Cats. 60-75, Eaton, RL, ed, Feline Res Group Woodland Park Zoo, Seattle, WA.
5849. Kleiman, DG. 1977. Monogamy in mammals. Q Rev Biol. 52:39-69.
5850. Kleiman, DG. 1977. Progress and problems in lion tamarin *Leontopithecus rosalia rosalia* reproduction. Int Zoo Yearb. 17:92-97.
5851. Kleiman, DG. 1977. Characteristics of reproduction and sociosexual interactions in pairs of lion tamarins (*Leontopithecus rosalia*) during the reproductive cycle. The Biology and Conservation of the Callitrichidae. 181-190, Kleiman, DG, ed, Smithsonian Inst Press, Washington, DC.
5852. Kleiman, DG. 1978. The development of pair preferences in the lion tamarin (*Leontopithecus rosalia*): Male competition or female choice. Biology and Behaviour of Marmosets. 205-207, Rothe, H, Wolters, JH, & Hearn, JP, eds, Eigenverlag Hartmut Rothe, Göttingen.
5853. Kleiman, DG. 1979. Parent-offspring conflict and sibling competition in a monogamous primate. Am Nat. 114:753-760.
5854. Kleiman, DG. 1981. *Leontopithecus rosalia*. Mamm Species. 148:1-7.
5855. Kleiman, DG. 1983. Ethology and reproduction of captive giant pandas (*Ailuropoda melanoleuca*). Z Tierpsychol. 62:1-46.
5856. Kleiman, DG, Ballou, JD, & Evans, RF. 1982. An analysis of recent reproductive trends in captive golden lion tamarins *Leontopithecus r rosalia* with comments on their future demographic management. Int Zoo Yearb. 22:94-101.

5857. Kleiman, DG & Davis, TM. 1979. Ontogeny and maternal care. Sp Publ Mus TX Tech Univ. 16(3):387-402.
5858. Kleiman, DG, Eisenberg, JF & Maliniak E. 1979. Reproductive parameters and productivity of caviomorph rodents. Vertebrate Ecology in the Northern Neotropics. 173-183, Eisenberg, JF, ed, Smithsonian Institution Press, Washington, DC.
5859. Kleiman, DG, Gracey, DW, & Hodgen, GD. 1978. Urinary chorionic gonadotropin levels in pregnant golden lion tamarins. J Med Primatol. 7:333-338.
5860. Kleiman, DG & Racey, PA. 1969. Observations on noctule bats (*Nyctalus noctula*) breeding in captivity. Lynx. 10:65-77.
5861. Klein, EH. 1982. Phenology of breeding and antler growth in white-tailed deer in Honduras. J Wildl Manage. 46:826-829.
5862. Klein, LL. 1971. Observations on copulation and seasonal reproduction of two species of spider monkeys, *Ateles belzebuth* and *A geoffroyi*. Folia Primatol. 15:233-248.
5863. Klima, M & Bangma, GC. 1987. Unpublished drawings of marsupial embryos from the Hill collection and some problems of marsupial ontogeny. Z Säugetierk. 52:201-211.
5864. Kline, PD. 1962. Vernal breeding of cottontails in Iowa. Proc IA Acad Sci. 69:244-252.
5865. Kline, PD. 1963. Notes on the biology of the jackrabbit in Iowa. Proc IA Acad Sci. 70:196-204.
5866. Kling, OR & Westfahl, PK. 1978. Steroid changes during the menstrual cycle of the baboon (*Papio cynocephalus*) and human. Biol Reprod. 18:392-400.
5867. Klingel, H. 1965. Notes on the biology of the plains zebra *Equus quagga boehmi* Matschie. E Afr Wildl J. 3:86-88.
5868. Klingel, H. 1969. Reproduction in the plains zebra, *Equus burchelli boehmi*: Behaviour and ecological factors. J Reprod Fert Suppl. 6:339-345.
5869. Klingel, H. 1969. The social organization and population ecology of the plains zebra (*Equus quagga*). Zool Afr. 4:249-263.
5870. Klingel, H. 1970. Zur Soziologie des Grevy-Zebra. Zool Anz Suppl. 33:311-316.
5871. Klingel, H. 1972. Social behaviour of African Equidae. Zool Afr. 7:175-185.
5872. Klingel, H & Klingel, U. 1966. Die Geburt eines Zebras (*Equus quagga boehmi* Matschie). Z Tierpsychol. 23:72-76.
5873. Klingener, D, Genoways, HH, & Baker, RJ. 1978. Bats from southern Haiti. Ann Carnegie Mus. 47:81-99.
5874. Klinger, SR, Robel, RJ, & Brown, BA. 1985. Morphological and reproductive characteristics of white-tailed deer from Fort Riley, Kansas. Southwest Nat. 30:589-596.
5875. Kliukhart, EG. 1967. Birth of a harbor seal pup. J Mamm. 48:677.
5876. Klockenhoff, H. 1969. Über die Kropfgazellen, *Gazella subgutturosa* (Guldenstaedt), 1780, Afgahistans und ihre Haltung im Zoologischen Garten Kabul. Fr Kölner Zoo. 12:91-96.
5877. Klopfer, PH & Dugard, J. 1976. Patterns of maternal care in lemurs III *Lemur variegatus*. Z Tierpsychol. 40:210-220.
5878. Klös, H & Klös, U. 1966. Bemerkungen zur künstlichen Aufzucht von Orang-Utans (*Pongo pygmaeus*). Sitzungsber Naturf Fr Berlin NF. 6:66-76.
5879. Klös, HG. 1966. A note on the birth of an elephant seal *Mirounga leonina* at West Berlin Zoo. Int Zoo Yearb. 6:193.
5880. Klös, HG. 1976. Breeding success with ruffled lemurs (*Lemur variegatus*) in the Berlin Zoo. Int Zoo News 23. 6(138):35.
5881. Klös, HG & Rahn, P. 1975. Über die Haltung und eine teilweise gelungene Aufzucht von Südlichen See-Elefanten (*Mirounga leonina*) im Zoo Berlin. Zool Beitr. 21:551-561.
5882. Klugh, AB. 1927. Ecology of the red squirrel. J Mamm. 8:1-32.
5883. Knaus, H. 1967. Zur Frage der Superfetation beim Feldhasen. Z Jagdwiss. 13:52-53.
5884. Knechtel, C & Piechoki, R. 1983. Zur Fortpflanzungsbiologie, Mortalität, Sexualität und Alter der Bisamratte (*Ondatra zibethica* L 1766). Hercynia. 20:259-278.
5884a. Knick, ST. 1990. Ecology of bobcats relative to exploitation and a prey decline in southeastern Idaho. Wildl Monogr. 108:1-42.
5885. Knick, ST, Brittell, JD, & Sweeney, SJ. 1984. Population characteristics of bobcats in Washington state. J Wildl Manage. 48:721-728.
5886. Knight, FM. 1987. The development of pelage insulation and its relationship to homeothermic ability in an altricial rodent, *Peromyscus leucopus*. Physiol Zool. 60:181-190.
5887. Knight, MH & Skinner, JD. 1981. Thermoregulatory, reproductive and behavioural adaptations of the big eared desert mouse, *Malacothrix typica* to its arid environment. J Arid Environ. 4:137-145.
5888. Knight, RR. 1970. The Sun River elk herd. Wildl Monogr. 23:1-66.
5889. Knipe, T. 1957. The javelina in Arizona. AZ Game Fish Wildl Bull. 2:1-96.
5890. Knobil, E. 1981. Patterns of hypohysiotropic signals and gonadotropin secretion in the rhesus monkey. Biol Reprod. 24:44-49.
5891. Knobil, E & Neill, JD, eds. 1988. The Physiology of Reproduction. Raven Press. New York.
5892. Knoch, HW. 1968. The eastern woodrat, *Neotoma floridana osagensis*: A laboratory colony. Trans KS Acad Sci. 71:361-372.
5893. Knopf, FL & Balph, DF. 1977. Annual periodicity of Uinta ground squirrels. Southwest Nat. 22:213-224.
5894. Knotts, LK & Glass, JD. 1988. Effects of photoperiod, β-endorphin, and naloxone on in vitro secretion of testosterone in white-footed mouse (*Peromyscus leucopus*). Biol Reprod. 39:205-212.
5895. Knowles, CJ. 1987. Reproductive ecology of black-tailed prairie dogs in Montana. Great Basin Nat. 47:202-206.
5896. Knowles, JM & Oliver, WLR. 1975. Breeding and husbandry of scimitar-horned oryx *Oryx dammah* at Marwell Zoo. Int Zoo Yearb. 15:228-229.
5897. Knowlton, FF & Mickael, ED. 1965. Placental ingestion and possible parturant mortality in wild white-tailed deer. J Mamm. 46:107.
5898. Knox, WE & Lister-Rosenoer, LM. 1978. Timing of gestation in rats by fetal and maternal weights. Growth. 42:43-53.
5899. Knox, WM, Miller, KV, & Marchinton, RL. 1988. Recurrent estrous cycles in white-tailed deer. J Mamm. 69:384-385.
5900. Kobayashi, K, Furihata, M, & Yamazaki, K. 1978. Breeding of the ocelot. Animals & Zoos. 30(8):6-9.
5901. Kobayashi, K, Onoda, H, & Kaneko, N. 1979. Reproduction of the white-tailed gnu. Animals & Zoos. 31(7):20-22.
5902. Koceva, V. 1970. A contribution to the study of the seasonal rhythm of the testis in the bat *Myotis oxygnatus*. God Zb Med Fak Skopje. 16:361-371. [Notes of translation provided by Dr AW Gustafson]
5903. Koch-Weser, S. 1984. Fledermäuse aus Obervolta, W-Afrika. Senckenbergia Biol. 64:255-311.

5904. Kodo, H, Hata, M, & Kijima, T. 1982. Handraising a douroucouli. Animals & Zoos. 34(11):8-11.
5905. Kodo, H, Otsuka, K, & Ishigami, M. 1979. The birth of gray jackals. Animals & Zoos. 31(3):6-9.
5906. Koefoed, A. 1923. Parringen hos Pindoinet. Flora og Fauna (Kopenhagen). 1923:105-107. [Cited by Herter, 1938]
5907. Koehler, CE & Richardson, PRK. 1990. *Proteles cristatus.* Mamm Species. 363:1-6.
5908. Koenig, L. 1970. Zur Fortpflanzung und Jugendentwicklung des Wüstenfuchses (*Fennecus zerda* Zimm 1780). Z Tierpsychol. 27:205-246.
5909. Koeppl, JW & Hoffmann, RS. 1981. Comparative postnatal growth of four ground squirrel species. J Mamm. 62:41-57.
5910. Koering, MJ. 1979. Folliculogenesis in primates: Process of maturation and atresia. Animal Models for Research on Contraception and Fertility. 187-199, Alexander, NJ, ed, Harper & Row, New York.
5911. Koering, MJ. 1983. Preantral follicle development during the menstrual cycle in the *Macaca mulatta* ovary. Am J Anat. 166:429-433.
5912. Koering, MJ, Baehler, EA, Goodman, AL, et al. 1982. Developing morphological asymmetry of ovarian follicular maturation in monkeys. Biol Reprod. 27:989-997.
5913. Koering, MJ, Wolf, RC, & Meyer, RK. 1973. Morphological and functional evidence for corpus luteum activity during late pregnancy in the rhesus monkey. Endocrinology. 93:686-693.
5914. Koffler, BR. 1972. *Meriones crassus.* Mamm Species. 9:1-4.
5915. Koford, CB. 1957. The vicuna and the puna. Ecol Monogr. 27:153-219.
5916. Koford, CB. 1958. Prairie dogs, white faces and blue grama. Wildl Monogr. 3:1-78.
5917. Koford, CB. 1961. The ecology and management of the vicuna in the Puna zone of Peru. Terre Vie. 16:342-353.
5918. Koford, CB. 1963. Rank of mothers and sons in bands of rhesus monkeys. Science. 141:356-357.
5919. Koford, CB. 1965. Population dynamics of rhesus monkeys on Cayo Santiago. Primate Behavior Field Studies of Monkeys and Apes. 160-174, DeVore, I, ed, Holt, Rinehart & Winston, New York.
5920. Koford, CB. 1966. Population changes in rhesus monkeys: Cayo Santiago, 1960-1964. Tulane Stud Zool. 13:1-7.
5920a. Koford, CB, Farber, PA, & Windle, WF. 1966. Twins and teratisms in rhesus monkeys. Folia Primatol. 4:221-?.
5921. Koford, CB & Koford, MR. 1948. Breeding colonies of bats, *Pipistrellus hesperus* and *Myotis subulatus melanorhinys.* J Mamm. 29:417-418.
5922. Koford, RR. 1982. Mating system of a territorial tree squirrel (*Tamiasciurus douglasii*) in California. J Mamm. 63:274-283.
5923. Kofron, CP. 1987. Seasonal reproduction of the springhare (*Pedetes capensis*) in southeastern Zimbabwe. Afr J Ecol. 25:185-194.
5924. Kohlbrugge, JHF. 1913. Befruchtung und Keimbildung bei der Fledermaus Xantharpya amiplexicaudata. Verh Kon Akad Wet Amsterdam Tweede Sectie. 17:3-37.
5925. Kohlbrugge, JHF. 1913. Die Verbreitung der Spematozoiden im weiblichen Körper und im befruchteten Ei. Arch Entw Mech Org. 35:165-188.
5926. Köhler-Rollefson, IU. 1991. *Camelus dromedarius.* Mamm Species. 375:1-8.
5927. Kohlkute, SD, Aitken, RJ, & Lunn, SF. 1983. Plasma testosterone response to hCG stimulation in the male marmoset monkey (*Callithrix jacchus jacchus*). J Reprod Fert. 67:457-463.
5928. Köhncke, M & Leonhardt, K. 1986. *Cryptoprocta ferox.* Mamm Species. 254:1-5.
5929. Köhncke, M & Schliemann, H. 1977. Über zwei Foeten von *Cryptoprocta ferox* Bennett, 1833. Mitt Hamburg Zool Mus Inst. 74:171-175.
5930. Koivisto, I, Wahlberg, C, & Muuronen, P. 1977. Breeding the snow leapard *Panthera uncia* at Helsinki Zoo 1967-1979. Int Zoo Yearb. 17:39-44.
5931. Kojima, T. 1951. On the brain of the sperm whale *Physeter catodon* L. Sci Rep Whales Res Inst. 6:49-72.
5932. Kojola, I. 1989. Mother's dominance status and differential investment in reindeer calves. Anim Behav. 38:177-185.
5933. Kojola, I & Nieminen, M. 1985. The mother-calf relationship in Cervidae. Suomen Riista. 32:74-89.
5934. Kojola, I & Nieminen, M. 1986. Behavioural ecology of wild forest reindeer and semidomesticated reindeer during the rut. Suomen Riista. 33:67-78.
5935. Kok, OB. 1982. Cannabilism and post-partum return to oestrus of a female cape giraffe. Afr J Ecol. 20:141-143.
5936. Kolar, K. 1965. Einige Mitteilungen über die Fortpflanzung und Jugendentwicklung von *Galaga senegalensis* in Gefangenschaft. Zool Garten NF. 31:109-117.
5937. Kolb, A. 1950. Beiträge zur Biologie einheimisher Fledermäuse. Zool Jb Syst. 78:547-572.
5938. Kolb, A. 1957. Au seiner Wochenstube des Mausohrs, *Myotis m myotis* (Borkhausen, 1797). Säugetierk Mitt. 5:10-18.
5939. Kolb, A. 1966. Geburtsvorgang bei *Myotis myotis* (Borkhausen, 1797) und anschlissendes Verhalten von Mutter und Jungen. Bijdr Dierkd. 36:69-74.
5940. Kolb, HH & Hewson, R. 1980. A study of fox populations in Scotland from 1971-1976. J Appl Ecol. 17:7-19.
5941. Kolb, HH & Hewson, R. 1980. The diet and growth of fox cubs in two regions of Scotland. Acta Theriol. 25:325-331.
5942. Kolenosky, G. 1977. Black bear. Can Wildl Serv.
5943. Kolenosky, GB. 1990. Reproductive biology of black bears in east-central Ontario. Int Conf Bear Res Manage. 8:385-391.
5944. Kolosov, AM. 1941. Reproductive biology of the common hare (*Lepus europaeus* Pall). Zool Zhur. 20:154-171. [Translation by Dept Lands Forests, Maples, ONT]
5945. Kolpovsky, VM. 1979. Postnatal changes of uterus and ovaries in *Mustela vison* (Carnivora, Mustelidae). Zool Zhurn. 58:409-418.
5946. Komarov, AB. 1965. Čislennosti i structura populjacii ondatry Problemy ondatrovodstva, Materialy naucno-proizoudstvennogo sovesianija po ondatra vostvu. Zentro-sojuz VNII2P Moskva. 24-26. [Cited by Knechtel and Piechoki, 1983]
5947. Komori, A. Survey on the breeding of Japanese serows, *Capricornis crispus* in captivity. unpublished. [Cited by Schaller 1977]
5948. Komori, A. 1966. The year of the serow. Animals & Zoos. 18(4):100-101.
5949. Kon, L. 1946. Contribution a l'étude de la néoformation d'ovules chez les Mammifères primitifs adultes. Rev Suisse Zool. 53:597-623.
5950. Kondo, T. 1971. Breeding of giraffes at Tama Zoological Park. Animals & Zoos. 23(10):330-333.
5951. Kondo, T. 1982. A population study of the Japanese wood mouse, *Apodemus speciosus* (Mammalia: Muridae), with special reference to its social behavior. Res Popul Ecol. 24:85-96.

5952. Konieczna, B. 1956. Sexual maturation and reproduction in the nutria (*Myocastor coypus*) Part 2: The ovary. Folia Biol. 4:139-150.
5952a. König, A, Radespiel, U, Siess, M, et al. 1990. Analysis of pairing-parturition- and interbirth-intervals in a colony of common marmosets (*Callithrix jacchus*). Z Säugetierkd. 55:308-314.
5953. König, B. 1985. Maternal activity budget during lactation in two species of Caviidae (*Cavia porcellus* and *Galea musteloides*). Z Tierpsychol. 68:215-230.
5954. Konnerup Madsen, H. 1950. Uddrag of beretningen for demonstrationsbrugene 1957. Dansk Pelsdyravl. 21:99-103.
5955. Koontz, FW & Roeper, NJ. 1983. *Elephantulus rufescens*. Mamm Species. 204:1-5.
5956. Koopman, KF. 1972. *Eudiscopus denticulus*. Mamm Species. 19:1-2.
5957. Kooy, J. Personal communication. [Cited by Naaktgeboren, C. 1964]
5958. Kooyman, GL. 1963. Milk analysis of the kangaroo rat, *Dipodomys merriami*. Science. 142:1467-1468.
5959. Kooyman, GL & Drabek, CM. 1968. Observation on milk, blood, and urine constituents of the weddell seal. Physiol Zool. 41:187-194.
5960. Kooyman, GL, Kerem, DH, Campbell, WB, et al. 1973. Pulmonary gas exchange in freely diving Weddell seals, *Leptonychotes weddelli*. Respir Physiol. 17:283-290.
5961. Kordek, WS & Lindzey, JS. 1980. Preliminary analysis of female reproductive tracts from Pennsylvania black bears. Bears: Their Biology and Management. 159-161, Martinka, CJ & McArthur, KL, eds, US Gov Printing Office, Washington, DC.
5962. Koritnik, DR, Laherty, RF, Rotten, D, et al. 1983. A radioimmunoassay for dehydroepiandrosterone sulfate in the circulation of rhesus monkeys. Steroids. 42:653-667.
5963. Korn, H & Tait, MJ. 1987. Initiation of early breeding in a population of *Microtus townsendi* (Rodentia) with the secondary plant compound 6-MBOA. Oecologia. 71:593-596.
5964. Korniljewa, LA. 1983. Einige Angaben zur Fortpflanzung turkmenischer Kulane (*Equus hemionus* Pall) im Leningrader Zoopark. Zool Garten NF. 53:320-322.
5965. Korniljewa, LA & Roshdestwenskaja, ID. 1975. Zur Zucht des groszen Ameisenbären, *Myrmecophaga tridactyla* L, im Leningrader Zoopark. Zool Garten NF. 45:377-384.
5966. Korolyev, VA. 1969. Peculiarities of the embryogenesis of the *Martes foina rosanovi*. Vestn Zool (Kiev). 5:26-30.
5967. Koshkina, TV & Khalansky, AS. 1962. On the reproduction of lemming (*Lemmus lemmus*) on Kola peninsula. Zool Zhurn. 41:604-615.
5968. Koskela, P & Vire, P. 1976. The abundance, autumn migration, population structure and body dimensions of the harvest mouse in northern Finland. Acta Theriol. 21:375-387.
5969. Kostian, E. 1970. Habitat requirements and breeding biology of the root vole, *Microtus oeconomus* (Pallas), on shore meadows in the Gulf of Bothina, Finland. Ann Zool Fennici. 7:329-340.
5970. Kostjan, EJ. 1934. Eisbären und ihr Wachstum. Zool Garten NF. 7:157-164.
5971. Kostron, K & Kukla, F. 1971. The seasonal changes of the mink's testicle volume. Acta Univ Agricult Fac Agron Brno. 19:171-178.
5972. Kosugi, K. 1978. Behavior and reproduction of the sable antelope. Animals & Zoos. 30(4):6-9.
5973. Kosugi, K & Furuya, T. 1980. Captive reproduction of the white rhinoceros. Animals & Zoos. 32(6):6-10.
5974. Kotera, S, Tanaka, T, Tajima, Y, et al. 1975. Establishment of an island breeding colony of Japanese monkeys as a laboratory animal. Lab Anim Handb. 6:95-105.
5975. Kotov, VA. 1966. The size and weight of *Capra caucasica* Guld. Zool Zhurn. 45:1229-1234.
5976. Kott, E & Robinson, WL. 1963. Seasonal variation in litter size of the meadow vole in southern Ontario. J Mamm. 44:467-470.
5977. Koudele, KA, Napolitano, AC, & Aulerich, RH. 1986. Inability to perceive photoperiod affects testes size and testosterone secretion in mink. Biol Reprod. 34(Suppl 1):66.
5978. Kovacs, KM & Lavigne, DM. 1985. Neonatal growth and organ allometry of northwest Atlantic harp seals (*Phoca groenlandica*). Can J Zool. 63:2793-2799.
5979. Kovacs, KM & Lavigne, DM. 1986. *Cystophora cristata*. Mamm Species. 258:1-9.
5980. Kovacs, KM & Lavigne, DM. 1986. Growth of grey seal (*Halichoerus grypus*) neonates: Differential maternal investment in the sexes. Can J Zool. 64:1937-1943.
5981. Kovlsisto, I, Wahlberg, C, & Muuronen, P. 1977. Breeding the snow leopard *Panthera uncia* at Helsinki Zoo. Int Zoo Yearb. 16:39.
5982. Kowalska-Dyrcz, A. 1967. Sexual maturation in young male of the common shrew. Acta Theriol. 12:172-173.
5983. Kowalska-Dyrcz, A. 1973. The structure of internal genital organs in Zapodidae (Rodentia). Acta Theriol. 18:107-116.
5984. Kowalski, K. 1957. *Microtus nivalis* (Martins, 1842) (Rodentia) in the Carpathians. Acta Theriol. 1:159-182.
5985. Kowalski, K. 1959. *Microtus socialis* (Pallas) (Rodentia) in the Lebanon mountains. Acta Theriol. 2:269-279.
5986. Kowalski, K. 1960. *Pitymys* Mc Murtrie 1831 (Microtidae, Rodentia) in the northern Carpathians. Acta Theriol. 4:81-91.
5987. Kowalski, K. 1984. Annual cycle of reproduction in *Apodemus sylvaticus* in Algeria. Acta Zool Fenn. 173:85-86.
5988. Kowalski, K, Gaisler, J, Bessam, H, et al. 1986. Annual life cycle of cave bats in northern Algeria. Acta Theriol. 31:185-206.
5989. Koyama, T, de la Pena, A, & Hagino, N. 1977. Plasma estrogen, progestin and luteinizing hormone during the normal menstrual cycle in the baboon: Role of luteinizing hormone. Am J Gyn Obs. 127:67-72.
5990. Kozlo, PG. 1980. Dynamics of the elk (*Alces alces* L) fecundity in Byelorussia. Zool Zhurn. 59:925-933.
5991. Kraemer, DC & Vera Cruz, NCV. 1972. Breeding baboons for laboratory use. Breeding Primates. 42-47, Beveridge, WIB, ed, Basel, Karger.
5992. Kraemer, HC, Horvat, JR, Doering, C, et al. 1982. Male chimpanzee development focusing on adolescence: Integration of behavioral with physiological changes. Primates. 23:393-405.
5993. Kraft, A. 1964. Management of moose in a Norwegian forest. Med Staten Viltunders 2nd serie. 16:1-61.
5994. Kraft, H. 1957. Das Verhalten von Muttertier und Neugeborenen bei Cameliden. Säugetierk Mitt. 5:174-175.
5995. Kralik, S. 1967. Breeding the caracal lynx *Felis caracal* at Brno Zoo. Int Zoo Yearb. 7:132.
5996. Kranz, KR & Lumpkin, S. 1982. Notes on the yellow-backed duiker *Cephalophus sylvicultor* in captivity with comments on its natural history. Int Zoo Yearb. 22:232-240.
5997. Kranz, KR, Xanten, WA, Jr, & Lumpkin, S. 1984. Breeding history of the dorcas gazelles *Gazella dorcas* at the National Zoological Park, 1961-1981. Int Zoo Yearb. 23:195-203.
5998. Krapp, F. 1974. Hohe Embryonenzahl auch bei *Neomys anomalus milleri* (Mottaz, 1907). Z Säugetierk. 39:201-203.

5999. Krasinski, Z & Raczynski, J. 1967. The reproduction biology of European bison living in reserves and in freedom. Acta Theriol. 12:407-444.
6000. Krátký, J. 1970. Postnatale Entwicklung des Grossmausohrs *Myotis myotis* (Borkhausen, 1797). Vestn Česk Spol Zool. 33:202-218.
6001. Krátký, J. 1971. Zur Ethologie des Mausohrs *Myotis myotis* Borkhousen, 1797. Zool Listy. 20:131-138.
6002. Kratochvíl, J. 1964. Über die männlichen Geschlechtsorgane von *Spalax leucodon hungaricus* Nehring. Acta Theriol. 8:189-206.
6003. Kratochvíl, J. 1968. Der Antritt des Vermehrungsprozesses der kleinen Erdsaugetiere in der Hohen Tatra. Zool Listy. 17:299-310.
6004. Kratochvíl, J. 1969. Der Geschlechtszyklus der Weibchen von *Pitymys subterraneus* und *P tatricus* (Rodenta) in der Hohen Tatra. Zool Listy. 18:99-120.
6005. Kratochvíl, J. 1970. Der Geschlechtszyklus der Männchen von *Pityms subterraneus* und *Pitymys tatricus* (Rodentia) in der Hohen Tatra. Zool Listy. 19:1-22.
6006. Kratochvíl, J. 1971. Die Hodengrösze als Kriterium der europaischen Arten der Gattung *Apodemus* (Rodentia, Muridae). Zool Listy. 20:293-306.
6007. Kratochvíl, J. 1974. Die Vermehrungsfähigkeit der Art *Arvicola terestris* (L) in der CSSR (Mammalia Microtidae). Zool Listy. 23:3-17.
6008. Kratochvíl, J, Pelikán, J, & Šebek, Z. 1956. Eine Anlyse vier populationen der Erdwuhlmaus aus der Tchechoslavakei. Zool Listy. 5:63-82.
6009. Krause, DB & Fadem, BH. 1987. Reproduction, development and physiology of the gray short-tailed opossum (*Monodelphis domestica*). Lab Anim Sci. 37:478-482.
6009a. Krause, WJ & Cutts, JH. 1983. Ultrastructural observations on the shell membrane of the north American opossum (*Didelphis virginiana*). Anat Rec. 207:335-338.
6009b. Krause, WJ & Cutts, JH. 1984. Scanning electron microscopic observations on the 9-day opossum (*Didelphis virginiana*) embryo. Acta Anat. 120:93-97.
6010. Krause, WJ & Cutts, JH. 1985. Placentation in the opossum, *Didelphis virginiana*. Acta Anat. 123:156-171.
6011. Krause, WJ & Cutts, JH. 1985. The allantois of the North American opossum (*Didelphis virginiana*) with preliminary observations on the yolk sac endoderm and trophectoderm. Anat Rec. 211:166-173.
6012. Krear, RH. 1966. An ecological and ethological study of the pika *Ochotona princeps saxatilis* Bangs in the Front Range of Colorado. Diss Abs Int. 26:6249.
6013. Krebs, CJ. 1964. The lemming cycle at Baker Lake, Northwest Territories, during 1959-1962. Tech Pap Arctic Inst N Am. 15:1-104.
6014. Krebs, CJ, Gilbert, BS, Boutin, S, et al. 1986. Population biology of snowshoe hares I Demography of food-supplemented populations in the southern Yukon, 1976-84. J Anim Ecol. 55:963-982.
6015. Krebs, CJ & Wingate, I. 1985. Population fluctuations in the small mammals of the Kluane Region, Yukon Territory. Can Field-Nat. 99:51-61.
6016. Krefting, LW, Stoeckeler, JH, Bradle, BJ, et al. 1962. Porcupine-timber relationships in the lake states. J Forestry. 60:325-339.
6017. Kreitmann, O, Bayard, F, & Hodgen, GD. 1980. 17β-hydroxysteroid dehydrogenase in monkey endometrium during the menstrual cycle and at the time of implantation. Steroids. 36:365-372. .
6018. Kreitmann-Gimbal, B, Bayard, F, Nixon, WE, et al. 1980. Patterns of estrogen and progesterone receptors in monkey endometrium during the normal menstrual cycle. Steroids. 35:471-479.
6019. Kress, A. 1984. Ultrastructural studies on oogenesis in the shrew (*Crocidura russula*) 1 The preantral follicle. J Morphol. 179:59-71.
6020. Krey, LC, Butler, WR, & Knobil, E. 1975. Surgical disconnection of the medial basal hypothalamus and pituitary function in the rhesus monkey I Gonadotropin Secretion. Endocrinology. 96:1073-1087.
6021. Krieg, H. 1924. Beobachtungen an argentinischen Beutelratten. Z Morphol Ökol Tiere. 1:637-659.
6022. Krieg, H. 1924. Chilenische Beutelratten. Z Morphol Ökol Tiere. 3:169-176.
6023. Krieg, H. 1929. Biologische Reisestudien in Südamerika IX Gürteltiere. Z Morphol Ökol Tiere. 14:166-190.
6024. Kriewaldt, FH & Hendrickx, AG. 1968. Reproductive parameters of the baboon. Lab Anim Care. 18:361-370.
6025. Krishna, A. 1984. Storage of spermatozoa in the female genital tract of the Indian pigmy pipistrelle bat, *Pipistrellus mimus* Wroughton. Arch Biol. 95:223-229.
6026. Krishna, A. 1985. Reproduction in the Indian pigmy pipistrelle bat *Pipistrellus mimus*. J Zool. 206A:41-51.
6027. Krishna, A & Dominic, CJ. 1978. Storage of spermatozoa in the female genital tract of the vespertilionid bat, (*Scotophilus heathi*). J Reprod Fert. 54:319-321.
6028. Krishna, A & Dominic, CJ. 1981. Reproduction in the vespertilionid bat, *Scotophilus heathi* Horsefield. Arch Biol. 92:247-258.
6029. Krishna, A & Dominic, CJ. 1982. Differential rates of fetal growth in two successive pregnancies in the emballonurid bat, *Taphozous longimanus* Hardwicke. Biol Reprod. 27:351-353.
6030. Krishna, A & Dominic, CJ. 1982. Reproduction in the Indian sheath-tailed bat. Acta Theriol. 27:97-106.
6031. Krishna, A & Dominic, CJ. 1983. Reproduction in the female short-nosed fruit bat, *Cynopterus sphinx* Vahl. Period Biol. 85:23-30.
6032. Krishne Gowda, CD. 1969. A brief note on breeding Indian elephants *Elephas maximus* at Mysore Zoo. Int Zoo Yearb. 9:99.
6033. Krishne Gowda, CD. 1969. Breeding the great Indian rhinoceros *Rhinoceros unicornis* at Mysore Zoo. Int Zoo Yearb. 9:101-102.
6034. Kristal, MB & Noonan, M. 1979. Perinatal maternal and neonatal behaviour in the captive reticulated giraffe. S Afr J Zool. 14:103-107.
6035. Kristiansson, H. 1981. Young production of European hedgehog in Sweden and Britain. Acta Theriol. 26:504-506.
6036. Krog, J, Wika, M, Lund-Larsen, T, et al. 1976. Spitsbergen reindeer, *Rangifer tarandus platyrhyncus* Vrolik: Morphology, fat storage and organ weights in late winter season. Norw J Zool. 24:407-417.
6037. Krogh, HJ and Cranjé, PA. 1975. A note on the birth of litters and post-natal evelopment in captive *Leggada minutoides*. Zool Afr. 10:107-108.
6038. Krohn, PL. 1960. The duration of pregnancy in rhesus monkeys *Macaca mulatta*. Proc Zool Soc London. 134:595-599.
6039. Krohn, PL & Zuckerman, S. 1950. The effect of androgen and oestrogen on the testes of immature monkeys. J Endocrinol. 6:256-260.
6040. Krohne, DT. 1980. Intraspecific litter size variation in *Microtus californicus* II Variation between populations. Evolution. 34:1174-1182.

6041. Krohne, DT. 1981. Intraspecific litter size variation in *Microtus californicus*: Variation within populations. J Mamm. 62:29-40.
6042. Krol, E. 1985. Reproductive energy budgets of hedgehogs during lactation. Zesz nauk Filii UW, 48, Biol. 10:105-117.
6043. Krölling, O. 1929. Über den Bau der Plazentome der Ziege und Gemse *Rupicapra rupicapra*. Baum Festschrift. [Cited by Krölling, 1931]
6044. Krölling, O. 1931. Über den Bau der Antilopenplazentome. Z Mikr Anat Forsch. 27:211-232.
6045. Kröning, F & Vorreyer, F. 1957. Untersuchungen über Vermehrungsraten und Körpergewichte beim weiblichen Rotwild. Z Jagdwiss. 3:145-153.
6046. Krott, P. 1959. Der Vielfrasz (*Gulo gulo*). VEB Gustav Fischer, Jena.
6047. Krott, P. 1973. Die Fortpflanzung des Edelmarders (*Martes martes* L) in freier Wildbahn. Z Jagdwiss. 19:113-117.
6048. Krott, P. 1982. Der Vielfrasz (*Gulo gulo* Linnaeus 1758) im Ökosystem. Säugetierk Mitt. 30:136-150.
6049. Krott, P & Gardner, C. 1984/85. Paarung des Vielfraszes, *Gulo gulo* (L, 1758) in freier Wildbahn. Saugetierk Mitt. 32:87.
6050. Krsmanovic, L, Mikeš, M, Habijan, V, et al. 1984. Reproductive activity of *Cricetus cricetus* L in Vojvodna, Yugoslavia. Acta Zool Fenn. 171:173-174.
6051. Kruckenberg, SM, Gier, HT, & Dennis, SM. 1973. Postnatal development of the prairie vole, *Microtus ochrogaster*. Lab Anim Sci. 23:53-55.
6052. Kruczek, M. 1986. Seasonal effects on sexual maturation of male bank voles (*Clethrionomys glareolus*). J Reprod Fert. 76:83-89.
6053. Kruczek, M. 1986. Seasonal variation of testicular activity in juvenile and mature bank voles (*Clethrionomys glareolus*). Folia Biol (Kraków). 34:373-380.
6054. Kruczek, M & Marchlewska-Koj, A. 1985. Androgen-dependent proteins in the urine of bank voles (*Clethrionomys glareolus*). J Reprod Fert. 75:189-192.
6055. Kruczek, M & Marchlewska-Koj, A. 1986. Puberty delay of bank vole females in high-density populations. Biol Reprod. 35:537-541.
6055a. Kruczek, M, Marchlewska-Koj, A, & Drickamer, LC. 1989. Social inhibition of sexual maturation in female and male bank voles (*Clethrionomys glareolus*). Acta Theriol. 34:479-485.
6056. Krüger, P. 1963. Paarung und Sexualzyklus als wahrscheinlichie Ursache der Spatherbst-und Winterwurfe beim ausgesetzten Kanada-Biber (*Castor canadensis* Kuhl, 1819) in Finland. Säugetierk Mitt. 11:16-20.
6057. Krumbiegel, I. 1943. Der Afrikanische Elefant. Verlag Dr Paul Schops, Leipzig.
6058. Krutyporoh, F & Treus, V. 1969. Domestication of eland. Moloch myas Skotov Mosk. 4:36-38. [An Breeding Abstr. 1970. 38:ref 778]
6059. Krutzsch, PH. 1946. Some observations on the big brown bat in San Diego County, California. J Mamm. 27:240-242.
6060. Krutzsch, PH. 1954. Notes on the habits of the bat, *Myotis californicus*. J Mamm. 35:539-545.
6061. Krutzsch, PH. 1955. Observations on the California mastiff bat. J Mamm. 36:407-414.
6062. Krutzsch, PH. 1955. Observations on the Mexican free-tailed bat, *Tadarida mexicana*. J Mamm. 36:236-242.
6062a. Krutzsch, PH. 1961. The reproductive cycle in the male vespertilionid bat *Myotis velifer*. Anat Rec. 139:309.
6063. Krutzsch, PH. 1962. Additional data on the os penis of Megachiroptera. J Mamm. 43:34-42.
6064. Krutzsch, PH. 1975. Reproduction of the canyon bat, *Pipistrellus hesperus*, in southwestern United States. Am J Anat. 143:163-200.
6065. Krutzsch, PH. 1979. Male reproductive patterns in nonhibernating bats. J Reprod Fert. 56:333-344.
6066. Krutzsch, PH. 1979. Personal communication. [Cited by Jerett, 1979]
6067. Krutzsch, PH & Crichton, EG. 1985. Observations on the reproductive cycle of female *Molossus fortis* (Chiroptera: Molossidae) in Puerto Rico. J Zool. 207A:137-150.
6068. Krutzsch, PH & Crichton, EG. 1986. Reproduction of the male eastern pipistrelle, *Pipistrellus subflavus*, in the north-eastern United States. J Reprod Fert. 76:91-104.
6069. Krutzsch, PH & Crichton, EG. 1987. Reproductive biology of the male little mastiff bat, *Mormopterus planiceps* (Chiroptera: Molossidae), in southeast Australia. Am J Anat. 178:352-368.
6069a. Krutzsch, PH & Crichton, EG. 1990. Reproductive biology of the male bent-winged bat, *Miniopterus schreibersii* (Vespertilionidae) in southeast South Australia. Acta Anat. 139:109-125.
6070. Krutzsch, PH, Crichton, EG, & Nagle, RB. 1982. Studies on prolonged spermatozoan survival in Chiroptera: A morphological examination of storage and clearance of intrauterine and cauda epididymal spermatozoa in the bats *Myotis lucifugus* and *M velifer*. Am J Anat. 165:421-434.
6071. Krutzsch, PH & Vaughan, TA. 1955. Additional data on the bacula of North American bats. J Mamm. 36:96-100.
6072. Krutzsch, PH, Watson, RH, & Lox, CD. 1976. Reproductive biology of the male leaf-nosed bat, *Macrotus waterhousii* in southwestern United States. Anat Rec. 184:611-636.
6073. Krutzsch, PH & Wells, WW. 1960. Androgenetic activity in the interscapular brown adipose tissue of the male hibernating bat (*Myotis lucifugus*). Proc Soc Exp Biol Med. 105:578-581.
6074. Kruuk, H. 1972. The Spotted Hyena: A Study of Predation and Social Behavior. Univ Chicago Press, Chicago.
6075. Kruuk, H, Conroy, JWH, & Moorhouse, A. 1987. Seasonal reproduction, mortality and food of otters (*Lutra lutra* L) in Shetland. Symp Zool Soc London. 58:263-278.
6076. Kruuk, H & Parish, T. 1982. Factors affecting population density, group size and territory size of the European badger, *Meles meles*. J Zool. 196:31-39.
6077. Krylov, VI. 1966. On sexual maturation of Pacific walrus females. Zool Zhurn. 45:919-927.
6078. Krzywinski, A, Krzywinska, K, Kisza, J, et al. 1980. Milk composition, lactation and the artificial rearing of red deer. Acta Theriol. 25:341-347.
6079. Krzywinski, A, Niedbalska, A, & Twardowski, L. 1984. Growth and development of hand reared deer fawns. Acta Theriol. 29:349-356.
6080. Kubota, K, Shimizu, T, Shibanai, S, et al. 1989. Histological properties and biological significance of pouch in red kangaroo, *Macropus rufus*. Anat Anz. 168:169-179.
6081. Kudolo, GB, Mbai, FN, & Eley, RM. 1986. Reproduction in the vervet monkey (*Cercopithecus aethiops*): Endometrial oestrogen and progesterone receptor dynamics during normal and prolonged menstrual cycles. J Endocrinol. 110:429-439.
6082. Kuehl, TJ & Dukelow, WR. 1975. Ovulation induction during an ovulatory season in *Saimiri sciurus*. J Med Primatol. 4:23-31.
6083. Kuehn, RE, Jensen, GD, & Morrill, RK. 1965. Breeding *Macaca nemestrina*: A program of birth engineering. Folia Primatol. 3:251-262.

6084. Kühlhorn, F. 1943. Beobachtungen über die Biologie von *Cebus apella*. Zool Garten NF. 15:221-234.
6084a. Kühlhorn, F. 1953. Säugetierkundliche Studien aus Süd-Mattogrosso. Säugetierk Mitt. 1:115-122.
6085. Kühn, E. 1953. Zum Wachstum männlicher Borstengürteltiere (*Chaetophractus villosus* Fisch). Zool Garten NF. 20:82-84.
6086. Kuhn, HJ. 1962. Zur Kenntnis der Microchiroptera Liberias. Zool Anz. 168:179-187.
6087. Kuhn, HJ & Schwaier, A. 1973. Implantation, early placentation, and chronology of embryogenesis in *Tupaia belangeri*. Z Anat Entwickl Gesch. 142:315-340.
6088. Kuhn, LW, Wick, WQ, & Pedersen, RJ. 1966. Breeding nests of Townsend's mole in Oregon. J Mamm. 47:239-249.
6089. Kukla, F. 1980. Copulation length related to fertility in mink females. Scientifur. 4(2):22-23.
6090. Kulichkov, AB. 1981. Polygamy of male sables and its relationship with the reproductive performance of females. Nauch Trudy Nauch-Issled Inst Pushn Zverov Krolik. 25:111-116. [Abstr in Scientifur. 1985. 9:305]
6091. Kulicke, H. 1960. Wintervermehrung von Rötelmaus (*Clethrionomys glareolus*), Erdmaus (*Microtus agrestis*) und Gelbhalsmaus (*Apodemus flavicollis*). Z Säugetierk. 25:89-91.
6092. Kulikov, AN & Ivashin, MV. 1958. On the problem of the reproduction cycle of *Balaenoptera physalis* L and *Megaptera nodosa* Bonn of the Atlantic sector of the Atlantic. Zool Zhurn. 38:123-125.
6093. Kulzer, E. 1958. Untersuchungen über die Biologie von Flughunden der Gattung *Rousettus* Gray. Z Morph Ökol Tiere. 47:374-402.
6094. Kulzer, E. 1966. Die Geburt bei Flughunden der Gattung *Rousettus* Gray (Megachiroptera). Z Säugetierk. 31:226-233.
6095. Kulzer, E. 1974. Jugendentwicklung und Temperaturregulation beim Chinchilla (*Chinchilla laniger* Molina, 1782). Z Säugetierk. 39:231-243.
6096. Kumar, A & Kurup, CU. 1985. Sexual behavior of the lion-tailed macaque, *Macaca silenus*. The Lion-Tailed Macaque: Status and Conservation. 109-130, Heltne, PG, ed, Alan R Liss, Inc, New York.
6097. Kumari, S & Prakash, I. 1983. Seasonal variation in the dimension of scent-marking gland of three desert rodents and its possible relationship with their reproductive performance. Proc Indian Acad Sci (Anim Sci). 92:299-304.
6098. Kumazawa, N. 1971. Birth of a Malayan tapir. Animals & Zoos. 23(12):402-405.
6099. Kumirai, A & Jones, JK, Jr. 1990. *Nyctinomops femorosaccus*. Mamm Species. 349:1-5.
6100. Kummer, H & Kurt, F. 1963. Social units of a free-living population of hamadryas baboons. Folia Primatol. 1:4-19.
6101. Kummer, J. 1970. Beobachtungen bei der Aufzucht und Haltung von Feldhasen (*Lepus europaeus* Pallas). Zool Garten NF. 38:138-139.
6102. Kunc, L. 1970. Breeding and rearing the northern lynx *Felis l lynx* at Ostrava Zoo. Int Zoo Yearb. 10:83-8.
6103. Kunc, L & Stehlik, J. 1968. Haltung und Zucht von Luchsen (*Lynx lynx*) im Zoologischen Garten Ostrava. Fr Kölner Zoo. 11:97-99.
6104. Kunkel, G & Toake, K-H. 1986. Beobachtugen zur Fortpflanzungsbiologie medititerraner Zwergfledermäuse (*Pipistrellus pipistrellus*). Z Säugetierk. 51:124-125.
6105. Kuns, ML & Tashian, RE. 1954. Notes on mammals from northern Chiapas, Mexico. J Mamm. 35:100-103.
6105a. Kunz, TH. 1971. Reproduction of some vespertilionid bats in central Iowa. Am Mid Nat. 86:477-486.
6106. Kunz, TH. 1973. Population studies of the cave bat (*Myotis velifer*): Reproduction, growth, and development. Occas Pap Mus Nat Hist Univ KS. 15:1-43.
6107. Kunz, TH. 1974. Reproduction, growth, and mortality of the vespertilionid bat, *Eptesicus fuscus*, in Kansas. J Mamm. 55:1-13.
6108. Kunz, TH. 1980. Daily energy budget of free-living bats. Int Bat Res Conf. 5:369-392.
6109. Kunz, TH. 1982. *Lasionycteris noctivagans*. Mamm Species. 172:1-5.
6110. Kunz, TH. 1987. Post-natal growth and energetics of suckling bats. Recent Advances in the Study of Bats. 395-420, Fenton, MB, Racey, PA, & Rayner, JMV, eds, Cambridge Univ Press.
6110a. Kunz, TH, ed. 1988. Ecological and Behavioral Methods for the Study of Bats. Smithsonian Inst Press, Washington, DC.
6111. Kunz, TH & Anthony, ELP. 1982. Age estimation and post-natal growth in the bat *Myotis lucifugus*. J Mamm. 63:23-32.
6112. Kunz, TH, August, PV, & Burnett, CD. 1983. Harem social organization in cave roosting *Artibeus jamaicensis* (Chiroptera: Phyllostomatidae). Biotropica. 15:133-138.
6113. Kunz, TH & Martin, RA. 1982. *Plecotus townsendii*. Mamm Species. 175:1-6.
6114. Kunz, TH, Stack, MH, & Jeness, R. 1983. A comparison of milk composition in *Myotis lucifugus* and *Eptesicus fuscus* (Chiroptera: Vestertilionidae). Biol Reprod. 28:229-234.
6115. Kuramoto, T. 1972. Studies on bats at the Akiyoshi-dai Plateau, with special reference to the ecological and phylogenetic aspects. Bull Akiyoshi-dai Sci Mus. 8:7-119.
6116. Kurnosov, KM. 1963. Peculiarities of the placenta formation in the elk (*Alces alces*). Zool Zhurn. 42:282-288.
6117. Kuroiwa, J & Imamichi, T. 1977. Growth and reproduction of the chinchilla: Age at vaginal opening, oestrous cycle, gestation period, litter size, sex ratio, and diseases frequently encountered. Exp Anim. 26:213-222.
6118. Kurt, F. 1968. Zusammenhänge zwischen Verhalten und Fortpflanzungsleistung beim Reh (*Capreolus capreolus* L). Z Jagdwiss. 14:97-106.
6119. Kurt, F. 1974. Remarks on the social structure and ecology of the Ceylon elephant in the Yala National Park. The Behavior of Ungulates and its Relation to Management. 618-634, Geist, V & Walther, F, eds, IUCN, Morges, Switzerland.
6119a. Kurta, A. 1980. Status of the Indiana bat, *Myotis sodalis*, in Michigan. MI Academician. 13:31-36.
6120. Kurta, A & Baker, RH. 1990. *Eptesicus fuscus*. Mamm Species. 356:1-10.
6121. Kurta, A, Bell, GP, Nagy, KA, et al. 1989. Water balance of free-ranging little brown bats (*Myotis lucifugus*) during pregnancy and lactation. Can J Zool. 67:2468-2472.
6122. Kurta, A, Bell, GP, Nagy, KA, et al. 1989. Energetics of pregnancy and lactation in free-ranging little brown bats (*Myotis lucifugus*). Physiol Zool. 62:804-818.
6122a. Kurta, A & Stewart, ME. 1990. Parturition in the silver-haired bat, *Lasionycteris noctivagans*, with a description of the neonates. Can Field-Nat. 104:598-600.
6123. Kürten, L. 1983. Haltung und Zucht der neotropischen Fledermaus *Carollia perspicillata*. Z Kölner Zoo. 26:53-57.
6124. Kurz, W. 1904. Der Uterus von *Tarsius spectrum* nach dem Wurf. Anat Hefte (Z Anat Entwicklungsgesch). 23:621-654.
6125. Kuschinski, L. 1974. Breeding binturongs *Arctictis binturong* at Glasgow Zoo. Int Zoo Yearb. 14:124-126.

6126. Kushner, H, Kraft-Schreyer, N, Angelakos, ET, et al. 1982. Analysis of reproductive data in a breeding colony of African green monkeys. J Med Primatol. 11:77-84.
6127. Kuster, J & Paul, A. 1984. Female reproductive characteristics in semifree-ranging barbary macaques (*Macaca sylvanus* L 1758). Folia Primatol. 43:69-83.
6128. Küsthardt, G. 1941. Weitere Beobachtungen an Schneemäusen. Z Säugetierk. 14:257-268.
6129. Kutzer, E. 1958. Untersuchungen über die Biologie von Flughunden der Gattung *Rousettus* Gray. Z Morphol Ökol Tiere. 47:374-402.
6130. Kutzer, E. 1966. Die Geburt bei Flughunden der Gattung *Rousettus* Gray (Megachiorptera). Z Säugetierk. 31:226-233.
6131. Kutzer, E & Frey, H. 1981. Effects of road traffic on hares. Proc World Lagomorph Conf, Univ Guelph, ONT. 1979:501-507.
6132. Kuvlesky, WP, Jr & Keith, LB. 1983. Demography of snowshoe hare populations in Wisconsin. J Mamm. 64:233-244.
6133. Kuwahara, H. 1974. A Great Indian rhinoceros gives birth to a baby. Animals & Zoos. 26(2):39-44.
6134. Kuwahata, T. 1973. On the breeding of cage reared red-backed vole, *Clethrionomys rufocanus bedfordiae* (Thomas). Exp Anim. 22 suppl:231-236.
6135. Kuyper, MA. 1979. A biological study of the golden mole *Amblysomus hottentotus*. M Sc Thesis. Univ Natal, Pietermaritzburg, South Africa.
6136. Kuyper, MA. 1985. The ecology of the golden mole *Amblysomus hottentotus*. Mamm Rev. 15:3-11.
6137. Kuyt, E. 1972. Food habits of wolves on barren ground caribou range. Can Wildl Ser Rep Ser. 21:1-36.
6137a. Kvam, T. 1990. Ovulation rates in European lynx, *Lynx lynx* (L), from Norway. Z Säugetierk. 55:315-320.
6137b. Kvam, T. 1991. Reproduction in European lynx, *Lynx lynx*. Z Säugetierk. 56:146-158.
6138. Kwiecinski, GG, Damassa, DA, & Gustafson, AW. 1986. Control of sex steroid-binding protein (SBP) in the male little brown bat: Relationship of plasma thyroxine levels to the induction of plasma SBP in immature males. J Endocrinol. 110:271-278.
6139. Kwiecinski, GG, Damassa, DA, Gustafson, AW, et al. 1987. Plasma sex steroid binding in Chiroptera. Biol Reprod. 36:628-635.
6140. Kwiecinski, GG, Krook, L, & Wimsatt, WA. 1987. Annual skeletal changes in the little brown bat, *Myotis lucifugus*, with particular reference to pregnancy and lactation. Am J Anat. 178:410-420.
6141. Kwiecinski, GG, Wimsatt, WA, & Krook, L. 1987. Morphology of thyroid C-cells and parathyroid glands in summer-active little brown bats, *Myotis lucifugus lucifugus*, with particular reference to pregnancy and lactation. Am J Anat. 178:421-427.
6142. La Gory, KE. 1980. Diurnal behavior of a white-tailed deer neonate. J Wildl Manage. 44:927-929.
6143. Labhsetwar, AP & Enders, AC. 1968. Progesterone in the corpus luteum and placenta of the armadillo, *Dasypus novemcinctus*. J Reprod Fert. 16:381-387.
6144. Labhsetwar, AP & Enders, AC. 1969. Pituitary LH content during delayed and post-implantation periods in the armadillo (*Dasypus novemcinctus*). J Reprod Fert. 18:383-390.
6144a. Lacey, EA & Sherman, PW. 1991. Social organization of naked mole-rat colonies: Evidence for divisions of labor. The Biology of the Naked Mole-Rat. 274-336 Sherman, PW, Jarvis, JUM, & Alexander, RD, eds, Princeton Univ Press, Princeton, NJ.
6145. Lacey, M. 1969. A note on breeding the nilgai *Boselaphus tragocamelus* at Stanley Zoo. Int Zoo Yearb. 9:115.
6146. Lacher, TE, Jr. 1979. Rates of growth in *Kerodon rupestris* and an assessment of its potential as a domesticated food source. Papers Avulsos Zool Šão Paulo. 33:67-76.
6147. Lacher, TE, Jr. 1981. The comparative social behavior of *Kerodon rupestris* and *Galea spixii* and the evolution of behavior in the Caviidae. Bull Carnegie Mus Nat Hist. 17:1-71.
6148. Lackey, JA. 1967. Growth and development of *Dipodomys stephensi*. J Mamm. 48:624-632.
6148a. Lackey, JA. 1970. Distributional records of bats from Veracruz. J Mamm. 51:384-385.
6149. Lackey, JA. 1976. Reproduction, growth and development in the Yucatan deer mouse, *Peromyscus yucatanicus*. J Mamm. 57:638-655.
6150. Lackey, JA. 1978. Reproduction, growth, and development in high-latitude and low-latitude populations of *Peromyscus leucopus* (Rodentia). J Mamm. 59:69-83.
6151. Lackey, JA. 1991. *Chaetodipus arenarius*. Mamm Species. 384:1-4.
6152. Lackey, JA. 1991. *Chaetodipus spinatus*. Mamm Species. 385:1-4.
6153. Lackey, JA, Huckaby, DG, & Ormiston, BG. 1985. *Peromyscus leucopus*. Mamm Species. 247:1-10.
6154. Ladman, AJ. 1967. The fine structure of the ductuli efferentes of the opossum. Anat Rec. 157:559-576.
6155. Ladovsky, W & Calixto, SL. 1984. Influence of estrogen and progesterone on the uterine sensitivity *in vitro* to neuropituitary hormones in the Brazilian marsupial *Didelphis albiventris*: Comparison with laboratory animals. Gen Comp Endocrinol. 53:69-77.
6156. Lagerkvist, G. 1981. Iller parning. Våra Pälsdjur. 52(3):79-81. [Abstract in Scientifur 6:37, 1982]
6157. Lahti, S & Helminen, M. 1974. The beaver *Castor fiber* (L) and *Castor canadensis* (Kuhl) in Finland. Acta Theriol. 19:177-189.
6158. Lair, H. 1985. Mating seasons and fertility of red squirrels in southern Quebec. Can J Zool. 63:2323-2327.
6159. Lait, H. 1985. Length of gestation in the red squirrel, *Tamiasciurus hudsonicus*. J Mamm. 66:809-810.
6160. Lajus, M, Canivenc, R, & Bonnin-Laffargue, M. 1964. Etude expérimentale de la répartition blastocytaire chez le Blaireau *Meles meles* L. CR Soc Biol. 158:1486-1488.
6161. Lall, SB. 1986. Folliculogenesis in *Rhinopoma kinneari* Wroughton (Microchiroptera: Mammalia). Myotis. 23-24:37-44.
6162. Lam, YM. 1983. Reproduction in the rice field rat, *Rattus argentiventer*. Malay Nat J. 36:249-282.
6163. Lambert, OG. 1962. Breeding the Indian palm squirrel *Funambulus palmarum* L. Int Zoo Yearb. 4:78-79.
6164. Lambiase, JT, Jr, Armann, RP, & Lindzey, JS. 1972. Aspects of reproductive physiology of male white-tailed deer. J Wildl Manage. 36:868-875.
6165. Lamming, GE, ed. 1984. Marshall's Physiology of Reproduction. Churchill Livingstone. London.
6166. Lampe, RP, Jones, JK, Jr, Hoffmann, RS, et al. 1974. The mammals of Carter County, southeastern Montana. Occas Pap Mus Nat Hist Univ KS. 25:1-39.
6167. Lamy, E & Canivenc, R. 1988. Photoperiodic reactivation of corpora lutea formed by gonadotrophic hormones in European badger *Meles meles* L. CR Hebd Séances Acad Sci Paris Ser III Sci Vie. 307:759-762.
6168. Lancaster, JB & Lee, RB. 1965. The annual reproductive cycle in monkeys and apes. Primate Behavior Field Studies of

Monkeys and Apes. 486-513, DeVore, I, ed, Holt, Rinehart & Winston, New York.
6169. Landau, R. 1938. Der ovariale und tubale Abschnitt des Genitaltraktus beim nicht-graviden und beim früh-graviden *Hemicentetes* Weibchen. Biomorphosis. 1:228-264.
6170. Lander, RH. 1979. Alaskan or northern fur seal. Mammals in the Seas, FAO Fish Ser 5. 2:19-27.
6171. Landowski, J. 1958. Erfahrungen in der Haltung von Schimpansen in Warschauer Zoo. Zool Garten NF. 24:256-264.
6171a. Landowski, J. 1972. Erste gelungene künstliche Aufzucht von Binturongs (*Arctictis binturong* Raffl) im Warschauer Zoo. Zool Garten. 42:38-50.
6172. Landry, SO, Jr. 1959. Lactation hair in the Asiatic squirrel and relationship of lactation hair to mammary hair. Science. 130:37.
6173. Landstrom, RE. 1971. Longevity of white-throated woodrat. J Mamm. 52:623.
6174. Lane, HH. 1909. Some observations on the habits and placentation of *Tatu novemcinctum*. State Univ OK (Norman) Res Bull. 1:5-18.
6175. Lane, HK. 1946. Notes on *Pipistrellus subflavus subflavus* (F Cuvier) during the season of parturition. PA Acad Sci. 20:57-61.
6176. Lang, CM. 1967. The estrous cycle of the squirrel monkey (*Saimiri sciureus*). Lab Anim Care. 17:442-451.
6177. Lang, CM. 1967. The estrous cycle of non human primates: A review of the literature. Lab An Care. 17:172-179.
6178. Lang, EM. 1955. Beobachtungen während zweier Giraffengeburten. Säugetierk Mitt. 3:1-5.
6179. Lang, EM. 1956. Haltung und Brunst von *Okapia* in Epulu. Säugetierk Mitt. 4:49-52.
6180. Lang, EM. 1956. Über das Steppenschuppentier (*Manis temminckii*). Zool Garten NF. 21:225-230.
6181. Lang, EM. 1957. Geburt eines Panzernashorns, *Rhinoceros unicornis* im Zoologischen Garten Basel. Säugetierk Mitt. 5:69-70.
6182. Lang, EM. 1958. Zur haltung des Strandwolfes (*Hyaena brunnea*). Zool Garten NF. 24:81-90.
6183. Lang, EM. 1959. The birth of a gorilla at Basle Zoo. Int Zoo Yearb. 1:3-7.
6184. Lang, EM. 1960. Notes on the birth of an okapi at Basle Zoo. Int Zoo Yearb. 2:94.
6185. Lang, EM. 1961. Beobachtungen am Indischen Panzernashorn (*Rhinoceros unicornis*). Zool Garten NF. 25:369-409.
6186. Lang, EM. 1961. Jambo, the second gorilla born at Basle Zoo. Int Zoo Yearb. 3:84-88.
6187. Lang, EM. 1967. The birth of an African elephant *Loxodonta africana* at Basle Zoo. Int Zoo Yearb. 7:154-157.
6188. Lang, EM. 1975. The Indian rhino in captivity. Breeding Endangered Species in Captivity. 293-307, Martin, RD, ed, Academic Press, New York.
6189. Lang, EM. 1976. Haltung und Zucht des kleinen Kudu (*Tragelaphus imberbis*). Zool Garten NF. 46:3-8.
6190. Lang, EM. 1977. The Indian rhino (*Rhinoceros unicornis*) in captivity. J Bombay Nat Hist Soc. 74:160.
6191. Lang, EM. 1978. Experiences with okapis in the Basle Zoo. Acta Zool Pathol Antverpiensia. 71:37-40.
6192. Lang, EM. 1979. Beobachtungen an der Sumpfantilope (*Tragelaphus spekei gratus*) im Zoologischen Garten Basel. Zool Garten NF. 49:8-16.
6193. Lang, EM. 1983. Die Somaliwildesel, *Equus asinus somalicus* im Basel. Zool Garten NF. 53:73-80.
6194. Lang, EM, Leutenegger, M, & Tobler, K. 1977. Indian rhinoceros, *Rhinoceros unicornis*, births in captivity. Int Zoo Yearb. 17:237-238.
6195. Lang, H & Chapin, JP. 1917. The American museum Congo Expedition collecton of bats, Part III, Field notes. Bull Am Mus Nat Hist. 37:497-563.
6196. Langdon, D & Schmidt, K. 1982. Hand-rearing a pygmy hippopotamus *Choeropsis liberiensis* at Melbourne Zoo. Int Zoo Yearb. 22:268-269.
6197. Lange, D. 1920. Notes on flying squirrels and gray squirrels. J Mamm. 1:243-244.
6198. Langevin, P & Barclay, MR. 1990. *Hypsignathus monstrosus*. Mamm Species. 357:1-4.
6199. Langenstein-Issel, B. 1950. Biologische und ökologische Untersuchungen über die Kurzohrmaus (*Pityms subterraneus* de Selys Longchamps). Z Pflanzenb u Pflanzensch. Ser 4 1:145-183.
6200. Langham, N. 1983. Distribution and ecology of small mammals in three rain forest localities of Peninsula Malaysia with particular reference to Kedah Peak. Biotropica. 15:199-206.
6201. Langham, NPE. 1982. The ecology of the common tree shrew: *Tupaia glis* in peninsular Malaysia. J Zool. 197:323-344.
6202. Langham, NPE & Ming, LY. 1977. Muscardi, Bonhote, 1902 a new mammal for peninsular Malaysia. Malay Nat J. 30:39-44.
6203. Langkavel, B. 1888. Hyrax. Zool Jahrb Syst. 3:336-347.
6204. Langley, CH & Giliomee, JH. 1974. Behaviour of the bontebok (*Damaliscus d dorcas*) in the Cape of Good Hope Nature Reserve. JS Afr Wildl Mgt Assoc. 4:117-121.
6205. Langley, R & Fairley, JS. 1982. Seasonal variations in infestations of parasites in a wood mouse *Apodemus sylvaticus* population in the west of Ireland. J Zool. 198:249-261.
6206. Langman, VA. 1977. Cow-calf relationships in giraffe (*Giraffa camelopardalis giraffa*). Z Tierpsychol. 43:264-286.
6207. Langtimm, CA & Dewsbury, DA. 1991. Phylogeny and evolution of rodent copulatory behaviour. Anim Behav. 41:217-225.
6208. Langvatn, R. 1986. Prostaglandin (PGF2α)-induced parturition in red deer (*Cervus elephus* L). Comp Biochem Physiol. 83C:19-22.
6209. Lanier, DL & Dewsbury, DA. 1977. Studies of copulatory behaviour in northern grasshopper mice (*Onychomys leucogaster*). Anim Behav. 25:185-192.
6210. Lanier, DL, Estep, DQ, & Dewsbury, DA. 1975. Copulatory behavior of golden hamsters: Effects on pregnancy. Physiol Behav. 15:209-212.
6211. Lanman, JT. 1977. Parturition in nonhuman primates. Biol Reprod. 16:28-38.
6212. Lanman, JT, Thau, R, Sundaram, K, et al. 1975. Ovarian and placental origins of plasma progesterone following fetectomy in monkeys (*Macaca mulatta*). Endocrinology. 96:591-597.
6213. Lantz, DE. 1910. Raising deer and other large game animals in the United States. USDA Biol Surv Bull. 36:1-61.
6214. Lantz, DE. 1917. The muskrat as a fur bearer. USDA Farm Bull. 869:1-22.
6215. Larkin, P & Roberts, M. 1979. Reproduction in the ring-tailed mongoose *Galidia elegans* at the National Zoological Park, Washington. Int Zoo Yearb. 19:188-193.
6216. Larkin, PA. 1948. The ecology of the mole (*Talpa europaea* L) populations. Ph D Diss. Univ Oxford, Oxford, England.
6217. Larsen, P. 1964. Some basic reproductive characteristics of pronghorn antelope in New Mexico. Proc W Assoc Game Fish Comm. 44:142-145.

6218. Larsen, F. 1984. Reproductive parameters of the minke whale, *Balaenoptera acutorostrata*, off west Greenland. Rep Int Whal Comm Sp Issue. 6:233-236.
6219. Larsen, F & Kapel, FO. 1983. Further biological studies of the west Greenland minke whale. Rep Int Whal Comm. 33:329-332.
6220. Larsen, T. 1984. Polar bear denning and cub production in Svalbard, Norway. J Wildl Manage. 48:320-326.
6221. Larson, LS. 1967. Age structure and sexual maturity within a western Maryland beaver (*Castor canadensis*) population. J Mamm. 48:408-413.
6222. Larsson, HO, Hagelin, M, & Hjern, M. 1982. Observations on a breeding group of pygmy marmosets *Cebuella pygmaea* at Skansen Aquarium. Int Zoo Yearb. 22:88-93.
6223. Larsson, TB, Hansson, L, & Hyholm, E. 1973. Winter reproduction in small rodents in Sweden. Oikos. 24:475-476.
6224. Larsson-Raznikiewicz, M & Mohamed, MA. 1986. Analysis of the casein content in camel *Camelus dromedarius* milk. Swed J Agric Res. 16:1318.
6225. Lasley, BL, Bogart, MH, & Shideler, SE. 1979. A comparison of lemur ovarian cycles. Animal Models for Research on Contraception and Fertility. 417-436, Alexander, NJ, ed, Harper & Row, New York.
6226. Lasley, BL, Czekala, NM, & Lindburg, DG. 1985. Urinary estrogen profiles in the lion-tailed macaque. The Lion-Tailed Macaque: Status and Conservation. 249-159, Hiltne, PG, ed, Alan R Liss, Inc, New York.
6227. Lasley, BL, Czekala, NM, Nakakura, KC, et al. 1979. Armadillos for studies on delayed implantation, quadruplets, uterus simplex, and fetal adrenal physiology. Animal Models for Research on Contraception and Fertility. 447-451, Alexander, NJ, ed, Harper & Row, New York.
6228. Lasley, BL, Hendrickx, AG, & Stabenfeldt, GH. 1974. Estradiol levels near the time of ovulation in the bonnet monkey (*Macaca radiata*). Biol Reprod. 11:237-244.
6229. Lasley, BL, Hodges, JK, & Czekala, NM. 1980. Monitoring the female reproductive cycle of great apes and other primate species by determination of estrogen and LH in small volumes of urine. J Reprod Fert Suppl. 28:121-129.
6230. Lasley, BL, Kasman, LH, Loskutoff, NM, et al. 1984. An update on endocrine monitoring of reproductive physiology. Am Assoc Zoo Vet Abstract Papers Ann Meeting. 1984:165-167.
6231. Lasley, BL, Presley, S, & Czekala, N. 1980. Monitoring ovulation in great apes: Urinary immunoreactive estrogen and bioactive luteinizing hormone. Proc Am Ass Zoo Vet. 1980:37-41.
6232. Lassen, RW, Ferrel, CM, & Leach, H. 1952. Food habits, productivity and condition of Doyle mule deer herd. CA Fish Game. 38:211-224.
6233. Lassieur, S & Wilson, DE. 1989. *Lonchorhina aurita*. Mamm Species. 347:1-4.
6234. Lataste, F. 1882. Sur le bouchon vaginal du *Pachyuromys duprasi* Lataste. Zool Anz. 5:235-239,.
6235. Lataste, F. 1883. Sur le bouchon vaginal des Rongeurs. J Anat Physiol. 19:144-171.
6236. Lataste, F. 1884. Description d'une espèce nouvelle de Gerbilline d'Arabie (*Meriones longifrons*). Proc Zool Soc London. 1884:88-109.
6237. Lataste, F. 1887. Notes prises au jour le jour sur différentes espèces de l'ordre des Rongeurs, observées en captivité. Actes Soc Linn Bordeaux. 40:1-659.
6238. Lataste, F. 1891. Des variations de durée de la gestation chez les Mammifères et des circonstances qui determinent ces variations: Théorie de la gestation retardée. CR Soc Biol. 43:21-31.
6239. Latimer, HB. 1950. Variation in the number and in the weights of the fetuses in each litter in a series of puppies. Growth. 14:107-110.
6240. Latimer, HB & Ibsen, HL. 1932. The postnatal growth in body weight of the cat. Anat Rec. 52:1-5.
6241. Latour, PB. 1987. Observations on demography, reproduction, and morphology of muskoxen (*Ovibos moschatus*) on Banks Islands, Northwest Territories. Can J Zool. 65:265-269.
6242. Lattanzio, RM & Chapman, JA. 1980. Reproductive and physiological cycles in an island population of Norway rats. Bull Chicago Acad Sci. 12:1-68.
6243. Lau, D. 1968. Beitrag zur Geweihentwicklung und Fortpflanzungsbiologie der Hirsche. Z Säugetierk. 33:194-214.
6244. Lauckhart, JB. 1948. Black-tailed deer in Western Washington. Proc W Assoc Game Fish Comm. 28:153-161.
6245. Lauer, BH & Baker, BE. 1969. Whale milk I Fin whale (*Balaenoptera physalus*) and beluga whale (*Delphinapterus leucas*) milk: Gross composition and fatty acid composition. Can J Zool. 47:95-97.
6246. Lauer, BH, Blood, DA, Pearson, AM, et al. 1969. Goat milk I Mountain goat (*Oreamnos americanus*) milk: Gross composition and fatty acid constitution. Can J Zool. 47:5-8.
6247. Lauer, BH, Kuyt, E, & Baker, BE. 1969. Wolf milk I Artic wolf (*Canis lupus arctos*) and husky milk: Gross composition and fatty acid constitution. Can J Zool. 47:99-102.
6248. Laufens, G. 1973. Beiträge zur Biologie der Fransenfledermäuse (*Myotis natteri* Kuhl, 1818). Z Säugetierk. 38:1-14.
6249. Laundre, JW & Keller, BL. 1984. Home range size of coyotes: a critical review. J Wildl Manage. 48:127-139.
6250. Laurie, A & Seidensticker, J. 1977. Behavioural ecology of the sloth bear (*Melursus ursinus*). J Zool. 182:187-204.
6251. Laurie, A. 1982. Behavioural ecology of the greater one-horned rhinoceros (*Rhinoceros unicornis*). J Zool. 196:307-341.
6252. Laurie, AH. 1937. The age of female blue whales and the effect of whaling on the stock. Discovery Rep. 15:223-284.
6253. Laurie, EMO. 1946. The coypu (*Myocastor coypus*) in Great Britain. J Anim Ecol. 15:22-34.
6254. Laurie, EMO. 1946. The reproduction of the house-mouse (*Mus musculus*) living in different environments. Proc R Soc. 133B:248-281.
6255. Laurie, WA, Lang, EM, & Groves, CP. 1983. *Rhinoceros unicornis*. Mamm Species. 211:1-6.
6256. Laursen, L & Bekoff, M. 1978. *Loxodonta africana*. Mamm Species. 92:1-8.
6257. LaVal, RK. 1967. Record of bats from the southeastern United States. J Mamm. 48:645-648.
6258. LaVal, RK. 1969. Records of bats from Honduras and El Salvador. J Mamm. 50:819-822.
6258a. LaVal, RK. 1972. Distributional records and band recoveries of bats from Puebla, Mexico. J Mamm. 16:449-451.
6259. LaVal, RK. 1973. Observations on the biology of *Tadarida brasiliensis cynocephala* in southeastern Louisiana. Am Mid Nat. 89:112-120.
6260. LaVal, RK. 1977. Notes on some Costa Rican bats. Brenesia. 10/11:77-83.
6261. LaVal, RK & Fitch, HS. 1977. Structure, movement and reproduction in three Costa Rican bat communities. Occas Pap Mus Nat Hist Univ KS. 69:1-28.

6262. LaVal, RK & LaVal, ML. 1977. Reproduction and behavior of the African banana bat, *Pipistrellus nanus*. J Mamm. 58:403-410.
6263. LaVal, RK & LaVal, ML. 1979. Notes on reproduction, behavior and abundance of the red bat, *Lasiurus borealis*. J Mamm. 60:209-212.
6264. Lavigne, D. 1979. Harp seal. Mammals in the Seas, FAO Fish Ser 5. 2:76-80.
6265. Lavigne, DM, Stewart, REA, & Fletcher, F. 1982. Changes in composition and energy content of harp seal milk during lactation. Physiol Zool. 55:1-9.
6266. Lavrov, NP. 1957. Akklimatrizatia ondatry v SSR. Isdatelstvo Centrosojuza, Moskva.
6267. Lavrow, N. 1933. To the biology of the musk-rat (Musquash) *Fiber zibethicus* L. Zool Zhur. 12:86-100.
6268. Law, G & Boyle, H. 1984. Breeding the Geoffroy's cat *Felis geoffroyi* at Glasgow Zoo. Int Zoo Yearb. 23:191-195.
6269. Lawlor, TE. 1965. The Yucatan deer mouse, *Peromyscus yucatanicus*. Univ KS Publ Mus Nat Hist. 16:421-438.
6270. Lawlor, TE. 1969. A systematic study of the rodent genus *Ototylomys*. J Mamm. 50:28-42.
6271. Lawlor, TE. 1982. *Ototylomys phyllotis*. Mamm Species. 181:1-3.
6272. Lawrence, B & Loveridge, A. 1953. Zoological results of a fifth expedition to East Africa 1 Mammals from Nyassaland and Tete with notes on the genus *Otomys*. Bull Mus Comp Zool. 110:1-80.
6273. Lawrow, H & Naumow, S. 1933. Die Verbreitung und Biologie des Wüstenziesels in Turkmenien ASSR (*Spermophilopsis leptodoctylus* Licht). Zool Zhurn. 12(2):80-116.
6274. Laws, RM. 1953. A new method of age determination in mammals with special reference to the elephant seal (*Mirounga leonina* Linn). Falkland Isl Depend Survey Sci Rep. 2:1-11.
6275. Laws, RM. 1953. The elephant seal (*Mirounga leonina* Linn) I Growth and age. Falkland Isl Depend Survey Sci Rep. 8:1-62.
6276. Laws, RM. 1956. The elephant seal (*Mirounga leonina* Linn) II General, social and reproductive behaviour. Falkland Isl Depend Survey Sci Rep. 13:1-88.
6277. Laws, RM. 1956. The elephant seal (*Mirounga leonina* Linn) III The physiology of reproduction. Falkland Isl Depend Survey Sci Rep. 15:1-66.
6278. Laws, RM. 1957. On the growth rates of the leopard seal, *Hydrurga leptonyx* (De Blainville, 1820). Säugetierk Mitt. 5:49-55.
6278a. Laws, RM. 1958. Growth rates and ages of crabeater seals, *Lobodon carcinophagus* Jacquinot & Pucheran. Proc Zool Soc London. 130:275-288.
6279. Laws, RM. 1959. The foetal growth rates of whales with special reference to the fin whale, *Balaenoptera physalus* Linn. Discovery Rep. 29:281-308.
6280. Laws, RM. 1960. Researches on the period of conception, duration of gestation and growth of the foetus of the fin whale, based on data from the International Whaling statistics; comments on C Naaktgeborens, EJ Slijper's and WL van Utrecht's paper. Norsk Hvalfangst Tid. 49:216-220.
6281. Laws, RM. 1961. Reproduction, growth and age of southern fin whales. Discovery Rep. 31:327-486.
6282. Laws, RM. 1967. Occurrence of placental scars in the uterus of the African elephant. J Reprod Fert. 14:445-449.
6283. Laws, RM. 1969. Aspects of reproduction in the African elephant, *Loxodonta africana*. J Reprod Fert Suppl. 6:193-217.
6284. Laws, RM. 1977. Seals and whales of the southern ocean. Phil Trans R Soc London Biol Sci. 279B:81-96.
6285. Laws, RM. 1979. Crabeater seal. Mammals in the Seas, FAO Fish Ser 5. 2:115-119.
6286. Laws, RM. 1979. Southern elephant seal. Mammals in the Seas, FAO Fish Ser 5. 2:106-109.
6287. Laws, RM. 1984. Seals. Antarctic Ecology. 2:621-715. Laws, RM, ed, Academic Press, New York.
6288. Laws, RM & Clough, G. 1966. Observations on reproduction in the hippopotamus *Hippopotamus amphibius* L. Symp Zool Soc London. 15:117-140.
6289. Laws, RM & Hofman, RJ. 1979. Ross seal. Mammals in the Seas, FAO Fish Ser 5. 2:120-124.
6290. Laws, RM & Parker, ISC. 1968. Recent studies on elephant populations in east Africa. Symp Zool Soc London. 21:319-359.
6291. Laws, RM, Parker, ISC, & Archer, AL. 1967. Estimating live weights of elephants from hind leg weights. E Afr Wildl J. 5:106-111.
6292. Laws, RM, Parker, ISC, & Johnstone, RCB. 1975. Elephants and Their Habits: The Ecology of Elephants in North Bunyhoro, Uganda. Clarendon Press, Oxford.
6293. Lay, DM. 1967. A study of the mammals of Iran: Resulting from the Street expedition of 1962-1963. Fieldiana Zool. 54:1-282.
6294. Lay, DM. 1978. Observations on reproduction in a population of pocket gophers, *Thomomys bottae*, from Nevada. Southwest Nat. 23:375-380.
6295. Lay, DW. 1942. Ecology of the opossum in eastern Texas. J Mamm. 23:147-159.
6296. Layne, J & Caldwell, DK. 1964. Behavior of the Amazon dolphin *Inia geoffrensis* (Blainville) in captivity. Zoologica. 49:81-108.
6297. Layne, JN. 1954. The biology of the red squirrel, *Tamiasciurus hudsonicus loquax* (Bangs), in central New York. Ecol Monogr. 24:227-267.
6298. Layne, JN. 1958. Notes on mammals of southern Illinois. Am Mid Nat. 60:219-254.
6299. Layne, JN. 1958. Reproductive characteristics of the grey fox in southern Illinois. J Wildl Manage. 22:157-163.
6300. Layne, JN. 1959. Growth and development of the eastern harvest mouse, *Reithrodontomys humulis*. Bull FL State Mus Biol Sci. 4:59-96.
6301. Layne, JN. 1966. Postnatal development and growth of *Peromyscus floridanus*. Growth. 30:23-45.
6302. Layne, JN. 1968. Ontogeny. Biology of *Peromyscus* (Rodentia). Sp Publ Am Soc Mamm. 2:148-253.
6303. Layne, JN. 1974. Ecology of small mammals in a flatwood habitat in north-central Florida, with emphasis on the cotton rat (*Sigmodon hispidus*). Am Mus Novit. 2544:1-48.
6304. Layne, JN & Hamilton, WJ, Jr. 1954. The young of the woodland jumping mouse, *Napaeozapus insignis insignis* (Miller). Am Mid Nat. 52:242-247.
6305. Layne, JN & McKeon, WH. 1956. Notes on the development of the red fox fetus. NY Fish Game J. 3:120-128.
6306. Layne, JN & McKeon, WH. 1956. Some aspects of red fox and gray fox reproduction in New York. NY Fish Game J. 3:44-74.
6307. Lazell, JD & Jarecki, L. 1985. Bats of Guana, British Virgin Islands. Bull Am Mus Nat Hist. 2819:1-7.
6307a. Lazenby-Cohen, KA & Cockburn, A. 1988. Lek promiscuity in a semelparous mammal, *Antechinus stuartii* (Marsupialia: Dasyuridae)?. Behav Ecol Sociobiol. 22:195-202.
6308. Le Boulenge, E. 1972. État de nos connaissances sur l'écologie du Rat musk *Ondatra zibethica* L. Terre Vie. 26:3-37.

6309. Le Count, AL. 1983. Evidence of wild black bears breeding while raising cubs. J Wildl Manage. 47:264-268.
6310. Le Gros Clark, WE. 1924. Notes on the living tarsier (*Tarsius spectrum*). Proc Zool Soc London. 1924:217-223.
6311. Le Gros Clark, WE. 1926. On the anatomy of the pen-tailed tree-shrew (*Ptilocercus lowii*). Proc Zool Soc London. 1926:1179-1309.
6312. Le Gros Clark, WE, McKeown, T, & Zuckerman, S. 1939. Visual pathways concerned in gonadal stimulation in ferrets. Proc R Soc London Biol Sci. 126B:449-468.
6312a. Le Louarn, H. 1968. Premiers résultats d'une étude sur les Micromammifères et les Oiseaux dans le Mélèzein du Briançonnais. Terre Vie. 22:326-342.
6313. Le Louarn, H & Janear, G. 1975. Repartition et biologie du Campagnol des neiges *Microtus nivalis* Martins dans la région de Briançon. Mammalia. 39:589-694.
6314. Le Louarn, H & Spitz, F. 1974. Biologie et écologie du Lérot *Eliomys quercinus* L dans les hautes-Alpes. Terre Vie. 28:544-563.
6315. le Riche, M. 1970. Birth of an oribi. Africana. 4(4):40-42.
6316. le Roux, P. 1966. The blue duiker (*Cephalophus monticola*). Unpublished review. Compiled during B Sc (Hons) Wildl Mgmt course. Pretoria Univ. [Cited by Mentis, 1972]
6317. Le Souef, AS. 1922. Notes on the nesting and breeding habits of the house building rat (*Conilurus conditor*) and Banfield's rat (*Uromys banfieldi*). Aust Zool. 3:15-16.
6318. Leader-Williams, N. 1979. Abnormal testes in reindeer, *Rangifer tarandus*. J Reprod Fert. 57:127-130.
6319. Leader-Williams, N. 1979. Age-related changes in the testicular and antler cycles of reindeer, *Rangifer tarandus*. J Reprod Fert. 57:117-126.
6320. Leader-Williams, N & Rosser, AM. 1983. Ovarian characteristics and reproductive performance of reindeer, *Rangifer tarandus*. J Reprod Fert. 67:247-256.
6321. Leamy, L. 1981. The effect of litter size on fertility in *Peromyscus leucopus*. J Mamm. 62:692-697.
6322. Leatherwood, S. 1977. Some preliminary impressions on the numbers and social behavior of free-swimming bottlenosed dolphin calves (*Tursiops truncatus*) in the northern Gulf of Mexico. Breeding Dolphins Present Status, Suggestions for the Future US Mar Mamm Comm Rep No MMC-76/07 NTIS No PB 273673. 143-167, Ridgeway, SH & Benirschke, K, eds, US Dept Commerce Nat Tech Info Serv, Arlington.
6323. Leatherwood, S & Walker, WA. 1979. The northern right whale dolphin *Lissodelphis borealis* Peale in the eastern North Pacific. Behaviour of Marine Animals. 3:85-141. Winn, HE & Olla, BL, eds, Academic Press, New York.
6324. Leavitt, WW & Blaha, GC. 1970. Circulating progesterone levels in the golden hamster during the estrous cycle, pregnancy, and lactation. Biol Reprod. 3:353-361.
6325. Lebas, F. 1987. Influence de la taille de la portée et de la production laitiere sur la quantité d'aliment ingerée par la Lapine allaitante. Reprod Nutr Dév. 27:207-208.
6326. Leberg, PL, Kennedy, ML, & van den Bussche, RA. 1983. Opossum demography and scent station visitation in western Tennessee. Proc Ann Conf SE Assoc Fish Wildl Agen. 37:34-40.
6327. LeBlanc, JS. 1956. Rearing of lemmings at Fort Churchill, Manitoba. J Mamm. 37:447-448.
6328. LeBoeuf, BJ. 1972. Sexual behavior in the northern elephant seal *Mirounga angustirostris*. Behaviour. 41:1-26.
6329. LeBoeuf, BJ. 1979. Northern elephant seal. Mammals in the Seas, FAO Fish Ser 5. 2:110-114.
6330. LeBoeuf, BJ & Briggs, KT. 1977. The cost of living in a seal harem. Mammalia. 41:167-195.
6331. LeBoeuf, BJ & Ortiz, CL. 1977. Composition of elephant seal milk. J Mamm. 58:683-685.
6332. LeBoeuf, BJ & Peterson, RS. 1969. Social status and mating activity in elephant seals. Science. 163:91-93.
6333. LeBoeuf, BJ, Whiting, RJ, & Gantt, RF. 1972. Perinatal behavior of northern elephant seal females and their young. Behaviour. 43:121-156.
6334. Lécaillon, A. 1908. Sur les cellules interstielles du testicule de la Taupe (*Talpa europaea*) considéré en dehors de la période de reproduction. CR Soc Biol. 66:599-601.
6335. Lechleitner, RR. 1959. Sex ratio, age classes and reproduction of the black-tailed jack rabbit. J Mamm. 40:63-81.
6336. Lecyk, M. 1958. The autumnal involution of the testes in the field vole (*Microtus arvalis* Pall). Zool Pol. 9:215-222.
6337. Lecyk, M. 1960. Seasonal changes in the epididymis of the field vole *Microtus arvalis* Pall. Zool Pol. 10:173-182.
6338. Lecyk, M. 1960. The possibility of superfetation in *Microtus arvalis* Pall. Zool Pol. 10:325-327.
6339. Lecyk, M. 1962. The effect of the length of daylight on reproduction in field vole (*Microtus arvalis* Pall). Zool Pol. 12:189-221.
6340. Lecyk, M. 1962. The dependence of breeding in the field vole (*Microtus arvalis* Pall) on light intensity and wave length. Zool Pol. 12:255-268.
6341. Lecyk, M. 1963. The effect of short daylight on sexual maturation in young individuals of the vole, *Microtus arvalis* Pall. Zool Pol. 13:77-86.
6342. Lederer, G. 1960. Ein Beitrag zur Ernährung, Haltung und Zucht pflanzenfressender Wildtiere in Gefangenschaft. Zool Garten NF. 25:215-234.
6343. Ledger, HP. 1963. Weights of some east African mammals. E Afr Wildl J. 1:123-124.
6344. Lee, AK & Cockburn, A. 1985. Evolutionary Ecology of Marsupials. Cambridge Univ Press, Cambridge.
6345. Lee, AK, Handasyde, KA, & Sanson, GD. 1991. Biology of the Koala. Surrey Beatty & Sons, NSW, Australia.
6346. Lee, AK, Woolley, P, & Braithwaite, RW. 1982. Life history strategies of dasyurid marsupials. Carnivorous Marsupials. 1:1-11. Archer, M, ed, R Zool Soc NSW, Mossman.
6347. Lee, C & Horvath, DJ. 1969. Management of the meadow vole (*Microtus pennsylvanicus*). Lab Anim Care. 19:88-91.
6348. Lee, C, Horvath, DJ, Metcalfe, RW, et al. 1970. Ovulation in *Microtus pennsylvanicus* in a laboratory environment. Lab Anim Care. 20:1098-1102.
6349. Lee, PC. 1987. Allomothering among African elephants. Anim Behav. 35:278-291.
6350. Lee, PC. 1987. Nutrition, fertility and maternal investment in primates. J Zool. 213:409-422.
6351. Lee, PC & Moss, CJ. 1986. Early maternal investment in male and female African elephant calves. Behav Ecol Sociobiol. 18:353-361.
6352. Lee, SY, Mossman, HW, Mossman, AS, et al. 1977. Evidence for a specific nidation site in ruminants. Am J Anat. 150:631-640.
6353. Lee, TG. 1902. On the early development of *Spermophilus tridecimlineatus*, a new type of mammalilan placentation. Science. 15:525-526.
6354. Lee, TG. 1903. Implantation of the ovum in *Spermophilus tridecemlineatus* Mitch. 417-435, EL Mark Anniversary Volume, Holt & Co, New York.
6355. Lee, TG. 1918. The implantation of the blastocyst and the formation of the decidual cavity in *Dipodomys*. Anat Rec. 14:43.

6356. Lee, TM, Smale, L, Zucker, I, et al. 1987. Role of photoperiod during pregnancy and lactation in the meadow vole, *Microtus pennsylvanicus*. J Reprod Fert. 81:343-350.
6357. Lee, TM, Smale, L, Zucker, I, et al. 1987. Influence of daylength experienced by dams on post-natal development of young meadow voles (*Microtus pennsylvanicus*). J Reprod Fert. 81:337-342.
6358. Lee, TM, Smale, L, Zucker, I, et al. 1987. Interaction of daylength and lactation in the control of pelage development and nest-building in female meadow voles (*Microtus pennsylvanicus*). J Reprod Fert. 81:351-356.
6359. Lee, TM, Spears, N, Tuthill, CR, et al. 1989. Maternal melatonin treatment influences rates of neonatal development of meadow vole pups. Biol Reprod. 40:495-502.
6359a. Lee, TM & Zucker, I. 1990. Photoperiod synchronizes reproduction and growth in the southern flying squirrel, *Glaucomys volans*. Can J Zool. 68:134-139.
6360. Leege, TA & Williams, RM. 1967. Beaver productivity in Idaho. J Wildl Manage. 31:326-332.
6361. Lefèvre, C. 1966. Etude de la croissance en élevage de *Pitymys duodecimcostatus* de Selys-Longchamps, originaires du Gard. Mammalia. 30:56-63.
6362. Legay, JM & Pontier, D. 1985. Age and fertility in domestic cats populations, *Felis catus*. Mammalia. 49:395-402.
6363. Lehman, LE. 1968. September birth of raccoons in Indiana. J Mamm. 49:126-127.
6364. Lehmann, I. 1938. Die Paravaginaldrüse von *Hemicentetes*. Biomorphosis. 1:202-227.
6365. Lehmkuhl, JF. 1980. Some aspects of the life history of blackbuck in Nepal. J Bombay Nat Hist Soc. 77:444-449.
6366. Leichter, J. 1986. Effect of paternal alcohol ingestion on fetal growth in rats. Growth. 50:228-233.
6367. Leiser, R & Enders, AC. 1980. Light-and-electron-microscopic study of the near-term paraplacenta of the domestic cat. Polar zone and paraplacental junctional areas. Acta Anat. 106:293-311.
6368. Leitch, I, Hytten, FE, & Billewicz, WZ. 1959. The maternal and neonatal weights of some mammalia. Proc Zool Soc London. 133:11-28.
6369. Lekagul, B & McNeely, JA. 1977. Mammals of Thailand. Kurusapha Ladprao Press, Bangkok.
6370. not used
6371. Lemke, CW. 1957. An unusually late pregnancy in a Wisconsin cottontail. J Mamm. 38:275.
6372. Lemke, TO, Cadena, A, Pine, RH, et al. 1982. Notes on opossums, bats, and rodents new to the fauna of Columbia. Mammalia. 46:225-234.
6372a. Lemke, TO & Tamsitt, JR. 1979. *Anoura cultrata* (Chiroptera: Phyllostomatidae) from Colombia. Mammalia. 43:579-581.
6373. Lemon, M. 1972. Peripheral plasma progesterone during pregnancy and the oestrous cycle in the tammar wallaby, *Macropus eugenii*. J Endocrinol. 55:63-71.
6374. Lemon, M & Bailey, LF. 1966. A specific protein difference in the milk from two mammary glands of a red kangaroo. Aust J Exp Biol Med Sci. 44:705-708.
6375. Lemon, M & Barker, S. 1967. Changes in milk composition of the red kangaroo, *Megaleia rufa* (Desmarest), during lactation. Aust J Exp Biol Med Sci. 45:213-219.
6376. Lensink, CJ. 1954. Occurrence of *Microtus xanthognathus* in Alaska. J Mamm. 35:259-260.
6377. Lent, PC. 1966. The caribou of northwestern Alaska. Environment of the Cape Thompson Region, Alaska. 481-517, Wilimovsky, NJ & Wolfe, JN, eds, US Atomic Energy Commission, Oak Ridge.
6378. Lent, PC. 1969. A preliminary study of the Okavango lechwe (*Kobus leche leche* Gray. E Afr Wildl J. 7:147-157.
6379. Lent, PC. 1974. A review of rutting behavior in moose. Nat Can. 101:307-323.
6380. Lent, PC. 1986. Observations on antler shedding by female barren-ground caribou. Can J Zool. 43:553-558.
6381. Lent, PC. 1988. *Ovibos moschatus*. Mamm Species. 302:1-9.
6381a. Lentfer, JW. 1955. A two-year study of the Rocky Mountain goat in the Crazy Mountains, Montana. J Wildl Manag. 19:417-429.
6382. Lentfer, JW & Sanders, DK. 1973. Notes on the captive wolf (*Canis lupus*) colony, Barrow, Alaska. Can J Zool. 51:623-627.
6383. Leon, JB, Smith, BB, Timm, KI, et al. 1990. Endocrine changes during pregnancy, parturition and the early post-partum period in the llama (*Lama glama*). J Reprod Fert. 88:503-511.
6384. Leonard, BV. 1976. The effect of fire on small mammals in southeastern Australia. Bull Aust Mamm Soc. 3:16.
6385. Leonard, RD. 1986. Aspects of reproduction of the fisher, *Martes pennanti*, in Manitoba. Can Field-Nat. 100:32-44.
6386. Leopold, AS, Riney, T, McCain, R, et al. 1951. The Jawbone deer herd. CA Div Fish Game, Game Bull. 4:1-139.
6387. Leopold, AS. 1959. Wildlife of Mexico. Univ CA Press, Berkeley.
6388. Lepri, JJ. 1983. Regulation of reproduction in an outbred colony of pine voles. Proc E Pine Meadow Vole Symp. 7:120-122.
6389. Lepri, JJ & Vandenbergh, JG. 1986. Puberty in pine voles, *Microtus pinetorum*, and the influence of chemosignals on female reproduction. Biol Reprod. 34:370-377.
6390. Lequin, RM, Elvers, LH, & Berlens, APMG. 1981. Early detection of pregnancy in rhesus and stump-tailed maceques (*Macaca mulatta* and *Macaca arctoides*). J Med Primatol. 10:189-198.
6391. Leraas, HJ. 1938. Observations on the growth and behavior of harvest mice. J Mamm. 19:441-444.
6392. Leslie, DM, Jr & Douglas, CL. 1979. Desert bighorn sheep on the River Mountains, Nevada. Wildl Monogr. 66:1-56.
6393. Leslie, G. 1970. Observations on the oriental short-clawed otter *Aonyx cinerea* at Aberdeen Zoo. Int Zoo Yearb. 10:79-81.
6394. Leslie, G. 1971. Note on the birth of a Malaysian fruit bat *Cynopterus brachyotis* at Aberdeen Zoo. Int Zoo Yearb. 11:79.
6395. Leslie, G. 1973. Breeding Scottish wild cats *Felis silvestris grampia* at Aberdeen Zoo. Int Zoo Yearb. 13:150.
6396. Leslie, G. 1974. Breeding the grey seal *Halinchoerus grypus* at Aberdeen Zoo. Int Zoo Yearb. 14:130-131.
6397. Leslie, PH & Ranson, RM. 1940. The mortality, fertility and rate of natural increase of the vole (*Microtus agrestis*) as observed in the laboratory. J Anim Ecol. 9:27-52.
6398. Leslie, PH, Tener, JS, Vizoso, M, et al. 1955. The longevity and fertility of the Orkney vole, *Microtus orcadenus*, as observed in the laboratory. Proc Zool Soc London. 125:115-125.
6399. Leslie, PH, Venables, VM, & Venables, LSV. 1952. The fertility and population structure of the brown rat (*Rattus norvegicus*) in corn-ricks and some other habitats. Proc Zool Soc London. 122:187-238.
6400. Leslie, PW & Fry, PH. 1989. Extreme seasonality of births among nomadic Turkana pastoralists. Am J Phys Anthropol. 79:103-115.

6401. Lessa, EP & Cook, JA. 1989. Interspecific variation in penial characters in the Genus *Ctenomys* (Rodentia: Octodontidae). J Mamm. 70:856-860.
6402. Letellier, F & Petter, F. 1962. Reproduction en captivité d'un Rongeur de Madagascar, *Macrotarsomys bastardi*. Mammalia. 26:132-133.
6403. Letizkaja, EP. 1984. Materials on reproduction and postnatal development of *Ellobius talpinus* (Rodentia, Cricetidae). Zool Zhurn. 63:1084-1089.
6404. Leung, K, Merkatz, I, & Solomon, S. 1972. Metabolism of ^{14}C-estradiol-17β injected intravenously into the pregnant baboon (*Papio cynocephalus*). Endocrinology. 91:523-528.
6405. Leutenegger, M. 1978. Pygmy hippopotamus *Choeropsis liberiensis* births in captivity. Int Zoo Yearb. 18:234.
6406. Leutenegger, W. 1979. Evolution of litter size in primates. Am Nat. 114:525-531.
6407. Leuthold, W. 1966. Variations in territorial behavior of Uganda kob *Adenota kob thomasi* (Neumann 1896). Behaviour. 27:215-258.
6408. Leuthold, W. 1971. Observation on the mother-young relationship in some antelopes. E Afr Wildl J. 9:153-4.
6409. Leuthold, W. 1974. Observations on home range and social organizatiodn of lesser kudu, *Tragelaphus imberbis* (Blyth, 1869). The Behaviour of Ungulates and Its Relation to Management. 206-234, Geist, V & Walther, F, eds, IUCN, Morges, Switzerland.
6410. Leuthold, W. 1978. On the ecology of the gerenuk *Litocranius walleri*. J Anim Ecol. 47:561-580.
6411. Leuthold, W. 1979. The lesser kudu, *Tragelaphus imberbis* (Blyth, 1869) Ecology and behaviour of an African antelope. Säugetierk Mitt. 27:1-75.
6412. Leuthold, W & Leuthold, BM. 1975. Parturition and related behaviour in the African elephant. Z Tierpsychol. 39:75-84.
6413. Leuthold, W & Leuthold, BM. 1975. Temporal patterns of reproduction in ungulates of Tsavo East National Park, Kenya. E Afr Wildl J. 13:159-169.
6414. Lewis, AW. 1972. Seasonal population changes in the cactus mouse, *Peromyscus eremicus*. Southwest Nat. 17:85-93.
6415. Lewis, JB. 1940. Mammals of Amelia County, Virginia. J Mamm. 21:422-428.
6416. Lewis, JG & Wilson, RT. 1979. The ecology of Swayne's hartebeest. Biol Cons. 15:1-12.
6417. Lewis, PR, Fletcher, TP, & Renfree, MB. 1986. Prostaglandin in the peripheral plasma of tammar wallabies during parturition. J Endocrinol. 111:103-109.
6418. Lewis, RE & Harrison, DL. 1962. Notes on bats from the Republic of Lebanon. Proc Zool Soc London. 138:473-486.
6419. Lewis, RE, Lewis, JH, & Atallah, SI. 1967. A review of Lebanese mammals: Lagomorpha and Rodentia. J Zool. 153:45-70.
6420. Lewis, RE, Lewis, JH, & Atallah, SI. 1968. A review of Lebanese mammals: Carnivora, Pinnipedia, Hyrocoidea, and Artiodactyla. J Zool. 154:517-531.
6421. Lewis, SE & Wilson, DE. 1987. *Vampyressa pusilla*. Mamm Species. 292:1-5.
6422. Leyhausen, P. 1966. Breeding the Brazilian ocelot-cat *Leopardus tigrinus* in captivity. Int Zoo Yearb. 6:176-178.
6423. Leyhausen, P & Tonkin, B. 1966. Breeding the black-footed cat *Felis nigripes* in captivity. Int Zoo Yearb. 6:178-182.
6424. Lhuillery, C, Martinet, L, Demarne, Y, et al. 1984. Food intake in captive leverets before weaning and the composition of the milk of the brown doe-hare (*Lepus europaeus*). Comp Biochem Physiol. 78A:73-76.
6425. not used
6426. not used
6427. not used
6428. Liché, H & Wodzicki, K. 1939. Vaginal smears and the oestrous cycle of the cat and lioness. Nature. 144:245-246.
6429. Licht, P, Zucker, I, Hubbard, G, et al. 1982. Circannual rhythms of plasma testosterone and luteinizing hormone levels in golden-mantled ground squirrels *Spermophilus lateralis*. Biol Reprod. 27:411-418.
6430. Lidicker, WZ, Jr. 1960. An analysis of intraspecific variation in the kangaroo rat *Dipodomys merriami*. Univ CA Publ Zool. 67:125-218.
6431. Lidicker, WZ, Jr. 1960. The baculum of *Dipodomys ornatus* and its implication for superspecific groupings of kangaroo rats. J Mamm. 41:495-499.
6432. Lidicker, WZ, Jr. 1966. Ecological observations on a feral house mouse population declining to extinction. Ecol Monogr. 36:27-50.
6433. Lidicker, WZ, Jr. 1973. Regulation of numbers in an island population of the California vole, a problem in community dynamics. Ecol Monogr. 43:271-302.
6434. Lidicker, WZ, Jr. 1976. Experimental manipulation of the timing of reproduction in the California vole. Res Pop Ecol. 18:14-27.
6435. Lidicker, WZ, Jr. 1980. The social biology of the California vole. Biologist. 62:46-55.
6436. Lidicker, WZ, Jr & Marlow, BJ. 1970. A review of the dasyurid marsupials genus *Antechinomys* Krefft. Mammalia. 34:212-227.
6436a. Lidicker, WZ, Jr & Ziegler, AC. 1968. Report on a collection of mammals from eastern New Guinea including species keys for fourteen genera. Univ CA Publ Zool. 87:1-64.
6437. Lie, G. 1955. Superfetation in cats, and some observations on the pubertal age of female cats. Nyt Mag Zool. 3:66-69.
6438. Lienhard, U. 1970. Beitrag zu-Wasserspitzmaus, *Neomys fodiens* (Pennant, 1771) mit hoher Embryonenzahl. Z Säugetierk. 35:106-107.
6439. Lienhardt, G. 1979. Beobachtungen zum Verhalten des Igels (*Erinaceus europaeus*) und seine Überlebensmöchlichkeit im heutigen Biotop. Zool Beitr. 25:447-484.
6440. Lienhardt, G. 1982. Beobachtungen zur Morphologie, Jugendentwicklung und zum Verhalten von Weiszbauchigeln *Erinaceus albiventris* (*Ateleris pruneri* [Wagner 1841]) in Gefangenschaft. Säugetierk Mitt. 30:251-259.
6441. Lienhart, R. 1940. A propos de la durée de la gestation chez le Lièvre. C R Soc Biol. 133:133-135.
6442. Liers, EE. 1951. Notes on the river otter (*Lutra canadensis*). J Mamm. 32:1-9.
6443. Liers, EE. 1958. Early breeding in the river otter. J Mamm. 39:438-439.
6444. Liggins, GC, France, JT, Knox, BS, et al. 1979. High corticosteroid levels in plasma of adult and foetal Weddell seals (*Leptonychotes weddelli*). Acta Endocrinol. 90:718-726.
6444a. Lilford, Lord. 1892. untitled. Proc Zool Soc London. 1892:542.
6445. Lim, BK. 1987. *Lepus townsendii*. Mamm Species. 288:1-6.
6445a. Lim, BL. 1970. Food habits and breeding cycle of the Malaysian fruit-eating bat, *Cynopterus brachyotis*. J Mamm. 51:174-177.
6446. Lim, BL. 1973. Breeding pattern, food habits and parasitic infestation of bats in Gunong Brinchang. Malay Nat J. 26:6-13.

6446a. Lim, BL. 1973. The banded linsang and the banded musang of west Malaysia. Malay Nat J. 26:105-111.
6447. Lim, BL, Chai, KS, & Muul, I. 1972. Notes on the food habit of bats from the fourth division, Sarawak, with special reference to a new record of Bornean bat. Sarawak Mus J. 20:351-357.
6448. Lim, BL & Muul, I. 1975. Notes on a rare species of arboreal rat, *Pithechir parvus* Kloss. Malay Nat J. 28:181-185.
6448a. Lim, BL & Omar, IARB. 1961. Observations on the habits in captivity of two species of wild cats, the leopard cat and the flat-headed cat. Malay Nat J. 15:48-51.
6449. Lincoln, DW & Renfree, MB. 1981. Mammary gland growth and milk ejection in the agile wallaby, *Macropus agilis*, displaying concurrent asynchronous lactation. J Reprod Fert. 63:193-203.
6450. Lincoln, GA. 1971. Puberty in a seasonally breeding male, the red deer stag *Cervus elaphus* L). J Reprod Fert. 25:41-54.
6451. Lincoln, GA. 1971. The seasonal reproductive changes in the red deer stag (*Cervus elaphus*). J Zool. 163:105-123.
6452. Lincoln, GA. 1974. Reproduction and "March madness" in the brown hare *Lepus europaeus*. J Zool. 174:1-14.
6453. Lincoln, GA. 1976. Seasonal changes in the pineal gland related to the reproductive cycle in the male hare, *Lepus europaeus*. J Reprod Fert. 46:489-491.
6454. Lincoln, GA. 1978. Plasma testosterone profiles in male macropodid marsupials. J Endocrinol. 77:347-351.
6455. Lincoln, GA. 1984. Antlers and their regeneration-A study using hummels, hinds and haviers. Proc R Soc Edinb. 82B:243-259.
6456. Lincoln, GA. 1987. Long-term stimulatory effects of a continuous infusion of LHRH agonist on testicular function in male red deer (*Cervus elaphus*). J Reprod Fert. 80:257-261.
6457. Lincoln, GA, Fraser, HM, & Fletcher, TJ. 1982. Antler growth in male red deer (*Cervus elaphus*) after active immunization against LH-RH. J Reprod Fert. 66:703-708.
6458. Lincoln, GA, Fraser, HM, & Fletcher, TJ. 1984. Induction of early rutting in male red deer (*Cervus elaphus*) by melatonin and its dependence on LHRH. J Reprod Fert. 72:339-343.
6459. Lincoln, GA & Guiness, FE. 1972. Effect of altered photoperiod on delayed implantation and moulting in roe deer. J Reprod Fert. 31:455-457.
6460. Lincoln, GA & Guiness, FE. 1973. The sexual significance of the rut in red deer. J Reprod Fert Suppl. 19:475-489.
6461. Lincoln, GA & Guiness, FE. 1977. Sexual selection in a herd of red deer. Proc Int Symp Comp Biol Reprod. 4:33-38.
6462. Lincoln, GA & Kay, RNB. 1979. Effects of season on the secretion of LH and testosterone in intact and castrated deer stags *Cervus elaphus*. J Reprod Fert. 55:75-80.
6463. Lincoln, GA, Youngson, RW, & Short, RV. 1970. The social and sexual behaviour of the red deer stag. J Reprod Fert Suppl. 11:71-103.
6464. Lindburg, DG & Lasley, BL. 1985. Strategies for optimizing the reproductive potential of lion-tailed macaque colonies in captivity. The Lion-Tailed Macaque: Status and Conservation. 343-356, Heltne, PG, ed, AR Liss, Inc, New York.
6465. Lindburg, DG, Shideler, S, & Fitch, H. 1985. Sexual behaviour in relation to time if ovulation in the lion-tailed macaque. Monogr Primatol. 7:131-148.
6466. Lindburg, DG. 1971. The rhesus monkey in north India: An ecological and behavioral study. Primate Behavior. 2-106, Rosenblum, LA, ed, Academic Press, New York.
6467. Lindburg, DG. 1983. Mating behavior and estrus in the Indian rhesus monkey. Perspectives in Primate Biology. 45-61, Seth, PK, ed, Today & Tomorrows Printers & Pub, New Delhi, India.
6468. Lindemann, W. 1953. Einiges über die Wildkatze der Ostkarpaten (*Felis s silvestris* Schreber, 1777). Säugetierk Mitt. 1:73-74.
6469. Lindemann, W. 1954. Zur Rassenfrage und Fortpflanzungsbiologie des karpatischen Braunbären, *Ursus arctos arctos* Linne, 1758. Säugetierk Mitt. 2:1-8.
6470. Lindeque, M & Skinner, JD. 1982. Aseasonal breeding in the spotted hyena (*Crocuta crocuta*, Erxleben), in southern Africa. Afr J Ecol. 20:271-278.
6471. Lindeque, M & Skinner, JD. 1982. Fetal androgens and sexual mimicry in spotted hyaenas (*Crocuta crocuta*). J Reprod Fert. 65:405-410.
6472. Lindeque, M, Skinner, JD, & Millar, RP. 1986. Adrenal and gonadal contribution to circulating androgens in spotted hyaenas (*Crocuta crocuta*) as revealed by LHRH, hCG and ACTH stimulation. J Reprod Fert. 78:211-217.
6473. Lindgren, E. 1975. An introduction to the biology of the rusa deer (*Cervus timorensis*) in the western district of Papua New Guinea. Fish Wildl Pap Vict. 8:36-51.
6474. Lindsay, DM. 1956. Some bat records from southeastern Indiana. J Mamm. 37:543-545.
6475. Lindsay, DM. 1960. Mammals of Ripley and Jefferson counties, Indiana. J Mamm. 41:253-262.
6476. Lindsay, N. 1983. Management of lemurs at Jersey Zoo. Management Prosimians and New World Primates, Proc Assn Br Wild An Keepers Symp. 8:29-34.
6477. Lindsay, NBD. 1979. A report on the field study of Geoffroy's tamarin *Saguinus oedipus geoffroyi*. Dodo, J Jersey Wildl Preserv Trust. 16:27-51.
6478. Lindsay, NBD. 1982. A second report on the management and breeding of the volcano rabbit *Romerolagus diazi* at the Jersey Wildlife Preservation Trust. Dodo, J Jersey Wildl Preserv Trust. 19:46-51.
6479. Lindsey, AA. 1937. The Weddell seal in the Bay of Whales, Antarctica. J Mamm. 18:127-144.
6480. Ling, JK. 1980. Sea lions breeding on north Fisherman Island, Western Australia. Western Aust Nat. 14:203-204.
6481. Ling, JK. 1987. Twin foetuses from an Australian sea lion, *Neophoca cinerea* (Pinnipedia: Otariidae). Aust Mamm. 9:59-60.
6481a. Ling, JK. 1992. *Neophoca cinerea*. Mamm Species. 392:1-7.
6482. Ling, JK & Bryden, MM. 1981. Southern elephant seal *Mirounga leonina* Linnaeus, 1758. Handbook of Marine Mammals. 2:297-327, Ridgway, SH & Harrison, RJ, eds, Academic Press, New York.
6482a. Ling, JK & Bryden, MM. 1992. *Mirounga leonina*. Mamm Species. 391:1-8.
6483. Ling, JK & Walker, GE. 1977. Seal studies in South Australia: Progress report for the period January 1976 to March 1977. S Aust Nat. 52:18-30.
6484. Ling, JK & Walker, GE. 1978. An 18-month breeding cycle in the Australian sea lion?. Search. 9:464-465.
6485. Ling, JK & Walker, GE. 1979. Seal studies in South Australia: Progress report for the period April 1977-July 1979. S Aust Nat. 54:68-78.
6486. Linsdale, J. 1927. Notes on the life history of *Synaptomys*. J Mamm. 8:51-54.
6487. Linsdale, J. 1928. Mammals of a small area along the Missouri River. J Mamm. 9:140-146.
6488. Linsdale, JM. 1938. Environmental responses of vertebrates in the Great Basin. Am Mid Nat. 19:1-206.

6489. Linsdale, JM. 1946. The California ground squirrel. Univ CA Press, Berkeley.
6490. Linsdale, JM & Tevis, LP, Jr. 1951. The Dusky-Footed Wood Rat VIII. Univ CA Press, Berkeley.
6491. Linsdale, JM & Tomich, PQ. 1953. A Herd of Mule Deer: A Record of Observations Made on the Hastings Natural History Reservation. Univ CA Press, Berkeley.
6492. Linton, RG. 1907. A contribution to the histology of the so-called Cowper's gland of the hedgehog (*Erinaceus europaeus*). Anat Anz. 31:61-70.
6493. Linzey, AV. 1970. Postnatal growth and development of *Peromyscus maniculatus nubiterrae*. J Mamm. 51:152-155.
6494. Linzey, AV. 1983. *Synaptomys cooperi*. Mamm Species. 210:1-5.
6495. Linzey, AV & Layne, JN. 1969. Comparative morphology of the male reproductive tract in the rodent genus *Peromyscus* (Muridae). Am Mus Novit. 2355:1-47.
6496. Linzey, DW. 1968. An ecological study of the golden mouse, *Ochrotomys nuttalli*, in the Great Smokey Mountains National Park. Am Mid Nat. 79:320-345.
6497. Linzey, DW & Linzey, AV. 1967. Growth and development of the golden mouse *Ochrotomys nuttalli nuttalli*. J Mamm. 48:445-448.
6497a. Linzey, DW & Linzey, AV. 1979. Growth and development of the southern flying squirrel (*Glaucomys volans volans*). J Mamm. 60:615-620.
6498. Linzey, DW & Packard, RL. 1977. *Ochrotomys nuttalli*. Mamm Species. 75:1-6.
6499. Lippert, W. 1973. Zum Fortpflanzungsverhalten des Mähnenwolfes, *Chrysocyon brachyurus* Illiger. Zool Garten NF. 43:225-247.
6500. Lippert, W. 1974. Observations on behavior during pregnancy and birth in orang-utan (*Pongo pygmaeus*) at Tierpark, Berlin (in German). Folia Primatol. 21:108-134.
6501. Lippert, W. 1977. Erfahrungen bei der Aufzucht von Orang-Utans (*Pongo pygmaeus*) im Tierpark Berlin. Zool Garten NF. 47:209-225.
6502. Lippold, LK. 1977. The Douc Langur: A time for conservation. Primate Conservation. 513-538, Prince Rainier III of Monaco & Bourne, GH, eds, Academic Press, New York.
6503. Lippold, LK & Brockman, DK. 1974. San Diego's douc langurs. Zoonoos (San Diego). 47(3):4-11.
6504. Lishak, RS. 1982. Vocalizations of nestling gray squirrels. J Mamm. 63:446-452.
6505. Liskop, KS, Sadleir, RMFS, & Saunders, BP. 1981. Reproduction and harvest of wolverine (*Gulo gulo*) in British Columbia. Proc Worldwide Furbearer Conf. 1:469-477.
6506. Liskowski, L & Wolf, RC. 1972. Urinary excretion of progesterone metabolites in pregnant rhesus monkeys. Proc Soc Exp Biol Med. 139:1123-1126.
6507. Little, GJ, Bryden, MM, & Barnes, A. 1987. Growth from birth to 20 days in the elephant seal, *Mirounga leonina*, at Macquarie Island. Aust J Zool. 35:307-312.
6508. Littlewood, A & Smith, J. 1979. Breeding and hand-rearing mandrills *Mandrillus sphinx* at Portland Zoo. Int Zoo Yearb. 19:161-165.
6509. Liu, X & Xiao, Q. 1986. Age composition, sex ratio and reproduction of bears in Heilongjaing Province, China. Acta Theriol Sinica. 6:161-170.
6510. Liversidge, R & de Jager, J. 1984. Reproductive parameters from hand-reared springbok *Antidorcas marsupialis*. S Afr J Wildl Res. 14:26-27.
6511. Llanos, AC & Crespo, JA. 1952. Ecologia de la vizcacha (*Lagostomus maximus maximus* Blainv) en el nordeste de la provincia de Entre Rios. Rev Inv Agric. 6:289-389.
6512. Llewellyn, JB. 1978. Reproductive patterns in *Peromyscus truei truei* in a pinyon-juniper woodland of western Nevada. J Mamm. 59:449-451.
6513. Llewellyn, LM. 1953. Growth rate of the raccoon fetus. J Wildl Manage. 17:320-321.
6514. Llewellyn, LM & Dale, FM. 1964. Notes on the ecology of the opossum in Maryland. J Mamm. 45:113-122.
6514a. Lllewellyn, LM & Enders, RK. 1954. Trans-uterine migration in the raccoon. J Mamm. 35:439.
6515. Llewellyn, LM & Handley, CO. 1945. The cottontail rabbits of Virginia. J Mamm. 26:379-390.
6516. Lloyd, HG. 1963. Intra-uterine mortality in the wild rabbit, *Oryctolagus cuniculus* (L) in populations of low density. J Anim Ecol. 32:549-563.
6517. Lloyd, HG. 1968. Observations on breeding in the brown hare (*Lepus europaeus*) during the first pregnancy of the season. J Zool. 156:521-528.
6518. Lloyd, HG. 1968. The control of foxes (*Vulpes vulpes* L). Ann Appl Biol. 61:334-345.
6519. Lloyd, HG. 1970. Variation and adaptation in reproductive performance. Symp Zool Soc London. 26:165-188.
6520. Lloyd, HG. 1981. The Red Fox. BT Batsford, London.
6521. Lloyd, HG & Englund, J. 1973. The reproductive cycle of the red fox in Europe. J Reprod Fert Suppl. 19:119-130.
6522. Lobanov, NV & Treus, VD. 1970. Antelopes of the genus *Connochaetes* in nature and in Askaniya-Nova. Vestnik Zool (Kiev). 6:55-62.
6523. Lobanov, NV & Treus, VD. 1971. Reproduction of *Amotragus lervia* Pall at the Zoo in Askaniya-Nova. Vestnik Zool (Kiev). 3:23-26.
6524. Lobanov, NV & Treus, VD. 1975. Protection and reproduction of *Bibos javanicus* D'Alton. Vistnek Zool (Kiev). 1:14-20.
6525. Lobao Tello, JLP & van Gelder, RG. 1975. The natural history of nyala, *Tragelaphus angasi* (Mammalia, Bovidae), in Mozambique. Bull Am Mus Nat Hist. 155:319-386.
6526. Lobatchew, SW. 1934. Zur Frage der vegetativ-geschlechtlichen Eunktion [*sic*] des Eierstockes bein [*sic*] dem Eichkörnchen. Zool Zhurn. 13(1):280-291.
6527. Lobos, AA. 1963. Observaciones sobre la madurez sexual del cachalote macho (*Physeter catodon* L), capturado en aguas Chilenas. Montemar. 11:99-125.
6528. Lochmiller, RL, Hellgren, EC, & Grant, WE. 1984. Selected aspects of collared peccary (*Dicotyles tajacu*) reproductive biology in a captive Texas herd. Zoo Biol. 3:145-149.
6529. Lochmiller, RL, Hellgren, EC, Grant, WE, et al. 1985. Description of collared peccary (*Tayassu tajacu*) milk composition. Zoo Biol. 4:375-379.
6530. Lochmiller, RL, Hellgren, EC, & Grant, WE. 1985. Relationships between internal morphology and body mass in the developing, nursling collared peccary, *Tayassu tajacu* (Tayassuidae). Growth. 49:154-166.
6531. Lochmiller, RL, Hellgren, EC, & Grant, WE. 1987. Physical characteristics of neonate, juvenile, and adult collared peccaries (*Tayassu tajacu angulatus*) from south Texas. J Mamm. 68:188-194.
6532. Lochmiller, RL, Hellgren, EC, Grant, WE, et al. 1985. Description of collared peccary (*Tayassu tajacu*) milk composition. Zoo Biol. 4:375-379.

6533. Lochmiller, RL, Whelan, JB, & Kirkpatrick, RL. 1982. Energetic cost of lactation in *Microtus pinetorium*. J Mamm. 63:475-481.
6534. Lockyer, C. 1972. The age of sexual maturity of the southern fin whale (*Balaenoptera physalus*) using annual layer counts in the ear plug. J Cons Perm Int Explor Mer. 34:276-294.
6535. Lockyer, C. 1977. Preliminary study of variations in age at sexual maturity of fin whales with year class, in six areas of the southern hemisphere. Rep Int Whal Commn. 27:141-147.
6536. Lockyer, C. 1981. Estimation of the energy cost of growth, maintenance and reproduction in the female minke whale (*Balaenoptera acutorostrata*), from the southern hemisphere. Rep Int Whal Comm. 31:337-343.
6537. Lockyer, C. 1981. The age at sexual maturity in fin whales off Iceland. Rep Int Whal Comm. 31:389-393.
6538. Lockyer, C, Gambell, R, & Brown, SG. 1977. Notes on age data of fin whales taken off Iceland 1967-1974. Rep Int Whal Comm. 27:427-450.
6539. Lockyer, C & Martin, AR. 1983. The sei whale off western Iceland II Age, growth and reproduction. Rep Int Whal Comm. 33:465-476.
6539a. Lockyer, C, Goodall, RNP, & Galeazzi, AR. 1988. Age and body-length characteristics of *Cephalorhynchus commersonii* from incidentally-caught specimens off Tierra del Fuego. Rep Int Whal Comm Sp Issue. 9:103-118.
6540. Lode, T. 1990. Reconnaissance du congenère et comportement sexuel chez un Mustelide: le Putois. Bull Soc Sc Nat Ouest France NS. 12:105-110.
6541. Loder, EG. 1894. Note on the period of gestation of the Indian antelope *Antelope cervicapra* (Linn). Proc Zool Soc London. 1894:476.
6542. Loeb, SC & Schwab, RG. 1987. Estimation of litter size in small mammals: Bias due to chronology of embryo resorption. J Mamm. 68:671-675.
6543. Lofts, B. 1960. Cyclical changes in the distribution of the testis lipids of a seasonal mammal (*Talpa europaea*). Q J Microsc Sci. 101:199-205.
6544. Logan, FD & Robson, FD. 1971. On the birth of a common dolphin (*Delphinus delphis* L) in captivity. Zool Garten NF. 40:115-124.
6545. Logan, KA, Irwin, LL, & Skinner, R. 1986. Characteristics of a hunted mountain lion population in Wyoming. J Wildl Manage. 50:648-654.
6546. Lohi, O & Valkosalo, K. 1982. Dräktighetstidens beroende av valpresultatet hos blårävar. Finsk Pälstidsk. 16:417-419. [Abstract: Scientifur, 8:316, 1984]
6547. Lohiya, NJ, Sharma, RS, Puri, CP, et al. 1988. Reproductive exocrine and endocrine profile of female langur monkeys, *Presbytis entellus*. J Reprod Fert. 82:485-492.
6548. Löhrl, H. 1938/39. Ökologische und physiologische Studien an einheimischen Muriden und Soriciden. Z Säugetierk. 13:114-160.
6549. Lombard, GL. 1961. The Cape fruit bat *Rousettus leachii*. Fauna Flora (Pretoria). 12:39-45.
6550. Lombardi, JR & Whitsett, JM. 1980. Effects of urine from conspecifics on sexual maturation in female prairie deermice, *Peromyscus maniculatus bairdii*. J Mamm. 61:766-768.
6551. Lombardo, DL & Terman, CR. 1980. The influence of the social environment on sexual maturation of female deer mice (*Peromyscus maniculatus bairdi*). Res Pop Ecol. 22:93-100.
6552. Long, CA. 1962. Records of reproduction for harvest mice. J Mamm. 43:103-104.
6553. Long, CA. 1965. The mammals of Wyoming. Univ KS Publ Mus Nat Hist. 14:493-758.
6554. Long, CA. 1972. Notes on habitat preference and reproduction in pigmy shrews, *Microsorex*. Can Field-Nat. 86:155-160.
6555. Long, CA. 1973. Reproduction in the white-footed mouse at the northern limits of its geographical range. Southwest Nat. 18:11-20.
6556. Long, CA. 1973. *Taxidea taxus*. Mamm Species. 26:1-4.
6557. Long, CA. 1974. *Microsorex hoyi* and *Microsorex thompsoni*. Mamm Species. 33:1-4.
6558. Long, CA & Frank, T. 1968. Morphometric variation and function in the baculum, with comments on correlation of parts. J Mamm. 49:32-43.
6559. Long, CA & Killingley, CA. 1983. The Badgers of the World. CC Thomas, Springfield, IL.
6560. Long, WS. 1940. Notes on the life histories of some Utah mammals. J Mamm. 21:170-180.
6561. Longhurst, W. 1944. Observations on the ecology of the Gunnison prairie dog in Colorado. J Mamm. 25:24-36.
6562. Longley, WH. 1963. Minnesota gray and fox squirrels. Am Mid Nat. 69:82-98.
6563. Lönnberg, E. 1902. On the female genital organs of *Cryptoprocta*. Bih K Svenska Akad Handl. 28(3):1-11.
6564. Lönnberg, E. 1928. Contributions to the biology and morphology of the badger, *Meles taxus*, and some other Carnivora. Ark Zool (Stockholm). 19A(26):1-11.
6565. Lönnberg, E & Gyldenstolpe, N. 1925. Zoological results of the Swedish expedition to Central Africa 1921 Vertebrata. Ark Zool (Stockholm). 17B(9):1-5.
6566. Lopez, MJ. 1981. Reproductive and temporal adaptations to seasonal change in the deer mouse, *Peromyscus maniculatus*. Ph D Diss. Univ TX, Austin.
6567. Lopez-Forment C, W. 1976. Registros de longevidad y estabilidad de las poblaciones del vampiro *Desmodus rotundus* en México. An Inst Biol Univ Nal Autón México Ser Zool. 47:197-198.
6568. Lopez-Forment, W. 1976. Some ecological aspects of the bat *Balantiopteryx plicata plicata* Peters, 1867 (Chiroptera: Emballonuridae) in Mexico. MS Thesis. Cornell Univ, Ithaca, New York.
6569. Lopez-Forment C, W & Cervantes, F. 1981. Preliminary observations on the ecology of *Romerolagus diazi* in México. Proc World Lagomorph Conf, Univ Guelph, ONT, Canada. 1979:949-955.
6570. Lopez-Forment C, W & Urbano V, G. 1979. Historia natural del zorillo manchado pigmeo, *Spilogale pygmaea*, con la descripcion de una nueva subespecie. An Inst Biol Univ Nal Auton México Ser Zool. 50:721-728.
6571. López-Fuster, MJ. 1984/85. Population structure of *Crocidura russula* Hermann, 1780 (Insectivora, Mammalia) in the Ebro Delta (Catalonia, Spain) throughout the year. Säugetierk Mitt. 32:21-25.
6572. López-Fuster, MJ. 1989. Reproductive strategy of the Millet's shrew (*Sorex coronatus* Millet, 1828) versus the common shrew (*Sorex araneus* L, 1758) in the northeast of the Iberian Peninsula (Mammalia, Insectivora: Soricidae). Zool Abhand Staat Mus Tierk Dresden. 44(13)143-148.
6573. López-Fuster, MJ, Castién, E, & Gosàlbez, J. 1988. Reproductive cycle and population structure of *Sorex coronatus* Millet, 1828 (Insectivora, Soricidae) in the northern Iberian Peninsula. Bonn Zool Beitr. 39:163-170.
6573a. López-Fuster, MJ, Gosàlbez, J, & Lluch, S. 1988. Characteristics of the reproductive cycle of the mole, *Talpa europaea*, in the northeast of the Iberian Peninsula. Acta Theriol. 33:131-137.

6574. López-Fuster, MJ, Gosàlbez, J, & San-Coma, V. 1985. Über die Fortpflanzang der Hausspitzmaus (*Crocidura russula* Hermann, 1780) im Ebro-Delta (Katalomien, Spanien). Z Säugetierk. 50:1-6.
6575. Lord, RD. 1959. Winter pregnancy of the cottontail. J Mamm. 40:443.
6576. Lord, RD, Jr. 1961. Magnitudes of reproduction in cottontail rabbits. J Wildl Manage. 25:28-33.
6577. Lorenz, R. 1972. Management and reproduction of the Goeldi's monkey *Callimico goeldii* (Thomas 1904), Primates. Saving the Lion Marmoset. 92-109, Bridgwater, DD, ed, The Wild Animal Propagation Trust, Wheeling, WV.
6578. Lorenz, R, Anderson, CO, & Mason, WA. 1973. Notes on reproduction in captive squirrel monkeys (*Saimiri sciureus*). Folia Primatol. 19:286-292.
6579. Lorenz, R & Heinemann, H. 1967. Beitrag zur morphologie und körperlichen jugendentwicklung der Springtamarin *Callimico goeldii* (Thomas, 1904). Folia Primatol. 6:1-27.
6580. Loskutoff, NM, Kasman, LH, Raphael, BL, et al. 1987. Urinary steroid evaluations to monitor ovarian function in exotic ungulates: IV Estrogen metabolism in the okapi (*Okapia johnstoni*). Zoo Biol. 6:213-218.
6581. Loskutoff, NM, Ott, JE, & Lasley, BL. 1982. Urinary steroid evaluations to monitor ovarian function in exotic ungulates I Pregnanediol-3-glucuronide immunoreactivity in the okapi (*Okapia johnstoni*). Zoo Biol. 1:45-53.
6581a. Loskutoff, NM, Raphael, BL, Nemec, LA, et al. 1990. Reproductive anatomy, manipulation of ovarian activity and non-surgical embryo recovery in suni (*Neotragus moschatus zuluensis*). J Reprod Fert. 88:521-532.
6582. Loskutoff, NM, Walker, L, Ott-Joslin, JE, et al. 1986. Urinary steroid evaluations to monitor ovarian function in exotic ungulates: II Comparison between the giraffe (*Giraffa camelopardalis*) and the okapi (*Okapia johnstoni*). Zoo Biol. 5:331-338.
6583. Lotshaw, R. 1971. Birth of two lowland gorillas *Gorilla g gorilla* at Cincinnati Zoo. Int Zoo Yearb. 11:84-87.
6584. Lott, DF. 1981. Sexual behavior and intersexual strategies in American bison. Z Tierpsychol. 56:97-114.
6585. Lott, DF & Galland, JC. 1985. Individual variation in fecundity in an American bison population. Mammalia. 49:300-302.
6586. Lott, DF & Galland, JC. 1985. Parturition in American bison: Precocity and systematic variation in cow isolation. Z Tierpsychol. 69:66-71.
6587. Lotze, JH & Anderson, S. 1979. *Procyon lotor*. Mamm Species. 119:1-8.
6588. Louch, CD, Ghosh, AK, & Pal, BC. 1965. Some aspects of the ecology of the Indian five-striped squirrel (*Funambulus pennanti* Wr) in Calcutta. J Bengal Nat Hist Soc. 34:5-21.
6589. Louch, CD, Ghosh, AK, & Pal, BC. 1966. Seasonal changes in weight and reproductive activity of *Suncus murinus* in west Bengal, India. J Mamm. 47:73-78.
6590. Loudon, ASI & Curlewis, JD. 1987. Refractoriness to melatonin and short day lengths in early seasonal quiescence in the Bennett's wallaby (*Macropus rufogriseus rufogriseus*). J Reprod Fert. 81:543-552.
6591. Loudon, ASI & Curlewis, JD. 1988. Cycles of antler and testicular growth in an aseasonal tropical deer (*Axis axis*). J Reprod Fert. 83:729-738.
6592. Loudon, ASI & Curlewis, JD. 1989. Evidence that the seasonally breeding Bennett's wallaby (*Macropus rufogriseus rufogriseus*) does not exhibit short-day photorefractoriness. J Reprod Fert. 87:641-648.
6593. Loudon, ASI, Curlewis, JD, & English, J. 1985. The effect of melatonin on the season embryonic diapause of the Bennett's wallaby (*Macropus rufogriseus rufogriseus*). J Zool. 206A:35-39.
6594. Loudon, ASI, McLeod, BJ, & Curlewis, JD. 1990. Pulsatile secretion of LH during the periovulatory and luteal phases of the oestrous cycle in the Père David's deer hind (*Elaphurus davidianus*). J Reprod Fert. 89:663-670.
6595. Loudon, ASI, McNeilly, AS, & Milne, JA. 1983. Nutrition and lactational control of fertility in red deer. Nature. 302:145-147.
6596. Loudon, ASI, Milne, JA, Curlewis, JD, et al. 1989. A comparison of the seasonal hormone changes and patterns of growth, voluntary food intake and reproduction in juvenile and adult red deer (*Cervus elaphus*) and Père David's deer (*Elaphurus davidianus*) hinds. J Endocrinol. 122:733-745.
6597. Loudon, ASI, Rothwell, N, & Stock, M. 1985. Brown fat, thermogenesis and physiological birth in a marsupial. Comp Biochem Physiol. 81A:815-819.
6598. Loughlin, TR & Perez, MA. 1985. *Mesoplodon stejnegeri*. Mamm Species. 250:1-6.
6599. Loughlin, TR, Perez, MA, & Merrick, RL. 1987. *Eumetopias jubatus*. Mamm Species. 283:1-7.
6600. Loughrey, AG. 1959. Preliminary investigation of the Atlantic walrus *Odobenus rosmarus rosmarus* (Linnaeus). Can Wildl Serv Wildl Manage Bull Ser 1. 14:1-123.
6601. Louis, TM, Somes, GW, Pryor, WH, et al. 1986. Estrogens and progesterone during gestation in *Dolichotis patagona*. Comp Biochem Physiol. 84A:295-298.
6601a. Loukashkin, AS. 1940. On the pikas of north Manchuria. J Mamm. 21:402-405.
6602. Loukashkin, AS. 1944. The giant rat-headed hamster, *Cricetulus triton nestor* Thomas, of Manchuria. J Mamm. 25:170-177.
6603. Louwman, JWW. 1970. Breeding the banded palm civet and the banded linsang *Hemigalus derbyanus* and *Prionodon linsang* at Wassenaar Zoo. Int Zoo Yearb. 10:81-82.
6604. Louwman, JWW. 1973. Breeding the tailless tenrec *Tenrec ecaudatus* at Wassenaar Zoo. Int Zoo Yearb. 13:125-126.
6605. Louwman, JWW & van Oyen, WG. 1968. A note on breeding Temminck's golden cat *Felis temmincki* at Wassenaar Zoo. Int Zoo Yearb. 8:47-49.
6606. Love, JA. 1978. A note on the birth of a baboon (*Papio anubis*). Folia Primatol. 29:303-306.
6607. Lovecky, DV, Estep, DQ, & Dewsbury, DA. 1979. Copulatory behaviour of cotton mice (*Peromyscus gossypinus*) and their reciprocal hybrids with white footed mice (*P leucopus*). Anim Behav. 27:371-375.
6608. Lovejoy, BP & Black, HC. 1974. Growth and weight of the mountain beaver, *Aplodontia rufa pacifica*. J Mamm. 55:364-369.
6609. Lovejoy, BP, Black, HC, & Hooven, EF. 1978. Reproduction, growth, and development of the mountain beaver (*Aplodontia rufa pacifica*). Northwest Sci. 52:323-328.
6610. Lovell, D, King, D, & Festing, M. 1972. Breeding performance of specific pathogen free guinea-pigs. Guinea-pig News Letter. 6:6-11.
6611. Loveridge, A. 1922. Notes on east African mammalia (other than horned ungulates) collected or kept in captivity 1915-1919. J E Afr Uganda Nat Hist Soc. 17:39-69.
6612. Loveridge, A. 1923. Notes on east African mammals, collected 1920-1923. Proc Zool Soc London. 1923:685-739.

6613. Low, WA. 1970. The influence of aridity on reproduction of the collared peccary (*Dicotyles tajacu*) (Linn) in Texas. Ph D Diss. Univ BC, Vancouver. [Diss Abst Int 31:7693B, 1971]
6614. Lowe, CE. 1958. Ecology of the swamp rabbit in Georgia. J Mamm. 39:116-127.
6615. Lowther, FD. 1940. A study of the activities of a pair of *Galago senegalensis moholi* in captivity, including the birth and postnatal development of twins. Zoologica. 25:433-462.
6616. Loy, J. 1971. Estrous behavior of free-ranging rhesus monkeys (*Macaca mulatta*). Primates. 12:1-32.
6617. Loy, J. 1975. The copulatory behaviour of adult male patas monkeys, *Erythrocebus patas*. J Reprod Fert. 45:193-195.
6618. Loy, J. 1981. The reproductive and heterosexual behaviours of adult patas monkeys in captivity. Anim Behav. 29:714-726.
6619. Loyd, M. 1965. The Arabian oryx in Muscat and Oman. E Afr Wildl J. 3:124-126.
6620. Lucas, NS, Hume, EM, & Smith, HH. 1927. On the breeding of the common marmoset (*Hapale jacchus* Linn) in captivity when irradiated with ultra-violet rays. Proc Zool Soc London. 1927:447-451.
6621. Lucas, NS, Hume, EM, & Smith, HH. 1937. On the breeding of the common marmoset (*Hapale jacchus* Linn) in captivity when irradiated with ultra-violet rays II A ten years' family history. Proc Zool Soc London. 107A:205-211.
6622. Luckett, WP. 1968. Morphogenesis of the placenta and fetal membranes of the tree shrews (family Tupaiidae). Am J Anat. 123:385-428.
6623. Luckett, WP. 1969. Evidence for the phylogenetic relationships of tree shrews (family Tupaiidae) based on the placental and foetal membranes. J Reprod Fert Suppl. 6:419-433.
6624. Luckett, WP. 1970. The fine structure of the placental villi of the rhesus monkey (*Macaca mulatta*). Anat Rec. 167:141-163.
6625. Luckett, WP. 1971. The development of the chorio-allantoic placenta of the African scaly-tailed squirrels (family Anamoluridae). Am J Anat. 130:159-178.
6626. Luckett, WP. 1976. Cladistic relationships among primate higher categories Evidence of the fetal membranes and placenta. Folia Primatol. 25:245-276.
6627. Luckett, WP. 1980. The use of fetal membrane data in assessing chiropteran phylogeny. Proc Int Bat Res Conf. 5:245-265.
6628. Luckett, WP & Mossman, HW. 1981. Development and phylogenetic significance of the fetal membranes and placenta of the African hystricognathous rodents *Bathyergus* and *Hystrix*. Am J Anat. 162:265-285.
6629. Luckey, TD, Mende, TJ, & Pleasants, J. 1955. The physical and chemical characterization of rat's milk. J Nutr. 54:345-359.
6630. Ludwig, DR. 1984. *Microtus richardsoni*. Mamm Species. 223:1-6.
6631. Ludwig, DR. 1984. *Microtus richardsoni* microhabitat and life history. Sp Publ Carnegie Mus Nat Hist. 10:319-331.
6632. Ludwig, DR. 1988. Reproduction and population dynamics of the water vole, *Microtus richardsoni*. J Mamm. 69:532-541.
6633. Ludwig, KS. 1961. Beitrag zum Bau der Gorilla-Placenta. Acta Anat. 45:110-123.
6634. Ludwig, KS. 1962. Beitrag zur Bau der Giraffenplacenta. Acta Anat. 48:206-223.
6635. Ludwig, KS. 1962. Zur Kenntnis der Geburtsplacenten der Ordnung Perissodactyla. Acta Anat. 49:154-167.
6636. Lueth, FX. 1967. Reproductive studies of some Alabama deer herds. Proc Annu Conf SE Assoc Game Fish Comm. 21:62-68.
6637. Lugg, DJ. 1966. Annual cycle of the Weddell seal in the Vestfold Hills, Antarctica. J Mamm. 47:317-322.
6638. Luick, JR, White, RG, Gau, AM, et al. 1974. Compositional changes in the milk secreted by grazing reindeer I Gross composition and ash. J Dairy Sci. 57:1325-1333.
6639. Lukens, PW, Jr & Davis, WB. 1957. Bats of the Mexican state of Guerrero. J Mamm. 38:1-14.
6640. Lumpkin, S & Koontz, FW. 1986. Social and sexual behavior of the rufous elephant-shrew (*Elephantulus rufescens*) in captivity. J Mamm. 67:112-119.
6641. Lumpkin, S, Koontz, F, & Howard, JG. 1982. The oestrous cycle of the rufous elephant-shrew, *Elephantulus rufescens*. J Reprod Fert. 66:671-673.
6642. Lumpkin, S & Kranz, KR. 1984. *Cephalophus sylvicultor*. Mamm Species. 225:1-7.
6643. Luna, ML. 1980. First field study of the yellow-tailed woolly monkey. Oryx. 15:386-389.
6644. Lundberg, K & Gerell, R. 1986. Territorial advertisement and mate attraction in the bat *Pipistrellus pipistrellus*. Ethology. 71:115-124.
6645. Lundquist, L, Nilsson, A, & Hansson, L. 1973. Winter breeding observed in northern red-backed mouse. Fauna. 26:216-217.
6646. Lunn, SF. 1978. Urinary oestrogen excretion in the common marmoset, *Callithrix jacchus*. Biology and Behaviour of Marmosets. 67-73, Rothe, E, Wolters, HJ, & Hearn, JP, eds, Eigenverlag Hartmut Rothe, Göttingen.
6647. Lunn, SF. 1983. Body weight changes throughout pregnancy in the common marmoset *Callithrix jacchus*. Folia Primatol. 41:204-217.
6648. Lunn, SF & McNeilly, AS. 1982. Failure of lactation to have a consistent effect on interbirth interval in the common marmoset, *Callithrix jacchus jacchus*. Folia Primatol. 37:99-105.
6649. Luong, AG & Galat, G. 1979. Consequences comportementales de pertubations sociales répétées sur une troupes de mones de Lowe *Cercopithecus campbelli lowei* de Cote d'Ivoire. Terre Vie. 33:49-58.
6650. Lusty, JA & Seaton, B. 1978. Oestrus and ovulation in the casiragua *Proechymis guairae* (Rodentia, Hystricomorpha). J Zool. 184:255-265.
6651. Lyapunova, EA, Vorantsov, NN, & Zakarjano, GG. 1975. Zygotic mortality in *Ellobius lutescens* (Rodentia: Microtinae). Experientia. 31:417-418.
6652. Lydekker, R. 1898. Wild Oxen, Sheep and Goats of all Lands. Rowland Ward, London.
6653. Lydekker, R. 1907. The Game Animals of India, Burma, Malaya, and Tibet. Rowland Ward, London.
6654. Lydekker, R. 1924. The Game Animals of India, Burma, Malaya, and Tibet; 2nd ed revised by JG Dollman. Rowland Ward, London.
6655. Lydersen, C & Gjertz, I. 1987. Population parameters of ringed seals (*Phoca hispida* Schreber, 1775) in the Svalbard area. Can J Zool. 65:1021-1027.
6656. Lyman, CP & O'Brien, RC. 1977. A laboratory study of the Turkish hamster, *Mesocricetus brandti*. Breviora. 442:1-27.
6657. Lynch, CD. 1974. A behavioural study of blesbok *Damaliscus dorcas phillipsi*, with special reference to territoriality. Mem Nasionale Mus Bloemfontein. 8:1-83.
6658. Lynch, CD. 1990. Reproduction in the lesser dwarf shrew, *Suncus varilla* (Mammalia: Soricidae). Navors Nas Mus (Bloemfontein). 7(3):45-59.
6659. Lynch, GR, Heath, HW, & Johnston, CM. 1981. Effect of geographical origin on the photoperiodic control of reproduction

in the white-footed mouse, *Peromyscus leucopus*. Biol Reprod. 25:475-480.
6660. Lynch, GR, Lynch, CB, Dube, M, et al. 1976. Early cold exposure: Effects on behavioral and physiological thermoregulation in the house mouse, *Mus musculus*. Physiol Zool. 49:191-199.
6661. Lynch, GR & Wichman, HA. 1981. Reproduction and thermoregulation in *Peromyscus*: Effects of chronic short days. Physiol Behav. 26:201-205.
6662. Lyne, AG. 1952. Notes on external characters of the pouch young of four species of bandicoot. Proc Zool Soc London. 122:625-649.
6663. Lyne, AG. 1964. Observations on the breeding growth of the marsupial *Perameles nasuata* (Geoffroy) with notes on other bandicoots. Aust J Zool. 12:322-339.
6664. Lyne, AG. 1967. Marsupials and Monotremes of Australia. Angus & Robertson. Sydney.
6665. Lyne, AG. 1971. Bandicoots in captivity. Int Zoo Yearb. 11:41-43.
6666. Lyne, AG. 1974. Gestation period and birth in the marsupial *Isoodon macrourus*. Aust J Zool. 22:303-309.
6667. Lyne, AG. 1975. Bandicoot birth. Aust Mamm. 1:398-399.
6668. Lyne, AG. 1976. Observations on oestrus and the oestrous cycle in the marsupials, *Isoodon macrourus* and *Perameles nasuta*. Aust J Zool. 24:513-521.
6669. Lyne, AG. 1976. Some observations on the oestrous cycle in bandicoots. Bull Aust Mamm Soc. 3:36.
6670. Lyne, AG. 1982. The bandicoots *Isoodon macrourus* and *Perameles nasuta*: Their maintenance and breeding in captivity. The Management of Australian Mammals in Captivity. 47-52, Evans, DD, ed, Zool Bd Victoria, Melbourne.
6671. Lyne, AG & Hollis, DE. 1979. Observations on the corpus luteum during pregnancy and lactation in the marsupials *Isooden macrourus* and *Perameles nasuta*. Aust J Zool. 27:881-899.
6672. Lyne, AG & Hollis, DE. 1982. Observations on the lateral vaginae and birth canals in the marsupials *Isoodon macrourus* and *Perameles nasuta* (Mammalia). J Zool. 198:263-277.
6673. Lyne, AG & Hollis, DE. 1983. Observations on Graafian follicles and their oocytes during lactation and after removal of pouch young in the marsupials *Isoodon macrourus* and *Perameles nasuta*. Am J Anat. 166:41-61.
6674. Lyne, AG, Pilton, PE, & Sharman, GB. 1959. Oestrous cycle, gestation period and parturition in the marsupial *Trichosurus vulpecula*. Nature. 183:622-623.
6675. Lyne, AG & Radford, HM. 1974. Some observations on reproduction in bandicoots. Aust Mamm. 1:293-294.
6676. Lyne, AG & Verhagen, AMW. 1957. Growth of the marsupial *Trichosurus vulpecula* and a comparison with some higher mammals. Growth. 21:167-195.
6677. Lyon, MW, Jr. 1903. Observations on the number of young of the lasiurine bats. Proc US Nat Mus. 26:425-426.
6678. Lyon, MW, Jr. 1913. Tree shrews: An account of the mammalian family Tupaiadae. Smithsonian Inst US Nat Mus Proc. 45:1-188.
6679. Lyons, PJ. 1979. Productivity and population structure of western Massachusetts beavers. Proc NE Fish Wildl Conf. 36:176-187.
6680. Maberly, CTA. 1969. The bushpig as a family man. Natal Wildl. 10(1):4-6.
6681. Maberry, MB. 1962. Breeding Indian elephants *Elephas maximus* at Portland Zoo. Int Zoo Yearb. 4:80-83.
6682. Maberry, MB & Dittebrandt, M. 1971. Note on mummified foetuses in a bactrian camel *Camelus bactrianus* at Portland Zoo. Int Zoo Yearb. 11:126-127.
6683. Macari, M & Machado, CR. 1978. Sexual maturity in rabbits defined by the physical and chemical characterictics of the semen. Lab Anim. 12:37-39.
6684. MacArthur, JA, Seamer, JH, & Veall, D. 1978. Establishment of a small breeding colony of rhesus monkeys. Lab Anim. 12:151-156.
6685. Macdonald, DW. 1979. 'Helpers' in fox society. Nature. 282:69-71.
6686. Macdonald, DW. 1980. Social factors affecting reproduction among red foxes (*Vulpes vulpes* L, 1758). The Red Fox; Symposium on Behavior and Ecology. 123-175, ed, Dr W Junk, Publ, The Hague.
6687. Macdonald, GJ. 1971. Reproductive patterns of three species of macaques. Fert Steril. 22:373-377.
6688. Macdonald, GJ, Yoshinaga, K, & Greep, RO. 1973. Progesterone values in monkeys near term. Am J Phys Anthropol. 38:201-206.
6689. MacDonald, SO & Jones, C. 1987. *Ochotona collaris*. Mamm Species. 281:1-4.
6690. Macêdo, RH & Mares, MA. 1988. *Neotoma albigula*. Mamm Species. 310:1-7.
6691. Macfarlane, JD & Taylor, JM. 1981. Sexual maturation in female Townsend's voles. Acta Theriol. 26:113-117.
6692. MacFarlane, JD & Taylor, JM. 1982. Nature of estrus and ovulation in *Microtus townsendii* (Bachman). J Mamm. 63:104-109.
6693. MacFarlane, JD & Taylor, JM. 1982. Pregnancy and reproductive performance in the Townsend's vole, *Microtus townsendii* (Bachman). J Mamm. 63:165-168.
6694. Machado, CRS, Calixto, SL, & Ladosky, W. 1982. Morphological and physiological factors involved in the contractility of the spermatic cord and ductus deferens of the opossum (*Didelphis albiventris*). J Reprod Fert. 65:275-280.
6695. Mack, D. 1979. Growth and development of infant red howling monkeys (*Alouatta seniculus*) in a free ranging population. pp 127-136, Vertebrate Ecology in the Northern Neotropics, Eisenberg, JF, ed. book. Smithsonian Inst Press, Washington, DC.
6696. Mack, D & Kafka, H. 1978. Breeding and rearing of woolly monkeys *Lagothrix lagotricha* at the National Zoological Park, Washington. Int Zoo Yearb. 18:117-122.
6697. Mack, G. 1966. Mammals from south-western Queensland. Mem QLD Mus. 13:213-228.
6698. Mackenzie, AR. 1985. Reproduction of farmed rusa deer (*Cervus timorensis*) in south-east Queensland, Australia. R Soc N Z Bull. 22:213-215.
6699. Mackerras, MJ & Smith, RH. 1960. Breeding the short-nosed marsupial bandicoot, *Isoodon macrourus* (Gould), in captivity. Aust J Zool. 8:371-382.
6700. Mackintosh, NA. 1942. The southern stocks of whalebone whales. Discovery Rep. 22:197-300.
6701. Mackintosh, NA. 1972. Biology of the populations of large whales. Sci Prog. 60:449-464.
6702. Mackintosh, NA & Wheeler, JFG. 1929. Southern blue and fin whales. Discovery Rep. 1:257-540.
6703. Mackler, SF & Buechner, HK. 1978. Play behavior and mother-young relationships in captive Indian rhinoceroses (*Rhinoceros unicornis*). Zool Garten NF. 48:177-186.
6704. Maclean, HIC. 1961. Four observations of killer whales with an account of the mating of these animals. Scottish Nat. 70:75-78.
6705. MacLeod, J. 1880. Contribution à l'étude de la structure de l'ovaire des Mammifères. Arch Biol. 1:241-278.

6706. MacLulich, DA. 1937. Fluctuations in numbers of the varying hare (*Lepus americanus*). Univ Toronto Stud Biol Ser. 43:1-136.
6707. MacLusky, NJ, Liederburg, I, Krey, LC, et al. 1980. Progestin receptors in the brain and pituitary of the bonnet monkey (*Macaca radiata*): Differences between the monkey and the rat in the distribution of progestin receptors. Endocrinology. 106:185-191.
6707a. Macmillan, L. 1955. The dugong. Walkabout. 21(Feb):17-20.
6708. Macmillen, RE. 1964. Population ecology, water relations, and social behavior of a southern California semidesert rodent fauna. Univ CA Publ Zool. 71:1-59.
6708a. Macnab, JA & Dirks, JC. 1941. The California red-backed mouse in the Oregon coast range. J Mamm. 22:174-180.
6709. MacNamara, M. 1980. Notes on the behavior and captive maintenance of maras, *Dolichotis patagonum*, at the Bronx Zoo. Zool Garten NF. 50:422-426.
6709a. MacNamara, M & Eldridge, W. 1987. Behavior and reproduction in captive pudu (*Pudu pudu*) and red brocket (*Mazama americana*): A descriptive and comparative analysis. Biology and Management of the Cervidae. 372-387, Wemmer, CM, ed, Smithsonian Inst Press, Washington, DC.
6710. MacPherson, AH & Manning, TH. 1959. The birds and mammals of Adelaide Peninsula, NWT. Bull Nat Mus Can. 161:1-63.
6711. MacRoberts, MH & MacRoberts, BB. 1966. The annual reproductive cycle of the barbary ape (*Macaca sylvana*) in Gibraltar. Am J Phys Anthrop. 25:299-304.
6712. Maddock, I. 1968. The care and breeding of leopards (*Panthera pardus*) at the Jersey Zoological Park. Ann Rep Jersey Wildl Preserv Trust. 5:42-43.
6713. Maddock, TH & McLeod, AN. 1974. Polyoestry in the little brown bat, *Eptesicus pumilus*, in central Australia. S Aust Nat. 48:50.
6714. Maddock, TH & McLeod, AN. 1976. Observations on the little brown bat, *Eptesicus pumilus caurinus* Thomas, in the Tennant creek area of the northern territory, part one: introduction and breeding biology. S Aust Nat. 50:42-50.
6715. Madhavan, A. 1971. Breeding habits in the Indian vespertilionid bat, *Pipistrellus ceylonicus chrysothrix* (Wroughton). Mammalia. 35:283-306.
6716. Madhavan, A. 1978. Breeding habits and associated phenomena in some Indian bats: Part V *Pipistrellus dormeri* (Dobson) - Vespertilionidae. J Bombay Nat Hist Soc. 75:426-433.
6717. Madhavan, A. 1980. Breeding habits and associated phenomena in some Indian bats: Part VI *Scotophilus heathi* (Horsfield) - Vespertilionidae. J Bombay Nat Hist Soc. 77:227-237.
6718. Madhavan, A, Patil, DR, & Gopalakrishna, A. 1978. Breeding habits and associated phenomena in some Indian bats: Part IV *Hipposideros fulvus fulvus* (Gray) - Hipposideridae. J Bombay Nat Hist Soc. 75:96-103.
6719. Madison, DM. 1984. Group nesting and its ecological and evolutionary significance in overwintering microtine rodents. Sp Publ Carnegie Mus Nat Hist. 10:267-274.
6720. Madkour, G. 1976. External genitalia and mammary glands of Egyptian female bats. Zool Anz. 197:62-66.
6721. Madkour, GA, Hammouda, EM, & Ibrahim, IG. 1983. Histomorphology of the male genitalia of two common Egyptian bats. Ann Zool (Agra, India). 20:1-24.
6722. Maeda, K. 1972. Growth and development of large noctule, *Nyctalus lasiopterus* Schreber. Mammalia. 36:269-278.
6723. Maeda, K & Dewa, H. 1982. Breeding habits of Japanese long-legged whiskered bats, *Myotis frater kaguyae* in Asahikawa, Japan. J Mamm Soc Japan. 9:82-87.
6723a. Maehr, DS, Land, ED, Roof, JC, et al. 1989. Early maternal behavior in the Florida panther (*Felis concolor coryi*). Am Mid Nat. 122:34-43.
6724. Magomedov, MRD. 1981. Effects of population density upon reproduction intensity in *Citellus pygmaeus* (Rodentia, Sciuridae). Zool Zhurn. 60:1048-1057.
6725. Magoun, AJ & Valkenburg, P. 1983. Breeding behavior of free-ranging wolverines (*Gulo gulo*). Acta Zool Fennici. 174:175-177.
6726. Maher, WJ. 1960. Recent records of the California grey whale (*Eschirichtius robustus*) along the north coast of Alaska. Arctic. 13:257-265.
6727. Maheswari, UK. 1982. Some observations on reproduction of the Indian hedgehog (*Hemiechinus auritus collaris* Gray, 1830). Säugetierk Mitt. 30:184-189.
6728. Mahmood, A, Baig, KJ, & Arslan, M. 1986. Studies on seasonal changes in the reproductive tract of female muskshrew, *Suncus murinus*. Pak J Zool. 18:263-272.
6729. Mahone, JP & Dukelow, WR. 1979. Seasonal variation of reproductive parameters in the laboratory-housed male cynomolgus macaque (*Macaca fascicularis*). J Med Primatol. 8:179-183.
6730. Mahoney, CJ. 1970. A study of menstrual cycle in *Macaca irus* with special reference to the detection of ovualtion. J Reprod Fert. 21:153-163.
6731. Mahoney, CJ. 1975. Aberrant menstrual cycles in *Macaca mulatta* and *M fascicularis*. Lab Anim Handb. 6:243-255.
6732. Mahoney, CJ. 1975. The accuracy of bimanual rectal palpation for determining the time of ovulation and conception in the rhesus monkey (*Macaca mulatta*). Lab Anim Handb. 6:127-138.
6733. Main, AR, Shield, JW, & Waring, H. 1959. Recent studies on marsupial ecology. Monogr Biol. 8:315-331.
6733a. Mainka, SA & Lothrop, CD, Jr. 1990. Reproductive and hormonal changes during the estrous cycle and pregnancy in Asian elephants (*Elephas maximus*). Zoo Biol. 9:411-419.
6734. Mainland, D. 1928. The pluriovular follicle, with reference to its occurrence in the ferret. J Anat. 62:139-158.
6735. Mainoya, JR. 1979. Spermatogenic and frontal sac gland activity in *Triaenops persicus* (Chiroptera: Hipposideridae). Afr J Ecol. 17:127-129.
6736. Maizumi, Y. 1972. Notes on the oestrous cycle of *Tamias sibiricus*. J Mamm Soc Japan. 5:186.
6737. Majeed, MA, Qureshi, AW, & Tayyab, M. 1966. Seasonal variation in calving frequency of west Pakistan buffalo and cow and their effect on reproductive efficiency. W Pak J Agric Res. 4:140-150.
6738. Majgier-Bobek, G. 1983. Ovarian histology and hormonal activity of roe deer fawns. Acta Theriol. 24:147-156.
6739. Mákelá, JF. 1985. Whelping results at the experimental farms in 1985. Finsk Pälstidskuft. 19:452. [Anim Breeding Abstr 54: ref 415, 1985]
6740. Malassine, A. 1970. Étude histologique et ultrastructurale du disque placentaire de Minioptère, *Miniopterus schreibersii* (Chiroptere): La nature endothéliochoriale, son caractère endocrine. Arch Anat Microsc Morph Exp. 59:99-112.
6741. Malbrant, R. 1936. Faune du Centre Africain Français. Paul Lechevalier, Paris.
6742. Malbrant, R & Maclatchy, A. 1949. Faune de l'Equateur Africain Français. Paul Lechevalier, Paris.

6743. Malcolm, JR. 1986. Socio-ecology of bat-eared foxes (*Otocyon megalotis*). J Zool. 208A:457-467.
6744. Maliniak, E & Taft, L. 1981. Husbandry and breeding protocol for *Monodelphis domestica*. unpublished mimeograph, Washington, DC, Nat Zool Park Dept Zool Res Bull. 1:1-4.
6745. Mallinson, JJC. 1967. Notes on the breeding of the African civet. Ann Rep Jersey Wildl Preserv Trust. 4:22-24.
6746. Mallinson, JJC. 1969. Notes on breeding the African civet *Viverra civetta* at Jersey Zoo. Int Zoo Yearb. 9:92-93.
6747. Mallinson, JJC. 1969. Observations on a breeding group of black-and-white colobus monkeys *Colobus polykomos*. Int Zoo Yearb. 9:79-81.
6748. Mallinson, JJC. 1969. Reproduction and development of Brazilian tapir *Tapirus terrestris*. Ann Rep Jersey Wildl Preserv Trust. 6:47-52.
6749. Mallinson, JJC. 1969. The breeding and maintenance of marmosets at Jersey Zoo. Ann Rep Jersey Wildl Preserv Trust. 6:5-10.
6750. Mallinson, JJC. 1970. Observations on the reproduction and development of vervet monkey with special reference to intersubspecific hybridization *Cercopithecus pygerythrus*, Cuvier 1821. Ann Rep Jersey Wildl Preserv Trust. 7:20-31.
6751. Mallinson, JJC. 1971. The breeding and maintenance of marmosets at Jersey Zoo. Int Zoo Yearb. 11:79-83.
6752. Mallinson, JJC. 1971. The pigmy hog *Sus salvanius* in northern Assam. Ann Rep Jersey Wildl Preserv Trust. 8:62-69.
6753. Mallinson, JJC. 1972. The reproduction of the African civet *Viverra civetta*. Ann Rep Jersey Wildl Preserv Trust. 9:35-38.
6754. Mallinson, JJC. 1973. The reproduction of the African civet *Viverra civetta* at Jersey Zoo. Int Zoo Yearb. 13:147-150.
6755. Mallinson, JJC. 1974. Notes on a breeding group of Sierra Leone striped squirrel *Funisciurus pyrrhopus leonis* (Thomas 1905) at the Jersey Zoological Park. Ann Rep Jersey Wildl Preserv Trust. 11:41-44.
6756. Mallinson, JJC. 1976. The breeding of the pygmy hog *Sus salvanius* (Hodgson) in northern Assam. Ann Rep Jersey Wildl Preserv Trust. 13:12-20.
6757. Mallinson, JJC. 1977. Breeding of the pigmy hog *Sus salvanius* (Hodgson) in northern Assam. J Bombay Nat Hist Soc. 74:288-298.
6758. Mallinson, JJC, Coffrey, P, & Usher-Smith, J. 1973. Maintenance, breeding and hand-rearing of lowland gorilla *Gorilla g gorilla* (Savage and Wyman, 1847), at the Jersey Zoological Park. Ann Rep Jersey Wildl Preserv Trust. 10:5-28.
6759. Mallory, FF & Brooks, RJ. 1978. Infanticide and other reproductive strategies in the collared lemming *Dicrostronyx groenlandicus*. Nature. 273:144-146.
6760. Mallory, FF & Brooks, RJ. 1980. Infanticide and pregnancy failure: Reproductive strategies in the female collared lemming (*Dicrostonyx groenlandicus*). Biol Reprod. 22:192-196.
6760a. Mallory, FF & Chulow, FV. 1977. Evidence of pregnancy failure in the wild meadow vole, *Microtus pennsylvanicus*. Can J Zool. 55:1-17.
6761. Mallory, FF, Elliott, JR, & Brooks, RJ. 1981. Changes in body size in fluctuating populations of the collared lemming: Age and photoperiod influences. Can J Zool. 59:174-182.
6762. Maloiy, GMO, Kamau, JMZ, Shkolnik, A, et al. 1982. Thermoregulation and metabolism in a small desert carnivore: The fennec fox (*Fennecus zerda*) (Mammalia). J Zool. 198:279-291.
6763. Maltz, E. 1976. Productivity in the desert: Bedouin goat, ibex and desert gazelle. Ph D Diss. Tel Aviv Univ, Tel Aviv. [Cited by Oftedal, 1984]
6764. Maltz, E & Shkolnik, A. 1984. Lactational strategies of desert ruminants: The bedouin goat, ibex and desert gazelle. Symp Zool Soc London. 51:193-213.
6765. Mambetzhumaev, AM. 1961. On the ecology of the deer *Cervus elaphus bactrianus* Lydekker. Zool Zhurn. 40:1725-1731.
6766. Mandal, AK. 1964. The behaviour of the rhesus monkeys (*Macaca mulatta* Zimmermann) in the Sundarbans. J Bengal Nat Hist Soc. 33:153-165.
6767. Manley, GH. 1966. Reproduction in lorisoid primates. Symp Zool Soc London. 15:493-509.
6768. Manley, GH. 1967. Gestation periods in the Lorisidae. Int Zoo Yearb. 7:80-81.
6769. Manly, DE. 1953. Parturition of a captive meadow vole. J Mamm. 34:130-131.
6770. Mann, DR, Blank, MS, Gould, KG, et al. 1982. Validation of radioimmunoassays for LH and FSH in the sooty mangabey (*Cercocebus atys*): Characteriztion of LH and the response to GnRH. Am J Primatol. 2:275-283.
6771. Mann, DR, Castracane, VD, McLaughlin, F, et al. 1983. Developmental patterns of serum luteinizing hormone, gonadal and adrenal steroids in the sooty mangabey (*Cercocebus atys*). Biol Reprod. 28:279-284.
6772. Mann, G. 1951. Aparato genital femennino de *Marmosa elegans*. Invest Zool Chilenas. 1(3):11-16.
6773. Mann, G. 1955. Monito del monte, *Dromiciops australis* Philippi. Invest Zool Chilenas. 2:159-166.
6774. Mann, GS & Bindra, OS. 1977. Reproductive activity of *Mus* spp in crop fields at Ludhiana. J Bombay Nat Hist Soc. 74:162-167.
6775. Mann, GS & Bindra, OS. 1979. Reproductive activity of four species of field-rats in crop fields at Ludhiana. J Bombay Nat Hist Soc. 76:501-507.
6776. Mann, T & Wilson, ED. 1962. Biochemical observations on the male accessory organs of nutria, *Myocastor coypus* (Molina). J Endocrinol. 25:407-408.
6777. Mann F, G. 1958. Reproduccion de *Dromiciops australis* (Marsupialia, Didelphydae). Invest Zool Chilenas. 4:209-213.
6778. Manniche, L. 1910. The terrestrial mammals and birds of northeast Greenland. Med om Grønland (Kopenhagen). 45:1-199. [Cited by Dittrich, 1961]
6779. Manning, RW & Jones, JK, Jr. 1988. *Perognathus fasciatus*. Mamm Species. 303:1-4.
6780. Manning, RW & Jones, JK, Jr. 1989. *Myotis evotis*. Mamm Species. 329:1-5.
6781. Manning, TH. 1954. Remarks on the reproduction, sex ratio, and life expectancy of the varying lemming, *Dicrostonyx groenlandicus*, in nature and captivity. Arctic. 7:36-48.
6782. Manning, TH. 1956. The northern red-backed mouse, *Clethrionomys rutilus* (Pallas) in Canada. Bull Nat Mus Can. 144:1-60.
6783. Mansell, WD. 1974. Productivity of white-tailed deer on the Bruce Peninsula, Ontario. J Wildl Manage. 38:808-814.
6784. Mansergh, I. 1984. *Burramys parvus* (Broom): A short review of the current state of knowledge concerning the species. Possums and Gliders. 413-416, Smith, AP & Hume, ID, eds, Surrey Beatty & Sons Pty Ltd, Sydney, Australia.
6785. Mansergh, I & Scotts, D. 1990. Aspects of the life history and breeding biology of the mountain pygmy-possum, *Burramys parvus*, (Marsupialia: Burramyidae) in alpine Victoria. Aust Mamm. 13:179-191.
6786. Mansfield, AW, 1958. The biology of the Atlantic walrus *Odobenus rosmarus rosmarus* (Linnaeus) in the eastern Canadian Arctic. Fish Res Bd Can Manu Rep Ser Biol. 653:1-146.

6787. Mansfield, AW. 1958. The breeding behaviour and reproductive cycle of the Weddell seal (*Leptonychotes weddelli* Lesson). Falkland Isl Depend Survey Sci Rep. 18:1-41.
6788. Mansfield, AW. 1967. Seals of arctic and eastern Canada. Bull Fish Res Bd Can. 137:1-30.
6789. Mansfield, AW. 1973. The Atlantic walrus *Odobenus rosmarus* in Canada and Greenland. IUCN Res Suppl Pap. 39:69-79.
6790. Mansfield, AW, Smith, TG, & Beck, B. 1975. The narwhal, *Monodon monoceros*, in eastern Canadian waters. J Fish Res Bd Can. 32:1041-1046.
6791. Manson, J. 1974. Aspekte van die Biologie en Gedrag van die Kaapse Grysbok *Raphiceros melanetis* Thunberg. MS Thesis. Univ Stellenbosch, Stellenbosch, S Africa. [Cited by Anderson, 1978]
6792. Manton, VJA. 1970. Breeding cheetahs *Acinonyx jubatus* at Whipsnade Park. Int Zoo Yearb. 10:85-86.
6793. Manton, VJA. 1971. A further report on breeding cheetahs *Acinonyx jubatus* at Whipshade Park. Int Zoo Yearb. 11:125-126.
6794. Manton, VJA. 1974. Birth of a cheetah *Acinonyx jubatus* to a captive-bred mother. Int Zoo Yearb. 14:126-129.
6795. Manton, VJA. 1975. Captive breeding of cheetahs. Breeding Endangered Species in Captivity. 337-344, Martin, RD, ed, Academic Press, New York.
6796. Manville, RH. 1949. A study of small mammal populations in northern Michigan. Misc Publ Mus Zool Univ MI. 73:1-83.
6797. Manville, RH. 1952. A late breeding cycle in *Peromyscus*. J Mamm. 33:389.
6798. Manville, RH. 1959. The Columbian ground squirrel in northwestern Montana. J Mamm. 40:26-45.
6799. Manville, RH. 1961. Notes on some mammals of the Gaspe Peninsula, Quebec. Can Field-Nat. 75:108-109.
6800. Maple, T, Erwin, J, & Mitchell, G. 1973. Age of sexual maturity in laboratory-born pairs of rhesus monkeys (*Macaca mulatta*). Primates. 14:427-428.
6801. Maple, TL, Lukas, J, Murdock, GK, et al. 1982. Notes on the births of two sable antelopes *Hippotragus niger*. Int Zoo Yearb. 22:218.
6802. Maple, TL, Murdock, GK, & Lukas, J. 1981. Birth weights of African antelopes on St Catherine's Island. Zool Garten NF. 51:388-390.
6803. Marafie, E, Nayak, R, & Al-Zaid, N. 1978. Breeding and reproductive physiology of the desert gerbil, *Meriones crassus*. Lab Anim Sci. 28:397-401.
6804. Marburger, RG, Robinson, RM, & Thomas, JU. 1966. A pseudohermaphroditic white-tailed deer. J Mamm. 47:711-712.
6805. Marcström, V. 1964. The muskrat *Ondatra zibethicus* in northern Sweden. Viltrevy. 2:329-407.
6806. Marcström, V. 1966. On the reproduction of the Norwegian lemming *Lemmus lemmus* L. Viltrevy. 4:311-342.
6807. Mares, MA. 1988. Reproduction, growth, and development in Argentine gerbil mice, *Eligmodontia typus*. J Mamm. 69:852-854.
6808. Mares, MA, Willig, MR, Streilein, KE, et al. 1981. The mammals of northeastern Brazil: A preliminary assessment. Ann Carnegie Mus. 50:81-137.
6809. Mares, MA & Wilson, DE. 1971. Bat reproduction during the Costa Rican dry season. BioScience. 21:471-477.
6810. Margolis, DJ & Lynch, GR. 1981. Effects of daily melatonin injections on female reproduction in the white-footed mouse, *Peromyscus leucopus*. Gen Comp Endocrinol. 44:530-537.
6811. Marie, M & Anouassi, A. 1986. Mating-induced luteinizing hormone surge and ovulation in the female camel (*Camelus dromedarius*). Biol Reprod. 35:792-798.
6812. Marie, M & Anouassi, A. 1987. Induction of luteal activity and progesterone secretion in the nonpregnant one-humped camel (*Camelus dromedarius*). J Reprod Fert. 80:183-192.
6813. Marik, M. 1931. Beobachtungen zur Fortpflanzungsbiologie der Uistiti (*Callithrix jacchus* L). Zool Garten NF. 4:347-349.
6814. Marinelli, L & Millar, JS. 1989. The ecology of beach-dwelling *Peromyscus maniculatus* on the Pacific Coast. Can J Zool. 67:412-417.
6814a. Marinkelle, CJ. 1970. *Vampyrops intermedius* sp n from Colombia. Rev Brasil Biol. 30:49-52.
6815. Mark-Siembida, J. 1974. Breeding viscachas *Lagostomus maximus* at Lodz Zoo. Int Zoo Yearb. 14:116-117.
6816. Markgren, G. 1964. Puberty, dentition and weight of yearling moose in a Swedish county. Viltrevy. 2:409-417.
6817. Markgren, G. 1969. Reproduction of moose in Sweden. Viltrevy. 6:129-301.
6818. Markgren, G. 1973. Factors affecting the reproduction of moose (*Alces alces*) in three different Swedish areas. Int Congr Game Biol. 11:67-70.
6819. Markham, J & Kirkwood, JK. 1988. Formula intake and growth of an addax hand-reared at the London Zoo. Int Zoo Yearb. 27:316-319.
6820. Markham, OD & Whicker, FW. 1973. Seasonal data on reproduction and body weights of pikas (*Ochotona princeps*). J Mamm. 54:496-498.
6821. Markley, MH & Bassett, CF. 1942. Habits of captive marten. Am Mid Nat. 28:604-616.
6822. Marks, KJ & Schadler, MH. 1979. Embryo rejection in the pine vole. Proc E Pine Meadow Vole Symp. 3:30-35.
6823. Marks, SA & Erickson, AW. 1966. Age determination in the black bear. J Wildl Manage. 30:389-410.
6824. Markus, CU. 1957. Vergleichende Untersuchungen über den Sexualcyclus Weiblicher Rotelmäuse (*Clethrionomys glareolus* Schreber 1780) und Feldmäuse (*Microtus arvalis* Pallas 1778). Wiss Z Univ Halle Math Nat. 6:1021-1032.
6824a. Marley, PB, Rodger, JC, White, IG, et al. 1981. 19-Hydroxylated prostaglandin in the semen of the marsupial *Trichosurus vulpecula* (brush-tailed possum). Comp Biochem Physiol. 70B:619-622.
6825. Marlow, BJ. 1961. Reproductive behavior of the marsupial mouse, *Antechinus flavipes* (Waterhouse) (Marsupialia) and the development of the pouch young. Aust J Zool. 9:203-218.
6826. Marlow, BJ. 1962. On the occurrence of *Antechinus muculatus* Gould and *Planigale ingrami* (Thomas) (Marsupialia, Dasyuridae) in Cape York Peninsula, Queensland. J Mamm. 43:433-434.
6827. Marlow, BJ. 1968. The sea lions of Dangerous Reef. Aust Nat Hist. 16:39-44.
6828. Marlow, BJ & King, JB. 1979. Hocker's (New Zealand) sea lions. Mammals in the Seas, FAO Fish Ser 5. 2:16-18.
6828a. Marlow, BJG. 1955. Observations on the herero musk shrew, *Crocidura flavescens herero* St Leger, in captivity. Proc Zool Soc London. 124:803-808.
6829. Marma, BB & Yunchis, VV. 1968. Observations on the breeding, management and physiology of snow leopards *Panthera u uncia* at Kaunas Zoo from 1962-1967. Int Zoo Yearb. 8:66-74.
6830. Marquette, WM. 1977. The 1976 catch of bowhead whales (*Balaena mysticetus*) by Alaskan eskimos, with a review of the fishery 1973-1976, and a biological summary of the species.

Process Report, US Dept Comm, Nat Oceanic Atmos Adm, Nat Mar Fish Serv NW, & AK Fisheries Center. Seattle.
6831. Marsden, HM & Conaway, CH. 1963. Behaviour and the reproductive cycle in the cottontail. J Wildl Manage. 27:161-170.
6832. Marsden, HM & Holler, NR. 1964. Social behavior in confined populations of the cottontail and the swamp rabbit. Wildl Monogr. 13:1-39.
6833. Marsh, CW. 1979. Comparative aspects of social organization in the Tana River red colobus *Colobus badius rufomitratus*. Z Tierpsychol. 51:337-362.
6834. Marsh, H. 1979. The life history parameters of the dugong and their implications for conservation. The Dugong Proc Seminar/Workshop, James Cook Univ N QLD, 1979, 53 Dept Zool James Cook Univ of N QLD, Townsville.
6835. Marsh, H. 1981. Preliminary description of the reproductive organs of the female dugong and suggested methods of specimen collection as part of a carcass salvage program. 141-144. The Dugong Proc Seminar/Workshop. James Cook Univ N QLD, 1979.
6836. Marsh, H. 1985. Observations on the ovaries of an isolated minke whale: Evidence for spontaneous sterile ovulation and structure of the resultant corpus. Sci Rep Whales Res Inst. 36:35-39.
6837. Marsh, H & Glover, TD. 1981. A preliminary description of the reproductive organs of the male dugong and suggested methods of specimen collection. The Dugong Proc Seminar/Workshop, James Cook Univ, N QLD, 1979. 149-154, J Cook Univ N QLD, Townsville.
6838. Marsh, H, Heinsohn, GE, & Channells, PW. 1984. Changes in the ovaries and uterus of the dugong, *Dugong dugon* (Sirenia: Dugongidae), with age and reproductive activity. Aust J Zool. 32:743-766.
6839. Marsh, H, Heinsohn, GE, & Glover, TD. 1984. Changes in the male reproductive organs of the dugong, *Dugong dugon* (Sirenia: Dugongidae), with age and reproductive activity. Aust J Zool. 32:721-742.
6840. Marsh, H, Heinsohn, GE, & Marsh, LM. 1984. Breeding cycle, life history and population dynamics of the dugong, *Dugong dugon* (Sirenia: Dugongidae). Aust J Zool. 32:767-788.
6841. Marsh, H & Kasuya, T. 1984. Changes in the ovaries of the short-finned pilot whale, *Globicephala macrorhynchus*, with age and reproductive activity. Rep Int Whal Comm Sp Issue. 6:311-335.
6842. Marsh, H, Spain, AV, & Heinsohn, GE. 1978. Physiology of the dugong. Comp Biochem Physiol. 61A:159-168.
6843. Marsh, RE & Howard, WE. 1968. Breeding ground squirrels, *Spermophilus beecheyi*, in captivity. J Mamm. 49:781-783.
6844. Marshall, AG & McWilliam, AN. 1982. Ecological observations on epomorphorine fruit-bats (Megachiroptera) in West African savanna woodland. J Zool. 198:53-67.
6845. Marshall, AJ. 1947. The breeding cycle of an equatorial bat (*Pteropus giganteus* of Ceylon). Proc Linn Soc London. 159:103-111.
6846. Marshall, AJ. 1948. Pre-gestational changes in the giant fruit bat (*Pteropus giganteus*) with special reference to an asymmetric endometrial reaction. Proc Linn Soc London. 161:26-36.
6847. Marshall, AJ. 1953. The unilateral endometrial reaction in the giant fruit-bat (*Pteropus giganteus*). J Endocrinol. 9:42-44.
6848. Marshall, AJ & Corbet, PS. 1959. The breeding biology of equatorial vertebrates: Reproduction of the bat *Chaerephon hindei* Thomas at latitude 0°26'N. Proc Zool Soc London. 132:607-616.
6849. Marshall, FHA. 1904. The oestrus cycle in the common ferret. Q J Microsc Sci. 48:323-345.
6850. Marshall, FHA. 1937. On the changeover in the oestrous cycle in animals after transferrence across the equator, with further observations on the incidence of the breeding seasons and the factors controlling sexual periodicity. Proc R Soc London Biol Sci. 122B:413-428.
6851. Marshall, FHA & Bowden, FP. 1934. The effect of irradiation with different wave-lengths on the oestrous cycle of the ferret, with remarks on the factors controlling sexual periodicity. J Exp Biol. 11:409-422.
6852. Marshall, FHA & Bowden, FP. 1936. The further effects of irradiation on the oestrous cycle of the ferret. J Exp Biol. 13:383-386.
6853. Marshall, FHA & Jolly, WA. 1905. Contributions to the physiology of mammalian reproduction Part I The oestrous cycle in the dog Part II The ovary as an organ of internal secretion. Proc R Soc London Biol Sci. 76B:395-398.
6854. Marshall, GR, Wickings, EJ, & Nieschlag, E. 1984. Testosterone can initiate spermatogenesis in an immature nonhuman primate, *Macaca fascicularis*. Endocrinology. 114:2228-2233.
6855. Marshall, JT, Jr. 1962. House mouse. Bull BP Bishop Mus. 225:241-246.
6856. Marshall, LG. 1977. *Lestodelphys halli*. Mamm Species. 81:1-3.
6857. Marshall, LG. 1978. *Lutreolina crassicaudata*. Mamm Species. 91:1-4.
6858. Marshall, LG. 1978. *Dromiciops australis*. Mamm Species. 99:1-5.
6859. Marshall, LG. 1978. *Glironia venusta*. Mamm Species. 107:1-3.
6860. Marshall, LG. 1978. *Chironectes minimus*. Mamm Species. 109:1-6.
6861. Marshall, LG. 1982. Evolution of South American Marsupialia. Mammalian Biology in South America. 251-272, Pymatuning Symposia in Ecology, Pittsburgh.
6862. Marshall, PJ & Sayer, JA. 1976. Population ecology and response to cropping of a hippopotamus population in eastern Zambia. J Appl Ecol. 13:391-403.
6863. Marshall, WH & Enders, RK. 1942. The blastocyst of the marten (*Martes*). Anat Rec. 84:307-310.
6864. Marsteller, FA & Lynch, CB. 1987. Reproductive responses to variation in temperature and food supply by house mice I Mating and pregnancy. Biol Reprod. 37:838-843.
6865. Marston, HR. 1926. The milk of the monotreme, *Echidna aculeata multi-aculeata*. Aust J Exp Biol Med Sci. 3:217-220.
6866. Marston, JH & Chang, MC. 1964. The time of ovulation, fertilization and cleavage in the Mongolian gerbil (*Meriones unguiculatus*). Anat Rec. 148:387.
6867. Marston, JH & Chang, MC. 1965. The breeding, management and reproductive physiology of the Mongolian gerbil (*Meriones unguiculatus*). Lab Anim Care. 15:34-48.
6868. Martan, J & Hruban, Z. 1970. Unusual spermatozoan formations in the epididymis of the flying squirrel (*Glaucomys volans*). J Reprod Fert. 21:167-170.
6869. Martan, JM, Adams, CS, & Perkins, BL. 1970. Epididymal spermatozoa of two species of squirrels. J Mamm. 51:376-378.
6870. Martel, D, Malet, C, Monier, M-N, et al. 1980. Nuclear receptor for oestrogen in the baboon endometrium: Detection, characterization and variation in its concentration during the menstrual cycle. J Endocrinol. 84:273-280.
6871. Martel, D, Malet, C, Olmedo, C, et al. 1980. Oestrogen receptor in the baboon endometrium: Cytosolic receptor,

detection, characterization and variation of its concentration during the menstrual cycle. J Endocrinol. 84:261-272.
6872. Martensz, ND & Everitt, BJ. 1982. Effects of passive immunization against testosterone on the sexual activity of female rhesus monkeys. J Endocrinol. 94:271-282.
6873. Martensz, ND & Herbert, J. 1982. Drug-induced hyperprolactinemia and discharges of luteinizing hormone evoked by oestrogen in ovariectomized rhesus monkeys. J Endocrinol. 92:111-122.
6874. Martensz, ND, Vellucci, SV, Fuller, LM, et al. 1987. Relation between aggressive behaviour and circadian rhythms in cortisol and testosterone in social groups of talapoin monkeys. J Endocrinol. 115:107-120.
6875. Martin, AR. 1982. Pregnancy rates of fin whales in the Icelandic catch. Rep Int Whal Comm. 32:325-329.
6876. Martin, DE. 1981. Breeding great apes in captivity. Reproductive Biology of the Great Apes. 343-373, Graham, CE, ed, Academic Press, New York.
6877. Martin, DE & Gould, KG. 1981. The male ape genital tract and its secretions. Reproductive Biology of the Great Apes. 127-162, Graham, CE, ed, Academic Press, New York.
6878. Martin, DE, Graham, CE, & Gould, KG. 1978. Successful artificial insemination in the chimpanzee. Symp Zool Soc London. 43:249-260.
6879. Martin, DE, Swenson, RB, & Collins, DC. 1977. Correlations of serum testosterone levels with age in male chimpanzees. Steroids. 29:471-481.
6880. Martin, DP. 1984. The effects of extended time in the laboratory on selected aspects of reproduction in the female rhesus monkey (*Macaca mulatta*). Am J Primatol. 7:39-55.
6881. Martin, EP. 1956. A population study of the prairie vole (*Microtus ochrogaster*) in north eastern Kansas. Univ KS Publ Mus Nat Hist. 8:361-416.
6882. Martin, IG. 1982. Maternal behavior of a short-tailed shrew (*Blarina brevicauda*). Acta Theriol. 27:153-156.
6883. Martin, JT & Baum, MJ. 1986. Neonatal exposure of female ferrets to testosterone alters sociosexual preferences in adulthood. Psychoneuroendocrinology. 11:167-176.
6884. Martin, K. 1943. The Colorado pika. J Mamm. 24:394-396.
6885. Martin, PA, Bevier, GW, & Dziuk, PJ. 1977. The effect of number of corpora lutea on the length of gestation in pigs. Biol Reprod. 16:633-637.
6886. Martin, PG. 1965. The potentialities of the fat-tailed marsupial mouse, *Sminthopsis crassicaudata* (Gould), as a laboratory animal. Aust J Zool. 13:559-562.
6887. Martin, R. 1976. Breeding great apes in captivity. New Scientist. 72:100-102.
6888. Martin, RA. 1967. Notes on the male reproductive tract of *Nectogale* and other soricid insectivores. J Mamm. 48:664-666.
6889. Martin, RD. 1966. Tree shrews: Unique reproductive mechanism of systemic importance. Science. 152:1402-1404.
6890. Martin, RD. 1968. Reproduction and ontogeny in tree-shrews (*Tupaia belangeri*) with reference to their general behaviour and taxonomic relationships. Z Tierpsychol. 25:409-495.
6891. Martin, RD. 1972. A laboratory breeding colony of the lesser mouse lemur. Breeding Primates. 161-171, Beveridge, WB, ed, Karger, Basel.
6892. Martin, RD. 1975. Breeding tree-shrews *Tupaia belangeri* and mouse lemurs *Microcebus murinus* in captivity. Int Zoo Yearb. 15:35-41.
6893. Martin, RD, Seaton, B, & Lasty, JA. 1975. Application of urinary hormone determinations in the management of gorillas. Ann Rep Wildl Preserv Trust. 12:61-70.
6894. Martin, RG & Lee, A. 1984. The koala, *Phascolarctos cinereus*, the largest marsupial folivore. Possums and Gliders. 463-467, Smith, AP & Hume, ID, eds, Aust Mamm Soc, Sydney.
6895. Martin, RJ. 1973. Growth curves for bushy-tailed woodrats based upon animals raised in the wild. J Mamm. 54:517-518.
6896. Martin, RL. 1971. The natural history and taxonomy of the rock vole, *Microtus chrotorrhinus*. Diss Abstr Int. 32(5):3079B.
6897. Martin, RL. 1979. Morphology, development, and adaptive values of the baculum of *Microtus chrotorrhinus* (Miller, 1894) and related forms. Säugetierk Mitt. 27:307-311.
6898. Martin, RW. 1981. Age-specific fertility in three populations of the koala, *Phascolarctos cinereus* Goldfuss, in Victoria. Aust Wildl Res. 8:275-283.
6899. Martin, RW & Lee, AK. 1984. The koala, *Phascolarctos cinereus*: The largest marsupial folivore. Possums and Gliders. 463-467, Smith, AP & Hume, ID, eds, Surrey Beatty & Sons Pty Ltd, Sydney, Australia.
6900. Martinet, L. 1963. Etablissement de la spermatogenèse chez le Campagnol des champs (*Microtus arvalis*) en fonction de la durée quotidienne d'éclairement. Ann Biol Anim Bioch Biophys. 3:343-352.
6901. Martinet, L. 1966. Modification de la spermatogenèse chez le Campagnol des champs (*Microtus arvalis*) en fonction de la durée quotidienne d'éclairement. Ann Biol Anim Bioch Biophys. 3:301-313.
6902. Martinet, L. 1967. Cycle saisonnier de reproduction du Campagnol des champs *Microtus arvalis*. Ann Biol Anim Bioch Biophys. 7:245-259.
6903. Martinet, L. 1968. Cycle saisonnier de reproduction du Campagnol des champs *Microtus arvalis*. Cycles Génitaux Saisonniers de Mammifères Sauvages. 67-78, Canivenc, R, ed, Masson et Cie, Paris.
6904. Martinet, L. 1980. Oestrous behaviour, follicular growth and ovulation during pregnancy in the hare (*Lepus europaeus*). J Reprod Fert. 59:441-445.
6905. Martinet, L, Allais, C, & Allain, D. 1981. The role of prolactin and LH in luteal function and blastocyst growth in mink (*Mustela vison*). J Reprod Fert Suppl. 29:119-130.
6906. Martinet, L, Allain, D, & Chabi, Y. 1985. Pineal denervation by cervical sympathetic ganglionectomy suppresses the role of photoperiod on pregnancy or pseudopregnancy, body weight and moulting periods in the mink (*Mustela vison*). J Endocrinol. 107:31-39.
6907. Martinet, L, Allain, D, & Meunier, M. 1983. Regulation in pregnant mink (*Mustela vison*) of plasma progesterone and prolactin concentrations and regulation of onset of the spring moult by daylight ratio and melatonin injections. Can J Zool. 61:1959-1963.
6908. Martinet, L & Demarne, Y. 1984. Nursing behaviour and lactation in the brown hare (*Lepus europaeus*) raised in captivity. Acta Zool Fennici. 171:187-190.
6909. Martinet, L, Legouis, JJ, & Moret, B. 1970. Quelques observations sur la reproduction du lièvre européen (*Lepus europaeus* Pallas) en captivité. Ann Biol Anim Bioch Biophys. 10:195-202.
6910. Martinet, L & Meunier, M. 1969. Influence des variations saisonnières de la luzerne sur la croissance, la mortalité et l'établissement de la maturité sexuelle chez le Campagnol des champs (*Microtus arvalis*). Ann Biol Anim Bioch Biophys. 9:451-462.

6911. Martinet, L & Meunier, M. 1975. Plasma and pituitary levels of LH in field voles, *Microtus arvalis*, reared under two different photoperiods. J Physiol (Paris). 70:539-547.
6912. Martinet, L, Ravault, JP, & Meunier, M. 1982. Seasonal variations in mink (*Mustela vison*) plasma prolactin measured by heterologous radioimmunoassay. Gen Comp Endocrinol. 48:71-75.
6913. Martinet, L & Raynaud, F. 1972. Mécanisme possible de la superfétation chez la Hase. CR Hebd Séances Acad Sci Sér D Sci Nat. 274:2683-2686.
6914. Martinet, L & Raynaud, F. 1975. Prolonged spermatozoan survival in the female hare uterus: Explanation of superfoetation. Biology of Spermatozoa INSERM Int Symposium Nouzilly, 1973. 134-144, ed, Karger, Basel.
6915. Martinez-Esteve, P. 1937. Le cycle sexuel vaginal chez le Marsupial *Didelphis ararae*. CR Soc Biol. 124:502-504.
6916. Martinez-Esteve, P. 1942. Observations on the histology of the opossum ovary. Contrib Embryol Carnegie Inst. 189:17-26.
6917. Martinka, CJ. 1974. Population characteristics of grizzly bears in Glacier National Park, Montana. J Mamm. 55:21-29.
6918. Martinsen, DL. 1968. Temporal patterns in the home ranges of chipmunks (*Eutamias*). J Mamm. 49:83-91.
6919. Martiroyan, BA. 1964. Propagation biology of the vole *Microtus* (*Chionomys*) *nivalis* in Armenian SSR. Zool Zhurn. 43:1552-1556.
6920. Marut, EL & Hodgen, GD. 1982. Asymmetric secretion of principal ovarian venous steroids in the primate luteal phase. Steroids. 39:461-469.
6921. Marut, EL, Williams, RF, Cowan, BD, et al. 1981. Pulsatile pituitary gonadotropin secretion during maturation of the dominant follicle in monkeys: Estrogen positive feedback enhances the biological activity of LH. Endocrinology. 109:2270-2272.
6922. Masake, RA. 1977. The ovary of the Thomson's gazelle. Z Säugetierk. 42:44-51.
6923. Masataka, N. 1981. A field study of the social behavior of Goeldi's monkeys (*Callimico goeldii*) in north Bolivia. Kyoto Univ Overseas Res Rep, New World Monkeys. 2:23-41.
6924. Maschkin, VI. 1982. New materials on ecology of *Marmota menzbieri* (Rodentia, Sciuridae). Zool Zhurn. 61:278-289.
6925. Maser, C. 1974. The sage vole, *Lagurus curtatus* (Cope, 1968), in the Crooked River National Grassland, Jefferson County, Oregon: A contribution to its life history and ecology. Säugetierk Mitt. 22:193-222.
6926. Maser, C & Towell, DE. 1984. Bacula of mountain lions, *Felis concolor*, and bobcat *F rufus*. J Mamm. 65:496-497.
6927. Mason, CF & Macdonald, SM. 1986. Otters: Ecology and Conservation. Cambridge Univ Press, Cambridge.
6928. Mason, DR. 1976. Observations on social organization, behaviour and distribution of impala in the Jack Scott Nature Reserve. S Afr J Wildl Res. 6:79-87.
6929. Mason, DR. 1982. Studies on the biology and ecology of the warthog *Phacochoerus aethiopicus sundevalli* Lonnberg 1908 in Zululand. D Sc Thesis. Univ Pretoria, Pretoria. [Cited by Skinner, 1984]
6930. Mason, DR. 1986. Reproduction in the male warthog *Phacochoerus aethiopicus* from Zululand, South Africa. S Afr J Zool. 21:39-47.
6931. Mason, K & Blackshaw, AW. 1973. The spermatogenic cycle of the bandicoot, *Isoodon macrourus*. J Reprod Fert. 32:307-308.
6932. Mason, WA. 1966. Social organization of the South American monkey, *Callicebus moloch*: A preliminary report. Tulane Stud Zool. 13:23-28.
6933. Mason, WA. 1968. Use of space by *Callicebus* groups. Primates: Studies in Adaptation and Variability. 200-216, Jay, PC, ed, Holt, Rinehart & Winston, New York.
6934. Masquelin, H & Swaen, A. 1880. Premières phases du développement du placenta maternel chez le Lapin. Arch Biol. 1:25-44.
6935. Massey, A & Vandenbergh, JG. 1980. Puberty delay by the urinary cue from female house mice in feral populations. Science. 209:821-822.
6936. Masui, M. 1967. Birth of a Chinese pangolin *Manis pentudactyla* at Ueno Zoo, Tokyo. Int Zoo Yearb. 7:114-116.
6937. Masure, AM & Bourlière, F. 1971. Surpeuplement, fecondité, mortalité et aggressivité dans une population captive de *Papio papio*. Terre Vie. 118:491-505.
6938. Mate, B. 1973. Population kinetics and related ecology of the northern sea lion, *Eumetopias jubatus* and the California sea lion, *Zalophus californianus*, along the Oregon coast. Ph D Diss. Univ OR, Eugene.
6939. Mate, B. 1979. California sea lion. Mammals in the Sea, FAO Fish Ser 5. 2:5-8.
6940. Mate, B & Gentry, RL. 1979. Northern (Steller) sea lion. Mammals in the Seas, FAO Fish Ser 5. 2:1-4.
6941. Mate, KE & Rodger, JC. 1991. Stability of the acrosome of the brush-tailed possum *Trichosurus vulpecula* and tammar wallaby *Macropus eugenii* in vitro and after exposure to conditions and agents known to cause capacitation or acrosome reaction of eutherian spermatozoa. J Reprod Fert. 91:41-48.
6942. Matharu, BS. 1966. Camel care. Indian Farming. 16(7):19-22.
6943. Mathews, M. 1973. Birth of a white rhinoceros in captivity. J Zoo Anim Med. 4:18.
6944. Mathisen, OA, Baade, RT, & Lopp, RJ. 1962. Breeding habits, growth and stomach contents of the Steller sea lion in Alaska. J Mamm. 43:469-476.
6945. Mathur, RS & Goyal, RP. 1974. Histomorphology of the male accessory reproductive organs of *Suncus murinus sindensis* Anderson, the common Indian shrew. Acta Anat. 89:610-615.
6945a. Matkin, CO & Leatherwood, S. 1986. General biology of the killer whale, *Orcinus orca*: A synopsis of knowledge. Behavioral Biology of Killer Whales. 35-68, AR Liss, ed, New York.
6946. Matomoras. Personal communication. [cited by Collett, F, 1981]
6947. Matschke, GH & Roughton, RD. 1977. Delayed antler development and sexual maturity among yearling male white-tailed deer. Proc Ann SE Conf Fish Wildl Agen. 31:51-56.
6948. Matson, JO. 1975. *Myotis planiceps*. Mamm Species. 60:1-2.
6949. Matson, JO & Abravaya, JP. 1977. *Blarinomys breviceps*. Mamm Species. 74:1-3.
6950. Matson, JR. 1952. Litter size in the black bear. J Mamm. 33:246-247.
6951. Matsushima, S, Sakai, Y, & Hira, Y. 1990. Effect of photoperiod on pineal gland volume and pinealocyte size in the Chinese hamster, *Cricetulus griseus*. Am J Anat. 187:32-38.
6952. Matsuura, Y. 1936. On the lesser rorqual found in the adjacent waters of Japan. Bull Jap Soc Sci Fish. 4:325-330.
6953. Matsuura, Y. 1942. On the northern porpoise whale *Berardius bairdii* Stejneger in the adjacent waters of "Bosyu". Zool Mag Tokyo. 54:466-473.

6953a. Matsuzaki, T, Kamiya, M, & Suzuki, H. 1985. Gestation period of the laboratory reared volcano rabbit (*Romerolagus diazi*). Exp Anim. 34:63-66.

6953b. Matsuzaki, T, Saito, M, & Kamiya, M. 1982. Breading and rearing of the volcano rabbit (*Romerolagus diazi*) in captivity. Exp Anim. 31:185-188.

6954. Matteri, RL, Baldwin, DM, Lasley, BL, et al. 1987. Biological and immunological properties of zebra pituitary gonadotropins: Comparison with horse and donkey gonadotropins. Biol Reprod. 36:113-111.

6955. Matthews, LH. 1935. The oestrous cycle and intersexuality of the female mole (*Talpa europaea* Linn). Proc Zool Soc London. 1935:347-383.

6956. Matthews, LH. 1937. The female sexual cycle in the British horse-shoe bats, *Rhinolophus ferrum-equinum insulanus* Barrett-Hamilton and *R hipposideros minutus* Montagu. Trans Zool Soc London. 23:224-266.

6957. Matthews, LH. 1937. The form of the penis in the British rhinolophid bats, compared with that in some vespertilionid bats. Trans Zool Soc London. 23:213-224.

6958. Matthews, LH. 1937. The humpback whale, *Megaptera nodosa*. Discovery Rep. 17:7-92.

6959. Matthews, LH. 1938. Notes on the southern right whale, *Eubalaena australis*. Discovery Rep. 17:169-182.

6960. Matthews, LH. 1938. The sei whale, *Balaenoptera borealis*. Discovery Rep. 17:183-290.

6961. Matthews, LH. 1938. The sperm whale *Physeter catodon*. Discovery Rep. 17:93-168.

6962. Matthews, LH. 1939. Post-partum oestrus in a bat. Nature. 143:643.

6963. Matthews, LH. 1939. Reproduction in the spotted hyena *Crocuta crocuta* (Erxleben). Phil Trans R Soc London Biol Sci. 230B:1-78.

6964. Matthews, LH. 1939. The bionomics of the spotted hyaena, *Crocuta crocuta* Erxl. Proc Zool Soc London. 109:43-56.

6965. Matthews, LH. 1941. Reproduction in the Scottish wild cat, *Felis silvestris grampia* Miller. Proc Zool Soc London. 111(B):59-77.

6966. Matthews, LH. 1942. Notes on the genitalia and reproduction of some African bats. Proc Zool Soc London. 111(B):289-346.

6967. Matthews, LH. 1946. Notes on the genital anatomy and physiology of the gibbon (*Hylobates*). Proc Zool Soc London. 116:339-364.

6968. Matthews, LH. 1947. A note on the female reproductive tract in the tree kangaroos (*Dendrolagus*). Proc Zool Soc London. 117:313-332.

6969. Matthews, LH. 1954. Placenta of the spotted hyaena, *Crocuta crocuta*. Proc Zool Soc London. 124:198.

6970. Matthews, LH. 1956. The sexual skin of the gelada baboon (*Theropithecus gelada*). Trans Zool Soc London. 28:543-552.

6971. Matthews, LH & Harrison, RJ. 1949. Subsurface crypts, oogenesis, and the corpus luteum in the ovaries of seals. Nature. 164:587-588.

6972. Matthey, R. 1954. Un cas nouveau de chromosomes sexuels multiples dans le genre *Gerbillus* (Rodentia Muridae: Gerbillinae). Experientia. 10:464-465.

6973. Matthey, R. 1958. Un nouveau type de détermination chromosomique du sexe chez les mammifères *Ellobius lutescens* Th et *Microtus* (*Chilotus*) *oregoni* Bachm (Murides: Microtines). Experientia. 14:240-241.

6974. Matthey, R. 1965. Le problème de la détermination du sexe chez *Acomys selousi* de Winton: Cytogénétique du genre *Acomys* (Rodentia: Murinae). Rev Suisse Zool. 72:119-144.

6975. Mattingly, DK & McClure, PA. 1982. Energetics of reproduction in large-littered cotton rats (*Sigmodon hispidus*). Ecology. 63:183-195.

6976. Mattingly, DK & McClure, PA. 1985. Energy allocation during lactation in cotton rats (*Sigmodon hispidus*) on a restricted diet. Ecology. 66:928-937.

6977. Mattlin, RH. 1981. Pup growth of the New Zealand fur seal *Arctocephalus forsteri* on the Open Bay Islands, New Zealand. J Zool. 193:303-314.

6978. Mauer, RA & Bradley, WG. 1966. Reproduction and food habits in Merriam's kangaroo rat. Desert Res Inst, Univ NV. 37:1-17.

6979. Mauget, R. 1972. Observations sur la reproduction du Sanglier (*Sus scrofa* L) à l'état sauvage. Ann Biol Anim Bioch Biophys. 12:195-202.

6980. Mauget, R. 1978. Seasonal reproductive activity of the European wild boar: Comparison with the domestic sow. Environmental Endocrinology. 79-80, Assenmacher, I & Farner, DS, eds, Springer, New York.

6981. Maurel, D. 1978. Seasonal changes of the testicular and thyroid functions in the badger, *Meles meles*. Environmental Endocrinology. 85-86, Assenmacher, I & Farner, DS, eds, Springer, New York.

6982. Maurel, D & Boissin, J. 1981. Plasma thyroxine and testosterone levels in the red fox (*Vulpes vulpes* L) during the annual cycle. Gen Comp Endocrinol. 43:402-404.

6983. Maurel, D & Boissin, J. 1982. Métabolisme périphérique de la testostérone en relation avec le cycle annuel de la testostérone et de la 5α-dihydrotestostérone plasmatiques chez le Blaireau européen (*Meles meles* L) et le Renard roux (*Vulpes vulpes* L). Can J Zool. 60:406-416.

6984. Maurel, D, Coutant, C, & Boissin, J. 1989. Effects of photoperiod, melatonin implants and castration on molting and on plasma thuroxine, testosterone and prolactin levels in the European badger (*Meles meles*). Comp Biochem Physiol. 93A:791-797.

6985. Maurel, D, Lacroix, A, & Boissin, J. 1984. Seasonal reproductive endocrine profiles in two wild mammals, the red fox *Vulpes vulpes* L and the European badger (*Meles meles* L). Acta Endocrinol. 105:130-138.

6986. Maurel, D, Laurent, AM, & Bossin, J. 1981. Short-term variations of plasma testosterone concentrations in the European badger (*Meles meles*). J Reprod Fert. 61:53-58.

6987. Maurel, D, Laurent, AM, Daniel, JY, et al. 1980. Étude de la capacité de liason de la protéine plasmatique liant la testostérone chez deux Mammifères sauvages à activité testiculaire cyclique, le Renard et le Blaireau. CR Hebd Séances Acad Sci. 291D:693-696.

6988. Maxwell, CS & Morton, ML. 1975. Comparative thermmoregulatory capabilities of neonatal ground squirrels. J Mamm. 56:821-828.

6989. Mayer, JJ & Brandt, PN. 1982. Identity, distribution, and natural history of the peccaries, Tayassuidae. Pymatuning Lab Ecol Sp Publ Ser. 6:433-455.

6990. Mayer, JJ & Wetzel, RM. 1986. *Catagonus wagneri*. Mamm Species. 259:1-5.

6991. Mayer, JJ & Wetzel, RM. 1987. *Tayassu pecari*. Mamm Species. 293:1-7.

6992. Mayer, WV & Boche, ET. 1954. Developmental patterns in the Barrow ground squirrel, *Spermophilus undulatus barrowensis*. Growth. 18:53-69.

6993. Mayer, WV. 1953. A preliminary study of the Barrow ground squirrel, *Citellus parryi barrowensis*. J Mamm. 34:334-345.

6994. Maynes, GM. 1972. Age estimation in the parma wallaby, *Macropus parma* Waterhouse. Aust J Zool. 20:107-118.
6995. Maynes, GM. 1972. Reproduction of the parma wallaby *Macropus parma*. Aust Mamm. 1:63-64.
6996. Maynes, GM. 1973. Aspects of reproduction in the whiptail wallaby, *Macropus parryi*. Aust Zool. 18:43-46.
6997. Maynes, GM. 1973. Reproduction in the parma wallaby, *Macropus parma* Waterhouse. Aust J Zool. 21:331-351.
6998. Maynes, GM. 1975. Breeding the parma wallaby in captivity. Breeding Endangered Species in Captivity. 167-170, Martin, RD, ed, Academic Press, New York.
6999. Maynes, GM. 1976. Growth of the parma wallaby, *Macropus parma* Waterhouse. Aust J Zool. 24:217-236.
7000. Maynes, GM. 1977. Distribution and aspects of the biology of the parma wallaby, *Macropus parma*, in New South Wales. Aust Wild Res. 4:109-126.
7001. Maynes, GM. 1977. Breeding and age structure of the population of *Macropus parma* on Kawau Island, New Zealand. Aust J Ecol. 2:207-214.
7002. Mazák, V. 1962. Zur Kenntnis der postnatalen Entwicklung der Rotelmaus *Clethrionomys glareolus* Schreber, 1780 (Mammalia, Microtidae). Vestn Čcsl Spol Zool. 26:77-104.
7003. Mazák, V. 1965. Hohes Alter einner Zwergmaus, *Micromys minutus*. Säugetierk Mitt. 13:161.
7004. Mazák, V. 1981. *Panthera tigris*. Mamm Species. 152:1-8.
7005. Mazin, VN. 1975. Data on reproduction of jerboas of the genus *Allactaga* in southeast Kazakhstan. Soviet J Ecol. 6:334-339.
7006. McAllan, BM & Dickman, CR. 1986. The role of photoperiod in the timing of reproduction in the dasyurid marsupial *Antechinus stuartii*. Oecologia. 68:259-264.
7007. McAllister, JA & Hoffmann, RS. 1988. *Phenacomys intermedius*. Mamm Species. 305:1-8.
7008. McArthur, JW, Beitins, IZ, Gorman, A, et al. 1981. The interrelationship between sex skin swellling and the urinary excretion of LH, estrone, and pregnanediol by the cycling female chimpanzee. Am J Primatol. 1:265-270.
7009. McArthur, JW, Ovadia, J, Smith, OW, et al. 1972. The menstrual cycle of the bonnet monkey (*Macaca radiata*). Folia Primatol. 17:107-121.
7010. McArthur, JW & Perley, R. 1969. Urinary gonadotropin excretion by infrahuman primates. Endocrinology. 84:508-513.
7011. McBee, K & Baker, RJ. 1982. *Dasypus novemcinctus*. Mamm Species. 162:1-9.
7012. McBride, AF & Kritzler, H. 1951. Observations on pregnancy, parturition, and postnatal behavior in the bottlenose dolphin. J Mamm. 32:251-266.
7013. McCabe, TT & Blanchard, BD. 1950. Three Species of *Peromyscus*. Rood Assoc, Santa Barbara, CA.
7013a. McCann, C. 1933. Notes on the colouration and habits of the white-browed gibbon or hoolock (*Hylobates hoolock* Harl). J Bombay Nat Hist Soc. 36:395-405.
7013b. McCann, C. 1941. Further observations on the flying-fox (*Pteropus giganteus* Brünn) and the fulvous fruit-bat (*Rousettus leschenaulti* Desm). J Bombay Nat Hist Soc. 42:587-592.
7014. McCann, G. 1975. A study of the genus *Berardius* Duvernoy. Sci Rep Whales Res Inst. 27:111-137.
7015. McCann, LJ. 1944. Notes on growth, sex and age ratios, and suggested management of Minnesota muskrats. J Mamm. 25:59-63.
7016. McCann, LJ. 1956. Ecology of the mountain sheep. Am Mid Nat. 56:297-324.
7017. McCann, TS. 1980. Population structure and social organization of southern elephant seals, *Mirounga leonina* (L). Biol J Linn Soc. 14:133-150.
7018. McCann, TS. 1981. Aggression and sexual activity of male southern elephant seals, *Mirounga leonina*. J Zool. 195:295-310.
7019. McCann, TS. 1981. The social organization and behaviour of the southern elephant seal, *Mirounga leonina* (L). Ph D Diss. Univ London, London.
7020. McCann, TS. 1982. Aggressive and maternal activities of female southern elephant seals (*Mirounga leonina*). Anim Behav. 30:268-276.
7021. McCann, TS. 1983. Activity budgets of southern elephant seals, *Mirounga leonina*, during the breeding season. Z Tierpsychol. 61:111-126.
7022. McCarley, H. 1958. Ecology, behavior and population dynamics of *Peromyscus nuttalli* in eastern Texas. TX J Sci. 10:147-171.
7023. McCarley, H. 1966. Annual cycle, population dynamics and adaptive behavior of *Citellus tridecemlineatus*. J Mamm. 47:294-316.
7024. McCarthy, TJ. 1982. Bat records from the Caribbean lowlands of El Peten Guatemala. J Mamm. 63:683-685.
7025. McCarty, R. 1975. *Onychomys torridus*. Mamm Species. 59:1-5.
7026. McCarty, R. 1978. *Onychomys leucogaster*. Mamm Species. 87:1-6.
7027. McCarty, R & Southwick, CH. 1977. Patterns of parental care in two cricetid rodents, *Onychomys torridus* and *Peromyscus leucopus*. Anim Behav. 25:945-948.
7028. McClenaghan, LR. 1984. Comparative ecology of sympatric *Dipodomys agilis* and *Dipodomys merriami* populations in southern California. Acta Theriol. 29:175-185.
7029. McClenaghan, LR, Jr & Gaines, MS. 1978. Reproduction in marginal populations of the hispid cotton rat (*Sigmodon hispidus*) in northeastern Kansas. Occas Pap Mus Nat Hist Univ KS. 74:1-16.
7030. McClure, PA & Randolph, JC. 1980. Relative allocation of energy to growth and development of homeothermy in the eastern wood rat (*Neotoma floridana*) and hispid cotton rat (*Sigmodon hispidus*). Ecol Monogr. 50:199-219.
7031. McConnell, SJ & Hinds, LA. 1985. Effect of pinealectomy on plasma melatonin, prolactin and progesterone concentrations during seasonal reproductive quiescence in the tammar, *Macropus eugenii*. J Reprod Fert. 75:433-440.
7032. McConnell, SJ, Tyndale-Biscoe, CH, & Hinds, LA. 1986. Change in duration of elevated concentrations of melatonin is the major factor in photoperiod response of the tammar, *Macropus eugenii*. J Reprod Fert. 77:623-632.
7033. McCoy, GW. 1912. Notes on bionomics of rats and ground squirrels. Public Health Rep. 27:1068-1072.
7034. McCracken, GF. 1978. Social organization and genetic studies in the polygynous bat, *Phyllostomus hastatus*. Bat Res News. 19:89.
7035. McCracken, GF. 1984. Communal nursing in Mexican free-tailed bat maternity colonies. Science. 223:1090-1091.
7036. McCracken, GF & Bradbury, J. 1977. Paternity and genetic heterogeneity in the polygamous bat, *Phyllostomus hastatus*. Science. 198:303-306.
7037. McCracken, HE. 1986. Observations on the oestrous cycle and gestation period of the greater bilby, *Macrotis lagotis* (Reid) (Marsupialia: Thylacomyidae). Aust Mamm. 9:5-16.
7038. McCrane, MP. 1966. Birth, behavior and development of a hand-reared two-toed sloth *Choloepus didactylus*. Int Zoo Yearb. 6:153-163.

7039. McCullagh, KG & Widdowson, EM. 1970. The milk of the African elephant. Br J Nutr. 24:109-117.
7040. McCulloch, CY. 1955. Breeding record of javelina, *Tayassu angulatus*, in southern Arizona. J Mamm. 36:146.
7041. McCulloch, CY & Inglis, JM. 1961. Breeding periods of the Ord kangaroo rat. J Mamm. 42:337-344.
7042. McCullough, DR. 1969. The Tule elk. Univ CA Publ Zool. 88:1-209.
7043. McCullough, RA. 1957. Possible occurrence of identical twins in the whitetail deer. J Mamm. 38:421-422.
7044. McCusker, JS. 1974. Breeding Malayan sun bears *Helarctos malayanus* at Fort Worth Zoo. Int Zoo Yearb. 14:118-119.
7045. McCusker, JS. 1985. Testicular cycles of the common long-nosed armadillo *Dasypus novemcinctus*, in north central Texas. The Evolution and Ecology of Armadillos, Sloths, and Vermilinguas. 255-261, Montgomery, GG, ed, Smithsonian Inst Press, Washington, DC.
7046. McDonald, EJ & Martell, AM. 1981. Twinning and postpartum activity in barren-ground caribou (*Rangifer tarandus*). Can Field-Nat. 95:354-355.
7046a. McDonald, IR, Lee, AK, Bradley, AJ, et al. 1981. Endocrine changes in dasyurid marsupial with differing mortality patterns. Gen Comp Endocrinol. 44:292-301.
7047. McDonald, IR, Lee, AK, Than, KA, et al. 1986. Failure of glucocorticoid feedback in males of a population of small marsupials (*Antechinus swainsonii*) during the period of mating. J Endocrinol. 108:63-68.
7048. McDonald, IR & Taitt, MJ. 1982. Steroid hormones in the blood plasma of Townsend's vole (*Microtus townsendii*). Can J Zool. 60:2264-2269.
7049. McDonald, IR & Waring, H. 1979. Hormones of marsupials and monotremes. Hormones and evolution. 873-924, Barrington, EJW, ed, Academic Press, London.
7050. McDonald, LE & Lasley, BL. 1978. Infertility in a captive group of white-lipped peccaries. J Zoo Anim Med. 9:90-95.
7051. McDougall, P. 1975. The feral goats of Kielderhead Moor. J Zool. 176:215-246.
7052. McDougall, WA. 1944. An investigation of the rat pest problem in Queensland cane fields: 2 Species and general habits. QLD J Agric Sci. 1:48-78.
7053. McDougall, WA. 1946. An investigation of the rat pest problem in Queensland cane fields: 4 Breeding and life histories. QLD J Agric Sci. 3:1-43.
7054. McEvoy, JS. 1970. Red-necked wallaby in Queensland. QLD J Agric J. 96:114-116.
7055. McEwan, EH. 1962. Reproduction of barren ground caribou *Rangifer tarandus groenlandicus* (Linnaeus) with relation to migration. Ph D Diss. McGill Univ, Montreal.
7056. McEwan, EH. 1970. Energy metabolism of barren ground caribou (*Rangifer tarandus*). Can J Zool. 48:391-392.
7057. McEwan, EH. 1971. Twinning in caribou. J Mamm. 52:479.
7058. McEwan, EH & Whitehead, PE. 1972. Reproduction in female reindeer and caribou. Can J Zool. 50:43-46.
7059. McEwen, EH & Scott, A. 1957. Pigmented areas in the uterus of the Arctic fox *Alopex lagopus innuitus* Merriam. Proc Zool Soc London. 128:347-348.
7059a. McFarlane, JR & Carrick, FN. 1992. Androgen concentrations in male platypuses. 69-74, in Platypus and Echidnas, Augee, ML, ed, Roy Zool Soc NSW, Sydney.
7059b. McFarlane, JR, Czekala, NM, & Papkoff, H. 1991. Zebra chorionic gonadotropin: Partial purification and characterization. Biol Reprod. 44:827-833.
7060. McGhee, ME & Genoways, HH. 1978. *Liomys pictus*. Mamm Species. 83:1-5.
7061. McGinnes, BS. 1972. Variation in peaks of fawning in Virginia. Proc Ann Conf SE Assoc Game Fish Comm. 26:23-27.
7062. McGinnis, SM & Schusterman, RJ. 1981. Northern elephant seal *Miroungo angustrostris* Gill, 1866. Handbook of Marine Mammals 2:329-349, Ridgeway, SH & Benirscheke, K, eds, Academic Press, New York.
7063. McGrew, JC. 1979. *Vulpes macrotis*. Mamm Species. 123:1-6.
7064. McGuckin, MA & Blackshaw, AW. 1987. Cycle of the seminiferous epithelium in the grey-headed fruit bat, *Pteropus polionotus*. Aust J Biol Sci. 40:203-210.
7065. McGuckin, MA & Blackshaw, AW. 1987. Seasonal changes in spermatogenesis (including germ cell degeneration) and plasma testosterone concentration in the grey-headed fruit bat, *Pteropus poliocephalus*. Aust J Biol Sci. 40:211-220.
7065a. McGuire, B, Heller, H, & Novak, M. 1988. Milk composition and volume in meadow voles and prairie voles. Acta Theriol. 33:537-544.
7066. McGuire, BA & Novak, MA. 1982. A comparison of maternal behavior in three species of voles (*Microtus pennsylvanicus*, *M pinetorum* and *M ochrogaster*) using a laboratory system. Proc E Pine Meadow Vole Symp. 6:139-145.
7067. McGuire, BA & Novak, MA. 1984. A comparison of maternal behaviour in the meadow vole (*Microtus pennsylvanicus*), prarie vole (*M ochrogaster*) and pine vole (*M pinetorum*). Anim Behav. 32:1132-1141.
7068. McHugh, T. 1958. Social behavior of the American buffalo (*Bison bison bison*). Zoologica. 43:1-40.
7069. McIlroy, JC. 1976. Aspects of the ecology of the common wombat, *Vombatus ursinus* I Capture, handling, marking, and radio-tracking techniques. Aust Wild Res. 3:105-116.
7070. McIlwaine, CP. 1962. Reproduction and body weights of the wild rabbit *Oryctolagus cuniculus* (L) in Hawke's Bay, New Zealand. NZ J Sci. 5:325-341.
7071. McIntosh, DL. 1963. Reproduction and growth of the fox in the Canberra district. CSIRO Wildl Res. 8:132-141.
7072. McIntosh, WB. 1967. The deer mouse. The UFAW Handbook of the Care and Management of Laboratory Animals, 3rd ed. 311-315, E & S Livingstone, Edinburgh.
7072a. McKay, GM. 1974. *Planigale ingrami* in captivity. Koolewong. 3(Mar):4-5.
7073. McKean, JL. 1972. Notes on some collections of bats (order Chiroptera) from Papua New Guinea and Bougainville Island. CSIRO Div Wildl Res Tech Pap. 26:1-5.
7074. McKean, JL & Hall, LS. 1964. Notes on microchiropteran bats. Vict Nat. 81:36-37.
7075. McKean, JL & Hamilton-Smith, E. 1967. Litter size and maternity sites in Australian bats (Chiroptera). Vict Nat. 84:203-206.
7076. McKean, JL & Price, WL. 1967. Notes on some Chiroptera from Queensland, Australia. Mammalia. 31:101-119.
7077. McKeever, S. 1958. Reproduction in the opossum in southwestern Georgia and northwestern Florida. J Wildl Manage. 22:303.
7078. McKeever, S. 1958. Reproduction in the raccoon in the southeastern United States. J Wildl Manage. 22:211.
7079. McKeever, S. 1963. Seasonal changes in body weight, reproductive organs, pituitary, adrenal glands, thyroid gland, and spleen of the Belding ground squirrel (*Citellus beldingi*). Am J Anat. 113:153-173.
7080. McKeever, S. 1964. The biology of the golden-mantled ground squirrel *Citellus lateralis*. Ecol Monogr. 34:383-401.

7081. McKeever, S. 1964. Variation in the weight of the adrenal, pituitary and thyroid gland of the white-footed mouse, *Peromyscus maniculatus*. Am J Anat. 114:1-15.
7082. McKeever, S. 1966. Reproduction in *Citellus beldingi* and *Citellus lateralis* in northeastern California. Symp Zool Soc London. 15:365-385.
7083. McKenna, J. 1974. Perinatal behaviour and parturition of a colobine, *Persbytis entellus entellus* (Hanuman langur). Lab Primate Newsl. 13:13-15.
7084. McKerrow, MJ. 1955. The lactation cycle of *Elephantulus myurus jamesoni* (Chubb). Phil Trans R Soc London Biol Sci. 238B:62-97.
7085. McKibbin, PE, Rajkumar, K, & Murphy, BD. 1984. Role of lipoproteins and prolactin in luteal function in the ferret. Biol Reprod. 30:160-166.
7086. McLaren, IA. 1958. Some aspects of growth and reproduction of the bearded seal, *Erignathus barbatus* (Erxleben). J Fish Res Bd Can. 15:219-227.
7087. McLaren, IA. 1958. The biology of the ringed seal (*Phoca hispida* Schreber) in the eastern Canadian arctic. Bull Fish Res Bd Can. 118:1-95.
7088. McLean, DD. 1954. Mountain lions in California. CA Fish Game. 40:147-166.
7089. McLean, IG. 1983. Paternal behavior and killing of young in arctic ground squirrels. Anim Behav. 31:32-44.
7090. McLean, IG & Towns, AJ. 1981. Differences in weight changes and the annual cycle of male and female arctic ground squirrels. Arctic. 34:249-254.
7091. McLean, PK & Pelton, MR. 1990. Some demographic comparisons of wild and panhandler bears in the Smoky Mountains. Int Conf Bear Res Manage. 8:105-112.
7092. McLellan, BN. 1989. Dynamics of a grizzly bear population during a period of industrial resource extraction III Natality and rate of increase. Can J Zool. 67:1865-1868.
7093. McLeod, JA. 1948. Preliminary studies on muskrat biology in Manitoba. Trans R Soc Can Third Ser Biol Sci. 42(Section V):81-96.
7094. McLeod, JA & Bondar, GF. 1952. Studies on the biology of the muskrat in Manitoba. Can J Zool. 30:243-253.
7095. McMahan, CA, Wigodsky, HS, & Moore, GT. 1976. Weight of the infant baboon (*Papio cynocephalus*) from birth to fifteen weeks. Lab Anim Sci. 26:928-931.
7096. McManus, JJ. 1971. Early postnatal growth and the development of temperature regulation in the Mongolian gerbil, *Meriones unguiculatus*. J Mamm. 52:782-792.
7097. McManus, JJ. 1974. *Didelphis virginiana*. Mamm Species. 40:1-6.
7098. McManus, JJ & Zurich, WM. 1972. Growth, pelage development and maturational molts of the Mongolian gerbil, *Meriones unguiculatus*. Am Mid Nat. 87:264-271.
7099. McManus, TJ, Wapstra, JE, Guiler, ER, et al. 1984. Cetacean strandings in Tasmania from February 1978 to May 1983. Pap Proc R Soc Tasmania. 118:117-135.
7100. McMillan, CA. 1981. Synchrony of estrus in macaque matrilines at Cayo Santigo. Am J Phys Anthrop. 54:251.
7101. McMillan, CA. 1986. Lineage-specific mating: Does it exist?. Cayo Santiago Macaques History, Behavior and Biology. 201-217, Rawlins, RG & Kessler, MJ, eds, State Univ NY Press, Albany.
7102. McMillin, JM, Seal, US, Keenlyne, KD, et al. 1974. Annual testosterone rhythm in the adult white-tailed deer (*Odocoileus virginianus borealis*). Endocrinology. 94:1034-1040.
7103. McMillin, JM, Seal, US, Rogers, L, et al. 1976. Annual testosterone rhythm in the black bear (*Ursus americanus*). Biol Reprod. 15:163-167.
7103a. McNab, BK. 1976. Seasonal fat reserves of bats in two tropical environments. Ecology. 57:332-338.
7104. McNab, AG & Crawley, MC. 1975. Mother and pup behaviour of the New Zealand fur seal, *Arctocephalus forsteri* (Lesson). Mauri Oro. 3:77-88.
7105. McNab, BK & Wright, PC. 1987. Temperature regulation and oxygen consumption in the Philippine tarsier *Tarsius syrichta*. Physiol Zool. 60:596-600.
7106. McNairn, IS & Fairall, N. 1979. Relationship between heart rate and metabolism in the hyrax (*Procavia capensis*) and guinea pig (*Cavia porcellus*). S Afr J Zool. 14:230-232.
7107. McNally, J. 1957. A field study of a koala population. Proc R Zool Soc NSW. 76:18-27.
7108. McNally, J. 1960. The biology of the water rat *Hydromys chrysogaster* Geoffroy (Muridae: Hydromyinae) in Victoria. Aust J Zool. 8:170-180.
7109. McNeely, JA. 1979. The world's smallest mammal. Tiger Paper. 6:47-48.
7110. McNeilly, AS, Abbott, DH, Lunn, SF, et al. 1981. Plasma prolactin concentrations during the ovarian cycle and lactation and their relationship to return to fertility *postpartum* in the common marmoset (*Callithrix jacchus*). J Reprod Fert. 62:353-360.
7111. McNeilly, AS, Martin, RD, Hodges, JK, et al. 1983. Blood concentrations of gonadotrophins, prolactin and gonadl steroids in males and non-pregnant and pregnant female African elephants (*Loxodonta africana*). J Reprod Fert. 67:113-120.
7112. McPhail, MK. 1934. Studies on the hypophysectomized ferret VII Inhibition of ovulation in the mated oestrous ferret. Proc R Soc London Biol Sci. 114B:124-128.
7113. McPhail, MK. 1935. Studies on the hypophysectomized ferret IX The effect of hypophysectomy on pregnancy and lactation. Proc R Soc London Biol Sci. 117B:34-45.
7114. McPhee, EC. 1988. Ecology and diet of some rodents from the lower montane region of Papua New Guinea. Aust Wildl Res. 15:91-102.
7115. McShea, WJ & Madison, DM. 1984. Communal nesting between reproductively active females in a spring population of *Microtus pennsylvanicus*. Can J Zool. 62:344-346.
7116. McShea, WJ & Madison, DM. 1989. Measurements of reproductive traits in a field population of meadow voles. J Mamm. 70:132-141.
7117. McToldridge, ER. 1969. A brief note on housing and breeding squirrel monkeys *Saimiri sciureus* at Santa Barbara Zoo. Int Zoo Yearb. 9:77.
7118. McToldridge, ER. 1969. Notes on breeding ring-tailed coatis *Nasua nasua* at Santa Barbara Zoo. Int Zoo Yearb. 9:89-90.
7119. McWilliam, AN. 1982. Adaptive responses to seasonality in four species of Microchiroptera in coastal Kenya. Ph D Diss. Univ Aberdeen, Aberdeen.
7120. McWilliam, AN. 1987. Polyestry and postpartum oestrus in *Tadarida* (*Chaerephon*) *pumila* (Chiroptera: Mollosidae) in northern Ghana, West Africa. J Zool. 213:735-739.
7121. McWilliam, AN. 1987. Territorial and pair behaviour of the African false vampire bat, *Cardioderma cor* (Chiroptera: Megadermatidae), in coastal Kenya. J Zool. 213:243-252.
7122. McWilliam, AN. 1987. The reproductive and social biology of *Coleura afra* in a seasonal environment. Recent Advances in the Study of Bats. 324-350, Fenton, MB, Racey, P, & Rayner, JMV, eds, Cambridge Univ Press, New York.

7123. Mead, JG, Odell, DK, Wells, RS, et al. 1980. Observations on a mass stranding of spinner dolphin, *Stenella longirostris*, from the west coast of Florida. Fish Bull USA. 78:353-360.
7124. Mead, JG, Walker, WA, & Houck, WJ. 1982. Biological observations on *Mesoplodon carlhubbsi* (Cetacea: Ziphiidae). Smithsonian Contrib Zool. 344:1-25.
7125. Mead, JI. 1989. *Nemorhaedus goral*. Mamm Species. 335:1-5.
7126. Mead, RA. 1968. Reproduction in eastern forms of the spotted skunk (genus *Spilogale*). J Zool. 156:119-136.
7127. Mead, RA. 1968. Reproduction in western forms of the spotted skunk (genus *Spilogale*). J Mamm. 49:373-390.
7128. Mead, RA. 1970. The reproductive organs of the male spotted skunk (*Spilogale putorius*). Anat Rec. 167:291-302.
7129. Mead, RA. 1971. Effects of light and blinding upon delayed implantation in the spotted skunk. Biol Reprod. 5:214-220.
7129a. Mead, RA. 1975. Effects of hypopyhysectomy on blastocyst survival, progesterone secretion and nidation in the spotted skunk. Biol Reprod. 12:526-533.
7130. Mead, RA. 1981. Delayed implantation in mustelids, with special emphasis on the spotted skunk. J Reprod Fert Suppl. 29:11-24.
7131. Mead, RA. 1981. Discussion. J Reprod Fert Suppl. 29:48-49.
7132. Mead, RA, Bremmer, S, & Murphy, BD. 1988. Changes in endometrial vascular permeability during the preiimplantation period in the ferret (*Mustela putorius*). J Reprod Fert. 82:293-298.
7132a. Mead, RA, Concannon, PW, & McRae, M. 1981. Effect of progestins on implantation in the western spotted skunk. Biol Reprod. 25:128-133.
7133. Mead, RA & Eik-Nes, KB. 1969. Seasonal variation in plasma levels of progesterone in the western forms of the spotted skunk. J Reprod Fert Suppl. 6:397-403.
7134. Mead, RA, Joseph, MM, Neirinckx, S, et al. 1988. Partial characterization of a luteal factor that induces implantation in the ferret. Biol Reprod. 38:798-803.
7135. Mead, RA & McRae, M. 1982. Is estrogen required for implantation in the ferret?. Biol Reprod. 27:540-547.
7136. Mead, RA & Neirinckx, S. 1989. Hormonal induction of oestrus and pregnancy in anoestrous ferrets (*Mustela putorius furo*). J Reprod Fert. 86:309-314.
7137. Mead, RA, Neirinckx, S, & Czekala, NM. 1990. Reproductive cycle of the steppe polecat (*Mustela eversmanni*). J Reprod Fert. 88:353-360.
7138. Mead, RA, Rector, M, Starypan, G, et al. 1991. Reproductive biology of captive wolverines. J Mamm. 72:807-814.
7139. Mead, RA & Rourke, AW. 1983. Protease inhibitor present in uterine fluid of the spotted skunk during delayed implantation. Biol Reprod. 28(Suppl):53.
7140. Mead, RA, Rourke, AW, & Swannack, A. 1979. Changes in uterine protein synthesis during delayed implantation in the western spotted skunk and its regulation by hormones. Biol Reprod. 21:39-46.
7141. Mead, RA & Swannack, A. 1978. Effects of hysterectomy on luteal function in the western spotted skunk (*Spilogale putorius latifrons*). Biol Reprod. 18:379-383.
7142. Mead, RA & Swannack, A. 1980. Aromatase activity in corpora lutea of the ferret. Biol Reprod. 22:560-565.
7143. Meagher, M. 1986. *Bison bison*. Mamm Species. 266:1-8.
7144. Meagher, MM. 1978. Bison. Big Game of North America. 123-133, Schmidt, JL & Gilbert, DL, eds, Stackpole Books, Harrisburg, PA.
7145. Mearns, EA. 1907. Mammals of the Mexican boundary of the United States. Bull US Nat Mus. 56:1-530.
7146. Measrock, V. 1954. Growth and reproduction in the females of two species of gerbil, *Tatera brantsi* (A Smith) and *Tatera afra* (Gray). Proc Zool Soc London. 124:631-658.
7147. Mech, LD. 1970. The Wolf: The Ecology and Behavior of an Endangered Species. The Natural History Press, Garden City, NY.
7148. Mech, LD. 1974. *Canis lupus*. Mamm Species. 37:1-6.
7149. Mech, LD. 1975. Disproportionate sex ratios of wolf pups. J Wildl Manage. 39:737-740.
7149a. Mech, LD & Turkowski, F. 1966. twenty-three raccoons in one winter den. J Mamm. 47:529-530.
7150. Meckley, PE & Ginther, OJ. 1972. Effects of litter and male on corpora lutea of the postpartum Mongolian gerbil. J Anim Sci. 34:297-301.
7151. Meckley, PE & Ginther, OJ. 1974. Occurrence of oestrus in the Mongolian gerbil after pairing with a male. Lab Anim. 8:93-97.
7152. Męczyński, S. 1974. Morphohistological structure of female genital organs in sousliks. Acta Theriol. 19:91-104.
7153. Medellín, RA. 1989. *Chrotoperus auritus*. Mamm Species. 343:1-5.
7154. Medellín, RA & Arita, HT. 1989. *Tonatia evotis* and *Tonatia silvicola*. Mamm Species. 334:1-5.
7154a. Medellín, RA, Navarro L, D, Davis, WB, et al. 1983. Notes on the biology of *Micronycteris brachyotis* (Dobson) (Chiroptera), in southern Veracruz, Mexico. Brenesia. 21:7-11.
7155. Medellín, RA, Wilson, DE, & Navarro L, D. 1985. *Micronycteris brachyotis*. Mamm Species. 251:1-4.
7156. Medin, DE. 1976. Modeling the dynamics of a Colorado mule deer population. Ph D Diss. CO State Univ, Fort Collins.
7157. Medin, DE & Anderson, AE. 1979. Modeling the dynamics of a Colorado mule deer population. Wildl Monogr. 68:1-77.
7158. Medjo, DC & Mech, LD. 1976. Reproduction activity in nine- and ten-month old wolves. J Mamm. 57:406-408.
7159. Medway, Lord. 1967. Observations on the breeding of the pencil-tailed tree mouse, (*Chiropodomys gliroides*). J Mamm. 48:20-26.
7160. Medway, Lord. 1962. The Malayan mole. Malay Nat J. 16:205-208.
7161. Medway, Lord. 1969. The Wild Mammals of Malaya and Offshore Islands including Singapore. Oxford Univ Press, London.
7162. Medway, Lord. 1970. Breeding of the silvered leaf monkey, *Presbytis cristata* in Malaya. J Mamm. 51:630-632.
7163. Medway, Lord. 1971. Observations of social and reproductive biology of the bent-winged bat *Miniopterus australis* in northern Borneo. J Zool. 165:261-273.
7164. Medway, Lord. 1972. Reproductive cycles of the flat-headed bats *Tylonycteris pachypus* and *T robustula* (Chiroptera: Vespertilioninae) in a humid equatorial environment. Zool J Linn Soc. 51:33-61.
7164a. Medway, Lord. 1978. The Wild Mammals of Malaya (Peninsular Malaysia) and Singapore, 2nd ed. Oxford Univ Press, Oxford.
7165. Meester, J. 1958. A litter of *Rattus namaquensis* born in captivity. J Mamm. 39:302-304.

7166. Meester, J. 1960. Early post-natal development of multimammate mice *Rattus* (*Mastomys*) *natalensis* (A Smith). Ann Transvaal Mus. 24:35-52.
7167. Meester, J. 1960. The dibatag, *Ammodorcas clarkei* (Thos) in Somalia. Ann Transvaal Mus. 24:53-60.
7167a. Meester, J. 1963. A systematic revision of the shrew genus *Crocidura* in southern Africa. Mem Transvaal Mus. 13:1-127.
7168. Meester, J & Hallett, AF. 1970. Notes on early postnatal development in certain southern African Muridae and Cricetidae. J Mamm. 51:703-711.
7169. Mehlhop, P & Lynch, JF. 1978. Population characteristics of *Peromyscus leucopus* introduced to islands inhabited by *Microtus pennsylvanicus*. Oikos. 31:17-26.
7170. Mehmood, A & Baig, KJ. 1989. Annual profiles of sex steroids in blood plasma and ovarian tissue of wild female musk shrew, *Suncus murinus*. Jap J Physiol. 39:767-771.
7171. Mehrer, CF. 1976. Gestation period in the wolverine *Gulo gulo*. J Mamm. 57:570.
7172. Mehta, RR, Jenco, JM, Gaynor, LV, et al. 1986. Relationships between ovarian morphology, vaginal cytology, serum progesterone, and urinary immunoreactive pregnanediol during the menstrual cycle of the cynomolgus monkey. Biol Reprod. 35:981-986.
7173. Mehta, VS, Prakash, A, & Singh, M. 1962. Gestation period in camels. Indian Vet J. 39:387-389.
7174. Meier, E. 1973. Beiträge zur Geburt des Damwildes (*Cervus dama* L). Z Säugetierk. 37:348-373.
7175. Meier, JE & Willis, MS. 1984. Techniques for hand-raising neonatal ruffed lemurs (*Varecia variegata*) and (*Varecia variegata rubra*) and a comparison of hand-raised and maternally-raised animals. J Zoo Anim Med. 15:24-31.
7176. Meier, PT & Svendsen, G. 1983. Alternative mating strategies of male woodchucks (*Marmota monax*). Am Zool. 23:932.
7177. Meile, P & Bubenik, A. 1979. Zur Bedeutung sozialer Auslöser für das Sozialverhalten der Gemse, *Rupicapra rupricapra* (Linne, 1758). Säugetierk Mitt, Sonderheft. 27:1-42.
7178. Meister, W & Davis, DD. 1953. Placentation of a primitive insectivore *Echinosorex gymnura*. Fieldiana Zool. 35:11-26.
7179. Meister, W & Davis, DD. 1956. Placentation of the pigmy treeshrew *Tupaia minor*. Fieldiana Zool. 35:73-84.
7180. Meister, W & Davis, DD. 1958. Placentation of the terrestial tree shrew (*Tupaia tana*). Anat Rec. 132:541-553.
7181. Meiszner, K & Gansloszer, U. 1985. Development of young in the kowari, *Dasyuroides byrnei* Spencer, 1896. Zoo Biol. 4:351-359.
7182. Melcher, JC, Armitage, KB, & Porter, WP. 1989. Energy allocation by yellow-bellied marmots. Physiol Zool. 62:429-448.
7183. Melchior, F. 1976. Künstliche Aufzucht von Wüstenfuchsen (*Fennecus zerda*). Zool Garten NF. 46:431-440.
7184. Melchior, G. 1977. Über die künstliche Aufzucht von Warzenschweinen (*Phacochoerus aethiopicus* Pallas). Zool Garten NF. 47:323-328.
7185. Mell, R. 1922. Beiträge zur Fauna Sinica I Die Vertebraten Sudchinas; Feldlisten und Feldnoten der Säuger, Vögel, Reptilien, Batrachier. Arch Naturgesch. 88A(10):1-134.
7186. Mellanby, K. 1974. The Mole. W Collins Sons & Co, London.
7187. Meller, RE, Keverne, EB, & Herbert, J. 1980. Behavioural and endocrine effects of naltrexone in male talapoin monkeys. Pharmacol Biochem Behav. 13:663-672.
7188. Mello, DA. 1977. Note on breeding of *Calomys expulsus*, Lund, 1841 (Rodentia, Cricetidae) under laboratory conditions. Rev Bras Pesqui Med Biol. 10:107.
7189. Mello, DA. 1978. Biology of *Calomys callosus* (Rengger, 1830) under laboratory conditions (Rodentia, Cricitinae). Rev Bras Biol. 38:807-811.
7190. Mello, DA. 1978. Some aspects of the biology of (*Oryzomys eliurus*) (Wagner, 1945) under laboratory conditions (Rodentia, Cricetidae). Rev Bras Biol. 38:293-295.
7191. Mello, DA. 1981. Studies on reproduction and longevity of *Calomys callosus* under laboratory conditions (Rodentia, Cricetidae). Rev Bras Biol. 41:841-843.
7192. Mello, DA. 1986. Breeding of wild-caught rodent Cricetidae *Holochilus brasiliensis* under laboratory conditions. Lab Anim. 20:195-196.
7193. Mello, DA & Mathias, CH. 1987. Criacao de *Akodon arviculoides* (Rodentia, Cricetidae) em laboratorio. Rev Bras Biol. 47:419-423.
7194. Melnick, DJ, Pearl, MC, & Richard, AF. 1984. Male migration and inbreeding avoidance in wild rhesus monkeys. Am J Primatol. 7:229-243.
7195. Melquist, WE & Hornocker, MG. 1983. Ecology of river otters in west central Idaho. Wildl Monogr. 83:1-60.
7196. Melton, DA. 1976. The biology of aardvark (Tubulidentata: Orycteropodidae). Mamm Review. 6:75-88.
7197. Melton, DA. 1983. Population dynamics of waterbuck (*Kobus ellipsiprymnus*) in the Umfolozi Game Reserve. Afr J Ecol. 21:77-91.
7198. Menard, N. 1982. Quelques aspects de la socioécologie de la Vigogne *Lama vicuna*. Terre Vie. 36:15-35.
7199. Menard, N, Vallet, D, & Gauthier-Hion, A. 1985. Démographie et reproduction de *Macaca sylvanus* dans differents habitats en Algérie. Folia Primatol. 44:65-81.
7200. Mendelsohn, JM. 1982. Notes on small mammals on the Springbok flats, Transvaal. S Afr J Zool. 17:197-201.
7201. Mendelssohn, H. 1965. Breeding the Syrian hyrax *Procavia capensis syriaca*. Int Zoo Yearb. 5:116-125.
7202. Mendoza, SP, Lowe, EL, Resko, JA, et al. 1978. Seasonal variations in gonadal hormones and social behavior in squirrel monkeys. Physiol Behav. 20:515-522.
7203. Mentis, MT. Personal communication. from Department of Game and Tsetse Control, N Rhodesia. [Mentis, 1972]
7204. Mentis, MT. 1972. A review of some life history features of the large herbivores. Lammergeyer. 16:1-89.
7205. Menzies, JI. 1957. Gene-controlled sterility in the African mouse (*Mastomys*). Nature. 179:1142.
7206. Menzies, JI. 1967. A preliminary note on the birth and development of a small-scaled tree pangolin *Manis tricuspis* at Ife University zoo. Int Zoo Yearb. 7:114.
7207. Menzies, JI. 1972. Notes on a hand-reared spotted cuscus, *Phalanger maculatus*. Int Zoo Yearb. 12:97-98.
7208. Menzies, JI. 1973. A study of leaf-nosed bats (*Hipposideros caffer* and *Rhinolophus landeri*) in a cave in northern Nigeria. J Mamm. 54:930-945.
7209. Menzies, JI & Dennis, E. 1979. Identification of New Guinea rodents. Handbook of New Guinea Rodents. Wau Ecol Inst Handb. 6:1-68.
7210. Menzies, JI & Pernetta, JC. 1986. A taxonomic revision of cuscuses allied to *Phalanger orientalis* (Marsupialia: Phalangeridae). J Zool. 1B:551-618.
7211. Mepham, TB & Beck, NFG. 1973. Variation in the yield and composition of milk throughout lactation in the guinea pig (*Cavia porcellus*). Comp Biochem Physiol. 45A:273-281.

7212. Merani, MS & Lizarralde, MS. 1980. *Akodon molinae* (Rodentia Cricetidae) a new laboratory animal: Breeding, management and reproductive performances. Lab Anim Sci. 14:129-131.
7213. Merani, MS, Vercellini, O, Acuña, AM, et al. 1983. Growth and reproduction of two species of *Akodon* and their hybrids. J Exp Zool. 228:527-535.
7214. Mercer, WE, Hearn, BJ, & Finlay, C. 1981. Arctic hare populations in insular Newfoundland. Proc World Lagomorph Conf, Univ Guelph, ONT, Canada. 1979:450-468.
7215. Merchant, J, Green, B, Messer, M, et al. 1989. Milk composition in the red-necked wallaby, *Macropus rufogriseus banksianus* (Marsupialia). Comp Biochem Physiol. 93A:483-488.
7216. Merchant, JC. 1976. Breeding biology of the agile wallaby, *Macropus agilis* (Gould) (Marsupialia: Macropodidae), in captivity. Aust Wildl Res. 3:93-103.
7217. Merchant, JC. 1979. The effect of pregnancy on the interval between one oestrus and the next in the tammar wallaby, *Macropus eugenii*. J Reprod Fert. 56:459-463.
7218. Merchant, JC. 1984. Energetics of growth in the eastern quoll, *Dasyurus viverrinus*. Bull Aust Mamm Soc. 8:148.
7219. Merchant, JC & Calaby, JH. 1981. Reproductive biology of the red-necked wallaby *Macropus rufogriseus banksianus* and Bennett's wallaby *M r rufogriseus* in captivity. J Zool. 194:203-217.
7220. Merchant, JC & Libke, JA. 1988. Milk composition in the northern brown bandicoot, *Isoodon macrourus* (Peramelidae, Marsupialia). Aust J Biol Sci. 41:495-505.
7221. Merchant, JC, Newgrain, K, & Green, B. 1983. Growth of the eastern quoll, *Dasyurus viverrinus*, (Shaw), (Marsupialia) in captivity. Aust Wildl Res. 11:21-29.
7222. Meritt, DA, Jr. 1971. The development of the La Plata three-banded armadillo *Tolypeutes matacus* at Lincoln Park Zoo, Chicago. Int Zoo Yearb. 11:195-196.
7223. Meritt, DA, Jr. 1973. In nature the behavior of armadillos. Yearb Am Philos Soc. 1972:383-384.
7224. Meritt, DA, Jr. 1976. The owl monkey *Aotus trivirgatus*, husbandry, behavior and breeding. Proc Nat Conf AAZPA. 1976:107-123.
7224a. Meritt, DA, Jr. 1976. The La Plata three-banded armadillo *Tolypeutes matacus* in captivity. Int Zoo Yearb. 16:153-156.
7225. Meritt, DA, Jr. 1980. Captive reproduction and husbandry of the douroucouli *Aotus trivirgatus* and the titi monkey *Callicebus* spp. Int Zoo Yearb. 20:52-59.
7226. Meritt, DA, Jr. 1981. Husbandry, reproduction and behaviour of the West African hedgehog *Erinaceus albiventris* at Lincoln Park Zoo. Int Zoo Yearb. 21:128-131.
7227. Meritt, DA, Jr. 1983. Preliminary observations on reproduction in the Central American agouti, *Dasyprocta punctata*. Zoo Biol. 2:127-131.
7228. Meritt, DA, Jr. 1984. The pacarana, *Dinomys branickii*. One Medicine. 154-161, Ryder, OA, & Byrd, ML, eds.
7229. Meritt, DA, Jr. 1985. Naked-tailed armadillos, *Cabassous* sp. The Evolution and Ecology of Armadillos, Sloths, and Vermilinguas. 389-391, Montgomery, GG, ed, Smithsonian Inst Press, Washington, DC.
7230. Meritt, DA, Jr. 1985. The two-toed Hoffmann's sloth, *Choloepus hoffmanni* Peters. The Evolution and Ecology of Armadillos, Sloths, and Vermilinguas. 333-341, Montgomery, GG, ed, Smithsonian Inst Press, Washington, DC.
7231. Meritt, DA, Jr. 1986. History of Geoffroy's tamarins, *Saguinus geoffroyi*, at Lincoln Park Zoological Gardens, 1974-1985. Primates: The Road to Self-Sustaining Populations. 261-267, Benirschke, K, ed, Springer-Verlag, New York.
7232. Meritt, DA, Jr & Meritt, GF. 1976. Sex ratios of Hoffmann's sloth, *Choloepus hoffmanni* Peters, and three-toed sloth, *Bradypus infuscatus* Wagler, in Panama. Am Mid Nat. 96:472-473.
7233. Merkatz, IR & Beling, CG. 1969. Urinary excretion of oestrogens and pregnanediol in the pregnant baboon. J Reprod Fert Suppl. 6:129-135.
7233a. Merkt, JR. 1987. Reproductive seasonality adn grouping patterns of the north Andean deer or taruca (*Hippocamelus antisensis*) in southern Peru. Biology and Management of the Cervidae, 388-401, Wemmer, CM, ed, Smithsonian Inst Press, Washington, DC.
7234. Merriam, CH. 1882. The vertebrates of the Adirondack region, northeastern New York. Trans Linn Soc NY. 1:6-168.
7235. Merriam, CH. 1884. The vertebrates of the Adirondack region, northeastern New York (Mammalia, concluded). Trans Linn Soc NY. 2:9-214.
7235a. Merriam, CH. 1898, Mammals of Tres Marias Islands, off western Mexico. Proc Biol Soc Wash. 12:13-19.
7236. Merriam, HR. 1964. The wolves of Coronation Island. Proc AK Sci Conf. 15:27-32.
7237. Merritt, JF. 1978. *Peromyscus californicus*. Mamm Species. 85:1-6.
7238. Merritt, JF. 1981. *Clethrionomys gapperi*. Mamm Species. 146:1-9.
7239. Merritt, JF. 1984. Winter Ecology of Small Mammals. Sp Publ Carnegie Mus Nat Hist. 10:1-380.
7240. Meschaks, P & Nordkvist, M. 1962. On the sexual cycle in the reindeer male. Acta Vet Scand. 3:151-162.
7241. Meserve, PL & Le Boulengé, E. 1987. Population dynamics and ecology of small mammals in the northern Chilian semiarid region. Fieldiana Zool NS. 39:413-431.
7242. Meserve, PL, Murua, R, Oscar-Lopetegui, N, et al. 1982. Observations on the small mammal fauna of a primary temperate rain forest in southern Chile. J Mamm. 63:315-317.
7243. Meslow, EC & Keith, LB. 1968. Demographic parameters of a snowshoe hare population. J Wildl Manage. 32:812-834.
7243a. Messer, M. 1974. Identification of N-acetyl-4-O-acetylneuralminyl-lactose in echidna milk. Biochem J. 139:415-420.
7244. Messer, M, Crisp, EA, & Newgrain, K. 1988. Studies on the carbohydrate content of milk of the crabeater seal (*Lobodon carcinophagus*). Comp Biochem Physiol. 90B:367-370.
7245. Messer, M & Elliott, C. 1987. Changes in α-lactalbumin, total lactose, UDP-galactose hydrolase and other factors in tammar wallaby (*Macropus eugenii*) milk during lactation. Aust J Biol Sci. 40:37-46.
7246. Messer, M, FitzGerald, PA, Merchant, JC, et al. 1987. Changes in milk carbohydrates during lactation in the eastern quoll, *Dasyurus viverrinus* (Marsupialia). Comp Biochem Physiol. 88B:1083-1086.
7247. Messer, M, Gadiel, PA, Ralston, GB, et al. 1983. Carbohydrates of the milk of the platypus. Aust J Biol Sci. 36:129-137.
7247a. Messer, M & Kerry, KR. 1973. Milk carbohydrates of the echidna and the platypus. Science. 180:201-202.
7247b. Messer, M & Mossop, GS. 1977. Milk carbohydrates of marsupials I Partial separation and characterization of neutral milk oligosaccharides of the eastern grey kangaroo. Aust J Biol Sci. 30:379-388.
7248. Messer, M & Nicholas, KR. 1991. Biosynthesis of marsupial milk oligosaccharides, characterization and

developmental changes of two galactosyltransferases in lactating mammary glands of the tammar wallaby *Macropus eugenii*. Biochim Biophys Acta. 1077:79-85.
7249. Messer, M & Trifonoff, E. 1980. Structure of a marsupial-milk trisaccharide. Carbohydrate Res. 83:327-334.
7250. Messick, JP & Hornocker, MG. 1981. Ecology of the badger in southwestern Idaho. Wildl Monogr. 76:1-53.
7251. Meurling, P, Andersson, B, Eriksson, M, et al. 1981. När smågnagarna blir fler. Forskning och Framsteg. 16(1):32-37. [Cited by Hansson, 1984]
7252. Meusy-Dessolle, N & Dang, DC. 1985. Plasma concentrations of testosterone, dihydro-testosterone, 4-androstenedione, dehydroepiandrosterone and oestradiol-17β in the crab-eating monkey (*Macaca fascicularis*) from birth to adulthood. J Reprod Fert. 74:347-359.
7253. Meyer, BJ & Meyer, RK. 1944. Growth and reproduction of the cotton rat, *Sigmodon hispidus hispidus*, under laboratory conditions. J Mamm. 25:107-129.
7254. Meyer, BJ & Meyer, RK. 1944. The effect of light on maturation and the estrous cycle of the cotton rat, *Sigmodon hispidus hispidus*. Endocrinology. 34:276-281.
7255. Meyer, MN. 1967. Peculiarities of the reproduction and development of *Phodopus sungorus* Pallas of different geographical populations. Zool Zhur. 46:604-613.
7256. Meyer, P. 1972. Zur Biologie und Ökologie des Atlashirsches *Cervus elaphus barbarus*, 1833. Z Säugetierk. 37:101-116.
7257. Meyer, RK. 1972. Chorionic gonadotrophin, corpus luteum function and embryo implantation in the rhesus monkey. Acta Endocrinol Suppl. 166:214-217.
7258. Meyer, RK, Wolf, RC, & Arslan, M. 1969. Implantation and maintenance of pregnancy in progesterone-treated ovariectomized monkeys (*Macaca mulatta*). Int Congr Primat. 2:30-35.
7259. Meyer-Holzapfel, M. 1968. Breeding the european wild cat *Felis s silvestris* at Berne Zoo. Int Zoo Yearb. 8:31-38.
7260. Meyers, K & Poole, WE. 1962. A study of the biology of the wild rabbit, *Oryctolagus cunniculus* L), in confined populations III Reproduction. Aust J Zool. 10:225-267.
7261. Meylan, A & Airoldi, J-P. 1975. Reproduction hivernale chez *Arvicola terrestris scherman* Shaw (Mammalia, Rodentia). Rev Suisse Zool. 82:689-694.
7262. Miall, LC & Greenwood, F. 1878. The anatomy of the Indian elephant. J Anat. 13:17-50.
7263. Michael, ED. 1964. Birth of white-tailed deer fawns. J Wildl Manage. 28:171-173.
7264. Michael, RP & Bonsall, RW. 1977. A 3-year study of an annual rhythm in plasma androgen levels in male rhesus monkeys (*Macaca mulatta*) in a constant laboratory environment. J Reprod Fert. 49:129-131.
7265. Michael, RP & Keverne, EB. 1971. An annual rhythm in the sexual activity of the male rhesus monkey, *Macaca mulatta*, in the laboratory. J Reprod Fert. 25:95-98.
7266. Michael, RP, Setchell, KDR, & Plant, TM. 1974. Diurnal changes in plasma testosterone and studies on plasma corticosteroids in non-anesthetized male rhesus monkeys (*Macaca mulatta*). J Endocrinol. 63:325-335.
7267. Michael, RP & Wilson, MI. 1975. Mating seasonality in castrated male rhesus monkeys. J Reprod Fert. 43:325-328.
7268. Michael, RP & Zumpe, D. 1977. Effects of androgen administration on sexually invitations by female rhesus monkeys (*Macaca mulatta*). Anim Behav. 25:936-944.
7269. Michael, RP & Zumpe, D. 1982. Influence of olfactory signals on the reproductive behaviour of social groups of rhesus monkeys (*Macaca mulatta*). J Endocrinol. 95:189-205.
7270. Michael, RP, Zumpe, D, & Bonsall, RW. 1984. Sexual behavior correlates with the diurnal plasma testosterone range in itact male rhesus monkeys. Biol Reprod. 30:652-657.
7271. Michaelis, H. 1966. Die weibliche Genitalcyklus der Rötelmaus *Clethrionomys glareolus* (Schreber 1780) und Zusammenhänge mit der vorgeburtlichen Sterblichkeit. Z Wissensch Zool. 174:290-376.
7272. Michalak, I. 1982. Reproduction and behaviour of the Mediterranean water shrew under laboratory conditions. Säugetierk Mitt. 30:307-310.
7273. Michalak, I. 1983. Reproduction, maternal and social behaviour of the European water shrew under laboratory conditions. Acta Theriol. 28:3-24.
7274. Michalak, I. 1986. Number and distribution of teats in *Neomys fodiens*. Acta Theriol. 31:119-127.
7275. Michalak, I. 1987. Growth and postnatal development of the European water shrew. Acta Theriol. 32:261-288.
7276. Michalak, I. 1987. Keeping and breeding the Eurasian water shrew *Neomys fodiens* under laboratory conditions. Int Zoo Yearb. 26:223-228.
7276a. Michalak, I. 1988. Behaviour of young *Neomys fodiens* in captivity. Acta Theriol. 33:487-504.
7277. Michalowski, DR. 1971. Hand-rearing a polar bear cub *Thalarctos maritimus* at Rochester Zoo. Int Zoo Yearb. 11:107-109.
7278. Michel, G, Elze, K, & Seifert, S. 1983. Zur Embryonalentwicklung des Bären unter besunderer Beachtung des Baues der Plazenta. Zool Garten NF. 53:290-294.
7279. Michener, DR. 1974. Annual cycle of activity and weight changes in Richardson's ground squirrel, *Spermophilus richardsonii*. Can Field-Nat. 88:409-413.
7280. Michener, GR. 1969. Notes on the breeding and young of the crest-tailed marsupial mouse, *Dasycercus cristicauda*. J Mamm. 50:633-635.
7281. Michener, GR. 1973. Climatic conditions and breeding in Richardson's ground squirrel. J Mamm. 54:499-503.
7282. Michener, GR. 1977. Gestation period and juvenile age at emergence in Richardson's ground squirrel. Can Field-Nat. 91:410-413.
7283. Michener, GR. 1979. Yearly variations in the population dynamics of Richardson's ground squirrels. Can Field-Nat. 93:363-370.
7284. Michener, GR. 1980. Differential reproduction among female Richardson's ground squirrels and its relation to sex ratio. Behav Ecol Sociobiol. 7:173-178.
7285. Michener, GR. 1980. Estrous and gestation periods in Richardson's ground squirrels. J Mamm. 61:531-534.
7286. Michener, GR. 1984. Copulatory plugs in Richardson's ground squirrel. Can J Zool. 62:267-270.
7287. Michener, GR. 1984. Sexual differences in body weight patterns in Richardson's ground squirrels during the breeding season. J Mamm. 65:59-66.
7288. Michener, GR. 1985. Chronology of reproductive events for female Richardson's ground squirrels. J Mamm. 66:280-288.
7289. Michener, GR & Koeppl, JW. 1985. *Spermophilus richardsonii*. Mamm Species. 243:1-8.
7289a. Middleton, AD. 1937. Whipsnade ecological survey. Proc Zool Soc London A. 1937:471-481.
7290. Midgley, EE. 1938. The visceral anatomy of the kangaroo rat. J Mamm. 19:304-317.

7291. Miegel, B. 1952. Die Biologie and Morphologie der Fortpflanzung der Bisamratte (*Ondatra zibethica* L). Z Mikr Anat Forsch. 58:531-598.
7292. Migula, P. 1969. Bioenergetics of pregnancy and lactation in European common vole. Acta Theriol. 14:167-179.
7293. Mihok, S. 1984. Life history of boreal meadow voles (*Microtus pennsylvanicus*). Sp Publ Carnegie Mus Nat Hist. 10:91-102.
7294. Mihok, S. 1987. Pregnancy test for meadow voles (*Microtus pennsylvanicus*) based on blood azurocyte counts. Can J Zool. 65:2830-2832.
7295. Mikhailova, OP. 1982. Determining oestrus in nutria. Krolikovodstvo i Zverovodstvo. 2:17-18. [Abstract: Scientifur, 9:308, 1985]
7296. Miles, P. 1967. Notes on the rearing and development of a hand-reared spider monkey *Ateles geoffoyi*. Int Zoo Yearb. 7:82-85.
7297. Millar, JCG. 1980. Aspects of the ecology of the American grey squirrel *Sciurus carolinsis* Gmelin in South Africa. MSC Thesis, Univ Stellenbosch, Stellenbosch.
7298. Millar, JS. 1970. Adrenal weight in relation to reproductive status in the pika, *Ochotona princeps* (Richardson). Can J Zool. 48:1137-1140.
7299. Millar, JS. 1970. Variations in fecundity of the red squirrel, *Tamiasciurus hudsonicus* (Erxleben). Can J Zool. 48:1055-1058.
7300. Millar, JS. 1972. Timing of breeding of pikas in southwestern Alberta. Can J Zool. 50:665-669.
7301. Millar, JS. 1973. Evolution of litter-size in the pika, *Ochotona princeps* (Richardson). Evolution. 27:134-143.
7302. Millar, JS. 1974. Success of reproduction in pikas, *Ochotona princeps* (Richardson). J Mamm. 55:527-542.
7303. Millar, JS. 1975. Tactics of energy partitioning in breeding *Peromyscus*. Can J Zool. 53:967-976.
7304. Millar, JS. 1978. Energetics of reproduction in *Peromyscus leucopus*: The cost of lactation. Ecology. 59:1055-1061.
7305. Millar, JS. 1979. Energetics of lactation in *Peromyscus maniculatus*. Can J Zool. 57:1015-1019.
7306. Millar, JS. 1983. Negative maternal effects on *Peromyscus maniculatus*. J Mamm. 64:540-543.
7307. Millar, JS. 1984. Reproduction and survival of *Peromyscus* in seasonal environments. Sp Publ Carnegie Mus Nat Hist. 10:253-266.
7308. Millar, JS, Burkholder, DAL, & Lang, TL. 1986. Estimating age at independence in small mammals. Can J Zool. 64:910-913.
7309. Millar, JS & Gyug, LW. 1981. Initiation of breeding by northern *Peromyscus* in relation to temperature. Can J Zool. 59:1094-1098.
7310. Millar, JS & Innes, DGL. 1985. Breeding by *Peromyscus maniculatus* over an elevational gradient. Can J Zool. 63:124-129.
7311. Millar, JS & Millar, WD. 1989. Effects of gestation on growth and development in *Peromyscus maniculatus*. J Mamm. 70:208-210.
7312. Millar, JS & Threadgill, DAL. 1987. The effect of captivity on reproduction and development in *Peromyscus maniculatus*. Can J Zool. 65:1713-1719.
7313. Millar, R & Fairall, N. 1976. Hypothalamic, pituitary and gonadal hormone production in relation to nutrition in the male hyrax (*Procavia capensis*). J Reprod Fert. 47:339-341.
7314. Millar, RP. 1971. Reproduction in the rock hyrax (*Procavia capensis*). Zool Afr. 6:243-261.
7315. Millar, RP & Glover, TD. 1970. Seasonal changes in the reproductive tract of the male rock hyrax, *Procavia capensis*. J Reprod Fert. 23:497-499.
7316. Millar, RP & Glover, TD. 1973. Regulation of seasonal sexual activity in an ascrotal mammal, the rock hyrax, *Procavia capensis*. J Reprod Fert Suppl. 19:203-220.
7316a. Miller, BJ & Anderson, SH. 1990. Comparison of black-footed ferret (*Mustela nigripes*) and domestic ferret (*M putorius furo*) courtship activity. Zoo Biol. 9:201-210.
7317. Miller, EH. 1974. A paternal role for Galapagos sea lions?. Evolution. 28:474-475.
7318. Miller, EH. 1975. Annual cycle of fur seals, *Arctocephalus forsteri* (Lesson) on the Open Bay Island, New Zealand. Pacific Sci. 29:139-152.
7319. Miller, EH. 1975. Social and evolutionary implications of territoriality in adult male New Zealand fur seals, *Arctocephalus forsteri* (Lesson, 1828), during the breeding season. Rapp P-V Réun Cons Int Explor Mer. 169:170-187.
7320. Miller, FL. 1965. Behavior associated with parturition in black-tailed deer. J Wildl Manage. 29:629-631.
7321. Miller, FL. 1970. Accidents to parturient black-tailed deer. Am Mid Nat. 83:303-304.
7322. Miller, FL & Gunn, A. 1982. Nursing and associated behavior of Peary caribou, *Rangifer tarandus pearyi*. Can Field-Nat. 96:200-202.
7323. Miller, FL & Parkes, GR. 1968. Placental remnants in the rumens of maternal caribou. J Mamm. 49:778-779.
7324. Miller, FW. 1930. Notes on some mammals of southern Matto Grosso, Brazil. J Mamm. 11:10-22.
7325. Miller, FW. 1930. A note on the pigmy vole in Colorado. J Mamm. 11:83-84.
7325a. Miller, FW. 1948. Early breeding of the Texas beaver. J Mamm. 29:419.
7326. Miller, GS, Jr. 1896. The beach mouse of Muskeget Island. Proc Boston Soc Nat Hist. 27:75-87.
7327. Miller, GS, Jr. 1900. A new mouse deer from lower Siam. Proc Biol Soc Wash. 13:185-186.
7328. Miller, GS, Jr. 1900. Mammals collected by Dr WL Abbott on Palo Lankawi and the Butang islands. Proc Biol Soc Wash. 13:187-193.
7328a. Miller, GS, Jr. 1904. Notes on the bats collected by William Palmer in Cuba. Proc US Nat Mus. 27:337-348.
7329. Miller, MA. 1946. Reproduction rates and cycles in the pocket gopher. J Mamm. 27:335-358.
7330. Miller, RE. 1939. The reproductive cycle in male bats of the species *Myotis lucifugus lucifugus* and *Myotis griscescens*. J Morphol. 64:267-295.
7331. Miller, RL & Pallotta, AJ. 1965. Comments on the maintenance of a small baboon colony. The Baboon in Medical Research. 111-124, Vagtborg, H, ed, Univ TX Press, Austin.
7332. Miller, RS. 1958. A study of a wood mouse population in Wytham Woods, Berkshire. J Mamm. 39:477-493.
7333. Miller-Ben-Shaul, D. 1962. Short-tailed shrews (*Blarina brevicauda*) in captivity. Int Zoo Yearb. 4:121-123.
7334. Milligan, SR. 1974. Social environment and ovulation in the vole, *Microtus agrestis*. J Reprod Fert. 41:35-47.
7335. Milligan, SR. 1975. Mating, ovulation and corpus luteum function in the vole, *Microtus agrestis*. J Reprod Fert. 42:35-44.
7336. Milligan, SR. 1975. Further observations on the influence of the social environment on ovulation in the vole, *Microtus agrestis*. J Reprod Fert. 44:543-544.
7337. Milligan, SR. 1976. Pregnancy blocking in the vole, *Microtus agrestis* I Effect of the social environment. J Reprod Fert. 46:91-95.

7338. Milligan, SR. 1976. Pregnancy blocking in the vole, *Microtus agrestis* II Ovarian, uterine and vaginal changes. J Reprod Fert. 46:97-100.
7339. Milligan, SR. 1978. The feedback of exogenous steroids on LH release and ovulation in the intact female vole (*Microtus agrestis*). J Reprod Fert. 54:309-311.
7340. Milligan, SR. 1979. Pregnancy blockage and the memory of the stud male in the vole (*Microtus agrestis*). J Reprod Fert. 57:223-225.
7341. Milligan, SR. 1980. Effect of bromocryptine treatment on the ovulatory response to oestradiol benzoate in the reflex ovulator, *Microtus agrestis*. J Endocrinol. 84:315-316.
7342. Milligan, SR. 1981. Analysis of the LH surge induced by mating and LH-RH in the vole, *Microtus agrestis*. J Reprod Fert. 63:39-45.
7343. Milligan, SR, Charlton, HM, & Versi, E. 1979. Evidence for a coitally induced 'mnemonic' involved in luteal function in the vole (*Microtus agrestis*). J Reprod Fert. 57:227-233.
7344. Milligan, SR & MacKinnon, PCB. 1976. Correlation of plasma LH and prolactin levels with the fate of the corpus luteum in the vole, *Microtus agrestis*. J Reprod Fert. 47:111-113.
7345. Mills, G. 1973. The brown hyaena. Afr Wild. 27:?.
7346. Mills, HB. 1937. A preliminary study of the bighorn of Yellowstone National Park. J Mamm. 18:205-212.
7347. Mills, JN, Ellis, BA, McKee, KT, et al. 1992. Reproductive characteristics of rodent assemblages in cultivated regions of central Argentina. J Mamm. 73:515-526.
7348. Mills, MGL. 1982. *Hyaena brunnea*. Mamm Species. 194:1-5.
7349. Mills, MGL. 1982. Notes on age determination, growth and measurements of brown hyaenas *Hyaena burnnea* from the Kalahari Gemsbok National Park. Koedoe. 25:55-61.
7350. Mills, MGL. 1982. The mating system of the brown hyaena, *Hyaena brunnea* in the southern Kalahari. Behav Ecol Sociobiol. 10:131-136.
7351. Mills, MGL. 1982. Factors affecting group size and territory size of the brown hyaena, *Hyaena brunnea* in the southern Kalahari. J Zool. 198:39-51.
7352. Mills, MGL. 1983. Mating and denning behaviour of the brown hyaena *Hyaena brunnea* and comparisons with other Hyaenidae. Z Tierpsychol. 63:331-342.
7353. Mills, RS. 1980. Parturition and social interaction among captive vampire bats, *Desmodus rotundus*. J Mamm. 61:336-337.
7354. Milner, J, Jones, C, & Jones, JK, Jr. 1990. *Nyctinomops macrotis*. Mamm Species. 351:1-4.
7355. Milton, K. 1981. Estimates of reproductive parameters for free-ranging *Ateles geoffroyi*. Primates. 22:574-579.
7356. Milton, K. 1985. Mating patterns of woolly spider monkeys, *Brachyteles arachnoides*: Implications for female choice. Behav Ecol Sociobiol. 17:53-59.
7357. Milyutin, NG. 1941. A note on the reproduction of the mole (*Talpa europaea brauneri* Satun). Zool Zhur. 20:482-484.
7358. Minchin, AK. 1937. Notes on the weaning of a young koala (*Phascolartos cinereus*). Rec S Aust Mus. 6:1-3.
7358a. Minoprio, JD. 1945. Sobre el *Chlamyphorus truncatus* Harlan. Acta Zool Lilloana. 3:5-58.
7359. Mirarchi, RE, Scanlon, PF, Kirkpatrick, RL, et al. 1977. Androgen levels and antler development in captive and wild whitetailed deer. J Wildl Manage. 41(2):178-183.
7360. Mirskaia, L & Crew, FAE. 1931. On the pregnancy rate in the lactating mouse and the effect of suckling on the duration of pregnancy. Proc R Soc Edinb. 51:1-7.
7360a. Mishra, HR & Wemmer, C. 1987. The comparative breeding ecology of four cervids in Royal Chitwan National Park, Nepal. Biology and Management of the Cervidae. 259-271, Wemmer, CM, ed, Smithsonian Inst Press, Washington, DC.
7361. Misonne, X. 1963. Les Rongeurs du Ruwenzori et des régions voisines. Expl Parc Nat Albert 2e Ser. 14:1-164.
7362. Misonne, X & Verschuren, J. 1976. Les Rongeurs du Nimba Libérien. Acta Zool Pathol Antverpiensia. 66:199-220.
7363. Mitani, JC. 1985. Mating behaviour of male orangutans in the Kutai Game Reserve, Indonesia. Anim Behav. 33:392-402.
7364. Mitchell, B. 1971. The weights of new-born to one-day-old red deer calves in Scottish moorland habitats. J Zool. 164:250-254.
7365. Mitchell, B. 1979. The reproductive performance of wild Scottish red deer, *Cervus elaphus*. J Reprod Fert Suppl. 19:271-285.
7366. Mitchell, B & Brown, D. 1973. The effects of age and body size on fertility of female red deer (*Cervus elaphus* L). Int Congr Game Bio. 11:89-98.
7367. Mitchell, B & Lincoln, GA. 1973. Conception dates in relation to age and condition in two populations of red deer in Scotland. J Zool. 171:141-152.
7368. Mitchell, BL & Uys, JMC. 1961. The problem of the lechwe (*Kobus leche*) on the Kafue Flats. Oryx. 6:171-183.
7369. Mitchell, BL. 1965. Breeding, growth and ageing criteia of Lichtenstein's hartebeest. Puku. 3:97-104.
7370. Mitchell, E. 1975. Review of biology and fisheries for smaller cetaceans. J Fish Res Bd Can. 32:891-983.
7371. Mitchell, E. 1977. Sperm whale maximum length limit: Proposed protection of 'harem masters'. Rep Int Whal Commn. 27:224-225.
7372. Mitchell, E & Kozicki, VM. 1975. Supplementary information on minke whale (*Balaenoptera acutorostrata*) from Newfoundland fishery. J Fish Res Bd Can. 32:985-994.
7373. Mitchell, E & Kozicki, VM. 1978. Sperm whale regional closed seasons: Proposed protection during mating and calving. Rep Int Whal Comm. 28:195-198.
7374. Mitchell, E & Kozicki, VM. 1984. Reproductive condition of male sperm whales, *Physeter macrocephalus*, taken off Nova Scotia. Rep Int Whal Comm Sp Issue. 6:243-252.
7375. Mitchell, GC. 1965. A natural history study of the funnel-eared Bat *Natalus stramineus*. MS thesis. Univ AZ, Tucson.
7376. Mitchell, GJ. 1967. Minimum breeding age of female pronghorn antelope. J Mamm. 48:489-490.
7377. Mitchell, MD, Clover, L, Thorburn, GD, et al. 1978. Specific change in the direction of prostaglandin synthesis by intra-uterine tissues of the rhesus monkey (*Macaca mulatta*) during late pregnancy. J Endocrinol. 78:343-350.
7378. Mitchell, MD, Hicks, BR, Thorburn, GD, et al. 1978. Production of thromboxane B2 by intra-uterine tissue from late pregnant rhesus monkeys (*Macaca mulatta*) *in vitro*. J Endocrinol. 79:103-106.
7379. Mitchell, MD, Hicks, BR, Thorburn, GD, et al. 1979. Intra-uterine tissues from late pregnant rhesus monkeys (*Macaca mulatta*) produce 6-oxo-prostaglandin $F_{1\alpha}$ *in vitro*. J Endocrinol. 81:339-343.
7380. Mitchell, MD, Mountford, LA, Natale, R, et al. 1980. Concentrations of oxytocin in the plasma and amniotic fluid of rhesus monkeys (*Macaca mulatta*) during the latter half of pregnancy. J Endocrinol. 84:473-478.
7381. Mitchell, OG. 1959. The reproductive cycle of the male arctic ground squirrel. J Mamm. 40:45-53.
7382. Mitchell, RM. 1977. Accounts of Nepalese mammals and analysis of the host-ectoparasite data by computer techniques. Ph D Thesis. IA State Univ.

7383. Mitchell, RM. 1979. The sciurid rodents (Rodentia: Sciuridae) of Nepal. J Asian Ecol. 1:21-28.
7384. Mitchell, SJ & Jones, SM. 1975. Diagnosis of pregnancy in marmosets (*Callithrix jacchus*). Lab Anim Sci. 9:49-56.
7385. Mitchell, WR, Loskutoff, NM, Czekala, NM, et al. 1982. Abnormal menstrual cycles in the female gorilla (*Gorilla gorilla*). J Zoo Anim Med. 13:143-147.
7386. Mitchell, WR, Presley, S, Czekala, NM, et al. 1982. Urinary immunoreactive estrogen and pregnanediol-3-glucuronide during the normal menstrual cycle of the female lowland gorilla (*Gorilla gorilla*). Am J Prim. 2:167-175.
7387. Miura, S. 1981. Behavior and social structure of captive muntjacs (2). Animals & Zoos. 33(4):12-17.
7388. Miura, S. 1984. Annual cycles of coat changes, antler growth, and reproductive behavior of sika deer in Nara Park, Japan. J Mamm Soc Japan. 10:1-7.
7389. Miura, S. 1986. Body and horn growth patterns in the Japanese serow, *Capricornis crispus*. J Mamm Soc Japan. 11:1-13.
7390. Miura, S, Kita, I, & Sugimura, M. 1987. Horn growth and reproductive history in female Japanese serow. J Mamm. 68:826-836.
7391. Miyaji, R. 1980. Reproduction of the common seal. Animals & Zoos. 32(12):6-10.
7392. Miyao, T, Morozumi, M, & Moruzumi, T. 1966. Small mammals of Mt Yatsugatake in Honshu VI Seasonal variation in sex ratio body weight and reproduction in the vole, *Microtus montebelli*. Zool Mag. 75:98-102.
7393. Miyao, T, Morozumi, T, Hanamura, H, et al. 1963. Small mammals on Mt Yatsugatake in Honshu II Seasonal differences of sex ratio, body weight and reproduction on *Apodemus argenteus* and *Clethrionomys andersoni* in the subalpine forest zone on Mt Yatsugatake. Zool Mag. 72:187-193.
7394. Miyao, T, Moruzumi, T, & Moruzumi, M. 1967. Small mammals of Mt Yatsugatake, Honshu VII Seasonal change in reproduction of the woodmouse, *Apodemus speciosus*. Zool Mag. 76:161-166.
7395. Miyashita, M & Nagase, K. 1981. Breeding the Mongolian gazelle *Procapra gutturosa* at Osaka Zoo. Int Zoo Yearb. 21:158-162.
7396. Miyazaki, N. 1977. Growth and reproduction of *Stenella coeruleoalba* off the Pacific coast of Japan. Sci Rep Whales Res Inst. 29:21-48.
7397. Miyazaki, N. 1977. School structure of *Stenella coeruleoalba*. Rep Int Whal Commn. 27:498-499.
7398. Miyazaki, N. 1980. Preliminary note on age determination and growth of the rough-toothed dolphin, *Steno bredanensis*, off the Pacific coast of Japan. Rep Int Whal Commn. 3:171-179.
7399. Miyazaki, N. 1984. Further analyses of reproduction in the striped dolphin, *Stenella coeruleoallaa*, off the Pacific coast of Japan. Rep Int Whal Comm Sp Issue. 6:343-353.
7400. Miyazaki, N, Fujise, Y, & Fujiyama, T. 1981. Body and organ weight of striped and spotted dolphins off the Pacific coast of Japan. Sci Rep Whales Res Inst. 33:27-67.
7401. Miyazaki, N & Wada, S. 1978. Observation of Cetacea during whale marking cruise in the western tropical Pacific, 1976. Sci Rep Whales Res Inst. 30:179-195.
7402. Mizroch, SA. 1981. Analyses of some biological parameters of the Antarctic fin whale (*Balaenoptera physalus*). Rep Int Whal Comm. 31:425-434.
7403. Mizue, K & Jimbo, H. 1950. Statistic study of foetuses of whales. Sci Rep Whales Res Inst. 3:119-131.
7404. Mizue, K & Murata, T. 1951. Biological investigation on the whales caught by the Japanese Antarctic whaling fleets season 1949-50. Sci Rep Whales Res Inst. 6:73-131.
7405. Mizuhara, H. 1982. Personal communication. [Cited by Lancaster & Lee, 1982]
7406. Mleczko, D. 1986. First rhino born at the New York Zoological Park. AAZPA Newsl. 27:16.
7407. Mlíkovsky, J. 1985. Sex ratio distribution in the Siberian tiger *Panthera tigris altaica* (Mammalia: Felidae). Z Säugetierk. 50:47-51.
7407a. Mlíkovsky, J. 1988. Secondary sex ratio in the Przewalski horse *Equus przewalskii* (Mammalia: Equidae). Z Säugetierk. 53:92-101.
7408. Mobarak, AM, ElWishy, AB, & Samira, MF. 1972. The penis and prepuce of the one-humped camel *C dromedarius*. Zbl Vet Med A. 19:787-795.
7409. Mock, OB. 1982. The least shrew (*Cryptotis parva*) as a laboratory animal. Lab Anim Sci. 32:177-179.
7410. Mock, OB & Conaway, CH. 1976. Reproduction of the least shrew (*Cryptotis parva*) in captivity. The Laboratory Animal in the Study of Reproduction, 6th Symp ICLA Tessaloniki 59-71, Ericksen, S, & Spiegel, AS, eds, G Fischer, New York.
7411. Modha, KL & Field, A. 1974. A record of twins in the Africa buffalo *Syncerus caffer* (Sparrmau). E Afr Wildl J. 12:319.
7412. Moehlman, PD. 1979. Jackal helpers and pup survival. Nature. 277:382-383.
7413. Moffatt, CB. 1922. The habits of the long-eared bat. Ir Nat. 31:105-111.
7414. Mogart, JR & Krausman, PR. 1982. Early breeding in bighorn sheep. Southwest Nat. 28:460-461.
7415. Moghe, MA. 1949. Interstitial cells of mammalian testes. J Zool Soc India. 1:101-106.
7416. Moghe, MA. 1951. Development and placentation of the Indian fruit bat, *Pteropus giganteus giganteus* (Brunnick). Proc Zool Soc London. 121:703-721.
7417. Moghe, MA. 1956. On the development and placentation of a megachiropteran bat *Cynopterus sphinx gangeticus* (Anderson). Proc Nat Inst Sci India. 22B:48-55.
7418. Moghe, MA. 1956. Some observations on the foetal membranes of the Indian palmcivet *Paradoxurus hermaphroditus hermaphroditus* (Schrater). Proc Nat Inst Sci India. 22B:41-47.
7419. Moghe, MA. 1957. The role of cytochemistry in the identification of the placenta of *Suncus murinus* [Insectivora: Soricidae]. Proc Zool Soc (Calcutta). HK Mookerjee Mem Vol:191-193.
7420. Mohanty, N & Chainy, GBN. 1985. Follicular growth in the musk shrew (*Suncus murinus*) ovary I Quantitative analysis and characterization of follicles. J Zool Soc India. 37:45-52.
7421. Mohapatra, S. 1978. A note on the breeding and longevity of the Indian pangolin (*Manis crassicaudata*) in captivity. J Bombay Nat Hist Soc. 75:921-923.
7422. Møhl-Hansen, U. 1954. Investigations on reproduction and growth of the porpoise (*Phocoena phocoena* L) from the Baltic. Vidensk Medd Dansk Naturh Foren. 116:368-396.
7423. Mohnot, SM. 1974. Ecology and behaviour of the common Indian langur *Presbytis intellus* Dufresne. Ph D Diss. Univ Jodpur, India. [Cited by Oppenheimer, 1977]
7424. Mohr, CE. 1933. Observations on the young of cave-dwelling bats. J Mamm. 14:49-53.
7425. Mohr, E. 1932. Materialien über die Hinzuchten des ehemaligen Hamburger Zoo. Zool Garten NF. 5:3-15.

7426. Mohr, E. 1936. Biologische Beobachtungen an *Solenodon paradoxus* Brandt in Gefangenschart I. Zool Anz. 113:177-188.
7427. Mohr, E. 1936. Biologische Beobachtungen an *Solenodon paradoxus* Brandt in Gefangenschart II. Zool Anz. 116:65-76.
7428. Mohr, E. 1937. Biologische Beobachtungen an *Solenodon paradoxus* Brandt in Gefangenschaft III. Zool Anz. 117:233-241.
7429. Mohr, E. 1938. Biologische Beobachtungen an *Solenodon paradoxus* Brandt in Gefangenschart IV. Zool Anz. 122:132-143.
7429a. Mohr, E. 1938. Vom järv (*Gulo gulo* L). Zool Garten. 10:14-21.
7430. Mohr, E. 1939. Die Baum-und Ferkelratten: Gattung *Capromys* Desmarest (sens ampl) und Plagiodontia Cuvier. Mitt Hamburg Zool Mus Inst. 48:48-118.
7431. Mohr, E. 1943. Einiges über die Saiga, *Saiga tatarica* L. Zool Garten NF. 15:175-185.
7432. Mohr, E. 1949. Einiges von Groszen und von Kleinen Mara (*Dolichotis patagonum* Zimm und *salinicola* Burm). Zool Garten NF. 16:111-132.
7433. Mohr, E. 1954. Die freilebende Nagetiere Deutchlands und der Nachbarländer Dritte Auflage. Gustav Fischer Verlag, Jena.
7434. Mohr, E. 1958. Zur Kenntnis des Hirschebers, *Babirussa babyrussa* Linné 1758. Zool Garten NF. 25:50-69.
7435. Mohr, E. 1963. Os penis uns Os clitoridis der Pinnipedia. Z Säugetierk. 28:19-37.
7436. Mohr, U, Schuller, H, Reznik, G, et al. 1973. Breeding of European hamsters. Lab Anim Sci. 23:799-802.
7437. Moisan, G. 1956. Late breeding in moose, *Alces alces*. J Mamm. 37:300.
7438. Mokkapati, S & Dominic, CJ. 1975. Sites of production of fructose and citric acid in the accessory sex glands of the male musk shrew, *Suncus murinus*. J Reprod Fert. 45:527-528.
7439. Mokkapati, S & Dominic, CJ. 1976. Sites of production of fructose and citric acid in the accessory reproductive glands of three species of male chiropterans. Biol Reprod. 14:627-629.
7440. Mokkapati, S & Dominic, CJ. 1977. Morphology of the accessory reproductive glands of some male Indian chiropterans. Anat Anz. 141:391-397.
7441. Mokkapati, S & Dominic, CJ. 1977. The accessory reproductive glands of five male mammals. Proc Zool Soc (Calcutta). 30:1-5.
7442. Mokkapati, S & Dominic, CJ. 1977. Accessory reproductive glands of the male indian mongoose, *Herpestes auropunctatus* Hodgson. J Mamm. 58:85-87.
7443. Molez, N. 1976. Adaptation alimentaire du Galago d'Allen aux milieu forestiers secondaires. Terre Vie. 30:210-228.
7444. Molinari, J & Soriano, PJ. 1987. *Sturnira bidens*. Mamm Species. 276:1-4.
7445. Mollaret, HH. 1962. Naissance de Damans en captivité. Mammalia. 26:530-532.
7446. Möller, D. 1971. Beitrag zur Reproduktion des Feldhasen (*Lepus europaeus* Pall) in der Deutschen Demokratischen Republik. Tag Ber Dt Akad Landwirtsch-Wiss Berlin. 113:191-202.
7447. Møller, OM. 1973. Progesterone concentrations in the peripheral plasma of the blue fox (*Alopex lagopus*) during pregnancy and the oestrous cycle. J Endocrinol. 59:429-438.
7448. Møller, OM. 1973. The progesterone concentrations in the peripheral plasma of the mink (*Mustela vison*) during pregnancy. J Endocrinol. 56:121-132.
7449. Møller, OM. 1974. The fine structure of the lutein cells in the blue fox (*Alopex lagopus*) with special reference to the secretory activity during pregnancy. Cell Tiss Res. 149:61-79.
7450. Møller, OM. 1974. Effects of ovariectomy on the plasma progesterone and maintenance of gestation in the blue fox, *Alopex lagopus*. J Reprod Fert. 37:141-143.
7451. Møller, OM. 1974. Plasma progesterone before and after ovariectomy in unmated and pregnant mink, *Mustela vison*. J Reprod Fert. 37:367-372.
7452. Møller, OM, Aursjø, JM, & Sjaastad, ØV. 1980. Oestradiol-17β concentrations in the peripheral plasma of the blue fox (*Alopex lagopus*) around oestrus. Acta Vet Scand. 21:140-142.
7453. Møller, OM, Mondain-Monval, M, Smith, A, et al. 1984. Temporal relationships between hormonal concentrations and the electrical resistance of the vaginal tract of blue foxes *Alopex lagopus* at pro-oestrus and oestrus. J Reprod Fert. 70:15-24.
7454. Mombaerts, J. 1944. Le sinus urogénital et les glandes sexuelles annexes du Hérisson (*Erinaceus europaeus* L). Arch Biol. 55:393-554.
7455. Monamy, V. 1991. An observation of free-living dusky antechinuses, *Antechinus swainsonii* (Marsupialia: Dasyuridae) during the breeding season. Aust Mamm. 14:23-24.
7456. Moncrief, ND. 1988. Absence of genic variation in a natural population of nine-banded armadillos, *Dasypus novemcinctus* (Dasypodidae). Southwest Nat. 33:229-231.
7457. Mondain-Monval, M, Bonnin, M, Canivenc, R, et al. 1980. Plasma estrogen level during delayed implantation in the European badger (*Meles meles* L). Gen Comp Endocrinol. 41:143-149.
7458. Mondain-Monval, M, Bonnin, M, Canivenc, R, et al. 1984. Heterologous radioimmunoassay of fox LH: Levels during the reproductive season and the anoestrus of the red fox (*Vulpes vulpes* L). Gen Comp Endocrinol. 55:125-132.
7459. Mondain-Monval, M, Bonnin, M, Scholler, R, et al. 1979. Androgens in peripheral blood of the red fox (*Vulpes vulpes* L) during the reproductive season and anoestrus. J Steroid Biochem. 11:1315-1322.
7460. Mondain-Monval, M, Bonnin, M, Scholler, R, et al. 1983. Plasma androgen pattern during delayed implantation in the European badger (*Meles meles* L). Gen Comp Endocrinol. 50:67-74.
7461. Mondain-Monval, M, Dutournet, B, Bonnin-Laffargue, M, et al. 1977. Ovarian activity during the anoestrus and the reproductive season of the red fox (*Vulpes vulpes* L). J Steroid Biochem. 8:761-769.
7462. Mondain-Monval, M, Møller, OM, Smith, AJ, et al. 1985. Seasonal variations of plasma prolactin and LH concentrations in the female blue fox (*Alopex lagopus*). J Reprod Fert. 74:439-448.
7463. Mondain-Monval, M, Smith, AJ, Simon, P, et al. 1988. Effect of melatonin implantation on the seasonal variation of FSH in the male blue fox (*Alopex lagopus*). J Reprod Fert. 83:345-354.
7464. Mondolfi, E. 1987. Baculum of the lesser andean coati, *Nasuella olivacea* (Gray), and of the larger grison, *Galictis vittata* (Schreber). Fieldiana Zool NS. 39:447-454.
7465. Mondolfi, E & Medina Padilla, G. 1957. Contribucion al conocimiento del "Perrito de Agua" (*Chironectes minimus* Zimmermann). Memoria-Soc Cienc Nat La Salle. 17:141-149.
7466. Mones, A & Ojasti, J. 1986. *Hydrochoerus hydrochaeris*. Mamm Species. 264:1-7.
7467. Monfort, A. 1974. Quelques aspects de la biologie des Phacocheres (*Phacochoerus aethiopicus*) au Parc National de l'Akagera, Rwanda. Mammalia. 38:177-200.
7468. Monfort, A & Monfort, N. 1974. Note sur l'écologie et le comportement des Oribis (*Ourebia ourebi*, Zimmerman 1783). Terre Vie. 28:169-208.

7469. Monfort, A & Monfort, N. 1977. L'"opération Éléphants" au Rawanda 1re Partie: Structure de la population du Bugesera et transferts de jeunes au Parc de l'Akagera. Terre Vie. 31:355-384.
7470. Monfort, A, Monfort, N, & Ruwet, JC. 1973. Éco-éthologie des Ongulés au Parc Nationale de l'Akagera (Rwanda). Ann Soc R Zool Belg. 103:177-208.
7471. Monfort, SL, Dahl, KD, Czekala, NM, et al. 1989. Monitoring ovarian function and pregnancy in the giant panda (*Ailuropoda melanoleuca*) by evaluating urinary bioactive FSH and steroid metabolites. J Reprod Fert. 85:203-212.
7472. Monfort, SL, Hess, DL, Shideler, SE, et al. 1987. Comparison of serum estradiol to urinary estrone conjugates in the rhesus macaque (*Macaca mulatta*). Biol Reprod. 37:832-837.
7473. Monfort, SL, Jayaraman, S, Shideler, SE, et al. 1986. Monitoring ovulation and implantation in the cynomolgus macaque (*Macaca fascucularis*) through evaluations of urinary estrone conjugates and progesterone metabolites: A technique for the routine evaluation of reproductive parameters. J Med Primatol. 15:17-26.
7474. Monfort, SL, Saranson, R, Hess, DL, et al. 1986. Comparison of serum estradiol to urinary estrone conjugates in rhesus monkeys. Biol Reprod. 34(Suppl):163.
7475. Monfort, SL, Wemmer, C, Kepler, TH, et al. 1990. Monitoring ovarian function and pregnancy in Eld's deer (*Cervus eldi thamin*) by evaluating urinary steroid metabolite excretion. J Reprod Fert. 88:271-281.
7476. Monfort-Braham, N. 1975. Variations dans la structure sociale du Topi, *Damaliscus korrigum* Ogilby, au Parc National de l'Akagera, Rwanda. Z Tierpsychol. 39:332-364.
7477. Monmignaut, C. 1964. Cycle oestral de quelques Muridés africains et d'un Cricétide malgache. Mammalia. 28:183-184.
7478. Monnett, C, Rotterman, LM, & Siniff, DB. 1991. Sex-related patterns of postnatal development of sea otters in Prince William Sound, Alaska. J Mamm. 72:37-41.
7479. Monroe, SE, Yamamoto, M, & Jaffe, RB. 1983. Changes in gonadotrope responsivity to gonadotropin releasing hormone during development of the rhesus monkey. Biol Reprod. 29:422-431.
7480. Montgomery, GG. 1964. Tooth eruption in preweaned raccoons. J Wildl Manage. 28:582-584.
7481. Montgomery, GG. 1969. Weaning of captive raccoons. J Wildl Manage. 33:154-159.
7482. Montgomery, GG. 1985. The Evolution and Ecology of Armadillos, Sloths, and Vermilinguas. Smithsonian Institution Press, Washington, DC.
7483. Montgomery, GG & Sunquist, ME. 1978. Habitat selection and use by two-toed and three-toed sloths. The Ecology of Arborial Folivores. 329-359, Montgomery, GG, ed, Smithsonian Inst Press, Washington, DC.
7484. Montgomery, PA, Patton, S, Huston, GE, et al. 1987. Gel electrophoretic analysis of proteins in human milk and colostrum. Comp Biochem Physiol. 86B:635-639.
7485. Moog, G. 1957. Geburt und Aufzucht einer gelbgrünen Meerkatze, *Cercopithecus callitrichus* H Geoffr. Zool Garten NF. 23:220-223.
7486. Mook Hong, S & Stetson, MH. 1988. Termination of gonadal refractoriness in Turkish hamsters, *Mesocricetus brandti*. Biol Reprod. 38:639-643.
7487. Moore, AW. 1929. Extra-uterine pregnancy in *Peromyscus*. J Mamm. 10:81.
7488. Moore, AW. 1929. Some notes upon Utah mammals. J Mamm. 10:259-260.
7489. Moore, AW. 1930. Six Utah mammal records. J Mamm. 11:87-88.
7490. Moore, AW. 1939. Notes on the Townsend mole. J Mamm. 20:499-501.
7491. Moore, AW. 1943. Notes on the sage mouse in eastern Oregon. J Mamm. 24:188-191.
7491a. Moore, CR. 1939. Modification of sexual development in the opossum by sex hormones. Proc Soc Exp Biol Med. 40:544-546.
7491b. Moore, CR. 1941. Embryonic differentiation of opossum prostate following castration and responses of the juvenile gland to hormones. Anat Rec. 80:315-327.
7491c. Moore, CR. 1941. On the role of sex hormones in sex differentiation in the opossum (*Didelphys virginiana*). Physiol Zool. 14:1-47.
7491d. Moore, CR. 1943. Sexual differentiation in the opossum after early gonadectomy. J Exp Zool. 94:415-462.
7492. Moore, CR & Bodian, D. 1940. Opossum pouch young as experimental material. Anat Rec. 76:319-327.
7492a. Moore, CR & Morgan, CF. 1942. Responses of the testis to androgenic treatments. Endocrinology. 30:990-999.
7492b. Moore, CR & Morgan, CF. 1943. First response of develping opossum gonads to equine gonadotrophic treatment. Endocrinology. 32:16-26.
7493. Moore, CR, Simmons, F, Wells, LJ, et al. 1934. On the control of reproductive activity in an annual-breeding mammal (*Citellus tridecemlineatus*). Anat Rec. 60:279-289.
7494. Moore, GT. 1975. The breeding and utilization of baboons for biomedical research. Lab Anim Sci. 25:798-801.
7495. Moore, HDM, Bonney, RC, & Jones, DM. 1981. Induction of oestrus and successful artificial insemination in the cougar, *Felis concolor*. Ann Proc Am Assoc Zoo Vet. 1981:141-142.
7496. Moore, HDM, Bonney, RC, & Jones, DM. 1981. Successful induced ovulation and artificial insemination in the puma (*Felis concolor*). Vet Rec. 108:282-283.
7497. Moore, HDM, Gems, S, & Hearn, JP. 1985. Early implantation stages in the marmoset monkey (*Callithrix jacchus*). Am J Anat. 172:265-278.
7498. Moore, JC. 1946. Mammals form Welaka, Putnam County, Florida. J Mamm. 27:49-59.
7498a. Moore, JC. 1949. Putnam County and other Florida mammal notes. J Mamm. 30:57-66.
7498b. Moore, JC. 1951. The status of the manatee in the Everglades National Park, with notes on its natural history. J Mamm. 32:22-36.
7498c. Moore, JC. 1953. Distribution of marine mammals to Florida waters. Am Mid Nat. 49:117-158.
7499. Moore, JC. 1956. Observations of manatees in aggregations. Am Mus Novit. 1811:1-24.
7500. Moore, JC. 1957. Newborn young of a captive manatee. J Mamm. 38:137-138.
7501. Moore, JC. 1957. The natural history of the fox squirrel, *Sciurus niger shermani*. Bull Am Mus Nat Hist. 113:1-71.
7502. Moore, JC. 1961. Geographic variation in some reproductive characteristics of diurnal squirrels. Bull Am Mus Nat Hist. 122:1-32.
7503. Moore, W, Jr. 1965. Observations on the breeding and care of the Chinese hamster, *Cricetulus griseus*. Lab Anim Care. 15:94-101.
7503a. Mooring, MS & Rubin, ES. 1991. Nursing behavior and early development of impala at San Diego Wild Animal Park. Zoo Biol. 10:329-339.
7504. Moors, PJ. 1975. The urogenital system and notes on the reproductive biology of the female rufous rat-kangaroo, *Aepyprymnus rufescens* (Gray) (Macropodidae). Aust J Zool. 23:355-361.

7505. Moran, RJ. 1970. Precocious antler development and sexual maturity in a captive elk. J Mamm. 51:812-813.
7506. Morejohn, GV. 1979. The natural history of Dall's porpoise in the north Pacific Ocean. Behavior of Marine Animals. 3:45-83, Winn, KE & Olla, BL, eds, Plenum Press, New York.
7507. Morejohn, GV & Baltz, DM. 1972. On the reproductive tract of the female dall porpoise. J Mamm. 53:606-608.
7507a. Moreno, LI, Salas, IC, & Glander, KE. 1991. Breech delivery and birth-related behaviors in wild mantled howling monkeys. Am J Primatol. 23:197-199.
7507b. Morgan, CF. 1943. The normal development of the ovary of the opossum from birth to maturity and its reactions to sex hormones. J Morphol. 72:27-85.
7508. Morgan, CF. 1946. Sexual rhythms in the reproductive tract of the adult female opossum and effects of hormonal treatments. Am J Anat. 78:411-463.
7509. Morgan, GS. 1989. *Geocapromys thoracatus*. Mamm Species. 341:1-5.
7510. Morgan, PR. 1974. Routine birth induction in rabbits using oxytocin. Lab Anim. 8:127-130.
7511. Mōri, J, Hafez, ESE, Jaszczak, S, et al. 1973. Serum LH during ovulatory and anovulatory menstrual cycles macaques. Acta Endocrinol. 73:751-758.
7512. Mōri, T, Son, SW, & Uchida, TA. 1986. Implications of prolonged sperm storage from the viewpoint of capacitation in the Japanese house-dwelling bat, *Pipistrellus abramus*. Dev Growth Diff Suppl. abs 94.
7513. Mōri, T & Uchida, TA. Personal communication. [Cited by Racey, 1979]
7514. Mōri, T & Uchida, TA. 1974. Electron microscopic analysis of the mechanism of fertilization in Chiroptera II Engulfment of spermatozoa by epithelial cells of the Fallopian tube in the Japanese bat, *Pipistrellus abramus*. Zool Mag. 83:163-170.
7515. Mōri, T & Uchida, TA. 1980. Sperm storage in the reproductive tract of the female Japanese long-fingered bat, *Miniopterus schreibersii fuliginosus*. J Reprod Fert. 58:429-433.
7516. Mōri, T & Uchida, TA. 1981. Ultrastructural observations of fertilization in the Japanese long-fingered bat, *Miniopterus schreibersii fuliginosus*. J Reprod Fert. 63:231-235.
7517. Mōri, T & Uchida, TA. 1981. Ultrastructural observations of ovulation in the Japanese long-fingered bat, *Miniopterus schreibersii fuliginosus*. J Reprod Fert. 63:391-395.
7518. Mōri, T & Uchida, TA. 1982. Changes in the morphology and behaviour of spermatozoa between copulation and fertilization in the Japanese long-fingered bat, *Miniopterus schreibersii fuliginosus*. J Reprod Fert. 65:23-28.
7519. Mōri, T, Oh, YK, & Uchida, TA. 1982. Sperm storage in the oviduct of the Japanese greater horseshoe bat, *Rhinolophus ferrumequinum nippon*. J Fac Agric Kyushu Univ. 27:47-53.
7520. Mori, U & Dunbar, RIM. 1985. Changes in the reproductive condition of female gelada baboons following the takeover of one-male units. Z Tierpsychol. 67:215-224.
7521. Morii, R. 1976. Biological study of the Japanese house bat *Pipistrellus abramus* (Temminck, 1840) in Kagawa prefecture. J Mamm Soc Japan. 6:248-258.
7522. Morii, R. 1980. External and cranial characters and the period of parturition in *Myotis macrodactylus*, Kagawa prefecture, Japan. Kagawa Seibutsu. 9:5-9.
7523. Morii, R. 1980. Postnatal development of external characters and behavior in young *Pipistrellus abramus*. J Mamm Soc Japan. 8:117-121.
7524. Morita, S. 1964. On the breeding season, litter size and gestation period of the Riukiu musk shrew, *Suncus murinus riukiuanus* Kuroda. Zool Mag. 73:196-201.
7525. Morrell, S. 1972. Life history of the San Joaquin kit fox. CA Fish Game. 58:162-174.
7526. Morris, B. 1958. The yolk-sac of the mole *Talpa europaea*. Proc Zool Soc London. 131:367-387.
7527. Morris, B. 1961. Some observations on the breeding season of the hedgehog and the rearing and handling of the young. Proc Zool Soc London. 136:201-206.
7528. Morris, B. 1963. Notes on the giant rat (*Cricetomys gambianus*) in Nyasaland. Afr Wild Life. 17:103-107.
7529. Morris, B. 1966. Breeding the European hedgehog *Erinaceus europaeus* in captivity. Int Zoo Yearb. 6:141-146.
7530. Morris, B. 1967. The European hedgehog (*Erinaceus europaeus* L). UFAW Handbook on the Care and Management of Lab Anim, 3rd ed. 478-488, E & S Livingstone, Edinburgh.
7531. Morris, D & Jarvis, C. 1959. Mammalian gestation periods. Int Zoo Yearb. 1:157-160.
7532. Morris, DJ & van Aarde, RJ. 1985. Sexual behavior of the female porcupine *Hystrix africaeaustralis*. Horm Behav. 19:400-412.
7533. Morris, DW. 1986. Proximate and ultimate controls on life-history variation: The evolution of litter size in white-footed mice (*Peromyscus leucopus*). Evolution. 40:169-181.
7534. Morris, J. 1975. Ovulation in raccoons and note on reproductive physiology of free-ranging raccoons. Trans MO Acad Sci. 7:261-262.
7535. Morris, JH, Negus, NC, & Spertzel, RO. 1967. Colonization of the tree shrew (*Tupaia glis*). Lab Anim Care. 17:514-520.
7536. Morris, M, Stevens, SW, & Adams, MR. 1980. Plasma oxytocin during pregnancy and lactation in the cynomologus monkey. Biol Reprod. 23:782-787.
7537. Morris, NE & Hanks, J. 1974. Reproduction in the bushbuck *Tragelaphus scriptus ornatus*. Arnoldia (Rhodesia). 7(1):1-8.
7538. Morris, P. 1977. Pre-weaning mortality in the hedgehog (*Erinaceus europaeus*). J Zool. 182:162-164.
7539. Morrison, JA. 1960. Characteristics of estrus in captive elk. Behaviour. 16:84-92.
7540. Morrison, JA. 1971. Morphology of corpora lutea in the Uganda kob antelope, *Adenota kob thomasi* (Neumann). J Reprod Fert. 26:297-305.
7541. Morrison, JA & Buechner, HK. 1971. Reproductive phenomena during the post-partum-preconception interval in the Uganda kob. J Reprod Fert. 26:307-317.
7542. Morrison, JA & Menzel, EW, Jr. 1972. Adaptation of a free-ranging rhesus monkey group to division and transplantation. Wildl Monogr. 31:1-78.
7543. Morrison, JA, Trainer, CE, & Wright, PL. 1959. Breeding season in elk as determined from known age embryos. J Wildl Manage. 23:27-34.
7544. Morrison, P, Dieterich, R, & Preston, D. 1976. Breeding and reproduction of fifteen wild rodents maintained as laboratory colonies. Lab Anim Sci. 26:237-243.
7545. Morrison, P, Dieterich, R, & Preston, D. 1977. Body growth in sixteen rodent species and subspecies maintained in laboratory colonies. Physiol Zool. 50:294-310.
7546. Morrison, PR, Ryser, PA, & Strecker, RL. 1954. Growth and the development of temperature regulation in the tundra redback vole. J Mamm. 35:376-386.

7546a. Morrissey, BL & Breed, WG. 1982. Variation in external morphology of the glans penis of Australian native rodents. Aust J Zool. 30:495-502.
7547. Morton, ML & Gallup, JS. 1975. Reproductive cycle of the Belding ground squirrel (*Spermophilus beldingi beldingi*): Seasonal and age differences. Great Basin Nat. 35:427-433.
7548. Morton, ML, Maxwell, CS, & Wade, CE. 1973. Body size, body composition, and behavior of juvenile Belding ground squirrels. Great Basin Nat. 34:121-134.
7549. Morton, ML & Parmer, RJ. 1975. Body size, organ size, and sex ratios in adult and yearling Belding ground squirrels. Great Basin Nat. 35:305-309.
7550. Morton, ML & Tung, HL. 1971. Growth and development in the Belding ground squirrel (*Spermophilus beldingi beldingi*). J Mamm. 52:611-616.
7551. Morton, SR. 1975. The life cycle of *Sminthopsis crassicaudata*. Aust Mamm. 1:398.
7552. Morton, SR. 1978. An ecological study of *Sminthopsis crassicaudata* (Marsupialia: Dasyuridae) III Reproduction and life history. Aust Wildl Res. 5:183-211.
7553. Morton, SR, Armstrong, MD, & Braithwaite, RW. 1987. The breeding season of *Sminthopsis virginiae* (Marsupialia: Dasyuridae) in the Northern Territory. Aust Mamm. 10:41-42.
7554. Morton, WRM. 1957. A chimpanzee placenta and foetus *in situ*. J Anat. 91:605.
7555. Morton, WRM. 1957. Placentation in the spotted hyena (*Crocuta crocuta* Erxleben). J Anat. 91:374-382.
7556. Morton, WRM. 1961. Observations on the full-term foetal membranes of three members of the Camelidae (*Camelus dromedarius* L, *Camelus bactrianus* L and *Lama glama* L). J Anat. 95:200-209.
7557. Moseley, EL. 1928. The number of young bats in one litter. J Mamm. 9:249.
7558. Moser, HG & Benirschke, K. 1962. Fetal zone of the adrenal gland in the nine-banded armadillo, *Dasypus novemcinctus*. Anat Rec. 143:47-59.
7559. Moshonkin, NN. 1983. The reproductive cycle in females of the European mink (*Lutreola lutreola*). Zool Zhurn. 62:1879-1883.
7560. Moss, CJ. 1983. Oestrous behaviour and female choice in the African elephant. Behaviour. 86:167-196.
7561. Mossman, AS. 1955. Reproduction of the brush rabbit in California. J Wildl Manage. 19:177-184.
7562. Mossman, AS & Mossman, HW. 1962. Ovulation, implantation, and fetal sex ratio in impala. Science. 137:869.
7563. Mossman, HW. 1937. Comparative morphogenesis of the fetal membranes and accessory uterine structures. Contrib Embryol Carnegie Inst Wash. 26(158):129-246.
7564. Mossman, HW. 1939. The epithelio-chorial placenta of an American mole, *Scalopus aquaticus*. Proc Zool Soc London. 109(B):373-375.
7565. Mossman, HW. 1957. Endotheliochorial placentation in the rodents, *Castor* and *Pedestes*. Proc Zool Soc (Calcutta). HK Mookerjee Mem Vol:183-189.
7566. Mossman, HW. 1957. The foetal membranes of the aard-vark. Mitt Natf Ges Bern NF. 14:119-127.
7567. Mossman, HW. 1967. Comparative biology of the placenta and fetal membranes. Fetal Homestasis. 13-97, Wynn, RM, ed, NY Acad Sci, New York.
7568. Mossman, HW & Fischer, TV. 1969. The preplacenta of *Pedetes*, the Träger and the maternal circulatory pattern in rodent placentae. J Reprod Fert Suppl. 6:175-184.
7569. Mossman, HW & Hisaw, FL. 1940. The fetal membranes of the pocket gopher, illustrating an intermediate type of rodent membrane formation. Am J Anat. 66:367-391.
7570. Mossman, HW, Hoffman, RA, & Kirkpatrick, CM. 1955. The accessory genital glands of male gray and fox squirrels correlated with age and reproductive cycles. Am J Anat. 97:257-301.
7571. Mossman, HW & Judas, I. 1949. Accessory corpoa lutea, lutein cell origin, and the ovarian cycle in the Canadian porcupine. Am J Anat. 85:1-39.
7572. Mossman, HW, Lawlah, JW, & Bradley, JA. 1932. The male reproductive tract of the Sciuridae. Am J Anat. 51:89-155.
7573. Mossman, HW & Owers, N. 1963. The shrew placenta: Evidence that it is endothelio-endothelial in type. Am J Anat. 113:245-271.
7574. Mossman, HW & Strauss, F. 1963. The fetal membranes of the pocket gopher illustrating an intermediate type of rodent membrane formation II From the beginning of the allantois to term. Am J Anat. 113:447-477.
7575. Mossman, HW & Weisenfeldt, LA. 1939. The fetal membranes of a primitive rodent, the thirteen-striped ground squirrel. Am J Anat. 64:59-109.
7576. Motta, M, Carreira, J, & Franco, A. 1983. A note on reproduction of *Didelphis marsupialis* in captivity. Mem Inst Oswaldo Cruz Rio de Janeiro. 78:507-509.
7577. Mottershead, GS. 1958. Interesting experiments at Chester Zoo. Zool Garten. 24:70-73.
7578. Mottershead, GS. 1963. The lesser panda in the Chester Zoological Gardens. Zool Garten. 27:300-302.
7579. Mottl, S. 1958. Der Geschlechtszyklus des Muffelwidders, *Ovis musimon* Schreber 1782. Zool Listy. 7:343-352.
7580. Moustgaard, J & Medarbegdere. 1957. Minkmaelkens sammensaetning og naeringsvaerdi. Forsogslab Årbog. 1957:11-12. [Cited by Venge, 1973]
7581. Mover, H, Ar, A, & Hellwing, S. 1989. Energetic costs of lactation with and without simultaneous pregnancy in the white-toothed shrew *Crocidura russula monacha*. Physiol Zool. 62:919-936.
7582. Mover, H, Hellwing, S, & Ar, A. 1988. Energetic cost of gestation in the white-tooth shrew *Crocidura russula monacha* (Soricidae, Insectivora). Physiol Zool. 61:17-25.
7583. Moynihan, M. 1970. Some behavior patterns of platyrrihine monkeys: *Saguinus geoffroyi* and some other tamarins. Smithsonian Contr Zool. 28:1-77.
7584. Moynihan, M. 1976. The New World Primates. Princeton Univ Press, Princeton.
7585. Muchlinski, AE. 1988. Population attributes related to the life-history strategy of hibernating *Zapus hudsonius*. J Mamm. 69:860-865.
7586. Mudar, KM & Allen, MS. 1986. A list of bats from northeastern Luzon, Phillippines. Mammalia. 219-225.
7587. Mueller, CC & Sadleir, RMFS. 1975. Attainment of early puberty in female black-tailed deer (*Odocoileus hemionus columbianus*). Theriogenology. 3:101-105.
7588. Mueller, CC & Sadleir, RMFS. 1977. Changes in the nutrient composition of milk of black-tailed deer during lactation. J Mamm. 58:421-423.
7589. Mueller, CC & Sadleir, RMFS. 1979. Age at first conception in black-tailed deer. Biol Reprod. 21:1099-1104.
7590. Mueller, CC & Sadleir, RMFS. 1980. Birth weights and early growth of captive mother-raised black-tailed deer. J Wildl Manage. 44:268-272.
7591. Mühlenberg, M & Roth, HH. 1985. Comparative investigations into the ecology of the kob antelope *Kobas kob*

kob (Erxleben 1777) in the Comoe National Park, Ivory Coast. S Afr J Wildl Res. 15:25-31.
7592. Muir, PD, Sykes, AR, & Barrell, Gk. 1988. Changes in blood content and histology during growth of antlers in red deer (*Cervus elaphus*) and their relationship to plasma testosterone levels. J Anat. 158:31-42.
7593. Mukku, VR, Murty, GSRC, Srinath, BR, et al. 1981. Regulation of testosterone rhythmicity by gonadotropins in bonnet monkeys (*Macaca radiata*). Biol Reprod. 24:814-819.
7593a. Mullally, DP. 1953. Hibernation in the golden-mantled ground squirrel, *Citellus lateralis bernardinus*. J Mamm. 34:65-73.
7594. Mullan, JM, Feldhamer, GA, & Morton, D. 1988. Reproductive characteristics of female sika deer in Maryland and Virginia. J Mamm. 69:388.
7595. Mullen, DA. 1968. Reproduction in brown lemmings (*Lemmus trimucronatus*) and its relevance to their cycle of abundance. Univ CA Publ Zool. 85:1-24.
7596. Mullen, RK. 1971. Energy metabolism and body water turnover rates of two species of freeliving kangaroo rats, *Dipodomys merriami* and *Dipodomys microps*. Comp Biochem Physiol. 39:379-390.
7597. Müller, G, Nicht, M, & Kuhne, H. 1969. Organgewichte von *Gerbillus pyramidum* Geoffrey, 1825. Z Versuchstierk. 11:123-135.
7598. Müller, H. 1954. Zur Fortpfanzungsbiologie des Hermelins (*Mustela erminea L*). Rev Suisse Zool. 61:451-453.
7599. Müller, H. 1970. Beiträge zur Biologie des Hermelins, *Mustela erminea* Linné, 1758. Säugetierk Mitt. 18:293-380.
7600. Müller, JP. 1977. Populationsökologie von *Arvicanthis abyssinicus* in der Grassteppe des Semien mountains National Park (Athiopien). Z Säugetierk. 42:145-172.
7601. Müller, S & Schlegel, H. 1839-42. Over de tot heden bekende eekhoorns (*Sciurus*) van den Indischen archipel. Verhandelingen over de Natuurlijke Geschiedenis der Nederlandsche Overzeesche Bezittingen. 85-101, ed, CI Temminck, Leiden.
7602. Müller, S & Schlegel, H. 1839-42. Beschrijving van een merkwaardig insectetend zoogdier *Hylomys suillus*. Verhandelingen over de Natuurlijke Geschiedenis der Nederlandsche Overzeesche Bezittingen. 153-159, CI Temminck, Leiden.
7603. Müller, S & Schlegel, H. 1839-42. Over de wilde zwijnen van den Indischen Archipel. Verhandelingen over de Natuurlijke Geschiedenis der Nederlandsche Overzeesche Bezittingen. 170-181, CI Temminck, Leiden.
7604. Müller, S & Schlegel, H. 1839-42. Over de herten van den Indischen archipel. Verhandelingen over de Natuurlijke Geschiedenis der Nederlandsche Overzeesche Bezittingen. 209-233, CI Temminck, Leiden.
7605. Müller-Using, D. 1957. Die Paarungsbiologie des Murmeltieres. Z Jagdwiss. 3:24-28.
7606. Mullican, TR & Keller, BL. 1986. Ecology of the sagebrush vole (*Lemniscus curtatus*) in southeastern Idaho. Can J Zool. 64:1218-1223.
7607. Mumford, RE. 1957. *Myotis occultus* and *Myotis yumanensis* breeding in New Mexico. J Mamm. 38:260.
7608. Mumford, RE. 1973. Natural history of the red bat (*Lasiurus borealis*) in Indiana. Period Biol. 75:155-158.
7609. Mumford, RE & Calvert, LL. 1960. *Myotis sodalis* evidently breeding in Indiana. J Mamm. 41:512.
7610. Mumford, RE & Handley, CO, Jr. 1956. Notes on the mammals of Jackson county, Indiana. J Mamm. 37:407-412.
7611. Mumford, RE, Oakley, LL, & Zimmerman, DA. 1964. June bat records from Guadalupe canyon, New Mexico. Southwest Nat. 9:43-45.
7612. Mumford, RE & Zimmerman, DA. 1962. Notes on *Choeronycteris mexicana*. J Mamm. 43:101-102.
7613. Munday, BL, Green, RH, & Obendorf, DL. 1982. A pygmy right whale *Caperea marginata* (Gray, 1846) stranded at Stanley, Tasmania. Pap Proc R Soc Tasmania. 116:1-3.
7614. Mundinger, JG. 1981. White-tailed deer reproductive biology in the Swan Valley, Montana. J Wildl Manage. 45:132-139.
7615. Mundy, KRD & Flook, DR. 1973. Background for managing grizzly bears in the national parks of Canada. Can Wildl Serv Rep Ser. 22:1-34.
7616. Mungall, EC. 1976. The Indian blackbuck antelope: A Texas review. Dept Wildl Fish Sci, TX Agric Exp Stat, TX A & M Univ.
7617. Mungall, EC. 1978. The Indian Blackbuck Antelope: A Texas View. Caesar Kleberg Res Prog Wildl Ecol, Dept Wild Fish Sci, TX Agric Exp Stn, TX A & M Univ, College Station.
7618. Munro, G. 1969. Breeding the lesser panda in Bremen Zoo Germany. Int Zoo News. 89:281-283.
7619. Munyer, EA. 1967. A parturition date for the hoary bat, *Lasiurus c cinereus*, in Illinois and notes on the newborn young. IL State Acad Sci. 60:95-97.
7620. Murakami, O. 1974. Growth and development of the Japanese wood mouse (*Apodemus speciosus*) I The breeding season in the field. Jap J Ecol. 24:194-206.
7621. Murariu, D. 1974. L'étude anatomo-histologique des glandes mammaires chez les Insectivores (Mammalia) de Roumanie. Trav Mus Hist Nat "Grigor Antipa". 14:431-438.
7622. Murata, K, Tanioka, M, & Murkami, N. 1986. The relationship between the pattern of urinary oestrogen and behavioural changes in the giant panda *Ailuropoda melanoleuca*. Int Zoo Yearb. 24/25:274-279.
7623. Murie, A. 1934. The moose of Isle Royale. Misc Publ Mus Zool Univ MI. 25:1-44.
7624. Murie, A. 1935. Mammals from Guatemala and British Honduras. Misc Publ Mus Zool Univ MI. 26:1-30.
7625. Murie, A. 1944. The Wolves of Mount McKinley. US Dept Interior Fauna Series. 5:1-238.
7626. Murie, J. 1872. On the female generative organs, viscera, and fleshy parts of *Hyaena brunnea*, Thunberg. Trans Zool Soc London. 7:503-512.
7627. Murie, J. 1872. Researches upon the anatomy of Pinnipedia Part 1 On the walrus (*Trichechus rosmarus* Linn). Trans Zool Soc London. 7:411-464.
7628. Murie, J. 1874. On the form and structure of the manatee *Manatus americanus*. Trans Zool Soc London. 8:127-202.
7629. Murie, JO. 1935. Alaska-Yukon Caribou. N Am Fauna. 54:1-90.
7630. Murie, JO. 1984. A comparison of life history traits in two populations of *Spermophilus columbianus* in Alberta, Canada. Acta Zool Fennici. 173:43-45.
7631. Murie, JO, Boag, DA, & Kivett, VK. 1980. Litter size in Columbian ground squirrels (*Spermophilus columbianus*). J Mamm. 61:237-244.
7632. Murie, JO & Harris, MA. 1982. Annual variation of spring emergence and breeding in Columbian ground squirrels (*Spermophilus columbianus*). J Mamm. 63:431-439.
7633. Murie, JO & McLean, IG. 1980. Copulatory plugs in ground squirrels. J Mamm. 61:355-356.
7634. Murie, OJ. 1940. Notes on the sea otter. J Mamm. 21:119-131.

7635. Murphy, BD. 1976. Effects of synthetic GnRH on litter size in ranch mink bred once or twice. Theriogenology. 6:463-467.
7636. Murphy, BD. 1979. Effects of GnRH on plasma LH and fertility in mink. Can J Anim Sci. 59:25-33.
7637. Murphy, BD. 1979. The role of prolactin in implantation and luteal maintenance in the ferret. Biol Reprod. 21:517-521.
7638. Murphy, BD. 1983. Effects of single or multiple injections of medroxyprogesterone acetate on the reproductive performance and gestation length of ranch mink. Can J Anim Sci. 63:989-991.
7639. Murphy, BD. 1983. Precocious induction of luteal activation and termination of delayed implantation in mink with the dopamine antagonist pimozide. Biol Reprod. 29:658-662.
7640. Murphy, BD, Concannon, PW, & Travis, HF. 1982. Effects of medroxyprogesterone acetate on gestation in mink. J Reprod Fert. 66:491-497.
7641. Murphy, BD, Concannon, PW, Travis, HF, et al. 1981. Prolactin: The hypophyseal factor that terminates embryonic diapause in mink. Biol Reprod. 25:487-491.
7642. Murphy, BD, DiGregorio, GB, Douglas, DA, et al. 1990. Interactions between melatonin and prolactin during gestation in mink (*Mustela vison*). J Reprod Fert. 89:423-429.
7643. Murphy, BD, Humphrey, WD, & Shepstone, SL. 1980. Luteal function in mink: The effects of hypophysectomy after the preimplantation rise in progesterone. Anim Reprod Sci. 3:225-232.
7643a. Murphy, BD & James, DA. 1974. Mucopolysaccharide histochemistry of the mink uterus during gestation. Can J Zool. 52:687-693.
7644. Murphy, BD, Mead, RA, & McKibbin, PE. 1983. Luteal contribution to the termination of preimplantation delay in mink. Biol Reprod. 28:497-503.
7645. Murphy, BD & Moger, WH. 1977. Progestins of mink gestation: The effects of hypophysectomy. Endocr Res Comm. 4:45-60.
7646. Murphy, CR & Smith, JR. 1970. Age determination of pouch young and juvenile Kangaroo Island wallabies. Trans R Soc S Aust. 94:15-20.
7647. Murphy, DA. 1963. A captive elk herd in Missouri. J Wildl Manage. 27:411-414.
7648. Murphy, ET. 1976. Breeding the clouded leopard *Neofelis nebulosa* at Dublin Zoo. Int Zoo Yearb. 16:122-124.
7649. Murphy, MF. 1952. Ecology and helminths of the Osage wood rat, *Neotoma floridana osagensis*, including the description of *Longistriata neotoma* n sp (Trichostrongylidae). Am Mid Nat. 48:204-218.
7650. Murphy, MF. 1982. Breeding statistics of western lowland gorilla *Gorilla g gorilla* in United States Zoological Parks. Int Zoo Yearb. 22:180-185.
7651. Murphy, RC & Nichols, JT. 1913. Long Island fauna and flora I The bats (order Chiroptera). Sci Bull Mus Brooklyn Inst Arts Sci NY. 2(1):1-15.
7652. Murr, E. 1929. Zur Erklärung der verlängerten Tragdauer bei Säugetieren. Zool Anz. 85:113-129.
7653. Murr, E. 1931. Experimentelle Abkurzung der Tragdauer beim Frettchen (*Putorius furo* L). Anz Akad Wiss Wien. 68:265-266.
7654. Murr, E. 1933. Experimentelle Abkurzung der Tragdauer durch Wärme. Z Vergl Physiol. 19:237-245.
7655. Murr, E. 1935. Die Fortpflanzung des Frettchens (*Putorius furo* L) Eine Zusammenfassung unserer Kenntnisse auf Grund eigener und fremder Untersuchungen, zugleich eine Einführung in die Fortpflanzungskunde der Säugetier. Z Züchtungsk. 32B:290, 385-408.
7656. Murray, BM & Murray, DF. 1969. Notes on mammals in alpine areas of the northern St Elias Mountains, Yukon Territory and Alaska. Can Field-Nat. 83:331-338.
7657. Murray, ER. Personal communication. [Cited by Mentis, 1972]
7658. Murray, GN. 1942. The gestation period of *Procavia capensis* (dassie). J S Afr Med Vet Assn. 13:27-28.
7659. Murray, JD, McKay, GM, & Sharman, GB. 1979. Studies on metatherian sex chromosomes IX Sex chromosomes of the greater glider (Marsupialia: Petauridae). Aust J Biol Sci. 32:375-386.
7660. Murray, MG. 1982. The rut of impala: Aspects of seasonal mating under tropical conditions. Z Tierpsychol. 59:319-337.
7661. Murtagh, CE & Sharman, GB. 1977. Chromosomal sex determination in monotremes. Proc Int Symp Comp Biol Reprod. 4:61-66.
7662. Murthy, KVR. 1969. Histophysiology of the penis in the Indian sheath-tailed bat *Taphozous longimanus* (Hardwicke), with notes on the morphology of the internal genitalia. J Zool Soc India. 21:149-159.
7663. Murthy, KVR. 1979. Studies on the male genitalia of Indian bats Part III Male genitalia of the Indian vampire bat *Megaderma lyra lyra* (Geoffroy). J Zool Soc India. 31:55-60.
7664. Murthy, KVR. 1981. Studies on the male genitalia of Indian bats Part V Male genitalia of the Indian vespertilianid bat *Pipistrellus ceylonicus chrysothrix* (Wroughton). J Zool Soc India. 33:53-61.
7665. Murthy, KVR & Vamburkar, SA. 1978. Studies on the male genitalia of Indian bats Part II Male genitalia of the giant Indian fruit bat *Pteropus giganteus giganteus* (Brunnich). J Zool Soc India. 30:47-55.
7666. Murua, R, Gonzalez, LA, & Meserve, PL. 1986. Population ecology of *Oryzomys longicaudatus philippii* (Rodentia: Cricetidae) in southern Chile. J Anim Ecol. 55:281-293.
7667. Musa, BE & Abusineia, ME. 1978. The oestrous cycle of the camel (*Camelus dromedarius*). Vet Rec. 103:556-557.
7668. Musa, BE & Sineina, MEA. 1976. Some observations on reproduction in the female camel (*Camelus dromedarius*). Acta Vet (Beograd). 26:63-67.
7669. Muse, PD & Louis, TM. 1980. Estrous cycle length of the mara, *Dolichotis patagona*. Anat Rec. 196:246.
7670. Musey, PI, Collins, DC, Gould, KG, et al. 1983. Differential oxidations of estradiol-17β by the chimpanzee in vivo. Am J Primatol. 5:271-275.
7671. Musser, GG. 1979. Results of the Archbold Expeditions No 102: The species of *Chiropodomys*, arboreal mice of Indochina and the Malay Archipelago. Bull Am Mus Nat Hist. 162:381-445.
7672. Musser, GG. 1981. A new genus of arboreal rat from West Java, Indonesia. Zool Verh. 189:1-35.
7673. Musser, GG. 1981. The giant rat of Flores and its relatives east of Borneo and Bali. Bull Am Mus Nat Hist. 169:69-175.
7674. Musser, GG. 1981. Results of the Archbold expeditions No 105 Notes on systematics of Indo-Malayan murid rodents, and descriptions of new genera and species from Ceylon, Sulawesi, and the Philippines. Bull Am Mus Nat Hist. 168:225-334.
7675. Musser, GG. 1982. Results of the Archbold expeditions No 110: *Crunomys* and the small-bodied shrew rats native to the Philippine Islands and Sulawesi (Celebes). Bull Am Mus Nat Hist. 174:1-95.
7675a. Musser, GG & Holden, ME. 1991. Sulawesi rodents (Muridae: Murinae): Morphological and geographical boundaries

of species in the *Rattus hoffmanni* group and a new species from Pulau Peleng. Bull Am Mus Nat Hist. 206:322-413.
7676. Mutere, FA. 1965. Delayed implantation in an equatorial fruit bat. Nature. 207:780.
7677. Mutere, FA. 1967. The breeding biology of equatorial vertebrates: Reproduction in the fruit bat, *Eidolon helvum*, at latitude 0°20'N. J Zool. 155:153-161.
7678. Mutere, FA. 1968. Breeding cycles in tropical bats in Uganda. J Appl Ecol. 5:8P-9P.
7679. Mutere, FA. 1968. The breeding biology of the fruit bat *Rousettus aegyptiacus* E Geoffroy living at 0°22'S. Acta Tropica. 25:97-108.
7680. Mutere, FA. 1970. Bat studies in Uganda. Stud Speleology. 2:61-68.
7681. Mutere, FA. 1970. The breeding biology of equatorial vertebrates: Reproduction in the insectivorous bat, *Hipposideros caffer*, living at 0°27'N. Bijdr Dierkd. 40:56-58.
7682. Mutere, FA. 1973. A comparative study of reproduction in two populations of the insectivorous bats, *Otomops martiensseni*, at latitudes 1°5'S and 2°30'S. J Zool. 171:79-92.
7683. Mutere, FA. 1973. Reproduction in two species of equatorial free-tailed bats (Molossidae). E Afr Wildl J. 11:271-280.
7684. Mutere, FA. 1980. *Eidolon helvum* revisited. Proc Int Bat Res Conf. 5:145-150.
7685. Muul, I. 1969. Mating behavior, gestation period, and development of *Glaucomys sabrinus*. J Mamm. 50:121.
7686. Muul, I. 1969. Photoperiod and reproduction in flying squirrels, *Glaucomys volans*. J Mamm. 50:542-549.
7687. Muul, I & Lim, LB. 1974. Reproductive frequency in Malaysian flying squirrels, *Hylopetes* and *Pteromyscus*. J Mamm. 55:393-400.
7688. Myers, D. 1982. Life cycle information: The mouse. The Guide for the Care and Use of Laboratory Animals. 78-23, DHEW Publ, Washington, DC.
7689. Myers, K. 1964. Influence of density on fecundity, growth rates, and mortality of the wild rabbit. CSIRO Wildl Res. 9:134-137.
7690. Myers, K & Poole, WE. 1958. Sexual behavior cycles in the wild rabbit, *Oryctolagus cuniculus* (L). CSIRO Wildl Res. 3:144-145.
7691. Myers, K & Poole, WE. 1962. A study of the biology of the wild rabbit, *Oryctolagus cuniculus* (L) in confined populations III Reproduction. Aust J Zool. 10:225-267.
7692. Myers, K & Poole, WE. 1962. Oestrous cycles in the rabbit *Oryctolagus cuniculus* (L). Nature. 195:358-359.
7693. Myers, K & Schneider, EC. 1964. Observations on reproduction, mortality, and behaviour in a small, free-living population of wild rabbits. CSIRO Wildl Res. 9:138-143.
7694. Myers, K, Bults, HG, & Gilbert, N. 1981. Stress in the rabbit. Proc World Lagomorph Conf, Univ Guelph, ONT, Canada. 1979:103-136.
7695. Myers, P. 1977. Patterns of reproduction of four species of vespertilionid bats in Paraguay. Univ CA Publ Zool. 107:1-41.
7696. Myers, P. 1981. Observations on *Pygoderma bilabiatum* (Wagner). Z Säugetierk. 46:146-151.
7697. Myers, P & Master, L. 1983. Reproduction by *Peromyscus maniculatus*: Size and compromise. J Mamm. 64:1-18.
7698. Myers, P & Master, LL. 1986. The interpretation of variation in litter size and weights of young: Three caveats. J Mamm. 67:572-575.
7699. Myers, P, Master, LL, & Garrett, RA. 1985. Ambient temperature and rainfall: An effect on sex ratio and litter size in deer mice. J Mamm. 66:289-298.
7699a. Myers, P & Wetzel, RM. 1979. New records of mammals from Paraguay. J Mamm. 60:638-641.
7700. Myers, P & Wetzel, RM. 1983. Systematics and zoogeography of the bats of the Chaco Boreal. MI Mus Zool Misc Publ. 165:1-59.
7700a. Myers, P, White, R, & Stallings, J. 1983. Additional records of bats from Paraguay. J Mamm. 64:143-145.
7701. Mykytowycz, R. 1959. Social behaviour of an experimental colony of wild rabbits, *Oryctolagus cuniculus* (L). CSIRO Wildl Res. 4:1-13.
7702. Mykytowycz, R & Fullagar, PJ. 1973. Effect of social environment on reproduction in the rabbit, *Oryctolagus cuniculus*. J Reprod Fert Suppl. 19:503-522.
7703. Myllymaki, A. 1977. Demographic mechanisms in the fluctuating populations of the field vole *Microtus agrestis*. Oikos. 29:468-493.
7704. Myrcha, A, Ryszkowski, L, & Walkowa, W. 1969. Bioenergetics of pregnancy and lactation in white mouse. Acta Theriol. 14:161-166.
7705. Mysterud, I. 1968. A third case of winter breeding in the wood lemming (*Myopus schisticolor*). Nytt Mag Zool. 16:22.
7706. Mysterud, I, Viitala, J, & Lahti, S. 1972. On winter breeding of the wood lemming (*Myopus schisticolor*). Norw J Zool. 20:91-92.
7707. Mystkowska, ET. 1975. Preimplantation development *in vivo* and *in vitro* in bank voles, *Clethrionomys glareolus*, treated with PMSG and HCG. J Reprod Fert. 42:287-292.
7708. Mystkowska, ET. 1980. The effect of litter size on body weight of young rats. Acta Theriol. 25:273-275.
7709. Naaktgeboren, C. 1961. Beobachtungen der Geburt des Frettchens (*Mustela furo* L). Bijdr Dierkd. 31:65-73.
7710. Naaktgeboren, C. 1965. Die Fortpflanzung des Rotfuchses, *Vulpes vulpes* (L), mit besonderer Berücksichtigung von Schwangerschaft und Geburt. Zool Anz. 175:235-263.
7711. Naaktgeboren, C. 1966. Notiz über einen trachtigen Uterus von *Okapia johnstoni* (Sclater 1901). Z Säugetierk. 31:171-176.
7712. Naaktgeboren, C. 1968. Some aspects of parturition in wild and domestic Canidae. Int Zoo Yearb. 8:8-13.
7713. Naaktgeboren, C, Slijper, EJ, & van Utrecht, WL. 1960. Research on the period of conception, duration of gestation and growth of the foetus in the fin whale, based on data from the International Whaling Statistics. Norsk Hvalfangst Tid. 49:113-119.
7714. Naaktgeboren, C & van Wagtendonk, A. 1966. Wahre Knoten in der Nabelschnuer nebst Bemerkungen über Plazentophagie bei Menschenaffen. Z Säugetierk. 31:376-382.
7714a. Nadeau, JH. 1985. Ontogeny. Sp Pub Am Soc Mamm. 8:254-285.
7715. Nadeau, JH, Lombardi, RT, & Tamarin, RH. 1981. Population structure and dispersal of *Peromyscus leucopus* on Muskeget Island. Can J Zool. 59:793-799.
7716. Nader, IA. 1968. Breeding records of the long-eared hedgehog *Hemiechinus auritus* (Gmelin). Mammalia. 32:528-529.
7717. Nader, IA. 1975. On the bats (Chiroptera) of the kingdom of Saudi Arabia. J Zool. 176:331-340.
7718. Nader, IA. 1982. New distributional records of bats from the Kingdom of Saudi Arabia (Mammalia: Chiroptera). J Zool. 198:69-82.
7719. Nader, IA & Kock, D. 1983. A new slit-faced bat from central Saudi Arabia. Senckenb Biol. 63:9-15.
7720. Nader, IA & Kock, D. 1983. Notes on some bats from the near east (Mammalia: Chiroptera). Int J Mamm Biol. 48:1-9.
7721. Nadler, RD. 1974. Periparturitional behavior of a primiparous lowland gorilla. Primates. 15:55-74.

7722. Nadler, RD. 1975. Cyclicity in tumescence of the perineal labia of female lowland gorillas. Anat Rec. 181:791-798.
7723. Nadler, RD. 1975. Second gorilla birth at the Yerkes Regional Primate Research Center. Int Zoo Yearb. 15:134-137.
7724. Nadler, RD. 1975. Sexual cyclicty in captive lowland gorilla. Science. 189:813-814.
7725. Nadler, RD. 1976. Sexual behavior of captive lowland gorillas. Arch Sex Behav. 5:487-502.
7726. Nadler, RD. 1977. Sexual behavior of captive orangutans. Arch Sex Behav. 6:457-475.
7727. Nadler, RD. 1980. Reproductive physiology and behaviour of gorillas. J Reprod Fert Suppl. 28:79-89.
7728. Nadler, RD. 1982. Reproductive behavior and endocrinology of organutans. The Orangutan Its Biology and Conservation. 232-248, DeBoer, LEM, ed, Dr W Junk, The Hague.
7729. Nadler, RD, Collins, DC, & Blank, MS. 1984. Luteinizing hormone and gonadal steroid levels during the menstrual cycle of orangutans. J Med Primatol. 13:305-314.
7730. Nadler, RD, Collins, DC, Miller, LC, et al. 1983. Menstrual cycle patterns of hormones and sexual behavior in gorillas. Horm Behav. 17:1-17.
7731. Nadler, RD, Graham, CE, Collins, DC, et al. 1979. Plasma gonadotropins, prolactin, gonadal steroids, and genital swelling during the menstrual cycle of lowland gorillas. Endocrinology. 105:290-296.
7732. Nadler, RD, Graham, CE, Collins, DC, et al. 1981. Portpartum amenorrhea and behavior of apes. Reproductive Biology of the Great Apes. 69-81, Graham, CE, ed, Academic Press, New York.
7733. Nadler, RD & Rosenblum, LA. 1972. Hormonal regulation of the "fatted" phenomenon in squirrel monkeys. Anat Rec. 173:181-188.
7734. Nadler, RN, Graham, CE, Gosselin, RE, et al. 1985. Serum levels of gonadotropins and gonadal steroids, including testosterone, during the menstrual cycle of the chimpanzee (*Pan troglodytes*). Am J Primatol. 9:273-284.
7735. Nagai, S & Tashiro, K. 1976. The mouflons of Tama Zoo. Animals & Zoos. 28(3):6-8.
7736. Nagai, Y & Ohno, S. 1977. Testis determining H-Y antigen in XO males of the mole-vole (*Ellobius lutescens*). Cell. 10:729-732.
7737. Nagasawa, H, Koshimizu, U, Watanabe, M, et al. 1989. Mammary gland growth and response to hormones in Mastomys compared with mice. Lab Ani Sci. 39:313-317.
7738. Nagel, A. 1977. Torpor in the European white-toothed shrews. Experientia. 33:1455-1456.
7739. Nagel, A. 1989. Development of temperature regulation in the common white-toothed shrew, *Crocidura russula*. Comp Biochem Physiol. 92A:409-413.
7740. Nagel, A. 1989. Paradoxical social effects on response to cold during ontogeny of the common white-toothed shrew, *Crocidura russula*. J Comp Physiol B. 159:301-304.
7741. Nagle, CA & Denari, JH. 1983. The Cebus monkey (*Cebus apella*). Reproduction in New World Primates. 39-67, Hearn, J, ed, MTP Press, Lancaster, England.
7742. Nagle, CA, Denari, JH, Quiroga, S, et al. 1979. The plasma pattern of ovarian steroids during the menstrual cycle in capuchin monkeys (*Cebus apella*). Biol Reprod. 21:979-983.
7743. Nagle, CA, Riarte, A, Quiroga, S, et al. 1980. Temporal relationship between follicular development, ovulation, and ovarian hormonal profile in the capuchin monkey (*Cebus apella*). Biol Reprod. 23:629-635.
7744. Nagorsen, D. 1985. *Kogia simus*. Mamm Species. 239:1-6.
7745. Nagorsen, D & Tamsitt, JR. 1981. Systematics of *Anoura cultrata, A brevirostrum*, and *A werckleae*. J Mamm. 62:82-100.
7746. Nagorsen, DW. 1987. *Marmota vancouverensis*. Mamm Species. 270:1-5.
7747. Nagy, F & Edmonds, RH. 1973. Morphology of the reproductive system of the armadillo: The spermatogonia. J Morphol. 140:307-320.
7748. Nagy, F & Edmonds, RH. 1973. Some observations on the fine structure of armadillo spermatozoa. J Reprod Fert. 34:551-553.
7749. Nagy, KA & Montgomery, GG. 1980. Field metabolic rate, water flux, and food consumption in three-toed sloths (*Bradypus variegatus*). J Mamm. 61:465-472.
7750. Nagy, KA, Seymour, RS, Lee, AK, et al. 1978. Energy and water budgets in free-living *Antechinus stuartii* (Marsupialia: Dasyuridae). J Mamm. 59:60-68.
7751. Nair, PV. 1989. Development of nonsocial behaviour in the Asiatic elephant. Ethology. 82:46-60.
7752. Naito, Y & Nishiwaki, M. 1972. The growth of two species of the harbour seal in the adjacent waters of Hokkaido. Sci Rep Whales Res Inst. 24:127-144.
7753. Naito, Y & Nishiwaki, M. 1973. Kurile harbour seal (*Phoca kurilensis*). IUCN Suppl Pap. 39:44-49.
7754. Nakai, K, Nimura, H, Tamura, M, et al. 1960. Reproduction and postnatal development of the colony bred *Meriones unguiculatus kurauchii* Mori. Exp Anim Tokyo. 9:157-159.
7755. Nakai, Y, Plant, TM, & Hess, DL. 1978. On the sites of the negative and positive feedback actions of estradiol in the control of gonadotropin secretion in the rhesus monkey. Endocrinology. 102:1008-1014.
7756. Nakakura, K, Czekala, NM, Lasley, BL, et al. 1982. Fetal-maternal gradients of steroid hormones in the nine-banded armadillo (*Dasypus novemcinctus*). J Reprod Fert. 66:635-643.
7757. Nakakura, KC & Lasley, BL. 1985. Fetal endocrinology of the common long-nosed armdillo *Dasypus novemcinctus*. The Evolution and Ecology of Armadillos, Sloths, and Vermilinguas. 247-253, Montgomery, GG, ed, Smithsonian Inst Press, Washington, DC.
7758. Nakamichi, M. 1983. Development of infant twin Japanese monkeys (*Macaca fuscata*) in a free-ranging group. Primates. 24:576-583.
7759. Nakano, O. 1928. Über die Verteilung des Glykogens beiden Zyklischen Veränderungen in den Geschlechtsorganen der Fledermaus und über die Nahrungsaufnahme der Spermien in den weiblichen Geschlechtswegen. Folia Anat Japan. 6:777-828.
7760. Nakata, K. 1984. Factors affecting litter size in the red-backed vole, *Clethrionomys rufocanus bedfordiae*, with special emphasis on population phase. Res Pop Ecol. 26:221-234.
7761. Nakata, K. 1986. Litter size of *Apodemus argenteus* in relation to the population cycle. J Mamm Soc Japan. 11:117-125.
7762. Nakatsu, A. 1977. Some observations on the Japanese field vole, *Microtus montebelli* (Milne-Edwards) in captivity II reproduction and growth of voles according to size of litter. Bull Gov For Exp Stn (Tokyo). 297:35-42.
7763. Nakazato, R, Nakayama, T, & Nakagawa, S. 1971. Hand-rearing agile wallabies *Protemnodon agilis* at Ueno Zoo, Tokyo. Int Zoo Yearb. 11:13-16.
7764. Nansen, F. 1961. Unter Robben und Eisbären. EA Brockhaus Verlag, Leipzig. [Cited by Dittrich, 1961]
7765. Napier, PH. 1976. Catalogue of Primates in the British Museum (Natural History) Part 1: Families Callitrichidae and Cebidae. Br Mus Nat Hist, London.

7766. Napier, PH. 1981. Catalogue of Primates in the British Museum (Natural History) and Elsewhere in the British Isles Part II: Family Cercopithecidae, subfamily Cercopithecinae. Br Mus Nat Hist, London.
7767. Nash, DJ & Seaman, RM. 1977. *Sciurus aberti*. Mamm Species. 80:1-5.
7768. Nash, LT. 1974. Parturition in a feral baboon (*Papio anubis*). Primates. 15:279-286.
7769. Nash, LT. 1983. Reproductive patterns in galagos (*Galago zanzibaricus* and *Galago garnettii*) in relation to climatic variability. Am J Primatol. 5:181-196.
7770. Nash, LT & Harcourt, CS. 1986. Social organization of galagos in Kenyan coastal forests: II *Galago garnettii*. Am J Primatol. 10:357-369.
7771. Nasir, UD, & Jeanloz, RD. 1983. Bonnet monkey cervical mucus glycoproteins: Study of the minor glycoprotein components of periovulatory phase mucus. Biol Reprod. 28:1189-1199.
7772. Nasledova, NI, Plotnikova, NS, & Ivanova, LN. 1984. The breeding of the water vole (*Arvicola terrestris*) under the controlled conditions. Zool Zhurn. 63:745-748.
7773. Nasu, K & Masaki, Y. 1970. Some biological parameters for stock assessment of the Antarctic sei-whale. Sci Rep Whales Res Inst. 22:63-74.
7774. Natal Park Board. Unpublished records. [Cited by Mentis, 1972]
7775. Nathanielz, PW. 1978. Endocrine mechanisms of parturition. Ann Rev Physiol. 40:411-445.
7776. Natoli, E & de Vito, E. 1991. Agonistic behaviour, dominance rank and copulatory success in a large multi-male feral cat, *Felis catus* L, colony in central Rome. An Beh. 42:227-241.
7777. Natuschke, G. 1960. Ergebnisse der Fledermausberingung und biologische Beobgachtungen an Fledermäusen in der Oberlausitz. Bonn Zool Bertr Sp Issue. 11:77-98.
7778. Naumov, NP & Lobachev, VS. 1975. Ecology of the desert rodents of the USSR (jerboas and gerbils). Rodents in Desert Environments. 465-598, Prakash, I & Ghosh, PK, eds, Dr W Junk, The Hague.
7779. Nauroz, MK. 1983. Die Waldmaus, *Apodemus sylvaticus* (Rodentia, Muridae) auf der Insel Mellum. Säugetierk Mitt. 31:141-159.
7780. Navarro, D & Wilson, DE. 1982. *Vampyrum spectrum*. Mamm Species. 184:1-4.
7781. Nawa, A. 1968. Observations on the breeding habits and the growth of *Prionailurus bengalensis manchurica*, in captivity. J Mamm Soc Japan. 4:12-19.
7782. Nawar, MM & Hafez, ESE. 1972. The reproductive cycle of the crab-eating macaque (*Macaca fascicularis*). Primates. 13:43-56.
7783. Nawar, SMA, Abul-Fadle, WS, & Mahmoud, SA. 1978. Studies on the ovarian activity of the dromedary (*Camelus dromedarius*). Z Mikr Anat Forsch. 92:385-408.
7784. Nawito, MF. 1967. Some reproductive aspects in the female camel. Thesis Faculty Veterinary Medicine. Warsaw Agric Univ, Warsaw. [Cited by Nawar et al, 1972]
7785. Nawito, MF, Shalash, MR, Hopper, R, et al. 1967. Reproduction in female camel. Bull Anim Sci Res Inst Cairo. 2:1-82.
7786. Nazarenko, YI. 1975. Sexual maturation, reproductive rate, and missed pregnancy, in female harp seals. Rapp P-V Réun Cons Int Expl Mer. 169:413-415.
7787. Neal, BJ. 1959. A contribution on the life history of the collared peccary in Arizona. Am Mid Nat. 61:177-190.
7788. Neal, BJ. 1965. Growth and development of the round-tailed and Harris antelope ground squirrels. Am Mid Nat. 73:479-489.
7789. Neal, BJ. 1965. Reproductive habits of round-tailed and Harris antelope ground squirrels. J Mamm. 46:200-206.
7790. Neal, BR. 1967. The ecology of small rodents in the grassland community of the Queen Elizabeth National Park, Uganda. Ph D Diss. Univ Southampton.
7791. Neal, BR. 1977. Reproduction of the multimammate rat, *Praomys* (*Mastomys*) *natalensis* (Smith), in Uganda. Z Säugetierk. 42:221-231.
7792. Neal, BR. 1977. Reproduction of the punctuated grass-mouse, *Lemniscomys striatus* in the Ruwenzori National Park, Uganda (Rodentia: Muridae). Zool Afr. 12:419-428.
7793. Neal, BR. 1981. Reproductive biology of the unstriped grass rat, *Arvicanthis*, in East Africa. Z Säugetierk. 46:174-189.
7794. Neal, BR. 1982. Reproductive biology of three species of gerbils (genus *Tatera*) in East Africa. Z Säugetierk. 47:287-296.
7795. Neal, BR. 1983. The breeding pattern of two species of spiny mice, *Acomys percivali* and *A wilsoni* (Muridae: Rodentia), in central Kenya. Mammalia. 47:311-321.
7796. Neal, BR. 1984. Relationship between feeding habits, climate and reproduction of small mammals in Meru National Park, Kenya. Afr J Ecol. 22:195-205.
7797. Neal, BR. 1986. Reproductive characteristics of African small mammals. Cimbebasia. 8:113-127.
7797a. Neal, BR. 1990. Observations on the early post-natal growth and development of *Tatera leucogaster*, *Aethomys chrysophilus* and *A namaquensis* from Zimbabwe, with a review of the pre- and post-natal growth and development of African muroid rodents. Mammalia. 54:245-270.
7797b. Neal, BR. 1991. Seasonal changes in reproduction and diet of the bushveld gerbil, *Tatera leucogaster* (Muridae: Rodentia), in Zimbabwe. Z Säugetierkd. 56:101-111.
7797c. Neal, BR & Alibhai, SK. 1991. Reproductive response of *Tatera leucogaster* (Rodentia) to supplemental food and 6-methoxybenzoxazolinone in Zimbabwe. J Zool. 223:469-473.
7798. Neal, J & Murphy, BD. 1977. Response of immature, mature nonbreeding and mature breeding ferret testis to exogenous LH stimulation. Biol Reprod. 16:244-248.
7799. Neal, J, Murphy, BD, Moger, WH, et al. 1977. Reproduction in the male ferret: Gonadal activity during the annual cycle; recrudescence and maturation. Biol Reprod. 17:380-385.
7800. Neas, JF & Hoffmann, RS. 1987. *Budorcas taxicolor*. Mamm Species. 277:1-7.
7801. Neaves, WB. 1973. Changes in testicular Leydig cells and plasma testosterone levels among seasonally breeding rock hyrax. Biol Reprod. 8:451-466.
7802. Neaves, WB, Griffin, JE, & Wilson, JD. 1980. Sexual dimorphism of the phallus in spotted hyaena (*Crocuta crocuta*). J Reprod Fert. 59:509-513.
7803. Nee, JA. 1969. Reproduction in a population of yellow-bellied marmots (*Marmota flaviventris*). J Mamm. 50:756-765.
7804. Negayama, K, Negayama, T, & Kondo, K. 1986. Behavior of Japanese monkey (*Macaca fuscata*) mothers and neonates at parturition. Int J Primatol. 7:365-378.
7805. Negus, NC. 1950. Breeding of three-year-old females in the Jackson Hole wildlife park buffalo herd. J Mamm. 31:463.
7806. Negus, NC. 1959. Breeding of subadult cottontail rabbits in Ohio. J Wildl Manage. 23:451-452.

7807. Negus, NC & Berger, PJ. 1977. Experimental triggering of reproduction in a natural population of *Microtus montanus*. Science. 196:1230-1231.
7808. Negus, NC, Berger, PJ, & Forslund, LG. 1977. Reproductive strategy of *Microtus montanus*. J Mamm. 58:347-353.
7809. Negus, NC & Findley, JS. 1959. Mammals of Jackson Hole, Wyoming. J Mamm. 40:371-381.
7810. Negus, NC, Gould, E, & Chipman, RK. 1961. Ecology of the rice rat, *Oryzomys palustris* (Harlan), on Breton Island, Gulf of Mexico, with a critique of the social stress theory. Tulane Stud Zool. 8:93-123.
7811. Negus, NC & Pinter, AJ. 1965. Litter sizes of *Microtus montanus* in the laboratory. J Mamm. 46:434-437.
7812. Negus, NC & Pinter, AJ. 1966. Reproductive responses of *Microtus montanus* to plants and plant extracts in the diet. J Mamm. 47:596-601.
7813. Nehring, A. 1893. Die Trächtigkeitsdauer des Dachses. Zool Garten. 34:107-110.
7814. Nehring, A. 1900. Die Zahl der Mammae by Cricetus, *Cricetulus* und *Mesocricetus*. Zool Anz. 23:572-573.
7815. Nehring, A. 1901. Die Zahl der Zitzen und der Embryonen bei *Mesocricetus* und *Cricetus*. Zool Anz. 24:130-131.
7816. Neill, JD, Johansson, EDB, & Knobil, E. 1967. Levels of progesterone in peripheral plasma during the menstrual cycle of the rhesus monkey. Endocrinology. 81:1161-1164.
7817. Neill, JD, Johansson, EDB, & Knobil, E. 1969. Patterns of circulating progesterone concentrations during the fertile menstrual cycle and the remainder of gestation in the rhesus monkey. Endocrinology. 84:45-48.
7818. Neill, JD, Johansson, EDB, & Knobil, E. 1969. Failure of hysterectomy to influence the normal pattern of cyclic progesterone secretion in the rhesus monkey. Endocrinology. 84:464-465.
7819. Neill, JD & Knobil, E. 1972. On the nature of the initial luteotropic stimulus of pregnancy in the rhesus monkey. Endocrinology. 90:34-38.
7820. Neill, JD, Patton, JM, Dailey, RA, et al. 1977. Luteinizing hormone releasing hormone (LHRH) in pituitary stalk blood of rhesus monkeys: Relationship to level of LH release. Endocrinology. 101:430-434.
7821. Nekipelow, V. 1940. Some new data on the Mongolian jerboa. Zool Zhurn. 19:313-320.
7822. Nel, JAJ. 1975. Aspects of the social ethology of some Kalahari rodents. Z Tierpsychol. 37:322-331.
7823. Nel, JAJ, Rautenbach, IL, Els, DA, et al. 1984. The rodents and other small mammals of the Kalahari Gemsbok National Park. Koedoe. 27(Suppl):195-220.
7824. Nel, JAJ & Stutterheim, CJ. 1973. Notes on early post-natal development of the Namaqua gerbil *Desmodillus auricularis*. Koedoe. 16:117-125.
7825. Nellis, CH. 1968. Productivity of mule deer on the National Bison range, Montana. J Wildl Manage. 32:344-349.
7826. Nellis, CH. 1969. Productivity of Richardson's ground squirrels near Rochester, Alberta. Can Field-Nat. 83:246-250.
7827. Nellis, CH, Thiessen, JL, & Prentice, CA. 1976. Pregnant fawn and quintuplet mule deer. J Wildl Manage. 40:795-796.
7827a. Nellis, DW. 1971. Additions to the natural history of *Brachyphylla* (Chiroptera). Carib J Sci. 11:91.
7828. Nellis, DW. 1989. *Herpestes auropunctatus*. Mamm Species. 342:1-6.
7828a. Nellis, DW & Ehle, CP. 1977. Observations on the behavior of *Brachyphylla cavernarum* (Chiroptera) in Virgin Islands. Mammalia. 41:403-409.
7829. Nellis, DW & Everard, COR. 1983. The biology of the mongoose in the Caribbean. Stud Fauna Curacao Other Carib I. 195:3-162.
7830. Nelsen, OE. 1944. Possible control of the lateral vaginal canal in the opossum during reproduction. Anat Rec. 89:563-564.
7831. Nelsen, OE & White, EL. 1941. A method for inducing ovulation in the anestrous opossum (*Didelphis virginiana*). Anat Rec. 81:529-535.
7832. Nelson, BB & Chapman, JA. 1982. Age determination and population characteristics of red foxes from Maryland. Z Saugetierk. 47:296-311.
7833. Nelson, JE. 1964. Notes on *Syconycteris australis* Peters, 1867 (Megachiroptera). Mammalia. 28:429-432.
7833a. Nelson, JE. 1965. Movements of Australian flying foxes (Pteropodidae: Megachiroptera). Aust J Zool. 13:53-73.
7834. Nelson, JE & Goldstone, A. 1986. Reproduction in *Peradorcas concinna* (Marsupialia: Macropodidae). Aust Wildl Res. 13:501-505.
7835. Nelson, JE & Hamilton-Smith, E. 1982. Some observations on *Notopteris macdonaldi* (Chiroptera: Pteropodidae). Aust Mamm. 5:247-252.
7836. Nelson, JE & Smith, G. 1971. Notes on growth rates in native cats of the family Dasyuridae. Int Zoo Yearb. 11:38-41.
7837. Nelson, ME & Mech, LD. 1981. Deer social organization and wolf predation in northeastern Minnesota. Wildl Monogr. 77:1-53.
7838. Nelson, ML & Inao, J. 1982. Seasonal changes in the pituitary gland of the feral Hawaiian mongoose (*Herpestes auropunctatus*). J Morphol. 174:133-140.
7839. Nelson, NS & Cooper, J. 1975. The growing conceptus of the domestic cat. Growth. 39:435-451.
7840. Nelson, RJ. 1985. Photoperiod influences reproduction in the prairie vole (*Microtus ochrogaster*). Biol Reprod. 33:596-602.
7841. Nelson, RJ. 1985. Photoperiodic regulation of reproductive development in male prairie voles: Influence of laboratory breeding. Biol Reprod. 33:418-422.
7842. Nelson, RJ, Dark, J, & Zucker, I. 1983. Influence of photoperiod, nutrition and water availability on reproduction of male California voles (*Microtus californicus*). J Reprod Fert. 69:473-477.
7843. Nelson, RJ & Desjardins, C. 1987. Water availability affects reproduction in deer mice. Biol Reprod. 37:257-260.
7844. Nelson, RJ, Frank, D, Smale, L, et al. 1989. Photoperiod and temperature affect reproductive and nonreproductive functions in male prairie voles (*Microtus ochrogaster*). Biol Reprod. 40:481-485.
7845. Nelson, TA & Woolf, A. 1985. Birth size and growth of white-tailed deer fawns in southern Illinois. J Wildl Manage. 49:374-377.
7846. Nerini, MK, Braham, HW, Marquette, WM, et al. 1984. Life history of the bowhead whale, *Balaena mysticetus* (Mammalia: Cetacea). J Zool. 204:443-468.
7847. Nero, RW. 1958. Hoary bat parturition date. Blue Jay. 16:130-131.
7848. Nero, RW & Wrigley, RE. 1977. Status and habits of the cougar in Manitoba. Can Field-Nat. 91:28-40.
7849. Nerquaye-Tetteh, JO & Clarke, JR. 1987. Hormonal requirements and the influence of age on the decidual cell reaction in the bank vole (*Clethrionomys glareolus*). J Reprod Fert. 79:575-579.
7850. Nerquaye-Tetteh, JO & Clarke, JR. 1990. Avoidance of used endometrial sites by blastocysts of the bank vole, *Clethrionomys glareolus*. J Reprod Fert. 89:729-734.

7851. Nesturch, MF. 1946. Beobachtungen über den Menstruationzyklus bei Cercopethicinen. Proc Moscow Zool Park. 3:83-94.
7852. Nette, T, Burles, D, & Hoefs, M. 1984. Observations of golden eagle, *Aquila chrysaetos*, predation on dall sheep, *Ovis dalli dalli*, lambs. Can Field-Nat. 98:252-254.
7853. Neugebauer, W. 1967. Breeding the southern elephant seal *Mirounga leonina* at Stuttgart Zoo. Int Zoo Yearb. 7:152-154.
7854. Neumann, AH. 1899. Burchell's and Grevy's zebras (*Equus burchelli* and *Equus grevyi*) in British east Africa. Great and Small Game of Africa. 84-89, Bryden, HA, ed, Rowland Ward, London.
7855. Neumann, AH. 1899. Thomson's gazelle (*Gazella thomsoni*). Great and Small Game of Africa. 352-354, Bryden, HA, ed, Rowland Ward, London.
7856. Neumann, AH. 1899. Grant's gazelle (*Gazella granti typica*) in east Africa. Great and Small Game of Africa. 355-360, Bryden, HA, ed, Rowland Ward, London.
7857. Neumann, AH. 1899. The inyala or nyala (*Tragelaphus angasi*) in Zululand. Great and Small Game of Africa. 460-467, Bryden, HA, ed, Rowland Ward, London.
7858. Neuville, H. 1935. De l'organe femelle de l'Hyène tachétée (*Hyaena crocuta* Erxt). Arch Mus Nat Hist Nat (Paris). 12:225-229.
7859. Neuweiler, G. 1969. Verhaltungsbeobachtungen an einer indischen Flughundkolonie (*Pteropus g giganteus* Brünn). Z Tierpsychol. 26:166-199.
7860. Nevo, E. 1961. Observations on Israeli populations of the mole rat *Spalax e ehrengergi* Nehring 1898. Mammalia. 25:129-143.
7861. Nevo, E. 1969. Mole rat *Spalax ehrenbergi*: Mating behavior and evolutionary significance. Science. 163:484-486.
7862. Nevo, E & Amir, E. 1961. Biological observations on the forest dormouse, (*Dryomys nitedula*) Pallas, in Israel (Rodentia: Muscardinidae). Bull Res Council Israel. 9B:200-201.
7863. Nevo, E, Guttman, R, Haber, M, et al. 1982. Activity patterns of evolving mole rats. J Mamm. 63:453-463.
7864. Newby, TC. 1973. Observations on the breeding behavior of the harbor seal in the state of Washington. J Mamm. 54:540-543.
7865. Newell-Morris, L, Orsini, J, & Seed, J. 1980. Age determination of the fetal and neonatal pigtailed macaque (*Macaca nemestrina*) from somatometric measurements. Lab Anim Sci. 30:180-187.
7866. Newman, HH. 1912. The ovum of the nine-banded armadillo: Growth of the ovocytes, maturation and fertilization. Biol Bull. 23:100-140.
7867. Newman, HH. 1913. Parthenogenetic cleavage of the armadillo ovum. Biol Bull. 25:52-78.
7868. Newman, HH & Patterson, JT. 1909. A case of normal identical quadruplets in the armadillo and its bearing on the problem of identical twins and sex determination. Biol Bull. 17:181-187.
7869. Newman, JR & Rudd, RL. 1978. Minimum and maximum metabolic rates of *Sorex sinuosus*. Acta Theriol. 23:371-380.
7870. Newman, LM & Hendrickx, AG. 1954. Fetal development in the normal thick-tailed bushbaby (*Galago crassicaudatus panganiensis*). Am J Primatol. 6:337-355.
7871. Newsome, AE. 1973. Cellular degeneration in the testis of red kangaroos during hot weather and drought in central Australia. J Reprod Fert Suppl. 19:191-201.
7872. Newson, J. 1964. Reproduction and prenatal mortality of snowshoe hares on Manitoulin Island, Ontario. Can J Zool. 42:987-1005.
7873. Newson, R. 1963. Differences in numbers, reproduction and survival between two neighboring populations of bank voles (*Clethrionomys glareolus*). Ecology. 44:110-120.
7874. Newson, R & de Vos, A. 1964. Population structure and body weights of snowshoe hares on Manitoulin Island, Ontario. Can J Zool. 42:975-986.
7875. Newson, RM. 1966. Reproduction in feral coypus (*Myocastor coypus*). Symp Zool Soc London. 15:323-334.
7876. Neyman, PF. 1977. Aspects of the ecology and social organization of free-ranging cotton-top tamarins (*Saguinus oedipus*) and the conservation status of the species. The Biology and Conservation of the Callitrichidae. 39-71, Kleiman, DG, ed, Smithsonian Inst Press, Washington, DC.
7877. Nicander, L & Glover, TD. 1973. Regional histology and fine structure of the epididymal duct in the golden hamster (*Mesocricetus auratus*). J Anat. 114:347-364.
7877a. Nice, MM, Nice, C, & Ewers, D. 1956. Comparison of behavior development in snowshoe hares and red squirrels. J Mamm. 37:64-74.
7878. Nichol, AA. 1938. Experimental feeding of deer. Univ AZ Coll Agr Exp Sta Tech Bull. 75:3-39.
7879. Nicholas, KR. 1988. Asynchronous dual lactation in a marsupial the tammar wallaby *Macropus eugenii*. Biochem Biophys Res Comm. 154:529-536.
7880. Nicholas, KR, Loughnan, M, Messer, M, et al. 1989. Isolation, partial sequence and asynchronous appearance during lactation of lysozyme and α-lactalbumin in the milk of a marsupial, the common ringtail possum (*Pseudocheirus peregrinus*). Comp Biochem Physiol. 94B:775-778.
7881. Nicholls, L. 1939. Period of gestation in *Loris*. Nature. 143:246.
7881a. Nichols, DG & Nichols, JT. 1934. Notes on Long Island, New York, bats. J Mamm. 15:156.
7882. Nichols, JD & Conley, W. 1982. Active season dynamics of a population of *Zapus hudsonius* in Michigan. J Mamm. 63:422-430.
7883. Nichols, L. 1978. Dall sheep reproduction. J Wildl Manage. 42:570-580.
7884. Nichols, L. 1978. Dall's sheep. Big Game of North America. 173-189, Schmidt, JL & Gilbert, DL, eds, Stackpole Books, Harrisburg, PA.
7885. Nicholson, AJ. 1941. The homes and social habits of the wood-mouse (*Peromyscus leucopus noveboracensis*) in southern Michigan. Am Mid Nat. 25:196-223.
7886. Nicholson, AJ & Warner, DW. 1953. The rodents of New Caledonia. J Mamm. 34:168-179.
7887. Nicholson, WS, Hill, EP, & Briggs, D. 1984. Denning, pup-rearing, and dispersal in the gray fox in east-central Alabama. J Wildl Manage. 48:33-37.
7888. Nicht, M. 1965. Ein Beitrag zur Kleinsäugerfauna der Hohen Tatra. Staat Mus (Dresden) Tierk Zool Abh. 28:57-63.
7889. Nicol, SC. 1982. Physiology of the Tasmanian devil *Sarcophilus harrisii* (Dasyuridae, Marsupialia). Carnivorous Marsupials. 1:273-278, Archer, M, ed, R Zool Soc NSW, Mossman, NSW.
7890. Nicoll, ME. 1983. Mechanisms and consequences of large litter production in *Tenrec ecaudatus* (Insectivora: Tenrecidae). Ann Mus R Afr Centr. 237:219-226.
7890a. Nicoll, ME. 1983. The biology of the giant otter shrew *Potamogale velox*. Res Rep Nat Geogr Soc. 21:331-337.
7891. Nicoll, ME. 1985. Responses to Seychelles tropical rain forest by a litter-foraging mammalian insectivore, *Tenrec ecaudatus*, native to Madagascar. J Anim Ecol. 54:71-88.

7892. Nicoll, ME & Racey, PA. 1985. Follicular development, ovulation, fertilization and fetal development in tenrecs (*Tenrec ecaudatus*). J Reprod Fert. 74:47-55.
7893. Nicoll, ME & Suttie, JM. 1982. The sheath-tailed bat *Coleura seychellensis* (Chiroptera: Emballonuridae) in the Seychelles Islands. J Zool. 197:421-426.
7894. Nielsen, HC & Torday, JS. 1983. Anatomy of fetal rabbit gonads and the sexing of fetal rabbits. Lab Anim. 17:148-150.
7895. Nielsen-Bondrup, S. 1987. Demography of *Clethrionomys gapperi* in different habitats. Can J Zool. 65:277-283.
7896. Nielson, PE. 1940. The fetal membranes of the kanagroo rat, *Dipodomys*, with a consideration of the phylogeny of the Geomyoidea. Anat Rec. 77:103-127.
7897. Nieminen, M. 1980. Nutritional and seasonal effects on the haematology and blood chemistry in reindeer (*Rangifer tarandus tarandus* L). Comp Biochem Physiol. 66A:399-413.
7898. Niethammer, G. 1950. Zur Jungenpflege und Orientierung der Hausspitzmaus (*Crocidura russula* Herm). Bonn Zool Beitr. 1:117-125.
7899. Niethammer, J & Winking, H. 1971. Die spanische Feldmaus (*Microtus arvalils asturianus* Miller, 1908). Bonn Zool Beitr. 22:220-235.
7900. Niethammer, J. 1956. Insektenfresser und Nager Spaniens. Bonn Zool Beitr. 7:249-295.
7901. Niethammer, J. 1960. Über die Säugetiere der Niedern Tauern. Mitt Zool Mus Berlin. 36:407-443.
7902. Niethammer, J. 1963. Notizen über den Maulwurf. Säugetierk Mitt. 11:79-80.
7903. Niethammer, J. 1972. Die Zahl der Mammae bei *Pitymys* und bei den Microtinen. Bonn Zool Beitr. 23:49-60.
7904. Niethammer, J. 1972. Zur Taxonomie und Biologie der Kurzohrmäusen. Bonn Zool Beitr. 23:290-309.
7905. Niethammer, J. 1973. Wurfgröszen griechischer Wühlmäuse (Microtinae). Bonn Zool Beitr. 24:361-365.
7906. Niethammer, J. 1976. Zur Fortpflanzung von Kleinsäugern in südwestlichen Marokko im Vorfrühling 1975. Säugetierk Mitt. 24:218-224.
7907. Niethammer, J. 1984. Die Zahl der Zitsen der kleinen Bandikuratte, *Bandicota bengalensis* Gray et Hardwicke, 1883. Z Säugetierk. 49:377-378.
7908. Nieuwenhuijsen, K, Bonke-Jansen, M, de Neef, KJ, et al. 1987. Physiological aspects of puberty in group-living stumptail monkeys (*Macaca arctoides*). Physiological & Behavior. 41:37-45.
7909. Nieuwenhuijsen, K, Lammers, AJJC, de Neef, KJ, et al. 1985. Reproduction and social rank in female stumptail macaques (*Macaca arctoides*). Int J Primatol. 6:77-99.
7910. Nievergelt, B. 1966. Der Alpensteinbock (*Capra ibex* L) in Seinem Lebensraum. Paul Parey, Hamburg.
7911. Nievergelt, B. 1966. Unterschiede in der Setzzeit beim Alpensteinbock (*Capra ibex* L). Rev Suisse Zool. 73:446-454.
7912. Nievergelt, B. 1974. A comparison of rutting behaviour and grouping in the Ethiopian and Alpine ibex. The Behaviour of Ungulates and its Relation to Management. 324-340, Geist, V & Walther, F, eds, IUCN, Morges, Switzerland.
7913. Niggol, K. 1960. Early mortality in fur seals according to sex. J Wildl Manage. 24:428-429.
7914. Niggol, K & Fiscus, CH, Jr. 1960. Northern fur seal twins. Mammalia. 24:457-458.
7915. Nigi, H. 1975. Menstrual cycle and some other related aspects of Japanese monkeys (*Macaca fuscata*). Primates. 16:207-216.
7916. Nigi, H. 1976. Some aspects related to conception of the Japanese monkey (*Macaca fuscata*). Primates. 17:81-88.
7917. Nigi, H, Tiba, T, Yamamoto, S, et al. 1980. Sexual maturation and seasonal changes in reproductive phenomena of male Japanese monkeys (*Macaca fuscata*) at Takasukiyama. Primates. 21:230-240.
7918. Nigi, H & Torii, R. 1983. Periovulatory time course of plasma estradiol and progesterone in the Japanese monkey (*Macaca fuscata*). Primates. 24:410-418.
7919. Nikiforov, LP. 1956. On the breeding of *Microtus* (*Stenocranius*) *gregalis* Pall in Kurgan forest-steppe in winter. Zool Zhurn. 35:464-466.
7920. Nikodémusz, E, Imre, R, Heltay, I, et al. 1985. The seasonal cycle of spermatogenesis in the roe deer, *Capreolus capreolus* L. Zool Jb Anat. 113:185-191.
7921. Nikolaevskii, LD. 1961. General outline of the anatomy of reindeer. Reindeer Husbandry, 2nd revised ed. 5-56, Zhigunov, PS, ed, US Dept Commerce, Springfield, VA.
7922. Nills, A. 1907. Die Fortpflanzung, des groszen Ameisenbären (*Myrmecophaga jubata*) in Nills Zoologischem Garten Stuttgart. Zool Beob. 48:145-151.
7923. Niort, PL. 1950. Une femelle en gestation de *Suncus etruscus*. Mammalia. 14:99-102.
7924. Nishida, T. 1979. The social structure of chimpanzees of the Mahale Mountains. The Great Apes. 73-121, Hamburg, DA & McCown, ER, ed, Benjamin/Cummings, Menlo Park, CA.
7925. Nishikata, S. 1978. Ecological studies on the population of *Apodemus argenteus argenteus* in Mt Kiyosumi, Chiba Pref I A life cycle and fluctuations of population size. J Mamm Soc Japan. 7:240-253.
7926. Nishiwaki, M. 1953. Hermaphroditism in a dolphin (*Prodelphinus caeruleo-albus*). Sci Rep Whales Res Inst. 8:215-218.
7927. Nishiwaki, M. 1955. On the sexual maturity of the Antarctic male sperm whale (*Physeter catodon* L). Sci Rep Whales Res Inst. 10:143-149.
7928. Nishiwaki, M. 1959. Humpback whales in Ryukyuan waters. Sci Rep Whales Res Inst. 14:49-87.
7929. Nishiwaki, M & Handa, C. 1958. Killer whales caught in the coastal waters off Japan for recent 10 years. Sci Rep Whales Res Inst. 13:85-96.
7930. Nishiwaki, M & Hayashi, K. 1950. Biological survey of fin and blue whales taken in the Antarctic season 1947-1948 by the Japanese fleet. Sci Rep Whales Res Inst. 3:132-190.
7931. Nishiwaki, M & Hibiya, T. 1951. On the sexual maturity of the sperm whale (*Physeter catodon*) found in the adjacent waters of Japan (I). Sci Rep Whales Res Inst. 6:153-165.
7932. Nishiwaki, M & Hibiya, T. 1952. On the sexual maturity of the sperm whale (*Physeter catodon*) found in the adjacent waters of Japan (II). Sci Rep Whales Res Inst. 7:121-124.
7933. Nishiwaki, M, Hibiya, T, & Kimura, S. 1954. On the sexual maturity of the sei whale of the Bonin Waters. Sci Rep Whales Res Inst. 9:165-177.
7934. Nishiwaki, M, Hibiya, T, & Kimura, S. 1956. On the sexual maturity of the sperm whale (*Physeter catodon*) found in the north Pacific. Sci Rep Whales Res Inst. 11:39-46.
7935. Nishiwaki, M & Kamiya, T. 1958. A beaked whale *Mesoplodon* stranded at Oiso beach, Japan. Sci Rep Whales Res Inst. 13:53-83.
7936. Nishiwaki, M & Marsh, H. 1985. Dugong *Dugong dugon* (Müller, 1776). 3:1-31. Handbook of Marine Mammals, Ridgway, SH & Harrison, RJ, eds. Academic Press, New York.
7937. Nishiwaki, M & Oye, T. 1951. Biological investigation on blue whales (*Balaenoptera musculus*) and fin whales (*Balaenoptera physalus*) caught by the Japanese Antarctic whaling fleets. Sci Rep Whales Res Inst. 5:91-167.

7938. Nissen, HW & Yerkes, RM. 1943. Reproduction in the chimpanzee: Report on forty-nine births. Anat Rec. 86:567-578.

7939. Niswender, GD & Spies, HG. 1973. Serum levels of luteinizing hormone, follicle-stimulating hormone and progesterone throughout the menstrual cycle of rhesus monkeys. J Clin Endocr Metab. 37:326-328.

7940. Nitikman, LZ. 1985. *Sciurus granatensis*. Mamm Species. 246:1-8.

7941. Nixon, CM. 1965. White-tailed deer growth and productivity in eastern Ohio. Game Res OH. 3:123-136.

7942. Nixon, CM. 1971. Productivity of white-tailed deer in Ohio. OH J Sci. 71:217-225.

7943. Nixon, CM, Hansen, LP, & Havera, SP. 1986. Demographic characteristics of an unexploited population of fox squirrels (*Sciurus niger*). Can J Zool. 64:512-521.

7943a. Nixon, CM, Hansen, LP, & Havera, SP. 1991. Growth patterns of fox squirrels in east-central Illinois. Am Mid Nat. 125:168-172.

7944. Nixon, CM & McClain, MW. 1975. Breeding seasons and fecundity of female gray squirrels in Ohio. J Wildl Manage. 39:426-438.

7945. Noback, CR. 1939. The changes in the vaginal smears and associated cyclic phenomena in the lowland gorilla (*Gorilla gorilla*). Anat Rec. 73:209-225.

7946. Noback, CR. 1946. Placentation and amniogenesis in the amnion of a baboon (*Papio papio*). Anat Rec. 94:553-567.

7947. Nogge, G. 1982. Jahresbericht 1981 der Aktiengesellschaft Zoologischer Garten Köln. Z Kölner Zoo. 25:3-25.

7948. Nogueira, JC, Campos, PA, & Ribeiro, MG. 1984. Histology, glycogen and mucosubstances histochemistry of the bulbo-urethral glands of the *Philander opossum* (Linnaeus, 1758). Anat Anz. 156:321-328.

7948a. Nogueira, JC, Ribeiro, MG, & Campos, PA. 1985. Histology and carbohydrate histochemistry of the prostate gland of the Brazilian four-eyed opossum (*Philander opossum* Linnaeus, 1758). Anat Anz. 159:241-252.

7949. Nogueira-Persona, L & Bustos-Obregon, E. 1983. The cycle of the seminiferous epithelium in the armadillo *Euphractus sexcinstus flavimanus*. Anat Rec. 205:143A.

7950. Nolf, P. 1895/96. Etude des modifications de la muquese utérine pendant la gestation chez le Murin *Vespertilio murinus*. Arch Biol. 14:561-693.

7951. Noll, UG. 1979. Postnatal growth and development of thermogenesis in *Rousettus aegyptiacus*. Comp Biochem Physiol. 63A:89-93.

7952. Noll-Banholzer, UG. 1979. Body temperature, oxygen consumption, evaporative water loss and heart rate in the fennec. Comp Biochem Physiol. 62A:585-592.

7953. Nomura, T & Ohsawa, N. 1976. The use and problems associated with non-human primates in the study of reproduction. The Laboratory Animal in the Study of Reproduction 6th Symposium ICLA Thessaloniki 9-11 July, 1975. 1-16, Antikatzides, T, Erichsen, S, & Spiegel, A, eds, Fischer, NY.

7954. Nomura, T, Ohsawa, N, Tagima, Y, et al. 1972. Reproduction of Japanese monkeys. Symp Use of Non-Human Primates for Research on Problems of Human Reproduction, Sukhumi, USSR, 13-17 Dec, 1971. Acta Endocrinol Suppl 166:473-482.

7955. Norbury, GL. 1987. Twins in the western grey kangaroo, *Macropus fuliginosus* (Marsupialia: Macropodidae), in northwestern Victoria. Aust Mamm. 10:33.

7956. Norbury, GL, Coulson, GM, & Walters, BL. 1988. Aspects of the demography of the western grey kangaroo, *Macropus fuliginosus melanops*, in semiarid north-west Victoria. Aust Wildl Res. 15:257-266.

7957. Norman, RL, Brandt, H, & van Horn, RN. 1978. Radioimmunoassay for luteinizing hormone (LH) in the ring-tailed lemur (*Lemur catta*) with antiovine LH and ovine ^{125}I-LH. Biol Reprod. 19:119-124.

7958. Norman, RL, Lindstrom, SA, Bangsberg, D, et al. 1984. Pulsatile secretion of luteinizing hormone during the menstrual cycle of rhesus macaques. Endocrinology. 115:261-266.

7959. Norman, RL & Spies, HG. 1979. Effect of luteinizing hormone-releasing hormone on the pituitary-gonadal axis in fetal and infant rhesus monkeys. Endocrinology. 105:655-659.

7960. Norris, KS & Prescott, JH. 1961. Observations on Pacific cetaceans of Californian and Mexican waters. Univ CA Publ Zool. 63:291-370.

7961. Norris, ML. 1984. Relationship between ovarian Δ^5-3β-hydroxysteroid dehydrogenase activity and implantation in non-lactating and lactating Mongolian gerbils. Reprod Nutr Dévelop. 24:953-961.

7962. Norris, ML. 1985. Disruption of pair bonding induces pregnancy failure in newly mated Mongolian gerbils (*Meriones unguiculatus*). J Reprod Fert. 75:43-47.

7963. Norris, ML & Adams, CE. 1972. Aggressive behaviour and reproduction in the Mongolian gerbil, *Meriones unguiculatus*, relative to age and sexual experience at pairing. J Reprod Fert. 31:447-450.

7964. Norris, ML & Adams, CE. 1972. The growth of the Mongolian gerbil, *Meriones unguiculatus*, from birth to maturity. J Zool. 166:277-282.

7965. Norris, ML & Adams, CE. 1974. Sexual development in the Mongolian gerbil, *Meriones unguiculatus*, with particular reference to the ovary. J Reprod Fert. 36:245-248.

7966. Norris, ML & Adams, CE. 1976. Incidence of pup mortality in the rat with particular reference to nesting material, maternal age and party. Lab Anim. 10:165-169.

7967. Norris, ML & Adams, CE. 1979. Exteroceptive factors and pregnancy block in the Mongolian gerbil, *Meriones unguiculatus*. J Reprod Fert. 57:401-404.

7968. Norris, ML & Adams, CE. 1979. The vaginal smear, mating, egg transport and preimplantation development in a wild guinea-pig, the cuis (*Galea musteloides*). J Reprod Fert. 55:457-461.

7969. Norris, ML & Adams, CE. 1979. Vaginal opening in the Mongolian gerbil, *Meriones unguiculatus*: Normal data and the influence of social factors. Lab Anim. 13:159-162.

7970. Norris, ML & Adams, CE. 1981. Mating post partum and length of gestation in the Mongolian gerbil (*Meriones unguiculatus*). Lab Anim. 15:189-191.

7971. Norris, ML & Adams, CE. 1981. Pregnancy concurrent with lactation in the Mongolian gerbil (*Meriones unguiculatus*). Lab Anim. 15:21-23.

7972. Norris, ML & Adams, CE. 1981. Time of mating and associated changes in the vaginal smear of the post-parturient Mongolian gerbil (*Meriones unguiculatus*). Lab Anim. 15:193-198.

7973. Norris, ML & Adams, CE. 1982. Lifetime reproductive performance of Mongolian gerbils (*Meriones unguiculatus*) with 1 or 2 ovaries. Lab Anim. 16:146-150.

7974. Norris, T. 1969. Ceylon sloth bear. Animals. 12:300-303.

7975. Notini, A. 1948. Biologiska Undersöknigar oner Grävlingen (*Meles meles*). Svenska Jägareforb Medd N:R. 13:1-256.

7975a. Nottingham, BG, Johnson, KG, Woods, JW, et al. 1982. Population characteristics and harvest relationships of a raccoon

population in east Tennessee. Proc Ann Conf SE Assoc Fish Wildl Agencies. 36:691-700.
7976. Novellie, PA. 1979. Courtship behaviour of blesbok (*Damaliscus dorcas phillipsi*). Mammalia. 43:263-274.
7977. Novellie, PA, Manson, J, & Bigalke, RC. 1984. Behavioural ecology and communication in the Cape grysbok. S Afr J Zool. 19:22-30.
7978. Novick, A. 1960. Successful breeding in captive *Artibeus*. J Mamm. 41:508-509.
7979. Novikov, GA. 1962. Carnivorous Mammals of the Fauna of the USSR. Israel Prog Sci Trans. Jerusalem.
7980. Novikov, GA, Airapetjants, AE, Pukinsky, YuB, et al. 1969. Some peculiarities of population of brown bears in the Leningrad district. Zool Zhurn. 48:885-901.
7981. Novikova, TG, Sobol, AV, & Diveea, GM. 1983. The concentration of sex hormones in female arctic foxes during the reproductive cycle. Nauch Trudy Nauchno-Issledo Vatelsku Inst Pushnogo Zverovodstva i Krolikovodstva. 29:190-197. [Anim Br Abst, 54:ref 1818, 1986]
7982. Novoa, C. 1970. Reproduction in Camelidae. J Reprod Fert. 22:3-20.
7983. Nowak, RM. 1991. Walker's Mammals of the World, 5th ed. Johns Hopkins Univ Press, Baltimore, MD.
7984. Nowak, RM & Paradiso, JL. 1983. Walker's Mammals of the World, 4th ed. John Hopkins Univ Press, Baltimore, MD.
7985. Nowosad, RF. 1973. Twinning in reindeer. J Mamm. 54:781.
7986. Noyes, M. 1968. A breeding pair of tree shrews. Ann Rep Jersey Wildl Preserv Trust. 5:39-41.
7987. Numerov, KD. 1966. On the sexual and age composition and on reproduction of the sable in Yenissei Siberia. Zool Zhurn. 45:421-429.
7988. Nyholm, NEI & Meurling, P. 1979. Reproduction of the bank vole, *Clethrionomys glareolus*, in northern and southern Sweden during several seasons and in different phases of the vole population cycle. Holarctic Ecol. 2:12-20.
7989. O, WS & Chow, PH. 1987. Asymmetry in the ovary and uterus of the golden hamster (*Mesocricetus auratus*). J Reprod Fert. 80:21-23.
7990. O, WS, Short, RV, Renfree, MB, et al. 1988. Primary genetic control of somatic sexual differentiation in a mammal. Nature. 331:716-717.
7991. O'Brien, SJ, Nash, WG, Wildt, DE, et al. 1985. A molecular solution to the riddle of the giant panda's phylogeny. Nature. 317:140-144.
7992. O'Brien, SJ, Roelke, ME, Marker, L, et al. 1985. Genetic basis for species vulnerability in the cheetah. Science. 227:1428-1434.
7993. O'Brien, SJ, Wildt, DE, Goldman, D, et al. 1983. The cheetah is depauperate in genetic variation. Science. 221:459-461.
7994. O'Connell, MA. 1979. Ecology of didelphid marsupials from northern Venezuela. Vertebrate Ecology in the Northern Neotropics. 73-87, Eisenberg, JF, ed, Smithsonian Inst, Washington, DC.
7995. O'Connell, MA. 1981. Population ecology of small mammals from northern Venezuela. Ph D Diss. TX Tech Univ, Lubbock, TX.
7996. O'Connell, MA. 1982. Population biology of North and South American grassland rodents: A comparative review. Spec Publ Pymatuning Lab Ecol. 6:167-185.
7997. O'Connell, MA. 1983. *Marmosa robinsoni*. Mamm Species. 203:1-6.
7998. O'Connell, MA. 1989. Population dynamics of neotropical small mammals in seasonal habitats. J Mamm. 70:532-548.
7999. O'Donoghue, CH. 1912. The corpus luteum in the non-pregnant *Dasyurus* and polyovular follicles in *Dasyurus*. Anat Anz. 41:353-368.
8000. O'Donoghue, CH. 1914. Über die Corpora lutea bei enigen Beuteltieren. Arch Mikr Anat Abt II. 84:1-47.
8001. O'Donoghue, ER. 1982. A resurgence in reproductive behavior in a previously inactive male orangutan (*Pongo pygmaeus abelii*). Zoo Biol. 1:157-159.
8002. O'Donoghue, PN. 1963. Reproduction in the female hyrax (*Dendrohyrax arborea rumenzorii*). Proc Zool Soc London. 141:207-239.
8003. O'Farrell, MJ & Blaustein, AR. 1974. *Microdipodops megacephalus*. Mamm Species. 46:1-3.
8004. O'Farrell, MJ & Blaustein, AR, 1974. *Microdipodops pallidus*. Mamm Species. 47:1-2.
8005. O'Farrell, MJ & Studier, EH. 1980. *Myotis thysanodes*. Mamm Species. 137:1-5.
8005a. O'Farrell, TP. 1965. Home range and ecology of snowshoe hares in interior Alaska. J Mamm. 46:406-418.
8006. O'Farrell, TP. 1975. Unusual fertilization of a grasshopper mouse, *Onychomys leucogaster*. Am Mid Nat. 93:255-256.
8007. O'Farrell, TP, Olson, RJ, Gilbert, RO, et al. 1975. A population of Great Basin pocket mice, *Perognathus parvus*, in the shrub-steppe of south-central Washington. Ecol Monogr. 45:1-28.
8008. O'Gara, BW. 1969. Unique aspects of reproduction in the female pronghorn *Antilocapra americana* Ord. Am J Anat. 125:217-232.
8009. O'Gara, BW. 1978. *Antilocapra americana*. Mamm Species. 90:1-7.
8010. O'Gara, BW, Moy, RF, & Bear, GD. 1971. The annual testicular cycle and horn casting in the pronghorn (*Antilocapra americana*). J Mamm. 52:537-544.
8011. O'Gorman, F. 1965. Mammals of Tory Island, Co Donegal, Ireland. Proc Zool Soc London. 145:155-158.
8012. O'Gorman, F & Fairley, FS. 1965. A colony of *Plecotus auritus* from Co Kilkenny. Proc Zool Soc London. 145:154-155.
8013. O'Gorman, FA. 1961. Fur seals breeding in the Falkland Island Dependencies. Nature. 192:914-916.
8014. O'Neal, GT, Flinders, JT, & Clary, WP. 1987. Behavioral ecology of the Nevada kit fox (*Vulpes macrotis nevadensis*) on a managed desert rangeland. Current Mammalogy. 443-481, Genoways, HH, ed, Plenum Press, NY, NY.
8015. O'Neil, T. 1959. The Muskrat in the Louisiana Coastal Marshes. LA Dept Wildl Fish, LA.
8016. O'Shea, TJ. 1980. Roosting, social organization and the annual cycle in a Kenya population of the bat *Pipistrellus nanus*. Z Tierpsychol. 53:171-195.
8017. O'Shea, TJ. 1991. *Xerus rutilus*. Mamm Species. 370:1-5.
8018. Oakey, RE. 1975. Serum cortisol binding capacity and cortisol concentration in the pregnant baboon and its fetus during gestation. Endocrinology. 97:1024-1029.
8019. Oaks, EC, Young, PJ, Kirkland, GL, Jr, et al. 1987. *Spermophilus variegatus*. Mamm Species. 272:1-8.
8020. Obenchain, JB. 1925. The brains of the South American marsupials *Caenolestes* and *Orolestes*. Field Mus Nat Hist. 4:175-232.
8020a. Obidina, VA. 1972. Winter breeding of *Alticola argentatus* under natural conditions. Soviet J Ecol. 3:567-568.
8020b. Oboussier, H & von Maydell, GA. 1960. Zur Kenntnis von *Presbytis entellus* (Dufresnes 1797). Zool Anz. 164:141-154.

8021. Odberg, FO. 1984. Some data on the fertility of bank voles (*Clethrionomys glareolus britannicus*) in the laboratory supporting the hypothesis of induced ovulation. Lab Anim. 18:33-35.
8022. Odell, D, Asper, ED, Baucom, J, et al. 1980. A recurrent mass stranding on the false killer whale, *Pseudorca crassidens*, in Florida. Fish Bull USA. 78:171-177.
8023. Odell, DK. 1975. Breeding biology of the California sea lion *Zalophus californianus*. Rapp PV Réun Cons Int Explor Mer. 169:374-378.
8024. Odell, DK. 1981. California sea lion *Zalophus californianus* (Lesson, 1828). Handbook of Marine Mammals, 1:67-97, Ridgway, SH & Harrison, RJ, eds, Academic Press, New York.
8025. Odor, DL, Gaddum-Rosse, P, Rumery, RE, et al. 1980. Cyclic variations in the oviductal ciliated cells during the menstrual cycle and after estrogen treatment in the pig-tailed monkey, *Macaca nemestrina*. Anat Rec. 198:35-57.
8026. Odum, EP. 1949. Small mammals of the Highlands (North Carolina) Plateau. J Mamm. 30:179-192.
8026a. Odum, EP. 1955. An eleven year history of a *Sigmodon* population. J Mamm. 36:368-378.
8027. Oduor-Okelo, D. 1978. A histological study on the ovary of the African cane rat (*Thryonomys swinderianus*). E Afr Wildl J. 16:257-264.
8028. Oduor-Okelo, D. 1985. Ultrastructural observations on the chorioallantoic placenta of the golden-rumped elephant shrew, *Rhynchocyon chrysopygus*. Afr J Ecol. 23:155-166.
8029. Oduor-Okelo, D & Gombe, S. 1982. Placentation in the cane rat (*Thryonomys swinderianus*). Afr J Ecol. 20:49-66.
8029a. Oduor-Okelo, D & Gombe, S. 1991. Development of the foetal membranes in the cane rat (*Thryonomys swinderianus*): A re-interpretation. Afr J Ecol. 29:157-167.
8030. Oduor-Okelo, D, Gombe, S, & Amoroso, EC. 1980. The placenta and fetal membranes of the short-nosed elephant shrew, *Elephantulus rufescens* (Peters, 1878). Säugetierk Mitt. 28:293-301.
8031. Oduor-Okelo, D & Neaves, WB. 1982. The chorioallantoic placenta of the spotted hyena (*Crocuta crocuta* Erxleben): An electron microscopic study. Anat Rec. 205:215-222.
8032. Oduor-Okelo, D, Musewe, VO, & Gombe, S. 1983. Electron microscopic study of the chorioallantoic placenta of the rock hyrax (*Heterohyrax brucei*). J Reprod Fert. 68:311-316.
8033. Oeming, A. 1965. A herd of musk-oxen *Ovibos moschatus* in captivity. Int Zoo Yearb. 5:58-65.
8034. Oettlé, AG. 1967. The multimammate mouse (*Praomys natalensis*; syn *Rattus natalensis*, *Mastomys coucha*). UFAW Handbook on the Care and Management of Laboratory Animals, 3rd ed. 468-477, E & S Livingstone, Edinburgh.
8035. Oftedal, OT. 1984. Milk composition, milk yield and energy output at peak lactation: A comparative review. Symp Zool Soc London. 51:33-85.
8036. Oftedal, OT, Boness, DJ, & Bower, WD. 1988. The composition of hooded seal (*Cystophora cristata*) milk: An adaptation for postnatal fattening. Can J Zool. 66:318-322.
8037. Oftedal, OT, Boness, DJ, & Tedman, RA. 1987. The behavior, physiology, and anatomy of lactation in the Pinnipedia. Current Mammalogy. 175-245, Genoways, HH, ed, Plenum Press, New York, New York.
8038. Oftedal, OT, Bowen, WD, Widdowson, EM, et al. 1989. Effects of suckling and the postsuckling fast on weights of the body and internal organs of harp and hooded seal pups. Biol Neonate. 56:282-300.
8039. Oftedal, OT, Hintz, HF, & Schryver, HF. 1983. Lactation in the horse: Milk composition and intake by foals. J Nutr. 113:2096-2106.
8040. Oftedal, OT, Iverson, SJ, & Boness, DJ. 1987. Milk and energy intakes of suckling California sea lion *Zalophus californianus* pups in relation to sex, growth, and predicted maintenance requirements. Physiol Zool. 60:560-575.
8041. Oftedal, OT & Jenness, R. 1988. Interspecies variation in milk composition among horses, zebras and asses (Perissodactyla: Equidae). J Dairy Res. 55:57-66.
8042. Ogden, J. 1983. Reproduction of wild-caught tamarins (*Saguinus mystax mystax*) under laboratory conditions. J Med Primatol. 12:343-345.
8043. Ogden, JD & Wolfe, LG. 1979. Reproduction of wild-caught marmosets (*Saguinus labiatus labiatus*) under laboratory conditions. Lab Anim Sci. 29:545-546.
8044. Ogden, JD, Wolfe, LG, & Deinhardt, FW. 1978. Breeding *Saguinus* and *Callithrix* species of marmosets under laboratory conditions. Marmosets in Experimental Medicine. 79-83, Gengozian, N & Deinhardt, F, eds, Karger, New York.
8045. Ogilvie, CS. 1949. Some notes on a Malayan bamboo rat. Malay Nat J. 4:24-28.
8046. Ogilvie, CS. 1958. The binturong or bear-cat. Malay Nat J. 13:1-3.
8047. Ogilvie, PW & Bridgwater, DD. 1967. Notes on the breeding of an Indian pangolin *Manis crassicaudata* at Oklahoma Zoo. Int Zoo Yearb. 7:116-118.
8048. Ogle, TF, Braach, HH, & Buss, IO. 1953. Fine structure and progesterone concentration in the corpus luteum of the African elephant. Anat Rec. 175:707-724.
8049. Ognev, SI. 1962. Mammals of Eastern Europe and Northern Asia Vol I Insectivora and Chiroptera. Israel Prog Sci Trans. Jerusalem.
8050. Ognev, SI. 1963. Mammals of the USSR and Adjacent Countries Vol V Rodents. Israel Prog Sci Trans. Jerusalem.
8051. Ognev, SI. 1963. Mammals of the USSR and Adjacent Countries Vol VI Rodents. Israel Prog Sci Trans. Jerusalem.
8052. Ognev, SI. 1964. Mammals of the USSR and Adjacent Countries Vol VII Rodents. Israel Prog Sci Trans. Jerusalem.
8053. Ognev, SI. 1966. Mammals of the USSR and Adjacent Countries Vol IV Rodents. Israel Prog Sci Trans. Jerusalem.
8054. Oh, YK. 1977. Periodic changes in the testis and ductus epididymis in Korean hibernating bats. Kor J Zool. 20:67-76.
8055. Oh, YK, Mōri, T, & Uchida, TA. 1983. Studies on the vaginal plug of the Japanese greater horseshoe bat, *Rhinolophus ferrumequinum nippon*. J Reprod Fert. 68:365-369.
8056. Oh, YK, Mōri, T, & Uchida, TA. 1985. Spermiogenesis in the Japanese greater horseshoe bat, *Rhinolophus ferrumequinum nippon*. J Fac Agr, Kyushu Univ. 29:203-209.
8057. Oh, YK, Mōri, T, & Uchida, TA. 1985. Prolonged survival of the Graafian follicle and fertilization in the Japanese greater horseshoe bat, *Rhinolophus ferrumequinum nippon*. J Reprod Fert. 73:121-126.
8058. Ohata, CA, Miller, K, & Irving, L. 1977. Metabolism of northern fur seals. Fed Proc. 36:546.
8058a. Ohdachi, S. 1992. Female reproduction in three species of *Sorex* in Hokkaido, Japan. J Mamm. 73:455-457.
8059. Ohlin, A. 1893. Some remarks on the bottlenose-whale (*Hyperoodon*). Acta Univ Lund. 29(8):1-13.
8060. Ohmart, RD. 1975. Reproduction in burros *Equus asinus* along the Colorado River during 1974. J Ariz Acad Sci. 10:22.
8061. Ohno, M & Fujino, K. 1952. Biological investigation of the whales caught by the Japanese Antarctic whaling fleets, season 1950/51. Sci Rep Whales Res Inst. 7:125-188.

8062. Ohno, S, Jainchill, J, & Stenius, C. 1963. The creeping vole (*Microtus oregoni*) as a gonosomic mosaic Part I: The OY/XY constitution of the male. Cytogenetics. 2:232-239.
8063. Ohsumi, S. 1960. Relative growth of the fin whale, *Balaenoptera physalus* (Linn). Sci Rep Whales Res Inst. 15:17-84.
8064. Ohsumi, S. 1964. Comparison of maturity and accumulation rate of corpora albicantia between left and right ovaries in Cetacea. Sci Rep Whales Res Inst. 18:123-148.
8065. Ohsumi, S. 1965. Reproduction of the sperm whale in the north-west Pacific. Sci Rep Whales Res Inst. 19:1-35.
8066. Ohsumi, S. 1966. Allomorphosis between body length at sexual maturity and body length at birth in the Cetacea. J Mamm Soc Japan. 3:3-7.
8067. Ohsumi, S. 1977. Bryde's whales in the pelagic whaling ground of the north Pacific. Rep Int Whal Commn Sp Issue. 1:140-149.
8068. Ohsumi, S. 1986. Yearly change in age and body length at sexual maturity of a fin whale stock in the eastern north Pacific. Sci Rep Whales Res Inst. 37:1-16.
8069. Ohsumi, S & Masaki, Y. 1975. Biological parameters of the Atlantic minke whale at the virginal population level. J Fish Res Bd Can. 32:995-1004.
8070. Ohsumi, S, Masaki, Y, & Kawamura, A. 1970. Stock of the Antarctic minke whale. Sci Rep Whales Res Inst. 22:75-125.
8071. Ohsumi, S, Nishiwaki, M, & Hibiya, T. 1958. Growth of fin whales in the northern Pacific. Sci Rep Whales Res Inst. 13:97-133.
8072. Ohta, K, Watarai, T, Oishi, T, et al. 1953. Composition of fin whale milk. Proc Japan Acad. 29:392-398.
8073. Ohta, K, Watarai, T, Ōishi, T, et al. 1955. Composition of fin whale milk. Sci Rep Whales Res Inst. 10:151-167.
8074. Ohtaishi, N & Yoneda, M. 1981. A thirty four year old male Kuril seal from Schiretoko Pen, Hokkaido. Sci Rep Whales Res Inst. 33:131-133.
8075. Ojasti, J. 1968. Notes on the mating behavior of the capybara. J Mamm. 49:534-535.
8076. Ojasti, J. 1973. Estudio Biologico del Chigiure o Capibara. Fondo Nacional de Investingaciones Agropecuiaras. [Cited by Kleiman et al, 1979]
8077. Ojeda, MM & Keith, LB. 1982. Sex and age composition and breeding biology of cottontail rabbit populations in Venezuela. Biotropica. 14:99-107.
8078. Oke, BO. 1985. Effect of season on the reproductive organs of the male African giant rat (*Cricetomys gambianus* Waterhouse) in Ibadan, Nigeria. Afr J Ecol. 23:67-70.
8079. Okhotina, MV. 1969. Some data on ecology of *Sorex (Ognevia) mirabilis* Ognev, 1937. Acta Theriol. 14:273-284.
8080. Okia, NO. 1973. The breeding pattern of the soft-furred rat, *Praomys morio* in an evergreen forest in southern Uganda. J Zool. 170:501-504.
8081. Okia, NO. 1974. Breeding in Franquet's bats, *Epomops franqueti* (Tomes), in Uganda. J Mamm. 55:462-465.
8082. Okia, NO. 1974. The breeding pattern of the eastern epauletted bat, *Epomophorus anurus* Heuglin, in Uganda. J Reprod Fert. 37:27-31.
8083. Okia, NO. 1976. The biology of the bush rat, *Aethomys hindei* Thomas in southern Uganda. J Zool. 180:41-56.
8084. Okia, NO. 1987. Reproductive cycles of east African bats. J Mamm. 68:138-141.
8085. Okulicz, WC, Darrow, JM, & Goldman, BD. 1988. Uterine steroid hormone receptors during the estrous cycle and during hibernation in the Turkish hamster (*Mesocricetus brandti*). Biol Reprod. 38:597-604.
8086. Okuzaki, M. 1979. Reproduction of raccoon dogs, *Nyctereutes procyonoides viverrinus*, Temminck, in captivity. Joshi Eiyo Dargaku Kiyo. 10:99-103.
8087. Olalla, AM. 1935. El genero *Sciurillus* representado en la amazoma y algunos observaciones sabre el mismo. Rev Mus Paulista. 19:425-430.
8088. Olds, N & Shoshani, J. 1982. *Procavia capensis*. Mamm Species. 171:1-7.
8089. Olds, TJ & Collins, LR. 1973. Breeding Matschie's tree kangaroo *Dendrolagus matschiei* in captivity. Int Zoo Yearb. 13:123-125.
8090. Olesen, CR & Thing, H. 1989. Guide to field classification by sex and age of the muskox. Can J Zool. 67:1116-1119.
8091. Oliff, WD. 1953. The mortality, fecundity and intrinsic rate of natural increase of the multimammate mouse, *Rattus (Mastomys) natalensis* (Smith) in the laboratory. J Anim Ecol. 22:217-226.
8092. Oliver, WLR. 1975. The Jamaican hutia (*Geocapromys brownii brownii*). Ann Rep Jersey Wildl Preserv Trust. 12:10-17.
8093. Oliver, WLR. 1976. The management of yapoks (*Chironectes minimus*) at Jersey Zoo, with observations on their behavior. Ann Rep Jersey Wildl Preserv Trust. 13:32-36.
8094. Oliver, WLR. 1977. The hutias Capromyidae of the West Indies. Int Zoo Yearb. 17:14-20.
8095. Oliver, WLR. 1980. The pygmy hog: The biology and conservation of the pygmy hog *Sus (Porcula) salvanius* and the hispid hare *Caprolagus hispidus*. Jersey Wildl Preserv Trust Sp Rep. 1:1-80.
8096. Olivera, J, Ramirez-Pulido, J, & Williams, SL. 1986. Reproduccion de *Peromyscus (Neotomodon) alstoni* (Mammalia: Muridae) en condiciones de laboratorio. Acta Zool Mex. 16:1-27.
8097. Oliveras, DM & Novak, MA. 1984. Paternal behavior as an indicator of social organization in *Microtus pennsylvanicus*, *M pinetorum* and *M ochrogaster*. Proc E Pine Meadow Vole Symp. 8:61-70.
8098. Oliveras, DM & Novak, MA. 1986. A comparison of paternal behaviour in the meadow vole *Microtus pennsylvanicus*, the pine vole *M pinetorum* and the prairie vole *M ochrogaster*. Anim Behav. 34:519-526.
8099. Olivier, G. 1952. Note sur la reproduction du Gibbon (*Hylobates concolor*). Terre Vie. 7:139-141.
8100. Olmedo, JE, Escos, J, & Gomedio, M. 1985. Reproduction de *Gazella cuvieri* en captivité. Mammalia. 49:501-507.
8101. Olsen, Ø. 1913. On the external characters and biology of Bryde's whale (*Balaenoptera brydei*), a new rorqual from the coast of South Africa. Proc Zool Soc London. 1913:1073-1090.
8102. Olsen, P. 1981. The stimulating effect of a phytohormone, gibberellic acid, on reproduction of *Mus musculus*. Aust Wildl Res. 8:321-325.
8103. Olsen, PD. 1982. Reproductive biology and development of the water rat, *Hydromys chrysogaster*, in captivity. Aust Wildl Res. 9:39-53.
8104. Olsen, PF. 1959. Muskrat breeding biology at Delta, Manitoba. J Wildl Manage. 23:40-53.
8105. Olsen, RW. 1968. Gestation period in *Neotoma mexicana*. J Mamm. 49:533-534.
8106. Omar, no initials. Personal communication. [Rowell, TE. 1970]
8107. Omar, A & DeVos, A. 1971. The annual reproductive cycle of an African monkey (*Cercopithecus mitis kolbi* Newman). Folia Primatol. 16:206-215.

8108. Ommanney, FD. 1932. The uro-genital system of the fin whale (*Balaenoptera physalus*). Discovery Rep. 5:363-466.
8109. Omura, H. 1950. On the body weight of sperm and sei whales located in the adjacent waters of Japan. Sci Rep Whales Res Inst. 4:1-13.
8110. Omura, H. 1950. Whales in the adjacent waters of Japan. Sci Rep Whales Res Inst. 4:27-113.
8111. Omura, H. 1953. Biological study on humpback whales in the Antarctic whaling areas IV and V. Sci Rep Whales Res Inst. 8:81-102.
8112. Omura, H. 1958. North Pacific right whale. Sci Rep Whales Res Inst. 13:1-52.
8113. Omura, H, Ohsumi, S, Nemoto, T, et al. 1969. Black right whales in the North Pacific. Sci Rep Whales Res Inst. 21:1-78.
8114. Omura, H & Sakiura, H. 1956. Studies on the little piked whale from the coast of Japan. Sci Rep Whales Res Inst. 11:1-38.
8115. Ondrias, JC. 1965. Contribution to the knowledge of *Microtus guentheri hartingi* from Thebes, Greece. Mammalia. 29:489-506.
8116. Onoyama, K & Haga, R. 1982. New record of four fetuses in a litter of the Yezo brown bear *Ursus arctos yesoensis*. J Mamm Soc Japan. 9:1-8.
8116a. Onyango, DW, Otianga-Owiti, GE, Odour-Okelo, D, et al. 1991. *In vitro* interstitial (Leydig) cell response to LH and concentrations of plasma testosterone and LH in the naked mole rat (*Heterocephalus glaber*, Ruppell). Afr J Ecol. 29:76-85.
8117. Oppenheimer, JR. 1969. Changes in forehead patterns and group composition of white-faced monkey (*Cebus capacinus*). Proc 2nd Int Congr Primatol, Atlanta 1968. 1:36-42.
8118. Oppenheimer, JR. 1976. *Presbytis entellus*: Birth in a free-ranging primate troop. Primates. 17:541-542.
8119. Oppenheimer, JR. 1977. *Presbytis entellus*, the Hanuman langur. Primate Conservation. 469-512, Prince Rainier III of Monaco & Bourne, GH, eds, Academic Press, New York.
8120. Orbell, E & Orbell, J. 1976. Hand-rearing saiga antelope *Saiga tatarica* at the Highland Wildlife Park. Int Zoo Yearb. 16:208-209.
8121. Øritsland, T. 1970. Sealing and seal research in the south-west Atlantic pack ice, Sept-Oct 1964. Antarctic Ecology. 1:367-376, Holgate, MW, ed. Academic Press, New York.
8122. Øritsland, T. 1975. Sexual maturity and reproductive performance of female hooded seals at Newfoundland. Res Bull Int Comm Northwest Atlantic Fish. 11:37-41.
8123. Øritsland, T & Benjaminsen, T. 1975. Sex ratio, age composition and mortality of hooded seals at Newfoundland. Res Bull Int Comm NW Atlantic Fish. 11:135-143.
8124. Ormrod, S. 1967. Milk analysis of the South American tapir *Tapirus terrestris*. Int Zoo Yearb. 7:157-158.
8125. Orr, RT. 1934. Description of a new snowshoe rabbit from eastern Oregon, with notes on its life history. J Mamm. 15:152-154.
8126. Orr, RT. 1940. The rabbits of California. Occ Pap CA Acad Sci. 19:1-207.
8127. Orr, RT. 1954. Natural history of the pallid bat, *Antrozous pallidus* (Le Conte). Proc CA Acad Sci. 28:165-246.
8128. Orr, RT & Poulter, TC. 1967. Some observations on reproduction, growth, and social behavior in the Steller sea lion. Proc CA Acad Sci. 35:193-226.
8129. Orsi, AM & Ferreira, AL. 1978. Definition of the stages of the cycle of the seminiferous epithelium of the opossum (*Didelphis azare* Temminck, 1825). Acta Anat. 100:153-160.
8130. Ortiz, CL, Le Boeuf, BJ, & Costa, DP. 1984. Milk intake of elephant seal pups: An index of parental investment. Am Nat. 124:416-422.
8131. Osadchuk, LV, Krass, PM, Trut, LN, et al. 1978. Change in the endocrine function of the ovaries of silver foxes during domestication. Dokl Akad Nauk SSSR Translation. 238:67-69.
8132. Osadchuk, LV, Krass, PM, Trut, LN, et al. 1978. Effect of selection for behavior on the endocrine function of the gonads in male silver-black foxes. Dokl Akad Nauk SSSR Translation. 240:200-202.
8133. Osborn, DJ. 1953. Age classes, reproduction, and sex ratios of Wyoming beaver. J Mamm. 34:27-44.
8134. Osborn, DJ & Helmy, I. 1980. The contemporary land mammals of Egypt (including Sinai). Fieldiana Zool NS. 5:1-579.
8134a. Osburn, W & Sclater, PL. 1865. Notes on the Cheiroptera of Jamaica. Proc Zool Soc London. 1865:61-85.
8135. Osgood, WH. 1912. Mammals from western Venezuela and eastern Columbia. Publ Field Mus Nat Hist Zool Ser. 10:33-66.
8136. Osgood, WH. 1921. A monographic study of the American marsupial, *Caenolestes*. Publ Field Mus Nat Hist Zool Ser. 14:3-156.
8137. Osgood, WH. 1936. New and imperfectly known small mammals from Africa. Publ Field Mus Nat Hist Zool Ser. 20:217-256.
8138. Osgood, WH. 1943. The mammals of Chile. Publ Field Mus Nat Hist Zool Ser. 30:1-268.
8139. Oshima, K, Hayashi, M, & Matsubayashi, K. 1977. Progesterone levels in the Japanese monkey (*Macaca fuscata fuscata*) during the breeding and nonbreeding season and pregnancy. J Med Primatol. 6:99-107.
8140. Osman, AM & El Azab, EA. 1974. Gonadal and epididymal sperm reserves in the camel, *Camelus dromedarius*. J Reprod Fert. 38:425-430.
8140a. Osman, DI. 1978. On the ultrastructure of modified Sertoli cells in the terminal segment of seminferous tubules in the boar. J Anat. 127:603-613.
8141. Osman, DI, Moniem, KA, & Tinegari, MD. 1979. Histological observations on the testis of the camel, with special emphasis on spermatogenesis. Acta Anat. 104:164-171.
8142. Osmaton, BB. 1950. The large red flying squirrel *Pteromys inornatus* Geoffroy. J Bombay Nat Hist Soc. 49:114-116.
8143. Osterud, EL, Lackey, S, & Wilson, ME. 1986. Estradiol increases somatomedin-C concentrations in adolescent rhesus macaques (*Macaca mulatta*). Am J Primatol. 11:53-62.
8144. Ostfeld, RS & Tamarin, RH. 1986. The role of seasonality in vole cycles. Can J Zool. 64:2871-2872.
8145. Ostwald, R, Wilken, K, Simons, J, et al. 1972. Influence of photoperiod and partial contact on estrus in the desert pocket mouse, *Perognathus penicillatus*. Biol Reprod. 7:1-8.
8146. Oswald, C & McClure, PA. 1985. Geographic variation in litter size in the cotton rat (*Sigmodon hispidus*): Factors influencing ovulation rate. Biol Reprod. 33:411-417.
8147. Oswald, C & McClure, PA. 1990. Energetics of concurrent pregnancy and lactation in cotton rats and woodrats. J Mamm. 71:550-509.
8148. Ottobre, JS, Nixon, WE, & Stouffer, RS. 1984. Induction of relaxin secretion in rhesus monkeys by human chorionic gonadotropin: Dependence on the age of the corpus luteum of the menstrual cycle. Biol Reprod. 31:1000-1006.
8149. Oud, JL & de Rooy, DG. 1977. Spermatogenesis in the Chinese hamster. Anat Rec. 187:113-124.

8150. Ovadia, J, McArthur, JW, Kopito, L, et al. 1971. The cervical mucus secretion of the bonnet monkey (*M radiata*): Anatomical basis and physiological regulation. Biol Reprod. 5:127-145.
8151. Ovadia, J, McArthur, JW, Smith, OW, et al. 1971. An individualized technique for inducing ovulation in the bonnet monkey, *Macaca radiata.* J Reprod Fert. 27:13-23.
8152. Ovaska, K & Herman, TB. 1987. Life history characteristics and movements of the woodland jumping mouse, *Napaeozapus insignis*, in Nova Scotia. Can J Zool. 66:1752-1762.
8153. Ovchinnikova, NA. 1971. Experimental study of the biological peculiarities and reproductive isolation of common and Transcaspian voles. Dokl Akad Nauk SSSR Translation. 200:570-572.
8154. Ovchinnikova, SL. 1969. Some peculiarities of the ecology of the Russian mole rat (*Spalax microphtalmus* Gould) in the Chernozem sone [*sic*]. Zool Zhurn. 48:1564-1570.
8155. Owen, JG. 1984. *Sorex fumeus.* Mamm Species. 215:1-8.
8156. Owen, JG & Hoffmann, RS. 1983. *Sorex ornatus.* Mamm Species. 212:1-5.
8157. Owen, R. 1832. On the mammary glands of the *Ornithorynchus paradoxus.* Phil Trans R Soc London. 1832:517-538.
8158. Owen, R. 1834. On the ova of the *Ornithorhynchus paradoxus.* Phil Trans R Soc London. 1834:555-566.
8159. Owen, R. 1834. On the young of the *Ornithorhynchus paradoxus*, Blum. Trans Zool Soc London. 1:221-228.
8160. Owen, R. 1841. Notes on the anatomy of the Nubian giraffe. Trans Zool Soc London. 2:217-248.
8161. Owen, R. 1849. Notes on the birth of the giraffe at the Zoological Society's gardens, and description of the foetal membranes and of some natural and morbid appearances observed in the dissection of the young animal. Trans Zool Soc London. 3:21-28.
8162. Owen, R. 1865. On the marsupial pouches, mammary glands, and mammary foetus of the *Echidna hystrix.* Phil Trans R Soc London. 155:671-686.
8163. Owen, R. 1866. On the aye-aye, (*Chiromys madagascariensis*, Desm; *Sciurus madagascariensis*, Gmel, Sonnerat; *Lemur psilodactylus*, Schreber, Shaw). Trans Zool Soc London. 5:33-101.
8164. Owen, R. 1880. On the ova of the *Echidna hystrix.* Phil Trans R Soc London. 1880:1051-1054.
8165. Owen, REA. 1970. Some observations on the sitatunga in Kenya. E Afr Wildl J. 8:181-196.
8166. Owen-Smith, RN. Personal communication. [Cited by Mentis, 1972]
8167. Owen-Smith, RN. 1974. The social system of the white rhinoceros. The Behaviour of Ungulates and its Relations to Management. 341-351, Geist, V & Walther, F, eds, IUCN, Morges, Switzerland.
8168. Owen-Smith, RN. 1975. The social ethology of the white rhinoceros *Ceratotherium simum* (Burchell, 1817). Z Tierpsychol. 38:337-384.
8169. Owens, DD & Owens, MJ. 1979. Communal denning and clan associations in brown hyenas (*Hyaena brunnea*, Thunberg) of the central Kalahari Desert. Afr J Ecol. 17:35-44.
8170. Owens, DD & Owens, MJ. 1984. Helping behaviour in brown hyenas. Nature. 308:843-845.
8171. Owers, NO. 1960. The endothelio-endothelial placenta of the Indian musk shrew, *Suncus murinus*: A new interpretation. Am J Anat. 106:1-25.
8172. Owiti, GEO, Cukierski, M, Tarara, RP, et al. 1986. Early placentation in the African green monkey (*Cercopithecus aethiops*). Acta Anat. 127:184-194.
8173. Owiti, GEO, Oduor-Okelo, P, & Gombe, S. 1985. Utrastructure of the chorioallantoic placenta of the springhare (*Pedetes capensis larvalis* Hollister). Afr J Ecol. 23:145-152.
8174. Owiti, GEO, Tarara, RP, & Hendrickx, AG. 1989. Fetal membranes and placenta of the African green monkey (*Cercopithecus aethiops*). Anatomy & Embryology. 179:591-604.
8175. Oxberry, BA. 1977. Ovarian morphology and steroidogenesis in the pallid bat, *Antrozous pallidus.* Anat Rec. 187:673.
8176. Oxberry, BA. 1979. Female reproductive patterns in hibernating bats. J Reprod Fert. 56:359-367.
8177. Oxnard, CE. 1981. The uniqueness of *Daubentonia.* Am J Phys Anthro. 54:1-21.
8178. Oyarzun, SE, Devison, D, & Mottram, K. 1984. Nutrition and early growth of twin hand-reared Nubian ibex *Capra ibex nebiana.* Int Zoo Yearb. 23:248-253.
8179. Oźdzeński, W & Mystkowski, ET. 1976. Stages of pregnancy of the bank vole. Acta Theriol. 21:279-286.
8180. Ozoga, JJ. 1987. Maximum fecundity in supplementally-fed northern Michigan white-tailed deer. J Mamm. 68:878-879.
8181. Ozoga, JJ & Verme, LJ. 1982. Physical and reproductive characteristics of a supplementally-fed white-tailed deer herd. J Wildl Manage. 46:281-301.
8182. Ozoga, JJ, Verme, LJ, & Bienz, CS. 1982. Parturition behavior and territoriality in white-tailed deer: Impact on neonatal mortality. J Wildl Manage. 46:1-11.
8183. Pacheco, J & Naranjo, CJ. 1978. Field ecology of *Dasypus sabanicola* in the flood savanna of Venezuela. Pan Am Health Org Sci Publ. 366:13-17.
8183a. Packard, JM, Babbitt, KJ, Hannon, PG, et al. 1990. Infanticide in captive collared peccaries (*Tayassu tajacu*). Zoo Biol. 9:49-53.
8184. Packard, JM, Dowdell, DM, Grant, WE, et al. 1987. Parturition and related behavior of the collared peccary (*Tayassu tajacu*). J Mamm. 68:679-681.
8185. Packard, JM & Mech, LD. 1980. Population regulation in wolves. Biosocial Mechanisms of Population Regulation. 135-150, Cohen, MN, Malpass, RS, & Klein, HG, eds, Yale Univ Press, New Haven.
8186. Packard, JM, Seal, US, Mech, LD, et al. 1985. Causes of reproductive failure in two family groups of wolves (*Canis lupus*). Z Tierpsychol. 68:24-40.
8187. Packard, RL. 1956. An observation on quadruplets in the red bat. J Mamm. 37:279-280.
8188. Packard, RL. 1956. The tree squirrels of Kansas: Ecology and economic importance. Univ KS Mus Nat Hist Misc Publ. 11:1-67.
8188a. Packard, RL. 1960. Speciation and evolution of the bygmy mice, genus *Baiomys.* Univ KS Publ Mus Nat Hist. 9:579-670
8189. Packard, RL. 1968. An ecological study of the fulvous harvest mouse in eastern Texas. Am Mid Nat. 79:68-88.
8189a. Packard, RL & Garner, H. 1964. Arboreal nests of the golden mouse in eastern Texas. J Mamm. 45:369-374.
8190. Packard, RL & Montgomery, JB, Jr. 1978. *Baiomys musculus.* Mamm Species. 102:1-3.
8191. Packer, C. 1979. Inter-troop transfer and inbreeding avoidance in *Papio anubis.* Anim Behav. 27:1-36.
8192. Packer, C. 1979. Male dominance and reproductive activity in *Papio anubis.* Anim Behav. 27:37-45.

8193. Packer, C. 1983. Demographic changes in a colony of Nile grassrats (*Arvicanthis niloticus*) in Tanzania. J Mamm. 64:159-161.
8194. Packer, C & Pusey, AE. 1982. Cooperation and competition within coalitions of male lions: Kin selection or game theory. Nature. 296:740-742.
8195. Padykula, HA & Taylor, JM. 1977. Uniqueness of the bandicoot chorioallantoic placenta (Marsupalia: Peramelidae) cytological and evolutionary interpretations. Proc Symp Comp Biol Reprod, 1976. 4:303-323.
8196. Pagels, JF & Aldeman, RG. 1971. A note on the cotton rat in central Virginia. VA J Sci. 22:195.
8197. Pagels, JF & Jones, C. 1974. Growth and development of the free-tailed bat, *Tadarida brasiliensis cynocephala* (LeConte). Southwest Nat. 19:267-276.
8198. Pagenstecher, no initials. 1859. Die Begattung von *Vesperugo pipistrellus*. Verhandl naturhist-mediz Verein Heidelberg. 1:194-195.
8199. Pagès, É. 1965. Notes sur les Pangolins du Gabon. Biol Gabon. 1:209-238.
8200. Pagès, É. 1968. Note sur la reproduction au Gabon de *Manis tricuspis*. Cycles Génitaux Saisonniers de Mammifères Sauvages. 151-164, Canivenc, R, ed, Masson et Cie, Paris.
8201. Pagès, É. 1972. Comportement aggrissif et sexuelle chez les Pangolins arboricoles (*Manis tricupis* et *M longicaudata*). Biol Gabon. 8:3-62.
8202. Pagès, É. 1972. Comportement maternel et développement du jeune chez un Pangolin arboricole (*M tricupis*). Biol Gabon. 8:63-120.
8203. Pagès, É. 1980. Ethoecology of *Microcebus coquereli* during the dry season. Nocturnal Malagasy Primates. 97-116, Charles-Dominique, P, Cooper, HM, Hladik, A, et al, eds, Academic Press, New York.
8204. Pahl, LI. 1987. Survival, age determination and population age structure of the common ringtail possum, *Pseudocheirus peregrinus*, in a *Eucalyptus* woodland and a *Leprospermum* thicket in southern Victoria. Aust J Zool. 35:625-639.
8205. Pahl, LI & Lee, AK. 1988. Reproductive traits of two populations of the common ringtail possum, *Pseudocheirus pereginus*, in Victoria. Aust J Zool. 36:83-97.
8206. Paino, G & Crasto, A. 1983. Sex hormone levels in the buffalo epididymis. Boll Zool. 50:101-103.
8207. Paintiff, JA & Anderson, DE. 1980. Breeding the margay *Felis wiedi* at New Orleans Zoo. Int Zoo Yearb. 20:223-224.
8208. Pal, AN. 1983. Male sex accessory glands of bat, *Miniopterus schreibersii fuliginosus* (Hodgson). Comp Physiol Ecol. 8:307-308.
8209. Palmeirim, JM & Hoffmann, RS. 1983. *Galemys pyrenaicus*. Mamm Species. 207:1-5.
8210. Palmer, LJ. 1934. Raising reindeer in Alaska. USDA Misc Publ. 207:1-40.
8211. Palmer, RS. 1951. The whitetail deer of Tom Logan Camps, Maine, with added notes on fecundity. J Mamm. 32:267-280.
8212. Palmer, SS & Bahr, NRA. 1986. Circannual serum concentrations of testosterone (T) in male and progesterone (P) in female black and polar bears. Biol Reprod. 34(Suppl):68.
8213. Palmer, SS, Nelson, RA, Ramsay, MA, et al. 1988. Annual changes in serum sex steroids in male and female black (*Ursus americanus*) and polar (*Ursus maritimus*) bears. Biol Reprod. 38:1044-1050.
8214. Palmer, WE, Hurst, GA, Leopold, BD, et al. 1991. Body weights and sex and age ratios for the swamp rabbit in Mississippi. J Mamm. 72:620-622.
8215. Palmieri, JR, van Dellen, AF, Tirtokusumo, S, et al. 1984. Trapping, care, and laboratory management of the silvered leaf monkey (*Presbytis cristatus*). Lab Anim Sci. 34:194-197.
8216. Palomo, LJ, Vargas, JM, & Antunez, A. 1989. Reproduction of *Microtus* (*Pitymys*) *duodecimcostatus* (Mammalia: Rodentia) in the south of Spain. Vie Milieu. 39:153-158.
8217. Pan, GW, Zhao, XX, Chen, BH, et al. 1986. Ovulation inducing effect of seminal plasma injected intramuscularly in the bactrian camel. Scientia Agric Sinica. 2:78-84.
8218. Panagis, K & Nel, JAJ. 1981. Growth and behavioural development in *Thamnomys dolichurus*. Acta Theriol. 26:381-392.
8219. Pandey, SD & Munshi, S. 1987. The genital system of the female large-eared hedgehog, *Hemiechinus auritus* Gmelin. Folia Biol (Kraków). 35:95-100.
8220. Panepinto, LM, Phillips, RW, Wheeler, LR, et al. 1978. The Yucatan miniature pig as a laboratory animal. Lab Anim Sci. 28:308-313.
8221. Panigel, M. 1970. Structure et ultrastructure comparées de la membrane placentaire chez certains Primates non humain (*Galago demidovii*, *Erythrocebus patas*, *Macaca irus* (*fascicularis*), *Macaca mulatta* et *Papio cynocyphelus*). CR Assn Anat. 145:319-337.
8222. Panigel, M. 1971. The electron microscopy of the placental villi in nonhuman primates *Galago demidovii*, *Erythrocebus patas*, *Macaca fascicularis*, *Macaca mulatta*, and *Papio cynocephalus*. Medical Primatology. 1970:536-552, Goldsmith, EI & Mooz-Jankowski, J. eds, Karger, Basel.
8223. Panouse, J. 1957. Les Mammifères du Maroc Primates Carnivores Pinnipedes, Artiodactyles. Trav Inst Sci Chérifien Sér Zool Rabat. 5:1-206.
8224. Panuska, JA. 1965. Delay of spring spermatogenesis in cold-exposed *Tamias striatus*. Am Zool. 5:739.
8225. Panuska, JA & Wade, NJ. 1957. Field observations on *Tamias striatus* in Wisconsin. J Mamm. 38:192-196.
8226. Panyutin, KK. 1963. Reproduction in the common noctule. Uchem Zap Mosk Obl Pedagog Inst. 126:63-66. [Biol Abst. 46:ref 96256]
8226a. Papke, RL, Concannon, PW, Travis, HF, et al. 1980. Control of luteal function and implantation in the mink by prolactin. J Anim Sci. 50:1102-1107.
8227. Paradiso, JL & Nowak, RM. 1972. *Canis rufus*. Mamm Species. 22:1-4.
8228. Parer, I. 1977. The population ecology of the wild rabbit, *Oryctolagus cuniculus* (L), in a Mediterranean-type climate in New South Wales. Aust Wildl Res. 4:171-205.
8229. Parer, I & Fullagar, PJ. 1986. Biology of rabbits, *Oryctolagus cuniculus* in subtropical Queensland. Aust Wildl Res. 13:545-557.
8229a. Parer, I & Libke, JA. 1991. Biology of the wild rabbit, *Oryctolagus cuniculus* (L), in the southern tablelands of New South Wales. Wildl Res. 18:327-341.
8230. Parer, JT. 1963. Vaginal contents and rectal temperature during the estrous cycle of the African dwarf goat. Am J Vet Rec. 24:1223-1226.
8231. Park, AW & Nowosielski-Slopowron, BJA. 1972. Biology of the rice rat (*Oryzomys palustris natator*) in a laboratory environment. Z Säugetierk. 37:42-51.
8232. Parker, C. 1979. Birth, care and development of Chinese hog badgers *Arctonyx collaris albogularis* at Metro Toronto Zoo. Int Zoo Yearb. 19:182-185.

8233. Parker, GR. 1977. Morphology, reproduction, diet, and behavior of the arctic hare (*Lepus arcticus monstrabilis*) on Axel Heiberg Island, Northwest Territories. Can Field-Nat. 91:8-18.
8234. Parker, GR. 1979. Unusually late pregnancy of a muskrat in southeastern New Bruswick. Can Field-Nat. 93:440-441.
8235. Parker, GR. 1981. Physical and reproductive characteristics of an expanding woodland caribou population (*Rangifer tarandus caribou*) in northern Labrador. Can J Zool. 59:1929-1940.
8236. Parker, GR & Maxwell, JW. 1980. Characteristics of a population of muskrats (*Ondatra zibethicus zibethicus*) in New Brunswick. Can Field-Nat. 94:1-8.
8237. Parker, GR & Maxwell, JW. 1984. An evaluation of spring and autumn trapping seasons for muskrats, *Ondatra zibethicus*, in eastern Canada. Can Field-Nat. 98:293-304.
8238. Parker, GR, Maxwell, JW, Morton, LD, et al. 1983. The ecology of the lynx (*Lynx canadensis*) on Cape Breton Island. Can J Zool. 61:770-786.
8239. Parker, GR & Smith, GEJ. 1983. Sex-and age-specific reproductive and physical parameters of the bobcat *Lynx rufus* on Cape Breton Island, Nova Scotia. Can J Zool. 61:1771-1782.
8240. Parker, KL & Wong, B. 1987. Raising black-tailed deer fawns at natural growth rates. Can J Zool. 65:20-23.
8241. Parker, PJ. 1977. An ecological comparison of marsupial and placental patterns of reproduction. The Biology of Marsupials. 273-286, Stonehouse, B & Gilmore, D, eds, Univ Park Press, Baltimore.
8242. Parker, SP, ed. 1990. Grzimek's Encyclopedia of Mammals. McGraw-Hill. NY, NY.
8243. Parkes, AS. 1931. The reproductive processes of certain mammals Part I The oestrous cycle of the Chinese hamster (*Cricetulus griseus*). Proc R Soc London Biol Sci. 108B:138-147.
8244. Parkes, AS. 1969. Multiple births in man. J Reprod Fert Suppl. 6:105-116.
8245. Parks, E. 1967. Second litters in the striped skunk. NY Fish Game J. 14:208-209.
8245a. Parodi, PW & Griffiths, M. 1983. A comparison of the positional distribution of fatty acids in milk triglycerides of the extant monotremes platypus (*Ornithorhynchus anatinus*) and echidna (*Tachyglossus aculeatus*). Lipids. 18:845-847.
8246. Parsons, FG. 1894. On the anatomy of *Atherurus africanus*. Proc Zool Soc London. 1894:675-692.
8247. Parsons, FG. 1898. On the anatomy of the African jumping-hare (*Pedetes caffer*) compared with that of the Dipodidae. Proc Zool Soc London. 1898:858-890.
8248. Parsons, G & Brown, MK. 1976. Yearling reproduction in beaver as related to population density in a portion of New York. Proc NE Fish Wildl Conf. 36:188-191.
8249. Parsons, HJ, Smith, DA, & Whittam, RF. 1986. Maternity colonies of silver-haired bats, *Lasionycteris noctivagans*, in Ontario and Saskatchewan. J Mamm. 67:598-600.
8250. Pascal, M. 1981. Le Dauphin de Commerson aux Iles Kerguelen. Terre Vie. 35:327-330.
8251. Pasley, JN & McKinney, TD. 1973. Grouping and ovulation in *Microtus pennsylvanicus*. J Reprod Fert. 34:527-530.
8252. Pastukhov, VD. 1968. On twins in *Pusa siberica* Gmel. Zool Zhurn. 47:479-482.
8253. Pasztor, LM & van Horn, RN. 1976. Twinning in prosimians. J Human Evol. 5:333-338.
8254. Patenaude, F. 1983. Care of the young in a family of wild beavers, *Castor canadensis*. Acta Zool Fenn. 174:121-122.
8255. Patenaude, F & Bovet, J. 1983. Parturition and related behavior in wild American beavers (*Castor canadensis*). Z Säugtierk. 48:136-145.
8256. Patric, EF. 1962. Reproductive characteristics of red-backed mouse during years of differing population densities. J Mamm. 43:200-205.
8257. Patterson, BD & Gallardo, MH. 1987. *Rhyncholestes raphanurus*. Mamm Species. 286:1-5.
8258. Patterson, JD. 1973. Ecologically different patterns of aggressive and sexual behavior in two troops of Ugandan baboons, *Papio anubis*. Am J Phys Anthropol. 38:641-647.
8259. Patterson, JT. 1912. A preliminary report on the demonstration of polyembryonic development in the armadillo (*Tatu novemcinctum*). Anat Anz. 41:369-381.
8260. Patterson, JT. 1918. Polyembryonic development in *Tatusia novemcincta*. J Morphol. 24:559-683.
8261. Patterson, JT & Hartman, CG. 1917. A polyembryonic blastocyst in the opossum. Anat Rec. 13:87-95.
8262. Pattie, D. 1973. *Sorex bendirii*. Mamm Species. 27:1-2.
8263. Patton, JL & Yang, SY. 1977. Genetic variation in *Thomomys bottae* pocket gophers: Macrogeographic patterns. Evolution. 31:697-720.
8264. Pauerstein, CJ, Eddy, CA, Croxatto, HD, et al. 1978. Temporal relationships of estrogen, progesterone, and luteinizing hormone levels to ovulation in women and infrahuman primates. Am J Obstet Gynecol. 130:876-886.
8265. Paul, A & Kuester, J. 1987. Sex ratio adjustment in a seasonally breeding primate species: Evidence from the Barbary macaque population at Affenberg Salem. Ethology. 74:117-132.
8265a. Paul, A & Kuester, J. 1990. Adaptive significance of sex ratio adjustment in semifree-ranging Barbary macaques (*Macaca sylvanus*) at Salem. Beh Ecol Sociobiol. 27:287-293.
8266. Paul, A & Thommen, D. 1984. Timing of birth, female reproductive success and infant sex ratio in semi free-ranging barbary macaques (*Macaca sylvanus*). Folia Primatol. 42:2-16.
8267. Paul, JR. 1967. Round-tailed muskrat in west central Florida. Q J FL Acad Sci. 30:227-229.
8268. Paul, JR. 1970. Observations on the ecology, populations and reproductive biology of the pine vole, *Microtus pinetorum*, in North Carolina. IL State Mus Rep Invest. 20:1-28.
8268a. Paul-Murphy, J, Tell, LA, Bravo, W, et al. 1991. Urinary steroid evaluations to monitor ovarian function in exotic ungulates: VIII Correspondence of urinary and plasma steroids in the llama (*Lama glama*) during nonconceptive and conceptive cycles. Zoo Biol. 10:225-236.
8269. Paulian, P. 1954. Sur quelques variations de la date de mise bas de l'Élephant de mer (*Mirounga leonina*). Mammalia. 18:375-379
8270. Paulraj, S. 1988. Breeding behaviour of the Malay giant squirrel *Ratufa bicolor* at Arignar Anna Zoological Park. Int Zoo Yearb. 27:279-282.
8271. Paulson, DD. 1988. *Chaetodipus baileyi*. Mamm Species. 297:1-5.
8272. Paulson, DD. 1988. *Chaetodipus hispidus*. Mamm Species. 320:1-4.
8273. Pavlenskii, LA. 1937. Materiali porazmnojeniiu i soderjaniiu enotovidnoi sobaki v nevole. Tr Novosibirsk Zoosada. 1:97-112. [Cited by Barbu, 1970]
8274. Pavlinin, VN. 1956. Reproduction and season of mole-trapping in the Ural. Zool Zhurn. 35:606-613.
8275. Pavlov, AN. 1959. Reproduction peculiarities of *Meriones meridianus* Pall and *M tamariscinus* Pall under the conditions of the north-western Caspian territory. Zool Zhurn. 38:1876-1885.
8276. Pawlinin, W. 1966. Der Zobel Neuer Brehm-Bücherei Bd 363. A Ziemsen, Wittenberg Lutherstadt.

8277. Payman, BC & Swanson, HH. 1980. Social influence on sexual maturation and breeding in the female Mongolian gerbil (*Meriones unguiculatus*). Anim Behav. 28:528-535.
8278. Payne, MR. 1977. Growth of a fur seal population. Phil Trans R Soc London Biol Sci. 279B:67-79.
8279. Payne, MR. 1979. Growth in the Antarctic fur seal *Arctocephalus gazella*. J Zool. 187:1-20.
8280. Payne, NF. 1984. Reproductive rates of beaver in Newfoundland. J Wildl Manage. 48:912-917.
8281. Payne, R. 1984. Reproduction rates and breeding area occupancy in the southern right whale, *Eubalaena australis*. Rep Int Whal Comm Sp Issue. 6:482.
8282. Payne, RL, Provost, EE, & Urbston, DF. 1966. Delineation of the period of rut and breeding season of a white-tailed deer population. Proc Ann Conf SE Assoc Game Fish Comm. 20:130-138.
8283. Peacock, EH. 1933. A Game-book for Burma. Witherby & Co, London.
8284. Peacock, LJ & Rogers, CM. 1959. Gestation period and twinning in chimpanzees. Science. 129:959.
8285. Peaker, M & Goode, JA. 1978. The milk of the fur-seal, *Arctocephalus tropicalis gazella* [*sic*] in particular the composition of aqueous phase. J Zool. 185:469-476.
8285a. Pearse, AS & Kellogg, R. 1938. Mammalia from Yucatan caves. Publ Carnegie Inst Washington. 491:301-304.
8286. Pearse, RJ. 1981. Notes on breeding, growth and longevity of the forester or eastern grey kangaroo, *Macropus giganteus* Shaw, in Tasmania. Aust Wildl Res. 8:229-235.
8287. Pearson, AM. 1962. Activity patterns, energy metabolism, and growth rate of the voles *Clethrionomys rufocanus* and *C glareolus* in Finland. Ann Zool Soc Zool Bot Fenn Vanamo. 24:1-57.
8288. Pearson, AM. 1972. Population characteristics of the northern interior grizzly in the Yukon Territory, Canada. Bears: Their Biology and Management: Int Conf Bear Res Manag, 2:32-35, Univ Alberta, Canada, IUCN, Morges, Switzerland.
8289. Pearson, AM. 1975. The northern interior grizzly bear, *Ursus arctos* L. Can Wildl Serv Rep Ser. 34:1-84.
8290. Pearson, H & Wright, AI. 1968. Some observations on the rearing of an okapi calf *Okapia johnstoni*. Int Zoo Yearb. 8:134-136.
8291. Pearson, J. 1944. The vaginal complex of the rat kangaroos. Aust J Sci. 7:80-83.
8292. Pearson, J. 1945. The female urogenital system of the Marsupialia with special reference to the vaginal complex. Pap Proc R Soc Tas. 1944:71-98.
8293. Pearson, J. 1946. The affinities of the rat-kangaroos (Marsupialia) as revealed by a comparative study of the female urogenital system. Pap Proc R Soc Tas. 1945:13-25.
8294. Pearson, J. 1949. Placentation of the Marsupialia. Proc Linn Soc London. 161:1-9.
8295. Pearson, J. 1950. A further note on the female urogenital system of *Hypsiprymnodon moschatus* (Marsupialia). Pap Proc R Soc Tas. 1949:203-210.
8296. Pearson, J. 1950. The relationship of the Potoridae to the Macropodidae (Marsupialia). Pap Proc R Soc Tas. 1949:211-230.
8297. Pearson, J & de Bavay, JM. 1951. The female urogenital system of *Antechinus* (Marsupialia). Pap Proc R Soc Tas. 85:137-142.
8298. Pearson, J & de Bavay, JM. 1953. The urogenital system of the Dasyurinae and the Thylacininae (Marsupialia, Dasyuridae). Pap Proc R Soc Tas. 87:175-199.
8299. Pearson, OP. 1944. Reproduction in the shrew (*Blarina brevicauda* Say). Am J Anat. 75:39-89.
8300. Pearson, OP. 1948. Life history of mountain viscachas in Peru. J Mamm. 29:345-374.
8301. Pearson, OP. 1949. Reproduction of a South American rodent, the mountain viscacha. Am J Anat. 84:143-171.
8302. Pearson, OP. 1951. Mammals in the highlands of southern Peru. Bull Mus Comp Zool. 106:117-204.
8303. Pearson, OP. 1952. Notes on a pregnant sea otter. J Mamm. 33:387.
8304. Pearson, OP. 1959. Biology of the subterraenean rodents, *Ctenomys* in Peru. Memorias del Museo de Historia Natural "Javier Prado". 9:1-56.
8305. Pearson, OP. 1967. La estructura por edades y la dinamica reproductiva en una poblacion de ratones de campo, *Akodon azarae*. Physis. 27:53-58.
8305a. Pearson, OP. 1983. Characteristics of a mammalian fauna from forests in Patagonia, southern Argentina. J Mamm. 64:476-492.
8305b. Pearson, OP. 1984. Taxonomy and natural history of some fossorial rodents of Patagonia, south Argentina. J Zool. 202:225-237.
8306. Pearson, OP. 1988. Biology and feeding dynamics of a South American herbivorous rodent, *Reithrodon*. Stud Neotropical Fauna Environ. 23:25-39.
8307. Pearson, O, Martin, S, & Bellati, J. 1987. Demography and reproduction of the silky desert mouse (*Eligmodontia*) in Argentina. Fieldiana Zool. 39:433-446.
8308. Pearson, OP & Baldwin, PH. 1953. Reproduction and age structure of a mongoose population in Hawaii. J Mamm. 34:436-447.
8309. Pearson, OP & Bassett, CF. 1944. Size of the vulva in the fox and its relation to fertility. Anat Rec. 89:455-459.
8310. Pearson, OP & Bassett, CF. 1946. Certain aspects of reproduction in a herd of silver foxes. Am Nat. 80:45-67.
8311. Pearson, OP & Enders, RK. 1943. Ovulation, maturation and fertilization in the fox. Anat Rec. 85:69-83.
8312. Pearson, OP & Enders, RK. 1944. Duration of pregnancy in certain mustelids. J Exp Zool. 95:21-35.
8313. Pearson, OP & Enders, RK. 1944. Time of ovulation and fertilization in the fox. Am Fur Breeder. 16(7):32-34.
8314. Pearson, OP, Koford, MR, & Pearson, AK. 1952. Reproduction of the lump nosed bat (*Corynorhinus rafinesquei*) in California. J Mamm. 33:273-320.
8315. Pearson, PG. 1952. Observations concerning the life history and ecology of the woodrat, *Neotoma floridana floridana* (Ord). J Mamm. 33:459-463.
8316. Pearson, PG. 1954. Mammals of Gulf Hammock, Levy County, Florida. Am Mid Nat. 51:468-480.
8317. Pechev, T. 1962. Etude du Mulot rupestre *Apodemus mystacinus* en Bulgarie. Mammalia. 26:293-310.
8318. Pechlaner, H & Thaler, E. 1983. Beitrag zur Fortpflanzungsbiologie des europaischen Fischotters (*Lutra lutra* L). Zool Garten NF. 53:49-58.
8319. Peddemors, VM, de Muelenaere, HJH, & Devchand, K. 1989. Comparative milk composition of the bottlenosed dolphin (*Tursiops truncatus*), humpback dolphin (*Sousa plumbea*) and common dolphin (*Delphinus delphis*) from southern African waters. Comp Biochem Physiol. 94A:639-641.
8320. Pedersen, A. 1945. Der Eisbär *Thalarctos maritimus* Phipps: Verbreitung und Lebensweise. Aktieselkabet E Bruun & Co, Kopenhagen. [Cited by Dittrich, 1961]
8321. Pedersen, T. 1952. A note on humpback oil and on the milk and milk fat from this species (*Megaptera nodosa*). Norsk Hvalf Tid. 41:375-378.

8322. Pehle, C. 1972. Statistisches zur Haltung vom Halbeseln *Equus hemionus*) in Tiergarten. Zool Garten NF. 42:189-203.
8323. Pehrson, Å & Lindlöf, B. 1984. Impact of winter nutrition on reproduction in captive mountain hares (*Lepus timidus*) (Mammalia: Lagomorpha). J Zool. 204:201-209.
8324. Peirce, EJ & Breed, WG. 1987. Cytological organization of the seminiferous epithelium in the Australian rodents *Pseudomys australis* and *Notomys alexis*. J Reprod Fert. 80:91-103.
8325. Peitz, B. 1978. Changes in water metabolism during the estrous cycle in three species of rodents. Physiol Zool. 51:256-266.
8326. Peitz, B. 1981. The oestrous cycle of the spiny mouse. J Reprod Fert. 61:453-459.
8327. Peitz, B, Foreman, D, & Schmitt, M. 1979. The reproductive tract of the male spiny mouse (*Acomys cahirinus*) and coagulation studies with other species. J Reprod Fert. 57:183-188.
8328. Pekárková, B. 1973. *Capra cylindricornis* Blyth, 1841 in Zoological garden Liberec. Lynx. 14:120-122.
8329. Pelikán, J. 1964. Vergleich einiger populationsdynamischer Faktoren bei *Apodemus sylvaticus* (L) und *A microps* Kr et Ros. Z Säugetierk. 29:242-251.
8330. Pelikán, J. 1965. Reproduction, population structure and elimination of males in *Apodemus agrarius* (Pall). Zool Listy. 14:317-332.
8331. Pelikán, J. 1966. Comparison of the birth rate in four *Apodemus* species. Zool Listy. 15:125-130.
8332. Pelikán, J. 1966. Notes on reproduction and sex ratio in *Apodemus agrarius* (Pall). Lynx. 6:121-123.
8333. Pelikán, J. 1967. Variability of body weight in three *Apodemus* species. Zool Listy. 16:199-220.
8334. Pelikán, J. 1967. Resorption rate in embryos of four *Apodemus* species. Zool Listy. 16:325-342.
8335. Pelikán, J. 1973. Notes on the reproduction of *Pitymys subterraneus* (de Sél Long). Zool Listy. 22:285-296.
8336. Pelikán, J & Holisova, V. 1986. Faunal and ecological characteristics of a marginal population of *Microtus agrestis*. Ann Naturhist Mus Wien. 88/89B:257-265.
8337. Peller, S. 1940. Growth, heredity and environment. Growth. 4:277-291.
8338. Pelletier, RM. 1986. Cyclic formation and decay of the blood-testis barrier in the mink (*Mustela vison*) a seasonal breeder. Am J Anat. 175:91-117.
8339. Pelt, FL. 1967. Remarks on breeding seasons of some Artiodactyla in captivity. Zool Garten NF. 34:293-296.
8340. Pelton, MR. 1968. Fluctuations in testicular condition of cottontail rabbits in Georgia. Proc Annu Conf SE Assoc Game Fish Comm. 22:198-202.
8341. Pelton, MR & Jenkins, JH. 1971. Productivity of Georgia cottontails. Proc Annu Conf SE Assoc Game Fish Comm. 25:261-268.
8342. Pelton, MR & Provost, EE. 1972. Onset of breeding and breeding synchrony by Georgia cottontails. J Wildl Manage. 36:544-549.
8343. Pelz, HJ. 1980. Populationsokölogie der Brandmaus, *Apodemus agrarius* (Pallas 1771), an ihrer westlichen Verbreitungsgrenze in Osthessen. Z Angew Zool. 67:179-209.
8344. Pemberton, JM & Balmford, AP. 1987. Lekking in fallow deer. J Zool. 213:762-765.
8345. Pemberton, JM & Dansie, O. 1983. Live weights of fallow deer (*Dama dama*) in British deer parks. J Zool. 199:171-177.
8346. Pembleton, EF & Williams, SL. 1978. *Geomys pinetis*. Mamm Species. 86:1-3.
8347. Peng, M, Lai, Y, Yang, C, et al. 1973. Reproductive parameters of the Taiwan monkey (*Macaca cyclopis*). Primates. 14:201-214.
8348. Pengelley, ET. 1966. Differential developmental patterns and their adaptive value in various species of the genus *Citellus*. Growth. 30:137-142.
8349. Pengelley, ET & Asmundson, SJ. 1975. Female gestation and lactation as zeitgebers for circannual rhythmicity in the hibernating ground squirrel, *Citellus lateralis*. Comp Biochem Physiol. 50A:621-625.
8350. Penrice, GW. 1899. Penrice's sing-sing (*Cobus defassa penricei*). Great and Small Game of Africa. 281-282, Bryden, HA, ed, Rowland Ward, London.
8351. Penrice, GW. 1899. The springbuck in West Africa. Great and Small Game of Africa. 340-343, Bryden, HA, ed, Rowland Ward, London.
8352. Penzhorn, BL. 1979. Social organization of the Cape mountain zebra *Equus z zebra* in the Mountain Zebra National Park. Koedoe. 22:115-156.
8353. Penzhorn, BL. 1982. Home range size of Cape mountain zebras *Equus zebra zebra* in the Mountain Zebra National Park. Koedoe. 25:103-108.
8354. Penzhorn, BL. 1985. Reproductive characteristics of a free-ranging population of Cape mountain zebra (*Equus zebra zebra*). J Reprod Fert. 73:51-57.
8355. Penzhorn, BL. 1988. *Equus zebra*. Mamm Species. 314:1-7.
8356. Penzhorn, BL & van der Merwe, NJ. 1988. Testis size and onset of spermatogenesis in Cape mountain zebras (*Equus zebra zebra*). J Reprod Fert. 83:371-375.
8357. Pepe, GJ & Albrecht, ED. 1980. The utilization of placental substrates for cortisol synthesis by the baboon fetus near term. Steroids. 35:591-597.
8358. Pepe, GJ, Titus, JA, & Townsley, JD. 1977. Increasing fetal adrenal formation of cortisol from pregnenolone during baboon (*Papio papio*) gestation. Biol Reprod. 17:701-705.
8359. Pépin, D. 1977. Phase finale du cycle de reproduction du Lièvre, *Lepus europaeus*. Mammalia. 41:221-230.
8360. Pépin, D, Spitz, F, Janeau, G, et al. 1987. Dynamics of reproduction and development of weight in the wild boar (*Sus scrofa*) in south-west France. Z Säugetierk. 52:21-30.
8361. Peppler, RD. 1979. Reproductive parameters in the nine-banded armadillo. Anat Rec. 193:649-650.
8362. Peppler, RD & Canale, J. 1980. Quantitative investigation of the annual pattern of follicular development in the nine-banded armadillo (*Dasypus novemcinctus*). J Reprod Fert. 59:193-197.
8363. Peppler, RD, Hossler, FE, & Stone, SC. 1986. Determination of reproductive maturity in the female nine-banded armadillo (*Dasypus novemcinctus*). J Reprod Fert. 76:141-146.
8364. Peppler, RD & Stone, SC. 1976. Plasma progesterone level in the female armadillo during delayed implantation and gestation: Preliminary report. Lab Anim Sci. 26:501-504.
8365. Peppler, RD & Stone, SC. 1980. Clomiphene-induced ovulation in the 9-banded armadillo (*Dasypus novemcinctus*). Lab Anim. 14:329-330.
8366. Peppler, RD & Stone, SC. 1980. Plasma progesterone level during delayed implantation, gestation and postpartum period in the armadillo. Lab Anim Sci. 30:188-191.
8367. Peppler, RD & Stone, SC. 1981. Annual pattern in plasma testosterone in the male armadillo, *Dasypus novemcinctus*. Anim Prod Sci. 4:49-53.

8368. Peracchi, AL & de Albuquerque, ST. 1971. Lista provisória dos quiropteros dos estatos do Rio de Janeiro e guanabara, Brasil (Mammalia, Chiroptera). Rev Brasil Biol. 31:405-413.
8369. Perachio, AA, Alexander, M, Marr, LD, et al. 1977. Diurnal variations of serum testosterone levels in intact and gonadectomized male and female rhesus monkeys. Steroids. 29:21-33.
8369a. Pereira, ME & Izard, MK. 1989. Lactation and care for unrelated infants in forest-living ringtailed lemurs. Am J Primatol. 18:101-108.
8370. Perevalov, AA. 1956. Some data on the biology of reproduction of the hare *Lepus tolai* Lehmani Sevrtz. Zool Zhurn. 35:141-154.
8371. Perez, GSA. 1972. Observations of Guam bats. Micronesica. 8:141-149.
8371a. Perez, LE, Czekala, NM, Weisenseel, KA, et al. 1988. Excretion of radiolabeled estradiol metabolites in the slow loris (*Mycticebus coucang*). Am J Primatol. 16:321-330.
8372. Pernetta, JC. 1977. Population ecology of British shrews in grassland. Acta Theriol. 22:279-296.
8373. Perret, M. 1977. Influence du groupement social sur l'activation sexuelle saisonniére [*sic*] chez le mâle de *Microcebus murinus* (Miller 1777). Z Tierpsychol. 43:159-179.
8374. Perret, M. 1980. Influence de la Captivité et du Groupement Social sur la Physiologie du Microcèbe (*Microcebus murimus*), Cheirogaleinae, Primates. Thèse de Docteur es Sciences, Centre d'Orsay, Paris. [Cited by Cross & Martin, 1981]
8375. Perret, M. 1982. Influence du groupement social sur la reproduction de la femelle de *Microcebus murinus* (Miller, 1777). Z Tierpsychol. 60:47-65.
8376. Perret, M. 1985. Diurnal variations in plasma testosterone concentrations in the male lesser mouse lemur (*Microcebus murinus*). J Reprod Fert. 74:205-213.
8377. Perret, M. 1985. Influence of social factors on seasonal variations in plasma testosterone levels of *Microcebus murinus*. Z Tierpsychol. 69:265-280.
8378. Perret, M. 1986. Plasma testosterone-binding globulin-binding capacity in the male lesser mouse lemur (*Microcebus murinus*): Relationship to seasonal and social factors. J Endocrinol. 110:169-175.
8379. Perret, M. 1986. Social influences on oestrous cycle length and plasma progesterone concentrations in the female lesser mouse lemur (*Microcebus murinus*). J Reprod Fert. 77:303-311.
8379a. Perret, M. 1990. Influence of social factors on sex ratio at birth, maternal investment and young survival ina aprosimian primate. Beh Ecol Sociobiol. 27:447-454.
8380. Perret, M & Atramentowics, M. 1989. Plasma concentrations of progesterone and testosterone in captive wooly opossums (*Caluromys philander*). J Reprod Fert. 85:31-41.
8380a. Perret, M & Ben M'Barek, S. 1991. Male influence on oestrous cycles in female woolly opossum (*Caluromys philander*). J Reprod Fert. 91:557-566.
8381. Perret, M & Predine, J. 1984. Effects of long-term grouping on serum cortisol levels in *Microcebus murinus* (Prosimii). Horm Behav. 18:346-358.
8382. Perret, M & Schilling, A. 1987. Role of prolactin in a pheromone-like sexual inhibition in the male lesser mouse lemur. J Endocrinol. 114:279-287.
8383. Perrin, MR. 1979. The roles of reproduction, survival and territoriality in the seasonal dynamics of *Clethrionomys gapperi* populations. Acta Theriol. 24:475-500.
8384. Perrin, MR. 1986. Some perspectives on the reproductive tactics of southern African rodents. Cimbebasia. 8:63-77.
8385. Perrin, MR & Clarke, JR. 1987. A preliminary investigation of the bioenergetics of pregnancy and lactation of *Praomys natalensis* and *Saccostomus campestris*. S Afr J Zool. 22:77-82.
8386. Perrin, MR & Swaenepoel, P. 1987. Breeding biology of the bushveld gerbil *Tatera leucogaster* in relation to diet, rainfall and life history theory. S Afr J Zool. 22:218-227.
8387. Perrin, WF, Coe, JM, & Zweifel, JR. 1976. Growth and reproduction of the spotted porpoise, *Stenella attenuata*, in the offshore eastern tropical Pacific. Fish Bull USA. 74:229-269.
8388. Perrin, WF & Henderson, JR. 1984. Growth and reproductive rates in two populations of spinner dolphins, *Stenella longirostris*, with different histories of exploitation. Rep Int Whal Comm Sp Issue. 6:417-430.
8389. Perrin, WF, Holts, DB, & Miller, RB. 1977. Growth and reproduction of the eastern spinner dolphin, a geographical form of *Stenella longirostris* in the eastern tropical Pacific. Fish Bull USA. 75:725-750.
8390. Perrin, WF, Miller, RB, & Sloan, PA. 1977. Reproductive parameters of the offshore spotted dolphin, a geographical form of *Stenella attenuata*, in the eastern tropical Pacific, 1973-75. Fish Bull USA. 75:629-633.
8390a. Perrin, WP & Reilly, SB. 1984. Reproductive parameters of dolphins and small whales of the family Delphinidae. Rep Int Whal Comm Sp Issue. 6:97-133.
8391. Perrotta, CA. 1959. Fetal membranes of the Canadian porcupine, *Erethizon dorsatum*. Am J Anat. 104:35-59.
8392. Perry, JS. 1942. Reproduction in the water-vole, *Arvicola amphibius* Linn. Proc Zool Soc London. 112A:118-130.
8393. Perry, JS. 1945. The reproduction of the wild brown rat (*Rattus norvegicus* Erxleben). Proc Zool Soc London. 115:19-46.
8394. Perry, JS. 1952. The growth and reproduction of elephants in Uganda. Uganda J. 16:51-66.
8395. Perry, JS. 1953. The reproduction of the African elephant, *Loxodonta africana*. Phil Trans R Soc London Biol Sci. 237B:93-149.
8396. Perry, JS. 1965. The structure and development of the reproducive organs of the female African elephant. Phil Trans R Soc London Biol Sci. 248B:35-51.
8397. Perry, JS. 1974. Implantation, foetal membranes and early placentation of the African elephant *Loxodonta africana*. Phil Trans R Soc London Biol Sci. 269B:109-135.
8398. Pervaiz, S & Brew, K. 1985. Composition of the milks of the bottlenose dolphin (*Tursiops truncatus*) and the Florida manatee (*Trichechus manatus latirostris*). Comp Biochem Physiol. 84A:357-360.
8399. Pervushin. 1966. Observations in delivery of sperm whales. Zool Zhurn. 45:1892-1893.
8400. Peters, DG & Rose, RW. 1979. The oestrous cycle and basal body temperature in the common wombat (*Vombatus ursinus*). J Reprod Fert. 57:453-460.
8401. Peters, JM, Maier, R, Hawthorne, BE, et al. 1972. Composition and nutrient content of elephant (*Elephas maximus*) milk. J Mamm. 53:717-724.
8402. Peters, H & Clarke, JR. 1974. The development of the ovary from birth to maturity in the bank vole (*Clethrionomys glareolus*) and the vole (*Microtus agrestis*). Anat Rec. 179:241-252.
8403. Peters, N. 1939. Über Grosze, Wachstum und Alter des Blauwales (*Balaenoptera musculus* (L) und Finnwales *Balaenoptera physalus* [L]). Zool Anz. 127:193-204.
8404. Peters, W. 1872. Contributions to the knowledge of *Pectinator* a genus of rodent Mammalia from north-eastern Africa. Trans Zool Soc London. 7:397-409.

8405. Petersen, KE & Yates, TL. 1980. *Condylura cristata.* Mamm Species. 129:1-4.
8406. Petersen, MK. 1977. Courtship and mating patterns of margays. The World's Cats. 3:22-35, Eaton, RL, ed, Carnivore Res Inst, Burke, Seattle.
8407. Petersen, MK & Petersen, MK. 1978. Growth rates and other post-natal developmental changes in margays. Carnivore. 1(1):87-92.
8408. Petersen, RP & Payne, NF. 1986. Productivity, size, age, and sex structure of nuisance beaver colonies in Wisconsin. J Wildl Manage. 50:265-268.
8409. Petersen, SL. 1986. Age-and hormone-related changes in vaginal smear patterns in the gray-tailed vole. J Reprod Fert. 78:49-56.
8410. Petersen, SL. 1986. Age-related changes in plasma oestrogen concentration, behavioural responsiveness to oestrogen, and reproductive success in female gray-tailed voles, *Microtus canicaudus.* J Reprod Fert. 78:57-64.
8410a. Peterson, GD. 1956. *Suncus murinus,* a recent introduction to Guam. J Mamm. 37:278-279.
8411. Peterson, RL. 1955. North American Moose. Univ Toronto Press, Toronto.
8412. Peterson, RL. 1965. The genus *Vampyressa* recorded from British Honduras. J Mamm. 46:676.
8413. Peterson, RL & Fenton, MB. 1969. A record of Siamese twinning in bats. Can J Zool. 47:154.
8414. Peterson, RM, Jr, Batzli, GO, & Banks, EM. 1976. Activity and energetics of the brown lemming in its natural habitat. Arctic Alp Res. 8:131-138.
8415. Peterson, RO, Woolington, JD, & Bailey, TN. 1984. Wolves of the Kenai Peninsula, Alaska. Wildl Monogr. 88:1-52.
8416. Peterson, RS. 1968. Social behavior in pinnipeds with particular reference to the northern fur seal. The Behavior and Physiology of Pinnipeds. 3-53, Rice, CE, ed, Meredith Corp, New York.
8417. Peterson, RS & Bartholomew, GA. 1967. The natural history and behavior of the California sealion. Sp Publ Am Soc Mamm. 1:1-79.
8418. Peterson, RS, Hubbs, CL, Gentry, RL, et al. 1968. The Guadalupe fur seal: Habitat, behavior, population size, and field identification. J Mamm. 49:665-675.
8419. Peterson, RS & Reeder, WG. 1966. Multiple birth in the northern fur seal. Z Säugetierk. 31:52-56.
8420. Pétra, PH & Schiller, HS. 1977. Sex steroid binding protein in the plasma of *Macaca nemestrina.* J Steroid Biochem. 8:655-661.
8421. Petric, A. Personal communication to V Hayssen. Brookfield Zoological Society hoofed stock records.
8422. Petrides, GA, Thomas, BO, & Davis, RB. 1951. Breeding of the ocelot in Texas. J Mamm. 32:116.
8423. Petrie, GF & Todd, RE. 1923. A report on plague investigations in Egypt. Report Public Health Laboratory, Cairo, Plague Report. [Cited by Buxton, 1936]
8424. Petrovsky, YT. 1961. Ecological peculiarities of *Citellus suslicus* Guld in Byelorussia. Zool Zhurn. 40:736-748.
8425. Petrusewicz, K, ed. 1983. Ecology of the bank vole. Acta Theriol Suppl 1. 28:1-242.
8426. Petter, F. 1957. La reproduction du Fennec. Mammalia. 21:307-309.
8427. Petter, F. 1959. Reproduction en captivité du Zorille du Sahara, *Poecilictis libyca.* Mammalia. 23:378-380.
8428. Petter, F. 1967. Gerbils. UFAW Handbook on the Care and Management of Lab Anim, 3rd ed. 449-451, E & S Livingstone, Edinburgh.
8429. Petter, F, Chippaux, A, & Monmignaut, C. 1964. Observations sur la biologie, la reproduction et la croissance de *Lemniscomys striatus* (Rongeus, Muridés). Mammalia. 28:620-627.
8430. Petter, F, Karimi, Y, & de Almeida, CR. 1967. Un nouveau Rongeur de laboratoire, le Cricitidé *Calomys callosus.* CR Hebd Séances Acad Sci Série D Sci Nat. 265:1974-1976.
8431. Petter, F, Quilici, M, Ranque, P, et al. 1969. Croisement d'*Arvicantyis niloticus* (Rongeurs, Murides) du Senegal et d'Ethiopie. Mammalia. 33:540-541.
8432. Petter, JJ. 1962. Recherches sur l'écologie et éthologie des Lémuriens malgache. Mém Mus Nat Hist Nat Paris (Nouv Ser) Sér A Zool. 27:1-146.
8433. Petter, JJ. 1965. The lemurs of Madagascar. Primate Behavior Field Studies of Monkeys and Apes. 292-319, DeVore, I, ed, Holt, Rinehart and Winston, New York.
8434. Petter, JJ. 1977. The aye-aye. Primate Conservation. 37-57, Prince Rainier III of Monaco & Bourne, GH, eds, Academic Press, New York.
8435. Petter, JJ, Albignac, R, & Rumpler, Y. 1977. Mammifères Lémuriens (Primates Prosimiens). Faune de Madagascar. 44:1-509.
8436. Petter, JJ & Petter-Rousseaux, A. 1963. Notes biologiques sur les Centetinae. Terre Vie. 17:66-80.
8437. Petter, JJ & Peyrieras, A. 1969. Nouvelle, contribution a l'etude d'un lemurien malgache, le aye-aye (*Daubentonia madagascariensis* E Geoffroy). Mammalia. 34:167-193.
8438. Petter, JJ & Peyrieras, A. 1970. Observations éco-éthologiques sur les Lémuriens malgaches du genre *Hapalemur.* Terre Vie. 117:356-382.
8439. Petter, JJ & Peyrieras, A. 1975. Preliminary notes on the behavior and ecology of *Hapalemur griseus.* Lemur Biology. 281-286, Tattersall, I & Sussman, RW, eds, Plenum Press, New York.
8440. Petter, JJ, Schilling, A, & Pariente, G. 1975. Observations on behavior and ecology of *Phaner lucifer.* Lemur Biology. 209-218, Tattersall, I & Sussman, RW, eds, Plenum Press, New York.
8441. Petter-Rousseaux, A. 1954. La périodicité sexuelle et les variations de génitalia femelles chez un Prosimien malgache, *Cheirogaleus major* (E Geoffroy). Compt Rend Hebd Séances Acad Sci. 239:1083-1085.
8442. Petter-Rousseaux, A. 1958. Variations cycliques de la morphologie des organes génitaux externes femelles chez certains Strepsirhini. CR Soc Biol. 152:951-953.
8443. Petter-Rousseaux, A. 1962. Recherches sur la biologie de la reproduction des Primates inferieurs. Mammalia Suppl 1. 26:1-88.
8444. Petter-Rousseaux, A. 1964. Reproductive physiology and behavior of the Lemuroidea. Evolutionary and Genetic Biology of Primates. 2:91-132. Buettner-Janusch, ed, Academic Press, New York.
8445. Petter-Rousseaux, A. 1968. Cycles génitaux saisonniers des Lémuriens malgaches. Cycles Génitaux Saisonniers de Mammifères Sauvages. 11-18, Canivenc, R, ed, Masson et Cie, Paris.
8446. Petter-Rousseaux, A. 1970. Observations sur l'influence de la photopériode sur l'activité sexuelle chez *Microcebus murinus* (Miller, 1777) en captivité. Ann Biol Anim Bioch Biophys. 10:203-208.
8447. Petter-Rousseaux, A. 1972. Application d'un système semestriel de variation de la photopériode chez *Microcebus murinus* (Miller, 1777). Ann Biol Anim Bioch Biophys. 12:367-375.

8448. Petter-Rousseaux, A. 1974. Photoperiod, sexual activity and body weight variations of *Microcebus murinus* (Miller, 1777). Prosimian Biology. 365-387, Martin, RD, Doyle, GA, & Walker, AC, eds, Univ Pittsburgh Press, Pittsburgh.
8449. Petter-Rousseaux, A. 1975. Activité sexuelle de *Microcebus murinus* (Miller, 1777) soumis à des régimes photopériodiques expérimentaux. Ann Biol Anim Bioch Biophys. 15:503-508.
8450. Petter-Rousseaux, A. 1979. Age of *Microcebus murinus* at the onset of testicular development: Preliminary observations on photoperiodic effect. Ann Biol Anim Bioch Biophys. 19:1801-1806.
8451. Petter-Rousseaux, A. 1980. Seasonal activity rhythms, reproduction, and body weight variations in five sympatric nocturnal prosimians, in simulated light and climatic conditions. Nocturnal Malagasy Primates. 137-152, Charles-Dominique, P, Cooper, HM, Hladik, A, et al eds, Academic Press, New York.
8452. Petter-Rousseaux, A. 1984. Annual variations in the plasma thyroxine levels in *Microcebus murinus*. Gen Comp Endocrinol. 55:405-409.
8453. Petter-Rousseaux, A & Bourlière, F. 1959. La périodicité sexuelle des femelles d'un Primate inférieur, le *Microcebus murinus* Effet d'une inversion des variations saisonnières de l'éclairement. CR Soc Biol. 153:226-228.
8454. Petter-Rousseaurux, A & Picon, R. 1981. Annual variation in the plasma testosterone in *Microcebus murinus*. Folia Primatol. 36:183-190.
8455. Petterborg, LF & Reiter, RJ. 1980. Effect of photoperiod and melatonin on testicular development in the white-footed mouse, *Peromyscus leucopus*. J Reprod Fert. 60:209-212.
8455a. Petterborg, LJ, Vaughan, MK, Johnson, LY, et al. 1984. Modification of testicular and thyroid function by chronic exposure to short photoperiod: A comparison in four rodent species. Comp Biochem Physiol. 78A:31-34.
8456. Pettigrew, BG & Sadleir, RMFS. 1974. The ecology of the deer mouse *Peromyscus maniculatus* in a costal coniferous forest I Population dynamics. Can J Zool. 52:107-118.
8457. Petzsch, H. 1937. Die Fortpflanzungsbiologie des Hamsters (*Cricetus cricetus* L). Naturf (Berlin). 13:337-340.
8458. Petzsch, H & Witstruk, KG. 1958. Beobachtungen an daghestanischen Turen (*Capra caucasica cylindricornis* Blyth) im Berg-Zoo Halle. Zool Garten NF. 25:6-29.
8459. Peyre, A. 1952. Note sur la structure histologique de l'ovaire du Desman des Pyrénées (*Galemys pyrenaicus* G). Bull Soc Zool France. 77:441-447.
8460. Peyre, A. 1955. Intersexualité du tractus génital femelle du Desman des Pyrénées (*Galemys pyrenaicus* G). Bull Soc Zool France. 80:132-138.
8461. Peyre, A. 1956. Ecologie et biogeographie du Desman (*Galemys pyrenaicus* G) dans les Pyrénées françaises. Mammalia. 20:405-418.
8462. Peyre, A. 1968. Cycles génitaux et correlations hypophysogénitales chez trois insectivores européens. Cycles Génitaux Saisonniers de Mammifères Sauvages. 133-142, Canivenc, R, ed, Masson et Cie, Paris.
8463. Peyre, A & Herlant, M. 1963. Ovo-implantation différée et correlations hypophysogénitales chez la femelle du Minioptère (*Miniopterus schreibersii* B). CR Hebd Séances Acad Sci Paris. 257:524-526.
8464. Peyre, A & Herlant, M. 1967. Ovo implantation différée et déterminism hormonal chez le Minioptère, *Miniopteres schreibersi* K (Chiroptère). CR Soc Biol. 161:1779-1782.
8465. Peyre, A & Malassine, A. 1969. L'équipement stéroidodeshydrogénasique et la fonction endocrine du placenta de Minioptère (Chiroptère). CR Soc Biol. 163:914-917.
8466. Pfeffer, P. 1961. L'écologie du Sanglier en Asie Centrale d'après les recherches d'A A Sloudsky. Terre Vie. 16:368-372.
8467. Pfeffer, P. 1967. Le Mouflon de Corse (*Ovis ammon musimon* Schreber, 1782); Position systématique, écologie et éthologie comparées. Mammalia, Suppl. 31:1-262.
8468. Pfeifer, S. 1980. Role of the nursing order in social development of mountain lion kittens. Dev Psychobiol. 13:47-53.
8469. Pfeifer, SR. 1982. Variability in reproductive output and success of *Spermophilus elegans* ground squirrels. J Mamm. 63:284-289.
8470. Pfeiffer, EW. 1954. Reproduction of a Primitive Rodent, *Alpodontia rufa*. Ph D Diss. Univ CA, Berkeley.
8471. Pfeiffer, EW. 1955. Hormonally induced "mammary hairs" of a primitive rodent *Aplodontia rufa*. Anat Rec. 122:241-255.
8472. Pfeiffer, EW. 1956. Notes on reproduction in the kangaroo rat, *Dipodomys*. J Mamm. 37:449-450.
8473. Pfeiffer, EW. 1956. The male reproductive tract of a primitive rodent, *Aplodontia rufa*. Anat Rec. 124:629-635.
8474. Pfeiffer, EW. 1958. The reproductive cycle of the female mountain beaver. J Mamm. 39:223-235.
8475. Pfeiffer, EW. 1960. Cyclic changes in the morphology of the vulva and clitoris of *Dipodomys*. J Mamm. 41:43-48.
8476. Pfister, H & Rimath, R. 1979. Das ASJV-Hasen project: Die schweizerische Hasenforschung. Schweiz Jagdztg. 2:3-53.
8477. Pflieger, R. 1982. Le Chamois. Gerfaut Club-Parnesse, Paris.
8478. Philippi, F. 1893. Un nuevo marsupial chileno. Anal Univ Chile. 86:31-34.
8479. Philippot, E, Goffart, M, & Dresse, A. 1965. Le systeme surrénalo-sympathique chez le paresseux (*Choloepus hoffmanni* Peters). Arch Int Physiol Bioch. 73:476-504.
8480. Phillippo, M, Lincoln, GA, & Lawrence, CB. 1972. The relationship between thyroidal calcitonin and seasonal and reproductive change in the stag (*Cervus elaphus* L). J Endocrinol. 53:xlviii-xlix.
8481. Phillips, CJ. 1968. Systematics of megachiropteran bats in the Solomon Islands. Univ KS Publ Mus Nat Hist. 16:777-837.
8482. Phillips, CJ & Jones, JK, Jr. 1968. Additional comments on reproduction in the woolly opossum (*Caluromys derbianus*) in Nicaragua. J Mamm. 49:320-321.
8483. Phillips, CJ & Jones, JK, Jr. 1969. Notes on reproduction and development in the four-eyed opossum, *Philander opossum*, in Nicaragua. J Mamm. 50:345-348.
8483a. Phillips, CJ & Jones, JK, Jr. 1971. A new subspecies of the long-nosed bat, *Hylonycteris underwoodi*, from Mexico. J Mamm. 52:77-80.
8484. Phillips, DM & Fadem, BH. 1987. The oocyte of a new world marsupial, *Monodelphis domestica*: Structure, formation, and function of the enveloping mucoid layer. J Exp Zool. 242:363-371.
8485. Phillips, GL. 1966. Ecology of the big brown bat (Chiroptera: Vespertilionidae) in northeastern Kansas. Am Mid Nat. 75:168-198.
8486. Phillips, IR. 1976. Skeletal development in the foetal and neonatal marmoset (*Callithrix jacchus*). Lab Anim. 10:317-333.
8487. Phillips, IR. 1976. The reproductive potential of the common cotton-eared marmoset (*Callithrix jacchus*) in captivity. J Med Primatol. 5:49-55.
8488. Phillips, IR & Grist, SM. 1975. The use of transabdominal palpation to determine the course of pregnancy in the marmoset (*Callithrix jacchus*). J Reprod Fert. 43:103-108.

8489. Phillips, JA. 1981. Growth and its relationship to the initial annual cycle of the golden-mantled ground squirrel, *Spermophilus lateralis*. Can J Zool. 59:865-871.
8490. Phillips, JFV. 1926. 'Wild pig' (*Potamocheorus choeropotamus*) at the Knysna: Notes by a naturalist. S Afr J Sci. 23:655-660.
8491. Phillips, MW, Stephens, MN, & Worden, AN. 1952. Observations on the rabbit in west Wales. Nature. 169:869-870.
8492. Phillips, WR & Inwards, SJ. 1985. The annual activity and breeding cycles of Gould's long-eared bat, *Nyctophilus gouldi* (Microchiroptera, Vespertilionidae). Aust J Zool. 33:111-126.
8493. Phillips, WWA. 1921. Notes on the habits of some Ceylon bats. J Bombay Nat Hist Soc. 28:448-452.
8494. Phillips, WWA. 1924. A guide to the mammals of Ceylon. Ceylon J Sci Series B Zool Geol. 13:1-64.
8495. Phillips, WWA. 1925. A guide to the mammals of Ceylon Part II Carnivora. Ceylon J Sci Series B Zool Geol. 13:143-184.
8496. Phillips, WWA. 1926. A guide to the mammals of Ceylon Part IV The Monkeys. Ceylon J Sci Series B Zool Geol. 13:261-284.
8497. Phillips, WWA. 1926. A guide to the mammals of Ceylon Part V The Pangolin. Ceylon J Sci Series B Zool Geol. 13:285-289.
8498. Phillips, WWA. 1927. A Guide to the mammals of Ceylon Part VI Ungulata. Ceylon J Sci Sect B Zool Geol. 14:1-50.
8499. Phillips, WWA. 1927. A Guide to the mammals of Ceylon Part VII Sirenia (The Dugong). Ceylon J Sci Sect B Zool Geol. 14:51-55.
8500. Phillips, WWA. 1928. A Guide to the mammals of Ceylon Part VIII Rodentia. Ceylon J Sci Sect B Geol Zool. 14:209-293.
8501. Phillips, WWA. 1935. Manual of the Mammals of Ceylon. Dulau & Co, London.
8502. Phillips, WWA. 1950. On the young of the Ceylon rusty-spotted cat (*Prionailurus rubiginosus phillipsi* Pocock). J Bombay Nat Hist Soc. 49:297-298.
8503. Phillips-Conroy, JE & Rogers, JA. 1985. Duration of the third stage of labor in a female yellow baboon from Mikumi National Park, Tanzania. Folia Primatol. 45:44-47.
8504. Phoenix, CH. 1976. Sexual behavior of castrated male rhesus monkeys treated with 19-hydroxytestosterone. Physiol Behav. 16:305-310.
8505. Phoenix, CH. 1980. Copulation, dominance, and plasma androgen levels in adult rhesus males born and reared in the laboratory. Arch Sex Behav. 9:149-167.
8506. Phoenix, CH & Chambers, KC. 1984. Sexual deprivation and its influence on testosterone levels and sexual behavior of old and middle-aged rhesus males. Biol Reprod. 31:480-486.
8507. Pi, JS. 1972. Notes on the ecology of five lorisiformes of Rio Muni. Folia Primatol. 18:140-151.
8508. Pi, JS. 1973. Contribution to the ecology of *Colobus polykomos satanas* (Waterhouse, 1838) of Rio Muni, Republic of Equatorial Guinea. Folia Primatol. 19:193-207.
8509. Pickering, DE. 1968. Reproduction characteristics in a colony of laboratory confined mulatta macaque monkeys. Folia Primatol. 8:169-179.
8510. Pickworth, S, Yerganian, G, & Chang, MC. 1968. Fertilization and early development in the Chinese hamster, *Cricetulus griseus*. Anat Rec. 162:197-208.
8511. Pidduck, ER & Falls, JB. 1973. Reproduction and emergence of juveniles in *Tamias striatus* (Rodentia: Sciuridae) at two localities in Ontario, Canada. J Mamm. 54:693-707.
8512. Piechocki, R. 1967. Die südostasiatischer Biber, *Castor fiber birulai* in der Mongolischen Volksrepublik. Arch Natursch Landschaftsforsch. 7:31-46.
8513. Pielowski, Z. 1971. The individual growth curve of the hare. Acta Theriol. 16:79-88.
8514. Pienaar, UdeV. 1963. The large mammals of the Kruger National Park: Their distribution and present day status. Koedoe. 6:1-37.
8515. Pienaar, UdeV. 1969. Observations on developmental biology, growth and some aspects of the population ecology of African buffalo, (*Syncerus caffer caffer* Sparrman) in the Kruger National Park. Koedoe. 12:29-52.
8516. Pieper, DR, Lobocki, CA, Thompson, M, et al. 1990. The olfactory bulbs tonically inhibit serum gonadotropin and prolactin levels in male hamsters on long or short photoperiods. J Neuroendo. 2:707-715.
8517. Pierce, JD, Jr, Ferguson, B, Salo, AL, et al. 1990. Patterns of sperm allocation across successive ejaculates in four species of voles (*Microtus*). J Reprod Fert. 88:141-149.
8518. Pietrzyk-Walknowska, J. 1956. Sexual maturation and reproduction in *Myocastor coypus* Part 3: The testicle. Folia Biol. 4:151-160.
8519. Pigozzi, G. 1989. Digging behaviour while foraging by the European badger, *Meles meles*, in a Mediterranean habitat. Ethology. 83:121-128.
8520. Pike, GC & MacAskie, IB. 1969. Marine mammals of British Columbia. Bull Fish Res Bd Can. 171:1-54.
8521. Pilawski, Z. 1969. Seasonal variation of ovulation response time after copulation in rabbits. Folia Biol. 17:211-217.
8522. Pilbeam, TE, Concannon, PW, & Travis, HF. 1979. The annual reproductive cycle of mink (*Mustela vison*). J Anim Sci. 48:578-584.
8523. Pilleri, G. 1959. Das Gehirn der Chinchillas und vergleichendanatomische Betrachtungen mit verwandten Nagerarten (Rodentia, Hystricomorpha). Acta Zool. 40:23-42.
8524. Pilleri, G. 1959. Das Gehirn von *Dolichotis patagona* und *Hydrochoerus hydrochoeris*, nebst Betrachtungen über die endocraniellen Verhältnisse (Rodentia, Hystricomorpha). Acta Zool. (Stockholm) 40:43-58.
8525. Pilleri, G. 1962. Zur Anatomy des Gehirnes von *Choeropsis liberiensis*. Acta Zool. (Stockholm) 43:229-246.
8526. Pilleri, G. 1964. Morphologie des Gehirnes des "southern right whale", *Eubales australis* Desmoulins 1822 (Cetacea, Mysticeti, Balaenidae). Acta Zool. (Stockholm) 45:245-272.
8527. Pilleri, G. 1970. Observations on the behaviour of *Platanista gangetica* in the Indus and Brahmaputra rivers. Invest Cetacea. 2:27-60.
8528. Pilleri, G. 1971. Beobachtungen über das Paarungsverhalten des Gangesdelphins, *Platanista gangetica*. Rev Suisse Zool. 78:231-234.
8529. Pilleri, G. 1971. Observations on the copulatory behaviour of the Gangetic dolphin, *Platanista gangetica*. Invest Cetacea. 3:31-33.
8530. Pilleri, G. 1972. Field observations carried out on the Indus dolphin *Platanista indi* in the winter of 1972. Invest Cetacea. 4:23-29.
8531. Pilleri, G. 1972. The cerebral anatomy of the Platanistidae (*Platanista gangetica, Platanista indi, Pontoporia plainvillei, Inia geoffrensis*). Invest Cetacea. 4:44-70.
8532. Pilleri, G. 1977. Hippomanes (allantoic 'calculi') in the Ganges dolphin, *Platanista gangetica*. Invest Cetacea. 8:121-122.
8533. Pilleri, G & Chen, P. 1982. The brain of the Chinese finless porpoise *Neophocaena asiaeorientalis* (Pilleri and Gihr, 1972): Macroscopic anatomy. Invest Cetacea. 13:27-68.
8534. Pilleri, G & Gihr, M. 1969. On the anatomy and behaviour of Risso's dolphin (*Grampus griseus* G Cuvier). Invest Cetacea. 1:74-93.

8535. Pilleri, G & Gihr, M. 1969. Zur Anatomie und Pathologie von *Inia geoffhensis* de Blainville 1817 (Cetacea, Susuidae) aus dem Beni, Bolivien. Invest Cetacea. 1:94-106.
8536. Pilleri, G & Gihr, M. 1970. The central nervous system of the mysticete and odontocete whales. Invest Cetacea. 2:89-128.
8537. Pilleri, G & Gihr, M. 1971. Brain-body weight ratio in *Pontoporia blainvillei.* Invest Cetacea. 3:69-73.
8538. Pilleri, G & Gihr, M. 1972. Contribution to the knowledge of the Cetaceans of Pakistan with particular reference to the genera *Neomeris, Sousa, Delphinus* and *Tursiops* and description of a new Chinese porpoise (*Neomeris asiadeorientalis*). Invest Cetacea. 4:107-162.
8539. Pilleri, G & Gihr, M. 1977. Observations on the Bolivian (*Inia bolissiensis* d'Orbigny, 1834) and the Amazonian bufeo (*Inia geoffrensis* de Blainville, 1817) with description of a new subspecies *Inia geoffrensis humboldtiana*. Invest Cetacea. 8:11-76.
8540. Pilleri, G & Gihr, M. 1982. The cepalization of *Cephalorhynchus commersoni.* Invest Cetacea. 13:79-85.
8541. Pilson, MEQ & Cooper, RW. 1967. Composition of milk from *Galago crassicaudatus.* Folia Primatol. 5:88-91.
8542. Pilson, MEQ & Kelly, AL. 1962. Composition of the milk from *Zalophus californianus*, the California sea lion. Science. 135:104-105.
8543. Pilson, MEQ & Waller, DW. 1970. Composition of milk from spotted and spinner porpoises. J Mamm. 51:74-49.
8544. Pilton, PE. 1961. Reproduction in the great grey kangaroo. Nature. 189:984-985.
8545. Pilton, PE & Sharman, GB. 1962. Reproduction in the marsupial *Trichosurus vulpecula.* J Endocrinol. 25:119-136.
8546. Pimentel, D. 1955. Biology of the Indian mongoose in Puerto Rico. J Mamm. 36:62-68.
8547. Pimlott, DH. 1959. Reproduction and productivity of Newfoundland moose. J Wildl Manage. 23:381-401.
8547a. Pimlott, DH, Shannon, JA, & Kolenosky, GB. 1969. The ecology of the timber wolf in Algonquin Provincial Park. Fish Wildl Res Branch Rep. 87:1-92.
8548. Pinder, L & Grosse, AP. 1991. *Blastocerus dichotomus.* Mamm Species. 380:1-4.
8549. Pine, RH. 1969. Reproduction in the gray-necked chipmunk, *Eutamias cinereicollis.* J Mamm. 50:642.
8550. Pine, RH. 1971. A review of the long-whiskered rice rat, *Oryzomys bombycinus* Goldman. J Mamm. 52:590-596.
8551. Pine, RH. 1972. The bats of the genus *Carollia.* TX A & M Univ Tech Monogr. 8:1-125.
8552. Pine, RH. 1979. Taxonomic notes on "*Monodelphis dimidiata itatiayae* (Miranda-Ribeiro)", *Monodelphis domestica* (Wagner) and *Monodelphis maraxina* Thomas (Mammalia: Marsupialia: Didelphidae). Mammalia. 43:495-499.
8553. Pine, RH. 1981. Reviews of the mouse opossums *Marmosa parvidens* Tate and *Marmosa invicta* Goldman (Mammalia: Marsupialia: Didelphidae) with description of a new species. Mammalia. 45:55-70.
8554. Pine, RH, Carter, DC, & LaVal, RK. 1971. Status of *Bauerus* van Gelder and its relationships to other nyctophiline bats. J Mamm. 52:663-669.
8555. Pine, RH, Dalby, PL, & Matson, JO. 1985. Ecology, postnatal development, morphometrics, and taxonomic status of the short-tailed opossum, *Monodelphis dimidiata*, an apparently semelparous annual marsupial. Ann Carnegie Mus. 54:195-231.
8556. Pine, RH, Miller, SD, & Schambezger, ML. 1979. Contributions to the mammalogy of Chile. Mammalia. 43:339-376.
8557. Pinkel, D, Gledhill, BL, Lake, S, et al. 1982. Sex preselection in mammals? Separation of sperm bearing Y and "O" chormosomes in the vole *Microtus oregoni.* Science. 218:904-906.
8558. Pinter, AJ. 1968. Effects of diet and light on growth, maturation, and adrenal size of *Microtus montanus.* Am J Physiol. 215:461-466.
8559. Pinter, AJ. 1970. Reproduction and growth for two species of grasshopper mice (*Onychomys*) in the laboratory. J Mamm. 51:236-243.
8560. Pinter, AJ. 1985. Effects of hormones and gonadal status on the midventral gland of the grasshopper mouse *Onychomys leucogaster.* Anat Rec. 211:318-322.
8561. Pinter, AJ. 1986. Population dynamics and litter size of the montane vole, *Microtus montanus.* Can J Zool. 64:1487-1490.
8562. Pinter, AJ & Negus, NC. 1965. Effects of nutrition and photoperiod on reproductive physiology of *Microtus montanus.* Am J Physiol. 208:633-637.
8562a. Pirlot, P. 1963. Algunas consideraciones sobre la ecologia de los mamiferos del oeste de Venezuela. Rev Univ Zulia. 23/24:55-100.
8563. Pirlot, P & Kamiya, T. 1982. Relative size of brain and brain components in three gliding placentals (Dermoptera, Rodentia). Can J Zool. 60:565-572.
8564. Pirlot, PL. 1954. Pourcentages de jeunes et périodes de reproduction chez quelques Rongeurs du Congo Belge. Ann Mus R Congo Belge NS Sci Zool. 1:41-46.
8565. Pirlot, PL. 1957. Données complementaires sur la reproduction de quelques Rongeurs d'Afrique. Rev Zool Bot Afr. 56:293-300.
8566. Pirta, RS & Singh, M. 1980. Changes in home ranges of rhesus monkeys (*Macaca mulatta*) groups living in natural habitats. Proc Indian Acad Sci (Anim Sci). 89:512-525.
8567. Pirta, RS & Singh, M. 1982. Differences in home ranges of rhesus monkeys (*Macaca mulatta*) groups living in three ecological habitats. Proc Indian Acad Sci (Anim Sci). 91:13-26.
8568. Pitcher, KW & Calkins, DG. 1981. Reproductive biology of Steller sea lions in the gulf of Alaska. J Mamm. 62:599-605.
8569. Pitman, JM, III & Bradley, EL. 1984. Hypothyroidism in reproductively inhibited prairie deer mice (*Peromyscus maniculatus bairdi*) from laboratory populations. Biol Reprod. 31:895-904.
8570. Pizzimenti, JJ. 1981. Increasing sexual dimorphism in prairie dogs: Evidence for changes during the past century. Southwest Nat. 26:41-47.
8571. Pizzimenti, JJ & Collier, GD. 1975. *Cynomys parvidens.* Mamm Species. 52:1-3.
8572. Pizzimenti, JJ & Hoffmann, RS. 1973. *Cynomys gunnisoni.* Mamm Species. 25:1-4.
8573. Pizzimenti, JJ & McClenaghan, LR, Jr. 1974. Reproduction, growth and development, and behavior in the Mexican prairie dog, *Cynomys mexicanus* (Merriam). Am Mid Nat. 92:130-145.
8574. Planel, H, Guilhem, A, & Soleilhavoup, J-P. 1961. Le cycle annuel du cortex surrénal d'un semi-hibernant: *Miniopterus schreiberii* [*sic*]. CR Assoc Anat. 47:620-633.
8575. Plant, TM. 1980. The effects of neonatal orchidectomy on the developmental pattern of gonadotropin secretion in the male rhesus monkey (*Macaca mulatta*). Endocrinology. 106:1451-1454.
8576. Plant, TM. 1981. Time courses of concentrations of circulating gonadotropin, prolactin, testosterone, and cortisol in adult male rhesus monkeys (*Macaca mulatta*) throughout the 24h light-dark cycle. Biol Reprod. 25:244-252.

8577. Plant, TM. 1982. Effects of orchidectomy and testosterone replacement treatment on pulsatile luteinizing hormone secretion in the adult rhesus monkey (*Macaca mulatta*). Endocrinology. 110:1905-1913.
8578. Plant, TM. 1982. Pulsatile luteinizing hormone secretion in the neonatal male rhesus monkey (*Macaca mulatta*). J Endocrinol. 93:71-74.
8579. Plant, TM. 1985. A study of the role of the postnatal testes in determining the ontogeny of gonadotropin secretion in the male rhesus monkey (*Macaca mulatta*). Endocrinology. 116:1341-1350.
8580. Plant, TM. 1986. Striking sex difference in the gonadotropin response to gonadectomy during infantile development in the rhesus monkey (*Macaca mulatta*). Endocrinology. 119:539-545.
8581. Plant, TM & Dubey, AK. 1984. Evidence from the rhesus monkey (*Macaca mulatta*) for the view that negative feedback control of luteinizing hormone secretion by the testis is mediated by a deceleration of hypothalamic gonadotropin-releasing hormone pulse frequency. Endocrinology. 115:2145-2153.
8582. Plant, TM, Hess, DL, Hotchkiss, J, et al. 1978. Testosterone and the control of gonadotropin secretion in the male rhesus monkey (*Macaca mulatta*). Endocrinology. 103:535-541.
8583. Plant, TM, James, VHT, & Michael, RP. 1969. Metabolism of [4-^{14}C] progesterone in the rhesus monkey (*Macaca mulatta*). J Endocrinol. 43:493-494.
8584. Plant, TM, James, VHT, & Michael, RP. 1971. Conversion of [4-^{14}C] progesterone to androsterone by female rhesus monkeys (*Macaca mulatta*). J Endocrinol. 51:751-761.
8585. Plant, TM & Michael, RP. 1974. Urinary excretion of androsterone throughout the menstrual cycle in the rhesus monkey (*Macaca mulatta*). J Reprod Fert. 41:205-209.
8586. Plant, TM, Moossy, J, Hess, DL, et al. 1979. Further studies on the effects of lesions in the rostral hypothalamus on gonadotropin secretion in the female rhesus monkey. Endocrinology. 105:465-473.
8587. Plant, TM, Nakai, Y, Belchetz, P, et al. 1978. The sites of action of estradiol and phentolamine in the inhibition of pulsatile, circhoral discharges of LH in the rhesus monkey (*Macaca mulatta*). Endocrinol. 102:1015-1018.
8588. Plant, TM, Schallenberger, E, & Hess, DL. 1980. Influence of suckling on gonadotropin secretion in the female rhesus monkey (*Macaca mulatta*). Biol Reprod. 23:760-766.
8589. Plant, TM & Zorub, DS. 1986. Pinealectomy in agonadal infantile male rhesus monkeys (*Macaca mulatta*) does not interrupt initiation of the prepubertal hiatus in gonadotropin secretion. Endocrinology. 118:227-232.
8590. Plant, TM, Zumpe, D, Sauls, M, et al. 1974. An annual rhythm in the plasma testosterone of adult male rhesus monkeys maintained in the laboratory. J Endocrinol. 62:403-404.
8591. Platt, NE, Jr, Prime, JH, & Witthames, SR. 1975. The age of the grey seal at the Farne Islands. Trans Nat Hist Soc Northumbria. 42:99-106.
8592. Platz, CC, Jr, Wildt, DE, Howard, JG, et al. 1983. Electroejaculation and semen analysis and freezing in the giant panda (*Ailuropoda melanoleuca*). J Reprod Fert. 67:9-12.
8593. Pleticha, P. 1969. Polarmoschusochsen in Norwegen. Fr Kölner Zoo. 12:125-129.
8594. Pleticha, P. 1973. Jugendentwicklung bei Alpensteinbocken *Capra ibex ibex* (Linne, 1758) im Zoo. Säugetierk Mitt. 21:297-307.
8595. Pleticha, PJR. 1972. Die Setzzeiten des Alpensteinbocks (*Capra ibex ibex*) in Gefangenschaft. Säugetierk Mitt. 20:354-359.
8596. Plotka, ED, Koller, DE, Letellier, MA, et al. 1983. Androgen receptors in prostate, neck and antler pedicle skin of fawn and adult intact and castrate white-tailed deer. Biol Reprod. 28(Suppl):82.
8597. Plotka, ED, Seal, US, & Verme, LJ. 1982. Morphologic and metabolic consequences of pinealectomy in deer. The Pineal Gland, Vol II Extra Reproductive Effects. 153-169, Reiter, RJ, ed, CRC Press, Boca Raton, FL.
8598. Plotka, ED, Seal, US, Letellier, MA, et al. 1979. Endocrine and morphologic effects of pinealectomy in white-tailed deer. Animal Models for Research on Contraception and Fertility. 452-466, Alexander, NJ, ed, Harper & Row, New York.
8599. Plotka, ED, Seal, US, Letellier, MA, et al. 1981. The effect of pinealectomy on seasonal phenotypic changes in white-tailed deer *Odocoileus virginianus borealis*. Pineal Function. 45-57, Matthews, CD & Seamark, RF, eds, Biomedical Press, Amsterdam, Elsevier.
8600. Plotka, ED, Seal, US, Letellier, MA, et al. 1984. Early effects of pinealectomy on LH and testosterone secretion in white-tailed deer. J Endocrinol. 103:1-7.
8601. Plotka, ED, Seal, US, Schmoller, GC, et al. 1977. Reproductive steroids in the white-tailed deer (*Odocoileus virginianus borealis*) I Seasonal changes in the female. Biol Reprod. 16:340-343.
8602. Plotka, ED, Seal, US, Schobert, EE, et al. 1975. Serum progesterone and estrogens in elephants. Endocrinology. 97:485-487.
8603. Plotka, ED, Seal, US, Verme, LJ, et al. 1977. Reproductive steroids in the white-tailed deer (*Odocoileus virginianus borealis*) II Progesterone and estrogen levels in peripheral plasma during pregnancy. Biol Reprod. 17:78-83.
8604. Plotka, ED, Seal, US, Verme, LJ, et al. 1980. Reproductive steroids in deer III Luteinizing hormone, estradiol and progesterone around estrus. Biol Reprod. 22:576-581.
8605. Plotka, ED, Seal, US, Verme, LJ, et al. 1982. Reproductive steroids in white-tailed deer IV Origin of progesterone during pregnancy. Biol Reprod. 26:258-262.
8606. Plotka, ED, Seal, US, Verme, LJ, et al. 1983. The adrenal gland in white-tailed deer: A significant source of progesterone. J Wildl Manage. 47:38-44.
8607. Plotka, ED, Seal, US, Zarembka, FR, et al. 1988. Ovarian function in the elephant: Luteinizing hormone and progesterone cycles in African and Asian elephants. Biol Reprod. 38:309-314.
8608. Plotka, ED, Witherspoon, DM, & Foley, CW. 1972. Luteal function in the mare as reflected by progesterone concentrations in peripheral blood plasma. Am J Vet Res. 33:917-920.
8609. Plotkin, VA. 1966. A rare instance of multifertility in *Balaenoptera borealis*. Zool Zhurn. 45:311-312.
8610. Poché, RM. 1975. Notes on reproduction in *Funisciurus anerythus* from Niger, Africa. J Mamm. 56:700-701.
8611. Poché, RM. 1975. The bats of National Park W, Niger, Africa. Mammalia. 39:39-50.
8612. Poché, RM. 1979. Notes on the big free-tailed bat (*Tadarida macrotis*) from south-west Utah, USA. Mammalia. 43:125-126.
8613. Poché, RM, Evans, SJ, Sultana, P, et al. 1987. Notes on the golden jackal (*Canis aureus*) in Bangladesh. Mammalia. 51:259-270.
8614. Pocock, RI. 1904. Exhibition of, and remarks upon, young examples of the Egyptian fat-tailed gerbile. Proc Zool Soc London. 1904:133-134.

8615. Pocock, RI. 1911. Exhibition of a newly born masked palm-civet (*Paradoxurus larvatus*), with an abnormal leg. Proc Zool Soc London. 1911:621-622.
8616. Pocock, RI. 1925. The external characters of an American badger (*Taxidea taxus*) and an American mink (*Mustela vison*), recently exhibited in the society's gardens. Proc Zool Soc London. 1925:17-25.
8616a. Pocock, RI. 1925. The external characters of the catarrine monkeys and ape. Proc Zool Soc London. 1925:1479-1579.
8617. Pocock, RI. 1926. The external characters of *Thylacinus*, *Sarcophilus*, and some related marsupials. Proc Zool Soc London. 1926:1037-1084.
8618. Pocock, RI. 1926. The external characters of the Patagonian weasel (*Lyncodon patagonicus*). Proc Zool Soc London. 1926:1085-1094.
8619. Pocock, RI. 1939. Fauna of British India Mammalia: Primates and Carnivora, 2nd ed. Taylor & Francis, London.
8620. Poduschka, W. 1974. Fortpflanzungseigenheiten und Jungenaufzucht des grossen Igel-Tenrek *Setifer setosus* (Froriep 1806). Zool Anz. 193:145-180.
8621. Poduschka, W. 1977. Das Paarungsvorspiel des Osteuropaischen Igels (*Erinaceus e roumanicus*) und theoretische Überlegungen zum Problem männlicher Sexualpheromone. Zool Anz. 199:187-208.
8622. Poduschka, W. 1981. Übertragung zweier Feten bei *Echinops telfari* (Insectivora: Tenrecidae). Zool Anz. 206:297-301.
8623. Poelker, RJ & Hartwell, HD. 1973. Black bear of Washington. WA State Game Dept Biol Bull. 14:1-180.
8624. Poglayen-Neuwall, I. 1962. Beiträge zu einem Ethogramm des Wickelbären *Poto flavus* Schreber. Z Säugetierk. 27:1-44.
8625. Poglayen-Neuwall, I. 1966. Notes on care, display, and breeding of olingos *Bassaricyon*. Int Zoo Yearb. 6:169-171.
8626. Poglayen-Neuwall, I. 1973. Observations on the birth of a kinkajou, *Potos flavus* (Schreber 1774). Zoologica. 58:41-42.
8627. Poglayen-Neuwall, I. 1975. Copulatory behavior, gestation and parturition of the tayra (*Eira barbara* L 1758). Z Säugetierk. 40:176-189.
8628. Poglayen-Neuwall, I. 1976. Fortpflanzung, Geburt und Aufzucht, nebst anderen Beobachtungen von Makibären (*Bassaricyon* Allen, 1876). Zool Beitr. 22:179-233.
8629. Poglayen-Neuwall, I. 1976. Zur Fortpflanzungsbiologie und Jugend-entwicklung von *Potos flavus* (Schreiber 1774). Zool Garten NF. 46:237-283.
8629a. Poglayen-Neuwall, I. 1977. Parturition in a hand-reared, primiparous gibbon (*Hylobates lar*). Zool Garten. 47:57-58.
8630. Poglayen-Neuwall, I. 1978. Breeding, rearing and notes on the behaviour of tayras *Eira barbara* in captivity. Int Zoo Yearb. 18:134.
8631. Poglayen-Neuwall, I & Poglayen-Neuwall, I. 1976. Postnatal development of tayras (Carnivora: *Eira barbara* L, 1758). Zool Beitr. 22:345-405.
8632. Poglayen-Neuwall, I & Poglayen-Neuwall, I. 1980. Gestation period and parturition of the ringtail *Bassariscus astatus* (Liechtenstein, 1830). Z Säugetierk. 45:73-81.
8633. Poglayen-Neuwall, I & Toweill, DE. 1988. *Bassariscus astutus*. Mamm Species. 327:1-8.
8634. Pogosianz, HE. 1967. The steppe lemming (*Lagurus lagurus* Pall). UFAW Handbook on the Care and Management of Lab Animals, 3rd ed. 452-456, E & S Livingstone, Edinburgh.
8635. Pohl, H. 1987. Control of annual rhythms of reproduction and hibernation by photoperiod and temperature in the Turkish hamster. J Therm Biol. 12:119-123.
8636. Pohl, L. 1908. Zur Naturgeschichte des kleinen Wiesels (*Ictis nivalis* L). Zool Anz. 33:264-267.
8637. Pohl, L. 1910. Wieselstudien. Zool Beob. 51:234-241.
8638. Pohle, C. 1972. Statistisches zur Haltung von Halbeseln (*Equus hemionus*) in Tiergarten. Zool Garten NF. 42:189-203.
8639. Pohle, C. 1973. Zur Zucht von Bengalkatzen (*Felis bengalensis*) in Tierpark Berlin. Zool Garten NF. 43:110-126.
8640. Pohle, C. 1974. Eine Zwillingsgeburt beim Burma-Leierhirsch (*Cervus eldi thamin*) im Tierpark Berlin und Notizen über Fortpflanzung und Geweihbildung bei diesen Hirschen. Zool Garten NF. 44:19-21.
8641. Pohle, C. 1974. Haltung und Zucht der Saiga-Antilope (*Saiga tatarica*) im Tierpark Berlin. Zool Garten NF. 44:387-409.
8642. Poiley, SM. 1949. Raising captive meadow voles (*Microtus p pennsylvanicus*). J Mamm. 30:317-318.
8643. Poiley, SM. 1972. Growth tables for 66 strains and stocks of laboratory animals. Lab Anim Sci. 22:759-778.
8644. Pojar, TM & Miller, LLW. 1984. Recurrent estrus and cycle length in pronghorn. J Wildl Manage. 48:973-979.
8645. Pokki, J. 1981. Distribution, demography and dispersal of the field vole *Microtus agrestis* (L), in the Tvärminne archipelago, Finland. Acta Zool Fenn. 164:1-48.
8646. Pokrovski, AV & Bolshakov, VN. 1968. Natural and potential fertility of two species of high mountain voles. Acta Theriol. 13:117-128.
8647. Pollack, EM. 1949. The Ecology of the Bobcat (*Lynx rufus rufus* Schreber) in the New England States. MS Thesis. Univ MA, Amherst.
8648. Pollock, JI. 1975. Field observations on *Indri indri*: A preliminary report. Lemur Biology. 287-311, Tattersall, I & Sussman, R, eds, Plenum Press, New York.
8649. Ponce de Lugo, CG. 1964. Length of menstrual cycles in reference to time of successful mating in *Macaca mulatta*. Biol Neonatorum. 6:104-111.
8650. Pond, CM. 1977. The significance of lactation in the evolution of mammals. Evolution. 31:177-199.
8651. Pook, AG. 1975. Breeding Goeldi's monkey (*Callimico goeldii*) at the Jersey Zoological Park. Ann Rep Jersey Wildl Preserv Trust. 12:17-20.
8652. Pook, AG. 1978. Breeding the Rodrigues fruit bat *Pteropus rodricensis* at the Jersey Zoological Park. Dodo J Jersey Wildl Preserv Trust. 14:30-33.
8653. Pook, AG. 1978. A comparison between the reproduction and parental behavior of the Goeldi's monkey (*Callimico goeldii*) and of the true marmosets (Callitrichidae). Biology and Behavior of Marmosets. 1-14, Rothe, H, Wolters, HJ, & Hearn, JP, eds, Eigenverlag Rothe, Göttingen.
8654. Pook, AG & Pook, G. 1979. A field study on the status and socioecology of the Goeldis monkey (*Callimico goeldii*) and other primates in northern Bolivia. Unpublished report. NY Zool Soc. [Cited by Heltne et al, 1981]
8655. Pook, AG & Pook, G. 1981. A field study of the socio-ecology of the Goeldi's monkey (*Callimico goeldii*) in northern Bolivia. Folia Primatol. 35:288-312.
8656. Poole, EL. 1932. Breeding of the hoary bat in Pennsylvania. J Mamm. 13:365-367.
8657. Poole, EL. 1936. Notes on the young of the Allegheny wood rat. J Mamm. 17:22-26.
8658. Poole, EL. 1938. Notes on the breeding of *Lasiurus* and *Pipistrellus* in Pennsylvania. J Mamm. 19:249.
8659. Poole, EL. 1940. A life history sketch of the Allegheny woodrat. J Mamm. 21:249-270.

8660. Poole, JH, Kasman, LH, Ramsay, EC, et al. 1984. Musth and urinary testosterone concentrations in the African elephant (*Loxodonta africana*). J Reprod Fert. 70:255-260.
8661. Poole, JH & Moss, CJ. 1981. Musth in the African elephant, *Loxodonta africana*. Nature. 292:830-831.
8662. Poole, TB. 1974. The effects of oestrous condition and familiarity on the sexual behaviour of polecats (*Mustela putorius* and *M furo* X *M putorius* hybrids). J Zool. 172:357-362.
8663. Poole, TB & Evans, RG. 1982. Reproduction, infant survival and productivity of a colony of common marmosets (*Callithrix jacchus jacchus*). Lab Anim. 16:88-97.
8664. Poole, TR & Evans, RG. 1982. Reproduction, infant survival and productivity of a colony of common marmosets (*Callithrix jacchus jacchus*). Lab Anim. 16:88-97.
8665. Poole, WE. 1973. A study of breeding in grey kangaroos *Macropus giganteus* Shaw and *Macropus fuliginosus* (Desmarest) in central New South Wales. Aust J Zool. 21:183-212.
8666. Poole, WE. 1975. Reproduction in the two species of grey kangaroos, *Macropus giganteus* Shaw and *M fuliginosus* (Desmarest) II Gestation, parturition and pouch life. Aust J Zool. 23:333-353.
8667. Poole, WE. 1976. Breeding biology and current status of the grey kangaroo *Macropus fuliginosus fuliginosus* of Kangaroo Island, South Australia. Aust J Zool. 24:169-187.
8668. Poole, WE. 1982. *Macropus giganteus*. Mamm Species. 187:1-8.
8669. Poole, WE. 1983. Breeding in the grey kangaroo, *Macropus giganteus*, from widespread locations in eastern Australia. Aust Wildl Res. 10:453-466.
8670. Poole, WE, Carpenter, SM, & Wood, JT. 1982. Growth of grey kangaroos and the reliability of age determination from body measurements I The eastern grey kangaroo, *Macropus giganteus*. Aust Wildl Res. 9:9-20.
8671. Poole, WE & Catling, PC. 1974. Reproduction in the two species of grey kangaroo *Macropus giganteus* Shaw and *Macropus fuliginosus* (Desmarest) I Sexual maturity and oestrus. Aust J Zool. 22:277-302.
8672. Poole, WE, Merchant, JC, Carpenter, SM, et al. 1985. Reproduction, growth and age determination in the yellow-footed rock-wallaby *Petrogale xanthopus* Gray, in captivity. Aust Wildl Res. 12:127-136.
8673. Poole, WE & Pilton, PE. 1964. Reproduction in the grey kangaroo, *Macropus canguru* in captivity. CSIRO Wildl Res. 9:218-234.
8674. Poole, WE, Sharman, GB, Scott, KJ, et al. 1982. Composition of milk from red and grey kangaroos with particular reference to vitamins. Aust J Biol Sci. 35:607-615.
8674a. Poole, WE, Westcott, M, & Simms, NG. 1992. Determination of oestrus in the female tammar, *Macropus eugenii*, by analysis of cellular composition of smears from the reproductive tract. Wildl Res. 19:35-46.
8675. Pope, NS. 1986. Seasonal differences in behavioral response to estradiol treatment in ovariectomized rhesus monkeys housed outdoors. Biol Reprod. 34(Suppl):381.
8676. Pope, NS, Gordon, TP, & Wilson, ME. 1986. Age, social rank and lactational status influence ovulatory patterns in seasonally breeding rhesus monkeys. Biol Reprod. 35:353-359.
8677. Pope, NS, Wilson, ME, & Gordon, TP. 1987. The effect of season on the induction of sexual behavior by estradiol in female rhesus monkeys. Biol Reprod. 36:1047-1054.
8677a. Popkin, BM, Bilsborrow, RE, & Akin, JS. 1982. Breast-feeding patterns in low-income countries. Science. 218:1088-1093.
8678. Popoff, N. 1911. Le tissu interstitiel et les corps jaunes de l'ovaire. Arch Biol. 26:403-556.
8679. Popov, L. 1979. Ladoga seal. Mammals in the Seas, FAO Fish Ser 5. 2:70-71.
8680. Popov, L. 1979. Baikal seal. Mammals in the Seas, FAO Fish Ser 5. 2:72-73.
8681. Popov, L. 1979. Caspian seal. Mammals in the Seas, FAO Fish Ser 5. 2:74-75.
8682. Popov, LA & Krylov, VI. 1978. On the period of reproduction in *Leptonychotes weddelli* on the King George Island (Waterloo), Antarctic. Zool Zhurn. 57:1250-1255.
8683. Porter, FL. 1979. Social behavior in the leaf-nosed bat, *Carollia perspicillata* I Social organization. Z Tierpsychol. 49:406-417.
8684. Porter, FL & McCracken, GF. 1983. Social behavior and allozyme variation in a captive colony of *Carollia perspicillata*. J Mamm. 64:295-298.
8685. Porter, G & Lacey, A. 1969. Breeding the Chinese hamster (*Cricetulus griseus*) in monogamous pairs. Lab Anim. 3:65-68.
8686. Porter, RH, Cavallaro, SA, & Moore, JD. 1980. Developmental parameters of mother-offspring interactions in *Acomys cahirinus*. Z Tierpsychol. 53:153-170.
8687. Porter, RH & Doane, HM. 1976. Maternal pheromone in the spiny mouse (*Acomys cahirinus*). Physiol Behav. 16:75-78.
8688. Portmann, A & Wirz, K. 1950. Die cerebralen Indices beim Okapi. Acta Tropica. 7:120-122.
8689. Porton, IJ, Kleiman, DG, & Rodden, M. 1987. Aseasonality of bush dog reproduction and the influence of social factors on the estrous cycle. J Mamm. 68:867-871.
8690. Posselt, J. 1963. The domestication of the eland. Rhod J Agr Res. 1:81-87.
8691. Potelov, VA. 1975. Reproduction of the bearded seal (*Erignathus barbatus*) in the Barents Sea. Rapp P V Réun Cons Int Explor Mer. 169:554.
8692. Potter, GE. 1930. *Marmosa* as a stowaway again. Science. 72:91.
8693. Potti, SP. 1966. A note on the breeding of the nilgiri tahr *Hemitragus hylocrius* at Trichuv Zoo. Int Zoo Yearb. 6:206.
8694. Pottinger, JA. 1911. Abnormal number of young in a markhor. J Bombay Nat Hist Soc. 20:1150.
8695. Potts, DM & Racey, PA. 1971. A light and electron microscope study of early development in the bat *Pipistrellus pipistrellus*). Micron. 2:322-348.
8695$_1$. Potvin, F. 1987. Wolf movements and population dynamics in Papineau-Labelle reserve, Quebec. Can J Zool. 66:1266-1273.
8695a. Poulet, AR. 1972. Recherches écologiques sur une savanne Sahelienne du Ferlo septentrionale Sénégal: Les Mammifères. Terre Vie. 26:440-472.
8695b. Poulet, AR. 1972. Evolution of the rodent population of a dry bush savanna in the Senegalese Sahel from 1969 to 1977. Bull Carnegie Mus Nat Hist. 6:113-117.
8696. Poulet, AR. 1984. Quelques observations sur la biologie de *Desmodiliscus braueri* Wettstein (Rodentia, Gerbillidae) dans le Sahel du Sénégal. Mammalia. 48:59-64.
8697. Poulter, TC, Pinney, TC, & Jennings, R. 1965. The rearing of Steller sea lions. Proc Ann Conf Bio Sonar Diving Mamm. 2:34-40.
8698. Pournelle, GH. 1952. Reproduction and early postnatal development of the cotton mouse, *Peromyscus gossypinus gossypinus*. J Mamm. 33:1-20.
8699. Pournelle, GH. 1955. Notes on the reproduction of a Baringo giraffe. J Mamm. 36:574.

8700. Pournelle, GH. 1960. Notes on the reproduction of the koala at San Diego Zoo. Int Zoo Yearb. 2:83.
8701. Pournelle, GH. 1960. Breeding notes on the east African colobus at San Diego Zoo. Int Zoo Yearb. 2:83-84.
8702. Pournelle, GH. 1960. San Diego's record of *Allenopithecus nigroviridis*. Int Zoo Yearb. 2:84.
8703. Pournelle, GH. 1960. Birth records of the pacific harbour seal at San Diego Zoo. Int Zoo Yearb. 2:89-90.
8704. Pournelle, GH. 1961. Notes on reproduction of the koala. J Mamm. 42:396.
8705. Pournelle, GH. 1962. Observations on the birth and early development of Allen's monkey. J Mamm. 43:265-266.
8706. Pournelle, GH. 1963. Eine Koala Kolonie im San Diego Zoo. Freunde der Kölner Zoo. 6:35-36.
8707. Pournelle, GH. 1964. A breeding note on the Asiatic ibex, *Capra ibex sibirica*. J Mamm. 45:629.
8708. Pournelle, GH. 1965. A breeding herd of lowland anoas *Anoa d depressicornis*. Int Zoo Yearb. 5:56-57.
8709. Pournelle, GH. 1965. Observations on birth and early development of the spotted hyena. J Mamm. 46:503.
8710. Pournelle, GH. 1967. Observations on reproductive behavior and early postnatal development of the proboscis monkey *Nasalis larvatus orientalis* at San Diego Zoo. Int Zoo Yearb. 7:90-92.
8711. Povilitis, A. 1985. Social behavior of the huemul (*Hippocamelus bisulcus*) during the breeding season. Z Tierpsychol. 68:261-286.
8712. Powell, N. Personal communication. [Cited by Mentis, 1972]
8713. Powell, RA. 1972. A comparison of populations of boreal red-backed vole (*Clethrionomys gapperi*) in tornado blowdown and standing forest. Can Field-Nat. 86:377-379.
8714. Powell, RA. 1981. *Martes pennanti*. Mamm Species. 156:1-6.
8715. Powell, RA. 1982. The Fisher: Life History, Ecology, and Behavior. Univ MN Press, Minneapolis.
8716. Powell, RA & Leonard, RD. 1983. Sexual dimorphism and energy expenditure for reproduction in female fisher *Martes pennanti*. Oikos. 40:166-174
8717. Powers, RA & Verts, BJ. 1971. Reproduction in the mountain cottontail rabbit in Oregon. J Wildl Manage. 35:605-613.
8718. Prakash, I. 1955. Notes on the desert hedgehog (*Hemiechinus auritus collaris* Gray. J Bombay Nat Hist Soc. 52:921-922.
8719. Prakash, I. 1958. The breeding season of the rhesus monkey *Macaca mulatta* (Zimmerman) in Rajasthan. J Bombay Nat Hist Soc. 55:154.
8720. Prakash, I. 1960. Breeding of mammals in Rajasthan desert, India. J Mamm. 41:386-389.
8721. Prakash, I. 1962. Ecology of the gerbils of Rajasthan desert, India. Mammalia. 26:311-331.
8722. Prakash, I. 1962. Group organization, sexual behavior and breeding season of certain Indian monkeys. Jap J Ecol. 12:83-86.
8723. Prakash, I. 1971. Breeding season and litter size of Indian desert rodents. Z Angew Zool. 58:441-454.
8724. Prakash, I. 1975. The population ecology of the rodents of the Rajasthan desert, India. Rodents in Desert Environments. 75-116, Prakash, I & Ghosh, PK, eds, Dr W Junk, The Hague.
8725. Prakash, I, Jain, AJ, & Purohit, KG. 1971. A note on the breeding and post-natal development of the Indian gerbil, *Tatera indica* (Hardwicke, 1807), in Rajastan desert. Säugetierk Mitt. 19:375-380.
8726. Prakash, I & Kametkar, LR. 1969. Body weight, sex and age factors in a population of northern palm squirrel, *Funambulus pennanti* Wroughton. J Bombay Nat Hist Soc. 66:99-115.
8727. Prakash, I & Purohit, KG. 1966. Some observations on the hairy-footed gerbille, *Gerbillus gleadowi* Munay, in the Rajasthan desert. J Bombay Nat Hist Soc. 63:431-434.
8728. Prakash, I & Rana, BD. 1970. A study of field population of rodents in the Indian desert. Z Angewandte Zoologie. 57:129-136.
8729. Prakash, I, Rana, BD, & Jain, AP. 1973. Reproduction in the cutch rock-rat, *Rattus cutchicus cutchicus*, in the Indian desert. Mammalia. 37:457-467.
8730. Prakash, I & Taneja, GC. 1969. Reproduction biology of the Indian desert hare, *Lepus nigricollis dayanus* Blandford. Mammalia. 33:102-117.
8731. Prange, HD, Schmidt-Nielsen, K, & Hackel, DB. 1968. Care and breeding of the fat sandrat (*Psammomys obesus* Crestzschmar). Lab Anim Care. 18:170-181.
8732. Prasad, MRN. 1951. Changes in the reproductive organs of the male palm squirrel *Funambulus palmarum* (Linn). Half Yrly J Mysore Univ. 12:89-105.
8733. Prasad, MRN. 1956. Reproductive cycle of the male Indian gerbile *Tateria indica Cavierii* (Waterhouse). Acta Zool (Stockholm). 37:87-122.
8734. Prasad, MRN. 1957. Male genital tract of the Indian and Ceylon palm squirrels and its bearing on the systematics of the Sciuridae. Acta Zool. (Stockholm) 38:1-26.
8735. Prasad, MRN, Dhaliwal, GK, Seth, P, et al. 1966. Biology of reproduction in the Indian palm squirrel, (*Funambulus pennanti*) (Wroughton). Symp Zool Soc London. 15:353-364.
8736. Prasad, MRN, Mossman, HW, & Scott, GL. 1979. Morphogenesis of the fetal membranes of an American mole, *Scalopus aquaticus*. Am J Anat. 155:31-68.
8737. Pratt, BL & Goldman, BD. 1986. Activity rhythms and photoperiodism of Syrian hamsters in a simulated burrow system. Physiol Behav. 36:83-89.
8738. Preble, NA. 1956. Notes on the life history of *Napaeozapus*. J Mamm. 37:196-200.
8739. Prejevalsky, N. 1876. Mongolia, the Tangut Country and the Solitudes of Northern Tibet, vol 1. Sampson, Low, Marston, Searle & Rivington, London.
8740. Prejevalsky, N. 1876. Mongolia, the Tangut Country and the Solitudes of Northern Tibet, vol 2. Sampson, Low, Marston, Searle & Rivington, London.
8741. Prell, H. 1930. Über doppelte Brunstzeit und verlängerte Tragzeit beiden europaischen Arten der Gattung *Ursus* Linné. Biol Zbl. 50:257-271.
8742. Prell, H. 1938. Die Tragzeiten der einheimischen Jagdtiere. Tharandter Forstl Jb. 89:696-701.
8743. Prelog, V & Meister, P. 1949. Untersuchungen, über Organextracte und Harn: Über die Isolierung von Progesteron aus dem Corpus luteum des Wales. Helv Chim Acta. 32:2435-2439.
8744. Preobrazhenskii, B. 1961. Management and breeding of reindeer. Reindeer Husbandry, 2nd revised ed, English trans. 78-128, Zhigunov, PS, ed, US Dept Commerce, Springfield, VA.
8745. Prescott, J & Ferron, J. 1978. Breeding and behaviour development of the American red squirrel *Tamaisciurus hudsonicus* in captivity. Int Zoo Yearb. 18:125-130.
8746. Presidente, PJA. 1982. Common brushtail possum *Trichosurus vulpecula*: Maintenance in captivity, blood values, diseases and parasites. The Management of Australian Mammals in Captivity. 55-66, Evans, DD, ed, Zool Bd Victoria, Melbourne.

8747. Presidente, PJA. 1982. Common wombat *Vombatus ursinus*: Maintenance in captivity, blood values, infections and parasitic diseases. The Management of Australian Mammals in Captivity. 133-143, Evans, DD, ed, Zool Bd Victoria, Melbourne.
8748. Preslock, JP, Hampton, SH, & Hampton, JK, Jr. 1973. Cyclic variations of serum progestins and immunoreactive estrogens in marmosets. Endocrinology. 92:1096-1101.
8749. Preslock, JP & Steinberger, E. 1976. Pathway of testosterone biosynthesis in the testis of the marmoset *Saguinus oedipus*. Steroids. 28:775-784.
8750. Preslock, JP & Steinberger, E. 1977. Testicular steroidogenesis in the common marmoset, *Callithrix jacchus*. Biol Reprod. 17:289-293.
8751. Preslock, JP & Steinberger, E. 1977. Testicular steroidogenesis in the mature and immature baboon *Papio anubis*. Gen Comp Endocrinol. 33:547-553.
8752. Preslock, JP & Steinberger, E. 1978. Substrate specificity for androgen biosynthesis in the primate testis. J Steroid Biochem. 9:163-167.
8753. Preslock, JP & Steinberger, E. 1978. Testicular steroidogenesis in the baboon *Papio anubis*. Steroids. 32:187-201.
8754. Preston, JR & Martin, RE. 1963. A grey shrew population in Harmon County, Oklahoma. J Mamm. 44:268-270.
8755. Price, DW, 1983. Gestation period and fetal growth in the gray whale. Rep Int Whal Comm. 33:539-544.
8756. Price, EO. 1966. Influence of light on reproduction in *Peromyscus maniculatus gracilis*. J Mamm. 47:343-344.
8757. Price, M. 1953. The reproductive cycle of the water shrew, *Neomys fodiens bicolor* Shaw. Proc Zool Soc London. 123:599-621.
8758. Pringle, JA & Pringle, VL. 1979. Observations on the lynx *Felis caracal* in the Bedford district. S Afr J Zool. 14:1-4.
8759. Prior, R. 1968. The Roe Deer of Cranborne Chase: An Ecological Survey. Oxford Univ Press, London.
8760. Procter, J. 1963. A contribution to the natural history of the spotted-necked otter (*Lutra maculicollis* Lichtenstein) in Tanganyika. E Afr Wildl J. 1:93-102.
8761. Proulx, G & Buckland, BML. 1985. Precocial breeding in a southern Ontario muskrat, *Ondatra zibethicus*, population. Can Field-Nat. 99:377-378.
8762. Proulx, G & Gilbert, FF. 1983. The ecology of the muskrat, *Ondatra zibethicus*, at Luther Marsh, Ontario. Can Field-Nat. 97:377-390.
8763. Provost, EE. 1962. Morphological characteristics of the beaver ovary. J Wildl Manage. 26:272-278.
8764. Provost, EE & Kirkpatrick, CM. 1952. Observations on the hoary bat in Indiana and Illinois. J Mamm. 33:110-113.
8765. Provost, EE, Nelson, CA, & Marshall, AD. 1973. Population dynamics and behavior in the bobcat. The World's Cats. 1:42-67, Eaton, RL, ed, World Wildlife, Safari, Winston, OR.
8766. Pruitt, WO, Jr. 1954. Notes on a litter of young masked shrews. J Mamm. 35:109-110.
8767. Pryce, CR, Abbott, DH, Hodges, JK, et al. 1988. Maternal behavior is related to prepartum urinary estradiol levels in red-bellied Tamarin monkeys. Physiol Behav. 44:717-726.
8768. Pryor, S & Bronson, FH. 1981. Relative and combined effects of low temperature, poor diet, and short daylength on the productivity of wild house mice. Biol Reprod. 25:734-743.
8769. Psenner, H. 1940. Beobachtungen an einem gefangenen groszen Wiesel (*Mustela erminea* L). Zool Garten NF. 12:315-322.
8770. Psenner, H. 1956. Neue Beobachtungen zur Fortpflanzungsbiologie des Murmeltieres. Z Jagdwiss. 2:148-152.
8771. Psenner, H. 1957. Neues vom Murmeltier, *Marmota m marmota* (Linné, 1758). Säugetierk Mitt. 5:4-10.
8772. Psenner, H. 1970. Über die Haltung von Luchsen (*Felis lynx*) in Innsbruker Zoo. Zool Garten NF. 39:232-239.
8773. Pucek, Z. 1960. Sexual maturation and variability of the reproductive system in young shrews (*Sorex* L) in the first calendar [sic] year of life. Acta Theriol. 3:269-296.
8774. Pucek, Z. 1964. The structure of the glans penis in *Neomys* Kaup, 1929 as a taxonomic character. Acta Theriol. 9:374-377.
8775. Pucek, Z. 1964. Morphological changes in shrews kept in captivity. Acta Theriol. 8:137-166.
8776. Pucek, Z. 1970. Seasonal and age change in shrews as an adaptive process. Symp Zool Soc London. 26:189-207.
8777. Pudney, J. 1976. Seasonal changes in the testis and epididymis of the American grey squirrel, *Sciurus carolinensis*. J Zool. 179:107-120.
8778. Pudney, J & Fawcett, DW. 1977. Putative steroidogenic cells associated with the ductuli efferentes of the ground squirrel (*Citellus lateralis*). Anat Rec. 188:453-476.
8779. Pudney, J & Lacy, D. 1977. Correlation between ultrastructure and biochemical changes in the testis of the American grey squirrel, *Sciurus carolinensis*, during the reproductive cycle. J Reprod Fert. 49:5-16.
8780. Puente, AE & Dewsbury, DA. 1976. Courtship and copulatory behavior of bottlenosed dolphins (*Tursiops truncatus*). Cetology. 21:1-9.
8781. Pugeat, M, Rockle, B, Chrousos, GP, et al. 1984. Plasma testosterone transport in primates. J Steroid Biochem. 20:473-478.
8782. Puget, A. 1966. Essai d'élevage en captivité étroite du Lièvre commun, *Lepus europaeus* Pallas, 1778. Bull Mus Nat Hist Natur 2 sér. 38:333-336.
8783. Puget, A. 1971. *Ochotona r rufescens* (Gray, 1842) en Afghanistan et son élevage en captivité. Mammalia. 35:25-37.
8784. Puget, A. 1973. The Afghan pika (*Ochotona rufescens refescens*): A new laboratory animal. Lab Anim Sci. 23:248-251.
8785. Puget, A. 1975. L'experience de trois années d'élevage d'*Ochotona rufescens rufescens* dans le vivarium. Zool Zhurn. 54:1079-1081.
8786. Puget, A. 1976. L'Ochotone Afghan: *Ochotona rufescens rufescens* (Mammalia Lagomorphe). Bull Soc Zool France. 101:203-207.
8787. Puget, A, Ballas, D, & Gouarderes, C. 1973. Essais de détermination d'un cycle sexuel chez l'Ochotone Afghan (*Ochotona rufescens rufescens*). Expérim Anim. 6:153-169.
8788. Puget, A & Gouarderes, C. 1974. Weight gain of the Afghan pika (*Ochotona rufescens rufescens*) from birth to 19 weeks of age, and during gestation. Growth. 38:117-129.
8789. Puisségur, C. 1935. Recherches sur le Desman des Pyrénées (*Galemys pyrenaicus* G). Bull Soc Hist Nat Toulouse. 67:163-227.
8790. Pull, H. 1938. Die Tragzeiten der einheimischen Jagdtiere. Tharandter Forstliches Jahrbuch. 89:696-701.
8791. Pulliainen, E. 1963. Occurrence and habits of the bear (*Ursus arctos*) in Finland. Suomen Riista. 16:23-30.
8792. Pulliainen, E. 1963. Occurrence and habits of the wolverine (*Gulo gulo*) in Finland. Suomen Riista. 16:109-119.
8793. Pulliainen, E. 1964. Studies on the wolf (*Canis lupus* L) in Finland. Ann Zool Fenn. 1:215-259.
8793a. Pulliainen, E. 1968. Breeding biology of the wolverine (*Gulo gulo* L.) in Finland. Ann Zool Fennici. 5:338-344.

8794. Puranik, PG & Patil, DR. 1984. Biology of the reproduction of the Indian mole rat, *Bandicota indica* (Rodentia; Muridae) I Breeding activity of the female. Indian J Zool. 12(1):63-74.
8795. Puri, CP, Puri, V, & Kumar, TCA. 1980. Bioactive luteinizing hormone in the plasma and cerebrospinal fluid of female rhesus monkeys. J Med Primatol. 9:39-49.
8796. Puri, V, Puri, CP, & Kumar, TCA. 1981. Serum levels of dihydrotestosterone in male rhesus monkeys estimated by a non-chromatographic radioimmunoassay method. J Steroid Biochem. 14:877-881.
8797. Purves, PE & Pilleri, G. 1978. The functional anatomy and general biology of *Pseudorca crassidens* (Owen) with a review of the hydrodynamics and acoustics in Cetacea. Invest Cetacea. 8:67-227.
8798. Puschmann, W. 1975. Wildtiere in Menschenhand. VEB Deutscher Landwirtschaftverlag, Berlin DDR.
8799. Puschmann, W, Schuppel, KF, & Kronberger, H. 1977. Detection of blastocyst in uterine lumen of Indian bear *Melursus u ursinus*). Sickness In Zoos. 389-391, Ippen, R & Schrader, MD, eds, Akademischer Verlag, Berlin, DDR.
8800. Pusey, AE. 1983. Mother-offspring relationships in chimpanzees after weaning. Anim Behav. 31:363-377.
8801. Pye, T. 1991. The New Holland mouse (*Pseudomys novaehollandiae*) (Rodentia: Muridae) in Tasmania: A field study. Wildl Res. 18:521-531.
8802. Pye, T & Bonner, WN. 1980. Feral brown rats, *Rattus norvegicus*, in South Georgia (South Atlantic Ocean). J Zool. 192:237-255.
8803. Quabbe, H-J, Bumke-Vogt, C, Gregor, M, et al. 1982. 24-Hour pattern of plasma prolactin in the male rhesus monkey and its relation to the sleep/wake cycle. Endocrinology. 110:969-975.
8804. Quadagno, DM. 1967. Litter size and implantation sites in feral house mice. J Mamm. 48:677.
8805. Quadagno, DM, Allin, JT, Brooks, RJ, et al. 1970. Some aspects of the reproductive biology of *Baiomys taylori ater*. Am Mid Nat. 84:550-551.
8806. Quadri, SK, Oyama, T, & Spies, HG. 1979. Effects of 17β-estradiol on serum prolactin levels and on prolactin responses to thyrotropin-releasing hormone in female rhesus monkeys. Endocrinology. 104:1649-1655.
8807. Quadri, SK & Spies, HG. 1976. Cyclic and diurnal patterns of serum prolactin in the rhesus monkey. Biol Reprod. 14:495-501.
8807a. Quay, WB. 1948. Notes on some bats from Nebraska and Wyoming. J Mamm. 29:181-182.
8808. Quay, WB. 1951. Observations on mammals of the Seward peninsula, Alaska. J Mamm. 32:88-99.
8809. Quay, WB. 1960. The reproductive organs of the collared lemming under diverse temperature and light conditions. J Mamm. 41:74-89.
8810. Quay, WB & Quay, JF. 1956. The requirements and biology of the collared lemming, *Dicrostonyx torquatus* Pallas, 1778, in captivity. Säugetierk Mitt. 4:174-180.
8811. Quilici, M, Ranque, P, & Camerlynck, P. 1969. Élevage au laboratoire d'*Arvicanthis niloticus* (Desmarest, 1822). Mammalia. 33:345-347.
8812. Quimby, DC. 1951. The life history and ecology of the jumping mouse, *Zapus hudsonius*. Ecol Monogr. 21:61-95.
8813. Quinn, NWS & Thompson, JE. 1987. Dynamics of an exploited Canada lynx population in Ontario. J Wildl Manage. 51:297-305.
8814. Quintanilla, R. 1973. Roedores perjudiciales para el agro de la Republica Argentina. Editorial Universitaria, Buenos Aires. [Cited by Dietrich, 1985]
8815. Quintero, F & Rasweiler, JJ, IV. 1973. The reproductive biology of the female vampire bat: *Desmodus rotundus*. Am Zool. 13:1284.
8816. Quintero, F & Rasweiler, JJ, IV. 1974. Ovulation and early embryonic development in the captive vampire bat, *Desmodus rotundus*. J Reprod Fert. 41:265-273.
8817. Qumsiyeh, MB. 1985. The bats of Egypt. Sp Publ Mus TX Tech Univ. 23:1-102.
8818. Qumsiyeh MB & Jones, JK, Jr. 1986. *Rhinopoma hardwickii* and *Rhinopoma muscatellus*. Mamm Species. 263:1-5.
8819. Qumsiyeh, MB & Schlitter, DA. 1982. The bat fauna of Jabal Al Akhdar, northeast Libya. Ann Carnegie Mus. 51:377-389.
8820. Quris, R. 1975. Ecologie et organisation sociale de *Cercocebus galeritus agilis* dans le nord-est du Gabon. Terre Vie. 29:337-398.
8821. Quris, R. 1976. Données comparatives sur la socio-écologie de huit espèces de Cercopithecidae vivant dans une même zone, due forêt primitive périodiquement inondée (nort-est du Gabon). Terre Vie. 30:193-209.
8822. Raak, G. 1940. Junge Sumpfluchse. Zool Garten NF. 12:198-200.
8822a. Rabb, GB. 1959. Reproductive and vocal behavior in captive pumas. J Mamm. 40:616-617.
8823. Rabb, GB. 1978. Birth, early behavior and clinical date [*sic*] on the okapi. Acta Zool Pathol Antverpiensia. 71:93-105.
8824. Rabb, GB, Ginsburg, BE, & Andrews, S. 1962. Comparative studies of canid behavior, IV Mating behavior in relation to social structure in wolves. Am Zool. 2:440.
8825. Rabb, GB & Rowell, JE. 1960. Notes on reproduction in captive marmosets. J Mamm. 41:401.
8825a. Racey, K. 1929. Observations on *Neurotrichus gibbsii gibbsii*. Murrelet. 10:61-62.
8826. Racey, PA. 1969. Diagnosis of pregnancy and experimental extension of gestation in the pipistrelle bat, *Pipistrellus pipistrellus*. J Reprod Fert. 19:465-474.
8827. Racey, PA. 1973. Environmental factors affecting the length of gestation in heterothermic bats. J Reprod Fert Suppl. 19:175-189.
8828. Racey, PA. 1973. The viability of spermatozoa after prolonged storage by male and female European bats. Period Biol. 75:201-205.
8829. Racey, PA. 1974. The reproductive cycle in male noctule bats, *Nyctalus noctula*. J Reprod Fert. 41:169-182.
8830. Racey, PA. 1975. The prolonged survival of spermatozoa in bats. The Biology of the Male Gamete. 385-416, Duckett, JG & Racey, PA, eds, Academic Press, London.
8831. Racey, PA. 1976. Induction of ovulation in the pipistrelle bat, *Pipistrellus pipistrellus*. J Reprod Fert. 46:481-483.
8832. Racey, PA. 1978. Seasonal changes in testosterone levels and androgen-dependent organs in male moles (*Talpa europaea*). J Reprod Fert. 52:195-200.
8833. Racey, PA. 1978. The effect of photoperiod on the initiation of spermatogenesis in pipistrelle bats, *Pipistrellus pipistrellus*. Proc Int Bat Res Conf. 4:255-258.
8834. Racey, PA. 1979. The prolonged storage and survival of spermatozoa in Chiroptera. J Reprod Fert. 56:391-402.
8835. Racey, PA. 1982. Ecology of bat reproduction. Ecology of Bats. 57-104, Kunz, TH, ed, Plenum Press, New York.
8836. Racey, PA & Kleiman, DG. 1970. Maintenance and breeding in captivity of some vespertilionid bats with special

reference to the noctule *Nyctalus noctula*. Int Zoo Yearb. 10:65-70.

8837. Racey, PA & Potts, DM. 1970. Relationship between stored spermatozoa and the uterine epithelium in the pipistrelle bat, *Pipistrellus pipistrellus*. J Reprod Fert. 22:57-64.

8838. Racey, PA & Skinner, JD. 1979. Endocrine aspects of sexual mimicry in spotted hyaenas *Crocuta crocuta*. J Zool. 187:315-326.

8839. Racey, PA, Suzuki, F, & Medway, Lord. 1975. The relationship between stored spermatozoa and the oviductal epithelium in bats of the genus *Tylonycteris*. The Biology of Spermatozoa. 123-133, Hafez, ESE & Thibault, CG, eds, Karger, Basel.

8840. Racey, PA & Swift, SM. 1981. Variations in gestation length in a colony of pipistrelle bats (*Pipistrellus pipistrellus*) from year to year. J Reprod Fert. 61:123-129.

8841. Racey, PA & Swift, SM. 1985. Feeding ecology of *Pipistrellus pipistrellus* (Chiroptera: Vespertilionidae) during pregnancy and lactation I Foraging behavior. J Anim Ecol. 54:205-215.

8842. Racey, PA & Tam, WH. 1974. Reproduction in male *Pipistrellus pipistrellus* (Mammalia: Chiroptera). J Zool. 172:101-122.

8843. Racey, PA, Uchida, TA, Mōri, T, et al. 1987. Sperm-epithelium relationships in relation to the time of insemination in little brown bats (*Myotis lucifugus*). J Reprod Fert. 80:445-454.

8844. Rachlow, JL & Bowyer, RT. 1991. Interannual variation in timing and synchrony of parturition in Dall's sheep. J Mamm. 72:487-492.

8845. Rachowiak, P. 1987. Some remarks on sex, age structure and reproductiveness of the root vole *Microtus oeconomus* (Pallas, 1776) in Chlebowo high peatbog. Przeglad Zool. 31:209-213.

8846. Raczyński, J. 1975. Progress in breeding European bison in captivity. Breeding Endangered Species in Captivity. 253-262, Martin, RD, ed, Academic Press, New York.

8847. Raczyński, J. 1980. Biologische Grundlagen der Züchtung und der Restitution des Wisents, *Bison bananus*. Zool Garten NF. 50:311-316.

8848. Radda, A, Pretzmann, G, & Steiner, HM. 1969. Bionomische und ökologische Studien an österreichischen Populationen der Gelbhalsmaus (*Apodemus flavicollis*, Melchior, 1834) durch Markierungsfang. Oecologia. 3:351-373.

8849. Raeside, JI & Ronald, K. 1981. Plasma concentrations of estrone, progesterone and corticosteroids during late pregnancy and after parturition in the harbour seal, *Phoca vitalina*. J Reprod Fert. 61:135-139.

8850. Rafferty-Machliss, G & Hartman, CG. 1953. Early death of the ovum in the opossum with observations on moribund mouse eggs. J Morphol. 92:455-485.

8851. Rahm, U. 1957. Der Baum-oder Waldschliefer, *Dendrohydrax dorsalis*. Zool Garten NF. 23:67-74.

8852. Rahm, U. 1961. Beobachtungen an der ersten in Gefangenschaft gehaltene *Mesopotamogale ruwenzorii* (Mammalia: Insectivora). Rev Suisse Zool. 68:73-90.

8853. Rahm, U. 1962. L'élevage et la reproduction en captivité de l'*Atherurus africanus* (Rongeurs, Hystricidae). Mammalia. 26:1-9.

8854. Rahm, U. 1966. Les Mammifères de la forêt équatoriale de l'est du Congo. Ann Mus R Afr Centr Sci Zool Sér 8. 149:39-121.

8855. Rahm, U. 1970. Note sur la reproduction des Sciuridés et Muridés dans la forêt équatoriale du Congo. Rev Suisse Zool. 77:635-646.

8856. Rahm, U & Christiaensen, A. 1963. Les Mammifères de la region occidentale du Lac Kivu. Ann Mus R Afr Centr Ann Sci Zool, Sér 8. 118:1-83.

8857. Rahm, UH. 1969. Gestation period and litter size of the mole rat, *Tachyoryctes ruandae*. J Mamm. 50:383-384.

8858. Raible, LH & Gorzalka, BB. 1986. Receptivity in Mongolian gerbils: dose and temporal parameters of ovarian hormone administration. Lab Anim. 20:109-113.

8859. Rainey, DG. 1956. Eastern woodrat, *Neotoma floridiana*: Life history and ecology. Univ KS Publ Mus Nat Hist. 8:535-546.

8860. Rainey, DG. 1965. Parturition in the desert wood rat. J Mamm. 46:340-341.

8861. Rajagopalan, PK. 1970. Breeding behavior and development of *Rattus rattus coroughtoni* Hinton, 1919 (Rodentia: Muridae) in the laboratory. J Bombay Nat Hist Soc. 67:552-558.

8862. Rajakaski, E & Konisto, D. 1966. Puberty of the female moose (*Alces alces*) in Finland. Suomen Riista. 18:157-162.

8863. Rajalakshmi, M & Prasd, MRN. 1970. Sites of formation of fructose, citric acid and sialic acid in the accessory glands of the giant fruit bat, *Pteropus giganteus giganteus* (Brünnich). J Endocrinol. 46:413-416.

8864. Rajasingh, GJ. 1983. A note on the longevity and fertility of the blackbuck, *Antilope cervicapra* Linnaeus. J Bombay Nat Hist Soc. 80:632-634.

8865. Rakkmatulina, IK. 1972. The breeding, growth, and development of pipistrelles in Azerbaidzhan. Sov J Ecol. 2:131-136.

8866. Ralls, K. 1973. *Cephalophus maxwellii*. Mamm Species. 31:1-4.

8867. Ralls, K. 1978. *Tragelaphus eurycerus*. Mamm Species. 111:1-4.

8868. Ralls, K, Barasch, C, & Minkowski, K. 1975. Behavior of captive mouse deer, *Tragulus napu*. Z Tierpsychol. 37:356-378.

8869. Ralph, CL, Young, S, Gettinger, R, et al. 1985. Does the manatee have a pineal body?. Acta Zool (Stokholm). 66:55-60.

8870. Ram, S, Singh, B, & Dhanda, OP. 1977. A note on genetic studies on gestation length, birth weight and intra-uterine development index in Indian camel (*Camelus dromedarius*) and factors affecting them. Indian Vet J. 54:953-955.

8871. Ramakrishna, PA. 1947. Post-partum oestrus in the Indian short-nosed fruit bat, *Cynopterus sphinx sphinx* (Vahl). Curr Sci. 16:186.

8872. Ramakrishna, PA. 1949. Gestation in the oriental vampires. Curr Sci. 18:307.

8873. Ramakrishna, PA. 1950. Parturition in certain Indian bats. J Mamm. 31:274-278.

8874. Ramakrishna, PA. 1950. Reproduction in *Cynopterus sphinx sphinx* (Vahl). Proc Nat Inst Sci India. 16:362.

8875. Ramakrishna, PA. 1950. Some aspects of reproduction in *Rhinolophus rouxi* (Temn). Proc Nat Inst Sci India. 16:360-362.

8876. Ramakrishna, PA. 1951. Some aspects of reproduction in the oriental vampires, *Lyroderma lyra lyra* Geoff and *Megaderma spasma* (Linn). J Mysore Univ. 11:107-118.

8877. Ramakrishna, PA. 1978. Parturition in the Indian rufous bat, *Rhinolophus rouxi* (Temminck). J Bombay Nat Hist Soc. 75:473-475.

8878. Ramakrishna, PA, Bhatia, D, & Gopalakrishna, A. 1981. Development of the corpus luteum in the Indian leaf-nosed bat *Hipposideros speoris* (Schneider). Curr Sci. 50:264-268.

8879. Ramakrishna, PA & Madhavan, A. 1977. Foetal mambranes and placentation in the vespertilionid bat, *Scotophilus heathi* (Horsefield). Proc Indian Acad Sci. 86B:117-126.
8880. Ramakrishna, PA & Prasad, MRN. 1962. Reproduction in the male slender loris, *Loris tardigrachus lydekkerianus* (Cabrera). Curr Sci. 31:468-469.
8881. Ramakrishna, PA & Prasad, MRN. 1967. Changes in the male reproductive organs of *Loris tardigradus lydekkerianus* (Cabrera). Folia Primatol. 5:176-189.
8882. Ramakrishna, PA & Rao, KVB. 1977. Reproductive adaptations in the Indian rhinolophid bat, *Rhinolophus rouxi* (Temminck). Curr Sci. 46:270-271.
8883. Ramaswami, LS. 1933. Some stages of placentation of *Vesperugo leiserli* (Kuhl). Half-yrly J Mysore Univ. 7:59-99.
8884. Ramaswami, LS. 1975. Some aspects of the reproductive biology of the langur monkey *Presbytis entellus entellus* Dufresne. Proc Indian Nat Sci Acad (Biol Sci). 41:1-30.
8885. Ramaswami, LS. 1986. Primate biology with special emphasis on reproduction. Proc Indian Nat Sci Acad. B52:315-332.
8886. Ramaswami, LS & Kumar, TCA. 1962. Reproductive cycle of the slender loris. Naturwissensch. 49:115-116.
8887. Ramaswami, LS & Kumar, TCA. 1963. Differential implantation of twin blastocysts in *Megaderma* (Microchiroptera). Experientia. 19:641-642.
8888. Ramaswami, LS & Kumar, TCA. 1965. Some aspects of reproduction of the female slender loris *Loris tardigradus* Cabr. Acta Zool (Stockholm). 46:257-274.
8889. Ramaswami, LS, Kurup, GU, & Gadgil, BA. 1982. Some aspects of the reproductive biology of the liontail macaque *Macaca silenus* (Linn) -- A zoo study. J Bombay Nat Hist Soc. 79:324-330.
8890. Ramaswamy, KR. 1961. Studies on the sex cycle of the Indian vampire bat, *Megaderma* (*Lyroderma*) *lyra lyra* (Geoffroy): I Breeding habits. Proc Nat Inst Sci India. 27B:287-301.
8891. Ramirez, MF, Freese, CH, & Revilla, CJ. 1977. Feeding ecology of the pygmy marmoset, *Cebuella pygmaea* in northeastern Peru. The Biology and Conservation of the Callitrichidae. 91-104, Kleiman, DG, ed, Smithsonian Inst Press, Washington, DC.
8892. Rammall, CG. 1964. Composition of thars' milk. NZ J Sci. 7:667-670.
8893. Ramsay, EC, Kasman, LH, & Lasley, BL. 1987. Urinary steroid evaluations to monitor ovarian function in exotic ungulates: V Estrogen and pregnanediol-3-glucuronide excretion in the black rhinoceros (*Diceros bicornis*). Zoo Biol. 6:275-282.
8894. Ramsay, EC, Lasley, BL, & Stabenfeldt, GH. 1981. Monitoring the estrous cycle of the Asian elephant (*Elephas maximus*), using urinary estrogens. Am J Vet Res. 42:256-260.
8895. Ramsay, M. 1987. Personal communication to V Hayssen.
8896. Ramsay, MA & Dunbrack, RL. 1986. Physiological constraints on life history phenomena: The example of small bear cubs at birth. Am Nat. 127:735-743.
8897. Ramsay, MA & Sadleir, RMFS. 1979. Detection of pregnancy in living bighorn sheep by progestin determination. J Wildl Manage. 43:970-973.
8898. Ramsay, MA & Stirling, I. Reproductive biology of female polar bears (*Ursus maritimus*). unpublished ms.
8899. Ramsay, MA & Stirling, I. 1982. Reproductive biology and ecology of female polar bears in western Hudson Bay. Nat Can. 109:941-946.
8900. Ramsay, MA & Stirling, I. 1986. On the mating system of polar bears. Can J Zool. 64:2142-2151.
8901. Ramsay, P. 1865. Description of a new genus and species of rat kangaroo allied to the genus *Hypsiprymnus*, proposed to be called *Hypsiprymnodon moschatus*. Proc Linn Soc NSW. 1:33-35.
8902. Rana, BD, Advani, R, & Soni, BK. 1983. Reproductive biology of *Rattus rattus rufescens* in the Indian desert. Z Angew Zool. 70:207-216.
8903. Rana, BD & Prakash, I. 1979. Reproductive biology and population structure of the house shrew, *Suncus murinus sindensis*, in western Rajasthan. Z Säugetierk. 44:333-343.
8904. Rana, BD & Prakash, I. 1984. Reproduction biology of the soft-furred field rat, *Rattus meltada pallidor* (Ryley, 1914) in the Rajasthan desert. J Bombay Nat Hist Soc. 81:59-70.
8905. Rana, BD, Soni, BK, & Jain, AP. 1985. Ecological note on *Bandicota bengalensis*. Mammalia. 49:578.
8906. Rand, AL. 1935. On the habits of some Madagascar mammals. J Mamm. 16:89-104.
8907. Rand, AL & Host, P. 1942. Results of the Archibold expeditions: Mammal notes from Highland county, Florida. Bull Am Mus Nat Hist. 80:1-21.
8908. Rand, RW. 1955. Reproduction in the female Cape fur seal, *Arctocephalus pusillus* (Schreber). Proc Zool Soc London. 124:717-740.
8909. Rand, RW. 1956. Notes on the Marion Island fur seal. Proc Zool Soc London. 126:65-82.
8910. Rand, RW. 1956. The Cape fur seal, *Arctocephalus pusillus* (Schreber) Its general characteristics and moult. Div Sea Fish S Afr Invest Rep. 21:1-52.
8911. Rand, RW. 1959. The Cape fur seal, *Arctocephalus pusillus*. Div Sea Fish S Afr Invest Rep. 34:1-75.
8912. Rand, RW. 1967. The Cape fur seal, *Arctocephalus pusillus* 3 General Behaviour on land and at sea. Div Sea Fish S Afr Invest Rep. 60:1-39.
8913. Randall, JA. 1985. Role of urine in coordinating reproduction in a desert rodent (*Dipodomys merriami*). Physiol Behav. 34:199-203.
8914. Randall, P, Taylor, P, & Banks, DR. 1984. Pregnancy and stillbirth in a lowland gorilla *Gorilla g gorilla*. Int Zoo Yearb. 23:183-185.
8915. Randolph, PA, Randolph, JC, Mattingly, K, et al. 1977. Energy costs of reproduction in the cotton rat, *Sigmodon hispidus*. Ecology. 58:31-45.
8916. Ransom, AB. 1966. Breeding seasons of white-tailed deer in Manitoba. Can J Zool. 44:59-62.
8917. Ransom, AB. 1967. Reproduction biology of white-tailed deer in Manitoba. J Wildl Manage. 31:114-123.
8918. Ransome, RD. 1973. Factors affecting the timing of births of the greater horse-shoe bat (*Rhinolophus ferrumequinum*). Period Biol. 75:169-175.
8919. Ransome, RD. 1978. Daily activity patterns of the greater horseshoe bat, *Rhinolophus ferrumequinum*, from April to September. Proc Int Bat Res Conf. 4:259-274.
8920. Ranson, RM. 1934. The field vole (*Microtus*) as a laboratory animal. J Anim Ecol. 3:70-76.
8921. Ranson, RM. 1941. Pre-natal and infant mortality in a laboratory population of voles (*Microtus agrestis*). Proc Zool Soc London. 111A:45-57.
8922. Ranson, RM. 1941. New laboratory animals from wild species: Breeding a laboratory stock of hedgehogs (*Erinaceus europaeus* L). J Hyg. 41:131-138
8923. Rao, AJ, Kotagi, SG, & Moudgal, NR. 1984. Serum concentrations of chorionic gonadotrophin, oestradiol-17β and progesterone during early pregnancy in the south Indian bonnet monkey (*Macaca radiata*). J Reprod Fert. 70:449-455.

8924. Rao, AMKM. 1981. Reproductive biology of the spiny field mouse, *Mus platythrix*. J Bombay Nat Hist Soc. 78:160-164.
8925. Rao, CRN. 1927. On the structure of the ovary and the ovarian ovum of *Loris lydekkerianus*, Cabr. Q J Microsc Sci. 71:57-74.
8926. Rao, LV & Pandey, RS. 1982. Seasonal changes in plasma progesterone concentrations in buffalo cows (*Bubalus bubalis*). J Reprod Fert. 66:57-61.
8927. Rao, LV & Pandey, RS. 1983. Seasonal variations in oestradiol-17β and luteinizing hormone in the blood of buffalo cows (*Bubalus bubalis*). J Endocrinol. 98:251-255.
8928. Raphael, MG. 1984. Late fall breeding of the northern flying squirrel, *Glaucomys sabrinus*. J Mamm. 65:138-139.
8929. Rapisarda, JJ, Bergman, KS, Steiner, RA, et al. 1983. Response to estradiol inhibition of tonic luteinizing hormone secretion decreases during the final stage of puberty in the rhesus monkey. Endocrinology. 112:1172-1179.
8930. Rasa, A. 1973. Intra-familial sexual repression in the dwarf mongoose *Helogale parvula*. Die Naturwissenschaften. 6:303-304.
8930a. Rasa, OAE. 1977. The ethology and sociology of the dwarf mongoose (*Helogale undulata rufula*). Z Tierpsychol. 43:337-406.
8931. Rasa, OAE. 1977. Differences in group member response to intruding conspecifics and frightening of potentially dangerous stimuli in dwarf mongooses (*Helogale undulata rufula*). Z Säugetierk. 42:108-112.
8932. Rasa, OAE. 1979. The effects of crowding on the social relationships and behaviour of the dwarf mongoose (*Helogale undulata rufula*). Z Tierpsychol. 49:317-329.
8933. Rashek, V. 1964. Reproduction of *Equus hemionus onager* Boddaert in Barsa-Kelmes Island (Aral sea). Vestn Čcsl Spol Zool. 28:89-95.
8934. Rasmussen, AT. 1917. Seasonal changes in the interstitial cells of the testis in the woodchuck (*Marmota monax*). Am J Anat. 22:475-515.
8935. Rasmussen, DR. 1981. Pair-bond strength and stability and reproductive success. Psychol Rev. 88:274-290.
8936. Rasmussen, DT. 1985. A comparative study of breeding seasonality and litter size in eleven taxa of captive lemurs (*Lemur* and *Varecia*). Int J Primatol. 6:501-517.
8937. Rasmussen, DT & Izard, MK. 1988. Scaling of growth and life history traits relative to body size, brain size, and metabolic rate in lorises and galagos (Lorisidae, Primates). Am J Phys Anthropol. 75:357-367.
8938. Rasmussen, KM, Ausman, LM, & Hayes, KC. 1980. Vital statistics from a laboratory breeding colony of squirrel monkeys (*Saimiri sciureus*). Lab Anim Sci. 30:99-106.
8939. Rasmussen, LE, Schmidt, MJ, Henneous, R, et al. 1982. Asian bull elephants: Flehmen-like responses to extratable components in female estrous urine. Science. 217:159-162.
8940. Rasmussen, LO, Buss, IO, Hess, DL, et al. 1984. Testosterone and dihydrotestosterone concentrations in elephant serum and temporal gland secretions. Biol Reprod. 30:352-362.
8941. Rasweiler, JJ, IV. 1972. The Laboratory Biology of the Long-tongued Bat, *Glossophaga soricina*: Maintenance Procedures, Estivation, the Menstrual Cycle, Histophysiology of the Oviduct and Intramural Implantation. Ph D Diss. Cornell Univ, Ithaca, NY.
8942. Rasweiler, JJ, IV. 1972. Reproduction in the long-tongued bat, *Glossaphaga soricina* I Preimplantation development and hystology of the oviduct. J Reprod Fert. 31:249-262.
8943. Rasweiler, JJ, IV. 1974. Reproduction in the long-tongued bat, *Glossophaga soricina* II Implantation and early embryonic development. Am J Anat. 139:1-35.
8944. Rasweiler, JJ, IV. 1977. Preimplantation development, fate of the zona pellucida, and observations on the glyogen-rich oviduct of the little bulldog bat, *Noctilio albiventris*. Am J Anat. 150:269-300.
8945. Rasweiler, JJ, IV. 1978. Unilateral oviductal and uterine reactions in the little bulldog bat, *Noctilio albiventris*. Biol Reprod. 19:467-492.
8946. Rasweiler, JJ, IV. 1979. Differential transport of embryos and degenerating ova by the oviducts of the long-tongued bat *Glossophaga soricina*. J Reprod Fert. 55:329-334.
8947. Rasweiler, JJ, IV. 1979. Early embryonic development and implantation in bats. J Reprod Fert. 56:403-416.
8948. Rasweiler, JJ, IV. 1982. The contribution of observations on early pregnancy in the little sac-winged bat, *Peropteryx kappleri*, to an understanding of the evolution of reproductive mechanisms in monovular bats. Biol Reprod. 27:681-702.
8949. Rasweiler, JJ, IV. 1984. Reproductive failure due to entrapment of oocytes in luteinized follicles of the little bulldog bat (*Noctilio albiventris*). J Reprod Fert. 71:95-101.
8950. Rasweiler, JJ, IV. 1987. Prolonged receptivity to the male and fate of spermatozoa in the female black mastiff bat, *Molossus ater*. J Reprod Fert. 79:643-654.
8951. Rasweiler, JJ, IV. 1988. Ovarian function in the captive black mastiff bat, *Molossus ater*. J Reprod Fert. 82:97-111.
8952. Rasweiler, JJ, IV. 1990. Implantation, development of the fetal membranes, and placentation in the captive black mastiff bat, *Molossus ater*. Am J Anat. 187:109-136.
8952a. Rasweiler, JJ, IV. 1991. Spontaneous decidual reactions and menstruation in the black mastiff bat, *Molossus ater*. Am J Anat. 191:1-22.
8952b. Rasweiler, JJ, IV. 1991. Development of the discoidal hemochorial placenta in the black mastiff bat, *Molossus ater*: Evidence for a role of maternal endothelial cells in the control of trophoblastic growth. Am J Anat. 191:185-207.
8953. Rasweiler, JJ, IV & Ishiyama, V. 1973. Maintaining frugivorous phyllostomatid bats in the laboratory: *Phyllostomus*, *Artibeus*, and *Sturnira*. Lab Anim Sci. 23:56-61.
8954. Raszynski, J. 1980. Biologische Grundlagen der Züchtung und der Restitution des Wisents, *Bison bonasus*. Zool Garten NF. 50:311-316.
8954a. Ratcliffe, FN. 1931. The flying fox (*Pteropus*) in Australia. Bull CSIRO. 53:1-81.
8955. Rathbun, GB. 1978. Evolution of the rump region in the golden rumped elephant-shrew. Bull Carnegie Mus Nat Hist. 6:11-19.
8956. Rathbun, GB. 1979. *Rhynchocyon chrysopygus*. Mamm Species. 117:1-4.
8957. Rathbun, GB, Beamar, P, & Maliniak, E. 1981. Capture, husbandry and breeding of rufous elephant-shrews *Elephantutus rufescens*. Int Zoo Yearb. 21:176-184.
8958. Rathman, NP & Mallory, FF. 1989. Normal and blocked pregnancy in the collared lemming: Morphometrical and histological changes. Can J Zool. 67:1363-1371.
8959. Ratomponirina, C, Andrianivo, J, & Rumpler, Y. 1982. Spermatogenesis in several intra- and interspecific hybrids of the lemur (*Lemur*). J Reprod Fert. 66:717-721.
8960. not used.
8961. Rauch, HG. 1957. Zum Verhalten von *Meriones tamariscinus* Pall (1778). Z Säugetierk. 22:218-240.
8962. Raun, GG. 1966. A population of woodrats (*Neotoma micropus*) in southern Texas. Bull TX Mem Mus. 11:1-62.

8963. Raun, GG & Wilks, BJ. 1964. Natural history of *Baiomys taylori* in southern Texas and competition with *Sigmodon hispidus* in mixed population. TX J Sci. 16:28-49.
8964. Rausch, RA. 1967. Some aspects of the population ecology of wolves, Alaska. Am Zool. 7:253-265.
8965. Rautenbach, IL. 1982. Mammals of the Transvaal. Ecoplan Monograph (Pretoria). 1:1-211.
8966. Rautenbach, IL & Nel, JAJ. 1975. Further records of smaller mammals from the Kalahari Gemsbok National Park. Koedoe. 18:195-198.
8967. Rauther, M. 1938. Über den männlichen Genitalapparat von *Solenodon paradoxus* Brandt (Mammalia, Insectivora). Zool Anz. 123:65-78.
8968. Raven, HC. 1936. Notes on the anatomy of the viscera of the giant panda (*Aileuropoda melanoleuca*). Am Mus Nov. 877:1-23.
8969. Ravindra, R, Bhatia, K, & Mead, RA. 1984. Steroid metabolism in corpora lutea of the western spotted skunk (*Spilogale putorius latifrons*). J Reprod Fert. 72:495-502.
8970. Ravindra, R & Mead, RA. 1984. Plasma estrogen levels during pregnancy in the western spotted skunk. Biol Reprod. 30:1153-1159.
8971. Ravindranath, N & Moudgal, NR. 1987. Use of tamoxifen, an antioestrogen, in establishing a need for oestrogen in early pregnancy in the bonnet monkey. J Reprod Fert. 81:327-336.
8972. Rawling, C. 1905. The Great Plateau. Edward Arnold, London.
8973. Rawlins, CGC. 1979. The breeding of white rhinos in captivity: A comparative survey. Zool Garten NF. 49:1-7.
8974. Rawlins, RG. 1979. Parturient and postpartum behaviour of a free-ranging Rhesus monkey (*Macaca mulatta*). J Mamm. 60:432-433.
8975. Rawlins, RG & Kessler, MJ. 1985. Climate and seasonal reproduction in the Cayo Santiago macaques. Am J Primatol. 9:87-99.
8976. Rawson, JMR & Dukelow, WR. 1973. Observation of ovulation in *Macaca fascicularis*. J Reprod Fert. 34:187-190.
8977. Rawson, JMR, Kuehl, TJ, & Dukelow, WR. 1973. Ovulation-menstrual cycle relationships in *Macaca fascicularis*. Fed Proc. 32(3):283.
8978. Ray, CC. 1981. Ross seal *Ommatophoca rossi* Gray, 1844. Handbook of Marine Mammals. Ridgway, SH & Harrison, RJ, eds, Academic Press, New York. 2:237-260
8978a. Raymond, MAV & Layne, JN. 1988. Aspects of reproduction in the southern flying squirrel in Florida. Acta Theriol. 33:505-518.
8979. Raynaud, A. 1944. Observations sur l'état de développement des glandes prostatiques des jeunes Mulots (*Apodemus sylvaticus* L) de sexe femelle. Bull Soc Zool France. 69:21-32.
8980. Raynaud, A. 1945. Existence constant de deux lobes prostatiques chez les Mulots (*Apodemus sylvaemus sylvaticus*). Bull Soc Zool France. 70:162-172.
8981. Raynaud, A. 1950. Action des facteurs externes sur le développement de l'appareil génital des Mulots (*Apodemus sylvaticus* L). CR Soc Biol. 144:945-948.
8982. Raynaud, A. 1950. État de développement de l'appareil génital des Mulots (*Apodemus sylvaticus* L) au cours des différentes saisons de l'année. CR Soc Biol. 144:938-940.
8983. Raynaud, A. 1950. Variations saisonnières des organes génitaux des Mulots (*Apodemus sylvaticus* L) de sex mâle Données pondérales et histologiques. CR Soc Biol. 144:941-945.
8984. Raynaud, A. 1951. Les glandes annexes du tractus uro-génital des Campagnols agrestes (*Microtus agrestis* L). Bull Biol France-Belg. 85:323-339.
8985. Raynaud, A. 1951. Recherches sur les variations saisonnières de l'activité génitale des Mulots (*Apodemus sylvaticus* L) du département de Tarn. Bull Soc Hist Nat Toulouse. 86:208-225.
8986. Raynaud, A. 1951. Reproduction en hiver, des Campagnols agrestes (*Microtus agrestis* L) dans le départment du Tarn. Bull Soc Zool France. 76:188-200.
8987. Raynaud, A. 1951. Spermatogenèse active, en hiver, chez les Mulots (*Apodemus sylvaticus* L) élevés dans une cage placée a l'exterieur, dans un jardin. CR Soc Biol. 145:1063-1069.
8988. Raynor, GS. 1960. Three litters in a pine mouse nest. J Mamm. 41:275.
8989. Rayor, LS. 1985. Effects of habitat quality on growth, age at first reproduction, and dispersal in Gunnison's prairie dogs (*Cynomys gunnisoni*). Can J Zool. 63:2835-2840.
8990. Read, B. 1986. Breeding and management of the Malayan tapir *Tapirus indicus* at St Louis Zoo. Int Zoo Yearb. 24/25:294-297.
8991. Read, B & Frueh, RJJ. 1980. Management and breeding of Speke's gazelle *Gazella spekei* at the St Louis Zoo, with a note on artificial insemination. Int Zoo Yearb. 20:99-104.
8992. Read, DG. 1984. Reproduction and breeding season of *Planigale gilesi* and *P tenuirostris* (Marsupialia: Dasyuridae). Aust Mamm. 7:161-173.
8993. Read, DG. 1987. A comparison of growth rates in dependent juveniles of *Planigale gilesi* and *P tenuirostris* (Marsupialia: Dasyuridae). Aust J Zool. 35:161-171.
8994. Read, DG, Fox, BJ, & Whitford, D. 1983. Notes on breeding in *Sminthopsis* (Marsupialia: Dasyuridae). Aust Mamm. 6:89-92.
8995. Reddi, AH & Prasad, MRN. 1967. Action of testosterone propionate and growth hormone on the os penis of the Indian palm squirrel, *Funambulus pennanti* (Wroughton). Gen Comp Endocrinol. 8:143-151.
8996. Reddi, AH & Prasad, MRN. 1968. The reproductive cycle of the male Indian palm squirrel, *Funambulus pennanti* Wroughton. J Reprod Fert. 17:235-245.
8997. Reddi, AH, Prasad, MRN, & Duraiswami, S. 1966. Fructose and citric acid in the male accessory glands of the Indian palm squirrel *Funambulus pennanti* (Wroughton). J Endocrinol. 35:193-197.
8998. Redenz, E. 1929. Das Verhalten der Säugetierspermatozoen zwischen Begattung und Befruchtung. Z Zellf Mikr Anat. 9:734-749.
8998a. Redfield, JA, Taitt, MJ, & Krebs, CJ. 1978. Experimental alterations of sex-ratios in populations of *Microtus oregonii*, the creeping vole. J Mamm. 47:55-69.
8999. Redford, KH & Wetzel, RM. 1985. *Euphractus sexcinctus*. Mamm Species. 252:1-4.
8999a. Redford, KH. 1987. The pampas deer (*Ozotoceros bezoarticus*) in central Brazil. 411-414, in Biology and Management of the Cervidae, Wemmer, CM, ed, Smithsonian Inst Press, Washington, DC.
9000. Reed, DF, Kincaid, KR, & Beck, TDI. 1979. Migratory mule deer fall from highway cliffs. J Wildl Manage. 43:272.
9001. Reed, EB. 1955. January breeding of *Peromyscus* in north central Colorado. J Mamm. 36:462-463.
9002. Reeder, EM. 1939. Cytology of the reproductive tract of the female bat *Myotis lucifugus lucifugus*. J Morphol. 64:431-451.

9002a. Reeder, WG & Noris, KS. 1954. Distribution, type locality, and habits of the fish-eating bat, *Pizonyx vivesi*. J Mamm. 35:81-87.
9003. Reeves, HM & Williams, RM. 1956. Reproduction, size, and mortality in the Rocky Mountain muskrat. J Mamm. 37:495-500.
9004. Reeves, RR & Leatherwood, S. 1985. Bowhead whale, *Balaena mysticetus* Linnaeus, 1758. Handbook of Marine Mammals. 3:305-344. Ridgway, SH & Harrison, RJ, eds, Academic Press, New York.
9005. Reeves, RR & Ling, JK. 1981. Hooded seal *Cystophora cristata* Erxleben, 1777. Handbook of Marine Mammals. 2:171-194. Ridgway, SH & Harrison, RJ, eds, Academic Press, New York.
9006. Reeves, RR & Tracey, S. 1980. *Monodon monoceros*. Mamm Species. 127:1-7.
9007. Regaud, C. 1904. État des cellules interstitielles du testicule chez la Taupe pendant la période de spermatogenèse et pendant l'état de repos des canalicules séminaux. CR Assoc Anat. 6:54-56.
9008. Reich, A. 1978. A case of inbreeding in the African wild dog *Lycaon pictus* in the Kruger National Park. Koedoe. 21:119-123.
9009. Reich, LM. 1981. *Microtus pennsylvanicus*. Mamm Species. 159:1-8.
9010. Reichman, OJ & van de Graaff, KM. 1973. Seasonal activity and reproductive patterns of five species of Sonoran desert rodents. Am Mid Nat. 90:118-126.
9010a. Reichman, OJ & van de Graaff, KM. 1975. Association between ingestion of green vegetation and desert rodent reproduction. J Mamm. 56:503-506.
9011. Reichstein, H. 1959. Populationsstudien an Erdmäusen, *Microtus agrestis* L (Markierungsversuche). Zool Jahrb Syst. 86:367-382.
9012. Reichstein, H. 1962. Beiträge zur Biologie eines Steppennagers, *Microtus* (*Phaeomys*) *brandti* (Raddle, 1861). Z Säugetierk. 27:146-163.
9013. Reichstein, H. 1964. Untersuchungen zum Körperwachstum und zum Reproduktionspotential der Feldmaus, *Microtus arvalis* (Pallas, 1779). Z Wissensch Zool. 170:112-222.
9014. Reichstein, H. 1967. Populationsstudien an steppenbewohnenden Nagetieren Ostafrikas. Z Säugetierkd. 32:309-313.
9015. Reig, OA. 1968. The chromosomes of the didelphid marsupial *Marmosa robinsoni* Bangs. Experientia. 25:185-186.
9016. Reig, OA. 1970. Ecological notes on the fossorial octodont rodent *Spalacopus cyanus* (Molina). J Mamm. 51:592-601.
9017. Reig, OA, Fernandez, DR, & Spotorno, OA. 1972. Further occurrence of a karyotype of 2N = 14 chromosomes in two species of Chilean didelphoid marsupials. Z Säugetierk. 37:37-42.
9018. Reilly, SB. 1984. Observed and maximum rates of increase in gray whales, *Eschrichtius robustus*. Rep Int Whal Comm Sp Issue. 6:389-399.
9019. Reimers, E. 1972. Growth in domestic and wild reindeer in Norway. J Wildl Manage. 36:612-619.
9020. Reimers, E. 1983. Reproduction in wild reindeer in Norway. Can J Zool. 61:211-217.
9021. Reimers, E, Klein, DR, & Sørumgård, R. 1983. Calving time, growth rate, and body size of Norwegian reindeer on different ranges. Arctic Alp Res. 15:107-118.
9022. Reimers, E & Ringberg, T. 1983. Seasonal changes in body weights of Svalbard reindeer from birth to maturity. Acta Zoologica Fennica. 175:69-72.
9023. Reinius, S, Fritz, GR, & Knobil, E. 1973. Ultrastructure and endocrinological correlates of an early implantation site in the rhesus monkey. J Reprod Fert. 32:171-173.
9024. Reiter, J, Panken, KJ, & LeBoeuf, BJ. 1981. Female competition and reproductive success in northern elephant seals. Anim Behav. 29:670-687.
9025. Reiter, J, Stinson, NL, & LeBoeuf, BJ. 1978. Northern elephant seal development: The transition from weaning to nutritional independence. Behav Ecol Sociobiol. 3:337-367.
9026. Reitz, FHH. 1972. Breeding the spotted hyaena *Crocuta crocuta* at Flamingo Park Zoo. Int Zoo Yearb. 12:118-119.
9027. Relexans, MC & Canivenc, R. 1967. Evolution pondérale du testicule du Blaireau européen *Meles meles* L au cours du cycle génital annuel. CR Soc Biol. 161:600-603.
9028. Rempe, U. 1957. Beobachtungen über Brunst, Paarung, Tragzeit, Geburt und Kreuzungen bei Mitgliedern der Untergattung *Putorius*. Säugetierk Mitt. 5:111-113.
9029. Rendall, P. 1897. Natural history notes from the West Indies. Zoologist Serie 4. I:341-345.
9030. Rendall, P. 1899. Lichtenstein's hartebeest (*Bubalis lichtensteini*). Great and Small Game of Africa. 160-165, Bryden, HA, ed, Rowland Ward, London.
9031. Renecker, LA. 1987. The composition of moose milk following late parturition. Acta Theriol. 32:129-133.
9032. Renfree, MB. 1972. Influence of the embryo on the marsupial uterus. Nature. 240:475-477.
9033. Renfree, MB. 1973. Proteins in the uterine secretions of the marsupial, *Macropus eugenii*. Dev Biol. 32:41-49.
9034. Renfree, MB. 1974. Ovariectomy during gestation in the American opossum, *Didelphis marsupialis virginiana*. J Reprod Fert. 39:127-130.
9035. Renfree, MB. 1975. Uterine proteins in the marsupial, *Didelphis marsupialis virginiana* during gestation. J Reprod Fert. 42:163-166.
9036. Renfree, MB. 1979. Initiation of development of diapausing embryos by mammary denervation during lactation in a marsupial. Nature. 278:549-551.
9037. Renfree, MB. 1980. Embryonic diapause in the honey possum *Tarsipes spencerae*. Search. 11:81.
9038. Renfree, MB. 1980. Endocrine activity in marsupial placentation. Endocrinology 1980. 83-86, Cumming, IA, Funder, JW, & Mendelsohn, FAO, eds, Aust Acad Sci, Canberra.
9039. Renfree, MB. 1981. Discussion. J Reprod Fert Suppl. 29:77.
9040. Renfree, MB. 1981. Embryonic diapause in marsupials. J Reprod Fert Suppl. 29:67-78.
9041. Renfree, MB. 1981. Marsupials: Alternative mammals. Nature. 293:100-101.
9041a. Renfree, MB. 1992. The role of genes and hormones in marsupial sexual differentiation. J Zool. 226:165-173.
9042. Renfree, MB, Flint, APF, Green, SW, et al. 1984. Ovarian steroid metabolism and oestrogens in the corpus luteum of the tammar wallaby. J Endocrinol. 101:231-240.
9043. Renfree, MB, Green, SW, & Young, IR. 1979. Growth of the corpus luteum and its progesterone content during pregnancy in the tammar wallaby, *Macropus eugenii*. J Reprod Fert. 57:131-136.
9044. Renfree, MB & Heap, RB. 1977. Steroid metabolism by the placenta, corpus luteum and endometrium during pregnancy in the marsupial *Macropus eugenii*. Theriogenology. 8:164.
9045. Renfree, MB, Lincoln, DW, Almeida, OFX, et al. 1981. Abolition of seasonal embryonic diapause in a wallaby by pineal denervation. Nature. 293:138-139.

9045a. Renfree, MB, Robinson, ES, Short, RV. et al. 1990. Mammary glands in male marsupials: 1 Primordia in neonatal opossums, *Didelphis virginiana* and *Monodelphis domestica*. Development. 110:385-390.
9046. Renfree, MB, Russell, EM, & Wooller, RD. 1984. Reproduction and life history of the honey possum, *Tarsipes rostratus*. Possums and Gliders. 427-437, Smith, AP & Hume, ID, eds, Aust Mamm Soc, Sydney.
9047. Renfree, MB & Tyndale-Biscoe, CH. 1973. Intrauterine development after diapause in the marsupial *Macropus eugenii*. Dev Biol. 32:28-40.
9048. Renfree, MB, Wallace, GI, & Young, IR. 1982. Effects of progesterone, oestradiol-17β and androstenedione on follicular growth after removal of the corpus luteum during lactational and seasonal quiescence in the tammar wallaby. J Endocrinol. 92:397-403.
9049. Reng, R. 1977. Die Placenta von *Microcebus murinus* Miller. Z Säugetierk. 42:201-214.
9049a. Renjun, L. 1988. Study on the regularity of reproduction in *Lipotes*. Aqua Mamm. 14:63-68.
9050. Renoir, JM, Mercier-Bodard, C, & Beaulieu, EE. 1980. Hormonal and immunological aspects of the phylogeny of sex steroid binding plasma protein. Proc Nat Acad Sci USA. 77:4578-4582.
9051. Rensenbrink, HP. 1971. Kleine kleine pandas. Artis. 16(5):162-167.
9052. Rensenbrink, HP. 1972. Weer jonge lippenberen. Artis. 18:40-47.
9053. Report on Plague Investigations in India. 1912. Observations on plague in eastern Bengal. J Hyg Plague Suppl. 1:157-192.
9054. Reppert, SM, Perlow, MJ, Tamarkin, L, et al. 1979. A diurnal melatonin rhythm in primate cerebrospinal fluid. Endocrinology. 104:295-301.
9055. Resko, JA. 1970. Androgen secretion by the fetal and neonatal rhesus monkeys. Endocrinology. 87:680-687.
9056. Resko, JA, Goy, RW, Robinson, JA, et al. 1982. The pubescent rhesus monkey: Some characteristics of the menstrual cycle. Biol Reprod. 27:354-361.
9057. Resko, JA, Malley, A, Begley, D, et al. 1973. Radioimmunoassay of testosterone during fetal development of the rhesus monkey. Endocrinology. 93:156-161.
9058. Resko, JA, Norman, RL, Niswender, GD, et al. 1974. The relationship between progestins and gonadotropins during the late luteal phase of the menstrual cycle in the rhesus monkey. Endocrinology. 94:128-135.
9059. Resko, JA, Quadri, SK, & Spies, HG. 1977. Negative feedback control of gonadotropins in male rhesus monkeys: Effects of time after castration and interactions of testosterone and estradiol-17β. Endocrinology. 101:215-224.
9060. Retfalvi, L. 1969. Sexual dimorphism in fetuses of the wapiti, *Cervus canadensis*. Can J Zool. 47:1418-1419.
9061. Retzius, G. 1900. Zur Kenntnis der Entwicklunggeschichte des Renntieres und des Rehes. Biol Untersuch. 9:109-117.
9062. Reuben, R. 1963. Note on the breeding season of *Rhinopoma hardwickei* Gray. J Bombay Nat Hist Soc. 60:722.
9063. Reuther, C. 1980. Der Fischotter, *Lutra lutra* L in Niedersachsen. Natursch Landschaftspf Niedersachsen. 11:1-182. [Cited by Mason & Macdonald, 1986]
9064. Reuther, RT. 1961. Breeding notes on mammals in captivity. J Mamm. 42:427-428.
9065. Reuther, RT. 1969. Growth and diet of young elephants in captivity. Int Zoo Yearb. 9:168-178.
9066. Reuther, RT & Doherty, J. 1968. Birth seasons of mammals at San Fransisco Zoo. Int Zoo Yearb. 8:97-101.
9067. Reventlow, A. 1943. Bericht über eine Faultiergeburt. Zool Garten NF. 15:22-26.
9068. Reventlow, A. 1949. The growth of our giraffes and giraffe-calves. Bijd Dierkd. 28:394-396.
9069. Reventlow, A. 1953. Remarks on American black bears. Zool Garten NF. 20:185-187.
9070. Revin, YV. 1968. A contribution to the biology of the northern pika (*Ochotona alpina* Pall) on the Olekmo-Charskoe Highlands (Yakutia). Zool Zhurn. 47:1075-1082.
9071. Revin, YV. 1983. Materials on ecology of *Lemmus amurensis* in south Yakutia. Zool Zhurn. 62:922-929.
9072. Reyes, FI, Winter, JSD, Faiman, C, et al. 1975. Serial serum levels of gonadotropins, prolactin and sex steroids in the nonpregnant and pregnant chimpanzee. Endocrinology. 96:1447-1455.
9073. Reynolds, HC. 1945. Some aspects of the life history and ecology of the opossum in central Missouri. J Mamm. 26:361-379.
9074. Reynolds, HC. 1952. Studies on reproduction in the opossum (*Didelphis virginiana virginiana*). Univ CA Publ Zool. 52:223-284.
9075. Reynolds, HG. 1958. The ecology of the Merriam kangaroo rat (*Dipodomys merriami* Mearns) on the grazing lands of southern Arizona. Ecol Monogr. 28:111-127.
9076. Reynolds, HG. 1960. Life history notes on Merriam's kangaroo rat in southern Arizona. J Mamm. 41:48-58.
9077. Reynolds, HG & Haskell, HS. 1949. Life history notes on Price and Bailey pocket mice of southern Arizona. J Mamm. 30:150-156.
9078. Reynolds, HG & Turkowski, F. 1972. Reproductive variations in the round-tailed ground squirrel as related to winter rainfall. J Mamm. 53:893-898.
9079. Reynolds, JE, III. 1977. Aspects of social behavior and ecology of a semi-isolated colony of Florida manatees, *Trichechus manatus*. MS Thesis. Univ Miami, Miami.
9080. Reynolds, JK & Stinson, RH. 1959. Reproduction in the European hare in southern Ontario. Can J Zool. 37:627-631.
9081. Reynolds, RL & van Horn, RN. 1977. Induction of estrus in intact *Lemur catta* under photoinhibition of ovarian cycles. Physiol Behav. 18:693-700.
9082. Reynolds, RL & Van Horn, RN. 1978. Exogenous estradiol and sexual receptivity in photoinhibited female *Lemur catta*. Physiol Behav. 21:383-385.
9083. Reynolds, TJ & Wright, JW. 1979. Early postnatal physical and behavioural development of degus (*Octodon degus*). Lab Anim. 13:93-99.
9084. Reynolds, V & Reynolds, F. 1965. Chimpanzees of the Budongo Forest. Primate Behavior Field Studies of Monkeys and Apes. 368-424, DeVore, I, ed, Holt, Rinehart & Winston, New York.
9085. Reznik-Schüller, H, Reznik, G, & Mohr, U. 1974. The European hamster (*Cricetus cricetus* L) as an experimental animal: Breeding methods and observations of their behaviour in the laboratory. Z Versuchstierk. 16:48-58.
9086. Rhine, RJ, Wasser, SK, & Norton, GW. 1988. Eight-year study of social and ecological correlates of mortality among immature baboons of Mikumi National Park, Tanzania. Am J Primatol. 16:199-212.
9087. Rhodes, DH. 1989. The influence of multiple photoperiods and pinealectomy on gonads, pelage and body weight in male meadow voles, *Microtus pennsylvanicus*. Comp Biochem Physiol. 93A:445-449.

9088. Rhodes, OE, Scribner, KT, Smith, MH, et al. 1985. Factors affecting the number of fetuses in a white-tailed deer herd. Proc Ann Conf SE Assn Fish Wildl Agen. 39:380-388.
9088a. Ribeiro, MG & Nogueira, JC. 1990. The penis morphology of the four-eyed opossum *Philander opossum*. Anat Anz. 171:65-72.
9089. Rice, DW. 1957. Life history and ecology of *Myotis austroriparius* in Florida. J Mamm. 38:15-32.
9090. Rice, DW. 1960. Population dynamics of the Hawaiian monk seal. J Mamm. 41:376-385.
9091. Rice, DW. 1963. Progress report on biological studies of the larger Cetacea in the waters off California. Norsk Hvalfangst-Tidende. 52:181-187.
9092. Rice, DW. 1977. Synopsis of biological data on the sei whale and Bryde's whale in the eastern north Pacific. Rep Int Whal Comm Sp Issue. 1:92-97.
9093. Rice, DW. 1983. Gestation period and fetal growth in the gray whale. Rep Int Whal Comm. 33:539-544.
9094. Rice, DW & Wolman, AA. 1971. The life history and ecology of the gray whale (*Eschrichtius robustus*). Am Soc Mamm Sp Publ. 3:1-42.
9095. Richard, A. 1974. Patterns of mating in *Propithecus verreauxi verreauxi*. Prosimian Biology. 49-74, Martin, RD, Doyle, GA, & Walker, AC, eds, Univ Pittsburg Press, Pittsburgh.
9096. Richard, AF. 1976. Preliminary observations on the birth and development of *Propithecus verreauxi* to the age of six months. Primates. 17:357-366.
9097. Richard, AF & Nicoll, ME. 1987. Female social dominance and basal metabolism in a Malagasy primate, *Propithecus verreauxi*. Am J Primatol. 12:309-314.
9098. Richard, PB. 1964. Notes sur la biologie de Daman des arbres (*Dendrohydrax dorsalis*). Biol Gabon. 1:73-84.
9099. Richard, PB. 1976. Determination de l'age et de la longevité chez le Desman des Pyrénées (*Galemys pyrenaicus*). Terre Vie. 30:181-192.
9100. Richards, C. 1983. Bats. The Australian Museum Complete Book of Australian Mammals. 332, Straham, R, ed, Angus & Roberts, Sydney.
9101. Richards, SH & Hine, RL. 1953. Wisconsin fox populations. Tech Wildl Bull WI Cons Dept. 6:1-78.
9102. Richardson, EG. 1977. The biology and evolution of the reproductive cycle of *Miniopterus schreibersii* and *M australis* (Chiroptera: Vespertilionidae). J Zool. 183:353-375.
9103. Richardson, PRK. 1985. The social behaviour and ecology of the aardwolf, *Proteles cristatus* (Spearman, 1783) in relation to its food resources. Ph D Diss. Zoology, Univ Oxford.
9104. Richardson, PRK. 1987. Aardwolf mating system: Overt cuckoldry in an apparently monogamous mammal. S Afr J Sci. 83:405-410.
9105. Richardson, WB. 1942. Ring-tailed cats (*Bassariscus astutus*): Their growth and development. J Mamm. 23:17-26.
9106. Richardson, WB. 1943. Woodrats (*Neotoma albigula*): Their growth and development. J Mamm. 24:130-143.
9107. Richins, GH, Smith, HD, & Jorgensen, CD. 1973. Growth and development of the western harvest mouse, *Reithrodontomys megalotis megalotis*. Great Basin Nat. 34:105-120.
9108. Richkind, M. 1977. Steroid hormone studies in pregnant and nonpregnant bottlenosed dolphins, *Tursiops truncatus*. Breeding Dolphins Present Status, Suggestions for the Future. 261-268, Ridgway, SH & Benirschke, K, eds, Nat Tech Inf Serv. Springfield, VA.
9109. Richkind, M & Ridgway, SH. 1975. Steroid hormone pattern in the pregnant and nonpregnant bottlenosed dolphin, *Tursiops truncatus* following the intramuscular and intravascular administration of NIH-FSH-Ovine-S9. 6th Ann Conf Workshop, Int Assoc Aquatic An Med.
9110. Richmond, M & Conaway, CH. 1969. Induced ovulation and oestrus in *Microtus ochrogaster*. J Reprod Fert Suppl. 6:357-376.
9111. Richmond, M & Conaway, CH. 1969. Management, breeding, and reproductive performance of the vole, *Microtus ochrogaster*, in a laboratory colony. Lab Anim Care. 19:80-87.
9111a. Richmond, ND & Grimm, WC. 1950. Ecology and distribution of the shrew *Sorex dispar* in Pennsylvania. Ecology. 31:279-282.
9112. Richter, AR & Labisky, RF. 1984. Reproductive dynamics among disjunct white-tailed deer herds in Florida. J Wildl Manage. 48:964-971.
9113. Rick, AM. 1968. Notes of bats from Tikal, Guatemala. J Mamm. 49:516-520.
9114. Rickart, EA. 1977. Reproduction, growth and development in two species of cloud forest *Peromyscus* from southern Mexico. Occas Pap Mus Nat Hist Univ KS. 67:1-22.
9115. Rickart, EA. 1986. Postnatal growth of the Piute ground squirrel (*Spermophilus mollis*). J Mamm. 67:412-415.
9116. Rickart, EA. 1987. *Spermophilus townsendii*. Mamm Species. 268:1-6.
9117. Rickart, EA. 1988. Population structure of the Piute ground squirrel (*Spermophilus mollis*). Southwest Nat. 33:91-96.
9118. Rickart, EA & Robertson, PB. 1985. *Peromyscus melanocarpus*. Mamm Species. 241:1-3.
9119. Rickart, EA & Yensen, E. 1991. *Spermophilus washingtoni*. Mamm Species. 371:1-5.
9120. Ride, WDL & Tyndale-Biscoe, CH. 1962. Mammals. W Aust Fish Bull. 2:54-97.
9121. Rideout, CB. 1978. Mountain Goat. Big Game of North America. 149-159, Schmidt, JL & Gilbert, DL, eds, Stackpole Books, Harrisburg, PA.
9122. Rideout, CB & Hoffmann, RS. 1975. *Oreamnos americanus*. Mamm Species. 63:1-6.
9123. Rider, V & Heap, RB. 1986. Heterologous anti-progesterone monoclonal antibody arrests early embryonic development and implantation in the ferret (*Mustela puttorius*). J Reprod Fert. 76:459-470.
9124. Ridgway, SH & Harrison, RJ. 1981. Handbook of Marine Mammals, Vol 1 & 2. Academic Press, New York.
9125. Rieck, W. 1956. Untersuchungen uber die Vermehrung des Feldhasen. Z Jagdwiss. 2:49-90.
9126. Riedman, M & Ortiz, CL. 1979. Changes in milk composition during lactation in the northern elephant seal. Physiol Zool. 52:240-249.
9127. Rieger, D & Murphy, BD. 1977. Episodic fluctuation in plasma testosterone and dihydrotestosterone in male ferrets during the breeding season. J Reprod Fert. 51:511-514.
9128. Rieger, I. 1979. A review of the biology of striped hyaenas, *Hyaena hyaena* (Linne, 1758). Säugetierk Mitt. 27:81-95.
9129. Rieger, I. 1979. Breeding the striped hyaena *Hyaena hyaena* in captivity. Int Zoo Yearb. 19:193-198.
9130. Rieger, I. 1981. *Hyaena hyaena*. Mamm Species. 150:1-5.
9131. Rieger, I & Peters, G. 1981. Einige Beobachtungen zum Paarungs-und Lautgebungsverhalten von Irbissen (*Uncia uncia*) im Zoologische Garten. Z Säugetierk. 46:35-48.
9132. Riesen, JE, Meyer, RK, & Wolf, RC. 1971. The effect of season on occurrence of ovulation in the rhesus monkey. Biol Reprod. 5:111-114.

9133. Rijksen, HD. 1978. A field study on Sumatran orang-utans (*Pongo pygmaeus abelli*, Lesson 1827): Ecology, behavior and conservation. Meded Landb Hoogesch Wageningen. 78(2):1-420.
9134. Riley, GA & McBride, RT. 1975. A survey of the red wolf (*Canis rufus*). The Wild Canids. 263-277, Fox, MW, ed, Van Nostrand Reinhold, New York.
9135. Riney, T & Child, G. 1960. Breeding season and the ageing criteria for the common duiker (*Sylvicapra grimmia*). Proc Fed Sci Congr Salisburg S Rhodesia. 1:291-299.
9136. Riney, T. 1956. Differences in proportion of fawns to hinds in red deer (*Cervus elaphus*) from several New Zealand environments. Nature. 177:488-489.
9136a. Rinker, GC. 1948. A bat (*Pipistrellus*) record from Honduras. J Mamm. 29:179-180.
9137. Riordan, DV. 1971. Reproductive variation in *Galago senegalensis* sub-species (lesser bush baby). Ann Rep Jersey Wildl Preserv Trust. 8:15-18.
9138. Riordan, DV. 1972. Reproduction in the spiny hedgehog tenrec (*Setifer setosus*) and the pigmy hedgehog tenrec *Echinops telfairi*. Ann Rep Jersey Wildl Preserv Trust. 9:18-25.
9139. Riordan, DV. 1973. Notes on the Cuban hutia *Capromys piloroides*. Ann Rep Jersey Wildl Preserv Trust. 10:38.
9140. Ripley, DS. 1952. Terriitorial and sexual behavior in the great Indian rhinoceros, a speculation. Ecology. 33:570-573.
9141. Risbridger, GP & Weiss, M. 1985. Gonadotrophin and steroid binding to adrenal cortex tissue of female brushtail possum, *Trichosurus vulpecula*. Aust J Zool. 33:831-835.
9142. Risler, L, Wasser, SK, & Sackett, GP. 1987. Measurement of excreted steroids in *Macaca nemestrina*. Am J Primatol. 12:91-100.
9142a. Rismiller, PD. 1992. Field observations on Kangaroo Island echidnas (*Tachyglossus aculeatus multiaculeatus*) during the breeding season. 101-105, in Platypus and Echidnas, Augee, ML, ed, R Zool Soc NSW, Sydney.
9143. Rissman, EF. 1987. Social variables influence female sexual behavior in the musk shrew (*Suncus murinus*). J Comp Psychol. 101:3-6.
9144. Rissman, EF. 1987. Gonadal influences on sexual beahvior in the male musk shrew (*Suncus murinus*). Horm & Behav. 21:132-136.
9145. Rissman, EF. 1991. Evidence that neural aromatization of androgen regulates the expression of sexual behaviour in female musk shrews. J Neuroendo. 3:441-448.
9146. Rissman, EF & Bronson, FH. 1987. Role of the ovary and adrenal gland in the sexual behavior of the musk shrew, *Suncus murinus*. Biol Reprod. 36:664-668.
9147. Rissman, EF, Clendenon, AL, & Krohmer, RW. 1990. Role of androgens in the regulation of sexual behavior in the female musk shrew. Neuroendocrinology. 51:468-473.
9148. Rissman, EF & Crews, D. 1988. Hormonal correlates of sexual behavior in the female musk shrew: The role of estradiol. Physiol Behav. 44:1-7.
9149. Rissman, EF & Crews, D. 1989. Effect of castration on epididymal sperm storage in male musk shrews (*Suncus murinus*) and mice (*Mus musculus*). J Reprod Fert. 86:219-222.
9150. Rissman, EF & Johnston, RE. 1985. Female reproductive development is not activated by male California voles exposed to family cues. Biol Reprod. 32:352-360.
9151. Rissman, EF, Nelson, RJ, Blank, JL, et al. 1987. Reproductive response of a tropical mammal, the musk shrew (*Suncus murinus*), to photoperiod. J Reprod Fert. 81:563-566.
9152. Rissman, EF, Sheffield, SD, Kretzmann, MB, et al. 1984. Chemical cues from families delay puberty in male California voles. Biol Reprod. 31:324-331.
9153. Rissman, EF, Silveira, J, & Bronson, FH. 1988. Patterns of sexual receptivity in the female musk shrew (*Suncus murinus*). Horm Behav. 22:186-193.
9154. Risting, S. 1928. Whales and whale foetuses statistics and catch measurements collected from the Norwegian Whalers Association. Rapp Réun Cons Perm Int Explor Mer. 50:5-122.
9155. Ritchie, ATA. 1963. The black rhinoceros (*Diceros bicornis* L). E Afr Wildl J. 1:54-62.
9155a. Ritke, ME. 1990. Quantitative assessment of variation in litter size of the raccoon *Procyon lotor*. Am Mid Nat. 123:390-398.
9155b. Ritke, ME. 1990. Sexual dimorphism in the raccoon (*Procyon lotor*): Morphological evidence for intrasexual selection. Am Mid Nat. 124:342-351.
9156. Rittner, M & Schmidt, U. 1982. The influence of the sexual cycle on the olfactory sensitivity of wild female house mice (*Mus musculus domesticus*). Z Säugetierk. 47:47-50.
9157. Robbel, H & Child, G. 1970. Notes on the 1969 rut in the Moremi. Botswana Notes & Records. 2:95-97. [Cited by Montis, 1972]
9158. Robbins, CT & Moen, AN. 1975. Uterine composition and growth in pregnant white-tailed deer. J Wildl Manage. 39:648-691.
9159. Röben, P. 1969. Die Spitzmäuse (*Soricidae*) der Heidelberger Umgebung. Säugetierk Mitt. 17:42-62.
9160. Robens, JF. 1968. Influence of maternal weight on pregnancy, number of corpora lutea, and implantation sites in the golden hamster (*Mesocricetus auratus*). Lab Anim Care. 18:651-653.
9161. Roberts, CM. 1973. The Embryology of Certain Hystricomorph Rodents London Hospital Medical College. Univ London, London.
9162. Roberts, CM & Perry, JS. 1974. Hystricomorph embryology. Symp Zool Soc London. 34:333-360.
9163. Roberts, CM & Weir, BJ. 1973. Implantation in the plains viscacha, *Lagostomus maximus*. J Reprod Fert. 33:299-307.
9164. Roberts, HA & Early, RC. 1952. Mammal survey of southeastern Pennsylvania. PA Game Comm. Harrisburg.
9165. Roberts, M. 1982. Demographic trends in a captive population of red pandas (*Ailurus fulgens*). Zoo Biol. 1:119-126.
9166. Roberts, M, Brand, S, & Maliniak, E. 1985. The biology of captive prehensile-tailed porcupines, *Coendou prehensilis*. J Mamm. 66:476-482.
9167. Roberts, M, Koontz, F, Phillips, L, et al. 1987. Management and biology of the prehensile-tailed porcupine *Coendou prehensilis* at Washington NZP and New York Zoological Park. Int Zoo Yearb. 26:265-275.
9168. Roberts, M, Maliniak, E, & Deal, M. 1984. The reproductive biology of the rock cavy, *Kerodon rupestris*, in captivity: A study of reproductive adaptation in a trophic specialist. Mammalia. 48:253-266.
9169. Roberts, M, Newman, L, & Peterson, G. 1982. The management and reproduction of the large hairy armadillo *Chaetophractus villosus* at the National Zoological Park. Int Zoo Yearb. 22:185-194.
9169a. Roberts, M, Phillips, L, & Kohn, F. 1990. Common ringtail possum (*Pseudocheirus peregrinus*) as a management model for the Pseudocheiridae: Reproductive scope, behavior, and biomedical values on a browse-free diet. Zoo Biol. 9:25-41.
9170. Roberts, MS. 1975. Growth and development of mother-reared red pandas *Ailurus fulgens*. Int Zoo Yearb. 15:57-63.
9171. Roberts, MS. 1978. The annual reproductive cycle of captive *Macaca sylvana*. Folia Primatol. 29:229-235.

9172. Roberts, MS. 1980. Breeding the red panda (*Ailurus fulgens*) at the National Zoological Park. Zool Garten NF. 50:253-263.
9173. Roberts, MS & Gittleman, JL. 1984. *Ailurus fulgens*. Mamm Species. 222:1-8.
9174. Roberts, MS & Kessler, DS. 1979. Reproduction in red pandas, *Ailurus fulgens* (Carnivora: Ailuropodidae). J Zool. 188:235-249.
9174a. Roberts, MS, Thompson, KV, & Cranford, JA. 1968. Reproduction and growth in captive punare (*Trichomys apereoides* Rodentia: Echimyidae) of the Brazilian Caatinga with reference to the reproductive strategies of Echimyidae. J Mamm. 69:542-551.
9175. Roberts, TJ. 1977. The Mammals of Pakistan. Ernest Benn Ltd, London.
9176. Robertson, PB. 1975. Reproduction and community structure of rodents over a transect in southern Mexico. Ph D Diss. Univ KS, Lawrence.
9177. Robertson, PB & Rickart, EA. 1975. *Cryptotis magna*. Mamm Species. 61:1-2.
9178. Robin, HA. 1881. Recherches anatomiques sur les Mammifères de l'ordre Chiroptères. Ann Sci Nat Zool Sér 6. 12:1-180.
9179. Robin, HA. 1881. Sur l'époque de l'accouplement des Chauves-souris. Bull Soc Philomatique Sér 7. 5:88-90.
9180. Robinette, WL & Archer, AL. 1971. Note on ageing criteria and reproduction of Thomson's gazelle. E Afr Wildl J. 9:83-98.
9181. Robinette, WL & Child, GFT. 1964. Notes on biology of the lechwe (*Kobus leche*). Puku. 2:84-117.
9182. Robinette, WL & Gashwiler, JS. 1950. Breeding season, productivity, and fawning period of the mule deer in Utah. J Wildl Manage. 14:457-469.
9183. Robinette, WL & Olsen, OA. 1944. Studies of the productivity of mule deer in central Utah. Trans N Am Wildl Conf. 9:156-161.
9184. Robinette, WL, Gashwiler, JS, Jones, DA, et al. 1955. Fertility of mule deer in Utah. J Wildl Manage. 19:115-136.
9185. Robinette, WL, Gashwiler, JS, & Morris, OW. 1961. Notes on caugar productivity and life history. J Mamm. 42:204-217.
9186. Robins, JP. 1954. Ovulation and pregnancy corpora lutea in the ovaries of the humpback whale. Nature. 173:201-203.
9187. Robins, JP. 1960. Age studies on the female humpback whale, *Megaptera nodosa* (Bonnaterre), in east Australian waters. Aust J Mar Freshwater Res. 11:1-13.
9188. Robinson, AC. 1988. The ecology of the bush rat, *Rattus fucipes* (Rodentia: Muridae) in Sherbrooke Forest, Victoria. Aust Mamm. 11:35-49.
9189. Robinson, AC, Robinson, JF, Watts, CHS, et al. 1976. The shark bay mouse *Pseudomys praeconis* and other mammals on Bernier Island, Western Australia. West Aust Nat. 13:149-155.
9190. Robinson, DJ & Cowan, IM. 1954. An introduced population of the gray squirrel (*Sciurus carolinensis* Gmelin) in British Columbia. Can J Zool. 32:261-282.
9190a. Robinson, ES, Renfree, MB, Short, RV, et al. 1991. Mammary glands in male marsupials: 2 Development of teat primordia in *Didelphis virginiana* and *Monodelphis domestica*. Reprod Fertil Dev. 3:295-301.
9191. Robinson, HGN, Gribble, WD, Page, WG, et al. 1965. Notes on the birth of a reticulated giraffe *Giraffa camelopardalis antiguorum*. Int Zoo Yearb. 5:49-52.
9192. Robinson, JA, Scheffler, G, Eisele, SG, et al. 1975. Effects of age and season on sexual behavior and plasma testosterone and dihydrotestosterone concentrations of laboratory-housed rhesus monkeys (*Macaca mulatta*). Biol Reprod. 13:203-210.
9193. Robinson, JA & Bridson, WE. 1978. Neonatal hormone patterns in the macaque I Steroids. Biol Reprod. 19:773-778.
9194. Robinson, JF, Robinson, AC, Watts, CHS, et al. 1978. Notes on rodents and marsupials and their ectoparasites collected in Australia in 1974-75. Tran R Soc S Aust. 102:59-70.
9195. Robinson, JS, Natale, R, Clover, L, et al. 1979. Prostaglandin E, thromboxane B2 and 6-oxo-prostaglandin F1α in amniotic fluid and maternal plasma of rhesus monkeys (*Macaca mulatta*) during the latter third of gestation. J Endocrinol. 81:345-349.
9196. Robinson, R. 1969. The breeding of spotted and black leopards. J Bombay Nat Hist Soc. 66:423-429.
9197. Robinson, R & Cox, HW. 1970. Reproductive performance in a cat colony over a 10-year period. Lab Anim. 4:99-112.
9198. Robinson, RM, Thomas, JW, & Marburger, RG. 1965. The reproductive cycle of male white-tailed deer in central Texas. J Wildl Manage. 29:53-59.
9199. Robinson, TS & Quick, FW. 1965. The cotton rat in Kentucky. J Mamm. 46:100.
9200. Roche, J. 1962. Nouvelles données sur la reproduction des Hyracoidés. Mammalia. 26:517-529.
9201. Rock, TW, Flood, PF, Rawlings, NC. 1991. Circannual changes in reproductive organs and spermatogenesis in male Saskatchewan beavers. J Mamm. 72:211-212.
9202. Rode, P. 1943. Faune de l'Empire Français, II Mammifères Ongulés de l'Afrique Noire, Part 1:1-123. Librairie La Rose, Paris.
9203. Rode, P. 1944. Faune de l'Empire Français, V Mammifères Ongulés de l'Afrique Noire, Part 2:125-209, Librairie La Rose, Paris.
9204. Rodger, JC. 1982. The testis and its excurrent ducts in American caenolestid and didelphid marsupials. Am J Anat. 163:269-282.
9205. Rodger, JC. 1990. Prospects for the artificial manipulation of marsupial reproduction and its application in research and conservation. Aust J Zool. 37:249-258.
9206. Rodger, JC & Bedford, JM. 1982. Induction of oestrus, recovery of gametes, and the timing of fertilization events in the opossum, *Didelphis virginiana*. J Reprod Fert. 64:159-169.
9207. Rodger, JC & Bedford, JM. 1982. Separation of sperm pairs and sperm-egg interaction in the opossum, *Didelphis virginiana*. J Reprod Fert. 64:171-179.
9208. Rodger, JC & White, IG. 1974. Carbohydrates of the prostate of two Australian marsupials, *Trichosurus vulpecula* and *Megaleia rufa*. J Reprod Fert. 34:267-274.
9209. Rodger, JC & White, IG. 1974. Sugars in semen of Australian marsupials. J Reprod Fert. 36:453.
9210. Rodger, JC & White, IG. 1975. Electroejaculation of Australian marsupials and analyses of the sugars in the seminal plasma from three macropod species. J Reprod Fert. 43:233-239.
9211. Rodgers, AR & Lewis, MC. 1986. Diet selection in arctic lemmings (*Lemmus sibiricus* and *Dicrostonyx groenlandicus*): Demography, home range, and habitat use. Can J Zool. 64:2717-2727.
9212. Rodgers, WA. 1984. Warthog ecology in south east Tanzania. Mammalia. 48:327-350.
9213. Rödl, P. 1974. Geburtsverlauf bei der Gelbhalsmaus, *Apodemus flavicollis* (Melchior, 1834). Säugetierk Mitt. 22:250-254.
9213a. Rodríguez-Durán, A & Kunz, TH. 1992. *Pteronotus quadridens*. Mamm Species. 395:1-4.

9214. Roecker, RM. 1950. The biology of the northern gray squirrel, *Sciurus carolinensis leucotis* (Gapper) in central New York. Ph D Diss. Cornell Univ, Ithaca.
9215. Roeder, JJ. 1978. Marking behaviour in genets (*G genetta* L): Seasonal variations and relation to social status in males. Behaviour. 67:149-156.
9216. Roeder, JJ. 1979. La reproduction de la Genette (*Genetta genetta* L) en captivité. Mammalia. 43:531-542.
9217. Roelvinck, W & Immendorf, M. 1962. Breeding the Bornean tupaia *Tupaia tana*. Int Zoo Yearb. 4:76.
9218. Roer, H. 1969. Das Alter der in vier Wochenstuben der Eifel ansassigen Weibchen des Mausohrs, *Myotis myotis* (Borkhausen, 1797). Säugetierk Mitt. 17:232-233.
9219. Roer, H. 1971. Zur Lebensweise einiger Microchiropteren der Namibwüste (Mammalia, Chiroptera). Staatl Mus Tierk (Dresden) Zool Abh. 32:43-55.
9220. Roer, H. 1973. Über die Ursachen hoher Jugendmortalität beim Mausohr, *Myotis myotis* (Chiroptera, Mamm). Bonn Zool Beitr. 24:332-341.
9221. Roest, AI. 1984. The Mono Bay kangaroo rat: A summary of current knowledge. Am Soc Mamm Poster. Humboldt State, 24-28 June.
9221a. Roest, AI. 1991. Captive reproduction in Heermann's kangaroo rat, *Dipodomys heermanni*. Zoo Biol. 127-137.
9222. Roff, C. 1960. Deer in Queensland. Queensl J Agr Sci. 17:43-58.
9223. Rogers, AL, Erickson, LF, Hoversland, AS, et al. 1969. Management of a colony of African pygmy goats for biomedical research. Lab Anim Care. 19:181-185.
9223a. Rogers, DS & Rogers, JE. 1992. *Heteromys oresterus*. Mamm Species. 396:1-3.
9223b. Rogers, DS & Rogers, JE. 1992. *Heteromys nelsoni*. Mamm Species. 397:1-2.
9224. Rogers, DS & Schmidly, DJ. 1982. Systematics of opening pocket mice (genus *Heteromys*) of the *desmarestianus* species group from Mexico and northern central America. J Mamm. 63:375-386.
9225. Rogers, JF & Dawson, WW. 1970. Foetal and placental size in a *Peromyscus* species cross. J Reprod Fert. 21:255-262.
9226. Rogers, JG & Beauchamp, CK. 1974. Relationships among three criteria of puberty in *Peromyscus leucopus noveboracensis*. J Mamm. 55:461-462.
9227. Rogers, LL. 1987. Effects of food supply and kinship on social behavior, movements and population growth of black bears in northeastern Minnesota. Wildl Monog. 97:1-72.
9228. Rogers, PM. 1981. The wild rabbit in the Camargue, southern France. Proc World Lagomorph Conf, Univ Guelph, ONT. 1979:587-599.
9228a. Rogowitz, GL & Wolfe, ML. 1991. Intraspecific variation in life-history traits of the white-tailed jackrabbit (*Lepus townsendii*). J Mamm. 72:796-806.
9229. Rohrbach, C. 1982. Investigation of the Bruce effect in the Mongolian gerbil (*Meriones unguiculatus*). J Reprod Fert. 65:411-417.
9230. Rojas, MA, Montenegro, MA, & Morales, B. 1982. Embryonic development of the degu, *Octodon degus*. J Reprod Fert. 66:31-38.
9231. Rojas-Mendoza, P. 1951. Estudio biólogico del conejo de los volcanes (género *Romerolagus*) (Mammalia, Lagomorpha). Tesis Prof Dept Biol. Fac Ciencias. Univ Nac Mexico.
9232. Rojo, MN & Sanchez-Cordero, V. 1984. Patrón del área de actividad de *Neotomodon alstoni alstoni* (Rodentia: Cricetinae). An Inst Biol Univ Nal Autón México, Ser Zool. 55:285-306.
9233. Röken, B & Röken, B. 1981. Künstliche Aufzucht eines neugeborenen männlichen Eisbären (*Thalarctos maritimus* Phipps, 1774) in Kolmårdens Djurpark. Zool Garten NF. 51:119-122.
9234. Rokitansky, G. 1953. Höchstlebensalter der Gelbhalsmaus, *Apodemus flavicollis* Melchior, 1834. Säugetierk Mitt. 1:29-30.
9235. Rolan, RG & Gier, HT. 1967. Correlation of embryo and placental scar counts of *Peromyscus maniculatus* and *Microtus ochrogaster*. J Mamm. 48:317-319.
9236. Rolleston, no initial. 1866. On the placental structures of the tenrec (*Centetes ecaudatus*), and those of certain other mammalia; with remarks on the value of the placental system of classification. Trans Zool Soc London. 5:285-316.
9237. Rollinat, R & Trouessart, E. 1895. Sur la reproduction des Chiroptères. CR Soc Biol Dixieme Sér. 2:53-54.
9238. Rollinat, R & Trouessart, E. 1895. Deuxieme note sur la reproduction des Chiroptères. CR Soc Biol Dixieme Sér. 2:534-536.
9239. Rollinat, R & Trouessart, E. 1895. Sur la reproduction des Chauves-Souris. Bull Soc Zool Fr. 20:25-28.
9240. Rollinat, R & Trouessart, E. 1896. Sur la reproduction des Chauves-souris Le Vespertillion murin (*Vespertilio murinas* Schreber). Mém Soc Zool Fr. 9(13):214-240.
9241. Rollinat, R & Trouessart, E. 1897. Sur la reproduction des Chauves-souris II Les Rhinolophes, et note sur leurs parasites épizootiques. Mém Soc Zool Fr. 10(8):114-138.
9242. Romer, JD. 1974. Milk analysis and weaning in the lesser Malay cheverotain *Tragulus javanicus*. Int Zoo Yearb. 14:179-180.
9243. Ronald, K. 1973. The Mediterranean monk seal, *Monachus monachus*. IUCN Suppl Pap. 39:30-41.
9244. Ronald, K & Dougan, JL. 1982. The ice lover: Biology of the harp seal (*Phoca groenlandica*). Science. 215:928-933.
9245. Ronald, K & Healey, PJ. 1981. Harp seal *Phoca groenlandica* Erxleben, 1777. Handbook of Marine Mammals, Ridgway, SH & Harrison, RJ, eds, Academic Press, New York. 2:55-87.
9246. Ronald, K & Thomson, CA. 1981. Parturition and postpartum behaviour of a captive harbour seal *Phoca vitulina*. Aquatic Mamm. 8:79-90.
9247. Roney, no initials. Personal communication. [Cited by Cary, 1976]
9248. Roney, E. 1975. Twin waterbucks at San Antonio Zoo. AAZPA Newsl. 16(10):14-15.
9249. Rongstad, OJ. 1965. A life history study of thirteen-lined ground squirrels in southern Wisconsin. J Mamm. 46:76-87.
9250. Rongstad, OJ. 1966. Biology of penned cottontail rabbits. J Wildl Manage. 30:312-319.
9251. Rongstad, OJ. 1969. Gross prenatal development of cottontail rabbits. J Wildl Manage. 33:164-168.
9252. Rood, JP. 1963. Observations on the behavior of the spiny rat *Heteromys melanoleucus* in Venezuela. Mammalia. 27:186-192.
9253. Rood, JP. 1965. Observations on population structure, reproduction and molt of the Scilly shrew. J Mamm. 46:426-433.
9254. Rood, JP. 1965. Observations on the life cycle and variation of the long-tailed field mouse *Apodemus sylvaticus* on the Isles of Scilly and Cornwall. J Zool. 147:99-107.
9255. Rood, JP. 1966. Observations on the reproduction of *Peromyscus* in captivity. Am Mid Nat. 76:496-503.
9256. Rood, JP. 1970. Ecology and social behavior of the desert cavy, *Microcavia australis*. Am Mid Nat. 83:415-454.
9257. Rood, JP. 1972. Ecological and behavioural comparisons of three genera of Argentine cavies. Anim Behav Monog. 5:1-83.

9258. Rood, JP. 1973. The banded mongoose. Africana. 5(2):14.
9259. Rood, JP. 1975. Population dynamics and food habits of the banded mongoose. E Afr Wildl J. 13:89-111.
9260. Rood, JP. 1978. Dwarf mongoose helpers at the den. Z Tierpsychol. 48:277-287.
9261. Rood, JP. 1980. Mating relationships and breeding suppression in the dwarf mongoose. Anim Beh. 28:143-150.
9262. Rood, JP. 1990. Group size, survival, reproduction, and routes to breeding in dwarf mongooses. Anim Beh. 39:566-572.
9263. Rood, JP & Waser, PM. 1978. The slender mongoose, *Herpestes sanguineus*, in the Serengeti. Carnivore. 1(3):54-58.
9264. Rood, JP & Weir, BJ. 1970. Reproduction in female wild guinea pigs. J Reprod Fert. 23:393-409.
9265. Roonwal, ML. 1949. Systematics, ecology and bionomics of mammals studied in connection with Tsutsugamushi disease (scrub typhus) in the Assam-Burma war theater during 1945. Trans Nat Inst Sci India. 3:67-122.
9265a. Roonwal, ML & Mohot, SM. 1977. Primates of South Asia. Harvard Univ Press, Cambridge, MA.
9266. Roos, TB & Shackelford, RM. 1955. Some observations on the gross anatomy of the genital system and two endocrine glands and body weights in the chinchilla. Anat Rec. 123:301-311.
9267. Root, DA & Payne, NF. 1984. Age-specific reproduction of gray foxes in Wisconsin. J Wildl Manage. 48:890-892.
9268. Roots, CG. 1966. Notes on the breeding of white-lipped peccaries *Tayassa albirostris* at Dudley Zoo. Int Zoo Yearb. 6:198-199.
9269. Rörig, G & Knoche, E. 1916. Beiträge zur Biologie der Feldmäuse. Arb Kais Biol Anst Land-Forstwirtsch. 9:333-420.
9270. Rosahn, PD & Greene, HS. 1936. The influence of intrauterine factors on the fetal weight of rabbits. J Exp Med. 63:901-921.
9271. Rose, J, Oldfield, JE, & Stormshak, F. 1986. Changes in serum prolactin concentrations and ovarian prolactin receptors during embryonic diapause in mink. Biol Reprod. 34:101-106.
9272. Rose, J, Stormshak, F, Adair, J, et al. 1983. Uterine progesterone metabolism in the mink during delayed implantation. Biol Reprod. 28(Suppl):137.
9273. Rose, RK & Gainies, MS. 1978. The reproductive cycle of *Microtus ochrogaster* in eastern Kansas. Ecol Monogr. 48:21-42.
9274. Rose, RM, Holadoy, JW, & Bernstein, IS. 1971. Plasma testosterone, dominance rank and aggressive behaviour in male rhesus monkeys. Nature. 231:366-368.
9275. Rose, RW. 1982. Tasmanian bettong *Bettongia gaimardi*: Maintenance and breeding in captivity. The Management of Australian Mammals in Captivity. 108-110, Evans, DD, ed, Zool Bd Victoria, Melbourne.
9276. Rose, RW. 1986. The control of pouch vacation in the Tasmanian bettong, *Bettongia gaimardi*. Aust J Zool. 34:485-491.
9277. Rose, RW. 1987. Reproductive biology of the Tasmanian bettong (*Bettongia gaimardi*: Macropodidae). J Zool. 212:59-67.
9278. Rose, RW. 1989. Age estimation of the Tasmanian bettong (*Bettongia gaimardi*) (Marsupialia: Potoroidae). Aust Wildl Res. 16:251-261.
9279. Rose, RW & Bradley, AJ. 1991. Testosterone levels in the Tasmanian bettong (*Bettongia gaimardi*). Gen Comp Endocrinol. 84:121-128.
9280. Rose, RW & McCartney, DJ. 1981. Reproduction of the red-bellied pademelon, *Thylogale billardierii* (Marsupialia). Aust Wildl Res. 9:27-32.
9281. Rose, RW & McCartney, DJ. 1981. Growth of the red-bellied pademelon, *Thylogale billardierii*, and age estimation of pouch young. Aust Wildl Res. 9:33-38.
9282. Roseberry, JL & Klimstra, WD. 1970. Productivity of white-tailed deer on Crab Orchard National Wildlife Refuge. J Wildl Manage. 34:23-28.
9283. Rosen, RC. 1975. Ontogeny of homeothermy in *Microtus pennsylvanicus* and *Octodon degus*. Comp Biochem Physiol. 52A:675-670.
9283a. Rosen, SI. 1972. Twin gorilla fetuses. Folia Primatol. 17:132-141.
9284. Rosenberg, H. 1971. Breeding the bat-eared fox *Otocyon megalotis* at Utica Zoo. Int Zoo Yearb. 11:101-102.
9285. Rosenberg, M. 1988. Birth weights in three Norwegian cities, 1860-1984 Secular trends and influencing factors. Ann Hum Biol. 15:275-288.
9286. Rosenblum, LA. 1968. Mother-infant relations and early behavioral development in the squirrel monkey. The Squirrel Monkey. 207-233, Rosenblum, LA & Cooper, RW, eds, Academic Press, New York.
9287. Rosenblum, LA. 1968. Some aspects of female reproductive physiology in the squirrel monkey. The Squirrel Monkey. 147-169, Rosenblum, LA & Cooper, RW, eds, Academic Press, New York.
9288. Rosenblum, LA. 1972. Reproduction of squirrel monkeys in the laboratory. Breeding Primates. 130-143, Beveridge, WIB, ed, Karger, Basel.
9289. Rosenblum, LA & Kaufman, IC. 1967. Laboratory observations of early mother-infant relations in pigtail and bonnet macaques. Social Communication Among Primates. 33-41, Altmann, SA, ed, Chicago Univ Press, Chicago.
9290. Rosenstrauch, A, Bedrak, E, & Friedlander, M. 1978. Testosterone production by testicular tissue of the camel (*Camelus dromedarius*) during the breeding season. J Steroid Biochem. 9:821.
9291. Rosenthal, MA. 1974. Hand-rearing patagonian cavies or maras *Dolichotis patagonum* at Lincoln Park, Chicago. Int Zoo Yearb. 14:214-215.
9292. Rosenthal, MA. 1975. Observations on the water opossum or yapok *Chironectes minimus* in captivity. Int Zoo Yearb. 15:4-6.
9293. Rosenthal, MA. 1981. Chimpanzee triplets born in captivity. Primates. 22:137-138.
9294. Rosenthal, MA & Meritt, DA, Jr. 1971. Hand-rearing Grant's gazelle *Gazella granti* at Lincoln Park Zoo, Chicago. Int Zoo Yearb. 11:130.
9295. Rosenthal, MA & Meritt, DA, Jr. 1973. Hand-rearing springhaas *Pedetes capensis* at Lincoln Park Zoo, Chicago. Int Zoo Yearb. 13:135-137.
9296. Roser, JF, Czekala, NM, Mortensen, RB, et al. 1986. Daily urinary hormone assays as a diagnostic tool to evaluate infertility in gorillas. Biol Reprod. 34(Suppl):131.
9296a. Roser, JR, Farmer, SW, Murthy, HMS, et al, 1984. Chemical, biological and immunological properties of pituitary gonadotropins from the donkey (*Equus asinus*): Comparison with the horse (*Equus caballus*). Biol Reprod. 30:1253-1262.
9297. Rosevear, DR. 1965. The Bats of West Africa. Trustees Brit Mus (Nat Hist), London.
9298. Rosner, W, Pugeat, MM, Chrousos, GP, et al. 1986. Steroid-binding proteins in primate plasma. Endocrinology. 118:513-517.
9299. Ross, GJB. 1969. Evidence for a southern breeding population of True's beaked whale. Nature. 222:585.

9300. Ross, GJB. 1977. The taxonomy of bottlenosed dolphins *Tursiops* species in South African waters, with notes on their biology. Ann Cape Prov Mus (Nat Hist). 11:135-194.
9301. Ross, GJB. 1979. Records of pygmy and dwarf sperm whales, genus *Kogia*, from southern Africa, with biological notes and some comparisons. Ann Cape Prov Mus (Nat Hist). 11:259-327.
9301a. Ross, GJB. 1984. The smaller cetaceans of the southeast coast of southern Africa. Ann Cape Prov Mus (Nat Hist). 15:173-410.
9302. Ross, GJB, Best, PB, & Donnelly, BG. 1975. New records of the pygmy right whale (*Caperea marginata*) from South Africa, with comments on distribution, migration, appearance, and behavior. J Fish Res Bd Can. 32:1005-1017.
9302a. Ross, RC. 1930. California Sciuridae in captivity. J Mamm. 11:76-78.
9303. Ross, RC. 1934. Age and fecundity of mule deer (*Odocoileus hemionus hemionus*). J Mamm. 15:72.
9303a. Rosser, AM. 1989. Environmental and reproductive seasonality of puku, *Kobus vardoni*, in Luangwa Valley, Zambia. Afr J Ecol. 27:77-88.
9304. Rostal, DC & Eaton, GG. 1983. Puberty in male Japanese macaques (*Macaca fuscata*): Social and sexual behavior in a confined troop. Am J Primatol. 4:135-141.
9305. Rotenberg, D. 1929. Notes on the male generative apparatus of *Tarsipes spenserae*. J R Soc W Aust. 15:9-14.
9306. Roth, CE. 1957. Notes on maternal care in *Myotis lucifugus*. J Mamm. 38:122-123.
9307. Roth, HH. 1962. Mitteilung über die Zwergluszpferdzucht in Gelsenkirchen. Zool Garten NF. 26:327-331.
9308. Roth, HH. 1964. Ein Beitrag zur Kenntnis von *Tremarctos ornatus* (Cuvier). Zool Garten NF. 29:107-129.
9309. Roth, HH. 1967. Über die Zusammensetzung von Warzenschwein-Milch. Zool Garten NF. 34:277-278.
9310. Roth, HH & Austen, B. 1966. Twin calves in elephants. Säugetierk Mitt. 14:342-345.
9311. Roth, HH, Kerr, MA, & Posselt, J. 1972. Studies on the utilization of semi-domesticated eland (*Taurotragus oryx*) in Rhodesia: Reproduction and herd increase. Z Tierzüchtung Züchtgsbiol. 89:69-83.
9312. Rothchild, I. 1981. The regulation of the mammalian corpus luteum. Rec Prog Hormone Res. 37:183-298.
9313. Rothe, H. 1974. Further observations on the delivery behaviour of the common marmoset (*Callithrix jacchus*). Z Säugetierk. 39:135-142.
9314. Rothe, H. 1975. Some aspects of sexuality and reproduction in groups of captive marmosets (*Callithrix jacchus*). Z Tierpsychol. 37:255-273.
9315. Rothe, H. 1977. Parturition and related behavior in *Callithrix jacchus* (Ceboidea, Callitrichidae). The Biology and Conservation of the Callitrichidae. 193-206, Kleiman, DG, ed, Smithsonian Inst Press, Washington, DC.
9316. Rothe, H, König, A, Darms, K, et al. 1987. Analysis of litter size in a colony of the common marmoset (*Callithrix jacchus*). Z Säugetierk. 52:227-235.
9317. Rothe, H & König, A. 1991. Sex ratio in newborn common marmosets (*Callithrix jacchus*): No indication for a functional germ cell chimerism. Z Säugetierk. 56:318-320.
9318. Rougeot, J. 1969. Accélération du rhythme de la reproduction chez le Mouflon de Corse (*Ovis ammon musimon* Schreber, 1782) au moyen de cycle photopériodiques semestriels. Ann Biol Anim Bioch Biophys. 9:441-443.
9318a. Rouk, CS & Carter, DC. 1972. A new species of *Vampyrops* (Chiroptera: Phyllostomatidae) from South America. Occas Pap Mus TX Tech Univ. 1:1-7.
9319. Rowan, W. 1945. Numbers of young of the common black and grizzly bears in western Canada. J Mamm. 26:197-199.
9320. Rowan, W. 1947. A case of six cubs in the common black bear. J Mamm. 28:404-405.
9321. Rowan, W & Keith, LB. 1956. Reproductive potential and sex ratios of snowshoe hares in northern Alberta. Can J Zool. 34:273-281.
9322. Rowbottom, MJ & Rowbottom, CH. 1980. Watching otters in the West Highlands of Scotland. Jotters. 6:2.
9323. Rowe, FP, Swinney, T, & Quy, RJ. 1983. Reproduction of the house mouse (*Mus musculus*) in farm buildings. J Zool. 199:259-269.
9324. Rowe, FP & Taylor, EJ. 1964. The numbers of harvest-mice (*Micromys minutus*) in corn-ricks. Proc Zool Soc London. 142:181-185.
9325. Rowe-Rowe, DT. 1972. The African weasel, *Poecilogale albinucha* Gray: Observations on behaviour and general biology. Lammergeyer. 15:39-58.
9326. Rowe-Rowe, DT. 1978. Reproduction and postnatal development of South African mustelines (Carnivora: Mustelidae). Zool Afr. 13:103-114.
9327. Rowe-Rowe, DT. 1978. The small carnivores of Natal. Lammergeyer. 25:1-48.
9328. Rowe-Rowe, DT & Bigalke, RC. 1972. Observations on the breeding and behaviour of blesbok. Lammergeyer. 15:1-14.
9329. Rowe-Rowe, DT & Meester, J. 1985. Biology of *Myosorex varius* in an African montane region. Acta Zool Fennici. 173:271-273.
9330. Rowell, J, Betteridge, KJ, Randall, GCB, et al. 1987. Anatomy of the reproductive tract of the female musk ox (*Ovibos moschatus*). J Reprod Fert. 80:431-444.
9331. Rowell, J & Flood, PF. 1988. Progesterone, oestradiol-17β and LH during the oestrous cycle of muskoxen (*Ovibos moschatus*). J Reprod Fert. 84:117-122.
9332. Rowell, TE. 1963. Behaviour and female reproductive cycles of rhesus macaques. J Reprod Fert. 6:193-204.
9333. Rowell, TE. 1967. Female reproductive cycles and the behavior of baboons and rhesus macaques. Social Communication among Primates. 15-32, Altmann, SA, ed, Univ Chicago Press, Chicago.
9334. Rowell, TE. 1970. Reproductive cycles of two *Cercopithecus* monkeys. J Reprod Fert. 22:321-338.
9335. Rowell, TE. 1972. Toward a natural history of the talapoin monkey in Cameroon. Ann Fac Sci Cameroon. 10:121-134.
9336. Rowell, TE. 1973. Social organization of wild talapoin monkeys. Am J Phys Anthrop. 38:593-598.
9337. Rowell, TE. 1977. Variation in age at puberty in monkeys. Folia Primatol. 27:284-296.
9338. Rowell, TE. 1977. Reproductive cycles of the talopoin monkey (*Miopithecus talopoin*). Folia Primatol. 28:188-202.
9339. Rowell, TE & Chalmers, NR. 1970. Reproductive cycles of the mangabey *Cercocebus albigena*. Folia Primatol. 12:264-272.
9340. Rowell, TE & Dixson, AF. 1975. Changes in social organization during the breeding season of wild talapoin monkeys. J Reprod Fert. 43:419-434.
9341. Rowell, TE & Richards, SM. 1979. Reproductive strategies of some African monkeys. J Mamm. 60:58-69.
9342. Rowlands, IW. 1936. Reproduction of the bank vole (*Evotomys glareolus* Schreber) II Seasonal changes in the

reproductive organs of the male. Phil Trans R Soc London. 226B:99-120.
9343. Rowlands, IW. 1938. Preliminary note on the reproductive cycle of the red squirrel (*Sciurus vulgaris*). Proc Zool Soc London. 108A:441-443.
9344. Rowlands, IW. 1967. The Ferret (*Mustela putorius furo* L). 582-593. UFAW Handbook on the Care and Management of Laboratory Animals, 3rd ed. E & S Livingstone, Edinburgh.
9345. Rowlands, IW. 1974. Mountain viscacha. Symp Zool Soc London. 34:131-141.
9346. Rowlands, IW & Heap, RB. 1966. Histological observations of the ovary and progesterone levels in the coypu (*Myocastor coypus*). Symp Zool Soc London. 15:335-352.
9347. Rowlands, IW & Parkes, AS. 1935. The reproductive processes of certain mammals VIII Reproduction in foxes (*Vulpes* spp). Proc Zool Soc London. 4:823-841.
9348. Rowlands, IW & Sadleir, RMFS. 1968. Induction of ovulation in the lion *Panthera leo*. J Reprod Fert. 16:105-114.
9349. Rowlands, IW, Tam, WH, & Kleiman, DG. 1970. Histological and biochemical studies on the ovary and of progesterone levels in the systemic blood of the green acouchi *Myoprocta pratti*. J Reprod Fert. 22:533-545.
9350. Rowlands, IW & Weir, BJ. 1984. Mammals: Non-primate eutherians. Marshall's Physiology of Reproduction, 455-658, Lamming, GE, ed, Churchill Livingstone, London.
9350a. Rowley, I & Mollison, BC. 1955. Copulation in the wild rabbit, *Oryctolagus cuniculus*. Behaviour. 8:81-84.
9350b. Rowley, J. 1929. Life history of the sea-lions on the California coast. J Mamm. 10:1-36.
9351. Rowsemitt, C, Kunz, TH, & Tamarin, RH. 1975. The timing and patterns of molt in *Microtus breweri*. Occas Pap Mus Nat Hist Univ KS. 34:1-11.
9352. Rowsemitt, CN & Berger, PJ. 1983. Diel plasma testosterone rhythms in male *Microtus montanus*, the montane vole, under long and short photoperiods. Gen Comp Endocrinol. 50:354-358.
9353. Rowsemitt, CN & O'Conner, AJ. 1989. Reproductive function in *Dipodomys ordii* stimulated by 6-methoxybenzoxazolinone. J Mamm. 70:805-809.
9354. Roy, MA, Wolf, RH, Martin, LN, et al. 1978. Social and reproductive behaviors in surrogate-reared squirrel monkeys (*Saimiri scuireus*). Lab Anim Sci. 28:417-421.
9355. Ru-Yung, S & Jinxiang, Z. 1987. Postnatal development of thermoregulation in the root vole (*Microtus oeconomus*) and the quantitative index of homeothermic ability. J Therm Biol. 12:267-272.
9355a. Rubin, D. 1943. Embryonic differentiation of Cowper's and Bartholin's glands of the opossum following castration and ovariectomy. J Exp Zool. 94:463-475.
9355b. Rubin, D. 1944. The relation of hormones to the development of Cowper's and Bartholin's glands in the opossum (*Didelphys virginiana*). J Morphol. 74:213-285.
9356. Ruch, W. 1967. Die Implantationszeit und deren Beeinflussung durch die Laktation bei *Acomys cahirinus dimidiatus*. Rev Suisse Zool. 74:566-569.
9357. Rudge, MR. 1969. Reproduction of feral goats *Capra hircus* L near Wellington New Zealand. NZ J Sci. 12:817-827.
9358. Rudnai, J. 1973. Reproductive biology of lions (*Panthera leo massaica* Neumann) in Nairobi National Park. E Afr Wildl J. 11:241-254.
9359. Rudnai, J. 1984. Suckling behaviour in captive *Dendrohyrax arboreus* (Mammalia: Hyracoidea). S Afr J Zool. 19:121-123.
9360. Rudnai, JA. 1973. The Social Life of the Lion. Medical & Technical Publ, Lancaster, UK.
9361. Rudran, R. 1973. The reproductive cycles of two subspecies of purple-faced langurs (*Presbytis senex*) with relation to environmental factors. Folia Primatol. 19:41-60.
9362. Rudran, R. 1978. Socioecology of the blue monkeys (*Cercopithecus mitis sticklmanni*) of the Kibale Forest, Uganda. Smithson Contr Zool. 249:1-88.
9363. Rudran, R. 1979. The demography and social mobility of a red howler (*Alouatta seniculus*) population in Venezuala. Vertebrate Ecology in the Northern Neotropics. 107-126, Eisenberg, JE, ed, Smithsonian Inst Press, Washington, DC.
9364. Ruedi, D. 1984. The great Indian rhinoceros (*Rhinoceros unicornis*). p 171-190 in One Medicine, Ryder, OA & Byrd, ML, ed. Springer-Verlag, NY.
9365. Ruempler, G. 1976. Zur Haltung von Flughunden (*Pteropus* spec) im Zoo. Fr Kölner Zoo. 19:3-8.
9366. Ruffer, DG. 1965. Sexual behaviour of the northern grasshopper mouse (*Onychomys leucogaster*). Anim Behav. 13:447-452.
9367. Ruffner, GA & Carothers, SW. 1982. Age structure, condition and reproduction of two *Equus asinus* (Equidae) populations from Grand Canyon National Park, Arizona. Southwest Nat. 27:403-411.
9368. Rühmekorf, E. 1961. Zucht von Fischkatzen (*Prionailurus viverinus*). Zool Garten NF. 26:121-122.
9369. Ruiz de Elvira, MC, Herndon, JG, & Wilson, ME. 1982. Influence of estrogen-treated females on sexual behavior and male testosterone levels of a social group of rhesus monkeys during the nonbreeding season. Biol Reprod. 26:825-834.
9370. Rukowski, NN. 1955. Zur Fortpflanzung der Wildkatze im Kaukasus. Bull Mosk Ges Naturforscher. 60(4):94-96. [Abstract: Zool Garten NF. 24:309, 1958]
9371. Rumbaugh, DM. 1967. "Avila" -- San Diego Zoo's captive-born gorilla *Gorilla g gorilla*. Int Zoo Yearb. 7:98-107.
9372. Rumery, RE, Gaddum-Rosse, P, Blandau, RJ, et al. 1978. Cyclic changes in ciliation of the oviductal epithelium in the pig-tailed macaque (*Macaca nemestrina*). Am J Anat. 153:345-366.
9373. Ruppenthal, GC, Boodlin, BL, & Sackett, GP. 1983. Perinatal hypothermia and maternal temperature declines during labor in pigtailed macaques (*Macaca nemestrina*). Am J Primatol. 4:81-92.
9374. Rusch, DA & Reeder, WG. 1978. Population ecology of Alberta red squirrels. Ecology. 59:400-420.
9375. Russell, AE & Zuckerman, S. 1935. A "sexual skin" in a marmoset. J Anat. 69:356-362.
9376. Russell, CP. 1932. Seasonal migration of mule deer. Ecol Monogr. 2:1-46.
9377. Russell, EM. 1974. The biology of kangaroos (Marsupialia: Macropodidae). Mamm Rev. 4:1-59.
9378. Russell, EM. 1982. Patterns of parental care and parental investment in marsupials. Biol Rev. 57:423-487.
9379. Russell, EM. 1982. Parental investment and desertion of young in marsupials. Am Nat. 119:744-748.
9380. Russell, EM. 1986. Observations on the behaviour of the honey possum, *Tarsipes rostratus* (Marsupialia: Tarsipedidae) in captivity. Aust J Zool Suppl. 121:1-63.
9381. Russell, EM & Richardson, BJ. 1971. Some observations on the breeding, age structure, dispersion and habitat of populations of *Macropus robustus* and *Macropus antilopinus* (Marsupialia). J Zool. 165:131-142.
9382. Russell, KR. 1966. Effects of a common environment on cottontail ovulation rates. J Wildl Manage. 30:819-827.

9383. Russell, R. 1984. Social behaviour of the yellow-bellied glider *Petaurus australis* in north Queensland. Possums and Gliders. 343-353, Smith, A & Hume, I, eds, Surrey Beatty & Sons Pty Ltd, Sydney, Australia.
9384. Russo, JP. 1956. The desert bighorn sheep in Arizona. AZ Game Fish Dept Wildl Bull. 1:1-153.
9385. Rust, HJ. 1946. Mammals of northern Idaho. J Mamm. 27:308-327.
9386. Rutberg, AT. 1984. Birth synchrony in American bison (*Bison bison*): A response to predation or season?. J Mamm. 65:418-423.
9387. Rutherford, WH. 1964. The beaver in Colorado: Its biology, ecology, management and economics. CO Game Fish Parks Dept, Game Res Div Tech Bull. 17:1-49.
9388. Ruud, JT. 1945. Further studies on the structure of the baleen plates and their application to age determination. Hvalråd Skr. 29:1-69.
9389. Ruud, JT, Jonsgard, Å, & Ottestad, P. 1950. Age-studies on blue whales. Hvalråd Skr. 33:1-66.
9390. Ryan, GE. 1976. Observations on the reproduction and age structure of the fox, *Vulpes vulpes* L, in New South Wales. Aust Wildl Res. 3:11-20.
9391. Ryan, KD. 1984. Hormonal correlates of photoperiod-induced puberty in a relfex ovulator, the female ferret (*Mustela furo*). Biol Reprod. 31:925-935.
9392. Ryan, KD. 1985. Maturation of ovarian function in female ferrets is influenced by photoperiod. J Reprod Fert. 74:503-507.
9393. Ryan, KD & Robinson, SL. 1985. A rise in tonic luteinizing hormone secretion occurs during photoperiod-stimulated sexual maturation of the female ferret. Endocrinology. 116:2013-2018.
9394. Ryan, KD & Robinson, SL. 1987. A study of spontaneous sexual maturation of the female ferret. Biol Reprod. 36:333-339.
9395. Ryan, KD, Siegel, SF, & Robinson, SL. 1983. Detailed patterns of LH secretion after ovariectomy in the female ferret. Biol Reprod. 28(Suppl):99.
9396. Ryan, KD, Siegel, SF, & Robinson, SL. 1985. Influence of day length and endocrine status on luteinizing hormone secretion in intact and ovariectomized adult ferrets. Biol Reprod. 33:690-697.
9397. Ryan, RM. 1963. Life history and ecology of the Australian lesser long-eared bat, *Nyctophillus geoffroyi* Leach. M Sc Thesis. Univ Melbourne, Melbourne. [Cited by McKean & Hamilton-Smith, 1960]
9398. Ryan, RM. 1966. Observations on the broad-nosed bat, *Scoteinus balstoni*, in Victoria. J Zool. 148:162-166.
9399. Ryberg, O. 1947. Studies on Bats and Bat Parasites. Bökfordaget Svensk Natur, Stockholm.
9400. Ryder, JA. 1888. The placentation of the two-toed ant-eater, *Cycloturus didactylus*. Proc Acad Nat Sci Philadelphia. 1887:115-118.
9400a. Ryg, M. 1984. Effects of nutrition on seasonal changes in testosterone levels in young male reindeer (*Rangifer tarandus tarandus*). Comp Biochem Physiol. 77A:619-621.
9401. Rylands, AB. 1981. Preliminary field observations on the marmoset *Callithrix humeralifer intermedius* (Hershkowitz, 1977) at Dardanelos, Rio Aripuana, Mato, Grosso. Primates. 22:46-59.
9402. Ryman, N, Beckman, G, Bruun-Petersen, G, et al. 1977. Variability of red cell enzymes and genetic implications of management policies in Scandinavian moose (*Alces alces*). Hereditas. 85:157-162.
9403. Ryman, N, Reuterwall, C, Nygrén, K, et al. 1980. Genetic variation and differentiation in Scandinavian moose (*Alces alces*): Are large mammals monomorphic?. Evolution. 34:1037-1049.
9404. Ryszkowski, L. 1971. Reproduction of bank voles and survival of juveniles in different pine forest ecosystems. Ann Zool Fennici. 8:85-91.
9405. Saad, MB & Baylé, JD. 1985. Seasonal changes in plasma testosterone, thyroxine, and cortisol levels in wild rabbits (*Oryctolagus cuniculus algirus*) of Zembra Island. Gen Comp Endocrinol. 57:383-388.
9406. Saayman, GS. 1973. Effects of ovarian hormones on the sexual skin and behavior of ovariectomized baboons (*Papio ursinus*) under free-ranging conditions. Symp 4th Int Congr Primatol, 1972. 2:64-98. S Karger, Basel.
9407. Saayman, GS & Tayler, CK. 1977. Observations on the sexual behavior of Indian Ocean bottlenosed dolphins (*Tursiops aduncus*). Breeding Dolphins Present Status, Suggestions for the Future. 113-129, Ridgway, SH & Benirschke, K, eds, US Mar Mamm Comm, Washington, DC.
9408. Saayman, GS & Tayler, CK. 1979. The socioecology of humpback dolphins (*Sousa* sp). Behaviour of Marine Animals. 3:165-226. Winn, HE & Olla, BL, eds, Plenum Press, New York.
9409. Sabilaev, AS. 1971. A contribution to the ecology of the northern three-toed jerboa (*Dipus sagitta*) in the north-west Kyzylkumy. Zool Zhurn. 50:1553-1563.
9410. Saboureau, M & Boissin, J. 1978. Seasonal changes and environmental control of testicular function in the hedgehog, *Erinaceus europaeus* L. Environmental Endocrinology. 111-112, Assenmacher, I & Farner, DS, eds, Springer, New York.
9411. Saboureau, M & Boissin, J. 1978. Variations saisonnières de la testostérone et de la thyroxinemie chez le Hérisson (*Erinaceus europaeus* L). CR Hebd Séances Acad Sci Paris Sér D Sci Nat. 286:1479-1482.
9412. Saboureau, M & Boissin, J. 1983. Peripheral metabolism of testosterone during the annual reproductive cycle in the male hedgehog, a hibernating mammal. Can J Zool. 61:2849-2855.
9413. Saboureau, M & Dutourné, B. 1981. The reproductive cycle in the male hedgehog (*Erinaceus europaeus* L): A study of endocrine and exocrine testicular functions. Reprod Nutr Dév. 21:109-126.
9414. Saboureau, M, Laurent, H-M, & Boissin, J. 1982. Plasma testosterone binding protein capacity in relation to the annual testicular cycle in hibernating mammal, the hedgehog (*Erinaceus europaeus* L). Gen Comp Endocrinol. 47:59-63.
9415. Sackmann, HJ, Telle, R, & Ring, I. 1979. Zur Ermittlung von Muffellammgewichten. Zool Garten NF. 49:127-130.
9416. Saddington, GM, Mrs. 1966. Notes on the breeding of the Siberian chipmunk *Tamias sibircus* in captivity. Int Zoo Yearb. 6:165-166.
9417. Sade, DS. 1964. Seasonal cycle in size of testes of free-ranging *Macaca mulatta*. Folia Primatol. 2:171-180.
9418. Sade, DS & Hildrech, RW. 1965. Notes on the green monkey (*Cereopithecus aethiops sabaeus*) on St Kitts, West Indies. Caribb J Sci. 5:67-81.
9419. Sadleir, RM & Shield, JW. 1960. Delayed birth in marsupial macropods -- The euro, the tammar and the marloo. Nature. 185:335.
9420. Sadleir, RM & Shield, JW. 1960. Delayed birth in a hill-kangaroo, the euro (*Macropus robustus*). Proc Zool Soc London. 135:642-643.
9421. Sadleir, RMFS. 1965. Reproduction in two species of kangaroo (*Macropus robustus* and *Megaleia rufa*) in the arid Pilbara region of Western Australia. Proc Zool Soc London. 145:239-261.
9422. Sadleir, RMFS. 1966. Notes on reproduction in the larger Felidae. Int Zoo Yearb. 6:184-187.

9423. Sadleir, RMFS. 1974. The ecology of the deer mouse *Peromyscus maniculatus* in a costal coniferous forest II Reproduction. Can J Zool. 52:119-131.
9423a. Sadleir, RMFS. 1987. Reproduction in female cervids. Biology and Management of the Cervidae. 124-144, Wemmer, CM, ed, Smithsonian Inst Press, Washington, DC.
9424. Sadleir, RMFS, Casperson, KD, & Harling, J. 1973. Intake and requirements of energy and protein for the breeding of wild deer mice, *Peromyscus maniculatus*. J Reprod Fert Suppl. 19:237-252.
9425. Sadleir, RMFS & Tyndale-Biscoe, CH. 1977. Photoperiod and the termination of embryonic diapause in the marsupial *Macropus eugenii*. Biol Reprod. 16:605-608.
9426. Sadowiak, M. 1971. Red-necked wallabies *Protemnodon rufogrisea* at Lody Zoo. Int Zoo Yearb. 11:22.
9427. Saether, BE & Haagenrud, H. 1983. Life history of the moose (*Alces alces*): Fecundity rates in relation to age and carcass weight. J Mamm. 64:226-232.
9428. Saez-Royuella, C & Telleria, JL. 1987. Reproductive trends of the wild boar (*Sus scrofa*) in Spain. Folia Zool (Praha). 36:21-25.
9428a. Safar-Hermann, N, Ismail, MN, Choi, HS, et al. 1987. Pregnancy diagnosis in zoo animals by estrogen determination in feces. Zoo Biol. 6:189-193.
9429. Sagar, P & Bindra, OS. 1978. Reproductive patterns of captive lesser bandicoot rat, *Bandicota bengalensis*, in Punjab (India). OH J Sci. 78:44-46.
9430. Sägesser, H. 1966. Über den Einfluss der Höhe auf einige biologische Erscheinungen beim Reh (*Capreolus. capreolus*) und bei der Gemse (*Rupicapra r rupicapra*). Rev Suisse Zool. 73:422-433.
9431. Sahu, A. 1983. Cell proliferation rate, renewal times and circadian mitotic rhythm in the genital tract during estrous cycle of the bandicoot rat, *Bandicota bengalensis*. Arch Biol. 94:155-169.
9432. Sahu, A. 1984. Histomorphic changes in the ovary during the estrous cycle of a wild rat, *Bandicota bengalensis*. Can J Zool. 62:1052-1058.
9433. Sahu, A & Chakraborty, S. 1983. Pineal cytology during estrous cycle of wild bandicoot rat, *Bandicota bengalensis*. Arch Biol. 94:399-411.
9434. Sahu, A & Chakraborty, S. 1986. Estradiol modulation of pineal gland activity in the wild bandicoot rat, *Bandicota bengalensis*. Acta Anat. 125:1-5.
9435. Sahu, A & Ghosh, A. 1982. Effect of grouping and sex on estrous regulation of a wild rat, *Bandicota bengalensis*. Biol Reprod. 27:1023-1025.
9436. Sahu, A & Maiti, BR. 1978. Estrous cycles of the bandicoot rat: A rodent pest. Zool J Linn Soc. 63:309-314.
9437. Saint Girons, H, Brosset, A, & Saint Girons, MC. 1969. Contribution á la connaissance du cycle annuel de la Chauvre-souris *Rhinolophus ferrumequinum* (Schreber, 1774). Mammalia. 33:357-470.
9438. Saint Girons, MC. 1955. Notes sur l'écologie des petits Mammifères du bocage Atlantique. Terre Vie. 10:4-41.
9439. Saint Girons, MC. 1962. Notes sur le date de reproduction en captivité du Fennec, *Fennecus zerda* (Zimmermann 1780). Z Säugetierk. 27:181-184.
9440. Saint Girons, MC, van Mourik, WR, & van Bree, PJH. 1968. La croissance ponderaie et le cycle annuel du Hamster, *Cricetus canescens* Nehring, 1899, en captivite. Mammalia. 32:577-602.
9441. Saint John, OB. 1891. Notes on *Herpestes mungo*. Proc Zool Soc London. 1891:245.
9442. Sairam, MR, Raj, HGM, & Moudgal, NR. 1966. Presence of a gonadotropin inhibitor in the urine of the bonnet monkey, *Macaca radiata*. Endocrinology. 78:923-928.
9443. Saito, T, Machida, K, Inoue, S, et al. 1980. Reproductive activity of *Microtus montebelli* at Okegawa City in Saitama Prefecture. J Mamm Soc Japan. 8:122-128.
9444. Sakamoto, K. 1982. Hand-rearing a striped hyena. Animals & Zoos. 34(9):8-11.
9445. Salama, A, Shalash, MR, & Hoppe, R. 1967. Physiology of reproduction in the buffalo cow. Bull Anim Sci Res Inst Cairo. 3:1-86.
9446. Saldarini, RJ, Spieler, JM, & Coppola, JA. 1972. Plasma estrogens, progestins and spinbarkeit characteristics during selected portions of the menstrual cycle of the cynomolgus monkey (*Macaca fuscicularis*). Biol Reprod. 7:347-355.
9447. Sale, JB. 1965. Gestation period and neonatal weight of the hyrax. Nature. 205:1240-1241.
9448. Sale, JB. 1965. Observations on parturition and related phenomena in the hyrax (Procaviidae). Acta Trop. 22:37-54.
9449. Sale, JB. 1969. Breeding season and litter size in Hyracoidea. J Reprod Fert Suppl. 6:249-263.
9450. Salter, RE & Hudson, RJ. 1982. Social organization of feral horses in western Canada. Appl Anim Ethology. 8:207-223.
9451. Salzmann, RC. 1963. Beiträge zur Fortpflanzungsbiologie von *Meriones shawi* (*Mammalis, Rodentia*). Rev Suisse Zool. 70:343-452.
9452. Samarsky, SL. 1962. On the reproduction of *Spalax leucodon* Nordmann on the territory of the Odessa region. Zool Zhurn. 41:1583-1584.
9453. Samarsky, SL. 1977. Reproduction of *Citellus pygmaeus* in the forrest-steppe Dnepr territory. Zool Zhurn. 56:113-119.
9454. Sampsell, RN. 1969. Hand rearing an aardvark *Orycteropus afer* at Crandon Park Zoo, Miami. Int Zoo Yearb. 9:97-99.
9455. Samuel, CA, Sumar, J, & Nathanielsz, PW. 1979. Histological observations of the adrenal glands of newborn alpacas (*Lama pacos*). Comp Biochem Physiol. 62A:387-396.
9456. Samuels, A, Silk, JB, & Rodman, PS. 1984. Changes in the dominance rank and reproductive behaviour of male bonnet macaques (*Macaca radiata*). Anim Behav. 32:994-1003.
9456a. Sanborn, CC. 1930. Distribution and habits of the three-banded armadillo (*Tolypeutes*). J Mamm. 11:61-68.
9457. Sanborn, CC & Nicholson, AJ. 1950. Bats from New Caledonia, the Solomon Islands and New Hebrides. Fieldiana Zool. 31:313-338.
9457a. Sanchez-Herrera, O, Tellez-Giron, G, Medellín, RA, et al. 1986. New records of mammals from Quintana Roo, Mexico. Mammalia. 50:275-278.
9458. Sandegren, FE. 1970. Breeding and maternal behavior in the Steller sea lion (*Eumetopias jubatus*) in Alaska. MS Thesis. Univ AK Coll. [Cited by Schusterman, 1981]
9459. Sandegren, FE. 1976. Courtship display, agonistic behavior and social dynamics in the Steller sea lion (*Eumetopias jubatus*). Behaviour. 57:159-172.
9460. Sandell, M. 1984. To have or not to have delayed implantation: The example of the weasel and the stoat. Oikos. 42:123-125.
9461. Sandell, M. 1990. The evolution of seasonal delayed implantation. Q Rev Biol. 65:23-42.
9462. Sanders, EH, Gardner, PD, & Berger, PJ. 1981. 6-methoxybenzoxazolinone: A plant derivative that stimulates reproduction in *Microtus montanus*. Science. 214:67-69.
9463. Sanderson, GC. 1950. Methods of measuring productivity in raccoons. J Wildl Manage. 14:389-402.

9464. Sanderson, GC. 1951. Breeding habits and a history of the Missouri raccoon population from 1941 to 1948. Trans N Am Wildl Conf. 16:445-460.
9465. Sanderson, GC. 1961. Estimating opossum populations by marking young. J Wildl Manage. 25:20-27.
9466. Sanderson, GC & Hubert, GF, Jr. 1981. Selected demographic characteristics of Illinois (USA) raccoons (*Procyon lotor*). Proc Worldwide Furbearer Conf. 1:487-513.
9467. Sanderson, GC & Nalbandov, AV. 1973. The reproductive cycle of the raccoon in Illinois. Bull IL Nat Hist Surv. 31:29-85.
9468. Sanderson, IT. 1940. The mammals of the north Cameroons forest area: Being the results of the Percy Sladen Expedition to the Mamfe Division of the British Cameroons. Trans Zool Soc London. 24:623-725.
9469. Sanderson, KJ & O'Driscoll, M. 1986. Breeding season of brushtail possums, *Trichosurus vulpecula* (Marsupialia: Phalangeridae) in Adelaide. Aust Mamm. 9:139-140.
9470. Sandes, JP. 1903. The corpus luteum of *Dasyurus viverrinus*, in the observations of the growth and atrophy of the Graafian follicle. Proc Linn Soc NSW. 28:364-405.
9471. Sandhu, S. 1984. Breeding biology of the Indian fruit bat, *Cynopterus sphinx* (Vahl) in central India. J Bombay Nat Hist Soc. 81:600-612.
9472. Sandifort, G. 1839-44. Ontleedkundige beschouwing van een volwassen orang-oetan (*Simia satyrus*, Linn) van het mannelijke geslacht. Verhandelingen over de Natuurlijke Geschiedenis der Nederlandsche Overzeesche Bezittingen. 29-56, CI Temminck, Leiden.
9473. Sangalang, GB & Freeman, HC. 1976. Steroids in the plasma of the gray seal, *Halichoerus grypus*. Gen Comp Endocrinol. 29:419-422.
9474. Sankhala, KS. 1967. Breeding behavior of the tiger *Panthera tigris* in Rajasthan. Int Zoo Yearb. 7:133-147.
9475. Sankhala, KS & Desai, JH. 1970. Reproductive behaviour of brow-antlered deer. J Bombay Nat Hist Soc. 67:561-565.
9476. Sano, Y & Miyake, T. 1978. Reproduction of the meerkat. Animals & Zoos. 30(7):6-9.
9477. Sans-Coma, V & Gosalbez, J. 1976. Sobre la reproducción de *Apdemus sylvaticus* L, 1758 en nordeste ibérico. Miscelanea Zool. 3:227-233.
9478. Sansom, GS. 1920. Parthenogenesis in the water vole, *Microtus amphibius*. J Anat. 55:68-77.
9479. Sansom, GS. 1922. Early development and placentation in *Arvicola* (*Microtus*) *amphibius*, with special reference to the origin of placental giant cells. J Anat. 56:333-365.
9480. Sansom, GS. 1927. The giant cells in the placenta of the rabbit. Proc R Soc London Biol Sci. 101B:354-368.
9481. Sansom, GS. 1937. The placentation of the Indian musk-shrew (*Crocidura caerulea*). Trans Zool Soc London. 23:267-314.
9482. Sansom, GS. 1932. Notes on some early blastocysts of the South American bat *Molossus*. Proc Zool Soc London. 1932:113-118.
9483. Sansom, GS. 1938. Early development and placentation of the Indian musk-shrew *Crocidura caerulea*. Biomorphosis. 1:321-322.
9484. Sansom, GS & Hill, JP. 1931. Observations on the structure and mode of implantation of the blastocyst of *Cavia*. Trans Zool Soc London. 21:295-340.
9485. Sanson, GD, Nelson, JE, & Fell, P. 1985. Ecology of *Peradorcas concinna* in Arnhemland in a wet and a dry season. Proc Ecol Soc Aust. 13:65-72.
9486. Santapau, H & Abdulali, H. 1960. How many young does a chital have?. J Bombay Nat Hist Soc. 57:653-654.
9487. Santiapilai, C & Chambers, MR. 1980. Aspects of the population dynamics of the wild pig (*Sus scrofa* Linnaeus, 1758) in the Ruhuna National Park, Sri Lanka. Spixiana. 3:239-250.
9488. Sapargeldyev, MS. 1984. A contribution to the ecology of the mouse-like hamster *Calomyscus mystax* (Rodentia, Cricetidae) in Turkmenistan. Zool Zhurn. 63:1388-1395.
9489. Sapkal, VM & Bhandarkar, WR. 1984. Breeding habits and associated phenomena in some Indian bats. J Bombay Nat Hist Soc. 81:380-386.
9490. Sapkal, VM & Gadegone, MM. 1981. Histochemical observations on the female reproductive tract of bats IV Mucopolysaccharides in the uterus of three species of bats. Z Mikrosk Anat Forsch. 95:93-107.
9491. Sapkal, VM & Khamre, KG. 1983. Breeding habits and associated phenomenon in some Indian bats, part VIII *Taphozous melanopogon* (Temminck): Emballonuridae. J Bombay Nat Hist Soc. 80:303-311.
9492. Sapolsky, RM. 1983. Endocrine aspects of social instability in the olive baboon (*Papio anubis*). Am J Primatol. 5:365-379.
9493. Sapolsky, RM. 1985. Stress-induced suppression of testicular function in the wild baboon: Role of glucocorticoids. Endocrinology. 116:2273-2278.
9494. Sapolsky, RM. 1986. Endocrine and behavioral correlates of drought in wild olive baboons (*Papio anubis*). Am J Primatol. 11:217-227.
9495. Sapolsky, RM. 1986. Stress-induced elevation of testosterone concentrations in high ranking baboons: Role of catecholamines. Endocrinology. 118:1630-1635.
9496. Sapozhenkov, YF. 1962. The ecology of the caracal (*Felis caracal* Mull) in the Karakumy. Zool Zhurn. 41:1111-1112.
9497. Sapozhenkov, YF. 1964. Ecology of the hare *Lepus tolai* Pall in Sand Karakumy. Zool Zhurn. 43:1382-1387.
9498. Sapozhenkov, YF. 1965. Ecology of the suslik *Spermophilopsis leptodactylus leptodactylus* Light in Sand Karakumy. Zool Zhurn. 44:1553-1557.
9499. Sapozhenkov, YF. 1965. Reproduction of mammals in the sand desert Karakumy. Zool Zhurn. 44:896-901.
9500. Sargeant, AB. 1978. Red fox prey demands and implications to prairie duck production. J Wildl Manage. 42:520-527.
9501. Sargeant, AB, Allen, SH, & Johnson, DH. 1981. Determination of age and whelping dates of live red fox pups. J Wildl Manage. 45:760-765.
9502. Sarker, NJ & Canivenc, R. 1982. Luteal vascularization in the European badger (*Meles meles* L). Biol Reprod. 26:903-908.
9503. Sather, JH. 1958. Biology of the great plains muskrat in Nebraska. Wildlife Monograph. 2:1-35.
9504. Sato, H, Paluch, HJ, & Kawakami, S. 1977. The birth of cheetahs. Animals & Zoos. 29(9):6-9.
9505. Sato, M & Yasui, K. 1978. Breeding of the European bison. Animals & Zoos. 30(2):6-9.
9506. Sauer, EGF. 1963. Courtship and copulation of the gray whale in the Bering Sea at St Lawrence Island, Alaska. Psychol Forsh. 27:157-174.
9507. Sauer, EGF. 1974. Zur Biologie der Zwerg-und Riesengalagos. Fr Kölner Zoo. 17:67-84.
9508. Sauer, EGF & Sauer, EM. 1972. Zur Biologie des kurzohrigen Elephantenspitzmaus, *Macroscelides proboscideus*. Fr Kölner Zoo. 15:119-139.
9509. Sauer, JR & Slade, NA. 1988. Body size as a demographic categorical variable: Ramifications for life history analysis of mammals. Evolution of Life Histories of Mammals: Theory and Pattern. 107-121, Boyce, MS, ed, Yale U Press, New Haven, CT.

9510. Saunders, JK, Jr. 1955. Fetus in yearling cow elk, *Cervus canadensis*. J Mamm. 36:145.
9510a. Saunders, JK, Jr. 1964. Physical characteristics of the Newfoundland lynx. J Mamm. 45:36-47.
9511. Sauther, ML & Nash, LT. 1987. Effect of reproductive state and body size on food consumption in captive (*Galago senegalensis braccatus*). Am J Phys Anthro. 73:81-88.
9511a. Savage, A & Snowdon, CT. 1982. Mental retardation and neurological deficits in a twin orangutan. Am J Primatol. 3:239-251.
9512. Savage, A, Ziegler, TE, & Snowdon, CT. 1988. Sociosexual development, pair bond formation, and mechanism of fertility suppression in female cotton-top tamarins (*Saguinus oedipus oedipus*). Am J Primatol. 14:345-359.
9513. Savage-Rumbaugh, ES & Wilkerson, BJ. 1978. Socio-sexual behavior in *Pan paniscus* and *Pan troglodytes*: A comparative study. J Hum Evol. 7:327-344.
9514. Savoy, JC. 1966. Breeding and hand-rearing of the giraffe *Giraffa camelopardalis* at Columbus Zoo. Int Zoo Yearb. 6:202-204.
9515. Sawina, NW & Opachowa, WR. 1981. Haltung und Zucht von Weiszhandgibbons *Hylobates lar* L, 1771) im Leningrader Zoo. Zool Garten NF. 51:343-352.
9516. Sawrey, DK, Baumgardner, DJ, Campa, MJ, et al. 1984. Behavioral patterns of Djungarian hamsters: An adaptive profile. Anim Learn Behav. 12:297-306.
9517. Sawrey, DK & Dewsbury, DA. 1985. Control of ovulation, vaginal estrus, and behavioral receptivity in voles (*Microtus*). Neurosci Biobehav Rev. 9:563-571.
9518. Sawrey, DK & Dewsbury, DA. 1991. Males accelerate reproductive development in female montane voles. J Mamm. 72:343-346.
9519. Saxton, GA, Jr & Servadda, DM. 1969. Human birth interval in east Africa. J Reprod Fert Suppl. 6:83-88.
9520. Sayer, JA & Rakha, AM. 1974. The age of puberty of the hippopotamus (*Hippopotamus amphibius* Linn) in the Luangwa river in eastern Zambia. E Afr Wildl J. 12:227-232.
9521. Sayler, A. 1969. The estrous cycle and estrous-related behavior in the collared lemming. Forma et Functio. 1:227-237.
9522. Scaling, A. 1970. Observations on a breeding pair of Sierra Leone striped squirrel, *Funisciurus pyrrhopus leonis*. Ann Rep Jersey Wildl Preserv Trust. 7:32-34.
9523. Scaling, A. 1971. Progress report on the breeding of pygmy hedgehog tenrec *Echinops telfairi* and spiny hedgehog tenrec *Setifer setosus*. Ann Rep Jersey Wildl Preserv Trust. 8:12-14.
9524. Scanlon, PF, Oelschlaeger, A, Berkaw, MN, et al. 1981. Seasonal influences on reproductive and endocrine characteristics of male cottontail rabbits. Proc World Lagomorph Conf, Univ Guelph, ONT, Canada. 1979:204-209.
9525. Scanlon, PF & Urbston, DF. 1978. Persistence of lactation in white-tailed deer. J Wildl Manage. 42:196-197.
9526. Scarlett, G & Wooley, PA. 1980. The honey possum, *Tarsipes spencerae* (Marsupialia: Tarsipedidae): A non-seasonal breeder?. Aust Mamm. 3:97-103.
9527. Schaaf, CD & Stuart, MD. 1983. Reproduction of the mongoose lemur (*Lemur mongoz*) in captivity. Zoo Biol. 2:23-38.
9528. Schaal, A. 1985. Observations préliminaires sur le cycle sexuel du Daim, *Cervus* (*Dama*) *dama* L. Mammalia. 49:288-291.
9529. Schacher, WH & Pelton, MR. 1975. Productivity of muskrats in east Tennessee. Proc Southeast Assoc Game Fish Comm. 29:594-608.
9530. Schadler, MH. 1978. Reproductive patterns in the pine vole. Proc E Pine Meadow Vole Symp. 2:82-87.
9531. Schadler, MH. 1981. Postimplantation abortion in pine voles (*Microtus pinetorum*) induced by strange males and pheromones of strange males. Biol Reprod. 25:295-297.
9532. Schadler, MH. 1981. Social organization and reproduction in freely reproducing colonies of pine voles in the laboratory. Proc E Pine Meadow Vole Symp. 5:120-123.
9533. Schadler, MH. 1982. Strange males block pregnancy in lactating pine voles, *Microtus pinetorum*, and effect survival and growth of nursing young. Proc E Pine Meadow Vole Symp. 6:132-138.
9534. Schadler, MH. 1983. Male siblings inhibit reproductive activity in female pine voles, *Microtus pinetorum*. Biol Reprod. 28:1137-1139.
9535. Schadler, MH. 1983. Social influences on reproduction in pine voles. Proc E Pine Meadow Vole Symp. 7:117-119.
9536. Schadler, MH & Butterstein, GM. 1979. Reproduction in the pine vole, *Microtus pinetorum*. J Mamm. 60:841-844.
9537. Schadler, MH & Butterstein, GM. 1987. Increase in serum levels of luteinizing hormone (LH) in pine voles, *Microtus pinetorum*, after induction of estrus and copulation. J Mamm. 68:410-412.
9538. Schadler, MH, Butterstein, GM, Faulkner, BJ, et al. 1988. The plant metabolite, 6-methoxybenzoxalinone, stimulates an increase in secretion of follicle-stimulating hormone and size of reproductive organs in *Microtus pinetorum*. Biol Reprod. 38:817-820.
9539. Schadler, MH & Gauger, BJ. 1980. Induction of abortion by strange males in pine vole females that are: 1) ten days pregnant or 2) pregnant and lactating. Proc E Pine Meadow Vole Symp. 4:24-27.
9540. Schäfer, no initials. 1950. Über den Schapi (*Hemitragus jemlaicus schaeferi*). Zool Anz. 145:247-260.
9541. Schäfer, E. 1937. Über das Zwergblauschaf (*Pseudois* spec nova) und das Groszblauschaf (*Pseudois najoor* Hdgs) in Tibet. Zool Garten NF. 9:263-278.
9542. Schagdasuren, O & Stubbe, M. 1974. Zur Säugetierfauna der Mongolei IV Der Ussurische Elch *Alces alces cameloides* (Milne-Edwards, 1867) in der Mongolei. Arch Natursch Lanschaftsforsch. 14:147-150.
9542a. Schaldach, WJ, Jr. 1966. New forms of mammals from southern Oaxaca, Mexico, with notes on some mammals of the coastal range. Säugetierk Mitt. 14:286-297.
9543. Schallenberger, E, Richardson, DW, & Knobil, E. 1981. Role of prolactin in the lactational amenorrhea of the rhesus monkey (*Macaca mulatta*). Biol Reprod. 25:370-374.
9544. Schaller, GB. 1965. The behavior of the mountain gorilla. Primate Behavior Field Studies of Monkeys and Apes. 324-367, DeVore, I, ed, Holt, Rinehart & Winston, New York.
9545. Schaller, GB. 1967. The Deer and the Tiger: A Study of Wildlife in India. Univ Chicago Press, Chicago.
9546. Schaller, GB. 1972. The Serengeti Lion: A Study of Predator-Prey Relations. Univ Chicago Press, Chicago.
9547. Schaller, GB. 1977. Mountain Monarchs. Univ Chicago Press, Chicago.
9548. Schaller, GB & Hamer, A. 1978. Rutting behavior of Père David's deer, *Elaphurus davidianus*. Zool Garten NF. 48:1-15.
9549. Schaller, GB, Jinchu, H, Wenshi, P, et al. 1985. The Giant Pandas of Wolong. Univ Chicago Press, Chicago.
9550. Schaller, GB & Mirza, ZB. 1974. On the behavior of the Punjab urial (*Ovis orientalis punjabiensis*). The Behaviour of Ungulates and its Relation to Management. 306-323, Geist, V & Walther, F, eds, IUCN, Morges, Switzerland.

9551. Schaller, GB & Vasconcelos, JMC. 1978. A marsh deer census in Brazil. Oryx. 14:345-351.
9552. Schally, AV, Itoh, Z, Carter, WH, et al. 1969. Effect of luteinizing hormone-releasing factor (LRF) on plasma LH levels in nutria (*Myocastor coypus*). Gen Comp Endocrinol. 12:176-179.
9553. Schams, D & Barth, D. 1982. Annual profiles of reproductive hormones in peripheral plasma of the male roe deer (*Capreolus capreolus*). J Reprod Fert. 66:463-468.
9554. Schams, D, Barth, D, & Karg, H. 1980. LH, FSH and progesterone concentrations in peripheral plasma of the female roe deer (*Capreolus capreolus*) during the rutting season. J Reprod Fert. 60:109-114.
9555. Scharlemann, E. 1953. Materialen zur Kenntnis und zur Erhaltung des ukran Bibers. Z Säugetierk. 18:146-162.
9556. Schaub, S. 1987. The world's first surviving *Macaca sylvanus* in a semifree-ranging colony. Folia Primatol. 49:106-110.
9557. Schauenberg, P. 1971. Le Chat des sables. Musées de Genève. 117:1-6.
9558. Schauenberg, P. 1972. Le Manul: Énigmatique Felide d'Asie. Musées de Genève. 125:5-9.
9559. Schauenberg, P. 1974. Données nouvelles sur le Chat de sables *Felis margarita* Loche, 1858. Rev Suisse Zool. 81:949-969.
9560. Schauenberg, P. 1974. Le Chat sauvage démystifié. Musées de Genève. 148:2-6.
9561. Schauenberg, P. 1976. Poid et taille de naissance du Chat forestier *Felis silvestris* Schreber. Mammalia. 40:687-689.
9562. Schauenberg, P. 1978. Note on Cuming's cloud rat, *Phloemys cumingi* Waterhouse 1839. Rev Suisse Zool. 83:341-347.
9563. Schauenberg, P. 1978. Note sur la reproduction du Manul *Otocolobus manul* (Pallas, 1776). Mammalia. 42:355-358.
9564. Schauenberg, P. 1979. Le Chat marbre. Musées de Genève. 190:6-11.
9565. Schauenberg, P. 1979. La reproduction du Chat des marais *Felis chaus* (Guldenstadt, 1776). Mammalia. 43:215-223.
9566. Schauenberg, P. 1979. Nôte sur la reproduction du Chat du Bengale *Prionailurus bengalensis* Kerr, 1792. Mammalia. 43:127-128.
9567. Schauenberg, P. 1979. Sa majesté l'Ocelot. Musées de Genève. 198:2-6.
9568. Schauenberg, P. 1979. Le baculum du Chat Forestier *Felis silvestris* Schreber, 1777. Rev Suisse Zool. 86:528-534.
9569. Schauenberg, P & Jotterfaud, M. 1975. Le Manul *Otocolobus manul* (Pallas,1776): Son caryotype et sa position dans la classification de Felides. Rev Suisse Zool. 82:425-429.
9570. Schaughnessy, PD. 1979. Cape (South African) fur seal. Mammals in the Seas, FAO Fish Ser 5. 2:37-40.
9571. Schaurte, WT. 1969. Über die Geburt eines Breitmaulnashorns, *Ceratotherium simum simum* (Burchell, 1817), in Wildschutzgebiet Krugerdorp in Transvaal. Säugetierk Mitt. 17:158-160.
9572. Scheck, SH. 1982. Development of thermoregulatory abilities in the neonatal hispid cotton rat, *Sigmodon hispidus texianus*, from northern Kansas and south-central Texas. Physiol Zool. 55:91-104.
9573. Scheffel, W. 1974. Notizen zur Haltung und Zucht der Sandkatze (*Felis margarita* Loche, 1858). Zool Garten NF. 44:338-348.
9574. Scheffel, W & Hemmer, H. 1975. Breeding Geoffroy's cat *Leopardus geoffroyi salinarum* in captivity. Int Zoo Yearb. 15:152-154.
9574a. Scheffer, TH. 1910. The pocket gopher. Bull KS State Agric Coll Exp Stn. 172:197-233.
9575. Scheffer, TH. 1924. Breeding habits of the mole and the gopher. Murrelet. 5:3-4.
9576. Scheffer, TH. 1924. Notes on the breeding of *Peromyscus*. J Mamm. 5:258-260.
9577. Scheffer, TH. 1925. Notes on the breeding of beavers. J Mamm. 6:129-130.
9578. Scheffer, TH. 1929. Mountain beavers in the pacific northwest: Their habits, economic status, and control. USDA Farmers' Bull. 1598:1-18.
9578a. Scheffer, TH. 1930. Bat matters. Murrelet. 11:11-13.
9579. Scheffer, TH. 1930. Determining the rate of replacement in a species. J Mamm. 11:466-469.
9579a. Scheffer, TH. 1933. Breeding of the Washington varying hare. Murrelet. 14:77-78.
9580. Scheffer, TH. 1938. Breeding records of Pacific coast pocket gophers. J Mamm. 19:220-224.
9581. Scheffer, TH. 1938. Pocket mice of Washington and Oregon in relation to agriculture. USDA Tech Bull. 608:1-15.
9582. Scheffer, TH. 1941. Ground squirrel studies in the four-rivers country, Washington. J Mamm. 22:270-279.
9583. Scheffer, TH. 1947. Ecological comparisons of the plains prairie-dog and the Zuni species. Trans KS Acad Sci. 49:401-406.
9584. Scheffer, VB. 1939. The os clitoridis of the Pacific otter. Murrelet. 20:20-21.
9585. Scheffer, VB. 1945. Growth and behavior of young sea lions. J Mamm. 26:390-392.
9586. Scheffer, VB. 1949. The clitoris bone in two pinnipeds. J Mamm. 30:269-270.
9587. Scheffer, VB. 1950. Growth of the testes and baculum in the fur seal, *Callorhinus ursinus*. J Mamm. 31:384-394.
9588. Scheffer, VB. 1950. Notes on the raccoon in southwest Washington. J Mamm. 31:444-448.
9589. Scheffer, VB. 1951. Measurements of sea otters from western Alaska. J Mamm. 32:10-14.
9590. Scheffer, VB & Slipp, JW. 1944. The harbor seal in Washington state. Am Mid Nat. 32:373-416.
9591. Scheffer, VB & Wilke, F. 1953. Relative growth in the northern fur seal. Growth. 17:129-145.
9592. Schenkel, R & Schenkel-Hulliger, L. 1969. Ecology and Behaviour of the Black Rhinoceros (*Diceros bicornis* L): A field study. Paul Parey, Hamburg.
9593. Scheurmann, E. 1975. Beobachtungen zur Fortpflanzung des Gayal, *Bibos frontalis* Lambert, 1837. Z Säugetierk. 40:113-127.
9594. Scheurmann, E, Senft, B, & Rietschel, W. 1977. Untersuchungen über die Zusammensetzung der Gayalmilch (*Bibos frontalis* Lambert 1837). Zool Garten NF. 47:24-32.
9595. Schiburr, R & Muschketat, L. 1983. Beutelteufel (*Sarcophilus harrisi* Boitard) im Zoo Rotterdam. Z Kölner Zoo. 26:84-89.
9596. Schierer, A, Mast, JC, & Hess, R. 1972. Contibution à l'étude éco-éthologique du grand Murin (*Myotis myotis*). Terre Vie. 26:38-53.
9597. Schladweiler, P & Stevens, DR. 1973. Reproduction of Shiras moose in Montana. J Wildl Manage. 37:535-544.
9598. Schlegel, H & Müller, S. 1839-42. Bijdragen tot de natuurlyke historie van den orang-oetan (*Simia satyrus*). Verhandelingen over de Natuurlijke Geschiedenis der Nederlandsche Overzeesche Bezittingen. 1-28, CI Temminck, Leiden.

9599. Schliemann, H & Maas, B. 1978. *Myzopoda aurita*. Mamm Species. 116:1-2.
9600. Schlott, M. 1950. Zur Kenntnis der Jugendentwicklung des Baribals (*Euarctos americanus* Pall). Zool Garten NF. 17:40-44.
9601. Schmidly, DJ. 1974. *Peromyscus attwateri*. Mamm Species. 48:1-3.
9602. Schmidly, DJ. 1974. *Peromyscus pectoralis*. Mamm Species. 49:1-3.
9603. Schmidly, DJ, Lee, MR, Modi, WS, et al. 1985. Systematics and notes on the biology of *Peromyscus hooperi*. Occas Pap Mus TX Tech Univ. 97:1-40.
9603a. Schmidly, DJ & Martin, CO. 1973. Notes on bats from the Mexican state of Querétaro. Bull S CA Acad Sci. 72:90-92.
9604. Schmidt, AM, Hess, DL, Schmidt, MJ, et al. 1988. Serum concentrations of oestradiol and progesterone, and sexual behaviour during the normal oestrous cycle in the leopard (*Panthera pardus*). J Reprod Fert. 82:43-49.
9605. Schmidt, AM, Nadal, LA, Schmidt, MJK, et al. 1979. Serum concentrations of oestradiol and progesterone during the normal oestrous cycle and early pregnancy in the lion (*Panthera leo*). J Reprod Fert. 57:267-272.
9606. Schmidt, CA, Engstrom, MD, & Genoways, HH. 1989. *Heteromys gaumeri*. Mamm Species. 345:1-4.
9607. Schmidt, CR. 1975. Captive breeding of the vicuna. Breeding Endangered Species in Captivity. 271-283, Martin, RD, ed, Academic Press, New York.
9607a. Schmidt, CR. 1990. Pigs. Grzimek's Encyclopedia of Mammals. 20-55, Packer, SP, ed, McGraw-Hill, New York.
9608. Schmidt, DV, Walker, LE, & Ebner, KE. 1971. Lactose synthetase activity in northern fur seal milk. Biochem Biophys Acta. 252:439-442.
9609. Schmidt, F. 1931. Erfahrungen und Beobachtungen bei Zuchtversuchen mit dem Marderhund. Deutche Pelztierz. 4:87-90.
9610. Schmidt, F. 1934. Über die Fortpflanzungsbiologie von sibirischem Zobel (*Martes zibellina* L) und europäischem Baummarder (*Martes martes* L). Z Säugetierk. 9:392-403.
9611. Schmidt, F. 1937. Über die Zucht des Marderhandes. Deutsche Pelztierzuchter. 12:235-237.
9612. Schmidt, FJW. 1931. Mammals of western Clark county, Wisconsin. J Mamm. 12:99-117.
9613. Schmidt, JL. 1984. Common duiker measurements in Natal and Zambia: An example of Bergmann's and Allen's rules. Lammergeyer. 32:8-10.
9614. Schmidt, U. 1974. Die Tragzeit der Vampirfledermäuse (*Desmodus rotundus*). Z Säugetierk. 39:129-132.
9615. Schmidt, U & Manske, U. 1973. Die Jugendentwicklung der Vampirfledermäuse (*Desmodus rotundus*). Z Säugetierk. 38:14-33.
9616. Schmidt-Nielsen, K, Dawson, TJ, Hammel, HT, et al. 1965. The jack-rabbit: A study in its desert survival. Hvalrjadet Skrifter. 48:125-142.
9617. Schmied, A. 1964. *Antilope cervicapra* (Bovidae) Mutter-Kind-Verhalten in den ersten Lebenstagen des Kitzes. Encyclopaedia cinematographica E. 1005:3-11.
9618. Schmitt, LH, Bradley, AJ, Kemper, CM, et al. 1989. Ecology and physiology of the northern quoll, *Dasyurus halluctatus* (Marsupialia, Dasyuridae), at Mitchell Plateau, Kimberly, Western Australia. J Zool. 217:539-558.
9619. Schneider, JE & Wade, GN. 1987. Body composition, food intake, and brown fat thermogenesis in pregnant Djungarian hamsters. Am J Physiol. 253:R314-R320.
9620. Schneider, KM. 1923. Beobachtungen aus dem Leipziger Zoologischen Garten über das Geschlechtsleben der Fleckenhyäne. Verh Deuts Zool Gesellsch E V. 28:78-79.
9621. Schneider, KM. 1926. Über Hyänenzucht. Pelztierzucht. 2(8):1-4.
9622. Schneider, KM. 1930. Einige Beobachtungen über das Geschlechtsleben des indischen Elephanten. Zool Garten NF. 3:305-314.
9623. Schneider, KM. 1933. Zur Fortpflanzung und Jugendentwicklung des kalifornischen Seelöwen. Zool Garten NF. 6:23-33.
9624. Schneider, KM. 1933. Zur Jugendentwicklung eines Eisbären. Zool Garten NF. 6:156-165.
9625. Schneider, KM. 1936. Zur Fortpflanzung, Aufzucht und Jugendentwicklung des Schabrackentapirs. Zool Garten NF. 8:84-90.
9626. Schneider, KM. 1950. Zur gewichtsmäsigen Jugendentwicklung gefangengehaltener Wildcaniden nebst einigen zeitlichen Bestimmungen über ihre Fortpflanzung Teil III 4: Der Marderhund (*Nyctereutes procyonoides* Gray). Zool Anz Suppl. 15:271-285.
9627. Schneider, KM. 1952. Einge Bilder zur Paarung der Fleckenhyäne, *Crocotta* (*sic*) *crocuta* Erxl. Zool Garten NF. 19:135-149.
9628. Schneider, KM. 1954. Vom Baumkanguruh (*Dendrolagus leucogenys* Matschie). Zool Garten NF. 21:63-106.
9629. Schneyer, A, Castro, A, & Odell, D. 1985. Radioimmunoassay of serum follicle-stimulating hormone and luteinizing hormone in the bottlenosed dolphin. Biol Reprod. 33:844-853.
9630. Schoemaker, AH. 1982. Fecundity in the captive howler monkey, *Alouatta caraya*. Zoo Biol. 1:149-156.
9631. Schoenfeld, H. 1904. La spermatogenèse chez la Noctule (*Vesperugo noctula*). Annl Bull Soc r Med Gand. 84:141-159.
9632. Schoenfeld, M & Yom-Tov, Y. 1985. The biology of two species of hedgehogs, *Erinaceus europaeus concolor* and *Hemiechinus auritus aegyptius*, in Israel. Mammalia. 49:339-356.
9633. Schofield, RD. 1958. Litter size and age ratios of Michigan red foxes. J Wildl Manage. 22:313-315.
9634. Schoknecht, PA. 1984. Growth and teat ownership in a litter of binturongs. Zoo Biol. 3:273-277.
9635. Schoknecht, PA, Cranford, JA, & Akers, RM. 1985. Variability in milk composition in the domestic ferret (*Mustela putorius*). Comp Biochem Physiol. 81A:589-591.
9636. Schomber, HW. 1963. Beiträge zur Kenntnis der Giraffengazelle *Litocranias walleri* Brooke, 1878). Säugetierk Mitt. 11(sp issue):1-44.
9637. Schomber, HW. 1964. Beiträge zur Kenntnis der Lamagazelle, *Ammodorcas clarkei* (Thomas, 1891). Säugetierk Mitt. 12:65-90.
9638. Schönberner, D. 1965. Beogbachtungen zur Fortpfanzungsbiologie des Wolfes, *Canis lupus*. Z Säugetierk. 30:171-178.
9639. Schoonmaker, JN, Bergman, KS, Steiner, RA, et al. 1982. Estradiol-induced luteal regression in the rhesus monkey: Evidence for an extraovarian site of action. Endocrinology. 110:1708-1715.
9639a. Schoonmaker, WJ. 1929. Weights of some New York mammals. J Mamm. 10:149-152.
9640. Schoonmaker, WJ. 1938. Notes on mating and breeding habits of foxes in New York state. J Mamm. 19:375-376.
9641. Schopper, D, Gaus, J, Claus, R, et al. 1984. Seasonal changes of steroid concentrations in seminal plasma of a European wild boar. Acta Endocrinol. 107:425-427.

9642. Schorger, AW. 1950. Harvest mouse in Dane County, Wisconsin. J Mamm. 31:363-364.
9643. Schowalter, DB & Gunson, JR. 1979. Reproductive biology of the big brown bat (*Eptesicus fuscus*) in Alberta. Can Field-Nat. 93:48-54.
9644. Schowalter, DB & Gunson, JR. 1982. Parameters of population and seasonal activity of striped skunks, *Mephitis mephitis*, in Alberta and Saskatchewan. Can Field-Nat. 96:409-420.
9645. Schowalter, DB, Gunson, JR, & Harder, LD. 1979. Life history characteristics of little brown bats (*Myotis lucifugus*) in Alberta. Can Field-Nat. 93:243-251.
9646. Schramm, P. 1961. Copulation and gestation in the pocket gopher. J Mamm. 42:167-170.
9647. Schreiber, GR. 1968. A note on keeping and breeding the Philippine tarsier *Tarsius syrichta* at Brookfield Zoo Chicago. Int Zoo Yearb. 8:114-115.
9648. Schroder, GD. 1979. Foraging behavior and home range utilization of the bannartail kangaroo rat (*Dipodomys spectabilis*). Ecology. 60:657-665.
9649. Schröder, L. 1957. Zur Anatomie und Topographie einiger Organe des Borstengürteltieres (*Chaetophractus villosus*). Zool Garten NF. 23:28-36.
9650. Schröder, W. 1971. Untersuchungen zur Ökologie des Gamswildes (*Rupicapra rupicapra* L) in einem Vorkommen der Alpen I Teil. Z Jagdwiss. 17:113-168.
9651. Schröder, W. 1971. Untersuchungen zur Ökologie des Gamswildes (*Rupicapra rupicapra* L) in einem Vorkommen der Alpen II Teil. Z Jagdwiss. 17:197-235.
9652. Schroder, W & Mensah, GA. 1987. Reproductive biology of *Thryonomys swinderianus* (Temminck). Z Säugetierk. 52:164-167.
9653. Schroeder, JP. 1990. Breeding bottlenose dolphins in captivity. The Bottlenose Dolphin. 435-446, Reeves, RR, ed, Academic Press, San Diego, CA.
9654. Schroeder, JP & Keller, KV. 1989. Seasonality of serum testosterone levels and sperm density in (*Tursiops truncatus*). J Exp Zool. 249:316-321.
9655. Schroeder, JP & Keller, KV. 1990. Artificial insemination of bottlenose dolphins. The Bottlenose Dolphin. 447-460, Reeves, RR, ed, Academic Press, San Diego, CA.
9656. Schröpfer, R. 1977. Die postnatale Entwicklung der Kleinwülmaus, *Pitymys subterraneus* De Selys-Longchamps, 1836 (Rodentia, Cricetidae). Bonn Zool Beitr. 28:249-268.
9657. Schryver, HF, Oftedal, OT, Williams, J, et al. 1986. A comparison of the mineral composition of milk of domestic and captive wild equids (*Equus przewalski, E zebra, E burchelli, E caballus, E assinus*). Comp Biochem Physiol. 85A:233-235.
9658. Schubart, O. 1930. Die nordamerikanische Seelöwen. Zool Garten NF. 3:173-183.
9659. Schuermann, E & Jainudeen, MR. 1972. "Musth" beim asiatischen Elefanten (*Elephas maximus*). Zool Garten NF. 42:131-142.
9660. Schuler, HM & Gier, HT. 1976. Duration of the cycle of the seminiferous epithelium in the prairie vole *Microtus ochrogaster* (*ochrogaster*). J Exp Zool. 197:1-12.
9661. Schulte, BA, Parsons, JA, Seal, US, et al. 1980. Heterologous radioimmunoassay for deer prolactin. Gen Comp Endocrinol. 40:59-68.
9662. Schulte, BA, Seal, US, Plotka, ED, et al. 1980. Seasonal changes in prolactin and growth hormone cells in the hypophyses of white tailed deer (*Odocoileus virginianus borealis*) studied by light microscopic immunocytochemistry and radioimmunoassay. Am J Anat. 158:369-377.
9663. Schulte, BA, Seal, US, Plotka, ED, et al. 1981. Characterization of seasonal changes in prolactin and growth hormone cells in the hypophyses of white-tailed deer (*Odocoileus virginianus borealis*) by ultrastructural and immunocytochemical techniques. Am J Anat. 160:277-284.
9664. Schulte, BA, Seal, US, Plotka, ED, et al. 1981. The effect of pinealectomy on seasonal changes in prolactin secretion in the white-tailed deer (*Odocoileus virginianus borealis*). Endocrinology. 108:173-178.
9665. Schulte, TL. 1937. The genito-urinary system of the *Elephas indicus* male. Am J Anat. 61:131-157.
9666. Schultz, AH. 1938. The relative weight of the testes in primates. Anat Rec. 72:387-394.
9666a. Schultz, AH. 1938. Genital swelling in the female orang-utan. J Mamm. 19:363-366.
9667. Schultz, AH. 1942. Growth and development of the proboscis monkey. Bull Mus Comp Zool. 89:279-314.
9667a. Schultz, SR & Johnson, MK. 1992. Breeding by male white-tailed deer fawns. J Mamm. 73:148-150.
9668. Schultze-Westrum, T. 1963. Die Wildziegen der agaischen Inseln. Säugetierk Mitt. 11:145-181.
9669. Schultze-Westrum, T. 1965. Innerartliche Verständigung durch Düfte beim Gleitbeutler *Petaurus breviceps papuanus* Thomas (Marsupialia, Phalangeridae). Z vergl Physiol. 50:151-220.
9670. Schulz, WC. 1966. Breeding and hand-rearing brown hyaenas *Hyaena brunnea* at Okahandja Zoopark. Int Zoo Yearb. 6:173-176.
9671. Schulze, W. 1970. Beiträge zum Vorkammen und zur Biologie der Haselmaus (*Muscardinus avellanarius* L) und des Siebenschläfers (*Glis glis*) im Südharz. Hercynia. 7:355-371.
9672. Schulze, W. 1973. Untersuchungen zur Biologie der Haselmaus (*Muscardinus avellanarius* L) im Südharz. Arch Natursch Landschaftsforsch. 13:107-121.
9673. Schürer, U. 1976. Beobachtungen an einem neugeborenen Flachlandtapir, *Tapirus terrestris* (Linne 1766). Zool Garten NF. 46:367-370.
9674. Schürer, U. 1978. Breeding black-footed cats in captivity. Carnivore. 1(2):109-111.
9675. Schürer, U. 1978. Haltung und Zucht von Schwartsfuszkatzen, *Felis nigripes* Burchell, 1822. Zool Garten NF. 48:385-400.
9676. Schuster, R. 1976. Lekking behavior in Kafue lechwe. Science. 192:1240-1242.
9677. Schusterman, RJ. 1981. Steller sea lion *Eumetopias jubatus* (Schreber, 1776). Handbook of Marine Mammals. 1:119-141, Ridgway, SH & Harrison, RJ, eds. Academic Press, New York.
9678. Schütz, H. 1953. Das Verhalten des Uterus und der Vagina des Igels (*Erinaceus europaeus* et *romanus* L) während des Oestrus, Anoestrus, der Gestation und der Lactation. Z Mikr Anat Forsch. 59:463-522.
9679. Schwab, H. 1952. Beobachtungen über die Begattung und die Spermakonservierung in den Geschlechtsorganen bei weiblichen Fledermäusen. Z Mikr Anat Forsch. 58:326-357.
9680. Schwagmeyer, PL. 1986. Effects of multiple mating on reproduction in female thirteen-lined ground squirrels. Anim Behav. 34:297-298.
9681. Schwagmeyer, PL & Brown, CH. 1983. Factors affecting male-male competition in thirteen-lined ground squirrels. Behav Ecol Sociobiol. 13:1-6.
9682. Schwaier, A. 1973. Breeding tupaias (*Tupaia belangeri*) in captivity. Z Versuchstierk. 15:255-271.
9683. Schwaier, A. 1975. The breeding stock of *Tupaia* at the Battelle Institut. Lab Anim Handb. 6:141-149.

9684. Schwaier, A & Kuhn, HJ. 1975. Chronology of the development of embryo and placenta of *Tupaia belangeri*. Lab Anim Handb. 6:257-258.
9685. Schwaier, A, Kuhn, HJ, & Hasan, SH. 1976. Serum progesterone levels during pregnancy in *Tupaia belangeri* correlated with ovarian and placental morphology. The Laboratory Animal in the Study of Reproduction, 6th Symposium ICLA, Thessaloniki 9-11 July, 1975. 75-82, Antikatzides, T, Ericksen, S, & Spiegel, A, eds, Fischer, NY.
9685a. Schwartz, A & Jones, JK, Jr. 1967. Bredin-Archbold-Smithsonian biological survey of Dominica 7 Review of bats of the endemic Antillean genus *Monophyllus*. Proc US Nat Mus. 124(3635):1-20.
9686. Schwartz, CC, Regelin, WL, & Franzmann, AW. 1982. Mule moose successfully breed as yearlings. J Mamm. 63:334-335.
9687. Schwartz, CW. 1942. Breeding season of the cottontail in central Missouri. J Mamm. 23:1-16.
9688. Schwartz, JE, II & Mitchell, GE. 1945. The Roosevelt elk on the Olympic Peninsula. J Wildl Manage. 9:295-324.
9689. Schwartz, OA & Bleich, VC. 1975. Comparative growth in two species of woodrats, *Neotoma lepida intermedia* and *Neotoma albigula renusta*. J Mamm. 56:653-666.
9690. Schwarz, SS & Smirnoff, VS. 1960. Zur Physiologie und Populationsdynamik der Bisamratte in der Walsteppe und im Hohen Norden. Zool Jb (Syst). 87:363-386.
9691. Schweers, S. 1984. Zur Fortpflanzungsbiologie des Zebraduckers *Cephalophus zebra* (Gray, 1838) im Vergleich zu anderen *Cephalophus-Arten*. Z Säugetierk. 49:21-36.
9692. Schwentker, V. 1963. The gerbil: A new laboratory animal. IL Vet. 6(4):5-9.
9693. Sclater, PL. 1883. Further notes on *Tragelaphus gratus*. Proc Zool Soc London. 1883:34-37.
9694. Sclater, PL. 1891. Breeding of *Tragelaphus gratus*. Proc Zool Soc London. 1891:213.
9695. Scoresby, W. 1820. An Account of the Arctic Regions, with a History and Description of the Northern Whale-Fishery, 2 volumes. Hurst, Robinson & Co, London.
9695a. Scott, EOG. 1949. Neonatal length as a linear function of adult length in Cetacea. Pap Proc R Soc Tasmania. 1948:75-93.
9696. Scott, EOG & Green, RH. 1975. Recent whale strandings in northern Tasmania. Pap Proc R Soc Tasmania. 109:91-96.
9697. Scott, GW & Fisher, KC. 1983. Sterility of two subspecies of the varying lemming *Dicrostonyx groenlandicus* bred in captivity. Can J Zool. 61:1182-1183.
9698. Scott, JN, Fritz, HI, & Nagy, F. 1979. Response to cryptorchidism of the testis and epididymis of the opossum (*Didelphis virginiana*). J Reprod Fert. 57:175-178.
9699. Scott, LM. 1984. Reproductive behavior of adolescent female baboons (*Papio anubis*) in Kenya. Female Primates: Studies by Women Primatologists. 77-100, Small, NF, ed, Alan R Liss, Inc, New York.
9699a. Scott, MP. 1987. The effect of mating and agonistic experience on adrenal function and mortality of male *Antechinus stuartii* (Marsupialia). J Mamm. 68:479-486.
9700. Scott, TG. 1939. Number of fetuses in the Hoy pigmy shrew. J Mamm. 20:251.
9701. Scruton, DM & Herbert, J. 1970. The menstrual cycle and its effect on behaviour in the talapoin monkey (*Miopithecus talapoin*). J Zool. 162:419-436.
9702. Scucchi, S. 1984. Interbirth intervals in a captive group of Japanese macaques. Folia Primatol. 42:203-208.
9703. Sculley, WC. 1898. Between Sun and Sand. Methuen and Co, London.
9704. Seal, US, Barton, R, Mather, L, et al. 1974. Long-term control of reproduction in female lions (*Panthera leo*) with implanted contraceptives. Proc Am Ass Zoo Vet. 1974:66-72.
9705. Seal, US & Plotka, ED. 1983. Age-specific pregnancy rates in feral horses. J Wildl Manage. 47:422-429.
9705a. Seal, US, Plotka, ED, Packard, JM, et al. 1979. Endocrine correlates of reproduction in the wolf I Serum progesterone, estradiol and LH during the estrous cycle. Biol Reprod. 21:1057-1066.
9706. Seal, US, Plotka, ED, Smith, JD, et al. 1985. Immunoreactive luteinizing hormone, estradiol, progesterone, testosterone, and androstonedione levels during the breeding season and anestrus in Siberian tigers. Biol Reprod. 32:361-368.
9707. Seal, US, Sinha, AA, & Doe, RP. 1972. Placental iron transfer: Relationship to placental anatomy and phylogeny of the mammal. Am J Anat. 134:263-269.
9708. Seal, US, Smith, JLD, Reindl, N, et al. 1983. Estrous cycles in Siberian tigers (*Panthera tigris altaica*). Biol Reprod. 28(Suppl):43.
9709. Sealander, JA & Price, JF. 1964. Free-tailed bat in Arkansas. J Mamm. 45:152.
9710. Sealander, JA, Jr & Walker, BQ. 1955. A study of the cotton rat in northwestern Arkansas. AR Acad Sci. 8:153-162.
9711. Sealy, SG. 1978. Litter size and nursery sites of the hoary bat near Delta, Manitoba. Blue Jay. 36:51-52.
9712. Searle, JB. 1984. Breeding the common shrew (*Sorex araneus*) in captivity. Lab Anim. 18:359-363.
9713. Sears, HS & Browman, LG. 1955. Quadruplets in a mule deer. Anat Rec. 122:335-339.
9714. Seaton, B. 1978. Patterns of oestrogen and testosterone during pregnancy in the gorilla. J Reprod Fert. 53:231-236.
9715. Sedlaczek, A. 1912. Über Plazentarbildung bei Antilopen. Anat Hefte. 46:577-597.
9716. Seebeck, JH. 1976. The broad toothed rat. Vict Nat. 93:56-58.
9717. Seebeck, JH. 1979. Status of the barred bandicoot, *Perameles gunnii*, in Victoria: With a note on husbandry of a captive colony. Aust Wildl Res. 6:255-264.
9718. Seebeck, JH. 1982. Long-nosed potoroos *Potorous tridactylus*: Husbandry and management of a captive colony. The Management of Australian Mammals in Captivity. 111-116, Evans, DD, ed, Zool Bd Victoria, Melbourne.
9719. Seebeck, JH, Brown, PR, Wallis, RL, et al. 1990. Bandicoots and Bilbies. Aust Mamm Soc, Surrey Beatty & Sons, NSW, Australia.
9720. Seebeck, JH & Johnson, PG. 1980. *Potorous longipes* (Marsupialia: Macropodidae): A new species from eastern Victoria. Aust J Zool. 28:119-134.
9721. Seegmiller, RF & Ohmart, RD. 1981. Ecological relationships of feral burros and desert bighorn sheep. Wildl Monogr. 78:1-58.
9722. Seeley, RR. 1983. Effects of indomethacin on reproduction under laboratory and field conditions in deer mice (*Peromyscus maniculatus*). Biol Reprod. 28:148-153.
9723. Segers, JC & Chapman, BK. 1984. Ecology of the spotted ground squirrel, *Spermophilus spilosoma* (Merriam) on Padre Island, Texas. Sp Publ Mus TX Tech Univ. 22:105-112.
9724. Seidel, DR & Booth, ES. 1960. Biology and breeding habits of the meadow mouse, *Microtus montanus*, in eastern Washington. Walla Walla College Publ Dept Biol Sci Biol Sta. 29:1-14.
9725. Seidensticker, J. 1977. Notes on early maternal behavior of the leopard. Mammalia. 41:111-113.

9726. Seidensticker, J, Sunquist, ME, & McDougal, C. 1990. Leopards living at the edge of the Royal Chitwan National Park, Nepal. Conservation in Developing Countries; Problems and Prospects. 415-423, Daniel, JC & Serrao, JS, eds. Bombay Nat Hist Soc, Oxford Univ Press, Bombay.
9727. Seidensticker, JC, IV, Hornocker, MG, Wiles, WV, et al. 1973. Mountain lion social organization in the Idaho Primitive Area. Wildl Monogr. 35:1-60.
9728. Seier, JV. 1986. Breeding vervet monkeys in a closed environment. J Med Primatol. 15:339-349.
9729. Seifert, S. 1970. Einige Ergebnisse aus dem Zuchtgeschehen bei Groszkatzen im Leipziger Zoo I Zum Sibirischen Tiger (*Panthera altaica* Temminck 1845). Zool Garten NF. 39:260-270.
9730. Seifert, S. 1978. Untersuchungen zur Fortpflanzungsbiologie der im Zoologischen Garten Leipzig gehaltenen Grosskatzen (*Panthera*, Oken, 1816) unter besonderer Berücksichtigung des Löwen *Panthera leo* (Linné, 1758). VEB Verlag Volk und Gesundheit, Berlin.
9731. Seip, DR. 1979. Energy intake in relation to puberty attainment in female black-tailed deer fauwn. MS Thesis. Simon Fraser Univ, Burnaby, BC.
9732. Seip, DR & Bunnell, FL. 1984. Body weights and measurements of Stone's sheep. J Mamm. 65:513-514.
9733. Seitz, A. 1952. Eisbärenzucht im Nürnberger Tiergarten. Zool Garten NF. 19:180-189.
9734. Seitz, A. 1957. Tragzeit, Fruchtbarkeit, Todesursachen beim Mähnenschaf (*Ammotragus lervia* Pallas). Zool Garten NF. 23:54-57.
9735. Seitz, A. 1967. Einige Feststellungen zur Lebensdauer der Elefanten in Zoologischen Gärten. Zool Garten NF. 34:31-55.
9736. Seitz, A. 1968/69. Einige Feststellungen zur Pflege und Aufzucht von Orang-Utans, *Pongo pygmaeus* Hoppius, 1763. Zool Garten NF. 36:225-245.
9737. Seitz, A. 1969. Notes on the body weights of new-born and young orang-utans *Pongo pygmaeus*. Int Zoo Yearb. 9:81-84.
9738. Seitz, A. 1970. Beitrag zur Haltung des Schabrakentapirs (*Tapirus indicus* Desmarest 1819). Zool Garten NF. 39:271-283.
9739. Seitz, A. 1973. Beitrag zur Zucht des Eisbären (*Thalarctos maritimus*). Zool Garten NF. 43:293-304.
9740. Sekii, S, Kondo, T, & Mezawa, Y. 1983. Reproduction of Grevy's zebra. Animals & Zoos. 35(8):6-10.
9741. Sekii, T. 1970. Birth of a hartebeest. Animals & Zoos. 22(6):6-9.
9742. Sekulic, R. 1978. Seasonality of reproduction in the sable antelope. E Afr Wildl J. 16:177-182.
9743. Sekulic, R. 1982. Birth in free-ranging howler monkeys *Alouatta seniculus*. Primates. 23:580-582.
9744. Sekulic, R. 1983. Male relationships and infant deaths in red howler monkeys (*Alouatta seniculus*). Z Tierpsychol. 61:185-202.
9745. Selenka, E. 1887. Studien über Entwicklungsgeschichte der Thiere, Das Opossum (*Didelphys virginiana*). 4 Heft:101-172. CW Kreidel, Wiesbaden.
9746. Selenka, E. 1892. Studien uber Entwicklungsgeschichte der Thiere 5 Heft. 173-233. CW Kreidel, Wiesbaden.
9747. Selle, RM. 1928. *Microtus californicus* in captivity. J Mamm. 9:93-98.
9748. Selous, FC. 1899. The Cape buffalo (*Bos caffer typicus*). Great and Small Game of Africa. 102-111, Bryden, HA, ed, Rowland Ward, London.
9749. Selous, FC. 1899. The pooko (or puku) kob (*Cobus vardoni typicus*). Great and Small Game of Africa. 294-299, Bryden, HA, ed, Rowland Ward, London.
9750. Selous, FC. 1899. The lechwe kob (*Cobus lechi*). Great and Small Game of Africa. 299-304, Bryden, HA, ed, Rowland Ward, London.
9751. Selous, FC. 1899. The sable antelope (*Hippotragus niger*) in south and south central Africa. Great and Small Game of Africa. 397-404, Bryden, HA, ed, Rowland Ward, London.
9752. Selous, FC. 1899. The roan antelope (*Hippotragus equinus*) in south and south central Africa. Great and Small Game of Africa. 406-411, Bryden, HA, ed, Rowland Ward, London.
9753. Selous, FC. 1899. The eland (*Taurotragus oryx* and *T oryx livingstoni*) in south and south central Africa. Great and Small Game of Africa. 421-432, Bryden, HA, ed, Rowland Ward, London.
9754. Selwood, L. 1980. A timetable of embryonic development of the dasyurid marsupial *Antechinus stuartii* (Macleay). Aust J Zool. 28:649-668.
9755. Selwood, L. 1981. Delayed embryonic development in the dasyurid marsupial, *Antechinus stuartii*. J Reprod Fert Suppl. 29:79-82.
9756. Selwood, L. 1982. A review of maturation and fertilization in marsupials with special reference to the dasyurid: *Antechinus stuartii*. Carnivorous Marsupials. 1:65-76. Archer, M, ed, R Zool Soc NSW, Mossman, NSW.
9757. Selwood, L. 1982. Brown antechinus *Antechinus stuartii*: Management of breeding colonies to obtain embryonic material and pouch young. The Management of Australian Mammals in Captivity. 31-37, Evans, DD, ed, Zool Bd Victoria, Melbourne.
9758. Selwood, L. 1983. Factors influencing pre-natal fertility in the brown marsupial mouse, *Antechinus stuartii*. J Reprod Fert. 68:317-324.
9759. Selwood, L. 1985. Synchronization of oestrus, ovulation and birth in female *Antechinus stuartii* (Marsupialia: Dasyuridae). Aust Mamm. 8:91-96.
9760. Selwood, L. 1987. Embryonic development in culture of two dasyurid marsupials, *Sminthopsis crassicaudata* (Gould) and *Sminthopsis macroura* (Spencer), during cleavage and blastocyst formation. Gamete Res. 16:355-370.
9761. Selwood, L & McCallum, F. 1987. Relationship between longevity of spermatozoa after insemination and the percentage of normal embryos in brown marsupial mice (*Antechinus stuartii*). J Reprod Fert. 79:495-503.
9761a. Selwood, L & Woolley, PA. 1991. A timetable of embryonic development, and ovarian and uterine changes during pregnancy, in the stripe-faced dunnart, *Sminthopsis macroura* (Marsupialia: Dasyuridae). J Reprod Fert. 91:213-227.
9762. Semb-Johansson, A, Wiger, R, & Engh, CE. 1979. Dynamics of freely growing, confined populations of the Norwegian lemming *Lemmus lemmus*. Oikos. 33:246-260.
9763. Semon, R. 1894. Zur Entwicklungsgeschichte der Monotremen. Zoologische Forschungsreisen in Australien Vol 2 Monotremen und Marsupalier Denkschr Naturwissensch Gesselsch Jena. 5:1894-1897.
9764. Sempéré, A. 1977. Plasma progesterone levels in the roe deer *Capreolus capreolus*. J Reprod Fert. 50:365-366.
9765. Sempéré, A. 1978. The annual cycle of plasma testosterone and territorial behavior in the roe deer. Environmental Endocrinology. 73-74, Assenmacher, I & Farner, DS, eds, Springer, New York.
9766. Sempéré, AJ & Boission, J. 1981. Relationship between antler development and plasma androgen concentrations in adult roe deer (*Capreolus capreolus*). J Reprod Endocrinol. 62:49-53.
9767. Sempéré, AJ, Boissin, J, Dutourne, B, et al. 1983. Variations de la concentration plasmatique en prolactine, LH, et

FSH et de l'activité testiculaire au cours de la premiére année de vie chez le Chevreuil (*Capreolus capreolus* L). Gen Comp Endocrinol. 52:247-254.
9768. Sempéré, AJ & Lacroix, A. 1982. Temporal and seasonal relationships between LH, testosterone and antlers in fawn and adult male roe deer (*Capreolus capreolus* L): A longitudinal study from birth to four years of age. Acta Endocrinol. 99:295-301.
9769. Senecal, D. 1977. Coyote. Hinterland Who's Who, Can Wildl Serv. 1-4.
9770. Senft, B & Rietschel, W. 1974. Untersuchungen über die Zusammensetzung der Milch des Okapis (*Okapia johnstoni*). Zool Garten NF. 44:175-179.
9771. Serafinski, W. 1955. Morphological and ecological investigations on Polish species of the genus *Sorex*. Acta Theriol. 1:27-86.
9772. Serebrennikov, MK. 1929. EVERMANNS Iltis (*Putorius eversmanni* Less) in den Wermut-Steppen des nordlichen Kazakstan. Z Säugetierk. 4:205-212.
9773. Serebrennikov, MK. 1931. Album einiger osteuropäischer, westsiberischer und turkestanischer Saugetiere II. Z Säugetierk. 6:160-163.
9774. Serebrennikov, MK. 1933. Album osteuropäischer, westsibirischer und turkestanischer Saugetiere III. Z Säugetierk. 8:33-39.
9775. Serena, M & Soderquist, TR. 1988. Growth and development of pouch young of wild and captive *Dasyurus geoffroii* (Marsupialia: Dasyuridae). Aust J Zool. 36:533-543.
9776. Sergeant, DE. 1962. On the external characters of the blackfish or pilot whales (genus *Globicephala*). J Mamm. 43:395-413.
9777. Sergeant, DE. 1962. The biology of the pilot or pothead whale *Globicephala melaena* (Traill) in Newfoundland waters. Bull Fish Res Bd Can. 132:1-84.
9778. Sergeant, DE. 1963. Minke whales, *Balaenoptera acutorostrata* Lacepede, of the western North Atlantic. J Fish Res Bd Can. 20:1489-1504.
9779. Sergeant, DE. 1965. Migrations of harp seals *Pagophilus groenlandicus* (Erxleben) in the north-west Atlantic. J Fish Res Bd Can. 22:433-464.
9780. Sergeant, DE. 1966. Reproductive rates of harp seals, *Pagophilus groenlandicus* (Erxleben). J Fish Res Bd Can. 23:757-766.
9781. Sergeant, DE. 1973. Biology of white whales (*Delphinapterus leucas*) in western Hudson Bay. J Fish Res Bd Can. 30:1065-1090.
9782. Sergeant, DE. 1973. Environment and reproduction in seals. J Reprod Fert Suppl. 19:555-561.
9783. Sergeant, DE. 1979. Hooded seal. Mammals in the Seas, FAO Fish Ser 5. 2:86-89.
9784. Sergeant, DE. 1982. Mass strandings of toothed whales (Odontoceti) as a population phenomenon. Sci Rep Whales Res Inst. 34:1-48.
9785. Sergeant, DE. 1986. Present status of white whales *Delphinapterus leucas* in the St Lawrence estuary. Nat Can. 113:61-81.
9786. Sergeant, DE & Brodie, PF. 1969. Body size in white whales, *Delphinapterus leucas*. J Fish Res Bd Can. 26:2561-2580.
9787. Sergeant, DE, Caldwell, DK, & Caldwell, MC. 1973. Age, growth and sexual maturity of bottlenosed dolphins (*Tursiops truncatus*) from Northeast Florida. J Fish Res Bd Can. 30:1009-1011.
9788. Sergeant, DE, Ronald, K, Boulva, J, et al. 1978. The recent status of *Monachus monachus*, the Mediterranean monk seal. Biol Conserv. 14:259-287.
9789. Sergeant, DE, St Aubin, DJ, & Geraci, JR. 1980. Life history and northwest Atlantic status of the Atlantic white-sided dolphin, *Lagenorhynchus acutus*. Cetology. 37:1-12.
9789a. Sernia, C. 1978. Steroid-binding proteins in the plasma of the echidna, *Tachyglossus aculeatus*, with comparative data for some marsupials and reptiles. Aust Zool. 20:87-97.
9790. Sernia, C, Bradley, AJ, & McDonald, IR. 1979. High affinity binding of adrenocortical and gonadal steroids by plasma proteins of Australian marsupials. Gen Comp Endocrinol. 38:496-503.
9791. Sernia, C, Garcia-Aragon, J, Thomas, WG, et al. 1990. Uterine oxytocin receptors in an Australian marsupial, the brushtail possum, *Trichosurus vulpecula*. Comp Biochem Physiol 95A:135-138.
9792. Sernia, C, Hinds, L, & Tyndale-Biscoe, CH. 1980. Progesterone metabolism during embryonic diapause in the tammar wallaby, *Macropus eugenii*. J Reprod Fert. 60:139-147.
9793. Sernia, C, Thomas, WG, & Gemmell, RT. 1991. Oxytocin receptors in the mammary gland and reproductive tract of a marsupial, the brushtail possum (*Trichosurus vulpecula*). Biol Reprod. 45:673-679.
9793a. Setchell, BP & Carrick, FN. 1973. Spermatogenesis in some Australian marsupials. Aust J Zool. 21:491-499.
9794. Setchell, KDR, Bull, R, & Adlercreutz, H. 1980. Steroid excretion during the reproductive cycle and pregnancy of the vervet monkey (*Cercopithecus aethiops pygerythrus*). J Steroid Biochem. 12:375-384.
9795. Setchell, KDR & Bonney, RC. 1979. The excretion of urinary steroids by the owl monkey (*Aotus trivirgatus*) studied using open tubular capillary column gas chromatography and mass spetrometry. J Steroid Biochem. 14:37-43.
9796. Setchell, KDR, Gosselin, SJ, Welsh, MB, et al. 1987. Dietary estrogens a probable cause of infertility and liver disease in captive cheetahs. Gastroenterology. 93:225-233.
9797. Seth, P & Prasad, MRN. 1967. Effect of bilateral adrenalectomy on the ovary of the five striped Indian palm squirrel, *Funambulus pennanti* (Wroughton). Gen Comp Endocrinol. 8:152-162.
9798. Seth, P & Prasad, MRN. 1967. Seasonal changes in the histochemical localization of δ-3β-hydroxysteroid dehydrogenase in the ovary of the five-striped Indian palm squirrel, *Funambulus pennanti* (Wroughton). Gen Comp Endocrinol. 9:383-390.
9799. Seth, P & Prasad, MRN. 1969. Reproductive cycle of the female five-striped palm squirrel *Funambulus pennanti* (Wroughton). J Reprod Fert. 20:211-222.
9800. Seth, PK & Seth, S. 1983. Population dynamics of free-ranging rhesus monkeys in different ecological conditions in India. Am J Primatol. 5:61-67.
9801. Seth-Smith, AMD & Parker, ISC. 1967. A record of twin foetuses in the African elephant. E Afr Wildl J. 5:167.
9802. Seton, ET. 1920. Notes on the breeding habits of captive deermice. J Mamm. 1:134-138.
9802a. Settle, G. 1976. Growth and behaviour of a litter of tiger cats (*Dasyurus maculatus*). Bull Aust Mamm Soc. 3:43.
9803. Settle, GA. 1978. The quiddity of tiger quolls. Aust Nat Hist. 19:164-165.
9804. Settle, GA & Croft, DB. 1982. Maternal behaviour of *Antechinus stuartii* (Dasyuridae, Marsupialia) in captivity. Carnivorous Marsupialia. 1:365-381, Archer, M, ed, R Zool Soc NSW, Mossman, NSW.

9805. Severaid, JH. 1942. The snowshoe hare its life history and artificial propagation. ME Dept Inland Fish Game.
9806. Severaid, JH. 1945. Breeding potential and artificial propagation of the snowshoe hare. J Wildl Manage. 9:290-295.
9807. Severaid, JH. 1950. The gestation period of the pika (*Ochotona princeps*). J Mamm. 31:356-357.
9808. Severaid, JH. 1955. The natural history of the pikas (mammalian genus *Ochotona*). Ph D Diss. Univ CA, Berkeley.
9809. Severinghaus, CW & Cheatum, EL. 1956. Life and times of the white-tailed deer. The Deer of North America. 57-186, Taylor, WP, ed, Wildl Manage Inst, Washington, DC.
9810. Seyfarth, RM. 1978. Social relationships among adult male and female baboons I Behaviour during sexual consortship. Behaviour. 64:204-226.
9811. Seymour, KL. 1989. *Panthera onca*. Mamm Species. 340:1-9.
9812. Shackelford, RM. 1952. Superfetation in the ranch mink. Am Nat. 86:311-319.
9813. Shackleton, CHL. 1974. Progesterone and oestrogen metabolism in the pregnant marmoset (*Callithrix jacchus*). J Steroid Biochem. 5:597-600.
9814. Shackleton, CHL & Mitchell, FL. 1975. The comparison of perinatal steroid endocrinology in simians with a view to finding a suitable animal model to study human problems. Lab Anim Handb. 6:159-181.
9815. Shackleton, DM. 1985. *Ovis canadensis*. Mamm Species. 230:1-9.
9816. Shackleton, DM, Patterson, RG, Haywood, J, et al. 1984. Gestation period in *Ovis canadensis*. J Mamm. 65:337-338.
9817. Shadle, AR. 1930. An unusual case of parturition in a beaver. J Mamm. 11:483-485.
9818. Shadle, AR. 1948. Gestation period in the porcupine, *Erethizon dorsatum dorsatum*. J Mamm. 29:162-164.
9819. Shadle, AR. 1951. Laboratory copulations and gestations of porcupine, *Erethizon dorsatum*. J Mamm. 32:219-221.
9820. Shadle, AR. 1952. Sexual maturity and first recorded copulation of a 16 month male porcupine, *Erethizon dorsatum dorsatum*. J Mamm. 33:239-241.
9821. Shadle, AR. 1953. Striped skunk produces two litters. J Wildl Manage. 17:388-389.
9822. Shadle, AR. 1956. Reproduction in a skunk, *Mephitus mephitus hudsonica*. J Mamm. 37:112-113.
9823. Shadle, AR & Ploss, WR. 1943. An unusual porcupine parturition and development of the young. J Mamm. 24:492-496.
9824. Shagaeva, VG & Kurnosov, KM. 1974. Morphogenesis of placenta and its patterns in *Camelus bactrianus*. Zool Zhurn. 53:1058-1065.
9825. Shaham, Y, Lelyveld, J, Marder, U, et al. 1978. Establishment of an albino sand rat (*Psammomys obesus*) colony and comparison with the natural coloured animal. Lab Anim. 12:13-17.
9826. Shaikh, AA, Shaikh, SA, Celaya, CL, et al. 1982. Ovulation pattern in successive cycles in the baboon. Primates. 23:592-595.
9827. Shaikh, AA, Naqvi, RH, & Shaikh, SA. 1978. Concentrations of oestradiol-17β and progesterone in the peripheral plasma of the cynomolgus monkey (*Macaca fascicularis*) in relation to the length of the menstrual cycle and its component phases. J Endocrinol. 79:1-7.
9828. Shalash, MR. 1965. Some reproductive aspects in the female camel. World Rev Anim Prod. 1(4):103-108.
9829. Shalash, MR. 1987. Reproduction in camels. Egypt J Vet Sci. 24:1-25.
9830. Shalash, MR & Nawito, M. 1963. Beitrag zur Sterilität des weiblichen Kamels. Dtsch Tierärzl Wchsch. 70:522.
9831. Shalash, MR & Nawito, M. 1964. Some reproductive aspects of the female camel. 5th Int Congr Anim Reprod Artif Insemin (Trento). 1:263-273.
9832. Shandilya, LN, Ramaswami, LS, & Shandilya, N. 1976. Oestrogen metabolites in urine during the menstrual cycle, pregnancy and puerperium in the Indian hanuman langur (*Presbytis entellus entullus*). J Reprod Fert. 47:7-11.
9833. Shandilya, LN, Ramaswami, LS, & Shandilya, N. 1977. Sialic acid concentration in the reproductive organs, pituitary gland and urine of the Indian langur monkey (*Presbytis entellus entellus*). J Endocrinol. 73:207-213.
9834. Shapiro, J. 1949. Ecological and life history notes on the porcupine in the Adirondacks. J Mamm. 30:247-257.
9835. Shapiro, LE, Austin, D, Ward, SE, et al. 1986. Familiarity and female mate choice in two species of voles (*Microtus ochrogaster* and *Microtus montanus*). Anim Behav. 34:90-97.
9836. Shaposhnikov, FD. 1956. Data on the ecology of musk-deer in northeastern Altai. Zool Zhurn. 35:1084-1093.
9837. Shaposhnikov, LV & Krushinskaya, ES. 1939. The Altaian *Marmota baibacina* Katsch, in the Daghestan ASSR. Zool Zhurn. 18:1048-1054.
9838. Shapovalov, AV. 1986. The performance of black nutria housed indoors. Nauch Trudy Nauchna Issedovatelskii Inst Pushnogo Zverovodstva i Krolikovodstva. 29:78-80. [An Br Abst. 54, ref 1827, 1986]
9839. Sharma, A & Mathur, RS. 1974. Histomorphological changes in the reproductive tract of female *Hemiechinus auritus collaris* Gray, in relation to the estrous cycle. Acta Zool (Stockholm). 55:235-243.
9840. Sharma, A & Mathur, RS. 1975. Cyclic changes in the uterine phosphatases of *Suncus murinus sindensis* Anderson, the Indian common house shrew. Acta Anat. 92:376-384.
9841. Sharma, A & Mathur, RS. 1976. Histomorphological changes in the female reproductive tract of *Suncus murinus sindensis* (Anderson) during the oestrous cycle. Folia Biol (Kraków). 24:277-283.
9842. Sharma, A, Mathur, RS, & Goyal, RP. 1977. Quantitative evaluation of phosphatases in the uteri of two Indian insectivores, *Hemiechinus auritis collaris* (Gray) and *Suncus murinus sindensis* (Anderson), in relation to the estrous cycle. Folia Biol (Kraków). 25:71-79.
9843. Sharma, IK. 1979. Habitats, feeding, breeding and reaction to man of the desert cat *Felis libyca* (Gray) in the Indian desert. J Bombay Nat Hist Soc. 76:498-499.
9844. Sharma, IK. 1982. Notes on the Indian pygmy pipistrelle (*Pipistrellus mimus* Wroughton) in the Thar desert. J Bombay Nat Hist Soc. 79:181-182.
9845. Sharma, KR, Muralidhar, K, & Moudgal, NR. 1978. Heterologous radioimmunoassay systems for measurement of follicle stimulating hormone (Follitropin) and luteinizing hormone (Lutropin) in the Bonnet monkey, *Macaca radiata*. Indian J Exp Biol. 16:153-156.
9846. Sharma, SS & Vyas, KK. 1970. Parturition in the camel (*Camelus dromedarius*). Ceylon Vet J. 18:7-9.
9847. Sharma, SS & Vyas, KK. 1971. Factors affecting gestation length in the Bikaneri camel (*Camelus dromedarius*). Ceylon Vet J. 19:67-68.
9848. Sharma, VD, Bhargava, KK, & Singh, M. 1963. Secondary sex ratio of normal births in Bikaneri camel. Indian Vet J. 40:561-563.
9849. Sharman, GB. 1955. Studies on marsupial reproduction II The oestrous cycle of *Setonix brachyurus*. Aust J Zool. 3:44-55.

9850. Sharman, GB. 1955. Studies on marsupial reproduction III Normal and delayed pregnancy in *Setonix brachyurus*. Aust J Zool. 3:56-70.
9851. Sharman, GB. 1956. Some aspects of marsupial reproduction. Proc Zool Soc London. 127:141-143.
9852. Sharman, GB. 1959. Marsupial reproduction. Monogr Biol. 8:332-368.
9853. Sharman, GB. 1961. The embryonic membranes and placentation in five genera of diprotodont marsupials. Proc Zool Soc London. 137:197-220.
9854. Sharman, GB. 1962. The initiation and maintenance of lactation in the marsupial, *Trichosurus vulpecula*. J Endocrinol. 25:375-385.
9855. Sharman, GB. 1963. Delayed implantation in marsupials. Delayed Implantation. 3-14, Enders, AC, ed, Univ Chicago Press, Chicago.
9856. Sharman, GB. 1964. The female reproductive sytem of the red kangaroo, *Megaleia rufa*. CSIRO Wildl Res. 9:50-57.
9857. Sharman, GB. 1965. The effects of suckling on normal and delayed cycles of reproduction in the red kangaroo. Z Säugetierk. 30:10-20.
9858. Sharman, GB & Calaby, JH. 1964. Reproductive behaviour in the red kangaroo, *Megaleia rufa*, in captivity. CSIRO Wildl Res. 9:58-85.
9859. Sharman, GB, Calaby, JH, & Poole, WE. 1966. Patterns of reproduction in female diprotodont marsupials. Symp Zool Soc London. 15:205-232.
9860. Sharman, GB & Clark, MJ. 1967. Inhibition of ovulation by the corpus luteum in the red kangaroo, *Megaleia rufa*. J Reprod Fert. 14:129-137.
9861. Sharman, GB & Pilton, PE. 1964. The life history and reproduction of the red kangaroo (*Megaleia rufa*). Proc Zool Soc London. 142:29-48.
9862. Sharman, GB, Frith, HJ, & Calaby, JH. 1964. Growth of the pouch young, tooth eruption, and age determination in the red kangaroo, *Megaleia rufa*. CSIRO Wildl Res. 9:20-49.
9863. Sharman, GB, Robinson, ES, Walton, SM, et al. 1970. Sex chromosomes and reproductive anatomy of some intersexual marsupials. J Reprod Fert. 21:57-68.
9864. Shaughnessy, PD. 1979. Cape fur seals in South West Africa. SW Afr Ann. 1979:101-103.
9865. Shaughnessy, PD. 1979. Cape (South African) fur seal. Mammals in the Sea, FAO Fish Ser 5. 2:37-40.
9866. Shaw, G. 1983. Effect of PGF-2α on uterine activity and concentrations of 13,14-dihydro-15-keto-$PGF_{2\alpha}$ in peripheral plasma during parturition in the tammar wallaby (*Macropus eugenii*). J Reprod Fert. 69:429-436.
9867. Shaw, G. 1990. Control of parturient behaviour by prostaglandin $F_{2\alpha}$ in the tammar wallaby (*Macropus eugenii*). J Reprod Fert. 88:335-342.
9868. Shaw, G & Renfree, MB. 1984. Concentrations of oestradiol-17β in plasma and corpora lutea throughout pregnancy in the tammar, *Macropus eugenii*. J Reprod Fert. 72:29-37.
9869. Shaw, G & Renfree, MB. 1986. Uterine and embryonic metabolism after diapause in the tammar wallaby, *Macropus eugenii*. J Reprod Fert. 76:339-347.
9870. Shaw, G & Rose, RW. 1979. Delayed gestation in the potoroo *Potorous tridactylus* (Kerr). Aust J Zool. 27:901-912.
9871. Shaw, WT. 1921. Moisture and altitude as factors in determining the seasonal activities of the Townsend ground squirrel in Washington. Ecology. 2:189-192.
9872. Shaw, WT. 1924. The home life of the Columbian ground squirrel. Can Field-Nat. 38:128-130, 151-153.
9873. Shaw, WT. 1925. Breeding and development of the Columbian ground squirrel. J Mamm. 6:106-113.
9874. Shaw, WT. 1925. Duration of the aestivation and hibernation of the Columbian ground squirrel (*Citellus columbianus*) and sex relation to the same. Ecology. 6:76-81.
9875. Shaw, WT. 1945. Seasonal and daily activities of the Columbian ground squirrel at Pullman, Washington. Ecology. 26:74-84.
9876. Sheets, RG, Linder, RL, & Dahlgren, RB. 1972. Food habits of two litters of black-footed ferret in South Dakota. Am Mid Nat. 87:249-251.
9877. Sheffer, DE. 1957. Cottontail rabbit propagation in small breeding pairs. J Wildl Manage. 21:90.
9878. Shehata, R. 1975. Female prostate in *Arvicanthis niloticus* and *Meriones libycus*. Acta Anat. 92:513-523.
9879. Shelden, RM. 1969. Reproduction in weasels (*Mustela erminea* and *Mustela frenata*): Some aspects of ovariectomy and ovarian steroid replacement. Diss Abstr. 30:444b.
9880. Shelden, RM. 1972. The fate of short-tailed weasel, *Musela erminea*, blastocysts following ovariectomy during diapause. J Reprod Fert. 31:347-352.
9881. Shelden, RM. 1973. Failure of ovarian steroids to influence blastocysts of weasels (*Mustela erminea*) ovariectomized during delayed implantation. Endocrinology. 92:638-641.
9882. Sheldon, C. 1930. Nova Scotia red-backed mice - *Clethreonomys gapperi ochraceus*. J Mamm. 11:318-320.
9883. Sheldon, C. 1934. Studies on the life histories of *Zapus* and *Napaeozapus* in Nova Scotia. J Mamm. 15:290-300.
9884. Sheldon, C. 1938. Vermont jumping mice of the genus *Napaeozapus*. J Mamm. 19:444-453.
9884a. Sheldon, WB. 1937. Notes on the giant panda. J Mamm. 18:13-19.
9885. Sheldon, WG. 1949. Reproductive behavior of foxes in New York state. J Mamm. 30:236-246.
9886. Sheldon, WG. 1959. A late breeding record of a bobcat in Massachusetts. J Mamm. 40:148.
9887. Shellhammer, H. 1982. *Reithrodontomys raviventris*. Mamm Species. 169:1-3.
9888. Shenbrot, GI. 1977. Age structure of the population and reproduction of *Alactagulus pygmaeus* (Rodentia, Dipopidae). Zool Zhurn. 56:1381-1388.
9889. Sheppard, DH. 1969. A comparison of reproduction in two chipmunk species (*Eutamias*). Can J Zool. 47:603-608.
9890. Sheppard, DH. 1972. Reproduction of Richardson's ground squirrel (*Spermophilus richardsonii*) in southern Sasketchewan. Can J Zool. 50:1577-1581.
9891. Sheppe, W. 1963. Population structure of deer mouse, *Peromyscus*, in the Pacific northwest. J Mamm. 44:180-185.
9892. Sheppe, W & Haas, P. 1981. The annual cycle of small mammal populations along the Chobe River, Botswana. Mammalia. 45:157-176.
9893. Sheppe, WA. 1973. Notes on Zambian rodents and shrews. Puku. 7:167-190.
9894. Sheridan, M & Tamarin, RH. 1988. Space use, longevity, and reproductive success in meadow voles. Behav Ecol Sociobiol. 22:85-90.
9895. Sherman, HB. 1930. Birth of the young of *Myotis austroriparius*. J Mamm. 11:495-503.
9896. Sherman, HB. 1937. Breeding habits of the free tailed bat. J Mamm. 18:176-187.
9896a. Sherman, HB. 1945. The Florida yellow bat, *Dasypterus floridanus*. Proc FL Acad Sci. 7:193-197.

9897. Sherman, PW. 1980. The limits of ground squirrel nepotism. Sociobiology: Beyond Nature/Nurture?. 505-544, Barlow, GW & Silverberg, J, eds, Westview Press, Boulder, Colorado.

9898. Sherman, PW. 1980. The meaning of nepotism. Am Nat. 116:604-606.

9899. Sherman, PW, Jarvis, JUM, & Alexander, RD. 1991. The Biology of the Naked Mole-Rat. Princeton Univ Press, Princeton, NJ.

9900. Sherman, PW & Morton, ML. 1984. Demography of Belding's ground squirrels. Ecology. 65:1617-1628.

9901. Sherry, BY. 1975. Reproduction of elephant in Gonarezhoa, south-eastern Rhodesia. Arnoldia (Rhodesia). 7(29):1-14.

9902. Sheth, AR, Shah, GV, Gadgil, BA, et al. 1975. Cyclic changes in the metabolism of glycogen in cervical mucus of bonnet monkeys (*Macaca radiata*). J Reprod Fert. 42:133-136.

9903. Shevchenko, VL. 1963. Reproduction and change in the numerousness of *Lagurus lagurus* Pall in the Ural region. Zool Zhurn. 42:114-125.

9903a. Shi, S, Dong, L, Chen, Y, et al. 1988. Reproductive endocrinological changes during the oestrus of the female giant panda. Acta Theriol Sinica. 8:1-6.

9904. Shideler, SE, Czekala, NM, Kasman, LH, et al. 1983. Monitoring ovulation and implantation in the lion-tailed macaque (*Macaca silenus*) through urinary estrone conjugate evaluations. Biol Reprod. 29:905-911.

9905. Shideler, SE, Czekela, NM, Benirschke, K, et al. 1983. Urinary estrogens during pregnancy of the ruffed lemur (*Lemur variegatus*). Biol Reprod. 28:963-969.

9906. Shideler, SE & Lindburg, DG. 1982. Selected aspects of *Lemur variegatus* reproductive biology. Zoo Biol. 1:127-134.

9907. Shideler, SE, Lindburg, DG, & Lasley, BL. 1983. Estrogen-behavior correlates in the reproductive physiology and behavior of the ruffed lemur (*Lemur variegatus*). Horm Behav. 17:249-263.

9908. Shideler, SE, Mitchell, WR, Lindburg, DG, et al. 1985. Monitoring luteal function in the lion-tailed macaque (*Macaca silenus*) through urinary progesterone metabolite measurements. Zoo Biol. 4:65-73.

9909. Shield, J. 1964. A breeding season difference in two populations of the Australian macropod marsupial (*Setonix brachyurus*). J Mamm. 45:616-625.

9910. Shield, J. 1968. Reproduction of the quokka, *Setonix brachyurus*, in captivity. J Zool. 155:427-444.

9911. Shield, JW & Woolley, P. 1960. Gestation time for delayed birth in the quokka. Nature. 188:163-164.

9912. Shield, JW & Woolley, P. 1963. Population aspects of delayed birth in the quokka (*Setonix brachyurus*). Proc Zool Soc London. 141:783-789.

9913. Shigehara, N. 1980. Epiphyseal union, tooth eruption, and sexual maturation in the common tree shrew, with reference to its systematic problem. Primates. 21:1-19.

9914. Shikawa, A & Namikawa, T. 1987. Postnatal growth and development in laboratory strains of large and small musk shrews (*Suncus murinus*). J Mamm. 68:766-774.

9915. Shiljaeva, LM. 1971. Structure of the arctic fox (*Alopex lagopus*) population and role of different generations in the control of its population density. Zool Zhurn. 50:1843-1852.

9916. Shillito (Babington), JF. 1963. Field observations on the growth, reproduction and activity of a woodland population of the common shrew *Sorex araneus*. Proc Zool Soc London. 140:99-114.

9916a. Shilo, RA & Tamarovskaya, MA. 1981. The growth and development of wolverines *Gulo gulo* at Novosibirsk Zoo. Int Zoo Yearb. 21:146-147.

9917. Shimada, A. 1973. Studies on *Tupaia glis* Diard as an experimental animal: Its breeding and growth. Jukken Dobutsu, Exp Anim. 22(Suppl):351-357.

9918. Shipp, E, Keith, K, Hughes, RL, et al. 1963. Reproduction in a free living population of domestic rabbits, *Oryctolagus cuniculus* (L) on sub-antarctic island. Nature. 200:858-860.

9919. Shiraishi, S. 1967. Some observations on the Japanese vole, *Microtus montebelli*, in the Chikugo river-bed, Kurume City I Sex ratio, breeding season, and embryo size. J Mamm Soc Japan. 3:57-63.

9920. Shively, C, Clarke, N, King, N, et al. 1982. Patterns of sexual behavior in male macaques. Am J Primatol. 2:373-384.

9921. Shoemaker, AH. 1979. Reproduction and development of the black howler monkey *Alouatta caraya* at Columbia Zoo. Int Zoo Yearb. 19:150-155.

9922. Shoemaker, AH. 1982. Fecundity in the captive howler monkey, *Alouatta caraya*. Zoo Biol. 1:149-156.

9923. Shoemaker, AH. 1982. Notes on the reproductive biology of the white-faced saki *Pithecia pithecia* in captivity. Int Zoo Yearb. 22:124-127.

9924. Sholl, SA, Goy, RW, & Uno, H. 1982. Differences in brain uptake and metabolism of testosterone in gonadectomized, adrenalectomized male and female rhesus monkeys. Endocrinology. 111:806-813.

9925. Sholl, SA, Toivola, PTK, & Robinson, JA. 1979. The dynamics of testosterone and dihydrotestosterone metabolism in the adult male rhesus monkey. Endocrinology. 105:402-405.

9926. Sholl, SA & Wolf, RC. 1974. The metabolic clearance rate of progesterone in the pregnant rhesus monkey. Endocrinology. 95:1287-1292.

9927. Sholl, SA & Wolf, RC. 1980. Progestin metabolism in the rhesus monkey corpus luteum. Steroids. 36:209-218.

9928. Sholl, SA, Wolf, RC, & Colas, AE. 1974. Δ^5-3β-hydroxysteroid dehydrogenase/δ^5-δ^4-isomerase activity in the corpus luteum of the pregnant rhesus monkey. Endocrinology. 94:908-910.

9929. Shorey, CD & Hughes, RL. 1973. Cyclical changes in the uterine endometrium and peripheral plasma concentrations of progesterone in the marsupial *Trichosurus vulpecula*. Aust J Zool. 21:1-19.

9930. Shorey, CD & Hughes, RL. 1973. Development, function, and regression of the corpus luteum in the marsupial *Trichosurus vulpecula*. Aust J Zool. 21:477-489.

9931. Short, C. 1970. Morphological development and aging of mule and white-tailed deer fetuses. J Wildl Manage. 34:383-388.

9932. Short, HL. 1961. Age at sexual maturity of Mexican free-tailed bats. J Mamm. 42:533-536.

9933. Short, HL. 1961. Growth and development of Mexican free-tailed bats. Southwest Nat. 6:156-163.

9934. Short, RV. Unpublished results. [Cited by Smith, et al, 1969]

9935. Short, RV. 1958. Progesterone in tissues and body fluids. Ph D Diss. Univ Cambridge, Cambridge. [Cited by Smith, et al, 1969]

9935a. Short, RV. 1966. Oestrous behaviour, ovulation and the formation of the corpus luteum in the African elephant, *Loxodonta africana*. E Afr Wildl J. 4:56-68.

9936. Short, RV & Buss, IO. 1965. Biochemical and histological observations on the corpora lutea of the African elephant, *Loxodonta africana*. J Reprod Fert. 9:61-67.

9937. Short, RV & Eckstein, P. 1961. Oestrogen and progesterone levels in pregnant rhesus monkeys. J Endocrinol. 22:15-22.
9938. Short, RV, Flint, APF, & Renfree, MB. 1985. Influence of passive immunization against GnRH on pregnancy and parturition in the tammar wallaby, *Macropus eugenii.* J Reprod Fert. 75:567-575.
9939. Short, RV & Hay, MF. 1966. Delayed implantation in the roe deer *Capreolus capreolus.* Symp Zool Soc London. 15:173-194.
9940. Short, RV & Mann, T. 1966. The sexual cycle of a seasonally breeding mammal, the roebuck (*Capreolus capreolus*). J Reprod Fert. 12:337-351.
9941. Short, RV, Mann, T, & Hay, MF. 1967. Male reproductive organs of the African elephant, *Loxodonta africana.* J Reprod Fert. 13:517-536.
9942. Short, RV & Sharman, GB. Unpublished results. [Cited by Smith et al, 1969]
9943. Shorten, Mrs Vizoso. 1951. Some aspects of the biology of the grey squirrel (*Sciurus carolinensis*) in Great Britain. Proc Zool Soc London. 121:427-459.
9944. Shortridge, GC. 1931. Field notes on two little-known antelopes the Damaraland dikdik (*Rhynchotragus damarensis*) and the Angolan impala. S Afr J Sci. 28:412-417.
9945. Shortridge, GC. 1934. The Mammals of South West Africa. William Heinemann Ltd, London.
9946. Shoshani, J & Eisenberg, JF. 1982. *Elephas maximus.* Mamm Species. 182:1-8.
9947. Shoshani, J, Goldman, CA, & Thewissen, JGM. 1988. *Orycteropus afer.* Mamm Species. 300:1-8.
9948. Shryer, J & Flath, DL. 1980. First record of the pallid bat (*Antrozous pallidus*) from Montana. Great Basin Nat. 40:115.
9949. Shubin, IG. 1965. Reproduction of *Ochotona pusilla* Pall. Zool Zhurn. 44:917-924.
9950. Shubin, IG. 1972. Reproduction and numbers of steppe lemming in northern Balkhash area. Soviet J Ecol. 3:450-452.
9951. Shubin, IG. 1974. Ecology of *Lagurus luteus* in the Zaisan hollow. Zool Zhurn. 53:272-277.
9952. Shubin, IG & Ismagilov, MI. 1969. A contribution to the ecology of the pygmy jerboa (*Salpingotus crassicauda*) in the Zaisansky hollow. Zool Zhurn. 48:1722-1726.
9953. Shubin, NG. 1964. The reproduction of chipmunk in the Tom River territory. Zool Zhurn. 3:910-917.
9954. Shubin, NG & Suchkova, HG. 1975. A contribution to the biology of *Sicista napaea.* Zool Zhurn. 54:475-479.
9955. Shul'gina, NK, Polyntsev, YV, Rozen, VB, et al. 1981. Content of sex steroids in the blood plasma of female sable at the initial stages of postnatal ontogenesis. Dokl Akad Nauk SSSR Translantion. 258:255-258.
9956. Shull, EM. 1958. Notes on the four-horned antelope *Tetracercus quadricornis* (Blainville). J Bombay Nat Hist Soc. 55:339-340.
9957. Shull, EM. 1962. Gestation period of the fourhorned antelope *Tetracercus quadricornis* (Blainville). J Bombay Nat Hist Soc. 59:945-947.
9958. Shump, AU, Aulerich, RJ, & Ringer, RK. 1976. Semen volume and sperm concentration in the ferret (*Mustela putorius*). Lab Ani Sci. 26:913-916.
9959. Shump, KA, Jr & Baker, RH. 1978. *Sigmodon alleni.* Mamm Species. 95:1-2.
9960. Shump, KA, Jr & Baker, RH. 1978. *Sigmodon leucotis.* Mamm Species. 96:1-2.
9961. Shump, KA, Jr & Shump, AU. 1982. *Lasiurus borealis.* Mamm Species. 183:1-6.
9962. Shump, KA, Jr & Shump, AU. 1982. *Lasiurus cinereus.* Mamm Species. 185:1-5.
9963. Shvareva, NV. 1982. Some peculiarities of morphofunctional activity of the ovarial gland in the dynamics of population cycle in *Dicrostonyx torquatus* (Rodentia) on the Wrangel Island. Zool Zhurn. 61:1740-1748.
9964. Siddiqi, MAH. 1937. The development of the penile urethra and the homology of Cowper's gland of male spermophile (*Citellus tridecemlineatus*) with a note on the prostatic utricle. J Anat. 72:109-115.
9965. Siddiqi, MAH. 1947. The prostate gland of *Herpestes edwardsii edwardsii.* J Anat. 81:123-126.
9966. Sidorov, GN & Botvinkin, AD. 1987. The corsac fox (*Vulpes corsac*) in southern Siberia. Zool Zh. 66:914-927.
9967. Siedel, DR & Booth, ES. 1960. Biology and breeding habits of the meadow mouse *Microtus montanus* in eastern Washington. Publ Walla Walla Coll. 29:1-12.
9968. Siegler, HR. 1954. Late-breeding snowshoe hare. J Mamm. 35:122.
9969. Sierts, W. 1950. Os clitoridis von *Zalophus californianus* Less und *Sciurus vulgaris fuscoater* Altum. Neue Ergebnisse und Probleme der Zoologie. 938-939, Herre, W, ed, Geest & Portig KG, Leipzig.
9969a. Sievert, J, Karesh, WB, & Sunde, V. 1991. Reproductive intervals in captive female western lowland gorillas with a comparison to wild mountain gorillas. Am J Primatol. 24:227-234.
9970. Sigg, H, Stolba, A, Abegglen, JJ, et al. 1982. Life history of Hamadryas baboons: Physical development, infant mortality, reproductive parameters and family relationships. Primates. 23:473-487.
9971. Sikes, SK. 1958. The calving of the hinds, *Sylvicapra grimmia* var *coronata*: The grey duiker. Niger Fld. 23:55-66. [Cited by Mentis, 1972]
9972. Sikes, SK. 1971. The Natural History of the African Elephant. Weidenfeld & Nicolson, London.
9973. Silk, JB. 1988. Social mechanisms of population regulation in a captive group of bonnet macaques (*Macaca radiata*). Am J Primatol. 14:111-124.
9974. Silk, JB, Clark-Wheatley, CB, Rodman, P, et al. 1981. Differential reproductive success and facultative adjustment of sex ratios among captive female bonnet macaques (*Macaca radiata*). Anim Behav. 29:1106-1120.
9975. Silver, H. 1961. Deer milk compared with substitute milk for fawns. J Wildl Manage. 25:66.
9976. Silver, H. 1965. An instance of fertility in a white-tailed buck fawn. J Wildl Manage. 29:634-636.
9977. Silverman, HB & Dunbar, MJ. 1980. Aggressive tusk use by the narwhal (*Monodon monoceros* L). Nature. 284:57-58.
9978. Simkin, DW. 1963. A study of moose reproduction and productivity in northwestern Ontario. MS Thesis. Cornell Univ, Ithaca, NY.
9979. Simkin, DW. 1965. Reproduction and productivity of moose in northwestern Ontario. J Wildl Manage. 29:740-750.
9980. Simmons, NM, Bayer, MB, & Sinkey, LO. 1984. Demography of Dall's sheep in the Mackensie Mountains, Northwest Territories. J Wildl Manage. 48:156-162.
9981. Simon, KJ. 1959. Preliminary studies on composition of milk of Indian elephants. Indian Vet J. 36:500-503.
9982. Simonds, PE. 1965. The bonnet macaque in south India. Primate Behavior Field Studies of Monkeys and Apes. 175-196, DeVore, I, ed, Holt, Rinehart & Winston, New York.

9983. Simons, LS. 1984. Seasonality of reproduction and dentinal structures in the harbor porpoise (*Phocoena phocoena*) of the North Pacific. J Mamm. 65:491-495.
9984. Simpson, CD. 1964. Notes on the banded mongoose *Mungos mungos* (Gmelin). Arnoldia (Rhodesia). 1(19):1-8.
9985. Simpson, CD. 1966. The banded mongoose. Animal Kingdom. 69:52-57.
9986. Simpson, CD. 1968. Reproduction and population structure in greater kudu in Rhodesia. J Wildl Manage. 32:149-161.
9987. Simpson, DA, Anthony, RG, Kelly, GM, et al. 1979. Dynamics of a pine vole population in a Pennsylvania orchard. Proc E Pine Meadow Vole Symp. 3:47-51.
9988. Simpson, JS & Jones, AC. 1982. Hybrid production in owl monkeys (*Aotus trivirgatus*). Lab Anim. 16:71-72.
9989. Simpson, KG. 1961. A rooftop breeding colony of Gould's wattled bat. Vict Nat. 78:325-327.
9990. Simpson, MJA & Simpson, AE. 1982. Birth sex ratios and social rank in rhesus monkeys mothers. Nature. 300:440-441.
9991. Sinclair, ARE. 1974. The natural regulation of buffalo populations in East Africa II Reproduction, recruitment and growth. E Afr Wildl J. 12:169-183.
9992. Sinclair, ARE. 1974. The natural regulation of buffalo populations in East Africa III Population trends and mortality. E Afr Wildl J. 12:185-200.
9993. Sinclair, ARE. 1977. Lunar cycle and timing of mating season in Serengeti wildebeest. Nature. 267:832-833.
9994. Sinclair, ARE. 1977. The African Buffalo: A Study of Resource Limitation of Populations. Univ Chicago Press, Chicago.
9995. Singh, SK & Dominic, CJ. 1981. Failure of 5-thio-D-glucose to induce antispermatogenic effects in the musk shrew, *Suncus murinus* L. Biol Reprod. 24:655-659.
9996. Singh, UB & Bharadwaj, MB. 1978. Morphological changes in the testis and epididymis of camels (*Camelus dromedarius*) Part I. Acta Anat. 101:275-279.
9997. Singh, UB & Bharadwaj, MB. 1978. Histological and histochemical studies on the testes of camel (*Camelus dromedarius*) Part II. Acta Anat. 101:280-288.
9998. Singh, V & Parkash, A. 1964. Mating behaviour in camel. Indian Vet J. 41:475-477.
9999. Singwi, MS & Lall, SB. 1983. Spermatogenesis in the non-scrotal bat, *Rhinopoma kinneari* Wroughton (Microchiroptera: Mammalia). Acta Anat. 116:136-145.
10000. Sinha, AA, Conaway, CH, & Kenyon, KW. 1966. Reproduction in the female sea otter. J Wildl Manage. 30:121-130.
10001. Sinha, AA & Erickson, AW. 1974. Ultrastructure of the placenta of Antarctic seals during the first third of pregnancy. Am J Anat. 141:263-279.
10002. Sinha, AA & Mead, RA. 1975. Ultrastructural changes in granulosa lutein cells and progesterone levels during preimplantation, implantation, and early placentation in western spotted skunk. Cell Tiss Res. 164:179-192.
10003. Sinha, AA & Mead, RA. 1976. Morphological changes in the trophoblast, uterus and corpus luteum during delayed implantation and implantation in the western spotted skunk. Am J Anat. 145:333-356.
10004. Sinha, AA & Mossman, HW. 1966. Placentation of the sea otter. Am J Anat. 119:521-554.
10005. Sinha, AA, Seal, US, Erickson, AW, et al. 1969. Morphogenesis of the white-tailed deer. Am J Anat. 126:201-241.
10006. Sinha, YP. 1980. Further observations on the field ecology of Rajasthan bats. J Bombay Nat Hist Soc. 77:465-470.
10007. Siniff, DB, DeMaster, DP, & Hofman, RJ. 1977. Analysis of the dynamics of a Weddell seal population. Ecol Monogr. 47:319-335.
10008. Siniff, DB, Stirling, I, Bengston, JL, et al. 1979. Social and reproductive behavior of crabeater seals (*Lobodon carcinophagus*) during the austral spring. Can J Zool. 57:2243-2255.
10009. Sirianni, JE, Swindler, DR, & Tarrant, LH. 1975. Somatometry of newborn *Macaca nemestrina*. Folia Primatol. 24:16-23.
10010. Sisk, CL. 1986. A decrease in responsiveness to testosterone negative feedback mediates the pubertal increase in luteinizing hormone (LH) pulse frequency in male ferrets. Biol Reprod. 34(Suppl):149.
10011. Sisk, CL & Desjardins, C. 1986. Pulsatile release of luteinizing hormone and testosterone in male ferrets. Endocrinology. 119:1195-1203.
10012. Sisk, CL & Ellis, GB. 1983. Episodic release of luteinizing hormone (LH) in intact and castrated male ferrets. Biol Reprod. 28(Suppl 1):66.
10013. Sivashankar, AK & Prasad, MRN. 1968. Seasonal variations in cholesterol concentration in the testis, adrenal, liver, and plasma of the Indian palm squirrel *Funambulus pennanti* (Wroughton). Gen Comp Endocrinol. 10:399-408.
10014. Sivertsen, E. 1941. On the biology of the harp seal, *Phoca groenlandica* Erxl, investigations carried out in the White Sea. Hvalråd Skr. 26:1-166.
10015. Siwela, AA & Tam, WH. 1981. Metabolism of androgens by the active and inactive prostate gland, and the seasonal change in systemic androgen levels in the grey squirrel (*Sciurus carolinensis* Gmelin). J Endocrinol. 88:381-391.
10016. Skarén, U. 1964. Zur Fortplanzungsbiologie des Waldlemmings. Arch Soc Zool Bot Fenn Vanamo. 18(Suppl):17-28.
10017. Skarén, U. 1973. Spring moult and onset of the breeding season of the commmon shrew (*Sorex araneus* L) in central Finland. Acta Theriol. 18:443-458.
10018. Skarén, U. 1978. Peltomyyran (*Microtus agrestis* L) talvili saantyminen Pohjois-savossa. Savon Luonto. 10:57-61. [Cited by Hansson, 1984]
10019. Skarén, U. 1979. Variation, breeding and moulting in *Sorex isodon* Turov in Finland. Acta Zool Fenn. 159:1-30.
10020. Skeldon, PC. 1973. Breeding cheetahs *Acinonyx jubatus* at Toledo Zoo. Int Zoo Yearb. 13:151-152.
10021. Skerten, R. 1972. Breeding black spider monkeys *Ateles paniscus paniscus* at Paignton Zoo. Int Zoo Yearb. 12:48-49.
10022. Skinner, JD. 1971. The sexual cycle of the impala ram *Aepyceros melampus* Lichtenstein. Zool Afr. 6:75-84.
10023. Skinner, JD. 1973. An appraisal of the status of certain antelope for game farming in South Africa. Z Tierzuchtg Züchtgsbiol. 90:263-277.
10024. Skinner, JD. 1976. Ecology of the brown hyaena *Hyaena brunnea* in the Transvaal with a distribution map for southern Africa. S Afr J Sci. 72:262-269.
10025. Skinner, JD. 1980. Productivity of mountain reedbuck *Redunca fulvorufula* (Afzelius, 1815) at the Mountain Zebra National Park. Koedoe. 23:123-130.
10026. Skinner, JD. 1984. Selected species of ungulates for game farming in South Africa. Acta Zool Fenn. 172:219-222.
10027. Skinner, JD. 1985. Wildlife management in practice: Conservation of ungulates through protection or utilization. Symp Zool Soc London. 54:25-46.
10028. Skinner, JD, Breytenbach, GJ, & Maberly, CTA. 1976. Observations on the ecology and biology of the bushpig

Potamochoerus porcus Linn in the northern Transvaal. S Afr J Wildl Res. 6:123-128.
10029. Skinner, JD, Dott, HM, de Vos, V, et al. 1980. On the sexual cycle of mature bachelor bontebok rams at the Bontebok National Park, Swellendam. S Afr J Zool. 15:117-120.
10030. Skinner, JD & Hall-Martin, AJ. 1975. A note on foetal growth and development of the giraffe *Giraffa camelopardalis giraffa*. J Zool. 177:73-79.
10031. Skinner, JD & Huntley, BJ. 1971. A report on the sexual cycle in the kudu bull *Tragelaphus strepsiceros* Pallas and a description of an intersex. Zool Afr. 6:293-299.
10032. Skinner, JD & Huntley, BJ. 1971. The sexual cycle of the blesbok ram *Damaliscus dorcas phillipsi*. Agroanimalia. 3:23-26.
10033. Skinner, JD & Ilani, G. 1979. The striped hyaena *Hyaena hyaena* of the Judean and Negev deserts and a comparison with the brown hyaena *H brunnea*. Israel J Zool. 28:229-232.
10034. Skinner, JD, Nel, JAJ, & Millar, RP. 1977. Evolution of time of parturition and differing litter sizes as an adaptation to changes in environmental conditions. Proc Int Symp Comp Biol Reprod. 4:39-44.
10035. Skinner, JD, Scorere, JA, & Millar, RP. 1975. Observations on the reproductive physiological status of mature herd bulls, bachelor bulls, and young bulls in the hippopotamus *Hippopotamus amphibius amphibius* Linnaeus. Gen Comp Endocrinol. 26:92-95.
10036. Skinner, JD, van Aarde, RJ, & van Jaarsveld, AS. 1984. Adaptations in three species of large mammals (*Antidorcas marsupialis*, (*Hystrix africaeaustralis*, *Hyaena brunnea*) to arid environments. S Afr J Zool. 19:82-86.
10037. Skinner, JD & van Zyl, JHM. 1969. Reproductive performance of the common eland, *Taurotragus oryx*, two environments. J Reprod Fert Suppl. 6:319-322.
10038. Skinner, JD & van Zyl, JHM. 1970. A study of growth of springbok ewes. African Wildlife. 24:149-154.
10039. Skinner, JD & van Zyl, JHM. 1970. The sexual cycle of the springbok ram (*Antidorcas marsupialis marsupialis*, Zimmermann). Proc S Afr Soc Anim Prod. 9:197-202.
10040. Skinner, JD & van Zyl, JHM. 1971. The post-natal development of the reproductive tract of the springbok ram lamb *Antidorcas marsupialis marsupialis* Zimmerman. Zool Afr. 6:301-311.
10041. Skinner, JD, van Zyl, JHM, & Oates, LG. 1974. The effect of season on the breeding cycle of plains antelope of the western Transvaal highveld. J S Afr Wildl Mgt Assoc. 4:15-23.
10042. Skinner, JD, van Zyl, JHM, & van Heerden, JHA. 1973. The effect of season on reproduction in the black wildebeest and red hartebeest in South Africa. J Reprod Fert Suppl. 19:101-110.
10043. Skinner, MP. 1922. The prong-horn. J Mamm. 3:82-105.
10044. Skira, IJ. 1978. Reproduction of the rabbit, *Oryctolagus cuniculus* (L), on Macquarie Island, subantarctic. Aust Wildl Res. 5:317-326.
10045. Sklenář, J. 1963. Rozmnozování netopýrū velkých (*Myotis myotis* Borkh). Lynx. 2:29-37.
10046. Skoczen, S. 1959. Determination of the sex of moles (*Talpa europaea* L) by means of their external features. Acta Theriol. 2:290-292.
10047. Skogland, T. 1984. The effects of food and maternal conditions on fetal growth and size in wild reindeer. Rangifer. 4:39-46.
10048. Skogland, T. 1986. Sex ratio variation in relation to maternal condition and parental investment in wild reindeer *Rangifer t tarandus*. Oikos. 46:417-419.
10049. Skowron-Cendrzak, A. 1956. Sexual maturation and reproduction in *Myocastor coypus* Part 1: The oestrous cycle. Folia Biol. 4:119-150.
10050. Skreb, N. 1954. Experimentelle Untersuchungen über die äusseren Ovulationsfaktoren bei der Fledermäusen *Nyctalus noctula*. Naturwissensch. 41:484.
10051. Skreb, N. 1955. Influence des gonadotrophines sur les Noctules (*Nyctalus noctula*) en hibernation. CR Soc Biol (Paris). 149:71-74.
10052. Skrivan, M, Stolc, L, & Louda, F. 1980. A study of reproduction factors in blue foxes. Scientifur. 4:25-31.
10053. Skryja, DD. 1974. Reproductive biology of the least chipmunk (*Eutamias minimus operarius*) in southeastern Wyoming. J Mamm. 55:221-224.
10054. Skryja, DD. 1978. Reproductive inhibition in female cactus mice (*Peromyscus eremicus*). J Mamm. 59:543-550.
10055. Skryja, DD & Clark, TW. 1970. Reproduction, seasonal changes in body weight, fat deposition, spleen and adrenal gland weight of the golden-mantled ground squirrel, *Spermophilus lateralis lateralis*, (Sciuridae) in the Laramie Mountains, Wyoming. Southwest Nat. 15:201-208.
10056. Skuncke, F. 1949. Ålgen. PA Norstedt & Søners Forlag, Stockholm. [Cited by Kraft, 1964]
10057. Slade, H. 1903. On the mode of copulation of the Indian elephant. Proc Zool Soc London. 1903:111-113.
10058. Slatkin, M & Hausfater, G. 1976. A note on the activities of a solitary male baboon. Primates. 17:311-322.
10059. Slaughter, L. 1971. Gestation period of the dorcus gazelle. J Mamm. 52:480-481.
10060. Sleptzov, MM. 1940. On some particularities of birth and nutrition of the young of the Black Sea porpoise *Delphinus delphis*. Zool Zhurn. 19:297-305.
10061. Sleptzov, MM. 1941. The biology of reproduction of the Black Sea dolphin *Delphinius delphis*. Zool Zhurn. 20:632-653.
10062. Sleptzov, MM. 1943. On the biology of reproduction of Pinnipedia of the Far East. Zool Zhurn. 22:109-128.
10063. Slijper, EJ. 1949. On some phenomena concerning pregnancy and parturition of the Cetacea. Bijdr Dierkd. 28:416-448.
10064. Slikker, W, Jr, Newport, GD, Slijper, J, et al. 1982. Placental transfer of synthetic and endogenous estrogen conjugates in the rhesus monkey (*Macaca mulatta*). Am J Primatol. 2:385-399.
10065. Slob, AK, Ooms, MP, & Vreeburg, JTM. 1979. Annual changes in serum testosterone in laboratory housed male stumptail macaques (*M arctoides*). Biol Reprod. 20:981-984.
10065a. Slooten, E. 1991. Age, growth, and reproduction in Hector's dolphin. Can J Zool. 69:1689-1700.
10065b. Slooten, E & Dawson, SM. 1988. Studies on Hector's dolphin, *Cephalorhynchus hectori*: A progress report. Rep Int Whal Comm Sp Issue. 9:325-338.
10066. Sludskij, AA. 1948. Ondatra i akklimatizatia ee v Kazachstane. Isdatelstvo akademic nauk Kazachsk, Alma-Ata. [Cited by Knechtel & Pieckocki, 1983]
10067. Sluiter, JW. 1954. Sexual maturity in bats of the genus *Myotis* II: Females of *M mystacinus* and supplementary data on female *M myotis* and *M emarginatus*. Proc K Ned Akad Wet Ser C. 57:696-700.
10068. Sluiter, JW. 1960. Reproductive rate of the bat *Rhinolophus hipposideros*. Proc K Ned Akad Wet Ser C. 63:383-393.
10069. Sluiter, JW & Bels, L. 1951. Follicular growth and spontaneous ovulation in captive bats during the hibernation period. Proc K Ned Akad Wet Ser C. 54:585-593.

10070. Sluiter, JW & Bouman, M. 1951. Sexual maturity in bats of the genus *Myotis*: I Size and histology of the reproductive organs during hibernation in connection with age and wear of teeth in female *Myotis myotis* and *Myotis emarginatus*. Proc K Ned Akad Wet Seri C. 54:594-601.
10071. Sluiter, JW & van Heerdt, PF. 1966. Seasonal habits of the noctuli bat (*Nyctalus noctula*). Arch Zool Nrl. 16:423-439.
10072. Sly, DL, Harbaugh, DW, London, WT, et al. 1983. Reproductive performance of a laboratory breeding colony of patas monkeys (*Erythrocebus patas*). Am J Primatol. 4:23-32.
10073. Smale, L. 1988. Influence of male gonadal hormones and familiarity on pregnancy interruption in prairie voles. Biol Reprod. 39:28-31.
10074. Smale, L, Nelson, RJ, & Zucker, I. 1988. Daylength influences pelage and plasma prolactin concentrations but not reproduction in the prairie vole, *Microtus ochrogaster*. J Reprod Fert. 83:99-109.
10075. Small, MF. 1982. Reproductive failure in macaques. Am J Primatol. 2:137-147.
10076. Small, MF. 1984. Aging and reproductive success in female *Macaca mulatta*. Female Primates: Studies by Women Primatologists. 249-259, Small, MF, ed, Alan R Liss, Inc, New York.
10077. Small, MF & Smith, DG. 1984. Sex differences in maternal investment by *Macaca mulatta*. Behav Ecol Sociobiol. 14:313-314.
10078. Small, MF & Smith, DG. 1986. The influence of birth timing upon infant growth and survival in captive rhesus macaques (*Macaca mulatta*). Int J Primatol. 7:289-304.
10078a. Smalley, AE. 1959. Pigmy sperm whale in Georgia. J Mamm. 40:452.
10079. Smiet, AC, Fulk, GW, & Lathiya, SB. 1979. Wild boar ecology in Thatta district: A preliminary study. Pak J Zool. 11:295-302.
10080. Smith, A. 1984. Demographic consequences of reproduction, dispersal and social interaction in a population of leadbeaters possum (*Gymnobelideus leadbeateri*). Possums and Gliders. 359-373, Smith, AP & Hume, ID, eds, Aust Mamm Soc, Sydney.
10081. Smith, A & Lee, AK. 1984. The evolution of strategies for survival and reproduction in possums and gliders. Possums and Gliders. 27-33, Smith, AP & Hume, ID, eds, Aust Mamm Soc, Sydney.
10082. Smith, AJ, Clausen, OPF, Kirkhus, B, et al. 1984. Seasonal changes in spermatogenesis in the blue fox (*Alopex lagopus*) quantified by DNA flow cytometry and measurement of soluble Mn^{2+}-dependent adenylate cyclase activity. J Reprod Fert. 72:453-461.
10083. Smith, AJ, Mondain-Monval, M, Andersen Berg, K, et al. 1987. Sexual development in the immature male blue fox (*Alopex lagopus*), investigated by testicular histology, DNA flow cytometry and measurement of plasma FSH, LH, testosterone and soluble testicular Mn^{2+}-dependent adenylate cyclase activity. J Reprod Fert. 81:505-515.
10084. Smith, AJ, Mondain-Monval, M, Andersen-Berg, K, et al. 1987. Effects of melatonin implantation on spermatogenesis, the moulting cycle and plasma concentrations of melatonin, LH prolactin and testosterone in the male blue fox (*Alopex lagopus*). J Reprod Fert. 79:379-390.
10085. Smith, AJ, Mondain-Monval, M, Andersen Berg, K, et al. 1987. Sexual development in the immatrue male blue fox (*Alopex lagopus*), investigated by testicular histology, DNA flow cytometry and measurement of plasma FSH, LH, testosterone and soluble testicular Mn^{2+}-dependent adenylate cyclase activity. J Reprod Fert. 81:505-515.
10086. Smith, AJ, Mondain-Monval, M, Møller, OM, et al. 1985. Seasonal variations of LH, prolactin, androstenedione, testosterone and testicular FSH binding in the male blue fox (*Alopex lagopus*). J Reprod Fert. 74:449-458.
10087. Smith, AJ, Mondain-Monval, M, Simon, P, et al. 1987. Preliminary studies of the effects of bromocriptine on testicular regression and the spring moult in a seasonal breeder, the male blue fox (*Alopex lagopus*). J Reprod Fert. 81:517-524.
10088. Smith, AP & Hume, ID. 1984. Possums and Gliders. Aust Mamm Soc, Surrey Beatty & Sons, NSW, Australia.
10088a. Smith, AT. 1974. The distribution and dispersal of pikas: Consequences of insular population structure. Ecology. 55:1112-1119.
10089. Smith, AT. 1978. Comparative demography of pikas (*Ochotona*): Effect of spatial and temporal age-specific mortality. Ecology. 59:133-139.
10090. Smith, AT. 1981. Population dynamics of pikas. Proc World Lagomorph Conf, Univ Guelph, ONT. 1979:572-586.
10091. Smith, AT. 1982. Population and reproductive trends of *Peromyscus gossypinus* in the Everglades of south Florida. Mammalia. 46:467-475.
10092. Smith, AT & Ivins, BL. 1983. Reproductive tactics of pikas: Why have two litters?. Can J Zool. 61:1551-1559.
10093. Smith, AT & Ivins, BL. 1984. Spatial relationships and social organization in adult pikas: A facultatively monogamous mammal. Z Tierpsychol. 66:289-308.
10094. Smith, AT & Weston, ML. 1990. *Ochotona princeps*. Mamm Species. 352:1-8.
10095. Smith, BW & McManus, JJ. 1975. The effects of litter size on the bioenergetics and water requirements of lactating *Mus musculus*. Comp Biochem Physiol. 51A:111-115.
10096. Smith, CC. 1965. Ageing criteria in the spring hare, *Pedetes capensis*, Forster. Arnoldia (Rhodesia). 1(26):1-6.
10097. Smith, CC. 1968. The adaptive nature of social organization in the genus of tree squirrels *Tamiasciurus*. Ecol Monogr. 38:31-63.
10098. Smith, CC. 1981. The invisible niche of *Tamiasciurus*: An example of nonpartitioning of resources. Ecol Monogr. 51:343-363.
10099. Smith, CF. 1936. Notes on the habits of the long-tailed harvest mouse. J Mamm. 17:274-278.
10100. Smith, CF. 1940. Weights of pocket gophers. J Mamm. 21:220.
10101. Smith, DA & Foster, JB. 1957. Notes on the small mammals of Churchill, Manitoba. J Mamm. 38:98-115.
10102. Smith, DA & Smith, LC. 1975. Oestrus, copulation, and related aspects of reproduction in female eastern chipmunks, *Tamias striatus* (Rodentia: Sciuridae). Can J Zool. 53:756-767.
10103. Smith, DG. 1980. Potential for cumulative inbreeding and its effects upon survival in captive groups of non-human primates. Primates. 21:430-436.
10104. Smith, DG. 1981. The association between rank and reproductive success of male rhesus monkeys. Am J Primatol. 1:83-90.
10105. Smith, DR. 1954. The bighorn sheep in Idaho. ID Dept Fish Game Wildl Bull. 1:1-154.
10106. Smith, E. 1956. Pregnancy in the little brown bat. Am J Physiol. 185:61-64.
10107. Smith, EW. 1951. Seasonal response of follicles in the ovaries of the bat *Myotis grisecens* to pregnancy urine gonadotrophin. Endocrinology. 49:67-72.

10108. Smith, GC. 1984. The biology of the yellow-footed antechinus, *Antechinus flavipes* (Marsupialia: Dasyuridae), in a swamp forest on Kiriaba Island, Cooloola, Queensland. Aust Wildl Res. 11:465-480.
10109. Smith, GC. 1985. Biology and habitat usage of sympatric populations of the fawn-footed melomys (*Melomys cervinipes*) and the grassland melomys (*M butoni*) (Rodentia: Muridae). Aust Zool. 21:551-563.
10110. Smith, GW & Johnson, DR. 1985. Demography of a Townsend ground squirrel population in southwestern Idaho. Ecology. 66:171-178.
10110a. Smith, HD & Jorgensen, CD. 1975. Reproductive biology of North American desert rodents. Monog Biol. 28:305-330.
10111. Smith, HD, Richins, GH, & Jorgensen, CD. 1978. Growth of *Dipodomys ordii* (Rodentia: Heteromyidae). Great Basin Nat. 38:215-221.
10112. Smith, HR, Sloan, RJ, & Walton, GS. 1981. Some management implications between harvest rate and population resiliency of the muskrat (*Ondatra zibethicus*). Proc Worldwide Furbearer Conf. 1:425-442.
10112a. Smith, JD & Genoways, HH. 1974. Bats of Margarita Island, Venezuela, with zoogeographic comments. Bull S CA Acad Sci. 73:64-79.
10113. Smith, JD & Hood, CS. 1980. Additional material of *Rhinolophus ruwenzorii* Hill, 1942, with comments on its natural history and taxonomic status. Proc Int Bat Res Conf. 5:163-171.
10114. Smith, JD & Hood, CS. 1981. Preliminary notes on bats from the Bismarck Archipelago (Mammalia: Chiroptera). Sci New Guinea. 8:81-121.
10115. Smith, JG. 1952. Food habits of mule deer in Utah. J Wildl Manage. 16:148-155.
10116. Smith, JG, Hanks, J, & Short, RV. 1969. Biochemical observation on the corpora lutea of the African elephant, *Loxodonta africana*. J Reprod Fert. 20:111-117.
10117. Smith, JR, Watts, CHS, & Crichton, EG. 1972. Reproduction in the Australian desert rodents *Notomys alexis* and *Pseudomys australis*. Aust Mamm. 1:1-7.
10118. Smith, LJ. 1968. A note on the birth of a white rhinoceros *Diceros simus* at Pretoria Zoo. Int Zoo Yearb. 8:134.
10119. Smith, MH. 1967. Mating behavior of *Peromyscus polionotus*. Q J FL Aca Sci. 30:230-240.
10120. Smith, MJ. 1971. Breeding the sugar glider *Petaurus breviceps* in captivity and growth of pouch young. Int Zoo Yearb. 11:26-28.
10121. Smith, MJ. 1973. *Petaurus breviceps*. Mamm Species. 30:1-5.
10122. Smith, MJ. 1979. Notes on reproduction and growth in the koala, *Phascolarctos cinereus* (Goldfuss). Aust Wildl Res. 6:5-12.
10123. Smith, MJ. 1984. Observations on the reproductive system and paracloacal glands of *Cercartetus lepidus* (Marsupialia: Burramyidae). Aust Mamm. 7:175-178.
10124. Smith, MJ. 1984. The reproductive system and paracloacal glands of *Petaurus breviceps* and *Gymnobelideus leadbeateri* (Marsupialia: Petauridae). Possums and Gliders. 321-330, Smith, AJ & Hume, ID, eds, Aust Mamm Soc, Sydney.
10125. Smith, MJ. 1989. Release of embryonic diapause in the brush-tailed bettong, *Bettongia penicillata*. Kangaroos, Wallabies and Rat-Kangaroos. 317-321, Grigg, G, Jarman, P, & Hume, I, Surrey Beatty & Sons, NSW, Australia.
10126. Smith, MJ, Bennett, JH, & Chesson, CM. 1978. Photoperiod and some other factors affecting reproduction in female *Sminthopsis crassicaudata* (Gould) (Marsupialia: Dasyuridae) in captivity. Aust J Zool. 26:449-463.
10127. Smith, MJ, Brown, BK, & Frith, HJ. 1969. Breeding of the brush tailed possum, *Trichosurus vulpecula* (Kerr) in New South Wales. CSIRO Wildl Res. 14:181-193.
10128. Smith, MH & McGinnis, JT. 1968. Relationships of latitude, altitude, and body size to litter size and mean annual production of offspring in *Peromyscus*. Res Pop Ecol. 10:115-126.
10129. Smith, MJ & Godfrey, GK. 1970. Ovulation induced by gonadotrophins in the marsupial, *Sminthopsis crassicaudata* (Gould). J Reprod Fert. 22:41-48.
10130. Smith, MJ & How, RA. 1973. Reproduction in the mountain possum, *Trichosurus caninus* (Ogilby), in captivity. Aust J Zool. 21:321-329.
10131. Smith, MJ & Sharman, GB. 1969. Development of dormant blastocysts induced by oestrogen in the ovariectomized marsupial *Macropus eugenii*. Aust J Biol Sci. 22:171-190.
10132. Smith, MSR. 1966. Studies on the Weddell seal in McMurdo Sound, Antarctica. Ph D Diss. Univ Canterbury. [Cited by Stirling, 1971]
10133. Smith, NB & Barkalow, FS. 1967. Precocious breeding in the gray squirrel. J Mamm. 48:328-330.
10134. Smith, NS & Buss, IO. 1973. Reproductive ecology of the female African elephant. J Wildl Manage. 37:524-534.
10135. Smith, NS & Buss, IO. 1975. Formation, function, and persistence of the corpora lutea of the African elephant (*Loxodonta africana*). J Mamm. 56:30-43.
10136. Smith, NS & Sowls, LK. 1975. Fetal development of the collared peccary. J Mamm. 56:619-625.
10137. Smith, PE. 1954. Continuation of pregnancy in rhesus monkeys (*Macaca mulatta*) following hypophysectomy. Endocrinology. 55:655-664.
10138. Smith, RA & Kennedy, ML. 1985. Demography of the raccoon (*Procyon lotor*) at Land Between The Lakes. Trans KE Acad Sci. 46:44-45.
10139. Smith, RE. 1958. Natural history of the prairie dog in Kansas. Univ KS Mus Nat Hist Misc Publ. 16:1-36.
10140. not used.
10141. Smith, RFC. 1969. Studies on the marsupial glider, *Schoinobates volans* (Kerr) 1 Reproduction. Aust J Zool. 17:625-636.
10142. Smith, RR. 1976. Observations on the Behaviour of a Small Captive Group of Gelada Baboons (*Theropithecus gelada*). Project for a Fellowship Inst Anim Tech. Unpublished.
10143. Smith, RR & Credland, PF. 1977. Menstrual and copulatory cycles in the gelada baboon *Theropithecus gelada*. Int Zoo Yearb. 17:183-185.
10144. Smith, RW. 1940. The land mammals of Nova Scotia. Am Mid Nat. 24:213-241.
10145. Smith, TG. 1973. Population dynamics of the ringed seal in the Canadian eastern Arctic. Bull Fish Res Bd Can. 181:1-55.
10146. Smith, TG & Hammill, MO. 1981. Ecology of the ringed seal, *Phoca hispida*, in its fast ice breeding habitat. Can J Zool. 59:966-981.
10147. Smith, VS, Ficken, R, Latchaw, P, et al. 1967. The influence of the mature male on the menstrual cycle of the female baboon. The Baboon in Medical Research. 2:621-624, Vagtborg, H, ed, Univ TX Press, Austin.
10148. Smith, WC & Reese, WC, Jr. 1968. Characteristics of a beagle colony I Estrous cycle. Lab Anim Care. 18:602-606.
10149. Smith, WJ, Smith, SL, Oppenheimer, EC, et al. 1973. Behavior of a captive population of black-tailed prairie dogs: Annual cycle of social behavior. Behaviour. 46:189-220.

10150. Smith, WP. 1991. *Odocoileus virginianus*. Mamm Species. 388:1-13.
10151. Smith, WW. 1954. Reproduction in the house mouse, *Mus musculus* L, in Mississippi. J Mamm. 35:509-515.
10152. Smithe, GE. 1939. Growth of fox foetus and length of gestation period. Canadian Silver Fox and Fur. 5:30.
10153. Smithers, RHN. 1966. The Mammals of Rhodesia, Zambia and Malawi. Collins, London.
10154. Smithers, RHN. 1971. The mammals of Botswana. Mus Mem Nation Mus Rhodesia. 4:1-340.
10155. Smithers, RHN. 1977. The serval-a spotted cat. Fauna Flora (Pretoria). 30:4-5.
10156. Smithers, RH. 1978. The regal caracal. Fauna Flora (Pretoria). 33:6-7.
10157. Smithers, RHN. 1978. The serval *Felis serval* Schreber, 1776. S Afr J Wildl Res. 8:29-37.
10158. Smithers, RHN. 1983. The Mammals of the Southern African Subregion. Univ Pretoria, Pretoria, S Africa.
10159. Smolen, MJ. 1981. *Microtus pinetorum*. Mamm Species. 147:1-7.
10160. Smolen, MJ, Genoways, HH, & Baker, RJ. 1980. Demographic and reproductive parameters of the yellow-cheeked pocket gopher (*Pappogeomys castanops*). J Mamm. 61:224-236.
10161. Smolen, MJ & Keller, BL. 1987. *Microtus longicaudus*. Mamm Species. 271:1-7.
10162. Smolenski, AJ & Rose, RW. 1988. Comparative lactation in two species of rat-kangaroo (Marsupialia). Comp Biochem Physiol. 90A:459-463.
10162a. Smuts, B & Nicholson, N. 1989. Reproduction in wild female olive baboons. Am J Primatol. 19:229-246.
10163. Smuts, GL. Personal communication. [Cited by Mentis, 1972]
10164. Smuts, GL. 1975. Home range sizes for Burchell's zebra *Equus burchelli antiquorum* from the Kruger National Park. Koedoe. 18:139-146.
10165. Smuts, GL. 1975. Reproduction and population characteristics of elephants in the Kruger National Park. J S Afr Wildl Mgmt Assoc. 5:1-10.
10166. Smuts, GL. 1975. Pre- and postnatal growth phenomena of Burchell's zebra *Equus burchelli antiquorum*. Koedoe. 18:69-102.
10167. Smuts, GL. 1976. Population characteristics of Burchell's zebra (*Equus burchelli antiquorum*, H Smith, 1841) in the Kruger National Park. S Afr J Wildl Res. 6:99-112.
10168. Smuts, GL. 1976. Reproduction in the zebra mare *Equus burchelli antiquorum* from the Kruger National Park. Koedoe. 19:89-132.
10169. Smuts, GL. 1976. Reproduction in the zebra stallion (*Equus burchelli antiquorum*) from the Kruger National Park. Zool Afr. 11:207-220.
10170. Smuts, GL, Hanks, J, & Whyte, IJ. 1978. Reproduction and social organization of lions from the Kruger National Park. Carnivore. 1(1):17-28.
10171. not used.
10172. Smuts, GL, Robinson, GA, & Whyte, IJ. 1980. Comparative growth of wild male and female lions (*Panthera leo*). J Zool. 190:365-373.
10173. Smuts, GL & Whyte, IJ. 1981. Relationships between reproduction and environment in the hippopotamus *Hippopotamus amphibius* in the Kruger National Park. Koedoe. 24:169-185.
10174. Smyth, DR & Philpott, CM. 1968. A field study of the rabbit bandicoot, *Macrotis lagotis*, Marsupialia, from Central Western Australia. Trans R Soc S Aust. 92:3-15.
10175. Smyth, M. 1966. Winter breeding in woodland mice, *Apodemus sylvaticus*, and voles, *Clethrionomys glareolus* and *Microtus agrestis*, near Oxford. J Anim Ecol. 35:471-485.
10176. Smythe, N. 1978. The natural history of the Central American agouti (*Dasyprocta punctata*). Smithsonian Contr Zool. 257:1-52.
10177. Snigirevskaya, EM. 1962. Biology of the *Eutamias sibericus* Lax on the Amur-Zeya plateau. Zool Zhurn. 41:1395-1401.
10178. Snow, CC. 1967. Some observations on the growth and development of the baboon. The Baboon in Medical Research. 2:187-189, Vagtborg, V, ed, Univ TX, Austin.
10179. Snow, JL, Jones, JK, Jr, & Webster, WD. 1980. *Centurio senex*. Mamm Species. 138:1-3
10180. Snowdon, CT, Savage, A, & McConnel, PB. 1985. A breeding colony of cotton-top tamarins (*Saguinus oedipus*). Lab Anim Sci. 35:477-480.
10181. Snyder, DP. 1956. Survival rates, longevity, and population fluctuations in the white-footed mouse, *Peromyscus leucopus*, in southeastern Michigan. Misc Publ Mus Zool Univ MI. 95:1-33.
10182. Snyder, DP. 1982. *Tamias striatus*. Mamm Species. 168:1-8.
10183. Snyder, LL. 1924. Some details on the life history and behavior of *Napaeozapus insignis abietorum* (Preble). J Mamm. 5:233-237.
10184. Snyder, PA. 1972. Behavior of *Leontopithecus rosalia* (the golden marmoset) and related species: A review. Saving the Lion Marmoset. 23-49, Bridgwater, DD, ed, The Wild Animal Propagation Trust, Wheeling, WV.
10185. Snyder, PA. 1974. Behavior of *Leontopithecus rosalia* (golden lion marmosets) and related species: A review. J Hum Evol. 3:109-122.
10186. Snyder, RL. 1985. The laboratory woodchuck. Lab Animal. 14(1):20-32.
10187. Snyder, RL & Christian, JJ. 1960. Reproductive cycle and litter size of the woodchuck. Ecology. 41:647-656.
10188. Snyder, RL, Davis, DE, & Christian, JJ. 1961. Seasonal changes in the weights of woodchucks. J Mamm. 42:297-312.
10189. Snytko, EG, Bernatskii, VG, & Nosova, NG. 1982. The study of mechanisms regulating ovulation in sables. Nauch Trudy Nauch: Issledo Inst Pushn Zverod Krolikov. 17:76-78. [Abstract in Scientifur 6:29, 1982]
10190. Soares, MJ & Hoffmann, JC. 1981. Seasonal reproduction in the mongoose, *Herpestes auropunctatus* I Androgen, luteinizing hormone, and follicle-stimulating hormone in the male. Gen Comp Endocrinol. 44:350-358.
10191. Soares, MJ & Hoffmann, JC. 1982. Seasonal reproduction in the mongoose, *Herpestes auropunctatus* II Testicular responsiveness to luteinizing hormone. Gen Comp Endocrinol. 47:226-234.
10192. Soares, MJ & Hoffmann, JC. 1982. Seasonal reproduction in the mongoose, *Herpestes auropunctatus* III Regulation of gonadotropin secretion in the male. Gen Comp Endocrinol. 47:235-242.
10193. Soares, MJ & Hoffmann, JC. 1982. Melatonin suppression of postcastration serum luteinizing hormone and follicle-stimulating hormone responses in the male mongoose, *Herpestes auropunctatus*. Gen Comp Endocrinol. 48:525-528.
10194. Soares, MJ & Hoffmann, JC. 1982. Role of daylength in the regulation of reproductive function in the male mongoose, *Herpestes auropunctatus*. J Exp Zool. 224:365-369.

10195. Sobolevsky, EI. 1982. Distribution, numbers and some ecological features of *Phoca vitulina inisularis* (Pinnepedia, Phocidae). Zool Zhurn. 61:1901-1908.
10196. Soderquist, TR & Serena, M. 1990. Occurrence and outcome of polyoestry in wild western quolls, *Dasyurus geoffroii* (Marsupialia: Dasyuridae). Aust Mamm. 13:205-208.
10197. Soholt, LF. 1976. Development of thermoregulation in Merriam's kangaroo rat, *Dipodomys merriami*. Physiol Zool. 49:152-157.
10197a. Sohlot, LF. 1977. Consumption of herbaceous vegetation and water during reproduction and development of Merriam's kangaroo rat, *Dipodomys merriami*. Am Mid Nat. 98:445-457.
10198. Soini, P. 1982. Ecology and population dynamics of the pigmy marmoset, *Cebuella pygmaea*. Folia Primatol. 39:1-21.
10199. Soini, P. 1987. Sociosexual behavior of a free-ranging *Cebuella pygmaea* (Callitrichidae, Platyrrhini) troop during post-partum estrus of its reproductive female. Am J Primatol. 13:223-230.
10200. Sokolov, VE. 1974. *Saiga tatarica*. Mamm Species. 38:1-4.
10201. Sokolov, VE, Isaev, SI, & Pavlova, EY. 1981. A study of mechanisms regulating growth and sexual maturation in great gerbils (*Rhombomys opimus*). Zool Zhurn. 60:579-586.
10202. Sokolov, VE, Isaev, SI, & Ratnikova, OA. 1981. Reproduction and marking behaviour of the great gerbils *Rhombomys opimus*. Zool Zhurn. 60:432-437.
10203. Sokolov, VE & Ovsyanikov, NJ. 1985. Dynamics of a local population of polar foxes in the absence of commercial trade. Proc Acad Sci USSR (Translations). 279:752-754.
10204. Sokolov, VE, Serbenyuk, MA, & Galanina, TM. 1983. Recognition by bank voles (*Clethrionomys glareolus*) Schreber, 1780) of phases of the female estrous cycle according to odor signals of excretions. Dokl Akad Nauk SSSR (Translation). 270:342-345.
10205. Sollberger, DE. 1940. Notes on the life history of the eastern flying squirrel. J Mamm. 21:282-293.
10206. Sollberger, DE. 1943. Notes on the breeding habits of the eastern flying squirrel (*Glaucomys volans volans*). J Mamm. 24:163-173.
10207. Solomatin, AO. 1964. Kulan, *Equus hemionus* Pallas in the USSR. Acta Soc Zool Bohemoslov. 28:178-192.
10208. Solomon, S & Leung, K. 1972. Steroid hormones in non-human primates during pregnancy. Acta Endocrinol Suppl. 166:178-190.
10209. Soma, H & Benirschke, K. 1977. Observation on the fetus and placenta of a proboscis monkey (*Nasalis larvatus*). Primates. 18:277-284.
10210. Somerville, TT. 1891. Notes on the lemming (*Myodes lemmus*). Proc Zool Soc London. 1891:655-658.
10211. Sonenshine, DE, Lauer, DM, Walker, TC, et al. 1979. The ecology of *Glaucomys volans* (Linnaeus, 1758) in Virginia. Acta Theriol. 24:363-377.
10212. Soo-Hoo, C. 1966. A brief note on the birth of a siamang *Hylobates syndactylus* at San Francisco Zoo. Int Zoo Yearb. 6:147.
10213. Sopelak, VM & Hodgen, GD. 1984. Transitory increases of luteinizing hormone and follicle-stimulating hormone following luteectomy in hemiovariectomized primates significantly increase progesterone secretion in vitro by the subsequent corpus luteum. Biol Reprod. 31:132-140.
10214. Sopelak, VM, Lynch, A, Williams, RF, et al. 1983. Maintenance of ovulatory menstrual cycles in chronically cannulated monkeys: A rest and mobile tether assembly. Biol Reprod. 28:703-706.
10215. Soper, JD. 1923. The mammals of Wellington and Waterloo counties, Ontario. J Mamm. 4:244-252.
10216. Soper, JD. 1941. History, range, and homelife of the northern bison. Ecol Monogr. 11:347-412.
10217. Soper, JD. 1942. Mammals of Wood Buffalo Park, northern Alberta and district of Mackenzie. J Mamm. 23:119-145.
10217a. Soper, JD. 1944. The mammals of southern Baffin Island, Northwest Territories, Canada. J Mamm. 25:221-254.
10218. Soper, JD. 1946. Mammals of the northern great plains along the international boundary in Canada. J Mamm. 27:127-153.
10219. Soper, JD. 1961. Field data on the mammals of southern Saskatchewan. Can Field-Nat. 75:23-41.
10220. Soper, JD. 1961. The mammals of Manitoba. Can Field-Nat. 75:171-219.
10221. Soppela, P, Nieminen, M, Saarela, S, et al. 1986. The influence of ambient temperature on metabolism and body temperature of newborn and growing reindeer calves (*Rangifer tarandus tarandus* L). Comp Biochem Physiol. 83A:371-386.
10222. Sorenson, MF, Rogers, JP, & Baskett, TS. 1968. Reproduction and development in confined swamp rabbits. J Wildl Manage. 32:520-531.
10223. Sorenson, MW & Conaway, CH. 1964. Observations of tree shrews in captivity. Sabah Soc J. 2:77-91.
10224. Sorenson, MW & Conaway, CH. 1968. The social and reproductive behavior of *Tupaia montana* in captivity. J Mamm. 49:502-512.
10225. Soriano, PJ & Molinari, J. 1987. *Sturnira aratathomasi*. Mamm Species. 284:1-4.
10226. Soriguer, RC & Rogers, PM. 1981. The European wild rabbit in Mediterranean Spain. Proc World Lagomorph Conf, Univ Guelph, ONT, Canada. 1979:600-613.
10227. Soroker, V, Hellwing, S, & Terkel, J. 1982. Parental behaviour in male and virgin white-toothed shrews, *Crocidura russula monacha* (Soricidae, Insectivora). Z Säugetierk. 47:321-324.
10228. Sosl, CL. 1987. Evidence that a decrease in testosterone negative feedback mediates the pubertal increase in luteinizing hormone pulse frequency in male ferrets. Biol Reprod. 37:73-81.
10229. Sosnovskii, IP. 1967. Breeding the red dog or dhole *Cuon alpinis* at Moscow Zoo. Int Zoo Yearb. 7:120-122.
10230. Souloumiac, J & Canivenc, R. 1976. Contrôle hormonal du fonctionnement des glandes odorantes de la Genetta mâle (*Genetta genetta* L). Arch Biol. 87:415-428.
10231. Southern, HN & Hook, O. 1963. Notes on breeding of small mammals in Uganda and Kenya. J Zool. 140:503-515.
10232. Southwell, CJ. 1984. Variability in grouping in the eastern grey kangaroo, *Macropus giganteus* I Group density. Aust Wild Res. 11:423-435.
10233. Southwick, CH. 1958. Population characteristics of house-mice living in English corn ricks: Density relationships. Proc Zool Soc London. 131:163-175.
10234. Southwick, CH, Beg, MA, & Siddiqi, MR. 1965. Rhesus monkeys in north India. Primate Behavior Field Studies of Monkeys and Apes. 111-159, DeVore, I, ed, Holt, Rinehart & Winston, New York.
10235. Sowls, LK. 1948. The Franklin ground squirrel, *Citellus franklinii* (Sabine), and its relationship to nesting ducks. J Mamm. 29:113-137.
10236. Sowls, LK. 1957. Reproduction in the Audubon cottontail in Arizona. J Mamm. 38:234-243.

10237. Sowls, LK. 1961. Gestation period of the collared peccary. J Mamm. 42:425-426.
10238. Sowls, LK. 1966. Reproduction in the collared peccary *Tayassu tajacu*. Symp Zool Soc London. 15:155-172.
10239. Sowls, LK & Phelps, RJ. 1968. Observations on the African bushpig *Potamochoerus porcus* Linn in Rhodesia. Zoologica. 53:75-84.
10240. Sowls, LK, Smith, NS, Holtan, DW, et al. 1976. Hormone levels and corpora lutea cell characteristics during gestation in the collared peccary. Biol Reprod. 14:572-578.
10241. Sowls, LK, Smith, VR, Jenness, R, et al. 1961. Chemical composition and physical properties of the milk of the collared peccary. J Mamm. 42:245-251.
10242. Spain, AV & Heinsohn, GE. 1975. Size and weight allometry in a north Queensland population of *Dugong dugon* (Muller) (Mammalia: Sirenia). Aust J Zool. 23:159-168.
10243. Spalding, DJ. 1966. Twinning in bighorn sheep. J Wildl Manage. 30:207.
10244. Spanel-Borowski, K, Bartke, A, & Petterborg, LJ. 1983. A possible mechanism of rapid luteolysis in white-footed mice, *Peromyscus leucopus*. J Morphol. 176:225-233.
10245. Spanel-Borowski, K, Petterborg, LJ, & Reiter, RJ. 1983. Morphological and morphometric changes in the ovaries of white-footed mice (*Peromyscus leucopus*) following exposure to long or short photoperiod. Anat Rec. 205:13-19.
10246. Sparks, DR. 1968. Occurrence of milk in stomach of young jackrabbits. J Mamm. 49:324-325.
10247. Speakman, JR & Racey, PA. 1986. The influence of body condition on sexual development of male brown long-eared bats (*Plecotus auritus*) in the wild. J Zool. 210A:515-525.
10248. Speakman, JR & Racey, PA. 1987. The energetics of pregnancy and lactation in the brown long-eared bat, *Plecotus auritus*. Recent Advances in the Study of Bats. 367-393, Fenton, MB, Racey, P, & Rayner, MNV, eds, Cambridge Univ Press, New York.
10249. Spears, N & Clarke, JR. 1986. Effect of male presence and of photoperiod on the sexual maturation of the field vole (*Microtus agrestis*). J Reprod Fert. 78:231-238.
10250. Spears, N & Clarke, JR. 1987. Effect of nutrition, temperature and photoperiod on the rate of sexual maturation of the field vole (*Microtus agrestis*). J Reprod Fert. 80:175-181.
10251. Spears, N & Clarke, JR. 1987. Comparison of the gonadal response of wild and laboratory field voles (*Microtus agrestis*) to different photoperiods. J Reprod Fert. 79:75-81.
10252. Spears, N, Finley, CM, Whaling, CS, et al. 1990. Sustained reproductive responses in Djungarian hamsters (*Phodopus sungorus*) exposed to a single long day. J Reprod Fert. 88:635-643.
10253. Speelman, SR, Dawson, WM, & Phillips, RW. 1944. Some aspects of fertility in horses raised under western range conditions. J Anim Sci. 3:233-241.
10254. Speller, SW. 1977. Arctic fox. Hinterland Who's Who, Can Wildl Serv. 1-3.
10255. Spencer, AW. 1984. Food habits, grazing activities, and reproductive development of long-tailed voles, *Microtus longicaudus* (Merriam) in relation to snow cover in the mountains of Colorado. Sp Publ Carnegie Mus Nat Hist. 10:67-90.
10256. Spencer, AW & Steinhoff, HW. 1968. An explanation of geographic variation in litter size. J Mamm. 49:281-286.
10257. Spencer, CC. 1943. Notes on the life history of Rocky Mountain bighorn sheep in the Tarryall Mountains of Colorado. J Mamm. 24:1-11.
10258. Spencer, DA. 1930. An interesting caesarean operation. J Mamm. 11:84-86.
10259. Spencer, DL & Lensink, CJ. 1970. The muskox of Nunivak Island, Alaska. J Wildl Manage. 34:1-15.
10260. Spencer, HE, Jr. 1955. The black bear and its status in Maine. ME Dept Int Fish Game Game Div Bull. 4:1-55.
10260a. Spencer, JA, Watson, JM, & Graves, JAM. 1991. The X chromosome of marsupials whares a highly conserved region with eutherians. Genomics. 9:598-604.
10261. Spencer, SR & Cameron, GN. 1982. *Reithrodontomys fulvescens*. Mamm Species. 174:1-7.
10261a. Spenrath, CA & LaVal, RK. 1970. Records of bats from Querétaro and San Luis Potosi, Mexico. J Mamm. 51:395-396.
10262. Speth, RL, Princhett, CL, & Jorgensen, CD. 1968. Reproductive activity of *Perognathus parvus*. J Mamm. 49:336-337.
10263. Spiegel, A. 1929. Biologische Beobachtungen an Javamakaken, *Macacus irus* F Cuv (*cynomolgus* L). Zool Anz. 81:45-65.
10264. Spiegel, A. 1954. Beobachtungen und Untersuchungen an Javamakaken. Zool Garten NF. 20:227-270.
10264a. Spies, HG & Chappel, SC. 1984. Mammals: Nonhuman primates. Marshall's Physiology of Reproduction. 659-712 Lamming, GE, ed, Churchill Livingstone, London.
10265. Spies, HG & Niswender, GD. 1972. Effect of progesterone and estradiol on LH release and ovulation in rhesus monkeys. Endocrinology. 90:257-261.
10266. Spies, HG, Norman, RL, & Buhl, AE. 1979. Twenty-four-hour patterns in serum prolactin and cortisol after partial and complete isolation of the hypothalamic-pituitary unit in rhesus monkeys. Endocrinology. 105:1361-1368.
10267. Spillett, JJ. 1966. Population studies of the lesser bandicoot rat in Calcutta. Indian Rodent Symp. 1:84-926.
10268. Spillett, JJ. 1966. Growth of three species of Calcutta rats. Indian Rodent Symp. 2:177-196.
10269. Spinage, CA. 1969. Naturalistic observations on the reproductive and maternal behaviour of the Uganda defassa waterbuck *Kobus defassa ugandae* Neumann. Z Tierpsychol. 26:39-47.
10270. Spinage, CA. 1969. Territoriality and social organization of the Uganda defassa waterbuck *Kobus defassa ugandae*. J Zool. 159:329-361.
10271. Spinage, CA. 1986. Maternal reproduction and health in the Grant's gazelle (*Gazella granti*). J Zool. 1B:461-520.
10272. Spinage, CA & Shelley, HJ. 1981. Tissue carbohydrate reserves in adult and foetal Grant's gazelles and wildebeest. Comp Biochem Physiol. 70A:87-89.
10273. Spitzenberger, F. 1976. Beiträge zur Kenntnis von *Dryomys laniger* Felten et Storch 1968 (Gliridae, Mammalia). Z Säugetierk. 41:237-249.
10274. Spitzenberger, F & Steiner, HM. 1962. Über Insektenfresser (Insectivora) und Wühlmäuse (Microtinae) der nordturkischen Feuchtwälder. Bonn Zool Beitr. 4:284-310.
10275. Spitzenberger, F & Steiner, HM. 1964. *Prometheomys schaposchnikovi* Satunin, 1901, in Nordost-Kleinasien. Z Säugetierk. 29:116-124.
10276. Spotte, S & Adams, G. 1981. Photoperiod and reproduction in captive female northern fur seals. Mammal Rev. 11:31-35.
10277. Spotte, S & Babus, B. 1980. Does a pregnant dolphin (*Tursiops truncatus*) eat more?. Cetology. 39:1-7.
10278. Spotte, S & Schneider, J. 1982. Early functional maturity of captive male northern elephant seals (*Mirounga angustirostris*). Zoo Biol. 1:355-358.

10279. Sprankel, H. 1959. Fortpflanzung von *Tupaia glis* Diard 1820 (Tupaiidae, Prosimiaes in Gefangenschaft). Naturwissensch. 46:338.
10280. Sprankel, H. 1960. Zucht von *Tupaia glis* Diard 1820 (Tupaiidae, Prosimiae) in Gefangenschaft. Naturwissensch. 47:213.
10281. Sprankel, H. 1961. Über Verhaltensweisen und Zucht von *Tupaia glis* (Diard 1820) in Gefangenschaft. Z Wissensch Zool. 165:186-220.
10282. Spühler, O. 1935. Genitalzyklus und Spermiogenese des Mausmaki (*Microcebus murinus* Miller). Z Zellf Mikr Anat. 23:442-463.
10283. Spurgeon, CH & Brooks, RJ. 1916. The implantation and early segmentation of the ovum of *Didelphis virginiana*. Anat Rec. 10:385-395.
10284. Squibb, RC, Kimball, JF, Jr, & Anderson, DR. 1986. Bimodal distribution of estimated conception dates in Rocky Mountain elk. J Wildl Manage. 50:118-121.
10285. Sreenivasan, MA, Bhat, HR, & Geevarghese, G. 1973. Breeding cycle of *Rhinolophus rouxi*, 1835 (Chiroptera: Rhinolophidae), in India. J Mamm. 54:1013-1017.
10286. Sreenivasan, MA, Bhat, HR, & Geevarghese, G. 1974. Observations on the reproductive cycle of *Cynopterus sphinx sphinx* Vahl, 1797 (Chiroptera: Pteropidae). J Mamm. 55:200-202.
10287. Srivastava, PK, Cavazos, F, & Lucas, FV. 1970. Biology of reproduction in the squirrel monkey (*Saimiri sciureus*): I The estrus (sic) cycle. Primates. 11:125-134.
10288. Srivastava, SC. 1952. Placentation in the mouse-tailed bat, *Rhinopoma kinneari* [Chiroptera]. Proc Zool Soc Bengal. 5:105-131.
10289. St John, OB. 1891. Note on *Herpestes mungo*. Proc Zool Soc London. 1891:245.
10290. Stabenfeldt, GH & Hendrickx, AG. 1972. Progesterone levels in the bonnet monkey (*Macaca radiata*) during the menstrual cycle and pregnancy. Endocrinology. 91:614-619.
10291. Stabenfeldt, GH & Hendrickx, AG. 1973. Progesterone levels in the sooty mangabey (*Cercocebus atys*) during the menstrual cycle, pregnancy and parturition. J Med Primatol. 2:1-10.
10292. Stabenfeldt, GH & Hendrickx, AG. 1973. Progesterone studies in the *Macaca fascicularis*. Endocrinology. 92:1296-1300.
10292a. Stacey, PJ & Baird, RW. 1991. Status of the false killer whale, *Pseudorca crassidens*, in Canada. Can Field-Nat. 105:189-197.
10293. Stachrowsky, WG. 1932. Zur Biologie des Eichhörnchens in der Gefangenschaft. Zool Zhurn. 11(1):82-104.
10294. Stachrowsky, WG. 1932. Zur Biologie des Schneehasens. Zool Zhurn. 11(2):70-78.
10295. Stack, H & Kunz, TH. 1981. Body composition and energy allocation during reproduction in adult female and juvenile *Eptesicus fuscus*. 12th N Am Symp Bat Res, 15-17 Oct 1981, Cornell Univ, Ithaca, NY.
10296. Stack, H & Kunz, TH. 1982. Milk composition and mammary gland development in *Myotis lucifugus* and *Eptesicus fuscus*. Bat Res News. 23:79.
10297. Stains, HJ. 1956. The raccoon in Kansas: Natural history, management, and economic importance. Univ KS Mus Nat Hist Misc Publ. 10:1-76.
10298. Stains, HJ. 1965. Female red bat carrying four young. J Mamm. 46:333.
10299. Stainthorpe, HL. 1970. Antelope milk analysis. Lammergeyer. 12:79.
10300. Stainthorpe, HL. 1972. Observations on captive eland in the Loteni Nature Reserve. Lammergeyer. 15:27-38.
10301. Stalling, DT. 1990. *Microtus ochrogaster*. Mamm Species. 355:1-9.
10302. Stallings, JR. 1986. Notes on the reproductive biology of the grey brocket deer (*Mazama gauazoubira*) in Paraguay. J Mamm. 67:172-174.
10303. Stanczyk, FZ, Hess, DL, Namkung, PC, et al. 1986. Alterations in sex steroid-binding protein (SBP), corticosteroid-binding globulin (CBG), and steroid hormone concentrations during pregnancy in rhesus macaques. Biol Reprod. 35:126-132.
10304. Standaert, TA, Truog, WE, Guthrie, RD, et al. 1984. Growth and development and the term and premature pigtailed macaque (*Macaca nemestrina*): Cardiovascular and respiratory changes during the first 3 weeks of life. Am J Primatol. 7:107-119.
10305. Stanford, JS. 1931. Notes on small mammals of Utah. J Mamm. 12:356-363.
10306. Stanley, WC. 1963. Habits of the red fox in northeastern Kansas. Univ KS Mus Nat Hist Misc Publ. 34:1-31.
10307. Stanton, TL, Craft, CM, & Reiter, RJ. 1986. Pineal melatonin: Circadian rhythm and variations during the hibernation cycle in the ground squirrel, *Spermophilus lateralis*. J Exp Zool. 239:247-254.
10308. Starck, D. 1956. Über den Reifegrad neugeborener Ursiden im Vergleich mit anderen Carnivoren. Sonderdruck Säugetierk Mitt. 4:21-27.
10309. Starck, D. 1957. Beobachtungen an *Heterocephalus glaber* Ruppel, 1842 (Rodentia, Bathyergidae in der Provinz Harrar). Z Säugetierk. 22:50-56.
10310. Starck, D. 1959. Ontogenie and Entwicklungsphysiologie der Säugetiere. Handbuch der Zoologie. 8(22):1-276.
10311. Starck, D & Frick, H. 1958. Beobachtungen an äthiopischen Primaten. Zool Jahrb Syst. 86:41-70.
10312. Starkov, ID. 1937. The sexual cycle of the silver fox. Usp Zooteh Nauk. 3:385-401. (An Breeding Abst 6:38,1937).
10313. Starrett, A. 1972. *Cyttarops alecto*. Mamm Species. 13:1-2.
10313a. Starrett, A & de la Torre, L. 1964. Notes on a collection of bats from Central America, with the third record for *Cyttarops alecto* Thomas. Zoologica. 49:53-63.
10313b. Starrett, A & Starrett, P. 1955. Observations on young blackfish, *Globicephala*. J Mamm. 36:424-429.
10314. Start, AN. 1972. Some bats of Bako National Park, Sarawak. Sarawak Mus J. 20:371-376.
10315. States, JB. 1976. Local adaptations in chipmunk (*Eutamias amoenus*) populations and evolutionary potential at species' borders. Ecol Monogr. 46:222-256.
10316. Stavy, M & Terkel, J. 1984. Plasma testosterone and estradiol levels during pregnancy in the hare (*Lepus capensis syriacus*). Acta Zool Fennici. 171:169-171.
10317. Stavy, M, Terkel, J, & Kohen, F. 1978. Plasma progesterone levels during pregnancy and pseudopregnancy in the hare (*Lepus europaeus syriacus*). J Reprod Fert. 54:9-14.
10318. Stavy, M, Terkel, J, & Marder, M. 1976. Aspects of rearing and reproduction of the hare, *Lepus europaeus syriacus*. Israel J Zool. 25:208.
10319. Stebbings, RE. 1966. Bechstein's bat, *Myotis bechsteini* in Dorset, 1960-1965. J Zool. 148:574-576.
10320. Stebbings, RE. 1967. Identification and distribution of bats in the genus *Plecotus* in England. J Zool. 153:291-310.

10321. Stebbings, RE. 1970. A bat new to Britain, *Pipistrellus mathusii*, with notes on its identification and distribution in Europe. J Zool. 161:282-286.
10322. Stebbins, LL. 1977. Energy requirements during reproduction of *Peromyscus maniculatus*. Can J Zool. 55:1701-1704.
10323. Stefan, Y. 1967. Morphologie et structure histologique du système génital mâle d'un Rongeur d'Iran: *Ellobius lutescens* (Thomas). Arch Anat Histol Embryol Norm Exp. 40:153-168.
10324. Stefan, Y. 1967. Hypogénitalism chez un Rongeur d'Iran: *Ellobius lutescens* (Thomas): Action comparée d'injections d'hypophyses de Rat et d'*Ellobius* sur le système génital de Ratons immatures. Comp Rend Hebd Séances Acad Sci Sér D Sci Nat. 264:2487-2489.
10325. Stefan, Y. 1976. Demonstration of the existence of a single morphological type of gonadotrophic cell in *Ellobius lutescens* (Microtinae) by an ultrastructural analysis of their development under various physiological and environmental conditions. Cell Tiss Res. 167:49-64.
10326. Stefan, Y & Medilanski, J. 1976. Development of rudimentary prostates of *Ellobius lutescens* (Microtinae) in organ culture: Effects of androgens. Experientia. 32:1074-1076.
10327. Stefan, Y & Steimer, T. 1974. The Leydig cell of a hypogonadal rodent (*Ellobius lutescens*, Th): Correlation between ultrastructure and biosynthetic activity. Biol Reprod. 19:913-921.
10328. Stefan, Y & Steimer, T. 1975. Analyse expérimental du cas d'hypogonadism présente par le mâle d'*Ellobius lutescens* (Th), rongeur d'Iran. CR Séances Soc Hist Nat Genève NS. 10:94-105.
10329. Stegeman, LC. 1930. Notes on *Synaptomys cooperi cooperi* in Washtenaw county, Michigan. J Mamm. 11:460-466.
10330. Stegeman, LC. 1937. Notes on young skunks in captivity. J Mamm. 18:194-202.
10331. Steger, RW, Huang, HH, Hodson, CA, et al. 1980. Effects of advancing age on hypothalamic-hypophysial-testicular functions in the male white-footed mouse (*Peromyscus leucopus*). Biol Reprod. 22:805-809.
10332. Stehlik, J. 1971. Breeding jaguars *Panthera onca* at Ostrava Zoo. Int Zoo Yearb. 11:121-123.
10332a. Stehn, RA & Jannett, FJ, Jr. 1981. Male induced abortion in various microtine rodents. J Mamm. 62:369-372.
10333. Stehn, R & Richmond, ME. 1975. Male-induced pregnancy termination in the prairie vole, *Microtus ochrogaster*. Science. 187:1211-1213.
10334. Steigers, WD, Jr & Flinders, JT. 1980. Mortality and movements of mule deer fawns in Washington. J Wildl Manage. 44:381-388.
10335. Steimer, T. 1976. Male hypogonadism in a rodent: In vitro testosterone synthesis. Experientia. 32:775.
10336. Steimer, T. 1977. Androgen production in the testis of *Ellobius lutescens* (Th) (Microtinae) in vitro: Correlation with plasma concentrations. Gen Comp Endocrinol. 34:83.
10337. Stein, FJ. 1978. Sex determination in the common marmoset (*Callithrix jacchus*). Lab Ani Sci. 28:75-80.
10338. Stein, G. 1929. Zur Kenntnis von *Erinaceus roumanicus* B Hamilt. Z Säugetierk. 4:240-250.
10339. Stein, GHW. 1938. Biologische Studien an deutschen Kleinsäugern. Arch Naturgesch NF. 7:477-513.
10340. Stein, GHW. 1950. Über Fortpflanzungszyklus, Wurfgrösze und Lebensdauer bei einigen kleinen Nagetieren. Schädlingsbekampfung. 42:122-131.
10341. Stein, GHW. 1950. Zur Biologie des Maulwurfs, *Talpa europaea* L. Bonn Zool Beitr. 1:97-116.
10342. Stein, GHW. 1951. Populationsanalytische Untersuchungen am europäischen Maulwurf II Über zeitliche Gröszenschwankungen. Zool Jb (Syst). 79:567-590.
10343. Stein, GHW. 1952. Über Massenvermehrung und Massenzusammenbruch bei der Feldmaus Populationsanalytische Untersuchungen an deutschen Kleinsäugetieren III *Microtus arvalis*. Zool Jb (Syst). 81:1-26.
10344. Stein, GHW. 1953. Über Umweltabhängigkeiten bei der Vermehrung der Feldmaus, *Microtus arvalis* Populationanalytische Untersuchungen an deutschen kleinen Säugetieren IV. Zool Jb Syst. 81:527-547.
10345. Stein, GHW. 1953. Über das Zahlenverhältnis der Geschlechter bei der Feldmaus, *Microtus arvalis* Populationsanalytische Untersuchungen an deutschen kleinen Säugetieren V. Zool Jb Syst. 82:137-156.
10346. Stein, GHW. 1961. Beziehungen zwischen Bestandsdichte und Vermehrung bei der Waldspitzmaus, *Sorex araneus*, und weiteren Rotzahnspitzmäusen. Z Säugetierk. 26:13-28.
10347. Stein, GHW. 1975. Über die Bestandsdichte und ihre Zusammenhänge bei der Wasserspitzmaus, *Neomys fodiens* (Pennant). Mitt Zool Mus Berlin. 51:187-198.
10348. Steinbacher, G. 1953. Wurfzahl beim Schwarzwild, *Sus s scrofa* (Linne, 1758). Säugetierk Mitt. 1:80.
10349. Steinbacher, G. 1958. Trächtigkeitsdauer beim Weiszbart-Gnu (*Connochaetes taurinus*). Säugetierk Mitt. 6:173.
10350. Steinbacher, G. 1958. Zur Fortpflanzungsbiologie des Braunbären (*Ursus a arctos*). Säugetierk Mitt. 6:27-28.
10351. Steinbacher, G. 1959. Trächtigkeitsdauer beim Weiszbartgnu. Säugetierk Mitt. 7:75.
10352. Steinbacher, G. 1963. Zur Trächtigkeitsdauer beim Braunbären (*Ursus arctos*). Säugetierk Mitt. 11:20-21.
10353. Steinbacher, G. 1966. Zum Lebensalter des Schnabeligels (*Tachyglossus*) und Langschnabeligels (*Zaglossus*). Säugetierk Mitt. 14:230.
10354. Steinbacher, G. 1978. Hohes Alter eines Dybowskyhirsch-♀ *Cervus nippon dybowskii*. Säugetierk Mitt. 26:240.
10355. Steinbacher, G. 1978. Hohes Alter eines Mantelpavian-♂ (*Papio hamadryas*). Säugetierk Mitt. 26:240.
10356. Steinlechner, S, Buchberger, A, & Heldmaier, G. 1987. Circadian rhythms of pineal N-acetyltransferase activity in the Djungarian hamster, *Phodopus sungorus*, in response to seasonal changes of natural photoperiod. J Comp Physiol A. 160:593-597.
10357. Steinemann, P. 1966. Künstliche Aufzucht eines Eisbären. Zool Garten NF. 32:129-145.
10358. Steiner, HM. 1968. Untersuchungen über die Variabilität und Bionomie der Gattung *Apodemus* (Muridae, Mammalia) der Donau-Auen von Stockeran (Niederösterreich). Z Wiss Zool. 177:1-96.
10359. Steiner, RA & Bremner, WJ. 1981. Endocrine correlates of sexual development in the male monkey, *Macaca fascicularis*. Endocrinology. 109:914-919.
10360. Steiner, RA, Clifton, DK, Spies, G, et al. 1976. Sexual differentiation and feedback control of luteinizing hormone secretion in the rhesus monkey. Biol Reprod. 15:206-212.
10361. Steiner, RA, Peterson, AP, Yu, JYL, et al. 1980. Ultradian luteinizing hormone and testosterone rhythms in the adult male monkey, *Macaca fascicularis*. Endocrinology. 107:1489-1493.
10362. Steiner, RA, Schiller, HS, Barber, J, et al. 1978. Luteinizing hormone regulation in the monkey (*Macaca nemestrina*): Failure of testosterone and dihydrotestosterone to block the estrogen-induced gonadotropin surge. Biol Reprod. 19:51-56.

10363. Steinhauf, D. 1959. Beobachtungen zum Brunstverhalten des Steinwildes (*Capra ibex*). Säugetierk Mitt. 7:5-9.
10364. Steiniger, B. 1976. Beiträge zum Verhalten und zur Sociologie des Bisams (*Ondatra zibethicus* L). Z Tierpsychol. 41:55-79.
10365. Steinmetz, H. 1937. Beobachtungen über die Entwicklung junger Zwergfluszpferde im Zoologischen Garten Berlin. Zool Garten NF. 9:255-263.
10366. Steklenev, EP. 1968. Des particularités anatomo-morphologiques de la structure et des fonctions physiologiqes des trompes de Fallope chez les Camelidés (genres *Lama* et *Camelus*). Int Cong Reprod Artif Insem Paris. 6:74.
10367. Steklenev, EP. 1969. Peculiarities of *Taurotragus oryx* Pall under conditions of half-free breeding in Askaniya-Nova. Vestnik Zool (Kiev). 1969:39-44.
10368. Steklenev, EP. 1978. Seasonal reproduction of *Cervus nippon hortulorum* SW under conditions of the Ukraine south. Vestnik Zool (Kiev). 1978:28-33.
10368a. Steklis, HD, Brammer, GL, Raleigh, MJ, et al. 1985. Serum testosterone, male dominance, and aggression in captive groups of vervet monkeys (*Cercopithecus aethiops sabaeus*). Horm Behav. 19:154-163.
10369. Steklis, HD, Raleigh, MJ, Kling, AS, et al. 1986. Biochemical and hormonal correlates of dominance and social behavior in all-male groups of squirrel monkeys (*Saimiri sciureus*). Am J Primatol. 11:133-145.
10370. Stempel, N. 1985. Fortpflanzung und Jugendentwicklung der Pestratte *Bandicota indica* (Bechstein). Bonn Zool Beitr. 36:9-36.
10371. Stenger, VG. 1974. Studies on reproduction in the stump-tailed macaque. Breeding Primates. 100-104, Beveridge, WIB, ed, Karger, Basel.
10371a. Stenlund, MH. 1955. A field study of the timber wolf (*Canis lupus*) on the Superior National Forest, Minnesota. Tech Bull MN Dept Conserv. 4:1-55
10372. Stenseth, NC. 1978. Is the female biased sex ratio in wood lemming *Myopus schisticolor* maintained by cyclic breeding?. Oikos. 30:83-89.
10373. Stenseth, NC, Gustafsson, TO, & Hansson, L. 1985. On the evolution of reproductive rates in microtine rodents. Ecology. 66:1795-1808.
10374. Stephens, RJ. 1962. Histology and histochemistry of the placenta and fetal membranes in the bat, *Tadarida brasiliensis cynocephala* (with notes on maintaining pregnant bats in captivity). Am J Anat. 111:259-285.
10375. Stephens, RJ. 1969. The development and fine structure of the allantoic placental barrier in the bat *Tadarida brasiliensis cynocephala*. J Ultrastruct Res. 28:371-398.
10376. Stephens, RJ & Cabral, LJ. 1972. The diffuse labryinthine endotheliochorial placenta of the free-tail bat: A light and electron microscope study. Anat Rec. 172:221-252.
10377. Stephens, RJ & Easterbrook, N. 1969. A new cytoplasmic organelle, related to both lipid and glycogen storage materials in the yolk sac of the bat, *Tadarida brasiliensis cynocephala*. Am J Anat. 124:47-56.
10378. Stephenson, RL. 1975. Abert's squirrel reproduction in central Arizona. J AZ Acad Sci. 10:22.
10379. Stĕrba, O, Hrabĕ, V, & Zima, J. 1986. Reproduction in a population of *Pitymys subterraneus* from the Tatras under conditions of high population density. Folia Zool. 35:215-228.
10380. Stĕrba, O & Klusak, K. 1984. Reproductive biology of fallow deer, *Dama dama*. Acta Sc Nat Brno. 18:1-46.
10381. Stern, BR & Smith, CG. 1984. Sexual behaviour and paternity in three captive groups of rhesus monkeys (*Macaca mulatta*). Anim Behav. 32:23-32.
10382. Stetson, MH, Sarafidis, E, & Rollag, MD. 1986. Sensitivity of adult male Djungarian hamsters (*Phodopus sangorus sangorus*) to melatonin injections throughout the day: Effects on reproductive system and the pineal. Biol Reprod. 35:618-623.
10383. Steven, DM. 1967. Voles (*Clethrionomys*). The UFAW Handbook on the Care and Management of Laboratory Animals, 3rd ed. 316-326, E & S Livingstone, Edinburgh.
10384. Stevens, VC. 1962. Regional varition in produtivity and reproduction physiology of the cottontail rabbit in Ohio. Trans N Am Wildl Nat Res Conf. 27:243-253.
10385. Stevens, VC, Sparks, SJ, & Powell, JE. 1970. Levels of estrogens, progestogens and luteinizing hormone during the menstrual cycle of the baboon. Endocrinology. 87:658-666.
10386. Stevens, WF & Weisbrod, AR. 1981. The biology of the European rabbit on San Juan Island, Washington, USA. Proc World Lagomorph Conf, Univ Guelph. ONT. 1979:870-879.
10387. Stevenson, M. 1973. Notes on pregnancy in the sooty mangabey *Cercocebus atys*. Int Zoo Yearb. 13:134-135.
10388. Stevenson, MF. 1976. Maintenance and breeding of the common marmoset *Callithrix jacchus* with notes on hand rearing. Int Zoo Yearb. 16:110-116.
10389. Stevenson, MF. 1976. Birth and perinatal behaviour in family groups of the common marmoset (*Callithrix jacchus jacchus*), compared to other primates. J Hum Evol. 5:365-381.
10390. Stevenson, MF. 1984. The captive breeding of marmosets and tamarins. Proc Symp Assoc Br Wild Anim Keepers. 8:49-67.
10391. Stevenson, MF & Sutcliffe, AG. 1978. Breeding a second generation of common marmosets *Callithrix jacchus* in captivity. Int Zoo Yearb. 18:109-114.
10392. Stevenson-Hamilton, J. 1913. Notes on albinism in the common reedbuck (*Cervicapra arundinum*), and on the habits and geographical distribution of Sharpe's steenbuck (*Raphiceros sharpei*). Proc Zool Soc London. 1913:537-541.
10393. Stewart, BE & Stewart, REA. 1989. *Delphinapterus leucas*. Mamm Species. 336:1-8.
10394. Stewart, BS & Leatherwood, S. 1985. Minke whale *Balaenoptera acutorostrata*. Handbook of Marine Mammals. 3:91-136, Ridgway, SH & Harrison, RJ, eds, Academic Press, New York.
10395. Stewart, DRM. 1963. The Arabian oryx (*Oryx leucoryx* Pallas). E Afr Wildl J. 1:103-118.
10395a. Stewart, F. 1982. Prolactin and luteinizing hormone receptors in marsupial corpora lutea: Relationship to control of luteal function. J Endocrinol. 92:63-72.
10396. Stewart, F. 1984. Mammogenesis and changing prolactin receptor concentrations in the mammary glands of the tammar wallaby (*Macropus eugenii*). J Reprod Fert. 71:141-148.
10397. Stewart, F & Tyndale-Biscoe, CH. 1982. Prolactin and luteinizing hormone receptors in marsupial corpora lutea: Relationship to control of luteal function. J Endocrinol. 92:63-72.
10398. Stewart, F & Tyndale-Biscoe, CH. 1983. Pregnancy and parturition in marsupials. Current Topics in Experimental Endocrinology Vol 4 The Endocrinology of Pregnancy and Parturition. 1-33, Martini, L & James, VHT, eds, Academic Press, New York.
10398a. Stewart, GR & Roest, AI. 1960. Distribution and habits of kangaroo rats at Morro Bay. J Mamm. 41:126-129.
10399. Stewart, KJ. 1977. The birth of a wild mountain gorilla (*Gorilla gorilla beringli*). Primates. 18:965-976.

10400. Stewart, KJ. 1984. Parturition in wild gorillas: Behavior of mothers, neonates, and others. Folia Primatol. 42:62-69.
10401. Stewart, R, Stewart, LC, & Haigh, JC. 1987. Gestation periods in two yearling captive moose, *Alces alces*, in Saskatchewan. Can Field-Nat. 101:103-104.
10402. Stewart, REA. 1986. Energetics of age-specific reproductive effort in female harp seals *Phoca groenlandica*. J Zool. 208A:503-517.
10403. Stewart, REA. 1987. Behavioral reproductive effort of nursing harp seals, *Phoca groenlandica*. J Mamm. 68:348-358.
10404. Stewart, REA & Lavigne, DM. 1980. Neonatal growth of northwest Atlantic harp seals, *Pagophilus groenlandicus*. J Mamm. 61:670-680.
10405. Stewart, REA & Murie, DJ. 1986. Food habits of lactating harp seals (*Phoca groenlandica*) in the gulf of St Lawrence in March. J Mamm. 67:186-188.
10406. Stewart, REA, Stewart, BE, Lavigne, DM, et al. 1989. Fetal growth of northwest Atlantic harp seals, *Phoca groenlandica*. Can J Zool. 67:1247-1257.
10407. Stewart, REA, Webb, BE, Lavigne, DM, et al. 1983. Determining lactose content of harp seal milk. Can J Zool. 61:1094-1100.
10408. Stewart, RM. Personal communication. [Cited by Mentis, 1972]
10409. Stewart, RR, Stewart, LMC, & Haigh, JC. 1985. Levels of some reproductive hormones in relation to pregnancy in moose: A preliminary report. Alces. 21:393-402.
10410. Stewart, RW & Bider, JR. 1974. Reproduction and survival of ditch-dwelling muskrats in southern Quebec. Can Field-Nat. 88:429-436.
10411. Stewart, ST. 1982. Late parturition in bighorn sheep. J Mamm. 63:154-155.
10412. Steyn, TJ. 1951. The breeding of lions in captivity. Fauna Flora (Pretoria). 2:37-55.
10413. Stickel, LF. 1979. Population ecology of house mice in unstable habitats. J Anim Ecol. 48:871-887.
10414. Stickley, AR, Jr. 1961. A black bear tagging study in Virginia. Proc Ann Conf SE Assoc Game Fish Comm. 15:43-54.
10415. Stieve, H. 1948. Der Bau der Primatenplazenta. Anat Anz. 96:299-329.
10416. Stieve, H. 1948. Zur Fortpflanzungsbiologie des Igels. Verh Deut Zool Kiel (Zool Anz Suppl 13). 13:253-256.
10417. Stieve, H. 1950. Anatomisch-biologische Untersuchungen über die Fortpflanzungstätigkeit des europäischen Rehes (*Capreolus capreolus capreolus* L). Z Mikrosk Anat Forsch. 55:427-530.
10418. Stieve, H. 1950. Die Fortpfanzungsbiologie des Rehwildes. Verh Deutschen Zoologen. 1949:348-354.
10419. Stieve, H. 1952. Die Paarungszeit des Daches (*Meles meles* L). Zool Garten NF. 19:126-133.
10420. Stieve, H. 1952. Zur Fortpflanzungsbiologie des europäischen Feldhasen (*Lepus europaeus* Pallas). Zool Anz. 148:101-114.
10421. Stieve-Miegel, B. 1955. Über Superfetation bei der Bisamratte (*Ondantra zibethica* L). Z Mikr Anat Forsch. 61:82-92.
10422. Stilmark, FR. 1963. Ecology of the chipmunk (*Eutamias sibericus* Laxm), in cedar forests of western Sayan. Zool Zhurn. 42:92-102.
10423. Stirling, EC. 1889. Preliminary notes on a new Australian mammal. Trans Proc Rep R Soc S Aust. 11:21-24.
10424. Stirling, EC. 1891. Description of a new genus and species of Marsupialia, *Notoryctes typhlops*. Trans R Soc S Aust. 14:154-187.
10425. Stirling, EC. 1891. Further notes on the habits and anatomy of *Notoryctes typhlops*. Trans R Soc S Aust. 14:283-291.
10426. Stirling, I. 1969. Birth of a Weddell seal pup. J Mamm. 50:155-156.
10427. Stirling, I. 1971. *Leptonychotes weddelli*. Mamm Species. 6:1-5.
10428. Stirling, I. 1971. Population dynamics of the Weddell seal (*Leptonychotes weddelli*) in McMurdo Sound, Antarctica, 1966-1968. Antarctic Res Ser. 18:141-161.
10429. Stirling, I. 1972. Observations on the Australian sea lion, *Neophoca cinerea* (Peron). Aust J Zool. 20:271-280.
10430. Stirling, I. 1979. Ribbon seal. Mammals in the Seas, FAO Fish Ser 5. 2:81-82.
10431. Stirling, I & Archibald, R. 1979. Bearded seal. Mammals in the Seas, FAO Fish Ser 5. 2:83-85.
10432. Stirling, I & Calvert, W. 1979. Ringed seal. Mammals in the Seas, FAO Fish Ser 5. 2:66-69.
10433. Stirling, I, Calvert, W, & Andriashek, D. 1980. Population ecology studies of the polar bear in the area of southeastern Baffin Island. Can Wildl Serv Occas Pap. 44:1-31.
10434. Stirling, I, Jonkel, C, Smith, P, et al. 1977. The ecology of the polar bear (*Ursus maritimus*) along the western coast of Hudson Bay. Can Wildl Serv Occas Pap. 33:1-69.
10435. Stoch, ZG. 1955. The male genital sytem and reproductive cycle of *Elephantulus myurus jamesoni* (Chubb). Phil Trans R Soc London Biol Sci. 238B:99-126.
10436. Stockard, AH. 1929. Observations on reproduction in the white-tailed prairie dog (*Cynomys leucurus*). J Mamm. 10:209-212.
10437. Stockard, AH. 1929. Observations on the seasonal activities of the white-tailed prairie-dog, *Cynomys leucurus*. Pap MI Acad Sci Art Let. 11:471-479
10438. Stockard, AH. 1934. Studies on the female reproductive system of the prairie dog (*Cynomys leucurus*) I Gross morphology. Pap MI Acad Sci Arts Let. 20:725-735.
10439. Stockard, AH. 1936. Studies on the female reproductive system of the prairie-dog, *Cynomys leucurus* II Normal cyclic phenomena of the ovarian follicles. Pap MI Acad Sci Art Let. 22:671-689
10440. Stockard, CR & Papanicolaou, GN. 1917. The existence of a typical oestrous cycle in the guinea-pig -- with a study of its histological and physiological changes. Am J Anat. 22:225-265.
10441. Stockrahm, DMB & Seabloom, RW. 1988. Comparative reproductive performance of black-tailed prairie dog populations in North Dakota. J Mamm. 69:160-164.
10442. Stodart, E. 1966. Observations on the behavior of the marsupial *Bettongia lesueri* in an enclosure. CSIRO Wildl Res. 11:91-101.
10443. Stodart, E. 1966. Management and behavior of breeding groups of the marsupial *Perameles nasua* Geoffroy in captivity. Aust J Zool. 14:611-623.
10444. Stodart, E & Myers, K. 1964. A comparison of behaviour, reproduction, and mortality of wild and domestic rabbits in confined populations. CSIRO Wildl Res. 9:144-159.
10445. Stoddard, HL. 1920. Nests of the western fox squirrel. J Mamm. 1:122-123.
10446. Stoddart, DM. 1970. Individual range, dispersion and dispersal in a population of water voles (*Arvicola terrestris* (L)). J Anim Ecol. 39:403-425.
10447. Stoddart, DM. 1971. Breeding and survival in a population of water voles. J Anim Ecol. 40:487-495.
10448. Stoddart, DM & Braithwaite, RW. 1979. A strategy for utilization of regenerating heathland habitat by the brown

bandicoot (*Isoodon obesulus*, Marsupialia, Peramelidae). J Anim Ecol. 48:165-179.

10449. Stokkan, KA, Hove, K, & Carr, WR. 1980. Plasma concentrations of testosterone and luteinizing hormone in rutting reindeer bulls (*Rangifer tarandrus*). Can J Zool. 58:2081-2083.

10450. Stolk, A. 1963. Ovulation, implantation and foetal sex ratio in the Uganda kob. Nature. 198:606.

10451. Stolzenberg, SJ, Jones, DCL, Kaplan, JN, et al. 1979. Studies with timed pregnant squirrel monkey (*Saimiri sciurus*). J Med Primatol. 8:29-38.

10452. Stone, WD. 1957. The gestation period of the two-toed sloth. J Mamm. 38:419.

10453. Stones, RC & Hayward, CL. 1968. Natural history of the desert woodrat *Neotoma lepida*. Am Mid Nat. 80:458-476.

10454. Storer, TI. 1930. Summer and autumn breeding of the California ground squirrel. J Mamm. 11:235-237.

10455. Storer, TI. 1931. A colony of Pacific pallid bats. J Mamm. 12:244-247.

10456. Storer, TI. 1953. Studies on rat reproduction in San Francisco. J Mamm. 34:365-373.

10457. Storey, AE. 1986. Influence of sires on male-induced pregnancy disruptions in meadow voles (*Microtus pennsylvanicus*) differs with stage of pregnancy. J Comp Psychol. 100:15-20.

10458. Storey, AE & Snow, DT. 1990. Postimplantation pregnancy disruptions in meadow voles: Relationship to variation in male sexual and aggressive behavior. Physiol Behav. 47:19-25.

10459. Storey, HE. 1945. The external genitalia and perfume gland in *Arctictis binturong*. J Mamm. 26:64-66.

10460. Storm, GL & Ables, ED. 1966. Notes on newborn and full-term wild red foxes. J Mamm. 47:116-118.

10461. Storm, GL, Andrews, RD, Phillips, RL, et al. 1976. Morphology, reproduction, dispersal, and mortality of midwestern red fox populations. Wildl Monogr. 49:1-82.

10462. Storm, GL, Andrews, RD, Phillips, RL, et al. 1976. Morphology, reproduction, dispersal, and mortality of midwestern red fox populations. Wildl Monogr. 49:1-82.

10463. Storm, GL & Sanderson, GC. 1968. Housing and reproductive performance of an outdoor colony of voles. J Mamm. 49:322-324.

10464. Storr, GM. 1964. Studies on marsupial nutrition 4 Diet of the quokka *Setonix brachyurus* (Quoy and Gaimard), on Rottnest Island. Aust J Biol Sci. 17:469-481.

10464a. Storrs, EE & Burchfield, HP. 1982. Gestation periods of 20 months in armadillos. Lab Anim Sci. 32:431.

10465. Storrs, EE, Burchfield, HP, & Rees, RJW. 1988. Superdelayed parturition in armadillos: A new mammalian survival strategy. Lepr Rev. 59:11-15.

10466. Storrs, EE & Williams, RJ. 1968. A study of monozygous quadruplet armadillos in relation to mammalian inheritance. Proc Nat Acad Sci. 60:910-914.

10467. Stott, K. Jr. 1946. Twins in green guenon. J Mamm. 27:394.

10468. Stott, K, Jr. 1953. Twinning in hooded capuchin. J Mamm. 34:385.

10469. Stoufflet, I, Mondain-Monval, M, Simon, P, et al. 1989. Patterns of plasma progesterone, androgen and oestrogen concentrations and in-vitro ovarian steriodogenesis during embryonic diapause and implantation in the mink (*Mustela vison*). J Reprod Fert. 87:209-221.

10470. Stout, GG. 1970. The breeding biology of the desert cottontail in the Phoenix region, Arizona. J Wildl Manage. 34:47-51.

10471. Stout, J & Sonenshine, DE. 1974. Ecology of an opossum population in Virginia, 1963-1969. Acta Theriol. 19:235-245.

10472. Stover, J. 1987. Variability of uterine cervical anatomy in the white-tailed gnu (*Connochaetes gnou*). Zoo Biol. 6:265-271.

10473. Stover, J, Evans, J, & Dolensek, EP. 1981. Inter species embryo transfer from the gaur to domestic Holstein. Proc Am Ass Zoo Vet. 1981:122-124.

10474. Strahan, R. 1983. The Australian Museum Complete Book of Australian Mammals. August Robertson Publ, Australia.

10474a. Strahan, R & Thomas, DE. 1975. Courtship of the platypus, *Ornithorhynchus anatinus*. Aust Zool. 18:165-178.

10475. Strahl, H. 1905. Doppelt-diskoidale Placenten bei amerikanischen Affen. Anat Anz. 26:429-430.

10476. Strahl, H. 1905. Eine Placenta mit einem Mesoplacentarium. Anat Anz. 26:524-528.

10477. Strahl, H. 1912. Zur Kenntnis der Wiederkäuerplazentome. Anat Anz. 40:257-264.

10478. Strahl, H. 1913. Über den Bau der Placenta von *Dasypus novemcinctus*. Anat Anz. 44:440-447.

10479. Straka, F. 1964. Beobachtungen über die Biologie der Blindmaus (*Spalax leucodon* Nordm) in Bulgarien. Zool Zhurn. 43:1539-1545.

10480. Straka, F. 1966. Zur Bionomie von *Apodemus microps* Krat et Ros in Bulgarien. Zool Listy. 15:97-104.

10481. Strasser, H. 1968. A breeding program for spontaneously diabetic experimental animals: *Psammomys obesus* (sand rat) and *Acomys cahirinus* (Spiny mouse). Lab Anim Care. 18:328-338.

10482. Stratz, KH. 1898. Der geschlechtsreife Säeugethiereierstock. Martinus Nijhof, Den Haag.

10483. Straus, WL, Jr. 1936. Electrical excitation of the cerebrum of the kangaroo rat. J Mamm. 17:374-382.

10484. Strauss, F. 1938. Die Befruchtung und der Vorgang der Ovulation bei *Ericulus* aus der Familie der Centetiden. Biomorphosis. 1:281-312.

10485. Strauss, F. 1939. Die Bildung des Corpus luteum by Centetiden. Biomorphosis. 1:489-544.

10486. Strauss, F. 1942. Vergleichende Beurteilung der Placentation bei den Insektivoren. Rev Suisse Zool. 49:269-282.

10487. Strauss, F. 1958. Erfahrungen mit einer Feldhasenzucht. Rev Suisse Zool. 65:434-441.

10488. Strauss, F. 1978. Eine Neuuntersuchung der Implantion und Placentation bei *Microcebus murinus*. Mitt Naturf Ges Bern NF. 35:107-119.

10489. Strauss, F. 1978. The ovoimplantation of *Microcebus murinus* Miller (Primates, Lemuroidea, Strepsirhini). Am J Anat. 152:99-110.

10490. Strautmann, EG. 1963. Ondatra v Kasachstane. Isdatelstvo akademii nauk Kazachkoj SSR, Alma-Ata. [Cited by Knechtel & Piechoki, 1983]

10491. Strazielle, L. 1980. Naissance d'un Elephant d'Asie au Parc Zoologique de Paris. Mammalia. 44:592-594.

10492. Strecker, JK. 1929. Notes on the Texas cotton and Attwater wood rats in Texas. J Mamm. 10:216-220.

10493. Streilein, KE. 1982. The Ecology of small mammals in the semiarid Brazilian Caatinga: I Climate and faunal composition. Ann Carnegie Mus. 51:79-107.

10494. Streilein, KE. 1982. The Ecology of small mammals in the semiarid Brazilian Caatinga: III Reproductive biology and population ecology. Ann Carnegie Mus. 51:251-269.

10495. Strelkov, P. 1960. The peculiarities of reproduction in bats (Vespertilionidae) near the northern border of their distribution. Symp Therilogicum Brno. 306-311, ed, Czechoslovak Acad Sci, Praha.

10496. Streubel, DP & Fitzgerald, JP. 1978. *Spermophilus spilosoma*. Mamm Species. 101:1-4.
10497. Streubel, DP & Fitzgerald, JP. 1978. *Spermophilus tridecemlineatus*. Mamm Species. 103:1-5.
10498. Stribley, JA, French, JA, & Inglett, BJ. 1987. Mating patterns in the golden lion tamarin (*Leontopithicus rosalia*): continuous receptivity and concealed estrus. Folia Primatol. 49:137-150.
10499. Stroganov, SU. 1969. Carnivorous Mammals of Siberia. Israel Prog Sci Trans, Smithsonian Inst, Nat Sci Foundation, Washington, DC.
10500. Stroman, HR & Slaughter, LM. 1972. The care and breeding of the pigmy hippopotamus *Choeropsis liberiensis* in captivity. Int Zoo Yearb. 12:126-131.
10501. Stroud, DC. 1982. Population dynamics of *Rattus rattus* and *R norvegicus* in a riparian habitat. J Mamm. 63:151-154.
10502. Struhsaker, TT. 1967. Social structure among vervet monkeys (*Cercopithecus aethiops*). Behaviour. 29:83-121.
10503. Struhsaker, TT. 1972. Behavior of vervet monkeys (*Cercopithecus aethiops*). Univ CA Publ Zool. 82:1-74.
10503a. Struhsaker, TT. 1975. The Red Colobus Monkey. Univ Chicago Press, Chicago, IL.
10504. Struhsaker, TT. 1976. A further decline in numbers of Ambolesi vervet monkeys. Biotropica. 8:211-214.
10505. Struhsaker, TT. 1977. Infanticide and social organization in the redtail monkey (*Cercopithecus ascianus schmidti*) in the Kibale Forest, Uganda. Z Tierpsychol. 45:75-84.
10506. Struhsaker, TT & Leland, L. 1985. Infanticide in a patrilineal society of red colobus monkeys. Z Tierpsychol. 69:89-132.
10507. Strum, SC & Western, JD. 1982. Variations in fecundity with age and environment in olive baboons (*Papio anubis*). Am J Primatol. 3:61-76.
10508. Struthers, PH. 1928. Breeding habits of the Canadian porcupine (*Erethizon dorsatum*). J Mamm. 9:300-308.
10509. Stuart, CT. 1981. Notes on the mammalian carnivores of the Cape Province, South Africa. Bontebok. 1:1-58. [Cited by Bakker & Meister, 1986]
10510. Stubbe, C, Stubbe, M, & Stubbe, I. 1982. Zur Reproduktion der Rehwildpopulation *Capreolus c capreolus* (L 1758): Des Wildforschungsgebiet Hakel. Hercynia. 19:97-109.
10511. Stubbe, M. 1967. Zur Populationsbiologie des Rotfuchses *Vulpes vulpes* (L). Hercynia. 4:1-10.
10512. Stubbe, M. 1969. Populationsbiologische Untersuchungen an *Mustela* marten. Hercynia. 6:306-318.
10513. Stubbe, M. 1969. Zur Biologie und zum Schutz des Fischotters *Lutra lutra* (L). Arch Naturschutz Landschaftsforsch. 9:315-324.
10514. Stubbe, M. 1977. Der Fischotter *Lutra lutra* (L 1758) in der DDR. Zool Anz. 199:265-285.
10515. Stubbe, M. 1980. Population ecology of the red fox (*Vulpes vulpes* L, 1758) in the GDR. The Red Fox; Symposium on Behavior and Ecology. 71-96, Zimen, E, ed, DR W Junk, The Hague.
10516. Stubbe, M & Chotolchu, N. 1968. Zur Säugetierfauna der Mongolei. Mitt Zool Mus Berlin. 44:5-121.
10517. Stubbe, M & Chotolchu, N. 1971. Zur Säugetierfauna der Mongolei-IIII Taiga-Pfeifhasen, *Ochotona alpina hyperborea* (Pallas 1881), aus dem Chenty. Mitt Zool Mus Berlin. 47:349-356.
10518. Stubbe, M & Stubbe, W. 1977. Zur Populationsbiologie des Rotfuchses *Vulpes vulpes* (L) III. Hercynia. 14:160-177.
10519. Stubbe, W & Stubbe, M. 1977. Vergleichende Beiträge zur Reproduktion und Geburtsbiologie von Wild-und Hausschwein *Sus scrofa* L. Jagd Wildforsch. 10:153-179.
10520. Studier, EH, Lysergen, VL, & O'Farrell, MJ. 1973. Biology of *Myotis thysanodes* and *M lucifugus* (Chiroptera: Vespertilionidae) II Bioenergetics of pregnancy and lactation. Comp Biochem Physiol. 44A:467-471.
10521. Studier, EH & O'Farrell, MJ. 1972. Biology of *Myotis thysanodes* and *M lucifugus* (Chiroptera: Vespertilionidae) I Thermoregulation. Comp Biochem Physiol. 41A:567-595.
10522. Studier, EH & O'Farrell, MJ. 1976. Biology of *Myotis thysanodes* and *M lucifugas* (Chiroptera: Vespertilionidae) III Metabolism, heart rate, breathing rate, evaporative water loss and general energetics. Comp Biochem Physiol. 54A:423-432.
10523. Studier, EH & O'Farrell, MJ. 1980. Physiological ecology of *Myotis*. Proc Int Bat Res Conf. 5:415-424.
10523a. Stuewer, FW. 1943. Raccoons: Their habits and management in Michigan. Ecol Monogr. 13:203-257.
10523b. Steuwer, FW. 1948. A record of red bats mating. J Mamm. 29:180-181.
10524. Stull, JW & Brown, WH. 1966. Fatty acid compositon of the milk fat of some desert mammals. J Mamm. 47:542.
10525. Stull, JW, Brown, WH, & Kooyman, GL. 1967. Lipids of the Weddell seal, *Leptonychotes weddelli*. J Mamm. 48:642-645.
10526. Stump, CW, Robins, JP, & Garde, ML. 1960. The development of the embryo and membranes of the humpback whale, *Megaptera nodosa* (Bonnaterre). Aust J Mar Freshwater Res. 11:365-386.
10527. Sturgess, I. 1948. The early embryology and placentation of *Procavia capensis*. Acta Zool (Stockholm). 29:393-497.
10528. Stutterheim, CJ & Skinner, JD. 1973. Preliminary notes on the behaviour and breeding of *Gerbillurus paeba paeba* (A Smith, 1834) in captivity. Koedoe. 16:127-148.
10529. Stüwe, M. 1985. Aspects of structure and reproduction of white-tailed deer populations, *Odocoileus virginianus*, in Venezuela and Virginia. Säugetierk Mitt. 32:137-141.
10530. Stüwe, M & Grodinsky, C. 1987. Reproductive biology of captive alpine ibex (*Capra i ibex*). Zoo Biol. 6:331-339.
10531. Su, JH, Aso, T, Motohashi, T, et al. 1980. Radioimmunoassay method for baboon plasma gonadotropins. Endocrinol Japan. 27:513-520.
10532. Subba Rau, A. 1927. Contributions to our knowledge of the placenta of Mustelidae, Ursidae, and Sciuridae. Proc Zool Soc London. 1927:1027-1069.
10533. Suckling, GC. 1984. Population ecology of the sugar glider, *Petaurus breviceps*, in a system of fragmented habitats. Aust Wildl Res. 11:49-75.
10534. Sudman, PD, Burns, JC, & Choate, JR. 1986. Gestation and postnatal development of the plains pocket gopher. TX J Sci. 38:91-94.
10535. Sugimura, M, Suzuki, Y, Kita, I, et al. 1983. Prenatal development of Japanese serows, *Capricornis crispus*, and reproduction in females. J Mamm. 64:302-304.
10536. Sugita, H. 1979. Reproduction of Père David's deer. Animals & Zoos. 31(9):6-9.
10537. Sugiyama, Y. 1984. Population dynamics of wild chimpanzees at Bassou, Guinea between 1976 and 1983. Primates. 25:391-400.
10538. Sugiyama, Y. 1984. Some aspects of infanticide and intermale competition among langurs, *Presbytis entellus* at Dharwar, India. Primates. 25:423-432.
10539. Sugiyama, Y, Yoshiba, K & Parthasarathy, MD. 1965. Home range, breeding season, male group, and inter-troop

relations in Hanuman langurs (*Presbytis entellus*). Primates. 6:73-106.
10540. Sulentich, JM, Williams, LR, & Cameron, GN. 1991. *Geomys breviceps*. Mamm Species. 383:1-4.
10541. Sullins, GL & Verts, BJ. 1978. Baits and baiting techniques for control of Belding's ground squirrels. J Wildl Manage. 42:890-896.
10542. Sullivan, EG. 1956. Gray fox reproduction, denning, range, and weights in Alabama. J Mamm. 37:346-351.
10543. Sullivan, TP & Sullivan, DS. 1982. Population dynamics and regulation of the Douglas squirrel (*Tamiasciurus dougalsii*) with supplemental food. Oecologica. 53:264-270.
10544. Sumangil, JP. 1966. Observations in the biology of Cotabato ricefield rats, problems and methods of control. Philippine Geographical J. 10:18-23.
10545. Sumangil, JP & Rosell, NC. 1966. Notes on the studies of Philippine species of field rats. Philippine Geographical J. 10:11-17.
10546. Sumar, J. 1983. Studies on Reproductive Pathology in Alpacas. Thesis MS Vet Sci. Swedish Univ Agric Sci, Uppsala, Sweden.
10547. Sumar, J, Smith, GW, Mayhua, E, et al. 1978. Adrenocorticul function in the fetal and newborn alpaca. Comp Biochem Physiol. 59A:79-84.
10548. Sumich, JL. 1986. Growth in young gray whales (*Eschrichtius robustus*). Mar Mamm Sci. 2:145-152.
10549. Summers, CF, Burton, RW, & Anderson, SS. 1975. Grey seal (*Halichoerus grypus*) pup production at North Rona: A study of birth and survival statistics collected in 1972. J Zool. 175:439-451.
10550. Summers, PM, Shephard, AM, Hodges, JK, et al. 1987. Successful transfer of the embryos of Prezewalski's horses (*Equus przewalskii*) and Grant's zebra (*E burchelli*) to domestic mares (*E caballus*). J Reprod Fert. 80:13-20.
10551. Summers, PM, Wennink, CJ, & Hodges, JK. 1985. Cloprostenol-induced luteolysis in the marmoset monkey (*Callithrix jacchus*). J Reprod Fert. 73:133-138.
10552. Sundaram, K, Tsong, YY, Hood, W, et al. 1973. Effect of immunization with estrone-protein conjugate in rhesus monkeys. Endocrinology. 93:843-847.
10553. Sundqvist, C, Lukola, A, & Valtonen, M. 1984. Relationship between serum testosterone concentrations and fertility in male mink (*Mustela vison*). J Reprod Fert. 70:409-412.
10554. Sung, KLP, Bradley, EL, & Terman, CR. 1977. Serum corticosterone concentrations in reproductively mature and unhibited deer mice *Peromyscus maniculatus bairdii*). J Reprod Fert. 49:201-206.
10555. Sunquist, ME. 1981. The social organization of tigers (*Panthera tigris*) in Royal Chitawan National Park, Nepal. Smithsonian Contr Zool. 336:1-98.
10556. Sunqvist, C, Amador, AG, & Bartke, A. 1989. Reproduction and fertility in the mink (*Mustela vison*). J Reprod Fert. 85:413-441.
10557. Sunqvist, C, Ellis, LC, & Bartke, A. 1988. Reproductive endocrinology of the mink (*Mustela vison*). Endocrine Rev. 9:247-266.
10558. Suprasert, A, Hirunagi, K, Fujioka, T, et al. 1989. Histochemistry of glycoconjugates in ovarian follicles of the adult house musk shrew, *Suncus murinus*. Acta Anat. 136:269-278.
10559. Surve, VG. 1970. The breeding of the Indian giant squirrel *Ratufa indica* in captivity. J Bombay Nat Hist Soc. 67:551-552.
10560. Sussman, RW. 1974. Ecological distinctions in sympatric species of *Lemur*. Prosimian Biology. 75-108, Martin, RD, Doyle, GA, & Walker, AC, eds, Univ Pittsburg Press, Pittsburgh.
10561. Sussman, RW. 1975. A preliminary study of the behavior and ecology of *Lemur fulvus rufus* Audebert, 1800. Lemur Biology. 237-258, Tattersall, I & Sussman, RW, eds, Plenum Press, New York.
10562. Sutherland, RL, Evans, SM, & Tyndale-Biscoe, CH. 1980. Macropodid marsupial luteinizing hormone: Validation of assay procedures and changes in concentrations in plasma during the oestrous cycle in the female tammar wallaby (*Macropus eugenii*). J Endocrinol. 86:1-12.
10562a. Suttie, JM, Fennessy, PF, Corson, ID, et al. 1989. LH and testosterone responses to GnRH in red deer (*Cervus elaphus*) stags kept in a manipulated photoperiod. J Reprod Fert. 85:213-219.
10563. Suttie, JM, Lincoln, GA, & Kay, RNB. 1984. Endocrine control of antler growth in red deer stags. J Reprod Fert. 71:7-15.
10564. Suttkus, RD & Jones, C. 1991. Observations on winter and spring reproduction in *Peromyscus leucopus* (Rodentia: Muridae) in southern Louisiana. TX J Sci. 43:179-189.
10565. Suttle, JM, Moore, HDM, Pierce, EJ, et al. 1988. Quantitative studies on variation in sperm head morphology of the hopping mouse, *Notomys alexis*. J Exp Zool. 247:166-171.
10565a. Sutton, DA. 1992. *Tamias amoenus*. Mamm Species. 390:1-8.
10566. Suzuki, F & Racey, PA. 1976. Fine structural changes in the epididymal epithelium of moles (*Talpa europaea*) throughout the year. J Reprod Fert. 47:47-54.
10567. Suzuki, F & Racey, PA. 1978. The organization of testicular interstitial tissue and changes in the fine structure of the Leydig cells of European moles (*Talpa europaea*) throughout the year. J Reprod Fert. 52:189-194.
10568. Svidenko, GD, Svidenko, AA, Nikitin, VP, et al. 1973. Reproduction of *Spermophilopsis leptodastylus* in Turkmenia. Zool Zhurn. 52:1836-1842.
10569. Svihla, A. 1929. Breeding habits and young of the red-backed mouse, *Evotomys*. Pap MI Acad Sci Arts Lett. 11:485-490.
10570. Svihla, A. 1929. Life history notes on *Sigmodon hispidus hispidus*. J Mamm. 10:352-353.
10571. Svihla, A. 1931. Habits of the Louisiana mink. J Mamm. 12:366-368.
10572. Svihla, A. 1931. Life history of the Texas rice rat (*Oryzomys palustris texensis*). J Mamm. 12:238-242.
10573. Svihla, A. 1932. A comparative life history study of the mice of the genus *Peromyscus*. Univ MI Mus Zool Misc Publ. 24:1-39.
10574. Svihla, A. 1933. Notes on the deer-mouse *Peromyscus maniculatus oreas* (Bangs). Murrelet. 14:13-14.
10575. Svihla, A. 1934. Development and growth of deermice (*Peromyscus maniculatus artemisiae*). J Mamm. 15:99-104.
10576. Svihla, A. 1935. Development and growth of the prairie deer mouse, *Peromyscus maniculatus bairdii*. J Mamm. 16:109-115.
10577. Svihla, A. 1936. Development and growth of *Peromyscus maniculatus oreas*. J Mamm. 17:132-137.
10578. Svihla, A. 1936. The Hawaiian rat. Murrelet. 17:3-14.
10579. Svihla, A. 1949. Notes on reproduction of the peninsula bear. Murrelet. 30:53-54.
10580. Svihla, A & Svihla, RD. 1931. The Louisiana muskrat. J Mamm. 12:12-28.

10581. Svihla, A & Svihla, RD. 1933. Notes on the life history of the wood rat, *Neotoma floridana rubida* Bangs. J Mamm. 14:73-75.
10582. Svihla, A & Svihla, RD. 1933. Notes on the jumping mouse *Zapus trinotatus trinotatus* Rhoads. J Mamm. 14:131-134.
10583. Svihla, RD. 1929. Habits of *Sylvilagus aquaticus littoralis*. J Mamm. 10:315-319.
10584. Svihla, RD. 1930. A family of flying squirrels. J Mamm. 11:211-213.
10585. Svihla, RD. 1930. Development of young red squirrels. J Mamm. 11:79-80.
10586. not used.
10587. Svihla, RD. 1930. Notes on the golden harvest mouse. J Mamm. 11:53-54.
10587a. Svihla, RD. 1931. Notes on desert and dusky harvest mice (*Reithrodontomys megalotis megalotis* and *R m nigrescens*). J Mamm. 12:363-365.
(*Reithrodontomys megalotis mgelaotis* and
10588. Svihla, RD. 1936. Breeding and young of the grasshopper mouse (*Onychomys leucogaster fuscogriseus*). J Mamm. 17:172-173.
10589. Sviridenko, PA. 1967. Propagation and variation in number of *Clethrionomys glareolus* Schreb in the Ukraine. Vestnik Zool (Kiev). 2:9-24.
10590. Svoboda, PL & Choate, JR. 1987. Natural history of the Brazilian free-tailed bat in the San Luis Valley of Colorado. J Mamm. 68:224-234.
10591. Swanepoel, CM. 1980. Some factors influencing the breeding season of *Praomys natalensis*. S Afr J Zool. 15:95-98.
10592. Swanepoel, P. 1976. An ecological study of rodents in northern Natal, exposed to dieldrin coverspraying. Ann Cape Prov Mus Nat Hist. 11(4):57-81.
10593. Swanepoel, P & Genoways, HH. 1983. *Brachyphylla cavernarum*. Mamm Species. 205:1-6.
10594. Swanepoel, P & Genoways, HH. 1983. *Brachyphylla nana*. Mamm Species. 206:1-3.
10595. Swank, WG. 1958. The mule deer in Arizona chaparral and an analysis of other important deer herds: A research and management study. AZ Game Fish Dept Wildl Bull. 3:1-109.
10596. Swanson, HH & Lockley, MR. 1978. Population growth and social structure of confined colonies of Mongolian gerbils: Scent gland size and marking behaviour as indices of social status. Aggressive Behav. 4:57-89.
10597. Swarth, HS. 1924. Birds and Mammals of the Sheena River region of northern British Columbia. Univ CA Publ Zool. 24:315-394.
10598. Swarth, HS. 1936. Mammals of the Atlin region, northwestern British Columbia. J Mamm. 17:398-405.
10599. Swayne, HCG. 1899. The elephant. Great and Small Game of Africa. 1-34, Bryden, HA, ed, Rowland Ward, London.
10600. Sweeney, JM, Sweeney, JR, & Provost, EE. 1979. Reproductive biology of a feral hog population. J Wildl Manage. 43:555-559.
10601. Sweet, G. 1907. Contributions to our knowledge of the anatomy of *Notoryctes typhlops* Stirling Pt IV. Q J Microsc Sci. 51:325-344.
10602. Swiezynski, K. 1968. The male reproductive organs of the European bison. Acta Theriol. 13:511-551.
10603. Swift, SM. 1980. Activity patterns of pipestrelle bats (*Pipestrellus pipestrellus*) in north-east Scotland. J Zool. 190:285-295.
10604. Swift, SM, Racey, PA, & Avery, MI. 1985. Feeding ecology of *Pipistrellus pipistrellus* (Chiroptera: Vespertilionidae) during pregnancy and lactation II Diet. J Anim Ecol. 54:217-225.
10605. Swihart, RK. 1984. Body size, breeding season length, and life history tactics of lagomorphs. Oikos. 43:282-290.
10606. Switzenberg, DF. 1950. Breeding productivity in Michigan red foxes. J Mamm. 31:194-195.
10607. Symes, CB. 1932. Notes on rats, fleas and plague in Kenya. Rec Med Res Lab Kenya. 3. [Cited by Buxton, 1936]
10608. Symington, MM. 1987. Ecological and social correlates of party size in the black spider monkey, *Ateles paniscus chamek*. Ph D Diss. Princeton Univ, NJ.
10609. Symington, MM. 1988. Demography, ranging patterns, and activity budgets of black spider monkeys (*Ateles paniscus chamek*) in the Manu National Park, Peru. Am J Primatol. 15:45-67.
10610. Symington, RB & Paterson, NJ. 1970. A preliminary report on the phenomenon of unilateral implantation in the right uterine horn of the common duiker, *Sylvicapra grimmia*. Arnoldia (Rhodesia). 4(32):1-5.
10611. Symons, HW & Weston, RS. 1958. Studies on the humpback whale, *Megaptera nodosa*, in the Bellingshausen Sea. Norsk Hvalf tid. 47:53-81.
10612. Szederjei, Á & Fabian, L. 1975. Giraffengeburten im Zoo Budapest. Zool Garten NF. 45:175-186.
10613. Taber, RD. 1953. Studies of black-tailed deer reproduction on three chaparral cover types. CA Fish & Game. 39(2):177-186.
10614. Taber, RD & Dasmann, RF. 1957. The dynamics of three natural populations of the deer *Odocoileus hemionus columbianus*. Ecology. 38:233-246.
10615. Taber, RD & Dasmann, RF. 1958. The black-tailed deer of the chaparral: Its life history and management in the north coast range of California. CA Fish Game Bull. 8:1-163.
10616. Taber, RD, Sheri, AN, & Ahmad, MS. 1967. Mammals of the Lyallpur region, West Pakistan. J Mamm. 48:392-407.
10617. Tacu, D. 1978. Resting metabolic rate of lactating and developing *Citellus citellus*. Acta Theriol. 18:297-301.
10618. Taddei, VA. 1976. The reproduction of some Phyllostomidae (Chiroptera) from the northwestern region of the state of São Paulo. Bol Zool Univ São Paulo. 1:313-330.
10618a. Taddei, VA, Vizotto, LD, & Martins, SM. 1976. Notas taxionômicas e biológicas sobre *Molossops brachymeles cerastes* (Thomas, 1901) (Chiroptera - Molossidae). Naturalia. 2:61-69.
10619. Taggart, DA & Temple-Smith, PD. 1990. Effects of breeding season and mating on total number and distribution of spermatozoa in the epididymis of the brown marsupial mouse, *Antechinus stuartii*. J Reprod Fert. 88:81-91.
10620. Taha, NM & Kielwein, G. 1990. Pattern of peptide-bound and free amino acids in camel, buffalo, and ass milk. Milchwissenschaft. 45:22-25.
10621. Tahiri-Zagret, C. 1969. Recherches sur le cycle oestrien de la femelle du Pangolin *Manis tricuspis* Rafinesquei d'après l'étude des frottis vaginaux. CR Séances Soc Biol. 163:2189-2193.
10622. Tähkä, KM. 1978. A histochemical study on the effects of photoperiod on gonadal and adrenal functions in the male bank vole (*Clethrionomys glareolus*). J Reprod Fert. 54:57-66.
10623. Tähkä, KM. 1980. A histochemical study on the effects of photoperiod on gonadal and adrenal functions in the female bank vole (*Clethrionomys glareolus*, Schreb). Gen Comp Endocrinol. 41:41-52.
10624. Tähkä, KM & Rajaniemi, H. 1985. Photoperiodic modulation of testicular LH receptors in the bank vole (*Clethrionomys glareolus*). J Reprod Fert. 75:513-519.
10625. Tähkä, KM, Ruokonen, A, Wallgren, H, et al. 1983. Temporal changes in testicular histology and steroidogenesis in

juvenile bank voles (*Clethrionomys glareolus*, Schreber) subjected to different photoperiods. Endocrinology. 112:1420-1426.
10626. Tähkä, KM, Teräväinen, T, Pankakoski, E, et al. 1989. The testes of moles (*Talpa europaea*) retain a considerable microsomal capacity for androgen synthesis during seasonal regression. Gen Comp Endocrinol. 76:301-309.
10627. Tähkä, KM, Teräväinen, T, & Wallgren, H. 1982. Effect of photoperiod on the bank vole (*Clethrionomys glareolus*, Schreber): An *in vitro* study. Gen Comp Endocrinol. 47:377-384.
10628. Taibel, AM. 1937. L'*Antelope cervicapra*: Osservazioni sul gruppo in allevamento presso la Stazione sperimentale di Avicultura di Rovigo. Rass Faun Rome. 4(2):3-19.
10629. Tait, AJ & Johnson, E. 1982. Spermatogenesis in the grey squirrel (*Sciurus carolinensis*) and changes during sexual regression. J Reprod Fert. 65:53-58.
10630. Tait, AJ, Pope, GS, & Johnson, E. 1981. Progesterone concentrations in peripheral plasma of non-pregnant and pregnant grey squirrels (*Sciurus carolinensis*). J Endocrinol. 89:107-116.
10631. Tait, WJE. 1969. The elusive and rare oribi. African Wild Life. 23:154-160.
10632. Takada, Y. 1983. Life history of small mammals in fallow fields 5 Reproduction in the large Japanese field mouse and the feral house mouse. J Mamm Soc Japan. 9:246-252.
10633. Takahata, Y. 1980. The reproductive biology of a free-ranging troop of Japanese monkeys. Primates. 21:303-329.
10634. Takahata, Y. 1982. The socio-sexual behavior of Japanese monkeys. Z Tierpsychol. 59:89-108.
10635. Takai, S & Yasui, K. 1980. Breeding and weaning the California sea lion. Animals & Zoos. 32(10):6-10.
10636. Takeshita, H. 1961. On the delivery behavior of squirrel monkeys (*Saimiri sciurea*) and a mona monkey (*Cercopithecus mona*). Primates. 3:59-72.
10636a. Talamantes, F & Ogren, L. 1988. The placenta as an endocrine organ: Polypeptides. The Physiology of Reproduction. 2093-2144, Knobil, E & Neill, JD, eds, Raven Press, New York.
10637. Talbot, LM & Talbot, MH. 1961. Preliminary observations on the population dynamics of the wildebeest in Narok District Kenya. E Afr Agric For J. 27:108-116.
10638. Talbot, LM & Talbot, MH. 1963. The wildebeest in western Masailand, East Africa. Wildl Monogr. 12:1-88.
10639. Talice, RV, Momigliano-Tedschi, E, Laffitte de Mosera, S, et al. 1959. Investigaciones sobre *Ctenomys torquatus*: Un roedor autoctono del Uruguay. Am Fac Med Univ Repub (Montiv). 44:452-462.
10640. Talmage, RV & Buchanan, GD. 1954. The armadillo (*Dasypus novemcinctus*). Rice Inst Pamphlet. 41(2):1-135.
10641. Talmage, RV, Buchanan, GD, Kraintz, FW, et al. 1954. The presence of a functional corpus luteum during delayed implantation in the armadillo. J Endocrinol. 11:44-49.
10642. Talpin, LE. 1980. Some observations on the reproductive biology of *Sminthopsis virginiae* (Tarragon), (Marsupialia: Dasyuridae). Aust Zool. 20:407-418.
10643. Tam, WH. 1971. The production of hormonal steroids by ovarian tissues of the chinchilla (*Chinchilla laniger*). J Endocrinol. 50:267-279.
10644. Tam, WH. 1972. Steroid metabolic pathways in the ovary of the chinchilla (*Chinchilla laniger*). J Endocrinol. 52:37-50.
10645. Tam, WH. 1973. Progesterone levels during the oestrous cycle and pregnancy in the cuis, *Galea musteloides*. J Reprod Fert. 35:105-114.
10646. Tam, WH. 1974. The synthesis of progesterone in some hystricomorph rodents. Symp Zool Soc London. 34:363-384.
10647. Tamarin, RH. 1977. Demography of the beach vole (*Microtus breweri*) and the meadow vole (*Microtus pennsylvanicus*) in southeastern Massachusetts. Ecology. 58:1310-1321.
10648. Tamarin, RH. 1977. Reproduction in the island beach vole, *Microtus breweri*, and the mainland meadow vole *Microtus pennsylvanicus*, in southeastern Massachusetts. J Mamm. 58:536-548.
10648a. Tamarin, RH, ed. 1985. Biology of New World *Microtus*. Sp Pub Am Soc Mamm. 8:1-893.
10649. Tamarin, RH & Kunz, TH. 1974. *Microtus breweri*. Mamm Species. 45:1-3.
10650. Tamarin, RH & Malecha, SR. 1972. Reproductive parameters in *Rattus rattus* and *Rattus exulans* of Hawaii, 1968 to 1970. J Mamm. 53:513-528.
10651. Tamarkin, L, Reppert, SM, Orloff, DJ, et al. 1980. Ontogeny of the pineal melatonin rhythm in the Syrian (*Mesocricetus auratus*) and Siberian (*Phodopus sungorus*) hamsters and in the rat. Endocrinology. 107:1061-1064.
10652. Tamsitt, JR. 1960. Some mammals of Riding Mountain National Park, Manitoba. Can Field-Nat. 74:147-150.
10652a. Tamsitt, JR. 1970. Comparative biochemistry and ecology of bats of the Puerto Rican region. Am Philos Soc Yearb. 1970:342-343.
10653. Tamsitt, JR & Häuser, C. 1985. *Sturnira magna*. Mamm Species. 240:1-4.
10654. Tamsitt, JR & Mejia, CA. 1962. The reproductive status of a population of the neotropical bat, *Artibeus jamaicensis*, at Providencia. Caribb J Sci. 2:139-144.
10655. Tamsitt, JR & Nagorsen, D. 1982. *Anoura cultrata*. Mamm Species. 179:1-5.
10656. Tamsitt, JR & Valdivieso, D. 1962. Le cycle oestral du Rat *Sigmodon bogotensis* de Colombie. Mammalia. 26:161-166.
10657. Tamsitt, JR & Valdivieso, D. 1963. Records and observations on Colombian bats. J Mamm. 44:168-179.
10658. Tamsitt, JR & Valdivieso, D. 1963. Reproductive cycle of the big fruit-eating bat, *Artibeus lituratus* Olfers. Nature. 198:104.
10659. Tamsitt, JR & Valdivieso, D. 1963. Condieron reproductora de una colonia Ecuatoriana del Murcielargo Myotis negro, *Myotis nigricans nigricans* (Familia Vespertilionidae). Caribb J Sci. 3:49-51.
10660. Tamsitt, JR & Valdivieso, D. 1964. Informations sur la reproduction des Cheiropteres Phyllostomides de Colombie. Mammalia. 28:397-402.
10661. Tamsitt, JR & Valdivieso, D. 1965. Reproduction of the female big fruit-eating bat, *Artibeus lituratus palmarum*, in Colombia. Caribb J Sci. 5:157-166.
10662. Tamsitt, JR & Valdivieso, D. 1965. The male reproductive cycle of the bat *Artibeus lituratus*. Am Mid Nat. 73:150-160.
10663. Tamsitt, JR & Valdivieso, D. 1966. Taxonomic comments on *Anoura caudifer*, *Artibeus lituratus*, and *Molossus molossus*. J Mamm. 47:230-238.
10664. Tamsitt, JR & Valdivieso, D. 1966. Parturition in the red fig-eating bat, *Stenoderma rufum*. J Mamm. 47:352-353.
10664a. Tamsitt, JR & Valdivieso, D, & Hernández C, J. 1965. Additional records of *Choeroniscus* in Colombia. J Mamm. 46:704.
10665. Tanaka, R. 1964. Population dynamics of the Smith's red-backed vole in highlands of Shikoku. Res Pop Ecol. 6:54-66.
10666. not used.

10667. Tanaka, T, Tokuda, K, & Kotera, S. 1970. Effects of infant loss on the interbirth interval of Japanese monkeys. Primates. 11:113-118.
10668. Tandler, J & Grosz, S. 1911. Über den Saisondimorphismus des Maulwurfhodens. Arch Entwicklungsmechank Org. 33:297-302.
10669. Tannenbaum, BR. 1975. Reproductive strategies in the white-lined bat. Ph D Diss. Cornell Univ, Ithaca, NY.
10670. Tappe, DT. 1941. Natural history of the Tulare kangaroo rat. J Mamm. 22:117-148.
10671. Tardif, SD. 1984. Social influences on sexual maturation of female *Saguinus oedipus oedipus*. Am J Primatol. 6:199-209.
10672. Tardif, SD, Carson, RL, & Clapp, NK. 1986. Breeding performance of captive-born cotton-top tamarin (*Saguinus oedipus*) females: Proposed explanations for colony differences. Am J Primatol. 11:271-275.
10673. Tardif, SD, Carson, RL, & Gangaware, BL. 1986. Comparison of infant care in family groups of the common marmoset (*Callithrix jacchus*) and the cotton-top tamarin (*Saguinus oedipus*). Am J Primatol. 11:103-110.
10674. Tardif, SD, Richter, CB, & Carson, RL. 1984. Effects of sibling-rearing experience on future success in two species of Callitrichidae. Am J Primatol. 6:377-380.
10675. Tardif, SD, Richter, CB, & Carson, RL. 1984. Reproductive performance of three species of Callitrichidae. Lab Anim Sci. 34:272-275.
10676. Tarkowski, AK. 1956. Studies on reproduction and prenatal mortality of the common-shrew (*Sorex araneus* L) I Foetal regression. Ann Univ Mariae Curie-Sklodowska Lublin Polinia. 9C:387-429.
10677. Tarkowski, AK. 1957. Studies on reproduction and prenatal mortality of the common shrew (*Sorex araneus* L) II Reproduction under natural conditions. Ann Univ Mariae Curie-Sklodowska. 10C:177-244.
10678. Tas'an (no initial) & Leatherwood, S. 1984. Cetaceans live-captured for Jaya Ancol Oceanarium, Djakarta, 1974-1982. Rep Int Whal Comm. 34:485-489.
10679. Tasaka, K & Yoshihara, K. 1982. The growth and development of chimpanzees. Animals & Zoos. 34(12):8-11.
10680. Tasse, J. 1986. Maternal and paternal care in the rock cavy, *Kerodon rupestris*, a South American hystricomorph rodent. Zoo Biol. 5:27-43.
10681. Tast, J. 1966. The root vole, *Microtus oeconomus* (Pallas), as an inhabitant of seasonally flooded land. Ann Zool Fenn. 3:127-171.
10682. Tate, GHH. 1931. Random observations on habits of South American mammals. J Mamm. 12:248-256.
10683. Tate, GHH. 1933. A systematic revision of the marsupial genus *Marmosa*. Bull Am Mus Nat Hist. 66:1-250.
10683a. Tate, GHH. 1935. Observations on the big-tailed shrew (*Sorex dispar* Batchelder). J Mamm. 16:213-215.
10684. Tate, GHH. 1945. Results of the Archbold Expeditions No 55, Notes on the squirrel-like and mouse-like possums (Marsupialia). Am Mus Novit. 1305:1-12.
10685. Tates, AD, Pearson, PL, & Geraedts, JPM. 1975. Identification of X and Y spermatozoa in the northern vole, *Microtus oeconomus*. J Reprod Fert. 42:195-198.
10686. Tattersall, I. 1976. Group structure and activity rhythm in *Lemur mongoz* (Primates, Lemuriformes) an Anjavan and Moheli Islands, Comoro Archipelago. Anthrop Pap Am Mus Nat Hist NY. 53:367-380.
10687. Taub, DM. 1980. Age at first pregnancy and reproductive outcome among colony-born squirrel monkeys (*Saimiri sciureus*, Brazilian). Folia Primatol. 33:262-272.
10688. Taub, DM. 1982. Sexual behavior of wild barbary macaque males (*Macaca sylvanus*). Am J Primatol. 2:109-113.
10689. Taub, DM, Adams, MR, & Auerbach, KG. 1978. Reproductive performance in a breeding colony of Brazilian squirrel monkeys (*Saimiri sciureus*). Lab Anim Sci. 28:562-566.
10690. Taverne, MAM & Bakker-Slotboom, MF. 1970. Observations on the delivered placenta and fetal membranes of the aardvark, *Orycteropus afer* (Pallas, 1776). Bijdr Dierkd. 40:154-162.
10691. Tavolga, MC & Essapian, FS. 1957. The behavior of the bottle-nosed dolphin (*Tursiops truncatus*): Mating, pregnancy, parturition and mother-infant behavior. Zoologica. 42:11-31.
10691a. Taya, K, Komura, H, Kondoh, M, et al. 1991. Concentrations of progesterone, testosterone and estradiol-17β in the serum during the estrous cycle of Asian elephants (*Elephas maximus*). Zoo Biol. 10:299-307.
10692. Tayes, MAF. 1948. Studies on the anatomy of the ovary and corpus luteum of the camel. Vet J. 104:179-186.
10693. Tayler, CK & Saayman, GS. 1972. The social organization and behaviour of dolphins (*Tursiops aduncus*) and baboons (*Papio ursinus*): Some comparisons and assessments. Ann Cape Prov Mus Nat Hist. 9:11-49.
10694. Taylor, I. 1972. *Ichneumia albicauda*. Mamm Species. 12:1-4.
10695. Taylor, JJ & Tomkinson, M. 1975. A comparative study of primate breast milk. Ann Rep Jersey Wildl Preserv Trust. 12:76-77.
10696. Taylor, JM. 1961. Reproductive biology of the Australian bush rat *Rattus assimilis*. Univ CA Publ Zool. 60:1-66.
10697. Taylor, JM. 1968. Reproductive mechanisms of the female southern grasshopper mouse, *Onychomys torridus longicaudus*. J Mamm. 49:303-309.
10698. Taylor, JM. 1984. The Oxford Guide to Mammals of Australia. Oxford Univ Press, Melbourne, Australia.
10699. Taylor, JM, Calaby, JH, & Redhead, TD. 1982. Breeding in wild populations of the marsupial-mouse *Planigale maculata sinualis* (Dasyuridae, Marsupialia). Carnivorous Marsupials. 1:83-87, Archer, M, ed, R Zool Soc NSW, Mossman, NSW.
10700. Taylor, JM & Calaby, JH. 1988. *Rattus fuscipes*. Mamm Species. 298:1-8.
10701. Taylor, JM & Calaby, JH. 1988. *Rattus lutreolus*. Mamm Species. 299:1-7.
10702. Taylor, JM, Calaby, JH, & Smith, SC. 1990. Reproduction in New Guinean *Rattus* and comparison with Australian *Rattus*. Aust J Zool. 38:587-602.
10703. Taylor, JM & Horner, BE. 1970. Gonadal activity in the marsupial mouse, *Antechinus bellus*, with notes on other species of the genus (Marsupialia: Dasyuridae). J Mamm. 51:659-668.
10704. Taylor, JM & Horner, BE. 1970. Observations on reproduction in *Leggadina* (Rodentia, Muridae). J Mamm. 51:10-17.
10705. Taylor, JM & Horner, BE. 1971. Reproduction in the Australian tree-rat *Conilurus penicillatus* (Rodentia: Muridae). CSIRO Wildl Res. 16:1-9.
10706. Taylor, JM & Horner, BE. 1971. Sexual maturation in the Australian rodent *Rattus fuscipes assimilis*. Aust J Zool. 19:1-17.
10707. Taylor, JM & Horner, BE. 1972. Breeding biology of three subspecies of the native Australian rat, *Rattus fuscipes*, in the laboratory. Aust Mamm. 1:8-13.
10708. Taylor, JM & Horner, BE. 1972. Observations on the reproductive biology of *Pseudomys* (Rodentia: Muridae). J Mamm. 53:318-328.

10709. Taylor, JM & Horner, BE. 1973. Reproductive characteristics of wild native Australian *Rattus* (Rodentia: Muridae). Aust J Zool. 21:437-476.
10710. Taylor, JM & Padykula, HA. 1978. Marsupial trophoblast and mammalian evolution. Nature. 271:588.
10711. Taylor, JM & Padykula, HA. 1982. Marsupial placentation and its evolutionary significance. J Reprod Fert Suppl. 31:95-105.
10712. Taylor, KD & Green, MG. 1976. The influence of rainfall on diet and reproduction in four African rodent species. J Zool. 180:367-389.
10713. Taylor, KM, Hungerford, DA, Snyder, RL, et al. 1968. Uniformity of karyotypes in the Camelidae. Cytogenetics. 7:7-15.
10714. Taylor, ME. 1969. Note on the breeding of two genera of viverrids, *Genetta* ssp and *Herpestes sanguineus*, in Kenya. E Afr Wildl J. 7:168.
10715. Taylor, ME. 1975. *Herpestes sanguineus*. Mamm Species. 65:1-5.
10716. Taylor, ME. 1987. *Bdeogale crassicauda*. Mamm Species. 294:1-4.
10717. Taylor, ME & Abrey, N. 1982. Marten, *Martes americana*, movements and habitat use in Algonquin Provincial Park, Ontario. Can Field-Nat. 96:439-447.
10718. Taylor, PJ, Jarvis, JUM, Crowe, TM, et al. 1985. Age determination in the Cape molerat *Georychus capensis*. S Afr J Zool. 20:261-267.
10719. Taylor, RJ. 1984. Foraging in the eastern grey kangaroo and the walleroo. J Anim Ecol. 53:65-74.
10720. Taylor, WP. 1918. Revision of the rodent genus *Aplodontia*. Univ Calif Publ Zool. 17:435-504.
10721. Taylor, WP. 1954. Food habits and notes on life history of the ring-tailed cat in Texas. J Mamm. 35:55-63.
10722. Tchernyavsky, FB. 1962. On the reproduction and growth of the bighorn (*Ovis nivicola* Esch). Zool Zhurn. 41:1556-1566.
10723. Teague, LG & Bradley, EL. 1978. The existence of a puberty accelerating pheromone in the urine of the male prairie deer mouse (*Peromyscus maniculatus bairdii*). Biol Reprod. 19:314-317.
10724. Teahan, CG, McKenzie, HA, & Griffiths, M. 1991. Some monotreme milk "whey" and blood proteins. Comp Biochem Physiol. 99B:99-118.
10725. Teahan, CG, McKenzie, HA, Shaw, DC, & Griffiths, M. 1991. The isolation and amino acid sequences of echidna (*Tachyglossus aculeatus*) milk lysozyme I and II. Biochem Int. 24:85-95.
10726. Teas, J, Taylor, HG, Richie, TL, et al. 1981. Parturition in rhesus monkeys (*Macaca mulatta*). Primates. 22:580-586.
10727. Tedman, RA. 1991. Morphology of the reproductive tract of a juvenile male ross seal, *Ommatophoca rossii* (Pinnipedia: Phocidae). Aust Mamm. 14:35-38.
10728. Tedman, RA. 1991. The female reproductive tract of the Australian sea lion, *Neophoca cinerea* (Peron, 1816) (Carnivora: Otariidae). Aust J Zool. 39:351-372.
10729. Tedman, RA & Green, B. 1987. Water and sodium fluxes and lactational energetics in suckling pups of Weddell seals (*Leptonychotes weddellii*). J Zool. 212:29-42.
10730. Teer, JG, Thomas, JW, & Walker, EA. 1965. Ecology and management of white-tailed deer in the Llano basin of Texas. Wildl Monogr. 15:1-62.
10731. Tefft, BC & Chapman, JA. 1983. Growth and development of nestling New England cottontails, *Sylvilagus transitionalis*. Acta Theriol. 28:317-320.
10732. Telfer, S & Breed, WG. 1976. The effect of age on the female reproductive tract of the hopping-mouse *Notomys alexis*. Aust J Zool. 24:533-540.
10733. Tembo, A. 1987. Population status of the hippopotamus on the Luangwa River, Zambia. Afr J Ecol. 25:71-77.
10734. Tembrock, G. 1957. Zur Ethologue des Rotfuchses (*Vulpes vulpes* [L]), unter besonderer Berücksichtigung der Fortpflanzung. Zool Garten NF. 23:289-532.
10735. Temple-Smith, PD. 1984. Reproductive structures and strategies in male possums and gliders. Possums and Gliders. 89-106, Smith, A & Hume, I, eds, Surrey Beatty & Sons Pty Ltd, Sydney, Australia.
10735a. Temple-Smith, PD. 1984. Phagocytosis of sperm cytoplasmic droplets by a specialized region in the epididymis of the brushtailed possum, *Trichosurus vulpecula*. Biol Reprod. 30:707-720.
10736. Temte, JL. 1985. Photoperiod and delayed implantation in the northern fur seal (*Callorhinus ursinus*). J Reprod Fert. 73:127-131.
10737. Ten Cate Hoedemaker, NJ. 1935. Mitteilung über eine reife Plazenta von *Phocoena phocoena* (Linnaeus). Arch Néerl Zool. 1:330-338.
10738. Tener, JS. 1954. A preliminary study of the musk-oxen of Fosheim Peninsula, Ellesmere Island, NWT. Can Wildl Serv Wildl Maneg Bull Ser 1. 9:1-34.
10739. Tener, JS. 1956. Gross composition of musk-ox milk. Can J Zool. 34:569-571.
10740. Tener, JS. 1965. Muskoxen in Canada. Queen's Printer, Ottawa.
10741. Terasawa, E. 1985. Developmental changes in the positive feedback effect of estrogen on luteinizing hormone release in ovariectomized female rhesus monkeys. Endocrinology. 117:2490-2497.
10742. Terasawa, E, Bridson, WE, Nass, TE, et al. 1984. Developmental changes in the luteinizing hormone secretory pattern in peripubertal female rhesus monkeys: Comparisons between gonadally intact and ovariectomized animals. Endocrinology. 115:2233-2240.
10743. Terasawa, E, Krook, C, Eman, S, et al. 1987. Pulsatile luteinizing hormone (LH) release during the progesterone-induced LH surge in the female rhesus monkey. Endocrinology. 120:2265-2271.
10744. Terasawa, E, Nass, TE, Yeoman, RR, et al. 1983. Hypothalamic control of puberty in the female rhesus macaque. Neuroendocrine Aspects of Reproduction. 149-182, Academic Press, New York.
10745. Terasawa, E, Noonan, JJ, Nass, TE, et al. 1984. Posterior hypothalamic lesions advance the onset of puberty in the female rhesus monkey. Endocrinology. 115:2241-2250.
10746. Terasawa, E, Yeoman, RR, & Schultz, NJ. 1984. Factors influencing the progesterone-induced luteinizing hormone surge in rhesus monkeys: Diurnal influence and time interval after estrogen. Biol Reprod. 31:732-741.
10747. Teräväinen, T & Saure, A. 1976. Changes in the testicular metabolism of dehydroepiandrosterone during the annual reproductive cycle of the hedgehog (*Erinaceus europaeus* L). Gen Comp Endocrinol. 29:328-332.
10748. Teräväinen, T & Tähkä, KM. 1985. Photoperiod-induced changes in the testicular metabolism of [4-^{14}C]17α-hydroxyprogesterone in the bank vole (*Clethrionomys glareolus*). J Reprod Fert. 74:625-630.
10749. Terborgh, J & Goldizen, AW. 1985. On the mating system of the cooperatively breeding saddle-backed tamarin (*Saguinus fuscicollis*). Behav Ecol Sociobiol. 16:293-299.

10750. Terman, CR. 1973. Recovery of reproductive function by prairie deer mice (*Peromyscus maniculatus bairdii*) from asymptotic populations. Anim Behav. 21:443-448.
10751. Terman, CR. 1973. Reproductive inhibition in asymptotic populations of prairie deer mice. J Reprod Fert Suppl. 19:457-463.
10751a. Terman, CR. 1992. Reproductive inhibition in female white-footed mice from Virginia. J Mamm. 73:443-448.
10752. Ternovsky, DV. 1983. Biology of reproduction and development of *Mustela erminea* (Carnivora, Mustelidae). Zool Zhurn. 62:1097-1105.
10753. Terrel, TL. 1972. The swamp rabbit, *Sylvilagus aquaticus*, in Indiana. Am Mid Nat. 87:283-295.
10754. Terwilliger, VJ. 1978. Natural history of Baird's tapir on Barro Colorado Island, Panama Canal Zone. Biotropica. 10:211-220.
10755. Tesh, RB & Cameron, RV. 1970. Laboratory rearing of the climbing rat, *Tylomys nudicaudatus*. Lab Anim Care. 20:93-96.
10756. Tesh, RB. 1970. Notes on the reproduction, growth, and development of echimyid rodents in Panama. J Mamm. 51:199-202.
10757. Tesh, RB. 1970. Observations on the natural history of *Diplomys darlingi*. J Mamm. 51:197-199.
10758. Teska, WR, Rybak, EN, & Baker, RH. 1981. Reproduction and development of the pygmy spotted skunk (*Spilogale pygmaea*). Am Mid Nat. 105:390-392.
10759. Testa, JW. 1987. Long-term reproductive patterns and sighting bias in Weddell seals (*Leptonychotes weddelli*). Can J Zool. 65:1091-1099.
10760. Tetley, H. 1941. On the Scottish wild cat. Proc Zool Soc London. 111(B):13-23.
10761. Tetley, H. 1941. On the Scottish fox. Proc Zool Soc London. 111(B):25-35.
10762. Tevis, L, Jr. 1955. Observations on chipmunks and mantled squirrels in northeastern California. Am Mid Nat. 53:71-77.
10763. Tham, P, Johnson, K, Murphy, B, et al. 1987. Predicting age from body weight of New Zealand white rabbit fetuses. Lab Anim Sci. 37:795-797.
10764. Thau, RB, Lanman, JT, & Brinson, AO. 1976. Declining plasma progesterone concentration with advancing gestation in blood from umbilical and uterine veins and fetal heart in monkeys. Biol Reprod. 14:507-510.
10765. Thau, RB, Lanman, JT, & Brinson, AO. 1977. Metabolic clearance rates, production rates and concentrations of progesterone in pregnant rhesus monkeys. Biol Reprod. 16:678-681.
10766. Thau, RB, Lanman, JT, & Brinson, AO. 1979. Observations on progesterone production and clearance in normal pregnant and fetectomized rhesus monkeys (*Macaca mulatta*). J Reprod Fert. 56:37-43.
10767. Thesiger, W. 1986. The Marsh Arabs. Longman, London. [Cited by Mason & Macdonald, 1986]
10768. Theuring, F, Ucer, U, & Hansmann, I. 1987. Absence of specific luteinizing hormone-releasing hormone (LHRH) receptors in the ovaries of Djungarian hamsters (*Phodopus sungorus*). Biol Reprod. 36:825-828.
10769. Thiessen, DD, Friend, HC, & Lindzey, G. 1968. Androgen control of territorial marking in the Mongolian gerbil. Science. 160:432-434.
10770. Thing, H, Klein, DR, Jingfors, K, et al. 1987. Ecology of muskoxen in Jameson Land, northeast Greenland. Holarctic Ecol. 10:95-103.
10771. Thomas, AD. 1946. The cape dassie *Procavia capensis*. African Wild Life. 1:64-68.
10772. Thomas, AD & Kolbe, FF. 1942. The wild pigs of South Africa. J S Afr Vet Assoc. 13:1-11.
10773. Thomas, DC. 1970. The Ovary, Reproduction, and Productivity of Female Columbian Black-tailed Deer. Ph D Diss. Univ British Columbia, Vancouver.
10774. Thomas, DC. 1982. The relationship between fertility and fat reserves of Peary caribou. Can J Zool. 60:597-602.
10775. Thomas, DC. 1983. Age-specific fertility of female Columbian black-tailed deer. J Wildl Manage. 47:501-506.
10776. Thomas, DC & Cowan, IM. 1975. The pattern of reproduction in female Columbian black-tailed deer, *Odocoileus hemionus columbianus*. J Reprod Fert. 44:261-272.
10777. Thomas, DC & Smith, ID. 1973. Reproduction in a wild black-tailed deer fawn. J Mamm. 54:302-303.
10778. Thomas, DW. 1983. The annual migrations of three species of west African fruit bats (Chiroptera: Pteropodidae). Can J Zool. 61:2266-2272.
10779. Thomas, DW, Fenton, MB, & Barclay, RMR. 1979. Social behavior of the little brown bat, *Myotis lucifugus*. Behav Ecol Sociobiol. 6:129-136.
10780. Thomas, DW & Marshall, AG. 1984. Reproduction and growth in three species of West African fruit bats. J Zool. 202:262-281.
10781. Thomas, J, DeMaster, D, Stone, S, et al. 1980. Observations of a newborn Ross seal pup (*Ommatophoca rossi*) near the Antarctic peninsula. Can J Zool. 58:2156-2158.
10782. Thomas, J, Pastukhov, V, Elsner, R, et al. 1982. *Phoca sibirica*. Mamm Species. 188:1-6.
10783. Thomas, JA & Birney, EC. 1979. Parental care and mating system of the prairie vole, *Microtus ochrogaster*. Behav Ecol Sociobiol. 5:171-186.
10784. Thomas, O. 1904. On the osteology and systematic position of the rare Malagasy bat *Myzopoda aurita*. Proc Zool Soc London. 1904:2-6.
10785. Thomas, ME. 1972. Preliminary study of the animal breeding patterns and population fluctuations of bats in three ecologically distinct habitats in southwestern Colombia. Ph D Diss. Tulane Univ, New Orleans.
10785a. Thomas, ME & McMurray, DN. 1974. Observations on *Sturnira aratathomasi* from Colombia. J Mamm. 55:834-836.
10786. Thomas, WD. 1958. Observations on the breeding in captivity of a pair of lowland gorillas. Zoologica. 43:95-104.
10787. Thomas, WD. 1962. Postnatal development of gemsbok. J Mamm. 43:98-101.
10788. Thomas, WD. 1965. Observations on a pair of cheetahs *Acinonyx jubatus* at Oklahoma City Zoo. Int Zoo Yearb. 5:114-116.
10788a. Thomas, WD. 1975. Observations on captive brockets *Mazama americana* and *M gouazoubira*. Int Zoo Yearb. 15:77-78.
10789. Thomé, H. 1980. Vergleichend-anatomishe Untersuchungen der prae-und postnatalen Entwicklung und der funktionellen Veränderungen des Uterus von Rotwild Cervus elaphus Linné, 1758) sowie Altersberechnungen an Feten dieser Art. Schr Arbeitskreis Wildbiol u Jagdwiss J-Liebig Univ Giessen. 6:1-119.
10790. Thompson, CB, Holter, JB, Hayes, HH, et al. 1973. Nutrition of white-tailed deer I Energy requirements of fawns. J Wildl Manage. 37:301-311.
10791. Thompson, DC. 1978. Regulation of a northern grey squirrel (*Sciurus carolinensis*) population. Ecology. 59:708-715.

10792. Thompson, JGE & Breed, WG. 1982. The effect of the social environment on vaginal perforation and oestrus in the hopping-mouse *Notomys alexis*. Aust J Zool. 30:169-173.
10793. Thompson, MA, Shideler, SE, & Czekala, NM. 1983. Detection of ovulation and implantation in representative species across the order Primates employing a rapid, simple RIA. Biol Reprod. 28(Suppl):141.
10794. Thompson, MJA. 1987. Longevity and survival of female pipistrelle bats (*Pipistrellus pipistrellus*) on the Vale of York, England. J Zool. 211:209-214.
10795. Thompson, MK & Bradley, EL. 1979. A study of the circadian rhythm and pre- and postpuberty concentrations of serum prolactin in male prairie deer mice (*Peromyscus maniculatus bairdii*). Gen Comp Endocrinol. 39:208-214.
10796. Thompson, RW & Tirmer, JC. 1982. Temporal geographic variation in the lambing season of bighorn sheep. Can J Zool. 60:1781-1793.
10797. Thompson, SD & Nicoll, ME. 1986. Basal metabolic rate and energetics of reproduction in therian mammals. Nature. 321:690-693.
10798. Thompson-Handler, N, Malenky, RK, & Badrian, N. 1984. Sexual behavior of *Pan paniscus* under natural conditions in the Lomako forest, Equatur, Zaire. The Pygmy Chimpanzee. 347-368, Susman, RL, ed, Plenum Press, New York.
10799. Thomson, AG. 1972. A short note on the reproduction of some Ghananian rodents. Rev Zool Bot Afr. 85:343-351.
10800. Thomson, APD. 1954. The onset of oestrus in normal and blinded ferrets. Proc R Soc London Biol Sci. 142B:126-135.
10801. Thomson, CE. 1982. *Myotis sodalis*. Mamm Species. 163:1-5.
10802. Thomson, JA & Owen, WH. 1964. A field study of the Australian ringtail possum *Pseudocheirus peregrinus* (Marsupialia: Phalangeridae). Ecol Monogr. 34:27-52.
10803. Thomson, PJ. 1973. Notes on the oribi (Mammalia, Bovidae) in Rhodesia. Arnoldia (Rhodesia). 6(21):1-5.
10804. Thorburn, GD & Challis, JRG. 1979. Endocrine control in parturition. Physiol Rev. 59:863-917.
10805. Thorburn, GD, Cox, RI, & Shorey, CD. 1971. Ovarian steroid secretion rates in the marsupial *Trichosurus vulpecula*. J Reprod Fert. 24:139-140.
10806. Thordarson, G, Holekamp, KE, & Talamantes, F. 1987. Development of an homologous radioimmunoassay for secreted prolactin from the California ground squirrel *Spermophilus beecheyi*). Biol Reprod. 36:1186-1190.
10807. Thorington, RW, Jr, Rudran, R, & Mack, D. 1979. Sexual dimorphisms of *Alouatta seniculus* and observations on capture techniques. Vertebrate Ecology in the Northern Neotropics. 97-106, Eisenberg, JE, ed, Smithsonian Inst Press, Washington, DC.
10808. Thorington, RW, Jr, Ruiz, JC, & Eisenberg, JF. 1984. A study of a black howling monkey (*Alouatta caraya*) population in northern Argentina. Am J Primatol. 6:357-366.
10809. Thornback, J. 1982. The IUCN Res Mammal Red Data Book, Part 1. IUCN, Gland, Switzerland.
10810. Thornber, EJ, Renfree, MB, & Wallace, GI. 1981. Biochemical studies of intrauterine components of the tammar wallaby *Macropus eugenii* during pregnancy. J Embryol Exp Morphol. 62:325-338.
10811. Thorne, ET, Dean, RE, & Hepworth, WG. 1976. Nutrition during gestation in relation to successful reproduction in elk. J Wildl Manage. 40:330-335.
10811a. Thornton, JE, Irving, S, & Goy, RW. 1991. Effects of prenatal antiandrogen treatment on masculinization and defeminization of guinea pigs. Physiol Behav. 50:471-475.
10812. Thorpe, PA & Herbert, J. 1976. Studies on the duration of the breeding season and photorefractoriness in female ferrets pinealectomized or treated with melatonin. J Endocrinol. 70:255-262.
10813. Thrasher, JD. 1969. Preliminary observations on mate compatibility in *Marmosa mitis*. Lab Anim Care. 19:67-70.
10814. Thrasher, JD, Barenfus, M, Rich, ST, et al. 1971. The colony management of *Marmosa mitis*, the pouchless opossum. Lab Anim Sci. 21:526-536.
10815. Tiba, T. 1981. Jahreszeitliche Schwankungen der Spermatogenese des Japanischen Makaken (*Macaca fuscata*) in Gefangenschaft insbesondere in Vergleich mit freilebenden Gruppen. Z Säugetierk. 46:352-363.
10815a. Tiba, T & Kita, I. 1990. Undifferentiated spermatogonia and their role in the seasonally fluctuating spermatogenesis in the ferret *Mustela putorius furo* (Mammalia). Zool Anz. 224:140-155.
10815b. Tiba, T & Kita, I. 1991. Undifferentiated spermatogonia and their role in the seasonally fluctuating spermatogenesis in the mink *Mustela vison*. Znat Histol Embryol. 20:118-128.
10816. Tiba, T & Nigi, H. 1980. Jahreszeitliche Schwankung in der Spermatogenese beim "free ranging" japanischen Makak (*Macaca fuscata*). Zool Anz. 204:371-387.
10817. Tiba, T, Sugimara, M, & Suzuki, Y. 1981. Kinetik der Spermatogenese bei der Wollhaargemse (*Capricornis crispus*) I Geschlechtsreife and jahrzeitliche Schwankung. Zool Anz. 207:16-24.
10818. Tiba, T, Sugimura, M, & Suzuki, Y. 1981. Kinetik der Spermatogenese bei der Wollhaargemse (*Capricornis crispus*) II Samenepithelzyklus and Samenepithelwelle. Zool Anz. 207:25-34.
10818a. Tiba, T, Takahashi, M, Igura, M, et al. 1992. Enhanced proliferation of undifferentiated spermatogonia after treatment of short photoperiod exposure in the Syrian hamster, *Mesocricetus auratus*. Anat Histol Embryol. 21:9-22.
10819. Tibbitts, FD & Hillemann, HH. 1959. The development and histology of the chinchilla placenta. J Morphol. 105:317-346.
10820. Tibbitts, FD & King, BF. 1975. The fine structure of the trophospongial layer of the kangaroo rat placenta. Anat Rec. 183:567-578.
10821. Tiehua, H, Chonghui, L, Yaoliang, Q, et al. 1980. Growth and development of the bandicoot rat, *Bandicota indica* (Bechstein). Acta Zoologica Sinica. 26:?-392.
10822. Tiemeier, OW. 1965. Bionomics in the black-tailed rabbit in Kansas. KS St Univ Agric Exp Sta Tech Bull. 140:5-40.
10823. Tien, D Van. 1961. Notes sur une collection de micro Mammifères de la région de Hon-Gay. Zool Anz. 166:290-298.
10824. Tien, D Van. 1963. Etude préliminaire de la faune des Mammifères de la région de Phu-Quy (Province NGHE-AN, centre Vietnam). Zool Anz. 171:448-456.
10825. Tien, D Van. 1966. Données sur la biologie du petit Écureuil raye de Hainan (*Callosciurus swinhoei hainanus*) au Nord-Vietnam. Z Säugetierk. 31:478-479.
10826. Tien, D Van. 1966. Sur une deuxième collection des Mammifères de la région de Yen-Bai (Nord-Vietnam). Staatl Mus Tierk (Dresden) Zool Abh. 28:285-292.
10827. Tien, D Van. 1968. Donnees sur l'écologie de l'Écureuil à ventre rouge de Manipur (*Callosciurus erythraeus erythrogaster* Blyth) au Nord-Vietnam. Zool Garten NF. 35:74-76.
10828. Tien, D Van. 1972. Données écologique sur l'Écureuil géant de McClelland (*Ratufa bicolor gigantea* (Rodentia, Sciuridae) au Vietnam. Zool Garten NF. 41:240-243.

10829. Tien, D Van. 1977. Données sur l'écologie et la biologie de l'Athérure malais (*Atherurus macrourus* L) (Rodentia: Hystricidae) au Nord-Vietnam. Zool Garten NF. 47:69-71.
10830. Tien, D Van & Sung, CV. 1971. Données écologiques sur les Rats de bambou (*Rhizomys pruinosus* Blyth et *Rhizomys sumatrensis cinerus* McClelland) au Vietnam. Zool Garten NF. 40:227-231.
10831. Tien, D Van & Sung, CV. 1975. Données sur la biologie et l'écologie du Porc-Épic (*Acanthion subcristatum*) (Swinhoe) (Rodentia, Hystricidae) au Nord-Vietnam. Zool Garten NF. 45:487-490.
10832. Tijskens, J. 1971. The oestrous cycle and gestation period of the mountain gorilla *Gorilla gorilla beringei*. Int Zoo Yearb. 11:181-183.
10833. Tikhomirov, EA. 1966. On the reproduction of seals belonging to the family Phocidae in the North Pacific. Zool Zhurn. 45:275-281.
10834. Tikhomirov, EA. 1975. Biology of the ice forms of seals in the Pacific section of the Antarctic. Rapp P-V Réun Cons Int Explor Mer. 169:409-412.
10835. Tileston, JV & Lechleitner, RR. 1966. Some comparisons of the black-tailed and white-tailed prairie dogs in north-central Colorado. Am Mid Nat. 75:292-316.
10836. Tillson, SA, Swisher, DA, Pharriss, BB, et al. 1976. Interrelationships between pituitary gonadotrophins and ovarian steroids in baboons during continuous intrauterine progesterone treatment. Biol Reprod. 15:291-296.
10836a. Tilson, RL. 1977. Social organization of simakobu monkeys (*Nasalis concolor*) in Siberut Island, Indonesia. J Mamm. 58:202-212.
10837. Timisjarvi, J, Nieminen, M, Roine, K, et al. 1982. Growth in the reindeer. Acta Vet Scand. 23:603-618.
10838. Timm, RM. 1974. Rediscovery of the rock vole (*Microtus chrotorrhinus*) in Minnesota. Can Field-Nat. 88:82-82.
10839. Timm, RM. 1975. Distribution, natural history, and parasites of mammals of Cook County, Minnesota. Occas Pap Bell Mus Nat Hist Univ Minn. 14:1-56.
10840. Timm, RM. 1982. *Ectophylla alba*. Mamm Species. 166:1-4.
10841. Timm, RM. 1985. *Artibeus phaeotis*. Mamm Species. 235:1-6.
10842. Timm, RM, Heaney, LR, & Baird, DD. 1977. Natural history of rock voles (*Microtus chrotorrhinus*) in Minnesota. Can Field-Nat. 91:177-181.
10843. Timm, RM & Kermott, LH. 1982. Subcutaneous and cutaneous melanins in *Rhabdomys*: Complementary ultraviolet radiation shields. J Mamm. 63:16-22.
10844. Timmermans, PJA, Schouten, WGP, & Krijnen, JCM. 1981. Reproduction of cynomolgus monkeys (*Macaca fascicularis*) in harems. Lab Anim. 15:119-123.
10845. Timofejew, WK. 1934. Materialen zur Biologie und Oekologie der Säugetiere auf der Insel Barsa-Kelmes im Aralsee im Zusammenhang mit der Akklimatisation der gelben Zieselmaus (*Citellus fulvus* Licht) auf dieser Insel. Zool Zhurn. 13:731-758.
10846. Tims, HWM. 1910. Report of a collection of seal-embryos (*Leptonychotes weddelli*) made during the voyage of the "Discovery" in the Antarctic seas, 1901-1904. Nat Antarctic Exp 1901-1904, Nat Hist Zool Bot. 5:1-23.
10847. Tingari, MD, Rahma, BA, & Saad, HM. 1984. Studies on the poll glands of the one-humped camel in relation to reproductive activity I Seasonal morphological and histochemical changes. J Anat. 138:193-205.
10848. Tingari, MD, Ramos, AS, Gail, ESE, et al. 1984. Morphology of the testis of the one-humped camel in relation to reproductive activity. J Anat. 139:133-143.
10849. Tinklepaugh, OL. 1930. Occurrence of vaginal plug in a chimpanzee. Anat Rec. 46:329-332.
10850. Tinklepaugh, OL. 1933. Sex cycles and other cyclic phenomena in a chimpanzee during adolescence, maturity and pregnancy. J Morphol. 54:521-547.
10851. Tinley, KL. 1969. Dikdik *Madoqua kirki* in South West Africa: Notes on distribution ecology and behavior. Madoqua. 1:7-33.
10852. Tobayama, T, Uchida, S, & Nishiwaki, M. 1969. A white bellied right whale dolphin caught in the waters of Ibaragi, Japan. J Mamm Soc Japan. 4:112-120.
10853. Tobayama, T, Uchida, S, & Nishiwaki, M. 1970. Twin foetuses from a blue whale dolphin. Sci Rep Whales Res Inst. 22:159-162.
10854. Tobet, SA & Baum, MJ. 1987. Role for prenatal estrogen in the deveolpment of masculine sexual behavior in the male ferret. Horm Behav. 21:419-429.
10855. Tobet, SA, Shim, JH, Osiecki, ST, et al. 1985. Androgen aromatization and 5α-reduction in ferret brain during perinatal development: Effects of sex and testosterone manipulation. Endocrinology. 116:1869-1877.
10856. Todd, AW. 1985. Demographic and dietary comparisons of forest and farmland coyote, *Canis latrans*, populations in Alberta. Can Field-Nat. 99:163-171.
10857. Todd, AW & Keith, LB. 1983. Coyote demography during a snowshoe hare decline in Alberta. J Wildl Manage. 47:394-404.
10858. Todhunter, R & Gemmell, RT. 1987. Seasonal changes in the reproductive tract of the male marsupial bandicoot, *Isoodon macrourus*. J Anat. 154:173-186.
10859. Toepfer, I. 1973. Milchzuzammensetzung vom Reh (*Capreolus c capreolus* L). Zool Garten NF. 43:147.
10860. Toivola, PTK, Bridson, WE, & Robinson, JA. 1978. Effect of luteinizing hormone-releasing hormone on the secretion of luteinizing hormone, follicle-stimulating hormone, and testosterone in adult male rhesus monkeys. Endocrinology. 102:1815-1821.
10861. Toll, JE, Baskett, TS, & Conaway, CH. 1960. Home range, reproduction, and foods of the swamp rabbit in Missouri. Am Mid Nat. 63:398-412.
10862. Tomich, PQ. 1962. The annual cycle of the California ground squirrel *Citellus beecheyi*. Univ CA Publ Zool. 65:213-282.
10863. Tomich, PQ & Devick, WS. 1970. Age criteria for the prenatal and immature mongoose in Hawaii. Anat Rec. 167:107-113.
10864. Tomilin, MI. 1936. Length of gestation period and menstrual cycle in the chimpanzee. Nature. 137:318-319.
10865. Tomilin, MI. 1940. Menstrual bleeding and genital swelling in *Miopithecus* (*Cercopithecus*) *talapoin*. Proc Zool Soc London. 110(A):43-45.
10866. Tomkins, IR. 1935. The marsh rabbit: An incomplete life history. J Mamm. 16:201-205.
10867. Tong, EH. 1958. Notes on the breeding of Indian rhinoceros, *Rhinoceros unicornis* at Whipsnade Park. Proc Zool Soc London. 130:296-299.
10868. Tong, EH. 1960. The breeding of the great Indian rhinoceros at Whipsnade park. Int Zoo Yearb. 2:12-15.
10869. Tong, JR. 1974. Breeding cheetahs, *Acinonyx jubatus*, at the Beekse Bergen Safari Park. Int Zoo Yearb. 14:129-130.

10870. Tonkin, BA & Kohler, E. 1978. Breeding the African golden cat *Felis* (*Profelis*) *auratus* in captivity. Int Zoo Yearb. 18:147-150.
10871. Tonkin, BA & Kohler, E. 1981. Observations on the Indian desert cat *Felis silvestris ornata*. Int Zoo Yearb. 21:151-154.
10872. Topal, G. 1974. Field observations on oriental bats: Sex ratio and reproduction. Vertebr Hung. 15:83-94.
10873. Torii, H & Miyake, T. 1986. Litter size and sex ratio of the masked palm civet, *Paguma larvata*, in Japan. J Mamm Soc Japan. 11:35-38.
10874. Torres, CN, Godinho, HP, & Setchell, BP. 1981. Frequency and duration of the stages of the cycle of the seminiferous epithelium of the nine-banded armadillo (*Dasypus novemcinctus*). J Reprod Fert. 61:355-340.
10875. Torres, S, Hulot, F, Meunier, M, et al. 1987. Comparative study of preimplantation development and embryonic loss in two rabbit strains. Reprod Nutr Dév. 27:707-714.
10876. Tovey, PE. 1981. Unpublished data, AK Coop Wildl Res Unit. [Cited by Keith, 1981]
10877. Toweill, DE & Toweill, DB. 1978. Growth and development of captive ringtails (*Bassariscus astutus flavus*). Carnivore. 1(3):46-53.
10878. Towers, PA & Martin, L. 1985. Some aspects of female reproduction in the grey-headed flying-fox, *Pteropus poliocephalus* (Megachiroptera: Pteropodidae). Aust Mamm. 8:257-263.
10879. Towers, PA & McGuckin, MA. 1984. Reproduction in the grey headed fruit bat (*Pteropus poliocephalus*). Bull Aust Mamm Soc. 8:86.
10880. Towers, PA, Shaw, G, & Renfree, MB. 1986. Urogenital vasculature and local steroid concentrations in the uterine branch of the ovarian vein of the female tammar wallaby (*Macropus eugenii*). J Reprod Fert. 78:37-47.
10881. Townsend, TW & Bailey, ED. 1975. Parturitional, early maternal, and neonatal behavior in penned white-tailed deer. J Mamm. 56:347-362.
10882. Townsley, JD. 1974. Estrogen excretion of the pregnant baboon (*Papio papio*). Endocrinology. 95:1759-1762.
10883. Toyama, Y, Calderon, FU, & Quesada, R. 1990. Ultrastructural study of crystalloids in Sertoli cells of the three-toed sloth (*Bradypus tridactylus*). Cell Tiss Res. 259:599-602.
10883a. Toyama, Y, Obinata, T, & Holtzer, H. 1979. Crystalloids of actin-like filaments in the sertoli cell of the swine testis. Anat Rec. 195:47-62.
10884. Trainer, CE. 1971. An investigation concerning the fertility of female Roosevelt elk (*Cervus canadensis roosevelti*) in Oregon. Proc West Assoc Game Fish Comm. 1971:378-385.
10885. Trapido, H. 1949. Gestation period, young, and maximum weight of the Isthmian capybara, *Hydrochoerus isthmius* Goldman. J Mamm. 30:433.
10886. Trapp, GR. 1962. Snowshoe hares in Alaska II Home range and ecology during early population increase. MS Thesis. Univ AK. [Cited by Keith, 1981]
10887. Trapp, GR. 1978. Comparative behavioral ecology of the ringtail and gray fox in southwestern Utah. Carnivore. 1(2):3-32.
10888. Tratz, EP. 1957. Beiträge zur Kenntnis der embryonalen Entwicklung der Gemse (*Rupicapra rupicapra*). Zool Garten NF. 23:194-220.
10889. Travis, TC & Holmes, WN. 1974. Some physiological and behavioural changes associated with oestrus and pregnancy in the squirrel monkey (*Saimiri sciureus*). J Zool. 174:41-66.
10890. Trebbau, P. 1972. Measurements and some observations on the freshwater dolphin *Inia geoffrensis* in the Apure River (Venezuela). Typewritten circular. 21:8. [Cited by Gilleri & Gihr, 1977]
10891. Trebbau, P. 1978. Some observations on the mating behavior of the Brazilian giant otter *Pteronura brasiliensis*. Zool Garten NF. 48:187-188.
10892. Trebbau, P. 1980. Some observations on the capybara (*Hydrochoerus hydrochaeris*). Zool Garten NF. 50:40-44.
10893. Tretheway, DEC & Verts, BJ. 1971. Reproduction in eastern cottontail rabbits in western Oregon. Am Mid Nat. 86:463-476.
10894. Treus, VD & Kravchenko, D. 1968. Methods of rearing and economic utilization of eland in the Askaniya-Nova Zoological Park. Symp Zool Soc London. 21:395-411.
10895. Treus, VD & Lobanov, NV. 1971. Acclimatisation and domestication of the eland *Taurotragus oryx* at Askanya-Nova Zoo. Int Zoo Yearb. 11:147-156.
10896. Treus, VD & Lobanov, NV. 1974. *Equus hemionus* Pall, 1775 and its reproduction. Vestnik Zool Kiev. 3:11-18.
10897. Treussier, M. 1977. Dynamics of a population of field mouse. Acta Theriol. 22:207-214.
10898. Trillmich, F. 1984. The Galapagos seals Part 2 Natural history of the Galapagos fur seal *Arctocephalus galapagoensis*, Heller. Key Evironments: Galapagos. 215-223, Perry, R, ed, Pergamon, Oxford, UK.
10899. Trillmich, F. 1986. Attendance behavior of Galapagos fur seals. Fur Seals Maternal Strategies on Land and at Sea. 168-185, Gentry, RL & Kooyman, GL, eds, Princeton Univ Press, Princeton.
10900. Trillmich, F. 1986. Attendance behavior of Galapagos sea lions. Fur Seals Maternal Strategies on Land and at Sea. 196-208, Gentry, RL & Kooyman, GL, eds, Princeton Univ Press, Princeton.
10901. Trillmich, F. 1986. Maternal investment and sex-allocation in the Galapagos fur seal, *Arctocephalus galapagoensis*. Behav Ecol Sociobiol. 19:157-164.
10902. Trillmich, F, Kirchmeier, D, Kirchmeier, O, et al. 1988. Characterization of proteins and fatty acid composition in Galapagos fur seal milk Occurence of whey and casein protein polymorphisms. Comp Biochem Physiol. 90B:447-452.
10903. Trillmich, F, Kooyman, GL, Majluf, P, et al. 1986. Attendance and diving behaviors of South American Fur Seals during El Niño in 1983. Fur Seals Maternal Strategies on Land and at Sea. 153-167, Gentry, RL & Kooyman, GL, eds, Princeton Univ Press, Princeton.
10904. Trillmich, F & Lechner, E. 1986. Milk of the Galapagos seal and sea lion, with a comparison of the milk of eared seals. J Zool. 209A:271-277.
10905. Trimberger, GW & Fincher, MG. 1956. Regularity of estrus, ovarian function, and conception rates in dairy cattle. Agric Exp Stn Bull Cornell Univ. 911:1-23.
10906. Tripp, HRH. 1971. Reproduction in elephant-shrews (Macroscelididae) with special reference to ovulation and implantation. J Reprod Fert. 26:149-159.
10907. Tripp, HRH. 1972. Capture, laboratory care and breeding of elephant-shrews (Macroscelididae). Lab Anim. 6:213-224.
10908. Trippensee, RE. 1936. The reproductive function in the cottontail rabbit (*Sylvilagus floridanus mearnsii* Allen) in southern Michigan. Proc N Am Wildl Conf. 1:344-350.
10909. Trodd, LL. 1962. Quadruplet fetuses in a white-tailed deer from Espanola, Ontario. J Mamm. 43:414.
10910. Troitzky, A. 1953. Contribution à l'étude des Pinnipèdes à propos de deux Phoques de le Mediterranée ramenés de croisière par S A S le Prince Rainier III de Monaco. Bull Inst Oceanog. 1032:1-46.

10911. Trojan, P & Wojciechowska, B. 1967. Resting metabolism rate during pregnancy and lactation in the European common vole - *Microtus arvalis* (Pall). Ekologia Polska - Seria A. 15:811-817.
10912. Trollope, J & Blurton-Jones, NG. 1970. Breeding the stump-tailed macaque, *Macaca arctoides*. Lab Anim. 4:161-169.
10913. Trollope, J & Blurton-Jones, NG. 1972. Age of sexual maturity in the stump-tailed macaque (*Macaca arctoides*): A birth from laboratory born parents. Primates. 13:229-230.
10914. Trollope, J & Blurton-Jones, NG. 1975. Aspects of reproduction and reproductive behavior in *Macaca arctoides*. Primates. 16:191-206.
10915. Trombulak, SC. 1987. Life history of the Cascade golden-mantled ground squirrel (*Spermophilus saturatus*). J Mamm. 68:544-554.
10916. Trombulak, SC. 1988. *Spermophilus saturatus*. Mamm Species. 322:1-4.
10917. Troughton, E. 1947. Furred Animals of Australia. C Scribner's Sons, New York.
10918. Trowbridge, BJ. 1983. Olfactory communication in the European otter (*Lutra lutra*). Ph D Diss, Univ Aberdeen, Aberdeen. [Cited by Mason & Macdonald, 1986]
10919. Troyer, JR. 1908. Neurosecretory material in the supraoptico-hypophyseal tract of the bat throughout the hibernating and summer periods. Anat Rec. 162:407-424.
10920. Trulio, LA, Loughry, WJ, Hennessy, DF, et al. 1986. Infanticide in California ground squirrels. An Behav. 34:291-294.
10921. Trupin, GL & Fadem, BH. 1982. Sexual behavior of the gray short-tailed opossum (*Monodelphis domestica*). J Mamm. 63:409-414.
10922. Tryon, CA, Jr. 1946. Montana records of beaver embryos. J Mamm. 27:396-397.
10923. Tryon, CA, Jr. 1947. Behavior and postnatal development of a porcupine. J Wildl Manage. 11:282-283.
10924. Tryon, CA, Jr. 1947. The biology of the pocket gopher (*Thomomys talpoides*) in Montana. MT State Coll Agric Exp Sta Tech Bull. 448:1-30.
10924a. Tryon, CA, Jr & Buck, PD. 1950. Montana records of antelope embryos and reproductive tracts. J Mamm. 31:192-193.
10925. Tsapljuk, OE. 1977. Age and seasonal patterns of reproductive biology in *Capreolus capreolus* in Kazakhstan. Zool Zhurn. 56:611-618.
10926. Tsellarius, YG, Semenova, LA, Tsimmerman, VG, et al. 1986. Influence of domestication on seasonal changes in stereological parameters of the testes in silver-back foxes. Proc Acad Sci USSR (Translations). 283:499-502.
10927. Tsonis, CG, Gaughwin, MD, & Breed, WG. 1981. Biochemistry of male accessory organs of conilurine rodents. Archs Androl. 6:239-242.
10927a. Tsubota, T. 1990. Studies on reproductive physiology of Hokkaido brown bears, *Ursus arctos yesoensis*. Jpn J Anim Reprod. 36:1P-10P.
10928. Tsubota, T & Kanagawa, H. 1989. Annual changes in serum testosterone levels and spermatogenesis in the Hokkaido brown bear, *Ursus arctos yesoensis*. J Mamm Soc Japan. 14:11-17.
10929. Tsubota, T, Kanagawa, H, Mano, T, & Aoi, T. 199?. Corpora albicantia and placental scars in the Hokkaido brown bear. Int Conf Bear Res Manage. 8:125-128.
10930. Tsui, HW, Tam, WH, Lofts, B, et al. 1974. The annual testicular cycle and androgen production *in vitro* in the masked civet cat *Paguma l larvata*. J Reprod Fert. 36:283-293.
10931. Tullner, WW. 1968. Urinary chorionic gonadotropin excretion in the monkey (*Macaca mulatta*): Early phase. Endocrinology. 82:874-875.
10932. Tullner, WW. 1972. Chorionic gonadotrophin in non-human primates. Acta Endocrinol Suppl. 166:200-213.
10933. Tullner, WW & Gray, CW. 1968. Chorionic gonadotropin excretion during pregnancy in a gorilla. Proc Soc Exp Biol Med. 128:954-956.
10934. Tullner, WW & Gray, CW. 1971. Unpublished results. [Cited by Tullner, 1972]
10935. Tullner, WW & Hertz, R. 1966. Chorionic gonadotropin levels in the rhesus monkey during early pregnancy. Endocrinology. 78:204-207.
10936. Tullner, WW & Hertz, R. 1966. Normal gestation and chorionic gonadotropin levels in the monkey after ovariectomy in early pregnancy. Endocrinology. 78:1076-1078.
10937. Tullner, WW & Hodgen, GD. 1974. Effects of fetectomy on plasma estrogens and progesterone in monkeys (*Macaca mulatta*). Steroids. 24:887-897.
10938. Tulloch, DG. Personal communication. [Cited by Sinclair, The African Buffalo, 1977]
10939. Tulloch, DG. 1979. The water buffalo, *Bubalus bubalis*, in Australia: Reproductive and parent-offspring behavior. Aust Wildl Res. 6:265-287.
10940. Tulloch, DG & Grassia, A. 1981. A study of reproduction in water buffalo in the Northern Territory of Australia. Aust Wildl Res. 8:335-348.
10941. Tumbleson, ME, Middleton, CC, Tinsley, OW, et al. 1969. Body weights and measurements of Hormel miniature swine from birth to nine months of age. Lab Anim Care. 19:596-601.
10942. Tumlison, R. 1987. *Felis lynx*. Mamm Species. 269:1-8.
10942a. Tumlinson, R. 1992. *Plecotus mexicanus*. 401:1-3.
10943. Tung, KSK, Ellis, LE, Childs, GV, et al. 1984. The dark mink: A model of male infertility. Endocrinology. 114:922-929.
10944. Tupikova, NV & Shvetzov, UG. 1956. Reproduction of the water-vole in the Volga-Akhtubinski region during the spring highwater period. Zool Zhurn. 35:130-140.
10945. Turcek, FJ. 1956. Über den Mufflon, *Ovis musimon* Schreber, 1782, in der Slowakei (CSR). Säugetierk Mitt. 4:167-171.
10946. Türcke, F & Schmincke, S. 1965. Das Muffelwild. Parey, Hamburg.
10947. Turek, FW, Desjardins, C, & Menaker, M. 1976. Differential effects of melatonin on the testes of photoperiodic and nonphotoperiodic rodents. Biol Reprod. 15:94-97.
10948. Turner, BN, Iverson, SL, & Severson, KL. 1976. Postnatal growth and development of captive Franklin's ground squirrels (*Spermophilus franklinii*). Am Mid Nat. 95:93-102.
10949. Turner, E. 1971. Birth of an Amazon dolphin, *Inia geoffrensis*, at Forth Worth zoological park. Zool Garten NF. 40:187-192.
10950. Turner, JC & Hansen, CG. 1980. Reproduction. The Desert Bighorn. 145-151, Monson, G & Sumner, L, eds, Univ AZ Press, Tucson.
10951. Turner, JJ & Herndon, JG. 1983. Seasonal changes in reproductive behavior in two ovariectomized female rhesus monkeys treated year-round with estradiol. Am J Primatol. 4:171-177.
10952. Turner, JW, Jr. 1972. Radioimmunoassay studies on plasma and pituitary LH levels and the half-life of circulating LH in intact and castrated gerbils. Endocrinology. 90:638-644.
10953. Turner, LW. 1972. Autecology of the Belding ground squirrel in Oregon. Ph D Diss. Univ AZ, Tucson.

10954. Turner, RW. 1974. Mammals of the Black Hills of South Dakota and Wyoming. Univ KS Mus Nat Hist Misc Publ. 60:1-178.
10954a. Turner, RW & Jones, JK, Jr. 1968. Additional notes on bats from western South Dakota. Southwest Nat. 13:444-447.
10955. Turner, RW, Padmowirjono, S, & Martoprawiro, S. 1975. Dynamics of the plague transmission cycle in central Java. Bull Penelitian Kesehatan. 3:41-71.
10956. Turner, W. 1871. On the gravid uterus and on the arrangements of the foetal membranes in the Cetacea. Trans R Soc Edinb. 26:467-504.
10957. Turner, (W). 1873. On the placentation of the sloths. J Anat. 7:302-303.
10958. Turner, (W). 1874. On the placentation of the sloths. J Anat. 8:362-376.
10959. Turner, W. 1875. Note on the placentation of hyrax (*Procavia capensis*). Proc R Soc London. 24:151-155.
10960. Turner, W. 1875-76. A further contribution to the placentation of the Cetacea (*Monodon monoceros*). Proc R Soc Edinb. 9:103-110.
10961. Turner, (W). 1876. On the placentation of the cape ant-eater (*Orycteropus capensis*). J Anat. 10:693-706.
10962. Turner, (W). 1876. On the placentation of the lemurs. Phil Trans R Soc London. 166:569-587.
10963. Turner, W. 1876. On the placentation of the sloths. Trans R Soc Edinb. 27:71-104.
10964. Turner, (W). 1878. On the placenta of the hog-deer (*Cervus porcinus*). J Anat. 13:94-98.
10965. Turner, (W). 1879. On the cotyledonary and diffused placenta of the Mexican deer (*Cervus mexicanus*). J Anat. 13:195-200.
10966. Turner, W. 1889. On the placentation of the Halicore dugong. Trans R Soc Edinb. 35:641-662.
10967. Turner, W. 1889. On the placentation of the Halicore dugong. Proc R Soc Edinb. 16:264.
10968. Turner, W. 1889. The placentation of the *Halicore dugong*. J Anat. 23:640-641.
10969. Turner, W. 1892. The lesser rorqual (*Balaenoptera rostrata*) in the Scottish seas, with observations on its anatomy. Proc R Soc Edinb. 19:36-74.
10970. Tutin, CEG. 1979. Mating patterns and reproductive strategies in a community of wild chimpanzees (*Pan troglodytes schweinfurthii*). Behav Ecol Sociobiol. 6:29-38.
10971. Tutin, CEG. 1980. Reproductive behaviour of wild chimpanzees in the Gombe National Park, Tanzania. J Reprod Fert. 28:43-57.
10972. Tutin, CEG & McGinnis, PR. 1981. Chimpanzee reproduction in the wild. Reproductive Biology of the Great Apes. 239-279, Graham, CE, ed, Academic Press, New York.
10973. Tutin, CEG & McGrew, WC. 1973. Sexual behavior of group-living adolescent chimpanzees. Am J Phys Anthrop. 38:195-200.
10974. Tuttle, MD. 1970. Distribution and zoogeography of Peruvian bats, with comments on natural history. Univ KS Sci Bull. 49:45-86.
10975. Tuttle, MD. 1975. Population ecology of the gray bat (*Myotis grisescens*): Factors influencing early growth and development. Occas Pap Mus Nat Hist Syst Ecol. 36:1-24.
10976. Tuttle, MD. 1976. Population ecology of the gray bat (*Myotis grisescens*): Philopatry, timing and patterns of movement, weight loss during migration, and seasonal adaptive strategies. Occas Pap Mus Nat Hist KS. 54:1-38.
10977. Tuttle, MD. 1976. Population ecology of the gray bat (*Myotis grisescens*): Factors influencing growth and survival of newly volant young. Ecology. 57:587-595.
10978. Tuttle, MD. 1979. Status, causes of decline and mangement of endangered gray bats. J Wildl Manage. 43:1-17.
10979. Tuttle, MD & Stevenson, D. 1982. Growth and survival of bats. Ecology of Bats. 105-150, Kunz, TH, ed, Plenum Press, New York.
10980. Twelves, J. 1986. Unpublished data. [Cited by Mason & Macdonald, 1986]
10981. Twichell, AR. 1939. Notes on the southern woodchuck in Missouri. J Mamm. 20:71-74.
10982. Twigg, GI. 1962. Notes on *Holochilus sciureus* in British Guiana. J Mamm. 43:369-374.
10983. Twigg, GI. 1965. Studies on *Holochilus sciureus berbicensis*, a cricetine rodent from the coastal region of British Guiana. Proc Zool Soc London. 145:263-283.
10984. Tyler, SJ. 1972. The behavior and social organization of the new forest ponies. An Behav Monogr. 5:87-196.
10985. Tyndale-Biscoe, CH. 1955. Observations on the reproduction and ecology of the brush-tailed possum, *Trichosurus vulpecula* Kerr (Marsupialia), in New Zealand. Aust J Zool. 3:162-184.
10986. Tyndale-Biscoe, CH. 1963. Effects of ovariectomy in the marsupial *Setonix brachyurus*. J Reprod Fert. 6:25-40.
10987. Tyndale-Biscoe, CH. 1963. Blastocyst transfer in the marsupial *Setonix brachyurus*. J Reprod Fert. 6:41-48.
10988. Tyndale-Biscoe, CH. 1963. The role of the corpus luteum in the delayed implantation of marsupials. Delayed Implantation. 15-28, Enders, AC, ed, Univ Chicago Press, Chicago.
10989. Tyndale-Biscoe, CH. 1965. The female urogenital system and reproduction of the marsupial *Lagostrophus fasciatus*. Aust J Zool. 13:255-267.
10990. Tyndale-Biscoe, CH. 1966. The marsupial birth canal. Symp Zool Soc London. 15:233-250.
10991. Tyndale-Biscoe, CH. 1968. Reproduction and postnatal development in the marsupial *Bettongia lesueur* (Quoy and Gaimard). Aust J Zool. 16:577-602.
10992. Tyndale-Biscoe, CH. 1969. Relaxin activity during the oestrous cycle of the marsupial, *Trichosurus vulpecula*. J Reprod Fert. 19:191-193.
10993. Tyndale-Biscoe, CH. 1970. Resumption of development by quiescent blastocysts transferred to primed, ovariectomized recipients in the marsupial, *Macropus eugenii*. J Reprod Fert. 23:25-32.
10994. Tyndale-Biscoe, CH. 1974. Reproduction in marsupials. Aust Mamm. 1:175-180.
10995. Tyndale-Biscoe, CH. 1979. Hormonal control of embryonic diapause and reactivation in the tammar wallaby. CIBA Found Symp 64 (Maternal Recognition of Pregnancy). 173-185, Excerpta Medica, Amsterdam.
10996. Tyndale-Biscoe, CH. 1980. Photoperiod and the control of seasonal reproduction in marsupials. Endocrinology 1980. 277-282, Cumming, IA, Funder, JW, & Mendelsohn, FAO, eds, Aust Acad Sci, Canberra.
10997. Tyndale-Biscoe, CH. 1981. Evidence for relaxin in marsupials. Relaxin. 225-230, Bryant-Greenwood, GD, Niall, HD, & Greenwood, FC, eds, Elsevier/North Holland, Amsterdam.
10998. Tyndale-Biscoe, CH. 1984. Reproductive physiology of possums and gliders. Possums and Gliders. 79-87, Smith, A & Hume, I, eds, Surrey Beatty & Sons Pty Ltd, Sydney, Australia.

10998a. Tyndale-Biscoe, CH. 1984. Mammals: Marsupials. Marshall's Physiology of Reproduction. 386-454, Lamming, GE, ed, Churchill Livingstone, London.
10999. Tyndale-Biscoe, CH. 1986. Embryonic diapause in a marsupial: Roles of the corpus luteum and pituitary in its control. Comparative Endocrinology: Developments and Directions. 137-155, Ralph, CL, ed, Alan R Liss, Inc, New York.
11000. Tyndale-Biscoe, CH & Hawkins, J. 1977. The corpora lutea of marsupials: Aspects of function and control. Proc 4th Symp Comp Biol Reprod, Canberra 1976. 245-251, Aust Acad Sci, Canberra.
11001. Tyndale-Biscoe, CH & Hearn, JP. 1981. Pituitary and ovarian factors associated with seasonal quiescence of the tammar wallaby, *Macropus eugenii.* J Reprod Fert. 63:225-230.
11001a. Tyndale-Biscoe, CH, Hearn, JP, & Renfree, MB. 1974. Control of reproduction in macropodid marsupials. J Endocrinol. 63:589-614.
11002. Tyndale-Biscoe, CH & Hinds, LA. 1981. Hormonal control of the corpus luteum and embryonic diapause in macropodid marsupials. J Reprod Fert Suppl. 29:111-117.
11003. Tyndale-Biscoe, CH & Hinds, LA. 1984. Seasonal pattern of circulating progesterone and prolactin and response to bromocriptine in the female tammar *Macropus eugenii.* Gen Comp Endocrinol. 53:58-68.
11004. Tyndale-Biscoe, CH & Hinds, LA. 1989. Influence of the immature testis on sexual differentiation in the tammar wallaby *Macropus eugenii* Macropodidae Marsupialia. Reprod Fert Dev. 1:243-254.
11005. Tyndale-Biscoe, CH, Hinds, LA, & Horn, CA. 1988. Fetal role in the control of parturition in the tammar, *Macropus eugenii.* J Reprod Fert. 82:419-428.
11006. Tyndale-Biscoe, CH, Hinds, LA, Horn, CA, et al. 1983. Hormonal changes at oestrus, parturition and post-partum oestrus in the tammar wallaby (*Macropus eugenii*). J Endocrinol. 96:155-161.
11007. Tyndale-Biscoe, CH & Mackenzie, RB. 1976. Reproduction in *Didelphis marsupialis* and *D albiventris* in Colombia. J Mamm. 57:249-265.
11008. Tyndale-Biscoe, CH & Renfree, M. 1987. Reproductive Physiology of Marsupials. Cambridge Univ Press. New York.
11009. Tyndale-Biscoe, CH & Rodger, JC. 1978. Differential transport of spermatozoa into the two sides of the genital tract of a monovular marsupial, the tammar wallaby (*Macropus eugenii*). J Reprod Fert. 52:37-43.
11010. Tyndale-Biscoe, CH & Smith, RCF. 1969. Studies on the marsupial glider, *Schoinobates volans* (Kerr) II Population structure and regulatory mechanisms. J Anim Ecol. 38:637-650.
11011. Tyser, RW. 1975. Taxonomy and reproduction of *Microtus canicaudus.* MS Thesis. OR State Univ, Corvallis.
11012. Tyson, E. 1698. Carigueya, seu Marsupiale Americanum; or the anatomy of an opossum, dissected at Gresham-College. Phil Trans R Soc London. 20:105-164.
11013. Uchida, T. 1951. Studies on the embryology of the Japanese house bat, *Pipistrellus tralatitius abramus* (Temminck) I On the period of gestation and the number of litter. Sci Bull Fac Agric Kyushu Univ. 12:11-14.
11014. Uchida, T. 1953. Studies on the embryology of the Japanese house bat, *Pipistrellus tralatitius abramus* (Temminck) II From the maturation of the ova to the fertilization, especially on the behaviour of follicle cells at the period of fertilization. Sci Bull Fac Agric Kyushu Univ. 14:153-168.
11015. Uchida, TA. 1957. Fertilization and hibernation in bats. Heridity, Tokyo. 11:14-17.
11016. Uchida, TA. 1966. Mammals of Japan (5): Order Chiroptera, genus *Pipistrellus.* Mamm Sci. 11:5-23. [Cited by Funakoski & Mehida, 1981]
11017. Uchida, TA, Inoue, C, & Kimura, K. 1984. Effects of elevated temperatures on the embryonic development and corpus luteum activity in the Japanese long-fingered bat, *Miniopterus schreibersi fuliginosus.* J Reprod Fert. 71:439-444.
11018. Uchida, TA & Mōri, T. 1987. Prolonged storage of spermatozoa in hibernating bats. Recent Advances in the Study of Bats. 351-365, Fenton, MB, Racey, P, & Rayner, JMV, eds, Cambridge Univ Press, New York.
11019. Uchida, TA, Mōri, T, & Oh, YK. 1984. Sperm invasion of the oviducal mucosa, fibroblastic phagocytosis and endometrial sloughing in the Japanese greater horseshoe bat, *Rhinolophus ferrumequinum nippon.* Cell Tiss Res. 236:327-331.
11020. Uchiyama, K. 1965. California sea-lion *Zalophus californicus* twins at Tokuyama Zoo. Int Zoo Yearb. 5:111.
11021. Udell, CC. 1981. Breeding the zebra duiker *Cephalophus zebra* at the Los Angeles Zoo. Int Zoo Yearb. 21:155-158.
11022. Udell, CC. 1984. Husbandry, breeding and behaviour of bongo *Tragelaphus eurycerus* at the Los Angeles Zoo. Int Zoo Yearb. 23:237-242.
11023. Ueckermann, E. 1956. Das Damwild. Paul Parey, Hamburg.
11024. Uhlig, HS. 1955. The determination of age of nestling and sub-adult gray squirrels in West Virginia. J Wildl Manage. 19:479-483.
11025. Uhlig, HG. 1956. Reproduction in the eastern flying squirrel in West Virginia. J Mamm. 37:295.
11026. Uieda, W, Sazima, I, & Filho, AS. 1980. Aspectos da biologia do morcego *Furipterus horrens* (Mammalia, Chiroptera, Furipteridae). Rev Brasil Biol. 40:59-66.
11027. Ulbrich, J. 1930. Die Bisamratte. Verlag Heinrich, Dresden.
11028. Ullmann, S. 1981. Observations on the primordial germ cells of bandicoots Peramelidae, Marsupalia. J Anat. 132:581-596.
11029. Ullmann, SL. 1984. Early differentiation of the testis in the native cat, *Dasyurus viverrinus* (Marsupialia). J Anat. 138:675-688.
11030. Ullmann, SL & Brown, R. 1983. Further observations on the potoroo (*Potorous tridactylus*) in captivity. Lab Anim. 17:133-137.
11031. Ullrey, DE, Schwartz, CC, Whetter, PA, et al. 1984. Blue-green color and composition of Stejneger's beaked whale (*Mesoplodon stejnegeri*) milk. Comp Biochem Physiol. 79B:349-352.
11032. Ullrich, W. 1961. Zur Biologie und Soziologie der Colobusaffen (*Colobus guereza caudatus* Thomas 1885). Zool Garten NF. 25:305-368.
11033. Ullrich, W. 1972. Zur Herdenordnung und Terriorialität des Barasingharhirsches (*Cervus duvauceli duvauceli*) in Assam. Zool Garten NF. 41:223-232.
11034. Ulmer, FA, Jr. 1957. Breeding of orang-utans. Zool Garten. 23:57-65.
11035. Ulmer, FA, Jr. 1961. Gestation period of the lion marmoset. J Mamm. 42:253-254.
11036. Ulmer, FA, Jr. 1963. Observations on the tarsier in captivity. Zool Garten NF. 27:106-121.
11037. Ulmer, FA, Jr. 1966. Gestation period of the Uganda kob antelope *Kobus kob thomasi.* Int Zoo Yearb. 6:204-205.
11038. Ulmer, FA, Jr. 1968. Breeding fishing cats *Felis viverrina* at Philadelphia Zoo. Int Zoo Yearb. 8:49-55.

11039. Underhill, JE. 1962. Notes on pika in captivity. Can Field-Nat. 76:177-178.
11040. Underwood, H, Whitsett, JM, & O'Brien, TG. 1985. Photoperiodic time measurement in the male deer mouse, *Peromyscus maniculatus*. Biol Reprod. 32:947-956.
11041. Underwood, LS. 1975. Notes on the arctic fox (*Alopex lagopus*) in the Prudhoe Bay area of Alaska. Biol Pap Univ AK Sp Rep. 2:145-149.
11042. Underwood, LS. 1983. Outfoxing the arctic cold. Nat Hist. 92(12):38-46.
11043. Underwood, R. 1978. Aspects of kudu ecology at Loskopdam Nature Reserve, eastern Transvaal. S Afr J Wildl Res. 8:43-47.
11044. Underwood, R. 1979. Mother-infant relationships and behavioural ontogeny in the common eland (*Taurotragus oryx oryx*). S Afr J Wildl Res. 9:27-45.
11044a. United States National Museum specimen tags. 1991. Examined by V Hayssen.
11045. Urbain, A. 1949. Sur la biologie du Castor du Rhône (*Castor fiber* L). Bijd Dierk. 28:472-476.
11046. Urbain, A. 1954. Biologie de l'Orycterope. Ann Mus R Congo Belge N S Sci Zool. 1:101-105.
11047. Urbain, A, Bullier, P, & Nouvel, J. 1945. Naissance d'un Élephant d'Asie (*Elephas maximus* L) au Parc zoologique du Bois de Vincinnes. Mammalia. 9:92-94.
11048. Urbain, A, Dechambre, E, & Rode, P. 1941. Observations faites sur une jeune Orang-utan né a la ménagerie du Jardin des Plantes. Mammalia. 5:83-85.
11048a. Urwin, VE & Allen, WR. 1982. Pituitary and chorionic gonadotrophic control of ovarian function during early pregnancy in equids. J Reprod Fert Suppl. 32:371-381.
11049. Uspenski, SM & Kistchinski, AA. 1972. New data on the winter ecology of the polar bear *Ursus maritimus* (Phipps) on Wrangel Island. Bears: Their Biology and Management, Proc Int Conf Bear Res Manage. 181-197, Calgary, ALB, IUCN, Morges, Switzerland.
11050. Usuki, H. 1967. Studies of the shrew mole (*Urotrichis talpoides*) II An ecological analysis of a small population. J Mamm Soc Japan. 3:47-52.
11051. Usuki, H. 1968. Studies of the shrew mole (*Urotrichis talpoides*) IV Seasonal reproductive conditions and some reflections on breeding cycle. J Mamm Soc Japan. 4:7-11.
11052. Utakoji, T. 1966. Chronology of nucleic acid synthesis in meiosis of the male Chinese hamster. Exp Cell Res. 42:585-596.
11053. Valdez, R. 1976. Fecundity of wild sheep (*Ovis orientalis*) in Iran. J Mamm. 57:762-763.
11054. Valdez, R, Alamia, LV, Bunch, TD, et al. 1977. Weights and measurements of Iranian wild sheep and wild goats. J Wildl Manage. 41:592-594.
11055. Valdez, R & LaVal, RK. 1971. Records of bats from Honduras and Nicaragua. J Mamm. 52:247-250.
11055a. Valdivieso, D, Conde, E, & Tamsitt, JR. 1968. Lactate dehydrogenase studies in Peurto Rican bats. Comp Biochem Physiol. 27:133-138.
11055b. Valdivieso, D & Tamsitt, JR. 1962. First records fo the pale spear-nosed bat in Colombia. J Mamm. 43:422-423.
11056. Valentincic, S, Bavdek, S, & Kusej, M. 1973. Gravidity of chamois females in Julian Alps. Int Congr Game Biol. 11:143-145.
11057. Valentincic, SI. 1958. Beitrag zur Kenntnis der Reproduktionserscheinungen beim Rotwild. Z Jagdwiss. 4:107-130.
11058. Valentincic Stane, SI. 1956. Resultate Zweijähriger Beobachtungen und Studien über den idealen Zuwachs bei Feldhasen auf der Insel, 'Biserni Otok'. Z Jagdwiss. 2:152-160.
11059. Valentine, GL & Kirkpatrick, RL. 1970. Seasonal changes in reproductive and related organs in the pine vole, *Microtus pinetorum*, in southwestern Virginia. J Mamm. 51:553-560.
11060. Valera, RB. 1955. Observations on the breeding habits, the shedding and the development of antlers of the Philippine deer (*Rusa* sp). Philliine J Forestry. 11:249-257.
11061. Valerio, DA, Courtney, KD, Miller, RL, et al. 1968. The establishment of *Macaca mulatta* breeding colony. Lab Anim Care. 18:589-594.
11062. Valerio, DA & Dalgard, DW. 1975. Experiences in the laboratory breeding of non-human primates. Lab Anim Handb. 6:49-62.
11063. Valerio, DA, Johnson, PT, & Thompson, GE. 1972. Breeding the greater bushbaby, *Galago crassicaudatus* in a laboratory environment. Lab Anim Sci. 22:203-206.
11064. Valerio, DA, Palotta, AJ, & Courtney, KD. 1969. Experiences in large-scale breeding of simians for medical experimentation. Ann NY Acad Sci. 162(1):282-296.
11065. Vallat, C. 1971. Birth of three cheetahs *Acinonyx jubatus* at Montpellier Zoo. Int Zoo Yearb. 11:124-125.
11066. Valtonen, MH & Mäkelä, JI. 1979. Monogamous raccoon dog (*Nyctereutes procyonoides*) and polygamous mating in fur farming. Scientifur. 3(1):21-22.
11067. Valtonen, MH & Mäkelä, JI. 1980. Reproduction and breeding of the raccoon dog. Scientifur. 4(1):18-19.
11068. Valtonen, MH, Rajakoski, EJ, & Lähteenmäki, P. 1978. Levels of oestrogen and progesterone in the plasma of the raccoon dog (*Nyctereutes procyonoides*) during oestrus and pregnancy. J Endocrinol. 76:549-550.
11069. Valtonen, MH, Rajakoski, EJ, & Mäkelä, JI. 1977. Reproductive features in the female raccoon dog (*Nyctereutes procyonoides*). J Reprod Fert. 51:517-518.
11070. Valverde, JA. 1957. Notes écologiques sur le Lynx d'Espagne *Felis lynx pardena* Temminck. Terre Vie. 12:51-67.
11071. Vamburkar, SA. 1958. The male genital tract of the Indian megachiropteran bat *Cynopterus sphinx gangeticus* And. Proc Zool Soc London. 130:57-77.
11072. van Aarde, RJ. 1978. Reproduction and population ecology in the feral house cat, *Felis catus*, on Marion Island. Carnivore Genetics Newsl. 3:288-316.
11073. van Aarde, RJ. 1980. Harem structure of the southern elephant seal *Mirounga leonina* at Kerguelen Island. Terre Vie. 34:31-44.
11074. van Aarde, RJ. 1983. Demographic parameters of the feral cat *Felis catus* population at Marion Island. S Afr J Wildl Res. 13:12-16.
11075. van Aarde, RJ. 1985. Circulating progesterone and oestradiol-17β concentrations in cyclic Cape porcupines, *Hystrix africaeaustralis*. J Reprod Fert. 75:583-591.
11076. van Aarde, RJ. 1985. Reproduction in captive female Cape porcupines (*Hystrix africaeaustralis*). J Reprod Fert. 75:577-582.
11077. van Aarde, RJ. 1987. Pre- and postnatal growth of the Cape porcupine *Hystrix africaeaustralis*. J Zool. 211:25-33.
11078. van Aarde, RJ. 1987. Reproduction in the Cape porcupine *Hystrix africaeaustralis*: An ecological perspective. S Afr J Sci. 83:605-607.
11079. van Aarde, RJ & Potgieter, HC. 1986. Circulating progesterone, progesterone-binding proteins and oestradiol-17β

concentrations in the pregnant Cape porcupine, *Hystrix africaeaustralis*. J Reprod Fert. 76:561-567.
11080. van Aarde, RJ & Skinner, JD. 1986. Functional anatomy of the ovaries of pregnant and lactating Cape porcupines, *Hystrix africaeaustralis*. J Reprod Fert. 76:553-559.
11081. van Aarde, RJ & Skinner, JD. 1986. Reproductive biology of the male Cape porcupine, *Hystrix africaeaustralis*. J Reprod Fert. 76:545-552.
11081a. van Aarde, RJ & van Wyk, V. 1991. Reproductive inhibition in the Cape porcupine, *Hystrix africaeaustralis*. J Reprod Fert. 92:13-19.
11082. van Arde, LMW. 1988. Reproduction in a laboratory colony of the pouched mouse, *Saccostomus campestris*. J Reprod Fert. 83:773-778.
11083. van Ballenberghe, V. 1983. Two litters raised in one year by a wolf pack. J Mamm. 64:171-172.
11084. van Bemmel, ACV. 1949/50. Revision of the rusine deer in the Indo-Australian Archipelago. Treubia. 20:191-261.
11085. van Bemmel, ACV. 1952. Contribution to the knowledge of the genera *Muntiacus* and *Arctogalidia* in the Indo-Australian Archipelago (Mammalia, Cervidae and Viverridae). Beaufortia. 2:1-50.
11086. van Bemmel, ACV. 1958. Birth of *Cebus capucinus hypoleucus*. Zool Garten NF. 24:246-247.
11087. van Bemmel, ACV. 1963. Das Züchten des Orang Utan im Zoo. Z Morph Anthrop. 53:65-71.
11088. van Bemmel, ACV. 1967. The banteng *Bos javanicus* in captivity. Int Zoo Yearb. 7:222-223.
11089. van Bemmel, ACV. 1968. Breeding tigers *Panthera tigris* at Rotterdam Zoo. Int Zoo Yearb. 8:60-63.
11090. van Bemmel, ACV. 1975. Breeding tigers as an aid to their survival. Breeding Endangered Species in Captivity. 329-335, Martin, RD, ed, Academic, New York.
11091. van Beneden, E. 1870. Recherches sur la composition et la signification de l'oeuf. Mém Couronnés Acad Belg. 34:1-279.
11092. van Beneden, E. 1875. La maturation de l'oeuf, la fécondation et les premières phases du dévelopement embryonnaires des mammifères d'après des recherches faites chez le lapin. Bull Acad R Belg 2ième series. 40:686-736.
11093. van Beneden, E. 1880. Contribution à la connaissance de l'ovaire des Mammifères. Arch Biol. 1:475-550.
11094. van Beneden, E. 1899. Recherches sur les premiers stades du developpement du Murin (*Vespertitlio murinus*). Anat Anz. 16:305-334.
11095. van Beneden, E & Julin, C. 1880. Observations sur la maturation, la fécondation et la sigmentation chez les Cheiroptères. Arch Biol. 1:551-572.
11096. van Bree, PJH, Collet, A, Desportes, G, et al. 1986. Le Dauphin de fraser, *Lagenodelphis hosei* (Cetacea, Odontoceti), espèce nouvelle pour la faune d'Europe. Mammalia. 50:57-86.
11097. Van de Graaf, K & Balda, RP. 1973. Importance of green vegetation for reproduction in the kangaroo rat, *Dipodomys merriami merriami*. J Mamm. 54:509-512.
11098. van den Broek, AJP. 1903. Untersuchungen über die weiblichen Geschlechtsorgane der Beuteltiere. Petrus Camper Ned Bijdr Anat. 3:221-346.
11099. van den Broek, AJP. 1904. Die Eihüllen und die Placenta von *Phoca vitulina*. Petrus Camper Ned Bijdr Anat. 2:546-570.
11100. van den Broek, AJP. 1910. Untersuchungen über den Bau der männlichen Geschlechtsorgane der Beuteltiere. Morphol Jb. 41:347-436.
11101. van den Broek, AJP. 1931. Einige Bemerkungen über der Bau der inneren Geschlechtsorganen der Monotremen. Morphol Jb Abt I. 67 II:134-156.
11102. van der Horst, CJ. 1935. On the reproduction of the springhare, *Pedestes caffer*. Pamph S Afr Biol Soc. 8:47.
11103. van der Horst, CJ. 1941. On the size of the litter and the gestation period of *Procavia capensis*. Science. 93:430-431.
11104. van der Horst, CJ. 1942. Early stages of the embryonic development of *Elephantulus*. S Afr J Med Sci Biol Suppl. 7:55-65.
11105. van der Horst, CJ. 1942. Some observations of the structure of the genital tract of *Elephantulus*. J Morphol. 70:403-429.
11106. van der Horst, CJ. 1943. The mechanism of egg transport from the ovary to the uterus in *Elephantulus*. S Afr J Med Sci. 8:41-49.
11106a. van der Horst, CJ. 1944. Remarks on the systematics of *Elephantulus*. J Mamm. 25:77-82.
11107. van der Horst, CJ. 1946. Some remarks on the biology of reproduction in the female of *Elephantulus*, the holy animal of Set. Trans R Soc S Afr. 31:181-199.
11108. van der Horst, CJ. 1948. Some early embryological stages of the golden mole, *Eremitalpa granti* (Broom). Sp Publ R Soc S Afr. Robert Bloom Commemorative Vol:225-229.
11109. van der Horst, CJ. 1949. An early stage of placentation in the aardvark, *Orycteropus*. Proc Zool Soc London. 119:1-18.
11110. van der Horst, CJ. 1949. Some natural experiments in the embryological development of *Elephantulus*. Arch Biol. 60:25-38.
11111. van der Horst, CJ. 1949. The placentation of *Tupaia javanica*. Proc K Ned Akad Wet. 52:1205-1213.
11112. van der Horst, CJ. 1950. The placentation of *Elephantulus*. Trans R Soc S Afr. 32:435-629.
11113. van der Horst, CJ. 1951. The post-partum involution of the uterus of *Elephantulus*. Acta Zool (Stockholm). 32:11-29.
11114. van der Horst, CJ. 1955. *Elephantulus* going into anoestrus; menstruation and abortion. Phil Trans R Soc London Biol Sci. 238B:27-61.
11115. van der Horst, CJ & Gillman, J. 1940. Mechanism of ovulation and corpus luteum formation in *Elephantulus*. Nature. 145:974.
11116. van der Horst, CJ & Gillman, J. 1940. Ovulation and corpus luteum formation in *Elephantulus*. S Afr J Med Sci. 5:73-91.
11117. van der Horst, CJ & Gillman, J. 1941. The menstrual cycle in *Elephantulus*. S Afr J Med Sci. 6:27-47.
11118. van der Horst, CJ & Gillman, J. 1941. The number of eggs and surviving embryos in *Elephantulus*. Anat Rec. 80:443-452.
11119. van der Horst, CJ & Gillman, J. 1942. The life history of the corpus luteum of menstruation in *Elephantulus*. S Afr J Med Sci. 7:21-41.
11120. van der Horst, CJ & Gillman, J. 1942. Pre-implantation phenomena in the uterus of *Elephantulus*. S Afr J Med Sci. 7:47-71.
11121. van der Horst, CJ & Gillman, J. 1942. Pre-implantation abortion in *Elephantulus*. S Afr J Med Sci. 7:120-126.
11122. van der Horst, CJ & Gillman, J. 1942. The spontaneous development of deciduomata in *Elephantulus*. S Afr J Med Sci. 7:127-133.
11123. van der Horst, CJ & Gillman, J. 1942. A critical analysis of the early gravid and premenstrual phenomena in the uterus of *Elephantulus*, *Macaca* and the human female. S Afr J Med Sci. 7:134-143.
11124. van der Horst, CJ & Gillman, J. 1945. The behavior of the Graafian follicle of *Elephantulus* during pregnancy, with

special reference to the hormonal regulation of ovarian activity. S Afr J Med Sci Biol Suppl. 10:1-14.
11125. van der Horst, CJ & Gillman, J. 1946. The corpus luteum of *Elephantulus* during pregnancy: Its form and its function. S Afr J Med Sci Biol Suppl. 11:87-101.
11126. van der Horst, CJ & Gillman, J. 1946. The reactions of the uterine blood vessels before, during and after pregnancy in *Elephantulus*. S Afr J Med Sci Biol Suppl. 11:103-110.
11127. van der Horst, G. 1972. Seasonal effects on the anatomy and histology of the reproductive tract of the male rodent mole *Bathyergus snillus snillus* (Schreber). Zool Afr. 7:491-520.
11128. van der Merwe, M. 1978. Delayed implantation in the Natal clinging bat *Miniopterus schreibersi natalensis* (A Smith 1834). Bat Res News. 19:110.
11129. van der Merwe, M. 1979. Foetal growth curves and seasonal breeding in the Natal clinging bat *Miniopterus schreibersi natalensis*. S Afr J Zool. 14:17-21.
11130. van der Merwe, M. 1979. Growth of ovarian follicles in the Natal clinging bat. S Afr J Zool. 14:111-117.
11131. van der Merwe, M. 1980. Delayed implantation in the Natal clinging bat *Miniopterus schreibersii natalensis* (A Smith, 1834). Proc Int Bat Res Conf. 5:113-123.
11132. van der Merwe, M. 1981. Fetal development of the bat *Miniopterus schreibersi natalensis*. S Afr J Zool. 16:172-182.
11133. van der Merwe, M. 1985. Histological study of implantation in the Natal clinging bat (*Miniopterus schreibersi natalensis*). J Reprod Fert. 65:319-323.
11134. van der Merwe, M. 1986. Reproductive strategy of *Miniopterus schreibersii natalensis*. Cimbebasia. 8:107-111.
11135. van der Merwe, M. 1987. Adaptive breeding strategies in some South African bats between 22°S and 28°S. S Afr J Sci. 83:607-609.
11136. van der Merwe, M, Giddings, SR, & Rautenbach, IL. 1987. Post-partum oestrus in the little free-tailed bat, *Tadarida* (*Chaerephon*) *pumila* (Microchiroptera: Molossidae) at 24°S. J Zool. 213:317-326.
11137. van der Merwe, M & Rautenbach, IL. 1986. Multiple births in Schlieffen's bat, *Nycticeius schlieffenii* (Peters, 1859) (Chiroptera: Vespertilionidae) from the southern African subregion. S Afr J Zool. 21:48-50.
11138. van der Merwe, M & Rautenbach, IL. 1987. Reproduction in Schlieffen's bat, *Nycticeius schlieffenii*, in the eastern Transvaal lowveld, South Africa. J Reprod Fert. 81:41-50.
11139. van der Merwe, M & Rautenbach, IL. 1990. Reproduction in the rusty bat, *Pipistrellus rusticus*, in the northern Transvaal bushveld, South Africa. J Reprod Fert. 89:537-542.
11140. van der Merwe, M, Rautenbach, IL, & van der Colf, WJ. 1986. Reproduction in females of the little free-tailed bat, *Tadarida* (*Chaerephon*) *pumilia*, in the eastern Transvaal, South Africa. J Reprod Fert. 77:355-364.
11141. van der Merwe, M & Skinner, JD. 1982. Annual reproductive pattern in the dassie *Procavia capensis*. S Afr J Zool. 17:130-135.
11142. van der Merwe, M, Skinner, JD, & Millar, RP. 1980. Annual reproductive pattern in the springhaas, *Pedetes capensis*. J Reprod Fert. 58:259-266.
11143. van der Merwe, M & van Aarde, RJ. 1989. Plasma progesterone concentrations in the female Natal clinging bat (*Miniopterus schreibersii natalensis*). J Reprod Fert. 87:665-669.
11144. van der Merwe, NJ. 1953. The jackal. Fauna Flora (Pretoria). 4:3-82.
11145. van der Merwe, NJ. 1959. Die wildehond (*Lycaon pictus*). Koedoe. 2:87-93.
11145a. van der Merwe, NJ & Rautenbach, IL. 1988. The placenta and foetal membranes of the lesser yellow house bat, *Scotophilus borbonicus* (E Geoffroy, 1803) (Chiroptera: Vespertilionidae). S Afr J Zool. 23:320-327.
11145b. van der Merwe, NJ, Rautenbach, IL, & Penzhorn, BL. 1988. A new pattern of early embryonic development in the seasonally breeding non-hibernating lesser yellow house bat, *Scotophilus borbonicus* (E Geoffroy, 1803) (Chiroptera: Vespertilionidae). Ann Transvaal Mus. 34:551-556.
11146. van der Spuy, CJ. Personal communication. [Cited by Dean, 1962]
11147. van der Stricht, O. 1911. Mécanisme de la fixation de l'oeuf de Chauve-souris (*V noctula*). CR Assoc Anat. 13:1-9.
11148. van der Stricht, O. 1912. Sur le processus de l'excrétion des glandes endocrines: Le corps jaune et la glande interstielle de l'ovaire. Arch Biol. 27:585-722.
11149. van der werff ten Bosch, JJ. 1969. Indices of human puberty. J Reprod Fert Suppl. 6:67-76.
11150. van der Westhuyzen, J. 1988. Haematology and iron status of the Egyptian fruit bat, *Rousettus aegyptaicus*. Comp Biochem Physiol. 90A:117-120.
11151. van Deusen, HM. 1971. *Zaglossus*: Egg-laying anteater. Fauna. 2:12-19.
11152. van Deusen, HM & George, GG. 1969. Results of the Archbold expeditions No 90 Notes on the echidnas (Mammalia, Tachyglossidae) of New Guinea. Am Mus Novit. 2383:1-23.
11153. van Doorn, C. 1968. Einiges über Fluszpferde: Ihre Zucht und Haltung in Rotterdammer Zoo. Fr Kölner Zoo. 11:85-90.
11154. van Doorn, C. 1969. Milus oder Davidhirsche (*Elaphurus davidianus* Milne-Edwards) im Rotterdamer Zoo. Fr Kölner Zoo. 12:97-103.
11155. van Doorn, C. 1971. Verzögerte Zyklus der Fortpflanzung bei Faultieren. Fr Kölner Zoo. 14:15-22.
11156. van Doorn, C. 1975. Der kanadische Elch in Blijdorp Zoo, Rotterdam. Fr Kölner Zoo. 18:117-125.
11157. van Doorn, C & Slijper, EJ. 1959. Some remarks on the birth of a father David's deer, *Elaphurus davidianus*, Milne Edw. Bijdr Dierkd. 29:73-74.
11158. van Dyck, S. 1979. Behaviour in captive individuals of the dasyurid marsupial *Planigale maculata* (Gould 1851). Mem QLD Mus. 19:413-429.
11159. van Dyck, S. 1979. Mating and other aspects of behaviour in wild striped possums. Vict Nat. 96:84-85.
11160. Van Dyck, S. 1980. The cinnamon antechinus, *Antechinus leo* (Marsupialia, Dasyuridae), a new species from the vine-forests of Cape York Peninsula. Aust Mamm. 3:5-17.
11161. Van Dyck, S. 1982. The status and relationships of the Atherton antechinus, *Antechinus godmani* (Marsupialia: Dasyuridae). Aust Mamm. 5:195-210.
11162. van Ee, CA. 1966. A note on breeding the Cape pangolin *Manis temmincki* at Bloemfontein Zoo. Int Zoo Yearb. 6:163-164.
11163. van Heerden, J & Kuhn, F. 1985. Reproduction in captive hunting dogs *Lycaon pictus*. S Afr J Wildl Res. 15:80-84.
11164. van Herwerden, M. 1908. Bijdrage tot de kennis van menstrueelen cyclus. Tijdschr Nederl Dierk Ver 2nd ser. 10:1-140.
11165. van Höffen, E. 1897. Die Fauna und Flora Grönlands Grönland expedition der Gesellschaft fur Erdkunde zu Berlin 1891-1893 Vol I. Berlin.
11167. van Hooff, JARAM. 1965. A large litter of lion cubs *Panthera leo* at Arnhem Zoo. Int Zoo Yearb. 5:116.

11168. van Horn, DR & Baker, BE. 1971. Seal milk II Harp seal (*Pagophilus groenlandicus*) milk: Effects of stage of lactation on the composition of the milk. Can J Zool. 49:1085-1088.
11169. van Horn, RN. 1980. Seasonal reproductive patterns in primates. Prog Reprod Biol. 5:181-221.
11170. van Horn, RN, Beamer, NB, & Dixson, AF. 1976. Diurnal variations of plasma testosterone in two prosimian primates (*Galago crassicaudatus crassicaudatus* and *Lemur catta*). Biol Reprod. 15:523-528.
11171. van Horn, RN & Eaton, GG. 1979. Reproductive physiology and behavior in prosimians. The Study of Prosimian Behavior. 79-122, Doyle, GA & Martin, RD, eds, Academic Press, New York.
11172. van Horn, RN & Resko, JA. 1977. The reproductive cycle of the ring-tailed lemur (*Lemur catta*): Sex steroid levels and sexual receptivity under controlled photoperiods. Endocrinology. 101:1579-1586.
11173. van Jaarsveld, AS & Skinner, JD. 1987. Spotted hyaena monomorphism: An adaptive 'phallusy'?. S Afr J Sci. 83:612-615.
11174. van Koersveld, E. 1953. The muskusrat *Ondatra zibethica* L in Nederland en zijn bestrijding. Jaarboek 1951-1952, Plantenziektenkundige Dienst Wageningen. 120:229-249.
11175. van Lavieren, LP & Bosch, ML. 1977. Evaluation des densités de grands Mammifères dans le Parc National de Bouba Ngida, Cameroun. Terre Vie. 31:3-32.
11176. van Lawick-Goodall, J. 1968. The behaviour of free-living chimpanzees in the Gombe Stream Reserve. Anim Behav Monogr. 1:161-311.
11177. van Lawick-Goodall, J. 1969. Some aspects of reproductive behaviour in a group of wild chimpanzees, *Pan troglodytes schweinfurthi*, at the Gombe Stream Chimpanzee Reserve, Tanzania, East Africa. J Reprod Fertil Suppl. 6:353-355.
11178. van Lennep, EW. 1950. Histology of the corpora lutea in blue and finwhale ovaries. Proc Kon Ned Akad Wetensch. 53:593-599.
11179. van Mourik, S. 1985. Embryonic and fetal devlelopment of rusa deer (*Cervus rusa timorensis*). Zool Anz. 215:159-167.
11180. van Mourik, S. 1986. Reproductive performance and maternal behavior of farmed rusa deer (*Cervus rusa timorensis*). Appl Anim Behav Sci. 15:147-159.
11181. van Mourik, S & Schurig, V. 1985. Hybridization between sambar (*Cervus rusa unicolor*) and rusa (*Cervus rusa timorensis*) deer. Zool Anz. 214:177-184.
11182. van Mourik, S & Stelmasiak, T. 1985. Seasonal variation in plasma prolactin concentrations in adult, male rusa deer (*Cervus rusa timorensis*). Comp Biochem Physiol. 82A:323-327.
11183. van Nieuwenhoven, PJ. 1956. Ecological observations in a hibernation-quarter of cave-dwelling bats in south-Limburg. Publ Natuurh Genootsch Limburg. 9:1-55.
11184. van Noordwyk, MA. 1985. Sexual behaviour of Sumatran long-tailed macaques (*Macaca fascicularis*). Z Tierpsychol. 70:277-296.
11185. van Oordt, GJ. 1921. Early developmental stages of *Manis javanica* Desm. Verh Kon Akad Wetensch Amsterdam Tweede Sectie. 2(3):1-102.
11186. van Peenen, PFD, Light, RH, Duncan, FJ, et al. 1974. Observations on *Rattus bartelsii* (Rodentia: Muridae). Treubia. 28:83-117.
11187. van Rompaey, H. 1988. *Osbornictis piscivora*. Mamm Species. 309:1-4.
11187a. Van Rompaey, H & Colyn, M. 1992. *Crossarchus ansorgei*. Mamm Species. 402:1-3.
11188. van Roosmalen, MGM. 1980. Habitat preferences, diet, feeding strategy and social organization of the black spider monkey (*Ateles paniscus paniscus* Linnaeus 1758) in Surinam. Rijksinstituut voor Natuurbeheer Rapport Leersum. [cited by Thornback, 1982]
11189. van Roosmalen, MGM, Mittermeier, RA, & Milton, K. 1981. The bearded sakis, genus *Chiropotes*. Ecology and Behavior of Neotropical Primates. 1:419-441. Coimbra-Filho, AF & Mittermeier, RA, eds, Acad Brasil Ciências, Rio de Jeneiro.
11190. van Soest, RWM & van Bree, PJH. 1970. Sex and age composition of a stoat population (*Mustela erminea* Linnaeus, 1758) from coastal dune region of the Netherlands. Beaufortia. 17:51-77.
11190a. van Tienhoven, A. 1983. Reproductive Physiology of Vertebrates. Cornell Univ Press, Ithaca, NY.
11191. van Utrecht, WL. 1978. Age and growth in *Phocoena phocoena* Linnaeus, 1758 (Cetacea, Odontoceti) from the North Sea. Bijdr Dierkd. 48:16-28.
11192. van Utrecht-Cock, CN. 1965. Age determination and reproduction of female fin whales (*Balaenoptera physalus*) (Linnaeus, 1758), with special regard to baleen plates and ovaries. Bijdr Dierkd. 35:39-100.
11193. van Vugt, D, Diefenbach, WD, Alston, E, et al. 1985. Gonadotropin-releasing hormone pulses in third ventricular cerebrospinal fluid of ovariectomized monkeys: Correlation with luteinizing hormone pulses. Endocrinology. 117:1550-1558.
11194. van Vugt, DA, Lam, NY, & Ferin, M. 1984. Reduced frequency of pulsatile luteinizing hormone secretion in the luteal phase of the rhesus monkey: Involvement of endogenous opiates. Endocrinology. 115:1095-1101.
11194a. Van Vuren, D & Armitage, KB. 1991. Duration of snow cover and its influence on life-hsitory variation in yellow-bellied marmots. Can J Zool. 69:1755-1758.
11195. van Wagenen, G. 1945. Mating in relation to pregnancy in the monkey. Yale J Biol Med. 17:745-760.
11196. van Wagenen, G. 1945. Optimal mating time for pregnancy in the monkey. Endocrinology. 37:307-312.
11197. van Wagenen, G. 1947. Early mating and pregnancy in the monkey. Endocrinology. 40:37-43.
11198. van Wagenen, G. 1949. Accelerated growth with sexual precocity in female monkeys receiving testosterone propionate. Endocrinology. 45:544-546.
11199. van Wagenen, G. 1972. Vital statistics from a breeding colony: Reproduction and pregnancy outcome in *Macaca mulatta*. J Med Primatol. 1:3-28.
11200. van Wagenen, G & Simpson, ME. 1952. Testicular development in the rhesus monkey. Anat Rec. 118:231-251.
11201. van Wagenen, G, Catchpole, HR, Negri, J, et al. 1965. Growth of the fetus and placenta of the monkey (*Macaca mulatta*). Am J Phys Anthrop. 23:23-34.
11202. van Weers, DJ. 1979. Notes on southeast Asia porcupines (Hystricidae, Rodentia) III on the taxonomy of the subgenus *Thecurus* Lyon, 1907 (genus *Hystrix* Linnaeus, 1758). Beaufortia. 28:17-33.
11203. van Weers, DJ. 1979. Notes on southeastern Asian porcupines (Hystricidae, Rodentia) IV on the taxonomy of the subgenus *Acanthion* F Cuvier, 1823 with notes on the other taxa of the family. Beaufortia. 29:215-272.
11204. van Wijngaarden, A. 1954. Biologie en bestrijding van de woelrat, *Arvicola terrestris terrestris* (L) in Nederland. Meded Plantenziektenkund Dienst Wageningen. 123:1-147.
11205. van Wijngaarden, A & van de Peppel, J. 1964. The badger, *Meles meles* (L), in the Netherlands. Lutra. 6:1-60.

11206. van Wijngaarden, A & van de Peppel, J. 1970. De otter, *Lutra lutra* (L) in Nederland. Lutra. 12:1-70.
11207. van Wormer, J. 1966. The World of the Black Bear. JB Lippincott Co, Philadelphia.
11208. van Wormer, J. 1969. The World of the Pronghorn. JB Lippincott Co, Philadelphia.
11208a. van Wyk, V & van Aarde, RJ. 1991. The stimulation of testosterone and LH secretion by synthetic GnRH in the male Cape porcupine. S Afr J Zool. 26:188-192.
11209. van Zyl, A. 1957. Serum protein-bound iodine and serum lipid changes in the baboon (*Papio ursinus*) I During the menstrual cycle. J Endocrinol. 14:309-316.
11210. van Zyl, A. 1957. Serum protein-bound iodine and serum lipid changes in the baboon (*Papio ursinus*) II During pregnancy, lactation and abortion. J Endocrinol. 14:317-324.
11211. van Zyl, JHM. 1966. Embryonic implantation in the springbok (*Antidorcas marsupialis marsupialis* (Zimmermann) and ovulation in the S A buffalo (*Syncerus caffer caffer* Sparrman). Fauna Flora (Pretoria). 17:37.
11212. van Zyl, JHM. 1977. The gemsbok on the S A Lombard Nature Reserve. Fauna Flora (Pretoria). 28:17.
11213. van Zyl, JHM & Skinner, JD. 1970. Growth and development of the springbok foetus. Afr Wildl. 24:309-316.
11214. van Zyl, JHM & Wehmeyer, AS. 1970. The composition of the milk of springbok (*Antidorcas marsupialis*), eland (*Taurotragus oryx*) and black wildebeest (*Connochaetes gnou*). Zool Afr. 5:131-133.
11215. VandeBerg, JL. 1983. The gray short-tailed opossum: A new laboratory animal. ILAR News. 26(3):9-12.
11216. Vandenbergh, JG. 1973. Environmental influences on breeding in rhesus monkeys. Symp 4th Int Congr Primat, 1972. 2:1-19.
11217. Vandenbergh, JG. 1987. Regulation of puberty and its consequences on population dynamics of mice. Am Zool. 27:891-898.
11218. Vandenbergh, JG & Drickamer, LC. 1974. Reproductive coordination among free-ranging rhesus monkeys. Physiol Behav. 13:373-376.
11219. Vanderbergh, JG & Vessey, S. 1968. Seasonal breeding of free-ranging rhesus monkeys and related ecological factors. J Reprod Fert. 15:71-79.
11220. Vania, J. 1965. Birth of a stellar sea lion pup. BioScience. 15:794-795.
11221. Vanoli, T. 1967. Beobachtungen an Pudus, *Mazama pudu* (Molina, 1782). Säugetierk Mitt. 15:155-163.
11222. Vargas, JM, España, M, Palomo, LJ, et al. 1984. Über die Geschlechtstätigkeit der *Mus spretus*: ♂♂ in Südspanien. Z Angew Zool. 71:257-273.
11222a. Vargas, JM, Palomo, LJ, & Palmqvist, P. 1991. Reproduction of the Algerian mouse (*Mus spretus* Lataste, 1883) in the south of the Iberian Peninsula. Bonn Zool Beitr. 42:1-10.
11223. Varona, LS. 1985. Modificaciones ontogenicas y dimorfismo sexual en *Mesoplodon gervaisi* (Cetacea; Ziphidae). Carib J Sci. 21:27-37.
11224. Várshavscky, SN & Krylova, KT. 1939. Some ecological particularities of the small ground squirrel (*Citellus pygmaeus* Pall) in various periods of its life. Zool Zhurn. 18:1026-1047.
11225. Vásárhelyi, I. 1929. Beitrage zur Kenntis der Lebensweise zweier Kleinsäuger (*Microtus arvalis* Pall). J trim Sect Soc R Sci Nat Hongrie. 26:84-91. [cited by Herfs, 1939]
11226. Vasilu, GD & Decei, P. 1964. Über den Luchs (*Lynx lynx*) der rumänischen Karpaten. Säugetierk Mitt. 12:155-183.
11227. Vaughan, MK, Vaughan, GM, & Reiter, RJ. 1973. Effect of ovariectomy and constant dark on the weight of reproductive and certain other organs in the female vole, *Microtus montanus*. J Reprod Fert. 32:9-14.
11228. Vaughan, MR & Keith, LB. 1980. Breeding by juvenile showshoe hares. J Wildl Manage. 44:948-951.
11228a. Vaughan, TA. 1954. Mammals of the San Gabriel Mountains of California. Univ KS Publ Mus Nat Hist. 7:513-582.
11229. Vaughan, TA. 1962. Reproduction in the plains pocket gopher in Colorado. J Mamm. 43:1-12.
11230. Vaughan, TA. 1969. Reproduction and population densities in a montane small mammal fauna. Univ KS Mus Nat Hist Misc Publ. 51:51-74.
11231. Vaughan, TA. 1982. Stephen's woodrat, a dietary specialist. J Mamm. 63:53-62.
11232. Vaughan, TA & Czaplewski, NJ. 1985. Reproduction in Stephen's woodrat: The wages of folivory. J Mamm. 66:429-443.
11233. Vaughan, TA & Vaughan, RP. 1986. Seasonality and the behavior of the African yellow-winged bat. J Mamm. 67:91-102.
11234. Vaughan, TA & Vaughan, RP. 1987. Parental behavior in the African yellow-winged bat (*Lavia frons*). J Mamm. 68:217-223.
11235. Vaughan, TA & Weil, WP. 1980. The importance of arthropods in the diet of *Zapus princeps* in a subalpine habitat. J Mamm. 61:122-124.
11235a. Vaughn, R. 1974. Breeding the tayra *Eira barbara* at the Antelope Zoo, Lincoln. Int Zoo Yearb. 14:120-122.
11236. Vaz-Ferreira, R. 1975. Behavior of the southern sea lion, *Otaria flavescens* (Shaw) in the Uruguayan islands. Rapp P-V Réun Cons Int Explor Mer. 169:219-227.
11237. Vaz-Ferreira, R. 1979. South American sea lion. Mammals in the Seas, FAO Fish Ser 5. 2:9-12.
11238. Vaz-Ferreira, R. 1979. South American fur seal. Mammals in the Seas, FAO Fish Ser 5. 2:34-36.
11239. Vaz-Ferreira, R. 1981. South American sea lion *Otaria flavescens* (Shaw, 1880). Handbook of Marine Mammals. 1:9-65. Ridgway, SH & Harrison, RJ, eds, Academic Press, New York.
11240. Veal, R & Caire, W. 1979. *Peromyscus eremicus*. Mamm Species. 118:1-6.
11240a. Vedder, L, 't Hart, L, & van Bree, PJH. 1992. Further notes on the pupping period in a recently founded colony of grey seals (*Halichoerus grypus*) in the Netherlands. Z Säugetierk. 57:116-117.
11241. Veenstra, AJF. 1958. The behaviour of the multimammate mouse *Rattus* (*Mestomys*) *natalensis* (A Smith). Anim Behav. 6:195-206.
11242. Veeraiah, DH. 1943. Some stages in the early development and placentation of *Funambulus palmarum* (Linn). Half-Yearly J Mysore Univ. 4B:63-110.
11243. Velhankar, DP, Hukeri, VB, Deshpande, RR, et al. 1973. Biometry of the genitalia and the spermatozoa of the giraffe. Indian Vet J. 50:789-792.
11244. Velich, R. 1958. Notes on mammals from eastern Nebraska. J Mamm. 39:147-148.
11245. Venables, UM & Venables, LSV. 1957. Mating behaviour of the seal *Phoca vitulina* in Shetland. Proc Zool Soc London. 128:387-396.
11246. Venge, O. 1956. Experiments on forced interruption of the copulation in mink. Acta Zool (Stockholm). 37:287-304.
11247. Venge, O. 1973. Reproduction in the mink. Årsskrift Kongelige Veterinaer: og Landbohøjskole. 1973:95-146.
11248. Ventura, J & Gonsálbez, J. 1987. Reproductive biology of *Arvicola sapidus* (Rodentia, Arvicolidae) in the Ebro delta (Spain). Z Säugetierk. 52:364-371.

11249. Ventura, J & Gonsálbez, J. 1990. Reproduction potential of *Arvicola terrestris* (Mammalia, Rodentia) in the northeast peninsula Iberian. Zool Anz. 225:45-54.
11250. Ventura, J & Gonsálbez, J. 1990. Reproductive cycle of *Arvicola terrestris* (Rodentia, Arvicolidae) in the Aran valley, Spain. Z Säugetierk. 55:383-391.
11251. Ventura, J, Gosálbez, J, & López-Fuster, MJ. 1991. Structure de population d'*Arvicola terrestris* (Linnaeus, 1758) (Rodentia, Arvicolidae) du Nord-Est Iberique. Mammalia. 55:85-90.
11252. Verevkin, MB. 1985. Biology of reproduction of *Meriones meridianus*. Zool Zhurn. 64:276-281.
11253. Verheyen, R. 1951. Contribution à l'étude éthologique des Mamifères du Parc National de l'Upemba. Institut des Parcs Nationaux du Congo Belge, Brussels.
11254. Vericad, JR. 1983. Estimación de la edad fetal y periodos de concepción y parto del jabalí (*Sus scrofa*) en los Pirineos Occidentales. Actas XV Congr Int Fauna Cinegetica y Silveste, Trujillo. 15. [Cited by Saez-Proyella & Talleria, 1981]
11255. Vericad-Corominas, JR. 1970. Estudio faunistico y biologico de los mamiferos montaraces del Pirineo. Publ Cent Pirenaico Biol Exp. 4:1-260.
11256. Verma, K. 1965. Notes on the biology and anatomy of the Indian treeshrew, *Anathana wroughtoni*. Mammalia. 29:289-330.
11257. Verme, LJ. 1961. Late breeding in northern Michigan deer. J Mamm. 42:426-427.
11258. Verme, LJ. 1962. Fecundity in a Michigan white-tailed deer. J Mamm. 43:112-113.
11259. Verme, LJ. 1963. Effect of nutrition on growth of white-tailed deer fawns. Trans N Am Wildl Nat Res Conf. 28:431-443.
11260. Verme, LJ. 1965. Reproduction studies on penned white-tailed deer. J Wildl Manage. 29:74-78.
11261. Verme, LJ. 1969. Reproductive patterns of white-tailed deer related to nutritional plane. J Wildl Manage. 33:881-887.
11262. Verme, LJ. 1970. Some characteristics of captive Michigan moose. J Mamm. 51:403-405.
11263. Verme, LJ. 1977. Assessment of natal mortality in upper Michigan deer. J Wildl Manage. 41:700-708.
11264. Verme, LJ. 1979. Influence of nutrition on fetal organ development in deer. J Wildl Manage. 43:791-796.
11265. Verme, LJ. 1984. Birth weights of fawns from doe fawn white-tailed deer. J Wildl Manage. 48:962-963.
11266. Verme, LJ. 1989. Maternal investment in white-tailed deer. J Mamm. 70:438-442.
11267. Verme, LJ, Ozoga, JJ, & Nellist, JT. 1987. Induced early estrus in penned white-tailed deer. J Wildl Manage. 51:54-56.
11267a. Vermeulen, HC & Nel, JAJ. 1988. The bush karoo rat *Otomys unisulcatus* on the Cape west coast. S Afr J Zool. 23:103-111.
11268. Vernon, MW, Dierschke, DJ, Sholl, SA, et al. 1983. Ovarian aromatase in granulosa and thecal cells of rhesus monkeys. Biol Reprod. 28:342-349.
11269. Verrill, AH. 1907. Notes on the habits and external characters of Solenodon of San Domingo (*Solenodon paradoxus*). Am J Sci 4th Ser. 24:55-57.
11270. Verschuren, J. 1957. Exploration du Parc National de la Garamba Vol 7: Ecolologie, Biologie et Systematique des Cheiroptères. Institut des Parcs Nationaux du Congo Belge. 1-465, Brussels.
11271. Verschuren, J. 1958. Exploration du Parc National de la Garamba Vol 9: Ecologie et Biologie des Grand Mammifères (Primates, Carnivores, Ongulés). Institut des Parcs Nationaux du Congo Belge, Brussels.
11272. Verschuren, J. 1966. Introduction è l'écologie et a la biologie des Cheiroptères. Explor Parc Nat Albert Mission, Bourlière, F & Verschuren, J, eds. 2:25-65.
11273. Verschuren, J. 1976. Les Cheiroptères du Mont Nimba (Liberia). Mammalia. 40:615-632.
11274. Versi, E, Chiappa, SA, Fink, G, et al. 1982. Effect of copulation on the hypothalamic content of gonadotrophic hormone-releasing hormone in the vole, *Microtus agrestis*. J Reprod Fert. 64:491-494.
11275. Versi, E, Chiappa, SA, Fink, G, et al. 1983. Pineal influences hypothalamic Gn-RH content in the vole, *Microtus agrestis*. J Reprod Fert. 67:365-368.
11276. Verts, BJ. 1967. Summer breeding of brush rabbits. The Murrelet. 48:19.
11277. Verts, BJ. 1967. The Biology of the Striped Skunk. Univ IL Press, Urbana.
11278. Verts, BJ & Carraway, LN. 1987. *Microtus canicaudus*. Mamm Species. 267:1-4.
11279. Verts, BJ & Carraway, LN. 1987. *Thomomys bulbivorus*. Mamm Species. 273:1-4.
11280. Verts, BJ & Carraway, LN. 1991. Summer breeding and fecundity in the camas pocket gopher (*Thomomys bulbivorus*). Northwest Nat. 72:61-65.
11281. Verts, BJ & Kirkland, GL, Jr. 1988. *Perognathus parvus*. Mamm Species. 318:1-8.
11282. Veselovský, Z. 1966. A contribution to the knowledge of the reproduction and growth of the two-toed sloth, *Choloepus didactylus* at Prague Zoo. Int Zoo Yearb. 6:147-153.
11283. Veselovský, Z. 1967. The Amur tiger *Panthera tigris altaica* in the wild and in captivity. Int Zoo Yearb. 7:210-215.
11284. Veselovský, Z. 1975. Notes on the breeding of cheetah (*Acinonyx jubatus* Schreber) at Prague Zoo. Zool Garten NF. 45:28-44.
11285. Veselovský, Z & Grundová, S. 1965. Beitrag zur Kenntnis des Dschungar-Hamsters, *Phodopus sungorus* (Pallas, 1773). Z Säugetierk. 30:305-311.
11286. Veselovský, Z & Volf, J. 1965. Breeding and care of rare Asian equids at Prague Zoo. Int Zoo Yearb. 5:28-37.
11287. Vessey, SH & Marsden, HM. 1975. Oviduct ligation in rhesus monkeys causes maladaptive epimelectic (care-giving) behavior. Int Congr Primatol. 5:321-325.
11288. Vestal, BM, Coleman, WC, & Chu, PR. 1980. Age of first leaving the nest in two species of deer mice (*Peromyscus*). J Mamm. 61:143-146.
11289. Vestal, EH. 1938. Biotic relations of the wood rat (*Neotoma fuscipes*) in the Berkeley hills. J Mamm. 19:1-36.
11290. Vevers, GM. 1926. Some notes on the recent birth of a hippopotamus (*H Amphibius*) in the gardens. Proc Zool Soc London. 1926:1097-1100.
11291. Vice, TE & Olin, FH. 1969. A note on the milk analysis and handrearing of the greater kudu *Tragelaphus strepsiceros* at San Antonio Zoo. Int Zoo Yearb. 9:114.
11292. Vidler, BO, Harthoorn, AM, Brocklesby, DW, et al. 1963. The gestation and parturition of the African buffalo (*Syncerus caffer caffer* Sparrman). E Afr Wildl J. 1:122-123.
11293. Vignoli, L. 1930. Il testiculo di *Rhinolophus ferrum equinum* nella sua maturazione sessuale e nelle sue variazioni cicliche annuali. Archs Ital Anat Embriol. 28:103-132.
11294. Viitala, J. 1975. Zur Soziologie subarktischer Populationen der Erdmaus, *Microtus agrestis* (Linne, 1761). Säugetierk Mitt. 23:1-9.

11295. Viljoen, S. 1975. Aspects of the ecology, reproductive physiology and ethology of the bush squirrel, *Paraxerus cepapi cepapi*. MS Thesis. Univ Pretoria, Pretoria. [Cited by Viljoen, 1977, J Reprod Fert]
11296. Viljoen, S. 1975. Aspects of reproduction of male bush squirrel, *Paraxerus cepapi cepapi* (A Smith, 1836) in the Transvaal. Publ Univ Pretoria, New Ser. 97:86-91.
11297. Viljoen, S. 1977. Behaviour of the bush squirrel *Paraxerus cepapi cepapi* (A Smith, 1836). Mammalia. 41:119-166.
11298. Viljoen, S. 1977. Factors affecting breeding synchronization in an African bush squirrel (*Paraxerus cepapi cepapi*). J Reprod Fert. 50:125-127.
11299. Viljoen, S. 1977. The yellow-footed squirrel of the bushveld. Fauna Flora (Pretoria). 28:15-16.
11300. Viljoen, S. 1978. Notes on the western striped squirrel *Funisciurus congicus* (Kuhl 1820). Madoqua. 2:119-128.
11301. Viljoen, S. 1980. Early postnatal development, parental care and interaction in the banded mongoose *Mungos mungo*. S Afr J Zool. 15:119-120.
11302. Viljoen, S & Du Toit, SHC. 1985. Postnatal development and growth of southern African tree squirrels in the genera *Funisciurus* and *Paraxerus*. J Mamm. 66:119-127.
11303. Villa R, B. 1953. Mamiferos silvestres del Valle de Mexico. An Inst Biol México D F. 23:269-492.
11304. Villa R, B & Villa-Cornejo, M. 1971. Observaciones acerca de algunos murciélagos del Norte de Argentina, especialmente de la biología del vampiro *Desmodus r rotundus*. An Inst Biol Univ Nal Autón México Ser Zool. 42:107-148.
11305. Villela, OMM & Alho, CJR. 1983. Postnatal development and growth of *Oryzomys subflavus* (Rodentia: Cricitidae) in laboratory setting. Rev Brasil Biol. 43:321-326.
11306. Vincent, J, Hitchins, PM, Bigalke, RC, et al. 1968. Studies on a population of nyala. Lammergeyer. 9:5-17.
11307. Vinogradov, BS & Argyropulo, AI. 1931. Zur Biologie der turkestanischen Springmäuse (Dipopidae). Z Säugetierk. 6:164-176.
11308. Vinson, GP. 1974. The control of the adrenocortical secretion in the brush-tailed possum, *Trichosurus vulpecula*. Gen Comp Endocrinol. 22:268-276.
11309. Vinson, GP & Renfree, MB. 1975. Biosynthesis and secretion of testosterone by adrenal tissue from the North American opossum, *Didelphis virginiana*, and the effects of tropic hormone stimulation. Gen Comp Endocrinol. 27:214-222.
11310. Viola, S. 1977. Observations on the brush tailed bettong (*Bettongia penicillata*), at the New York Zoological Park. Int Zoo Yearb. 17:156-157.
11311. Visser, J. 1979. The small cats. Afr Wild Life. 31:26-28.
11312. Vladykov, VD. 1944. Etudes sur les Mammifères aquatiques III Chasse, biologie et valeur économique du Marsouin Blanc ou Beluga (*Delphinapterus leucas*) du fleuve et du golfe Saint-Laurent. Contr Inst Biol Univ Montréal. 15:5-194.
11313. Vlasák, P. 1972. The biology of reproduction and post-natal development of *Crocidura suaveolens* Palls, 1811 under laboratory conditions. Acta Univ Carolinae Biol. 1970:207-292.
11314. Vlasák, P. 1973. Vergleich der postnatalen Entwicklung der Arten *Sorex araneus* L und *Crocidura suaveolens* (Pall) mit Bemerkungen zur Methodik der Laborzucht (Insectivora: Soricidae). Vestn Čcsl Spol Zool. 37:222-233.
11315. Vogel, P. 1970. Biologische Beobachtungen an Etruskerspitzmausen (*Suncus etruscus* Savi, 1832). Z Säugetierk. 35:173-185.
11316. Vogel, P. 1972. Beitrag zur Fortpflanzungsbiologie der Gattungen *Sorex*, *Neomys* und *Crocidura* (Soricidae). Verh Naturf Ges Basel. 82:165-192.
11317. Vogel, P. 1981. Occurrence and interpretation of delayed implantation in insectivores. J Reprod Fert Suppl. 29:51-60.
11318. Vogt, C. 1881. Recherches sur l'embryogénie des Chauves-Souris (Chiroptères). C R Assn Franc Sci la 10 session Alger. 1881:655-661.
11319. Vohralík, V. 1974. Biology of the reproduction of the common hamster *Cricetus cricetus* (L). Vestn Čcsl Spol Zool. 38:228-240.
11320. Voisin, J. 1972. Notes on the behaviour of the killer whale, *Orcinus orca* (L). Norw J Zool. 20:93-96.
11321. Volčanezkij, I. 1934. Gefangenschaftsbeobachtungen am Steppeniltis (*Putorius eversmanni* Less). Zool Garten NF. 7:262-273.
11322. Volčanezkij, I & Furssajev, A. 1934. Über die Ökologie von *Citellus pygmaeus* Pall im pestendemischen Gebiete des westlichen Kasakstan. Z Säugetierk. 9:404-423.
11323. Volcani, R. 1953. Seasonal variation in spermatogenesis of some farm animals under the climatic conditions of Israel. Bull Res Council Israel. 3:123-126.
11324. Volcani, R, Zisling, R, Sklan, D, et al. 1973. The composition of chinchilla milk. Br J Nutr. 29:121-125.
11325. Volf, J. 1957. A propos de la reproduction du Fennec. Mammalia. 21:454-455.
11326. Volf, J. 1959. La reproduction des Genettes au zoo de Prague. Mammalia. 23:168-171.
11327. Volf, J. 1963. Bemerkungen zur Fortpflanzungsbiologie der Eisbären, *Thalarctos maritimus* (Phipps) in Gefangenschaft. Z Säugetierk. 28:163-166.
11328. Volf, J. 1963. Einige Bemerkungen zur Aufzucht von Eisbären (*Thalarctos maritinus*) in Gefangenschaft. Zool Garten NF. 28:97-108.
11329. Volf, J. 1964. Trente-deux jeunes de la Genette. Mammalia. 28:658-659.
11330. Volf, J. 1966. Zucht und Fortpflanzangsbiologie der Braunbären (*Ursus arctos* L) in Gefangenschaft. Lynx. 6:185-189.
11331. Volf, J. 1968. Breeding the european wild cat *Felis s silvestris* at Prague Zoo. Int Zoo Yearb. 8:38-40.
11332. Volf, J. 1974. Auszerordentlich frühe Geschlechtsreife beim Fluszpferd (*Hippopotamus amphibius* Linné 1758). Zool Garten NF. 44:385-386.
11333. Volf, J. 1974. Breeding of cheetah (*Acinonyx jubatus* Schreber, 1776) in Prague Zoological Garden. Lynx. 15:45-49.
11334. Volf, J. 1974. Einfluss der Gefangenschaft auf die Geschechtszyklus der Przewalski-Wildpferde (*Equus przewalskii* Polj, 1881). Lynx. 16:78-83.
11335. Volf, J. 1975. Breeding of Przewalski wild horses. Breeding Endangered Species in Captivity. 263-270, Martin, RD, ed, Academic Pres, New York.
11336. Volf, J. 1977. Fruchtbarkeit, Ende der Fruchtbarkeitsperiode und Langlebigkeit der Przewalski-Wildpferde. Lynx. 19:81-86.
11337. Volf, J. 1981. Lebensdauer der Przwalski-Wildpferde (*Equus przewalskii* Polj, 1881) in Gefangenschaft. Zool Garten NF. 51:385-387.
11338. Volf, J & Brodský, O. 1970. Die Fortpflanzung und künstliche Aufzucht des Pumas. Lynx. 11:124-130.
11339. Volleth, M & Yong, HS. 1987. *Glischropus tylopus*, the first known old-world bat with a X- autosome translocation. Experientia. 43:922-924.

11340. von der Borgh, S. 1963. Unilateral hormone effect in the marsupial *Trichosurus vulpecula*. J Reprod Fert. 5:447-449.
11341. von Fischer, B. 1981. Die Rolle des Prolaktins bei der Trächtigkeit des Rehes (*Capreolus c capreolus*). Z Säugetierk. 46:259-264.
11342. von Gadow, K. 1972. Personal communication. [Cited by Mentis, 1972]
11343. von Holst, D. 1969. Sozialer Stress by Tupaias (*Tupaia belangeri*). Z vergl Physiol. 63:1-58.
11344. von Holst, D. 1974. Social stress in the tree-shrew: Its causes and physiological and ecological consequences. Prosimian Biology. 389-411, Martin, RD, Doyle, GA, & Walker, AC, eds, Duckworth & Co, London.
11345. von Jhering, H. 1885. Über die Fortpflanzung der Gürteltiere. Sitzingsberichte Preusz Akad Wissensch. 1885:1051-1053.
11346. von Ketelhodt, HF. 1973. Breeding notes in blue duiker. Zool Afr. 8:138.
11347. von Ketelhodt, HF. 1976. Observations on the lambing interval of the Cape bushbuck, *Tragelaphus scriptus sylvaticus*. Zool Afr. 11:221-225.
11348. von Ketelhodt, HF. 1977. Observations on the lambing interval of the grey duiker, *Sylivicapra grimmia grimmia*. Zool Afr. 12:232-233.
11349. von Ketelhodt, HF. 1977. The lambing interval of the blue duiker, *Cephalophus monticola* Gray, in captivity, with observations on its breeding and care. S Afr J Wildl Res. 7:41-43.
11350. von Ketelhodt, HF. 1977. The composition of the milk of the African dwarf goat, springbok and blue duiker. Zool Afr. 12:232.
11351. von Kölliker, ni. 1889. On the placenta of the genus *Tragulus*. J Anat. 14:373-374.
11352. von Richter, W. 1966. Untersuchungen über angeborene Verhaltensweise des Schabrackentapirs (*Tapirus indicus*) und des Flachlandtapirs (*Tapirus terrestris*). Zool Beitr NF. 1966:67-159.
11353. von Richter, W. 1969. Observations on the biology and behaviour of two black wildebeest (*Connochaetes gnou*) populations. Roneoed Progr Rep. [Cited by Mentis, 1972]
11354. von Richter, W. 1970. Observations on the biology and behaviour of the black wildebeest *Connochaetes gnou*. Roneoed Progr Rep. [Cited by Mentis, 1972]
11355. von Richter, W. 1971. Observations on the biology and ecology of the black wildebeest (*Connochaetes gnou*). J S Afr Wildl Mgmt Assoc. 1:3-16.
11356. von Richter, W. 1974. *Connochaetes gnou*. Mamm Species. 50:1-6.
11357. von Schantz, T. 1981. Female cooperation male competition, and dispersal in the red fox *Vulpes vulpes*. Oikos. 37:63-68.
11358. von Schantz, T. 1984. "Non-breeders" in the red fox *Vulpes vulpes*: A case of resource surplus. Oikos. 42:59-65.
11359. von Vietinghoff-Riesch, A. 1952. Beiträge zur Biologie des Siebenschläfers (*Glis glis* L). Bonn Zool Beitr. 3:167-186.
11360. von Vietinghoff-Riesch, A. 1955. Neuere Untersuchungen über die Biologie des Siebenschläfers, *Glis glis glis* (Linné, 1758) auf Grund von Freilandmarkierungen im Deister, Niedersachsen, und Beobachtungen in Tierhaus in Hannoversch-Munden. Säugetierk Mitt. 3:113-121.
11361. von W Schulte, H. 1916. Monographs of the Pacific Cetacea II The sei whale (*Balaenoptera borealis* Lesson) 2 Anatomy of a foetus of *Balaenoptera borealis*. Mem Mus Nat Hist. 1(5):394-502.
11362. von Wrangel, H. 1939/42. Beiträge zur Biologie der Rötelmaus, *Clethrionomys glareolus* Schreb. Z Säugetierk. 14:52-112.
11363. Vorhies, CT. 1921. Caesarian operation on *Lepus alleni*, and notes on the young. J Mamm. 2:114-116.
11364. Vorhies, CT & Taylor, WP. 1922. Life history of the kangaroo rat *Dipodomys spectabilis spectabilis* Merriam. USDA Bull. 1091:1-40.
11365. Vorhies, CT & Taylor, WP. 1933. The life histories and ecology of jackrabbits, *Lepus alleni* and *Lepus californicus* ssp, in relation to grazing in Arizona. Univ AZ Agric Exp Sta Tech Bull. 49:471-587.
11366. Voronov, GA. 1984. The ecology and postnatal development in *Microtus sachalensis* (Rodentia, Cricetidae). Zool Zhurn. 63:1693-1704.
11367. Vorontsov, NM. 1973. The evolution of the sex chromosome. Cytotaxonomy and Vertebrate Evolution. 619-657, Chiarelli, AB & Capanna, E, eds, Academic Press, New York.
11368. Voss, G. 1964. Notizen zur Fortpflanzungs-Rhythmik. Zool Garten NF. 28:263-265.
11369. Voss, G. 1965. Zweillingsgeburt beim Groszohr-Hirsch, *Odocoileus hemionus* Raf. Z Säugetierk. 30:20-24.
11370. Voss, G. 1969. Breeding the pronghorn antelope and the saiga antelope *Antelocapra americana* and *Saiga tatarica* Winnipeg Zoo. Int Zoo Yearb. 9:116-118.
11371. Voss, G. 1970. On the gestation of the white-handed gibbon or lar, *Hylobates lar* L. Zool Garten NF. 39:295-296.
11372. Voss, R. 1979. Male accessory glands and the evolution of copulatory plugs in rodents. Occas Pap Mus Zool Univ MI. 689:1-17.
11372a. Voss, RS. 1988. Systematics and ecology of ichthyomyine rodents (Muroidea): Patterns of morphological evolution on a small adaptive radiation. Bull Am Mus Nat Hist. 188:259-493.
11373. Voss, RS, Heideman, PD, Mayer, VL, et al. 1992. Husbandry, reproduction, and postnatal development of the neotropical muroid rodent *Zygodontomys brevicauda*. Lab Anim. 26:38-46.
11374. Vrolik, W. 1854. Bijdrage tot de Natuur-en ontleedkundige kennis van den *Manatanus americanus*. Bijdr Dierkd. 6:53-80.
11375. Wachtel, SS, Koo, GC, Ohno, S, et al. 1976. H-Y antigen and the origin of XY female wood lemmings (*Myopus schisticolor*). Nature. 264:638-639.
11376. Wachtendorf, W. 1951. Beiträge zur Ökologie und Biologie der Haselmaus (*Muscardinus avellanarius*) im Alpenvorland. Zool Jb Syst. 80:189-204.
11377. Wackernagel, H. 1965. Grant's zebra *Equus burchelli boehmi* at Basle Zoo: A contribution to breeding biology. Int Zoo Yearb. 5:38-41.
11378. Wackernagel, H. 1968. A note on breeding the serval cat *Felis serval* at Basle Zoo. Int Zoo Yearb. 8:46-47.
11379. Wade, O. 1927. Breeding habits and early life of the thirteen-striped ground squirrel, *Citellus tridecemlineatus* (Mitchell). J Mamm. 8:269-276.
11380. Wade, O. 1928. Notes on the time of breeding and the number of young of *Cynomys ludovicianus*. J Mamm. 9:149-151.
11381. Wade-Smith, J & Richmond, ME. 1978. Induced ovulation, development of the corpus luteum, and tubal transport in the striped skunk *Mephitis mephitis*. Am J Anat. 153:123-142.
11382. Wade-Smith, J & Richmond, ME. 1978. Reproduction in captive striped skunks (*Mephitis mephitis*). Am Mid Nat. 100:452-455.

11383. Wade-Smith, J, Richmond, ME, Mead, RA, et al. 1980. Hormonal and gestational evidence for delayed implantation in the striped skunk, *Mephitis mephitis*. Gen Comp Endocrinol. 42:509-515.
11384. Wade-Smith, J & Verts, BJ. 1982. *Mephitis mephitis*. Mamm Species. 173:1-7.
11385. Wadsworth, CE. 1969. Reproduction and growth of *Eutamias quadrivittatus* in southeastern Utah. J Mamm. 50:256-261.
11386. Waechter, A. 1975. Ecologie de la Fouine en Alsace. Terre Vie. 29:399-457.
11387. Wagner, HO. 1960. Beitrag zur Biologie des mexikanischen Spieszhirsches *Mazama sartorii* (Saussure). Z Tierpsychol. 17:358-363.
11388. Wagner, HO. 1961. Die Nagetiere einer Gebirgsabdachung in Südmexiko und ihre Beziehungen zur Umwelt. Zool Jahrb (Syst). 89:177-242.
11389. Wahby, A. 1932. Vie et moeurs des *Capra aegagrus* (Pallas) des Mts Taurus (région d'Alaya). Arch Zool Ital. 16:545-549.
11390. Wahlström, A. 1928. Beiträge zur Biologie von *Sorex vulgaris* L. Z Säugetierk. 3:284-294.
11391. Wahlström, A. 1929. Beiträge zur Biologie von *Crocidura leucodon* (ERM). Z Säugetierk. 4:157-185.
11392. Wainer, JW. 1976. Studies of an island population of *Antechinus minimus* (Marsupialia, Dasyuridae). Aust Zool. 19:1-7.
11393. Wakefield, NA. 1963. The Australian pigmy-possums. Vict Nat. 80:99-116.
11394. Wakefield, NA. 1970. Notes on Australian pigmy possums (*Cercartetus*, Phalangeridae, Marsupialia). Vict Nat. 87:11-18.
11395. Wakefield, NA & Warnecke, RM. 1963. Some revision in *Antechinus* (Marsupialia) 1. Vict Nat. 80:194-219.
11396. Wakefield, NA & Warnecke, RM. 1967. Some revision in *Antechinus* (Marsupialia) 2. Vict Nat. 84:69-99.
11397. Waldschmidt, A & Müller, EF. 1988. A comparison of postnatal thermal physiology and energetics in an altricial (*Gerbillus perpallidus*) and a precocial (*Acomys cahirinus*) rodent species. Comp Biochem Physiol. 90A:169-181.
11398. Walhovd, H. 1965. Alderbestemmelse av hare (*Lepus timmidus* L) med data om alders-og kjonnsfordeling vekst ogvekt. Med Statens Viltsunders 2nd serie. 22:1-57.
11399. Walhovd, H. 1984. The breeding habits of the European hedgehog (*Erinaceus europaeus* L) in Denmark. Z Säugetierk. 49:269-277.
11400. Walin, T, Soivio, A, & Kristofferson, R. 1968. Histological changes in the reproductive system of female hedgehogs during the hibernation season. Ann Zool Fennici. 5:227-229.
11401. Walker, A. 1968. A note on hand-rearing a potto *Perodicticus potto*. Int Zoo Yearb. 8:110-111.
11402. Walker, EP. 1929. Evidence on the gestation period of martens. J Mamm. 10:206-209.
11403. Walker, GE & Ling, JK. 1981. Australian sea lion *Neophoca cinerea* (Peron, 1816). Handbook of Marine Mammals. 1:99-118. Ridgway, SH & Harrison, RJ, eds, Academic Press, New York.
11404. Walker, GE & Ling, JK. 1981. New Zealand sea lion *Phocarctos hookeri* (Gray, 1844). Handbook of Marine Mammals. 1:25-38. Ridgway, SH & Harrison, RJ, eds, Academic Press, New York.
11405. Walker, LA, Czekala, NM, Cornell, LH, et al. 1987. Analysis of the ovarian cycle and pregnancy of the killer whale by urinary hormone measurement. Biol Reprod Suppl 1. 36:130.
11406. Walker, LC, Kaack, B, Brizzee, KR, et al. 1981. Prenatal ionizing irradiaton and early postnatal growth of Columbian and Bolivian squirrel monkeys (*Saimiri sciureus*). Am J Primatol. 1:379-387.
11407. Walker, LI, Spotorno, AE, & Arrau, J. 1984. Cytogenetic and reproductive studies of two nominal subspecies of *Phyllotis darwini* and their experimental hybrids. J Mamm. 65:220-230.
11408. Walker, ML, Gordon, TP, & Wilson, ME. 1982. Reproductive performance in capture-acclimated female rhesus monkeys (*Macaca mulatta*). J Med Primatol. 11:291-302.
11409. Walker, ML, Gordon, TP, & Wilson, ME. 1983. Menstrual cycle characteristics of seasonally breeding rhesus monkeys. Biol Reprod. 29:841-848.
11410. Walker, ML, Wilson, ME, & Gordon, TP. 1983. Female rhesus monkey aggression during the menstrual cycle. Anim Behav. 31:1047-1054.
11411. Walker, MT & Gemmell, RT. 1983. Organogenesis of the pituitary, adrenal and lung at birth in the wallaby, *Macropus rufogriseus*. Am J Anat. 168:331-344.
11412. Walker, MT & Gemmell, RT. 1983. Plasma concentrations of progesterone, oestradiol-17β and 13,14-dihydro-15-oxoprostaglandin F-2α in the pregnant wallaby (*Macropus rufogriseus rufogriseus*). J Endocrinol. 97:369-377.
11413. Walker, MT & Hughes, RL. 1981. Ultrastructural changes after diapause in the uterine glands, corpus luteum and blastocyst of the red-necked wallaby, *Macropus rufogriseus banksianus*. J Reprod Fert Suppl. 29:151-158.
11414. Walkinshaw, LH. 1947. Notes on the arctic hare. J Mamm. 28:353-357.
11415. Wall, F. 1908. Birth of Himalayan cat-bears (*Ailurus fulgens*) in captivity. J. Bombay Nat Hist Soc. 18:903-904.
11416. Wallace, C & Fairall, N. 1967. Chromosome polymorphism in the impala *Aepyceros melampus melampus*. S Afr J Sci. 63:482-486.
11417. Wallace, GI. 1981. Uterine factors and delayed implantation in macropodid marsupials. J Reprod Fert Suppl. 29:173-181.
11418. Wallace, GI. 1984. A histological study of the early stages of pregnancy in the bent-winged bat (*Miniopterus schreibersii*) in north-eastern New South Wales, Australia (30° 27'S). J Zool. 185:519-537.
11419. Wallace, HF. 1913. The Big Game of Central and Western China. John Murray, London.
11420. Wallace, MC, Jojolla, J, & Krausman, PR. 1988. Unusual reproduction in a Arizona elk. Southwest Nat. 33:249.
11421. Wallach, EE, Virutamasen, P, & Wright, KH. 1973. Menstrual cycle characteristics and side of ovulation in the rhesus monkey. Fert Steril. 24:715-721.
11422. Wallage-Drees, JM. 1983. Effects of food on onset of breeding in rabbits, *Oryctolagus cuniculus* (L) in a sand dune habitat. Acta Zool Fennica. 174:57-59.
11423. Wallen, K, Winston, LA, Gaventa, S, et al. 1984. Periovulatory changes in female sexual behavior and patterns of ovarian steroid secretion in group-living rhesus monkeys. Horm Behav. 181:431-450.
11424. Wallis, J. 1982. Sexual behavior of captive chimpanzees (*Pan troglodytes*): Pregnant versus cycling females. Am J Primatol. 3:77-88.
11425. Wallis, R & Baxter, G. 1980. The swamp antechinus (*Antechinus minimus maritimus*): Notes on a captive specimen. Vict Nat. 97:211-213.

11426. Wallis, RL & Maynes, GM. 1973. Ontogeny of thermoregulation in *Macropus parma* (Marsupialia, Macropodidae). J Mamm. 54:278-281.
11427. Wallis, SJ. 1981. Notes on the ecology of the Orkney vole (*Microtus arvalis orcadensis*). J Zool. 195:532-536.
11428. Wallis, SJ. 1983. Sexual behavior and reproduction of *Cercocebus albigena johnstonii* in Kibale Forest, western Uganda. Int J Primatol. 4:153-166.
11429. Walsh, SW, Meyer, RK, Wolf, RC, et al. 1977. Corpus luteum and fetoplacental functions in monkeys hypophysectomized during late pregnancy. Endocrinology. 100:845-850.
11430. Walsh, SW, Resko, JA, Grumbach, MM, et al. 1980. In utero evidence for a functional fetoplacental unit in rhesus monkey. Biol Reprod. 23:264-270.
11431. Walsh, SW, Wolf, RC, Meyer, RK, et al. 1977. Chorionic gonadotropin, chorionic somatomammotropin and prolactin in the uterine vein and peripheral plasma. Endocrinology. 100:851-855.
11432. Walsh, SW, Wolf, RC, Meyer, RK, et al. 1979. Estrogens in the uteroovarian, uterine and peripheral plasma in pregnant rhesus monkeys. Biol Reprod. 20:606-610.
11433. Walther, FR. 1965. Verhaltensstudien an der Grantgazelle (*Gazella granti* Brooke, 1872) im Ngorogoro-Krater. Z Tierpsychol. 22:167-208.
11434. Walton, DW, Brooks, JE, Thinn, KK, et al. 1980. Reproduction in *Rattus exulans* in Rangoon, Burma. Mammalia. 44:349-360.
11435. Walton, DW, Brooks, JE, Tun, UMM, et al. 1977. The status of *Rattus norvegicus* in Rangoon, Burma. Jap J Sanitary Zool. 28:363-366.
11436. Walton, DW, Brooks, JE, Tun, UMM, et al. 1978. Observations on reproductive activity among female *Bandicota bengalensis* in Rangoon. Acta Theriol. 23:489-501.
11437. Walton, DW, King, RE, Brooks, JE, et al. 1980. Observations on reproduction in *Mus musculus* L in Rangoon. Z Säugetierk. 45:57-60.
11438. Walton, GM & Walton, DW. 1973. Notes on hedgehogs of the lower Indus Valley. Korean J Zool. 16:161-170.
11439. Walton, KC. 1976. The reproductive cycle in the male polecat *Putorius putorius* in Britain. J Zool. 180:498-503.
11440. Walton, SM. 1971. Sex-chromosome mosaicism in pouch young of marsupials *Perameles* and *Isoodon*. Cytogenetics. 10:115-120.
11441. Wan, P. 1968. Notes on the first propagating record of the Palaearctic flying squirrel, *Pteromys volans aluco* (Thomas) from Korea. J Mamm Soc Japan. 4:40-43.
11442. Wandeler, A & Huber, W. 1969. Gewichtswachstum und jahreszeithche Gewichtsschwankungen bei Reh und Gemse. Rev Suisse Zool. 76:686-694.
11443. Wandeler, AI. 1972-1974. Die Fortpflanzungsleistung des Rehs (*Capreolus capreolus* L) im Berner Mittelland. Jahrb Naturhist Mus Bern. 5:245-301.
11444. Wandeler, AI & Graf, M. 1982. Der Geschlechtszyklus weiblicher Dachse (*Meles meles* L) in der Schweiz. Rev Suisse Zool. 89:1009-1016.
11445. Wandrey, R. 1975. Contribution to the study of the social behaviour of captive golden jackals (*Canis aureus* L). Z Tierpsychol. 39:365-402.
11446. Wandrey, R. 1978. Ein Fall von Superfötation bei Robben. Fr Kölner Zoo. 21:53.
11447. Wang, F. 1985. Preliminary study on the ecology of *Trogopterus xanthipes*. Acta Theriol Sinica. 5:103-110.
11448. Wang, F. 1985. Preliminary study on the ecology of the complex-toothed flying squirrel. Contemporary Mammalogy in China and Japan. 67-69, Kawamichi, T, ed, Mamm Soc Japan.
11448a. Wang, P, Lu, H, & Zhao, W. 1988. Energy metabolism of *Mustela vison* during pregnancy and lactation. Acta Theriol Sinica. 8:139-145.
11449. Wang, X & Hoffmann, RS. 1987. *Pseudois nayaur* and *Pseudois schaeferi*. Mamm Species. 278:1-6.
11450. Ward, HL. 1905. The number of young of the red bat. Bull WI Nat Hist Soc. 3:181-182.
11451. Ward, KL & Renfree, MB. 1984. Effects of progesterone on parturition in the tammar, *Macropus eugenii*. J Reprod Fert. 72:21-28.
11451a. Ward, OG & Wurster-Hill, DH. 1989. Ecological studies of Japanese raccoon dogs, *Nyctereutes procyonoides viverrinus*. J Mamm. 70:330-334.
11452. Ward, OG & Wurster-Hill, DH. 1990. *Nyctereutes prycyonoides*. Mamm Species. 358:1-5.
11453. Ward, SJ. 1990. Life history of the eastern pygmy-possum, *Cercartetus nanus* (Burramyidae: Marsupialia), in south-eastern Australia. Aust J Zool. 38:287-304.
11454. Ward, SJ. 1990. Life history of the feathertail glider, *Acrobates pygmaeus* (Acrobatidae: Marsupialia) in south-eastern Australia. Aust J Zool. 38:503-517.
11455. Ward, SJ. 1990. Reproduction in the western pgymy-possum, *Cercartetus concinnus* (Marsupialia: Burramyidae), with notes on reproduction of some other small possum species. Aust J Zool. 38:423-438.
11455a. Ward, SJ. 1992. Life history of the little pygmy-possum, *Cercartetus lepidus* (Marsupialia: Burramyidae), in the Big Desert, Victoria. Aust J Zool. 40:43-55.
11456. Ward, SJ & Renfree, MB. 1988. Reproduction in females of the feathertail glider *Acrobates pygmaeus* (Marsupialia). J Zool. 216:225-239.
11457. Ward, SJ & Renfree, MB. 1988. Reproduction in males of the feathertail glider *Acrobates pygmaeus* (Marsupialia). J Zool. 216:241-251.
11458. Warembourg, M. 1983. Progestagen-concentrating cells in the brain, uterus, vagina and mammary glands of the galago (*Galago senegalensis*). J Reprod Fert. 68:189-193.
11459. Waring, J, Moir, RJ, & Tyndale-Biscoe, CH. 1966. Comparative physiology of marsupials. Adv Comp Physiol Biochem. 2:237-376.
11460. Warkentin, MJ. 1968. Observations on the behavior and ecology of the nutria in Louisiana. Tulane Stud Zool. 15:10-17.
11461. Warneke, R. 1979. Australian fur seal. Mammals in the Seas, FAO Fish Ser 5. 2:41-44.
11462. Warren, ER. 1926. Notes on the breeding of wood rats of the genus *Neotoma*. J Mamm. 7:97-101.
11463. Warren, HB. 1974. Aspects of the behaviour of the impala male *Aepyceros melampus* during the rut. Arnoldia (Rhodesia). 6(27):1-9.
11464. Warren, RJ, Vogelsang, RW, Kirkpatrick, RL, et al. 1978. Reproductive behaviour of captive white-tailed deer. Anim Behav. 26:179-183.
11465. Warner, RM. 1982. *Myotis auriculus*. Mamm Species. 191:1-3.
11466. Warner, RM & Czaplewski, NJ. 1984. *Myotis volans*. Mamm Species. 224:1-4.
11467. Warwick, T. 1940. A contribution to the ecology of the musk-rat (*Ondatra zibetheca*) in the British Isles. Proc Zool Soc London. 110(A):165-201.

11468. Wassif, K. 1953. On the occurrence of hedgehogs of the genus *Paraechinus* in the El Tahreer Province of Egypt. Bull Zool Soc Egypt. 11:40-47.
11469. Waterman, JM. 1986. Behaviour and use of space by juvenile Columbian ground squirrels (*Spermophilus columbianus*). Can J Zool. 64:1121-1127.
11470. Watkins, LC. 1972. *Nycticeius humeralis*. Mamm Species. 23:1-4.
11471. Watkins, LC. 1977. *Euderma maculatum*. Mamm Species. 77:1-4.
11471a. Watkins, LC, Jones, JK, Jr, & Genoways, HH. 1972. Bats of Jalisco, Mexico. Sp Publ Mus TX Tech Univ. 1:1-44.
11472. Watkins, LC & Shump, KA, Jr. 1981. Behavior of the evening bat *Nycticeius humeralis* at a nursery roost. Am Mid Nat. 105:258-268.
11473. Watson, CRB & Watson, RT. 1986. Observations on the post-natal development of the tiny musk shrew, *Crocidura bicolor*. S Afr J Zool. 21:352-354.
11474. Watson, H. 1978. Coastal Otters (*Lutra lutra* L) i. Shetland. Vincent Wildlife Trust, London. [Cited by Mason & MacDonald, 1986]
11474a. Watson, JM & Graves, JAM. 1992. Clues about the evolution of mammalian sex chromosomes from monotreme gene mapping. 35-43, in Platypus and Echidnas, Augee, ML, ed, R Zool Soc NSW, Sydney.
11474b. Watson, JM, Spencer, JA, Riggs, AD, et al. 1990. The X chromosome of monotremes shares a highly conserved region with the eutherian and marsupial X chromosome despite the absence of X chromosome inactivation. Proc Nat Acad Sci. 87:7125-7129.
11474c. Watson, JM, Spencer, JA, Riggs, AD, et al. 1991. Sex chromosome evolution: Platypus gene mapping suggests that part of the human X chromosome was originally autosomal. Proc Nat Acad Sci. 88:11256-11260.
11475. Watson, JS. 1950. Some observations on the reproduction of *Rattus rattus* L. Proc Zool Soc London. 120:1-12.
11475a. Watson, JS. 1951. The rat problem in Cyprus. Col Res Publ. 9:1-66.
11476. Watson, JS. 1954. Breeding season of the wild rabbit in New Zealand. Nature. 174:608-609.
11477. Watson, JS. 1956. The present distribution of *Rattus exulans* (Peale) in New Zealand. NZ J Sci Tech B. 37:560.
11478. Watson, JS. 1957. Reproduction of the wild rabbit, *Oryctolagus cuniculus* (L) in Hawke's Bay, New Zealand. NZ J Sci Tech. 38:451-482.
11479. Watson, M. 1872. Contributions to the anatomy of the Indian elephant (*Elephas indicus*), Part II Urinary and generative organs. J Anat. 7:60-74.
11480. Watson, M. 1878. On the male generative organs of *Chlamydophorus truncatus* and *Dasypus sexcinctus*. Proc Zool Soc London. 1878:673-681.
11481. Watson, M. 1878. On the male generative organs of *Hyaena crocuta*. Proc Zool Soc London. 1878:416-428.
11482. Watson, M. 1881. Additional observations on the anatomy of the spotted hyaena. Proc Zool Soc London. 1881:516-521.
11483. Watson, M. 1881. On the female organs and placentation of the raccoon (*Procyon lotor*). Proc R Soc London. 32:272-298.
11484. Watson, M. 1883. Additional observations on the structures of the female organs of the Indian elephant (*Elephas indicus*). Proc Zool Soc London. 1883:517-521.
11485. Watson, M. 1885. On the anatomy of the female organs of the Proboscidae. Trans Zool Soc London. 11:111-130.
11486. Watson, M, Clulow, FV, & Mariotti, F. 1983. Influence of olfactory stimuli on pregnancy of the meadow vole, *Microtus pennsylvanicus*, in the laboratory. J Mamm. 64:706-708.
11487. Watson, PF & D'Souza, F. 1975. Detection of oestrus in the African elephant (*Loxodonta africana*). Theriogenology. 4:203-209.
11488. Watson, RM. 1969. Reproduction of wildebeest, *Connochaetes taurinus albojubatus* Thomas, in the Serengeti region: Its significance to conservation. J Reprod Fert Suppl. 6:287-310.
11489. Watson, RM. 1970. Generation time and intrinsic rates of natural increase in wildebeeste (*Connochaetes taurinus albojubatus* Thomas). J Reprod Fert. 22:557-561.
11490. Watt, LJ. 1934. Frequency distribution of litter size in mice. J Mamm. 15:185-189.
11491. Watts, CHS. 1969. Distribution and habits of the rabbit bandicoot. Trans R Soc S Aust. 93:135-141.
11492. Watts, CHS. 1974. The Nuyts Islands bandicoot (*Isoodon obsculus nauticus*). S Aust Nat. 49:20-24.
11493. Watts, CHS. 1982. Australian hydromyine rodents: Maintenance of captive colonies. The Management of Australian Mammals in Captivity. 180-184, Evans, DD, ed, Zool Bd VIC, Melbourne.
11494. Watts, CHS. 1982. The husbandry of Australian *Rattus*. The Management of Australian Mammals in Captivity. 177-179, Evans, DD, ed, Zool Bd VIC, Melbourne.
11495. Watts, CHS & Aslin, HJ. 1974. Notes on the small mammals of north-eastern South Australia and south-western Queensland. Trans R Soc S Aust. 98:61-69.
11495a. Watts, DP. 1990. Mountain gorilla life histories, reproductive competition, and sociosexual behavior and some implications for captive husbandry. Zoo Biol. 9:185-200.
11495b. Watts, DP. 1991. Mountain gorilla reproduction and sexual behavior. Am J Primatol. 24:211-225.
11496. Watts, ES. 1985. Adolescent growth and development of monkeys, apes and humans. Nonhuman Primate Models for Human Growth and Development. 1-65, Alan R Liss, Inc, New York.
11497. Watzka, M. 1940. Mikroskopisch-anatomische Untersuchungen über die Ranzzeit und Tragdauer des Hermelins (*Putorius ermineus*). Z Mikr Anat Forsch. 48:359-374.
11498. Wauchope, J. 1984. The narrow-nosed planigale, *Planigale tenuirostris* from the southern Flinders ranges. S Aust Nat. 58:40-44.
11499. Wayne, NL & Rissman, EF. 1990. Effects of photoperiod and social variables on reproduction and growth in the male musk shrew (*Suncus murinus*). J Reprod Fert. 89:707-715.
11500. Wayre, P. 1969. Breeding roe deer *Capreolus capreolus* at Norfolk Wildlife Park. Int Zoo Yearb. 9:112-114.
11501. Wayre, P. 1969. Breeding the European lynx *Felis lynx* at the Norfolk Wildlife Park. Int Zoo Yearb. 9:95-96.
11502. Wayre, P. 1979. The Private Life of the Otter. Batsford, London. [Cited by Mason & Macdonald, 1986]
11503. Weaker, FJ. 1980. Morphology of the prostate gland in the nine-banded armadillo. Acta Anat. 106:405-414.
11504. Weaver, RL. 1940. Notes on a collection of mammals from the southern coast of the Labrador penninsula. J Mamm. 21:417-422.
11505. Webb, BE, Stewart, REA, & Lavigne, DM. 1984. Mineral constituents of harp seal milk. Can J Zool. 62:831-833.
11506. Webb, J. 1974. Observations on the birth of a nyala *Tragelaphus angasi* at Marwell Zoo. Int Zoo Yearb. 14:132-133.

11507. Webb, JW & Nellis, DW. 1981. Reproductive cycle of white-tailed deer of St Croix, Virgin Islands. J Wildl Manage. 45:253-258.
11507a. Webb, RG, Baker, RH, & Dalby, PL. 1967. Vertebrados de la Isla del Toro, Veracruz. An Inst Biol Univ Mal Autón México Ser Zool. 1:1-8.
11507b. Webb-Peploe, CG. 1946/7. Field notes on the mammals of south Tinnevelly, south India. J Bombay Nat Hist Soc. 46:629-644.
11508. Weber, B. 1970. Wasserspitzmaus, *Neomys fodiens*, mit 11 Embryonen. Hercynia. 7:372-373.
11509. Weber, BJ & Wolfe, ML. 1982. Use of serum progesterone levels to detect pregnancy in elk. J Wildl Manage. 46:835-838.
11510. Weber, E. 1972. Breeding cotton-headed tamarins *Saguinus oedipus* at Melbourne Zoo. Int Zoo Yearb. 12:49-50.
11511. Weber, E. 1974. Breeding the eastern native-cat *Dasyurus viverrinus* at Melbourne Zoo. Int Zoo Yearb. 14:106-107.
11512. Weber, E. 1975. Note on the breeding of the eastern native cat at Melbourne Zoo. Breeding Endangered Species in Captivity. 183-186, Martin, RD, ed, Academic Press, New York.
11513. Webley, GE & Johnson, E. 1982. Effect of ovariectomy on the course of gestation in the grey squirrel (*Sciurus carolinensis*). J Endocrinol. 93:423-426.
11514. Webley, GE & Johnson, E. 1983. Reproductive physiology of the grey squirrel (*Sciurus carolinensis*). Mamm Rev. 13:149-154.
11515. Webley, GE, Pope, GS, & Johnson, E. 1984. Testosterone, 17β-hydroxy-5α-androstan-3-one and 4-androstene-3,17-dione in the plasma of male and female grey squirrels (*Sciurus carolinensis*). J Steroid Biochem. 20:1207-1209.
11516. Webley, GE, Pope, GS, & Johnson, E. 1985. Seasonal changes in the testis and accessory reproductive organs and seasonal and circadian changes in plasma testosterone concentrations in the male grey squirrel (*Sciurus carolinensis*). Gen Comp Endocrinol. 59:15-23.
11517. Webley, GE, Richardson, MC, Summers, PM, et al. 1989. Changing responsiveness of luteal cells of the marmoset monkey (*Callithrix jacchus*) to luteotrophic and luteolytic agents during normal and conception cycles. J Reprod Fert. 87:301-310.
11518. Webster, AB, Gartshore, RG, & Brooks, RJ. 1981. Infanticide in the meadow vole *Microtus pennsylvanicus*: Significance in relation to social system and population cycling. Behav Neural Biol. 31:342-347.
11519. Webster, DG, Evans, RL, & Dewsbury, DA. 1980. Behavioral patterns of round-tailed muskrats (*Neofiber alleni*). FL Scientist. 43:1-6.
11520. Webster, JR & Barrell, GK. 1985. Advancement of reproductive activity, seasonal reduction in prolactin secretion and seasonal pelage changes in pubertal red deer hinds (*Cervus elaphus*) subjected to artificially shortened daily photoperiod or daily melatonin treatments. J Reprod Fert. 73:255-260.
11520a. Webster, JR, Suttie, JM, & Corson, ID. 1991. Effects of melatonin implants on reproductive seasonality of male red deer (*Cervus elaphus*). J Reprod Fert. 92:1-11.
11521. Webster, WD & Jones, JK, Jr. 1982. *Reithrodontomys megalotis*. Mamm Species. 167:1-5.
11522. Webster, WD & Jones, JK, Jr. 1982. *Artibeus aztecus*. Mamm Species. 177:1-3.
11523. Webster, WD & Jones, JK, Jr. 1982. *Artibeus toltecus*. Mamm Species. 178:1-3.
11524. Webster, WD & Jones, JK, Jr. 1983. *Artibeus hirsutus* and *Artibeus inopinatus*. Mamm Species. 199:1-3.
11525. Webster, WD & Jones, JK, Jr. 1984. *Glossophaga leachii*. Mamm Species. 226:1-3.
11526. Webster, WD & Jones, JK, Jr. 1984. Notes on a collection of bats from Amazonian Ecuador. Mammalia. 48:247-252.
11527. Webster, WD & Jones, JK, Jr. 1985. *Glossophaga mexicana*. Mamm Species. 245:1-2.
11528. Webster, WD, Jones, JK, Jr, & Baker, RJ. 1980. *Lasiurus intermedius*. Mamm Species. 132:1-3.
11529. Webster, WD & Owen, RD. 1984. *Pygoderma bilabiatum*. Mamm Species. 220:1-3.
11530. Wedemeyer, KO. 1941. Beiträge zur Kleinnsäugerfauna Luneburgs. Z Säugetierk. 16:271-288.
11531. Wegge, P. 1973. Reproductive rates of red deer (*Cervus elaphus atlanticus* L) in Norway. Int Cong Game Biol. 11:79-87.
11532. Wegge, P. 1975. Reproduction and early calf mortality in Norwegian red deer. J Wildl Manage. 39:92-100.
11533. Wehrenberg, WB, Chaichareon, DP, Dierschke, DJ, et al. 1977. Vascular dynamics of the reproductive tract in the female rhesus monkey: Relative contribution of ovarian and uterine arteries. Biol Reprod. 17:148-153.
11534. Wehrenberg, WB, Dierschke, DJ, & Wolf, RC. 1979b. Uteroovarian pathways and maintenance of early pregnancy in rhesus monkeys. Biol Reprod. 20:601-605.
11535. Wehrenberg, WB, Dierschke, DJ, Wolf, RC, et al. 1979. The effect of ligating the ovarian and uterine arteries on ovarian function in cyclic rhesus monkeys. Biol Reprod. 20:596-600.
11536. Wehrenberg, WB, Dierschke, DJ, Wolf, RC, et al. 1980. Effect of intrauterine and intramuscular administration of human chorionic gonadotropin on corpus luteum function in cyclic rhesus monkey. Biol Reprod. 23:10-14.
11537. Wehrenberg, WB & Dyrenfurth, I. 1983. Photoperiod and ovulatory mentrual cycles in female macaque monkeys. J Reprod Fert. 68:119-122.
11538. Wehrenberg, WB, Dyrenfurth, I, & Ferin, M. 1980. Endocrine characteristics of the menstrual cycle in the Assamese monkey (*Macaca assamensis*). Biol Reprod. 23:522-525.
11539. Wehrenberg, WB & Ferin, M. 1982. Regulation of pulsatile prolactin secretion in primates. Biol Reprod. 27:99-103.
11540. Wehrenberg, WB & Wardlaw, SL, Frantz, AG, et al. 1982. β-Endorphin in hypophyseal portal blood: Variations throughout the menstrual cycle. Endocrinology. 111:879-881.
11541. Weick, RF. 1981. Induction of the luteinizing hormone surge by intrahypothalamic application of estrogen in the rhesus monkey. Biol Reprod. 24:415-422.
11542. Weick, RF, Dierschke, DJ, Karsch, FJ, et al. 1972. The refractory period following estrogen-induced LH surges in the rhesus monkey. Endocrinology. 91:1528-1530.
11543. Weick, RF, Dierschke, DJ, Karsch, FJ, et al. 1973. Periovulatory time courses of circulating gonadotropic and ovarian hormones in the rhesus monkey. Endocrinology. 93:1140-1147.
11544. Weideman, W. 1970. Vergleichende Untersuchungen an Gehirnen südamerikanischer Nagetiere. Z Wissensch Zool. 181:66-139.
11545. Weinbach, AP. 1941. The human growth curve: I Prenatal. Growth. 5:217-233.
11546. Weinbach, AP. 1941. The human growth curve: II Birth to puberty. Growth. 5:235-255.
11547. Weiner, C, Schlechter, N, & Zucker, I. 1984. Photoperiodic influences on testicular development of deer mice from two diffeent altitudes. Biol Reprod. 30:507-513.
11548. Weiner, J. 1987. Limits to energy budget and tactics in energy investment during reproduction in the Djungarian hamster

(*Phodopus sungorus sungorus* Pallas 1770). Symp Zool Soc London. 57:167-187.
11549. Weinmann, E & Mauler, R. 1975. Breeding *Macaca fascicularis* under laboratory conditions. Lab Anim Handb. 6:89-93.
11550. Weir, BJ. 1960?. Aspects of reproduction in chinchilla. J Reprod Fert. 12:410-411.
11551. Weir, BJ. 1967. The care and management of laboratory hystricomorph rodents. Lab Anim. 1:95-104.
11552. Weir, BJ. 1969. The induction of ovulation in the chinchilla. J Endocrinol. 43:55-60.
11553. Weir, BJ. 1970. The management and breeding of some more hystricomorph rodents. Lab Anim. 4:83-97.
11554. Weir, BJ. 1971. Some notes on reproduction in the Patagonian mountain viscacha, *Lagidium boxi* (Mammalia: Rodentia). J Zool. 164:463-467.
11555. Weir, BJ. 1971. Some observations on reproduction in the female green acouchi, *Myoprocta pratti*. J Reprod Fert. 24:193-201.
11556. Weir, BJ. 1971. Some observations on reproduction in the female agouti, *Dasyprocta aguti*. J Reprod Fert. 24:203-211.
11557. Weir, BJ. 1971. The reproductive physiology of the plains viscacha, *Lagostomus maximus*. J Reprod Fert. 25:355-363.
11558. Weir, BJ. 1971. The reproductive organs of the female plains viscacha, *Lagostomus maximus*. J Reprod Fert. 25:365-373.
11559. Weir, BJ. 1971. The evocation of oestrus in the cuis, *Galea musteloides*. J Reprod Fert. 26:405-408.
11560. Weir, BJ. 1973. Another hystricomorph rodent: Keeping casiragua (*Proechimys guairae*) in captivity. Lab Anim. 7:125-134.
11561. Weir, BJ. 1973. The role of the male in the evocation of oestrus in the cuis, *Galea musteloides* (Rodentia: Hystricomorpha). J Reprod Fert Suppl. 19:421-432.
11562. Weir, BJ. 1974. The tuco-tuco and plains viscacha. Symp Zool Soc London. 34:113-128.
11563. Weir, BJ. 1974. Reproductive characteristics of hystricomorph rodents. Symp Zool Soc London. 34:265-301.
11564. Weir, BJ & Rowlands, IW. 1973. Reproductive strategies of mammals. Ann Rev Ecol Syst. 4:139-163.
11565. Weir, BJ & Rowlands, IW. 1974. Functional anatomy of the hystricomorph ovary. Symp Zool Soc London. 34:303-332.
11566. Weiss, G, Butler, WR, Hotchkiss, J, et al. 1976. Periparturitional serum concentrations of prolactin, the gonadotropins, and the gonadal hormones in the rhesus monkey. Proc Soc Exp Biol Med. 151:113-116.
11567. Weiss, G, Dierschke, DJ, Karsch, FJ, et al. 1973. The influence of lactation on luteal function in the rhesus monkey. Endocrinology. 93:954-959.
11568. Weiss, G, Hotchkiss, J, Dierschke, DJ, et al. 1974. Metabolic clearance rate of progesterone during lactation in the rhesus monkey. Proc Soc Exp Biol Med. 146:901-903.
11569. Weiss, G, Rifkin, I, & Atkinson, LE. 1976. Induction of ovulation in premenarchial rhesus monkeys with human gonadotropins. Biol Reprod. 14:401-404.
11570. Weiss, G, Steinetz, BG, Dierschke, DJ, et al. 1981. Relaxin secretion in the rhesus monkey. Biol Reprod. 24:565-567.
11571. Weiss, M. 1986. Differences in steroid metabolism by testis, prostrate and epididymis in the immatrue and adult possum (*Trichosurus vulpecula*). Comp Biochem Physiol. 84B:571-574.
11572. Weiss, M. 1988. Factors influencing prostatic 5α-reductase in possum (*Trichosurus vulpecula*). Comp Biochem Physiol. 89B:21-26.
11573. Weiss, M & Carson, RS. 1987. Induction of adrenocortical special zone in the male possum (*Trichosurus vulpecula*). Comp Biochem Physiol. 86A:361-365.
11574. Weixin, L. 1988. Litter size and survival rate in captive giant pandas. Int Zoo Yearb. 27:304-307.
11575. Welker, C. 1972. Breeding of the prosimian primate *Galago crassicaudatus* E Geoffroy, 1812 in a family group. Folia Primatol. 18:379-389.
11576. Welker, C. 1982. On the life history of a female greater galago, *Galago agisymbanus* Coquerel, 1859, in captivity. Folia Primatol. 38:136-137.
11577. Welker, C, Röber, J, & Lührmann, B. 1981. Zum Anteil der einzelnen Gruppenmitglieder an der Jungenaufzucht beim Weiszbüscheläffchen, *Callithrix jacchus*, beim Lisztäffchen *Saguinus oedipus* und beim Springaffen *Callicebus moloch*. Zool Anz. 207:201-209.
11578. Welles, R & Welles, F. 1961. The Bighorn of Death Valley. Fauna of the National Parks, US Fauna Series. 6:1-242.
11579. Wells, LJ. 1935. Seasonal sexual rhythm and its experimental modification in the male of the thirteen-lined ground squirrel (*Citellus tridecemlineatus*). Anat Rec. 62:409-447.
11580. Wells, LJ. 1938. Gonadotropic potency of the hypophysis in a wild male rodent with annual rut. Endocrinology. 22:588-594.
11581. Wells, LJ & Moore, CR. 1936. Hormonal stimulation of spermatogenesis in the testis of the gound squirrel. Anat Rec. 66:181-200.
11582. Wells, LJ & Overholser, MD. 1938. Sperm formation and growth of accessory reproductive organs in hypophysectomized ground squirrels in response to substances from blood and human urine. Anat Rec. 72:231-247.
11583. Wells, LJ & Zalesky, M. 1940. Effects of low environmental temperature on the reproductive organs of male mammals with annual aspermia. Am J Anat. 66:429-447.
11584. Wells, ME. 1968. A comparison of the reproductive tracts of *Crocuta crocuta*, *Hyaena hyaena* and *Proteles cristatus*. E Afr Wildl J. 6:63-70.
11585. Wells, NM & Giacalone, J. 1985. *Syntheosciurus brochus*. Mamm Species. 249:1-3.
11586. Wells, RS. 1984. Reproductive behavior and hormonal correlates in Hawaiian spinner dolphins, *Stenella longirostris*. Rep Int Whal Comm Sp Issue. 6:465-472.
11587. Wells, RT. 1978. Field observations of the hairy-nosed wombat, *Lasiorhinus latifrons* (Owen). Aust Wildl Res. 5:299-303.
11588. Wells-Gosling, N & Heaney, LR. 1984. *Glaucomys sabrinus*. Mamm Species. 229:1-8.
11588a. Welter, WA & Sollberger, DE. 1939. Notes on the mammals of Rowan and adjacent counties in eastern Kentucky. J Mamm. 20:77-81.
11589. Wemmer, C. 1971. Birth, development and behavior of a fanaloka *Fossa fossa* at the National Zoological Park, Washington DC. Int Zoo Yearb. 11:113-115.
11590. Wemmer, C & Grodinsky, C. 1988. Reproduction in captive female brow-antlered deer (*Cervus eldi thamin*). J Mamm. 69:389-392.
11590a. Wemmer, C, Halverson, T, Rodden, M, et al. 1989. The reproductive biology of female Père David's deer (*Elaphurus davidianus*). Zoo Biol. 8:49-55.

11591. Wemmer, C & Murtaugh, J. 1981. Copulatory behavior and reproduction in the binturong, *Arctictis binturong*. J Mamm. 62:342-352.
11592. Wemmer, CW. 1983. Sociology and management. The Biology and Management of an Extinct Species, Père David's Deer. 126-132, Beck, BB & Wemmer, CW, eds, Noyes Publications, Park Ridge, NJ.
11593. Wendland, V. 1963. Entststehen und Vergehen einer Waldpopulation der Feldmaus (*Microtus arvalis*) im Berliner Grunewald. Sitzungsber Gesellsch Naturforsch Freunde Berlin NF. 3:56-66.
11594. Wendt, H. 1964. Erfolgreiche Zucht des Baumwollköpfchens oder Pincheäffchens *Leontocebus* (*Oedipomidas*) *oedipus* (Linné, 1758), in Gefangenschaft. Säugetierk Mitt. 12:49-52.
11595. Werner, RM, Montrey, RD, Roberts, CR, et al. 1980. Establishment of a cynomolgus monkey (*Macaca fascicularis*) breeding colony in Malaysia: A feasability study. Lab Anim Sci. 30:571-574.
11596. Wesley, DE, Knox, KL, & Nagy, JG. 1973. Energy metabolism of pronghorn antelopes. J Wildl Manage. 37:563-573.
11597. West, NB & Brenner, RM. 1983. Estrogen receptor levels in the oviducts and endometria of cynomolgus macaques during the menstrual cycle. Biol Reprod. 29:1303-1312.
11598. West, NO. 1968. The length of the estrous cycle in the Colombian black-tailed deer or coast deer (*Odocoileus lennionus columbianus*). BS Thesis. Univ BC, Vancouver.
11599. Westfahl, PK & Kling, OR. 1982. Relationship of estradiol to lutal function in the cycling baboon. Endocrinology. 110:64-69.
11600. Westfahl, PK & Resko, JA. 1983. Effects of clomiphene on luteal function in the nonpregnant cynomolgus macaque. Biol Reprod. 29:963-969.
11601. Westfahl, PK, Stadelman, HL, Horton, LE, et al. 1984. Experimental induction of estradiol positive feedback in intact male monkeys: Absence of inhibition by physiologic concentrations of testosterone. Biol Reprod. 31:856-862.
11602. Westlin, LM. 1982. Increased fertility in young primiparous female bank voles, *Clethrionomys glareolus*, treated with prolactin or progesterone after mating. J Reprod Fert. 66:113-115.
11603. Westlin, LM. 1982. Sterile matings at the beginning of the breeding season in *Clethrionomys rufocanus* and *Microtus agrestis*. Can J Zool. 60:2568-2571.
11604. Westlin, LM & Gustafsson, TO. 1983. Influence of sexual experience and social environment on fertility and incidence of mating in young female bank vole (*Clethrionomys glareolus*). J Reprod Fert. 69:173-177.
11605. Westlin, LM & Gustafsson, TO. 1984. Influence of age and artificial vaginal stimulation on fertility in female bank voles (*Clethrionomys glareolus*). J Reprod Fert. 71:103-106.
11606. Westlin, LM & Niklasson, M. 1983. Activity in the corpora lutea resulting from sterile matings in *Clethrionomys glareolus*: A histochemical study. J Endocrinol. 99:9-12.
11607. Westlin-van Aarde, LM. 1988. Reproduction in a laboratory colony of the pouched mouse, *Saccostomus campestris*. J Reprod Fert. 83:773-778.
11608. Westlin-van Aarde, LM & Gaskin, DE. 1989. Social environment and reproduction in female pouched mice, *Saccostomus campestris*. J Reprod Fert. 86:367-372.
11609. Wetter, WA & Sallberger, DE. 1939. Notes on the mammals of Rowan and adjacent counties in eastern Kentucky. J Mamm. 20:77-81.
11610. Wetzel, RM. 1985. Taxonomy and distribution of armidillos, Dasypodidae. The Evolution and Ecology of Armadillos, Sloths, and Vermilinguas. 23-46, Montgomery, GG, ed, Smithsonian Inst Press, Washington, DC.
11611. Wharton, CH. 1950. Notes on the life history of the flying lemur. J Mamm. 31:269-273.
11612. Wharton, CH. 1950. Notes on the Philippine tree shrew, *Urogale everetti* Thomas. J Mamm. 31:352-354.
11613. Wharton, CH. 1950. The tarsier in captivity. J Mamm. 31:260-268.
11614. Wharton, DC. 1986. Management procedures for the successful breeding of the striped grass mouse *Lemniscomys striatus*. Int Zoo Yearb. 24/25:260-263.
11614a. Wharton, DC. 1987. Captive management of the large Malayan chevrotain (*Tragulus napu*) at New York Zoological Park. Biology and Management of the Cervidae. 418-421, Wemmer, CM, ed, Smithsonian Inst Press, Washington, DC.
11615. Whateley, A. 1980. Comparative body measurements of male and female spotted hyaenas from Natal. Lammergeyer. 28:40-43.
11616. Whateley, A & Brooks, PM. 1985. The carnivores of Hluhluwe and Umfolozi Game Reserves 1973-1982. Lammergeyer. 35:1-27.
11617. Wheatley, BP. 1982. Energetics of foraging in *Macaca fascicularis* and *Pongo pygmaeus* and selective advantage of large body size in the orang-utan. Primates. 23:348-363.
11618. Wheeler, AG, Hurst, PR, Poyser, NL, et al. 1983. Uterine histology and prostaglandin concentrations and utero-ovarian venous steroid and prostaglandin concentrations during the luteal phase of the menstrual cycle in baboons (*Papio* spp) with or without an IUD. J Reprod Fert. 67:35-46.
11619. Wheeler, JFG. 1930. The age of fin whales at physical maturity with a note on multiple ovulations. Discovery Rep. 2:403-434.
11620. Wheeler, SH & King, DR. 1985. The European rabbit in south-western Australia II Reproduction. Aust Wildl Res. 12:197-212.
11621. Whitaker, A. 1905. Notes on the breeding habits of bats. Naturalist. 30:325-330.
11622. Whitaker, A. 1907. Notes on the breeding habits of bats. Naturalist. 32:74-83.
11623. Whitaker, A. 1909. Notes on bats. Naturalist. 34:71-77.
11624. Whitaker, JO, Jr. 1963. A study of the meadow jumping mouse, *Zapus hudsonius* (Zimmerman), in central New York. Ecol Monogr. 33:215-254.
11625. Whitaker, JO, Jr. 1972. *Zapus hudsonius*. Mamm Species. 11:1-7.
11626. Whitaker, JO, Jr. 1974. *Cryptotis parva*. Mamm Species. 43:1-8.
11627. Whitaker, JO, Jr & Mumford, RE. 1972. Notes on occurrence and reproduction of bats in Indiana. Proc IN Acad Sci. 81:376-383.
11628. Whitaker, JO, Jr & Sly, GR. 1970. First record of *Reithrodontomys megalotis* in Indiana. J Mamm. 51:381.
11629. Whitaker, JO, Jr & Wrigley, RE. 1972. *Napaeozapus insignis*. Mamm Species. 14:1-6.
11630. White, CM & Moore, SA. 1977. Intrauterine distribution and attachment of fetuses in white-tailed deer. J Mamm. 58:668-670.
11631. White, F. 1986. Census and preliminary observations on the ecology of the black-faced black spider monkey (*Ateles paniscus chamek*) in Manu National Park, Peru. Am J Primatol. 11:125-132.

11632. White, FN. 1914. Variations in the sex ratio of *Mus rattus* associated with an unusual mortality of adult females. Proc R Soc London. 87:335-344.
11633. White, IG. 1981. Epididymal compounds and their influence on the metabolism and survival of spermatozoa. Am J Primat. 1:143-156.
11634. White, JCD. 1953. Composition of whale's milk. Nature. 171:612.
11635. White, JM, Williams, G, Samour, JH, et al. 1985. The composition of milk from captive aardvark (*Orycteropus afer*). Zoo Biol. 4:245-251.
11636. White, JR. 1984. Born captive, released in the wild. Sea Frontiers. 30:369-375.
11637. White, M. 1973. The white-tailed deer of the Arkansas National Wildlife Refuge. TX J Sci. 24:457-489.
11638. White, M, Knowlton, FF, & Glazener, WC. 1972. Effects of dam-newborn fawn behavior on capture and mortality. J Wildl Manage. 36:897-906.
11639. White, RG, Holleman, DF, & Tiplady, BA. 1989. Seasonal body weight, body condition, and lactational trends in muskoxen. Can J Zool. 67:1125-1133.
11640. White, RG & Luick, JR. 1984. Plasticity and constraints in the lactational strategy of reindeer and caribou. Symp Zool Soc London. 51:215-232.
11641. White, RJ, Blaine, CR, & Blakley, GA. 1973. Detecting ovulation in *Macaca nemestrina* by correlation of vaginal cytology, body temperature and perineal tumescence with laparoscopy. Am J Phys Anthrop. 38:189-194.
11642. Whitehead, GK. 1972. Deer of the World. Constable, London.
11643. Whitehead, H & Arnblom, T. 1987. Social organization of sperm whales off the Galapagos Islands, February-April 1985. Can J Zool. 65:913-919.
11644. Whitehead, PE & McEwan, EH. 1973. Seasonal variation in the plasma testosterone concentration of reindeer and caribou. Can J Zool. 51:651-658.
11645. Whitehead, PE & McEwan, EH. 1980. Progesterone levels in peripheral plasma of rocky mountain bighorn ewes (*Ovis canadensis*) during the estrous cycle and pregnancy. Can J Zool. 58:1105-1108.
11645a. Whiteman, EE. 1940. Habits and pelage changes in captive coyotes. J Mamm. 21:435-438.
11646. Whitford, D, Fanning, FD, & White, AW. 1982. Some information on reproduction, growth and development in *Planigale gilesi* (Dasyuridae, Marsupialia). Carnivorous Marsupials. 1:77-81, Archer, M, ed, R Zool Soc NSW, Mossman, NSW.
11646a. Whitlow, WB & Hall, ER. 1933. Mammals of the Pocatello region of southeastern Idaho. Univ CA Publ Zool. 40:235-276.
11647. Whitney, R & Burdick, HO. 1966. Observations on Russian steppe lemmings (*Lagurus lagurus*). Symp Zool Soc London. 15:311-322.
11648. Whitsett, JM, Lawton, AD, & Miller, LL. 1984. Photosensitive stages in pubertal development of male deer mice (*Peromyscus maniculatus*). J Reprod Fert. 72:269-276.
11649. Whitsett, JM & Miller, LL. 1982. Photoperiod and reproduction in female deer mice. Biol Reprod. 26:296-304.
11650. Whitsett, JM, Noden, PF, Cherry, J, et al. 1984. Effect of transitional photoperiods on testicular development and puberty in male deer mice (*Peromyscus maniculatus*). J Reprod Fert. 72:277-286.
11651. Whitsett, JM, Underwood, H, & Cherry, J. 1983. Photoperiodic stimulation of pubertal development in male deer mice: Involvement of the circadian system. Biol Reprod. 28:652-656.
11652. Whitsett, JM, Underwood, H, & Cherry, J. 1984. Influence of melatonin on pubertal development in male deer mice (*Peromyscus maniculatus*). J Reprod Fert. 72:287-293.
11653. Whitten, JEJ. 1981. Ecological separation of three diurnal squirrels in tropical rain forest on Siberut Island, Indonesia. J Zool. 193:405-420.
11654. Whitten, PL. 1984. Competition among female vervet monkeys. Female Primates: Studies by Women Primatologists. 127-140, Small, SF, ed, Alan R Liss, Inc, New York.
11655. Whitworth, MR. 1984. Maternal care and behavioural development in pikas *Ochotona princeps*. Anim Behav. 32:743-752.
11656. Whitworth, MR & Southwick, CH. 1981. Growth of pika in laboratory confinement. Growth. 45:66-72.
11657. Whitworth, MR & Southwick, CH. 1984. Sex differences in the ontogeny of social behavior in pikas: Possible relationships to dispersal and territoriality. Behav Ecol Sociobiol. 15:175-182.
11658. Wick, WQ & Penttila, HE. 1957. Beaver litter of seven from Skagit County, Washington. Murrelet. 38:7.
11659. Wickings, EJ, Hanker, JP, & Nieschlag, E. 1980. Serum levels of biologically active luteinizing hormone following pituitary stimulation with luteinizing hormone releasing hormone and two analogues in male rhesus monkeys (*Macaca mulatta*): Effect of season. J Endocrinol. 85:12P.
11660. Wickings, EJ & Nieschlag, E. 1978. The effects of active immunization with testosterone on pituitary-gonadal feedback in the male rhesus monkey (*Macaca mulatta*). Biol Reprod. 18:602-607.
11661. Wickings, EJ, Usadel, KH, Dathe, G, et al. 1980. The role of follicle stimulating hormone in testicular function of the mature rhesus monkey. Acta Endocrinol. 95:117-128.
11662. Wickler, W. 1967. Socio-sexual signals and their intra-specific imitation among primates. Primate Ethology. 69-147, Morris, D, ed, Weidenfeld & Nicolson, London.
11663. Wickler, W & Uhrig, D. 1969. Verhalten und Ökologische Nische der Gelbflugelfledermaus *Lavia frons* (Geoffroy) (Chiroptera, Megadermatidae). Z Tierpsychol. 26:726-736.
11664. Widowski, TM., Ziegler, TE, Elowson, AM, et al. The role of males in the stimulation of reproductive function in female cotton-top tamarins, *Saguinus o oedipus*. Anim Behav. 40:731-741.
11665. Wiebe, RH, Diamond, E, Akesel, S, et al. 1984. Diurnal variations of androgens in sexually mature male Bolivian squirrel monkeys (*Saimiri sciureus*) during the breeding season. Am J Primatol. 7:291-297.
11666. Wiebe, RH, Williams, LE, Abee, CR, et al. 1988. Seasonal changes in serum dehydroepiandrosterone, androstenedione, and testosterone levels in the squirrel monkey (*Saimiri boliviensis boliviensis*). Am J Primat. 14:285-291.
11667. Wiegert, RG. 1961. Nest construction and oxygen consumption of *Condylura*. J Mamm. 42:528-529.
11668. Wieland, H. 1973. Beiträge zur Biologie und zum Massenwechsel der groszen Wühlmaus (*Arvicola terrestris* L). Zool Jahrb Syst. 100:351-428.
11669. Wigal, RA & Chapman, JA. 1983. Age determination, reproduction, and mortality of the gray fox (*Urocyon cinereoargenteus*) in Maryland, USA. Z Säugetierk. 48:226-245.
11670. Wiger, R. 1979. Demography of a cyclic population of the bank vole *Clethrionomys glareolus*. Oikos. 33:373-385.

11671. Wight, HM. 1918. The life-history and control of the pocket gopher in the Willamette Valey. OR Agric Coll Exp Sta Bull. 153:1-55.
11672. Wight, HM. 1922. The Willamette Valley pocket gopher. Murrelet. 3(3):6-8.
11672a. Wight, HM. 1925. Notes on the tree mouse, *Phenacomys silvicola*. J Mamm. 6:282-283.
11673. Wight, HM. 1930. Breeding habits and economic relations of the Dallas pocket gopher. J Mamm. 11:40-48.
11674. Wight, HM. 1931. Reproduction in the eastern skunk (*Mephitis mephitis nigra*). J Mamm. 12:42-47.
11675. Wigley, TB, Roberts, TH, & Arner, DH. 1983. Reproductive characteristics of beaver in Mississippi. J Wildl Manage. 47:1172-1177.
11676. Wigley, TB, Roberts, TH, & Arner, DH. 1984. Methods of determining litter size in beaver. Proc Ann Conf SE Assn Fish Wildl Ag. 38:197-200.
11677. Wikramanayake, ED & Dryden, GL. 1986. Scanning electron microscopy of reproductive epithelia in the female musk shrew, *Suncus murinus*. Acta Theriol. 31:129-136.
11678. Wilcox, DE & Mossman, HW. 1945. The common occurrence of "testis" cords in the ovaries of the shrew (*Sorex nagrans*, Baird). Anat Rec. 92:183-195.
11679. Wildhagen, A. 1953. On the Reproduction of Voles and Lemming in Norway. Statens Viltunderspkelser, Oslo.
11680. Wildt, DE, Bush, M, Howard, JG, et al. 1983. Unique seminal quality in the South African cheetah and a comparative evaluation in the domestic cat. Biol Reprod. 29:1019-1025.
11681. Wildt, DE, Bush, M, O'Brien, SJ, et al. 1991. Semen characteristics in free-living koalas (*Phascolarctos cinereus*). J Reprod Fert. 92:99-107.
11682. Wildt, DE, Doyle, LL, Stone, S, et al. 1977. Correlation of perineal swelling with serum ovarian hormanal levels, vaginal cytology, and ovarian follicular development during the baboon reproductive cycle. Primates. 18:261-270.
11683. Wildt, DE, Guthrie, SC, & Seager, SW. 1978. Ovarian and behavioral cyclicity of the laboratory maintained cat. Horm Behav. 10:251-257.
11684. Wildt, DE, O'Brien, SJ, Howard, JG, et al. 1987. Similarity in ejaculate-endocrine characteristics in captive versus free-ranging cheetahs of two subspecies. Biol Reprod. 36:351-360.
11685. Wildt, DE, O'Brien, SJ, Packer, C, et al. 1986. Reproductive and genetic consequences of founding an isolated population of east African lions. Biol Reprod. 34(Suppl):203.
11686. Wildt, DE, Phillips, LG, Simmons, LG, et al. 1988. A comparative analysis of ejaculate and hormonal characteristics of the captive male cheetah, tiger, leopard and puma. Biol Reprod. 38:245-255.
11687. Wildt, DE, Platz, CC, Chakraborty, PK, et al. 1979. Oestrus and ovarian activity in a female jaguar (*Panthera onca*). J Reprod Fert. 56:555-558.
11688. Wildt, DE, Platz, CC, Seager, SWJ, et al. 1981. Induction of ovarian activity in the cheetah (*Acinonyx jubatus*). Biol Reprod. 24:217-222.
11689. Wildt, L, Hausler, A, & Hutchison, JS. 1981. Estradiol as a gonadotropin releasing hormone in the rhesus monkey. Endocrinology. 108:2011-2013.
11690. Wildt, L, Häusler, A, Marshall, G, et al. 1981. Frequency and amplitude of gonadotropin-releasing hormone stimulation and gonadotrpin secretion in the rhesus monkey. Endocrinology. 109:376-385.
11691. Wildt, L, Hutchison, JS, Marshall, G, et al. 1981. On the site of action of progesterone in the blockade of the estradiol-induced gonadotropin discharge in the rhesus monkey. Endocrinology. 109:1293-1294.
11692. Wilen, R, Goy, RW, Resko, JA, et al. 1977. Pubertal body weight and growth in the female rhesus pseudohermaphrodite. Biol Reprod. 16:470-473.
11693. Wilen, R & Naftolin, F. 1976. Age, weight and weight gain in the individual pubertal female rhesus monkey (*Macaca mulatta*). Biol Reprod. 15:356-360.
11694. Wilen, R & Naftolin, F. 1978. Pubertal age, weight, and weight gain in the individual female New World monkey (*Cebus albifrons*). Primates. 19:769-774.
11695. Wiley, RW. 1980. *Neotoma floridana*. Mamm Species. 139:1-7.
11696. Wiley, RW. 1984. Reproduction in the southern plains woodrat (*Neotoma micropus*) in western Texas. Sp Publ Mus TX Tech Univ. 22:137-164.
11697. Wilhelm, JH. 1933. Das Wild des Okawangogebietes und des Caprivizipfels. J S W Afr Sci Soc. 6:51-74.
11697a. Wilk, RJ, Solberg, JW, Berns, VD, et al 1988. Brown bear, *Ursus arctos*, with six young. Can Field-Nat. 102:541-543.
11698. Wilke, F. 1958. Fat content of fur-seal milk. Murrelet. 39:40.
11698a. Wilke, F, Taniwaki, T, & Kuroda, N. 1953. *Phocoenoides* and *Lagenorhynchus* in Japan, with notes on hunting. J Mamm. 34:488-497.
11699. Wilkes, GE & Janssens, PA. 1986. Physiological and metabolic changes associated with weaning in the tammar wallaby, *Macropus eugenii*. J Comp Physiol B. 156:829-837.
11700. Wilkins, KT. 1986. *Reithrodontomys montanus*. Mamm Species. 257:1-5.
11701. Wilkins, KT. 1987. *Lasiurus seminolus*. Mamm Species. 280:1-5.
11702. Wilkins, KT. 1989. *Tadarida brasiliensis*. Mamm Species. 331:1-10.
11703. Wilkinson, GS. 1985. The social organization of the common vampire bat II Mating system, genetic structure, and relatedness. Behav Ecol Sociobiol. 17:123-134.
11704. Wilkinson, JF & de Fremery, P. 1940. Gonadotropic hormones in the urine of the giraffe. Nature. 46:491.
11705. Wilkinson, PF. 1971. The first verified occurrence of twinning in the musk ox. J Mamm. 52:238.
11706. Wilks, BJ. 1963. Some aspects of the ecology and population dynamics of the pocket gopher (*Geomys bursarius*) in southern Texas. TX J Sci. 15:241-283.
11707. Wilks, JW. 1977. Endocrine characterization of the menstrual cycle of the stumptailed monkey (*Macaca arctoides*). Biol Reprod. 16:474-478.
11708. Wilks, JW, Marciniak, RD, Hildebrand, DL, et al. 1980. Periovulatory endocrine events in the stumptailed monkey (*Macaca arctoides*). Endocrinology. 107:237-244.
11709. Wilks, JW & Noble, AS. 1983. Steroidogenic responsiveness of the monkey corpus luteum to exogenous chorionic gonadotropin. Endocrinology. 112:1256-1266.
11710. Will, U & Reichstein, H. 1972. Erfolgreiche Gefangenschaftzucht bei Brandmäusen, *Apodemus agrarius* (Pallas, 1771). Z Säugetierk. 37:359-362.
11710a. Willan, K. 1990. Reproductive biology of the southern AAfrican ice rat. Acta Theriol. 35:39-51.
11711. Willan, K & Meester, J. 1978. Breeding biology and postnatal development of the African dwarf mouse. Acta Theriol. 23:55-73.
11712. Williams, CE & Caskey, AL. 1965. Soil fertility and cottontail fecundity in southeastern Missouri. Am Mid Nat. 74:211-224.

11713. Williams, CF. 1986. Social organization of the bat, *Carollia perspicillata* (Chiroptera: Phyllostomidae). Ethology. 71:265-282.
11714. Williams, DF & Findley, JS. 1979. Sexual size dimorphism in vespertilionid bats. Am Mid Nat. 102:113-126.
11715. Williams, DF & Kilburn, KS. 1991. *Dipodomys ingens*. Mamm Species. 377:1-7.
11716. Williams, ES, Thorne, ET, Kwiatkowski, DR, et al. 1991. Reproductive biology and management of captive black-footed ferrets (*Mustela nigripes*). Zoo Biol. 10:383-398.
11716a. Williams, ES, Thorne, ET, Kwiatkowski, DR, et al. 1992. Comparative vaginal cytology of the estrous cycle of black-footed ferrets (*Mustela nigripes*), Siberian polecats (*M eversmanni*), and domestic ferrets (*M putorius furo*). J Vet Diagn Invest. 4:38-44.
11716b. Williams, G & Crawford, MA. 1987. The transfer of radiolabelled n-6 esential fatty acids by the mammary system of the tree-shrew (*Tupia* [*sic*] *tana*). Comp Biochem Physiol. 86B:575-580.
11717. Williams, L. 1967. Breeding Humbolt's woolly monkey *Lagothrix lagothrica* at Murrayton Woolly Monkey Sanctuary. Int Zoo Yearb. 7:86-89.
11718. Williams, L. 1968. Man and Monkey. JB Lippincott & Co, Philadelphia.
11719. Williams, L, Vituli, W, McElhinney, T, et al. 1986. Male behavior through the breeding season in *Saimiri boliviensis boliviensis*. Am J Primatol. 11:27-35.
11719a. Williams, LR & Cameron, GN. 1990. Intraspecific response to variation in food resources by Attwater's pocket gopher. Ecology. 71:797-810.
11720. Williams, LR & Cameron, GN. 1991. *Geomys attwateri*. Mamm Species. 382:1-5.
11721. Williams, R & Williams, A. 1982. The life cycle of *Antechinus swainsonii* (Dasyuridae, Marsupialia). Carnivorous Marsupials. 1:89-95, Archer, M, ed, R Zool Soc NSW, Mossman, NSW.
11722. Williams, RF & Hodgen, GD. 1980. Reinitiation of diurnal rhythm of prolactin secretion in postpartum rhesus monkeys. Biol Reprod. 23:276-280.
11723. Williams, RF, Johnson, DK, & Hodgen, GD. 1978. Ovarian estradiol secretion during early pregnancy in monkeys: Luteal versus extra-luteal secretion and effect of chorionic gonadotropin. Steroids. 32:539-545.
11724. Williams, RF, Johnson, DK, & Hodgen, GD. 1979. Resumption of estrogen-induced gonadotropin surges in postpartum monkeys. J Clin Endocrinol Metab. 49:422-428.
11725. Williams, RG, Golley, FB, & Carmon, JL. 1965. Reproductive performance of a laboratory colony of *Peromyscus polionotus*. Am Mid Nat. 73:101-110.
11726. Williams, SL. 1982. *Geomys personatus*. Mamm Species. 170:1-5.
11727. Williams, SL & Baker, RJ. 1974. *Geomys arenarius*. Mamm Species. 36:1-3.
11728. Williams, SL & Genoways, HH. 1980. Morphological variation in the southeastern pocket gopher, *Geomys pinetis* (Mammalia: Rodentia). Ann Carnegie Mus Nat Hist. 49:405-453.
11729. Williams, SL & Genoways, HH. 1981. Systematic review of the Texas pocket gopher, *Geomys personatus* (Mammalia: Rodentia). Ann Carnegie Mus Nat Hist. 50:435-473.
11730. Williams, SL, Ramímrez-Pulido, J & Baker, RJ. 1985. *Peromyscus alstoni*. Mamm Species. 242:1-4.
11731. Williamson, BR. 1976. Reproduction in female African elephant in the Wankie National Park, Rhodesia. S Afr J Wild Res. 6:89-93.
11731a. Williamson, DT. 1991. Condition, growth and reproduction in female red lechwe (*Kobus leche leche* Gray 1850). Afr J Ecol. 29:105-117.
11732. Williamson, P, Fletcher, TP, & Renfree, MB. 1990. Testicular development and maturation of the hypothalamic-pituitary-testicular axis in the male tammar, *Macropus eugenii*. J Reprod Fert. 88:549-557.
11733. Willig, A & Wendt, S. 1970. Aufzucht und Verhalten des Geoffroyi-Perücken-äffchens, *Oedipomidas geoffroyi* Pucheran, 1845. Säugetierk Mitt. 18:117-122.
11734. Willig, MR. 1985. Ecology, reproductive biology, and systematics of *Neoplatymops maltogrossensis* (Chiorptera: Molossidae). J Mamm. 66:618-628.
11735. Willig, MR. 1985. Reproductive activity of female bats from northeast Brazil. Bat Res News. 26:17-20.
11736. Willig, MR. 1985. Reproductive patterns of bats from Catingas and Cerrado biomes in northeast Brazil. J Mamm. 66:668-681.
11737. Willig, MR & Hollander, RR. 1987. *Vampyrops lineatus*. Mamm Species. 275:1-4.
11738. Willig, MR & Jones, JK, Jr. 1985. *Neoplatymops mattogrossensis*. Mamm Species. 244:1-3.
11739. Willis, R. 1983. Management of white-faced saki monkeys (*Pithecia pithecia*) at the Zoological Society of London. 24-28, Management of Prosimians and New World Primates Proc Symp Assoc Br Wild Animal Keepers.
11740. Willis, KB, Willig, MR, & Jones, JK, Jr. 1990. *Vampyrodes caraccioli*. Mamm Species. 359:1-4.
11741. Willis, RB. 1980. Breeding the Malayan giant squirrel *Ratufa bicolor* at London Zoo. Int Zoo Yearb. 20:218-220.
11742. Willner, GR, Chapman, JA, & Pursley, D. 1979. Reproduction, physiological responses, food habits, and abundance of nutria in Maryland marshes. Wildl Monogr. 65:1-43.
11743. Willner, GR, Dixon, KR, & Chapman, JA. 1983. Age determination and mortality of the nutria (*Myocastor coypus*) in Maryland USA. Z Säugetierk. 48:19-34.
11744. Willner, GR, Dixon, KR, Chapman, JA, et al. 1980. A model for predicting age-specific body weights of nutria without age determination. J Appl Ecol. 7:343-347.
11745. Willner, GR, Feldhamer, GA Zucker, EE, et al. 1980. *Ondatra zibethicus*. Mamm Species. 141:1-8.
11746. Wilson, BA. 1986. Reproduction in the female dasyurid *Antechinus minimus maritimus* (Marsupialia: Dasyuridae). Aust J Zool. 34:189-198.
11747. Wilson, BA & Bourne, AR. 1984. Reproduction in the male dasyurid *Antechinus minimus maritimus* (Marsupialia: Dasyuridae). Aust J Zool. 32:311-318.
11748. Wilson, BA & Bourne, AR. 1984. Reproduction in the marsupial mouse *Antechinus minimus*. Acta Zool Fennica. 171:165-167.
11749. Wilson, CG. 1977. Gestation and reproduction in golden lion tamarins. The Biology and Conservation of the Callitrichidae. 191-192, Kleiman, DG, ed, Smithsonina Inst Press, Washington, DC.
11750. Wilson, CG. 1980. The breeding and management of Nilgiri tahr *Hemitragus hylocrius* at Memphis Zoo. Int Zoo Yearb. 20:104-106.
11751. Wilson, DE. 1970. Opossum predation: *Didelphis* on *Philander*. J Mamm. 51:386-387.
11752. Wilson, DE. 1971. Ecology of *Myotis nigricans* (Mammalia: Chiroptera) on Barro Colorado Island, Panama Canal Zone. J Zool London. 163:1-13.

11753. Wilson, DE. 1978. *Thyroptera discifera*. Mamm Species. 104:1-3.
11754. Wilson, DE. 1979. Reproductive patterns. Sp Publ Mus TX Tech Univ. 16:317-378.
11755. Wilson, DE & Findley, JS. 1970. Reproductive cycle of a neotropical insectivorous bat, *Myotis nigricans*. Nature. 225:1155.
11756. Wilson, DE & Findley, JS. 1971. Spermatogenesis in some neotropical species of *Myotis*. J Mamm. 52:420-426.
11757. Wilson, DE & Findley, JS. 1977. *Thyroptera tricolor*. Mamm Species. 71:1-3.
11758. Wilson, DE & Hirst, SN. 1977. Ecology and factors limiting roan and sable antelope populations in South Africa. Wildl Monogr. 54:1-111.
11759. Wilson, DE & LaVal, RS. 1974. *Myotis nigricans*. Mamm Species. 39:1-3.
11760. Wilson, ED & Dewees, AA. 1962. Body weights, adrenal weights and oestrous cycles of nutria. J Mamm. 43:362-364.
11761. Wilson, JT & Hill, JP. 1908. Observations on the development of *Ornithorhynchus*. Phil Trans R Soc London Biol Sci. 199B:31-168.
11762. Wilson, KA. 1954. Litter production of coastal North Carolina muskrats. Proc SE Assoc Game Fish Comm. 8:13-19.
11763. Wilson, KA. 1959. The otter in North Carolina. Proc SE Assoc Game Fish Comm. 13:267-277.
11764. Wilson, M, Daly, M, & Behrends, P. 1985. The estrous cycle of two species of kangaroo rats (*Dipodomys microps* and *D merriami*). J Mamm. 66:726-732.
11765. Wilson, M, Plant, TM, & Michael, RP. 1972. Androgens and the sexual behaviour of male rhesus monkeys. J Endocrinol. 52:ii.
11766. Wilson, ME. 1981. Social dominance and female reproductive behaviour in rhesus monkeys (*Macaca mulatta*). Anim Behav. 29:472-482.
11767. Wilson, ME, Gordon, TP, & Bernstein, IS. 1978. Timing of births and reproductive success in rhesus monkey social groups. J Med Primatol. 7:202-212.
11768. Wilson, ME, Gordon, TP, Blank, MS, et al. 1984. Timing of sexual maturity in female rhesus monkeys (*Macaca mulatta*) housed outdoors. J Reprod Fert. 70:625-633.
11769. Wilson, ME, Gordon, TP, & Chikazawa, D. 1982. Female mating relationships in rhesus monkeys. Am J Primatol. 2:21-27.
11770. Wilson, ME, Gordon, TP, & Collins, DC. 1982. Variation in ovarian steroids associated with the annual mating period in female rhesus monkeys (*Macaca mulatta*). Biol Reprod. 27:530-539.
11771. Wilson, ME, Gordon, TP, & Collins, DC. 1986. Ontogeny of luteinizing hormone secretion and first ovulation in seasonal breeding rhesus monkeys. Endocrinology. 118:293-301.
11772. Wilson, ME, Pope, NS, & Gordon, TP. 1986. Seasonal modulation of luteinizing hormone (LH) secretion by estradiol (E_2) in rhesus monkey. Biol Reprod. 34(Suppl 1):185.
11773. Wilson, ME, Pope, NS, & Gordon, TP. 1987. Seasonal modulation of luteinizing-hormone secretion in female rhesus monkeys. Biol Reprod. 36:975-984.
11774. Wilson, ME, Walker, ML, & Gordon, TP. 1984. Effects of age, lactation, and repeated cycles on rhesus monkey copulatory intervals. Am J Primatol. 7:21-26.
11775. Wilson, ME, Walker, ML, Schwartz, SM, et al. 1985. Gonadal status influences developmental patterns of serum prolactin in female rhesus monkeys housed outdoors. Endocrinology. 116:640-645.
11776. Wilson, MI. 1977. Characterization of the oestrous cycle and mating season of squirrel monkeys from copulatory behaviour. J Reprod Fert. 51:57-63.
11777. Wilson, MI, Brown, GM, & Wilson, D. 1978. Annual and diurnal changes in plasma androgen and cortisol in adult male squirrel monkeys (*Saimiri sciureus*) studied longitudinally. Acta Endocrinol. 87:424-433.
11778. Wilson, P. 1984. Aspects of reproductive behaviour of bharal (*Pseudois nayaur*) in Nepal. Z Säugetierk. 49:36-42.
11779. Wilson, P & Franklin, WL. 1985. Male group dynamics in inter-male aggression of guanacos in southern Chile. Z Tierpsychol. 69:305-328.
11779a. Wilson, RT. 1986. Reproductive performance and survival of young one-humped camels on Kenya commercial ranches. Anim Prod. 42:375-380.
11780. Wilson, SN & Sealander, JA. 1971. Some characteristics of white-tailed deer reproduction in Arkansas. Proc Ann Conf SE Assoc Game Fish Comm. 25:53-65.
11781. Wilson, VJ. 1965. Observations on the greater kudu *Tragelaphus strepsiceros* Pallas from a tsetse control hunting scheme in northern Rhodesia. E Afr Wildl J. 3:27-36.
11782. Wilson, VJ. 1966. Observations on Lichtenstein's hartebeest, *Alcelaphus lichtensteini* over a three-year period and their response to various tsetse control measures in eastern Zambia. Arnoldia (Rhodesia). 2(15:1-13.
11783. Wilson, VJ. 1968. Weights of some mammals from eastern Zambia. Arnoldia (Rhodesia). 3(32):1-20.
11784. Wilson, VJ. 1969. Eland, *Taurotragus oryx*, in eastern Zambia. Arnoldia (Rhodesia). 4(12):1-9.
11785. Wilson, VJ. 1969. The large mammals of the Matopos National Park. Arnoldia (Rhodesia). 4(13):1-32.
11786. Wilson, VJ. 1970. Data from the culling of kudu, *Tragelaphus strepsiceros* Pallas in the Kyle National Park, Rhodesia. Arnoldia (Rhodesia). 4(36):1-26.
11787. Wilson, VJ & Child, G. 1964. Notes on bushbuck (*Tragelaphus scriptus*) from a tsetse fly control area in northern Rhodesia. Puku. 2:118-128.
11788. Wilson, VJ & Child, G. 1965. Notes on klipspringer from tsetse fly control areas in eastern Zambia. Arnoldia (Rhodesia). 1(35):1-9.
11789. Wilson, VJ & Clarke, JE. 1962. Observations on the common duiker, *Sylvicapra grimmia* Linn, based on material collected from a tsetse control game elimination scheme. Proc Zool Soc London. 138:487-497.
11790. Wilson, VJ & Kerr, MA. 1969. Brief notes on reproduction in steenbok *Raphicerus campestris* Thunberg. Arnoldia (Rhodesia). 4(23):1-5.
11791. Wilsson, L. 1971. Observations and experiments on the ethology of the European beaver (*Castor fiber* L). Viltrevy. 8:115-260.
11792. Wimsatt, WA. 1942. Survival of spermatozoa in the female reproductive tract of the bat. Anat Rec. 83:299-305.
11792a. Wimsatt, WA. 1944. Further studies on the survival of spermatozoa in the female reproductive tract of the bat. Anat Rec. 88:193-204.
11793. Wimsatt, WA. 1944. An analysis of implantation in the bat, *Myotis lucifugus lucifugus*. Am J Anat. 74:355-411.
11794. Wimsatt, WA. 1944. Growth of the ovarian follicle and ovulation in *Myotis lucifugus lucifugus*. Am J Anat. 74:129-173.
11795. Wimsatt, WA. 1945. Notes on breeding behavior, pregnancy, and parturition in some vespertilionid bats of the eastern United States. J Mamm. 26:23-33.
11796. Wimsatt, WA. 1945. The placentation of a vespertilionid bat, *Myotis lucifugus lucifugus*. Am J Anat. 77:1-52.

11797. Wimsatt, WA. 1950. New histological observations on the placenta of sheep. Am J Anat. 87:391-436.
11798. Wimsatt, WA. 1954. The fetal membranes and placentation of the tropical American vampire bat *Desmodus rotundus murinus*. Acta Anat. 21:285-341.
11799. Wimsatt, WA. 1958. The allantoic placental barrier in Chiroptera: A new concept of its organization and histochemistry. Acta Anat. 32:141-186.
11800. Wimsatt, WA. 1960. Some problems of reproduction in relation to hibernation in bats. Bull Mus Comp Zool. 124:249-269.
11801. Wimsatt, WA. 1960. An analysis of parturition in Chiroptera, including new observations on *Myotis l lucifugus*. J Mamm. 41:183-200.
11802. Wimsatt, WA. 1963. Delayed implantation in the Ursidae, with particular reference in the black bear (*Ursus americanus* Pallas). Delayed Implantation. 49-70, Enders, AC, ed, Univ Chicago Press, Chicago.
11803. Wimsatt, WA. 1975. Some comparative aspects of implantation. Biol Reprod. 12:1-40.
11804. Wimsatt, WA. 1979. Reproductive asymetry in unilateral pegnancy in Chiroptera. J Reprod Fert. 56:345-357.
11805. Wimsatt, WA & Enders, AC. 1978. Morphology of the uterus, placenta and paraplacental organs in the disc-winged bat *Thyroptera tricolor* Spix. Bat Res News. 19:30-31.
11806. Wimsatt, WA & Enders, AC. 1980. Structure and morphogenesis of the uterus, placenta, and paraplacental organs of the neotropical disc-winged bat *Thyroptera tricolor spix* [sic] (Microchiroptera: Thyropteridae). Am J Anat. 159:209-243.
11807. Wimsatt, WA, Enders, AC, & Mossman, HW. 1973. A reexamination of the chorioallantoic placental membrane of a shrew, *Blarina brevicauda*: Resolution of a controversy. Am J Anat. 138:207-233.
11808. Wimsatt, WA & Gopalakrishna, A. 1958. Occurrence of a placental hematoma in the primitive sheath-tailed bats (Emballonuridae), with observations on its structure, develoment and histochemistry. Am J Anat. 103:35-67.
11809. Wimsatt, WA & Kallen, FC. 1952. Anatomy and histophysiology of the penis of a vespertilionid bat, *Myotis lucifugus lucifugus*, with particular reference to its vascular organization. J Morphol. 90:415-465.
11810. Wimsatt, WA & Kallen, FC. 1957. The unique maturation response of the Graafian follicles of hibernation of vespertilionid bats and the question of its significance. Anat Rec. 129:115-131.
11811. Wimsatt, WA, Krutzsch, PH, & Napolitano, L. 1966. Studies on sperm survival mechanisms in the female reproductive tract of hibernating bats I Cytology and ultrastructure of intra-uterine spermatozoa in *Myotis lucifugus*. Am J Anat. 119:25-59.
11812. Wimsatt, WA & Parks, HF. 1966. Ultrastructure of the surviving follicle of hibernation and of the ovum-follicle cell relationship in the vespertilionid bat *Myotis lucifugus*. Symp Zool Soc London. 15:419-454.
11813. Wimsatt, WA & Trapido, H. 1952. Reproduction and the female reproductive cycle in the tropical American vampire bat, *Desmodus rotundus murinus*. Am J Anat. 91:415-445.
11814. Wimsatt, WA & Wislocki, GB. 1947. The placentation of the American shrews *Blarina brevicauda* and *Sorex fumeus*. Am J Anat. 80:361-435.
11815. Winegarner, CE & Winegarner, MS. 1982. Reproductive history of a bobcat. J Mamm. 63:680-682.
11816. Wing, ES. 1960. Reproduction in the pocket gopher in north-central Florida. J Mamm. 41:35-43.
11817. Wing, LD & Buss, IO. 1970. Elephants and forests. Wildl Monogr. 19:1-92.
11818. Wingate, LR. 1986. Age of sexual maturity in female *Rhinolophus clivosus* Cretzschmar 1828 and *Myotis tricolor* (Temminck 1832). Cimbebasia. 8A:155-159.
11819. Winkler, P, Loch, H, & Vogel, C. 1984. Life history of Hanuman langurs (*Presbytis entellus*): Reproductive parameters, infant mortality, and troop development. Folia Primatol. 43:1-23.
11820. Winkling, H. 1976. Karyologie und Biologie der beiden iberischen Wuhlmausarten *Pitymys mariae* und *Pitymys duodecimocostatus*. Z Zool Syst Evolut-forsch. 14:104-129.
11821. Winn, RM. 1989. The aye-ayes, *Daubentonia madagascariensis*, at the Paris Zoological Garden: Maintenance and preliminary behavioural observations. Folia Primatol. 52:109-123.
11822. Winowgradow, B. 1928. Über eine neue Springmaus (*Scriptopoda lichtensteini* sp n) aus der Karokorum-Wuste, Russich-Turkestan. Z Säugetierk. 2:92-101.
11823. Winter, JSD, Faiman, C, Hobson, WC, et al. 1975. Pituitary gonadal relations in infancy: I Patterns of serum gonadotropin concentrations from birth to four years of age in man and chimpanzee. J Clin Endocrinol Metab. 40:545-551.
11824. Winter, JSD, Faiman, C, Hobsen, WC, et al. 1980. The endocrine basis of sexual development in the chimpanzee. J Reprod Fert Suppl. 28:131-138.
11825. Winterer, J, Merriam, GR, Gross, E, et al. 1984. Idiopathic precocious puberty in the chimpanzee: A case report. J Med Primatol. 13:73-79.
11826. Winterer, J, Palmer, AE, Cicmanec, J, et al. 1985. Endocrine profile of pregnancy in the patas monkey (*Erythrocebus patas*). Endocrinology. 116:1090-1093.
11827. Winters, GH. 1978. Production, mortality, and sustainable yield of northwest Atlantic harp seals (*Pagophilus groenlandicus*). J Fish Res Bd Can. 1978:1249-1261.
11828. Winters, SJ, Troen, P, & Plant, TM. 1981. Relationship between testosterone binding globulin and the failure of androgens to suppress serum gonadotropin concentrations in long-term castrated adult male rhesus monkeys. J Steroid Biochem. 14:1223-1227.
11829. Wipper, E. 1975. Ökologie des Seehundes, *Phoca vitulina* (Linné, 1758), an der niedersächsischen Nordseeküste. Säugetierk Mitt. 23:32-63.
11830. Wirtz, JH. 1954. Reproduction in the pocket gopher *Thomomys talpoides rostralis*, Hall and Montague. J CO WY Acad Sci. 4:62.
11831. Wirtz, WO, III. 1968. Reproduction, growth and development, and juvenile mortality in the Hawaiian monk seal. J Mamm. 49:229-238.
11832. Wirtz, WO, III. 1972. Population ecology of the Polynesian rat, *Rattus exulans*, on Kure Atoll, Hawaii. Pacific Sci. 26:433-464.
11833. Wirtz, WO, III. 1973. Growth and development of *Rattus exulans*. J Mamm. 54:189-202.
11833a. Wischusen, EW, Ingle, NR, & Richmond, ME. ms. Observations on the reproductive biology and social behaviour of the Phillippine flying lemur (*Cynocephalus volans*). Malay Nat J.
11834. Wiseman, GL & Hendrickson, GO. 1950. Notes on the life history and ecology of the opossum in southeast Iowa. J Mamm. 31:331-337.
11835. Wishart, WD. 1981. January conception in an elk in Alberta. J Wildl Manage. 45:544.
11836. Wislocki, GB. 1925. On the placentation of the sloth (*Bradypus griseus*). Contrib Embryol Carnegie Inst. 16(78):5-22.

11837. Wislocki, GB. 1926. Remarks on the placentation of a platyrrhine monkey (*Ateles geoffroyi*). Am J Anat. 36:467-487.
11838. Wislocki, GB. 1927. On the placentation of the tridactyl sloth (*Bradypus griseus*) with a description of some characters of the fetus. Contrib Embryol Carnegie Inst. 19(105):209-228.
11839. Wislocki, GB. 1928. Further observations upon the minute structure of the labyrinth in the placenta of the sloths. Anat Rec. 40:385-395.
11840. Wislocki, GB. 1928. Observations on the gross and microscopic anatomy of the sloths (*Bradypus griseus griseus* Gray and *Choloepus hoffmanni* Peters). J Morphol. 46:317-397.
11841. Wislocki, GB. 1928. On the placentation of the two-toed anteater *Cyclopes didactylus*. Anat Rec. 39:69-83.
11842. Wislocki, GB. 1928. The placentation of hyrax (*Procavia capensis*). J Mamm. 9:117-126.
11843. Wislocki, GB. 1929. The placentation of primates, with a consideration of the phylogeny of the placenta. Contrib Embryol Carnegie Inst. 20(111):51-80.
11844. Wislocki, GB. 1930. On an unusual placental form in the Hyracoidea: Its bearing on the theory of the phylogeny of the placenta. Contrib Embryol Carnegie Inst. 21(122):83-95.
11845. Wislocki, GB. 1930. On a series of placental stages of a platyrrhine monkey (*Ateles geoffroyi*) with some remarks upon age, sex, and breeding period in platyrrhines. Contrib Embryol Carnegie Inst. 22(133):173-192.
11846. Wislocki, GB. 1931. Notes on the female reproductive tract (ovaries, uterus and placenta) of the collared peccary (*Pecari angulatus bangsi* Goldman). J Mamm. 12:143-149.
11847. Wislocki, GB. 1932. On the female reproductive tract of the gorilla, with a comparison of that of other primates. Contr Embryol Carnegie Inst. 23(135):163-204.
11848. Wislocki, GB. 1932. Placentation in the marmoset (*Oedipomas geoffoyi*), with remarks of twinning in monkeys. Anat Rec. 52:381-399.
11849. Wislocki, GB. 1933. On the placentation of the harbor porpoise (*Phocoena phocoena* Linnaeus). Biol Bull. 65:80-98.
11850. Wislocki, GB. 1935. The placentation of the manatee (*Trichechus latirostris*). Mem Mus Comp Zool. 54:159-178.
11851. Wislocki, GB. 1936. The external genitalia of the simian primates. Human Biol. 8:309-347.
11852. Wislocki, GB. 1939. Observations on twinning in marmosets. Am J Anat. 64:445-483.
11853. Wislocki, GB. 1940. The placentation of *Solenodon paradoxus*. Am J Anat. 66:497-531.
11854. Wislocki, GB. 1941. The placentation of an antelope (*Rhynchotragus kirkii nyikae* Heller). Anat Rec. 81:221-235.
11855. Wislocki, GB. 1943. Hemopoeisis in the chorionic villi of the placenta of platyrrhine monkeys. Anat Rec. 85:349-363.
11856. Wislocki, GB. 1943. Studies on the growth of deer antlers II Seasonal changes in the male reproductive tract of the Virginia deer (*Odocoileus virginianus borealis*) with a discussion of the factors controlling the antler-gonad periodicity. Essays in Biology in Honor of Herbert M Evans. 631-653, Farquhar, ST, Leake, CD, Lyons, WB, et al, eds, Univ CA Press, Berkeley.
11857. Wislocki, GB & Amoroso, EC. 1956. The placenta of the wolverine (*Gulo gulo luscus* Linnaeus). Bull Mus Comp Zool. 114:91-100.
11858. Wislocki, GB & Enders, RK. 1941. The placentation of the bottle-nosed porpoise (*Tursiops truncatus*). Am J Anat. 68:97-125.
11859. Wislocki, GB & Fawcett, DW. 1941. The placentation of the Jamaican bat (*Artibeus jamaicensis parvipes*). Anat Rec. 81:307-331.
11860. Wislocki, GB & Fawcett, DW. 1949. The placentation of the pronghorned antelope (*Antilocapra americana*). Bull Mus Comp Zool. 101:545-558.
11861. Wislocki, GB & Streeter, GL. 1938. On the placentation of the macaque (*Macaca mulatta*) from the time of implantation until the formation of the definitive placenta. Contr Embryol Carnegie Inst. 27(160):1-66.
11862. Wislocki, GB & van der Westhuysen, OP. 1940. The placentation of *Procavia capensis* with a discussion of the placental affinities of the Hyracoidea. Contr Embryol Carnegie Inst. 28:65-88.
11863. Withers, PC. 1979. Ecology of a small mammal community on a rocky outcrop in the Namib Desert. Madoqua. 2:229-246.
11864. Withers, PC. 1983. Seasonal reproduction by small mammals of the Namib desert. Mammalia. 47:195-204.
11865. Witkin, JW. 1985. Luteinizing hormone-releasing hormone in olfactory bulb of primates. Am J Primatol. 8:309-315.
11866. Włodek, K & Krzywiński, A. 1987. Zu Biologie und Verhalten des Marderhundes (*Nyctereutes procyonoides*) in Polen. Z Jagdwissensch. 32:203-215.
11867. Wojciechowska, B. 1970. The growth and net production in the common vole during postnatal period. Acta Theriol. 15:81-88.
11868. Wolf, RC, O'Connor, RF, & Robinson, JA. 1977. Cyclic changes in plasma progestins and estrogens in squirrel monkeys. Biol Reprod. 17:228-231.
11869. Wolf, RH, Harrison, RM, & Martin, TW. 1975. A review of reproductive patterns in New World monkeys. Lab Anim Sci. 25:814-821.
11870. Wolfe, JL. 1982. *Oryzomys palustris*. Mamm Species. 176:1-5.
11871. Wolfe, JL. 1985. Population ecology of the rice rat (*Oryzomys pallustris*) in a coastal marsh. J Zool. 205A:235-244.
11872. Wolfe, JL & Linzey, AV. 1977. *Peromyscus gossypinus*. Mamm Species. 70:1-5.
11873. Wolfe, LD. 1984. Japanese macaque female sexual behavior. Female Primates: Studies by Women Primatologists. 141-157, Small, MF, ed, Alan R Liss, Inc, New York.
11874. Wolfe, LG, Deinhardt, F, Ogden, JD, et al. 1975. Reproduction of wild-caught and laboratory-born marmoset species used in biomedical research (*Saguinus* sp, *Callithrix jacchus*). Lab Anim Sci. 25:802-813.
11875. Wolfe, LG, Ogden, JD, Deinhardt, JB, et al. 1972. Breeding and hand-rearing marmosets for viral oncogenesis studies. Breeding Primates. 145-157, Beveridge, WIB, ed, Karger, Basel.
11876. Wolff, JO. 1980. Social organization of the taiga vole (*Microtus xanthognathus*). Biologist. 62:34-45.
11877. Wolff, JO. 1985. Comparative population ecology of *Peromyscus leucopus* and *Peromyscus maniculatus*. Can J Zool. 63:1548-1555.
11878. Wolff, JO & Lidicker, WZ, Jr. 1980. Population ecoloy of the taiga vole, *Microtus xanthognathus*, in interior Alaska. Can J Zool. 58:1800-1812.
11879. Woliński, Z. 1980. Einflusz von Zoo-und Reservatsbedingungen auf den Geschlechtszyklus des Wisents (*Bison bonasus* L, 1758). Zool Garten NF. 50:321-326.
11880. Wolman, AA. 1985. Gray whale *Eschrichtius robustus* (Lilljeborg, 1861). Handbook of Marine Mammals. 3:67-90. Ridgway, SH & Harrison, RJ, eds. Academic Press, New York.
11881. Wolton, RJ, Arak, PA, Godfray, CJ, et al. 1982. Ecological and behavioural studies of the Megachiroptera at

Mount Nimba, Liberia, with notes on Microchiroptera. Mammalia. 46:419-448.
11882. Wong, B & Parker, KL. 1988. Estrus in black-tailed deer. J Mamm. 69:168-171.
11883. Wong, YC, Breed, WG, & Chow, PH. 1988. Ultrastructural features of the ventral prostate epithelial cells in the Australian plains rat, *Pseudomys australis*). Acta Anat. 133:289-296.
11884. Wood, AK, Short, RE, Darling, AE, et al. 1986. Serum assays for detecting pregnancy in mule and white-tailed deer. J Wildl Manage. 50:684-687.
11885. Wood, BJ & Liau, SS. 1984. A long-term study of *Rattus tiomanicus* populations in an oil plantation in Johore, Malaysia III Bionomics and natural regulation. J Appl Ecol. 21:473-495.
11886. Wood, DH. 1970. An ecological study of *Antechinus stuartii* (Marsupialia) in a south-east Queensland rain forest. Aust J Zool. 18:185-207.
11887. Wood, DH. 1980. The demography of a rabbit population in an arid region of New South Wales, Australia. J Anim Ecol. 49:55-79.
11888. Wood, FD. 1935. Notes on the breeding behavior and fertility of *Neotoma fuscipes macrotis* in captivity. J Mamm. 16:105-109.
11889. Wood, FG. 1977. Birth of porpoises at Marineland, Florida, 1939 to 1969, and comments on problems involved in captive breeding of small Cetacea. Breeding Dolphins Present Status, Suggestions for the Future. US Mar Mamm Comm Rep No MMC-76/07 NTIS no PB 273673 US Dept Comm Nat Tech Info Serv. 47-60, Ridgway, SH & Benirschke, K, eds, Arlington.
11890. Wood, JE. 1949. Reproductive pattern of the pocket gopher (*Geomys breviceps brazensis*). J Mamm. 30:36-44.
11891. Wood, JE. 1955. Notes on young pocket gophers. J Mamm. 36:143-144.
11892. Wood, JE. 1955. Notes on reproduction and rate of increase of raccoons in the Post Oak region of Texas. J Wildl Manage. 19:409-410.
11893. Wood, JE. 1958. Age structure and productivity of a gray fox population. J Mamm. 39:74-86.
11894. Wood, TJ. 1967. Ecology and population dynamics of the red squirrel (*Tamiasciurus hudsonicus*) in Wood Buffalo National Park. MS Thesis, Univ SAS, Saskatoon. [not seen]
11895. Wood, TJ & Munroe, SA. 1977. Dynamics of snowshoe hare populations in the Maritime Provinces of Canada. Can Wildl Serv Occ Pap. 30:1-19.
11896. Wood, TJ & Tessier, GD. 1974. First records of eastern flying squirrel (*Glaucomys volans*) from Nova Scotia. Can Field-Nat. 88:83-84.
11897. Wood Jones, F. 1914. Some phases in the reproductive history of the female mole (*Talpa europea*). Proc Zool Soc London. 1914:191-216.
11898. Wood Jones, F. 1916. The genitalia of the Chiroptera. J Anat. 51:36-60.
11899. Wood Jones, F. 1917. The genitalia of *Tupaia*. J Anat. 51:118-126.
11900. Wood Jones, F. 1921. On the habits of *Trichosurus vulpecula*. J Mamm. 2:18-93.
11901. Wood Jones, F. 1923. The Mammals of South Australia Part I: The Monotremes and the Carnivorous Marsupials. Government Printer, Adelaide.
11902. Wood Jones, F. 1924. The Mammals of South Australia Part II: The Bandicoots and the Herbivorous Marsupials. Government Printer, Adelaide.
11903. Wood Jones, F. 1949. The study of a generalized marsupial (*Dasycercus cristicauda* Krefft). Trans Zool Soc London. 26:409-501.
11904. Woodall, PF. 1982. An index of male fecundity in live water voles (*Arvicola terrestris*). J Zool. 197:292-295.
11905. Woodcock, WH. 1955. Management and breeding of a new laboratory species, *Meriones libycus*. J Anim Techn Assn. 6:10-14.
11906. Woods, CA. 1973. *Erethizon dorsatum*. Mamm Species. 29:1-6.
11907. Woods, CA & Boraker, DK. 1975. *Octodon degus*. Mamm Species. 67:1-5.
11907a. Woods, CA, Contreras, L, Willner-Chapman, G. et al. 1992. *Myocastor coypus*. Mamm Species. 398:1-8.
11908. Woodward, SL. 1979. The social system of feral asses (*Equus asinus*). Z Tierpsychol. 49:304-316.
11909. Woolf, A & Harder, JD. 1979. Population dynamics of a captive white-tailed deer herd with emphasis on reproduction and mortality. Wildl Monogr. 67:1-53.
11910. Woolf, A. 1971. Influence of lambing and morbidity on weights of captive Rocky Mountain bighorns. J Mamm. 52:242-243.
11910a. Woollard, HH. 1925. The anatomy of *Tarsius spectrum*. Proc Zool Soc London. 1925:1071-1184.
11910b. Woollard, P. 1971. Differential mortality of *Antechinus stuartii* (Macleay): Nitrogen balance and somatic changes. Aust J Zool. 19:347-353.
11911. Wooller, RD, Renfree, MB, Russell, EM, et al. 1981. Seasonal changes in a population of the nectar-feeding marsupial *Tarsipes spencerae* (Marsupalia: Tarsipedidae). J Zool. 195:267-279.
11912. Woolley, P. 1966. Reproductive biology of *Antechinus stuartii* Macleay (Marsupialia: Dasyuridae). Ph D Diss. Aust Nat Univ, Canberra.
11913. Woolley, P. 1966. Reproduction in *Antechinus* spp and other dasyurid marsupials. Symp Zool Soc London. 15:281-294.
11914. Woolley, P. 1971. Maintenance and breeding of laboratory colonies of *Dasyuroides byrnei* and *Dasycercus cristicauda*. Int Zoo Yearb. 11:351-354.
11915. Woolley, P. 1971. Observations on the reproductive biology of the dibbler, *Antechinus apicalis* (Marsupialia: Dasyuridae). J Proc R Soc West Aust. 54:99-102.
11915a. Woolley, P. 1973. Breeding patterns and the breeding and laboratory maintenance of dasyurid marsupials. Jikken Dobutsu (Exp Anim). 22:161-172.
11916. Woolley, P. 1974. The pouch of *Planigale subtilissima* and other dasyurid marsupials. J R Soc West Aust. 57:11-15.
11917. Woolley, P. 1975. The seminiferous tubules in dasyurid marsupials. J Reprod Fert. 45:255-261.
11917a. Woolley, P. 1976. An accessory erectile structure on the penis of some dasyurid marsupials. Bull Aust Mamm Soc. 3:25.
11917b. Woolley, PA. 1981. *Antechinus bellus*, another dasyurid marsupial with post-mating mortality of males. J Mamm. 62:381-382.
11918. Woolley, PA. 1984. Reproduction in *Antechinomys laniger* ("spenceri" form) (Marsupialia: Dasyuridae): Field and laboratory investigations. Aust Wildl Res. 11:481-489.
11919. Woolley, PA. 1988. Reproduction in the ningbing antechinus (Marsupialia: Dasyuridae): Field and laboratory observations. Aust Wildl Res. 15:149-156.
11920. Woolley, PA. 1990. Reproduction in *Sminthopsis macroura* (Marsupialia: Dasyuridae) I The female. 1990. Aust J Zool. 38:187-205.

11921. Woolley, PA. 1990. Reproduction in *Sminthopsis macroura* (Marsupialia: Dasyuridae) II The male. 1990. Aust J Zool. 38:207-217

11922. Woolley, PA. 1991. Reproduction in *Pseudantechinus macdonnellensis* (Marsupialia: Dasyuridae): Field and laboratory observations. Wildl Res. 18:13-25.

11922a. Woolley, PA. 1991. Reproduction in *Dasykaluta rosamondae* (Marsupialia: Dasyuridae): Field and laboratory observations. Aust J Zool. 39:549-568.

11923. Woolley, PA & Ahern, LD. 1983. Observations on the ecology and reproduction of *Sminthopsis leucopus* (Marsupialia: Dasyuridae). Proc R Soc Vict. 95:169-180.

11924. Woolley, PA & Allison, A. 1982. Observations on the feeding and reproductive states of captive feather-tailed possums, *Distoechurus pennatus* (Marsupialia: Burramyidae). Aust Mamm 5:285-287.

11925. Woolley, PA & Gilfillan, SL. 1991. Confirmation of polyoestry in captive white-footed dunnarts, *Sminthopsis leucopus* (Marsupialia: Dasyuridae). Aust Mamm. 14:137-138.

11926. Woolley, PA & Scarlet, G. 1984. Observations on the reproductive anatomy of male *Tarsipes rostratus* (Marsupialia: Tarsipedidae). Possums and Gliders. 445-450, Smith, AP & Hume, ID, eds, Aust Mamm Soc, Sydney.

11927. Woolley, PA & Valente, A. 1986. Reproduction in *Sminthopsis longicaudata* (Marsupialia: Dasyuridae): Laboratory observations. Aust Wildl Res. 13:7-12.

11928. Woolley, PA & Watson, MR. 1984. Observations on a captive outdoor breeding colony of a small dasyurid marsupial, *Sminthopsis crassicaudata*. Aust Wildl Res. 11:249-254.

11929. Woolpy, JH. 1968. The social organization of wolves. Nat Hist. 77(5):46-55.

11930. Worth, CB. 1950. Observations on the behavior and breeding of captive rice rats and woodrats. J Mamm. 31:421-426.

11931. Worth, CB. 1967. Reproduction, development and behavior of captive *Oryzomys laticeps* and *Zygodontomys brevicauda* in Trinidad. Lab Anim Care. 17:355-361.

11932. Worth, RW, Charlton, HM, & Mackinnon, PCB. 1973. Field and laboratory studies of the control of luteinizing hormone secretion and gonadal activity in the vole, *Microtus agrestis*. J Reprod Fert Suppl. 19:89-99.

11933. Worthy, GAJ & Lavigne, DM. 1987. Mass loss, metabolic rate, and energy utilization by harp and gray seal pups during the postweaning fast. Physiol Zool. 60:352-364.

11934. Wragg, LE. 1953. Notes on the life history of the muskrat in southern Ontario. Can Field-Nat. 67:174-177.

11935. Wrangham, RW. 1969. Captivity behaviour and post-natal development of the cape pouched rat, *Saccostomus campestris* Peters. Puku. 5:207-210.

11936. Wright, BS. 1983. Cougar. Can Wildl Serv, Hinterlands Who's Who. 1983:1-4.

11937. Wright, EM, Jr & Bush, DE. 1977. The reproductive cycle of the capuchin (*Cebus apella*). Lab Anim Sci. 27:651-654.

11938. Wright, JW, Niles, DM, Wheeler, GG, et al. 1965. Northernmost records of some neotropical bat genera. J Mamm. 46:330-331.

11939. not used.

11940. Wright, PC. 1981. The night monkeys, genus *Aotus*. Ecology and Behavior of Neotropical Primates. 1:211-240. Coimbra-Filho, AF & Mittemermeier, RA, eds, Academia Brassileira Ciêcias, Rio de Janeiro.

11941. Wright, PC. 1984. Biparental care in *Aotus trivirgatus* and *Callicebus moloch*. Female Primates: Studies by Women Primatologists. 59-75, Small, MF, ed, Alan R Liss, Inc, New York.

11942. Wright, PC, Izard, MK, & Simons, EL. 1986. Reproductive cycles in *Tarsius bancanus*. Am J Primatol. 11:207-215.

11943. Wright, PC, Toyama, LM, & Simons, EL. 1986. Courtship and copulation in *Tarsius bancanus*. Folia Primatol. 46:142-148.

11944. Wright, PL. 1942. A correlation between the spring molt and spring changes in the sexual cycle in the weasel. J Exp Zool. 91:103-110.

11945. Wright, PL. 1942. Delayed implantation in the long-tailed weasel (*Mustela frenata*), the short-tailed weasel (*Mustela cicognani*) and the marten (*Martes americana*). Anat Rec. 83:341-353.

11946. Wright, PL. 1947. Preimplantation stages in the long-tailed weasel (*Mustela frenata*). Anat Rec. 100:593-603.

11947. Wright, PL. 1947. The sexual cycle of the male long-tailed weasel (*Mustela frenata*). J Mamm. 28:343-352.

11948. Wright, PL. 1948. Breeding habits of captive long-tailed weasels (*Mustela frenata*). Am Mid Nat. 39:338-344.

11949. Wright, PL. 1963. Variations in reproductive cycles in North American mustelids. Delayed Implantation. 77-95, Enders, AC, ed, Univ Chicago Press, Chicago.

11950. Wright, PL. 1966. Observations on the reproductive cycle of the American badger (*Taxidea taxus*). Symp Zool Soc London. 15:27-45.

11951. Wright, PL. 1969. The reproductive cycle of the male American badger, *Taxidea taxus*. J Reprod Fert Suppl. 6:435-445.

11952. Wright, PL. 1981. Discussion. J Reprod Fert Suppl. 29:49-50.

11953. Wright, PL & Coulter, MW. 1967. Reproduction and growth in Maine fishers. J Wildl Manage. 31:70-87.

11954. Wright, PL & Dow, SA, Jr. 1962. Minimum breeding age in pronghorn antelope. J Wildl Manage. 26:100-101.

11955. Wright, PL & Rausch, R. 1955. Reproduction in the wolverine *Gulo gulo*. J Mamm. 36:346-355.

11956. Wrigley, RE. 1969. Ecological notes on the mammals of southern Quebec. Can Field-Nat. 83:201-211.

11957. Wrigley, RE. 1972. Systematics and biology of the woodland jumping mouse, *Napaeozapus insignis*. IL Biol Monogr. 47:1-117.

11957a. Wrigley, RE, Dubois, JE, & Copland, HWR. 1991. Distribution and ecology of six rare species of prairie rodents in Manitoba. Can Field-Nat. 105:1-12.

11958. Wright, S. 1922. The effects of inbreeding and crossbreeding on guinea pigs. USDA Tech Bull. 1090:1-63.

11959. Wroot, AJ, Wroot, SA, & Murie, JO. 1987. Intraspecific variation in postnatal growth of Columbian ground squirrels (*Spermophilus columbianus*). J Mamm. 68:395-398.

11960. Wu, JT. 1975. Time of implantation in the Mongolian gerbil (*Meriones unguiculatus*) and its hormonal requirements. Biol Reprod. 13:298-303.

11961. Wu, JT & Chang, MC. 1972. Effects of progesterone and estrogen on the fate of blastocysts in ovariectomized pregnant ferrets: A preliminary study. Biol Reprod. 7:231-237.

11962. Wu, JT & Chang, MC. 1973. Hormonal requirements for implantation and embryonic development in the ferret. Biol Reprod. 9:350-355.

11963. Wüst, G. 1976. Geburt und perinatales Verhalten beim Steppenzebra, *Equus quagga*. Zool Garten NF. 46:305-352.

11964. Wyatt, JM & Vevers, GM. 1935. On the birth of a chimpanzee recently born in the society's gardens. Proc Zool Soc London. 1935:195-197.

11965. Wyman, J. 1967. The jackals of the Serengeti. Animals. 10:79-83.
11966. Wynn, RM. 1973. Fine structure of the placenta. Handbk Physiol, Section 7, Endocrinol, Vol II, Female Reproductive System, Part 2. 261-276, Greep, RO & Astwood, EB, eds, Am Physiol Soc, Washington, DC.
11967. Wynn, RM & Amoroso, EC. 1964. Placentation in the spotted hyena (*Crocuta crocuta* Erxleben), with particular reference to the circulation. Am J Anat. 115:327-361.
11968. Wynn, RM & Davies, J. 1965. Comparative electron microscopy of the hemochorial placenta. Am J Obs Gyn. 91:533-549.
11969. Wynne-Edwards, KE & Lisk, RD. 1984. Djungarian hamsters fail to conceive in the presence of multiple males. Anim Behav. 32:626-628.
11970. Wynne-Edwards, KE, Huck, UW, & Lisk, RD. 1987. Influence of pre- and post-copulatory pair contact on pregnancy success in Djungarian hamsters, *Phodopus campbelli.* J Reprod Fert. 80:241-249.
11971. Wynne-Edwards, KE, Terranova, PF, & Lisk, RD. 1987. Cyclic Djungarian hamsters, *Phodopus campbelli*, lack the progesterone surge normally associated with ovulation and behavioral receptivity. Endocrinology. 120:1308-1316.
11972. Xanten, WA, Jr. 1972. Gestation period in the bongo (*Boocercus eurycerus*). J Mamm. 53:232.
11973. Xanten, WA, Jr, Collins, LR, & Connery, MM. 1973. Breeding and birth of a bongo *Boocercus eurycerus* at the National Zoological Park, Washington. Int Zoo Yearb. 13:152-153.
11974. Xanten, WA, Jr, Kafka, H, & Olds, E. 1976. Breeding the binturong, *Arctictis binturong* at the National Zoological Park Washington. Int Zoo Yearb. 16:117-119.
11975. Xanten, WA, Jr, Kessler, DS, & Grumm, J. 1988. Breeding and management of Prevost's squirrel *Callosciurus prevosti* at the National Zoological Park. Int Zoo Yearb. 27:283-286.
11976. Xia, X & Millar, JS. 1986. Sex-related dispersion of breeding deer mice in the Kananaskis valley, Alberta. Can J Zool. 64:933-936.
11977. Ximenez, A. 1975. *Felis geoffroyi.* Mamm Species. 54:1-4.
11978. Xu, YS, Wang, HY, Zeng, GA, et al. 1985. Hormone concentrations before and after semen-induced ovulation in the bactrian camel (*Camelus bactrianus*). J Reprod Fert. 74:341-346.
11979. Yablokov, AV & Bogoslovskaya, LS. 1984. A review of Russian research on the biology and commercial whaling of the gray whale. The Gray Whale *Eschrichtius robustus*. 465-485, Jones, ML, Swartz, SL, & Leatherwood, S, eds, Academic Press, New York.
11980. Yablokov, AV, Bel'kovich, VM, & Borisov, VI. 1972. Whales and Dolphins. Nauka, Moscow. [Translation by Nat Techn Inf Service, Springfield, VA]
11981. Yadav, M. 1972. Characteristics of blood in the pouch young of a marsupial (*Setonix brachyurus*). Aust J Zool. 20:249-263.
11982. Yadav, M. 1973. Spontaneous development of quiescent blastocyst in a marsupial, *Setonix brachyurus*, carrying a suckling pouch young. Lab Anim. 7:89-92.
11982a. Yadav, M & Eadie, M. 1973. Passage of maternal immunoglobins to the pouch young of a marsupial, *Setonix brachyurus*. Aust J Zool. 21:171-181.
11983. Yadav, M, Stanley, NF, & Waring, H. 1972. The microbial flora of the gut of the pouch-young and the pouch of a marsupial, *Setonix brachyurus*. J Gen Microbiol. 70:437-442.
11984. Yadav, RN. 1968. Notes on breeding the Indian wolf *Canis lupus pallipes* at Jaipur Zoo. Int Zoo Yearb. 8:17-18.
11985. Yadav, RN. 1971. Breeding the stump-tailed macaque *Macaca arctoides* at Jaipur Zoo. Int Zoo Yearb. 11:83-84.
11986. Yagil, R & Etzion, Z. 1980. Hormonal and behavioural patterns in the male camel (*Camelus dromedarius*). J Reprod Fert. 58:61-65.
11987. Yagil, R & Etzion, Z. 1984. Enhanced reproduction in camels (*Camelus dromedarius*). Comp Biochem Physiol. 79A:201-204.
11988. Yagil, R, Saran, A, & Etzion, Z. 1984. Camels' milk: For drinking only?. Comp Biochem Physiol. 78A:263-266.
11989. Yahner, RH. 1978. The adaptive nature of the social system and behavior in the eastern chipmunk, *Tamias striatus*. Behav Ecol Sociobiol. 3:397-427.
11990. Yahr, P & Kessler, S. 1975. Suppression of reproduction in water-deprived Mongolian gerbils (*Meriones unguiculatus*). Biol Reprod. 12:249-254.
11991. Yalden, DW. 1985. *Tachyoryctes macrocephalus*. Mamm Species. 237:1-3.
11991a. Yamada, JK & Durrant, BS. 1989. Reproductive parameters of clouded leopards (*Neofelis nebulosa*). Zoo Biol. 8:223-231.
11992. Yamada, M. 1954. Some remarks on the pygmy sperm whale, *Kogia*. Sci Rep Whales Res Inst. 9:37-57.
11993. Yamaji, T, Dierschke, DJ, Bhattacharya, AN, et al. 1972. The negative feedback control by estradiol and progesterone of LH secretion in the ovariectomized rhesus monkey. Endocrinology. 90:771-777.
11994. Yamaji, T, Dierschke, DJ, Hotchkiss, J, et al. 1971. Estrogen induction of LH release in the rhesus monkey. Endocrinology. 89:1034-1041.
11995. Yamaji, T, Peckham, WD, Atkinson, LE, et al. 1973. Radioimmunoassay of rhesus monkey follicle stimulating hormone (RhFSH). Endocrinology. 92:1652-1659.
11996. Yamamoto, S. 1967. Breeding Japanese serows *Capricornis crispus* in captivity. Int Zoo Yearb. 7:174-175.
11997. Yamamoto, S. 1967. Notes on hand-rearing chimpanzee twins *Pan troglodytes* at Kobe Zoo. Int Zoo Yearb. 7:97-98.
11998. Yamamoto, S. 1967. Notes on the breeding of black rhinoceroses *Diceros bicornis* at Kobe Zoo. Int Zoo Yearb. 7:163.
11999. Yamashita, T. 1971. Birth of Chinese water deer. Animals & Zoos. 23(5):150-153.
11999a. Yang, S & Zhuge, Y. 1989. Studies on reproduction and population age structure of *Suncus murinus*. Acta Theriol Sinica. 9:195-210.
12000. Yang, CS, Kuo, CC, Del Favero, JE, et al. 1968. Care and raising of newborn Taiwan monkeys (*Macaca cyclopis*) for virus studies. Lab Anim Care. 18:536-543.
12001. Yankell, SL, Schwartzman, RM, & Resnick, B. 1970. Care and breeding of the Mexican hairless dog. Lab Anim Care. 20:940-945.
12002. Yanushko, PA. 1958. Population dynamics of *Cervus elaphus brauneri* (Scharlemagne). Zool Zhurn. 37:1228-1235.
12003. Yardeni-Yaron, E. 1951. Reproductive behaviour of the Levante vole (*Microtus guentheri* D A). Bull Res Council Israel. 1(4):96-99.
12004. Yasui, K. 1972. Raising and breeding of Japanese serow. Animals & Zoos. 24(6):186-189. (Abst Zool Garten 50:207, 1980).
12005. Yasukawa, N, Michael, SD, & Christian, JJ. 1978. Estrous cycle regulation in the whitefooted mouse (*Peromyscus*

leucopus) with special reference to vaginal cast formation. Lab Ani Sci. 28:46-50.

12006a. Yates, TL & Schmidly, DJ. 1977. Systematics of *Scalopus aquaticus* (Linneaus) in Texas and adjacent states. Occas Pap Tx Tech Univ. 45:1-36.

12006. Yates, TL & Schmidly, DJ. 1978. *Scalopus aquaticus.* Mamm Species. 105:1-4.

12007. Yatsenko, EN. 1959. Reproduction of the vole *Prometheomys schaposchnikovi satunin.* Zool Zhurn. 38:916-919.

12007a. Ye, R & Liang, J. 1989. Study on the growth and development of plateau pika under the condition of artificial feeding. Acta Theriol Sinica. 9:110-118.

12007b. Ye, R, Zhou, W, Bai, Q, et al. 1990. Breeding of plateau pika under artificial conditions. Acta Theriol Sinica. 10:287-293.

12008. Yellon, SM & Goldman, BD. 1984. Photoperiod control of reproductive development in the male Djungarian hamster (*Phodopus sungorus*). Endocrinology. 114:664-670.

12009. Yellon, SM & Goldman, BD. 1987. Influence of short days on diurnal patterns of serum gonadotrophins and prolactin concentrations in the male Djungarian hamster, *Phodopus sungorus.* J Reprod Fert. 80:167-174.

12009a. Yen, SSC & Lein, A. 1984. Mammals: Man. Marshall's Physiology of Reproduction. 713-788, Lamming, GE, ed, Churchill Livingstone, London.

12010. Yenikoye, A. 1984. Variations annuelles du comportement d'oestrus, du taux des possibilités d'ovulations chez la Brebis Peuhl du Niger. Reprod Nutr Dév. 24:11-19.

12011. Yenikoye, A, Andre, D, Ravault, J, et al. 1981. Étude de quelques characteristiques de reproduction chez la Brebis Peulh, du Niger. Reprod Nutr Dév. 21:937-951.

12012. Yeoman, RR, Aksel, S, Hazelton, JM, et al. 1988. In vitro bioactive luteinizing hormone assay shows cyclical seasonal hormonal changes and response to luteinizing-hormone releasing hormone in the squirrel monkey (*Saimiri boliviensis boliviensis*). Am J Primatol. 14:167-175.

12013. Yeoman, RR & Terasawa, E. 1984. An increase in single unit activity of the medial basal hypothalamus occurs during the progesterone-induced luteinizing hormone surge in the female rhesus monkey. Endocrinology. 115:2445-2452.

12014. Yerger, RW. 1955. Life history notes on the eastern chipmunk, *Tamias striatus lysteri* (Richardson), in central New York. Am Mid Nat. 53:312-323.

12015. Yerkes, RM. 1943. Chimpanzees: A Laboratory Colony. Yale Univ Press, New Haven, CT.

12016. Yerkes, RM & Elder, JH. 1936. Oestrus, receptivity, and mating in chimpanzees. Comp Psychol Monogr. 13(5):1-39.

12017. Yerkes, RM & Elder, JH. 1936. The sexual and reproductive cycles of chimpanzee. Proc Nat Acad Sci US. 22:276-283.

12018. Yin, T. 1962. Twin elephant calves and interval between births of successive elephant calves. J Bombay Nat Hist Soc. 69:643-644.

12019. Yoakum, JD. 1978. Pronghorn. Big Game of North America. 103-121, Schmidt, JL & Gilbert, DL, eds, Stackpole Books, Harrisburg, PA.

12020. Yocom, CF. 1967. Ecology of feral goats in Haleakala National Park, Maui, Hawaii. Am Mid Nat. 77:418-451.

12021. Yokoyama, K, Ohtsu, R, & Uchida, TA. 1979. Growth and LDH isozyme patterns in the pectoral and cardiac muscles of the Japanese lesser horseshoe bat, *Rhinolophus cornutus cornutus* from the standpoint of adaptation for flight. J Zool. 187:85-96.

12021a. Yonekura, M, Matsui, S, & Kasuya, T. 1980. On the external characters of *Globicephala macrorhynchus* off Taiji, Pacific coast of Japan. Sci Rep Whales Res Inst. 32:61-95.

12022. Yoshida, H. 1968. Notes on the Japanese water shrew, *Chimarrogale platycephala platycephala* (Temminck, 1842), from Kyushu. J Mamm Soc Japan. 4:26-28.

12023. Yoshida, H. 1970. Small mammals of Mt Kiyomizo, Fukuoka Pref 2 Reproduction in the shrew mole, *Urotrichis talpoides.* J Mamm Soc Japan. 5:85-90.

12024. Yoshida, H. 1971. Small mammals of Mt Kiyomizu, Fukouka Pref 3 Reproduction in the Japanese wood mouse. J Mamm Soc Japan. 5:123-129.

12025. Yoshida, H. 1972. Small mammals of Mt Kiyomizu, Fukouka Pref 4 Reproduction in the Japanese long-tailed field mouse, *Apodemus argenteus.* J Mamm Soc Japan. 5:170-177.

12026. Yoshida, H. 1973. Small mammals of Mt Kiyomizu, Fukuoka Pref 5 Reproduction of the Smilth's red-backed vole, *Anteliomys smithi.* J Mamm Soc Japan. 5:206-212.

12027. Yoshida, T, Nakajima, M, Hiyaoka, A, et al. 1982. Menstrual cycle lengths and the estimated time of ovulation in the cynomolgus monkey (*Macaca fascicularis*). Jukken Dubutsu (Experimental Animals). 31:165-174.

12028. Yoshida, T, Suzuki, K, Cho, F, et al. 1987. Serum chorionic gonadotropin levels determined by radioreceptor assay and early diagnosis of pregnancy in the cynomolgus monkey (*Macaca fascicularis*). Am J Primatol. 12:101-106.

12029. Yoshida, Y, Baba, N, Oya, M, et al. 1977. On the formation and regression of corpus luteum in the northern fur seal ovaries. Sci Rep Whales Res Inst. 29:12-128.

12030. Yoshihara, M & Miura, S. 1984. Birth records of captive Reeve's muntjacs. J Mamm Soc Japan. 10:35-36.

12031. Yoshinaga, K, Hess, DL, Hendrickx, AG, et al. 1988. The development of the sexually indifferent gonad in the prosimian, *Galago crassicaudatus crassicaudatus.* Am J Anat. 181:89-105.

12032. Yoshino, S. 1970. Birth and growth of a South American tapir. Animals & Zoos. 22(10):6-9.

12033. Yoshiyuki, M, Iijima, M, & Ogawara, Y. 1970. The embryo-size in a population of *Pipistrellus abramus* in Saitama, Japan. J Mamm Soc Japan. 5:74-75.

12034. Yosida, TH. 1978. Experimental breeding and cytogenetics of the soft-furred rat, *Millardia meltada.* Lab Anim. 12:73-77.

12035. Young, CJ & Jones, JK, Jr. 1982. *Spermophilus mexicanus.* Mamm Species. 164:1-4.

12036. Young, CJ & Jones, JK, Jr. 1983. *Peromyscus yucatanicus.* Mamm Species. 196:1-3.

12037. Young, CJ & Jones, JK, Jr. 1984. *Reithrodontomys gracilis.* Mamm Species. 218:1-3.

12038. Young, DAB. 1976. Breeding and fertility of the Egyptian spiny mouse, *Acomys cahirinus*: Effect of different environments. Lab Anim. 10:15-24.

12039. Young, E. 1972. Observations on the movement and daily home range size of impala, *Aepyceros melampus* (Lichtenstein) in the Kruger National Park. Zool Afr. 7:187-195.

12040. Young, EG & Grant, GA. 1931. The composition of vixen milk. J Biol Chem. 93:805-810.

12041. Young, IR & Renfree, MB. 1979. The effects of corpus luteum removal during gestation on parturition in the tammar wallaby (*Macropus eugenii*). J Reprod Fert. 56:249-254.

12042. Young, RA. 1979. Observations on parturition, litter size, and foetal development at birth in the chocolate wattled bat, *Chalinolobus morio* (Vespertilionidae). Vict Nat. 96:90-91.

12043. Young, RA & Sims, EAH. 1979. The woodchuck, *Marmota monax*, as a laboratory animal. Lab Anim Sci. 29:770-780.
12044. Young, SP. 1958. The Bobcat of North America. Wildl Manage Inst, Washington.
12045. Young, SP & Goldman, EA. 1946. The Puma: Mysterious American Cat. Am Wildl Inst, Washington.
12046. Young, WA. 1961. Rearing an American tapir (*Tapirus terrestris*). Int Zoo Yearb. 3:94-95.
12047. Young, WC & Yerkes, RM. 1943. Factors influencing the reproductive cycle in the chimpanzee: The period of adolescent sterility and related problems. Endocrinology. 33:121-154.
12048. Youngman, PM. 1956. A population of the striped field mouse, *Apodemus agrarius coreae*, in central Korea. J Mamm. 37:1-10.
12049. Youngman, PM. 1975. Mammals of the Yukon Territory. Nat Mus Can Publ Zool. 10:1-192.
12050. Youngman, PM. 1990. *Mustela lutreola*. Mamm Species. 362:1-3.
12051. Yousef, MK, Robertson, WD, Dill, DB, et al. 1970. Energy expenditure of running kangaroo rats, *Dipodomys merriami*. Comp Biochem Physiol. 36:387-393.
12052. Zaidi, P, Wickings, EJ, & Nieschlag, E. 1982. The effects of ketamine HCl and barbiturate anesthesia on the metabolic clearance and production rates of testosterone in the male rhesus monkey. J Steroid Biochem. 16:463-466.
12053. Zajaczek, S & Kaminski, Z. 1947. Réactions des glandes endocrines du Hérisson sous l'influence d'injections hormonales. CR Séances Soc Biol. 141:1104-1105.
12054. Zalesky, M. 1934. A study of the seasonal changes in the adrenal gland of the thirteen-lined ground squirrel (*Citellus tridecemlineatus*), with particular reference to its sexual cycle. Anat Rec. 60:291-316.
12055. Zalesky, M. 1935. A study of the seasonal changes in the thyroid gland of the thirteen-lined ground squirrel (*Citellus tridecemlineatus*), with particular reference to its sexual cycle. Anat Rec. 62:109-137.
12056. Zalesky, M & Wells, LJ. 1937. Experimental studies of the thyroid I Effects of thyroidectomy on reproductive organs in males of an annual breeding ground squirrel. Anat Rec. 69:79-94.
12057. Zalkin, V. 1936. On the biology of the white bear of the Franz-Joseph Archipelago. Bull Soc Nat Moscow Sect Biol. 45:355-363.
12058. Zalkin, VI. 1940. Some materials on the biology of the porpoise of the Azov and Black Seas. Zool Zhurn. 19:160-170.
12059. Zambatis, N. 1976. The honey badger: A tough customer. Fauna Flora (Pretoria). 27:1-3.
12060. Zamboni, l, Conaway, CH, & van Pelt, L. 1974. Seasonal changes in production of semen in free-ranging rhesus monkeys. Biol Reprod. 11:251-267.
12061. Zammuto, RM. 1983. Effect of a climatic gradient on Columbian ground squirrel (*Spermophilus columbianus*) life history. Ph D Diss. Univ W ONT, London, ONT, Canada.
12062. Zammuto, RM. 1984. Relative abilities of three tests to detect variance heterogeneity among mammalian litter sizes. Can J Zool. 62:2287-2289.
12063. Zammuto, RM & Millar, JS. 1985. Environmental predictability, variability, and *Spermophilus columbianus* life history over an elevational gradient. Ecology. 66:1784-1794.
12064. Zammuto, RM & Sherman, PW. 1986. A comparison of time-specific and cohort-specific life tables for Belding's ground squirrels, *Spermophilus beldingi*. Can J Zool. 64:602-605.
12065. Zaniewski, L. 1965. Sex determination in the European beaver, *Castor fiber* Linnaeus, 1758. Acta Theriol. 10:297-301.
12066. Zara, JL. 1973. Breeding and husbandry of the capybara *Hydrochaeris hydrochaeris* at Evansville Zoo. Int Zoo Yearb. 13:137-139.
12067. Zarrow, MX, Eleftheriou, BE, Whitecotten, GL, et al. 1961. Separation of the pubic symphysis during pregnancy and after treatment with relaxin in two subspecies of *Peromyscus maniculatus*. Gen Comp Endocrinol. 1:386-391.
12068. Zarrow, MX & Wilson, ED. 1963. Hormonal control of the pubic symphysis of the Skomer bank vole (*Clethrionomys skomerensis*). J Endocrinol. 28:103-106.
12069. Zavala, DC, Farber, JP, Rhodes, MC, et al. 1974. Pulmonary oxygen toxicity in opossum pouch young, weanlings and mothers: Preliminary observations. J Lab Clin Med. 84:206-217.
12070. Zbytovský, P. 1974. Beitrag zur Biologie der Erdmaus (*Microtus agrestis* L). Lynx. 15:50-57.
12071. Zegers, DA. 1984. *Spermophilus elegans*. Mamm Species. 214:1-7.
12072. Zeigler, DL. 1978. The Okanogan mule deer. WA Dept Game Biol Bull. 15:1-106.
12073. Zeiller, W. 1977. Miami Seaquarium dolphin breeding program. Breeding Dolphins Present Status, Suggestions for the Future. US Mar Mamm Comm Rep No MMC-76/07 NTIS no PB 273673, US Dept Comm Nat Tech Info Serv. 61-65, Ridgway, SH & Benirschke, K, eds, Arlington, VA.
12074. Zejda, J. 1955. Die Analyse der Frühlingspopulation der Rötelmaus in der Hohen Tatra. Zool Entomol Listy. 4:313-328.
12075. Zejda, J. 1961. Age structure in populations of the bank vole, *Clethrionomys glareolus* Schreber, 1780. Zool Listy. 10:249-264.
12076. Zejda, J. 1962. Winter breeding in the bank vole, *Clethrionomys glareolus* Schreb. Zool Listy. 11:309-321.
12077. Zejda, J. 1964. Development of several populations of the bank vole, *Clethrionomys glareolus* Schreb, in a peak year. Zool Listy. 13:15-30.
12078. Zejda, J. 1965. Das Gewicht, das Alter un die Geschlechtsaktivität bei der Rötelmaus (*Clethrionomys glareolus* Schreb). Z Säugetierk. 29:1-9.
12079. Zejda, J. 1966. Litter size in *Clethrionomys glareolus* Schreber 1780. Zool Listy. 15:193-206.
12080. Zejda, J. 1968. A study on embryos and newborns of *Clethrionomys glareolus* Schreb. Zool Listy. 17:115-126.
12081. Zejda, J & Hrabě, V. 1973. The baculum and the duration of the period of reproductive activity in *Clethrionomys glareolus*. Zool Listy. 22:201-212.
12082. Zejda, J & Pelikán, J. 1984. Influence of radiation on the litter size of two microtine rodents. Sp Publ Carnegie Mus Nat Hist. 10:235-241.
12083. Zelenka, G. 1965. Observations sur l'écologie de la Marmotte des Alpes. Terre Vie. 19:238-256.
12084. Zeleznik, AJ. 1981. Premature elevation of systemic estradiol reduces serum levels of follicle-stimulating hormone and lengthens the follicular phase of the menstrual cycle in rhesus monkeys. Endocrinology. 109:352-355.
12085. Zeleznik, AJ, Hutchison, JS, & Schuler, HM. 1985. Interference with the gonadotropin-suppressing actions of estradiol in macaques overrides the selection of a single preovulatory follicle. Endocrinology. 117:991-999.
12086. Zeleznik, AJ, Hutchison, JS, & Schuler, HM. 1987. Passive immunization with anti-estradiol antibodies during the luteal phase of the menstrual cycle potentiates the perimenstrual rise in serum gonadotrophin concentrations and stimulates

follicular growth in the cynomolgus monkey (*Macaca facicularis*). J Reprod Fert. 80:403-410.
12087. Zeleznik, AJ & Resko, JA. 1980. Progesterone does not inhibit gonadotropin-induced follilcular maturation in the female rhesus monkey (*Macaca mulatta*). Endocrinology. 106:1820-1826.
12088. Zellmer, G. 1960. Hand-rearing of giraffe at Bristol Zoo. Int Zoo Yearb. 2:90-93.
12089. Zeng, Z & Brown, JH. 1987. Population ecology of a desert rodent: *Dipodomys merriami* in the Chihuahuan desert. Ecology. 68:1328-1340.
12090. Zenkovich, BA. 1938. Milk of large-sized cetaceans. Dokl Akad Nauk SSSR. 20:203-205.
12091. Zetterberg, H. 1945. Tva Fredlösa. Lindblad, Uppsala.
12092. Zeveloff, SI & Doerr, PD. 1981. Reproduction of raccoons in North Carolina. J Elisha Mitchell Sci Soc. 97:194-199.
12093. Zhang, X. 1989. Protein secretion by the endometrium during pregnancy in the ewe. Reprod Fertil Dév. 1:15-30.
12093a. Zhao, QK & Deng, ZY. 1988. *Macaca thibetana* at Mt Emei, China: I A cross-sectional study of growth and development. Am J Primatol. 16:251-260.
12093b. Zhao, QK & Deng, ZY. 1988. *Macaca thibetana* at Mt Emei, China: II Birth seasonality. Am J Primatol. 16:261-268.
12094. Zherebtzova, OV. 1982. Materials on biology of *Erinaceus* (*Hemiechinus*) *auritus* in Zaunguzskiye Karakumy. Zool Zhurn. 61:411-418.
12095. Zhigal'skii, OA & Bernshtein, AD. 1987. Population factors regulating reproduction in the red-backed vole. Proc Acad Sci USSR (Translations). 291:643-645.
12096. Ziegler, L. 1843. Beobachtungen über die Brunst und den Embryo der Rehe. Hellwingsche Hofbücherei, Hannover.
12097. Ziegler, TE. 1989. Postpartum ovulation and conception in Goeldi's monkey, *Callimico goeldii*. Folia Primatol. 52:206-210.
12098. Ziegler, TE, Bridson, WE, Snowdon, CT, et al. 1987. Urinary gonadotropin and estrogen excretion during the postpartum estrus, conception, and pregnancy in the cotton-top tamarin (*Saguinus oedipus oedipus*). Am J Primatol. 12:127-140.
12099. Ziegler, TE, Savage, A, Scheffler, G, et al. 1987. The endocrinology of puberty and reproductive functioning in female cotton-top tamarins (*Saguinus oedipus*) under varying conditions. Biol Reprod. 37:618-627.
12100. Ziegler, TE, Sholl, SA, Scheffler, G, et al. 1989. Excretion of estrone, estradiol, and progesterone in the urine and feces of the female cotton-top tamarin (*Saguinus oedipus oedipus*). Am J Primatol. 17:185-195.
12100a. Ziegler, TE, Snowdon, CT, & Bridson, WE. 1990. Reproductive performance and excretion of urinary estrogens and gonadotropins in the female pygmy marmoset (*Cebuella pygmaea*). Am J Primatol. 22:191-203.
12101. Ziegler, TE, Snowdon, CT, Warneke, M, et al. 1990. Urinary excretion of oestrone conjugates nad gonadotrophins during pregnancy in the Goeldi's monkey (*Callimico goeldii*). J Reprod Fert. 89:163-168.
12102. Ziegler, TE & Stein, FJ. 1981. Reproductive data from a colony of common marmosets (*Callithrix jacchus*). Am J Primatol. 1:336.
12103. Zimen, E. 1976. On the regulation of pack size in wolves. Z Tierpsychol. 40:300-341.
12104. Zimina, RP. 1962. The ecology of *Ochotona macrotis* Gunther dwelling in the area of the Tersky-Alatau mountain range. Bull Mnskovskogo Obshchestvo Ispytotele i Privody. 67(3):5-12.
12104a. Zimmerman, EG. 1970. Additional records of bats from Utah. Southwest Nat. 15:263-264.
12105. Zimmermann, K. 1964. Zur Säugetier-Faun Chinas. Mitt Zool Mus Berlin. 40:87-140.
12106. Zimmermann, RR. 1969. Early weaning and weight gain in infant rhesus monkeys. Lab Anim Care. 19:644-647.
12107. Zimmermann, W. 1980. Zur Haltung und Zucht von Saiga-Antilopen (*Saiga tatarica tatarica*) im Kölner Zoo. Z Kölner Zoo. 23:120-127.
12108. Zimmy, ML. 1965. Thirteen-lined ground squirrels born in captivity. J Mamm. 46:521-522.
12109. Zinchenko, VL & Ivashin, MV. 1987. Siamese twins of minke whales of the southern hemisphere. Sci Rep Whales Res Inst. 38:165-169.
12110. Zippelius, H-M. 1958. Zur Jugendentwicklung der Waldspitzmaus, *Sorex armeus*. Bonn Zool Beitr. 9:120-129.
12111. Zubko, YP & Ostryakov, SI. 1961. On the reproduction of *Ellobius talpinus* Pallas in the south of the Ukraine. Zool Zhurn. 40:1577-1579.
12112. Zucker, EE & Chapman, JA. 1984. Morphological and physiological characteristics of muskrats from three different physiographic regions of Maryland, USA. Z Säugetierk. 49:90-104.
12113. Zucker, EL & Kaplan, JR. 1981. A reinterpretation of sexual harassment in patas monkeys. Anim Behav. 29:957-958.
12114. Zucker, I, Johnston, PG, & Frost, D. 1980. Comparative, physiological and biochronometric analysis of rodent seasonal reproductive cycles. Prog Reprod Biol. 5:102-133.
12115. Zucker, I & Licht, P. 1983. Circannual and seasonal variations in plasma luteinizing hormone levels of ovariectomized ground squirrels (*Spermophilus lateralis*). Biol Reprod. 28:178-185.
12116. Zucker, I & Licht, P. 1983. Seasonal variations in plasma luteinizing hormone levels of gonadectomized male ground squirrels (*Spermophilus lateralis*). Biol Reprod. 29:278-285.
12117. Zuckerman, S. 1930. The menstrual cycle of the primates Part I General nature and homology. Proc Zool Soc London. 1930:691-754.
12118. Zuckerman, S. 1931. The menstrual cycle of the primates Part III The alleged breeding-season of primates, with special reference to the chacma baboon (*Papio porcarius*). Proc Zool Soc London. 1931:325-343.
12119. Zuckerman, S. 1931. The menstrual cycle of the primates Part IV Observations on the lactation period. Proc Zool Soc London. 1931:593-602.
12120. Zuckerman, S. 1932. The menstrual cycle of the primates Part VI Further observations on the breeding of primates, with special reference to the suborders Lemuroidea and Tarsioidea. Proc Zool Soc London. 1932:1059-1075.
12121. Zuckerman, S. 1935. The menstrual cycle of the primates Part VIII The oestrin-withdrawal theory of menstruation. Proc R Soc London. 118B:13-21.
12122. Zuckerman, S. 1935. The Aschheim-Zondek diagnosis of pregnancy in the chimpanzee. Am J Physiol. 110:597-601.
12123. Zuckerman, S. 1937. The duration and phases of the menstrual cycle in primates. Proc Zool Soc London. 107A:315-329.
12124. Zuckerman, S. 1937. The interaction of ovarian hormones in the menstrual cycle. Les Hormones Sexuelles, CR, Colloque Int Collège France. 121-135, Brouha, L, ed, Hermann & Cie, Paris.
12125. Zuckerman, S. 1938. The 'female prostate' in the green monkey (*Cercopithecus aethiops sabeus*). J Anat. 72:472.

12126. Zuckerman, S. 1940/41. Periodic uterine bleeding in spayed rhesus monkeys injected with a constant threshold dose of oestrone. J Endocrinol. 2:263-267.
12127. Zuckerman, S. 1948. Duration of reproduction life in the baboon. J Endocrinol. 5:220-221.
12128. Zuckerman, S. 1953. The breeding seasons of mammals in captivity. Proc Zool Soc London. 122:827-950.
12129. Zuckerman, S & Parkes, AS. 1932. The menstrual cycle of the primates Part V The cycle of the baboon. Proc Zool Soc London. 1932:139-191.
12130. Zuckerman, S, van Wagenen, G, & Gardiner, RH. 1938. The sexual skin of the rhesus monkey. Proc Zool Soc London. 108:385-401.
12131. Zukowsky, L. 1961. Zuchterfolg bei einem Kapuzineraffen im Münsterchen Zoo. Zool Garten NF. 26:104-106.
12132. Zumpe, D & Michael, RP. 1970. Ovarian hormones and female sexual invitations in captive rhesus monkeys (*Macaca mulatta*). Anim Behav. 18:293-301.
12133. Zumpe, D & Michael, RP. 1977. Effects of ejaculations by males on the sexual invitations of female rhesus monkeys (*Macaca mulatta*). Behaviour. 60:260-277.
12134. Zumpt, I. 1970. The ground squirrel. Afr Wildl. 24:115-121.
12135. Żurowski, W & Doboszynská, T. 1975. Superfoetation in European beaver. Acta Theriol. 20:97-104.
12136. Żurowski, W & Kasperczyk, B. 1986. Characteristics of a European beaver population in the Suwalki lakeland. Acta Theriol. 31:311-325.
12137. Żurowski, W, Kisza, J, & Kruk, A. 1971. Composition of milk of European beavers, *Castor fiber* Linnaeus, 1758. Acta Theriol. 16:405-408.
12138. Zurowski, W, Kisza, J, Kruk, A, et al. 1974. Lactation and chemical composition of the milk of the European beaver (*Castor fiber* L). J Mamm. 55:847-850.
12139. Zwahlen, R. 1975. Zur Fortpflanzung und Jungenentwicklung des Eichhörnchens, *Sciurus vulgaris* (Linne, 1758). Säugetierk Mitt. 23:9-13.

Appendix 1
Mammalian Species Accounts 1-402

Published by the American Society of Mammalogists[1]

Order Monotremata
- Family Tachyglossidae
  - Ornithorhynchidae

Order Marsupialia
- Family Didelphidae
  - *Caluromys derbianus*, 140 (1423)
  - *Chironectes minimus*, 109 (6860)
  - *Didelphis virginiana*, 40 (7097)
  - *Glironia venusta*, 107 (6859)
  - *Lestodelphys halli*, 81 (6856)
  - *Lutreolina crassicaudata*, 91 (6857)
  - *Marmosa robinsoni*, 203 (7997)
  - *Monodelphis kunsi*, 190 (281)
- Microbiotheriidae
  - *Dromiciops australis*, 99 (6858)
- Caenolestidae
  - *Rhyncholestes raphanurus*, 286 (8257)
- Dasyuridae
- Myrmecobiidae
- Thylacinidae
- Notoryctidae
- Peramelidae
- Thylacomyidae
- Phalangeridae
- Burramyidae
- Petauridae
  - *Petaurus breviceps*, 30 (10121)
- Macropodidae
  - *Macropus giganteus*, 187 (8668)
- Phascolarctidae
- Vombatidae
- Tarsipedidae

Order Edentata
- Family Myrmecophagidae
- Bradypodidae
- Choloepidae
- Dasypodidae
  - *Dasypus novemcinctus*, 162 (7011)
  - *Euphractus sexcinctus*, 252 (8999)

Order Insectivora
- Family Solenodontidae
- Tenrecidae
- Chrysochloridae
- Erinaceidae
- Soricidae
  - *Blarina brevicauda*, 261 (3725)
  - *Cryptotis goodwini*, 44 (1812); *magna*, 61 (9177); *mexicana*, 28 (1811); *parva*, 43 (11626)
  - *Megasorex gigas*, 16 (368)
  - *Microsorex hoyi*, 33 (6557); *thompsoni*, 33 (6557)
  - *Notiosorex crawfordi*, 17 (369)
  - *Sorex bendirii*, 27 (8262); *dispar*, 155 (5795); *fumeus*, 215 (8155); *gaspensis*, 155 (5795); *longirostris*, 143 (3511); *merriami*, 2 (367); *nanus*, 131 (4894); *ornatus*, 212 (8156); *pacificus*, 231 (1613); *palustris*, 296 (760); *tenellus*, 131 (4894); *trowbridgii*, 337 (3724)
- Talpidae
  - *Condylura cristata*, 129 (8405)
  - *Galemys pyrenaicus*, 207 (8209)
  - *Neurotrichus gibbsii*, 387 (1617)
  - *Parascalops breweri*, 98 (4247)
  - *Scalopus aquaticus*, 105 (12006)
  - *Scapanus orarius*, 253 (4456)

Order Scandentia
- Family Tupaiidae

Order Dermoptera
- Family Cynocephalidae

Order Chiroptera
- Family Pteropodidae
  - *Eidolon helvum*, 312 (2430)
  - *Epomophorus gambianus*, 344 (1139); *wahlbergi*, 394 (47a)
  - *Hypsignathus monstrosus*, 357 (6198)
- Rhinopomatidae
  - *Rhinopoma hardwickei*, 263 (8818); *muscatellum*, 263 (8818)
- Emballonuridae
  - *Balantiopteryx io*, 313 (384); *infusca*, 313 (384); *plicata*, 301 (383)
  - *Cyttarops alecto*, 13 (10313)
  - *Diclidurus albus*, 316 (1671)
- Craseonycteridae
  - *Craseonycteris thonglongyai*, 160 (4748)
- Nycteridae
- Megadermatidae
  - *Macroderma gigas*, 260 (5067)
- Rhinolophidae
- Noctilionidae
  - *Noctilio albiventris*, 197 (4956); *leporinus*, 216 (4955)
- Mormoopidae
  - *Pteronotus davyi*, 346 (62); *parnellii*, 209 (4658); *quadridens*, 395 (9213a)
- Phyllostomidae
  - *Anoura cultrata*, 179 (10655)
  - *Ardops nichollsi*, 24 (5419)
  - *Artibeus aztecus*, 177 (11522); *hirsutus*, 199 (11524); *inopinatus*, 199 (11524); *phaeotis*, 235 (10841); *toltecus*, 178 (11523)
  - *Brachyphylla cavernarum*, 205 (10593); *nana*, 206 (10594)
  - *Centurio senex*, 138 (10179)
  - *Chiroderma improvisum*, 134 (5412)
  - *Choeronycteris mexicana*, 291 (385)
  - *Chrotoperus auritus*, 343 (7153)
  - *Desmodus rotundus*, 202 (4049)
  - *Diphylla ecaudata*, 227 (4050)

---

[1] The taxonomy in this appendix follows that of individual Mammalian Species accounts. Square brackets [] provide the name used in this compendium.

Ectophylla alba, 166 (10840)
Erophylla sezekorni, 115 (526)
Glossophaga leachii, 226 (11525); *mexicana*, 245 (11527); *soricina*, 379 (216)
Hylonycteris underwoodi, 32 (5424)
Leptonycteris nivalis, 307 (4642)
Lonchorhina aurita, 347 (6233)
Macrophyllum macrophyllum, 62 (4401)
Macrotus waterhousii, 1 (280)
Micronycteris brachyotis, 251 (7155); *megalotis*, 376 (203)
Monophyllus plethodon, 58 (4942); *redmani*, 57 (4941)
Platyrrhinus [*Vampyrops*] *helleri*, 373 (3236)
Pygoderma bilabiatum, 220 (11529)
Stenoderma rufum, 18 (3699)
Sturnira aratathomasi, 284 (10225); *bidens*, 276 (7444); *lilium*, 333 (3599); *magna*, 240 (10653); *thomasi*, 68 (5422)
Tonatia evotis, 334 (7154); *silvicola*, 334 (7154)
Uroderma bilobatum, 279 (527)
Vampyressa pusilla, 292 (6421)
Vampyrodes caraccioli, 359 (11740)
Vampyrops lineatus, 275 (11737)
Vampyrum spectrum, 184 (7780)

Natalidae
Natalus major, 130 (5036); *micropus*, 114 (5700)

Furipteridae

Thyropteridae
Thyroptera discifera, 104 (11753); *tricolor*, 71 (11757)

Myzopodidae
Myzopoda aurita, 116 (9599)

Vespertilionidae
Antrozous pallidus, 213 (4668)
Bauerus [*Antrozous*] *dubiaquercus*, 282 (3045)
Eptesicus fuscus, 356 (6120)
Euderma maculatum, 77 (11471)
Eudiscopus denticulus, 19 (5956)
Idionycteris phyllotis, 208 (2196)
Lasionycteris noctivagans, 172 (6109)
Lasiurus borealis, 183 (9961); *cinereus*, 185 (9962); *intermedius*, 132 (11528); *seminolus*, 280 (11701)
Myotis auriculus, 191 (11465); *austroriparius*, 332 (5392); *evotis*, 329 (6780); *keenii*, 121 (3287); *lucifugus*, 142 (3216); *nigricans*, 39 (11759); *planiceps*, 60 (6948); *sodalis*, 163 (10801); *thysanodes*, 137 (8005); *velifer*, 149 (3288); *volans*, 224 (11466)
Nycticeius humeralis, 23 (11470)
Pipistrellus subflavus, 228 (3549)
Plecotus mexicanus, 401 (10942a); *rafinesquii*, 69 (5397); *townsendii*, 175 (11870)
Rhogeessa gracilis, 76 (5406)

Mystacinidae

Molossidae
Neoplatymops [*Molossops*] *mattogrossensis*, 244 (11738)
Nyctinomops aurispinosus, 350 (5411); *femorosaccus*, 349 (6099); *macrotis*, 351 (7354)
Tadarida brasiliensis, 331 (11702)

Order Primates
Family Cheirogaleidae
Lemuridae
Indriidae
Daubentoniidae
Lorisidae
Galagidae
Tarsiidae
Callithricidae
Leontopithecus rosalia, 148 (5854)
Callimiconidae
Cebidae
Callicebus moloch, 112 (5390)
Cercopithecidae
Hylobatidae
Pongidae
Pongo pygmaeus, 4 (4106)
Hominidae

Order Carnivora
Family Canidae
Atelocynus [*Dusicyon*] *microtis*, 256 (849)
Canis latrans, 79 (744); *lupus*, 37 (7148); *rufus*, 22 (8227)
Cerdocyon [*Dusicyon*] *thous*, 186 (848)
Chrysocyon brachyurus, 234 (2578)
Cuon alpinus, 100 (1961)
Nyctereutes procyonoides, 358 (11452)
Urocyon cinereoargenteus, 189 (3529)
Vulpes macrotis, 123 (7063); *velox*, 122 (2913)

Ursidae
Ailuropoda melanoleuca, 110 (1816)
Ursus maritimus, 145 (2475)

Procyonidae
Ailurus fulgens, 222 (9173)
Bassariscus astutus, 327 (8633)
Potos flavus, 321 (3398)
Procyon lotor, 119 (6587)

Mustelidae
Enhydra lutris, 133 (3098)
Martes americana, 289 (1869); *pennanti*, 156 (8714)
Mephitis mephitis, 173 (11384)
Mustela erminea, 195 (5741); *lutreola*, 362 (12050); *nigripes*, 126 (4802)
Taxidea taxus, 26 (6556)

Viverridae
Cryptoprocta ferox, 254 (5928)
Osbornictis piscivora, 309 (11187)

Herpestidae
Bdeogale crassicauda, 294 (10716)
Crossarchus ansorgei, 402 (11187a); *obscurus*, 290 (3866)
Herpestes auropunctatus, 342 (7828); *sanguineus*, 65 (10715)
Ichneumia albicauda, 12 (10694)
Liberiictis kuhni, 348 (3867)

Protelidae
Proteles cristatus, 363 (5907)

Hyaenidae
Hyaena brunnea, 194 (7348); *hyaena*, 150 (9130)

Felidae
Felis concolor, 200 (2186); *geoffroyi*, 54 (11977); [*Lynx*] *lynx*, 269 (10942)
Panthera onca, 340 (9811); *tigris*, 152 (7004)
Uncia [*Panthera*] *uncia*, 20 (4617)

Appendix 2
Core Journals

The following core journals were reviewed volume by volume minimally from 1964 to 1988, but usually from volume 1 through 1991. Other journals were examined as citations to them arose. Journal abbreviations used in the bibliography are parenthetically included. Other abbreviations in the bibliography are those used by the BioSciences Information Service.

Acta Anatomica
(Acta Anat)
Acta Endocrinologica
(Acta Endocrinol)
Acta Theriologica
(Acta Theriol)
African Journal of Ecology
(Afr J Ecol)
African Wildlife
(Afr Wildl)
American Journal of Anatomy
(Am J Anat)
American Journal of Physiology
(Am J Physiol)
American Journal of Primatology
(Am J Primatol)
American Midland Naturalist
(Am Mid Nat)
American Museum of Natural History, Bulletin
(Bull Am Mus Nat Hist)
American Mus Novitates
(Am Mus Novit)
American Naturalist
(Am Nat)
American Zoologist
(Am Zool)
Anatomischer Anzeiger
(Anat Anz)
Anatomical Record
(Anat Rec)
Animal Behaviour
(Anim Behav)
Animal Behaviour Monographs
(Anim Behav Monogr)
Annales de Biologie Animale, Biochimie et Biophysique
(Ann Biol Anim Biochim Biophys)
Annales Zoologici Fennici
(Ann Zool Fenn)
Annales Sociéte Royale Zoologique de Belgique
(Ann Soc R Zool Belg)
Annales Musée Royal de l'Afrique Central Série Quarto Zoologique
(Ann Mus R Afr Cent Ser Quarto Zool)
Annals of the Cape Provincial Museums Natural History
(Ann Cape Prov Mus Nat Hist)
Annals of the Transvaal Museum
(Ann Transvaal Mus)
Annotations Zoologicae Japonensis
(Annot Zool Japon)
Aquatic Mammals
(Aquat Mamm)
Archives d'Anatomie Microscopique et de Morphologie Expérimentale
(Arch Anat Microsc Morphol Exp)
Arctic
(Arctic)
Arkiv fur Zoologi (Stockholm)
(Ark Zool (Stockholm))
Arnoldia (Rhodesia)
(Arnoldia (Rhodesia))
Arnoldia (Zimbabwe)
(Arnoldia (Zimbabwe))
Australian Journal of Biological Science
(Aust J Biol Sci)
Australian Journal of Experimental Biology and Medical Science
(Aust J Exp Biol Med Sci)
Australian Journal of Marine and Freshwater Research
(Aust J Mar Freshwater Res)
Australian Journal of Science
(Aust J Sci)
Australian Journal of Zoology
(Aust J Zool)
Australian Mammalogy
(Aust Mammal)
Australian Wildlife Research
(Aust Wildl Res)
Australian Zoology
(Aust Zool)
Behaviour
(Behaviour)
Behavioral Ecology and Sociobiology
(Behav Ecol Sociobiol)
Bibliotheca Primatologica
(Bibl Primatol)
Bijragen tot di Dierkunde
(Bijdr Dierk)
Biologia Gabonica
(Biol Gabon)
Biological Bulletin
(Biol Bull)
Biology of the Neonate
(Biol Neonate)
Biology of Reproduction
(Biol Reprod)
Biotropica
(Biotropica)
Bonner Zoologische Beitraege
(Bonn Zool Beitr)
Bulletin Fisheries Research Board of Canada
(Bull Fish Res Bd Can)
Bulletin Museum of Comparative Zoology at Harvard University
(Bull Mus Comp Zool)
Bulletin Societé de Zoologique de France
(Bull Soc Zool France)
Canadian Field-Naturalist
(Can Field-Nat)
Canadian Journal of Fisheries and Aquatic Sciences
(Can J Fish Aquatic Sci)
Canadian Journal of Zoology
(Can J Zool)
Canadian Wildlife Service Reports
(Can Wildl Serv Rept)
Carnivore
(Carnivore)
Cimbebasia
(Cimbebasia)
Comparative Biochemistry and Physiology
(Comp Biochem Physiol)
Contributions to Primatology
(Contrib Primatol)
CSIRO Wildlife Research
(CSIRO Wildl Res)
Discovery Reports
(Discovery Rep)
East African Wildlife Journal
(E Afr Wildl J)
Ecology
(Ecology)
Ecological Monographs
(Ecol Monogr)
Endocrinologica Japan
(Endocrinol Jpn)
Endocrinology
(Endocrinology)
Evolution
(Evolution)
Fertility and Sterility
(Fertil Steril)
Fieldiana Zoology
(Fieldiana Zool)
Folia Primatologica
(Folia Primatol)

Freunde der Kölner Zoo
(Fr Köln Zoo)
Gamete Research
(Gamete Res)
General and Comparative Endocrinology
(Gen Comp Endocrinol)
Great Basin Naturalist
(Great Basin Nat)
Growth
(Growth)
Handbuch der Zoologie
(Handb Zool)
Hercynia
(Hercynia)
Hormones and Behavior
(Horm Behav)
Hvalradets Skrifter
(Hvalradets Skr)
International Journal of Primatology
(Int J Primatol)
International Zoo Yearbook
(Int Zoo Yearb)
Israel Journal of Zoology
(Israel J Zool)
Journal of Agricultural Research
(J Agr Res)
Journal of Anatomy
(J Anat)
Journal of Animal Ecology
(J Anim Ecol)
Journal of Bombay Natural History Society
(J Bombay Nat Hist Soc)
Journal of Comparative Physiology
(J Comp Physiol)
Journal of Comparative Physiology and Psychology
(J Comp Physiol Psychol)
Journal of Endocrinology
(J Endocrinol)
Journal of Experimental Biology
(J Exp Biol)
Journal of Experimental Zoology
(J Exp Zool)
Journal of Mammalogy
(J Mamm)
Journal of the Mammal Society of Japan
(J Mamm Soc Japan)
Journal of Medical Primatology
(J Med Primatol)
Journal of Morphology
(J Morphol)
Journal of Physiology
(J Physiol)
Journal of Reproduction and Fertility
(J Reprod Fertil)
Journal of Steroid Biochemistry
(J Steroid Biochem)
Journal of Wildlife Management
(J Wildl Manage)
Journal of Zoo Animal Medicine
(J Zoo Anim Med)
Journal of Zoology
(J Zool)
Koedoe
(Koedoe)
Lab Animals
(Lab Anim)
Laboratory Animal Care
(Lab Anim Care)
Laboratory Anim Science
(Lab Anim Sci)
Lammergeyer
(Lammergeyer)
Mammal Review
(Mammal Rev)
Mammalia
(Mammalia)
Mammal Species
(Mamm Species)
Marine Mammal Science
(Mar Mamm Sci)
Medical Primatology
(Med Primatol)
Mitteilungen der Naturforschenden Gesellschaft in Bern
(Mitt Naturforsch Ges Bern)
Mitteilungen aud dem Hamburgischen Zoologischen Museum und Institut
(Mitt Hamb Zool Mus Inst)
New Zealand Journal of Science
(N Z J Sci)
Netherlands Journal of Zoology
(Neth J Zool)
Norwegian Journal of Zoology
(Norw J Zool)
Oecologia
(Oecologia)
Oikos
(Oikos)
Philosophical Transactions of the Royal Society of London
(Phil Trans R Soc Lond)
Pakistan Journal of Zoology
(Pakistan J Zool)
Periodicum Biologorum
(Per Biol)
Physiology and Behavior
(Physiol Behav)
Physiological Zoology
(Physiol Zool)
Placenta
(Placenta)
Primates
(Primates)
Primates in Medicine
(Primat Med)
Proceedings of the Indian Academy of Sciences (Animal Science)
(Proc Indian Acad Sci (Anim Sci))
Proceedings of the Koninklijke Nederlandse Akademie van Wetenschappen
(Proc K Ned Akad Wetensch)
Proceedings of the National Institute of Sciences, India
(Proc Nat Inst Sci India)
Proceedings of the Royal Society of Edinburgh
(Proc R Soc Edinburgh)
Proceedings of the Royal Society of London Series B
(Proc R Soc London B)
Puku
(Puku)
Rapports et Process-Verbaux des Réunions Conseil International pour l'Exploration de la Mer
(Rapp P-V Réun Cons Int Explor Mer)
Report International Whaling Commission
(Rep Int Whal Comm)
Reproduction, Nutrition, Dévellopement
(Reprod Nutr Dév)
Respiration
(Respiration)
Respiration Physiology
(Respir Physiol)
Revista Brasileira Biologia
(Rev Bras Biol)
Revue de Zoologie et de Botanique Africaine
(Rev Zool Afr)
Revue Suisse de Zoologie
(Rev Suisse Zool)
Säugetierkundliche Mitteilungen
(Säugetierk Mitt)
Smithsonian Contributions to Zoology
(Smithson Contrib Zool)
South African Journal of Science
(S Afr J Sci)
South African Journal of Zoology
(S Afr J Zool)
Southwest Naturalist
(Southwest Nat)
Terre Vie
(Terre Vie)
Transactions North American Wildlife Conference
(Trans N Am Wildl Conf)
Tulane Studies in Zoology
(Tulane Stud Zool)
University of California Contributions to Zoology
(Univ CA Contrib Zool)
University of Kansas Publications Museum of Natural History
(Univ KS Publ Mus Nat Hist)
Vestnik Ceskoslovenske Spoleanosti Zoologicki
(Vest Ccsl Spol Zool)

Vestnik Zoologii (Kiev)
(Vestn Zool (Kiev))
Victorian Naturalist
(Victorian Nat)
Viltrevy
(Viltrevy)
Wildlife Monographs
(Wildl Monogr)
Wildlife Research
(Wildl Res)
Zeitschrift der Kölner Zoo
(Z Köln Zoo)
Zeitschrift für Jagdwissenschaft
(Z Jagdwiss)
Zeitschrift für Mikroskopisch-Anatomische Forschung
(Z Mikrosk Anat Forsch)
Zeitschrift für Säugetierkunde
(Z Säugetierk)
Zeitschrift für Tierpsychologie
(Z Tierpsychol)
Zeitschrift für du Wissenschartliche Zoologie
(Z Wiss Zool)
Zentralblatt für das Gesamte Anatomie
(Zentralbl Gesamte Anat)
Zoo Biology
(Zoo Biol)
Zoologica
(Zoologica)
Zoologicheskii Zhurnal
(Zool Zh)
Zoologicki Listy
(Zool Listy)
Zoologische Garten
(Zool Gart)
Zoologische Jahrbucher
(Zool Jb)

Index to Popular or Well-Known Mammals

Index of Species

The page number listed refers the reader to the first listing of information for the famlily to which that species belongs. An asterisk (*) indicates we found no reproductive data for that species.